GATE PREP SERIES

GATE

Graduate Aptitude Test in Engineering

AF646232

2022

Electronics & Communication Engineering

Solved Papers

(2000-2021)

G. K. PUBLICATIONS (P) LTD.

Title : GATE 2022 : Electronics & Communication Engineering Solved Papers (2000-2021)

Language : English

Editor's Name : Vinit Garg

Typeset & Published by :

Career Launcher Infrastructure (P) Ltd.

A-45, Mohan Cooperative Industrial Area, Near Mohan Estate Metro Station, New Delhi - 110044

Marketed by :

G.K. Publications (P) Ltd.

Plot No. 9A, Sector-27A, Mathura Road, Faridabad, Haryana-121003

ISBN : **978-93-90820-00-9**

Printer's Details : Made in India, New Delhi.

For product information :

Visit ***www.gkpublications.com*** or email to ***gkp@gkpublications.com***

Contents

GATE

Graduate Aptitude Test in Engineering

IIT Institutes

GATE 2021 will be conducted by
Indian Institute of Technology, Mumbai

Indian Institute of Technology, Kharagpur

GATE 2020 conducted by
Indian Institute of Technology, Delhi

Indian Institute of Technology, Chennai

Indian Institute of Technology, Guwahati

Indian Institute of Technology, Roorkee

Indian Institute of Technology, Kanpur

Preface

The Graduate Aptitude Test in Engineering (GATE) is an online exam conducted by the IITs for admissions to PG courses in IITs, IISc Bangalore, NITs and many state run universities as well as private universities. Also there are more than 37 PSUs that use GATE score for recruitments. A large number of corporates are also using GATE score as a tool to screen students for placements.

GK Publications is well known as the "publisher of choice" to students preparing for GATE and other technical examinations in the country. We published the first set of books in 1994 when GATE exam, both objective and conventional, was conducted in the paper and pencil environment, and used as a check point for entry to post graduate courses in IITs and IISCs. At that time, students had little access to technology and relied mainly on instructor led learning followed by practice with books available for these examinations.

A lot has changed since then!

Today, GATE is conducted in an online only mode with multiple choice and numerical based questions. The score is valid for three years and is used not only for post graduate courses but is also used by major PSUs for recruitment. Today's students have easy access to technology and the concept of a monologue within the classroom has changed to dialogue where students come prepared with concepts and then discuss topics. They learn a lot of things on the go with their mobile devices and practice for mock tests online.

We, as a leading publisher of GATE books, have also embraced change. Today, our books are no more guides and papers only but come with a fully supported mobile app and a web portal. The mobile app provides access to video lectures, short tests and regular updates about the exam. The web portal in additional to what is available on the app provides full length mock tests to mimic the actual exam and help you gauge your level of preparedness. The combination of practice content in print, video lectures, and short and full length tests on mobile and web makes this product a complete courseware for GATE preparation.

This book includes previous years GATE questions along with detailed solution of each question for better understanding. It will help the GATE aspirants to know an idea about the pattern of questions asked in GATE examination.

We also know that improvement is a never ending process and hence we welcome your suggestions and feedback or spelling and technical errors if any. Please write to us at gkp@gkpublications.com

We hope that our small effort will help you prepare well for the examination.

We wish you all the best!

GK Publications Pvt. Ltd.

About GATE

The Graduate Aptitude Test in Engineering (GATE) conducted by IISc and IITs has emerged as one of the bench mark tests for engineering and science aptitude in facilitating admissions for higher education (M.Tech./Ph.D.) in IITs, IISc and various other Institutes/Universities/Laboratories in India. With the standard and high quality of the GATE examination in 27 disciplines of engineering and science, Humanities and Social Sciences subjects, it identifies the candidate's understanding of a subject and aptitude and eligibility for higher studies. During the last few years, GATE score is also being used as one of the criteria for recruitment in Government Organizations such as Cabinet Secretariat, and National/State Public Sector Undertakings in India. Because of the importance of the GATE examination, the number of candidates taking up GATE exams has increased tremendously. GATE exams are conducted by the IITs and IISc as a computer based test having multiple choice questions and numerical answer type questions. The questions are mostly fundamental, concept based and thought provoking. From 2017 onwards GATE Exam is being held in Bangladesh, Ethiopia, Nepal, Singapore, Sri Lanka and United Arab Emirates. An Institute with various nationalities in its campus widens the horizons of an academic environment. A foreign student brings with him/her a great diversity, culture and wisdom to share. Many GATE qualified candidates are paid scholarships/assistantship, especially funded by Ministry of Human Resources Development, Government of India and by other Ministries. Indian Institute of Technology Bombay is the Organizing Institute for GATE 2021.

Why GATE?

Admission to Post Graduate and Doctoral Programmes

Admission to postgraduate programmes with MHRD and some other government scholarships/ assistantships in engineering colleges/institutes is open to those who qualify through GATE. GATE qualified candidates with Bachelor's degree in Engineering/Technology/Architecture or Master's degree in any branch of Science/Mathematics/Statistics/Computer Applications are eligible for admission to Master/Doctoral programmes in Engineering/Technology/Architecture as well as for Doctoral programmes in relevant branches of Science with MHRD or other government scholarships/ assistantships. Candidates with Master's degree in Engineering/Technology/Architecture may seek admission to relevant Ph.D programmes with scholarship/assistantship without appearing in the GATE examination.

Financial Assistance

A valid GATE score is essential for obtaining financial assistance during Master's programs and direct Doctoral programs in Engineering/Technology/Architecture, and Doctoral programs in relevant branches of Science in Institutes supported by the MHRD or other Government agencies. As per the directives of the MHRD, the following procedure is to be adopted for admission to the post-graduate programs (Master's and Doctoral) with MHRD scholarship/assistantship. Depending upon the norms adopted by a specific institute or department of the Institute, a candidate may be admitted directly into a course based on his/her performance in GATE only **or** based on his/her performance in GATE **and** an admission test/interview conducted by the department to which he/she has applied **and/or** the candidate's academic record. If the candidate is to be selected through test/interview for post-graduate programs, a minimum of 70% weightage will be given to the performance in GATE and the remaining 30% weightage will be given to the candidate's performance in test/interview and/or academic record, as per MHRD guidelines. The admitting institutes could however prescribe a minimum passing percentage of marks in the test/interview. Some colleges/institutes specify GATE qualification as the mandatory requirement even for admission without MHRD scholarship/assistantship.

To avail of the financial assistance (scholarship), the candidate must first secure admission to a program in these Institutes, by a procedure that could vary from institute to institute. Qualification in GATE is also a minimum requirement to apply for various fellowships awarded by many Government organizations. Candidates are advised to seek complete details of admission procedures and availability of MHRD scholarship/assistantship from the concerned admitting institution. The criteria for postgraduate admission with scholarship/assistantship could be different for different institutions. The management of the post-graduate scholarship/assistantship is also the responsibility of the admitting institution. Similarly, reservation of seats under different categories is as per the policies and norms prevailing at the admitting institution and Government of India rules. *GATE offices will usually not entertain any enquiry about admission, reservation of seats and / or award of scholarship / assistantship.*

PSU Recruitments

As many as 37 PSUs are using GATE score for recruitment. It is likely that more number of PSUs may start doing so by next year. Below is the list of PSUs:

MDL, BPCL, GAIL, NLC LTD, CEL, Indian Oil, HPCL, NBPC, NECC, BHEL, WBSEDCL, NTPC, ONGC, Oil India, Power Grid, Cabinet Secretariat, Govt. of India, BAARC, NFL, IPR, PSPCL, PSTCL, DRDO, OPGC Ltd., THDC India Ltd., BBNL, RITES, IRCON, GHECL, NHAI, KRIBHCO, Mumbai Railway Vikas Corporation Ltd. (MRVC Ltd.), National Textile Corporation, Coal India Ltd., BNPM, AAI, NALCO, EdCIL India.

Important :

1. Admissions in IITs/IISc or other Institutes for M.Tech./Ph.D. through GATE scores shall be advertised separately by the Institutes and GATE does not take the responsibility of admissions.
2. Cabinet Secretariat has decided to recruit officers for the post of Senior Field Officer (Tele) (From GATE papers of EC, CS, PH), Senior Research Officer (Crypto) (From GATE papers of EC, CS, MA), Senior Research Officer (S&T) (From GATE papers EC, CS, CY, PH, AE, BT) in the Telecommunication Cadre, Cryptographic Cadre and Science & Technology Unit respectively of Cabinet Secretariat. The details of the scheme of recruitment shall be published in National Newspaper/Employment News by the concerned authority.
3. Some PSUs in India have expressed their interest to utilize GATE scores for their recruitment purpose. The Organizations who intend to utilize GATE scores shall make separate advertisement for this purpose in Newspapers etc.

Who Can Appear for GATE?

Eligibility for GATE

Before starting the application process, the candidate must ensure that he/she meets the eligibility criteria of GATE given in Table.

Eligibility Criteria for GATE 2021

Degree/Program	Qualifying Degree/Examination	Description of Eligible Candidates	Expected Year of Completion
B.E. / B.Tech. / B.Pharm.	Bachelor's degree in Engineering / Technology (4 years after 10+2 or 3 years after B.Sc. / Diploma in Engineering / Technology)	Currently in the 3rd year or higher grade or already completed	2022
B. Arch.	Bachelor's degree of Architecture (5- year course) / Naval Architecture (4- year course) / Planning (4- year course)	Currently in the 3rd year or higher grade or already completed	2023 (for 5-year program), 2022 (for 4-year program)
B.Sc. (Research) / B.S.	Bachelor's degree in Science (Post-Diploma/4 years after 10+2)	Currently in the 3rd year or higher grade or already completed	2022
Pharm. D. (after 10+2)	6 years degree program, consisting of internship or residency training, during third year onwards	Currently in the $3^{rd}/4^{th}/5^{th}/6^{th}$ year or already completed	2024
M.B.B.S.	Degree holders of M.B.B.S. and those who are in the $5^{th}/6^{th}/7^{th}$ semester or higher semester of such programme.	5th, 6th, 7th or higher semester or already completed	2022
M. Sc. / M.A. / MCA or equivalent	Master's degree in any branch of Arts/Science/Mathematics/Statistics/ Computer Applications or equivalent	Currently in the first year or higher or already Completed	2022
Int. M.E./ M.Tech. (Post-B.Sc.)	Post-B.Sc Integrated Master's degree programs in Engineering/ Technology (4-year program)	Currently in the 1st/ $2^{nd}/3^{rd}/4^{th}$ year or already completed	2024
Int. M.E./ M.Tech. or Dual Degree (after Diploma or 10+2)	Integrated Master's degree program or Dual Degree program in Engineering/Technology (5-year program)	Currently in the 3rd $/4^{th}/5^{th}$ year or alreadycompleted	2023
B.Sc. / B.A. / B.Com.	Bachelor degree in any branch of Science / Arts / Commerce (3 years program)	Currently in the 3rd year or already completed	2021
Int. M.Sc. / Int. B.S. / M.S.	Integrated M.Sc. or 5-year integrated B.S.-M.S. program	Currently in the 3rd year or higher or already completed	2022
Professional Society Examinations (equivalent to B.E. / B.Tech. / B.Arch.)	B.E./B.Tech./B.Arch. equivalent examinations of Professional Societies, recognized by MoE/UPSC/AICTE (e.g. AMIE by Institution of Engineers-India, AMICE by the Institute of Civil Engineers-India and so on)	Completed Section A or equivalent of such professional courses	NA

In case a candidate has passed one of the qualifying examinations as mentioned above in 2020 or earlier, the candidate has to submit the degree certificate / provisional certificate / course completion certificate / professional certificate / membership certificate issued by the society or institute. In case, the candidate is expected to complete one of the qualifying criteria in 2021 or later as mentioned above, he/she has to submit a certificate from Principal or a copy of marks card for section A of AMIE.

Certificate From Principal

Candidates who have to submit a certificate from their college Principal have to obtain one from his/her institution beforehand and upload the same during the online submission of the application form.

Candidates With Backlogs

Candidates, who have appeared in the final semester/year exam in 2021, but with a backlog (arrears/failed subjects) in any of the papers in their qualifying degree should upload a copy of any one of the mark sheets of the final year,

OR

obtain a declaration from their Principal along with the signature and seal beforehand and upload the same during the online submission of the application form.

GATE Structure

Structure of GATE

GATE 2021 will be conducted on 27 subjects (papers). Table below shows the list of papers and paper codes for GATE 2021. A candidate is allowed to appear in ANY ONE or UP TO TWO papers of the GATE examination. However, note that the combination of TWO papers in which a candidate can appear MUST be from the pre-defined list as given in Table. Also note that for a paper running in multiple sessions, a candidate will be mapped to appear for the examination in one of the sessions ONLY.

List of GATE Papers and Corresponding Codes

GATE Paper	Code	GATE Paper	Code
Aerospace Engineering	AE	Instrumentation Engineering	IN
Agricultural Engineering	AG	Mathematics	MA
Architecture and Planning	AR	Mechanical Engineering	ME
Bio-medical Engineering	BM	Mining Engineering	MN
Biotechnology	BT	Metallurgical Engineering	MT
Civil Engineering	CE	Petroleum Engineering	PE
Chemical Engineering	CH	Physics	PH
Computer Science and Information Technology	CS	Production and Industrial Engineering	PI
Chemistry	CY	Statistics	ST
Electronics and Communication Engineering	EC	Textile Engineering and Fibre Science	TF
Electrical Engineering	EE	Engineering Sciences	XE*
Environmental Science & Engineering	ES	Humanities & Social Sciences	XH**
Ecology and Evolution	EY	Life Sciences	XL***
Geology and Geophysics	GG		

Note: *Environmental Science and Engineering (ES) and Humanities and Social Sciences (XH) are two new papers introduced in GATE 2021.*

*XE Paper Sections	Code	**XH Paper Sections	Code	***XL Paper Sections	Code
Engineering Mathematics (Compulsory) (15 marks)	A	Reasoning and Comprehension (Compulsory) (25 marks)	B1	Chemistry (Compulsory) (25 marks)	P
Any TWO optional Sections (2x35 = 70 marks)		Any ONE optional Section (60 marks)		Any TWO optional Sections (2x30 = 60 marks)	
Fluid Mechanics	B	Economics	C1	Biochemistry	Q
Materials Science	C	English	C2	Botany	R
Solid Mechanics	D	Linguistics	C3	Microbiology	S
Thermodynamics	E	Philosophy	C4	Zoology	T
Polymer Science and Engineering	F	Psychology	C5	Food Technology	U
Food Technology	G	Sociology	C6		
Atmospheric and Oceanic Sciences	H				

*XE (Engineering Sciences), **XH (Humanities & Social Sciences), ***XL (Life Sciences), papers are of general nature and will be comprised of Sections listed in the above table

Note: *Each subject/paper is of total 100 marks. General Aptitude (GA) section of 15 marks is common for all papers. Hence remaining 85 marks are for the respective subject/paper code.*

Combination of Two Papers Allowed to Appear in GATE 2021 (subject to availability of infrastructure and schedule)

Code of The First (Primary) Paper	Codes of Papers Allowed as The Second Paper
AE	XE
AG	ES
AR	CE
BM	BT / XL
BT	BM / XL
CE	AR / ES
CH	CY / PE / XE
CS	MA
CY	CH / XL
EC	IN / PH
EE	IN
ES	AG / CE
EY	XL
GG	MN / PE / PH
IN	EC / EE / PH
MA	CS / PH / ST
ME	XE
MN	GG / XE
MT	PH / XE
PE	CH / GG / XE
PH	EC / GG / IN / MA / MT / ST
PI	XE
ST	MA / PH
TF	XE
XE	AE / CH / ME / MN / MT / PE / PI / TF
XH	----
XL	BM / BT / CY / EY

General Aptitude Questions

All the papers will have a few questions that test the General Aptitude (Language and Analytical Skills), apart from the core subject of the paper.

Duration and Examination Type

All the papers of the GATE 2021 examination will be for 3 hours duration and they consist of 65 questions for a total of 100 marks. Since the examination is an ONLINE computer based test (CBT), at the end of the stipulated time (3-hours), the computer screen will automatically close the examination inhibiting any further action.

Candidates will be permitted to occupy their allotted seats 40 minutes before the scheduled start of the examination. Candidates can login and start reading the instructions 20 minutes before the start of examination. The late login time (if any) recorded by the computer system MUST NOT be beyond 30 minutes from the actual starting time of the examination. Under NO circumstances, will a candidate be permitted to login after 30 minutes from the actual examination starting time. Candidates will NOT be permitted to leave the examination hall before the end of the examination.

Pattern of Question Papers

GATE 2021 may contain questions of THREE different types in all the papers:

(i) **Multiple Choice Questions (MCQ)** carrying 1 or 2 marks each, in all the papers and sections. These questions are objective in nature, and each will have choice of four answers, out of which ONLY ONE choice is correct.

Negative Marking for Wrong Answers: For a wrong answer chosen in a MCQ, there will be negative marking. For 1-mark MCQ, 1/3 mark will be deducted for a wrong answer. Likewise, for 2-mark MCQ, 2/3 mark will be deducted for a wrong answer.

(ii) **Multiple Select Questions (MSQ)** carrying 1 or 2 marks each in all the papers and sections. These questions are objective in nature, and each will have choice of four answers, out of which ONE or MORE than ONE choice(s) is / are correct.

Note: There is NO negative marking for a wrong answer in MSQ questions. However, there is NO partial credit for choosing partially correct combinations of choices or any single wrong choice.

(iii) **Numerical Answer Type (NAT)** Questions carrying 1 or 2 marks each in most of the papers and sections. For these questions, the answer is a signed real number, which needs to be entered by the candidate using the virtual numeric keypad on the monitor (keyboard of the computer will be disabled). No choices will be shown for these types of questions. The answer can be a number such as 10 or –10 (an integer only). The answer may be in decimals as well, for example, 10.1 (one decimal) or 10.01 (two decimals) or –10.001 (three decimals). These questions will be mentioned with, up to which decimal places, the candidates need to present the answer. Also, for some NAT type problems an appropriate range will be considered while evaluating these questions so that the candidate is not unduly penalized due to the usual round-off errors. Candidates are advised to do the rounding off at the end of the calculation (not in between steps). Wherever required and possible, it is better to give NAT answer up to a maximum of three decimal places.

Note: *There is NO negative marking for a wrong answer in NAT questions.*
Also, there is NO partial credit in NAT questions.

Marking Scheme – Distribution of Marks and Questions

General Aptitude (GA) Questions

In all papers, GA questions carry a total of 15 marks. The GA section includes 5 questions carrying 1-mark each (sub-total 5 marks) and 5 questions carrying 2-marks each (sub-total 10 marks).

Question Papers other than GG, XE, XH and XL

These papers would contain 25 questions carrying 1-mark each (sub-total 25 marks) and 30 questions carrying 2-marks each (sub-total 60 marks) consisting of some MCQ type questions, while the remaining may be MSQ and / or NAT questions.

GG (Geology and Geophysics) Paper

Apart from the General Aptitude (GA) section, the GG question paper consists of two parts: Part A and Part B. Part A is compulsory for all the candidates. Part B contains two sections: Section 1 (Geology) and Section 2 (Geophysics). Candidates will have to attempt questions in Part A and questions in either Section 1 or Section 2 of Part B. The choice of Section 1 OR Section 2 of Part B has to be made at the time of online registration in GOAPS. At the examination hall, candidate cannot request for change of section.

Part A consists of 25 questions carrying 1-mark each (sub-total 25 marks and some of these will be MSQ and/or numerical answer type questions while remaining questions will be MCQ type). Either section of Part B (Section 1 and Section 2) consists of 30 questions carrying 2-marks each (sub-total 60 marks and some of these will be MSQ and/or numerical answer type questions while remaining questions will be MCQ type).

XE Paper (Engineering Sciences)

A candidate appearing in the XE paper has to answer the following:

- GA – General Aptitude carrying a total of 15 marks.
- Section A – Engineering Mathematics (Compulsory): This section contains 11 questions carrying a total of 15 marks: 7 questions carrying 1-mark each (sub-total 7 marks), and 4 questions carrying 2-marks each (sub-total 8 marks). Some questions will be of numerical answer type while remaining questions will be MCQ type.
- Any two of XE Sections B to H: The choice of two sections from B to H can be made during the examination after viewing the questions. Only TWO optional sections can be answered at a time. A candidate wishing to change midway of the examination to another optional section must first choose to deselect one of the previously chosen optional sections (B to H). Each of the optional sections of the XE paper (Sections B through H) contains 22 questions carrying a total of 35 marks: 9 questions carrying 1-mark each (sub-total 9 marks) and 13 questions carrying 2-marks each (sub-total 26 marks). Some questions will be of MSQ and/or numerical answer type while remaining questions will be MCQ type.

XH Paper (Humanities and Social Sciences)

A candidate appearing in the XH paper has to answer the following:

- **GA – General Aptitude** carrying a total of 15 marks.
- **Section B1 – Reasoning and Comprehension (Compulsory):** This section contains 15 questions carrying a total of 25 marks: 5 questions carrying 1-mark each (sub-total 5 marks) and 10 questions carrying 2-marks each (sub-total 20 marks). Some questions will be of MSQ and/or numerical answer type while remaining questions will be MCQ type.
- **Any ONE of XH Sections C1 to C6:** The ONE choice of section from C1 to C6 has to be made at the time of online registration in GOAPS. At the examination hall, candidate cannot request for change of section. Each of the optional sections of the XH paper (Sections C1 through C6) contains 40 questions carrying a total of 60 marks: 20 questions carrying 1-mark each (sub-total 20 marks) and 20 questions carrying 2-marks each (sub-total 40 marks). Some questions will be of MSQ and/or numerical answer type while remaining questions will be MCQ type.

XL Paper (Life Sciences)

A candidate appearing in the XL paper has to answer the following:

- **GA – General Aptitude** carrying a total of 15 marks.
- **Section P–Chemistry (Compulsory):** This section contains 15 questions carrying a total of 25 marks: 5 questions carrying 1-mark each (sub-total 5 marks) and 10 questions carrying 2-marks each (sub-total 20 marks). Some questions will be of MSQ and/or numerical answer type while remaining questions will be MCQ type.
- **Any two of XL Sections Q to U:** The choice of two sections from Q to U can be made during the examination after viewing the questions. Only TWO optional sections can be answered at a time. A candidate wishing to change midway of the examination to another optional section must first choose to deselect one of the previously chosen optional sections (Q to U). Each of the optional sections of the XL paper (Sections Q through U) contains 20 questions carrying a total of 30 marks: 10 questions carrying 1-mark each (sub-total 10 marks) and 10 questions carrying 2-marks each (sub-total 20 marks). Some questions will be of MSQ and/or numerical answer type while remaining questions will be MCQ type.

GATE Score

After the evaluation of the answers, the actual (raw) marks obtained by a candidate will be considered for computing the GATE Score. For multi-session papers (subjects), raw marks obtained by the candidates in different sessions will be converted to Normalized marks for that particular subject. Thus, raw marks (for single session papers) or normalized marks (for multi-session papers) will be used for computing the GATE Score, based on the qualifying marks.

Calculation of Normalized Marks for Multi-Session Papers

In GATE 2021 examination, some papers may be conducted in multi-sessions. Hence, for these papers, a suitable normalization is applied to take into account any variation in the difficulty levels of the question papers across different sessions. The normalization is done based on the fundamental assumption that "in all multi-session GATE papers, the distribution of abilities of candidates is the same across all the sessions". This assumption is justified since the number of candidates appearing in multi-session papers in GATE 2021 is large and the procedure for allocation of session to candidates is random. Further, it is also ensured that for the same multi-session paper, the number of candidates allotted in each session is of the same order of magnitude.

Based on the above, and considering various normalization methods, the committee arrived at the following formula for calculating the normalized marks for the multi-session papers.

Nsormalization mark of jth candidate in the ith session $\widehat{M_{ij}}$ is given by

$$\widehat{M_{ij}} = \frac{\bar{M}_t^g - M_q^g}{\bar{M}_{ti} - M_{iq}} (M_{ij} - M_{iq}) + M_q^g$$

where

M_{ij} : is the actual marks obtained by the jth candidate in i^{th} session

$\bar{M}_t^g$: is the average marks of the top 0.1% of the candidates considering all sessions

M_q^g : is the sum of mean and standard deviation marks of the candidates in the paper considering all sessions

$\bar{M}_{ti}$: is the average marks of the top 0.1% of the candidates in the ith session

M_{iq} : is the sum of the mean marks and standard deviation of the ith session

Calculation of GATE Score for All Papers

For all papers for which there is only one session, actual marks obtained by the candidates will be used for calculating the GATE 2021 Score. For papers in multi-sessions, normalized marks will be calculated corresponding to the raw marks obtained by a candidate and the GATE 2021 Score will be calculated based on the normalized marks.

The GATE 2021 score will be computed using the formula given below.

$$\text{GATE Score} = S_q + (S_t - S_q)\frac{(M - M_q)}{(\bar{M}_t - M_q)}$$

where

- M : marks obtained by the candidate (actual marks for single session papers and normalized marks for multi-session papers)
- M_q : is the qualifying marks for general category candidate in the paper
- $\bar{M}_t$: is the mean of marks of top 0.1% or top 10 (whichever is larger) of the candidates who appeared in the paper (in case of multi-session papers including all sessions)
- S_q : 350, is the score assigned to M_q
- S_t : 900, is the score assigned to $\bar{M}_t$

In the GATE 2021 the qualifying marks (M_q) for general category student in each subject will be 25 marks (out of 100) or $\mu + \sigma$, whichever is larger. Here μ is the mean and σ is the standard deviation of marks of all the candidates who appeared in the paper.

After the declaration of results, GATE Scorecards can be downloaded by the GATE qualified candidates ONLY.

The GATE 2021 Committee has the authority to decide the qualifying mark/score for each GATE paper. In case of any claim or dispute with respect to GATE 2021 examination or score, the Courts and Tribunals in Mumbai alone will have the exclusive jurisdiction to entertain and settle them.

GATE Syllabus

Section 1: Engineering Mathematics

Linear Algebra: Vector space, basis, linear dependence and independence, matrix algebra, eigenvalues and eigenvectors, rank, solution of linear equations- existence and uniqueness.

Calculus: Mean value theorems, theorems of integral calculus, evaluation of definite and improper integrals, partial derivatives, maxima and minima, multiple integrals, line, surface and volume integrals, Taylor series.

Differential Equations: First order equations (linear and nonlinear), higher order linear differential equations, Cauchy's and Euler's equations, methods of solution using variation of parameters, complementary function and particular integral, partial differential equations, variable separable method, initial and boundary value problems.

Vector Analysis: Vectors in plane and space, vector operations, gradient, divergence and curl, Gauss's, Green's and Stokes' theorems.

Complex Analysis: Analytic functions, Cauchy's integral theorem, Cauchy's integral formula, sequences, series, convergence tests, Taylor and Laurent series, residue theorem.

Probability and Statistics: Mean, median, mode, standard deviation, combinatorial probability, probability distributions, binomial distribution, Poisson distribution, exponential distribution, normal distribution, joint and conditional probability.

Section 2: Networks, Signals and Systems

Circuit analysis: Node and mesh analysis, superposition, Thevenin's theorem, Norton's theorem, reciprocity.

Sinusoidal steady state analysis: phasors, complex power, maximum power transfer.

Time and frequency domain analysis of linear circuits: RL, RC and RLC circuits, solution of network equations using Laplace transform.

Linear 2-port network parameters, wye-delta transformation.

Continuous-time signals: Fourier series and Fourier transform, sampling theorem and applications.

Discrete-time signals: DTFT, DFT, z-transform, discrete-time processing of continuous-time signals. LTI systems: definition and properties, causality, stability, impulse response, convolution, poles and zeroes, frequency response, group delay, phase delay.

Section 3: Electronic Devices

Energy bands in intrinsic and extrinsic semiconductors, equilibrium carrier concentration, direct and indirect band-gap semiconductors.

Carrier transport: diffusion current, drift current, mobility and resistivity, generation and recombination of carriers, Poisson and continuity equations.

P-N junction, Zener diode, BJT, MOS capacitor, MOSFET, LED, photo diode and solar cell.

Section 4: Analog Circuits

Diode circuits: clipping, clamping and rectifiers.

BJT and MOSFET amplifiers: biasing, ac coupling, small signal analysis, frequency response.

Current mirrors and differential amplifiers.

Op-amp circuits: Amplifiers, summers, differentiators, integrators, active filters, Schmitt triggers and oscillators.

Section 5: Digital Circuits

Number representations: binary, integer and floating-point- numbers.

Combinatorial circuits: Boolean algebra, minimization of functions using Boolean identities and Karnaugh map, logic gates and their static CMOS implementations, arithmetic circuits, code converters, multiplexers, decoders.

Sequential circuits: latches and flip-flops, counters, shift-registers, finite state machines, propagation delay, setup and hold time, critical path delay.

Data converters: sample and hold circuits, ADCs and DACs.

Semiconductor memories: ROM, SRAM, DRAM.

Computer organization: Machine instructions and addressing modes, ALU, data-path and control unit, instruction pipelining.

Section 6: Control Systems

Basic control system components; Feedback principle; Transfer function; Block diagram representation; Signal flow graph; Transient and steady-state analysis of LTI systems; Frequency response; Routh-Hurwitz and Nyquist stability criteria; Bode and root-locus plots; Lag, lead and lag-lead compensation; State variable model and solution of state equation of LTI systems.

Section 7: Communications

Random processes: autocorrelation and power spectral density, properties of white noise, filtering of random signals through LTI systems.

Analog communications: amplitude modulation and demodulation, angle modulation and demodulation, spectra of AM and FM, superheterodyne receivers.

Information theory: entropy, mutual information and channel capacity theorem.

Digital communications: PCM, DPCM, digital modulation schemes (ASK, PSK, FSK, QAM), bandwidth, inter-symbol interference, MAP, ML detection, matched filter receiver, SNR and BER.

Fundamentals of error correction, Hamming codes, CRC.

Section 8: Electromagnetics

Maxwell's equations: differential and integral forms and their interpretation, boundary conditions, wave equation, Poynting vector.

Plane waves and properties: reflection and refraction, polarization, phase and group velocity, propagation through various media, skin depth.

Transmission lines: equations, characteristic impedance, impedance matching, impedance transformation, S-parameters, Smith chart.

Rectangular and circular waveguides, light propagation in optical fibers, dipole and monopole antennas, linear antenna arrays.

Chapter-Wise Analysis

GATE PAPERS (Electronics & Communication Engineering)

Subject	2011	2012	2013	2014	2015	2016	2017	2018	2019	2020	2021
1. Engineering Maths											
1 Mark Questions	3	4	4	3	4	5	4	5	6	4	5
2 Marks Questions	4	7	6	4	3	4	5	3	3	3	3
Total Marks	**11**	**18**	**16**	**11**	**10**	**13**	**14**	**11**	**12**	**10**	**11**
2. Network Theory											
1 Mark Questions	2	1	3	3	3	5	1	1	1	3	1
2 Marks Questions	3	4	2	2	3	3	2	3	4	2	5
Total Marks	**8**	**9**	**7**	**7**	**9**	**11**	**5**	**7**	**9**	**7**	**11**
3. Signals & Systems											
1 Mark Questions	1	2	0	4	2	1	3	2	3	2	2
2 Marks Questions	4	3	1	4	4	2	3	2	1	2	3
Total Marks	**9**	**8**	**2**	**12**	**10**	**5**	**9**	**6**	**5**	**6**	**8**
4. Electronic Devices & Circuits											
1 Mark Questions	2	4	6	4	4	3	5	3	3	2	3
2 Marks Questions	4	4	4	2	3	3	3	2	2	4	2
Total Marks	**10**	**12**	**14**	**8**	**10**	**9**	**11**	**7**	**7**	**10**	**7**
5. Analog Circuits											
1 Mark Questions	3	0	1	2	2	2	1	2	3	3	3
2 Marks Questions	3	2	6	5	3	3	4	5	5	5	5
Total Marks	**9**	**4**	**13**	**12**	**8**	**8**	**9**	**12**	**13**	**13**	**13**
6. Digital Circuits & Microprocessor											
1 Mark Questions	2	5	3	3	4	2	3	3	2	3	3
2 Marks Questions	3	1	3	5	4	3	4	4	1	3	2
Total Marks	**8**	**7**	**9**	**13**	**12**	**8**	**11**	**11**	**4**	**9**	**7**
7. Control Systems											
1 Mark Questions	6	1	4	3	2	2	3	3	1	2	2
2 Marks Questions	4	3	4	4	3	3	4	5	4	4	1
Total Marks	**14**	**7**	**12**	**11**	**8**	**8**	**11**	**13**	**9**	**10**	**4**

Subject	2011	2012	2013	2014	2015	2016	2017	2018	2019	2020	2021
8. Communication systems 1 Mark Questions	1	2	3	2	2	2	4	4	3	3	4
2 Marks Questions	1	2	0	2	4	4	2	3	6	4	5
Total Marks	**3**	**6**	**3**	**6**	**10**	**10**	**8**	**10**	**15**	**11**	**14**
9. Electromagnetic Theory 1 Mark Questions	3	6	1	1	2	3	1	2	3	3	2
2 Marks Questions	4	4	3	2	3	5	3	3	4	3	3
Total Marks	**11**	**14**	**7**	**5**	**8**	**13**	**7**	**8**	**11**	**9**	**8**
10. Computer-organization 1 Mark Questions	–	–	–	–	–	–	–	–	–	–	0
2 Marks Questions	–	–	–	–	–	–	–	–	–	–	1
Total Marks	–	–	–	–	–	–	–	–	–	–	**2**
11. General Aptitude 1 Mark Questions	5	5	5	5	5	5	5	5	5	5	5
2 Marks Questions	5	5	5	5	5	5	5	5	5	5	5
Total Marks	**15**	**15**	**15**	**15**	**15**	**15**	**15**	**15**	**15**	**15**	**15**

SOLVED PAPER - 2021

INSTRUCTIONS

1. Total of 65 questions carrying 100 marks, out of which 10 questions carrying a total of 15 marks are in General Aptitude (GA)
2. The Engineering Mathematics will carry around **15% of the total marks,** the General Aptitude section will carry **15% of the total marks** and the **remaining 70% of the total marks.**
3. **Types of Questions**
   (*a*) **Multiple Choice Questions (MCQ)** carrying 1 or 2 marks each in all papers and sections. These questions are objective in nature, and each will have a choice of four options, out of which the candidate has to mark the correct answer(s).
   (*b*) **Numerical Answer Questions** of 1 or 2 marks each in all papers and sections. For these questions the answer is a real number, to be entered by the candidate using the virtual keypad. No choices will be shown for these type of questions.
   (*c*) **Multiple Select Questions (MSQ)** carrying 1 or 2 marks each in all the papers and sections. These questions are objective in nature, and each will have choice of four answers, out of which ONE or MORE than ONE choice(s) is / are correct.
4. For **1-mark** multiple-choice questions, **1/3 marks** will be deducted for a wrong answer. Likewise, for **2-marks** multiple-choice questions, **2/3** marks will be deducted for a wrong answer. There is no negative marking for numerical answer type questions & multiple selection questions.

Chapter-Wise Analysis

#	Chapters	Marks		
		1	2	Total
1	General Aptitude	5	5	15
2	Engineering Mathematics	5	3	11
3	Network Theory	1	5	11
4	Signals and Systems	2	3	8
5	Electronic Devices & Circuits	3	2	7
6	Analog Circuits	3	5	13

#	Chapters	Marks		
		1	2	Total
7	Digital Circuits & Microprocessor	3	2	7
8	Control Systems	2	1	4
9	Communication Systems	4	5	14
10	Electromagnetic Theory	2	3	8
11	Computer-organization	-	1	2

GENERAL APTITUDE (GA)

Q.1 to Q.5: Multiple Choice Questions (MCQ), carry ONE mark each.

1. The current population of a city is 11,02,500. If it has been increasing at the rate of 5% per annum, what was its population 2 years ago?

(a) 9,92,500 (b) 9,95,006

(c) 10,00,000 (d) 12,51,506

2. p and q are positive integers and $\frac{p}{q}+\frac{q}{p}=3$, then,

$\frac{p^2}{q^2}+\frac{q^2}{p^2}=$

(a) 3 (b) 7

(c) 9 (d) 11

3.

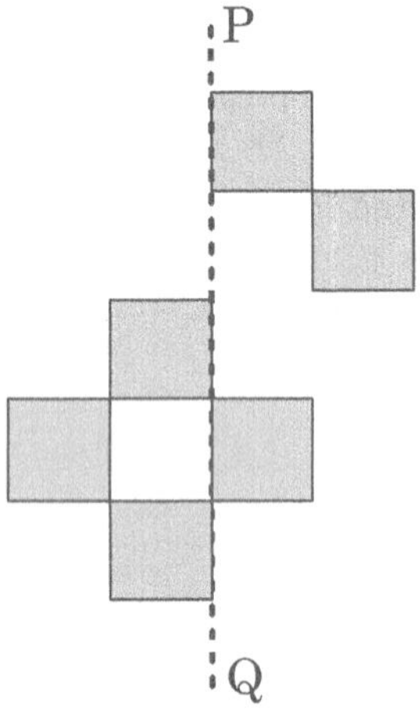

The least number of squares that must be added so that the line P-Q becomes the line of symmetry is ________

(a) 4 (b) 3

(c) 6 (d) 7

4. *Nostalgia* is to *anticipation* as _____ is to ______

Which one of the following options maintains a similar logical relation in the above sentence?

(a) Present, past (b) Future, past

(c) Past, future (d) Future, present

5. Consider the following sentences:

(i) I woke up from sleep.

(ii) I woked up from sleep.

(iii) I was woken up from sleep.

(iv) I was wokened up from sleep.

Which of the above sentences are grammatically CORRECT?

(a) (i) and (ii) (b) (i) and (iii)

(c) (ii) and (iii) (d) (i) and (iv)

Q.6 to Q.10: Multiple Choice Question (MCQ), carry TWO marks each.

6. Given below are two statements and two conclusions.

Statement 1 : All purple are green.

Statement 2 : All black are green.

Conclusion I : Some black are purple.

Conclusion II : No black is purple.

Based on the above statements and conclusions, which one of the following options is logically CORRECT?

(a) Only conclusion I is correct.

(b) Only conclusion II is correct.

(c) Either conclusion I or II is correct.

(d) Both conclusion I and II are correct.

7. Computers are ubiquitous. They are used to improve efficiency in almost all fields from agriculture to space exploration. Artificial intelligence (AI) is currently a hot topic. AI enables computers to learn, given enough training data. For humans, sitting in front of a computer for long hours can lead to health issues.

Which of the following can be deduced from the above passage?

(i) Nowadays, computers are present in almost all places.

(ii) Computers cannot be used for solving problems in engineering.

(iii) For humans, there are both positive and negative effects of using computers.

(iv) Artificial intelligence can be done without data.

(a) (ii) and (iii)

(b) (ii) and (iv)

(c) (i), (iii) and (iv)

(d) (i) and (iii)

8. Consider a square sheet of side 1 unit. In the first step, it is cut along the main diagonal to get two triangles. In the next step, one of the cut triangles is revolved about its short edge to form a solid cone. The volume of the resulting cone, in cubic units, is ________

(a) $\frac{\pi}{3}$

(b) $\frac{2\pi}{3}$

(c) $\frac{3\pi}{2}$

(d) 3π

9.

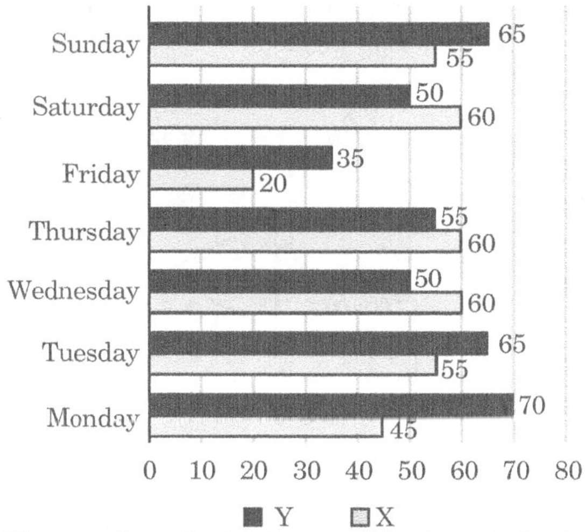

The number of minutes spent by two students, X and Y, exercising every day in a given week are shown in the bar chart above.

The number of days in the given week in which one of the students spent a minimum of 10% more than the other student, on a given day, is

(a) 4 (b) 5

(c) 6 (d) 7

10.

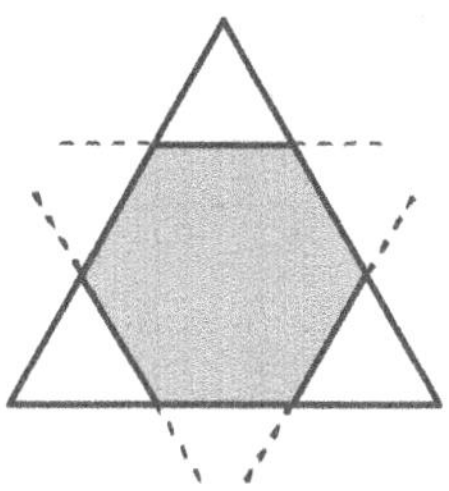

Corners are cut from an equilateral triangle to produce a regular convex hexagon as shown in the figure above.

The ratio of the area of the regular convex hexagon to the area of the original equilateral triangle is

(a) 2 : 3 (b) 3 : 4

(c) 4 : 5 (d) 5 : 6

ELECTRONICS & COMMUNICATION ENGINEERING

Q.1 - Q-15: Multiple Choice Questions (MCQ), carry ONE mark each.

1. The vector function $F(r) = -x\hat{i} + y\hat{j}$ is defined over a circular arc C shown in the figure.

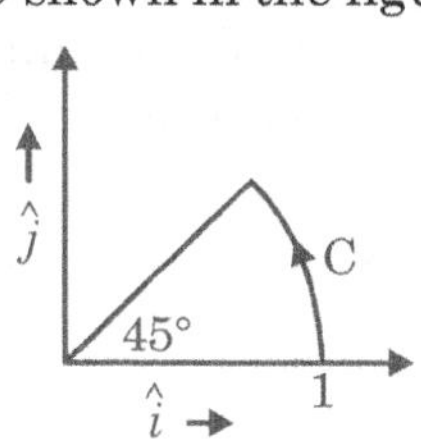

The line integral of $\int_C F(r).\,dr$ is

(a) $\frac{1}{2}$ (b) $\frac{1}{4}$

(c) $\frac{1}{6}$ (d) $\frac{1}{3}$

2. Consider the differential equation given below.

$$\frac{dy}{dx} + \frac{x}{1-x^2}y = x\sqrt{y}$$

The integrating factor of the differential equation is

(a) $\left(1-x^2\right)^{-\frac{3}{4}}$ (b) $\left(1-x^2\right)^{-\frac{1}{4}}$

(c) $\left(1-x^2\right)^{-\frac{3}{2}}$ (d) $\left(1-x^2\right)^{-\frac{1}{2}}$

3. Two continuous random variables X and Y are related as

$$Y = 2X + 3$$

Let σ_X^2 and σ_Y^2 denote the variances of X and Y, respectively. The variances are related as

(a) $\sigma_Y^2 = 2\sigma_X^2$ (b) $\sigma_Y^2 = 4\sigma_X^2$

(c) $\sigma_Y^2 = 5\sigma_X^2$ (d) $\sigma_Y^2 = 25\sigma_X^2$

4. Consider a real-valued base-band signal $x(t)$, band limited to 10 kHz. The Nyquist rate for the signal $y(t) = x(t)\, x\left(1+\frac{t}{2}\right)$ is

(a) 15 kHz (b) 30 kHz

(c) 60 kHz (d) 20 kHz

5. Consider two 16-point sequences $x[n]$ and $h[n]$. Let the linear convolution of $x[n]$ and $h[n]$ be denoted by $y[n]$, while $z[n]$ denotes the 16-point inverse discrete Fourier transform (IDFT) of the product of the 16-point DFTs of $x[n]$ and $h[n]$. The value(s) of k for which $z[k] = y[k]$ is/are

(a) $k = 0, 1, 2, ..., 15$ (b) $k = 0$

(c) $k = 15$ (d) $k = 0$ and $k = 15$

6. A bar of silicon is doped with boron concentration of 10^{16} cm^{-3} and assumed to be fully ionized. It is exposed to light such that electron-hole pairs are generated throughout the volume of the bar at the rate of 10^{20} cm^{-3}s^{-1}. If the recombination lifetime is 100 μs, intrinsic carrier concentration of silicon is 10^{10} cm^{-3} and assuming 100% ionization of boron, then the approximate product of steady-state electron and hole concentrations due to this light exposure is

(a) 10^{20} cm^{-6} (b) 2×10^{20} cm^{-6}

(c) 10^{32} cm^{-6} (d) 2×10^{32} cm^{-6}

7. The energy band diagram of a p-type semiconductor bar of length L under equilibrium condition (*i.e.,* the Fermi energy level E_F is constant) is shown in the figure. The valance band E_V is sloped since doping is non-uniform along the bar. The difference between the energy levels of the valence band at the two edges of the bar is Δ.

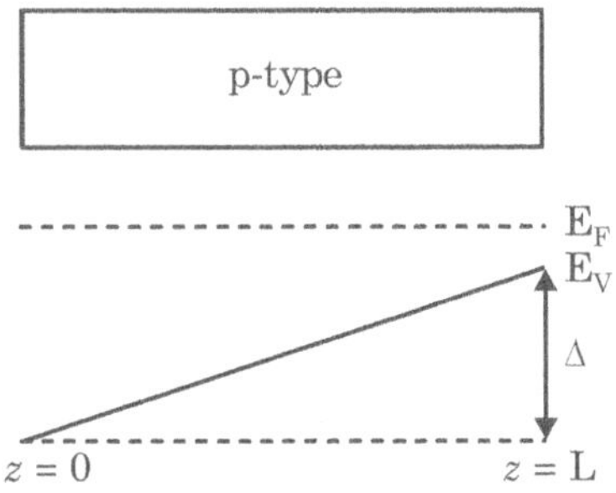

If the charge of an electron is q, then the magnitude of the electric field developed inside this semiconductor bar is

(a) $\frac{\Delta}{qL}$ (b) $\frac{2\Delta}{qL}$

(c) $\frac{\Delta}{2qL}$ (d) $\frac{3\Delta}{2qL}$

8. In the circuit shown in the figure, the transistors M_1 and M_2 are operating in saturation. The channel length modulation coefficients of both the transistors are non-zero. The transconductance of the MOSFETs M_1 and M_2 are g_{m1} and g_{m2}, respectively, and the internal resistance of the MOSFETs M_1 and M_2 are r_{01} and r_{02}, respectively.

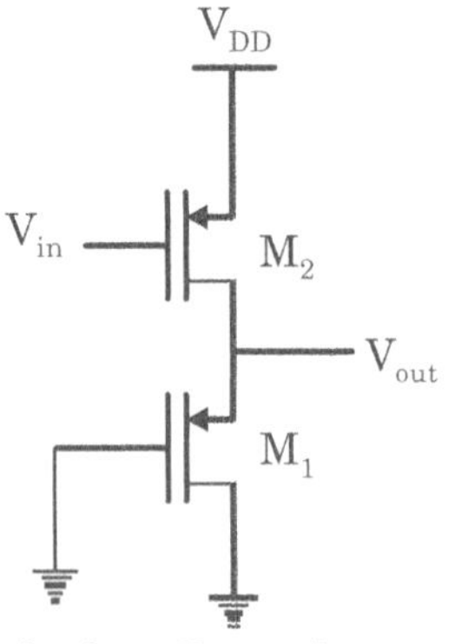

Ignoring the body effect, the ac small signal voltage gain $\left(\frac{\partial V_{out}}{\partial V_{in}}\right)$ of the circuit is

(a) $-g_{m2}(r_{01} \| r_{02})$

(b) $-g_{m2}\left(\frac{1}{g_{m1}} \| r_{02}\right)$

(c) $-g_{m1}\left(\frac{1}{g_{m2}} \| r_{01} \| r_{02}\right)$

(d) $-g_{m2}\left(\frac{1}{g_{m1}} \| r_{01} \| r_{02}\right)$

9. For the circuit with an ideal OPAMP shown in the figure, V_{REF} is fixed.

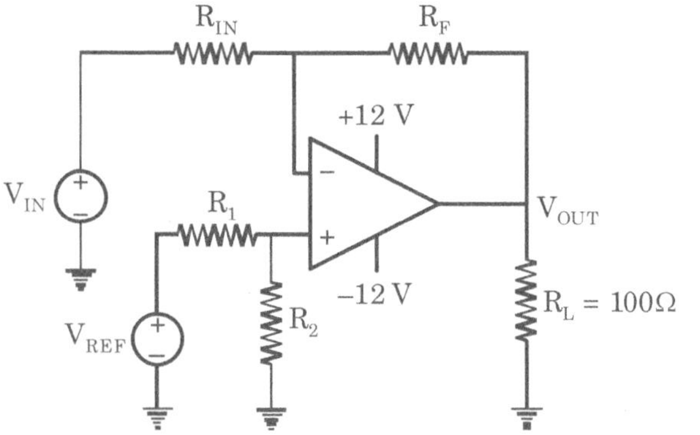

If $V_{OUT} = 1$ volt for $V_{IN} = 0.1$ volt and $V_{OUT} = 6$ volt for $V_{IN} = 1$ volt, where V_{OUT} is measured across R_L connected at the output of this OPAMP, the value of $\frac{R_F}{R_{IN}}$ is

(a) 3.285 (b) 2.860

(c) 3.825 (d) 5.555

10. Consider the circuit with an ideal OPAMP shown in the figure.

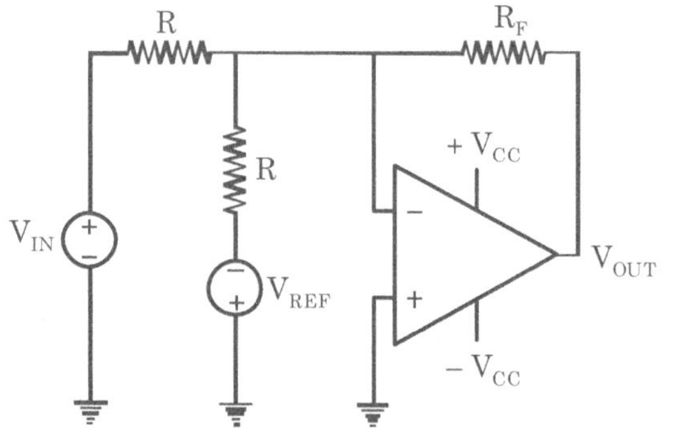

Assuming $|V_{IN}| << |V_{CC}|$ and $|V_{REF}| << |V_{CC}|$, the condition at which V_{OUT} equals to zero is

(a) $V_{IN} = V_{REF}$ (b) $V_{IN} = 0.5\ V_{REF}$

(c) $V_{IN} = 2\ V_{REF}$ (d) $V_{IN} = 2 + V_{REF}$

11. If $(1235)_x = (3033)_y$, where x and y indicate the bases of the corresponding numbers, then

(a) $x = 7$ and $y = 5$ (b) $x = 8$ and $y = 6$

(c) $x = 6$ and $y = 4$ (d) $x = 9$ and $y = 7$

12. Addressing of a 32K × 16 memory is realized using a single decoder. The minimum number of AND gates required for the decoder is

(a) 2^8

(b) 2^{32}

(c) 2^{15}

(d) 2^{19}

13. The block diagram of a feedback control system is shown in the figure.

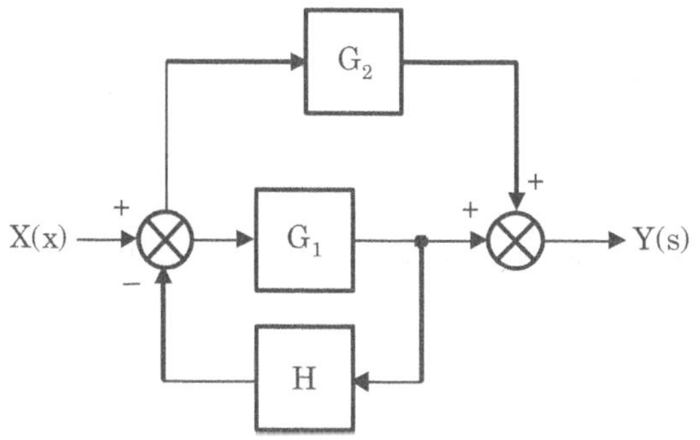

The transfer function $\frac{Y(s)}{X(s)}$ of the system is

(a) $\frac{G_1 + G_2 + G_1G_2H}{1 + G_1H}$ (b) $\frac{G_1 + G_2}{1 + G_1H + G_2H}$

(c) $\frac{G_1 + G_2}{1 + G_1H}$ (d) $\frac{G_1 + G_2 + G_1G_2H}{1 + G_1H + G_2H}$

14. The complete Nyquist plot of the open-loop transfer function G(*s*)H(*s*) of a feedback control system is shown in the figure.

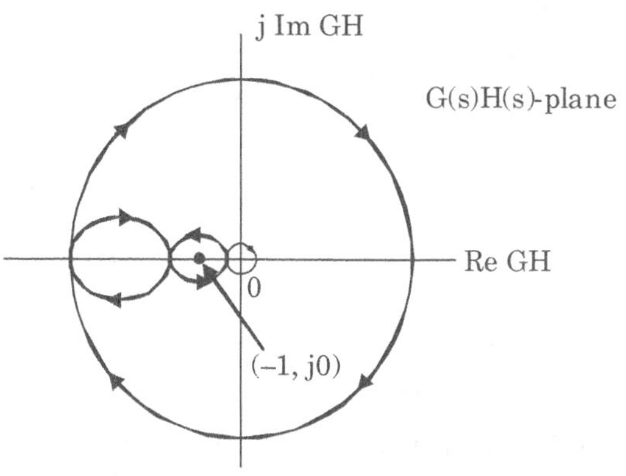

If G(*s*)H(*s*) has one zero in the right-half of the *s*-plane, the number of poles that the closed-loop system will have in the right-half of the s-plane is

(a) 0 (b) 1

(c) 4 (d) 3

15. Consider a rectangular coordinate system (*x, y, z*) with unit vectors a_x, a_y, and a_z. A plane wave traveling in the region $z \geq 0$ with electric field vector $E = 10 \cos(2 \times 10^8 t + \beta z) a_y$ is incident normally on the plane at $z = 0$, where β is the phase constant. The region $z \geq 0$ is in free space and the region $z < 0$ is filled with a lossless medium (permittivity $\varepsilon = \varepsilon_0$, permeability $\mu = 4\mu_0$, where $\varepsilon_0 = 8.85 \times 10^{-12}$ F/m and $\mu_0 = 4\pi \times 10^{-7}$ H/m). The value of the reflection coefficient is

(a) $\frac{1}{3}$ (b) $\frac{3}{5}$

(c) $\frac{2}{5}$ (d) $\frac{2}{3}$

Q. 16 to Q. 25 : Numerical Answer Type (NAT), carry ONE mark each.

16. If the vectors (1. 0, –1. 0, 2. 0), (7. 0, 3. 0, *x*) and (2. 0, 3. 0, 1. 0) in $\mathbb{R}^3$ are linearly dependent, the value of *x* is ________

17. Consider the vector field $F = a_x(4y - c_1 z) + a_y(4x + 2z) + a_z(2y + z)$ in a rectangular coordinate system (*x, y, z*) with unit vectors a_x, a_y, and a_z. If the field F is irrotational (conservative), then the constant c_1 (in integer) is ________

18. Consider the circuit shown in the figure.

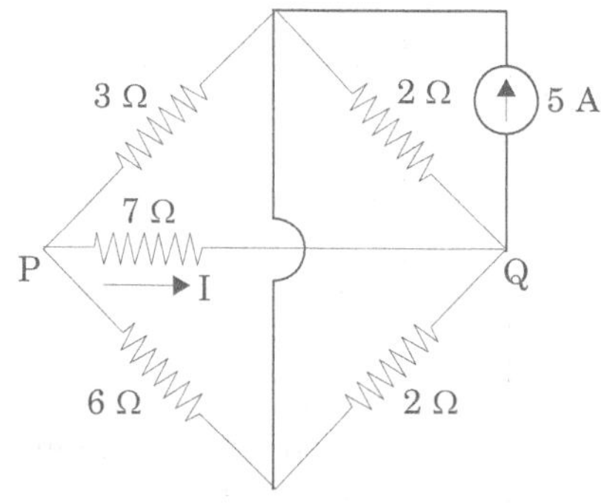

The current I flowing through the 7Ω resistor between P and Q (*rounded off to one decimal place*) is __________ A.

19. Consider the circuit shown in the figure.

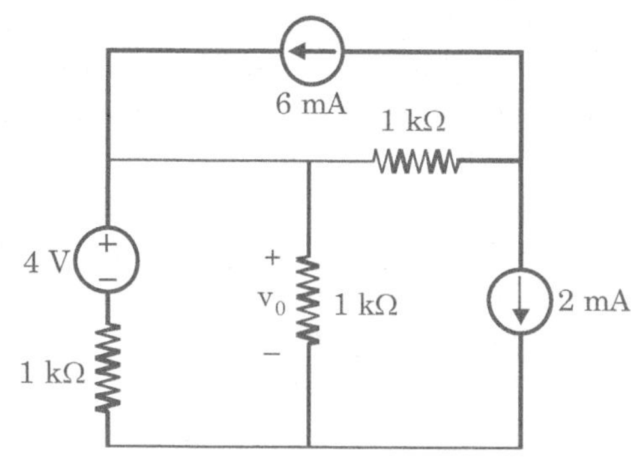

The value of v_0 (rounded off to one decimal place) is __________ V.

20. An 8-bit unipolar (all analog output values are positive) digital-to-analog converter (DAC) has a full-scale voltage range from 0 V to 7. 68 V. If the digital input code is 10010110 (the leftmost bit is MSB), then the analog output voltage of the DAC (*rounded off to one decimal place*) is ________V.

21. The autocorrelation function $R_X(\tau)$ of a wide-sense stationary random process X(*t*) is shown in the figure.

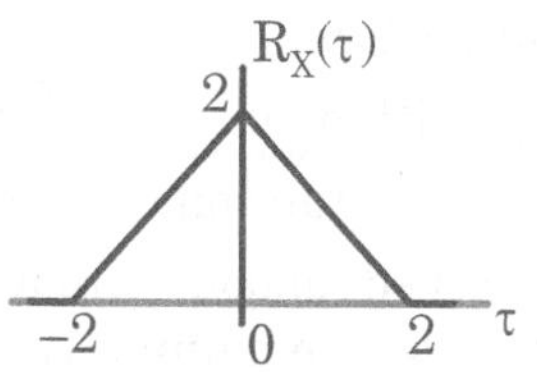

The average power of X(*t*) is __________

22. Consider a carrier signal which is amplitude modulated by a single-tone sinusoidal message signal with a modulation index of 50%. If the carrier and one of the sidebands are suppressed in the modulated signal, the percentage of power saved (*rounded off to one decimal place*) is ______

23. A speech signal, band limited to 4 kHz, is sampled at 1.25 times the Nyquist rate. The speech samples, assumed to be statistically independent and uniformly distributed in the range – 5 V to + 5 V, are subsequently quantized in an 8-bit uniform quantizer and then transmitted over a voice-grade AWGN telephone channel. If the ratio of transmitted signal power to channel noise power is 26 dB, the minimum channel bandwidth required to ensure reliable transmission of the signal with arbitrarily small probability of transmission error (*rounded off to two decimal places*) is ________ kHz.

24. A 4 kHz sinusoidal message signal having amplitude 4 V is fed to a delta modulator (DM) operating at a sampling rate of 32 kHz. The minimum step size required to avoid slope overload noise in the DM (*rounded off to two decimal places*) is ________ V.

25. The refractive indices of the core and cladding of an optical fiber are 1.50 and 1. 48, respectively. The critical propagation angle, which is defined as the maximum angle that the light beam makes with the axis of the optical fiber to achieve the total internal reflection, (*rounded off to two decimal places*) is ________ degree.

Q. 26 to Q. 35 : Multiple Choice Question (MCQ), carry TWO marks each.

26. Consider the integral

$$\oint_C \frac{\sin(x)}{x^2(x^2+4)}dx$$

where C is a counter-clockwise oriented circle defined as $|x - i| = 2$. The value of the integral is

(a) $-\frac{\pi}{8}\sin(2i)$ (b) $\frac{\pi}{8}\sin(2i)$

(c) $-\frac{\pi}{4}\sin(2i)$ (d) $\frac{\pi}{4}\sin(2i)$

27. A box contains the following three coins.

I. A fair coin with head on one face and tail on the other face.

II. A coin with heads on both the faces.

III. A coin with tails on both the faces.

A coin is picked randomly from the box and tossed. Out of the two remaining coins in the box, one coin is then picked randomly and tossed. If the first toss results in a head, the probability of getting a head in the second toss is

(a) $\frac{2}{5}$ (b) $\frac{1}{3}$

(c) $\frac{1}{2}$ (d) $\frac{2}{3}$

28. The switch in the circuit in the figure is in position P for a long time and then moved to position Q at time t = 0.

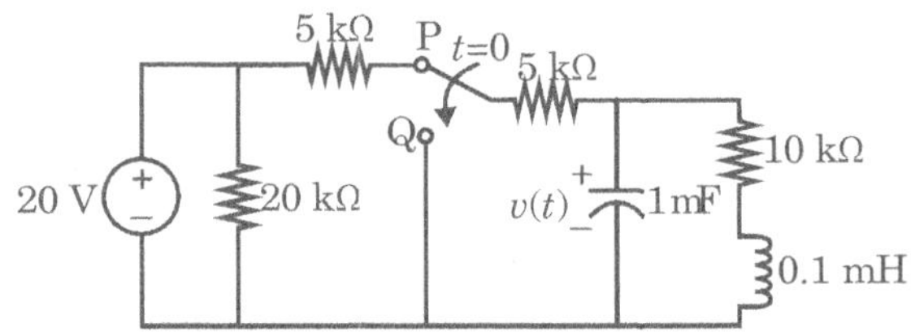

The value of $\frac{dv(t)}{dt}$ at $t = 0^+$ is

(a) 0 V/s (b) 3 V/s

(c) – 3 V/s (d) – 5 V/s

29. Consider the two-port network shown in the figure.

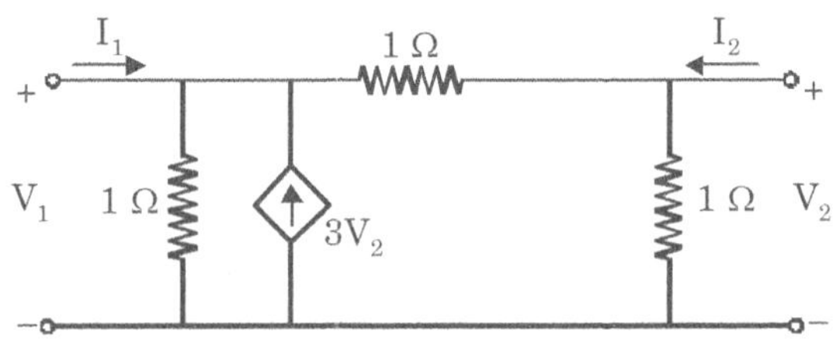

The admittance parameters, in siemens, are

(a) $y_{11} = 2, y_{12} = -4, y_{21} = -4, y_{22} = 2$

(b) $y_{11} = 1, y_{12} = -2, y_{21} = -1, y_{22} = 3$

(c) $y_{11} = 2, y_{12} = -4, y_{21} = -1, y_{22} = 2$

(d) $y_{11} = 2, y_{12} = -4, y_{21} = -4, y_{22} = 3$

30. For an n-channel silicon MOSFET with 10 nm gate oxide thickness, the substrate sensitivity $\frac{\partial V_T}{\partial |V_{BS}|}$ is found to be 50 mV/V at a substrate voltage $|V_{BS}| = 2$ V, where V_T is the threshold voltage of the MOSFET. Assume that, $|V_{BS}| >> 2\Phi_B$, where $q\Phi_B$ is the separation between the Fermi energy level E_F and the intrinsic level E_i in the bulk. Parameters given are

Electron charge (q) = 1.6×10^{-19} C

Vacuum permittivity (ε_0) = 8.85×10^{-12} F/m

Relative permittivity of silicon (ε_{Si}) = 12

Relative permittivity of oxide (ε_{ox}) = 4

The doping concentration of the substrate is

(a) 7.37×10^{15} cm^{-3} (b) 4.37×10^{15} cm^{-3}

(c) 2.37×10^{15} cm^{-3} (d) 9.37×10^{15} cm^{-3}

31. The propagation delays of the XOR gate, AND gate and multiplexer (MUX) in the circuit shown in the figure are 4 ns, 2 ns and 1 ns, respectively.

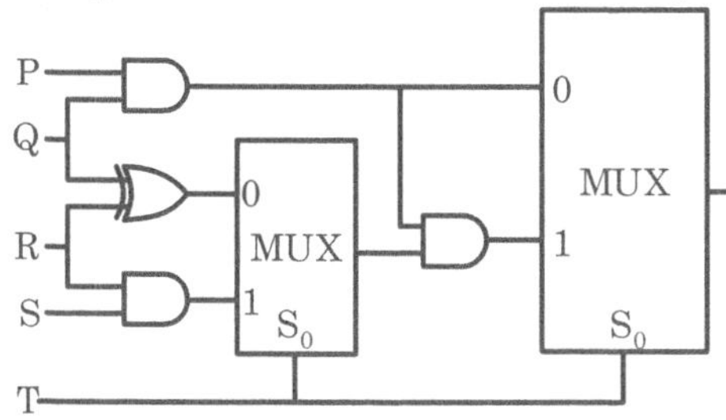

If all the inputs P, Q, R, S and T are applied simultaneously and held constant, the maximum propagation delay of the circuit is

(a) 3 ns (b) 5 ns

(c) 6 ns (d) 7 ns

32. The content of the registers are R_1 = 25 H, R_2 = 30 H and R_3 = 40 H. The following machine instructions are executed.

PUSH{R_1}
PUSH{R_2}
PUSH{R_3}
POP{R_1}
POP{R_2}
POP{R_3}

After execution, the content of registers R_1, R_2, R_3 are

(a) R_1 = 40H, R_2 = 30H, R_3 = 25H

(b) R_1 = 25H, R_2 = 30H, R_3 = 40H

(c) R_1 = 30H, R_2 = 40H, R_3 = 25H

(d) R_1 = 40H, R_2 = 25H, R_3 = 30H

33. The electrical system shown in the figure converts input source current $i_s(t)$ to output voltage $v_0(t)$.

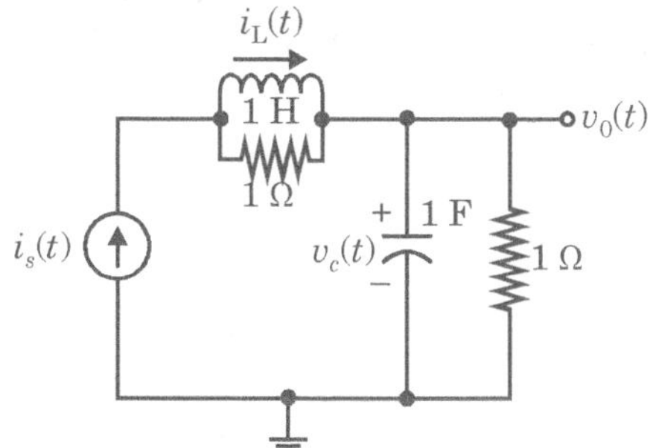

Current $i_L(t)$ in the inductor and voltage $v_C(t)$ across the capacitor are taken as the state variables, both assumed to be initially equal to zero, i.e., $i_L(0) = 0$ and $v_C(0) = 0$. The system is

(a) completely state controllable as well as completely observable

(b) completely state controllable but not observable

(c) completely observable but not state controllable

(d) neither state controllable nor observable

34. A digital transmission system uses a (7, 4) systematic linear Hamming code for transmitting data over a noisy channel. If three of the message-codeword pairs in this code $(m_i; c_i)$, where c_i is the codeword corresponding to the ith message m_i, are known to be (1 1 0 0 ; 0 1 0 1 1 0 0), (1 1 1 0 ; 0 0 1 1 1 1 0) and (0 1 1 0 ; 1 0 0 0 1 1 0), then which of the following is a valid codeword in this code?

(a) 1 1 0 1 0 0 1

(b) 1 0 1 1 0 1 0

(c) 0 0 0 1 0 1 1

(d) 0 1 1 0 1 0 0

35. The impedance matching network shown in the figure is to match a lossless line having characteristic impedance Z_0 = 50 Ω with a load impedance Z_L. A quarter–wave line having a characteristic impedance Z_1 = 75 Ω is connected to Z_L. Two stubs having characteristic impedance of 75 Ω each are connected to this quarter–wave line. One is a short-circuited (S.C.) stub of length 0.25λ connected across PQ and the other one is an open–circuited (O.C.) stub of length 0.5λ connected across RS.

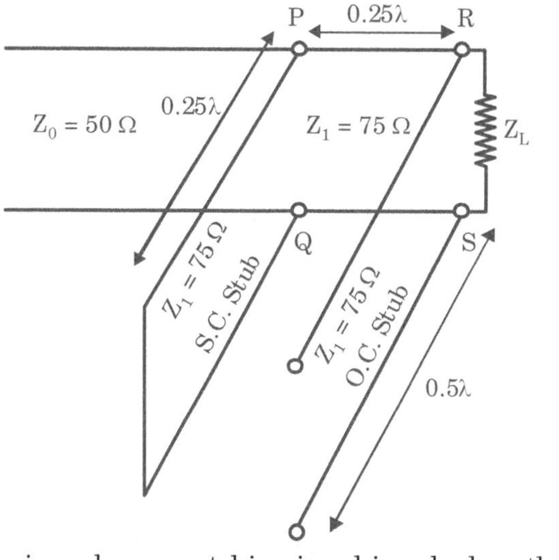

The impedance matching is achieved when the real part of Z_L is

(a) 112.5 Ω (b) 75.0 Ω

(c) 50.0 Ω (d) 33.3 Ω

Q. 36 to Q. 55 Numerical Answer Type (NAT), carry TWO marks each.

36. A real 2 × 2 non-singular matrix A with repeated eigenvalue is given as

$$A = \begin{bmatrix} x & -3.0 \\ 3.0 & 4.0 \end{bmatrix}$$

where x is a real positive number. The value of x (*rounded off to one decimal place*) is ________

37. For a vector field $D = \rho \cos^2 \varphi a_\rho + z^2 \sin^2 \varphi\, a_\varphi$ in a cylindrical coordinate system (ρ, φ, z) with unit vectors a_ρ, a_φ and a_z, the net flux of D leaving the closed surface of the cylinder $(\rho = 3,\ 0 \le z \le 2)$ (*rounded off to two decimal places*) is ________

38. In the circuit shown in the figure, the switch is closed at time t = 0, while the capacitor is initially charged to – 5V (*i.e.,* $v_c(0) = -5$ V).

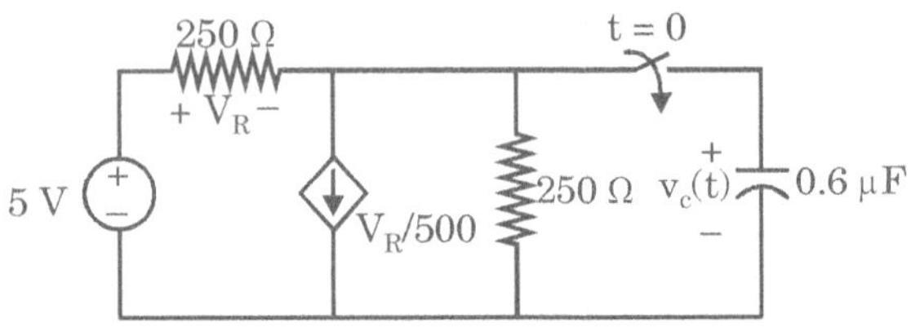

The time after which the voltage across the capacitor becomes zero (*rounded off to three decimal places*) is ________ ms.

39. The exponential Fourier series representation of a continuous-time periodic signal $x(t)$ is defined as

$$x(t) = \sum_{k=-\infty}^{\infty} a_k e^{jk\omega_0 t}$$

where ω_0 is the fundamental angular frequency of $x(t)$ and the coefficients of the series are ak. The following information is given about $x(t)$ and a_k.

I. $x(t)$ is real and even, having a fundamental period of 6

II. The average value of $x(t)$ is 2

III. $a_k = \begin{cases} k, & 1 \le k \le 3 \\ 0, & k > 3 \end{cases}$

The average power of the signal $x(t)$ (*rounded off to one decimal place*) is ________

40. For a unit step input $u[n]$, a discrete-time LTI system produces an output signal $(2\delta[n+1] + \delta[n] + \delta[n-1])$. Let $y[n]$ be the output of the system for an input $\left(\left(\frac{1}{2}\right)^n u(n)\right)$. The value of $y[0]$ is ________

41. Consider the signals $x[n] = 2^{n-1} u[-n+2]$ and $y[n] = 2^{-n+2} u[n+1]$, where $u[n]$ is the unit step sequence. Let $X(e^{j\omega})$ and $Y(e^{j\omega})$ be the discrete-time Fourier transform of $x[n]$ and $y[n]$, respectively. The value of the integral

$$\frac{1}{2\pi}\int_0^{2\pi} X(e^{j\omega})Y(e^{-j\omega})d\omega$$

(*rounded off to one decimal place*) is ________

42. A silicon P-N junction is shown in the figure. The doping in the P region is 5×10^{16} cm^{-3} and doping in the N region is 10×10^{16} cm^{-3}. The parameters given are

Built-in voltage $(\Phi_{bi}) = 0.8$ V

Electron charge $(q) = 1.6 \times 10^{-19}$ C

Vacuum permittivity $(\varepsilon_0) = 8.85 \times 10^{-12}$ F/m

Relative permittivity of silicon $(\varepsilon_{Si}) = 12$

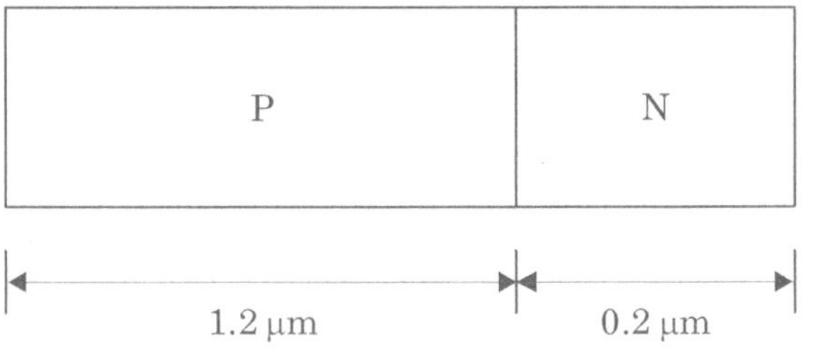

The magnitude of reverse bias voltage that would completely deplete one of the two regions (P or N) prior to the other (*rounded off to one decimal place*) is ________ V.

43. An asymmetrical periodic pulse train v_{in} of 10 V amplitude with on-time $T_{ON} = 1$ ms and off-time $T_{OFF} = 1$ μs is applied to the circuit shown in the figure. The diode D_1 is ideal.

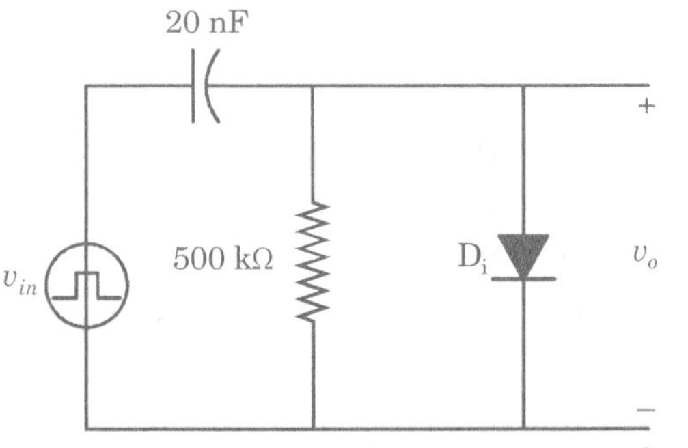

The difference between the maximum voltage and minimum voltage of the output waveform v_o (*in integer*) is ________ V.

44. For the transistor M_1 in the circuit shown in the figure, $\mu_n C_{ox} = 100$ μA/V^2 and (W/L) = 10, where μ_n is the mobility of electron, C_{ox} is the oxide capacitance per unit area, W is the width and L is the length.

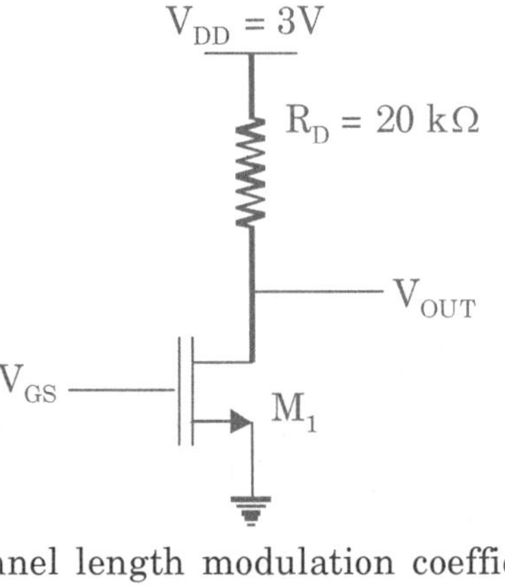

The channel length modulation coefficient is ignored. If the gate-to-source voltage V_{GS} is 1 V to keep the transistor at the edge of saturation, then the threshold voltage of the transistor (*rounded off to one decimal place*) is ________ V.

45. A circuit with an ideal OPAMP is shown in the figure. A pulse V_{IN} of 20 ms duration is applied to the input. The capacitors are initially uncharged.

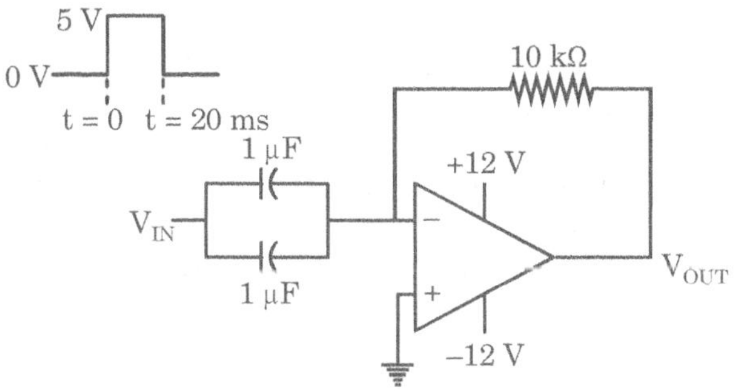

The output voltage V_{OUT} of this circuit at $t = 0^+$ (*in integer*) is ________ V.

46. The propagation delay of the exclusive-OR (XOR) gate in the circuit in the figure is 3 ns. The propagation delay of all the flip-flops is assumed to be zero. The clock (Clk) frequency provided to the circuit is 500 MHz.

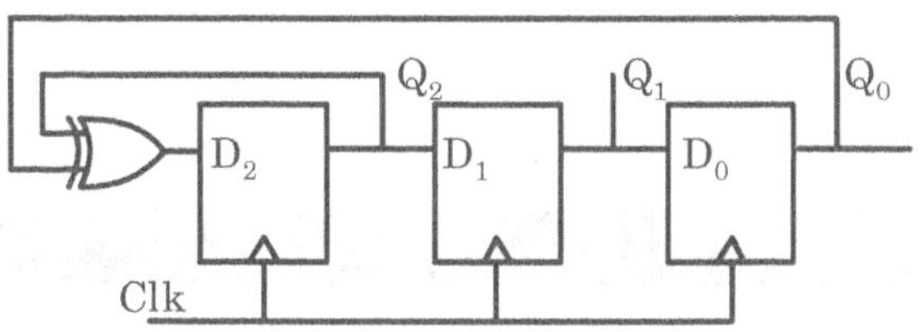

Starting from the initial value of the flip-flop outputs $Q_2Q_1Q_0 = 1\ 1\ 1$ with $D_2 = 1$, the minimum number of triggering clock edges after which the flip-flop outputs $Q_2Q_1Q_0$ becomes 1 0 0 (*in integer*) is ________

47. The circuit in the figure contains a current source driving a load having an inductor and a resistor in series, with a shunt capacitor across the load. The ammeter is assumed to have zero resistance. The switch is closed at time t = 0.

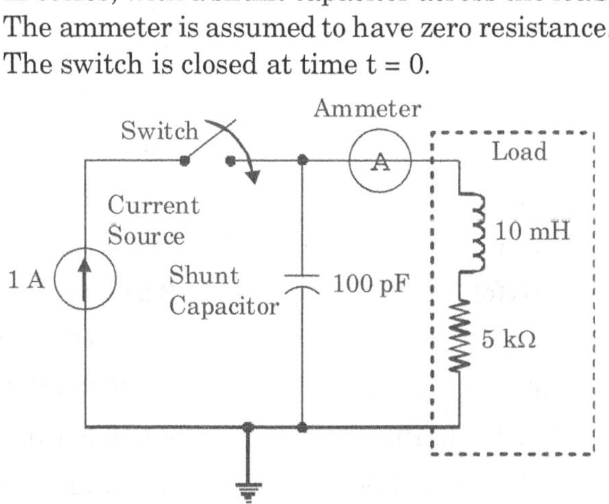

Initially, when the switch is open, the capacitor is discharged and the ammeter reads zero ampere. After the switch is closed, the ammeter reading keeps fluctuating for some time till it settles to a final steady value. The maximum ammeter reading that one will observe after the switch is closed (*rounded off to two decimal places*) is ________ A.

48. A unity feedback system that uses proportional-integral (PI) control is shown in the figure.

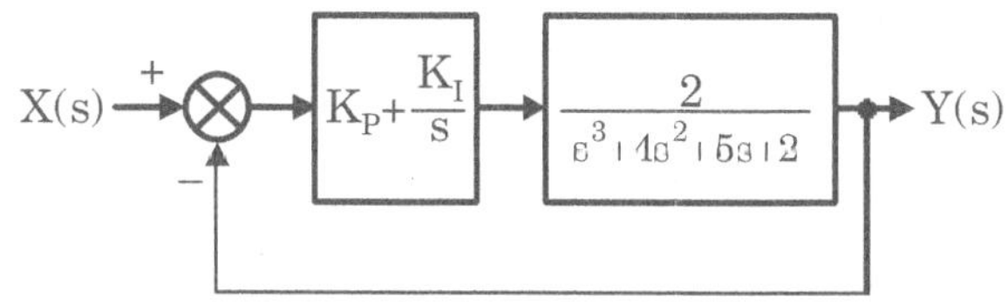

The stability of the overall system is controlled by tuning the PI control parameters K_P and K_I. The maximum value of K_I that can be chosen so as to keep the overall system stable or, in the worst case, marginally stable (*rounded off to three decimal places*) is ________

49. A sinusoidal message signal having root mean square value of 4 V and frequency of 1 kHz is fed to a phase modulator with phase deviation constant 2 rad/volt. If the carrier signal is $c(t) = 2\cos(2\pi 10^6 t)$, the maximum instantaneous frequency of the phase modulated signal (*rounded off to one decimal place*) is ________ Hz.

50. Consider a superheterodyne receiver tuned to 600 kHz. If the local oscillator feeds a 1000 kHz signal to the mixer, the image frequency (*in integer*) is ________ kHz.

51. In a high school having equal number of boy students and girl students, 75% of the students study Science and the remaining 25% students study Commerce. Commerce students are two times more likely to be a boy than are Science students. The amount of information gained in knowing that a randomly selected girl student studies Commerce (*rounded off to three decimal places*) is ________ bits.

52. A message signal having peak-to-peak value of 2 V, root mean square value of 0.1 V and bandwidth of 5 kHz is sampled and fed to a pulse code modulation (PCM) system that uses a uniform quantizer. The PCM output is transmitted over a channel that can support a maximum transmission rate of 50 kbps. Assuming that the quantization error is uniformly distributed, the maximum signal to quantization noise ratio that can be obtained by the PCM system (*rounded off to two decimal places*) is ________

53. Consider a polar non-return to zero (NRZ) waveform, using + 2V and – 2 V for representing binary '1' and '0' respectively, is transmitted in the presence of additive zero–mean white Gaussian noise with variance 0. 4 V^2. If the *a priori* probability of transmission of a binary '1' is 0. 4, the optimum threshold voltage for a maximum *a posteriori* (MAP) receiver (*rounded off to two decimal places*) is _________ V.

54. A standard air–filled rectangular waveguide with dimensions a = 8 cm, b = 4 cm, operates at 3.4 GHz. For the dominant mode of wave propagation, the phase velocity of the signal is v_p. The value (*rounded off to two decimal places*) of $\frac{v_p}{c}$, where c denotes the velocity of light, is _________

55. An antenna with a directive gain of 6 dB is radiating a total power of 16 kW. The amplitude of the electric field in free space at a distance of 8 km from the antenna in the direction of 6 dB gain (*rounded off to three decimal places*) is _________ V/m.

ANSWERS

General Aptitude (GA)

1. (c)	**2.** (b)	**3.** (c)	**4.** (c)
5. (b)	**6.** (c)	**7.** (d)	**8.** (a)
9. (c)	**10.** (a)		

Electronics & Communication Engineering

1. (a)	**2.** (b)	**3.** (b)	**4.** (b)
5. (c)	**6.** (d)	**7.** (a)	**8.** (d)
9. (d)	**10.** (a)	**11.** (b)	**12.** (c)
13. (c)	**14.** (d)	**15.** (a)	**16.** (8.0 to 8.0)
17. (0 to 0)	**18.** (0.5 to 0.5)	**19.** (1.0 to 1.0)	**20.** (4.5 to 4.5)
21. (2.0 to 2.0)	**22.** (94.2 to 94.6)	**23.** (9.24 to 9.28)	**24.** (3.12 to 3.16)
25. (9.30 to 9.44)	**26.** (a)	**27.** (b)	**28.** (c)
29. (c)	**30.** (a)	**31.** (c)	**32.** (a)
33. (d)	**34.** (d)	**35.** (a)	**36.** (10.0 to 10.0)
37. (56.50 to 56.60)	**38.** (0.132 to 0.146)	**39.** (31.9 to 32.1)	**40.** (0.0 to 0.0)
41. (7.9 to 8.1)	**42.** (8.1 to 8.4)	**43.** (10 to 10)	**44.** (0.5 to 0.5)
45. (–12 to –12)	**46.** (5 to 5)	**47.** (1.40 to 1.50)	**48.** (3.125 to 3.125)
49. (1011313.5 to 1011313.9)	**50.** (1400 to 1400)	**51.** (3.320 to 3.325)	**52.** (30.70 to 30.74)
53. (0.03 to 0.05)	**54.** (1.15 to 1.25)	**55.** (0.224 to 0.264)	

EXPLANATIONS

GENERAL APTITUDE

1. As given that,

Current population of a city = 11, 02, 500

Increasing rate = 5%/Annum

The population 2 years ago = ?

Let us Consider, the population 2 years ago = x

Using compound interest formula, we have

$$A = P\left(1 + \frac{R}{100}\right)^n$$

$$\Rightarrow \quad 11{,}02{,}500 = x\left(1 + \frac{5}{100}\right)^2$$

$$\Rightarrow \quad 11{,}02{,}500 \times \frac{100 \times 100}{105 \times 105} = x$$

$$\Rightarrow \quad \boxed{x = 10{,}00{,}000}$$

Hence, the population 2 years ago = 10,00,000.

2. As Given that, $\frac{p}{q} + \frac{q}{p} = 3$

By Squaring on both sides;

$$\left(\frac{p}{q} + \frac{q}{p}\right)^2 = 9$$

$$\Rightarrow \frac{p^2}{q^2} + \frac{q^2}{p^2} + 2 = 9$$

Hence, $\boxed{\frac{p^2}{q^2} + \frac{q^2}{p^2} = 7}$

3.

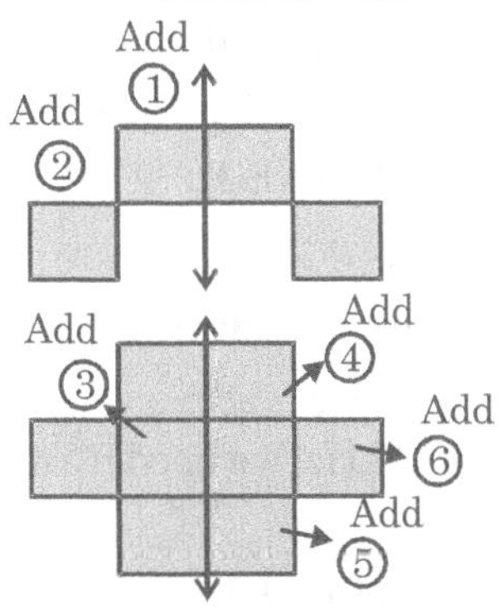

Hence, the least number of squares added = 6.

4. Nostalgia (means excessively sentimental yearning for return to or of some past period) is to anticipation. (means visualization of a future event or state)

5. Wake pastense is woke and third form is woken so, the verb forms are:

V1	V2	V3
Wake	Woke	Woken

6.

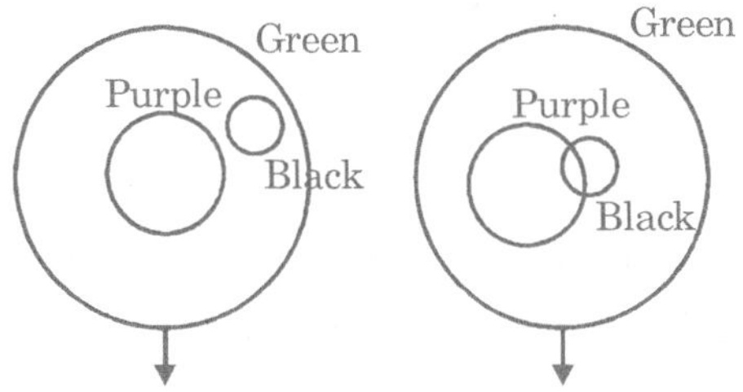

No black is purple   Some black are purple

7. The passage deduces that computers are present every where and they have both positive and negative effects on humans.

8. Square:

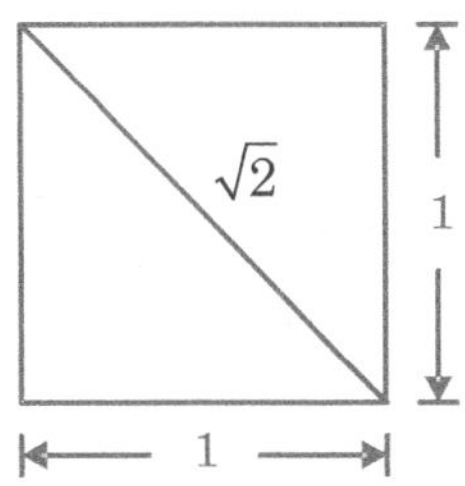

Solid core formed by revolving tringle about its short edge.

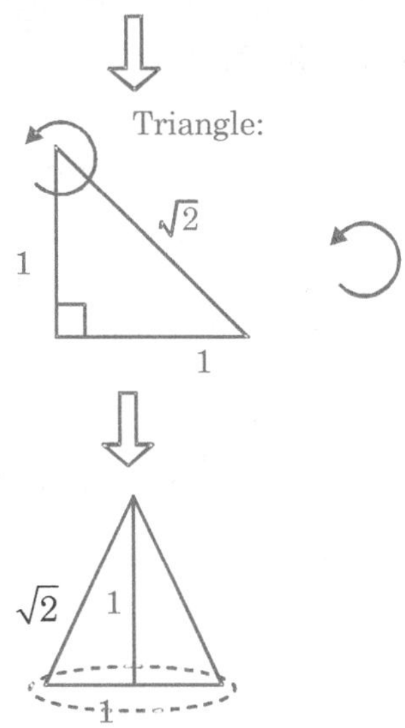

Radius of base circle(r) = 1 unit

Height of cone(h) = 1 unit

Volume of cone $= \frac{\pi}{3} \times r^2 h = \frac{\pi}{3} \times 1 \times 1 = \frac{\pi}{3}$ unit^3

9. From the bar graph, it is clear that except Thursday, total 6 days one of the students spent a minimum of 10% more than the other student.

10. In regular hexagon, each interior angle = 120°,

So, if we cut corners of an equilateral triangle, then the removed triangles are also equilateral triangles.

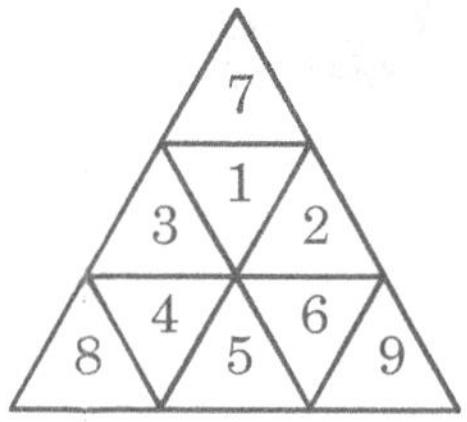

$\therefore$ Let, side of regular hexagon = x units

Then, side of original equilateral triangle = 3x

$\therefore$ Area of regular hexagon $= \frac{3\sqrt{3}}{2}x^2$

Area of equilateral triangle $= \frac{\sqrt{3}}{4}(3x)^2$

$$\therefore \quad \text{Required ratio} = \frac{3\sqrt{3}}{2}x^2 : \frac{\sqrt{3}}{4}(3x)^2$$

$$= \frac{3\sqrt{3}}{2}x^2 : \frac{9\sqrt{3}}{4}x^2$$

$$= 1 : \frac{3}{2} = 2 : 3$$

ELECTRONICS & COMMUNICATION ENGINEERING

1. Parametric equations of the curve C (circular arc) x = cos t; y = sin t where t is parameter, and radius = 1

$$\int_C F(r)\,dr = \int(-x\hat{i} + y\hat{j})(dx\hat{i} + dy\hat{j})$$

$$\therefore \int_C F(r).dr = \int_C(-xdx + ydy)$$

$$= \int_C\{(-\cos t)(-\sin t\, dt) + (\sin t)(\cos t\, dt)\}$$

where $(t \to 0 \text{ to } \frac{\pi}{4})$

$$= \int_0^{\frac{\pi}{4}} 2\sin t \cos t\, dt$$

$$= \int_0^{\frac{\pi}{4}} \sin 2t\, dt$$

$$= \left(\frac{-\cos 2t}{2}\right)_0^{\frac{\pi}{4}}$$

$$= \frac{-1}{2}(0-1) = \frac{1}{2}$$

Hence option (a) is correct answer.

2. As given differential equation $x(t) \times x(\frac{t}{2})$

$$\frac{dy}{dx} + \underbrace{\left(\frac{x}{1-x^2}\right)}_{P} y = \underbrace{(x)}_{Q} y^{1/2} \quad \text{...(i)}$$

is a Bernoulli's equation in y.

Dividing both sides by $y^{1/2}$

$$y^{-1/2}\frac{dy}{dx} + \left(\frac{x}{1-x^2}\right)y^{1/2} = x \quad \text{...(ii)}$$

Let $y^{1/2} = t$

$$\Rightarrow \quad \frac{1}{2}y^{-\frac{1}{2}}\frac{dy}{dx} = \frac{dt}{dx}$$

$$\therefore \quad \text{(2) becomes } 2\frac{dt}{dx} + \left(\frac{x}{1+x^2}\right)t = x$$

$$\Rightarrow \quad \frac{dt}{dx} + \left(\frac{x}{2(1+x^2)}\right)t = \frac{x}{2} \quad \text{...(iii)}$$

is a linear equation in t.

$$\therefore \quad I.F = e^{\int \frac{x}{2(1-x^2)}dx}$$

$$= e^{-\frac{1}{4}\int \frac{-2x}{1-x^2}dx}$$

$$= e^{\ell n\left((1-x^2)^{-\frac{1}{4}}\right)}$$

$$\Rightarrow \boxed{I \cdot F = (1-x^2)^{-\frac{1}{4}}}$$

Hence, option (b) is correct answer.

3. As Given that, Y = 2X + 3

$$Var(Y) = E(Y^2) - (E(Y))^2$$
$$= E[(2X+3)^2] - [E(2X+3)]^2$$
$$= 4[E(X^2) - (E(X))^2]$$
$$\sigma_Y^2 = 4\sigma_X^2$$

Hence, option (b) is correct answer.

4. As is given that

x(t) is band limited to 10 kHz, we need to compute Nyquist rate of

$$y(t) = x(t)x\left(1+\frac{t}{2}\right)$$

x(t) having maximum frequency $f_{m_1} = 10$ kHz

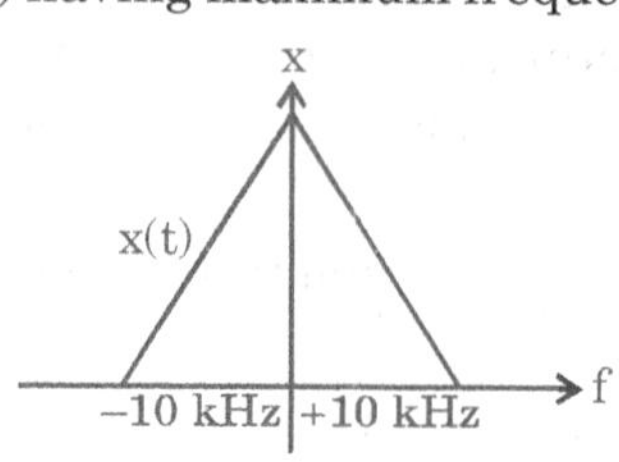

$x\left(\frac{t}{2}\right)$ having maximum frequency $f_{m_2} = 5$ kHz

$x\left(\frac{t+2}{2}\right) = x\left(\frac{1}{2}+1\right) = x\left(1+\frac{t}{2}\right)$ will have

maximum frequency same as $x\left(\frac{t}{2}\right)$ i.e., 5 kHz,

since time shifting does not affect the frequency limit of original signal.

$x(t).x\left(1+\frac{t}{2}\right)$ will have maximum frequency.

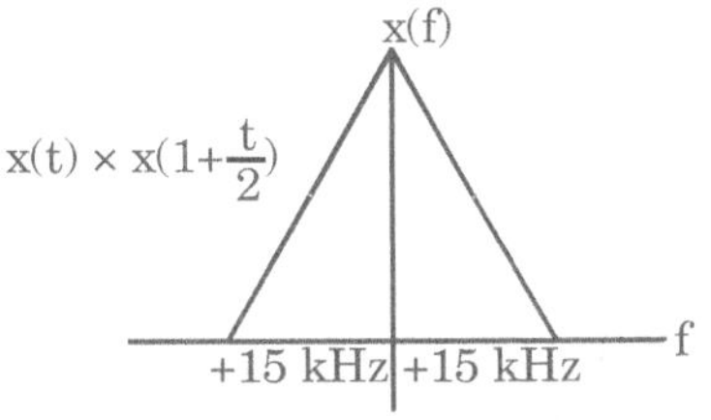

$$f_{m_1} + f_{m_2} = 10 + 5 = 15 \text{ kHz}$$

Since in frequency domain it will lead to convolution and in convolution upper limit is sum of upper limit.

So, hence, Nyquist rate = $2\left(f_{m_1} + f_{m_2}\right) = 2 \times 15$

= 30 kHz

Hence, option (b) is correct answer.

5. As given that

$y(n) = x(n) * h(n)$

where, * represents linear convolution

$Z(n) = \text{IDFT } [X(k). H(k)]$

16 point DFT is considered

We need to find for what value of k, y(k) = z(k)

$\text{IDFT}[X(k). H(k)] = x(n) \circledast h(n)$ [$\circledast$: circular convolution]

where,

x(n) have samples

h(n) have 16 samples

As we know that the length of linear convolution is = $N_1 + N_2 - 1$

Then x(n) * h(n) wil have 16 + 16 – 1 = 31 number of samples x(n) $\circledast$ h(n) will have 16 number of samples.

While computing circular convolution through liner convolution, we used to add last (k – 1) number of samples of linear convolution in beginning without disturbing the kth sample of linear and circular convolution. So the kth samples of linear convolution and circular convolution same.

Now, take an example, $x[n] = \left[\underset{\uparrow}{1}, 2, 3\right]$

$h(n) = \left[\underset{\uparrow}{4}, 5, 6\right]$

x(n) * h(n)	1	2	3		
	4	5	6		
	4	8	12		
		5	10	15	
			6	12	18
	4	13	28	27	18
	27	28			
	31	41	28		

So, $y(n) = x(n) * h(n) = \left[\underset{\uparrow}{4}, 13, 28, 27, 18\right]$

$z(n) = x(n) \circledast h(n) = [31, 31, 28]$

Here we have taken 3 samples in both x(n), h(n), and can see the 3rd samples of y(n), z(n) are same i.e., 28. [3rd sample is equivalent to n = 2 or k = 2]

Instead of 3 samples, if we take 16 samples then 16th samples will appear same in both y(n), z(n) [16th samples is equivalent to n = 15 or k = 15 sincestarting value of n/k is 0 not 1].

Hence, option (c) is correct answer.

6. As Given that: $P_o = 10^{16}$ cm^{-3}, $\tau = 100$ μ sec, $n_i = 10^{10}$ cm^{-3}, $g_o = 10^{20}$ cm^{-3}s^{-1}

Now using mass action law

$$n_o = \frac{n_i^2}{p_o} = \frac{10^{20}}{10^{16}} = 10^4 \text{ cm}^{-3}$$

As we know that

Excess charge carrier:

$$\Delta P = \Delta n$$
$$= 10^{20} \text{ cm}^{-3}\text{s}^{-1} \times 100 \times 10^{-6}\text{s}$$
$$= 10^{16} \text{ cm}^{-3}$$
$$\text{Product} = (P_o + \Delta P)(n_o + \Delta n)$$
$$= (10^{16} + 10^{16}) \times (10^4 + 10^{16})$$
$$= 2 \times 10^{32} \text{ cm}^{-6}$$

Hence option (d) is correct answer.

7. This is the case of non-uniform doping.

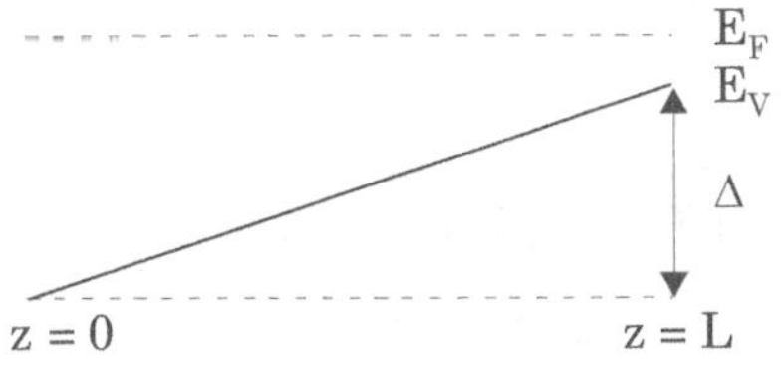

In a semi conductor result in built in electric filed so the relationship between the electric field and slope of energy-level is given by

$$E_x = \frac{-d\psi}{dx}$$

$$E_x = \frac{-d}{dx}\left[\frac{-E_x}{q}\right]$$

$$|E_x| = \left|\frac{1}{q}\frac{dE_v}{dx}\right|$$

$$\Rightarrow \quad E_x = \frac{1}{q}\left(\frac{\Delta - 0}{L - 0}\right)$$

$$\boxed{\therefore\ E = \frac{1}{q}.\frac{\Delta}{L}}$$

Hence, option (a) is corrct answer.

8. A.C equivalent of M_1 transistor ($\lambda \neq 0$)

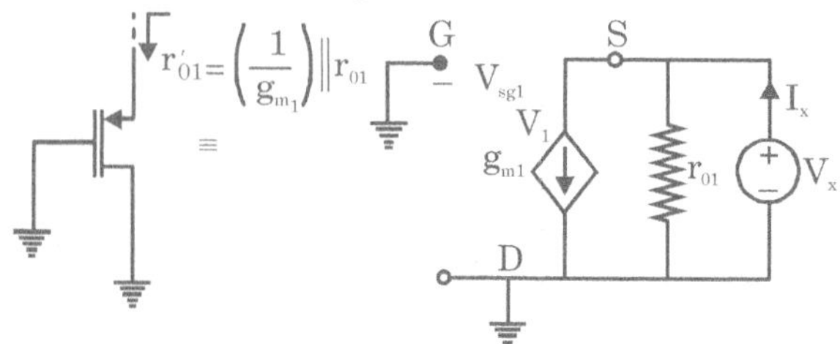

Apply KCL at source node

$$i_x = \frac{V_x}{r_{01}} + g_{m1}V_{sg1}$$

$$= \frac{V_x}{r_{01}} + g_{m1}V_x \qquad (\because V_{sg1} = V_x)$$

$$= \left(\frac{1}{r_{01}} + g_{m1}\right)V_x$$

$$r'_{01} = \frac{V_x}{i_x} = \frac{1}{\frac{1}{r_{01}} + g_{m1}} = \frac{1}{g_{m1}} \| r_{01}$$

A.c equivalent of complete amplifier circuit

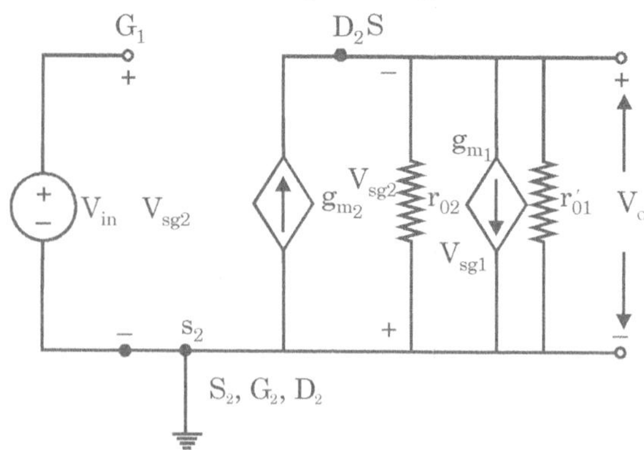

Now, Apply KCL at output node

$$g_{m_2}V_{sg2} = \frac{V_0}{r_{02}} + \frac{V_0}{r'_{01}}$$

$$g_{m_2}V_{sg2} = V_0\left(\frac{1}{r_{02}} + \frac{1}{r'_{01}}\right)$$

$$g_{m_2}V_{sg2} = V_0\frac{1}{r_{eq}}$$

$$V_{sg2} = V_{s2} - V_{g2}$$

$$= 0 - V_{g2} - V_{in} \qquad (\because V_{in} = V_{g2})$$

$$-g_{m_2}V_{in} = V_{out}r_{eq}$$

$$\frac{V_{out}}{V_{in}} = -g_{m_2}r_{eq}$$

$$= -g_{m_2}\left(r_{02} \| r'_{01}\right)$$

$$= -g_{m_2}\left(r'_{01} \| r_{02}\right)$$

$$A_9 = \frac{\partial V_{out}}{\partial V_{in}}$$

$$= -g_{m_2}\left(\frac{1}{g_{m_1}} \| r_{01} \| r_{02}\right)$$

Hence, option (d) is the correct answer.

9. According to the Question:

For ideal op-amp and virtual short concept

$$V_+ = V_-$$

$$= \frac{R_2}{R_1 + R_2}.V_{REF}$$

Apply KCL at node V_-, then

$I_1 = I_2$ ($\because$ Input current of ideal op-amp is zero)

$$\frac{V_{IN} - V_-}{R_{IN}} = \frac{V_- - V_{OUT}}{R_F}$$

$$\frac{R_F}{R_{IN}} = \frac{V_- - V_{OUT}}{V_{IN} - V_-} \qquad \text{...(i)}$$

If $V_{OUT} = 1$ volt for $V_{IN} = 0.1$ volt then equation (i) becomes

$$\frac{R_F}{R_{IN}} = \frac{V_- - 1}{0.1 - V_-} \qquad \text{...(ii)}$$

For $V_{OUT} = 6V$ for $V_{IN} = 1$ volt, then equation (i) becomes

$$\frac{R_F}{R_{IN}} = \frac{V_- - 6}{1 - V_-} \qquad \text{...(iii)}$$

By Compare equation (i) and (iii)

$$\frac{V_- - 1}{0.1 - V_-} = \frac{V_- - 6}{1 - V_-}$$

$$V_- = V_+ = -0.097 \text{ volt}$$

Substitute the value of V_- in equation (iii), then

$$\frac{R_F}{R_{IN}} = \frac{-0.097 - 6}{1 - (-0.097)} = -5.55$$

$$\boxed{\therefore \frac{R_F}{R_{IN}} = 5.55}$$

10. According to the question given circuit is:

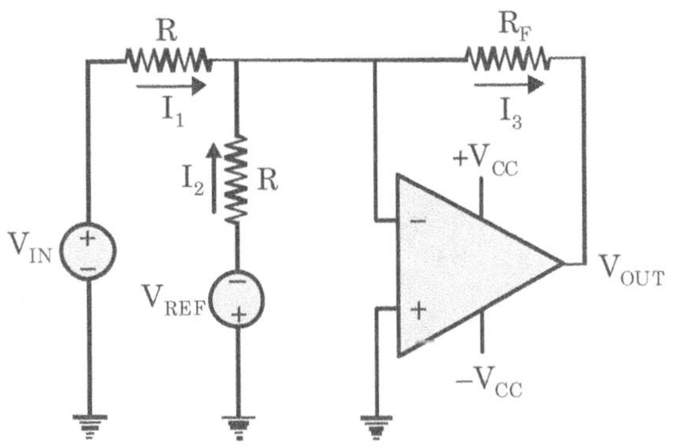

As per the virtual short concept of an ideal op-amp

$$V_+ = V_- = 0V$$

Apply KCL at (c)

$$I_1 + I_2 = I_3$$

$$\frac{V_{IN} - 0}{R} + \left(\frac{-V_{REF} - 0}{R}\right) = \frac{0 - V_{OUT}}{R_f}$$

$[\therefore V_{OUT} = 0V]$

$$\frac{V_{IN}}{R} - \frac{V_{REF}}{R} = 0$$

$$\boxed{\therefore V_{IN} = V_{REF}}$$

Hence, option (a) is correct answer.

11. As given that

$(1235)_x = (3033)_y$

By converting both side into base 10.

$(1 \times x^3) + (2 \times x^2) + (3 \times x^1) + (5 \times x^0)$

$= (3 \times y^3) + (0 \times y^2) + (3 \times y^1) + (3 \times y^0)$

$\Rightarrow x^3 + 2x^2 + 3x + 5 = 3y^3 + 3y + 3$

Let Checking with option B i.e., x = 8, y = 6

$8^3 + (2 \times 8^2) + (3 \times 8) + 5 = (3 \times 6^3) + (3 \times 6) + 3$

$\Rightarrow 669 = 699$ (valid)

12. Total of memory:

$$32k \times 16 = 2^5 \times 2^{10} \times 16 = 2^{15} \times 16 = 2^P \times n$$

Number of address line = p = 15

15 address line can generate 2^{15} number of different address, and each of these address will be accessed through output of decoder, so the size of decoder needed is 15 to 2^{15} which needs 2^{15} number of AND gates.

In a 2 × 4 Decoder the internal structure is

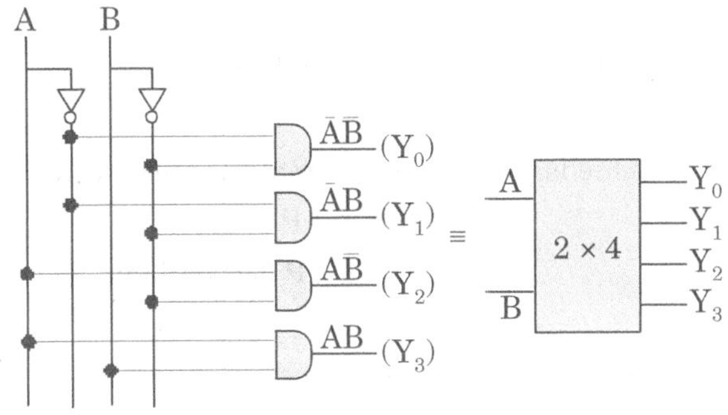

So, a 2 × 4 decoder need 4 AND gates.

2 × 2^{15} decoders need 2^{15} AND gates.

Basically, number of AND gate needed for a decoder is equal to number of output lines of decoder.

Hence, option (c) is correct answer.

13.

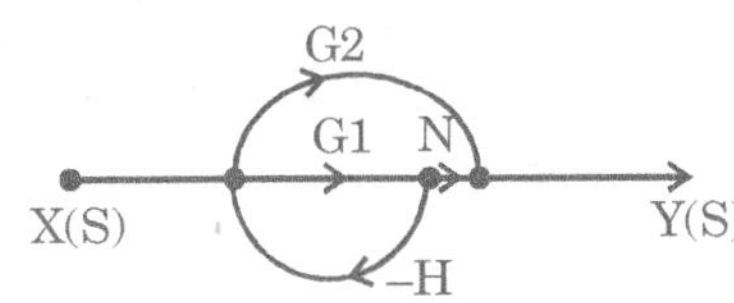

$$Y(s) = [X(s) - NH]G_2 + N$$

Where $N = [X(s) - NH]G_1$

$$N = X(s)\frac{G_1}{1 + G_1H}$$

$$Y(s) = G_2X(s) - NG_2H + N$$

$$Y(s) = G_2X(s) - \frac{X(s)G_1G_2H}{1 + G_1H} \times X(s)\frac{G_1}{1 + G_1H}$$

$$Y(s) = X(s)\left[\left[\frac{G_2 + G_1G_2H - G_1G_2H}{1 + G_1H}\right] + \frac{G_1}{1 + G_1H}\right]$$

So, Transfer function $= \dfrac{Y(s)}{X(s)} = \dfrac{G_1 + G_2}{1 + G_1H}$

Hence, option (c) is correct answer.

14. As given figure,

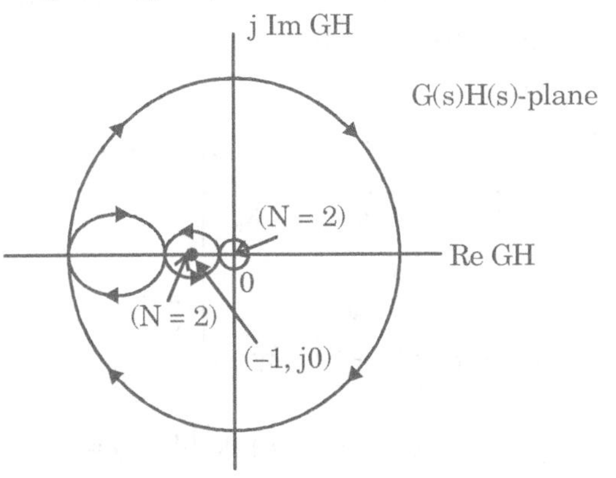

W.K. T, N = Z – P

GH have no poles on the right half(R.H)s-plane

$$P = 0$$

When (–1, + j0) lies at P then

$$N = -1 - 1 = -2$$

[One is clockwise and another anti-clock wise]

$$N = Z - P$$

$$-2 = 1 - P \text{ Closed loop poles}$$

$$P = 3$$

Hence option 'd' is correct answer.

15. Considering two medium

Z < 0 | Z = 0 | Z > 0

$\text{impedance}(\eta_1) = 120\pi\sqrt{\frac{\mu}{\epsilon}}$ | $\text{impedance}(\eta_1) = 120\pi\sqrt{\frac{\mu}{e}}$

$= 120\pi\sqrt{\frac{4}{1}} = 240\pi$ | $= 120\pi\sqrt{\frac{1}{1}} = 120\pi$

Lossless medium | Free space

with $\varepsilon = \varepsilon_0$, $\mu = 4\mu_0$

As we know that,

$$\Rightarrow \text{Reflection coefficient } \frac{\eta_2 - \eta_1}{\eta_2 + \eta_1} = \frac{240\pi - 120\pi}{240\pi + 120\pi}$$

$$= \frac{120\pi}{360\pi} = \frac{1}{3}$$

Hence, option (a) is the correct answer.

16. As given that, $X_1 = (1.0, -1.0, 2.0)$, $X_2 = (7.0, 3.0, x)$ and $X_3 = (2.0, 3.0, 1.0)$ be three linearly dependent vectors. Consider these vectors as rows in a matrix A. For linearly dependent vectors, we have

$$|A| = 0$$

$$\Rightarrow \begin{vmatrix} 1 & -1 & 2 \\ 7 & 3 & x \\ 2 & 3 & 1 \end{vmatrix} = 0$$

$$\Rightarrow 1 \times (3 - 3x) + 1 \times (7 - 2x) + 2 \times (21 - 6) = 0$$

$$\Rightarrow -5x + 10 + 30 = 0$$

$$\Rightarrow x = 8$$

17. For irrotational field, we have curl $(\nabla \times \vec{F}) = 0$

$$\Rightarrow \begin{vmatrix} \hat{i}_x & \hat{j}_y & \hat{k}_z \\ \frac{\partial}{\partial x} & \frac{\partial}{\partial y} & \frac{\partial}{\partial z} \\ 4y - c_1 z & 4x + 2z & 2y + z \end{vmatrix} = 0$$

$$\Rightarrow \hat{i}_x(2 - z) - \hat{j}_y(0 + c_1) + \hat{k}_z(4 - 4) = 0$$

$$\Rightarrow c_1 = 0$$

18. From the given circuit

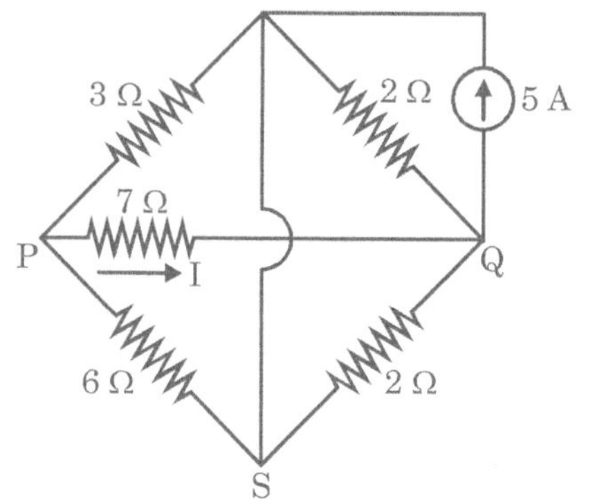

Now redraw the network we have,

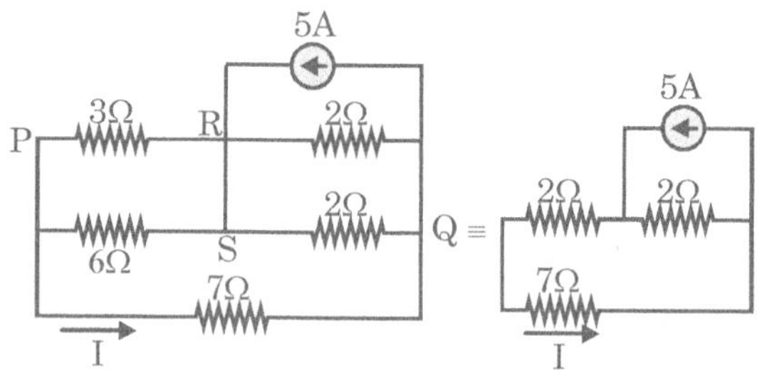

Now, by current division rule (CDR):

$$(I) = \frac{5 \times 1}{1 + 2 + 7} = \frac{5}{10} = 0.5 \text{ Amps}$$

19. From the given circuit

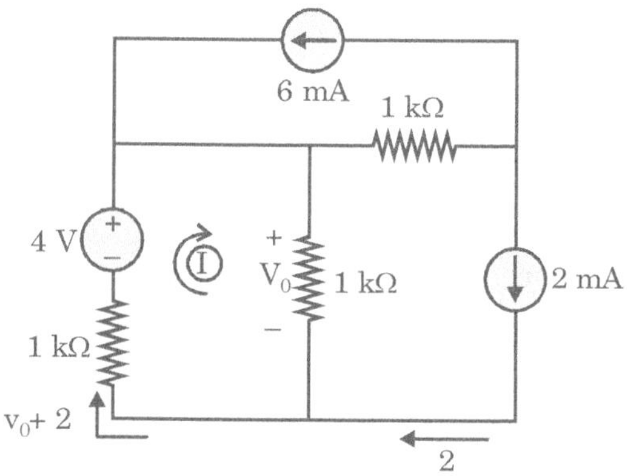

By KVL

$$4 - V_0 - (V_o + 2) = 0$$

$$V_0 + V_0 = 4 - 2$$

$$2V_0 = 2$$

$$\Rightarrow V_0 = 1 \text{ Volt}$$

20. As given that

8-bit DAC have F.S.O = 7.68V

We need to obtain analog output voltage for input (10010110).

$$\text{Resolution} = \frac{V_{F.S.O}}{2^n - 1} = \frac{7.68}{2^8 - 1} = 0.03011$$

So, Analog output Voltage

= Resolution × Decimal equivalent of $(10010110)_2$

$$= 0.030 \times 150 = 4.5\ 17 \text{ V}$$

21. As we know that,

Average power of $x(t) = E[x^2(t)]$

$= R_x(0) = 2$

22. As we know that,

% power saving

$$= \frac{\text{Power Saved}}{\text{Total power}} \times 100$$

Total power in amplitude modulated wave by a single tune sinusoidal signal:

$$= \underbrace{P_C}_{\text{carrier}} + \underbrace{P_C \frac{\mu^2}{4}}_{\text{sideband}} + \underbrace{P_C \frac{\mu^2}{4}}_{\text{sideband}}$$

After suppressing carrier and one sideband

$$\text{Power} = \frac{P_C \mu^2}{4}$$

$\Rightarrow$ Savings in power $= P_C + P_C \frac{\mu^2}{4}$

$$\%\text{ Power saved} = \frac{P_C + P_C \frac{\mu^2}{4}}{P_C + P_C \frac{\mu^2}{4} + P_C \frac{\mu^2}{4}} \times 100$$

$$= \frac{4+\mu^2}{2(2+\mu^2)} \times 100$$

For 50% modulation this value is $(\because \mu = 0.5)$

$$\frac{4+(0.5)^2 \times 100}{2\left(2+(0.5)^2\right)} = \frac{4.25}{4.5} \times 100 = 94.4\%$$

23. As we know that, Channel capacity relation for an AWGN channel

$$C = B \log_2[1 + \text{SNR}]$$

Data rate $= n.F_s$

$= 8 \times 1.25 \times 2 \times 4000$

$= 80 \times 10^3$ bits per record

$\Rightarrow$ Data rate $(R) \leq C$

$$R \leq B \log_2[1 + \text{SNR}]$$

$$\Rightarrow \quad 80 \times 10^3 \leq B \log_2[1 + 400]$$

$$\Rightarrow \quad B \geq \frac{80 \times 10^3}{\log_2[401]}$$

$$\geq \frac{80 \times 10^3}{8.64} \geq 9.259 \text{ kHz}$$

$B \geq 9.26$ KHz

24. As given that

$F_S = 32$ KHz Δ to avoid slope-overload error.

To avoid slope overload in DM

$$\Delta.F_S \geq = \left|\frac{d}{dt} m(t)\right|_{max}$$

$$\therefore m(t) = 4 \sin(2\pi \times 4 \times 10^3 t)$$

$$\Rightarrow \quad \Delta.32 \times 10^3 \geq 4 \times 4 \times 2\pi \times 10^3 t$$

$$\Rightarrow \quad \Delta \geq \frac{32\pi}{32}$$

$$\Rightarrow \quad \Delta \geq 3.14$$

25. As we know that

The total interval neglection angle of incidence

$$(\theta_C) = \sin^{-1}\left[\frac{\eta_2}{\eta_1}\right]$$

$$\theta_c = \sin^{-1}\left(\frac{1.48}{1.5}\right) = \sin^{-1}(0.9866) = 80.63$$

$\Rightarrow$ Angle it should make with axis is

$\theta_1 = 90° - \theta_c$

$\theta_1 = 90 - 80.63 = 9.36°$

26. C(curve) is $|x - i \Rightarrow = 2$.

$x^2(x^2 + 4) = 0$ gives $x = 0, 2i$, $2i$ are the singular points.

Only $x = 0$, $2i$ lies inside C.

$$\text{Let } f(x) = \frac{\sin x}{x^2(x^2+4)} = \left[\frac{\sin x}{4x^2}\right] - \left[\frac{\sin x}{4(x^2+4)}\right]$$

$$\therefore \oint_C \frac{\sin x}{x^2(x^2+4)} dx = \frac{1}{4}\oint_C \frac{\sin x}{(x-0)^2} dx - \frac{1}{4}\oint_C \frac{\left(\frac{\sin x}{x+2i}\right)}{x-2i} dx$$

$$\int_C f(z)\,dz = 2\pi i \left[\underset{x=0}{\text{Res}} + \underset{x=2i}{\text{Res}}\right]$$

$$= \frac{1}{4} \times \frac{2\pi i}{1!}\left[\frac{d}{dx}(\sin x)\right]_{x=0} - \frac{1}{4} \times 2\pi i \left[\frac{\sin x}{x+2i}\right]_{x=2i}$$

($\because$ Using Cauchy's integral formula)

$$= 2\pi i \left[\frac{1}{4} - \frac{\sin 2i}{16i}\right]$$

$$= \frac{\pi i}{2} - \frac{\pi}{8}\sin(2i)$$

Hence, no option matched.

27. As givne that,

A box contains the three coins such as,(i) HT (ii) HH (iii) TT.

So, probability of head in first toss $P(H_1) = \frac{1}{2}$

As we know that $P[H_2/H_1] = P[H_2 \cap H_1]/P(H_1)$

$$\text{So, } P(H_2 \cap H_1) = \left(\frac{1}{12} + \frac{1}{12}\right) = \frac{1}{6}$$

$$\text{Therefore, } P\left(\frac{H_2}{H_1}\right) = \frac{1}{6} \times \frac{2}{1} = \frac{1}{3}$$

Hence, option (b) is correct answer.

28. As given network is

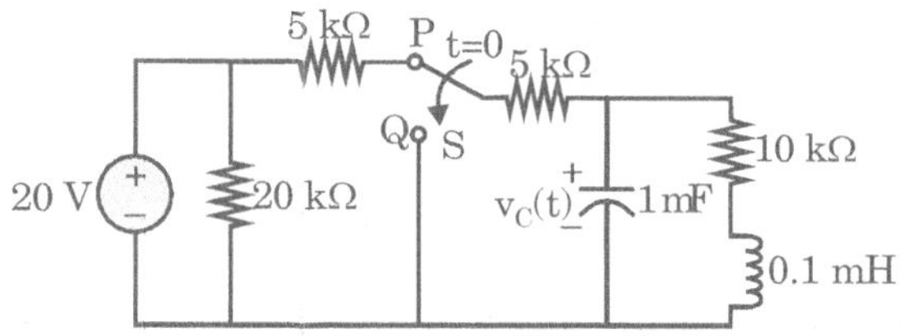

→ For t<0, 'S' is in position (P)

At, t = 0 – (S.S), L → S.C, C → O.C

We need to obtain $\left.\dfrac{dV_C(t)}{dt}\right|_{t=0^+}$

at $t = 0^-$
- → Switch at P
- → Capacitor open circuit
- → Inductor short circuit
- → Network in steady state

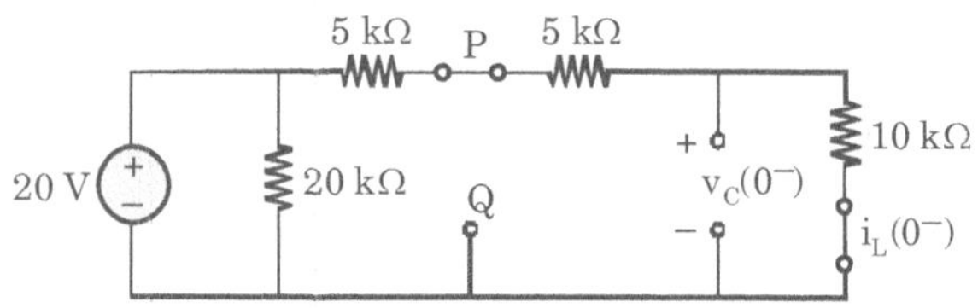

at $t = 0^-$

$$i_L(0^-) = \frac{20V}{(5+5+10)k\Omega}$$

$$= \frac{20V}{20k\Omega} = 1 \text{ mA}$$

$$V_C(0^-) = (10\ k\Omega)[i_L(0^-)]$$

$$= [10\ k\Omega][1 \text{ mA}] = 10 \text{ V}$$

at $t = 0^+$
- → Switch at Q
- → Capacitor replaced by ideal voltage source of value $V_C(0^-)$
- → Inductor replaced by ideal current source of value $i_L(0^-)$

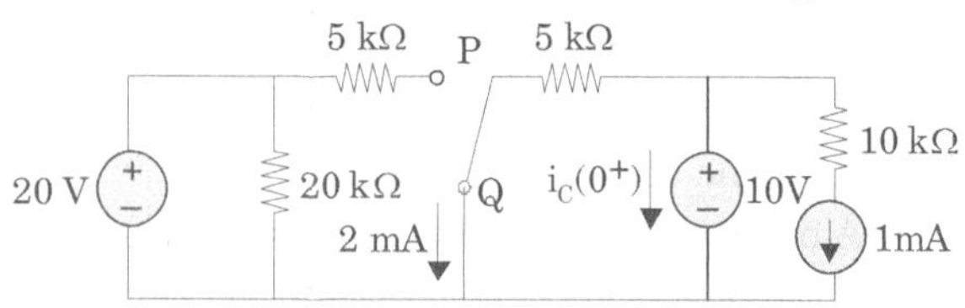

at $t = 0^-$

Now By KCL $i_C(0^+) = -3$ mA

$$i_C(t) = C\frac{dV_C(t)}{dt}$$

$$\Rightarrow \quad \frac{dV_C(t)}{dt} = \frac{i_C(t)}{C}$$

$$\Rightarrow \quad \left.\frac{dV_C(t)}{dt}\right|_{t=0^+} = \frac{i_C(0^+)}{C}$$

$$= \frac{-3 \text{ mA}}{1 \text{ mF}} = -3 \text{ V/sec}$$

Hence, option (c) is correct answer.

29. As given network

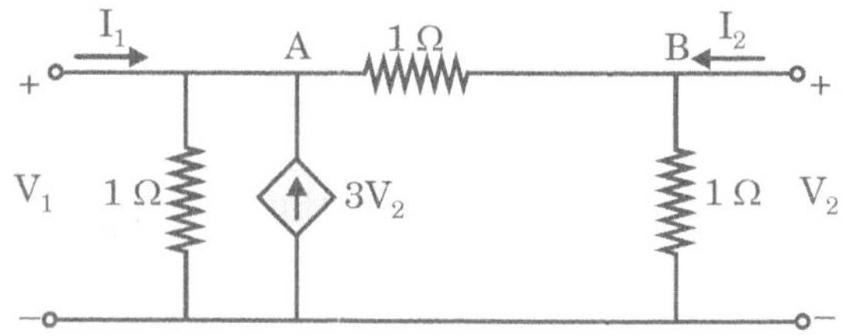

By nodal analysis at node A!

Node A	Node B
$-I_1 - 3V_2 + \frac{V_1}{1} + \frac{V_1 - V_1}{1} = 0$	$-I_2 + \frac{V_1}{1} + \frac{V_2 - V_1}{1} = 0$
$\Rightarrow I_1 = V_1 + V_1 - V_3 - 3V_2$	$\Rightarrow I_2 = -V_1 + V_2 + V_2$
$\Rightarrow I_1 = 2V_1 - 4V_2$ ....(i)	$\Rightarrow I_2 = -V_1 + 2V_2$ ....(ii)
$\Rightarrow I_1 = Y_{11}V_1 + Y_{12}V_2$	$\Rightarrow I_2 = Y_{21}V_1 + Y_{22}V_2$
So $Y_{11} = 2, Y_{21} = -4$	So $Y_{21} = -1, Y_{22} = 2$

From, equation (i) & (ii)

$$Y = \begin{bmatrix} 2 & -4 \\ -1 & 2 \end{bmatrix}$$

Hence, option (c) is the correct answer.

30. As we know that

For N-channel MOSFET their threshold voltage (V_T) is given by

$$V_T = V_{T_0} + \gamma\left(\sqrt{|V_{BS}| + |2\phi_B|} - \sqrt{|2\phi_B|}\right) \quad ....(i)$$

Where, V_{T_0} threshold voltage no body-bias i.e., $|V_{BS}| = 0V$

$$\gamma = \frac{\sqrt{2\epsilon_s qN_a}}{\varepsilon_{ox}} \quad ...(ii)$$

(∵ N-channel MOSFET is having P-type susbstrate)

Differentiate equation (i) w.r.t V_{BS}

$$\frac{\partial V_T}{\partial |V_{BS}|} = \frac{\gamma}{2\sqrt{|V_{BS}| + |2\phi_B|}}$$

$$= 50 \text{ mV/V} \quad ...(iii)$$

$$50 \times 10^{-3} \text{ v/v} = \frac{\gamma}{2\sqrt{2}} \quad (\because |V_{BS}| >> |2\phi_B|)$$

$$\gamma = 2\sqrt{2} \times 50 \times 10^{-3} = 0.141$$

From equation (iii)

$$\gamma = \sqrt{\frac{2\in_s qN_a}{\frac{\in_{ox}}{t_{ox}}}}$$

$$= \frac{t_{ox}\sqrt{2\in_s qN_a}}{\in_{ox}}$$

$$\gamma = \frac{t_{ox}\sqrt{2\in_s qN_a}}{\in_{ox}} \quad ...(iv)$$

Now, Square on both the sides

$$\gamma^2 = \left(\frac{t_{ox}}{\epsilon_{ox}}\right)^2 \times 2\epsilon_s q N_a$$

So, $$N_a = \gamma^2\left(\frac{\epsilon_{ox}}{t_{ox}}\right)^2 \times \frac{1}{2\epsilon_s q}$$

By putting respective values

$$= (0.14)^2\left(\frac{4\times 8.854\times 10^{-14}}{10\times 10^{-9}\times 10^2}\right)$$

$$\times \frac{1}{2\times 12\times 8.854\times 10^{-4}\times 1.6\times 10^{-19}}$$

$$= \frac{2.491\times 10^{-27}}{3.3984\times 10^{-43}} = 7.329\times 10^{15}$$

$$\boxed{\text{So}, N_a = 7.37\times 10^{15}\,\text{cm}^{-3}}$$

Hence, option (a) is the correct answer.

31. As given combination circuit:

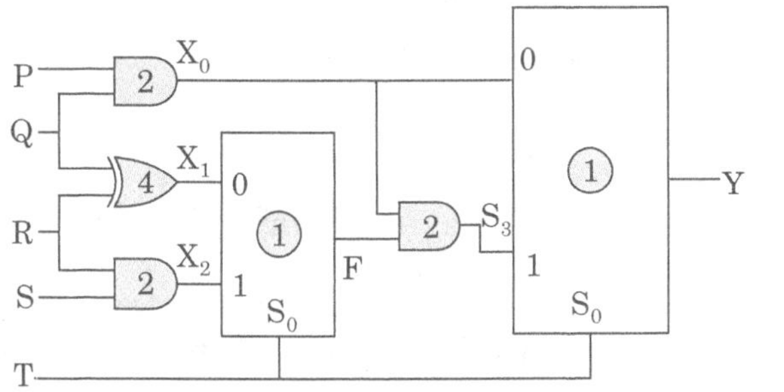

When T = 0, to get output only signal x_0 is essential for second MUX, so the delay involved in this case is 2n sec for x_0, MUX delay 1 nsec, total 2 + 1 = 3 nsec.

When T = 1, to get output Y we need X_2 and X_3, X_2 will be available at 2n sec (delay of AND gate)

F will be available after 1 n-sec of X_2 2 + 1 = 3 n sec (i.e., delay of MUX with X_2),

X_3 will available after 2 n sec of F i.e., 3 + 2 = 5 n sec

Y will be available after 1 n sec of X_3 i.e., 5 + 1 = 6 n sec

When T = 0, propagation delay (Y) at 3 n sec

T = 1, propagation delay Y at 6 n sec

Hence, Worst case delay is 6 n sec.

32. As given that

The content of Registers

R_1 = 25 H, R_2 = 30 H, R_3 = 40 H

PUSH $\{R_1\} \to$ content of R_1 will be pushed to top 2 bytes of stack

PUSH $\{R_2\} \to$ content of R_2 will be pushed to top 2 bytes of stack

PUSH $\{R_3) \to$ content of R_3 will be pushed to top 2 bytes of stack

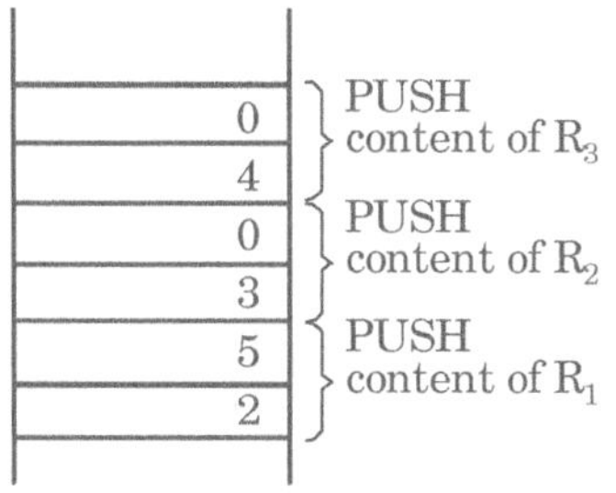

POP $R_1 \to$ Top 2 types of stack will be popped to R_1, So R_1 = 40H

POP $R_2 \to$ Top 2 bytes of stack will be popped to R_2, So R_2 = 30 H

POP $R_3 \to$ Top 2 bytes of stack will be popped R_3, So R_3 = 25H

Hence, after execution of above 6 instructions

R_1 = 40H, R_2 = 30H, R_3 = 25H.

Hence, option (a) is the correct answer.

33. As given that:

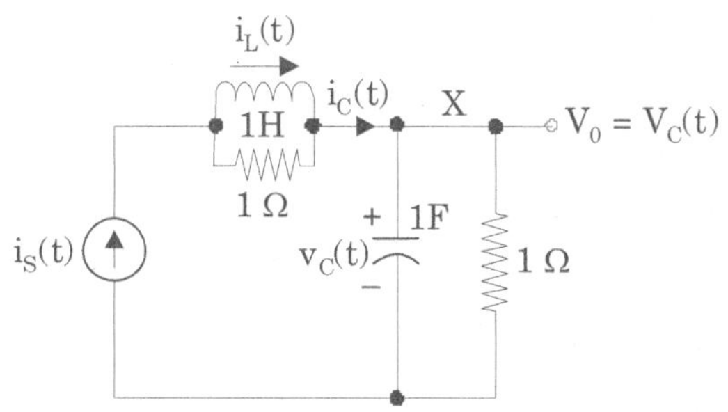

Here, $i_L(t)$ and $V_C(t)$ are state variables

At node X by KCL

$$-i_S(t) + 1\frac{dV_C(t)}{dt} + \frac{V_C(t)}{1} = 0$$

$$V_C'(t) = \frac{dV_C(t)}{dt}$$

$$= i_S(t) - V_C(t)$$

Now, by current-division principle

$$i_L(s) = \frac{i(s)\times 1}{s+1}$$

$$i_S(s) = \frac{di_L(t)}{dt} + i_L(t)$$

$$i_L(s) = \frac{di_L(t)}{d(t)} = i_S(t) - i_L(t)$$

So, state matrix A can be obtained by

$$\begin{bmatrix} i_L(t) \\ V_C(t) \end{bmatrix} = \begin{bmatrix} -1 & 0 \\ 0 & -1 \end{bmatrix}\begin{bmatrix} i_L(t) \\ V_C(t) \end{bmatrix} + \begin{bmatrix} 1 \\ 1 \end{bmatrix} i_S(t)$$

$$A = \begin{bmatrix} -1 & 0 \\ 0 & -1 \end{bmatrix} \quad B = \begin{bmatrix} 1 \\ 1 \end{bmatrix}$$

C = controllability = [B AB]

$$AB = \begin{bmatrix} -1 & 0 \\ 0 & -1 \end{bmatrix}\begin{bmatrix} 1 \\ 1 \end{bmatrix} = \begin{bmatrix} -1 \\ -1 \end{bmatrix}$$

$$C = \begin{bmatrix} 1 & -1 \\ 1 & -1 \end{bmatrix}$$

$$|C| = \begin{bmatrix} 1 & -1 \\ 1 & -1 \end{bmatrix} = -1 + 1 = 0$$

$$\text{Observability} = 0 = \begin{bmatrix} C \\ CA \end{bmatrix}$$

$$\boxed{V_0(t) = V_C(t) + i_S(t)}$$

$$O = \begin{bmatrix} 0 & 1 \\ 0 & -1 \end{bmatrix}$$

$|O| = 0$

C = [0 1]

$$C_A = [0\ \ 1]\begin{bmatrix} -1 & 0 \\ 0 & -1 \end{bmatrix} = [0\ \ 1]$$

Hence, option(d) system is neither state controllable non observable.

34. As we know that

Hamming distance : Distance between two code words is the number of places the codeword differ.

Hamming weight : of a codeword is equal to the numbers of non-zero components.

[If hamming weight of the code words formed by module-2]

As Given that, $[m_1, c_1]$ = [1100; 0101100]

$[m_2, c_2]$ = [1110,0011110]

$[m_3, c_3]$ = [0110; 1000110]

c_1 = 0101100

∴ c_2 = 0011110

c_3 = 1000110

Now by taking option 4 code word 0110 100 = c_4

$d(c_1, c_2) = 3 = W(c_1 \oplus c_2) = W(0110010) = 3$

$d(c_1, c_3) = 4 = W(c_1 \oplus c_3) = W(1101010) = 4$

$d(c_1, c_4) = 2 = W(c_1 \oplus c_4) = W(0011000) = 2$

Hence option (d) is valid code word.

35.

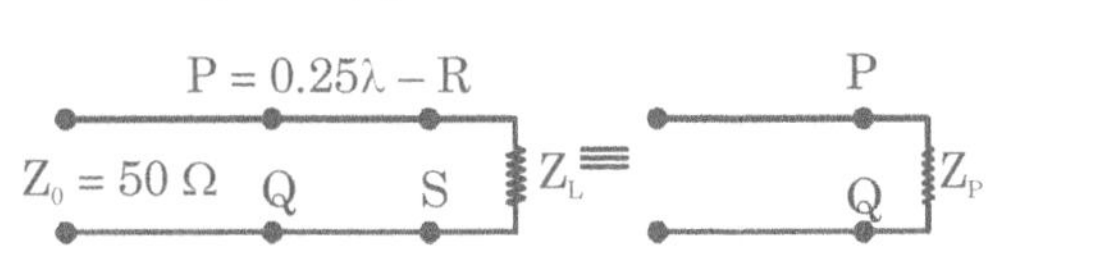

Impedance across RS = $Z_L \parallel \infty = Z_L$

Load impedance across PQ

$$= \frac{Z_0^2}{0} \parallel \frac{(75)^2}{Z_L} = \infty \parallel \frac{(75)^2}{Z_L} \parallel \frac{(75)^2}{Z_L}$$

For matching $50\ \Omega = \frac{(75)^2}{Z_L} \Rightarrow Z_L = 112.5$

Hence, option (a) is the correct answer.

36. As we know that

The characteristic equation is $|A - \lambda I| = 0$

$$\begin{bmatrix} x-\lambda & -3 \\ 3 & 4-\lambda \end{bmatrix} = 0$$

$\Rightarrow (x - \lambda)(4 - \lambda) + 9 = 0$

$\Rightarrow \lambda^2 + (-4 - x)\lambda + (4_x + 9) = 0$, a quadratic equatin in λ

For repeated eigen values, we have

$B^2 - 4AC = 0$, where A = 1, B = – 4 – x, C = 4x + 9

$(-4 - x)^2 - 4(1)(4x + 9) = 0$

$16 + x^2 + 8x - 16x - 36 = 0$

$x^2 - 8x - 20 = 0$

$(x - 10)(x + 2) = 0$

$\Rightarrow$ x = 10 or – 2

∵ x is positive real number.

∴ x = 10

37. As given that

$D = (\rho \cos^2 \phi)a_\rho + (z^2 \sin^2 \phi)a_\phi$ be a vector in cylindrical system then

$$\text{div}(D) = \frac{1}{\rho}\left\{\frac{\partial}{\partial \rho}(\rho D_\rho) + \frac{\partial}{\partial \phi}(D_\phi) + \frac{\partial}{\partial z}(D_z)\right\}$$

Where $D_\rho = \rho \cos^2\phi$; $D_\phi = z^2 \sin^2\phi$; $D_z = 0$

$$\Rightarrow \quad \text{div}(D) = \frac{1}{\rho}\{2\rho \cos^2\phi;\ 2z^2 \sin\phi \cos\phi\}$$

$$= (1 + \cos 2\phi) + \frac{z^2 \sin 2\phi}{\rho}$$

∴ The net Electric flux of D leaving the closed surface of the cylinder is

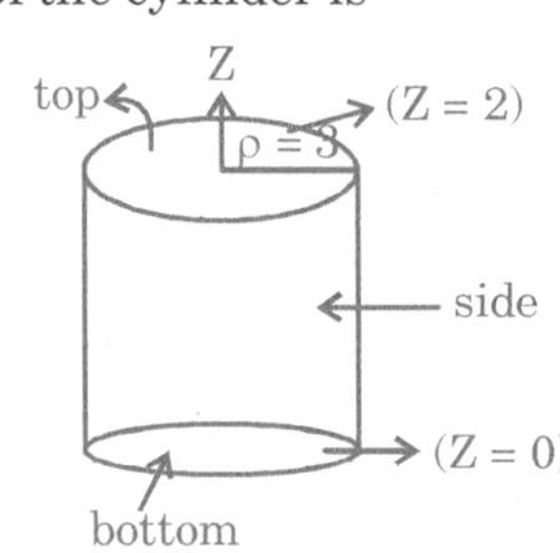

$$\Psi_{net} = \Psi_{bottom} + \Psi_{top} + \Psi_{side}$$

$$\Psi_{net} = \oiint_S D.d\vec{s} = \oiiint_V \text{div}(D)dV$$

(Using Gauss divergence theorem)

$$\Psi_{net} = \int_{\rho=0}^{3}\int_{\phi=0}^{2\pi}\int_{z=0}^{2}(1 + \cos 2\phi)\rho d\rho d\phi dz$$

$$+\int_{\rho=0}^{3}\int_{\phi=0}^{2\pi}\int_{z=0}^{2}\left(\frac{z^2\sin 2\phi}{\rho}\right)$$

$$\left(\because dV = \rho d\rho d\phi dz \text{ and } \begin{array}{r}'\rho' \text{ limits; 0 to 3,}\\ \phi \text{ limits; 0 to } 2\pi,\\ z \text{ limits; 0 to 2}\end{array}\right)$$

$$= \left(\frac{\rho^2}{2}\right)_0^3 \left(\phi + \frac{\sin 2\phi}{2}\right)_0^{2\pi} \times (z)_0^2 + (\ell n\rho)_0^3$$

$$\times\left(\frac{-\cos 2\phi}{2}\right)_0^{2\pi} \times \left(\frac{z^3}{3}\right)_0^2 = \left(\frac{9}{2}\right) \times (2\pi + 0) \times 2 + 0$$

$$\Psi_{net} = 18\pi \approx 56.55$$

38. As given Circuit:

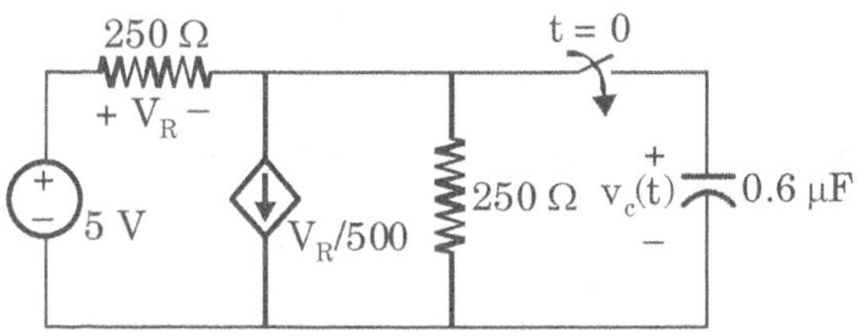

$V(0) = -5V = V_0$

for t < 0, is closed

For final value at t

= ∞ → Switch is closed → Capacitor is open circuit → Network in steady state

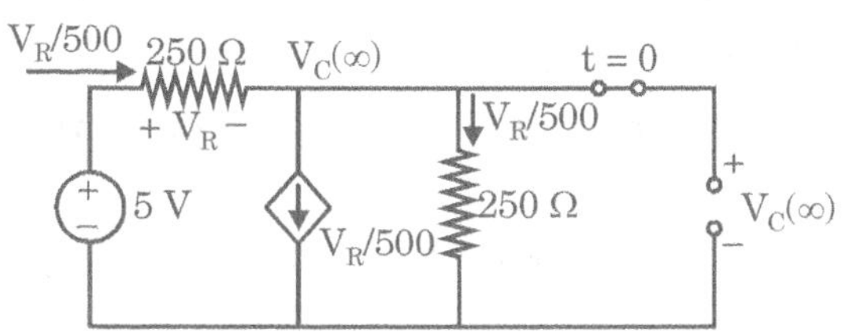

By KVL in outer loop

$$\Rightarrow 5 - V_R\left(\frac{V_R}{500}250\right) = 0$$

$$\Rightarrow 5 - V_R - \frac{V_R}{2} = 0$$

$$\Rightarrow \frac{3V_R}{2} = 0 \Rightarrow V_R = \frac{10}{3}V$$

$$\Rightarrow V_C(\infty) = \frac{V_R}{500} \times 250 = \frac{V_R}{2} = \frac{10}{6} = \frac{5}{3}V = 1.66\text{ V}$$

Time constant $\tau = R_{th}C$

$$R_{th} = \frac{V_{th}}{I_N}$$

- In the above circuit V_{th} and $V_C(\infty)$ are same

  So $V_{th} = \frac{5}{3}V$

- Computation of I_N

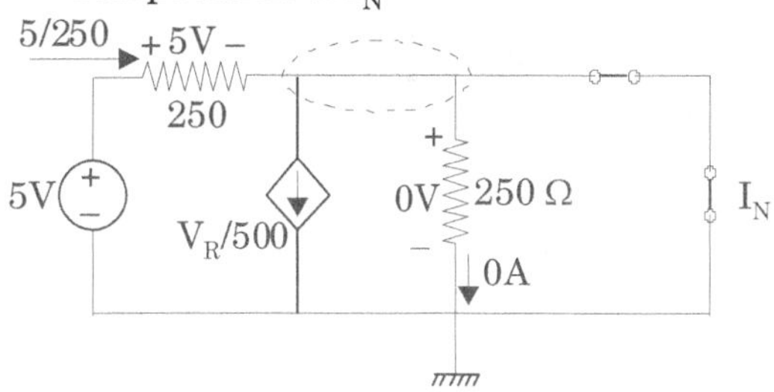

$$\text{By KCL } \frac{5}{250} = \frac{5}{500} + 0 + I_N$$

$$\Rightarrow I_N = \frac{5}{250} - \frac{5}{500} = \frac{1}{50} - \frac{1}{100} = \frac{2-1}{100} = \frac{1}{100}A$$

$$R_{th} = \frac{V_{th}}{I_N} = \frac{\frac{5}{3}}{\frac{1}{100}} = \frac{500}{3}\Omega$$

$$\tau = R_{th}C = \frac{500}{3} \times 0.6 \times 10^{-6} = 1 \times 10^{-4}\text{ sec}$$

$$V_C(t) = V_C(\infty) + \left[V_C(0^-) - V_C(\infty)\right]e^{-\frac{t}{\tau}}$$

$$= 1.66 + [-5 - 1.66]e^{-10^4 t}$$

$$\Rightarrow 0 = 1.66 - 6.66\, e^{-10^4 t}$$

$$\Rightarrow e^{-10^4 t} = \frac{1.66}{6.66}$$

$$\Rightarrow -10^4 t = \ell n\left(\frac{1.66}{6.66}\right)$$

$$\Rightarrow t = -10^{-4}\ell n\left(\frac{1.66}{6.66}\right)$$

$$= (-10^4)(1.389)$$

$$= 1.389 \times 10^{-4}\text{ sec} = 0.138\text{ msec}$$

39. As given that

For a periodic signal x(t) is

- Its period is 6, it is real, even.
- Average value of x(t) is 2 i.e., $a_0 = 2$.
- $a_k = \begin{cases} k; & k \le k \le 3 \\ 0; & k > 3 \end{cases} \Rightarrow \begin{array}{l} a_1 = 1 \\ a_2 = 2 \\ a_3 = 3 \end{array}$

If $|a_1| = 1$ then $|a_{-1}| = 1$

$|a_2| = 2$ then $|a_{-2}| = 2$ etc.

$|a_3| = 3$ then $|a_{-3}| = 3$ etc.

We need to compute its average power.

Now, by Parseval's theorem

$$P = \frac{1}{T}\int_0^T |x(t)|^2\,dt = \sum_{k=-\infty}^{\infty} |a_k|^2$$

$$= a_0^2 + 2\sum_{k=1}^{\infty} |a_k|^2$$

$$= a_0^2 + 2\sum_{k=1}^{3} |a_k|^2 \quad (\because\ a_k = 0 \text{ for } k > 3)$$

$$= a_0^2 + 2\left[|a_1|^2 + |a_2|^2 + |a_3|^3\right]$$

$$= 2^2 + 2[1^2 + 2^2 + 3^2]$$

$$P = 4 + 28 = 32 \text{ watt}$$

40. As given that

For unit step input u(n), the output is $2\delta(n + 1) + \delta(n) + \delta(n - 1)$. We need to obtain the system response for input $\left(\frac{1}{2}\right)^n u(n)$.

$$s(n) = 2\delta(n + 1) + \delta(n) + \delta(n - 1) \quad [s(n) : \text{step response}]$$

$$h(n) = s(n) - s(n - 1) \quad (h(n) : \text{impulse response})$$

$$= 2\delta(n + 1) + \delta(n) + \delta(n - 1) - [2\delta[(n-1)+1] + \delta(n-1) + \delta(n-1) - 1]$$

$$= 2\delta(n + 1) + \delta(n) + \delta(n - 1) - 2\delta(n) - \delta(n-1) - \delta(n-2)$$

$$= 2\delta(n + 1) - \delta(n) - \delta(n - 2)$$

If the input is $x(n) = \left(\frac{1}{2}\right)^n u(n)$ then its response is: $y(n) = x(n)*h(n)$

$$= \left[\left(\frac{1}{2}\right)^n u(n)\right] * [2\delta(n + 1) - \delta(n) - \delta(n - 2)]$$

$$= \left[\left(\frac{1}{2}\right)^n u(n) * 2\delta(n+1)\right] - \left[\left(\frac{1}{2}\right)^n u(n) * \delta(n)\right] - \left[\left(\frac{1}{2}\right)^n u(n) * \delta(n-2)\right]$$

$$y(n) = 2\left(\frac{1}{2}\right)^{n+1} u(n+1) - \left(\frac{1}{2}\right)^n u(n) - \left(\frac{1}{2}\right)^{n-2} u(n-2)$$

As we want Y(0), it is due to first 3 terms of y(n)

$$\text{So, } y(0) = 2\left(\frac{1}{2}\right)^1 u(1) - \left(\frac{1}{2}\right)^0 u(0) - \left(\frac{1}{2}\right)^2 u(-2)$$

$$= \left(2\times\frac{1}{2}\times 1\right) - (1\times 1) - \left[\left(\frac{1}{2}\right)^{-2} \times 0\right]$$

$$= 1 - 1 - 0 = 0$$

41. As given that

$$x(n) = 2^{n-1} u(-n + 2)$$
$$y(n) = 2^{-n+2} u(n + 1)$$

$$I = \frac{1}{2\pi}\int_0^{2\pi} X(e^{j\omega})Y(e^{-j\omega})d\omega = ?$$

Let, $P(e^{j\omega}) = X(e^{j\omega})Y(e^{-j\omega})$

Then $\frac{1}{2\pi}\int_0^{2\pi} P(e^{j\omega})d\omega = P(0) = P(n)|_{n=0}$

$$P(n) = \frac{1}{2\pi}\int_0^{2\pi} P(e^{j\omega})e^{-j\omega n}d\omega$$

If we make n = 0 in above equation then

$$P(0) = \frac{1}{2\pi}\int_0^{2\pi} P(e^{j\omega})d\omega$$

If $P(e^{j\omega}) = X(e^{j\omega})Y(e^{-j\omega})$

Then $P(n) = x(n) * y(-n) \quad [\because\ y(-n) \leftrightarrow Y(e^{-j\omega})]$

$P(n) = x(n) * g(n) \quad$ [Where $g(n) = y(-n)$]

$$= \sum_{k=-\infty}^{\infty} x(k)g(n - k) \quad [\because g(n) = 2^{n+2}u(-n + 1)]$$

$$= \sum_{k=-\infty}^{\infty} \left[2^{k-1}u(-k + 2)\right]\left[2^{n-k+2}u[-(n - k) + 1]\right]$$

$$= \sum_{k=-\infty}^{\infty} (2^{k-1})(2^{n-k+2})u(-k + 2)u(k - n + 1)$$

$$\Rightarrow P(0) = \sum_{k=-\infty}^{\infty} 2^{k-1}.2^{-k+2}\left[u(-k + 2).u(k + 1)\right]$$

$$\left[\begin{array}{l} u(-k + 2) = 1; -\infty \le k \le 2 \\ u(k + 1) = 1; -1 \le k \le \infty \\ u(-k + 2), u(k + 1) = \begin{cases} 1; & -1 \le k \le 2 \\ 0; & \text{else} \end{cases} \end{array}\right]$$

$$\Rightarrow \quad P(0) = \sum_{k=-1}^{2} 2^{k-1+(-k+2)}$$

$$= \sum_{k=-1}^{2} 2$$

$$= \sum_{k=-1}^{2} 2(1)^k$$

$$= 2(1)^{-1} + 2(1)^0 + 2(1)^1 + 2(1)^2$$

$$= (2 \times 1) + (2 \times 1) + (2 \times 1) + (2 \times 1)$$

$$= 2 + 2 + 2 + 2 = 8$$

42. According to the question

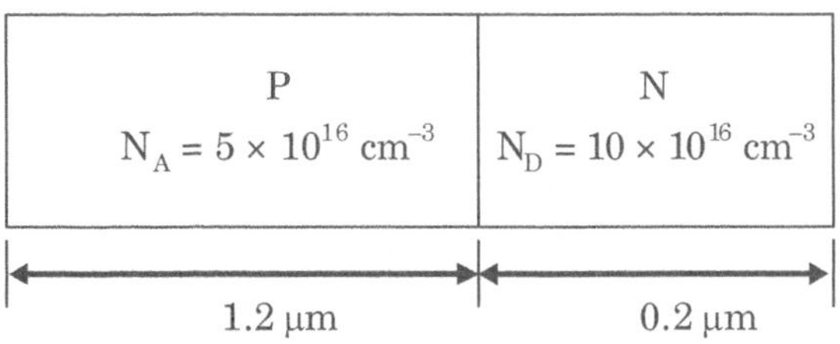

According to the junction law

$$\frac{W_P}{W_n} = \frac{N_D}{N_A} = \frac{10 \times 10^{16}}{5 \times 10^{16}} = 2$$

$$W_P = 2W_n$$

This indicates that, N-region will depleted entirely at first;

$$W = W_P + W_n$$
$$= 2W_n + W_n$$
$$W = 3W_n = 3 \times 0.2\ \mu m$$

$$\therefore \quad W = 0.6\ \mu m = 0.6 \times 10^{-4}\ cm \qquad (\because \in - \in_0 \in_r)$$

$$W = \sqrt{\left(\frac{2\in}{q}\right)\left(\frac{1}{N_A} + \frac{1}{N_D}\right).V_J}$$

(where $V_J = V_{bi} + V_{RB}$)

$$= \sqrt{\left(\frac{2\in}{q}\right)\left(\frac{1}{N_A} + \frac{1}{N_D}\right)(V_{bi} + V_{RB})}$$

$$V_{RB} = \frac{W_q^2 \times N_A.N_D}{2\in(N_A + N_D)} - V_{bi}$$

$$V_{RB} = \frac{(0.6 \times 10^{-4})^2 \times 1.6 \times 10^{-19} \times 5 \times 10^{16} \times 10 \times 10^{16}}{2 \times 12 \times 8.85 \times 10^{-14}(5 \times 10^{16} + 10 \times 10^{16})} - 0.8V$$

$$= 9.095\ V - 0.8\ V = 8.24\ V$$

$$\boxed{\text{Hence, } V_{RB} = 8.24\ V}$$

43. As given that,

$V_{in} = 10V$

$T_{on} = 1m$ sec,

$T_{off} = 1\ \mu$ sec

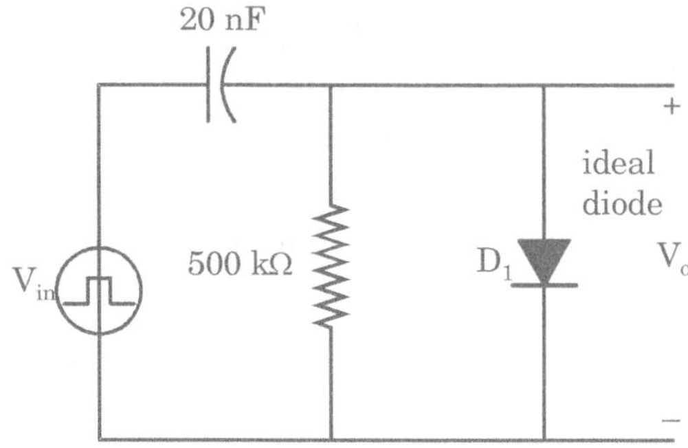

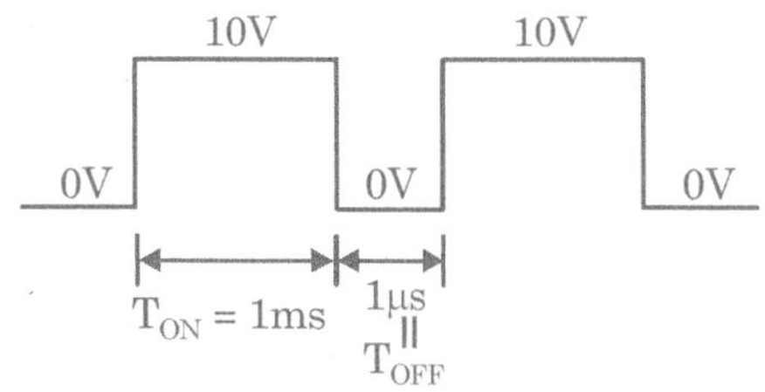

$$T = T_{ON} + T_{OFF} = 1.001\ m\ sec.$$

$$\tau = RC = 20 \times 10^{-9} \times 500 \times 10^3 = 10\ ms$$

So, $\tau >> T$

In T_{ON} duration diode be in conduction and their equivalent circuit is

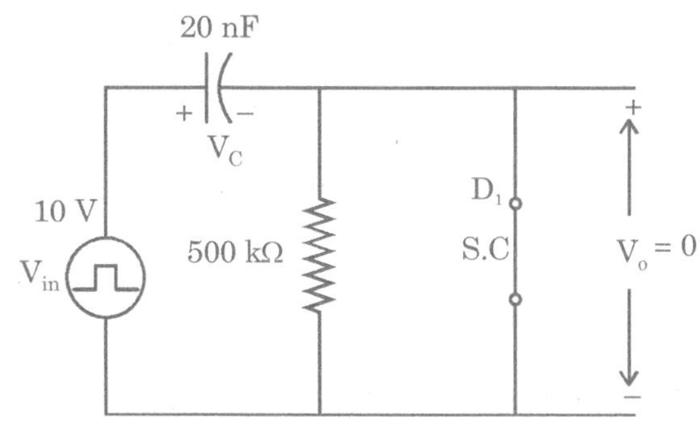

$$+V_{in} - V_C = 0V$$

$$\Rightarrow \quad \boxed{V_C = V_{in} = +10V}$$

$$\boxed{V_0 = 0V} \qquad ....(i)$$

Once the capacitor in fully charge and $\tau >> T$; so diode will not conduct further and their equivalent circuit is

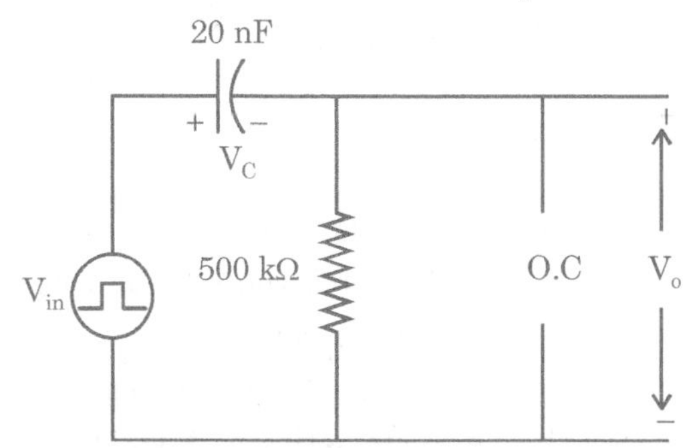

$$+V_{in} - V_C - V_0 = 0V$$

$$V_0 = V_{in} - V_C$$

In steady state: when $V_{in} = 10V$

$$V_0 = 10V - 10V = 0V$$

When $V_{in} = 0V$, then

$$V_0 = 0 - 10V$$

$$V_0 = -10\ V \qquad ....(ii)$$

So, the output waveform in steady state is

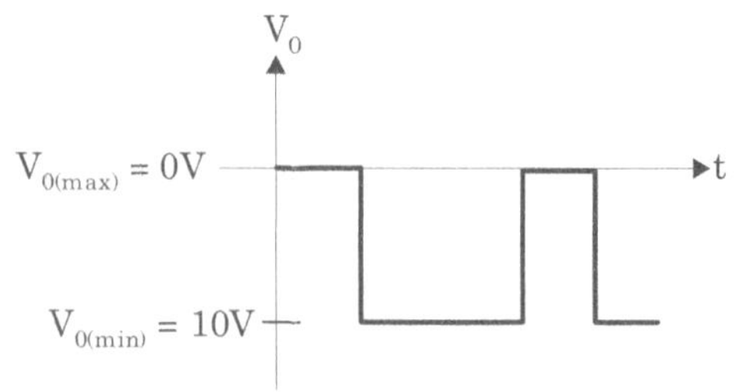

Hence, the difference between maximum and minimum voltage of the output wave form.

$$V_{0(differential)} = V_{0(max)} - V_{0(min)} = 0V - (-10V) = 10V$$

44. As given circuit:

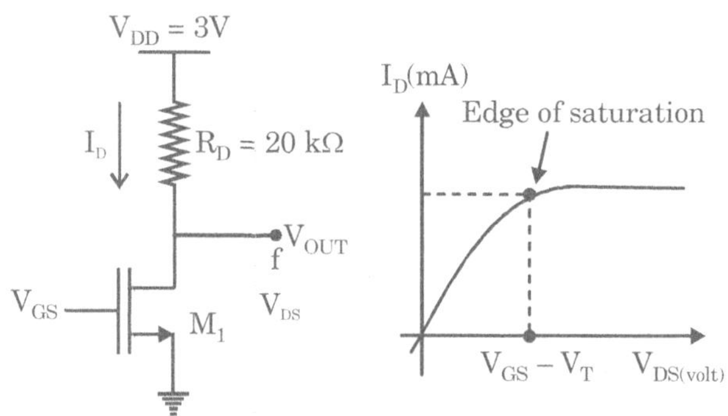

Edge of saturation $V_{DS} = (V_{GS} - V_T)$

$$I_{D(sat)} = \frac{1}{2}\mu_n C_{ox}\frac{W}{L}(V_{DS})^2$$

$$\frac{3-1}{20}\text{mA} = \frac{1}{2} \times 100\ \mu\text{A/v}^2 \times 10(V_{GS} - V_T)^2$$

$$\frac{1}{10}\text{mA} = \frac{1}{2} \times \frac{\text{mA}}{V^2} \times (1 - V_T)^2$$

$$(1 - V_T)^2 = \frac{2}{10}V^2 = \frac{1}{5}V^2$$

$$1 - V_T = \pm\ 0.447\ \text{volt}$$

$$V_T = 1 - 0.447\ \text{volt} = 0.552\ \text{volt}$$

$$1 - V_T = -\ 0.447\ \text{volt}$$

$$V_T = 1 + 0.447 = 1.447\ \text{volt}$$

As per the characteristics of MOS transition $V_G - V_T > 0V$

Hence, $\boxed{V_T = 0.55\ \text{volt}}$ is accepted value.

45. As given circuit:

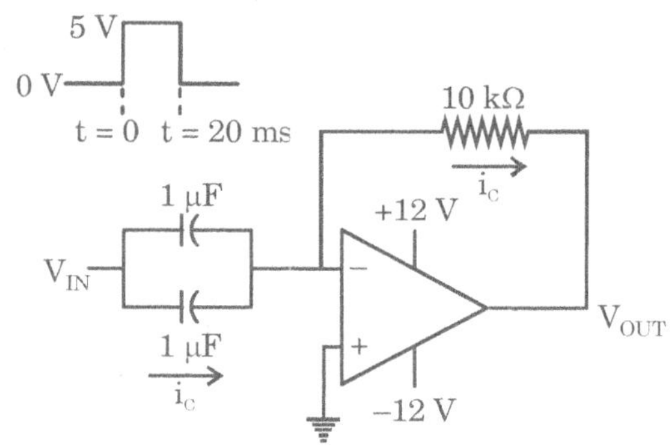

Initially capacitors are uncharged, therefore $V_{C1} = V_{C2} = 0$ means capacitors will behave as short cirict.

$$i_C = \left[\frac{0 - V_0}{R}\right] = C\frac{dV_{in}}{dt}$$

$$V_0 = -RC\frac{d}{dt}V_{in}$$

$$= -RC\ [\text{Slope of input } V_{in}]$$

$$= -RC(\infty)$$

Capacitors doesn't allow sudden change in voltage, so at $t = 0^+$, $V_{c1} = V_{c2} = 0V$

So, the voltage at $V_- = 5V$ att = 0 and $V_+ = +V$

$V_- > V_+ \Rightarrow V_+ - V_- < 0V$, so op-amp will produce, saturated negative output voltage.

Hence $\boxed{V_{out} = -12V}$

46. As given circuit:

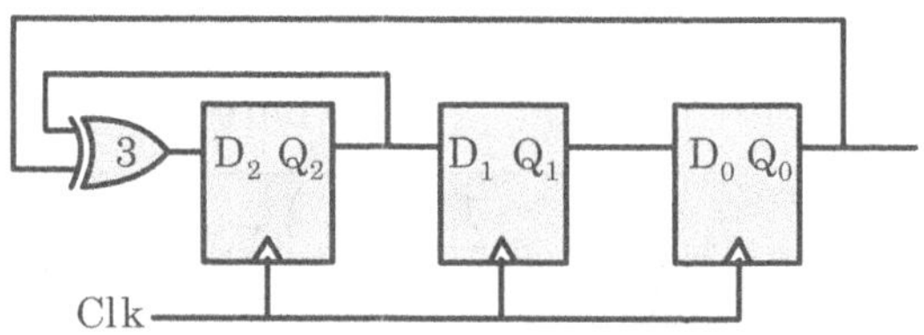

$f_{clk} = 500$ MHz, $T_{clk} = 2$n secm $T_{XOR} = 3$ns,

Initially $Q_2\ Q_1\ Q_0 = 111$, $D_2 = 1$

Delay of XOR gate is 3 n sec.

We need to obtain after how many clock contents of $Q_2\ Q_1\ Q_0$ becomes100

Its clear from above circuit that

$$Q_2^+ = D_2,\ Q_1^+ = D_1,\ Q_0^+ = D_0$$

Now, draw the waveform's which will give a clear picture.

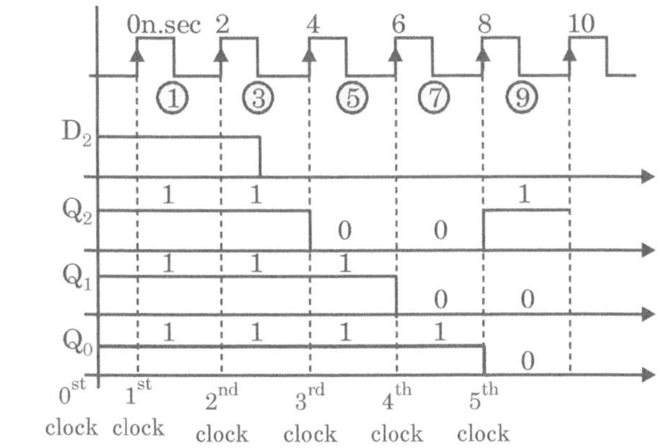

$Q_2Q_1Q_0$: 111 111 011 001 100

We can see after 5th clock content $Q_2\ Q_1\ Q_0$ is100

Here, XOR gate will change its output after 3n sec of its new input, which is happening here at 3nsec and 7n sec

47. As given circuit:

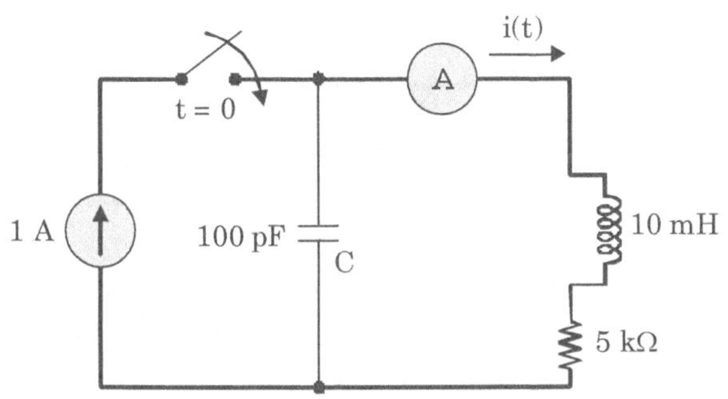

As given that $V_C(0^-) = 0V$

$$i_L(0^-) = 0A$$

In Laplace domain the above circuit for t > 0 becomes

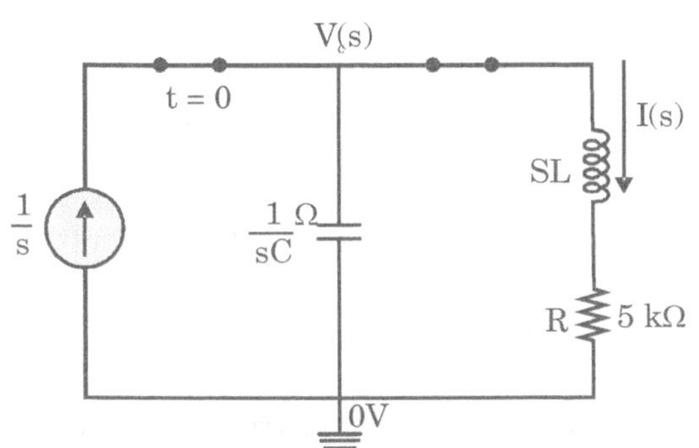

Here, ammeter replaced by short circuit as its internal resistance given is 0Ω.

Now, by current division rule

$$I(s) = \frac{1}{s}\frac{\frac{1}{sC}}{R+SL+\frac{1}{sC}}$$

$$= \frac{1}{s(s^2LC+SCR+1)}$$

$$= \frac{\frac{1}{LC}}{s\left(s^2+\frac{R}{L}s+\frac{1}{LC}\right)}$$

$$i(\infty) = \lim_{t\to\infty} SI(s) = 1$$

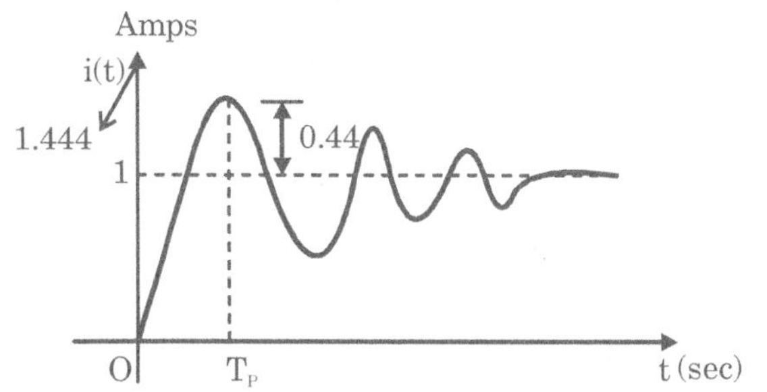

The expression of I(S) is identical to unit step response of a standard 2nd order closed loop system i.e.,

$$Y(s) = \frac{\omega_n^2}{s\left(s^2+2\xi w_n s+\omega_n^2\right)}$$

Now, by comparison $\omega_n = \frac{1}{\sqrt{LC}}$

$$= \frac{1}{\sqrt{10\times10^{-3}\times100\times10^{-12}}}$$

$$= 10^6 \text{ r/sec}$$

$$2\xi\omega_n = \frac{R}{L}$$

$$\Rightarrow \quad \xi = \frac{R}{2L\omega_n}$$

$$= \frac{5\times10^3}{2\times10\times10^{-3}\times10^6}$$

$$\xi = \frac{1}{4} = 0.25$$

$$\xi = 0.25$$

underdamped response:

$$\text{Overshoot} = e^{\frac{-\pi\xi}{\sqrt{1-\xi^2}}} = e^{\frac{-0.25A}{\sqrt{1-0.25^2}}} = 0.4443$$

Hence, the maximum ammeter reading just after the switch closed is:

$$\frac{i_L(t)}{\max} = i(\infty) + 0.444 = 1 + 0.444 = 1.444 \text{ A}$$

48. As given block diagram

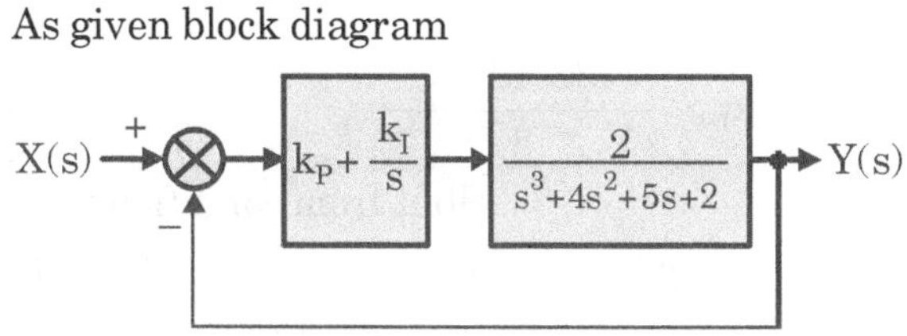

The characteristic equation of the system is

$$1 + G(s) = 0$$

$$1+\left(k_p+\frac{k_I}{s}\right)\left(\frac{2}{s^3+4s^2+5s+2}\right)=0$$

$$\Rightarrow 1+\left(\frac{k_p s+k_I}{s}\right)\left(\frac{2}{s^3+4s^2+5a+2}\right)=0$$

$\Rightarrow s(s^3 + 4s^2 + 5s + 2) + 2k_ps + 2k_I = 0$

$\Rightarrow s^4 + 4s^3 + 5s^2 + 2s + 2k_ps + 2k_I = 0$

$\Rightarrow s^4 + 4s^3 + 5s^2 + (2 + 2k_p)s + 2k_I = 0$

From necessary condition $\Rightarrow K_p > -1, k_1 > 0$

For stability all roots of characteristic equation must be negative i.e., the closed loop poles should lie on left half of s-plane.

By R-H criterion.

$$\begin{array}{c|ccc} s^4 & 1 & 5 & 2k_I \\ s^3 & 4 & 2+2k_p & 0 \\ s^2 & \dfrac{18-2k_p}{4} & 2k_I & 0 \\ s^1 & \dfrac{\left(\dfrac{18-2k_p}{4}\right)(2+2k_p)-8k_p}{\dfrac{18-2k_p}{4}} & & 0 \\ s^0 & 2k_I & & \end{array}$$

For stability all element of 1st column must be positive.

Condition 1 : $2k_I > 0 \Rightarrow k_I > 0$

Condition 2 : $\left[\dfrac{18-2k_p}{4}\right] > 0 \Rightarrow k_p < 9$

Condition 3 : $\left[\dfrac{\left(\dfrac{18-2k_p}{4}\right)(2+2k_p)-8k_1}{\dfrac{18-2k_p}{4}}\right] > 0$

$$\Rightarrow (2+2k_p)-\left(\frac{8k_I\times 4}{18-2k_p}\right)>0$$

$$\Rightarrow 1+k_p>\frac{16k_I}{18-2k_p}$$

$$\Rightarrow 1+k_p>\frac{8k_I}{9-k_p}$$

$$\Rightarrow k_I>\frac{(1+k_p)(9-k_p)}{8}$$

[but from condition 2 : $k_p < 9$]

If we try to plot the graph between k_I and k_P k_I

$$= \frac{(1+k_p)(9-k_p)}{8}$$

For maxima or minima $\dfrac{dk_I}{dk_P} = 0$

$\Rightarrow \dfrac{1}{8}[1 + k_P(-1) + (9 - k_P)] = 0$

$\Rightarrow 9 - k_P = 1 + k_P$

$\Rightarrow k_P = 4$

If, $k_P = 0$, $k_I = \dfrac{9}{8} = 1.125$

$k_P = 4, k_I = 3.125$

$k_P = 9, k_I = 0$

So, the plot between k_P and k_I will be

The standard part is satisfying the condition $(k_I > 0)$ and $(k_P < 9)$

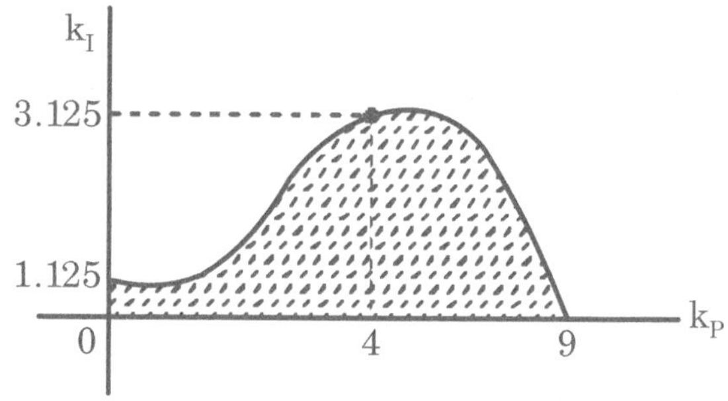

But in the range of k_P (i.e.,) $(0 < k_P < 9)$, the maximum possible value of k_I is 3.125 when $k_P = 4$.

Hence, $(k_I)_{max}$: 3.125

49. As given that, $V_{rms} = 4$

$f_m = 1$ KHz

$K_P = 2$ rad/Volt

$\therefore Vrms = \dfrac{V_m}{\sqrt{2}}$

$V_m = V_{rms}\sqrt{2} = 4\sqrt{2}$

$x_{pm}(t) = A_C \cos(2\pi f_c t + k_p m(t))$

Instantaneous phase $\phi_1(t) = 2\pi f_c t + k_p m(t)$

Instantaneous frequency $\omega_1(t) = 2\pi f_c + k_p \dfrac{d}{dt} m(t)$

Maximum instantaneous frequency (in Hz)

$$f_{i\,max} = f_C + \frac{1}{2\pi}k_p\left(\frac{d}{d+}m(t)\right)_{max}$$

$$= 1000\times 10^3 + \frac{1}{2\pi}\times 2\times 4\sqrt{2}\times 10^3$$

$$= 1000 \times 10^3 + 11.312 \times 10^3$$

$m(t) = 4\sqrt{2}\sin\left(2\pi\times 10^3 t\right)$

Hence maximum instantaneous frequency is 1011312 kHz.

50. As given that

$f_S = 600$ kHz

$f_{LO} = 1000$ kHz

f_{SI} = Image frequency

$F_{SI} = F_S + 2f_{IF}$

$\therefore f_{IF} = 1000k - 600k = 400$ k

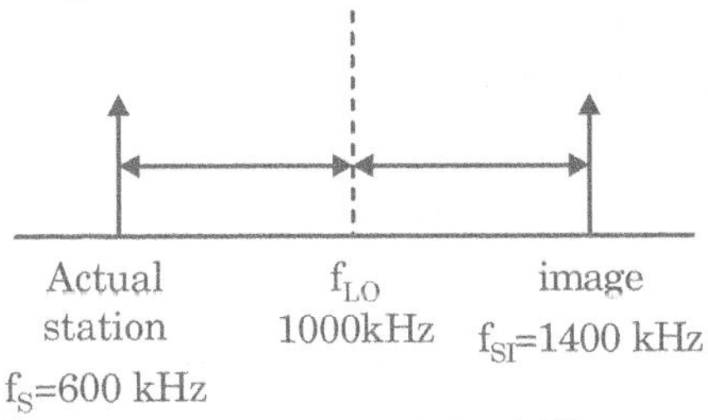

$f_{SI} = 600 + 2 \times 400 = 1400$ kHz

51. As given probabilities:

Students: Boys 50%, $P(B) = \frac{1}{2}$; Girls 50%, $P(G) = \frac{1}{2}$

Subjects: Science 75%, $P(S) = \frac{3}{4}$; Commerce 25%, $P(C) = \frac{1}{4}$

It is given that conditional probability of Boy being science student is half for boy being commerce student.

Let us consider

$$P\left\{\frac{\text{Boy}}{\text{Science}}\right\} = x$$

$$\Rightarrow P\left\{\frac{\text{Boy}}{\text{Commerce}}\right\} = 2x$$

$$\Rightarrow P\{\text{Boy}\} = P\left\{\frac{\text{Boy}}{\text{Science}}\right\} P\{\text{Science}\} + P\left\{\frac{\text{Boy}}{\text{Commerece}}\right\} P\{\text{Commerce}\}$$

$$\Rightarrow \frac{1}{2} = x\frac{3}{4} + 2x.\frac{1}{4}$$

$$\Rightarrow x = \frac{2}{5}$$

$$\Rightarrow P\left\{\frac{\text{Boy}}{\text{Commerece}}\right\} = \frac{4}{5}$$

$$\Rightarrow P\left\{\frac{\text{Girl}}{\text{Commerce}}\right\} = 1 - P\left\{\frac{\text{Boy}}{\text{Commerece}}\right\} = \frac{1}{5}$$

Now,

$$P\left\{\frac{\text{Commerce}}{\text{Girl}}\right\} = \left[\frac{P\left\{\frac{\text{Girl}}{\text{Commerce}}\right\}.P\{\text{Commerce}\}}{P\{\text{Girl}\}}\right]$$

$$= \frac{\frac{1}{5}\times\frac{1}{4}}{\frac{1}{2}} = \frac{1}{10}$$

$$\Rightarrow I\left(\frac{\text{Commerce}}{\text{Girl}}\right) = \log_2\left(\frac{1}{\frac{1}{110}}\right)$$

$$= \log_2 10 = 3.32 \text{ bits}$$

52. As we know that

Signal to quantization noise ratio $(SNR)_q$

$$= \frac{\text{signal power}}{\text{Quantization noise power}}$$

Signal power = $(RMS)^2 = (0.1)^2 = 0.01$

Quantization step size Δ

$$= \frac{x(t)_{max} - x(t)_{min}}{L} = \frac{2}{L}$$

Data rate = n.Fs

Where 'n' in number of encoder bits.

F_S is sampling frequency.

As given that

$n.F_S \le 50 \times 10^3$ bits per second

Minimum $F_S = 2 \times 5$ kHz = 10 kHz

$\Rightarrow n \times 10 \times 10^3 \le 50 \times 10^3$

$\Rightarrow n \le 5$, taking maximum n = 5

$$\text{Quantization step size } \Delta = \frac{\text{Peak to peak Voltage}}{2^n}$$

$= \frac{2}{2^5} = \frac{1}{2^4}$, quantization noise power $= \frac{\Delta^2}{12}$ to maximum $(SNR)_q$, quantization noise power should be minimized.

$$\Rightarrow (SNR)_{q\,max} = \frac{\text{Signal power}}{\text{Noise power}} = \frac{(0.01)}{\frac{(1/24)^2}{12}}$$

$$= \frac{0.01 \times 12}{\left(\frac{1}{16}\right)^2} = 30.72$$

53. Optimum threshold for a maximum a posteriori (MAP) receiver under AWGN channel is

$$V_{th} = \left(\frac{a_1 + a_2}{2}\right) + \frac{\sigma_n^2}{a_1 - a_2}\ell_n\left[\frac{P_0}{P_1}\right]$$

$(\because a_1 = 2, a_2 = -2)$

$$= 0 + \frac{0.4}{2-(-2)}\ell_n\left(\frac{0.6}{0.4}\right)$$

$$= 0.1 \times 0.4 = 0.04 \text{ V}$$

54. As we know that

$$V_P = \frac{C}{\sqrt{1-\left(\frac{f_c}{f}\right)^2}},$$

Since a > b, dominant mode $f_C\,(TE)_{10}$:

$$f_C\left(TE_{10}\right) = \frac{C}{2a}$$

$$= \frac{3\times 10^8}{2\times 8\times 10^{-2}} = 1.875 \text{ GHz}$$

$$\Rightarrow \quad \frac{V_P}{C} = \frac{1}{\sqrt{1-\left(\frac{1.875}{3.4}\right)^2}}$$

$$= \frac{1}{\sqrt{0.695}} = 1.198$$

Hence, the factor $\left[\frac{VP}{C}\right]$ is 1.198.

55. As given that

For directive gain:

10 log D = 6

$D = 10^{0.6}$

= 3.98

P_{rad} = 16 kw

Distance (r) = 8 km

Since antenna gain is 6 dB

Total effective Isotropic, Radiated power

EIRP = 16 × 4 = 64 kW

Power density at a distance of 8 km

$$= \frac{64\times 10^3}{4\pi\times(8000)^2}$$

$$\Rightarrow \quad \frac{|E_0|^2}{2\times 120\pi} = \frac{64\times 10^3}{4\pi\times(8000)^2}$$

$$\Rightarrow \quad E_0 = \sqrt{\frac{64\times 60\times 10^3}{8000\times 8000}} = 0.245 \text{ V/m}$$

SOLVED PAPER – 2020

ELECTRONICS & COMMUNICATION ENGINEERING

No. of Questions : 65 *Maximum Marks : 100*

INSTRUCTIONS

1. Total of 65 questions carrying 100 marks, out of which 10 questions carrying a total of 15 marks are in General Aptitude (GA)
2. The Engineering Mathematics will carry around **15% of the total marks,** the General Aptitude section will carry **15% of the total marks** and the **remaining 70% of the total marks.**
3. **Types of Questions**
   (*a*) **Multiple Choice Questions (MCQ)** carrying 1 or 2 marks each in all papers and sections. These questions are objective in nature, and each will have a choice of four options, out of which the candidate has to mark the correct answer(s).
   (*b*) **Numerical Answer Questions** of 1 or 2 marks each in all papers and sections. For these questions the answer is a real number, to be entered by the candidate using the virtual keypad. No choices will be shown for these type of questions.
4. For **1-mark** multiple-choice questions, **1/3 marks** will be deducted for a wrong answer. Likewise, for **2-marks** multiple-choice questions, 2/3 marks will be deducted for a wrong answer. There is no negative marking for numerical answer type questions.

Chapter-Wise Analysis

#	Chapters	Marks		
		1	2	Total
1	General Aptitude	5	5	15
2	Engineering Mathematics	4	3	10
3	Network Theory	3	2	7
4	Signals and Systems	2	2	6
5	Electronic Devices & Circuits	2	4	10

#	Chapters	Marks		
		1	2	Total
6	Analog Circuits	3	5	13
7	Digital Circuits & Microprocessor	3	3	9
8	Control Systems	2	4	10
9	Communication Systems	3	4	11
10	Electromagnetic Theory	3	3	9

GENERAL APTITUDE (GA)

Q. 1 to Q. 5 : Carry One Mark Each.

1. The untimely loss of life is a cause of serious global concern as thousands of people get killed ____ accidents every year while many other die ____ diseases like cardio vascular disease, cancer, etc.

(a) in, of (b) from, of

(c) during, from (d) from, from

2. He was not only accused of theft ___ of conspiracy.

(a) rather (b) but also

(c) but even (d) rather than

3. Select the word that fits the analogy:

Explicit : Implicit :: Express : _____

(a) Impress (b) Repress

(c) Compress (d) Suppress

4. The Canadian constitution requires that equal importance be given to English and French. Last year, Air Canada lost a lawsuit, and had to pay a six- figure fine to a French-speaking couple after they filed complaints about formal in-flight announcements in English lasting 15 seconds, as opposed to informal 5 second messages in French.

The French-speaking couple were upset at ____.

(a) the in-flight announcements being made in English.

(b) the English announcements being clearer than the French ones.

(c) the English announcements being longer than the French ones.

(d) equal importance being given to English and French.

5. A superadditive function f(.) satisfies the following property

$$f(x_1 + x_2) \geq f(x_1) + f(x_2)$$

Which of the following functions is a superadditive function for x > 1 ?

(a) e^x (b) $\sqrt{x}$

(c) $\frac{1}{x}$ (d) e^{-x}

Q. 6 to Q. 10 : Carry Two Marks Each.

6. The global financial crisis in 2008 is considered to be the most serious world-wide financial crisis, which started with the sub-prime lending crisis in USA in 2007. The sub-prime lending crisis led to the banking crisis in 2008 with the collapse of Lehman Brothers in 2008. The sub-prime lending refers to the provision of loans to those borrowers who may have difficulties in repaying loans, and it arises because of excess liquidity following the East Asian crisis.

Which one of the following sequences shows the correct precedence as per the given passage?

(a) East Asian crisis — Sub-prime lending crisis — Banking crisis — Global financial crisis

(b) Sub-prime lending crisis — Global financial crisis — Banking crisis — East Asian crisis

(c) Banking crisis — Sub-prime lending crisis — Global financial crisis — East Asian crisis

(d) Global financial crisis — East Asian crisis — Banking crisis — Sub-prime lending crisis

7. It is quarter past three in your watch. The angle between the hour hand and the minute hand is ___

(a) 0° (b) 7.5°

(c) 15° (d) 22.5°

8. A circle with centre O is shown in the figure. A rectangle PQRS of maximum possible area is inscribed in the circle. If the radius of the circle is a, then the area of the shaded portion is _____.

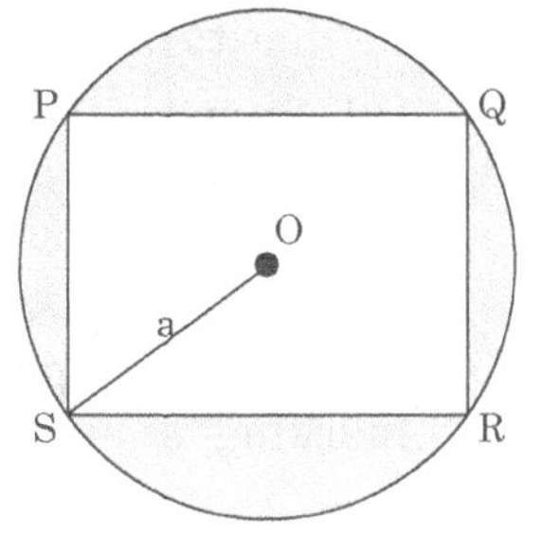

(a) $\pi a^2 - a^2$

(b) $\pi a^2 - \sqrt{2}a^2$

(c) $\pi a^2 - 2a^2$

(d) $\pi a^2 - 3a^2$

9. a, b, c are real numbers. The quadratic equation $ax^2 - bx + c = 0$ has equal roots, which is β, then

(a) $\beta = b/a$ (b) $\beta^2 = ac$

(c) $\beta^3 = bc/(2a^2)$ (d) $b^2 \neq 4ac$

10. The following figure shows the data of students enrolled in 5 years (2014 to 2018) for two schools P and Q. During this period, the ratio of the average number of the students enrolled in school P to the average of the difference of the number of students enrolled in schools P and Q is ____

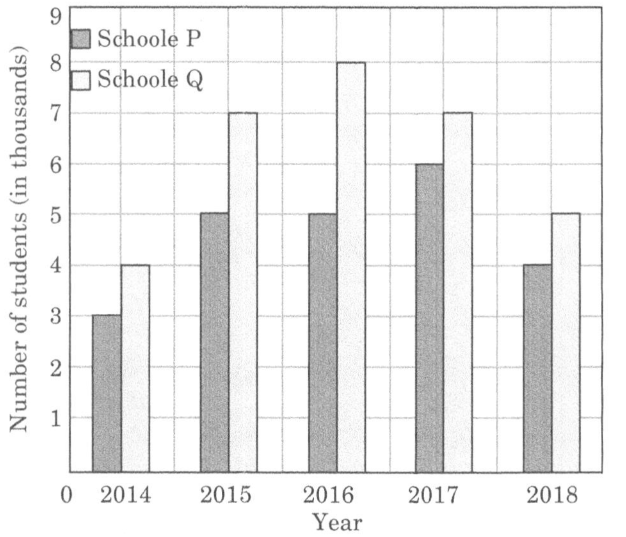

(a) 8 : 23 (b) 23 : 8

(c) 23 : 31 (d) 31 : 23

ELECTRONICS & COMMUNICATION ENGINEERING

Q. 1 to Q. 25 : Carry One Mark Each.

1. If v_1 , v_2, ..., v_6 are six vectors in $\mathbb{R}^4$, which one of the following statements is FALSE ?

(a) It is not necessary that these vectors span $\mathbb{R}^4$

(b) These vectors are not linearly independent

(c) Any four of these vectors form a basis for $\mathbb{R}^4$

(d) If $\{v_1, v_3, v_5, v_6\}$ spans $\mathbb{R}^4$, then it forms a basis for $\mathbb{R}^4$

2. For a vector field $\vec{A}$, which one of the following is FALSE ?

(a) $\vec{A}$ is solenoidal if $\nabla . \vec{A} = 0$

(b) $\nabla \times \vec{A}$ is another vector field.

(c) $\vec{A}$ is irrotational if $\nabla^2 \vec{A} = 0$.

(d) $\nabla \times (\nabla \times \vec{A}) = \nabla(\nabla . \vec{A}) - \nabla^2 \vec{A}$

3. The partial derivative of the function

$$f(x, y, z) = e^{1-x\cos y} + xze^{-\frac{1}{(1+y^2)}}$$

with respect to x at the point (1, 0, e) is

(a) – 1 (b) 0

(c) 1 (d) $\frac{1}{e}$

4. The general solution of $\frac{d^2y}{dx^2} - 6\frac{dy}{dx} + 9y = 0$ is

(a) $y = C_1e^{3x} + C_2e^{-3x}$

(b) $y = (C_1 + C_2x)e^{-3x}$

(c) $y = (C_1 + C_2x)e^{3x}$

(d) $y = C_1e^{3x}$

5. The output y[n] of a discrete-time system for an input x[n] is

$$y[n] = \max_{-\infty \le k \le n} |x[k]|$$

The unit impulse response of the system is

(a) 0 for all n

(b) 1 for all n

(c) unit step signal u[n]

(d) unit impulse signal δ[n]

6. A single crystal intrinsic semiconductor is at a temperature of 300K with effective density of states for holes twice that of electrons. The thermal voltage is 26mV. The intrinsic Fermi level is shifted from mid-bandgap energy level by

(a) 18.02 meV (b) 9.01 meV

(c) 13.45 meV (d) 26.90 meV

7. Consider the recombination process via bulk traps in a forward biased pn homojunction diode. The maximum recombination rate is U_{max}. If the electron and the hole capture cross-sections are equal, which one of the following is FALSE ?

(a) With all other parameters unchanged, U_{max} decreases if the intrinsic carrier density is reduced.

(b) U_{max} occurs at the edges of the depletion region in the device

(c) U_{max} depends exponentially on the applied bias

(d) With all other parameters unchanged, U_{max} increases if the thermal velocity of the carriers increases.

8. The components in the circuit shown below are ideal. If the op-amp is in positive feedback and the input voltage V_i is a sine wave of amplitude 1 V, the output voltage V_0 is

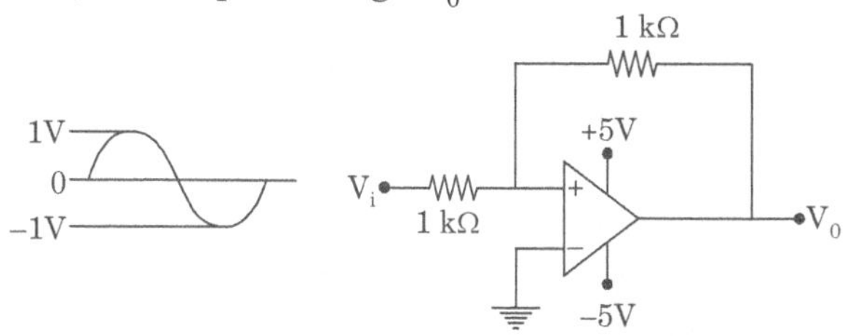

(a) a non-inverted sine wave of 2V amplitude.

(b) an inverted sine wave of 1V amplitude.

(c) a square wave of 5V amplitude.

(d) a constant of either +5 V or – 5V.

9. In the circuit shown below, the Thevenin voltage V_{TH} is

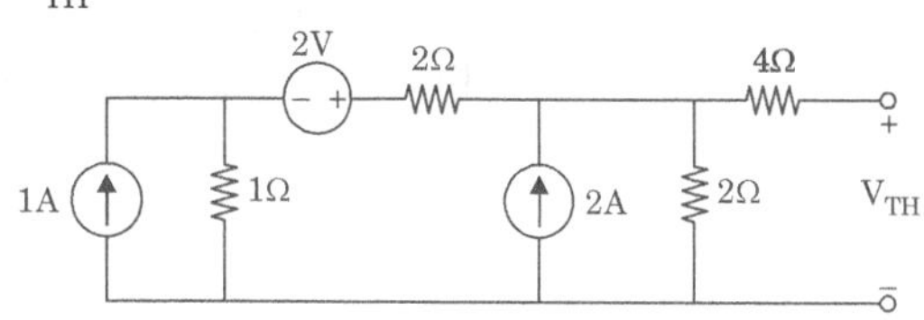

(a) 2.4 V (b) 2.8 V

(c) 3.6 V (d) 4.5 V

10. The figure below shows a multiplexer where S_1 and S_0 are the select lines, I_0 to I_3 are the input data lines, EN is the enable line, and F(P, Q, R) is the output. F is

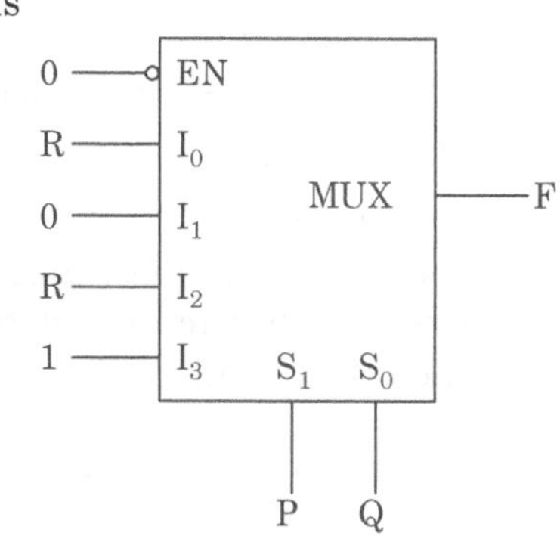

(a) $PQ + \bar{Q}R$

(b) $P + Q\bar{R}$

(c) $P\bar{Q}R + \bar{P}Q$

(d) $\bar{Q} + PR$

11. The pole-zero map of a rational function G(s) is shown below. When the closed contour Γ is mapped into the G(s)-plane, then the mapping encircles.

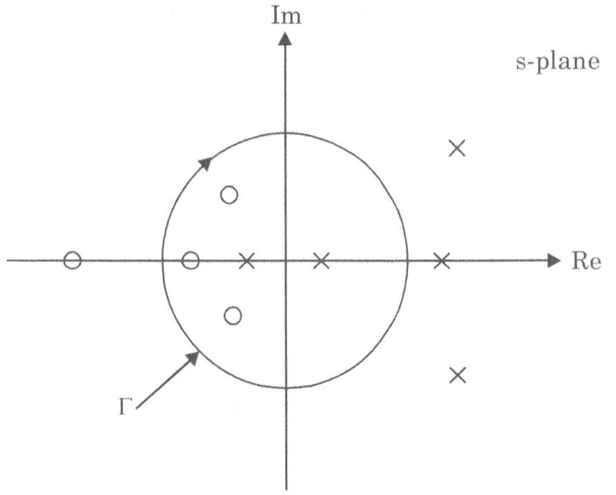

(a) the origin of the G(s)-plane once in the counter-clockwise direction.

(b) the origin of the G(s)-plane once in the clockwise direction.

(c) the point $-1 + j0$ of the G(s)-plane once in the counter-clockwise direction.

(d) the point $-1 + j0$ of the G(s)-plane once in the clockwise direction.

12. A digital communication system transmits a block of N bits. The probability of error in decoding a bit is α. The error event of each bit is independent of the error events of the other bits. The received block is declared erroneous if at least one of its bits is decoded wrongly. The probability that the received block is erroneous is

(a) $N(1 - \alpha)$

(b) α^N

(c) $1 - \alpha^N$

(d) $1 - (1 - \alpha)^N$

13. The impedances $Z = jX$, for all X in the range $(-\infty, \infty)$, map to the Smith chart as

(a) a circle of radius 1 with centre at (0, 0)

(b) a point at the centre of the chart.

(c) a line passing through the centre of the chart.

(d) a circle of radius 0.5 with centre at (0.5, 0)

14. Which one of the following pole-zero plots corresponds to the transfer function of an LTI system characterized by the input-output difference equation given below?

$$y[n] = \sum_{k=0}^{3} (-1)^k x[n-k]$$

(a)

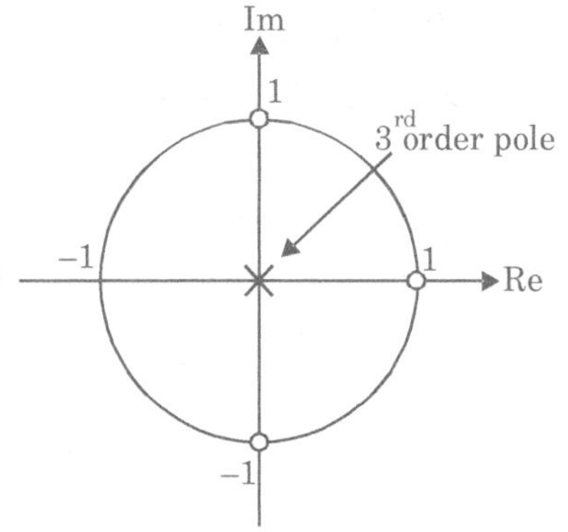

(b)

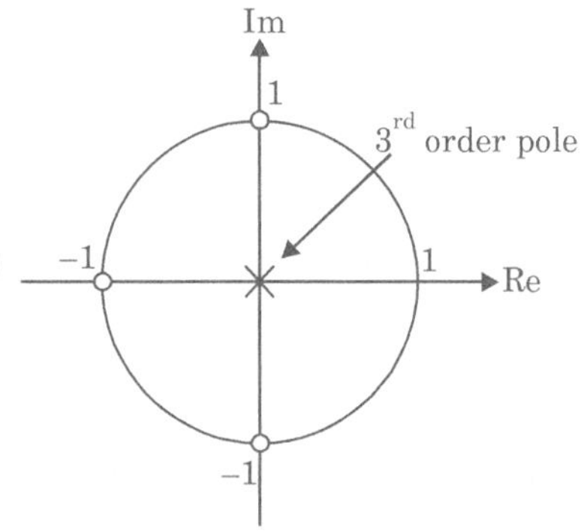

(c)

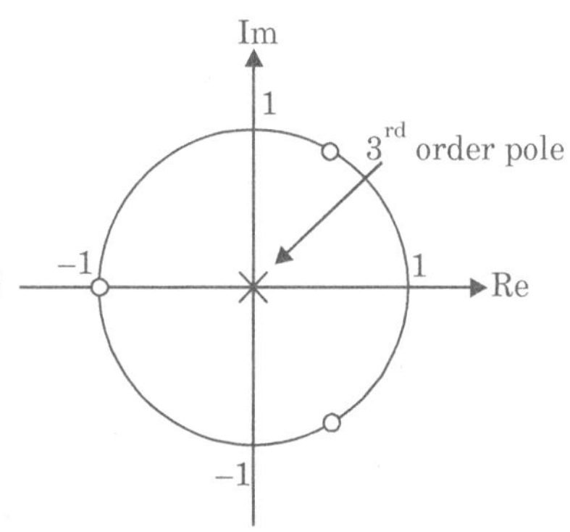

(d)

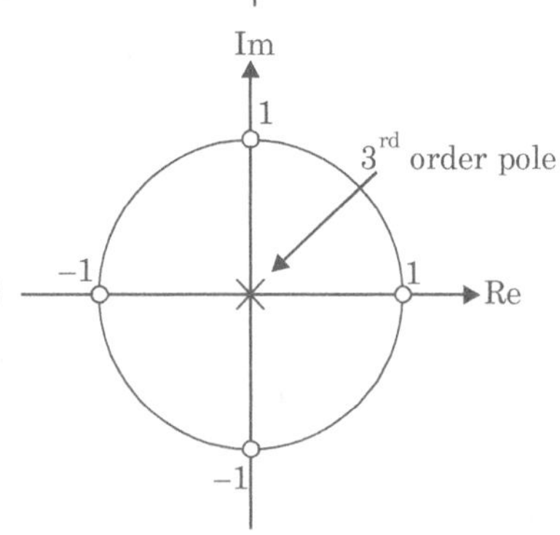

15. In the given circuit, the two-port network has the impedance matrix $Z = \begin{bmatrix} 40 & 60 \\ 60 & 120 \end{bmatrix}$. The value of Z_L for which maximum power is transferred to the load is ____ Ω.

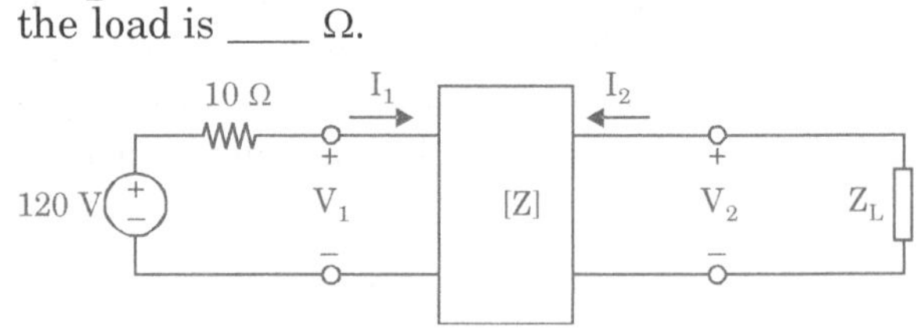

16. The current in the RL-circuit shown below is $i(t) = 10\cos(5t - \pi/4)$A.

The value of the inductor **(rounded off to two decimal places)** is ____ H.

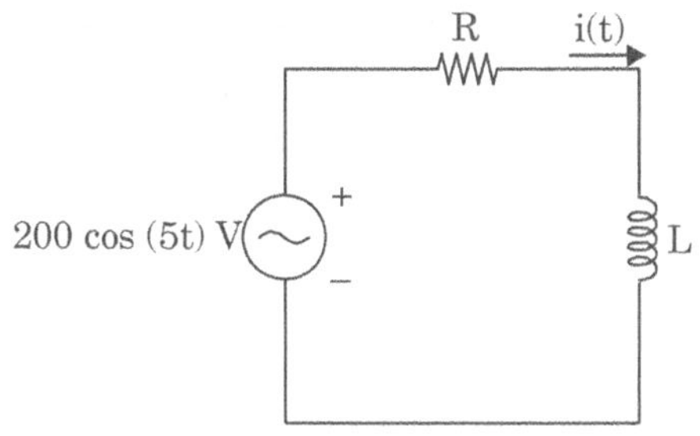

17. In the circuit shown below, all the components are ideal and the input voltage is sinusoidal. The magnitude of the steady-state output V_0 **(rounded off to two decimal places)** is ______ V.

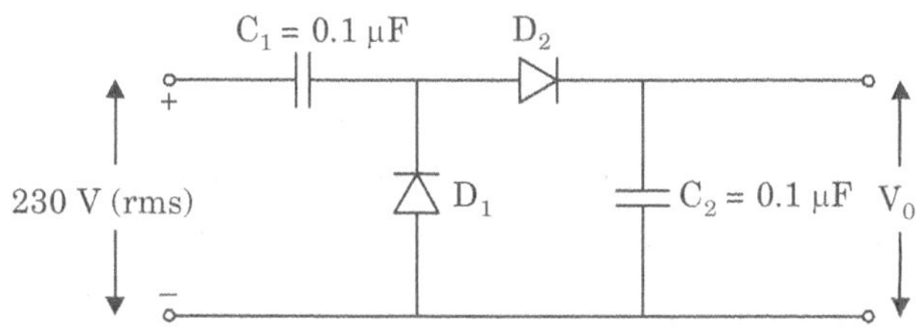

18. In the circuit shown below, all the components are ideal. If V_i is +2V, the current I_0 sourced by the op-amp is _____ mA.

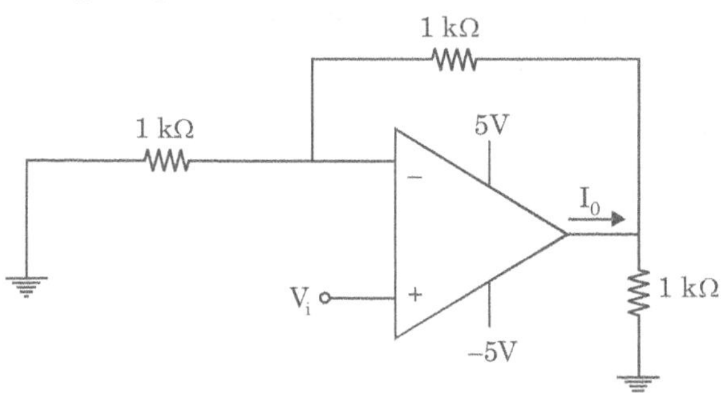

19. In an 8085 microprocessor, the number of address lines required to access a 16K byte memory bank is ______.

20. A 10-bit D/A converter is calibrated over the full range from 0 to 10V. If the input to the D/A converter is 13A (in hex), the output **(rounded off to three decimal places)** is ____ V.

21. A transmission line of length $3\lambda/4$ and having a characteristic impedance of 50Ω is terminated with a load of 400 Ω. The impedance **(rounded off to two decimal places)** seen at the input end of the transmission line is _____ Ω.

22. A binary random variable X takes the value +2 or –2. The probability $P(X = +2) = \alpha$. The value of α **(rounded off to one decimal place)**, for which the entropy of X is maximum, is ____.

23. The loop transfer function of a negative feedback system is

$$G(s)H(s) = \frac{K(s + 11)}{s(s + 2)(s + 8)}$$

The value of K, for which the system is marginally stable, is ______.

24. The random variable

$$Y = \int_{-\infty}^{\infty} W(t)\,\phi(t)dt$$

$$\text{Where } \phi(t) - \begin{cases} 1; & 5 \le t \le 7 \\ 0; & \text{otherwise} \end{cases}$$

and W(t) is a real white Gaussian noise process with two-sided power spectral density $S_W(f) = 3$ W/Hz, for all f. The variance of Y is ____.

25. The two sides of a fair coin are labelled as 0 and 1. The coin is tossed two times independently. Let M and N denote the labels corresponding to the outcomes of those tosses. For a random variable X, defined as X = min(M, N), the expected value E(X) **(rounded off to two decimal places)** is _____.

Q. 26 to Q. 55 : Carry Two Marks Each.

26. Consider the following system of linear equations.

$x_1 + 2x_2 = b_1$; $2x_1 + 4x_2 = b_2$; $3x_1 + 7x_2 = b_3$; $3x_1 + 9x_2 = b_4$

Which one of the following conditions ensures that a solution exists for the above system?

(a) $b_2 = 2b_1$ and $6b_1 - 3b_3 + b_4 = 0$

(b) $b_3 = 2b_1$ and $6b_1 - 3b_3 + b_4 = 0$

(c) $b_2 = 2b_1$ and $3b_1 - 6b_3 + b_4 = 0$

(d) $b_3 = 2b_1$ and $3b_1 - 6b_3 + b_4 = 0$

27. Which one of the following options contains two solutions of the differential equation $\frac{dy}{dx} = (y - 1)x$?

(a) $\ln|y - 1| = 0.5x^2 + C$ and $y = 1$

(b) $\ln|y - 1| = 2x^2 + C$ and $y = 1$

(c) $\ln|y - 1| = 0.5x^2 + C$ and $y = -1$

(d) $\ln|y - 1| = 2x^2 + C$ and $y = -1$

28. The current I in the given network is

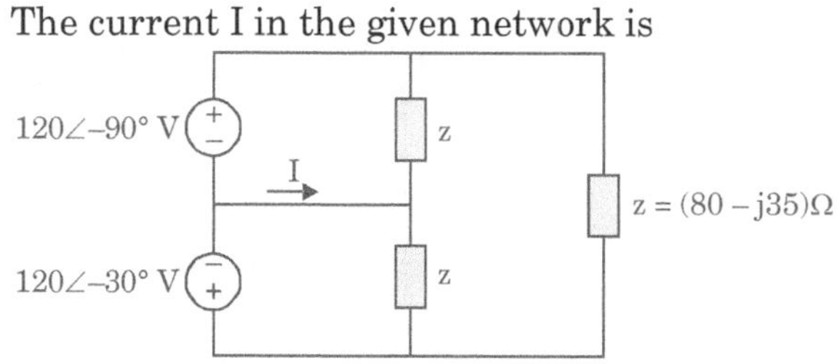

(a) 0A (b) 2.38 ∠–96.37° A

(c) 2.38 ∠143.63° A (d) 2.38 ∠– 23.63° A

29. A finite duration discrete time signals x[n] is obtained by sampling the continuous-time signal $x(t) = \cos(200\pi t)$ of sampling instants $t = \frac{n}{400}$, n = 0, 1, 2, ... 7. The 8-point discrete Fourier transform (DFT) of x[n] is defined as

$$X[k] = \sum_{n=0}^{7} x[n]e^{-j\frac{\pi kn}{4}}, k = 0, 1, ..., 7$$

Which one of the following statements is TRUE?

(a) All X[k] are non-zero

(b) Only X[4] is non-zero

(c) Only X[2] and X[6] are non-zero

(d) Only X[3] and X[5] are non-zero

30. For the given circuit, which one of the following is the correct state equation?

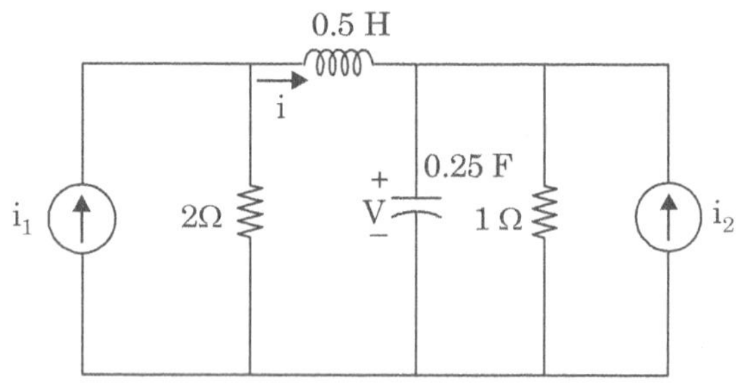

(a) $\frac{d}{dt}\begin{bmatrix} v \\ i \end{bmatrix} = \begin{bmatrix} -4 & 4 \\ -2 & -4 \end{bmatrix}\begin{bmatrix} v \\ i \end{bmatrix} + \begin{bmatrix} 0 & 4 \\ 4 & 0 \end{bmatrix}\begin{bmatrix} i_1 \\ i_2 \end{bmatrix}$

(b) $\frac{d}{dt}\begin{bmatrix} v \\ i \end{bmatrix} = \begin{bmatrix} -4 & -4 \\ -2 & 4 \end{bmatrix}\begin{bmatrix} v \\ i \end{bmatrix} + \begin{bmatrix} 4 & 4 \\ 4 & 0 \end{bmatrix}\begin{bmatrix} i_1 \\ i_2 \end{bmatrix}$

(c) $\frac{d}{dt}\begin{bmatrix} v \\ i \end{bmatrix} = \begin{bmatrix} 4 & -4 \\ -2 & -4 \end{bmatrix}\begin{bmatrix} v \\ i \end{bmatrix} + \begin{bmatrix} 0 & 4 \\ 4 & 4 \end{bmatrix}\begin{bmatrix} i_1 \\ i_2 \end{bmatrix}$

(d) $\frac{d}{dt}\begin{bmatrix} v \\ i \end{bmatrix} = \begin{bmatrix} -4 & -4 \\ -2 & -4 \end{bmatrix}\begin{bmatrix} v \\ i \end{bmatrix} + \begin{bmatrix} 4 & 0 \\ 0 & 4 \end{bmatrix}\begin{bmatrix} i_1 \\ i_2 \end{bmatrix}$

31. A one-sided abrupt pn junction diode has a depletion capacitance C_D of 50 pF at a reverse bias of 0.2V. The plot of $1/C_D^2$ versus the applied voltage V for this diode is a straight line as shown in the figure below. The slope of the plot is _____ $\times 10^{20}$ $F^{-2}V^{-1}$.

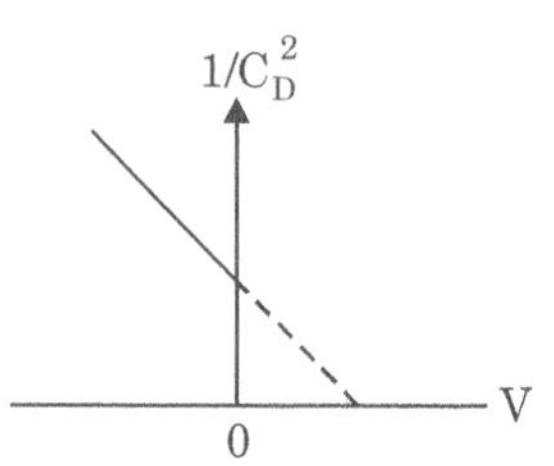

(a) – 5.7 (b) – 3.8

(c) – 1.2 (d) – 0.4

32. The band diagram of a p-type semiconductor with a band-gap of 1eV is shown. Using this semiconductor, a MOS capacitor having V_{TH} of – 0.16 V, C'_{ox} of 100 nF/cm² and a metal work function of 3.87 eV is fabricated. There is no charge within the oxide. If the voltage across the capacitor is V_{TH}, the magnitude of depletion charge per unit area (in C/cm²) is

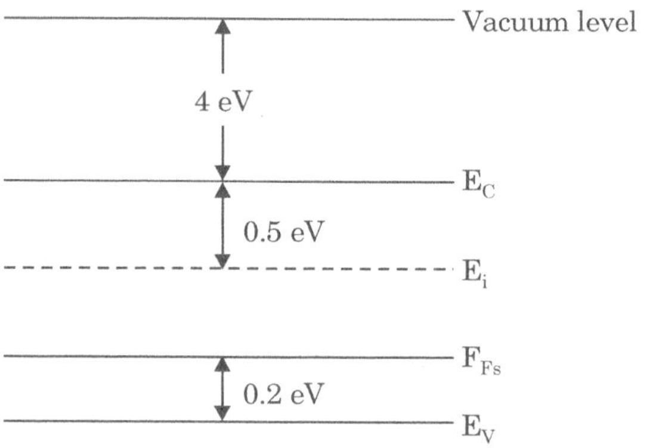

(a) 1.70×10^{-8} (b) 0.52×10^{-8}

(c) 1.41×10^{-8} (d) 0.93×10^{-8}

33. The base of an npn BJT T1 has a linear doping profile $N_B(x)$ as shown below. The base of another npn BJT T2 has a uniform doping N_B of 10^{17} cm^{-3}. All other parameters are identical for both the devices. Assuming that the hole density profile is the same as that of doping, the common-emitter current gain of T2 is

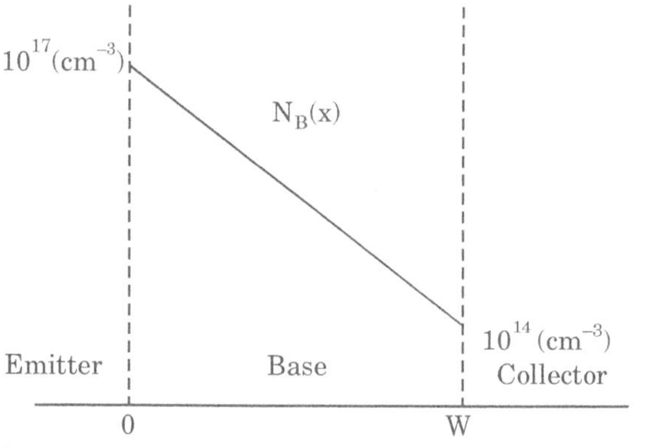

(a) approximately 2.0 times that of T1.

(b) approximately 0.3 times that of T1.

(c) approximately 2.5 times that of T1.

(d) approximately 0.7 times that of T1.

34. A pn junction solar cell of area 1.0 cm², illuminated uniformly with 100 mW cm^{-2}, has the following parameters: Efficiency = 15%, open circuit voltage = 0.7V, fill factor = 0.8, and thickness = 200 μm. The charge of an electron is 1.6×10^{-19} C. The average optical generation rate (in cm^{-3} s^{-1}) is

(a) 0.84×10^{19} (b) 5.57×10^{19}

(c) 1.04×10^{19} (d) 83.60×10^{19}

35. For the BJT in the amplifier shown below, V_{BE} = 0.7 V, kT/q = 26 mV. Assume that BJT output resistance (r_o) is very high and the base current is negligible. The capacitors are also assumed to be short circuited at signal frequencies. The input v_i is direct coupled. The low frequency voltage gain v_o/v_i of the amplifier is

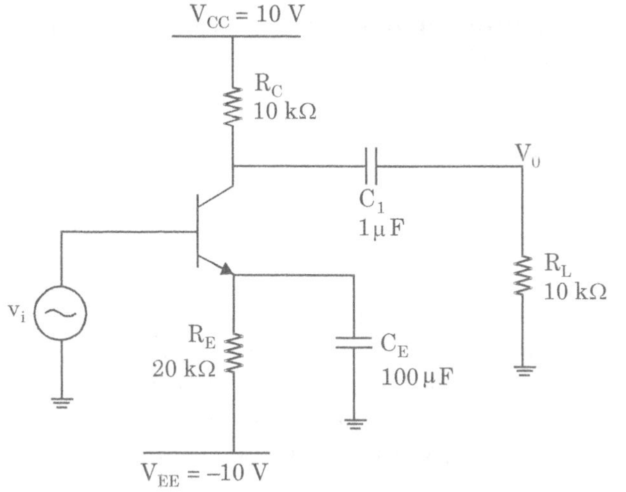

(a) – 89.42 (b) – 128.21
(c) – 178.85 (d) – 256.42

36. An ehhancement MOSFET of threshold voltage 3 V is being used in the sample and hold circuit given below. Assume that the substrate of the MOS device is connected to –10 V. If the input voltage V_1 lies between ±10 V, the minimum and the maximum values of V_G required for proper sampling and holding respectively, are

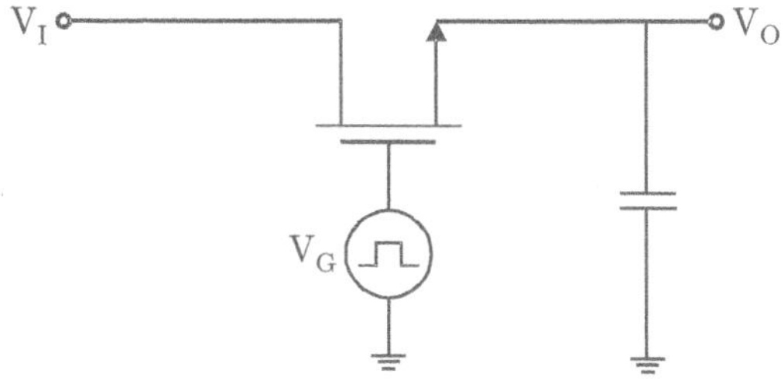

(a) 3 V and –3 V. (b) 10 V and –10 V.
(c) 13 V and –7V. (d) 10 V and –13 V.

37. Using the incremental low frequency small-signal model of the MOS device, the Norton equivalent resistance of the following circuit is

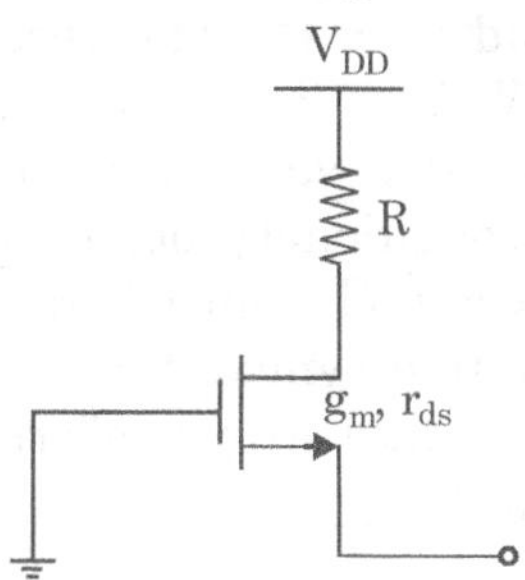

(a) $r_{ds} + R + g_m r_{ds} R$ (b) $\frac{r_{ds} + R}{1 + g_m r_{ds}}$
(c) $r_{ds} + \frac{1}{g_m} + R$ (d) $r_{ds} + R$

38. P, Q, and R are the decimal integers corresponding to the 4-bit binary number 1100 considered in signed magnitude, 1's complement and 2's complement representations, respectively. The 6-bit 2's complement representation of (P + Q + R) is

(a) 110101 (b) 110010
(c) 111101 (d) 111001

39. The state diagram of a sequence detector is shown below. State S_0 is the initial state of the sequence detector. If the output is 1, then

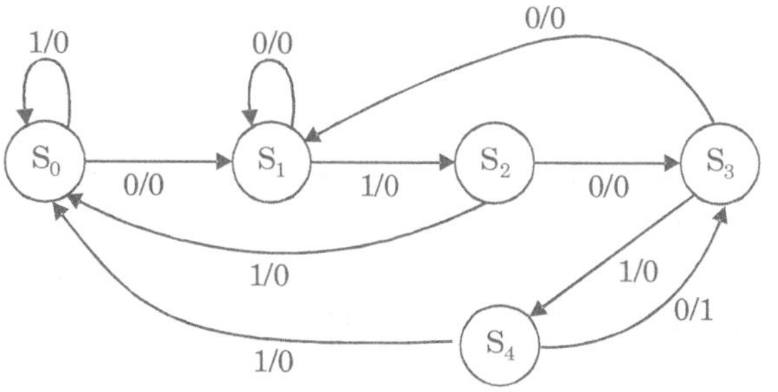

(b) the sequence 01010 is detected.
(d) the sequence 01011 is detected.
(a) the sequence 01110 is detected.
(c) the sequence 01001 is detected.

40. The characteristic equation of a system is

$$s^3 + 3s^2 + (K + 2)s + 3K = 0.$$

In the root locus plot for the given system, as K varies from 0 to ∞, the break-away or break-in point(s) lie within

(a) (–1, 0) (b) (– 2, – 1)
(c) (– 3, – 2) (d) (– ∞, – 3)

41. The components in the circuit given below are ideal. If R = 2 kΩ and C = 1 μF, the –3 dB cut-off frequency of the circuit in Hz is

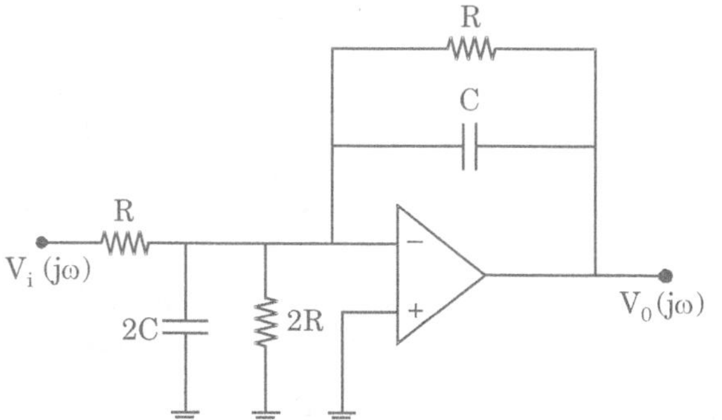

(a) 14.92 (b) 34.46
(c) 59.68 (d) 79.58

42. For the modulated signal $x(t) = m(t)\cos(2\pi f_c t)$, the message signal $m(t) = 4\cos(1000\pi t)$ and the carrier frequency f_c is 1 MHz. The signal x(t) is passed through a demodulator, as shown in the figure below. The output y(t) of the demodulator is

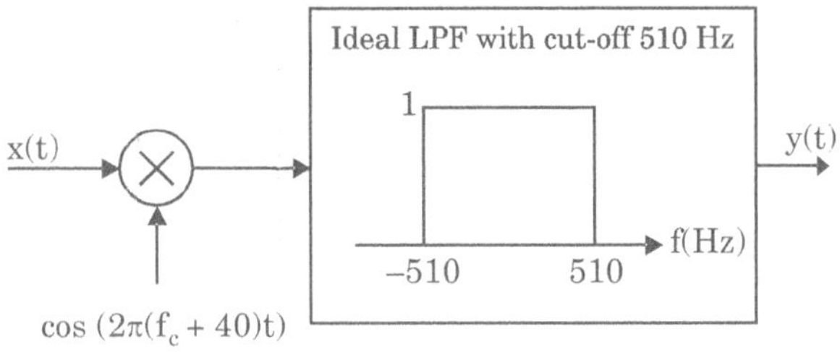

(a) $\cos(460\pi t)$ (b) $\cos(920\pi t)$

(c) $\cos(1000\pi t)$ (d) $\cos(540\pi t)$

43. For an infinitesimally small dipole in free space, the electric field E_θ in the far field is proportional to $\left(\frac{e^{-jkr}}{r}\right)\sin\theta$, where $k = \frac{2\pi}{\lambda}$. A vertical infinitesimally small electric dipole ($\delta l << \lambda$) is placed at a distance $h(h > 0)$ above an infinite ideal conducting plane, as shown in figure. The minimum value of 'h', for which one of the maxima in the far field radiation pattern occurs at $\theta = 60°$, is

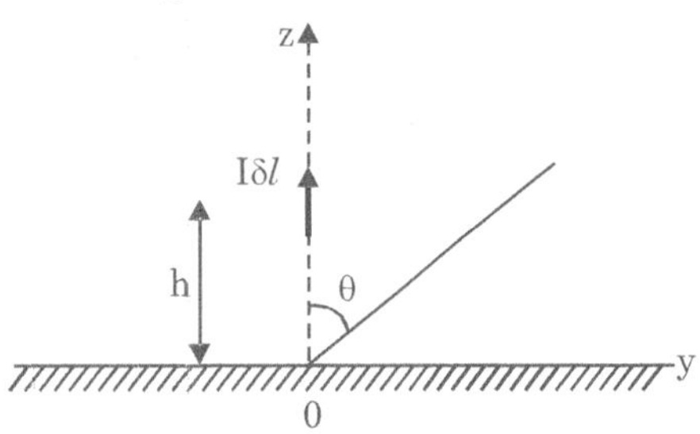

Infinite conducting plane

(a) λ (b) 0.5λ

(c) 0.25λ (d) 0.75λ

44. In the voltage regulator shown below, V_I is the unregulated input at 15 V. Assume $V_{BE} = 0.7$ V and the base current is negligible for both the BJTs. If the regulated output V_0 is 9 V, the value of R_2 is ______ Ω.

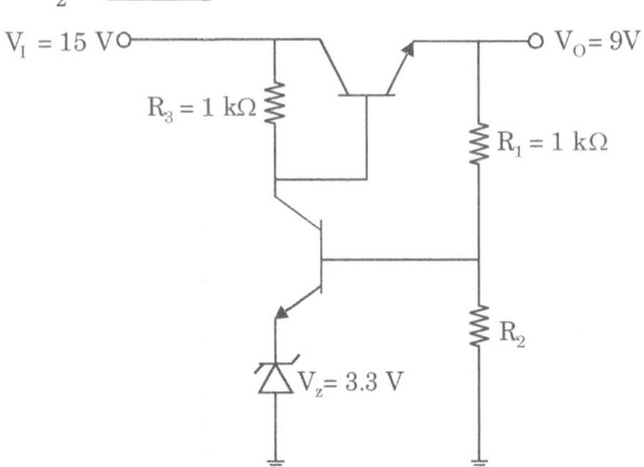

45. The magnetic field of a uniform plane wave in vacuum is given by

$$\vec{H}(x,y,z,t) = (\hat{a}_x + 2\hat{a}_y + b\hat{a}_z)\cos(\omega t + 3x - y - z)$$

The value of b is _____.

46. For a 2-port network consisting of an ideal lossless transformer, the parameter S_{21} (**rounded off to two decimal places**) for a reference impedance of 10Ω, is _______.

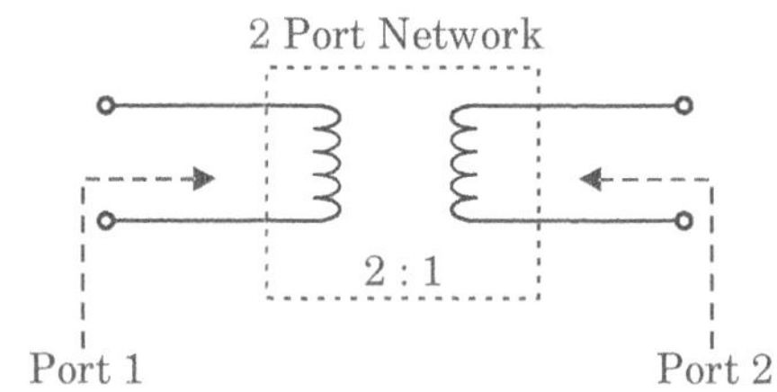

47. $S_{PM}(t)$ and $S_{FM}(t)$ as defined below, are the phase modulated and the frequency modulated waveforms, respectively, corresponding to the message signal m(t) shown in the figure.

$$S_{PM}(t) = \cos(1000\pi t + K_p m(t))$$

and $S_{PM}(t) = \cos\left(100\pi t + K_f \int_{-\infty}^{t} m(\tau)d\tau\right)$

where K_p is the phase deviation constant in radians/volt and K_f is the frequency deviation constant in radians/second/volt. If the highest instantaneous frequencies of $S_{PM}(t)$ and $S_{FM}(t)$ are same, then the value of the ratio $\frac{K_p}{K_f}$ is ___ seconds.

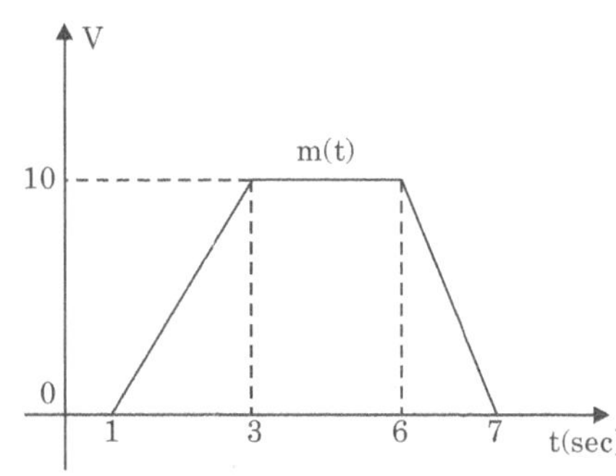

48. In a digital communication system, a symbol S randomly chosen from the set $\{s_1, s_2, s_3, s_4\}$ is transmitted. It is given that $s_1 = -3$, $s_2 = -1$, $s_3 = +1$ and $s_4 = +2$. The received symbol is $Y = S + W$. W is a zero-mean unit-variance Gaussian random variable and is independent of S. P_i is the conditional probability of symbol error for the maximum likelihood (ML) decoding when the transmitted symbol $S = s_i$. The index i for which the conditional symbol error probability P_i is the highest is _____.

49. A system with transfer function $G(s) = \frac{1}{(s+1)(s+a)}$, a > 0 is subjected to an input 5cos3t. The steady state output of the system is $\frac{1}{\sqrt{10}}\cos(3t - 1.892)$. The value of a is _____.

50. For the components in sequential circuit shown below, t_{pd} is the propagation delay, t_{setup} is the setup time, and t_{hold} is the hold time. The maximum clock frequency (**rounded off to the nearest integer**), at which the given circuit can operate reliably is _____ MHz.

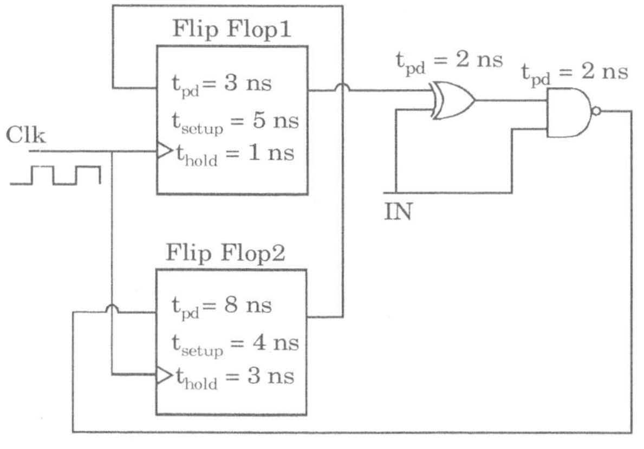

51. For the solid S shown below, the value of $\iiint_S x\,dx\,dy\,dz$ (**rounded off to two decimal places**) is

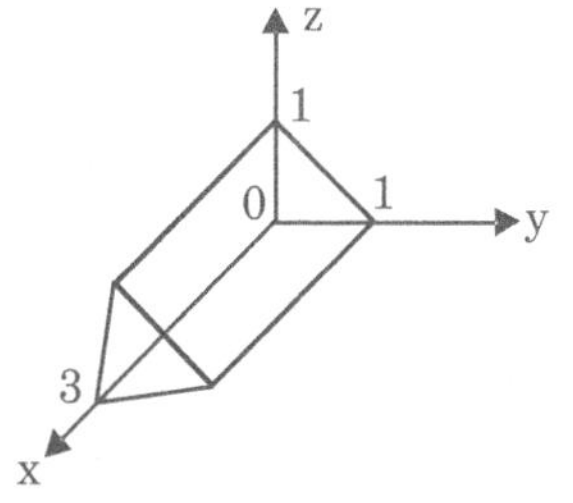

52. X(ω) is Fourier transform of x(t) shown below. The value of $\int_{-\infty}^{\infty} |X(\omega)|^2$ (**rounded off to two decimal places**) is _____.

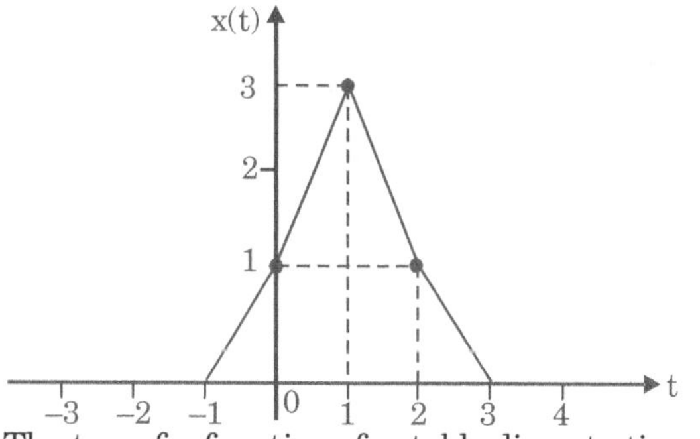

53. The transfer function of a stable discrete-time LTI system is $H(z) = \frac{K(z-\alpha)}{(z+0.5)}$, where K and α are real numbers. The value of α (**rounded off to one decimal place**) with |α| > 1, for which the magnitude response of the system is constant over all frequencies, is _____.

54. X is a random variable with uniform probability density function in the interval [– 2, 10]. For Y = 2X – 6, the conditional probability $P(Y \le 7 \mid X \ge 5)$ (**rounded off to three decimal places**) is _____.

55. Consider the following closed loop control system

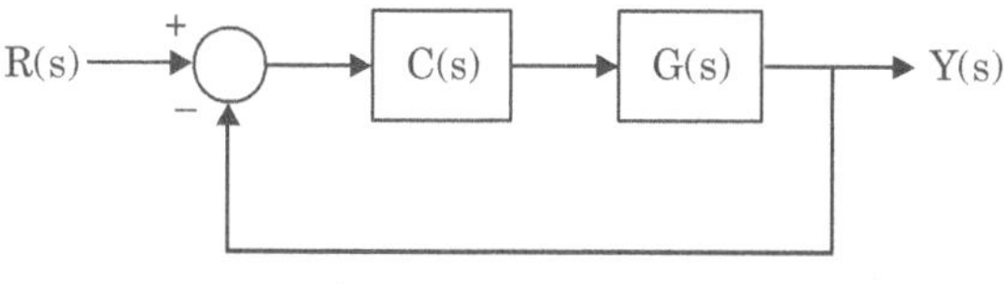

Where $G(s) = \frac{1}{s(s+1)}$ and $C(s) = K\frac{s+1}{s(s+3)}$

If the steady state error for a unit ramp input is 0.1, then the value of K is _____.

ANSWERS

General Aptitude (GA)

1. (a)	**2.** (b)	**3.** (b)	**4.** (c)	**5.** (a)	**6.** (a)
7. (b)	**8.** (c)	**9.** (c)	**10.** (b)		

Electronics & Communication Engineering

1. (c)	**2.** (c)	**3.** (b)	**4.** (c)	**5.** (c)	**6.** (b)
7. (b)	**8.** (d)	**9.** (c)	**10.** (a)	**11.** (b)	**12.** (d)
13. (a)	**14.** (a)	**15.** (48 to 48)	**16.** (2.80 to 2.85)	**17.** (644 to 657)	**18.** (6 to 6)
19. (14 to 14)	**20.** (3.050 to 3.080)	**21.** (6.25 to 6.25)	**22.** (0.5 to 0.5)	**23.** (160 to 160)	**24.** (6 to 6)
25. (0.25 to 0.25)	**26.** (a)	**27.** (a)	**28.** (c)	**29.** (c)	**30.** (a)
31. (a)	**32.** (a)	**33.** (a)	**34.** (a)	**35.** (a)	**36.** (c)
37. (b)	**38.** (a)	**39.** (a)	**40.** (a)	**41.** (d)	**42.** (b)
43. (a)	**44.** (800 to 800)	**45.** (1 to 1)	**46.** (0.8 to 0.8)	**47.** (2 to 2)	**48.** (3 to 3)
49. (4 to 4)	**50.** (76 to 78)	**51.** (2.25 to 2.25)	**52.** (58.50 to 58.80)	**53.** (–2 to –2)	**54.** (0.3 to 0.3)
55. (30 to 30)					

EXPLANATIONS

GENERAL APTITUDE

5. x^m is super additive function, if $x > 1$.
example: $2^3 > 2^1 + 2^2$
$f(x_1 + x_2) > f(x_1) + f(x_2)$ Satisfying
Hence, e^x is super additive fucntion.

7. As given that, Time: 03 : 15
So, h = 3, m = 15

Then, the angle between hands $(\theta) = \left|30h - \frac{11m}{2}\right|$

Hence, $\theta = \left|30 \times 3 - \frac{165}{2}\right| = 75°$

8. From given figure:
Area of shaded portion
= Area of circle – Area of rectangle
As we know that,
Area of the circle = πa^2 And
The maximum possible, area of rectangle inscribed in the circle, when it becomes square.

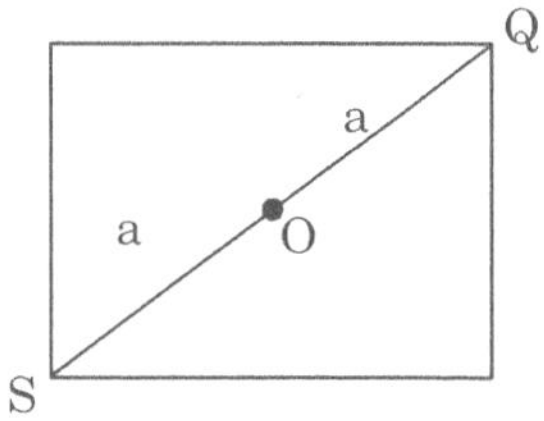

Diagonal = 2a

Area of square = $\frac{1}{2}(\text{diagonal})^2 = 2a^2$

So, required shaded Area.
= Area of circle – Area of square = $\pi a^2 - 2a^2$

9. As given that, the quadratic equation:
$ax^2 - bx + c = 0$
According to question it has equal roots (β).

So, sum of roots = $\beta + \beta = \frac{b}{a} \Rightarrow 2\beta = \frac{b}{a}$ ...(1)

And, product of roots = $\beta * \beta = \frac{c}{a} \Rightarrow \beta^2 = \frac{c}{a}$ ...(2)

By multiplying (1) & (2) equation:
$\Rightarrow (2\beta)(\beta^2) = (b/a)(c/a)$
$(2b^3) = bc/a^2$
$b^3 = bc/(2a^2)$

10. Average number of students in school (P) in 5 years

$$= \frac{3000 + 5000 + 5000 + 6000 + 4000}{5} = 4600$$

Average number of students in school (Q) in 5 years

$$= \frac{4000 + 7000 + 8000 + 7000 + 5000}{5} = 6200$$

The difference of the number of students enrolled in school P and Q = (6200 – 4600) = 1600

The ratio of the average number of the student enrolled in school P to the average of the diffrence of the number of student enrolled in school P and Q then,

Required ratio $= \frac{4600}{1600} = \frac{23}{8} = 23 : 8$

ELECTRONICS & COMMUNICATION ENGINEERING

1. Option (d) is a wrong statement.
To form a basis, the set $\{v_1, v_3, v_5, v_6\}$ should be linearly independent set.
If $\{v_1, v_3, v_5, v_6\}$ spans R^4 then it may (or) may not be a basis for R^4.

2. For a vector field $\vec{A}$
$\vec{A}$ is irrotational (or) conservative only if $\nabla \times \vec{A} = 0$
So, $\vec{A}$ is irrotational if $\nabla^2\vec{A} = 0$ is false.

3. As given that $f(x, y, z) = e^{1-x\cos y} + xze^{\left(\frac{-1}{1+y^2}\right)}$

$$f_x = \frac{\partial f}{\partial x} = e^{1-x\cos y}(0 - \cos y) + z.e^{\frac{-1}{1+y^2}}$$

$$\therefore \left(\frac{\delta f}{\delta x}\right)_{(1,0),e} = e^{1-1}(-\cos 0) + (e).e^{\left(\frac{-1}{1+0}\right)}$$

$$= (-1 + 1) = 0$$

4. Let us Consider $\frac{d}{dx} = D$
$f(D) = D^2 - 6D + 9 = 0$
⇒ Auxiliary equation
⇒ $D^2 - 6D + 9 = 0$
⇒ $(D - 3)^2 = 0$
⇒ D = 3, 3 are real and equal roots
So, the general solution of given differential equation is $y = (C_1 + C_2x)e^{3x}$.
Hence, option (d) is correct.

5. The output of a discrete time system
$y(n) = \max|x(k)|; \quad -\infty \le k \le n$
The unit impulse response is
Now, apply $x(n) = \delta(n) \Rightarrow X(k) = \delta(k)$
So, $y(n) = \max|\delta(k)| = 1, n \ge 0 = u(n)$

6. As given that
$T = 300K, N_V = 2N_C\ V_T = 26$ mV
As we know that,

$$E_F = \frac{E_C + E_V}{2} - \frac{kT}{2}\ln\left(\frac{N_C}{N_V}\right)$$

If $N_C = N_V$ then $E_{fi} = \frac{E_C + E_V}{2}$
⇒ Midband gap energy $\quad (\because N_V = 2N_C)$

$$E_F = E_{fi} - \frac{kT}{2}\ln\left(\frac{N_C}{2N_C}\right)$$

$$(E_{fi} - E_F) = \frac{kT}{2}\ln\left(\frac{1}{2}\right) = \frac{0.026}{2}\ln\left(\frac{1}{2}\right)$$

$$= -9.01 \times 10^{-3} = -9.01 \text{ meV}$$

7. With all other parameters unchanges U_{max}, decreases if the intrinsic carrier density reduced is false because intrinsic carrier density will not affect generation and recombination rate. so, statement (c) is false.

8.

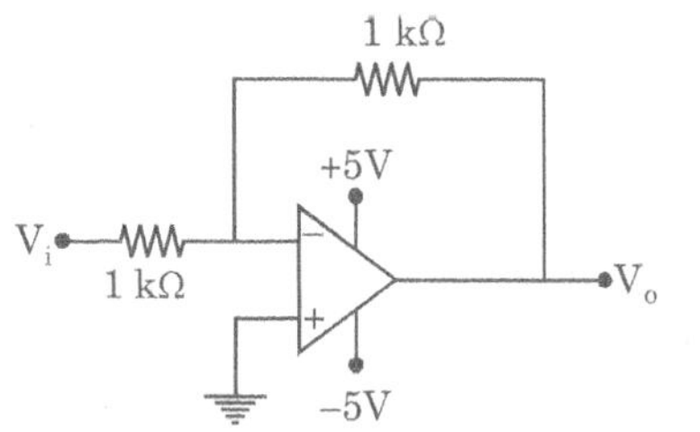

The given circuit is a schmittrigger of non-inverting type $V_0 = \pm 5V$

$$V^+ = \left[\frac{V_o \times 1 + V_i \times 1}{1+1}\right] = \left[\frac{V_o + V_i}{2}\right]$$

Consider $V_o = \pm 5V$, $V^+ = \left[\frac{\pm 5 + V_i}{2}\right]$

If $V_{in} > +5V$ then V_o changes from –5V to +5V

If $V_{in} < -5V$ then V_o changes from +5V to –5V

But the given input has maximum peak value $V_{in} = \pm 1V$, therefore the input cannot flip the output. Hence, output stays or cannot change from +5V (or) – 5V

Transfer characteristics

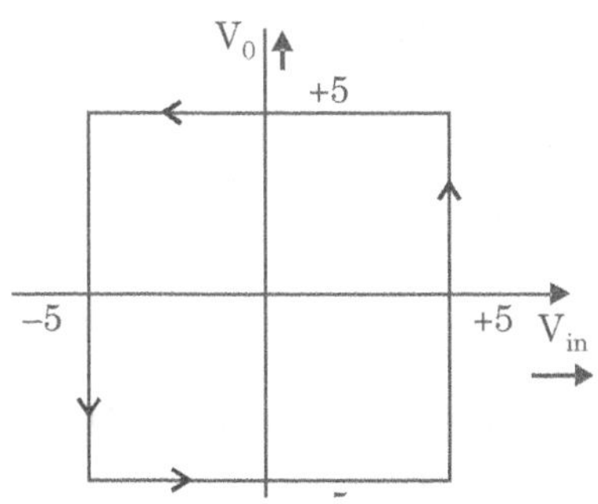

9. By Applying source transformation:

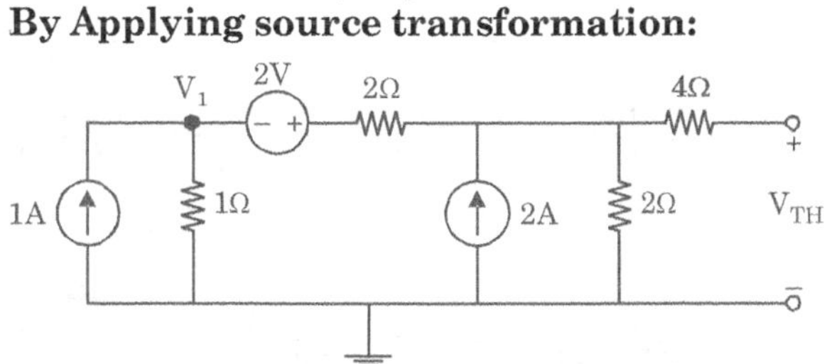

Now, By KCL:

$$-1 + \frac{V_1}{1} + \frac{(V_1 - V_{TH} + 2)}{2} = 0$$

$$3V_1 - V_{TH} = 0 \quad \text{...(1)}$$

$$\frac{V_{TH}}{2} - 2 + \frac{(V_{TH} - V_1 - 2)}{2} = 0$$

$$-V_1 + 2V_{TH} = 6 \quad \text{...(2)}$$

Now, equation (2) multiply by 3,

$$-3V + 6V_{TH} = 18 \quad \text{...(3)}$$

And, By adding equation (1) with equation (3),

$$5V_{TH} = 18$$

$$V_{TH} = \frac{18}{5} = 3.6V$$

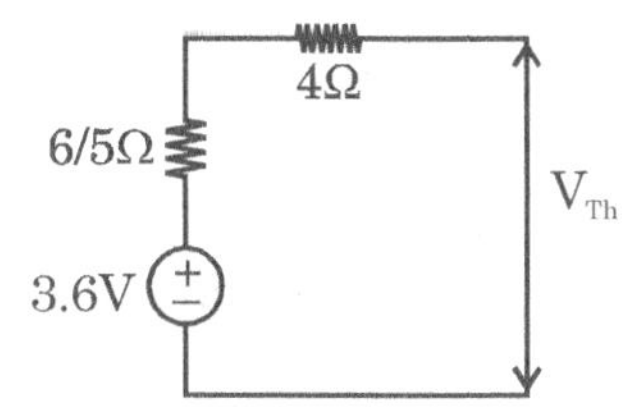

$V_{Th} = 3.6V$

10. If EN = 0 then the multiplexer is Enabled.

P \ QR	00	01	11	10
0	0	1	0	0
1	0	1	1	1

So, the output

$$F = \bar{P}\bar{Q}.R + P.\bar{Q}.R + P.Q.1 = \bar{Q}R + PQ.$$

11. Here, S-plane contour is encircling 2-poles and 3-zeros in clockwise direction. So, the corresponding G(s) plane contour encircle origin 2 times in anticlockwise direction and 3 times in clockwise direction.

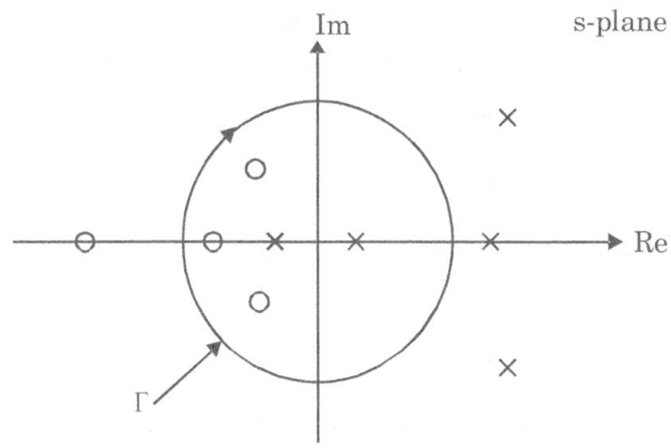

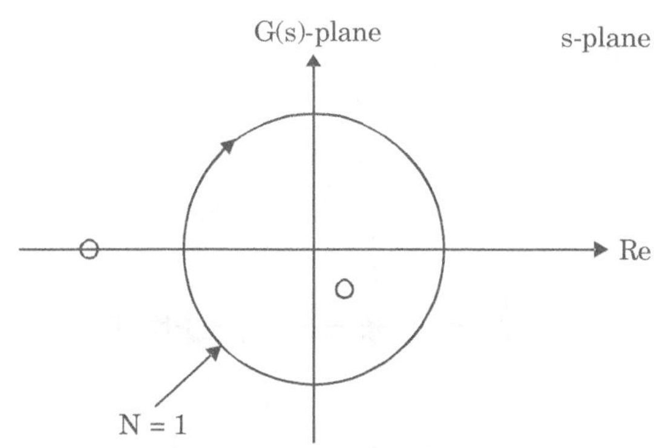

Effectively once encircles the origin in the clock wise direction.

12. The probability of error:

For; P(1 bit error) = α,

P(1 bit correct or Noerror) = $(1 - \alpha)$

P(N bit error) = α^N,

P(N bit correct or No error in received block)

$$= (1 - \alpha)^N$$

The received block declared erroneous if atleast one of bits decoding wrongly (i.e correct probability means all bits should be correct)

P(No error in received block is erroneous)

$$= 1 - (1 - \alpha)^N$$

13. The normalized impedance for the given impedance

$$\Rightarrow \left[\frac{Z}{Z_0}\right] = \left[\frac{jx}{Z_0}\right] \quad [\because \text{Normalized resistance}]$$

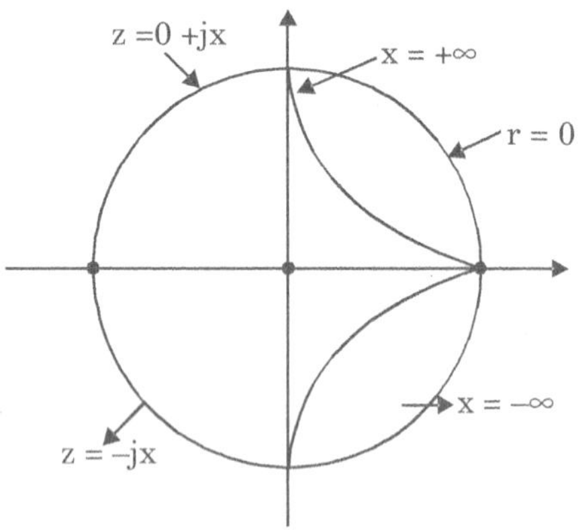

When r = 0 and x lies from $-\infty$ to ∞ is a unit Circle of (radius 1) with centre at (0, 0), on a complex reflection coefficient plane. That indicates pure induction and capacitive load.

14. As given that,

$$y(n) = \sum_{k=0}^{3} (-1)^k x(n-k)$$

$$y(n) = x(n) - x(n-1) + x(n-2) - x(n-3)$$

$$H(z) = \frac{Y(z)}{X(z)} = 1 - z^{-1} + z^{-2} + z^{-3}$$

$$= \frac{z^3 - z^2 + z - 1}{z^3} = \frac{(z-1)(z^2+1)}{z^3}$$

Pole zero plot:

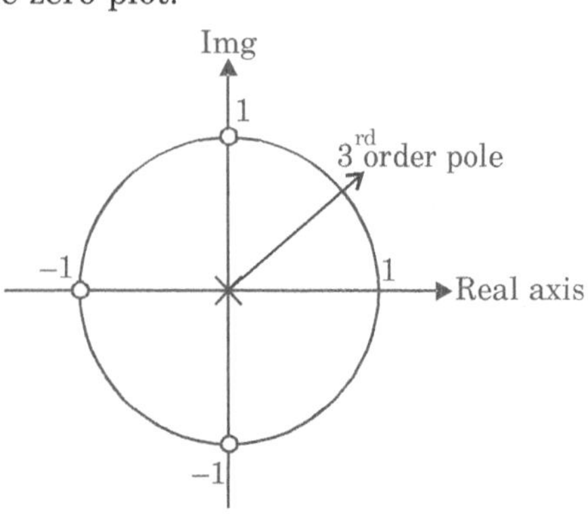

one zero at z = 1 & 2 zeros at z = ±j

3 poles at z = 0

Hence, option (c) is correct.

15. $V_1 = 40I_1 + 60I_2$ ...(1)

$V_2 = 60I_1 + 120I_2$ ...(2)

By compairing eq(i) and (ii) by standard Z-transform equation

$V_1 = Z_{11}I_1 + Z_{12}I_2$

$V_2 = Z_{21}I_1 + Z_{22}I_2$

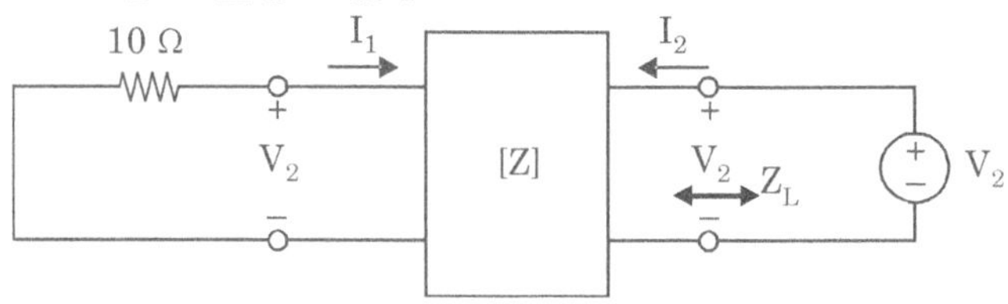

$Z_{11} = 40, Z_{12} = 60, Z_{21} = 60, Z_{22} = 120,$

$R_S = 10\Omega$(given)

From maximum power transfer theorem:-

$$Z_L = Z_{th} = Z_{22} - \left[\frac{Z_{12} \times Z_{21}}{R_S + Z_{11}}\right] = 120 - \left[\frac{60 \times 60}{10 + 40}\right]$$

$Z_L = 48\Omega$

16. As given that

$\bar{V} = 200\angle 0°$

$\bar{I} = 10\angle -45°$

$$Z = \frac{V}{I} = \frac{200\sqrt{0°}}{10\sqrt{-45°}} = 20\angle + 45°$$

$$Z = \left(10\sqrt{2} + j10\sqrt{2}\right)$$

$$Z = 10\sqrt{2} + j\ 10\sqrt{2}$$

$$\downarrow$$

$$X_L$$

$$X_L = \omega_L = 10\sqrt{2} \Rightarrow L = \frac{10\sqrt{2}}{\omega} = \frac{10\sqrt{2}}{5}$$

So, L = 2.828H

17. As given that

$V_{in} = 230 \times \sqrt{2} = V_m$

During negative cycle C_1 charges [D_1 forward biased]

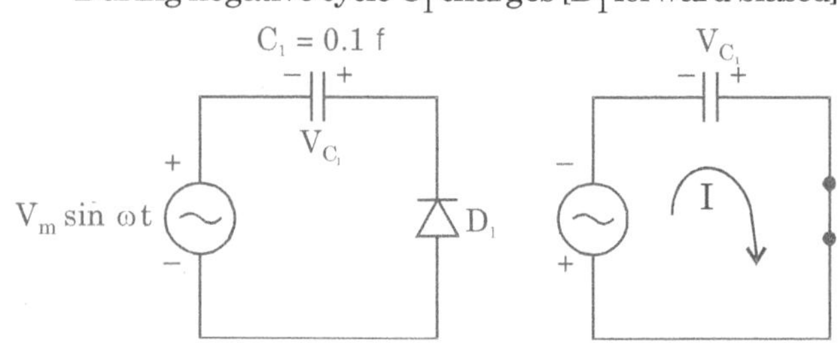

By KVL, $-V_m + V_{C_1} = 0$

$$V_{C_1} = V_m$$

During positive cycle D_2 forward biased

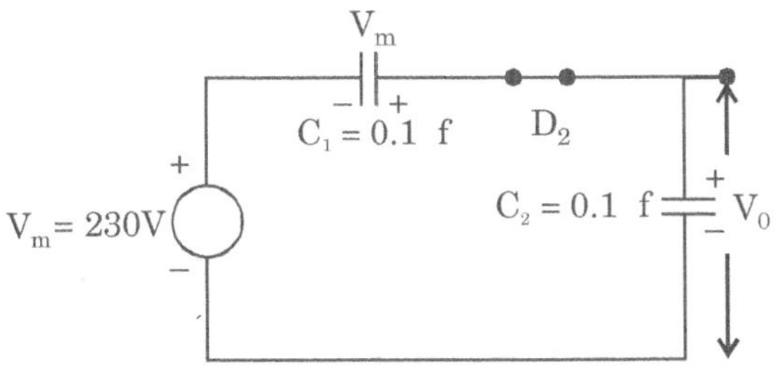

By KVL; $-V_m - V_m + V_o = 0$

$\because$ Voltage doubles $(V_o) = 2Vm$

$$= 2 \times 230 \times \sqrt{2} = 650.440V$$

$$V_0 = 650.44V$$

18.

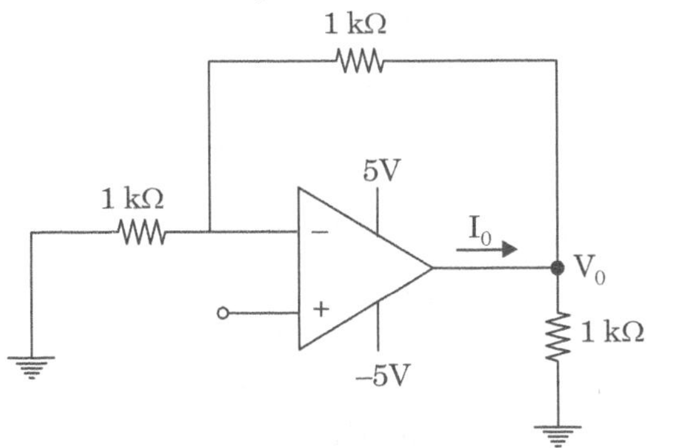

Given op-amp is in non-inverting mode, then output

$$V_0 = \left(1 + \frac{R_f}{R_i}\right) V_{in} = \left(1 + \frac{1k}{1k}\right) \times 2V = 4V$$

By appling KCL at node V_0

$$\frac{V_i - V_0}{1k\Omega} + \frac{0 - V_0}{2k\Omega} + I_0 = 0$$

$$\frac{2-4}{1k\Omega} + \frac{0-4}{1k\Omega} + I_0 = 0$$

$$-2 + I_0 - 4 = 0$$

$$I_o = 6mA$$

19. In 8085 microprocessor, the number of address lines(n) required to a access a 16KB number of memory locations(N) is $2^n = 16KB = 2^4 \times 2^{10}$

$$(\because 1KB = 2^{10})$$

$$\Rightarrow \quad 2^n = 2^{14}$$

Hence, the number of adder lines required are n = 14.

20. The Number of steps for digital input is $(13A)_{16}$

$$\Rightarrow \quad 1 \times 16^2 + 3 \times 16^1 + 10 \times 16^0 = 314$$

$$\text{step size} = \frac{10}{2^{10-1}} = \frac{10}{2^9}$$

As we know that

DAC output

= Number of steps for given digital input × step size

$$= 314 \times \frac{10}{2^9} = \frac{3140}{512} = 6.132 \text{ volts.}$$

21. As given that :

Length of tranmission line :

$$(\ell) = \frac{3\lambda}{4}$$

$$Z_o = 50\Omega$$

$$Z_R = 400\Omega \qquad \left(\because l = \frac{\lambda}{4}\right)$$

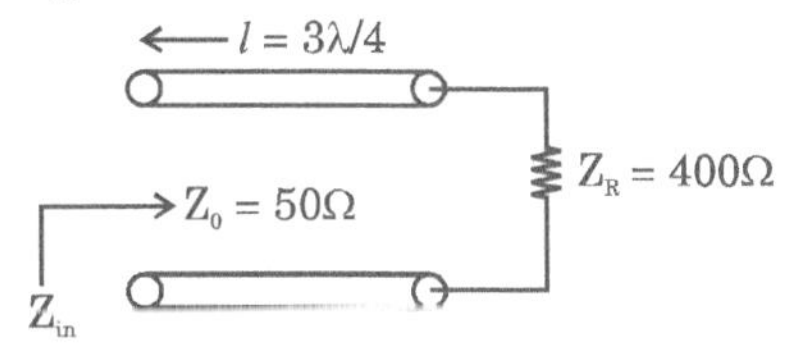

As length of the transmission line is $\frac{3\lambda}{4}$, (odd multiples of quarter wavelengths like $\frac{\lambda}{4}, \frac{3\lambda}{4}, \frac{5\lambda}{4}$....) and hence this line must be a quarter wave transmission line.

Assume this quarter wave transmission line is lossless then $Z_{in} = \frac{Z_o^2}{Z_R} = \frac{50^2}{400} = 6.25\Omega$

22. As given $p(x = 2) = \alpha$

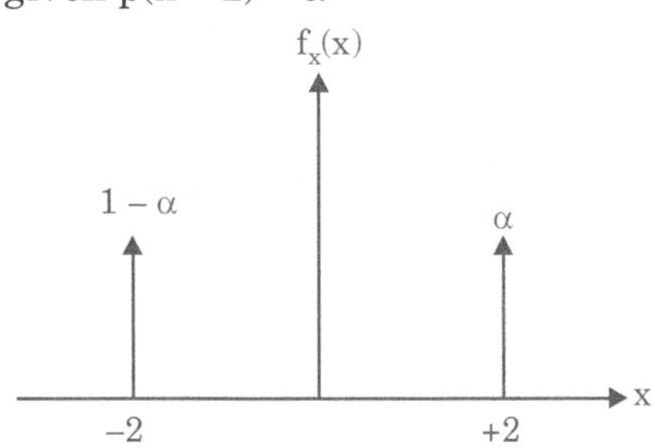

Entropy will be maximum when the provided probabilities are equal

i.e. $P(X = +2) = P(x = -2) = \alpha = 1/2 = 0.5$

23. Characteristic equation q(s) for the given open loop system will be: $1 + G(s)H(s) = 0$

$$q(s) = s(s + 2)(s + 8) + K(s + 11) = 0$$

$$q(s) = s^3 + 10s^2 + 16s + Ks + 11K = 0$$

$$q(s) = s^3 + 10s^2 + (16 + K)s + 11K = 0$$

Now, By using R–H criteria:

s^3	1	16 + K
s^2	10	11 K
s^1	$\frac{10(16+K) - 11K}{10}$,	0
s^0	11 K	

For system to be marginal stability,

$$\left(\frac{10(K + 16) - 11K}{10}\right) = 0$$

$$\text{or, } 10K + 160 - 11K = 0$$

Hence, K = 160

24. As given that

$$Y = \int_{-\infty}^{\infty} W(t)\phi(t)dt$$

where, $\phi(t) = \begin{Bmatrix} 1; 5 \le t \le 7, \\ 0; \text{otherwise} \end{Bmatrix}$

let $\phi(t) = h(-(t-\tau)) \Rightarrow$ where τ is the symbol duration

$$Y = \int_{-\infty}^{\infty} W(t)h(-(t-\tau))dt \text{ Let } t = \tau \Rightarrow dt = d\tau$$

$$Y = \int_{-\infty}^{\infty} W(\tau)h((\tau - t))d\tau$$

$$Y(t) = W(t) * h(t)$$

$$\Rightarrow \quad S_Y(f) = |H(f)|^2 s_w(f)$$

$$\delta_Y^2 = E[Y^2] - E[Y]^2$$

$$E[Y^2] = \int_{-\infty}^{\infty} S_Y(f)df = \int_{-\infty}^{\infty} |H(f)|^2 S_W(f)df$$

$$E[Y^2] = E\left[\int_{-\infty}^{\infty} W(t)\phi(t)\right]dt$$

$$= E\left[\int_{-\infty}^{\infty} W(t)\phi(t)dt\right] = 0$$

$$S_w(f) = \frac{N_0}{2} = \frac{N_0}{2}\int_{-\infty}^{\infty} |H(f)|^2 df$$

$$= \frac{N_0}{2}\int_{-\infty}^{\infty} |\phi(f)|^2 df \quad (\because \phi(t) = h(-(t-\tau)))$$

$$|\phi(f)^2| = |H(f)|2 = \frac{N_0}{2}\int_{-\infty}^{\infty} |\phi(t)|^2 dt$$

(from the Rayleigh's theorem)

$$E[y^2] = 3 \times 2 = 6, \quad (\because E_\phi = 2)$$

25. Coin tossed two times:

S = {(H, H), (H, T), (T, H), (T, T)}

M Outcomes		X = min (M, N)
0(T),	0(T)	0(T)
0(T),	1(H)	0(T)
1(H),	0(T)	0(T)
1(H),	1(H)	1(H)

P(X = 1) = 1/4, P(X = 0) = 3/4

As we know that,

$$E(X) = \sum_i x_i P(X = x_i) \quad (\because i = 0, 1)$$

$$\text{So, } E(X) = 0 \times \frac{3}{4} + 1 \times \frac{1}{4} = 0.25$$

26. As given that

$X_1 + 2X_2 = b_1$;

$2X_1 + 42X_2 = b_2$;

$3X_1 + 7X_2 = b_3$;

$3X_1 + 9X_2 = b_4$;

So, Consider augmented matrix

$$(A|B) = \left[\begin{array}{cc|c} 1 & 2 & b_1 \\ 2 & 4 & b_2 \\ 3 & 7 & b_3 \\ 3 & 9 & b_4 \end{array}\right]$$

By : $R_2 \to R_2 - 2R_1$; $R_3 \to R_3 - 3R_1$; $R_4 \to R_4 - 3R_1$

$$(A|B) \sim \left[\begin{array}{cc|c} 1 & 2 & b_1 \\ 0 & 0 & b_2 - 2b_1 \\ 0 & 1 & b_3 - 3b_1 \\ 0 & 3 & b_4 - 3b_1 \end{array}\right]$$

By : $R_2 \leftrightarrow R_3$

$$(A|B) \sim \left[\begin{array}{cc|c} 1 & 2 & b_1 \\ 0 & 1 & b_3 - 3b_1 \\ 0 & 0 & b_2 - 2b_1 \\ 0 & 3 & b_4 - 3b_1 \end{array}\right]$$

By : $R_4 \to R_4 - 3R_2$

$$(A|B) \sim \left[\begin{array}{cc|c} 1 & 2 & b_1 \\ 0 & 1 & b_3 - 3b_1 \\ 0 & 0 & b_2 - 2b_1 \\ 0 & 0 & 6b_1 - 3b_1 + b_4 \end{array}\right]$$

Here, $\rho(A) = 2$

To have a solution, $\rho(A) = 2 = \rho(A|B)$

if $b_2 - 2b_1 = 0$ and $6b_1 - 3b_3 + b_4 = 0$ then

$$\rho(A|B) = 2$$

$\therefore$ solution exists if $b_2 = 2b_1$

and $\quad 6b_1 - 3b_3 + b_4 = 0$

27. As given that

$$\frac{dy}{dx} = (y-1)x$$

By variable separable method:

$$\Rightarrow \int \frac{1}{y-1}dy = \int x\,dx + c$$

$$\Rightarrow \log|(y-1)| = \frac{x^2}{2} + C \quad \text{(where, } y \ne 1)$$

$$\left(\because \int \frac{f'(x)dx}{f(x)}dx = \log|f(x)| + C\right)$$

So, $\log|y-1| = (0.5)x^2 + C$ for $y \ne 1$ & $y = -1$ is a general solution.

Hence, option (a) is correct.

28. By applying super-position theorem

Step-I

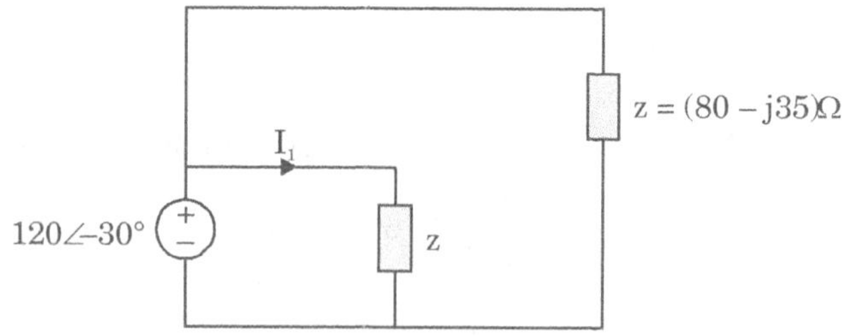

$$I_1 = -\frac{[120\angle -30°]}{Z} = \frac{-[120\angle -30°]}{80 - j35}$$

Step-II

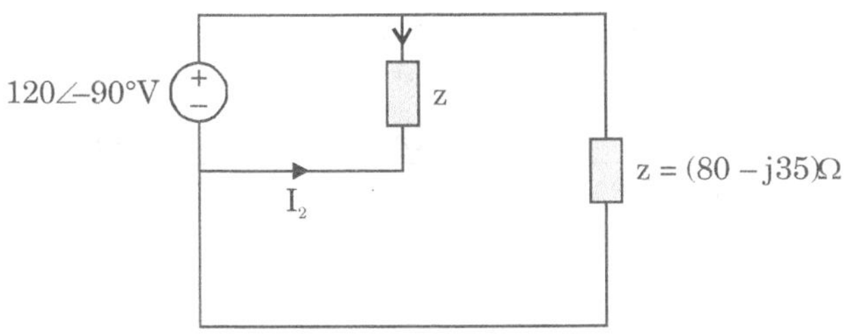

$$I_2 = -\frac{[120\angle -90°]}{Z} = -\frac{[-j120]}{80 - j35} = \frac{j120}{80 - j35}$$

$$I = I_1 + I_2 = \frac{1}{(80 - j35)}[-103.92 + j60 + j120]$$

$$I = \frac{(-103.92 + j180)}{(80 - j35)} = \left[\frac{207.34\angle 120°}{87.32\angle 23.63°}\right]$$

$$I = 2.380\angle 143.63° A$$

29. As given that

$$x(t) = \cos(200\pi t)$$

$$t = \frac{n}{400}$$

$$x(n) = \cos\left(\frac{200\pi n}{400}\right) \Rightarrow x(n) = \cos\left(\frac{\pi n}{2}\right)$$

$$(\because n = 0, 1, \ldots, 7)$$

$x(n) = \{1, 0, -1, 0, 1, 0, -1, 0\}$

When we look at x(n), it contains only even samples.

Let us consider $z(n) = \{1 -1\ 1 -1\} \xleftrightarrow{\text{DFT}} Z(k)$

$$Z(k) = Wn|_{(n=4)} Z(n) = \begin{bmatrix} Z(0) \\ Z(1) \\ Z(2) \\ Z(3) \end{bmatrix}$$

$$= \begin{bmatrix} 1 & 1 & 1 & 1 \\ 1 & -j & 1 & j \\ 1 & -1 & 1 & -1 \\ 1 & j & -1 & -j \end{bmatrix} \begin{bmatrix} 1 \\ -1 \\ 1 \\ -1 \end{bmatrix} = \begin{bmatrix} 0 \\ 0 \\ 4 \\ 0 \end{bmatrix}$$

$N = 4$

We know zero interpolation in time domain corresponds to replication of DFT spectrum

$$x(n) = z\left(\frac{n}{2}\right) \xleftrightarrow{\text{DFT}} X(k) = \{Z(k), Z(k)\}$$

$$= \{0,0,4,0,0,0,4,0\}$$

(or)

$$\cos\left(\frac{2\pi}{N}k_0 n\right) \leftrightarrow \frac{N}{2}[\delta(k - k_0) + \delta(k + k_0)]$$

$$\cos\left(\frac{n\pi}{2}\right) = \cos\left(\frac{2\pi}{8}2n\right) \leftrightarrow \frac{8}{2}[\delta(k-2) + \delta(k+2)]$$

So, non zero samples are –2, 2. From periodicity property non zero samples are –2 + 8, 2 = 6, 2.

30. From source transformation:

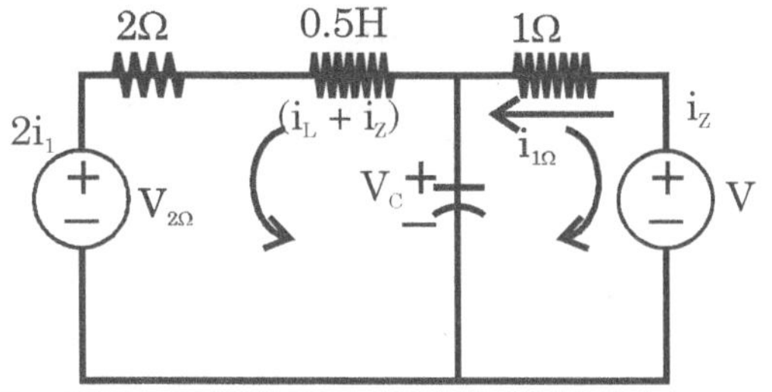

By KCL,

$$i_L + i_Z = C\frac{dV_C}{dt} + i_{1\Omega} \quad \ldots(1)$$

By KVL,

$$-V_{2\Omega} + L\frac{di_L}{dt} + V_C = 0 \quad \ldots(2)$$

Now, voltage across 2Ω, current theory 1Ω

$$i_1 = \frac{V_{2\Omega}}{2} + i_L \Rightarrow V_{2\Omega} = 2[i_1 - 1_L] \quad \ldots(3)$$

$$1.\ i_{1\Omega} - V_C = 0 \Rightarrow i_{1\Omega} = V_C \quad \ldots(4)$$

$$i_L + i_2 = \frac{1}{4}\left[\frac{dV_C}{dt}\right] + V_C$$

$$-[2(i_1 - i_L)] + \frac{1}{2}\frac{di_L}{dt} + V_C = 0$$

$$\frac{1}{4}\frac{dV_C}{dt} = -V_C + i_L + i_2$$

$$\frac{1}{2}\frac{di_L}{dt} = -V_C + i_L + i_2$$

$$\frac{dV_C}{dt} = -4V_C + 4i_L + 4i_2$$

$$\frac{di_L}{dt} = -2V_C - 4i_L + 4i_1$$

$$\begin{bmatrix} \frac{dV_C}{dt} \\ \frac{di_L}{dt} \end{bmatrix} = \frac{d}{dt}\begin{bmatrix} V_c \\ i_L \end{bmatrix} = \begin{bmatrix} -4 & 4 \\ -2 & -4 \end{bmatrix}\begin{bmatrix} V_C \\ i_L \end{bmatrix} + \begin{bmatrix} 0 & 4 \\ 4 & 0 \end{bmatrix}\begin{bmatrix} i_1 \\ i_2 \end{bmatrix}$$

31. If we assume $V_0 = 0.5$ V

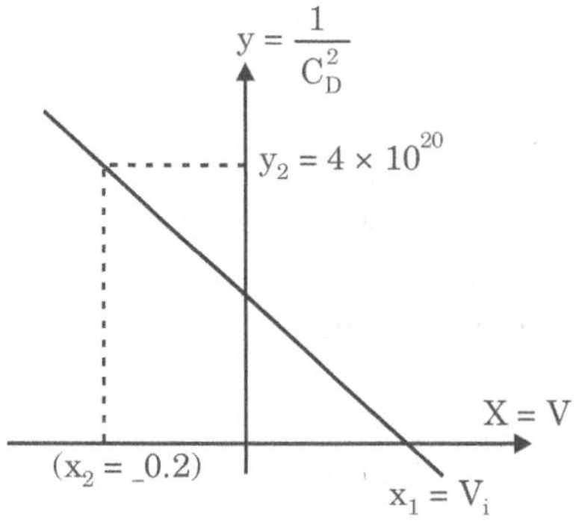

$$y_2 = \frac{1}{C_D^2}$$

$$= \left[\frac{1}{(50 \times 10^{-12})^2}\right]$$

$$= 4 \times 10^{20}$$

$$\because \quad C_D = \sqrt{\frac{q\varepsilon}{2(V_0 + V_R)}\left(\frac{N_A N_D}{N_A + N_D}\right)}$$

$$C_D^2 \propto \frac{1}{(V_o + V_R)}$$

or

$$\frac{1}{C_D^2} \propto (V_0 + V_R)$$

$$k = \frac{\frac{1}{C_D^2}}{V_0 + V_R} = \text{slope}$$

$$= \frac{\left(\frac{1}{50pF}\right)^2}{0.5 + 0.2}$$

$$= \frac{1}{(50pF)^2 \times 0.7}$$

$$= 5.71 \times 10^{20}\ F^{-2}\ V^{-1}$$

Hence, option (a) is correct.

32. For Mos Capacitance:

As given that

$$f_S = 4eV + 0.5eV + 0.3eV = 4.8eV$$

$$V_{TH} = -0.16V,\ C_{ox} = 100nF/cm^2$$

$$\phi_m = 3.87\ eV$$

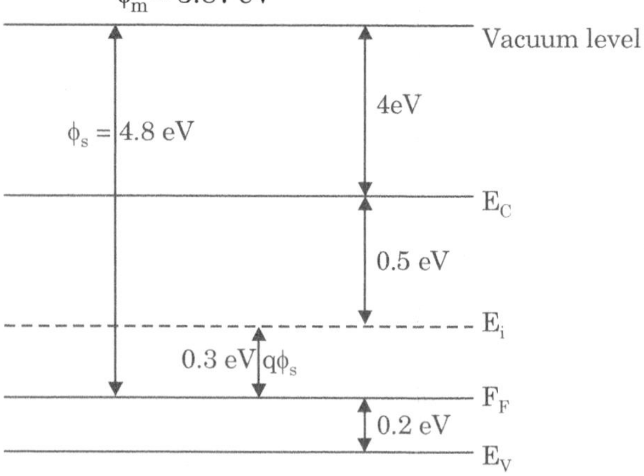

$\therefore \quad V_{FB} = \phi_{ms}$ $\quad$ ($\because$ No oxide charge)

$\therefore \quad V_{FB} = \phi_{ms} = \phi_m - \phi_s = 3.87 - 4.8 = -0.93$

$$\phi_t = 2\,\phi_F,\ \phi_F = 0.3V$$

$$\phi_t = 2 \times 0.3 = 0.6V$$

As we know that,

$$V_T - V_{FB} = \frac{|Q_{dinv}|}{C_{ox}} + \phi_t$$

$$-0.16 - (-0.93) = \frac{|Q_{dinv}|}{100n} + 0.6$$

$$|Q_{dinv}| = 0.17 \times 100 \times 10^{-9} = 17 \times 10^{-9}$$

$$= 1.7 \times 10^{-8}\ C/cm^2$$

33. As we know that

$$\frac{\beta_{T2}}{\beta_{T1}} = \frac{\text{Area under the carrier profile of base of T1}}{\text{Area under the carrier profile of base of T2}}$$

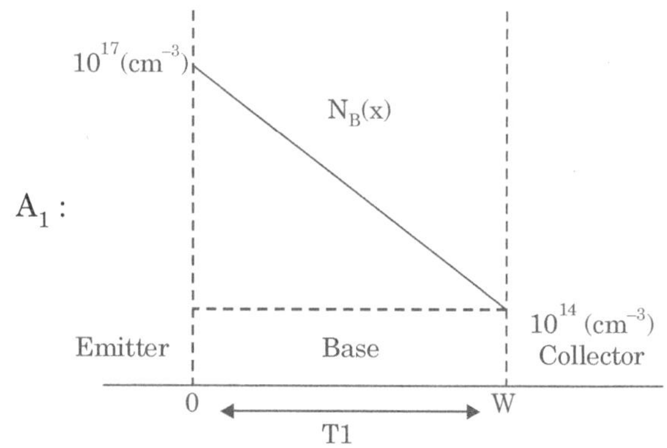

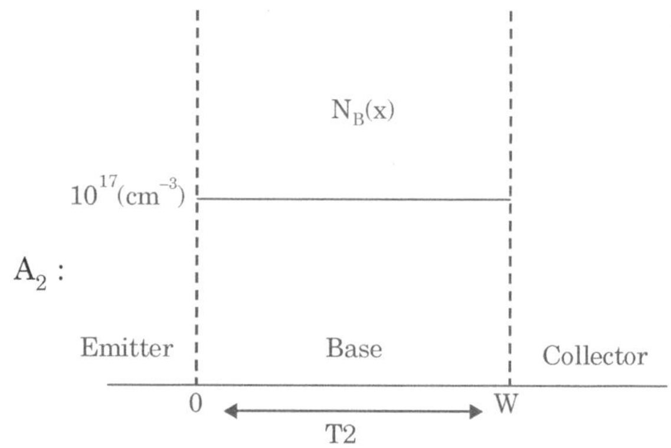

$$\frac{\beta_{T2}}{\beta_{T1}} = \frac{\int_0^W N_{A_2}(X)dx}{\int_0^W N_{A_1}(X)dx}$$

$$= \frac{10^{17} \times W}{\frac{1}{2}(10^{17} - 10^{14})W + 10^{14}W}$$

$$= \frac{10^{17}}{\frac{1}{2}(10^{17} - 10^{14}) + 10^{14}}$$

$$\Rightarrow \quad \beta_{T2} = 2\beta_{T1}$$

34. As given that,

$$A = 1\ cm^2$$

$$P_{in}\ (/cm^2) = 100\ mW/cm^2$$

$$\eta = 0.15,\ V_{oc} = 0.7,\ \text{Fill factor}(F) = 0.8$$

Thickness = 200 μm

$$\text{Power }(/cm^2) = \frac{100mW/cm^2}{200 \times 10^{-4}} = 5W/cm^3$$

As we know that,

$$\eta = \frac{V_m I_m}{P_{in}}, \; F = \frac{V_m I_m}{V_{OC} I_{SC}}$$

$$V_m I_m = \eta \, P_{in} = 0.15 \times 100\text{mw}$$

$$I_{SC} = \frac{V_m I_m}{V_{OC} \times F} = \frac{0.15 \times 100\,\text{mw}}{0.7 \times 0.8}$$

$$= \frac{15}{0.56}\ \text{mA}$$

$$G_L = \frac{I_{SC}}{q \times \text{Area} \times \text{thickness}}$$

$$= \frac{15 \times 10^{-3}}{0.56 \times 1.6 \times 10^{-19} \times 200 \times 10^{-4}}$$

$$= \frac{15 \times 10^{19}}{0.56 \times 32}$$

$$G_L = 0.84 \times 10^{19}/\text{cm}^3$$

Hence, Optical generation rate = $0.84 \times 10^{19}/\text{cm}^3\text{-s}$

35. DC Analysis: $I_E = \dfrac{10 - 0.7}{20} = 0.465\text{mA} = I_{C_{DC}}$

$$g_m = \frac{I_{C_{DC}}}{V_t} = \frac{0.465}{26}\ \text{A/V} = 0.0178$$

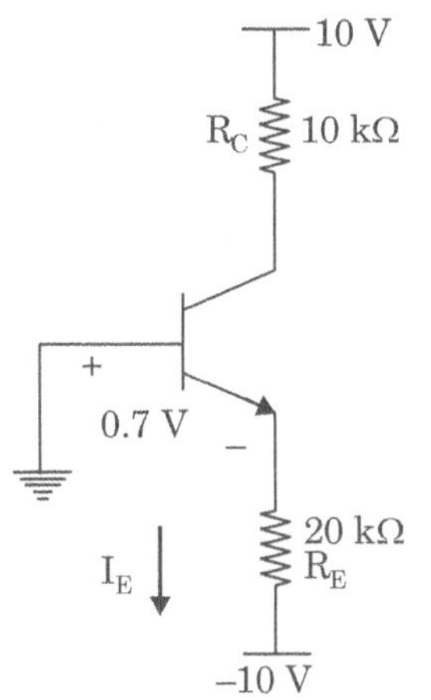

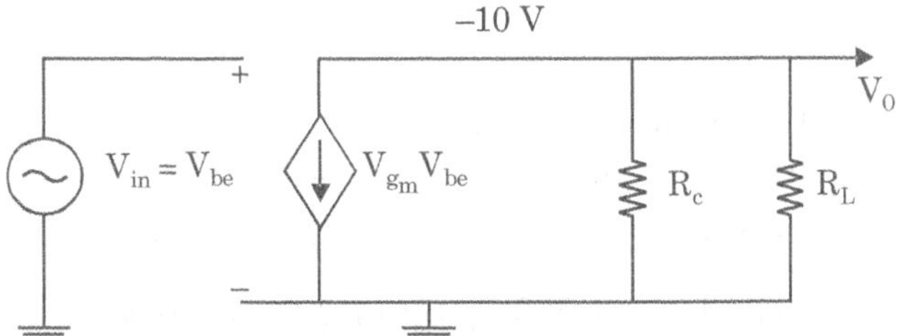

$$V_{out} = -g_m (R_C \parallel R_L)$$

$$V_{in} = V_{be}$$

$$\frac{V_{out}}{V_{in}} = \frac{-g_m (R_C \parallel R_L) V_{be}}{V_{be}}$$

$$\frac{V_{out}}{V_{in}} = -g_m (R_C \parallel R_L)$$

$$= -0.0178(5000) = -89.423$$

36.

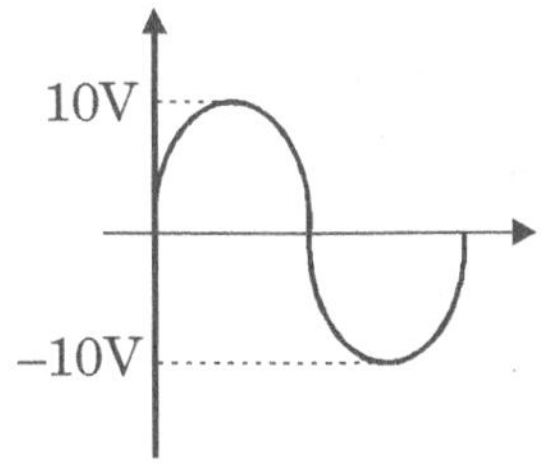

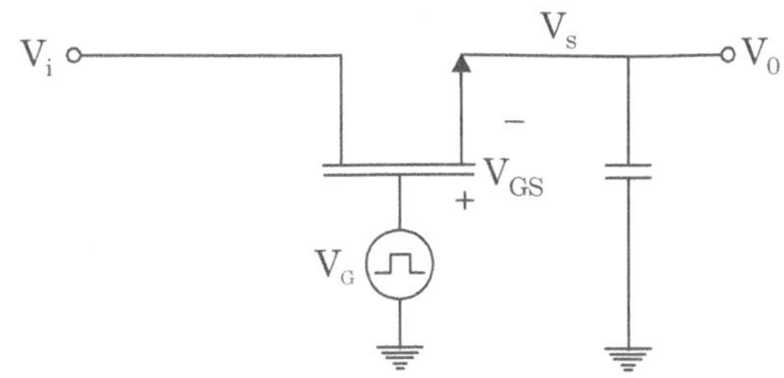

$$V_T = 3V$$

$$V_i = \pm 10V$$

During sampling (i.e when input ±10V)

MOSFET must be ON:

when $V_i = 10V \;\Rightarrow\; V_G > 10V + 3V$ [when, $V_{i\,max} = 10V$]

$$V_G > 13\ V$$

For holding MOSFET should be off

when $V_i = -10V$

$\Rightarrow$ $V_G > V_i + V_T$ [when $V_{i\,min} = -10V$]

$$V_G > -10 + 3V$$

$$V_G > -7V$$

37.

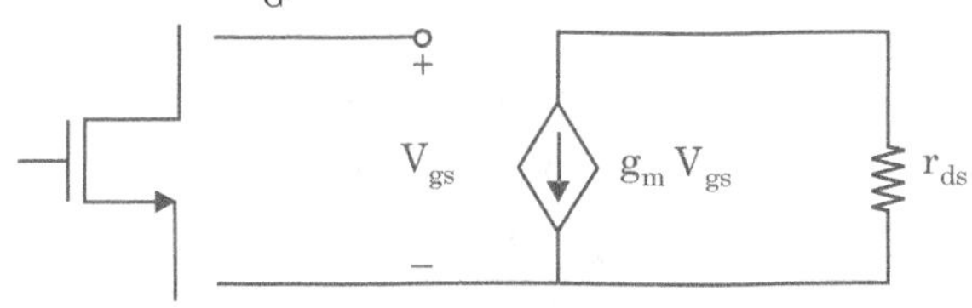

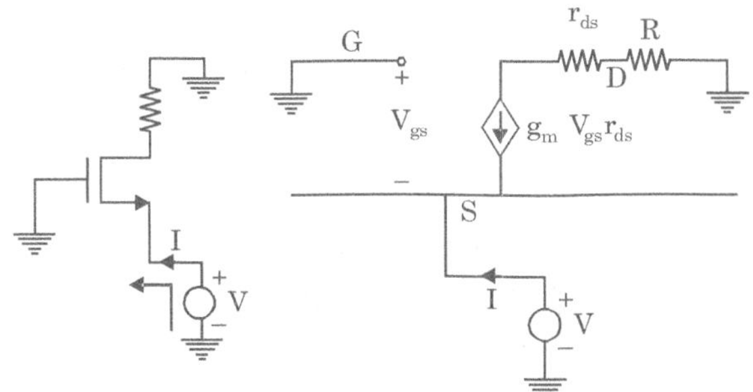

By Applying KVL:

$$V_{gs} + V = 0 \;\Rightarrow\; V = -V_{gs}$$

$$I_x = \frac{V_x}{r_{ds} + R} = \frac{-g_m r_{ds} V_{gs} + V}{r_{ds} + R}$$

$$I_x = \frac{g_m r_{ds} V + V}{r_{ds} + R}$$

$$= (r_{ds} + R) I_x = V(1 + g_m r_{ds})$$

$$R_N = \frac{V_x}{I_x} = \frac{r_{ds} + R}{1 + g_m r_{ds}}$$

38. As given that:

the binary number is 1100

So, sign magnitude of 1100

$= -4 = P$

1's complement of 1100

$= -3 = Q$

2's complement of 1100

$= -4 = R$

Thus, $P + Q + R = (-4) + (-3) + (-4)$

$P + Q + R = -11$

$= $ 2's complement of $(+11)$

or

$= $ 2's complement of (001011)

$= (110101)$

Hence, P + Q + R in 6-digit 2's complement representation is 110101.

39. As Given that S_0 is initial state of the sequence detector

If the output to be 1, the sequence of states traversed are

$S_0,\ S_1,\ S_2,\ S_3,\ S_4,\ S_3$

So, the sequence detected is $\rightarrow$ 0 1 0 1 0

40. As given that, the characteristic Equation of a system is:

$Q(s) = 1 + G(s).H(s) = 0$

$s^3 + 3s^2 + (K + 2)s + 3K = 0$

$s^3 + 3s^2 + 2s + Ks + 3K = 0$

$s(s^2 + 3s + 2) + K(s + 3) = 0$

$$\left[-K = \frac{S^3 + 3S^2 + 2S}{(S+3)}\right]$$

$$\frac{-dK}{dS} = \left[\frac{(S+3)(3S^2+6S+2)-(S^3+3S^2+2S)}{(S+3)^2}\right]$$

$= 0$

$3S^3 + 6S^2 + 2S + 9S^2 + 18S + 6 - S^3 - 3S^2 - 2S = 0$

$2S^3 + 12S^2 + 18S + 6 = 0$

$S = (-0.46, -3.87, -1.65)$

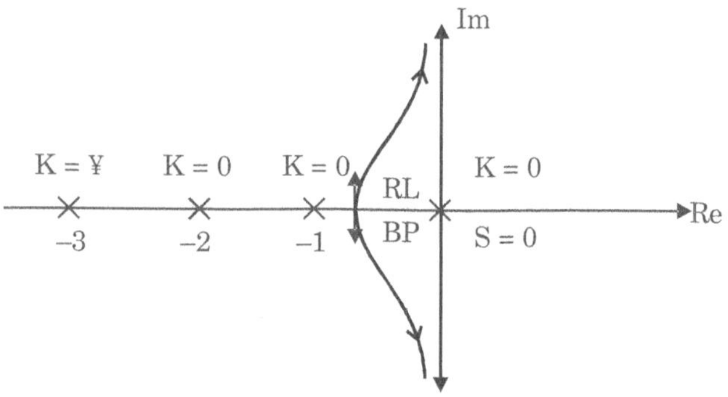

Break-away point exist between (–1, 0)

41.

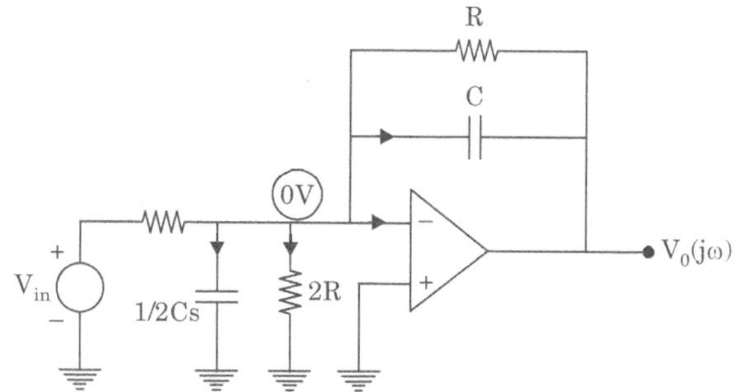

If no current flows through 2C and 2R in the given above Ckt.

Now, By using KCL

$$\frac{V_{in} - 0}{R} = \frac{0-0}{1/2sC} + \frac{0-0}{2R} + 0 + \frac{0 - V_o}{1/sC} + \frac{0 - V_o}{R}$$

$$\frac{V_{in}}{R} = -V_o\left[\frac{1}{R} + sC\right]$$

$$\left|\frac{V_o}{V_{in}}\right| = \frac{1}{1 + sCR} = \frac{1}{1 + s\tau}$$

$$\omega_{3dB} = \frac{1}{\tau} = \frac{1}{RC}$$

So, op-amp active filter (LPF) inverting type 3dB cut-off frequency.

$$f_{3dB} = \frac{1}{2\pi RC} = \frac{1}{2\pi(2k)(1\mu)}$$

$$= \frac{1}{2\pi \times 2 \times 10^3 \times 10^{-6}} = 79.58 \text{ Hz}$$

42.

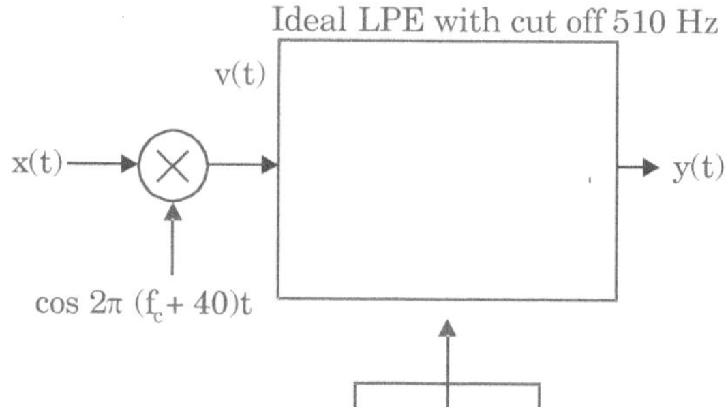

The out put of the multiplex:

$v(t) = m(t)\cos(2\pi f_c t)\cos(2\pi(f_c + 40)t)$

$v(t) = 4\cos(2\pi 500t)\cos(2\pi.10^6 t)\cos(2\pi(10^6 + 40)t)$

$v(t) = 4\cos(2\pi 500t)\frac{1}{2}\cos(4\pi 10^6 + 2\pi 40)t + \cos(2\pi 40t)$

$v(t) = 2\cos(2\pi.500t)[\cos(2\pi.40t) + \cos(2\pi(2.10^6 + 40)t]$

$v(t) = \cos(2\pi.540t) + \cos(2\pi.460t) + \cos(2\pi(2.10^6 + 540)t) + \cos(2\pi(2.10^6 + 460t))$

So, output of low pass filter:

$y(t) = \cos(2\pi 460t)$

$y(t) = \cos(920\pi t)$

43. Consider an infinitesimal electric dipole ($dl << \lambda$), which is vertically placed at height 'h' above an infinite grounded conductor.

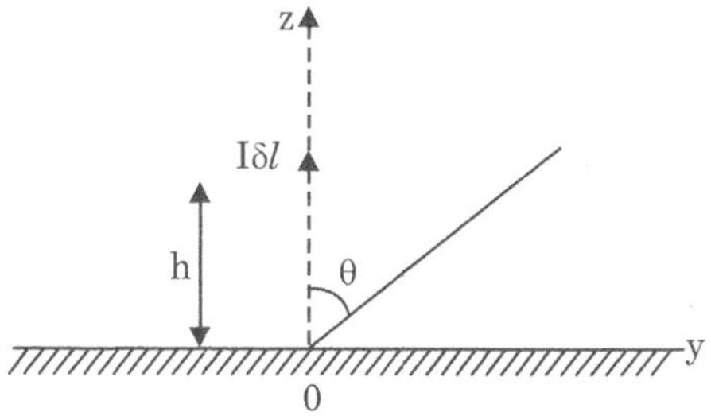

Infinite conduction plate

The resultant electric field intensity is given by

$$E_\theta = \frac{j\eta I_0 \beta dl \sin\theta}{4\pi r} e^{-j\beta r} 2\cos(\beta h \cos\theta) \; ; z \geq 0$$
$$= 0 \qquad ; z < 0$$

(or)

$$E_\theta = \frac{j\eta I_0 k dl \sin\theta}{4\pi r} e^{-j\beta r} 2\cos(kh\cos\theta) \; ; z \geq 0$$
$$= 0 \qquad ; z < 0$$

where k (or) β is phase shift constant (or) wave number.

$$\beta = \frac{2\pi}{\lambda}$$

Given $\theta_{max} = 60°$

Field will have maxima,

when $|\cos(kh\cos\theta_{max})| = 1$

$$\Rightarrow \quad kh\cos\theta_{max} = \pi$$
$$\frac{2\pi}{\lambda} h\cos 60° = \pi$$
$$\frac{2\pi}{\lambda} .h. \frac{1}{2} = \pi$$
$$\therefore \quad h = \lambda$$

(or) $$\frac{\partial E_0}{\partial h} = 0$$

Consider, $$\frac{j\eta k I_0 dl e^{-jkr} \sin\theta}{4\pi r} = P$$

$$\frac{\partial}{\partial h}\{2P\cos(kh\cos\theta_{max})\} = 0$$
$$-2pk\cos\theta_{max}\sin(kh\cos\theta_{max}) = 0$$
$$\sin(kh\cos\theta_{max}) = 0$$
$$kh\cos\theta_{max} = n\pi$$

For, $n = 1$

$$kh\cos 60 = \pi$$
$$\frac{2\pi}{\lambda} \times h. \frac{1}{2} = \pi$$

$\therefore$ So, minimum $h = \lambda$

44. $$V_{R2} = 0.7 + 3.3 = 4V$$
$$V_o = IR_1 + IR_2$$
$$9 = I(R_1 + R_2) \quad \text{[Given Vo = 9V]}$$

Now, By Neglecting base current

So, $$9 = \frac{4}{R_2}(1k + R_2)$$

$$\frac{9 \times R_2}{R_2 + 1k\Omega} = 4$$
$$5R_2 = 4k$$
$$R_2 = \frac{4000}{5}$$
$$R_2 = 800\ \Omega$$

45. As given that,

$$\bar{H} = (a_{\hat{x}} + 2a_{\hat{y}} + ba_{\hat{z}})\cos(\omega t + 3x - y - z)$$
$$\bar{H} = \bar{H}_0 \cos(\omega t - \bar{k}.\bar{r})$$
$$= \bar{H}_0 \cos\left[\omega t - (-3a_{\hat{x}} + a_{\hat{y}} + a_{\hat{z}})\cdot(xa_{\hat{x}} + ya_{\hat{y}} + za_{\hat{z}})\right]$$

So $$\bar{H}_0 = a_{\hat{x}} + 2a_{\hat{y}} + ba_{\hat{z}}$$
$$\hat{H}_0 = \left[\frac{(1a_{\hat{x}} + 2a_{\hat{y}} + ba_{\hat{z}})}{\sqrt{1^2 + 2^2 + b^2}}\right]$$
$$\bar{k} = -3a_{\hat{x}} + a_{\hat{y}} + a_{\hat{z}}$$
$$\hat{k} = \frac{-3a_{\hat{x}} + a_{\hat{y}} + a_{\hat{z}}}{\sqrt{(3)^2 + 1^2 + 1^2}}$$

For uniform plane wave $\hat{H}_0.\hat{k} = 0$

$$\frac{a\hat{x} + 2\hat{y} + b\hat{z}}{\sqrt{1^2 + 2^2 + b^2}} . \frac{(-3a_{\hat{x}} + a_{\hat{y}} + a_{\hat{z}})}{\sqrt{9 + 1 + 1}} = 0$$
$$-3 + 2 + b = 0$$
$$b = 1$$

46. For ideal Tranformer of n : 1, the scattering matrix is:

$$\begin{bmatrix} S_{11} & S_{12} \\ S_{21} & S_{22} \end{bmatrix} = \begin{bmatrix} \frac{n^2 - 1}{n^2 + 1} & \frac{2n}{n^2 + 1} \\ \frac{2n}{n^2 + 1} & \frac{1 - n^2}{n^2 + 1} \end{bmatrix} \quad ...(1)$$

$$\therefore \quad n = 2$$
$$\Rightarrow S_{12} = S_{21} = \left[\frac{2n}{n^2 + 1}\right]$$
$$= \frac{2 \times 2}{2^2 + 1}$$
$$= \frac{4}{5} = 0.8$$

47. As given that

$$S(t)_{pm} = A_c \cos(2\pi f_c t + k_p m(t)$$

$$S(t)_{fm} = A_c \cos[2\pi f_c t + k_f \int_0^{\infty} m(t)dt]$$

Highest instantaneous frequencies are same.

$$f_i = \frac{1}{2\pi}\frac{d}{dt}\theta(t)$$

$$\because \; f_{i,\,PM} = f_{i,\,FM}$$

$$f_c + \frac{K_p}{2\pi}\frac{dm(t)}{dt}\bigg|_{max} = f_c + \frac{K_f}{2\pi} m(t)\bigg|_{max}$$

$$K_p \frac{dm(t)}{dt}\bigg|_{max} = K_f m(t)\big|_{max}$$

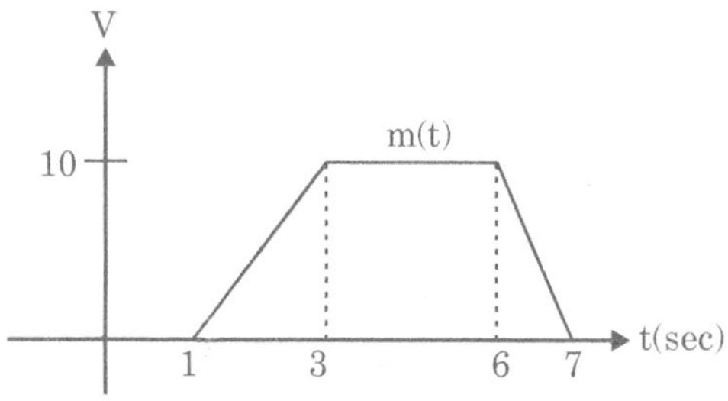

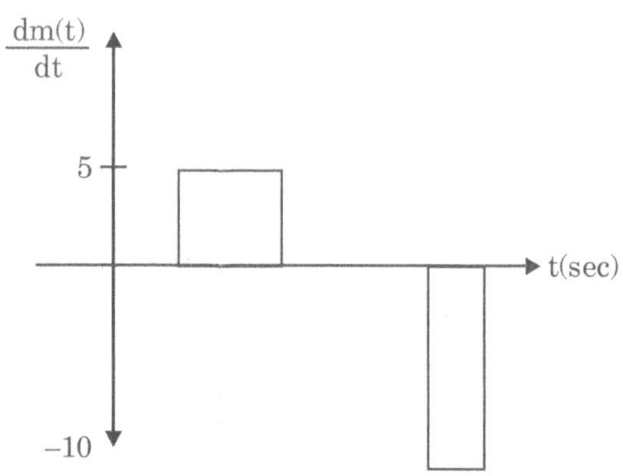

$$K_p.5 = k_f.10$$

$$\frac{K_p}{K_f} = \frac{10}{5} = 2 \text{ sec}$$

48. $S_1 = -3, S_2 = -1, S_3 = +1, S_4 = +2$

$Y = S + W$

$$W = N(0, 1) = \frac{1}{\sqrt{2\pi}} e^{-\omega^2/2}$$

$f_{s1}(s_1) = -3 + W = S_1$

Since the noise variable is gaussian with zero mean and ML decoding is used, the decision boundary between two adjacent signal points will be their arithmetic mean.

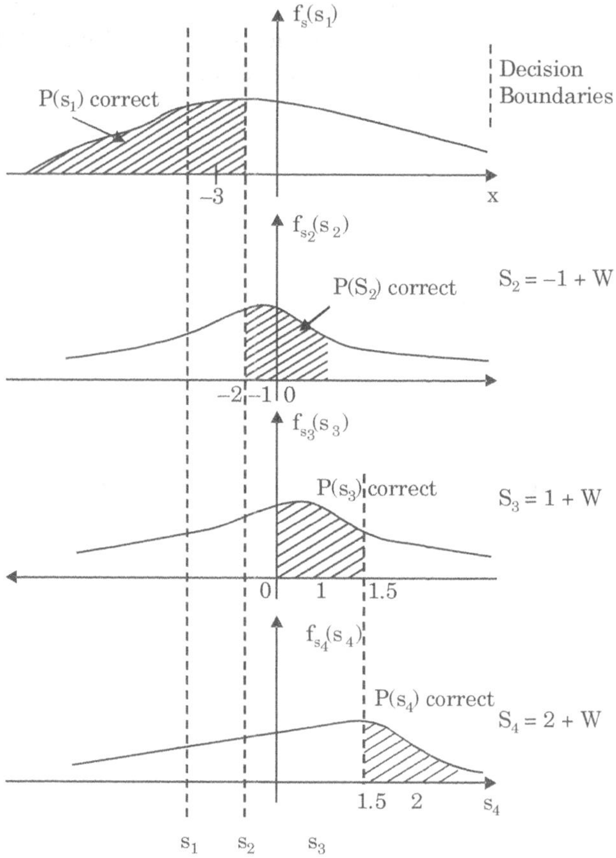

In all above graphs, the shaded area indicates the conditional probability of decoding a symbol correctly when it is transmited.

Since all symmbols are equally likely, decision boundaries are mid points.

i.e., for S_1, decision boundary – 2 V i.e., anything below – 2 V considered as S_1.

For S_2 decision boundaries are – 2 & 0V, i.e., any thing – 2V to 0V considered as S_2.

For S_3 decision boundaries are 0V & 1.5V.

For S_4 decision boundaries is 1.5V i.e., 1.5 V to ∞ considered as S_4.

P(Correct symbol receiveing is less for S_3 only therefore S_3 has highest conditional symbol error probability).

49. As given that,

$$G(s) = \frac{1}{(s+1)(s+a)} \quad a > 0$$

$$x(t) = 5 \cos 3t$$

$$y_{ss}(t) = \frac{1}{\sqrt{10}} \cos(3t - 1.892)$$

$$x(t) = A\cos(\omega_0 t + \theta)$$

$$H(s)|_{s\, j\omega_0} = |H(\omega_0)| \angle H(\omega_0)$$

$Y^{ss}(t) = A\,|H(\omega_0)|\cos(\omega_0 t + \theta + \angle H(\omega_0))$

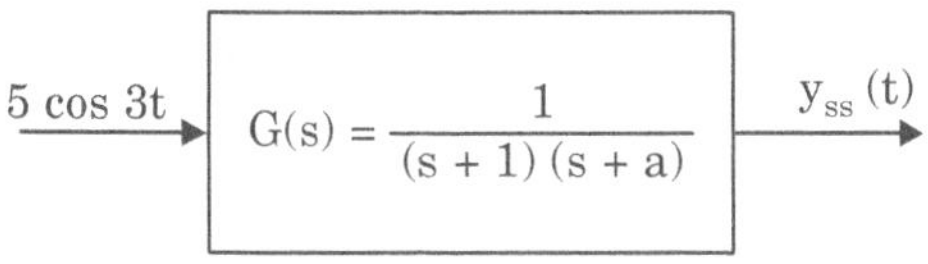

$$(\because S = j\omega_0)$$

$$G(S) = \frac{1}{(1 + j\omega_0)(a + j\omega_0)}$$

$$|G(j\omega_0)| = \frac{1}{\sqrt{(\omega_0^2 + 1)(\omega_0^2 + a^2)}} \qquad (\because \omega_0 = 3)$$

So, $\quad |G(j3)| = \dfrac{1}{\sqrt{a^2 + 9}\sqrt{1 + 9}}$

From the above expression

$$A|G(j3)| = \frac{1}{\sqrt{10}} \qquad (\because A = 5)$$

$$\frac{5}{\sqrt{10}\sqrt{a^2 + 9}} = \frac{1}{\sqrt{10}} \Rightarrow 5 = \sqrt{a^2 + 9}$$

$$25 - a^2 + 9 = 0$$

So, $\quad a = 4$

50. In any sequential circuit, the condition for proper operation is

Clock period (T) $\geq (t_{pd} + t_{comb\ logic} + t_{setup})$

So, total propagation delay:

For Flipflop 1 $\Rightarrow T \geq (8 + 5) \Rightarrow T \geq 13$ns

For Flipflop 2 $\Rightarrow T \geq (3 + 2 + 2 + 4) \Rightarrow T \geq 11$ns

Thus, $T \geq 13$ns

$\Rightarrow$ frequency of operation (f) = $\dfrac{1000}{13}$ MHz

$f_{max} = 76.92$ MHz.

51. From the given figures, the limits of te integralare given by

$$\text{limits}: \begin{cases} x : 0 \text{ to } 3 \\ y : 0 \text{ to } 1 \\ z : 0 \text{ to } 1 - y \end{cases}$$

$$I = \int_{y=0}^{1}\int_{z=0}^{1-y} xdz \int_{x=0}^{3} x\,dx\,dy\,dz$$

$$I = \int_{x=0}^{3}\left[\int_{y=0}^{1}\left\{\int_{z=0}^{1-y} xdz\right\} dy\right] dx$$

$$= \int_{x=0}^{3}\left[\int_{y=0}^{1} x(z)_0^{1-y}\,dy\right] dx$$

$$= \int_{x=0}^{3}\left[\int_{y=0}^{1} x.(1 - y)dy\right] dx$$

$$= \left(\frac{x^2}{2}\right)_0^3 . \left(y - \frac{y^2}{2}\right)_0^1$$

$$= \frac{9}{2}\left(1 - \frac{1}{2}\right) = \frac{9}{4} = 2.25$$

52. As given figure:

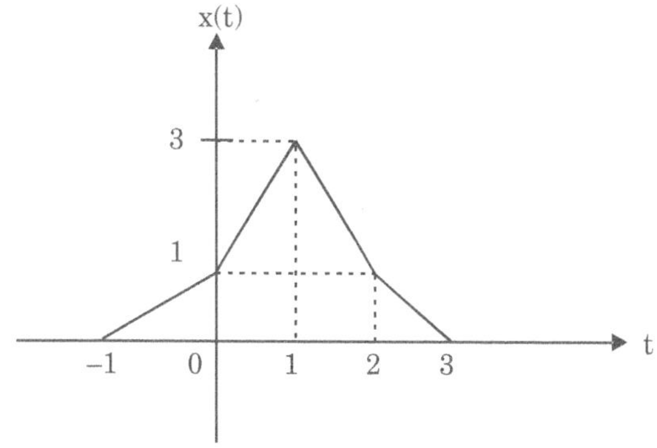

Because of symmetry of the signal w.r.t to t = 1

From praseval's theorem

$$\int_{-\infty}^{\infty} |X(\omega)|^2\,d\omega = 2\pi\int_{-\infty}^{\infty} |X(t)|^2\,dt$$

$$\int_{-\infty}^{\infty} |X(\omega)|^2\,d\omega = 2\pi\left[\int_{-1}^{3} |X(t)|^2\,dt\right]$$

$$= 2(2\pi)\left[\int_{-1}^{0} |(t + 1)|^2\,dt + \int_{0}^{1}(2t + 1)^2\,dt\right]$$

$$2(2\pi)\left[\int_{-1}^{0} |(t + 1)|^2\,dt + \int_{0}^{1}(4t^2 + 4t + 1)\,dt\right]$$

$$= 4\pi\left[\left[\frac{(t + 1)^3}{3}\right]_{-1}^{0} + \left[\frac{4t^3}{3} + \frac{4t^2}{2} + t\right]_0^1\right]$$

$$= 4\pi\left[\frac{1}{3} + \frac{4}{3} + 2 + 1\right] = 4\pi\left[\frac{5}{3} + 2\right]$$

$$= \frac{56\pi}{3}$$

$$= 58.64$$

53. As given that T/f of LTI system is:

$$H(z) = \frac{K(z - \alpha)}{z \mid 0.5}, |\alpha| > 1$$

$$|H(e^{j\omega})| = 1 \ \forall\ \omega$$

For an all pass filter poles & zeros are reciprocal to each other.

To get constant magnitude for all frequency pole located at z = – 0.5

$\therefore$ zero is located at $= \dfrac{1}{\text{Pole}} = \dfrac{1}{-0.5} = -2$

$\therefore$ zero $= \alpha = -2$

So, $H(z) = \dfrac{k(z + 2)}{z + 0.5}$

54. X follows uniform distribution over [–2, 10]

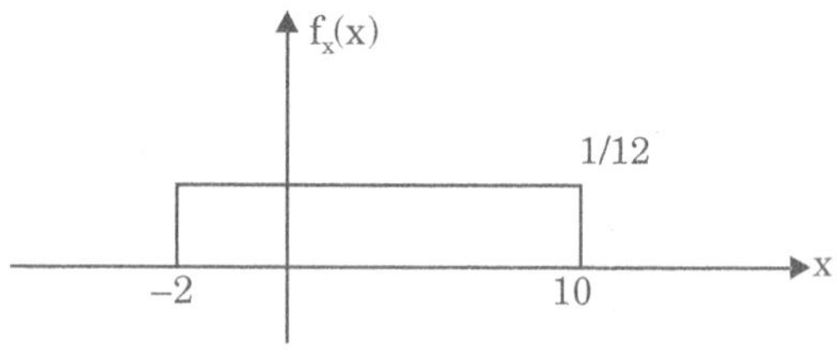

$$\left[f(x) = \left[\frac{1}{b-a}\right] = \left[\frac{1}{10-(-2)}\right]\right] = \frac{1}{12}$$

As given that,

Y = 2x – 6

$$X = \frac{y+6}{2}$$

for, Y = 7, X = 6.5

$$P(Y \le 7/X \ge 5) = P(Y \le 6.5/Y \ge 5)$$

$$\frac{P[X > 5 \text{ and } X < 6.5]}{P[X > 5]}$$

$$= \frac{\int_5^{6.5} 1/12\, dy}{\int_5^{10} 1/12\, dy} = \frac{1.5}{5} = 0.3$$

55.

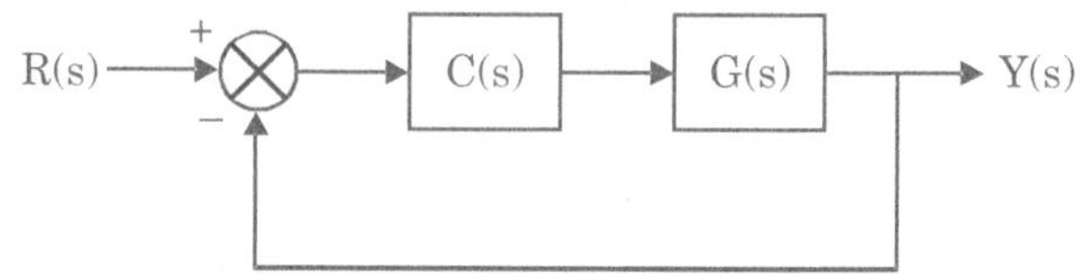

The open loop transfer function for the

$$\text{System} = C(s)G(s) = \frac{k(s+1)}{s(s+1)(s+3)} = \frac{k}{s(s+3)}$$

Since the system is type–1, so far a given Unit ramp input steady state.

$$e_{ss}\,(\text{ramp}) = \frac{1}{k_v} = 0.1$$

where, $k_v = \lim_{s\to 0} C(s)G(s) = \lim_{s\to 0} s.\frac{k}{s(s+3)} = \frac{k}{3}$

$$e_{ss} = \frac{1}{k_V} = \frac{3}{k} = 0.1$$

$$k = \frac{3}{0.1}$$

$$k = 30$$

SOLVED PAPER – 2019

ELECTRONICS & COMMUNICATION ENGINEERING

No. of Questions : 65 *Maximum Marks : 100*

INSTRUCTIONS

1. Total of 65 questions carrying 100 marks, out of which 10 questions carrying a total of 15 marks are in General Aptitude (GA)
2. The Engineering Mathematics will carry around **15% of the total marks,** the General Aptitude section will carry **15% of the total marks** and the **remaining 70% of the total marks.**
3. **Types of Questions**
   (*a*) **Multiple Choice Questions (MCQ)** carrying 1 or 2 marks each in all papers and sections. These questions are objective in nature, and each will have a choice of four options, out of which the candidate has to mark the correct answer(s).
   (*b*) **Numerical Answer Questions** of 1 or 2 marks each in all papers and sections. For these questions the answer is a real number, to be entered by the candidate using the virtual keypad. No choices will be shown for these type of questions.
4. For **1-mark** multiple-choice questions, **1/3 marks** will be deducted for a wrong answer. Likewise, for **2-marks** multiple-choice questions, **2/3** marks will be deducted for a wrong answer. There is no negative marking for numerical answer type questions.

Chapter-Wise Analysis

#	Chapters	Marks		
		1	2	Total
1	General Aptitude	5	5	15
2	Engineering Mathematics	6	3	12
3	Networks	1	4	9
4	Signals and Systems	3	1	5
5	Electronic Devices	3	2	7

#	Chapters	Marks		
		1	2	Total
6	Analog Circuits	3	5	13
7	Digital Circuits & Microprocessor	2	1	4
8	Control Systems	1	4	9
9	Communications	3	6	15
10	Electromagnetics	3	4	11

GENERAL APTITUDE (GA)

Q. 1 to Q. 5 : Carry One Mark Each.

1. Five different books (P, Q, R, S, T) are to be arranged on a shelf. The books R and S are to be arranged first and second, respectively from the right side of the shelf. The number of different order in which P, Q and T may be arranged is ________.

   (*a*) 2 (*b*) 120
   (*c*) 6 (*d*) 12

2. The boat arrived ________ dawn.

   (*a*) on (*b*) at
   (*c*) under (*d*) in

3. It would take one machine 4 hours to complete a production order and another machine 2 hours to complete the same order. If both machines work simultaneously at their respective constant rates, the time taken to complete the same order is _______ hours.

   (*a*) 2/3 (*b*) 7/3
   (*c*) 4/3 (*d*) 3/4

4. When he did not come home, she _______ him lying dead on the roadside somewhere

   (*a*) concluded (*b*) pictured
   (*c*) notice (*d*) looked

5. The strategies that the company _______ to sell its products ________ house-to-house marketing.

   (*a*) uses, include
   (*b*) use, includes
   (*c*) uses, including
   (*d*) used, includes

Q. 6 to Q. 10 : Carry Two Marks Each.

6. "Indian history was written by British historians– extremely well documented and researched, but not always impartial. History had to serve its purpose: Everything was made subservient to the glory of the Union Jack. Latter-day Indian scholar presented a contrary picture."

   From the text above, we can infer that :

   Indian history written by British historians ____.

   (*a*) was well documented and not researched but was always biased
   (*b*) was not well documented and researched and was sometimes biased
   (*c*) was well documented and researched but was sometimes biased
   (*d*) was not well documented and researched and was always biased

7. Two design consultants, P and Q, started working from 8 AM for a client. The client budgeted a total of USD 3000 for the consultants. P stopped working when the hour hand moved by 210 degrees on the clock. Q stopped working when the hour hand moved by 240 degrees. P took two tea breaks of 15 minutes each during her shift, but took no lunch break. Q took only one lunch break for 20 minutes, but no tea breaks. The market rate for consultants is USD 200 per hour and breaks are not paid. After paying the consultants, the client shall have USE _____ remaining in the budget.

   (*a*) 000.00 (*b*) 433.33
   (*c*) 166.67 (*d*) 300.00

8. Five people P, Q, R, S and T work in a bank. P and Q don't like each other but have to share an office till T gets a promotion and moves to the big office next to the garden. R, who is currently sharing an office with T wants to move to the adjacent office with S, the handsome new intern.

   Given the floor plan, what is the current location of Q, R and T?

   (O = Office, WR = Washroom)

   (*a*)
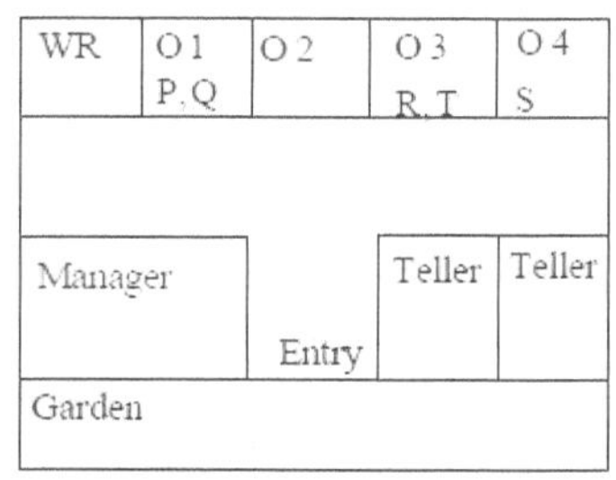

   (*b*)
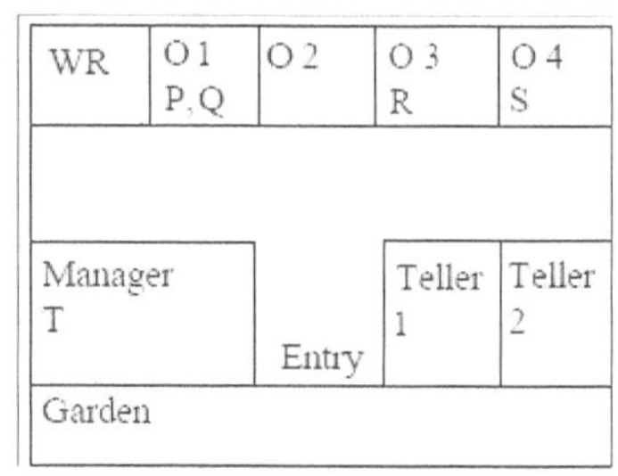

   (*c*)
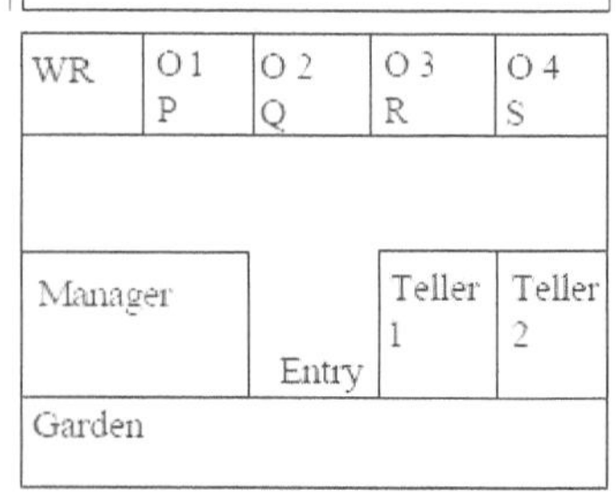

   (*d*)
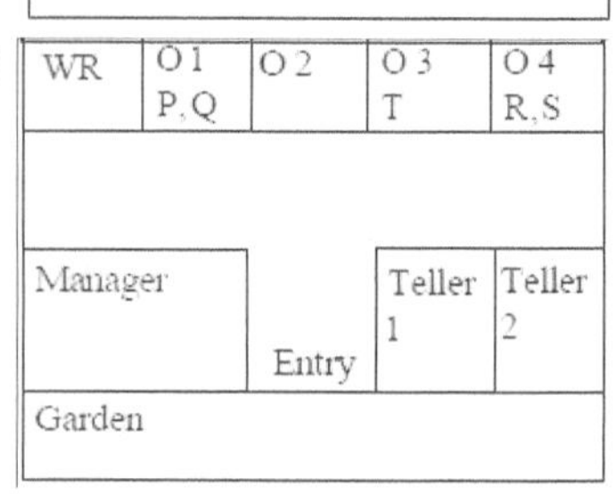

9. Four people are standing in a line facing you. They are Rahul, Mathew, Seema and Lohit. One is an engineering, one is a doctor, one a teacher and another a dancer. You are told that:

1. Mathew is not standing next to Seema
2. There are two people standing beween Lohit and the engineer
3. Rahul is not a doctor
4. The teacher and the dancer are standing next to each other.
5. Seema is turning to her right to speak to the doctor standing next to her.

Who amongst them is an engineer?

(*a*) Rahul
(*b*) Mathew
(*c*) Seema
(*d*) Lohit

10. The bar graph in Panel (a) shows the proportion of male and female illiterates in 2001 and 2011. The proportions of males and females in 2001 and 2011 are given in Panel (b) and (c), respectively. The total population did not change during this period.

The percentage increase in the total number of literate from 2001 to 2011 is ________.

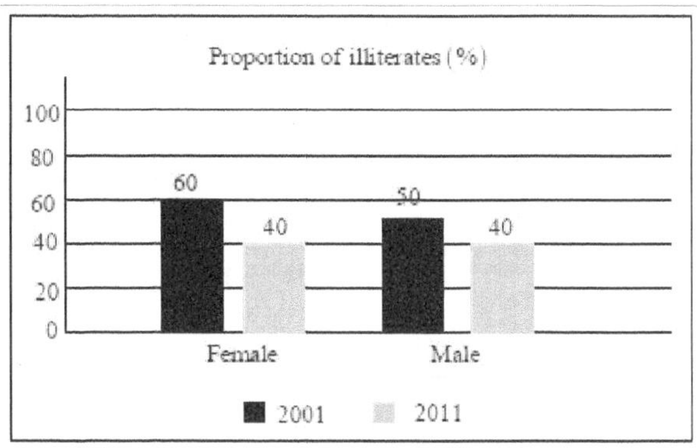

Panel (a)

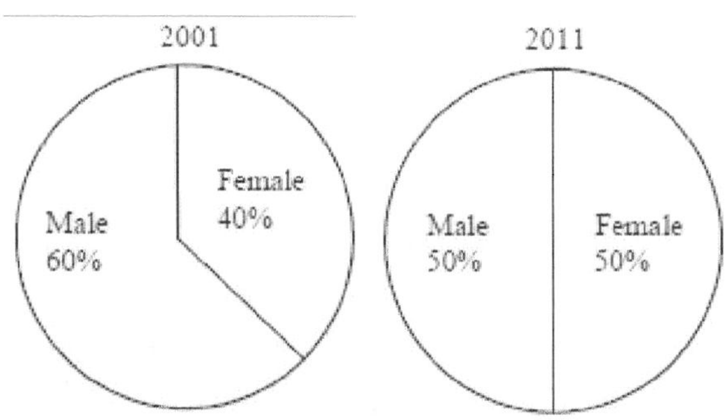

Panel (b) Panel (c)

(*a*) 33.43 (*b*) 35.43
(*c*) 34.43 (*d*) 30.43

ELECTRONICS & COMMUNICATION ENGINEERING

Q. 1 to Q. 25 : Carry One Mark Each.

1. Radiation resistance of a small dipole current element of length l at a frequency of 3 GHz is 3 ohms. If the length is changed by 1%, then the percentage change in the radiation resistance, rounded off two decimal places, is ________%.

2. Which one of the following functions is analytic over the entire complex plane?

(*a*) $\ell n(z)$
(*b*) $\cos(z)$
(*c*) $e^{1/z}$
(*d*) $\dfrac{1}{1-z}$

3. The value of integral $\int_0^{\pi}\int_y^{\pi}\frac{\sin x}{x}dx\,dy$, is equal to ________.

4. In the circuit shown, V_S is a square wave of period T with maximum and minimum values of 8V and – 10 V, respectively. Assume that the diode is ideal and $R_1 = R_2 = 50\ \Omega$. The average value of V_L is ________ volts (rounded off to 1 decimal place).

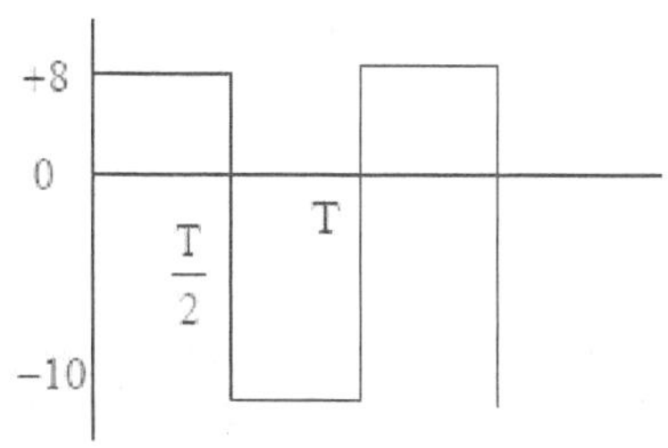

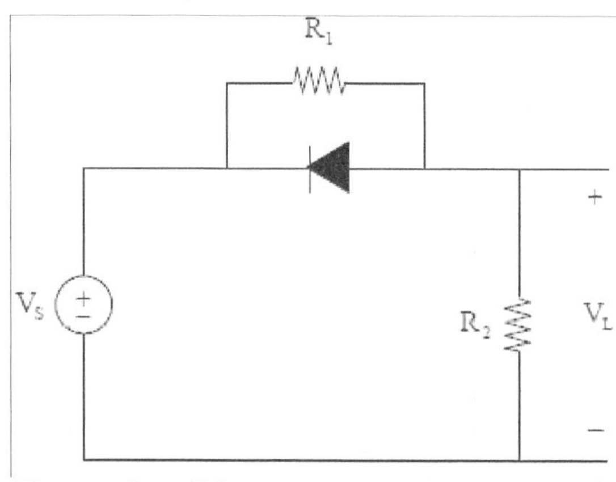

5. The number of distinct eigen values of the matrix

$$A = \begin{bmatrix} 2 & 2 & 3 & 3 \\ 0 & 1 & 1 & 1 \\ 0 & 0 & 3 & 3 \\ 0 & 0 & 0 & 2 \end{bmatrix}$$ is equal to ________.

6. The families of curves represented by the solution of the equation $\frac{dy}{dx} = -\left(\frac{x}{y}\right)^n$

For n = – 1 and n = + 1, respectively, are

(*a*) Hyperbolas and Circles

(*b*) Circles and Hyperbolas

(*c*) Hyperbolas and Parabolas

(*d*) Parabolas and Circles

7. In the table shown, List-I and List-II, respectively, contain terms appearing on the left-hand side and the right-hand side of Maxwell's equations (in their standard form). Match the left-hand side with the corresponding right-hand side.

List - I	List - II
1. $\nabla . D$	(P) 0
2. $\nabla \times E$	(Q) ρ
3. $\nabla . B$	(R) $-\frac{\partial B}{\partial t}$
4. $\nabla \times H$	(S) $J + \frac{\partial D}{\partial t}$

(*a*) 1-Q, 2-R, 3-P, 4-S (*b*) 1-Q, 2-S, 3-P, 4-R

(*c*) 1-P, 2-R, 3-Q, 4-S (*d*) 1-R, 2-Q, 3-S, 4-P

8. Consider the signal $f(t) = 1 + 2\cos(\pi t) + 3\sin\left(\frac{2\pi}{3}t\right) + 4\cos\left(\frac{\pi}{2}t + \frac{\pi}{4}\right)$, where t is in seconds.

Its fundamental time period, in seconds, is _____.

9. If X and Y are random variables such that E[2X + Y] = 0 and E[X + 2Y] = 33, the E[X] + E[Y] = _____.

10. In the circuit shown, what are the values of F for EN = 0 and EN = 1, respectively?

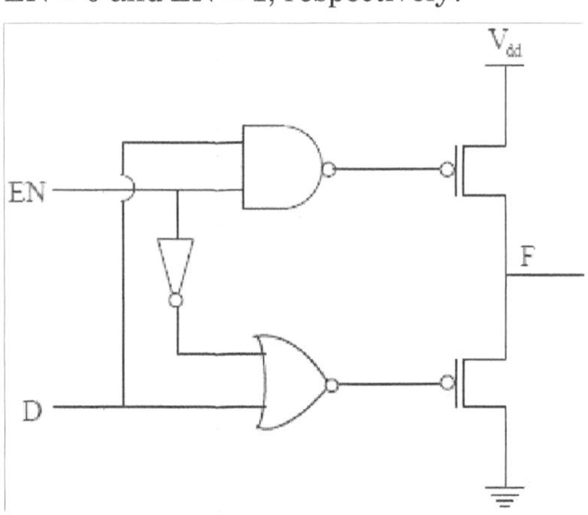

(A) 0 and 1 (B) Hi – Z and $\bar{D}$

(C) 0 and D (D) Hi – Z and D

11. Let H(z) be the z-transform of a real-valued discrete time signal h[n]. If $P(z) = H(z)H\left(\frac{1}{z}\right)$ has a zero $z = \frac{1}{2} + \frac{1}{2}j$, and P(z) has a total of four zeros, which one of the following plots represents all the zeros correctly?

(A)

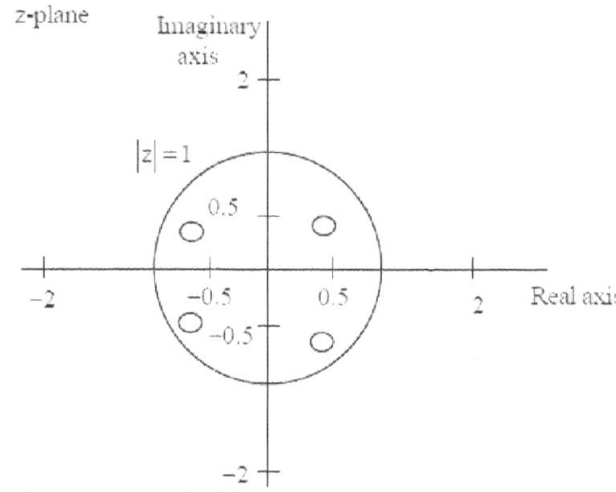

(B)

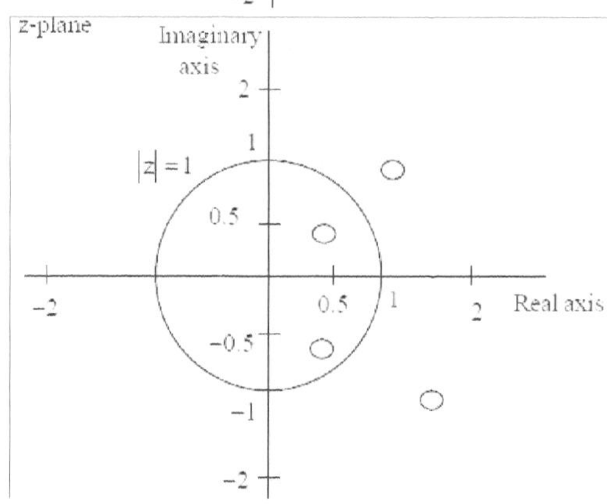

(C)

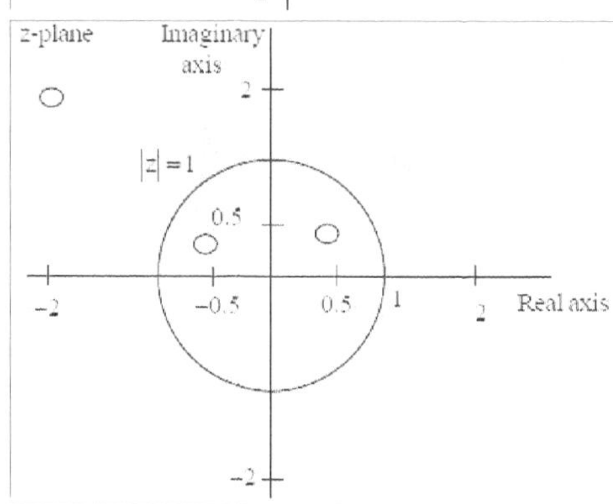

(D)

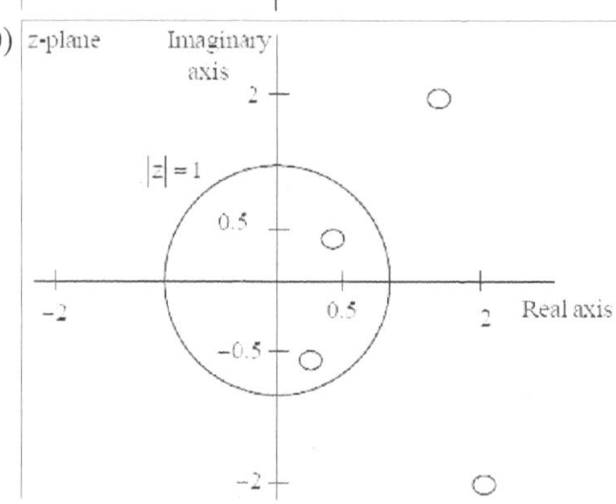

12. Which one of the following options describes correctly the equilibrium band diagram at T = 300 K of a Silicon pnn⁺p⁺⁺ configuration shown in the figure?

p	n	n^+	p^{++}

(A)

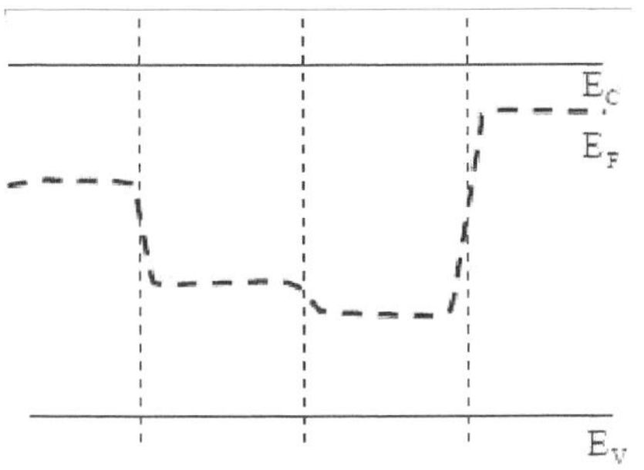

(B)

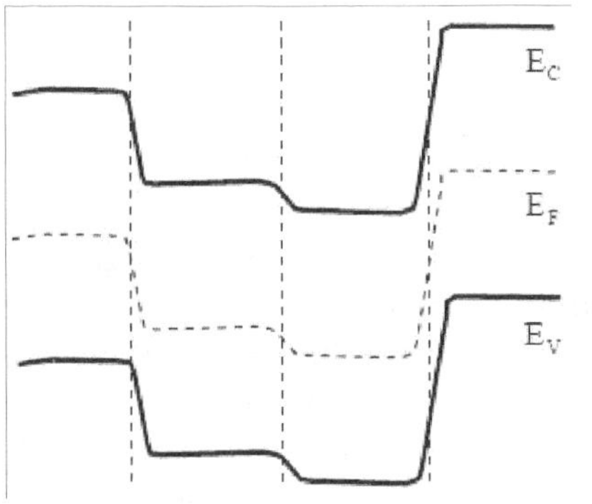

(C)

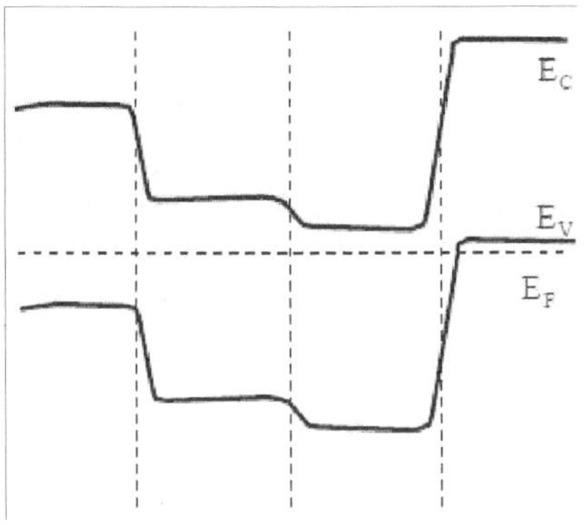

(D)

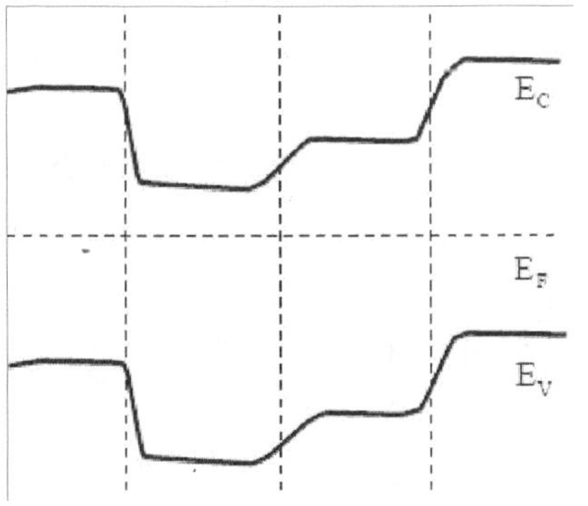

13. In the circuit shown, A and B are the inputs and F is the output. What is the functionality of the circuit?

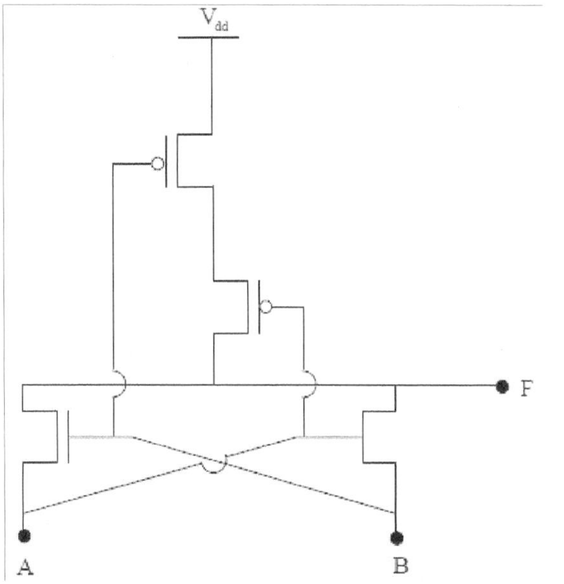

(A) XOR (B) XNOR

(C) Latch (D) SRAM cell

14. The correct circuit representation of the structure shown in the figure is

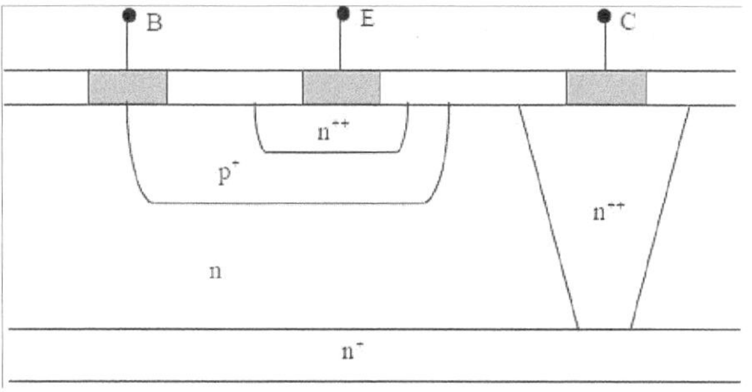

(A)

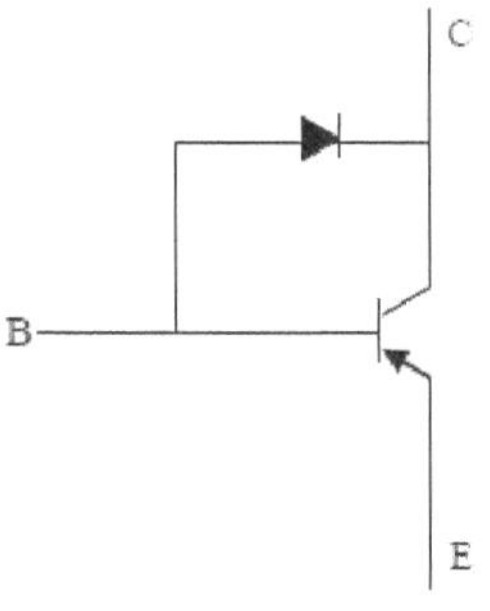

(B)

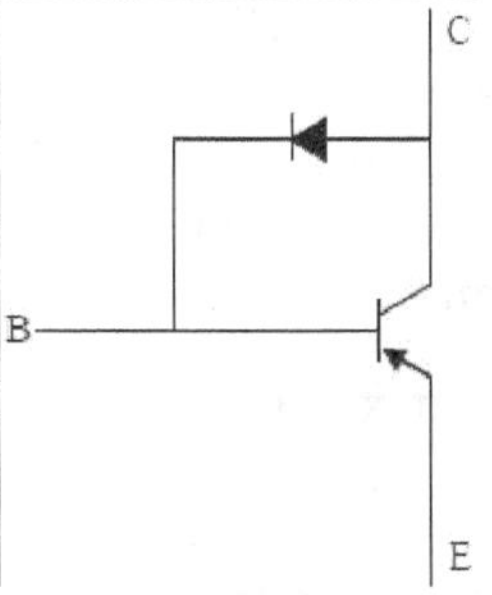

(C) 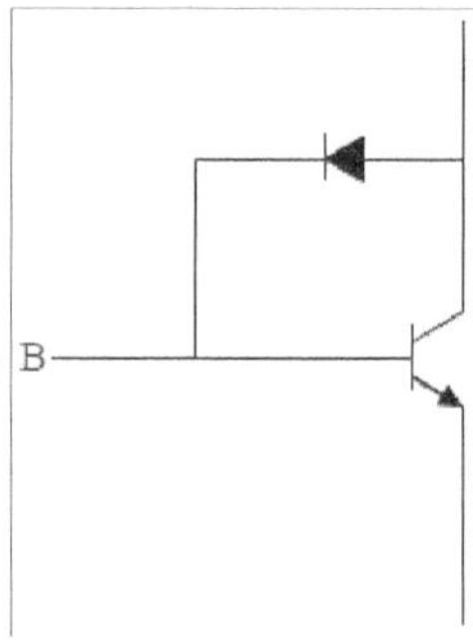

(D) 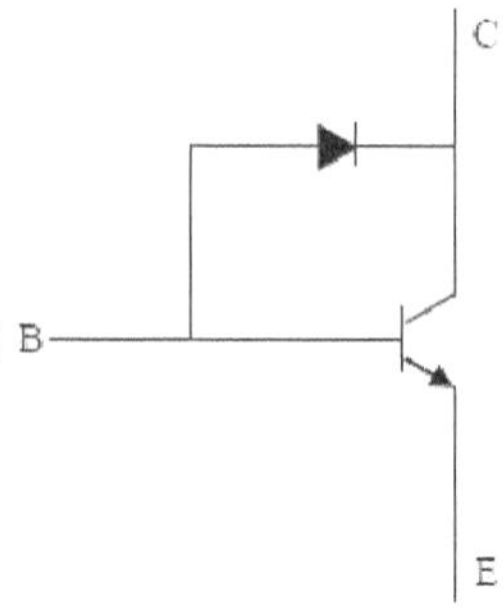

15. What is the electric flux $\left(\int \vec{E}.d\hat{a}\right)$ through a quarter-cylinder of height H (as shown in the figure) due to an infinitely long the line charge along the axis of the cylinder with a charge density of Q?

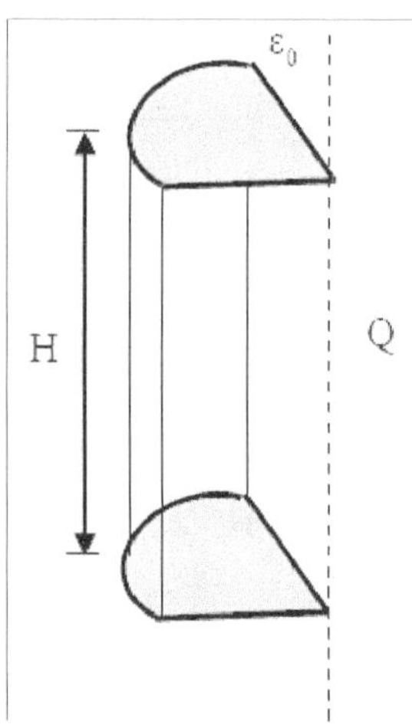

(A) $\frac{4H}{Q\varepsilon_0}$ (B) $\frac{HQ}{4\varepsilon_0}$

(C) $\frac{HQ}{\varepsilon_0}$ (D) $\frac{H\varepsilon_0}{4Q}$

16. Let Z be an exponential random variable with mean 1. That is, the cumulative distribution function of Z is given by

$$F_z(x) = \begin{cases} 1 - e^{-x} & \text{if } x \geq 0 \\ 0 & \text{if } x < 0 \end{cases}$$

The $\Pr(Z > 2 \mid Z > 1)$, rounded off two decimal places, is equal to ______.

17. A linear Hamming code is used to map 4-bit messages to 7-bit codewords. The encoder mapping is linear. If the message 0001 is mapped to the codeword 0000111, and the message 0011 is mapped to the codeword 1100110, then the message 0010 is mapped to

(A) 0010011

(B) 1111111

(C) 1111000

(D) 1100001

18. In the circuit shown, the clock frequency, i.e., the frequency of the Clk signal, is 12 kHz. The frequency of the signal at Q_2 is _____ kHz.

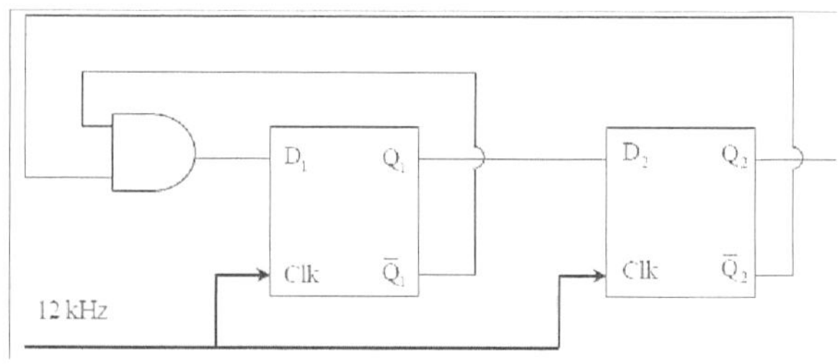

19. A standard CMOS inverter is designed with equal rise and fall times ($\beta_n = \beta_p$). If the width of the pMOS transistor in the inverter is increased, what would be the effect on the LOW noise margin (NM_L) and the HIGH noise margin NM_H?

(A) NM_L increases and NM_H decrease

(B) Both NM_L and NM_H increase

(C) No change in the noise margins

(D) NM_L decreases and NM_H increases

20. For an LTI system, the Bode plot for its gain is as illustrated in the figure shown. The number of system poles N_p and the number of system zeros N_z in the frequency range $1\text{Hz} \leq f \leq 10^7$ Hz is

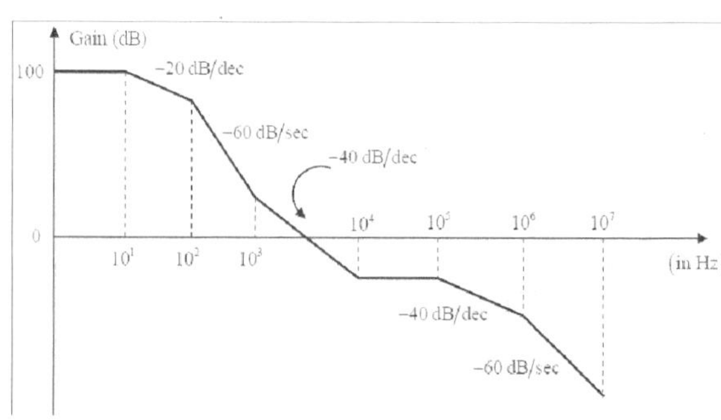

(A) $N_p = 4$, $N_z = 2$

(B) $N_p = 7$, $N_z = 4$

(C) $N_p = 6$, $N_z = 3$

(D) $N_p = 5$, $N_z = 2$

21. The figure shows the high-frequency C-V curve of a MOS capacitor (at T = 300 K) with $\Phi_{ms} = 0$ V and no oxide charges. The flat-band, inversion, and accumulation conditions are represented, respectively, by the points.

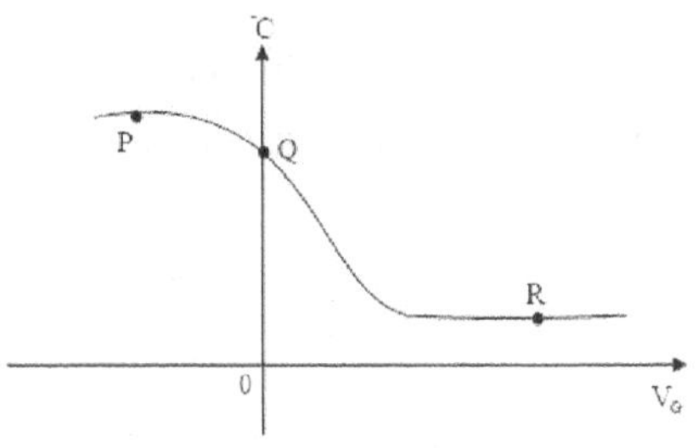

(A) R, P, Q (B) Q, P, R
(C) P, Q, R (D) Q, R, P

22. Consider the two-port resistive network shown in the figure. When an excitation of 5V is applied across Port 1, and Port 2 is shorted, the current through the short circuit at Port 2 is measured to be 1 A (see (a) in the figure).

Now, if an excitation of 5 V is applied across port 2, and port 1 is shorted (see (b) in the figure), what is the current through the short circuit at port 1?

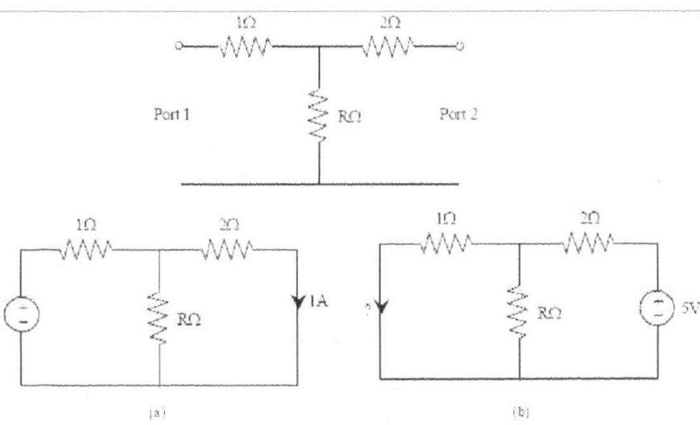

(A) 1 A
(B) 2 A
(C) 2.5 A
(D) 0.5 A

23. Let Y(s) be the unit-step response of a causal system having a transfer function

$$G(s) = \frac{3 - s}{(s+1)(s+3)}$$

That is, $Y(s) = \frac{G(s)}{s}$. The forced response of the system is

(A) $u(t) - 2e^{-t}u(t) + e^{-3t}u(t)$
(B) $2u(t)$
(C) $u(t)$
(D) $2u(t) - 2e^{-t}u(t) + e^{-3t}u(t)$

24. The baseband signal m(t) shown in the figure is phase-modulated to generate the PM signal $\varphi(t) = \cos(2\pi f_c t + km(t))$.

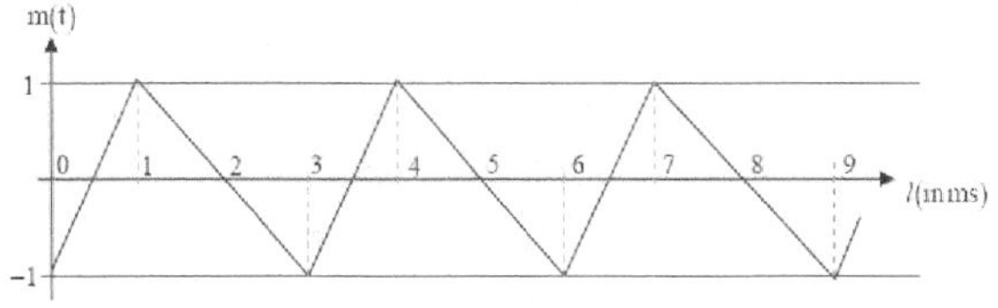

The time t on the x-axis in the figure is in milliseconds. If the carrier frequency is $f_c = 50$ kHz and $k = 10\pi$, the ratio of the minimum instantaneous frequency (in kHz) to the maximum instantaneous frequency (in kHz) is ______ (rounded off to 2 decimal places).

25. The value of the contour integral

$$\frac{1}{2\pi j}\oint\left(z + \frac{1}{z}\right)^2 dz$$

Evaluated over the unit circle $|Z| = 1$ is ______.

Q. 26 to Q. 55 : Carry Two Marks Each.

26. In the circuit shown, the threshold voltages of the pMOS($|V_{tp}|$) and nMOS (V_{tn}) transistors are both equal to 1 V. All the transistors have the same output resistance r_{ds} of 6MΩ. The other parameters are listed below:

$$\mu_n C_{ox} = 60\mu A / V^2; \left(\frac{W}{L}\right)_{nMOS} = 5$$

$$\mu_p C_{ox} = 30\mu A / V^2; \left(\frac{W}{L}\right)_{pMOS} = 10$$

μ_n and μ_p are the carrier mobilities, and C_{ox} is the oxide capacitance per unit area. Ignoring the effect of channel length modulation and body bias, the gain of the circuit is ______ (rounded off to 1 decimal place).

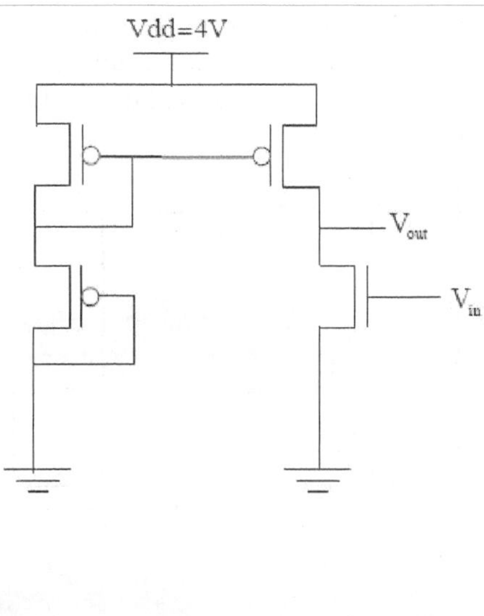

27. It is desired to find three-tap causal filter which gives zero signal as an output to and input of the form

$$x[n] = c_1 \exp\left(-\frac{j\pi n}{2}\right) + c_2 \exp\left(\frac{j\pi n}{2}\right),$$

Where c_1 and c_2 are arbitrary real numbers. The desired three-tap filter is given by

h[0] = 1, h[1] = a, h[2] = b and

h[n] = 0 for n < 0 or n > 2.

What are the values of the filter taps a and b if the output is y[n] = 0 for all n, when x[n] is as given above?

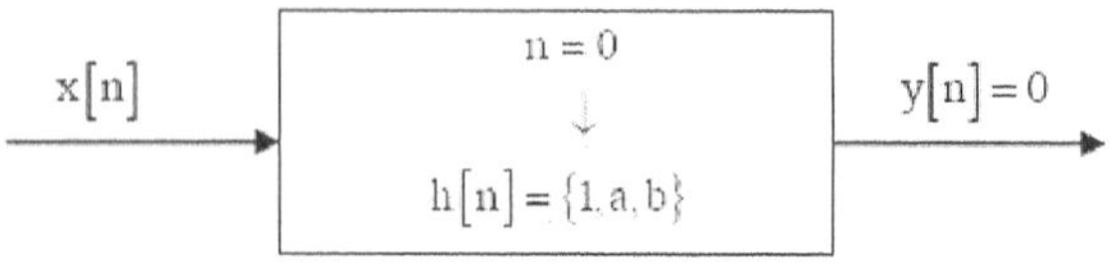

(A) a = –1, b = 1 (B) a = 0, b = 1

(C) a = 1, b = 1 (D) a = 0, b = – 1

28. Let h[n] be length-7 discrete-time finite impulse response filter, given by

h[0] = 4, h[1] = 3, h[2] = 2, h[3] = 1

h[–1] = –3, h[–2] = –2, h[–3] = –1,

and h[n] is zero for $|n| \geq 4$. A length-3 finite impulse response approximation g[n] of h[n] has to be obtained such that

$E(h, g) = \int_{-\pi}^{\pi} \left|H(e^{j\omega}) - G(e^{j\omega})\right|^2 d\omega$ is minimized,

where $H(e^{j\omega})$ and $G(e^{j\omega})$ are the discrete-time Fourier transforms of h[n] and g[n], respectively. For the filter that minimizes E(h, g), the value of 10g[–1] + g[1], rounded off to 2 decimal places, is_________.

29. Consider the line integral $\int_C (xdy - ydx)$. The integral being taken in a counterclockwise direction over the closed curve C that forms the boundary of the region R shown in the figure below. The region R is the area enclosed by the union of a 2 × 3 rectangle and a semi-circle of radius 1. The line integral evaluates to

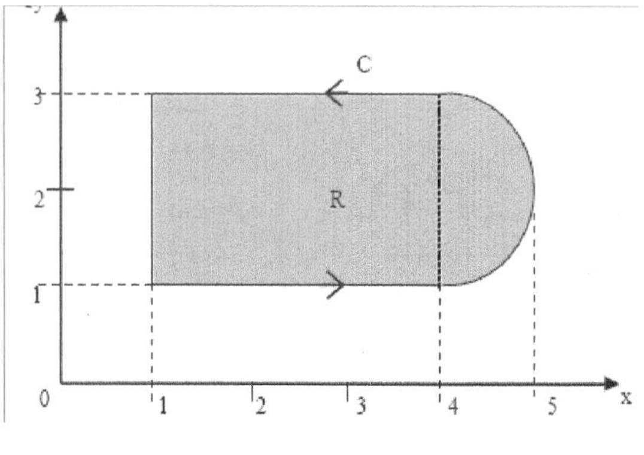

(A) $16 + 2\pi$

(B) $6 + \pi/2$

(C) $12 + \pi$

(D) $8 + \pi$

30. A rectangular waveguide of width W and height h has cut-off frequencies for TE_{10} and $T_{E_{11}}$ modes in the ratio 1 : 2. The aspect ratio w/h, rounded off to two decimal places, is ______.

31. A Germanium sample of dimension 1 cm × 1 cm is illuminated with a 20 mW, 600 nm laser light source as shown in the figure. The illuminated sample surface has a 100 nm of loss-less Silicon dioxide layer that reflects one-fourth of the incident light. From the remaining light, one-third of the power is reflected form the silicon dioxide-Germanium interface, one-third is absorbed in the Germanium layer, and one-third is transmitted through the other side of the sample. If the absorption coefficient of Germanium at 600 nm is 3×10^4 cm^{-1} and the bandgap is 0.66 eV, the thickness of the Germanium layer, rounded off to 3 decimal places, is _____ μm.

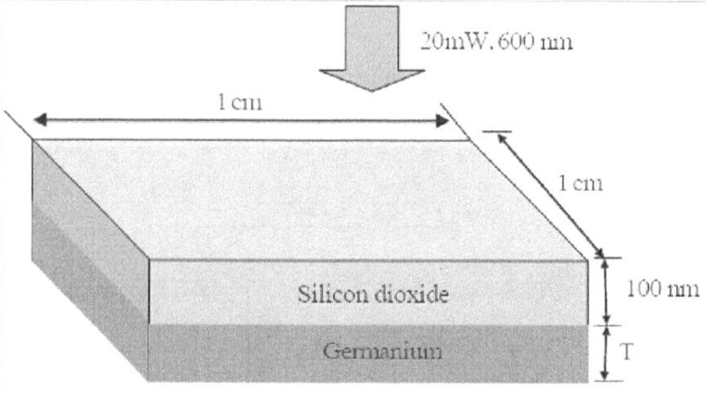

32. In the circuit shown, the breakdown voltage and the maximum current of the Zener diod are 20 V and 60 mA, respectively. The values of R_1 and R_L are 200 Ω and 1k Ω, respectively. What is the range of V_i that will maintain the Zener diode in the 'n' state?

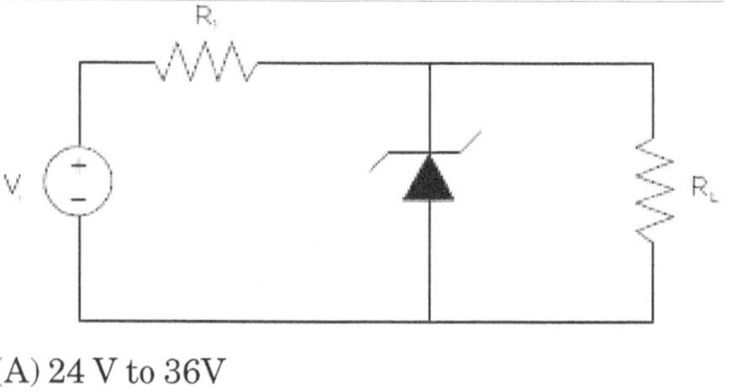

(A) 24 V to 36V

(B) 22 V to 34 V

(C) 20 V to 28 V

(D) 18 V to 24

33. A single bit, equally likely to be 0 and 1, is to be sent across an additive white Gaussian noise (AWGN) channel with power spectral density $N_0/2$. Binary signaling, with $0 \to p(t)$ and $I \to q(t)$, is used for the transmission, along with an optimal receiver that minimizes the bit-error probability.

Let $\varphi_1(t), \varphi_2(t)$ form and orthonormal signal set.

If we choose $p(t) = \varphi_1(t)$ and $q(t) = -\varphi_1(t)$, we would obtain a certain bit-error probability P_b.

If we keep $p(t) = \varphi_1(t)$, but take $q(t) = \sqrt{E}\varphi_2(t)$, for what value of E would we obtain the same bit-error probability P_b?

(A) 3 (B) 1

(C) 2 (D) 0

34. Consider a six-point decimation-in-time Fast Fourier Transform (FFT) algorithm, for which the signal-flow graph corresponding to X[I] is shown in the figure. Let $W_6 = \exp\left(-\frac{j2\pi}{6}\right)$. In the figure, what should be the values of the coefficients a_1, a_2, a_3 in terms of W_6 so that X[I] is obtained correctly?

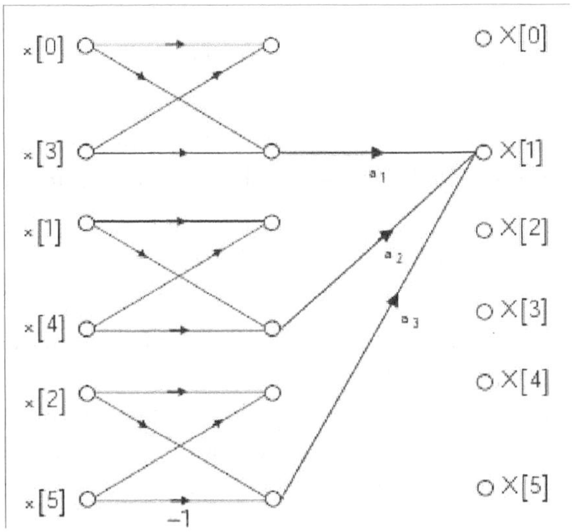

(A) $a_1 = 1, a_2 = W_6^2, a_3 = W_6$

(B) $a_1 = -1, a_2 = W_6^2, a_3 = W_6$

(C) $a_1 = -1, a_2 = W_6, a_3 = W_6^2$

(D) $a_1 = 1, a_2 = W_6, a_3 = W_6^2$

35. The quantum efficiency (η) and responsivity (R) at a wavelength λ (in μ m) in a p-i-n photo detector are related by

(A) $R = \frac{\eta \times \lambda}{1.24}$ (B) $R = \frac{\lambda}{\eta \times 1.24}$

(C) $R = \frac{1.24 \times \lambda}{\eta}$ (D) $R = \frac{1.24}{\eta \times \lambda}$

36. Consider a long-channel MOSFET with a channel length 1 μ m and width 10μm. The device parameters are acceptor concentration $N_A = 5 \times 10^{16}$ cm^{-3}, electron mobility μ_n = 800 cm^2/V–s, oxide capacitance/area C_{ox} = 3.45 $\times 10^{-7}$F/cm^2, threshold voltage V_T = 0.7V. The drain saturation current $(I_{D_{sat}})$ for a gate voltage of 5V is ______ mA (rounded off to two decimal places).

$$\left[\varepsilon_0 = 8.854 \times 10^{-14} \text{F/cm}, \varepsilon_{Si,} = 11.9\right]$$

37. A voice signal m (t) is in the frequency range 5 kHz to 15 kHz. The signal is amplitude modulated to generate an AM signal f(t) = A (1 + m (t)) cos $2\pi f_c t$, where f_c = 600kHz.

The Am signal f(t) is to be digitized and archived. This is done by first sampling f(t) at 1.2 times the Nyquist frequency, and then quantizing each sample using a 256-level quantizer. Finally, each quantized sample is binary coded using K bits, where K is the minimum number of bits required for the encoding. The rate, in Megabits per second (rounded off to 2 decimal places), of the resulting stream of coded bits is _____ Mbps.

38. A random variable X takes values – 1 and + 1 with probabilities 0.2 and 0.8, respectively. It is transmitted across a channel which adds noise N, so that the random variable at the channel output is Y = X + N. The noise N is independent of X, and is uniformly distributed over the interval [–2, 2]. The receiver makes a decision

$$X = \begin{cases} -1, \text{ if } Y \le \theta \\ +1, \text{ if } Y > \theta \end{cases}$$

Where the threshold $\theta \in [-1, 1]$ is chosen so as to minimize the probability of error, rounded off to 1 decimal place, is ________.

39. Let the state-space representation of an LT1 system be $\dot{x}(t) = A_x(t) + Bu(t)$,

y(t) = Cx(t) + du (t) where A,B,C are matrices, d is a scalar, u(t) is the input to the system, and y (t) is its output. Let $B = [0\ 0\ 1]^T$ and d = 0. Which one of the following options for A and C will ensure that the transfer function of this LTI system is

$$H(s) = \frac{1}{S^3 + 3s^2 + 2s + 1}?$$

(A) $A = \begin{bmatrix} 0 & 1 & 0 \\ 0 & 0 & 1 \\ -1 & -2 & -3 \end{bmatrix}$ and $C = [1\ \ 0\ \ 0]$

(B) $A = \begin{bmatrix} 0 & 1 & 0 \\ 0 & 0 & 1 \\ -1 & -2 & -3 \end{bmatrix}$ and $C = [0\ \ 0\ \ 1]$

(C) $A = \begin{bmatrix} 0 & 1 & 0 \\ 0 & 0 & 1 \\ -3 & -2 & -1 \end{bmatrix}$ and $C = [0\ \ 0\ \ 1]$

(D) $A = \begin{bmatrix} 0 & 1 & 0 \\ 0 & 0 & 1 \\ -3 & -2 & -1 \end{bmatrix}$ and $C = [1\ \ 0\ \ 0]$

40. Two identical copper wires W1 and W2, placed in parallel as shown in the figure, carry currents I and 2I, respectively, in opposite directions. If the two wires are separated by a distance of 4r, then the magnitude of the magnetic field $\vec{B}$ between the wires at a distance r form W1 is

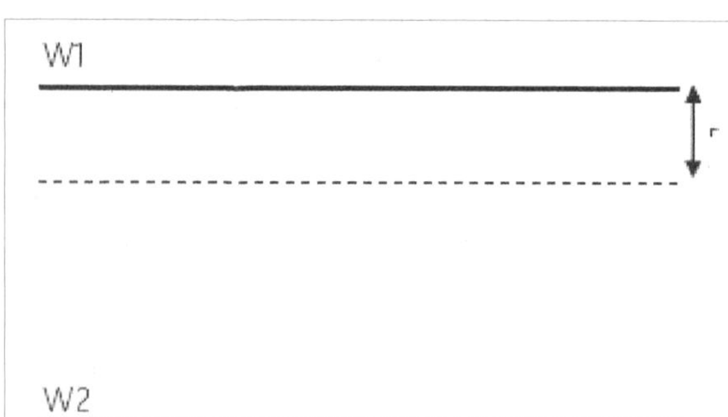

(A) $\frac{5\mu_0 I}{6\pi r}$ (B) $\frac{\mu_0 I}{6\pi r}$

(C) $\frac{6\mu_0 I}{5\pi r}$ (D) $\frac{\mu_0^2 I^2}{2\pi r^2}$

41. In the circuit shown, V_s is a 10V square wave of period, T = 4 ms with R = 500Ω and C = 10 μF.The capacitor is initially unchanged at t = 0, and the diode is assumed to be ideal. The voltage across the capacitor (V_c) at 3 ms is equal to ______ volts (rounded off to one decimal place).

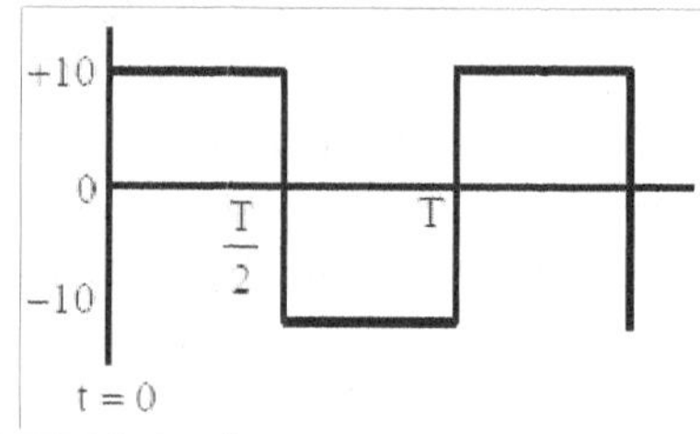

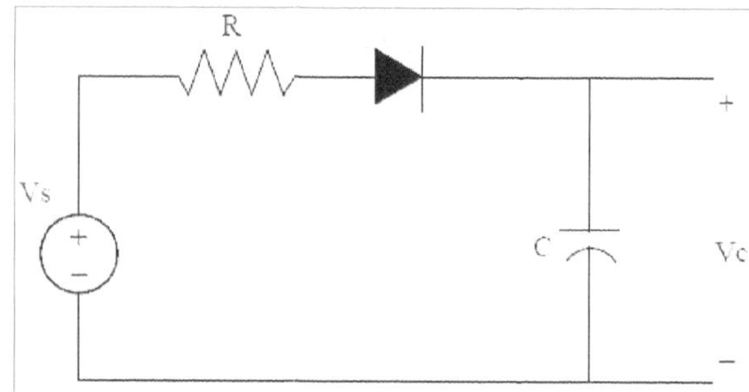

42. Consider a causal second-order system with the transfer function $G(s) = \frac{1}{1 + 2s + s^2}$

With a until-step $R(s) = \frac{1}{s}$ as an input. Let C(s) be the corresponding output. The time taken by the system output C(t) to reach 94% of its steady-state value $\lim_{t \to \infty} c(t)$ rounded off to two decimal places, is

(A) 5.25 (B) 2.81

(C) 4.50 (D) 3.89

43. The RC circuit shown below has a variable resistance R(t) given by the following expression:

$$R(t) = R_0\left(t - \frac{t}{T}\right) \text{ for } 0 \le t < T$$

Where $R_0 = 1\Omega$, C = 1F. We are also given that T = $3R_0C$ and the source voltage is V_s = 1 V. If the current at time t = 0 is 1A, then the current I(t), in amperes, at time t = T/2 is ______ (rounded off to 2 decimal places).

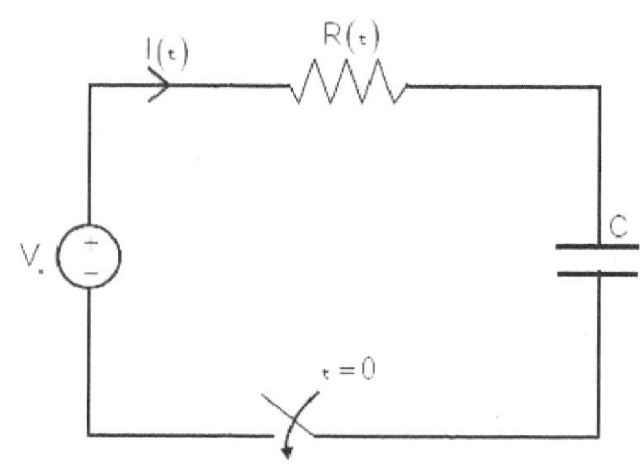

44. In an ideal p-n junction with an ideality factor of 1 at T = 300 K, the magnitude of the reverse-bias voltage required to reach 75% of its reverse saturation current, rounded off to 2 decimal places, is ______mV. [$k = 1.38 \times 10^{-23}\ JK^{-1}$, $h = 6.625 \times 10^{-34}\ J-s$, $q = 1.602 \times 10^{-19} C$]

45. The dispersion equation of a waveguide, which relates the wave number k to the frequency ω, is

$$k(\omega) = (1/c)\sqrt{\omega^2 - \omega_0^2}$$

Where the speed of light $c = 3 \times 10^8$ m/s. and ω_0 is a constant. If the group velocity is 2×10^8 m/s, then the phase velocity is

(A) 2×10^8 m/s (B) 1.5×10^8 m/s

(C) 3×10^8 m/s (D) 4.5×10^8 m/s

46. The state transition diagram for the circuit shown is

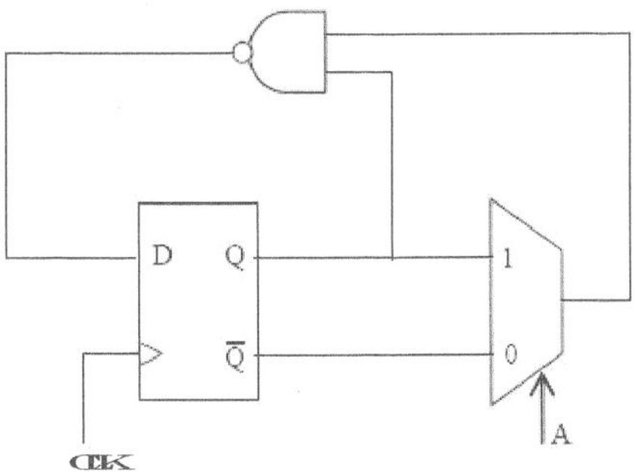

(A)
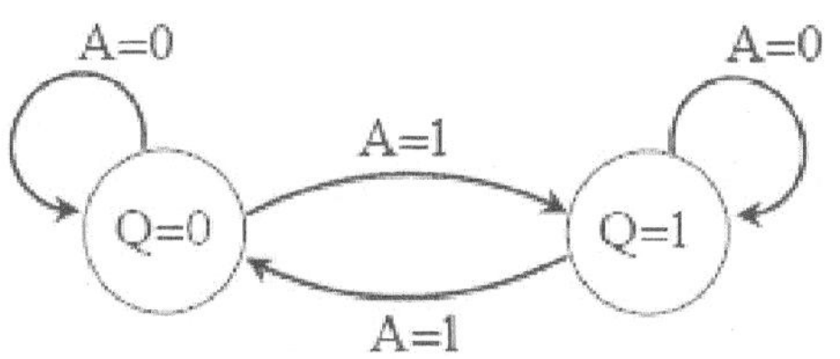

(B)
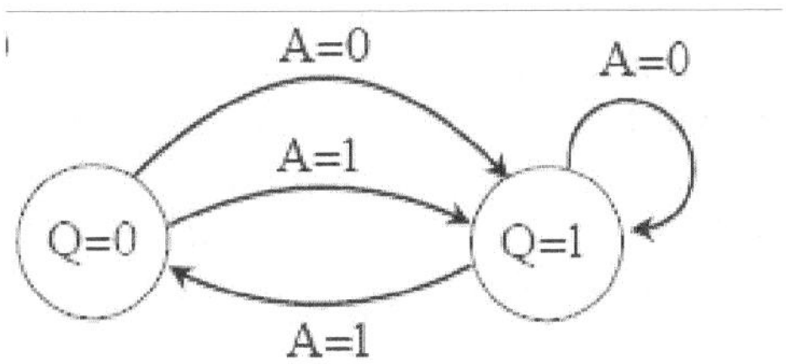

(C)
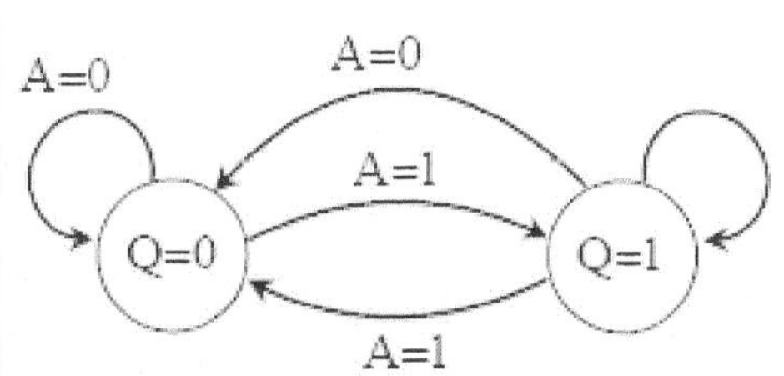

(D)
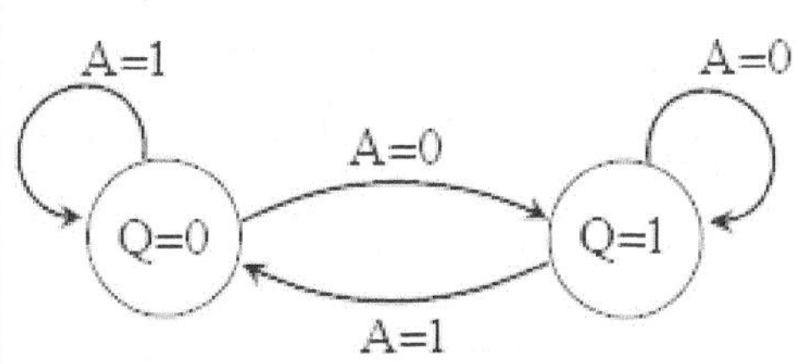

47. Consider a unity feedback system, as in the figure shown, with an integral compensator $\frac{k}{s}$ and open-loop transfer function

$$G(s) = \frac{1}{s^2 + 3s + 2}$$

Where K > 0. The positive value of K for which there are exactly two poles of the unity feedback system on the jω axis is equal to ______ (rounded off to two decimal places).

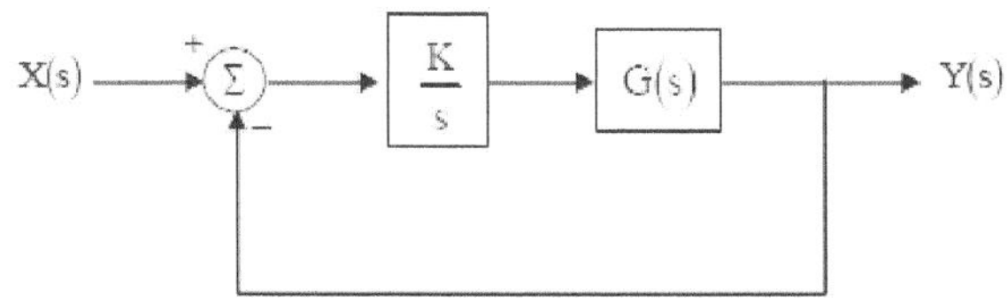

48. Consider a differentiable function f(x) on the set of real numbers such that $f(-1) = 0$ and $|f'(x)| \le 2$. Given these conditions, which one of the following inequalities is necessarily true for all $x \in [-2, 2]$?

(A) $f(x) \le 2|x+1|$ (B) $f(x) \le 2|x|$

(C) $f(x) \le \frac{1}{2}|x+1|$ (D) $f(x) \le \frac{1}{2}|x|$

49. Consider the homogeneous oridinary differential equation $x^2\frac{d^2y}{dx^2} - 3x\frac{dy}{dx} + 3y = 0,\ x > 0$ with y(x) as a general solution. Given that y(1) = 1 and y(2) = 14 the value of y(1.5), (rounded off of to two decimal places), is ______.

50. Let a random process Y(t) be described as $Y(t) = h(t) \times X(t) + Z(t)$, where X(t) is a white noise process with power spectral density $S_x(f) = 5$W/Hz. The filter h(t) has a magnitude response given by $|H(f)| = 0.5$ for $-5 \le f \le 5$, and zero elsewhere. Z(t) is a stationary random process, uncorrelated with X(t), with power spectral density as shown in the figure.

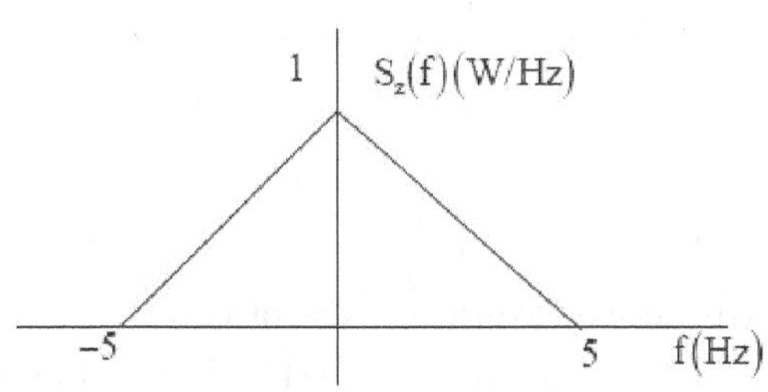

The power in Y(t), in watts, is equal to ______W. (rounded off to two decimal places).

51. In the circuits shown, the threshold voltage of each nMOS transistor is 0.6V. Ignoring the effect of channel length modulation and body bias, the values of V_{out1} and V_{out2}, respectively, in volts, are

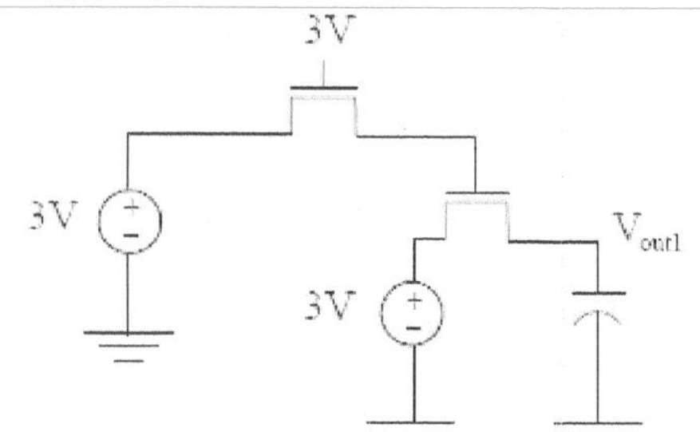

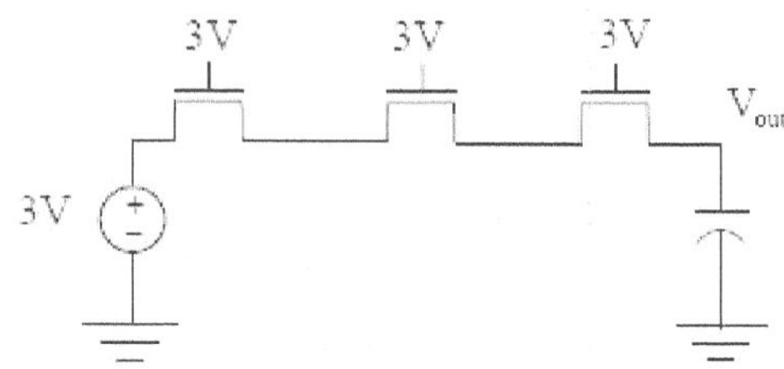

(A) 2.4 and 1.2

(B) 2.4 and 2.4

(C) 1.8 and 1.2

(D) 1.8 and 2.4

52. A CMOS inverter, designed to have a mid-point voltage V_I equal to half of V_{dd}, as shown in the figure, has the following parameters:

$V_{dd} = 3V$

$\mu_n C_{ox} = 100\mu A/V^2$; $V_{tn} = 0.7V$ for nMOS

$\mu_p C_{ox} = 40\mu A/V^2$; $|V_{tp}| = 0.9V$ for pMOS

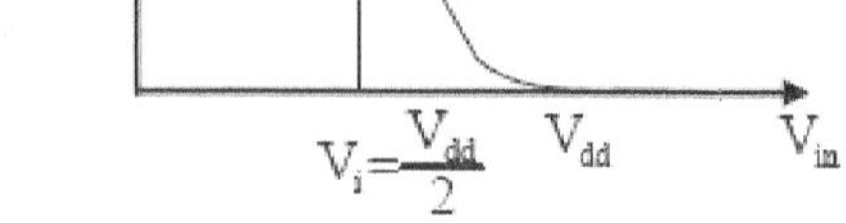

The ratio of $\left(\frac{W}{L}\right)_n$ to $\left(\frac{W}{L}\right)_p$ is equal to _____. (rounded off to three decimal places).

53. In the circuit shown, if $v(t) = 2\sin(1000t)$ volts, $R = 1k\Omega$, and $C = 1\mu F$, then the steady-state current $i(t)$, in milliamperes (mA) is

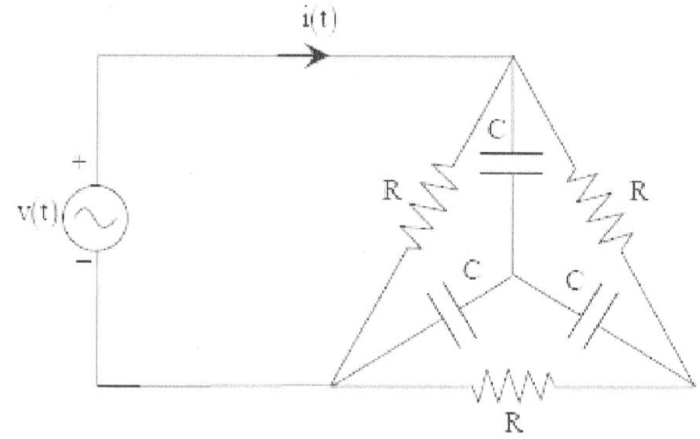

(A) $\sin(1000t) + \cos(1000t)$

(B) $\sin(1000t) + 3\cos(1000t)$

(C) $2\sin(1000t) + 2\cos(1000t)$

(D) $3\sin(1000t) + \cos(1000t)$

54. The block diagram of a system is illustrated in the figure shown, where X(s) is the input and Y(s) is the output. The transfter function $H(s) = \frac{Y(s)}{X(s)}$ is

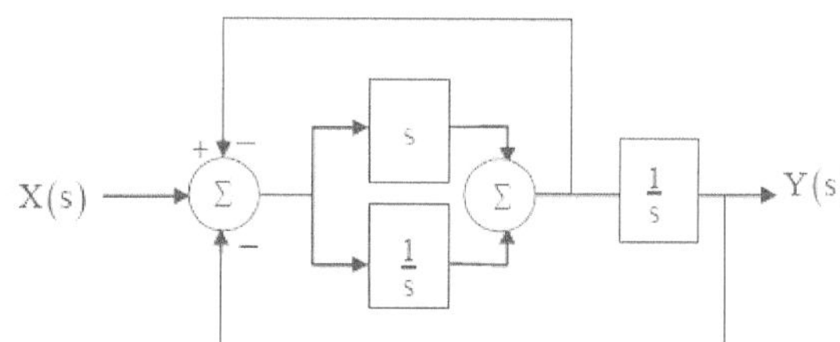

(A) $H(s) = \frac{s^2+1}{2s^2+1}$ (B) $H(s) = \frac{s^2+1}{s^3+2s^2+s+1}$

(C) $H(s) = \frac{s+1}{s^2+s+1}$ (D) $H(s) = \frac{s^2+1}{s^3+s^2+s+1}$

55. In the circuit shown, $V_1 = 0$ and $V_2 = V_{dd}$. The other relevant parameters are mentioned in the figure. Ignoring the effect of channel length modulation and the body effect, the value of I_{out} is ____mA (rounded off to one decimal place).

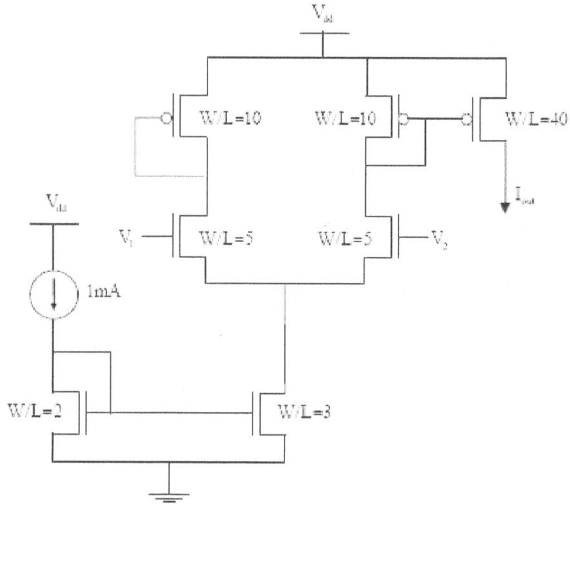

ANSWERS

General Aptitude (GA)

1. (C)	**2.** (B)	**3.** (C)	**4.** (B)	**5.** (A)	**6.** (C)	**7.** (C)	**8.** (A)
9. (B)	**10.** (D)						

Electronics & Communication Engineering

1. (2)	**2.** (B)	**3.** (2)	**4.** (–3)	**5.** (3)	**6.** (A)	**7.** (A)	**8.** (12)
9. (11)	**10.** (D)	**11.** (B)	**12.** (C)	**13.** (B)	**14.** (D)	**15.** (B)	**16.** (0.37)
17. (D)	**18.** (4)	**19.** (B)	**20.** (C)	**21.** (D)	**22.** (A)	**23.** (A)	**24.** (0.75)
25. (0)	**26.** (–900)	**27.** (B)	**28.** (–27)	**29.** (C)	**30.** (1.732)	**31.** (0.115)	**32.** (A)
33. (A)	**34.** (D)	**35.** (A)	**36.** (25.51)	**37.** (0.192)	**38.** (0.1)	**39.** (B)	**40.** (A)
41. (3.31)	**42.** (C)	**43.** (0.10)	**44.** (35.87)	**45.** (D)	**46.** (B)	**47.** (6)	**48.** (A)
49. (5.25)	**50.** (17.5)	**51.** (D)	**52.** (0.225)	**53.** (D)	**54.** (B)	**55.** (6)	

EXPLANATIONS

GENERAL APTITUDE

1. ∴ The number of different orders in which P, Q and T arranged = 3! × 2 × 1 = 6

3. Time taken $= \left(\frac{4 \times 2}{4+2}\right)$ hr $= \frac{4}{3}$ hours

7. Given, P and Q started working from 8 A.M. for a client

   Total budget = USD 3000

   P worked exactly 7 hours and took 30 min break in between.

   ⇒ 'P' worked number of hours = 6.5 hours

   'Q' worked exactly 8 hours and took 20 min break in between.

   ⇒ 'Q' worked number of hours ≈ 7.67 hours

   Client amount paid by for both P and Q

   = USD 200/hr

   ∴ Total USD Paid = (6.5 × 200 + 7.67 × 200)

   = (1300 + 1534) = 2834

   Remaining amount left

   = 3000 – 2834 ≈ 166 (approx)

9. According to the given data;

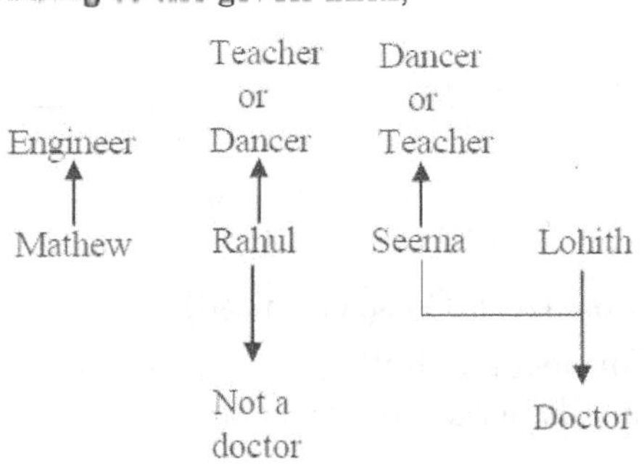

   ∴ Mathew must be an Engineer.

10. Let us assume, that population = 100 [2001 – 2011]

From panel (a);	From panel (b & c)
Percentage of male literates in 2001 = 50%	Number of males in 2001 = 60
Percentage of female literates in 2001 = 40%	Number of females in 2001 = 40
Percentage of male literates in 2011 = 60%	Number of males in 2011 = 50
Percentage of female literates in 2011 = 60%	Number of females in 2011 = 50

∴ Number of male literates in 2001

$$= 60 \times \frac{50}{100} = 30$$

Number of male literates in 2011

$$= 50 \times \frac{60}{100} = 30$$

Number of female literates in 2001

$$= 40 \times \frac{40}{100} = 16$$

Total number of literates in 2001

= 30 + 16 = 46

Number of female literates in 2011

$$= 50 \times \frac{60}{100} = 30$$

Total number of literates in 2011

= 30 + 30 = 60

∴ Percentage increase in the total number of literates from 2001–2011

$$= \left[\frac{60-46}{46} \times 100\right]\% = \left[\frac{14}{46} \times 100\right]\% = 30.43\%$$

ELECTRONICS & COMMUNICATION ENGINEERING

1. Given, $R_{rad} = 3\Omega$, and f = 3 GHz

   % change in length $\frac{d\ell}{\ell} \times 100\%$

   $\frac{dR_{rad}}{R_{rad}} \times 100\% = ?$

   $$R_{rad} = \frac{80\pi^2}{\lambda^2}\ell^2 \quad ...(i)$$

   Change in resistance with respect to change in length is

   $$\frac{dR_{rad}}{d\ell} = 2\ell\left(\frac{80\pi^2}{\lambda^2}\right) \quad ...(ii)$$

   dividing eq. (i) and (ii) we get

   $$= \frac{\frac{dR_{rad}}{d\ell}}{R_{rad}} = \frac{2\ell}{\ell^2} = \frac{2}{\ell}$$

   $$\therefore \frac{dR_{rad}}{R_{rad}} = 2 \times \frac{d\ell}{\ell}$$

   $$\therefore \frac{dR_{rad}}{R_{rad}} \times 100\% = 2 \times \left(\frac{\partial\ell}{\ell} \times 100\%\right)$$

   $$\left[\because \frac{d\ell}{\ell} \times 100\% = 1\%\right]$$

   $\therefore$ percent change in radiation resistance = 2%

2. $\ell n(z)$ is not analytic at z = 0;

   and $e^{\frac{1}{z}}$ is also not analytic at z = 0 and

   $\frac{1}{1-z}$ is also not analytic at z = 0

   But cos (z) is analytic over the entire complex plane,

   Since $\cos z = \cos(x + iy) = \cos x \cos(iy) - \sin x \sin(iy)$

   $= \cos x \cos h\, y - i \sin x \sin h\, y;$

   Now comparing with u + iv we get

   Where $u(x, y) = \cos x \cos h\, y;$

   $v(x,y) = -\sin x \sin h\, y$

$u_x = -\sin x.\cosh y$	$v_x = -\cos x.\sinh y$
$u_y = \cos x \sinh y$	$v_y = -\sin x \cosh y$

3. 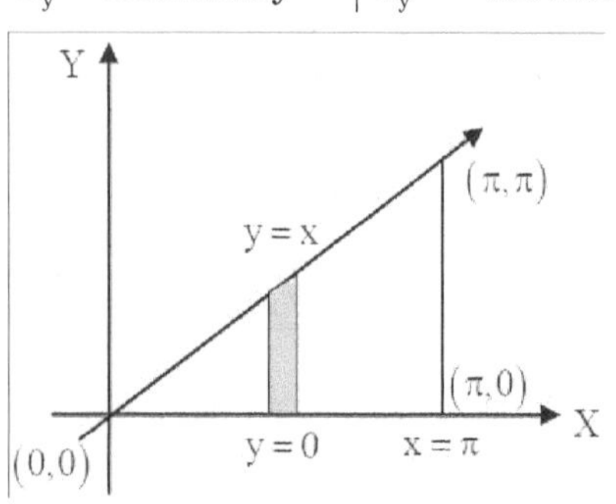

$$\int_0^{\pi}\int_0^{\pi} \frac{\sin x}{x} dx\, dy$$

Given limits are x = y → x = π & y = 0 → y = π.

For change of order of integration.

Consider a strip parallel to Y-axis. Then

Limits of y are : y = 0 → y = x and

Limits of x : x = 0 → x = π

$$\therefore \int_0^{\pi}\int_0^{\pi} \frac{\sin x}{x} dx\, dy = \int_{x=0}^{\pi}\left[\int_{y=0}^{\pi} \frac{\sin x}{x}.dy\right] dx$$

$$= \int_{x=0}^{\pi} \frac{\sin x}{x}[y]_0^x .\, dx$$

$$= \int_{x=0}^{\pi} \sin x.dx$$

$$= [-\cos x]_0^{\pi}$$

$$= -[-1-1] = 2$$

$$\Rightarrow \int_0^{\pi}\int_y^{\pi} \frac{\sin y}{x} dx\, dy = 2$$

4. 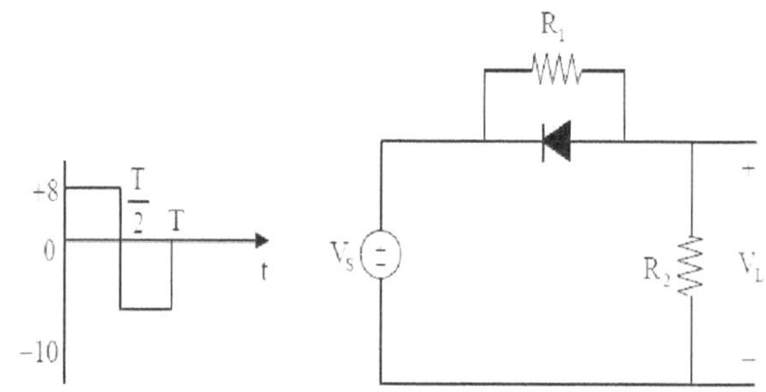

Case (i) :

For +ve half cycle diode is reverse bias : (open circuited)

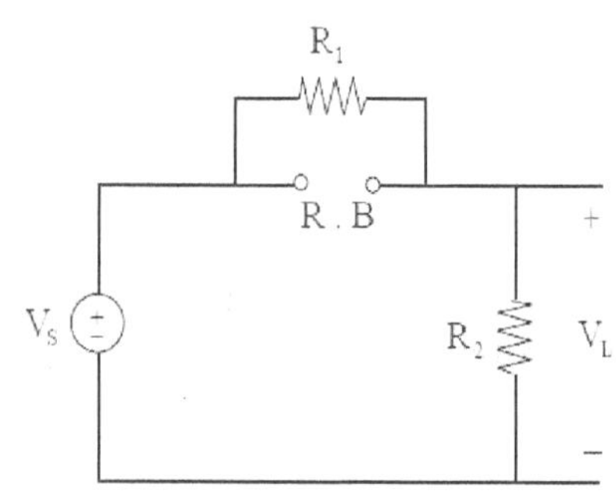

$$V_L = \frac{R_2}{R_1 + R_2}.V_S = \frac{50}{100} \times 8V = 4V$$

Case (ii) : (Colse circuited)

For negative half cycle diode is forward bias and ideally act as short circuit.

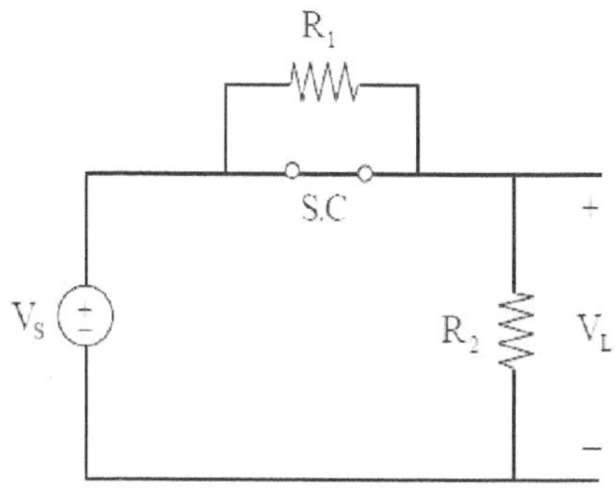

$V_L = V_S = -10$ V.

Output Waveform

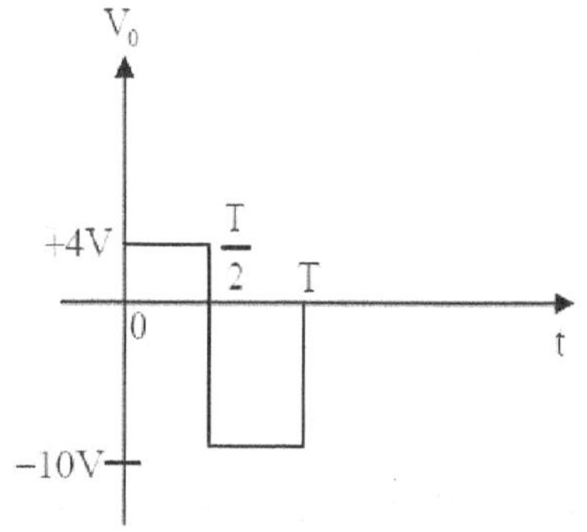

$$V_{avg} = \frac{4V \times \frac{T}{2} + \left(-10V \times \frac{T}{2}\right)}{T} = 2V - 5V = -3V$$

$\therefore\ V_{avg} = -3$ V

5. Given,

$$\text{Matrix A} = \begin{bmatrix} 2 & 2 & 3 & 3 \\ 0 & 1 & 1 & 1 \\ 0 & 0 & 3 & 3 \\ 0 & 0 & 0 & 2 \end{bmatrix}$$

$\rightarrow$ Upper triangular matrix

$\therefore$ Eigen values of A are $\lambda = 2, 1, 3, 2$ [diagonal elements]

$\therefore$ The number of distinct eigen values of the matrix = 3. i.e. 1, 2, 3.

6. Given differential equation is

$$\frac{dy}{dx} = -\left(\frac{x}{y}\right)^n$$

$$\therefore\ \frac{dy}{dx} = -\frac{x^n}{y^n} \quad ...(1)$$

For n = –1; we have

$$\frac{dy}{dx} = -\frac{x^{-1}}{y^{-1}}$$

$$\Rightarrow\ \frac{dy}{dx} = -\frac{y}{x}$$

By method of variable separable form.

$$\Rightarrow\ \frac{1}{y}dy = \frac{-1}{x}dx$$

Integrative on both sides we get,

$$\ell ny = -\ell nx + \ell nc \Rightarrow \ell n(xy) = \ell nc$$

$$\Rightarrow\ xy = c^2$$

Which represents family of hyperbolas.

For n = 1; we have

$$\frac{dy}{dx} = -\frac{x}{y} \quad [\because (1)]$$

By method of variable separable

$$\Rightarrow\ ydy = -xdx$$

Integrating on bothsides get

$$\frac{y^2}{2} = -\frac{x^2}{2} + c \Rightarrow \frac{x^2}{2} + \frac{y^2}{2} = c$$

$$\Rightarrow\ x^2 + y^2 = c$$

$\Rightarrow\ x^2 + y^2 = (\sqrt{c})^2$ which represents family of circles.

7. $\nabla.D = \rho$

$$\nabla \times \bar{E} = \frac{-\partial \bar{B}}{\partial t}$$

$$\nabla.\bar{B} = 0$$

$$\nabla \times \bar{H} = \bar{J} + \frac{\partial \bar{D}}{\partial t}$$

$\therefore$ 1 – Q, 2 – R, 3 – P, 4 – S

Option (a) is correct.

8. It is given that,

$$f(t) = 1 + 2\cos \pi t + 3\sin \frac{2\pi}{3}t + 4\cos\left(\frac{\pi}{2}t + \frac{\pi}{4}\right)$$

Its fundamental frequency (ω_0)

$$\omega_0 = \frac{\text{HCF of } (\pi, 2\pi, \pi)}{\text{LCM of } (1,3,2)} = \frac{\pi}{6}$$

$$\text{Now, } T_0 = \frac{2\pi}{\omega_0} = \frac{2\pi \times 6}{\pi} = 12 \text{ sec}$$

9. Given, E[2x + y] = 0

and E[x + 2y] = 33

$$\Rightarrow\ 2E(x) + E(y) = 0 \quad ...(i)$$

$$\Rightarrow\ E(x) + 2E(y) = 33 \quad ...(ii)$$

Solving eq. (i) and (ii) we get

E(y) = 22 and E(x) = – 11

$\therefore\ E(x) + E(y) = -11 + 22 = 11$

$\Rightarrow\ E(x) + E(y) = 11$

10.

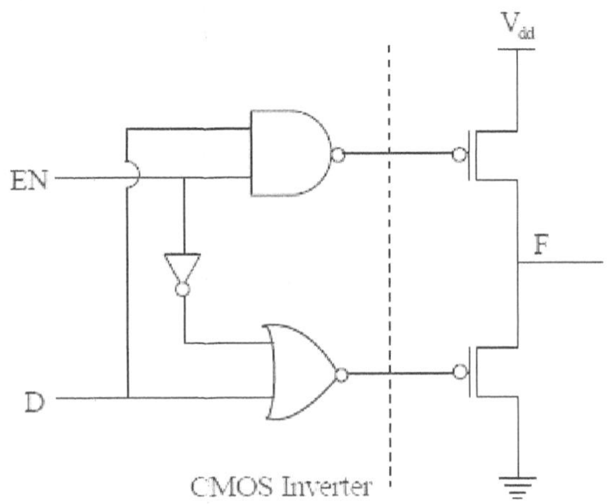

→ NAND gate enabled, when their enable input is "1" and NOR gate enabled, when their enable input is "0".

Case (i) :

When EN = 0, then both the logic gates NAND and NOR disabled, so CMOS inverter input is floating. So, output is also high impedance state.

F = Hi – z

Case (ii) :

When EN = 1, then both the logic gates NAND and NOR are enabled with output $\overline{D}$ that is input of CMOS inverter.

So, $F = \overline{\overline{D}} = D$

11. It is given that H(z) is z-transform of a real-valued signal h(n).

$P(z) = H(z)H\left(\frac{1}{z}\right)$ and P(z) has 4 zeros ⇒ P(z) are,

sum of zeros of H(z) and zeros of $H\left(\frac{1}{z}\right)$.

If $z_1 = \frac{1}{2} + j\frac{1}{2}$ is one zero then there must be a zero at $z_1^* = \frac{1}{2} - j\frac{1}{2}$

Let z_1, z_1^* represent zeros of H(z) then the zeros of $H\left(\frac{1}{z}\right)$ will be $\frac{1}{z_1}$ and $\left(\frac{1}{z_1}\right)^*$

$$\frac{1}{z_1} = \frac{1}{\frac{1}{2} + j\frac{1}{2}} = 1 - i$$

$$\left(\frac{1}{z_1}\right)^* = 1 + j$$

Hence zeros of P(z) are $\left(\frac{1}{2} \pm j\frac{1}{2}\right)$ and $(1 \pm j)$

or $[0.707\underline{|\pm 45°}]$ and $[\sqrt{2}\underline{|\pm 45°}]$

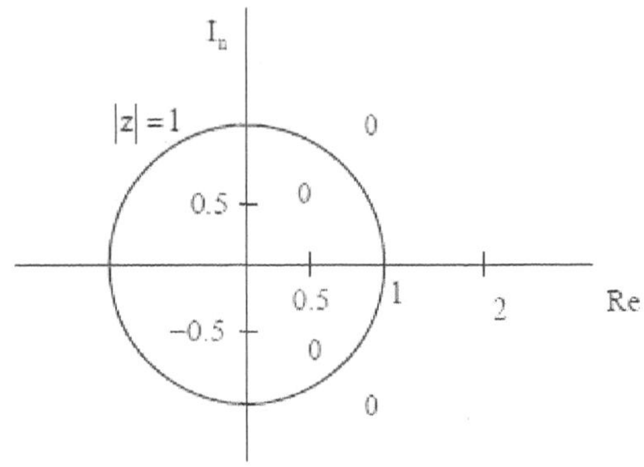

Even option D looks like similar but in option B, the zeros that are outside the unit circle have real part 2, but we need 1.

12.

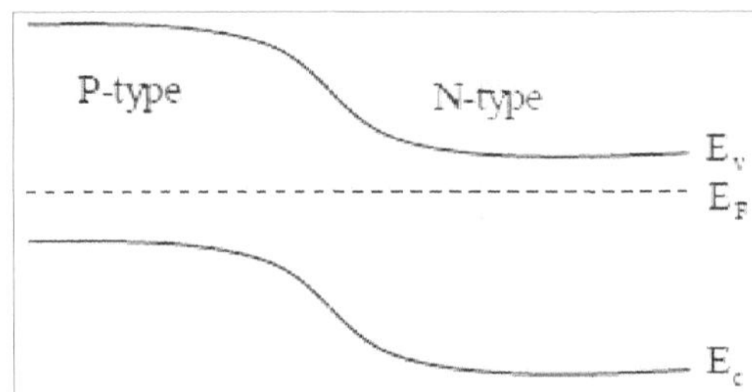

Energy 'band diagram of PN junction diode.

- For pnn⁺p⁺⁺ configuration, Fermi-energy level (E_F) is more closer to E_V compare to E_C.
- For P-type of semiconductor

$$E_F = E_V - kT\ell n\left(\frac{N_A}{N_V}\right) \quad [\text{For } N_A < N_V]$$

13. PMOS behaves as ON switch for logic 0.

and NMOS behaves as ON switch for logic 1.

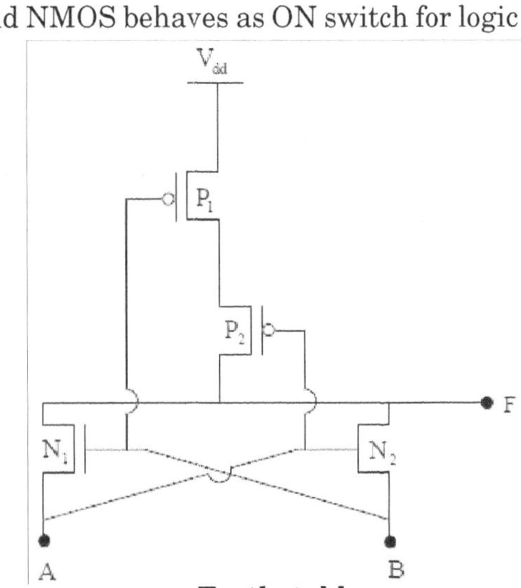

Truth table

A	B	P_1	N_1	P_2	N_2	F
0	0	0	0	0	0	1
0	1	1	1	0	0	0
1	0	0	0	1	1	0
1	1	1	1	1	1	1

So from the above truth table we can say $F = A \odot B$.

14. Aluminium metal in contact with a lightly doped n type silicon forms a non ohmic rectifying contact because the trivalent aluminium easily dissolves into silicon and convert the part of n type semiconductor into p type.

15. The total electric flux leaving the cylinder of height "H" is

$$= \oiint \bar{E}.d\bar{A} = \frac{QH}{\epsilon_0}$$

∴ The electric flux through a quarter cylinder of height "H" is

$$= \frac{\oiint \bar{E}.d\bar{A}}{4} = \frac{QH}{4\epsilon_0}$$

∴ Option (B) is correct.

16. Given,

Cumulative distribution function,

$$F_z(x) = \begin{cases} 1 - e^{-x} & \text{if } x \geq 0 \\ 0 & \text{if } x < 0 \end{cases}$$

→ probability density function.

$$\Rightarrow F_z'(x) = f_z(x) = \begin{cases} e^{-x}; & x \geq 0 \\ 0; & x < 0 \end{cases}$$

Where Z be an exponential R.V with mean '1'

Using conditional probability

$$\therefore \Pr\left[\frac{z>2}{z>1}\right] = \frac{\Pr[z>2 \cap z>1]}{\Pr[z>1]}$$

$$= \frac{\Pr[z>2]}{\Pr[z>1]} = \frac{\int_2^\infty f(x)dx}{\int_1^\infty f(x)dx} = \frac{\int_0^\infty e^{-x}dx}{\int_1^\infty e^{-x}dx}$$

$$\Rightarrow \Pr\left[\frac{z>2}{z>1}\right] = \frac{\left[-e^{-x}\right]_2^\infty}{\left[-e^{-x}\right]_1^\infty} = \frac{\left[0+e^{-2}\right]}{\left[0+e^{-1}\right]}$$

$$= \frac{e^{-2}}{e^{-1}} = \frac{1}{e} \approx 0.37$$

17. Message (1) ⇒ 0001

Message (2) ⇒ 0011

0000111 ⇒ Codeword (1)

1100110 ⇒ Codeword (2)

Since it is a linear hamming code.

Message (1) + Message (2) results in codeword (1) + codeword (2)

⇒ Addition of binary is logical XOR

∴

Message	Codeword
0001	0000111
⊕ 0011	⊕ 1100110
0010	1100001
↗ Message (3)	↗ Codeword (3)

18. 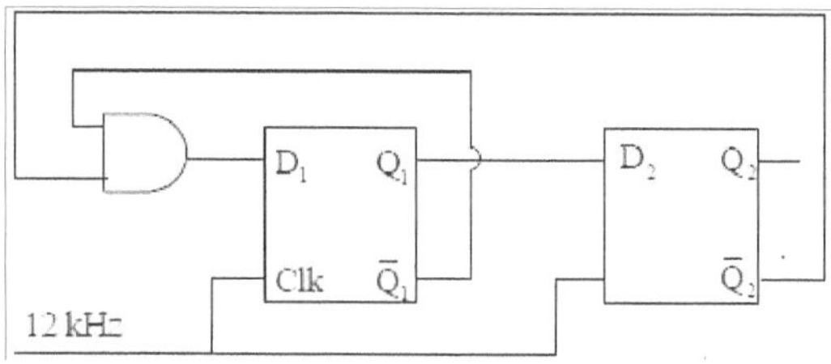

$D_2(Q)_1$	$D_1(\bar{Q}_1\bar{Q}_2)_2$	Q_2	Q_1	
–	–	0	0	← Initially
0	1	0	1	1st clock pulse
1	0	1	0	2nd clock pulx
0	0	0	0	3rd clock puls

$$Q_2Q_1 \rightarrow 00 \rightarrow 01 \rightarrow 10 \rightarrow 00$$

It is a MOD-3 synchronous counter

$$\text{So, } f_{Q_2} = \frac{f_i}{3} = \frac{12\text{kHz}}{3} = 4\text{ kHz}$$

19. B $NM_L = V_{IL} - V_{OLU}$ = Increase

$NM_H = V_{OHU} - V_{IH}$ = Increase

$$V_{IH} = V_{TN} + \frac{(V_{DD} + V_{TP} - V_{TN})}{\left(\frac{k_n}{k_p} - 1\right)} \left[\frac{2\frac{k_n}{k_p}}{\sqrt{\frac{34n}{k_p} + 1}} - 1 \right] > 1$$

$$k_p \alpha \left(\frac{W}{L}\right)_p, k_n \alpha \left(\frac{W}{L}\right)_n$$

Here, pMOS width is increased $\frac{k_p}{k_n} > 1 \Rightarrow \frac{k_n}{k_p} < 1$

$$V_{OH} = \frac{1}{2}\left[\left(1 + \frac{k_n}{k_p}\right)V_1 + V_{DD} - \left(\frac{k_n}{k_p}\right)V_{TN} - V_{TP}\right] < 1$$

Similarly,

$$V_{IL} = V_{TN} + \frac{(V_{DD} + V_{TP} - V_{TN})}{\left(\frac{k_n}{k_p} - 1\right)}\left[2\sqrt{\frac{\frac{k_n}{k_p}}{\frac{k_n}{k_p} + 3}} - 1\right] > 1$$

$$V_{OLU} = \frac{VI\left(1 + \frac{k_n}{k_p}\right) - V_{DD} - \left(\frac{k_n}{k_p}\right)V_{TN} - V_{TP}}{2\left(\frac{k_n}{k_p}\right)} < 1$$

20. From the given bode plot

$$N_p = 6, N_z = 3$$

21.

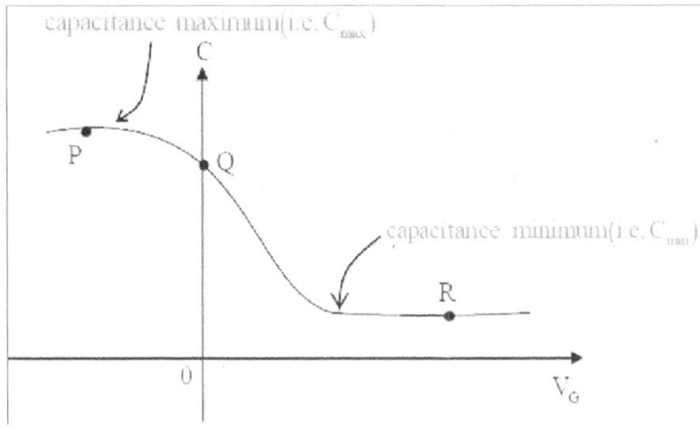

At higher frequency C_{min} minimum capacitance obtained in inversion-regime so point R-belongs to inversion regime.

Maximum capacitance obtained in accumulation regime so point p-line is accumulation regime.

For $\phi_m = 0V$, flat band occurs at $V_G = 0V$, so point ϕ lies at flat-band regime.

22.

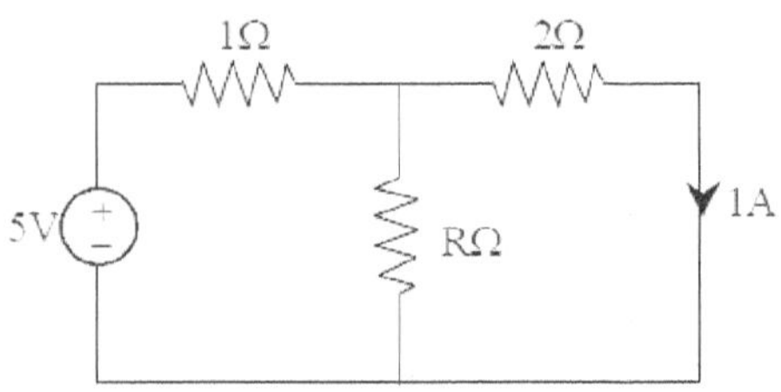

(figure a)

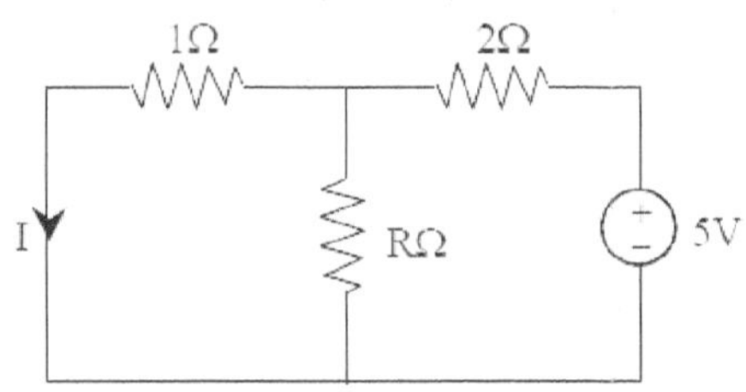

(figure b)

By reciprocity theorem,

Current I = 1 A

$$\left[\frac{V}{I} = k, \frac{V_1}{V_2} = \frac{I_1}{I_2}\right]$$

23. Forced response is the response, due to external input signal

$$Y(s) = \frac{G(s)}{s} = \frac{3 - s}{s(s+1)(s+3)}$$

By partial fraction

$$= \frac{1}{s} + \frac{-2}{s+1} + \frac{1}{s+3}$$

Taking increase lap case transform.

$$\Rightarrow y(t) = u(t) - 2e^{-t}u(t) + e^{-3t}u(t)$$

24. Instantaneous frequency $(f_1) = \frac{1}{2\pi}\frac{d\theta(t)}{dt}$

where $\theta(t) = 2\pi f_c t + km(t)$

$$\therefore f_1 = \frac{1}{2\pi}\frac{d}{dt}\left[2\pi f_c t + km(t)\right] = f_c + \frac{K}{2\pi}\frac{dm}{dt}$$

$$f_{i_{max}} = f_c + \frac{K}{2\pi}\frac{dm}{dt}\bigg|_{max}^{1} \text{ and } f_{i_{min}} = f_c + \frac{K}{2\pi}\frac{dm}{dt}\bigg|_{min} \quad \text{(ii)}$$

$$\text{Now } \frac{dm}{dt}\bigg|_{max} = 2\frac{dm}{dt}\bigg|_{min} = -1$$

Replacing the value of

$$\frac{dm}{dt}\bigg|_{max} \text{ and } \frac{dm}{dt}\bigg|_{min}$$

In eq. (i) and (ii) we get.

$$f_{i_{max}} = 50\text{ K} + (5)(2)\text{K} = 60\text{K}$$

$$f_{i_{min}} = (50\text{K}) + (5)(-1)\text{K} = 45\text{K}$$

$$\text{Now Required Ratio} = \frac{f_{i_{min}}}{f_{i_{max}}}$$

$$= \frac{45\text{K}}{60\text{K}}$$

$$= \frac{3}{4} = 0.75$$

25. $$\frac{1}{2\pi j}\oint\left(z + \frac{1}{z}\right)^2 dz = \frac{1}{2\pi j}\oint\frac{(z^2+1)^2}{z^2}dz \quad \text{...(i)}$$

∴ Singular point z = 0; which lies inside unit circle $|z| = 1$

∴ Using Cauchy's generalization integration formula, we get.

$$\oint_c \frac{f(z)}{(z - z_0)^n}dz = 2\pi j\frac{\left[f^{n-1}(z_0)\right]}{(n-1)!}$$

$$\Rightarrow \therefore \oint_C \frac{(z^2+1)^2}{(z-0)^2}dz = 2\pi j\frac{\left[(z^2+1)^2\right]'}{1!}\Bigg/ z = 0;$$

$$= 2\pi j[2(z^2 + 1)(2z)]_{z=0}$$

$$= 2\pi j[0] = 0.$$

From eq. (i) we get,

$$\frac{1}{2\pi j}\oint\left(z + \frac{1}{z}\right)^2 dz = \frac{1}{2\pi j}\oint\frac{(z^2+1)^2}{(z-0)^2} = \frac{1}{2\pi j}(0) = 0$$

26. 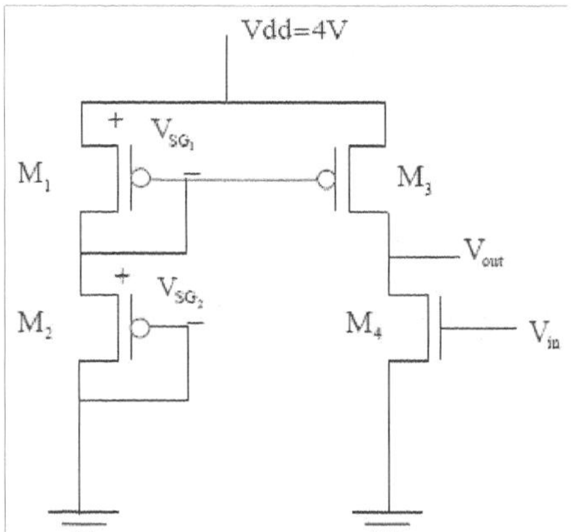

M_1 and M_2 will have equal current flowing also since they are identical $\left(\frac{W}{L}\right)_1 = \left(\frac{W}{L}\right)_2$

$\therefore \; V_{SG_1} = V_{SG_2}$ also, By KVL in loop

$4 = V_{SG_1} + V_{SG_2}$

$\therefore V_{SG_1} = V_{SG_2} = 2V$

$\therefore$ Current through M_1 and M_2 is given by

$$I = \left(\frac{1}{2}\right)(\mu_p C_{ox})\left(\frac{W}{L}\right)\left[V_{SG} - |V_T|\right]^2$$

$$= \left(\frac{1}{2}\right)(30)(10)(2-1)^2$$

$$= 15(10)(1)^2 = 0 = 150\mu A$$

M_1 and M_3 are matched with same $\left(\frac{W}{L}\right)$ and same V_{SG}

$\therefore \; I_{M_3} = I_{M_4} = 150\mu A$

Now for MOSFET M_4,

$$g_m = \sqrt{2\mu_n C_{ox} \frac{W}{L} I_D}$$

$$= \sqrt{2 \times 60\mu \times 5 \times 150\mu} = 300\mu℧$$

Now voltage gain (A_v)

$$= -g_m(r_d \parallel r_d) = \left(-300\frac{\mu A}{V}\right)(6M\Omega \parallel 16M\Omega)$$

$$= \left(-300\frac{\mu A}{V}\right)(3M\Omega) = -900V/V$$

27. It is given that
$h(n) = [1, a, b]$

and $x(n) = C_1 e^{-j\frac{\pi}{2}n} + C_2 e^{j\frac{\pi}{2}n}$

$\Rightarrow y(n) = 0$

Now If $h(n) = [1, a, b]$

$\therefore \; H(e_{j\omega}) = 1 + ae^{-j\omega} + be^{-j2\omega}$

$\rightarrow$ when $x(n) = C_1 e^{-j\frac{\pi}{2}n} + C_2 e^{j\frac{\pi}{2}n}$ then exiression of y(n)

Since the input x(n) contain 2 frequencies $\pm\frac{\pi}{2}$,

Let evaluate $\left|H(e^{j\omega})\right|$ at f nil 2 frequency

$$H(e^{j-\pi/2}) = 1 + ae^{-j\left(\frac{-\pi}{2}\right)} + be^{-j2\left(\frac{-\pi}{2}\right)}$$

$$= 1 + ae^{+j\pi/2} + be^{j\pi}$$

$$= 1 + [a(j)] + [b(-1)]$$

$$= (1 - b) + j(a)$$

$$\therefore \quad H(e^{j\pi/2}) = (1 - b) - ja$$

$$\left|H(e^{j\pi/2})\right| = \left|H\left(e^{-j\frac{\pi}{2}}\right)\right| = \sqrt{(1-b)^2 + a^2}$$

So the expression of y(n) is

$$y(n) = \left[(1-b)^2 + a^2\right]^{1/2} C_1 e^{-j\left(\frac{1}{2}n + \phi_1\right)}$$

$$+ \left[(1-b)^2 + a^2\right]^{1/2} C_2 e^{j\left(\frac{\pi}{2}n + \phi_2\right)}$$

for $y(n) = 0$

$\Rightarrow \; k = \sqrt{(1-b)^2 + a^2} = 0$

Now from option

from option (a) $a = -1, b = 1$, then

$k = \sqrt{0^2 + 1^2} = 1$(not correct)

from option (b) $a = 0, b = 1$ then

$k = \sqrt{0^2 + 0^2} = 0$(correct)

from option (c) $a = 1, b = 1$, then

$k = \sqrt{0^2 + 1^2} = 1$(not correct)

from option (d) $a = 0, b = 1$, then

$k = \sqrt{2^2 + 0^2} = 2$(not correct)

28. It is given that

$$h(n) = [-1, -2, -3, 4, 3, 2, 1] \quad \text{...(i)}$$

$$g(n) = [a, \underset{\uparrow}{b}, c] \quad \text{...(ii)}$$

It is mentioned that

$E(h,g)]\int_{-x}^{\pi}\left|H(e^{j\omega}) - G(e^{j\omega})\right|^2 d\omega$, is minimised,

If h(n) and g(n) represent IDTFT of $H(e^{j\omega})$, $G(e^{j\omega})$

then $E(h, g) = 2\pi\sum|h(n) - g(n)|^2$ (by Parseval theorem)

$h(n) \leftrightarrow H(e^{j\omega})$

$g(n) \leftrightarrow G(e^{j\omega})$

Now, $h(n) - g(n) \leftrightarrow H(e^{j\omega}) - G(e^{j\omega})$

Energy of

$$[h(n) - g(n)] = \frac{1}{2\pi}\int_{-\pi}^{\pi}\left|H(e^{j\omega}) - G(e^{j\omega})\right|^2 d\omega$$

$\Rightarrow 2\pi$ energy of

$$[h(n) - g(n)] = \int_{-\pi}^{\pi}\left|H(e^{j\omega}) - G(e^{j\omega})\right|^2 d\omega$$

$$\Rightarrow E(h, g) = \int_{-x}^{\pi}\left|H(e^{j\omega}) - G(e^{j\omega})\right|^2 d\omega$$

$$= 2\pi\sum_{n=\infty}^{\infty}\left|h(n) - g(n)\right|^2$$

We want to minimize E (h, g)

Using equation (i) and equation (ii) we get

$h(n) - g(n) = [-1, -2, -3 - a, 4 - b, 3 - c, 2, 1]$

$$\therefore \quad E(h,g) = 2\pi\sum\left|h(n) - g(n)\right|^2$$

$$= 2\pi\sum\left[h(n) - g(n)\right]^2$$

$$\therefore \quad E(h, g) = 2\pi[(-1)^2 + (-2)^2 + (-3 - a)^2 + (4 - b)^2 + (3 - c)^2 + 2^2 + 1^2]$$

$$= 2\pi[10 + (-3 - a)^2 + (4 - b)^2 + (3 - 1)^2]$$

To minimize the value of E(h,g)

Equate, $-3 - a = 0 \Rightarrow a = -3$

and $4 - b = 0 \Rightarrow b = 4$

and $3 - c = 0 \Rightarrow c = 3$

$\rightarrow g(n) = [a, b, c] = [-3, 4, 3]$

So $10g(-1) + g(1) = 10a + c$

$= (10(-3) + 3) = -30 + 3 = -27$

29.

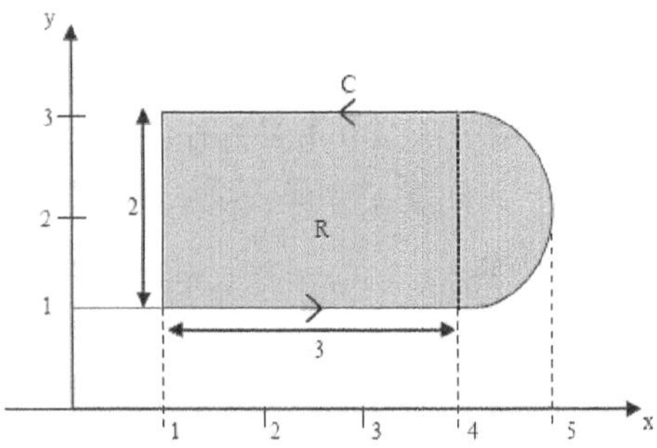

$\int_C (xdy - ydx) = \int_C (-y\,dx + x\,dy)$; where C is a

Closed figure formed by rectangle and a semi circle having radius 1.

To countert line integral to area integral.

Now Using Green's theorem, we have

$$\oint_C\left(-\underset{M}{y}\,dx + \underset{N}{y}\,dy\right) = \int_R (1+1)\,dx\,dy$$

$$= 2\iint_R dx\,dy \rightarrow \text{Area of the region R.}$$

$= 2$[Area of rectangle + Area of semi-circle]

$$= 2\left[(3 \times 2) + \frac{1}{2}\pi(1^2)\right]$$

$$= 2\left[6 + \frac{\pi}{2}\right] = (12 + \pi)$$

30. Cutoff frequency of TE_{10} mode is $f_{C_{10}} = \dfrac{C}{2W}$

Cutoff frequency of TE_{11} mode

$$f_{C_{11}} = \frac{C}{2W}\sqrt{1^2 + \left(\frac{W}{h}\right)^2}$$

Given $\dfrac{f_{C_{10}}}{f_{C_{11}}} = \dfrac{1}{2} \Rightarrow \dfrac{(C/2w)}{\left(\dfrac{C}{2W}\right)\sqrt{1^2 + \left(\dfrac{W}{h}\right)^3}} = \dfrac{1}{2}$

$$\frac{1}{\sqrt{1 + \frac{\omega^2}{h^2}}} = \frac{1}{2}$$

Now squaring both the side,

$$\therefore \quad 1^2 + \left(\frac{W}{h}\right)^2 = 4$$

$$\therefore \quad \frac{W}{h} = \sqrt{3}$$

$\therefore$ Required ratio $= \dfrac{W}{h} = \sqrt{3} = 1.732.$

31.

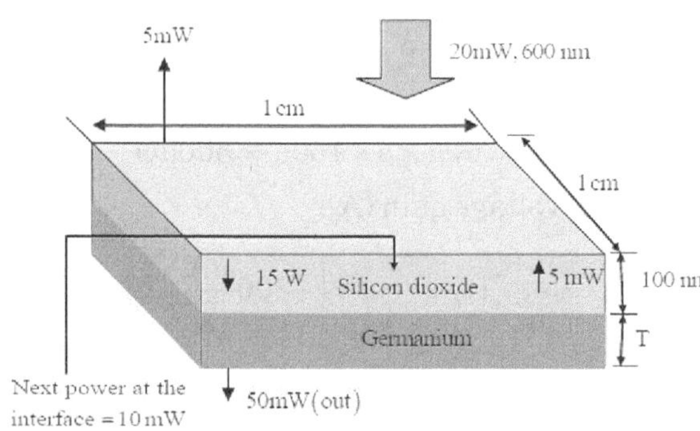

Given

$$\alpha = 3 \times 10^4 \text{ cm}$$

$$2\alpha = 6 \times 10^4 \text{ cm}$$

Now, $\quad P(L) = P(0)e^{-2\alpha l.}$

$$L = -\frac{1}{2\alpha}\ln\frac{P(2)}{P(0)}$$

$$L = -\frac{1}{6\times10^4}\ln\frac{5}{10}$$

$$= 0.11 \times 10^{-6}\,\text{m}$$

$$= 0.115\,\mu_m$$

32. 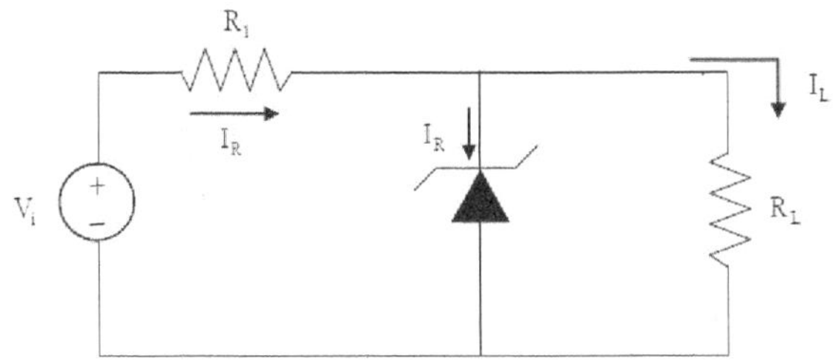

$$I_L = \frac{20-0}{1}\text{mA} = 20\text{mA}$$

At node A

Now, $I_R = I_Z + I_L$

Now, $V_{i\,min} = I_{R_{min}} R_1 + V_Z$

$I_{R_{min}} = 0 + I_2 = 20\text{mA}.$

$V_{imin} = (20\times10^{-3}\times200) + 20 = 4V + 20v = 24v$

and $V_{imax} = I_{R_{max}} + R_1 + V_Z$

Now, $I_{R_{max}} = I_{z\,max} + I_L = 60\text{mA} + 20\text{mA} = 80\text{mA}$

$\therefore\ V_{i\,max} = 80 \times 10^{-3} \times 200 + 20 = 16V + 20V = 36V$

$\therefore\ V_{i\,max} = 24V$ and $V_{i\,max} = 36V$

33. **Case 1 :**

$P(t) = \phi_1\ (t),\ q(t) = -\phi_1\ (t)$

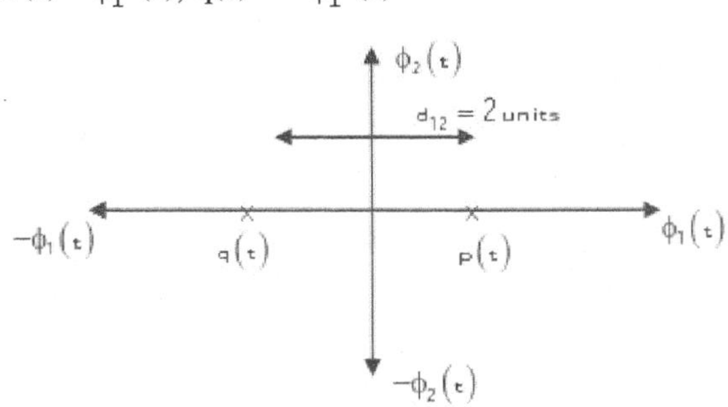

Case 2:

$P(t) = \phi_1(t)\ q(t) = \sqrt{E}\ \phi_2(t)$

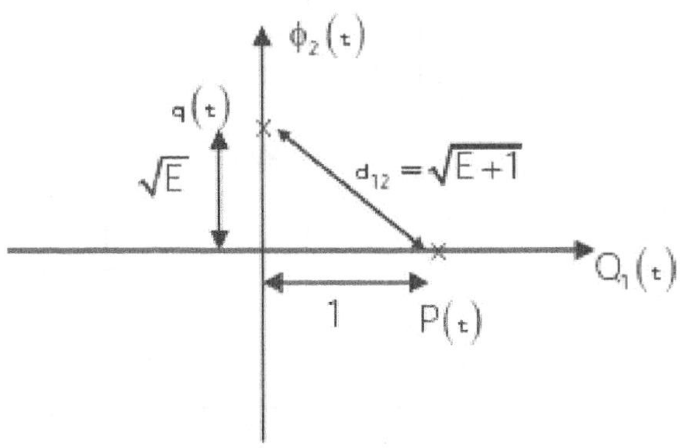

For same probability of error distance between points should be same for both cases

$\therefore \sqrt{E+1} = 2$

$\Rightarrow\ E = 3$

34. Using DIT algorithm are can obtain FFT coefficient [X(I)].

The given butterfly structure is a standard structure where

$a_1 = W_6^0 = 1$

$a_2 = W_6^1 = W_6$

$a_3 = W_6^2$

35. Responsivity $(R) = \frac{e\eta}{hv} = \frac{e\eta s}{hc} = \frac{\eta\times\lambda}{1.24}(A/W)$

36. Given data; $L = 1\,\mu m\ \mu m,\ W = 10\,\mu m$

$NA = 5 \times 10^{16}\ cm^{-3},\ \mu_n = 800\frac{cm^2}{v-sec}$

$C_{ox} = 3.45 \times 10^{-7}$ F/cm, $V_T = 0.7V$

$\varepsilon_0 = 8.854 \times 10^{-14}$ F/cm, $\varepsilon_{si} = 11.9$ and $V_G = 5V$

Now, $I_{D_{sat}} = \frac{1}{2}\mu_n C_{ox}\frac{W}{L}\left[V_G - V_T\right]^2$

$= \frac{1}{2}\times800\times3.45\times10^{-7}\times\frac{10}{1}\left[5-0.7\right]^2 = 0.0255A$

$\approx 25.51\text{mA}$

37. 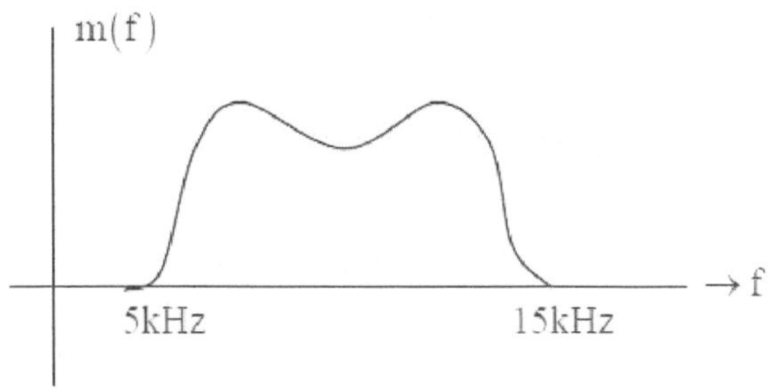

Modulation done using amplitade modulation technique

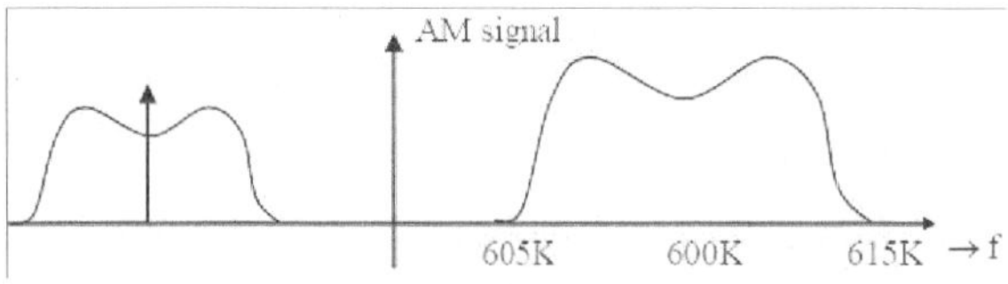

Now by sampling

$$\frac{2f_{+1}}{n} \le f_s \le \frac{2f_1}{n-1}$$

Here $f_H = 615K,\ f_L = 605$ K.

$\therefore\ 1 \le n \le \frac{f_H}{B} : B = f_H - f_L$

$\therefore\ n \le 61.5,\ n = 61.$

$\therefore \; f_s \geq \frac{2\times 615}{61}$

Minimum sampling is 20 K.

$R_b = 1.2 f_s \times 8 = 0.192$ M bit/s

38. $\frac{\text{When } X = -1 \text{ is transmitted}}{P(X=-1) = 0.2}$

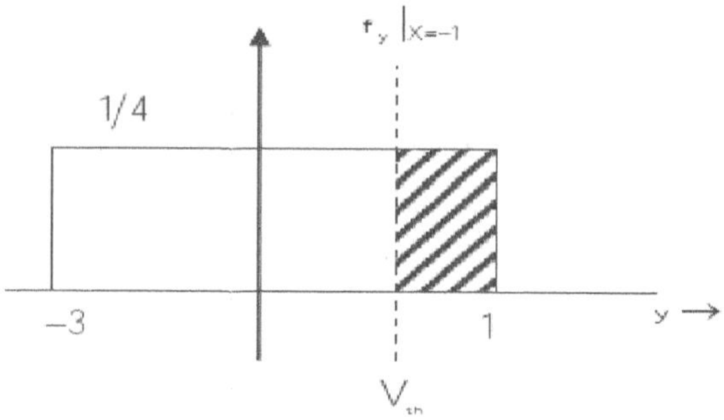

$P_{e-1} = (0.2)\left(\frac{1}{4}\right)(1 - V_{th})$

When X = 1 is transmitted P[x = 1] = 0.8

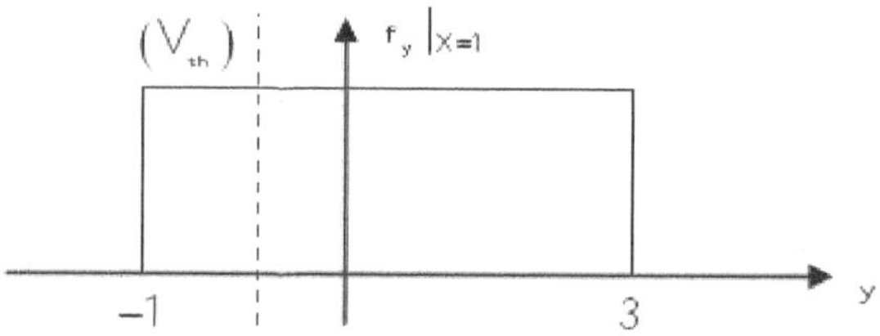

$P_{e_1} = (0.8)\left(\frac{1}{4}\right)(V_{th}+1)$

$P_{e_1} = \frac{(0.2)(1)(1-V_{th})+(0.8)(1)(V_{th}+1)}{4}$

$P_e = \frac{0.2(1-V_{th})+0.8(V_{th}+1)}{4}$

$= \frac{(0.2+0.8)+0.6V_{th}}{4} = \frac{1+0.6V_{th}}{4}$

Now, $-1 \leq V_{th} \leq 1$

$\therefore P_{e_{min}}$ when $V_{th} = -1$

$\therefore P_{e_{min}} = \frac{1+(0.6)(-1)}{4} = \frac{0.4}{4} = 0.1$

39. Given,

C matrix start from right to left same sign.

$H(S)\frac{Y(s)}{V(s)} = \frac{1}{s^3+3s^2+2s+1}$

A matrix element start from right to left and take sopposite sign.

Controllable.

$A\begin{bmatrix} 0 & 1 & 0 \\ 0 & 0 & 1 \\ -1 & -2 & -3 \end{bmatrix}$ and $C = \begin{bmatrix} 0 & 0 & 1 \end{bmatrix}$

40.

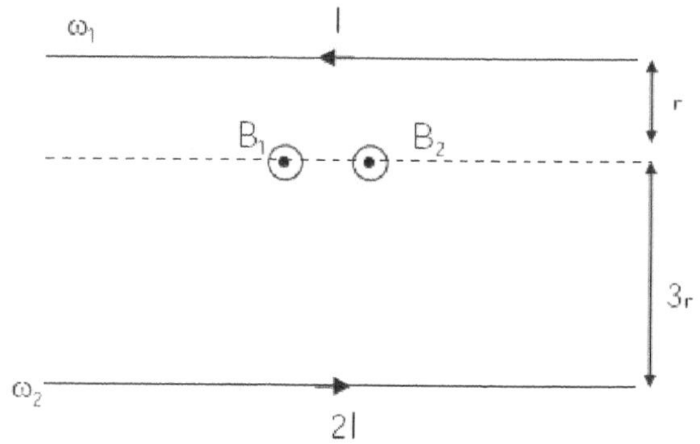

Between the wires ω_1 and ω_2 the $\overline{B}$ – fields due to I and 2I will gets added up.

$\therefore$ The B – field due to ω_1 at a distance 'r' from the ω_1 is

$\therefore$ The B – field due to ω_1 at a distance 'r' from the ω_1 is

$$B_1 = \frac{\mu_0 I}{2\pi r} \rightarrow (1)$$

The B – field dut to ω_2 at a distance "3r" from the ω_2 is given by

$$B_2 = \frac{\mu_0(2I)}{2\pi(3r)} \rightarrow (2)$$

$\therefore$ Now, total magnetic field (B)

$$= B_1 + B_2 = \frac{\mu_0 I}{2\pi}\left[\frac{1}{r}+\frac{2}{3r}\right]$$

$$B = \frac{5\mu_0 I}{6\pi r}\,\omega b/m^2.$$

41. Given: T = 4ms, R = 500 Ω, C = 10μF

Now, $\tau = RC = 500 \times 10 \times 10^{-6}$ sec = 5 ms

and $\frac{T}{2} = \frac{4ms}{2} = 2ms$

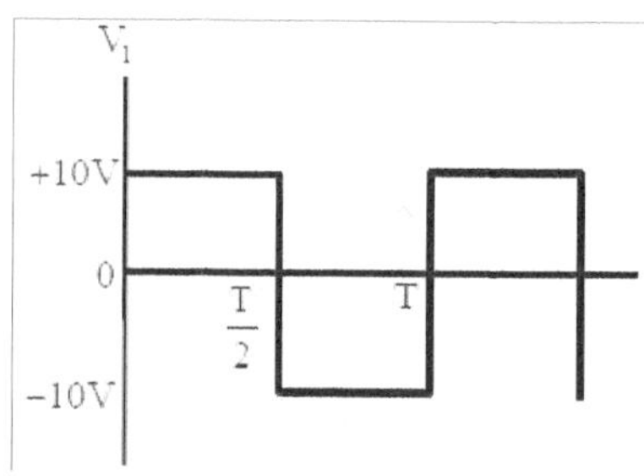

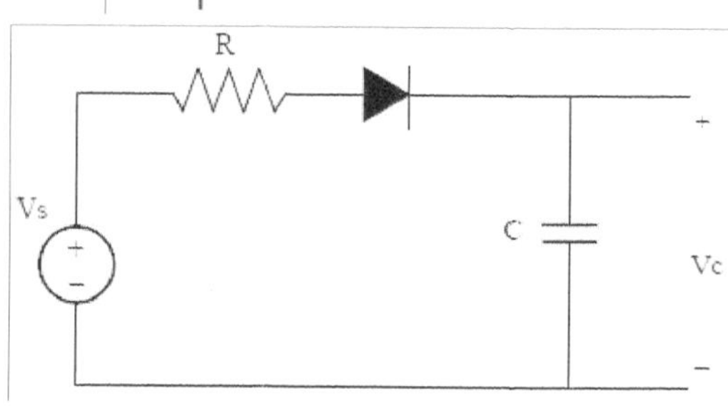

For positive half cycle, diode will be is forward biased and capacitor start to charge

$$\therefore \quad V_C = V_C(\infty)\left[1 - e^{\frac{-t}{RC}}\right] = 10\left[1 - e^{\frac{2ms}{5ms}}\right]$$

$$= 10\left[1 - e^{\frac{2}{5}}\right] = 4.51V$$

$\therefore \quad V_C = 3.31V$

42. Given, $G(s) = \dfrac{1}{1 + 2s + s^2}$

$\therefore C(s) = G(s)R(s) = \dfrac{1}{s(s+1)^2}$

Taking increase Laplace transform,

$C(t) = 1 - e^{-1} - te^{-1}$

Now checking from options,

From Option (A)

$0.94 = 1 - e^{-5.25} - 5.25\, e^{-5.25} = 0.89$

option A is wrong

From option (C)

$\therefore \quad 0.94 = 1 - e^{-4.50} - 4.50\, e^{-4.50} = 0.94$

43. For t > 0, Redrawing the given network.

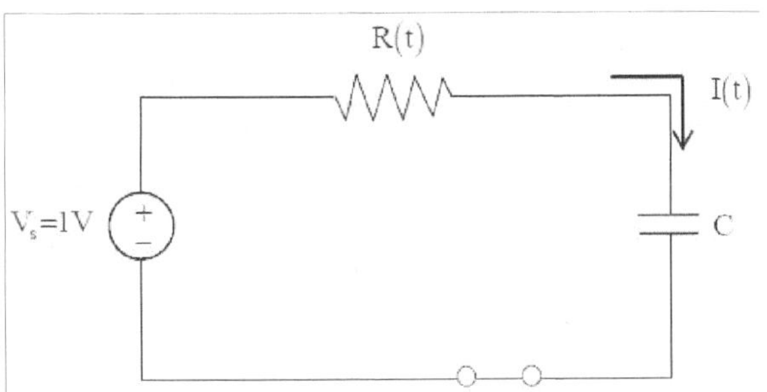

Given data

$R_0 = 1\Omega$, $C = 1F$, $T = 3R_0C$

Now $R(t) = R_0\left(1 - \dfrac{t}{T}\right) 1 \quad 0 \le t \le T$

$$R_0\left(1 - \frac{t}{3R_0C}\right) = R_0\left(1 - \frac{t}{3}\right) = (1 - t/3)$$

$\therefore \quad R(0) = R_0 - 1\Omega$

Redrawing the network at t = 0,

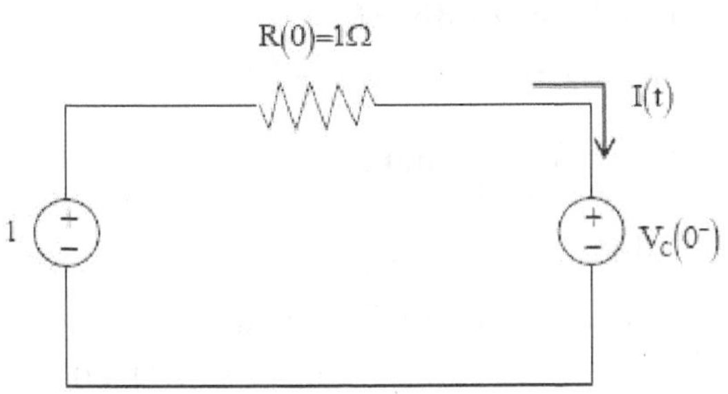

Now $I(0) = \dfrac{1 - V_c(0^-)}{R(0)}$

$\therefore \quad 1 = \dfrac{1 - V_C(0^-)}{1} \qquad (\because I(0) = 1A)$

$\therefore \quad 1 = 1 - V_C(0) \Rightarrow V_C(0^-) = 0V$

Now fort $= \infty$, capacitor will be open circuit

$V_C(\infty) = 1V$

Now, $\tau = RC = R(t)C = R_0\left(1 - \dfrac{t}{3}\right)$

Now $V_C(t) = 1 + [0 - 1]\, e^{\frac{-t}{(1-t/3)}}$

$$\therefore \quad V_C\left(\frac{T}{2}\right) = 1 - e^{-\left(\frac{T/2}{1-T/6}\right)}$$

$$= 1 - e^{-\left(\frac{3/2}{1-3/6}\right)} \quad (\because T = 3)$$

$$= 1 - e^{-\left(\frac{1.5}{0.5}\right)} = 1 - e^{-3} = 0.95$$

Now $I(t) = \dfrac{V_s(t) - V_C(t)}{R(t)}$

$$\therefore \quad I\left(\frac{T}{2}\right) = \frac{V_s\left(\frac{T}{2}\right) - V_C\left(\frac{T}{2}\right)}{R\left(\frac{T}{2}\right)} = \frac{1 - 0.95}{0.5} = 0.10$$

44. $I = I_0\left[\exp\left(-\dfrac{V_R}{V_T}\right) - 1\right]$

$$\therefore \quad \frac{-3}{4} I_0 = I_0\left[\exp\left(\frac{V_R}{V_T}\right) - 1\right]$$

$$\therefore \quad \exp\left[\frac{-V_R}{V_T}\right] = \frac{-3}{4} + 1 = \frac{1}{4}$$

$$\therefore \quad \frac{-V_R}{V_T} = \ell n\left(\frac{1}{4}\right)$$

$$\therefore \quad -V_R = V_T \ell n\left(\frac{1}{4}\right) = 25.9 \times \ell n\left(\frac{1}{4}\right) mV$$

$\therefore \quad |-V_R| = 35.87$ mV

45. Given data $C = 3 \times 10^8$ m/sec

Group velocity $v_g = 2 \times 10^8$ m/sec

Now $v_p v_g = C^2$

$$\therefore \quad \boxed{P = \frac{C^2}{v_g}}$$

$\therefore$ Phase velocity $(v_p) = \dfrac{C^2}{v_g}$

$$= \frac{(3 \times 10^8)}{2 \times 10^8} = 4.5 \times 10^8 \text{ m/sec}$$

46. The given circuit is

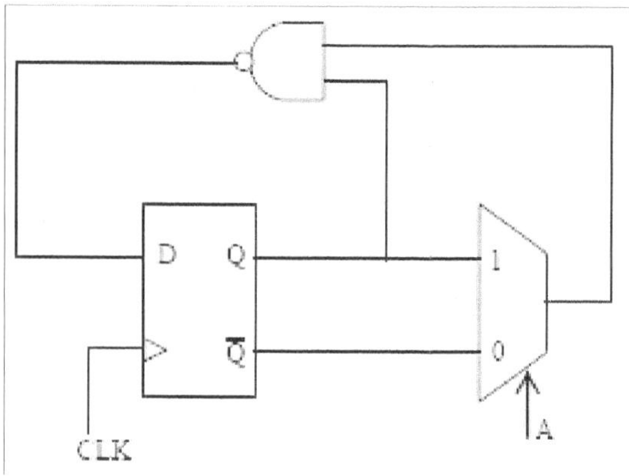

For Q = 0, state diagram

When $Q = 0, \overline{Q} = 1$ and $A = 0$, then $Y = \overline{Q} = 1$, so $D = \overline{QY} = \overline{0.1} = \overline{0} = 1$ after 1 clock $Q^+ = 1$

When $Q = 0, \overline{Q} = 1$ and $A = 1$, then $Y = Q = 1$, so $D = \overline{QY} = \overline{0.0} = 1$ after 1 clock $Q^+ = 1$

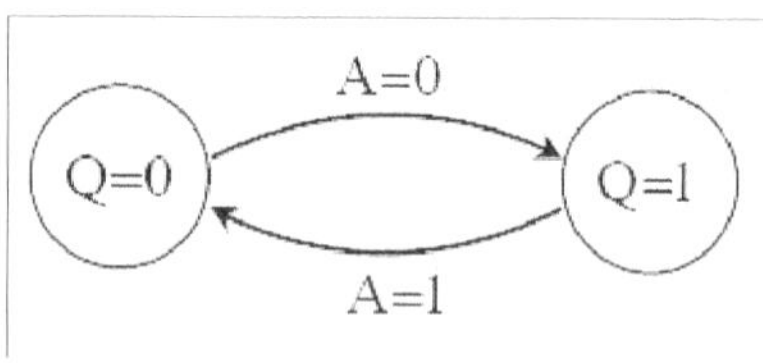

[State Diagram for Q = 0]

For Q = 1

→ When $Q = 1, \overline{Q} = 0$ and $A = 0$, then $Y = \overline{Q} = 0$, so $D = \overline{QY} = \overline{1.0} = \overline{0} = 1$ after 1 clock $Q^+ = 1$

→ When $Q = 1, \overline{Q} = 0$ and $A = 1$, then $Y = Q = 1$, so $D = \overline{QY} = \overline{1.1} = 0$ after 1 clock $Q^+ = 0$

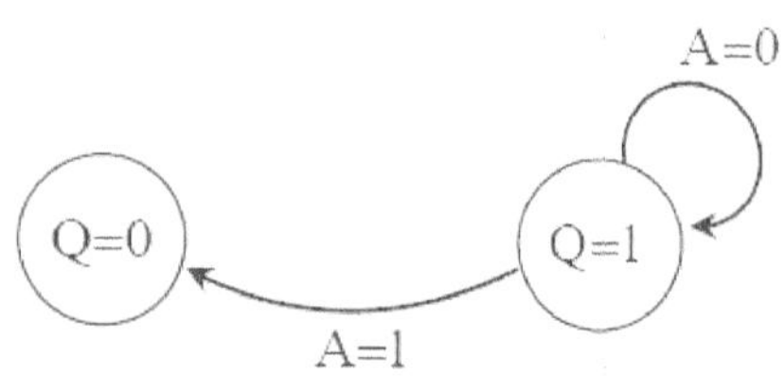

[State Diagram for Q = 1]

By combining both we can draw a single state diagram

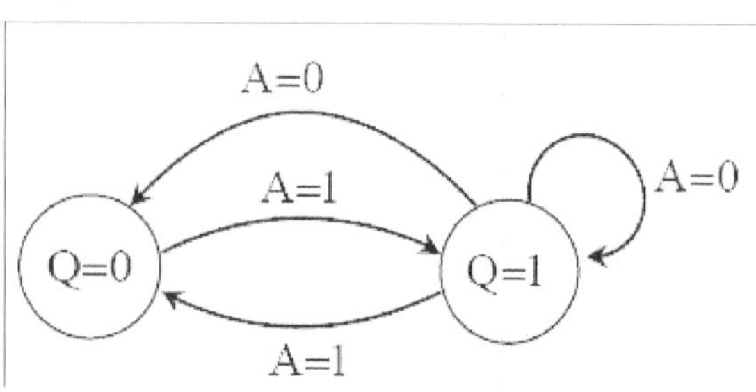

[Combined State Diagram]

47. Given, A unit feedback system as shown in figure

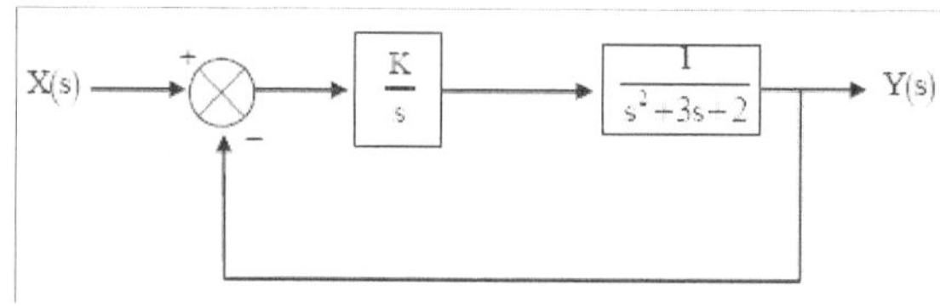

$CE = s(s^2 + 3s + 2) + k = 0$

$\Rightarrow s^3 + 3s^2 + 5 + k = 0$

Rowth Hurwitz criteria.

s^3	1	2
s^2	3	k
s^1	$\frac{6-k}{3}$	0
s^0	k	0

If this row will be zero then two poles will be on jω axis

$\therefore\ 6 - K = 0 \Rightarrow K = 6$

Lets cross check : $3s^2 + 6 = 0 \Rightarrow s^2 + 2 = 0$

$s = \pm j2$ ⟶ Two poles on jω

48. Satisfy the given condition

Given $|f'(x)| \le 2; f(-1) = 0$

$\Rightarrow -2 \le f'(x) \le 2$; where $x \in [-2, 2]$

Now applying Lagrange's Mean Value Theorem over interval [–1, 2]

$$\Rightarrow -2 \le f'(x) \le 2 \Rightarrow -2 \le \frac{f(2) - f(-1)}{2 - (-1)} \le 2$$

$$\left[\because f'(c) = \frac{f(b) - f(a)}{b - a}\right]$$

$$\Rightarrow -2 \le \frac{f(2) - 0}{3} \le 2$$

$\Rightarrow -2 \le f(2) \le 6$ ...(2)

Hence option (a) satisfies the equation.

49. Given differential equatin,

$$x^2 \frac{d^2y}{dx^2} - 3x\frac{dy}{dx} + 3y = 0, \qquad ...(i)$$

$x > 0$ & $y(1) = 1$ and $y(2) = 14$

(Cauchy-Euler Linear D. E.)

Equation (i) can be written as

$[x^2D^2 - 3xD + 3]y = 0$

Let $xD = \theta$; $x^2D^2 = \theta(\theta - 1)$;

where $\theta = \frac{d}{dz}$ and $x = e^z$

$\therefore [\theta(\theta - 1) - 3\theta + 3]y = 0 \Rightarrow [\theta^2 - 4\theta + 3]y = 0$

Consider the Auxillary equation as

$\theta^2 - 4\theta + 3 = 0 \Rightarrow (\theta - 3)(\theta - 1) = 0$

$\Rightarrow \theta = 1, 3 \rightarrow$ (Roots are real and distinct)

$\therefore$ The solution is,

$y = C_1 e^{1.z} + C_2 e^{3.z} \Rightarrow y = C_1 x + C_2 x^3 [\because x = e^z]$ ...(ii)

Given y = 1at x = 1

$\therefore\ C_1 + C_2$ ...(iii)

and y = 14 at x = 2

$\therefore\ 14 = 2C_1 + 8C_2$

$\therefore\ C_1 + 4C_2 = 7$ ...(iv)

Solving eq. (iii) and (iv) we get

$\boxed{C_2 = 2 \text{ and } C_1 = 1}$

Now from eq. (ii),

$y = (-1)x + (2)x^3$

$\therefore\ y(1.5) = (-1)(1.5) + 2(1.5)^3$

$\therefore\ y(1.5) = 5.25$

50. Power in y(t) $= \begin{bmatrix} \text{Power in} \\ h(t) \times X(t) \end{bmatrix} + [\text{Power in } Z(t)]$

$$\text{Power in } h(t) \times X(t) = \int_{-\infty}^{\infty} |H(f)|^2 S_{xx}(f)df$$

$$= \int_{-\infty}^{\infty} |H(f)|^2 (5)df = \int_{-5}^{5} (025)(.5)df$$

$= (10)(1.25) = 12.5\text{W}$

Power in $S_2(f)$ = Area under power spectral density

$$P_{z(t)} = \left(\frac{1}{2}\right)(10)(1) = 5\text{W}$$

$\therefore$ Power in y(t) = (12.5 + 5)W = 17.5W

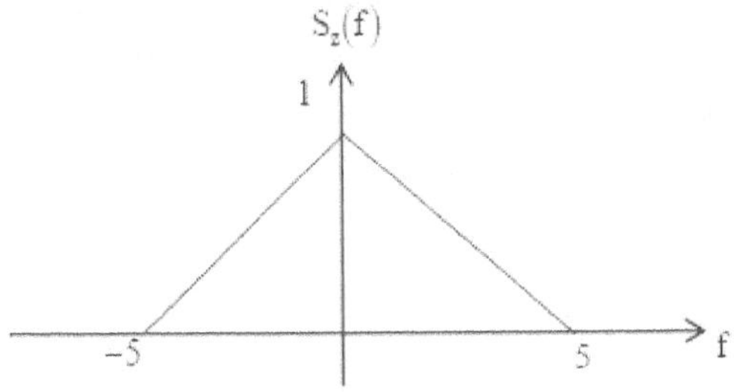

51.

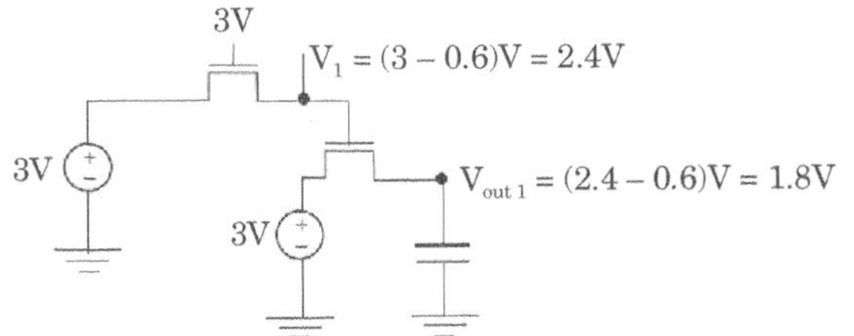

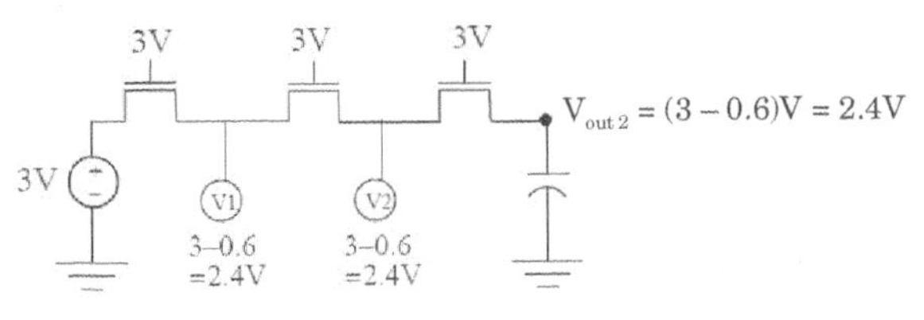

52.

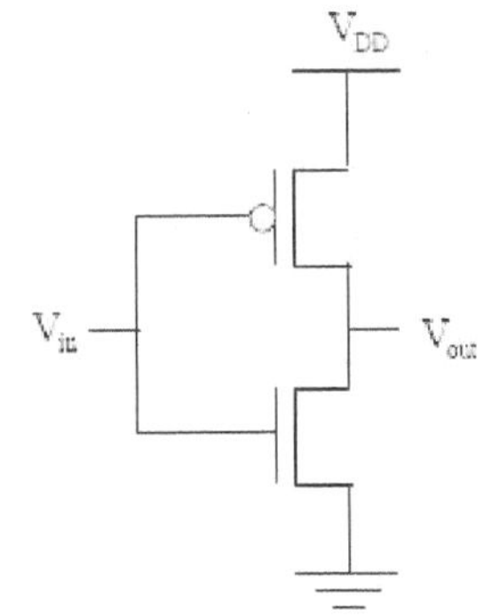

At $V_i = \frac{V_{DD}}{2}$, both the MOSFETs are in saturation and both MOSFETs have the same current.

$$\left(\frac{V_{dd}}{2}\right)$$
$$\uparrow$$

$$\therefore \left(\frac{1}{2}\right)(100\mu A / V^2)\left(\frac{W}{L}\right)_n [1.5 - 0.7]^2$$

$$= \left(\frac{1}{2}\right)(40\mu A / V^2)\left(\frac{W}{L}\right)_P (3 - 1.5 - 0.9)^2$$

$$\frac{\left(\frac{W}{L}\right)_n}{\left(\frac{W}{L}\right)_p} = \frac{(40)(0.6)^2}{(100)(0.8)^2} = \frac{40 \times 0.36}{100 \times 0.64} = 0.225$$

53.

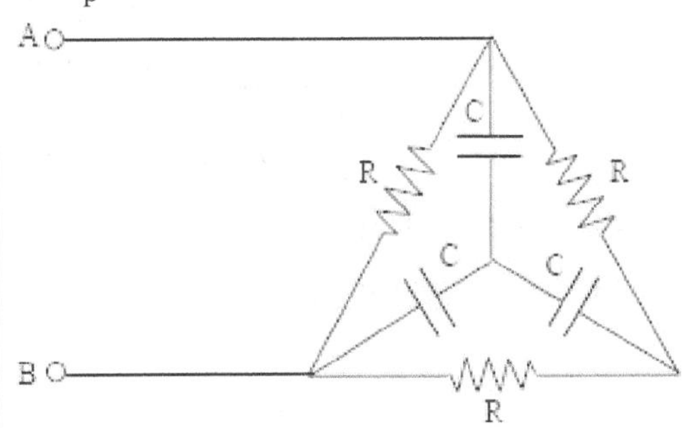

It is given that $V(t) = 2\sin 100t = \overline{V} = 2\angle 0°$

$R = 1k\Omega$, $C = 1\mu F$

By observity the circuit we can say

$$\bar{I} = \frac{\overline{V}}{Z}$$

$\rightarrow$ When each element of star network are same then its corresponding delta element are same and it becomes

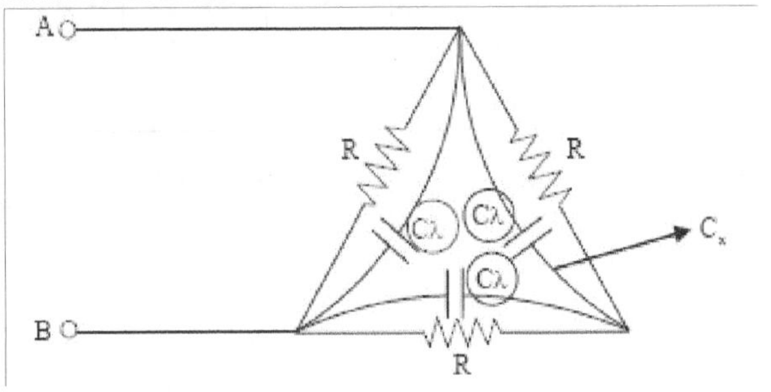

$Z_\Delta = 3Z^*$, but in capacitor case $Z = \frac{1}{j\omega C}$, So if the capacitor of star network are C each then in its delta equivalent it becomes $C/3 = C_x$

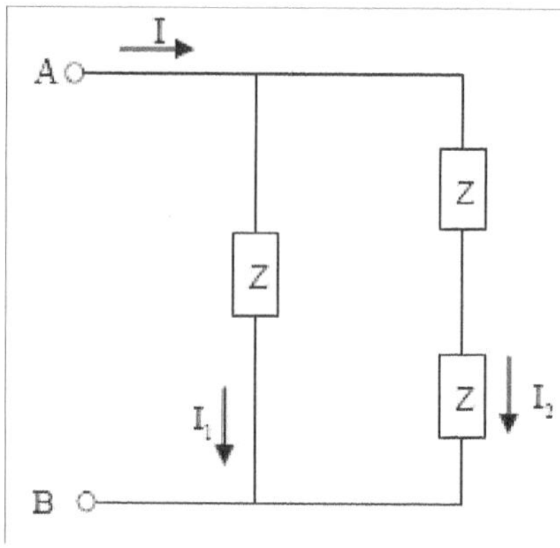

Further redrawing the network.

$$\text{Where } Z = R \,||\, \frac{1}{j\omega C_x} = \frac{R/jwC_x}{R + \frac{1}{j\omega C_x}} = \frac{R}{1 + j\omega RC_x}$$

$$\frac{R}{1 + j\omega R\frac{C}{3}} = \frac{3R}{3 + j\omega RC} = \frac{3R}{3 + j}$$

$$\text{Now } \bar{I} = \bar{I}_1 + \bar{I}_2 = \frac{\bar{V}}{Z} + \frac{\bar{V}}{2Z} = \frac{2\underline{|0}}{(3R/3 + j)} + \frac{2\underline{|0}}{(65R/3 + j)}$$

$$= \frac{3 + j}{3R}[2\underline{|0} + 1\underline{|0}] = \frac{3 + j}{R}(1\underline{|0}) = \frac{[3\underline{|0}] + [1\underline{|90}]}{R}$$

$$= 3\underline{|0} + 1\underline{|90}\text{mA} \;\; (\because 12 - 1000\Omega)$$

$$= 3\sin 1000t + \sin(1000t + 90°)$$

54. Given block diagram

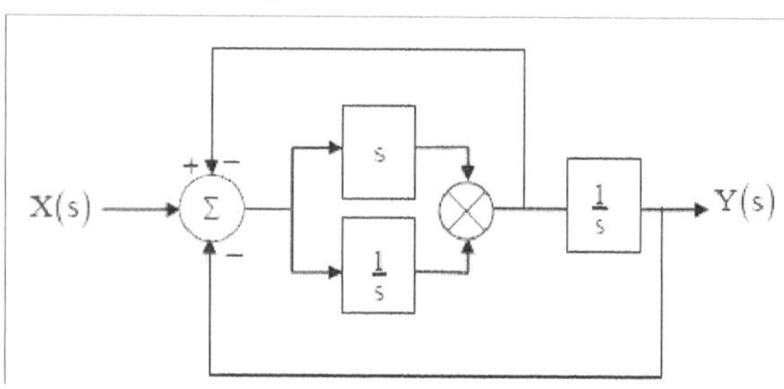

It can be reduced to

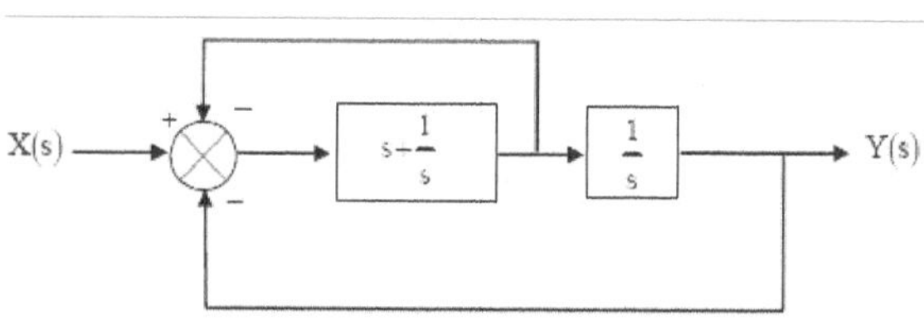

$$H(S) = \frac{Y(s)}{X(s)} = \frac{\frac{(s^2 + 1)}{s} \times \frac{1}{s}}{1 + \frac{s^2 + 1}{s} + \frac{s^2 + 1}{s^2}}$$

$$= \frac{\frac{(s^2 + 1)}{s^2}}{\frac{s^2 + (s^2 + 1)s + s^2 + 1}{s^2}}$$

Hence, the transfer function

$$H(s) = \frac{Y(s)}{X(s)} = \frac{(s^2 + 1)}{(s^3 + 2s^2 + s + 1)}$$

55.

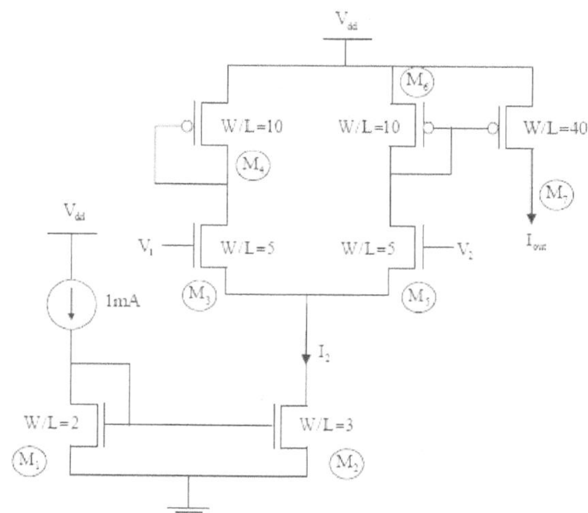

M_1 and M_2 have the same V_{gs}

∴ Current flows in the ratio of W/L

$$\therefore I_2 = \left(\frac{3}{2}\right)^{(1mA)} = 1.5mA$$

$V_1 = 0$; Therefore M_3 is in cut off and entire I_2 current flows through M_5 branch.

$$\therefore I_5 = 1.5mA$$

$$I_{out} = \left(\frac{40}{10}\right)^{[I_5]}$$

$$\therefore I_{out} = 4 \times 1.5mA = 6mA$$

SOLVED PAPER – 2018

ELECTRONICS & COMMUNICATION ENGINEERING

No. of Questions : 65 *Maximum Marks : 100*

INSTRUCTIONS

1. Total of 65 questions carrying 100 marks, out of which 10 questions carrying a total of 15 marks are in General Aptitude (GA)
2. The Engineering Mathematics will carry around 15% **of the total marks,** the General Aptitude section will carry **15% of the total marks** and the **remaining 70% of the total marks.**
3. **Types of Questions**
   (*a*) **Multiple Choice Questions (MCQ)** carrying 1 or 2 marks each in all papers and sections. These questions are objective in nature, and each will have a choice of four options, out of which the candidate has to mark the correct answer(s).
   (*b*) **Numerical Answer Questions** of 1 or 2 marks each in all papers and sections. For these questions the answer is a real number, to be entered by the candidate using the virtual keypad. No choices will be shown for these type of questions.
4. For **1-mark** multiple-choice questions, **1/3 marks** will be deducted for a wrong answer. Likewise, for **2-marks** multiple-choice questions, 2/3 marks will be deducted for a wrong answer. There is no negative marking for numerical answer type questions.

Chapter-Wise Analysis

#	Chapters	Marks		
		1	2	Total
1	General Aptitude	5	5	15
2	Engineering Mathematics	5	3	11
3	Networks	1	3	7
4	Signals and Systems	2	2	6
5	Electronic Devices	3	2	7

#	Chapters	Marks		
		1	2	Total
6	Analog Circuits	2	5	12
7	Digital Circuits & Microprocessor	3	4	11
8	Control Systems	3	5	13
9	Communications	4	3	10
10	Electromagnetics	2	3	8

GENERAL APTITUDE

1. "By giving him the last _______ of the cake, you will ensure lasting _______ in our house today."

   The words that best fill the blanks in the above sentence are

   (*a*) peas, piece (*b*) piece, peace
   (*c*) peace, piece (*d*) peace, peas

2. "Even though there is a vast scope for its ________, tourism has remained a/an ________ area."

   The words that best fill the blanks in the above sentence are

   (*a*) improvement, neglected
   (*b*) rejection, approved
   (*c*) fame, glum
   (*d*) interest, disinterested

3. If the number 715 ? 423 is divisible 3 (? denotes the missing digit in the thousandths place), then the smallest whole number in the place of ? is _________

   (*a*) 0 (*b*) 2
   (*c*) 5 (*d*) 6

4. What is the value of $1+\frac{1}{4}+\frac{1}{16}+\frac{1}{64}+\frac{1}{256}+....?$

   (*a*) 2 (*b*) $\frac{7}{4}$
   (*c*) $\frac{3}{2}$ (*d*) $\frac{4}{3}$

5. As 1.5 m in tall person is standing at a distance of 3 m from a lamp post. The light from the lamp at the top of the post casts her shadow. The length of the shadow is twice her height. What is the height of the lamp post in meters?

   (*a*) 1.5 (*b*) 3
   (*c*) 4.5 (*d*) 6

6. Leila aspires to buy a car worth Rs. 10,00,000 after 5 years. What is the minimum amount in Rupees that she should deposit now in a bank which offers 10% annual rate of interest, if the interest was compounded annually?

   (*a*) 5,00,000 (*b*) 6,21,000
   (*c*) 6,66,667 (*d*) 7,50,000

7. Two alloys A and B contain gold and copper in the ratios of 2 : 3 and 3 : 7 by mass, respectively. Equal masses of alloys A and B are melted to make an alloy C. The ratio of gold to copper in alloy C is ______.

   (*a*) 5 : 10 (*b*) 7 : 13
   (*c*) 6 : 11 (*d*) 9 : 13

8. The Cricket Board has long recognized John's potential as a leader of the team. However, his on-field Temper has always been a matter of concern for them since his junior days. While this aggression has filled stadia with die-hard fans, it has taken a toll on his own batting. Until recently, it appeared that he found it difficult to convert his agression into big scores. Over the past three seasons though, that picture of John has been replaced by a cerebral, calculative and successful batsman-captain. After many years, it appears that the team has finally found a complete captain.

   Which of the following statements can be logically inferred from the above paragraph?

   (i) Even as a junior cricketer, John was considered a good captain.
   (ii) Finding a complete captain is a challenge.
   (iii) Fans and the Cricket Board have differing views on what they want in a captain.
   (iv) Over the past three seasons John has accumulated big scores.

   (*a*) (i), (ii) and (iii) only (*b*) (iii) and (iv) only
   (*c*) (ii) and (iv) only (*d*) (i), (ii), (iii) and (v)

9. A cab was involved in a hit and run accident at right you are given the following data about the cabs in the city and the accident.

   (i) 85% of cabs in the city are green and the remaining cabs are blue.
   (ii) A witness identified the cab involved in the accident as blue.
   (iii) It is known that a witness can correctly identify the cab colour only 80% of the time.

   Which of the following options is closest to the probability that the accident was caused by a blue cab?

   (*a*) 12%
   (*b*) 15%
   (*c*) 41%
   (*d*) 80%

10. A coastal region with unparalleled beauty is home to many species of animals. It is dotted with coral reefs and unspoilt white sandy beaches. It has remained inaccessible to tourists due to poor connectivity and lack of accommodation. A company has spotted the opportunity and is planning to develop a luxury resort with helicopter service to the nearest major city airport. Environmentalists are upset that this would lead to the region becoming crowded and polluted like any other major beach resorts.

Which one of the following statements can be logically inferred from the information given in the above paragraph?

(*a*) The culture and tradition of the local people will be influenced by the tourists.

(*b*) The region will become crowded and polluted due to tourism.

(*c*) The coral reefs are on the decline and could soon vanish.

(*d*) Helicopter connectivity would lead to an increase in tourists coming to the region.

TECHNICAL SECTION

1. Two identical nMOS transistors M_1 and M_2 are connected as shown below. The circuit is used as an amplifier with the input connected between G and S terminals and the output taken between D and S terminals, V_{bias} and V_D are so adjusted that both transistors are in saturation. The transconductance of this combination is defined as $g_m = \frac{\partial i_D}{\partial v_{GS}}$ while the output resistance is $r_o = \frac{\partial v_{DS}}{\partial i_D}$, where i_D is the current flowing into the drain of M_2. Let g_{m1}, g_{m2} be the transconductances and r_{o1}, r_{o2} be the output resistance of transistors M_1 and M_2, respectively

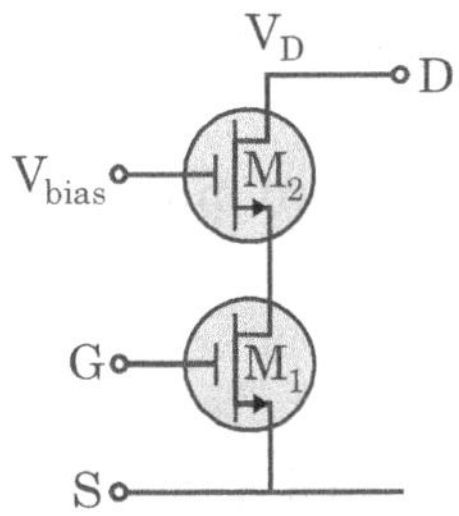

Which of the following statements about estimates for g_m and r_o is correct?

(*a*) $g_m \approx g_{m1} \cdot g_{m2} \cdot r_{o2}$ and $r_o \approx r_{o1} + r_{o2}$

(*b*) $g_m \approx g_{m1} + g_{m2}$ and $r_o \approx r_{o1} + r_{o2}$

(*c*) $g_m \approx g_{m1}$ and $r_o \approx r_{o1} \cdot g_{m2} \cdot r_{o2}$

(*d*) $g_m \approx g_{m1}$ and $r_o \approx r_{o2}$

2. In the circuit shown below, the op-amp is deal and Zener voltage of the diode is 2.5 volts. At the input, unit step voltage is applied i.e. $v_{IN}(t) = u(t)$ volts. Also at t = 0, the voltage across each of the capacitors is zero.

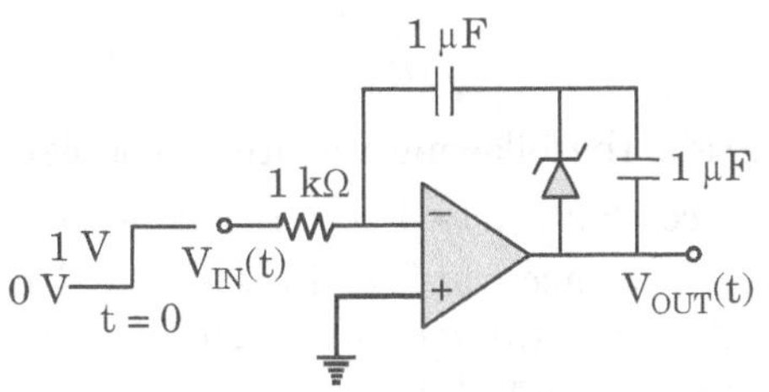

The time t, in milliseconds, at which the output voltage v_{OUT} crosses –10 V is

(*a*) 2.5

(*b*) 5

(*c*) 7.5

(*d*) 10

3. A good transimpedance amplifier has

(*a*) low input impedance and high output impedance

(*b*) high input impedance and high output impedance

(*c*) high input impedance and low output impedance

(*d*) low input impedance and low output impedance

4. Let the input be u and the output be y of a system, and the other parameters are real constants. Identify which among the following systems is not a linear system :

(*a*) $$\frac{d^3y}{dt^3} + a_1\frac{d^2y}{dt^2} + a_2\frac{dy}{dt} + a_3y = b_3u + b_2\frac{du}{dt} + b_1\frac{d^2u}{dt^2}$$

(with initial rest conditions)

(*b*) $y(t) = \int_0^t e^{\alpha(t-\tau)}\beta u(\tau)\, d\tau$

(*c*) $y = au + b,\ b \neq 0$

(*d*) $y = au$

5. The Nyquist stability criterion and the Routh criterion both are powerful analysis tools for determining the stability of feedback controllers. Identify which of the following statements is FALSE :

(*a*) Both the criteria provide information relative to the stable gain range of the system.

(*b*) The general shape of the Nyquist plot is readily obtained from the Bode magnitude plot for all minimum-phase systems.

(*c*) The Routh criterion is not applicable in the condition of transport lag, which can be readily handled by the Nyquist criterion.

(*d*) The closed-loop frequency response for a unity feedback system cannot be obtained from the Nyquist plot.

6. Consider $p(s) = s^3 + a_2s^2 + a_1s + a_0$ with all real coefficients. It is known that its derivative $p'(s)$ has no real roots. The number of real roots of p(s) is

7. In a p-n junction diode at equilibrium, which one of the following statements is NOT TRUE?
   (*a*) The hole and electron diffusion current components are in the same direction.
   (*b*) The hole and electron drift current components are in the same direction.
   (*c*) On an average, holes and electrons drift in opposite direction.
   (*d*) On an average, electrons drift and diffuse in the same direction.

8. The logic *f*(X, Y) realized by the given circuit is

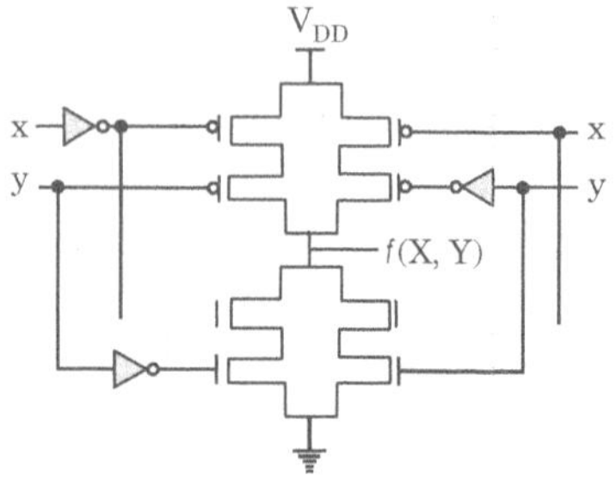

   (*a*) NOR (*b*) AND
   (*c*) NAND (*d*) XOR

9. A function F(A, B, C) defined by three Boolean variables A, B and C when expressed as sum of products is given by

$$F = \bar{A}\cdot\bar{B}\cdot\bar{C} + \bar{A}\cdot B\cdot\bar{C} + A\cdot\bar{B}\cdot\bar{C}$$

where, $\bar{A}, \bar{B}$ and $\bar{C}$ are the complements of the respective variables. The product of sums (POS) form of the function F is
   (*a*) $F = (A + B + C)\cdot(A + \bar{B} + C)\cdot(\bar{A} + B + C)$
   (*b*) $F = (\bar{A} + \bar{B} + \bar{C})\cdot(\bar{A} + B + \bar{C})\cdot(A + \bar{B} + \bar{C})$
   (*c*) $F = (A + B + \bar{C})\cdot(A + \bar{B} + \bar{C})\cdot(\bar{A} + B + \bar{C})\cdot(\bar{A} + \bar{B} + C)\cdot(\bar{A} + \bar{B} + \bar{C})$
   (*d*) $F = (\bar{A} + \bar{B} + C)\cdot(\bar{A} + B + C)\cdot(A + \bar{B} + C)\cdot(A + B + \bar{C})\cdot(A + B + C)$

10. The points P, Q and R shows on the Smith chart (normalized impedance chart) in the following figure represent :

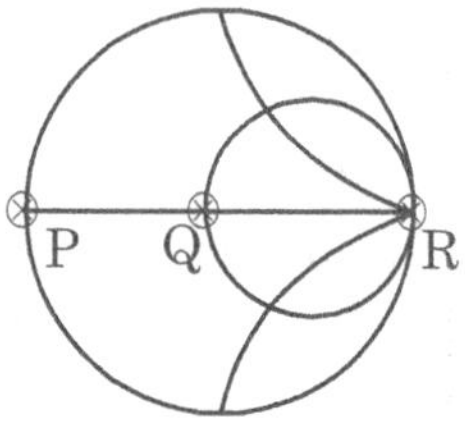

   (*a*) P : Open Circuit, Q : Short Circuit, R : Matched : Load
   (*b*) P : Open Circuit, Q : Matched Load, R : Short Circuit
   (*c*) P : Short Circuit, Q : Matched Load, R : Open Circuit
   (*d*) P : Short Circuit, Q : Open Circuit, R : Matched Load

11. Let M be a real 4 × 4 matrix. Consider the following statements :

S1 : M has 4 linearly independent eigenvectors.

S2 : M has 4 distinct eigenvalues.

S3 : M is non-singular (invertible).

Whict one among the following is TRUE?
   (*a*) S1 implies S2 (*b*) S2 implies S1
   (*c*) S1 implies S3 (*d*) S3 implies S2

12. Let $f(x,y) = \dfrac{ax^2 + by^2}{xy}$, where a and b are constant. If $\dfrac{\partial f}{\partial x} = \dfrac{\partial f}{\partial y}$ at x = 1 and y = 2, then the relation between a and b is
   (*a*) $a = \dfrac{b}{4}$ (*b*) $a = \dfrac{b}{2}$
   (*c*) a = 2b (*d*) a = 4b

13. A discrete-time all-pass system has two of its poles at $0.25\angle 0°$ and $2\angle 30°$. Which one of the following statements about the system is TRUE?
   (*a*) It has two more poles at $0.5\angle 30°$ and $4\angle 0°$.
   (*b*) It is stable only when the impulse response is two-sided.
   (*c*) It has constant phase response over all frequencies.
   (*d*) It has constant phase response over the entire z-plane.

14. Let x(t) be a periodic function with period T = 10. The Fourier series coefficients for this series are denoted by a_k, that is

$$x(t) = \sum_{k=-\infty}^{\infty} a_k e^{jk\frac{2\pi}{T}t}$$

The same function x(t) can also be considered as a periodic function with period T' = 40. Let b_k be the Fourier series coefficients when period is taken as T'. If $\sum_{k=-\infty}^{\infty} |a_k| = 16$, then $\sum_{k=-\infty}^{\infty} |b_k|$ is equal to
   (*a*) 256 (*b*) 64
   (*c*) 16 (*d*) 4

15. Consider the following amplitude modulated signal:

s(t) = cos(2000 πt) + 4cos(2400 πt) + cos(2800 πt)

The ratio (accurate to three decimal places) of the power of the message signal to the power of the carrier signal is _________.

16. Consider a binary channel code in which each codeword has a fixed length of 5 bits. The Hamming distance between any pair of distinct codewords in this code is at least 2. The maximum number of codewords such a code can contain is ___________.

17. A binary source generates symbols $X \in \{-1, 1\}$ which are transmitted over a noisy channel. The probability of transmitting X = 1 is 0.5. Input to the threshold detector is R = X + N. The probability density function $f_N(n)$ of the noise N is shown below.

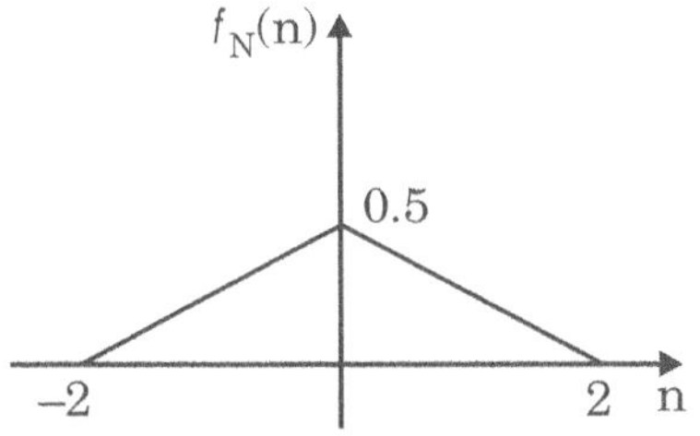

If the detection thershold is zero, then the probability of error (correct to two decimal places) is ___________.

18. A p-n step junction diode with a contact potential of 0.65 V has a depletion width of 1μm at equilibrium. The forward voltage (in volts, correct to two decimal places) at which this width reduces to 0.6 μm is __________.

19. A traffic signal cycles from GREEN to YELLOW, YELLOW to RED and RED to GREEN. In each cycle, GREEN is turned on for 70 seconds, YELLOW is turned on for 5 seconds and the RED is turned on for 75 seconds. This traffic light has to be implemented using a finite state machine (FSM). The only input to this FSM is a clock of 5 second period. The minimum number of flip-flops required to implement this FSM is ________.

20. There are two photolithography system : one with light source of wavelength λ_1 = 156 nm (System 1) and another with light source of wavelength λ_2 = 325 nm (System 2). Both photolithography systems are otherwise identical. If the minimum feature sizes that can be realized using System 1 and System 2 are L_{min1} and L_{min2} respectively, the ratio L_{min1}/L_{min2} (correct to two decimal places) is _________.

21. A lossy transmission line has resistance per unit length R = 0.05 Ω/m. The line is distortionless and has characteristic impedance of 50 Ω. The attenuation constant (in Np/m correct to three decimal places) of the line is _________.

22. Consider matrix $A = \begin{bmatrix} k & 2k \\ k^2 - k & k^2 \end{bmatrix}$ and vector $X = \begin{bmatrix} x_1 \\ x_2 \end{bmatrix}$. The number of distinct real values of k for which the equation AX = 0 has infintely many solution is _______.

23. Let X_1, X_2, X_3 and X_4 be independent normal random variables with zero mean and unit variance. The probability that X_4 is the smallest among the four is ________.

24. Taylor series expansion of $f(x) = \int_0^x e^{-\left(\frac{t^2}{2}\right)} dt$ around x = 0 has the form

$$f(x) = a_0 + a_1x + a_2x^2 + ...$$

The coefficient a_2 (correct to two decimal places) is equal to _________.

25. The ABCD matrix for a two-port network is defined by :

$$\begin{bmatrix} V_1 \\ I_1 \end{bmatrix} = \begin{bmatrix} A & B \\ C & D \end{bmatrix} \begin{bmatrix} V_2 \\ -I_2 \end{bmatrix}$$

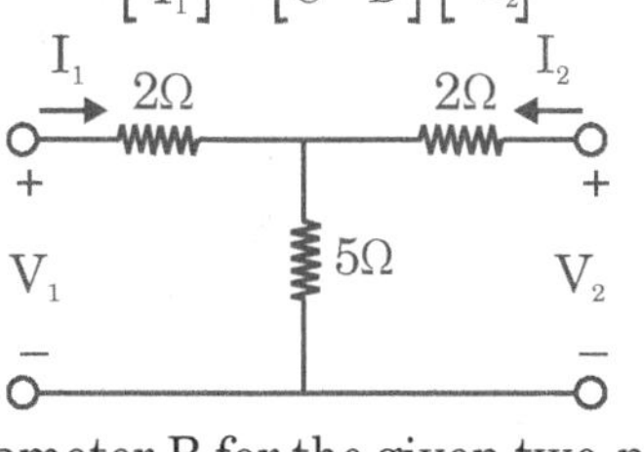

The parameter B for the given two-port network (in ohms, correct to two decimal places) is ______.

26. The circuit shown in the figures is used to provide regulated voltage (5 V) across the 1 kΩ resistor. Assume that the Zener diode has a constant reverse breakdown voltage for a current range, starting from a minimum required Zener current, I_{Zmin} = 2 mA to its maximum allowable current. The input voltage V_1 may vary by 5% from its nominal value of 6 V. The resistance of the diode in the breakdown region in negligible.

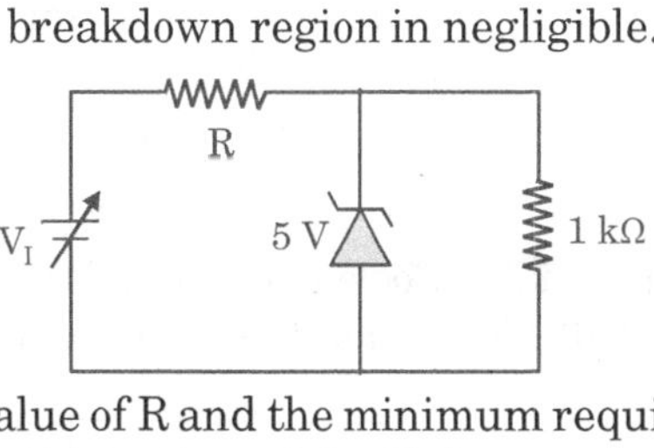

The value of R and the minimum required power dissipation rating of the diode, respectively, are

(*a*) 186 Ω and 10 mW

(*b*) 100 Ω and 40 mW

(*c*) 100 Ω and 10 mW

(*d*) 186 Ω and 40 mW

27. Let $c(t) = A_c\cos(2\pi f_c t)$ and $m(t) = \cos(2\pi f_m t)$. It is given that $f_c >> 5f_m$. The signal $c(t) + m(t)$ is applied to the input of a non-linear device, whose output $v_0(t)$ is related to the input $v_i(t)$ as $v_0(t) = av_1(t) + bv_i^2(t)$, where a and b are positive constants. The output of the non-linear device is passed through an ideal band-pass filter with center frequency f_c and bandwidth $3f_m$, to produce an amplitude modulated (AM) wave. If it is desired to have the sideband power of the AM wave to be half of the carrier power, then a/b is

(*a*) 0.25 (*b*) 0.5
(*c*) 1 (*d*) 2

28. Consider a white Gaussian noise process N(t) with two-sided power spectral density $S_N(f)$ = 0.5 W/HZ as input to a filter with impulse response $0.5e^{-t^2/2}$ (where t is in seconds) resulting in output Y(t). The power in Y(t) in watts is

(*a*) 0.11 (*b*) 0.22
(*c*) 0.33 (*d*) 0.44

29. The state equation and the output equation of a control system are given below :

$$\dot{x} = \begin{bmatrix} -4 & -1.5 \\ 4 & 0 \end{bmatrix} x + \begin{bmatrix} 2 \\ 0 \end{bmatrix} u,$$

$$y = [1.5 \quad 0.625]x$$

The transfer function representation of the system is

(*a*) $\dfrac{3s+5}{s^2+4s+6}$ (*b*) $\dfrac{3s-1.875}{s^2+4s+6}$

(*c*) $\dfrac{4s+1.5}{s^2+4s+6}$ (*d*) $\dfrac{6s+5}{s^2+4s+6}$

30. Red (R), Green (G) and Blue (B) Light Emitting Diodes (LEDs) were fabricated using p-n junctions of three different inorganic semiconductors having different band-gaps. The built-in voltages of red, green and blue diodes are V_R, V_G and V_B, respectively. Assume donor and acceptor doping to be the same (N_A and N_D respectively) in the p and n sides of all the three diodes.

Which one of the following relationships about the built-in voltages is TRUE?

(*a*) $V_R > V_G > V_B$ (*b*) $V_R < V_G < V_B$
(*c*) $V_R = V_G = V_B$ (*d*) $V_R > V_G < V_B$

31. A four-variable Boolean function is realized using 4 × 1 multiplexers as shown in the figure.

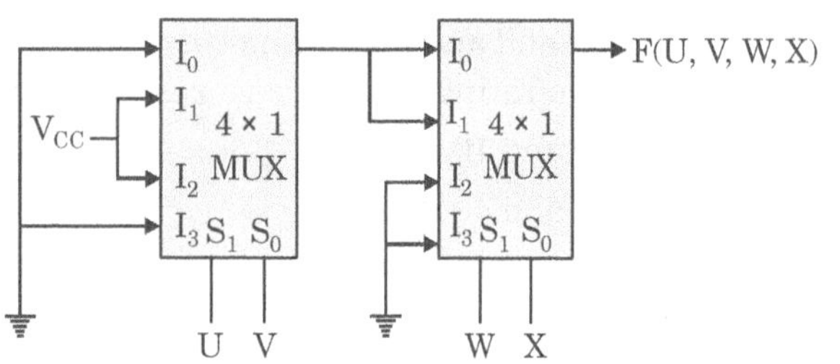

The minimized expression for F(U, V, W, X) is

(*a*) $(UV + \bar{U}\bar{V})\bar{W}$

(*b*) $(UV + \bar{U}\bar{V})(\bar{W}\bar{X} + \bar{W}X)$

(*c*) $(U\bar{V} + \bar{U}V)\bar{W}$

(*d*) $(U\bar{V} + \bar{U}V)(\bar{W}\bar{X} + \bar{W}X)$

32. A 2 × 2 ROM array is built with the help of diodes as shown in the circuit below. Here W_0 and W_1 are signals that select the word lines and B_0 and B_1 are signals that are output of the sense amps based on the stored data corresponding to the bit lines during the read operation.

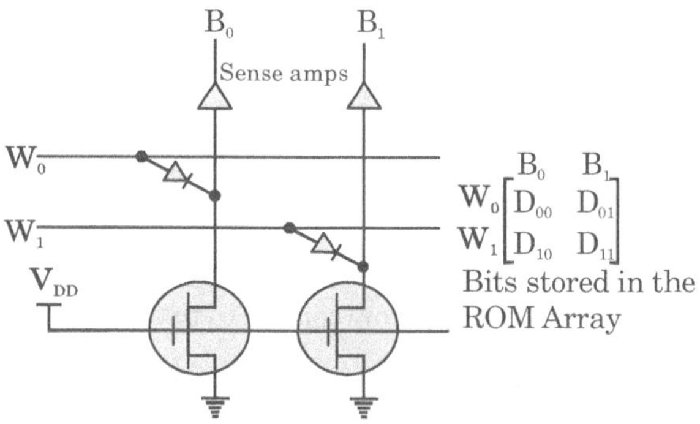

During the read operation, the selected word line goes high and the other word line is in a high impedance state. As per the implementation shown in the circuit diagram above, what are the bits corresponding to D_{ij} (where i = 0 or 1 and j = 0 or 1) stored in the ROM?

(*a*) $\begin{bmatrix} 1 & 0 \\ 0 & 1 \end{bmatrix}$ (*b*) $\begin{bmatrix} 0 & 1 \\ 1 & 0 \end{bmatrix}$

(*c*) $\begin{bmatrix} 1 & 0 \\ 1 & 0 \end{bmatrix}$ (*d*) $\begin{bmatrix} 1 & 1 \\ 0 & 0 \end{bmatrix}$

33. The distance (in meters) a wave has to propagate in a medium having a skin depth of 0.1 m so that the amplitude of the wave attenuates by 20 dB, is

(*a*) 0.12 (*b*) 0.23
(*c*) 0.46 (*d*) 2.3

34. A curve passes through the point (x = 1, y = 0) and satisfies the differential equation $\dfrac{dy}{dx} = \dfrac{x^2+y^2}{2y} = \dfrac{y}{x}$. The equation that describes the curve is

(*a*) $In\left(1+\dfrac{y^2}{x^2}\right) = x-1$ (*b*) $\dfrac{1}{2}In\left(1+\dfrac{y^2}{x^2}\right) = x-1$

(*c*) $In\left(1+\dfrac{y}{x}\right) = x-1$ (*d*) $\dfrac{1}{2}In\left(1+\dfrac{y}{x}\right) = x-1$

35. For the circuit given in the figure, the voltage V_C (in volts) across the capacitor is

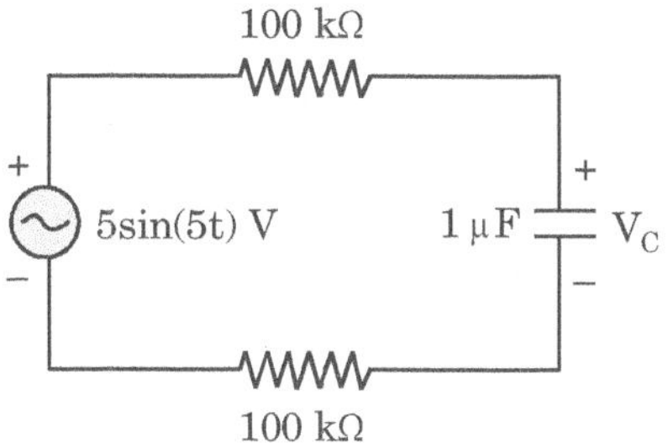

(*a*) $1.25\sqrt{2}\sin(5t - 0.25\pi)$

(*b*) $1.25\sqrt{2}\sin(5t - 0.125\pi)$

(*c*) $2.5\sqrt{2}\sin(5t - 0.25\pi)$

(*d*) $2.5\sqrt{2}\sin(5t - 0.125\pi)$

36. For the circuit given in the figure, the magnitude of the loop current (in amperes, correct to three decimal places) 0.5 second after closing the switch is ________.

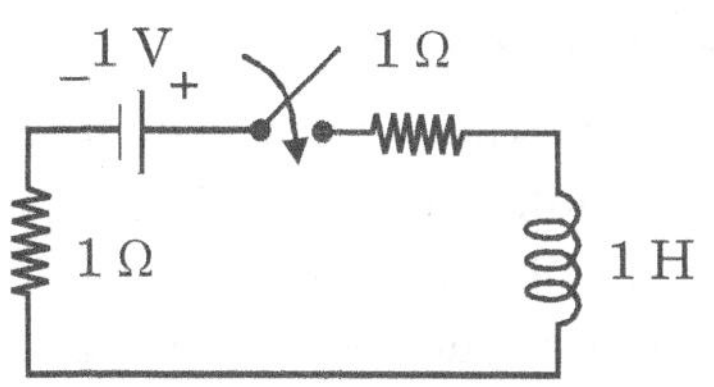

37. A DC current of 26 μA flows through the circuit shown. The diode in the circuit is forward biased and it has an ideality factor of one. At the quiescent point, the diode has a junction capacitance of 0.5 nF. Its neutral region resistances can be neglected. Assume that the room temperature thermal equivalent voltage is 26 mV.

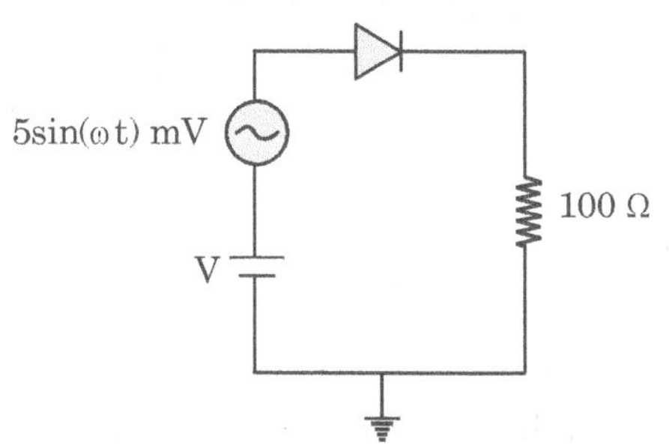

For $\omega = 2 \times 10^6$ rad/s, the amplitude of the small-signal component of diode current (in μA, correct to one decimal place) is ________.

38. An op-amp based circuit is implemented as shown below.

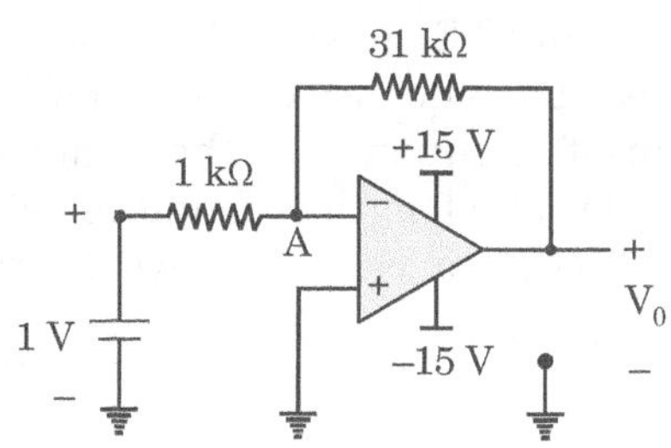

In the above circuit, assume the op-amp to be ideal. The voltage (in volts, correct to one decimal place) at node A, connected to the negative input of the op-amp as indicated in the figure is ________.

39. The input 4sinc(2t) is fed to a Hilbert transformer to obtain y(t), as shown in the figure below :

Here $\sin c(x) = \dfrac{\sin(\pi, x)}{\pi x}$. The value (accurate to two decimal places) of $\int_{\infty}^{\infty} |y(t)|^2\, dt$ is ________.

40. A random variable X takes values –0.5 and 0.5 with probabilities $\frac{1}{4}$ and $\frac{3}{4}$, respectively. The noisy observation of X is Y = X + Z, where Z has uniform probability density over the interval (–1, 1). X and Z are independent. If the MAP rule based detector outputs $\hat{X}$ as

$$\hat{X} = \begin{cases} -0.5, & Y < \alpha \\ 0.5, & Y \geq \alpha, \end{cases}$$

then the value of α (accurate to two decimal places) is ________.

41. For a unity feedback control system with the forward path transfer function $G(s) = \dfrac{K}{s(s+2)}$. The peak resonant magnitude M_r of the closed-loop frequency response is 2. The corresponding value of the gain K (correct to two decimal places) is ________.

42. The figure below shows the Bode magnitude and phase plots of a stable transfer function

$$G(s) = \frac{n_0}{s^3 + d_2 s^2 + d_1 s + d_0}.$$

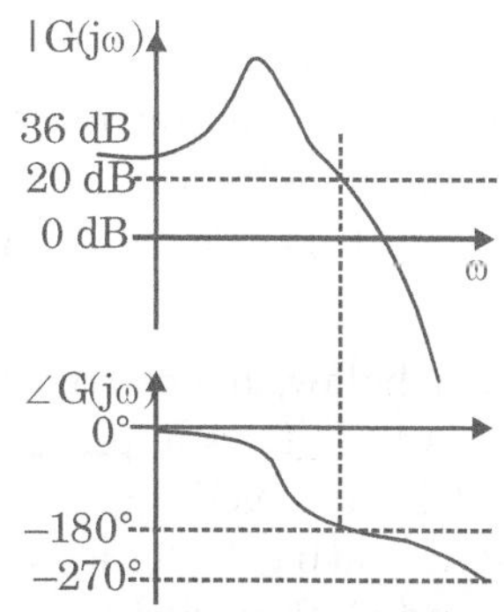

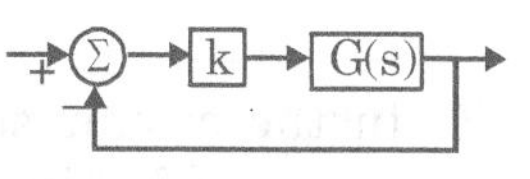

Consider the negative unity feedback configuration with gain k in the feedforward path. The closed loop is stable for $k < k_0$. The maximum value of k_0 is ________.

43. In the circuit shown below, the (WIL) value for M_2 is twice that for M_1. The two nMOS transistors are otherwise identical. The threshold voltage V_T for both transistors is 1.0 V. Note that V_{GS} for M_2 must be > 1.0 V.

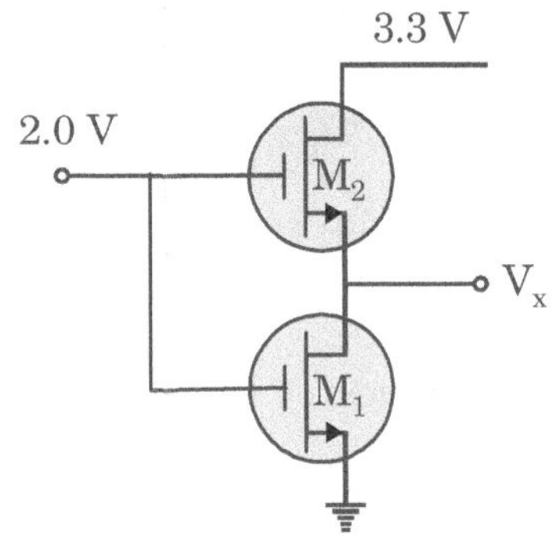

Current through the nMOS transistors can be modeled as

$$I_{DS} = \mu C_{ox}\left(\frac{W}{L}\right)\left((V_{GS} - V_T)V_{DS} - \frac{1}{2}V_{DS}^2\right) \quad \text{for}$$

$V_{DS} \leq V_{GS} - V_T$

$$I_{DS} = \mu C_{ox}\left(\frac{W}{L}\right)(V_{GS} - V_T)^2/2 \text{ for } V_{DS} \geq V_{GS} - V_T$$

The voltage (in volts, accurate to two decimal places) at V_x is ________.

44. A solar cell of area 1.0 cm^2, operating at 1.0 sun intensity, has a short circuit current of 20 MA, and an open circuit voltage of 0.65 V. Assuming room temperature operation and thermal equivalent voltage of 26 mV, the open circuit voltage (in volts, correct to two decimal places) at 0.2 sun intensity is ________.

45. A junction is made between p^- Si with doping density N_{A1} = 10^{15} cm^{-3} and p Si with doping density N_{A2} = 10^{17} cm^{-3}.

Given : Boltzmann constant k = 1.38 × 10^{-23} J . K^{-1}, electronic charge q = 1.6 . 10^{-19} C. Assume 100% acceptor ionization.

At room temperature (T = 300K), the magnitude of the built-in potential (in volts, correct to two decimal places) across this junction will be ________.

46. In the circuit shown below, a positive edge-triggered D flip-flop is used for sampling input data D_{IN} using clock CK. The XOR gate outputs 3.3 volts for logic HIGH and 0 volts for logic LOW levels. The data bit and clock periods are equal and the value of $\Delta T/T_{CK}$ = 0.15, where the parameters ΔT and T_{CK} are shown in the figure. Assume that the Flip-Flop and the XOR gate are ideal.

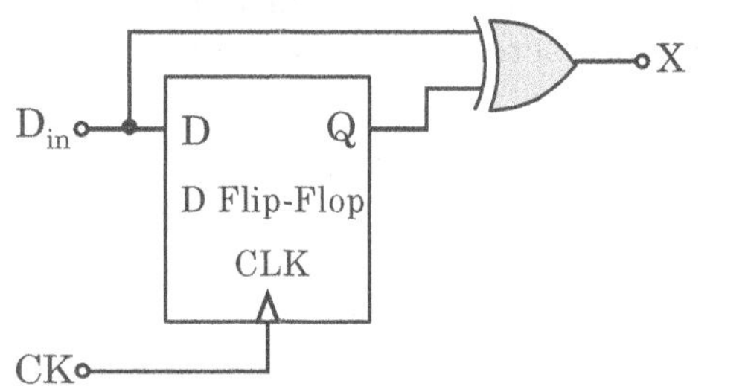

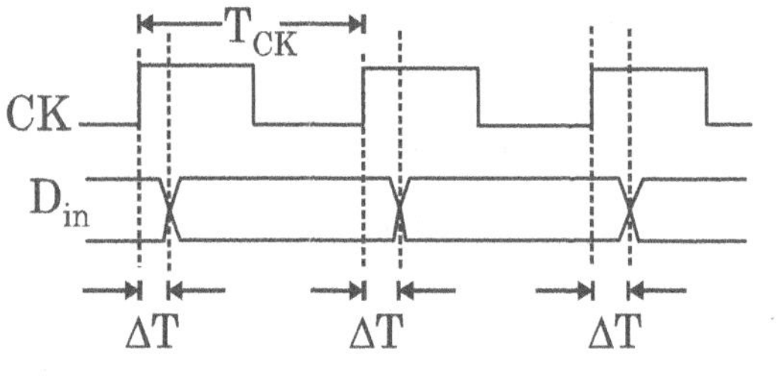

If the probability of input data bit (D_{in}) transition in each clock period is 0.3, the average value (in volts, accurate to two decimal places) of the voltage at node X, is ________.

47. The logic gates shown in the digital circuit below use strong pull-down nMOS transistors for LOW logic level at the outputs. When the pull-downs are off, high-value resistors set the output logic levels to HIGH (i.e. the pull-ups are weak). Note that some nodes are intentionally shorted to implement "wired logic". Such shorted nodes will be HIGH only if the outputs of all the gates whose outputs are shorted are HIGH.

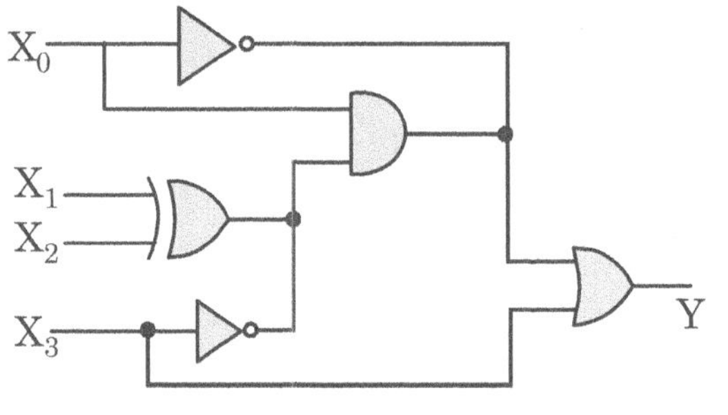

The number of distinct values of X_3 X_2 X_1 X_0 (out of the 16 possible values) that give Y = 1 is ________.

48. The cut-off frequency of TE_{01} mode of an air filled rectangular waveguide having inner dimensions a cm × b cm (a > b) is twice, that of the dominant TE_{10} mode. When the waveguide is operated at a frequency which is 25% higher than the cut-off frequency of the dominant mode, the guide wavelength is found to be 4 cm. The value of b (in cm, correct to two decimal places) is ________.

49. A uniform plane wave travelling in free space and having the electric field

$$\vec{E} = \left(\sqrt{2}\hat{a}_x - \hat{a}_z\right)\cos\left[6\sqrt{3}\pi \times 10^8 t - 2\pi\left(x + \sqrt{2}z\right)V/m\right]$$

is incident on a dielectric medium (relative permittivity > 1, relative permeability = 1) as shown in the figure and there is no reflected wave.

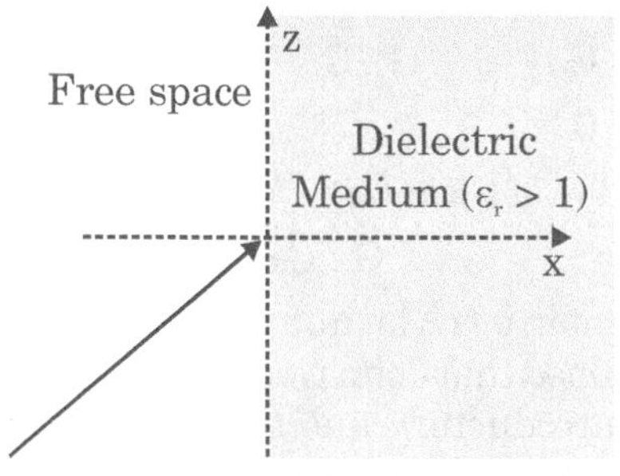

The relative permittivity (correct to two decimal places) of the dielectric medium is ________.

50. The position of a particle y(t) is described by the differential equation :

$$\frac{d^2y}{dt^2} = -\frac{dy}{dt} - \frac{5y}{4}$$

The initial conditions are $y(0) = 1$ and $\left.\frac{dy}{dt}\right|_{t=0} = 0$.

The position (accurate to two decimal places) of the particle at $t = \pi$ is ________.

51. The contour C given below is on the complex plane $z = x + jy$, where $j = \sqrt{-1}$.

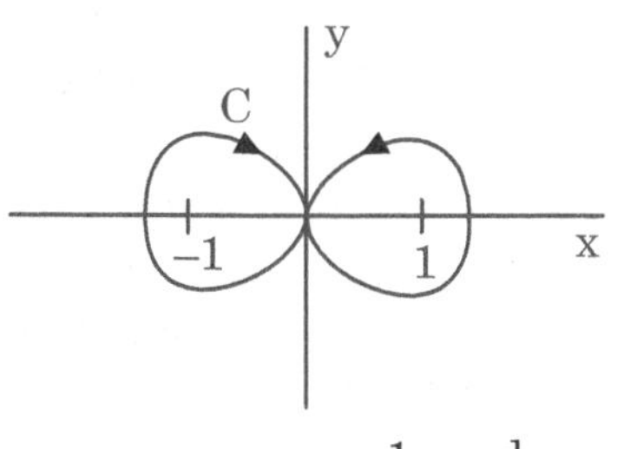

The value of the integral $\frac{1}{\pi j}\oint_c \frac{dz}{z^2 - 1}$ is ________.

52. Let $r = x^2 + y - z$ and $z^3 - zy + yz + y^3 = 1$. Assume that x and y are independent variables. At $(x, y, z) = (2, -1, 1)$, the value (correct to two decimal places) of $\frac{\partial r}{\partial x}$ is ________.

53. Consider the network shown below with $R_1 = 1\ \Omega$, $R_2 = 2\ \Omega$ and $R_3 = 3\ \Omega$. The network is connected to a constant voltage source of 11 V.

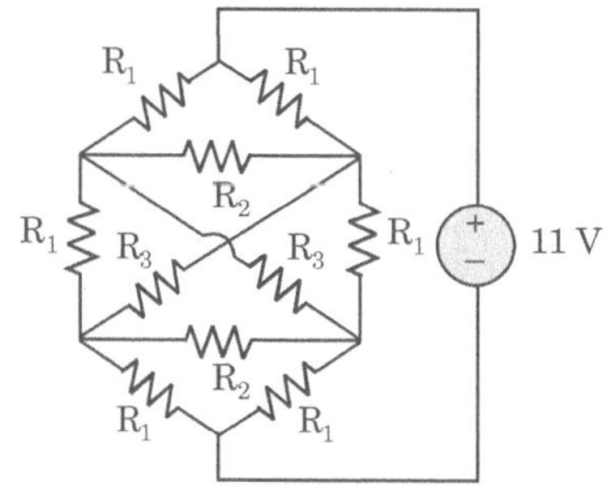

The magnitude of the current (in amperes, accurate to two decimal places) through the source is ______.

54. A band limited low-pass signal x(t) of bandwidth 5 kHz is sampled at a sampling rate f_s. The signal x(t) is reconstructed using the reconstruction filter $H(f)$ whose magnitude response is shown below:

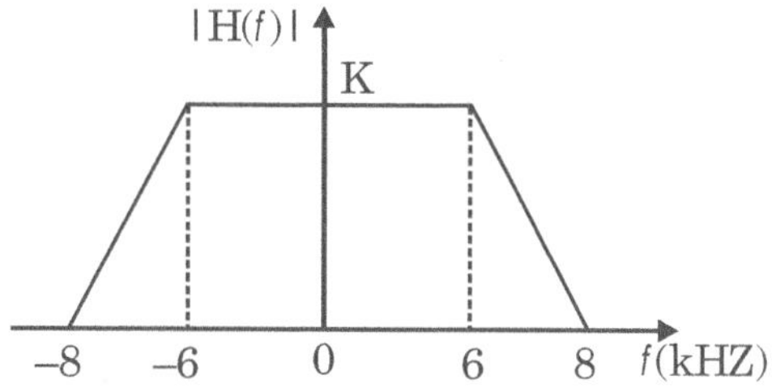

The minimum sampling rate f_s (in kHz) for perfect reconstruction of x(t) is ________.

55. Let $X[k] = k + 1,\ 0 \le k \le 7$ be 8-point DFT of a sequence x[n], where $X[k] = \sum_{n=0}^{N-1} x[n]e^{-j2\pi nk/N}$.

The value (correct to two decimal places) of $\sum_{n=0}^{3} x[2n]$ is ________.

ANSWERS

General Aptitude (GA)

1. (*b*)	2. (*a*)	3. (*b*)	4. (*d*)	5. (*b*)	6. (*b*)	7. (*b*)	8. (*c*)	9. (*a*)	10. (*b*)

Technical Section

1. (*c*)	2. (*c*)	3. (*d*)	4. (*c*)	5. (*b*)	6. (*b*)	7. (*d*)	8. (*d*)
9. (*c*)	10. (*c*)	11. (*c*)	12. (*d*)	13. (*b*)	14. (*c*)	15. 0.25	16. 16
17. 0.125	18. 0.416	19. 5	20. 0.48	21. 0.001	22. 2	23. 0.25	24. 0
25. 4.80	26. (*b*)	27. (*d*)	28. (*b*)	29. (*a*)	30. (*b*)	31. (*c*)	32. (*a*)
33. (*b*)	34. (*a*)	35. (*c*)	36. 0.316	37. 6.40	38. 0.5	39. 8	40. –0.50
41. 14.928	42. 0.10	43. 0.4226	44. 0.608	45. 0.1192	46. 0.8415	47. 8	48. 0.75
49. 2	50. –0.21	51. 2	52. 4.50	53. 8	54. 13	55. 3	

EXPLANATIONS

GENERAL APTITUDE

3. Rule of divisibility by 3 : Sum of all digits should be divisible by 3

715 ? 423

$7 + 1 + 5 + 4 + 2 + 3 = 22$

So, the next number after sum 22 which are divisible by 3 are 24, 27, 30 etc.

So, $22 + 2 = 24$

$22 + 5 = 27$

But according to given condition 2 is right answer.

4. The given series is in G.P,

where $a = 1$ and $r = \frac{1}{4}$

Now, in G.P, $s_n = \frac{a}{1-r}$

$$\frac{1}{1-\frac{1}{4}} = \frac{1}{3/4} = \frac{4}{3}$$

5.

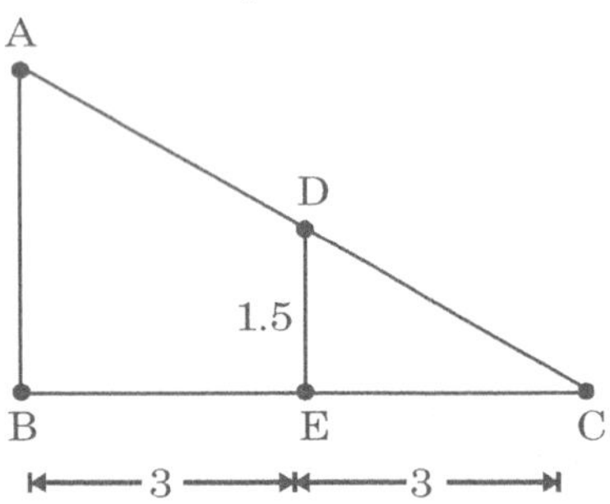

Let 'AB' be the height of lamp post

Here ΔABC and ΔDEC are similar triangles,

$\therefore \quad \frac{AB}{DE} = \frac{BC}{EC}$

$\therefore \quad \frac{AB}{1.5} = \frac{6}{3}$

$\therefore \quad \frac{AB}{1.5} = 2$

AB = 3 meter

Hence height of lamp post is 3 m.

6. P : A

$(10)^5 : (11)^5$

$$\therefore \quad P = \frac{(10)^6 \times (10)^5}{(11)^5}$$

$$= \frac{(10)^{11}}{(11)^5}$$

$= 620921.323 \approx 621000$

7.

A	B
G : C	G : C
2 : 3	3 : 7

Now making the mass equal of alloy A and B.

A	B
G : C	G : C
4 : 6	3 : 7

$\therefore$ Alloy of C contain

$G : C = 4 + 3 : 6 + 7 = 7 : 13$

8. Statement (i) is not true as nowhere it is mentioned that John was a captain in junior team. The introductory line emphasizes on the board recognizing John's potential (Latent quality/ possibility) as leader of the team.

Statement (iii) also manipulates the facts mentioned in the argument.

The 3rd statement of the argument while this aggression has filled stadia with die-hard fans does not indicate fans expectations from John as a caption.

Statement (ii) The concluding statement of the para suggests that finding a completer captain is a tough task as it look John many years to become a successful and calculative batsman - captain.

Statement (iv) can be explicity concluded from the last 4 lines of the para.

9. Probability that accident was caused by blue cab

$= 0.15 \times 0.8 = 0.12 = 12\%$

10. (A) is beyond the scope of given information option (c) can also be discarded on the same grounds.

The argument deals with the coastal region becoming crowded and polluted because of the upcoming luxury resort. Option (b) precisely underlines the theme of the para.

TECHNICAL SECTION

1.

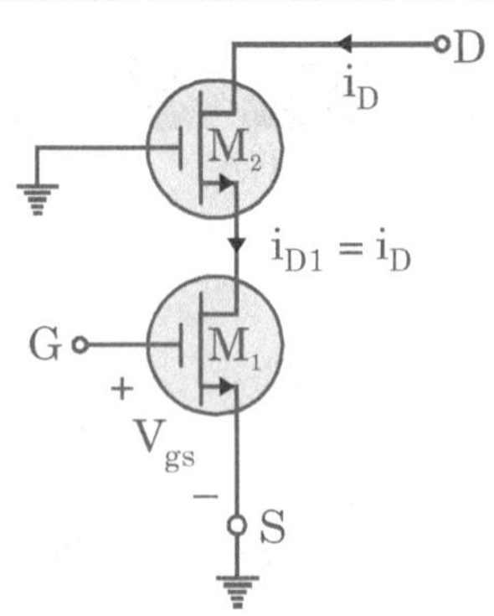

$$g_m = \frac{\Delta I_D}{\Delta V_{in}} = \frac{i_D}{v_{gs}} = \frac{i_{D1}}{v_{gs}} = g_{m1}$$

To calculate r_0 :

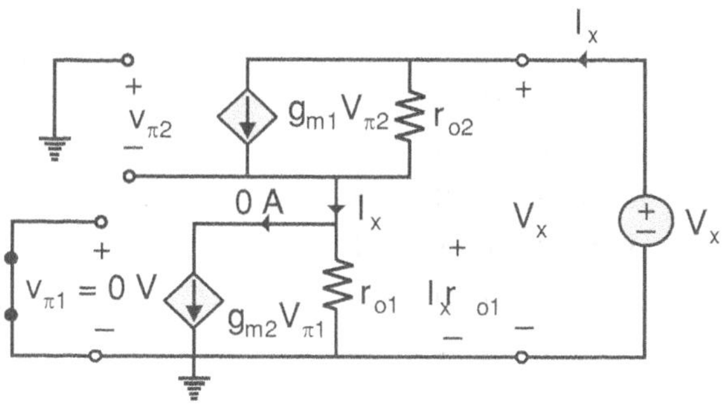

$V_{\pi2} = -I_x r_{o1}$

$$I_x = g_{m2}V\pi_2 + \frac{(V_x - I_x r_{o1})}{r_{o2}}$$

$$\therefore \quad I_x = -g_{m2}\, r_{o1} I_x + \frac{V_x}{r_{o2}} - I_x\frac{r_{o1}}{r_{o2}}$$

$$\text{Now, } V_x = r_{o2}\left[1 + r_{o1}gm_2 + \frac{r_{o1}}{r_{o2}}\right]I_x$$

$$r_o = \frac{V_x}{I_x} = r_{o1} + r_{o2} + r_{o1}\, r_{o2}\, g_{m2} = r_{o1}\, r_{o2}\, g_{m2}$$

2. For t > 0,

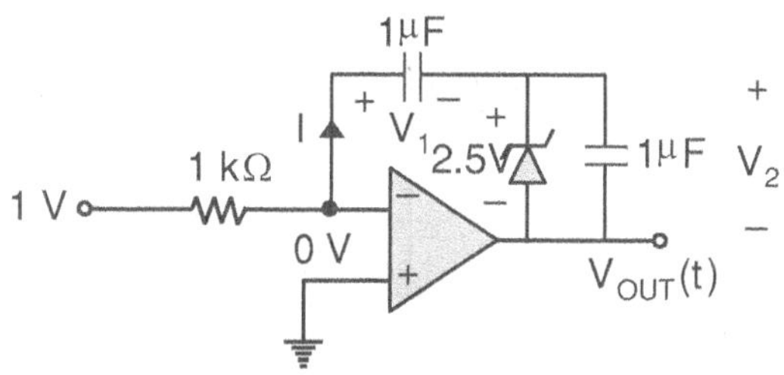

$$I = \frac{1V}{1k\Omega} = 1mA$$

Till t = 2.5 msec, both V_1 and V_2 will increase and after t = 2.5 msec, V_1 = 2.5 V and V_2 increases with time.

when $v_{out}(t) = -10$ V, $V_1 = 7.5$ V

$$\text{So,} \quad \frac{1}{1\mu F}\int_0^t (1mA)\,dt = 7.5V$$

$$10^3 t = 7.5$$

$$t = 7.5 \text{ msec}$$

3. A good transimpedance amplifier should have low input impedance and low output impedance.

5. For Bode magnitude plotting only magnitude information is required for plotting. While in Nyquist plotting we need both magnitude and phase information.

6. Here degree of p(s) is '3', so it will have three roots.

$p'(s) = 3s^2 + 2a_2s + a_1$ will have two roots and has no real root

$\therefore$ p(s) will have one real root.

8. From pull-down network,

$$\overline{f(X,Y)} = \overline{X}\overline{Y} + XY = X \odot Y$$

$$f(X,Y) = \overline{X \odot Y} = X \oplus Y$$

9.

C \ AB	$\overline{A}\overline{B}$ 00	$\overline{A}B$ 01	AB 11	$A\overline{B}$ 10
$\overline{C}$ 0	1	1		1
C 1				

$$F(A,B,C,D) = \overline{A}\overline{B}\overline{C} + \overline{A}B\overline{C} + A\overline{B}\overline{C}$$

$$= \Sigma M(0, 2, 4) = \Pi m(1, 3, 5, 6, 7)$$

$$= (A + B + \overline{C})(A + \overline{B} + \overline{C})(\overline{A} + B + \overline{C})$$

$$(\overline{A} + \overline{B} + C)(\overline{A} + \overline{B} + \overline{C})$$

10. For Short circuit, r = x = 0 ⇒ Point "P"
For Open circuit, r = x = ∞ ⇒ Point "R"
For Matched load, r = 1 and x = 0 ⇒ Point "Q"
P : Short Circuit, Q : Matched Load, R : Open Circuit

11. Eigen vectors corresponding to distinct eigen values are linearly independent.

12. Here, $f(x,y) = \dfrac{ax^2 + by^2}{xy} = a\left(\dfrac{x}{y}\right) + b\left(\dfrac{y}{x}\right)$

Now $\dfrac{\partial f}{\partial x} = \dfrac{a}{y} - \dfrac{by}{x^2}$

Now $\dfrac{\partial f}{\partial x}$ at x = 1 and y = 2

$$\left.\frac{\partial f}{\partial x}\right|_{(1,2)} = \left[\frac{a}{y} - \frac{by}{x^2}\right]_{(1,2)} \left[\frac{a}{2} - \frac{b(2)}{1}\right] = \frac{a}{2} - 2b$$

Now $\dfrac{\partial f}{\partial y} = \left(-\dfrac{ax}{y^2} + \dfrac{b}{x}\right)$

Now $\dfrac{\partial f}{\partial y}$ at x = 1 and y = 2

$$\therefore \quad \left.\frac{\partial f}{\partial y}\right|_{(1,2)} = \left[-\frac{ax}{y^2} + \frac{b}{x}\right]_{(1,2)} = -\frac{a}{4} + b$$

Now according to question,

$$\frac{\partial f}{\partial x} = \frac{\partial f}{\partial y}$$

$$\therefore \quad \frac{a}{2} - 2b = -\frac{a}{4} + b$$

$$\therefore \quad \frac{3a}{4} = 3b$$

$$\therefore \quad a = 4b$$

13. 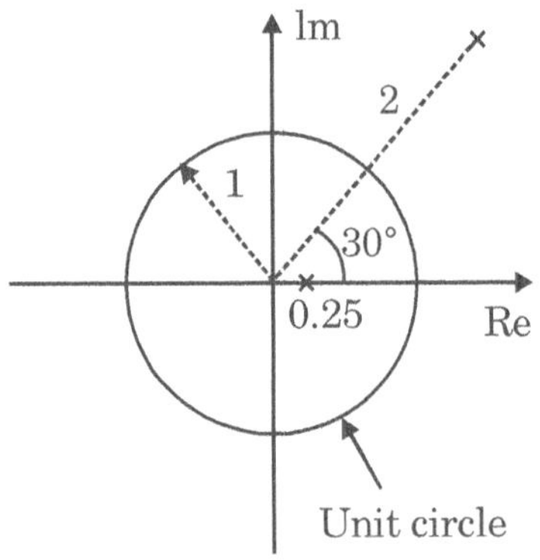

The ROC should encircle unit circle to make the system stable. From the given pole pattern it is clear that, to make the system stable, the ROC should be two-sided. Thus impulse response for the system should be two-sided.

14. Change in only time period or frequency does not change in the value of Fourier series coefficients.

So, $b_k = a_k$

$$\sum_{k=-\infty}^{\infty} |b_k| = \sum_{k=-\infty}^{\infty} |a_k| = 16$$

15. $$s(t) = \cos(2000\pi t) + 4\cos(2400\pi t) + \cos(2800\pi t) \quad ...(i)$$

Standard form of the AM signal is given below,

$$s(t) = \frac{\mu A_c}{2}\cos\left[2\pi(f_c - f_m)t\right] + A_c\cos(2\pi f_c t) + \frac{\mu A_c}{2}\cos\left[2\pi(f_c + f_m)t\right]$$

Now comparing equation (*i*) with standard equation we get

$$A_c = 4$$

and $$\frac{\mu A_c}{2} = 1$$

$$\therefore \quad \mu = \frac{2}{A_c}$$

Now $$\frac{A_m}{A_c} = \frac{2}{A_c} \Rightarrow A_m = 2$$

Now $$\frac{P_m}{P_c} = \frac{\frac{1}{2}A_m^2}{\frac{1}{2}A_c^2} = \frac{A_m^2}{A_c^2} = \frac{(2)^2}{(4)^2} = \frac{1}{4} = 0.25$$

16. For, n = 5 and $d_{min} = 2$ (given)

For $d_{min} = 2$, the codewords can be formed as follows:

0	0	0	0	1
0	0	0	1	0
0	0	1	0	0
0	0	1	1	1
0	1	0	0	0
0	1	0	1	1
0	1	1	0	1
0	1	1	1	0
1	0	0	0	0
1	0	0	1	1
1	0	1	0	1
1	0	1	1	0
1	1	0	0	1
1	1	0	1	0
1	1	1	0	0
1	1	1	1	1

$\Rightarrow$ Total 16 codewords are possible

17. Let s_0 and s_1 be the transmitted symbols representing the transmitted value {–1, 1} respectively and let r_0 and r_1 be the received symbols.

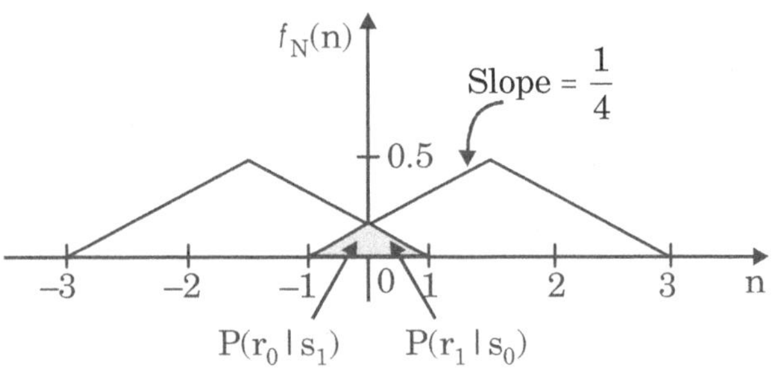

Now Probability of error, (P_e)

$$= P(s_1)P(r_0|s_1) + P(s_0)P(r_1|s_0)$$

$$P(r_0|s_1) = P(r_1|s_0)$$

$$= \frac{1}{2} \times 1 \times \frac{1}{4} = \frac{1}{8}$$

$$P(s_0) = P(s_1) = \frac{1}{2} \text{ (given)}$$

$$\therefore \quad P_e = \left(\frac{1}{2} \times \frac{1}{8} + \frac{1}{2} \times \frac{1}{8}\right)$$

$$= \frac{1}{2}\left(\frac{1}{8} + \frac{1}{8}\right)$$

$$= \frac{1}{8} = 0.125$$

18. $$W_{dep} = \sqrt{\frac{2\varepsilon}{q}\left(\frac{1}{N_A}+\frac{1}{N_D}\right)(V_{bi}-V_{AK})}$$

$$\frac{\sqrt{(0.65-V_{AK})}}{\sqrt{0.65}} = \frac{0.6\mu m}{1\mu m}$$

$$\therefore \quad \sqrt{\frac{0.65-V_{AK}}{0.65}} = 0.6$$

$$\therefore \quad \sqrt{\frac{1-V_{AK}}{0.65}} = 0.6$$

Now squaring on both side

$$\therefore \quad 1-\frac{V_{AK}}{0.65} = 0.36$$

$$\therefore \quad V_{AK} = 0.65\,(1-0.36) = 0.65\times 0.64 = 0.416\text{ V}$$

19. GREEN → 70 seconds

YELLOW → 5 seconds

RED → 75 seconds

Clock period → 5 seconds

Total number of unique states

$$= \frac{70+5+75}{5} = \frac{150}{5} = 30$$

∴ No of flipflop required (n) = $\log_2^{30} \approx 5$

20. $\therefore \quad L = \dfrac{k\lambda}{NA}$

$$L_{min} \propto \lambda$$

$$\frac{L_{min1}}{L_{min2}} = \frac{\lambda_1}{\lambda_2} = \frac{156\text{ nm}}{325\text{ nm}} = 0.48$$

21. For a distortionless transmission line,

$$\frac{L}{R} = \frac{C}{G}$$

Propagation constant,

$$\gamma = \alpha + j\beta = \sqrt{(R+j\omega L)(G+j\omega C)}$$

$$= \sqrt{RG}\left(1+j\omega\frac{L}{R}\right)$$

Attenuation constant, $\alpha = \sqrt{RG}$

Characteristic impedance,

$$Z_0 = \sqrt{\frac{(R+j\omega L)}{(G+j\omega C)}} = \sqrt{\frac{R}{G}}$$

$$\sqrt{G} = \frac{\sqrt{R}}{Z_0}$$

So,

$$\alpha = \sqrt{R}\cdot\frac{\sqrt{R}}{Z_0} = \frac{R}{Z_0} = \frac{0.05}{50} = \frac{0.01}{10} = 0.001\text{Np/m}$$

22. AX = 0 has infinitely many solutions (given)

$$\Rightarrow \quad |A| = 0$$

$$\therefore \quad \begin{vmatrix} k & 2k \\ k^2-k & k^2 \end{vmatrix} = 0$$

$$\therefore \quad k(k^2) - 2k(k^2-k) = 0$$

$$\therefore \quad k^3 - 2k^3 + 2k^2 = 0$$

$$\therefore \quad k^2(2-k) = 0$$

∴ k = 0, 2 ⇒ "two" distinct values of k

23. $P(X_4 \text{ is smallest}) = \dfrac{3!}{4!} = \dfrac{3\times 2}{4\times 3\times 2} = \dfrac{1}{4} = 0.25$

24. $$f(x) = \int_0^x e^{-t^2/2}dt$$

Differentiating $f(x)$ w.r.t. 'x' we get

$$f'(x) = e^{-x^2/2} \text{ and } f''(x) = e^{-x^2/2}(-x)$$

Now $f''(0) = 0$

$$a_2 = \frac{f''(0)}{2!} = 0$$

25. $$V_1 = AV_2 - BI_2 \quad ...(i)$$

and $$I_1 = CV_2 - DI_2 \quad ...(ii)$$

Now inorder to find B, put $V_2 = 0$ (i.e. port 2 is short circuited) in equation (i) we get

$$V_1 = -BI_2$$

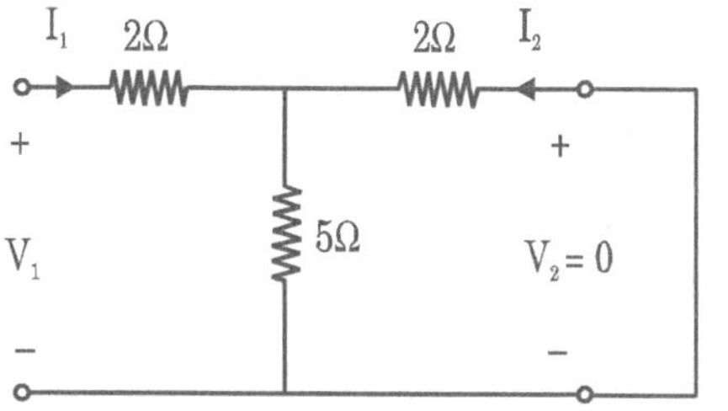

$$\therefore \quad I_1 = \frac{V_1}{2\Omega + (5\Omega \,||\, 2\Omega)}$$

$$= \frac{V_1}{\left(2+\frac{10}{7}\right)}$$

$$= \frac{7V_1}{24\Omega} \quad \left[\because 5\Omega \,||\, 2\Omega = \frac{5\times 2}{5+2} = \frac{10}{7}\Omega\right]$$

Now, $$I_2 = -I_1 \times \frac{5\Omega}{5\Omega+2\Omega} = \frac{-5V_1}{24\Omega}$$

$$\therefore \quad B = -\frac{V_1}{I_2}$$

$$= \frac{24}{5}\Omega = 4.80\Omega$$

26. $V_I = 6\text{ V} \pm 5\% = 6\text{ V} \pm 0.3\text{ V} = 5.7\text{ V to } 6.3\text{ V}$

$$I_L = \frac{5V}{1k\Omega} = 5mA$$

$$I_{s(min)} = I_L + I_{z(min)}$$

$$= 5\text{ mA} + 2\text{ mA} = 7\text{ mA}$$

Now $\quad I_s = \frac{V_I - V_z}{R}$

$$\therefore \quad I_{s(min)} = \frac{V_{I(min)} - V_z}{R_{max}} = 7mA$$

$$\therefore \quad R_{max} = \frac{V_{I(min)} - V_z}{I_s(min)}$$

$$= \frac{5.7-5}{7}k\Omega = \frac{700}{7}\Omega = 100\Omega$$

Now, when, R = 100 Ω,

$$I_{s(max)} = \frac{6.3-5}{100}A = 13mA$$

$$I_{z(max)} = I_{s(max)} - I_L = 13\text{ mA} - 5\text{ mA} = 8\text{ mA}$$

$$P_{z(min)} = V_z\, I_{z(max)} = (5 \times 8)\text{mW} = 40\text{ mW}$$

27. $v_i(t) = A_c\cos(2\pi f_c t) + \cos(2\pi f_m t)$

$$v_o(t) = av_i(t) + bv_i^2(t)$$

$$\left[aA_c\cos(2\pi f_c t) + a\cos(2\pi f_m t)\right]$$

$$+ b[A_c^2\cos^2(2\pi f_c t) + \cos^2(2\pi f_m t)$$

$$+ 2A_c\cos(2\pi f_c t)\cos(2\pi f_m t)]$$

Now passing above signal through band pass filter we get,

$$y(t) = aA_c\cos 2\pi f_c t + 2bA_c\cos(2\pi f_c t)\cos(2\pi f_m t)$$

$$= aA_c\left[1 + \frac{2b}{a}\cos(2\pi f_m t)\right]\cos(2\pi f_c t)$$

Now, modulation index, $(\mu) = \frac{2b}{a}$

Power of side band $(P_{SB}) = \frac{\mu^2}{2}P_c = \frac{1}{2}P_c$

$$\therefore \quad \mu^2 = 1 \Rightarrow \mu = 1$$

$$\therefore \quad \frac{2b}{a} = 1$$

$$\therefore \quad \frac{a}{b} = 2$$

29. Transfer function,

$$T(s) = \frac{Y(s)}{U(s)} = C[sI - A]^{-1}B$$

$$A = \begin{bmatrix} -4 & -1.5 \\ 4 & 0 \end{bmatrix}$$

$$B = \begin{bmatrix} 2 \\ 0 \end{bmatrix}$$

$$C = [1.5 \quad 0.625]$$

$$[sI - A] = \begin{bmatrix} s+4 & 1.5 \\ -4 & s \end{bmatrix}$$

$$[sI - A]^{-1} = \frac{1}{(s^2+4s+6)}\begin{bmatrix} s & -1.5 \\ 4 & s+4 \end{bmatrix}$$

$$[sI - A]^{-1}B = \frac{1}{s^2+4s+6}\begin{bmatrix} 2s \\ 8 \end{bmatrix}$$

$$C[sI - A]^{-1}B = \frac{1}{s^2+4s+6}[1.5 \quad 0.625]\begin{bmatrix} 2s \\ 8 \end{bmatrix}$$

$$T(s) = \frac{3s+5}{s^2+4s+6}$$

30. $\lambda_R > \lambda_G > \lambda_B$

Energy gap, $\quad E_g \propto \frac{1}{\lambda}$

So, $\quad E_{gR} < E_{gG} < E_{gB}$

Materials with high energy gap will have high built-in voltages, when doping concentrations are same.

So, $\quad V_R < V_G < V_B$

31.

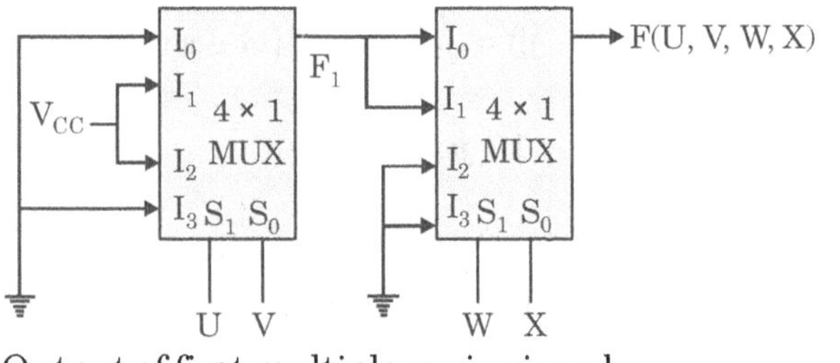

Output of first multiplexer is given by

$$F_1 = \bar{U}V + U\bar{V}$$

and Output of the second multiplexer is given by

$$F = \bar{W}\bar{X}F_1 + \bar{W}XF_1 = \bar{W}F_1 = (\bar{U}V + U\bar{V})\bar{W}$$

$$(\because F_1 = \bar{U}V + U\bar{V})$$

32.

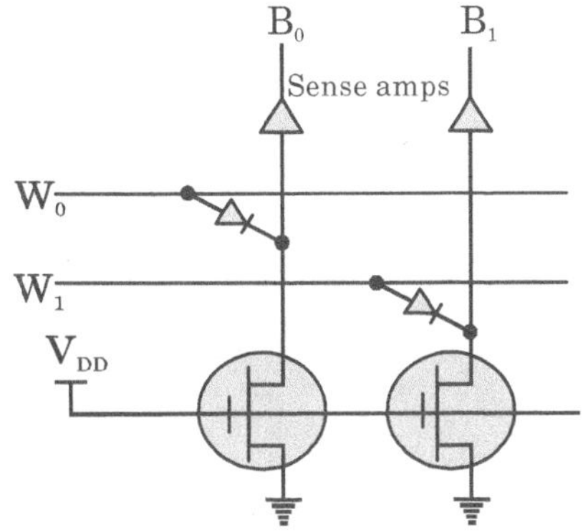

When, $W_0 = V_{DD}$, $B_0 = V_{DD}$; otherwise $B_0 = 0$ ...(*i*)

When $W_1 = V_{DD}$, $B_1 = V_{DD}$; otherwise $B_1 = 0$ ...(*ii*)

From (*i*) and (*ii*) we get

$\Rightarrow$ $B_0 = W_0$ and $B_1 = W_1$

Hence, $\begin{matrix} & B_0 & B_1 \\ W_0 & 1 & 0 \\ W_1 & 0 & 1 \end{matrix}$

33. Attenuation constant, (α)

$$= \frac{1}{\text{skin depth}} = 10\text{Np/m}$$

$$\therefore \quad 20\log_{10}\left(\frac{E_0}{E_x}\right) = 20 \text{ dB}$$

$$\therefore \quad \frac{E_0}{E_x} = 10^1$$

$$\therefore \quad \frac{E_0}{E_x} = 10$$

$$\therefore \quad E_x = \frac{E_0}{10}$$

$$\text{Now } E_x = E_0 e^{-\alpha x} = E_0 e^{-10x} = \frac{E_0}{10}$$

$$\therefore \quad e^{-10x} = \frac{1}{10}$$

$$x = \frac{1}{10}\text{In}(10) = 0.23\text{m}$$

34. $$\frac{dy}{dx} = \frac{x^2}{2y} + \frac{y}{2} + \frac{y}{x}$$

Let $y = xt$

$$\therefore \quad \frac{dy}{dx} = t + x\frac{dt}{dx}$$

$$\therefore \quad t + x\frac{dt}{dx} = \frac{x}{2t} + \frac{tx}{2} + t$$

$$\therefore \quad x\frac{dt}{dx} = x\left(\frac{1}{2t} + \frac{t}{2}\right)$$

$$\therefore \quad x\frac{dt}{dx} = x\left(\frac{1+t^2}{2t}\right)$$

$$\text{Now } \int \frac{2t}{1+t^2}dt = \int dx + C$$

$$In(1 + t^2) = x + C$$

$$\therefore \quad t = \frac{y}{x}$$

$$\text{So, In}\left(1 + \frac{y^2}{x^2}\right) = x + C$$

At $x = 1$, $y = 0$

$$\text{In}\left(1 + \frac{0}{1}\right) = \text{In}(1) = 0 = 1 + C$$

$$C = -1$$

$$\text{So, In}\left(1 + \frac{y^2}{x^2}\right) = x - 1$$

35. $$\frac{1}{\omega C} = \frac{1}{5 \times 10^{-6}} = 200 \text{ k}\Omega$$

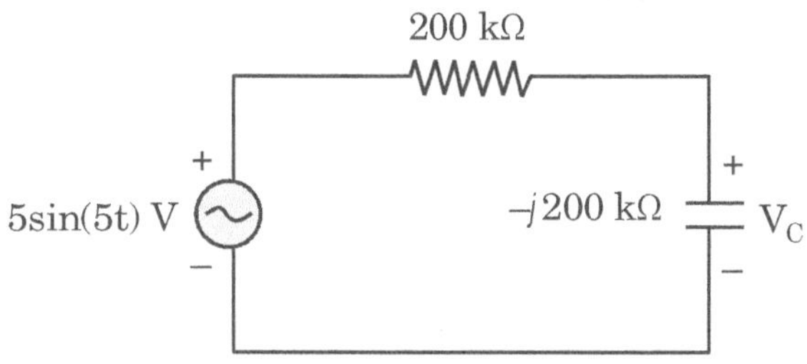

$$V_C = \frac{5\angle 0°}{200 - j200} \times (-j200)\text{V} = \frac{5\angle 0° \times 1\angle -90°}{\sqrt{2}\angle -45°}\text{V}$$

$$= \frac{5}{\sqrt{2}}\angle -45°\text{V} = 2.5\sqrt{2}\sin\left(5t - \frac{\pi}{4}\right)\text{V}$$

$$= 2.5\sqrt{2}\sin(5t - 0.25\pi)\text{V}$$

36. Loop current,

$$i(t) = \frac{1}{1+1}\left(1 - e^{-1/\tau}\right)\text{A} \;;\; t > 0$$

$$\text{Now } \tau = \frac{L}{R_{eq}} = \frac{1}{1+1} = \frac{1}{2}\text{sec}$$

$$\therefore \quad i(t) = \frac{1}{2}\left(1 - e^{-2t}\right)\text{A} \;;\; t > 0 \qquad \left(\because \tau = \frac{1}{2}\right)$$

Now, at t = 0.5 sec, $i(t) = \frac{1}{2}\left(1 - e^{-1}\right)\text{A} = 0.316\text{A}$

37. The small-signal equivalent model of the given circuit is given below.

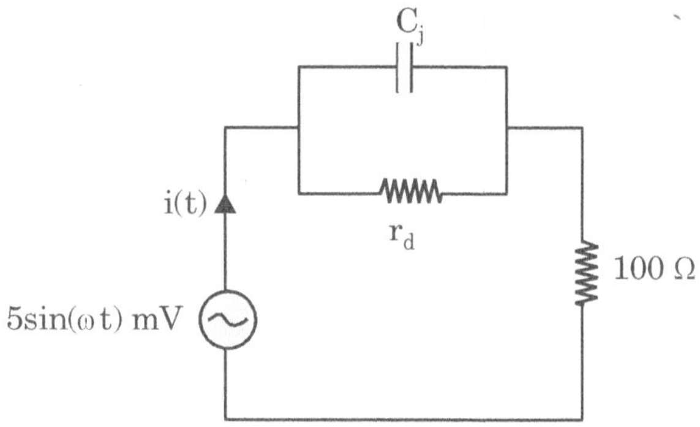

Given data, $\omega = 2 \times 10^6$ rad/sec
$C_i = 0.5$ nF
$I_{DC} = 26\ \mu A$
$V_T = 26$ mV
$\eta = 1$

Now,

$$r_d = \frac{\eta V_T}{I_{DC}} = \frac{26mV}{26\mu A} = 1k\Omega$$

$$\frac{1}{\omega Cj} = \frac{1}{2 \times 10^6 \times 0.5 \times 10^{-9}}\Omega = 1k\Omega$$

∴ Total impedance of the circuit

$$Z = \left(r_d \parallel \frac{1}{j\omega Cj}\right) + 100\Omega$$

$$\left(r_d \parallel \frac{1}{j\omega Cj}\right) = \frac{(1000)(-j1000)}{1000 - j1000}\Omega = \frac{-j(1+j)}{2}k\Omega$$

$$= \frac{1}{2}(1-j)k\Omega = (500 - j500)\Omega$$

$$\therefore\ Z = [(500 + j500) + 100]\Omega$$

$$\therefore\ Z = 600 - j\,500\ \Omega$$

$$\sqrt{(600)^2 + (500)^2}$$

Now magnitude

$$|Z| = 100\sqrt{36+25} = 100\sqrt{61}\ \Omega$$

$$\text{Now, } I_m = \frac{V_m}{|Z|} = \frac{5mV}{100\sqrt{61}\Omega} = \frac{50}{\sqrt{61}}\mu A = 6.40\mu A$$

38. Applying the concept of virtual ground, we get,

$$V_o = \frac{R_2}{R_1}\cdot V_{in} \qquad [\because \text{non-inverting amplifier}]$$

$$\therefore\ V_0 = \frac{31k\Omega}{1k\Omega} \times 1V$$

$$V_0 = -31\ V < -15\ V \qquad [\text{Not possible}]$$

Hence, the output voltage of the op-amp is equal to –15 V.

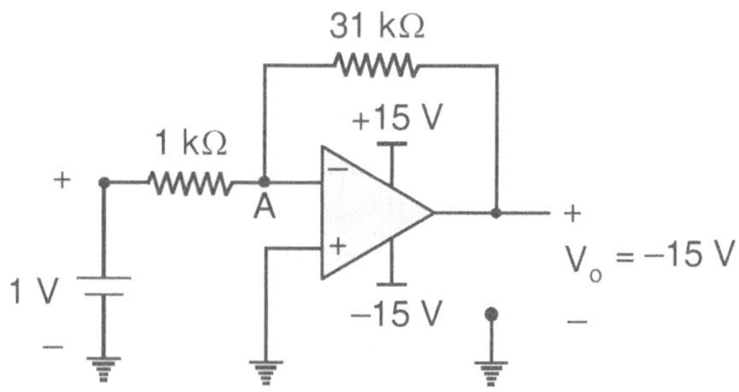

Now applying KCL at node 'A', we get,

$$\frac{V_A - 1}{1k\Omega} + \frac{V_A - (-15)}{31k\Omega} = 0$$

$$\frac{V_A}{1k\Omega} + \frac{V_A}{31k\Omega} = \frac{-15}{31k\Omega} + \frac{1}{1k\Omega}$$

$$\therefore\ V_A\left[\frac{1}{1} + \frac{1}{31}\right] = -\frac{15}{31} + 1$$

$$V_A = 0.5\ V$$

39. Hilbert transorm does not alter the amplitude spectrum of the signal.

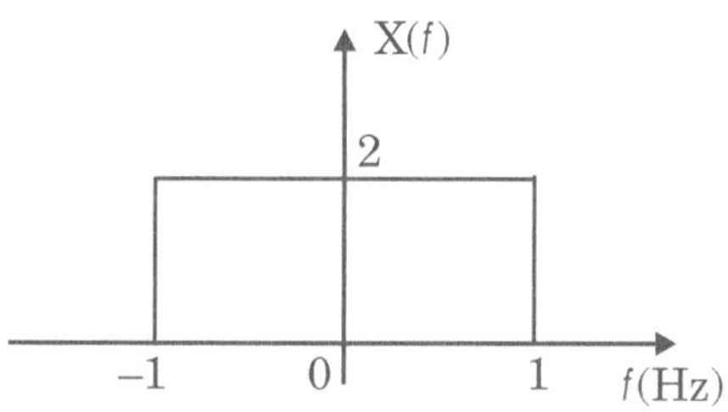

$$\text{So, } \int_{-\infty}^{\infty}|y(t)|^2 dt = \int_{-\infty}^{\infty}|x(t)|^2 dt = \int_{-\infty}^{\infty}|X(f)|^2 df$$

$$x(t) = 4\text{sinc}(2t)$$

$$\text{sinc}(t) \xleftrightarrow{\text{CTFT}} \text{rect}(f)$$

$$4\,\text{sinc}(2t) \xleftrightarrow{\text{CTFT}} \frac{4}{2}\text{rect}\left(\frac{f}{2}\right) = 2\ \text{rect}\left(\frac{f}{2}\right)$$

$$\int_{-\infty}^{\infty}|X(f)|^2\, df = 2 \times (2)^2 = 8$$

$$\text{So, } \int_{-\infty}^{\infty}|y(t)|^2\, dt = 8$$

40.

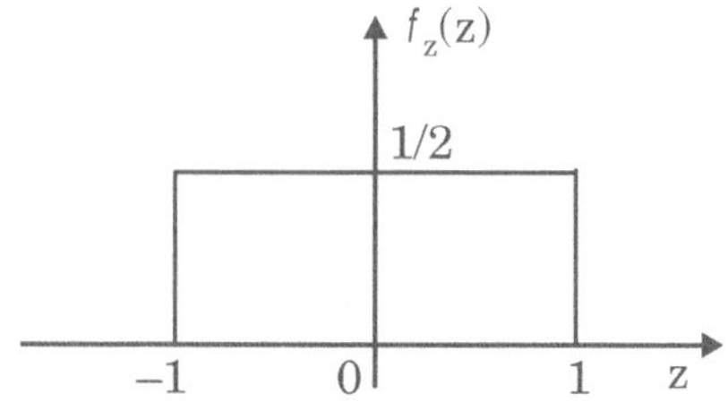

$$P(x_0) = \frac{1}{4}$$

$$P(x_1) = \frac{3}{4}$$

MAP criteria,

$$f_\gamma(y \mid x_0)P(x_0) \underset{x_1}{\overset{x_0}{\gtrless}} f_\gamma(y \mid x_1)P(x_1)$$

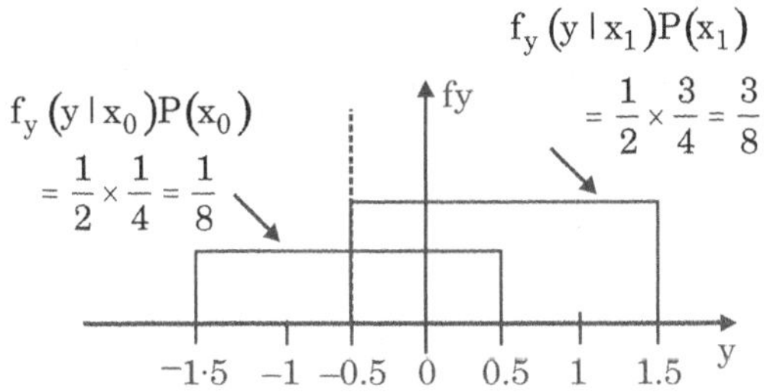

Optimum threshold exists here

So, $\alpha = -0.50$

41. Maximum resonant peak,

$$M_r = \frac{1}{2\xi\sqrt{1-\xi^2}} = 2$$

$$2\xi\sqrt{1-\xi^2} = \frac{1}{2}$$

$$\xi^2(1-\xi^2) = \frac{1}{16}$$

$$\xi^4 - \xi^2 + \frac{1}{16} = 0$$

$$\therefore \quad \xi^2 = 1 \pm \sqrt{1 - 4(1)\left(\frac{1}{16}\right)}$$

$$= \frac{1}{2} \pm \frac{\frac{2}{\sqrt{\frac{3}{4}}}}{2} = \frac{1}{2} \pm \frac{\sqrt{3}}{4}$$

As $M_r = 2 > 1$, $\xi < \frac{1}{\sqrt{2}}$ and $\xi^2 < \frac{1}{2}$

$$\therefore \quad \xi^2 = \frac{1}{2} - \frac{\sqrt{3}}{4}$$

Given, $G(s) = \frac{K}{s(s+2)} = \frac{\omega_n^2}{s(s+2\xi\omega_n)}$

So, $\omega_n = \sqrt{K}$

$$2\xi\sqrt{K} = 2$$

$$\sqrt{K} = \frac{1}{\xi}$$

$$K = \frac{1}{\xi^2} = \frac{1}{\left(\frac{1}{2} - \frac{\sqrt{3}}{4}\right)} = \frac{4}{2-\sqrt{3}} = 14.928$$

42. For G(s), $M_{dB}(\omega_{pc}) = 20\text{ dB}$

When cascaded with k,

$$GM_{dB} = -20\text{ dB} - 20\log_{10}(k) > 0\text{ dB}$$

$$\therefore \quad 20 + 20\log_{10}(k) < 0$$

$$\therefore \quad 20\log_{10}(k) < -20$$

$$\therefore \quad \log_{10}k < -1$$

$$k < 10^{-1} = 0.10$$

$$\therefore \quad k_0 = 0.10$$

43. Let, $K_n = \frac{\mu_n C_{ox}}{2}\left(\frac{W}{L}\right)$

Given than $\left(\frac{W}{L}\right)_2 = 2\left(\frac{W}{L}\right)_1$

So, $K_{n_2} = 2K_{n_1}$

For M_1, $V_{GS1} - V_T = (2-1)V = 1\text{ V}$

For M_2, $V_{GS2} - V_T = 2 - V_x - 1 = 1\text{ V} - V_x < 1\text{ V}$

$$V_{DS2} = (3.3 - V_x) > (V_{GS2} - V_T)$$

So, M_1 will be in linear region and M_2 will be in saturation region.

$$I_{D1} = I_{D2}$$

$$\therefore \quad K_{n1}\left[2(V_{GS1} - V_T)V_{DS1} - V_{DS1}^2\right] = K_{n2}(V_{GS2} - V_T)^2$$

$$K_{n1}\left[2(2-1)V_x - V_x^2\right] = 2K_{n1}(2 - V_x - 1)^2$$

$$(\because K_{n_2} = 2K_{n_1})$$

$$\therefore \quad 2V_x - V_x^2 = 2(1 + V_x^2 - 2V_x) = 2V_x^2 - 4V_x + 2$$

$$\therefore \quad 3V_x^2 - 6V_x + 2 = 0$$

$$\therefore \quad V_x = 1 \pm \sqrt{\frac{4 - \frac{8}{3}}{4}} = 1 \pm \sqrt{\frac{1}{3}}V$$

$$V_{GS2} = (2 - V_x) \geq V_T \Rightarrow (1 - V_x) \geq 0$$

$$\therefore \quad V_x = 1 - \sqrt{\frac{1}{3}} = 0.4226V$$

44. For solar cell, $V_{OC} = V_T \ln\left(\frac{I_{SC}}{I_o}\right)$

$$V_{OC2} - V_{OC1} = V_T \ln\left(\frac{I_{SC2}}{I_{SC1}}\right) = V_T \ln\left(\frac{0.20}{1.0}\right)$$

$$V_{OC2} = V_{OC1} - 0.026\ln(5)$$

$$= 0.65 - 0.041845 = 0.608\text{ V}$$

46.

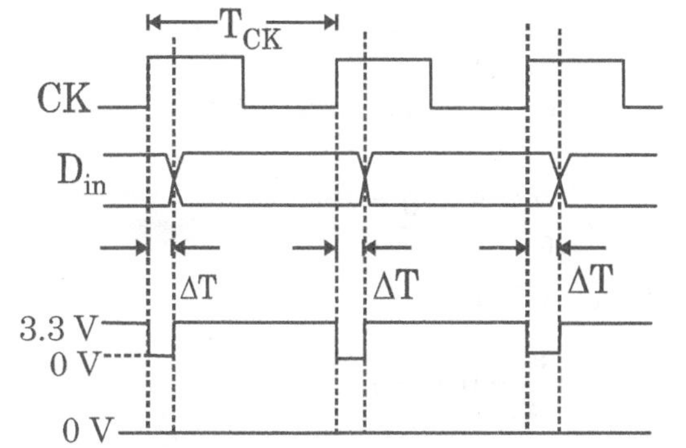

When input is changed during clock period (with probability 0.3)

When input is not changed during clock period (with probability 0.7)

$$V_{X(avg)} = \left[0.3 \times 3.3\left(1 - \frac{\Delta T}{T_{CK}}\right)\right] + [0.7 \times 0]V$$

$$\left(\because \frac{\Delta T}{T_{CK}} = 0.15V\right)$$

$$= 0.3 \times 3.3 \times (1 - 0.15)\ V$$

$$= (0.3 \times 3.3 \times 0.85)\ V = 0.8415\ V$$

47.

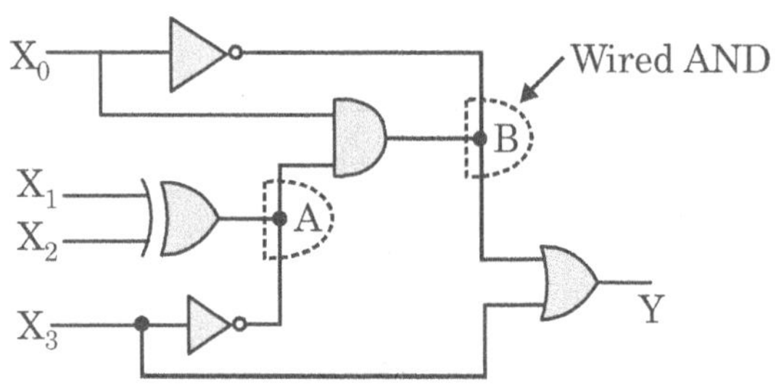

$$A = (X_1 \oplus X_1)\ \bar{X}_3$$

$$B = \left[(X_1 \oplus X_2)\bar{X}_3\ X_0\right] \cdot \bar{X}_0 = 0$$

$$Y = B + X_3 = 0 + X_3 = X_3$$

Out of 16 possible combinations of $X_3\ X_2\ X_1\ X_0$ Output will be high for 8 combinations.

$\therefore$ Y will be high for 8 combinations.

48. $$f_{c(01)} = 2 f_{c(10)} = \frac{2c}{2a} = \frac{c}{a}$$

Now $\frac{c}{2b} = \frac{c}{a} \Rightarrow a = 2b$

$\therefore\ b = \frac{a}{2}$

Operating frequency, $f = 1.25 f_{c(10)}$

$$f_{c(10)} < 1.25 f_{c(10)} < [f_{c(01)} = 2 f_{c(10)}]$$

So, for the given frequency, the waveguide will work in TE_{10} mode.

$$\text{So, } \lambda_g = \frac{\lambda_0}{\sqrt{1 - \left(\frac{f_{c(10)}}{f}\right)^2}} = \frac{c/f}{\sqrt{1 - \left(\frac{1}{1.25}\right)^2}} = \frac{c/f}{0.6}$$

$$\frac{C}{(1.25) f_{c(10)} (0.6)} = \lambda_g = 4\text{cm}$$

$$\frac{C}{f_{c(10)}} = 3 \times 10^{-2} = 2a$$

$\therefore\ a = 1.5$ cm

$\Rightarrow b = \frac{a}{2} = 0.75$ cm

49.

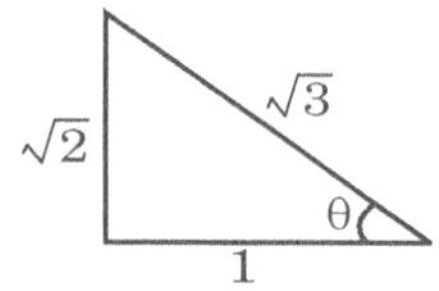

$$\vec{K}_i = 2\pi\left(\hat{x} + \sqrt{2}\hat{z}\right) = 2\pi\sqrt{3}\left(\frac{1}{\sqrt{3}}\hat{x} + \sqrt{\frac{2}{3}}\hat{z}\right)$$

$$\cos\theta_{ix} = \frac{1}{\sqrt{3}}$$

$$\Rightarrow \quad \tan\theta_{ix} = \sqrt{2}$$

Now as there is no reflected wave,

$\therefore\ \theta_{ix} = \theta_B$ = Brewester angle

As the wave is parallelly polarized,

$$\therefore\ \tan\theta_B = \sqrt{\frac{\epsilon_r \epsilon_0}{\epsilon_0}} = \sqrt{\epsilon_r} = \sqrt{2}$$

So, $\epsilon_r = 2$

50. $$\frac{d^2y}{dt^2} = -\frac{dy}{dt} - \frac{5y}{4}$$

$$\therefore\ \frac{d^2y}{dt^2} + \frac{dy}{dt} + \frac{5y}{4} = 0 \qquad \text{...(}i\text{)}$$

Now, $y(0) = 1$

and $y'(0) = 0$

By applying Laplace transform, in equation (*i*) we get,

$$s^2Y(s) - s(1) + sY(s) - 1 + \frac{5}{4}Y(s) = 0$$

$$\therefore\ Y(s)\left[s^2 + s + \frac{5}{4}\right] = s + 1$$

$$\therefore\ Y(s) = \frac{s+1}{s^2 + s + \frac{5}{4}} = \frac{s+1}{\left(s + \frac{1}{2}\right)^2 + 1}$$

$$= \frac{\left(s + \frac{1}{2}\right)}{\left(s + \frac{1}{2}\right)^2 + 1} + \frac{\frac{1}{2}}{\left(s + \frac{1}{2}\right)^2 + 1}$$

By taking inverse Laplace transform, we get,

$$y(t) = e^{-t/2}\left[\cos(t) + \frac{1}{2}\sin(t)\right];\ t > 0$$

Now, At $t = \pi$,

$$y(t = \pi) = e^{-\pi/2}\left[(-1) + (0)\right] = -e^{-\pi/2}$$

$$= -0.2078 \simeq -0.21$$

51.

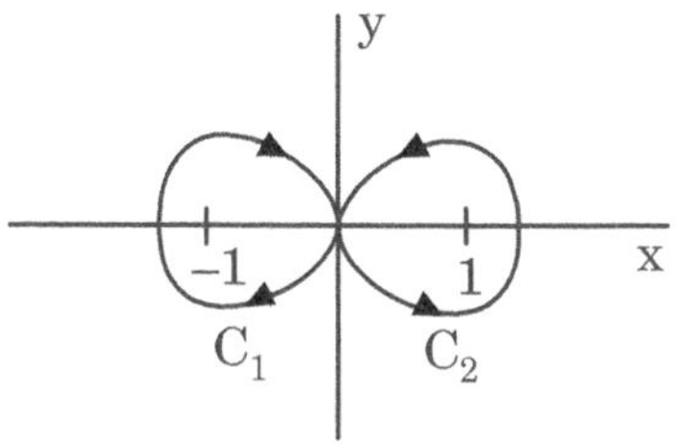

$$\frac{1}{\pi i}\oint_C \frac{dz}{z^2-1}$$

$$= 2\left[\frac{1}{2\pi i}\oint_{C_1}\frac{dz}{(z+1)(z-1)} + \frac{1}{2\pi i}\oint_{C_2}\frac{dz}{(z+1)(z-1)}\right]$$

$$= 2\left[-\left(\frac{1}{z-1}\right)\bigg|_{z=-1} + \left(\frac{1}{z+1}\right)\bigg|_{z=1}\right]$$

$$= 2\left[-\left(-\frac{1}{2}\right) + \left(\frac{1}{2}\right)\right] = 2$$

52.

$$r = x^2 + y - z \qquad ...(i)$$

$$\therefore \quad \frac{\partial r}{\partial x} = 2x - \frac{\partial z}{\partial x} \qquad ...(ii)$$

Now, $z^3 - xy + yz + y^3 = 1$ ...(*iii*)

$$\therefore \quad 3z^2\frac{\partial z}{\partial x} - y + y\frac{\partial z}{\partial x} = 0$$

$$\therefore \quad \frac{\partial z}{\partial x} = \frac{y}{3z^2 + y}$$

By substituting the value of $\frac{\partial z}{\partial x}$ in equation (ii), we get,

$$\frac{\partial r}{\partial x} = 2x - \frac{y}{3z^2 + y}$$

$$\text{Now} \quad \frac{\partial r}{\partial x}\bigg|_{(2,-1,1)} = 2(2) - \frac{(-1)}{3(1)^2 + (-1)}$$

$$= 4 + \frac{1}{3-1} = 4.50$$

53.

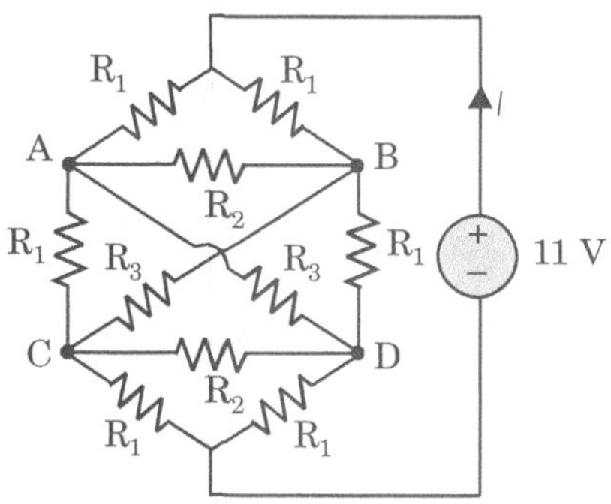

As the network is symmetric,

$V_A = V_B$ and $V_C = V_D$

So, current through R_2 resistors is zero and as $V_A = V_B$ and $V_C = V_D$, electrically the circuit can be reduced as,

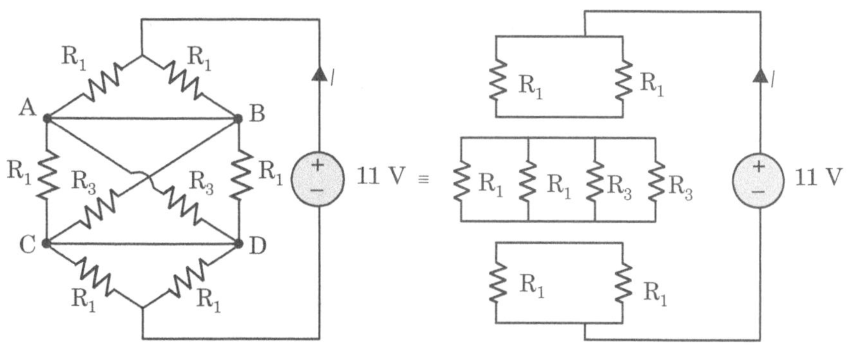

Given data, $R_1 = 1\ \Omega$ and $R_3 = 3\ \Omega$

Now Total resistance, $R_T = 2(R_1 \| R_1) + (R_1 \| R_1 \| R_3 \| R_3) = R_1 + \left(\frac{R_1}{2}\bigg\|\frac{R_3}{2}\right)$

$$\therefore \quad R_T = 1 + \left(\frac{1}{2}\bigg\|\frac{3}{2}\right)\Omega = 1 + \frac{3/2}{4} = \frac{11}{8}\Omega$$

$$\text{Now} \quad I = \frac{V}{R_T}$$

$$\therefore \quad I = \frac{11V}{R_T} = \frac{11}{(11/8)} = 8A$$

54. Let an arbitrary spectrum for x(t) as shown below:

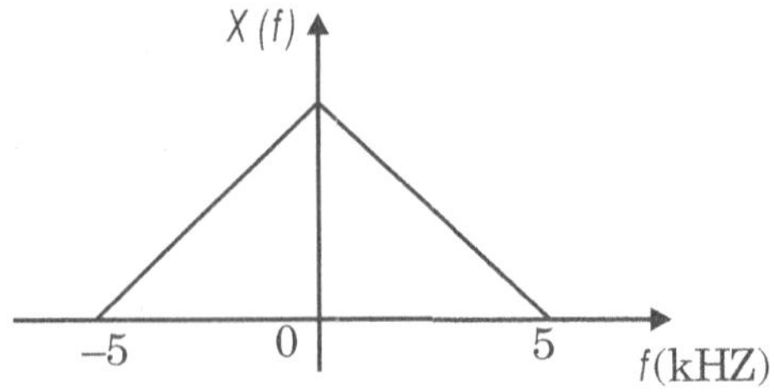

The spectrum of the sampled signal can be given as,

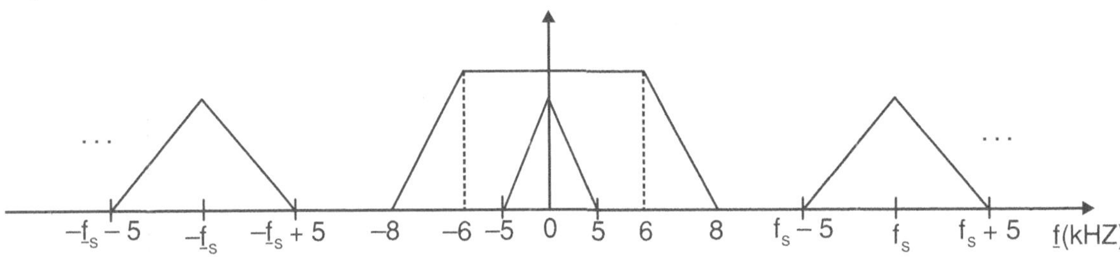

For proper reconstruction of the signal,

$$f_s - 5 \geq 8$$

$$\therefore \quad f_s \geq 8 + 5 = 13 \text{ kHz}$$

$$\therefore \quad f_{s(\min)} = 13 \text{ kHz}$$

55.

$$X(k) = \{1, 2, 3, 4, 5, 6, 7, 8\}$$

$$\sum_{n=0}^{3} x[2n] = x[0] + x[2] + x[4] + x[6]$$

$$= 4.5 - 0.5 - 0.5j - 0.5 - 0.5 + 0.5j$$

$$= 4.5 - 1.5 = 3$$

SOLVED PAPER – 2017

ELECTRONICS & COMMUNICATION ENGINEERING

No. of Questions : 65 *Maximum Marks : 100*

INSTRUCTIONS

1. Total of 65 questions carrying 100 marks, out of which 10 questions carrying a total of 15 marks are in General Aptitude (GA)
2. The Engineering Mathematics will carry around **15% of the total marks,** the General Aptitude section will carry **15% of the total marks** and the **remaining 70% of the total marks.**
3. **Types of Questions**
   (*a*) **Multiple Choice Questions (MCQ)** carrying 1 or 2 marks each in all papers and sections. These questions are objective in nature, and each will have a choice of four options, out of which the candidate has to mark the correct answer(s).
   (*b*) **Numerical Answer Questions** of 1 or 2 marks each in all papers and sections. For these questions the answer is a real number, to be entered by the candidate using the virtual keypad. No choices will be shown for these type of questions.
4. For **1-mark** multiple-choice questions, **1/3 marks** will be deducted for a wrong answer. Likewise, for **2-marks** multiple-choice questions, 2/3 marks will be deducted for a wrong answer. There is no negative marking for numerical answer type questions.

Chapter-Wise Analysis

#	Chapters	Marks		
		1	2	Total
1	General Aptitude	5	5	15
2	Engineering Mathematics	4	5	14
3	Networks	1	2	5
4	Signals and Systems	3	3	9
5	Electronic Devices	5	3	11

#	Chapters	Marks		
		1	2	Total
6	Analog Circuits	1	4	9
7	Digital Circuits & Microprocessor	3	4	11
8	Control Systems	3	4	11
9	Communications	4	2	8
10	Electromagnetics	1	3	7

TECHNICAL SECTION

(Q.1 – 25) : Carry One Mark Each

1. The voltage of an electromagnetic wave propagating in a coaxial cable with uniform characteristic impedance is $V(l) = e^{-\gamma l + j\omega t}$ Volts, where l is the distance along the length of the cable in metres, $\gamma = (0.1 + j40)\text{m}^{-1}$ is the complex propagation constant, and $\omega = 2\pi \times 10^9$ rad/s is the angular frequency. The absolute value of the attenuation in the cable in dB/metre is

2. In the circuit shown, the positive angular frequency ω (in radians per second) at which the magnitude of the phase difference between the voltages V_1 and V_2 equals $\frac{\pi}{4}$ radians, is ______

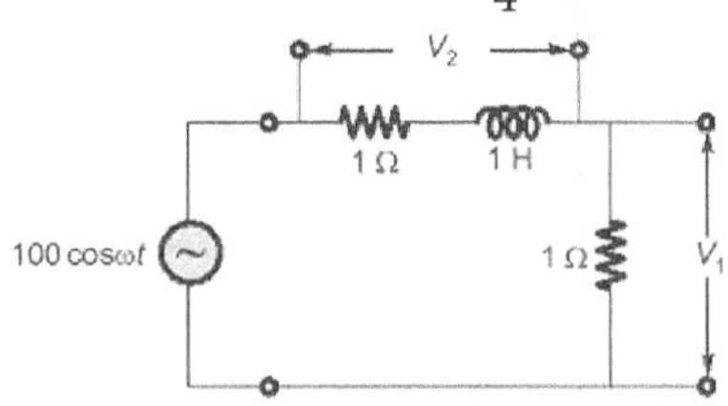

3. The open loop transfer function

$$G(s) = \frac{(s+1)}{s^p(s+2)(s+3)}$$

where p is an integer, is connected in unity feedback configuration as shown in the figure.

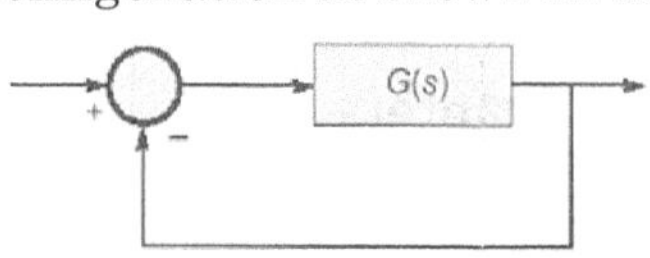

Given that the steady state error is zero for unit step input and is 6 for unit ramp input, the value of the parameter p is ________.

4. In a digital communication system, the overall pulse shape $p(t)$ at the receiver before the sampler has the Fourier transform P(*f*). If the symbols are transmitted at the rate of 2000 symbols per second, for which of the following cases is the inter symbol interference zero?

(*a*)

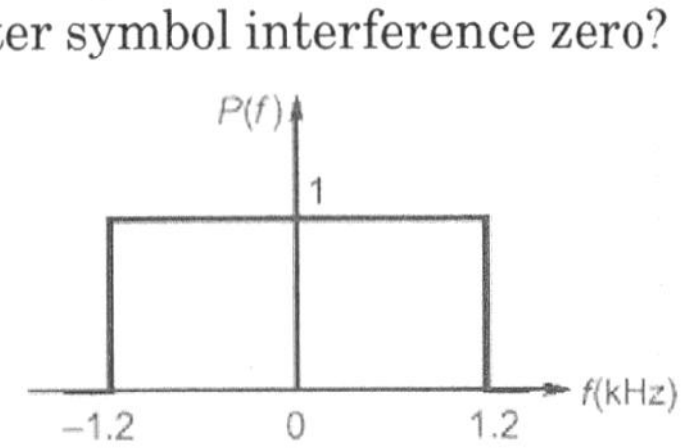

(*b*)

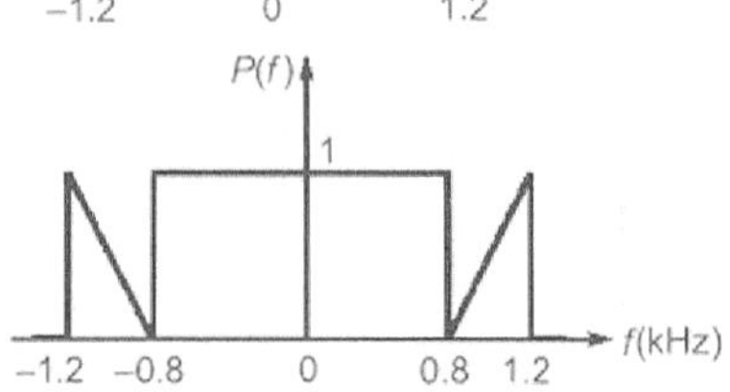

(*c*)

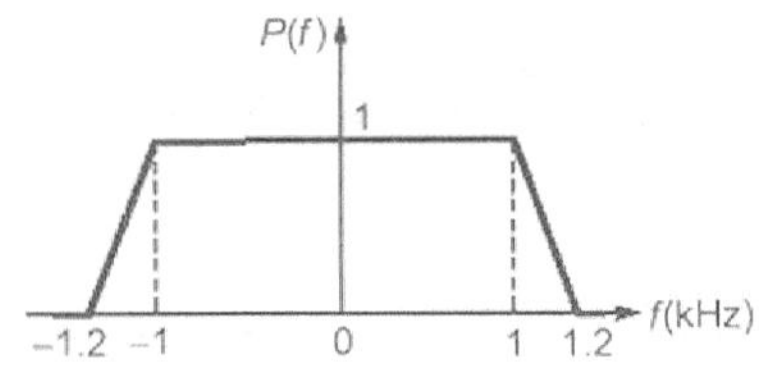

(*d*)

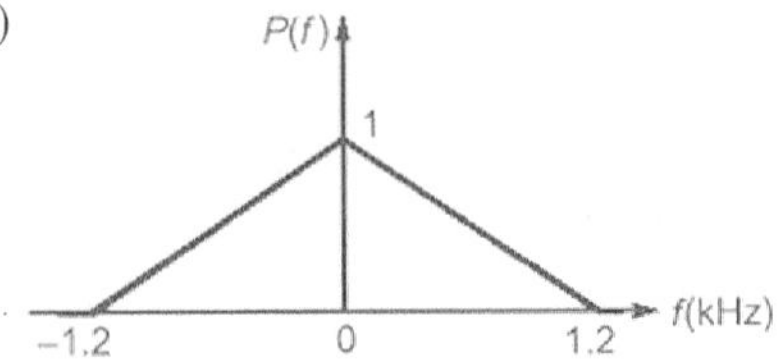

5. The Miller effect in the context of a common Emitter amplifier explains

(*a*) an increase in the low-frequency cutoff frequency

(*b*) an increase in the high-frequency cutoff frequency

(*c*) a decrease in the low-frequency cutoff frequency

(*d*) a decrease in the high-frequency cutoff frequency

6. For the operational amplifier circuit shown, the output saturation voltages are ± 15 V. The upper and lower threshold voltages for the circuit are, respectively,

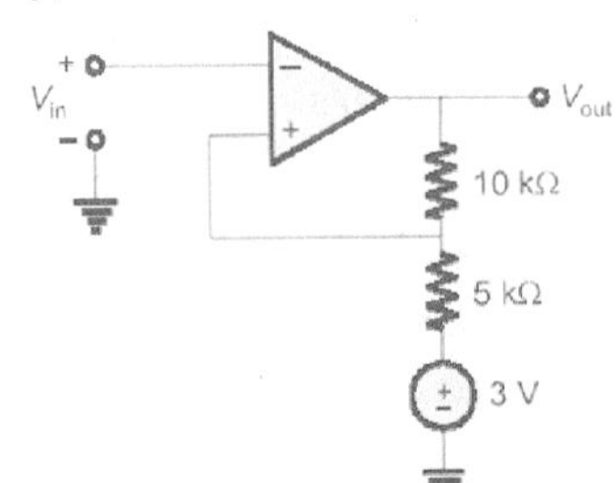

(*a*) + 5V and – 5 V (*b*) + 7V and – 3 V

(*c*) + 3V and – 7 V (*d*) + 3V and – 3 V

7. Consider the 5 × 5 matrix

$$A = \begin{bmatrix} 1 & 2 & 3 & 4 & 5 \\ 5 & 1 & 2 & 3 & 4 \\ 4 & 5 & 1 & 2 & 3 \\ 3 & 4 & 5 & 1 & 2 \\ 2 & 3 & 4 & 5 & 1 \end{bmatrix}$$

It is given that A has only one real eigenvalue. Then the real eigenvalue of A is

(*a*) – 2.5

(*b*) 0

(*c*) 15

(*d*) 25

8. Consider a single input single output discrete-time system with $x[n]$ as input and $y[n]$ as output, where the two are related as

$$y[n] = \begin{cases} n|x[n]|, & \text{for } 0 \le n \le 10 \\ x[n] - x[n-1], & \text{otherwise} \end{cases}$$

Which one of the following statements is true about the system?

(*a*) It is causal and stable

(*b*) It is causal but not stable

(*c*) It is not causal but stable

(*d*) It is neither causal nor stable

9. Let (X_1, X_2) be independent random variables. X_1 has mean 0 and variance 1, while X_2 has mean 1 and variance 4. The mutual information $I(X_1 : X_2)$ between X_1 and X_2 in bits is ________

10. Consider the following statements about the linear dependence of the real valued functions $y_1 = 1$, $y_2 = x$ and $y_3 = x^2$, over the field of real numbers.

I. y_1, y_2 and y_3 are linearly independent on $-1 \le x \le 0$.

II. y_1, y_2 and y_3 are linearly dependent on $0 \le x \le 1$.

III. y_1, y_2 and y_3 are linearly independent on $0 \le x \le 1$.

IV. y_1, y_2 and y_3 are linearly dependent on $-1 \le x \le 0$

Which one among the following is correct?

(*a*) Both I and II are true

(*b*) Both I and III are true

(*c*) Both II and IV are true

(*d*) Both III and IV are true

11. A good transconductance amplifier should have

(*a*) high input resistance and low output resistance

(*b*) low input resistance and high output resistance

(*c*) high input and output resistances

(*d*) low input and output resistances

12. Consider a wireless communication link between a transmitter and a receiver located in free space, with finite and strictly positive capacity. If the effective areas of the transmitter and the receiver antennas, and the distance between them are all doubled, and everything else remains unchanged, the maximum capacity of the wirelss link

(*a*) increases by a factor of 2

(*b*) decreases by a factor of 2

(*c*) remains unchanged

(*d*) decreases by a factor of $\sqrt{2}$

13. Consider a stable system with transfer function

$$G(s) = \frac{s^p + b_1 s^{p-1} + \ldots + b_p}{s^q + a_1 s^{q-1} + \ldots + a_q}$$

where $b_1, \ldots, b_p$ and $a_1, \ldots, a_q$ are real valued constants. The slope of the Bode log magnitude curve of G(*s*) converges to – 60 dB/decade as $\omega \to \infty$. A possible pair of values for p and q is

(*a*) $p = 0$ and $q = 3$

(*b*) $p = 1$ and $q = 7$

(*c*) $p = 2$ and $q = 3$

(*d*) $p = 3$ and $q = 5$

14. The rank of the matrix $M = \begin{bmatrix} 5 & 10 & 10 \\ 1 & 0 & 2 \\ 3 & 6 & 6 \end{bmatrix}$ is

(*a*) 0 (*b*) 1

(*c*) 2 (*d*) 3

15. Consider the following statements for continuous-time linear time invariant (LTI) systems.

I. There is no bounded input bounded output (BIBO) stable system with a pole in the right half of the complex plane.

II. There is no causal and BIBO stable system with a pole in the right half of the complex plane.

Which one among the following is correct?

(*a*) Both I and II are true

(*b*) Both I and II are not true

(*c*) Only I is true

(*d*) Only II is true

16. In the latch circuit shown, the NAND gates have non-zero, but unequal propagation delays. The present input condition is : P = Q = '0'. If the input condition is changed simultaneously to P = Q = '1', the outputs X and Y are

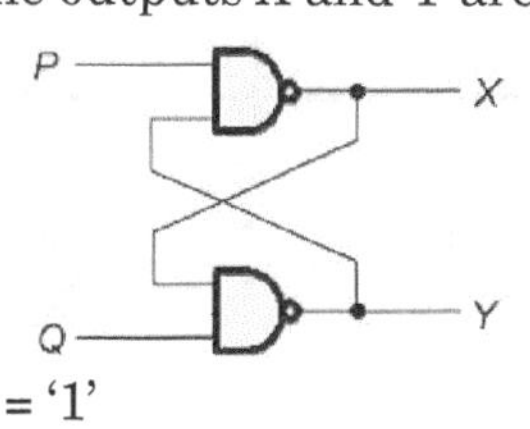

(*a*) X = '1', Y = '1'

(*b*) either X = '1', Y = '0' or X = '0', Y = '1'

(*c*) either X = '1', Y = '1' or X = '0', Y = '0'

(*d*) X = '0', Y = '0'

17. The clock frequency of an 8085 microprocessor is 5 MHz. If the time required to execute an instruction is 1.4 μs, then the number of T-states needed for executing the instruction is

(a) 1 (b) 6

(c) 7 (d) 8

18. Which one of the following statements about differential pulse code modulation (DPCM) is true?

(a) The sum of message signal sample with its prediction is quantized.

(b) The message signal sample is directly quantized, and its prediction is not used.

(c) The difference of message signal sample and a random signal is quantized.

(d) The difference of message signal sample with its prediction is quantized.

19. Consider the D-Latch shown in the figure, which is transparent when its clock input CK is high and has zero propagation delay. In the figure, the clock signal CLK1 has a 50% duty cycle and CLK2 is a one-fifth period delayed version of CLK1. The duty cycle at the output of the latch in percentage is ________

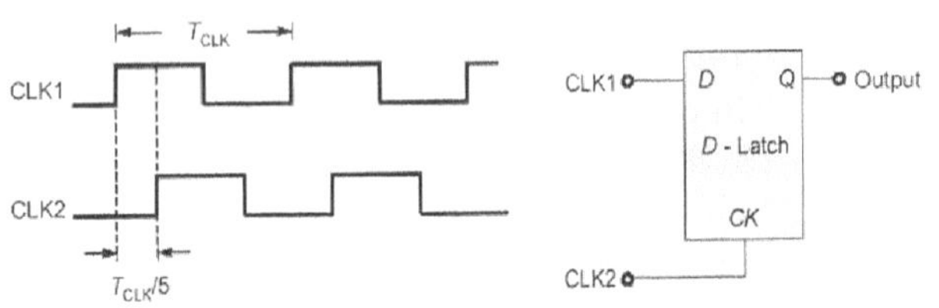

20. Which of the following can be the pole-zero configuration of a phase-lag controller (lag compensator)?

(a)

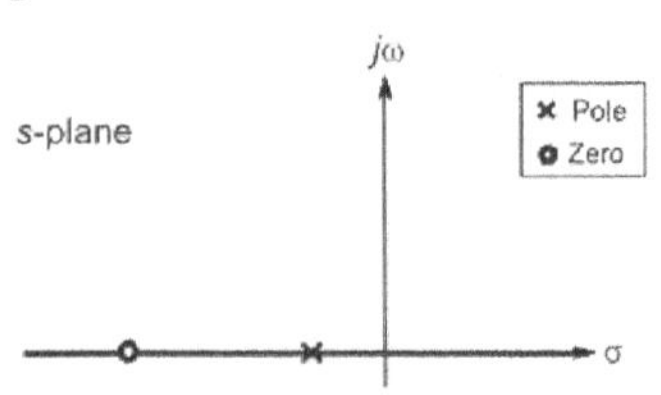

(b)

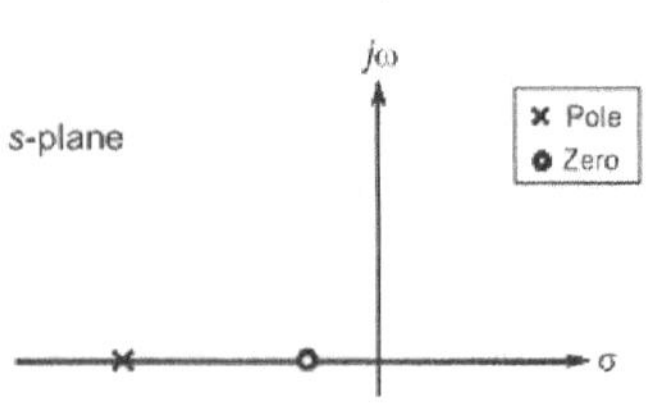

(c)

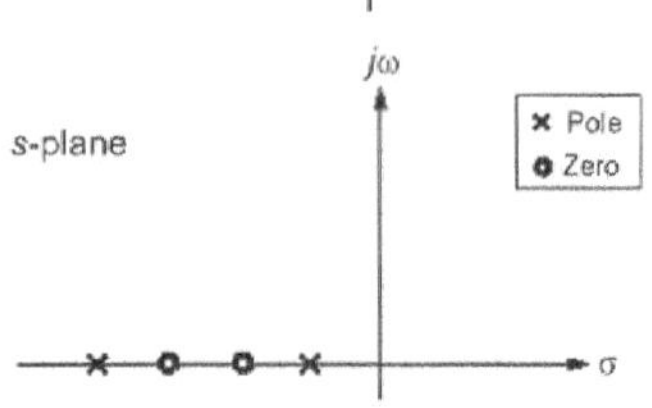

(d)

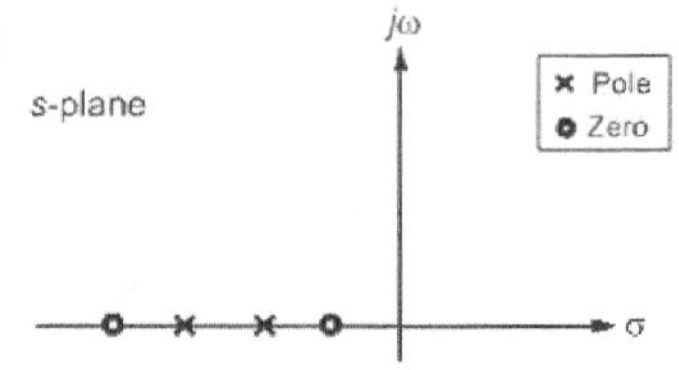

21. A periodic signal $x(t)$ has a trigonometric Fourier series expansion

$$x(t) = a_0 + \sum_{n=1}^{\infty}\left(a_n \cos n\omega_0 t + b_n \sin n\omega_0 t\right)$$

If $x(t) = -x(-t) = -x(t - \pi/\omega_0)$, we can conclude that

(a) a_n are zero for all n and b_n are zero for n even

(b) a_n are zero for all n and b_n are zero for n odd

(c) a_n are zero for n even and b_n are zero for n odd

(d) a_n are zero for n odd and b_n are zero for n even

22. For a narrow base PNP BJT, the excess minority carrier concentrations (Δn_E for emitter, Δp_B for base, Δn_C for collector) normalized to equilibrium minority carrier concentrations (n_{E0} for emitter, p_{B0} for base, $^{n}C_0$ for collector) in the quasi-neutral emitter, base and collector regions are shown below. Which one of the following biasing modes is the transistor operating in?

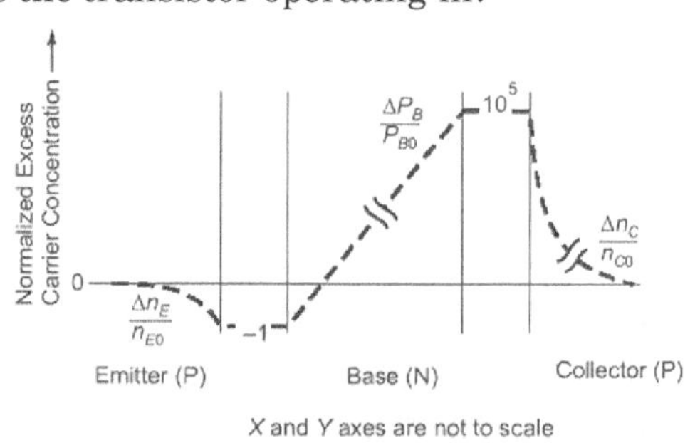

(a) Forward active

(b) Saturation

(c) Inverse active

(d) Cutoff

23. A bar of Gallium Arsenide (GaAs) is doped with Silicon such that the Silicon atoms occupy Gallium and Arsenic sites in the GaAs crystal. Which one of the following statements is true?

(a) Silicon atoms act as p-type dopants in Arsenic sites and n-type dopants in Gallium sites.

(b) Silicon atoms act as n-type dopants in Arsenic sites and p-type dopants in Gallium sites.

(c) Silicon atoms act as p-type dopants in Arsenic as well as Gallium sites.

(d) Silicon atoms act as n-type dopants in Arsenic as well as Gallium sites.

24. An $n^+ - n$ Silicon device is fabricated with uniform and non-degenerate donor doping concentrations of $N_{D1} = 1 \times 10^{18}$ cm^{-3} and $N_{D2} = 1 \times 10^{15}$ cm^{-3} corresponding to the n^+ and n regions respectively. At the operational temperature T, assume complete impurity ionization, $kT/q = 25$ mV, and intrinsic carrier concentration to be $n_i = 1 \times 10^{10}$ cm^{-3}. What is the magnitude of the built-in potential of this device?

(*a*) 0.748 V (*b*) 0.460 V

(*c*) 0.288 V (*d*) 0.173 V

25. Three fair cubical dice are thrown simultaneously. The probability that all three dice have the same number of dots on the faces showing up is (up to third decimal place) _______.

(Q.26 – 55) : Carry Two Marks Each

26. Let $h[n]$ be the impulse response of a discrete-time linear time invariant (LTI) filter. The impulse response is given by

$$h[0] = \frac{1}{3};$$

$$h[1] = \frac{1}{3};$$

$$h[2] = \frac{1}{3};$$

and $h[n] = 0$

for $n < 0$

and $n > 2$

Let $H(\omega)$ be the discrete-time Fourier transform (DTFT) of $h[n]$, where ω is the normalized angular frequency in radians. Given that $H(\omega_0) = 0$ and $0 < \omega_0 < \pi$, the value of ω_0 (in radians) is equal to _________.

27. Two discrete-time signals $x[n]$ and $h[n]$ are both non-zero only for $n = 0, 1, 2$ and are zero otherwise. It is given that

$$x[0] = 1,$$

$$x[1] = 2,$$

$$x[2] = 1,$$

$$h[0] = 1$$

Let $y[n]$ be the linear convolution of $x[n]$ and $h[n]$. Given that $y[1] = 3$ and $y[2] = 4$, the value of the expression $(10y[3] + y[4])$ is ________.

28. In the circuit shown, the voltage $V_{IN}(t)$ is described by :

$$V_{IN}(t) = \begin{cases} 0, & \text{for } t < 0 \\ 15 \text{ Volts}, & \text{for } t \geq 0 \end{cases}$$

where t is in seconds. The time (in seconds) at which the current I in the circuit will reach the value 2 Amperes is ________.

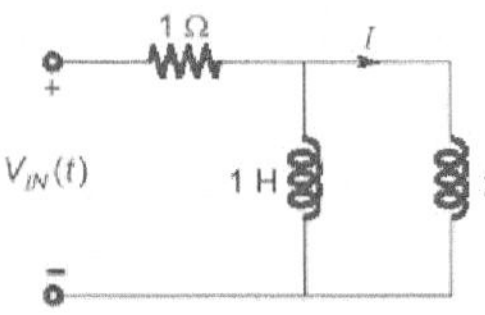

29. As shown, two Silicon (Si) abrupt p-n junction diodes are fabricated with uniform donor doping concentrations of $N_{D1} = 10^{14}$ cm^{-3} and $N_{D2} = 10^{16}$cm^{-3} in the n-regions of the diodes, and uniform acceptor doping concentrations of $N_{A1} = 10^{14}$ cm^{-3} and $N_{A2} = 10^{16}$ cm^{-3} in the p-regions of the diodes, respectively. Assuming that the reverse bias voltage is >> built-in potentials of the diodes, the ratio C_2/C_1 of their reverse bias capacitances for the same applied reverse bias, is ________.

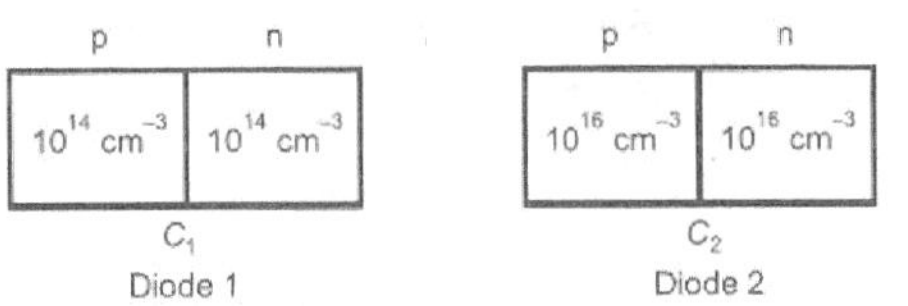

30. As shown, a uniformly doped Silicon (Si) bar of length L = 0.1 μm with a donor concentration $N_D = 10^{16}$ cm^{-3} is illuminated at $x = 0$ such that electron and hole pairs are generated at the rate of $G_L = G_{L0}\left(1 - \frac{x}{L}\right), 0 \leq x \leq L,$

where $G_{L0} = 10^{17}$ cm^{-3} s^{-1}. Hole lifetime is 10^{-4} s, electronic charge $q = 1.6 \times 10^{-19}$ C, hole diffusion coefficient
$D_p = 100$ cm^2/s and low level injection condition prevails. Assuming a linearly decaying steady state excess hole concentration that goes to 0 at $x = L$, the magnitude of the diffusion current density at $x = L/2$, in A/cm^2, is ________

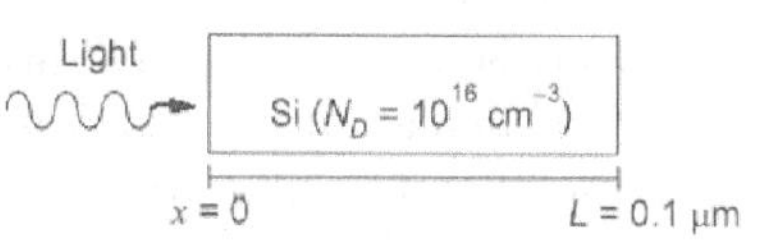

31. A half wavelength dipole is kept in the x-y plane and oriented along 45° from the x-axis. Determine the direction of null in the radiation pattern for $0 \leq \phi \leq \pi$. Here the angle $\theta (0 \leq \theta \leq \pi)$ is measured from the z-axis, and the angle $\phi (0 \leq \phi \leq 2\pi)$ is measured from the x-axis in the x-y plane.

(*a*) $\theta = 90°, \phi = 45°$ (*b*) $\theta = 45°, \phi = 90°$

(*c*) $\theta = 90°, \phi = 135°$ (*d*) $\theta = 45°, \phi = 135°$

32. Which one of the following options correctly describes the locations of the roots of the equation $s^4 + s^2 + 1 = 0$ on the complex plane?

(*a*) Four left half plane (LHP) roots.

(*b*) One right half plane (RHP) root, one LHP root and two roots on the imaginary axis.

(*c*) Two RHP roots and two LHP roots.

(*d*) All four roots are on the imaginary axis

33. For the DC analysis of the Common-Emitter amplifier shown, neglect the base current and assume that the emitter and collector currents are equal. Given that $V_T = 25$ mV, $V_{BE} = 0.7$ V, and the BJT output resistance r_o is practically infinite. Under these conditions, the midband voltage gain magnitude, $A_V = |v_o/v_i|$V/V, is ______

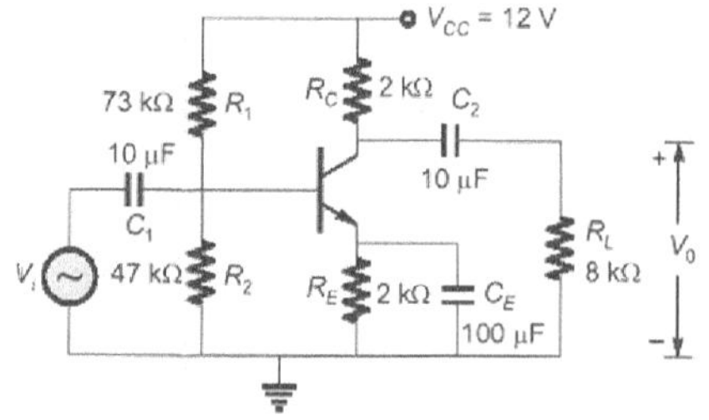

34. For the circuit shown, assume that the NMOS transistor is in saturation. Its threshold voltage $V_{tn} = 1$ V and its transconductance parameter $\mu_n C_{ox}\left(\frac{W}{L}\right) = 1 \text{mA/V}^2$. Neglect channel length modulation and body bias effects. Under these conditions, the drain current I_D in mA is ______.

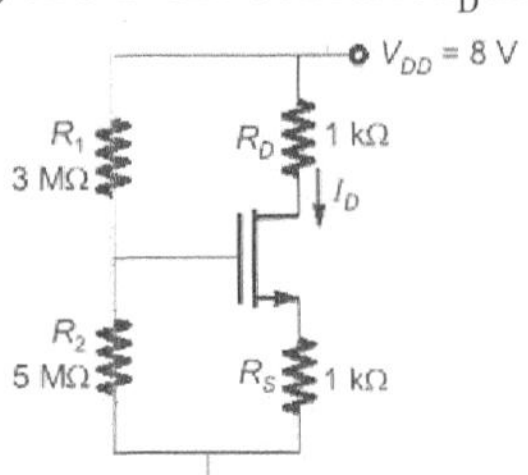

35. The dependence of drift velocity of electrons on electric field in a semiconductor is shown below. The semiconductor has a uniform electron concentration of $n = 1 \times 10^{16}$ cm^{-3} and electronic charge $q = 1.6 \times 10^{-19}$ C. If a bias of 5 V is applied across a 1 m region of this semiconductor, the resulting current density in this region, in kA/cm², is ______.

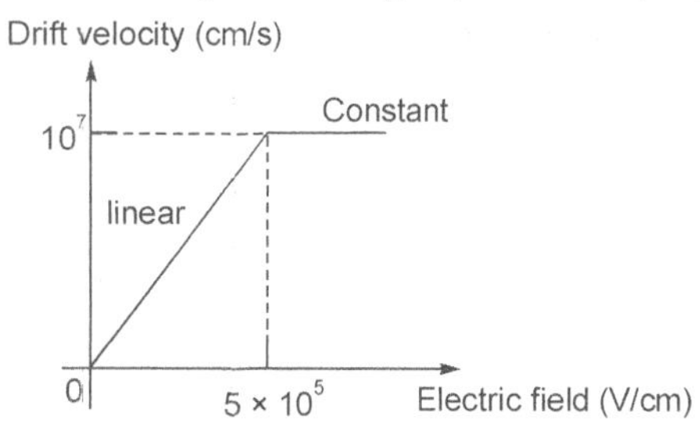

36. A linear time invariant (LTI) system with the transfer function

$$G(s) = \frac{K(s^2 + 2s + 2)}{(s^2 - 3s + 2)}$$

is connected in unity feedback configuration as shown in the figure.

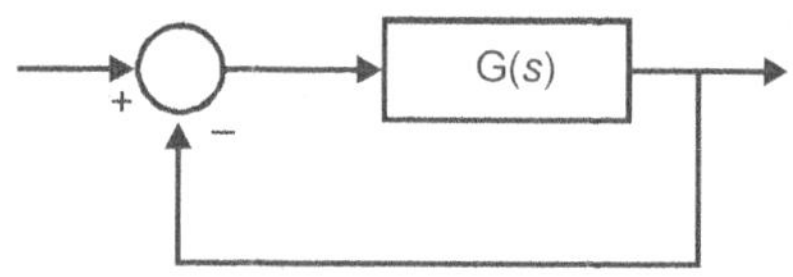

For the closed loop system shown, the root locus for $0 < K < \infty$ intersects the imaginary axis for K = 1.5. The closed loop system is stable for

(*a*) K > 1.5 (*b*) 1 < K < 1.5

(*c*) 0 < K < 1.5 (*d*) no positive value of K

37. The amplifier circuit shown in the figure is implemented using a compensated operational amplifier (op-amp), and has an open-loop voltage gain, $A_0 = 10^5$ V/V and an open-loop cut-off frequency, $f_c = 8$ Hz. The voltage gain of the amplifier at 15 kHz, in V/V is ______.

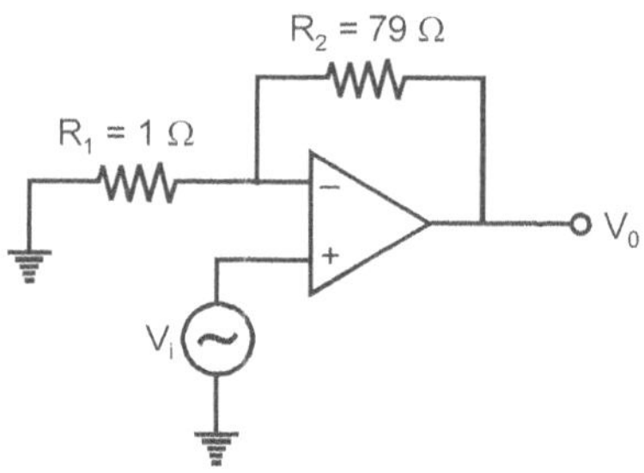

38. The expression for an electric field in free space is $E = E_0(\hat{x} + \hat{y} + j2\hat{z})e^{-j(\omega t - kx + ky)}$, where x, y, z represent the spatial coordinates, t represents time, and ω, k are constants. This electric field

(*a*) does not represent a plane wave.

(*b*) represents a circularly polarized plane wave propagating normal to the z-axis.

(*c*) represents an elliptically polarized plane wave propagating along the *x-y* plane.

(*d*) represents a linearly polarized plane wave.

39. Starting with $x = 1$, the solution of the equation $x^3 + x = 1$, after two iterations of Newton-Raphson's method (up to two decimal places) is ______.

40. Let X(t) be a wide sense stationary random process with the power spectral density $S_X(f)$ as shown in figure (a), where f is in Hertz (Hz). The random process X(t) is input to an ideal lowpass filter with the frequency response

$$H(f) = \begin{cases} 1, & |f| \le \frac{1}{2}\text{Hz} \\ 0, & |f| > \frac{1}{2}\text{Hz} \end{cases}$$

as shown in figure (*b*). The output of the lowpass filter is $Y(t)$.

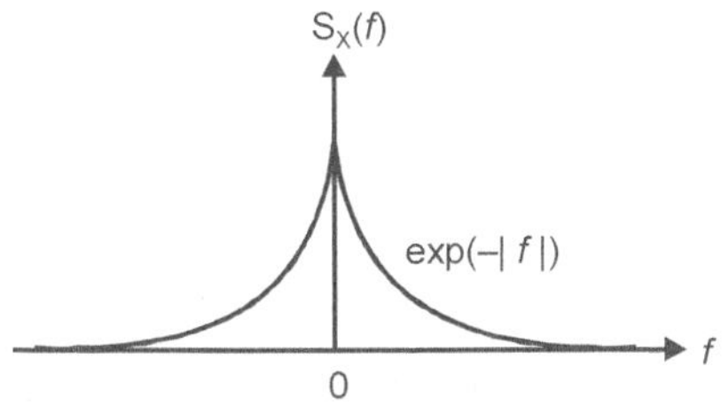

(a)

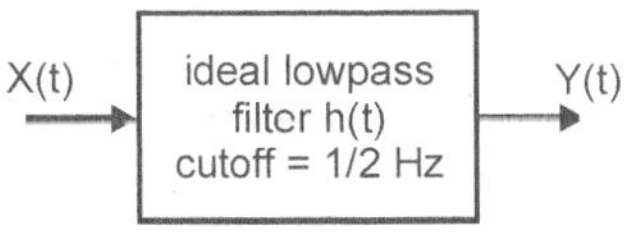

(b)

Let E be the expectation operator and consider the following statements :

I. $E(X(t)) = E(Y(t))$

II. $E(X^2(t)) = E(Y^2(t))$

III. $E(Y^2(t)) = 2$

Select the correct option :

(*a*) only I is true

(*b*) only II and III are true

(*c*) only I and II are true

(*d*) only I and III are true

41. Let $I = \int_C (2zdx + 2ydy + 2xdz)$ where x, y, z are real, and let C be the straight line segment from point A : (0, 2, 1) to point B : (4, 1, – 1). The value of I is _________.

42. The Nyquist plot of the transfer function

$$G(s) = \frac{K}{(s^2 + 2s + 2)(s + 2)}$$

does not encircle the point $(-1 + j0)$ for K = 10 but does encircle the point $(-1 + j0)$ for K = 100. Then the closed loop system (having unity gain feedback) is

(*a*) stable for K = 10 and stable for K = 100

(*b*) stable for K = 10 and unstable for K = 100

(*c*) unstable for K = 10 and stable for K = 100

(*d*) unstable for K = 10 and unstable for K = 100

43. In the figure shown, the npn transistor acts as a switch

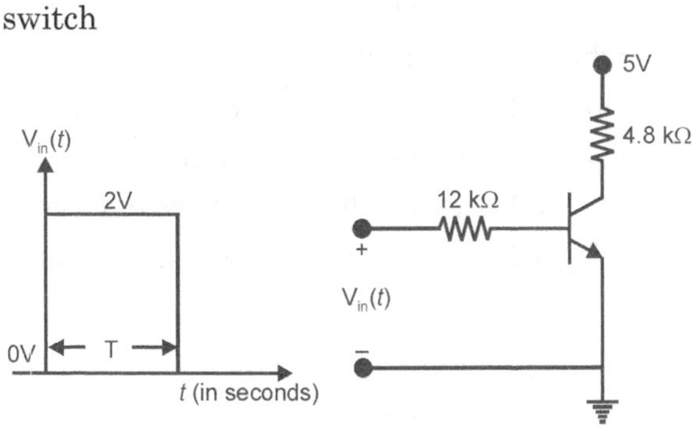

For the input $V_{in}(t)$ as shown in the figure, the transistor switches between the cut-off and saturation regions of operation, when T is large. Assume collector-to-emitter voltage at saturation $V_{CE(sat)} = 0.2$ V and base-to-emitter voltage $V_{BE} = 0.7$ V. The minimum value of the common-base current gain (α) of the transistor for the switching should be _________.

44. The figure shows an RLC circuit excited by the sinusoidal voltage $100\cos(3t)$ Volts, where t is in seconds. The ratio $\dfrac{\text{amplitude of } V_2}{\text{amplitude of } V_1}$ is _______.

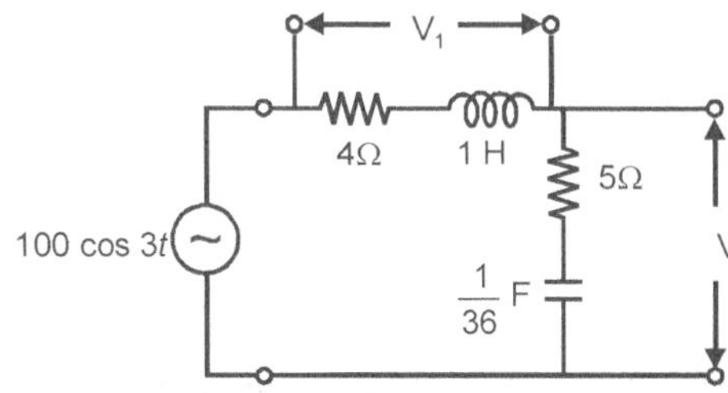

45. An optical fiber is kept along the $\hat{z}$ direction. The refractive indices for the electric fields along $\hat{x}$ and $\hat{y}$ directions in the fiber are $n_x = 1.5000$ and $n_y = 1.5001$, respectively ($n_x \neq n_y$ due to the imperfection in the fiber cross-section). The free space wavelength of a light wave propagating in the fiber is 1.5 m. If the lightwave is circularly polarized at the input of the fiber, the minimum propagation distance after which it becomes linearly polarized, in centimetres, is _______.

46. A 4-bit shift register circuit configured for right-shift operation, i.e. $D_{in} \rightarrow A$, $A \rightarrow B$, $B \rightarrow C$, $C \rightarrow D$, is shown. If the present state of the shift register ABCD = 1101, the number of clock cycles required to reach the state ABCD = 1111 is _______.

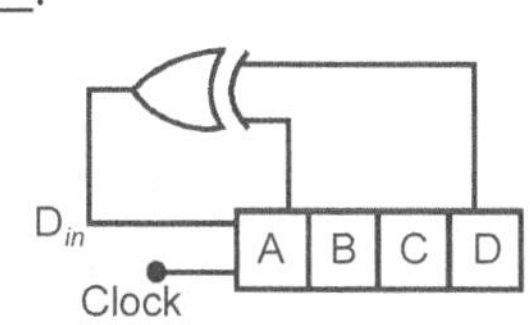

47. Which one of the following gives the simplified sum of products expression for the Boolean function $F = m_0 + m_2 + m_3 + m_5$, where m_0, m_2, m_3 and m_5 are minterms corresponding to the inputs A, B and C with A as the MSB and C as the LSB?

(*a*) $\bar{A}B + \bar{A}\bar{B}\bar{C} + A\bar{B}C$

(*b*) $\bar{A}\bar{C} + \bar{A}B + A\bar{B}C$

(*c*) $\bar{A}\bar{C} + A\bar{B} + A\bar{B}C$

(*d*) $\bar{A}BC + \bar{A}\bar{C} + A\bar{B}C$

48. Let $x(t)$ be a continuous time periodic signal with fundamental period T = 1 seconds. Let $\{a_k\}$ be the complex Fourier series coefficient of $x(t)$, where k is integer valued. Consider the following statements about $x(3t)$:

I. The complex Fourier series coefficients of $x(3t)$ are $\{a_k\}$ where k is integer valued.

II. The complex Fourier series coefficients of $x(3t)$ are $\{3a_k\}$ where k is integer valued.

III. The fundamental angular frequency of $x(3t)$ is 6π rad/s.

For the three statements above, which one of the following is correct?

(*a*) only II and III are true

(*b*) only I and III are true

(*c*) only III is true

(*d*) only I is true

49. A continuous time signal $x(t) = 4\cos(200\pi t) + 8\cos(400\pi t)$, where t is in seconds, is the input to a linear time invariant (LTI) filter with the impulse response

$$h(t) = \begin{cases} \dfrac{2\sin(300\pi t)}{\pi t}, & t \neq 0 \\ 600, & t = 0 \end{cases}$$

Let $y(t)$ be the output of this filter. The maximum value of $|y(t)|$ is _________.

50. The following FIVE instructions were executed on an 8085 microprocessor.

MVI A, 33 H

MVI B, 78 H

ADD B

CMA

ANI 32 H

The accumulator value immediately after the execution of the fifth instructions is

(*a*) 00 H (*b*) 10 H

(*c*) 11 H (*d*) 32 H

51. A finite state machine (FSM) is implemented using the D flip-flops A and B, and logic gates, as shown in the figure below. The four possible states of the FSM are Q_AQ_B = 00, 01, 10 and 11.

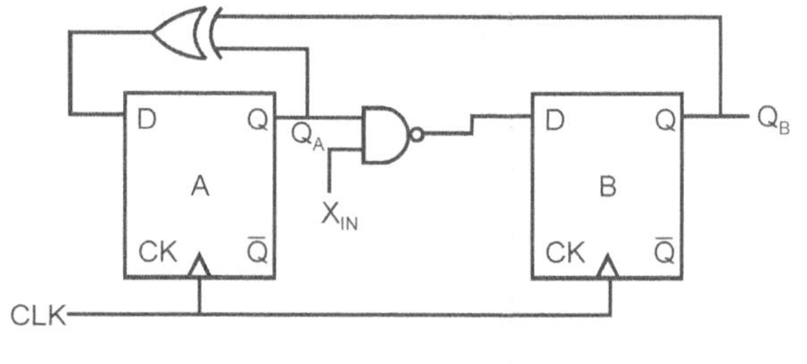

Assume that X_{IN} is held at constant logic level throughout the operation of the FSM. When the FSM is initialized to the Q_AQ_B = 00 and clocked, after a few clock cycles, it starts cycling through

(*a*) all of the four possible states if $X_{IN} = 1$

(*b*) three of the four possible states if $X_{IN} = 0$

(*c*) only two of the four possible states if $X_{IN} = 1$

(*d*) only two of the four possible states if $X_{IN} = 0$

52. Let $f(x) = e^{x + x^2}$ for real x. From among the following, choose the Taylor series approximation of $f(x)$ around $x = 0$, which includes all powers of x less than or equal to 3,

(*a*) $1 + x + x^2 + x^3$ (*b*) $1 + x + \frac{3}{2}x^2 + x^3$

(*c*) $1 + x + \frac{3}{2}x^2 + \frac{7}{6}x^3$ (*d*) $1 + x + 3x^2 + 7x^3$

53. A three dimensional region R of finite volume is described by

$$x^2 + y^2 \le z^3 \,;\, 0 \le z \le 1,$$

where x, y, z are real. The volume of R (up to two decimal places) is _________.

54. In binary frequency shift keying (FSK), the given signal waveforms are

$u_0(t) = 5\cos(20000\pi t)$; $0 \le t \le$ T. and

$u_1(t) = 5\cos(22000\pi t)$; $0 \le t \le$ T,

where T is the bit-duration interval and t is in seconds. Both $u_0(t)$ and $u_1(t)$ are zero outside the interval $0 \le t \le$ T. With a matched filter (correlator) based receiver, the smallest positive value of T (in milliseconds) required to have $u_0(t)$ and $u_1(t)$ uncorrelated is

(*a*) 0.25 ms

(*b*) 0.5 ms

(*c*) 0.75 ms

(*d*) 1.0 ms

55. Which one of the following is the general solution of the first order differential equation

$$\frac{dy}{dx} = (x + y - 1)^2,$$

where x, y are real?

(*a*) $y = 1 + x + \tan^{-1}(x + c)$, where c is a constant.

(*b*) $y = 1 + x + \tan(x + c)$, where c is a constant.

(*c*) $y = 1 - x + \tan^{-1}(x + c)$, where c is a constant.

(*d*) $y = 1 - x + \tan(x + c)$, where c is a constant.

GENERAL APTITUDE (GA)

(Q.1 – 5) : Carry One Mark Each

1. In the summer, water consumption is known to decrease overall by 25%. A Water Board official states that in the summer household consumption decreases by 20%, while other consumption increases by 70%.

   Which of the following statements is correct?

   (*a*) The ratio of household to other consumption is 8/17.

   (*b*) The ratio of household to other consumption is 1/17.

   (*c*) The ratio of household to other consumption is 17/8.

   (*d*) There are errors in the official's statement.

2. Some tables are shelves. Some shelves are chairs. All chairs are benches. Which of the following conclusions can be deduced from the preceding sentences?

   (*i*) At least one bench is a table

   (*ii*) At least one shelf is a bench

   (*iii*)At least one chair is a table

   (*iv*)All benches are chairs

   (*a*) only (*i*) (*b*) only (*ii*)

   (*c*) only (*ii*) and (*iii*) (*d*) only (*iv*)

3. I ______ made arrangements had I ______ informed earlier.

   (*a*) could have, been (*b*) would have, being

   (*c*) had, have (*d*) had been, been

4. 40% of deaths on city roads may be attributed to drunken driving. The number of degrees needed to represent this as a slice of a pie chart is

   (*a*) 120 (*b*) 144

   (*c*) 160 (*d*) 212

5. She has a sharp tongue and it can occasionally turn ________.

   (*a*) hurtful (*b*) left

   (*c*) methodical (*d*) vital

(Q.6 – 10) : Carry Two Marks Each

6. Trucks (10 m long) and cars (5 m long) go on a single lane bridge. There must be a gap of at least 20 m after each truck and a gap of at least 15 m after each car. Trucks and cars travel at a speed of 36 km/h. If cars and trucks go alternately. What is the maximum number of vehicles that can use the bridge in one hour?

   (*a*) 1440 (*b*) 1200

   (*c*) 720 (*d*) 600

7. "If you are looking for a history of India, or for an account of the rise and fall of the British Raj, or for the reason of the cleaving of the subcontinent into two mutually antagonistic parts and the effects this mutilation will have in the respective sections, and ultimately on Asia, you will not find it in these pages; for though I have spent a lifetime in the country, I lived too near the seat of events, and was too intimately associated with the actors, to get the perspective needed for the impartial recording of these matters".

   Here, the word 'antagonistic' is closest in meaning to

   (*a*) impartial (*b*) argumentative

   (*c*) separated (*d*) hostile

8. A contour line joins locations having the same height above the mean sea level. The following is a contour plot of a geographical region. Contour lines are shown at 25 m intervais in this plot.

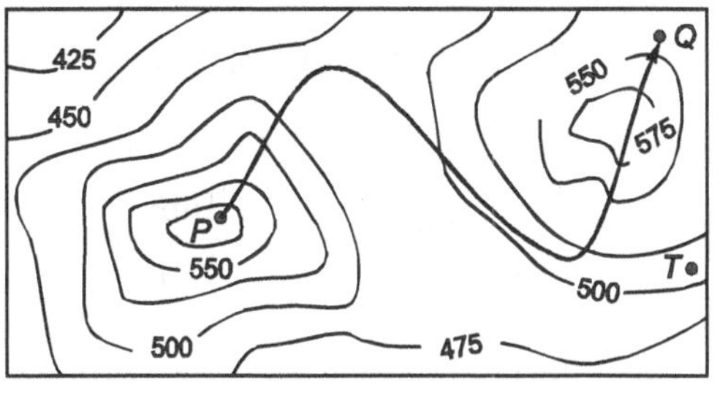

   The path from P to Q is best described by

   (*a*) Up-Down-Up-Down

   (*b*) Down-Up-Down-Up

   (*c*) Down-Up-Down

   (*d*) Up-Down-Up

9. There are 3 Indians and 3 Chinese in a group of 6 people. How many subgroups of this group can we choose so that every subgroup has at least one Indian?

   (*a*) 56 (*b*) 52

   (*c*) 48 (*d*) 44

10. S, T, U, V, W, X, Y and Z are seated around a circular table. T's neighbours are Y and V. Z is seated third to the left of T and second to the right of S. U's neighbours are S and Y; and T and W are not seated opposite each other. Who is third to the left of V?

    (*a*) X

    (*b*) W

    (*c*) U

    (*d*) T

ANSWERS

Technical Section

1. 0.868	2. 1	3. 1	4. (*c*)	5. (*d*)	6. (*b*)	7. (*c*)
8. (*a*)	9. 0	10. (*b*)	11. (*c*)	12. (*c*)	13. (*a*)	14. (*c*)
15. (*d*)	16. (*b*)	17. (*c*)	18. (*d*)	19. 30	20. (*a*)	21. (*a*)
22. (*c*)	23. (*a*)	24. (*d*)	25. 0.028	26. 2.10	27. 31	28. 0.34
29. 10	30. 16	31. (*a*)	32. (*c*)	33. 128	34. 2	35. 1.6
36. (*a*)	37. 44.4	38. (*c*)	39. 0.68	40. (*a*)	41. –11	42. (*b*)
43. 0.902	44. 2.6	45. 0.375	46. 10	47. (*b*)	48. (*b*)	49. 8
50. (*b*)	51. (*d*)	52. (*c*)	53. 0.7853	54. (*b*)	55. (*d*)	

General Aptitude (GA)

1. (*d*)	2. (*b*)	3. (*a*)	4. (*b*)	5. (*a*)	6. (*a*)	7. (*d*)	8. (*c*)	9. (*a*)	10. (*a*)

EXPLANATIONS

TECHNICAL SECTION

1. $$V(l) = V_0 e^{-\alpha l} e^{-j\beta l} e^{j\omega t}$$

Now, $$\text{Attenuation} = \frac{|\text{Input}|}{|\text{Output}|}$$

$$\text{Attenuation per meter} = \frac{|V_0|}{|V_0(1m)|} = e^{\alpha}$$

$$\text{Attenuation in } dB/m = (20 \log e^{\alpha})dB/m$$
$$= 20(0.1) \log e$$
$$= 0.868\ dB/m$$

2. 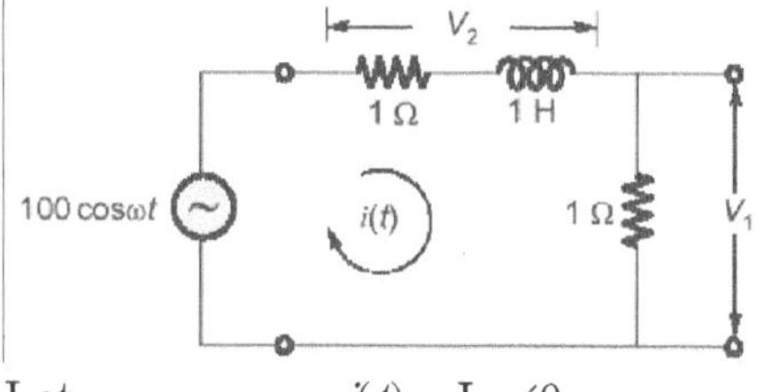

Let, $i(t) = I_m \angle \theta_i$

and $Z_2 = 1 + j\omega = \sqrt{1+\omega^2} \angle \theta_2$

Now, $\theta_2 = \tan^{-1}\left(\frac{\omega}{1}\right)$

$V_1 = i(t)\,(1\ \Omega) = I_m \angle \theta_i$

and $V_2 = i(t) Z_2 = I_m \sqrt{1+\omega^2} \angle \theta_2 + \theta_i$

Now from given data,

$$(\theta_i + \theta_2) - (\theta_i) = \frac{\pi}{4}$$

$$\therefore \quad \theta_2 = \frac{\pi}{4}$$

and $$\tan^{-1}\left(\frac{\omega}{1}\right) = \frac{\pi}{4}$$

$$\therefore \quad \omega = \tan\frac{\pi}{4}$$

$$\therefore \quad \omega = 1 \text{ rad/sec}$$

3. Steady state error for type 1 system, for step input is zero and for ramp input is $\frac{1}{K_V}$ and for parabolic input is infinity.

Hence, the given system must be of type-1 and $p = 1$.

4. From given data,

symbol rate $(R_b) = 2000$ symbols/sec

For zero ISI, raised cosine pulses will be used.

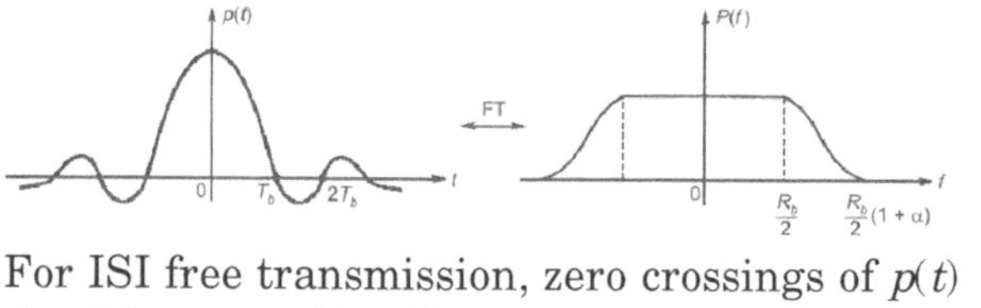

For ISI free transmission, zero crossings of $p(t)$ should occur at T_b, $2T_b$,.....

From option (*c*)

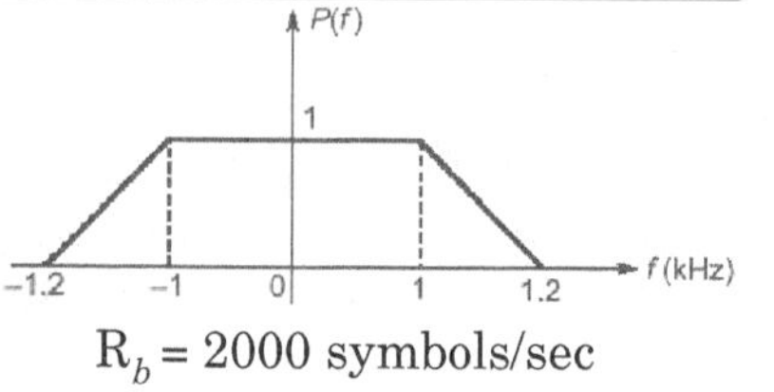

$$R_b = 2000 \text{ symbols/sec}$$

$$\therefore \quad \frac{R_b}{2} = \frac{2000}{2} = 1 \text{ kHz}$$

Hence, option (*c*) is correct.

5. Miller effect increases input capacitance and hence decreases the higher cut-off frequency.

6. 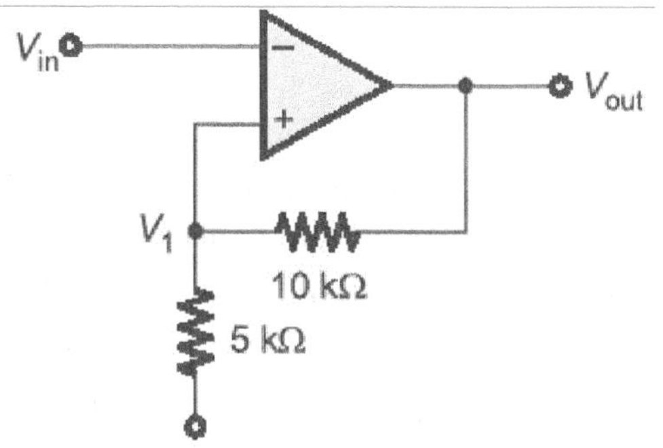

At node V_1,

$$V_1 = \frac{(3\times 10)+(V_0\times 5)}{15} = \frac{30+5V_0}{15} = \frac{5}{15}(6+V_0) = \frac{(6+V_0)}{3}$$

Now, $V_{UT} = \frac{6+15}{3} = 7V$

and $V_{LT} = \frac{6-15}{3} = -3V$

7. $|A-\lambda I| = 0$

$$\therefore \begin{bmatrix} 1 & 2 & 3 & 4 & 5 \\ 5 & 1 & 2 & 3 & 4 \\ 4 & 5 & 1 & 2 & 3 \\ 3 & 4 & 5 & 1 & 2 \\ 2 & 3 & 4 & 5 & 1 \end{bmatrix} - \lambda \begin{bmatrix} 1 & 0 & 0 & 0 & 0 \\ 0 & 1 & 0 & 0 & 0 \\ 0 & 0 & 1 & 0 & 0 \\ 0 & 0 & 0 & 1 & 0 \\ 0 & 0 & 0 & 0 & 1 \end{bmatrix} = 0$$

$$\begin{bmatrix} 1-\lambda & 2 & 3 & 4 & 5 \\ 5 & 1-\lambda & 2 & 3 & 4 \\ 4 & 5 & 1-\lambda & 2 & 3 \\ 3 & 4 & 5 & 1-\lambda & 2 \\ 2 & 3 & 4 & 5 & 1-\lambda \end{bmatrix} = 0$$

Sum of all elements in any one row must be zero,

i.e., $15-\lambda = 0$

$\therefore$ $\lambda = 15$

8. Since present output does not depend upon future values of input hence system is causal and also every bounded input produces bounded output, So we can say that system is stable.

9. Mutual information of two random variables is a measure of the mutual dependence of the two variables.

Given that, X and Y are independent.

Hence, $I(X : Y) = 0$

10. Any of the given three functions cannot be written as the linear combination of other two functions. Hence, the statements I and III are correct.

11. A good transconductance amplifier should have very high input resistance and very high output resistance.

12. As per Friis free space propagation equation,

$$W_r = \frac{W_t A_{er} A_{et}}{(\lambda d)^2}$$

where A_{er} and A_{et} are doubled and d is also doubled, and W_r is same. Hence capacity is also same.

Hence maximum capacicty of wireless link remains unchanged

13. Final slope = – 60 dB/decade,

$\Rightarrow$ $q - p = 3$

Now among options, option (a) satisfies the above condition.

14. $$M = \begin{bmatrix} 5 & 10 & 10 \\ 1 & 0 & 2 \\ 3 & 6 & 6 \end{bmatrix}$$

Now applying, $R_1 \leftrightarrow R_2$ in above matrix we get

$$\begin{bmatrix} 1 & 0 & 2 \\ 5 & 10 & 10 \\ 3 & 6 & 6 \end{bmatrix}$$

Applying $R_2 \leftarrow R_2 - 5R_1$ and $R_3 \leftarrow R_3 - 3R_1$, we get

$$\begin{bmatrix} 1 & 0 & 2 \\ 0 & 10 & 0 \\ 0 & 6 & 0 \end{bmatrix}$$

Now applying, $R_3 \leftarrow R_3 - \frac{6}{10}R_2$, we get

$$\begin{bmatrix} 1 & 0 & 2 \\ 0 & 10 & 0 \\ 0 & 0 & 0 \end{bmatrix}$$

The above matrix is in Echelon form

Rank of matrix M is given by

$$\rho(M) = 2$$

15. A BIBO stable system can have poles in right half of complex plane, if it is a non-causal system. So, statement-I is wrong.

A causal and BIBO stable system should have all poles in the left half of complex plane. So, statement-II is correct.

$\therefore$ option (*d*) is correct.

16. 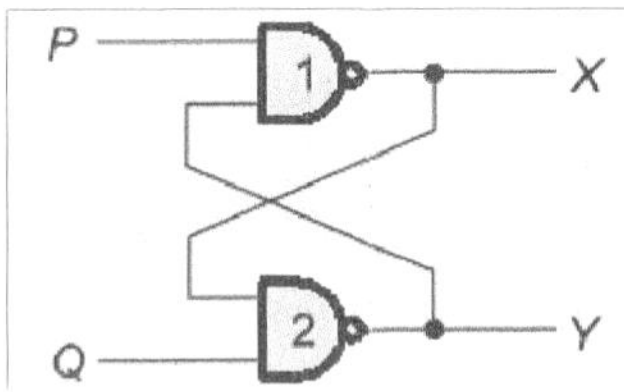

Present input condition : P = Q = 0

$\Rightarrow$ Corresponding outputs are X = Y = 1

When input condition is changed to

P = Q = 1 from P = Q = 0:

Possibility - 1 :

Let gate-2 is faster than gate-1, then the possible outputs are X = 1, Y = 0

Possibility - 2 :

Let gate-1 is faster than gate-2, then the possible outputs are X = 0, Y = 1

Hence, option (*b*) is correct.

17. From given data,

$$f_{CLK} = 5 \text{ MHz}$$

$$\text{Execution time} = 1.4\ \mu s$$

$$\text{Execution time} = n(\text{T} - \text{state})$$

where n is number of T-states required to execute the instruction

$$\therefore \quad T_{CLK} = \frac{1}{f_{CLK}} = 0.2\ \mu s$$

Hence $$n = \frac{1.4\ \mu s}{T_{CLK}} = \frac{1.4}{0.2} = 7$$

18. In DPCM, difference of message signal sample with its prediction is quantized.

19. 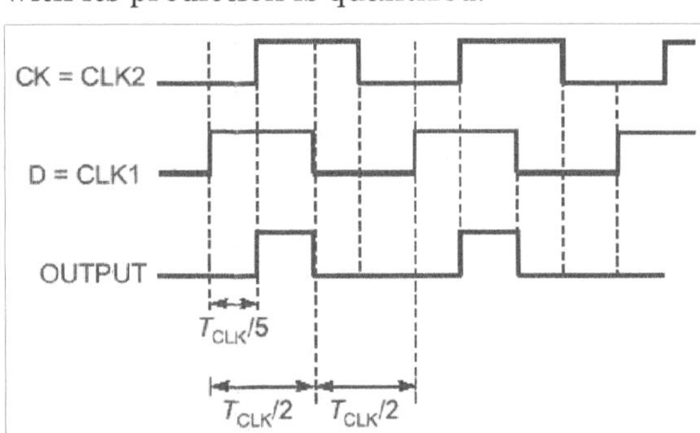

Duty cycle at the output

$$= \left[\frac{\frac{T_{CLK}}{2} - \frac{T_{CLK}}{5}}{T_{CLK}} \times 100\right]\%$$

$$= \left(\frac{3}{10} \times 100\right)\% = 30\%$$

20. Phase lag controller transfer function is given by

$$G_C(s) = \frac{s+Z}{s+P}; \text{ where } |Z| > |P|$$

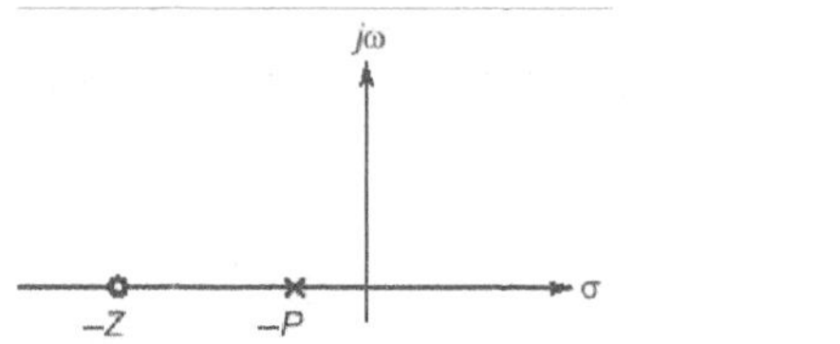

21. As Signal has odd and half wave symmetries. Hence, all a_n are zero and b_n are zero for n even.

22. Emitter-base junction (J_E) is in reverse biased.

Collector-base junction (J_C) is in Forward biased.

Hence, transistor is working in inverse active mode.

24. $$V_0 = V_T \ln \frac{n_1}{n_2}$$

$$= V_T \ln \frac{N_{D1}}{N_{D2}} \quad \left(\begin{array}{l} \because \quad n_1 = N_{D1} \\ \text{and } n_2 = N_{D2} \end{array}\right)$$

$$= 0.25 \ln \frac{10^{18}}{10^{15}}$$

$$= 0.025 \ln(1000)$$

$$= 0.173 \text{ V}$$

25. When three dice are thrown

Total number of possible cases

$$= (6)^3 = 216$$

Outcomes for all three dice having same number are,

$$\left\{\begin{array}{ccc} (1,1,1) & (2,2,2) & (3,3,3) \\ (4,4,4) & (5,5,5) & (6,6,6) \end{array}\right\}$$

Hence number of favourable cases = 6

$\therefore$ Required probability

$$= \frac{\text{Number of favourable cases}}{\text{Total number of outcomes}}$$

$$= \frac{6}{216} = \frac{1}{36} = 0.028$$

26. Since $$h[n] = \frac{1}{3}\delta[n] + \frac{1}{3}\delta[n-1] + \frac{1}{3}\delta[n-2]$$

$$\therefore \quad H(e^{j\omega}) = \frac{1}{3} e^{j\omega}[1 + 2\cos\omega]$$

$$\therefore \quad H(e^{j\omega_0}) = 0,$$

$$\therefore \quad (1 + 2\cos\omega_0) = 0$$

$$\therefore \quad \cos\omega_0 = -\frac{1}{2}$$

$$\therefore \quad \omega_0 = \frac{2\pi}{3} = 2.10 \text{ radians}$$

27.
$$x[n] = \{1, 2, 1\}$$
$$h[n] = \{1, a, b\}$$
$$y[n] = [A, 3, 4, B, C\}$$

x \ h	1	a	b
1	1	a	b
2	2	2a	2b
1	1	a	b

$y[0] = 1$; $y[1] = 2 + a$; $y[2] = 1 + 2a + b$; $y[3] = a + 2b$; $y[4] = b$

Given, $y[1] = 2 + a = 3 \Rightarrow a = 1$

and $y[2] = 1 + 2a + b = 4 \Rightarrow b = 1$

$\therefore$ $y[3] = a + 2b = 1 + 2(1) = 3$

and $y[4] = b = 1$

$\therefore$ $10y[3] + y[4] = (10 \times 3) + 1 = 31$

28.
$$i_s(t) = \frac{V}{R}\left[1 - e^{\frac{-Rt}{L}}\right]$$
$$i_s(t) = \frac{15}{1}\left[1 - e^{\frac{-3t}{2}}\right] \text{A}$$

Now current through 2 H,
$$i(t) = i_s(t)\frac{1}{1+2}$$
$$\therefore \quad i(t) = 5\left[1 - e^{\frac{-3t}{2}}\right] \text{A}$$

Now for $i(t) = 2$A,
$$\therefore \quad 2 = 5\left[1 - e^{\frac{-3t}{2}}\right]$$
$$\therefore \quad \frac{2}{5} = 1 - e^{-\frac{3t}{2}}$$
$$\therefore \quad 1 - \frac{2}{5} = e^{-\frac{3t}{2}}$$
$$\therefore \quad \frac{3}{5} = e^{-\frac{3t}{2}}$$
$$\therefore \quad t = 0.3405 \text{ sec}$$

29.
$$W = \sqrt{\frac{2\varepsilon V_R}{e}\left(\frac{1}{N_A} + \frac{1}{N_D}\right)} \quad \because V_R >> V_{bi}$$

Under reverse biased condition,
$$C = \frac{\varepsilon A}{W}$$
$$\Rightarrow \quad C \propto \frac{1}{W}$$

Hence
$$\frac{C_2}{C_1} = \frac{W_1}{W_2}$$
$$= \sqrt{\frac{\frac{2\varepsilon V_R}{e}\left(\frac{2}{10^{14}}\right)}{\frac{2\varepsilon V_R}{e}\left(\frac{2}{10^{16}}\right)}} = 10$$

30. Net hole density varying in the direction of x is,
$$p_n(x) = p_{no} + \Delta p$$
$$= p_{no} + G_L\tau_p \qquad (\because \Delta p = G_L\tau_p)$$
$$= p_{no} + G_L\tau_p\left(1 - \frac{x}{L}\right)$$
$$\left(\because G_L = G_{L0}\left(1 - \frac{x}{L}\right) \text{given}\right)$$

Now,
$$J_{p.diff} = -eD_p\frac{dp}{dx}$$
$$= -eD_p\left[\frac{-G_{L0}\tau_p}{L}\right]$$
$$= \frac{1.6 \times 10^{-19} \times 100 \times 10^{17} \times 10^{-4}}{0.1 \times 10^{-4}} \text{ A/cm}^2$$
$$= 16 \times 10^{-19+2+17-4+4} = 16 \text{ A/cm}^2$$

31.

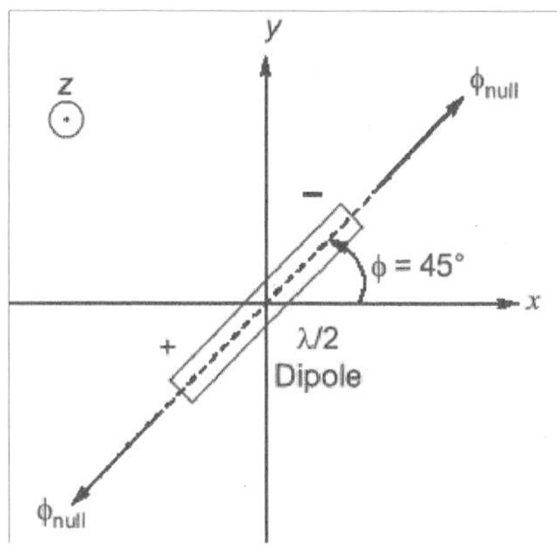

As the dipole is in X-Y plane, the radiation pattern has nulls for $\theta = 90°$, $\phi = 45°$ and $225°$

for $0 \le \phi \le \pi$, $\phi = 45°$

32. Rouths Hurwitz criteria
$$f(s) = s^4 + s^2 + 1 = 0$$

$$\begin{array}{c|ccc} s^4 & 1 & 1 & 1 \\ s^3 & \not{0}4 & \not{0}2 & 0 \\ s^2 & 0.5 & 1 & 0 \\ s^1 & -6 & 0 & 0 \\ s^0 & 1 & 0 & 0 \end{array} \qquad \frac{dA(s)}{ds} = 4s^3 + 2s^1$$

There are two sign changes in the first column of the R-H table and the order of auxillary equation is 4. Hence, four poles are symmetric about origin.

$\therefore$ 2 poles are in right half of S plane and 2 poles are in left half of S plane

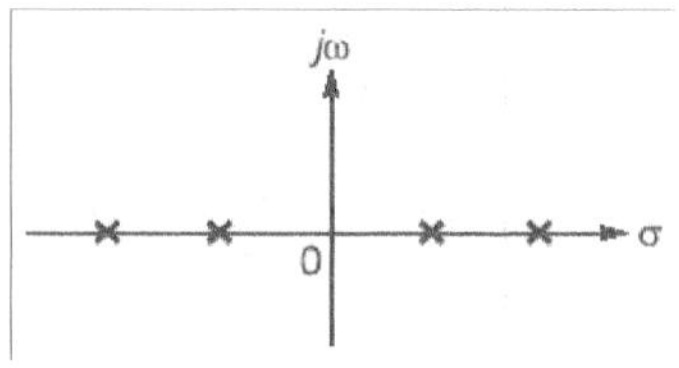

33. $V_{Th} = \dfrac{12 \times 47}{73 + 47} = 4.7\text{ V}$

$I_C = \dfrac{V_{Th} - V_{BE}}{R_E}$

$= \dfrac{4.7 - 0.7}{2} = \dfrac{4}{2} = 2\text{ mA}$

Now $g_m = \dfrac{I_C}{V_T} = \dfrac{2}{25} = 80\text{ m℧}$

Now $A_V = -g_m R'_L = -80 \times \dfrac{2 \times 8}{10} = -128$

$\therefore$ $|A_V| = 128$

34. $V_{GS} = \dfrac{8 \times 5}{8} - 1 \times I_D = 5 - I_D$

$\therefore$ $I_D = 5 - V_{GS}$ ...(*i*)

Now, $I_D = \dfrac{\mu_0 C_{ox}}{2}\dfrac{W}{L}(V_{GS} - V_T)^2$

$\therefore$ $5 - V_{GS} = \dfrac{1}{2}(V_{GS} - 1)^2$

$\therefore$ $10 - 2V_{GS} = V_{GS}^2 + 1 - 2V_{GS}$

$\therefore$ $V_{GS}^2 = 9 \Rightarrow V_{GS} = 3\text{ V}$

Hence $I_D = 5 - V_{GS} = 5 - 3 = 2\text{ mA}$

35. $E = \dfrac{V}{d} = \dfrac{5}{10^{-4}} = 5 \times 10^4\text{ V/cm}$

slope of the curve, (m)

$= \dfrac{10^7 - 0}{5 \times 10^5} = 20 \quad \left(\begin{array}{l} \because y = mx \\ \Rightarrow m = \dfrac{y}{x} \end{array}\right)$

Now $v_d = 20 \times E = 20 \times 5 \times 10^4 = 10^6\text{ V/cm}$

and $J = nev_d = 1 \times 10^{16} \times 1.6 \times 10^{-19} \times 10^6$

$= 1.6 \times 10^3\text{ A/cm}^2 = 1.6\text{ kA/cm}^2$

36. Given that, $G(s) = \dfrac{K(s^2 + 2s + 2)}{s^2 - 3s + 2}$

The root locus plot for the given system

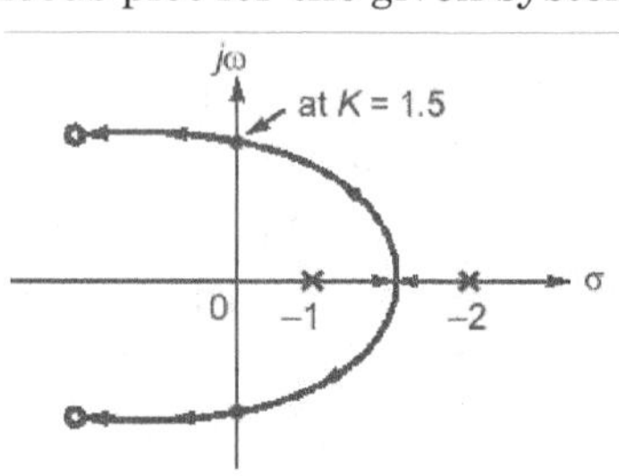

$\therefore$ System is stable for $K > 1.5$.

37. In the given circuit,

Feedback factor $(\beta) = \dfrac{R_1}{R_1 + R_2} = \dfrac{1}{80}$

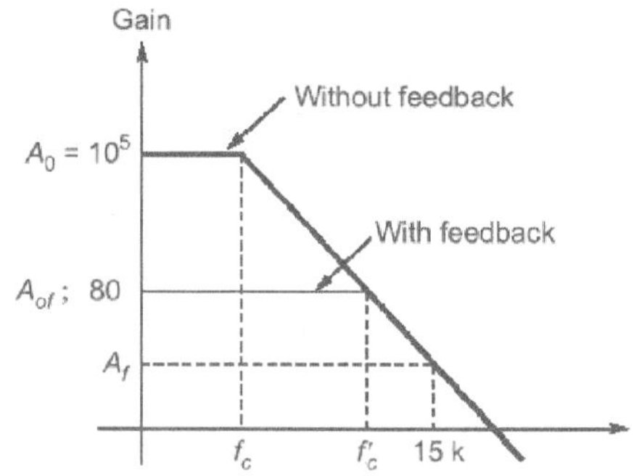

$\therefore$ $A_{nf} = \dfrac{A_0}{1 + A_0\beta} \simeq 80$

and $f'_c = f_c(1 + A_0\beta)$

$= 8\left(1 + \dfrac{10^5}{80}\right)\text{ Hz} = 10{,}008\text{ Hz}$

Now Gain at $f = 15$ kHz given by

$A_f = \dfrac{A_{0f}}{\sqrt{1 + \left(\dfrac{f}{f'_c}\right)}}$

$= \dfrac{80}{\sqrt{\left(1 + \left(\dfrac{15000}{10008}\right)^5\right)}} \simeq 44.4$

39. $f(x) = x^3 + x -$

Now $f(1) = (1)^3 + 1 - 1 = 1$

and $f'(x) = \dfrac{d}{dx}(x^3 + x - 1) = 3x^2 + 1$

$\therefore$ $f'(1) = 4$

Now by Newton-Raphson method,

$x_{i+1} = x_i - \dfrac{f(x_i)}{f'(x_i)}$

For $x_0 = 1$, $x_1 = x_0 - \dfrac{f(x_0)}{f'(x_0)}$

$= 1 - \dfrac{1}{4} = 0.75$

and for $x_1 = 0.75$

$x_2 = x_1 - \dfrac{f(x_1)}{f'(x_1)}$

$= 0.75 - \dfrac{f(0.75)}{f'(0.75)}$

$= 0.75 - \dfrac{0.171875}{2.6875}$

$= 0.686 \approx 0.68$

40. The given input power spectral density is as follows:

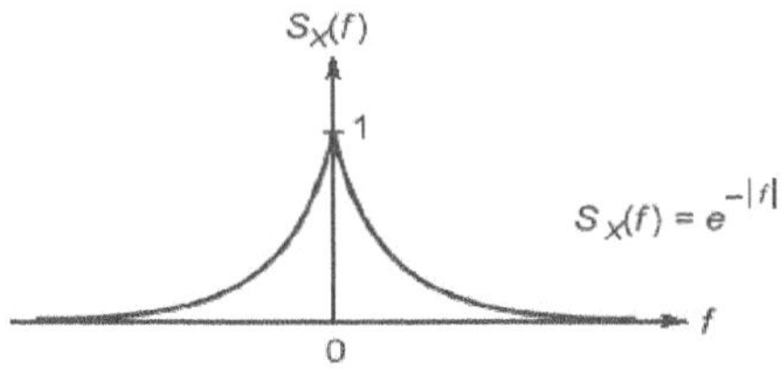

Now frequency response of the low pass filter is as follows:

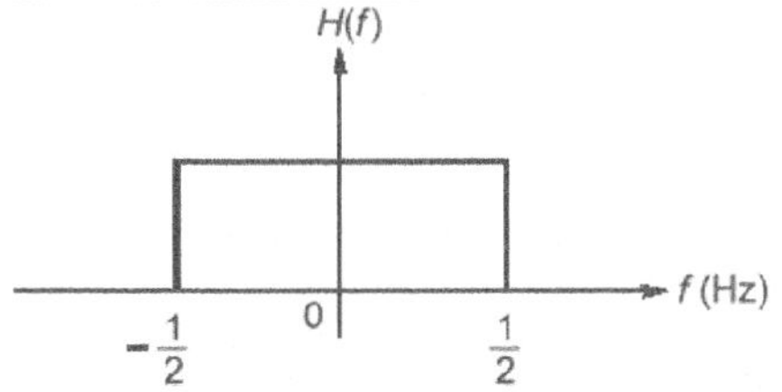

Now, $E[Y(t)] = H(0)\,E[X(t)]$

and $H(0) = 1$

$\therefore$ $E[Y(t)] = E[X(t)]$

Now $E[Y^2(t)] \neq E[X^2(t)]$

Since, LPF does not allow total power from input to output.

$$E[X^2(t)] = \int_0^{\infty} S_x(f)\,df = 2\text{ W}$$

As $E[Y^2(t)] \neq E[X^2(t)]$.

$E[Y^2(t)] \neq 2$

Hence only statement - I is correct.

41. A(0, 2, 1) and 6(4, 1, –1)

The equation of the line AB is

$$\frac{x-0}{4-0} = \frac{y-2}{1-2} = \frac{z-1}{-1-1} = t(\text{say})$$

$\therefore$ $x = 4t;$ $y = -t + 2;$ $z = -2t + 1$

$\therefore$ $dx = 4dt;$ $dy = -dt;$ $dz = -2dt$

Also t varies from 0 to 1

$$\therefore I = \int_0^1 2(-2t+1)4\,dt + 2(-t+2)(-dt) + 2(4t)(-2\,dt)$$

$$= \int_0^1 (-16t + 8 + 2t - 4 - 16t)\,dt = \int_0^1 (-30t + 4)\,dt$$

$$= \left(-30\frac{t^2}{2} + 4t\right)\Bigg|_0^1 = (-15 + 4) = -11$$

42. Given that,

$$G(s) = \frac{K}{(s^2 + 2s + 2)(s + 2)}$$

For K = 10; N = P – Z = 0 $\Rightarrow$ Z = 0 : stable

For K = 100; N = P – Z = 1 $\Rightarrow$ Z $\neq$ 0 : unstable

Where P is open loop poles in RHS of s-plane

Z is closed loop poles in RHS of s-plane

N = number of encirclements about the point $(-1 + j0)$

43.

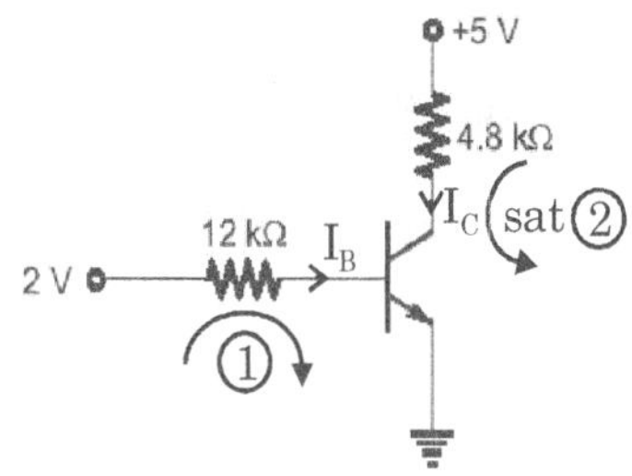

From Loop 1

$$\therefore \quad I_B = \frac{2-0.7}{12} = 0.10833\text{ mA}$$

From loop 2

$$\therefore \quad I_{C(sat)} = \frac{5-0.2}{4.8} = 1\text{ mA}$$

Now $$I_B \geq I_{B(min)} = \frac{I_{C(sat)}}{\beta}$$

$$\therefore \quad I_B \geq \frac{1\text{ mA}}{\beta}$$

$$\therefore \quad \beta \geq \frac{1}{0.10833} \text{ and } \beta_{min} = 9.23$$

Now $$\alpha_{min} = \frac{\beta_{min}}{1+\beta_{min}} = \frac{9.23}{1+9.23} = 0.902$$

44.

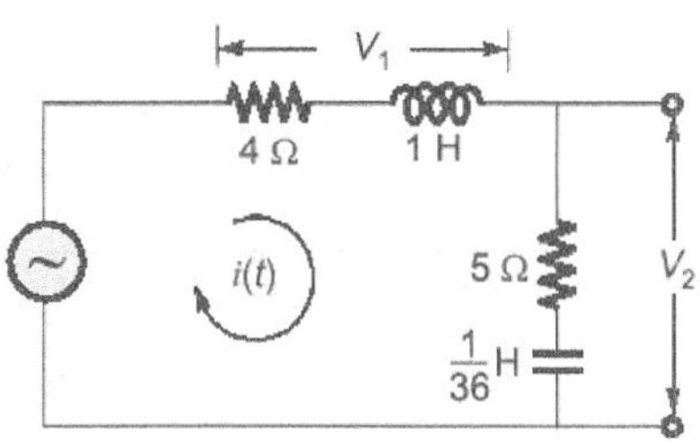

$\omega = 3$ rad/sec

and $Z_1 = (4 + j3)\Omega$

and $Z_2 = (5 - j12)\,\Omega$

$|V_1| = |i|\,|z_1| = |i|\sqrt{4^2 + 3^2} = 5|i|$

and $|V_2| = |i|\,|z_2| = |i|\sqrt{5^2 + 12^2} = 13|i|$

Now, $$\frac{|V_2|}{|V_1|} = \frac{13|i|}{5|i|} = \frac{13}{5} = 2.6$$

45. Initially the wave is circularly polarized. So, the initial phase difference between field components in $\hat{a}_x$ direction and $\hat{a}_y$ direction is $\frac{\pi}{2}$.

To become linearly polarized, the wave must travel a minimum distance, such that, the phase difference at that point between the field components in $\hat{a}_x$ direction and $\hat{a}_y$ direction is π (i.e., the travel of this minimum distance should

provide an additional phase difference of $\frac{\pi}{2}$ between $\hat{a}_x$ and $\hat{a}_y$ field components).

$$\therefore \quad z_{min}k_x \sim z_{min}k_y = \frac{\pi}{2}$$

$$\therefore \quad z_{min}\left[\frac{\omega}{v_{px}} \sim \frac{\omega}{v_{py}}\right] = \frac{\pi}{2}$$

$$\left[\because k_x = \frac{\omega}{v_{px}} \text{ and } k_y = \frac{\omega}{v_{py}}\right]$$

$$\therefore \quad 2\pi z_{min}\left[\frac{f}{c}\sqrt{\varepsilon_{rx}} \sim \frac{f}{c}\sqrt{\varepsilon_{rx}}\right] = \frac{\pi}{2}$$

$$\therefore \quad \frac{4z_{min}}{\lambda_0}[n_x \sim n_y] = 1$$

$$\therefore \quad z_{min} = \frac{\lambda_0}{4(n_x \sim n_y)}$$

$$= \frac{1.5}{4(1.5 \sim 1.5001)}\mu m$$

$$= \frac{1.5}{4(0.0001)}\mu m$$

$$= \frac{1.5}{4}\text{ cm}$$

$$\therefore \quad z_{min} = 0.375 \text{ cm}$$

46.

Clock Number	$D_{in} = A \oplus D$ Just before clock	A	B	C	D
Initial	-	1	1	0	1
1	0	0	1	1	0
2	0	0	0	1	1
3	1	1	0	0	1
4	0	0	1	0	0
5	0	0	0	1	0
6	0	0	0	0	1
7	1	1	0	0	0
8	1	1	1	0	0
9	1	1	1	1	0
10	1	1	1	1	1

Hence inorder to reach the state ABCD = 1111, total 10 clock pulse is required.

47. The given Boolean function is,

$$F = m_0 + m_2 + m_3 + m_5$$

Now from x-map

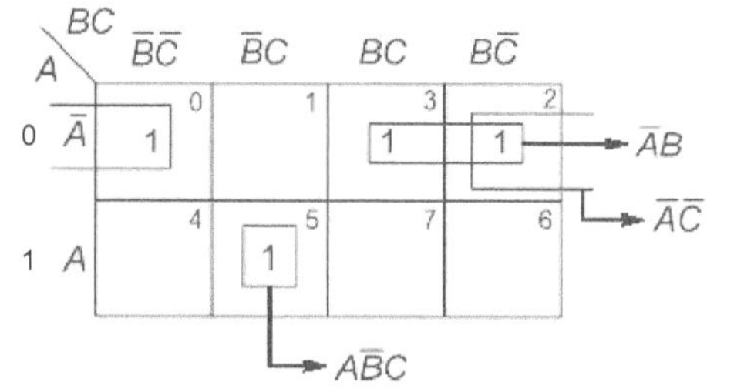

$$F = \overline{A}\overline{C} + \overline{A}B + A\overline{B}C$$

48. Initially T = 1 sec, so $\omega_0 = 2\pi$ rad/sec. When $x(t)$ is compressed by 3, frequency will expand by same factor but there is no change in values of a_k, a_k remains constant.

So, both statement I and III are correct.

49. $$x(t) = 4\cos 200\pi t + 8\cos 400\pi t$$

Taking fourier transform of $h(t)$ we get

$$h(t) = \frac{2\sin 300\pi t}{\pi t} \xleftrightarrow{\text{F.T}}$$

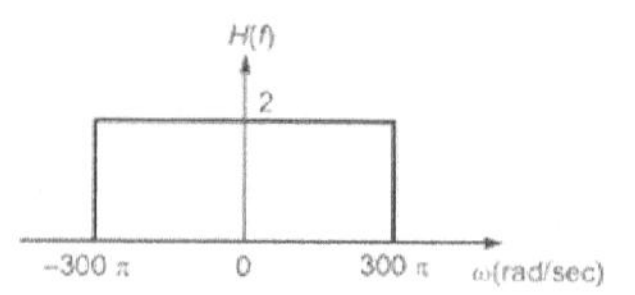

So, $y(t) = 8\cos 200\pi t$

Hence $|y(t)|_{max} = 8$

50. MVI A, 33 H : (A) = 33 H

MVI B, 78 H : (B) = 78 H

ADD B : (A) ← (A) + (B)

= 33 H + 78 H

= AB H

⇒ (A) = AB H

(A) = 1010 1011

CMA : Complement Accumulator

⇒ (A) = 0101 0100

ANI 32 H : (A) ← (A) AND 32 H

0101 0100

0011 0010

0001 0000

⇒ (A) = 10 H

51. From the given diagram

$$D_A = Q_A \oplus Q_B$$

and $$D_B = \overline{Q_A X_{IN}}$$

For $$X_{IN} = 0$$

$$D_A = Q_A \oplus Q_B$$

and $$D_B = 1$$

Present State				Next State	
Q_A	Q_B	$D_A = Q_A \oplus Q_B$	D_B	Q_A^-	Q_B^-
0	0	0	1	0	1
0	1	1	1	1	1
1	1	0	1	0	1
0	0	1	1	1	1

So, for $X_{IN} = 0$,

Number of possible states = 2

For $X_{IN} = 1$:

$$D_A = Q_A \oplus Q_B$$

and $$D_B = \overline{Q_A}$$

Present State				Next State	
Q_A	Q_B	$D_A = Q_A \oplus Q_B$	D_B	Q_A^-	Q_B^-
0	0	0	1	0	1
0	1	1	1	1	1
1	1	0	0	0	0
0	0	0	1	0	1

So, for $X_{IN} = 1$,

Number of possible states = 3

So, the option (*d*) is correct.

52. At $x = 0$,

$$f(x) = e^x e^{x^2}$$

$$= \left(1 + x + \frac{x^2}{2!} + \frac{x^3}{3!} + \ldots\right)\left(1 + x^2 + \frac{x^4}{2!} + \frac{x^6}{3!} + \ldots\right)$$

$$= 1 + x^2 + \frac{x^4}{2} + \frac{x^6}{6} + \ldots + x + x^3 + \frac{x^5}{2} + \frac{x^7}{6} + \ldots$$

$$+ \frac{x^2}{2} + \frac{x^4}{2} + \frac{x^6}{4} + \frac{x^8}{12} \ldots + \frac{x^3}{6} + \frac{x^5}{6} + \frac{x^7}{12} + \frac{x^9}{36} + \ldots$$

$$= 1 + x + \frac{3}{2}x^2 + \frac{7}{6}x^3$$

53. Let $x^2 + y^2 = t^2$

and $t^2 = z^3$

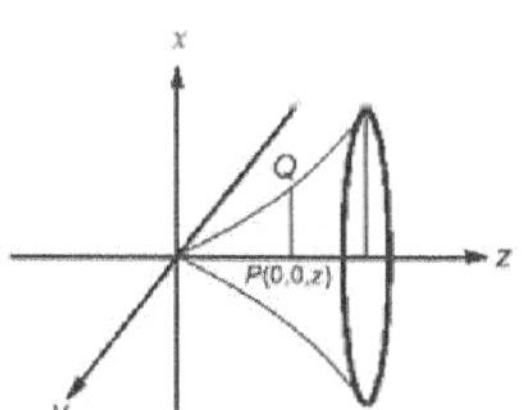

Here revolution is about *z* axis

Now volume of region R = $\int_0^1 \pi(PQ)^2\, dz$

Where PQ is radius of circle at some z.

Now $PQ = \sqrt{x^2 + y^2}$

$\therefore$ $(PQ)^2 = x^2 + y^2 = z^3$

$\therefore$ Volume of region R = $\int_0^1 \pi t^2 dz = \int_0^1 \pi z^3 dz$

$$= \frac{\pi}{4} = 0.7853$$

54. $u_0(t) = 5 \cos 20000\, \pi t$

and $u_1(t) = 5 \cos 22000\, \pi t$

Also $f_1 = 11000$ Hz

and $f_2 = 10000$ Hz

Now for FSK waveforms to be uncorrelated,

$$f_1 - f_2 = n\frac{R_b}{2};$$

$$n = 1, 2, 3, \ldots..$$

$\therefore$ $R_b = \dfrac{2(f_1 - f_2)}{n} = \dfrac{2000}{n}$ bits/sec

$\Rightarrow$ $R_{b(\max)} = 2000$ bits/sec

($\because$ $n = 1$ is minimum value)

Now $T_{b(\min)} = \dfrac{1}{R_{b(\max)}} = 0.5$ ms

55. $$\frac{dy}{dx} = (x + y - 1)^2 \quad \ldots(i)$$

Let $x + y - 1 = t$ ...(*ii*)

Now differiating equation (*ii*) w.r.t to '*x*' we get

$$1 + \frac{dy}{dx} = \frac{dt}{dx}$$

$\therefore$ $$\frac{dy}{dx} = \frac{dt}{dx} - 1$$

Substituting equations (*ii*) and (*iii*) in equation (*i*) we get

$$\frac{dt}{dx} - 1 = t^2$$

$$\frac{dt}{dx} = t^2 + 1$$

$$\frac{dt}{t^2 + 1} = dx$$

Now integrating on both side we get,

$$\int \frac{1}{t^2 + 1} dt = \int dx$$

$\therefore$ $\tan^{-1} t = x + c$

$\therefore$ $x + y - 1 = \tan(x + c)$ ($\because$ $t = x + y - 1$)

$\therefore$ $y = 1 - x + \tan(x + c)$

GENERAL APTITUDE (GA)

1. Let H is household consumption and P be other consumption

   Now according to question,

   $$H \times 0.8 + P \times 1.7 = (H + P) \times 0.75$$

   From this it is concluded that ratio is negative.

2. From given condition

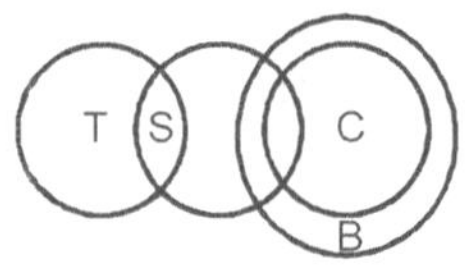

   Only conclusion (*ii*) follows

3. Use of conditional sentence based on past participle form.

   I could have made arrangements had I been informed earlier.

4. $$40\% \text{ of } 360° = \frac{40}{100} \times 360° = 144°$$

   So, the angle subtended on pie chart will be 144°

5. Hurtful means causing pain or suffering or something that is damaging or harmful. The expression 'sharp tongue' defines a bitter or critical manner of speaking.

6. According to question,

   Length of Truck + gap required = 10 + 20 = 30 m

   and

   Length of Car + gap required = 5 + 15 = 20 m

   Alternative pairs of Truck and Car needs

   30 + 20 = 50 m

   Let '*n*' be the number of repetition of (Truck + Car) in 1 hour (3600 secs)

   Given speed = 36 km/hr

   = 10 m/sec

   $$\therefore \quad \frac{50m \times n}{3600 \text{ secs}} = 36 \text{ km/hr}$$

   $$\therefore \quad \frac{50n}{3600} \text{ m/sec} = 10 \text{ m/sec}$$

   $$\therefore \quad n = \frac{36000}{50}$$

   = 720 (Trucks + Cars)

   So, 720 (Trucks + Cars) passes

   = (720 × 2) = 1440 vehicles

7. Antagonist is a adversary or one who opposes/contends against another. Hence the closest meaning to word "antagonistic" is hostile.

8. Contour lines can be observed to cross region with height

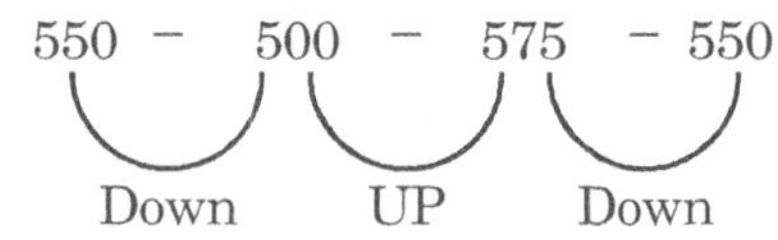

   Down-Up-Down satisfies.

9. Subgroups containing only Indians

   $$= {}^3C_1 + {}^3C_2 + {}^3C_3 = 3 + 3 + 1 = 7$$

   and Subgroups containing one Indian and rest Chinese

   $$= {}^3C_1[{}^3C_1 + {}^3C_2 + {}^3C_3]$$
   $$= 3[3 + 3 + 1] = 21$$

   and Subgroups containing two Indian and remaining Chinese

   $$= {}^3C_2[{}^3C_1 + {}^3C_2 + {}^3C_3] = 21$$

   and Subgroups containing three Indian and remaining Chinese

   $$= {}^3C_3[{}^3C_1 + {}^3C_2 + {}^3C_3] = 7$$

   Hence total number of subgroups

   $$= 7 + 21 + 21 + 7 = 56.$$

10. Following circular seating arrangement can be drawn from the given data

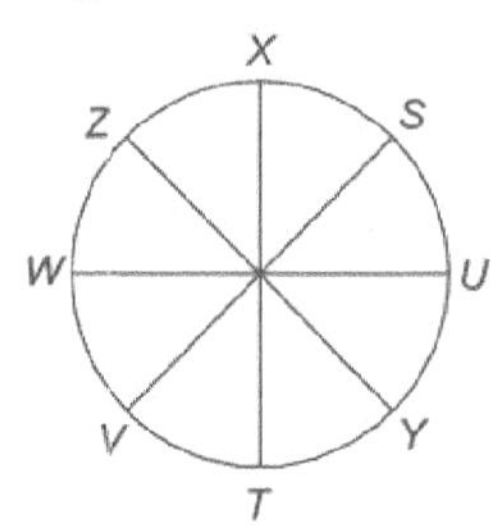

   From the given arrangement 'X' is the third to the left of 'V'.

SOLVED PAPER – 2016

ELECTRONICS & COMMUNICATION ENGINEERING
Set - 1

No. of Questions : 65 *Maximum Marks : 100*

INSTRUCTIONS

1. Total of 65 questions carrying 100 marks, out of which 10 questions carrying a total of 15 marks are in General Aptitude (GA)
2. The Engineering Mathematics will carry around **15% of the total marks,** the General Aptitude section will carry **15% of the total marks** and the **remaining 70% of the total marks.**
3. **Types of Questions**
   (*a*) **Multiple Choice Questions (MCQ)** carrying 1 or 2 marks each in all papers and sections. These questions are objective in nature, and each will have a choice of four options, out of which the candidate has to mark the correct answer(s).
   (*b*) **Numerical Answer Questions** of 1 or 2 marks each in all papers and sections. For these questions the answer is a real number, to be entered by the candidate using the virtual keypad. No choices will be shown for these type of questions.
4. For **1-mark** multiple-choice questions, **1/3 marks** will be deducted for a wrong answer. Likewise, for **2-marks** multiple-choice questions, 2/3 marks will be deducted for a wrong answer. There is no negative marking for numerical answer type questions.

GENERAL APTITUDE (GA)

(Q.1 – 5) : Carry One Mark Each

1. Which of the following is CORRECT with respect to grammar and usage?
   Mount Everest is ____________.
   (*a*) the highest peak in the world
   (*b*) highest peak in the world
   (*c*) one of highest peak in the world
   (*d*) one of the highest peak in the world
2. The policeman asked the victim of a theft, "What did you ?"
   (*a*) loose (*b*) lose
   (*c*) loss (*d*) louse
3. Despite the new medicine's ______________ in treating diabetes, it is not ______________widely.
   (*a*) effectiveness --- prescribed
   (*b*) availability --- used
   (*c*) prescription --- available
   (*d*) acceptance --- proscribed
4. In a huge pile of apples and oranges, both ripe and unripe mixed together, 15% are unripe fruits. Of the unripe fruits, 45% are apples. Of the ripe ones, 66% are oranges. If the pile contains a total of 5692000 fruits, how many of them are apples?
   (*a*) 2029198 (*b*) 2467482
   (*c*) 2789080 (*d*) 3577422
5. Michael lives 10 km away from where I live. Ahmed lives 5 km away and Susan lives 7 km away from where I live. Arun is farther away than Ahmed but closer than Susan from where I live. From the information provided here, what is one possible distance (in km) at which I live from Arun's place?
   (*a*) 3.00
   (*b*) 4.99
   (*c*) 6.02
   (*d*) 7.01

(Q.6 – 10) : Carry Two Marks Each

6. A person moving through a tuberculosis prone zone has a 50% probability of becoming infected. However, only 30% of infected people develop the disease. What percentage of people moving through a tuberculosis prone zone remains infected but does not show symptoms of disease?
   (*a*) 15
   (*b*) 33
   (*c*) 35
   (*d*) 37
7. In a world filled with uncertainty, he was glad to have many good friends. He had always assisted them in times of need and was confident that they would reciprocate. However, the events of the last week proved him wrong.
   Which of the following inference(s) is/are logically valid and can be inferred from the above passage?

(*i*) His friends were always asking him to help them.

(*ii*) He felt that when in need of help, his friends would let him down.

(*iii*) He was sure that his friends would help him when in need.

(*iv*) His friends did not help him last week.

(*a*) (*i*) and (*ii*) (*b*) (*iii*) and (*iv*)

(*c*) (*iii*) only (*d*) (*iv*) only

8. Leela is older than her cousin Pavithra. Pavithra's brother Shiva is older than Leela. When Pavithra and Shiva are visiting Leela, all three like to play chess. Pavithra wins more often than Leela does. Which one of the following statements must be TRUE based on the above?

(*a*) When Shiva plays chess with Leela and Pavithra, he often loses.

(*b*) Leela is the oldest of the three.

(*c*) Shiva is a better chess player than Pavithra.

(*d*) Pavithra is the youngest of the three.

9. If $q^{-a} = \frac{1}{r}$ and $r^{-b} = \frac{1}{s}$ and $s^{-c} = \frac{1}{q}$, the value of abc is .

(*a*) $(rqs)^{-1}$ (*b*) 0

(*c*) 1 (*d*) r^+q^+s

10. P, Q, R and S are working on a project. Q can finish the task in 25 days, working alone for 12 hours a day. R can finish the task in 50 days, working alone for 12 hours per day. Q worked 12 hours a day but took sick leave in the beginning for two days. R worked 18 hours a day on all days. What is the ratio of work done by Q and R after 7 days from the start of the project?

(*a*) 10:11 (*b*) 11:10

(*c*) 20:21 (*d*) 21:20

TECHNICAL SECTION

(Q.1 – 25) : Carry One Mark Each

1. Let $M^4 = I$, (where I denotes the identity matrix) and $M \neq I$, $M^2 \neq I$ and $M^3 \neq I$. Then, for any natural number k, M^{-1} equals:

(*a*) M^{4k+1} (*b*) M^{4k+2}

(*c*) M^{4k+3} (*d*) M^{4k}

2. The second moment of a Poisson-distributed random variable is 2. The mean of the random variable is _______.

3. Given the following statements about a function $f: \mathbb{R} \to \mathbb{R}$, select the right option:

P : If $f(x)$ is continuous at $x = x_0$, then it is also differentiable at $x = x_0$.

Q : If $f(x)$ is continuous at $x = x_0$, then it may not be differentiable at $x = x_0$.

R : If $f(x)$ is differentiable at $x = x_0$, then it is also continuous at $x = x_0$.

(*a*) P is true, Q is false, R is false

(*b*) P is false, Q is true, R is true

(*c*) P is false, Q is true, R is false

(*d*) P is true, Q is false, R is true

4. Which one of the following is a property of the solutions to the Laplace equation: $\nabla^2 f = 0$?

(*a*) The solutions have neither maxima nor minima anywhere except at the boundaries.

(*b*) The solutions are not separable in the coordinates.

(*c*) The solutions are not continuous.

(*d*) The solutions are not dependent on the boundary conditions.

5. Consider the plot of $f(x)$ versus x as shown below.

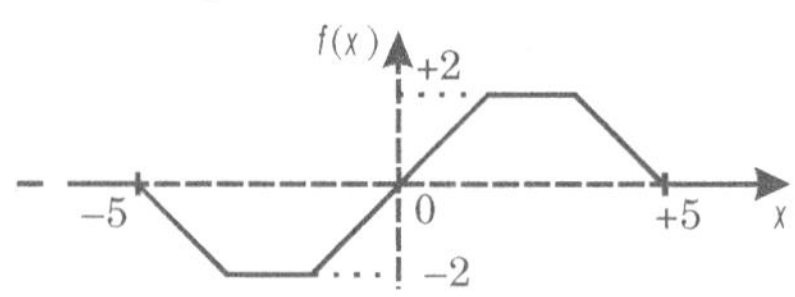

Suppose $F(x) = \int_{-5}^{x} f(y)dy$. Which one of the following is a graph of $F(x)$?

(*a*) 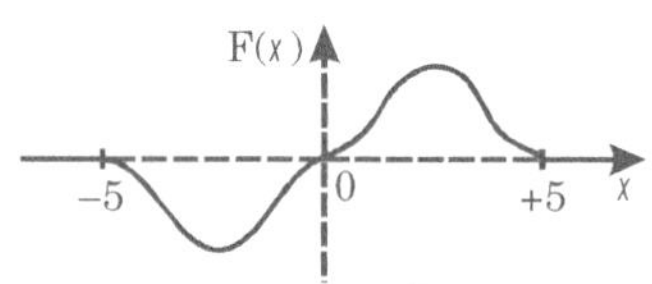

(*b*) 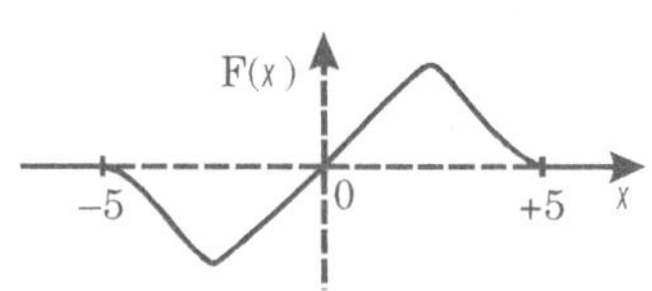

(*c*) 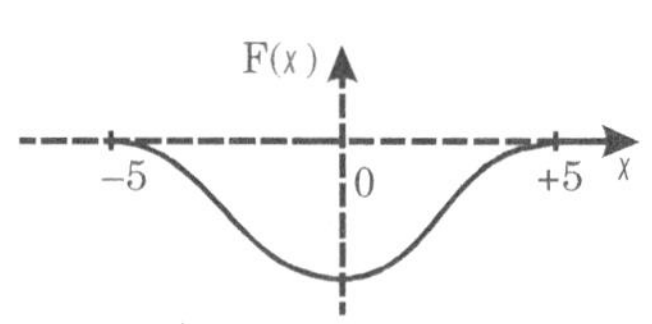

(*d*) 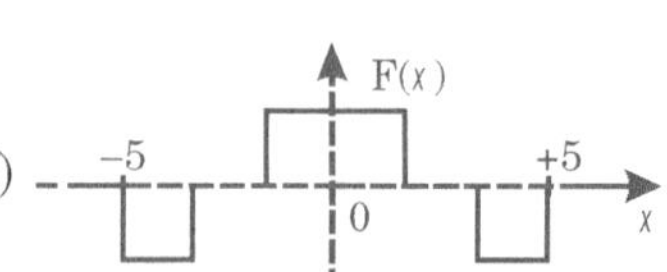

6. Which one of the following is an eigen function of the class of all continuous-time, linear, time-invariant systems ($u(t)$ denotes the unit-step function)?

(*a*) $e^{j\omega_0 t}u(t)$ (*b*) $\cos(\omega_0 t)$

(*c*) $e^{j\omega_0 t}$ (*d*) $\sin(\omega_0 t)$

7. A continuous-time function $x(t)$ is periodic with period T. The function is sampled uniformly with a sampling period T_s. In which one of the following cases is the sampled signal periodic?

(a) $T = \sqrt{2}\, T_2$

(b) $T = 1.2\, T_s$

(c) Always

(d) Never

8. Consider the sequence $x[n] = a^n u[n] + b^n u[n]$, where $u[n]$ denotes the unit-step sequence and $0 < |a| < |b| < 1$. The region of convergence (ROC) of the z-transform of $x[n]$ is

(a) $|z| > |a|$

(b) $|z| > |b|$

(c) $|z| < |a|$

(d) $|a| < |z| < |b|$

9. Consider a two-port network with the transmission matrix: $T = \begin{pmatrix} A & B \\ C & D \end{pmatrix}$. If the network is reciprocal, then

(a) $T^{-1} = T$

(b) $T^2 = T$

(c) Determinant (T) = 0

(d) Determinant (T) = 1

10. A continuous-time sinusoid of frequency 33 Hz is multiplied with a periodic Dirac impulse train of frequency 46 Hz. The resulting signal is passed through an ideal analog low-pass filter with a cutoff frequency of 23 Hz. The fundamental frequency (in Hz) of the output is ________.

11. A small percentage of impurity is added to an intrinsic semiconductor at 300 K. Which one of the following statements is true for the energy band diagram shown in the following figure?

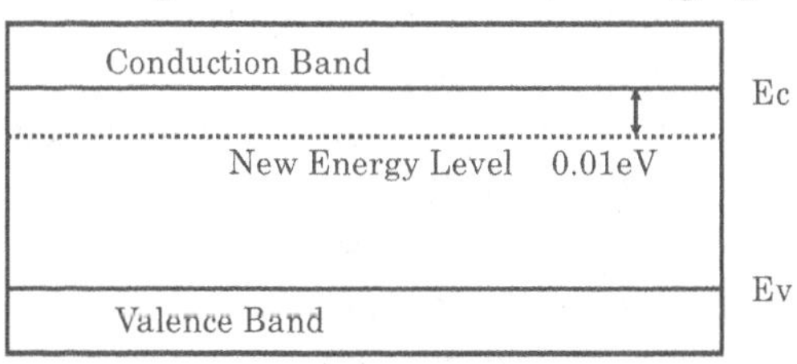

(a) Intrinsic semiconductor doped with pentavalent atoms to form n-type semiconductor

(b) Intrinsic semiconductor doped with trivalent atoms to form n-type semiconductor

(c) Intrinsic semiconductor doped with pentavalent atoms to form p-type semiconductor

(d) Intrinsic semiconductor doped with trivalent atoms to form p-type semiconductor

12. Consider the following statements for a metal oxide semiconductor field effect transistor (MOSFET):

P : As channel length reduces, OFF-state current increases.

Q : As channel length reduces, output resistance increases.

R : As channel length reduces, threshold voltage remains constant.

S : As channel length reduces, ON current increases.

Which of the above statements are INCORRECT?

(a) P and Q

(b) P and S

(c) Q and R

(d) R and S

13. Consider the constant current source shown in the figure below. Let β represent the current gain of the transistor.

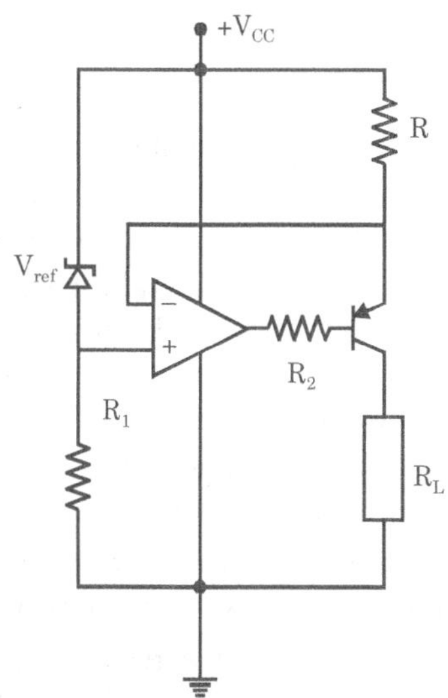

The load current I_0 through R_L is

(a) $I_0 = \left(\frac{\beta+1}{\beta}\right)\frac{V_{ref}}{R}$ (b) $I_0 = \left(\frac{\beta}{\beta+1}\right)\frac{V_{ref}}{R}$

(c) $I_0 = \left(\frac{\beta+1}{\beta}\right)\frac{V_{ref}}{2R}$ (d) $I_0 = \left(\frac{\beta}{\beta+1}\right)\frac{V_{ref}}{2R}$

14. The following signal V_i of peak voltage 8 V is applied to the non-inverting terminal of an ideal opamp. The transistor has $V_{BE} = 0.7$ V, $\beta = 100$; $V_{LED} = 1.5$ V, $V_{CC} = 10$ V and $-V_{CC} = -10$ V.

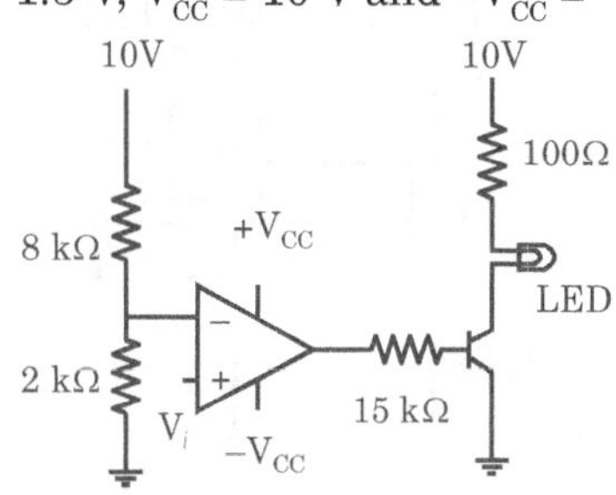

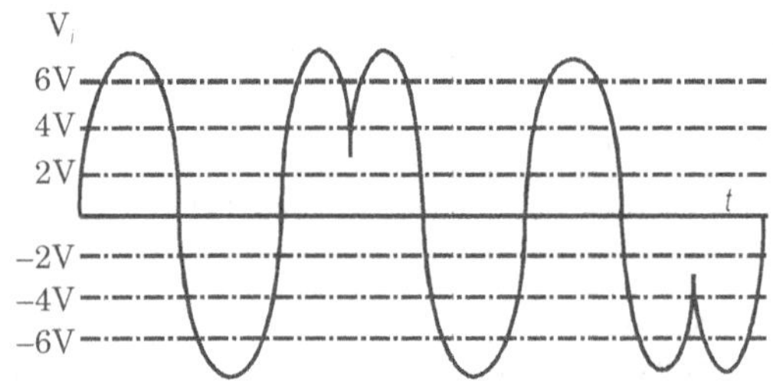

The number of times the LED glows is ________.

15. Consider the oscillator circuit shown in the figure. The function of the network (shown in dotted lines) consisting of the 100 kΩ resistor in series with the two diodes connected back-to-back is to:

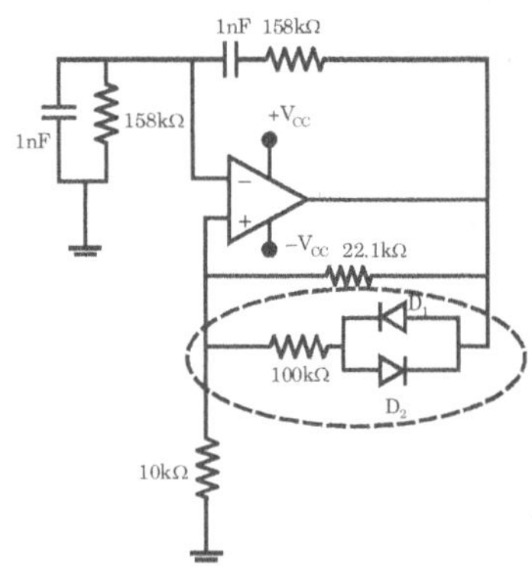

(*a*) introduce amplitude stabilization by preventing the op amp from saturating and thus producing sinusoidal oscillations of fixed amplitude

(*b*) introduce amplitude stabilization by forcing the opamp to swing between positive and negative saturation and thus producing square wave oscillations of fixed amplitude

(*c*) introduce frequency stabilization by forcing the circuit to oscillate at a single frequency

(*d*) enable the loop gain to take on a value that produces square wave oscillations

16. The block diagram of a frequency synthesizer consisting of a Phase Locked Loop (PLL) and a divide-by-N counter (comprising ÷ 2, ÷ 4, ÷ 8, ÷ 16 outputs) is sketched below. The synthesizer is excited with a 5 kHz signal (Input 1). The free-running frequency of the PLL is set to 20 kHz. Assume that the commutator switch makes contacts repeatedly in the order 1-2-3-4.

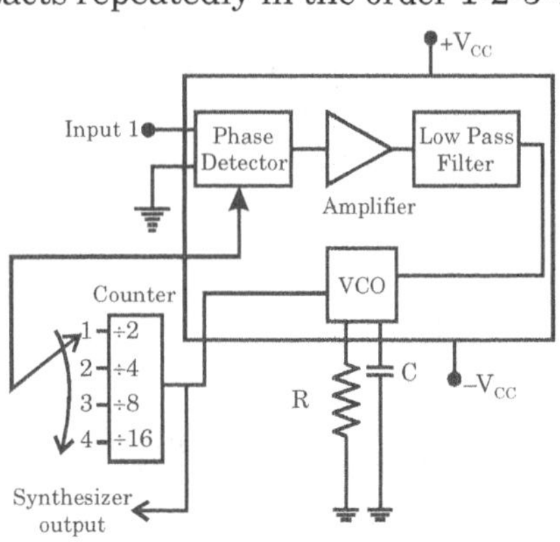

The corresponding frequencies synthesized are:

(*a*) 10 kHz, 20 kHz, 40 kHz, 80 kHz

(*b*) 20 kHz, 40 kHz, 80 kHz, 160 kHz

(*c*) 80 kHz, 40 kHz, 20 kHz, 10 kHz

(*d*) 160 kHz, 80 kHz, 40 kHz, 20 kHz

17. The output of the combinational circuit given below is

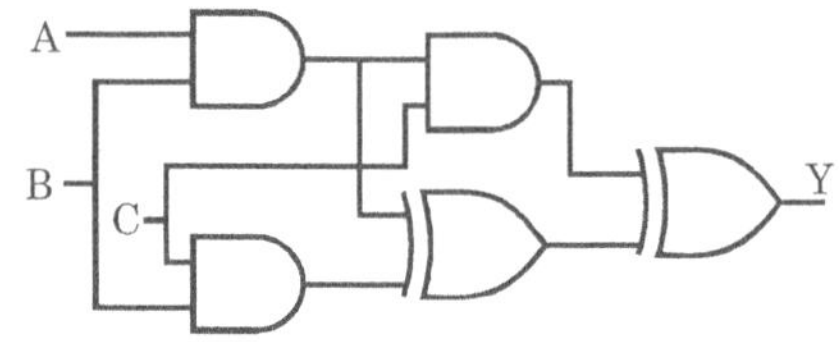

(*a*) A + B + C (*b*) A(B + C)

(*c*) B(C + A) (*d*) C(A + B)

18. What is the voltage V_{out} in the following circuit?

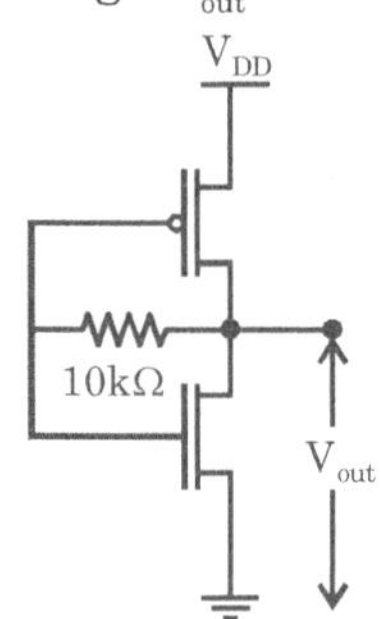

(*a*) 0 V

(*b*) ($|V_T$ of PMOS$|$ + V_T of NMOS) / 2

(*c*) Switching threshold of inverter

(*d*) V_{DD}

19. Match the inferences X, Y, and Z, about a system, to the corresponding properties of the elements of first column in Routh's Table of the system characteristic equation.

X : The system is stable ...

Y : The system is unstable ...

Z : The test breaks down ...

P : ... when all elements are positive

Q : ... when any one element is zero

R : ... when there is a change in sign of coefficients

(*a*) X → P, Y → Q, Z → R

(*b*) X → Q, Y → P, Z → R

(*c*) X → R, Y → Q, Z → P

(*d*) X → P, Y → R, Z → Q

20. A closed-loop control system is stable if the Nyquist plot of the corresponding open-loop transfer function

(*a*) encircles the s-plane point (–1 + *j*0) in the counterclockwise direction as many times as the number of right-half s-plane poles.

(*b*) encircles the s-plane point (0 – *j*1) in the clockwise direction as many times as the number of right-half s-plane poles.

(*c*) encircles the s-plane point $(-1 + j0)$ in the counterclockwise direction as many times as the number of left-half s-plane poles.

(*d*) encircles the s-plane point $(-1 + j0)$ in the counterclockwise direction as many times as the number of right-half s-plane zeros.

21. Consider binary data transmission at a rate of 56 kbps using baseband binary pulse amplitude modulation (PAM) that is designed to have a raised-cosine spectrum. The transmission bandwidth (in kHz) required for a roll-off factor of 0.25 is ________.

22. A superheterodyne receiver operates in the frequency range of 58 MHz – 68 MHz. The intermediate frequency f_{IF} and local oscillator frequency f_{LO} are chosen such that $f_{IF} \le f_{LO}$. It is required that the image frequencies fall outside the 58 MHz – 68 MHz band. The minimum required f_{IF} (in MHz) is ________.

23. The amplitude of a sinusoidal carrier is modulated by a single sinusoid to obtain the amplitude modulated signal $s(t) = 5\cos 1600\pi t + 20\cos 1800\pi t + 5\cos 2000\pi t$. The value of the modulation index is ________

24. Concentric spherical shells of radii 2 m, 4 m, and 8 m carry uniform surface charge densities of 20 nC/m², –4 nC/m² and ρ_s, respectively. The value of ρ_s (nC/m²) required to ensure that the electric flux density $\vec{D} = \vec{0}$ at radius 10 m is ________.

25. The propagation constant of a lossy transmission line is $(2 + j5)\text{m}^{-1}$ and its characteristic impedance is $(50 + j0)\ \Omega$ at $\omega = 10^6$ rad s^{-1}. The values of the line constants L, C, R, G are, respectively,

(*a*) L = 200 μH/m, C = 0.1 μF/m, R = 50 Ω/m, G = 0.02 S/m

(*b*) L = 250 μH/m, C = 0.1 μF/m, R = 100 Ω/m, G = 0.04 S/m

(*c*) L = 200 μH/m, C = 0.2 μF/m, R = 100 Ω/m, G = 0.02 S/m

(*d*) L = 250 μH/m, C = 0.2 μF/m, R = 50 Ω/m, G = 0.04 S/m

(Q.26 – 55) : Carry Two Marks Each

26. The integral $\frac{1}{2\pi}\iint_D (x+y+10)\,dx\,dy$, where D denotes the disc: $x^2 + y^2 \le 4$, evaluates to______.

27. A sequence x[n] is specified as

$$\begin{bmatrix} x[n] \\ x[n-1] \end{bmatrix} = \begin{bmatrix} 1 & 1 \\ 1 & 0 \end{bmatrix}^n \begin{bmatrix} 1 \\ 0 \end{bmatrix}, \text{ for } n \ge 2.$$

The initial conditions are $x[0] = 1$, $x[1] = 1$, and $x[n] = 0$ for $n < 0$. The value of $x[12]$ is ______.

28. In the following integral, the contour C encloses the points $2\pi j$ and $-2\pi j$

$$-\frac{1}{2\pi}\oint_C \frac{\sin z}{(z-2\pi j)^3}\,dz$$

The value of the integral is ________.

29. The region specified by $\{(\rho, \varphi, z): 3 \le \rho \le 5, \frac{\pi}{8} \le \varphi \le \frac{\pi}{4}, 3 \le z \le 4.5\}$ in cylindrical coordinates has volume of _______.

30. The Laplace transform of the causal periodic square wave of period T shown in the figure below is

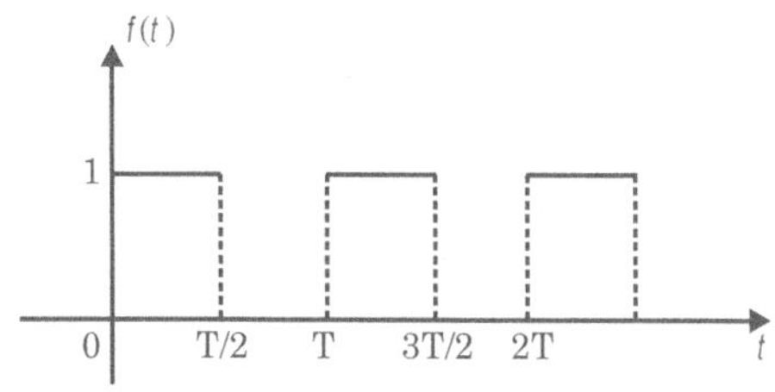

(*a*) $F(s) = \dfrac{1}{1+e^{-sT/2}}$ (*b*) $F(s) = \dfrac{1}{s\left(1+e^{-\frac{sT}{2}}\right)}$

(*c*) $F(s) = \dfrac{1}{s(1-e^{-sT})}$ (*d*) $F(s) = \dfrac{1}{1-e^{-sT}}$

31. A network consisting of a finite number of linear resistor (R), inductor (L), and capacitor (C) elements, connected all in series or all in parallel, is excited with a source of the form

$$\sum_{k=1}^{3} a_k \cos(k\omega_0 t), \text{ where } a_k \ne 0,\ \omega_0 \ne 0.$$

The source has nonzero impedance. Which one of the following is a possible form of the output measured across a resistor in the network?

(*a*) $\sum_{k=1}^{3} b_k \cos(k\omega_0 t + \phi_k)$, where $b_k \ne a_k, \forall k$

(*b*) $\sum_{k=1}^{4} b_k \cos(k\omega_0 t + \phi_k)$, where $b_k \ne 0, \forall k$

(*c*) $\sum_{k=1}^{3} a_k \cos(k\omega_0 t + \phi_k)$

(*d*) $\sum_{k=1}^{2} a_k \cos(k\omega_0 t + \phi_k)$

32. A first-order low-pass filter of time constant T is excited with different input signals (with zero initial conditions up to $t = 0$). Match the excitation signals X, Y, Z with the corresponding time responses for $t \ge 0$:

X : Impulse P : $1 - e^{-t/T}$
Y : Unit step Q : $t - T(1 - e^{-t/T})$
Z : Ramp R : $e^{-t/T}$

(a) X → R, Y → Q, Z → P
(b) X → Q, Y → P, Z → R
(c) X → R, Y → P, Z → Q
(d) X → P, Y → R, Z → Q

33. An AC voltage source V = 10 sin(t) volts is applied to the following network. Assume that R_1 = 3 kΩ, R_2 = 6 kΩ and R_3 = 9 kΩ, and that the diode is ideal.

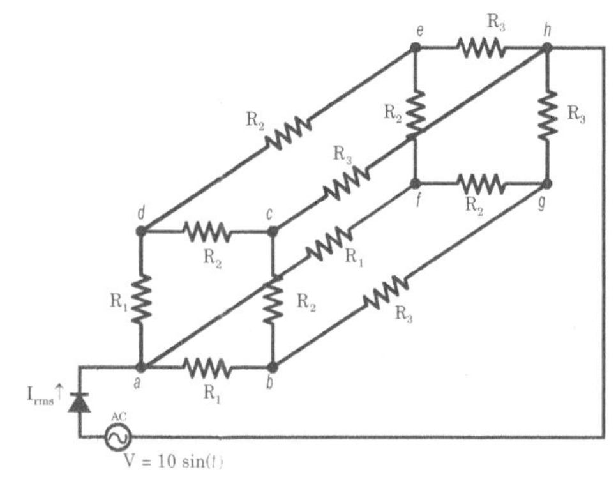

RMS current I_{rms} (in mA) through the diode is ________.

34. In the circuit shown in the figure, the maximum power (in watt) delivered to the resistor R is ________.

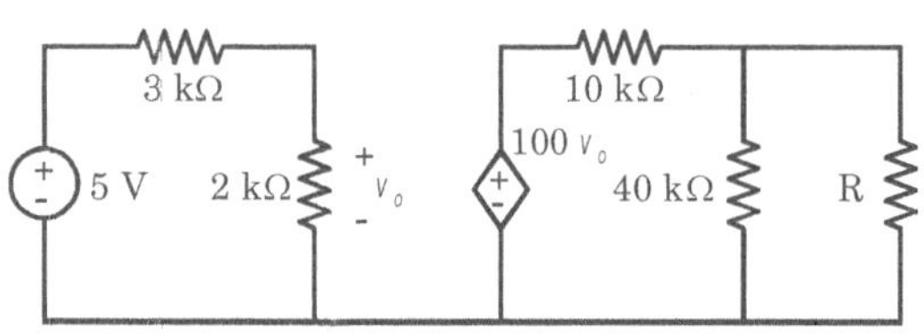

35. Consider the signal

$$x[n] = 6\,\delta[n+2] + 3\delta[n+1] + 8\delta[n] + 7\delta[n-1] + 4\delta[n-2].$$

If $X(e^{j\omega})$ is the discrete-time Fourier transform of x[n], then $\frac{1}{\pi}\int_{-\pi}^{\pi} X(e^{j\omega})\sin^2(2\omega)\,d\omega$ is equal to ________.

36. Consider a silicon p-n junction with a uniform acceptor doping concentration of 10^{17} cm^{-3} on the p-side and a uniform donor doping concentration of 10^{16} cm^{-3} on the n-side. No external voltage is applied to the diode. Given: kT/q = 26 mV, $n_i = 1.5 \times 10^{10}$ cm^{-3}, $\varepsilon_{Si} = 12\varepsilon_0$, $\varepsilon_0 = 8.85 \times 10^{-14}$ F/m, and q = 1.6×10^{-19} C.

The charge per unit junction area (nC cm^{-2}) in the depletion region on the p-side is ________.

37. Consider an n-channel metal oxide semiconductor field effect transistor (MOSFET) with a gate-to-source voltage of 1.8 V. Assume that $\frac{W}{L} = 4$, $\mu_N C_{ox} = 70 \times 10^{-6}$ AV^{-2}, the threshold voltage is 0.3V, and the channel length modulation parameter is 0.09 V^{-1}. In the saturation region, the drain conductance (in micro seimens) is ________.

38. The figure below shows the doping distribution in a p-type semiconductor in log scale.

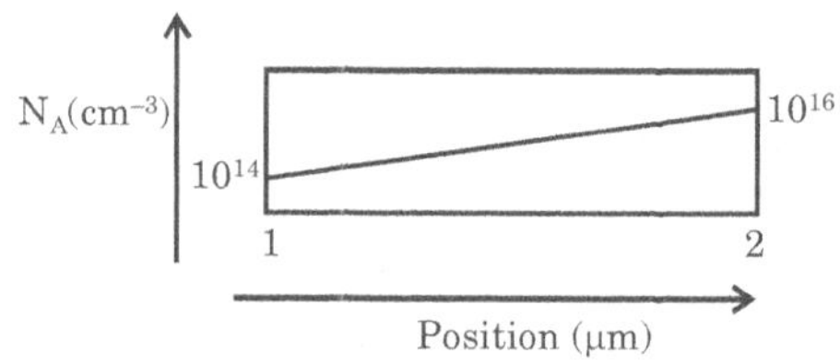

The magnitude of the electric field (in kV/cm) in the semiconductor due to non uniform doping is ________.

39. Consider a silicon sample at T = 300 K, with a uniform donor density $N_d = 5 \times 10^{16}$ cm^{-3}, illuminated uniformly such that the optical generation rate is $G_{opt} = 1.5 \times 10^{20}$ cm^{-3} s^{-1} throughout the sample. The incident radiation is turned off at t = 0. Assume low-level injection to be valid and ignore surface effects. The carrier lifetimes are τ_{p0} = 0.1 μs and τ_{n0} = 0.5 μs.

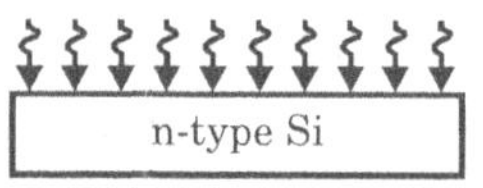

The hole concentration at t = 0 and the hole concentration at t = 0.3 μs, respectively, are

(a) 1.5×10^{13} cm^{-3} and 7.47×10^{11} cm^{-3}
(b) 1.5×10^{13} cm^{-3} and 8.23×10^{11} cm^{-3}
(c) 7.5×10^{13} cm^{-3} and 3.73×10^{11} cm^{-3}
(d) 7.5×10^{13} cm^{-3} and 4.12×10^{11} cm^{-3}

40. An ideal opamp has voltage sources $V_1, V_3, V_5, \ldots, V_{N-1}$ connected to the non-inverting input and $V_2, V_4, V_6, \ldots, V_N$ connected to the inverting input as shown in the figure below ($+V_{CC}$ = 15 volt, $-V_{CC}$ = −15 volt). The voltages $V_1, V_2, V_3, V_4, V_5, V_6, \ldots$ are 1, − 1/2, 1/3, −1/4, 1/5, −1/6, ... volt, respectively. As N approaches infinity, the output voltage (in volt) is ________.

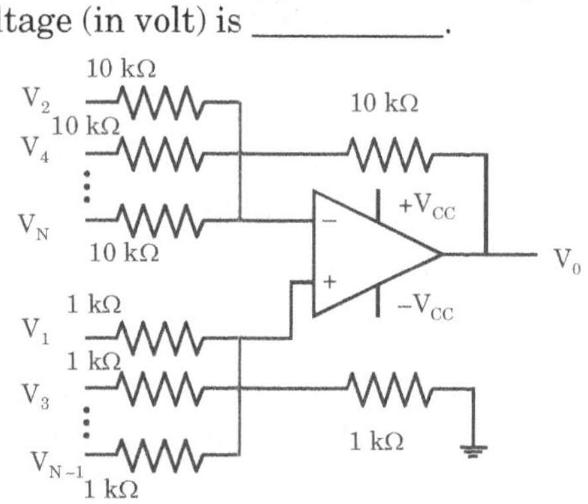

41. A *p-i-n* photodiode of responsivity 0.8A/W is connected to the inverting input of an ideal opamp as shown in the figure, $+V_{cc}$ = 15 V, $-V_{cc}$ = –15V, Load resistor R_L = 10 kΩ. If 10 μW of power is incident on the photodiode, then the value of the photocurrent (in μA) through the load is ______

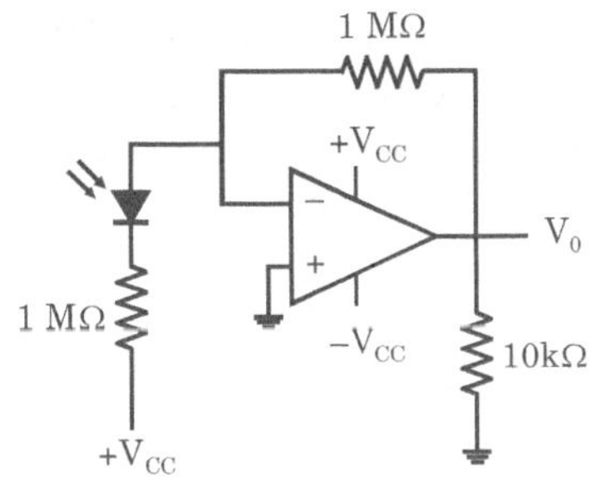

42. Identify the circuit below.

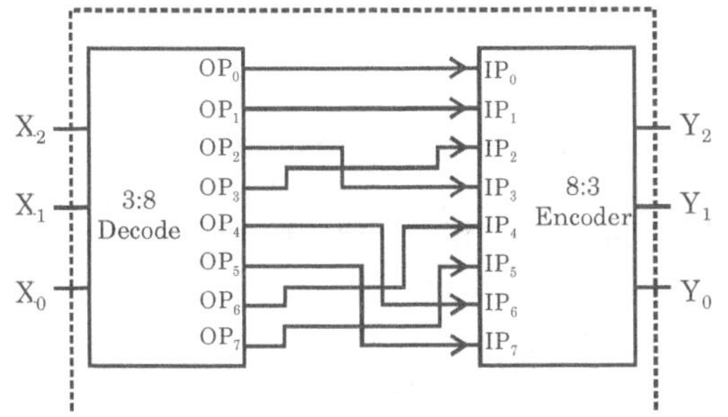

(*a*) Binary to Gray code converter
(*b*) Binary to XS3 converter
(*c*) Gray to Binary converter
(*d*) XS3 to Binary converter

43. The functionality implemented by the circuit below is

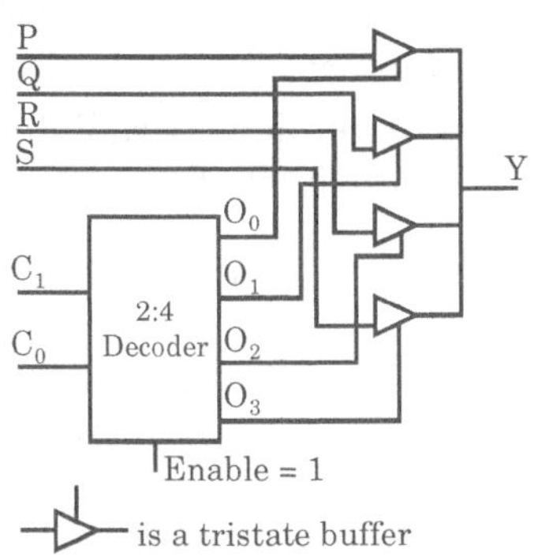

(*a*) 2-to-1 multiplexer
(*b*) 4-to-1 multiplexer
(*c*) 7-to-1 multiplexer
(*d*) 6-to-1 multiplexer

44. In an 8085 system, a PUSH operation requires more clock cycles than a POP operation. Which one of the following options is the correct reason for this?

(*a*) For POP, the data transceivers remain in the same direction as for instruction fetch (memory to processor), whereas for PUSH their direction has to be reversed.

(*b*) Memory write operations are slower than memory read operations in an 8085 based system.

(*c*) The stack pointer needs to be pre-decremented before writing registers in a PUSH, whereas a POP operation uses the address already in the stack pointer.

(*d*) Order of registers has to be interchanged for a PUSH operation, whereas POP uses their natural order.

45. The open-loop transfer function of a unity-feedback control system is

$$G(s) = \frac{K}{s^2 + 5s + 5}$$

The value of K at the breakaway point of the feedback control system's root-locus plot is________.

46. The open-loop transfer function of a unity-feedback control system is given by

$$G(s) = \frac{K}{s(s+2)}$$

For the peak overshoot of the closed-loop system to a unit step input to be 10%, the value of K is ____________.

47. The transfer function of a linear time invariant system is given by

$$H(s) = 2s^4 - 5s^3 + 5s - 2$$

The number of zeros in the right half of the s-plane is ________.

48. Consider a discrete memoryless source with alphabet S = $\{s_0, s_1, s_2, s_3, s_4, ...\}$ and respective probabilities of occurrence $P = \left\{\frac{1}{2}, \frac{1}{4}, \frac{1}{8}, \frac{1}{16}, \frac{1}{32}, ...\right\}$.

The entropy of the source (in bits) is _______.

49. A digital communication system uses a repetition code for channel encoding/decoding. During transmission, each bit is repeated three times (0 is transmitted as 000, and 1 is transmitted as 111). It is assumed that the source puts out symbols independently and with equal probability. The decoder operates as follows: In a block of three received bits, if the number of zeros exceeds the number of ones, the decoder decides in favor of a 0, and if the number of ones exceeds the number of zeros, the decoder decides in favor of a 1. Assuming a binary symmetric channel with crossover probability p = 0.1, the average probability of error is ________.

50. An analog pulse $s(t)$ is transmitted over an additive white Gaussian noise (AWGN) channel. The received signal is $r(t) = s(t) + n(t)$, where $n(t)$ is additive white Gaussian noise with power spectral density $\frac{N_0}{2}$. The received signal is passed through a filter with impulse response $h(t)$. Let E_s and E_h denote the energies of the pulse $s(t)$ and the filter $h(t)$, respectively. When the signal-to-noise ratio (SNR) is maximized at the output of the filter (SNR_{max}), which of the following holds?

(*a*) $E_s = E_h; SNR_{max} = \frac{2E_s}{N_0}$

(*b*) $E_s = E_h; SNR_{max} = \frac{E_s}{2N_0}$

(*c*) $E_s > E_h; SNR_{max} = \frac{2E_s}{N_0}$

(*d*) $E_s < E_h; SNR_{max} = \frac{2E_h}{N_0}$

51. The current density in a medium is given by

$$\vec{j} = \frac{400 \sin\theta}{2p(r^2+4)} \hat{a}_r \ \text{Am}^{-2}$$

The total current and the average current density flowing through the portion of a spherical surface $r = 0.8$ m, $\frac{\pi}{12} \le \theta \le \frac{\pi}{4}, 0 \le \phi \le 2\pi$ are given, respectively, by

(*a*) 15.09 A, 12.86 Am^{-2}

(*b*) 18.73 A, 13.65 Am^{-2}

(*c*) 12.86 A, 9.23 Am^{-2}

(*d*) 10.28 A, 7.56 Am^{-2}

52. An antenna pointing in a certain direction has a noise temperature of 50 K. The ambient temperature is 290 K. The antenna is connected to a pre-amplifier that has a noise figure of 2 dB and an available gain of 40 dB over an effective bandwidth of 12 MHz. The effective input noise temperature T_e for the amplifier and the noise power P_{ao} at the output of the preamplifier, respectively, are

(*a*) $T_e = 169.36$ K and $P_{ao} = 3.73 \times 10^{-10}$ W

(*b*) $T_e = 170.8$ K and $P_{ao} = 4.56 \times 10^{-10}$ W

(*c*) $T_e = 182.5$ K and $P_{ao} = 3.85 \times 10^{-10}$ W

(*d*) $T_e = 160.62$ K and $P_{ao} = 4.6 \times 10^{-10}$ W

53. Two lossless X-band horn antennas are separated by a distance of 200λ. The amplitude reflection coefficients at the terminals of the transmitting and receiving antennas are 0.15 and 0.18, respectively. The maximum directivities of the transmitting and receiving antennas (over the isotropic antenna) are 18 dB and 22 dB, respectively. Assuming that the input power in the lossless transmission line connected to the antenna is 2 W, and that the antennas are perfectly aligned and polarization matched, the power (in mW) delivered to the load at the receiver is ________.

54. The electric field of a uniform plane wave travelling along the negative z direction is given by the following equation:

$$\vec{E}_w^i = (\hat{a}_x + j\hat{a}_y)E_0 e^{jkz}$$

This wave is incident upon a receiving antenna placed at the origin and whose radiated electric field towards the incident wave is given by the following equation:

$$\vec{E}_a = (\hat{a}_x + 2\hat{a}_y)E_1 \frac{1}{r} e^{-jkr}$$

The polarization of the incident wave, the polarization of the antenna and losses due to the polarization mismatch are, respectively,

(*a*) Linear, Circular (clockwise), –5dB

(*b*) Circular (clockwise), Linear, –5dB

(*c*) Circular (clockwise), Linear, –3dB

(*d*) Circular (anti clockwise), Linear, –3dB

55. The far-zone power density radiated by a helical antenna is approximated as:

$$\vec{W}_{rad} = \vec{W}_{average} \approx \hat{a}_r C_0 \frac{1}{r_2} \cos^4\theta$$

The radiated power density is symmetrical with respect to ϕ and exists only in the upper hemisphere: $0 \le \theta \le \frac{\pi}{2}; 0 \le \phi \le 2\pi$; C_0 is a constant.

The power radiated by the antenna (in watts) and the maximum directivity of the antenna, respectively, are

(*a*) 1.5 C_0, 10 dB

(*b*) 1.256 C_0, 10 dB

(*c*) 1.256 C_0, 12 dB

(*d*) 1.5 C_0, 12 dB

ANSWERS

GENERAL APTITUDE

1. (*a*)	2. (*b*)	3. (*a*)	4. (*a*)	5. (*c*)	6. (*c*)	7. (*b*)	8. (*d*)	9. (*c*)	10. (*c*)

TECHNICAL SECTION

1. (*c*)	2. (1)	3. (*b*)	4. (*a*)	5. (*c*)	6. (*c*)
7. (*b*)	8. (*b*)	9. (*d*)	10. (13)	11. (*a*)	12. (*c*)
13. (*b*)	14. (3)	15. (*a*)	16. (*a*)	17. (*c*)	18. (*b*)
19. (*d*)	20. (*a*)	21. (35)	22. (5)	23. (0.5)	24. (–0.25)
25. (*b*)	26. (20)	27. (233)	28. (– 133.87)	29. (4.714)	30. (*b*)
31. (*a*)	32. (*c*)	33. (1)	34. (0.8)	35. (8)	36. (4.836)
37. (28.35)	38. (0.0133)	39. (*c*)	40. (15)	41. (–8)	42. (*a*)
43. (*b*)	44. (*c*)	45. (1.25)	46. (2.87)	47. (3)	48. (2)
49. (0.028)	50. (*a*)	51. (*)	52. (*a*)	53. (2.99)	54. (*c; d*)
55. (*b*)					

* *None of the given option is correct*

EXPLANATIONS

GENERAL APTITUDE

2. Lose (verb)
3. Here 'effectiveness' is noun and 'prescribed' is verb. So these words are apt and befitting with the given word 'medicine'.
4. Given,

Total no. of fruits = 5692000

Unripe type of apples = 45% of 15% of 5692000

$$= \frac{45}{100} \times \frac{15}{100} \times 5692000$$

$$= 384210$$

Ripe type of apples = (100 – 66)% × (100 – 15)% × 5692000

$$= \frac{34}{100} \times \frac{85}{100} \times 5692000$$

$$= 1644988$$

∴ Total no. of apples = 384210 + 1644988

= 2029198

5. From given data, the following diagram can be possible.

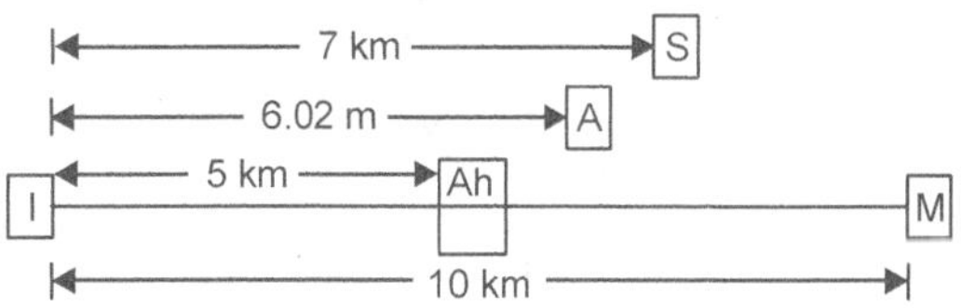

Here S → Susan

A → Arun

Ah → Ahmed

M → Michael

From the above diagram, Arun lives farthest away than Ahmed means more than 5 km but closer than Susan means less than 7 km, from the given alternatives.

So, only option 'c' is possible.

6. The percentage of people moving through a tuberculosis prone zone remains infected but does not show symptoms of disease.

$$= (100 - 30)\% \times 50\%$$

$$= \frac{70}{100} \times \frac{50}{100} = \frac{35}{100} = 35\%$$

8. From the given question two statements will be followed.

For statment I

Arrange the given data according to their ages

$$\text{Age} \uparrow \quad \text{Pavithra} \rightarrow \text{Leela} \rightarrow \text{Shiva}$$

(Shiva ↑ Leela ↑ Pavithra)

For statement II

Arrange the given data according to their winning

(Pavithra ↑ Leela)

So, from statement I and II, it is clear that only option (*d*) is possible (i.e., statement I).

9. Given $q^{-a} = \frac{1}{r}$

$\Rightarrow \quad \frac{1}{q^a} = \frac{1}{r}$

$\Rightarrow \quad q^a = r \quad ...(i)$

$r^{-b} = \frac{1}{s}$

$\Rightarrow \quad \frac{1}{r^b} = \frac{1}{s}$

$\Rightarrow \quad s = r^b \quad ...(ii)$

$s^{-c} = \frac{1}{q}$

$\Rightarrow \quad \frac{1}{s^c} = \frac{1}{q}$

$\Rightarrow \quad s^c = q \quad ...(iii)$

From equation (*i*),

$q^a = r$

$(s^c)^a = r \quad$ (from eq. (*iii*))

$S^{ac} = r$

$s^{ac} = (s)^{\frac{1}{b}} \quad$ (from eq. (*ii*))

$s^{abc} = s = s^1$

$\therefore \quad abc = 1$

So the value of *abc* is 1

10. Q can finish 1 day's 1 hour's work

$$= \frac{1}{25 \times 12} = \frac{1}{300}$$

R can finish 1 day's 1 hour's work

$$= \frac{1}{50 \times 12} = \frac{1}{600}$$

Now Q working hours

$$= (7 - 2) \times 12 = 60 \text{ hr}$$

$\therefore$ R working hours $= 7 \times 18 = 126$

After 7 days, the ratio of work done by Q & R is

$$Q : R$$

$$\frac{60}{300} : \frac{126}{600}$$

$$Q : R = 120 : 126 = 20 : 21$$

TECHNICAL SECTION

1. Given, $M^4 = I$

$\Rightarrow \quad M^8 = M^4 = I$

$\Rightarrow \quad M^7 = M^{-1}$

$\Rightarrow \quad M^{12} = M^8 = I$

$\Rightarrow \quad M^{11} = M^{-1}$

$\Rightarrow \quad M^{16} = M^{12} = I$

$\Rightarrow \quad M^{15} = M^{-1}$

....

$\therefore \quad M^{-1} = M^{4k+3}$

where *k* is a natural number.

2. Here, $E(x^2) = 2$

$\therefore \quad V(x) = E(x^2) - (E(x))^2$

Let mean of the poisson random variable be *x*.

$$x = 2 - x^2$$

$$\left(\text{Since } V(x) = x \text{ and } (E(x))^2 = x^2\right)$$

$$x^2 + x - 2 = 0$$

$$x^2 + 2x - x - 2 = 0$$

$$x(x + 2) - 1\,(x + 2) = 0$$

$$(x - 1)\,(x + 2) = 0$$

$$\therefore \quad x = 1, -2$$

Mean of random variable $(\lambda) = 1$

3. Since continuous function may not be differentiable at $x = x_0$, but differentiable function is always continuous.

5. From the given question integration of ramp is parabolic, integration of step is ramp.

So, the required graph of $f(x)$ is shown below.

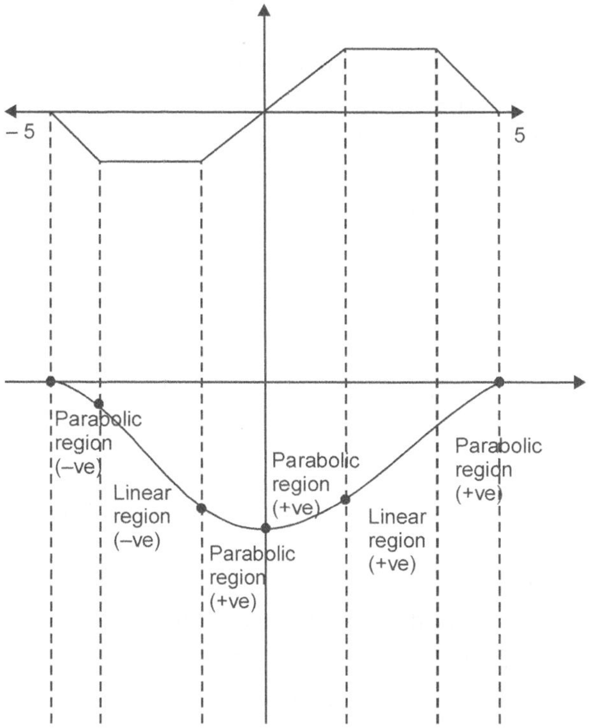

7. If $\frac{\omega_0}{2\pi}$ is a rational number, then discrete time signal $x(n) = \cos(\omega_0 n)$ will be periodic.

8. Given sequence

$$x(n) = (a)^n x(n) + (b)^n x(n),$$

Also given $\quad 0 < |a| < |b| < 1$

$\therefore$ The region of convergence (ROC)

$$= (|z| > |a|) \cap (|z| > |b|)$$

$$= |z| > |b|$$

9. A two port network is reciprocal in transmission parameters if (AD – BC) = 1

   i.e., the value of Determinant (T) = 1

10. Given, $\quad f_m = 33$ Hz,

    $f_s = 46$ Hz

    The frequency in sampled singal = ± 33, 13, 79, 59, 125. .... The above frequencies are passed to a LPF of cut-off frequency 23 Hz. The output frequency = 13 Hz.

11. Donor energy level close to conduction band.

12. As per the given statement

    P : TRUE

    Q : FALSE, As channel length reduces, output resistance reduces.

    R : FALSE, As channel length reduces, threshold voltage reduces

    S : TRUE

    So, the given statement Q and R are incorrect.

13.

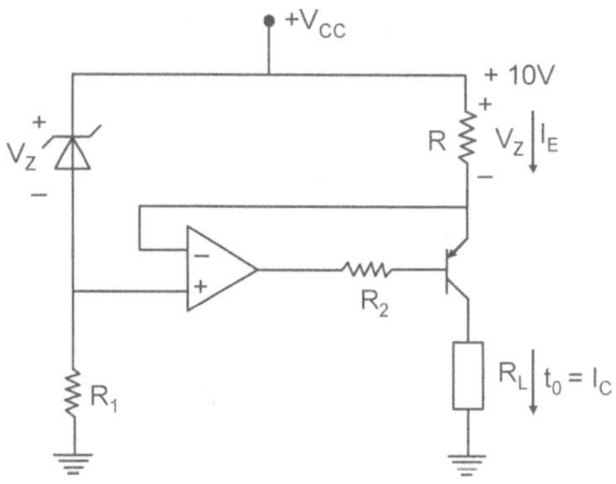

From the given figure, we have

$$V_P = V_N \text{ (Virtual short)}$$

$$I_0 = I_C = \left(\frac{\beta}{\beta+1}\right) I_E = \left(\frac{\beta}{\beta+1}\right)\frac{V_Z}{R}$$

$$= \left(\frac{\beta}{\beta+1}\right)\frac{V_{ref}}{R}$$

14.

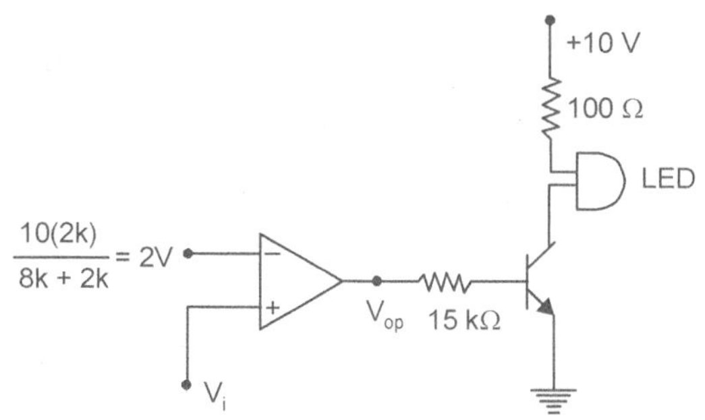

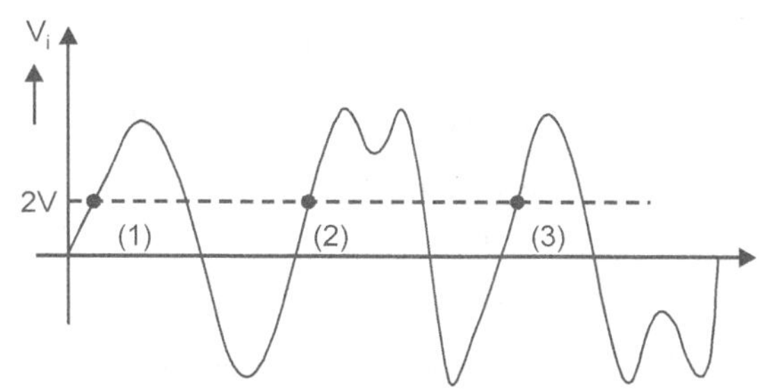

V_i crosses 2V, 3 times

So, the LED glows 3 times.

15. The circuit shown (in fig.) is a wein bridge oscillator. So the amplitude of oscillations can be determined and stabilized by using a nonlinear control network. As the oscillations grow, the diodes start to conducting that causes the effective resistance in the feedback to decrease. Therefore equilibrium will be reached at the output amplitude that causes the loop gain to be exactly unity.

16. Here the required diagram of frequency synthesizer consisting of phase locked loop (PPL) and a divide by-N counter is shown below.

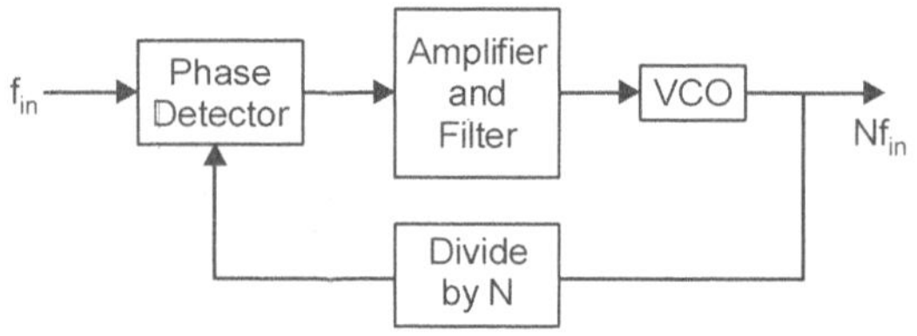

f_{in}	Divide by N	VCO output (Nf_{in})
5kHz	2	10kHz
5kHz	4	20kHz
5kHz	8	40kHz
5kHz	16	80kHz

So, the corresponding frequencies synthesizer are respectively 10 kHz, 20kHz, 40kHz, 80kHz.

17. From the given combinational circuit output (Y)

$$= ABC \oplus AB \oplus BC$$
$$= AB(C \oplus 1) \oplus BC$$
$$= AB\bar{C} \oplus BC \quad (\text{Since } C \oplus 1 = C \cdot \bar{1} + 1 \cdot \bar{C})$$
$$= C \cdot 0 + \bar{C}$$
$$= 0 + \bar{C}$$
$$= \bar{C}$$
$$= B(A\bar{C} \oplus C)$$
$$= B\left[A\bar{C} \cdot \bar{C} + C\,\overline{A\bar{C}}\right]$$
$$= B\left[A\bar{C} + C\left(\bar{A} + \bar{\bar{C}}\right)\right]$$
$$= B\left[A\bar{C} + \bar{A}C + C\right] \qquad \left(\text{Since } \bar{\bar{C}} = C\right)$$
$$= B\left[A\bar{C} + C(\bar{A} + 1)\right]$$
$$= B\left[C + A\bar{C}\right] \qquad \left(\text{Since } \bar{A} + 1 = 1\right)$$
$$= B[C + A]$$
$$\therefore \quad Y = B(C + A)$$

19. P | + + + — Stable (when all elements are positive)

Q | + 0 + — Roath array breaks (when any on element is zero)

R | + – + — Un-stable (when there is a change in sign of coefficients)

$\therefore$ X – P, Y – R, Z – Q

Here X = The system is stable
Y = The system is unstable
Z = The test breaks down

20. Here the Nyquist plot of the corresponding open-loop transfer function is

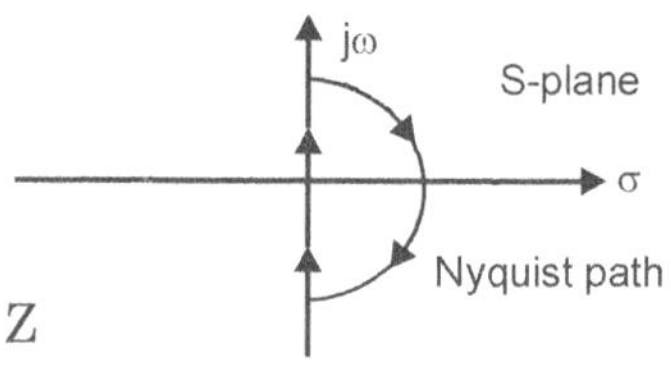

$$N = P - Z$$

For closed loop stability

$$Z = 0, N = P$$

So, the point (– 1, j0) encircles the s-plane is the counter clockwise direction equaling P poles of the right-half s-plane.

21. From given data

binary data transmission (R_b) = 56 kbps

Roll-off factor (α) = 0.25

$\therefore$ Transmission bandwidth

$$(BW) = \frac{R_b}{2}[1+\alpha]$$
$$= \frac{56}{2}[1+0.25]\text{kHz}$$
$$= 28 \times 1.25 \text{ kHz}$$
$$= 35 \text{ kHz}$$

22. From given data

The frequency range of superheterodyne receiver

$$(f_s) = 58 \text{ MHz - } 68\text{MHz}$$

When $f_s = 58$ MHz

$$f_{si} = f_s + 21F > 68 \text{ MHz}$$
$$21F > 10 \text{ MHz}$$
$$\therefore \quad 1F \geq 5 \text{ MHz}$$

So, the minimum required F_{1F} is 5 MHz.

23. Here the amplitude modulated signal

$$s(t) = 5\cos 1600\pi t + 20\cos 1800\pi t + 5\cos 2200\pi t$$

$$s(t) = \frac{A_c\mu}{2}\cos 2\pi(f_c - f_m)t + A_c\cos 2\pi f_c t + \frac{A_c\mu}{2}\cos 2\pi(f_c + f_m)t$$

So, $A_c = 20$ and $A_c\mu = 10$

or $\dfrac{A_c\mu}{2} = 5$

$$\therefore \quad \mu = \frac{10}{20} = \frac{1}{2} = 0.5$$

24.

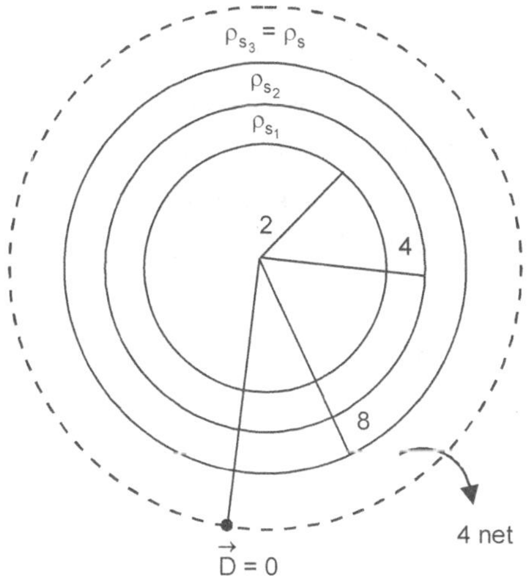

Given uniform change densities are

$$\rho_{S_1} = 20 \text{ nc/m}^2$$

$$\rho_{S_2} = -4 \text{ nc/m}^2$$

$$\therefore \quad \rho_{S_3} = ? \text{ (unknown)}$$

Now electric flux density at $r = 10$ m is given by

$$\vec{D} = \left(\frac{\psi \text{ net leaving the sphere of radius } r = 10 \text{ m}}{\text{Area of sphere of radius } r = 10 \text{ m}}\right)\hat{a}_r$$

but $\vec{D} = 0$

$$\psi_{net}\big|_{at\ r=10} = 0$$

$$20 \times 10^{-9} \times 4\pi(2)^2 + (-4 \times 10^{-9}) \times 4\pi(4)^2 + \rho_{S_3} 4\pi(8)^2 = 0$$

$$\therefore \quad \rho_{S_3} = \rho_S = -0.25 \text{ nc/m}^2$$

25. Given,

Propagation contact of a lossy transmission line

$$(P) = (2 + j5) \text{ m}^{-1},$$

characteristic impedance

$$(z_0) = 50\,\Omega,$$

angular frequency $(\omega) = 10^6$ rad/sec,

Now $\quad P = \sqrt{(R + j\omega L)(G + j\omega C)}$

$$z_0 = \sqrt{\frac{(R + j\omega L)}{(G + jwC)}}$$

$$Pz_0 = R + j\omega L$$

$$\Rightarrow \quad R + j\omega L = (100 + j250)$$

$$\therefore \quad R = 100\ \Omega/\text{m}$$

$$L = \frac{250}{\omega}$$

$$L = \frac{250}{10^6} = 250\ \mu\text{H}/m$$

(Since $\omega = 10^6$ rad/s)

$$\therefore \quad \frac{P}{z_0} = G + j\omega C$$

$$G + j\omega C = \left(\frac{2}{50} + j\frac{5}{50}\right)$$

$$\therefore \quad G = \frac{2}{50} = \frac{1}{25}$$

$$= 0.04 \text{ s/m}$$

and

$$C = \frac{5}{50 \times \omega}$$

$$= \frac{5}{50 \times 10^6} \text{ F/m}$$

$$= \frac{1}{10} \times 10^{-6}$$

$$= 0.1\ \mu \text{ F/m}$$

Hence line constant L, C, R & G are respectively

$$L = 250\ \mu\text{H/m}$$

$$C = 0.1\ \mu\text{F/m}$$

$$R = 100\ \Omega/\text{m}$$

and $\quad G = 0.04$ s/m

26. Given integral is converted into polar coordinates, we get

$$\frac{1}{2\pi}\iint_D (x + y + 10)\,dxdy$$

$$= \frac{1}{2\pi}\int_{r=0}^{2}\int_{\theta=0}^{2\pi} (r\cos\theta + r\sin\theta + 10)\,pdrd\theta$$

$$= \frac{1}{2\pi}\int_{r=0}^{2}\int_{\theta=0}^{2\pi} (r^2\cos\theta + r^2\sin\theta + 10r)dr\,d\theta$$

$$= \frac{1}{2\pi}\int_{0=0}^{2\pi}\left\{\frac{r^3}{3}\cos\theta + \frac{r^3}{3}\sin\theta + 5r^2\right\}_0^2 d\theta$$

$$= \frac{1}{2\pi}\int_0^{\pi}\left\{\frac{8}{3}\cos\theta + \frac{8}{3}\sin\theta + 20\right\}d\theta$$

$$= \frac{1}{2\pi}\left[\frac{8}{3}\sin\theta - \frac{8}{3}\cos\theta + 20\,\theta\right]_0^{2\pi}$$

$$= \frac{1}{2\pi}\left[\left(\frac{8}{3}\sin 2\pi - \frac{8}{3}\cos 2\pi + 40\pi\right) - \left(\frac{8}{3}\sin 0 - \frac{8}{3}\cos 0\right)\right]$$

$$= \frac{1}{2\pi}\left[\left(0 - \frac{8}{3} + 40\pi\right) - \left(-\frac{8}{3}\right)\right]$$

$$= \frac{1}{2\pi}\left\{\left(-\frac{8}{3} + 40\pi\right) - \left(-\frac{8}{3}\right)\right\}$$

$$= \frac{1}{2\pi}\left(-\frac{8}{3}+40\pi+\frac{8}{3}\right)$$

$$= \frac{1}{2\pi}\times 40\pi$$

$$= 20$$

27. Given sequence $x[n]$ is given by

$$\begin{bmatrix} x(n) \\ x(n-1) \end{bmatrix} = \begin{bmatrix} 1 & 1 \\ 1 & 0 \end{bmatrix}^n \begin{bmatrix} 1 \\ 0 \end{bmatrix}, n \geq 2$$

Put $n = 2$, we get

$$\begin{bmatrix} x(2) \\ x(1) \end{bmatrix} = \begin{bmatrix} 1 & 1 \\ 1 & 0 \end{bmatrix}^2 \begin{bmatrix} 1 \\ 0 \end{bmatrix}$$

$$= \begin{bmatrix} 2 & 1 \\ 1 & 1 \end{bmatrix}\begin{bmatrix} 1 \\ 0 \end{bmatrix} = \begin{bmatrix} 2 \\ 1 \end{bmatrix}$$

$x(2) = 2$

$x(1) = 1$

$n = 3$

$$\begin{bmatrix} x(3) \\ x(2) \end{bmatrix} = \begin{bmatrix} 1 & 1 \\ 1 & 0 \end{bmatrix}^3 \begin{bmatrix} 1 \\ 0 \end{bmatrix}$$

$$= \begin{bmatrix} 3 & 2 \\ 2 & 1 \end{bmatrix}\begin{bmatrix} 1 \\ 0 \end{bmatrix} = \begin{bmatrix} 3 \\ 2 \end{bmatrix}$$

$x(3) = 3$

$x(2) = 2$

From the above values we can write the recursive relation is as

$$x(n) = x(n-1) + x(n-2)$$
$$x(2) = x(1) + x(0)$$
$$= 1 + 1 = 2$$
$$x(3) = x(2) + x(1)$$
$$= 2 + 1 = 3$$
$$x(4) = x(3) + x(2)$$
$$= 3 + 2 = 5$$
$$x(5) = x(4) + x(3)$$
$$= 5 + 3 = 8$$
$$x(6) = x(5) + x(4)$$
$$= 8 + 5 = 13$$
$$x(7) = x(6) + x(5)$$
$$= 13 + 8 = 21$$
$$x(8) = x(7) + x(6)$$
$$= 21 + 13 = 34$$
$$x(9) = x(8) + x(7)$$
$$= 34 + 21 = 55$$
$$x(10) = x(9) + x(8)$$
$$= 55 + 34 = 89$$
$$x(11) = 89 + 55 = 144$$
$$x(12) = 144 + 89 = 233$$

So, the required value of $x(12)$ is 233.

28. Given integral

$$-\frac{1}{2\pi}\oint_C \frac{\sin z}{(Z-2\pi j)^3}\, dz = -\frac{1}{2\pi}\times 2\pi i \cdot \frac{f''(2\pi j)}{2!}$$

Now $f(z) = \sin z$

$f'(z) = \cos z$

$\therefore$ $f''(z) = -\sin z$

$\therefore$ $f''(z_0) = -\sin 2\pi j$

Now $\dfrac{1}{2\pi}\oint_c \dfrac{\sin z}{(z-2\pi j)^3}\, dz = -\dfrac{1}{2\pi}\times 2\pi j\left(\dfrac{-\sin(2\pi j)}{2}\right)$

$$= j \times j\frac{\sin h\, 2\pi}{2}$$

$$= -\frac{1}{2}(\sin h\, 2\pi) = -133.87$$

So, the value of integral is – 133.87

29. The region of cylinder is given by

$$3 \leq \rho \leq 5$$

$$\frac{\pi}{8} \leq \phi \leq \frac{\pi}{4}$$

$$3 \leq z \leq 4.5$$

Now the differential volume of cylinder is

$$dv = \rho d\rho\, \rho\phi\, dz$$

$$\text{Volume (V)} = \int_{\rho=3}^{5}\int_{\phi=\frac{\pi}{8}}^{\pi/4}\int_{z=3}^{4.5} \rho d\rho\, d\phi\, dz$$

$$= \left.\frac{\rho^2}{2}\right|_3^5 \times \phi\Big|_{\frac{\pi}{8}}^{\frac{\pi}{4}} \times z\Big|_3^{4.5}$$

$$= \frac{1}{2}(25-9)\times\left(\frac{\pi}{4}-\frac{\pi}{8}\right)\times(4.5-3)$$

$$= \frac{1}{2}\times 16\times\frac{\pi}{8}\times\frac{3}{2}$$

$\therefore$ $V = 4.714\ m^3$

30. From the given figure.

One period of signal

$$x_1(t) = u(t) - u(t-T/2)$$

$\therefore$ $X_1(s) = \dfrac{1}{s} - \dfrac{e^{-sT/2}}{s}$

$$= \frac{1-e^{-sT/2}}{s}$$

$$X(s) = \frac{1}{1-e^{-sT}} X_1(s)$$

$$\left(\text{Here } X_1(s) = \frac{1-e^{-sT/2}}{s}\right)$$

$$\therefore \quad X(s) = \frac{1}{s\left(1+e^{-sT/2}\right)}$$

31.

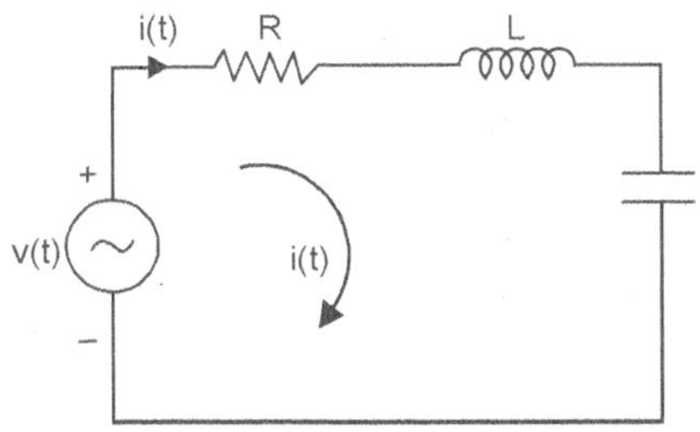

Let us consider a series RLC-circuit (with voltage source)

Here

$V(t) = a_1 \cos \omega_0 t + a_2 \cos 2\omega_0 t + a_3 \cos 3\,\omega_0 t$

$i(t) = b_1 \cos(\omega_0 t + \phi_2) + b_3 \cos(2\omega_0 t + \phi_2)$
$+ b_3 \cos(3\omega_0 t + \phi_3)$

$$i(t) = \sum_{k=1}^{3} b_k \cos\left(k\omega_0 t + \phi_k\right)$$

Where $b_k \neq a_k, \ \forall\, k$

32. Transfer function

$$H(s) = \frac{1}{1+s\tau}$$

$$\therefore \quad V_0(s) = H(s)\cdot V_I(s)$$

(*a*) if $V_i(t) = \delta(t)$

$$V_I(s) = 1$$

$$V_0(s) = H(s)\cdot V_I(s) = \frac{1}{1+s\tau}$$

$$\upsilon_0(t) = \frac{1}{\tau} e^{-\frac{t}{\tau}} \qquad \text{(impulse)}$$

(*b*) if $v_i(t) = u(t)$

$$V_I(s) = \frac{1}{s}$$

$$V_0(s) = \frac{1}{s(1+s\tau)} = \frac{1}{s} - \frac{1}{s+\frac{1}{\tau}}$$

$$\upsilon_0(t) = \left(1-e^{-t/\tau}\right) \qquad \text{(Unit step)}$$

(*c*) $v_i(t) = r(t)$

$$\Rightarrow \quad V_I(s) = \frac{1}{s^2}$$

$$V_0(s) = H(s)\cdot V_I(s)$$

$$= \frac{1}{s^2(1+s\tau)}$$

$$= \frac{1}{s^2} - \frac{\tau}{s} + \frac{t}{s+\frac{1}{\tau}}$$

$$\upsilon_0(t) = t - \tau\left(1-e^{-t/\tau}\right) \qquad \text{(Ramp)}$$

33. For diode is ideal the circuit network is given by

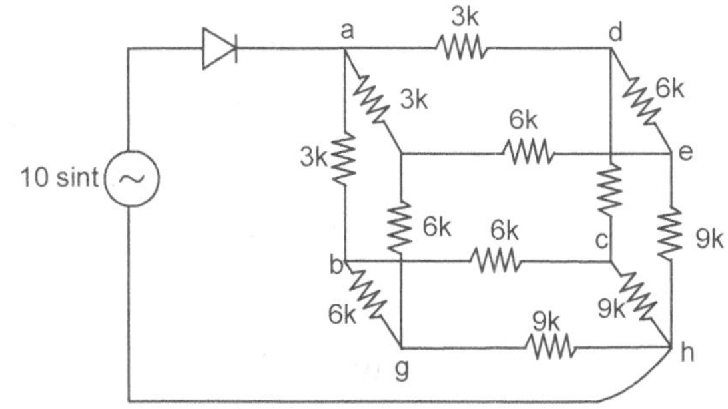

At same potential we can join nodes, therefore network becomes

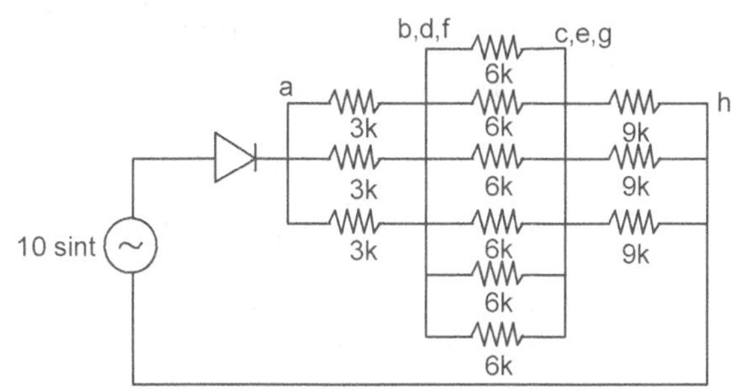

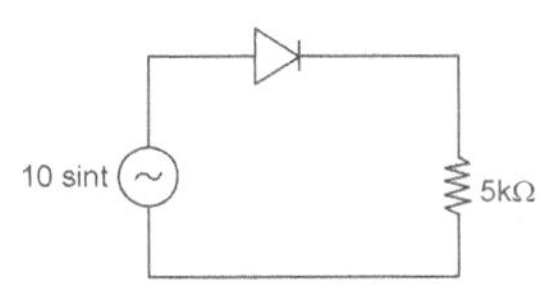

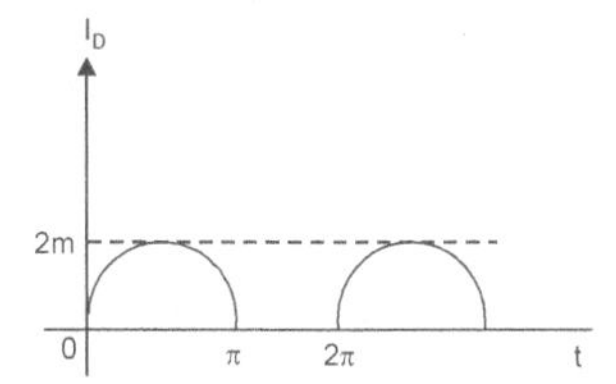

$$I_{D(RMS)} = \frac{2m}{2} = 1\text{ mA}$$

So, the RMS current (I_{RMS}) through the diode is 1 mA.

34. From the given circuit

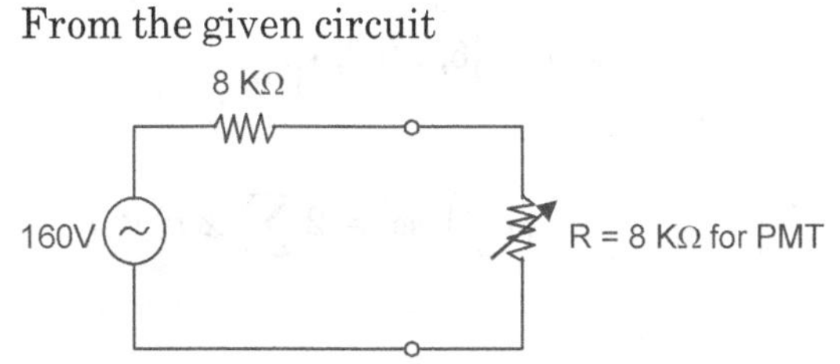

$$V_0 = \left(\frac{5}{5k}\right)\times 2k = 2V$$

$$V_{OC} = V_{th} = V_{40k}$$
$$= i_{40k} \times 40k$$
$$= \left(\frac{200}{50k}\right) \times 40k$$
$$= 160 \text{ V}$$

$$\Rightarrow \quad I_{sc} = \frac{200}{10k} = 20 \text{ mA}$$

So, $$R_{th} = \frac{V_{th}}{I_{sc}}$$
$$= \frac{160\text{v}}{20 \text{ mA}} = 8 \ k\Omega$$

The maximum power delivered to the resister R

$$(P_{max}) = \frac{v_s^2}{4 R_L}$$
$$W = \frac{(160)^2}{4 \times 8k}$$
$$= \frac{160 \times 160}{4 \times 8 \times 1000} = 0.8 \text{ w}$$

35. Plancheral's relation is given by

$$\frac{1}{2\pi}\int_{-\pi}^{\pi} X\left(e^{j\omega}\right)Y\left(e^{j\omega}\right)d\omega = \sum_{n=-\infty}^{\infty} x(n)y(n)$$

$$Y(e^{j\omega}) = \sin^2(2\omega) = \frac{1-\cos(4\omega)}{2}$$
$$= \frac{1}{2} - \frac{1}{4}e^{j4\omega} - \frac{1}{4}e^{-j4\omega}$$
$$y(n) = \frac{1}{2}\delta(n) - \frac{1}{4}\delta(n+4) - \frac{1}{4}\delta(n-4)$$
$$y(n) = \left\{-\frac{1}{4}, 0, 0, 0, \underset{\uparrow}{\frac{1}{2}}, 0, 0, 0, -\frac{1}{4}\right\}$$
$$x(n) = \left\{6, 3, \underset{\uparrow}{8}, 7, 4\right\}$$

$$\frac{1}{\pi}\int_{-\pi}^{\pi} X\left(e^{j\omega}\right)\cdot Y\left(e^{j\omega}\right)d\omega = 2\sum_{n=-\infty}^{\infty} x(n)y(n)$$

$$2\sum_{n=-\infty}^{\infty} x(n)y(n) = 2 \times 8 \times \frac{1}{2} = 8$$

36. From given data

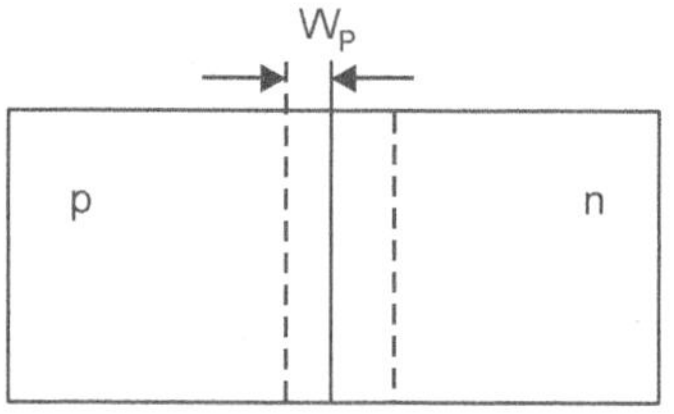

$$\varepsilon_{si} = 12\ \varepsilon_0$$
$$= 12 \times 8.85 \times 10^{-14} \text{ F/m}$$
$$N_D = 10^{16} \text{ cm}^{-3}$$
$$= 10^{22} \text{ m}^{-3}$$
$$N_A = 10^{17} \text{ cm}^{-3}$$
$$= 10^{23} \text{ m}^{-3}$$

$$V_0 = \frac{kT}{q}\ell n\left[\frac{N_A N_D}{ni^2}\right]$$
$$= 0.026\ell n\left[\frac{10^{23} \times 10^{22}}{\left(1.5 \times 10^{16}\right)^2}\right]$$
$$= 0.757 \text{ V}$$

$$W = \sqrt{\frac{2\varepsilon}{q}V_0\left(\frac{1}{N_A} + \frac{1}{N_D}\right)}$$
$$= \sqrt{\frac{2 \times 12 \times 8.85 \times 10^{-14}}{1.6 \times 10^{-19}} \times 0.757\left(\frac{1}{10^{23}} + \frac{1}{10^{22}}\right)}$$
$$= 3.325 \times 10^{-8} \text{ m}$$
$$= 3.325 \times 10^{-6} \text{ cm}$$

$$W_p = \frac{N_D}{N_A + N_D}\omega$$
$$= \frac{10^{22}}{10^{22} + 10^{23}} \times 3.325 \times 10^{-8}$$
$$= 3.023 \times 10^{-9} \text{ m}$$
$$= 3.023 \times 10^{-7} \text{ cm}$$
$$Q = W_P N_A eA$$
$$\Rightarrow \quad \frac{Q}{A} = W_P N_A$$
$$e = 3.023 \times 10^{-7} \times 10^{17} \times 1.6 \times 10^{-19}$$
$$= 4.836 \times 10^{-9} \text{ cm}^{-2}$$
$$= 4.836 \text{ nC-cm}^{-2}$$

Hence, the charge per unit function area (nC-cm^{-2}) in the depletion region on the *p*-side is 4.836 nC-cm^{-2}.

37. Here Drain conductance (in saturation region),

$$g_d = \frac{1}{r_d} = \lambda.I_D$$

$$I_D = \frac{1}{2}\mu_m C_{ox}\frac{W}{L}\left[V_{gs} - V_T\right]^2$$

$$= \frac{1}{2}\times 70\times 10^{-6}\times 4\left[1.8V - 0.3V\right]^2$$

$$= \frac{1}{2}\times 7\times 10^{-5}\times 4\times (1.5)^2$$

$$= 14\times 10^{-5}\times 2.25$$

$$= 31.5\times 10^{-5} A$$

$$= 0.315 \text{ mA}$$

$$g_d = 0.09 \times 0.315 \text{ mA/V}$$

$$g_d = 28.35 \times 10^{-6} \text{ A/v}$$

$$= 28.35\ \mu \text{ Seimens}$$

So, the drain conductance is 28.35 μ seimens.

38. The doping distribution in P-type semiconductor (in log scale) is shown below in the figures.

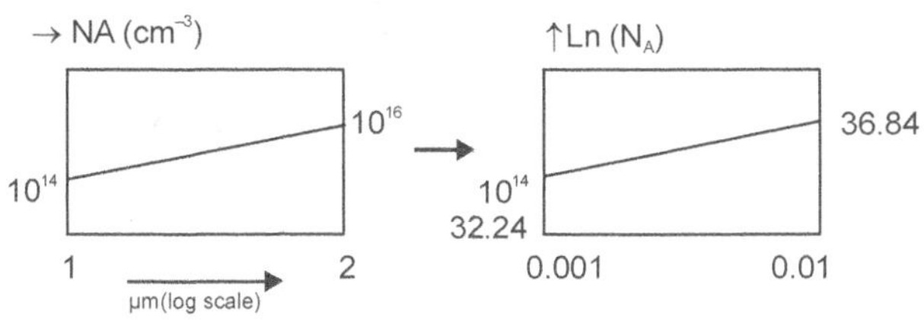

$$qD_p\frac{dp}{dx} = q\mu_p p\varepsilon$$

$$\mu_p V_T\frac{dp}{dx} = \mu_p p\varepsilon$$

$$\varepsilon = \frac{V_T}{p}\frac{dp}{dx}$$

$$p \cong N_A$$

$$\varepsilon = \frac{V_T}{N_A}\frac{dN_A}{dx}$$

$$\Rightarrow \quad \varepsilon = V_T\frac{d}{dx}\ln\left[N_A(x)\right]$$

$$\log_{10} x_1 = 1\ \mu m$$

$$\Rightarrow \quad x_1 = 10^1\ \mu m$$

$$= 0.001 \text{ cm}$$

$$\log_{10} x_2 = 2\mu m$$

$$\Rightarrow \quad x_2 = 10^2\ \mu m$$

$$= 0.01 \text{ cm}$$

$$\ln(10^{14}) = 32.23$$

$$\ln(10^{16}) = 36.84$$

$$\varepsilon = 0.026\left[\frac{36.84 - 32.23}{0.01 - 0.001}\right]$$

$$= 0.0133 \text{ /cm}$$

Hence, the magnitude of electric field in the semiconductor due to non-uniform doping is 0.0133 / cm.

39. The hole Concentration

$$P_n(t) = P_{n_0} + P_n(0)e^{-t/\tau_p}$$

at low level injuction

$\Rightarrow P_{n_0}$ (neglective)

$$\therefore \quad GR = \frac{P_n(0)}{\tau_{n_0}}$$

$$\Rightarrow \quad P_{n_0}(0) = GR \times \tau_{n_0}$$

$$= 1.5 \times 10^{20} \times 0.5 \times 10^{-6}$$

$$= 7.5 \times 10^{13}/\text{cm}^3$$

At $t = 0$

$$\Rightarrow \quad P(t) = P_n(0)\cdot e^0$$

$$= 7.5 \times 10^{13} \text{ /cm}^3$$

At $t = 0.3$ μs

$$\Rightarrow \quad P(t) = P_n(0)e^{\frac{0.3}{0.1}}$$

$$= 3.73 \times 10^{11} \text{ /cm}^3$$

So, the hole concentration at $t = 0$ and at $t = 0.3$ μs are 7.5×10^{13} cm^{-3} and 3.73×10^4 cm^{-3} respectively.

40. Here an ideal opamp has voltage source V_1, V_3, V_5, -------, V_{N-1} connected to the non-inverting input and V_2, V_4, V_6, -------, V_N connected to the inverting input as shown in the fig. below.

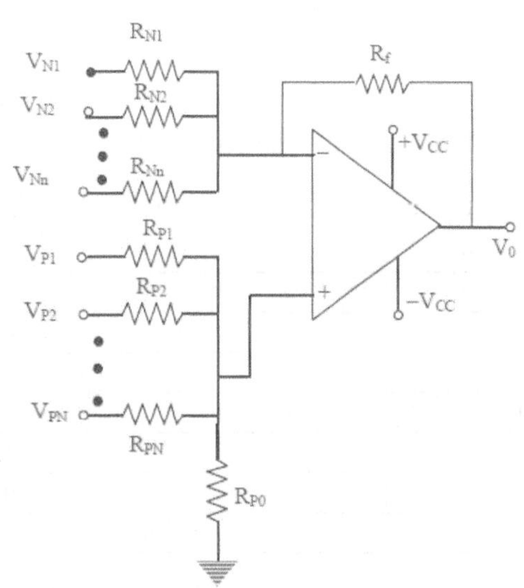

Using superposition, the output can be shown by

$$V_0 = \left[1+\frac{R_f}{R_N}\right]$$

$$\left[\frac{R_p}{R_{p1}}V_{p1}+\frac{R_p}{R_{p2}}V_{p2}+\ldots\ldots\ldots\ldots\frac{R_P}{R_{PN}}V_{pn}\right]$$

$$-\left[\frac{R_f}{R_{N1}}V_{N1}+\frac{R_f}{R_{N2}}V_{N2}+\ldots\ldots\ldots\ldots\frac{R_f}{R_{Nn}}V_{Nn}\right]$$

Where $R_N = R_{N1}\|R_{N2}\|\|\ldots\ldots\|R_{Nn}$

and $R_p = R_{p1}\|R_{p2}\ldots\ldots R_{PN}\|R_{PO}$

In the problem given

$$R_f = R_{N1} = R_{N2} = \ldots\ldots$$
$$= R_{Nn} = 10\text{ k}\Omega$$
$$R_{p1} = R_{p2} = R_{p3} = \ldots\ldots$$
$$= R_{PN} = R_{PO} = 1\text{ k}\Omega$$

$$\therefore \quad V_0 = \left[1+\frac{10k}{\left(\frac{10k}{n}\right)}\right]$$

$$\left[\frac{\frac{1k}{(1+n)}}{1k}V_{P1}+\frac{\left(\frac{1k}{1+n}\right)}{1k}V_{P2}+\ldots\ldots\right]$$

$$-\left[\frac{10k}{10k}V_{N1}+\frac{10k}{10k}V_{N2}+\ldots\ldots\right]$$

$$\therefore \quad V_0 = (V_{P1}+V_{P2}+\ldots\ldots V_{Pn}) - (V_{N1}+V_{N2}+\ldots\ldots V_{Nn})$$

If the series approaches ∞ then

$$V_0 = \left(1+\frac{1}{3}+\frac{1}{5}+\frac{1}{7}\ldots\ldots\right)-\left(\frac{-1}{2}\frac{-1}{4}\frac{-1}{6}\cdot-\ldots\ldots\right)$$

$$= 1+\frac{1}{2}+\frac{1}{3}+\frac{1}{4}+\frac{1}{5}+\ldots\ldots = \infty$$

This series is called harmonic series which is a divergent infinite series

$$\therefore \quad V_0 = +\infty = +V_{sat}$$
$$= +V_{CC} = 15V$$

Hence, the output voltage is 15 V (positive).

41. From given data

The photo diode with responsivity 0.8 A/W

$$\therefore \quad \text{Diode current} = 0.8\text{A/W}[10\ \mu\text{W}]$$
$$= 8\times 10^{-6}\text{ A}$$
$$= 8\ \mu\text{A}$$
$$V_0 = -8\mu\ (1\text{M})$$
$$= -8\times 10^{-6}\times 10^{-6}\text{ V}$$
$$= -8\text{V}$$

$$I_L = \frac{-8}{10k} = \frac{-8}{10\times 10^3}$$
$$= \frac{-8}{10^4} = -8\times 10^{-4}\text{ A}$$
$$= -800\times 10^{-6}\text{ A}$$
$$= -800\ \mu\text{A}$$

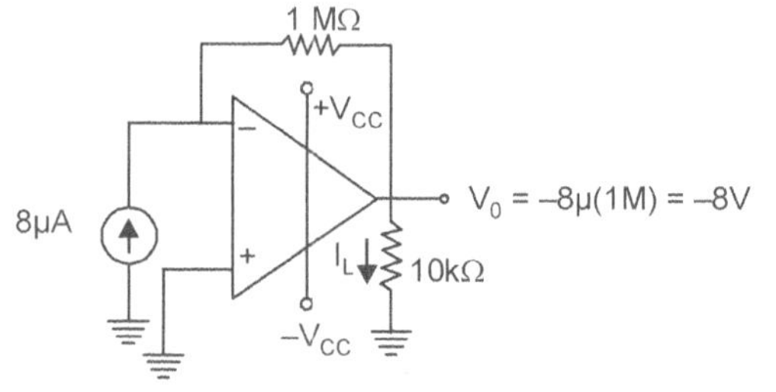

Hence, the value of photo current (throughput the load) is – 800 μA.

42. Here the truth table for the given circuit is shown below.

X_2	X_1	X_0	OP_0	OP_1	OP_2	OP_3	OP_4	OP_5	OP_6	OP_7	IP_0	IP_1	IP_2	IP_3	IP_4	IP_5	IP_6	IP_7	Y_2	Y_1	Y_0
0	0	0	1	0	0	0	0	0	0	0	1	0	0	0	0	0	0	0	0	0	0
0	0	1	0	1	0	0	0	0	0	0	0	1	0	0	0	0	0	0	0	0	1
0	1	0	0	0	1	0	0	0	0	0	0	0	0	1	0	0	0	0	0	1	1
0	1	1	0	0	0	1	0	0	0	0	0	0	1	0	0	0	0	0	0	1	0
1	0	0	0	0	0	0	1	0	0	0	0	0	0	1	0	0	0	0	–	–	–
1	0	1	0	0	0	0	0	1	0	0	0	0	0	0	1	0	0	0	–	–	–
1	1	0	0	0	0	0	0	0	1	0	0	0	0	0	0	1	0	0	1	0	1
1	1	1	0	0	0	0	0	0	0	1	0	0	0	0	1	0	0	0	1	0	0

Hence it will be binary to Gray Code Converter.

43.

C_1	C_0	Y
0	0	P
0	1	Q
1	0	R
1	1	S

Thus it is "4 to 1 multiplexer" (it means 4 inputs and 1 output)

44. Push takes 12T states due to pre decrement and pop takes 10T states.

45. The open-loop transfer function of a unity feedback control system is

$$G(s) = \frac{K}{s^2+5s+5}$$

For break away point $\frac{dk}{ds} = 0$

$$\frac{d}{ds}\left(\frac{1}{s^2+5s+5}\right) = 0$$

$$\frac{\left(s^2+5s+5\right)\cdot\frac{d(1)}{ds} - 1\cdot\frac{d}{ds}\left(s^2+5s+5\right)}{\left(s^2+5s+5\right)^2} = 0$$

$$0-(2s+5) = 0$$

$$2s+5 = 0$$

$\therefore$ $s = -2.5$ is a break away point

The value of K is Obtain from magnitude condition

$$\left|\frac{K}{s^2+5s+5}\right|_{s=-2.5} = 1$$

$$\left|\frac{K}{(-2.5)^2+5(-2.5)+5}\right| = 1$$

$$\Rightarrow \quad \left|\frac{K}{6.25-12.5+5}\right| = 1$$

$$\therefore \quad K = 1.25$$

So, the value of K at the breakaway point is 1.25.

46. Given the peak overshoot of the closed-loop system to a unit step input

$$\%M_p = 10\%$$

$$M_p = 0.1$$

$$\Rightarrow \quad M_p = e^{-\pi\xi/\sqrt{1-\xi^2}}$$

$$0.1 = e^{-\pi\xi/\sqrt{1-\xi^2}}$$

$$\Rightarrow \quad \ln(0.1) = \frac{-\pi\xi}{\sqrt{1-\xi^2}}$$

$$\Rightarrow \quad 2.3 = \frac{\pi\xi}{\sqrt{1-\xi^2}}$$

$$\xi = 0.59$$

Also given $\quad G(s) = \frac{K}{s(s+2)}$

CE:-

$$1 + G(s) = 0$$

$$\Rightarrow \quad s^2 + 2s + K = 0$$

$$2\,\varepsilon\omega_n = 2$$

$$2 \times 0.59 \times \omega_n = 2$$

$$\omega_n = 1.69 \text{ r/sec}$$

$$K = \omega_n^2 = 1.69 \times 1.69$$

$$= 2.87 \text{ r/sec}$$

Hence the value of K is 2.87 r/sec.

47. The transfer function of a linear time invariant system

$$H(s) = 2s^4 - 5s^3 + 5s - 2$$

Now, RH – Criteria

①	$+s^4$	2	0	–2
	$-s^3$	– 5	+ 5	
③	$+s^2$	2	– 2	
	$+s^1$	0(2)		
③	$-s^0$	– 2		

3 Sign Changes

3 Roots (Zeros) in the right half of the *s*-Plane.

48. The entropy of the source

$$(H) = \frac{1}{2}\log_2(2) + \frac{1}{4}\log_2(4) + \frac{1}{8}\log_2(8) + \frac{1}{16}\log_2(16) + ...$$

$$= \frac{1}{2}\cdot 1 + \frac{1}{4}\log_2(2^2) + \frac{1}{8}\log_2(2^3) + \frac{1}{16}\log_2(2^4) + ...$$

$$= \frac{1}{2}\cdot 1 + \frac{1}{4}\log_2(2) + \frac{1}{8}\times 3\log_2(2) + \frac{1}{16} \times 4\log_2(2) + ...$$

$$H = \frac{1}{2}\times 1 + \frac{1}{4}\times 2 + \frac{1}{8}\times 3 + \frac{1}{16}\times 4 + ...$$

$$= \sum_{n=0}^{\infty} n\left(\frac{1}{2}\right)^n = \frac{\frac{1}{2}}{\left(1-\frac{1}{2}\right)^2} = 2$$

49. Probability of error

$$P_e = P^3 + 3P^2(1-P)$$
$$P = 0.1$$
$$P_e = (0.1)^3 + 3 \times (0.1)^2 (1 - 0.1)$$
$$= 0.001 + 3 \times 0.01 \times 0.9$$
$$= 0.001 + 0.027$$
$$= 0.028$$

Thus the average probability of error is 0.028.

50. The impulse response of the filter is same on the signal, so we have $E_s = E_h$

Signal-to-noise ratio

$$SNR = \frac{2E_s}{N_o}$$

$$E_s = E_s$$

$$SNR = \frac{2E_s}{N_o}$$

Hence, the output of the fitter (SNR_{max}) is $\frac{2E_s}{N_o}$.

51. Correct option is not given

Current density in a medium is given by

$$\vec{J} = \frac{400\sin\theta}{2\pi(r^2+4)} \hat{a}_r \text{ A/m}^{-2}$$

Now current passing through the portion of sphere of radius (r) = 0.8 m is given by

$$I = \int_s \vec{J}.d\vec{s} \quad (r = \text{constant})$$

$$d\vec{s} = r^2 \sin\theta \, d\theta \, d\phi \, \hat{a}rd \quad (\because r = 0.8 \text{ m})$$

$$I = \int_{\theta=\frac{\pi}{2}}^{\frac{\pi}{4}} \int_{\phi=0}^{2\pi} \frac{400\sin\theta}{2\pi(r^2+4)} r^2 \sin\theta \, d\theta \, d\phi$$

$$= \frac{400(0.8)^2}{2\pi(0.8^2+4)}\left[\left(\frac{\pi}{4}-\frac{\pi}{12}\right) - \left(\sin\left(\frac{\pi}{2}\right)-\sin\left(\frac{\pi}{6}\right)\right)\right] \times (2\pi)$$

$\therefore$ I = 7.45 A

The average current density through the given sphere surface is given by

$$J = \frac{1}{\text{Area of } r = 0.8\text{m sphere}}$$

$$= \frac{7.45}{(0.8)^2 \int_{\theta=\frac{\pi}{2}}^{\frac{\pi}{4}} \int_{\phi=0}^{2\pi} \sin\theta \, d\theta \, d\phi}$$

$$= \frac{7.45}{1.04}$$

$\therefore$ J = 7.15 A/m^2

Hence the total current and average current density are 7.45 A and 7.15 A/m^2.

52.

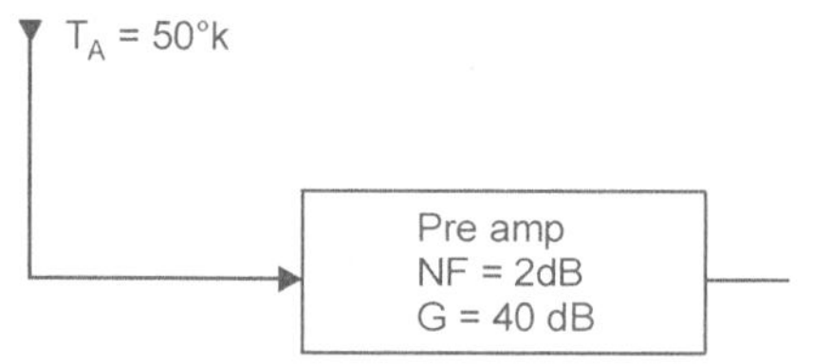

$$10 \log_{10} NF = 2dB$$
$$\log_{10} NF = 0.2$$
$$NF = 10^{0.2}$$

Noise temperature = $(NF - 1)\, T_o$
$= (10^{0.2} - 1)\, 290°$
$= 169.36$ K

Noise power (i/p)
$= k T_e B$
$= 1.38 \times 10^{-23} \times (169.36 + 50) \times 12 \times 10^6$

Noise power at (o/p) = $(3.632 \times 10^{-14}) \times 10^4$
$= 3.73 \times 10^{-10}$ watts

53. For lossless horn antennas

$$\eta_T = \eta_R = 1$$

Power Gain = Directivity

Directivity of Transmitting antenna,

$$D_T = 18 \text{ dB}$$
$$10 \log D_T = 18$$
$$G_T \text{ (or) } D_T = 63.09$$

Directivity of Receiving antenna,

$$D_R = 22 \text{ dB}$$
$$10 \log D_R = 22$$
$$G_R \text{ (or) } D_R = 158.48$$

input power P_{in} = 2 W

Spacing, $r = 200\,\lambda$

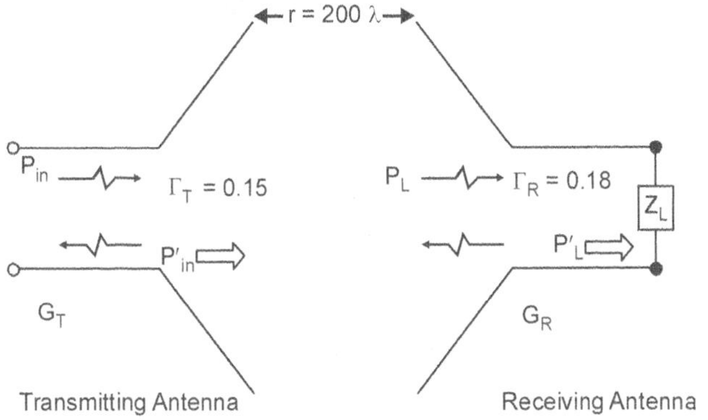

Friis transmission formula in given by

$$P_L = G_T G_R \left[\frac{\lambda}{4\pi r}\right]^2 P'_{in}$$

where :

P'_{in} : Input power (prime indicates power due to reflection)

$$P'_{in} = \left|1 - \Gamma_T^2\right| P_{in}$$

$$= \left|1 - (0.15)^2\right| \times 2$$

$$P'_{in} = 1.955 \text{ W}$$

$$P_L = 63.09 \times 158.48 \left[\frac{\lambda}{4\pi \times 200\lambda}\right]^2 \times 1.955$$

$$= 3.1 \times 10^{-3}$$

As there is a reflection at the terminals of receiving antenna power delivered to the load in given by

$$P'_L = \left\{1 - \left|\Gamma_R^2\right|\right\} \times P_L$$

$$= \left\{\left|1 - (0.18)^2\right|\right\} \times 3.1 \times 10^{-3}$$

$$\therefore \quad P'_L = 2.99 \text{ mW}$$

Hence, the power delivered to the load at the receiver is 2.99 mW.

54. Given

Electric field of incident wave is given by

$$E_W^i = (\hat{a}_x + j\hat{a}_y) E_0 e^{jkz}$$

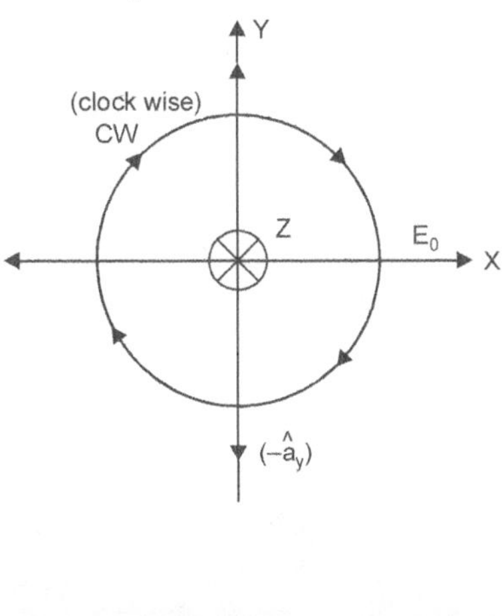

at $z = 0$;

$$\vec{E}_w^i = E_0 \cos\omega t \hat{a}_x - E_0 \sin\omega t \hat{a}_y$$

(in time varying form)

at $\omega t = 0$

$$\vec{E}_w^i = E_0 \hat{a}_x$$

at $\omega t = \frac{\pi}{2}$

$$\vec{E}_w^i = E_0(-\hat{a}_y)$$

As a tip of electric field intensity is tracing a circle when time varies, hence the wave is said to be circularly polarized in clockwise direction (or) RHCP. Polarizing vector of incident wave is given by,

$$\hat{P}_i = \frac{\hat{a}_x + j\hat{a}_y}{\sqrt{2}}$$

radiated electric field from the antenna is

$$\vec{E}_a = (\hat{a}_x + 2\hat{a}_y) E_I \frac{1}{\gamma} e^{-jk\gamma}$$

at $r = 0$

$$\vec{E}_a = E_I \cos\omega t \, \hat{a}_x + 2E_I \cos\omega t \hat{a}_y$$

(in time varying form)

As both x and y components are in-phase hence the wave is said to be linear polarized. Polarizing vector of radiated field is $\hat{P}_a = \frac{(\hat{a}_x + 2\hat{a}_y)}{\sqrt{5}}$

polarizing mismatch; The polarizing mismatch is said to have, if the polarization of receiving antenna is not same on the polarization of the incident wave. The polarization loss factor (PLF) characterizes the loss of EM power due to polarization mismatch.

$$\text{PLF} = \left|\hat{P}_i . \hat{P}_a\right|^2$$

in dB; PLF (dB) = 10 log (PLF)

$$\text{PLF} = \left|\left(\frac{\hat{a}_x + j\hat{a}_y}{2}\right) . \left(\frac{\hat{a}_x + 2\hat{a}_y}{\sqrt{5}}\right)\right|$$

$$= \left|\frac{1 + j2}{\sqrt{2}\sqrt{5}}\right|^2$$

$$= \frac{1}{2} (\text{or}) 0.5$$

$$\text{PLF (dB)} = 10 \log 0.5$$

$$= -3.0102$$

$$= -3 \text{ dB(approx)}$$

So, the loss due to the polarization is – 3 dB.

55. Given

Power density radiated by the antenna is given by

$$\vec{W}_{rad} = \frac{C_0'}{r^2}\cos^4\theta\ \hat{a}_r\ \text{W/m}^2$$

Power radiated (or) average power radiated by the antenna in given by

$$P_{rad} = \oint_s \vec{W}_{rad}\cdot d\vec{s}$$

$$= \int_{\theta=0}^{\pi/2}\int_{\phi=0}^{2\pi} \frac{C_0'}{r^2}\cos^4\theta\, r^2\sin\theta\, d\theta\, d\phi$$

($\because$ radiated only in the upper hemisphere)

$$= C_0'\,(2\pi)\frac{1}{5}$$

$$\therefore\quad P_{rad} = 1.256\ C_0'\ \text{Watt}$$

Now Maximum directivity of the antenna in given by

$$D = 4\pi\frac{U_{max}}{P_{rad}}$$

$$U = r^2\, W_{rad}$$

$$U = r^2 \times \frac{C_0'}{r^2}\cos^4\theta$$

$$U = C_0'\ \cos^4\theta$$

$$U_{max} = C_0'$$

$$\therefore\quad D = \frac{4\pi C_0'}{1.256\ C_0} = 10$$

$$D_{(dB)} = 10\log 10$$

$$\therefore\quad D = 10\ \text{dB}$$

Thus the power radiated by the antenna and the maximum directivity of the the antenna are $1.256\ C_0'$ Watt and 10 dB respectively.

■■

SOLVED PAPER – 2016

ELECTRONICS & COMMUNICATION ENGINEERING
Set - 2

No. of Questions : 65 | *Maximum Marks : 100*

INSTRUCTIONS

1. Total of 65 questions carrying 100 marks, out of which 10 questions carrying a total of 15 marks are in General Aptitude (GA)
2. The Engineering Mathematics will carry around **15% of the total marks,** the General Aptitude section will carry **15% of the total marks** and the **remaining 70% of the total marks.**
3. **Types of Questions**
   (*a*) **Multiple Choice Questions (MCQ)** carrying 1 or 2 marks each in all papers and sections. These questions are objective in nature, and each will have a choice of four options, out of which the candidate has to mark the correct answer(s).
   (*b*) **Numerical Answer Questions** of 1 or 2 marks each in all papers and sections. For these questions the answer is a real number, to be entered by the candidate using the virtual keypad. No choices will be shown for these type of questions.
4. For **1-mark** multiple-choice questions, **1/3 marks** will be deducted for a wrong answer. Likewise, for **2-marks** multiple-choice questions, **2/3** marks will be deducted for a wrong answer. There is no negative marking for numerical answer type questions.

GENERAL APTITUDE (GA)

(Q.1 – 5) : Carry One Mark Each

1. Based on the given statements, select the appropriate option with respect to grammar and usage.

   Statements :

   I. The height of Mr. X is 6 feet.

   II. The height of Mr. Y is 5 feet.

   (*a*) Mr. X is longer than Mr. Y.
   (*b*) Mr. X is more elongated than Mr. Y
   (*c*) Mr. X is taller than Mr. Y
   (*d*) Mr. X is lengthier than Mr. Y

2. The students _____ the teacher on teachers' day for twenty years of dedicated teaching.

   (*a*) facilitated
   (*b*) felicitated
   (*c*) fantasized
   (*d*) facillitated

3. After India's cricket world cup victory in 1985, Shrotria who was playing both tennis and cricket till then, decided to concentrate only on cricket. And <u>the rest is history</u>.

   What does the underlined phrase mean in this context?

   (*a*) history will rest in peace
   (*b*) rest is recorded in history books
   (*c*) rest is well known
   (*d*) rest is archaic

4. Given (9 inches)$^{1/2}$ = (0.25 yards)$^{1/2}$, which one of the following statements is TRUE?

   (*a*) 3 inches = 0.5 yards
   (*b*) 9 inches = 1.5 yards
   (*c*) 9 inches = 0.25 yards
   (*d*) 81 inches = 0.0625 yards

5. S, M, E and F are working in shifts in a team to finish a project. M works with twice the efficiency of others but for half as many days as E worked. S and M have 6 hour shifts in a day whereas E and F have 12 hours shifts. What is the ratio of contribution of M to contribution of E in the project?

   (*a*) 1 : 1 (*b*) 1 : 2
   (*c*) 1 : 4 (*d*) 2 : 1

(Q.6 – 10) : Carry Two Marks Each

6. The Venn diagram shows the preference of the student population for leisure activities.

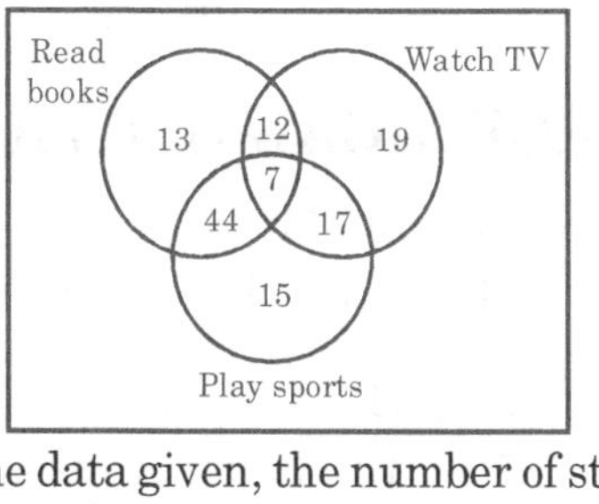

   From the data given, the number of students who like to read books or play sports is ____.

   (*a*) 44 (*b*) 51
   (*c*) 79 (*d*) 108

7. Social science disciplines were in existence in an amorphous form until the colonial period when they were institutionalized. In varying degrees, they were intended to further the colonial interest. In the time of globalization and the economic rise of postcolonial countries like India, conventional ways of knowledge production have become obsolete.

Which of the following can be logically inferred from the above statements?

I. Social science disciplines have become obsolete.

II. Social science disciplines had a pre-colonial origin.

III. Social science disciplines always promote colonialism.

IV. Social science must maintain disciplinary boundaries.

(*a*) II only (*b*) I and III only
(*c*) II and IV only (*d*) III and IV only

8. Two and a quarter hours back, when seen in a mirror, the reflection of a wall clock without number markings seemed to show 1 : 30. What is the actual current time shown by the clock?

(*a*) 8:15 (*b*) 11:15
(*c*) 12:15 (*d*) 12:45

9. M and N start from the same location. M travels 10 km East and then 10 km North-East. N travels 5 km South and then 4 km South-East. What is the shortest distance (in km) between M and N at the end of their travel?

(*a*) 18.60 (*b*) 22.50
(*c*) 20.61 (*d*) 25.00

10. A wire of length 340 mm is to be cut into two parts. One of the parts is to be made into a square and the other into a rectangle where sides are in the ratio of 1:2. What is the length of the side of the square (in mm) such that the combined area of the square and the rectangle is a MINIMUM?

(*a*) 30 (*b*) 40
(*c*) 120 (*d*) 180

TECHNICAL SECTION

(Q.1 – 25) : Carry One Mark Each

1. The value of x for which the matrix

$$A = \begin{bmatrix} 3 & 2 & 4 \\ 9 & 7 & 13 \\ -6 & -4 & -9+x \end{bmatrix}$$ has zero as an eigen value

is ______.

2. Consider the complex valued function $f(z) = 2z^3 + b|z|^3$ where z is a complex variable. The value of b for which the function $f(z)$ is analytic is ______

3. As x varies from –1 to +3, which one of the following describes the behaviour of the function $f(x) = x^3 - 3x^2 + 1$?

(*a*) $f(x)$ increases monotonically.

(*b*) $f(x)$ increases, then decreases and increases again.

(*c*) $f(x)$ decreases, then increases and decreases again.

(*d*) $f(x)$ increases and then decreases

Then the given function $f(x)$ increases, then decreases and increase again.

4. How many distinct values of x satisfy the equation $\sin(x) = x/2$, where x is in radians?

(*a*) 1 (*b*) 2
(*c*) 3 (*d*) 4 or more

5. Consider the time-varying vector

$I = \hat{x}15\cos(\omega t) + \hat{y}5\sin(\omega t)$ in Cartesian coordinates, where $\omega > 0$ is a constant.

When the vector magnitude $|I|$ is at its minimum value, the angle θ that I makes with the x axis (in degrees, such that $0 \le \theta \le 180°$) is ______.

6. In the circuit shown below, V_s is constant voltage source and I_L is a constant current load.

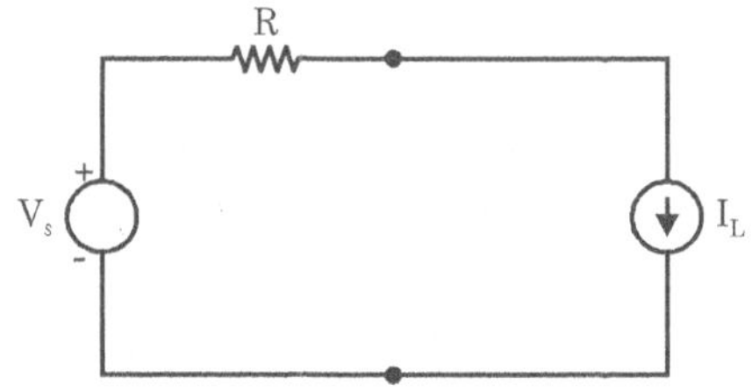

The value of I_L that maximizes the power absorbed by the constant current load is

(*a*) $\frac{V_s}{4R}$ (*b*) $\frac{V_s}{2R}$
(*c*) $\frac{V_s}{R}$ (*d*) ∞

7. The switch has been in position 1 for a long time and abruptly changes to position 2 at $t = 0$.

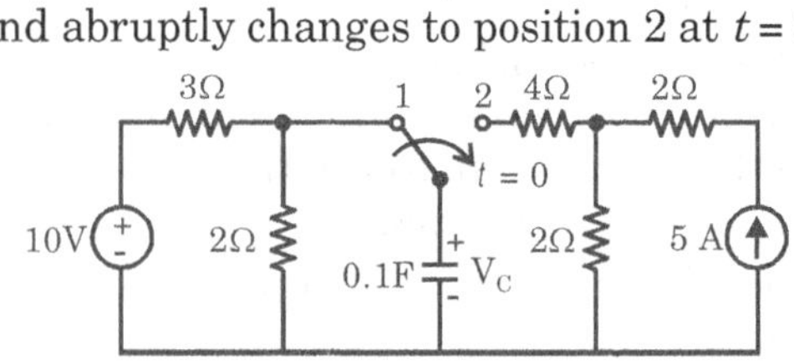

If time t is in seconds, the capacitor voltage V_C (in volts) for $t > 0$ is given by

(*a*) $4(1 - \exp(-t/0.5))$ (*b*) $10 - 6\exp(-t/0.5)$
(*c*) $4(1 - \exp(-t/0.6))$ (*d*) $10 - 6\exp(-t/0.6)$

8. The figure shows an RLC circuit with a sinusoidal current source.

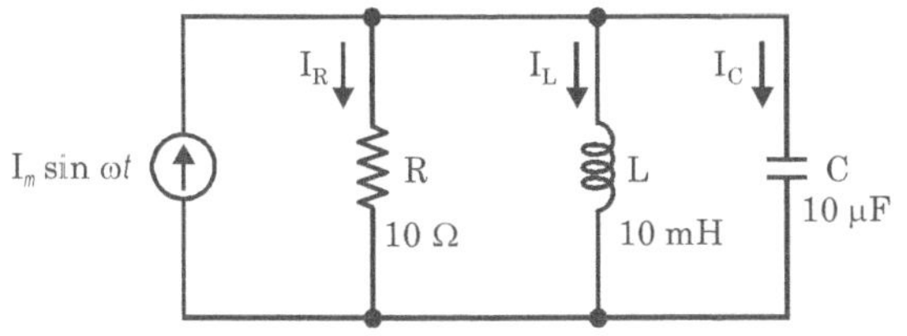

At resonance, the ratio $\left|\frac{I_L}{I_R}\right|$, i.e., the ratio of the magnitudes of the inductor current phasor and the resistor current phasor, is_____

9. The Z-parameter matrix for the two-port network shown is

$$\begin{bmatrix} 2j\omega & j\omega \\ j\omega & 3+2j\omega \end{bmatrix},$$

Where the entries are in Ω.

Suppose $Z_b(j\omega) = R_b + j\omega$

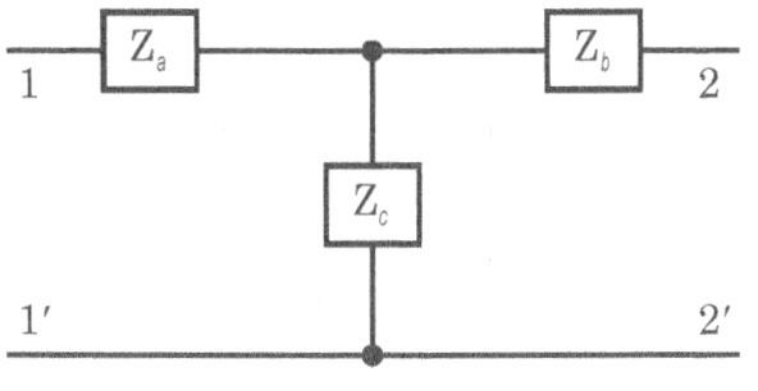

Then the value of R_b (in Ω) equals _____

10. The energy of the signal $x(t) = \frac{\sin(4\pi t)}{4\pi t}$ is ________.

11. The Ebers-Moll model of a BJT is valid
   (*a*) only in a active mode
   (*b*) only in active and saturation modes
   (*c*) only in active and cut-off modes
   (*d*) in active, saturation and cut-off modes

12. A long-channel NMOS transistor is biased in the linear region with V_{DS} = 50 mV and is used as a resistance. Which one of the following statements is NOT correct?
   (*a*) If the device width W is increased, the resistance decreases.
   (*b*) If the threshold voltage is reduced, the resistance decreases.
   (*c*) If the device length L is increased, the resistance increase.
   (*d*) If V_{GS} is increased, the resistance increases.

13. Assume that the diode in the figures has V_{on} = 0.7 V, but is otherwise ideal.

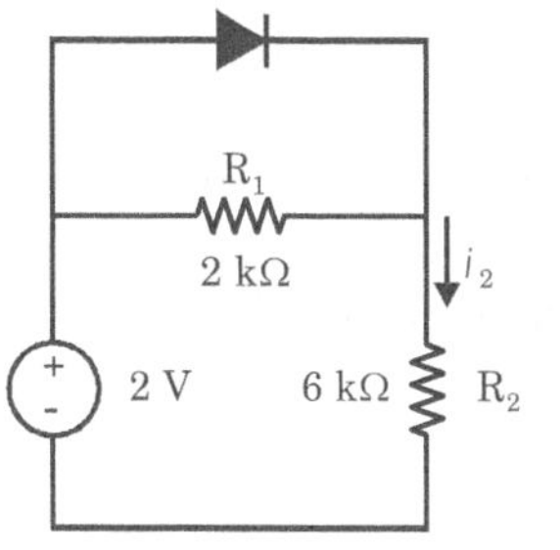

The magnitude of the current i_2 (in mA) is equal to_____

14. Resistor R_1 in the circuit below has been adjusted so that I_1 = 1 mA. The bipolar transistors Q_1 and Q_2 are perfectly matched and have very high current gain, so their base currents are negligible. The supply voltage V_{CC} is 6 V. The thermal voltage kT/q is 26 mV.

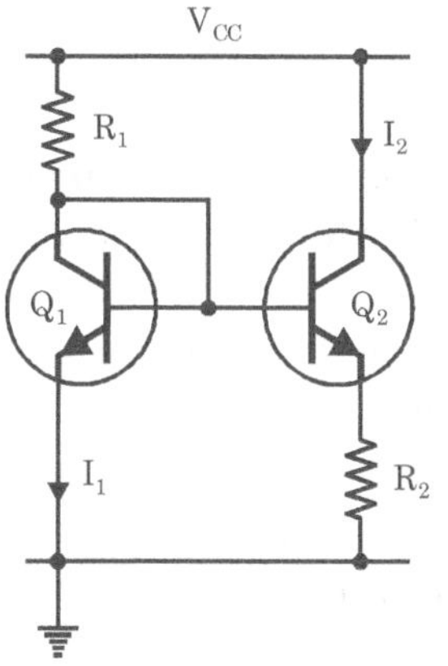

The value of R_2 (in Ω) for which I_2 = 100 μA is __________.

15. Which one of the following statements is correct about an ac-coupled commonemitter amplifier operating in the mid-band region?
   (*a*) The device parasitic capacitances behave like open circuits, whereas coupling and bypass capacitances behave like short circuits.
   (*b*) The device parasitic capacitances, coupling capacitances and bypass capacitances behave like open circuits.
   (*c*) The device parasitic capacitances, coupling capacitances and bypass capacitances behave like short circuits.
   (*d*) The device parasitic capacitances behave like short circuits, whereas coupling and bypass capacitances behave like open circuits.

16. Transistor geometries in a CMOS inverter have been adjusted to meet the requirement for worst case charge and discharge times for driving a load capacitor C. This design is to be converted to that of a NOR circuit in the same technology, so that its worst case charge and discharge times while driving the same capacitor are similar. The channel lengths of all transistors are to be kept unchanged. Which one of the following statements is correct?

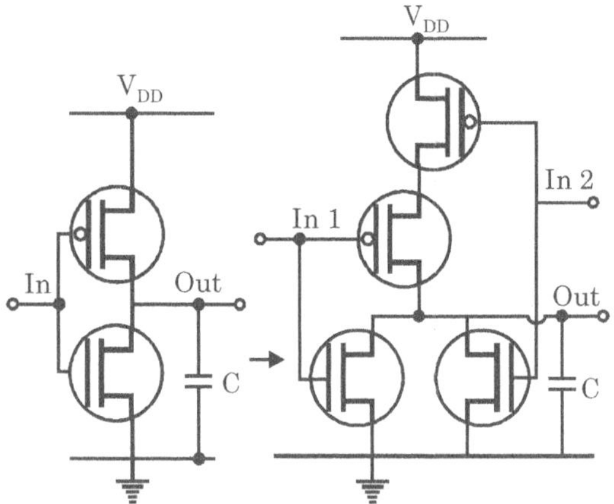

(*a*) Widths of PMOS transistors should be doubled, while widths of NMOS transistors should be halved.

(*b*) Widths of PMOS transistors should be doubled, while widths of NMOS transistors should not be changed.

(*c*) Widths of PMOS transistors should be halved, while widths of NMOS transistors should not be changed.

(*d*) Widths of PMOS transistors should be unchanged, while widths of NMOS transistors should be halved.

17. Assume that all the digital gates in the circuit shown in the figure are ideal, the resistor R = 10 kΩ and the supply voltage is 5 V. The D flip-flops D_1, D_2, D_3, D_4 and D_5 are initialized with logic values 0, 1, 0, 1 and 0, respectively. The clock has a 30% duty cycle.

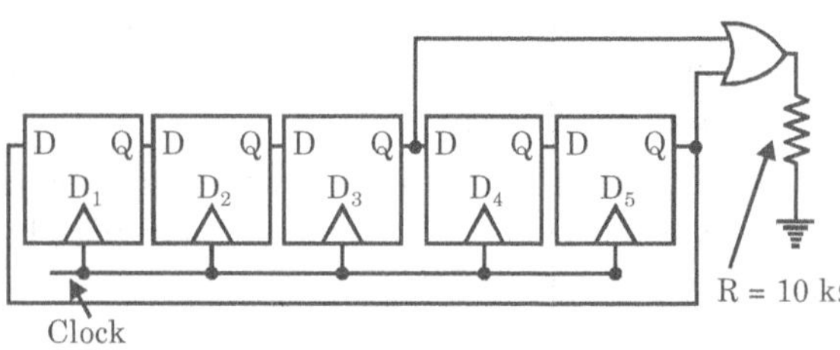

The average power dissipated (in mW) in the resistor R is _____.

18. A 4 :1 multiplexer is to be used for generating the output carry of a full adder. A and B are the bits to be added while C_{in} is the input carry and C_{out} is the output carry. A and B are to be used as the select bits with A being the more significant select bit.

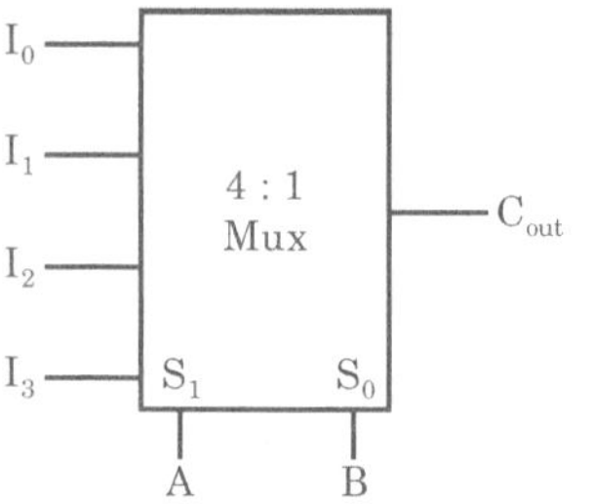

Which one of the following statements correctly describes the choice of signals to be connected to the inputs I_0, I_1, I_2 and I_3 so that the output is C_{out}?

(*a*) $I_0 = 0$, $I_1 = C_{in}$, $I_2 = C_{in}$ and $I_3 = 1$

(*b*) $I_0 = 1$, $I_1 = C_{in}$, $I_2 = C_{in}$ and $I_3 = 1$

(*c*) $I_0 = C_{in}$, $I_1 = 0$, $I_2 = 1$ and $I_3 = C_{in}$

(*d*) $I_0 = 0$, $I_1 = C_{in}$, $I_2 = 1$ and $I_3 = C_{in}$

19. The response of the system $G(s) = \dfrac{s-2}{(s+1)(s+3)}$ to the unit step input $u(t)$ is $y(t)$. The value of $\dfrac{dy}{dt}$ at $t = 0^+$ is ______.

20. The number and direction of encirclements around the point $-1 + j0$ in the complex plane by the Nyquist plot of $G(s) = \dfrac{1-s}{4+2s}$ is

(*a*) zero

(*b*) one, anti-clockwise

(*c*) one, clockwise

(*d*) two, clockwise

21. A discrete memoryless source has an alphabet $\{a_1, a_2, a_3, a_4\}$ with corresponding probabilities $\left\{\frac{1}{2}, \frac{1}{4}, \frac{1}{8}, \frac{1}{8}\right\}$. The minimum required average codeword length in bits to represent this source for error-free reconstruction is________.

22. A speech signal is sampled at 8 kHz and encoded into PCM format using 8 bits/sample. The PCM data is transmitted through a baseband channel via 4-level PAM. The minimum bandwidth (in kHz) required for transmission is _____.

23. A uniform and constant magnetic field $B = \hat{z}B$ exists in the $\hat{z}$ direction in vacuum. A particle of mass m with a small charge q is introduced into this region with an initial velocity $v = \hat{x}v_z + \hat{z}v_z$. Given that B, m, q, v_x and v_z are all nonzero, which one of the following describes the eventual trajectory of the particle?

(*a*) Helical motion in the $\hat{z}$ direction.

(*b*) Circular motion in the xy plane.

(*c*) Linear motion in the $\hat{z}$ direction.

(*d*) Linear motion in the $\hat{x}$ direction.

24. Let the electric field vector of a plane electromagnetic wave propagating in a homogenous medium be expressed as $E = \hat{x}E_x e^{-j(\omega t - \beta z)}$, where the propagation constant β is a function of the angular frequency ω. Assume that $\beta(\omega)$ and E_x are known and are real. From the information available, which one of the following CANNOT be determined?

(*a*) The type of polarization of the wave.

(*b*) The group velocity of the wave.

(*c*) The phase velocity of the wave.

(*d*) The power flux through the $z = 0$ plane.

25. Light from free space is incident at an angle θ_i to the normal of the facet of a step-index large core optical fibre. The core and cladding refractive indices are $n_1 = 1.5$ and $n_2 = 1.4$, respectively.

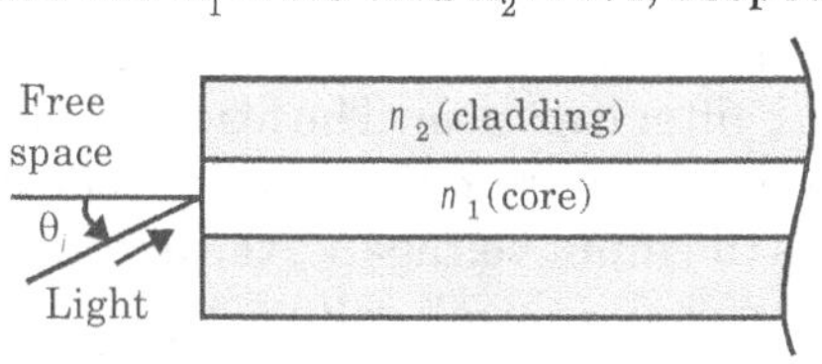

The maximum value of θ_i (in degrees) for which the incident light will be guided in the core of the fibre is ________.

(Q.26 – 55) : Carry Two Marks Each

26. The ordinary differential equation

$$\frac{dy}{dx} = -3y + 2, \text{ with } y(0) = 1$$

is to be solved using the forward Euler method. The largest time step that can be used to solve the equation without making the numerical solution unstable is ________.

27. Suppose C is the closed curve defined as the circle $x^2 + y^2 = 1$ with C oriented anti clockwise. The value of $\oint (xy^2 dx + x^2 y dy)$ over the curve C equals ________.

28. Two random variables X and Y are distributed according to

$$f_{X,Y}(x, y) = \begin{cases} (x + y) & 0 \le x \le 1,\ 0 \le y \le 1 \\ 0 & \text{otherwise} \end{cases}$$

The probability $P(X + Y \le 1)$ is ________

29. The matrix $A = \begin{bmatrix} a & 0 & 3 & 7 \\ 2 & 5 & 1 & 3 \\ 0 & 0 & 2 & 4 \\ 0 & 0 & 0 & b \end{bmatrix}$ has det(A)=100 and trace (A)=14. The value of $|a - b|$ is ________.

30. In the given circuit, each resistor has a value equal to 1 Ω.

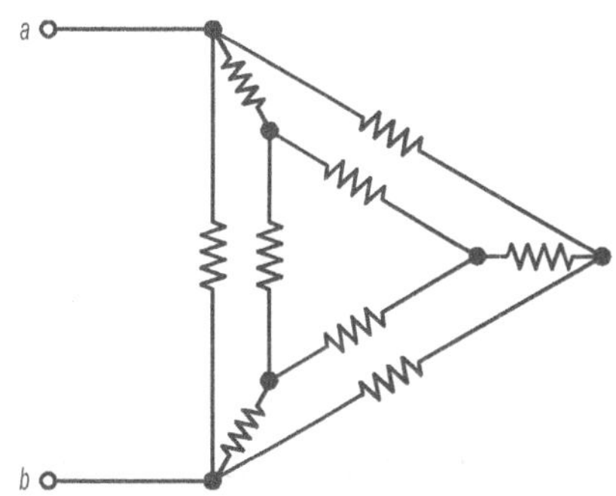

What is the equivalent resistance across the terminals a and b?

(*a*) 1/6 Ω (*b*) 1/3 Ω

(*c*) 9/20 Ω (*d*) 8/15 Ω

31. In the circuit shown in the figure, the magnitude of the current (in amperes) through R_2 is ___

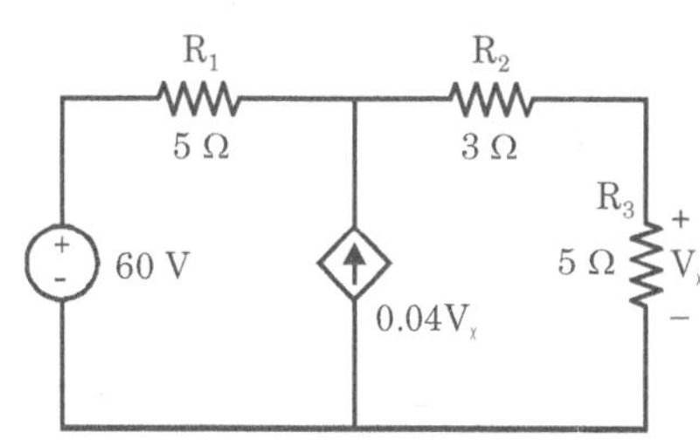

32. A continuous-time filter with transfer function $H(s) = \frac{2s + 6}{s^2 + 6s + 8}$ is converted to a discrete time filter with transfer function $G(s) = \frac{2z^2 - 0.5032z}{z^2 - 0.5032z + k}$ so that the impulse response of the continuous-time filter, sampled at 2 Hz, is identical at the sampling instants to the impulse response of the discrete time filter. The value of k is _____.

33. The Discrete Fourier Transform (DFT) of the 4-point sequence

$x[n] = \{x[0], x[1], x[2], x[3]\} = \{3, 2, 3, 4\}$ is

$X[k] = \{X[0], X[1], X[2], X[3]\} = \{12, 2j, 0, -2j\}$.

If $X_1[k]$ is the DFT of the 12-point sequence $x_1[n] = \{3, 0, 0, 2, 0, 0, 3, 0, 0, 4, 0, 0\}$, the value of

$\left|\dfrac{X_1[8]}{X_1[11]}\right|$ is ______.

34. The switch S in the circuit shown has been closed for a long time. It is opened at time $t = 0$ and remains open after that. Assume that the diode has zero reverse current and zero forward voltage drop.

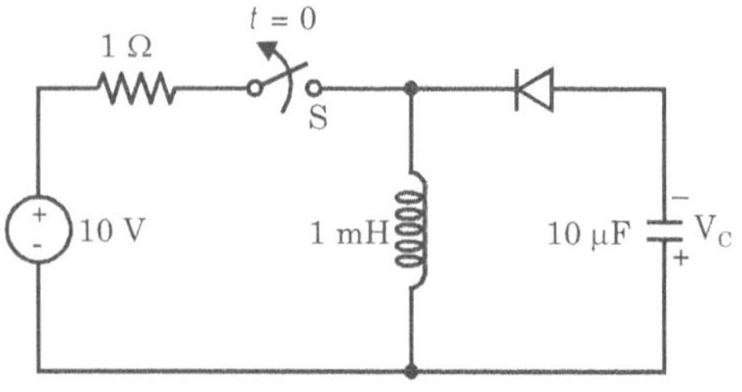

The steady state magnitude of the capacitor voltage VC (in volts) is ______.

35. A voltage V_G is applied across a MOS capacitor with metal gate and *p*-type silicon substrate at T = 300 K. The inversion carrier density (in number of carriers per unit area) for $V_G = 0.8$ V is 2×10^{11} cm^{-2}. For $V_G = 1.3$ V, the inversion carrier density is 4×10^{11} cm^{-2}. What is the value of the inversion carrier density for $V_G = 1.8$ V?

(*a*) 4.5×10^{11} cm^{-2}

(*b*) 6.0×10^{11} cm^{-2}

(*c*) 7.2×10^{11} cm^{-2}

(*d*) 8.4×10^{11} cm^{-2}

36. Consider avalanche breakdown in a silicon p^+n junction. The *n*-region is uniformly doped with a donor density N_D. Assume that breakdown occurs when the magnitude of the electric field at any point in the device becomes equal to the critical field E_{crit}. Assume E_{crit} to be independent of N_D. If the built-in voltage of the p^+n junction is much smaller than the breakdown voltage, V_{BR}, the relationship between V_{BR} and N_D is given by

(*a*) $V_{BR} \times \sqrt{N_D}$ = constant

(*b*) $N_D \times \sqrt{V_{BR}}$ = constant

(*c*) $N_D \times V_{BR}$ = constant

(*d*) N_D / V_{BR} = constant

37. Consider a region of silicon devoid of electrons and holes, with an ionized donor density of $N_d^+ = 10^{17}$ cm^{-3}. The electric field at $x = 0$ is 0 V/cm and the electric field at $x = L$ is 50 kV/cm in the positive x direction. Assume that the electric field is zero in the y and z directions at all points.

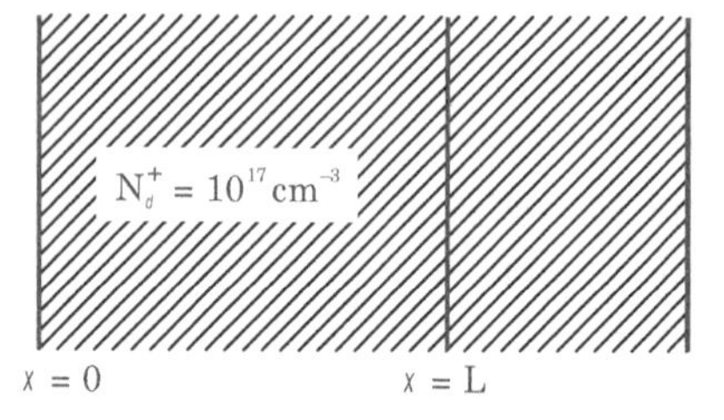

Given $q = 1.6 \times 10^{-19}$ coulomb, $\epsilon_0 = 8.85 \times 10^{-14}$ F/cm, $\epsilon_r = 11.7$ for silicon, the value of L in nm is _____

38. Consider a long-channel NMOS transistor with source and body connected together. Assume that the electron mobility is independent of V_{GS} and V_{DS}. Given,

$g_m = 0.5$ μA/V for $V_{DS} = 50$ mV and $V_{GS} = 2$ V,

$g_d = 8$ μA/V for $V_{GS} = 2$ V and $V_{DS} = 0$ V,

where $g_m = \dfrac{\partial I_D}{\partial V_{GS}}$ and $g_d = \dfrac{\partial I_D}{\partial V_{DS}}$

The threshold voltage (in volts) of the transistor is ________.

39. The figure shows a half-wave rectifier with a 475 μF filter capacitor. The load draws a constant current $I_0 = 1$ A from the rectifier. The figure also shows the input voltage V_i, the output voltage V_C and the peak-to-peak voltage ripple u on V_C. The input voltage V_i is a triangle-wave with an amplitude of 10 V and a period of 1 ms.

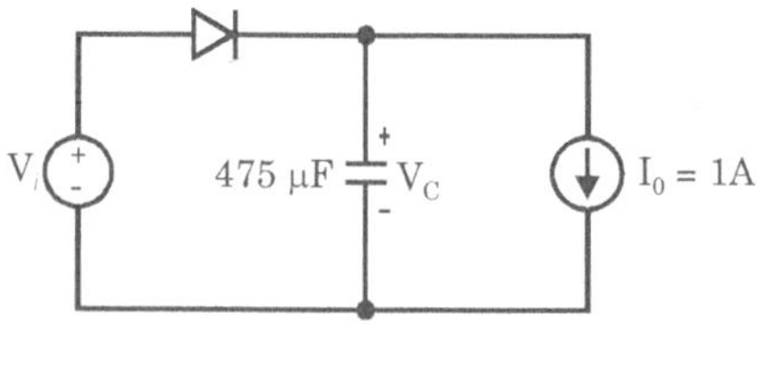

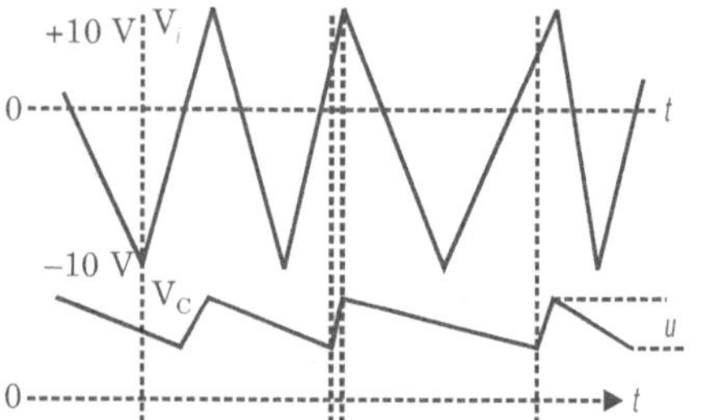

The value of the ripple u (in volts) is ______

40. In the opamp circuit shown, the Zener diodes Z_1 and Z_2 clamp the output voltage V_0 to + 5 V or –5 V. The switch S is initially closed and is opened at time t = 0.

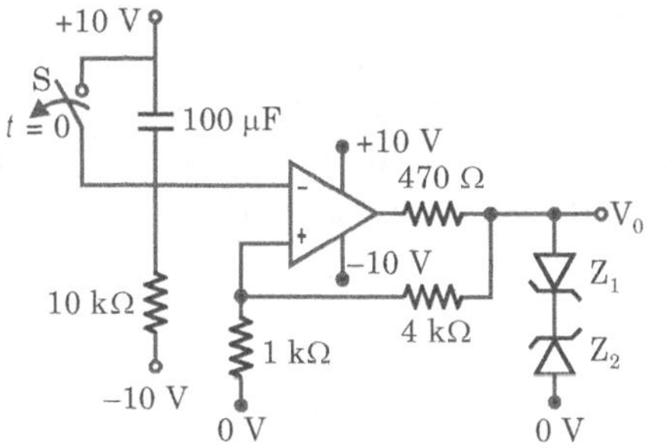

The time $t = t_1$ (in seconds) at which V_0 changes state is ________

41. An opamp has a finite open loop voltage gain of 100. Its input offset voltage V_{ios} (= + 5 mV) is modeled as shown in the circuit below. The amplifier is ideal in all other respects. V_{input} is 25 mV.

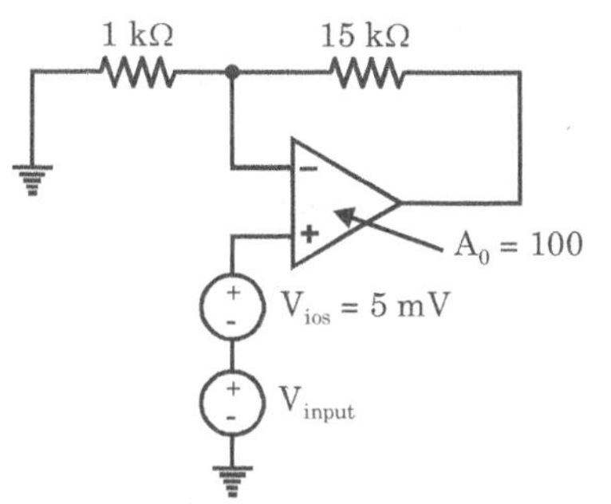

The output voltage (in millivolts) is ________

42. An 8 Kbyte ROM with an active low Chip Select input ($\overline{CS}$) is to be used in an 8085 microprocessor based system. The ROM should occupy the address range 1000 H to 2 FFFH. The address lines are designated as A_{15} to A_0, where A_{15} is the most significant address bit.

Which one of the following logic expressions will generate the correct $\overline{CS}$ signal for this ROM?

(*a*) $A_{15} + A_{14} + (A_{13} . A_{12} + \bar{A}_{13} . \bar{A}_{12})$

(*b*) $A_{15} . A_{14} . (\bar{A}_{13} + \bar{A}_{12})$

(*c*) $\bar{A}_{15} . \bar{A}_{14} . (A_{13} + \bar{A}_{14} + \bar{A}_{13} . \bar{A}_{12})$

(*d*) $\bar{A}_{15} + \bar{A}_{14} + A_{13} . A_{12}$

43. In an N bit flash ADC, the analog voltage is fed simultaneously to $2^N - 1$ comparators. The output of the comparators is then encoded to a binary format using digital circuits. Assume that the analog voltage source V_{in} (whose output is being converted to digital format) has a source resistance of 75 Ω as shown in the circuit diagram below and the input capacitance of each comparator is 8 pF.

The input must settle to an accuracy of 1/2 LSB even for a full scale input change for proper conversion. Assume that the time taken by the thermometer to binary encoder is negligible.

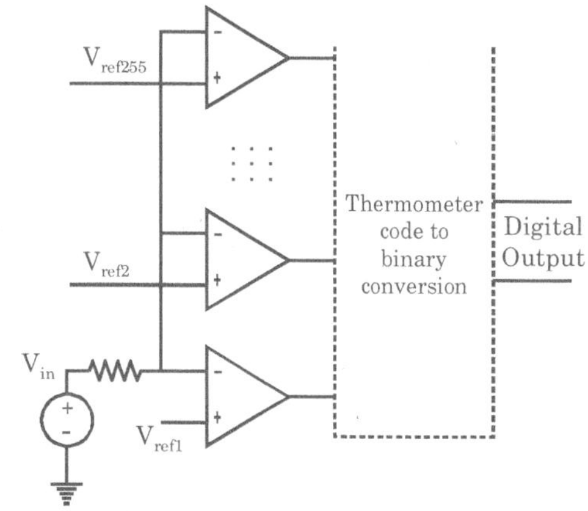

If the flash ADC has 8 bit resolution, which one of the following alternatives is closest to the maximum sampling rate ?

(*a*) 1 mega samples per second

(*b*) 6 mega samples per second

(*c*) 64 mega samples per second

(*d*) 256 mega samples per second

44. The state transition diagram for a finite state machine with states A, B and C, and binary inputs X, Y and Z, is shown in the figure.

Which one of the following statements is correct?

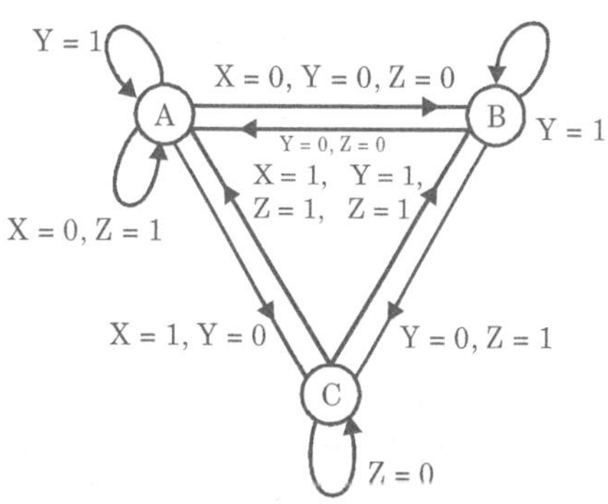

(*a*) Transitions from State A are ambiguously defined.

(*b*) Transitions from State B are ambiguously defined.

(*c*) Transitions from State C are ambiguously defined.

(*d*) All of the state transitions are defined unambiguously.

45. In the feedback system shown below $G(s) = \frac{1}{(s^2 + 2s)}$.

The step response of the closed-loop system should have minimum settling time and have no overshoot.

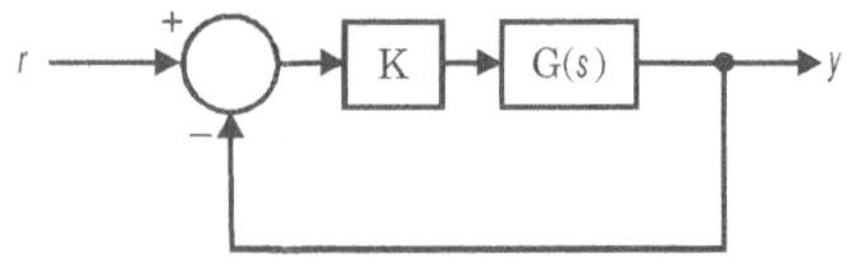

The required value of gain k to achieve this is ________.

46. In the feedback system shown below

$$G(s) = \frac{1}{(s+1)(s+2)(s+3)}$$

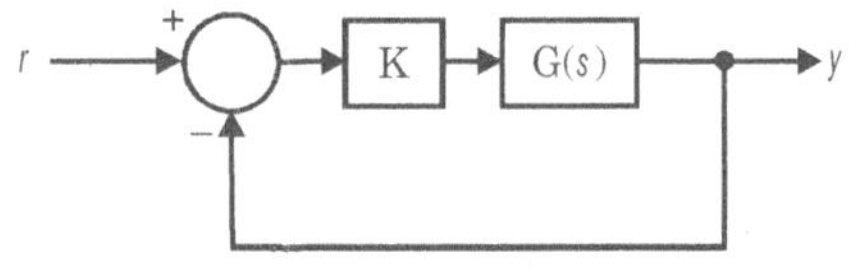

The positive value of k for which the gain margin of the loop is exactly 0 dB and the phase margin of the loop is exactly zero degree is ________

47. The asymptotic Bode phase plot of

$G(s) = \frac{k}{(s+0.1)(s+10)(s+p_1)}$, with k and p_1 both positive, is shown below.

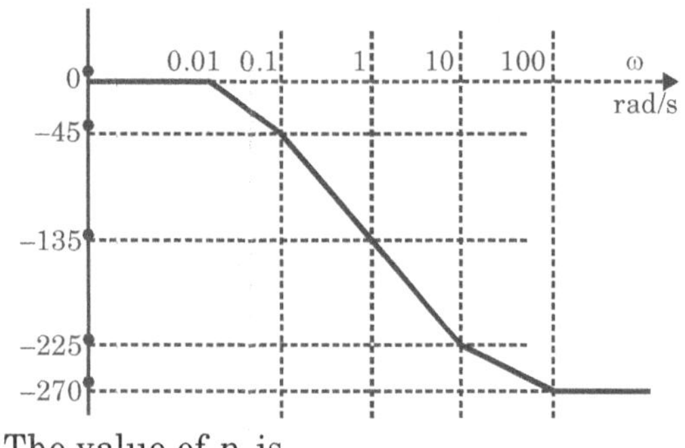

The value of p_1 is ________

48. An information source generates a binary sequence $\{\alpha_n\}$. α_n can take one of the two possible values −1 and +1 with equal probability and are statistically independent and identically distributed. This sequence is precoded to obtain another sequence $\{\beta_n\}$, as $\beta_n = \alpha_n + k\alpha_{n-3}$. The sequence $\{\beta_n\}$ is used to modulate a pulse $g(t)$ to generate the baseband signal

$$X(t) = \sum_{n=-\infty}^{\infty} \beta_n g(t - nT),$$

where $g(t) = \begin{cases} 1, & 0 \le t \le T \\ 0, & \text{otherwise} \end{cases}$

If there is a null at $f = \frac{1}{3T}$ in the power spectral density of $X(t)$, then k is ________

49. An ideal band-pass channel 500 Hz- 2000 Hz is deployed for communication. A modem is designed to transmit bits at the rate of 4800 bits/s using 16-QAM. The roll-off factor of a pulse with a raised cosine spectrum that utilizes the entire frequency band is ________

50. Consider a random process $X(t) = 3V(t) - 8$, where V(t) is a zero mean stationary random process with autocorrelation $R_v(\tau) = 4e^{-5|\tau|}$. the power in $X(t)$is ________

51. A binary communication system makes use of the symbols "zero" and "one". There are channel errors. Consider the following events:

x_0 : a "zero" is transmitted
x_1 : a "one" is transmitted
y_0 : a "zero" is received
y_1 : a "one" is received

The following probabilities are given:

$$P(x_0) = \frac{1}{2},\ P(y_0 | x_0) = \frac{3}{4},$$

and $P(y_0 | x_0) = \frac{1}{2}$.

The information in bits that you obtain when you learn which symbol has been received (while you know that a "zero" has been transmitted) is ________.

52. The parallel-plate capacitor shown in the figure has movable plates. The capacitor is charged so that the energy stored in it is E when the plate separation is d. The capacitor is then isolated electrically and the plates are moved such that the plate separation becomes $2d$.

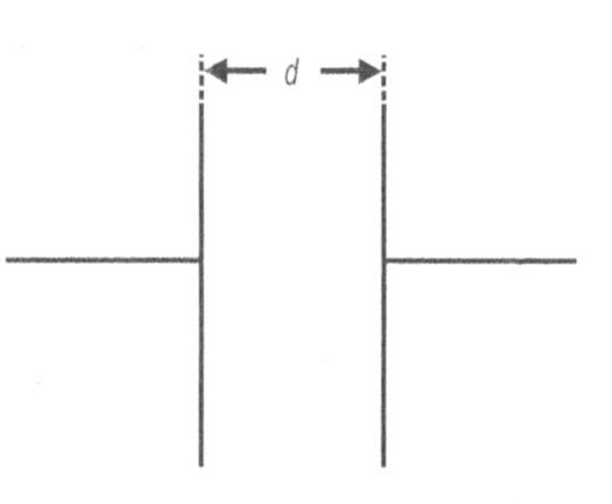

At this new plate separation, what is the energy stored in the capacitor, neglecting fringing effects?

(*a*) 2E (*b*) $\sqrt{2}$E
(*c*) E (*d*) E/2

53. A lossless microstrip transmission line consists of a trace of width w. It is drawn over a practically infinite ground plane and is separated by a dielectric slab of thickness t and relative permittivity $\varepsilon_r > 1$. The inductance per unit length and the characteristic impedance of this line are L and Z_0, respectively.

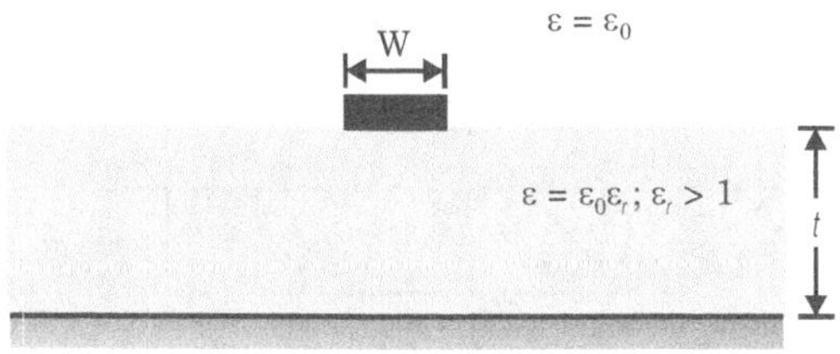

Which one of the following inequalities is always satisfied?

(a) $Z_0 > \sqrt{\frac{Lt}{\varepsilon_0 \varepsilon_r w}}$ (b) $Z_0 < \sqrt{\frac{Lt}{\varepsilon_0 \varepsilon_r w}}$

(c) $Z_0 > \sqrt{\frac{Lw}{\varepsilon_0 \varepsilon_r t}}$ (d) $Z_0 < \sqrt{\frac{Lw}{\varepsilon_0 \varepsilon_r t}}$

54. A microwave circuit consisting of lossless transmission lines T_1 and T_2 is shown in the figure. The plot shows the magnitude of the input reflection coefficient Γ as a function of frequency f. The phase velocity of the signal in the transmission

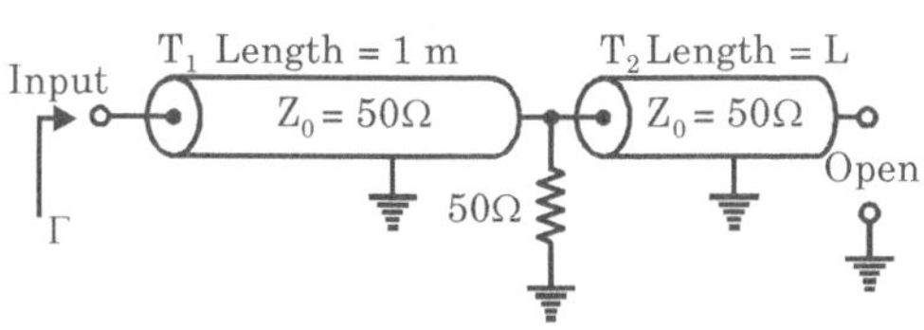

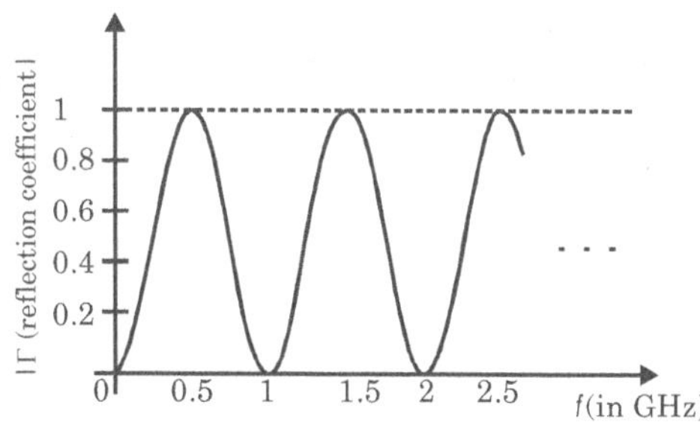

The length L (in meters) of T_2 is ________

55. A positive charge q is placed at $x = 0$ between two infinite metal plates placed at $x = -d$ and at $x = +d$ respectively. The metal plates lie in the yz plane.

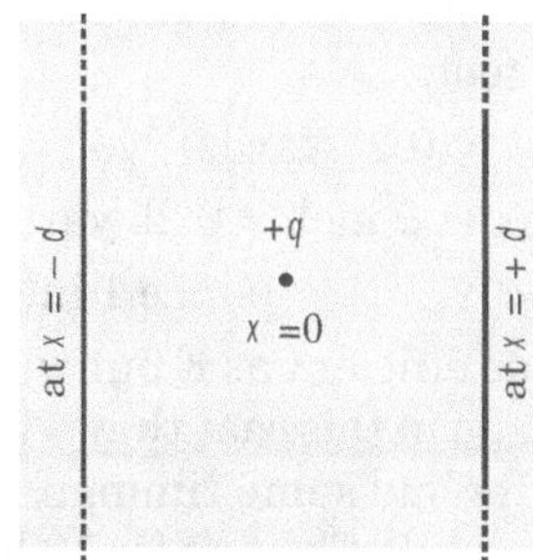

The charge is at rest at $t = 0$, when a voltage +V is applied to the plate at – d and voltage –V is applied to the plate at $x = +d$. Assume that the quantity of the charge q is small enough that it does not perturb the field set up by the metal plates. The time that the charge q takes to reach the right plate is proportional to

(a) d/V (b) $\sqrt{d}$/V

(c) $d/\sqrt{V}$ (d) $\sqrt{d/V}$

ANSWERS

GENERAL APTITUDE

1. (c)	2. (b)	3. (c)	4. (c)	5. (b)	6. (d)	7. (a)	8. (d)	9. (c)	10. (b)

TECHNICAL SECTION

1. (1)	2. (0)	3. (b)	4. (c)	5. (90°)	6. (b)	7. (d)	8. (0.316)
9. (3)	10. (0.25)	11. (d)	12. (d)	13. (0.25)	14. (598.67)	15. (a)	16. (b)
17. (1.5)	18. (a)	19. (1)	20. (a)	21. (1.75)	22. (16)	23. (a)	24. (d)
25. (32.58)	26. (0.66)	27. (0)	28. (0.33)	29. (3)	30. (d)	31. (5)	32. (0.049)
33. (6)	34. (100)	35. (b)	36. (c)	37. (32.36)	38. (1.2)	39. (2.105)	40. (0.7985)
41. (413.8)	42. (a)	43. (a)	44. (c)	45. (1)	46. (60)	47. (1)	48. (–1)
49. (0.25)	50. (100)	51. (0.405)	52. (a)	53. (a)	54. (0.1)	55. (c)	

EXPLANATIONS

GENERAL APTITUDE

1. 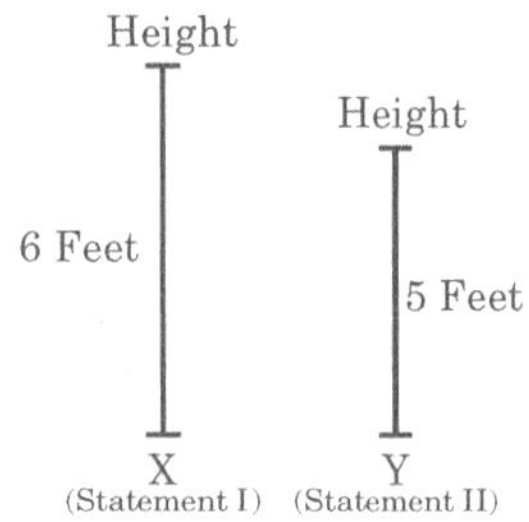

Hence from the given figure Mr. X is taller than Mr. Y by 1 foot.

4. (9 inches)$^{1/2}$ = (0.25 yards)$^{1/2}$,
Solving we get 9 inch = 0.25 yards
(since 1 inch = 0.028 yard)

5. M is twice as efficient as E but worked for half as many days. So in this case they will do equal work if their shifts had same timings. But M's shift is for 6 hours, while E's shift for 12 hours. Hence, E will do twice the work as M.
Ratio of contribution of M : E in work is 1 : 2.

6. Given Venn diagram is

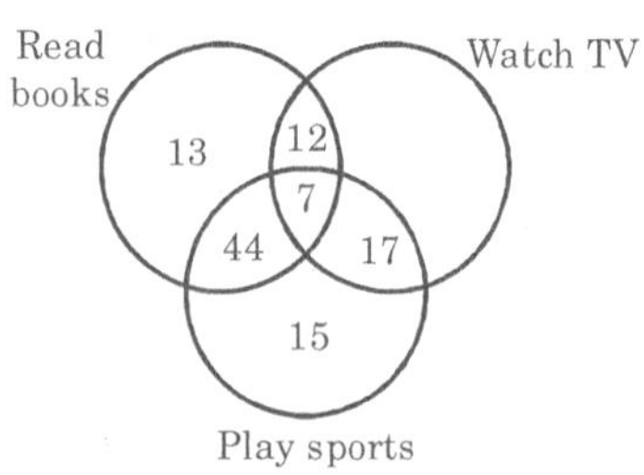

The number of students who like to read books or play sports
= 13 + 12 + 44 + 7 + 15 + 17
= 108

7. Social science disciplines had a pre-colonial origin.

8. 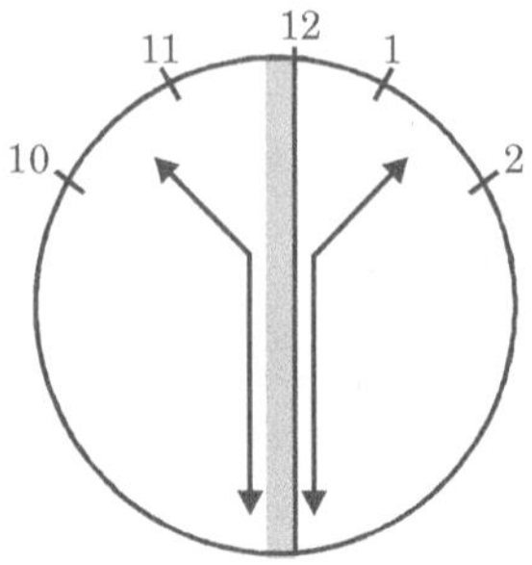

Mirror image of 1 : 20 is 10 : 30
10 : 30 was the time two and quarter hour back so time now will be 12 : 45

9. 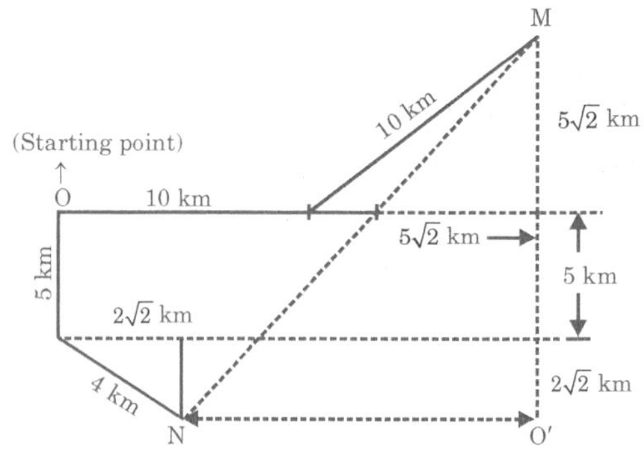

From the given figure

$$MN = \sqrt{(O'M)^2 + (O'N)^2}$$

$$O'M = 5\sqrt{2} + 5 + 2\sqrt{2} = 5 + 7\sqrt{2}$$

$$O'N = 10 + 5\sqrt{2} - 2\sqrt{2} = 10 + 3\sqrt{2}$$

$$MN = \sqrt{(5 + 7\sqrt{2})^2 + (10 + 3\sqrt{2})^2}$$

$$= \sqrt{25 + 98 + 70\sqrt{2} + 100 + 18 + 60\sqrt{2})}$$

$$\approx 20.61 \text{ km}$$

TECHNICAL SECTION

1. From the given question,
A has an eigen value is zero

$$|A| = 0$$

$$\begin{vmatrix} 3 & 2 & 4 \\ 9 & 7 & 13 \\ -6 & -4 & -9+x \end{vmatrix} = 0$$

$$3(-63 + 7x + 52) - 2(-81 + 9x + 78) + 4(-36 + 42) = 0$$
$$3(7x - 11) - 2(9x - 3) + 4(6) = 0$$
$$21x - 33 - 18x + 6 + 24 = 0$$
$$3x - 3 = 0$$
$$x = 1$$

2. From the question, the complex variable function is:

$$f(z) = 2z^3 + b_1 |z|^3$$

Given $f(z)$ is analytic.
which is possible only when $b = 0$
since $|z^3|$ is differentiable at the origin but not analytic.
$2z^3$ is analytic everywhere
$\therefore$ $f(z) = 2z^3 + b|z^3|$ is analytic
only when $b = 0$

3. Given $f(x) = x^3 - 3x^2 + 1$

$$f'(x) = 3x^2 - 6x$$

$$f'(x) = 0$$

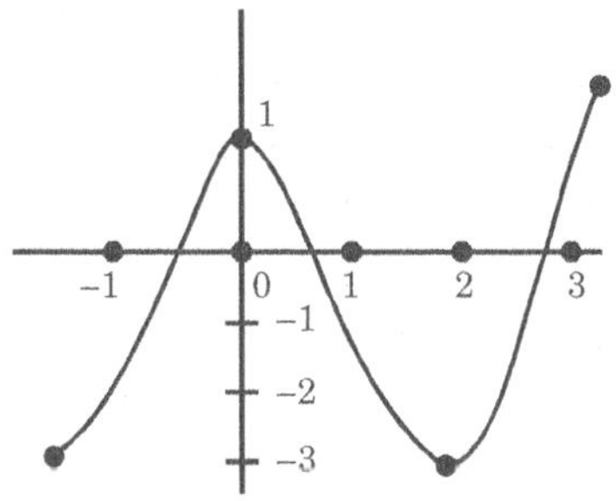

$$3x^2 - 6x = 0$$

$$3x(x-2) = 0$$

$$x = 0, 2$$

$$f''(x) = 6x - 6$$

At $x = 0$ $f''(0) = -6$ maxima

$x = 2$ $f''(2) = 6$ minima

Then the given function $f(x)$ increases, then decreases and increase again.

4. Here the given equation is

$$\sin(x) = \frac{x}{2} \text{ (where } x \text{ is in radians)}$$

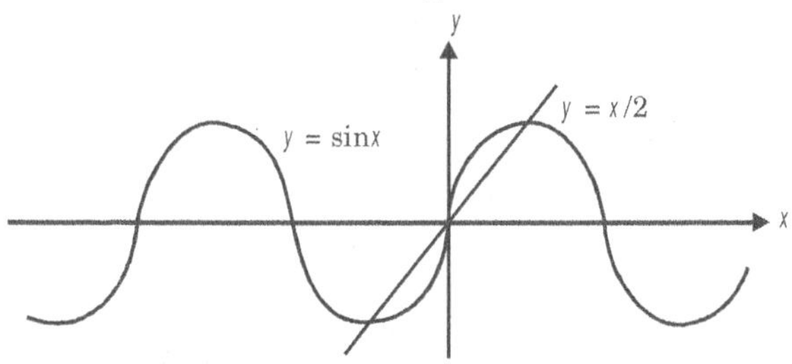

Hence there are 3 distinct values.

5. Here the given time-varying vector

$$I = \hat{x}15\cos\omega t + \hat{y}5\sin\omega t$$

$$|I| = \sqrt{(15\cos\omega t)^2 + (5\sin\omega t)^2}$$

$$= \sqrt{225\cos^2\omega t + 25\sin^2\omega t}$$

$$= \sqrt{225\cos^2\omega t + 25(1-\cos^2\omega t)}$$

$$= \sqrt{25 + 200\cos^2\omega t}$$

$|I|$ is minimum when $\cos^2\omega t = 0$

$$\cos^2\omega t = \cos 90°$$

or $\theta = \omega t = 90°$

So, the value of θ is 90°

6. In maximum power transformation, half of the voltage drop across source resistance, remaining half across the load.

$\therefore$ Voltage across source (R)

$$I_L R = \frac{V_s}{2} \text{ (from the given figure)}$$

$$I_L = \frac{V_s}{2R}$$

7. At $t = 0^-$, Switch is at position-1

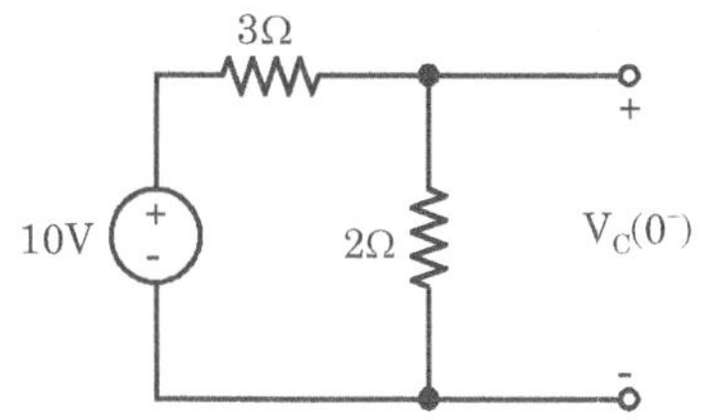

where, $V_c(0^-) = \frac{10 \times 2}{2+3} = 4\text{ V}$ ...(1)

$$V_c(0^-) = V_c(0^+) = 4\text{ V}$$

At $t = \infty$

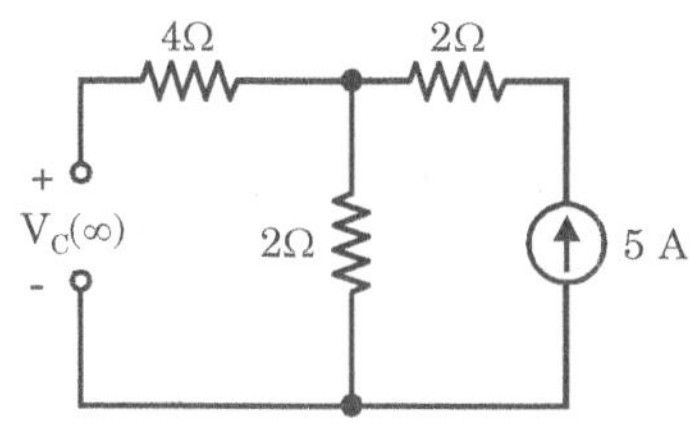

$V_c(\infty) = 5 \times 2 = 10\text{ V}$ ...(2)

The time constant of the circuit is

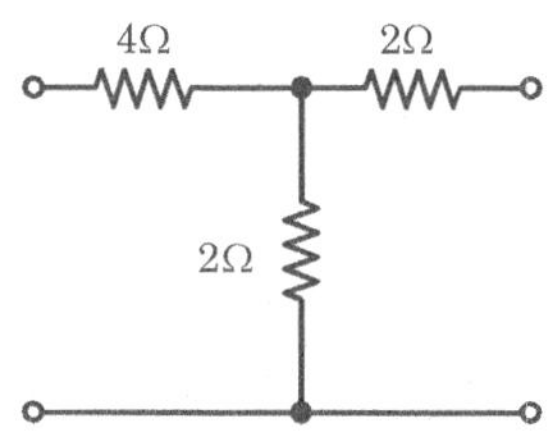

$$\tau = R_{eq}C_{eq}$$

$$= (4 + 2) \times 0.1$$

$$= 6 \times 0.1 = 0.6\text{ sec}$$

Therefore $V_c(t) = V_c(\infty) + [V_c(0^+) - V_c(\infty)]e^{-t/\tau}$

$$= 10 + (4 - 10)e^{-t/0.6}$$

$$V_c(t) = (10 - 6e^{-t/0.6})\text{ V}$$

8. At resonance (for parallel RLC circuit)

$$I_R = I$$

$$I_L = QI\angle{-90°}$$

$$I_C = QI\angle 90°$$

For parallel RLC circuit

$$\left|\frac{I_L}{I_R}\right| = \frac{IQ}{I} = Q = R\sqrt{\frac{C}{L}}$$

$$= 10\sqrt{\frac{10\times10^{-16}}{10\times10^{-3}}} = 10\sqrt{10^{-3}}$$

$$= 0.316$$

9. For T - network

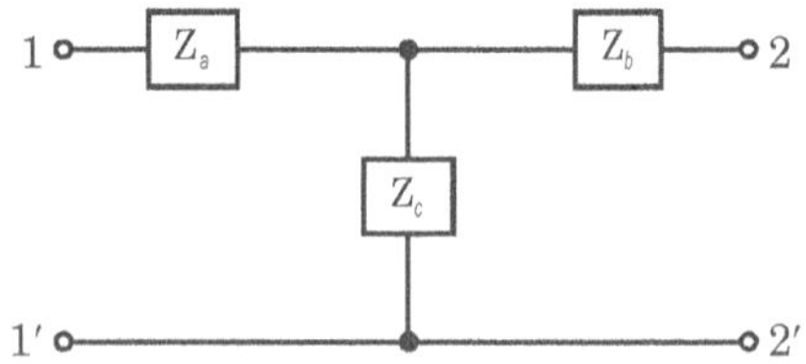

$$Z_{11} = Z_a + Z_c$$
$$Z_{22} = Z_b + Z_c$$
and $$Z_{12} = Z_{21} = Z_c$$

Given $$[Z] = \begin{bmatrix} 2j\omega & j\omega \\ j\omega & 3+2j\omega \end{bmatrix}$$

Therefore $$Z_{12} = j\omega$$
and $$Z_{22} = 3 + 2j\omega$$
$$= 3 + j\omega + j\omega$$
$$= Z_b + Z_c$$
$$= R_b + j\omega + Z_c$$
$\therefore$ $$R_b = 3\ \Omega$$

Hence the value of R_b is 3Ω.

10. Given the energy of the signal

$$x(t) = \frac{\sin(4\pi t)}{4\pi t}$$

$$\frac{\sin at}{\pi t} \leftrightarrow$$

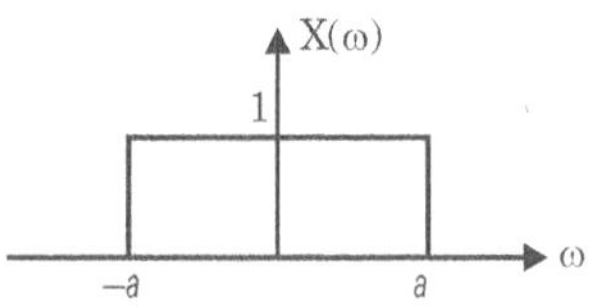

$$\frac{\sin 4\pi t}{4\pi t} = \frac{1}{4} \times \frac{\sin 4\pi t}{\pi t} \leftrightarrow$$

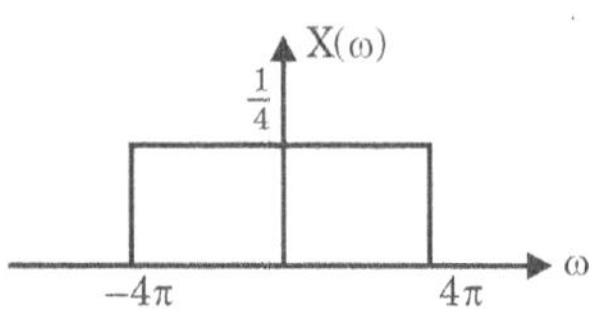

Energy, $$E_{x(t)} = \frac{1}{2\pi}\int_{-\infty}^{\infty} |X(\omega)|^2\, d\omega$$
$$= \frac{1}{2\pi}\int_{-4\pi}^{4\pi} \left(\frac{1}{4}\right)^2 d\omega$$
$$= \frac{1}{2\pi} \times \frac{1}{16}\left[\omega\right]_{-4\pi}^{4\pi}$$
$$= \frac{1}{2\pi} \times \frac{1}{16}\left[8\pi\right]$$
$$= \frac{1}{4} = 0.25\ \text{J}$$

11. Ebers-Moll model of a BJT is valid for all the region of operation.

12. Here $$r_{ds} = \frac{1}{\mu_n C_{OX} \frac{W}{L}(V_{GS} - V_T)}$$

r_{ds} = channel resistance

W ↑ r ↓ A → correct

V_T ↓ r ↓ B → correct

L ↑ r ↑ C → correct

V_{GS} ↑ r ↓

$\therefore$ D → Wrong statement

So, statement (*d*) is not correct.

13. Here V_{on} = 0.7 V (assume the diode in the figure)

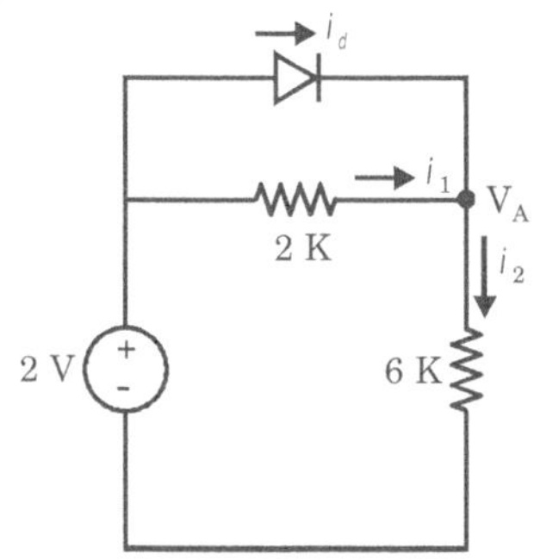

Let diode : ON

$\Rightarrow$ $$V_A = 2 - 0.7 = 1.3\ \text{V}$$

$$i_2 = \frac{1.3}{6\text{K}};$$

$$i_1 = \frac{0.7}{2\text{K}}$$

$$i_d = i_2 - i_1$$
$$= \frac{1.3}{6}\text{mA} - \frac{0.7}{2}\text{mA}$$
$$= \frac{1.3 - 2.1}{6} = -\frac{0.8}{6} = -\text{ve}$$

Which is not possible

When diode is OFF

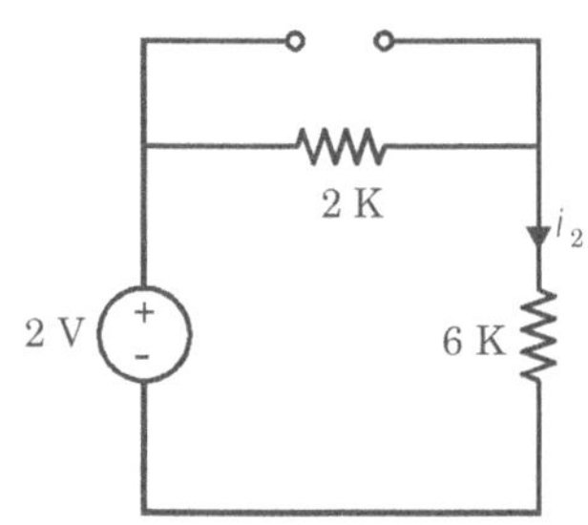

$$i_2 = \frac{2\text{V}}{2\text{K} + 6\text{K}} = 0.25\ \text{mA}$$

Hene the magnitude of current i_2 is 0.25 mA.

14. From the given figure

$$R_2 = \frac{V_T}{I_2}\ln\left(\frac{I_1}{I_2}\right)$$

$$= \frac{26\times10^{-3}}{100\times10^{-6}}\left(\frac{1\times10^{-3}}{100\times10^{-6}}\right)$$

$$= 598.67\ \Omega$$

17.

CLK	Q_1	Q_2	Q_3	Q_4	Q_5	$Y = Q_3 + Q_5$
0	0	1	0	1	0	0
1	0	0	1	0	1	1
2	1	0	0	1	0	0
3	0	1	0	0	1	1
4	1	0	1	0	0	1
5	0	1	0	1	0	0

The waveform of the gate output

$$Y = Q_3 + Q_5$$

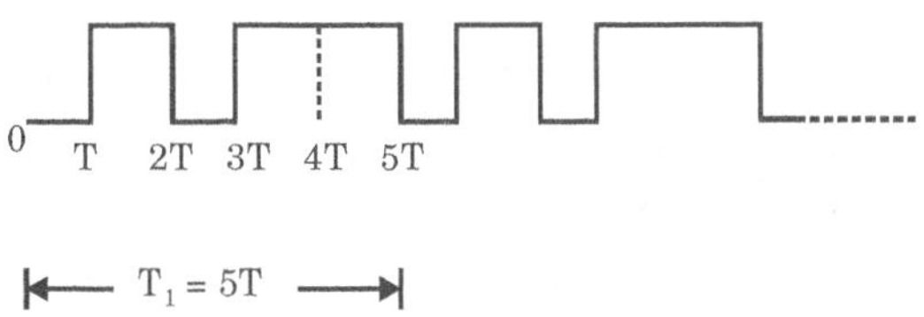

Average power dissipated

$$P = \frac{V^2}{R}\times\frac{T_{ON}}{T}$$

$$= \frac{5^2}{10}\times\frac{3T}{5T}$$

$$= 1.5\ \text{mW}$$

18. In case of a full adder,

$$C_{out}(A, B, C_{in}) = \Sigma(3, 5, 6, 7)$$

A, B: Applied at select input of MUX; C_{in}: Applied at data input of MUX

	I_0	I_1	I_2	I_3
$\overline{C_{in}}$	0	2	4	(6)
C_{in}	1	(3)	(5)	(7)
	0	C_{in}	C_{in}	1

$\therefore$

$$I_0 = 0$$
$$I_1 = C_{in}$$
$$I_2 = C_{in}$$
$$I_3 = 1$$

19. Given the response of the system

$$G(s) = \frac{s-2}{(s+1)(s+3)}$$

Now $\quad L\left(\frac{dy}{dt}\right) = (sY(s) - y(0))$

$$Y(s) = G(s)\times\frac{1}{s} = \frac{s-2}{s(s+1)(s+3)}$$

$$y(0) = \underset{s\to\infty}{\text{Lt}}\ sY(s)$$

(Applying initial value theorem)

$$= \underset{s\to\infty}{\text{Lt}}\ \frac{s-2}{(s+1)(s+3)}$$

$$= \frac{\left(1-\frac{2}{s}\right)}{s\left(1+\frac{1}{s}\right)\left(1+\frac{3}{s}\right)}$$

$$y(0) = 0$$

$$L\left(\frac{dy}{dt}\right) = sY(s) = \frac{s\times(s-2)}{s(s+1)(s+3)}$$

$$= \frac{s-2}{(s+1)(s+3)}$$

$$\left.\frac{dy}{dt}\right|_{t=0} = \underset{s\to\infty}{\text{Lt}}\ sL\left(\frac{dy}{dt}\right)$$

$$= \underset{s\to\infty}{\text{Lt}}\ \frac{s\times(s-2)}{(s+1)(s+3)}$$

$$= \frac{s\times s\left(1-\frac{2}{s}\right)}{s\left(1+\frac{1}{s}\right)s\left(1+\frac{3}{s}\right)}$$

$$= \frac{1-0}{(1+0)(1+0)} = 1$$

20. Here the Nyquist plot is

$$G(s) = \frac{1-s}{4+2s}$$

and $\quad \angle G(s)| = -\tan^{-1}\omega - \tan^{-1}\frac{\omega}{2}$

at $s = 0$, $\quad |G(s)| = \frac{1-0}{4+0} = \frac{1}{4} = 0.25$

and $\quad \angle G(s)|_{\omega=0} = -\tan^{-1}0° - \tan^{-1}\frac{0°}{2} = 0°$

$$\text{at } s=\infty, \quad |G(s)| = \frac{s\left(\frac{1}{s}-1\right)}{s\left(\frac{4}{s}+2\right)} = \frac{\left(\frac{1}{s}-1\right)}{\left(\frac{4}{s}+2\right)}$$

$$\frac{(0+1)}{(0+2)} = -\frac{1}{2} = -0.5$$

and $\angle G(s)|_{\omega=\infty} = -\tan^{-1} 180° - \tan^{-1} 90° = -180°$

Nyquist plot is

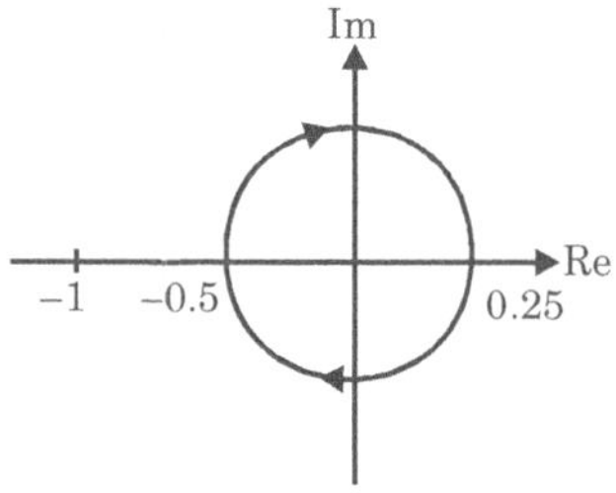

Hence the number of encirclements of $(-1 + j0) = 0$

21. The minimum required average codeword length in bits for error free reconstruction

$$L_{min} = H \text{ (Entropy)}$$

$$H = \frac{1}{2}\log_2 2 + \frac{1}{4}\log_2 4 + \frac{1}{8}\log_2 8 + \frac{1}{8}\log_2 8$$

$$= \frac{1}{2}\log_2 2 + \frac{1}{4}\log_2 2^2 + \frac{1}{8}\log_2 2^3 + \frac{1}{8}\log_2 2^3$$

$$= \frac{1}{2}\log_2 2 + \frac{1}{4}\times 2\log_2 2 + \frac{1}{8}\times 3\log_2 2 + \frac{1}{8}\times 3\log_2 2$$

$$= \frac{1}{2}+\frac{1}{2}+\frac{3}{8}+\frac{3}{8} = 1.75$$

$\Rightarrow \quad L_{min} = 1.75$ bits/word

Hence the minimum required average codeword length is 1.75 bits/word.

22. From the given data

$f_s = 8$ kHz (speech signal)

$n = 8$ bits/sample ; $M = 4$

The minimum band width

$$BW_{min} = \frac{R_b}{2\log_2 M} = \frac{R_b}{2\log_2 4} = \frac{f_s \times n}{2\log_2 2^2}$$

$$= \frac{f_s \times n}{4\log_2 2} = \frac{f_s \times n}{4} = \frac{8\times 8}{4} = 16 \text{ kHz}$$

Hence the minimum band width is 16 kHz.

23. Ba_z magnetic field (magnetic field exists in the $\hat{z}$ direction in vacuum).

$v_x a_x + v_z a_z$ velocity

$$F = Q\,(v \times B) \text{ by Lorent's law}$$

$$= Q\,(v_x a_x + v_z a_z) \times Ba_z$$

$$F_y = Q\, v_x \,.\, B\,(-a_y)$$

This results in a circular path in the XY plane with $v_z a_z$ component causing a linear path.

Both result in a helical path in Z axis.

24. $v_p = \omega/\beta$ can be calculated. Polarization can be identified.

μ_r and ϵ_r cannot be found, due to which power flux cannot be calculated as power flux

$$P = \frac{1}{2}\frac{|E|^2}{\eta}, \text{ where } \eta = 120\pi \times \sqrt{\frac{\mu_r}{\epsilon_r}}$$

Hence, the power flux through the z = 0 plane.

25. $$\sin\alpha_{max} = \sin\theta_i = \sqrt{n_1^2 - n_2^2} = \sqrt{1.5^2 - 1.4^2}$$

(Here θ_i be the incident angle)

$\Rightarrow \; a_{max} = \sin\theta_i = \sin^{-1}(0.5385) = 32.58°$

$\therefore \; \theta_i = 32.58°$

Hence, the maximum value of θ_i is 32.58°.

26. Given differential equations

$$\frac{dy}{dx} = -3y + 2, \; y(0) = 1$$

If $|1 - 3h| < 1$, then solution of differential equation is stable.

$$-1 < 1 - 3h < 1$$

$$-2 < -3h < 0$$

$$0 < h < 2/3$$

$$h_{max} = \frac{2}{3} = 0.66$$

Hence, the largest time step that can be used to solve the equation is 0.66.

27. By Green's theorem

$$C = \oint xy^2 dx + x^2 y dy \quad \text{(where C = Closed curve)}$$

$$= \iint_R \left(\frac{d}{dx}(x^2 y) - \frac{d}{dy}(xy^2)\right) dxdy$$

$$= \iint_R (2xy - 2xy) = 0$$

So, the value of closed curve C is 0.

28. The probability $P(X + Y \le 1)$

$$= \int_{x=0}^{1}\int_{y=0}^{(1-x)} f_{xy}(x, y)\,dx\,dy$$

$$= \int_{x=0}^{1}\int_{y=0}^{(1-x)} (x + y)\,dx\,dy = \int_{x=0}^{1}\left(xy + \frac{y^2}{2}\right)_0^{1-x}$$

$$= \int_{x=0}^{1}\left(x(1-x) + \frac{(1-x)^2}{2}\right)dx$$

$$= \int_{x=0}^{1}(x - x^2)\,dx + \int_{x=0}^{1}\frac{(1-x)^2}{2}\,dx$$

$$= \int_{x=0}^{1}\left(\frac{1}{2} - \frac{x^2}{2}\right)dx = \left(\frac{x}{2} - \frac{x^3}{6}\right)_0^1$$

$$= \frac{1}{2} - \frac{1}{6} = \frac{1}{3} = 0.33$$

29. Trace of A = 14

$a + 5 + 2 + b = 14$

(Taking the diagonal element and then adding)

$$a + b = 7 \quad \ldots(i)$$

$$\det(A) = 100$$

$$5\begin{vmatrix} a & 3 & 7 \\ 0 & 2 & 4 \\ 0 & 0 & b \end{vmatrix} = 100$$

$$5 \times 2 \times a \times b = 100$$

$$10\,ab = 100$$

$$ab = 10 \quad \ldots(ii)$$

From equation (*i*) and (*ii*)

either $a = 5,\ b = 2$

or $a = 2\ b = 5$

$$|a - b| = |5 - 2| = 3$$

30. $R_{eq} \Rightarrow$

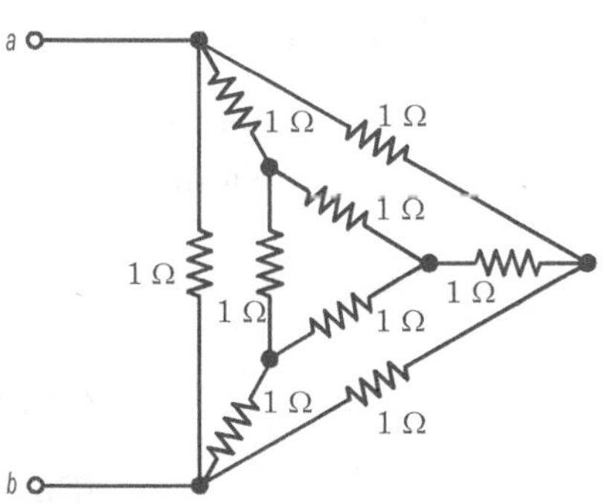

By using delta to star conversion

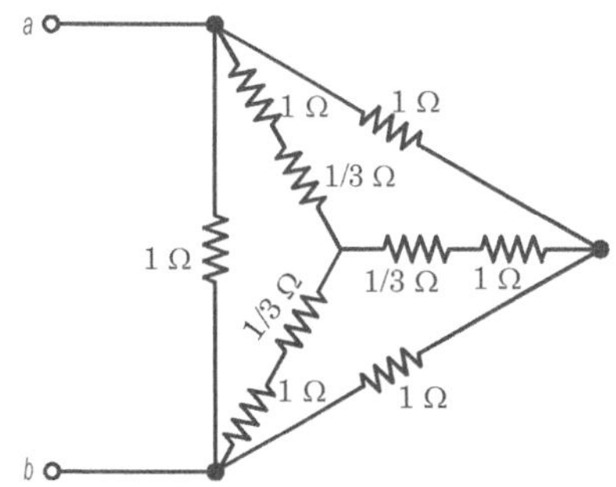

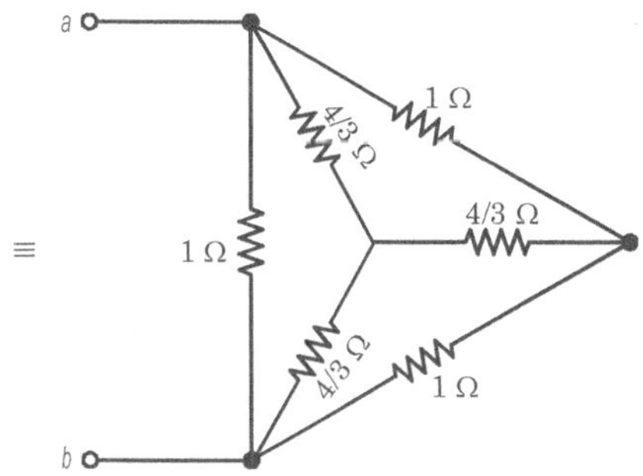

Again by star to delta conversion

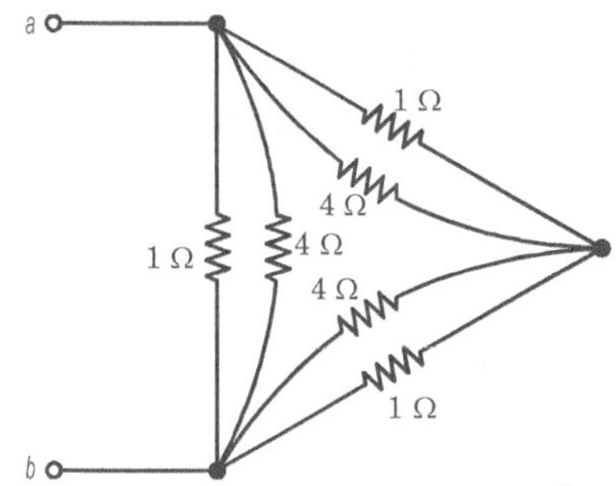

$$R_{ab} = \{4 \parallel 1) + (4 \parallel 1)\} \parallel \{1 \parallel 4)\} \quad \left(\because \frac{R_1 \cdot R_2}{R_1 + R_2}\right)$$

$$= \left(\frac{4}{5} + \frac{4}{5}\right) \Big\| \frac{4}{5} = \frac{8}{15}\Omega$$

So, the equivalent resistance across the terminals a and b is $\frac{8}{15}\Omega$.

31. From the given figure

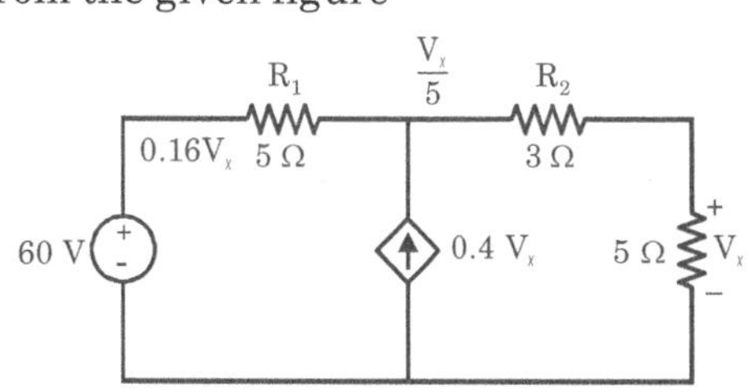

Using KVL (kirchhoff's voltage law) in the outer loop

$$60 - 5(0.16\,V_x) - \frac{V_x}{5} \times 3 - V_x = 0$$

or $V_x = 25$ V

$\therefore$ The current flowing through

$$R_2 = \frac{V_x}{5} = \frac{25}{5} = 5\text{ A}$$

So, the magnitude of the current is 5 A.

32. Given a continuous-time filter with transfer function

$$H(s) = \frac{2s+6}{s^2+6s+8} = \frac{1}{s+2} + \frac{1}{s+4}$$

$$h(t) = e^{-2t}\,u(t) + e^{-4t}\,u(t)$$

Given sampling frequency $(f_s) = 2$ Hz

For discrete time,

$$t = nT_s = \frac{n}{2}$$

$$h[n] = (e^{-n} + e^{-2n})\,u[n]$$

$$H(z) = \frac{1}{1-e^{-1}Z^{-1}} + \frac{1}{1-e^{-2}Z^{-1}}$$

$$= \frac{Z}{Z-e^{-1}} + \frac{Z}{Z-e^{-2}}$$

$$= \frac{2Z^2 - 0.5032Z}{Z^2 - 0.5032Z + 0.049}$$

$\therefore \quad k = 0.049$

Hence the value of k is 0.049.

33. Discrete Fourier Transform (DFT) of the 4-point sequence is

$$x_1[n] = x\left[\frac{n}{3}\right]$$

$X_1[K] = \{12, 2j, 0, -2j, 12, 2j, 0, -2j, 12, 2j, 0, -2j\}$

$X_1[8] = 12; \quad X_1(11) = -2j$

$$\left|\frac{X_1[8]}{X_1[11]}\right| = \left|\frac{12}{-2j}\right| = 6$$

34. At $t = 0^-$

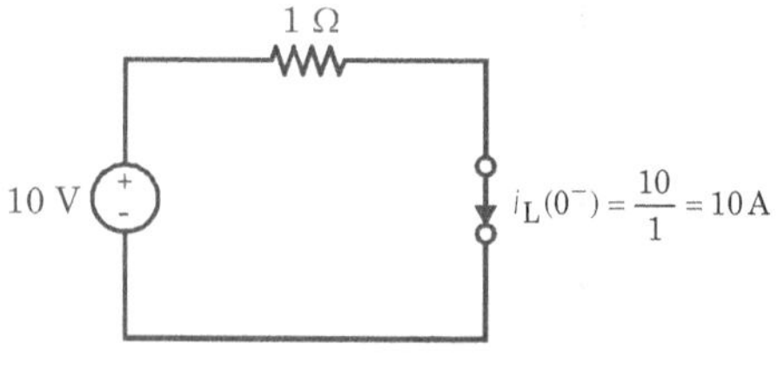

$$i_L(0^-) = \frac{10}{1} = 10 \text{ A}$$

for $t > 0$ (using Laplace transform)

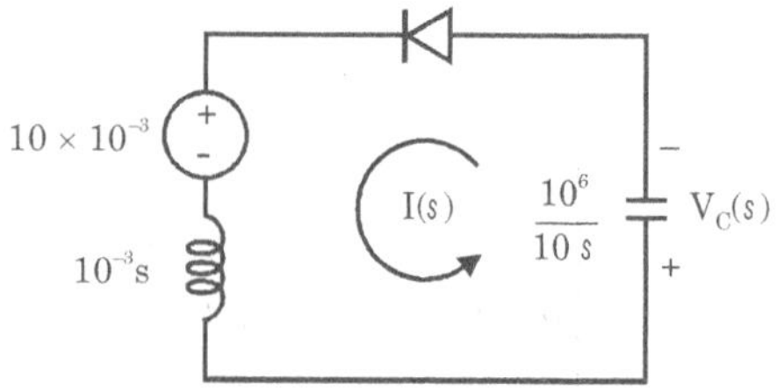

$$I(s) = \frac{10\times10^{-3}}{10^{-3}s + \frac{10^6}{10s}}$$

$$V_c(s) = I(s)\times\frac{10^6}{10s}$$

$$V_c(s) = \frac{10^6}{s^2+10^8}$$

Taking inverse Laplace transform, we get

$V_c(t) = 100\sin 10^4 t$ V ($\because V_c(t) = V_0 \sin\omega t$)

$\therefore$ Steady state magnitude voltage across capacitor is 100 V.

35. In a MOS - Capacitor

$$(V_G - V_T) \propto Q$$

Where V_G = voltage applied across a MOS capacitor with metal gate and p-type silicon

Q = Inversion carrier density

$$\frac{(V_{G1}-V_T)}{(V_{G2}-V_T)} = \frac{Q_1}{Q_2}$$

$$\frac{0.8-V_T}{1.3-V_T} = \frac{2\times10^{11}}{4\times10^{11}}$$

On solving, we get $\quad V_T = 0.3$ V

Now Consider $\quad \frac{V_{G2}-V_T}{V_{G3}-V_T} = \frac{Q_2}{Q_3}$

$$\frac{1.3-0.3}{1.8-0.3} = \frac{4\times10^{11}}{Q_3}$$

Inversion carrier density with $V_G = 1.8$ V

$$Q_3 = 6\times10^{11}/\text{cm}^2$$

Hence, the value of inversion carrier density is $6\times10^{11}/\text{cm}^2$

36. In any type of PN junction

$$V_{BR} \propto \frac{1}{\text{Doping Concentration}}$$

i.e. $V_{BR} \propto \frac{1}{N_D}$ (or) $V_{BR} = \frac{\varepsilon E^2}{2qN_D}$

(Here V_{BR} = break down voltage, N_D = donor density)

$\therefore \quad V_{BR}\times N_D \rightarrow$ is a constant

37. $|\varepsilon| = \frac{q}{\varepsilon_0}N_D X_N$

$$50\times10^3 = \frac{1.6\times10^{-19}}{8.85\times10^{-14}\times11.7}\times10^{17}\times X_N$$

(Since $\epsilon = \epsilon_0\,\epsilon_r$)

On solving $\quad X_N = 3.2356\times10^{-6}$ cm

$= 3.236\times10^{-8}$ m

$\therefore \quad X_N = L = 32.36$ nm

Hence, the value of L is 32.36 nm.

38. Since $V_{GS} > V_{DS}$, MOSFET is in linear operation

$$I_D = K'_N [V_{GS} - V_T] V_{DS}$$

$$\frac{dI_D}{dV_{GS}} = K'_N V_{DS}$$

$$g_m = K'_N V_{DS}$$

$$0.5 \times 10^{-6} = K'_N [50 \times 10^{-3}]$$

$$\therefore \quad K'_N = 10^{-5}$$

$$I_D = K'_N [V_{GS} - V_T] V_{DS}$$

$$\frac{dI_D}{dV_{dS}} = K'_N [V_{GS} - V_T]$$

$$g_d = K'_N [V_{GS} - V_T]$$

$$8 \times 10^{-6} = 10^{-5} [2 - V_T]$$

$$\frac{8 \times 10^{-6}}{10^{-5}} = 2 - V_T$$

$$8 \times 10^{-1} = 2 - V_T$$

$$0.8 = 2 - V_T$$

$$\therefore \quad V_T = 1.2 \text{ V}$$

Hence threshold voltage of the transistor is 1.2 V.

39.
$$V_{ripple} = \frac{I_{DC}.T}{C}$$

Here T = Time period

C = Filter capacitor

$$\therefore \quad V_{ripple} = \frac{1 \times 1 \times 10^{-3}}{475 \times 10^{-6}}$$

$$= 2.105 \text{ Volts}$$

40. Initially switch is closed and $V_B = 10$ V

$\Rightarrow V_{01} = -10$ V

$\Rightarrow V_0 = -V_2 = -5$ V

$$\Rightarrow V_A = \frac{V_0}{4K + 1K} \times 1K = \frac{-5}{5K} \times K = -1 \text{ V}$$

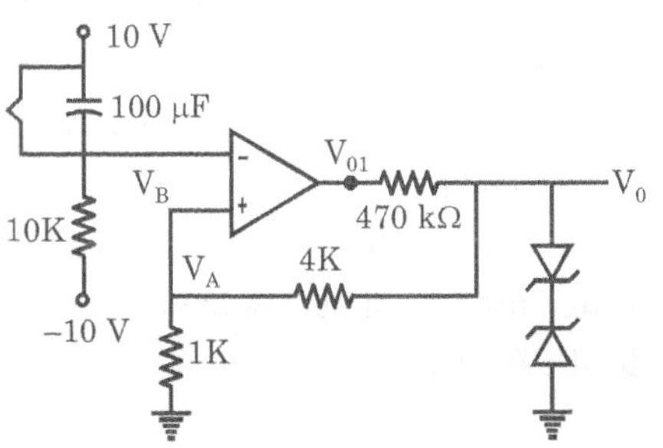

At $t = 0$;

The switch is opened and as $t \to \infty$, V_B approaches –10 V.

Let at $t = T_1$,

V_B exceeds V_A (–1 V) so that V_{01} changes from –10 V to 10 V

$\Rightarrow V_0$ changes from –5 V to 5 V

$$V_B = V_f + (V_i - V_f)e^{-t/\tau}$$

$$= -10 + [10 - 1 - 10]e^{-t/RC}$$

(Since $\tau = RC$)

At $t = T_1$ $\quad V_B = -1$

$$-1 \text{ V} = -10 + 20e^{-T_1/RC}$$

$$\Rightarrow \quad T_1 = RC \ln \frac{20}{9}$$

$$= 10 \times 10^3 \times 100 \times 10^{-6} \times 0.798$$

$$= 0.798 \text{ sec}$$

Hence, the required time at which V_0 changes state is 0.798 s.

41. Overall input = $V_{ios} + V_{input}$

(Here V_{ios} = input offset voltage)

= 5 mV + 25 mV = 30 mV

$$V_0 = \frac{\left(1 + \frac{R_F}{R_1}\right)}{1 + \frac{1}{A_{OL}}\left(1 + \frac{R_F}{R_1}\right)} \times \text{Overall input}$$

$$= \frac{1 + \frac{15K}{1K}}{1 + \frac{1}{100}\left(1 + \frac{15K}{1K}\right)} \times 30 \times 10^{-3}$$

$$= \frac{\frac{16K}{1K}}{1 + \frac{1}{100}\left(1 + \frac{16K}{K}\right)} \times 30 \times 10^{-3}$$

$$= \frac{16}{116} \times 30 \times 10^{-3} \times 100$$

$$= \frac{48000}{116} \times 10^{-3} \text{ V}$$

$$= 413.79 \text{ mV}$$

Hence, the output V_0 is 413.79 mV.

42. Here 8 kbyte ROM is given

$\therefore 2^n = 8 \text{ kB} = 2^3(2^{10}) = 2^{13} \quad (a^m \times a^n = a^{m+n})$

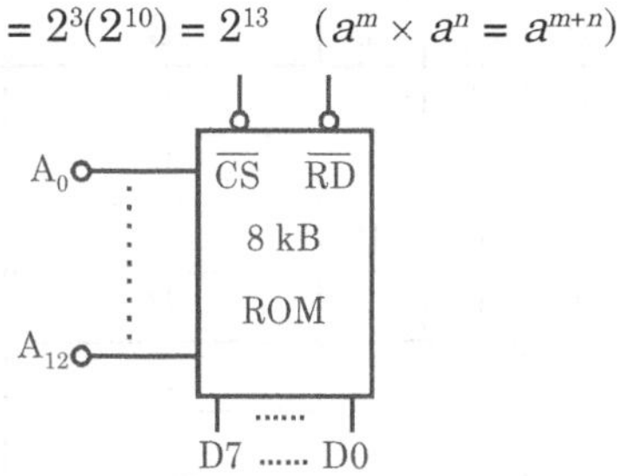

∴ 13 Address lines are required for memory chip.

But the address range given as 1000 H - 2FFF H

A_{15}	A_{14}	A_{13}	A_{12}	A_{11}	A_{10}	A_9	A_8	A_7	A_6	A_5	A_4	A_3	A_2	A_1	A_0
0	0	0	1	0	0	0	0	0	0	0	0	0	0	0	0
0	0	1	0	1	1	1	1	1	1	1	1	1	1	1	1

In order to get $\overline{CS}$ as low, the condition is

$A_{15} = A_{14} = 0$ and $A_{13} = 0/1$, $A_{12} = 1/0$.

43. The total capacitance $= (2^N - 1) \times C = (2^8 - 1) \times 8$ pF $= 2.04$ nF (Here the input capacitance of each comparator = 8 pF)

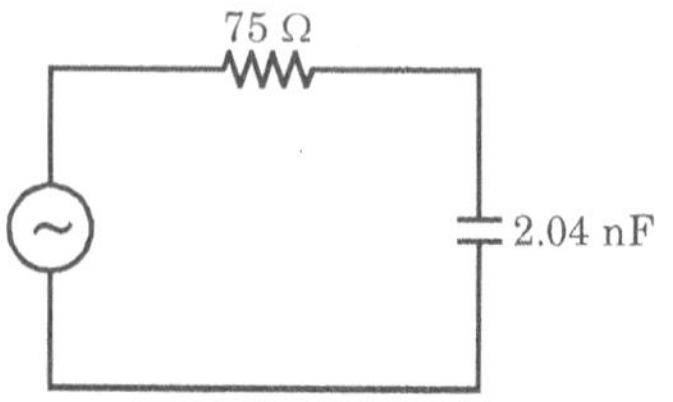

The time constant $= \tau = RC = 153$ ns

∵ Settling time $(T_s) = 5\tau = 5\,RC = 765$ ns

$$\therefore \text{ Sampling rate } = \frac{1}{\text{Settling time } (T_s)}$$

$$\approx 1 \text{ M samples/sec}$$

Hence, the maximum sampling rate is 1 mega samples per second.

44. **For State A**

Binary input			**Present State**	**Next State**
X	**Y**	**Z**		
0	0	0	A	B
0	0	1	A	A
0	1	0	A	A
0	1	1	A	A
1	0	0	A	C
1	0	1	A	C
1	1	0	A	A
1	1	1	A	A

Present State B

Binary input			**Present State**	**Next State**
X	**Y**	**Z**		
0	0	0	B	A
0	0	1	B	C
0	1	0	B	B
0	1	1	B	B
1	0	0	B	A
1	0	1	B	B
1	1	0	B	B
1	1	1	B	B

Present State C

Binary input			**Present State**	**Next State**
X	**Y**	**Z**		
0	0	0	C	C
0	0	1	C	C
0	1	0	C	C
0	1	1	C	B
1	0	0	C	C
1	0	1	C	A
1	1	0	C	C
1	1	1	C	AB

In state 'C' when XYZ = 111; the ambiguity occurs. Because from state 'C'

When, X = 1, Z = 1 ⇒ Next state = A

When, Y = 1, Z = 1 ⇒ Next state = B

So, transitions from state 'C' are ambiguous defined.

45.
$$G(s) = \frac{1}{s^2 + 2s}$$

$$\frac{Y(s)}{R(s)} = \frac{\frac{K}{s^2+2s}}{1 + \frac{K}{s^2+2s}} = \frac{K}{s^2 + 2s + K}$$

Minimum settling time and no overshoot implies

$$\xi = 1$$

$$\omega_n = \sqrt{K}$$

$$2 \times \xi \times \omega_n = 2$$

$$\omega_n = 1$$

$$\Rightarrow \quad \sqrt{K} = 1 \text{ or } K = 1$$

Hence, the required value of gain K to achieve this is 1.

46.
$$1 + G(s)H(s) = 1 + \frac{K}{(s+1)(s+2)(s+3)} = 0$$

$$(s + 1)(s^2 + 5s + 6) + K = 0$$

$$s^3 + 5s^2 + 6s + s^2 + 5s + 6 + K = 0$$

$$\Rightarrow s^3 + 6s^2 + 11s + 6 + K = 0$$

Gain margin = 0 dB and phase margin = 0°

It implies marginal stable system

By Routh Array

s^3	1	11
s^2	6	$(6 + K)$
s	$\frac{66 - 6 - K}{6}$	0
s^0	$6 + K$	

For marginal stable system,

$$60 - K = 0$$

$$\Rightarrow \quad K = 60$$

Hence, the positive value of K is 60.

47. $G(s) = \dfrac{K}{(s+0.1)(s+10)(s+p_1)}$

$$\angle G(s) = -\tan^{-1}\frac{\omega}{0.1} - \tan^{-1}\frac{\omega}{10} - \tan^{-1}\frac{\omega}{p_1}$$

$$\angle G(s)|_{\omega=1} = -\tan^{-1}\frac{1}{0.1} - \tan^{-1}\frac{1}{10} - \tan^{-1}\frac{1}{p_1}$$

$$= -135°$$

$$-\tan^{-1} 10 - \tan^{-1} 0.1 - \tan^{-1}\frac{1}{p_1} = -135°$$

$$-84.28° - 5.72° - \tan^{-1}\frac{1}{p_1} = -135°$$

$$-\tan^{-1}\frac{1}{p_1} = -135° + 90°$$

$$\tan^{-1}\frac{1}{p_1} = 45°$$

$$\frac{1}{p_1} = 1 \Rightarrow p_1 = 1$$

Hence, the value of p_1 is 1.

48. Power spectral density of $X(t) = S_X(f)$

$$S_X(f) = \frac{|G(f)|^2}{T}\sum_{n=-\infty}^{\infty} R_b(\tau)e^{j2\pi fn\tau},$$

$$R_b(\tau) = E[\alpha_n\beta_{n-\tau}]$$

$$= E[(\alpha_n + k\alpha_{n-3})(\alpha_{n-\tau} + k\,\alpha_{n-\tau-3})]$$

$$= E[\alpha_n\alpha_{n-\tau}] + kE[\alpha_{n-3}\,\alpha_{n-\tau}] + kE[\alpha_n\alpha_{n-\tau-3}] + k^2E[\alpha_{n-3}\,\alpha_{n-\tau-3}]$$

$$= E[\alpha_n\alpha_{n-\tau}] + kE[\alpha_{n-3}\,\alpha_{n-\tau-3+3}] + kE[\alpha_n\alpha_{n-\tau-3}] + k^2E[\alpha_{n-3}\,\alpha_{n-\tau-3}]$$

$$= R(\tau) + kR(\tau - 3) + kR(\tau + 3) + k^2R(\tau)$$

$$= (1 + k^2)R(\tau) + kR(\tau + 3) + kR(\tau - 3)$$

Hence autocorrelation function can be defined as:

$$R_b(\tau) = \begin{Bmatrix} 1+k^2 & \tau = 0 \\ k & \tau = \pm 3 \\ 0 & \text{otherwise} \end{Bmatrix}$$

Power spectral density

$$s_b(f) = 1 + k^2 + 2k\cos(2\pi f3T)$$

Null will occur at $\quad f = \dfrac{1}{3T}$

At $\quad f = \dfrac{1}{3T}$

$$s_b(f) = 1 + k^2 + 2k\cos(2\pi) = 0$$

$$\Rightarrow 1 + k^2 + 2k = 0$$

$$\Rightarrow (k + 1)^2 = 0$$

$$\therefore \quad k = -1$$

Hence the value of k is – 1.

49. Here the range of Channel spectrum 500 Hz – 2000 Hz (Here R_b = rate of transmission)

Hence Bandwidth (BW) = 1500 Hz

$$BW = \frac{R_b}{\log_2 M}(1+\alpha)$$

$$1500 = \frac{R_b}{\log_2 16}(1+\alpha)$$

$$1500 = \frac{4800}{\log_2 16}(1+\alpha)$$

$$1500 = 1200\,(1 + \alpha)$$

$$\Rightarrow \quad \alpha = 0.25$$

Hence, the roll-off factor of a-pulse is 0.25.

50. Given random process

$$X(t) = 3V(t) - 8 \quad \text{and} \quad E[V(t)] = 0$$

$$R_v(\tau) = 4e^{-5|\tau|}$$

Power of $X(t) = E[X^2(t)]$

$$= E\,[9V^2(t)] + 64 - 48\,E\,[V(t)]$$

$$= 9E\,[V^2(t)] + 64 - 48\,E\,[V(t)]$$

$$E\,[V^2(t)] = R_v(0) = 4$$

Power of $X(t) = ((9 \times 4) + 64)$

$$= 100$$

Hence the power in $X(t)$ is 100.

51.

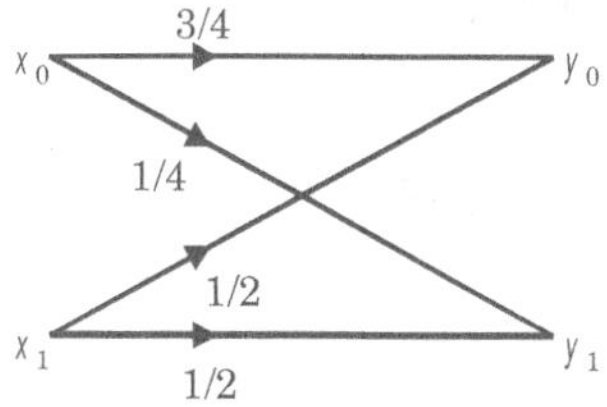

$$P(x_0) = \frac{1}{2}; \qquad P(x_1) = \frac{1}{2}$$

$$P\left(\frac{y_0}{x_0}\right) = \frac{3}{4}; \qquad P\left(\frac{y_0}{x_1}\right) = \frac{1}{2}$$

$$[P(x, y)] = [P(x)]_d\,[P(y|x)]$$

$$= \begin{bmatrix} 1/2 & 0 \\ 0 & 1/2 \end{bmatrix}\begin{bmatrix} 3/4 & 1/4 \\ 1/2 & 1/2 \end{bmatrix}$$

$$[P(x,y)] = \begin{bmatrix} 3/8 & 1/8 \\ 1/4 & 1/4 \end{bmatrix}$$

$$P(y \mid x_0) = -\sum_{k=0}^{1} P(x_0, y_k) \log_2 P\left(\frac{y_k}{x_0}\right)$$

$$= -\{P(x_0, y_0)\log_2 P(y_0 \mid x_0) + P(x_0, y_1)\log_2 P(y_1 \mid x_0)\}$$

$$= -\left\{\frac{3}{8}\log_2 \frac{3}{4} + \frac{1}{8}\log_2 \frac{1}{4}\right\}$$

$$= 0.405$$

52. Let $E = E_1$,

$$\therefore \quad \text{Energy } E_1 = \frac{Q_1^2}{2C_1}$$

(Here Q = charge across capacitor)

For electrically isolated

$$\Rightarrow \quad Q_2 = Q_1$$
$$d_2 = 2d_1$$

$$\Rightarrow \quad C_2 = \frac{C_1}{2} \quad (\text{Here } d_1 = d_2 = \text{plate separation})$$

$$E_2 = \frac{Q_2^2}{2C_2} = \frac{Q_1^2}{\frac{2C_1}{2}} = 2\left(\frac{Q_1^2}{2C_1}\right)$$

$$= 2E_1 = 2E$$

Hence the energy stored in the capacitor is 2E.

53. $$Z_0 = \sqrt{\frac{L}{C}}$$

Here Z_0 = Characteristics impedance
L = Inductance per unit length

$$\Rightarrow \quad Z_0 = \sqrt{\frac{Lt}{\epsilon_{eff} W}}$$

as $\quad \epsilon_{eff} < \epsilon_0 \epsilon_r$

$$Z_0 > \sqrt{\frac{Lt}{\epsilon_0 \epsilon_r W}}$$

54. At frequency of 1 GHz the line behaves as $\frac{\lambda}{2}$ length there by giving open circuit at the input of T_2 line. (From figure)

With 50 Ω termination at T_1 the load is matched and $|\Gamma| = 0$ as seen in the graph.

Here $|\Gamma|$ = input reflection coefficient

$T_1 - T_2$ = lossless transmission line

$$f = 1 \text{ GHz}$$

$$\lambda = \frac{2 \times 10^{10}}{1 \times 10^9} = 20 \text{ cm}$$

Now $$L = \frac{\lambda}{2} = 10 \text{ cm} = 0.1 \text{ m}$$

Hence, the length (L) of T_2 is 0.1 m.

55. For velocity being free,

Since $\quad K.E. = \frac{1}{2}mV^2$

$$V = \frac{E}{q}$$

E = Energy
V = Applied voltage
q = electric charge of a metal plate

$$\frac{1}{2}mV^2 = qV$$

$$v = \frac{d}{t} = \sqrt{\frac{2qV}{m}}$$

$$\frac{d}{t} = \sqrt{\frac{2qV}{m}}$$

$$t = \frac{d}{\sqrt{\frac{2qV}{m}}}$$

$$\Rightarrow \quad t \propto \frac{d}{\sqrt{V}}$$

■■

SOLVED PAPER – 2016

ELECTRONICS & COMMUNICATION ENGINEERING

Set - 3

No. of Questions : 65 | *Maximum Marks : 100*

INSTRUCTIONS

1. Total of 65 questions carrying 100 marks, out of which 10 questions carrying a total of 15 marks are in General Aptitude (GA)
2. The Engineering Mathematics will carry around **15% of the total marks,** the General Aptitude section will carry **15% of the total marks** and the **remaining 70% of the total marks.**
3. **Types of Questions**
   (*a*) **Multiple Choice Questions (MCQ)** carrying 1 or 2 marks each in all papers and sections. These questions are objective in nature, and each will have a choice of four options, out of which the candidate has to mark the correct answer(s).
   (*b*) **Numerical Answer Questions** of 1 or 2 marks each in all papers and sections. For these questions the answer is a real number, to be entered by the candidate using the virtual keypad. No choices will be shown for these type of questions.
4. For **1-mark** multiple-choice questions, **1/3 marks** will be deducted for a wrong answer. Likewise, for **2-marks** multiple-choice questions, **2/3** marks will be deducted for a wrong answer. There is no negative marking for numerical answer type questions.

GENERAL APTITUDE (GA)

(Q.1 – 5) : Carry One Mark Each

1. An apple costs Rs. 10. An onion costs Rs. 8.
Select the most suitable sentence with respect to grammar and usage.
(*a*) The price of an apple is greater than an onion.
(*b*) The price of an apple is more than onion.
(*c*) The price of an apple is greater than that of an onion.
(*d*) Apples are more costlier than onions.

2. The Buddha said, "Holding on to anger is like grasping a hot coal with the intent of throwing it at someone else; you are the one who gets burnt."
Select the word below which is closest in meaning to the word underlined above.
(*a*) burning (*b*) igniting
(*c*) clutching (*d*) flinging

3. M has a son Q and a daughter R. He has no other children. E is the mother of P and daughter-inlaw of M. How is P related to M?
(*a*) P is the son-in-law of M.
(*b*) P is the grandchild of M.
(*c*) P is the daughter-in-law of M.
(*d*) P is the grandfather of M.

4. The number that least fits this set: (324, 441, 97 and 64) is ________.
(*a*) 324 (*b*) 441
(*c*) 97 (*d*) 64

5. It takes 10 s and 15 s, respectively, for two trains travelling at different constant speeds to completely pass a telegraph post. The length of the first train is 120 m and that of the second train is 150 m. The magnitude of the difference in the speeds of the two trains (in m/s) is ________.
(*a*) 2.0
(*b*) 10.0
(*c*) 12.0
(*d*) 22.0

(Q.6 – 10) : Carry Two Marks Each

6. The velocity V of a vehicle along a straight line is measured in m/s and plotted as shown with respect to time in seconds. At the end of the 7 seconds, how much will the odometer reading increase by (in m) ?

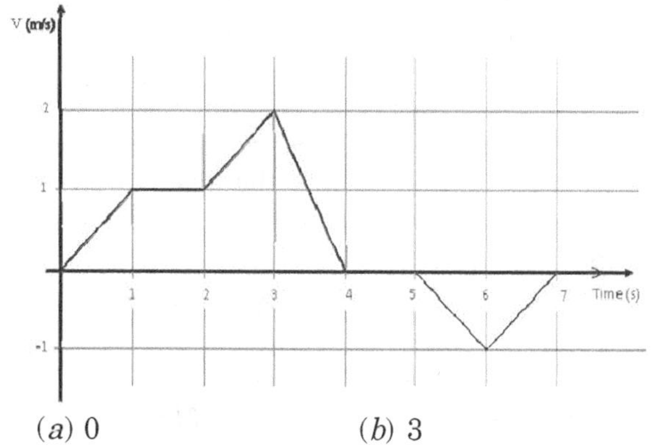

(*a*) 0 (*b*) 3
(*c*) 4 (*d*) 5

7. The overwhelming number of people infected with rabies in India has been flagged by the World Health Organization as a source of concern. It is estimated that inoculating 70% of pets and stray dogs against rabies can lead to a significant reduction in the number of people infected with rabies.

Which of the following can be logically inferred from the above sentences ?

(*a*) The number of people in India infected with rabies is high

(*b*) The number of people in other parts of the world who are infected with rabies is low

(*c*) Rabies can be eradicated in India by vaccinating 70% of stray dogs

(d) Stray dogs are the main source of rabies worldwide

8. A flat is shared by four first year undergraduate students. They agreed to allow the oldest of them to enjoy some extra space in the flat. Manu is two months older than Sravan, who is three months younger than Trideep. Pavan is one month older than Sravan. Who should occupy the extra space in the flat ?

(*a*) Manu (*b*) Sravan

(*c*) Trideep (*d*) Pavan

9. Find the area bounded by the lines $3x + 2y = 14$, $2x - 3y = 5$ in the first quadrant

(*a*) 14.95 (*b*) 15.25

(*c*) 15.70 (*d*) 20.35

10. A straight line is fit to a data set ($\ln x$, y). This line intercepts the abscissa at $\ln x = 0.1$ and has a slope of –0.02. What is the value of y at $x = 5$ from the fit?

(*a*) –0.030 (*b*) –0.014

(*c*) 0.014 (*d*) 0.030

TECHNICAL SECTION

(Q.1 – 25) : Carry One Mark Each

1. Consider a 2 × 2 square matrix

$$A = \begin{bmatrix} \sigma & x \\ \omega & \sigma \end{bmatrix},$$

where x is unknown. If the eigenvalues of the matrix A are $(\sigma + j\omega)$ and $(\sigma - j\omega)$, then x is equal to

(*a*) $+j\omega$ (*b*) $-j\omega$

(*c*) $+\omega$ (*d*) $-\omega$

2. For $f(z) = \dfrac{\sin(z)}{z^2}$, the residue of the pole at z = 0 is _______.

3. The probability of getting a "head" in a single toss of a biased coin is 0.3. The coin is tossed repeatedly till a "head" is obtained. If the tosses are independent, then the probability of getting "head" for the first time in the fifth toss is _______

4. The integral $\int_0^1 \dfrac{dx}{\sqrt{(1-x)}}$ is equal to_______

5. Consider the first order initial value problem

$$y' = y + 2x - x^2,\ y(0) = 1,\ (0 \le x < \infty)$$

with exact solution $y(x) = x^2 + e^x$. For $x = 0.1$, the percentage difference between the exact solution and the solution obtained using a single iteration of the second-order Runge-Kutta method with step-size $h = 0.1$ is _______.

6. Consider the signal $x(t) = \cos(6\pi t) + \sin(8\pi t)$, where t is in seconds. The Nyquist sampling rate (in samples/second) for the signal $y(t) = x(2t + 5)$ is

(*a*) 8 (*b*) 12

(*c*) 16 (*d*) 32

7. If the signal $x(t) = \dfrac{\sin(t)}{\pi t} * \dfrac{\sin(t)}{\pi t}$ with * denoting the convolution operation, then $x(t)$ is equal to

(*a*) $\dfrac{\sin(t)}{\pi t}$ (*b*) $\dfrac{\sin(2t)}{2\pi t}$

(*c*) $\dfrac{2\sin(t)}{\pi t}$ (*d*) $\left(\dfrac{\sin(t)}{\pi t}\right)^2$

8. A discrete-time signal $x[n] = \delta[n-3] + 2\,\delta[n-5]$ has z-transform $X(z)$. If $Y(z) = X(-z)$ is the z-transform of another signal $y[n]$, then

(*a*) $y[n] = x[n]$ (*b*) $y[n] = x[-n]$

(*c*) $y[n] = -x[n]$ (*d*) $y[n] = -x[-n]$

9. In the RLC circuit shown in the figure, the input voltage is given by

$$v_i(t) = 2\cos(200t) + 4\sin(500t).$$

The output voltage $v_o(t)$ is

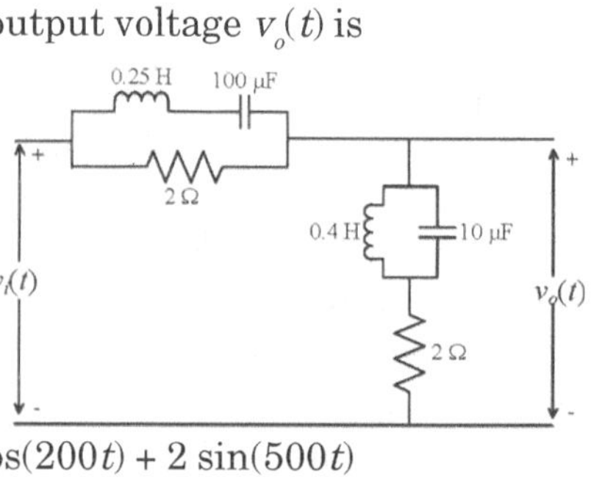

(*a*) $\cos(200t) + 2\sin(500t)$

(*b*) $2\cos(200t) + 4\sin(500t)$

(*c*) $\sin(200t) + 2\cos(500t)$

(*d*) $2\sin(200t) + 4\cos(500t)$

10. The I-V characteristics of three types of diodes at the room temperature, made of semiconductors X, Y and Z, are shown in the figure. Assume that the diodes are uniformly doped and identical in all respects except their materials. If E_{gx}, E_{gY} and E_{gZ} are the band gaps of X, Y and Z, respectively, then

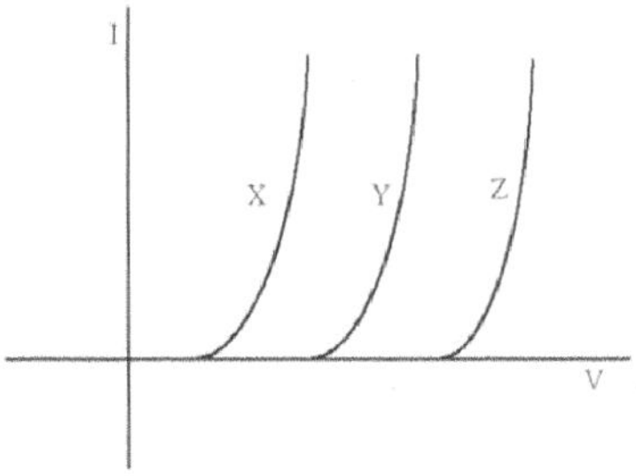

(*a*) $E_{gX} > E_{gY} > E_{gZ}$
(*b*) $E_{gx} = E_{gY} = E_{gz}$
(*c*) $E_{gx} < E_{gY} < E_{gZ}$
(*d*) no relationship among these band gaps exists.

11. The figure shows the band diagram of a Metal Oxide Semiconductor (MOS). The surface region of this MOS is in

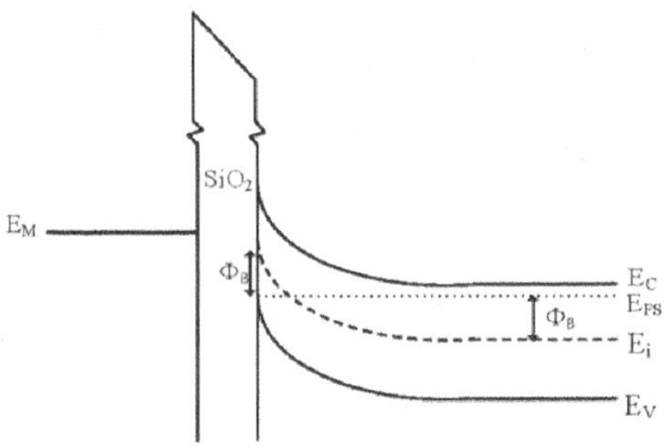

(*a*) inversion (*b*) accumulation
(*c*) depletion (*d*) flat band

12. The figure shows the I-V characteristics of a solar cell illuminated uniformly with solar light of power 100 mW/cm². The solar cell has an area of 3 cm² and a fill factor of 0.7. The maximum efficiency (in %) of the device is _______

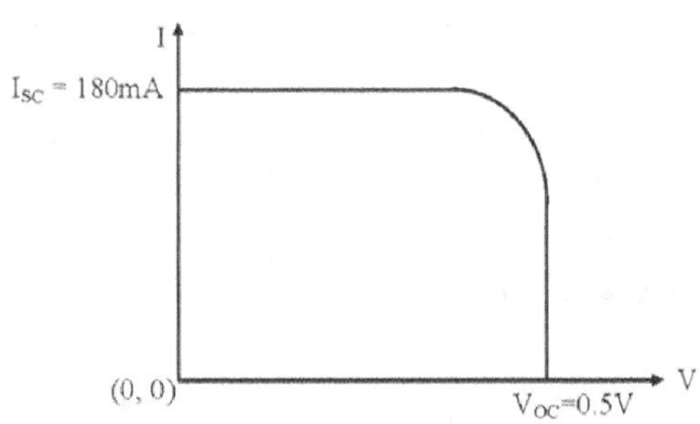

13. The diodes D1 and D2 in the figure are ideal and the capacitors are identical. The product RC is very large compared to the time period of the ac voltage. Assuming that the diodes do not breakdown in the reverse bias, the output voltage V_o (in volt) at the steady state is _______

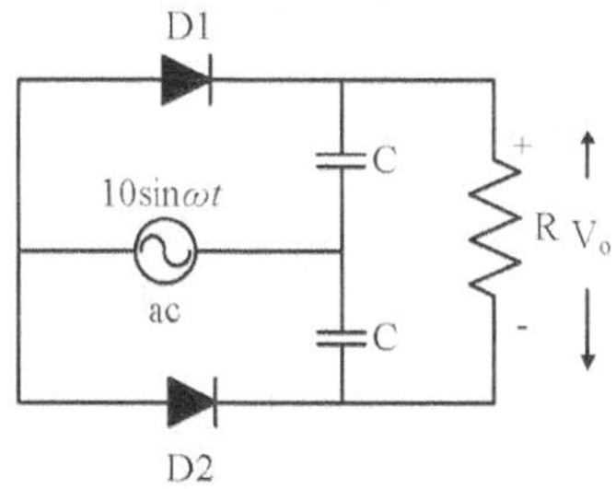

14. Consider the circuit shown in the figure. Assuming $V_{BE1} = V_{EB2} = 0.7$ volt, the value of the dc voltage V_{C2} (in volt) is _______

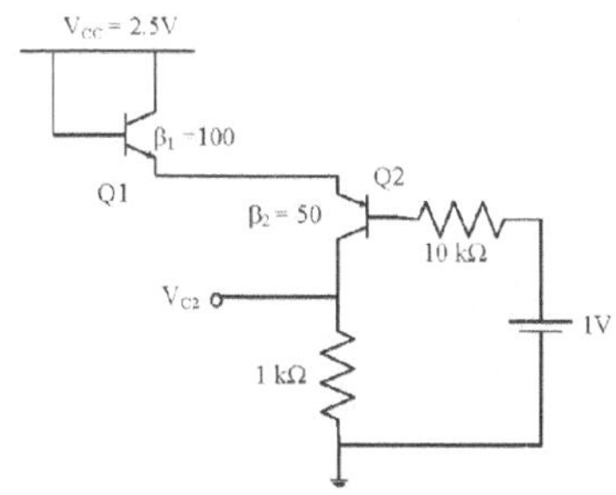

15. In the astable multivibrator circuit shown in the figure, the frequency of oscillation (in kHz) at the output pin 3 is _______

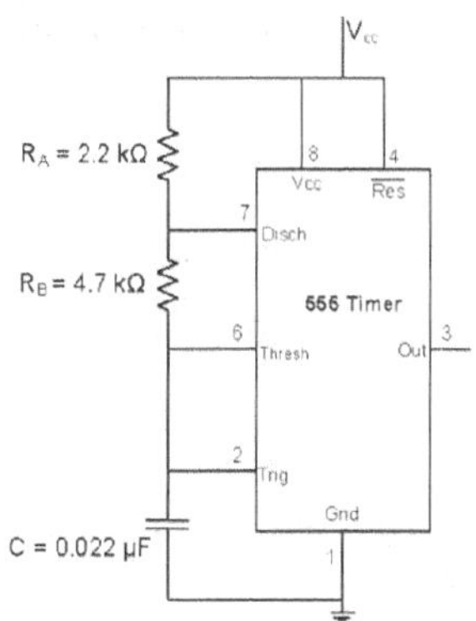

16. In an 8085 microprocessor, the contents of the accumulator and the carry flag are A7 (in hex) and 0, respectively. If the instruction RLC is executed, then the contents of the accumulator (in hex) and the carry flag, respectively, will be

(*a*) 4E and 0 (*b*) 4E and 1
(*c*) 4F and 0 (*d*) 4F and 1

17. The logic functionality realized by the circuit shown below is

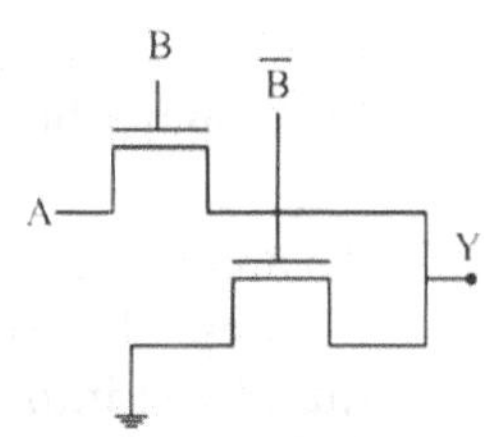

(*a*) OR
(*b*) XOR
(*c*) NAND
(*d*) AND

18. The minimum number of 2-input NAND gates required to implement a 2-input XOR gate is

(*a*) 4 (*b*) 5
(*c*) 6 (*d*) 7

19. The block diagram of a feedback control system is shown in the figure. The overall closed-loop gain G of the system is

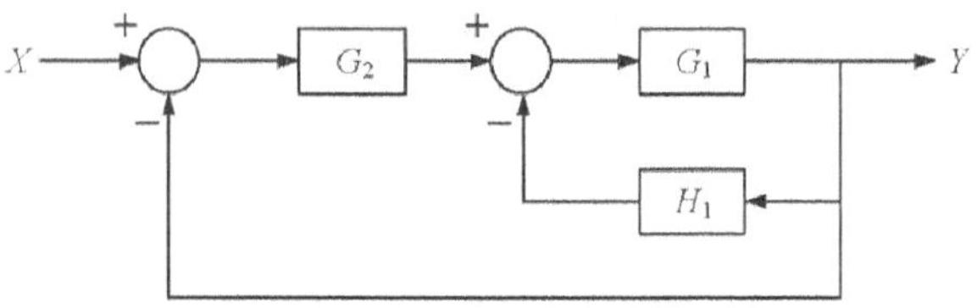

(*a*) $G = \dfrac{G_1G_2}{1+G_1H_1}$

(*b*) $G = \dfrac{G_1G_2}{1+G_1G_2+G_1H_1}$

(*c*) $G = \dfrac{G_1G_2}{1+G_1G_2H_1}$

(*d*) $G = \dfrac{G_1G_2}{1+G_1G_2+G_1G_2H_1}$

20. For me unity feedback control system shown in the figure, the open-loop transfer function G(s) is given as

$$G(s) = \frac{2}{s(s+1)}.$$

The steady state error e_{ss} due to a unit step input is

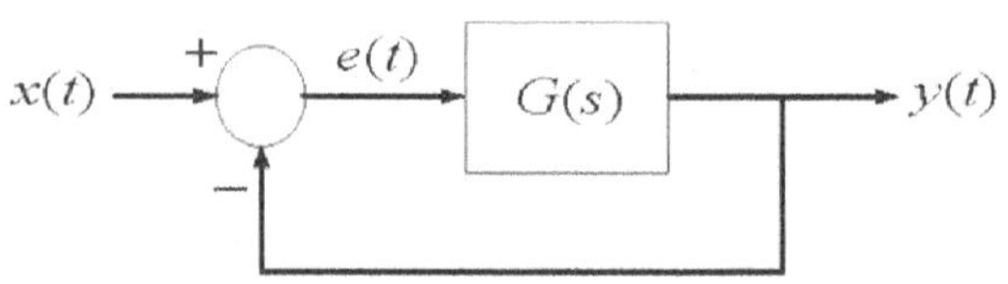

(*a*) 0

(*b*) 0.5

(*c*) 1.0

(*d*) ∞

21. For a superheterodyne receiver, the intermediate frequency is 15 MHz and the local oscillator frequency is 3.5 GHz. If the frequency of the received signal is greater than the local oscillator frequency, then the image frequency (in MHz) is_______.

22. An analog baseband signal, bandlimited to 100 Hz, is sampled at the Nyquist rate. The samples are quantized into four message symbols that occur independently with probabilities $p_1 = p_4 = 0.125$ and $p_2 = p_3$. The information rate (bits/sec) of the message source is _______.

23. A binary baseband digital communication system employs the signal

$$p(t) = \begin{cases} \dfrac{1}{\sqrt{T_S}}, & 0 \le t \le T_S \\ 0, & \text{otherwise} \end{cases}$$

for transmission of bits. The graphical representation of the matched filter output y(t) for this signal will be

(*a*)

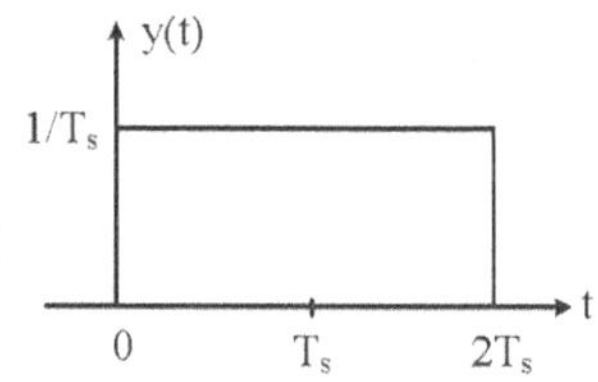

(*b*)

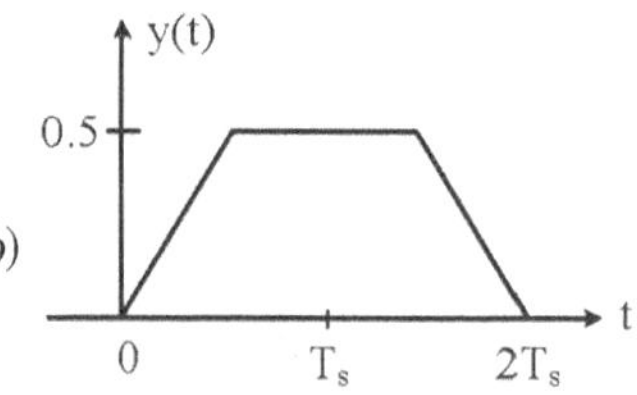

(*c*)

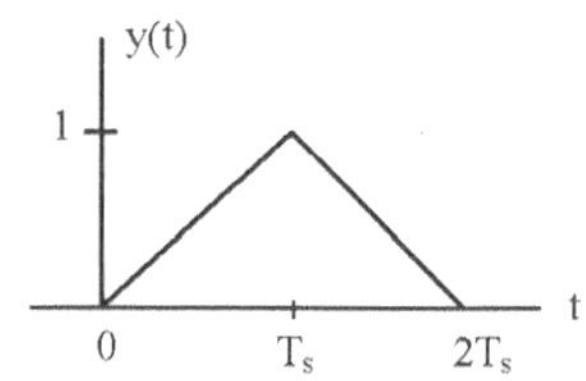

(*d*)

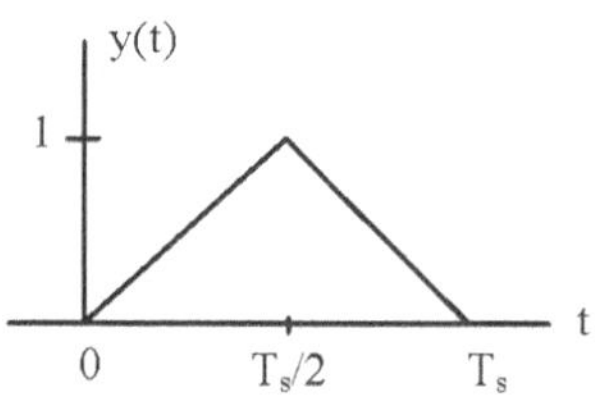

24. If a right-handed circularly polarized wave is incident normally on a plane perfect conductor, then the reflected wave will be

(*a*) right-handed circularly polarized

(*b*) left-handed circularly polarized

(*c*) elliptically polarized with a tilt angle of 45°

(*d*) horizontally polarized

25. Faraday's law of electromagnetic induction is mathematically described by which one of the following equations?

(*a*) $\nabla \bullet \vec{B} = 0$

(*b*) $\nabla \bullet \vec{D} = \rho_V$

(*c*) $\nabla \times \vec{E} = -\dfrac{\partial \vec{B}}{\partial t}$

(*d*) $\nabla \times \vec{H} = \sigma\vec{E} + \dfrac{\partial \vec{D}}{\partial t}$

(Q.26 – 55) : Carry Two Marks Each

26. The particular solution of the initial value problem given below is

$$\frac{d^2y}{dx^2} + 12\frac{dy}{dx} + 36y = 0 \text{ with } y(0) = 3$$

and $\left.\frac{dy}{dx}\right|_{x=0} = -36$

(*a*) $(3 - 18x)\, e^{-6x}$ (*b*) $(3 + 25x)\, e^{-6x}$

(*c*) $(3 + 20x)\, e^{-6x}$ (*d*) $(3 - 12x)\, e^{-6x}$

27. If the vectors $e_1 = (1, 0, 2)$, $e_2 = (0, 1, 0)$ and $e_3 = (-2, 0, 1)$ form an orthogonal basis of the three-dimensional real space $\mathbb{R}^3$, then the vector $u = (4, 3, -3) \in \mathbb{R}^3$ can be expressed as

(*a*) $u = -\frac{2}{5}e_1 - 3e_2 - \frac{11}{5}e_3$

(*b*) $u = -\frac{2}{5}e_1 - 3e_2 + \frac{11}{5}e_3$

(*c*) $u = -\frac{2}{5}e_1 + 3e_2 + \frac{11}{5}e_3$

(*d*) $u = -\frac{2}{5}e_1 + 3e_2 - \frac{11}{5}e_3$

28. A triangle in the xy-plane is bounded by the straight lines $2x = 3y$, $y = 0$ and $x = 3$. The volume above the triangle and under the plane $x + y + z = 6$ is ________.

29. The values of the integral $\frac{1}{2\pi j}\oint_c \frac{e^z}{z-2}\,dz$ along a closed contour c in anti-clockwise direction for

(*i*) the point $z_0 = 2$ inside the contour *c*, and

(*ii*) the point $z_0 = 2$ outside the contour *c*,

respectively, are

(*a*) (*i*) 2.72, (*ii*) 0

(*b*) (*i*) 7.39, (*ii*) 0

(*c*) (*i*) 0, (*ii*) 2.72

(*d*) (*i*) 0, (*ii*) 7.39

30. A signal $2\cos\left(\frac{2\pi}{3}t\right) - \cos(\pi t)$ is the input to an LTI system with the transfer function $H(s) = e^{s} + e^{-s}$.

If C_k denotes the kth coefficient in the exponential Fourier series of the output signal, then C_3 is equal to

(*a*) 0

(*b*) 1

(*c*) 2

(*d*) 3

31. The ROC (region of convergence) of the z-transform of a discrete-time signal is represented by the shaded region in the *z*-plane. If the signal $x|n| = (2.0)^{|n|}$, $-\infty < n < +\infty$, then the ROC of its z-transform is represented by

(*a*)

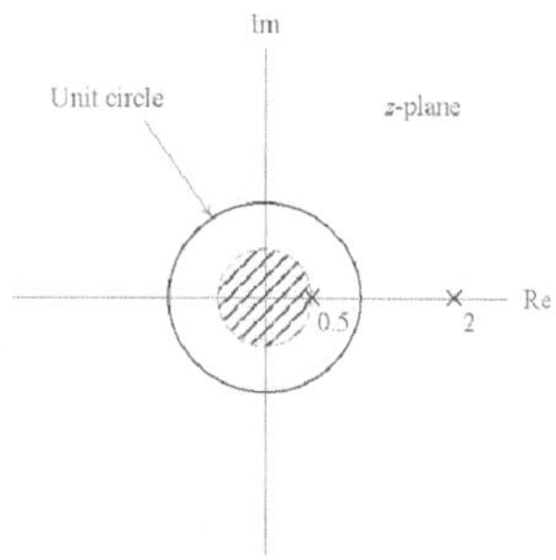

(*b*)

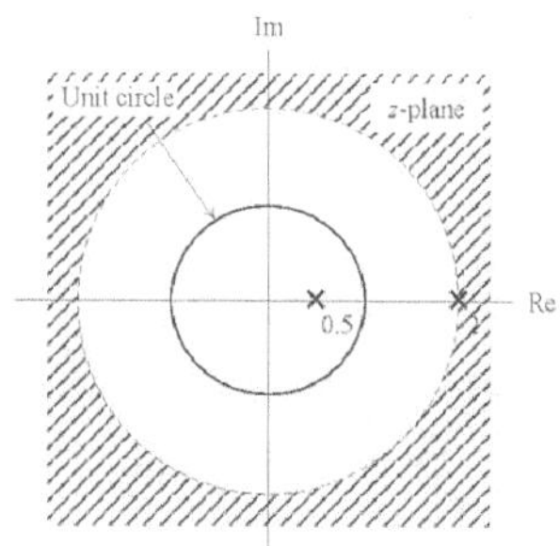

(*c*)

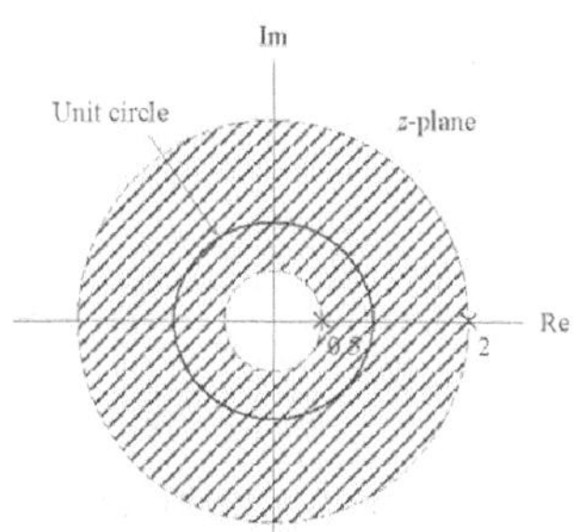

(*d*)

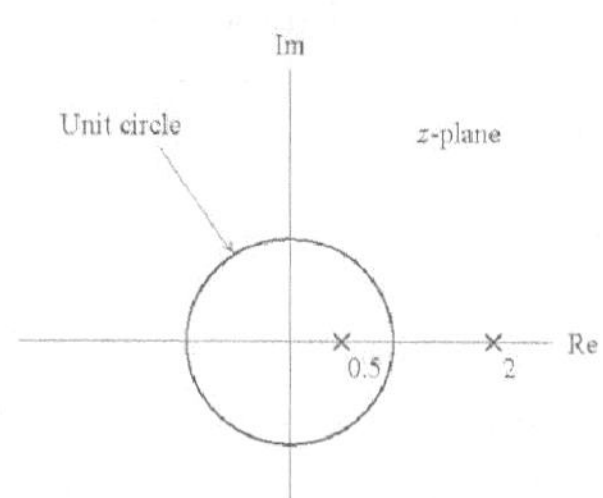

(ROC does not exist)

32. Assume that the circuit in the figure has reached the steady state before time $t = 0$ when the 3 Ω resistor suddenly burns out, resulting in an open circuit. The current $i(t)$ (in ampere) at $t = 0^+$ is________.

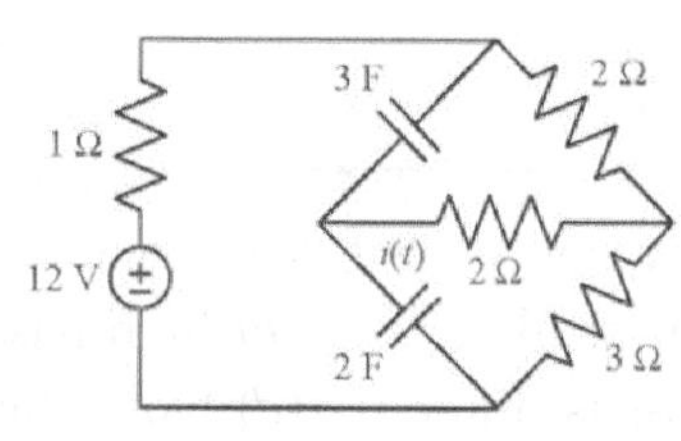

33. In the figure shown, the current i (in ampere) is________.

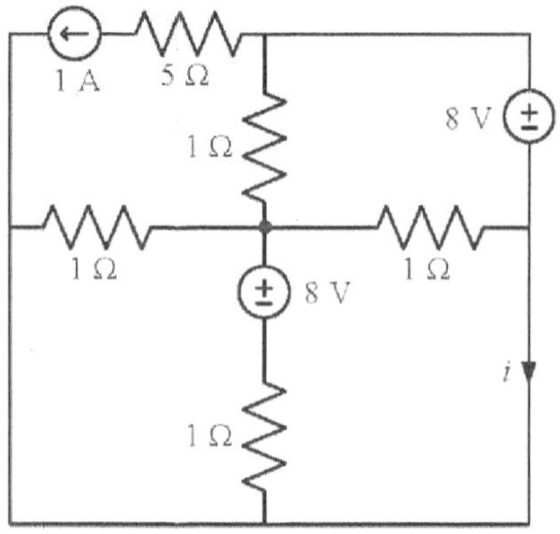

34. The z-parameter matrix $\begin{bmatrix} z_{11} & z_{12} \\ z_{21} & z_{22} \end{bmatrix}$ for the two-port network shown is

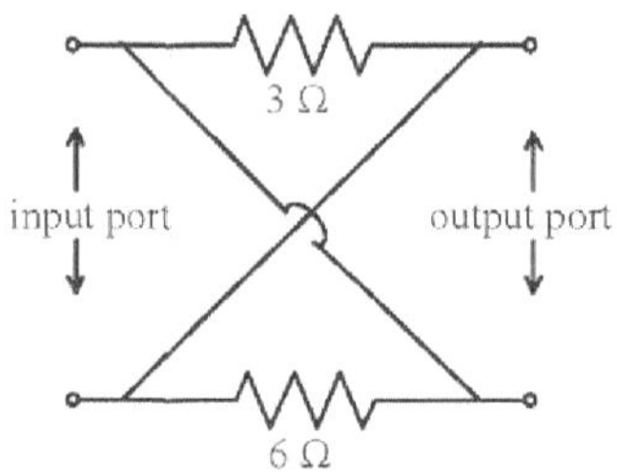

(*a*) $\begin{bmatrix} 2 & -2 \\ -2 & 2 \end{bmatrix}$ (*b*) $\begin{bmatrix} 2 & 2 \\ 2 & 2 \end{bmatrix}$

(*c*) $\begin{bmatrix} 9 & -3 \\ 6 & 9 \end{bmatrix}$ (*d*) $\begin{bmatrix} 9 & 3 \\ 6 & 9 \end{bmatrix}$

35. A continuous-time speech signal $x_a(t)$ is sampled at a rate of 8 kHz and the samples are subsequently grouped in blocks, each of size N. The DFT of each block is to be computed in real time using the radix-2 decimation-in-frequency FFT algorithm. If the processor performs all operations sequentially, and takes 20 (μs for computing each complex multiplication (including multiplications by 1 and -1) and the time required for addition/subtraction is negligible, then the maximum value of N is_______.

36. The direct form structure of an FIR (finite impulse response) filter is shown in the figure.

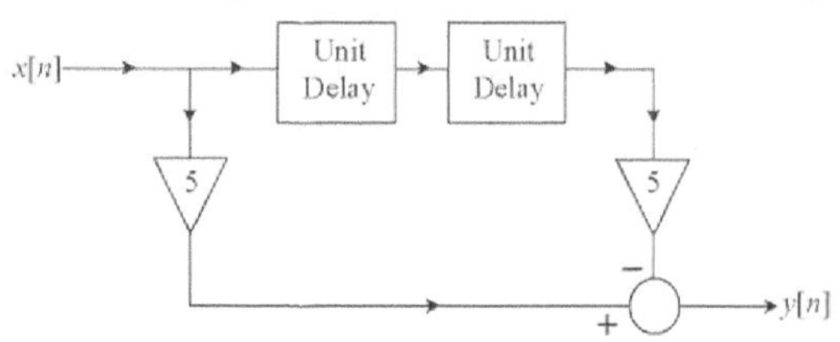

The filter can be used to approximate a

(*a*) low-pass filter (*b*) high-pass filter
(*c*) band-pass filter (*d*) band-stop filter

37. The injected excess electron concentration profile in the base region of an *npn* BJT, biased in the active region, is linear, as shown in the figure. If the area of the emitter-base junction is 0.001 cm^2, μ_n = 800 cm^2/(V-s) in the base region and depletion layer widths are negligible, then the collector current Ic (in mA) at room temperature is______.
(Given: thermal voltage V_T = 26 mV at room temperature, electronic charge $q = 1.6 \times 10^{-19}$ C)

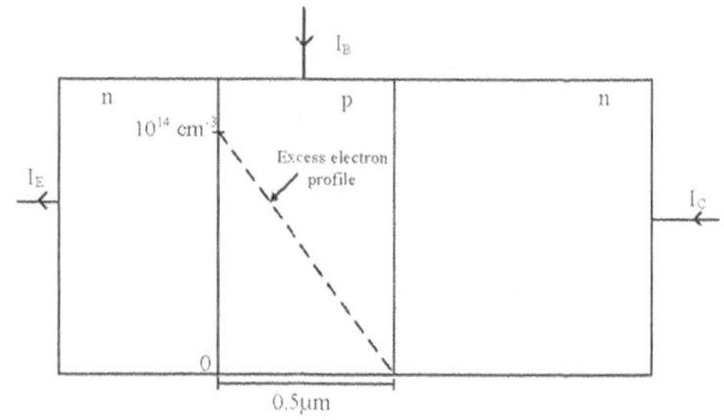

38. Figures I and II show two MOS capacitors of unit area. The capacitor in Figure I has insulator materials X (of thickness t_1 = 1 nm and dielectric constant ε_1 = 4) and Y (of thickness t_2 = 3 nm and dielectric constant ε_2 = 20). The capacitor in Figure II has only insulator material X of thickness t_{Eq}. If the capacitors are of equal capacitance, then the value of t_{Eq} (in nm) is______

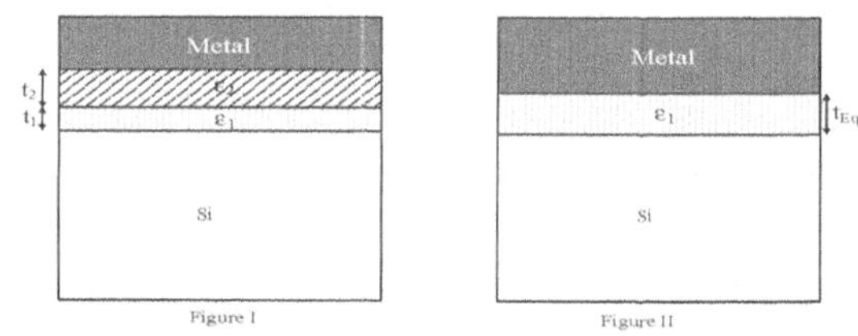

39. The I-V characteristics of the zener diodes Dl and D2 are shown in Figure I. These diodes are used in the circuit given in Figure II. If the supply voltage is varied from 0 to 100 V, then breakdown occurs in

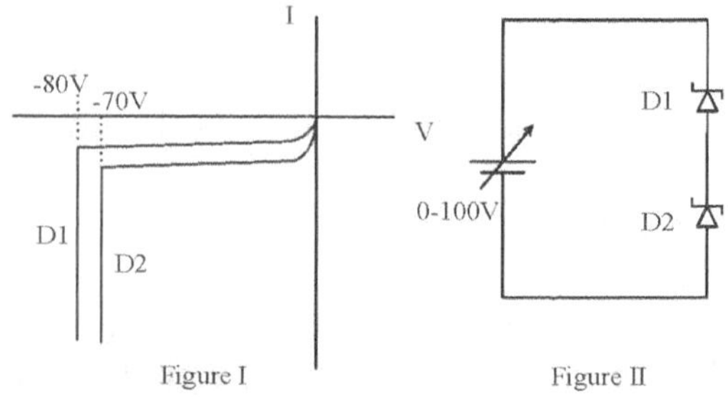

(*a*) D1 only
(*b*) D2 only
(*c*) both D1 and D2
(*d*) none of D1 and D2

40. For the circuit shown in the figure, $R_1 = R_2 = R_3 = 1\ \Omega$, $L = 1\ \mu H$ and $C = 1\ \mu F$. If the input $V_{in} = \cos(10^6 t)$, then the overall voltage gain (V_{out}/V_{in}) of the circuit is _______.

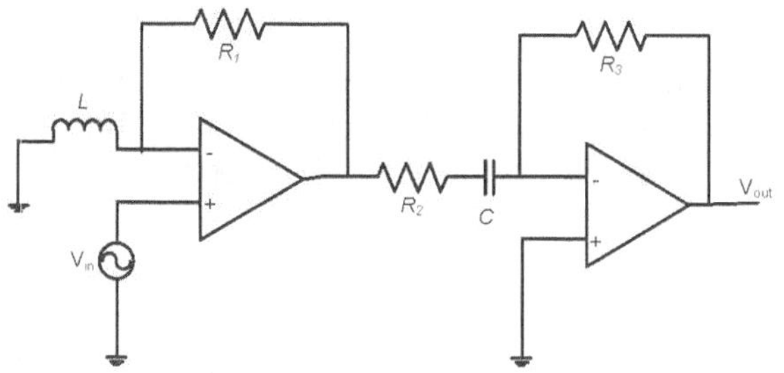

41. In the circuit shown in the figure, the channel length modulation of all transistors is non-zero $(\lambda \neq 0)$. Also, all transistors operate in saturation and have negligible body effect. The ac small signal voltage gain (V_o/V_{in}) of the circuit is

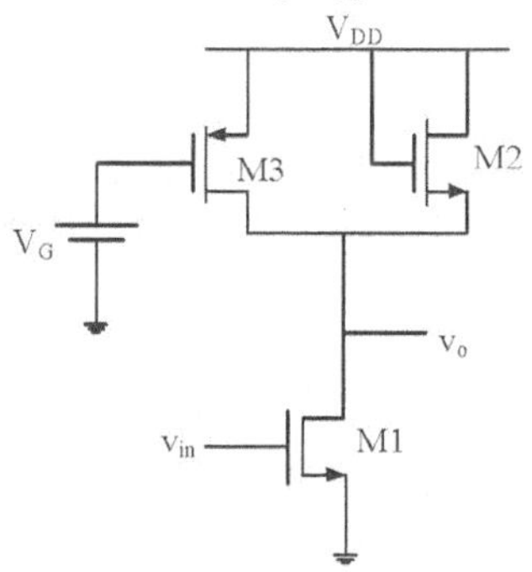

(*a*) $-g_{m1}(r_{o1} \parallel r_{o2} \parallel r_{o3})$

(*b*) $-g_{m1}(r_{o1} \parallel \frac{1}{g_{m3}} \parallel r_{o3})$

(*c*) $-g_{m1}(r_{o1}(\parallel(\frac{1}{g_{m3}} \parallel r_{o2}) \parallel r_{o3})$

(*d*) $-g_{m1}(r_{o1} \parallel (\frac{1}{g_{m3}} \parallel r_{o3}) \parallel r_{o2})$

42. In the circuit shown in the figure, transistor M1 is in saturation and has transconductance $g_m = 0.01$ siemens. Ignoring internal parasitic capacitances and assuming the channel length modulation λ to be zero, the small signal input pole frequency (in kHz) is _______.

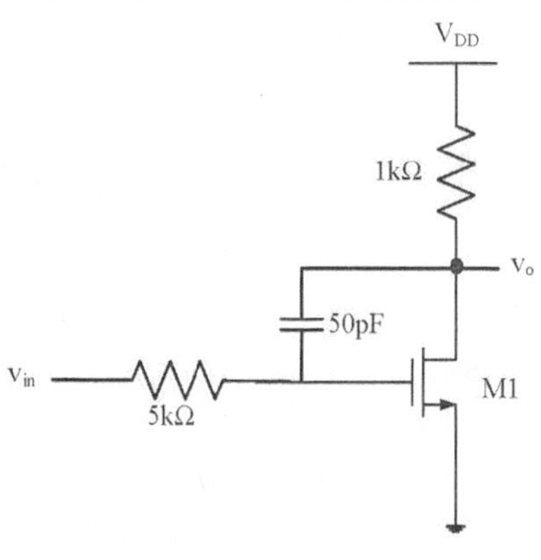

43. Following is the K-map of a Boolean function of five variables P, Q, R, S and X. The minimum sum-of-product (SOP) expression for the function is

RS \ PQ	00	01	11	10
00	0	0	0	0
01	1	0	0	1
11	1	0	0	1
10	0	0	0	0

X=0

RS \ PQ	00	01	11	10
00	0	1	1	0
01	0	0	0	0
11	0	0	0	0
10	0	1	1	0

X=1

(*a*) $PQS\bar{X} + P\bar{Q}S\bar{X} + Q\bar{R}\bar{S}X + QR\bar{S}X$

(*b*) $\bar{Q}S\bar{X} + Q\bar{S}X$

(*c*) $\bar{Q}SX + Q\bar{S}\bar{X}$

(*d*) $\bar{Q}S + Q\bar{S}$

44. For the circuit shown in the figure, the delays of NOR gates, multiplexers and inverters are 2 ns, 1.5 ns and 1 ns, respectively. If all the inputs P, Q, R, S and T are applied at the same time instant, the maximum propagation delay (in ns) of the circuit is _______.

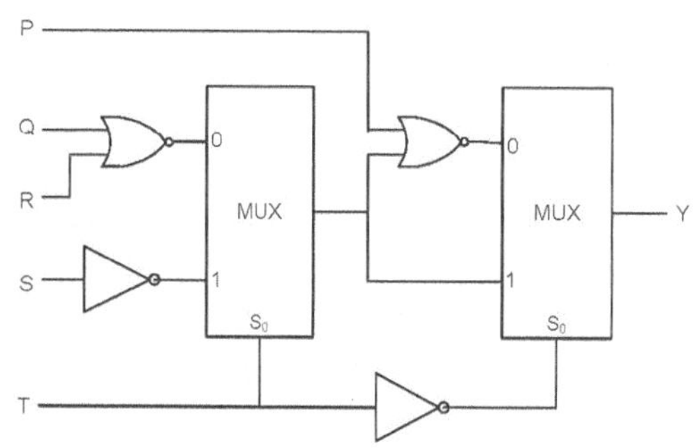

45. For the circuit shown in the figure, the delay of the bubbled NAND gate is 2 ns and that of the counter is assumed to be zero.

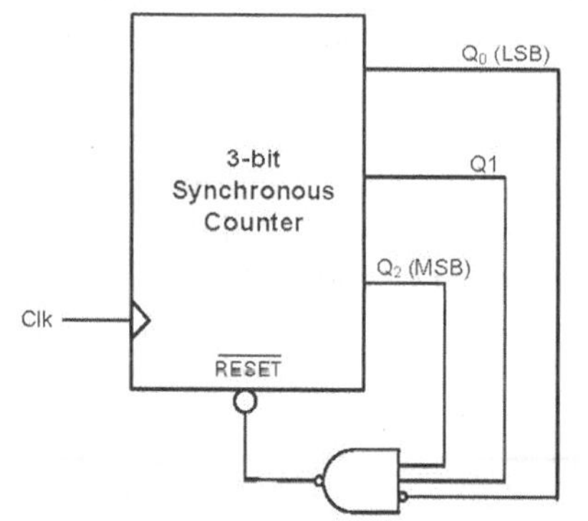

If the clock (Clk) frequency is 1 GHz, then the counter behaves as a

(*a*) mod-5 counter

(*b*) mod-6 counter

(*c*) mod-7 counter

(*d*) mod-8 counter

46. The first two rows in the Routh table for the characteristic equation of a certain closed-loop control system are given as

s^3	1	$(2K + 3)$
s^2	$2K$	4

The range of K for which the system is stable is

(*a*) $-2.0 < K < 0.5$

(*b*) $0 < K < 0.5$

(*c*) $0 < K < \infty$

(*d*) $0.5 < K < \infty$

47. A second-order linear time-invariant system is described by the following state equations

$$\frac{d}{dt}x_1(t) + 2x_1(t) = 3u(t)$$

$$\frac{d}{dt}x_2(t) + x_2(t) = u(t)$$

where $x_1(t)$ and $x_2(t)$ are the two state variables and $u(t)$ denotes the input. If the output $c(t) = x_1(t)$, then the system is

(*a*) controllable but not observable

(*b*) observable but not controllable

(*c*) both controllable and observable

(*d*) neither controllable nor observable

48. The forward-path transfer function and the feedback-path transfer function of a single loop negative feedback control system are given as

$$G(s) = \frac{K(s+2)}{s^2 + 2s + 2} \text{ and } H(s) = 1,$$

respectively. If the variable parameter K is real positive, then the location of the breakaway point on the root locus diagram of the system is _______.

49. A wide sense stationary random process X(*t*) passes through the LTI system shown in the figure. If the autocorrelation function of X(*t*) is $R_X(\tau)$, then the autocorrelation function $R_Y(\tau)$ of the output Y(*t*) is equal to

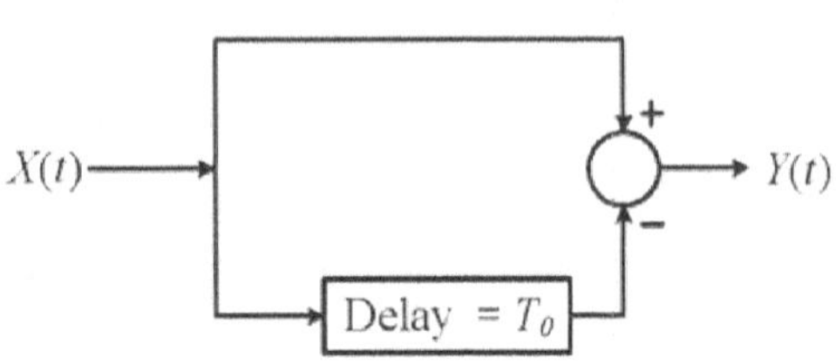

(*a*) $2R_X(\tau) + R_X(\tau - T_0) + R_X(\tau + T_0)$

(*b*) $2R_X(\tau) - R_X(\tau - T_0) - R_X(\tau + T_0)$

(*c*) $2R_X(\tau) + 2R_X(\tau - 270)$

(*d*) $2R_X(\tau) - 2R_X(\tau - 270)$

50. A voice-grade AWGN (additive white Gaussian noise) telephone channel has a bandwidth of 4.0 kHz and two-sided noise power spectral density $\frac{\eta}{2} = 2.5 \times 10^{-5}$ Watt per Hz. If information at the rate of 52 kbps is to be transmitted over this channel with arbitrarily small bit error rate, then the minimum bit-energy E_b (in mJ/bit) necessary is _______.

51. The bit error probability of a memoryless binary symmetric channel is 10^{-5}. If 10^5 bits are sent over this channel, then the probability that not more than one bit will be in error is _______.

52. Consider an air-filled rectangular waveguide with dimensions a = 2.286 cm and b = 1.016 cm. At 10 GHz operating frequency, the value of the propagation constant (per meter) of the corresponding propagating mode is _______

53. Consider an air-filled rectangular waveguide with dimensions a = 2.286 cm and b = 1.016 cm. The increasing order of the cut-off frequencies for different modes is

(*a*) $TE_{01} < TE_{10} < TE_{11} < TE_{20}$

(*b*) $TE_{20} < TE_{11} < TE_{10} < TE_{01}$

(*c*) $TE_{10} < TE_{20} < TE_{01} < TE_{11}$

(*d*) $TE_{10} < TE_{11} < TE_{20} < TE_{01}$

54. A radar operating at 5 GHz uses a common antenna for transmission and reception. The antenna has a gain of 150 and is aligned for maximum directional radiation and reception to a target 1 km away having radar cross-section of 3 m². If it transmits 100 kW, then the received power (in μW) is _______.

55. Consider the charge profile shown in the figure. The resultant potential distribution is best described by

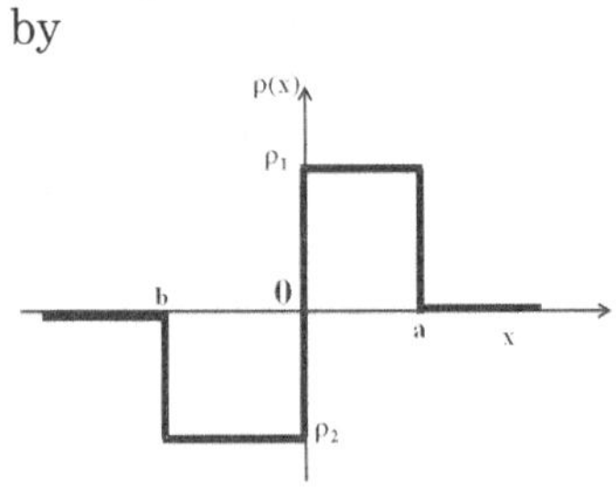

(*a*) 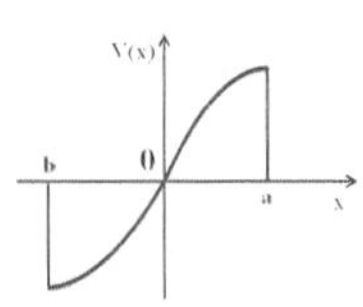

(*b*) 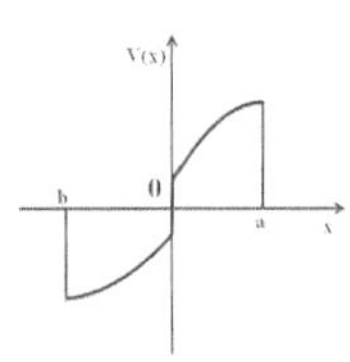

(*c*) 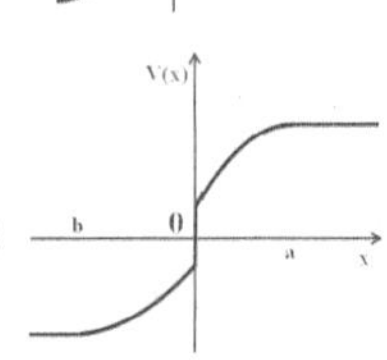

(*d*) 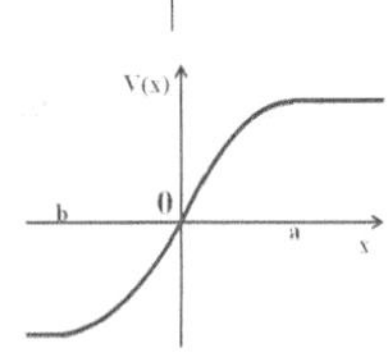

ANSWERS

GENERAL APTITUDE

1. (*c*)	2. (*c*)	3. (*b*)	4. (*c*)	5. (*a*)	6. (*d*)	7. (*a*)	8. (*c*)	9. (*b*)	10. (*a*)

TECHNICAL SECTION

1. (*d*)	2. (1)	3. (0.07203)	4. (2)	5. (0.06)	6. (*c*)
7. (*a*)	8. (*c*)	9. (*b*)	10. (*c*)	11. (*a*)	12. (21)
13. (0)	14. (0.5)	15. (5.65)	16. (*d*)	17. (*d*)	18. (*a*)
19. (*b*)	20. (*a*)	21. (3485)	22. (362.255)	23. (*c*)	24. (*b*)
25. (*c*)	26. (*a*)	27. (*d*)	28. (10)	29. (*b*)	30. (*b*)
31. (*d*)	32. (–1)	33. (–1)	34. (*a*)	35. (8)	36. (*c*)
37. (6.656)	38. (1.6)	39. (*a*)	40. (–1)	41. (*c*)	42. (57.8745)
43. (*b*)	44. (7)	45. (*d*)	46. (*d*)	47. (*a*)	48. (–3.41)
49. (*b*)	50. (31.503)	51. (0.735)	52. (158.07)	53. (*c*)	54. (0.0122)
55. (*c*)					

EXPLANATIONS

GENERAL APTITUDE

2. The meaning of underlined word grasping means clutching (or holding something tightly).

3. 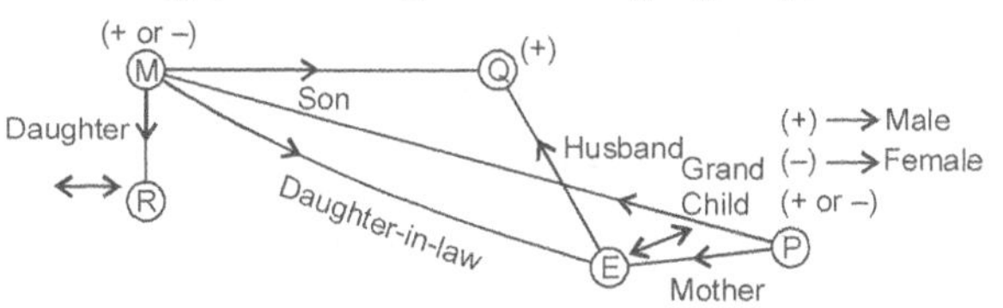

So, from the above relation diagram it is clear that P is the grandchild of M.

5. Here the speed of first train $(S_1) = \frac{120}{10} = 12$ m/s

and the speed of second train $(S_2) = \frac{150}{15} = 10$ m/s

$\therefore$ The magnitude of difference in the speeds of two trains $= (S_1 - S_2) = (12 - 10)$ m/s $= 2$ m/s

6. From the given figure.

The odometer reading increases from starting point to end point.

Odometer reading = Area of the given diagram

Area of the velocity and time graph per second

1st sec $\Rightarrow$ triangle

$$= \frac{1}{2} \times 1 \times 1 = \frac{1}{2}$$

2nd sec $\Rightarrow$ square

$$= 1 \times 1 = 1$$

3rd sec $\Rightarrow$ square + triangle

$$= 1 \times 1 + \frac{1}{2} \times 1 \times 1 = 1\frac{1}{2}$$

4th sec $\Rightarrow$ triangle

$$= \frac{1}{2} \times 1 \times 2 = 1$$

5th sec $\Rightarrow$ straight line

$$= 0$$

6th sec $\Rightarrow$ triangle

$$= \frac{1}{2} \times 1 \times 1 = \frac{1}{2}$$

7th sec $\Rightarrow$ triangle

$$= \frac{1}{2} \times 1 \times 1 = \frac{1}{2}$$

Total Odometer reading at 7 seconds

$$= \left(\frac{1}{2} + 1 + 1\frac{1}{2} + 1 + 0 + \frac{1}{2} + \frac{1}{2}\right) \text{m}$$

$$= 5 \text{ m}$$

So, the odometer reading is increased by 5 m.

8. Manu age = sravan age + 2 months

Manu age = Trideep age – 3 months

Pavan age = Sravan's age + 1 month

From the above statement

Trideep age > Man > Pavan > Sravan

Hence, Trideep can occupy the extra space in the flat.

9. 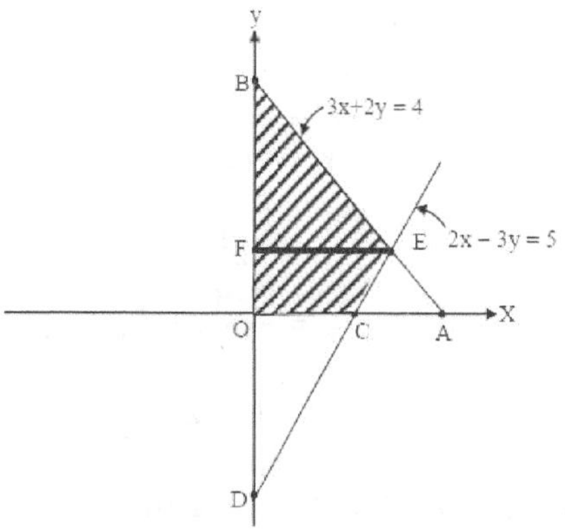

$$A = \left[\frac{14}{3}, 0\right]$$
$$B = [0, 7],$$
$$C = \left[\frac{5}{2}, 0\right]$$
$$D = \left[0, \frac{-5}{3}\right]$$
$$E = [4, 1],$$
$$F = [0, 1]$$

Required area = Area of Δ BFE + Area of quadrilateral FEOC

$$= \frac{1}{2} \times 4 \times 6 + \frac{1}{2} \times (4 + 2.5) \times 1$$
$$= 2 \times 6 + \frac{1}{2} \times 6.5$$
$$= 12 + 3.25$$
$$= 15.25 \text{ sq. units}$$

So, the area bounded by the given lines is 15.25 sq. units.

10. The equation of a straight line is

$$y = mx + c$$
$$m = \text{slope} = -0.02$$

set $(\log x, y)$

Now, the line intercepts the abscissa at

$$\log x = 0.1$$

So, at abscissa $y = 0$

$\therefore$ $y = mx + c$

$0 = -0.02 \times 0.1 + c$

$\therefore$ $c = 0.002$

$y = mx + c$

$y = -0.02 \times \log x + c$

At $x = 5$

$\therefore$ $y = -0.02 \times \log 5 + 0.002 = -0.030$

So, the value of y is -0.030.

TECHNICAL SECTION

1. From the given square matrix

$$\det(A) = \sigma^2 - \omega x$$
$$(\sigma + j\omega)(\sigma - j\omega) = \sigma^2 - \omega x$$
$$(\sigma)^2 - (j\omega)^2 = \sigma^2 - \omega x$$
$$\sigma^2 - (j^2)(\omega^2) = \sigma^2 - \omega x$$
$$\sigma^2 + \omega^2 = \sigma^2 - \omega x \quad (\text{since } j^2 = -1)$$
$$\omega^2 = -\omega x$$

$\therefore$ $x = -\omega$

2. Given $f(z) = \dfrac{\sin z}{z^2}$

$$= \frac{1}{z^2}\left\{z - \frac{z^3}{3!} + \frac{z^5}{5!} - \dots\dots\dots\right\}$$
$$= \frac{1}{z} - \frac{z}{3!} + \frac{z^3}{5!} - \dots\dots\dots$$

Residue of $f(z) = 1$ (at $z = 0$)

3. The probability of getting "head" for the first time in fifth loss

$$(P) = (0.7)^4 (0.3)$$
$$= 0.07203$$

4. Let $I = \displaystyle\int_0^1 \frac{dx}{\sqrt{1-x}}$

$$= \left[-2\sqrt{1-x}\right]_0^1$$
$$= -2[(0) - 1]$$

$\therefore$ $I = 2$

5. Given the first order initial value problem is

$$\frac{dy}{dx} = y + 2x - x^2$$
$$y(0) = 1$$
$$0 \le x < \infty$$

Given $f(x, y) = y + 2x - x^2$,

$x_o = 0,$

$y_o = 1,$

$h = 0.1$

$$k_1 = hf(x_o, y_o)$$
$$= 0.1(1 + 2(0) - 0^2)$$
$$= 0.1$$
$$k_2 = hg(x_0 + h, y_0 + k_1)$$
$$= 0.1((y_0 + k_1) + 2(x_0 + h) - (x_0 + h)^2)$$

(since $h = 0.1$)

$$= 0.1((1 + 0.1) + 2(0.1) - (0.1)^2)$$
$$= 0.1(1.1 + 0.2 - 0.01)$$
$$= 0.129$$

$\therefore$ $y_1 = y_0 + \dfrac{1}{2}(k_1 + k_2)$

$$= 1 + \frac{1}{2}(0.1 + 0.129)$$

(since $k_1 = 0.1$ and $k_2 = 0.129$)

$$= 1 + 0.1145 = 1.1145$$

For exact solution,

$$y(x) = x^2 + e^x$$
$$y(0.1) = (0.1)^2 + e^{0.1}$$
$$= 0.01 + 1.1050$$
$$= 1.1152$$
$$\text{error} = 1.1152 - 1.1145$$
$$= 0.00069$$

$$\text{Relative error} = \frac{\text{error}}{y(x)} = \frac{0.00069}{1.1152}$$
$$= 0.00062$$

Percentage Error = $(0.00062 \times 100)\% = 0.06\%$

6. Given signal

$$x(t) = \cos(6\pi t) + \sin(8\pi t)$$

Where t is in seconds

and $$y(t) = x(2t+5)$$

$$y(t) = \cos(12\pi t + 30\pi) + \sin(16\pi t + 40\pi)$$

$$f_{m_1} = 6$$

$$f_{m_2} = 8$$

$$\therefore \quad f_m = 8 \text{ Hz}$$

$$(f_S)_{min} = 2f_m = 16 \text{ Hz}$$

Thus the Nyquist sampling rate is 16 Hz.

7. Here the convolution of two sinc pulses is sinc pulse.

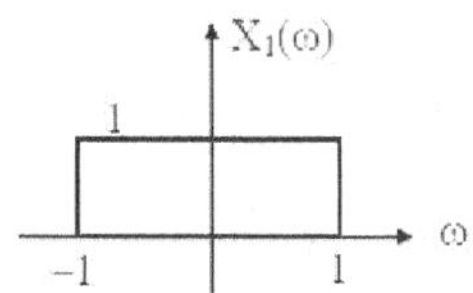

So, $$x_1(t) = \frac{\sin t}{\pi t}$$

Now $$x(t) = x_1(t) * x_1(t)$$

$$X(\omega) = X_1(\omega)\cdot X_1(\omega) = X_1(\omega)$$

$$\therefore \quad x(t) = x_1(t) = \frac{\sin t}{\pi t}$$

So, the value of $x(t)$ is $\frac{\sin t}{\pi t}$

8. Here $(a)^n x(n) \leftrightarrow X(z/a)$

$$a = -1$$

$$(-1)^n x(n) \leftrightarrow X(-z)$$

but $$x(n) = \delta[n-3] + 2\delta[n-5]$$

$$y(n) = (-1)^n x(n)$$

$$= (-1)^n [\delta(n-3) + 2\delta[n-5]$$

$$\therefore \quad y(n) = -\delta(n-3) - 2\delta(n-5)$$

$$= -x(n)$$

Hence, the value of signal $y(n)$ is $-x(n)$.

9. The input voltage is given by

$$V_i(t) = 2\cos 200t + 4\sin 500t$$

Let us apply Superposition theorem only consider 2cos 200*t*, then circuit becomes

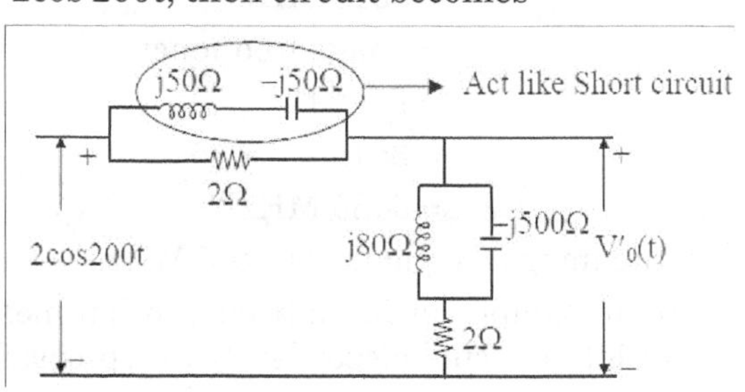

So, $$V_0'(t) = 2\cos 200t$$

Now consider only 4 sin 500*t*, then circuit becomes

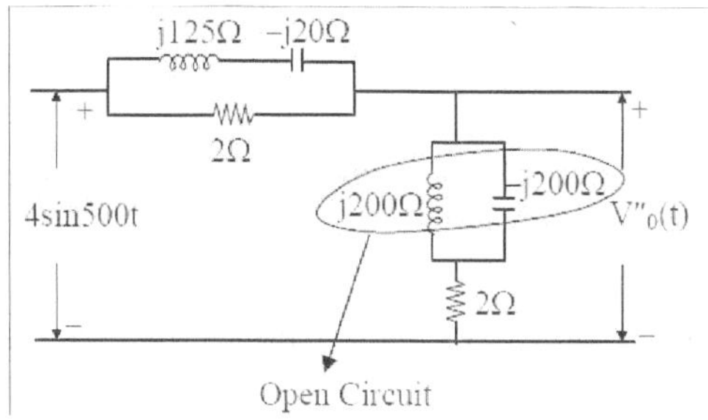

So, again $V_0''(t) = 4\sin 500t$

Finally (according to superposition theorem)

$$V_0(t) = V_0'(t) + V_0''(t)$$

$$\therefore \quad V_0(t) = 2\cos(200t) + 4\sin(500t)$$

10. From the given fig :

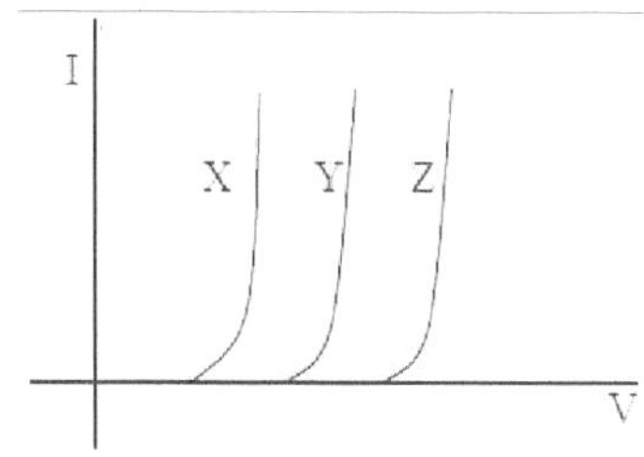

$$V_{r_3} > V_{r_2} > V_{r_1}$$

$$V_r \propto E_g$$

Where V_r is cut-in voltage

So, $$E_{gZ} > E_{gY} > E_{gX}$$

12. $$\text{Fill factor} = \frac{P_{max}}{P_T} = \frac{P_{max}}{I_{SC}\cdot V_{OC}}$$

where P_{max} = Maximum power

P_T = Theoretical power $= I_{SC}\cdot V_{OC}$

$$0.7 = \frac{P_{max}}{180\times 10^{-3}\times 0.5}$$

$$\therefore \quad P_{max} = 63\times 10^{-3} \text{ W}$$

Now Maximum efficiency

$$(\eta_{max}) = \frac{P_{max}}{P_{in}}$$

$$= \left(\frac{63\times 10^{-3}\,\text{W}}{100\times 10^{-3}\,\dfrac{\text{W}}{\text{cm}^2}\times 3\text{cm}^2}\times 100\right)\%$$

$$= 21\%$$

Thus the maximum efficiency (in %) of the device is 21%.

13. 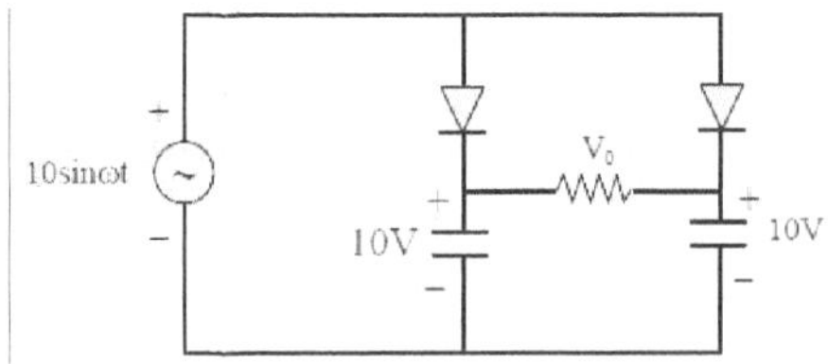

Diodes are ideal therefore during Positive cycle of output voltage

$$(V_0) = 10 - 10 = 0 \text{ V}$$

During Negative cycle, the diodes are Reverse biased so output

$$(V_0) = 0 \text{ V}$$

$$\therefore \quad V_0 = 0 \text{ V (always)}$$

So, the output voltage V_0 (in volt) at steady state is 0 V.

14. Here

$$V_{E_1} = (2.5 - 0.7) = 1.8 \text{ V}$$

$$V_{B_2} = V_{E1} - V_{EB2} = (1.8 - 0.7) = 1.1 \text{ V}$$

$$I_{B_2} = \frac{V_{B_2} - 1}{10k} = \frac{1.1 - 1}{10k} = \frac{0.1}{10k}$$

$$I_{C2} = \beta I_{B2} = 50\left[\frac{0.1}{10k}\right]$$

$$\therefore \quad V_{C2} = I_{C2}(1K) = \frac{50(0.1)}{10k}(1k) = 0.5 \text{ V}$$

Hence, the value of the dc voltage V_{C2} is 0.5V.

15. 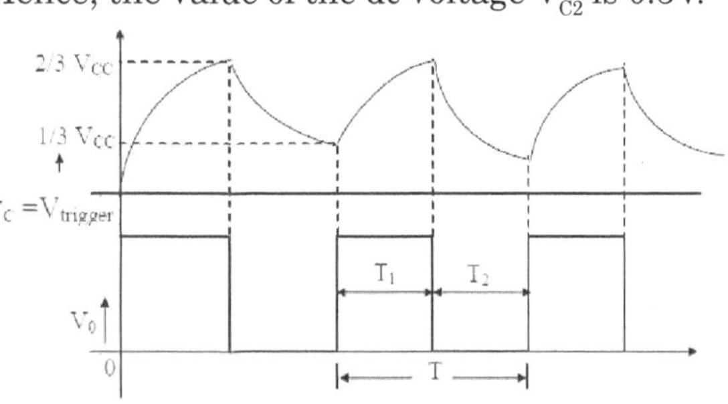

From the above figure

$$T_1 = 0.693\,(R_A + R_B)C$$
$$= 0.693\,(2.2k + 4.7k)\,0.022\,\mu$$
$$= 0.693 \times 6.9 \times 10^3 \times 0.022 \times 10^{-6}$$
$$= 0.1052 \text{ m sec}$$

$$T_2 = 0.693\,R_B C$$
$$= 0.693\,(4.7\,k)\,0.022\,\mu$$
$$= 0.693 \times 4.7 \times 10^3 \times 0.022 \times 10^{-6}$$
$$= 0.0716562 \text{ m sec}$$

Total period $T = T_1 + T_2$

$$= 0.1768562 \text{ m sec}$$

Frequency of oscillations

$$(f) = \frac{1}{T} = 5.65 \text{ kHz}$$

16. Given

		CY
$A = A7_H =$	1010 0111	0

After executing instruction

RLC ⇒ A =	0100 1111	1

$A = 4F_H$ and CY = 1

17. From given fig. it is a AND gate

Input		Output
A	B	Y
0	0	0
0	1	0
1	0	0
1	1	1

18. The minimum no. of NAND gates required for 2-input EX-OR gate = 4

19. From the given block diagram, we have

$$\frac{Y(s)}{X(s)} = G(s) = \frac{G_1G_2}{1 + G_1H_1 + G_2G_2}$$

20. Given $G(s) = \dfrac{2}{s(s+1)}$ and $H(s) = 1$

⇒ Type-1 System, to the unit step input the $e_{ss} = 0$

So, the steady state error e_{ss} due to a unit step input is 0.

21. 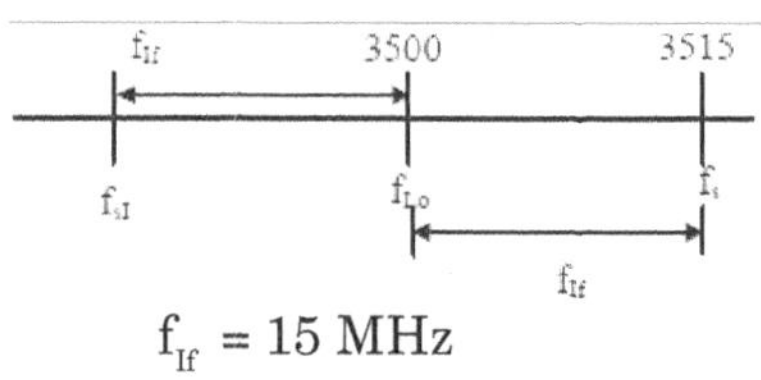

$$f_{If} = 15 \text{ MHz}$$
$$f_{Lo} = 3500 \text{ MHz}$$
$$f_s - f_{Lo} = f_{If}$$
$$f_s = f_{Lo} + f_{If} = 3515 \text{ MHz}$$
$$f_{si} = \text{image frequency} = f_s - 2f_{If} = 3515 - 2 \times 15 = 3485 \text{ MHz}$$

So, the image frequency is 3485 MHz.

22. Here the samples are quantized into four message symbols that occur independently with probabilities are given by

$P_1 = 0.125$
$P_4 = 0.125$
$P_2 = 0.375$
$P_3 = 0.375$

$$H = 0.125\log_2 \frac{1}{0.125} + 0.125\log_2 \frac{1}{0.125} + 0.375\log_2 \frac{1}{0.375} + 0.375 \times \log_2 \frac{1}{0.375}$$

$$= 0.125\log_2 8 + 0.125\log_2 8 + 0.375\log_2\left(\frac{8}{3}\right) + 0.375\log_2\left(\frac{8}{3}\right)$$

Information rate = H × 200
= 1.811 × 200
= 362.255

23. The graphical representation of the matched filter output $y(t)$ for this signal is given by

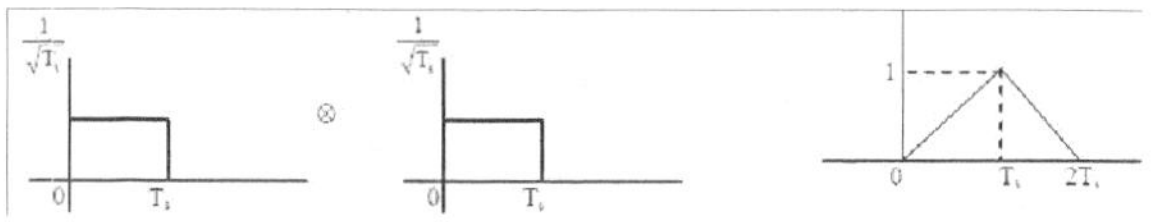

24.

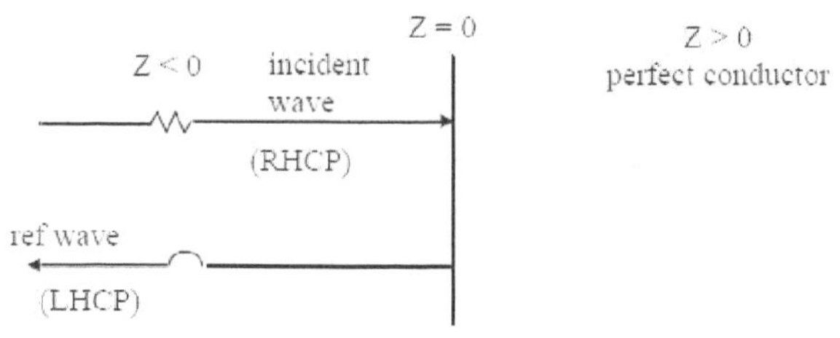

Here RHCP → right handed circularly polarized
LHCP → Left-handed circularly polarized

If the wave is incident on perfect conductor then reflection coefficient is given by

$$\Gamma \equiv \frac{E_{r_0}}{E_{i_0}} = -1$$

$$E_{r_0} = E_{i_0} \angle 180°$$

If incident wave is traveling along + Z direction then the reflected wave will be traveling along – Z direction.

Thus, the reflected wave is left hand circularly polarized (LHCP).

25. Faraday's law of electromagnetic induction is mathematically described by the equation is given by

$$\nabla \times \vec{E} = -\frac{\partial \vec{B}}{\partial t}$$

or

$$\nabla \times \vec{E} = -\mu\frac{\partial \vec{H}}{\partial t}$$

26. The differential equation is given by

$$\frac{d^2y}{dx^2} + 12\frac{dy}{dx} + 36y = 0$$

$\Rightarrow D^2 + 12D + 36 = 0$

$(D+6)^2 = 0$

$\therefore \quad D = -6, -6$

The solution is given by

$$y = C_1e^{-6x} + C_2xe^{-6x} \quad \ldots(i)$$

$y(0) = 3$

$\Rightarrow \quad 3 = C_1$

$C_1 = 3$

$(i) \Rightarrow \quad y = 3e^{-6x} + C_2xe^{-6x}$

$$\frac{dy}{dx} = -18e^{-6x} + C_2\left\{-6xe^{-6x} + e^{-6x}\right\}$$

$$\Rightarrow \quad \left.\frac{dy}{dx}\right|_{x=0} = -18 + C_2$$

$\Rightarrow \quad -36 = -18 + C_2$

$\Rightarrow \quad C_2 = -18$

$\therefore$ The solution of $y = 3\,e^{-6x} - 18\,x\,e^{-6x}$

(Since $C_1 = 3$ & $C_2 = -18$)

27. The equation is given by

$$u = x_1e_1 + x_2e_2 + x_3e_3 \quad \ldots(i)$$

Putting the value of u, e_1, e_2 & e_3 in the given equation (i), we get

$$(4, 3, -3) = x_1(1, 0, 2) + x_2(0, 1, 0) + x_3(-2, 0, 1)$$

$x_1 - 2x_3 = 4 \rightarrow (a)$

$x_2 = 3 \rightarrow (b)$

$2x_1 + x_3 = -3 \rightarrow (c)$

On solving these equations, we get

$$x_1 = -\frac{2}{5},$$

$$x_2 = 3,$$

$$x_3 = \frac{-11}{5}$$

$$\therefore \quad u = -\frac{2}{5}e_1 + 3e_2 - \frac{11}{5}e_3$$

28.

$$\text{Volume} = \iint z\,dx\,dy$$

$$= \int_{x=0}^{3}\int_{y=0}^{\frac{2}{3}x}(6 - x - y)\,dx\,dy = 10$$

$$\left(\text{since } 2x = 3y,\ y = \frac{2}{3}x\ \&\ x = 3\right)$$

$x + y + z = 6$

$\therefore \quad z = 6 - x - y$

29. (*i*) When the point $z_0 = 2$ inside the contour *c*.

$$\frac{1}{2\pi j}\oint_c \frac{e^z}{(z-2)}dz = \frac{1}{2\pi j}2\pi j f(2)$$

$\Rightarrow \quad e^2 = 7.39$

(*ii*) When the point $z_0 = 2$ outside the contour *c*.

$$\frac{1}{2\pi j}\oint_c \frac{e^z}{(z-2)}dz = 0 \; (\because \; z = 2 \text{ lies out side } c)$$

30. The transfer function

$H(e^{j\omega}) = e^{j\omega} + e^{-j\omega} = 2\cos\omega$

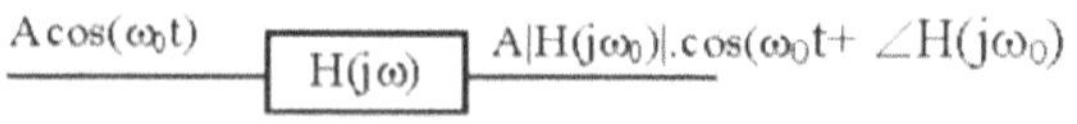

Here $\quad x(t) = 2\cos\left(\frac{2\pi}{3}t\right)$

$$\omega_0 = \frac{2\pi}{3}$$

$$H(j\omega_0) = 2\cos\left(\frac{2\pi}{3}\right) = 2\left(\frac{-1}{2}\right) = -1$$

$$y(t) = 2\cos\left(\frac{2\pi}{3}t + 180°\right)$$

$$x(t) = \cos \pi t$$

$$\omega_0 = \pi$$

$$H(j\omega_0) = 2\cos(\pi) = -2$$

$$y(t) = 2\cos(\pi t + 180°)$$

$$y(t) = 2\cos\left(\frac{2\pi}{3}t + \pi\right) - 2\cos(\pi t + \pi)$$

$$\omega_1 = \frac{2\pi}{3} \qquad \omega_2 = \pi$$

$$T_1 = 3 \qquad T_2 = 2$$

$$T_0 = 6$$

$$\omega_0 = \frac{2\pi}{T_0} = \frac{\pi}{3}$$

$$y(t) = 2\cos(2\omega_0 t + \pi) - 2\cos(3\omega_0 t + \pi)$$

$$y(t) = e^{j(2\omega_0 t+\pi)} + e^{-j(2\omega_0 t+\pi)} - e^{j(3\omega_0 t+\pi)} - e^{-j(3\omega_0 t+\pi)}$$

$$y(t) = -e^{j(2\omega_0 t)} - e^{-j(2\omega_0 t)} + e^{j(3\omega_0 t)} + e^{-j(3\omega_0 t)}$$

$$C_3 = 1$$

Thus the value of C_3 is 1.

31. $x(y) = (2)^n u(n) + \left(\frac{1}{2}\right)^n u(-n-1)$

ROC = $(|z| > 2) \cap (|z| > ½) = \phi$

So, the ROC of z-tansform is null.

32. Circuit at $t = 0^-$, the direction of current and correct component was not given.

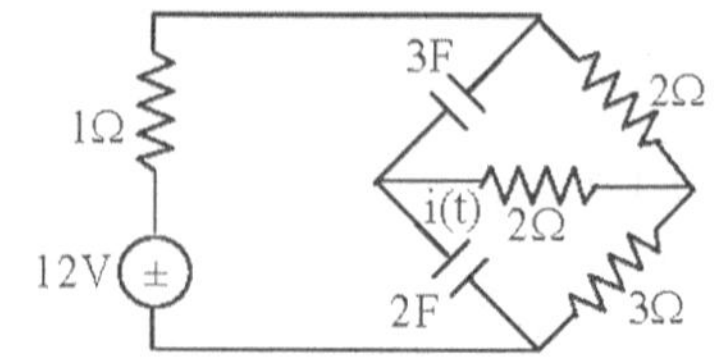

Not $t = 0^+$

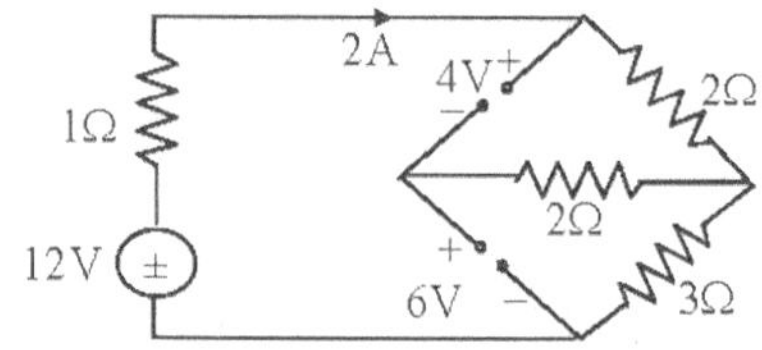

So, $\quad i(0^+) = \frac{-4}{4} = -1A$

∴ The magnitude of he current is 1 ampere (negative).

33.

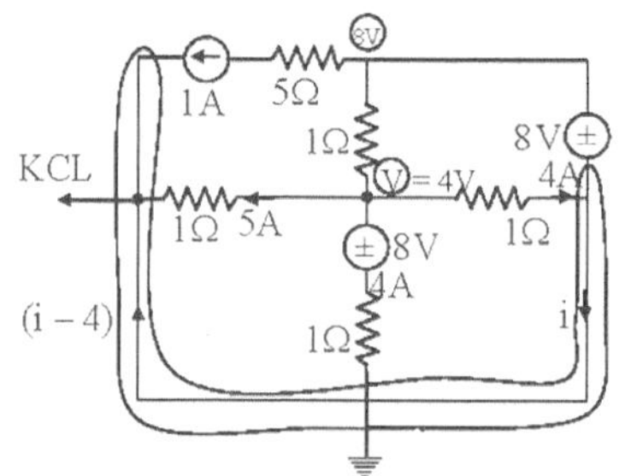

From above fig.

Nodal Analysis

$$\frac{(V-8)}{1} + \frac{V}{1} + \frac{(V-8)}{1} + \frac{V}{1} = 0$$

$$4V = 16$$

$$V = 4 \text{ Volts}$$

Now from KCL

$$i - 4 + 4 + 1 = 0$$

$$i = -1A$$

Hence the current (*i*) in the given circuit is –1 Amp.

34. Given network is in lattice form

Where $\quad Z_a = 3\Omega$

$Z_b = 0\Omega$

$Z_c = 0\Omega$

$Z_d = 6\Omega$

But it is not symmetrical & balanced

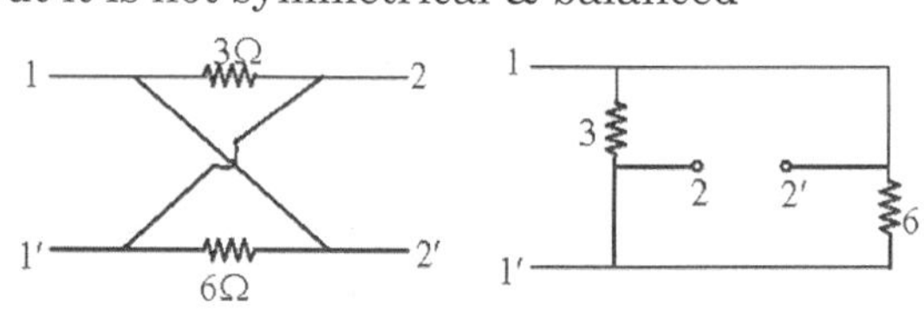

So, $Z_{11} = \left.\frac{V_1}{I_1}\right|_{I_2=0} = 3//6 = 2\Omega$

$Z_{21} = \left.\frac{V_2}{I_1}\right|_{I_2=0}$

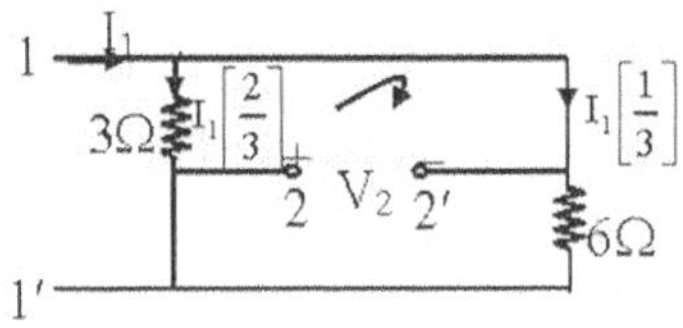

KVL

$-V_2 - 2I_1 + 0 = 0$

$V_2 = -2I_1$

$\Rightarrow \quad Z_{21} = -2\Omega$

Also

$Z_{22} = \left.\frac{V_2}{I_2}\right|_{I_1=0} = 3//6 = 2\Omega$

$Z_{21} = \left.\frac{V_1}{I_2}\right|_{I_1=0}$

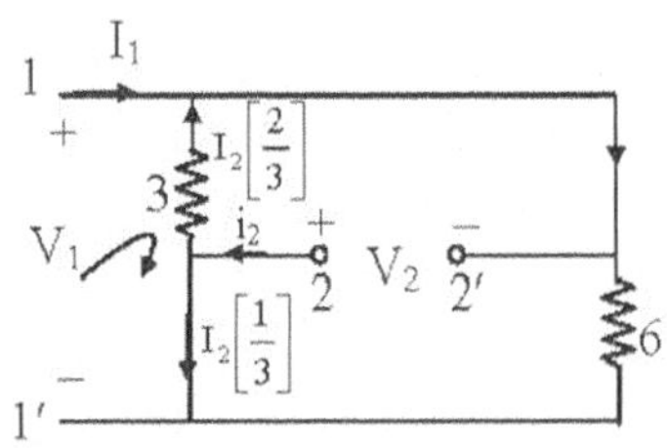

KVL

$-V_1 - 2I_2 = 0$

$V_1 = -2I_2$

$\Rightarrow \quad Z_{12} = -2\Omega$

So, the Z-parameter matrix for the two port network is given by

$$|Z| = \begin{bmatrix} 2 & -2 \\ -2 & 2 \end{bmatrix}\Omega$$

35. The number of complex multiplications required

for DIF-FFT = $\left(\frac{N}{2}\log_2 N\right)$

$\therefore \quad \left(\frac{N}{2}\log_2 N\right)(20\ \mu\text{ sec}) = 125\ \mu\text{sec}$

Solving the above equation, we get N = 8

36. From the given figure

$$y(n) = 5[x(n) - x(n-2)]$$
$$Y(e^{j\omega}) = 5[1 - e^{-2j\omega}]X(e^{j\omega})$$
$$H(e^{j\omega}) = 5[1 - e^{-2j\omega}]$$

ω	$\lvert H(e^{j\omega})\rvert$
0	0
$\frac{\pi}{2}$	10
π	0

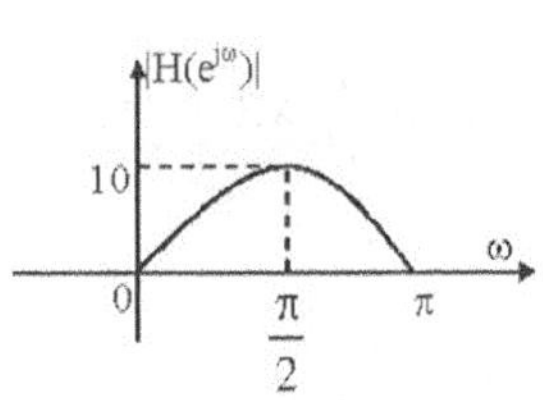

Thus, it is a Band pass filter

37. We know that

Collector current $(I_C) = AeD_n\frac{dn}{dx} = Ae\mu_n V_T\frac{dn}{dx}$

Where

A = Area for the emitter-base function

e = Electronic charge

V_T = Thermal voltage

$$I_C = 0.001 \times 1.6 \times 10^{-19} \times 800 \times 0.026 \times \left(\frac{10^4 - 0}{0.5 \times 10^{-4}}\right)$$

$I_C = 6.656$ mA

Hence the collector current (I_C) is 6.656 mA.

38. We known that

Now $\quad C = \frac{\varepsilon A}{d}$

Where C = Capacitance

A = Area of a capacitor

ε = dielectric constant

Now $\quad C = \frac{C_1C_2}{C_1 + C_2}$

$$= \left[\frac{\frac{4}{1\times10^{-9}} \times \frac{20}{3\times10^{-9}}}{\frac{4}{1\times10^{-9}} + \frac{10}{3\times10^{-9}}}\right] 8.8521\times10^{-12}$$

$= 2.5 \times 10^9\ \varepsilon_0$

$C = \frac{\varepsilon_r\varepsilon_0}{t_{eq}}$

$$t_{eq} = \frac{\varepsilon_r\varepsilon_0}{2.5\times10^9\ \varepsilon_0} = \frac{4\times\varepsilon_0}{2.5\times10^9\ \varepsilon_0}$$

$= 1.6 \times 10^{-9}$ m

$= 1.6$ nm

So, the value of thickness (t_{eq}) is 1.6 nm.

39.

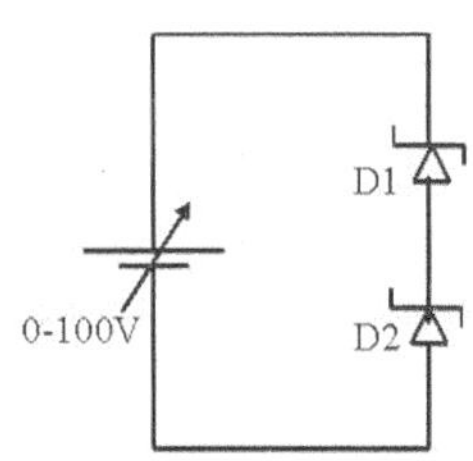

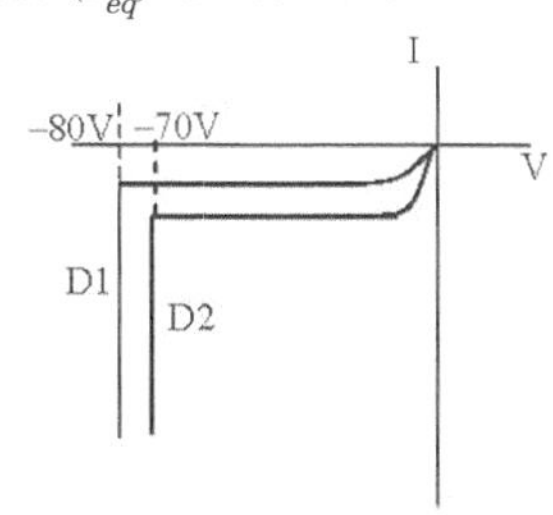

From the above fig., it is clear that both zener diodes are in reverse biased

$V_{BZ_1} = 80$ V

$V_{BZ_2} = 70$ V

D_1 have list saturation current.

When we will vary the voltage above 80V

D_1 get breaks down & will replaced by 80V. & through it '∞' current can flow through it.

But because of D_2 we will take minimum current i.e. net current equals to reverse saturation current of D_2 as we know.

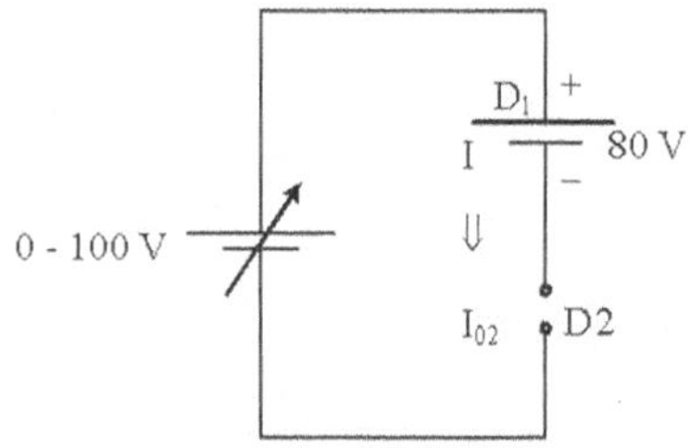

The diode have least saturation will break down first and it will replaced by its break down voltage and the net current equal upto other diode reverse saturation current.

40.

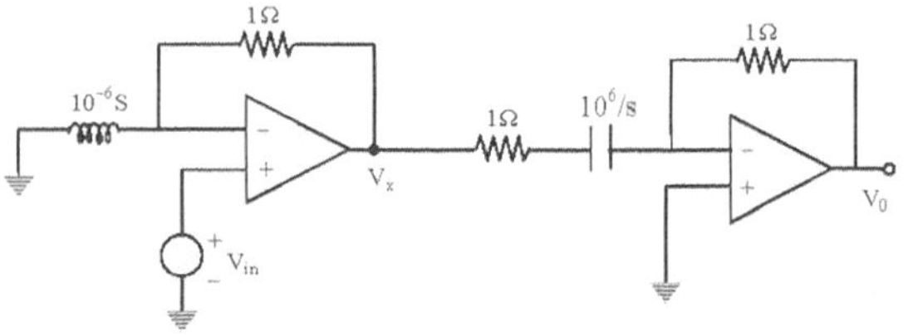

From above fig.

$$\frac{V_x}{V_{in}} = 1 - \frac{1}{10^{-6}s} = 1 + \frac{10^6}{s} = \frac{s+10^6}{s}$$

$$V_0 = \left[\frac{-1}{1+\frac{10^6}{s}}\right]V_x = \frac{V_0}{V_x} - \frac{-s}{s+10^6}$$

$$\therefore \quad \frac{V_0}{V_{in}} = \frac{V_0}{V_x}\cdot\frac{V_x}{V_{in}} = \left[\frac{-s}{s+10^6}\right]\left[\frac{s+10^6}{s}\right] = -1$$

$\therefore$ overall voltage gain $\left(\frac{V_0}{V_{in}}\right) = -1$

41. The ac small signal voltage gain

$$\frac{V_0}{V_{in}} = g_{m1}\left[r_{01} \,||\left(\frac{1}{g_{m2}} \,||\, r_{02}\right)||\, r_{03}\right]$$

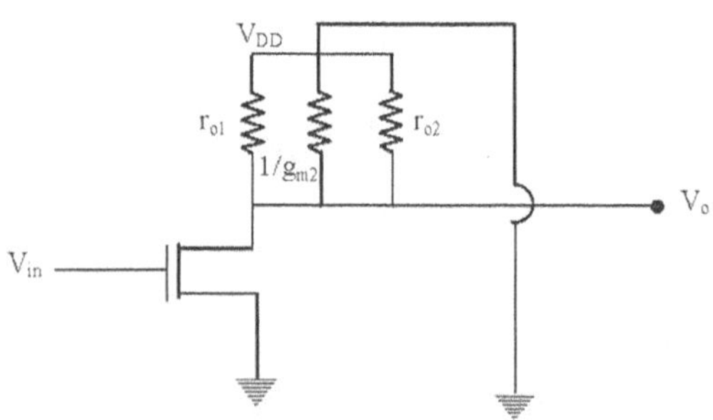

42.

$$C_{M1} = 50PF\,[1 - A_V]$$
$$A_V = -g_m R_D$$
$$= -0.01 \times 1$$
$$A_V = -10$$
$$C_{Mi} = 50PF\,[1 + 10]$$
$$= 0.55 \times 10^{-9}\,F$$
$$= 0.55\,nF$$

$$f_p = \frac{1}{2\pi R_i C_{mi}}$$

$$f_p = \frac{1}{2\pi \times 5K \times 0.55\,mF}$$

$$= \frac{1}{2\pi \times 5 \times 10^3 \times 0.55 \times 10^{-9}}$$

$$= 57.8745\text{ kHz}$$

So, the small signal input pole frequency is 57.8745 kHz.

43. The minimum sum-of-product (SOP) expression for the function

$$= \bar{Q}.S.\bar{X} + Q.\bar{S}.X$$

44. The maximum propagation delay of the circuit

$= t_{pd,\,NOR} + t_{pd,\,mux} + t_{pd,\,NOR} + t_{pd,mux}$

$= (2 + 1.5 + 2 + 1.5)$ ns

$= 7$ ns

45. At 6th Clock pulse

$Q_2Q_1Q_0 = 110$

It makes NAND gate output '0' at 8 ns due to its delay. At that time counter receives 7th, 8th Clock pulse and counts 111, 000. Hence, it is mod–8 counter.

46.

s^3	1	$2k+3$
s^2	$2k$	4
s^1	$\frac{2k(2k+3)-4}{2k}$	>0 for stability
s^0	4	

$4k^2 + 6k - 4 > 0$

$4k^2 + 8k - 2k - 4 > 0$

$4k(k+2) - 2(k+2) > 0$

$(4k-2)(k+2) > 0$

$(4k-2) > 0,\ k+2 > 0$

$k > -2,\ k > 0.5$

$0.5 < k < \infty$

So the range of k for which the system is stable is $0.5 < k < \infty$.

47. Given $\frac{d}{dt}x_1(t) + 2x_1(t) = 3u(t)$

$\therefore \quad \dot{x}_1 = -2x_1 + 3U$

and $\frac{d}{dt}x_2(t) + x_2(t) = u(t)$

$\therefore \quad \dot{x}_2 = -x_2 + U$

$c = x_1$

$$\begin{bmatrix} \dot{x}_1 \\ \dot{x}_2 \end{bmatrix} = \begin{bmatrix} -2 & 0 \\ 0 & -1 \end{bmatrix}\begin{bmatrix} x_1 \\ x_2 \end{bmatrix} + \begin{bmatrix} 3 \\ 1 \end{bmatrix}U$$

$$[c] = [1 \quad 0]\begin{bmatrix} x_1 \\ x_2 \end{bmatrix}$$

Hence by applying Gilbert's test, the system is controllable but not observable.

48. Given transfer function

$$G(s) = \frac{k(s+2)}{(s^2+2s+2)}, \quad H(s) = 1$$

At break away point $\Rightarrow \frac{dk}{ds} = 0$

$$\frac{d}{ds}\left(\frac{s+2}{s^2+2s+2}\right) = 0$$

$$\left[\frac{(s^2+2s+2)\frac{d}{ds}(s+2)-(s+2)\frac{d}{ds}(s^2+2s+2)}{(s^2+2s+2)^2}\right]$$

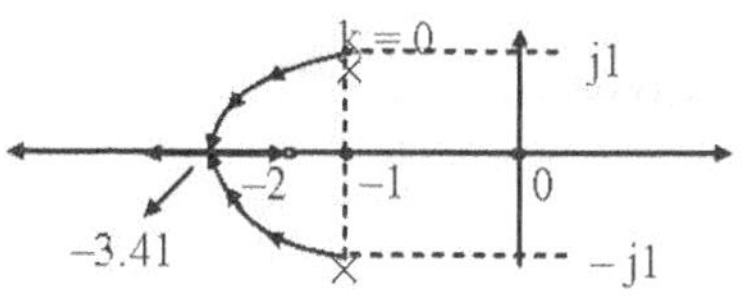

$$\Rightarrow \left[\frac{1(s^2+2s+2)-(s+2)(2s+2)}{(s^2+2s+2)^2}\right] = 0$$

$\Rightarrow -s^2 - 4s - 2 = 0$

$\Rightarrow -0.58, -3.41$

Valid BAP is –3.41

49. $Y(t) = X(t) - X(t - T_0)$

Autocorrelation function for o/p

$$= R_y(\tau) = E[y(t)\, Y(t+\tau)]$$

$$R_y(\tau) = E\,[(X(t) - X(t-T_0)]\,[X(t+\tau) - X(t+\tau-T_0)]$$

$$R_y(\tau) = E\,[(X(t)\,X(t+\tau) - X(t)\,X(t+\tau-T_0) - X(t-T_0)\,X(t+\tau) + X(t-T_0)\,X(t+\tau-T_0)]$$

$$R_y(\tau) = [R_x(\tau) - R_x(\tau - T_0) - R_x(\tau + T_0) + R_x(\tau)]$$

$$R_y(\tau) = 2\,R_x(\tau) - R_x(\tau - T_0) - R_x(\tau + T_0)$$

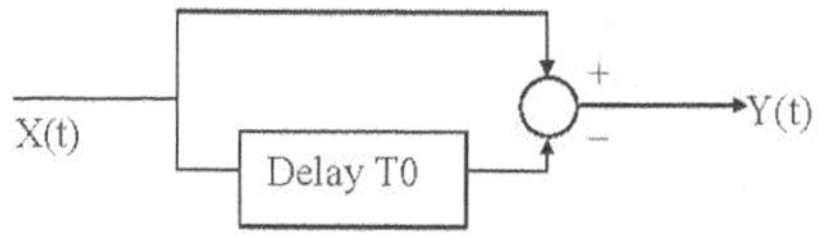

So, the autocorrection function of $R_y(\tau)$ of the output $Y(t)$ is $2R_x(\tau) - R_x(\tau - T_0) - R_x(\tau + T_0)$.

50. From given data

Channel transmission rate (C) = 52 kbps

Channel band width B = 4 kHz

$$\frac{\eta}{2} = 2.5 \times 10^{-5}$$

$$N = 4 \times 10^3 \times 2.5 \times 10^{-5} \times 2$$

$$C = B\log_2\left[1 + \frac{S}{N}\right]$$

$$S = 1638.2$$

$$E_b = \frac{S}{R_b} = \frac{J/\text{sec}}{\text{bits/sec}} = 31.503$$

$$\frac{C}{B} = \log_2(1 + S/N)$$

$\Rightarrow \log_2(1 + S/N) = \frac{C}{B}$

$\Rightarrow (1 + S/N) = 2^{C/B} = 2^{52/4} = 2^{13} = 8192$

$\Rightarrow S/N = 8191$

$\therefore\ S = 8191 \times N$

$\Rightarrow S = 8191 \times 4 \times 10^3 \times 2.5 \times 10^{-5} \times 2$

$\quad = 819.1 \times 2$

$$E_b = \frac{819.1 \times 2}{R_b} = 31.503$$

So the minimum bit energy (E_b) is 31.503 mJ/bit.

51. $P = 10^{-5}\ N = 10^5$

Given question can be solved by two methods.

Method 1 :

Binomial : ${}^nC_x\, p^x\, q^{n-x}$

$$P[x=0] + P[x=1] = {}^{10^5}C_0(10^{-5})^0(1-10^{-5})^{10^5} + {}^{10^5}C_1(10^{-5})^1(1-10^{-5})^{10^5-1}$$

$$= (1)\,(1) \times 0.367 + 0.367 = 0.735$$

Method 1 :

$$\text{Poisson} = \frac{e^{-\lambda}\lambda^x}{x!}$$

$$\lambda = np = 10^{-5} \times 10^5 = 1$$

(since $n = 10^5$, $p = 10^{-5}$)

$$= \frac{e^{-1}(1)^1}{1!} + e^{-1}$$

$\Rightarrow \quad 2 \times e^{-1} = 0.735$

So, the probability that not more than one bit will be in error is 0.735.

52. Given

Air filled rectangular waveguide

a = 2.286 cm

b = 1.016 cm

f = 10 GHz

Assume dominant mode (TE_{10}) is propagating in the waveguide,

So cut-off frequency of TE_{10} mode is given by

$$f_c(TE_{10}) = \frac{c}{2a} = \frac{3 \times 10^{10}}{2 \times 2.286}$$

$$f_c = 6.56 \text{ GHz}$$

propagation constant $\bar{\gamma}$ is given by

$$\bar{\gamma} = j\beta$$

$$= i\omega\sqrt{\mu_0 \epsilon_0}\sqrt{1-\left(\frac{f_c}{f}\right)^2}$$

$$= j2\pi\times10\times10^9\times\frac{1}{3\times10^8}\sqrt{1-\left(\frac{6.56}{10}\right)^2}$$

$$\bar{\gamma} = j158.07 \text{ m}^{-1}$$

Therefore the value of propagation constant is given by

$$|\bar{\gamma}| = 158.07 \text{ m}^{-1}$$

53. Given

a = 2.286 cm

b = 1.016 cm

air filled rectangular waveguide

$$f_{c(TE_{11})} = \frac{c}{2}\sqrt{\frac{1}{a^2}+\frac{1}{b^2}} \quad (\text{Since } m = 1, n = 1)$$

$$= \frac{3\times10^{10}}{2}\sqrt{\left(\frac{1}{(2.216)^2}+\frac{1}{1.016^2}\right)}$$

$$f_{c(TE_{11})} = 16.15 \text{ GHz}$$

$$f_{c(TE_{01})} = \frac{c}{2b} = \frac{3\times10^{10}}{2\times1.016} = 14.76 \text{ GHz}$$

$$f_{c(TE_{20})} = \frac{c}{a} = \frac{3\times10^{10}}{2.286} = 13.12 \text{ GHz}$$

$$f_{c(TE_{10})} = \frac{c}{2a} = \frac{3\times10^{10}}{2\times2.286} = 6.56 \text{ GHz}$$

∴ Increasing order of the cut-off frequency is given by

$$TE_{10} < TE_{20} < TE_{01} < TE_{11}$$

54. Given

frequency, $f = 5$ GHz $= 5\times10^9$ Hz

wave length, $\lambda = \frac{c}{f} = \frac{3\times10^8}{5\times10^9} = 0.06$m

gain of antenna, G = 150

Range of target, $R_{max} = 1$ km $= 10^3$ m,

radar cross-section, $\sigma = 3\text{m}^2$,

transmitted power, $P_t = 100$ kW

Radar range equation is given by

$$R_{max} = \left[\frac{P_t\times G\times\frac{\lambda^2}{4\pi}\times G\times\sigma}{(4\pi)^2\times P_R}\right]^{\frac{1}{4}} \quad \left(\text{Since } A_e = \frac{\lambda^2}{4\pi}G\right)$$

The received power, P_R is given by

$$P_R = \frac{100\times10^3\times150\times150\times(0.06)^2\times3}{(4\pi)^3\times(10^3)^4}$$

$$= 1.22\times10^{-8} = 0.0122\ \mu\text{W}$$

Hence, the received power is 0.0122 μW.

55. Let us consider $a = 1$ and $b = -1$

For line (1) :

Now the equation of the line to point (–1, 0) and (0, –1) is given by

$$y - 0 = \frac{-1}{1}(t-1)$$

$$y = -t-1$$

$$\int_{-1}^{t} -(t+1)\,dt = \frac{-(t+1)^2}{2} \quad 1 < t < 0$$

For line (2) :

The equation of the line to the point (0, –1) and (1, 0) is given by

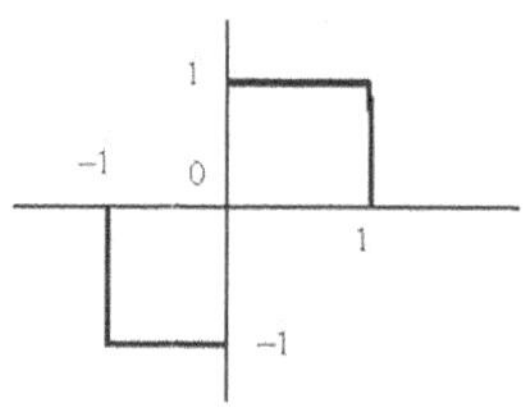

$$y-(-1) = \frac{1}{1}(t-0)$$

$$y+1 = t$$

$$y = t-1$$

$$\int_{0}^{t}(t-1)\,dt = \frac{(t-1)^2}{2} \quad 0 < t < 1$$

At $t = 0$:

$$y = \frac{-(t+1)^2}{2}$$

$$y = \frac{-1}{2}$$

After Ist integration

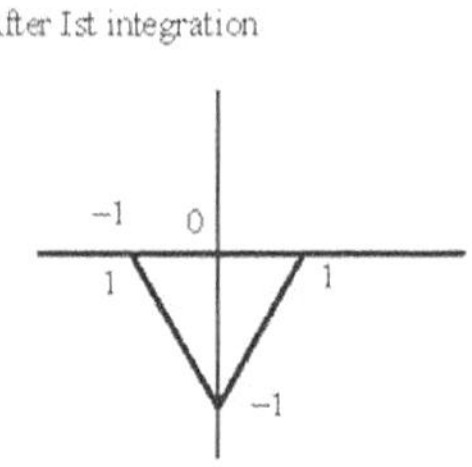

At $t = 0^+$:

$$y = \frac{(t-1)^2}{2}$$

$$y = \frac{1}{2}$$

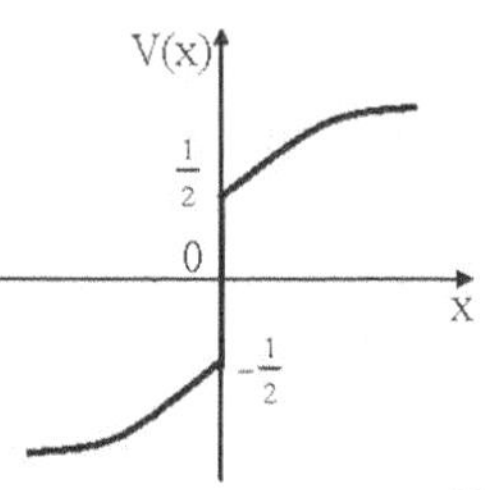

SOLVED PAPER – 2015

ELECTRONICS & COMMUNICATION ENGINEERING

Set - 1

No. of Questions : 65 *Maximum Marks : 100*

GENERAL APTITUDE (GA)

Q. (1–5) Carry One Mark Each

1. Choose the word most similar in meaning to the given word :

   Educe

   (*a*) Exert (*b*) Educate

   (*c*) Extract (*d*) Extend

2. If $\log_x\left(\frac{5}{7}\right) = -\frac{1}{3}$, then the value of x is

   (*a*) $\frac{343}{125}$ (*b*) $\frac{125}{343}$

   (*c*) $-\frac{25}{49}$ (*d*) $-\frac{49}{25}$

3. Operators $\square, \diamond$ and $\rightarrow$ are defined by

   $$a \square b = \frac{a-b}{a+b};\ a \diamond b = \frac{a+b}{a-b};\ a \rightarrow b = ab.$$

   Find the value $(66 \square 6) \rightarrow (66 \diamond 6)$

   (*a*) -2 (*b*) -1

   (*c*) 1 (*d*) 2

4. Choose the most appropriate word from the options given below to complete the following sentence.

   The principal presented the chief guest with a ____________, as token of appreciation.

   (*a*) momento

   (*b*) memento

   (*c*) momentum

   (*d*) moment

5. Choose the appropriate word/phrase, out of the four options given below, to complete the following sentence :

   Frogs _________.

   (*a*) Croak

   (*b*) Roar

   (*c*) Hiss

   (*d*) Patter

(Q. 6–10) : Carry Two Marks Each

6. A cube of side 3 units is formed using a set of smaller cubes of side 1 unit. Find the proportion of the number of faces of the smaller cubes visible to those which are NOT visible.

   (*a*) 1 : 4

   (*b*) 1 : 3

   (*c*) 1 : 2

   (*d*) 2 : 3

7. Fill in the missing value

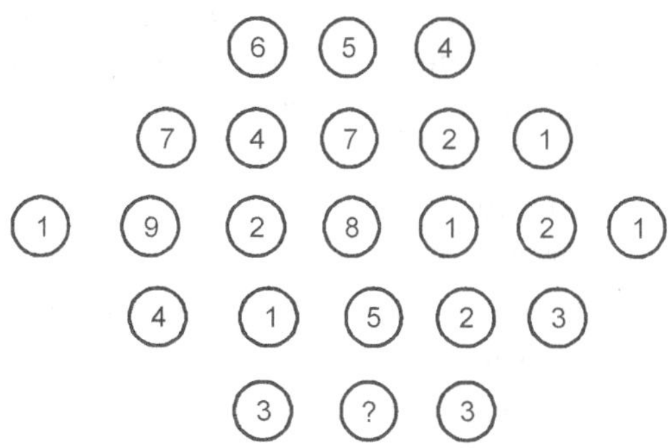

8. Humpty Dumpty sits on a wall every day while having lunch. The wall sometimes breaks. A person sitting on the wall falls if the wall breaks.

   Which one of the statements below is logically valid and can be inferred from the above sentences?

   (*a*) Humpty Dumpty always falls while having lunch

   (*b*) Humpty Dumpty does not fall sometimes while having lunch

   (*c*) Humpty Dumpty never falls during dinner

   (*d*) When Humpty Dumpty does not sit on the wall, the wall does not break

9. The following question presents a sentence, part of which is underlined. Beneath the sentence you find four ways of phrasing the underline part. Following the requirements of the standard written English, select the answer that produces the most effective sentence.

Tuberculosis, together with its effects, ranks one of the leading causes of death in India.

(*a*) ranks as one of the leading causes of death

(*b*) rank as one of the leading causes of death

(*c*) has the rank of one of the leading causes of death

(*d*) are one of the leading causes of death

10. Read the following paragraph and choose the correct statement.

Climate change has reduced human security and threatened human well being. An ignored reality of human progress is that human security largely depends upon environmental security. But on the contrary, human progress seems contradictory to environmental security. To keep up both at the required level is a challenge to be addressed by one and all. One of the ways to curb the climate change may be suitable scientific innovations, while the other may be the Gandhian perspective on small scale progress with focus on sustainability.

(*a*) Human progress and security are positively associated with environmental security.

(*b*) Human progress is contradictory to environmental security.

(*c*) Human security is contradictory to environmental security.

(*d*) Human progress depends upon environmental security.

TECHNICAL SECTION

Q. (1–25) Carry One Mark Each

1. A region of negative differential resistance is observed in the current voltage characteristics of a silicon PN junction if

(*a*) Both the P-region and the N-region are heavily doped

(*b*) The N-region is heavily doped compared to the P-region

(*c*) The P-region is heavily doped compared to the N-region

(*d*) An intrinsic silicon region is inserted between the P-region and the N-region

2. A silicon sample is uniformly doped with donor type impurities with a concentration of $10^{16}/cm^3$.

The electron and hole mobilities in the sampleare 1200 $cm^2/V - s$ and 400 $cm^2/V - s$ respectively.

Assume complete ionization of impurities.

The charge of an electron is 1.6×10^{-19}C. The resistivity of the sample (in Ω – cm) is ______.

3. A unity negative feedback system has the open-loop transfer function $G(s) = \frac{k}{s(s+1)(s+3)}$.

The value of gain K (> 0), at which root locus crosses the imaginary axis is ________.

4. Suppose A and B are two independent events with probabilities P(A) ≠ 0 and P(B) ≠ 0. Let $\bar{A}$ and $\bar{B}$ be their complements. Which one of the following statements is FALSE ?

(*a*) $P(A \cap B) = P(A)P(B)$

(*b*) $P(A/B) = P(A)$

(*c*) $P(A \cup B) = P(A) + P(B)$

(*d*) $P(\bar{A} \cap \bar{B}) = P(\bar{A})\,P(\bar{B})$

5. The waveform of a periodic signal $x(t)$ is shown in the figure.

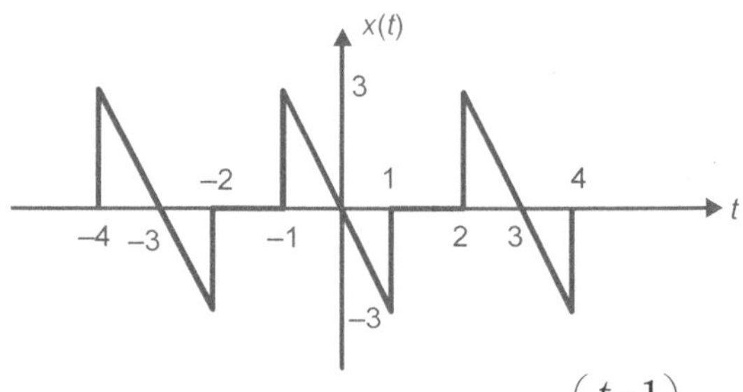

A signal $g(t)$ is defined by $g(t) = x\left(\frac{t-1}{2}\right)$. The average power of $g(t)$ is ________.

6. In the network shown in the figure, all resistors are identical with R = 300 Ω. The resistance R_{ab} (in Ω) of the network is ________.

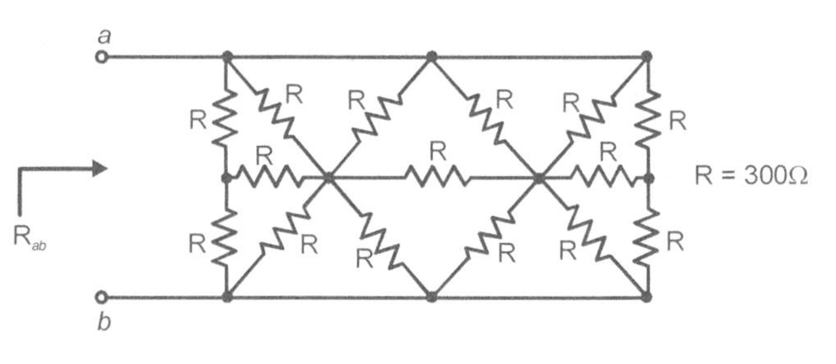

7. Consider a system of linear equation :

$x - 2y + 3z = -1$

$x - 3y + 4z = 1$ and

$-2x + 4y - 6z = k$

The value of k for which the system has infinitely many solutions is ________.

8. The polar plot of transfer function

$$G(s) = \frac{10(s+1)}{s+10} \text{ for } 0 \le \omega \le \infty \text{ will be in the}$$

(a) First quadrant

(b) Second quadrant

(c) Third quadrant

(d) Fourth quadrant

9. Let $z = x + iy$ be a complex variable. Consider that contour integration is performed along the unit circle in anticlockwise direction. Which one of the following statements is NOT TRUE ?

(a) The residue of $\frac{z}{z^2-1}$ at $z = 1$ is $\frac{1}{2}$

(b) $\oint_c z^2 dz = 0$

(c) $\frac{1}{2\pi i}\oint_c \frac{1}{z} dz = 1$

(d) $\bar{z}$ (complex conjugate of z) is an analytical function

10. Consider the signal

$s(t) = m(t)\cos(2\pi f_c t) + \hat{m}(t)(2\pi f_c t)$ where $\hat{m}(t)$ denotes the Hilbert transform of $m(t)$ and the bandwidth of $m(t)$ is very small compared to f_c. The signal $s(t)$ is a

(a) High-pass signal

(b) Low-pass signal

(c) Band-pass signal

(d) Double sideband suppressed carrier signal

11. For the circuit with ideal diodes shown in the figure, the shape of the output (v_{out}) for the given sine wave input (v_{in}) will be

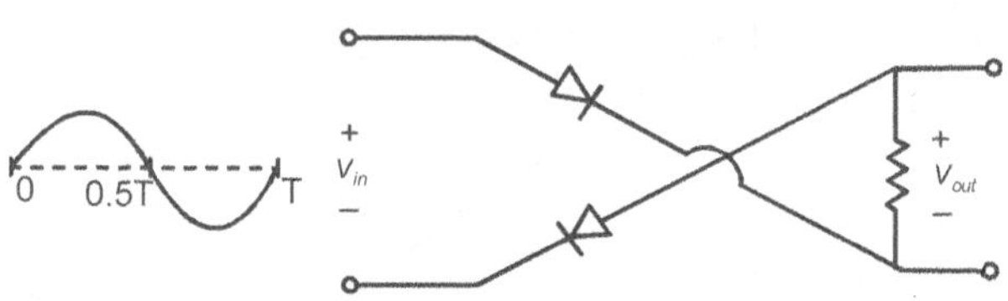

(a)

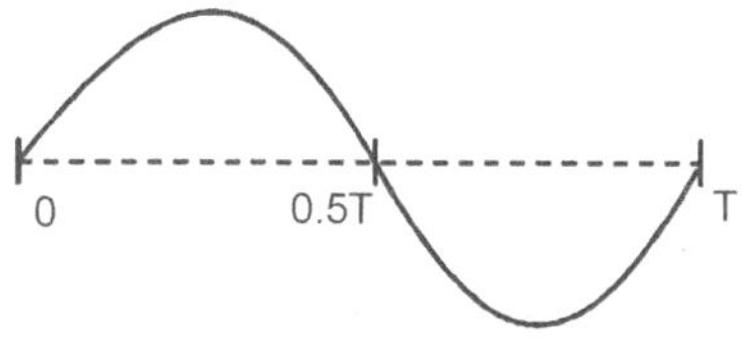

(b)

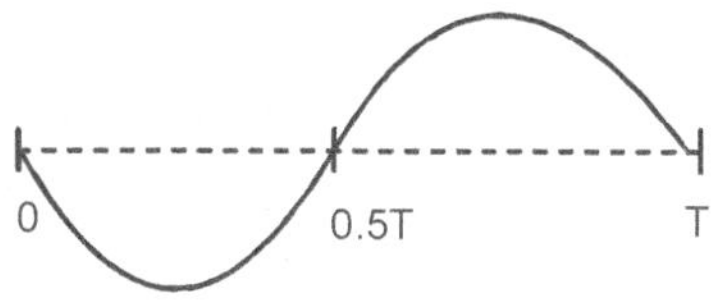

(c)

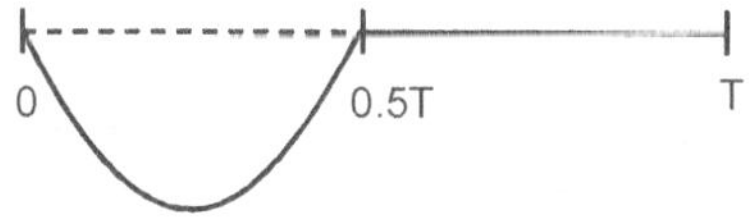

(d)

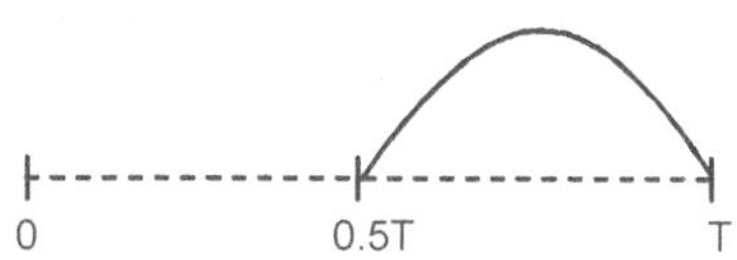

12. The result of convolution $x(-t) * \delta(-t - t_0)$ is

(a) $x(t + t_0)$

(b) $x(t - t_0)$

(c) $x(-t + t_0)$

(d) $x(-t - t_0)$

13. The value of p such that the vector $\begin{bmatrix}1\\2\\3\end{bmatrix}$ is an eigenvector of the matrix $\begin{bmatrix}4 & 1 & 2\\ P & 2 & 1\\ 14 & -4 & 10\end{bmatrix}$, is____.

14. In the given circuit, the values of V_1 and V_2 respectively are

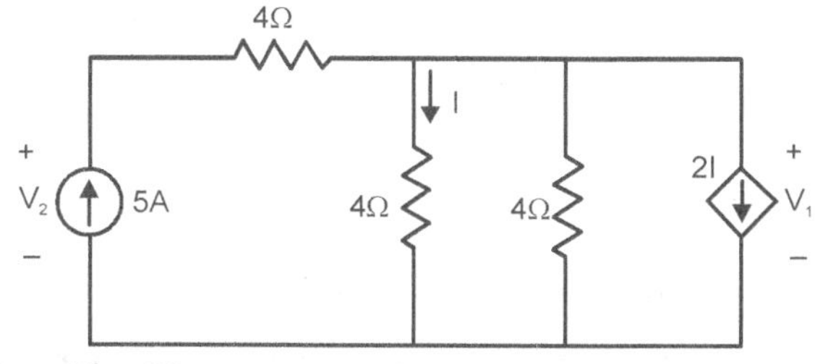

(a) 5V, 25V (b) 10V, 30V

(c) 15V, 35V (d) 0V, 20V

15. In an 8085 microprocessor, the shift registers which store the result of an addition and the overflow bit are, respectively

(a) B and F (b) A and F

(c) H and F (d) A and C

16. Negative feedback in a closed-loop control system DOES NOT

(*a*) reduce the overall gain

(*b*) reduce bandwidth

(*c*) improve disturbance rejection

(*d*) reduce sensitivity to parameter variation

17. A 16 kb (=16,384 bit) memory array is designed as a square with an aspect ratio of one (number of rows is equal to the number of columns). The minimum number of address lines needed for the row decoder is _______.

18. A function $f(x) = 1 - x^2 + x^3$ is defined in the closed interval [– 1, 1]. The value of x, in the open interval (– 1, 1) for which the mean value theorem is satisfied, is

(*a*) $-\frac{1}{2}$ (*b*) $-\frac{1}{3}$

(*c*) $\frac{1}{3}$ (*d*) $\frac{1}{2}$

19. The electric field component of a plane wave traveling in a lossless dielectric medium is given by

$$\vec{E}(z, t) = \hat{a}_y 2\cos\left(10^8 t - \frac{z}{\sqrt{2}}\right) \text{V/m.}$$

The wavelength (in m) for the wave is _______.

20. In the circuit shown, the switch SW is thrown from position A to position B at time $t = 0$. The energy (in μJ) taken from the 3V source to charge the 0.1 μF capacitor from 0V to 3V is

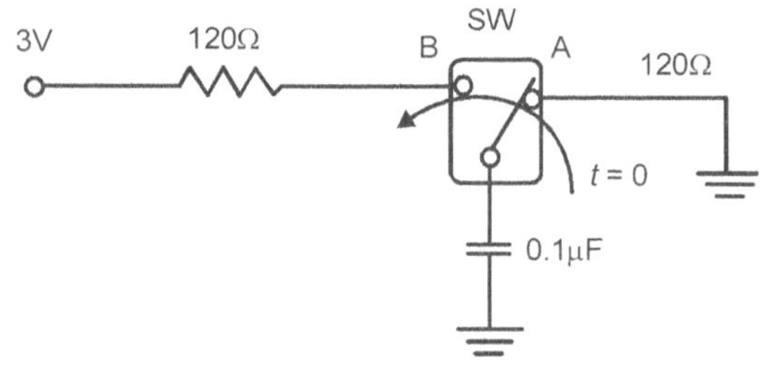

(*a*) 0.3 (*b*) 0.45

(*c*) 0.9 (*d*) 3

21. In the circuit shown below, the Zener diode is ideal and the Zener voltage is 6V. The output voltage V_0 (in volts) is _________.

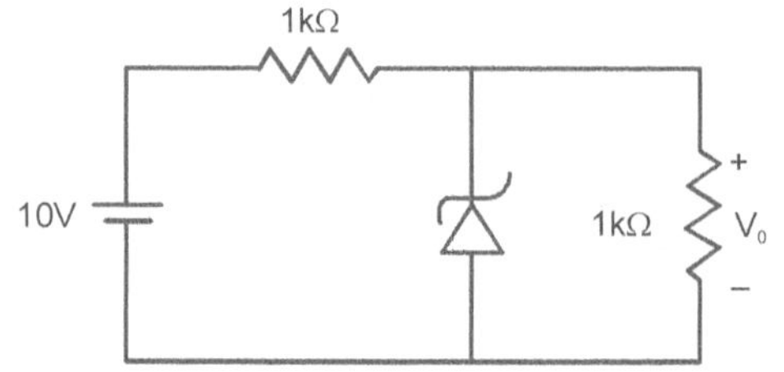

22. Consider a four bit D to A converter. The analog value corresponding to digital signals of values 0000 and 0001 are 0V and 0.0625V respectively. The analog value (in Volts) corresponding to the digital signal 1111 is _________.

23. In the circuit shown, at resonance, the amplitude of the sinusoidal voltage (in Volts) across the capacitor is _________.

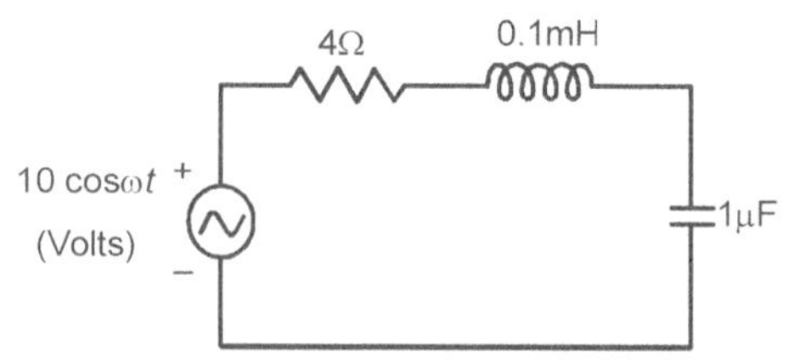

24. A sinusoidal signal of 2 kHz frequency is applied to a delta modulator. The sampling rate and step-size Δ of the delta modulator are 20,000 samples per second and 0.1V, respectively. To prevent slope overload, the maximum amplitude of the sinusoidal signal (in Volts) is

(*a*) $\frac{1}{2\pi}$ (*b*) $\frac{1}{\pi}$

(*c*) $\frac{2}{\pi}$ (*d*) π

25. Consider a straight, infinitely long, current carrying conductor lying on the z-axis. Which one of the following plots (in linear scale) qualitatively represents the dependence of H_ϕ on r, where H_ϕ is the magnitude of the azimuthal component of magnetic field outside the conductor and r is the radial distance from the conductor ?

(*a*)
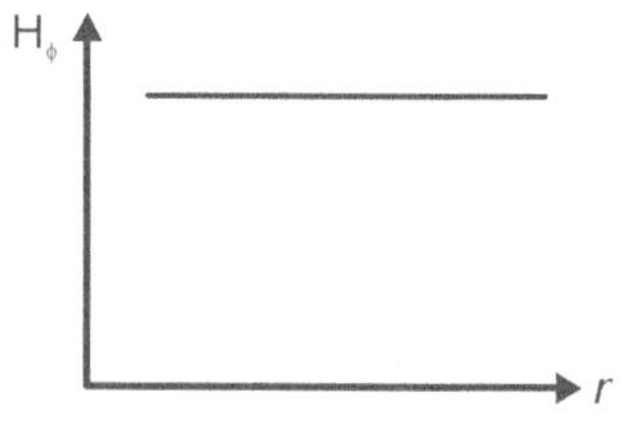

(*b*)
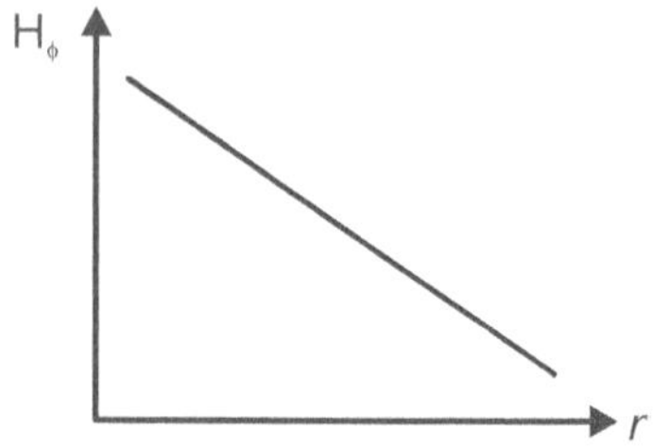

(c) 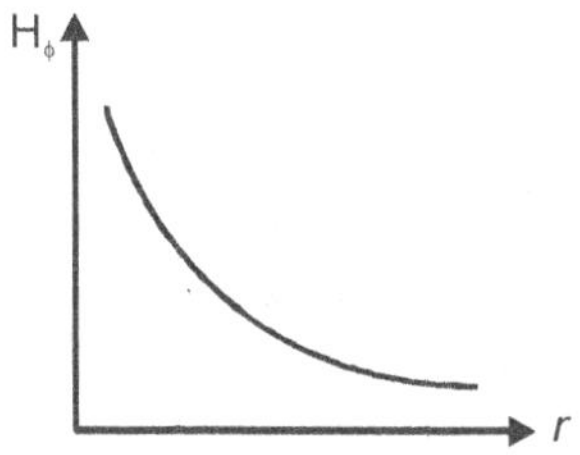

(d) 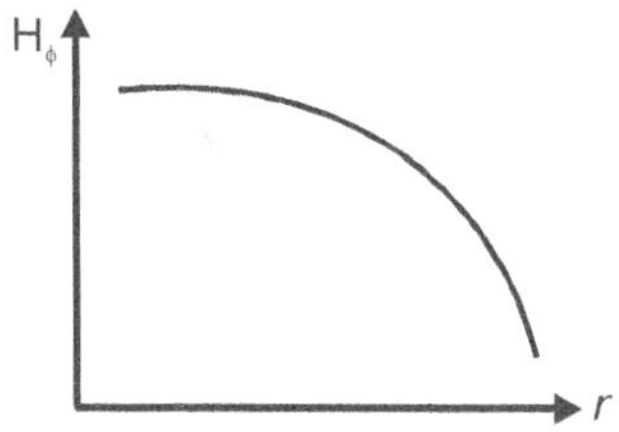

Q. (26–55) : Carry Two Marks Each

26. The input X to the Binary Symmetric Channel (BSC) shown in the figure is '1' with probability 0.8. The cross-over probability is 1/7. If the received bit Y = 0, the conditional probability that '1' was transmitted is ________.

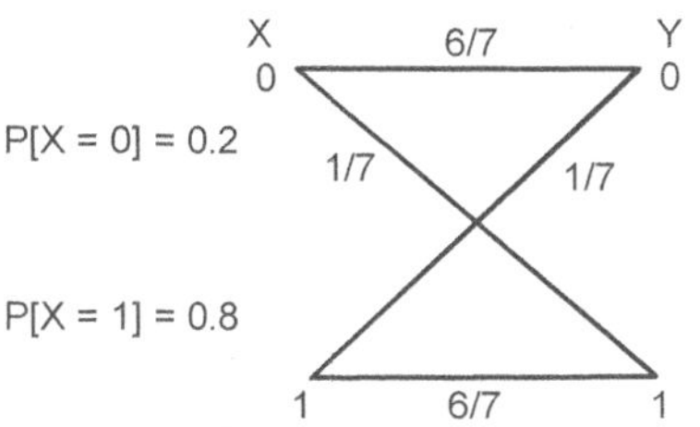

27. The transmitted signal in a GSM system is of 200 kHz bandwidth and 8 users share a common bandwidth using TDMA. It at a given time 12 users are talking in a cell, the total bandwidth of the signal received by the base station of the cell will be at least (in kHz) ________.

28. For the discrete-time system shown in the figure, the poles of the system transfer function are located at

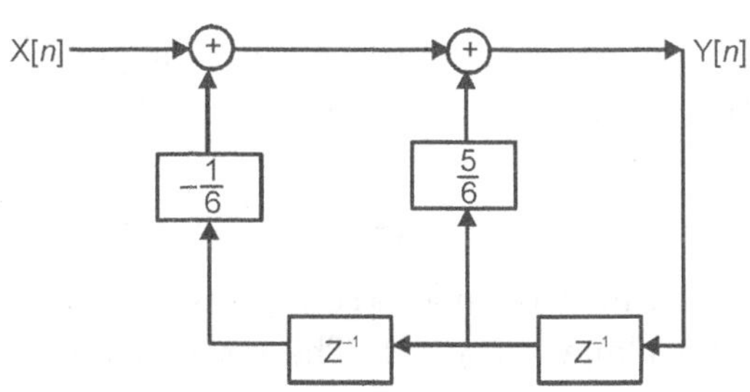

(a) 2, 3 (b) $\frac{1}{2}, 3$

(c) $\frac{1}{2}, \frac{1}{3}$ (d) $2, \frac{1}{3}$

29. The circuit shown in the figure has an ideal opamp. The oscillation frequency and the condition to sustain the oscillations, respectively, are

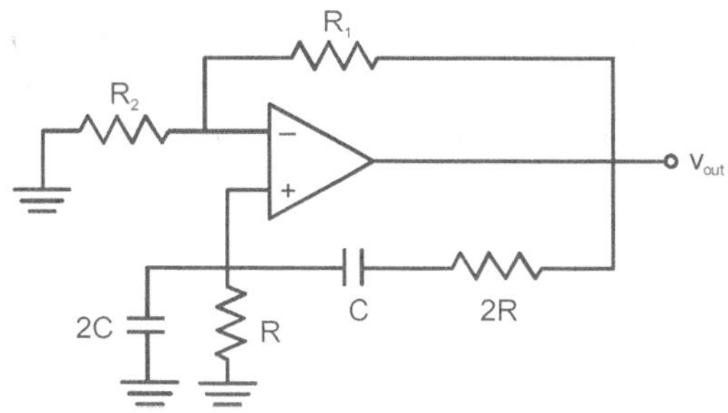

(a) $\frac{1}{CR}$ and $R_1 = R_2$ (b) $\frac{1}{CR}$ and $R_1 = 4R_2$

(c) $\frac{1}{2CR}$ and $R_1 = R_2$ (d) $\frac{1}{2CR}$ and $R_1 = 4R_2$

30. A source emits bit 0 with probability $\frac{1}{3}$ and bit 1 with probability $\frac{2}{3}$. The emitted bits are communicated to the receiver. The receiver decides for either 0 or 1 based on the received value R. It is given that the conditional density functions of R are as

$$f_{R|0}(r) = \begin{cases} \frac{1}{4}, -3 \le x \le 1, \\ 0, \text{otherwise}, \end{cases} \text{ and } f_{R|1}(r) = \begin{cases} \frac{1}{6}, -1 \le x \le 5, \\ 0, \text{otherwise}. \end{cases}$$

The minimum decision error probability is

(a) 0

(b) $\frac{1}{12}$

(c) $\frac{1}{9}$

(d) $\frac{1}{6}$

31. For a silicon diode with long P and N regions, the accepter and donor impurity concentrations are 1×10^{17} cm^{-3} and 1×10^{15} cm^{-3}, respectively. The lifetimes of electrons in P region and holes in N region are both 100 μs. The electron and hole diffusion coefficients are 49 cm^2/s and 36 cm^2/s, respectively. Assume $kT/q = 26$ mV, the intrinsic carrier concentration is 1×10^{10} cm^{-3} and $q = 1.6 \times 10^{-19}$C. When a forward voltage of 208 mV is applied across the diode, the hole current density (in nA/cm^2) injected from P region to N regions is ________.

32. For the NMOSFET in the circuit shown, the threshold voltage is V_{th}, where $V_{th} > 0$. The source voltage V_{ss} is varied from 0 to V_{DD}. Neglecting the channel length modulation, the drain current I_D as a function V_{ss} is represented by.

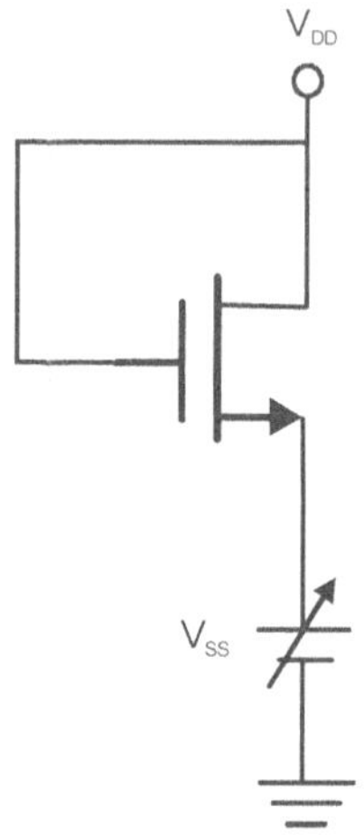

(a) 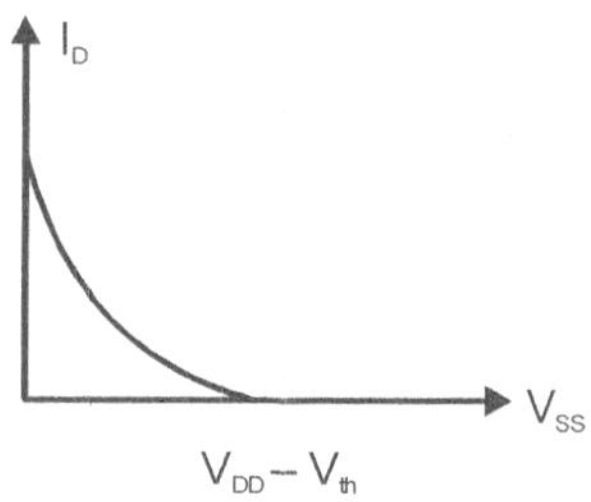

(b) 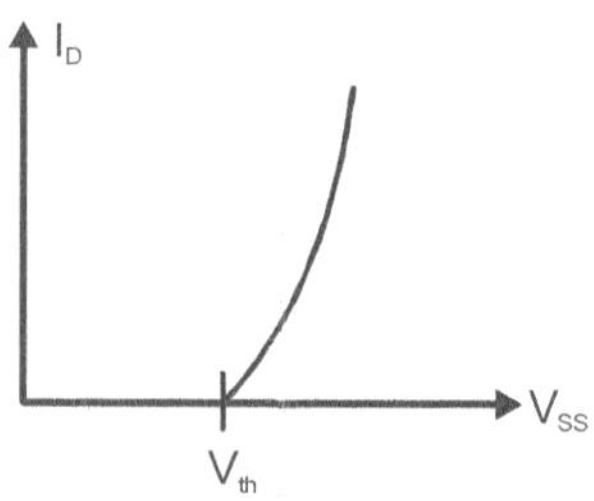

(c) 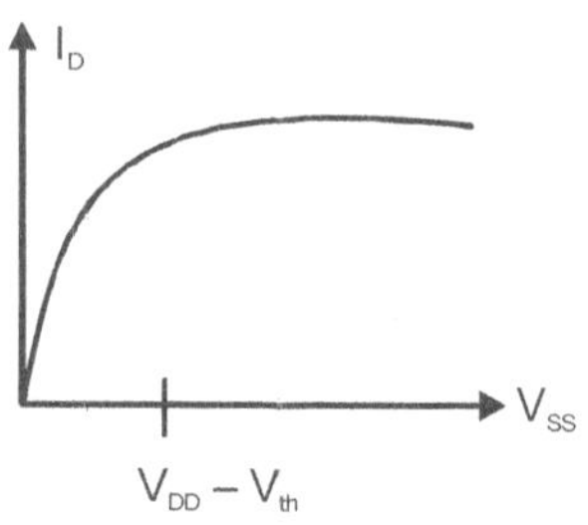

(d) 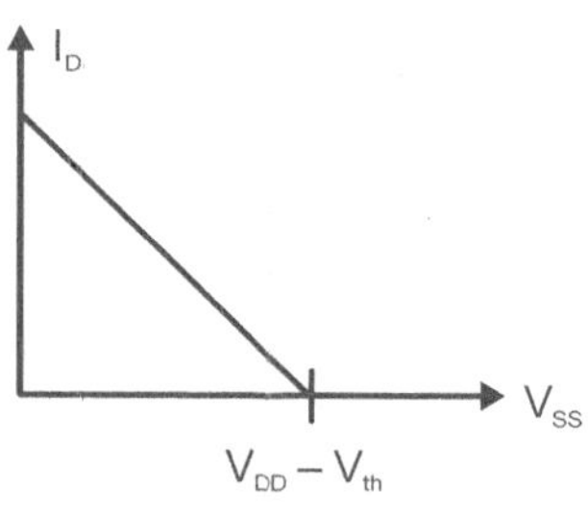

33. In the system shown in figure (*a*), $m(t)$ is a low-pass signal with bandwidth W Hz. The frequency response of the band-pass filter $H(f)$ is shown in figure (*b*). If it is desired that the output signal $z(t) = 10x(t)$, the maximum value of W (in Hz) should be strictly less than ________.

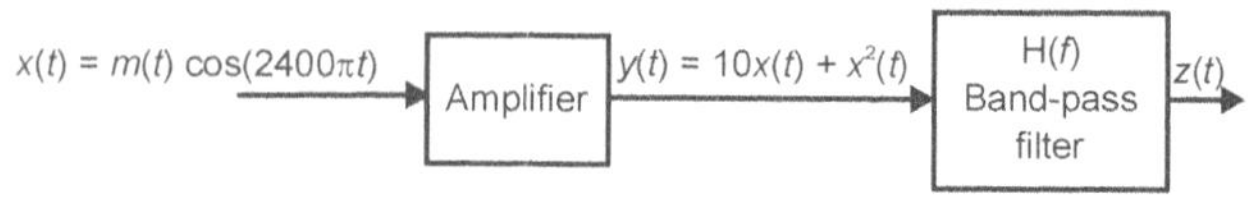

(*a*)

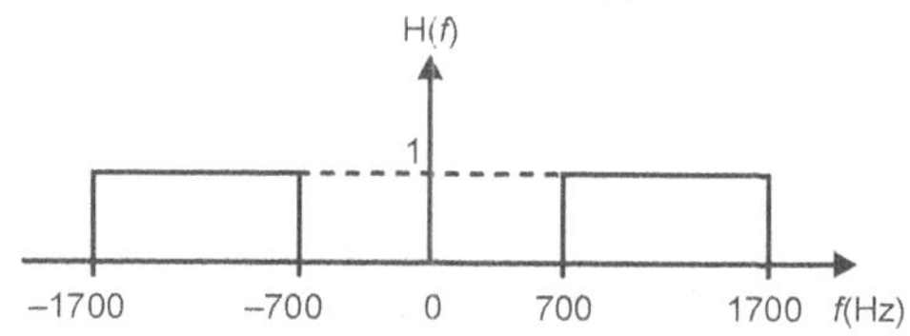

34. In the circuit shown, $I_1 = 80$ mA and $I_2 = 4$ mA. Transistors T_1 and T_2 are identical. Assume that the thermal voltage V_T is 26 mV at 27°C. At 50°C, the value of the voltage $V_{12} = V_1 - V_2$ (in mV) is ________.

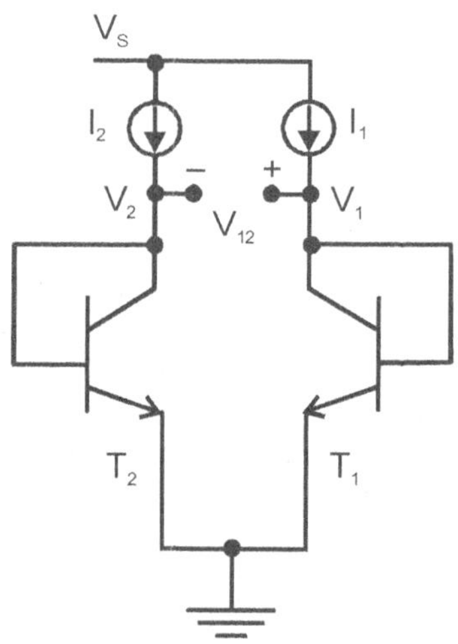

35. The electric field intensity of a plane wave traveling in free space is given by the following expression

$$E(x, t) = a_y 24\pi \cos(\omega t - k_0 x)\ (\text{V/m})$$

In this field, consider a square area 10 cm × 10 cm on a plane $x + y = 1$. The total time averaged power (in mW) passing through the square area is ________.

36. The maximum area (in square units) of a rectangle whose vertices lie on the ellipse $x^2 + 4y^2 = 1$ is _____.

37. In the given circuit, the maximum power (in Watts) that can be transferred to the load R_L is ________.

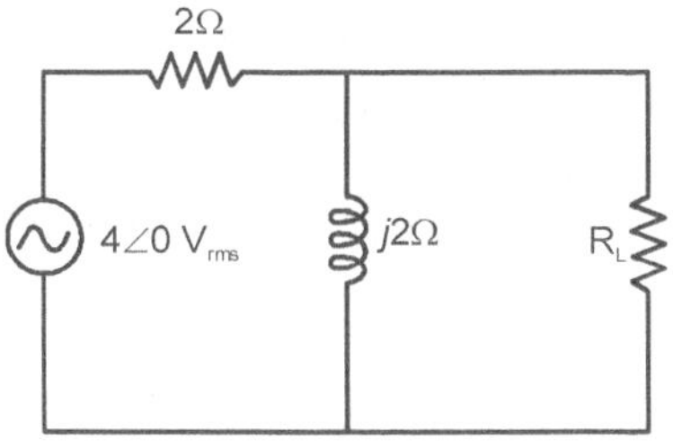

38. A lead compensator network includes a parallel combination of R and C in the feed-forward path. If the transfer function of the compensator is $G_c(s) = \frac{s+2}{s+4}$. The value of RC is _____.

39. A MOSFET in saturation has a drain current of 1mA for V_{DS} = 0.5V. If the channel length modulation coefficient is $0.05V^{-1}$, the output resistance (in kΩ) of the MOSFET is ________.

40. The built in potential of an abrupt p-n junction is 0.75V. If its junction capacitance (C_J) at a reverse bias (V_R) of 1.25 V is 5 pF, then the value of C_J (in pf) when V_R = 7.25V is _____.

41. In the circuit shown, assume that the opamp is ideal. The bridge output voltage V_0 (in mV) for δ = 0.05 is ________.

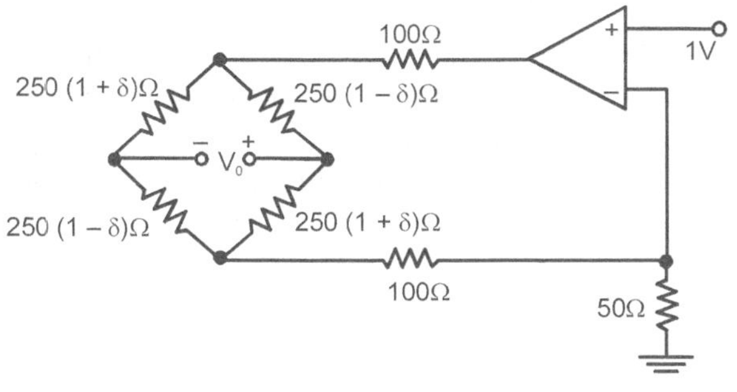

42. The damping ratio of a series RLC circuit can be expressed as

(a) $\frac{R^2C}{2L}$

(b) $\frac{2L}{R^2C}$

(c) $\frac{R}{2}\sqrt{\frac{C}{L}}$

(d) $\frac{2}{R}\sqrt{\frac{L}{C}}$

43. In the circuit shown, switch SW is closed at $t = 0$. Assuming zero initial conditions, the value of $v_c(t)$ (in Volts) at t = 1 sec is ________.

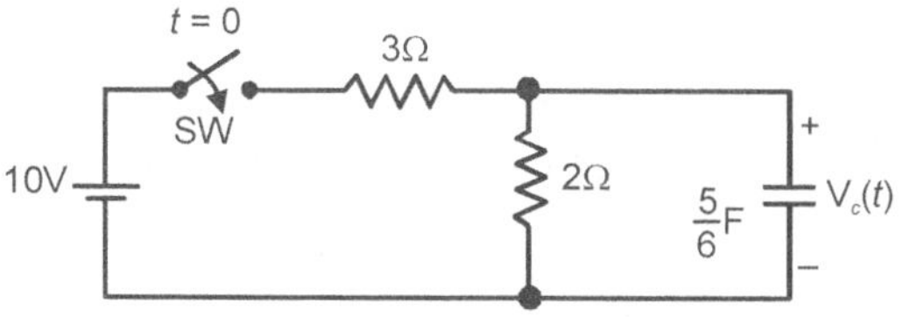

44. Consider a uniform plane wave with amplitude (E_0) of 10V/ m and 1.1 GHz frequency travelling in air, and incident normally on a dielectric medium with complex relative permittivity (ε_r) and permeability (μ_r) as shown in the figure.

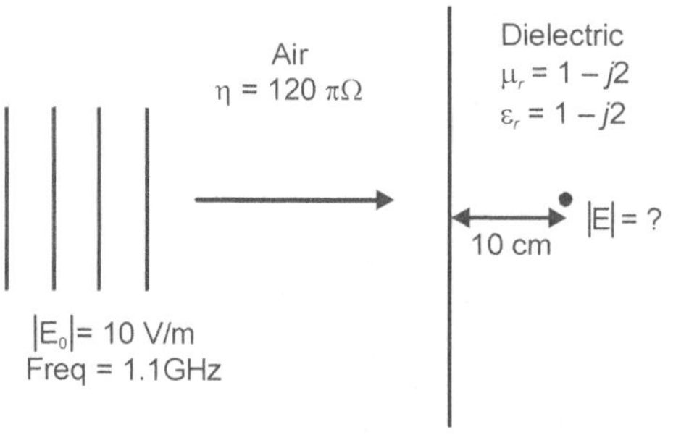

The magnitude of the transmitted electric field component (in V/m) after it has travelled a distance of 10 cm inside the dielectric region is ________.

45. A vector $\vec{P}$ is given by $\vec{P} = x^3y\vec{a}_x - x^2y^2\vec{a}_y - x^2yz\vec{a}_z$.

Which one of the following statements is TRUE?

(a) $\vec{P}$ is solenoidal, but not irrotational

(b) $\vec{P}$ is irrotational, but not solenoidal

(c) $\vec{P}$ is neither solenoidal nor irrotational

(d) $\vec{P}$ is both solenoidal and irrotational

46. All the logic gates shown in the figure have a propagation delay of 20 ns. Let A = C = 0 and B = 1 until time t = 0. At t = 0, all the inputs flip (i.e., A = C = 1 and B = 0) and remain in that state. For t > 0, output Z = 1 for a duration (in ns) of

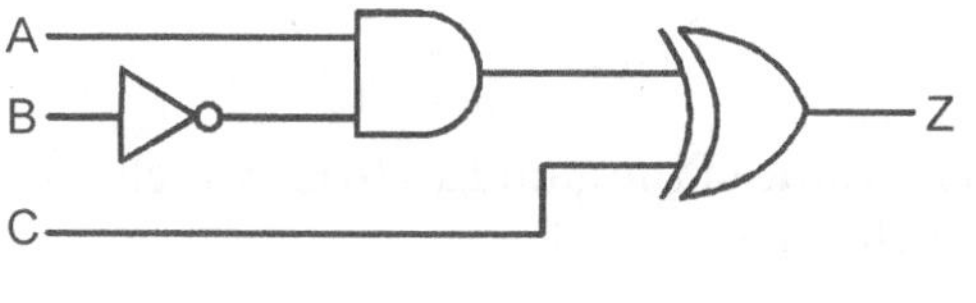

47. A plant transfer function is given as

$$G(s) = \left(K_p + \frac{K_1}{s}\right)\frac{1}{s(s+2)}.$$

When the plant operates in a unity feedback configuration, the condition for the stability of the closed loop system is

(a) $K_p > \frac{K_1}{2} > 0$
(b) $2K_1 > K_p > 0$
(c) $2K_1 < K_p$
(d) $2K_1 > K_p$

48. A 3-input majority gate is defined by the logic function M(a, b, c) = ab + bc + ca. Which one of the following gates is represented by the function

$M\left(\overline{M(a, b, c)}, M(a, b, \bar{c}), c\right)$?

(a) 3-input NAND gate
(b) 3-input XOR gate
(c) 3-input NOR gate
(d) 3-input XNOR gate

49. The longitudinal component of the magnetic field inside an air-filled rectangular waveguide made of a perfect electric conductor is given by the following expression

$H_z(x, y, z, t) = 0.1 \cos(25\pi x) \cos(30.3\pi y) \cos(12\pi \times 10^9 t - \beta z)$ (A/m)

The cross-sectional dimensions of the waveguide are given as a = 0.08 m and b = 0.033 m. The mode of propagation inside the waveguide is

(a) TM_{12}
(b) TM_{21}
(c) TE_{21}
(d) TE_{12}

50. The open-loop transfer function of a plant in a unity feedback configuration is given as

$$G(s) = \frac{K(s+4)}{(s+8)(s^2-9)}.$$

The value of the gain K(> 0) for which $-1 + j2$ lies on the root locus is ________.

51. Two sequences [a, b, c] and [A, B, C] are related as,

$$\begin{bmatrix} A \\ B \\ C \end{bmatrix} = \begin{bmatrix} 1 & 1 & 1 \\ 1 & W_3^{-1} & W_3^{-2} \\ 1 & W_3^{-2} & W_3^{-4} \end{bmatrix}\begin{bmatrix} a \\ b \\ c \end{bmatrix} \text{ where } W_3 = e^{j\frac{2\pi}{3}}$$

If another squence [p, q, r] is derived as,

$$\begin{bmatrix} p \\ q \\ r \end{bmatrix} = \begin{bmatrix} 1 & 1 & 1 \\ 1 & W_3^{1} & W_3^{2} \\ 1 & W_3^{2} & W_3^{4} \end{bmatrix}\begin{bmatrix} 1 & 0 & 0 \\ 0 & W_3^{2} & 0 \\ 0 & 0 & W_3^{4} \end{bmatrix}\begin{bmatrix} A/3 \\ B/3 \\ C/3 \end{bmatrix},$$

then the relationship between the sequences [p, q, r] and [a, b, c] is

(a) $[p, q, r] = [b, a, c]$
(b) $[p, q, r] = [b, c, a]$
(c) $[p, q, r] = [c, a, b]$
(d) $[p, q, r] = [c, b, a]$

52. The solution of the differential equation

$$\frac{d^2y}{dt^2} + 2\frac{dy}{dt} + y = 0 \text{ with } y(0) = y'(0) = 1 \text{ is}$$

(a) $(2 - t)e^t$
(b) $(1 + 2t)e^{-t}$
(c) $(2 + t)e^{-t}$
(d) $(1 - 2t)e^t$

53. The pole-zero diagram of a causal and stable discrete-time system is shown in the figure. The zero at the origin has multiplicity 4. The impulse response of the system is $h[n]$. If $h[0] = 1$, we can conclude

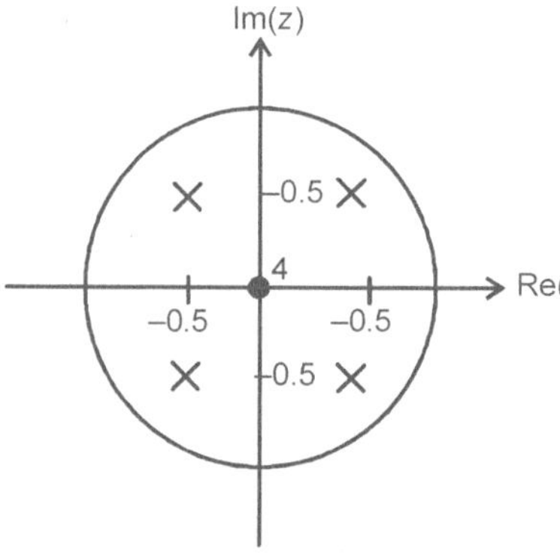

(a) $h[n]$ is real for all n
(b) $h[n]$ is imaginary for all n
(c) $h[n]$ is real for only even n
(d) $h[n]$ is purely imaginary for only odd n

54. Which one of the following graphs describes the function $f(x) = e^{-x}(x^2 + x + 1)$

(a)

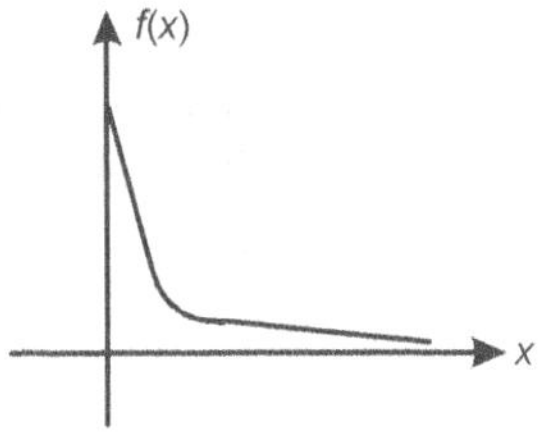

(b)

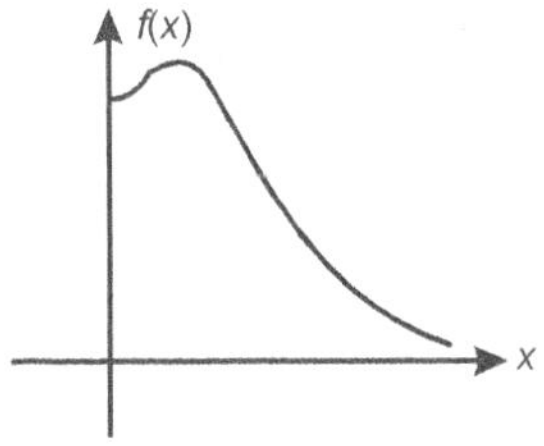

(c)

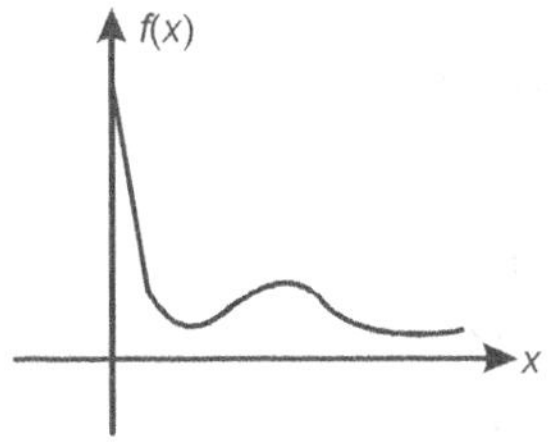

(d)

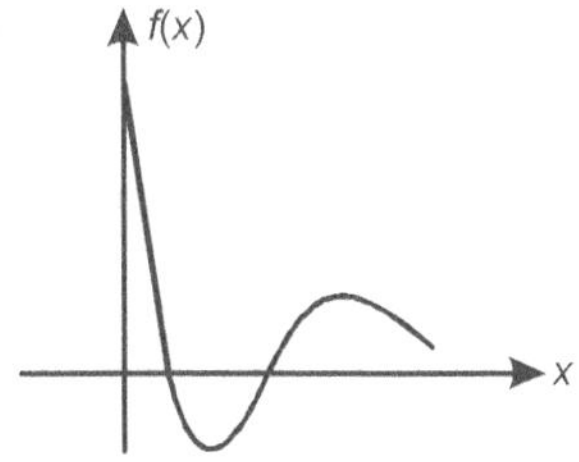

55. The Boolean expression

$$F(X, Y, Z) = \bar{X}Y\bar{Z} + X\bar{Y}\bar{Z} + XY\bar{Z} + XYZ$$

converted into the canonical product of sum (POS) form is

(a) $(X+Y+Z)(X+Y+\bar{Z})(X+\bar{Y}+\bar{Z})(\bar{X}+Y+\bar{Z})$

(b) $(X+\bar{Y}+Z)(\bar{X}+Y+\bar{Z})(\bar{X}+\bar{Y}+Z)(\bar{X}+\bar{Y}+\bar{Z})$

(c) $(X+Y+Z)(\bar{X}+Y+\bar{Z})(X+\bar{Y}+Z)(\bar{X}+\bar{Y}+\bar{Z})$

(d) $(X+\bar{Y}+\bar{Z})(\bar{X}+Y+Z)(\bar{X}+\bar{Y}+Z)(X+Y+Z)$

ANSWERS

GENERAL APTITUDE

1. (c)	2. (a)	3. (c)	4. (b)	5. (a)	6. (c)
7. (3)	8. (b)	9. (a)	10. (b)		

TECHNICAL SECTION

1. (a)	2. (0.52)	3. (12)	4. (c)	5. (2)	6. (100)
7. (2)	8. (a)	9. (d)	10. (c)	11. (c)	12. (d)
13. (17)	14. (a)	15. (b)	16. (b)	17. (7)	18. (b)
19. (8.885)	20. (c)	21. (5)	22. (0.93 – 0.94)	23. (25)	24. (a)
25. (c)	26. (0.4)	27. (400)	28. (c)	29. (d)	30. (d)
31. (28-30)	32. (a)	33. (350)	34. (83.5 – 84.0)	35. (53.3)	36. (1)
37. (1.66)	38. (0.5)	39. (20)	40. (2.5)	41. (250)	42. (c)
43. (2.528)	44. (0.1)	45. (a)	46. (40)	47. (a)	48. (b)
49. (c)	50. (25.5)	51. (c)	52. (b)	53. (a)	54. (b)
55. (a)					

EXPLANATIONS

GENERAL APTITUDE

1. The word similar in meaning to Educe is Extract.

2. $$\frac{5}{7} = x^{-1/3}$$

$$\Rightarrow \quad \frac{7}{5} = x^{1/3}$$

$$\Rightarrow \quad \left(\frac{7}{5}\right)^3 = x$$

$$\Rightarrow \quad x = \frac{343}{125}$$

3. $$66\square 6 = \frac{66-6}{66+6} = \frac{60}{72} = \frac{5}{6}$$

$$66\diamond 6 = \frac{66+6}{66-6} = \frac{72}{60} = \frac{6}{5}$$

$$(66\square 6)\rightarrow(66\diamond 6) = \frac{5}{6}\times\frac{6}{5} = 1$$

4. The principal presented the chief guest with a memento, as token of appreciation.

5. Frogs make 'croak' sound.

6. 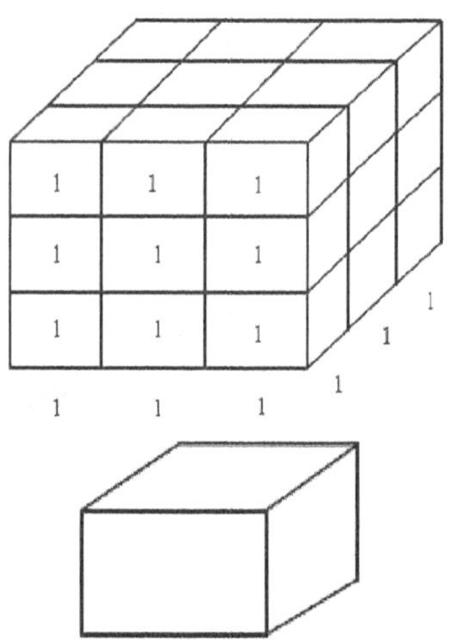

Number of faces per cube = 6

Total number of cubes = 9 × 3 = 27

∴ Total number of faces = 27 × 6 = 162

∴ Total number of non visible faces = 162 – 54 = 108

$$\therefore \frac{\text{Number of visible faces}}{\text{Number of non visible faces}} = \frac{54}{108} = \frac{1}{2}$$

7. Middle number is the average of number on both sides

$$\therefore \text{ Average of 3 and 3 is } \frac{3+3}{2} = \frac{6}{2} = 3$$

TECHNICAL SECTION

1. In Tunnel diode, both the P region and N region are heavily doped. It shows the negative differential resistance in the current (–) voltage characterstics.

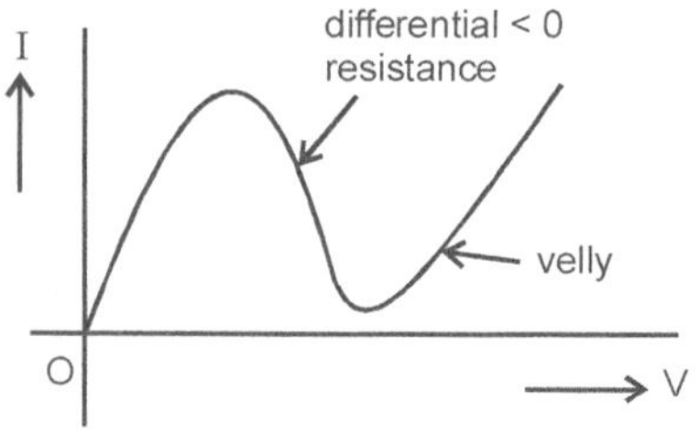

2. For donar type impurities,

Conductivity

$$\sigma_N = N_D q\mu_n$$

$$= 10^{16} \times 1.6 \times 10^{-19} \times 1200 = 1.92 \ \mho/\text{cm}$$

Resistivity

$$\rho = \frac{1}{\sigma_N} = \frac{1}{1.92}$$

$$\rho = 0.52\ \Omega - \text{cm}$$

3. The open loop transfer function with unity feedback system

$$G(s) = \frac{K}{s(s+1)(s+3)} \qquad \because H(s) = 1$$

We know that

Closed loop characteristic equation of the s/s

$$1 + G(s)H(s) = 0$$

$$1 + \frac{K}{s(s+1)(s+3)} = 0$$

$$s(s+1)(s+3) + K = 0$$

$$s^3 + 4s^2 + 3s + K = 0$$

Routh array corresponding to characteristic equation

s^3	1	3
s^2	4	K
s^1	$\frac{12-K}{4}$	0
s^0	K	0

For Root Locus crosses the imaginary axis, all the elements of a row must be zero.

$$\frac{12-K}{4} = 0$$

$$\therefore \quad 12 - K = 0$$

$$\therefore \quad K = 12$$

4. Here A and B are two independent events

$$\therefore \quad P(A \cap B) = P(A)\,P(B)$$

and $\quad P(A/B) = P(A)$

and $\quad P(B/A) = P(B)$

Also if A and B are independent event then $\bar{A}$ and $\bar{B}$ are also independent event

$$\therefore \quad P(\bar{A} \cap \bar{B}) = P(\bar{A})P(\bar{B})$$

Hence option (*a*), (*b*) and (*d*) are correct

Now $\quad P(A \cup B) = P(A) + P(B) - P(A \cap B)$

$$= P(A) + P(B) - P(A)\,P(B)$$

$\therefore$ Option (*c*) is false.

6.

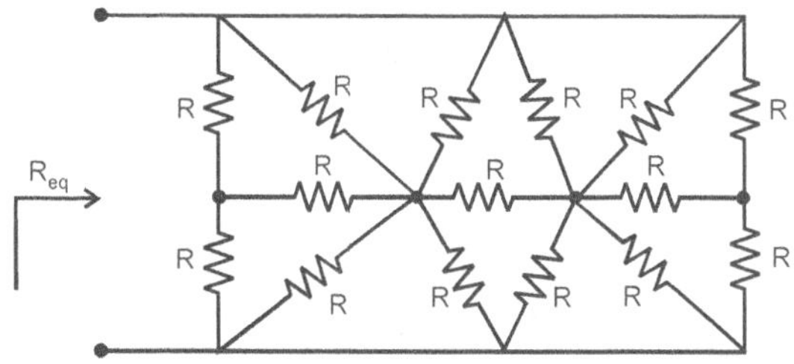

R = 300 Ω

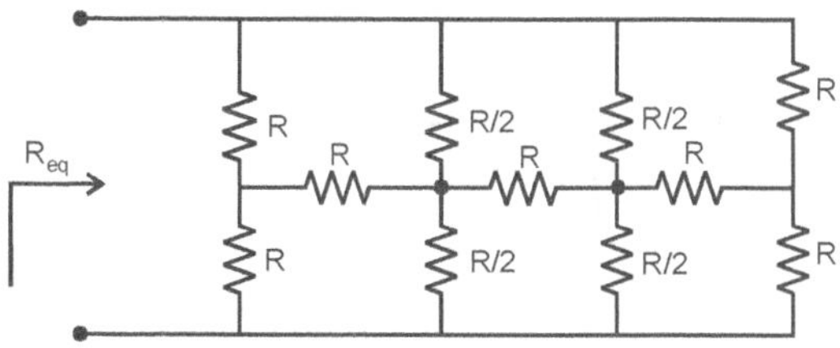

These are balanced wheatstone bridges.

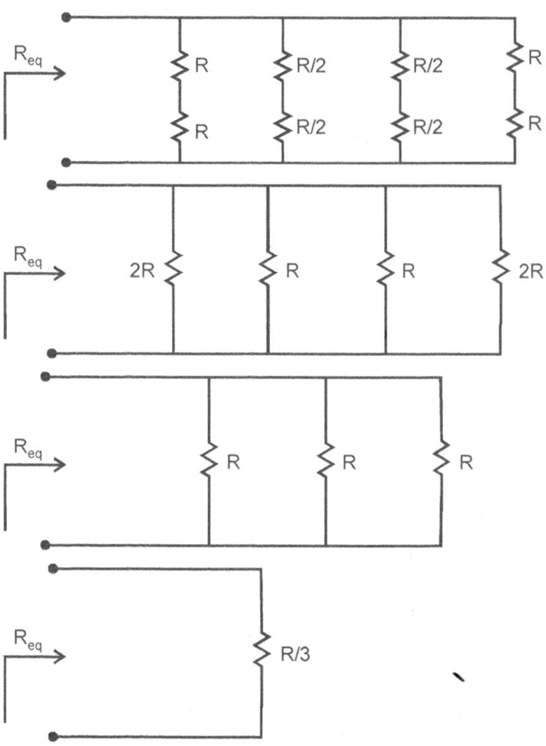

$$R_{eq} = \frac{R}{3} = \frac{300}{3} = 100\ \Omega$$

$$R_{eq} = 100\Omega$$

7.
$$x - 2y + 3z = -1$$
$$x - 3y + 4z = 1$$
$$-2x + 4y - 6z = k$$

Arugmented matrix (A/B) is given by

$$(A/B) = \begin{bmatrix} 1 & -2 & 3 & -1 \\ 1 & -3 & 4 & 1 \\ -2 & 4 & -6 & k \end{bmatrix}$$

Now applying row reduction technique—

$$R_2 \to R_2 - R_1$$
$$R_3 \to R_3 + 2R_1$$

We get

$$\begin{bmatrix} 1 & -2 & 3 & -1 \\ 0 & -1 & 1 & 2 \\ 0 & 0 & 0 & k-2 \end{bmatrix}$$

For infinite solution

$$\rho(A/B) = \rho(A) = r$$

$$\therefore \quad k - 2 = 0$$

$$\therefore \quad \boxed{k = 2}$$

8. The polar plot of the transfer function

$$G(s) = \frac{10(s+1)}{(s+10)} = \frac{10(1+s)}{10\left(1+\dfrac{s}{10}\right)}$$

$$G(s) = \frac{(1+s)}{(1+0.1s)}$$

for frequency response

$$s = j\omega$$

$$G(j\omega) = \frac{(1+j\omega)}{(1+j0.1\omega)}$$

$$|G(j\omega)| = \frac{\sqrt{1+\omega^2}}{\sqrt{1+(0.1\omega)^2}} \quad \begin{cases} |G(j\omega)|_{\omega=0} = 1 \\ |G(j\omega)|_{\omega=\infty} = 10 \end{cases}$$

$$\phi = \angle G(j\omega)$$
$$= \tan^{-1}(\omega) - \tan^{-1}(0.1\omega)$$

for $(0 \le \omega \le \infty)$

$$\phi = +\text{ve}$$

So it is always in first quadrant.

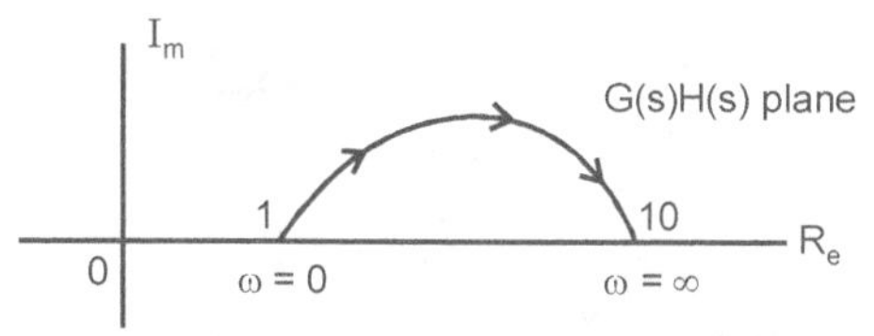

9.
$$z = x + iy$$

Let $\quad F(z) = \bar{z} = x - iy$

Here $\quad u = x$ and $v = -y$

$\Rightarrow \quad u_x = 1$ and $v_x = 0$

$u_y = 0$ and $v_y = -1$

As C – R equation is not satisfied.

$\therefore \quad u_x \neq v_y$

$\therefore$ $\bar{z}$ is not analytic.

From option (*a*) $z = 1$

Residue $\left(\dfrac{z}{z^2-1}\right)$ at $z = 1$

$$= \lim_{z\to 1}(z-1)\frac{z}{z^2-1}$$

$$= \lim_{z\to 1}\frac{(z-1)z}{(z+1)(z-1)}$$

$$= \lim_{z\to 1}\frac{z}{z+1}$$

$$= \frac{1}{1+1} = \frac{1}{2}$$

From option (*b*)

As z^2 is analytic everywhere

Now by using Cauchy's theorem

$$\oint_c z^2 dz = 0\ y$$

11.

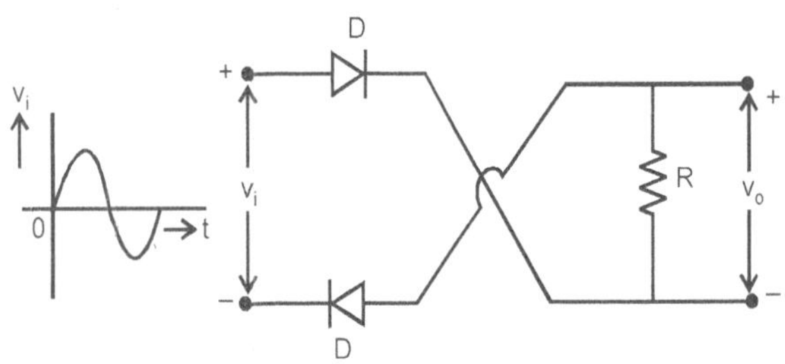

Consider + ve half cycle,

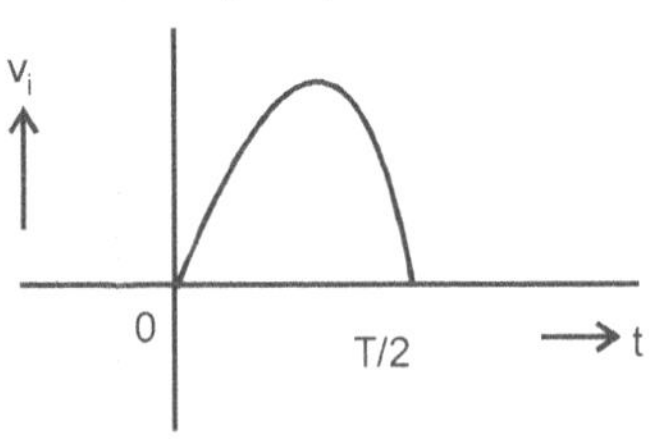

Both the diodes are forward bias.

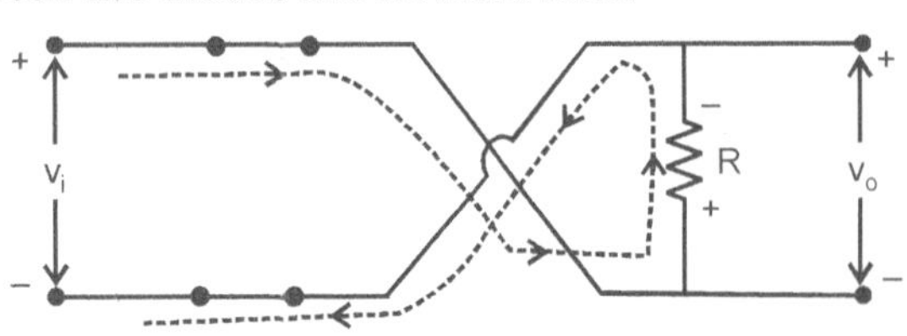

$\boxed{V_o = -V_i}$

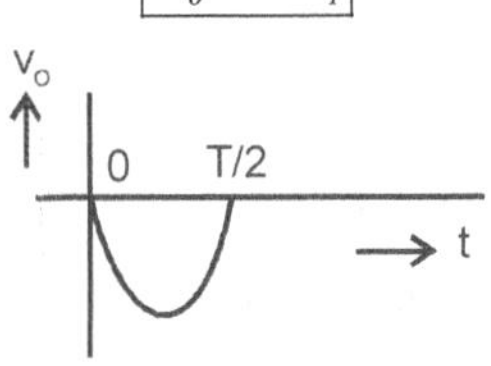

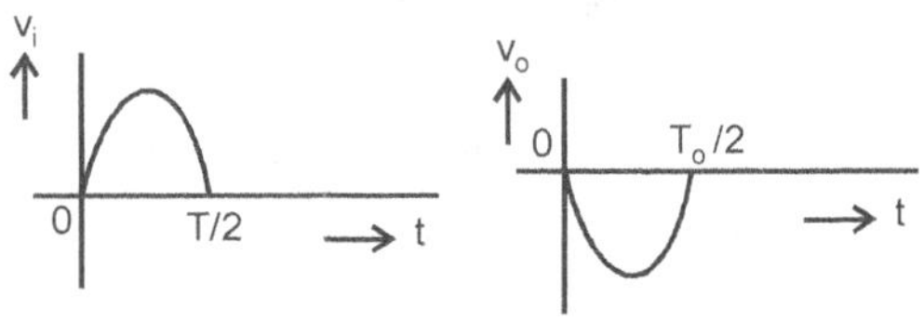

Consider –ve half cycle

Both the diodes are reverse bias.

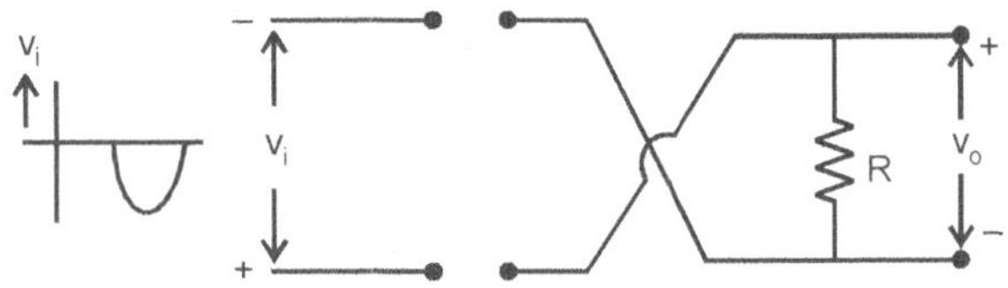

$\boxed{V_o = 0}$ volt

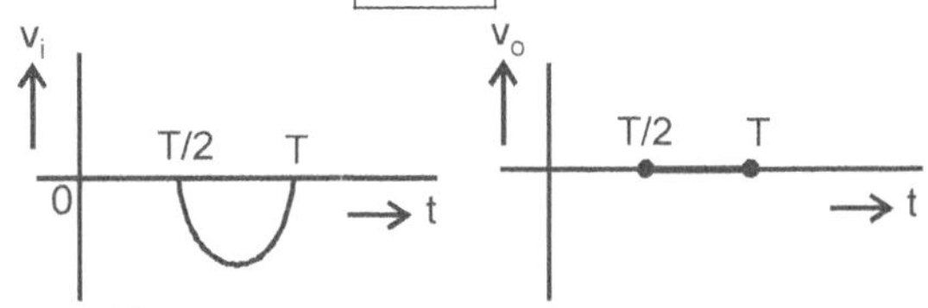

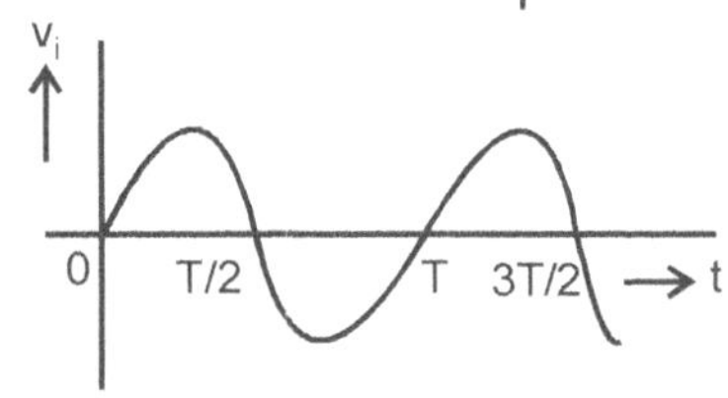

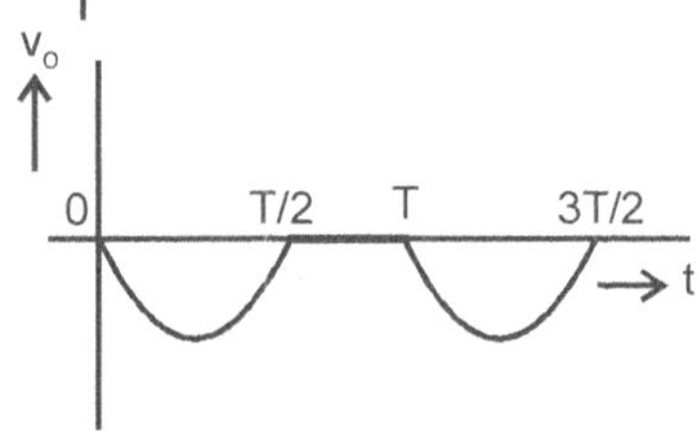

12. $x(-t) \otimes \delta(-t-t_0) = ?$

$$\delta(-t-t_0) = \begin{cases} 0, & -t-t_0 \neq 0 \\ 1, & -t-t_0 = 0 \end{cases}$$

$$\Rightarrow \quad t = -t_0$$

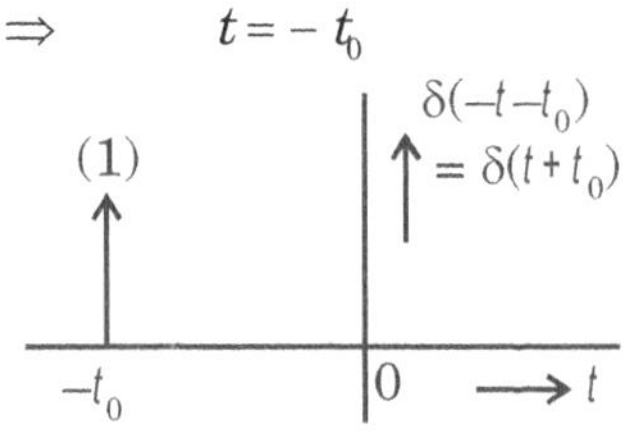

$$\delta(-t-t_0) = \delta(t+t_0)$$

We know that

$$x(t) \otimes \delta(t-t_0) = x(t-t_0)$$

$$x(t) \otimes \delta(t+t_0) = x(t+t_0)$$

$$x(-t) \otimes \delta(t+t_0) = x(-t-t_0)$$

or $\boxed{x(-t) \otimes \delta(-t-t_0) = x(-t-t_0)}$

13. $AX = \lambda X$

$$\Rightarrow \begin{bmatrix} 4 & 1 & 2 \\ P & 2 & 1 \\ 14 & -4 & 10 \end{bmatrix} = \lambda \begin{bmatrix} 1 \\ 2 \\ 3 \end{bmatrix}$$

$$\Rightarrow \quad \begin{bmatrix} 12 \\ P+7 \\ 36 \end{bmatrix} = \begin{bmatrix} \lambda \\ 2\lambda \\ 3\lambda \end{bmatrix}$$

$$\Rightarrow \quad \lambda = 12 \quad \text{...(1)}$$

$$2\lambda = P + 7 \quad \text{...(2)}$$

and $3\lambda = 36$

i.e., $\lambda = 12$

$\therefore$ Equation (2) gives $P + 7 = 24 \Rightarrow P = 17$

14. By nodal analysis

$$-5 + I + I + 2I = 0$$

$$4I = 5$$

$$I = \frac{5}{4} A$$

$$V_1 = 4I = 5 \text{ volts}$$

$$V_2 = 4(5) + V_1$$

$$= 20 + V_1$$

$$= 25 \text{ volts}$$

15. In 8085 μP,

Result of addition → Accumulator overflow bit is carry which is stored in flag register

Shift Registers are

A (accumulator)

F (flag register)

16. Negative feedback in a closed loop control system

→ reduce overall gain

→ Increase sand width

→ Improve disturbance rejection

→ Reduce sensitivity to parameter variation

Therefore, negative feedback in a closed Loop control system doesn't reduce bandwidth.

17. Structure of memory chip

= number of row × number of column

= M × N

Number of row = number of Column

M × M = 16 KB

$M^2 = 16 \times 1024 = 2^{14}$

$M = 2^7$

Number of Address lines = $\log_2 M$

$$n = \log_2 2^7$$

$$n = 7$$

18. The given function is

$$F(x) = 1 - x^2 + x^3,$$

where $x \in [-1, 1]$

$$\Rightarrow \quad F'(x) = -2x + 3x^2$$

By mean value theorem

$$F'(x) = \frac{F(1) - F(-1)}{1-(-1)}$$

Now $F(1) = 1 - (1)^2 + (1)^3 = 1$

and $F(-1) = 1 - (-1)^2 + (-1)^3$

$= 1 - 1 - 1 = -1$

$$\therefore \quad F'(x) = \frac{1-(-1)}{1-(-1)} = \frac{2}{2} = 1$$

$\therefore \quad F'(x) = 1$

$\therefore \quad -2x + 3x^2 = 1$

$\therefore \quad 3x^2 - 2x - 1 = 0$

$\therefore \quad 3x^2 - 3x + x - 1 = 0$

$\therefore \quad 3x(x-1) \quad + 1\,(x-1) = 0$

$\therefore \quad (3x + 1) = 0$

or $\quad x - 1 = 0$

$\therefore \quad x = -\frac{1}{3}$

or $\quad x = 1$

Now $-\frac{1}{3}$ lies between (– 1, 1)

$\therefore \quad x = -\frac{1}{3}$

19. $$\vec{E} = 2\cos\left(10^8 t - \frac{z}{\sqrt{2}}\right)\hat{a}_x \text{ volt/met}$$

Compairing it with equation

$$\vec{E} = E_m \cos(\omega t - \beta z)\,\hat{a}_x \text{ volt/met}$$

$$\omega = 10^8 \text{ rad/sec}$$

$$\beta = \frac{1}{\sqrt{2}} \text{ rad/met}$$

$$\lambda = \frac{2\pi}{\beta} \text{met}$$

$$\lambda = \frac{2\pi}{1/\sqrt{2}}$$

$$= 2\sqrt{2}\pi \text{ met}$$

$$\boxed{\lambda = 8.886 \text{ met}}$$

21.

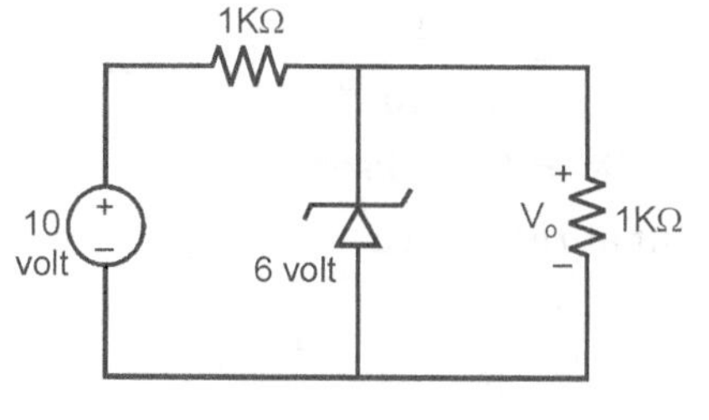

Using voltage divider Rule

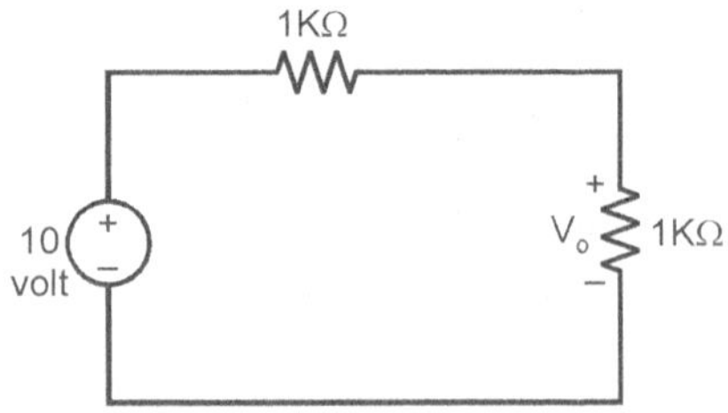

$$V_o = 10 \times \frac{1}{1+1}$$
$$= 5 \text{ volt}$$

since $V_o < V_z$

Zener wouldn't breakdown

Output voltage

$$V_o = 5 \text{ volt}$$

22. We know that for n bit D/A converter

$$V_0 = K\{2^{n-1} b_{n-1} + 2^{n-2} b_{n-2} + ... + 2^0 b_0\}$$

For 4 bit D/A converter

$n = 4$

$$V_0 = K\{2^3 b_3 + 2^2 b_2 + 2^1 b_1 + 2^0 b_0\}$$

when digital signal is 0001

$V_0 = 0.0615$ volt

$$0.0615 = K\{2^3 \times 0 + 2^2 \times 0 + 2^1 \times 0 + 2^0 \times 1\}$$

$K = 0.0615$

when digital signal is 1111

$V_0 = ?$

$$V_0 = 0.0615\{2^3 \times 1 + 2^2 \times 1 + 2^1 \times 1 + 2^0 \times 1\}$$
$$= 0.0615 \times 15$$

$\boxed{V_0 = 0.93}$ Volt

23.

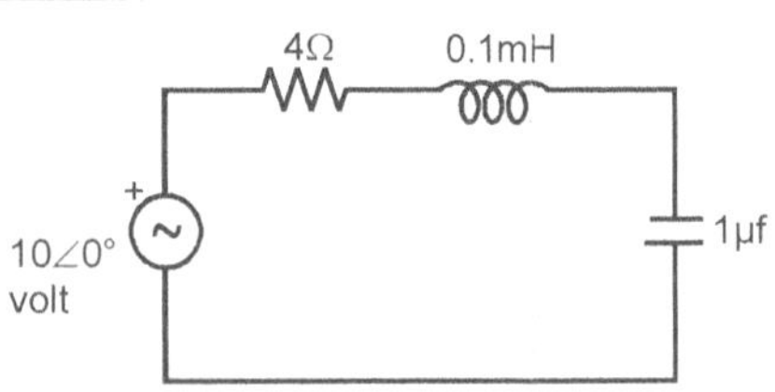

Resonant frequency

$$\omega_r = \frac{1}{\sqrt{LC}}$$
$$= \frac{1}{\sqrt{0.1 \times 10^{-3} \times 1 \times 10^{-6}}}$$
$$= \frac{1}{\sqrt{1 \times 10^{-10}}} = 10^5 \text{ rad/sec}$$

$\omega_r = 10^5$ rad/sec

Current at resonance

$$I = \frac{V}{R} = \frac{10}{4} = 2.5 \text{ Ap}$$

$$X_L = W_L = 10^5 \times 0.1 \times 10^{-3} = 10\ \Omega$$
$$V_L = IX_L = 2.5 \times 10$$
$$= 25 \text{ Volt}$$

At Resonance

$$V_C = V_L$$
$$V_C = 25 \text{ volt}$$

25.

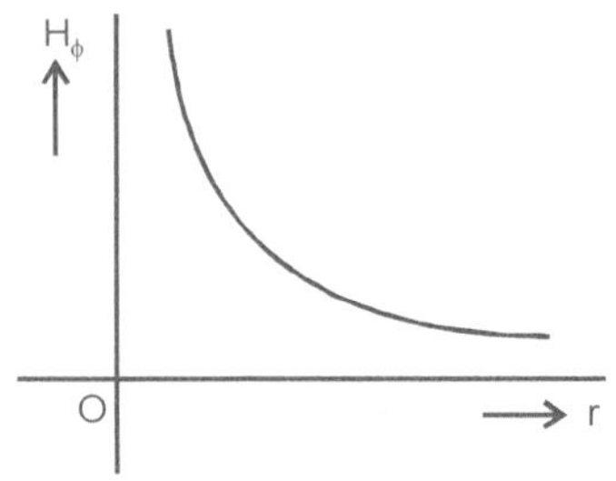

We know that magnetic field around a current carrying conductor

$$H_\phi = \frac{I}{2\pi r} i\phi$$
$$|H_\phi| = \frac{I}{2\pi r}$$
$$|H_\phi|\ \alpha\ \frac{1}{r}$$

28.

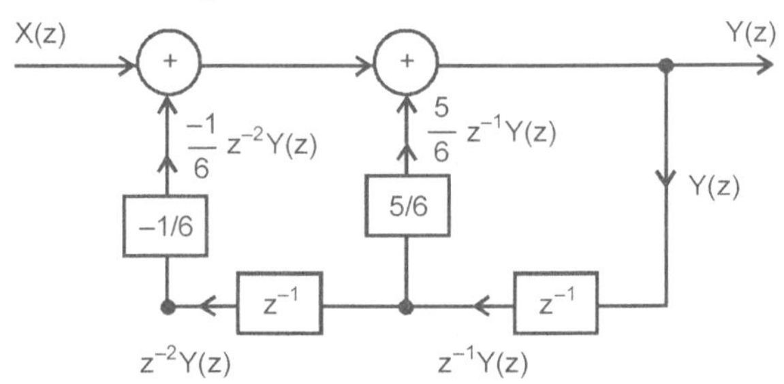

$$X(z) - \frac{1}{6} z^{-2}Y(z) + \frac{5}{6} z^{-1}Y(z) = Y(z)$$
$$Y(z) - \frac{5}{6} z^{-1}Y(z) + \frac{1}{6} z^{-2}Y(z) = X(z)$$
$$Y(z)\left\{1 - \frac{5}{6} z^{-1} + \frac{1}{6} z^{-2}\right\} = X(z)$$

T/F of the S/S $\quad H(z) = \dfrac{Y(z)}{X(z)}$

$$= \frac{1}{\left(1 - \frac{5}{6} z^{-1} + \frac{1}{6} z^{-2}\right)}$$
$$H(z) = \frac{z^2}{\left(z^2 - \frac{5}{6} z + \frac{1}{6}\right)}$$
$$H(z) = \frac{z^2}{\left(z - \frac{1}{2}\right)\left(z - \frac{1}{3}\right)}$$

Pole location

$$z-\frac{1}{2}=0 \quad \Rightarrow \quad Z=\frac{1}{2}$$

$$z-\frac{1}{3}=0 \quad \Rightarrow \quad Z=\frac{1}{3}$$

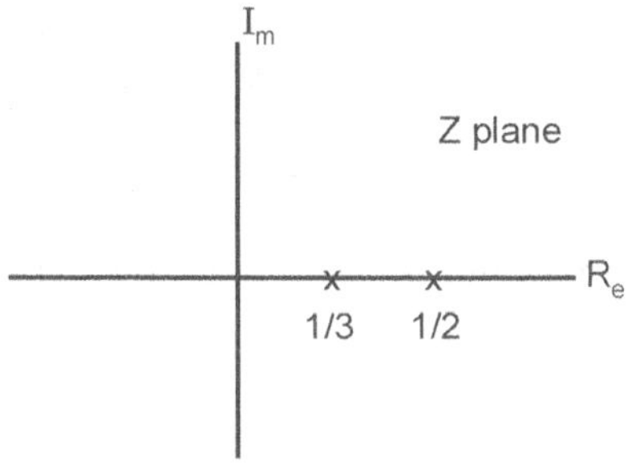

32. $V_{GS} = V_{DS}$

Hence MOS Transistor is in saturation.

In saturation,

$$I_D = k(V_{GS} - V_r)^2$$
$$= k(V_{DD} - V_{SS} - V_T)^2$$

As $V_{SS} \uparrow I_D \downarrow$ (Not linearly because square factor)

Hence option (*a*) correct

33.

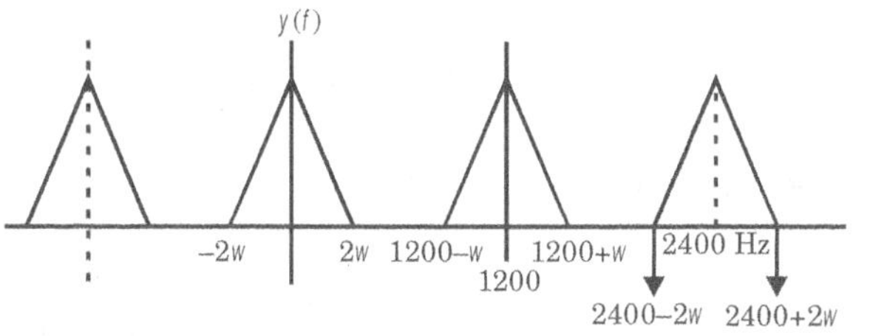

When output is 10 $x(t)$

It must be $1200 - w > 2w$ ...(1)

$2400 - 2w > 1700$ ...(2)

From (1) and (2)

$w < 400$ ...(3)

$w < 350$ ...(4)

So from (3) and (4)

$w < 350$

35. $E = 24\,\pi \cos(\omega t - \beta z)\,\bar{a}x\, \text{V/m}$

$\vec{P}_{avg}$ = average Poynting vector

$$= \frac{(24\pi)^2}{2\eta}\bar{a}_z$$

Surface over the plane

$$X + Y = 1(x + y + 0z = 1)$$

Normal to this plane is

$$\hat{n} = (1, 1, 0)$$

$$\text{Power crossing} = \int_S \bar{P}_{avg}.d\bar{S}$$
$$= \int_S \frac{(24\pi)^2}{2\eta}\bar{a}_z.(1,1,0)dS$$
$$= 0 \text{ mW}$$

36.

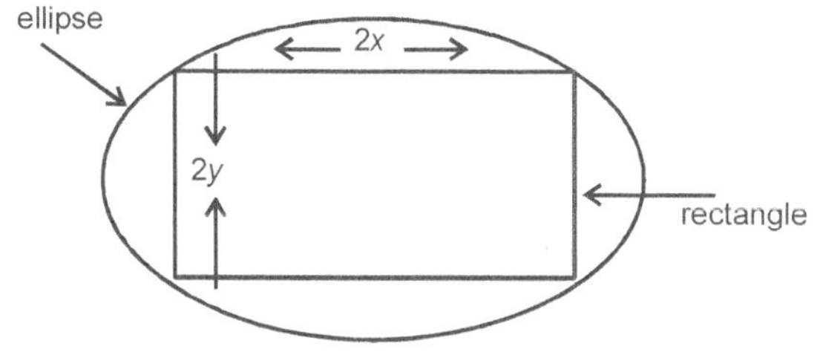

Here rectangle is inscribed in ellipse.

Let the length and breadth of rectangle be $2x$ and $2y$ respectively.

Now area of rectangle = $(2x) \times (2y)$

$= 4xy$

Let F = (Area of rectangle)^2

$= (4xy)^2$

$= 16x^2y^2$

$$= 16x^2\left(\frac{1-x^2}{4}\right)$$

$$\left[\because y^2 = \frac{1-x^2}{4} \text{ from equation of ellipse}\right]$$

$= 4x^2(1 - x^2)$

Now $F'(x) = 0$

$\therefore$ $x(1 - 2x^2) = 0$

$\therefore$ $x = 0$

or $1 - 2x^2 = 0$

$\Rightarrow$ $2x^2 = 1$

$\Rightarrow$ $x^2 = \frac{1}{2}$

$\Rightarrow$ $x = \frac{1}{\sqrt{2}}$

For $x = 0$,

$y^2 = \frac{1}{4}$

$\Rightarrow$ $y = \pm\frac{1}{2}$

and for $x = \frac{1}{\sqrt{2}}$

$$y^2 = \frac{1-\frac{1}{2}}{4} = \frac{1}{8}$$

$$\Rightarrow \quad y = \frac{1}{\sqrt{8}}$$

Now $\quad F''(x) = 8 - 48\,x^2$

$\therefore \quad F''(x) < 0$ for

$$x = \frac{1}{\sqrt{2}}$$

$\Rightarrow$ F(x) is maximum at $x = \frac{1}{\sqrt{2}}$

The maximum area of rectangle inscribed in ellipse is

$$4xy = 4\left(\frac{1}{\sqrt{2}}\right)\left(\frac{1}{\sqrt{8}}\right)$$

$$= \frac{4}{4} = 1$$

37.

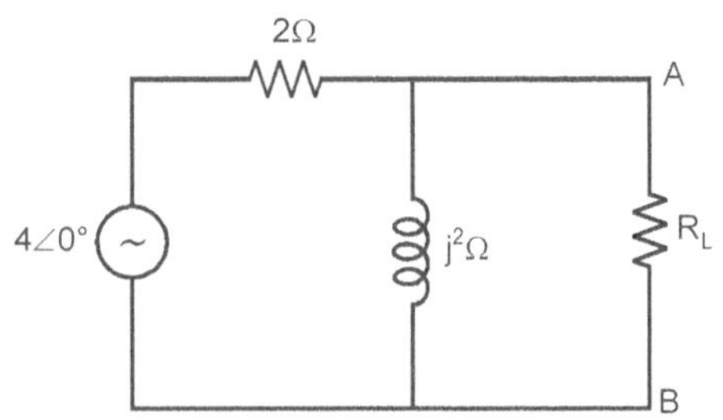

Thevenins equivalent

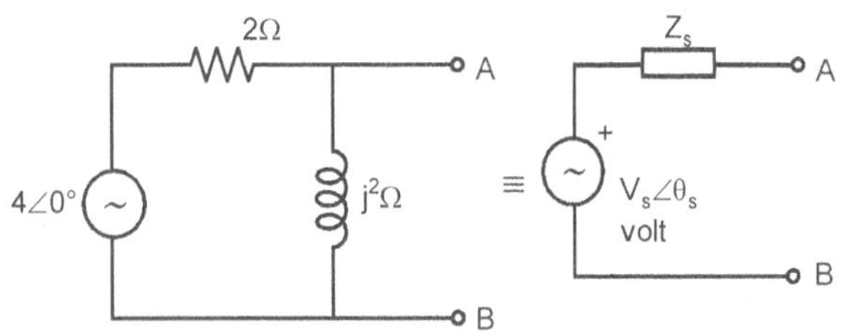

$$V_S\angle\theta_S = 4\angle 0^\circ \times \frac{j2}{(2+j2)}$$

$$= 2.828\ \angle 45^\circ \text{ volt}$$

$$= 2\sqrt{2}\ \angle 45^\circ \text{ volt}$$

$$Z_S = \frac{2\times(j2)}{(2+j2)}$$

$$= (1 + 1j)\Omega$$

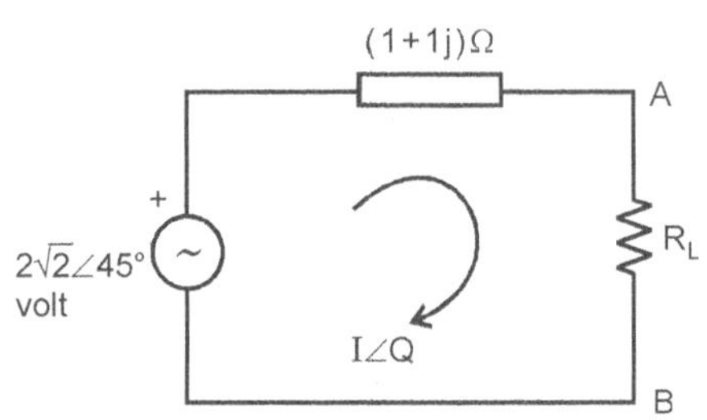

$$I\angle\theta = \frac{2\sqrt{2}\angle 45^\circ}{(1+R_L)+j1}$$

$$P_L = I^2R_L$$

where $\quad I = |I\angle\theta|$

$$= \frac{2\sqrt{2}}{\sqrt{(1+R_L)^2+(1)^2}}$$

$$P_L = \frac{8.\,R_L}{\{(R_L+1)^2+1\}}$$

For maximum power transfer to the load

$$\frac{dP_L}{dR_L} = 0$$

$$\frac{dP_L}{dR_L} = 8\left\{\frac{((R_L+1)^2+1).1 - R_L(2(R_L+1))}{\{(R_L+1)^2+1\}^2}\right\} = 0$$

$$= 8\{(R_L+1)^2 + 1 - 2R_L^2 - 2R_L\} = 0$$

$$8\{2 - R_L^2\} = 0$$

$$\boxed{R_L = \sqrt{2}\ \Omega}$$

$$P_L = \frac{8}{(\sqrt{2}+1)^2+1} \times \sqrt{2}$$

$$= 1.66 \text{ watts}$$

$$\boxed{P_L = 1.66 \text{ watts}}$$

38. The transfer function of a phase lead N/W is given as

$$G(s) = \frac{(s+2)}{(s+4)} = \frac{2\left(1+\frac{s}{2}\right)}{4\left(1+\frac{s}{4}\right)}$$

$$G(s) = \frac{0.5(1+0.5s)}{(1+0.25s)}$$

Phase lead N/W is

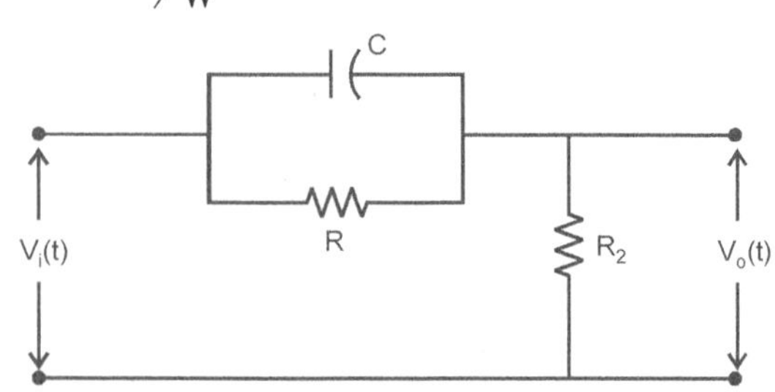

T/F of phase lead N/W is

$$G(s) = \frac{V_o(s)}{V_i(s)} = \frac{\alpha(1+\tau s)}{\alpha(1+\alpha\tau s)}$$

Where $\quad \alpha = \frac{R_2}{R+R_2}, \ \tau = RC$

compairing with standard T/F

$$\tau = RC = 0.5$$

$$\boxed{RC = 0.5}$$

39. Under channel length modulation

$$I_D = I_{Dsat}(1 + \lambda V_{DS})$$

$$\frac{dI_D}{dV_{DS}} = \frac{I}{r_0} = \lambda I_{Dsat}$$

$$r_0 = \frac{1}{\lambda I_{Dsat}} = \frac{1}{0.05 \times 10^{-3}} = 20k\Omega$$

40.

$$C_j \propto \frac{1}{\sqrt{V_{b1} + V_R}}$$

$$\frac{C_{2j}}{C_{1j}} = \sqrt{\frac{V_{b1} + V_{R_1}}{V_{b1} + V_{R_2}}}$$

$$C_{2j} = C_{1j}\sqrt{\frac{2}{8}} = \frac{C_{1j}}{2} = 2.5\ pF$$

41.

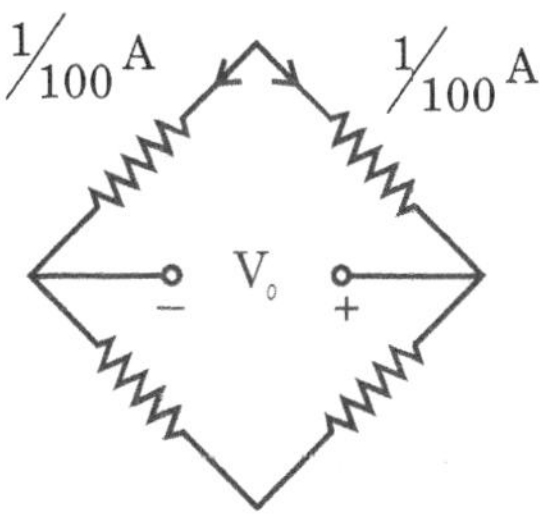

$$I_{50\Omega} = \frac{1}{50}A = I_{100\Omega}$$

$$V_o = \frac{1}{100}[250(1+\delta) - 250(1-\delta)]$$

$$= \frac{1}{100} \times 250 \times 2\delta$$

$$\doteq \frac{1}{1000} \times 250 = 0.25\ V$$

$$= 250\ mV$$

42. Consider series R-L-C network circuit

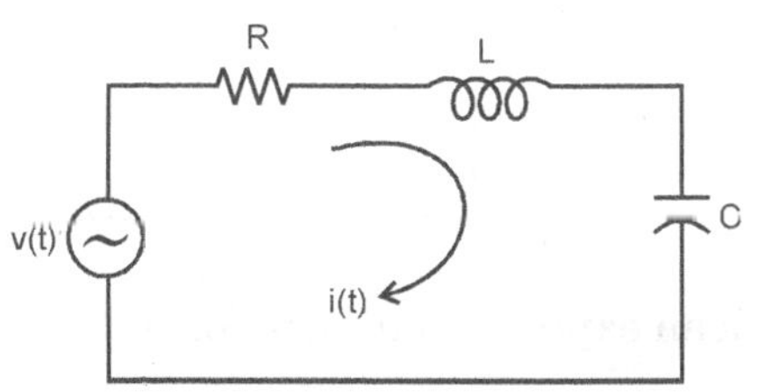

$$V(t) = Ri(t) + L\frac{di(t)}{dt} + \frac{1}{C}\int i(t)dt$$

Taking Laplace Transformation both sides assuming initial conditions are zero.

$$V(s) = RI(s) + LsI(s) + \frac{1}{Cs}I(s)$$

$$V(s) = \left(R + Ls + \frac{1}{Cs}\right)I(s)$$

T/F of the system

$$\frac{I(s)}{V(s)} = \frac{Cs}{(LCs^2 + RCs + 1)}$$

$$= \frac{\frac{1}{L}s}{s^2 + \frac{R}{L}s + \frac{1}{LC}}$$

Characteristic equation of the s/s

$$s^2 + \frac{R}{L}s + \frac{1}{LC} = 0 \quad ...(i)$$

Comparing with standard characteristic equation

$$s^2 + 2\xi\omega_n s + \omega_n^2 = 0$$

$$\omega_n^2 = \frac{1}{LC}$$

$$\omega_n = \frac{1}{\sqrt{LC}}\ \text{rad/sec}$$

$$2\xi\omega_n = \frac{R}{L}$$

$$2\xi \cdot \frac{1}{\sqrt{LC}} = \frac{R}{L}$$

$$\xi = \frac{R}{2L} \times \sqrt{LC} = \frac{R}{2}\sqrt{\frac{C}{L}}$$

Damping ratio $\boxed{\xi = \frac{R}{2}\sqrt{\frac{C}{L}}}$

47. $G(s) = \left(K_P + \frac{K_1}{s}\right)\frac{1}{s(s+2)}$

$$G(s) = \frac{sK_P + K_1}{s^2(s+2)}$$

Closed Loop Characterstic equation

$$1 + G(s) = 0$$

$$1 + \frac{sK_P + K_1}{s^2(s+2)} = 0$$

$$s^3 + 2s^2 + sK_P + K_1 = 0$$

Routh Array

s^3	1	K_P
s^2	2	K_1
s	$\frac{2K_P - K_1}{2}$	0
s^0	K_1	0

For closed loop stability

$$K_1 > 0$$

$$2K_P - K_1 > 0$$
$$K_1 < 2K_P$$
$$\frac{K_1}{2} < K_P$$
$$K_P > \frac{K_1}{2} > 0$$

50. The open loop transfer function with unity feed back

$$G(s) = \frac{K(s+4)}{(s+8)(s^2-9)}$$
$$= \frac{K(s+4)}{(s+8)(s-3)(s+3)}$$

for a point lie on root locus,

Two criteria must be fulfilled.

(*i*) Magnitude criteria $|G(s)\,H(s)| = 1$

(*ii*) Angle criteria $\angle G(s)H(s) = \pm 180°$

For determining the value of K

Magnitude criteria must be fulfilled

$$|G(s)H(s)|_{s=s_1} = 1$$
$$\left|\frac{K(s+4)}{(s+8)(s+3)(s-3)}\right|_{s=-1+j2} = 1$$
$$\frac{|K||(-1+j2+4)|}{|(-1+j2+8)||(-1+j2+3)||(-1+j2-3)|} = 1$$
$$\frac{K|(3+2j)|}{|(7+2j)||(2+2j)||(-4+2j)|} = 1$$
$$\frac{K(\sqrt{13})}{(\sqrt{53})(\sqrt{8})(\sqrt{20})} = 1$$
$$K = \frac{(\sqrt{53})(\sqrt{8})(\sqrt{20})}{(\sqrt{13})}$$
$$K = 25.54$$

52. The given differential equation is of the form

$$(D^2 + 2D + 1)y = 0$$
$$\therefore \quad D^2 + 2D + 1 = 0$$
$$\therefore \quad (D + 1)^2 = 0$$
$$\therefore \quad D = -1, -1$$

$\therefore$ Solution of above differential equation is

$$y(t) = (c_1 + c_2t)e^t$$

(Complimentary function)

$$\therefore \quad y'(t) = c_2e^{-t} + (c_1 + c_2t)(-e^{-t})$$
$$\therefore \quad y(0) = 1,\ y'(0) = 1$$

gives $\quad c_1 = 1$

and $\quad c_2 + c_1(-1) = 1$

$\therefore \quad c_2 = 2$

Hence $\quad y(t) = (1 + 2t)e^{-t}$

54. $f(x) = e^{-x}(x^2 + x + 1)$

function at origin

$$f(0) = e^{-0}(0 + 0 + 1) = 1$$
$$\boxed{f(0) = 1}$$

for maximum or minimum value of function

$$\frac{df(x)}{dx} = 0$$
$$\frac{df(x)}{dx} = e^{-x}(2x + 1) + (x^2 + x + 1)\,e^{-x}(-1) = 0$$
$$= e^{-x}(x^2 + x + 1 - 2x - 1) = 0$$
$$= e^{-x}(x^2 - x) = 0$$
$$e^{-x}.x(x-1) = 0$$
$$x = 1$$
$$f(1) = e^{-1}(1 + 1 + 1) = 3e^{-1}$$
$$\boxed{f(1) = 1.10}$$

$$\frac{d^2 f(x)}{dx^2} < 0 \text{ (It is maximum)}$$
$$f(x) = (x^2 + x + 1)e^{-x}$$
$$= \left(x^2 + x + \frac{1}{4} + \frac{3}{4}\right)e^{-x}$$
$$= \left\{\left(x + \frac{1}{2}\right)^2 + \frac{3}{4}\right\}e^{-x}$$

Value of $f(x)$ can't be negative for all positive value of x.

at $x \to \infty$

$e^{-x} \to 0$

$$f(x)\big|_{x\to\infty} \to 0$$

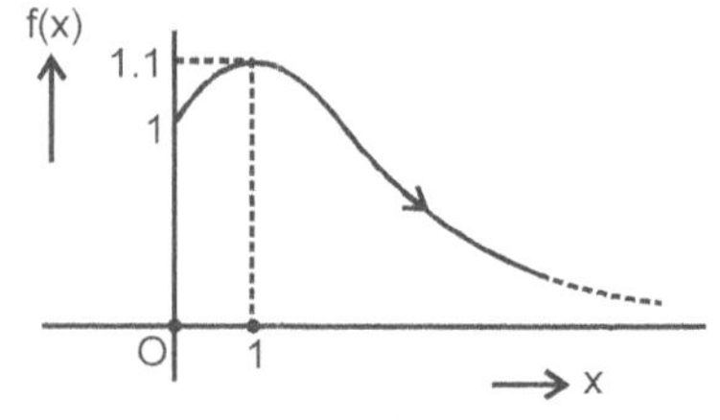

55. Boolean expression in SOP form

$$F(X, Y, Z) = \bar{X}Y\bar{Z} + X\bar{Y}\bar{Z} + XY\bar{Z} + XYZ$$
$$= \sum m\,(2, 4, 6, 7)$$
$$= \pi m\,(0, 1, 3, 5)$$
$$= (X + Y + Z)\cdot(X + Y + \bar{Z})$$
$$\cdot\,(X + \bar{Y} + \bar{Z})\cdot(\bar{X} + Y + \bar{Z})$$

SOLVED PAPER – 2015

ELECTRONICS & COMMUNICATION ENGINEERING

Set - 2

No. of Questions : 65 *Maximum Marks : 100*

GENERAL APTITUDE (GA)

Q. No. 1 – 5 Carry One Mark Each

1. What is the adverb for the given word below ?
   Misogynous
   (*a*) Misogynousness (*b*) Misogynity
   (*c*) Misogynously (*d*) Misogynous

2. Choose the appropriate word–phrase out of the four options given below, to complete the following sentence.
   Dhoni, as well as the other ream members of Indian team ________ present on the occasion
   (*a*) Were (*b*) Was
   (*c*) Has (*d*) Have

3. Ram and Ramesh appeared in an interview for two vacancies in the same department. The probability of Ram's selection is 1/6 and that of Ramesh is 1/8. What is the probability that only one of them will be selected ?
   (*a*) $\frac{47}{18}$ (*b*) $\frac{1}{4}$
   (*c*) $\frac{13}{48}$ (*d*) $\frac{35}{48}$

4. Choose he word most similar in meaning to the given word :
   Awkward
   (*a*) Inept (*b*) Graceful
   (*c*) Suitable (*d*) Dreadful

5. An electric bus has onboard instruments that report the total electricity consumed since the start of the trip as well as the total distance covered. During a single day of operation, the bus travels on stretches M, N, O and P, in that order. The cumulative distances traveled and the corresponding electricity consumption are shown in the Table below

Stretch	Cumulative distance (km)	Electricity used (kWh)
M	20	12
N	45	25
O	75	45
P	100	57

The stretch where the electricity consumption per km is minimum is
(*a*) M (*b*) N
(*c*) O (*d*) P

Q. No. 6 – 10 Carry Two Marks Each

6. Given below are two statements followed by two conclusions. Assuming these statements to be true, decide which one logically follows.
   Statements :
   I. All film stars are playback singers.
   II. All film directors are film stars
   Conclusions :
   I. All film directors are playback singers.
   II. Some film stars are film directors.
   (*a*) Only conclusion I follows
   (*b*) Only conclusion II follows
   (*c*) Neither conclusion I nor II follows
   (*d*) Both conclusions I and II follow

7. Lamenting the gradual sidelining of the arts ill school curricula, a group of prominent artists wrote to the Chief Minister last year, asking him to allocate rnore funds to support arts education in schools. However, no such increase has been announced in this year's Budget. The artists expressed their deep anguish at their request not being approved, but many of them remain optimistic about funding in the future
   Which of the statement(s) below is/are logically valid and can be inferred from the above statements ?
   (*i*) The artists expected funding for the arts to increase this year
   (*ii*) The Chief Minister was receptive to the idea of increasing funding for the arts
   (*iii*) The Chief Minister is a prominent artist
   (*iv*) Schools are giving less importance to arts education nowadays
   (*a*) (*iii*) and (*iv*) (*b*) (*i*) and (*iv*)
   (*c*) (*i*), (*ii*) and (*iv*) (*d*) (*i*) and (*iii*)

8. A tiger is 50 leaps of its own behind a deer. The tiger takes 5 leaps per minute to the deer's 4. If the tiger and the deer cover 8 metre and 5 metre per leap respectively. What distance in metres will be tiger have to run before it catches the deer?

9. If $a^2 + b^2 + c^2 = 1$, then $ab + bc + ac$ lies in the interval

(a) $\left[1, \frac{2}{3}\right]$ (b) $\left[-\frac{1}{2}, 1\right]$

(c) $\left[-1, \frac{1}{2}\right]$ (d) [2, – 4]

10. In the following sentence certain parts are underlined and marked P, Q and R. One of the parts may contain certain error or may not be acceptable in standard written communication. Select the part containing an error. Choose D as your answer if there is no error.

The student corrected $\underset{\text{P}}{\underline{\text{all the errors}}}$ that $\underset{\text{Q}}{\underline{\text{the instructor marked}}}$ on the $\underset{\text{R}}{\underline{\text{answer book}}}$

(a) P (b) Q

(c) R (d) No error

ELECTRONICS AND COMMUNICATION ENGINEERING

Q. No. 1 – 25 Carry One Mark Each

1. Let the signal $f(t) = 0$ outside the interval T_1, T_2, where T_1 and T_2 are finite. Furthermore, $|f(t)| < \infty$. The region of convergence (RoC) of the signal's bilateral Laplace transform F(s) is

1. A parallel strip containing the $j\Omega$ axis
2. A parallel strip not containing the $j\Omega$ axis
3. The entire s–plane
4. A half plane containing the $j\Omega$ axis

(a) 1 (b) 2

(c) 3 (d) 4

2. A unity negative feedback system has an open–loop transfer function $G(s) = \frac{K}{s(s+10)}$. The gain K for the system to have a damping ratio of 0.25 is ________.

3. A mod-n counter using a synchronous binary up–counter with synchronous clear input is shown in the figure. The value of n is ________.

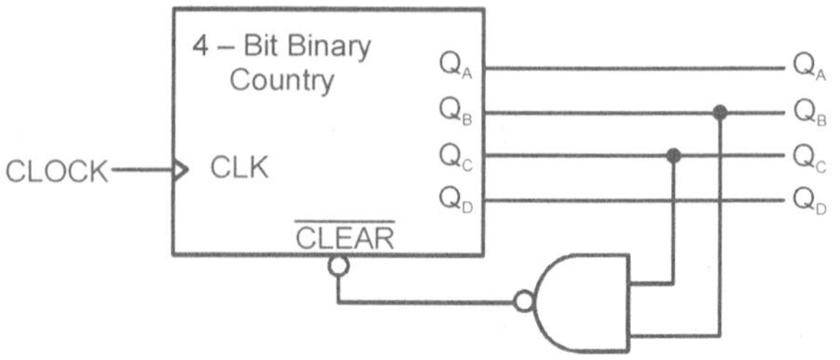

4. By performing cascading and/or summing/differencing operations using transfer function blocks $G_1(s)$ and $G_2(s)$, one **CANNOT** realize a transfer function of the form

(a) $G_1(s)G_2(s)$

(b) $\frac{G_1(s)}{G_2(s)}$

(c) $G_1(s)\left(\frac{1}{G_1(s)} + G_2(s)\right)$

(d) $G_1(s)\left(\frac{1}{G_1(s)} - G_2(s)\right)$

5. The electric field of a uniform plane electromagnetic wave is

$$\vec{E} = \left(\vec{a}_x + j4\vec{a}_y\right)\exp\left[j\left(2\pi \times 10^7 t - 0.2z\right)\right]$$

The polarization of the wave is

(a) Right handed circular

(b) Right handed elliptical

(c) Left handed circular

(d) Left handed elliptical

6. A piece of silicon is doped uniformly with phosphorous with a doping concentration of $10^{16}/cm^3$. The expected value of mobility versus doping concentration for silicon assuming full dopant ionization is shown below. The charge of an electron is 1.6×10^{-19}C. The conductivity (in S cm^{-1}) of the silicon sample at 300 K is ________

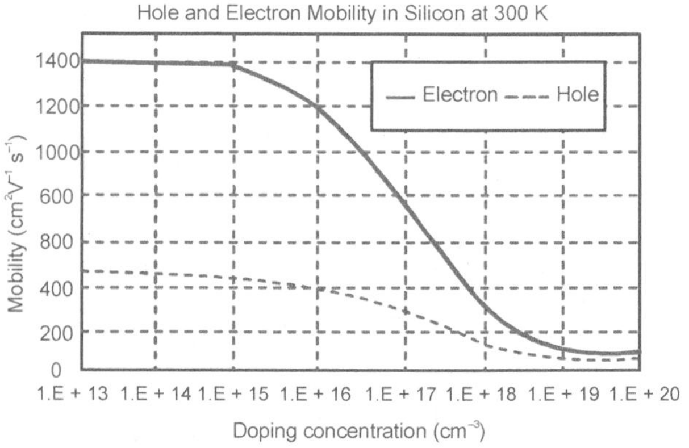

7. In the figure shown, the output Y is required to be $Y = AB + \overline{C}\overline{D}$. The gates G_1 and G_2 must be, respectively,

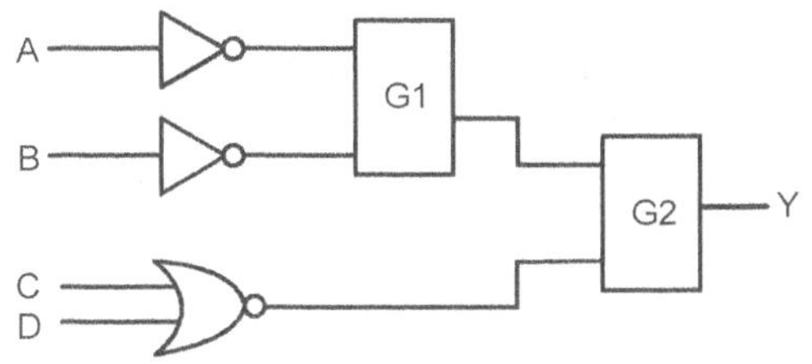

(*a*) NOR, OR (*b*) OR, NAND

(*c*) NAND, OR (*d*) AND, NAND

8. In the bistable circuit shown, the ideal opamp has saturation level of ± 5V. The value of R_1 (in kΩ) that gives a hysteresis width of 500 mV is __________.

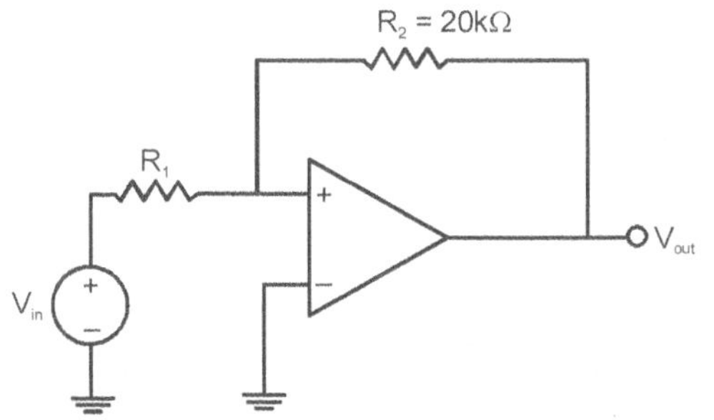

9. Two causal discrete–time signal $x[n]$ and $y[n]$ are related as $y[n] = \sum_{m=0}^{n} x[m]$

It the *z*–transform of $y[n]$ is $\dfrac{2}{z(z-1)^2}$, the value of $x[2]$ is __________.

10. The bilateral Laplace transform of a function

$$f(t) = \begin{cases} 1 \text{ if } a \le t \le b \\ 0 \text{ otherwise} \end{cases}$$

(*a*) $\dfrac{a-b}{s}$ (*b*) $\dfrac{e^s(a-b)}{s}$

(*c*) $\dfrac{e^{-as} - e^{-bs}}{s}$ (*d*) $\dfrac{e^{s(a-b)}}{s}$

11. The 2–port admittance matrix of the circuit shown is given by

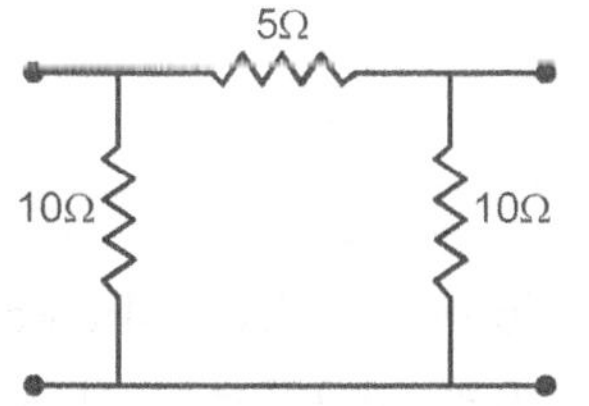

(*a*) $\begin{bmatrix} 0.3 & 0.2 \\ 0.2 & 0.3 \end{bmatrix}$ (*b*) $\begin{bmatrix} 15 & 5 \\ 5 & 15 \end{bmatrix}$

(*c*) $\begin{bmatrix} 3.33 & 5 \\ 5 & 3.33 \end{bmatrix}$ (*d*) $\begin{bmatrix} 0.3 & 0.4 \\ 0.4 & 0.3 \end{bmatrix}$

12. The value of *x* for which all the eigen–values of the matrix given below are real is

$$\begin{bmatrix} 10 & 5+j & 4 \\ x & 20 & 2 \\ 4 & 2 & -10 \end{bmatrix}$$

(*a*) $5 + j$ (*b*) $5 - j$

(*c*) $1 - 5j$ (*d*) $1 + 5j$

13. The signal $\cos\left(10\pi t + \frac{\pi}{4}\right)$ is ideally sampled at a sampling frequency of 15 Hz. The sampled signal is passed through a filter with impulse response $\left(\frac{\sin(\pi t)}{\pi t}\right)\cos\left(10\pi t - \frac{\pi}{2}\right)$. The filter output is

(*a*) $\frac{15}{2}\cos\left(40\pi t - \frac{\pi}{4}\right)$

(*b*) $\frac{15}{2}\left(\frac{\sin(\pi t)}{\pi t}\right)\cos\left(10\pi t + \frac{\pi}{4}\right)$

(*c*) $\frac{15}{2}\cos\left(10\pi t - \frac{\pi}{4}\right)$

(*d*) $\frac{15}{2}\left(\frac{\sin(\pi t)}{\pi t}\right)\cos\left(40\pi t - \frac{\pi}{2}\right)$

14. A sinusoidal signal of amplitude A is quantized by a uniform quantizer Assume that the signal utilizes all the representation levels of the quantizer. If the signal to quantization noise ratio is 31.8 dB, the number of levels in the quantizer is __________.

15. The magnitude and phase of the complex Fourier series coefficients a_k of a periodic signal $x(t)$ are shown in the figure. Choose the correct statement form the four choices given.

Notation : C is the set of complex numbers, R is the set of purely real numbers, and P is the set purely imaginary numbers.

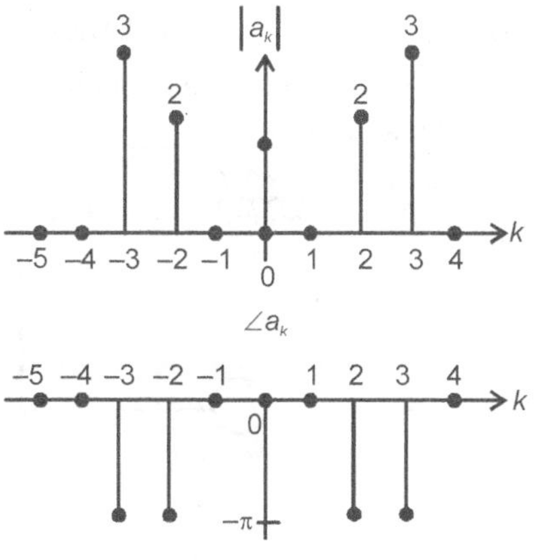

(a) $x(t) \in R$

(b) $x(t) \in P$

(c) $x(t) \in (C - R)$

(d) The information given is not sufficient to draw any conclusion about $x(t)$

16. The general solution of the differential equation

$\frac{dy}{dx} = \frac{1+\cos 2y}{1-\cos 2x}$ is

(a) $\tan y - \cot x = c$ (c is a constant)

(b) $\tan x - \cot y = c$ (c is a constant)

(c) $\tan y + \cot x = c$ (c is a constant)

(d) $\tan x + \cot y = c$ (c is a cons tant)

17. An n–type silicon sample is uniformly illuminated with light which generates 10^{20} electron hole pairs per cm^3 per second. The minority carrier lifetime in the sample is $1\mu s$. Int eh steady state, the hole concentration in the sample is approximately 10^x, where x is an integer. The value of x is ________

18. If the circuit shown has to function as a clamping circuit, which one of the following conditions should be satisfied for sinusoidal signal of period T ?

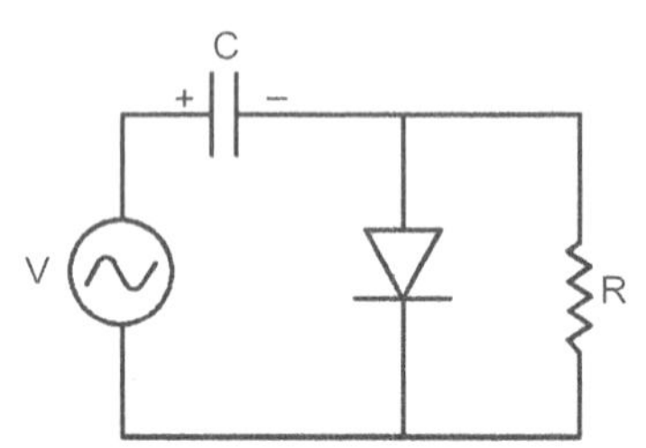

(a) RC << T (b) RC = 0.35 T

(c) RC ≈ T (d) RC >> T

19. In a source free region in vacuum, if the electrostatic potential $\varphi = 2x^2 + y^2 + cz^2$, the value of constant c must be ____________

20. In an 8085 microprocessor, which one of the following instructions changes the content of the accumulator ?

(a) MOV B, M (b) PCHL

(c) RNZ (d) SBI BEH

21. The voltage (V_C) across the capacitor (in Volts) in the network shown is ________

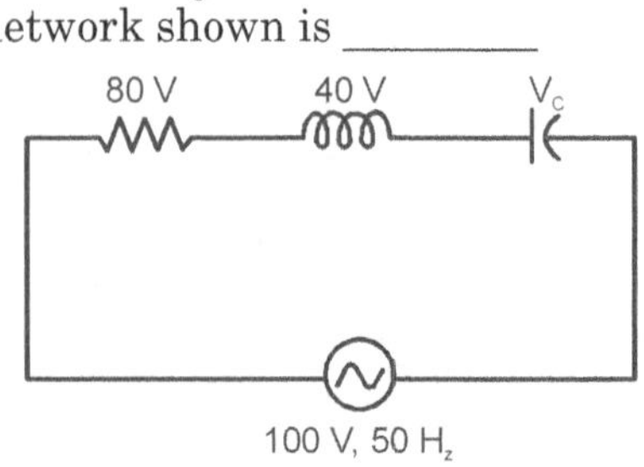

22. Let $f(z) = \frac{az+b}{cz+d}$. If $f(z_1) = f(z_2)$ for all $z_1 \neq z_2$, $a = 2$, $b = 4$ and $c = 5$, then d should be equal to ______.

23. In the circuit shown the average value of the voltage V_{ab} (in Volts) in steady state condition is __________.

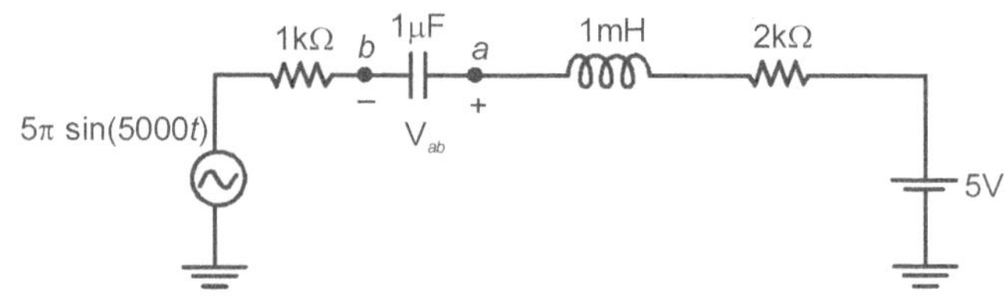

24. For the signal flow graph shown in the figure, the value of $\frac{C(s)}{R(s)}$ is

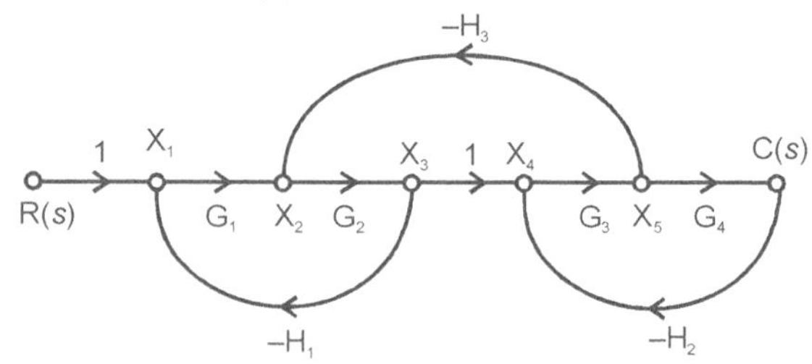

(a) $\frac{G_1G_2G_3G_4}{1-G_1G_2H_1-G_3G_4H_2-G_2G_3H_3+G_1G_2G_3G_4H_1H_2}$

(b) $\frac{G_1G_2G_3G_4}{1+G_1G_2H_1+G_3G_4H_2+G_2G_3H_3+G_1G_2G_3G_4H_1H_2}$

(c) $\frac{1}{1+G_1G_2H_1+G_3G_4H_2+G_2G_3H_3+G_1G_2G_3G_4H_1H_2}$

(d) $\frac{1}{1-G_1G_2H_1-G_3G_4H_2-G_2G_3H_3+G_1G_2G_3G_4H_1H_2}$

25. In the circuit shown, $V_0 = V_{0A}$ for switch SW in position A and $V_0 = V_{0B}$ SW in position B. Assume that the opamp is ideal. The value of $\frac{V_{0B}}{V_{0A}}$ is _____

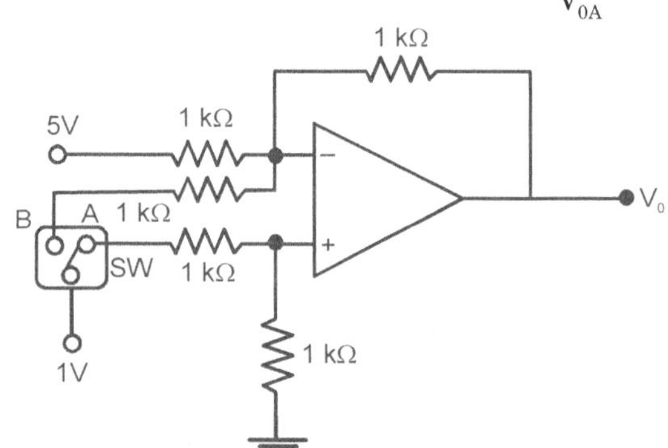

Q. No. 26 – 55 carry Two Marks Each

26. Let X ∈ {0, 1) and Y ∈ {0, 1} be two independent binary random variables. If P (X = 0) = p and P(Y = 0) = q, then P(X + Y ≥ 1) is equal to

(a) $pq + (1 - p)(1 - q)$ (b) pq

(c) $p(1 - q)$ (d) $1 - pq$

27. An LC tank circuit consists of an ideal capacitor C connected in parallel with a coil of inductance L having an internal resistance R. The resonant frequency of the tank circuit is

(*a*) $\frac{1}{2\pi\sqrt{LC}}$ (*b*) $\frac{1}{2\pi\sqrt{LC}}\sqrt{1-R^2\frac{C}{L}}$

(*c*) $\frac{1}{2\pi\sqrt{LC}}\sqrt{1-\frac{L}{R^2C}}$ (*d*) $\frac{1}{2\pi\sqrt{LC}}\left(1-R^2\frac{C}{L}\right)$

28. $\{X_n\}_{n=-\infty}^{n=\infty}$ is an independent and identically distributed (*i, i, d,*) random process with X_n equally likely to be +1 or – 1. $\{Y_n\}_{n=-\infty}^{n=\infty}$ is another random process obtained as $Y_n = X_n + 0.5\,X_{n-1}$. The autocorrelation function of $\{Y_n\}_{n=-\infty}^{n=\infty}$ denoted by $R_Y[k]$ is

(*a*)

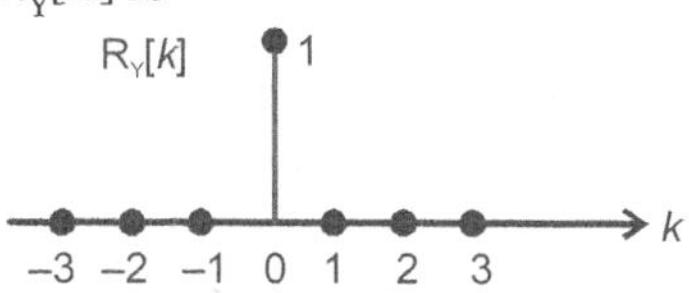

(*b*)

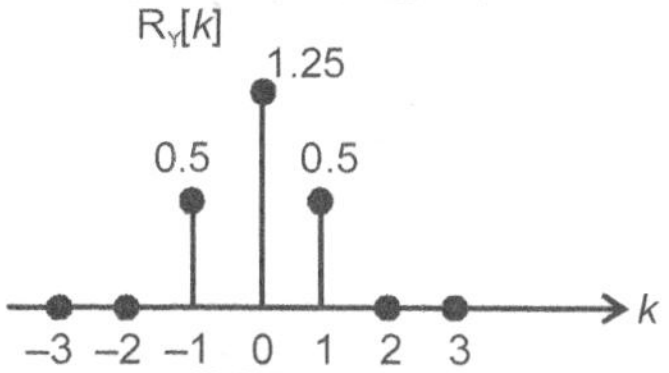

(*c*)

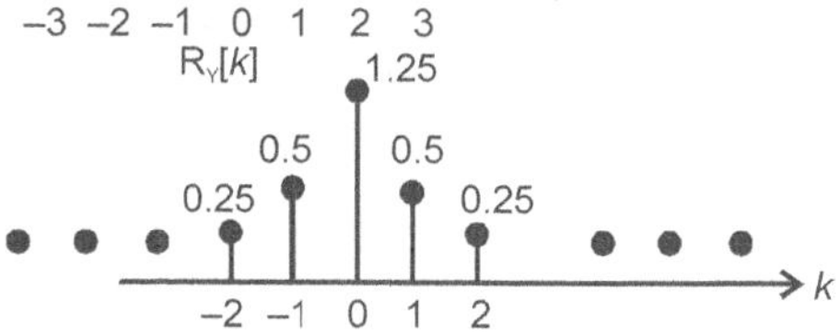

(*d*)

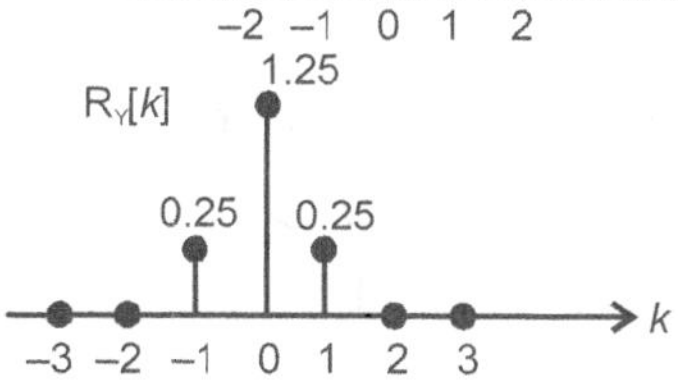

29. In a MOS capacitor with an oxide layer thickness of 10 nm, the maximum depletion layer thickness is 100 nm. The permittivities of the semiconductor and the oxide layer are ε_s and ε_{ox} respectively. Assuming $\varepsilon_s/\varepsilon_{ox} = 3$, the ratio of the maximum capacitance to the minimum capacitance of this MOS capacitor is ____________

30. Let the random variable X represent the number of times a fair coin needs to be tossed till two consecutive heads appear for the first time. The expectation of X is ____________

31. In the circuit shown, the Norton equivalent resistance (in Ω) across terminals $a - b$ is ________.

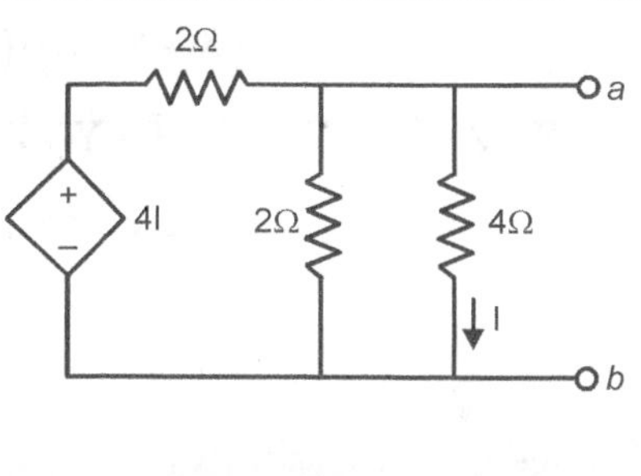

32. The figure shows a binary counter with synchronous clear input. With the decoding logic shown, the counter works as a

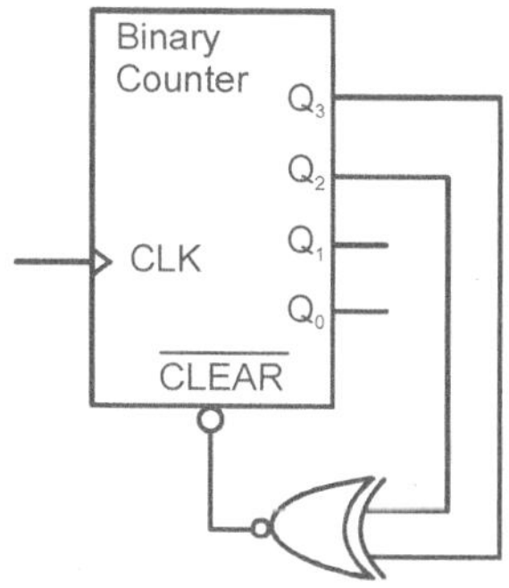

(*a*) mod–2 counter (*b*) mod–4 counter
(*c*) mod–5 counter (*d*) mod–6 counter

33. In the ac equivalent circuit shown, the two BJTs are bbiased in active region and have identical parameters with β >> 1. The open circuit small signal voltage gain is approximately ____________.

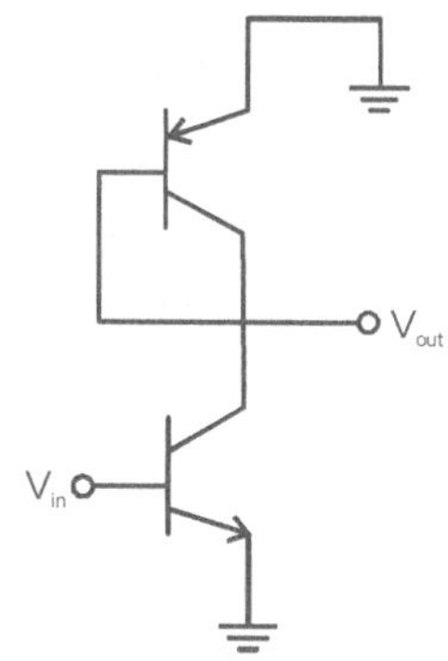

34. The state variable representation of a system is given as

$$\dot{x}=\begin{bmatrix}0 & 1\\ 0 & -1\end{bmatrix}x;\ \ x(0)=\begin{bmatrix}1\\ 0\end{bmatrix}$$

$$y=\begin{bmatrix}0 & 1\end{bmatrix}x$$

The response $y(t)$ is

(*a*) $\sin(t)$ (*b*) $1 - e^t$
(*c*) $1 - \cos(t)$ (*d*) 0

35. Consider the differential equation $\frac{dx}{dt} = 10 - 0.2x$ with initial condition $x(0) = 1$. The response $x(t)$ for $t > 0$ is

(*a*) $2 - e^{-0.2t}$ (*b*) $2 - e^{0.2t}$
(*c*) $50 - 49e^{-0.2t}$ (*d*) $50 - 49\,e^{0.2t}$

36. For the voltage regulator circuit shown, the input voltage (V_{in}) is 20V ± 20% and the regulated output volage (V_{out}) is 10V. Assume the opamp to be ideal. For a load R_L drawing 200 mA, the maximum power dissipation in Q_1 (in Watts) is ____________.

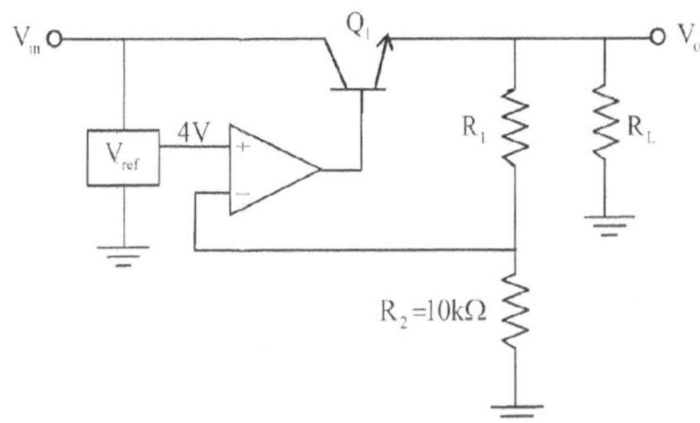

37. Input $x(t)$ and output $y(t)$ of an LTI system are related by the differential equation $y''(t) - y'(t) - 6y(t) = x(t)$. If the system is neither causal nor stable, the impulse response $h(t)$ of the system is

(a) $\frac{1}{5}e^{3t}u(-t)+\frac{1}{5}e^{-2t}u(-t)$

(b) $-\frac{1}{5}e^{3t}u(-t)+\frac{1}{5}e^{-2t}u(-t)$

(c) $\frac{1}{5}e^{3t}u(-t)-\frac{1}{5}e^{-2t}u(t)$

(d) $-\frac{1}{5}e^{3t}u(-t)-\frac{1}{5}e^{-2t}u(t)$

38. The diode in the circuit given below has $V_{ON} = -0.7$ V but is ideal otherwise. The current (in mA) in the 4kΩ resistor is ________.

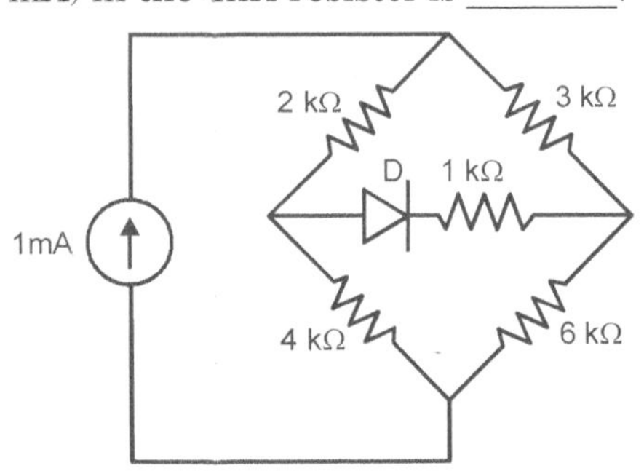

39. A zero mean white Gaussian noise having power spectral density $\frac{N_0}{2}$ is passed through an LTI filter whose impulse response $h(t)$ is shown in the figure. The variance of the filtered noise at $t = 4$ is

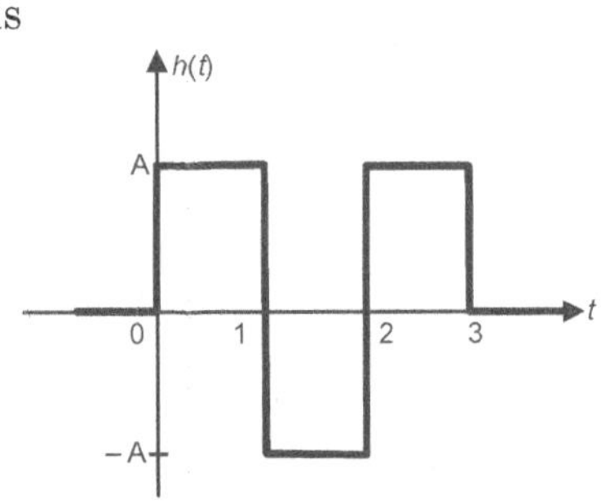

(a) $\frac{3}{2}A^2N_0$ (b) $\frac{3}{4}A^2N_0$

(c) A^2N_0 (d) $\frac{1}{2}A^2N_0$

40. Assuming that the opamp in the circuit shown below is ideal, the output voltage V_0 (in volts) is ___________

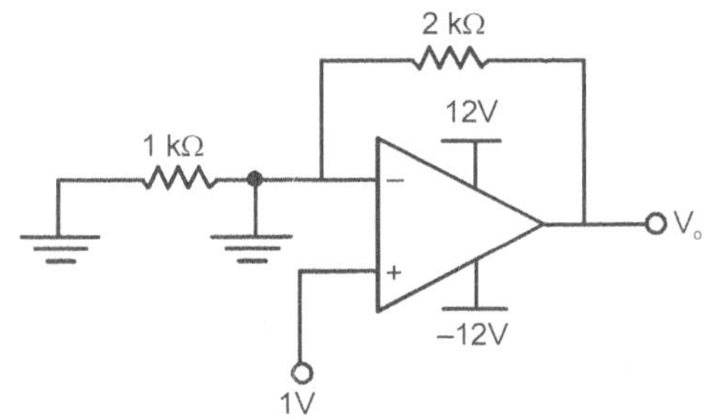

41. A 1 – to – 8 demultiplexer with data input in D_{in}, address inputs S_0, S_1, S_2 (with S_0 as the LSB) and $\overline{Y}_0$ to $\overline{Y}_7$ as the eight demultiplexed output, is to be designed using two 2 – to – 4 decoders (with enable input $\overline{E}$ and address input A_0 and A_1) as shown in the figure D_{in}, S_0, S_1 and S_2 are to be connected to P, Q, R and S, but not necessarily in this order. The respective input connections to P, Q, R and S terminals should be

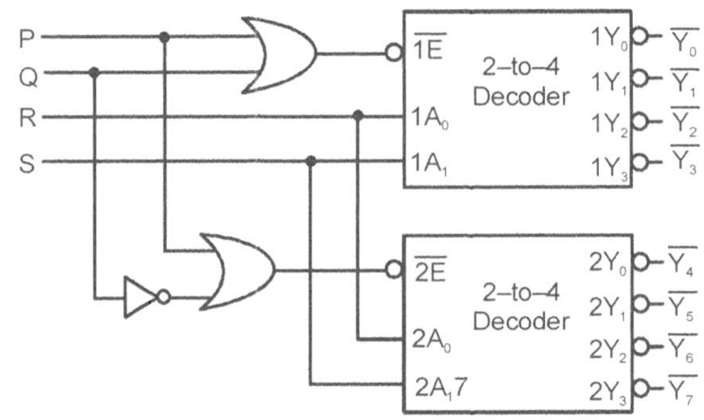

(a) S_2, D_{in}, S_0, S_1 (b) S_1, D_{in}, S_0, S_2

(c) D_{in}, S_0, S_1, S_2 (d) D_{in}, S_2, S_0, S_1

42. The value of the integral

$$\int_{-\infty}^{\infty} 12\cos(2\pi t)\frac{\sin(4\pi t)}{4\pi t}\,dt \text{ is } \underline{\hspace{3cm}}$$

43. A function of Boolean variables X, Y and Z is expressed in terms of the min–terms as F(X, Y, Z) = Σ(1, 2, 5, 6, 7)

Which one of the product of sums given below is equal to the function F(X, Y, Z) ?

(a) $(\overline{X}+\overline{Y}+\overline{Z})\cdot(\overline{X}+Y+Z)\cdot(X+\overline{Y}+\overline{Z})$

(b) $(X+Y+Z)\cdot(X+\overline{Y}+\overline{Z})\cdot(\overline{X}+Y+Z)$

(c) $(\overline{X}+\overline{Y}+Z)\cdot(\overline{X}+Y+\overline{Z})\cdot(X+\overline{Y}+Z)\cdot(X+Y+\overline{Z})\cdot(X+Y+Z)$

(d) $(X+Y+\overline{Z})\cdot(\overline{X}+Y+Z)\cdot(\overline{X}+Y+\overline{Z})\cdot(\overline{X}+\overline{Y}+Z)\cdot(\overline{X}+\overline{Y}+\overline{Z})$

44. The transfer function of a mass–spring damper system is given by

$$G(s) = \frac{1}{Ms^2 + Bs + K}$$

The frequency response data for the system are given in the following table.

ω in rad/s	$\lvert G(j\omega)\rvert$ in dB	$\arg(G(j\omega))$ in deg
0.01	–18.5	–0.2
0.1	–18.5	–1.3
0.2	–18.4	–2.6
1	–16	–16.9
2	–11.4	–89.4
3	–21.5	–151
5	–32.8	–167
10	–45.3	–174.5

The unit step response of the system approaches a steady state value of ________

45. Two half–wave dipole antennas placed as shown in the figure are excited with sinusoidally varying currents of frequency 3 MHz and phase shift of $\pi/2$ between them (the element at the origin leads in phase). If the maximum radiated E–field at the point P in the x–y plane occurs at an azimuthal angle of 60° the distance d (in meters) between the antennas is ________.

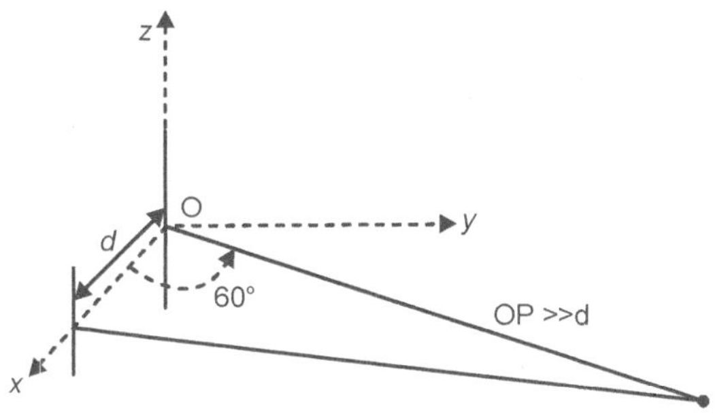

46. An air–filled rectangular waveguide of internal dimensions a cm × bcm ($a > b$) has a cutoff frequency of 6 GHz for the dominant TE_{10} mode. For the same waveguide, if the cutoff frequency of the TM_{11} mode is 15 GHz, the cutoff frequency of the TE_{01} mode in GHz is ________.

47. Consider two real sequences with time–origin marked by the bold value

$x_1[n] = \{1, 2, 3, 0\}$, $x_2[n] = \{1, 3, 2, 1\}$

Let $X_1(k)$ and $X_2(k)$ be 4–point DFTs of $x_1[n]$ and $x_2[n]$, respectively

Another sequence $x_3[n]$ is derived by taking 4–point inverse DFT of $x_3(k) = x_1(k)\, x_2(k)$.

The value of $x_3[2]$ is ________

48. Let $x(t) = as(t) + s(-t)$ with $s(t) = \beta e^{-4t}u(t)$, where $u(t)$ is unit step function. If the bilateral Lapalce transform of $x(t)$ is

$$X(s) = \frac{16}{s^2 - 16} \qquad -4 < \text{Re}\{s\} < 4$$

Then the value of β is ________

49. Consider a binary, digital communication system which used pulses g(t) and – g(t) for transmitting bits over an AWGN channel. If the receiver uses a matched filter, which one of the following pulses will give the minimum probability of bit error ?

(*a*)

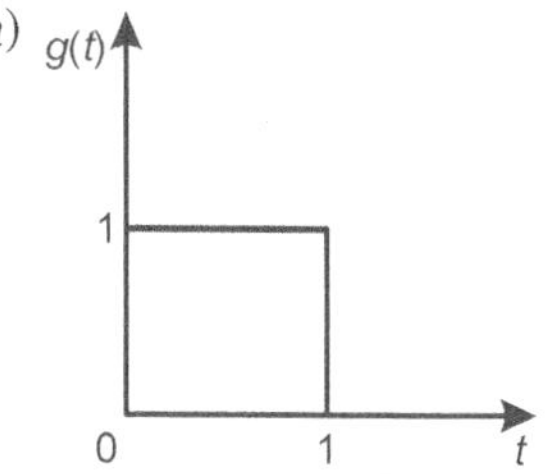

(*b*)

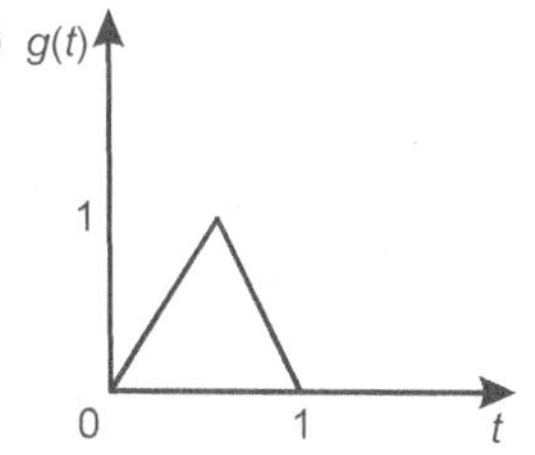

(*c*)

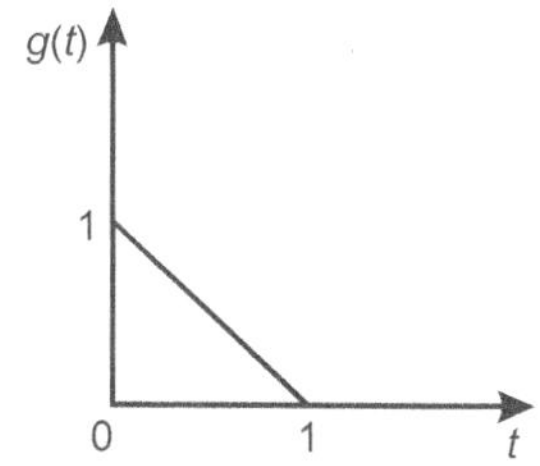

(*d*)

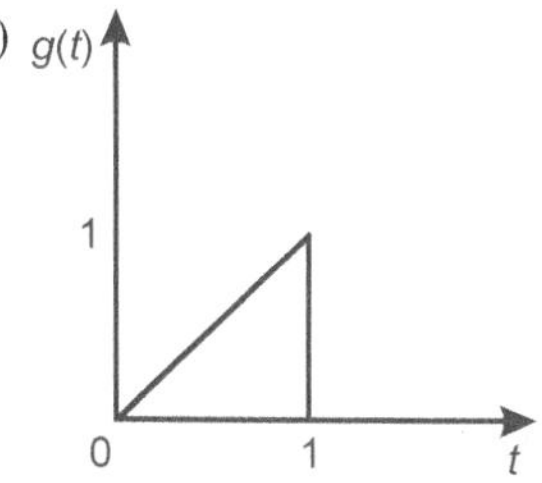

50. The electric field of a plane wave propagating in a lossless non–magnetic medium is given by the following expression

$$E(z, t) = a_x 5\cos\left(2\pi \times 10^9 t + \beta z\right) + a_y 3\cos\left(2\pi \times 10^9 t + bz - \frac{\pi}{2}\right)$$

The type of the polarization is

(*a*) Right Hand Circular

(*b*) Left Hand Elliptical

(*c*) Right Hand Elliptical

(*d*) Linear

51. The energy band diagram and electron density profile $n(x)$ in a semiconductor are shown in the figure. Assume that $n(x) = 10^{15} e^{\left(\frac{q\alpha x}{kT}\right)} \text{cm}^{-3}$, with $\alpha = 0.1$ V/cm and x expressed in cm. Given, $\frac{kT}{q} = 0.026\text{V}, D_n = 36\text{cm}^2 s^{-1}$, and $\frac{D}{\mu} = \frac{kT}{q}$.

The electron current density (in A/cm²) at $x = 0$ is

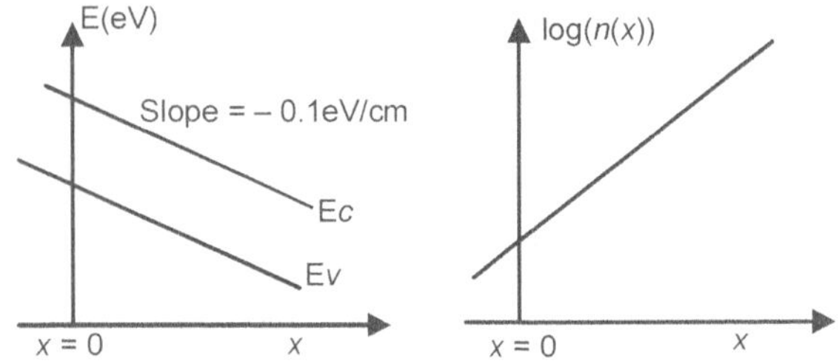

(a) -4.4×10^{-2} (b) -2.2×10^{-2}

(c) 0 (d) -4.4×10^{-2}

52. A dc voltage of 10V is applied across an n–type silicon bar having a rectangular cross–section and a length of 1cm as shown in figure. The donor doping concentration N_D and the mobility of electrons μ_n are 10^{16} and 1000 $\text{cm}^2\text{V}^{-1}s^{-1}$, respectively. The average time (in ms) taken by the electrons to move from one end of the bar to other end is ____________.

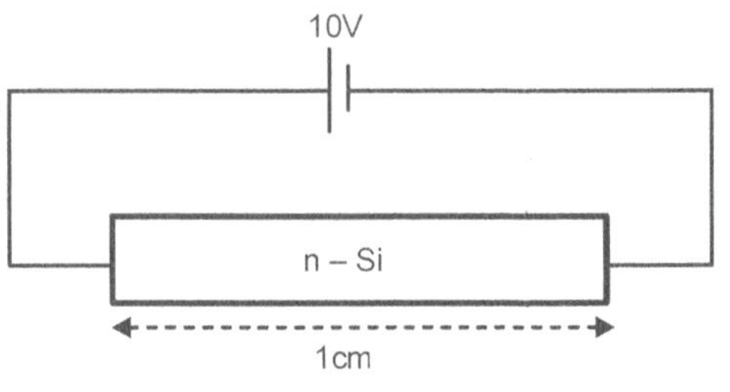

53. In the circuit shown, the initial voltages across the capacitors C_1 and C_2 are 1V and 3V, respectively. The switch is closed at time $t = 0$. The total energy dissipated (in Joules) in the resistor R until steady state is reached is ______

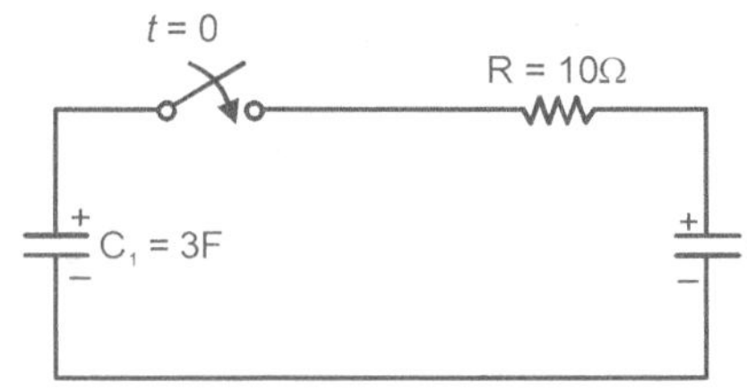

54. The output of a standard second–order system for a unit step input is given as

$$y(t) = 1 - \frac{2}{\sqrt{3}} e^{-t} \cos\left(\sqrt{3}t - \frac{\pi}{6}\right)$$

The transfer function of the system is

(a) $\frac{2}{(s+2)(s+\sqrt{3})}$

(b) $\frac{1}{s^2+2s+1}$

(c) $\frac{3}{s^2+2s+3}$

(d) $\frac{4}{s^2+2s+4}$

55. If C denotes the counterclockwise unit circle, the value of the contour integral $\frac{1}{2\pi j}\oint_C \text{Re}\{z\}\, dz$ is __________.

ANSWERS

GENERAL APTITUDE

1. (c)	2. (b)	3. (b)	4. (a)	5. (d)	6. (d)	7. (b)
8. (800)	9. (b)	10. (b)				

ELECTRONICS AND COMMUNICATION

1. (c)	2. (400)	3. (7)	4. (b)	5. (d)	6. (1.92)	7. (a)
8. (1)	9. (0)	10. (c)	11. (a)	12. (b)	13. (a)	14. (32)
15. (a)	16. (d)	17. (14)	18. (d)	19. (–3.1)–(–2.9)	20. (d)	21. (100)
22. (10)	23. (5)	24. (b)	25. (1.5)	26. (d)	27. (b)	28. (b)
29. (4.33)	30. (1.5)	31. (1.333)	32. (c)	33. (–1)	34. (d)	35. (c)
36. (2.8056)	37. (b)	38. (0.6)	39. (a)	40. (11–12)	41. (d)	42. (3)
43. (b)	44. (0.12)	45. (50)	46. (13.7)	47. (11)	48. (–2)	49. (a)
50. (b)	51. (c)	52. (100)	53. (1.5)	54. (d)	55. (0.5)	

EXPLANATIONS

GENERAL APTITUDE

3. $P(\text{Ram}) = \frac{1}{6}$; $\quad p(\text{Ramesh}) = \frac{1}{8}$

p(only at) = p(Ram) × p(not ramesh) + p(Ramesh) × p(n_0 × Ram)

$$= \frac{1}{6} + \frac{7}{8} + \frac{1}{8} \times \frac{5}{6}$$

$$\Rightarrow \quad \frac{12}{40} = \frac{1}{4}$$

5. For $M \Rightarrow \frac{12}{20} = 0.6$

$N \Rightarrow \frac{13}{25} = 0.52$

$O \Rightarrow \frac{20}{30} = 0.66$

$P \Rightarrow \frac{12}{25} = 0.48$

8. Tiger–1 leap ⇒ 8 meter

Speed = 5 leap/hr = 40m min

Deer → 1 leap = 5 meter

speed = 4hr = 20m/min

Let at time '*t*' the tiger catches the deer.

∴ Distance travelled by deer + initial distance between them

50 × 8 ⇒ 400m = distance covered by tiger.

$\Rightarrow 40 \times t = 400 + 20t$

$$\Rightarrow t = \frac{400}{200} = 20 \text{ min}$$

⇒ total distance ⇒ 400 + 20 × t = 800 ms

10. The is not required in 'Q'

ELECTRONICS AND COMMUNICATION

1. For a finite duration time domain signal ROC is entire s-plane.

2.
$$G(s) = \frac{k}{s^2 + 10s}$$

$$\text{CLTF} = \frac{k}{s^2 + 10s + k}$$

$$\xi = 0.25$$

$$k = \omega_n^2$$

$$\omega_n = \sqrt{k}$$

$$\xi = \frac{10}{2\sqrt{k}} = 0.25$$

$$\sqrt{k} = \frac{10}{0.5}$$

$$k = (20)^2$$

$$k = 400$$

3. To find the modulus of the counter, consider the status of the inputs (Q_B, Q_C) as 1.

So, $Q_AQ_BQ_CQ_D$ = 0110

So, it is a MOD-7 counter

5. $E = \left(a_x + 4ja_y\right)e^{j(2\pi\times10^7 t - 0.2z)}$ $\qquad \omega = 2\pi \times 10^7$

$E_z = \cos \omega t$ $\qquad \beta = 0.2$

$E_y = 4\cos(\omega + \pi/2) = -4\sin \omega t$

So, it left hand elliptical polrization

6. As per the graph mobility of electrons at the concentration $10^{16}/\text{cm}^3$ is $1200 \frac{\text{cm}^2}{\text{V}-s}$

So, $\quad \mu_n = 1200 \frac{\text{cm}^2}{\text{V}-s}$

$$\sigma_N = N_D q_n$$
$$= 10^{16} \times 1.6 \times 10^{-19} \times 1200$$
$$= 1.92 \text{ S cm}^{-1}$$

7. Given expression is $Y = AB + \bar{C}\bar{D}$

The first term can be obtained by considering G_1 as NOR gate, and second term $\left(\bar{C}\bar{D}\right)$ is obtained from another lower NOR-Gate. So, final expression can be implemented by considering G_2 as OR-Gate.

8.

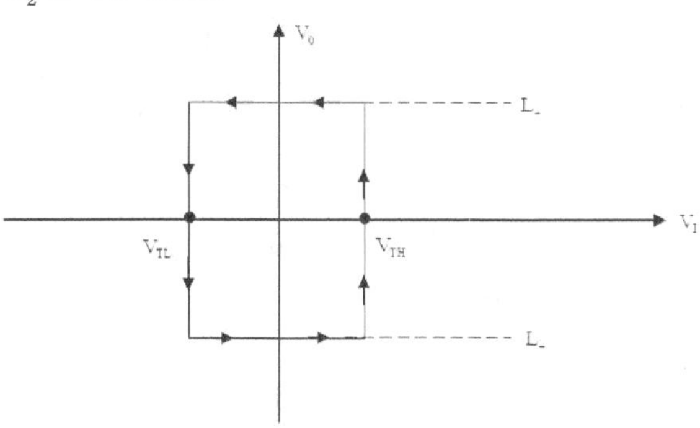

$$\text{Hysteresis} = V_{TH} - V_{TL}$$

$$= -L_-\left(\frac{R_1}{R_2}\right) + L_+\left(\frac{R_1}{R_2}\right)$$

$$500\text{mV} = -(-5)\left(\frac{R_1}{20k}\right) + 5\left(\frac{R_1}{20k}\right)$$

$$= \frac{R_1}{2k}$$

$$\therefore \quad R_1 = 500 \times 2 \times 10^3 \times 10^{-3}$$

$$= 1000\ \Omega = 1\text{k}\Omega$$

9. $$y[n] = \sum_{m=0}^{n} x[m]$$

According to accumulation property of z-transform,

$$Y[z] = \frac{X(z)}{(1-z^{-1})}$$

$$\Rightarrow \quad \frac{2}{z(z-1)} = \frac{zX(z)}{(z-1)}$$

$$\therefore \quad X[z] = \frac{2z^{-2}}{(z-1)} = \frac{2z^{-3}}{(1-z^{-1})}$$

$$\therefore \quad x[n] = 2u[n-3] \text{ thus } x[2] = 0$$

10. Given $$f(t) = \begin{cases} 1 & a \le t \le b \\ 0 & \text{otherwise} \end{cases}$$

$$L\{f(t)\} = \int_0^\infty e^{-st} f(t)dt$$

$$= \int_0^\infty e^{-st} f(t) + \int_a^\infty e^{-st} f(t)dt + \int_b^\infty e^{-st} f(t)dt$$

$$= 0 + \int_a^b e^{-st} dt + 0$$

$$= \left.\frac{e^{-st}}{-s}\right|_a^b$$

$$= \frac{-1}{s}\left[e^{-bs} - e^{-as}\right]$$

$$= \frac{e^{-as} - e^{-bs}}{s}$$

12. Let $$A = \begin{bmatrix} 10 & 5+J & 4 \\ x & 20 & 2 \\ 4 & 2 & -10 \end{bmatrix}$$

Given that all eigen values of A are real.

$\Rightarrow$ A is Hermitian

$\Rightarrow A^\theta = A$ i.e. $(\bar{A})^T = A$

$$\begin{bmatrix} 10 & \bar{x} & 4 \\ 5-j & 20 & 2 \\ 4 & 2 & -10 \end{bmatrix} = \begin{bmatrix} 10 & 5+J & 4 \\ x & 20 & 2 \\ 4 & 2 & -10 \end{bmatrix}$$

$\Rightarrow x = 5 - j$

13. Given signal is $x(t) = \cos\left(10\pi t + \frac{\pi}{4}\right)$

Neglect the phase-shift $\frac{\pi}{4}$ and it can be inserted at the end result.

$$\therefore \text{ If } \quad x_1(t) = \cos 10\pi t \xrightarrow{L} X_1(f)$$

$$= \frac{1}{2}[\delta(f-5) + \delta(f+5)]$$

Given filter impulse response is,

$$h(t) = \left(\frac{\sin \pi t}{\pi t}\right)\cos\left(40\pi t - \frac{\pi}{2}\right)$$

$$= (\sin ct) \sin (40\pi t)$$

$$\therefore \quad H(f) = \text{rect } f * \frac{1}{2j}[\delta(f-20) - \delta(f+20)]$$

$$= \frac{1}{2j}[\text{rect}(f-20) - \text{rect}(f+20)]$$

$X_1(f)$ repeats with a value $f_o = 15$Hz and each impulse value is $\frac{15}{2}$

Thus the sampled signal spectrum and the spectrum of the filter are as follows:

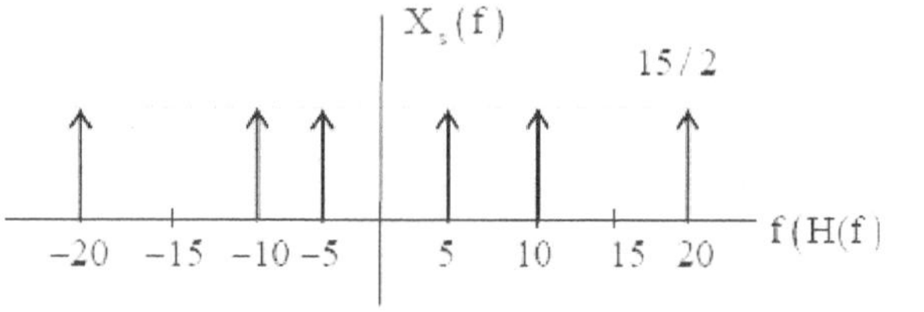

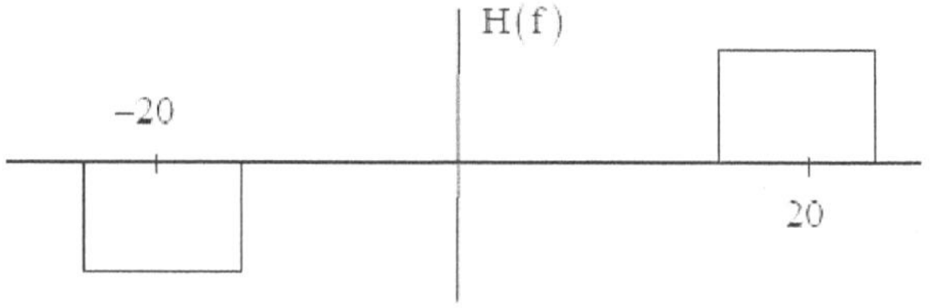

$$\therefore \quad X_s(f)\,H(f) = \frac{15}{4j}[\delta(f-20) - \delta(f+20)]$$

$$x_r(t) = \frac{15}{2}\sin(40\pi t) \rightarrow \text{recovered signal}$$

$$= \frac{15}{2}\cos\left(40\pi t - \frac{\pi}{2}\right)$$

Insert the neglected phase shift $\frac{\pi}{4}$

$$\therefore \quad x_r(t) = \frac{15}{2}\cos\left(40\pi t - \frac{\pi}{2} + \frac{\pi}{4}\right)$$

$$= \frac{15}{2}\cos\left(40\pi t - \frac{\pi}{4}\right)$$

14. Signal power = $\frac{A^2}{2}$

Quantization step size, $\Delta = \frac{2A}{L}$

Quantization noise power = $\frac{\Delta^2}{12}$

$$= \frac{4A^2}{12L^2}$$

$$= \frac{A^2}{3L^2}$$

$\Rightarrow$ Signal to quantiation noise ratio = $\frac{3}{2}L^2$

Given signal to quantization noise ratio = 31.8dB or 1513.56

$$\Rightarrow \quad \frac{3}{2}L^2 = 1513.56$$

$$\Rightarrow \quad L = 31.76$$

$$\approx 32$$

15. $\angle a_k = -\pi$ only changes the sign of the magnitude $|a_k|$. Since the magnitude spectrum $|a_k|$ is even the corresponding time-domain signal is real.

16. Given $\frac{dy}{dx} = \frac{1-\cos 2y}{1+\cos 2x}$

$\Rightarrow \frac{dy}{1-\cos 2y} = \frac{dx}{1+\cos 2x}$ Variable - Separable

$\Rightarrow \frac{dy}{2\sin^2 y} = \frac{dx}{2\cos^2 x}$

$\Rightarrow \int \operatorname{cosec}^2 y\,dy = \int \sec^2 x\,dx$

$\Rightarrow -\cot y = \tan x + k$

$\Rightarrow -\tan x - \cot y = k$

$\Rightarrow \tan x + \cot y = c$ where $c = -k$

17. The concentration of hole-electron pair in 1 sec = $10^{20} \times 10^{-6} = 10^{14}/\text{cm}^3$

So, the power of 10 is 14.

$$x = 14$$

20. Generally arithmetic or logical instructions update the data of accumulator and flags. So, in the given option only SBT BE H is arithmetic instruction.

SBI BEH→ Add the content of accumulator with immediate data BE H and store the result in accumulator.

21.

$$V = \sqrt{V_R^2 + (V_c + V_L)^2}$$

$$(100)^2 = (80)^2 + (V_c - 40)^2$$

$$(V_c - 40)^2 = (180)(20)$$

$$(V_c - 40) = \pm\sqrt{2 \times 90 \times 20}$$

$$V_c - 40 = \pm 60$$

$$V_c = \pm 60 + 40$$

$$V_c = 60 + 40$$

$$V_c = 100 \text{ V}$$

22. $f(z) = \frac{az+b}{cz+d}$ if $f(z_1) = f(z_2)$ for $z_1 \neq z_2$

$a = 2$, $b = 4$ and $c = 5$

$$f(z) = \frac{2z+4}{5z+d}$$

$$f(z_1) = f(z_2)$$

$$\Rightarrow \quad \frac{2z_1+4}{5z_1+d} = \frac{2z_2+4}{5z_2+d}$$

$\Rightarrow 10z_1 z_2 + 20z_2 + 2dz_1 + 4d$

$= 10z_1 z_2 + 20z_1 + 2dz_2 + 4d$

$\Rightarrow \quad 20(z_2 - z_1) = 2d(z_2 - z_1)$

$\Rightarrow \quad d = 10$

24. Using mason gain formula we get it directly.

25.

$$V_{OB} = -5\left(\frac{1\Omega}{1k\Omega}\right) - 1\left(\frac{1\Omega}{1k\Omega}\right)$$

$$= -6\text{V}$$

$$V_{OA} = -5\left(\frac{1k\Omega}{1k\Omega}\right) + \frac{1}{2}\left(1 + \frac{1k\Omega}{1k\Omega}\right)$$

$$= -4\text{V} \qquad \left(\because V_+ = \frac{1}{2}\text{V}\right)$$

$$\therefore \quad \frac{V_{OB}}{V_{OA}} = 1.5$$

26. P{$x = 0$} = P ⇒ P{$x = 1$} = $1 - p$

P{$y = 0$} = q ⇒ P{$y = 1$} = $1 - q$

Let Z = X + Y

X	Y	Z
0	0	0
0	1	1
1	0	1
1	1	2

From above table,

P{X + Y + Z} ⇒ P < Z ≥ B

P{Z ≥ 1} = P{X = 0 and Y = 1} + P{X = 1 and Y = 0} + P{X = 1 and Y = 1}

= 1 – P{X = 0 and Y = 0}

= 1 – pq

27.

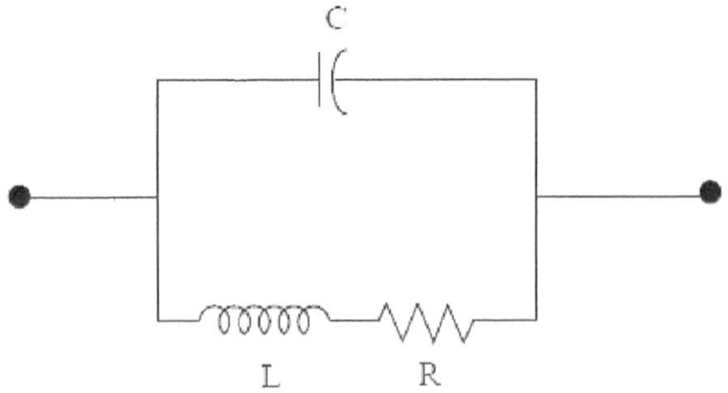

$$Y = Y_c + Y_{LR}$$

$$Y = j\omega C + \frac{1}{(j\omega L + R)}$$

$$= j\omega C + \frac{(R - j\omega L)}{(R^2 - \omega^2 L^2)}$$

Placing Imaginary part to zero we get option (*b*).

28.
$$R_Y(k) = R_y(n, n+k)$$
$$= E[Y(n).Y(n+k)]$$
$$Y(n) = x(n) + 0.5x(n-1)$$
$$R_Y(k) = E[(x[n]+0.5x[n-1])(x(n+k)+0.5x(n+k-1))]$$
$$= E[(x[n].x(n+k)+x(n)0.5x(n+k-1)+0.5x(n-1).x(n+k) +0.25x(n-1)x(n+k-1)$$
$$= E[x[n].x(n+k)+0.5E[x(n)x(n+k-1)] +0.5E[(x(n-1)x(n+k))] +0.25E[x(n-1)x(n+k-1)]]$$
$$= R_X(k) + 0.5R_x(k-1) + 0.5R_x(k+1) + 0.25R_x(k)$$
$$R_y(k) = 1.25R_x(k) + 0.5R_x(k-1) + 0.5R_x(k+1)$$
$$R_x(k) = E[x(n).x(n+k)]$$

if $k = 0$

$$R_x(0) = E[x^2(n)]$$
$$= 1^2.\frac{1}{2} + (-1)^2 \times \frac{1}{2}$$
$$= 1$$

if $k \neq 0$,

$$R_x(k) = E[x(n)].E[x(n+k)] = 0$$

$$\left\{\begin{array}{l} \because E[x(n)] = 0 \\ E[x(n+k)] = 0 \end{array}\right\}$$

$$\Rightarrow R_y(0) = 1.25R_x(0) + 0.5R_x(-1) + 0.5R_x(1)$$
$$= 1.25$$
$$R_y(1) = 1.25R_x(1) + 0.5R_x(0) + 0.5R_x(2)$$
$$= 0.5$$
$$R_y(-1) = 1.25R_x(-1) + 0.5R_x(-2) + 0.5R_x(0)$$
$$= 0.5$$

$R_y(k)$ for k other than 0, 1 and –1 = 0

$\Rightarrow R_y(k)$

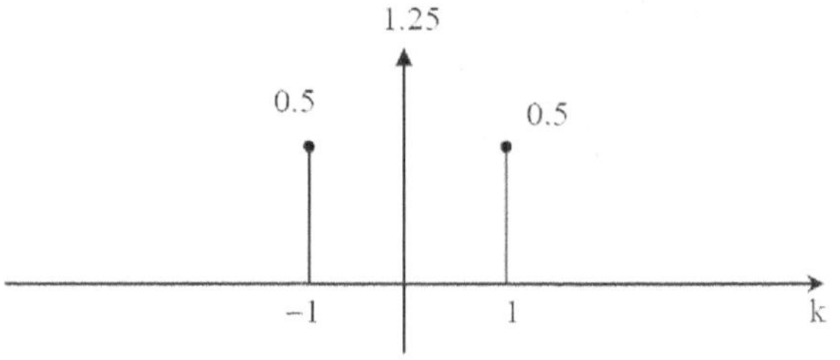

29.
$$\frac{C_{max}}{C_{min}} = \frac{\frac{\epsilon_{ox}}{t_{ox}}}{\frac{\frac{\epsilon_{ox}}{t_{ox}} \times \frac{\epsilon_s}{X_{dmax}}}{\frac{\epsilon_{ox}}{t_{ox}} + \frac{\epsilon_s}{X_{dmax}}}}$$
$$= \left[1 + \frac{X_{dmax}}{t_{ox}} \times \frac{\epsilon_{ox}}{\epsilon_s}\right]$$
$$= \left[1 + \frac{100}{10} \times \frac{1}{3}\right]$$
$$= 4.33$$

30. Let x be a random variable which denotes number of tosses to get two heads.

$$P(x = 2) = HH = \frac{1}{2} \times \frac{1}{2}$$
$$P(x = 3) = THH = \frac{1}{2} \times \frac{1}{2} \times \frac{1}{2}$$
$$P(x = 4) = TTHH = \frac{1}{2} \times \frac{1}{2} \times \frac{1}{2} \times \frac{1}{2}$$

....

$$E(x) = 2\left(\frac{1}{2}\times\frac{1}{2}\right) + 3\times\left(\frac{1}{2}\times\frac{1}{2}\times\frac{1}{2}\right) + 4\left(\frac{1}{2}\times\frac{1}{2}\times\frac{1}{2}\times\frac{1}{2}\right) + ...$$
$$= 2 \times \frac{1}{2^2} + 3 \times \frac{1}{2^3} + 4 \times \frac{1}{2^4} + ...$$
$$= \frac{1}{2}\left[2.\frac{1}{2^2} + 3.\frac{1}{2^2} + 4 \times \frac{1}{2^3} + ...\right]$$
$$= \frac{1}{2}\left[\left(1 + 2.\frac{1}{2} + 3\frac{1}{2^2} + ...\right) - 1\right]$$
$$= \frac{1}{2}\left[\left(1 - \frac{1}{2}\right)^{-2} - 1\right]$$
$$= \frac{1}{2}[4 - 1]$$
$$= \frac{3}{2}$$

31.

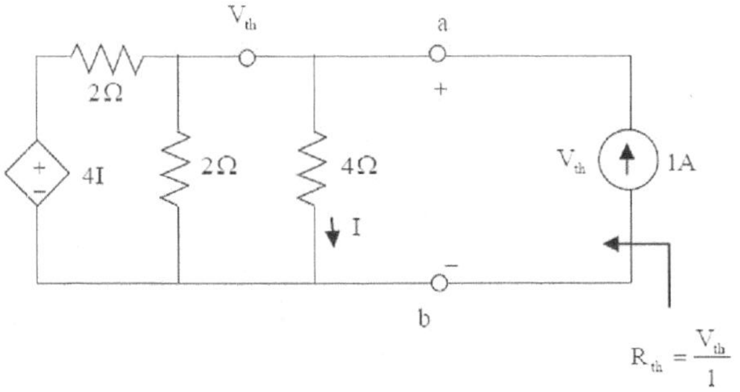

Nodal @ '*a*'

$$\frac{V_{th}-4I}{2}+\frac{V_{th}}{2}+\frac{V_{th}}{4}-1 = 0$$

Also $\quad V_{th} = 4I$

$\Rightarrow \quad V_{th} = \frac{4}{3}$

$$R_{th} = \frac{4}{3} = 1\frac{1}{3} = 1.333$$

32.

Q_3	Q_2	Q_1	Q_0
0	0	0	0
0	0	0	1
0	0	1	0
0	0	1	1
0	1	0	0

Once the output of Ex-NOR gate is 0 then counter will be RESET. So, Ex-NOR-gate will produce logic 0 for $Q_3 = 0$, $Q_2 = 1$. So, the counter will show the sequence like:

$$0 \rightarrow 1 \rightarrow 2 \rightarrow 3 \rightarrow 4$$

So, it is MOD-5 counter.

33.

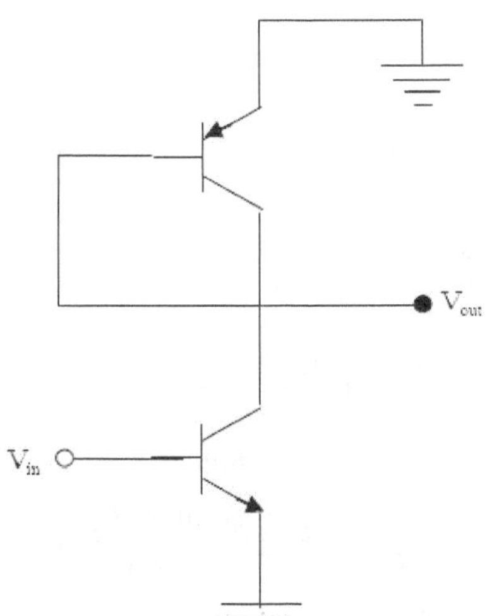

When Base and collector is shorted, it act as a diode.

So, $\quad V_0 = -0.7V.$

Gain, $\quad \frac{V_0}{V_{in}} = -\frac{0.7V}{0.7V} = -1$

34. $\quad \dot{X} = AX$

$$X(s) = (sI - A)^{-1} X(0)$$

$$X(s) = \begin{bmatrix} s & -1 \\ 0 & s+1 \end{bmatrix}^{-1} \begin{bmatrix} 1 \\ 0 \end{bmatrix}$$

$$X(s) = \begin{bmatrix} 1/s \\ 0 \end{bmatrix}$$

$$x(t) = \begin{bmatrix} 1 \\ 0 \end{bmatrix}$$

$$y(t) = [0 \;\; 1]\begin{bmatrix} 1 \\ 0 \end{bmatrix} = 0$$

35. Given $\quad$ D.E $\frac{dx}{dt} = 10 - 0.2\,x$

$$x(0) = 1$$

$$\Rightarrow \quad \frac{dx}{dt} + (0.2)\,x = 10$$

Auxiliary equation is $m + 0.2 = 0$

$$m = -0.2$$

Complementary solution

$$x_c = C\,e^{(-0.2)t}$$

$$x_p = \frac{1}{D+(0.2)} 10\,e^{0t} = \frac{10\,e^{0t}}{0.2}$$

$$= 50e^{0t} = 50$$

$$x = x_c + x_p$$

$$= C\,e^{(-0.2)t} + 50$$

Given $\quad x(0) = 1$

$\Rightarrow \quad C + 50 = 1$

$\Rightarrow \quad C = -49$

$$x = 50 - 49\,e^{(-0.2)t}$$

36.

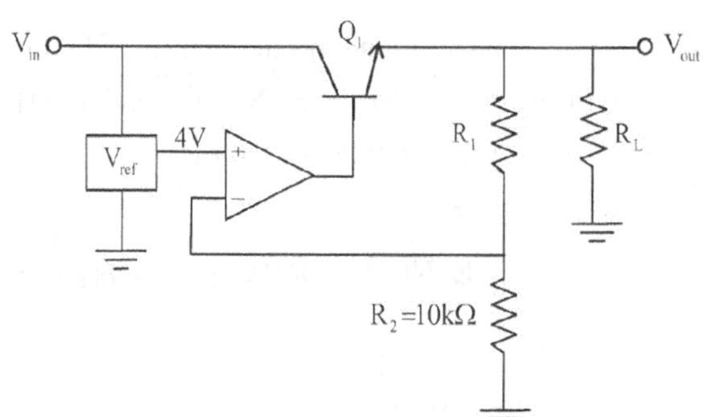

$$P_{Q_1(max)} = V_{CE(max)} \times I_{c\,max} \quad ...(i)$$
$$V_{CE(max)} = (24 - 10)V = 14V$$
$$I_{c\,max} = (200 + 0.4)mA$$
$$I_E = I_c = 200mA + 0.4mA$$
$$= 200.4mA \left(\because I_{R_2} = I_{R_1} = \frac{4-0}{10} mA\right)$$

Put values in equation (i), we get

$$P_{Q_1(max)} = 14 \times 200.4 \times 10^{-3} \text{ Watt}$$
$$= 2.8056 \text{ Watt}$$

37. The given differential equation is,

$$y''(t) - y'(t) - 6y(t) = x(t)$$

On applying Laplace transform on both sides,

$$s^2y(s) - sy(0) - y(0) - [sy(s) - y(0)] - 6y(s) = x(s)$$

To calculate the transfer function all initial conditions are taken as '0'.

$$\therefore \quad (s^2 - s - 6)y(s) = x(s)$$
$$H(s) = \frac{1}{(s^2 - s - 6)}$$
$$= \frac{1}{(s-3)(s+2)}$$
$$= \frac{1}{5}\left[\frac{1}{s-3} - \frac{2}{s+2}\right]$$

It is given that $h(t)$ is non-casual and un-stable.

To satisfy both the conditions ROC should be left of the left most pole.

Using the following standard pair

$$\frac{1}{s+a} \longleftrightarrow -e^{-at}u(-t); \sigma < -a$$
$$\frac{1}{s-a} \longleftrightarrow -e^{at}u(-t); \sigma < a$$
$$H(s) = \frac{1}{5}\left[\frac{1}{s-3} - \frac{1}{s+2}\right]$$
$$= \frac{1}{5}\left[-e^{3t}u(-t) + e^{-2t}u(-t)\right]$$
$$= \frac{-1}{5}e^{-3t}u(-t) + \frac{1}{5}e^{-2t}u(-t)$$

So option (b) is correct.

39. Let $N(t)$ be the noise at the output of filter.

Variance of $N(t) = E(N^2(t)) - E(N(+1))^2$

Since the input noise is zero mean,

Output noise mean is also zero

$$E(N(t)) = E(W(t)) \cdot \left(\int_{-\infty}^{\infty} h(t)dt\right)$$
$$E(W(t)) = 0$$

$W(t)$ is white noise

$$\Rightarrow \quad \text{var}(N(t)) = E(N^2(t)) = R_N(0)$$

Since $R_N(\tau) = h(\tau) * h(-\tau) * R_\omega(\tau)$

Since $R_N(\tau) = \frac{N_o}{2} \cdot \delta(\tau)$

$$R_N(\tau) = [h(\tau) * h(-\tau)] \cdot \frac{N_o}{2}$$
$$R_N(\tau) = \frac{N_o}{2}\int_{-\infty}^{\infty} h(k).h(\tau + k)\,dk$$
$$R_N(0) = \frac{N_o}{2}\int_{-\infty}^{\infty} h^2(k)dk = \frac{N_o}{2}(3A^2)$$
$$= \frac{3}{2} \cdot A^2.N_o$$

41.

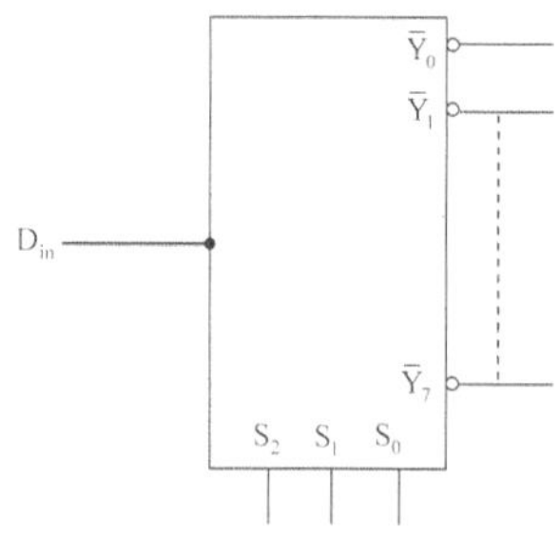

We need to implement 1 : 8 DEMUX

Select lines of DEMUX should be mapped to address lines of decoder. So, LSB of DEMUX should be connected to LSB of address lines of decoder.

$R \rightarrow S_0$

$S \rightarrow S_1$

Input to both the decoder should be same so

$P \rightarrow D_{in}$

NOT gate along with OR gate in case to select one decoder at a time so $Q \rightarrow S_a$.

$P \rightarrow D_{in}$

$Q \rightarrow S_2$

$R \rightarrow S_0$

$S \rightarrow S_1$

42. $\int_{-\infty}^{\infty} 12\cos 2\pi t \frac{\sin 4\pi t}{4\pi t}dt$

$$\frac{12}{4\pi}\int_0^{\infty}\frac{2\cos 2\pi t \sin 4\pi t}{t}dt$$
$$\frac{3}{\pi}\left[\int_0^{\infty}\frac{\sin 6\pi t\,dt}{t} + \int_0^{\infty}\frac{\sin 2\pi t\,dt}{t}\right]$$

($\because$ sin A – cos B = sin (A + B) + sin (A – B))

$$= \frac{3}{\pi}\left[\int_0^{\infty} e^{\theta t}\frac{6\sin 6\pi t}{t}dt + \int_0^{\infty} e^{\theta t}\frac{\sin 2\pi t}{t}dt\right]$$
$$= \frac{3}{\pi}\left[L\left\{\frac{\sin 6\pi t}{t}\right\} + L\left\{\frac{\sin 2\pi t}{t}\right\}\right] \text{with } s = 0$$

$$= \frac{3}{\pi}\left[\int_s^\infty \frac{6\pi}{s^2+36\pi^2}ds + \int_s^\infty \frac{2\pi}{s^2+4\pi^2}ds\right] \text{with } s=0$$

$$= \frac{3}{\pi}\left[6\pi.\frac{1}{6\pi}\tan^{-1}\left(\frac{s}{6\pi}\right) + 2\pi.\frac{1}{2\pi}\tan^{-1}\left(\frac{s}{2\pi}\right)\Big|_s^\infty\right] \text{with } s=0$$

$$= \frac{3}{\pi}\left[\tan^{-1}\infty - \tan^{-1}\left(\frac{s}{6\pi}\right) + \tan^{-1}(\infty) - \tan^{-1}\left(\frac{s}{2\pi}\right)\right]$$

$$\Rightarrow \frac{3}{\pi}\left[\frac{\pi}{2} - \tan^{-1}0 + \frac{\pi}{2} - \tan^{-1}0\right]$$

$$\Rightarrow \frac{3}{\pi}\left[\frac{\pi}{2} - 0 + \frac{\pi}{2} - 0\right] = \frac{3}{\pi}\times\pi = 3$$

43. Given minterm is :

$F(X, Y, Z) = \Sigma(1, 2, 5, 6, 7)$

So, maxterm is : $F(X, Y, Z) = \pi M(0, 3, 4)$

$POS = (X+Y+Z)(X+\overline{Y}+\overline{Z})(\overline{X}+Y+Z)$

45. $\psi = \delta + \beta d\cos\theta$

For maximum field, $\psi = 0$ $\Big|$ $\lambda = \frac{3\times10^8}{3\times10^6} = 100$ m

$$\delta + \beta\cos\theta = 0$$

$$-\frac{\pi}{2} + \frac{2\pi}{\lambda}d\cos 60 = 0$$

$$\frac{\pi}{2} = \frac{2\pi}{100}(d)\frac{1}{2}$$

$$d = 50 \text{ m}$$

46. TE_{01}

$$f_c = \frac{3\times10^8}{2}.\frac{1}{b}$$

$$f_c = 13.7 \text{ GHz}$$

TE_{10}

$$f_c = 6 \text{ GHz}$$

$$f_c = \frac{1}{2\sqrt{\mu\epsilon}}\sqrt{\left(\frac{m}{a}\right)^2 + \left(\frac{n}{b}\right)^2}$$

$$a = \frac{1}{40}$$

TM_{11}

$$15\times10^9 = \frac{3\times10^8}{2}\sqrt{\frac{1}{a^2}+\frac{1}{b^2}}$$

$$= \frac{1}{b} = 91.65$$

47. $x_1[n] = \{1, 2, 3, 0\}$, $x_2[n] = \{1, 3, 2, 1\}$

$X_3(k) = X_1(k)\,X_2(k)$

Based on the properties of DFT,

$x_1[n] \otimes x_2[n] = X_1(k)\,X_2(k) = x_3[n]$

Circular convolution between two 4-point signals is as follows :

$$\begin{bmatrix}1&0&3&2\\2&1&0&3\\3&2&1&0\\0&3&2&1\end{bmatrix}\begin{bmatrix}1\\3\\2\\1\end{bmatrix} = \begin{bmatrix}9\\8\\11\\14\end{bmatrix}$$

$\therefore$ $x_3[2] = 11$

48. $x(t) = \alpha s(t) + s(-t)$ & $s(t) = \beta e^{-4t}u(t)$

$x(t) = \alpha\beta e^{-4t}u(t) + \beta e^{4t}u(-t)$

$$\alpha\beta e^{-4t}u(t) \xrightarrow{L} \frac{\alpha\beta}{s+4}$$

$$\beta e^{4t}u(-t) \xrightarrow{L} \frac{\beta}{s-4}$$

$$\therefore\ X(s) = \frac{\alpha\beta}{s+4} - \frac{\beta}{s-4}$$

$$\beta\left[\frac{\alpha(s-4)-(s+4)}{s^2-16}\right] = \frac{16}{s^2-16};\ -4 < \sigma < +4$$

On solving the numerator $\beta = -2$

49. Optimum receiver for AWGN channel is given by matched filter.

In case of matched filter receiver,

$$\text{Probability of error} = Q\left(\sqrt{\frac{2E}{N_u}}\right)$$

$\Rightarrow$ Probability of error is minimum for which E is maximum

Now looking at options

Energy in option (*a*) = $1^2 = 1$

Energy in option (*c*) and (*d*) is same = 1/3

$$\text{Energy in option } (b) = 2\left[\int_0^{1/2}(2t)^2\,dt\right]$$

$$= 2\left[\int_0^{1/2}4t^2\,dt\right]$$

$$= 2.4\left(\frac{t^3}{3}\right)\Big|_0^{1/2} = 1/3$$

Thus option (*a*) is correct answer.

50. $E_x = 5\cos(\omega t + \beta z)$

$$E_y = 3\cos\left(\omega t + \beta z - \frac{\pi}{2}\right)$$

$$\phi = -\frac{\pi}{2}$$

But the wave is propagating along negative z-direction

So, left hand elliptical (LED)

52.
$$\varepsilon = \frac{V}{d}$$
$$= \frac{10}{1} = 10 \text{ V/m}$$
$$V_d = \mu\varepsilon$$
$$= 1000 \times 10 = 10^4 \text{cm/s}$$
$$V_d = \frac{L}{T}$$
$$\Rightarrow \quad T = \frac{L}{V_d}$$
$$= \frac{1 \times 100}{10^4 \times 10^2} = 100\mu s$$

53.

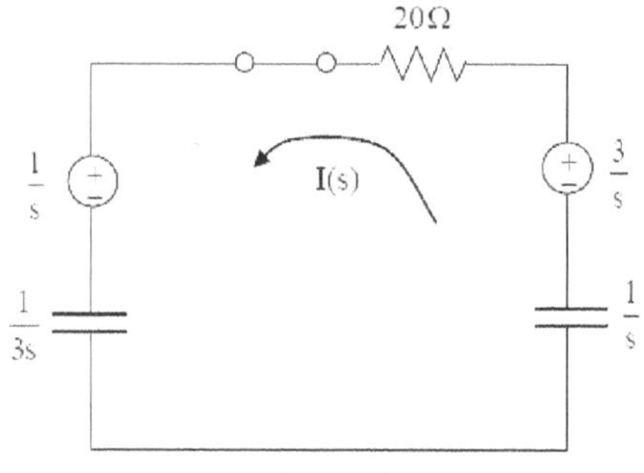

$$I(s) = \frac{\left(\frac{3}{s} - \frac{1}{s}\right)}{\left(10 + \frac{1}{3s} + \frac{3}{3s}\right)}$$
$$I(s) = \frac{2}{\left(10s + \frac{4}{3}\right)} = \frac{2}{10\left(s + \frac{4}{30}\right)}$$
$$i(t) = \frac{1}{5} e^{-\frac{4}{30}t} ; \; t \geq 0$$

$$E_R = \int_0^\infty i^2(t) 10\, dt$$
$$= \left(\frac{10}{25}\right) \int_0^\infty e^{\frac{-4}{15}t} dt$$
$$= \frac{10}{25} \cdot \left.\frac{e^{\frac{4}{15}t}}{\frac{-4}{15}}\right]_0^\infty$$
$$= 0 - \frac{10}{25} \times \frac{15}{-4} = 1.5 \text{ J}$$

54. Here $\xi\omega_n = 1$
$$\sqrt{1 - \xi^2} = \frac{\sqrt{3}}{2}$$
$$\xi = \frac{1}{2}$$
$$\omega_n = 2$$

55. $\frac{1}{2\pi J} \oint_e \text{Re}(z)\, dz$ where C is $|z| = 1$

Put $z = e^{j\theta}$ $\quad \Rightarrow d\theta = je^{j\theta} d\theta$
$$\frac{1}{2\pi j} \int_0^{2\pi} \text{Re}\left(e^{j\theta}\right) je^{j\theta} d\theta$$
$$= \frac{1}{2\pi j} \int_0^{2\pi} \cos\theta . j(\cos\theta + j\sin\theta)\, d\theta$$
$$= \frac{j}{2\pi j}\left[\int_0^{2\pi} \cos^2\theta\, d\theta - \int_0^{2\pi} \cos\theta \sin\theta\, d\theta\right]$$
$$= \frac{j}{2\pi j}[\pi - 0] = \frac{1}{2}$$

■■

SOLVED PAPER – 2015

ELECTRONICS & COMMUNICATION ENGINEERING
Set - 3

No. of Questions : 65 *Maximum Marks : 100*

GENERAL APTITUDE (GA)

Q. No. 1 – 5 Carry One Mark Each

1. Find the missing sequence in the letter series below :

A, CD, GHI, ?, UVWXY

(*a*) LMN (*b*) MNO

(*c*) MNOP (*d*) NOPQ

2. Choose the correct verb to fill in the blank below :

Let us _________.

(*a*) Introvert (*b*) alternate

(*c*) atheist (*d*) altruist

3. Choose the most appropriate word from the options given below to complete the following sentence ?

If the athlete had wanted to come first in the race, he ________ several hours every day.

(*a*) Should practice (*b*) Should have practised

(*c*) Practised (*d*) Should be practicing

4. Choose the most suitable one word substitute for the following expression

Connotation of a road or way

(*a*) Pertinacious (*b*) Viaticum

(*c*) Clandestine (*d*) Ravenous

5. If $x > y >$ I, which of the following must be true?

(*i*) $\ln x > \ln y$ (*ii*) $e^x > e^y$

(*iii*) $y^x > x^y$ (*iv*) $\cos x > \cos y$

(a) (*i*) and (*ii*) (*b*) (*i*) and (*iii*)

(*c*) (*iii*) and (*iv*) (*d*) (*ii*) and (*iv*)

Q. No. 6 – 10 Carry Two Marks Each

6. From a circular sheet of paper of radius 30 cm, a sector of 10% area is removed. If the remaining part is used to make a conical surface, then the ratio of the radius and height of the cone is ______.

7. In the following question, the first and the last sentence of the passage are in order and numbered 1 and 6. The rest of the passage is split into 4 parts and numbered as 2, 3, 4, and 5. These 4 parts are not arranged in proper order. Read the sentences and arrange them in a logical sequence to make a passage and choose the correct sequence from the given options.

1. One Diwali, the family rises early in the morning.
2. The whole family, including the young and the old enjoy doing this,
3. Children let off fireworks later in the night with their friends.
4. At sunset, the lamps are lit and the family performs various rituals
5. Father, mother, and children visit relatives and exchange gifts and sweets.
6. Houses look so pretty with lighted lamps all around.

(*a*) 2, 5, 3, 4 (*b*) 5, 2, 4, 3

(*c*) 3, 5, 4, 2 (*d*) 4, 5, 2, 3

8. Ms. X will be in Bagdogra from 01/05/2014 to 20/05/2014 and from 22/05/2014 to 31/05/2014. On the morning of 21/05/2014, she will reach Kochi via Mumbai.

Which one of the statements below is logically valid and can be inferred from the above sentences ?

(*a*) Ms. X will be in Kochi for one day, only in May

(*b*) Ms. X will be in Kochi for only one day in May

(*c*) Ms. X will be only in Kochi for one day in May

(*d*) Only Ms. X will be in Kochi for one day in May.

9. $\log \tan 1° + \log \tan 2° + ... + \log \tan 89°$ is ....

(*a*) 1 (*b*) $\frac{1}{\sqrt{2}}$

(*c*) 0 (*d*) – 1

10. Ram and Shyam shared a secret and promised to each other that it would remain between them. Ram expressed himself in one of the following ways as given in the choices below.

Identify the correct way as per standard English.

(*a*) It would remain between you and me.

(*b*) It would remain between I and you

(*c*) It would remain between you and I

(*d*) It would remain with me.

ELECTRONICS AND COMMUNICATION ENGINEERING

Q. No. 1 – 25 Carry One Mark Each

1. A coaxial cable is made of two brass conductors. The spacing between the conductors is filled with Teflon ($\varepsilon_r = 2.1$, $\tan\delta = 0$). Which one of the following circuits can represent the lumped element model of a small piece of this cable having length Δz?

(*a*)
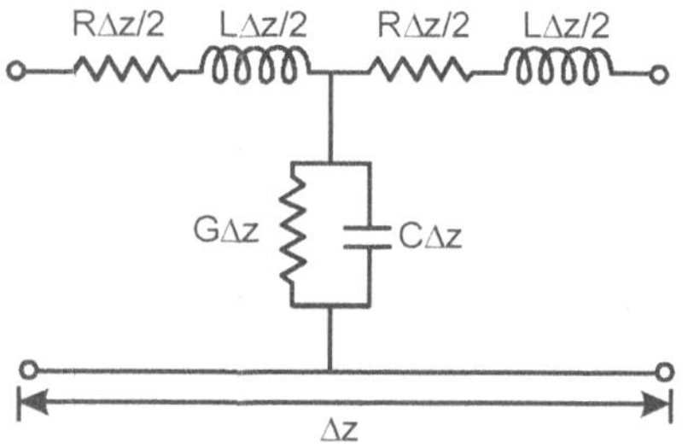

(*b*)
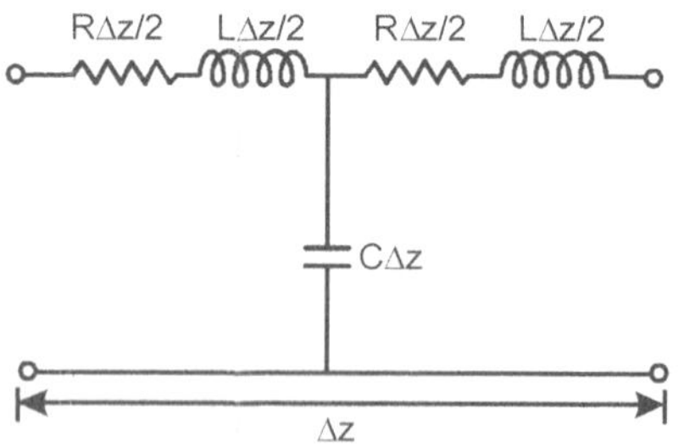

(*c*)
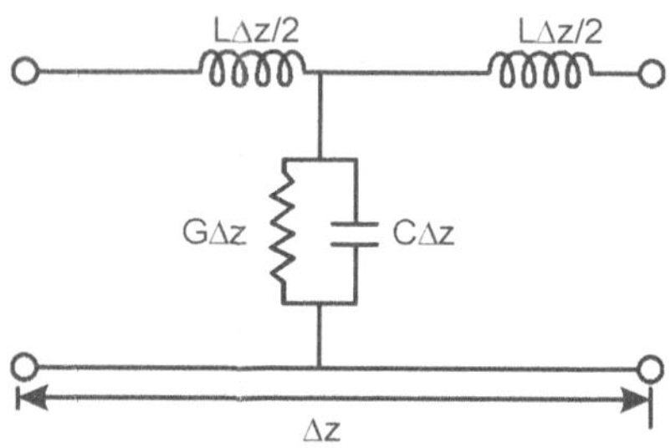

(*d*)
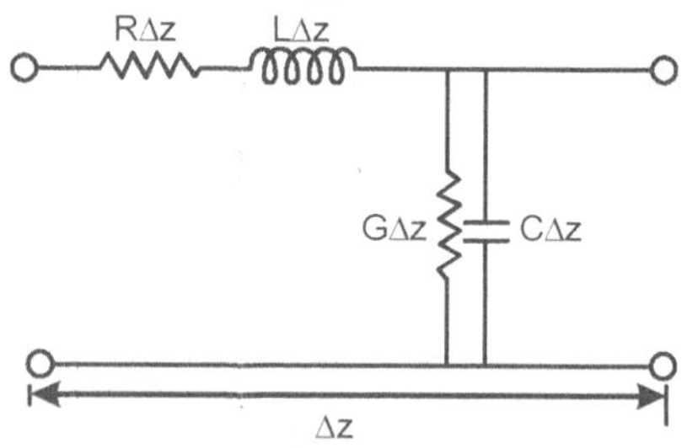

2. The phase margin (in degrees) of the system

$$G(s) = \frac{10}{s(s+10)} \text{ is } \_\_\_\_\_\_.$$

3. In the circuit shown, diodes D_1, D_2 and D_3 are ideal, and the inputs E_1, E_2 and E_3 are '0 V' for logic '0' and '10 V' for logic '1'. What logic gate does the circuit represent?

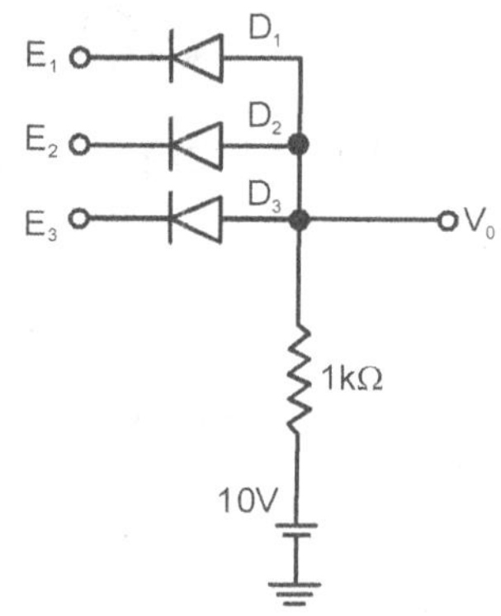

(*a*) 3 input OR gate

(*b*) 3 input NOR gate

(*c*) 3 input AND gate

(*d*) 3 input XOR gate

4. In the circuit shown in the figure, the BJT has a current gain (β) of 50. For an emitter base voltage V_{EB} = 600 mV, the emitter collector voltage V_{EC} (in Volts) is ______.

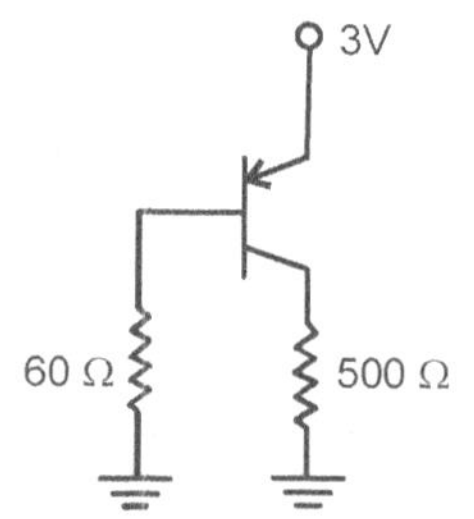

5. The contour on the x-y plane, where the partial derivative of $x^2 + y^2$ with respect to y is equal to the partial derivative of $6y + 4x$ with respect to x, is

(*a*) $y = 2$ (*b*) $x = 2$

(*c*) $x + y = 4$ (*d*) $x - y = 0$

6. For $A = \begin{bmatrix} 1 & \tan x \\ -\tan x & 1 \end{bmatrix}$ the determinant of $A^T A^{-1}$ is

(*a*) $\sec^2 x$ (*b*) $\cos 4x$

(*c*) 1 (*d*) 0

7. In the circuit shown, the voltage Vx (in Volts) is ______.

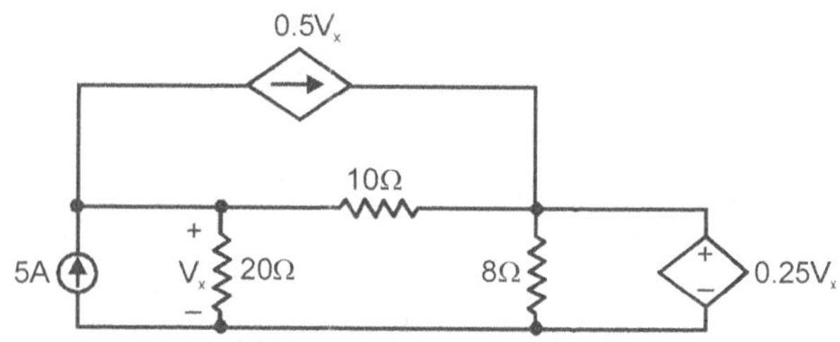

8. Which one of the following 8085 microprocessor programs correctly calculates the product of two 8-bit numbers stored in registers B and C ? (Options)

(*a*)
```
MVI A, 00 H
JNZ LOOP
CMP C
LOOP DCR B
HLT
```

(*b*)
```
MVI, A, 00H
CMP C
LOOP DCR B
JNZ LOOP
HLT
```

(*c*)
```
MVI A, 00H
LOOP ADD C
```

(*d*)
```
MVI A, 00H
ADD C
```

9. Consider the function $g(t) = e^{-t}\sin(2\pi t)\,u(t)$ where $u(t)$ is the unit step function. The area under $g(t)$ is _______.

10. In the circuit shown using an ideal opamp, the 3-dB cut-off frequency (in Hz) is _______.

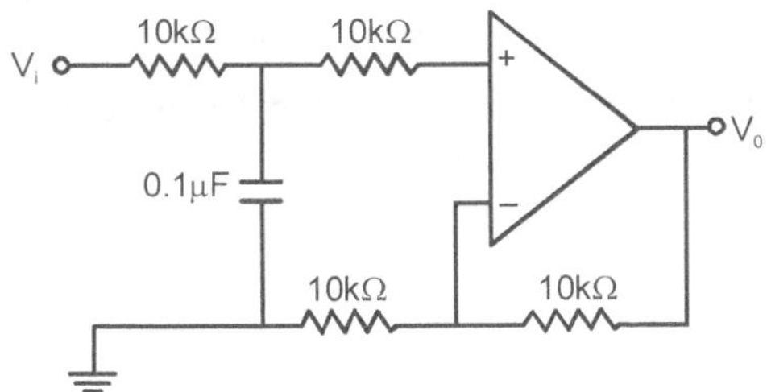

11. The modulation scheme commonly used for transmission from GSM mobile terminals is

(*a*) 4-QAM
(*b*) 16-PSK
(*c*) Walsh-Hadamard orthogonal codes
(*d*) Gaussian Minimum Shift Keying (GMSK)

12. Which one of the following processes is preferred to form the gate dielectric (SiO_2) of MOSFETs ?

(*a*) Sputtering
(*b*) Molecular beam epitaxy
(*c*) Wet oxidation
(*d*) Dry oxidation

13. Consider the Bode plot shown in figure. Assume that all the poles and zeros are real valued.

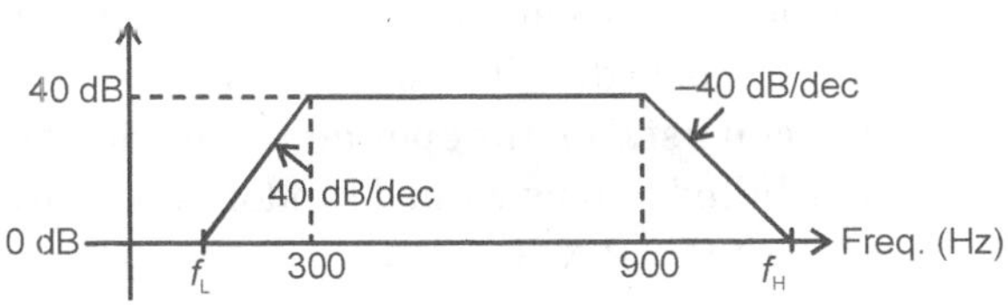

The value of $f_H - f_L$ (in Hz) is _______.

14. In the circuit shown, assume that diodes D_1 and D_2 are ideal. In the steady-state condition the average voltage V_{ab} (in Volts) across the 0.5 μF capacitor is _______.

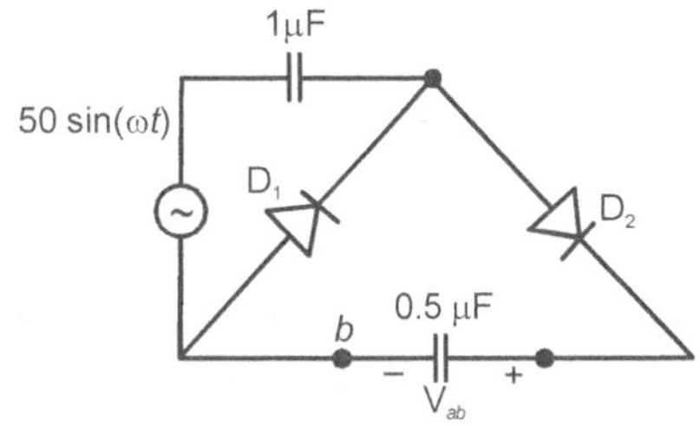

15. The transfer function of a first order controller is given as $G_C(s) = \frac{K(s+a)}{s+b}$ where, K, a and b are positive real numbers. The condition for this controller to act as a phase lead compensator is

(*a*) $a < b$ (*b*) $a > b$
(*c*) $K < ab$ (*d*) $K > ab$

16. A message signal $m(t) = A_m \sin(2\pi f_m t)$ is used to modulate the phase of a carrier $A_c \cos(2\pi f_c t)$ to get the modulated signal $y(t) = A_c \cos(2\pi f_c t + m(t))$. The bandwidth of $y(t)$

(*a*) depends on A_m but not on f_m
(*b*) depends on f_m but not on A_m
(*c*) depends on both A_m and f_m
(*d*) does not depends on A_m or f_m

17. The directivity of an antenna array can be increased by adding more antenna elements, as a larger number of elements

(*a*) improves the radiation efficiency
(*b*) increases the effective area of the antenna
(*c*) results in a better impedance matching
(*d*) allows more power to be transmitted by the antenna

18. For the circuit shown in the figure, the Thevenin equivalent voltage (in Volts) across terminals a-b is _______.

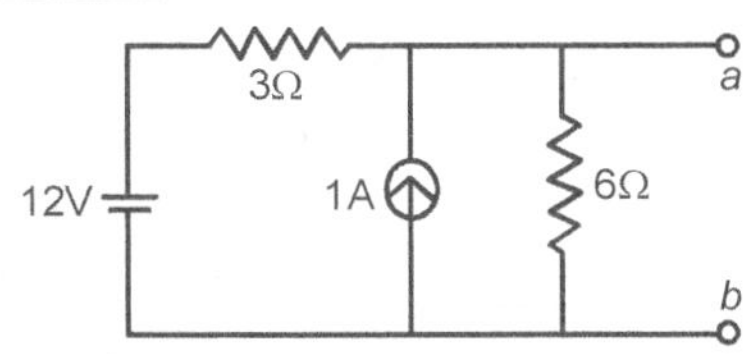

19. The impulse response of an LTI system can be obtained by

(*a*) differentiating the unit ramp response
(*b*) differentiating the unit step response
(*c*) integrating the unit ramp response
(*d*) integrating the unit step response

20. Consider a four point moving average filter defined by the equation $y[n] = \sum_{i=0}^{3} \alpha_i [n-i]$

The condition on the filter coefficients that results in a null at zero frequency is

(*a*) $\alpha_1 = \alpha_2 = 0;\ \alpha_0 = -\alpha_3$
(*b*) $\alpha_1 = \alpha_2 = 1;\ \alpha_0 = -\alpha_3$
(*c*) $\alpha_0 = \alpha_3 = 0;\ \alpha_1 = \alpha_2$
(*d*) $\alpha_1 = \alpha_2 = 0;\ \alpha_0 = \alpha_3$

21. If C is a circle of radius r with centre z_0, in the complex z-plane and if n is a non-zero integer, then $\oint \frac{dz}{(z - z_0)^{n+1}}$ equals

(*a*) $2\pi nj$ (*b*) 0
(*c*) $\frac{nj}{2\pi}$ (*d*) $2\pi n$

22. At very high frequencies, the peak output voltage V_0 (in Volts) is ________.

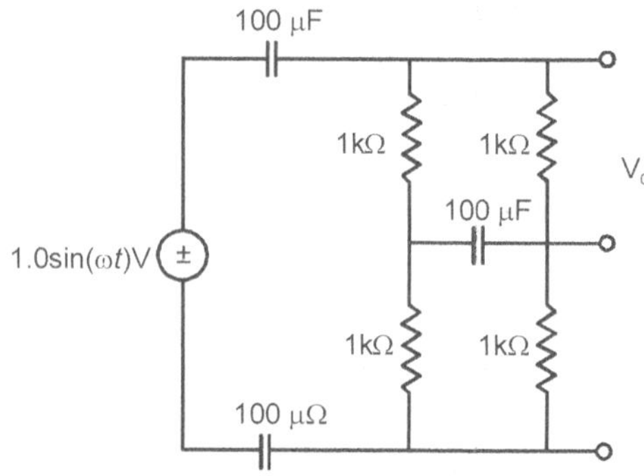

23. If the base width in a bipolar junction transistor is doubled, which one of the following statements will be TRUE ?

(*a*) Current gain will increase
(*b*) Unity gain frequency will increase
(*c*) Emitter base junction capacitance will increase
(*d*) Early voltage will increase

24. The value of $\sum_{n-0}^{\infty} n\left(\frac{1}{2}\right)^n$ is ________.

25. The circuit shown consists of J-K flip-flops, each with an active low asynchronous reset $(\overline{R}_d \text{ input})$. The counter corresponding to this circuit is

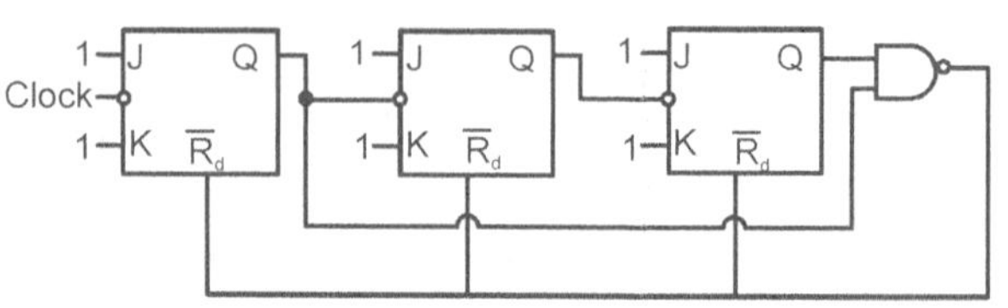

(*a*) a modulo-5 binary up counter
(*b*) a modulo-6 binary down counter
(*c*) a modulo-5 binary down counter
(*d*) a modulo-6 binary up counter

Q. No. 26 – 55 Carry Two Marks Each

26. A 200 m long transmission line having parameters shown in the figure is terminated into a load R_L. The line is connected to a 400 V source having source resistance R_S through a switch which is closed at $t = 0$. The transient response of the circuit at the input of the line ($z = 0$) is also drawn in the figure. The value of R_L (in Ω) is ______

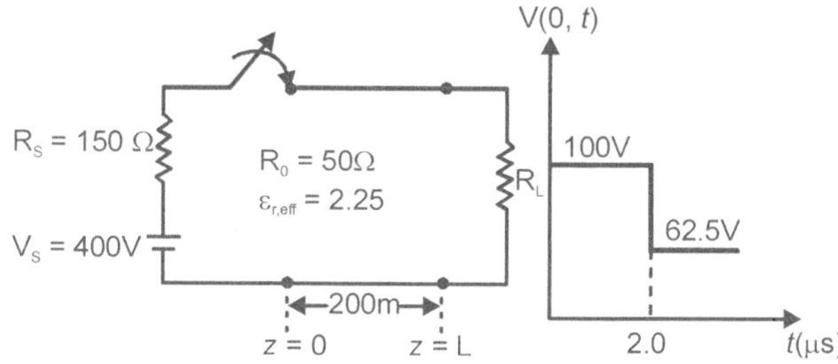

27. A coaxial capacitor of inner radius 1 mm and outer radius 5 mm has a capacitance per unit length of 172 pF/m. If the ratio of outer radius to inner is doubled, the capacitance per unit length (in pF/m) is ________.

28. A universal logic gate can implement any Boolean function by connecting sufficient number of them appropriately. Three gates are shown.

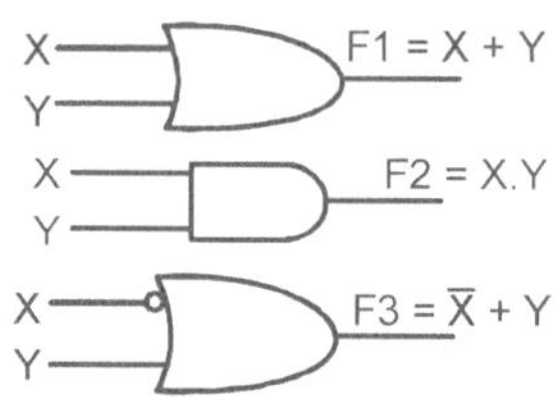

Which one of the following statements is TRUE ?

(*a*) Gate 1 is a universal gate.
(*b*) Gate 2 is a universal gate.
(*c*) Gate 3 is a universal gate.
(*d*) None of the gates shown is a universal gate.

29. The Newton-Raphson method is used to solve the equation $f(x) = x^3 - 5x^2 + 6x - 8 = 0$. Taking the initial guess as $x = 5$, the solution obtained at the end of the first iteration is ________.

30. A random binary wave $y(t)$ is given by

$$y(t) = \sum_{n=-\infty}^{\infty} X_n p(t - nT - \phi).$$

where $p(t) = u(t) - u(t - T)$, $u(t)$ is the unit step function and φ is an independent random variable with uniform distribution in [0, T]. The sequence $\{X_n\}$ consists of independent and identically distributed binary valued random variables with $P\{X_n = +1\} = P\{X_n = -1\} = 0.5$ for each n.

The value of the auto correlation $R_{yy}\left(\frac{3T}{4}\right) \triangleq E\left[y(t)y\left(t - \frac{3T}{4}\right)\right]$ equal ________.

31. A three bit pseudo random number generator is shown. Initially the value of output $Y = Y_2\,Y_1\,Y_0$ is set to 111. The value of output Y after three clock cycles is

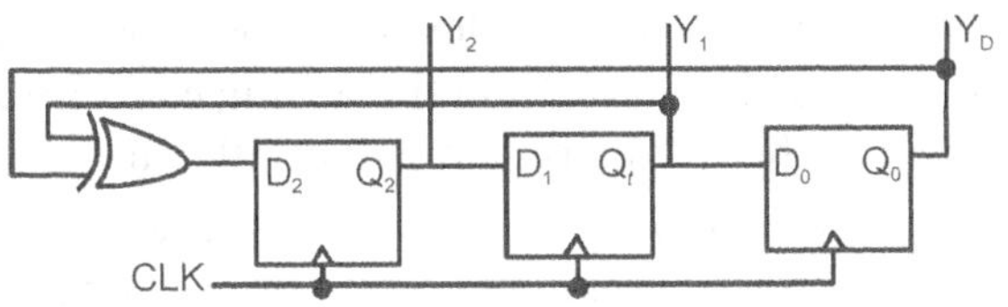

(*a*) 000 (*b*) 001

(*c*) 010 (*d*) 100

32. In the circuit shown, assume that the opamp is ideal. If the gain (V_o/V_{in}) is –12, the value of R (in kΩ) is ______

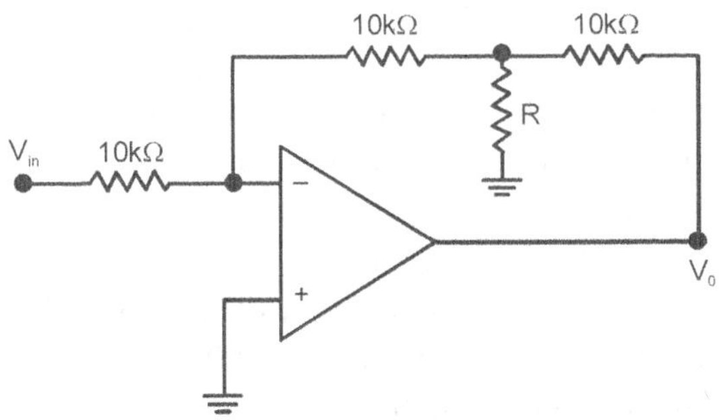

33. Two sequences $x_1[n]$ and $x_2[n]$ have the same energy. Suppose $x_1[n] = \alpha\, 0.5^n\, u[n]$, where α is a positive real number and $u[n]$ is the unit step sequence. Assume

$$x_2[n] = \begin{cases} \sqrt{1.5} & \text{for } n = 0, 1 \\ 0 & \text{otherwise} \end{cases}$$

Then the value of α is ________.

34. The ABCD parameters of the following 2-port network are

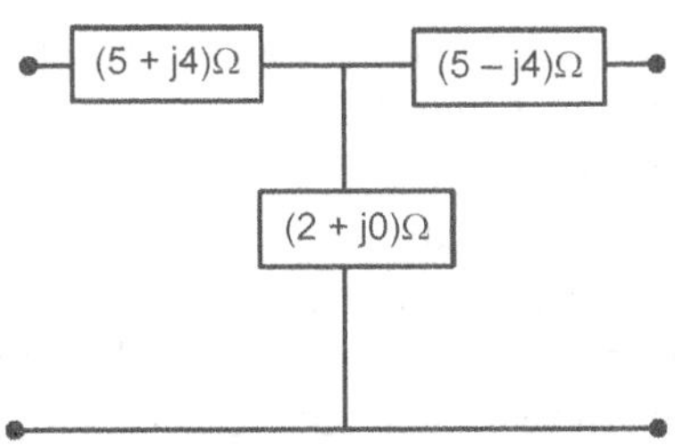

(*a*) $\begin{bmatrix} 3.5 + j2 & 20.5 \\ 20.5 & 3.5 - j2 \end{bmatrix}$

(*b*) $\begin{bmatrix} 3.5 + j2 & 0.5 \\ 0.5 & 3.5 - j2 \end{bmatrix}$

(*c*) $\begin{bmatrix} 10 & 2 + j0 \\ 2 + j0 & 10 \end{bmatrix}$

(*d*) $\begin{bmatrix} 7 + j4 & 0.5 \\ 30.5 & 7 - j4 \end{bmatrix}$

35. A network is described by the state model as

$$\dot{x}_1 = 2x_1 - x_2 + 3u$$
$$\dot{x}_2 = -4x_2 - u$$
$$y = 3x_1 - 2x_2$$

The transfer function $H(s) = \left(\dfrac{Y(s)}{U(s)}\right)$ is

(*a*) $\dfrac{11s+35}{(s-2)(s+4)}$ (*b*) $\dfrac{11s-35}{(s-2)(s+4)}$

(*c*) $\dfrac{11s+38}{(s-2)(s+4)}$ (*d*) $\dfrac{11s-38}{(s-2)(s+4)}$

36. In the circuit shown, the current I flowing through the 50 Ω resistor will be zero if the value of capacitor C (in μF) is ______.

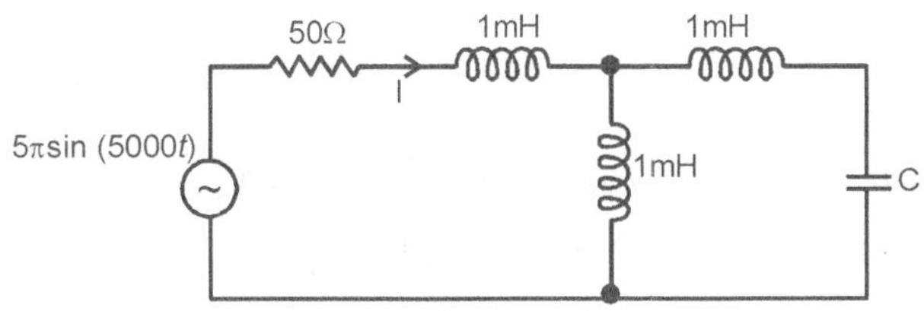

37. A realization of a stable discrete time system is shown in figure. If the system is excited by a unit step sequence input $x[n]$, the response $y[n]$ is

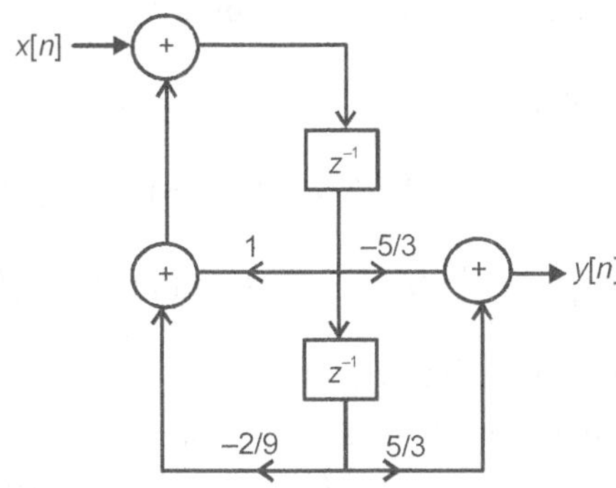

(*a*) $4\left(-\frac{1}{3}\right)^n u(n) - 5\left(-\frac{2}{3}\right)^n u[n]$

(*b*) $5\left(-\frac{2}{3}\right)^n u(n) - 3\left(-\frac{1}{3}\right)^n u[n]$

(*c*) $5\left(\frac{1}{3}\right)^n u(n) - 5\left(\frac{2}{3}\right)^n u[n]$

(*d*) $5\left(\frac{2}{3}\right)^n u(n) - 5\left(\frac{1}{3}\right)^n u[n]$

38. The complex envelope of the bandpass signal

$$x(t) = -\sqrt{2}\left(\frac{\sin\left(\frac{\pi t}{5}\right)}{\frac{\pi t}{5}}\right)\sin\left(\pi t - \frac{\pi}{4}\right),$$

centered about $f = \frac{1}{2}$ Hz, is

(a) $\left(\dfrac{\sin\left(\frac{\pi t}{5}\right)}{\frac{\pi t}{5}}\right)e^{j\frac{\pi}{4}}$ (b) $\left(\dfrac{\sin\left(\frac{\pi t}{5}\right)}{\frac{\pi t}{5}}\right)e^{-j\frac{\pi}{4}}$

(c) $\sqrt{2}\left(\dfrac{\sin\left(\frac{\pi t}{5}\right)}{\frac{\pi t}{5}}\right)e^{j\frac{\pi}{4}}$ (d) $\sqrt{2}\left(\dfrac{\sin\left(\frac{\pi t}{5}\right)}{\frac{\pi t}{5}}\right)e^{j\frac{\pi}{4}}$

39. In the circuit shown, assume that the diodes D_1 and D_2 are ideal. The average value of voltage V_{ab} (in Volts), across terminals '*a*' and '*b*' is ______.

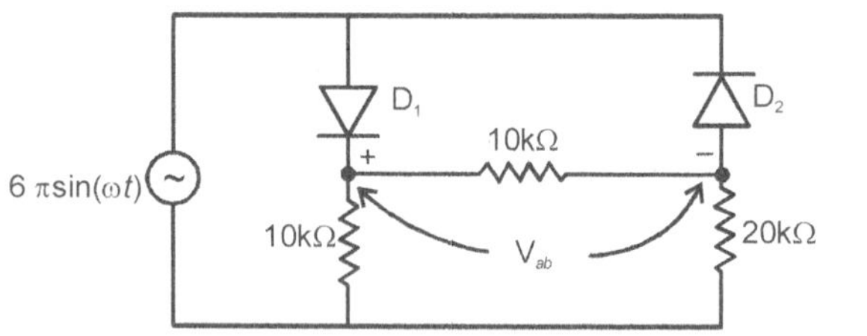

40. Consider the differential equation

$$\frac{d^2x(t)}{dt^2}+3\frac{dx(t)}{dt}+2x(t)=0.$$

Given $x(0) = 20$ and $x(1) = \dfrac{10}{e}$,

where $e = 2.71$. the value of $x(2)$ is ________.

41. Let $\bar{x}[n]=1+\cos\left(\frac{\pi n}{8}\right)$ be periodic signal with period 16. Its DFS coefficient are defined by $\alpha_k=\frac{1}{16}\sum_{x=0}^{15}\bar{x}[n]\exp\left(-j\frac{\pi}{8}kn\right)$ for all k. The value of the coefficient α_{31} is __________.

42. A fair die with faces {1, 2, 3, 4, 5, 6} is thrown repeatedly till '3' is observed for the first time. Let X denote the number of times the die is thrown. The expected value of X is ________.

43. The electric field profile in the depletion region of a *p-n* junction in equilibrium is shown in the figure. Which one of the following statements is **NOT TRUE** ?

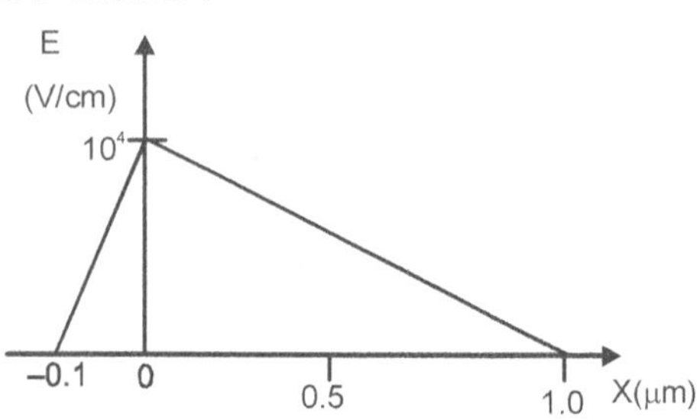

(a) The left side of the junction is *n*-type and the right side is *p*-type

(b) Both the *n*-type and *p*-type depletion regions are uniformly doped

(c) The potential difference across the depletion region is 700 mV

(d) If the *p*-type region has a doping concentration of 10^{15} cm^{-3}, then the doping concentration in the *n*-type region will be 10^{16} cm^{-3}

44. A vector field $D = 2\rho^2 a_\rho + z\, a_z$ exists inside a cylindrical region enclosed by the surfaces $\rho = 1$, $z = 0$ and $z = 5$. Let S be the surface bounding this cylindrical region. The surface integral of this field on S $\left(\oiint_s D.ds\right)$ is ________.

45. An npn BJT having reverse saturation current $I_s = 10^{-15}$ A is biased in the forward active region with $V_{BE} = 700$ mV. The thermal voltage (V_T) is 25 mV and the current gain (β) may vary from 50 to 150 due to manufacturing variations. The maximum emitter current (in μA) is ________.

46. Consider the 3 m long lossless air-filled transmission line shown in the figure. It has a characteristic impedance of 120πΩ, is terminated by a short circuit, and is excited with a frequency of 37.5 MHz. What is the nature of the input impedance (Z_{in})?

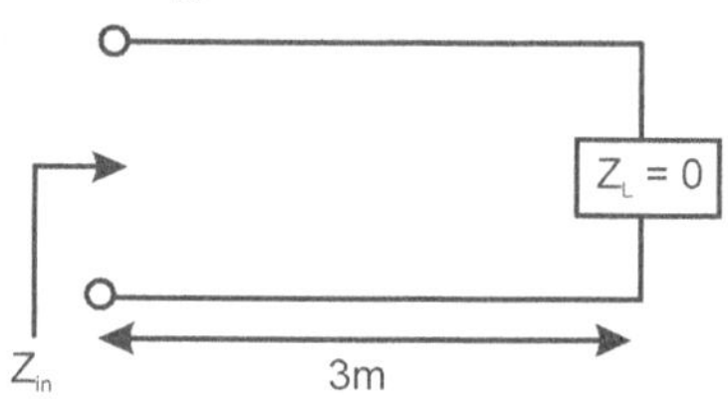

(a) Open (b) Short

(c) Inductive (d) Capacitive

47. The current in an enhancement mode NMOS transistor biased in saturation mode was measured to be 1 mA at a drain-source voltage of 5 V. When the drain-source voltage was increased to 6 V while keeping gate-source voltage same, the drain current increased to 1.02 mA. Assume that drain to source saturation voltages is much smaller than the applied drain-source voltage. The channel length modulation parameter λ (in V^{-1}) is __________.

48. For the system shown in figure, $s = -2.75$ lies on the root locus if K is _______.

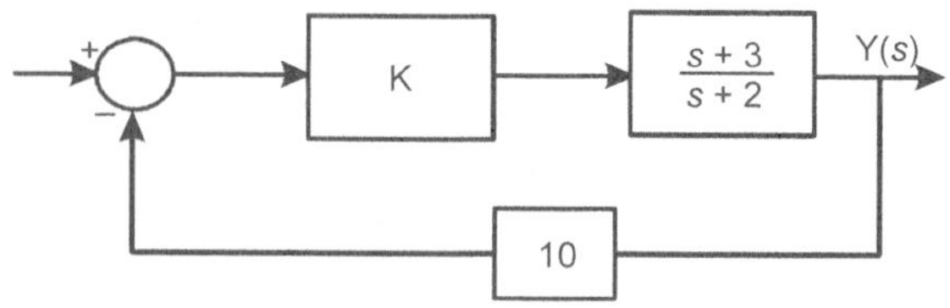

49. An SR latch is implemented using TTL gates as shown in the figure. The set and reset pulse inputs are provided using the push-button switches. It is observed that the circuit fails to work as desired. The SR latch can be made functional by changing

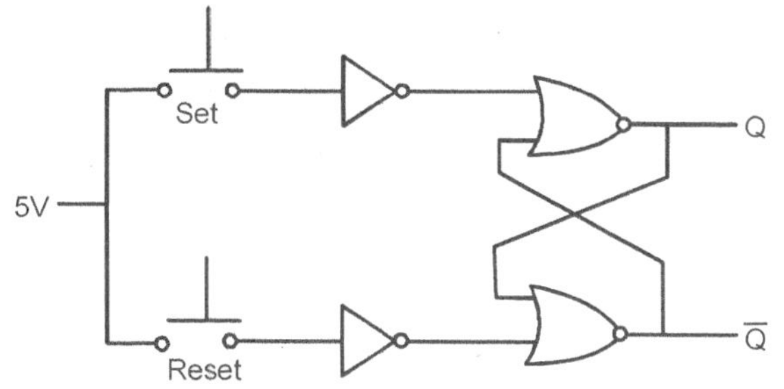

(a) NOR gates to NAND gates

(b) inverters to buffers

(c) NOR gates to NAND gates and inverters to buffers

(d) 5 V to ground

50. The variance of the random variable X with probability density function

$$f(x) = \frac{1}{2}|x|e^{-|x|} \quad \text{is \_\_\_\_\_}$$

51. Consider a continuous-time signal defined as

$$x(t) = \left(\frac{\sin\left(\frac{\pi t}{2}\right)}{\left(\frac{\pi t}{2}\right)}\right) * \sum_{n=-\infty}^{\infty} \delta(t - 10n)$$

Where '*' denotes the convolution operation and t is in seconds. The Nyquist sampling rate (in samples / sec) for $x(t)$ is ________.

52. In the circuit shown, the both the enhancement mode NMOS transistors have the following characteristics : $k_n = \mu_n C_{ox}$ (W/L) = 1 mA/V^2; V_{TN} = 1V. Assume that the channel length modulation parameter λ is zero and body is shorted to source. The minimum supply voltage V_{DD} (in volts) needed to ensure that transistor M_1 operates in saturation mode of operation is ________.

53. The position control of a DC servo-motor is given in the figure. The values of the parameters are K_T = 1 N-m A, R_a = 1 Ω, L_a = 0.1H. J = 5 kg-m^2, B = 1 N-m (rad/sec) and K_b = 1 V/(rad/sec). The steady-state position response (in radians) due to unit impulse disturbance torque T_d is ________.

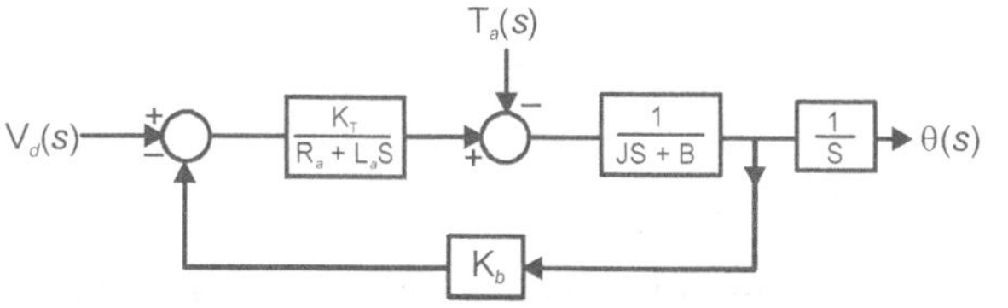

54. The characteristic equation of an LTI system is given by $F(s) = s^5 + 2s^4 + 3s^3 + 6s^2 - 4s - 8 = 0$. The number of roots that lie strictly in the left half s-plane is ________.

55. Suppose $x[n]$ is an absolutely summable discrete-time signal. Its z-transform is a rational function with two poles and two zeroes. The poles are at $z = \pm 2j$. Which one of the following statements is TRUE for the signal $x[n]$?

(a) It is a finite duration signal.

(b) It is a causal signal.

(c) It is a non-causal signal.

(d) It is a periodic signal

ANSWERS

GENERAL APTITUDE

1. (c) 2. (b) 3. (b) 4. (b) 5. (a) 6. (1.9-2.2) 7. (b) 8. (b) 9. (c) 10. (a)

ELECTRONICS AND COMMUNICATION ENGINEERING

1. (b) 2. (84.32) 3. (c) 4. (2) 5. (a) 6. (c) 7. (8) 8. (c) 9. (0.155) 10. (159.15)

11. (d) 12. (d) 13. (8970) 14. (100) 15. (a) 16. (c) 17. (b) 18. (10) 19. (b) 20. (a)

21. (b) 22. (0.5) 23. (d) 24. (2) 25. (a) 26. (30) 27. (120.22) 28. (c) 29. (4.2903) 30. (0.25)

31. (d) 32. (1) 33. (1.49-1.51) 34. (b) 35. (a) 36. (20) 37. (c) 38. (c) 39. (5) 40. (0.8556)

41. (0.5) 42. (6) 43. (c) 44. (78.53) 45. (1475) 46. (d) 47. (0.022) 48. (0.3) 49. (d) 50. (6)

51. (0.4) 52. (3) 53. (– 0.5) 54. (2) 55. (c)

EXPLANATIONS

GENERAL APTITUDE (GA)

1.

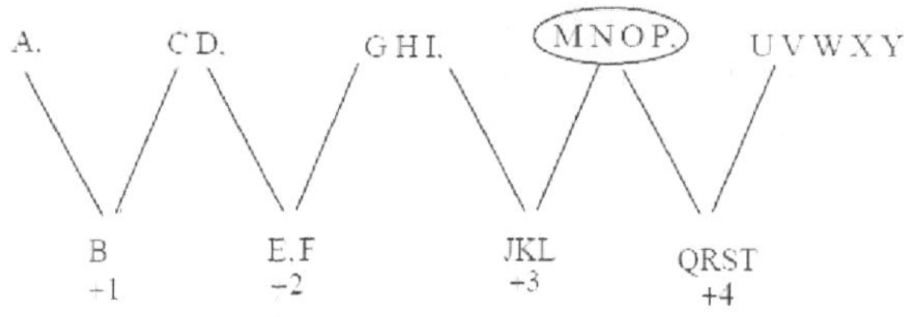

3. For condition regarding something which already happened, should have practiced is the correct choice.

5. For whole numbers, greater the value greater will be its log.

   Same logic for power of *e*.

6. 90% of area of sheet = Cross sectional area of cone

   $\Rightarrow\ 0.9 \times \pi \times 30 \times 30 = \pi \times r_1 \times 30$

   $\Rightarrow\ 27 \text{ cm} = r_1$

   $\therefore$ height of the cone = $\sqrt{30^2 - 27^2}$ = 13.08 cm

8. Second sentence says that Ms. X reaches Kochi on 21/05/2014. Also she has to be in Bagdogora on 22/05/2014.

   $\therefore$ She stays in Kochi for only one day in may.

9. $\log \tan 1° + \log \tan 89° = \log(\tan 1° \times \tan 89°)$

   $= \log(\tan 1° \times \cot 1°)$

   $= \log 1 = 0$

   Using the same logic total sum is '0'.

ELECTRONICS AND COMMUNICATION ENGINEERING

1. Loss tangent $\tan\delta = 0 = \dfrac{\sigma}{\omega \in}$

   $\sigma = 0$

   G $\rightarrow$ Conductivity of the dielectric material

   So, $\sigma = 0 = G$

3. **Case (i)** : If any input is logic 0 (i.e., 0V) then corresponding diode is 'ON' and due to ideal diode output voltage $V_0 = 0$ as well as if there is any input logic 1 (i.e., 10V) corresponding diode will be OFF.

   Case (ii) : If all the inputs are high (i.e., 10V) then all the diodes are R.B (OFF) and output voltage $V_0 = 10V$.

   So, it is a positive logic 3-inputs AND gate.

4. $V_{EB} = 0.7$ V

   $I_B = 0.0383$ mA

   $I_C = 1.916$ mA

   $V_{EC} = 3 - I_C R_C = 3 - (1.916 \times 0.5) = 2.04$ V

5. The partial derivative of x^2y^2 with respect to y is $0 + 2y \Rightarrow 2y$. The partial derivate of $6y + 4x$ with respect x is $0 + 4 = 4$. Green that both are equal.

   $\Rightarrow\quad 2y = 4$

   $\Rightarrow\quad y = 2$

6. $$A = \begin{bmatrix} 1 & \tan x \\ -\tan x & 1 \end{bmatrix}$$

   $$A^T = \begin{bmatrix} 1 & -\tan x \\ \tan x & 1 \end{bmatrix}$$

   Now $$A^{-1} = \frac{1}{|A|}(\text{adj }(A))$$

   $$A = \frac{1}{\sec^2 x}\begin{bmatrix} 1 & -\tan x \\ \tan x & 1 \end{bmatrix}$$

   $\therefore\quad |A^T.A^{-1}| = 1$

7.

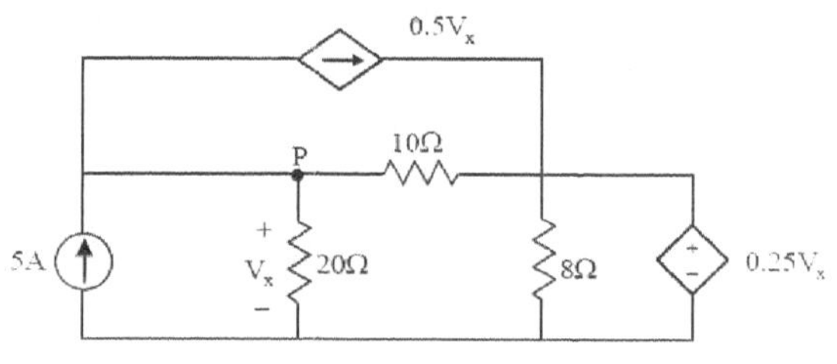

   Apply KCL at point P

   $$\frac{V_x}{20} + \frac{V_x - 0.25V_x}{10} + 0.5V_x = 5$$

   $$V_x\left(\frac{1}{20} + \frac{0.75}{10} + 0.5\right) = 5$$

   $$V_x\left(\frac{5}{8}\right) = 5$$

   $\Rightarrow V_x = 8V$

8. MVI A, 00H $\leftarrow$ Load accumulator by 00H

   Loop : ADDC $\leftarrow$ Add the content of accumulator with content of C register and store result in accumulator.

   This will continue till B register reaches to 004.

   DCRB

   JNZ LOOP

   HLT

   So, repetitive addition of a number as many times will give the product of these two numbers.

9. $g(t) = e^{-t}\sin(2\pi t).u(t)$

Let $y(t) = \sin(2\pi t).u(t)$

then $Y(s) = \dfrac{2\pi}{s^2+(2\pi)^2}$

$g(t) = e^{-1}y(t)$

$\Rightarrow$ $G(s) = Y(s+1)$

$$G(s) = \frac{2\pi}{(s+1)^2+(2\pi)^2}$$

$$G(s) = \int_{-\infty}^{\infty} g(t).e^{-st}dt$$

$$G(0) = \int_{-\infty}^{\infty} g(t).dt$$

$$\Rightarrow \text{Area under g(t)} = \frac{2\pi}{1+(2\pi)^2} = \frac{6.28}{40.438} = 0.155$$

10. $f_{3dB} = \dfrac{1}{2\pi RC}$

$$= \frac{1}{2\pi\times 10\times 10^3\times 0.1\times 10^{-6}}$$

$= 159.15$ Hz

13. $40 = \dfrac{40-0}{\log_{10}(300)-\log_{10}(f_L)}$

$$\log_{10}\left(\frac{300}{f_L}\right) = 1$$

$300 = 10\, f_L$

$f_L = 30$ Hz ...(*i*)

$$-40 = \frac{0-40}{\log_{10} f_H - \log(900)}$$

$$\log_{10}\left(\frac{f_H}{900}\right) = 1$$

$f_H = 900\times 10 = 9000$ ...(*ii*)

$f_H - f_L = 9000 - 30 = 8970$ Hz

15. For phase lead compensator

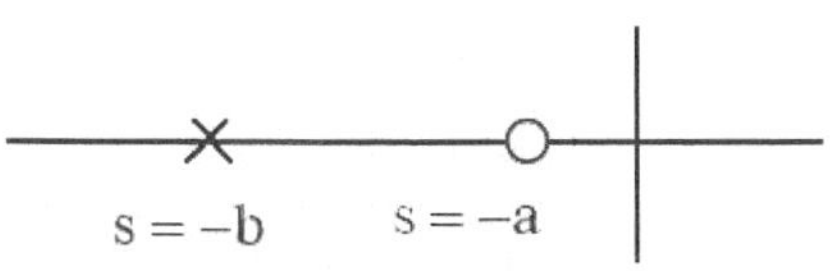

a < b

16. $y(t) = A_c\cos[2\pi f_c t + m(t)]$

$m(t) = A_m\sin(2\pi f_m t)$

Since $y(t)$ is phase modulated signal,

$\phi(t) = 2\pi f_c t + m(t)$

Bandwidth $= 2[\Delta f + f_m]$

$$\Delta f = \frac{1}{2\pi}\frac{d}{dt}m(t)$$

$\Rightarrow$ Δf depends on A_m as well as f_m. Thus Bandwidth depends on both A_m and f_m.

17. $D = \dfrac{4\pi}{\lambda^2}Ae$

$D\uparrow \rightarrow Ae\uparrow$

18.

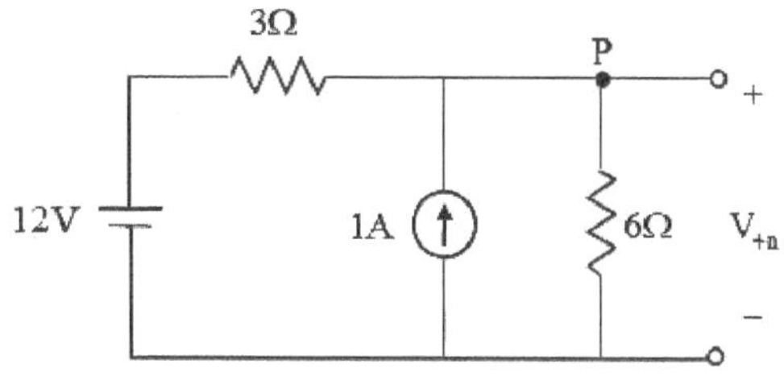

Apply nodal equation at point P

$$V_{+n}\left(\frac{1}{3}+\frac{1}{6}\right) = \frac{12}{3}+1$$

$$\frac{V_{+n}}{2} = 5$$

$\Rightarrow$ $V_{+n} = 10$ V

19. Let $h(t)$ be the impulse response of the system

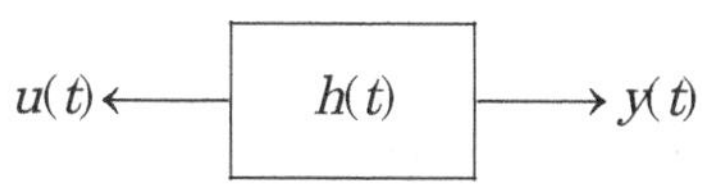

$y(t)$ is unit step response of the system

$$y(t) = \int_{-\infty}^{t} h(\tau)d\tau$$

If we need to get $h(t)$, then we have to differentiate $y(t)$.

Thus differentiating the unit-step response gives impulse response for LTI system.

20. Given $y[n] = \sum_{i=0}^{2}\alpha_i x(n-i)$

$\Rightarrow$ $y[n] = \alpha_0 x[n] + \alpha_1 x[n-1] + \alpha_2 x[n-2] + \alpha_3 x[n-3]$

Getting a null at zero frequency implies that given filter can be high pass filter but it cannot be low pass filter.

High pass filter is possible if we have negative coefficients.

Let say, $\alpha_1 = \alpha_2 = 0,\ \alpha_0 = -\alpha_3$

$\Rightarrow$ $y[n] = -\alpha_3 x[n] + a_3 x[n-3]$

$H(z) = -\alpha_3[1-z^{-3}]$

$\Rightarrow$ $H(e^{j\Omega}) = -\alpha_3[1-e^{-j3\Omega}]$

$$= -\alpha_3 e^{\frac{-j3\Omega}{2}}\left[\frac{e^{j\frac{3\Omega}{2}} - e^{-j\frac{3\Omega}{2}}}{2j}\right]\times 2j$$

$$= -\alpha_3 2j\sin\left(\frac{3\Omega}{2}\right).e^{-j\frac{3\Omega}{2}}$$

$$= -\alpha_3 2.\sin\frac{3\Omega}{2}.e^{-j\frac{3\Omega}{2}}.e^{j\frac{\pi}{2}}$$

$\Rightarrow$ $H(e^{j\Omega})|_{\Omega=0} = 0$

In other cases it in not possible.

21. By Cauchy's Integral formula,

$$\oint_C \frac{f(z)}{(z-z_0)^{n+1}}dz = \frac{2\pi i - f^n(z_0)}{n!}$$

$$\oint_C \frac{dz}{(z-z_0)^{n+1}} = \frac{2\pi i}{n!}\times 0$$

$$= 0 \qquad \begin{pmatrix} \because f(z) = 1 \\ f^n(z) = 0 \text{ at any } z_0 \end{pmatrix}$$

22. For capacitor $Z_C = \frac{1}{j\omega C}$

Very high frequency means $\omega \to \infty \Rightarrow Z_C \to 0$

So, all capacitors are replaced by short circuit.

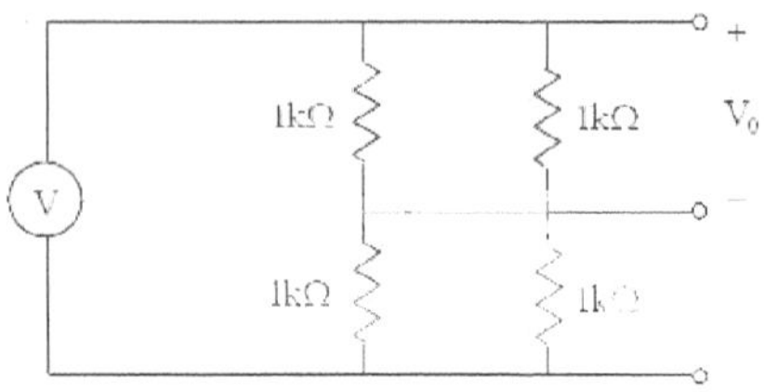

By voltage division,

$$V_0 = \frac{V_i}{2}$$

$$V_0 = \frac{1.\sin\omega t}{2} = 0.5 \sin \omega t$$

Thus, Peak voltage = 0.5

23. W_B doubled (increased) → early effect is still present but its effect less severe relative to previous W_B. Slope I_C Vs V_{CE} decreases

24. $$\sum_{n=0}^{\infty} h.\left(\frac{1}{2}\right)^n = ?$$

This can be solved using Z-transform and also using algebra as arithmetic-geometric series. We will use algebra.

$$\sum_{n=0}^{\infty} h.\left(\frac{1}{2}\right)^n = 0.1 + 1.\frac{1}{2} + 2.\left(\frac{1}{2}\right)^2 + \ldots$$

[$a + (a+d)r + (a+2d)r^2 + \ldots$ infinite terms]

$$= \frac{a}{1-r} + \frac{rd}{(1-r)^2}$$

For above series, $a = 0$, $d = 1$, $r = \frac{1}{2}$.

$$\Rightarrow \quad \sum_{n=0}^{\infty}\left(\frac{1}{2}\right)^n.n = \frac{0}{1-\frac{1}{2}} + \frac{1(\frac{1}{2})}{\left(1-\frac{1}{2}\right)^2} = 2$$

25. 1. Clock is taken from normal output and it is –ve edge triggering. So, it is UP-counter.

2. Input of the NAND-gate is taken from Q_2 and Q_0. So $Q_2 = 1$ and $Q_0 = 1$.

3. To find the modulus

$(Q_2, Q_1, Q_0) = (1\ 0\ 1)$

So, it is MOD – 5 binary UP-counter.

26.

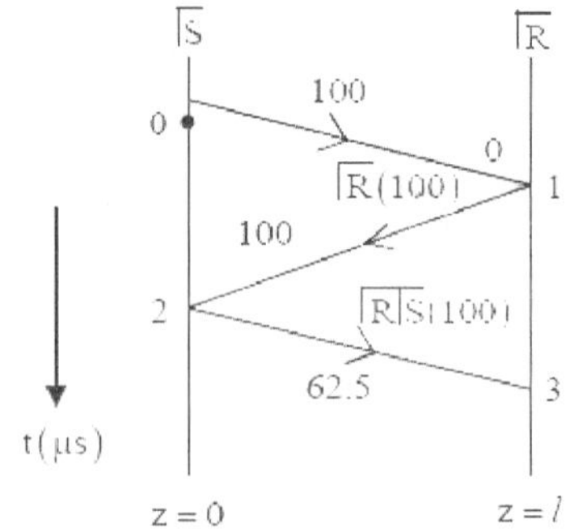

Given $V(t = 2\mu s, Z = 0) = 62.5$

$$62.5 = V(t = 0, z = 0) + V(t = 1, z = 0) + V(t = 2, z = 0)$$

$$62.5 = 100 + \Gamma_R(100) + \Gamma_R\Gamma_S(100)$$

$$\Gamma_R = \frac{R_L - 50}{R_L + 50}, \Gamma_S = \frac{1}{2}$$

So, $$\Gamma_R = \frac{-1}{4}$$

$$R_L = 30\ \Omega$$

27. $$C = \frac{2\pi \in l}{\ln(b/a)}$$

$$\frac{C}{l} = \frac{2\pi\in}{\ln(b/a)} = C_1$$

$$\frac{C_1}{C_2} = \frac{\ln(b_2/a_2)}{\ln(b_1/a_1)}$$

$$\frac{172\text{pF}}{C_2} = \frac{\ln(10/1)}{\ln(5)}$$

$$C_2 = \frac{\ln(5)}{\ln(10)}172\text{pF}$$

$$C_2 = 120.22 \text{ pF}$$

29. $$f(x) = x^3 - 5x^2 + 6x - 8$$

$$x_0 = 5$$

$$f'(x) = 3x^2 - 10x + 6$$

By Newton-Raphson method.

$$x_1 = x_0 - \frac{f(x_0)}{f'(x_0)}$$

$$= 5 - \frac{f(5)}{f'(5)}$$

$$= 5 - \frac{22}{31}$$

$$= 5 - 0.7097$$

$$= 4.2903$$

30.
$$y(t) = \sum_{n=-\infty}^{\infty} X_n P(t - nT - \phi)$$

$$R_{yy(z)} = \left[1 - \frac{|\tau|}{T}\right]$$

Derivation of above autocorrelation function can be found in any book dealing with random process.

[B.P. Lathi, Simon, Haykin, Schaum series].

$$R_{yy}\left(\frac{3T}{4}\right) = \left[1 - \frac{3\pi/4}{\pi}\right] = \frac{1}{4} = 0.25$$

31.

	$D_2(Q_1 \oplus Q_0)$	$D_1(Q_2)$	$D_0(Q_1)$	$Y_2(Q_2)$	$Q_1(Y_1)$	Q_1Y_0
	–	–	–	1	1	1
1st clcok	0	1	1	0	1	1
2nd clcok	0	0	1	0	0	1
3rd clcok	1	0	0	1	0	0

After three clock pulses output $Y_2Y_1Y_0 = 100$

34. For the standard 'T' network, obtain the Z-matrix first and then convert it into T-matrix

$$Z = \begin{bmatrix} 7+j4 & 2 \\ 2 & 7-j4 \end{bmatrix}$$

$$\Delta Z = [(7+j4)(7-j4)] - 4$$
$$= 49 + 16 - 4 = 16$$

$$A = \frac{Z_{11}}{Z_{21}} = \frac{7+j4}{2} = 3.5 + j2$$

$$B = \frac{\Delta Z}{Z_{21}} = \frac{61}{2} = 30.5$$

$$C = \frac{1}{Z_{21}} = \frac{1}{2} = 0.5$$

$$D = \frac{Z_{22}}{Z_{21}} = \frac{7-j4}{2} = 3.5 - j2$$

$$T = \begin{bmatrix} 3.5+j2 & 30.5 \\ 0.5 & 3.5-j2 \end{bmatrix}$$

35.
$$A = \begin{bmatrix} 2 & -1 \\ 0 & -4 \end{bmatrix} \quad B = \begin{bmatrix} 3 \\ -1 \end{bmatrix}$$

$$C = [3 - 2]$$

$$H(s) = C(sI - A)^{-1}B$$

$$= [3\ \ -2]\begin{bmatrix} s-2 & 1 \\ 0 & s+4 \end{bmatrix}\begin{bmatrix} 3 \\ -1 \end{bmatrix}$$

$$= [3\ \ -2]\frac{1}{(s-2)(s+4)}\begin{bmatrix} s+4 & -1 \\ 0 & s-2 \end{bmatrix}\begin{bmatrix} 3 \\ -1 \end{bmatrix}$$

$$= [3\ \ -2]\frac{1}{s^2 - 2s - 8}\begin{bmatrix} 3s+12+1 \\ -s+2 \end{bmatrix}$$

$$= \frac{95+39+25+4}{(s-2)(s+4)} = \frac{11s+35}{(s-2)(s+4)}$$

36.

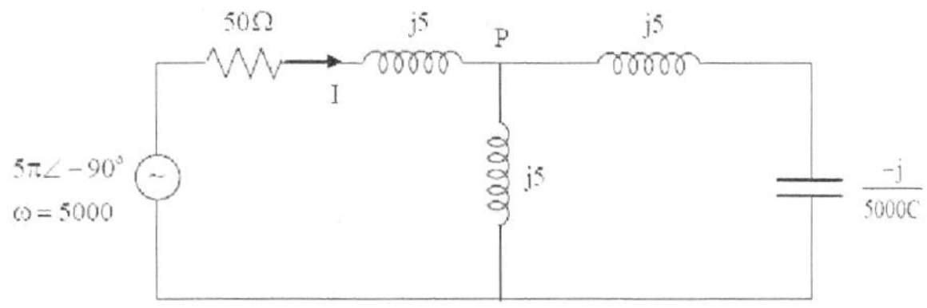

If $I = 0, V_p = V_i$,

$$V_i\left(\frac{1}{j5} + \frac{1}{j\left(5 - \frac{1}{5000C}\right)}\right) = 0$$

$$\Rightarrow \quad \frac{1}{j5} = \frac{-1}{j\left(5 - \frac{1}{5000C}\right)} = 0$$

$$\frac{1}{5} = \frac{1}{\frac{1}{5000C} - 5}$$

$$\Rightarrow \quad 5 = \frac{1}{5000C} - 5$$

$$\Rightarrow \quad \frac{1}{5000C} = 10$$

$$\Rightarrow \quad C = \frac{1}{5 \times 10^4} = 20\ \mu F$$

37.

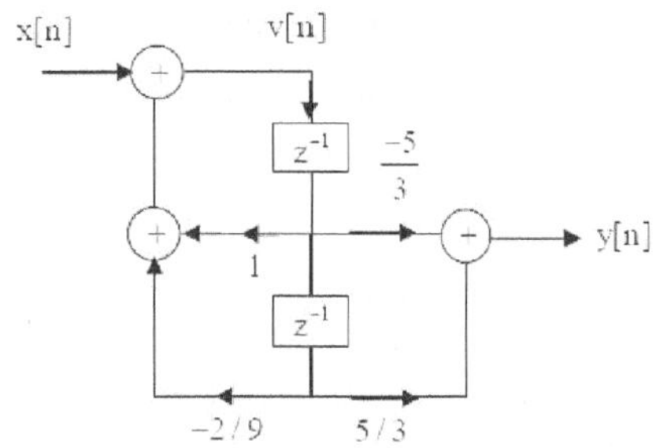

From the graph

$$v[n] = x[n] + v[n-1] - \frac{2}{9}v[n-2]$$

$$y[n] = -\frac{5}{3}v[n-1] + \frac{5}{3}v[n-2]$$

$$V(z) = \left[1 - z^{-1} + \frac{2}{9}z^{-2}\right] = X(z)$$

$$\Rightarrow \quad \frac{V(z)}{X(z)} = \frac{1}{1 - z^{-1} + \frac{2}{9}z^{-2}} \quad \ldots(i)$$

$$\frac{Y(z)}{V(z)} = \frac{-5}{3}z^{-1} + \frac{5}{3}z^{-2} \quad \ldots(ii)$$

Multiplying (*i*) and (*ii*) we get

$$\Rightarrow \quad \frac{Y(z)}{X(z)} = \frac{-\frac{5}{3}z^{-1}[1 - z^{-1}]}{1 - z^{-1} + \frac{2}{9}z^{-2}}$$

For unit step response,

$$X(z) = \frac{1}{1 - z^{-1}}$$

$$\Rightarrow \quad Y(z) = \frac{-\frac{5}{3} z^{-1}}{1 - z^{-1} + \frac{2}{9} z^{-2}}$$

$$= \frac{A}{1 - \frac{1}{3} z^{-1}} + \frac{B}{1 - \frac{2}{3} z^{-1}}$$

On solving,

A = 5; B = –5

$$\Rightarrow \quad y[n] = 5\left(\frac{1}{3}\right)^n u[n] - 5\left(\frac{2}{3}\right)^n u[n]$$

38. $$x(t) = \sqrt{2}\left(\frac{\sin(\pi t/5)}{\pi t/5}\right)\sin\left(\pi t - \frac{\pi}{4}\right)$$

we can write above expression as

$$x(t) = -\sqrt{2}\left(\frac{\sin(\pi t/5)}{\pi t/5}\right)\left[\cos\frac{\pi}{4}\sin\pi t - \sin\frac{\pi}{4}\cos\pi t\right]$$

$$= \left(\frac{\sin(\pi t/5)}{\pi t/5}\right)\cos\pi t - \frac{\sin(\pi t/5)}{(\pi t/5)}\sin(\pi t)$$

Also

$x(t) = x_c(t)\cos 2\pi f_c t - x_s(t)\sin(2\pi f_c t)$

[Low pass representation of Bandpass signals]

$$x_c(t) = \frac{\sin(\pi t/5)}{\pi t/5}$$

$$x_s(t) = \frac{\sin(\pi t/5)}{\pi t/5}$$

$x_{ce}(t)$ is the complex envelope of $x(t)$

$$x_{ce}(t) = x_c(t) + jx_s(t)$$

$$= \frac{\sin(\pi t/5)}{\pi t/5}[1 + j]$$

$$= \frac{\sqrt{2}\sin(\pi t/5)}{(\pi t/5)} e^{j\pi/4}$$

40. Given $\frac{d^2x(t)}{dt^2} + \frac{3dx(t)}{dt} + 2x(t) = 0$

$x(0) = 20$

$x(1) = \frac{10}{e}$

$x(2) = $ ------

Auxillary equation is

$m^2 + 3m + 2 = 0$

$m = 1, -2$

Complementary solution $x_e = c_1e^{-t} + c_2e^{-2t}$

Particular solution $x_p = 0$

Finally $\quad x = x_c + x_p = c_1e^{-t} + c_2e^{-2t}$

$x(0) = 20$

$\Rightarrow \quad 20 = c_1 + c_2 \rightarrow (a)$

$x(1) = \frac{10}{e}$

$\Rightarrow \quad \frac{10}{e} = c_1e^{-1} + c_2e^{-2}$

$\Rightarrow \quad 10 = c_1 + c_2e^{-1} \rightarrow (b)$

From (a) $\quad c_1 = 20 - c_2$

Now $\quad 10 = (20 - c_2) + c_2e^{-1}$

$10 = c_2(e^{-1} - 1) + 20$

$$c_2 = \frac{10 - 20}{e^{-1} - 1}$$

$$= -\frac{10}{e^{-1} - 1}$$

$$= \frac{10}{1 - e^{-1}} = \frac{10e}{e - 1}$$

$$\Rightarrow \quad c_1 = 20 - \frac{10e}{e - 1}$$

$$= \frac{20e - 20 - 10e}{e - 1}$$

$$= \frac{10e - 20}{e - 1}$$

$$x(t) = \frac{10e - 20}{e - 1}e^{-1} + \frac{10e}{e - 1}e^{-2t}$$

$$x(2) = \left(\frac{10e - 20}{e - 1}\right)^{e^{-2}} + \left(\frac{10e}{e - 1}\right)e^{-4}$$

$= 0.8556$

41. $$x[n] = 1 + \cos\left(\frac{\pi}{8}n\right)$$

N = 16

$$x[n] = 1 + \frac{1}{2}e^{j\frac{2\pi n}{16}} + \frac{1}{2}e^{-j\frac{2\pi n}{16}}$$

$a_{-1} = \frac{1}{2}$, $a_1 = \frac{1}{2}$, $a_0 = 1$

$a_1 = a_{-1+16}$

$\Rightarrow \quad a_{-1} = a_{15} = \frac{1}{2}$

$\Rightarrow \quad a_0 = 1,\ a_1 = \frac{1}{L},\ a_2$ to $a_{14} = 0,\ a_{15} = \frac{1}{2}$

DFS coefficients are also periodic with period 16.

$a_{31} = a_{16+15}$

$a_{31} = a_{15}$

$\Rightarrow \quad a_{31} = \frac{1}{2}$

43. Built in potential

$$\psi_o = \frac{1}{2}\times\left(10^6\ V/_m\right)\times(1.1\times10^{-6}\,m)$$
$$= 0.55 \text{ volts}$$

But in Question (option (*c*)) is given as 700 mV.

44. $$D = 2\rho^2 a\rho + z a_z$$

$$\oint_Z D.ds = \int_V (\nabla.D)dv$$

$$\nabla.D = \frac{1}{\rho}\frac{\partial}{\partial\rho}(\rho D_\rho) + \frac{1}{\rho}\frac{\partial D\phi}{\partial\phi} + \frac{\partial D_z}{\partial z}$$

$$= \frac{1}{\rho}\frac{\partial}{\partial\rho}(\rho 2\rho^2) + 0 + 1$$

$$= \frac{1}{\rho}2(3)\rho^2 + 1 = 6\rho + 1$$

$$\int_V(\nabla.D)dv = \int_{\rho=0}^{1}\int_{\phi=0}^{2\pi}\int_{z=0}^{5}(6\rho+1)\rho\ d\rho d\phi dz$$

$$= \left(\frac{6\rho^3}{3} + \frac{\rho^2}{2}\right)\Bigg|_0^1 (2\pi)(5)$$

$$= \left(2 + \frac{1}{2}\right)10\pi$$

$$\int_V(\nabla.D)dv = 78.53$$

45. $$I_B = \frac{I_C}{\beta} = \frac{I_S}{\beta}e^{V_{BE}/V_T}$$

$$I_E = (\beta+1)I_B$$

$$= \frac{\beta+1}{\beta}I_S.e^{V_{BE}/V_T}$$

$$= (1.02)(10^{-9}\times10^{-6})e^{\frac{700\times10^{-3}}{25\times10^{-3}}}$$

$$= 1475\ \mu A$$

46. $$Z_{in} = JZ_0 \tan \beta l$$

$$\beta l = \frac{2\pi}{\lambda}.l = \frac{2\pi}{8}(3) = \frac{3\pi}{4}$$

$$\lambda = \frac{3\times10^8}{37.5\times10^6} = 8$$

Short circuited line

$$0 < \beta l < \frac{\pi}{2} \rightarrow \text{Inductor}$$

$$\frac{\pi}{2} < \beta l < \pi \rightarrow \text{Capacitor}$$

47. NMOS SATURATION

$$I_D = 1mA\ @\ V_{DS} = 5V$$
$$I_D = 1.02mA\ @\ V_{DS} = 6V$$
$$I_{D\,Sat} << V_{DS}$$
$$I_D = k(V_{GS} - V_T)^2(1 + \lambda V_{DS})$$
$$I_D = k'(1 + \lambda V_{DS})$$
$$10^{-3} = k'(1 + 5\lambda)$$
$$1.02\times10^{-3} = k'(1 + 6\lambda)$$
$$1.02 = \frac{1+6\lambda}{1+5\lambda}$$
$$1.02 + 5.1\lambda = 1 + 6\lambda$$
$$0.02 = 0.9\lambda$$
$$\lambda = 0.022\ V^{-1}$$

48.

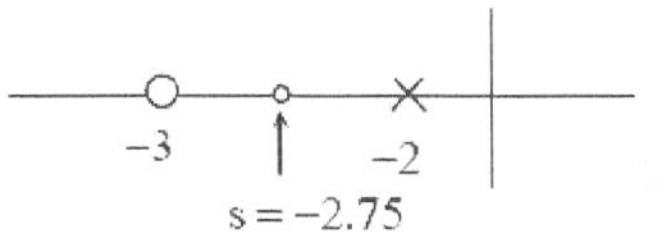

$$G(s)H(s) = \frac{10k(s+3)}{(s+2)}$$
$$= \frac{k(s+3)}{(s+2)}$$

For k' $$k' = \frac{\text{Poles length}}{\text{zero length}}$$
$$= \frac{0.75}{0.25} = 3$$
$$10k = 3$$
$$k = 0.3$$

50. $$Var(x) = E(x^2) - (E(x))^2$$

$$E(x) = \int_{-\infty}^{\infty} x\frac{1}{2}|x|e^{-|x|}dx$$

$$= \frac{-1}{2}\int_{-\infty}^{0} x^2e^x dx + \frac{1}{2}\int_0^{\infty} x^2.e^{-x}dx$$
$$= 0$$

$$E(x)^2 = \int_{-\infty}^{\infty} x^2.\frac{1}{2}|x|e^{-|x|}dx$$

$$= 2\left[\frac{1}{2}\int_0^{\infty} x^2.xe^{-x}dx\right]$$

$$= \int_0^{\infty} x^3.e^{-x}dx$$

$$= x^3.\frac{e^{-x}}{-1}\Bigg|_0^{\infty} - (3x^2)(e^{-x})\Big|_0^{\infty} + 6x(-e^{-x})\Big|_0^{\infty} - 6e^{-x}\Big|_0^{\infty}$$
$$= 6$$

$$\Rightarrow Var(x) = 6 - 0 = 6$$

51. $$x(t) = \frac{\sin(\pi t/2)}{(\pi t/2)} * \sum_{n=-\infty}^{\infty} \delta(t-10n)$$

Convolution in time domain becomes multiplication in frequency domain.

$$\frac{1}{10}\sum_{n=-\infty}^{\infty} \delta(f-kf_s)$$

$$F_s = \frac{1}{T_s} = 0.1$$

$$\frac{\sin(\pi\, t/2)}{(\pi\, t/2)} \longleftrightarrow$$

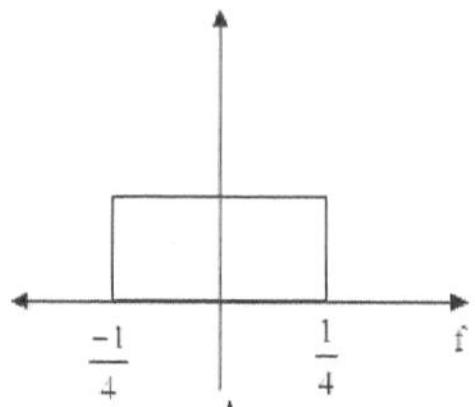

$$\sum_{n=-\infty}^{\infty} \delta(t-10n) \longleftrightarrow$$

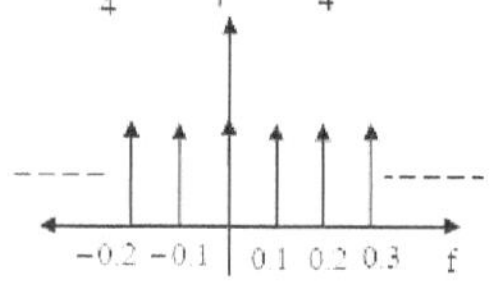

Multiplication in frequency domain will result maximum frequency is 0.2.

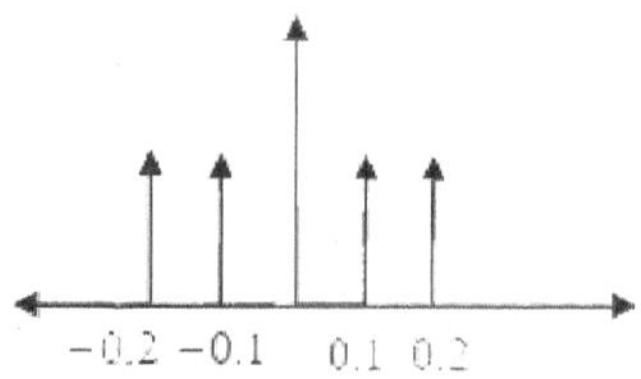

Thus Nyquist rate= 0.4 samples/sec

52.

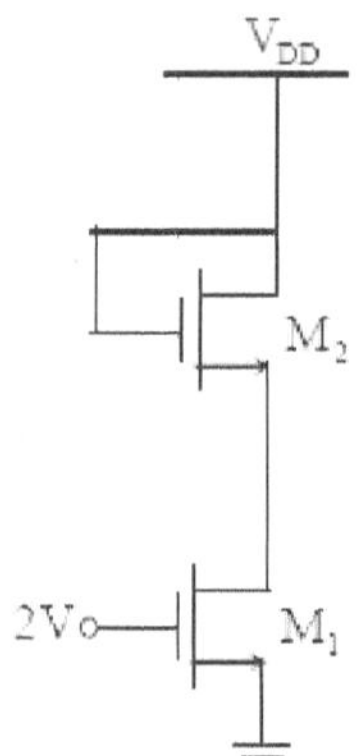

Lower transistor (M_1) to work in saturation

$$V_{DS1} \geq V_{GS1} - V_+$$

So, for minimum V_{DD}

$$V_{DS1} = V_{GS1} - V_+$$

$$V_{DS1} = 2 - 1 = 1V$$

$$V_{DS1} = V_{D1} - V_{S1}$$

$$1V = V_{D1} - 0$$

$$\therefore \quad V_{D1} = 1V$$

and $$I_{D1} = K'(V_{GS1} - V_+)^2$$

$$V_{D1} = \frac{1\text{ mA}}{V^2} \times (2-1)^2 = 1\text{ mA}$$

Now transistor M_2, $V_{DG} = 0V$

So, it will work into saturation region and same current will flow

$$I_{D2} = I_{D1} = K'(V_{GS2} - V_+)^2$$

$$1\text{ mA} = \frac{1\text{ mA}}{V^2} \times (V_{DD}-1-1)^2 \quad (\because V_{S2} = V_{D1})$$

$$\therefore \quad V_{DD} = 3V$$

53. $$T_d(s) = 1$$

$$\theta(s) = -\frac{1}{s\left[(Js+B) + \dfrac{K_bK_T}{R_a + L_as}\right]}$$

Steady State Value is $\lim_{s\to 0} s\theta(s) = -0.5$

54. $5^5 + 2s^4 + 3s^3 + 6s^2 - 4s - 8 = 0$

S^5	1	3	−4	
S^4	2	6	−8	
S^3	8	12	0	
S^2	3	−8	0	$\rightarrow$ as $\in \rightarrow 0$
S^1	−9.33	0	0	
S^0	−8			

$$2s^4 + 6s^2 - 8 = 0$$

Let $x = s^2$, then

$$2x^2 + 6x - 8$$

$$\Rightarrow \quad x = 1, -4$$

$$s^2 = 1, -4$$

$$\Rightarrow \quad s = \pm 1, \pm 2$$

Lies on RHS $\rightarrow s = 1, 2$

There are only two poles on LHS.

55. Since $x[n]$ in absolutely summable thus its ROC must include unit circle.

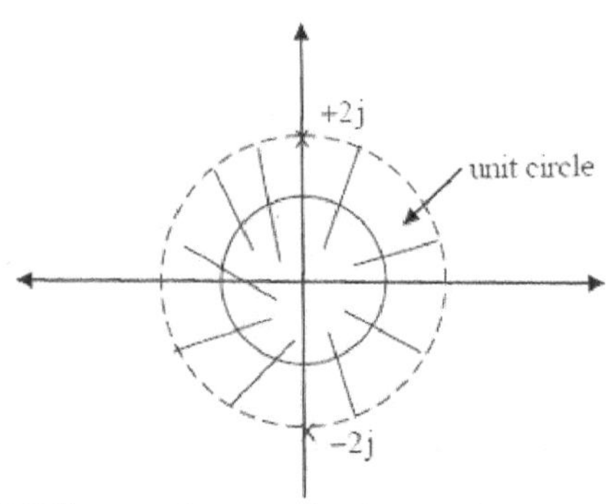

Thus ROC must be inside the circling radius 2.

$x[n]$ must be a non-causal signal.

SOLVED PAPER – 2014

ELECTRONICS AND COMMUNICATION ENGINEERING

Set -1

No. of Questions : 65 *Maximum Marks : 100*

GENERAL APTITUDE

(Q. 1 – 5) Carry One Mark Each.

1. Choose the most appropriate phrase from the options given below to complete the following sentence.

   The aircraft ________ take off as soon as its flight plan was filed.

   (*a*) is allowed to

   (*b*) will be allowed to

   (*c*) was allowed to

   (*d*) has been allowed to

2. Read the statements:

   All women are enterpreneurs.

   Some women are doctors.

   Which of the following conclusions can be logically inferred from the above statements?

   (*a*) All women are doctors

   (*b*) All doctors are enterpreneurs

   (*c*) All enterpreneurs are women

   (*d*) Some enterpreneurs are doctors

3. Choose the most appropriate word from the options given below to complete the following sentence.

   Many ancient cultures attributed disease to supernatural causes. However, modern science has largely helped ________ such notions.

   (*a*) impel (*b*) dispel

   (*c*) propel (*d*) repel

4. The statistics of runs scored in a series by four batsmen are provided in the follwing table. Who is the most consistent batsman of these four?

Batsman	Average	Standard deviation
K	31.2	5.21
L	46.0	6.35
M	54.4	6.22
N	17.9	5.90

   (*a*) K (*b*) L

   (*c*) M (*d*) N

5. What is the next number in the series?

   12 35 81 173 357 ___

(Q. 6 – 10) Carry Two Marks Each.

6. Find the odd one from the following

   (*a*) W,E,K,O (*b*) I,Q,W,A

   (*c*) F,N,T,X (*d*) N,V,B,D

7. For submitting tax returns, all resident males with annual income below ₹ 10 lakh should fill up Form P and all resident females with income below ₹ 8 lakh should fill up Form Q. All people with incomes above ₹ 10 lakh should fill up Form R, except non residents with income above ₹ 15 lakhs, who should fill up Form S. All others should fill Form T, An example of a person who should fill From T is

   (*a*) a resident male with annual income ₹ 9 lakh

   (*b*) a resident female with annual income ₹ 9 lakh

   (*c*) a non-resident male with annual income ₹ 16 lakh

   (*d*) a non-resident female with annual income ₹ 16 lakh

8. A train that is 280 metres long, travelling at a uniform speed, crosses a platform in 60 seconds and passes a man standing on the platform in 20 seconds. What is the length of the platform in metres?

9. The exports and imports (in crores of ₹) of a country from 2000 to 2007 are given in the following bar chart. If the trade deficit is defined as excess of imports over exports, in which year is the trade deficit 1/5th of the exports?

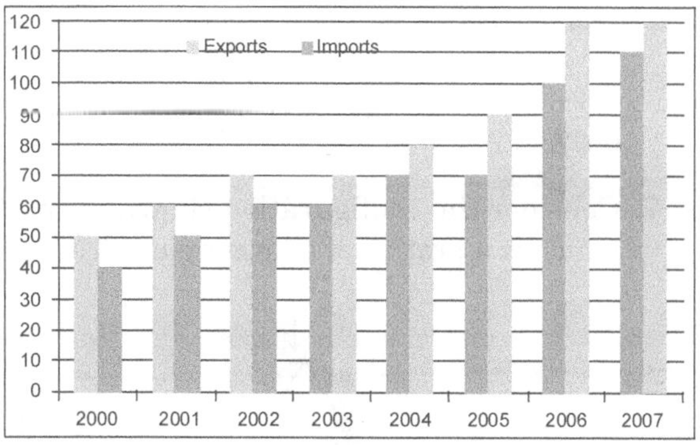

   (*a*) 2005 (*b*) 2004

   (*c*) 2007 (*d*) 2006

10. You are given three coins: on has heads on both faces, the second has tails on both faces, and the third has a head on one face and a tail on the other. You choose a coin at random and toss it, and it comes up heads. The probability that the other face is tails is

(*a*) 1/4 (*b*) 1/3

(*c*) 1/2 (*d*) 2/3

ELECTRONICS AND COMMUNICATION

(Q. 1 – 25) Carry One Mark Each.

1. For matrices of same dimension M, N and scalar c, which one of these properties DOES NOT ALWAYS hold?

(*a*) $(M^T)^T = M$ (*b*) $(cM)^T = c(M)^T$

(*c*) $(M + N)^T = M^T + N^T$ (*d*) MN = NM

2. In a housing society, half of the families have a single child per family, while the remaining half have two children per family. The probability that a child picked at random, has a sibling is _____

3. C is a closed path in the z- plane given by $|z| = 3$. The value of the integral $\oint_C \left(\frac{z^2 - z + 4j}{z + 2j}\right) dz$ is

(*a*) $-4\pi(1 + j2)$ (*b*) $4\pi(3 - j2)$

(*c*) $-4\pi(3 + j2)$ (*d*) $4\pi(1 - j2)$

4. A real (4 × 4) matrix A satisfies the equation $A^2 = I$, where I is the (4 × 4) identity matrix. The positive eigen value of A is _______.

5. Let X_1, X_2 and X_3 be independent and identically distributed random variables with the uniform distribution on [0, 1]. The probability P {X_1 is the largest} is ______

6. For maximum power transfer between two cascaded sections of an electrical network, the relationship between the output impedance Z_1 of the first section to the input impedance Z_2 of the second section is

(*a*) $Z_2 = Z_1$ (*b*) $Z_2 = -Z_1$

(*c*) $Z_2 = Z_1^*$ (*d*) $Z_2 = -Z_1^*$

7. Consider the configuration shown in the figure which is a protion of a larger electrical network

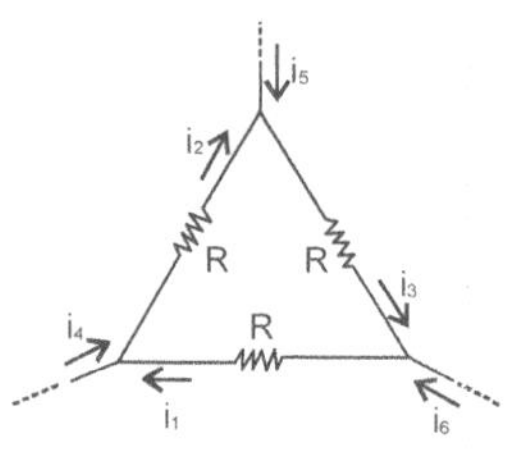

For R = 1 Ω and currents $i_1 = 2$ A, $i_4 = -1$ A, $i_5 = -4$ A, which one of the following is **TRUE?**

(*a*) $i_6 = 5$ A

(*b*) $i_3 = -4$ A

(*c*) Data is sufficient to conclude that the supposed currents are impossible

(*d*) Data is insufficient to identify the currents i_2, i_3, and i_6

8. When the optical power incident on a photodiode is 10μW and the responsivity is 0.8 A/W, the photocurrent generated (in μA) is ________.

9. In the figure, assume that the forward voltage drops of the PN diode D_1 and Schottky diode D_2 are 0.7 V and 0.3 V, respectively. If ON denotes conducting state of the diode and OFF denotes non-conducting state of the diode, then in the circuit,

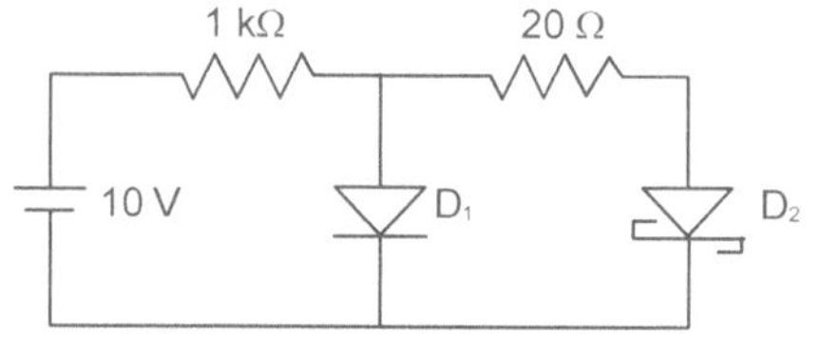

(*a*) both D_1 and D_2 are ON

(*b*) D_1 is ON and D_2 is OFF

(*c*) both D_1 and D_2 are OFF

(*d*) D_1 is OFF and D_2 is ON

10. If fixed positive charges are present in the gate oxide of an n-channel enhancement type MOSFET, it will lead to

(*a*) a decrease in the threshold voltage

(*b*) channel length modulation

(*c*) an increase in substrate leakage current

(*d*) an increase in accumulation capacitance

11. A good current buffer has

(*a*) low input impedance and low output impedance

(*b*) low input impedance and high output impedance

(*c*) high input impedance and low output impedance

(*d*) high input impedance and high output impedance

12. In the ac equivalent circuit shown in the figure, if i_{in} is the input current and R_F is very large, the type of feedback is

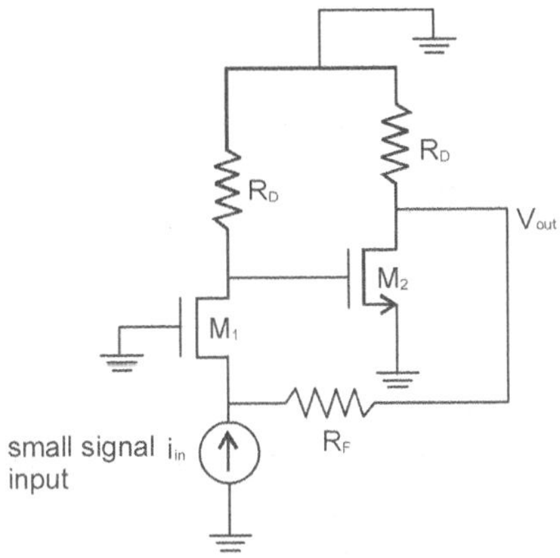

(*a*) voltage-voltage feedback

(*b*) voltage-current feedback

(*c*) current-voltage feedback

(*d*) current-current feedback

13. In the low-pass filter shown in the figure, for a cut-off frequency of 5 kHz, the value of R_2 (in kΩ) is ________.

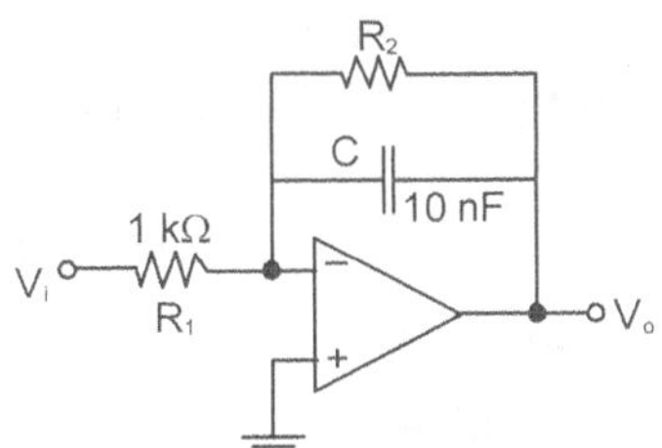

14. In the following circuit employing pass transistor logic, all NMOS transistors are identical with a threshold voltage of 1 V. Ignoring the body-effect, the output voltages at P, Q and R are,

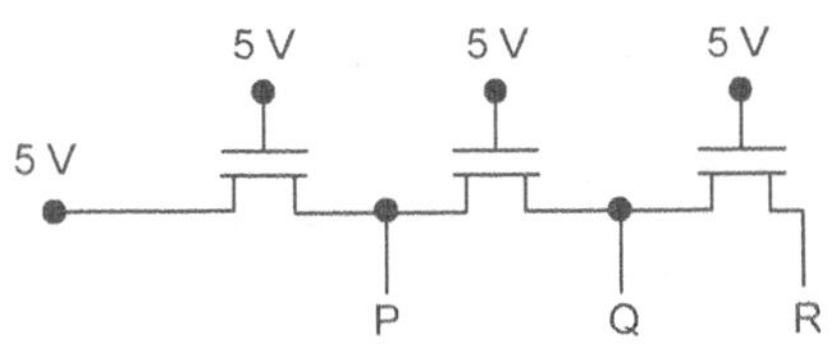

(*a*) 4V, 3V, 2V (*b*) 5V, 5V, 5V

(*c*) 4V, 4V, 4V (*d*) 5V, 4V, 3V

15. The Boolean expression $(X + Y)(X + \overline{Y}) + \overline{(X\overline{Y}) + \overline{X}}$ simplifies to

(*a*) X (*b*) Y

(*c*) XY (*d*) X+Y

16. Five JK flip-flops are cascaded to form the circuit shown in figure. Clock pulses at a frequency of 1 MHz are applied as shown. The frequency (in kHz) of the waveform at Q3 is ________.

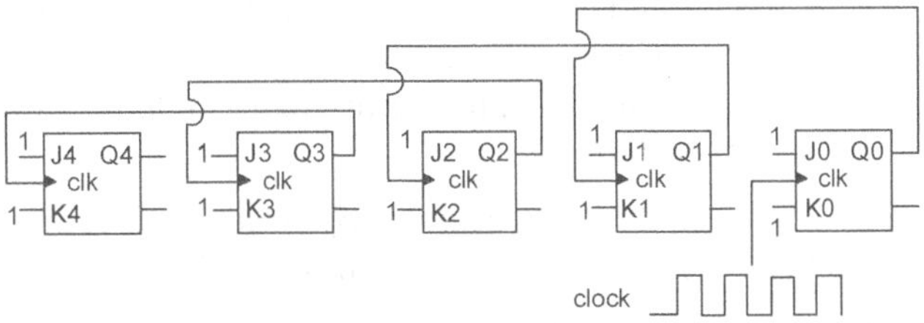

17. A discrete-time signal $x[n] = \sin(\pi^2 n)$, n being an interger, is

(*a*) periodic with period π.

(*b*) periodic with period π^2.

(*c*) periodic with period π/2.

(*d*) not periodic.

18. Consider two real valued signals, $x(t)$ band-limited to [– 500 Hz, 500 Hz] and $y(t)$ band-limited to [–1 kHz, 1 kHz]. For $z(t) = x(t) \cdot y(t)$, the Nyquist sampling frequency (in kHz) is _____.

19. A continuous, linear time-invariant filter has an impulse reponse $h(t)$ described by

$$h(t) = \begin{cases} 3 \text{ for } 0 \le t \le 3 \\ 0 \text{ otherwise} \end{cases}$$

When a constant input of value 5 is applied to this filter, the steady state outout is _____.

20. The forward path transfer function of a unity negative feedback system is given by

$$G(s) = \frac{K}{(s+2)(s-1)}$$

The value of K which will place both the poles of the closed-loop system at the same location, is ______.

21. Consider the feedback system shown in the figure. The Nyquist plot of G (s) is also shown. Which one of the following conclusions is correct?

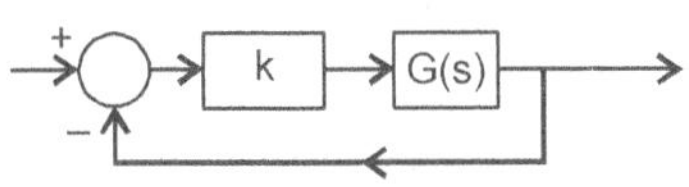

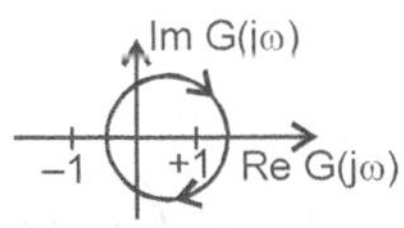

(*a*) G(*s*) is an all-pass filter

(*b*) G(*s*) is a strictly proper transfer function

(*c*) G(*s*) is a stable and minimum-phase transfer function

(*d*) The closed-loop system is unstable for sufficiently large and positive *k*

22. In a code-division multiple access (CDMA) system with N = 8 chips, the maximum number of users who can be assigned mutually orthogonal signature sequences is ______

23. The capacity of a Binary Symmetric Channel (BSC) with cross-over probability 0.5 is _______

24. A two-port network has scattering parameters given by $[S]=\begin{bmatrix} s_{11} & s_{12} \\ s_{21} & s_{22} \end{bmatrix}$. If the port-2 of the two-port is short circuited, the s_{11} parameter for the resultant one-port network is

(*a*) $\dfrac{s_{11}-s_{11}s_{22}+s_{12}s_{21}}{1+s_{22}}$ (*b*) $\dfrac{s_{11}+s_{11}s_{22}-s_{12}s_{21}}{1+s_{22}}$

(*c*) $\dfrac{s_{11}+s_{11}s_{22}+s_{12}s_{21}}{1-s_{22}}$ (*d*) $\dfrac{s_{11}-s_{11}s_{22}+s_{12}s_{21}}{1-s_{22}}$

25. The force on a point charge $+q$ kept at a distance d from the suface of an infinite grounded metal plate in a medium of permittivity ε is

(*a*) 0

(*b*) $\dfrac{q^2}{16\pi \in d^2}$ away from the plate

(*c*) $\dfrac{q^2}{16\pi \in d^2}$ towards the plate

(*d*) $\dfrac{q^2}{4\pi \in d^2}$ towards the plate

(Q. 26 – 55) Carry Two Marks Each.

26. The Taylor series expansion of $3 \sin x + 2 \cos x$ is

(*a*) $2 + 3x - x^2 - \dfrac{x^3}{2} + ...$

(*b*) $2 - 3x + x^2 - \dfrac{x^3}{2} + ...$

(*c*) $2 + 3x + x^2 + \dfrac{x^3}{2} + ...$

(*d*) $2 - 3x - x^2 + \dfrac{x^3}{2} + ...$

27. For a function $g(t)$, it is given that $\int_{-\infty}^{+\infty} g(t)\, e^{-j\omega t}\, dt = \omega e^{-2\omega^2}$ for any real value ω. If $y(t) = \int_{-\infty}^{+\infty} g(\tau)\, d\tau$, then $\int_{-\infty}^{+\infty} y(t)\, dt$ is

(*a*) 0 (*b*) $-j$

(*c*) $-\dfrac{j}{2}$ (*d*) $\dfrac{j}{2}$

28. The volume under the surface $z(x, y) = x + y$ and above the triangle in the x-y plane defined by $\{0 \le y \le x \text{ and } 0 \le x \le 12\}$ is _______.

29. Consider the matrix

$$j_6 = \begin{vmatrix} 0 & 0 & 0 & 0 & 0 & 1 \\ 0 & 0 & 0 & 0 & 1 & 0 \\ 0 & 0 & 0 & 1 & 0 & 0 \\ 0 & 0 & 1 & 0 & 0 & 0 \\ 0 & 1 & 0 & 0 & 0 & 0 \\ 1 & 0 & 0 & 0 & 0 & 0 \end{vmatrix}$$

Which is obtained by reversing the order of the columns of the identity matrix I_6. Let $P = I_6 + \alpha J_6$, where α is a non-negative real number. The value of α for which det(P) = 0 is ______.

30. A Y-network has resistances of 10Ω each in two of its arms, while the third arm has a resistance of 11Ω. In the equivalent Δ-network, the lowest value (in Ω) among the three resistances is _______.

31. A 230 V rms source supplies power to two loads connected in parallel. The first load draws 10 kW at 0.8 leading power factor and the second one draws 10 kVA at 0.8 lagging power factor. The complex power delivered by the source is

(*a*) (18 + j 1.5) kVA (*b*) 18 – j 1.5) kVA

(*c*) (20 + j 1.5) kVA (*d*) (20 – j 1.5) kVA

32. A periodic variable x is shown in the figure as a function of time. The root-mean-square (rms) value of x is _______.

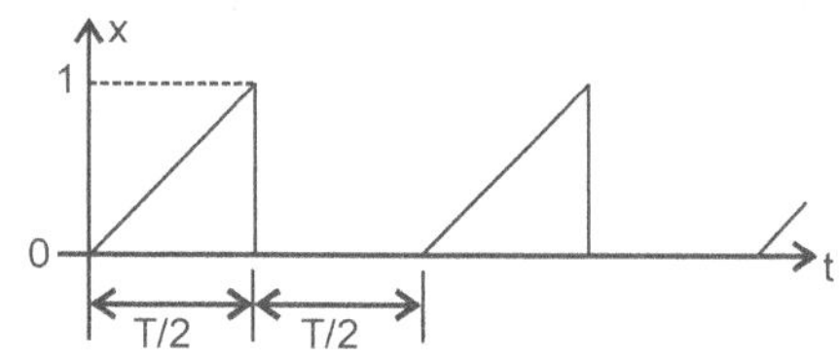

33. In the circuit shown in the figure, the value of capacitor C (in mF) needed to have critically damped response $i(t)$ is _______.

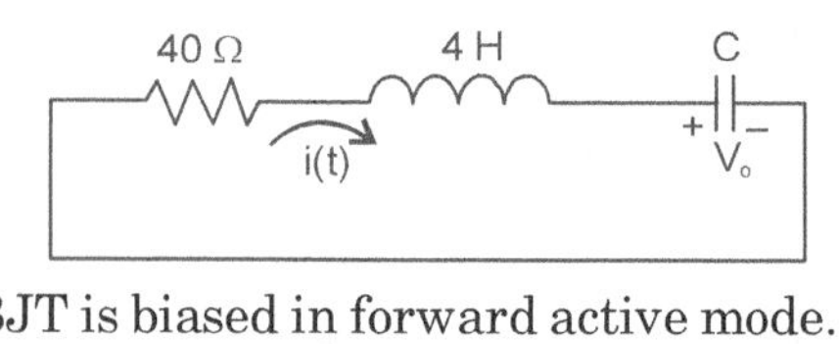

34. A BJT is biased in forward active mode. Assume V_{BE} = 0.7 V, kT/q = 25 mV and reverse saturation current Is = 10^{-13} A. The transconductance of the BJT (in mA/V) is ________.

35. The doping concentrations on the p-side and n-side of a silicon diode are 1×10^{16} cm^{-3} and 1×10^{17} cm^{-3}, respectively. A forward bias of 0.3 V is applied to the diode. At T = 300 K, the intrinsic carrier concentration of silicon $n_i = 1.5 \times 10^{10}$ cm^{-3} and $\frac{kT}{q} = 26$ mV. The electron concentration at the edge of the depletion on the p-side is

(*a*) 2.3×10^{9} cm^{-3}

(*b*) 1×10^{16} cm^{-3}

(*c*) 1×10^{17} cm^{-3}

(*d*) 2.25×10^{6} cm^{-3}

36. A depletion type N-channel MOSFET is biased in its linear region for use as a voltage controlled resistor. Assume threshold voltage $V_{TH} = -0.5$ V, $V_{GS} = 2.0$V, $V_{DS} = 5$V, W/L = 100, $C_{OX} = 10^{-8}$ F/cm^2 and μ_n 800cm^2/V$_{-s}$. The value of the resistance of the voltage controlled resistor (in Ω) is ______.

37. In the voltage regulator circuit shown in the figure, the op-amp is ideal. The BJT has $V_{BE} = 0.7$ V and β = 100, and the zener voltage is 4.7 V. For a regulated output of 9 V, the value of R (in Ω) is ____.

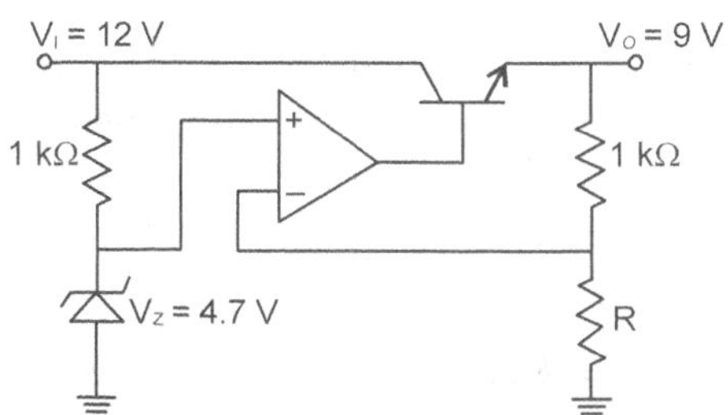

38. In the circuit shown, the op-amp has finite input impedance, infinite voltage gain and zero input offset voltage. The output voltage V_{out} is

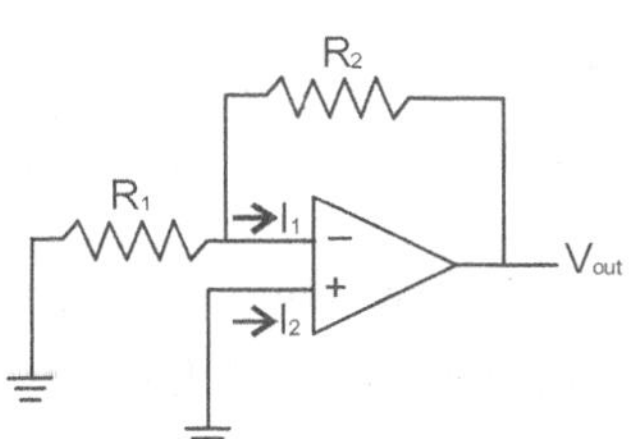

(*a*) $-I_2(R_1 + R_2)$ (*b*) I_2R_2

(*c*) I_1R_2 (*d*) $-I_1(R_1 + R_2)$

39. For the amplifier shown in the figure, the BJT parameters are $V_{BE} = 0.7$ V, β = 200, and thermal voltage $V_T = 25$ mV. The voltage gain (v_o/v_i) of the amplifier is ______.

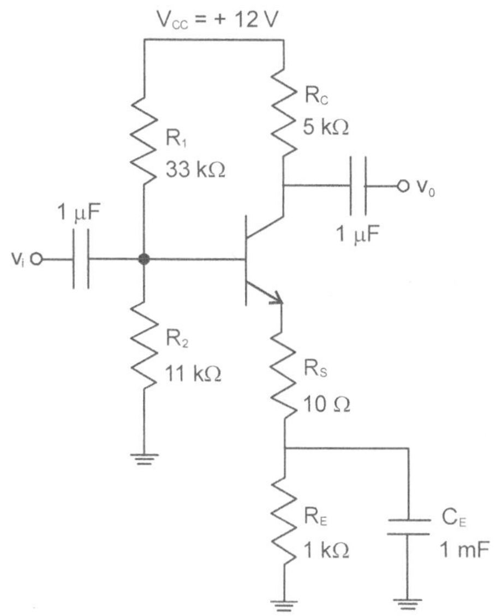

40. The output F in the digital logic circuit shown in the figure is

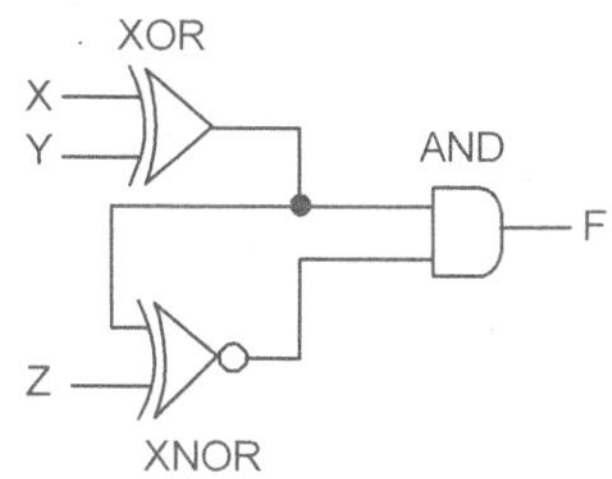

(*a*) $F = \bar{X}YZ + X\bar{Y}Z$

(*b*) $F = \bar{X}Y\bar{Z} + X\bar{Y}\bar{Z}$

(*c*) $F = \bar{X}\bar{Y}Z + XYZ$

(*d*) $F = \bar{X}\bar{Y}\bar{Z} + XYZ$

41. Cosider the Boolean function, $F(w,x,y,z) = wy + xy + \bar{w}xyz + \bar{w}\bar{x}y + xy + \bar{x}\bar{y}\bar{z}$. Which one of the following is the complete set of essential prime implicants?

(*a*) $w, y, xz, \bar{x}\bar{z}$ (*b*) w, y, xz

(*c*) $y, \bar{x}\bar{y}\bar{z}$ (*d*) $y, xz, \bar{x}\bar{z}$

42. The digital logic shown in the figure satisfies the given state diagram when Q1 is connected to input A of the XOR gate.

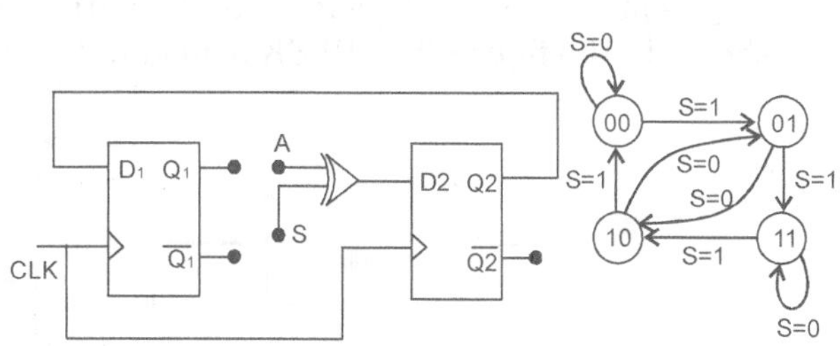

Suppose the XOR gate is replaced by an XNOR gate. Which one of the following options preserves the state diagram?

(*a*) Input A is connected to $\overline{Q2}$

(*b*) Input A is connected to Q_2

(*c*) Input A is connected to $\overline{Q1}$ and S is complemented

(*d*) Input A is connected to $\overline{Q1}$

43. Let $x[n] = \left(-\frac{1}{9}\right)^n u(n) - \left(-\frac{1}{3}\right)^n u(-n-1)$. The Region of Convergence (ROC) of the z-transform of $x[n]$

(*a*) is $|z| > \frac{1}{9}$ (*b*) is $|z| < \frac{1}{3}$

(*c*) is $\frac{1}{3} > |z| > \frac{1}{9}$ (*d*) does not exist

44. Consider a discrete time periodic signal $x[n] = \sin\left(\frac{\pi n}{5}\right)$. Let a_k be the complex Fourier series coefficients of $x[n]$. The cofficients $\{a_k\}$ are non-zero where k = Bm ± 1, when m is any integer. The value of B is _____.

45. A system is described by the following differential equation, where $u(t)$ is the input to the system and $y(t)$ is the output of the system.

$$\dot{y}(t) + 5y(t) = u(t)$$

When $y(0) = 1$ and $u(t)$ is a unit step function, $y(t)$ is

(*a*) $0.2 + 0.8e^{-5t}$ (*b*) $0.2 - 0.2e^{-5t}$

(*c*) $0.8 + 0.2e^{-5t}$ (*d*) $0.8 - 0.8e^{-5t}$

46. Consider the state space model of a system, as given below

$$\begin{bmatrix}\dot{x}_1\\ \dot{x}_2\\ \dot{x}_3\end{bmatrix} = \begin{bmatrix}-1 & 1 & 0\\ 0 & -1 & 0\\ 0 & 0 & -2\end{bmatrix}\begin{bmatrix}x_1\\ x_2\\ x_3\end{bmatrix} + \begin{bmatrix}0\\ 4\\ 0\end{bmatrix}u; y = \begin{bmatrix}1 & 1 & 1\end{bmatrix}\begin{bmatrix}x_1\\ x_2\\ x_3\end{bmatrix}.$$

The system is

(*a*) controllable and observable

(*b*) uncontrollable and observable

(*c*) uncontrollable and unobservable

(*d*) controllable and unobservable

47. The phase margin in degrees of $G(s) = \frac{10}{(s+0.1)(s+1)(s+10)}$ calculated using the asymptotic Bode plot is _____.

48. For the following feedback system $G(s) = \frac{1}{(s+1)(s+2)}$. The 2%-settling time of the step response is required to be less than 2 seconds.

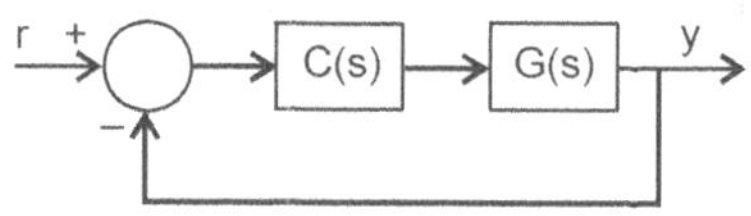

Which one of the following compensators C(*s*) achieves this?

(*a*) $3\left(\frac{1}{s+5}\right)$ (*b*) $5\left(\frac{0.03}{s}+1\right)$

(*c*) $2(s + 4)$ (*d*) $4\left(\frac{s+8}{s+3}\right)$

49. Let X be a real-valued random variable with $E[X]$ and $E[X^2]$ denoting the mean value of X and X^2, respectively. The relation which always holds true is

(*a*) $(E[X])^2 > E[X^2]$ (*b*) $E[X^2] \geq (E[X])^2$

(*c*) $E[X^2] = (E[X])^2$ (*d*) $E[X^2] > (E[X])^2$

50. Consider a random process $X(t) = \sqrt{2}\sin(2\pi t + \varphi)$, where the random phase φ is uniformly distributed in the interval $[0, 2\pi]$. The auto-correlation $E[X(t_1)X(t_2)]$ is

(*a*) $\cos(2\pi(t_1 + t_2))$ (*b*) $\sin(2\pi(t_1 - t_2))$

(*c*) $\sin(2\pi(t_1 + t_2))$ (*d*) $\cos(2\pi(t_1 - t_2))$

51. let $Q(\sqrt{\gamma})$ be the BER of a BPSK system over an AWGN channel with two-sided noise power spectral density $N_0/2$. The parameter γ is a function of bit energy and noise power spectral density.

A system with two independent and identical AWGN channels with noise power spectral density $N_0/2$ is shown in the figure. The BPSK demodulator receives the sum of outputs of both the channels.

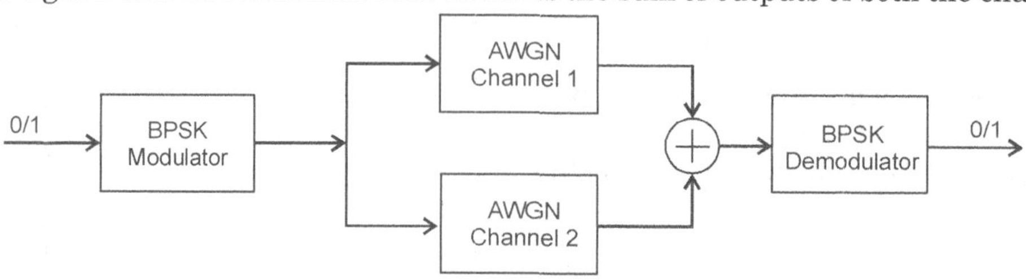

If the BER of this system is $Q(b\sqrt{\gamma})$, then the value of b is _____.

52. A fair coin is tossed repeatedly until a 'Head' appears for the first time. Let L be the number of tosses to get this first 'Head'. The entropy H(L) in bits is _____.

53. In spherical coordinates, let $\hat{a}_\theta, \hat{a}_\phi$ denote unit vectors the θ, ϕ directions.

$$E = \frac{100}{r} \sin\theta \cos(\omega t - \beta r)\hat{a}_\theta \; V/m$$

and $$H = \frac{0.265}{r} \sin\theta \cos(\omega t - \beta r)\hat{a}_\phi \; A/m$$

represent the electric and magnetic field components of the EM wave at large distances r from a dipole antenna, in free space. The average power (W) crossing hemispherical shell located at r = 1km, $0 \le \theta \le \pi/2$ is _____

54. For a parallel plate transmission line, let v be the speed of propagation and Z be the characteristic impedance. Neglecting fringe effects, a reduction of the spacing between the plates by a factor of two results in

(*a*) halving of v and no change in Z

(*b*) no changes in v and halving of Z

(*c*) no change in both v and Z

(*d*) halving of both v and Z

55. The input impedance of a $\frac{\lambda}{8}$ section of a lossless transmission line of characteristic impedance 50Ω is found to be real when the other end is terminated by a load Z_L (= R + *j*X)Ω. If X is 30Ω, the value of R (in Ω) is ____

ANSWERS

GENERAL APTITUDE

1. (*c*) **2.** (*d*) **3.** (*b*) **4.** (*a*) **5.** (725 to 725) **6.** (*d*) **7.** (*b*) **8.** (560 to 560) **9.** (*d*) **10.** (*b*)

ELECTRONICS AND COMMUNICATION

1. (*d*) **2.** (0.65 to 0.68) **3.** (*c*) **4.** (0.99 to 1.01) **5.** (0.32 to 0.34)

6. (*c*) **7.** (*a*) **8.** (7.99 to 8.01) **9.** (*d*) **10.** (*a*)

11. (*b*) **12.** (*b*) **13.** (3.1 to 3.26) **14.** (*c*) **15.** (*a*)

16. (62.4 to 62.6) **17.** (*d*) **18.** (2.99 to 3.01) **19.** (44 to 46) **20.** (2.24 to 2.26)

21. (*d*) **22.** (7.99 to 8.01) **23.** (– 0.01 to 0.01) **24.** (*b*) **25.** (*c*)

26. (*a*) **27.** (*b*) **28.** (862 to 866) **29.** (0.99 to 1.01) **30.** (29.08 to 29.10)

31. (*b*) **32.** (0.39 to 0.42) **33.** (9.99 to 10.01) **34.** (5.7 to 5.9) **35.** (*a*)

36. (499 to 501) **37.** (1092 to 1094) **38.** (*c*) **39.** (– 240 to – 230) **40.** (*a*)

41. (*d*) **42.** (*d*) **43.** (*c*) **44.** (9.99 to 10.01) **45.** (*a*)

46. (*b*) **47.** (42 to 48) **48.** (*c*) **49.** (*b*) **50.** (*d*)

51. (1.4 to 1.42) **52.** (1.99 to 2.01) **53.** (55.4 to 55.6) **54.** (*b*) **55.** (39 to 41)

EXPLANATIONS

GENERAL APTITUDE

4. If the standard deviation is less, there will be less deviation or batsman is more consistent.

5. The given series is

   12, 35, 81, 173, 357, ...

   The given series follows the following pattern

   $$12 \times 2 + 11 = 35$$
   $$35 \times 2 + 11 = 81$$
   $$81 \times 2 + 11 = 173$$
   $$173 \times 2 + 11 = 357$$
   $$357 \times 2 + 11 = 725$$

6. W, E, K, O,   I, Q, W, A,   F, N, T, X,   N, V, B, D,

   +8 +6 +4   +8 +6 +4   +8 +6 +4   +8 +6 (+2)

   −2 −2   −2 −2   −2 −2   −2 −4

   Hence the odd one from the following group is N, V, B, D.

7. By reading the instructions, it is clear that person who fills form T is a resident female with annual income of ₹ 9 Lakh.

8. Let the length of platform = x.m

   and train length = 280 m (given)

   According to Question

   $$\frac{x+280}{60} = \frac{280}{20}$$

   $$\therefore \quad x = 560 \text{ m.}$$

9. **For 2005,**

   trade deficit = (90 – 70) crores
   = 20 crores

   Now $\frac{1}{5}$th of export = $\frac{1}{5}$(70) crores
   = 14 crores
   ≠ 20 crores.

   Hence option (a) is wrong.

   For 2004,

   Trade deficit = (80 – 70) crores
   = 10 crores

   Now $\frac{1}{5}$th of export = $\frac{1}{5}$(70) crores
   = 14 crores
   ≠ 20 crores.

   Hence option (b) is wrong.

   For 2007,

   Trade deficit = (120 – 110) crores
   = 10 crores

   Now $\frac{1}{5}$th of export = $\frac{1}{5}$(110) crores
   = 22 crores
   ≠ 10 crores.

   Hence option (c) is wrong.

   For 2006

   Trade deficit = (120 – 100) crores
   = 20 crores

   Now $\frac{1}{5}$th of export = $\frac{1}{5}$(100) crores
   = 20 crores
   = Trade deficits

   Hence option (d) is correct.

10. The probaility that the other face is tail, is $\frac{1}{3}$.

ELECTRONICS AND COMMUNICATION

1. Matrix multiplication is not commutative in general.

2. Let the number of families in housing society be x.

   $\therefore$ $\frac{x}{2}$ family have single child.

   $\therefore$ Total $\frac{x}{2}$ children

   Now remaining $\frac{x}{2}$ families have 2 children.

   $\therefore$ Total $\frac{x}{2} \times 2 = x$, sibling children

   Probability that a child picked at random has a

   $$\text{sibling} = \frac{x}{x+\frac{x}{2}}$$

   $$= \frac{x \times 2}{3x} = \frac{2}{3} = 0.667$$

3. $Z = -2j$ is a singularity lies inside C : $|Z| = 3$

   $\therefore$ By Cauchy's integral formula,

   $$\oint_C \frac{Z^2 - Z + 4j}{Z + 2j} dz = 2\pi j.[Z^2 - Z + 4j]_{Z=-2j}$$

   $$= 2\pi j[-4 + 2j + 4j]$$

   $$= -4\pi[3 + j2]$$

4. A real (4 × 4) matrix A satisfies the equation $A^2 = I$ where I is the (4 × 4) identity matrix. The positive eigen value of A is 1.

5. X_1, X_2, X_3 are independent and identically distributed random variables.

So P{X_1 is the largest} = 1/3 = 0.33

6. Two cascaded sections

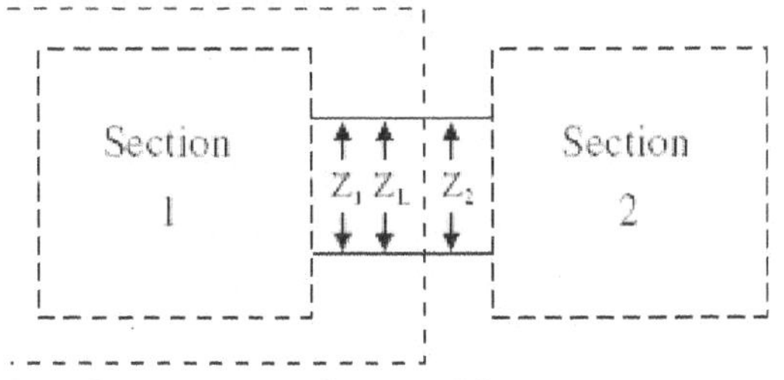

Z_1 = Output impedance of first section

Z_2 = Input impedance of second section

For maximum power transfer, upto 1st section is

$Z_L = Z_1^*$

$Z_L = Z_2 \Rightarrow Z_1^*$

7. Using KVL at all the three nodes.

$$i_4 + i_1 - i_2 = 0$$
$$i_6 + i_3 - i_1 = 0$$
$$i_5 + i_2 - i_3 = 0$$

By solving the equation, $i_6 = 5A$

8. Responsivity $(R) = \dfrac{I_P}{P_0}$

where I_P = photo current

P_0 = Incident power

$I_P = R \times P_0$

$= 8\ A$

9. Assume both the diode ON.

Then circuit will be as per figure (2)

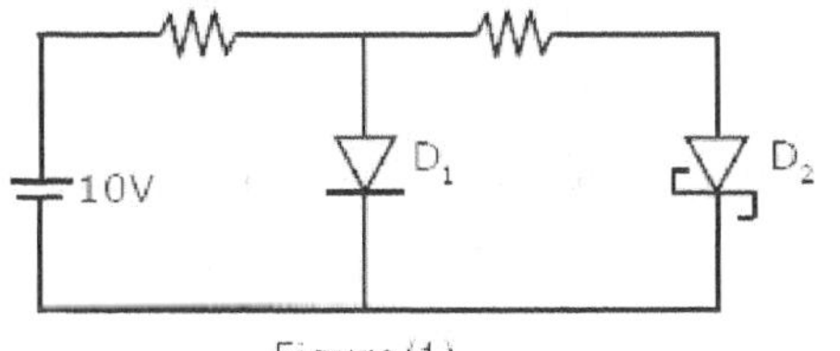

Figure (1)

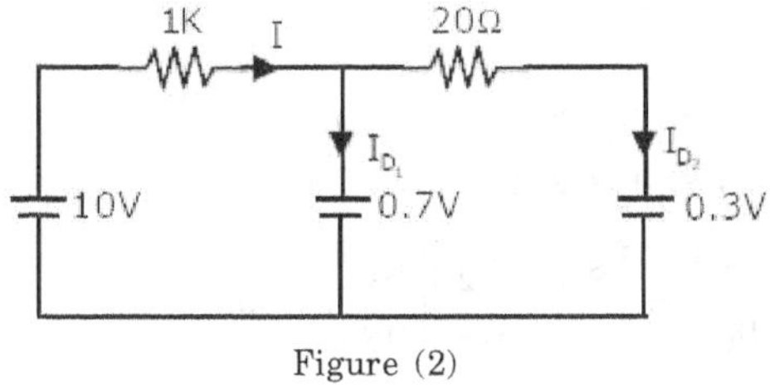

Figure (2)

$$\therefore\ I = \frac{10 - 0.7}{1k} = 9.3 \text{ mA}$$

$$I_{D_2} = \frac{0.7 - 0.3}{20} = 20 \text{ mA}$$

Now, $I_{D_1} = I - I_{D_2}$

$= -10.7$ mA (Not Possible)

$\therefore$ D_1 is OFF and hence D_2 – ON

10. If fixed positive charges are present in the gate oxide of an n-channel enhancement type MOSFETS it will lead to a decrease in the threshold voltage.

11. Ideal current Buffer has $Z_i = 0$

$Z_0 = \infty$

12. Output sample is voltage and is added at the input or current

$\therefore$ It is voltage – shunt negative feedback i.e., voltage-current negative feedback

13. f = 5KHz

$$\text{Cut off frequency (LPF)} = \frac{1}{2\pi R_2 C} = 5 \text{ KHz}$$

$$\Rightarrow R_2 = \frac{1}{2\pi \times 5 \times 10^3 \times 10 \times 10^{-9}} = 3.18 \text{ k}\Omega$$

14. Assume al NMOS are in saturation

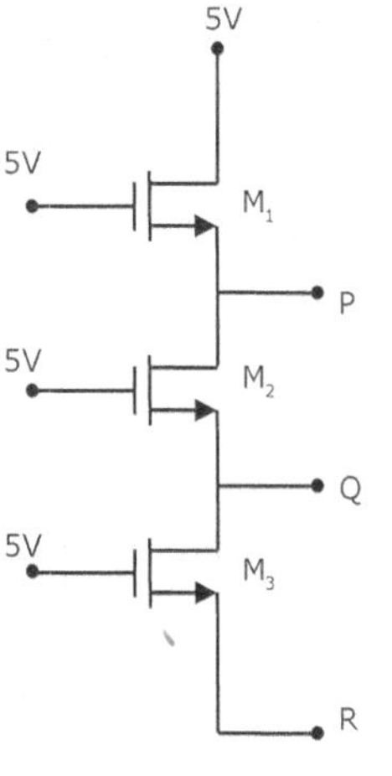

$\therefore$ $V_{DS} \geq (V_{GS} - V_T)$

For m_1,

$(5 - Vp) \geq (5 - V_p - 1)$

$(5 - Vp) > (4 - Vp) \Rightarrow$ Sat

$\therefore$ $I_{D_1} = k(V_{GS} - V_T)^2$

$I_{D_1} = K(4 - V_P)^2$ ...(1)

For m_2,

$I_{D_1} = K(5 - V_Q - 1)^2$

$I_{D_2} = K(4 - V_Q)^2$ ...(2)

$\therefore$ $I_{D_1} = I_{D_2}$

$(4 - V_P)^2 = (4 - V_Q)^2$

$\Rightarrow V_P = V_Q$ & $V_P + V_Q = 8$

$\Rightarrow V_P = V_Q = 4V$

For m_3,

$I_{D_3} = K(5 - V_R - 1)^2$

$\therefore$ $I_{D_2} = I_{D_3}$

$(4 - V_Q)^2 = (4 - V_R)^2$

$\Rightarrow V_R = V_Q = 4V$

$\therefore$ $V_P = V_Q = V_R = 4V$

15. $f = (x+y)(x+\bar{y}) + \overline{(x\bar{y})+\bar{x}}$

$= x + xy + x\bar{y} + (\overline{x\bar{y}})\cdot x$

$= x + xy + x\bar{y} + (\bar{x}+y)\cdot x$

$= x(1+y+\bar{y})$

$= x$

16. $f_{Q_3} = \dfrac{\text{input frequency}}{\text{modulus of counter}} = \dfrac{1\text{MHz}}{16}$

$f_{Q_3} = 62.5$ kHz.

17. Time period of a discrete signal,

$$\frac{w_0}{2\pi} = \frac{K}{N}$$

$$N = \frac{2\pi K}{w_0}$$

$$= \frac{2\pi K}{\pi^2} = \frac{2K}{\pi}$$

N is a irrational number so signal is not periodic.

18. Multiplication in the time domain

= Convolution in frequency domain

$x_1(t)\cdot x_2(t) = X_1(j\omega)X_2(j\omega)$

Fundamental frequencies = $f_1, f_1 \pm f_2, f_1 \pm 2f_2 \ldots$

= 500, 1500....

Nyquist rate = 2 × 1500

= 3000 Hz

= 3 kHz

19. 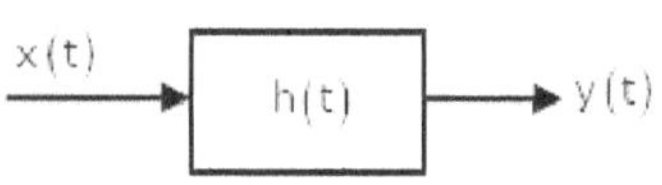

y(t) = x (t) * h(t)

x(t) = 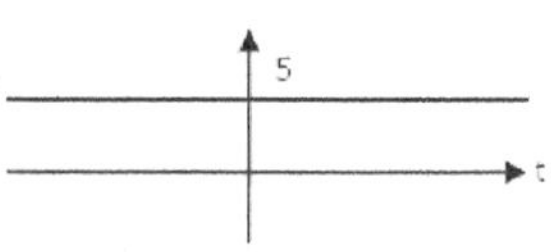

h (t) = 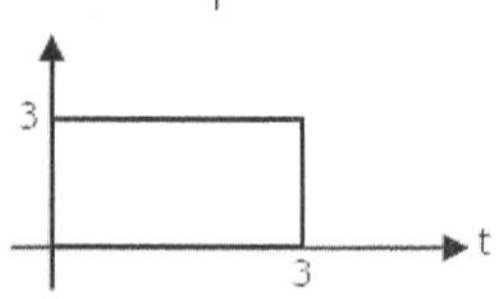

$$y(t) = \int_0^3 3.5\, d\tau = 45 \text{(steady state output)}$$

20. Using root locus,

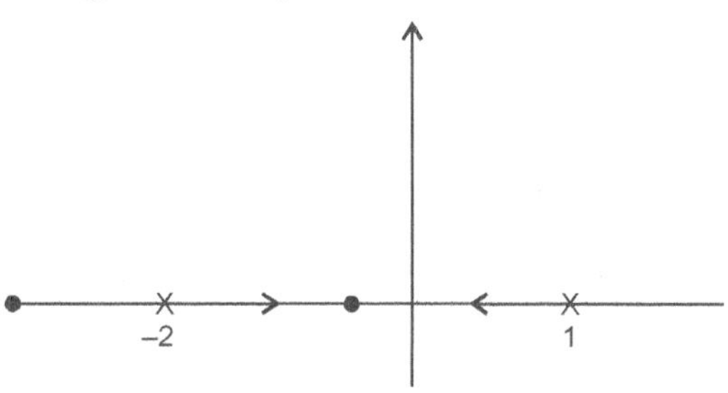

Break away point,

$$1 + \frac{K}{(s+2)(s-1)} = 0$$

$$K = -(s+2)(s-1)$$

$$\frac{dK}{ds} = -2s - 1 = 0$$

$$s = -0.5$$

$$|G(s)|_{s=-0.5} = 1$$

$$\therefore \quad \left|\frac{K}{(s+2)(s-1)}\right|_{s=-0.5} = 1$$

21. For all pass system the pole zero pair must be symmetrical about imaginary axis with zero on the RHS and pole on the LHS of s-plane.

This is not minimum phase system.

Encirclement to the critical point (–1, 0) = N = 0

Open loop pole at RHS = P = 1

$\therefore \quad N = P - z$

$z = 1$ (Close loop pole at RHS)

Thus system is unstable.

22. Spreading factor(SF) = $\dfrac{\text{chip rate}}{\text{symbol rate}}$

This if a single symbol is represented by a code of 8 chips

Chip rate = 80 × symbol rate

S.F (Spreading Factor)= $\dfrac{8 \times \text{symbol rate}}{\text{symbol rate}} = 8$

Spread factor (or) process gain and determine to a certain extent the upper limit of the total number of uses supported simultaneously by a station.

23. Channel capacity of BSC is

$$C = P\log_2 P + (1-P)\log_2(1-P) + 1$$

$$= 0.5 \log_2 0.5 + 0.5 \log_2 0.5 + 1$$

$$= 0 \qquad [\because \log_2 0.5 = 1]$$

It is the case of channel with independent input and output, hence C = 0

24. 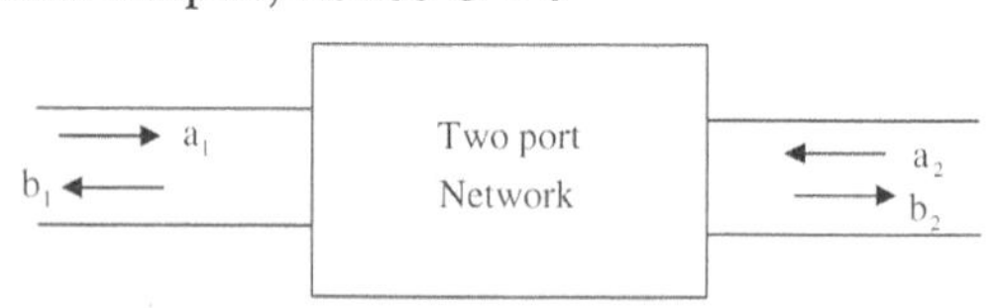

$b_1 = s_{11}a_1 + s_{12}a_2$

$b_2 = s_{21}a_1 + s_{22}a_2$

$$\begin{bmatrix} b_1 \\ b_2 \end{bmatrix} = \begin{bmatrix} s_{11} & s_{12} \\ s_{21} & s_{22} \end{bmatrix}\begin{bmatrix} a_1 \\ a_2 \end{bmatrix}; \; s_1 = \left.\frac{b_1}{a_1}\right|_{a_2=0}$$

By verification Answer (b) satisfies.

25.

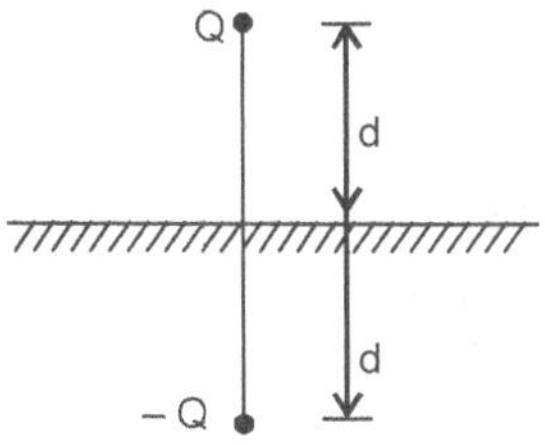

$$F = \frac{Q_1 Q_2}{4\pi \in R^2}$$

$$= \frac{q^2}{4\pi \in (2d)^2} = \frac{q^2}{16\pi \in d^2}$$

26. Taylor series expansion of 3 sin x + 2 cos x

Taylor series is given by

$$f(x) = f(x_0) + f'(x_0)(x - x_0) + \frac{f''(x_0)}{2!}(x - x_0)^2 + \frac{f'''(x_0)}{3!}(x - x_0)^3 + ...$$

Here $x_0 = 0$

$$\therefore \quad f(x) = f(0) + f'(0)(x) + \frac{f''(0)}{2}x^2 + \frac{f'''(0)}{6}(x)^3 + ... \quad ...(i)$$

Now $f(x) = 3 \sin x + 2 \cos x$

$\Rightarrow$ $f(0) = 2$

Now $f'(x) = 3 \cos x + 2(- \sin x)$

$= 3 \cos x - 2 \sin x$

$\Rightarrow$ $f'(0) = 3$

$f''(x) = -3 \sin x - 2 \cos x$

$\Rightarrow$ $f''(0) = -2$

$f'''(x) = -3 \cos x + 2 \sin x$

$\Rightarrow$ $f'''(0) = -3$

From equation (i),

$$f(x) = f(0) + f'(0)\,x + \frac{f''(0)}{2}x^2 + \frac{f'''(0)}{6}(x)^3 + ...$$

$$= 2 + 3x + \frac{(-2)}{2}x^2 + \left(\frac{-3}{6}\right)x^3 + ...$$

$$= 2 + 3x - x^2 - \frac{1}{2}x^3 + ...$$

27. Given,

$$\int_{-\infty}^{\infty} g(t).e^{-jwt}dt = \omega.e^{-2w^2} \text{ (let } G(j\omega))$$

$$\Rightarrow \int_{-\infty}^{\infty} g(t)dt = 0$$

$$y(t) = \int_{-\infty}^{t} g(z).dz$$

$\Rightarrow$ y(t) = g(t) * u(t) [u(t) in unit step function]

$$\Rightarrow \quad Y(j\omega) = G(j\omega).\,U(j\omega)$$

$$Y(j\omega) = \int_{-\infty}^{\infty} y(t).e^{-j\omega t}dt$$

$$\Rightarrow \quad Y(j0) = \int_{-\infty}^{\infty} y(t)\,dt$$

$$= \left[\omega.e^{-2w^2}\left[\frac{1}{j\omega} + \pi\delta(\omega)\right]\right]_{\omega = 0}$$

$$= \frac{1}{j} = -j$$

28. The volume under the surface $z(x, y) = x + y$ and above the triangle in the $x - y$ plane is given

$$\text{Volume} = \int_0^{12}\left[\int_0^x (x + y)dy\right]dx$$

$$= \int_0^{12}\left[xy + \frac{y^2}{2}\right]_0^x dx$$

$$= \int_0^{12}\left[x^2 + \frac{x^2}{2}\right]dx$$

$$\therefore \quad \text{Volume} = \left[\frac{x^3}{3} + \frac{x^3}{6}\right]_0^{12}$$

$$= \left[\frac{2x^3 + x^3}{6}\right]_0^{12}$$

$$= \left[\frac{x^3}{2}\right]_0^{12}$$

$$= \frac{1}{2}(12)^3$$

$$= 864.$$

29.
$$J_6 = \begin{bmatrix} 0 & 0 & 0 & 0 & 0 & 1 \\ 0 & 0 & 0 & 0 & 1 & 0 \\ 0 & 0 & 0 & 1 & 0 & 0 \\ 0 & 0 & 1 & 0 & 0 & 0 \\ 0 & 1 & 0 & 0 & 0 & 0 \\ 1 & 0 & 0 & 0 & 0 & 0 \end{bmatrix}$$

$$I_6 = \begin{bmatrix} 1 & 0 & 0 & 0 & 0 & 0 \\ 0 & 1 & 0 & 0 & 0 & 0 \\ 0 & 0 & 1 & 0 & 0 & 0 \\ 0 & 0 & 0 & 1 & 0 & 0 \\ 0 & 0 & 0 & 0 & 1 & 0 \\ 0 & 0 & 0 & 0 & 0 & 1 \end{bmatrix}$$

Now $P = I_6 + \alpha J_6$

$$= \begin{bmatrix} 1 & 0 & 0 & 0 & 0 & 0 \\ 0 & 1 & 0 & 0 & 0 & 0 \\ 0 & 0 & 1 & 0 & 0 & 0 \\ 0 & 0 & 0 & 1 & 0 & 0 \\ 0 & 0 & 0 & 0 & 1 & 0 \\ 0 & 0 & 0 & 0 & 0 & 1 \end{bmatrix} + \begin{bmatrix} 0 & 0 & 0 & 0 & 0 & \alpha \\ 0 & 0 & 0 & 0 & \alpha & 0 \\ 0 & 0 & 0 & \alpha & 0 & 0 \\ 0 & 0 & \alpha & 0 & 0 & 0 \\ 0 & \alpha & 0 & 0 & 0 & 0 \\ \alpha & 0 & 0 & 0 & 0 & 0 \end{bmatrix}$$

$$\therefore \; P = \begin{bmatrix} 1 & 0 & 0 & 0 & 0 & \alpha \\ 0 & 1 & 0 & 0 & \alpha & 0 \\ 0 & 0 & 1 & \alpha & 0 & 0 \\ 0 & 0 & \alpha & 1 & 0 & 0 \\ 0 & \alpha & 0 & 0 & 1 & 0 \\ \alpha & 0 & 0 & 0 & 0 & 1 \end{bmatrix}$$

$$|P| = 1\begin{vmatrix} 1 & 0 & 0 & \alpha & 0 \\ 0 & 1 & \alpha & 0 & 0 \\ 0 & \alpha & 1 & 0 & 0 \\ \alpha & 0 & 0 & 1 & 0 \\ 0 & 0 & 0 & 0 & 1 \end{vmatrix} - \alpha \begin{vmatrix} 0 & 1 & 0 & 0 & \alpha \\ 0 & 0 & 1 & \alpha & 0 \\ 0 & 0 & \alpha & 1 & 0 \\ 0 & \alpha & 0 & 0 & 1 \\ \alpha & 0 & 0 & 0 & 0 \end{vmatrix}$$

$$= \begin{vmatrix} 1 & \alpha & 0 & 0 \\ \alpha & 1 & 0 & 0 \\ 0 & 0 & 1 & 0 \\ 0 & 0 & 0 & 1 \end{vmatrix} + \alpha \begin{vmatrix} 0 & 1 & \alpha & 0 \\ 0 & \alpha & 1 & 0 \\ 0 & 0 & 0 & 1 \\ \alpha & 0 & 0 & 0 \end{vmatrix}$$

$$= \begin{vmatrix} 1 & 0 & 0 \\ 0 & 1 & 0 \\ 0 & 0 & 1 \end{vmatrix} - \alpha \begin{vmatrix} \alpha & 0 & 0 \\ 0 & 1 & 0 \\ 0 & 0 & 1 \end{vmatrix} - \alpha \begin{vmatrix} 0 & 1 & 0 \\ 0 & 0 & 1 \\ \alpha & 0 & 0 \end{vmatrix}$$

$$+ \alpha^2 \begin{vmatrix} 0 & \alpha & 0 \\ 0 & 0 & 1 \\ \alpha & 0 & 0 \end{vmatrix}$$

$$|P| = 1(1) - \alpha(\alpha) - \alpha((-1)(-\alpha)) + \alpha^2[(-\alpha)(-\alpha)] = 0$$

$$\therefore \quad 1 - \alpha^2 - \alpha^2 + \alpha^4 = 0$$

$$\alpha^4 - 2\alpha^2 + 1 = 0$$

$$(\alpha^2 - 1)^2 = 0$$

$$\therefore \quad \alpha^2 = 1$$

$$\therefore \quad \alpha = \pm\sqrt{1}$$

$\therefore \; \alpha = 1$ [$\because$ α is non-negative real number]

30.

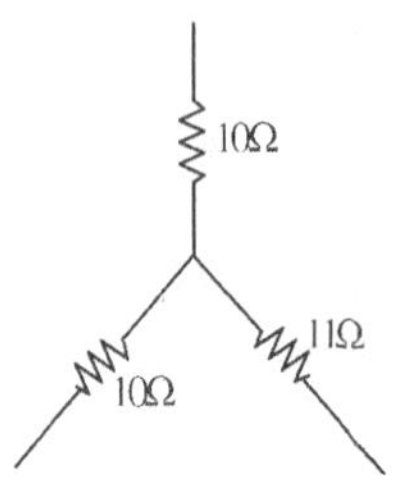

Star Connection

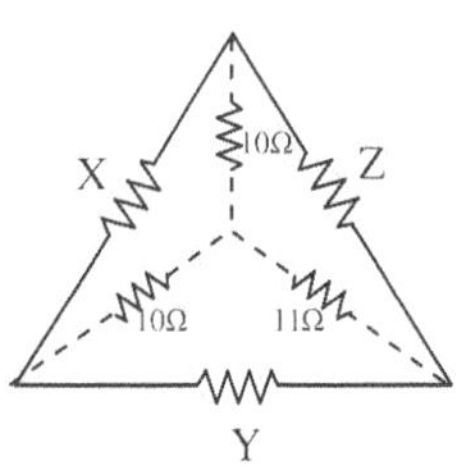

Delta Connection

X = 29.09Ω

y = 32Ω

z = 32Ω

$$X = \frac{(10)(10) + (10)(11) + (10)(11)}{11}\,\Omega$$

$$y = \frac{(10)(10) + (10)(11) + (10)(11)}{10}\,\Omega$$

$$z = \frac{(10)(10) + (10)(11) + (10)(11)}{10}\,\Omega$$

i.e, lowest value among three resistances is 29.09Ω

31.

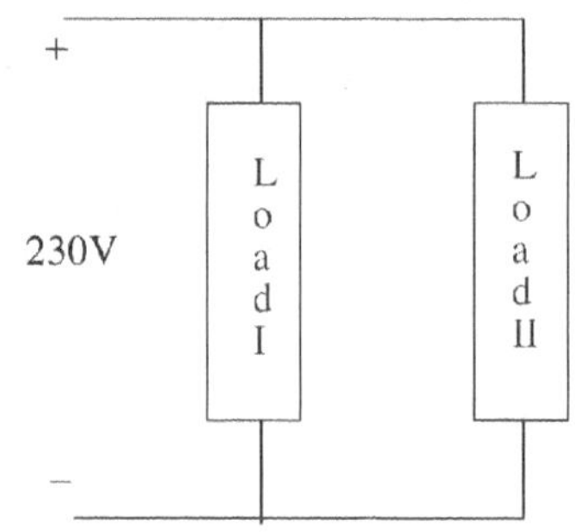

Load I:

P = 10kw

$\cos\phi = 0.8$

$Q = P\tan\phi = 7.5$KVAR

$\Big\}\; S_1 = P - jQ = 10 - j7.5$ KVA

Load II : S = 10 KVA

$\cos\phi = 0.8 \quad \sin\phi = \dfrac{Q}{S}$

$\cos\phi = \dfrac{P}{S}$

$0.8 = \dfrac{P}{10} \rightarrow P = 8$kw Q = 6 KVAR

$S_1 = P + jQ = 8 + j6$

Complex power delivered by the source is $S_I + S_{II} = 18 - j1.5$ KVA

32. RMS value $= \sqrt{\dfrac{1}{T}\displaystyle\int_0^T f^2(t)\,dt}$

where T is time period

$$= \sqrt{\frac{1}{T}\left[\int_0^{T/2}\left(\frac{2}{T}t\right)^2 dt + \int_{T/2}^{T}(0)^2 dt\right]}$$

$$= \sqrt{\frac{1}{T}\left[\int_0^{T/2}\frac{4}{T^2}t^2 dt\right]}$$

$$= 0.408$$

33. For ciritical damping,

$$\xi = \frac{1}{2Q} = 1$$

where Q = Quality factor

For series circuit,

$$Q = \frac{1}{R}\sqrt{\frac{L}{C}}$$

$$\frac{1}{\frac{2}{R}\sqrt{\frac{L}{C}}} = 1$$

$$C = \left(\frac{2}{R}\right)^2 \times L = \left(\frac{2}{40}\right)^2 \times 4 = 10 \text{ mF.}$$

34. $$g_m = \frac{I_C}{V_T}$$

$$I_C = I_S\left[e^{\frac{V_{BE}}{V_T}} - 1\right]$$

$$I_C = 10^{-13} \times \left[e^{\frac{0.7}{0.025}} - 1\right]$$

$$I_C = 144.6257 \text{ mA}$$

$$g_m = \frac{0.144}{0.025}$$

$$= 5.78 \frac{\text{mA}}{\text{V}}$$

35. Electron concentration,

$$n \simeq \frac{n_i^2}{N_A} e^{V_{in}/V_T}$$

$$= \frac{(1.5 \times 10^{10})^2}{1 \times 10^{16}} e^{0.3/26\text{mV}}$$

$$= 2.3 \times 10^9/\text{ cm}^3$$

36. $$r_{DS} = \frac{1}{(\mu_n C_{on})\left(\frac{W}{L}\right)(V_{GS} - V_T)}$$

$$r_{DS} = \frac{1}{800 \times 10^{-4} \times 10^8 \times 100 \times (20 - 0.5)}$$

$$r_{DS} = 500\ \Omega$$

37. Given circuit is an op-amp series regulator, V_0 is given by

$$V_0 = \left(1 + \frac{R_1}{R_2}\right)V_z$$

$$9V = \left(1 + \frac{1k\Omega}{R_2}\right)4.7$$

$$R_2 = 1093.02\ \Omega$$

38. Given, $Z_i = \infty$

$$A_{0_L} = \infty$$

$$V_{i_0} = 0$$

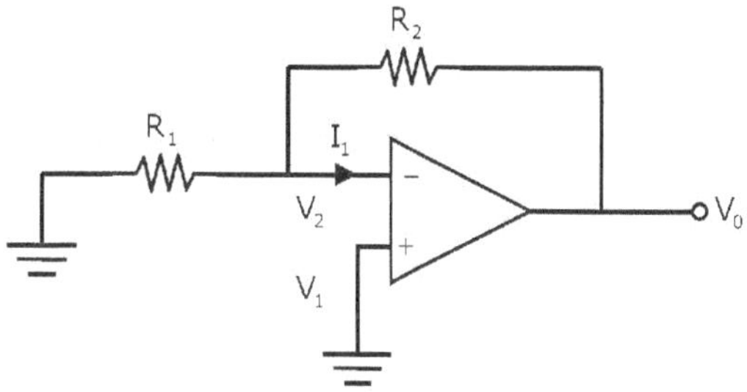

$$V_2 = (R_1 // R_2)\, I_1$$

$$= \frac{R_1 R_2}{R_1 + R_2} I_1 \quad ...(1)$$

KCL at inverting node

$$\frac{V_2}{R_1} + \frac{V_2 - V_0}{R_2} = 0 \quad (\because Z_i = \infty)$$

$$\frac{V_0}{R_2} = V_2\left[\frac{1}{R_1} + \frac{1}{R_2}\right]$$

$$\frac{V_0}{R_2} = \left(\frac{R_1 R_2}{R_1 + R_2}\right) I_1 \left[\frac{R_2 + R_1}{R_1 R_2}\right]$$

$$\Rightarrow \quad V_0 = I_1 R_2$$

39. $V_{BE} = 0.7V$, $\beta = 200$, $V_T = 25mV$

DC Analysis:

$$V_B = 12 \times \frac{11k}{11k + 33K} = 3V$$

$$V_E = 3 - 0.7 = 2.3V$$

$$I_E = \frac{2.3}{10 + 1K} = 2.277mA$$

$$I_B = 11.34\ \ A$$

$$I_C = 2.26 \text{ mA}$$

$$r_e = \frac{25\,\text{mV}}{2.277\,\text{mA}} = 10.98\,\Omega$$

$$A_V = \frac{V_0}{V_i} = \frac{-\beta R_C}{\beta r_e + (1+\beta)(R_s)}$$

$$= \frac{-200 \times 5K}{200 \times 10.98 + (201)10}$$

$$A_V = -237.76$$

40. 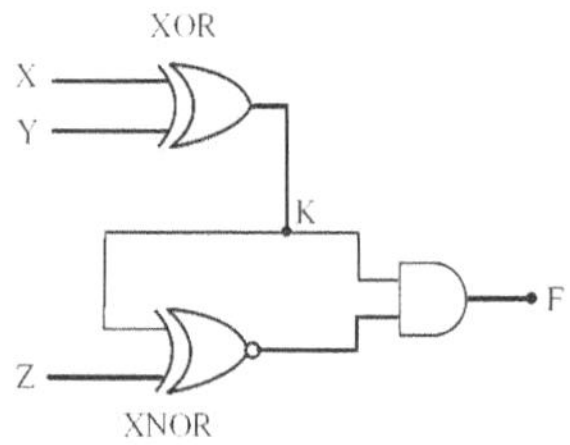

Assume dummy variable K as a output of XOR gate $K = X \oplus Y = \bar{X}Y + X\bar{Y}$

$$F = K.(K \odot Z)$$
$$= (\bar{K}\bar{Z} + K.Z)$$
$$= K.\bar{K}\bar{Z} + K.K.Z)$$
$$= 0 + K.Z \quad (\because K.\bar{K} = 0 \text{ and } K.K = K)$$

Put the value of K in above expression

$$F = (\bar{X}Y + X\bar{Y})\,Z$$
$$= \bar{X}YZ + X\bar{Y}Z$$

41. 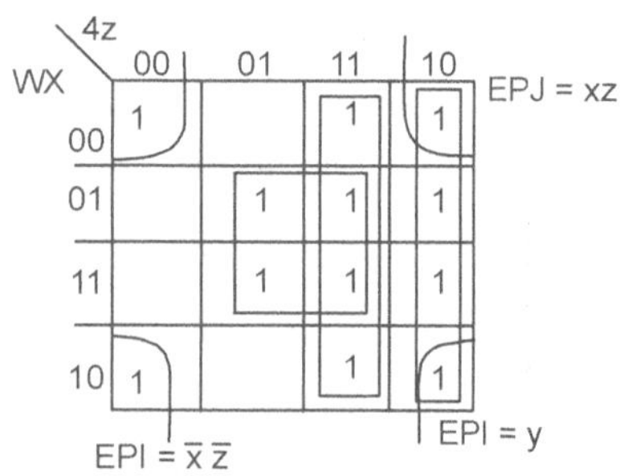

42.

$$D_2 = A \oplus S$$
$$D_2 = Q_1 \oplus S$$

But $D_2 = A \odot S$

if $A = \bar{Q}_1$

then $D_2 = \bar{Q}_1 \odot S$

$$D_2 = \bar{Q}_1 \oplus S$$

So input A is connected to $\bar{Q}_1$

43. $x(n) = \underbrace{\left[-\frac{1}{9}\right]4(x)}_{\text{Right side Signal}} - \underbrace{\left(\frac{1}{3}\right)^n 4(-n-1)}_{\text{Left side signal}}$

ROC is $|z| > \frac{1}{9}$ ROC is $|z| < \frac{1}{3}$

So ROC is $\frac{1}{3} > |z| > \frac{1}{9}$

45.

$$\frac{dy}{dt} + 5y(t) = u(t)$$
$$y(0) = 1$$
$$sy(s) - y(0) + 5y(s) = \frac{1}{s}$$
$$5y(s) - 1 + sy(s) = \frac{1}{s}$$
$$y(s)[s+5] = \left[\frac{1}{s} + 1\right]$$
$$y(s) = \frac{s+1}{s(s+5)}$$
$$y(s) = \frac{1}{5s} + \frac{4}{5(s+5)}$$

Applying inverse laplace transform

$$y(t) = \frac{1}{5}u(t) + \frac{4}{5}e^{-5t}u(t)$$
$$y(t) = (0.2 + 0.8\,e^{-5t})$$

46. From the given state model.

$$A = \begin{bmatrix} -1 & 1 & 0 \\ 0 & -1 & 0 \\ 0 & 0 & -2 \end{bmatrix} \quad B = \begin{bmatrix} 0 \\ 4 \\ 0 \end{bmatrix} \quad c = [1\,1\,1]$$

Controllable : $Q_c = c = [B \quad AB \quad A^2B]$

if $|Q_c| \neq 0 \rightarrow$ controllable

$$Q_c = \begin{bmatrix} 0 & 4 & -8 \\ 4 & -4 & 4 \\ 0 & 0 & 0 \end{bmatrix} \Rightarrow |Q_c| = 0$$

$\therefore$ Uncontrollable

Observable : $Q_0 = \begin{bmatrix} C \\ CA \\ CA^2 \end{bmatrix}$

if $|Q_0| \neq 0 \rightarrow$ observable

$$Q_0 = \begin{bmatrix} 1 & 1 & 1 \\ -1 & 0 & -2 \\ 1 & 1 & 4 \end{bmatrix} \Rightarrow |Q_0| = 1$$

$\therefore$ Observable

The system is uncontrollable and observable.

47. $G(S) = \dfrac{10}{(s+0.1)(s+1)(s+10)}$

$$G(S) = \frac{10}{0.1\left[1 + \frac{s}{0.1}\right][1+s]\left[1 + \frac{s}{10}\right].10}$$

$$G(S) = \frac{10}{[1+10s][1+s][1+0.1s]}$$

By Approximation, $G(S) = \dfrac{10}{[10s+1]}$

48.
$$G(s) = \frac{2(s+4)}{(s+1)(s+2)}$$
$$T(s) = \frac{G(S)}{1+G(S)+H(S)}$$
$$= \frac{2(S+4)}{(S+1)(S+2)+2(S+4)}$$

Compare with $S^2 + 2\zeta W_n S + w_n^2 = 0$
$$\zeta W_n = 2.5$$
$$\tau_{sett} = \frac{2}{\zeta w_n} = 0.8 \text{ sec.}$$

49.
$$x^2 = E(x^2) - [E(x)]^2$$

σ_x^2 can never be negative
$$E[x^2] \geq [E(x)]^2$$

50. Given $X(t) = \sqrt{2}\sin(2\pi t + \phi)$

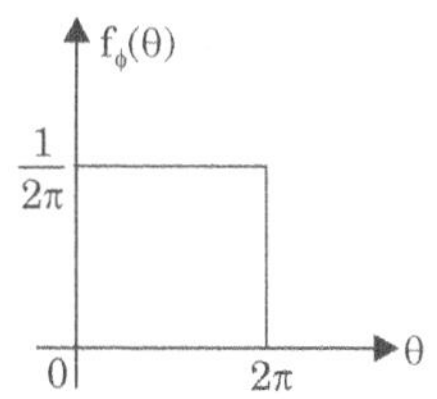

ϕ in uniformly distributed in the interval $[0, 2\pi]$
$E[X(t_1)X(t_2)]$
$$= \int_0^{2\pi} \sqrt{2}\sin(2\pi t_1 + \theta)\sqrt{2}\sin(2\pi t_2 + \theta) f_\phi(\theta)\, d\theta$$
$$= 2\int_0^{2\pi} \sin(2\pi t_1 + \theta)\sin(2\pi t_2 + \theta).\frac{1}{2\pi}.d\theta$$
$$= \frac{1}{2\pi}\int_0^{2\pi} \sin(2\pi(t_1 + t_2) + 2\theta)\, d\theta$$
$$+ \frac{1}{2\pi}\int_0^{2\pi} \cos(2\pi(t_1 - t_2))\, d\theta$$

First integral will result into zero as we are integrating from 0 to 2π.
Second integral result into $\cos[2\pi(t_1 - t_2)]$
$\Rightarrow E[X(t_1)X(t_2)] = \cos(2\pi(t_1 - t_2))$

51. Bit error rate for BPSK $= Q\left(\sqrt{\frac{2E}{NO}}\right) \cdot \left\{Q\left(\sqrt{\frac{E}{N_O/2}}\right)\right\}$

$$\Rightarrow Y = \frac{2E}{N_O}$$

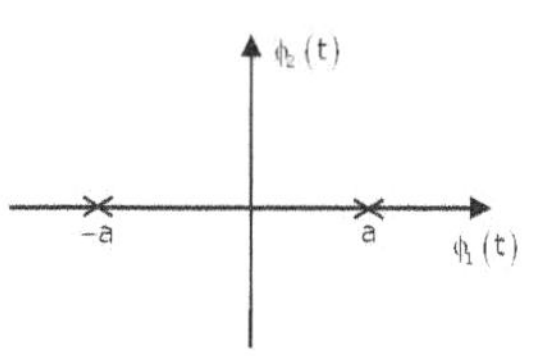

Function of bit energy and noise $P_{SD}\ \frac{N_O}{2}$

Counterllation diagram of BPSK
Channel is A_{WGN} which implies noise sample as independent
Let $2x + n_1 + n_2 = x^1 + n^1$
where $x^1 = 2x$
$n^1 = n_1 + n_2$

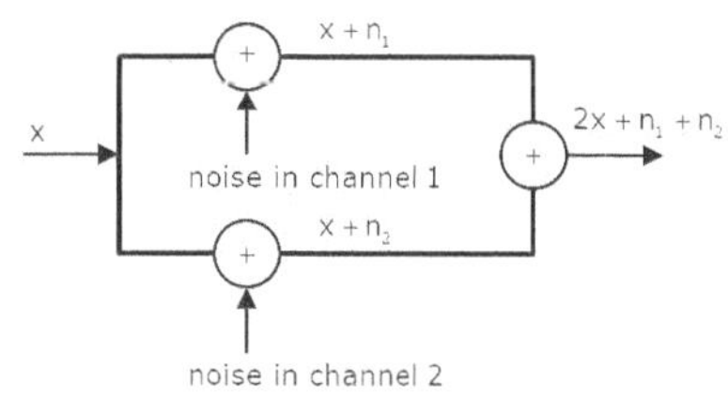

Now Bit error rate $= Q\left(\sqrt{\frac{2E^1}{N_O^1}}\right)$

E^1 is energy in x^1
N_O^1 is PSD of h^1
$E^1 = 4E$ [as amplitudes are getting doubled]
$N_O^1 = N_O$ [independent and identical channel]

$$\Rightarrow \text{Bit error rate} = Q\left(\sqrt{\frac{4E}{N_O}}\right) = Q\left(\sqrt{2}\sqrt{\frac{2E}{N_O}}\right)$$
$$\Rightarrow b = \sqrt{2} \text{ or } 1.414$$

52. In this problem random variable is L
L can be 1, 2, ...

$$P\{L = 1\} = \frac{1}{2}$$
$$P\{L = 2\} = \frac{1}{4}$$
$$P\{L = 3\} = \frac{1}{8}$$
$$H\{L\} = \frac{1}{2}\log_2\frac{1}{1/2} + \frac{1}{4}\log_2\frac{1}{1/4} + \frac{1}{8}\log_2\frac{1}{1/8} + \ldots$$
$$= 0 + 1.\frac{1}{2} + 2.\frac{1}{4} + 3.\frac{1}{8} + \ldots$$

[Arithmatic geometric series summation]
$$= \frac{2}{1-\frac{1}{2}} + \frac{\frac{1}{2}\cdot 1}{\left(1-\frac{1}{2}\right)^2} = 2$$

53. $E_\theta = \frac{100}{r}\sin\theta e^{-j\beta r}$

$H_Q = \frac{0.265}{r}\sin\theta\, e^{-j\beta r}$

$P_{avg} = \frac{1}{2}\int_s E_\theta H_Q^*.\, ds$

$= \frac{1}{2}\int_s \frac{100(0.265)}{r^2}\sin^2\theta\, r^2 \sin\theta\, d\theta\, d\phi$

$P_{aet} = \frac{1}{2}\int_s (26.5)\sin^2 d\theta\, d\phi$

$= 13.25\int_{\theta=0}^{\pi/2}\sin^3\theta\, d\theta \int_{Q=0}^{2\pi} d\phi = 13.25\left(\frac{2}{3}\right)(2\pi)$

P = 55.5 w

54. $Z_o = \frac{276}{\sqrt{\epsilon_r}}\log\left(\frac{d}{r}\right)$

d → distance between the two plates

so, z_o – changes, if the spacing between the plates changes.

$V = \frac{1}{\sqrt{LC}}$ → independent of spacing between the plates

55. Given, $\ell = \lambda/s$

$Z_o = 50\Omega$

$Z_{in}\left(\ell = \lambda/8\right) = Z_o\left[\frac{Z_L + JZ_o}{Z_o + KZ_L}\right]$

$Z_{in} = 50\left[\frac{Z_L + J50}{50 + JZ_L}\right] = 50\left[\frac{Z_L + J50}{50 + JZ_L}\times\frac{50 - JZ_L}{50 - JZ_L}\right]$

$Z_{in} = 50\left[\frac{50Z_L + 50Z_L + J(50^2 - Z_L^2)}{50^2 + Z_L^2}\right]$

Given, Z_{in} → Real

So, $I_{mg}(Z_{in}) = 0$

$50^2 - Z_L^2 = 0$

$Z_L^2 = 50^2$

$R^2 + X^2 = 50^2$

$R^2 = 50^2 - X^2 = 50^2 - 30^2$

$R = 40\Omega$

SOLVED PAPER – 2014

ELECTRONICS AND COMMUNICATION ENGINEERING

Set - 2

No. of Questions : 65 *Maximum Marks : 100*

GENERAL APTITUDE

Q. No. 1 – 5 Carry One Mark Each

1. Choose the most appropriate word from the options given below to complete the following sentence.

   Communication and interpersonal skills are _____ important in their own ways.

   (*a*) each (*b*) both
   (*c*) all (*d*) either

2. Which of the options given below best completes the following sentence?

   She will feel much better if she ________.

   (*a*) will get some rest
   (*b*) gets some rest
   (*c*) will be getting some rest
   (*d*) is getting some rest

3. Choose the most appropriate pair of words from the options given below to complete the following sentence.

   She could not _____ the thought of ________ the election to her bitter rival.

   (*a*) bear, loosing (*b*) bare, loosing
   (*c*) bear, losing (*d*) bare, losing

4. A regular die has six sides with numbers 1 to 6 marked on its sides. If a very large number of throws show the following frequencies of occurrence: $1 \rightarrow 0.167$; $2 \rightarrow 0.167$; $3 \rightarrow 0.152$; $4 \rightarrow 0.166$; $5 \rightarrow 0.168$; $6 \rightarrow 0.180$. We call this die

   (*a*) irregular
   (*b*) biased
   (*c*) Gaussian
   (*d*) insufficient

5. Fill in the missing number in the series.

   2 3 6 15 _____ 157.5 630

Q. No. 6 – 10 Carry Two Marks Each

6. Find the odd one in the following group :

   Q,W,Z,B B,H,K,M W,C,G,J M,S,V,X

   (*a*) Q,W,Z,B (*b*) B,H,K,M
   (*c*) W,C,G,J (*d*) M,S,V,X

7. Lights of four colors (red, blue, green, yellow) are hung on a ladder. On every step of the ladder there are two lights. If one of the lights is red, the other light on that step will always be blue. If one of the lights on a step is green, the other light on that step will always be yellow. Which of the following statements is not necessarily correct?

   (*a*) The number of red lights is equal to the number of blue lights
   (*b*) The number of green lights is equal to the number of yellow lights
   (*c*) The sum of the red and green lights is equal to the sum of the yellow and blue lights
   (*d*) The sum of the red and blue lights is equal to the sum of the green and yellow lights

8. The sum of eight consecutive odd numbers is 656. The average of four consecutive even numbers is 87. What is the sum of the smallest odd number and second largest even number?

9. The total exports and revenues from the exports of a country are given in the two charts shown below. The pie chart for exports shows the quantity of each item exported as a percentage of the total quantity of exports. The pie chart for the revenues shows the percentage of the total revenue generated through export of each item. The total quantity of exports of all the items is 500 thousand tonnes and the total revenues are 250 crore rupees. Which item among the following has generated the maximum revenue per kg?

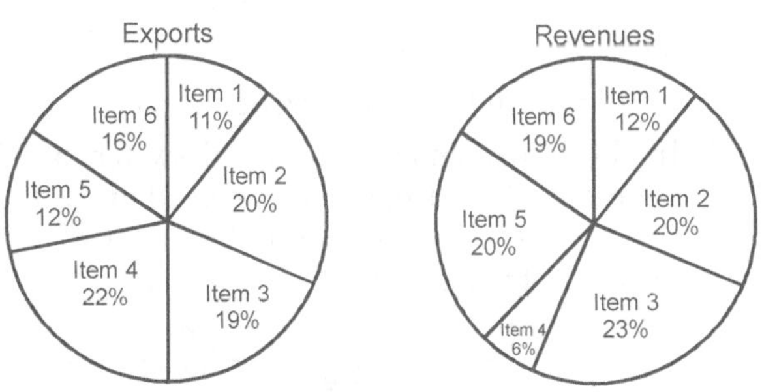

   (*a*) Item 2 (*b*) Item 3
   (*c*) Item 6 (*d*) Item 5

10. It takes 30 minutes to empty a half-full tank by draining it at a constant rate. It is decided to simultaneously pump water into the half-full tank while draining it. What is the rate at which water has to be pumped in so that it gets fully filled in 10 minutes?

(*a*) 4 times the draining rate

(*b*) 3 times the draining rate

(*c*) 2.5 times the draining rate

(*d*) 2 times the draining rate

ELECTRONICS AND COMMUNICATION

Q. No. 1 – 25 Carry One Mark Each

1. The determinant of matrix A is 5 and the determinant of matrix B is 40. The determinant of matrix AB is ________.

2. Let X be a random variable which is uniformly chosen from the set of positive odd numbers less than 100. The expectation E[X] is ________.

3. For $0 \le t < \infty$, the maximum value of the function $f(t) = e^{-t} - 2e^{-2t}$ occurs at

(*a*) $t = \log_e 4$ (*b*) $t = \log_e 2$

(*c*) $t = 0$ (*d*) $t = \log_e 8$

4. The value of $\lim_{x \to \infty} \left(1 + \frac{1}{x}\right)^x$ is

(*a*) ln2 (*b*) 1.0

(*c*) e (*d*) ∞

5. If the characteristic equation of the differential equation $\frac{d^2y}{dx^2} + 2\alpha\frac{dy}{dx} + y = 0$ has two equal roots, then the values of α are

(*a*) ± 1 (*b*) 0,0

(*c*) $\pm j$ (*d*) $\pm 1/2$

6. Norton's theorem states that a complex network connected to a load can be replaced with an equivalent impedance

(*a*) in series with a current source

(*b*) in parallel with a voltage source

(*c*) in series with a voltage source

(*d*) in parallel with a current source

7. In the figure shown, the ideal switch has been open for a long time. If it is closed at $t = 0$, then the magnitude of the current (in mA) through the 4kΩ resistor at $t = 0^+$ is ________.

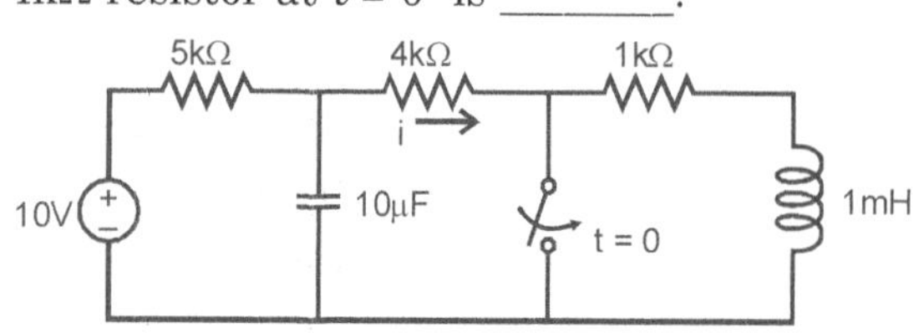

8. A silicon bar is doped with donor impurities $N_D = 2.25 \times 10^{15}$ atoms/cm^3. Given the intrinsic carrier concentration of silicon at T = 300 K is $n_i = 1.5 \times 10^{10}$ cm^{-3}. Assuming complete impurity ionization, the equilibrium electron and hole concentrations are

(*a*) $n_0 = 1.5 \times 10^{16}$ cm^{-3}, $p_0 = 1.5 \times 10^5$ cm^{-3}

(*b*) $n_0 = 1.5 \times 10^{10}$ cm^{-3}, $p_0 = 1.5 \times 10^{15}$ cm^{-3}

(*c*) $n_0 = 2.25 \times 10^{15}$ cm^{-3}, $p_0 = 1.5 \times 10^{10}$ cm^{-3}

(*d*) $n_0 = 2.25 \times 10^{15}$ cm^{-3}, $p_0 = 1 \times 10^5$ cm^{-3}

9. An increase in the base recombination of a BJT will increase

(*a*) the common emitter dc current gain β

(*b*) the breakdown voltage BV_{CEO}

(*c*) the unity-gain cut-off frequency f_T

(*d*) the transconductance g_m

10. In CMOS technology, shallow P-well or N-well regions can be formed using

(*a*) low pressure chemical vapour deposition

(*b*) low energy sputtering

(*c*) low temperature dry oxidation

(*d*) low energy ion-implantation

11. The feedback topology in the amplifier circuit (the base bias circuit is not shown for simplicity) in the figure is

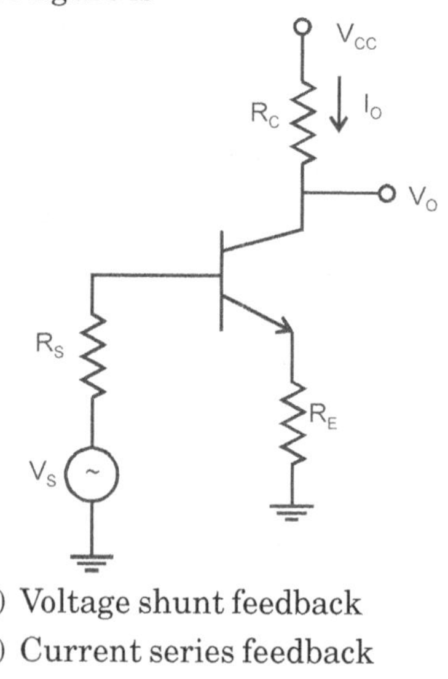

(*a*) Voltage shunt feedback

(*b*) Current series feedback

(*c*) Current shunt feedback

(*d*) Voltage series feedback

12. In the differential amplifier shown in the figure, the magnitudes of the common-mode and differential-mode gains are A_{cm} and A_d, respectively. If the resistance R_E is increased, then

(*a*) A_{cm} increases

(*b*) common-mode rejection ratio increases

(*c*) A_d increases

(*d*) common-mode rejection ratio decreases

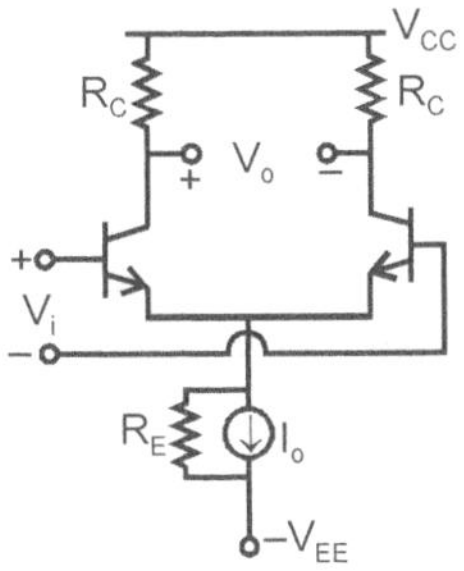

13. A cascade connection of two voltage amplifiers A1 and A2 is shown in the figure. The open-loop gain A_{v0}, input resistance R_{in}, and output resistance R_0 for A1 and A2 are as follows:

A1 : $A_{v0} = 10$, $R_{in} = 10\text{k}\Omega$, $R_0 = 1\text{k}\Omega$

A2 : $A_{v0} = 5$, $R_{in} = 5\text{k}\Omega$, $R_0 = 200\Omega$

The approximate overall voltage gain V_{out}/V_{in} is ________.

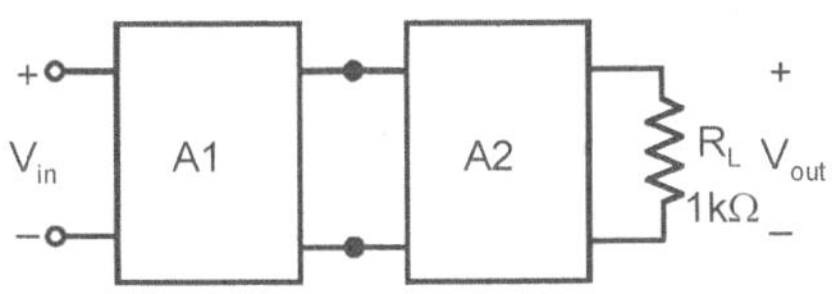

14. For an *n*-variable Boolean function, the maximum number of prime implicants is

(*a*) 2(*n*-1) (*b*) *n*/2

(*c*) 2^n (*d*) $2^{(n-1)}$

15. The number of bytes required to represent the decimal number 1856357 in packed BCD (Binary Coded Decimal) form is ________.

16. In a half-subtractor circuit with X and Y as inputs, the Borrow (M) and Difference (N = X – Y) are given by

(*a*) $M = X \oplus Y$, $N = XY$

(*b*) $M = XY$, $N = X \oplus Y$

(*c*) $M = \overline{X}Y$, $N = X \oplus Y$

(*d*) $M = X\overline{Y}$, $N = \overline{X \oplus Y}$

17. An FIR system is described by the system function $H(z) = 1 + \frac{7}{2}z^{-1} + \frac{3}{2}z^{-2}$. The system is

(*a*) maximum phase

(*b*) minimum phase

(*c*) mixed phase

(*d*) zero phase

18. Let $x[n] = x[-n]$. Let X(*z*) be the *z*-transform of *x*[*n*]. If $0.5 + j0.25$ is a zero of X(*z*), which one of the following must also be a zero of X(*z*).

(*a*) $0.5 - j0.25$ (*b*) $1/(0.5 + j0.25)$

(*c*) $1/(0.5 - j0.25)$ (*d*) $2 + j4$

19. Consider the periodic square wave in the figure shown.

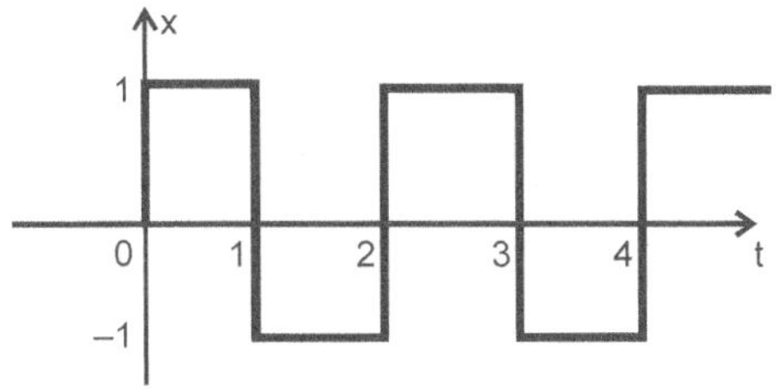

The ratio of the power in the 7th harmonic to the power in the 5th harmonic for this waveform is closest in value to ________.

20. The natural frequency of an undamped second-order system is 40 rad/s. If the system is damped with a damping ratio 0.3, the damped natural frequency in rad/s is ________.

21. For the following sytem,

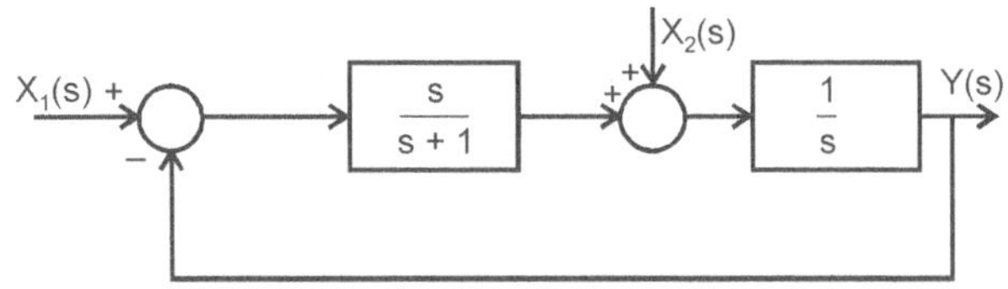

When $X_1(s) = 0$, the transfer function $\frac{Y(s)}{X_2(s)}$ is

(*a*) $\frac{s+1}{s^2}$ (*b*) $\frac{1}{s+1}$

(*c*) $\frac{s+2}{s(s+1)}$ (*d*) $\frac{s+1}{s(s+2)}$

22. The capacity of a band-limited additive white Gaussian noise (AWGN) channel is given by $C = W\log_2\left(1+\frac{P}{\sigma^2 W}\right)$ bits per second (bps), where W is the channel bandwidth, P is the average power received and σ^2 is the one-sided power spectral density of the AWGN. For a fixed $\frac{P}{\sigma^2} = 1000$, the channel capacity (in kbps) with infinite bandwidth ($W \to \infty$) is approximately

(*a*) 1.44 (*b*) 1.08

(*c*) 0.72 (*d*) 0.36

23. Consider sinusoidal modulation in an AM system. Assuming no overmodulation, the modulation index (μ) when the maximum and minimum values of the envelope, respectively, are 3 V and 1 V, is ________.

24. To maximize power transfer, a lossless transmission line is to be matched to a resistive load impedance via a $\lambda/4$ transformer as shown.

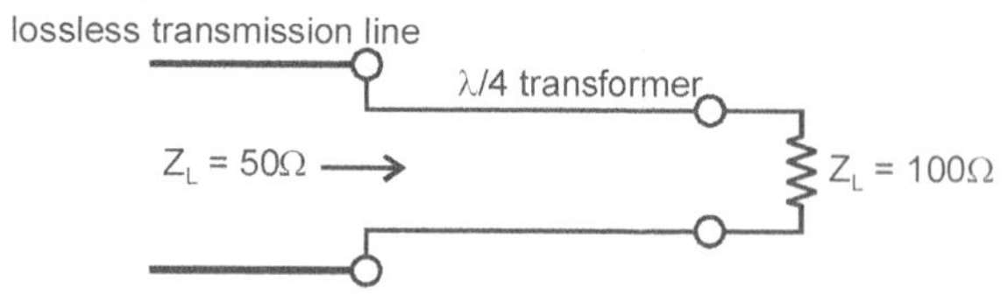

The characteristic impedance (in Ω) of the $\lambda/4$ transformer is ________.

25. Which one of the following field patterns represents a TEM wave travelling in the positive x direction?

(*a*) $E = +8\hat{y}$, $H = -4\hat{z}$ (*b*) $E = -2\hat{y}$, $H = -3\hat{z}$

(*c*) $E = +2\hat{z}$, $H = +2\hat{y}$ (*d*) $E = -3\hat{y}$, $H = +4\hat{z}$

Q. No. 26 – 55 Carry Two Marks Each

26. The system of linear equations

$$\begin{pmatrix} 2 & 1 & 3 \\ 3 & 0 & 1 \\ 1 & 2 & 5 \end{pmatrix}\begin{pmatrix} a \\ b \\ c \end{pmatrix} = \begin{pmatrix} 5 \\ -4 \\ 14 \end{pmatrix} \text{ has}$$

(*a*) a unique solution

(*b*) infinitely many solutions

(*c*) no solution

(*d*) exactly two solutions

27. The real part of an analytic function $f(z)$ where $z = x + jy$ is given by $e^{-y}\cos(x)$. The imaginary part of $f(z)$ is

(*a*) $e^{y}\cos(x)$ (*b*) $e^{-y}\sin(x)$

(*c*) $-e^{y}\sin(x)$ (*d*) $-e^{-y}\sin(x)$

28. The maximum value of the determinant among all 2 × 2 real symmetric matrices with trace 14 is ________.

29. If $\vec{r} = x\hat{a}_x + y\hat{a}_y + z\hat{a}_z$ and $|\vec{r}| = r$,

then div $(r^2\nabla(\ln r))$ = ________.

30. A series LCR circuit is operated at a frequency different from its resonant frequency. The operating frequency is such that the current leads the supply voltage. The magnitude of current is half the value at resonance. If the values of L, C and R are 1 H, 1 F and 1Ω, respectively, the operating angular frequency (in rad/s) is ________.

31. In the h-parameter model of the 2-port network given in the figure shown, the value of h_{22} (in S) is ________.

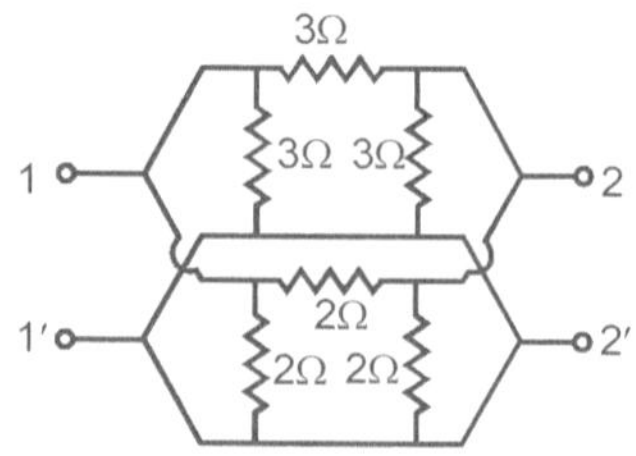

32. In the figure shown, the capacitor is initially uncharged. Which one of the following expressions describes the current I(t) (in mA) for $t > 0$?

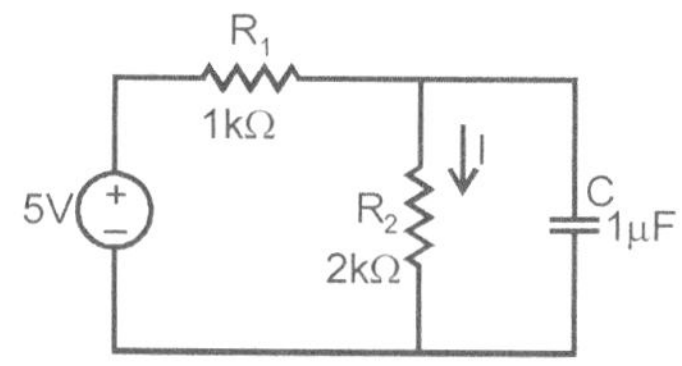

(*a*) $I(t) = \frac{5}{3}(1 - e^{-t/\tau}), \tau = \frac{2}{3}$ msec

(*b*) $I(t) = \frac{5}{2}(1 - e^{-t/\tau}), \tau = \frac{2}{3}$ msec

(*c*) $I(t) = \frac{5}{3}(1 - e^{-t/\tau}), \tau = 3$ msec

(*d*) $I(t) = \frac{5}{2}(1 - e^{-t/\tau}), \tau = 3$ msec

33. In the magnetically coupled circuit shown in the figure, 56 % of the total flux emanating from one coil links the other coil. The value of the mutual inductance (in H) is ________.

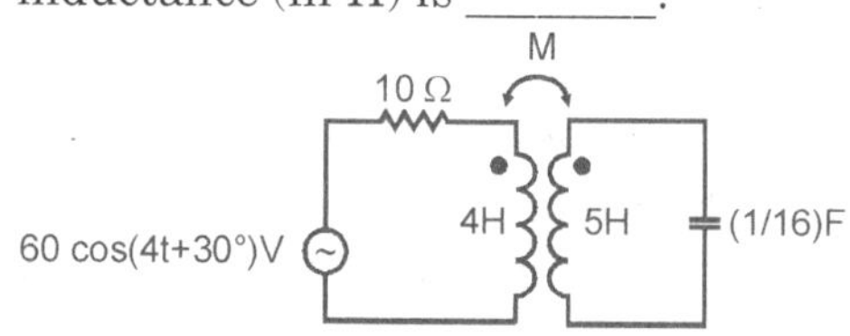

34. Assume electronic charge $q = 1.6 \times 10^{-19}$ C, kT/q = 25 mV and electron mobility μ_n = 1000 cm²/V-s. If the concentration gradient of electrons injected into a P-type silicon sample is 1×10^{21}/cm⁴, the magnitude of electron diffusion current density (in A/cm²) is ________.

35. Consider an abrupt PN junction (at T = 300 K) shown in the figure. The depletion region width X_n on the N-side of the junction is 0.2 m and the permittivity of silicon (ε_{si}) is 1.044×10^{-12} F/cm. At the junction, the approximate value of the peak electric field (in kV/cm)is ________.

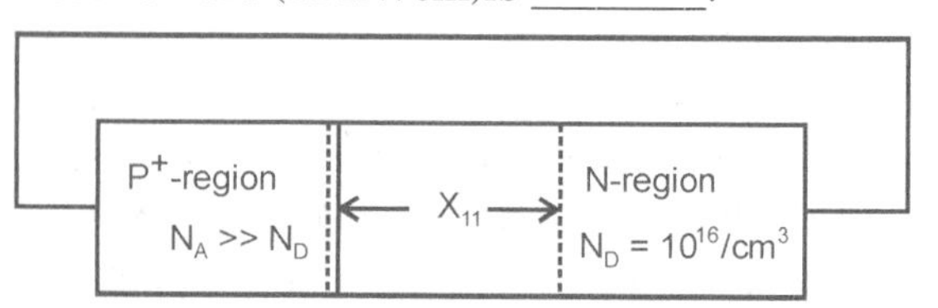

36. When a silicon diode having a doping concentration of $N_A = 9 \times 10^{16}$ cm^{-3} on p-side and $N_D = 1 \times 10^{16}$ cm^{-3} on n-side is reverse biased, the total depletion width is found to be 3 m. Given that the permittivity of silicon is 1.04×10^{-12} F/cm, the depletion width on the p-side and the maximum electric field in the depletion region, respectively, are

(*a*) 2.7 m and 2.3×10^5 V/cm

(*b*) 0.3 m and 4.15×10^5 V/cm

(*c*) 0.3 m and 0.42×10^5 V/cm

(*d*) 2.1 m and 0.42×10^5 V/cm

37. The diode in the circuit shown has $V_{on} = 0.7$ Volts but is ideal otherwise. If $V_i = 5\sin(\omega t)$ Volts, the minimum and maximum values of V_O (in Volts) are, respectively,

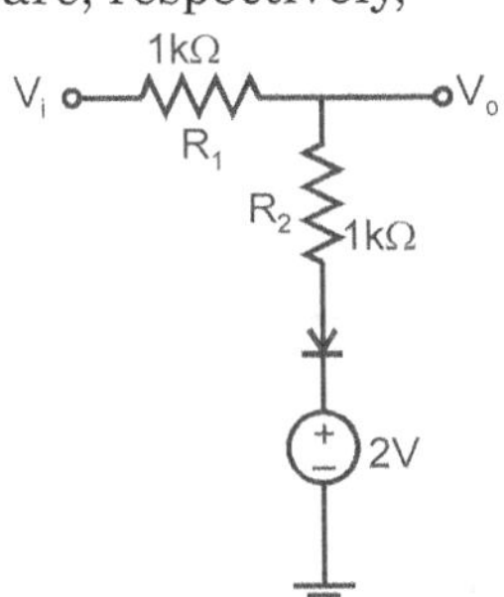

(*a*) – 5 and 2.7 (*b*) 2.7 and 5

(*c*) – 5 and 3.85 (*d*) 1.3 and 5

38. For the n-channel MOS transistor shown in the figure, the threshold voltage V_{Th} is 0.8 V. Neglect channel length modulation effects. When the drain voltage $V_D = 1.6$ V, the drain current I_D was found to be 0.5 mA. If V_D is adjusted to be 2 V by changing the values of R and V_{DD}, the new value of I_D (in mA) is

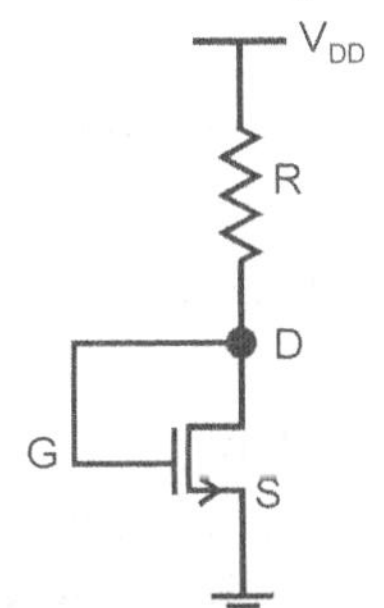

(*a*) 0.625 (*b*) 0.75

(*c*) 1.125 (*d*) 1.5

39. For the MOSFETs shown in the figure, the threshold voltage $|V_t| = 2$V and $K = \frac{1}{2} C_{ox}\left(\frac{W}{L}\right) = 0.1$ mA/V^2. The value of I_D (in mA) is ________.

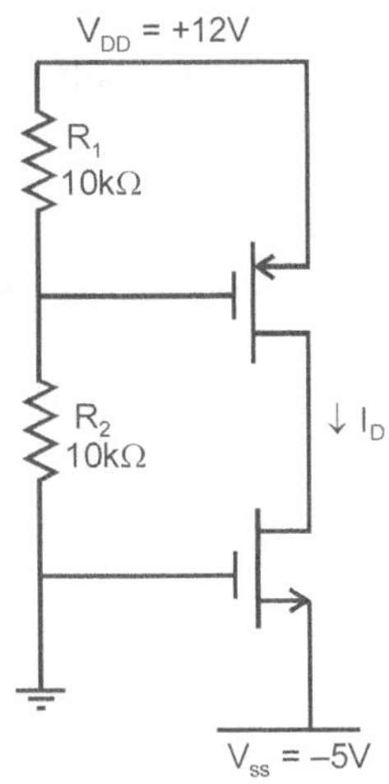

40. In the circuit shown, choose the correct timing diagram of the output (y) from the given waveforms W1, W2, W3 and W4.

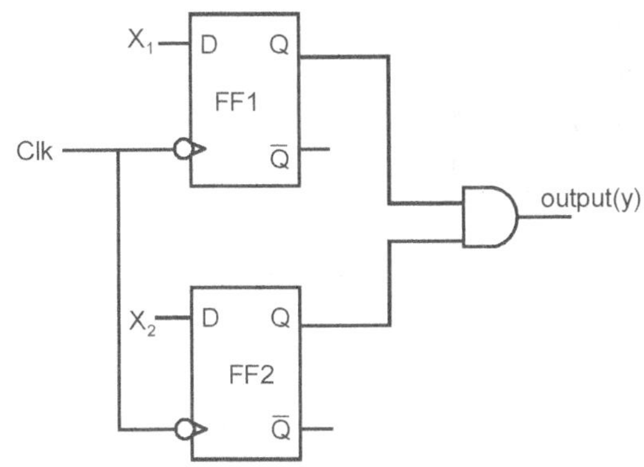

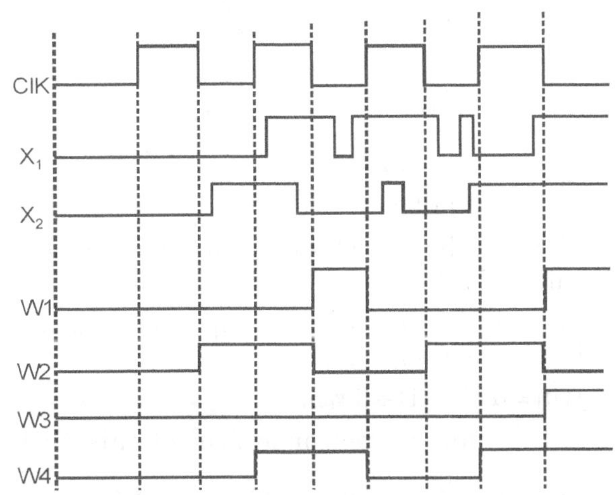

(*a*) W1 (*b*) W2

(*c*) W3 (*d*) W4

41. The outputs of the two flip-flops Q1, Q2 in the figure shown are initialized to 0,0. The sequence generated at Q1 upon application of clock signal is

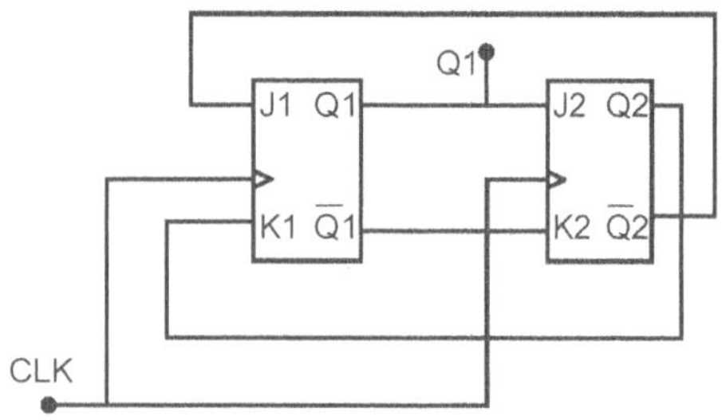

(*a*) 01110... (*b*) 01010...

(*c*) 00110... (*d*) 01100...

42. For the 8085 microprocessor, the interfacing circuit to input 8-bit digital data ($DI_0 - DI_7$) from an external device is shown in the figure. The instruction for correct data transfer is

(*a*) MVI A, F8H

(*b*) IN F8H

(*c*) OUT F8H

(*d*) LDA F8F8H

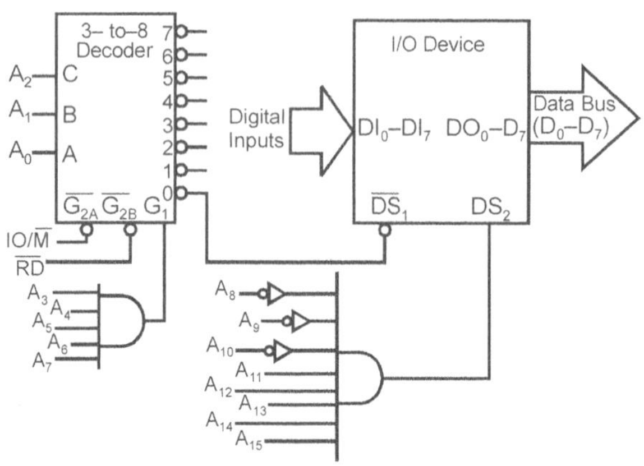

43. Consider a discrete-time signal

$$x[n] = \begin{cases} n \text{ for } 0 \le n \le 10 \\ 0 \text{ otherwise} \end{cases}$$

If $y[n]$ is the convolution of $x[n]$ with itself, the value of $y[4]$ is ________.

44. The input-output relationship of a causal stable LTI system is given as

$y[n] = \alpha\, y[n-1] + \beta\, x[n]$

If the impulse response $h[n]$ of this system satisfies the condition $\sum_{n=0}^{\infty} h[n] = 2$, the relationship between α and β is

(*a*) $\alpha = 1-\beta/2$ (*b*) $\alpha = 1+\beta/2$

(*c*) $\alpha = 2\beta$ (*d*) $\alpha = -2\beta$

45. The value of the integral $\int_{-\infty}^{\infty} \text{sinc}^2(5t)\, dt$ is ________.

46. An unforced liner time invariant (LTI) system is represented by

$$\begin{bmatrix} \dot{x}_1 \\ \dot{x}_2 \end{bmatrix} = \begin{bmatrix} -1 & 0 \\ 0 & -2 \end{bmatrix} \begin{bmatrix} x_1 \\ x_2 \end{bmatrix}$$

If the initial conditions are $x_1(0) = 1$ and $x_2(0) = -1$, the solution of the state equation is

(*a*) $x_1(t) = -1,\ x_2(t) = 2$

(*b*) $x_1(t) = -e^{-t},\ x_2(t) = 2e^{-t}$

(*c*) $x_1(t) = e^{-t},\ x_2(t) = -e^{-2t}$

(*d*) $x_1(t) = -e^{-t},\ x_2(t) = -2e^{-t}$

47. The Bode asymptotic magnitude plot of a minimum phase system is shown in the figure.

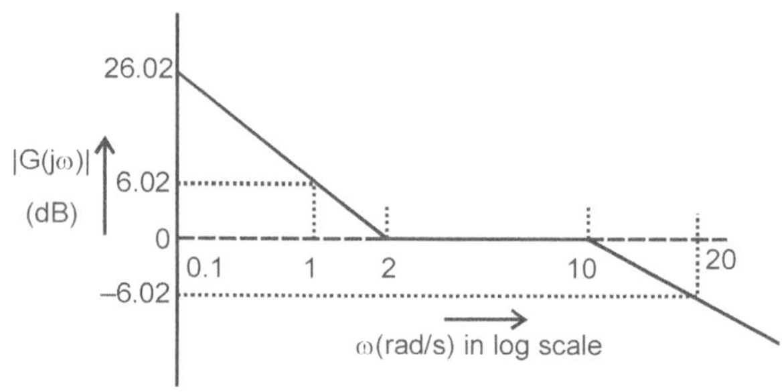

If the system is connected in a unity negative feedback configuration, the steady state error of the closed loop system, to a unit ramp input, is ________.

48. Consider the state space system expressed by the signal flow diagram shown in the figure.

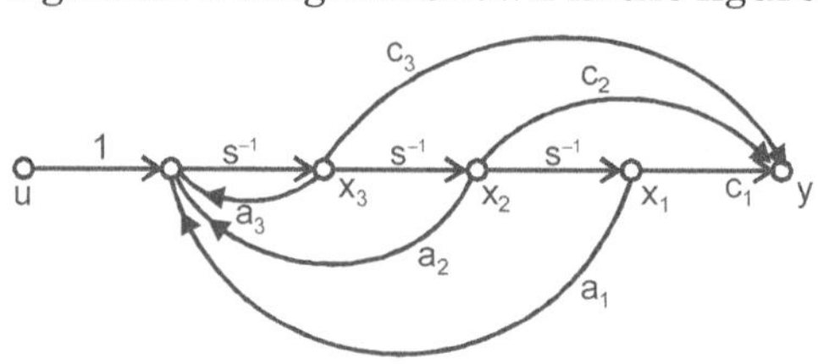

The corresponding system is

(*a*) always controllable

(*b*) always observable

(*c*) always stable

(*d*) always unstable

49. The input to a 1-bit quantizer is a random variable X with pdf $f_X(x) = 2e^{-2x}$ for $x \ge 0$ and $f_X(x) = 0$ for $x < 0$. For outputs to be of equal probability, the quantizer threshold should be ________.

50. Coherent orthogonal binary FSK modulation is used to transmit two equiprobable symbol waveforms $s_1(t) = \alpha \cos 2\pi f_1 t$ and $s_2(t) = \alpha \cos 2\pi f_2 t$, where α = 4mV. Assume an AWGN channel with two-sided noise power spectral density

$\frac{N_0}{2} = 0.5 \times 10^{-12}$ W/Hz. Using an optimal receiver and the relation $Q(v) = \frac{1}{\sqrt{2\pi}} \int_v^{\infty} e^{-u^2/2}\, du$, the bit error probability for a data rate of 500 kbps is

(*a*) $Q(2)$ (*b*) $Q(2\sqrt{2})$

(*c*) $Q(4)$ (*d*) $Q(4\sqrt{2})$

51. The power spectral density of a real stationary random process $X(t)$ is given by

$$S_X(f) = \begin{cases} \frac{1}{W}, & |f| \le W \\ 0, & |f| > W \end{cases}$$

The value of the expectation $E\left[\pi X(t)\, X\left(t - \frac{1}{4W}\right)\right]$ is ________.

52. In the figure, $M(f)$ is the Fourier transform of the message signal $m(t)$ where A = 100 Hz and B = 40Hz. Given $v(t) = \cos(2\pi f_c t)$ and $w(t) = \cos(2\pi(f_c + A)t)$, where $f_c > A$. The cutoff frequencies of both the filters are f_c.

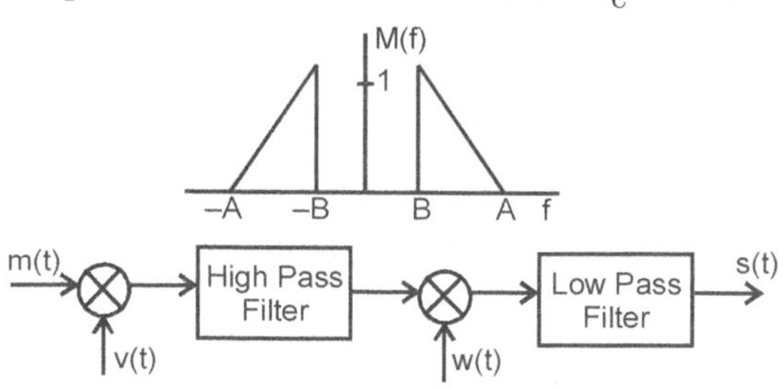

The bandwidth of the signal at the output of the modulator (in Hz) is ________.

53. If the electric field of a plane wave is
$\vec{E}(z, t) = \hat{x}3\cos(\omega t - kz + 30°) - \hat{y}4\sin(\omega t - kz + 45°)$ (mV/m), the polarization state of the plane wave is

(*a*) left elliptical

(*b*) left circular

(*c*) right elliptical

(*d*) right circular

54. In the transmission line shown, the impedance Z_{in} (in ohms) between node A and the ground is ________.

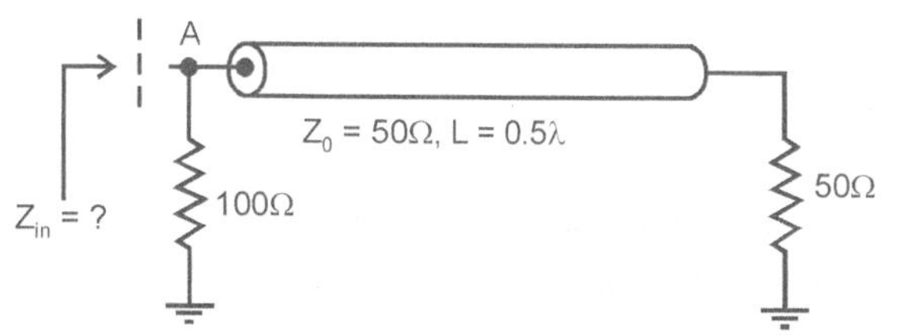

55. For a rectangular waveguide of internal dimensions $a \times b (a > b)$, the cut-off frequency for the TE_{11} mode is the arithmetic mean of the cut-off frequencies for TE_{10} mode and TE_{20} mode. If $a = \sqrt{5}$ cm, the value of b (in cm) is ________.

ANSWERS

GENERAL APTITUDE

1. (*b*) 2. (*b*) 3. (*c*) 4. (*b*) 5. (45) 6. (*c*) 7. (*d*) 8. (163) 9. (*d*)
10. (*a*)

ELECTRONICS AND COMMUNICATION

1. (200) 2. (50) 3. (*a*) 4. (*c*) 5. (*a*) 6. (*d*) 7. (1.2mA) 8. (*d*) 9. (*b*)
10. (*d*) 11. (*b*) 12. (*b*) 13. (34.722) 14. (*d*) 15. (4) 16. (*c*) 17. (*c*) 18. (*b*)
19. (0.5) 20. (38.15rad/sec) 21. (*d*) 22. (*a*) 23. (0.5) 24. (70.7Ω) 25. (*b*) 26. (*b*)
27. (*b*) 28. (49) 29. (3) 30. (0.45 r/sec) 31. (1.24) 32. (*a*) 33. (2.49 Henry)
34. (4000) 35. (30.66) 36. (*b*) 37. (*c*) 38. (*c*) 39. (0.9) 40. (*c*) 41. (*d*) 42. (*d*)
43. (10) 44. (*a*) 45. (0.2) 46. (*c*) 47. (0.50) 48. (*a*) 49. (0.35) 50. (*c*) 51. (4)
52. (60) 53. (*a*) 54. (33.33Ω) 55. (2)

EXPLANATIONS

4. For a very large number of throws, the frequency should be same for unbiased throw. As it not same, then the die is baised.

5. 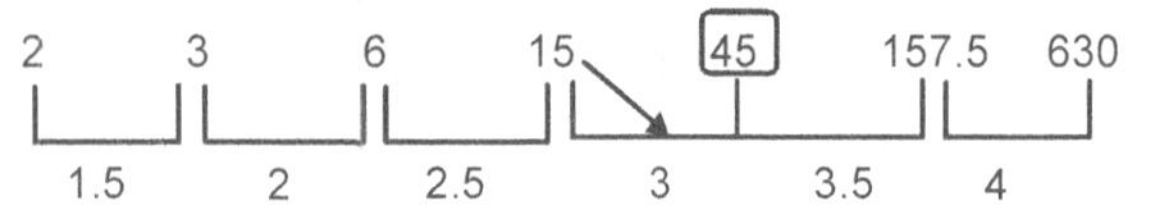

$\frac{\text{2nd number}}{\text{1st number}}$ is in increasing order as shown above

6. 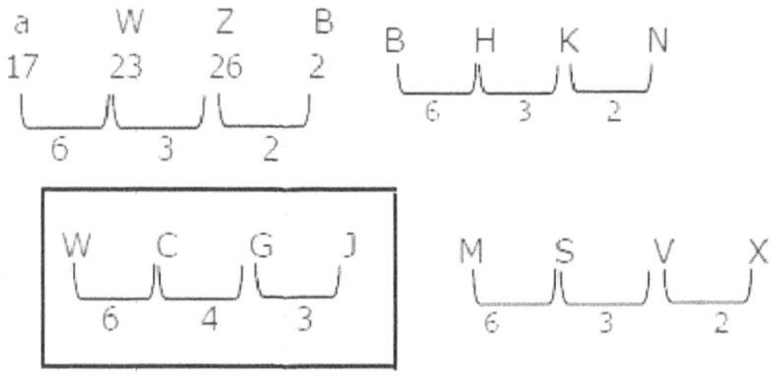

8. Eight consecutive odd number = 656

a-6, a-1, a, $a+2$, $a+4$, $a+6$

$$a + 8 = 656$$

$$a = 81$$

Smallest $m = 75$ ...(1)

Average consecutive even numbers

$$\Rightarrow \frac{a-2+a+a+2+a+4}{4} = 87$$

$$\Rightarrow a = 86$$

Second largest number = 88

$$1 + 2 = 163$$

9. **Item : 2**

$$\frac{\frac{20}{100}\times 250\times 10^7}{\frac{20}{100}\times 500\times 10^3}$$

$0.5 \times 10^4 = 5 \times 10^3$ $\boxed{1}$

= Item 2

Item : 3

$$\frac{23\times 250\times 10^7}{19\times 500\times 10^3}$$

1.2 = Item 3

Item : 6

$\frac{19}{16}$ = 1.18 = Item 6

Item : 5

$\frac{20}{12} = \frac{5}{3} = 1.6$

$\Rightarrow$ $\boxed{1.6 = \text{Item 5}}$

10. 

V_{half} = 30(s) drawing rate = s

$$\text{Total volume} = 60\text{ S tank}$$

$$(s^1)(10) - (s)10 = 30s$$

$$s^1(s) - s = 3s$$

$$s1 = 4s$$

s^1 = 4 drawing rate

1. $|AB| = |A|\cdot|B| = (5)\cdot(40) = 200$

2. $X = 1, 3, 5, \ldots\ldots, 99$

$\Rightarrow$ $n = 50$ (number of observations)

$$\therefore \quad E(x) = \frac{1}{n}\sum_{i=1}^{n} X_i$$

$$= \frac{1}{50}[1+3+5+\ldots+99]$$

$$= \frac{1}{50}(50)^2 = 50$$

3. $f'(t) = -e^{-t} + 4e^{-2t} = 0$

$$\Rightarrow e^{-t}\left[4e^{-t} - 1\right] \Rightarrow e^{-t} = \frac{1}{4}$$

$$\Rightarrow t = \log_e^4$$

and $f''(t) < 0$

at $t = \log_e^4$

4. $\lim_{x\to\infty}\left(1+\frac{1}{x}\right)^x = e$ (standard limit)

5. For equal roots, Discriminant

$$B^2 - 4AC = 0$$

$$\Rightarrow 4\alpha^2 - 4 = 0$$

$$\Rightarrow \alpha = \pm 1$$

6. Norton's theorm

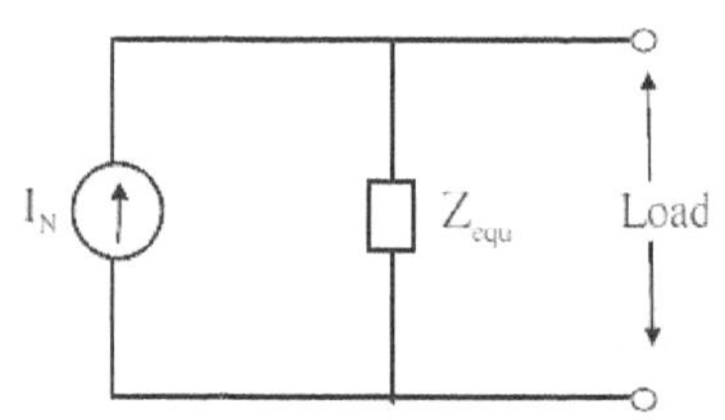

7. 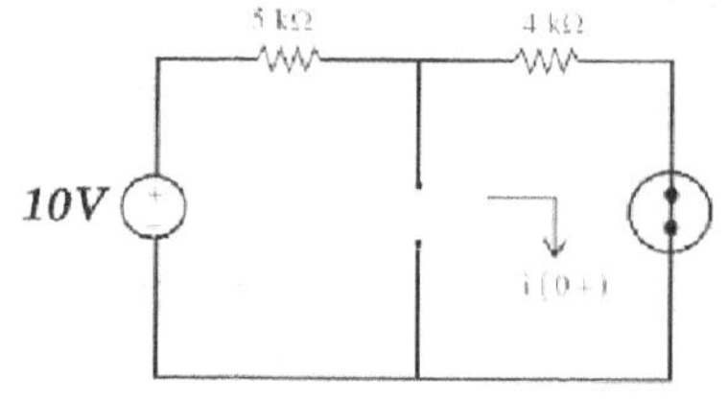

For $t = o+$

$$i(o+) = \frac{10}{9K}$$

$$\Rightarrow 1.11 \text{ mA}$$

$$i(o+) \simeq 1.2 \text{ mA}$$

8.
$$N_D = 2.25 \times 10^{15} \text{ Atom/cm}^3$$
$$h_i = 1.5 \times 10^{10} \text{ /cm}^3$$

Since complete ionization taken place,

$$h_0 = N_D = 2.25 \times 10^{15} \text{ /cm}^3$$

$$P_0 = \frac{n_i^2}{n_0} = \frac{(1.5\times10^{10})^2}{2.25\times10^{15}}$$
$$= 1 \times 10^5 \text{ /cm}^3$$

12.

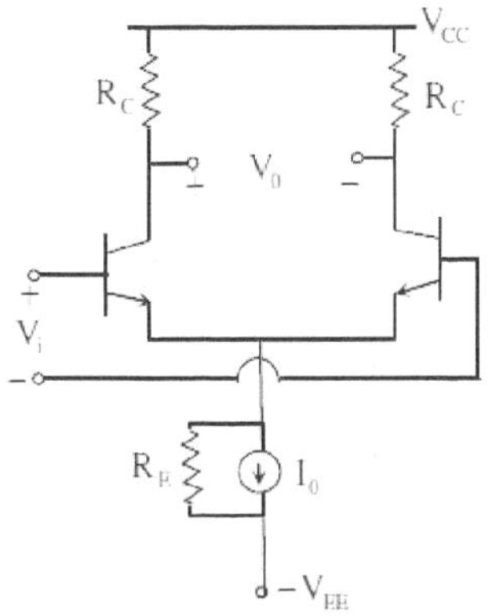

A_d does not depend on R_E

A_{cm} decreases as R_E is increased

$$\therefore \quad \text{CMRR} = \frac{A_d}{A_{cm}} = \text{Increases}$$

13. Overall voltage gain,

$$A_V = \frac{V_0}{V_i}$$
$$= A_{V_1} A_{V_2}\left[\frac{Z_{i_2}}{Z_{i_2}+Z_{0_1}}\right]\left[\frac{R_L}{R_L+Z_{0_2}}\right]$$
$$= 10\times5\left[\frac{5k}{5k+1k}\right]\left[\frac{1k}{1k+200}\right]$$
$$A_V = 34.722$$

14. For an n-variable Boolean function, the maximum number of prime implicants = $2^{(n-1)}$

15. In packed BCD (Binary Coded Decimal) typically encoded two decimal digits within a single byte by taking advantage of the fact that four bits are enough to represent the range 0 to 9.

So, 1856357 is required 4-byte to stored these BCD digits.

16. Function Table for Half-subtractor is

X	Y	Difference (N)	Borrow (M)
0	0	0	0
Hence, N = X ⊕ Y and m = XY			
0	1	1	1
1	0	1	0
1	1	0	0

Hence, $N = X \oplus Y$ and $m = \bar{X}Y$

17. Minimum phase system has all zeros inside unit circle maximum phase system has all zeros outside unit circle mixed phase system has some zero outside unit circle and some zeros inside unit circle.

for
$$H(s) = 1+\frac{7}{2}Z^{-1}+\frac{3}{2}Z^{-2}$$

One zero is inside and one zero outside unit circle hence mixed phase system

18. Given $\quad x[n] = x[-n]$

$\Rightarrow \quad x(z) = x(z^{-1})$

[Time reversal property in z – transform]

$\Rightarrow$ if one zero is $0.5 + j0.25$

then other zero will be $\dfrac{1}{0.5+j0.25}$

19. For a periodic sequence wave, nth harmonic component is $\alpha\dfrac{1}{n}$

$\Rightarrow$ power in n[th] harmonic component is $\alpha\dfrac{1}{n^2}$

$\Rightarrow$ Ratio of the power in 7th harmonic to power in 5th harmonic for given waveform is

$$\frac{\frac{1}{7^2}}{\frac{1}{5^2}} = \frac{25}{49} \approx 0.5$$

20. Given
$$\omega_n = 40 \text{ r/sec}$$
$$\xi = 0.3$$
$$\omega_d = \omega_n\sqrt{1-\xi^2}$$
$$\omega_d = 40\sqrt{1-(0.3)^2}$$
$$\boxed{\omega_d = 38.15 \text{ r/sec}}$$

21.

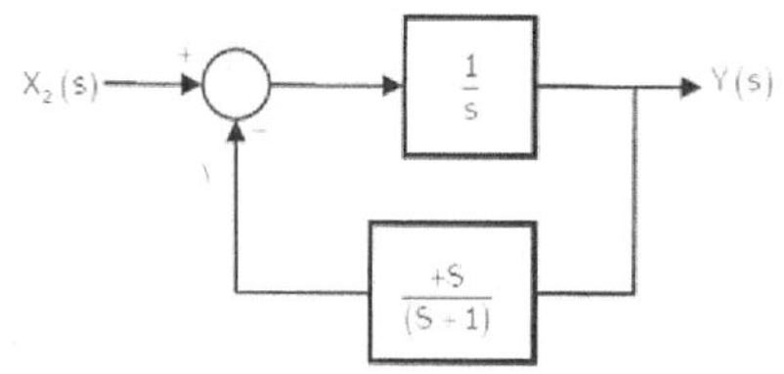

If $\quad X_1(s) = 0$

$\dfrac{Y(s)}{X_2(s)}$; The block diagram becomes

$$\frac{Y(s)}{X_2(s)} = \frac{\frac{1}{s}}{1+\frac{1}{s}\cdot\frac{s}{(s+1)}}$$
$$= \frac{\frac{1}{s}}{\frac{(s+2)}{s+1}} \Rightarrow \frac{(s+1)}{s(s+2)}$$

22.

$$C = \lim_{w\to\infty} \omega \log_2\left[1+\frac{P}{\sigma^2\omega}\right]$$

$$= \lim_{w\to\infty} \frac{\omega \ln\left[1+\frac{P}{\sigma^2\omega}\right]}{\ln 2}$$

$$= \frac{1}{\ln 2}\lim_{w\to\infty} \frac{\ln\left[1+\frac{P}{\sigma^2\omega}\right]}{\frac{P}{\sigma^2\omega}}\cdot\frac{P}{\sigma^2}$$

$$= \frac{P}{\sigma^2 \ln_2}\lim_{w\to\infty} \frac{\ln\left[1+\frac{P}{\sigma^2\omega}\right]}{\frac{P}{\sigma^2\omega}}$$

This limit is equivalent to

$$\lim_{w\to\infty} \frac{\ln[1+x]}{x} = 1$$

$$= \frac{P}{\sigma^2 \cdot \ln_2}$$

$$= \ln_2 e \frac{P}{\sigma^2}$$

$= 1.44$ KGpa

23.

$$\mu = \frac{A(t)_{max} - A(t)min}{A(t)_{max} + A(t)min}$$

$$\mu = \frac{3-1}{3+1}$$

$$= \frac{1}{2} = 0.5$$

24. Here impedance is matched by using QWT $\left(\frac{\lambda}{4}\right)$

$$\therefore \quad Z_0' = \sqrt{Z_L Z_{in}}$$

$$= \sqrt{100\times 50}$$

$$= 50\sqrt{2}$$

$$= Z_0' = 70.7\ \Omega$$

25. For TEM wave

Electric field (E), Magnetic field (H) and

Direction of propagation (P) are orthogonal to each other.

Here $P = +a_x$

By verification

$$E = -2a_y,$$

$$H = -3a_z$$

$$E \times H = -a_y \times -a_z$$

$$= +a_x \rightarrow P$$

26.

$$\left[\frac{A}{B}\right] = \left[\begin{array}{ccc|c} 2 & 1 & 3 & 5 \\ 3 & 0 & 1 & -4 \\ 1 & 2 & 5 & 14 \end{array}\right]$$

$$\begin{array}{l} R_2 \rightarrow 2R_2 - 3R_1 \\ R_3 \rightarrow 2R_3 - R_1 \end{array} \left[\begin{array}{ccc|c} 2 & 1 & 3 & 5 \\ 0 & -3 & -7 & -23 \\ 0 & 3 & 7 & 23 \end{array}\right] \xrightarrow{R_3 + R_2}$$

$$\left[\begin{array}{ccc|c} 2 & 1 & 3 & 5 \\ 0 & -3 & -7 & -23 \\ 0 & 0 & 0 & 0 \end{array}\right]$$

Since, rank (A) = rank $\left(\frac{A}{B}\right)$ < number of unknowns

$\therefore$ Equations have infinitely many solutions.

27. real part $u = e^{-y}\cos x$

and $V = ?$

$$dv = \frac{\partial v}{\partial x}dx + \frac{\partial v}{\partial y}dy$$

$$= -\frac{\partial u}{\partial y}dx + \frac{\partial u}{\partial x}dy$$

(Using C – R equations)

$$= e^{-y}\cos x\, dx - e^{-y}\sin x\, dy$$

$$= d\left[e^{-y}\sin x\right]$$

Integrating, we get $V = e^{-y}\sin x$

28. General 2×2 real symmetric matrix is $\begin{bmatrix} y & x \\ x & z \end{bmatrix}$

$\Rightarrow$ $\det = yz - x^2$

and trace is $y + z = 14$ (given)

$\Rightarrow$ $z = 14 - y$ ...(*)

Let $f = yz - x^2(\det)$

$= -x^2 - y^2 + 14y(u\sin g^*)$

Using maxima and minima of a function of two variables, we have f is maximum at $x = 0$, $y = 7$ and therefore, maximum value of the determinant is 49

29.

$$\nabla(\ln r) = \frac{\vec{r}}{r^2}$$

$$\Rightarrow \operatorname{div}\left(r^2\nabla(\ln r)\right) = \operatorname{div}(\vec{r}) = 3$$

$$\left[\nabla(\ln r) = \sum \hat{a}_x \frac{\partial}{\partial x}(\ln r) = \sum \hat{a}_x\left(\frac{1}{r}\right)\left(\frac{x}{r}\right)\right.$$

$$\left. = \frac{1}{r^2}\sum \hat{a}_x x = \frac{\vec{r}}{r^2}\right]$$

30. The operating frequency (w_x), at which current leads the supply.

i.e., $\omega_x < \omega_r$

again magnitude of current is half the value at resonance

i.e., at $\omega = \omega_x$

$$\Rightarrow \quad I_x = \frac{V}{|Z|}$$

at $\omega = \omega_x$

$$\Rightarrow \quad I_{resonance} = \frac{V}{R}$$

$$I_x = \frac{I_{resonance}}{2}$$

i.e.,
$$\frac{V}{|Z|} = \frac{V}{2R} = |Z| = 2R$$

Given, $R = 1\Omega$;
$L = 1H$;
$C = 1F$

$$|Z| = \sqrt{R^2 + \left(\frac{1}{\omega_c} - \omega L\right)^2} = 2$$

$$= R^2 + \left(\frac{1}{\omega_c} - \omega L\right)^2 = 4$$

By substituting R, L & C values,

$$\Rightarrow \quad 1 + \left(\frac{1}{\omega} - \omega\right)^2 = 4$$

$$\Rightarrow \quad \omega^2 + \frac{1}{\omega^2} = 5$$

Assume $\omega^2 = x$,

then, $x + \frac{1}{x} = 5$

$$\Rightarrow \quad x^2 - 5x + 1 = 0$$

$$x_{1,2} = 4.791, 0.208$$

if $x = 4.791$

$\Rightarrow \quad \omega = 2.18$ r/sec

if $x = 0.208$

$\Rightarrow \quad \omega = 0.45$ r/sec

But $\omega_x < \omega_r$

So, operating frequency

$\omega_x = 0.45$ r/sec

31. If two, $\frac{\pi - n}{ws}$ are connected in parallel,

The y-parameter are added

i.e., $y_{equ} = y_1 + y_1$

$$y_1 = \begin{bmatrix} \frac{2}{3} & -\frac{1}{3} \\ -\frac{1}{3} & \frac{2}{3} \end{bmatrix}$$

$$y_2 = \begin{bmatrix} 1 & -\frac{1}{2} \\ -\frac{1}{2} & 1 \end{bmatrix}$$

$$y_{equ} = \begin{bmatrix} \frac{5}{3} & -\frac{5}{6} \\ -\frac{5}{6} & \frac{5}{3} \end{bmatrix}$$

$$h = \begin{bmatrix} \frac{1}{y_{11}} & -\frac{y_{12}}{y_{11}} \\ \frac{y_{21}}{y_{11}} & \frac{\Delta y}{y_{11}} \end{bmatrix}$$

where $\Delta y = y_{11}\,y_{22} - y_{12} - y_{21}$

The value of $h_{22} = \Delta y$

$$= \left[\left(\frac{5}{3}\right)\left(\frac{5}{3}\right)\right] - \left[\left(\frac{-5}{6}\right)\left(\frac{-5}{6}\right)\right]$$

$$\Delta y = 2.0833$$

$$y_{11} = \frac{5}{3}$$

$$\therefore \quad h_{22} = 1.24$$

32. $v_c(t) = V_{R2}(t)$

$$= V_{final} + [V_{final} - V_{final}]\,e^{-t/\tau}$$

$$\boxed{\tau = R_{equ} \cdot C_{equ}} \Rightarrow \frac{2}{3} \times 10^3 \times 10^{-6}$$

$$R_{equ} = 2K \| 1K \Rightarrow \frac{2}{3} K\Omega$$

$$C_{equ} = 1\mu F$$

$$\boxed{\tau = \frac{2}{3}\text{ msec}}$$

$$V_{initial} = 0 \text{ volts}$$

$$V_{final} = V_{s.s}$$

$$= 5 \cdot \frac{2}{3} = \frac{10}{3}\text{volts}$$

$$v_{R2}(t) = \frac{10}{3} - \frac{10}{3} e^{-t/\tau}$$

$$v_{R2}(t) = \frac{10}{3}\left[1 - e^{-t/\tau}\right]\text{volts}$$

$$\Rightarrow \quad i_{R2}(t) = \frac{v_{R2}(t)}{2K}$$

$$= \frac{5}{3}\left[1 - e^{-t/\tau}\right]\text{mA}$$

33. Given 56% of the total flux emanating from one coil links to other coil.

i.e., $K = 56\% \Rightarrow 0.56$

We have, $K = \dfrac{M}{\sqrt{L_1L_2}}$

$L_1 = 4H;$

$L_2 = 5H$

$M = (0.56)\sqrt{20}$

$\Rightarrow \quad m = 2.50\text{ H}$

34. Given $q = 1.6 \times 10^{-19};$

$\dfrac{kJ}{q} = 2.5\text{ mV},$

$\mu_n = 1000\text{ cm}^2/\text{v} - \text{s}$

From Einstein relation,

$$\frac{D_n}{\mu_n} = \frac{kJ}{q}$$

$\Rightarrow \quad D_n = 25\text{ mV} \times 1000\text{ cm}^2/\text{v} - \text{S}$

$\Rightarrow \quad = 25\text{ cm}^2/\text{s}$

Diffusion current Density

$$J = qD_n\frac{dn}{dx}$$

$= 1.6 \times 10^{-19} \times 25 \times 1 \times 10^{21}$

$= 4000\text{ A/cm}^2$

35. Given $x_n = 0.2\ \mu m,$

$\in_{Si} = 1.044 \times 10^{-12}\text{ F}/\mu_n$

$N_D = 10^{16}/\text{cm}^3$

Peak Electric field,

$$E = \frac{qN_Dx_n}{\in}$$

$$= \frac{1.6\times10^{-19}\times10^{16}\times0.00002}{1.044\times10^{-12}}$$

$= 30.66\text{ KV/cm}$

36. Given $N_A = 9 \times 10^{16}/\text{cm}^3;$

$N_D = 1 \times 10^{16}/\text{cm}^3$

Total depletion width

$x = x_n + x_p = 3\ \mu\text{m}$

$\in = 1.04 \times 10^{-12}\text{ F/cm}$

$$\frac{x_n}{x_p} = \frac{N_A}{N_D}$$

$$= \frac{9\times10^{16}}{1\times10^{16}}$$

$x_n = 9x_p \quad \text{...(1)}$

Total Depletion width,

$x_n + x_p = 3\ \mu\text{m}$

$9x_p + x_p = 3\ \mu\text{m}$

$x_p = 0.3\ \mu\text{m}$

Maximum Electric field,

$$E = \frac{qN_Ax_p}{\in}$$

$$= \frac{1.6\times10^{-19}\times9\times10^{16}\times0.3\ \mu\text{m}}{1.04\times10^{-12}}$$

$= 4.15 \times 10^5\text{ V/cm}$

37. When Vi makes Diode 'D' OFF,

$V_0 = V_i$

$\therefore \quad V_0(\text{min}) = -5V$

When V_i makes diode 'D' ON,

$$V_0 = \frac{(V_i - 0.7 - 2)}{R_1 + R_2} + V_{on} + 2V$$

$$\therefore \quad V_0(\text{max}) = \frac{(5-0.7-2)1k}{1k+1k} + 0.7 + 2V$$

$= 3.85\text{ V}$

38. Given $V_{Th} = 0.8\text{ V}$

When $V_D = 1.6\text{ V},$

$I_D = 0.5\text{ mA}$

$$= \frac{1}{2}\mu_n \cos\frac{W}{L}(V_{DS} - V_{Th})^2$$

[∵ Device is in sat]

$$\Rightarrow \quad \frac{1}{2}\mu_n \cos\frac{W}{L} = 0.78125\times10^{-3}\text{ A/V}^2$$

When $V_D = 2V$

$$I_D = \frac{1}{2}\mu_n \cos\frac{\omega}{L}(V_{DS} - V_{Th})^2$$

$= 078125 \times 10^{-3}\ (2 - 0.8)1.125\text{ mA}$

39.

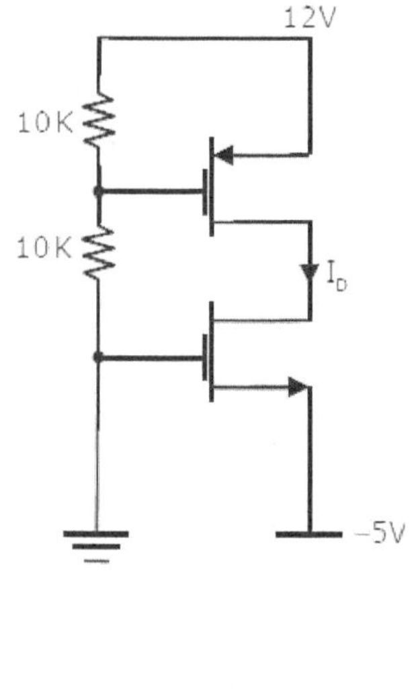

Given $|V_t| = 2V,$

$$K = \frac{1}{2}\mu\cos\frac{W}{L}$$
$$= 0.1\ A/V^2$$
$$I_{D_1} = I_{D_2}$$
$$= \frac{1}{2}\mu_n\cos\frac{W}{L}\left(V_{GS_1} - V_t\right)^2$$
$$= 0.1\ mA/V^2(5-2)^2$$
$$= 0.9\ mA$$

40. This circuit has used negative edge triggered, so output of the D-flip flop will changed only when CLK signal is going from HIGH to LOW (1 to 0)

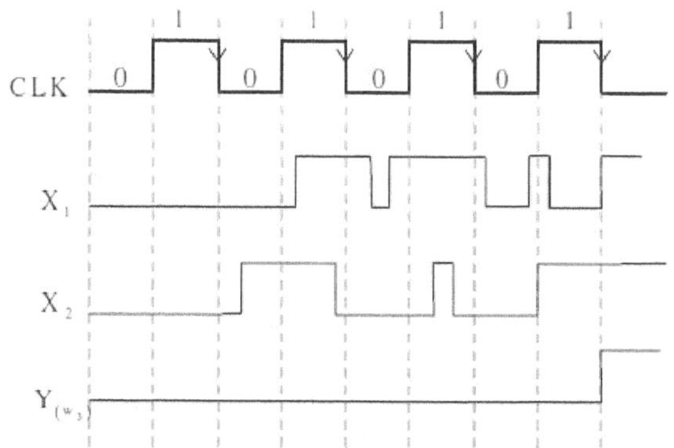

This is a synchronous circuit, so both the flip flops will trigger at the same time and will respond on falling edge of the Clock. So, the correct output (Y) waveform is associated to w_3 waveform.

41.

Clock	$J_1(\bar{Q}_2)$	$K_1(Q_2)$	$J_2(Q_1)$	$K_2(\bar{Q}_1)$	Q_1	Q_2
Initial →	–	–	–	–	0	0
1st CP →	1	0	0	1	1	0
2nd CP →	1	0	1	0	1	1
3rd CP →	0	1	1	0	0	1
4th CP →	0	1	0	1	0	0

So, the output sequence generated at Q_1 is 01100 .....

42. This circuit diagram indicating that it is memory mapped I/O because to enable the 3-to-8 decoder $\overline{G_{2A}}$ is required active low signal through $(I_o/\overline{m})$ and $\overline{G_{2B}}$ is required active low through $(\overline{R_D})$ it means I/o device read the status of device LDA instruction is appropriate with device address.

Again to enable the decoder o/p of AND gate must be 1 and Ds_2 signal required is 1 which is the o/p of multi-i/p AND gate to enable I/O device.

So,

A_{15}	A_{14}	A_{13}	A_{12}	A_{11}	A_{10}	A_9	A_8
1	1	1	1	1	0	0	0
F				8			

A_7	A_6	A_5	A_4	A_3	A_2	A_1	A_0
1	1	1	1	1	0	0	0
F				8			

Device address = F8F8H

The correct instruction used → LDA F8F8H

43. Given
$$x[n] = \begin{cases} n & \text{for } 0 \le n \le 10 \\ 0 & \text{elsewhere} \end{cases}$$
$$y[n] = x[n] * x[n]$$
$$y[n] = \sum_{k=0}^{n} x[k]\cdot x[n-k]$$
$$\Rightarrow \quad y[4] = \sum_{k=0}^{4} x[k]\cdot x[G-k]$$
$$= x(0)\cdot x(4) + x(1)x(3) + x(2)x(2) + x(3)x(1) + x(4)\cdot x(0)$$
$$= 0 + 3 + 4 + 3 + 0 = 10$$

44. Given system equation as
$$y[n] = \alpha y[n-1] + \beta x[n]$$
$$\Rightarrow \quad \frac{y(z)}{x(z)} = \frac{\beta}{1-\alpha z^{-1}}$$
$$\Rightarrow \quad H(z) = \frac{\beta}{1-\alpha z^{-1}}$$
$$h[n] = \beta(\alpha)^h u[n] \quad \text{[causal system]}$$

Also given that $\sum_{h=0}^{\infty} h[n] = 2$
$$\beta\left[\frac{1}{1-\alpha}\right] = 2$$
$$1-\alpha = \frac{\beta}{2}$$
$$\alpha = 1-\frac{\beta}{2}$$

45. We can use pasrevalis theorem

Let $x(t)\sin(5t) = \dfrac{\sin 5\pi t}{5\pi t}$

⇒ in frequency domain

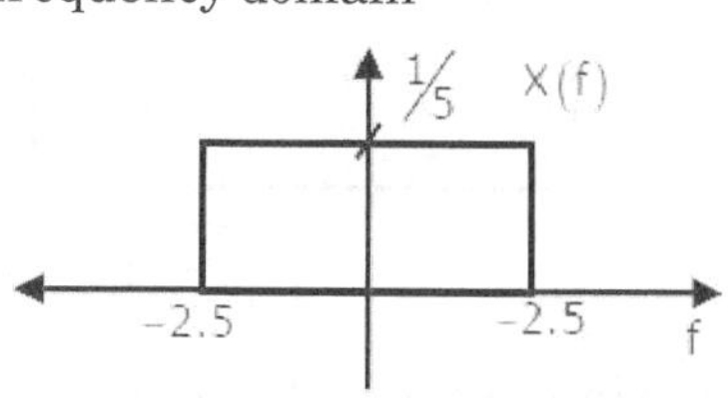

Now, $\displaystyle\int_{-\infty}^{\infty} x^2(t)dt = \int_{-\infty}^{\infty} x^2(t)df$

$$= \int_{-2.5}^{2.5}\left(\frac{1}{5}\right)^2 = \frac{1}{25}\times 5 = \frac{1}{5} = 0.2$$

46. Solution of state equation of

$$X(t) = L^{-1}\left[SI - A^{-1}\right]\cdot X(0)$$

$$X(0) = \begin{bmatrix}1\\-1\end{bmatrix} A = \begin{bmatrix}-1 & 0\\0 & -2\end{bmatrix}$$

$$[SI - A]^{-1} = \begin{bmatrix}S+1 & 0\\0 & S+2\end{bmatrix}^{-1}$$

$$= \frac{1}{(S+1)(S+2)}\begin{bmatrix}S+2 & 0\\0 & S+1\end{bmatrix}$$

$$[SI - A]^{-1} = \begin{bmatrix}\frac{1}{S+1} & 0\\0 & \frac{1}{S+2}\end{bmatrix}$$

$$L^{-1}\left[(SI - A)^{-1}\right] = \begin{bmatrix}L^{-1}\left[\frac{1}{S+1}\right] & 0\\0 & L^{-1}\left[\frac{1}{S+2}\right]\end{bmatrix}$$

$$L^{-1}\left[(SI - A)^{-1}\right] = \begin{bmatrix}e^{-t} & 0\\0 & e^{-2t}\end{bmatrix}$$

$$\begin{bmatrix}X_1(t)\\X_2(t)\end{bmatrix} = \begin{bmatrix}e^{-t} & 0\\0 & e^{-2t}\end{bmatrix}\begin{bmatrix}1\\-1\end{bmatrix}$$

$$\begin{bmatrix}X_1(t)\\X_2(t)\end{bmatrix} = \begin{bmatrix}-e^{t}\\-e^{-2t}\end{bmatrix}$$

$\therefore$ $$\boxed{\begin{matrix}X_1(t) = e^{-t}\\X_2(t) = -e^{-2t}\end{matrix}}$$

47.

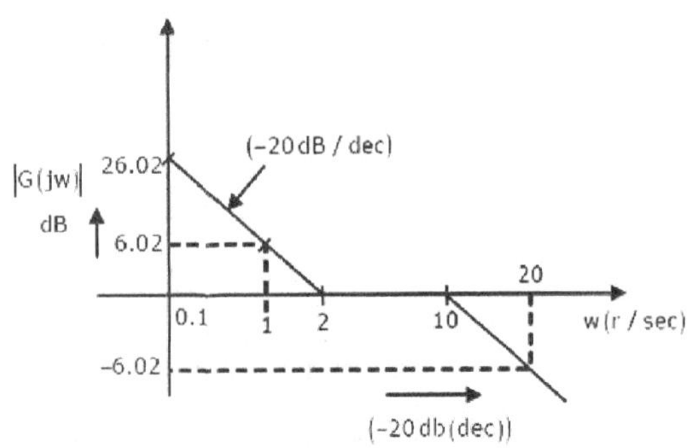

→ Due to initial slope, it is a type-1 system, and it has non zero velocity error coefficient (KV)

→ The magnitude plot is giving 0dB at 2r/sec.

Which gives k_v

$\therefore$ $$k_v = 2$$

The steady state error $e_{ss} = \dfrac{A}{k_v}$

given unit ramp input; A = 1

$$e_{ss} = \frac{1}{2}$$

$$\boxed{e_{ss} = 0.50}$$

48. From the given signal flow graph, the state model is

$$\begin{bmatrix}\dot{X}_1\\\dot{X}_2\\\dot{X}_3\end{bmatrix} = \begin{bmatrix}0 & 1 & 0\\0 & 0 & 1\\a_3 & a_2 & a_1\end{bmatrix}\begin{bmatrix}X_1\\X_2\\X_3\end{bmatrix} + \begin{bmatrix}0\\0\\1\end{bmatrix}u$$

$$Y = \begin{bmatrix}C_1 & C_2 & C_3\end{bmatrix}\begin{bmatrix}X_1\\X_2\\X_3\end{bmatrix}$$

$$A = \begin{bmatrix}0 & 1 & 0\\0 & 0 & 1\\a_3 & a_2 & a_1\end{bmatrix};$$

$$B = \begin{bmatrix}0\\0\\1\end{bmatrix};$$

$$C = [C_1\, C_2\, C_3]$$

Controllability :

$$Q_c = \begin{bmatrix}B & AB & A^2B\end{bmatrix}$$

$$Q_C = \begin{bmatrix}0 & 0 & 1\\0 & 1 & a_1\\1 & a_1 & a_2 + a_1^2\end{bmatrix}$$

$$|Q_C| = 1 \neq 0$$

Observability

$$Q_O = \begin{bmatrix}C\\CA\\CA^2\end{bmatrix} \Rightarrow$$

$$\begin{bmatrix}C_1 & C_2 & C_3\\a_3c_3 & c_1 + a_2c_3 & c_2 + a_1c_3\\c_2a_3 + c_3(a_1a_3) & a_2c_2 + c_3(a_1a_2 + a_3) & c_1 + a_1c_2 + c_3(a_1^2 + a_2)\end{bmatrix}$$

$|Q_0| \Rightarrow$ depends on a_1, a_2, a_3 & c_1 & c_2 & c_3

It is always controllable

49.

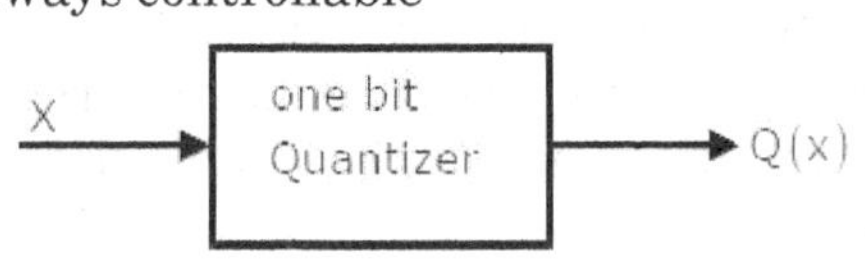

One bit quantizer will give two levels.

Both levels have probability of $\frac{1}{2}$

Pd of input X is

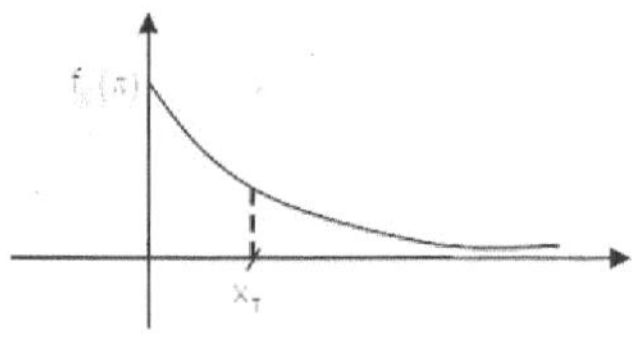

Let x_T be the thsuhold

$$Q(x) = \begin{bmatrix} x_1 & x \geq x_T \\ x_2 & x < x_T \end{bmatrix}$$

Where x_1 and x_2 are two levels

$$P\{Q(r) = x_1\} = \frac{1}{2}$$

$$\Rightarrow \quad \int_{x_T}^{\infty} 2.e^{-2x} dx = \frac{1}{2}$$

$$2 \cdot \frac{e^{-2x}}{-2}\Bigg|_{x_T}^{\infty} = \frac{1}{2}$$

$$-e^{-2\infty} + e^{-2x_T} = \frac{1}{2}$$

$$e^{-2x_T} = \frac{1}{2}$$

$$-2x_T = \ln\frac{1}{2}$$

$$-2x_T = -0.693$$

$$x_T = 0.35$$

50. For Binary F_{SK}

Bit error probability = $Q\left(\sqrt{\frac{E}{N_o}}\right)$

$E \rightarrow$ Energy per bit [no. of symbols = No. of bits]

$$E = \frac{A^2T}{2},$$

$$A = 4 \times 10^{-3},$$

$$T = \frac{1}{500 \times 10^3} \text{[inverse of data rate]}$$

$$\Rightarrow \quad E = \frac{16 \times 10^{-6} \times 2 \times 10^{-6}}{2}$$

$$= 16 \times 10^{-12}$$

$$N_0 = 1 \times 10^{-12}$$

$$\Rightarrow \quad P_e = Q\left(\sqrt{\frac{16 \times 10^{-12}}{1 \times 10^{-12}}}\right)$$

$$= Q(4)$$

51. Given
$$S_x(f) = \begin{cases} \frac{1}{w}, & |f| \leq w \\ 0, & |f| \geq w \end{cases}$$

$$R_x(t) = \int_{-w}^{w} \frac{1}{w} \cdot e^{j2\pi ft} df$$

$$= \frac{1}{w}\frac{e^{j2\pi wt} - e^{-j2\pi wt}}{j2\pi t}$$

$$= \frac{1}{w}\left(\frac{\sin(2\pi wt)}{\pi t}\right)$$

Now, $E\left[\pi \times (t) \cdot x\left(t - \frac{1}{4w}\right)\right] = \pi R_x\left(\frac{1}{4w}\right)$

$$\Rightarrow \quad \pi \cdot \frac{1}{w} \cdot \frac{\sin\left(2\pi w \cdot \frac{1}{4w}\right)}{\pi \cdot \frac{1}{4w}} = \frac{4}{1}$$

52. $m(t) \leftrightarrow M(f)$

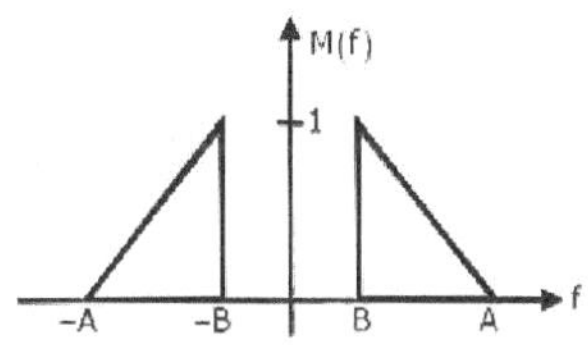

After multiplication with

$$V(t) = \cos(2\pi f_c t)$$

Let $\quad w^1(t) = m(t).V(t)$

$\Rightarrow$ $W^1(f)$ (spectrum of $w^1(t)$) is

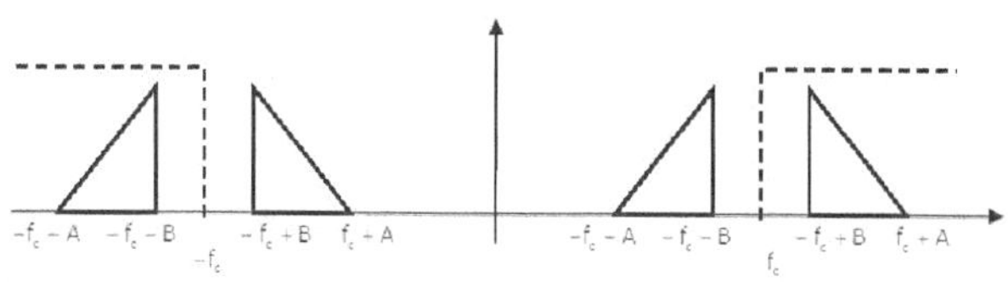

After his pass filter

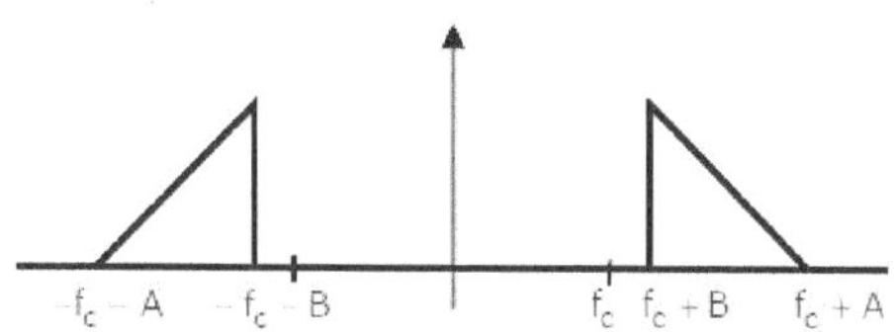

After multiplication with $\cos(2\pi(f_c + A)t)$ and low pass filter of cut off f_c

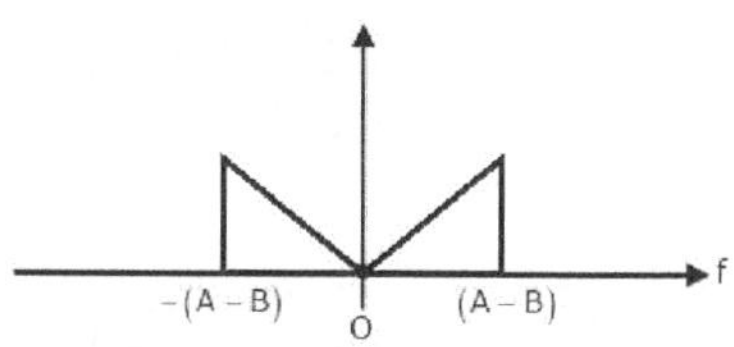

Bandwidth $= A - B$

$= 100 - 40$

$= 60$

53.
$$E(z_1 t) = 3\cos(\text{cot} - kz + 30°)a_x - 4$$
$$- \sin(\omega t - kz + 45°)a_y$$
$$E_x = 3\cos(\omega t - kz + 30°)$$
$$E_y = -4\cos(\omega t - kz + 45°)$$
At $z = 0$
$$E_x = 3\cos(\omega t + 30°)$$
$$E_y = -4\sin(\omega t + 45°)$$
$$|E_x| \neq |E_y|$$
→ so Elliptical polarization
$$Q = 30° - 135° = -105°$$
∴ left hand elliptical (LEP)

54. Here $\ell = \dfrac{\lambda}{2}$
$$Z_{in}\left(\ell = \frac{\lambda}{2}\right) = Z_L$$
$$= 50\ \Omega$$
$$\therefore \quad Z_{in} = (100 \| 50)$$
$$= \frac{100}{3}$$
$$= 33.33\ \Omega$$

55.
$$t_{c10} = \frac{C}{2}\sqrt{\left(\frac{1}{a}\right)^2}$$
$$t_{c10} = K\left(\frac{1}{a}\right);$$
$$t_{c20} = \text{sss } K\left(\frac{2}{a}\right)$$
$$t_{c11} = K\sqrt{\frac{1}{a^2} + \frac{1}{b^2}}$$
Given
$$t_{c11} = \frac{f_{c10} + f_{c20}}{2}$$
$$K\sqrt{\frac{1}{a^2} + \frac{1}{b^2}} = \frac{K}{2}\left[\frac{1}{a} + \frac{2}{a}\right]$$
$$\sqrt{\frac{1}{a^2} + \frac{1}{b^2}} = \frac{3}{2a}$$
$$\frac{1}{5} + \frac{1}{b^2} = \frac{9}{4(5)}$$
$$\Rightarrow \quad -\frac{1}{5} + \frac{9}{20} = \frac{1}{b^2}$$
$$-0.2 + 0.45 = \frac{1}{b^2}$$
$$\therefore \quad \frac{1}{b^2} = \frac{1}{2^2}$$
$$\Rightarrow \quad b = 2 \text{ cm}$$

■■

SOLVED PAPER – 2014

ELECTRONICS AND COMMUNICATION ENGINEERING

Set - 3

No. of Questions : 65 *Maximum Marks : 100*

GENERAL APTITUDE

Q. No. 1 – 5 Carry One Mark Each

1. "India is a country of rich heritage and cultural diversity." Which one of the following facts best supports the claim made in the above sentence?
   (*a*) India is a union of 28 states and 7 union territories.
   (*b*) India has a population of over 1.1 billion.
   (*c*) India is home to 22 official languages and thousands of dialects.
   (*d*) The Indian cricket team draws players from over ten states.

2. The value of one U.S. dollar is 65 Indian Rupees today, compared to 60 last year. The Indian Rupee has _______.
   (*a*) Depressed (*b*) Depreciated
   (*c*) Appreciated (*d*) Stabilized

3. 'Advice' is _______.
   (*a*) a verb (*b*) a noun
   (*c*) an adjective (*d*) both a verb and a noun

4. The next term in the series 81,54,36,24... is _______

5. In which of the following options will the expression P < M be definitely true?
   (*a*) M < R > P > S
   (*b*) M > S < P < F
   (*c*) Q < M < F = P
   (*d*) P = A < R < M

Q. No. 6 – 10 Carry Two Marks Each

6. Find the next term in the sequence: 7G, 11K, 13M, _______
   (*a*) 15Q (*b*) 17Q
   (*c*) 15P (*d*) 17P

7. The multi-level hierarchical pie chart shows the population of animals in a reserve forest. The correct conclusions from this information are:

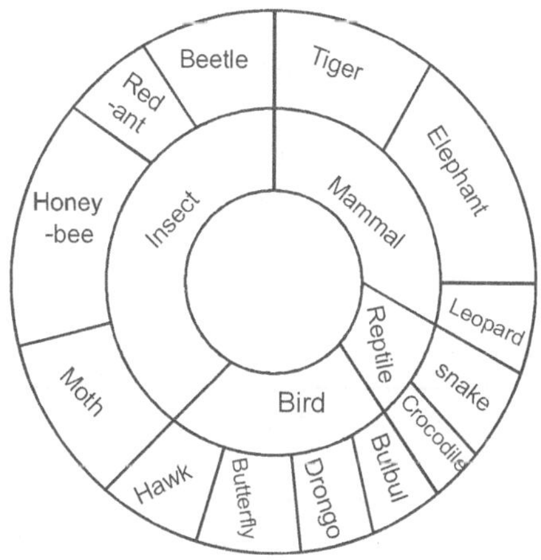

   (*i*) Butterflies are birds
   (*ii*) There are more tigers in this forest than red ants
   (*iii*) All reptiles in this forest are either snakes or crocodiles
   (*iv*) Elephants are the largest mammals in this forest
   (*a*) (*i*) and (*ii*) only
   (*b*) (*i*), (*ii*), (*iii*) and (*iv*)
   (*c*) (*i*), (*iii*) and (*iv*) only
   (*d*) (*i*), (*ii*) and (*iii*) only

8. A man can row at 8 km per hour in still water. If it takes him thrice as long to row upstream, as to row downstream, then find the stream velocity in km per hour.

9. A firm producing air purifiers sold 200 units in 2012. The following pie chart presents the share of raw material, labour, energy, plant and machinery, and transportation costs in the total manufacturing cost of the firm in 2012. The expenditure on labour in 2012 is ₹ 4,50,000. In 2013, the raw material expenses increased by 30% and all other expenses increased by 20%. If the company registered a profit of ₹ 10 lakhs in 2012, at what price (in ₹) was each air purifier sold?

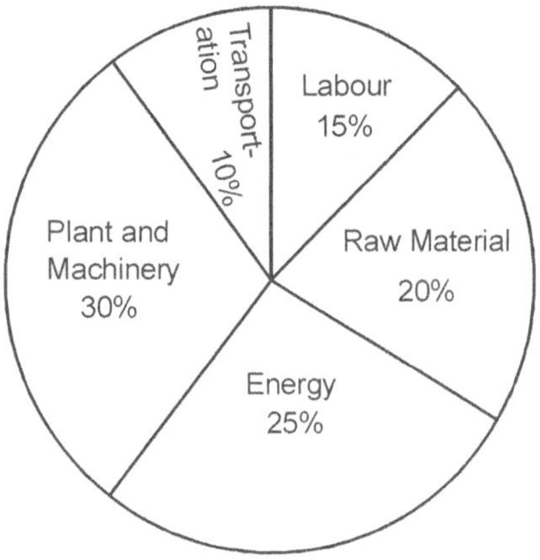

10. A batch of one hundred bulbs is inspected by testing four randomly chosen bulbs. The batch is rejected if even one of the bulbs is defective. A batch typically has five defective bulbs.

The probability that the current batch is accepted is ________

ELECTRONICS AND COMMUNICATION

Q. No. 1 – 25 Carry One Mark Each

1. The maximum value of the function $f(x) = \ln(1 + x) - x$ (where $x > -1$) occurs at $x =$ ________.

2. Which ONE of the following is a linear non-homogeneous differential equation, where x and y are the independent and dependent variables respectively?

(*a*) $\frac{dy}{dx} + xy = e^{-x}$ (*b*) $\frac{dy}{dx} + xy = 0$

(*c*) $\frac{dy}{dx} + xy = e^{-y}$ (*d*) $\frac{dy}{dx} + e^{-y} = 0$

3. Match the application to appropriate numerical method.

Application

P1 : Numerical integration

P2 : Solution to a transcendental equation

P3 : Solution to a system of linear equations

P4 : Solution to a differential equation

Numerical Method

M1 : Newton-Raphson Method

M2 : Runge-Kutta Method

M3: Simpson's 1/3-rule

M4: Gauss Elimination Method

(*a*) P1—M3, P2—M2, P3—M4, P4—M1

(*b*) P1—M3, P2—M1, P3—M4, P4—M2

(*c*) P1—M4, P2—M1, P3—M3, P4—M2

(*d*) P1—M2, P2—M1, P3—M3, P4—M4

4. An unbiased coin is tossed an infinite number of times. The probability that the fourth head appears at the tenth toss is

(*a*) 0.067 (*b*) 0.073

(*c*) 0.082 (*d*) 0.091

5. If $z = xy \ln(xy)$, then

(*a*) $x\frac{\partial z}{\partial x} + y\frac{\partial z}{\partial y} = 0$ (*b*) $y\frac{\partial z}{\partial x} = x\frac{\partial z}{\partial y}$

(*c*) $x\frac{\partial z}{\partial x} = y\frac{\partial z}{\partial y}$ (*d*) $y\frac{\partial z}{\partial x} + x\frac{\partial z}{\partial y} = 0$

6. A series RC circuit is connected to a DC voltage source at time $t = 0$. The relation between the source voltage V_S, the resistance R, the capacitance C, and the current $i(t)$ is given below:

$$V_s = Ri(t) + \frac{1}{C}\int_0^t i(u)du$$

Which one of the following represents the current $i(t)$?

(*a*)

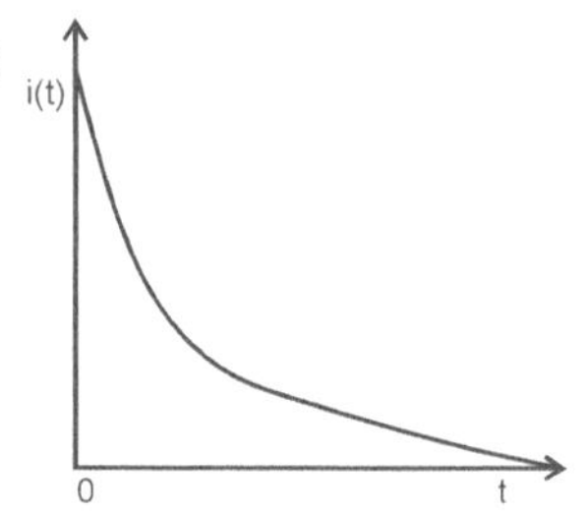

(*b*)

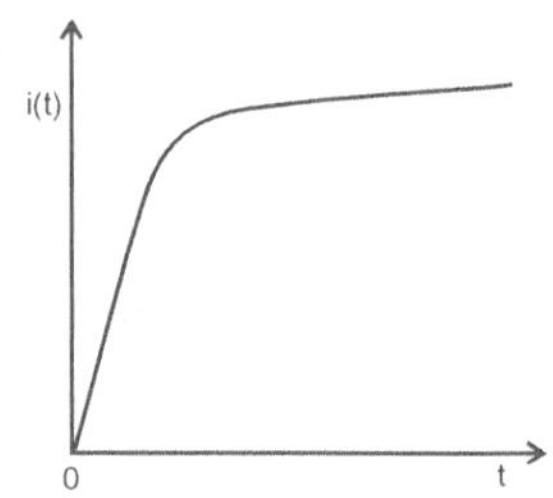

(*c*)

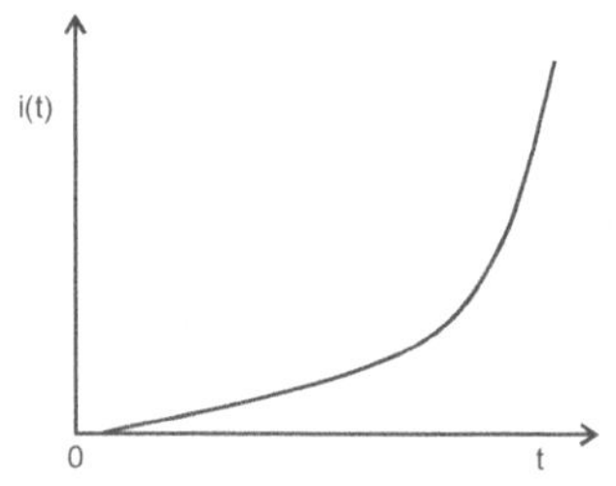

(*d*)

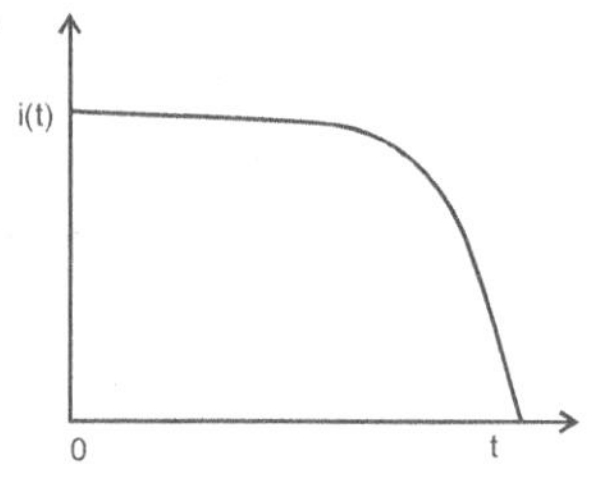

7. In the figure shown, the value of the current I (in Amperes) is ________.

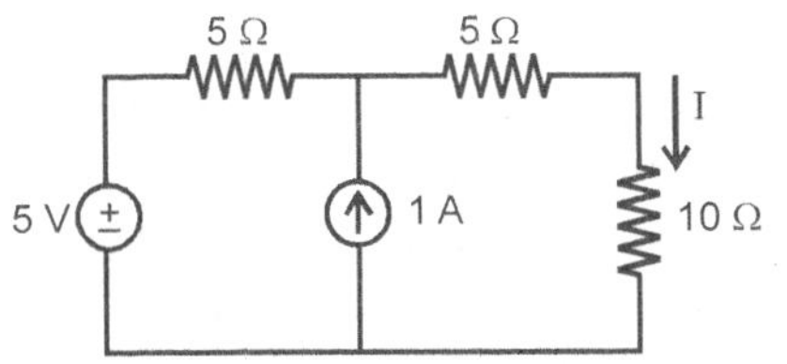

8. In MOSFET fabrication, the channel length is defined during the process of
   (*a*) Isolation oxide growth
   (*b*) Channel stop implantation
   (*c*) Poly-silicon gate patterning
   (*d*) Lithography step leading to the contact pads

9. A thin P-type silicon sample is uniformly illuminated with light which generates excess carriers. The recombination rate is directly proportional to
   (*a*) The minority carrier mobility
   (*b*) The minority carrier recombination lifetime
   (*c*) The majority carrier concentration
   (*d*) The excess minority carrier concentration

10. At T = 300 K, the hole mobility of a semiconductor $\mu_p = 500\ \text{cm}^2/\text{V}-\text{s}$ and $\frac{kT}{q} = 26\ m\text{V}$.

    The hole diffusion constant D_P in cm²/s is ________

11. The desirable characteristics of a transconductance amplifier are
    (*a*) High input resistance and high output resistance
    (*b*) High input resistance and low output resistance
    (*c*) Low input resistance and high output resistance
    (*d*) Low input resistance and low output resistance

12. In the circuit shown, the PNP transistor has $|V_{BE}| = 0.7$ and $\beta = 50$. Assume that $R_B = 100\text{k}\Omega$. For V_0 to be 5 V, the value of R_C (in kΩ) is ________

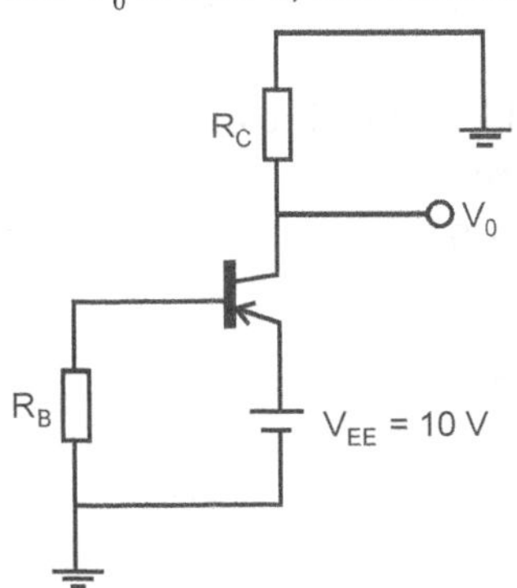

13. The figure shows a half-wave rectifier. The diode D is ideal. The average steady-state current (in Amperes) through the diode is approximately ________.

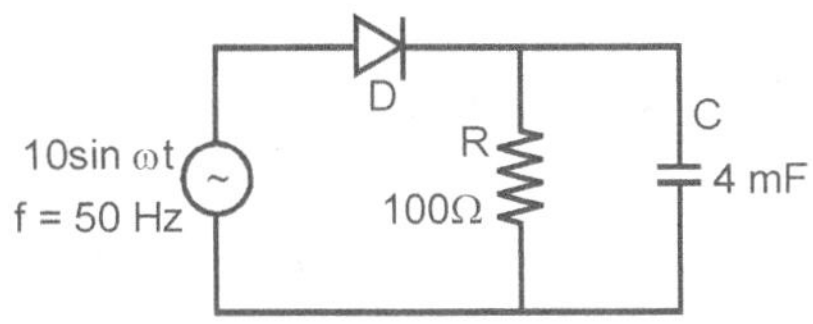

14. An analog voltage in the range 0 to 8 V is divided in 16 equal intervals for conversion to 4-bit digital output. The maximum quantization error (in V) is ________

15. The circuit shown in the figure is a

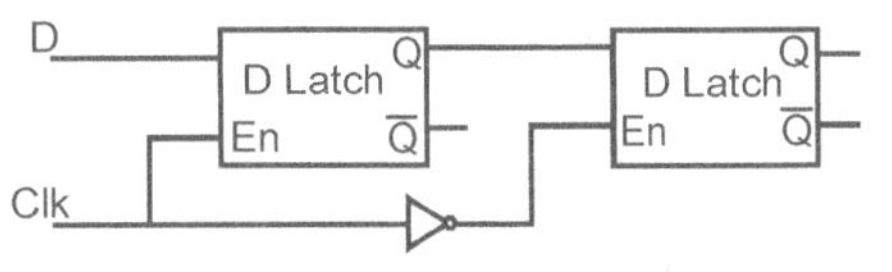

    (*a*) Toggle Flip Flop
    (*b*) JK Flip Flop
    (*c*) SR Latch
    (*d*) Master-Slave D Flip Flop

16. Consider the multiplexer based logic circuit shown in the figure.

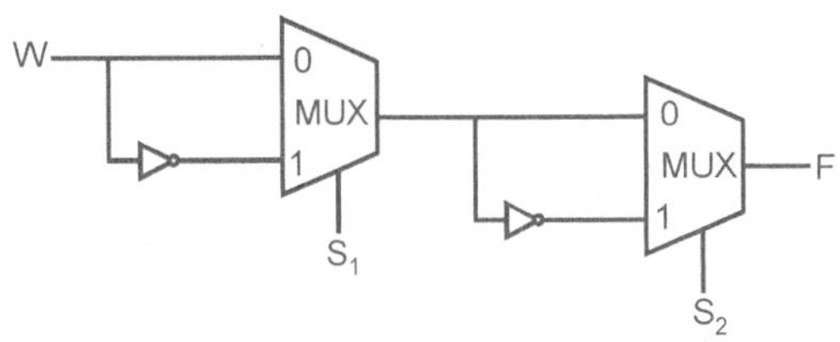

    Which one of the following Boolean functions is realized by the circuit?
    (*a*) $F = W\overline{S}_1\overline{S}_2$
    (*b*) $F = WS_1 + WS_2 + S_1S_2$
    (*c*) $F = \overline{W} + S_1 + S_2$
    (*d*) $F = W \oplus S_1 \oplus S_2$

17. Let $x(t) = \cos(10\pi t) + \cos(30\pi t)$ be sampled at 20 Hz and reconstructed using an ideal low-pass filter with cut-off frequency of 20 Hz. The frequency/frequencies present in the reconstructed signal is/are
    (*a*) 5 Hz and 15 Hz only
    (*b*) 10 Hz and 15 Hz only
    (*c*) 5 Hz, 10 Hz and 15 Hz only
    (*d*) 5 Hz only

18. For an all-pass system $H(z)=\dfrac{(z^{-1}-b)}{(1-az^{-1})}$, where $|H(e^{-jw})| = 1$, for all ω. If Re(a) ≠ 0, Im(a) ≠ 0, then b equals

(a) a (b) a^*

(c) $1/a^*$ (d) $1/a$

19. A modulated signal is $y(t) = m(t)\cos(40000\pi t)$, where the baseband signal $m(t)$ has frequency components less than 5 kHz only. The minimum required rate (in kHz) at which $y(t)$ should be sampled to recover $m(t)$ is ________.

20. Consider the following block diagram in the figure.

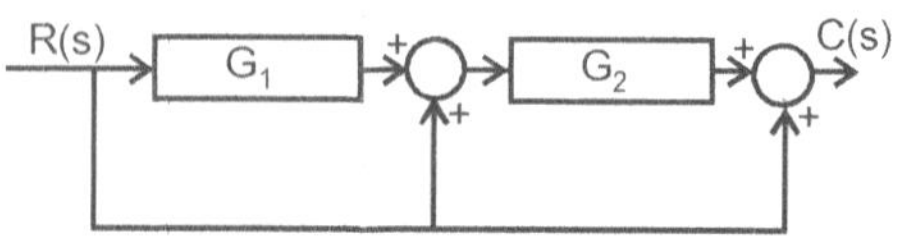

The transfer function $\dfrac{C(s)}{R(s)}$ is

(a) $\dfrac{G_1G_2}{1+G_1G_2}$ (b) $G_1G_2 + G_1 + 1$

(c) $G_1G_2 + G_2 + 1$ (d) $\dfrac{G_1}{1+G_1G_2}$

21. The input $-3e^{2t}u(t)$, where $u(t)$ is the unit step function, is applied to a system with transfer function $\dfrac{s-2}{s+3}$. If the initial value of the output is –2, then the value of the output at steady state is ________.

22. The phase response of a passband waveform at the receiver is given by

$$\varphi(f) = -2\pi\alpha(f - f_c) - 2\pi\beta f_c$$

Where f_c is the centre frequency, and α and β are positive constants. The actual signal propagation delay from the transmitter to receiver is

(a) $\dfrac{\alpha-\beta}{\alpha+\beta}$ (b) $\dfrac{\alpha\beta}{\alpha+\beta}$

(c) α (d) β

23. Consider an FM signal $f(t) = \cos[2\pi f_c t + \beta_1 \sin 2\pi f_1 t + \beta_2 \sin 2\pi f_2 t.]$. The maximum deviation of the instantaneous frequency from the carrier frequency f_c is

(a) $\beta_1 f_1 + \beta_2 f_2$ (b) $\beta_1 f_2 + \beta_2 f_1$

(c) $\beta_1 + \beta_2$ (d) $f_1 + f_2$

24. Consider an air filled rectangular waveguide with a cross-section of 5 cm × 3 cm. For this waveguide, the cut-off frequency (in MHz) of TE_{21} mode is ________.

25. In the following figure, the transmitter Tx sends a wideband modulated RF signal via a coaxial cable to the receiver Rx. The output impedance Z_T of Tx, the characteristic impedance Z_0 of the cable and the input impedance Z_R of Rx are all real.

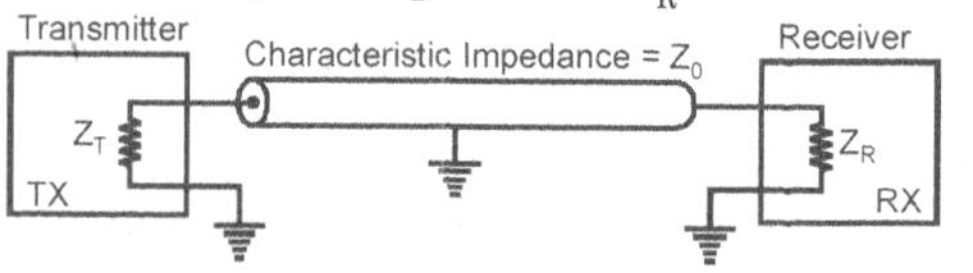

Which one of the following statements is TRUE about the distortion of the received signal due to impedance mismatch?

(a) The signal gets distorted if $Z_R \neq Z_0$, irrespective of the value of Z_T

(b) The signal gets distorted if $Z_T \neq Z_0$, irrespective of the value of Z_R

(c) Signal distortion implies impedance mismatch at both ends: $Z_T \neq Z_0$ and $Z_R \neq Z_0$

(d) Impedance mismatches do not result in signal distortion but reduce power transfer efficiency

Q. No. 26 – 55 Carry Two Marks Each

26. The maximum value of $f(x) = 2x^3 - 9x^2 + 12x - 3$ in the interval $0 \le x \le 3$ is ________.

27. Which one of the following statements is NOT true for a square matrix?

(a) If A is upper triangular, the eigenvalues of A are the diagonal elements of it

(b) If A is real symmetric, the eigenvalues of A are always real and positive

(c) If A is real, the eigenvalues of A and A^T are always the same

(d) If all the principal minors of A are positive, all the eigenvalues of A are also positive

28. A fair coin is tossed repeatedly till both head and tail appear at least once. The average number of tosses required is ________.

29. Let X_1, X_2, and X_3 be independent and identically distributed random variables with the uniform distribution on [0, 1]. The probability $P\{X_1 + X_2 \le X_3\}$ is ________.

30. Consider the building block called 'Network N' shown in the figure.

Let C = 100 F and R = 10kΩ

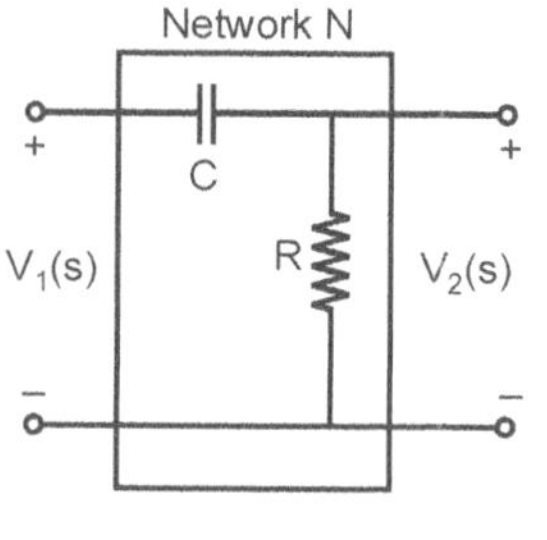

Two such blocks are connected in cascade, as shown in the figure.

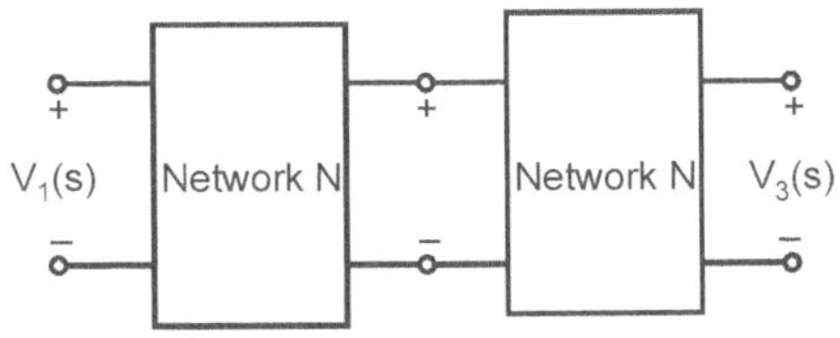

The transfer function $\frac{V_3(s)}{V_1(s)}$ of the cascaded network is

(*a*) $\frac{s}{1+s}$ (*b*) $\frac{s^2}{1+3s+s^2}$

(*c*) $\left(\frac{s}{1+s}\right)^2$ (*d*) $\frac{s}{2+s}$

31. In the circuit shown in the figure, the value of node voltage V_2 is

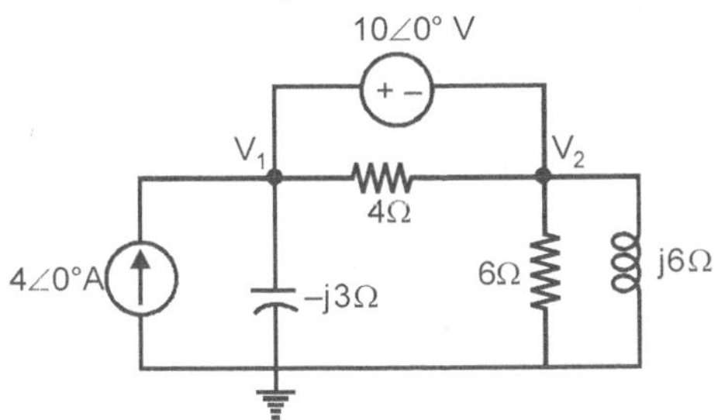

(*a*) $22 + j2$ V (*b*) $2 + j22$ V

(*c*) $22 - j2$ V (*d*) $2 - j22$ V

32. In the circuit shown in the figure, the angular frequency ω (in rad/s), at which the Norton equivalent impedance as seen from terminals b-b′ is purely resistive, is_______.

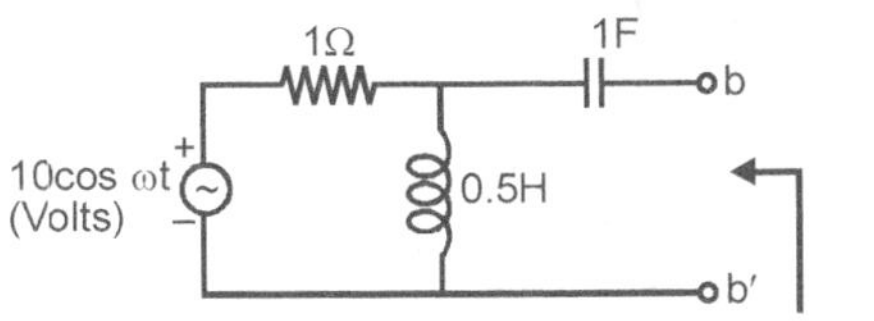

33. For the Y-network shown in the figure, the value of R_1 (in Ω) in the equivalent Δ-network is _______.

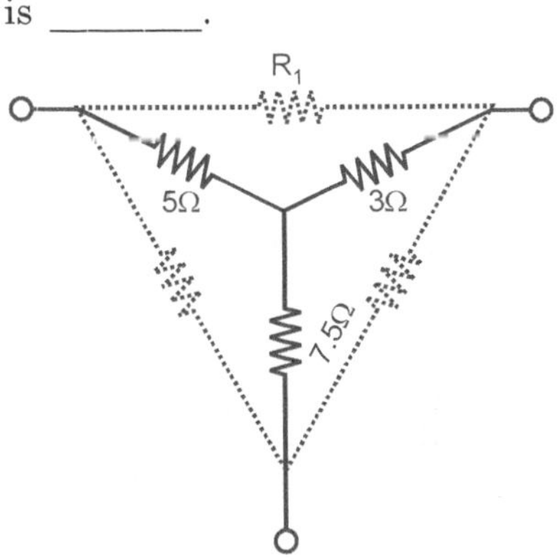

34. The donor and accepter impurities in an abrupt junction silicon diode are 1×10^{16} cm^{-3} and 5×10^{18} cm^{-3}, respectively. Assume that the intrinsic carrier concentration in silicon $n_i = 1.5 \times 10^{10}$ cm^{-3} at 300 K, $\frac{kT}{q} = 26$ mV and the permittivity of silicon $\varepsilon_{si} = 1.04 \times 10^{-12}$ F/cm. The built-in potential and the depletion width of the diode under thermal equilibrium conditions, respectively, are

(*a*) 0.7 V and 1×10^{-4} cm

(*b*) 0.86 V and 1×10^{-4} cm

(*c*) 0.7 V and 3.3×10^{-5} cm

(*d*) 0.86 V and 3.3×10^{-5} cm

35. The slope of the I_D vs. V_{GS} curve of an n-channel MOSFET in linear regime is $10^{-3}\ \Omega^{-1}$ at $V_{DS} = 0.1$V. For the same device, neglecting channel length modulation, the slope of the $\sqrt{I_D}$ vs V_{GS} curve (in$\sqrt{A}$/V) under saturation regime is approximately _______.

36. An ideal MOS capacitor has boron doping-concentration of 10^{15} cm^{-3} in the substrate. When a gate voltage is applied, a depletion region of width 0.5 m is formed with a surface (channel) potential of 0.2 V. Given that $\varepsilon_0 = 8.854 \times 10^{-14}$ F/cm and the relative permittivities of silicon and silicon dioxide are 12 and 4, respectively, the peak electric field (in V/ m) in the oxide region is _______.

37. In the circuit shown, the silicon BJT has β = 50. Assume $V_{BE} = 0.7$ V and $V_{CE(sat)} = 0.2$ V. Which one of the following statements is correct?

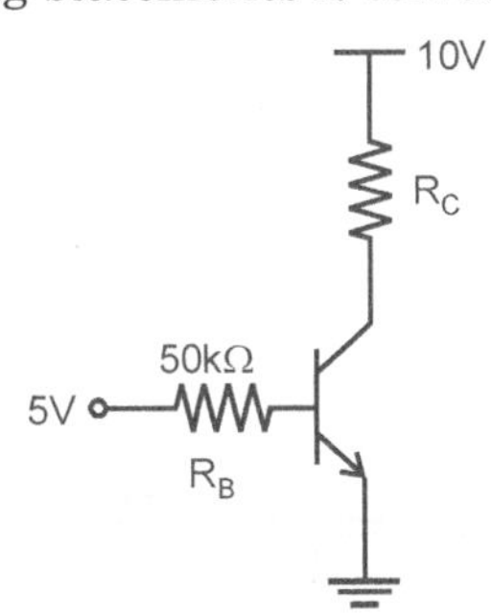

(*a*) For $R_C = 1$ kΩ, the BJT operates in the saturation region

(*b*) For $R_C = 3$ kΩ, the BJT operates in the saturation region

(*c*) For $R_C = 20$ kΩ, the BJT operates in the cut-off region

(*d*) For $R_C = 20$ kΩ, the BJT operates in the linear region

38. Assuming that the Op-amp in the circuit shown is ideal, V_O is given by

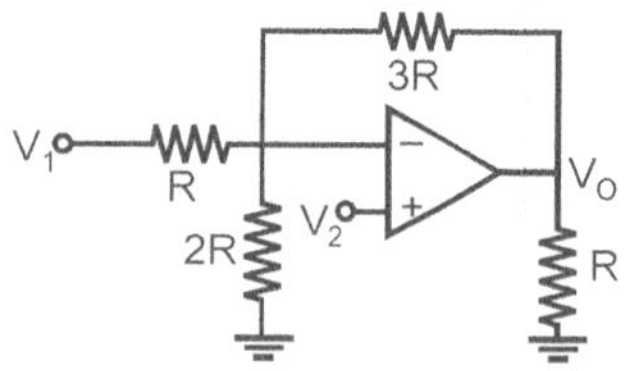

(*a*) $\frac{5}{2}V_1 - 3V_2$ (*b*) $2V_1 - \frac{5}{2}V_2$

(*c*) $-\frac{3}{2}V_1 + \frac{7}{2}V_2$ (*d*) $-3V_1 + \frac{11}{2}V_2$

39. For the MOSFET M_1 shown in the figure, assume $W/L = 2$, $V_{DD} = 2.0$ V, ${}_nC_{ox} = 100$ A/V² and $V_{TH} = 0.5$ V. The transistor M_1 switches from saturation region to linear region when V_{in} (in Volts) is ________.

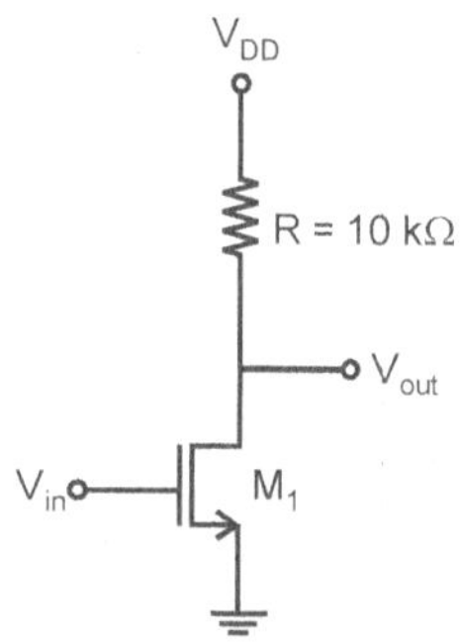

40. If WL is the Word Line and BL the Bit Line, an SRAM cell is shown in

(*a*)

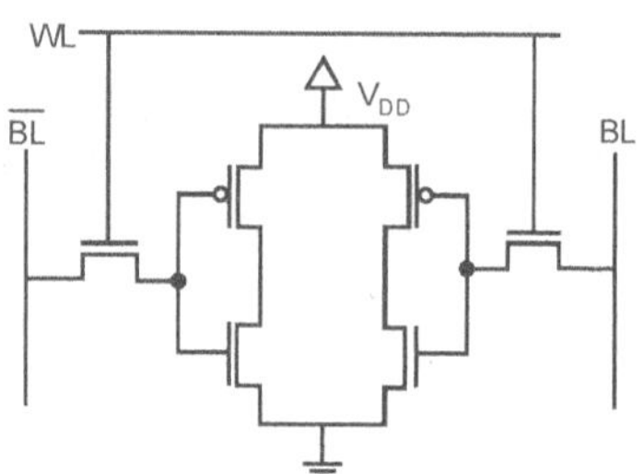

(*b*)

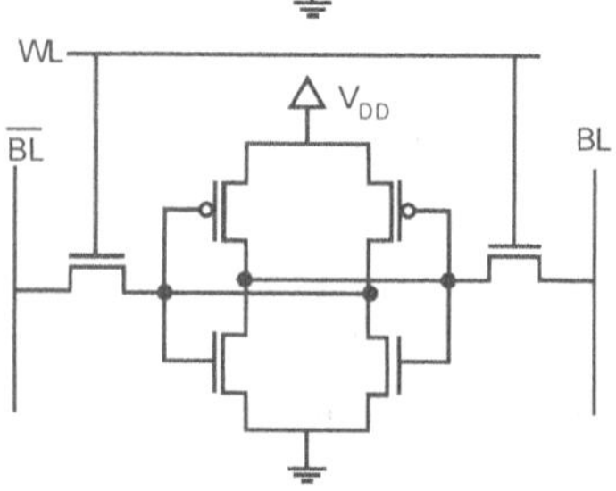

(*c*)

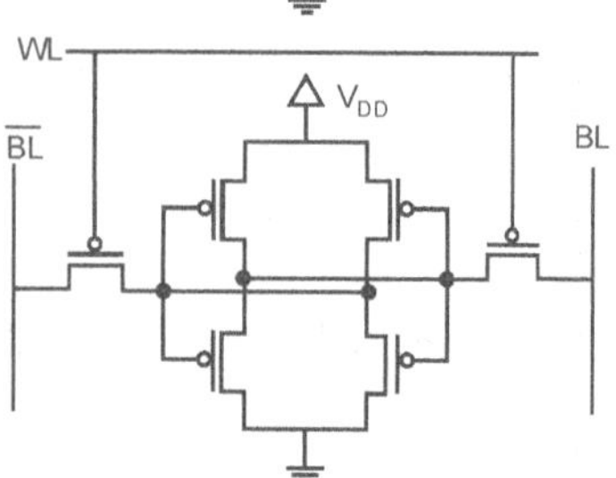

(*d*)

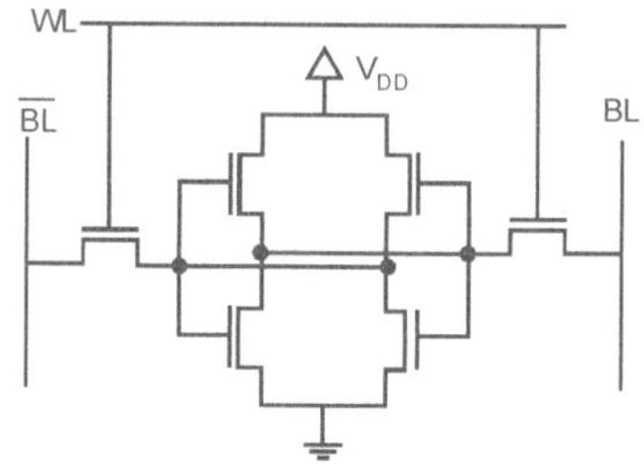

41. In the circuit shown, W and Y are MSBs of the control inputs. The output F is given by

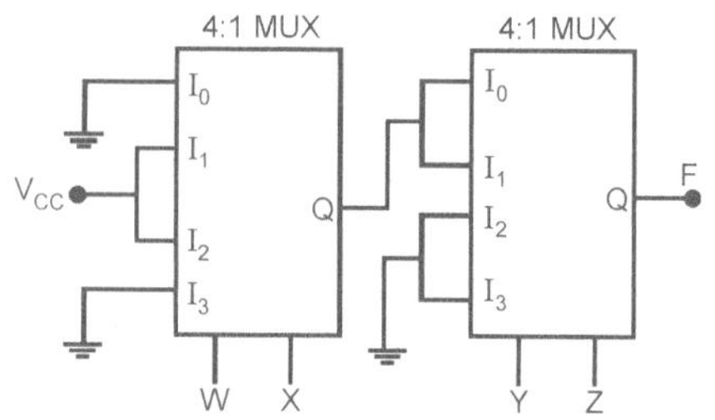

(*a*) $F = W\overline{X} + \overline{W}X + \overline{Y}\overline{Z}$

(*b*) $F = W\overline{X} + \overline{W}X + \overline{YZ}$

(*c*) $F = W\overline{X}\overline{Y} + \overline{W}X\overline{Y}$

(*d*) $F = \left(\overline{W} + \overline{X}\right)\overline{Y}\,\overline{Z}$

42. If X and Y are inputs and the Difference ($D = X - Y$) and the Borrow (B) are the outputs, which one of the following diagrams implements a half-subtractor?

(*a*)

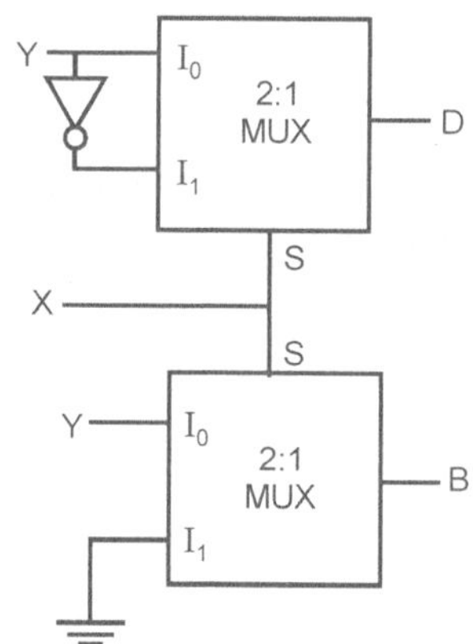

(*b*)

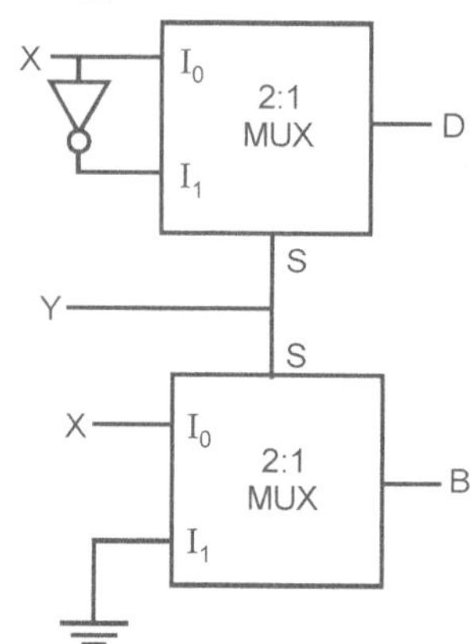

(c)

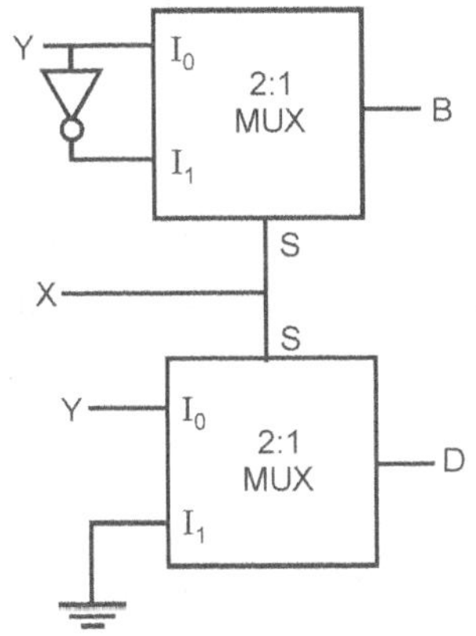

(d)

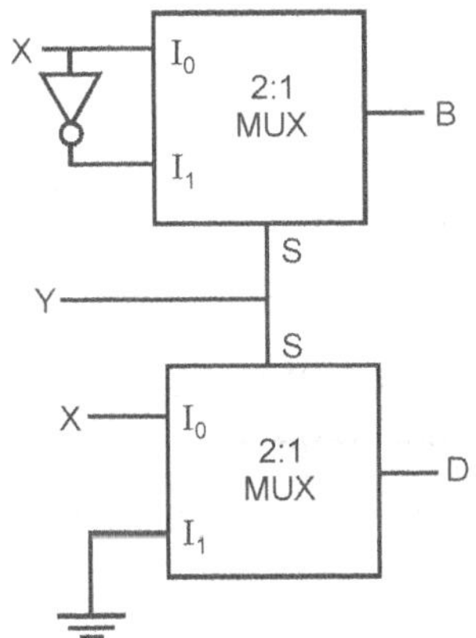

43. Let $H_1(z) = (1 - pz^{-1})^{-1}$, $H_2(z) = (1-qz^{-1})^{-1}$, $H(z) = H_1(z) + rH_2(z)$. The quantities p, q, r are real numbers. Consider $p=\frac{1}{2}, q=\frac{1}{4}, |r|<1$. If the zero of $H(z)$ lies on the unit circle, then $r =$ ________

44. Let $h(t)$ denote the impulse response of a causal system with transfer function $\frac{1}{s+1}$. Consider the following three statements.

S1: The system is stable.

S2: $\frac{h(t+1)}{h(t)}$ is independent of t for $t>0$.

S3: A non-causal system with the same transfer function is stable.

For the above system,

(a) Only S1 and S2 are true

(b) Only S2 and S3 are true

(c) Only S1 and S3 are true

(d) S1, S2 and S3 are true

45. The z-transform of the sequence $x[n]$ is given by $X(z) = \frac{1}{(1-2z^{-1})^2}$, with the region of convergence $|z| > 2$. Then, $x[2]$ is ________.

46. The steady state error of the system shown in the figure for a unit step input is ________.

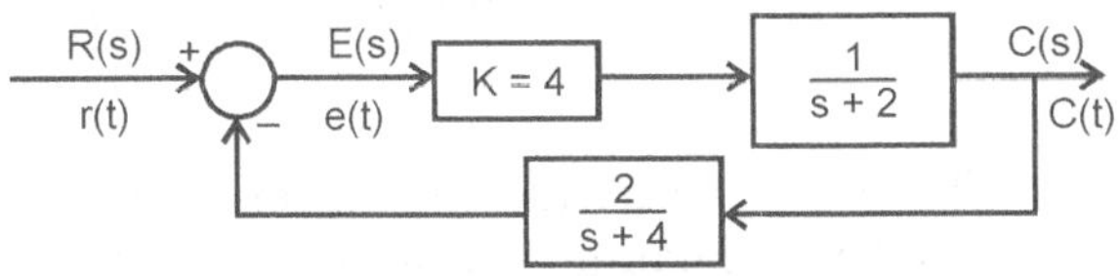

47. The state equation of a second-order linear system is given by

$$\dot{x}(t) = Ax(t),\ x(0) = x_0$$

For $x_0 = \begin{bmatrix} 1 \\ -1 \end{bmatrix}$, $x(t) = \begin{bmatrix} e^{-t} \\ -e^{-t} \end{bmatrix}$ and for $x_0 = \begin{bmatrix} 0 \\ 1 \end{bmatrix}$,

$x(t) = \begin{bmatrix} e^{-t} - e^{-2t} \\ -e^{-t} + 2e^{-2t} \end{bmatrix}$

When $x_0 = \begin{bmatrix} 3 \\ 5 \end{bmatrix}$, $x(t)$ is

(a) $\begin{bmatrix} -8e^{-t} + 11e^{-2t} \\ 8e^{-t} - 22e^{-2t} \end{bmatrix}$ (b) $\begin{bmatrix} 11e^{-t} - 8e^{-2t} \\ -11e^{-t} + 16e^{-2t} \end{bmatrix}$

(c) $\begin{bmatrix} 3e^{-t} - 5e^{-2t} \\ -3e^{-t} + 10e^{-2t} \end{bmatrix}$ (d) $\begin{bmatrix} 5e^{-t} - 3e^{-2t} \\ -5e^{-t} + 6e^{-2t} \end{bmatrix}$

48. In the root locus plot shown in the figure, the pole/zero marks and the arrows have been removed. Which one of the following transfer functions has this root locus?

(a) $\frac{s+1}{(s+2)(s+4)(s+7)}$

(b) $\frac{s+4}{(s+1)(s+2)(s+7)}$

(c) $\frac{s+7}{(s+1)(s+2)(s+4)}$

(d) $\frac{(s+1)(s+2)}{(s+7)(s+4)}$

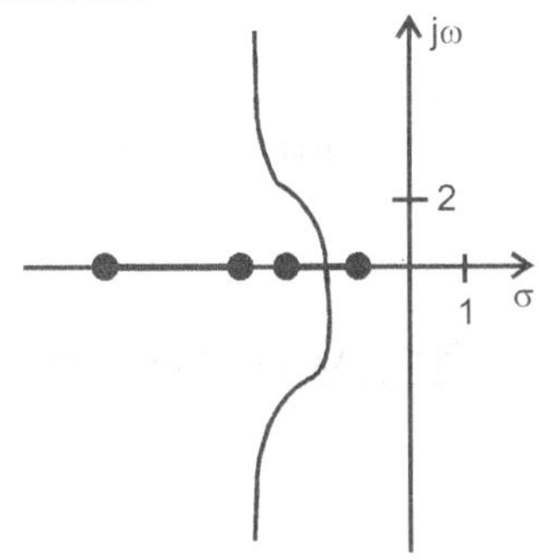

49. Let $X(t)$ be a wide sense stationary (WSS) random process with power spectral density $S_X(f)$. If $Y(t)$ is the process defined as $Y(t) = X(2t-1)$, the power spectral density $S_Y(f)$ is

(a) $S_Y(f) = \frac{1}{2}S_X\left(\frac{f}{2}\right)e^{-j\pi f}$

(b) $S_Y(f) = \frac{1}{2}S_X\left(\frac{f}{2}\right)e^{-j\pi f/2}$

(c) $S_Y(f) = \frac{1}{2}S_X\left(\frac{f}{2}\right)$

(d) $S_Y(f) = \frac{1}{2}S_X\left(\frac{f}{2}\right)e^{-j2\pi f}$

50. A real band-limited random process X(*t*) has two-sided power spectral density

$$S_x(f) = \begin{cases} 10^{-6}(3000-|f|)\text{Watts/Hz} & \text{for}|f| \le 3\,k\text{Hz} \\ 0 & \text{otherwise} \end{cases}$$

Where *f* is the frequency expressed in Hz. The signal X(*t*) modulates a carrier cos16000 πt and the resultant signal is passed through an ideal band-pass filter of unity gain with centre frequency of 8 kHz and band-width of 2 kHz. The output power (in Watts) is ________.

51. In a PCM system, the signal $m(t) = \{\sin(100\pi t) + \cos(100\pi t)\}$ V is sampled at the Nyquist rate. The samples are processed by a uniform quantizer with step size 0.75 V. The minimum data rate of the PCM system in bits per second is ________.

52. A binary random variable X takes the value of 1 with probability 1/3. X is input to a cascade of 2 independent identical binary symmetric channels (BSCs) each with crossover probability 1/2. The outputs of BSCs are the random variables Y_1 and Y_2 as shown in the figure.

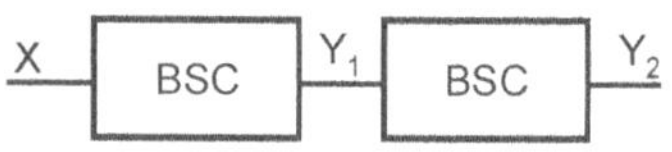

The value of $H(Y_1) + H(Y_2)$ in bits is ________.

53. Given the vector A = (cos *x*) (sin *y*) $\hat{a}_x$ + (sin x) (cos *y*) $\hat{a}_y$, where $\hat{a}_x$, $\hat{a}_y$ denote unit vectors along *x*, *y* directions, respectively. The magnitude of curl of A is ________

54. A region shown below contains a perfect conducting half-space and air. The surface current $\vec{K}_s$ on the surface of the perfect conductor is $\vec{K}_s = \hat{x}2$ amperes per meter. The tangential $\vec{H}$ field in the air just above the perfect conductor is

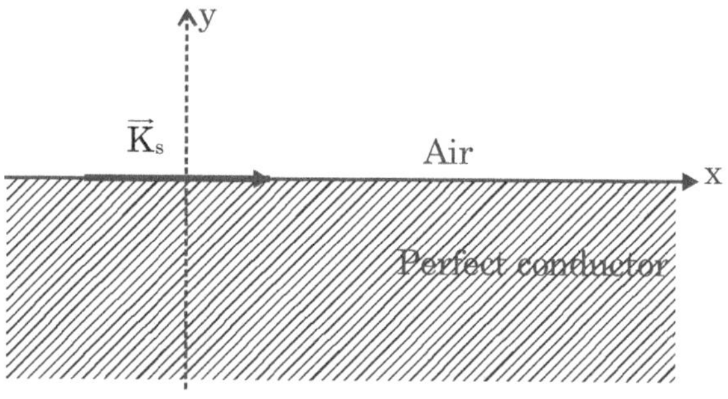

(*a*) $(\hat{x}+\hat{z})2$ amperes per meter

(*b*) $\hat{x}2$ amperes per meter

(*c*) $-\hat{z}2$ amperes per meter

(*d*) $\hat{z}2$ amperes per meter

55. Assume that a plane wave in air with an electric field $\vec{E} = 10\cos\left(\omega t - 3x - \sqrt{3z}\right)\hat{a}_y$ V/m is incident on a non-magnetic dielectric slab of relative permittivity 3 which covers the region Z > 0. The angle of transmission in the dielectric slab is ________ degrees.

ANSWERS

GENERAL APTITUDE

1. (*c*) 2. (*b*) 3. (*b*) 4. (16) 5. (*d*) 6. (*b*) 7. (*d*) 8. (4) 9. (20,000) 10. (0.8145)

ELECTRONICS AND COMMUNICATION

1. (0)	2. (*a*)	3. (*b*)	4. (*c*)	5. (*c*)	6. (*a*)	7. (0.5)
8. (*c*)	9. (*d*)	10. (13)	11. (*a*)	12. (1.075)	13. (0.09)	14. (0.25)
15. (*d*)	16. (*d*)	17. (*a*)	18. (*b*)	19. (10 KHz)	20. (*c*)	21. (0)
22. (*c*)	23. (*a*)	24. (7810MHz)	25. (*c*)	26. (6)	27. (*b*)	28. (3)
29. (0.16)	30. (*b*)	31. (*d*)	32. (*2 r/sec*)	33. (10Ω)	34. (*d*)	35. (0.07)
36. (2.4)	37. (*b*)	38. (*d*)	39. (1.5)	40. (*b*)	41. (*c*)	42. (*a*)
43. (–0.5)	44. (*a*)	45. (12)	46. (0.5)	47. (*b*)	48. (*b*)	49. (*c*)
50. (2.5)	51. (200)	52. (2)	53. (0)	54. (*d*)	55. (30)	

EXPLANATIONS

GENERAL APTITUDE

1. Diversity is shown in terms of difference language.

4.
$$81 - 54 = 27;$$
$$27 \times \frac{2}{3} = 18$$
$$54 - 36 = 18;$$
$$18 \times \frac{2}{3} = 12$$
$$36 - 24 = 12;$$
$$12 \times \frac{2}{3} = 8$$
$$\therefore \quad 24 - 8 = 6$$

7. It is not mentioned that elephant is the largest animal.

8. 4 km/hr

Speed of man = 8

Left distance = d

Time taken = $\frac{d}{8}$

Upstream :

Speed of stream = s

⇒ speed upstream = S′ = $(8 - s)$

$$t' = \left(\frac{d}{8-s}\right)$$

Downstream :

Given speed downstream

$$= t'' = \frac{d}{8+s}$$
$$\Rightarrow \quad 3t' = t''$$
$$\Rightarrow \quad \frac{3d}{8-s} = \frac{d}{8+s}$$
$$\Rightarrow \quad \frac{3d}{8-s} = \frac{d}{8+s}$$
$$\Rightarrow \quad s = 4 \text{ km/hr}$$

9. Total expenditure = $\frac{15}{100}x$

$= 4,50,000$

$x = 3 \times 10^6$

Profit = 10 lakhs

So, total selling price = 40,00,000 ...(1)

Total purifies = 200 ...(2)

S.P. of each purifier = $\frac{(1)}{(2)}$ = 20,000

10. Probability for one bulb to be non defective is $\frac{95}{100}$

∴ Probabilities that none of the bulbs is defectives

$$\left(\frac{95}{100}\right)^4 = 0.8145$$

ELECTRONICS AND COMMUNICATION

1.
$$f'(x) = 0$$
$$\Rightarrow \quad \frac{1}{1+x} - 1 = 0$$
$$\Rightarrow \quad \frac{-x}{1+x} = 0$$
$$\Rightarrow \quad x = 0$$
and
$$f^{11}(x) = \frac{-1}{(1+x)^2} < 0 \text{ at } x = 0$$

2. (*a*) $\frac{dy}{dx} + xy = e^{-x}$ is a first order linear equation (non-homogeneous)

(*b*) $\frac{dy}{dx} + xy = 0$ is a first order linear equation (homogeneous)

(*c*), (*d*) are non linear equations

3. P1 – M3, P2 – M1, P3 – M4, P4 – M2

4. P [fourth head appears at the tenth toss] = P [getting 3 heads in the first 9 tosses and one head at tenth toss]

$$= \left[{}^9C_3 \cdot \left(\frac{1}{2}\right)^9\right] \times \left[\frac{1}{2}\right]$$
$$= \frac{21}{256} = 0.082$$

5.
$$\frac{\partial z}{\partial x} = y\left[x \times \frac{1}{xy} \times y + \ln xy\right]$$
$$= y[1 + \ln xy]$$
and
$$\frac{\partial z}{\partial y} = x(1 + \ln xy)$$
$$\Rightarrow \quad x\frac{\partial z}{\partial x} = y\frac{\partial z}{\partial y}$$

6. In a series RC circuit,

→ Initially at $t = 0$, capacitor charges with a current of $\frac{V_S}{R}$ and in steady state at $t = \infty$, capacitor behaves like open circuit and no current flows through the circuit

→ So the current $i(t)$ represents an exponential decay function

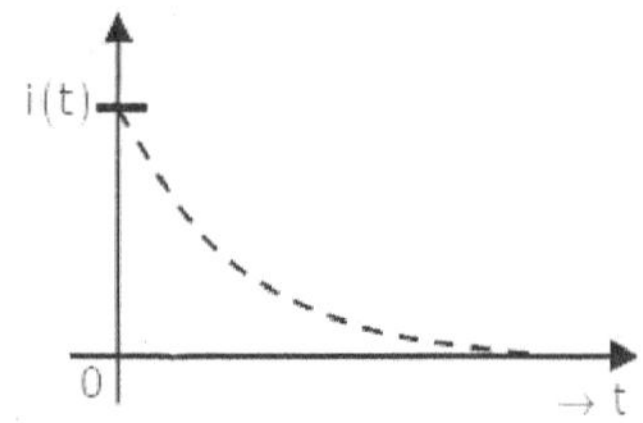

7. 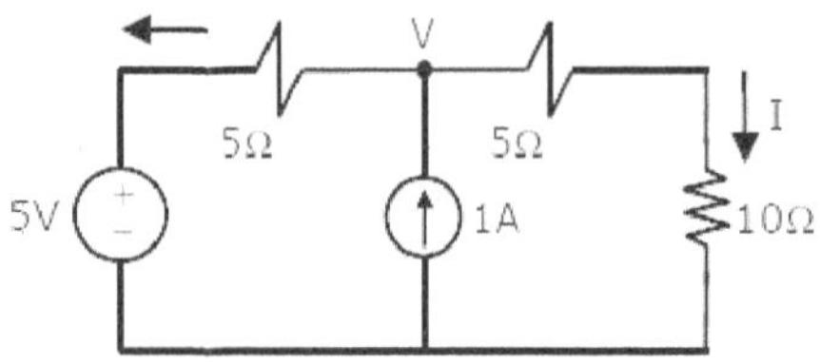

Apply KCL at node V,

$$\frac{V-5}{5} - 1 + \frac{V}{15} = 0$$

$$\Rightarrow \quad V = \frac{30}{4} \text{ volts}$$

$$\Rightarrow \quad \text{current } I = \frac{V}{15}$$

$$\Rightarrow \quad \frac{2}{4} = 0.50 \text{ Amperes}$$

9. Recombination rate,

$$R = B\left(n_{n_o} + n_n'\right)\left[P_{n_o} + P_n'\right]$$

n_{n_o} & P_{n_o} = Electron and hole concentrations respectively under thermal equilibrium

n_n' & p_n' = Excess elements and hole concentrations respectively

10. From Einstein relation,

$$\frac{D_P}{\mu_P} = \frac{kJ}{q}$$

$$\Rightarrow \quad D_P = 26 \text{ mV} \times 500 \text{ cm}^2/v-s$$

$$= 13 \text{ cm}^2/\text{s}$$

11. Transconductance amplifier must have

$$z_i = \infty$$

and $\quad z_0 = \infty$ ideally

12. KVL in base loop gives,

$$I_B = \frac{10-0.7}{100K} = 93\ \mu A$$

$$\Rightarrow \quad I_C = \beta I_B = 50 \times 93\ \mu A$$

$$= 4.65 \text{ mA}$$

from figure, $\quad V_0 = I_C R_C$

$$\Rightarrow \quad R_C = \frac{V_0}{I_C} = \frac{5V}{4.65mA}$$

$$= 1.075\ \Omega$$

13.

$$V_{dc} = V_m - \frac{I_{dc}}{4fc}$$

$$I_{dc}R_L = V_m - \frac{I_{dc}}{4fc}$$

$$I_{dc}\left[R_L + \frac{1}{4fc}\right] = V_m$$

$$\Rightarrow \quad I_{dc} = \frac{10}{100 + \frac{1}{4 \times 50 \times 4 \times 10^{-3}}}$$

$$= 0.09 \text{ A}$$

14. Maximum quantization error is $\frac{\text{step} - \text{size}}{2}$

$$\text{step} - \text{size} = \frac{8-0}{16}$$

$$= \frac{1}{2} = 0.5\text{V}$$

Quantization error = 0.25 V

15. Latches are used to construct Flip-Flop. Latches are level triggered, so if you use two latches in cascaded with inverted clock, then one latch will behave as master and another latch which is having inverted clock will be used as a slave and combined it will behave as a flip-flop. So given circuit is implementing Master-Slave D flip-flop

16. 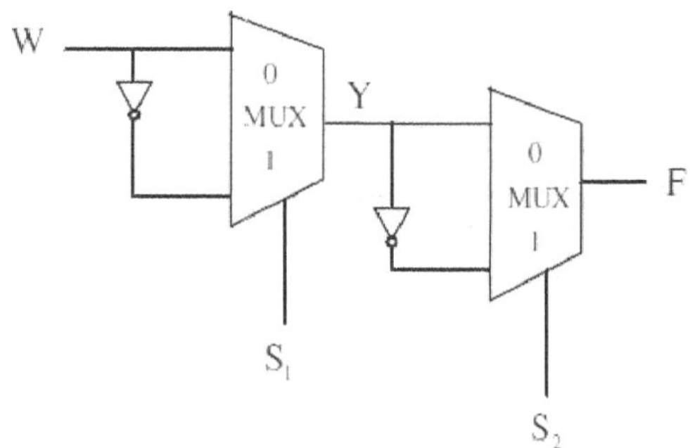

Output of first MUX = $w\bar{s}_1 + \bar{w}s_1 = w \oplus s_1$

Let $\quad Y = w \oplus s_1$

Output of second MUX = $Y\bar{s}_2 + \bar{Y}s_2$

$$= Y \oplus s_2$$

$$= w \oplus s_1 + s_2$$

17.

$$x(t) = \cos(10\,\pi t) + \cos(30\,\pi t),$$

$$F_s = 20 \text{ Hz}$$

Spectrum of $x(t)$

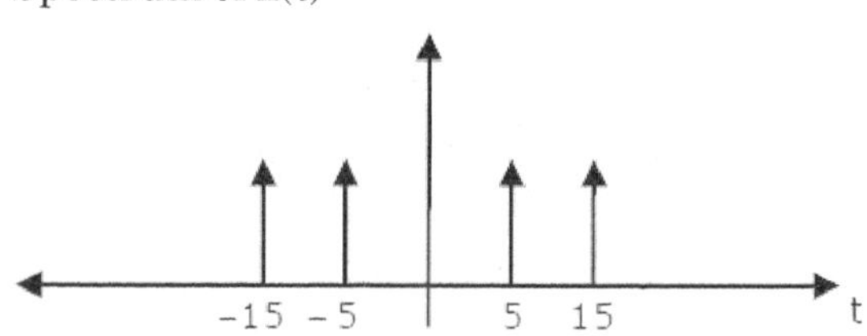

Spectrum of sampled version of $x(t)$

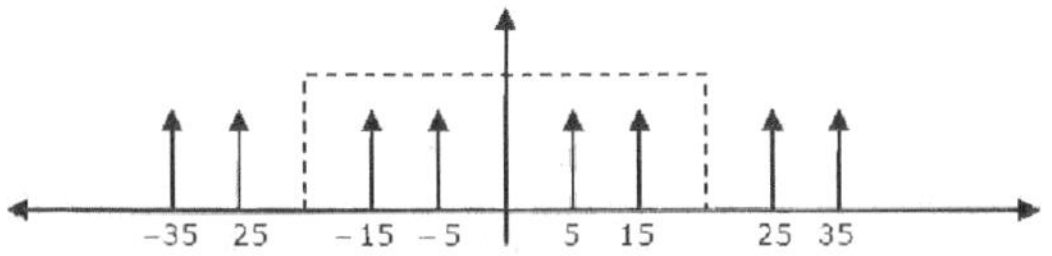

After LPF, signal will contain 5 and 15 Hz component only

18. For an all pass system,

$$\text{pole} = \frac{1}{\text{zero}^*}$$

or $$\text{zero} = \frac{1}{\text{pole}^*}$$

$$\text{pole} = a$$

$$\text{zero} = \frac{1}{b}$$

$$\Rightarrow \quad \frac{1}{b} = \frac{1}{a^*}$$

or $$b = a^*$$

19. Since $m(t)$ is a base band signal with maximum frequency 5 KHz, assumed spreads as follows :

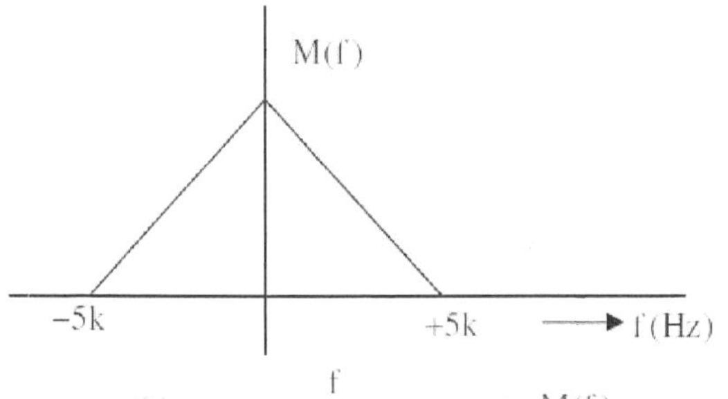

$$m(t) \xrightarrow{f} M(f)$$

$$\because \; y(t) = m(t)\cos(40000\pi t) \xrightarrow{7}$$

$$m(f)\frac{*1}{2}[\delta(f-20k)+\delta(f+20k)]$$

$$\therefore \; y(f) = \frac{1}{2}\left[M(f-20k)+M(f+20k)\right]$$

Thus the spectrum of the modulated signal is as follows :

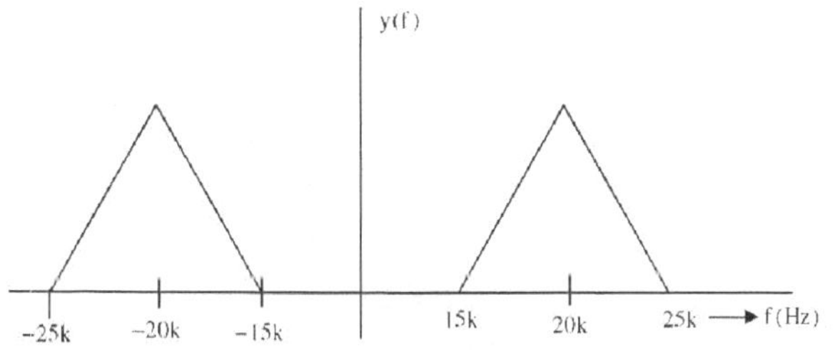

If $y(t)$ is sampled with a sampling frequency 'f_s' then the resultant signal is a periodic extension of successive replica of $y(f)$ with a period 'fs'.

It is observed that 10 KHz and 20 KHz are the two sampling frequencies which causes a replica of $M(f)$ which can be filtered out by a LPF.

Thus the minimum sampling frequency (f_s) which extracts $m(t)$ from $g(f)$ is 10 KHz.

20. By drawing the signal flow graph for the given block diagram

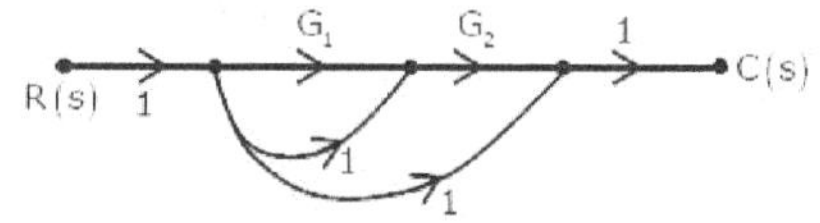

Number of parallel paths are three

Gains $\quad P_1 = G_1G_2,$

$P_2 = G_2,$

$P_3 = 1$

By mason's gain formula,

$$\frac{C(s)}{R(s)} = P_1 + P_2 + P_3$$

$$\Rightarrow \quad \boxed{G_1G_2 + G_2 + 1}$$

21. 1. $$\frac{Y(s)}{X(s)} = \frac{S-2}{S+3}$$

$$\Rightarrow \quad SY(s) + 3Y(s) = S \times (s) - 2X(s)$$

Due to initial condition, we can write above equation as

$$Sy(s) - y(0) + 3y(s) = sx(s) - x(0^-) - 2x(s)$$

$$y(0^-) = -2,$$

$$x(0^-) = 0$$

$$\left[x(t) = 3e^{2t}u(t)\right]$$

$$\Rightarrow \quad Sy(s) + 2 + 3y(s) = (s-2)\left(\frac{-3}{s-2}\right)$$

$$(s+3)y(s) = -3-2$$

$$\Rightarrow \quad y(s) = \frac{-5}{5+3}$$

$$\Rightarrow \quad y(t) = -5e^{-3t}u(t)$$

$y(\infty)$ (steady sate) = 0

2. $$H(s) = \frac{s-2}{s+3};$$

$$X(t) = -3e^{2t}.u(t)$$

$$\therefore \quad X(s) = \frac{-3}{s-2}$$

$$\Rightarrow \quad Y(s) = \frac{-3}{s+3}$$

$$y(t)\big|_{\text{at } t=\infty} \Rightarrow y(\infty) = \lim_{s\to 0} S.y(s)$$

$$= \lim_{s\to 0}\frac{-3s}{s+3}$$

$$y(\infty) = 0$$

22. Phase response of pass band waveform

$$\phi(f) = -2\pi\alpha(f - f_c) - 2\pi\beta f_c$$

Group delay $t_y = \frac{-d\phi(f)}{2\pi\, df} = \alpha$

Thus 'α' is actual signal propagation delay from transmitter to receiver.

23. Instantaneous phase

$$\phi_1(t) = 2\pi f_c t + \beta_1 \sin 2\pi f_1 + \beta_2 \sin 2\pi f_2 t$$

Instantaneous frequency

$$f_i(t) = \frac{d}{dt}\phi_1(t) \times \frac{1}{2\pi}$$

$$= f_c + \beta_1 f_1 \cos 2\pi f_1 t + \beta_2 f_2 \cos 2\pi f_2 t$$

Instantaneous frequency deviation

$$= \beta_1 f_1 \cos 2\pi f_1 t + \beta_2 f_2 \cos 2\pi f_2 t$$

Maximum $\Delta f = \beta_1 f_1 + \beta_2 f_2$

24.

$$f_c(TE_{21}) = \frac{C}{2}\sqrt{\left(\frac{2}{9}\right)^2 + \left(\frac{1}{b}\right)^2}$$

$$= \frac{3\times10^{10}}{2}\sqrt{\left(\frac{2}{5}\right)^2 + \left(\frac{1}{3}\right)^2}$$

$$= 1.5\times10^{10}\sqrt{0.16+0.111}$$

$$= 0.52 \times 1.5 \times 10^{10}$$

$$= 7.81 \text{ GHz}$$

$$= 7810 \text{ MHz}$$

25. Signal distortion implies impedance mismatch at both ends. i.e.,

$$Z_T \neq Z_0$$

$$Z_R \neq Z_0$$

26.

$$f'(x) = 6x^2 - 18x + 12 = 0$$

$\Rightarrow$ $x = 1, 2 \in [0, 3]$

Now $f(0) = -3$;

$f(3) = 6$

and $f(1) = 2$;

$f(2) = 1$

Hence, $f(x)$ is maximum at $x = 3$ and the maximum value is 6.

27. Consider,

$A\begin{bmatrix} -1 & 1 \\ 1 & -1 \end{bmatrix}$ which is real symmetric matrix

Characteristic equation is

$$|A - \lambda I| = 0$$

$\Rightarrow$ $(1+\lambda)^2 - 1 = 0$

$\Rightarrow$ $\lambda + 1 = \pm 1$

$\therefore$ $\lambda = 0, -2$ (not positive)

(*b*) is not true

(*a*), (*c*), (*d*) are true using properties of eigen values

28. Let the first toss be Head.

Let x denotes the number of tosses (after getting first head) to get first tail.

We can summarize the even as :

Event (After getting first H)	x	Probability $(p(x))$
T	1	$\frac{1}{2}$
HT	2	$\frac{1}{2} * \frac{1}{2} = \frac{1}{4}$
HHT	3	$\frac{1}{8}$

and so on ......

$$E(x) = \sum_{x=1}^{\infty} x p(x)$$

$$= 1x\frac{1}{2} + 2x\frac{1}{4} + 3x\frac{1}{8}$$

Let, $S = 1x\frac{1}{2} + 2x\frac{1}{4} + 3x\frac{1}{8}$ ...(I)

$\Rightarrow$ $\frac{1}{2}S = \frac{1}{4} + 2x\frac{1}{8} + 3x\frac{1}{16}$ ...(II)

(I – II) gives

$$\left(1 - \frac{1}{2}\right)S = \frac{1}{2} + \frac{1}{4} + \frac{1}{8} + \frac{1}{16} + \ldots$$

$\Rightarrow$ $\frac{1}{2}S = \frac{\frac{1}{2}}{1 - \frac{1}{2}} = 1$

$\Rightarrow$ $S = 2$

$\Rightarrow$ $E(x) = 2$

i.e. The expected number of tosses (after first head) to get first tail is 2 and same can be applicable if first toss results in tail.

Hence the average number of tosses is

$$1 + 2 = 3$$

29. Given $x_1 x_2$ and x_3 be independent and identically distributed with uniform distribution on [0, 1]

Let $z = x_1 + x_2 - x_3$

$$\Rightarrow P\{x_1 + x_2 \le x_3\} = P\{x_1 + x_2 - x_3 \le 0\}$$

$$= P\{z \le 0\}$$

Let us find probability density function of random variable z.

Since Z is summation of three random variable x_1, x_2 and $-x_3$

Overall pdf of z is convolution of the pdf of $x_1 x_2$ and $-x_3$

pdf of $\{x_1 + x_2\}$ is

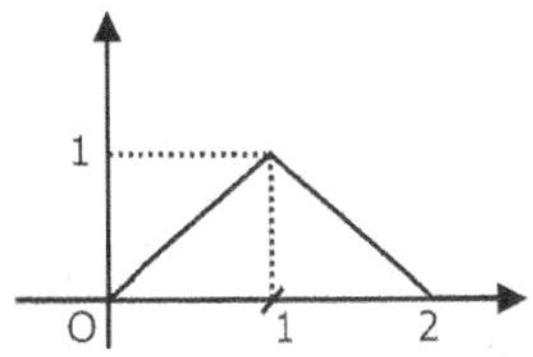

pdf of $-x_3$ is

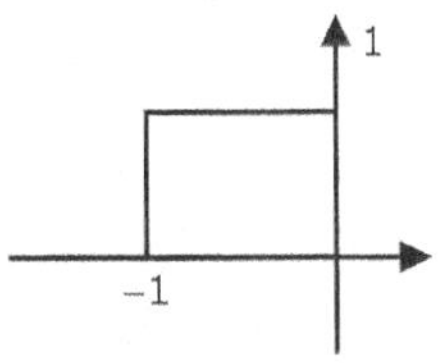

$$P\{z \le 0\} = \int_{-1}^{0} \frac{(z+1)^2}{2} dz$$

$$= \left.\frac{(z+1)^3}{6}\right|_{-1}^{0}$$

$$= \frac{1}{6} = 0.16$$

30. Two blocks are connected in cascade, Represent in s-domain,

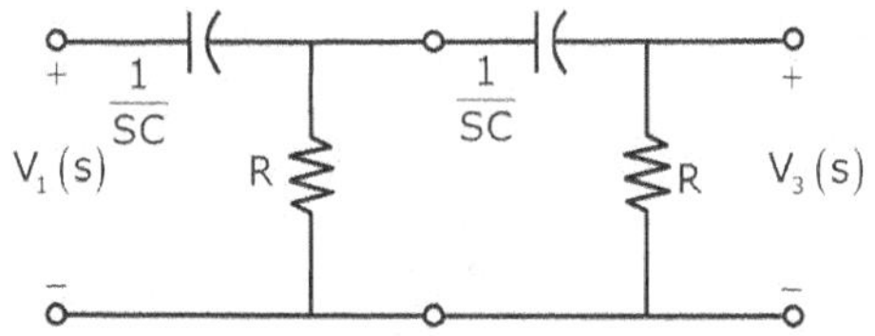

$$\frac{V_3(s)}{V_1(s)} = \frac{R \cdot R}{\frac{1}{sc}\left[R + R + \frac{1}{SC}\right] + R\left[\frac{1}{SC} + R\right]}$$

$$= \frac{R \cdot R}{\frac{1}{SC} \cdot \frac{1}{SC}\left[2R(SC) + 1\right] + \frac{R}{SC}\left[1 + RSC\right]}$$

$$= \frac{S^2C^2.R.R}{\left[1 + 2R(SC)\right] + RSC + R^2S^2C^2}$$

$$= \frac{S^2.100 \times 100 \times 10^{-6} \times 10^{-6} \times 10 \times 10 \times 10^3 \times 10^3}{S^2 \times 100 \times 10^6 \times 10^4 \times 10^{-12} + 3S + 100 \times 10^{-6} \times 10^4 + 1}$$

$$\frac{V_3(s)}{V_1(s)} = \frac{S^2}{1 + 3S + S^2}$$

31.

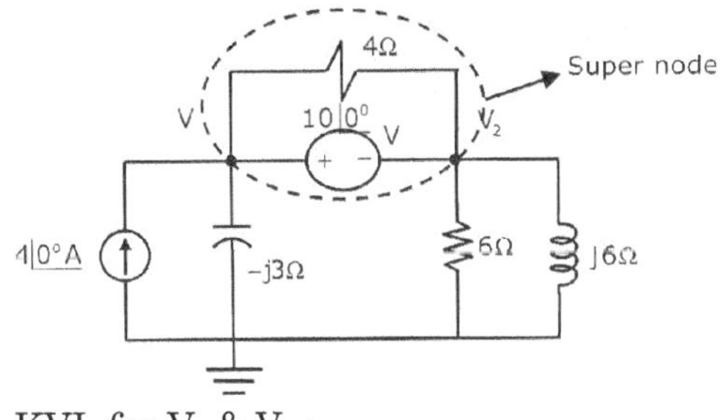

KVL for V_1 & V_2 :

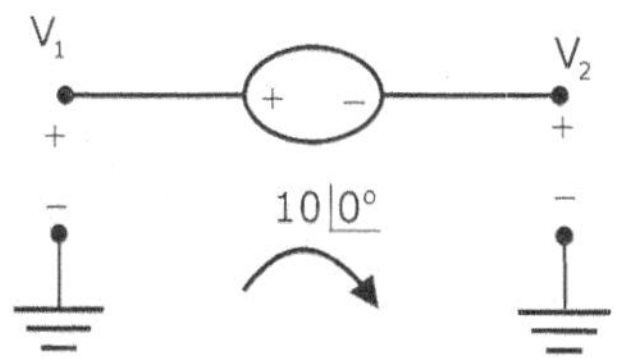

$$V_1 - V_2 = 10\angle 0^\circ \quad ...(1)$$

$$\boxed{V_1 = V_2 + 10\angle 0^\circ}$$

KCL at super node :

$$-4\angle 0^\circ + \frac{V_1}{-j3} + \frac{V_2}{6} + \frac{V_2}{j6} = 0 \quad ...(2)$$

$$\frac{V_1}{-j3} + \frac{V_2}{6} + \frac{V_2}{j6} = 4\angle 0^\circ$$

from (1) & (2),

$$\frac{V_2 + 10\angle 0^\circ}{-j3} + \frac{V_2}{6} + \frac{V_2}{j6} = 4\angle 0^\circ$$

$$V_2\left[\frac{1}{-j3} + \frac{1}{6} + \frac{1}{j6}\right] = 4\angle 0^\circ + \frac{10}{j3}$$

$\therefore$ $V_2 = (2 - j22)$ Volts

32. Norton's equivalent impedance

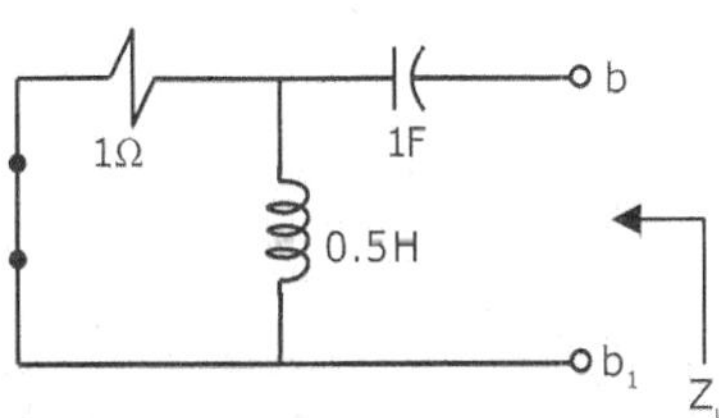

$$Z_N = \frac{1 * j\omega \cdot \frac{1}{2}}{1 + j\omega \cdot \frac{1}{2}} + \frac{1}{j\omega.1} = \frac{j\omega}{2 + j\omega} + \frac{1}{j\omega}$$

$$Z_N = \frac{(2 - \omega^2) + j\omega}{\left[2j\omega - \omega^2\right]}$$

$$\Rightarrow \quad Z_N = \frac{\left[\left(\omega^2-2\right)-j\omega\right]\cdot\left[\omega^2+2j\omega\right]}{\left[\omega^4+4\omega^2\right]}$$

Equating imaginary term to zero

i.e., $\omega^3 - 4\omega = 0$

$$\Rightarrow \quad \omega\left(\omega^2-4\right) = 0$$

$$\Rightarrow \quad \boxed{\omega = 2r/\sec}$$

33.

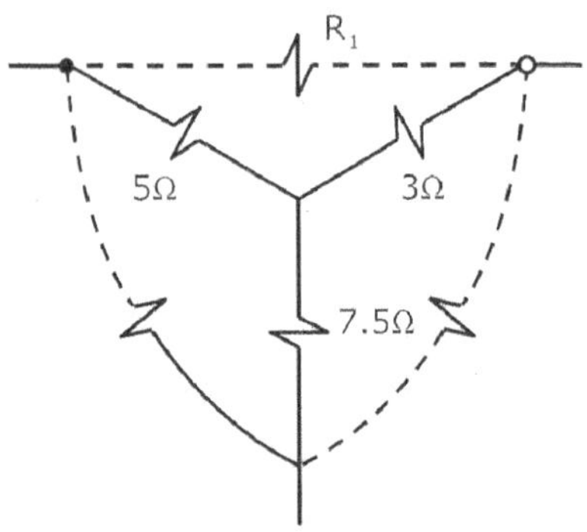

$$R_1 = \frac{(7.5)(5)+(3)(5)+(7.5)(3)}{7.5}\Omega$$

$$R_1 = 10\Omega$$

34. $$V_{bi} = V_T \ln\frac{N_A N_D}{n_i^2}$$

$$= 26 \text{ mv } \ln\left[\frac{5\times10^{18}\times1\times10^{16}}{\left(1.5\times10^{10}\right)^2}\right]$$

$$= 0.859 \text{ V}$$

$$W = \sqrt{\frac{2\varepsilon_S V_{bi}}{q}\left[\frac{N_A+N_D}{N_A N_D}\right]}$$

$$= 3.34\times10^{-5} \text{ cm}$$

35. In linear region,

$$I_D = k\left[(V_{GS}-V_T)V_{DS} - \frac{V_{DS}^2}{2}\right]$$

$$\frac{\partial I_D}{\partial V_{GS}} = 10^{-3}$$

$$= kV_{DS}$$

$\because V_{DS}$ is small, $\frac{V_{DS}^2}{2}$ is neglected

$$\Rightarrow \quad K = \frac{10^{-3}}{0.1}$$

$$= 0.01$$

In saturation region,

$$I_D = \frac{1}{2}k(V_{GS}-V_T)^2$$

$$\sqrt{I_D} = \sqrt{\frac{k}{2}}(V_{GS}-V_T)$$

$$\frac{\partial\sqrt{I_D}}{\partial V_{GS}} = \sqrt{\frac{k}{2}}$$

$$= \sqrt{\frac{0.01}{2}}$$

$$= 0.07$$

36. $$E_s = \frac{2\times0.2}{0.5}$$

$$= 0.8 \text{ v}/\mu\text{m}$$

$$E_{ox} = \frac{E_s}{E_{ox}}E_s$$

$$= 2.4 \text{ v}/\mu\text{m}$$

37. KVL in base loop,

$$5 - I_B(50k) - 0.7 = 0$$

$$I_B = \frac{5-0.7}{50\,k}$$

$$= 80 \text{ A}$$

$$\Rightarrow \quad I_C = \beta I_B$$

$$= 50\times86 \text{ A}$$

$$= 4.3 \text{ mA}$$

$$\therefore \quad R_C = \frac{10-V_{CE}(\text{sat})}{I_C}$$

$$= \frac{10-0.2}{4.3\text{mA}}$$

$$R_C = 2279\Omega$$

and the BJT is in saturation

38. Virtual ground and KCL at inverting terminal gives

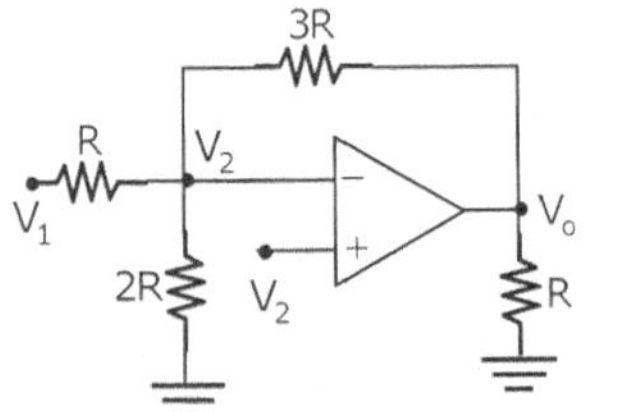

$$\frac{V_2-V_1}{R}+\frac{V_2}{2R}+\frac{V_2-V_0}{3R} = 0$$

$$\frac{V_0}{3R} = \frac{V_2}{R}+\frac{V_2}{3R}+\frac{V_2}{2R}-\frac{V_1}{R}$$

$$V_0 = -3V_1 + \frac{11}{2}V_2$$

39. Transistor m_1 switch from saturation to linear

$\Rightarrow \quad V_{DS} = V_{GS} - V_T;$

where $\quad V_{DS} = V_0$

and $\quad V_{GS} = V_i$

$\therefore \quad V_{DS} = V_0 = V_i - V_T$

Drain current $\quad I_D = \frac{1}{2}\mu_n \cos\frac{w}{L}(V_{GS} - V_T)^2$

$$\frac{V_{DD} - V_o}{10K} = \frac{1}{2} \times 100 \times 10^{-6} \times 2(V_{GS} - 0.5)^2$$

$$\frac{2 - (V_i - 0.5)}{10K} = 100 \times 10^{-6}(V_i - 0.5)^2$$

$\Rightarrow \quad V_i = 1.5V$

40. For an SRAM construction four MOSFETs are required (2-PMOS and 2-NMOS) with interchanged outputs connected to each CMOS inverter. So option (*b*) is correct.

41.

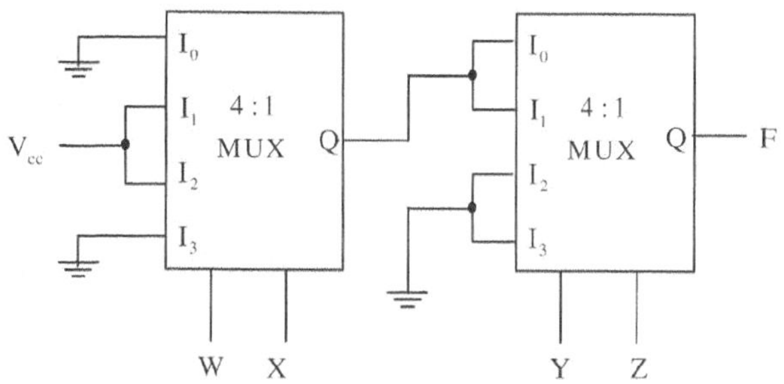

The output of the first MUX

$$= \overline{W} \times V_{cc} + W\overline{X}.V_{cc}$$

$$= \overline{W}X \times W\overline{X} \quad (\because V_{cc} = \text{logic } 1)$$

$$= W \oplus X$$

Let $\quad Q = W \oplus X$

The output of the second MUX

$$= Q.\overline{Y}\overline{Z} + Q.\overline{Y}Z$$

$$= Q.\overline{Y}(\overline{Z} + Z)$$

$$= Q.\overline{Y}.1 = Q.\overline{Y}$$

Put the value of Q in above expression

$$= (\overline{W}X + W\overline{X}).\overline{Y}$$

$$= \overline{W}X.\overline{Y} + W\overline{X}.\overline{Y}$$

42.

X	Y	D	B
0	0	0	0
0	1	1	1
1	0	1	0
1	1	0	0

So, $\quad D = X \oplus Y$

$$= \overline{X}Y + X\overline{Y}$$

and $\quad B = \overline{X}.Y$

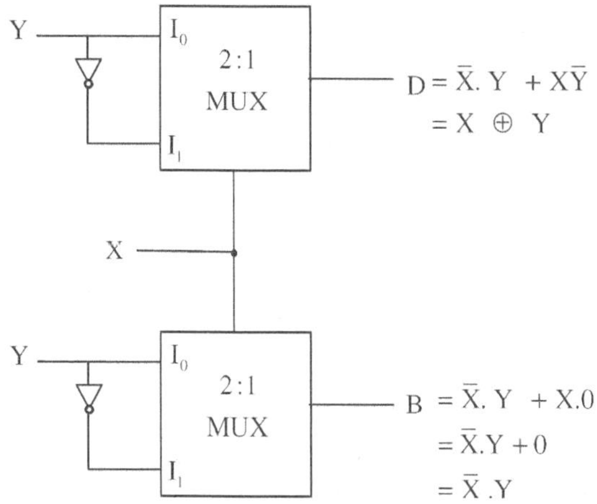

43. $\quad H_1(z) = (1 - Pz^{-1})^{-1}$

$\quad H_2(z) = (1 - qz^{-1})^{-1}$

$$H(z) = \frac{1}{1 - Pz^{-1}} + r\frac{1}{(1 - qz^{-1})}$$

$$= \frac{1 - qz^{-1} + r(1 - Pz^{-1})}{(1 - Pz^{-1})(1 - Pz^{-1})}$$

$$= \frac{(1 + r) - (q + rp)z^{-1}}{(1 - Pz^{-1})(1 - Pz^{-1})}$$

zero of $\quad H(z) = \frac{q + rp}{1 + r}$

Since zero is existing on unit circle

$$\Rightarrow \quad \frac{q + rp}{1 + r} = 1$$

or $$\frac{q + rp}{1 + r} = -1$$

$$\frac{-\frac{1}{4} + \frac{r}{2}}{1 + r} = 1$$

or $$\frac{-\frac{1}{4} + \frac{r}{2}}{1 + r} = -1$$

$$-\frac{1}{4} + \frac{r}{2} = 1 + r$$

or $$-\frac{1}{4} + \frac{r}{2} = -1 - r$$

$\Rightarrow \quad r = -\frac{5}{2}$

$\Rightarrow \quad \frac{r}{2} = -\frac{5}{4}$

or $\quad \frac{3}{4} = \frac{-3r}{2}$

$r = -\frac{1}{2}$

$\Rightarrow \quad r = -0.5$

$r = -\frac{5}{2}$ is not possible

44. $h(t) \leftrightarrow H(s) = \frac{1}{s+1}$

$\Rightarrow \quad h(t) = e^{-t}u(t)$

S_1 : System is stable (TRUE)

Because $h(t)$ absolutely integrable

S_2 : $\frac{h(t+1)}{h(t)}$ is independent of time (TRUE)

$\frac{e^{-(t+1)}}{e^{-t}} \Rightarrow e^{-1}$ (independent of time)

S_3 : A non-causal system with same transfer function is stable

$\frac{1}{s+1} \leftrightarrow -e^{-t}u(-t)$ (a non-causal system) but this is not absolutely integrable thus unstable.

Only S_1 and S_2 are TRUE

45. $X(z) = \frac{1}{(1-2z^{-1})^2}$

$= \frac{1}{(1-2z^{-1})}\frac{1}{(1-2z^{-1})}$

$x[n] = 2^n u[n] * 2^n u[n]$

$x[n] = \sum_{k=0}^{n} 2^k . 2^{(n-k)}$

$\Rightarrow \quad x[2] = \sum_{k=0}^{2} 2^k . 2^{(2-k)} \; 2^0.2^2 + 2^1.2^1 + 2^2.2^0$

$= 2^0.2^2 + 2^1.2^1 + 2^2.2^0$

$= 4 + 4 + 4 = 12$

46. Given $\quad G(s) = \frac{4}{s+2};$

$H(s) = \frac{2}{s+4}$

For unit step input,

$k_p = \lim_{s\to 0} G(s)H(s)$

$k_p = \lim_{s\to 0}\left(\frac{4}{s+2}\right)\left(\frac{2}{s+4}\right)$

$\boxed{k_p = 1}$

Steady state error $e_{ss} = \frac{A}{1+k_p}$

$e_{ss} = \frac{1}{1+1}$

$e_{ss} = \frac{1}{2} \Rightarrow 0.50$

47. Apply linearity principle,

$$\begin{bmatrix}3\\5\end{bmatrix} = a\begin{bmatrix}1\\-1\end{bmatrix} + b\begin{bmatrix}0\\1\end{bmatrix} s$$

$a = 3;\ b = 8$

$$\Rightarrow \quad x(t) = 3\begin{bmatrix}e^{-t}\\-e^{-t}\end{bmatrix} + \begin{bmatrix}e^{-t} - e^{-2t}\\-e^{-t} + 2e^{-2t}\end{bmatrix}$$

$$\Rightarrow \quad x(t) = \begin{bmatrix}11e^{-t} - 8e^{-2t}\\-11e^{-t} + 16e^{-2t}\end{bmatrix}$$

48. For transfer function $\frac{(s+4)}{(s+1)(s+2)(s+3)}$

From pole zero plot

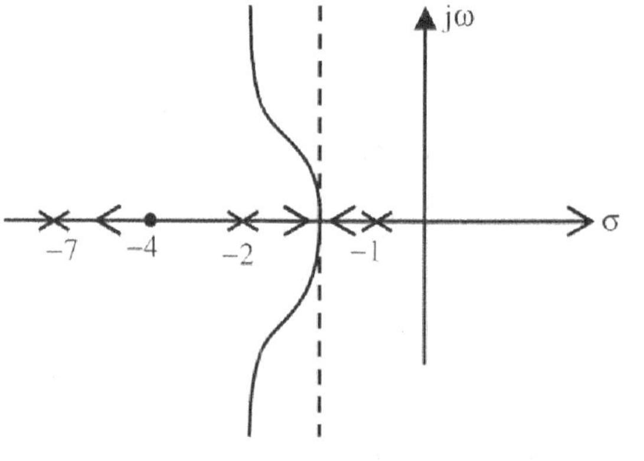

49. Shifting in time domain does not change PSD. Since PSD is Fourier transform of autocorrelation function of WSS process, autocorrelation function depends on time difference.

$X(t) \leftrightarrow R_x(z) \leftrightarrow S_x(f)$

$Y(t) = X(2t-1) \leftrightarrow R_y(2\zeta) \leftrightarrow \frac{1}{2}S_x\left(\frac{f}{2}\right)$

[time scaling property of Fourier transform]

50.

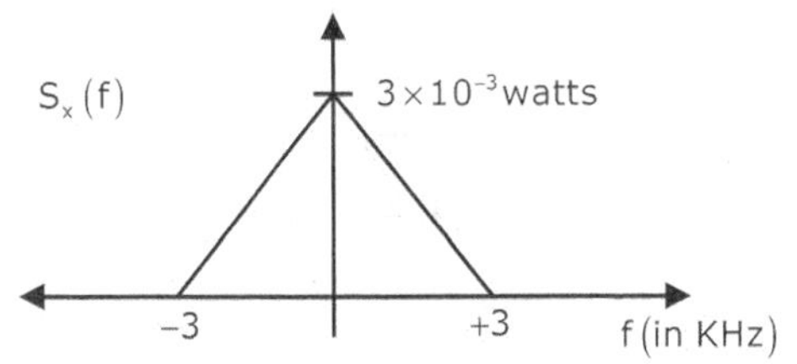

After modulation with cos (16000 πt)

$$S_y(f) = \frac{1}{4}[s_x(f - f_c) + S_x(f + f_c)]$$

This is obtain the power spectral density Random process $y(t)$, we shift the given power spectral density random process $x(t)$ to the right by f_c shift it to be the left by f_c and the two shifted power spectral and divide by 4.

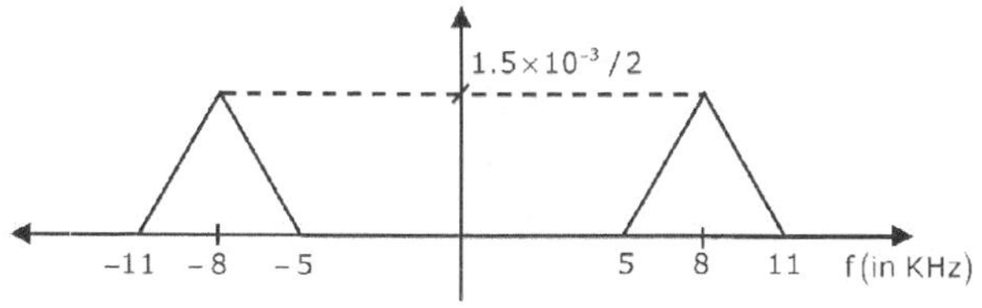

After band pass filter of center frequency 8 KHz and BW of 2 KHz

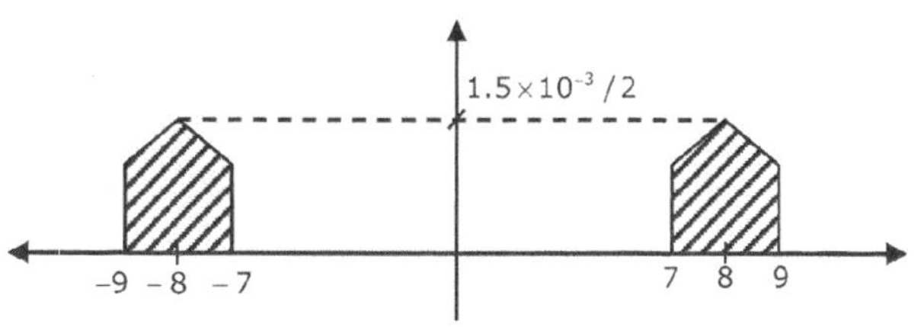

Total output power is area of shaded region

= 2[Area of one side portion]

= 2[Area of triangle + Area of rectangle]

$$= \frac{2\left[-\frac{1}{2}\times2\times10^{3}\times0.5\times10^{-3}+2\times10^{3}\times1\times10^{-3}\right]}{2}$$

= [0.5 + 2] = 2.5 watts

51. Nyquist rate = 2 × 50Hz

= 100 samples/sec

$$\Delta = \frac{m(t)_{\max} - m(t)_{\min}}{L}$$

$$\Rightarrow \quad L = \frac{\sqrt{2}-(-\sqrt{2})}{0.75}$$

$$L = \frac{2\sqrt{2}}{0.75}$$

= 3.77 = 4

Number of bits required to encode '4' levels

= 2 bits/level

Thus data rate = 2 × 100 = 200 bits/sec

52. Let $P\{x = 2\} = \frac{1}{3}$,

$$P\{x = 0\} = \frac{2}{3}$$

to find $H(Y_1)$ we need to know

$P\{y_1 = 0\}$ and $P\{y_2 = 1\}$

$P\{Y_1 = 0\} = P\{Y_1 = 0/x_1 = 0\}$

$P\{x_1 = 0\} + P\{y_1 = 0/x_1 = 1\}\ P\{x_1 = 1\}$

$$= \frac{1}{2}.\frac{1}{3}+\frac{1}{2}\times\frac{2}{3}=\frac{1}{2}$$

$$P\{y_1 = 1\} = \frac{1}{2}$$

$$\Rightarrow \quad H(y_1) = \frac{1}{2}\log_2^2+\frac{1}{2}\log_2^2 = 1$$

Similarly $P(y_2 = 0\} = \frac{1}{2}$

and $P\{y_2 = 1\} = \frac{1}{2}$

$\Rightarrow \quad H\{y_2\} = 1$

$\Rightarrow \quad H\{y_1\} + H\{y_2\} = 2$ bits

53. (1)

$$\text{Curl } \vec{A} = \begin{vmatrix} \hat{a}_x & \hat{a}_y & \hat{a}_z \\ \frac{\partial}{\partial x} & \frac{\partial}{\partial y} & \frac{\partial}{\partial z} \\ \cos x \sin y & \sin x \cos y & 0 \end{vmatrix}$$

$$= \vec{0}$$

$$\therefore \quad \left|\text{Curl } \vec{A}\right| = 0$$

(2) Given $A = \cos x \sin y\hat{a}_x + \sin x \cos y\hat{a}_y$

$$\nabla\times A = \begin{vmatrix} a_x & a_y & a_z \\ \frac{\partial}{\partial x} & \frac{\partial}{\partial y} & \frac{\partial}{\partial z} \\ \cos x \sin y & \sin x \cos y & 0 \end{vmatrix}$$

$$= a_x(0) - a_y(0) + a_z(\cos x\cos y - \cos x\cos y) = 0$$

$$\therefore \quad |\nabla\times A| = 0$$

54. Given medium (1) is perfect conductor Medium (2) is air

$\therefore \quad H_1 = 0$

From boundary conditions

$$(H_1 - H_2) \times a_n = K_S$$

$$\left.\begin{matrix} H_1 = 0 \\ a_n = a_y \end{matrix}\right| K_S = 2\hat{a}_x$$

$$-H_2 \times a_y = 2\hat{a}_x$$

$$-(H_x a_x + H_y a_y + H_z a_z) \times a_y = 2a_x$$

$$-H_x a_z + H_z a_x = 2a_x$$

$\therefore \quad H_z = 2$

$$\boxed{H = 2a_z}$$

55. Given $\quad E = 10\cos(\omega t - 3x - \sqrt{3}z)\, a_y$

$$E = E_0 e^{-J\beta(x\cos\theta_x + y\cos\theta_y + z\cos\theta_z)}$$

So

$$\beta_x = \beta\cos\theta_x = 3$$

$$\beta_y = \beta\cos\theta_y = 0$$

$$\beta_z = \beta\cos\theta_z = \sqrt{3}$$

$$\beta_x^2 + \beta_y^2 + \beta_z^2 = \beta^2$$

$$9 + 3 = \beta^2$$

$$\Rightarrow \quad \beta = \sqrt{13}$$

$$\beta\cos\theta_z = \sqrt{3}$$

$$\Rightarrow \quad \cos\theta_z = \sqrt{\frac{3}{13}}$$

$$\Rightarrow \quad \theta_z = 61.28 = \theta_i$$

$$\frac{\sin\theta_i}{\sin\theta_t} = \sqrt{\frac{E_2}{E_1}}$$

$$\Rightarrow \quad \frac{\sin 61.28}{\sin\theta_t} = \sqrt{\frac{3}{1}}$$

$$\Rightarrow \quad \frac{0.8769}{\sqrt{3}} = \sin\theta_t$$

$$\theta_t = 30.4$$

$$\Rightarrow \quad \theta_t \simeq 30^\circ$$

■■

SOLVED PAPER – 2014

ELECTRONICS AND COMMUNICATION ENGINEERING

Set - 4

No. of Questions : 65 | *Maximum Marks : 100*

GENERAL APTITUDE

Q. No. 1 – 5 Carry One Mark Each

1. Which of the following options is the closest in meaning to the word underlined in the sentence below?

   In a democracy, everybody has the freedom to <u>disagree</u> with the government.

   (*a*) Dissent (*b*) Descent
   (*c*) Decent (*d*) Decadent

2. After the discussion, Tom said to me, 'Please revert!' He expects me to ________.

   (*a*) Retract (*b*) Get back to him
   (*c*) Move in reverse (*d*) Retreat

3. While receiving the award, the scientist said, "I feel vindicated". Which of the following is closest in meaning to the word 'vindicated'?

   (*a*) Punished (*b*) Substantiated
   (*c*) Appreciated (*d*) Chastened

4. Let $f(x, y) = x^n y^m = P$. If x is doubled and y is halved, the new value of f is

   (*a*) 2^{n-m} P (*b*) 2^{m-n} P
   (*c*) $2(n - m)$P (*d*) $2(m - n)$P

5. In a sequence of 12 consecutive odd numbers, the sum of the first 5 numbers is 425. What is the sum of the last 5 numbers in the sequence?

Q. No. 6 – 10 Carry Two Marks Each

6. Find the next term in the sequence:

   13M, 17Q, 19S, ___

   (*a*) 21W (*b*) 21V
   (*c*) 23W (*d*) 23 V

7. If 'KCLFTSB' stands for 'best of luck' and 'SHSWDG' stands for 'good wishes', which of the following indicates 'ace the exam'?

   (*a*) MCHTX (*b*) MXHTC
   (*c*) XMHCT (*d*) XMHTC

8. Industrial consumption of power doubled from 2000-2001 to 2010-2011. Find the annual rate of increase in percent assuming it to be uniform over the years.

   (*a*) 5.6 (*b*) 7.2
   (*c*) 10.0 (*d*) 12.2

9. A firm producing air purifiers sold 200 units in 2012. The following pie chart presents the share of raw material, labour, energy, plant and machinery, and transportation costs in the total manufacturing cost of the firm in 2012. The expenditure on labour in 2012 is ₹ 4,50,000. In 2013, the raw material expenses increased by 30% and all other expenses increased by 20%. What is the percentage increase in total cost for the company in 2013?

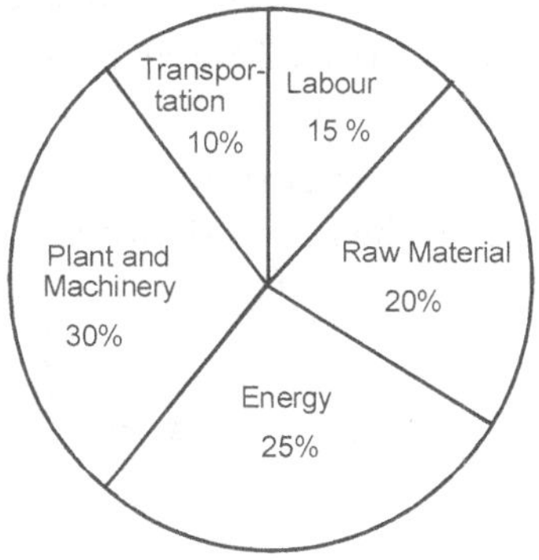

10. A five digit number is formed using the digits 1, 3, 5, 7 and 9 without repeating any of them. What is the sum of all such possible five digit numbers?

    (*a*) 6666660
    (*b*) 6666600
    (*c*) 6666666
    (*d*) 6666606

ELECTRONICS AND COMMUNICATION

Q. No. 1 – 25 Carry One Mark Each

1. The series $\sum_{n=0}^{\infty} \frac{1}{n!}$ converges to

   (*a*) $2 \ln 2$ (*b*) $\sqrt{2}$
   (*c*) 2 (*d*) e

2. The magnitude of the gradient for the function $f(x, y, z) = x^2 + 3y^2 + z^3$ at the point (1,1,1) is ________.

3. Let X be a zero mean unit variance Gaussian random variable. E[|X|] is equal to _____

4. If a and b are constants, the most general solution of the differential equation

$$\frac{d^2x}{dt^2}+2\frac{dx}{dt}+x=0 \text{ is}$$

(a) ae^{-t} (b) $ae^{-t}+bte^{-t}$
(c) $ae^{t}+bte^{-t}$ (d) ae^{2t}

5. The directional derivative of $f(x, y)=\frac{xy}{\sqrt{2}}(x+y)$ at (1, 1) in the direction of the unit vector at an angle of $\frac{\pi}{4}$ with y-axis, is given by _____ .

6. The circuit shown in the figure represents a

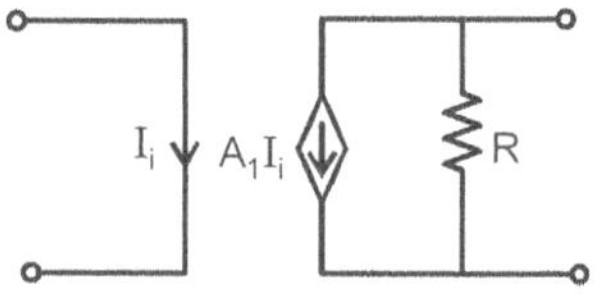

(a) Voltage controlled voltage source
(b) Voltage controlled current source
(c) Current controlled current source
(d) Current controlled voltage source

7. The magnitude of current (in mA) through the resistor R_2 in the figure shown is_______.

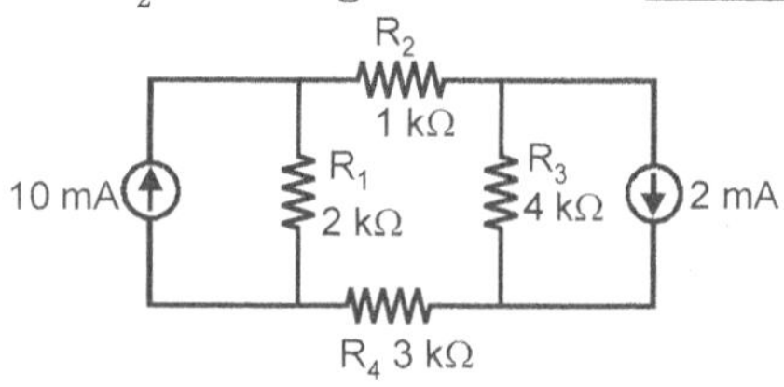

8. At T = 300 K, the band gap and the intrinsic carrier concentration of GaAs are 1.42 eV and 10^6 cm^{-3}, respectively. In order to generate electron hole pairs in GaAs, which one of the wavelength (λ_C) ranges of incident radiation, is most suitable? (Given that: Plank's constant is 6.62×10^{-34} J-s, velocity of light is 3×10^{10} cm/s and charge of electron is 1.6×10^{-19} C)

(a) 0.42 μm < λ_C < 0.87 μm
(b) 0.87 μm < λ_C < 1.42 μm
(c) 1.42 μm < λ_C < 1.62 μm
(d) 1.62 μm < λ_C < 6.62 μm

9. In the figure ln (ρ_i) is plotted as a function of 1/T, where ρ_i is the intrinsic resistivity of silicon, T is the temperature, and the plot is almost linear.

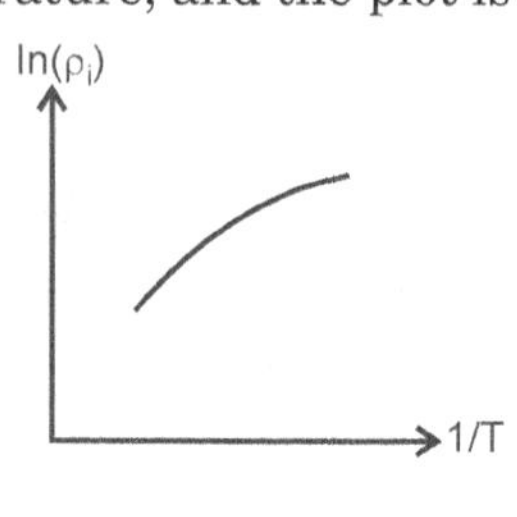

The slope of the line can be used to estimate

(a) Band gap energy of silicon (E_g)
(b) Sum of electron and hole mobility in silicon ($\mu_n + \mu_p$)
(c) Reciprocal of the sum of electron and hole mobility in silicon $(\mu_n + \mu_p)^{-1}$
(d) Intrinsic carrier concentration of silicon (n_i)

10. The cut-off wavelength (in μm) of light that can be used for intrinsic excitation of a semiconductor material of bandgap E_g = 1.1 eV is ________.

11. If the emitter resistance in a common-emitter voltage amplifier is not bypassed, it will

(a) Reduce both the voltage gain and the input impedance
(b) Reduce the voltage gain and increase the input impedance
(c) Increase the voltage gain and reduce the input impedance
(d) Increase both the voltage gain and the input impedance

12. Two silicon diodes, with a forward voltage drop of 0.7 V, are used in the circuit shown in the figure. The range of input voltage V_i for which the output voltage $V_0 = V_i$, is

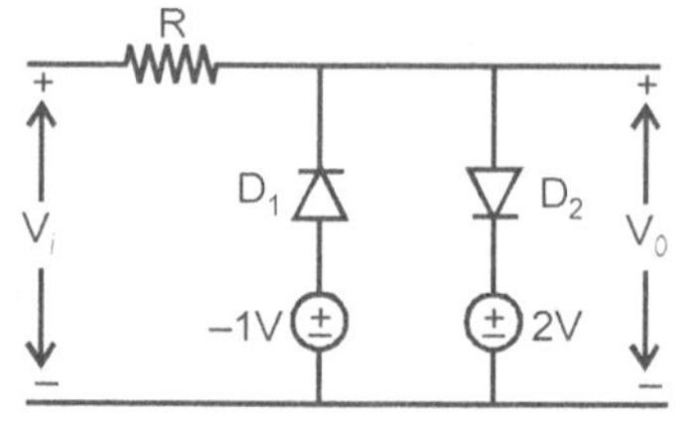

(a) – 0.3V < V_i < 1.3V
(b) – 0.3V < V_i < 2V
(c) – 1.0V < V_i < 2.0V
(d) – 1.7V < V_i < 2.7V

13. The circuit shown represents:

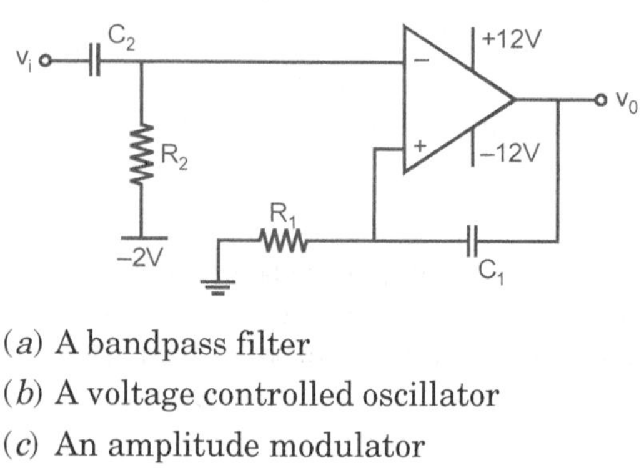

(a) A bandpass filter
(b) A voltage controlled oscillator
(c) An amplitude modulator
(d) A monostable multivibrator

14. For a given sample-and-hold circuit, if the value of the hold capacitor is increased, then

(*a*) Droop rate decreases and acquisition time decreases

(*b*) Droop rate decreases and acquisition time increases

(*c*) Droop rate increases and acquisition time decreases

(*d*) Droop rate increases and acquisition time increases

15. In the circuit shown in the figure, if C = 0, the expression for Y is

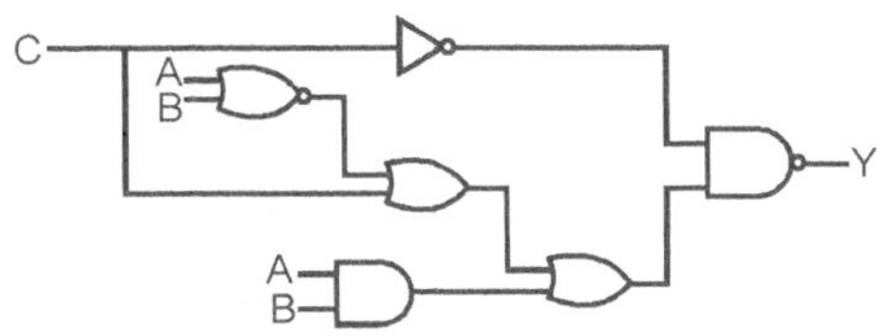

(*a*) $Y = A\bar{B} + \bar{A}B$ (*b*) $Y = A + B$

(*c*) $Y = \bar{A} + \bar{B}$ (*d*) $Y = AB$

16. The output (Y) of the circuit shown in the figure is

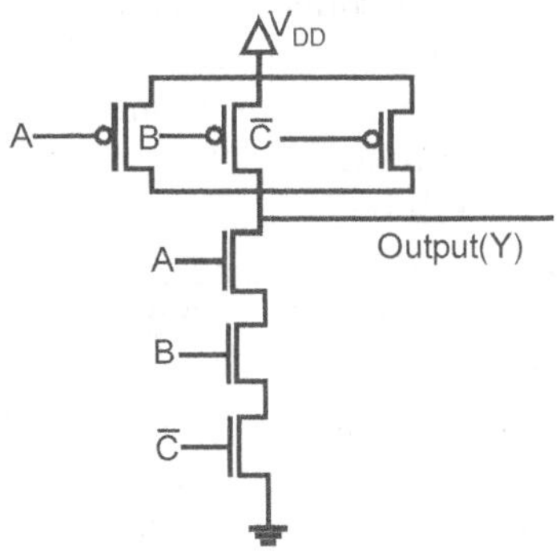

(*a*) $\bar{A} + \bar{B} + C$ (*b*) $A + \bar{B}.\bar{C} + A.\bar{C}$

(*c*) $\bar{A} + B + \bar{C}$ (*d*) $A.B.\bar{C}$

17. A Fourier transform pair is given by

$$\left(\frac{2}{3}\right)^n u[n+3] \overset{FT}{\Leftrightarrow} \frac{Ae^{-j6\pi f}}{1-\left(\frac{2}{3}\right)e^{-j2\pi f}}$$

where $u[n]$ denotes the unit step sequence. The value of A is ________.

18. A real-valued signal $x(t)$ limited to the frequency band $|f| \le \frac{W}{2}$ is passed through a linear time invariant system whose frequency response is

$$H(f) = \begin{cases} e^{-j4\pi f}, & |f| \le \frac{W}{2} \\ 0, & |f| > \frac{W}{2} \end{cases}$$

The output of the system is

(*a*) $x(t+4)$ (*b*) $x(t-4)$

(*c*) $x(t+2)$ (*d*) $x(t-2)$

19. The sequence $x[n] = 0.5^n u[n]$, where $u[n]$ is the unit step sequence, is convolved with itself to obtain $y[n]$. Then $\sum_{n=-\infty}^{+\infty} y[n]$ is________.

20. In a Bode magnitude plot, which one of the following slopes would be exhibited at high frequencies by a 4th order all-pole system?

(*a*) – 80 dB/decade

(*b*) – 40 dB/decade

(*c*) + 40 dB/decade

(*d*) + 80 dB/decade

21. For the second order closed-loop system shown in the figure, the natural frequency (in rad/s) is

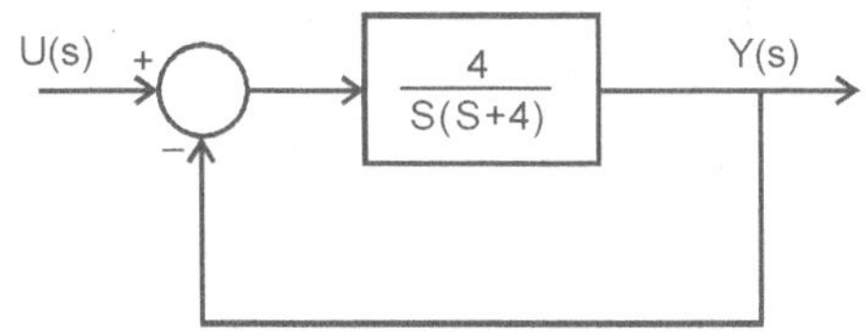

(*a*) 16 (*b*) 4

(*c*) 2 (*d*) 1

22. If calls arrive at a telephone exchange such that the time of arrival of any call is independent of the time of arrival of earlier or future calls, the probability distribution function of the total number of calls in a fixed time interval will be

(*a*) Poisson (*b*) Gaussian

(*c*) Exponential (*d*) Gamma

23. In a double side-band (DSB) full carrier AM transmission system, if the modulation index is doubled, then the ratio of total sideband power to the carrier power increases by a factor of____________.

24. For an antenna radiating in free space, the electric field at a distance of 1 km is found to be 12 mV/m. Given that intrinsic impedance of the free space is 120π Ω, the magnitude of average power density due to this antenna at a distance of 2 km from the antenna (in nW/m²) is____________.

25. Match column A with column B.

	Column A	Column B
P.	Point electromagnetic source	1. Highly directional
Q.	Dish antenna	2. End free
R.	Yagi-Uda antenna	3. Isotropic

	P	Q	R
(*a*)	1	2	3
(*b*)	3	1	2
(*c*)	2	1	3
(*d*)	3	2	1

Q. No. 26–55 Carry Two Marks Each

26. With initial values $y(0) = y'(0) = 1$ the solution of the differential equation $\frac{d^2y}{dx^2} + 4\frac{dy}{dx} + 4y = 0$ at $x = 1$ is ______

27. Parcels from sender S to receiver R pass sequentially through two post-offices. Each post-office has a probability $\frac{1}{5}$ of losing an incoming parcel, independently of all other parcels. Given that a parcel is lost, the probability that it was lost by the second post-office is________.

28. The unilateral Laplace transform of $f(t)$ is $\frac{1}{s^2+s+1}$. Which one of the following is the unilateral Laplace transform of $g(t) = t.\ f(t)$?

(*a*) $\frac{-s}{(s^2+s+1)^2}$ (*b*) $\frac{-(2s+1)}{(s^2+s+1)^2}$

(*c*) $\frac{s}{(s^2+s+1)^2}$ (*d*) $\frac{2s+1}{(s^2+s+1)^2}$

29. For a right angled triangle, if the sum of the lengths of the hypotenuse and a side is kept constant, in order to have maximum area of the triangle, the angle between the hypotenuse and the side is

(*a*) 12° (*b*) 36°

(*c*) 60° (*d*) 45°

30. The steady state output of the circuit shown in the figure is given by $y(t) = A(\omega)\sin(\omega t + \phi(\omega))$. If the amplitude $|A(\omega)| = 0.25$, then the frequency ω is

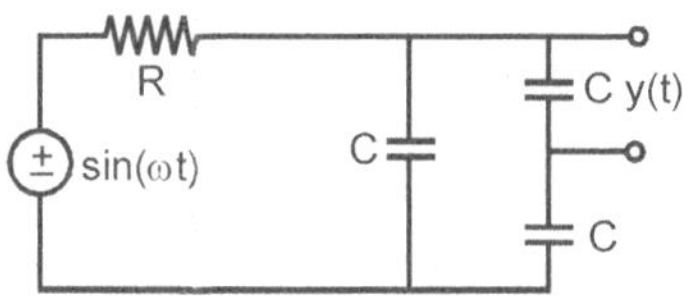

(*a*) $\frac{1}{\sqrt{3}RC}$ (*b*) $\frac{2}{\sqrt{3}RC}$

(*c*) $\frac{1}{RC}$ (*d*) $\frac{2}{RC}$

31. In the circuit shown in the figure, the value of $V_0(t)$ (in Volts) for $t \to \infty$ is ______.

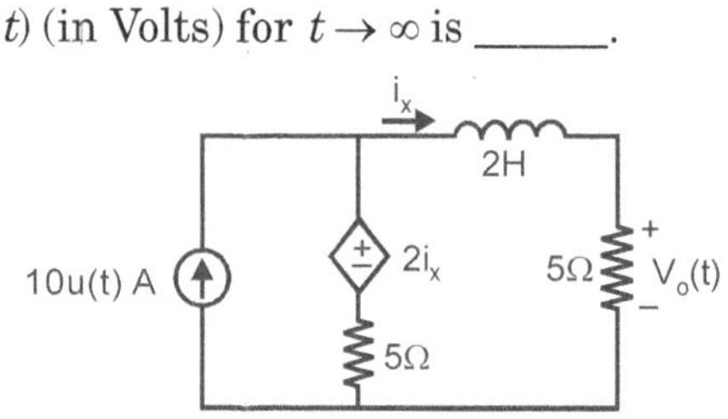

32. The equivalent resistance in the infinite ladder network shown in the figure is R_e.

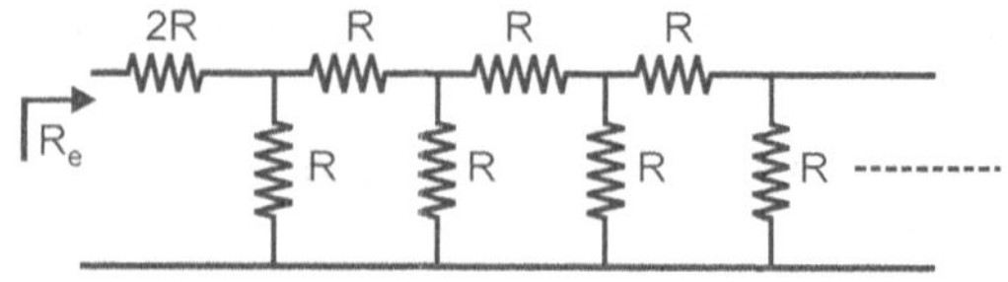

The value of R_e/R is ________

33. For the two-port network shown in the figure, the impedance (Z) matrix (in Ω) is

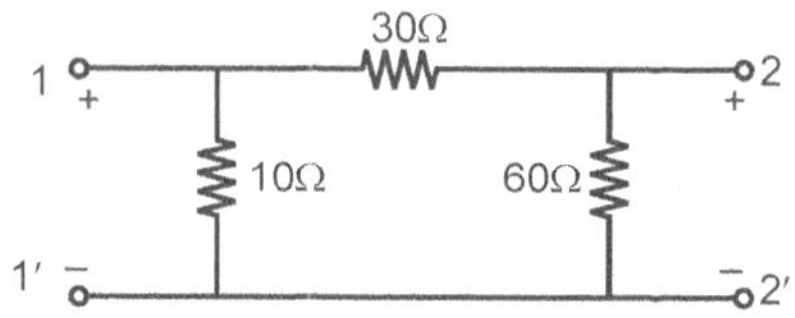

(*a*) $\begin{bmatrix} 6 & 24 \\ 42 & 9 \end{bmatrix}$ (*b*) $\begin{bmatrix} 9 & 8 \\ 8 & 24 \end{bmatrix}$

(*c*) $\begin{bmatrix} 9 & 6 \\ 6 & 24 \end{bmatrix}$ (*d*) $\begin{bmatrix} 42 & 6 \\ 6 & 60 \end{bmatrix}$

34. Consider a silicon sample doped with $N_D = 1 \times 10^{15}/cm^3$ donor atoms. Assume that the intrinsic carrier concentration $n_i = 1.5 \times 10^{10}/cm^3$. If the sample is additionally doped with $N_A = 1 \times 10^{18}/cm^3$ acceptor atoms, the approximate number of electrons/cm³ in the sample, at T = 300 K, will be ______________.

35. Consider two BJTs biased at the same collector current with area A_1 = 0.2 μm × 0.2 μm and A_2 = 300 μm × 300 μm. Assuming that all other device parameters are identical, kT/q = 26 mV, the intrinsic carrier concentration is 1×10^{10} cm^{-3}, and $q = 1.6 \times 10^{-19}$ C, the difference between the base-emitter voltages (in mV) of the two BJTs (i.e., $V_{BE1} - V_{BE2}$) is ______________.

36. An N-type semiconductor having uniform doping is biased as shown in the figure.

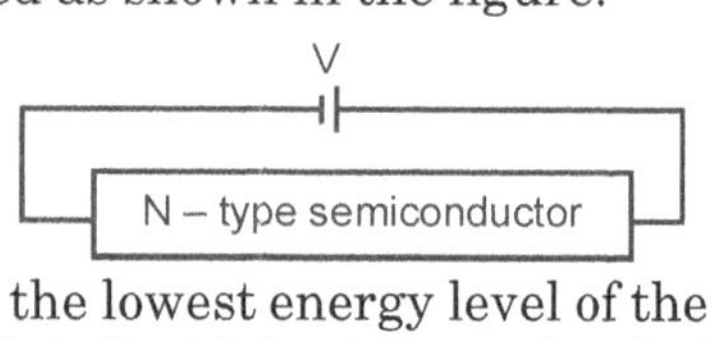

If E_C is the lowest energy level of the conduction band, E_V is the highest energy level of the valance band and E_F is the Fermi level, which one of the following represents the energy band diagram for the biased N-type semiconductor?

(*a*)
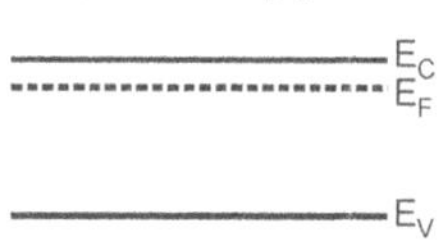

(*b*)
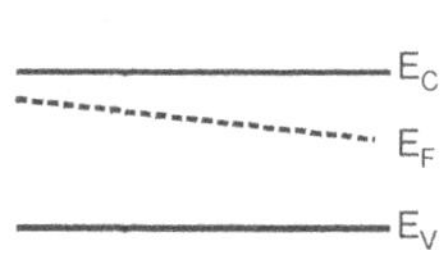

(*c*)
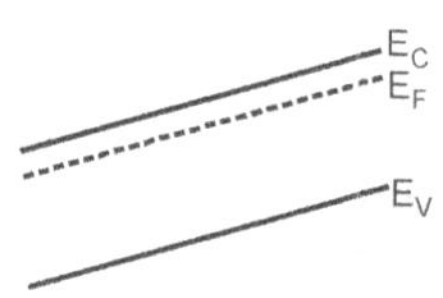

(*d*)
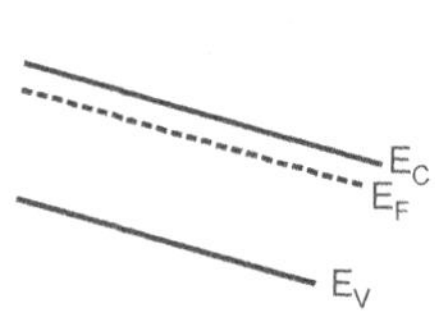

37. Consider the common-collector amplifier in the figure (bias circuitry ensures that the transistor operates in forward active region, but has been omitted for simplicity). Let I_C be the collector current, V_{BE} be the base-emitter voltage and V_T be the thermal voltage. Also, g_m and r_0 are the small-signal transconductance and output resistance of the transistor, respectively. Which one of the following conditions ensures a nearly constant small signal voltage gain for a wide range of values of R_E?

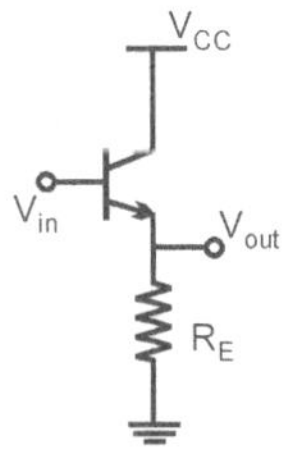

(*a*) $g_m R_E << 1$

(*b*) $I_C R_E >> V_T$

(*c*) $g_m r_0 >> 1$

(*d*) $V_{BE} >> V_T$

38. A BJT in a common-base configuration is used to amplify a signal received by a 50Ω antenna. Assume kT/q = 25 mV. The value of the collector bias current (in mA) required to match the input impedance of the amplifier to the impedance of the antenna is________.

39. For the common collector amplifier shown in the figure, the BJT has high β, negligible $V_{CE(sat)}$, and V_{BE} = 0.7 V. The maximum undistorted peak-to-peak output voltage v_o (in Volts) is______.

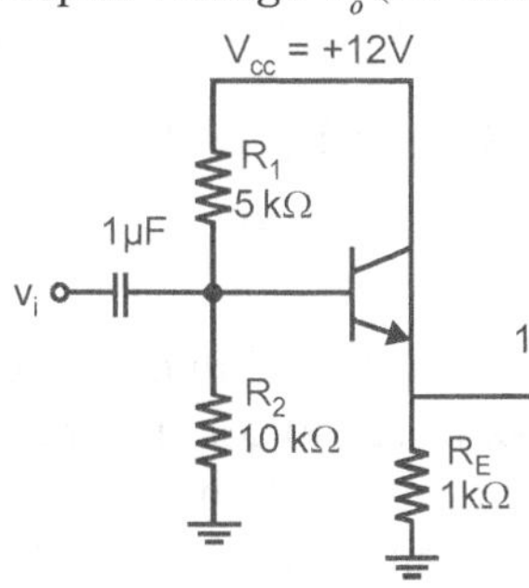

40. An 8-to-1 multiplexer is used to implement a logical function Y as shown in the figure. The output Y is given by

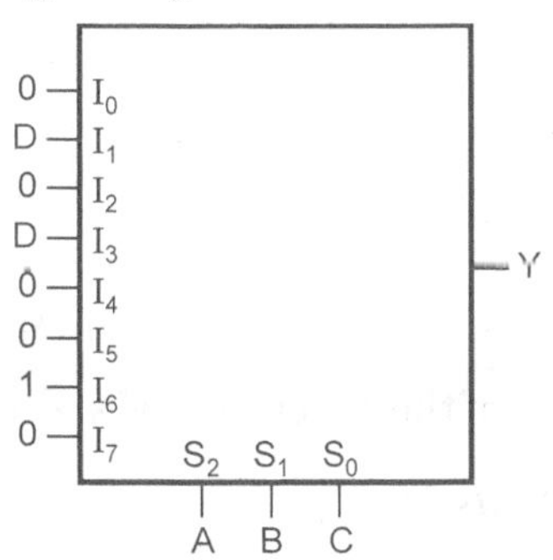

(*a*) $Y = A\bar{B}C + A\bar{C}D$

(*b*) $Y = \bar{A}BC + A\bar{B}D$

(*c*) $Y = AB\bar{C} + \bar{A}CD$

(*d*) $Y = \bar{A}\bar{B}D + A\bar{B}C$

41. A 16-bit ripple carry adder is realized using 16 identical full adders (FA) as shown in the figure. The carry-propagation delay of each FA is 12 ns and the sum-propagation delay of each FA is 15 ns. The worst case delay (in ns) of this 16-bit adder will be __________.

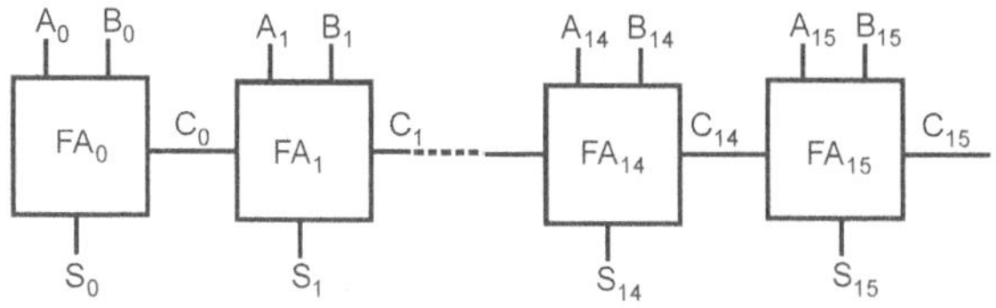

42. An 8085 microprocessor executes "STA 1234H" with starting address location 1FFEH (STA copies the contents of the Accumulator to the 16-bit address location). While the instruction is fetched and executed, the sequence of values written at the address pins A_{15}-A_8 is

(*a*) 1FH, 1FH, 20H, 12H

(*b*) 1FH, FEH, 1FH, FFH, 12H

(*c*) 1FH, 1FH, 12H, 12H

(*d*) 1FH, 1FH, 12H, 20H, 12H

43. A stable linear time invariant (LTI) system has a transfer function $H(s) = \dfrac{1}{s^2 + s - 6}$. To make this system causal it needs to be cascaded with another LTI system having a transfer function $H_1(s)$. A correct choice for $H_1(s)$ among the following options is

(*a*) $s + 3$ (*b*) $s - 2$

(*c*) $s - 6$ (*d*) $s + 1$

44. A causal LTI system has zero initial conditions and impulse response $h(t)$. Its input $x(t)$ and output $y(t)$ are related through the linear constant-coefficient differential equation

$$\frac{d^2 y(t)}{dt^2} + \alpha \frac{dy(t)}{dt} + \alpha^2 y(t) = x(t)$$

Let another signal $g(t)$ be defined as

$$g(t) = \alpha^2 \int_0^t h(\tau)d\tau + \frac{dh(t)}{dt} + \alpha h(t)$$

If G(*s*) is the Laplace transform of $g(t)$, then the number of poles of G(*s*) is _______.

45. The N-point DFT X of a sequence $x[n]$, $0 \le n \le N-1$ is given by

$$X[k] = \frac{1}{\sqrt{N}} \sum_{n=0}^{N-1} x[n] e^{-j\frac{2\pi}{N}nk}, 0 \le k \le N-1$$

Denote this relation as X = DFT(*x*). For N = 4, which one of the following sequences satisfies DFT(DFT (*x*)) = *x*?

(*a*) $x = [1\ 2\ 3\ 4]$ (*b*) $x = [1\ 2\ 3\ 2]$

(*c*) $x = [1\ 3\ 2\ 2]$ (*d*) $x = [1\ 2\ 2\ 3]$

46. The state transition matrix $\phi(t)$ of a system $\begin{bmatrix} \dot{x}_1 \\ \dot{x}_2 \end{bmatrix} = \begin{bmatrix} 0 & 1 \\ 0 & 0 \end{bmatrix} \begin{bmatrix} x_1 \\ x_2 \end{bmatrix}$ is

(a) $\begin{bmatrix} t & 1 \\ 1 & 0 \end{bmatrix}$ (b) $\begin{bmatrix} 1 & 0 \\ t & 1 \end{bmatrix}$

(c) $\begin{bmatrix} 0 & 1 \\ 1 & t \end{bmatrix}$ (d) $\begin{bmatrix} 1 & t \\ 0 & 1 \end{bmatrix}$

47. Consider a transfer function

$$G_p(s) = \frac{ps^2 + 3ps - 2}{s^2 + (3+p)s + (2-p)}$$

with p a positive real parameter. The maximum value of p until which G_p remains stable is ________.

48. The characteristic equation of a unity negative feedback system is $1 + KG(s) = 0$. The open loop transfer function $G(s)$ has one pole at 0 and two poles at –1. The root locus of the system for varying K is shown in the figure.

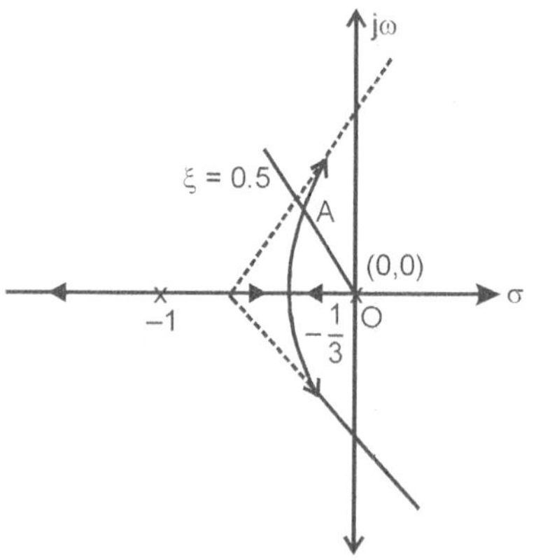

The constant damping ratio line, for $\xi = 0.5$, intersects the root locus at point A. The distance from the origin to point A is given as 0.5. The value of K at point A is ________.

49. Consider a communication scheme where the binary valued signal X satisfies $P\{X = +1\} = 0.75$ and $P\{X = -1\} = 0.25$. The received signal $Y = X + Z$, where Z is a Gaussian random variable with zero mean and variance σ^2. The received signal Y is fed to the threshold detector. The output of the threshold detector $\hat{X}$ is:

$$\hat{X} = \begin{cases} +1, & Y > \tau \\ -1, & Y \le \tau. \end{cases}$$

To achieve a minimum probability of error $P\{\hat{X} \neq X\}$, the threshold τ should be

(a) Strictly positive

(b) Zero

(c) Strictly negative

(d) Strictly positive, zero, or strictly negative depending on the nonzero value of σ^2

50. Consider the Z-channel given in the figure. The input is 0 or 1 with equal probability.

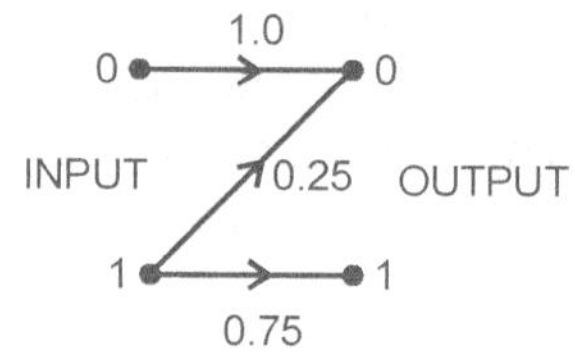

If the output is 0, the probability that the input is also 0 equals ____________

51. An M-level PSK modulation scheme is used to transmit independent binary digits over a band-pass channel with bandwidth 100 kHz. The bit rate is 200 kbps and the system characteristic is a raised-cosine spectrum with 100% excess bandwidth. The minimum value of M is ________.

52. Consider a discrete-time channel $Y = X + Z$, where the additive noise Z is signal-dependent. In particular, given the transmitted symbol $X \in \{-a, +a\}$ at any instant, the noise sample Z is chosen independently from a Gaussian distribution with mean βX and unit variance. Assume a threshold detector with zero threshold at the receiver.

When $\beta = 0$, the BER was found to be $Q(a) = 1 \times 10^{-8}$

$\left(Q(v) = \frac{1}{\sqrt{2\pi}} \int_v^{\infty} e^{-u^2/2} du, \text{ and for } v > 1, \text{ use } Q(v) \approx e^{-v^2/2} \right)$

When $\beta = -0.3$, the BER is closest to

(a) 10^{-7} (b) 10^{-6}

(c) 10^{-4} (d) 10^{-2}

53. The electric field (assumed to be one-dimensional) between two points A and B is shown. Let ψ_A and ψ_B be the electrostatic potentials at A and B, respectively. The value of $\psi_B - \psi_A$ in Volts is ________.

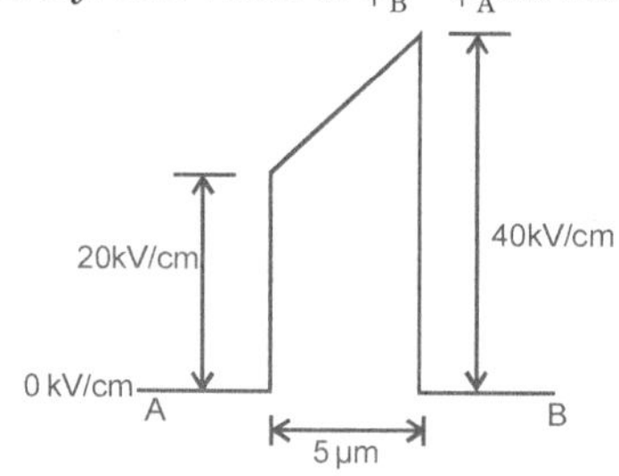

54. Given $\vec{F} = z\hat{a}_x + x\hat{a}_y + y\hat{a}_z$. If S represents the portion of the sphere $x^2 + y^2 + z^2 = 1$ for $z \ge 0$, then $\int_S \nabla \times \vec{F} \cdot \vec{ds}$ is ________.

55. If $E = -(2y^3 - 3yz^2)\,\hat{x} - (6xy^2 - 3xz^2)\,\hat{y} + (6xyz)\,\hat{z}$ is the electric field in a source free region, a valid expression for the electrostatic potential is

(a) $xy^3 - yz^2$ (b) $2xy^3 - xyz^2$

(c) $y^3 + xyz^2$ (d) $2xy^3 - 3xyz^2$

ANSWERS

GENERAL APTITUDE (GA)

1. (*a*)	2. (*b*)	3 . (*b*)	4. (*a*)	5. (495)	6. (*c*)	7. (*b*)
8. (*b*)	9. (22%)	10. (*b*)				

ELECTRONICS AND COMMUNICATION

1. (*d*)	2. (7)	3. (0.8)	4. (*b*)	5. (3)	6. (*c*)	7. (2.8)
8. (*a*)	9. (*a*)	10. (1.125)	11. (*b*)	12. (*d*)	13. (*d*)	14. (*b*)
15. (*a*)	16. (*a*)	17. (3.375)	18. (*d*)	19. (4)	20. (*a*)	21. (*c*)
22. (*a*)	23. (4)	24. (47.7)	25. (*b*)	26. (0.54)	27. (0.44)	28. (*d*)
29. (*c*)	30. (*b*)	31. (31.25)	32. (2.618)	33. (*c*)	34. (225.2)	35. (381)
36. (*d*)	37. (*b*)	38. (0.5)	39. (9.4)	40. (*c*)	41. (195)	42. (*a*)
43. (*b*)	44. (1)	45. (*b*)	46. (*d*)	47. (2)	48. (0.375)	49. (*c*)
50. (0.8)	51. (16)	52. (*c*)	53. (–15)	54. (3.14)	55. (*d*)	

EXPLANATIONS

GENERAL APTITUDE (GA)

1. Dissent is to disagree
2. Revert means set back
3. Vindicate has 2 meanings
   1. Clear of blain
   2. Substantiate, justify
4. $(2x)^n \times \left(\frac{y}{2}\right)^m = 2^{n-m} \times x^n\, y^m$
5. Let consecutive odd numbers be a-10, a-8, a-4, a-2, a, ... a + 12

   Sum of 1st 5 number = $5a - 30 = 425 \Rightarrow a = 91$

   Last 5 numbers = $(a + 4) + (a + 6) + ... + (a + 12)$

   $$= (95 + 97 + 99 + 101 + 103) = \boxed{495}$$
6. 13 M

   17(13 + 4) Q(M + 4)

   19(17 + 2) S(Q + 2)

   32(19 + 4) W = (s + 4)

   ⇒ 23 W
7. KCLFTSB SHSWDG

   Reverse order: Reverse order:

   BCS TOF LUCK GO OD W IS HES

   Ace the exam

   Reverse order should be

   MAXE EHT ECA

   Looking at the options we have M X H T C
8. $$A = P\left(1 + \frac{r}{100}\right)^n$$

   $$A = 2P$$

   $$2 = \left(1 + \frac{r}{100}\right)^{10}$$

   $$\therefore \quad r = 7.2$$
9. Let total cost in 2012 is 100

   Raw material increases in 2013 to 1.3 × 20 = 26

   Other Expenses increased in 2013 to 1.2 × 80 = 96

   Total Cost in 2013 = 96 + 26 = 122

   Total Cost increased by 22%

   Hint:Labour cost (i.e, 4,50,000) in 2012 is redundant data.
10. 1 appears in units place in 4! Ways

    Similarly all other positions in 4! Ways

    Same for other digits.

    Sum of all the numbers

    = (11111) X 4! (1 + 3 + 5 + 7 + 9) = 6666600

ELECTRONICS AND COMMUNICATION

1. $$\sum_{n=0}^{\infty} \frac{1}{n!} = 1 + \frac{1}{1!} + \frac{1}{2!} + \ldots$$

   $$= e \text{ as } e^x = 1 + \frac{x}{1!} + \frac{x^2}{2!} + \ldots \ \forall\ x \text{ in R}$$
2. $$(\nabla f)_{P(1,1,1)} = \left(\vec{i}(2x) + \vec{j}(6y) + \vec{k}(3z^2)\right)_{P(1,1,1)}$$

   $$= 2\vec{i} + 6\vec{j} + 3\vec{k}$$

   $$|(\nabla f)_P| = \sqrt{4 + 36 + 9} = 7$$
3. X ~ N(0, 1)

   $$\Rightarrow \quad f(x) = \frac{1}{\sqrt{2\pi}} e^{-x^2/2}, -\infty < x < \infty$$

   $$\therefore \quad E\{|x|\} = \int_{-\infty}^{\infty} |x| . f(x)\, dx$$

   $$= \frac{1}{\sqrt{2\pi}} x^2 \int_0^{\infty} x e^{-x^2/2}\, dx$$

   $$= \frac{2}{\sqrt{2\pi}} \int_0^{\infty} e^{-u}\, du = \sqrt{\frac{2}{\pi}} = 0.797 \simeq 0.8$$

4. A.E. : $-m^2 + 2m + 1 = 0 \Rightarrow m = -1, -1$

$\therefore$ general solution is $x = (a + bt)e^{-t}$

5. $$f = \frac{1}{\sqrt{2}}(x^2 y + xy^2)$$

$$\Rightarrow \quad \nabla f = \vec{i}\left[\frac{2xy + y^2}{\sqrt{2}}\right] + \vec{j}\left[\frac{x^2 + 2xy}{\sqrt{2}}\right]$$

at (1, 1), $\nabla f = \frac{3}{\sqrt{2}}\vec{i} + \frac{3}{\sqrt{2}}\vec{j}$

$\hat{e}$ = unit vector in the direction i.e., making an angle of $\frac{\pi}{4}$ with y-axis

$$= \left(\sin\frac{\pi}{4}\right)\vec{i} + \left(\cos\frac{\pi}{4}\right)\vec{j}$$

$\therefore$ Directional derivative = $\hat{e}\nabla f = 2\left(\frac{3}{\sqrt{2}}\right)\left(\frac{1}{\sqrt{2}}\right) = 3$

6. 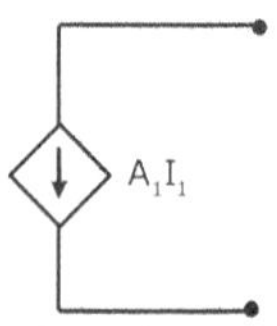

The dependent source represents a current controlled current source

7. By source transformation

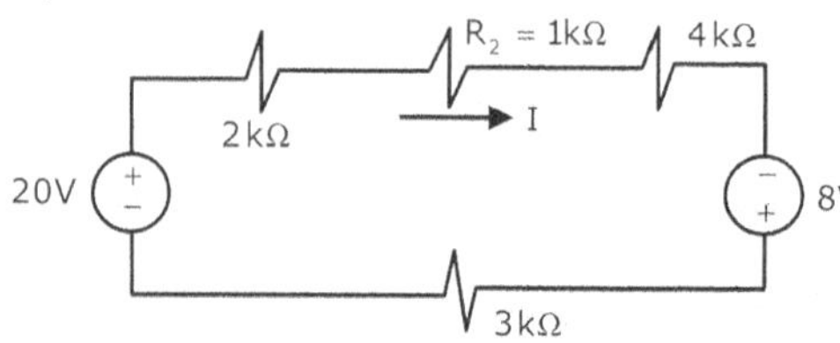

By KVL,

$20 - 10k.I + 8 = 0$

$$\Rightarrow I = \frac{28}{10k}$$

$\Rightarrow I = 2.8$ mA

8. $E = \frac{hC}{\lambda} \Rightarrow \lambda = \frac{6.62 \times 10^{-34} \times 3 \times 10^8}{1.42 \times 1.6 \times 10^{-19}} = 0.87\ \mu m$

9. $n_i \alpha T^{3/2} e^{-Eg/kT}$ and $\rho_i \alpha \frac{1}{\eta_i}$

$\therefore$ From the graph, Energy graph of S_i can be estimated

10. $E = \frac{hC}{\lambda}$

$$\Rightarrow \lambda = \frac{6.6 \times 10^{-34} \times 3 \times 10^8}{1.1 \times 1.6 \times 10^{-19}} = 1.125\ \mu m$$

11. When a CE amplifier's emitter resistance is not by passed, due to the negative feedback the voltage gain decreases and input impedance increases

12. When $V_i < -1.7V$; D_1 – ON and D_2 – OFF

$\therefore V_0 = -1.7$ V

When $V_i > 2.7$ V; D_1 – OFF & D_2 – ON

$\therefore V_0 = 2.7V$

When $-1.7 < V_i < 2.7$ V, Both D_1 & D_2 OFF

$\therefore V_0 = V_i$

14. Capacitor drop rate = $\frac{dv}{dt}$

For a capacitor, $\frac{dv}{dt} \propto \frac{1}{c}$

$\therefore$ Drop rate decreases as capacitor value is increased

For a capacitor, $Q = cv = i \times t \Rightarrow t \propto c$

$\therefore$ Acquisition time increases as capacitor value increased

15. 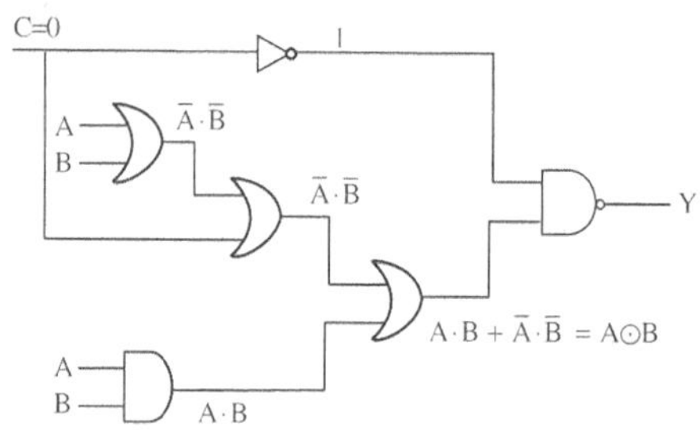

$Y = \overline{1.A \odot B}$

$= \overline{A \odot B} = A \oplus B = \bar{A}B + A\bar{B} + \bar{A}B$

16. 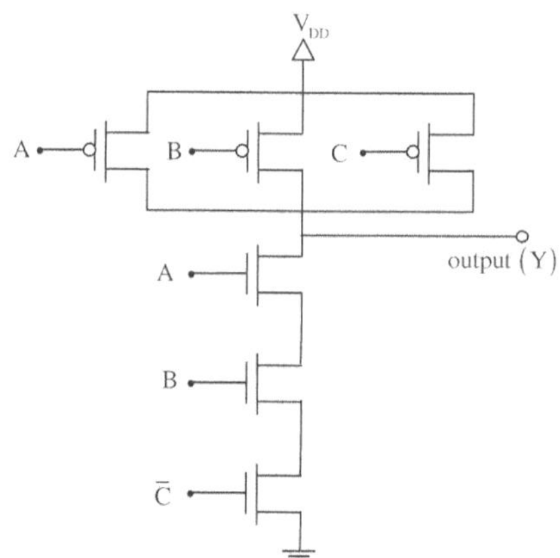

This circuit is CMOS implementation

If the NMOS is connected in series, then the output expression is product of each input with complement to the final product.

So, $Y = \overline{A.B.\bar{C}} = \bar{A} + \bar{B} + C$

17. $$x[n] = \left(\frac{2}{3}\right)^n u[n+3]$$

$$X(e^{j\Omega}) = \sum_{n=-3}^{\infty}\left(\frac{2}{3}\right)^n . e^{-j\Omega n} = \frac{\left(\frac{2}{3}\right)^{-3} . e^{-j3\Omega}}{1 - \frac{2}{3}e^{-j\Omega}}$$

$$\Rightarrow \quad A = \left(\frac{3}{2}\right)^3 = \frac{27}{8} = 3.375$$

18. Let $x(t)$ Fourier transform be $x(t)$

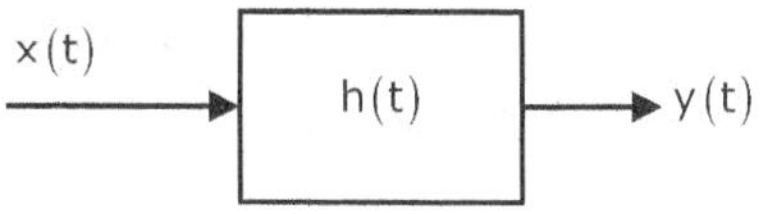

$y(t) = x(t) * h(t)$ (convolution)

$\Rightarrow$ $Y(f) = X(f).\ H(f)$

$\Rightarrow$ $Y(f) = e^{-j4\pi f}.\ X(f)$

$\Rightarrow$ $y(t) = x(t-2)$

19. $y[n] = x[n] * x[n]$

Let $Y(e^{j\Omega})$ is F.T. pair with $y[n]$

$\Rightarrow$ $Y(e^{j\Omega}) - X(e^{j\Omega})\ .\ X(e^{j\Omega})$

$$Y(e^{j\Omega}) = \frac{1}{1-0.5e^{-j\Omega}}.\frac{1}{1-0.5e^{-j\Omega}}$$

also $Y(e^{j\Omega}) = \sum_{h=-\infty}^{\infty} y[n].e^{-j\Omega n}$

$$\Rightarrow \sum_{h=-\infty}^{\infty} y[n] = Y(e^{j0}) = \frac{1}{0.5}.\frac{1}{0.5} = 4$$

20. → In a BODE diagram, in plotting the magnitude with respect to frequency, a pole introduce a line 4 slope -20dB / dc

→ If 4th order all-pole system means gives a slope of (–20)* 4 dB / dec i.e. – 80dB / dec

21. Transfer function $\frac{Y(s)}{U(s)} = \frac{4}{S^2 + 4s + 4}$

If we compare with standard 2nd order system transfer function

i.e., $\frac{w_n^2}{s^2 + 2\xi w_n s + w_n^2}$

$w_n^2 = 4 \Rightarrow w_n = 2$ rad/sec

22. Poisson distribution: It is the property of Poisson distribution.

23. $\frac{\text{Ratio of total side band power}}{\text{Carrier power}} \alpha\ \mu^2$

If it in doubled, this ratio will be come 4 times

24. Electric field of an antenna is

$$E_0 = \frac{\eta I_0 dl}{4\pi}\sin\theta\left[\underset{\text{Radiation field}}{\frac{j\beta}{r}} + \underset{\text{inductive field}}{\frac{1}{r^2}} + \underset{\text{Electrostatic field}}{\frac{J}{\beta r^3}}\right]$$

$\therefore$ $E\ \alpha\ \frac{1}{r}$ $\qquad \frac{E_1}{E_2} = \frac{r_2}{r_1}$

$\Rightarrow$ $E_2 = 6$ mv/m

$$P = \frac{1}{2}\frac{E^2}{\eta} = \frac{1}{2}\frac{36\times10^{-8}}{120\pi} = 47.7 \text{ nw/m}^2$$

25. 1. Point electromagnetic source, can radiate fields in all directions equally, so isotropic.

2. Dish antenna → highly directional

3. Yagi – uda antenna → End fire

Figure: Yagi-uda antenna

26. A.E : $m^2 + 4m + 4 = 0$

$\Rightarrow$ $m = -2, -2$

$\therefore$ solutions is $y = (a + bx)e^{-2x}$ ...(1)

$y = (a + bx)(-2e^{-2x}) + e^{-2x}(b)$ ...(2)

using $y(0) = 1$; $y(0) = 1$, (1) and (2) gives

$a = 1$ and $b = 3$

$\therefore$ $y = (1 + 3x)e^{-2x}$

at $x = 1$, $y = 4e^{-2} = 0.541 \simeq 0.54$

27. Parcel will be lost if

a. it is lost by the first post office

b. it is passed by first post office but lost by the second post office

Prob(parcel is lost) = $\frac{1}{5} + \frac{4}{5}\times\frac{1}{5} = \frac{9}{25}$

P (Parcel lost by second post if it passes first post office) = P (Parcel passed by first post office) × P(Parcel lost by second post office)

$$= \frac{4}{5}\times\frac{1}{5} = \frac{4}{25}$$

Prob (parcel lost by 2nd post office | parcel lost)

$$= \frac{\frac{4}{25}}{\frac{9}{25}} = \frac{4}{9} = 0.44$$

28. **Exp : (1)**

If $f(t) \leftrightarrow F(s)$

Then $\quad tf(t) \leftrightarrow \frac{-d}{ds}F(s)$

$$= \frac{-d}{ds}\left(\frac{1}{s^2+s+1}\right) = \frac{-(2s+1)}{(s^2+s+1)^2}$$

$$= \frac{2s+1}{(s^2+s+1)^2}$$

Exp : (2)

$$F(s) = \frac{1}{s^2+s+1}$$

$$L\,[g(t) = t.f(t)] = -\frac{d}{ds}[F(s)]$$

(using multiplication by t)

29. Let x (opposite side), y (adjacent side) and z (hypotenuse) of a right angled triangle.

Given Z + y = K(constant) ......(1) and angle between them say 'θ' then Area,

$$A = \frac{1}{2}xy = \frac{1}{2}(z\sin\theta)(z\cos\theta) = \frac{z^2}{4}\sin 2\theta$$

Now (1)

$$\Rightarrow \quad z + z\sin\theta = k$$

$$\Rightarrow \quad z = \frac{k}{1+\sin\theta}$$

$$= \quad A = \frac{k^2}{4}\left[\frac{\sin 2\theta}{(1+\sin\theta)^2}\right]$$

In order to have maximum area, $\frac{dA}{d\theta} = 0$

$$\Rightarrow \frac{k^2}{4}\left[\frac{(1+\sin\theta)^2(2\cos 2\theta) - \sin 2\theta(\cos\theta).2(1+\sin\theta)}{(1+\sin\theta)^4}\right] = 0$$

$\Rightarrow \theta = \frac{\pi}{6} = 30°$, Answer obtained is different than official key

30.

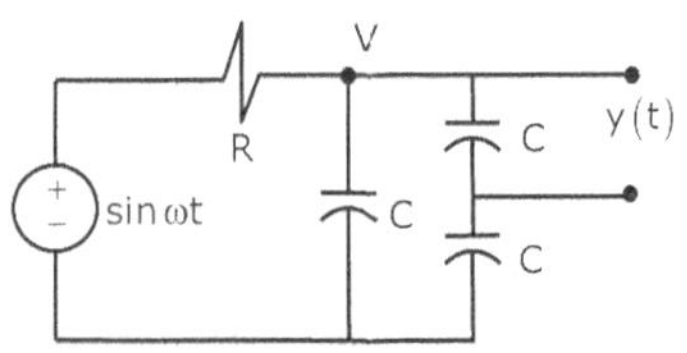

By nodal method,

$$\frac{V - 1\angle 0°}{R} + \frac{V}{\left(\frac{1}{j\omega c}\right)} + \frac{V}{\left(\frac{2}{j\omega c}\right)} = 0$$

$$V\left[\frac{1}{R} + j\omega c + \frac{j\omega c}{2}\right] = \frac{1\angle 0°}{R}$$

$$V = \frac{2}{2 + 3j\omega RC}$$

$$Y = \frac{V}{2} \Rightarrow \frac{1}{2 + j\omega 3RC}$$

$$\text{given } |A(\omega)| = \frac{1}{4} \Rightarrow \frac{1}{\sqrt{4 + 9R^2c^2.\omega^2}}$$

$$\Rightarrow \omega = \frac{2}{\sqrt{3}RC}$$

31.

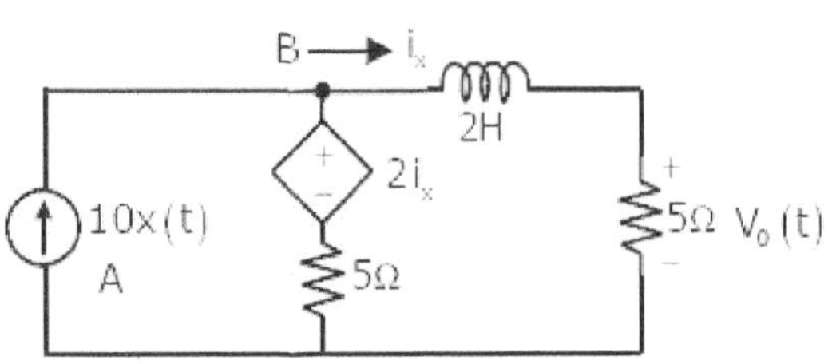

For $t \to \infty$, i.e., at steady state, inductor will behave as a shot circuit and hence $V_B = 5.i_x$

By KCL at node B, $-10 + V_B - 2i_x + i_x = 0$

$$\Rightarrow \quad i_x = \frac{50}{8}$$

$$V_0(t) = 5i_x(t)$$

$$\Rightarrow \quad V_0(t) = \frac{250}{8} = 31.25 \text{ volts}$$

32.

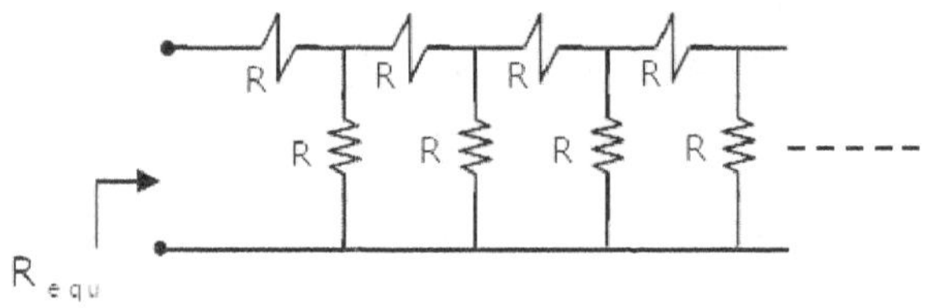

→ For an infinite ladder network, if all resistance are having same value of R

Then equivalent resistance is $\left(\frac{1+\sqrt{5}}{2}\right)$.R

→ For the given network, we can split in to R is in series with $R_{equivalent}$

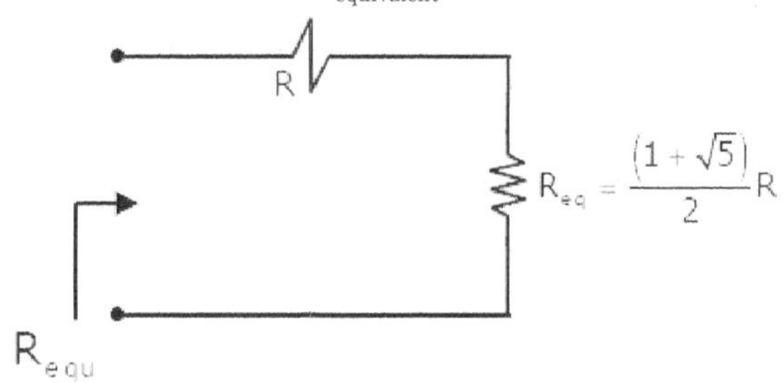

$$\Rightarrow \quad R_{equ} = R + 1.618R$$

$$\Rightarrow \quad \frac{R_{equ}}{R} = 2.618$$

33. For the two-part network

$$Y \text{ matrix} = \begin{bmatrix} \frac{1}{30} + \frac{1}{10} & -\frac{1}{30} \\ \frac{-1}{30} & \frac{1}{60} + \frac{1}{30} \end{bmatrix}$$

$$Z_{matrix} = [Y]^{-1}$$

$$Z = \begin{bmatrix} 0.1333 & -0.0333 \\ -0.0333 & 0.05 \end{bmatrix}$$

$$Z = \begin{bmatrix} 9 & 6 \\ 6 & 24 \end{bmatrix}$$

34.

$$P = N_A - N_D$$

$$= 1 \times 10^{18} - 1 \times 10^{13}$$

$$= 9.99 \times 1017$$

$$\eta = \frac{\eta_i^2}{P} = \frac{(1.5 \times 10^{10})}{9.99 \times 10^{17}} = 225.2/\text{cm}^3$$

35.
$$I_{C_1} = I_{C_2} \text{ (Given)}$$
$$I_{S_1} e^{\frac{V_{BE_1}}{V_T}} = I_{S_2} e^{\frac{V_{BE_2}}{V_T}}$$
$$e^{\frac{(V_{BE_1} - V_{BE_2})}{V_T}} = \frac{I_{S_2}}{I_{S_1}}$$
$$V_{BE_1} - V_{BE_2} = V_T \ln \frac{I_{S_2}}{I_{S_1}}$$
$$= 26 \times 10^{-3} \ln\left[\frac{300 \times 300}{0.2 \times 0.2}\right] \because I_S \alpha A$$
$$\left(V_{BE_1} - V_{BE_2}\right) = 381 \text{ mV}$$

37.
$$A_V = \frac{R_E}{r_e + R_E}$$
$$= \frac{R_E}{\frac{V_T}{I_E} + R_E} = \frac{I_E R_E}{V_T + I_E R_E}$$
$$\therefore \quad A_V \simeq \frac{I_C R_E}{V_T + I_C R_E} \qquad (\because I_C \simeq I_E)$$

$\therefore$ $I_C R_E >> U_T \Rightarrow A_V$ in almost constant

38. Input impedance of CB amplifier,
$$z_i = r_e = \frac{V_I}{I_E}$$
$$\Rightarrow \quad 50 = \frac{25\text{mV}}{I_E}$$
($\because$ signal is received from 50Ω antenna and V_T = 25mV)
$$\Rightarrow \quad I_E = \frac{25\text{mV}}{50\Omega} = 0.5 \text{ mA}$$

39. $\because$ β = high, I_B is neglected
$$\therefore \quad V_B = 12 \times \frac{10\,k}{10\,k + 5\,k} = 8\text{V}$$
$$V_E = V_B - 0.7 = 7.3\text{V}$$
$$\therefore \quad V_{CE} = 12 - 7.3 = 4.7\text{V}$$
$\therefore$ Maximum undistorted
$$V_0(p-p) = 2 \times 4.7\text{V}$$
$$= 9.4\text{V}$$

40.

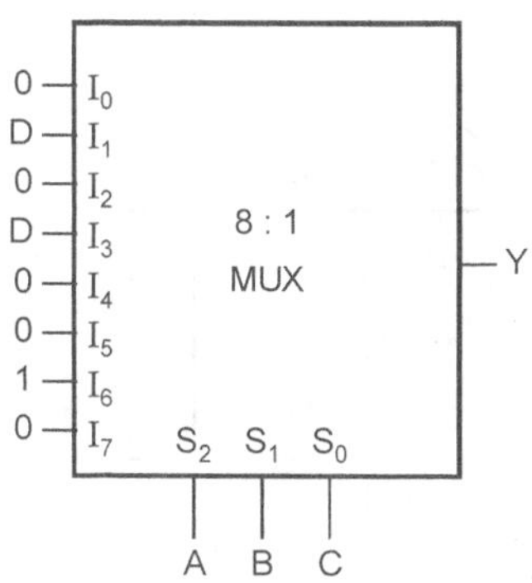

$$Y = \bar{A}\bar{B}CD + \bar{A}BCD + AB\bar{C}$$
Remaining combinations of the select lines will produce output 0.

So,
$$Y = \bar{A}CD(\bar{B} + B) + AB\bar{C}$$
$$= \bar{A}CD + AB\bar{C}$$
$$= AB\bar{C} + \bar{A}CD$$

41.

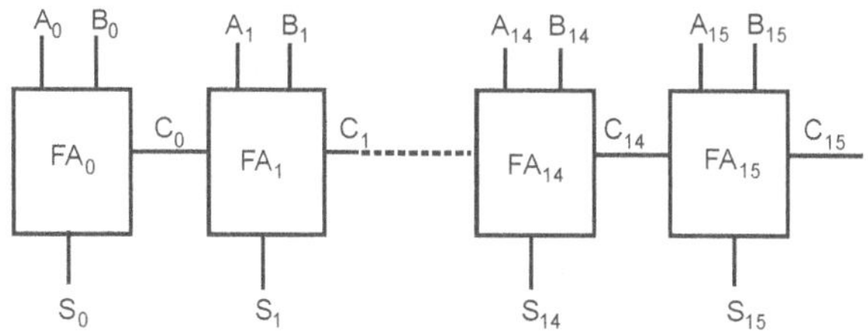

This is 16-bit ripple carry adder circuit, in their operation carry signal is propagating from 1st stage FA0 to last state FA15, so their propagation delay is added together but sum result is not propagating. We can say that next stage sum result depends upon previous carry.

So, last stage carry (C_{15}) will be produced after 16 × 12ns = 192ns

Second last stage carry (C_{14}) will be produced after 180 ns.

For last stage sum result (S_{15}) total delay
= 180ns + 15ns = 195ns

So, worst case delay = 195 ns

42. Let the opcode of STA is XXH and content of accumulator is YYH.

Instruction: STA 1234 H

Starting address given = 1FFEH

So, the sequence of data and addresses is given below:

Address (in hex) $A_{15}-A_8$	A_7-A_0	: Data (in hex)
1F	FE H	→ XXH
1F	FF H	→ 34H
20	00 H	→ 12 H
12	34 H	→ YYH

43. Given, $\quad H(s) = \frac{1}{s^2 + s - 6} = \frac{1}{(s+3)(s-2)}$

It is given that system is stable thus its ROC includes $j\omega$ axis. This implies it cannot be causal, because for causal system ROC is right side of the rightmost pole.

⇒ Poles at $s = 2$ must be removes so that it can be become causal and stable simultaneously.

$$\Rightarrow \quad \frac{1}{(s+3)(s-2)}(s-2) = \frac{1}{s+3}$$

Thus $H_1(s) = s - 2$

44. Given differential equation

$$s^2 y(s) + \alpha\, sy(s) + a^2 y(s) = x(s)$$

$$\Rightarrow \frac{y(s)}{x(s)} = \frac{1}{s^2 + \alpha s + \alpha^2} = \text{H}(s)$$

$$g(t) = \alpha^2 \int_0^t h(z)\,dz + \frac{d}{dt} h(t) + \alpha h(t)$$

$$= \alpha^2 \frac{\text{H}(s)}{s} + \text{SH}(s) + \alpha \text{H}(s)$$

$$= \alpha^2 \frac{1}{s(s^2 + \alpha s + \alpha^2)} + s \frac{1}{(s^2 + 2s + \alpha^2)} + \frac{\alpha}{s^2 + \alpha s + \alpha^2}$$

$$= \frac{\alpha^2 + \alpha s + s^2}{s(s^2 + \alpha s + \alpha^2)} = \frac{1}{s}$$

No. of poles = 1

45. This can be solved by directly using option and satisfying the condition given in question

$$\text{X} = \text{DFT}\,(x)$$

$$\text{D}_{\text{FT}}(\text{D}_{\text{FT}}(x)) = \text{DFT(X)} = \frac{1}{\sqrt{\text{N}}} \sum_{n=0}^{\text{N}-1} \text{X}[n] e^{-j\frac{2\pi}{\text{N}} nk}$$

DFT $y\,[1\ 2\ 3\ 4]$

$$\text{X} = \frac{1}{\sqrt{4}} \begin{bmatrix} 1 & 1 & 1 & 1 \\ 1 & -j & -1 & j \\ 1 & -1 & 1 & -1 \\ 1 & +j & -1 & -j \end{bmatrix} \begin{bmatrix} 1 \\ 2 \\ 3 \\ 4 \end{bmatrix} = \frac{1}{\sqrt{4}} \begin{bmatrix} 10 \\ 2+2j \\ 2 \\ -2-j2 \end{bmatrix}$$

DFT of (x) will not result in $[1\ 2\ 3\ 4]$

Try with DFT of $y\,[1\ 2\ 3\ 2]$

$$\text{X} = \frac{1}{\sqrt{4}} \begin{bmatrix} 1 & 1 & 1 & 1 \\ 1 & -j & -1 & j \\ 1 & -1 & 1 & -1 \\ 1 & +j & -1 & -j \end{bmatrix} \begin{bmatrix} 1 \\ 2 \\ 3 \\ 2 \end{bmatrix}$$

$$= \frac{1}{\sqrt{4}} \begin{bmatrix} 8 \\ -2 \\ 0 \\ -2 \end{bmatrix} = \begin{bmatrix} 4 \\ -1 \\ 0 \\ -1 \end{bmatrix}$$

$$\text{DFT of} \begin{bmatrix} 4 \\ -1 \\ 0 \\ -1 \end{bmatrix} = \frac{1}{\sqrt{4}} \begin{bmatrix} 1 & 1 & 1 & 1 \\ 1 & -j & -1 & j \\ 1 & -1 & 1 & -1 \\ 1 & +j & -1 & -j \end{bmatrix} \begin{bmatrix} 4 \\ -1 \\ 0 \\ -1 \end{bmatrix}$$

$$= \frac{1}{2} \begin{bmatrix} 2 \\ 4 \\ 6 \\ 4 \end{bmatrix} = \begin{bmatrix} 1 \\ 2 \\ 3 \\ 2 \end{bmatrix}$$

Same as x

Then 'b' is right option

46. Given state model,

$$\begin{bmatrix} \dot{x}_1(t) \\ \dot{x}_2(t) \end{bmatrix} = \begin{bmatrix} 0 & 1 \\ 0 & 0 \end{bmatrix} \begin{bmatrix} x_1(t) \\ x_2(t) \end{bmatrix}$$

$$\text{A} = \begin{bmatrix} 0 & 1 \\ 0 & 0 \end{bmatrix}$$

$\phi(t) \Rightarrow$ state transistion matrix

$$\phi(t) = \text{L}^{-1}\left[(\text{SI} - \text{A})^{-1}\right]$$

$$[\text{SI} - \text{A}]^{-1} = \begin{bmatrix} s & -1 \\ 0 & s \end{bmatrix}^{-1}$$

$$\Rightarrow \frac{1}{s^2} \begin{bmatrix} s & 1 \\ 0 & s \end{bmatrix}$$

$$\phi(t) = \text{L}^{-1} \begin{bmatrix} 1/s & 1/s^2 \\ 0 & 1/s \end{bmatrix}$$

$$\phi(t) = \begin{bmatrix} 1 & t \\ 0 & 1 \end{bmatrix}$$

47. Given $\quad \text{G}_p(s) = \dfrac{ps^2 + 3ps - 2}{s^2 + (3+p)s + (2-p)}$

By R – H criteria

The characteristic equation is

$$s^2 + (3 + p)\, s + (2 - p) = 0$$

i.e. $s^2 + (3 + p)\, s + (2 - p) = 0$

By forming R-H array,

s^2	1	$(2-p)$
s^1	$(3+\phi)$	0
s^0	$(2-p)$	

For stability, first column elements must be positive and non-zero

i.e. $(1)(3 + p) > 0 \Rightarrow \boxed{p > -3}$

and $(2)(2 - p) > 0 \Rightarrow \boxed{p < 2}$

i.e. $\boxed{-3 < p < 2}$

The maximum value of p unit which G_p remains stable is 2

48.

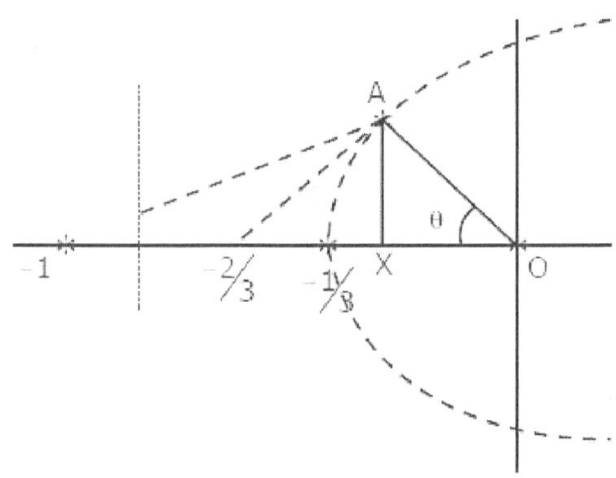

We know that the co-ordinate of point A of the given root locus i.e., magnitude condition

$$|G(s)H(s)| = 1$$

Here, the damping factor $\xi = 0.5$ and the length of OA = 5

$$\xi = 0.5$$

Then in the right angle triangle

$$\cos\theta = \frac{OX}{OA}$$

$$\Rightarrow \quad \cos 60 = \frac{OX}{0.5}$$

$$\Rightarrow \quad OX = \frac{1}{4}$$

$$\Rightarrow \quad \sin\theta = \frac{AX}{OA}$$

$$\Rightarrow \quad \sin 60 = \frac{AX}{0.5}$$

$$\Rightarrow \quad AX = \frac{\sqrt{3}}{4}$$

So, the co-ordinate of point A is $-1/4 + j\sqrt{3}/4$

Substituting the above value of A in the transfer function and equating to 1 i.e. by magnitude condition,

$$\left|\frac{k}{s(s+1)^2}\right|_{s=-1/4+j\sqrt{3}/4} = 1$$

$$k = \sqrt{\frac{1}{16}+\frac{3}{16}}\cdot\left(\sqrt{\frac{9}{16}+\frac{3}{16}}\right)^2$$

$$k = 0.375$$

49. $H_1 : x = +1$; $H_0 : x = -1$

$P(H_1) = 0.75$; $P(H_2) = 0.25$

Received signal $\gamma = X + Z$

Where $Z \sim N(0, -2)$; $f_z(z) = \frac{1}{\sigma\sqrt{2\pi}} e^{-z^2/2\sigma^2}$

$$\text{Received signal } \gamma = \begin{cases} 1+Z & \text{if} \quad X = 1 \\ -1+Z & \text{if} \quad X = -1 \end{cases}$$

$$f_y(y/H_1) = \frac{1}{\sigma\sqrt{2\pi}} e^{-\frac{1}{2\sigma^2}(\gamma-1)^2}$$

$$f_y(y/H_0) = \frac{1}{\sigma\sqrt{2\pi}} e^{-\frac{1}{2\sigma^2}(\gamma+1)^2}$$

At optimum threshold y_{opt}: for minimum probability of error

$$\left.\frac{f_y(y/H_1)}{f_y(y/H_0)}\right|_{y=y_{opt}} = \frac{P(H_0)}{P(H_1)}$$

$$\left. e^{-\frac{1}{2\sigma^2}\left[(\gamma-1)^2-(\gamma+1)^2\right]}\right|_{y_{opt}} = \frac{P(H_0)}{P(H_1)}$$

$$e^{+2y_{opt}/\sigma^2} = \frac{P(H_0)}{P(H_1)}$$

$$y_{opt} = \frac{\sigma^2}{2} 1_n \left(\frac{P(H_0)}{P(H_1)}\right) = \frac{-1.1\sigma^2}{2} = -0.55\sigma^2$$

y_{opt} = Optimum threshold

$y_{opt} < 0 \therefore$ Threshold is negative

50. Given channel

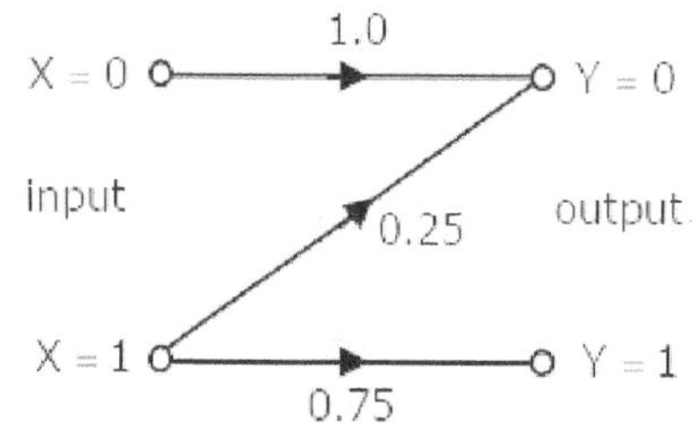

We have to determine, $P\{x = 0 / y = 0\}$

$$P\{x = 0 / y = 0\} = \frac{P\{y=0/x=0\}P\{x=0\}}{P\{y=0\}}$$

$$= \frac{1\cdot 1/2}{1\cdot 1/2 + 0.25\times\frac{1}{2}}$$

$$= \frac{4}{5} = 0.8$$

51. Bandwidth requirement for m-level PSK

$$= \frac{1}{T}(1+\alpha)$$

[Where T is symbol duration. α is roll of factor]

$$\Rightarrow \quad \frac{1}{T}(1+\alpha) = 100 \times 10^3$$

$$\alpha = 1 \quad \text{[100\% excess bandwidth]}$$

$$\Rightarrow \quad \frac{1}{T}(2) = 100 \times 10^3$$

$$\Rightarrow \quad T = \frac{2}{100\times 10^3} = 20 \quad \text{sec}$$

$$\text{Bit duration} = \frac{1}{200\times 10^3}$$

$$= 0.5 \times 10^{-5} = 5 \times 10^{-6} \text{ sec}$$

$$\text{Bit duration} = \frac{\text{Symbol duration}}{\log_2 m}$$

$$\Rightarrow \quad \log_2 m = \frac{20\times 10^{-6}\text{ sec}}{5\times 10^{-6}} = 4$$

$$\Rightarrow \quad M = 16$$

52. $X \in [-a, a]$ and $P(x = -a) = P(x = a) = \frac{1}{2}$

$$\gamma = X + Z \rightarrow \text{Received signal}$$

$$Q(a) = 1 \times 10^{-8}$$

$$Q(a) \approx e^{-v^2/2}$$

$$Z \sim N(\beta, X, 1)$$

$$f_z(z) = \frac{1}{\sqrt{2\pi}} e^{-\frac{1}{2}(Z - \beta x)}$$

$$\gamma = \begin{cases} -a + z & \text{if } x = -a \\ a + z & \text{if } x = +a \end{cases}$$

$$H_1 : x = +a$$

$$H_0 : x = -a$$

and Threshold = 0

$$f_y(y/H_1) = \frac{1}{\sqrt{2\pi}} e^{-\frac{1}{2}(y - a(1+\beta))^2}$$

$$f_y(y/H_0) = \frac{1}{\sqrt{2\pi}} e^{-\frac{1}{2}(y + a(1+\beta))^2}$$

BER :

$$P_e = P(H_1)P(e/H_1) + P(H_0)P(e/H_0)$$

$$= \frac{1}{2}\int_{-\infty}^{0} \frac{1}{\sqrt{2\pi}} e^{-\frac{1}{2}(y - a(1+\beta))^2} dy + \frac{1}{2}\int_{0}^{\infty} \frac{1}{\sqrt{2\pi}} e^{-\frac{1}{2}(y + a(1+\beta))^2}$$

$$dy = Q(a(1 + \beta))$$

$$\beta = 0$$

$$P_e = Q(a) = 1 \times 10^{-8} = e^{-a^2/2}$$

$$\Rightarrow \quad a = 6.07$$

$$\beta = -0.3$$

$$P_e = Q(6.07\,(1 - 0.3))$$

$$= Q(4.249)$$

$$P_e = e^{-(4.249)^2/2}$$

$$= 1.2 \times 10^{-4}$$

$$P_e \simeq 10^{-4}.$$

53. A (0kV/cm, 20kV/cm) B (5×10^{-4}kV/cm, 40kV/cm)

$$E - 20 = \frac{40 - 20}{5 \times 10^{-4}}(x - 0)$$

$$\Rightarrow \quad E = 4 \times 10^4\, x + 20$$

$$V_{AB} = -\int_A^B E.dl$$

$$= -\int_0^{5\times10^{-4}/\text{cm}} (4 \times 10^4\, x + 20)dx$$

$$= -\left(4 \times 10^4 \frac{x^2}{2} + 20x\right)\Bigg|_0^{5\times10^{-4}}$$

$$= -(2 \times 10^4 \times 25 \times 10^{-8} + 20 \times 5 \times 10^{-4})$$

$$= -(50 \times 10^{-4} + 100 \times 10^{-4})$$

$$= -150 \times 10^{-4}\text{ kV}$$

$$\Rightarrow \quad V_{AB} = -15\text{V}$$

54. $\int_s \nabla \times \vec{F}.d\vec{s} = \oint_C \vec{F}.d\vec{r}$ (using stoke's theorem and C is closed curve i.e.,

$$x^2 + y^2 = 1,\ z = 0$$

$$\Rightarrow \quad x = \cos\theta,\ y = \sin\theta \text{ and } \theta = 0 \text{ to } 2\theta$$

$$= \oint_C zdx + xdy + ydz$$

$$= \oint_C xdy = \int_0^{2\pi} \cos\theta(\cos\theta\ d\theta)$$

$$= \frac{1}{2}\left(\theta + \frac{\sin 2\theta}{2}\right)_0^{2\pi} = \pi = 3.14$$

55. Given $E = -\left(2y^3 - 3yz^2\right)a_x - \left(6xy^2 - 3xz^2\right)a_y + 6xyza_z$

By verification option (*d*) satisfy

$$E = -\nabla V$$

■■

SOLVED PAPER – 2013

ELECTRONICS AND COMMUNICATION ENGINEERING

No. of Questions : 65 | *Maximum Marks : 100*

Q. (1–25) carry one mark each.

1. A bulb in a staircase has two switches, one switch being at the ground floor and the other one at the first floor. The bulb can be turned ON and also can be turned OFF by any one of the switches irrespective of the state of the other switch. The logic of switching of the bulb resembles
   (*a*) an AND gate (*b*) an OR gate
   (*c*) an XOR gate (*d*) a NAND gate

2. Consider a vector field $\vec{A}(\vec{r})$. The closed loop line integral $\oint \vec{A} \bullet d\vec{l}$ can be expressed as
   (*a*) $\oiint (\nabla \times \vec{A}) \bullet d\vec{s}$ over the closed surface bounded by the loop
   (*b*) $\oiiint (\nabla \bullet \vec{A})dv$ over the closed volume bounded by the loop
   (*c*) $\iiint (\nabla \bullet \vec{A})dv$ over the open volume bounded by the loop
   (*d*) $\iint (\nabla \times \vec{A}) \bullet d\vec{s}$ over the open surface bounded by the loop

3. Two systems with impulse responses $h_1(t)$ and $h_2(t)$ are connected in cascade. Then the overall impulse response of the cascaded system is given by
   (*a*) product of $h_1(t)$ and $h_2(t)$
   (*b*) sum of $h_1(t)$ and $h_2(t)$
   (*c*) convolution of $h_1(t)$ and $h_2(t)$
   (*d*) subtraction of $h_2(t)$ from $h_1(t)$

4. In a forward biased *pn* junction diode, the sequence of events that best describes the mechanism of current flow is
   (*a*) injection, and subsequent diffusion and recombination of minority carriers
   (*b*) injection, and subsequent drift and generation of minority carriers
   (*c*) extraction, and subsequent diffusion and generation of minority carriers
   (*d*) extraction, and subsequent drift and recombination of minority carriers

5. In IC technology, dry oxidation (using dry oxygen) as compared to wet oxidation (using steam or water vapor) produces
   (*a*) superior quality oxide with a higher growth rate
   (*b*) inferior quality oxide with a higher growth rate
   (*c*) inferior quality oxide with a lower growth rate
   (*d*) superior quality oxide with a lower growth rate

6. The maximum value of θ until which the approximation sin θ ≈ 0 holds to within 10% error is
   (*a*) 10° (*b*) 18°
   (*c*) 50° (*d*) 90°

7. The divergence of the vector field
   $$\vec{A} = x\hat{a}_x + y\hat{a}_y + z\hat{a}_z \text{ is}$$
   (*a*) 0 (*b*) $\frac{1}{3}$
   (*c*) 1 (*d*) 3

8. The impulse response of a system is $h(t) = tu(t)$. For an input $u(t-1)$, the output is
   (*a*) $\frac{t^2}{2}u(t)$ (*b*) $\frac{t(t-1)}{2}u(t-1)$
   (*c*) $\frac{(t-1)^2}{2}u(t-1)$ (*d*) $\frac{t^2-1}{2}u(t-1)$

9. The Bode plot of a transfer function G(s) is shown in the figure below.

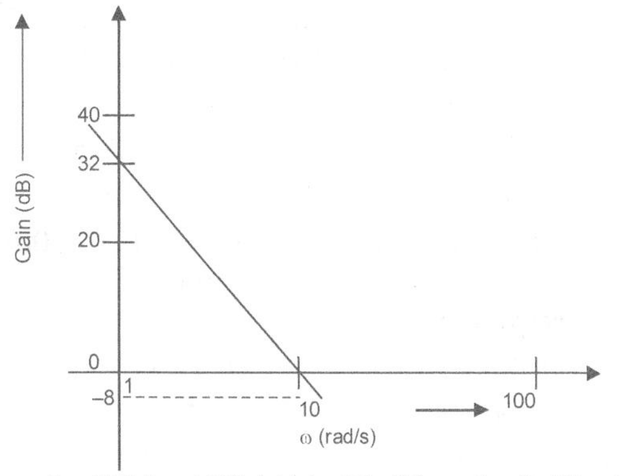

   The gain (20 log | G(s) |) is 32 dB and –8 dB at 1 rad/s and 10 rad/s respectively. The phase is negative for all ω. Then G (s) is
   (*a*) $\frac{39.8}{s}$ (*b*) $\frac{39.8}{s^2}$
   (*c*) $\frac{32}{s}$ (*d*) $\frac{32}{s^2}$

10. In the circuit shown below what is the output voltage (V_{out}) if a silicon transistor Q and an ideal op-amp are used?
   (*a*) –15 V
   (*b*) –0.7 V
   (*c*) +0.7 V
   (*d*) +15 V

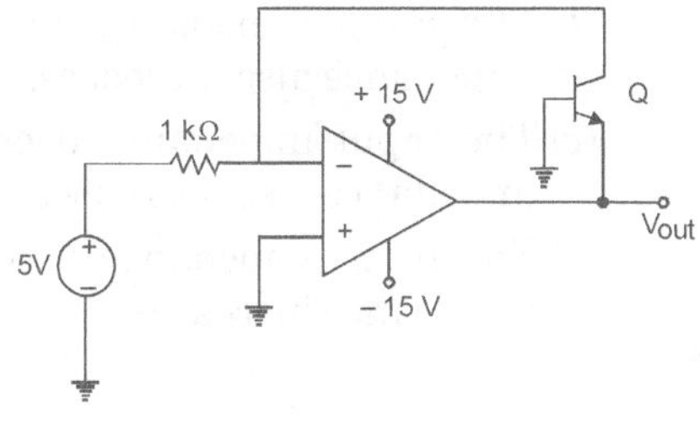

11. Consider a delta connection of resistors and its equivalent star connection as shown below. If all elements of the delta connection are scaled by a factor K, K > O, the elements of the corresponding star equivalent will be scaled by a factor of

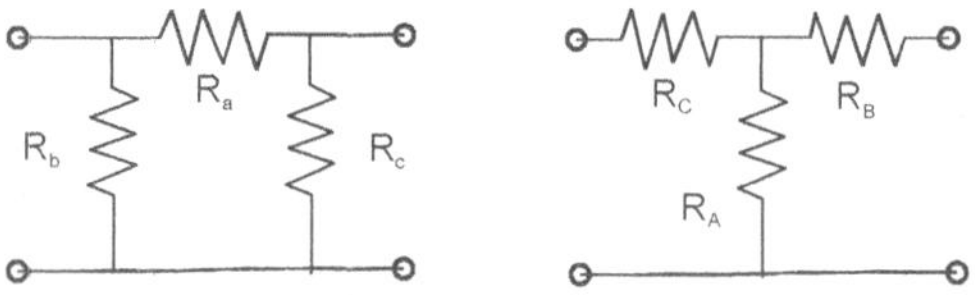

(a) k^2 (b) k
(c) $1/k$ (d) $\sqrt{k}$

12. For 8085 microprocessor, the following program is executed.

```
    MVI A, 05H;
    MVI B, 05H;
PTR: ADD B;
    DCR B;
    JNZ PTR;
    ADI 03H;
    HLT;
```

At the end of program, accumulator contains
(a) 17H (b) 20H
(c) 23H (d) 05H

13. The bit rate of a digital communication system is R kbits/s. The modulation used is 32-QAM. The minimum bandwidth required for ISI free transmission is
(a) $R/10$ Hz (b) $R/10$ kHz
(c) $R/5$ Hz (d) $R/5$ kHz

14. For a periodic signal $v(t) = 30 \sin 100t + 10 \cos 300t + 6 \sin (500t + \pi/4)$, the fundamental frequency in rad/s is
(a) 100 (b) 300
(c) 500 (d) 1500

15. In a voltage-voltage feedback as shown below, which one of the following statements is TRUE if the gain k is increased?

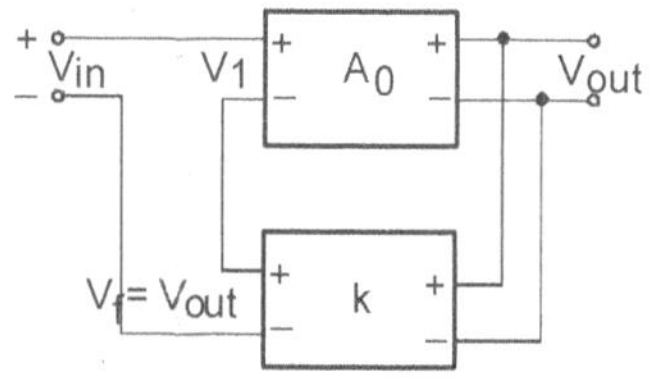

(a) The input impedance increases and output impedance decreases.
(b) The input impedance increases and output impedance also increases.
(c) The input impedance decreases and output impedance also decreases.
(d) The input impedance decreases and output impedance increases.

16. A band-limited signal with a maximum frequency of 5kHz is to be sampled. According to the sampling theorem, the sampling frequency which is not valid is
(a) 5 kHz (b) 12 kHz
(c) 15 kHz (d) 20 kHz

17. In a MOSFET operating in the saturation region, the channel length modulation effect causes
(a) an increase in the gate-source capacitance
(b) a decrease in the transconductance
(c) a decrease in the unity-gain cutoff frequency
(d) a decrease in the output resistance

18. Which one of the following statements is NOT TRUE for a continuous time causal and stable LTI system?
(a) All the poles of the system must lie on the left side of the $j\omega$ axis.
(b) Zeros of the system can lie anywhere in the s-plane.
(c) All the poles must lie within $|s| = 1$.
(d) All the roots of the characteristic equation must be located on the left side of the $j\omega$ axis.

19. The minimum eigen value of the following matrix is $\begin{bmatrix} 3 & 5 & 2 \\ 5 & 12 & 7 \\ 2 & 7 & 5 \end{bmatrix}$
(a) 0 (b) 1
(c) 2 (d) 3

20. A polynomial $f(x) = a_4x^4 + a_3x^3 + a_2x^2 + a_1x - a_0$ with all coefficients positive has
(a) no real roots
(b) no negative real root
(c) odd number of real roots
(d) at least one positive and one negative real root

21. Assuming zero initial condition, the response $y(t)$ of the system given below to a unit step input $u(t)$ is

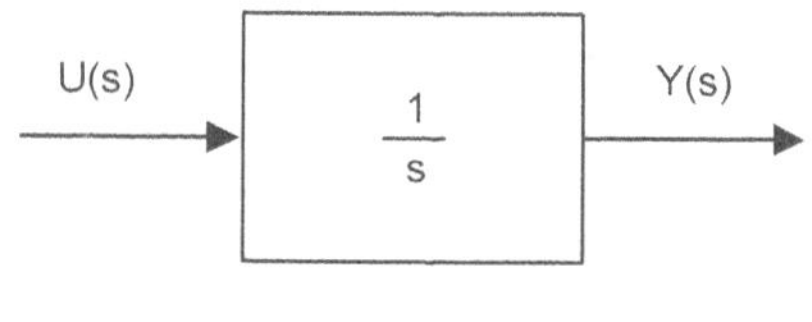

(a) $u(t)$
(b) $tu(t)$
(c) $\frac{t^2}{2}u(t)$
(d) $e^{-1}\,u(t)$

22. The transfer function $\frac{V_2(s)}{V_1(s)}$ of the circuit shown below is

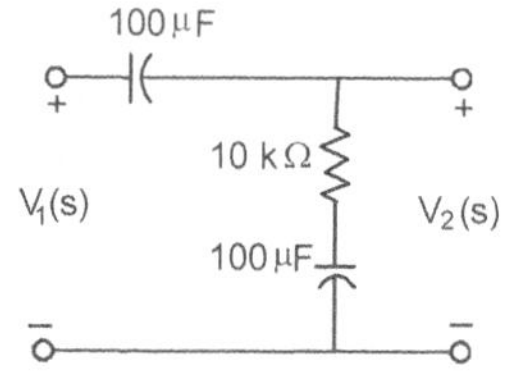

(a) $\frac{0.5s+1}{s+1}$
(b) $\frac{3s+6}{s+2}$
(c) $\frac{s+2}{s+1}$
(d) $\frac{s+1}{s+2}$

23. A source $v_s(t) = V \cos 100\pi t$ has an internal impedance of $(4 + j3)\ \Omega$. If a purely resistive load connected to this source has to extract the maximum power out of the source, its value in Ω should be

(a) 3 (b) 4

(c) 5 (d) 7

24. The return loss of a device is found to be 20 dB. The voltage standing wave ratio (VSWR) and magnitude of reflection coefficient are respectively

(a) 1.22 and 0.1 (b) 0.81 and 0.1

(c) –1.22 and 0.1 (d) 2.44 and 0.2

25. Let $g(t) = e^{-\pi t^2}$, and $h(t)$ is a filter matched to $g(t)$. If $g(t)$ is applied as input to $h(t)$, then the Fourier transform of the output is

(a) $e^{-\pi f^2}$ (b) $e^{-\pi f^2/2}$

(c) $e^{-\pi |f|}$ (d) $e^{-2\pi f^2}$

Q. (26–55) carry two marks each.

26. Let U and V be two independent zero mean Gaussian random variables of variances $\frac{1}{4}$ and $\frac{1}{9}$ respectively. The probability $P(3V \geq 2U)$ is

(a) $\frac{4}{9}$ (b) $\frac{1}{2}$

(c) $\frac{2}{3}$ (d) $\frac{5}{9}$

27. Let A be an $m \times n$ matrix and B an $n \times m$ matrix. It is given that determinant $(I_m + AB)$ = determinant $(I_n + BA)$, where I_k is the $k \times k$ identity matrix. Using the above property, the determinant of the matrix given below is

$$\begin{bmatrix} 2 & 1 & 1 & 1 \\ 1 & 2 & 1 & 1 \\ 1 & 1 & 2 & 1 \\ 1 & 1 & 1 & 2 \end{bmatrix}$$

(a) 2 (b) 5

(c) 8 (d) 16

28. In the circuit shown below, if the source voltage $V_a = 100\angle 53{:}13°$ V then the Thevenin's equivalent voltage in Volts as seen by the load resistance R_L is

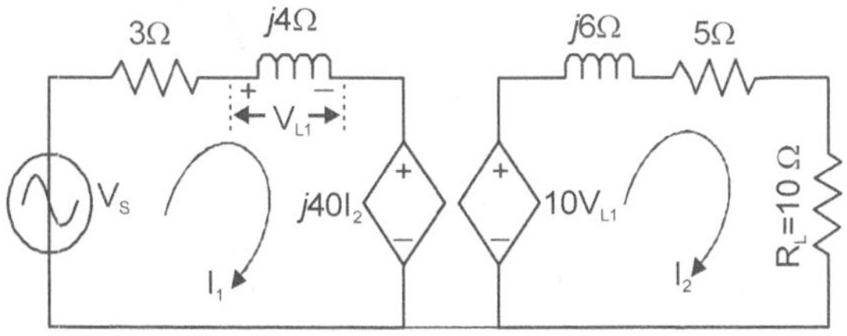

(a) $100\angle 90°$ (b) $800\angle 0°$

(c) $800\angle 90°$ (d) $100\angle 60°$

29. The open-loop transfer function of a dc motor is given as $\frac{w(s)}{V_a(s)} = \frac{10}{1+10s}$. When connected in feedback as shown below, the approximate value of K_a that will reduce the time constant of the closed loop system by one hundred times as compared to that of the open-loop system is

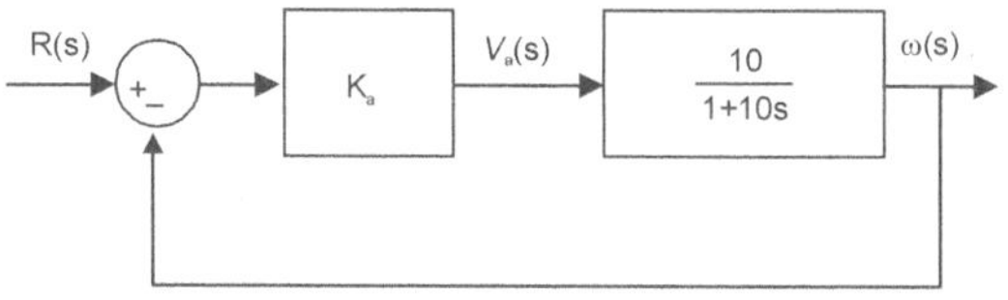

(a) 1 (b) 5

(c) 10 (d) 100

30. In the circuit shown below, the knee current of the ideal Zener diode is 10 mA. To maintain 5 V across R_L, the minimum value of R_L in Ω and the minimum power rating of the Zener diode in mW, respectively, are

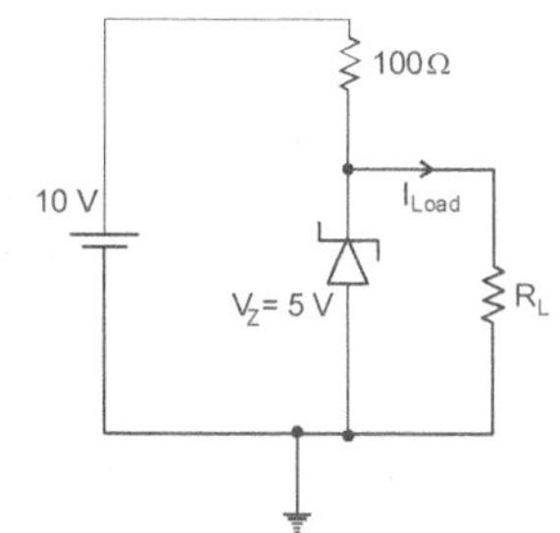

(a) 125 and 125

(b) 125 and 250

(c) 250 and 125

(d) 250 and 250

31. The following arrangement consists of an ideal transformer and an attenuator which attenuates by a factor of 0.8. An ac voltage V_{wx1} = 100V is applied across WX to get an open circuit voltage V_{yz1} across YZ. Next, an ac voltage V_{yz2} = 100V is applied across YZ to get an open circuit voltage V_{wx2} across WX. Then, V_{yz1} / V_{wx1}, V_{wx2} / V_{yz2} are respectively,

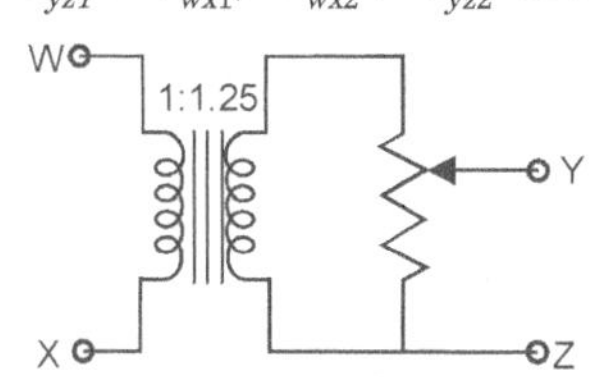

(a) 125/100 and 80/100

(b) 100/100 and 80/100

(c) 100/100 and 100/100

(d) 80/100 and 80/100

32. Two magnetically uncoupled inductive coils have Q factors q_1 and q_2 at the chosen operating frequency. Their respective resistances are R_1 and R_2. When connected in series, their effective Q factor at the same operating frequency is

(a) $q_1 + q_2$

(b) $(1/q_1)+(1/q_2)$

(c) $(q_1R_1 + q_2R_2) / (R_1 + R_2)$

(d) $(q_1 R_2 + q_2 R_1) / (R_1 + R_2)$

33. The impulse response of a continuous time system is given by $h(t) = \delta(t - 1) + \delta(t - 3)$. The value of the step response at $t = 2$ is

(a) 0 (b) 1

(c) 2 (d) 3

34. The small-signal resistance (i.e., dV_B/dI_D) in kΩ offered by the n-channel MOSFET M shown in the figure below, at a bias point of V_B = 2 V is (device data for M: device transconductance parameter $k_n = \mu_A C_{ox}'$ (W/L) = 40 μA/V², threshold voltage V_{TN} = 1 V, and neglect body effect and channel length modulation effects)

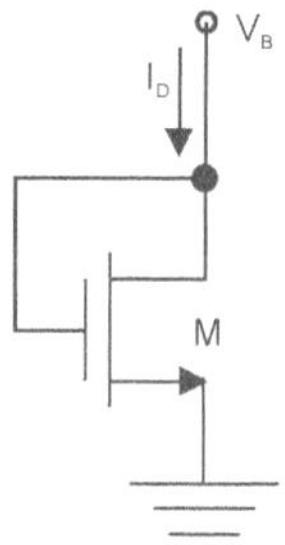

(*a*) 12.5

(*b*) 25

(*c*) 50

(*d*) 100

35. The ac schematic of an NMOS common-source stage is shown in the figure below, where part of the biasing circuits has been omitted for simplicity. For the *n*-channel MOSFET M, the transconductance g_m = 1 mA/V, and body effect and channel length modulation effect are to be neglected. The lower cutoff frequency in Hz of the circuit is approximately at

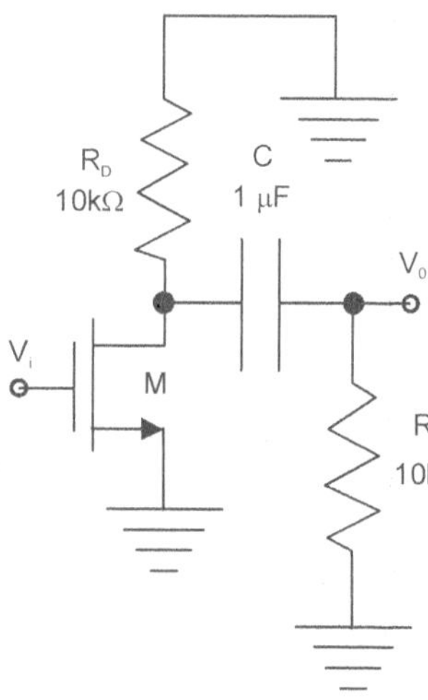

(*a*) 8 (*b*) 32

(*c*) 50 (*d*) 200

36. A system is described by the differential equation

$$\frac{d^2y}{dt^2} + 5\frac{dy}{dt} + 6y(t) = x(t).$$

Let $x(t)$ be a rectangular pulse given by

$$x(t) = \begin{cases} 1 & 0 < t < 2 \\ 0 & \text{otherwise} \end{cases}$$

Assuming that $y(0) = 0$ and $\frac{dy}{dt} = 0$ at $t = 0$, the Laplace transform of $y(t)$ is

(*a*) $\frac{e^{-2s}}{s(s+2)(s+3)}$ (*b*) $\frac{1-e^{-2s}}{s(s+2)(s+3)}$

(*c*) $\frac{e^{-2s}}{(s+2)(s+3)}$ (*d*) $\frac{1-e^{-2s}}{(s+2)(s+3)}$

37. A system described by a linear, constant coefficient, ordinary, first order differential equation has an exact solution given by $y(t)$ for $t > 0$, when the forcing function is $x(t)$ and the initial condition is $y(0)$. If one wishes to modify the system so that the solution becomes $-2y(t)$ for $t > 0$, we need to

(*a*) change the initial condition to $-y(0)$ and the forcing function to $2x(t)$

(*b*) change the initial condition to $2y(0)$ and the forcing function to $-x(t)$

(*c*) change the initial condition to $j\sqrt{2}y(0)$ and the forcing function to $j\sqrt{2}x(t)$

(*d*) change the initial condition to $-2y(0)$ and the forcing function to $-2x(t)$

38. Consider two identically distributed zero-mean random variables U and V. Let the cumulative distribution functions of U and 2V be F(*x*) and G(*x*) respectively. Then, for all values of *x*

(*a*) $F(x) - G(x) \le 0$ (*b*) $F(x) - G(x) \ge 0$

(*c*) $(F(x) - G(x)) \cdot x \le 0$ (*d*) $(F(x) - G(x)) \cdot x \ge 0$

39. The DFT of a vector $[a \ b \ c \ d]$ is the vector $[\alpha \ \beta \ \gamma \ \delta]$. Consider the product

$$[p \ q \ r \ s] = [a \ b \ c \ d]\begin{bmatrix} a & b & c & d \\ d & a & b & c \\ c & d & a & b \\ b & c & d & a \end{bmatrix}$$

The DFT of the vector $[p \ q \ r \ s]$ is a scaled version of

(*a*) $[\alpha^2 \ \beta^2 \ \gamma^2 \ \delta^2]$

(*b*) $[\sqrt{\alpha} \ \sqrt{\beta} \ \sqrt{\gamma} \ \sqrt{\delta}]$

(*c*) $[\alpha+\beta \ \ \beta+\delta \ \ \delta+\gamma \ \ \gamma+\alpha]$

(*d*) $[\alpha \ \beta \ \gamma \ \delta]$

40. The signal flow graph for a system is given below. The transfer function $\frac{Y(s)}{U(s)}$ for this system is

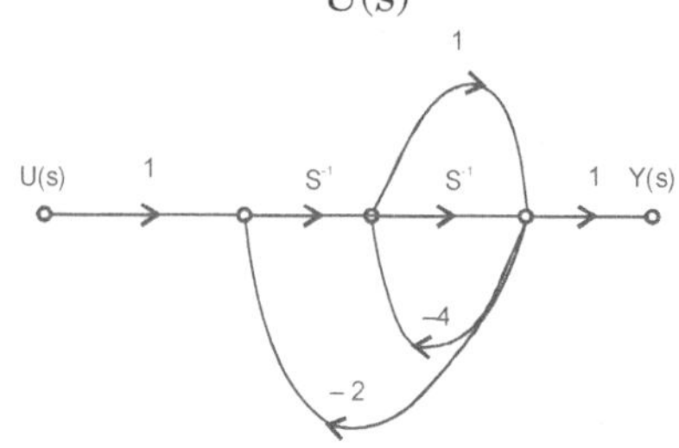

(*a*) $\frac{s+1}{5s^2+6s+2}$ (*b*) $\frac{s+1}{s^2+6s+2}$

(*c*) $\frac{s+1}{s^2+4s+2}$ (*d*) $\frac{1}{5s^2+6s+2}$

41. In the circuit shown below the op-amps are ideal. Then V_{out} in Volts is

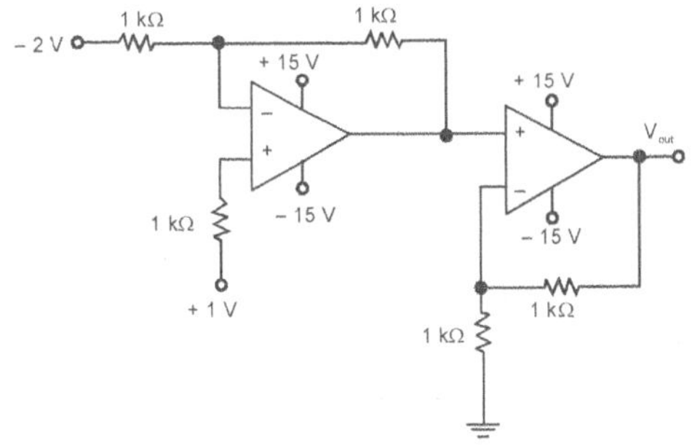

(a) 4 (b) 6

(c) 8 (d) 10

42. In the circuit shown below, Q_1 has negligible collector-to-emitter saturation voltage and the diode drops negligible voltage across it under forward bias. If V_{cc} is +5 V, X and Y are digital signals with 0 V as logic 0 and V_{cc} as logic 1, then the Boolean expression for Z is

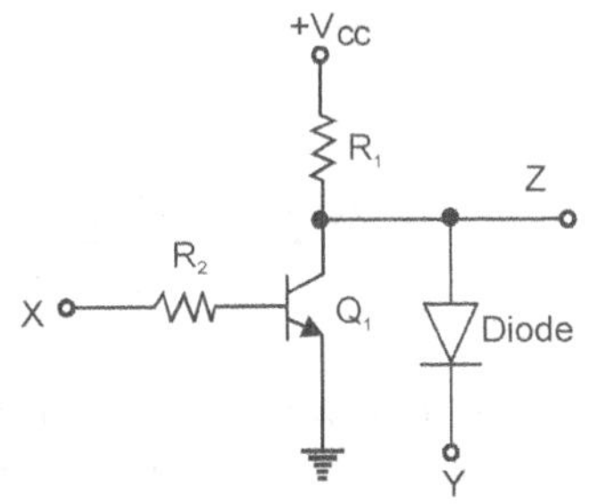

(a) XY (b) $\bar{X}Y$

(c) $X\bar{Y}$ (d) $\overline{XY}$

43. A voltage 1000 sin ωt Volts is applied across YZ. Assuming ideal diodes, the voltage measured across WX in Volts, is

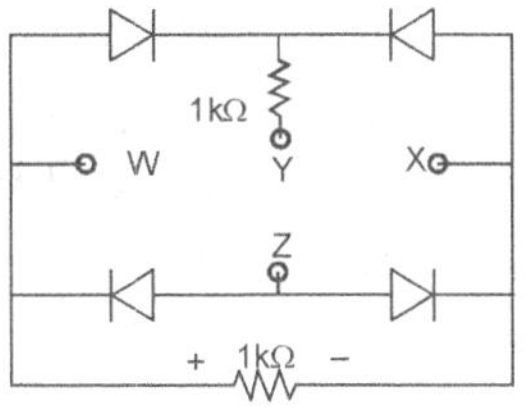

(a) sinωt

(b) (sin ωt + | sin ωt |)/2

(c) (sin ωt − | sin ωt |)/2

(d) 0 for all t

44. Three capacitors C_1, C_2 and C_3 whose values are 10μF, 5μF, and 2μF respectively, have breakdown voltages of 10V, 5V, and 2V respectively. For the interconnection shown below, the maximum safe voltage in Volts that can be applied across the combination, and the corresponding total charge in μC stored in the effective capacitance across the terminals are respectively,

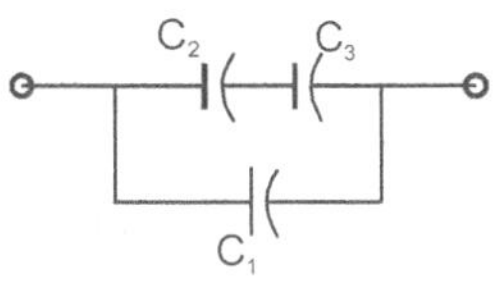

(a) 2.8 and 36

(b) 7 and 119

(c) 2.8 and 32

(d) 7 and 80

45. There are four chips each of 1024 bytes connected to a 16 bit address bus as shown in the figure below. RAMs 1, 2, 3 and 4 respectively are mapped to addresses

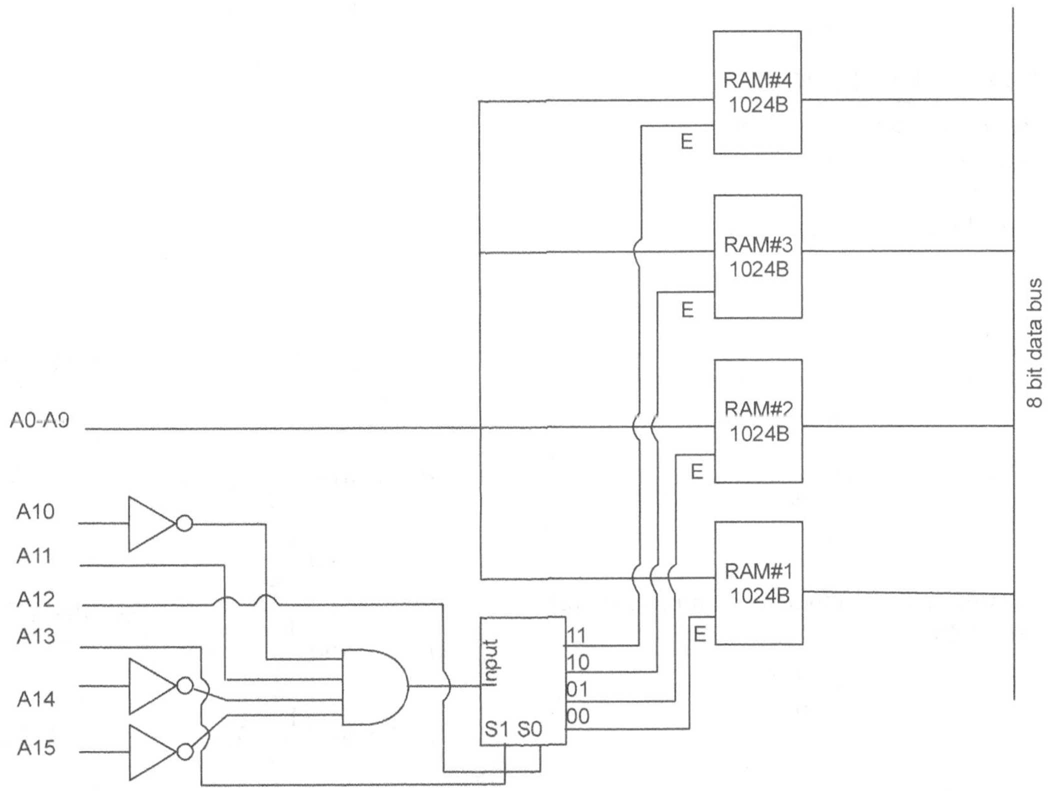

(*a*) 0C00H-0FFFH, 1C00H-1 FFFH, 2C00H-2FFFH, 3C00H-3FFFH

(*b*) 1800H-1FFFH, 2800H-2FFFH, 3800H-3FFFH,4800H-4FFFH

(*c*) 0500H-08FFH, 1500H-18FFH, 3500H-38FFH, 5500H-58FFH

(*d*) 0800H-0BFFH, 1800H-1BFFH, 2800H-2BFFH, 3800H-3BFFH

46. In the circuit shown below, the silicon *npn* transistor Q has a very high value of β. The required value of $R1_2$ in kΩ to produce I_C = 1 *m*A is

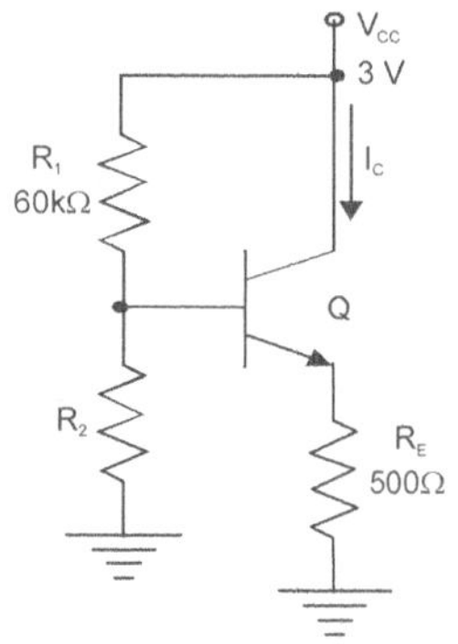

(*a*) 20 (*b*) 30

(*c*) 40 (*d*) 50

47. Let U and V be two independent and identically distributed random variables such that P(U = +1) = P(U = –1) = $\frac{1}{2}$. The entropy H (U +V) in bits is

(*a*) $\frac{3}{4}$ (*b*) 1

(*c*) $\frac{3}{2}$ (*d*) $\log_2 3$

COMMON DATA QUESTIONS

Common Data for Questions 48 and 49:

Bits 1 and 0 are transmitted with equal probability. At the receiver, the pdf of the respective received signals for both bits are as shown below.

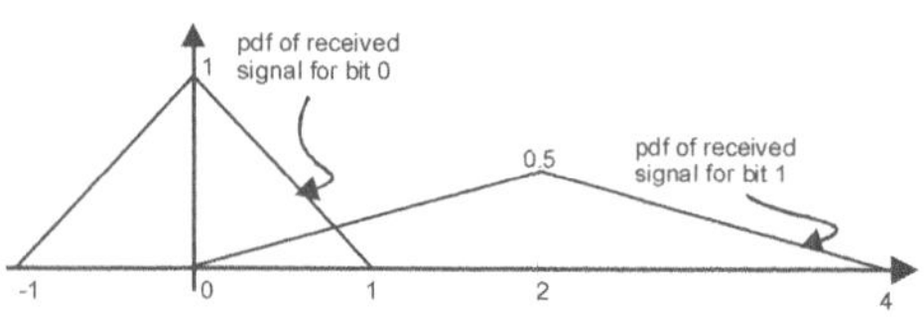

48. If the detection threshold is 1, the BER will be

(*a*) $\frac{1}{2}$ (*b*) $\frac{1}{4}$

(*c*) $\frac{1}{8}$ (*d*) $\frac{1}{16}$

49. The optimum threshold to achieve minimum bit error rate (BER) is

(*a*) $\frac{1}{2}$ (*b*) $\frac{4}{5}$

(*c*) 1 (*d*) $\frac{3}{2}$

Common Data for Questions 50 and 51:

Consider the following figure

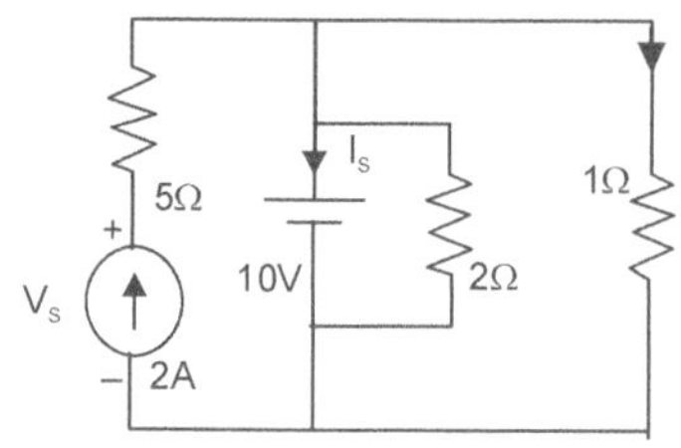

50. The current I_s in Amps in the voltage source, and voltage V_s in Volts across the current source respectively, are

(*a*) 13, –20

(*b*) 8, –10

(*c*) –8, 20

(*d*) –13, 20

51. The current in the 1 Ω resistor in Amps is

(*a*) 2

(*b*) 3.33

(*c*) 10

(*d*) 12

LINKED ANSWER QUESTIONS

Statement for Linked Answer Questions 52 and 53 :

A monochromatic plane wave of wavelength λ = 600μm is propagating in the direction as shown in the figure below. $\vec{E}_i, \vec{E}_r$ and $\vec{E}_t$ denote incident, reflected, and transmitted electric field vectors associated with the wave.

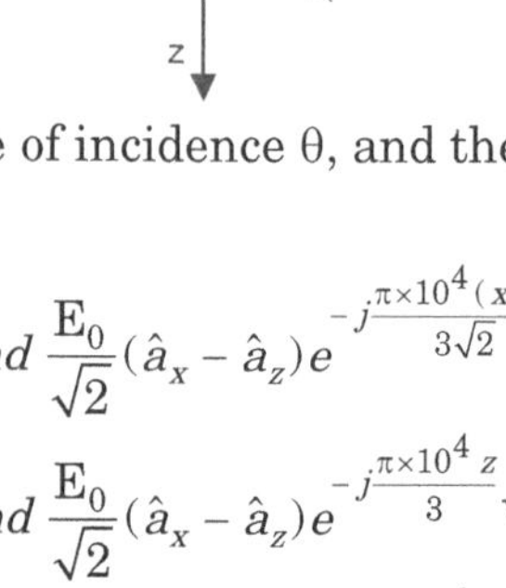

52. The angle of incidence θ_i and the expression for $\vec{E}_i$ are

(*a*) 60° *and* $\frac{E_0}{\sqrt{2}}(\hat{a}_x - \hat{a}_z)e^{-j\frac{\pi\times10^4(x+z)}{3\sqrt{2}}}$ V / *m*

(*b*) 45° *and* $\frac{E_0}{\sqrt{2}}(\hat{a}_x - \hat{a}_z)e^{-j\frac{\pi\times10^4 z}{3}}$ V / *m*

(*c*) 45° *and* $\frac{E_0}{\sqrt{2}}(\hat{a}_x - \hat{a}_z)e^{-j\frac{\pi\times10^4(x+z)}{3\sqrt{2}}}$ V / *m*

(*d*) 60° *and* $\frac{E_0}{\sqrt{2}}(\hat{a}_x - \hat{a}_z)e^{-j\frac{\pi\times10^4 z}{3}}$ V / *m*

53. The expression for $\vec{E}_r$ is

(a) $0.23\frac{E_0}{\sqrt{2}}(\hat{a}_x - \hat{a}_z)e^{-j\frac{\pi\times10^4(x-z)}{3\sqrt{2}}}$ V / m

(b) $-\frac{E_0}{\sqrt{2}}(\hat{a}_x - \hat{a}_z)e^{j\frac{\pi\times10^4 z}{3}}$ V / m

(c) $0.44\frac{E_0}{\sqrt{2}}(\hat{a}_x - \hat{a}_z)e^{-j\frac{\pi\times10^4(x-z)}{3\sqrt{2}}}$ V / m

(d) $\frac{E_0}{\sqrt{2}}(\hat{a}_x - \hat{a}_z)e^{-j\frac{\pi\times10^4(x+z)}{3}}$ V / m

Statement for Linked Answer Questions 54 and 55 :

The state diagram of a system is shown below. A system is described by the state-variable equations

$$\dot{X} = AX + Bu;$$
$$y = CX + Du$$

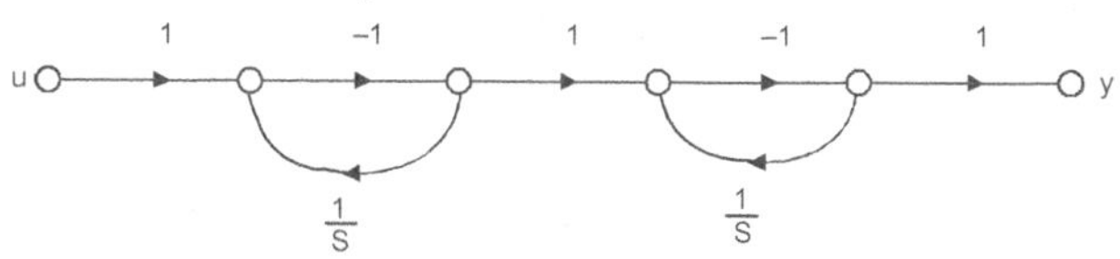

54. The state-variable equations of the system shown in the figure above are

(a) $\dot{X} = \begin{bmatrix} -1 & 0 \\ 1 & -1 \end{bmatrix}X + \begin{bmatrix} -1 \\ 1 \end{bmatrix}u$, $y = [1 \ \ -1]X + u$

(b) $\dot{X} = \begin{bmatrix} -1 & 0 \\ -1 & -1 \end{bmatrix}X + \begin{bmatrix} -1 \\ 1 \end{bmatrix}u$, $y = [-1 \ \ -1]X + u$

(c) $\dot{X} = \begin{bmatrix} -1 & 0 \\ -1 & -1 \end{bmatrix}X + \begin{bmatrix} -1 \\ 1 \end{bmatrix}u$, $y = [-1 \ \ -1]X - u$

(d) $\dot{X} = \begin{bmatrix} -1 & -1 \\ 0 & -1 \end{bmatrix}X + \begin{bmatrix} -1 \\ 1 \end{bmatrix}u$, $y = [1 \ \ -1]X - u$

55. The state transition matrix e^{At} of the system shown in the figure above is

(a) $\begin{bmatrix} e^{-t} & 0 \\ te^{-t} & e^{-t} \end{bmatrix}$

(b) $\begin{bmatrix} e^{-t} & 0 \\ -te^{-t} & e^{-t} \end{bmatrix}$

(c) $\begin{bmatrix} e^{-t} & 0 \\ e^{t} & e^{t} \end{bmatrix}$

(d) $\begin{bmatrix} e^{-t} & -te^{-t} \\ 0 & e^{-t} \end{bmatrix}$

GENERAL APTITUDE (GA) QUESTIONS

(Q. 56–60) carry one mark each.

56. Choose the grammatically CORRECT sentence:

(a) Two and two add four.

(b) Two and two become four.

(c) Two and two are four.

(d) Two and two make four.

57. **Statement** : You can always give me a ring whenever you need.

Which one of the following is the best inference from the above statement?

(a) Because I have a nice caller tune.

(b) Because I have a better telephone facility.

(c) Because a friend in need is a friend indeed.

(d) Because you need not pay towards the telephone bills when you give me a ring.

58. In the summer of 2012, in New Delhi, the mean temperature of Monday to Wednesday was 41°C and of Tuesday to Thursday was 43°C. If the temperature on Thursday was 15% higher than that of Monday, then the temperature in °C on Thursday was

(a) 40 (b) 43

(c) 46 (d) 49

59. Complete the sentence : Dare _____ mistakes.

(a) commit (b) to commit

(c) committed (d) committing

60. They were requested not to **quarrel** with others.

Which one of the following options is the closest in meaning to the word **quarrel**?

(a) make out (b) call out

(c) dig out (d) fall out

(Q. 61–65) carry two marks each.

61. A car travels 8 km in the first quarter of an hour, 6 km in the second quarter and 16 km in the third quarter. The average speed of the car in km per hour over the entire journey is

(a) 30 (b) 36

(c) 40 (d) 24

62. Find the sum to n terms of the series 10 + 84 + 734 + ...

(a) $\frac{9(9^n+1)}{10}+1$ (b) $\frac{9(9^n-1)}{8}+1$

(c) $\frac{9(9^n-1)}{8}+n$ (d) $\frac{9(9^n-1)}{8}+n^2$

63. **Statement** : There were different streams of freedom movements in colonial India carried out by the moderates, liberals, radicals, socialists, and so on.

Which one of the following is the best inference from the above statement?

(a) The emergence of nationalism in colonial India led to our Independence.

(b) Nationalism in India emerged in the context of colonialism.

(c) Nationalism in India is homogeneous.

(d) Nationalism in India is heterogeneous.

64. The set of values of p for which the roots of the equation $3x^2 + 2x + p(p-1) = 0$ are of opposite sign is

(*a*) $(-\infty, 0)$

(*b*) $(0, 1)$

(*c*) $(1, \infty)$

(*d*) $(0, \infty)$

65. What is the chance that a leap year, selected at random, will contain 53 Saturdays?

(*a*) $\frac{2}{7}$ (*b*) $\frac{3}{7}$

(*c*) $\frac{1}{7}$ (*d*) $\frac{5}{7}$

ANSWERS

1. (*c*)	2. (*d*)	3. (*c*)	4. (*a*)	5. (*d*)	6. (*b*)	7. (*d*)	8. (*c*)	9. (*b*)	10. (*b*)
11. (*b*)	12. (*a*)	13. (*b*)	14. (*a*)	15. (*a*)	16. (*a*)	17. (*d*)	18. (*c*)	19. (*a*)	20. (*c*)
21. (*b*)	22. (*d*)	23. (*c*)	24. (*a*)	25. (*d*)	26. (*c*)	27. (*b*)	28. (*c*)	29. (*c*)	30. (*b*)
31. (*b*)	32. (*c*)	33. (*b*)	34. (*b*)	35. (*a*)	36. (*b*)	37. (*d*)	38. (*d*)	39. (*a*)	40. (*a*)
41. (*c*)	42. (*b*)	43. (*d*)	44. (*c*)	45. (*d*)	46. (*c*)	47. (*c*)	48. (*c*)	49. (*b*)	50. (*d*)
51. (*c*)	52. (*c*)	53. (*c*)	54. (*a*)	55. (*a*)	56. (*d*)	57. (*c*)	58. (*c*)	59. (*b*)	60. (*b*)
61. (*c*)	62. (*d*)	63. (*d*)	64. (*b*)	65. (*a*)					

EXPLANATIONS

1. Let us consider the switches A and B and bulb Y. Switches can be 2 positions up (0) or down (1)

Starting with both A and B in up position. Let the bulb be OFF. Now since B can operate independently when B goes down, the bulb goes ON

A	B	Y
up (0)	up (0)	OFF
up (0)	down (1)	ON

Now keeping A in down position when B goes down, the bulb will go OFF.

A	B	Y
down (1)	up (0)	ON
down (1)	down (1)	OFF

find truth table corresponds to XOR gate.

2. Hence, for a vector field $\vec{A}(\vec{r})$

$$\oint \vec{A} \cdot \vec{de} = \iint \left(\vec{\nabla} \times \vec{A}\right) \cdot \vec{ds}$$

→ according to Stokes theorem.

3.

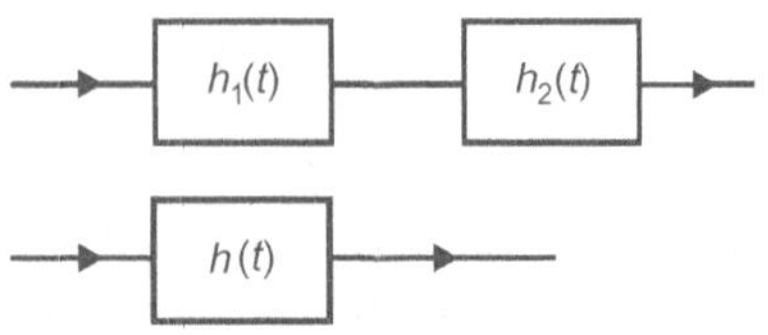

$$h(t) = h_1(t) \oplus h_2(t)$$

4. Due to application of voltage (forward bias minority carrier are injected from either side of diode one subsequent diffusion takes place and finally recombination.

Injection and subsequent diffusion and recombination of minority carriers

5. As during dry oxidation, quality of oxide is superior as it does not contain ullter which is responsible for containmation/impurity but simultaneously lower the growth rate

6.
$$\lim_{\theta \to 0} \frac{\sin\theta}{\theta} = 1$$
$$\lim_{\theta \to 0} \sin\theta = \theta$$

as within 10% of error. $= 180° \times 1 = 18°$

7.
$$\bar{A} = xa_x + ya_y + za_z$$
$$\nabla \cdot A = \frac{\partial A_x}{\partial x} + \frac{\partial A_y}{\partial y} + \frac{\partial A_z}{\partial z} = 1 + 1 + 1 = 3$$

8. As $h(t) = t\,a(f)$

input response

$\delta(t) \to t\,a(t)$

$$u(t) \to \int_{-\infty}^{t} +a(I)dt = \int_{0}^{t} t dt = \frac{t^2}{2}a(t)$$

$$u(t-1) \to \frac{(t-1)^2}{2}a(t-1)$$

9. $20 \log k = 32$

$k = 10^{1.6}$

$= 39.8$

$$T(s) = \frac{39.8}{s^2}$$

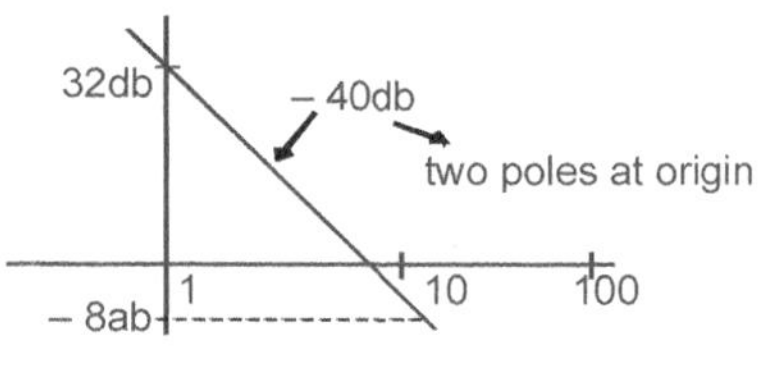

10. Using the concept of "virtual ground" in an operational amplifier, we can set the voltage at the point to zero volts since the non inverting terminal is grounded.

Once $V_A = 0$, V_C will also be zero.

We know that for a silicon *n–p–n* transistor,

$V_{BE} = V_B - V_E = 0.7$ V

Since, $V_B = 0 \Rightarrow V_E = -0.7$ V

Hence the output voltage is the same as the emitter voltage

so, $V_{out} = -0.7$ V

11.

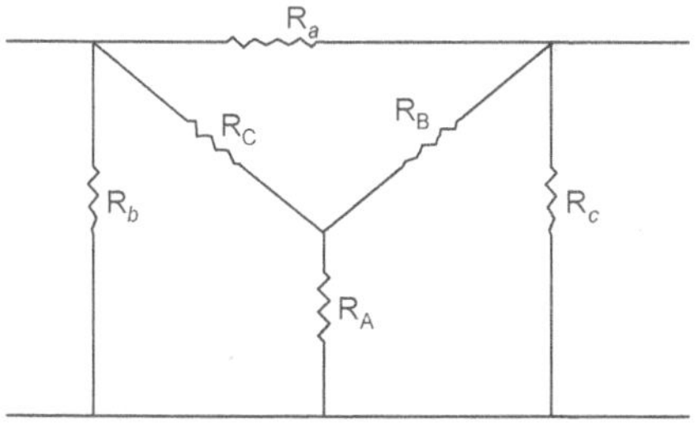

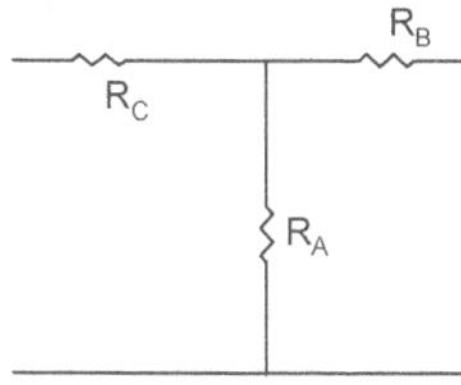

$$R_c = \frac{R_a \cdot R_b}{R_a + R_b + R_c} \text{ as } R_a \text{ is scaled by facts } k$$

$$R_c' = \frac{R_a' \cdot R_b'}{R_a' + R_b + R_c} = \frac{k^2 R_a \cdot R_b}{k(R_a + R_b + R_c)}$$

$$= k \cdot \frac{R_a \times R_b}{R_a + R_b + c}$$

so elements corresponding to star equivalence will be seated by facts k.

12. Accumulator changes as follows

(05 + 05 + 04 + 03 + 02 + 01)H

At the end of Loop accumulator contains = 14H

ADI 03H $\rightarrow$ A = (14 + 03) = 17H

13. Bit rate given = R Kbits/second

Modulation = 32-QAM

No. of bits/symbol = 5[$\log_2 32$]

Symbol rate = $\frac{R}{5}$ k symbols/second

Finally we are transmitting symbols.

$E_T \rightarrow$ transmission bandwidth

$$B_T = \frac{R(\text{symbol rate})}{(1+\alpha)} = \frac{R}{5(1+\alpha)}$$

For B_T to be minimum, α has to be maximum

$$\Rightarrow B_T = \frac{R}{5\times 2} = \frac{R}{10}$$

Maximum value of α is '1' which is a roll off factor

14. LCM of T_1, T_2, T_3 will be:

$$= \frac{\text{LCM of numerator}}{\text{HCF of denominator.}} = \frac{2\pi}{100}$$

$\therefore$ Overall time period = $\frac{2\pi}{100}$ sec.

Harmonic frequency = 100 rad/sec.

15. Input Independence of a voltage-voltage feedback circuit $= 2i(1 + A_0 k)$

Z_i = initial input impedance (without feed output Impedance of a voltage-voltage feedback circuit

$$= Z_0/(1 + A_0 k)$$

Z_0 = initial output impedance (without feedback)

Hence, As K is increased, the input impedance will increase and output impedance will decrease.

16. Here $f_m = 5$KHZ

$\therefore$ $f_s \geq 2f_m = 10$ KHZ

B,C,D options are greater than 10 KHZ

17. $I_D = K_n\left(\frac{\omega}{2}\right)(V_{GS} - V_t)^2(1 + \lambda V_{DS})$ output resistance is given by$\rightarrow$

$$r_0 = \frac{1}{\partial I_D / \partial V_{DS}} \quad ...(1)$$

$$\frac{\partial I_D}{\partial V_{DS}} = \lambda k' n \frac{\omega}{2}(V_{GS} - V_t)^2 \quad ...(2)$$

$$r_0 = \frac{1}{\partial I_D / dV_{DS}} = \frac{1}{\lambda k' n \frac{\omega}{L}(V_{Gs} - V_t)^2}$$

For Ideal cases

$$I_D = k' n\left(\frac{\omega}{L}\right)(V_{Gs} - V_t)^2 \quad \frac{\partial ID}{\partial V_{DS}} = 0$$

$$r = \infty$$

18. **Consider option A:** In which all the pols lie on the left of $j\omega$ axis which satisfy casual stable LT1 system.

Option B: For a stable casual system, there are no restriction for the position of zeroes on s plane.

Option C: text true.

Option D: Roots of characteristic equation are all closed loop poles and they all line on the left side of the $j\omega$ axis.

19. $|A| = 3[60–49] – 5[25 – 14] + 2[35 – 24] = 0$

$|A| = a_1 \times a_2 \times a_3 = 0$

which, implies either of the above eigen values equal to zero. It may be one or two negative eigen values.

20. $f(n) = a_4 x^4 + a_3 x^3 + a_2 x^2 + a_1 x - a_0$

If complex roots are in even no. (in pair) then the real roots will also be even.

$\rightarrow$ option (c) is wrong.

from the equation,

Product of roots = $\frac{-a_0}{a_4}(< 0)$

As no. of roots = 4

$\therefore$ Products of 4 nos . < 0

$\Rightarrow$ either 1 or 3 nos < 0.

$\therefore$ B is

Product of roots:

$Z_1 Z_1^*, Z_2 Z_2^* < 0$

$|Z_1|^2 |Z_2|^2 < 0$ which is not possible

21. $H(s) = \frac{1}{s}$

$h(t) = u(t)$

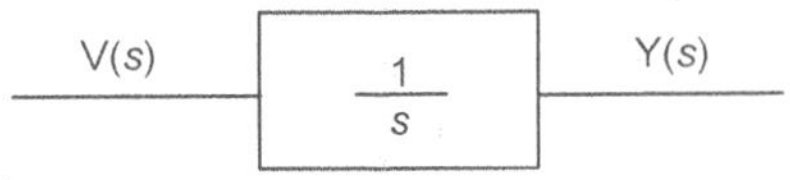

$\underline{\underline{u}}(t)$ = input

output $= u(t) \oplus h(t) = u(t) \oplus u(t) = r(t) = +u(t)$

22. Taking Laplace transformation of the circuit,

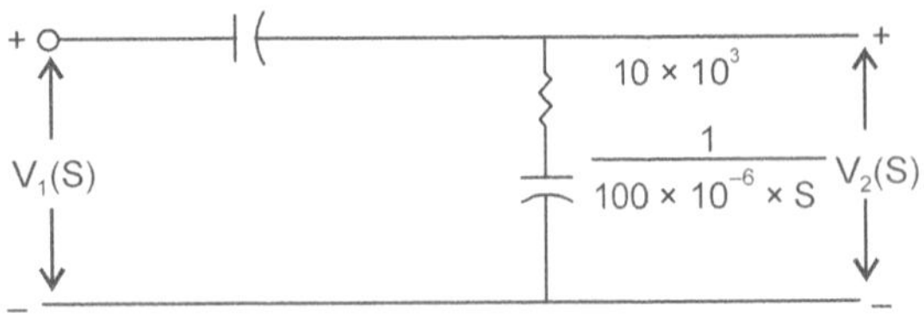

By applying voltage devider rule:

$$V_2(s) = \frac{10 \times 10^3 + \frac{10^4}{s}}{10 \times 10^3 + \frac{10^4}{s} + \frac{10^4}{s}} \times V_1(s)$$

$$\frac{V_2(s)}{V_1(s)} = \frac{1 + \frac{1}{s}}{1 + \frac{2}{s}} = \frac{s+1}{s+2}$$

23. $Z_i = (4 + j3)\ \Omega$

$Z_L = \sqrt{4^2 + 3^2}\ \Omega = 5\ \Omega$

24. The reflection co-efficient is $-20\log\Gamma = 20\text{dB}$

$\Rightarrow \log\Gamma = -1\text{dB}; \quad \Rightarrow \Gamma = 10^{-1} \Rightarrow \Gamma = 0.1$

Relation between Γ and VSWR is

$$S = \frac{1 + |\Gamma|}{1 - |\Gamma|} = \frac{1 + 0.1}{1 - 0.1} = \frac{1.1}{0.9} = 1.22$$

25. $g(t) = e^{-\pi t^2}$

$h(t) = g(-t) = e^{-(-t)^2} = e^{-t^2}$

As we know

$$F[r(t)] = F[h(t) \oplus g(t)] \qquad \Big| \; e^{-\pi t^2} \rightleftharpoons e^{-\pi f^2}$$
$$= H(s) \cdot G(s)$$
$$= e^{-\pi f^2} \cdot e^{-\pi f^2}$$
$$= e^{-2\pi f^2}$$

26. Given:

U & V be two independent & identical, distributed random variable

U & V are $\xleftrightarrow{\text{I.I.D}}$

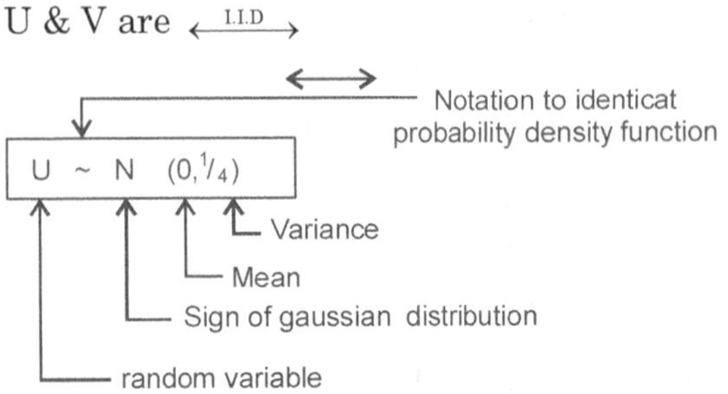

$\boxed{V \sim N(0, 1/9).}$

To find: $P(3V \geq 2U)$

Solution: Rearrange the problem

$$P\left[\left(\frac{V-O}{1/3}\right) \geq \left(\frac{U-O}{1/2}\right)\right]$$

Let denote

$$X = \frac{V-O}{\frac{1}{3}} \quad \left\{ \begin{array}{l} E[X] = E\left[\frac{V}{\frac{1}{3}}\right] = \frac{1}{\frac{1}{3}} E[V] = O \\ E\left[1 \times 1^2\right] = E\left[\frac{|V|^2}{\frac{1}{9}}\right] = \frac{1}{\frac{1}{9}} E[|V|] \\ = 9 \times \frac{1}{9} \\ = 1. \end{array} \right.$$

$$Y = \frac{U-O}{\frac{1}{2}}$$

then X & Y become

$X \sim N(0, 1)$.

normalised distribution (Gaussian)

$Y \sim N(0, 1)$

normalised Gaussian distributed.

You can see clearly. How they became normalised distribution, as we know that normalised distribution has zero-mean & variance '1'.

$$E[|y|] = E\left[\frac{|U|}{\frac{1}{2}}\right] = \frac{1}{\frac{1}{2}} E[|U|] = 0$$

$$E\left[|y|^2\right] = E\left[\frac{|U|^2}{\frac{1}{4}}\right] = \frac{1}{\frac{1}{4}} E\left[|U|^2\right] = 4 \times \frac{1}{4} = 1.$$

Question reduce to:

$$P(X \geq Y) = P(X - Y \geq O)$$

As let, $Z = X - Y$, it is just linear combination e random variable which is gaussion distribution with mean – O & variance – 1.

$\therefore$ their Linear combination is also gaussian $= P(Z \geq O)$

Note: By using central limit theorm, note here mean will be $\boxed{E[|Z|] = 0}$ as mean remain same & variance become 'n' times of resulting random variable, so figure will became

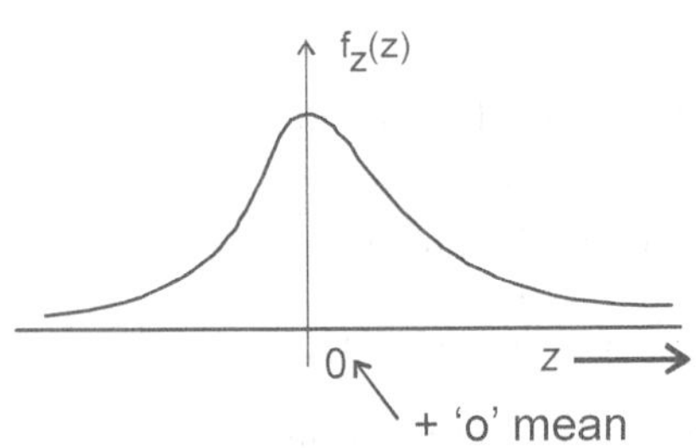

27. Given:

A : $m \times n$ matrix

B : $n \times m$ matrix

det $(I_m + AB) = \det(I_n + BA)$

I_K : K × K identify matrix

To find:

$$\det\begin{bmatrix}2&1&1&1\\1&2&1&1\\1&1&2&1\\1&1&1&2\end{bmatrix}=\det(M) \qquad \{\because say\}$$

Analysis: We will break matrix m to match $(I_m + AB)$

Plan:

1. As per analysis part we will break matrix m into sum of I_m and AB
2. Then use $\det(I_m + AB) = \det(I_m + BA)$

Carrying out plan:

$$\begin{bmatrix}2&1&1&1\\1&2&1&1\\1&1&2&1\\1&1&1&2\end{bmatrix}=\begin{bmatrix}1&0&0&0\\0&1&0&0\\0&0&1&0\\0&0&0&1\end{bmatrix}+\begin{bmatrix}1&1&1&1\\1&1&1&1\\1&1&1&1\\1&1&1&1\end{bmatrix} \quad ...(1)$$

Now we will break second matrix in RHS of above as follows

$$\begin{bmatrix}1&1&1&1\\1&1&1&1\\1&1&1&1\\1&1&1&1\end{bmatrix}_{4\times4}=\begin{bmatrix}1\\1\\1\\1\end{bmatrix}_{4\times1}\begin{bmatrix}1&1&1&1\end{bmatrix}_{1\times4} \quad ...(2)$$

Using (2) into (1), we get

$$\begin{bmatrix}2&1&1&1\\1&2&1&1\\1&1&2&1\\1&1&1&2\end{bmatrix}=I_4+\begin{bmatrix}1\\1\\1\\1\end{bmatrix}\begin{bmatrix}1&1&1&1\end{bmatrix} \quad ...(3)$$

Let $A=\begin{bmatrix}1\\1\\1\\1\end{bmatrix}_{4\times1}$ and $B=\begin{bmatrix}1&1&1&1\end{bmatrix}_{1\times4}$

$$\therefore \begin{bmatrix}2&1&1&1\\1&2&1&1\\1&1&2&1\\1&1&1&2\end{bmatrix}=I_4+A_{4\times1}B_{1\times4}\Rightarrow \boxed{\begin{matrix}m=4\\n=1\end{matrix}}$$

But we are given that $\det(I_m + AB) = \det(I_n + BA)$

$$\therefore \det(I_4+A_{4\times1}B_{1\times4}) = \det\left(I_1+\begin{bmatrix}1&1&1&1\end{bmatrix}\begin{bmatrix}1\\1\\1\\1\end{bmatrix}\right)$$

$= \det(1 + 4)$ $\{\because I_1 = 1\}$

$= \det(5)$

$= 5$ {∵ determinant of a scalar is the same scalar}

28. Given: $V_s = 100\angle 53.13°V$

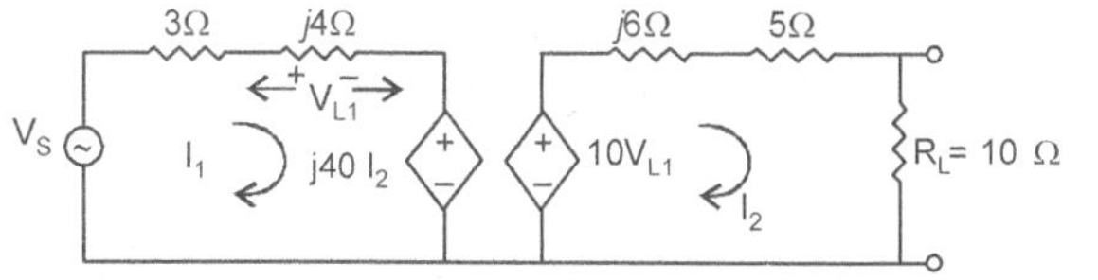

To find: Thevenin's voltage across Load resistonce

Solution

* For V_{th}' open it.

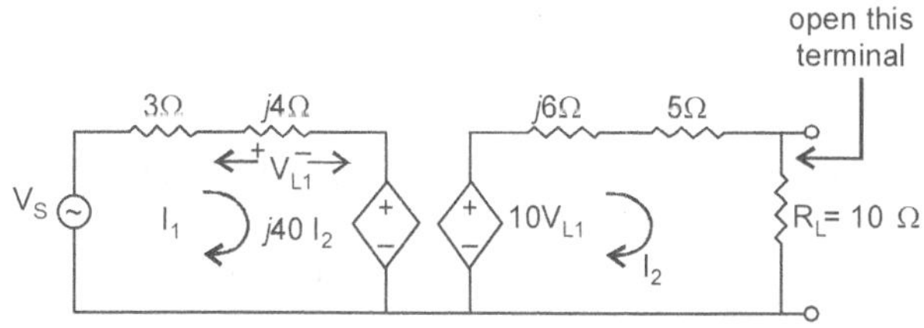

* Opening, then $I_2 = 0$

* When $I_2 = 0$, then $j40I_2 = 0$ {voltage source will short circuit)

∴ Circuit became

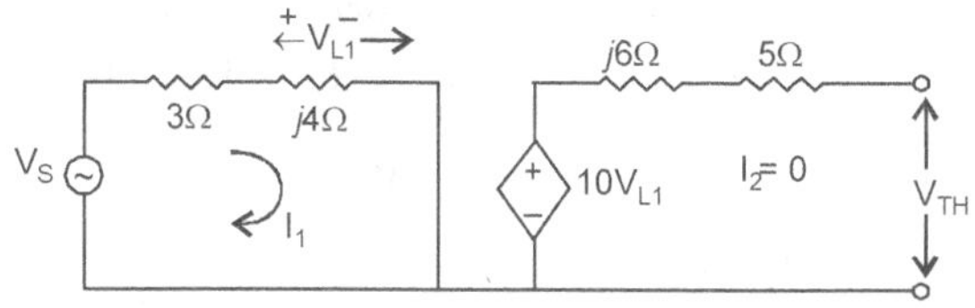

$\therefore \boxed{I_1=\frac{V_s}{3+4j}}$ $\qquad \therefore \boxed{V_{L1}=j4\times\frac{V_s}{3+4j}}$

* $V_{TH} = 10V_{L1}$ because no-current flowing through circuit.

$$V_{TH} = \frac{10\times j4\times V_s}{3+4j} = \frac{40j\,V_s}{3+4j}$$

From rectangular domain to polar domain.

$$=\frac{40\underline{|90°}}{5\underline{|53\cdot13}}\times100\underline{|53\cdot13°} \qquad \boxed{V_{TH}=800\underline{|90°}}$$

29. Given:

Open loop transfer function of a dc motor as

$$\frac{w(s)}{V_a(s)} = \frac{10}{1+10s}$$

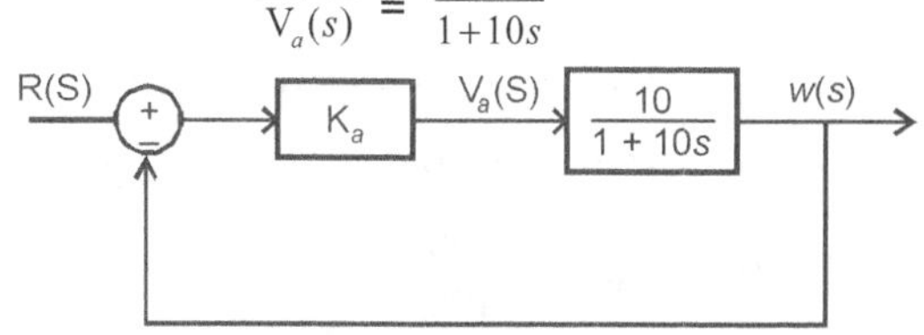

Topic : P controller with unity feed back

Formula: For first order system loop transfer function is $\frac{C(s)}{R(s)}=\frac{K}{1+sT}$ comparing with $\frac{w(s)}{V_a(s)}=\frac{10}{1+10s}$

$T_{\text{open loop}}=10$

Now for closed loop overall transfer function is given by

$$\frac{w(s)}{R(s)} = \frac{K_a\left(\frac{10}{1+10s}\right)}{1+K_a\left(\frac{10}{1+10s}\right)}$$

$$= \frac{K_a 10}{1+10s+K_a 10} = \frac{10K_a}{10s+(10K_a+1)}$$

Dividing numerator and denominator by $10K_a + 1$

Now $$\frac{w(s)}{R(s)} = \frac{\frac{10K_a}{10K_a+1}}{1+\left(\frac{10}{10K_a+1}\right)s}$$

So $T_{closed\ loop} = \frac{10}{10K_a+1}$ (By comparing from formula)

In Question given that time constant of closed loop system is $\frac{1}{100}$ times of time constant of open loop system

so $\frac{10}{10K_a+1} = \frac{1}{100}10\left(T_{closed\ loop} = \frac{1}{100}T_{openloop}\right)$

$10\ K_a + 1 = 100$; $10\ K_a = 99$; $K_a = 9.9 \simeq 10$

$\therefore \boxed{K_a = 10}$ approximate value

30. $I_s = I_z + I_L$

$I_s - I_z = I_L$

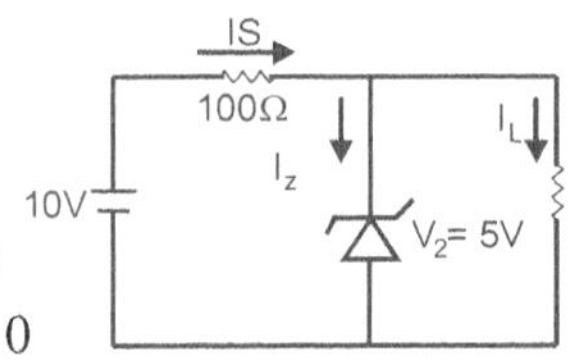

Two extreme condition:

If I_z(min),then I_L (max)

If I_z(max)then I_L (min) = 0

$I_z(\text{max}) = I_s = \frac{10-5}{10} = 50$ mA

$I_z(\text{min}) = I_s - I_L(\text{max})$

$I_L(\text{max}) = I_s - I_z(\text{min}) = I_s - I_z = (50 - 10) = 40$mA

$R_L(\text{min}) = \frac{V}{I_L(\text{max})} = \frac{5}{40}K = 125\Omega$

$P_z = V_z \times I_z(\text{max}) = 5 \times 50 \text{ mA} = 250$ mw

31. 1st case :

$V_{wx1} = 100$ V

So, $Vy'z_1 = \frac{M_2}{M_1}Vwx_1 = 1.25 \times 100 = 125V$

$\therefore$ $Vy'z_1 = Vy'z_1 \times x = 125 \times 0.8 = 100$ v.

$\therefore$ $Vyz_1/Vwx_1 = \frac{100}{100}$.

2nd case:

$Vyz_2 = 100$ V

$\therefore$ $Vy't_2 = \frac{100}{\alpha} = \frac{100}{0.8} = 125v$

Now, $Vwx_2 = \frac{M_1}{M_2}Vy'z_2 = \frac{1}{1.25} \times 125 = 100v$

$\therefore$ $Vwx_2/Vyz_2 = \frac{100}{100}$

32. Let the effective Q factor is q_1 then it can be written using inductance and resistance of equivalent circuit.

$$q = \frac{\omega Leq\cdot}{Req\cdot} = \frac{\omega(L_1+L_2)}{R_1+R_2}$$

Now we substitute the value of L_1 and L_2 in terms of q_1 and q_2

$\therefore$ $$q = \omega\left[\frac{q_1R_1}{\omega} + \frac{q_2R_2}{\omega}\right] = (q_1R_1 + q_2R_2)/(R_1+R_2)$$

33. As $h(t) = \delta(t-1) + \delta(t-3)$

$r(t) = h(t) \oplus u(t)$

$= [\delta(t-1) + \delta(t-3)] \oplus u(t)$

$= u(t-1) + u(t-3)$

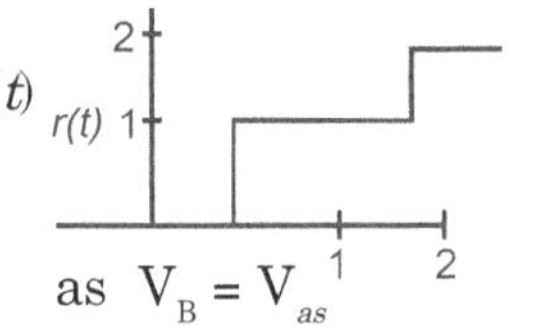

34. $I_D = \frac{\mu}{2}\cos\frac{w}{L}(V_{as} - V_T)^2$ as $V_B = V_{as}$

$\Rightarrow$ $$\frac{\partial I_D}{\partial V_{as}} = \frac{\partial ID}{\partial V_{as}} = \frac{\mu}{2}\cos\frac{w}{L} \times 2(V_{as} - V_T)$$

$$= \mu\cos\frac{W}{L}(V_{as} - V_T)$$

$$= 40\times10^{-6}(2-1) = 40\times10^{-6}$$

$\Rightarrow$ $$\frac{dV_B}{\partial I_D} = \frac{\partial V_{as}}{\partial I_D} = \frac{1}{40\times10^{-6}} = 25K$$

35. $$w_L = \frac{1}{\tau} = \frac{1}{CR_e q}$$

$$w_L = \frac{1}{1\times10^{-6}\times(10k+10k)}$$

$$2\pi f_L = \frac{1}{1\times10^{-6}\times20\times10^3}$$

$$f_L = \frac{1}{2\pi\times1\times10^{-6}\times10^3} \cong 8$$

36. Since $n(t)$ can be written in function of t using step function $x(t) = 4(t) = 4(t-L)$

we need to $x(s)$ laplace transform $x(t)$

$$x(s) = \frac{1}{s} - \frac{1}{s}e^{-2s}$$

$\therefore$ $$L\left(\frac{d^2y(t)}{d(t)}\right) = s^2Y(s)$$

$\therefore$ $$L\left[\frac{d^2y(t)}{dt} + s\frac{dy}{dt} + 6y = x\right]$$

$\Rightarrow$ $$s^2y(s) + ssy(s) + 6y(s) = x(s)$$

$\therefore$ $$y(s) = \frac{x(s)}{s^2+5s+6} = \frac{1-e^{-2s}}{s(s+2)(s+3)}$$

37. Given:

$\Rightarrow$ A linear constant coefficient, first order differential equation of $y(t)$

$\Rightarrow$ forcing function – $x(t)$

$\Rightarrow$ initial condition : $y(0)$

$\Rightarrow$ $y(t) > 0$

Find out:

$-2y(t)$, by changing the value of $x(t)$ and $y(0)$

Analysis:

Lets the differential equation is

$$\frac{dy}{dt} + py = x(t)$$

$\because$ $y(t)$ = function of $x(t)$ and $y(0) = f(x(t), y(0))$

Let find the relation between $y(t)$ and $x(t)$ define integrating factor IF.

$$\text{I.F.} = e^{\int Pdt}$$

Then $y(t) = \frac{1}{\text{I.F.}}\int x(t)\text{I.F.}\,dt + y(0)$

I.F. is independent of $n(t)$ and $y(0)$

So for find $-2y(t)$

$$-2y(t) = -2\frac{1}{\text{I.F.}}\int n(t)\text{I.F.}\,dt + (-2)y(0)$$

$$-2y(t) = \frac{1}{\text{I.F.}}\int[-2n(t)]\text{I.F.}\,dt + [-2y(0)]$$

So for get $-2y(t)$, change the $n(t)$ by $-2x(t)$, and $y(0)$ with $-2y(0)$

38. Given:

Two random variables (R.V.) U and V; Identically distributed.

Mean:

E (U) = 0 and E(V) = 0

Cumulative distribution function (CDF):

$F_U(x) = F(x)$

$F_{2V}(x) = G(x)$

To find: $F(x) - G(x)$ and $(F(x) - G(x))\,x$

Analysis:

Identically distributed R.V. U and V means their probability density function(pol) will be some. And CDF is the integration of polf. So their CDF will also be some

Hence $F_U(x) = F_V(x)$

Plan:

1. First find $F(x) - G(x)$; Then check its positivity or negativity.
2. Secondly find $(F(x) - G(x))x$; Then check its positivity or negativity.

Carrying out plan:

We know that $F_U(x) = F(x) = Pr(U \le x)$ ...(1)

and $F_{2V} = G(x) = Pr(2V \le x)$ ...(2)

Now, consider $F_V(x) - F_{2V}(x)$ {From (1) and (2)}

$= F(x) - G(x)$ {From (1) and (2)}

$= Pr(U \le x) - Pr(2V \le x)$ {From (1) and (2)}

Now, $P(2V \le x) = P\left(V \le \frac{x}{2}\right)$ ...(3)

Use this result into above expression

$\Rightarrow$ $F(x) - G(x) = Pr(U \le x) - Pr\left(V \le \frac{x}{2}\right)$ ....(4)

We know that $Pr(U \le x) = F_U(x)$ ...(5)

and $Pr\left(V \le \frac{x}{2}\right) = F_V\left(\frac{x}{2}\right)$ ...(6)

From analysis section we have $F_U(x) = F_V(x)$...(7)

From (6) and (7) $F_V\left(\frac{x}{2}\right) = F_U\left(\frac{x}{2}\right)$ ...(8)

Using (5) and (6) into (4)

$F(x) - G(x) = F_u(x) - F_v\left(\frac{x}{2}\right)$ ....(9)

From (8) and (9), we have

$\Rightarrow$ $F(x) - G(x) = F_U(x) - F_U\left(\frac{x}{2}\right)$

Using (1) into above

$\Rightarrow$ $F(x) - G(x) = F(x) - F\left(\frac{x}{2}\right)$ ...(10)

Property of CDF: CDF is a non-decreasing function.

Case 1: x is +ve $(x \ge 0)$ $\Rightarrow x \ge \frac{x}{2}$

$\Rightarrow$ $F(x) \ge F\left(\frac{x}{2}\right)$ {$\because$ Non-decreasing function}

$\Rightarrow$ $F(x) - F\left(\frac{x}{2}\right) \ge 0$

So from (10) $F(x) - G(x) \ge 0$ ...(11)

Case 2: x is –ve $(x \le 0)$ $\Rightarrow x \le \frac{x}{2}$

$\Rightarrow$ $F(x) \le F\left(\frac{x}{2}\right) \Rightarrow F(x) - F\left(\frac{x}{2}\right) \le 0$

{$\because$ Non-Negative function}

So from (10) $F(x) - G(x) \le 0$ ...(12)

So from case 1 and case 2, result our answer depends on positivity and negativity of x. So for $x \ge 0$ our answer is option (B); But for $x \le 0$ our answer is option (A). But not always (A) and (B) so we reject these two options.

Now we have to find some compact form for our answer which will be true for any value of x. *i.e.* for positive and negative x.

Case 3: $x \ge 0$

From (11) $F(x) - G(x) \ge 0$

Multiply by x, we get

$(F(x) - G(x))x \ge 0$ ...(13)

Case 4: $x \le 0$

From (12) $F(x) - G(x) \le 0$

Multiply by x, we get

$F(x) - G(x)x \ge 0$ ...(14)

So from (13) and (14) we have

$(F(x) - G(x))x \ge 0$ for all x

$\therefore$ Option (D) is correct answer.

Example: A Gaussion random variable with mean ZERO has following probability density function (pelf)

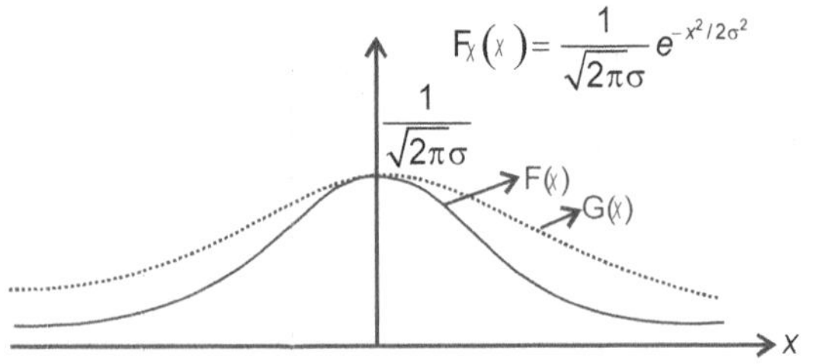

The CDF will be (approximately)

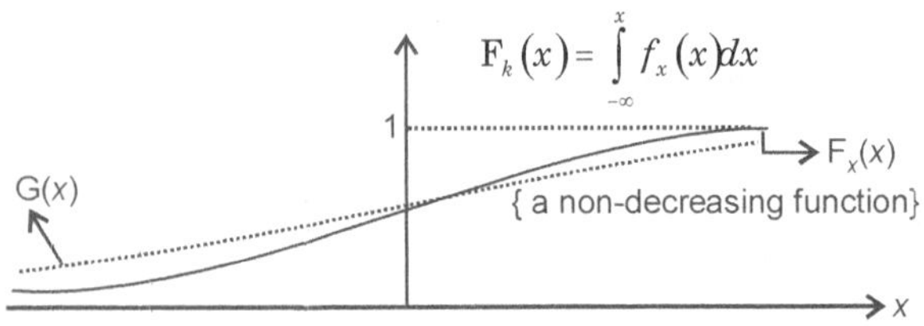

Conclusion:

The important point to remember in question is that "CDF function is a non-decreasing function."

39. Given:

DFT [a b c d] = [α β η δ] ...(1)

$$[p \quad q \quad r \quad s] = [a \quad b \quad c \quad d]\begin{bmatrix} a & b & c & d \\ d & a & b & c \\ c & d & a & b \\ b & c & d & a \end{bmatrix} \quad ...(2)$$

To find:

DFT [*p q r s*]?

Analysis: We know that

DFT $[x_1(n) \otimes x_2(n)] = x_1(K)\, x_2(K)$ ...(3)

where ⊗ represent circular convolution

and $x_1(k) = \text{DFT}[x_1(n)]$ ...(4)

$x_2(k) = \text{DFT}[x_2(n)]$ ...(5)

we know that circular convolution can be found by "MATRIX METHOD"

MATRIX METHOD: Let two sequences

$x_1(n) = \{x_1(0), x,(1), \ldots x_1(N)\}^T$ {a colum vector}

$x_2(n) = \{x_2(0), x_2(1), \ldots\ldots x_2(N)\}^T$ {a colum vector}

Design a matrix say by using sequence $x_2(n)$

$$M = \begin{bmatrix} x_2(0) & x_2(N) & x_2(N-1) & \ldots & x_2(1) \\ x_2(1) & x_2(0) & x_2(N) & & . \\ . & x_2(1) & x_2(0) & & . \\ . & . & x_2(1) & & . \\ . & . & . & & . \\ . & . & . & . & . \\ . & . & . & . & . \\ . & . & . & & x_2(N-1) \\ x_2(N-1) & x_2(N-2) & x_2(N-3) & & x_2(N) \\ x_2(N) & x_2(N-1) & x_2(N-2) & \ldots\ldots & x_2(0) \end{bmatrix} \quad ...(6)$$

Now circular convolution is given by

$x_1(n) * x_2(n) = M \times x_1(n)$ ...(7)

Plan:

1. Design matrix M by using vactor [a b c d]
2. Find circular convolution using equation 7

Carrying our plan:

Using vector [a b c d] design matrix M

$$M = \begin{bmatrix} a & d & c & b \\ b & a & d & c \\ c & b & a & d \\ d & c & b & a \end{bmatrix}$$

Now circular convolution of [a b c d] with it self.

$$[a \quad b \quad c \quad d]^T \otimes [a \quad b \quad c \quad d]^T = \begin{bmatrix} a & d & c & b \\ b & a & d & c \\ c & b & a & d \\ d & c & b & a \end{bmatrix}\begin{bmatrix} a \\ b \\ c \\ d \end{bmatrix}$$

Take transpose on both sides

$$\begin{bmatrix} a \\ b \\ c \\ d \end{bmatrix} \otimes \begin{bmatrix} a \\ b \\ c \\ d \end{bmatrix} = [a \quad b \quad c \quad d]\begin{bmatrix} a & d & c & b \\ b & a & d & c \\ c & b & a & d \\ d & c & b & a \end{bmatrix} \quad ...(8)$$

⇒from 8 and 2, we get

$$[p \quad q \quad r \quad s] = \begin{bmatrix} a \\ b \\ c \\ d \end{bmatrix} \otimes \begin{bmatrix} a \\ b \\ c \\ d \end{bmatrix}$$

Take DFT both sides

$$\text{DFT}\ [p\ q\ r\ s] = \text{DFT}\left\{\begin{bmatrix} a \\ b \\ c \\ d \end{bmatrix} \oplus \begin{bmatrix} a \\ b \\ c \\ d \end{bmatrix}\right\}$$

= DFT [a b c d] · DFT [a b c d] = [α β γ δ] · [α β γ δ]

$$\boxed{\text{DFT}[p \quad q \quad r \quad s] = [\alpha^2 \quad \beta^2 \quad \gamma^2 \quad \delta^2]} \begin{Bmatrix} \text{Element wise} \\ \text{multiplication} \end{Bmatrix} \quad ...(9)$$

∴ Option (A) is the correct answer

Conclusion:

The important point to be noted is that sequences can be written as vectors. So vector and sequences were same thing.

40.

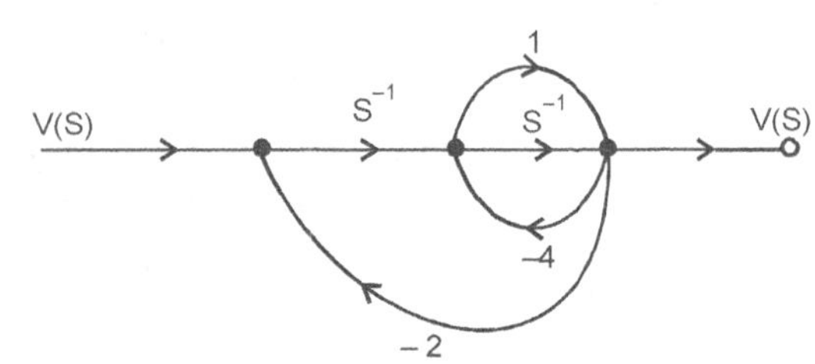

Forward path =

$P_1 = S^{-1}, S^{-1} = S^{-2}$ | $\Delta_1 = 1$

$P_2 = S^{-1}$ $\Delta_2 = 1$

Loop =

$$L_1 = -4S^{-1}$$
$$L_2 = -2S^{-1}S^{-1} = -2S^{-2}$$
$$L_3 = -2S^{-1}$$
$$L_4 = -4$$

$$T(S) = \frac{S^{-2}+S^{-1}}{1+4+2S^{-1}+2S^{-2}+4S^{-1}} = \frac{S^{-1}+S^{-2}}{S+6S^{-1}+2S^{-2}}$$

$$= \frac{S+1}{5S^2+6S+2}$$

41.

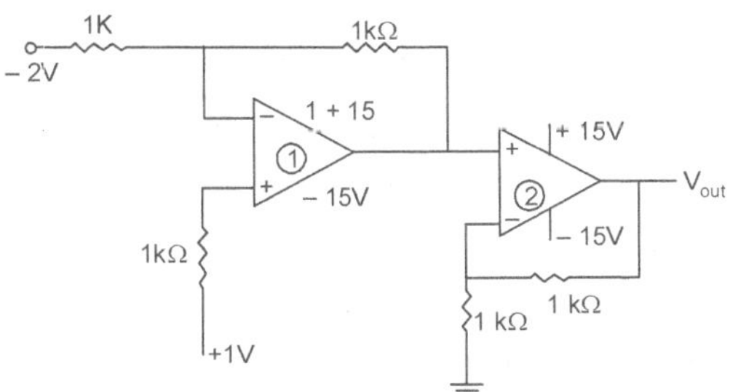

$$V\text{ out} = \left[+1v\left(1+\frac{1}{1}\right) - 2v\left(\frac{-1}{1}\right)\right]\left[1+\frac{1}{1}\right]$$

⇑ Gain of non inverty amp (1) ⇑ Gain of inverty amp (1) ⇑ Gain of non inverty amp (2)

$$= [1 \times 2 + 2] \times 2\ v = 8v$$

42.

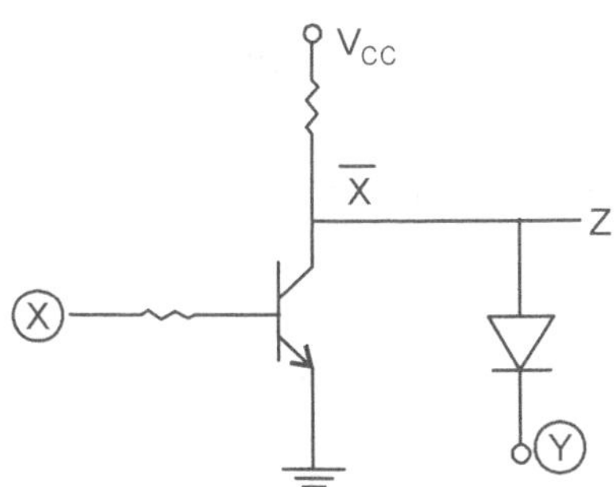

$$z = \bar{x}.y$$

43. 'D' 0 for all +

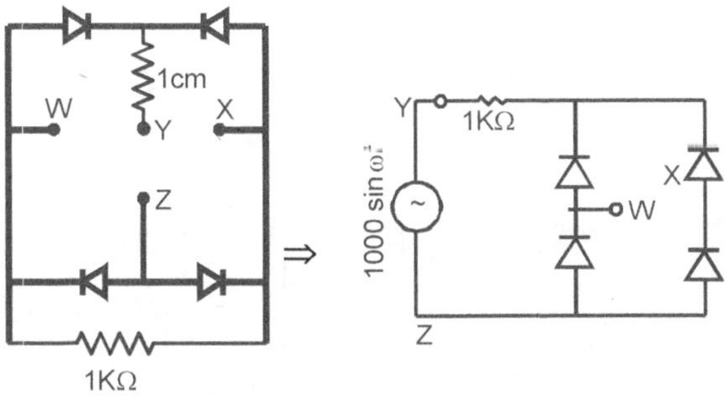

Note:

⇒ All diode conducts only done negative half.

⇒ XW is at symmetrical point so voltage across XW is zero for all time.

44. **Given values:**

	Capacitance Value	Voltage Breakdown
C_1	10 F	10V
C_2	5 F	5V
C_3	2 F	2V.

Given circuit:

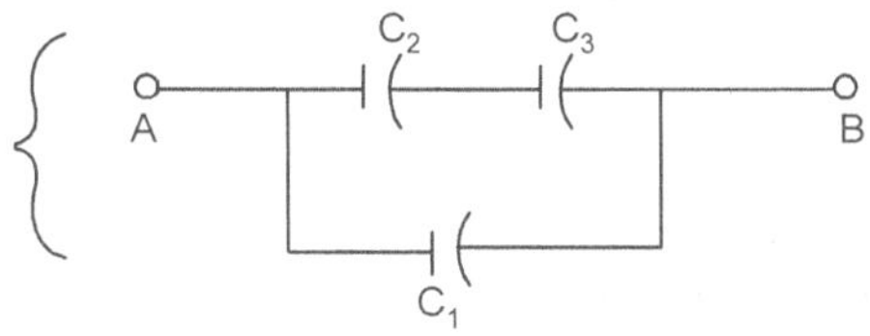

To find:

Max. safe voltage in *volts* that can be applied across the combination and Corresponding *total charge* is *C.*

Devising a plan:

(*a*) We will first calculate equivalent capacitance of given figure.

(*b*) By calculating capacitance, we can check which answer can be the solution, it reduce your effort in solving problem.

(*c*) Then apply the remaining answer for remaining options.

Solving:

Equivalent capacitance, as C_2 & C_3 are in series and we know that when capacitance are in series equivalent capacitance = $\frac{C_2C_3}{C_2+C_3}$ & when two capacitor are in parallel then their addition

$$= C_1 + \frac{C_2C_3}{C_2+C_3}$$

$$* \quad C = C_1 + \frac{C_2C_3}{C_2+C_3} = 10\mu F + \frac{5\mu F \times 2\mu F}{5\mu F + 2\mu F}. = 11.4285\ \ F$$

Now check the option by using formula $CV = \theta$

(*a*) $2.8 \times 11.4285 = 32\ \mu C \neq 36\ \mu C$ (option (*a*) is wrong)

hence '*c*' option can be right

Now when 7V

(*b*) $7V \times 11.4285\ \ F = 80\ \ c \neq 119\ \ C$ { B is

hence '*d*' option is right.

only '*c*' & '*d*' are left.

⇒ Now apply 7V first in given circuit,

when voltage across AB is 7V then C_1 can handle it broz max breakdown voltage across is 10Y.

and across

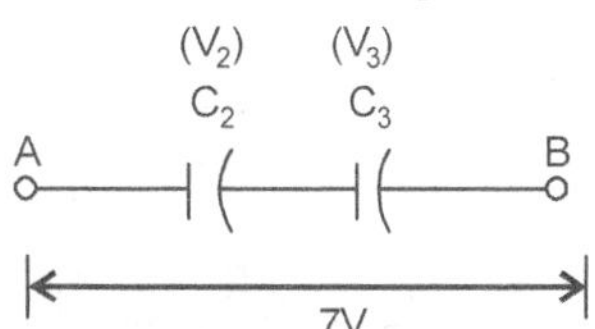

as they are in series so charge across them will be same.

$$Q_2 = Q_3$$
$$C_2V_2 = C_3V_3$$
$$C_2V_2 = C_3V_3 \quad [V_2 + V_3 = 7V]$$

$$C_2V_2 = C_3(7 - V_2)$$
$$C_2V_2 = 7C_3 - C_3V_2$$
$$V_2 = \frac{7C_3}{C_2 + C_3}$$
$$V_2 = \frac{7 \times 2}{2+5} = 2V \qquad \left\{\boxed{V_2 = 2V}\right.$$

and $V_3 = 7 - V_2 = 7 - 2 \quad \boxed{V_3 = 5V}$

As when we take '7V' then for that V_2 = 2V & V_3 = 5V, but max voltage across V_3 can be 2V [because above which it breakdown].

as $\boxed{V_3 \le 2V}$

But, when we take 7V then V_3 have to be 5V which is not possible hence '*d*' is also wrong.

⇒ Only option left is '*c*'

But, I will show that it is also right.

When $V_{AB} = 2.\delta$ V

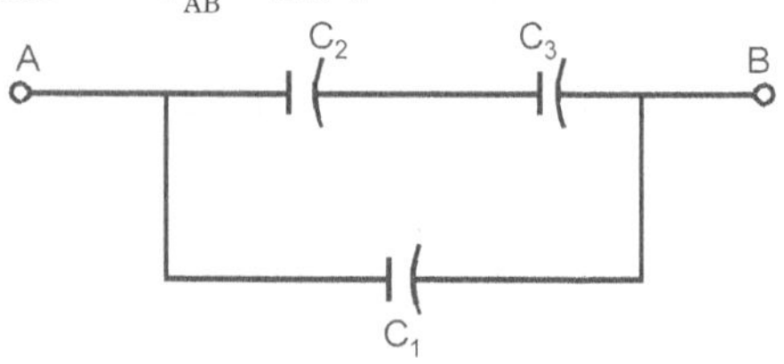

voltage across C_1 = 2.8 V [possible as breakdown voltage is 10 V].

$$V_2 = \frac{C_3 \times 2.8}{C_2 + C_3}$$

$\boxed{V_2 = \cdot 8V}$ [possible as breakdown voltage across C_2 is 5V)]

$$V_3 = \left(\frac{C_2}{C_2 + C_3}\right) \times 2.8$$

= 2 V [possible as breakdown voltage across C_3 is 2V]

hence option 'D' is correct

***Verification**: As I already showed that only '*s*' is true no one else.

Conclusion: voltage across method capacitance is not same as voltage across method resistance. For determine voltage across resistance we just do as we want to determine across R_1 & R_2

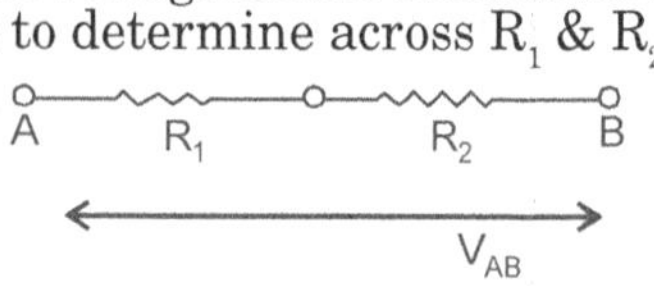

$$\boxed{V_{(R_1)} = \frac{R_1}{R_1 + R_2} V_{AB}}$$

But in capacitor

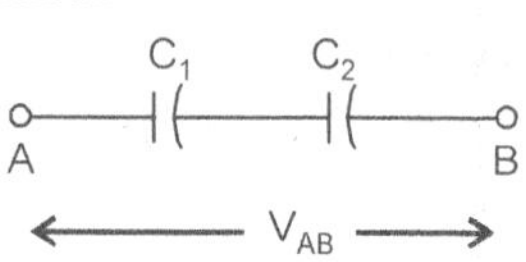

$$\boxed{V_{(C_1)} = \frac{C_1}{C_1 + C_2} V_{AB}}$$

So, don't apply resistance voltage method in to capacitor one, if you do that then you will obtain 'D' as answer, but which is wrong.

45. s1 s0

A_{15}	A_{14}	A_{13}	A_{12}	A_{11}	A_{10}	A_9	A_8	A_7	A_6	A_5	A_4	A_3	A_2	A_1	A_0	
0	0	0	0	1	0	×	×	×	×	×	×	×	×	×	×	RAM 1
0	0	0	1	1	0	×	×	×	×	×	×	×	×	×	×	RAM 2
0	0	1	0	1	0	×	×	×	×	×	×	×	×	×	×	RAM 3
0	0	1	1	1	0	×	×	×	×	×	×	×	×	×	×	RAM 4

RAM1 – 0800H–0BFF H

RAM2 – 1800H–1BFF H

RAM3 – 2800H – 23FF H

RAM4 – 3800H–3BFF H

Let's plan to solve problem

47. U X V are two independent and identically distribution random variable

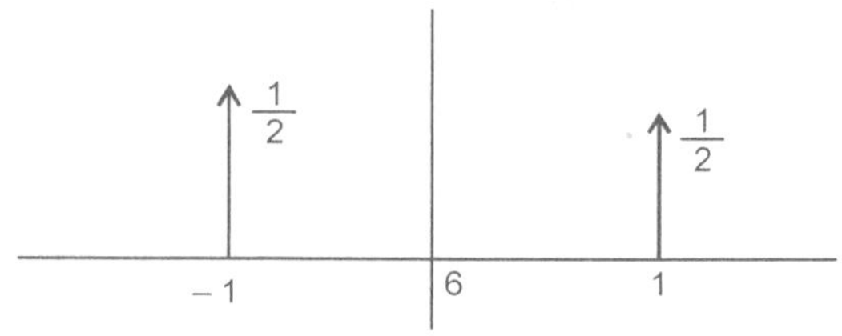

so $p_u(u) = \frac{1}{2}\delta(u+1) + \frac{1}{2}\delta(u-1)$

similarly $p_v(v) = \frac{1}{2}\delta(v+1) + \frac{1}{2}\delta(v-1)$

$$Z = (U + V)$$
$$P_z(z) = P_u(u) \oplus P_v(u)$$
$$= \frac{1}{4}\delta(z+2) + \frac{1}{2}\delta(z) + \frac{1}{4}\delta(z-2)$$

So entropy H(*z*) = H (U + V)

$$= P_1\log + P_2\log\frac{1}{P_2} + P_3 4\frac{1}{P_3}$$
$$= \frac{1}{4}\log_2 4 + \frac{1}{2}\log 2 + \frac{1}{4}\log_2 4 = \frac{1}{2} + \frac{1}{2} + \frac{1}{2} = \frac{3}{2}$$

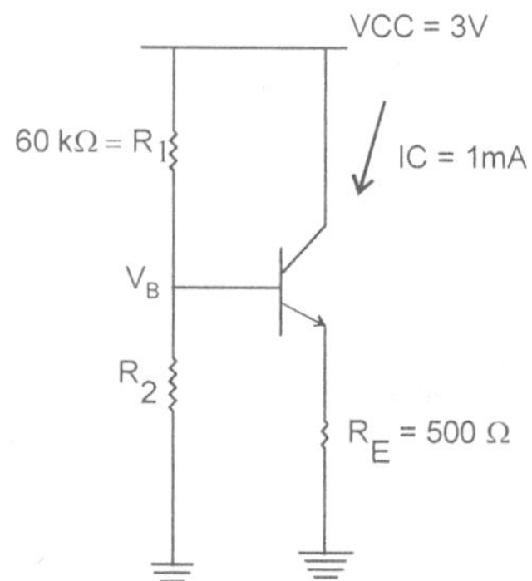

As IC = I_E

since B → ∞

IB = 0

$$VB = V_{BE} + R_E I_E = V_{BE} + R_E I_C$$

$$V_B = .7 + 1 \times 10^{-3} \times 500 = 1.2$$

$$V_B = 1.2V = \frac{R_2}{R_1+R_2} = \frac{R_2}{60k+R_2} \times 3V$$

$$1.2\,(60 + R_2) = 3R_2$$

$$R_2 = \frac{1.2 \times 60}{1.8} = 40k$$

48.

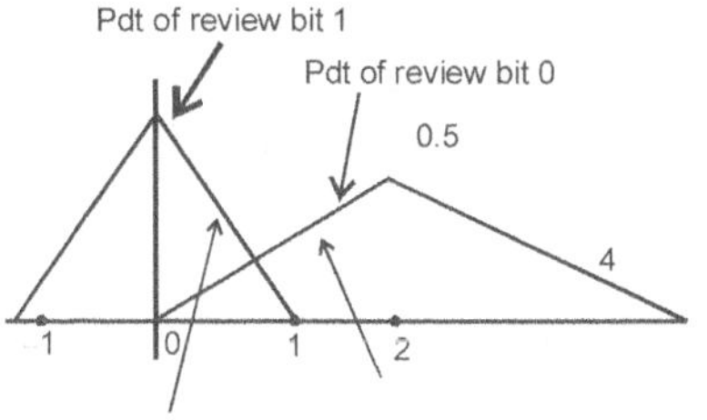

$$y = -x + 1 \quad ...(i)$$

$$y = \frac{.5}{2} \times x\ ; \quad y = \frac{x}{4} \quad ...(ii)$$

Solving (*i*) and (*ii*)

$$-x+1 = \frac{x}{4} \Rightarrow 1 = \frac{x}{4}+1 = \frac{5x}{4} \Rightarrow x = \frac{4}{5}$$

$$\therefore \quad y = \frac{1}{5}$$

So optimum ratio to achieve minimum bit rate (BER) $\frac{4}{5}$

49.

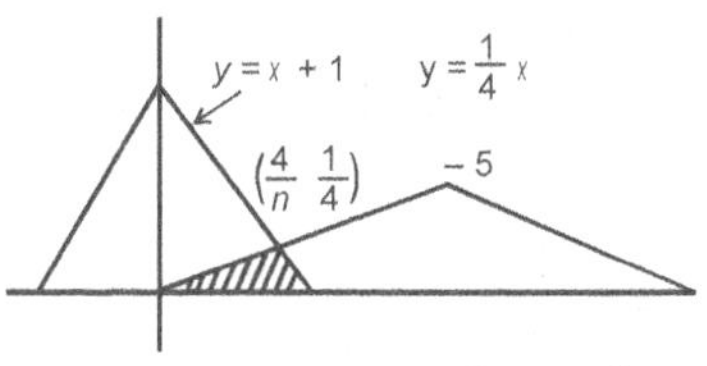

$$\text{Bit error rate} = \frac{1}{2} \times 1 \times \frac{1}{4} = \frac{1}{8}$$

50–51

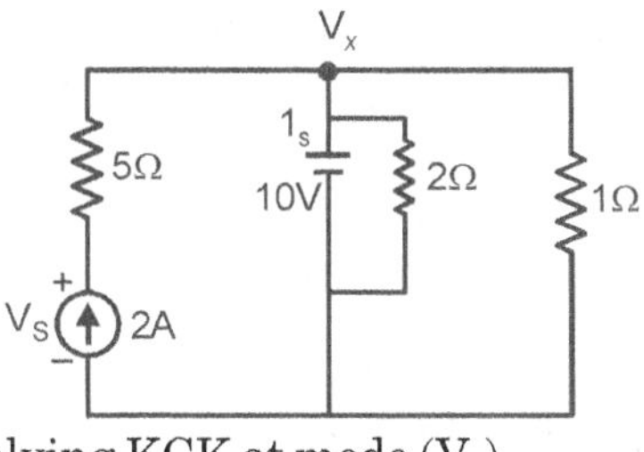

⇒ Applying KCK at mode (V_x)

$$2 - I_S - \frac{10}{2} - \frac{10}{1} = 0$$

$$I_S = -5 - 10 + 2 = -13$$

$$\Rightarrow \quad \text{Current through } 1\Omega = \frac{10}{1} = 10\text{ A}$$

$$\Rightarrow \quad V_s - 5 \times 2 - 10V = 0$$

$$V_s = 20V$$

50. (C) Current through 1Ω = 10A]

51. (D) [$I_s = -13A$, $V_s = 20$]

52. **Given:**

Electric field vectors of a monochromatic plane wave.

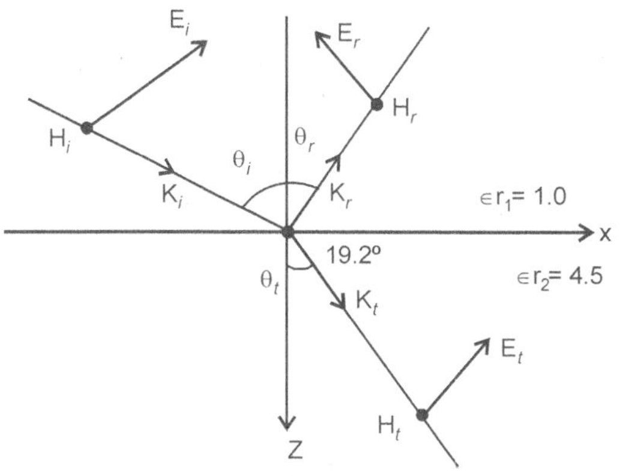

$$\lambda = 600 \quad m$$

$$K_0 = \frac{2\pi}{\lambda} = \frac{\pi}{3} \times 10^4 \text{ rad/mt}$$

Topic:

Plane wave propagation (Electro magnetics)

Formula:

Snell's law, $\sqrt{Er_1}\sin\theta_i = -\sqrt{Er_2}\sin\theta_t$

Solution:

To find the angle of incidence use snell's law.

$$\sqrt{Er_1}\sin\theta_i = \sqrt{Er_2}\sin\theta_t$$

$$\Rightarrow \quad \sqrt{1.0}\sin\theta_i = \sqrt{4.5}\sin 19.2$$

⇒ $\theta i = 44.3 \simeq 45°$, is the desired incidence angle.

To find the incidence Electric filed vector, we need to find the propagation constant first.

We can resolve K_i into two components as shown in the figure.

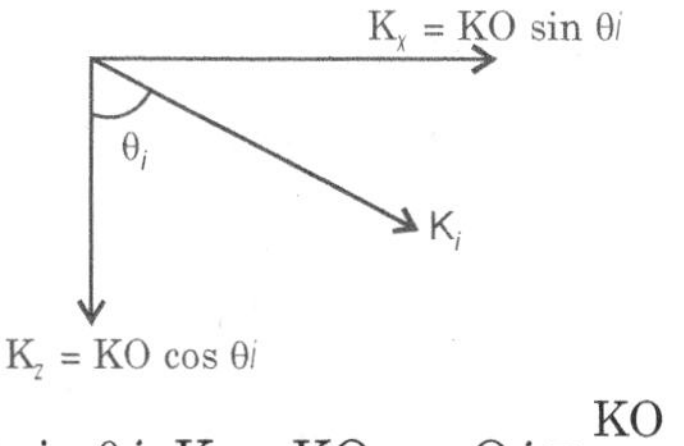

$$Kx = KO \sin\theta i;\ K_w = KO \cos Qi = \frac{KO}{\sqrt{2}} = \frac{KO}{\sqrt{2}}$$

Now the phase term can be given by

$$e^{-j(\beta_x x + \beta_z z)} = e^{-j\left(\frac{2\pi}{\lambda}\cos\theta_i x + \frac{2\pi}{\lambda}\sin\theta_i z\right)} = e^{-j\frac{\pi \times 10^4}{3\sqrt{2}}(x+z)} \quad ...(1)$$

The components consituting the amplitude part of E*i* are

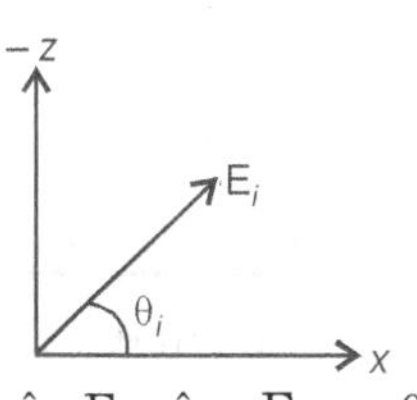

$$Eio = Ex a\hat{x} - Ez a\hat{z} = E_0 \cos\theta_i a\hat{x} - E_0 \sin\theta_i a\hat{z}$$

$$= \frac{E_0}{\sqrt{2}}(\hat{a}_x - \hat{a}_z) \text{ v/m} \quad ...(2)$$

Therefore, the expression for $\bar{E}_i$ can be obtained from (1) and (2) equations

$$Ei = \frac{E_0}{\sqrt{2}}(\hat{a}_x - \hat{a}_z)e^{-j\frac{\pi \times 10^4}{3\sqrt{2}}(x+z)} \text{ v/mt}$$

53. Given:

Electric field vectors of a monochromatic plane wave.

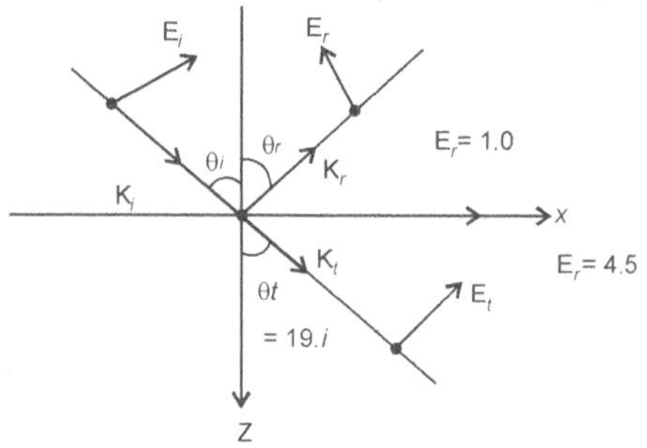

$$\lambda = 600 \quad \text{m}$$

$$K_0 = \frac{2\pi}{\lambda} = \frac{\pi}{3} \times 10^4 \text{ rad/m}$$

Topic:

Plane wave propagation (Electromagnetics)

Formula:

For an oblique incidence the reflection coefficient is

$$T = \frac{\eta_2 \cos\theta_t - \eta_1 \cos\theta_i}{\eta_2 \cos\theta_t + \eta_1 \cos\theta_i} = \frac{\sqrt{\frac{\mu_2}{\epsilon_2}}\cos\theta_t - \sqrt{\frac{\mu_1}{\epsilon_1}}\cos\theta_i}{\sqrt{\frac{\mu_2}{\epsilon_2}}\cos\theta_t - \sqrt{\frac{\mu_1}{\epsilon_1}}\cos\theta_i}$$

Here $\quad {}_1 = {}_2 = {}_0 = 4\pi \quad 10^{-7}$ H/mt

$$\epsilon_1 = \epsilon_0;\ \epsilon_2 = 4.5\ \epsilon_0$$

Solution:

First we find the angle of incidnece θi from snell's law as follows

$$\sqrt{\epsilon r_1}\sin\theta_i = \sqrt{\epsilon r_2}\sin\theta_t$$

On substituting the values we get $\theta_i \simeq 45°$

The reflection coefficient is,

$$T = \frac{\frac{1}{\sqrt{4.5}}\cos 19.2 - \frac{1}{\sqrt{1}}\cos 45°}{\frac{1}{\sqrt{4.5}}\cos 19.2 + \frac{1}{\sqrt{1}}\cos 45°} = -0.23$$

To find the reflected field vector $\bar{E}_r$, we need to find the propagation constant, for reflected wave in medium.

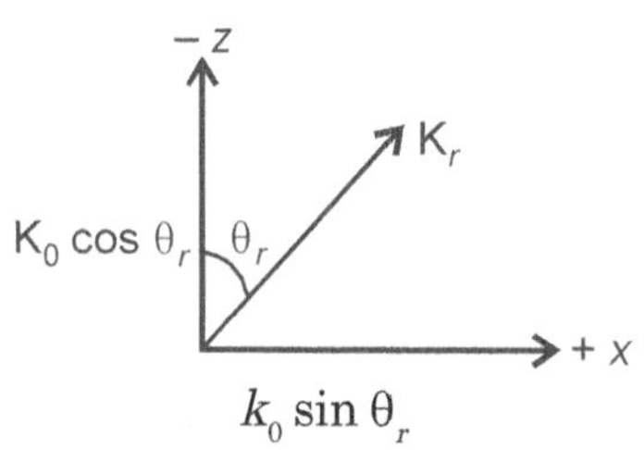

$$\boxed{\theta_i + \theta_r = 90°}$$

The phase term will be

$$e^{-j(\beta_x x + \beta_z(-z))} = e^{-j\left(\frac{2\pi}{\lambda}\cos\theta_r x - \frac{2\pi}{\lambda}\sin\theta_r z\right)} = e^{-j\frac{\pi\times 10^4}{3\sqrt{2}}(x-z)} \quad ...(1)$$

The amplitude term can be obtained by resolving $\bar{E}_r$ into two components

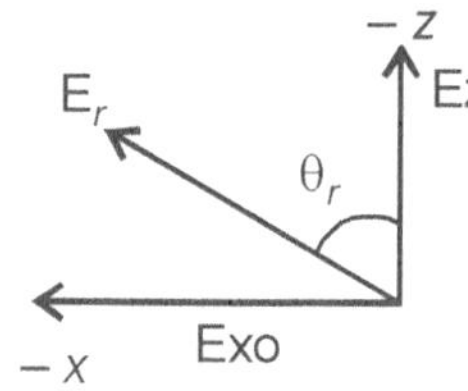

$$E_{r0} = \left(-Ex_0 a\hat{x} - Ez_0 a\hat{z}\right) \times \text{reflection coefficient}$$

$$= \left(-E_0 \sin\theta_r a\hat{x} - E_0 \cos\theta_r a\hat{z}\right) \times T$$

$$= \frac{0.23E_0}{\sqrt{2}}(\hat{a}_x + \hat{a}_z) \text{ v/m} \quad ...(2)$$

Thus the expression for $\bar{E}_r$ can be obtained form (1) and (2) as,

$$\bar{E}_r = 0.23\frac{E_0}{\sqrt{2}}(\hat{a}_x + \hat{a}_z)e^{-j\frac{\pi\times 10^4}{3\sqrt{2}}(x-z)} \text{ V/m}$$

55. Given:

Using data of previous question

State transtition matrix:

$$A = \begin{bmatrix} -1 & 0 \\ 1 & -1 \end{bmatrix}$$

Find out:

e^{At}, which is given as

$$L^{-1}[[SI - A]^{-1}]$$

Solution

$$SI - A = S\begin{bmatrix} 1 & 0 \\ 0 & 1 \end{bmatrix} - \begin{bmatrix} -1 & 0 \\ 1 & -1 \end{bmatrix}$$

$$= \begin{bmatrix} S+1 & 0 \\ -1 & S+1 \end{bmatrix}$$

$$[SI - A]^{-1} = \frac{1}{(S+1)^2}\begin{bmatrix} (S+1) & 0 \\ +1 & (S+1) \end{bmatrix}$$

$$= \begin{bmatrix} \frac{1}{S+1} & 0 \\ \frac{1}{(S+1)^2} & \frac{1}{S+1} \end{bmatrix}$$

taking Laplace Inverse

$$L^{-1}[SI - A]^{-1} = L^{-1}\begin{bmatrix} \frac{1}{S+1} & 0 \\ \frac{1}{(S+1)^2} & \frac{1}{S+1} \end{bmatrix}$$

$$= \begin{bmatrix} L^{-1}\left(\frac{1}{S+1}\right) & 0 \\ L^{-1}\left(\frac{1}{(S+1)^2}\right) & L^{-1}\left(\frac{1}{S}+1\right) \end{bmatrix}$$

$$= \begin{bmatrix} e^{-t} & 0 \\ te^{-t} & e^{-t} \end{bmatrix}$$

■■

SOLVED PAPER – 2012

ELECTRONICS AND COMMUNICATION ENGINEERING

Time : 3 Hours | Maximum Marks : 100

Q.1-25 Carry one mark each.

1. The current i_b through the base of silicon npn transistor is $1 + 0.1 \cos (10000\pi t)$ mA. At 300 K, the r_π in the small signal model of the transistor is

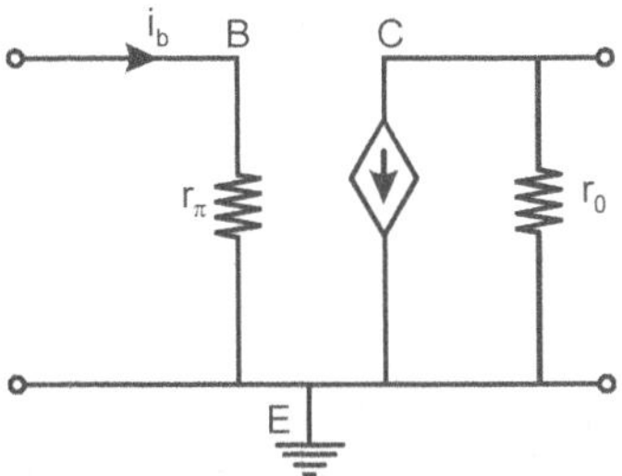

(*a*) 250 Ω (*b*) 27.5 Ω

(*c*) 22.5 Ω (*d*) 25 Ω

2. The power spectral density of a real process X(t) for positive frequencies is shown below. The values of $E[X^2(t)]$ and $|E[X(t)]|$, respectively, are

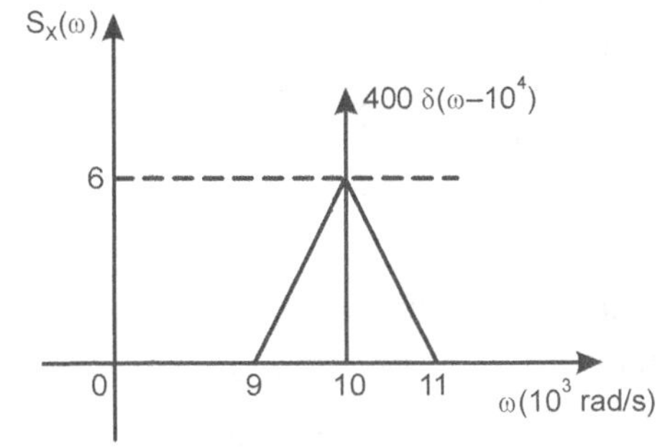

(*a*) $\frac{6000}{\pi}, 0$ (*b*) $\frac{6400}{\pi}, 0$

(*c*) $\frac{6400}{\pi}, \frac{20}{(\pi\sqrt{2})}$ (*d*) $\frac{6000}{\pi}, \frac{20}{(\pi\sqrt{2})}$

3. In a baseland communication link frequencies upto 3500 Hz are used for signaling. Using a raised cosine pulse with 75% excess bandwidth dnd for not inter-symbol interference, the maximum possible signaling rate in symbols per second is

(*a*) 1750 (*b*) 2526

(*c*) 4000 (*d*) 5250

4. A plane wave propagating in air with $\vec{E} = (8\hat{a}_x + 6\hat{a}_y + 5\hat{a}_z)\, e^{j(\omega t+3x-4y)}$ V/m is incident on a perfectly condicting slab positioned at $x \le 0$. the $\vec{E}$ field of the reflected wave is

(*a*) $(-8a_x - 6\hat{a}_y - 5\hat{a}_z)e^{j(\omega t+3x+4y)}$ V/m

(*b*) $(-8\hat{a}_x + 6\hat{a}_y - 5\hat{a}_z)e^{j(\omega t+3x+4y)}$ V/m

(*c*) $(-8\hat{a}_x - 6\hat{a}_y - 5\hat{a}_z)e^{j(\omega t-3x-4y)}$ V/m

(*d*) $(-8\hat{a}_x + 6\hat{a}_y - 5\hat{a}_z)e^{j(\omega t-3x-4y)}$ V/m

5. The electric field of a uniform plane electromagnetic wave in free space, along the positive x direction is given $\vec{E} = 10(\hat{a}_y + \hat{a}_z)e^{-j25x}$. The frequency and polarization of the wave, respectively are

(*a*) 1.2 GHz and left circular

(*b*) 4 Hz and left circular

(*c*) 1.2 GHz and right circular

(*d*) 4 Hz and right circular

6. Consider the given circuit.

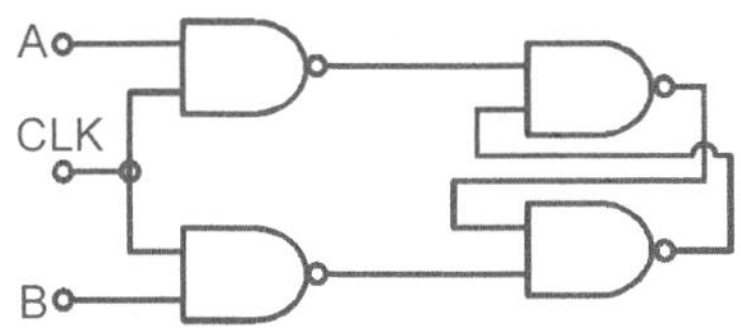

In this circuit, the race around

(*a*) does not occur

(*b*) occurs when CLK = 0

(*c*) occurs when CLK = 1 and A = B = 1

(*d*) occurs when CLK = 1 and A = B = 0

7. The output Y of a 2-bit comparator is logic 1 whenever the 2-bit input A is greater than the 2-bit input B. The number of combinations for which the output is logic 1, is

(*a*) 4 (*b*) 6

(*c*) 8 (*d*) 10

8. The i-v characteristics of the diode in the circuit given below are

$$i = \begin{cases} \frac{v - 0.7}{500} \text{A}, & v \ge 0.7\text{V} \end{cases}$$

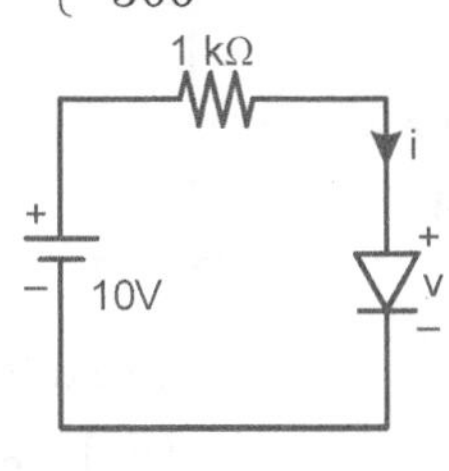

The current in the circuit is

(*a*) 10 mA (*b*) 9.3 mA

(*c*) 6.67 mA (*d*) 6.2 mA

9. In the following figure, C_1 and C_2 are ideal capacitors. C_1 has been charged to 12 V before the ideal switch S is closed at t = 0. The current i(t) for all t is

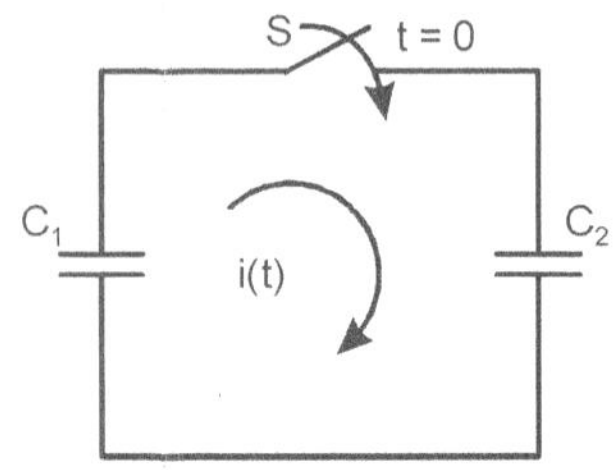

(*a*) zero

(*b*) a step function

(*c*) an exponentially decaying function

(*d*) an impulse function

10. The average power delivered to an impedance $(4 - j3)\Omega$ by a currrent $5\cos(100\ \pi t + 100)$ A is

(*a*) 44.2 W (*b*) 50 W

(*c*) 62.5 W (*d*) 125 W

11. The unilateral laplace transform of f(t) is $\frac{1}{s^2+s+1}$. The unilateral laplace transform of t f(t) is

(*a*) $-\frac{s}{\left(s^2+s+1\right)^2}$ (*b*) $-\frac{2s+1}{\left(s^2+s+1\right)^2}$

(*c*) $\frac{s}{\left(s^2+s+1\right)^2}$ (*d*) $\frac{2s+1}{\left(s^2+s+1\right)^2}$

12. With initial condition x(1) = 0.5, the solution of the differential equation,

$t\frac{dx}{dt}+x=t$ is

(*a*) $x = t - \frac{1}{2}$ (*b*) $x = t^2 - \frac{1}{2}$

(*c*) $x = \frac{t^2}{2}$ (*d*) $x = \frac{t}{2}$

13. The diodes and capacitors in the circuits shown are ideal. The voltage v(t) across the diode D1 is

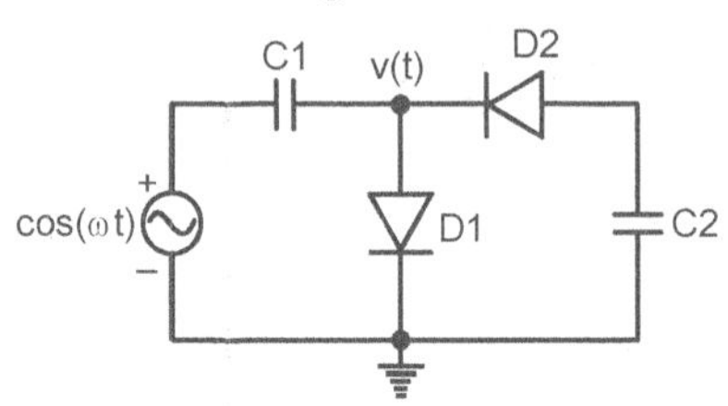

(*a*) $\cos(\omega t) - 1$ (*b*) $\sin(\omega t)$

(*c*) $1 - \cos(\omega t)$ (*d*) $1 - \sin(\omega t)$

14. In the circuit shown

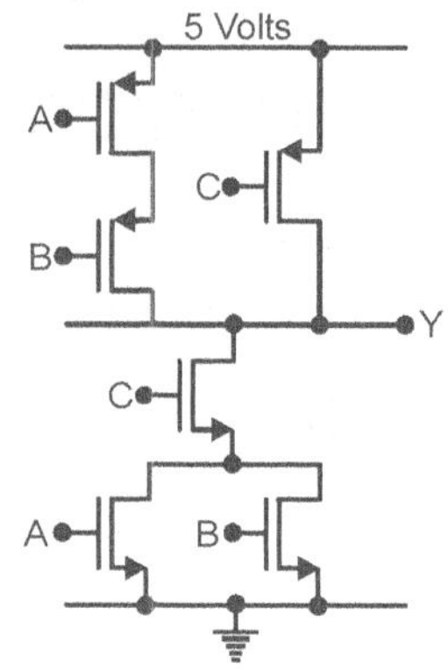

(*a*) $Y = \overline{A}\overline{B} + \overline{C}$ (*b*) Y = (A+B)C

(*c*) $Y = \left(\overline{A} + \overline{B}\right) + \overline{C}$ (*d*) Y = AB+C

15. A source alphabet consists of N symbols with the probability of the first two symbols being the same. A source encoder increases the probability of the first symbols by a small amount ε and decreases that of the second by ε. After encoding the entropy of the source.

(*a*) increases

(*b*) remains the same

(*c*) increases only if N = 2

(*d*) decreases

16. A coaxial-cable with an inner diameter of 1 mm and outer diameter of 2.4 mm is filled with a dielectric of relative permitivity 10.89.

Given $\mu_0 = 4\pi \times 10^{-7}$ H/m. $\varepsilon_0 = \frac{10^{-9}}{36\pi}$ F/m, the characteristic impedance of the cable is

(*a*) 330 Ω (*b*) 100 Ω

(*c*) 143.3 Ω (*d*) 43.4 Ω

17. The radiation pattern of an antenna in spherical co-ordinates is given by

$$F(\theta) = \cos^4\theta,\ 0 \le \theta \le \pi/2$$

The directivity of the antenna is

(*a*) 10 dB (*b*) 12.6 dB

(*c*) 11.5 dB (*d*) 18 dB

18. If $x[n] = (1/3)^{|n|} - (1/2)^n\ u[n]$, then the region of convergence (ROC) of its Z-transform in the Z-plane will be

(*a*) $\frac{1}{3} < |z| < 3$ (*b*) $\frac{1}{3} < |z| < \frac{1}{2}$

(*c*) $\frac{1}{2} < |z| < 3$ (*d*) $\frac{1}{3} < |z|$

19. In the sum of products functions

$f(X,Y,Z) = \sum(2, 3, 4, 5)$, the prime implicants are

(*a*) $\overline{X}Y, X\overline{Y}$ (*b*) $\overline{X}Y, X\overline{Y}\overline{Z}, X\overline{Y}Z$

(*c*) $\overline{X}Y\overline{Z}, \overline{X}YZ, X\overline{Y}$ (*d*) $\overline{X}Y\overline{Z}, \overline{X}YZ, X\overline{Y}\overline{Z}, X\overline{Y}Z$

20. A system with transfer function is excited by $\sin(\omega t)$. The steady-state output of the system is zero at.

$$G(s) = \frac{(s^2+9)(s+2)}{(s+1)(s+3)(s+4)}$$

Find ω at which steady state output is zero

(a) $\omega = 1$ rad/s (b) $\omega = 2$ rad/s

(c) $\omega = 3$ rad/s (d) $\omega = 4$ rad/s

21. The impedance looking into nodes 1 and 2 in the given circuit is

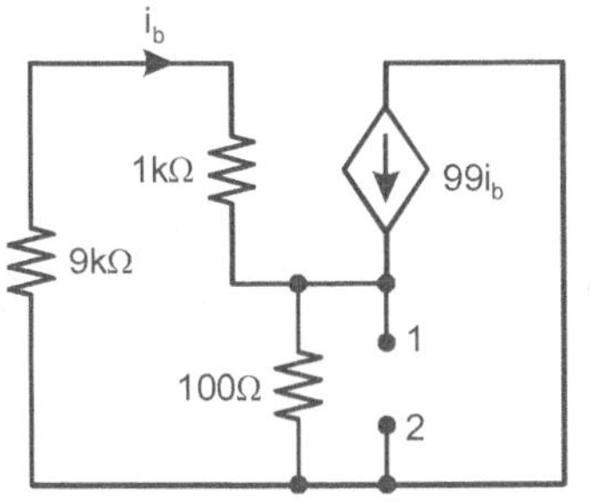

(a) 50 Ω (b) 100 Ω

(c) 5 kΩ (d) 10.1 kΩ

22. In the circuit shown below, the current through the inductor is

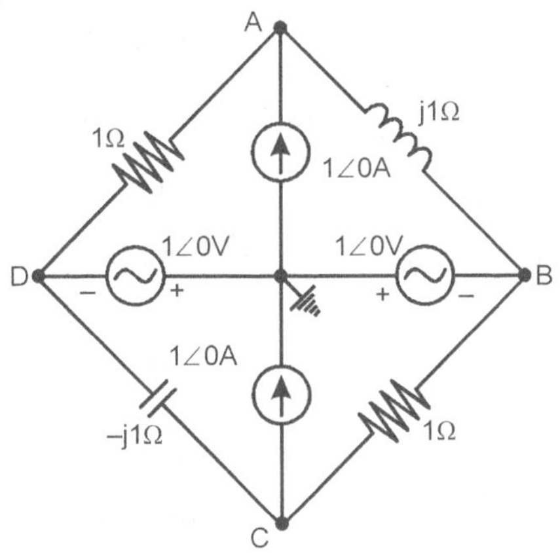

(a) $\frac{2}{1+j}$A (b) $\frac{-1}{1+j}$A

(c) $\frac{1}{1+j}$A (d) 0 A

23. Given

$f(z) = \frac{1}{z+1} - \frac{2}{z+3}$. If C is a counterclockwise path in the z-plane such that $|z+1| = 1$, the value of $\frac{1}{2\pi}\oint_c f(z)dz$ is

(a) –2 (b) –1

(c) 1 (d) 2

24. Two independent random variables X and Y are uniformly distributed in the interval [–1, 1]. The probability that max [X, Y] is less than $\frac{1}{2}$ is

(a) $\frac{3}{4}$ (b) $\frac{9}{16}$

(c) $\frac{1}{4}$ (d) $\frac{2}{3}$

25. If $x = \sqrt{-1}$, then the value of x^x is

(a) $e^{-\pi/2}$ (b) $e^{\pi/2}$

(c) x (d) 1

Q.26-55 carry two marks each.

26. The source of a silicon ($n_i = 10^{10}$ per cm^3) n-channel MOS transistor has an area of 1 sq μm and a depth of 1 μm. If the dopant density in the source is 10^{19}/cm^3, the number of hole in the source region with the above volume is approximately

(a) 10^7 (b) 100

(c) 10 (d) 0

27. A BPSK scheme operating over an AWGN channel with noise power spectral density of $N_0/2$, uses equiprobable signals $s_1(t) = \sqrt{\frac{2E}{T}}\sin(\omega_0 t)$ and $s_2(t) = -\sqrt{\frac{2E}{T}}\sin(\omega_0 t)$ over the symbol interval (0, T). If the local oscillator in the coherent receiver is ahead in pahse by 45° with respect to the received signal, the probability of error in the resulting system is

(a) $Q\left(\sqrt{\frac{2E}{N_0}}\right)$ (b) $Q\left(\sqrt{\frac{E}{N_0}}\right)$

(c) $Q\left(\sqrt{\frac{E}{2N_0}}\right)$ (d) $Q\left(\sqrt{\frac{E}{4N_0}}\right)$

28. A transmission line with a characteristic impedance of 100Ω is used to match a 50Ω section to a 200Ω section. if matching is to be done both at 429MHz and 1GHz, the length of the transmisson line can be approximately

(a) 82.5 cm (b) 1.05 m

(c) 1.58 m (d) 1.75 m

29. The input x(t) and output y(t) of s system are related as $y(t) = \int_{-\infty}^{t} x(\tau)\cos(3\tau)d\tau$. The system is

(a) time-invariant and stable

(b) stable and not time-invariant

(c) time-invariant and not stable

(d) not time-invariant and not stable

30. The feedback system shown below oscillates at 2 rad/s when

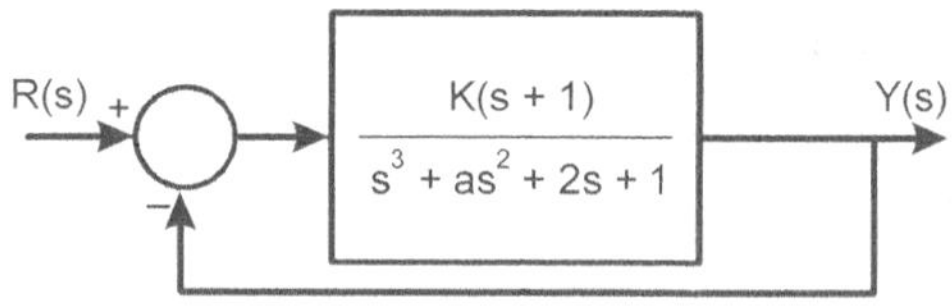

(*a*) K = 2 and a = 0.75 (*b*) K = 3 and a = 0.75

(*c*) K = 4 and a = 0.5 (*d*) K = 2 and a = 0.5

31. The Fourier transform of a signal h(t) is $H(j\omega) = 2(\cos\omega)\dfrac{(\sin 2\omega)}{\omega}$. The value of h(0) is

(*a*) $\frac{1}{4}$ (*b*) $\frac{1}{2}$

(*c*) 1 (*d*) 2

32. The state variable description of an LTI system is given by

$$\begin{pmatrix} \dot{x}_1 \\ \dot{x}_2 \\ \dot{x}_3 \end{pmatrix} = \begin{pmatrix} 0 & a_1 & 0 \\ 0 & 0 & a_2 \\ a_3 & 0 & 0 \end{pmatrix}\begin{pmatrix} x_1 \\ x_2 \\ x_3 \end{pmatrix} + \begin{pmatrix} 0 \\ 0 \\ 1 \end{pmatrix}u$$

$$y = \begin{pmatrix} 1 & 0 & 0 \end{pmatrix}\begin{pmatrix} x_1 \\ x_2 \\ x_3 \end{pmatrix}$$

where y is the output and u is the input. The system is controllable for

(*a*) $a_1 \neq 0, a_2 = 0, a_3 \neq 0$ (*b*) $a_1 = 0, a_2 \neq 0, a_3 \neq 0$

(*c*) $a_1 = 0, a_2 \neq 0, a_3 = 0$ (*d*) $a_1 \neq 0, a_2 \neq 0, a_3 = 0$

33. Assuming both the voltage source are in phase, the value of R for which maximum power is transferred from circuit A to circuit B is

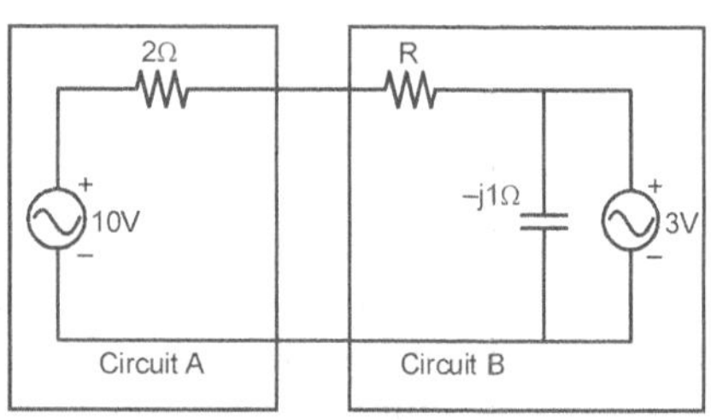

(*a*) 0.8Ω (*b*) 1.4Ω

(*c*) 2Ω (*d*) 2.8Ω

34. Consider the differential equation

$$\frac{d^2y(t)}{dt^2} + 2\frac{dy(t)}{dt} + y(t) = \delta(t) \text{ with } y(t)\big|_{t=0^+} = -2$$

and $\left.\dfrac{dt}{dt}\right|_{t=0^+} = 0$

The numerical value of $\left.\dfrac{dt}{dt}\right|_{t=0}$ is

(*a*) –2 (*b*) –1

(*c*) 0 (*d*) 1

35. The direction of vector A is radially outward from the origin with $|A| = kr^n$ where $r^2 = -x^2 + y^2 + z^2$ and k is a constant. The value of n for which $\nabla.A = 0$ is

(*a*) –2 (*b*) 1

(*c*) 1 (*d*) 0

36. A fair coin is tossed till a head appears for the first time. The probability that the number of required tosses is odd, is

(*a*) $\frac{1}{3}$ (*b*) $\frac{1}{2}$

(*c*) $\frac{2}{3}$ (*d*) $\frac{3}{4}$

37. In the CMOS circuit shown, electron and hole mobilities are equal, and M1 and M2 are equally sized. The device M1 is in the lienar region if

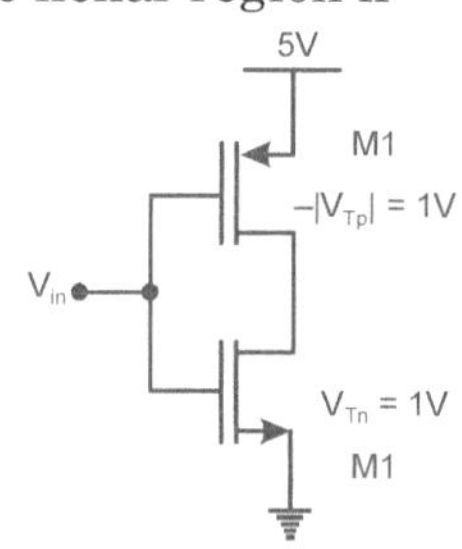

(*a*) $V_{in} < 1.875$ V

(*b*) 1.875 V $< V_{in} <$ 3.125 V

(*c*) $V_{in} > 3.125$ V

(*d*) $0 < V_{in} < 5$V

38. A binary symmetric chanel (BSC) has a transition probability of $\frac{1}{8}$. If the binary transmit symbol X is such that $P(X = 0) = \frac{9}{10}$, then the probability of error for an optimum receiver will be

(*a*) $\frac{7}{80}$ (*b*) $\frac{63}{80}$

(*c*) $\frac{9}{10}$ (*d*) $\frac{1}{10}$

39. The signal m(t) as shown is applied both to a phase modulator (with k_p) as the phase constant) and a frequency modulator (with k_f as the frequency constant) having the same carrier frequency.

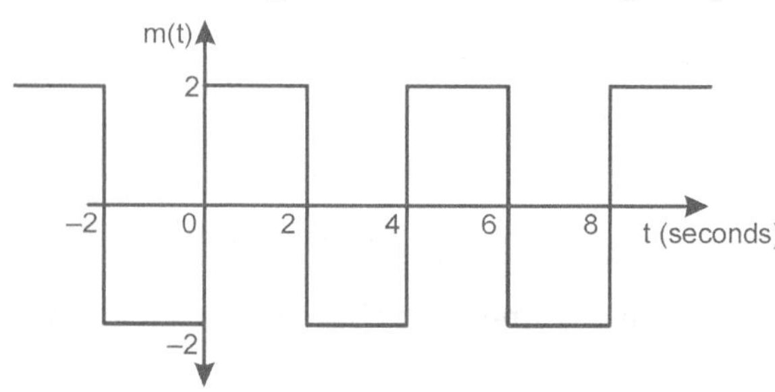

The ratio k_p/k_f (in rad/Hz) for the same maximum phase deviation is

(*a*) 8π (*b*) 4π

(*c*) 2π (*d*) π

40. The magnetic field along the propagation direction inside a rectangular waveguide with the cross section shown in the figure is

$H_z = 3\cos(2.094 \times 10^2 x)\cos(2.618 \times 10^2 y)\cos(6.283 \times 10^{10} t - \beta z)$

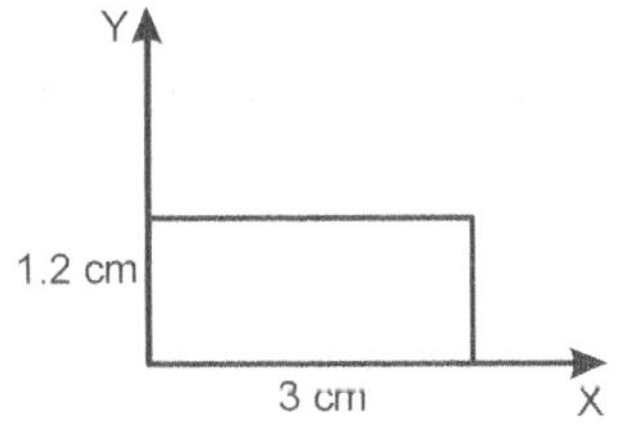

The phase velocity v_p of the wave inside the waveguide satisfies

(a) $v_p > c$ (b) $v_p = c$
(c) $0 = v_p < c$ (d) $v_p = 50$

41. The circuit shown is a

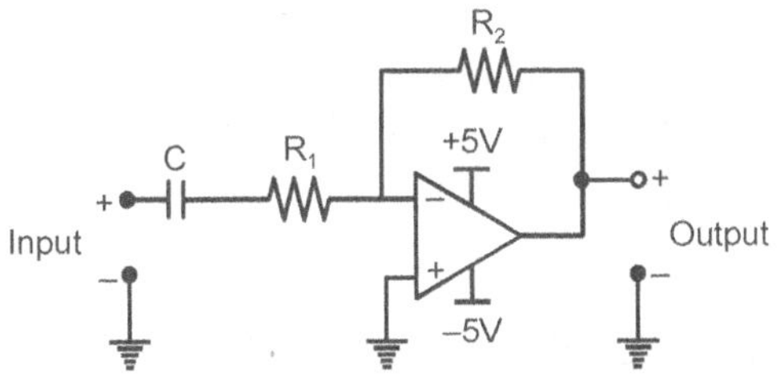

(a) low pass filter with $f_{3dB} = \dfrac{1}{(R_1 + R_2)C}$ rad/s

(b) high pass filter with $f_{3dB} = \dfrac{1}{R_1 C}$ rad/s

(c) low pass filter with $f_{3dB} = \dfrac{1}{R_1 C}$ rad/s

(d) high pass filter with $f_{3dB} = \dfrac{1}{(R_1 + R_2)C}$ rad/s

42. Let y[n] denote the convolution of h[n] and g[n], where $h[n] = (1/2)^n u[n]$ and g[n] is a causal sequence. If y[0] = 1 and y[1] = 1/2, then g[1] equals

(a) 0 (b) $\dfrac{1}{2}$

(c) 1 (d) $\dfrac{3}{2}$

43. The state transition diagram for the logic circuit shown is

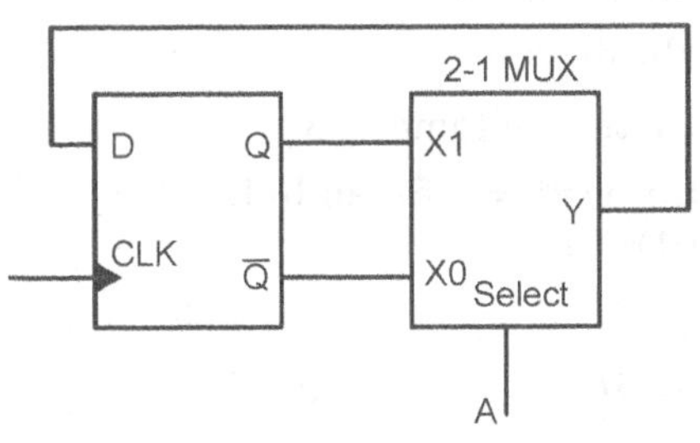

(a)
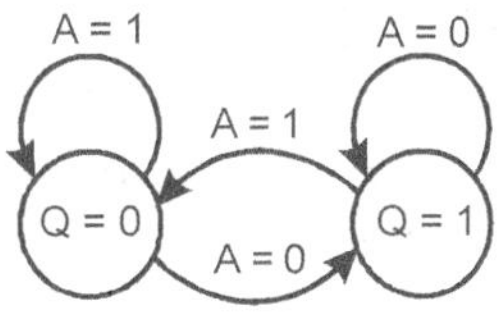

(b)
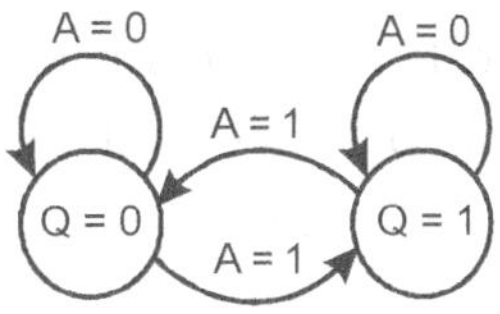

(c)
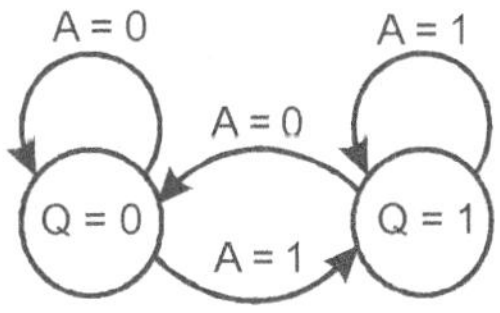

(d)
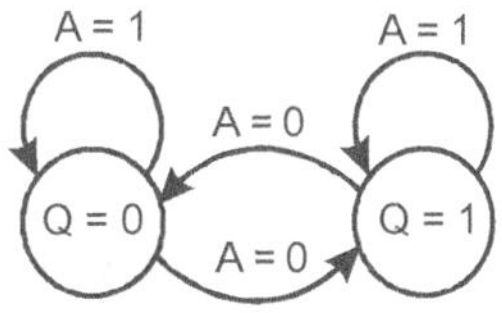

44. The voltage gain A_e of the circuit shown below is

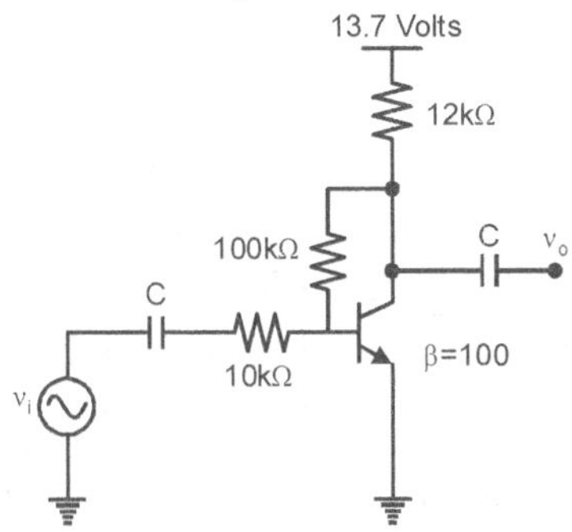

(a) $|A_v| = 200$ (b) $|A_v| = 100$
(c) $|A_v| = 20$ (d) $|A_v| = 10$

45. If $V_A - V_B = 6$ V, then $V_C - V_D$ is

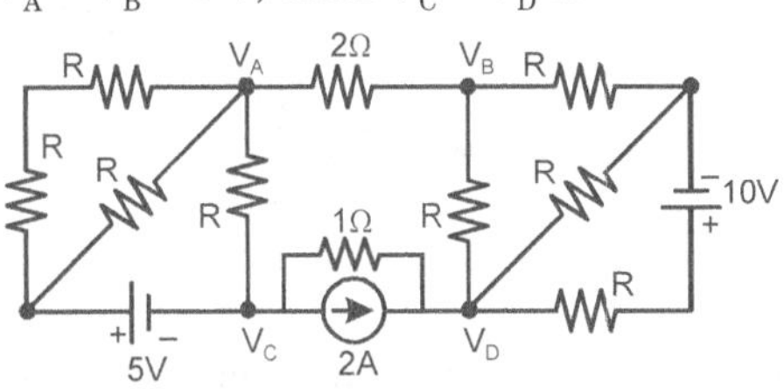

(a) –5V (b) 2V
(c) 3V (d) 6V

46. The maximum value of $f(x) = x^3 - 9x^2 + 24x + 5$ in the interval [1, 6] is

(a) 21 (b) 25
(c) 41 (d) 46

47. Given that

$A = \begin{bmatrix} -5 & -3 \\ 2 & 0 \end{bmatrix}$ and $I = \begin{bmatrix} 1 & 0 \\ 0 & 1 \end{bmatrix}$, the value of A^3 is

(a) $15A + 12I$ (b) $19A + 30I$
(c) $17A + 15I$ (d) $17A + 21I$

COMMON DATA QUESTIONS

Common Data for Q.48-49

With 10V dc commented at port A in the linear nonreciprocal two-port network shown below, the following were observed :

(*i*) 1Ω connected at port B draws a current of 3 A

(*ii*) 2.5Ω connected at port B draws a current of 2 A

48. With 10V dc connected at port A, the current drawn by 7Ω connected at port B is

(*a*) $\frac{3}{7}$ A (*b*) $\frac{5}{7}$ A

(*c*) 1A (*d*) $\frac{9}{7}$ A

49. For the same network, with 6V dc connected at port A, 1Ω connected at port B draws $\frac{7}{3}$ A. If 8V dc is connected to port A, the open circuit voltage at port B is

(*a*) 6V (*b*) 7V

(*c*) 8A (*d*) 9V

Common Data for Q.50-51

In the three dimensional view of a silicon n-channel MOS transistor shown below, δ = 20 nm. The transistor is of width 1 mm. The depletion width formed at every p-n junction is 10 nm. The relative permitivities of Si and SiO_2, respectively, are 11.7 and 3.9, and $\varepsilon_0 = 8.9 \times 10^{-12}$ F/m.

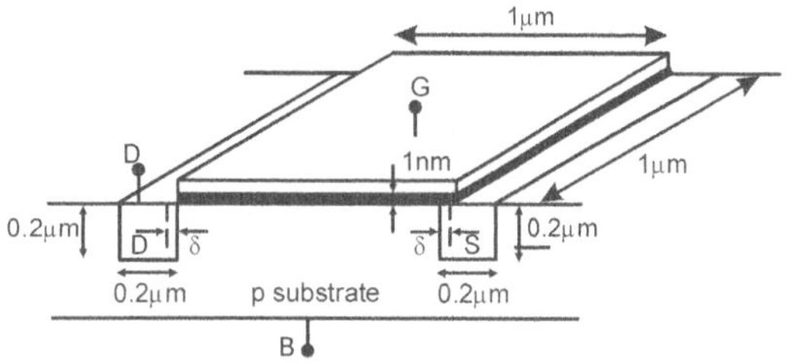

50. The gate source overlap capacitance is approximately

(*a*) 0.7 *f*F (*b*) 0.7 pF

(*c*) 0.35 *f*F (*d*) 0.24 pF

51. The source body junction capacitance is approximately

(*a*) 2 *f*F (*b*) 7 *f*F

(*c*) 2 pF (*d*) 7 pF

LINKED ANSWER QUESTIONS

Statement for Linked Answer Q.52-53

An infinitely long uniform solid wire of radius a carries a uniform dc current of density $\vec{j}$

52. The magnetic field at a distance r from the center of the wire is proportional to

(*a*) r for r < a and $1/r^2$ for r > a

(*b*) 0 for r < a and 1/r for r > a

(*c*) r for r < a and 1/r for r > a

(*d*) 0 for r < a and $1/r^2$ for r > a

53. A hole of radius b(b < a) is now drilled along the length of the wire at a distance d from the center of the wire shown below.

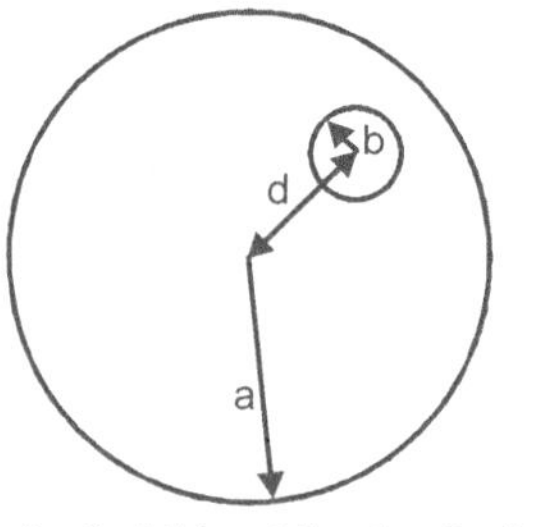

The magnetic field inside the hole is

(*a*) uniform and depends only on d

(*b*) uniform and depends only on b

(*c*) uniform and depends on both b and d

(*d*) non uniform

Statement for Linked Answer Q.54-55

The transfer function of a compensator is given as

$$G_c(s) = \frac{s+a}{s+b}$$

54. $G_c(s)$ is a lead compensator if

(*a*) a = 1, b = 2 (*b*) a = 3, b = 2

(*c*) a = –3, b = –1 (*d*) a = 3, b = 1

55. The phase of the above lead compensator is maximum at

(*a*) $\sqrt{2}$ rad/s (*b*) $\sqrt{3}$ rad/s

(*c*) $\sqrt{6}$ rad/s (*d*) $1/\sqrt{3}$ rad/s

Q. 56 – 60 carry one mark each.

56. One of the parts (A, B, C, D) in the sentence given below contains an ERROR. Which one of the following is **INCORRECT** ?

I requested that he should be given the driving test today instead of tomorrow.

(*a*) requested that

(*b*) should be given

(*c*) the driving test

(*d*) instead of tomorrow

57. If $(1.001)^{1259} = 3.52$ and $(1.001)^{2062} = 7.85$, then $(1.001)^{3321} =$

(*a*) 2.23 (*b*) 4.33

(*c*) 11.37 (*d*) 27.64

58. Choose the most appropriate alternative from the options given below to complete the following sentence :

If the tired soldier wanted to lie down, he ___ the mattress out on the balcony.

(*a*) should take (*b*) shall take

(*c*) should have taken (*d*) will have taken

59. Choose the most appropriate word from the options given below to complete the following sentence :

Given the seriousness of the situation that he had to face, his ___ was impressive.

(*a*) beggary (*b*) nomenclature

(*c*) jealousy (*d*) nonchalance

60. Which one of the following options is the closest in meaning to the word given below ?

Latitude

(*a*) Eligibility (*b*) Freedom

(*c*) Coercion (*d*) Meticulousness

Q. 61 – 65 carry two marks each.

61. A and B are friends. They decide to meet between 1 PM and 2 PM on a given day. There is a condition that whoever arrives first will not wait for the other for more than 15 minutes. The probability that they will meet on that day is

(*a*) $\frac{1}{4}$ (*b*) $\frac{1}{16}$

(*c*) $\frac{7}{16}$ (*d*) $\frac{9}{16}$

62. **One of the legacies of the Roman legions was discipline. In the legions, military law prevailed and discipline was brutal. Discipline on the battlefield kept units obedient, intact and fighting, even when the odds and conditions were against them.**

Which one of the following statements best sums up the meaning of the above passage ?

(*a*) Thorough regimentation was the main reason for the efficiency of the Roman legions even in adverse circumstances.

(*b*) The legions were treated inhumanly as if the men were animals.

(*c*) Discipline was the armies' inheritance from their seniors.

(*d*) The harsh discipline to which the legions were subjected to led to the odds and conditions being against them.

63. Raju has 14 currency notes in his pocket consisting of only Rs. 20 notes and Rs. 10 notes. The total money value of the notes is Rs. 230. The number of Rs. 10 notes that Raju has is

(*a*) 5 (*b*) 6

(*c*) 9 (*d*) 10

64. There are eight bags of rice looking alike, seven of which have equal weight and one is slightly heavier. The weighing balance is of unlimited capacity. Using this balance, the minimum number of weighings required to identify the heavier bag is

(*a*) 2 (*b*) 3

(*c*) 4 (*d*) 8

65. The data given in the following table summarizes the monthly budget of an average household.

Category	Amount (Rs.)
Food	4000
Clothing	1200
Rent	2000
Savings	1500
Other expenses	1800

The approximate percentage of the monthly budget **NOT** spent on savings is

(*a*) 10% (*b*) 14%

(*c*) 81% (*d*) 86%

ANSWERS

1. (*d*)	2. (*b*)	3. (*c*)	4. (*c*)	5. (*a*)	6. (*a*)	7. (*b*)	8. (*d*)	9. (*d*)	10. (*b*)
11. (*d*)	12. (*d*)	13. (*a*)	14. (*a*)	15. (*d*)	16. (*)	17. (*a*)	18. (*c*)	19. (*a*)	20. (*c*)
21. (*b*)	22. (*c*)	23. (*c*)	24. (*b*)	25. (*a*)	26. (*d*)	27. (*b*)	28. (*c*)	29. (*d*)	30. (*a*)
31. (*c*)	32. (*d*)	33. (*a*)	34. (*d*)	35. (*a*)	36. (*c*)	37. (*a*)	38. (*a*)	39. (*b*)	40. (*c*)
41. (*b*)	42. (*a*)	43. (*d*)	44. (*d*)	45. (*a*)	46. (*c*)	47. (*b*)	48. (*c*)	49. (*b*)	50. (*a*)
51. (*a*)	52. (*c*)	53. (*a*)	54. (*a*)	55. (*a*)	56. (*b*)	57. (*d*)	58. (*a*)	59. (*d*)	60. (*b*)
61. (*c*)	62. (*a*)	63. (*a*)	64. (*a*)	65. (*d*)					

EXPLANATIONS

1. For the given signal model of transistor,

$$g_m = \frac{I_C}{V_T}, \; V_T = 25\ mV \text{ at } 300\ K$$

$$\beta = g_m r_\pi$$

or $$r_\pi = \frac{\beta}{g_m} = \frac{\beta V_T}{I_C} = \frac{V_T}{I_b}$$

DC value of $I_b = 1mA$

then, $$r_\pi = \frac{25 \times 10^{-3}}{1 \times 10^{-3}} = 25\Omega$$

2. PSD of $x(t)$

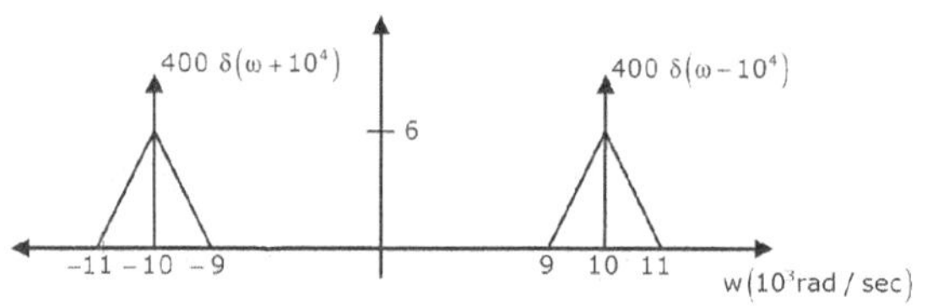

$$E[x^2(t)] = R_{xx}(0)$$

$$R_{xx}(0) = \frac{1}{2\pi}\int_{-\infty}^{\infty} S_{xx}(\omega)\, d\omega$$

$$R_{xx}(\tau) \leftrightarrow S_{xx}(\omega)$$

Fourier transform pair

$$= \frac{1}{2\pi}\left[\frac{1}{2} \times 2 \times 10^3 \times 6 + \frac{1}{2} \times 2 \times 10^3 \times 6 + 400 + 400\right]$$

$$= \frac{6400}{\pi}$$

Since PSD of $x(t)$ does not contain any DC component, hence mean value of $x(t)$ is zero.

$$E[x(t)] = 0$$

3. The frequency response for the raised cosine pulse is shown below.

α = roll — off factor

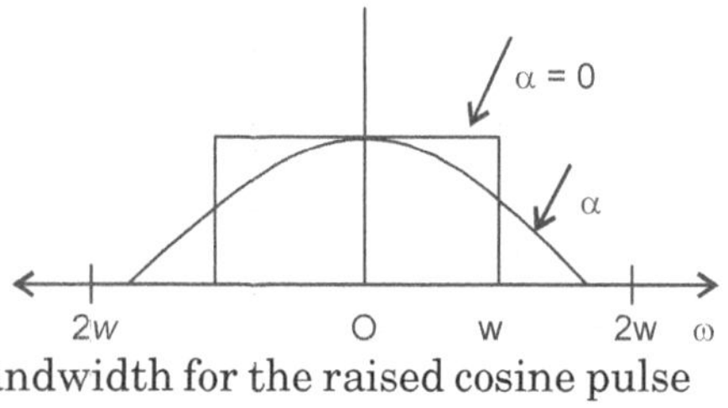

Bandwidth for the raised cosine pulse

$$f_B' = \frac{1+\alpha}{2T} H_z$$

given, $$f_B' = f_B(1.75) = \frac{1}{2T}(1.75)$$

then, $$\frac{1}{2T}(1+\alpha) = \frac{1}{2T}(1.75)$$

$$\alpha = 0.75$$

Then, maximum possible signaling rate,

$$= \frac{2 f_B}{1+\alpha} = \frac{2 \times 3500}{1+0.75} = 4000 \text{ symbols/sec.}$$

4. $E_i = (8a_x + 6a_y + 5a_z)\, e^{j(\omega t + 3x - 4y)} V/m$

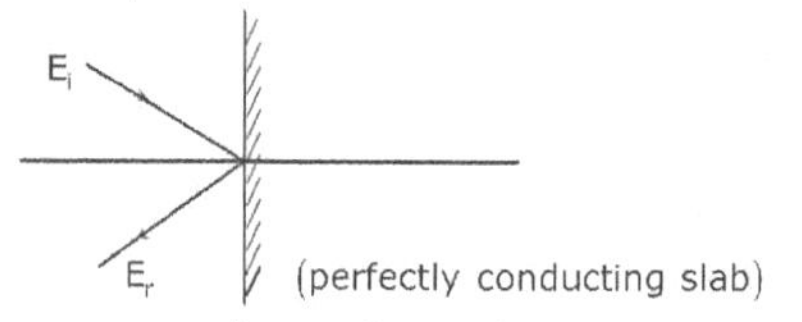

$x > 0 \quad x = 0 \quad x < 0$

I. Electric field inside a perfect conductor = 0

$\therefore \; E_{transmitted} = 0$

II. For E_1 and E_r y direction is same since slab is positioned at $x = 0$ and only x is reversed.

$\therefore \; E_i + E_r = 0$

$$E_r = -E_i$$

$$= \left[-\left(8\bar{a}_x + 6\bar{a}_y + 5\bar{a}_2\right)\right] \cdot e^{j(\omega t - 3x - 4y)}$$

5. Instantaneous form of field vector $\bar{E}$,

$$E_{ins} = R_e\left(\bar{E} e^{j\omega t}\right)$$

$$= 10\left[R_e\left(\bar{a}_y + j\bar{a}_z\right) \cdot e^{-j25x} \cdot e^{j\omega t}\right]$$

$$= 10\left[R_e \cdot \left(\bar{a}_y + j\bar{a}_z\right) e^{j(\omega t - 25x)}\right]$$

$$= 10\left[\cos(\omega t - 25x) + \sin(\omega t - 25x)\right]$$

$$\bar{E}_{ins} = 10\left[\cos(\omega t - 25x)\bar{a}_y + \cos\left(\omega t - 25x - \frac{\pi}{2}\right) a_z\right] \quad \text{...(A)}$$

From the expression,

$$\beta = 25 m^{-1}$$

Then, $$f = \frac{C\beta}{2\pi} = \frac{3 \times 10^8 \times 25}{2\pi} = 1.25 \text{ GHz}$$

As seen from equation. (A)

The z-component of $\bar{E}_{ins}$ lags by $-\frac{\pi}{2}$ to y-component, which suggets left circularised wave.

6.

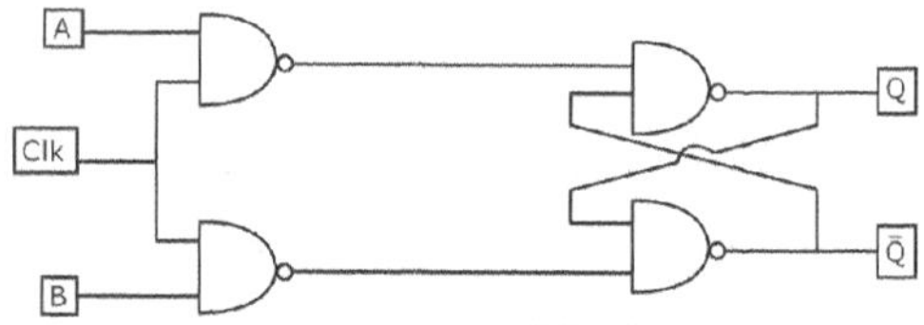

$$Q_{next} = \overline{\overline{A.CLK}.\overline{Q}}$$

$$= A.CLK + Q$$

$$\overline{Q}_{next} = A.CLK + \overline{Q}$$

If CLK =1 and A and B = 1, then

$$\left.\begin{aligned} Q_{next} &= 1 \\ \overline{Q}_{next} &= 1 \end{aligned}\right\} \text{No race around}$$

If CLK = 1 and A = B = 0, then

$$\left.\begin{aligned} Q_{next} &= Q \\ \overline{Q}_{next} &= \overline{Q} \end{aligned}\right\} \text{No race around}$$

Thus race around does not occur in the circuit

7. Truth table for the comparator is given below

A	B	Y
00	00	0
01	00	1
10	00	1
11	00	1
00	01	0
01	01	0
10	01	1
11	01	1
00	10	0
01	10	0
10	10	0
11	10	1
00	11	0
01	11	0
10	11	0
11	11	0

It can be seen from the truth-table that

Y = 1 occurs for six number of combinations where A > B.

8. Using KVL around the loop, we get

$$10 - 1000i - V = 0 \quad \text{...(1)}$$

Given $\quad i = \dfrac{V - 0.7}{500}$

or, $\quad V = 500i + 0.7 \quad \text{...(2)}$

Putting equation (2) in equation (1), we have

$$10 - 1000i - (500i + 0.7) = 0$$

or, $\quad i = \dfrac{10 - 0.7}{1500} = 6.2 \times 10^{-3}\ \text{A} = 6.2\text{mA}$

9. When switch in closed at $t = 0$, capacitor C_1 will discharge and C_2 will get charge since both C_1 and C_2 are ideal and there is no-resistance in the circuit charging and discharing time constant will be zero.

Thus current will exist like an impulse function.

10. $z = 4 - j3 = R_L - jX_C$; $R_L = 4$;

$I = 5\cos(100\pi t + 100) = I_m \cos(\omega t + \alpha)$

$$P = \frac{1}{2} I_m^2 R_L = \frac{1}{2} \times 5^2 \times 4 = 50\text{W}$$

11. If $f(t) \leftrightarrow F(s)$, then $tf(t) \leftrightarrow -\dfrac{d}{ds}F(s)$

Thus if $\quad F(s) = \dfrac{1}{s^2 + s + 1}$

$$tf(t) \rightarrow \frac{-d}{ds}\left(\frac{1}{s^2 + s + 1}\right) = \frac{2s + 1}{(s^2 + s + 1)^2}$$

12. Given differential equation is

$$t\frac{dx}{dt} + x = t$$

$$\Rightarrow \quad \frac{dx}{dt} + \frac{x}{t} = 1$$

$$\text{IF} = e^{\int \frac{1}{t} dt} = e^{\log t} = t;$$

solution is $\quad x(\text{IF}) = \int (\text{IF})\, dt$

$$xt = \int t\, dt$$

$$\Rightarrow \quad xt = \frac{t^2}{2} + c;$$

Given: $\quad x(1) = 0.5$

$$\Rightarrow \quad 0.5 = \frac{1}{2} + c$$

$$\Rightarrow \quad c = 0$$

Hence required solution $xt = \dfrac{t^2}{2}$

$$\Rightarrow \quad x = \frac{t}{2}$$

13.

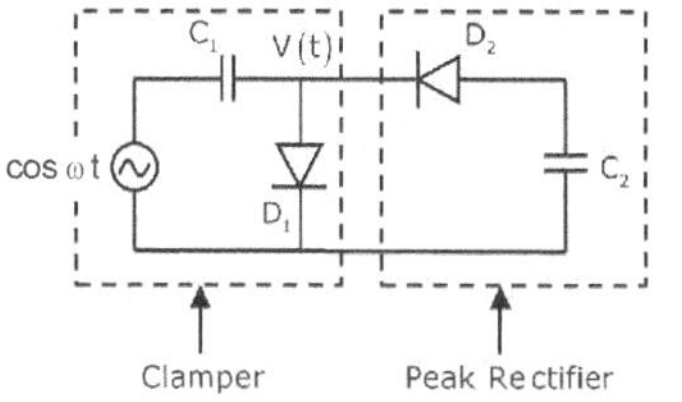

When excited by $\cos(\omega t)$, clamping section clamp the positive peak to 0 volts and negative peak to –2 volts. So whole $\cos(\omega t)$ is lower by –1 volt.

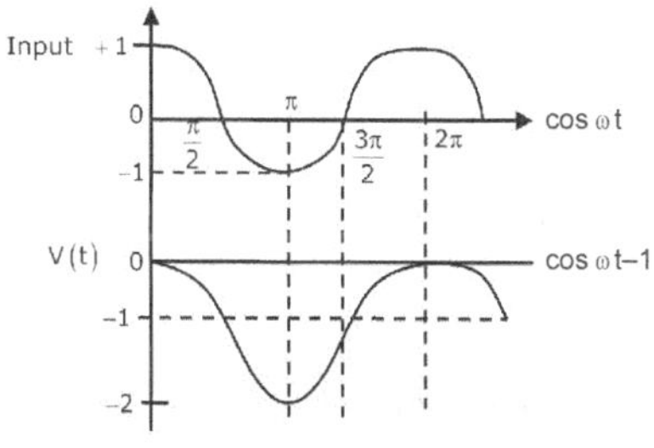

14. $$Y = \overline{(A + B)C} = \overline{(A + B)} + \overline{C} = \overline{A}.\overline{B} + \overline{C}$$

15. Entropy is maximum when all symbols are equiprobable

If probability of symbols at different then entropy in going to decrease

16. Characteristic impedance

$$Z_0 = \frac{1}{2\pi}\sqrt{\frac{\mu}{\in}}\ln\left(\frac{b}{a}\right)$$

$$= \frac{1}{2\pi}\sqrt{\frac{\mu}{\in_0 \in_r}}\ln\left(\frac{2.4}{1}\right)$$

Substitute the values, we get $Z_0 = 15.3\Omega$

17. Directivity of an antenna is, $D = \frac{4\pi U_{max}}{P_{rad}}$

Where U_{max} is the maximum radiation intensity and P_{rad} is radiated power

Now, $$P_{rad} = \int_0^{2\pi} d\phi$$

$$= \int_0^{2\pi}\int_0^{\pi/2} F(\theta)\cdot\sin\theta\cdot d\theta\cdot d\phi$$

$$= \int_0^{2\pi}\int_0^{\pi/2} \cos^4\theta\cdot\sin\theta\cdot d\theta\cdot d\phi$$

$$= 2\pi\int_0^{\pi/2} \cos^4\theta\cdot\sin\theta\cdot d\theta$$

$$= (-2\pi)\left[\frac{\cos^5\theta}{5}\right]_0^{\pi/2} = \frac{2\pi}{5}$$

Now, $$D = \frac{4\pi U_{max}}{P_{rad}}$$

[Here, $U_{max} = |\cos^4\theta|_{max} = 1$]

$$= \frac{4\pi}{2\pi/5} = 10$$

In dB, directivity, $D = 10\log_{10}D = 10dB$

18. For $n > 0$,

$$X[n] = \left(\frac{1}{3}\right)^n u[n] - \left(\frac{1}{2}\right)^n u[n]$$

For $n < 0$

$$X[n] = \left(\frac{1}{3}\right)^{-n} u[-n-1]$$

Hence, $x[n] = \left(\frac{1}{3}\right)^n u(n) + \left(\frac{1}{3}\right)^{-n} u[-n-1] - \left(\frac{1}{2}\right)^n u[n]$

For, $\left(\frac{1}{3}\right)^n u(n)$ to converage,

$$\sum_{n=-\infty}^{\infty}\left(\frac{1}{3}\right)^n z^{-n} < \infty$$

or, $$\left|\frac{1}{3}z^{-1}\right| < 1$$

or, $$|z| > \frac{1}{3} \quad \text{...(A)}$$

For $\left(\frac{1}{3}\right)^{-n} u(-n-1)$ to converge,

$$\sum_{n=-\infty}^{-1}\left(\frac{1}{3}\right)^{-n} z^{-n} < \infty$$

or, $|3z^{-1}| < \infty$ or, $|z| > 3$ ...(B)

For $\left(\frac{1}{2}\right)^n u(n)$ to converge, $\sum_{n=-\infty}^{\infty}\left(\frac{1}{2}\right)^n z^n < \infty$, 1

or, $\left|\frac{1}{2}z^{-1}\right| < 1$ or, $|z| > \frac{1}{2}$ ...(C)

By inequalities (A), (B) and (C), it can be seen that common region of covergence is,

$$\frac{1}{2} < |z| < 3$$

19.

x \ yz	00	01	11	10
0	0	0	1	1
1	1	1	0	0

Implicates are $\bar{x}yz$, $\bar{x}y\bar{z}$, $x\bar{y}\bar{z}$, $x\bar{y}z$

Prime implicants are $\bar{x}y$ and $x\bar{y}$

20. $$X(t) = \sin(\omega t)$$

$$X(s) = \mathcal{L}\sin(\omega t) = \frac{\omega}{s^2+\omega^2}$$

Output,

$$Y(s) = X(s)\cdot G(s)$$

$$= \frac{\omega}{s^2+\omega^2}\cdot\frac{(s^2+9)(s+2)}{(s+1)(s+3)(s+4)}$$

Now, $$\underset{t\to\infty}{Lt}\, y(t) = \underset{s\to 0}{Lt}\, y(s)$$

$$= \underset{s\to 0}{Lt}\frac{s\omega}{s^2+\omega^2}\cdot\frac{(s^2+9)(s+2)}{(s+1)(s+3)(s+4)}$$

cancellation of terms $s^2 + 9$ and $s^2 + \omega^2$ will led to applicability of final value theorem as presence of pairs of complex conjugate poles on the imaginary axis will led to $x(t)$ containing dinusoidal components and $y(\infty)$ will be defined.

For this to be hold, $\omega^2 = 9$

$\Rightarrow$ $\omega = 3$ rad/sec

21. Open-circuiting terminal 1-2.

$$V_{oc} = (99\,i_b + i_b)\cdot 100$$
$$= 10^4 i_b$$

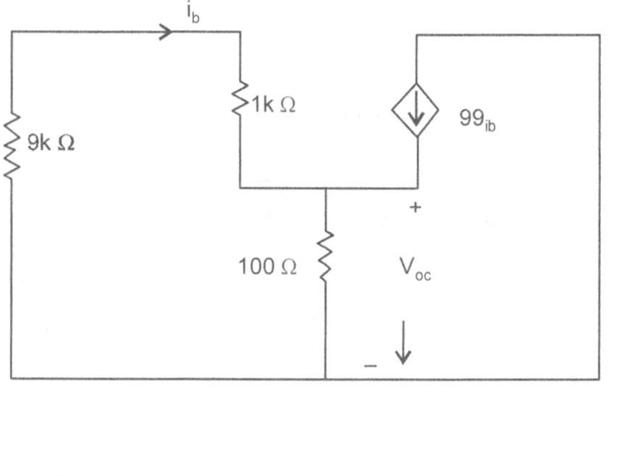

Short-circuiting the terminal, we get

$$I_{sc} = i_b + +99i_b$$

$$I_{sc} = 100i_b$$

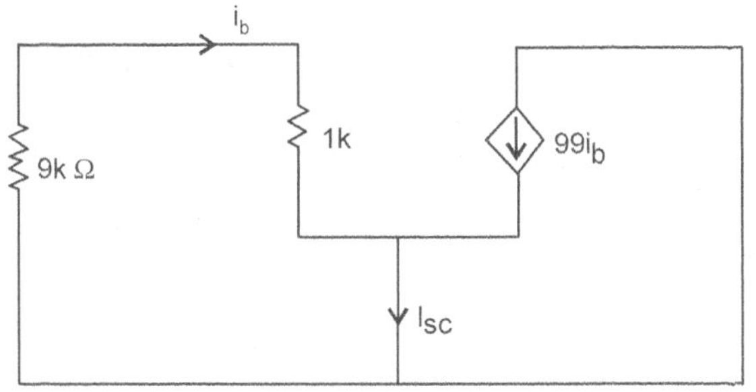

Therefore, impedance working into nodes 1 and 2,

$$Z_{th} = \frac{V_{oc}}{I_{sc}} = 100\Omega$$

22.

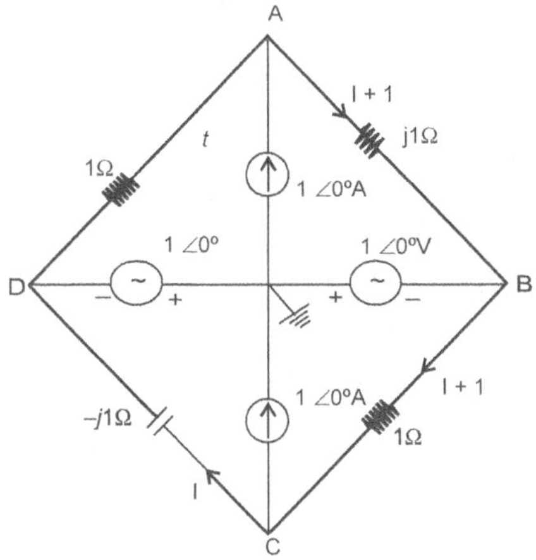

As, $V_B + 1\angle 0° - 1\angle 0° \quad = V_D$

or $\qquad V_B = V_D$

Then, voltage drop across BD,

$$V_{BD} = 0$$

and therefore no current flows through it.

Now, using KVL along the loop ABCD we get,

or, $-I.1 - j(I+1) - (I+1) + Ij = 0$

or, $\qquad -2I - j - 1 = 0$

or, $\qquad I = -\frac{j+1}{2}$

Current through Inductor

$$I + 1 - -\left(\frac{j+1}{3}\right) + 1 - \frac{1}{2} - \frac{1}{2}$$

$$= \frac{1-j}{2} = \frac{1-j}{(1-j)(1+j)} = \frac{1}{1+j}$$

23. $$f(z) = \frac{1}{z+1} - \frac{2}{z+3}$$

$$f(z) = \frac{-z+1}{(z+1)(z+3)}$$

Pole at $z = -1$ lies within the circle $|z+1| = 1$ and pole at $z = -3$ lies outside the circle $|z+1| = 1$

$\oint_c f(z).dz = 2\pi i$(Sum of the residues at the singular points within C)

Or, $\frac{1}{2\pi i}\oint_c f(z)\cdot dz = \text{Res } f(-1) = \underset{z\to -1}{\text{Lt}} \frac{-z+1}{z+3} = 1$

24. Distribution of two random variables X and Y are illustrated below :

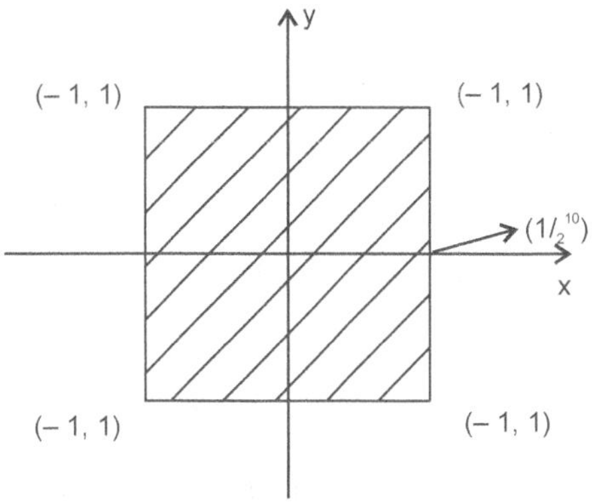

The probability that max [X, Y] is less than $\frac{1}{2}$ is the shaded region, divided by the total area enclosed by the rectangle,

$$P\left(\max[X,Y] < \frac{1}{2}\right) = \frac{\text{Shaded Area}}{\text{Total Area}}$$

$$= \frac{\frac{3}{2}\times\frac{3}{2}}{4} = \frac{9}{16} = 9$$

25. *Given*: $x = \sqrt{-1};\ x^x = \left(\sqrt{-1}\right)^{\sqrt{-1}} = i^i$

We know, $e^{i\theta} = \cos\theta + i\sin\theta$

$$\Rightarrow \qquad e^{i\frac{\pi}{2}} = \cos\frac{\pi}{2} + i\sin\frac{\pi}{2} = i$$

$$\therefore \qquad (i)^i = \left(e^{\frac{i\pi}{2}}\right)^i = \left(e^{-\frac{\pi}{2}}\right)^i$$

26. $$p = \frac{n_i^2}{10^{19}} = 10\text{ cm}^{-3}$$

Volume $= 10^{-6} \times 10^{-6}\text{ m}^3 = 10^{-12}\text{ cm}^3$

$\therefore$ Total holes $= 10^{-11} \approx 0$

27. $$s_1(t) = \sqrt{\frac{2E}{T}}\sin\omega_c t$$

$$s_2(t) = -\sqrt{\frac{2E}{T}}\sin\omega_c t$$

As local oscillator in the coherent receiver is ahead in phase by 45° with respect to the received signal the amplitude decreases by a factor of

$$\sin 45° = \frac{1}{\sqrt{2}},\ s_1'(t) = \frac{s_1(t)}{\sqrt{2}},\ s_1'(t) = \frac{s_2(t)}{\sqrt{2}}$$

then,

Energy of the difference signal at the output,

$$E_d = \frac{1}{T}\int_0^T \left[s_1'(t) - s_2'(t)\right]^2 \cdot dt$$

$$= \frac{E}{T}\int_0^T (2\sin\omega_c t)^2 \cdot dt$$

$$= 4\frac{E}{T}\int_0^T \frac{(1-\cos 2\omega_c t)}{2} \cdot dt$$

$$E_d = 2E$$

Now, probability of error,

$$P_e = Q\left(\sqrt{\frac{E}{2N_0}}\right)$$

$$P_e = Q\left(\sqrt{\frac{E}{2N_0}}\right)$$

29. Given, $y = \int_{-\infty}^{t} x(J)\cos(3J)\cdot dJ$

$$y(t-t_0) = \int_{-\infty}^{t-t_0} x(J)\cos(3J)\cdot dJ$$

For input, $x(t-t_0)$

Output $y'(t) = \int_{-\infty}^{t} x(J-t_0)\cos 3J \cdot dJ$

$$= \int_{-\infty}^{t-t_0} x(J)\cdot \cos 3(J+t_0)\cdot dJ$$

Since, $y'(t) \neq y(t-t_0)$, the system is not time-invariant.

For bounded input, $x(J) = \cos(3J)[-1 \le \cos 3J \le 1]$

Output, $y(t) = \int_{-\infty}^{t} \cos^2 3J$

$$= \left.\frac{1+\cos 6J}{2}\right|_{-\infty}^{t} = \frac{1+\cos 6t}{2}$$

for $t \to \infty$,

$y(t) \to \infty$ [$y(\infty)$ not defined]

Hence, system is not stable.

30. Characteristics equation is given by

$$1 + G(s)H(s) = 0$$

or, $1 + \dfrac{k(s+1)}{s^3 + as^2 + 2s + 1} = 0$

or, $s^3 + as^2 + (2+k)s + k + 1 = 0$

Constructing Routh-array, we have

s^3	1	$2+k$
s^2	a	$k+1$
s^1	$\dfrac{a(2+k)-(1+k)}{a}$	
s^0	$1+k$	

For oscillations to occur, $\dfrac{a(2+k)-(1+k)}{a} = 0$

or, $a = \dfrac{1+k}{2+k}$ ...(1)

And, the frequency of oscillation is given by,

$as^2 + (k+1) = 0$

or, $a(j\omega)^2 + (k+1) = 0$

or, $(k+1) = 4a$ [given $\omega = 2$ rad/sec] ...(2)

From equation (1) and equation (2), we get

$k = -1, 2$

For $k = -1$, $a = 0$ [System will not oscillate]

For $k = 2$, $a = \dfrac{1+k}{2+k} = \dfrac{3}{4} = 0.75$

31. $H(j\omega) = 2\cos\omega \dfrac{\sin 2\omega}{\omega}$

putting, $H_1(j\omega) = 2\cos\omega = e^{j\omega} + e^{-j\omega}$

$$H_2(j\omega) = \frac{2\sin 2\omega}{2\omega} = \frac{2\cdot\sin(2\omega)}{2\omega}$$

Inverse fourier transform of $H_2(j\omega)$ is,

$$h_2(t) = \begin{cases} \frac{1}{2}, & -2 \le + \le 2 \\ 0, & \text{otherwise} \end{cases}$$

as illustrated below,

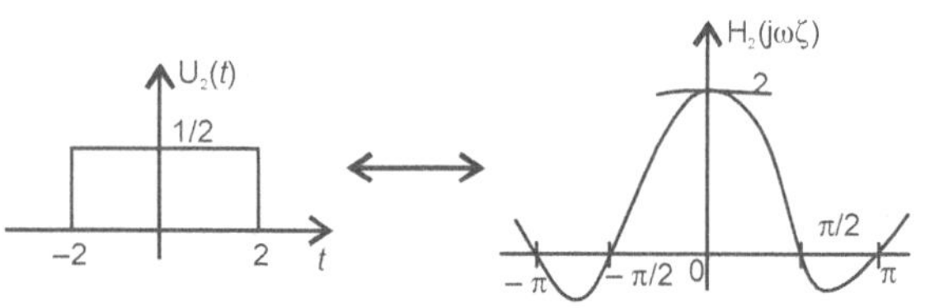

Then, using $e^{j\omega}\cdot x(\omega) \leftrightarrow x(t+1)$ (time-shifting property).

We get, $h(t) = \mathcal{L}^{-1}4(j\omega) = h_2(t-1) + h_2(t+1)$

Which is drawn below

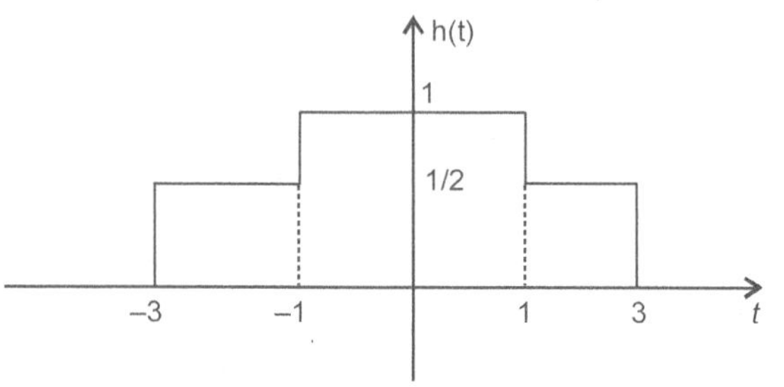

At $t = 0$, $h(t) = h(0) = 1$

32. Controllability matrix is given by, $Q_c = [B\ AB\ A^2B]$

$$AB = \begin{bmatrix} 0 & a_1 & 0 \\ 0 & 0 & a_2 \\ a_3 & 0 & 0 \end{bmatrix}\begin{bmatrix} 0 \\ 0 \\ 1 \end{bmatrix} = \begin{bmatrix} 0 \\ a_2 \\ 0 \end{bmatrix}$$

$$A^2B = AAB = \begin{bmatrix} 0 & a_1 & 0 \\ 0 & 0 & a_2 \\ a_3 & 0 & 0 \end{bmatrix}\begin{bmatrix} 0 \\ a_2 \\ 0 \end{bmatrix} = \begin{bmatrix} a_1a_2 \\ 0 \\ 0 \end{bmatrix}$$

Then, $Q_c = \begin{bmatrix} 0 & 0 & a_1a_2 \\ 0 & a_2 & 0 \\ 1 & 0 & 0 \end{bmatrix}$

For the system to be controllable, the rank of Q_c must be 3. and $|Q_c| = 0$

So satisfy this condition,

$$a_1a_2 \neq 0$$

$$a_2 \neq 0$$

and, $1(0 - a_1a_2^2) \neq 0$

or $a_1a_2^2 \neq 0$

Hence $a_1 \neq 0,\ a_2 \neq 0,\ a_3 = 0$

33. Power transferred from circuit A to circuit B

$$I = \frac{10-3}{2+R} = \frac{7}{2+R}$$

$$V = 3 + IR = 3 + \frac{7R}{2+R}$$

$$= \left(\frac{6+10R}{2+R}\right)$$

$$P = VI = \left(\frac{7}{R+2}\right)\left(\frac{6+10R}{R+2}\right)$$

$$P = \frac{42+70R}{(R+2)^2}$$

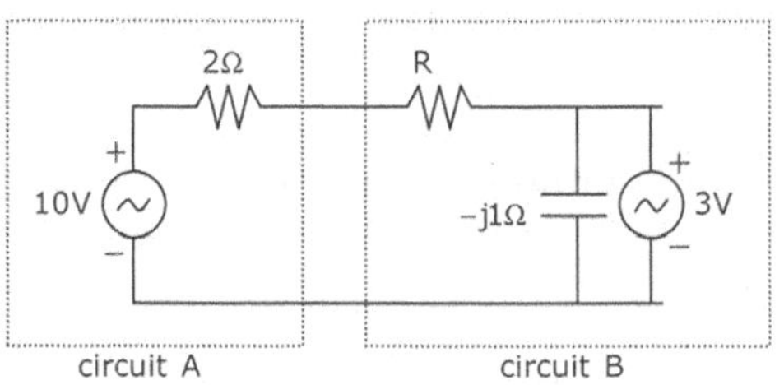

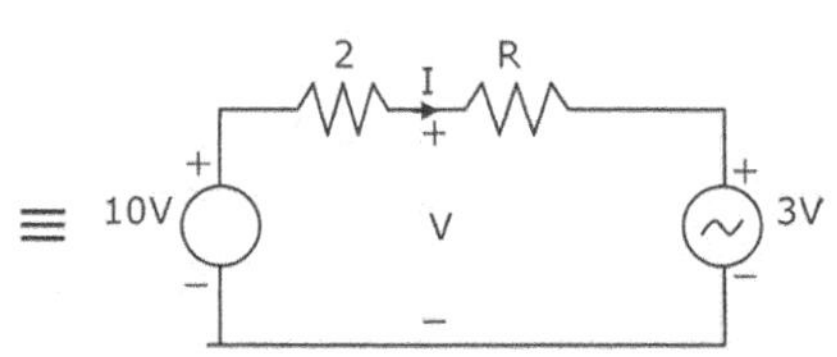

$$\frac{dP}{dR} = \frac{(R+2)^2(70) - (42+70R)2(R+2)}{(R+2)^4} = 0$$

$\Rightarrow$ $70(R+2)^2 = (42 + 70R)\,2\,(R+2)$

$\Rightarrow$ $5(R + 2) = 2(3 + 5R)$

$\Rightarrow$ $5R + 10 = 6 + 10R$

$\Rightarrow$ $4 = 5R$

$\Rightarrow$ $R = 0.8\Omega$

34. $\dfrac{d^2y(t)}{dt^2} + \dfrac{2dy(t)}{dt} + y(t) = \delta(t)$

Converting to s-domain,

$$s^2y(s) - sy(0) - y'(0) + 2[sy(s) - y(0)] + y(s) = 1$$

$$[s^2 + 2s + 1]y(s) + 2s + 4 = 1$$

$$y(s) = \frac{-3-2s}{(s^2+2s+1)}$$

Find inverse Lapalce transform

$$y(t) = [-2e^{-t} - te^{-t}]u(t)$$

$$\frac{dy(t)}{dt} = 2e^{-t} + te^{-t} - e^{-t}$$

$$\left.\frac{dy(t)}{dt}\right|_{t=0^+} = 2 - 1 = 1$$

35. In spherical co-ordinate,

$$\nabla\cdot\vec{A} = \frac{1}{r^2}\frac{\partial}{\partial r}(r^2A_r) + \frac{1}{r\sin\theta}\cdot\frac{\partial}{\partial r}(A_\theta\sin\theta) + \frac{1}{r\sin\theta}\cdot\frac{\partial}{\partial\phi}A_\phi$$

Given, $\bar{A} = kr^n\cdot\bar{a}_r$

then, $\bar{\nabla}\cdot\bar{A} = \dfrac{1}{r^2}\dfrac{\partial}{\partial r}(r^2kr^n) + 0$

[There is no component of $\bar{A}$ along $\bar{A}_\theta$ and $\bar{A}_\phi$]

$$= \frac{1}{r^2}\frac{\partial}{\partial r}(r^{n+2})$$

For $\bar{\nabla}\cdot\bar{A} = 0$, $\dfrac{\partial}{\partial r}(r^{n+2}) = 0$

or, $(n+2)\,r^{n+1} = 0$

or, $n = -2$

36. P(odd tosses) = P(H) + P(TTH) + P(TTTTH) + ...

$$= \frac{1}{2} + \left(\frac{1}{2}\right)^3 + \left(\frac{1}{2}\right)^5 + \ldots$$

$$= \frac{1}{2}\left(1 + \left(\frac{1}{2}\right)^2 + \left(\frac{1}{2}\right)^4 + \ldots\right)$$

$$= \frac{1}{2}\left[1 + \left(\frac{1}{4}\right) + \left(\frac{1}{4}\right)^2 + \ldots\right]$$

$$= \frac{1}{2}\left(\frac{1}{1-\frac{1}{4}}\right) = \frac{1}{2}\times\frac{4}{3} = \frac{2}{3}$$

37. This inverter given is CMOS inverter and since both NMOS and PMOS inverter have equal values of Houshold voltage, it is also known as symmetric inverter.

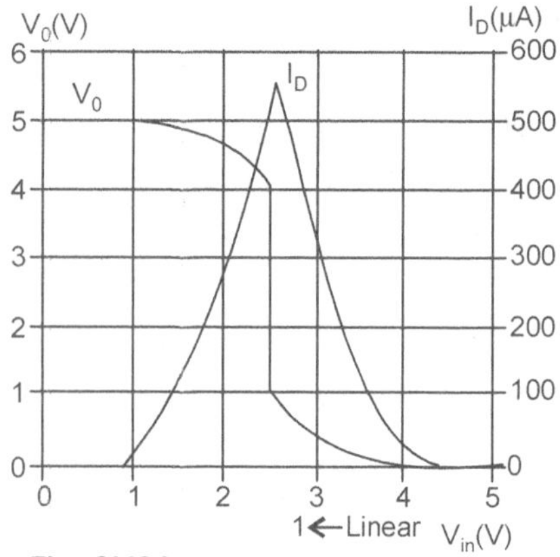

Fig. *CMOS inverter characteristics*

The characteristic of this inverter is illustrated above, it can be seen here that for V_{IN} slightly greater than V_T. PMOS (M1) stays in trade region (linear region) and NMOS (M2) stays in saturation. As V_{IN} increases, the drain current increase, voltage drop across M1 increase, and the output voltage reduces. At some point later, PMOS (M1) will enter into saturation region.

Hence right option for M1 to be in linear region is $V_{in} < 1.875$ volts.

38. $$P[X = 0] = \frac{0}{10}$$

$$P[X = 1] = 1 - P[X = 0] = \frac{1}{10}$$

Probability of error for an optimum receiver,

$$P_e = P[X = 1]\,(1 - \text{transmition prbability})$$

$$= \frac{1}{10}\left(1 - \frac{1}{8}\right) = \frac{7}{80}$$

39. For phase modulation,

Instantaneous angle $Q_i(t)$ is given by,

$$Q_i(t) = 2\pi f_c t + k_p m(t)$$

Phase deviation, $\phi(t) = Q_i(t) - 2\pi f_c t = k_p m(t)$

and $|\phi(t)|_{max} = |k_p m(t)|_{max} = k_p.2$ ...(1)

For frequency modulation,

Instantaneous frequency $f_i(t)$ is given by,

$$f_i(t) = f_c + k_f m(t)$$

and $$Q_i(t) = 2\pi \int_{-\infty}^{t} t_i(J).dJ$$

$$Q_i(t) = 2\pi f_c t + 2\pi k_f \int_{-\infty}^{t} m(J).dJ$$

Phase deviation. $\phi(t) = Q_i(t) - 2\pi f_c t$

$$= 2\pi k_f \int_{-\infty}^{t} m(J).dJ$$

$$|\phi(t)|_{max} = \left| 2\pi k_f \right| \left| \int_{-\infty}^{t} m(J).dJ \right|_{max}$$

$$= 2\pi k_f \int_0^2 2.dJ = 8\pi k_f \quad ...(2)$$

Given that maximum phase deviation is same for both FM and PM

then, $8\pi k_f = k_p 2$

or, $$\frac{k_p}{k_f} = 4\pi$$

40. $$H_z = H_0 \cos\left(\frac{m\pi x}{a}\right)\cos\left(\frac{n\pi y}{\beta}\right)\cos(\omega t - \beta z)$$

Given :

$H_z = 3\cos(2.094 \times 10^2 x)\cos(2.618 \times 10^2 y)$
$\cos(6.283 \times 10^{10} - \beta z)$

Comparing two equations, we get $m = 2$, $n = 1$

$f = 10$ GHz

$$f_C = \frac{C}{2}\sqrt{\left(\frac{m}{a}\right)^2 + \left(\frac{n}{b}\right)^2}$$

$$= \frac{3 \times 10^{10}}{2}\sqrt{\left(\frac{2}{3}\right)^2 + \left(\frac{1}{1.2}\right)^2} = 1.5 \text{ GHz}$$

Thus wave will not propagate inside waveguide Hence $V_p = 0$

41. $$V_0(s) = -\left(\frac{R_2}{R_1 + \frac{1}{CS}}\right)V_i(s)$$

$$V_0(s) = -\frac{R_2 CS}{(R_1 CS + 1)}V_i(s)$$

Thus cut-off frequency is $\frac{1}{R_1 C}$ and filter in high pass filter

42. $$y[n] = \sum_{k=0}^{\infty}\left(\frac{1}{2}\right)^k g(n-k)$$

$$y[0] = \sum_{k=0}^{\infty}\left(\frac{1}{2}\right)^k g(-k) = 1$$

$$\Rightarrow \quad \left(\frac{1}{2}\right)^0 g(0) = 1$$

$$\Rightarrow \quad g(0) = 1$$

Since $g(n)$ is Causal sequence

$$g(-1), g(-2), \ldots = 0$$

$$y[1] = \sum_{k=0}^{\infty}\left(\frac{1}{2}\right)^k g[1-k]$$

$$\Rightarrow \quad \left(\frac{1}{2}\right)^0 g[1] + \left(\frac{1}{2}\right)^1 g(0) = \frac{1}{2}$$

$$g[1] = 0$$

43. $\left.\begin{array}{l} A = 0, \quad y = Q \\ A = 1, \quad y = \overline{Q} \end{array}\right\}$

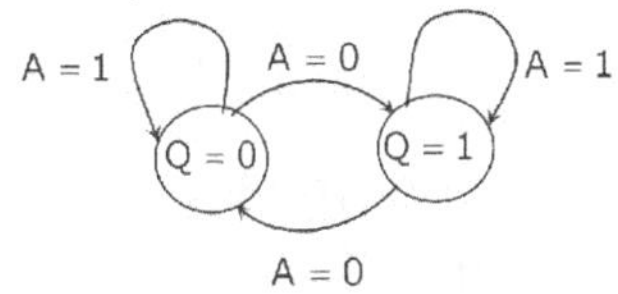

When ever A = 1, output gets into same state
When ever A = 0, output gets toggled

44. Drawing the simplified model of the given circuit,

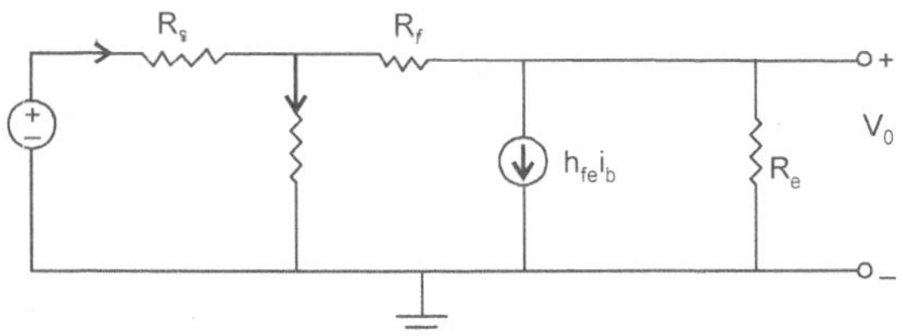

Using KVL for the base and collector nodes, we have

$$\frac{V_b}{h_{ie}} + \frac{V_b - V_s}{R_s} + \frac{V_b - V_0}{R_f} = 0 \qquad ...(i)$$

and $$h_{fe}\frac{V_b}{h_{ie}} + \frac{V_0 - V_b}{R_f} + \frac{V_0}{R_c} = 0 \qquad ...(ii)$$

from equation (*i*), $$V_b = \frac{\left[\dfrac{V_s}{R_s} + \dfrac{V_0}{R_f}\right]}{\left[\dfrac{1}{h_{ie}} + \dfrac{1}{R_s} + \dfrac{1}{R_f}\right]}$$

and from equation (*ii*) $$V_o = \frac{\left(\dfrac{1}{R_f} - \dfrac{h_{fe}}{h_{ie}}\right)}{\dfrac{1}{R_f} + \dfrac{1}{R_c}} \cdot V_b$$

Then,

$$\text{Output Voltage, } V_o = \frac{\left(\dfrac{1}{R_f} - \dfrac{h_{fe}}{h_{ie}}\right)\left(\dfrac{V_s}{R_s} + \dfrac{V_0}{R_f}\right)}{\left(\dfrac{1}{R_f} + \dfrac{1}{R_c}\right)\left(\dfrac{1}{h_{ie}} + \dfrac{1}{R_s} + \dfrac{1}{R_f}\right)}$$

or, $$V_o = \frac{\left(\dfrac{1}{100k} - \dfrac{100}{1.1k}\right)\left(\dfrac{V_s}{10k} + \dfrac{V_o}{100k}\right)}{\left(\dfrac{1}{100k} + \dfrac{1}{12k}\right)\left(\dfrac{1}{1.1k} + \dfrac{1}{10k} + \dfrac{1}{100k}\right)}$$

$(h_{ie} = 1.1k\Omega)$

$$= \frac{(-90.89)}{(0.093)(1.01)}\left[\frac{V_s}{10} + \frac{V_o}{100}\right] = -96.76\left[V_s + \frac{V_o}{10}\right]$$

On $|A_{vs}| \approx 10$

45. $I = \dfrac{V_A - V_B}{2} = \dfrac{6}{2} = 3A;$

Since current entering any network is same as leaving in $V_C - V_D$ branch. Also it is I = 3A

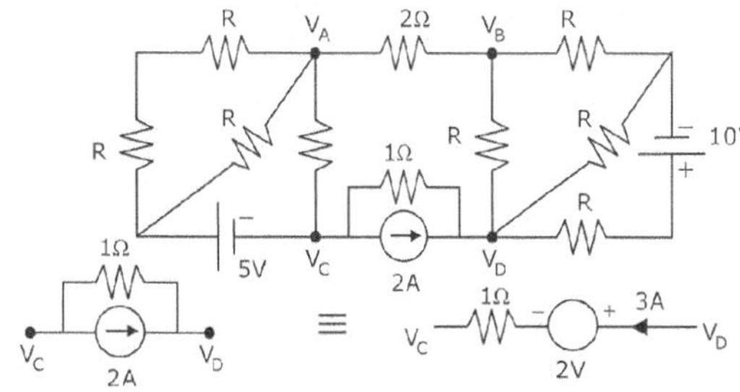

$$V_D = 2 + 3 + V_C$$
$$= 5 + V_C;$$
$$V_C - V_D = -5V$$

46. $$f(x) = x^3 - 9x^2 + 24x + 5$$

To find maximum value of $f(x)$, putting $f'(x) = 0$

or, $3x^2 - 18x + 24 = 0$

or, $x^2 - 6x + 8 = 0$

$x = 2,4$

Now, at $x = 2$, $f''(x) = (2x - 6)|_{x=2} = -2 < 0$

$x = 4$, $f''(x) = (2x - 6)|_{x=4} = 2 > 0$

Hence, a maxima occurs at $x = 2$ and a minima occurs at $x = 4$.

But at the end of interval, $x = 6$,

$$f(x) = 6^3 - 9 \times 6^2 + 24 \times 6 + 5 = 41$$

Which infers that maximum value of a function is simply the greatest value in the immediate neighbourhood of the maxima point ($x = 2$, in this case). In fact there may be several maximum and minimum values of a funcation is an interval same is also true for the minimum value of a function.

47. Characteristic equation of A is

$$|A - \lambda I| = 0$$

or, $$\begin{vmatrix} -5-\lambda & -3 \\ 2 & 0-\lambda \end{vmatrix} = 0$$

or, $(5 + \lambda)\lambda + 6 = 0$

or, $\lambda^2 + 5\lambda + 6 = 0 \qquad ...(1)$

Now, CAYLEY-HAMILTON Theorm states that every square matrix satisfies its own characteristic equation; i.e. if the characteristic equation for the nth order square matrix A is,

$$|A - \lambda I| = (-1)^n \lambda^n + k_1\lambda^{n-1} + ... + k_n = 0$$

then, $(-1)^n A^n + k_1 A^{n-1} + ... + k_n = 0$

Then, by using this theorem to equation (1), we get

$$A^2 + 5A + 6I = 0$$

or, $A^2 = -5A - 6I \qquad ...(2)$

Multiplying by A to equation (2) both sides, we get

$$A^3 = -5A^2 - 6A$$

or $A^3 = -5[-5A - 6I] - 6A = 19A + 30I$

or, $A^3 = 19A + 30I$

48.

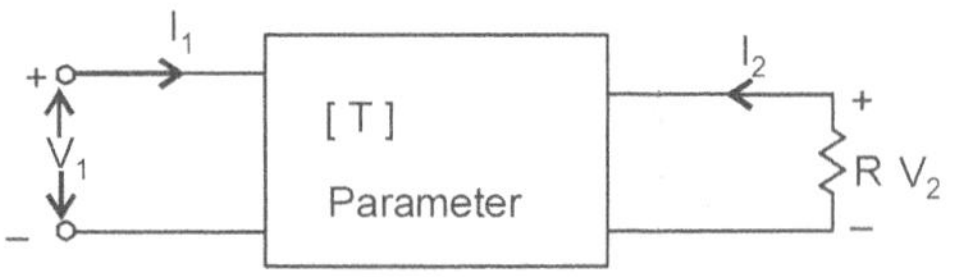

By definition

$$\begin{bmatrix} V_1 \\ I_1 \end{bmatrix} = \begin{bmatrix} A & B \\ C & D \end{bmatrix} \begin{bmatrix} V_2 \\ -I_2 \end{bmatrix}$$

$$V_1 = AV_2 - BI_2$$
$$I_1 = CV_2 - DI_2$$

Given, $V_1 = 10$ volts, $V_2 = -I_2R = 3$ volts, $I_2 = -3A$

Then, $10 = 3A + 3B$ ...(1)

and $V_1 = 10V$,

$R' = 2.5\Omega,\ I_2' = -2A, V_2' = -I_2'R' = 5V$

then, $10 = 5A + 2B$ ...(2)

From equation (1) and equation (2), we get

$$A = \frac{10}{9}, B = \frac{20}{9}$$

Now, For $V_1 = 10V,\ R = 7\Omega$

We get, $10 = A(-I_2R) - BI_2$

or, $$10 = \frac{10}{9}(-7I_2) - \frac{20}{9}I_2$$

$\Rightarrow$ $I_2 = -1$ amp

It means current being drawn is 1 amp.

50. Gate-source capacitance C_g is,

$$C_g = \frac{\in_1 A_1}{d_1} = \frac{\in_0 \in_{r_1} A_1}{d_1}$$

Since there is layer of S_iO_2 between gate and source, therefore,

$\in_{r_1} = 3.9,\quad A_1 = l \times \delta$

$$= 1 \times 10^{-6} \times (20 \times 10^{-9})$$
$$= 20 \times 10^{-15}\ m$$
$$d_1 = 1nm = 10^{-9}m$$

Then, $$C_g = \frac{(8.9 \times 10^{-12}) \times 3.9 \times 2 \times 10^{-14}}{10^{-9}}$$

$$= 0.69 \times 10^{-15} \approx 0.7 \text{ femco Farad}$$
$$= 0.7\ fF$$

51. There is a layer of S_i between source and body, then, source- body junction capacitance is,

$$C_j = \frac{\in_0 \in_r A}{d}$$

$A = (0.2\quad m) \times (1\quad m) = 0.2 \times 10^{-12} m^2$

$d = 10\quad m = 10 \times 10^{-9}$ m

[Width of depletion layer

$\in_r = 11.7$ (for silicon) formed at $p - n$ function]

then, $$C_j = \frac{8.9 \times 10^{-12} \times 11.7 \times 0.2 \times 10^{-12}}{10^{-8}}$$

$$= 2.08 \times 10^{-15}F$$
$$\cong 2fF$$

52. For region $0 \le r \le a$

Using Ampere's circuital law

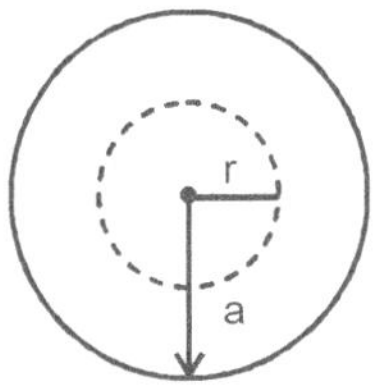

$$\int H.dl = I_{enc} = \int J.ds$$
$$I_{cnc} = (J.\ \pi r^2)$$
$$H.(2\pi r) = J\pi r^2$$
$$H = \frac{J.r}{2}$$

Thus magnetic increases with r for region $r > a$

$$H_\phi(2\pi r) = J.(\pi a^2)$$
$$H_\phi = \frac{J.\pi a^2}{2\pi r}$$
$$= \frac{Ja^2}{2r}$$

Magnetic field is proportional to $\frac{1}{r}$

54. $$\phi = \tan^{-1}\frac{\omega}{a} - \tan^{-1}\frac{\omega}{b}$$

For phase lead ϕ should be + ve

$\Rightarrow$ $\tan^{-}\frac{\omega}{a} > \tan^{-1}\frac{\omega}{b}$

$\Rightarrow$ $a < b$

both options (a) and (c) satisfier but option (c) will put pole and zero as RHS of s-plane thus not possible, Option (a) is right.

55. $$G_c(j\omega) = \frac{j\omega + a}{j\omega + b}$$

$$\phi = \angle G_c(j\omega) = \tan^{-1}\frac{\omega}{a} - \tan^{-1}\frac{\omega}{b}$$

For, ϕ to be maximum, $\frac{d\phi}{a\omega} = 0$

or, $$\frac{\frac{1}{a}}{1 + \left(\frac{\omega}{a}\right)^2} - \frac{\frac{1}{b}}{1 + \left(\frac{\omega}{b}\right)^2} = 0$$

or, $$b + \frac{\omega^2}{b} = a + \frac{\omega^2}{a}$$

or, $$\omega^2\left[\frac{1}{b} - \frac{1}{a}\right] = a - b$$

or, $$\omega = \sqrt{ab}\ \text{rad/sec}$$

For $a = 1$, $b = 2$

We have, $\omega = \sqrt{2}$ rad/sec

57. Let $1.001 = x$

$x^{1259} = 3.52$; and $x^{2062} = 7.85$

$$x^{3321} = x^{1259} \quad x^{2062} = 3.52 \times 7.85$$
$$= 27.65$$

61.

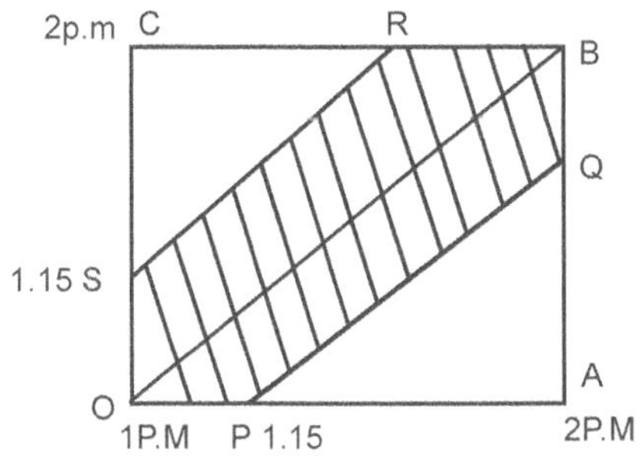

OB is the line when both A and B arrive at same time.

Total sample space = $60 \times 60 = 3600$

Favourable cases = Area of OABC – Area of PQRS

$$= 3600 - 2 \times \left(\frac{1}{2} \times 45 \times 45\right)$$
$$= 1575$$

$\therefore$ Required probability = $\frac{1575}{3600} = \frac{7}{16}$

63. Let the number of Rs. 20 notes be x and Rs. 10 notes be y

$$20x + 10y = 230$$
$$x + y = 14$$
$$x = 9$$
and $$y = 5$$

Hence numbers of 10 rupee notes are 5.

65. Total budget = 10,500

Expenditure other than savings = 9000

Hence, $\frac{9000}{10500} = 86\%$

SOLVED PAPER – 2011

ELECTRONICS AND COMMUNICATION ENGINEERING

Time : 3 Hours | Maximum Marks : 100

Q.(1–25) carry one mark each.

1. The modes in a rectangular waveguide are denoted by TE_{mn}/TM_{mn} where m and n are the eigen numbers along the larger and smaller dimensions of the waveguide respectively. Which one of the following statements is **TRUE**?
   (*a*) The TM_{10} mode of the waveguide does not exist
   (*b*) The TE_{10} mode of the waveguide does not exist
   (*c*) The TM_{10} and the TE_{10} modes both exist and have the same cut-off frequencies
   (*d*) The TM_{10} and the TM_{01} modes both exist and have the same cut-off frequencies

2. The **Column-1** lists the attributes and the **Column-2** lists the modulation systems. Match the attribute to the modulation system that best meets it.

   Column-1
   A. Power efficient tranmission of signals
   B. Most bandwidth efficient transmission of voice signals
   C. Simplest receiver structure
   D. Bandwidth efficient transmission of signals with significant *dc* component

   Column-2
   1. Conventional AM
   2. FM
   3. VSB
   4. SSB-SC

	A	B	C	D
(*a*)	4	2	1	3
(*b*)	2	4	1	3
(*c*)	3	2	1	4
(*d*)	2	4	3	1

3. The differential equation

$$100\frac{d^2y}{dt^2} - 20\frac{dy}{dt} + y = x(t)$$

describes a system with an input $x(t)$ and an output $y(t)$. The system, which is initially relaxed, is excited by a unit step input. The output $y(t)$ can be represented by the waveform

(*a*)
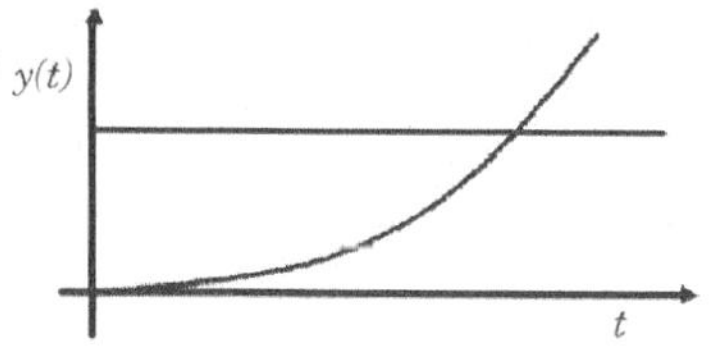

(*b*)
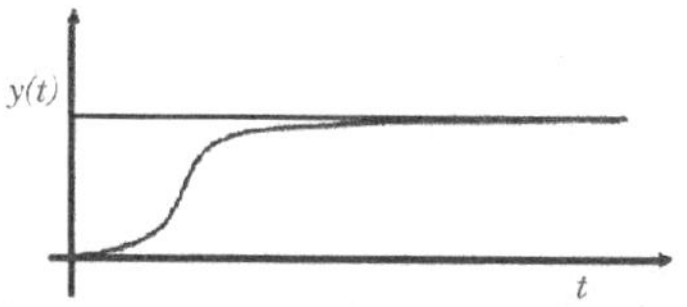

(*c*)
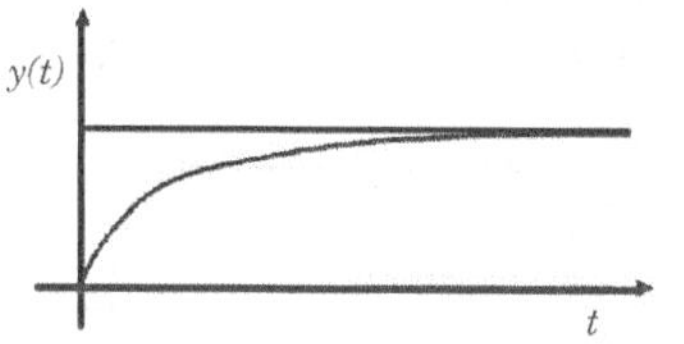

(*d*)
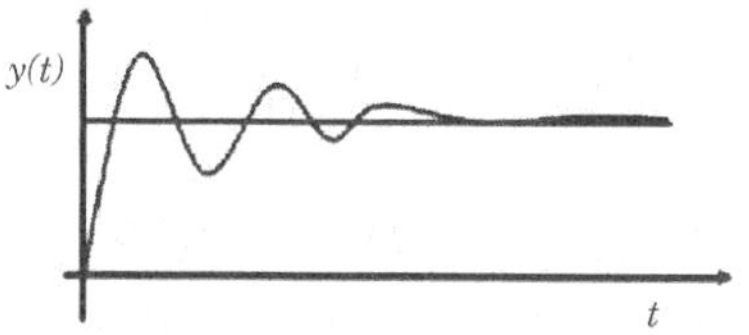

4. For the transfer function $G(j\omega) = 5 + j\omega$, the corresponding Nyquist plot for positive frequency has the form

(*a*)
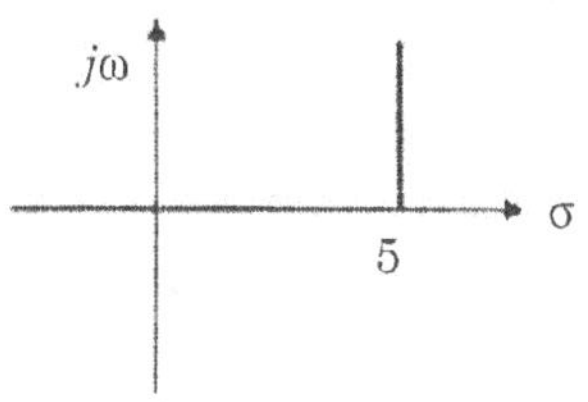

(*b*)
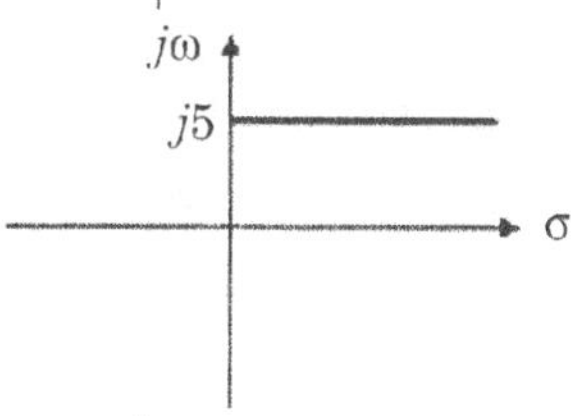

(*c*)
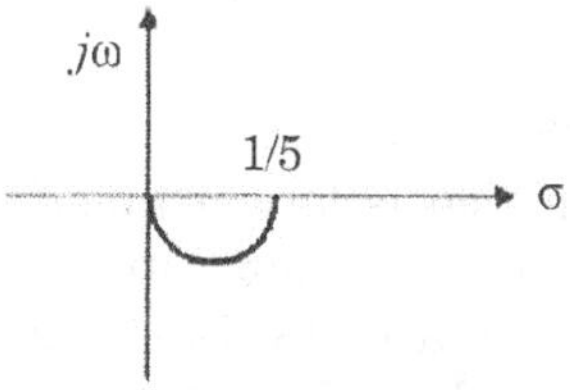

(*d*)
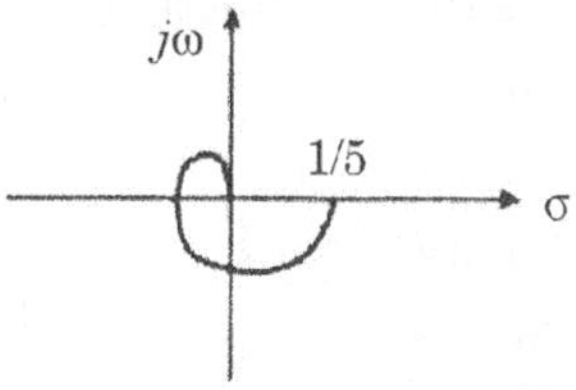

5. The trigonometric Fourier series of an even function does not have the

(*a*) dc term (*b*) cosine terms

(*c*) sine terms (*d*) odd harmonic terms

6. When the output Y in the circuit below is "1", it implies that data has

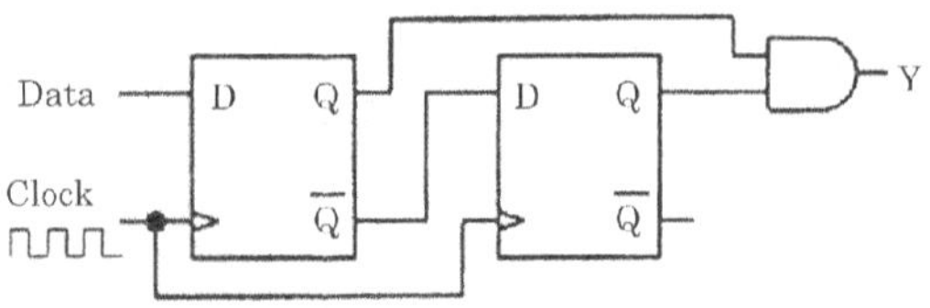

(*a*) changed from "0" to "1"

(*b*) changed from "1" to "0"

(*c*) changed in either direction

(*d*) not changed

7. The logic function implemented by the circuit below is (ground implies a logic "0").

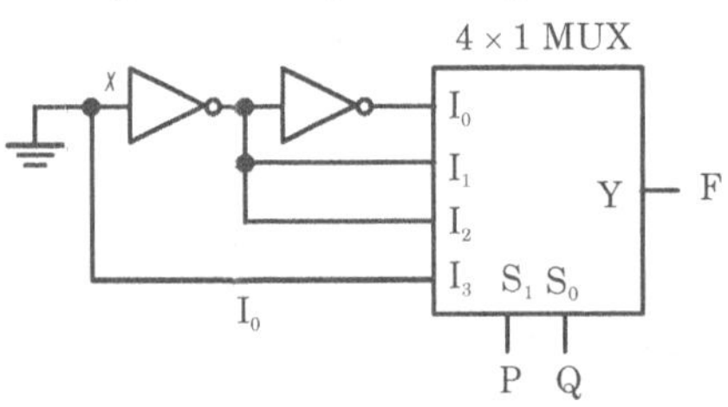

(*a*) F = AND (P, Q) (*b*) F = OR (P, Q)

(*c*) F = XNOR (P, Q) (*d*) F = XOR (P, Q)

8. The circuit below implements a filter between the input current i and the output voltage v_0. Assume that the opamp is ideal. The filter implemented is a.

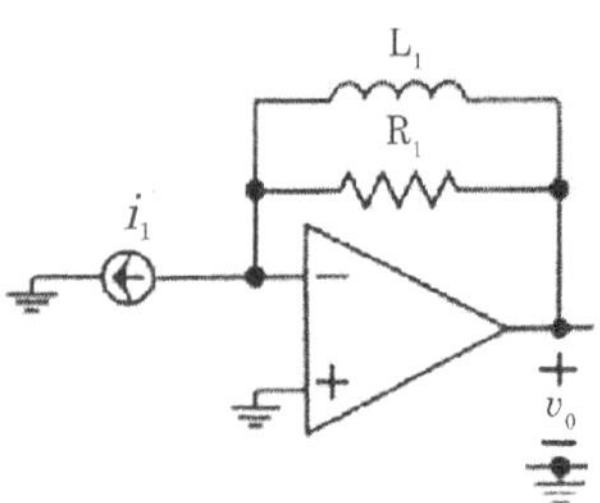

(*a*) low pass filter

(*b*) band pass filter

(*c*) band stop filter

(*d*) high pass filter

9. A silicon PN junction is forward biased with a constant current at room temperature. When the temperature is increased by 10°C, the forward bias voltage across the PN junction

(*a*) increases by 60 mV

(*b*) decreases by 60 mV

(*c*) increases by 25 mV

(*d*) decreases by 25 mV

10. In the circuit shown below, the Norton equivalent current in amperes with respect to the terminals P and Q is

(*a*) $6.4 - j4.8$

(*b*) $6.56 - j7.87$

(*c*) $10 + j0$

(*d*) $16 + j0$

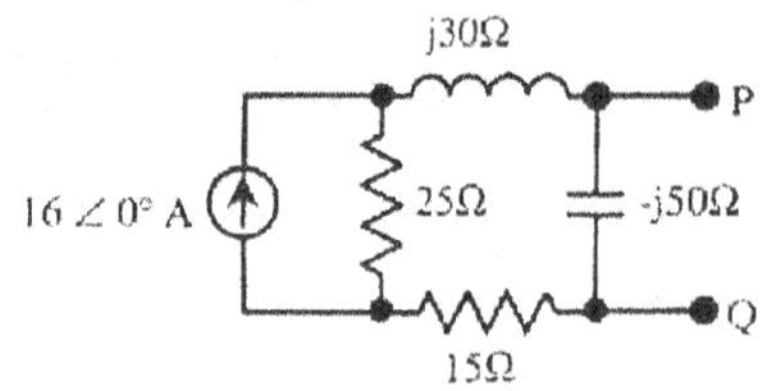

11. In the circuit shown below, the value of R_L such that the power transferred to R_L is maximum is

(*a*) 5 Ω

(*b*) 10 Ω

(*c*) 15 Ω

(*d*) 20 Ω

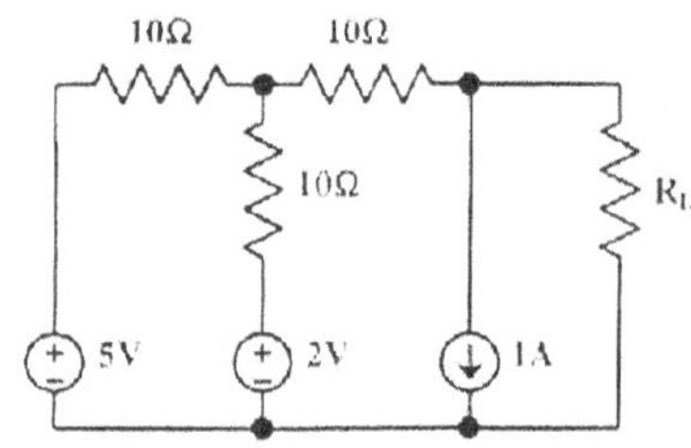

12. The value of the integral $\oint_C \frac{-3z+4}{(z^2+4z+5)} dz$ where c is the circle $|z| = 1$ is given by

(*a*) 0 (*b*) $\frac{1}{10}$

(*c*) $\frac{4}{5}$ (*d*) 1

13. A transmission line of characteristic impedance 50 Ω is terminated by a 50 Ω load. When excited by a sinusoidal voltage source at 10 GHz, the phase difference between two point spaced 2 mm apart on the line is found to be $\frac{\pi}{4}$ radians. The phase velocity of the wave along the line is

(*a*) 0.8×10^8 m/s (*b*) 1.2×10^8 m/s

(*c*) 1.6×10^8 m/s (*d*) 3×10^8 m/s

14. Consider the following statements regarding the complex Poynting vector $\vec{P}$ for the power radiated by a point source in an infinite homogeneous and lossless medium. Re($\vec{P}$) denotes the real part of $\vec{P}$, S denotes a spherical surface whose centre is at the point source, and $\hat{n}$ denotes the unit surface normal on S. Which of the following statements is TRUE?

(*a*) Re($\vec{P}$) remains constant at any radial distance from the source

(*b*) Re($\vec{P}$) increases with increasing radial distance from the source

(*c*) $\oiint_S \text{Re}(\vec{P}).\hat{n}\, dS$ remains constant at any radial distance from the source.

(*d*) $\oiint_S \text{Re}(\vec{P}).\hat{n}\, dS$ remains with increasing radial distance from the source

15. An analog signal is band-limited to 4 kHz, sampled at the Nyquist rate and the samples are quantized into 4 levels. The quantized levels are assumed to be independent and equally probable. If we transmit two quantized samples per second, the information rate is

(*a*) 1 bit/sec (*b*) 2 bits/sec

(*c*) 3 bits/sec (*d*) 4 bits/sec

16. The root locus plot for a system is given below. The open loop transfer function corresponding to this plot is given by

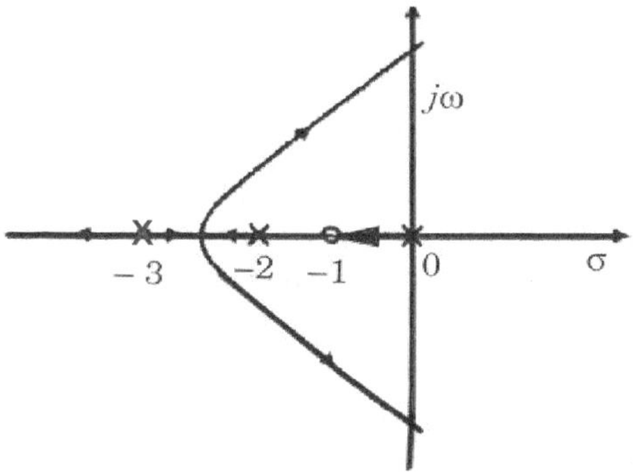

(*a*) $G(s)\,H(s) = k\dfrac{s(s+1)}{(s+2)(s+3)}$

(*b*) $G(s)\,H(s) = k\dfrac{(s+1)}{s(s+2)(s+3)^2}$

(*c*) $G(s)\,H(s) = k\dfrac{1}{s(s-1)(s+2)(s+3)}$

(*d*) $G(s)\,H(s) = k\dfrac{(s+1)}{s(s+2)(s+3)}$

17. A system is defined by its impulse response $h(n) = 2^n u(n-2)$. The system is

(*a*) stable and causal

(*b*) causal but not stable

(*c*) stable but not causal

(*d*) unstable and noncausal

18. If the unit step response of a network is $(1 - e^{-at})$, then its unit impulse response is

(*a*) $a\,e^{-at}$ (*b*) $a^{-1}\,e^{-at}$

(*c*) $(1 - a^{-1})\,e^{-at}$ (*d*) $(1 - a)\,e^{-at}$

19. The output Y in the circuit below is always "1" when

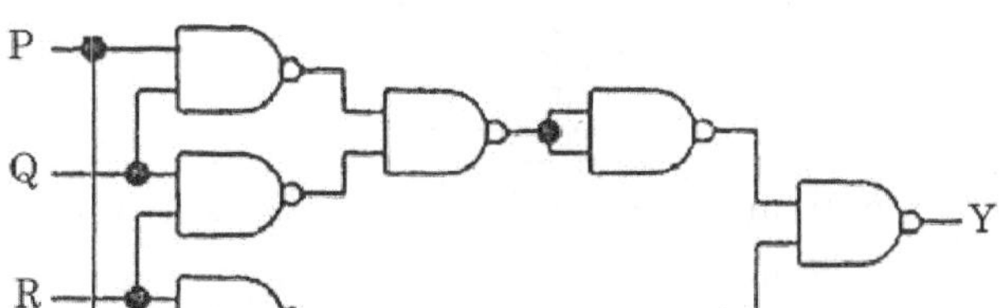

(*a*) two or more of the inputs, P, Q, R are "0"

(*b*) two or more of the inputs P, Q, R are "1"

(*c*) any odd number of the inputs P, Q, R is "0"

(*d*) any odd number of the inputs P, Q, R is "1"

20. In the circuit shown below, capacitors C_1 and C_2 are very large and are shorts at the input frequency, v_i is a small signal input. The gain magnitude $|v_a / v_i|$ at 10 Mrad/s is

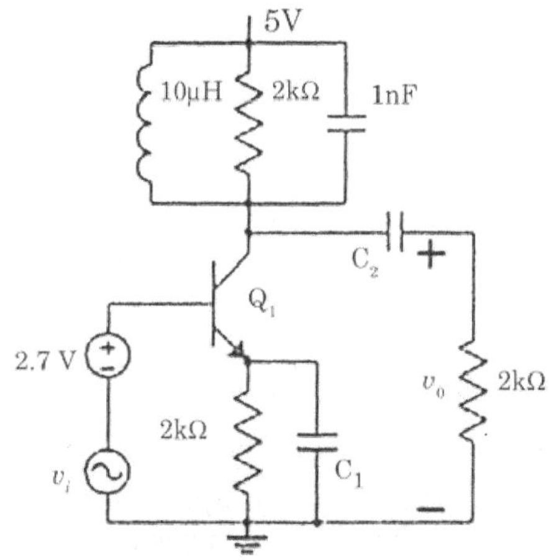

(*a*) maximum (*b*) minimum

(*c*) unity (*d*) zero

21. Drift current in semiconductors depends upon

(*a*) only the electric field

(*b*) only the carrier concentration gradient

(*c*) both the electric field and the carrier concentration

(*d*) both the electric field and the carrier concentration gradient

22. A Zener diode, when used in voltage stabilization circuits, is biased in

(*a*) reverse bias region below the breakdown voltage

(*b*) reverse breakdown region

(*c*) forward bias region

(*d*) forward bias constant current mode

23. The circuit shown below is driven by a sinusoidal input $v_i = V_P \cos(t/RC)$. The steady output v_0 is

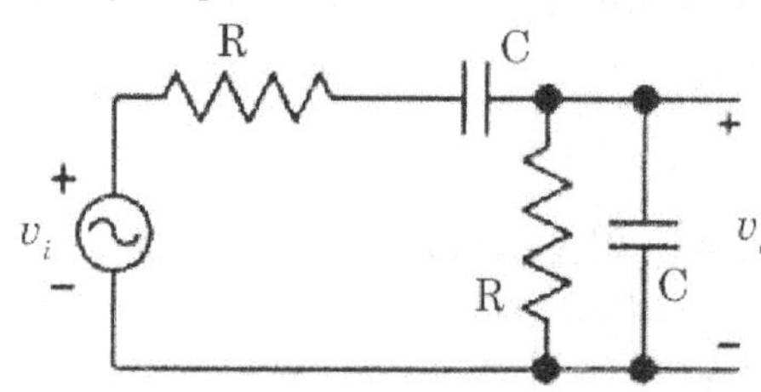

(*a*) $(V_P/3)\cos(t/RC)$ (*b*) $(V_P/3)\sin(t/RC)$

(*c*) $(V_P/2)\cos(t/RC)$ (*d*) $(V_P/2)\sin(t/RC)$

24. Consider a closed surface S surrounding a volume V. If $\vec{r}$ is the position vector of a point inside S with $\hat{n}$ the unit normal on S, the value of the integral $\oiint_S 5\vec{r}\cdot\hat{n}\, dS$ is

(*a*) 3 V (*b*) 5 V

(*c*) 10 V (*d*) 15 V

25. The solution of the differential equation $\frac{dy}{dx} = ky$, $y(0) = c$ is

(*a*) $x = ce^{-ky}$ (*b*) $x = ke^{cy}$

(*c*) $y = ce^{kx}$ (*d*) $y = ce^{-kx}$

Q.(26 – 55) carry two marks each.

26. The electric and magnetic fields for a TEM wave of frequency 14 GHz in a homogeneous medium of relative permittivity ε_r and relative permeability $\mu_r = 1$ are given by

$$\vec{E} = E_P\, e^{j(\omega t - 280\pi y)}\, \hat{u}_z \text{ V/m}$$

$$\vec{H} = 3\, e^{j(\omega t - 280\pi y)}\, \hat{u}_x \text{ A/m}$$

Assuming the speed of light in free space to be 3×10^8 m/s, the initrinsic impedance of free space to be 120π, the relative permittivity ε_r of the medium and the electric field amplitude E_P are

(*a*) $\varepsilon_r = 3$, $E_P = 120\pi$ (*b*) $\varepsilon_r = 3$, $E_P = 360\pi$

(*c*) $\varepsilon_r = 9$, $E_P = 360\pi$ (*d*) $\varepsilon_r = 9$, $E_P = 120\pi$

27. A message signal $m(t) = \cos 2000\,\pi\, t + 4\cos 4000\,\pi t$ modulates the carrier $c(t) = \cos 2\pi f_c t$ where $f_c = 1$ MHz to produce an AM signal. For demodulating the generated AM signal using an envelope detector, the time constant RC of the detector circuit should satisfy

(*a*) 0.5 ms < RC < 1 ms (*b*) 1 μs << RC < 0.5 ms

(*c*) RC << 1 μs (*d*) RC >> 0.5 ms

28. The block diagram of a system with one input u and two outputs y_1 and y_2 is given below.

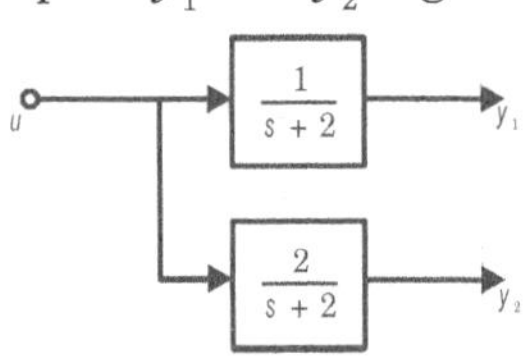

A state space model of the above system in terms of the state vector $\underline{x}$ and the output vector $\underline{y} = [y_1\ y_2]^T$ is

(*a*) $\dot{\underline{x}} = [2]\,\underline{x} + [1]u$; $\underline{y} = [1\ \ 2]\,\underline{x}$

(*b*) $\dot{\underline{x}} = [-2]\,\underline{x} + [1]u$; $\underline{y} = \begin{bmatrix}1\\2\end{bmatrix}\underline{x}$

(*c*) $\dot{\underline{x}} = \begin{bmatrix}-2 & 0\\0 & -2\end{bmatrix}\underline{x} + \begin{bmatrix}1\\1\end{bmatrix}u$; $\underline{y} = [1\ \ 2]\,\underline{x}$

(*d*) $\dot{\underline{x}} = \begin{bmatrix}2 & 0\\0 & 2\end{bmatrix}\underline{x} + \begin{bmatrix}1\\1\end{bmatrix}u$; $\underline{y} = \begin{bmatrix}1\\2\end{bmatrix}\underline{x}$

29. Two systems $H_1(z)$ and $H_2(z)$ are connected in cascade as shown below. The overall output $y(n)$ is the same as the input $x(n)$ with one unit delay. The transfer function of the second system $H_2(z)$ is

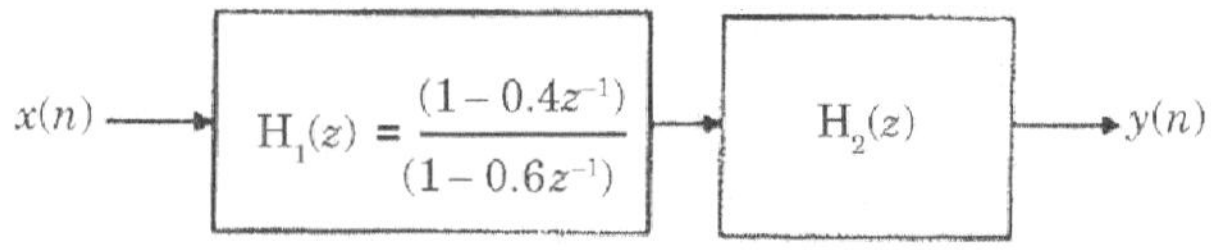

(*a*) $\dfrac{(1-0.6z^{-1})}{z^{-1}(1-0.4z^{-1})}$ (*b*) $\dfrac{z^{-1}(1-0.6z^{-1})}{(1-0.4z^{-1})}$

(*c*) $\dfrac{z^{-1}(1-0.4z^{-1})}{(1-0.6z^{-1})}$ (*d*) $\dfrac{(1-0.4z^{-1})}{z^{-1}(1-0.6z^{-1})}$

30. An 8085 assembly language program is given below. Assume that the carry flag is initially unset. The content of the accumulator after the execution of the program is

```
MVI A, 07 H
RLC
MOV B, A
RLC
RLC
ADD B
RRC
```

(*a*) 8 CH (*b*) 64 H

(*c*) 23 H (*d*) 15 H

31. The first points of the 8-point DFT of a real valued sequence are 5, $1 - j3$, 0, $3 - j4$, 0 and $3 + j4$. The last two points of the DFT are respectively

(*a*) 0, $1 - j3$ (*b*) 0, $1 + j3$

(*c*) $1 + j3$, 5 (*d*) $1 - j3$, 5

32. For the BJT Q_1 in the circuit shown below, $\beta = \infty$, $V_{BEon} = 0.7$V. The switch is initially closed. At time $t = 0$, the switch is opened. The time t at which Q_1 leaves the active region is

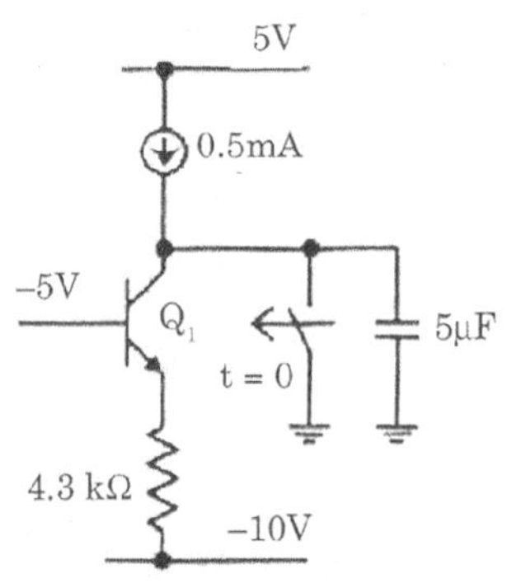

(*a*) 10 ms

(*b*) 25 ms

(*c*) 50 ms

(*d*) 100 ms

33. In the circuit shown below, the network N is described by the following Y matrix :

$$Y = \begin{bmatrix} 0.1S & -0.01S \\ 0.01S & 0.1S \end{bmatrix}.$$ The voltage gain $\frac{V_2}{V_1}$ is

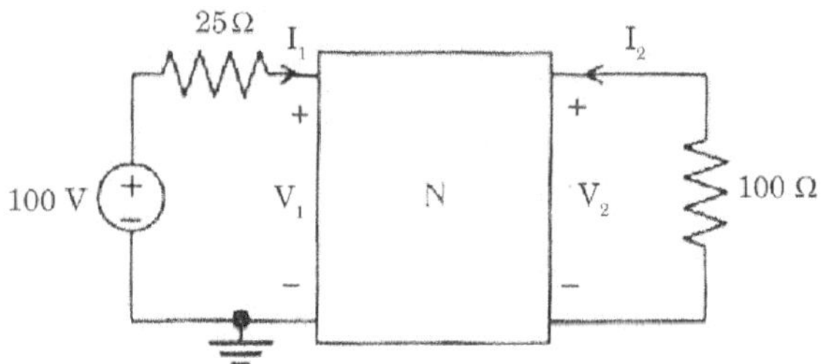

(a) $\frac{1}{90}$ (b) $-\frac{1}{90}$

(c) $-\frac{1}{99}$ (d) $-\frac{1}{11}$

34. In the circuit shown below, the initial charge on the capacitor is 2.5 mC, with the voltage polarity as indicated. The switch is closed at time $t = 0$. The current $i(t)$ at a time t after the switch is closed is

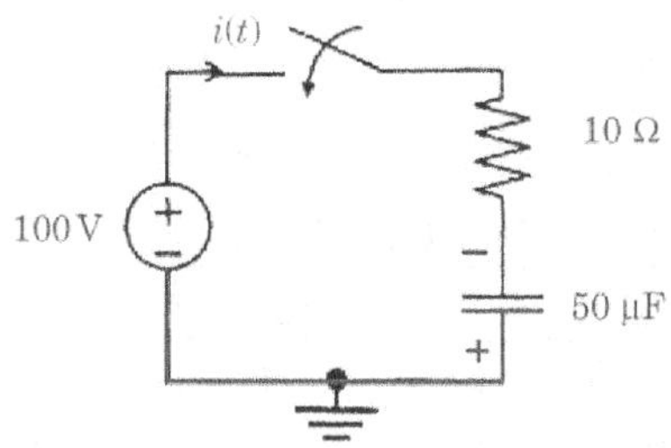

(a) $i(t) = 15 \exp(-2 \times 10^3\, t)$A

(b) $i(t) = 5 \exp(-2 \times 10^3\, t)$A

(c) $i(t) = 10 \exp(-2 \times 10^3\, t)$A

(d) $i(t) = -5 \exp(-2 \times 10^3\, t)$A

35. The system of equations

$x + y + z = 6$

$x + 4y + 6z = 20$

$x + 4y + \lambda z = \mu$

has NO solution for values of λ and μ given by

(a) $\lambda = 6, \mu = 20$

(b) $\lambda = 6, \mu \neq 20$

(c) $\lambda = 6, \mu = 20$

(d) $\lambda \neq 6, \mu \neq 20$

36. A fair dice is tossed two times. The probability that the second toss results in a value that is higher than the first toss is

(a) $\frac{2}{36}$ (b) $\frac{2}{6}$

(c) $\frac{5}{12}$ (d) $\frac{1}{2}$

37. A current sheet $\vec{j} = 10\hat{u}_y$ A/m lies on the dielectric interface $x = 0$ between two dielectric media with $\varepsilon_{r1} = 5$, $\mu_{r1} = 1$ in Region-1 ($x < 0$) and $\varepsilon_{r2} = 2$, $\mu_{r2} = 2$ in Region-2 ($x > 0$). If the magnetic field in Region-1 at $x = 0^-$ is $\overline{H}_1 = 3\hat{u}_x + 30\hat{u}_y$ A/m, the magnetic field in region-2 at $x = 0^+$ is

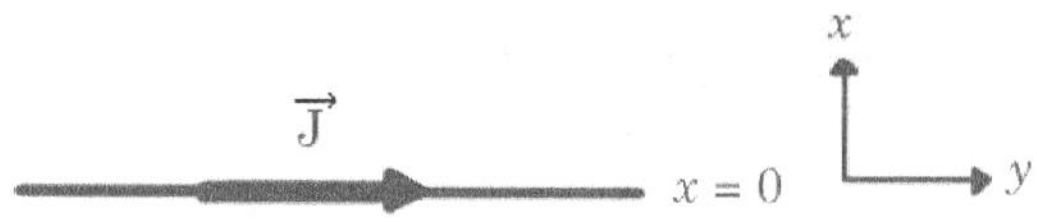

$x > 0$ (Region –1) : $\varepsilon_{r1} = 5$, $\mu_{r1} = 1$

(a) $\vec{H}_2 = 1.5\hat{u}_x + 30\hat{u}_y - 10\hat{u}_z$ A/m

(b) $\vec{H}_2 = 3\hat{u}_x + 30\hat{u}_y - 10\hat{u}_z$ A/m

(c) $\vec{H}_2 = 1.5\hat{u}_x + 40\hat{u}_y$ A/m

(d) $\vec{H}_2 = 3\hat{u}_x + 30\hat{u}_y + 10\hat{u}_z$ A/m

38. A transmission line of characteristic impedance 50 Ω is terminated in a load impedance Z_L. The VSWR of the line is measured as 5 and the first of the voltage maxima in the line is observed at a distance of λ/4 from the load. The value of Z_L is

(a) 10 Ω (b) 250 Ω

(c) $(19.23 + j46.15)$ Ω (d) $(19.23 - j46.15)$ Ω

39. X(t) is a stationary random process with autocorrelation function $R_x(\tau) = \exp(-\pi\tau^2)$. This process is passed through the system shown below. The power spectral density of the output process Y(t) is

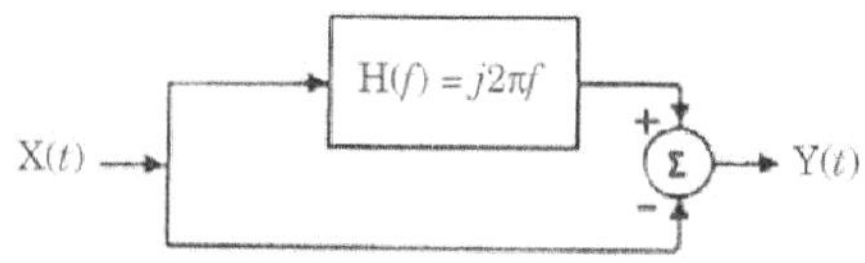

(a) $(4\pi^2 f^2 + 1) \exp(-\pi f^2)$

(b) $(4\pi^2 f^2 - 1) \exp(-\pi f^2)$

(c) $(4\pi^2 f^2 + 1) \exp(-\pi f)$

(d) $(4\pi^2 f^2 - 1) \exp(-\pi f)$

40. The output of a 3-stage Johnson (twisted-ring) counter is fed to a digital-to-analog (D/A) converter as shown in the figure below. Assume all states of the counter to be unset initially. The waveform which represents the D/A converter output V_0 is

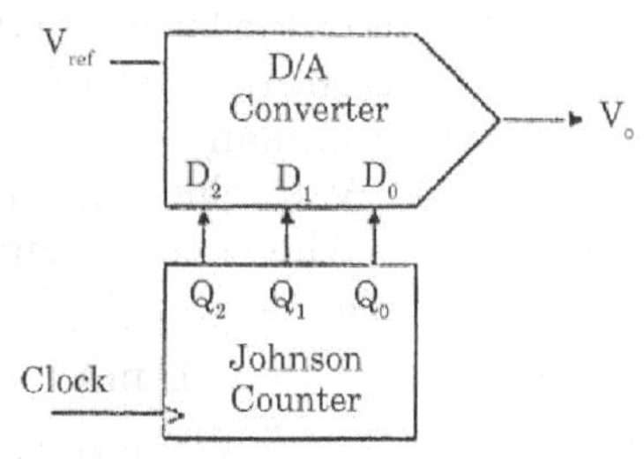

(a) 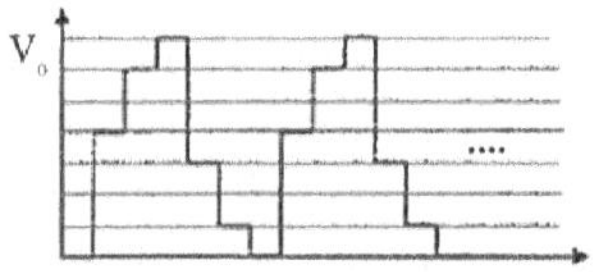

(b) 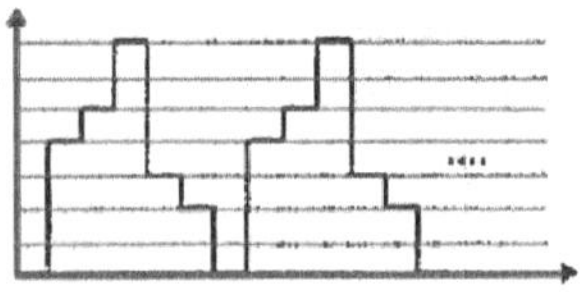

(c) 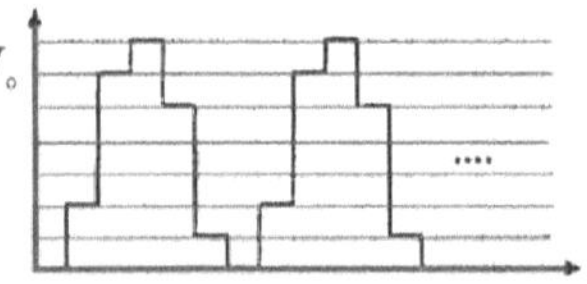

(d) 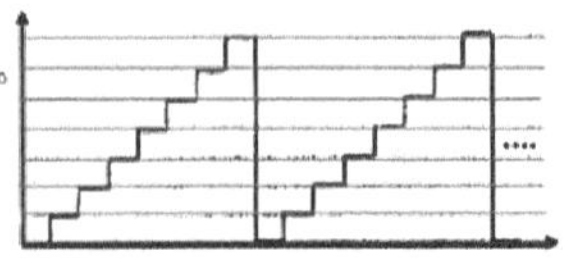

41. Two D flip-flops are connected as a synchronous counter that goes through the following $Q_B\ Q_A$ sequence $00 \rightarrow 11 \rightarrow 01 \rightarrow 10 \rightarrow 00 \rightarrow ...$

The connections to the inputs D_A and D_B are

(a) $D_A = Q_B,\ D_B = Q_A$

(b) $D_A = \vec{Q}_A,\ D_B = \vec{Q}_B$

(c) $D_A = (Q_A \vec{Q}_B + \vec{Q}_A Q_B),\ D_B = Q_A$

(d) $D_A = (Q_A Q_B + \vec{Q}_A \vec{Q}_B),\ D_B = \vec{Q}_B$

42. In the circuit shown below, for the MOS transistors, $\mu_c C_{ox} = 100\ \text{mA/V}^2$ and the threshold voltage $V_T = 1\text{V}$. The voltage V_x at the source of the upper transistor is

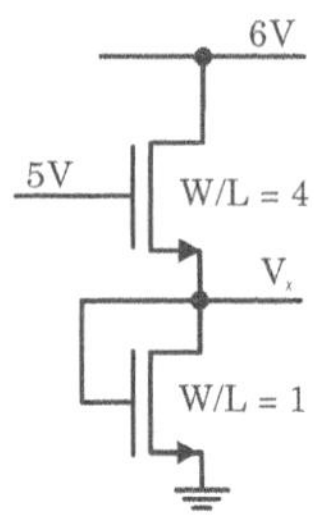

(a) 1 V

(b) 2 V

(c) 3 V

(d) 3.67 V

43. An input $x(t) = \exp(-2t)u(t) + \delta(t-6)$ is applied to an LTI system with impulse response $h(t) = u(t)$. The output is

(a) $[1 - \exp(-2t)]\ u(t) + u(t+6)$

(b) $[1 - \exp(-2t)]\ u(t) + u(t-6)$

(c) $0.5[1 - \exp(-2t)]\ u(t) + u(t+6)$

(d) $0.5[1 - \exp(-2t)]\ u(t) + u(t-6)$

44. For a BJT, the common-base current gain $\alpha = 0.98$ and the collector base junction reverse bias saturation current $I_{CO} = 0.6\ \mu\text{A}$. This BJT is connected in the common emitter mode and operated in the active region with a base drive current $I_B = 20\ \mu\text{A}$. The collector current I_C for this mode of eperation is

(a) 0.98 mA (b) 0.99 mA

(c) 1.0 mA (d) 1.01 mA

45. If $F(s) = L[f(t)] = \dfrac{2(s+1)}{s^2+4s+7}$ then the initial and final values of $f(t)$ are respectively

(a) 0, 2 (b) 2, 0

(c) $0, \dfrac{2}{7}$ (d) $\dfrac{2}{7}, 0$

46. In the circuit shown below, the current I is equal to

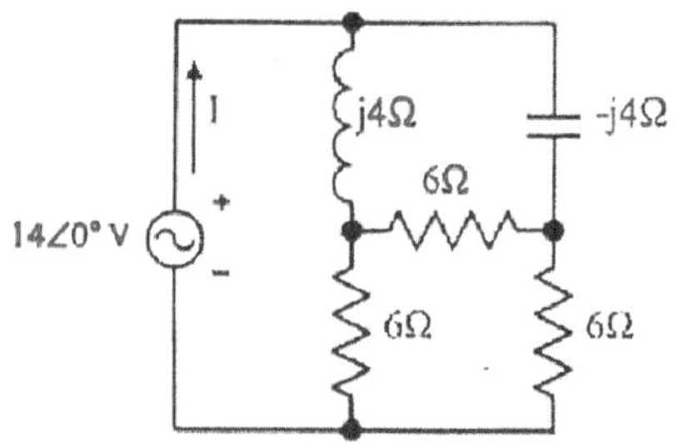

(a) $1.4\angle 0°$ A (b) $2.0\angle 0°$ A

(c) $2.8\angle 0°$ A (d) $3.2\angle 0°$ A

47. A numerical solution of the equation $f(x) = x + \sqrt{x} - 3 = 0$ can be obtained using Newton-Raphson method. If the starting value is $x = 2$ for the iteration, the value of x that is to be used in the next step is

(a) 0.306 (b) 0.739

(c) 1.694 (d) 2.306

COMMON DATA QUESTIONS

Common Data for Questions (48-49):

The channel resistance of an N-channel JFET shown in the figure below is 600 Ω when the full channel thickness (t_{ch}) of 10 μm is available for conduction. The built-in voltage of the gate P^+ N junction (V_{in}) is –1 V. When the gate to source voltage (V_{GS}) is 0 V, the channel is depleted by 1 μm on each side due to the built-in voltage and hence the thickness available for conduction is only 8 μm.

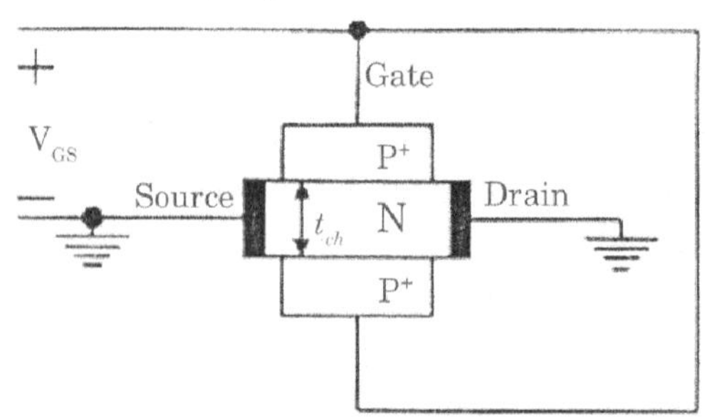

48. The channel resistance when $V_{GS} = 0$ V is

(a) 480 Ω (b) 600 Ω

(c) 750 Ω (d) 1000 Ω

49. The channel resistance when $V_{GS} = -3$ V is

(a) 360 Ω (b) 917 Ω

(c) 1000 Ω (d) 3000 Ω

Common Data for Questions (50-51):

The input-output transfer function of a plant $H(s) = \frac{100}{s(s+10)^2}$. The plant is placed in a unity negative feedback configuration as shown in the figure below.

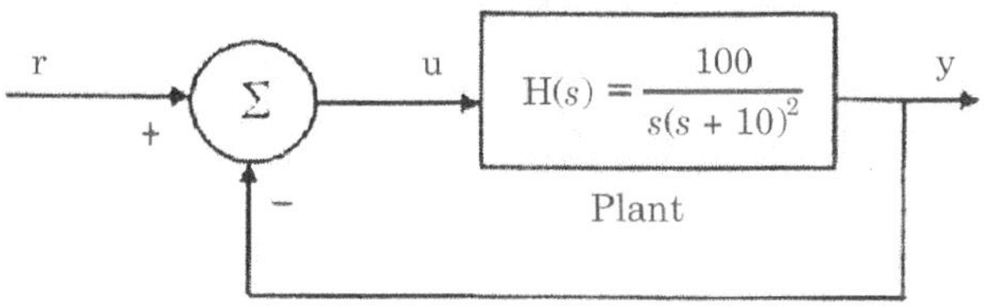

50. The signal flow graph that DOES NOT model the plant transfer function H(*s*) is

(*a*) 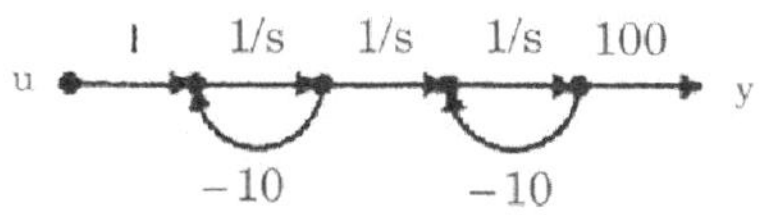

(*b*) 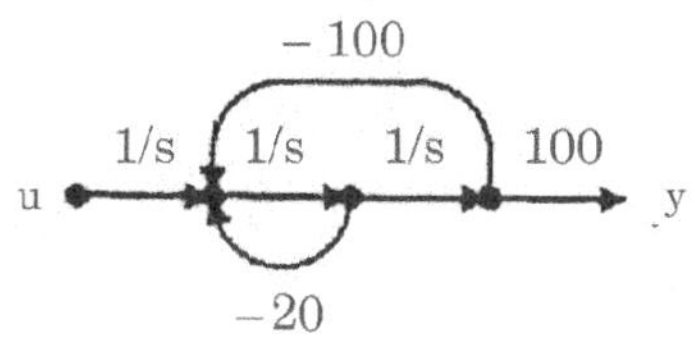

(*c*) 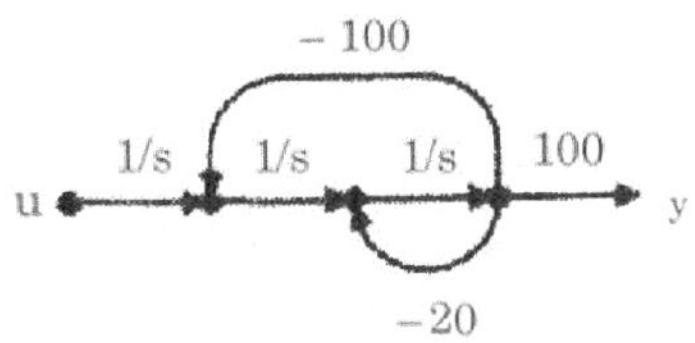

(*d*) 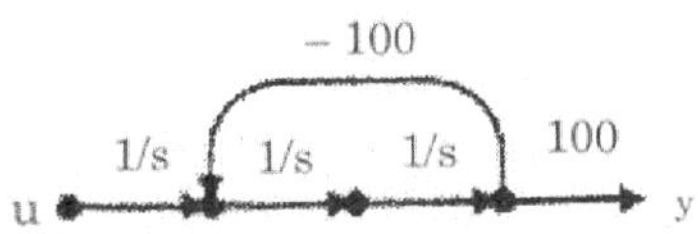

51. The gain margin of the system under closed loop unity negative feedback is

(*a*) 0 dB (*b*) 20 dB

(*c*) 26 dB (*d*) 46 dB

LINKED ANSWER QUESTIONS

Statement for Linked Answer Questions (52-53):

A four-phase and an eight-phase signal constellation are shown in the figure below.

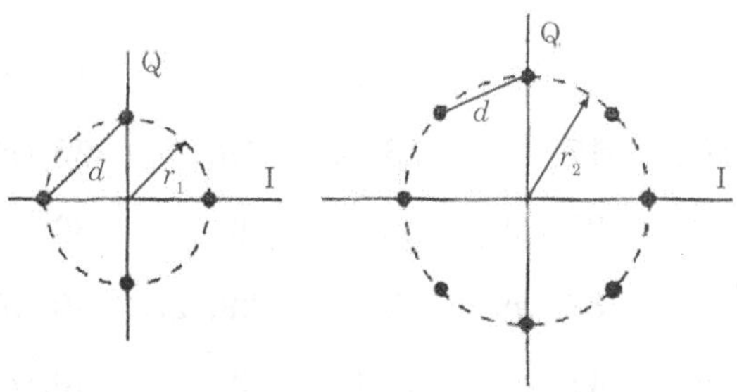

52. For the constraint that the minimum distance between pairs of signal points be *d* for both constellations, the radii r_1 and r_2 of the circles are

(*a*) $r_1 = 0.707d$, $r_2 = 2.782d$

(*b*) $r_1 = 0.707d$, $r_2 = 1.932d$

(*c*) $r_1 = 0.707d$, $r_2 = 1.545d$

(*d*) $r_1 = 0.707d$, $r_2 = 1.307d$

53. Assuming high SNR and that all signals are equally probable, the additional average transmitted signal energy required by the 8-PSK signal to achieve the same error probability as the 4-PSK signal is

(*a*) 11.90 dB (*b*) 8.73 dB

(*c*) 6.79 dB (*d*) 5.33 dB

Statement for Linked Answers Question (54-55):

In the circuit shown below, assume that the voltage drop across a forward biased diode is 0.7 V. The thermal voltage $V_t = kT/q = 25$ mV. The small signal input $v_i = V_P \cos(\omega t)$ where $V_P = 100$ mV.

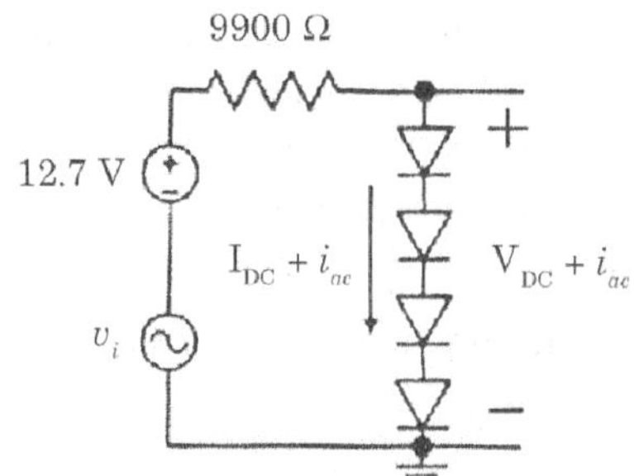

54. The bias current I_{DC} through diodes is

(*a*) 1 mA (*b*) 1.28 mA

(*c*) 1.5 mA (*d*) 2 mA

55. The ac output voltage v_{ac} is

(*a*) $0.25 \cos(\omega t)$ mV (*b*) $1 \cos(\omega t)$ mV

(*c*) $2 \cos(\omega t)$ mV (*d*) $22 \cos(\omega t)$ mV

GENERAL APTITUDE (GA) QUESTIONS

Q.(56 – 60) carry one mark each.

56. The question below consists of a pair of related words followed by four pairs of words. Select the pair that best expresses the relation in the original pair :

Gladiator : Arena

(*a*) dance : stage (*b*) commuter : train

(*c*) teacher : classroom (*d*) lawyer : courtroom

57. There are two candidates P and Q in an election. During the campaign, 40% of the voters promised to vote for P, and rest for Q. However, on the day of election 15% of the voters went back on their promise to vote for P and instead voted for Q. 25% of the voters went back on their promise to vote for Q and instead voted for P. Suppose, P lost by 2 votes, then what was the total number of voters?

(*a*) 100 (*b*) 110

(*c*) 90 (*d*) 95

58. Choose the most appropriate word from the options given below to complete the following sentence :

It was her view that the country's problems had been______ by foreign technocrats, so that to invite them to come back would be counter-productive.

(*a*) identified (*b*) ascertained

(*c*) exacerbated (*d*) analysed

59. Choose the word from the options given below that is most nearly opposite in meaning to the given word:

Frequency

(*a*) periodicity (*b*) rarity

(*c*) gradualness (*d*) persistency

60. Choose the most appropriate word from the options given below to complete the following sentence:

Under ethical guidelines recently adopted by the Indian Medical Association, human genes are to be manipulated only to correct diseases for which ______ treatments are unsatisfactory.

(*a*) similar (*b*) most

(*c*) uncommon (*d*) available

Q.(61-65) carry two marks each.

61. The horse has played a little known but very important role in the field of medicine. Horses were injected with toxins of diseases until their blood built up immunities. Then a serum was made from their blood. Serums to fight with diphtheria and tetanus were developed this way.

It can be inferred from the passage, that horses were

(*a*) given immunity to diseases

(*b*) generally quite immune to diseases

(*c*) given medicines to fight toxins

(*d*) given diphtheria and tetanus serums

62. The fuel consumed by a motorcycle during a journey while traveling at various speeds is indicated in the graph below.

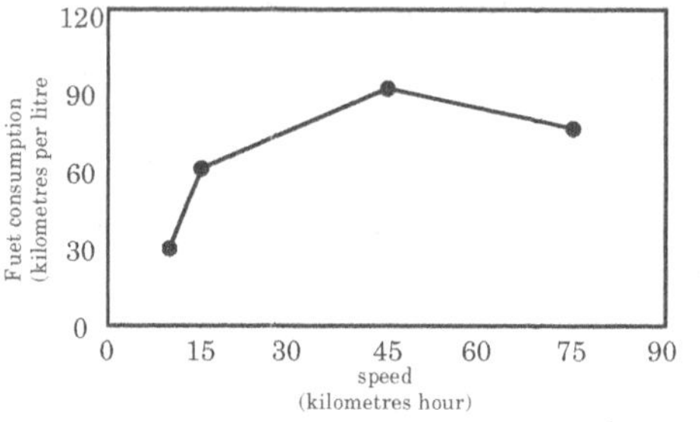

The distances covered during four laps of the journey are listed in the table below

Lap	Distance (kilometres)	Average speed (kilometres per hour)
P	15	15
Q	75	45
R	40	75
S	10	10

From the given data, we can conclude that the fuel consumed per kilometre was least during the lap

(*a*) P (*b*) Q

(*c*) R (*d*) S

63. Three friends, R, S and T shared toffee from a bowl, R took 1/3rd of the toffees, but returned four to the bowl. S took 1/4th of what was left but returned three toffees to the bowl. T took half of the remainder but returned two back into the bowl. If the bowl had 17 toffees left, how many toffees were originally there in the bowl?

(*a*) 38 (*b*) 31

(*c*) 48 (*d*) 41

64. Given that $f(y) = |y|/y$, and q is any non-zero real number, the value of $|f(q) - f(-q)|$ is

(*a*) 0 (*b*) –1

(*c*) 1 (*d*) 2

65. The sum of n terms of the series 4 + 44 + 444 + ... is

(*a*) $(4/81)[10^{n+1} - 9n - 1]$

(*b*) $(4/81)[10^{n-1} - 9n - 1]$

(*c*) $(4/81)[10^{n+1} - 9n - 10]$

(*d*) $(4/81)[10^{n} - 9n - 10]$

ANSWERS

1. (*a*)	2. (*b*)	3. (*a*)	4. (*a*)	5. (*c*)	6. (*a*)	7. (*d*)	8. (*d*)	9. (*d*)	10. (*a*)
11. (*c*)	12. (*a*)	13. (*c*)	14. (*d*)	15. (*d*)	16. (*b*)	17. (*b*)	18. (*a*)	19. (*b*)	20. (*a*)
21. (*c*)	22. (*b*)	23. (*a*)	24. (*d*)	25. (*c*)	26. (*d*)	27. (*b*)	28. (*b*)	29. (*b*)	30. (*c*)
31. (*c*)	32. (*c*)	33. (*d*)	34. (*a*)	35. (*b*)	36. (*c*)	37. (*a*)	38. (*a*)	39. (*a*)	40. (*a*)
41. (*d*)	42. (*c*)	43. (*d*)	44. (*d*)	45. (*b*)	46. (*b*)	47. (*c*)	48. (*c*)	49. (*c*)	50. (*d*)
51. (*c*)	52. (*d*)	53. (*d*)	54. (*a*)	55. (*b*)	56. (*d*)	57. (*a*)	58. (*c*)	59. (*b*)	60. (*d*)
61. (*c*)	62. (*a*)	63. (*c*)	64. (*d*)	65. (*b*)					

EXPLANATIONS

1. For TM_{modes} in rectangular waveguide,

$$\bar{E}_z = E_{02} \sin\left(\frac{m\pi x}{a}\right)\sin\left(\frac{n\pi y}{b}\right) \cdot e^{-j\beta_g z}$$

If either $m = 0$ or $n = 0$, the field, intensities all vaminto So, there is no TM_{01} or TM_{10} mode in rectangular waveguide.

2. Power efficient transmission → FM
Most bandwidth efficient → SSB-SC
Transmission of voice signal
Simplest receives structure → conventional AM
Bandwidth efficient transmission of → VSB
Signals with significant DC component

3. Given : $\frac{100 d^2 y}{dt^2} - \frac{20 dy}{dt} + y = x(t)$
Apply L.T, we get

$$(100s^2 - 20s + 1)\,Y(s) = \frac{1}{s}$$

$$\left[\text{since} \quad x(t) = u(t)\,x(s) = \frac{1}{s}\right]$$

$$\therefore \quad Y(s) = \frac{1}{s(100s^2 - 20s + 1)}$$

Hence poles with positive real part. So, system is unstable.

4. $$|G(jw)|_{w=0} = 5, \quad \angle G(jw)|_{w=0} = 0$$

$$|G(jw)|_{w=\infty} = \infty, \quad \angle G(jw)|_{w=\infty} = \frac{\pi}{2}$$

Hence, the correct option is (*a*)

5. $f(t)$ is even function, hence
where, b_k = coefficient of sine terms, $b_k = 0$

6. When data is 0, Q is 0.
And Q′ is 1 first flip flop
Data is changed to 1
Q is 1 → first 'D′
Q′ is connected to 2nd flip flop
$\therefore \quad Q_2 = 1$
So, inputs of AND gate is 1
$\Rightarrow \quad y = 1$

7. Truth-table for the given circuit is,

$S_1(P)$	$S_0(Q)$	Y
0	0	$I_0 = 0$
0	1	$I_1 = 1$
1	0	$I_2 = 1$
1	1	$I_3 = 0$

k-map for the above

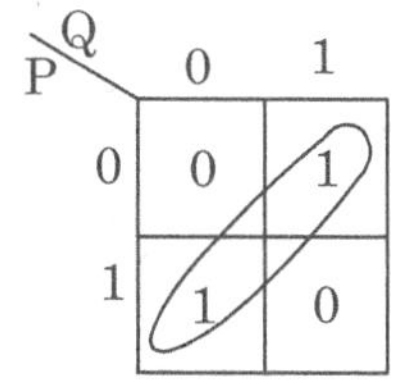

From the *k*-map,

$$Y = P\bar{Q} + \bar{P}Q = X\,OR(PQ)$$

8. When $\omega = 0$; inductor acts as a short circuit so, $V_0 = 0$
When $\omega = \infty$, inductor acts as open circuit
$\Rightarrow \quad V_0 = i_1 R_1$
So, it acts as a high pass filter.

9. For Si forward bias voltage change by – 2.5mV/°C
For 10°C increases,

$$\text{change} = -2.5 \times 10 = -25\text{mV}$$

10. When terminals P and Q are short circuit.
Then circuit becomes

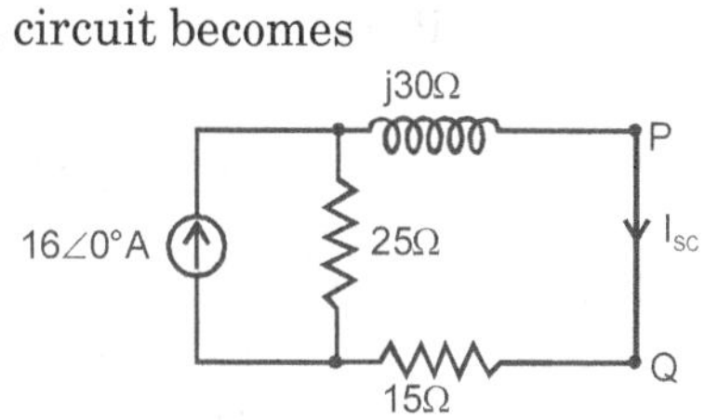

From current division rules

$$I_{SC} = \frac{16(25)}{25 + 15 + j30} = \frac{(16)(25)}{40 + j30}$$

$$= \frac{(16)25)}{10(4 + j3)} = 6.4 - j4.8$$

11. For maximum power transmission

$$R_L = R^*_{TH}$$

The equivalent circuit is drawn here after short-circuting voltage source and open-circuting current source

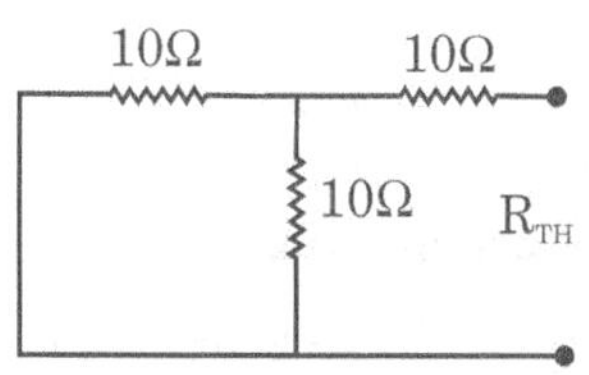

$$\therefore \quad R_{TH} = 10 + \frac{10 \times 10}{20}$$

$$\text{so,} \quad R_{TH} = R_L = 15\Omega$$

12. $$\oint_c \frac{-3z + 4}{z^2 + 4z + 5}\,dz = 0$$

Since $z^2 + 4z + 5 = (z + 2)^2 + 1 = 0$

$z = -2 \pm j$ will be outside the unit circle.

Since the poles lies outside the circle $|z| = 1$, $f(z)$ is analytic every where and $\oint_c f(z).dz = 2\pi i(0) = 0$.

13. $\phi(\text{Phare difference}) = \frac{2\pi}{\lambda} \cdot \Delta x$

Δx = path difference

or, $\frac{\pi}{4} = \frac{2\pi}{\lambda} \cdot \Delta x = \frac{2\pi}{\lambda}[2 \times 10^{-3}]$

or, $\lambda = 16 \times 10^{-3}$m

Then, phase velocity

$$V_P = f\lambda = 10 \times 10^9 \times 16 \times 10^{-3}$$
$$= 1.6 \times 10^8 \text{ m/s}$$

14. $\oiint_s \text{Re}(\vec{P}) \cdot \hat{n} ds$ gives average power and it decreases with increasing radial distance from the source.

15. Entropy, $4(x) = \sum_{i=1}^{4} P_i \log_2 \left(\frac{1}{P_i}\right)$

As quantized level are independent and equiprobable

$$P_1 = P_2 = P_3 = P_4 = \frac{1}{4}$$

then, $H(x) = 4\left[\frac{1}{4}\log_2 4\right] = 2$ bits/sample

And, Information rate, R = Ω4

= 2 sample/sec × 2 bits/sample

= 4 bits/sec.

16. As seen in the roof-locus plot,

zero is at $s = -1$

and poles are at $s = 0, -2, -3$

Angle of Asymptote $= \frac{(2q+1)\pi}{m-n}$

m = No. of poles

n = No. of zeros

$\Rightarrow \frac{\pi}{3} = \frac{\pi}{(m-1)}$

$\Rightarrow m = 4$

Hence, No. of poles at $s = -3$ is two, and both terminates at ∞.

The correct option is (*b*).

17. $h(n) = 2^n u(n-2)$

$h(n)$ is existing for $n > 2$; thus

$h(n) = 0$; $n < 0$; hence causal

$$\sum_{n=-\infty}^{\infty} |h(n) = \sum_{n=\infty}^{\infty} 2^n u(n-2)$$

$$\sum_{n=2}^{\infty} 2^n = \infty\text{; hence system is unstable}$$

18. $S(t) \rightarrow$ step response

Impulse response, $h(t) = \frac{d}{dt}(S(t))$

$$= \frac{d}{dt}\left(1 - e^{-at}\right) = ae^{-at}$$

19. In the circuit

output Y = PQ + PR + RQ

Hence, two or more inputs are '1', Y is always '1'.

20. In the parallel RLC circuit,

L = 10μH and C = 1nF

$$\therefore \quad \omega_g = \frac{1}{\sqrt{LC}} = \frac{1}{\sqrt{10 \times 10^{-6} \times 10^{-9}}}$$

$= 10^7$ rad/s = 10Mrad/s

Hence for a tuned amplifier, gain is maximum at resonant frequency

21. Drift current, $J = \sigma E$

$$J = (n\mu_n + p\mu_p) qE$$

Hence it depends on carrier concentration and electric field.

22. Zener Diode I-V Characteristics

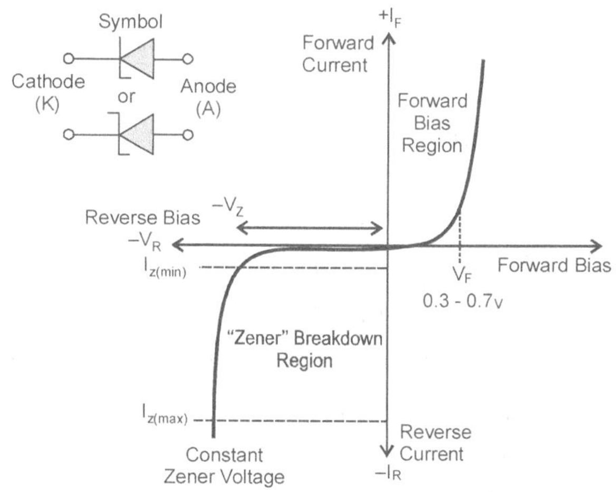

From the I-V characteristics curve above, we can see that the zener diode has a region in its reverse bias characteristics of almost a constant negative voltage regardless of the value of the current flowing through the diode and remains nearly constant even with large changes in current as long as the zener diodes current remains between the breakdown current $I_{z(min)}$ and the maxium current rating $I_{z(max)}$.

This ability to control itself can be used to great effect to regulate or stabilise a voltage source against supply or load variations and this is accomplished when using zener diode in reverse-bias region below the breakdown voltage.

23. $$\frac{V_0}{V_i} = \frac{z_2}{z_1 + z_2}$$

where, $z_2 = R \| \frac{1}{j\omega C}$

$$z_1 = R + \frac{1}{j\omega C}$$

$$z_2 = \frac{R}{R(j\omega C) + 1}$$

Given: $\omega = \frac{1}{RC}$ $\left[\text{since } v_i = v_p \cos\left(\frac{t}{RC}\right)\right]$

$\Rightarrow$ $$z_2 = \frac{R}{1+j}$$

$$z_1 = R + \frac{1}{j\omega c} = R + \frac{R}{j} = R(1-j)$$

$$\frac{v_o}{v_1} = \frac{\frac{R}{1+j}}{\frac{R}{1+j} + R(1-j)} = \frac{1}{1+2} = \frac{1}{3}$$

24. Apply Divergence theorem, we have

$$\oiint_S 5\vec{r}.\vec{n}.dx = \iiint_v 5\nabla.\vec{r}dV$$

$$= 5(3)\iiint_v dv = 15\text{ V}$$

(since $\nabla.\vec{r} = 3$ and $\vec{r}$ are position vector)

25. Given : $y(0) = C$

and $\frac{dy}{dx} = ky$

$\Rightarrow$ $\frac{dy}{y} = kdx$

Taking log, $\ln y = kx + c$

$\Rightarrow$ $y = e^{kx} e^c$

When $y(0) = C$,

$y = k_1 e^0$

$\therefore$ $y = ce^{kx}$ ...($\because k_1 = C$)

26. From the given expressions for $\bar{E}$ and $\bar{H}$

$\beta = 280\pi$

$\Rightarrow$ $\frac{2\pi}{\lambda} = 280\pi, \lambda = \frac{1}{140}m.$

$$v = f\lambda = 14 \times 10^9 \times \frac{1}{140}$$

$$= 1 \times 10^8 \text{ m/s}$$

Now, $v = \frac{C}{\sqrt{\mu_r t_r}} = \frac{C}{\sqrt{t_r}}$

or, $10^8 = \frac{3 \times 10^8}{\sqrt{t_e}}$

$\Rightarrow$ $t_r = 9$

and, $\frac{E_p}{H_p} = \eta = \sqrt{\frac{\mu}{t}} = \sqrt{\frac{\mu_0\mu_2}{t_0 t_r}} = \frac{1}{3}\sqrt{\frac{\mu_0}{t_0}}$

$\Rightarrow$ $E_p = 3 \times \frac{1}{3} \times 120\pi = 120\pi$

27. 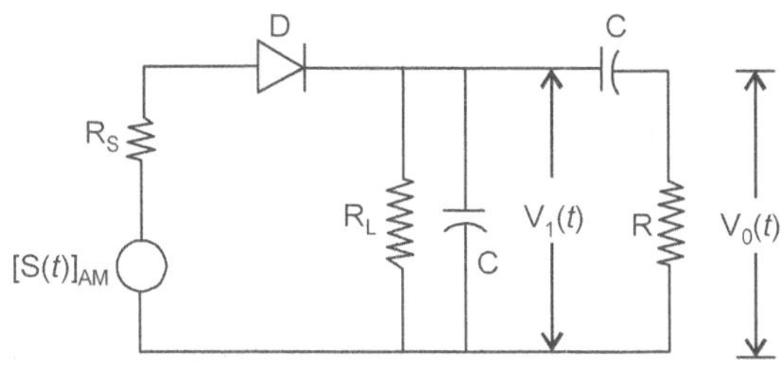

Circuit diagram of Envelope detection

Modulating signals are normaly restricted to a specific frequency range which limits the maximum rate of fall of AM's have amplitude. This rate of range of $m(t)$ depends upon the triphent frequency in $M(f)$.

During discharging f capacitor C, time constant should be large enough so that C does not discharge too much between the positive peaks of the carrier, it should be rather small enough to follow the maximum rate of charge of $m(t)$.

that is, $R_L C << \frac{1}{W}$

Now, during charging, capacitor should be charged in a smallest time possible, for this to be hold,

$$R_s C << \frac{1}{F_C}$$

Hence, the condition for satisfactory operation

$$\frac{1}{f_c} << R_L C << \frac{1}{W}$$

W(highest frequency in $m(t)$) = 2000 Hz

$f_c = 1$ MHz

The, $1\ \mu s << R_L C << 6.5$ ms

28. From the given figure,

$$\frac{Y_1(s)}{U(s)} = \frac{1}{s+2}$$

$\Rightarrow$ $$\frac{X_1(s)}{U(s)} \cdot \frac{Y_1(s)}{X_1(s)} = \frac{1}{s+2} \quad ...(1)$$

and $$\frac{Y_2(s)}{U(s)} = \frac{2}{s+2}$$

$\Rightarrow$ $$\frac{Y_2(s)}{X_2(s)} \cdot \frac{X_2(s)}{U(s)} = \frac{2}{s+2} \quad ...(2)$$

From equation (1), we get

$$\frac{X_1(s)}{U(s)} = \frac{1}{s+2}$$

or, $SX_1(s) + 2X_1(s) = U(s)$

or, $\dot{X}_1(t) + 2X_1(t) = U(t)$ ...(3)

and $\dfrac{Y_1(s)}{X_1(s)} = 1$

or, $Y_1(s) = X_1(s)$

or, $Y_1(t) = X_1(t)$ ...(4)

from equation (2),

$$\frac{X_2(s)}{U(s)} = \frac{L}{s+2}$$

or, $SX_2(s) + 2X_2(s) = 1\ U(s)$

or, $\dot{X}_2(t) + 2X_2(t) = 1\ U(t)$ ...(5)

and $\dfrac{Y_2(s)}{U(s)} = 2$

$Y_2(s) = 2U(s)$

or $X_2(t) = 2U(t)$ ...(6)

Combining equation (3), (4), (5) and (6), we get

$$\begin{bmatrix} \dot{x}_1(t) \\ \dot{x}_2(t) \end{bmatrix} = \begin{bmatrix} -2 & 0 \\ -2 & 0 \end{bmatrix} \begin{bmatrix} x_1(t) \\ x_2(t) \end{bmatrix} + \begin{bmatrix} 1 \\ 1 \end{bmatrix} u(t)$$

or $X_1 = [1\ 2]\ u(t)$

29. Overall transfer function $= z^{-1}$

(since unit delay T.F $= z^{-1}$)

$$H_1(z)\,H_2(z) = z^{-1}$$

$$H_2(z) = \frac{z^{-1}}{H_1(z)} = z^{-1}\frac{(1-0.6z^{-1})}{(1-0.4z^{-1})}$$

30. MVI A, 07 H ⇒ 0000 0111 ⇒ content of A

RLC ⇒ 0000 1110 ⇒ content of A

MOV B, A ⇒ 0000 1110 ⇒ content of B

RLC ⇒ 0001 1100 ⇒ content of B

RLC ⇒ 011 1000 ⇒ content of B

ADD B

A 0000 1110

+

B 0011 1000

0100 0110

RRC → $\dfrac{0010}{2}\ \dfrac{0011}{3}$ 23 H

31. DFT points are complex conjugates of each other and they one symmetric to the middle point.

$x(0) = x^*(7)$

$x(1) = x^*(6)$

$x(2) = x^*(5)$

$x(3) = x^*(4)$

Hence Last two points will be

$x^*(0)$ and $x^*(1) = 1+j3$ and 5

32. Apply KVL at the BE junction

$$I_E = \frac{-5-0.7+10}{4.3k\Omega} = \frac{4.3}{4.3k\Omega} = 1\text{ mA}$$

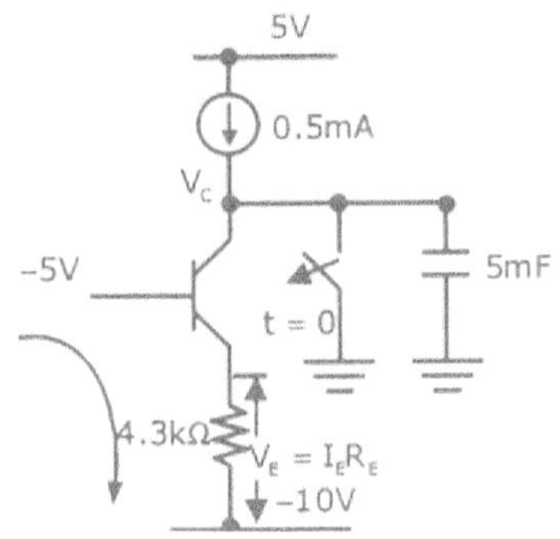

Always $I_E = 1\text{mA}$;

At collector junction

$I_{cap} + (0.5\text{mA}) = 1\text{mA}$ (since $\beta = \infty$; $I_E = I_C$)

$I_{cap} = 1 - 0.5 = 0.5$ mA always constant

$V_{CE} = V_C - V_E$

⇒ $V_C = V_{CE} + V_E$

$= 0.7 + (4.3) \times 10^3 \times 10^{-3}$

$= 0.7 + 4.3$ ...(since $V_E = I_E R_E$)

$V_C = 5V = V_{cap}$

$$V_{cap} = I_{cap}\frac{t}{C}$$

or $$t = \frac{V_{cap}(C)}{I_{cap}} = \frac{(5)\times 5\times 10^{-6}}{0.5\times 10^{-3}} = 50\text{ ms}$$

33. $V_1 = 100V + 25I_1$, $V_2 = -I_2 R_L$

$V_2 = -100I_2$

$I_2 = Y_{z1}V_1 + Y_{z2}V_2$

∴ $-0.01V_2 = 0.01\ V_1 + 0.1V_2$

⇒ $$\frac{V_2}{V_1} = \frac{-1}{11}$$

34.

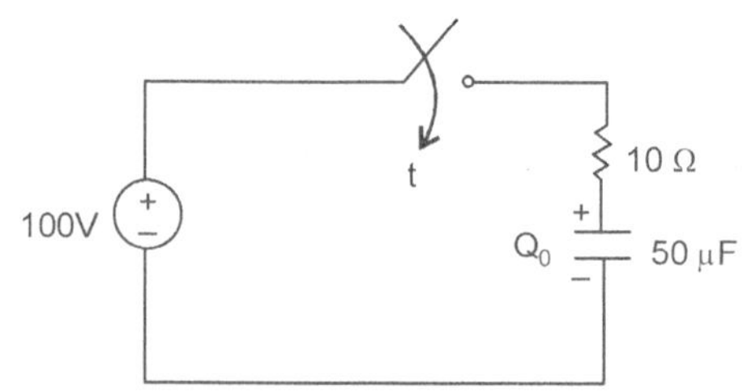

Q_0(Initial charge on capacita)

$= CV_c(0-) = CV_c(0+)$

⇒ $$V_c(0-) = V_c(0+) = \frac{Q_0}{C} = \frac{2.5\times 10^{-3}}{50\times 10^{-6}}$$

$= -50$ volts

The capacitor voltage at time t,

$$V_c(t) = \left[V_c(0+) - V_c(\infty)\right]e^{-t/RC} + V_c(\infty)$$

$V_c(\infty) = 100$ volts

$$V_C(t) = [-50-100]e^{-2\times10^3 t}+100$$
$$= -150e^{-2\times10^3 t}+100$$

[The sign is –ve because the capacitor voltage is in positive direction of the current.]

then, $i(t) = C\dfrac{dv_c(t)}{dt}$

$$= (50\times10^{-6})\times(-150\times-2\times10^3)e^{-2\times10^3 t}$$
$$= +15e^{-2\times10^3 t}$$

35. Given equations are

$$x+y+z=6$$
$$x+4y+6z=20$$

and $x+4y+\lambda z=\mu$

If $\lambda = 6$ and $\mu = 20$, then

$$x+4y+6z=20$$
$$x+4y+6z=20 \text{ infinite solution}$$

If $\lambda = 6$ and $\mu \neq 20$, then

$$x+4y+6z=20$$
$$(\mu \neq 20) \text{ no solution}$$

If $\lambda \neq 6$ and $\mu = 20$, then

$$x+4y+6z=20 \text{ will have solution}$$
$$x+4y+\lambda z=20$$

$\lambda \neq 6$ and $\mu \neq 20$ will also give solution

36. Total number of cause = 36

Total number of favorable causes

$$= 5+4+3+2+1 = 15$$

$\therefore$ Probability $= \dfrac{15}{36} = \dfrac{5}{12}$

(1, 1) (2, 1) (3, 1) (4, 1) (5, 1) (6, 1)

37. $$\overline{H}_{t2} - \overline{H}_{t1} = \overline{J}\times\overline{a}_n$$

$\Rightarrow$ $$\overline{H}_{t2} = \overline{H}_{t1} - 10\hat{u}_z = 30\hat{u}_y - 10\hat{u}_z$$

And $$\overline{B}_{n1} = \overline{B}_{n2}$$

$\therefore$ $$\mu_1\overline{H}_1 = \mu_2\overline{H}_2$$

$\Rightarrow$ $$\overline{H}_2 = \frac{\mu_1}{\mu_2}\overline{H}_1$$

Normal component in x direction

$$\overline{H}_2 = \frac{1}{2}(3)\hat{u}_x = 1.5\hat{u}_x;$$
$$\overline{H}_2 = 1.5\hat{u}_x + 30\hat{u}_y - 10\hat{u}_z \text{ A/m}$$

38. The reflection coefficient at the receiving end,

$$J_l = \frac{Z_L - Z_0}{Z_L + Z_0},\ J_l = |J_l|\cdot e^{J\phi_l}$$

Q_l is the phase-angle between incident & reflected wave.

and standing-wave ratio, $P = \dfrac{1+|J_l|}{1-|J_l|}$

or $$5 = \frac{1+|J_l|}{1-|J_l|}, |J_x| = \frac{2}{3}$$

Now, voltage expression at distance d from the load,

$$V(d) = V_T e^{\beta l}\left[1+|J_l|e^{j(\phi_l-2\beta d)}\right]$$

Given, $V(d)$ is maximum at $d = \dfrac{\lambda}{4}$

then, $$\phi_l - 2\beta\cdot\frac{\lambda}{4} = 0 \text{ for voltage maximum}$$

or, $$\phi_l = \frac{\beta\lambda}{2} = \frac{2\pi}{\lambda}\cdot\frac{\lambda}{2} = \pi$$

then, $$J_l = |J_l|\cdot e^{j\pi} = \frac{-2}{3}$$

and, $$\frac{-2}{3} = \frac{Z_L - Z_0}{Z_L + Z_0}$$

or, $$\frac{-2}{3} = \frac{Z_L - 50}{Z_L + 50}$$

or, $$Z_L = 10\Omega$$

39. In terms of fourier transform,

$$\mathcal{L}[x(t)] = x(f)$$
$$\mathcal{L}[y(t)] = y(f)$$

then, $$y(f) = x(f)H(f) - x(f)$$
$$y(f) = x(f)[H(f)-1] = x(f) - [j2\pi f - 1]$$

Power spectial dencity of $y(f)$,

$$S_y(f) = [j2\pi f - 1]^2\cdot S_u(f)$$

[Using $S_y(f) = |H(f)^2|\cdot S_x f$]

Hence $H(f)$ is the transfer characteristic of the filter.

Given, $$R_x(T) = e^{-(\pi J^2)}$$
$$S_x(f) = e^{-(f^2)} \quad \left[e^{-\pi t^2} \longleftrightarrow e^{-\pi f^2}\right]$$

as $e^{-\pi t^2}$ is a Gaussian function.

then, $$S_y(f) = e^{-\pi f^2}\cdot\left[1+4\pi^2 f^2\right]$$

40. For Johnson counter sequence

D_2	D_1	D_0	V_0
0	0	0	+ 1
1	0	0	+ 4
1	1	0	+ 6
1	1	1	+ 7
0	1	1	+ 3
0	0	1	+ 1
0	0	0	+ 0

41.

Q(Present)		Q(next)			
Q_B	Q_A	Q_B^1	Q_A^1	D_B	D_A
0	0	1	1	1	1
1	1	0	1	0	1
0	1	1	0	1	0
1	0	0	0	0	0

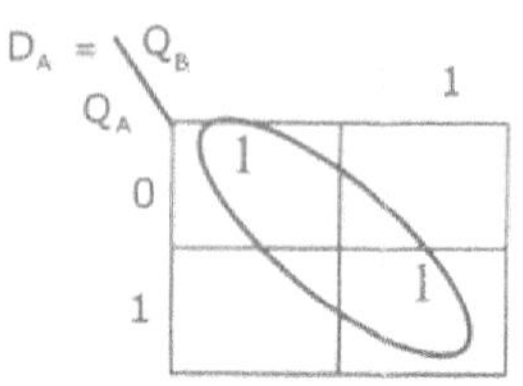

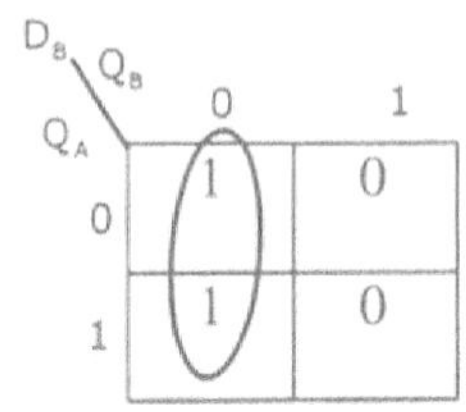

42. Transistor having $\frac{W}{L} = 4$

$$V_{DS} = 6 - V_x \text{ and } V_{GS} = 5 - V_x$$
$$V_{GS} - V_T = 5 - V_x - 1 = 4 - V_x$$
$$V_{DS} > V_{GS} - V_T$$

Hence transistor in saturation region.

Transistor having $\frac{W}{L} = 1$

Drain is connected to gate

Thus transistor in saturation

$$V_{DS} > V_{GS} > V_T \quad (\because V_{DS} = V_{GS})$$

Current flowing in both transistors is same

$$\mu_n C_{0x}\left(\frac{W}{L}\right)_1 \left(\frac{(V_{GS})_1 - V_T}{2}\right)^2$$
$$= \mu_n C_{0x}\left(\frac{W}{L}\right)_2 \cdot \left(\frac{(V_{GS})_2 - V_T}{2}\right)^2$$
$$\Rightarrow 4\frac{(5 - V_x - 1)^2}{2} = 1\frac{(V_x - 1)^2}{2} \quad ...(\because V_{GS} = V_x - 0)$$
$$\Rightarrow 4\left(V_x^2 - 8V_x + 16\right) = V_x^2 - 2V_x + 1$$
$$\Rightarrow 3V_x^2 - 30V_x + 63 = 0$$
$$\Rightarrow V_x = 3V$$

43. Here, $x(s) = \frac{1}{s+2} + e^{-6s}$ and $H(s) = \frac{1}{s}$

$$Y(s) = H(s) \times (s)$$
$$= \frac{1}{s(s+2)} + \frac{e^{-6s}}{s}$$
$$= \frac{1}{2}\frac{1}{s} - \frac{1}{2(s+2)} + \frac{e^{-6s}}{s}$$
$$\Rightarrow y(t) = 0.5\,(1 - e^{-2t})\,u(t) + u(t-6)$$

44. We know $I_C = \beta I_B + (1 + \beta) I_{CBO}$

where, $\beta = \frac{\alpha}{1-\alpha} = \frac{0.98}{1-0.98} = 49$

$$I_B = 20\mu A,\ I_{CBO} = 0.6\mu A$$
$$\therefore \quad I_C = 1.01 \text{ mA}$$

45. Initial value of $f(t)$

$$f(0) = \lim_{t\to 0} f(t) = \lim_{s\to\infty} sf(s) = \lim_{s\to\infty} \frac{2s(s+1)}{s^2+4s+7}$$
$$= \lim_{s\to\infty} \frac{2\left(1+\frac{1}{s}\right)}{1+\frac{4}{s}+\frac{7}{s^2}} = 2$$

final vlaue of $f(t)$

$$f(\infty) = \lim_{t\to\infty} f(t) = \lim_{s\to 0} sf(s)$$
$$= \lim_{s\to 0} \frac{2s(s+1)}{s^2+4s+7} = 0$$

46. Applying delta – to – star conversion circuit becomes

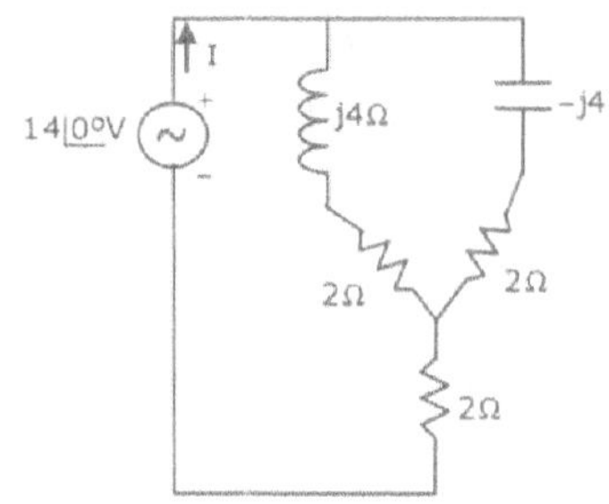

Net impedance = $(2 + j4) \parallel (2 - j4) + 2$

$$= \frac{4+16}{4} + 2 = 7\Omega$$
$$I = \frac{14\angle 0^\circ}{7} = 2\angle 0^\circ A$$

47.
$$X_{n+1} = x_n - \frac{f(x_n)}{f'(x_n)}$$
$$f(x_0) = f(2) = \left(2 + \sqrt{2} - 3\right) = \sqrt{2} - 1$$

and, $f'(x_0) = f'(2) = 1 + \frac{1}{2\sqrt{2}} = \frac{2\sqrt{2}+1}{2\sqrt{2}}$,

$$X_1 = X_0 - \frac{f(X_0)}{f'(X_0)}$$
$$\Rightarrow \quad X_1 = 2 - \frac{\left(\sqrt{2}-1\right)}{\frac{2\sqrt{2}+1}{2\sqrt{2}}} = 1.694$$

48.

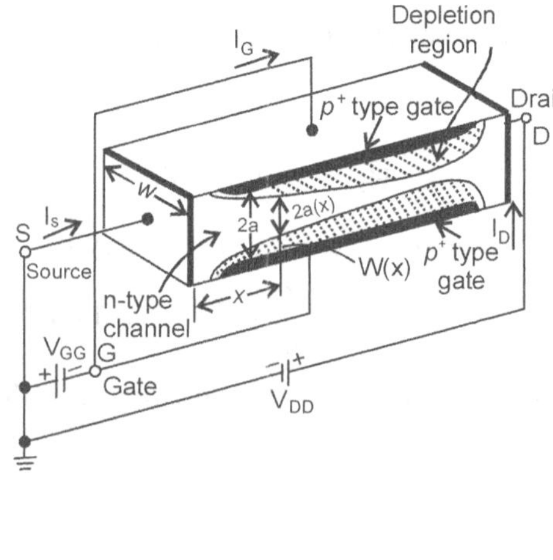

Depletion layer width, $W(x) = a - b(x)$

$$= \frac{2t}{qN_0}\left[V_0 - V(x)\right]^{1/2}$$

V_0 is the build in voltage.

or, $W(x) \alpha [V_0 - V(x)]^{1/2}$

$V(x)$ is the applied voltage and is a negative number for an applied reverse bias.

For $V_{as} = 0$ volts,

$W(x) = W_1$

and, For $V_{as} = 3$ volts,

$W(x) = W_2$

then, $\frac{W_1(x)}{W_2(x)} = \sqrt{\frac{1}{1-(-3)}} = \sqrt{\frac{1}{4}} = \frac{1}{2}$

$\Rightarrow$ $W_2(x) = 2W_1(x)$

(*a*) For, $V_{as} = 0$,

Channel width = 10 m – 2 m = 8 m

and, as channel resistance $\alpha = \frac{1}{\text{channel width}}$

$\Rightarrow$ $\frac{r_{d_1}}{r_{d_2}} = \frac{8\mu m}{10\mu m}$

$\Rightarrow$ $r_{d_2} = \frac{10}{8} \times 600 = 750\ \Omega$

(*b*) For, $V_{as} = -3$,

$W_2(x) = 4$ m

Channel width = (10 – 4) m = 6 m

then, $\frac{r_{d_3}}{r_{d_4}} = \frac{10}{6}$

or, $r_{d_3} = \frac{10}{6} \times 600 = 1000\Omega$

49. Width of depletion large,

$$W \propto \sqrt{V_{bi} + V_{GS}}$$

$\therefore$ $\frac{W_2}{W_1} = \sqrt{\frac{-1-3}{-1}}$

$\Rightarrow$ $W_2 = 2W_1 = 2(1\mu m) = 2\mu m$

$\therefore$ Channel thickness = 10 – 4 = 6μm

Now 8μm $\rightarrow$ 750Ω

Hence for 6 μm, $r_d = \frac{8}{6} \times 750 = 1000\ \Omega$

50.

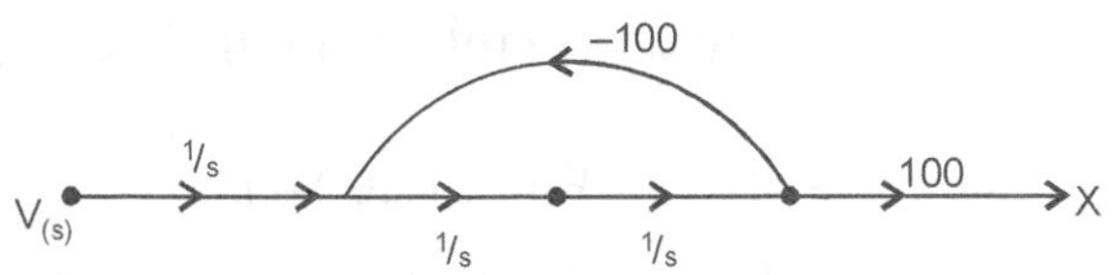

Forward path, $P_1 = \frac{100}{s^3}$

Individual loop gain, $P_{11} = \frac{-100}{s^2}$

then, transfer f^n, $T(s) = \frac{P_1\Delta_1}{1-P_{11}} = \frac{\frac{100}{s^3}}{1+\frac{100}{s^2}}$

[$\Delta = 1$ as forward path touches the loop]

$$T(s) = \frac{100}{s(s^2+100)}$$

51. $H(s) = \frac{100}{s(s+10)^2}$

Phase cross over frequency

$$= -90 - 2\tan^{-1}\left(\frac{\omega}{10}\right) = -180^\circ$$

$\Rightarrow$ $-2\tan^{-1}\left(\frac{\omega}{10}\right) = -90^\circ$

$\Rightarrow$ $\tan^{-1}\left(\frac{\omega}{10}\right) = 45^\circ$

$\Rightarrow$ $\omega = 10$ rad/sec

$$|H(jw)| = \frac{100}{|j10|(j10+10)^2} = \frac{1}{10.2} = \frac{1}{20}$$

$$GM = 20\log\frac{1}{1/20} = 20\log 20 = 26\text{ dB}$$

52. For M-ary signal constellator;

$$d = 2\sin\left(\frac{\pi}{M}\right)\cdot\sqrt{E_s}$$

(VE_s is the distance of any point from origin)

For, 4-ary; $d_1 = 2\sin\left(\frac{\pi}{4}\right)\cdot r_1$

and for 8-ary, $d_2 = 2\sin\left(\frac{\pi}{4}\right)\cdot r_2$

given, $d_1 = d_2 = d$

$$r_1 = \frac{d}{2\sin\frac{\pi}{4}} = 0.707d$$

and $$r_2 = \frac{d}{2\sin\frac{\pi}{8}} = 1.307d$$

53. Probability error, $P_e \alpha \sqrt{E_s}$

Then, $\frac{P_{e_1}}{P_{e_2}} = \frac{\sqrt{E_{s_1}}}{\sqrt{E_{s_2}}} = \frac{r_1}{r_2}$

$\Rightarrow$ $\frac{E_{s_1}}{E_{s_2}} = \left(\frac{r_1}{r_2}\right)^2 = \left(\frac{1.307}{6.707}\right)^2 = 3.42$

For having same error, probability,

P signal energy for 4-PSK signal should be 3.42 greater that of 8-PSK signal.

In $dB = 10\log 3.42 = 5.33\ dB$

54. $I_{DC} = \frac{12.7-(0.7+0.7+0.7+0.7)}{9900} = 1\text{mA}$

For DC biasing, considering only DC voltage source.

55. AC dynamic resistance,

$$r_d = \frac{V_T}{I_D} = \frac{25mV}{1mA} = 25\Omega$$

Assume $\eta = 1$ (since forward drop = 0.7V)

Ac dynamic resistance offered by each diode = 25Ω

$$\therefore \quad V_{ac} = V_i(ac)\left[\frac{4 \times 25\Omega}{9900 + 100}\right]$$

$$= 100 \times 10^{-3} \cos \omega t \left[\frac{100}{10000}\right]$$

$$= 1\cos(\omega t) \text{ mV}$$

56. Given relationship is worker: workplace.

A gladiator is

(*i*) a person, usually a professional combatant trained to entertain the public by engaging in mortal combat with another person or a wild.

(*ii*) A person engaged in a controversy or debate, especially in public.

57.

P	Q
40%	60%
–6%	+6%
+15%	–15%
49%	51%

since 2% = 2

∴ 100% = 100

58. Clues in the question are -- foreign technocrats did something negatively to the problems – so it is counter-productive to invite them. All other options are non-negative. The best choice is exacerbated which means aggravated or worsened.

59. The best antonym here is rarity which means shortage or scarcity.

60. The context seeks to take a deviation only when the existing / present / current / alternative treatments are unsatisfactory. So the word for the blank should be a close synonym of existing / present/current/ alternative. Available is the closest of all.

61. From the passage it cannot be inferred that horses are given immunity as in (*a*), since the aim is to develop medicine and in turn immunize humans. (*b*) is correct since it is given that horses develop immunity after some time. Refer "until their blood built up immunities". Even (*c*) is invalid since medicine is not built till immunity is developed in the horses. (*d*) is incorrect since specific examples are cited to illustrate and this cannot capture the essence.

62.

Fuel	consumption	Actual
P	60 km/*l*	$\frac{15}{60} = \frac{1}{4}l$
Q	90 km/*l*	$\frac{75}{90} = \frac{5}{6}l$
R	75 km/*l*	$\frac{40}{75} = \frac{8}{15}l$
S	30 km/*l*	$\frac{10}{30} = \frac{1}{3}l$

63. Let total number of toffees in bowl be x

R took $\frac{1}{3}$ of toffees and returned 4 to the bowl

∴ Number of toffees with R = $\frac{1}{3}x - 4$

Remaining of toffees in bowl = $\frac{2}{3}x + 4$

Number of toffees with S = $\frac{1}{4}\left(\frac{2}{3}x+4\right) - 3$

Remaining toffees in bowl = $\frac{3}{4}\left(\frac{2}{3}x+4\right) + 3$

Number of toffees with T = $\frac{1}{2}\left(\frac{3}{4}\left(\frac{2}{3}x+4\right)+3\right) - 2$

Remaining toffees in bowl = $\frac{1}{2}\left[\frac{3}{4}\left(\frac{2}{3}x+4\right)+4\right]+2$

$$\text{Given :} \quad \frac{1}{2}\left[\frac{3}{4}\left(\frac{2}{3}x+4\right)+3\right] + 2 = 17$$

$$\Rightarrow \quad \frac{3}{4}\left(\frac{2}{4}x+4\right) = 27$$

$$\Rightarrow \quad x = 48$$

64. Given: $f(y) = \frac{|y|}{y}$

$$\Rightarrow \quad f(q) = \frac{|q|}{q};$$

$$f(-q) = \frac{|-q|}{-q} = \frac{-|q|}{q}$$

$$|f(q) - f(q)| = \frac{|q|}{q} + \frac{|q|}{q} = \frac{2|q|}{q} = 2$$

65. Let $S = 4(1 + 11 + 111 + ...) = \frac{4}{9}(9 + 99 + 999 + ...)$

$$= \frac{4}{9}\{(10-1) + (10^2-1) + (10^3-1) + \text{.........}\}$$

$$= \frac{4}{9}\{(10 + 10^2 + ...+ 10^n) - n\}$$

$$= \frac{4}{9}\left\{10\frac{(10^n-1)}{9} - n\right\}$$

$$= \frac{4}{81}\{10^{n+1} - 9n - 10\}$$

SOLVED PAPER – 2010

ELECTRONICS AND COMMUNICATION ENGINEERING

Time : 3 Hours **Maximum Marks : 100**

Q.(1 – 25) carry one mark each.

1. The eigenvalues of a skew-symmetric matrix are
   (a) always zero
   (b) always pure imaginary
   (c) either zero or pure imaginary
   (d) always real

2. The trigonometric Fourier series for the waveform $f(t)$ shown below contains

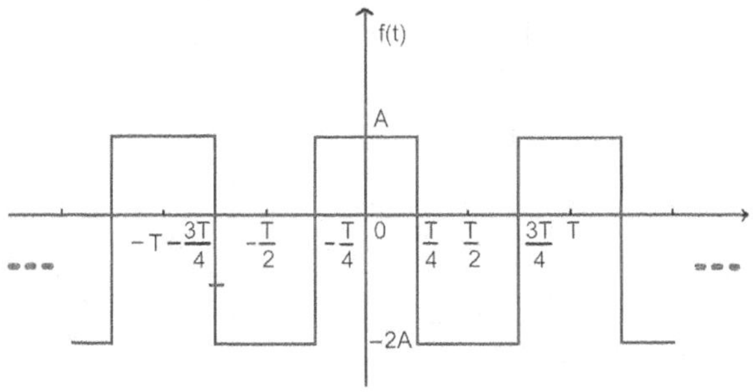

   (a) only cosine terms and zero value for the dc component
   (b) only cosine terms and a positive value for the dc component
   (c) only cosine terms and a negative value for the dc component
   (d) only sine terms and a negative value for the dc component

3. A function $n(x)$ satisfies the differential equation

$$\frac{d^2n(x)}{dx^2} - \frac{n(x)}{L^2} = 0$$ where L is a constant. The boundary conditions are $n(0) = K$ and $n(\infty) = 0$. The solution to this equation is
   (a) $n(x) = K \exp(x/L)$
   (b) $n(x) = K \exp(-x/\sqrt{L})$
   (c) $n(x) = K^2 \exp(-x/L)$
   (d) $n(x) = K \exp(-x/L)$

4. For the two-part network shown below, the short-circuit admittance parameter matrix is

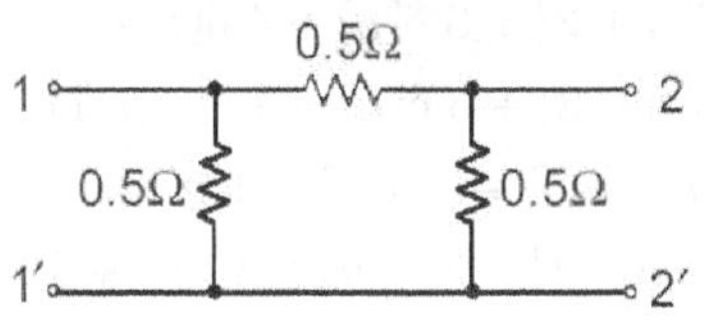

   (a) $\begin{bmatrix} 4 & -2 \\ -2 & 4 \end{bmatrix} S$ (b) $\begin{bmatrix} 1 & -0.5 \\ -0.5 & 1 \end{bmatrix} S$
   (c) $\begin{bmatrix} 1 & 0.5 \\ 0.5 & 1 \end{bmatrix} S$ (d) $\begin{bmatrix} 4 & 2 \\ 2 & 4 \end{bmatrix} S$

5. For a parallel RLC circuit, which one of the following statements is not correct?
   (a) The bandwidth of the circuit decreases if R is increased
   (b) The bandwidth of the cicuit remains same if L is increased
   (c) At resonance, input impedance is a real quantity
   (d) At resonance, the magnitude of input impedance attains its minimum value

6. At room temperature, a possible value for the mobility of electrons in the inversion layer of a silicon n-channel MOSFET is
   (a) 450 cm^2/V-s (b) 1350 cm^2/V-s
   (c) 1800 cm^2/V-s (d) 3600 cm^2/V-s

7. Thin gate oxide in a CMOS process is preferably grown using
   (a) wet oxidation (b) dry oxidation
   (c) epitaxial deposition (d) ion implantation

8. In the silicon BJT circuit shown below, assume that the emitter area of transistor Q1 is half that of transistor Q2.

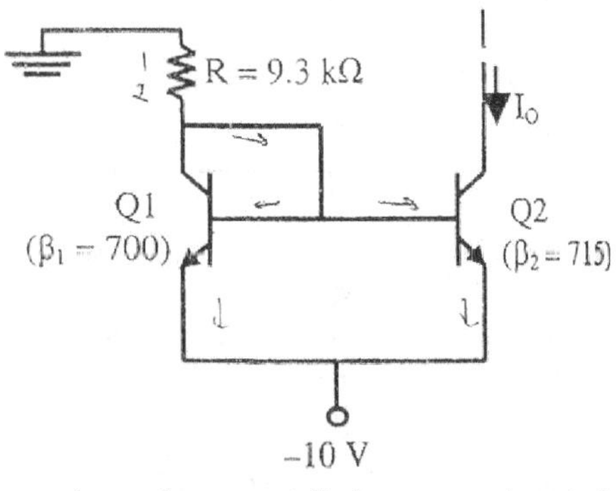

The value of current I_0 is approximately
   (a) 0.5 mA
   (b) 2 mA
   (c) 9.3 mA
   (d) 15 mA

9. The amplifier circuit shown below uses a silicon transistor. The capacitors C_C and C_E can be assumed to be short at signal frequency and the effect of output resistance r_0 can be ignored. If C_E is disconnected from the circuit, which one of the following statements is TRUE?

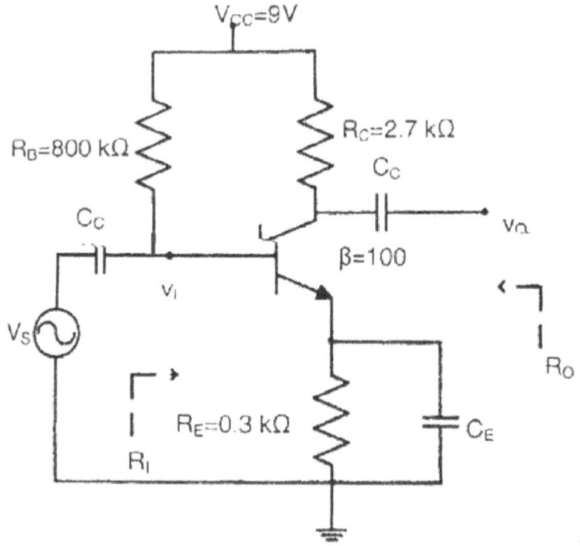

(*a*) The input resistance R_i increases and the magnitude of voltage gain A_v decreases

(*b*) The input resistance R_i decreases and the magnitude of voltage gain A_v increases

(*c*) Both input resistance R_i and the magnitude of voltage gain A_v decrease

(*d*) Both input resistance R_i and the magnitude of voltage gain A_v increase

10. Assuming the OP–AMP to be ideal, the voltage gain of the amplifier shown below is

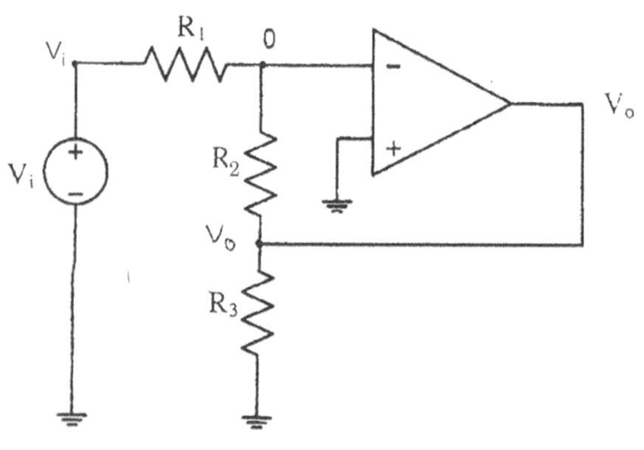

(*a*) $-\frac{R_2}{R_1}$ (*b*) $-\frac{R_3}{R_1}$

(*c*) $-\left(\frac{R_2 \parallel R_3}{R_1}\right)$ (*d*) $-\left(\frac{R_2 + R_3}{R_1}\right)$

11. Match the logic gates in **Column A** with their equivalents in **Column B**.

Column A Column B

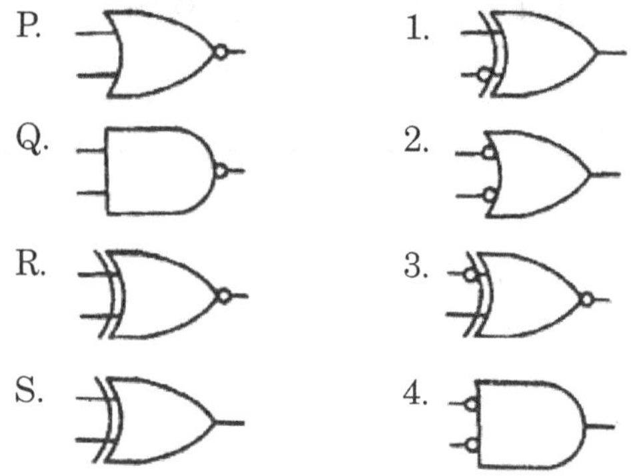

(*a*) P-2, Q-4, R-1, S-3 (*b*) P-4, Q-2, R-1, S-3

(*c*) P-2, Q-4, R-3, S-1 (*d*) P-4, Q-2, R-3, S-1

12. For the output F to be 1 in the logic ciruit shown, the input combination should be

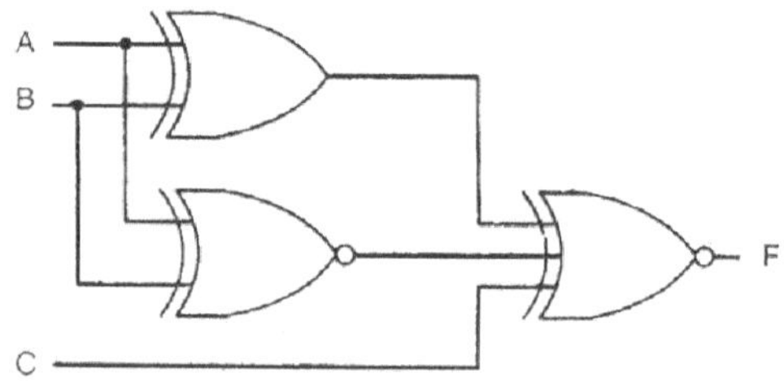

(*a*) A = 1, B = 1, C = 0 (*b*) A = 1, B = 0, C = 0

(*c*) A = 0, B = 1, C = 0 (*d*) A = 0, B = 0, C = 1

13. In the circuit shown, the device connected to Y5 can have address in the range

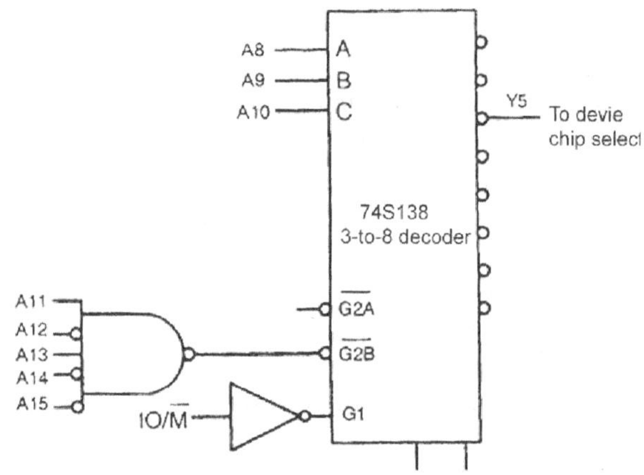

(*a*) 2000 – 20FF (*b*) 2D00 – 2DFF

(*c*) 2E00 – 2EFF (*d*) FD00 – FDFF

14. Consider the *z*-transform

$$X(z) = 5z^2 + 4z^{-1} + 3;\ 0 < |z| < \infty.$$

The inverse z-transform $x[n]$ is

(*a*) $5\ \delta[n + 2] + 3\ \delta[n] + 4\ \delta[n - 1]$

(*b*) $5\ \delta[n - 2] + 3\ \delta[n] + 4\ \delta[n + 1]$

(*c*) $5\ u[n + 2] + 3\ u[n] + 4\ u[n - 1]$

(*d*) $5\ u[n - 2] + 3\ u[n] + 4\ u[n + 1]$

15. Two discrete time systems with impulse responses $h_1[n] = \delta[n-1]$ and $h_2[n] = \delta[n-2]$ are connected in cascade. The overall impulse response of the cascaded system is

(a) $\delta[n-1] + \delta[n-2]$ (b) $\delta[n-4]$

(c) $\delta[n-3]$ (d) $\delta[n-1]\,\delta[n-2]$

16. For an N-point FFT algorithm with $N = 2^m$, which one of the following statements is TRUE?

(a) It is not possible to construct a signal flow graph with both input and output in normal order

(b) The number of butterflies in the mth stage is N/m

(c) In-place computation requires storage of only 2N node data

(d) Computation of a butterfly requires only one complex multiplication

17. The transfer function Y(s)/R(s) of the system shown is

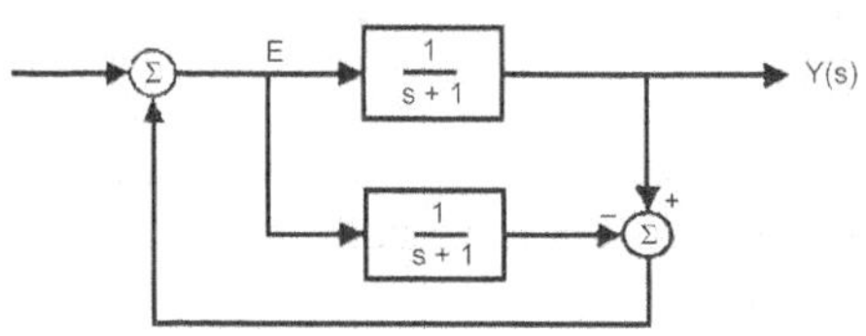

(a) 0 (b) $\frac{1}{s+1}$

(c) $\frac{2}{s+1}$ (d) $\frac{2}{s+3}$

18. A system with the transfer function $\frac{Y(s)}{X(s)} = \frac{s}{s+p}$ has an output $y(t) = \cos\left(2t - \frac{\pi}{3}\right)$ for the input signal $x(t) = p\cos\left(2t - \frac{\pi}{2}\right)$. Then, the system parameter 'p' is

(a) $\sqrt{3}$ (b) $\frac{2}{\sqrt{3}}$

(c) 1 (d) $\frac{\sqrt{3}}{2}$

19. For the asymptotic Bode magnitude plot shown below, the system transfer function can be

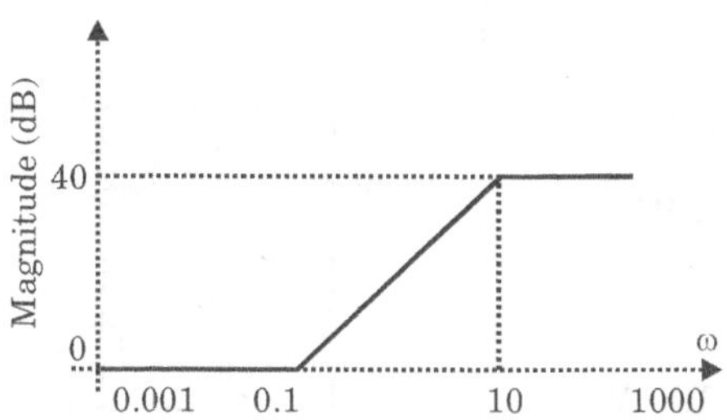

(a) $\frac{10s+1}{0.1s+1}$ (b) $\frac{100s+1}{0.1s+1}$

(c) $\frac{100s}{10s+1}$ (d) $\frac{0.1s+1}{10s+1}$

20. Suppose that the modulating signal is $m(t) = 2\cos(2\pi f_m t)$ and the carrier signal is $x_c(t) = A_c \cos(2\pi f_c t)$. Which one of the following is a conventional AM signal without over-modulation ?

(a) $x(t) = A_c m(t) \cos(2\pi f_c t)$

(b) $x(t) = A_c[1 + m(t)] \cos(2\pi f_c t)$

(c) $x(t) = A_c \cos(2\pi f_c t) + \frac{A_c}{4} m(t) \cos(2\pi f_c t)$

(d) $x(t) = A_c \cos(2\pi f_m t) \cos(2\pi f_c t) + A_c \sin(2\pi f_m t) \sin(2\pi f_c t)$

21. Consider an angle modulated signal

$x(t) = 6\cos[2\pi \times 10^6 t + 2\sin(8000\pi t) + 4\cos(8000\pi t)]$ V.

The average power of $x(t)$ is

(a) 10 W (b) 18 W

(c) 20 W (d) 28 W

22. If the scattering matrix [S] of a two port network is

$$[S] = \begin{bmatrix} 0.2\angle 0^\circ & 0.9\angle 90^\circ \\ 0.9\angle 90^\circ & 0.1\angle 90^\circ \end{bmatrix}$$

then the network is

(a) lossless and reciprocal

(b) losseless but not reciprocal

(c) not lossless but reciprocal

(d) neither lossless nor reciprocal

23. A transmission line has a characteristic impedance of 50 Ω and a resistance of 0.1 Ω/m. If the line is distortionless, the attenuation constant (in Np/m) is

(a) 500 (b) 5

(c) 0.014 (d) 0.002

24. Consider the pulse shape $s(t)$ as shown. The impulse response $h(t)$ of the filter matched to this pulse is

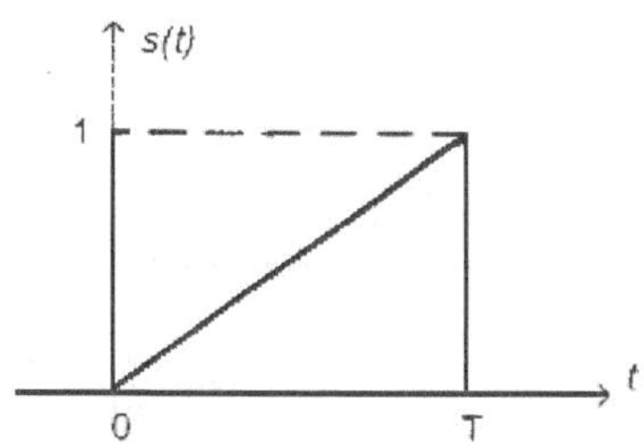

(a) 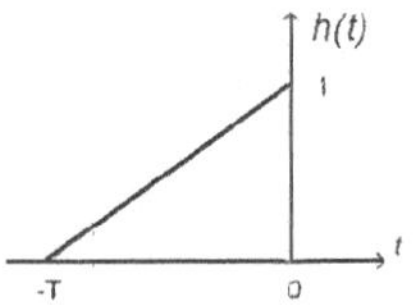

(b) 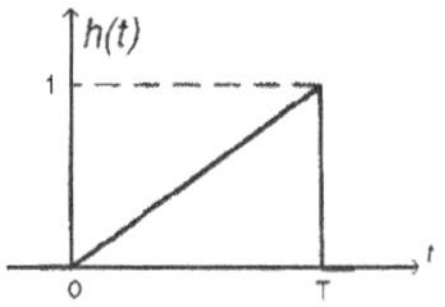

(c) 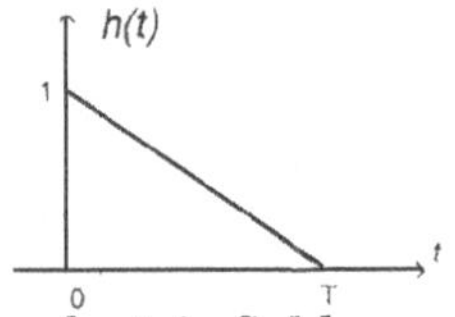

(d) 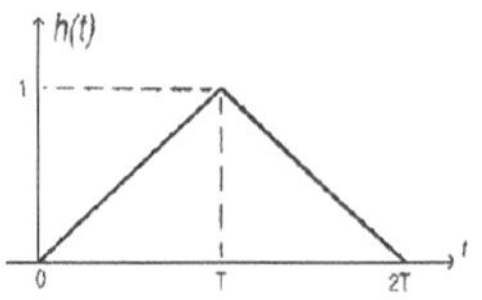

25. The electric field component of a time harmonic plane EM wave travelling in a nonmagnetic lossless dielectric medium has an amplitude of 1V/m. If the relative permittivity of the medium is 4, the magnitude of the time-average power density vector (in W/m^2) is

(a) $\frac{1}{30\pi}$ (b) $\frac{1}{60\pi}$

(c) $\frac{1}{120\pi}$ (d) $\frac{1}{240\pi}$

Q.(26 – 55) carry two marks each.

26. If $e^y = x^{\frac{1}{x}}$, then y has a

(a) maximum at $x = e$

(b) minimum at $x = e$

(c) maximum at $x = e^{-1}$

(d) minimum at $x = e^{-1}$

27. A fair coin is tossed independently four times. The probability of the event "the number of times heads show up is more than the number of times tails show up" is

(a) $\frac{1}{16}$ (b) $\frac{1}{8}$

(c) $\frac{1}{4}$ (d) $\frac{5}{16}$

28. If $\vec{A} = xy\,\hat{a}_x + x^2\,\hat{a}_y$, then $\oint_C \vec{A} \cdot d\vec{l}$ over the path shown in the figure is

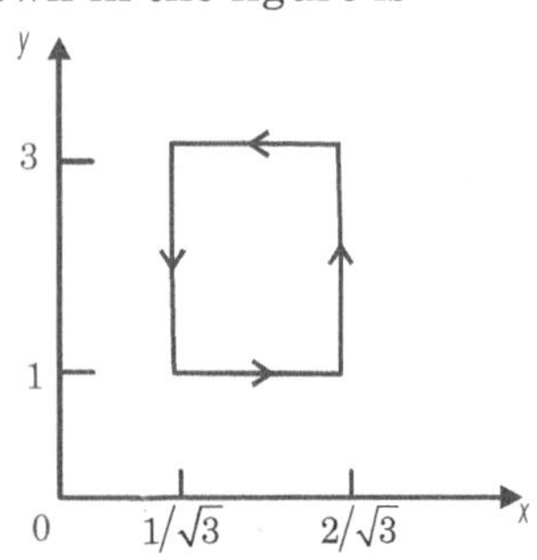

(a) 0 (b) $\frac{2}{\sqrt{3}}$

(c) 1 (d) $2\sqrt{3}$

29. The residues of a complex function $X(z) = \frac{1-2z}{z(z-1)(z-2)}$ at its poles are

(a) $\frac{1}{2}, -\frac{1}{2}$ and 1 (b) $\frac{1}{2}, \frac{1}{2}$ and –1

(c) $\frac{1}{2}$, 1 and $-\frac{3}{2}$ (d) $\frac{1}{2}$, –1 and $\frac{3}{2}$

30. Consider a differential equation $\frac{dy(x)}{dx} - y(x) = x$ with the initial conditions $y(0) = 0$. Using Euler's first order method with a step size of 0.1, the value of $y(0.3)$ is

(a) 0.01 (b) 0.031

(c) 0.0631 (d) 0.1

31. Given $f(t) = \mathcal{L}^{-1}\left[\frac{3s+1}{s^3 + 4s^2 + (K-3)s}\right]$.

If $\lim_{t\to\infty} f(t) = 1$, then the value of K is

(a) 1 (b) 2

(c) 3 (d) 4

32. In the circuit shown, the switch S is open for a long time and is closed at $t = 0$. The current $i(t)$ for $t \geq 0^+$ is

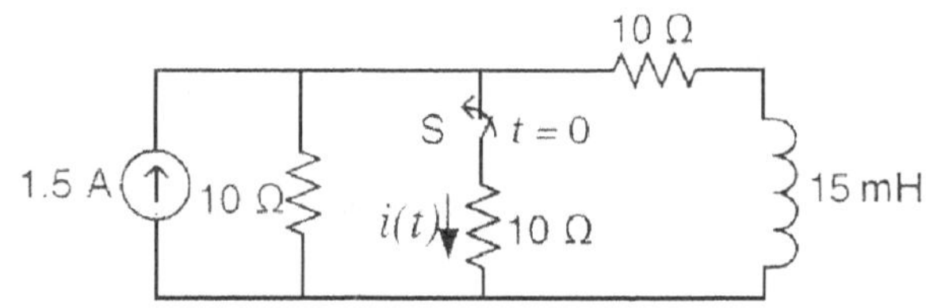

(a) $i(t) = 0.5 - 0.125e^{-1000t}$ A

(b) $i(t) = 1.5 - 0.125e^{-1000t}$ A

(c) $i(t) = 0.5 - 0.5e^{-1000t}$ A

(d) $i(t) = 0.375e^{-1000t}$ A

33. The current I in the circuit shown is

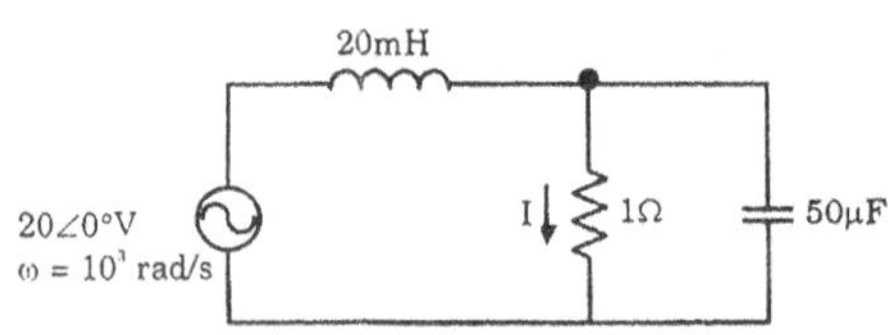

(a) $-j1$ A (b) $j1$ A

(c) 0 A (d) 20 A

34. In the circuit shown, the power supplied by the voltage source is

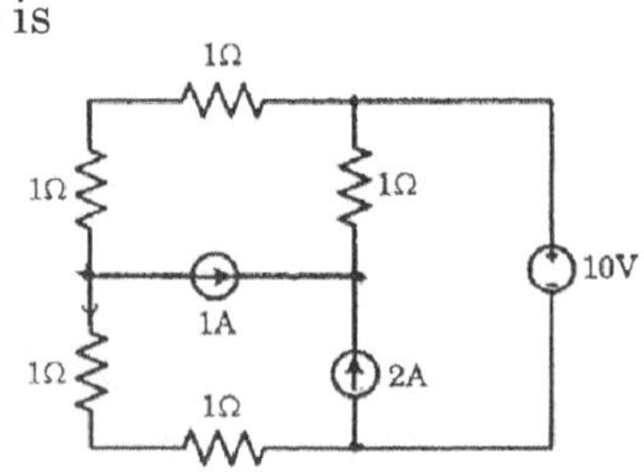

(a) 0 W

(b) 5 W

(c) 10 W

(d) 100 W

35. In a uniformly doped BJT, assume that N_E, N_B and N_C are the emitter, base and collector dopings in atoms/cm^3, respectively. If the emitter injection effeciency of the BJT is close to unity, which one of the following conditions is TRUE?

(*a*) $N_E = N_B = N_C$

(*b*) $N_E >> N_B$ and $N_B > N_C$

(*c*) $N_E = N_B$ and $N_B < N_C$

(*d*) $N_E < N_B < N_C$

36. Compared to a p-n junction with $N_A = N_D = 10^{14}/\text{cm}^3$, which one of the following statements is TRUE for a p-n junction with $N_A = N_D = 10^{20}/\text{cm}^3$?

(*a*) Reverse breakdown voltage is lower and depletion capacitance is lower

(*b*) Reverse breakdown voltage is higher and depletion capacitance is lower

(*c*) Reverse breakdown voltage is lower and depletion capacitance is higher

(*d*) Reverse breakdown voltage is higher and depletion capacitance is higher

37. Assuming that all flip-flops are in reset condition initially, the count sequence observed at Q_A in the circuit shown is

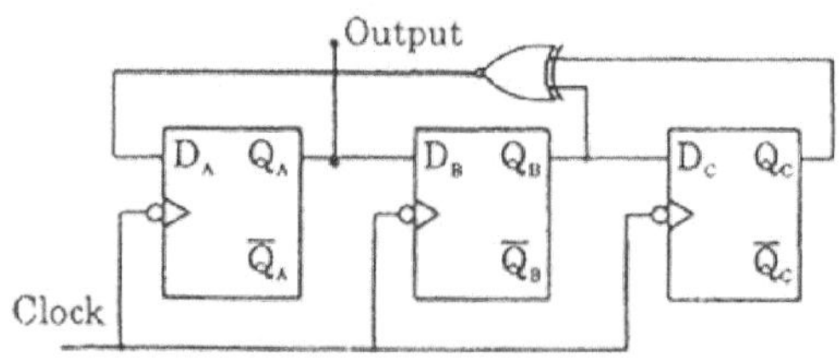

(*a*) 0010111... (*b*) 0001011...

(*c*) 0101111... (*d*) 0110100...

38. The transfer characteristic for the precision rectifier circuit shown below is (assume ideal OP-AMP and practical diodes)

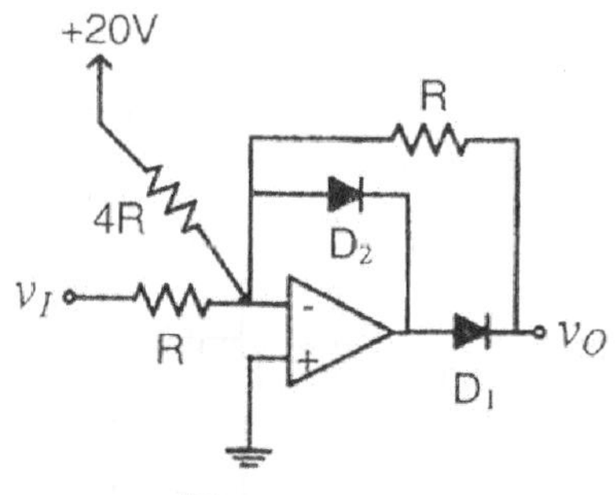

(*a*)

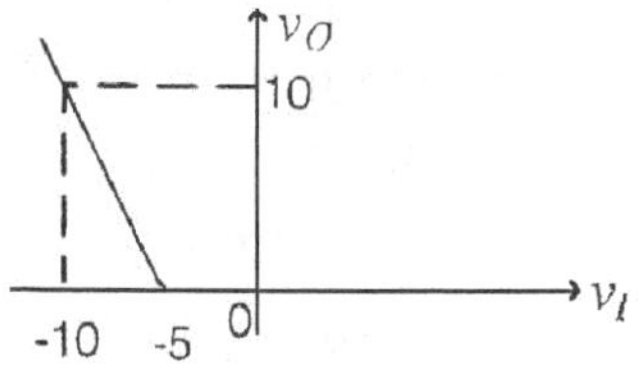

(*b*)

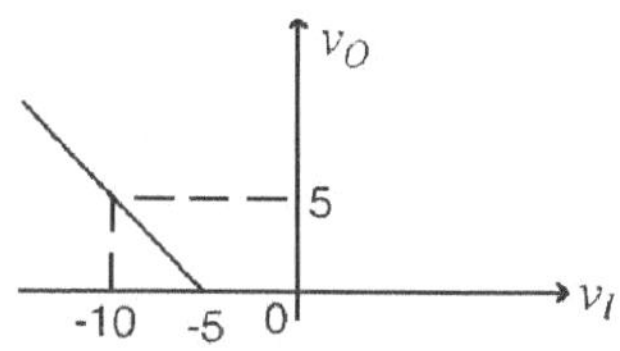

(*c*)

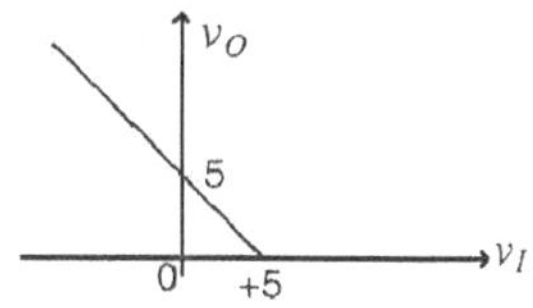

(*d*)

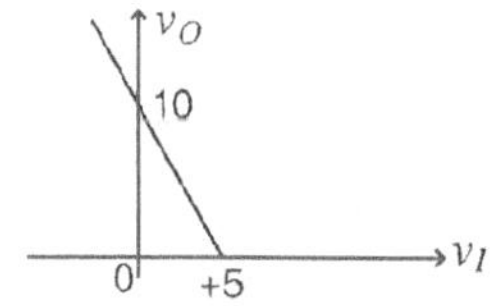

39. The Boolean function realized by the logic circuit shown is

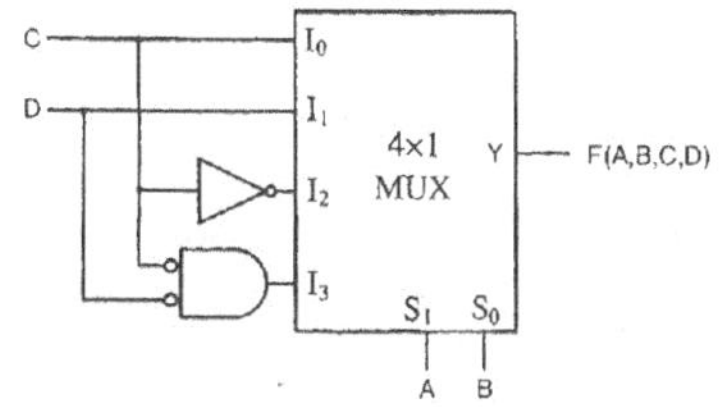

(*a*) F = Σm (0, 1, 3, 5, 9, 10, 14)

(*b*) F = Σm (2, 3, 5, 7, 8, 12, 13)

(*c*) F = Σm (1, 2, 4, 5, 11, 14,15)

(*d*) F = Σm (2, 3, 5, 7, 8, 9, 12)

40. For the 8085 assembly language program given below, the content of the accumulator after the execution of the program is

3000	MVI	A, 45H
3002	MOV	B, A
3003	STC	
3004	CMC	
3005	RAR	
3006	XRA	B

(*a*) 00H (*b*) 45H

(*c*) 67H (*d*) E7H

41. A continuous time LTI system is described by

$$\frac{d^2y(t)}{dt^2} + 4\frac{dy(t)}{dt} + 3\,y(t) = 2\frac{dx(t)}{dt} + 4x(t)$$

Assuming zero initial conditions, the response $y(t)$ of the above system for the input $x(t) = e^{-2t}\,u(t)$ is given by

(*a*) $(e^t - e^{3t})\,u(t)$ (*b*) $(e^{-t} - e^{-3t})\,u(t)$

(*c*) $(e^{-t} + e^{-3t})\,u(t)$ (*d*) $(e^t + e^{3t})\,u(t)$

42. The transfer function of a discrete time LTI system is given by

$$H(z) = \frac{2 - \frac{3}{4}z^{-1}}{1 - \frac{3}{4}z^{-1} + \frac{1}{8}z^{-2}}$$

Consider the following statements:

S1 : The system is stable but causal for ROC: $|z| > \frac{1}{2}$

S2 : The system is stable but not causal for ROC: $|z| < \frac{1}{4}$

S3 : The system is neither stable nor causal for ROC: $\frac{1}{4} < |z| < \frac{1}{2}$

Which one of the following statements is valid ?

(*a*) Both S1 and S2 are true

(*b*) Both S2 and S3 are true

(*c*) Both S1 and S3 are true

(*d*) S1, S2 and S3 are all true

43. The Nyquist sampling rate for the signal

$$s(t) = \frac{\sin(500\pi t)}{\pi t} \times \frac{\sin(700\pi t)}{\pi t}$$ is given by

(*a*) 400 Hz (*b*) 600 Hz

(*c*) 1200 Hz (*d*) 1400 Hz

44. A unity negative feedbook closed loop system has a plant with the transfer function $G(s) = \frac{1}{s^2+2s+2}$ and a controller $G_c(s)$ in the feed forward path. For a unit step input, the transfer function of the controller that gives minimum steady state error is

(*a*) $G_c(s) = \frac{s+1}{s+2}$

(*b*) $G_c(s) = \frac{s+2}{s+1}$

(*c*) $G_c(s) = \frac{(s+1)(s+4)}{(s+2)(s+3)}$

(*d*) $G_c(s) = 1 + \frac{2}{s} + 3s$

45. X(*t*) is a stationary process with the power spectral density $S_x(f) > 0$ for all *f*. The process is passed through a system shown below.

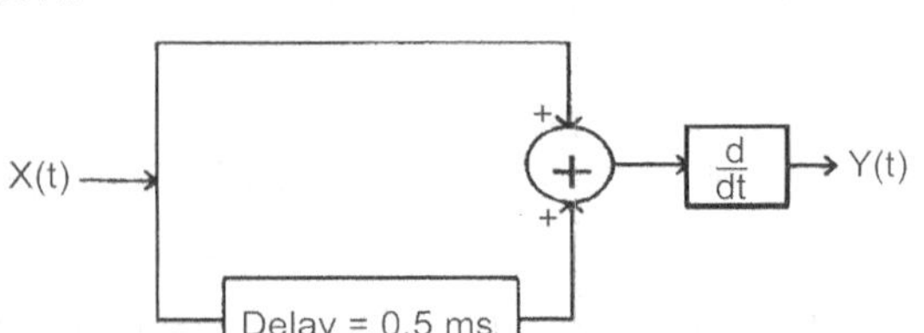

Let $S_Y(f)$ be the power spectral density of Y(*t*). Which one of the following statements is correct?

(*a*) $S_Y(f) > 0$ for all *f*

(*b*) $S_Y(f) = 0$ for $|f| > 1$ kHz

(*c*) $S_Y(f) = 0$ for $f = nf_0$, $f_0 = 2$ kHz, *n* any integer

(*d*) $S_Y(f) = 0$ for $f = (2n+1)f_0$, $f_0 = 1$ kHz, *n* any integer

46. A plane wave having the electric field component $\vec{E}_i = 24\cos(3 \times 10^8 t - \beta y)\hat{a}_z$ V/m and traveling in free space is incident normally on a lossless medium with $\mu = \mu_0$ and $\varepsilon = 9\varepsilon_0$ which occupies the region $y \geq 0$. The reflected magnetic field component is given by

(*a*) $\frac{1}{10\pi}\cos(3 \times 10^8 t + y)\,\hat{a}_x$ A/m

(*b*) $\frac{1}{20\pi}\cos(3 \times 10^8 t + y)\,\hat{a}_x$ A/m

(*c*) $-\frac{1}{20\pi}\cos(3 \times 10^8 t + y)\,\hat{a}_x$ A/m

(*d*) $-\frac{1}{10\pi}\cos(3 \times 10^8 t + y)\,\hat{a}_x$ A/m

47. In the circuit shown, all the transmission line sections are lossless. Voltage Standing Wave Ratio (VSWR) on the 60Ω line is

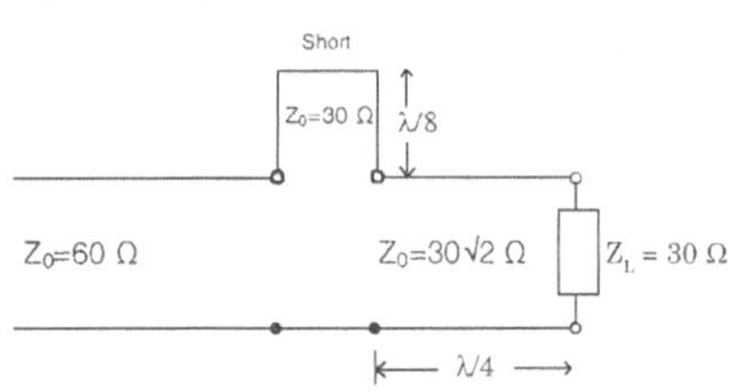

(*a*) 1.00 (*b*) 1.64

(*c*) 2.50 (*d*) 3.00

Common Data Questions

Common Data for Q. (48 – 49) :

Consider the common emitter amplifier shown below with the following circuit parameters:

$\beta = 100$, $g_m = 0.3861$ A/V, $r_o = \infty$, $r_\pi = 259\ \Omega$, $R_S = 1\ k\Omega$, $R_B = 93\ k\Omega$, $R_C = 250\ \Omega$, $R_L = 1\ k\Omega$, $C_1 = \infty$ and $C_2 = 4.7\ \mu F$.

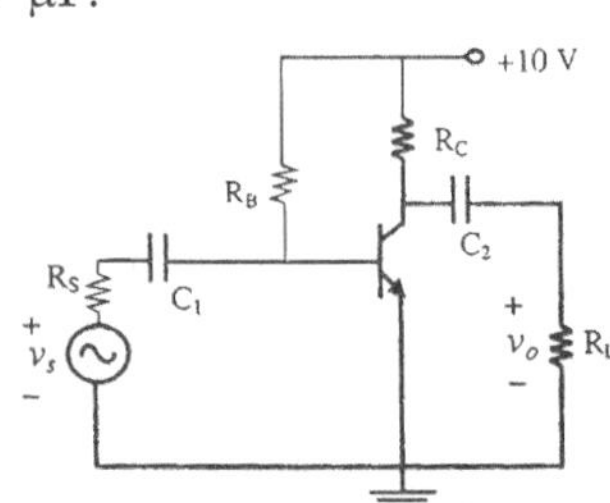

48. The resistance seen by the source v_s is

(*a*) 258 Ω (*b*) 1258 Ω

(*c*) 93 kΩ (*d*) ∞

49. The lower cut-off frequency due to C_2 is

(*a*) 33.9 Hz (*b*) 27.1 Hz

(*c*) 13.6 Hz (*d*) 16.9 Hz

Common data for Q. (50 – 51) :

The signal flow graph of a system is shown below.

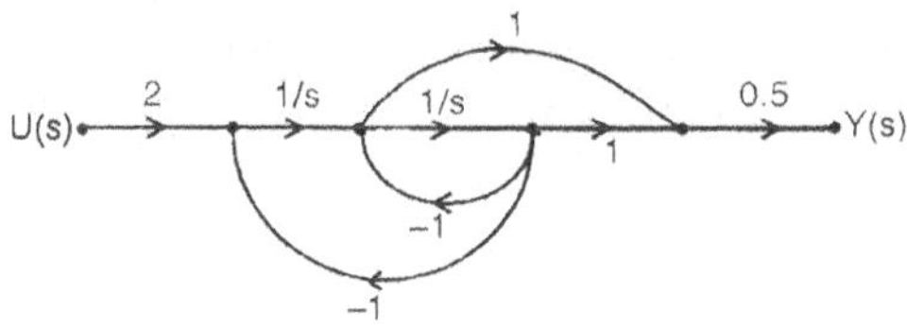

50. The state variable representation of the system can be

(*a*) $\dot{x} = \begin{bmatrix} 1 & 1 \\ -1 & 0 \end{bmatrix} x + \begin{bmatrix} 0 \\ 2 \end{bmatrix} u, \; y = \begin{bmatrix} 0 & 0.5 \end{bmatrix} x$

(*b*) $\dot{x} = \begin{bmatrix} -1 & 1 \\ -1 & 0 \end{bmatrix} x + \begin{bmatrix} 0 \\ 2 \end{bmatrix} u, \; y = \begin{bmatrix} 0 & 0.5 \end{bmatrix} x$

(*c*) $\dot{x} = \begin{bmatrix} 1 & 1 \\ -1 & 0 \end{bmatrix} x + \begin{bmatrix} 0 \\ 2 \end{bmatrix} u, \; y = \begin{bmatrix} 0.5 & 0.5 \end{bmatrix} x$

(*d*) $\dot{x} = \begin{bmatrix} -1 & 1 \\ -1 & 0 \end{bmatrix} x + \begin{bmatrix} 0 \\ 2 \end{bmatrix} u, \; y = \begin{bmatrix} 0.5 & 0.5 \end{bmatrix} x$

51. The transfer function of the system is

(*a*) $\dfrac{s+1}{s^2+1}$ (*b*) $\dfrac{s-1}{s^2+1}$

(*c*) $\dfrac{s+1}{s^2+s+1}$ (*d*) $\dfrac{s-1}{s^2+s+1}$

Linked Answer Questions

Statement for linked Answer Q. (52 – 53) :

The silicon sample with unit cross-sectional area shown below is in thermal equilibrium. The following information is given : T = 300 K, electronic charge = 1.6×10^{-19} C, thermal voltage = 26 mV and electron mobility = 1350 cm²/V-s

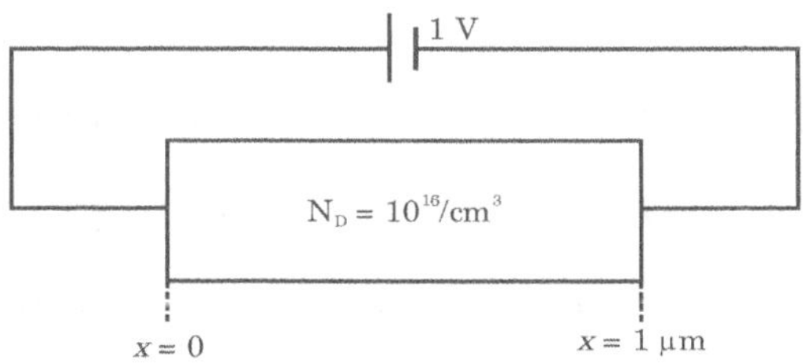

52. The magnitude of the electric field at $x = 0.5$ μm is

(*a*) 1 kV/cm (*b*) 5 kV/cm

(*c*) 10 kV/cm (*d*) 26 kV/cm

53. The magnitude of the electron drift current density at $x = 0.5$ μm is

(*a*) 2.16×10^4 A/cm² (*b*) 1.08×10^4 A/cm²

(*c*) 4.32×10^3 A/cm² (*d*) 6.48×10^2 A/cm²

Statement for linked Answer Q. (54 – 55) :

Consider a baseband binary PAM receiver shown below. The additive channel noise $n(t)$ is white with power spectral density $S_N(f) = N_0/2 = 10^{-20}$ W/Hz. The low-pass filter is ideal with unity gain and cut-off frequency 1 MHz. Let Y_k represent the random variable $y(t_k)$.

$Y_k = N_k$ if transmitted bit $b_k = 0$

$Y_k = a + N_k$ if transmitted bit $b_k = 1$

where N_k represents the noise sample value. The noise sample has a probability density function,

$P_{N_k}(n) = 0.5\alpha\, e^{-\alpha|n|}$ (This has mean zero and variance $2/\alpha^2$).

Assume transmitted bits to be equiprobable and threshold z is set to $a/2 = 10^{-6}$ V.

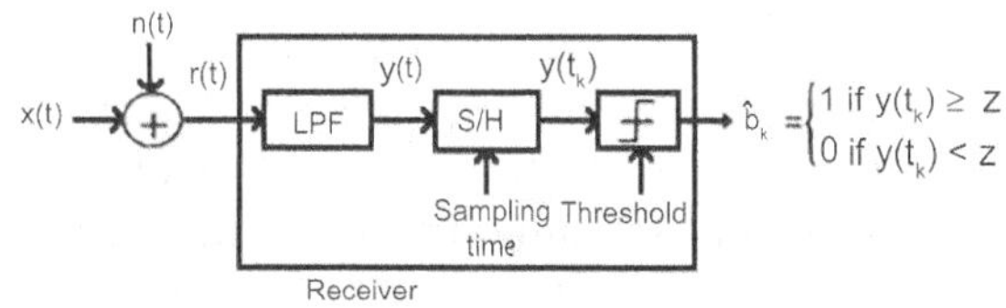

54. The value of the parameter α (in V^{-1}) is

(*a*) 10^{10} (*b*) 10^7

(*c*) 1.414×10^{-10} (*d*) 2×10^{-20}

55. The probability of bit error is

(*a*) $0.5 \times e^{-3.5}$ (*b*) $0.5 \times e^{-5}$

(*c*) $0.5 \times e^{-7}$ (*d*) $0.5 \times e^{-10}$

General Aptitude (GA) Questions

Q.(56 – 60) carry one mark each.

56. Which of the following options is the closest in meaning to the word below:

Circuitous

(*a*) cyclic (*b*) indirect

(*c*) confusing (*d*) crooked

57. The question below consists of a pair of related words followed by four pairs of words. Select the pair that best expresses the relation in the original pair.

Unemployed : Worker

(*a*) fallow : land (*b*) unaware : sleeper

(*c*) wit : jester (*d*) renovated : house

58. Choose the most appropriate word from the options given below to complete the following sentence:

If we manage to__________our natural resources, we would leave a better planet for our children.

(*a*) uphold (*b*) restrain

(*c*) cherish (*d*) conserve

59. Choose the most appropriate word from the options given below to complete the following sentence:

His rather casual remarks on politics ________ his lack of seriousness about the subject.

(*a*) masked (*b*) belied

(*c*) betrayed (*d*) suppressed

60. 25 persons are in a room. 15 of them play hockey, 17 of them play football and 10 of them play both hockey and football. Then the number of persons playing neither hockey nor football is

(*a*) 2 (*b*) 17

(*c*) 13 (*d*) 3

Q.(61 – 65) carry two marks each.

61. Modern warfare has changed from large scale clashes of armies to suppression of civilian populations. Chemical agents that do their work silently appear to be suited to such warfare; and regretfully, there exist people in military establishments who think that chemical agents are useful tools for their cause.

Which of the following statements best sums up the meaning of the above passage:

(*a*) Modern warfare has resulted in civil strife

(*b*) Chemical agents are useful in modern warfare

(*c*) Use of chemical agents in warfare would be undesirable

(*d*) People in military establishments like to use chemical agents in war

62. If 137 + 276 = 435 how much is 731 + 672?

(*a*) 534 (*b*) 1403

(*c*) 1623 (*d*) 1513

63. 5 skilled workers can build a wall in 20 days; 8 semi-skilled workers can build a wall in 25 days; 10 unskilled workers can build a wall in 30 days. If a team has 2 skilled, 6 semi-skilled and 5 unskilled workers, how long will it take to build the wall?

(*a*) 20 days (*b*) 18 days

(*c*) 16 days (*d*) 15 days

64. Given digits 2, 2, 3, 3, 3, 4, 4, 4, 4 how many distinct 4 digit numbers greater than 3000 can be formed?

(*a*) 50 (*b*) 51

(*c*) 52 (*d*) 54

65. Hari (H), Gita (G), Irfan (I) and Saira (S) are siblings (i.e., brothers and sisters). All were born on Ist January. The age difference between any two successive siblings (that is born one after another) is less than 3 years. Given the following facts:

I. Hari's age + Gita's age > Irfan's age + Saira's age

II. The age difference between Gita and Saira is 1 year. However, Gita is not the oldest and Saira is not the youngest.

III. There are no twins.

In what order were they born (oldest first)?

(*a*) HSIG (*b*) SGHI

(*c*) IGSH (*d*) IHSG

ANSWERS

1. (*c*)	2. (*c*)	3. (*d*)	4. (*d*)	5. (*d*)	6. (*b*)	7. (*b*)	8. (*b*)	9. (*a*)	10. (*a*)
11. (*d*)	12. (*d*)	13. (*b*)	14. (*a*)	15. (*c*)	16. (*d*)	17. (*b*)	18. (*b*)	19. (*a*)	20. (*c*)
21. (*b*)	22. (*c*)	23. (*d*)	24. (*c*)	25. (*c*)	26. (*a*)	27. (*d*)	28. (*c*)	29. (*c*)	30. (*b*)
31. (*d*)	32. (*a*)	33. (*a*)	34. (*a*)	35. (*b*)	36. (*c*)	37. (*d*)	38. (*b*)	39. (*d*)	40. (*c*)
41. (*b*)	42. (*c*)	43. (*c*)	44. (*d*)	45. (*d*)	46. (*a*)	47. (*b*)	48. (*b*)	49. (*b*)	50. (*d*)
51. (*c*)	52. (*c*)	53. (*a*)	54. (*b*)	55. (*d*)	56. (*b*)	57. (*a*)	58. (*d*)	59. (*c*)	60. (*d*)
61. (*c*)	62. (*c*)	63. (*d*)	64. (*b*)	65. (*b*)					

EXPLANATIONS

1. Skew symmetric matrix, $A = -A^T$

Since magnitude of diagonal element is same only sign is changed, **so eigen value** must be either zero **or** pure imaginary to satisfy this relation.

2. As $f(t) = f(-t)$ [$f(t)$ is an even function]

Fourier sine series coefficients are zero: $b_n = 0$ for all $n \geq 1$.

Now, DC value

$$a_0 = \frac{1}{T}\int_{-T/2}^{T/2} f(t)\cdot dt'$$

$$= \frac{1}{T}\int_{-T/2}^{-T/4} (-2A)\cdot dt + \frac{1}{T}\int_{-T/4}^{0} A\cdot dt + \frac{1}{T}\int_{0}^{T/4} A\cdot dt + \frac{1}{T}\int_{T/4}^{T/2} (-2A)\cdot dt$$

$$= \frac{1}{T}[(-2A)\cdot t]_{-T/2}^{-T/4} + \frac{1}{T}[(A)\cdot t]_{-T/4}^{0} + \frac{1}{T}\cdot(A\cdot t)_0^{T/4} + \frac{1}{T}[(-2A)\cdot t]_{T/4}^{T/2}$$

$$= \frac{-A}{2}[-\text{ve}]$$

Therefore, the fourier series for given waveform will contain negative DC term and cosine terms.

3. Given : $\dfrac{d^2 n(x)}{dx^2} - \dfrac{n(x)}{L^2} = 0,$

where, L = constant

Boundary conditions are : $n(0) = K$

and $n(\infty) = 0$

$$\therefore \quad \left(p^2 - \frac{1}{L^2}\right) = 0$$

$$\Rightarrow \quad p = +\frac{1}{L}$$

$$\therefore \quad n(x) = A_1 e^{-\frac{1}{L}x} + A_2 e^{\frac{1}{L}x}$$

Applying *Boundary conditions*:

$$n(0) = Ae^0 = A_1 + A_2 = k$$

and $n(\infty) = 0$

$$\therefore \quad n(\infty) = A_1\cdot 0 + A_2 e^{\infty} = 0$$

$$\Rightarrow \quad A_2 = 0$$

$$\therefore \quad A_1 = k$$

Hence solution is $n(x) = ke^{-\frac{1}{L}x}$

4. Given circuit is

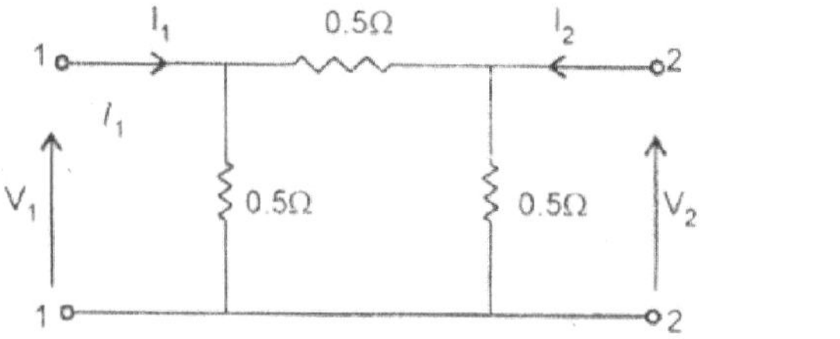

$$\begin{bmatrix} I_1 \\ I_2 \end{bmatrix} = \begin{bmatrix} Y_{11} & Y_{12} \\ Y_{21} & Y_{22} \end{bmatrix}\begin{bmatrix} V_1 \\ V_2 \end{bmatrix}$$

$$Y_{11} = \left.\frac{I_1}{V_1}\right|_{V_2=0} = \frac{1}{0.5110\times 5} = \frac{1}{0.25} = 4$$

$$Y_{12} = \left.\frac{I_1}{V_2}\right|_{V_1=0} = \frac{1}{0.5} = 2$$

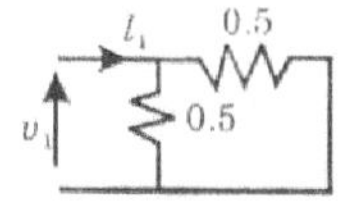

$$Y_{21} = \left.\frac{I_2}{V_1}\right|_{V_2=0} = \frac{1}{0.5} = 2$$

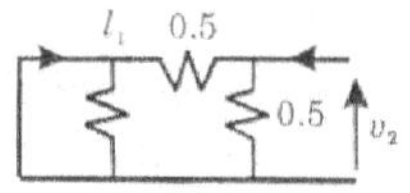

$$Y_{22} = \left.\frac{I_2}{V_2}\right|_{V_1=0} = \frac{1}{0.5110\times 5} = 4$$

Hence short-circuit admittance parameter matrix is

$$Y = \begin{bmatrix} 4 & 2 \\ 2 & 4 \end{bmatrix} \text{S.}$$

5. $I = I_m \text{Sin } wt$

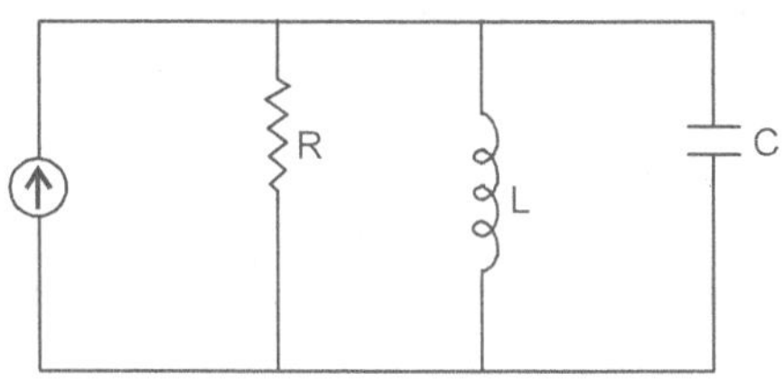

Parallel RCL circuit

Admittance,

$$Y = \frac{1}{R} + j\omega c + \frac{1}{j\omega L} = \frac{1}{R} + J\left(\omega c - \frac{1}{\omega L}\right)$$

Frequency response of voltage amplitude

$$|V| = \frac{I_m}{\sqrt{\left(\frac{1}{R}\right)^2 + \left(\omega c - \frac{1}{1L}\right)^2}}$$

At resonance, $\omega_o C - \dfrac{1}{\omega L} = 0,\ \omega = \dfrac{1}{\sqrt{LC}}$

Impedance, $Z = \dfrac{1}{Y} = R$

[Real quality and having maximum]

Half-power frequencies,

$$\omega_1, \omega_2 = -\frac{1}{2RC} \pm \sqrt{\left(\frac{1}{2RC}\right)^2 + \frac{1}{LC}}$$

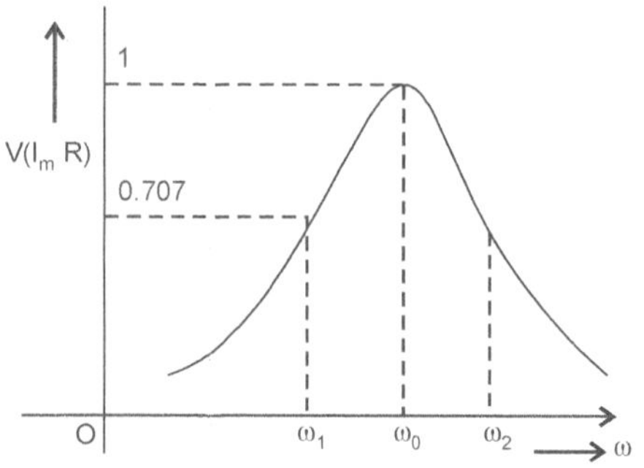

Then, BW (Bandwidth)

$$= \omega_2 - \omega_1 = \frac{1}{RC}$$

BW decreases mean R is increased

Also, BW is independent of any variation in L.

6. For n-channel MOSFET,
mobility of inversion layer = 1350 cm²/V-s

8.

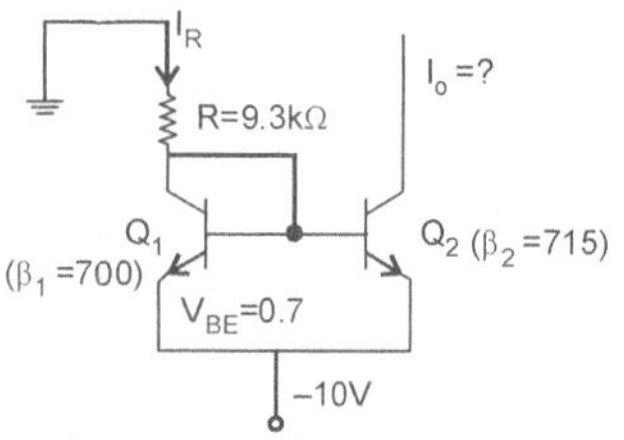

Let both transistors are in active region, therefore voltage at Q_1 base

$(V_{Base})_{Q1} = 0.7\text{-}10 = -9.3V$

Current through R,

$$I_R = \frac{9.3V}{9.3k\Omega} = 1 \text{ mA} = I_C$$

Since emitter area of $Q_1 = \frac{1}{2}$ [emitter area of Q_2]

i.e. $A_{Q1} = \frac{A_{Q2}}{2}$

$\therefore (\beta_2)_{effective} = 2 \times \beta_2 = 1430$

Since effective β of Q_2 is double of Q_1, so collector current also will be double nearly.

$$I_0 = I_{C2} \simeq 2 \times I_{C1} \simeq 2 \text{ mA}$$

9. Drawing low-frequency simplified h-Parameter model of the given circuit,

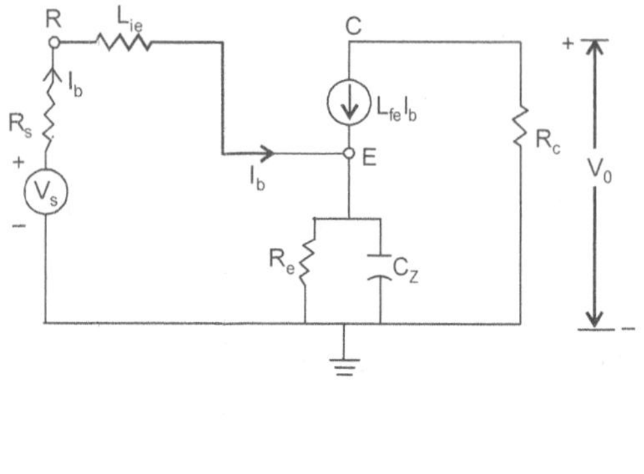

$$V_0 = -I_b h_{fe} Rc$$

$$= -h_{fe} Rc \frac{V_s}{R_s + h_{ie} + Z'_e}$$

or, $$A_v = \frac{-V_0}{V_s}$$

$$= \frac{-h_{fe} Rc}{R_s + h_{ie} + Z'_e}$$

$$Z'_e = \frac{(1 + h_{fe}) R_e}{1 + fwc_2 R_e}$$

Input impedance,

$$Z'_i = R_s + h_{ie} + Z'_e$$

When C_2 is disconnected,

$$A'_v = \frac{V_0}{V_s}$$

$$= \frac{-h_{ef} Rc}{R_s + h_{ie} + R_e(1 + h_{fe})} \qquad \text{[Decrease]}$$

and Input impedance,

$$Z'_i = R_j + h_{ie} + R_e(1 + h_{fe}) \qquad \text{[Increase]}$$

10.

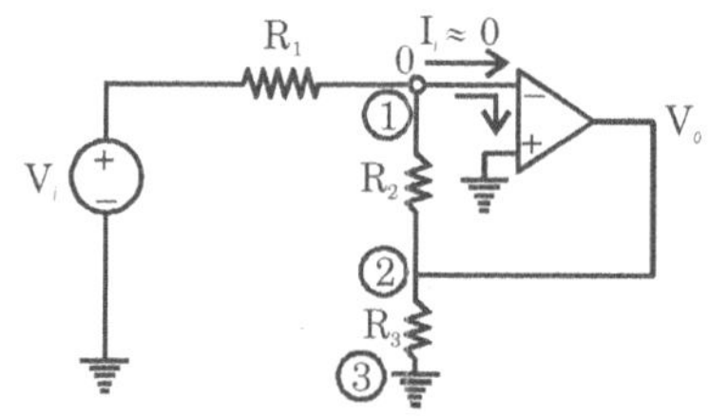

As V2 – V1 = V2 – V3

R2 and R3 in parallel.

Using KCL at mode (1), we have

$$\Rightarrow \quad \frac{0 - v_i}{R_1} + 0 - v_0\left[\frac{1}{R_2} + \frac{1}{R_3}\right] = 0$$

or, $$\frac{v_o}{v_i} = -\left(\frac{R_2 \| R_3}{R_1}\right)$$

11.

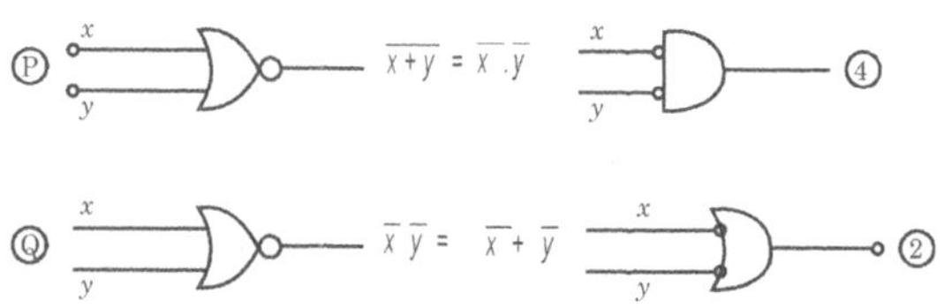

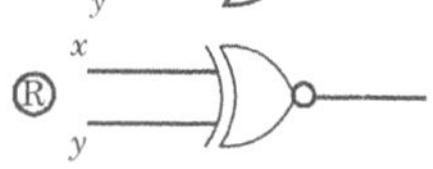

$$= \overline{x \oplus y} = \overline{x\bar{y} + y\bar{x}}$$

$$= \overline{(x\bar{y})} \cdot \overline{(y\bar{x})} = \overline{(\bar{x} + y) \cdot (\bar{y} + x)}$$

$$= \overline{\bar{x}\bar{y} + xy + x\bar{x} + y\bar{y}}$$

$$= \overline{\bar{x}\bar{y} + xy} \equiv$$

$$= x \oplus y = \bar{x}\bar{y} + xy$$

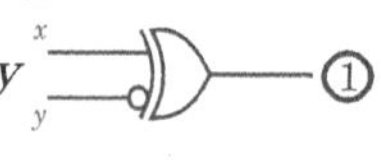

12. Output $F = (A \oplus B) \odot (A \odot B) \odot C$

$$= \{(A \oplus B)(A \odot B) + (A \odot B)(A \oplus B)\} \odot C$$

$$= \{(A \oplus B)(A \odot B)\} \odot C$$

$$= 0 \odot C = \bar{C}$$

It can be inferred that output F is only completed of c function.

Therefore, for C = 0, F =1

13.

A_{15}	A_{14}	A_{13}	A_{12}	A_{11}	A_{10}	A_9	A_8
0	0	1	0	1	1	0	1
To enable chip					for y_5		

A_7	A_6	A_5	A_4	A_3	A_2	A_1	A_0
0	0	0	0	0	0	0	0
⋮	⋮	⋮	⋮	⋮	⋮	⋮	⋮
⋮	⋮	⋮	⋮	⋮	⋮	⋮	⋮
1	1	1	1	1	1	1	1

Hence range of the address is

2DOO – 2DFF

Alternately

$A_{15}\ A_{14}\ \ldots\ \ldots\ \ldots\ \ldots\ \ldots\ A_8\ A_7\ \ldots\ \ldots\ \ldots\ \ldots\ \ldots\ A_0$

0 0 1 0 1 1 0 1 0 0 0 0 0 0 0 0

2 D 0 0

= 2D00

Maximum address

$A_{15}\ A_{14}\ \ldots\ \ldots\ \ldots\ \ldots\ \ldots\ A_8\ A_7\ \ldots\ \ldots\ \ldots\ \ldots\ \ldots\ A_0$

0 0 1 0 1 1 0 1 1 1 1 1 1 1 1 1

2 D F F

= 2DFF

Hence range of the address is 2D00 – 2DFF

14. From, Z-transform property of

$$Z^{m} \leftrightarrow \delta[n - m]$$

We has$X(Z) \leftrightarrow 5\delta[n + 2] + 4\delta[n - 1] + 3\delta[n]$

15. $h_1[n] = \delta(n - 1)$ $\quad h_1(z) = z^{-1}$

$h_2(n) = \delta(n - 2)$ $\quad h_2(z) = z^{-2}$

Hence in cascade, overall z-transform of impulse response,

$$H(z) = h_1(z) \cdot h_2(z)$$

$$= z^{-1} \cdot z^{-2} = z^{-3}$$

$$\therefore \quad h(n) = \delta(n - 3)$$

16. For N-point Fast Fourier Transform (FFT), with $N = 2^m$.

we need only one complex multiplication to compute a butterfly.

17. 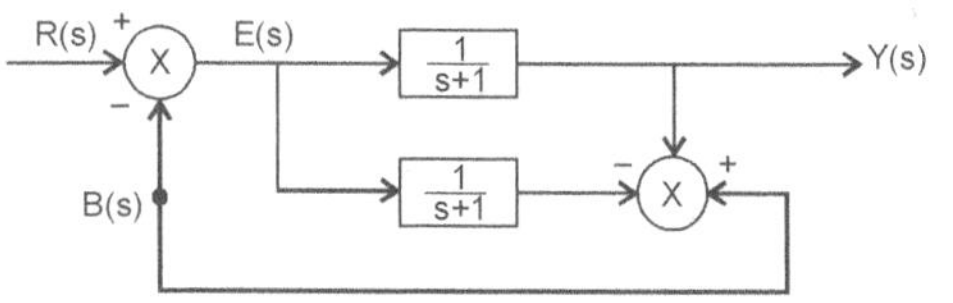

$$Y(s) = E(s) \cdot \frac{1}{(s+1)} \quad \ldots(i)$$

$$E(s) = R(s) - B(s)$$

$$= R(s) - \left\{Y(s) - \frac{1}{(s+1)} \cdot E(s)\right\}$$

$$E(s) = R(s) - \{Y(s) - Y(s)\}$$

$$E(s) = R(s)$$

From equation (*i*),

$$Y(s)(s + 1) = R(s)$$

$$\therefore \quad \frac{Y(s)}{R(s)} = \frac{1}{(s+1)}$$

Alternately

Signal flow graph of given system

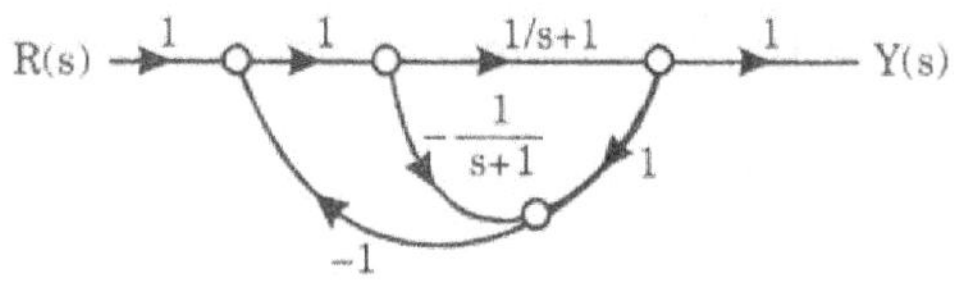

$$P_1 = \frac{1}{s+1};\ \Delta_1 = 1$$

$$L_1 = \frac{1}{s+1};\ \Delta = 1 - \{L_1 + L_2]$$

$$L_2 = -\frac{1}{s+1}$$

$$\therefore \quad \frac{Y(s)}{R(s)} = \frac{P_1\Delta_1}{\Delta} = \frac{\left(\frac{1}{s+1}\right)(1)}{(1)} = \frac{1}{s+1}$$

18. Given : $\dfrac{Y(s)}{X(s)} = \dfrac{s}{s+p}$

Output, $y(t) = \cos\left(2t - \frac{\pi}{3}\right)$

$$= \cos 2t \cdot \frac{1}{2} + \sin 2t \cdot \frac{\sqrt{3}}{2}$$

Input, $x(t) = p\cos\left(2t - \frac{\pi}{2}\right)$

$$= p\cos 2t \cdot 0 + p \cdot \sin 2t \cdot 1 = p\sin 2t$$

$$\therefore \quad Y(s) = \frac{p \cdot 2}{s^2 + 4};$$

$$Y(s) = \frac{1}{2}\frac{s}{s^2+4} + \frac{\sqrt{3}}{2}\frac{2}{s^2+4}$$

$$\frac{Y(s)}{X(s)} = \frac{\frac{1}{2(s^2+4)}\left\{s + 2\sqrt{3}\right\}}{\frac{2p}{s^2+4}} = \frac{s + 2\sqrt{3}}{4p}$$

$$\therefore \quad \frac{s+2\sqrt{3}}{4p} = \frac{s}{s+p}$$

$$\frac{j\omega+2\sqrt{3}}{4p} = \frac{j\omega}{\omega j+p}$$

$$\Rightarrow \quad \frac{\sqrt{\omega^2+12}}{4p} = \frac{\omega}{\sqrt{\omega^2+p^2}}$$

$$\Rightarrow \quad (\omega^2+12)(\omega^2+p^2) = 16p^2\omega^2$$

But $\omega = 2$

$$\Rightarrow \quad \omega^2 = 4$$

$$\therefore \quad 16(4+p^2) = 16\cdot 4\cdot p^2$$

$$\Rightarrow \quad 4+p^2 = 4p^2$$

$$\Rightarrow \quad p = \frac{2}{\sqrt{3}}$$

19. At $\omega_1 = 0.1$, slope changes from 0 *dB* to + 20 *dB*

$$\therefore \quad \text{Factor} = \left(1+\frac{s}{0.1}\right)$$

At $\omega_2 = 10$, slope changes from + 20 *dB* to 0 *dB*

$$\therefore \quad \text{Factor} = \frac{1}{\left(1+\frac{s}{10}\right)}$$

So,
$$T(s) = \frac{k\left(1+\frac{s}{0.1}\right)}{\left(1+\frac{s}{10}\right)}$$

$$= \frac{k(10s+1)}{0.1s+1}$$

At $\omega = 0.1$, magnitude = 0 *dB*,

$$\therefore \quad 20\log k = 0$$

$$\Rightarrow \quad k = 1$$

$$\therefore \quad T(s) = \frac{1+10s}{1+0.1s}$$

20. Conventional AM wave is represented as,

$$x(t) = A_C[1+k_a m(t)]\cos(2\pi f_c t) \quad \text{...(A)}$$

For over modulation, $|k_a m(t)|_{\max} > 1$

Hence, only signals given in option (*b*) and (*c*) are conventional signal.

For, $x(t) = A_c\cos(2\pi f_c t) + \frac{A_c}{4} m(t)\cos(2\pi f_c t)$

$$|k_a m(t)|_{\max} = \left|\frac{1}{4} m(t)\right|_{\max} = \left|\frac{1}{4}\cdot 2\right| = \frac{1}{2} < 1$$

(Under modulated)

For, $x(t) = A_c[1+m(t)]\cos(2\pi f_c t)$

$$|k_a m(t)|_{\max} = |1\cdot m(t)|_{\max} = 2 > 1$$

(Over modulated)

21. The terms $[2\sin(8000\pi t) + 4\cos(8000\pi t)]$ can be expressed as,

$$[\cos\alpha\sin(8000\pi t) + \sin\alpha\cdot\cos(8000\pi t)]$$

Hence $\cos\alpha = \frac{2}{\sqrt{20}} = \frac{1}{\sqrt{5}}, \sin\alpha = \frac{2}{\sqrt{5}}$

or, $\sqrt{20}[\sin(8000\pi t+\phi)]$,

where $\tan^{-1}\phi = 2$

Hence, $x(t) = 6\cos[2\pi\times 10^6 t + \sqrt{20}\cdot\sin(8000\pi t+\phi)]$

For angle-modulated signal hainvg constant amplitude A_c.

The average power is given by,

$$P_{av} = \frac{A_c^2}{2} = \frac{6^2}{2} = 18 \text{ watts}$$

22. Given : $[S] = \begin{bmatrix} 0.2\angle 0^\circ & 0.9\angle 90^\circ \\ 0.9\angle 90^\circ & 0.1\angle 90^\circ \end{bmatrix} = \begin{bmatrix} S_{11} & S_{12} \\ S_{21} & S_{22} \end{bmatrix}$

For lossless reciprocal networks,

$$|S_{11}| = |S_{22}|$$

and $|S_{11}|^2 + |S_{22}|^2 = 1$

For reciprocal, $S_{12} = S_{22}$, so system is reciprocal.

23. Given : $Z_0 = 50\ \Omega$, $R = 0.1\ \Omega/m$

For distortion loss, $\frac{R}{L} = \frac{G}{C}$

Attenuation in (Np/m), $\alpha = \sqrt{RG}$

$$Z_0 = \sqrt{\frac{R}{G}}$$

$$\Rightarrow \quad G = \frac{R}{Z_0^2}$$

$$\therefore \quad \alpha = \sqrt{\frac{R\cdot R}{Z_0^2}}$$

$$= \frac{R}{Z_0} = \frac{0.1}{50} = 0.002 \text{ Np/m}$$

24. Impulse response of the matched filter,

$$h(t) = s(T-t)$$

Given,

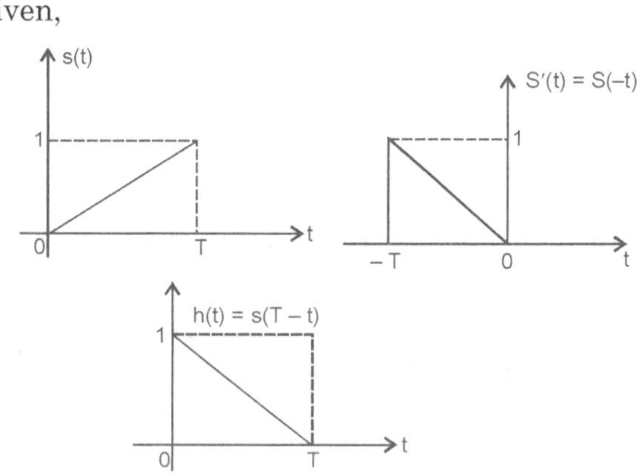

25. Given : E = 1 V/m, $\varepsilon_r = 4$

Average power density,

$$P_{av} = \frac{E^2}{2\eta}$$

$$= \frac{-1}{2\times 120\pi\sqrt{\frac{\mu_r}{\varepsilon_r}}}$$

$$= \frac{1}{2\times 120\pi\times\sqrt{\frac{1}{4}}} = \frac{1}{120\pi}$$

26. Given : $e^y = x^{\frac{1}{x}}$

Taking log, $\ln e^y = \ln x^{\frac{1}{x}}$

$\Rightarrow \quad y = \frac{1}{x}\ln x$

Differentiating

$$\frac{dy}{dx} = \frac{-1}{x^2}(\ln x) + \left(\frac{1}{x^2}\right) = 0 \quad \ldots(i)$$

$\Rightarrow \quad \frac{1}{x^2}\{-\ln x + 1\} = 0$

$\Rightarrow \quad \frac{1}{x^2} = 0 \text{ or } \ln x = 1$

$\Rightarrow \quad x = \infty \text{ or } x = e^1 = e$

Differentiating equation (*i*), we have

$$\frac{d^2y}{dx^2} = \frac{-2}{x^3} - \left\{\frac{1}{x^3} - \ln x \cdot \frac{2}{x^3}\right\}$$

For $x = e$, $\frac{d^2y}{dx^2} = -\frac{2}{e^3} - \left\{\frac{1}{e^3} - \frac{2}{e^3}\right\} = -\frac{1}{e^3}$ (–ve)

Hence maxima at $x = e$.

27. Required probability

$$= {}^4C_4\left(\frac{1}{2}\right)^4\left(\frac{1}{2}\right)^0 + {}^4C_3\left(\frac{1}{2}\right)^3\left(\frac{1}{2}\right)^1$$

$$= \left(\frac{1}{2}\right)^4 \cdot 1 + 4\cdot\left(\frac{1}{2}\right)^4$$

$$= 5\left(\frac{1}{2}\right)^4 = \frac{5}{16}$$

28. Given : $\vec{A} = xy\hat{a}_x + x^2\hat{a}_y$

$\oint \vec{A}\cdot\vec{dl} = ?$

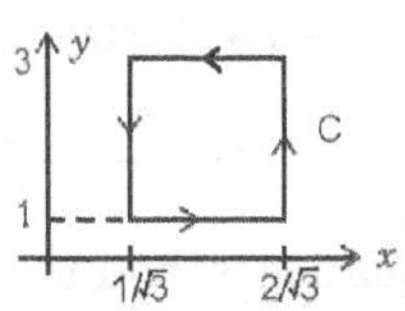

Line integral $= \oint (xy\hat{a}_x + x^2\hat{a}_y)\cdot(dx\hat{a}_x + dy\hat{a}_y)$

$= \oint (xydx + x^2dy)$

$= \oint xydx + \oint x^2dy$

$$= \int_{\frac{2}{\sqrt{3}}}^{\frac{2}{\sqrt{3}}} x\cdot 1dx + \int_{\frac{2}{\sqrt{3}}}^{-\frac{1}{\sqrt{3}}} x\cdot 3dx + \int_3^1\left(\frac{1}{\sqrt{3}}\right)^2\cdot dy + \int_1^3\left(\frac{2}{\sqrt{3}}\right)^2\cdot dy$$

$$= \left\{\frac{\left(\frac{2}{\sqrt{3}}\right)^2 - \left(\frac{1}{\sqrt{3}}\right)^2}{2}\right\} + \left[\frac{3}{2}\left\{\left(\frac{1}{\sqrt{3}}\right)^2 - \left(\frac{2}{\sqrt{3}}\right)^2\right\}\right]$$

$$+\frac{1}{3}(1-3) + \frac{4}{3}(3-1)$$

$$= \frac{1}{2} + \left(-\frac{3}{2}\right) + \left(-\frac{2}{3}\right) + \frac{8}{3}$$

$= -1 + 2 = 1$

29. Given : $x(z) = \frac{1-2z}{z(z-1)(z-2)}$

Residue at $z = a$ of G(*z*)

$$= \ln_{z\to a} (z-a)\,G(z)$$

Pole 0, 1, 2

At $P_1 = 0$,

Residue: $R_1 = \lim_{z\to 0}\frac{(1-2z)}{(z-1)(z-2)} = \frac{1}{2}$

At $P_2 = 1$,

Residue : $R_2 = \lim_{z\to 1}\frac{(1-2z)}{z(z-2)} = 1$

At $P_3 = 0$,

Residue : $R_3 = \lim_{z\to 2}\frac{1-2z}{z(z-1)} = -\frac{3}{2}$

30. Given : $\frac{dy(x)}{dx} - y(x) = x; \quad y(0) = 0$

$h = 0.1 \qquad y(0.3) = ?$

$x_1 = 0$

$y_1 = 0$

$\therefore \left(\frac{dy}{dx}\right)_1 = 0 - 0 = 0$

$y_2 = y_1 + h\left(\frac{dy}{dx}\right)_1 = 0 + 0.1\times 0 = 0$

$x_2 = x_1 + h = 0.1$

$y_2 = 0$

$= x_2 + y_2(x)$

$\therefore \left(\frac{dy}{dx}\right)_2 = 0.1 + 0 = 0.1$

$y_3 = y_2 + h\left(\frac{dy}{dx}\right)_2 = 0 + 0.1\times 0.1 = 0.01$

$x_3 = x_2 + \Delta = 0.2$

$y_3 = 0.01$

$\therefore \left(\frac{dy}{dx}\right)_3 = x_3 + y_3(x)$

$= 0.2 + 0.01$

$= 0.21$

$\therefore \quad y_4 = y_3 + h\cdot\left(\frac{dy}{dy}\right)_3$

$$= 0.01 + 0.1 * 0.21$$
$$= 0.01 + 0.021$$
$$= 0.031$$
$$x_4 = x_3 + \Delta = 0.3,$$
$$y_4 = 0.31$$
$$\therefore \quad y(0.3) = 0.031$$

31. Given : $f(t) = \mathcal{L}^{-1}\left[\dfrac{(3s+1)}{s^3+4s^2+(k-3)s}\right]$

$$\Rightarrow \quad F(s) = \mathcal{L}\{F(t)\} = \frac{3s+1}{s^3+4s^2+(k-3)s}$$

$$\lim_{t\to\infty} f(t) = \lim_{s\to 0} F(s) = 1 \text{ (given)}$$

$$\therefore \quad \lim_{s\to 0} \frac{s\cdot(3s+1)}{s^3+4s^2+(k-3)s} = 1$$

$$\Rightarrow \quad \lim_{s\to 0} \frac{3s+1}{s^2+4s+k-3} = 1$$

$$\Rightarrow \quad \frac{1}{k-3} = 1$$

$$\Rightarrow \quad k-3 = 1$$

$$\Rightarrow \quad k = 4$$

32.

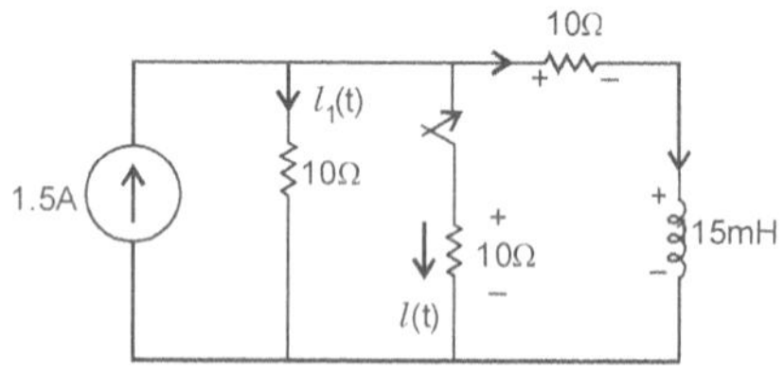

$$i(0^+) = \frac{0.75}{2} = 0.375 \text{ A}$$

$$i(\infty) = 0.5 \text{ A}$$

$$i(t) = i(\infty) - \{i(\infty) - i(0^+)\}\, e^{-\frac{Rt}{L}}$$

where, R is equivalent resistance seen across L with current source opened.

$$R = 10 + (10 \| 10)$$
$$= 15\ \Omega$$

$$i(t) = 0.5 - \{0.5 - 0.375\}e\, e^{-\frac{5t}{15\times 10^{-3}}}$$

$$i(t) = 0.5 - 0.125\, e^{-1000\,t} \text{ A}$$

33.

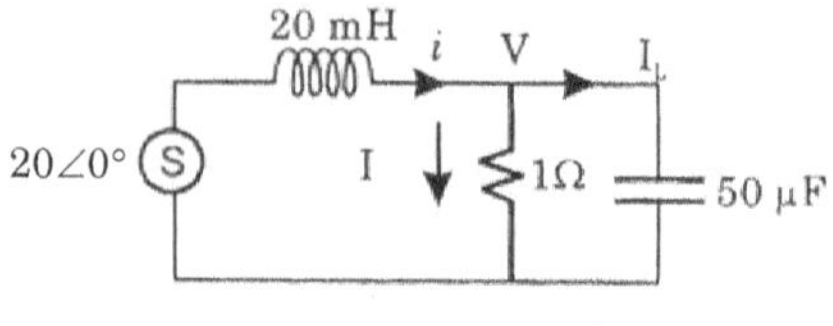

$$V = 20\angle 0°$$
$$\omega = 10^3 \text{ rad/ s}$$
$$i = I + I_L$$
$$i = \frac{V}{X_L + R \| X_C}$$
$$\therefore \quad I = \frac{X_c}{R + X_C}(i)$$

But $\quad X_C = \dfrac{-j}{\omega C} = \dfrac{-j}{10^3 \times 50 \times 10^{-6}} = -j20$

$$X_L = j\omega L = j10^3 \times 20\text{m} = 20j$$
$$R = 1\ \Omega$$
$$V = 20$$

$$\therefore \quad i = \frac{20}{20j + \dfrac{1-(-j20)}{1-j20}}$$

$$= \frac{20(1-j20)}{20j + 400 - j20} = \frac{1}{20}(1-j20)$$

$$\therefore \quad I = \frac{-j20}{1-j20}\cdot\frac{1}{20}(1-j20) = -j1 \text{ Amp}$$

34.

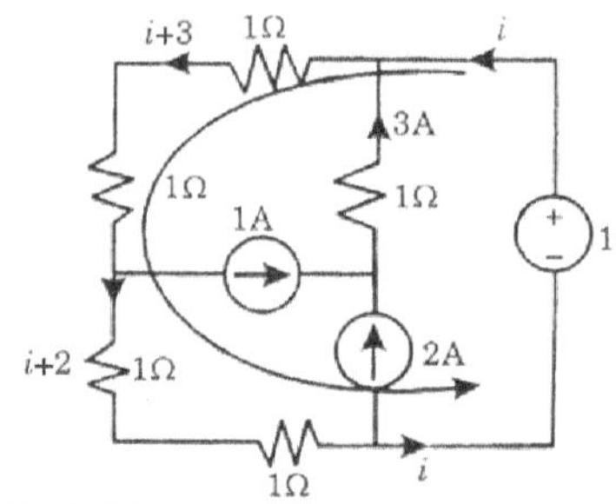

By KVL

$$10 = (i+3)*(1+1) + (i+2)*(1+1)$$

$$\Rightarrow \quad 10 = 2i + 6 + 2i + 4$$

$$\Rightarrow \quad 4i = 0$$

$$\Rightarrow \quad i = 0$$

Hence power drawn from supply,

$$P = Vi = 10 \times 0 = 0 \text{ W}$$

Alternately

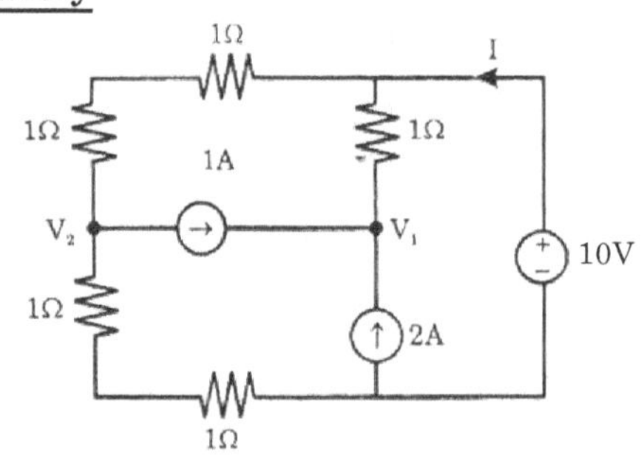

By node analysis :

$$\frac{10 - V_2}{2} - I + 10 - V_1 = 0 \qquad \ldots(i)$$

$$10 - V_1 + I + 2 = 0 \qquad \ldots(ii)$$

and $\quad \dfrac{10 - V_2}{2} + (-1) - \dfrac{V_2}{2} = 0 \qquad \ldots(iii)$

From three equation (s) and three variables I, V_1, V_2

$$I = 0,\ V_2 = 4,\ V_1 = 13$$

Hence power drawn from the supply,

$$P = 10 \times I = 0$$

35. N_E, N_B, N_C

Since emitter injection efficiency of the BJT is close to unity.

$\therefore \quad N_E > N_B > N_C$

and $\quad N_E >> N_B$

36. Given : p–n junction with $N_A = N_D = 10^4/cm^3$

For $N_A = N_D = 10^{20}/cm^3$

Since doping increases, so reverse breakdown voltage decreases and depletion capacitor increases.

37.

CLK	$\frac{D_A}{(Q_B \oplus Q_C)}$	DB (Q_A)	DC(Q_B)	Q_A	Q_B	Q_C
0				0	0	0
1	1	0	0	1	0	0
2	1	1	0	1	1	0
3	0	1	1	0	1	1
4	1	0	1	1	0	1
5	0	1	0	0	1	0
6	0	0	1	0	0	1
7	0	0	0	0	0	0

38.

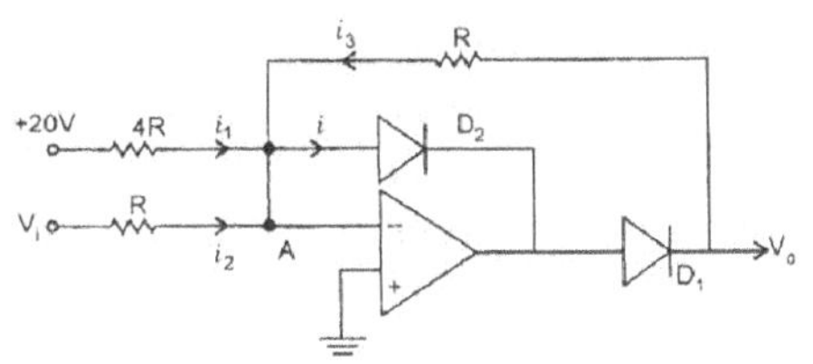

Since input is connected to negative terminal, so output always (+ve)

Since $\quad v_A = 0$ (virtually grounded)

Case 1 : $i_1 + i_2 + i_3 = i$

$$\therefore \quad \frac{20}{4R} + \frac{V_i}{R} + \frac{V_0}{R} = i$$

If $v_0 = +$ ve, diode not conduct, so $i = 0$

$$\therefore \quad \frac{20}{4R} + \frac{V_i}{R} + \frac{V_0}{R} = 0$$

$\Rightarrow \quad 5 + v_i + v_0 = 0$

$\Rightarrow \quad v_0 = -5 - v_i \geq 0$

$\Rightarrow \quad -v_i \geq 5$

$\Rightarrow \quad v_i \leq -5$

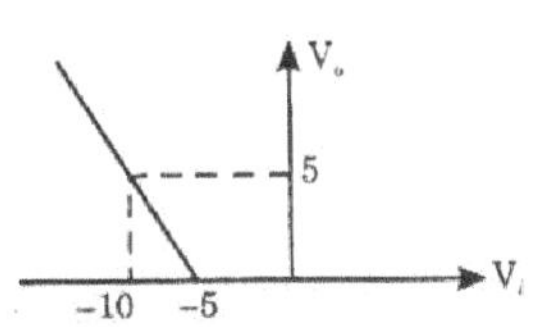

Case 2 : Since diode will conduct, hence

For $v_i > -5$, $\quad v_0 = 0$

For input $v_i = -10$,

$$v_0 = -5 + 10$$
$$= +5 \text{ volts}$$

<u>Alternately</u>

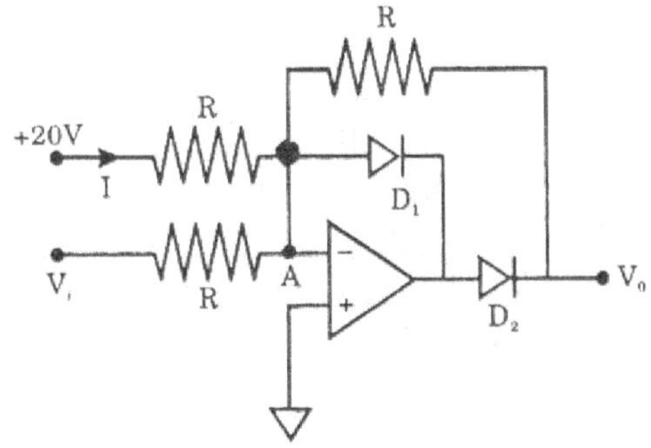

Point A will be at zero volt, so 20 V source supply current,

$$I = \frac{20}{R}$$

Case 1 : $v_i \geq 0$,

v_i will supply current

D_1 and D_2 will be in forward bias, so

$$v_0 = 0$$

Case 2 : $-5 \leq v_i < 0$,

+ 20 V source will supply current to both v_i source and circuit and diodes D_1 and D_2 will still in forward bias. So

$\therefore \quad v_0 = 0$

Case 3 : $\quad v_i < -5$,

So v_i will sink current more than $\frac{20}{R} = I$

So, D_1 and D_2 will be in reverse bias

$\therefore \quad v_0 \propto v_i$

So

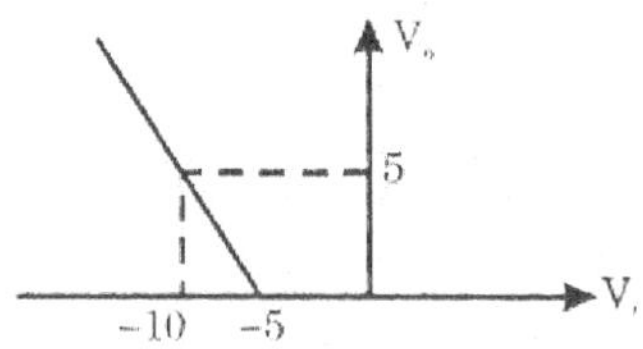

39. For 4 × 1 MUX,

$$F = \bar{S}_1\bar{S}_0 I_0 + \bar{S}_1 S_0 I_1 + S_1\bar{S}_0 I_2 + S_1 S_0 I_3$$

Given : $S_1 = A$, $S_0 = B$, $I_0 = C$, $I_1 = D$, $I_2 = \bar{C}$,

$$I_3 = \bar{C}\cdot\bar{D}$$

$\therefore$ F (A, B, C, D)

$= \bar{A}\bar{B}C + \bar{A}BD + A\bar{B}\bar{C} + AB\bar{C}\bar{D}$

$= \bar{A}\bar{B}C(D + \bar{D}) + \bar{A}B(C + \bar{C})D + A\bar{B}\bar{C}(D+\bar{D}) + AB\bar{C}\bar{D}$

$$= \underbrace{\bar{A}\bar{B}CD}_{3} + \underbrace{\bar{A}\bar{B}C\bar{D}}_{2} + \underbrace{\bar{A}BCD}_{7} + \underbrace{\bar{A}B\bar{C}D}_{5} + \underbrace{A\bar{B}\bar{C}D}_{9} + \underbrace{A\bar{B}\bar{C}\bar{D}}_{8} + \underbrace{AB\bar{C}\bar{D}}_{12}$$

$= \Sigma m\,(2, 3, 5, 7, 8, 9, 12)$

40. MVI A, 45H A = 45H
MOV B, A B = 45 H
STC set carry ⇒ carry flag = 1
CMC compliment carry ⇒ carry Flag = 0
RAR Rotate with carry

0	1	0	0	0	1	0	1	0

⇒ carry Flag = 1

$$A = \underbrace{0010}_{2}\ \underbrace{0010}_{2} = 22H$$

XRA B → XOR with 45H
22H = 00100010
45 H = 01000101
Output 01100111
⇒ A = 67 H

41. Zero initial conditions : $y(t) = ?$ for $x(t) = e^{-2t}\,u(t)$

Given equation is

$$\ddot{y}(t) + 4\dot{y}(t) + 3y(t) = 2\dot{x}(t) + 4x(t)$$

Taking Laplace

$s^2Y(s) + 4sY(s) + 3Y(s) = 2sX(s) + 4X(s)$

$$\Rightarrow \quad \frac{Y(s)}{X(s)} = \frac{2s+4}{s^2+4s+3} \qquad ...(i)$$

Given : $X(s) = \dfrac{1}{(s+2)}$

$$\therefore \quad Y(s) = \frac{(2s+4)}{(s^2+4s+3)(s+2)}$$

$$= \frac{2}{(s+1)(s+3)}$$

Taking inverse laplace transform

$$\Rightarrow \quad Y(s) = \left\{\frac{1}{s+1} - \frac{1}{s+3}\right\}$$

$$\therefore \quad y(t) = \left(e^{-t} - e^{-3t}\right)u(t)$$

42. A discrete-time LT1 system is B1B0 (Bounded input bounded output) stable if and only if its impulse response $h[n]$ is absolutely summable, that is

$$\sum_{n=-\infty}^{\infty} |h[n]| < \infty$$

Now, $\quad H(z) = \sum_{n=-\infty}^{\infty} h[n]\, z^{-n}$

let $\quad Z = e^{j\Omega}$ so that

$|z| = |e^{-1\Omega}| = 1$ then

$$|H/(e^{j\Omega})| = \frac{N \qquad 0}{2}$$

$$= \sum_{n=-\infty}^{\infty} |h[n]| < \infty$$

It can be inferred that if the system is stable, then H(z) converges for $z = e^{j\Omega}$. hence, for a stable discrete-time LT1 system, the ROC of H(z) must contain the unit circle $|z| = 1$.

Further, for the system having impulse response

$$h[n] = \alpha^n u[n],\ H(z) = \frac{z}{z-\alpha},\quad |z| > |\alpha|$$

For $|z| > |\alpha|$, ROC of H(z) includes $z = \infty$, therefore $h(n)$ is a causal sequence and thus the system is causal.

Now, Given, $$H(z) = \frac{2 - \frac{3}{4}Z^{-1}}{1 - \frac{3}{4}Z^{-1} + \frac{1}{8}Z^{-2}}$$

$$= \frac{1}{1-\frac{1}{4}Z^{-1}} + \frac{1}{1-\frac{1}{2}Z^{-1}}$$

For, S1 : ROC, $|z| > \frac{1}{2}$, H(z) contain unit circle and includes $z = \infty$, the system is stable and causal.

For, S2 : ROC $|z| < \frac{1}{4}$, H(z) does not contain unit circle and excludes $z = \infty$, the system is neither stable nor causal.

For, S3 : ROC $\frac{1}{4} < |z| < \frac{1}{2}$, H($z$) does not contain unit circle and excludes $z = \infty$, the system is neither stable nor causal.

43. $$s(t) = \frac{\sin 500\pi t}{\pi t} \times \frac{\sin 700\pi t}{\pi t}$$

Assume, $s_1(t) = \dfrac{\sin 500\pi t}{\pi t}$

and $\quad s_2(t) = \dfrac{\sin 700\pi t}{\pi t}$

$s_1(t)$ and its fourier transform is shown below :

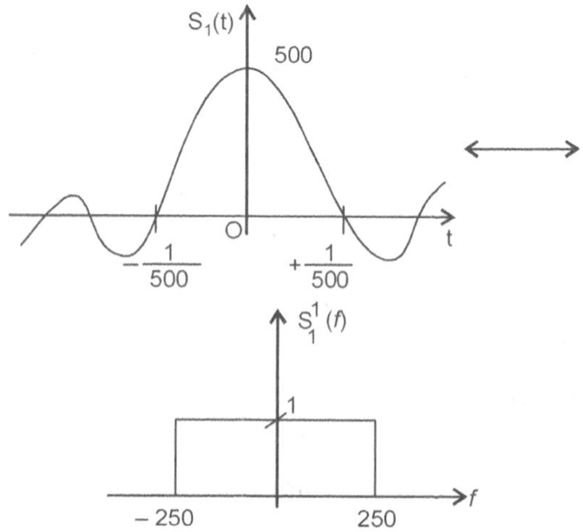

$s_2(t)$ and its fouriar transfroum is shown below :

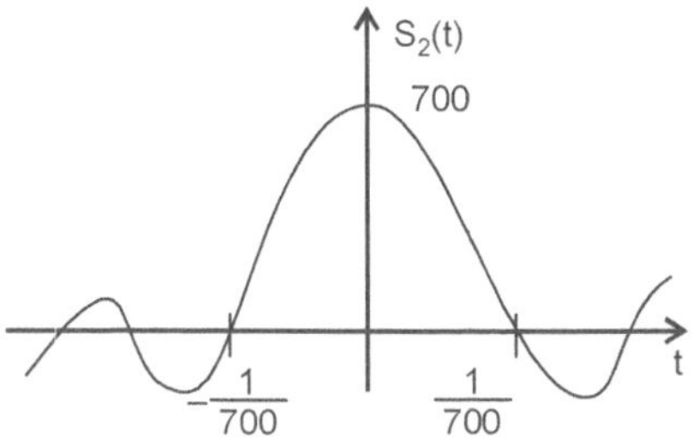

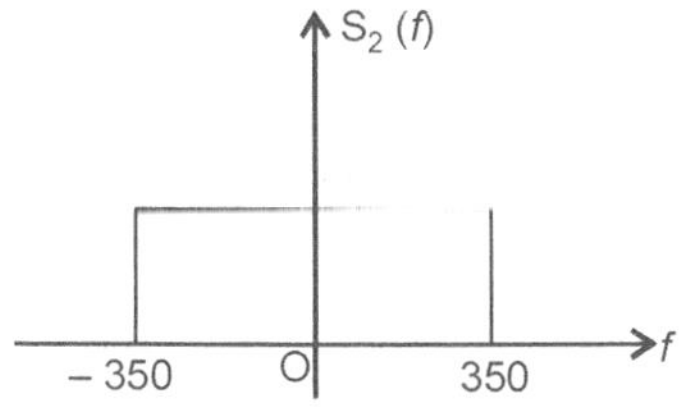

From frequency convolution theorem, $x_1(t)x_2(t) \leftrightarrow 1\, x_1(f) \cdot x_2(f)$, we find that $s(t)$ is band-limited signal and its bandwidth is equal to the sum of bandwidths of $s_1(t)$ and $s_2(t)$, i.e., 600 Hz. Thus, Nyquist rate is 1200 Hz.

44. $E(s) = R(s) - C(s)$

$E(s) = R(s) - E(s) \cdot C(s)\, G_c(s)$

$$E(s) = \frac{R(s)}{1+G(s)G_c(s)}$$

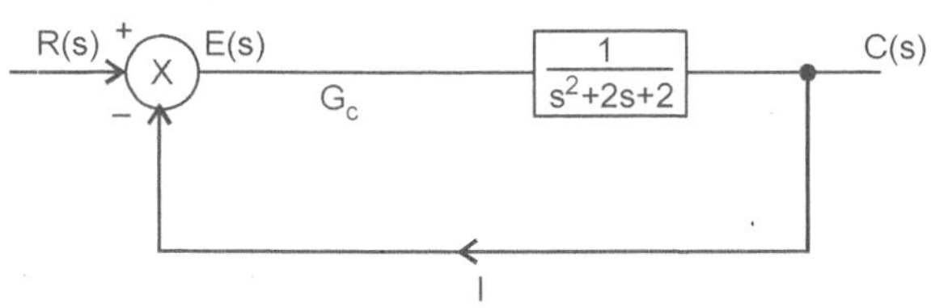

Here, $R(s) = \frac{1}{s}$

$$\lim_{s\to 0} G(s) = \frac{1}{0+0+2} = \frac{1}{2} = 0.5$$

$\therefore$ Steady state error, $e_{ss} = \lim_{s\to 0} sE(s)$

$$= \lim_{s\to 0} \frac{1}{1+G(s)G_c(s)}$$

For unit step input,

$$k(s) = \frac{1}{s}$$

$$\therefore \quad e_{ss} = \lim_{s\to 0} \frac{1}{1+G(s)G_c(s)}$$

$$= \frac{1}{1+\lim_{s\to 0} G(s) \lim_{s\to 0} G_c(s)}$$

$$= \frac{1}{1+0.5\lim_{s\to 0} G_c(s)}$$

e_{ss} = minimum

if $\quad \underset{s\to 0}{G_c(s)}$ = maximum

Hence for option (*d*) : $\underset{s\to 0}{G_c(s)} = \infty$

$\therefore \quad e_{ss} = 0$

45.

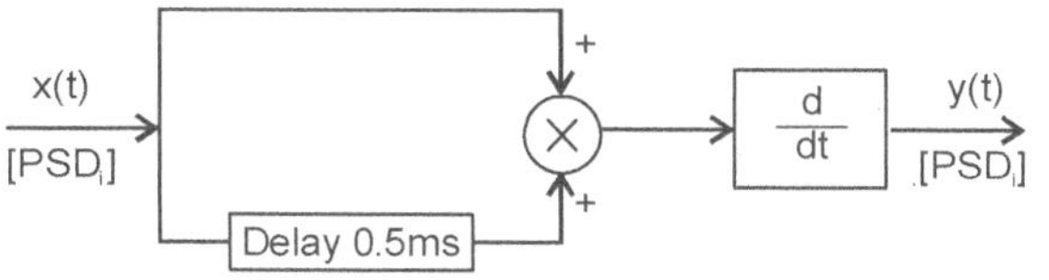

$$y'(t) = \frac{d}{dt}[x(t) + x(t-T)]$$

$$y'(f) = \left[x(f) + e^{-j2\pi fT} \times (f)\right] j\omega$$

$$H'(f) = j\omega\left[1 + e^{-j2\pi fT}\right]$$

$$= [1 + \cos 2\pi fT - j\sin 2\pi fT] j\omega$$

$$|H'(f)|^2 = \omega^2.2.(1 + \cos 2\pi fT)$$

$\therefore \quad PSD_o = |H(f)|^2 PSD_i$

$$= 2\omega^2[1 + \cos(\pi.2fT)]\, PSD_i$$

$\therefore \; [1 + \cos(\pi 2fT)] = 0$; for $f = \frac{(2n+1)}{2T}$

46. $\bar{E}_i = 24\cos(3\times 10^8 t - \beta y)\bar{a}_z V/m$

Comparing $\bar{E}_i$ with $\bar{E} = E_m \cos(wt - \beta y) \cdot \bar{a}_z V/m$ it can be inferred that the plane wave given by $\bar{E} \times \bar{H}$ is travelling in the y direction having electric field along the $+\bar{a}_z$ direction and magnetic field along $+\bar{a}_x$ direction.

Now,

as, $\frac{|\bar{E}_i|}{|\bar{H}_i|} = \eta$

$$\bar{H}_i = \frac{\bar{E}_i}{\eta}$$

$$= \frac{1}{5\pi}\cos(3\times 10^8 t - \beta y) \cdot \bar{a}_x A/m$$

Also, $\frac{\bar{H}_r}{\bar{H}_i} = \frac{\eta_1 - \eta_2}{\eta_1 + \eta_2}$

Given, $u_{r_1} = u_{r_2} = 1$

$E_{r_1} = E_{r_2} = 9$

$$\frac{\sqrt{\frac{\mu_0 \mu_{r_1}}{E_{r_1} E_0}} - \sqrt{\frac{\mu_0}{E_0} \cdot \frac{\mu_{r_2}}{E_{r_2}}}}{\sqrt{\frac{\mu_0}{E_0} \cdot \frac{\mu_{r_1}}{E_{r_1}}} \sqrt{\frac{\mu_0}{E_0} \cdot \frac{\mu_{r_2}}{E_{r_2}}}} = \frac{\frac{1}{\sqrt{E_{r_1}}} - \frac{1}{\sqrt{E_{r_2}}}}{\frac{1}{\sqrt{E_{r_1}}} + \frac{1}{\sqrt{E_{r_2}}}}$$

$$= \frac{\sqrt{9} - \sqrt{1}}{\sqrt{9} + \sqrt{1}} = \frac{1}{2}$$

then, $\bar{H}_r = \frac{1}{10\pi}\cos(3\times 10^8 t - \beta y)\bar{a}_x A/m$

47.

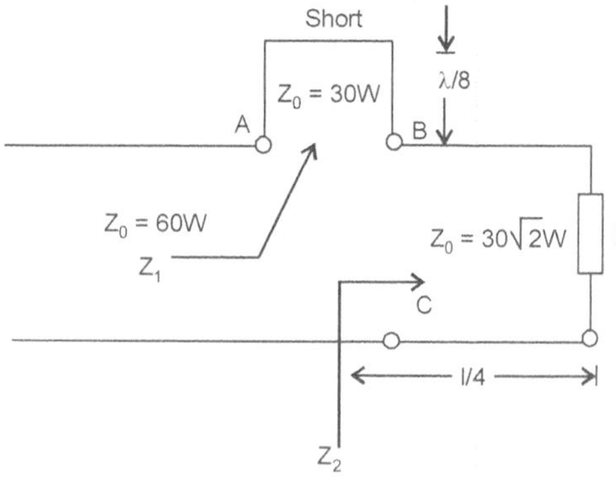

In input impedance is given by,

$$Z_{in} = Z_0\left(\frac{Z_L + jZ_0 + \tan\beta l}{Z_0 + jZ_L \tan\beta l}\right)$$

Where Z_0 = characteristics impedance,

Z_L = load impedance

l = length of the line,

$$\beta = \frac{2\pi}{\lambda}$$

then input impedance looking into terminal AB,

$$Z_1 = 30\left[\frac{0 + 30j\tan\left(\frac{2\pi}{\lambda}\cdot\frac{\lambda}{8}\right)}{30+0}\right] = 30j$$

Input impedance looking into terminal BC,

$$Z_2 = 30\sqrt{2}\left[\frac{30 + j30\sqrt{2}\tan\left(\frac{2\pi}{\lambda}\cdot\frac{\lambda}{4}\right)}{30\sqrt{2} + 30j\tan\left(\frac{2\pi}{\lambda}\cdot\frac{\lambda}{4}\right)}\right] = 60$$

$$= 30\sqrt{2}\left[\frac{\frac{30}{\tan\frac{\pi}{2}} + \frac{j30\sqrt{2}}{\frac{30\sqrt{2}}{\tan\frac{\pi}{2}}} + 30j}{}\right] = 60$$

Total load impedance looking into terminal AC,

$$Z_L = Z_1 + Z_2 = 60 + 30j$$

Reflection coefficient,

$$T_e = \frac{Z_L - Z_0}{Z_L + Z_0}$$

$$= \left[\frac{60 + 30j - 60}{60 + 30j + 60}\right]$$

$$= \left[\frac{j}{4+j}\right] = \frac{1}{\sqrt{17}}$$

Hence, VSWR, of $= \frac{1 + |j_e|}{1 - |j_e|}$

$$= \frac{1 + \frac{1}{\sqrt{17}}}{1 - \frac{1}{\sqrt{17}}} = 1.64$$

48. Small-signal equivalent model of the given circuit,

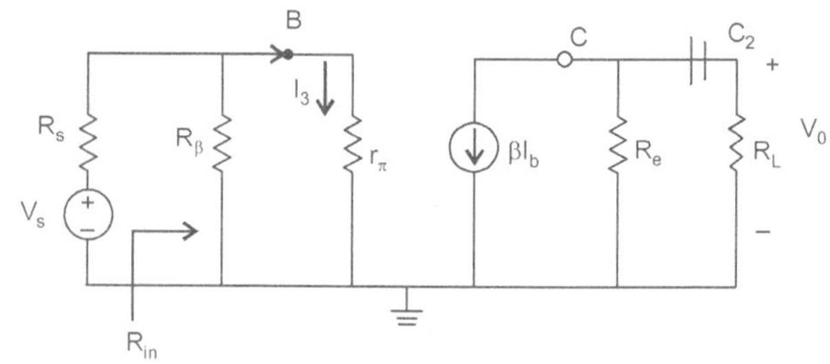

$$C_1 \to \infty$$

$\frac{1}{WC_1} \to 0$ [capacitor C_1 is short circuited]

Input initance,

$$R_{in} = r_\pi \,||\, R_B$$
$$= (259 \,||\, 93K)\Omega$$
$$= 258\Omega$$

Input resitance seen by source

$$R_{in} = R_{in} + R_s$$
$$= 1000 + 258$$
$$= 1258\Omega$$

49. Lower cut-off frequency,

$$f_L = \frac{1}{2\pi(R_L + R_C)C_2} = \frac{1}{2\pi \times 1250 \times 4.7 \times 10^{-6}}$$
$$= 27.1 \text{ Hz}$$

50.

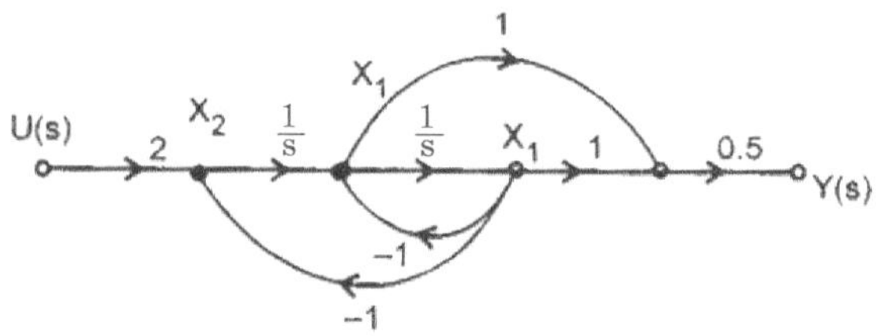

From signal flow graph

$$\dot{x}_2 = 2\,u(t) - x_1$$
$$\dot{x}_1 = x_2 - x_1$$
$$y = (x_1 + x_2) \times 0.5$$

$$\Rightarrow \quad [Y] = [0.5 \;\; 0.5]\begin{bmatrix} x_1 \\ x_2 \end{bmatrix}$$

$$\therefore \quad \dot{x}_1 = -x_1 + x_2$$
$$\dot{x}_2 = -x_1 + 2u(s)$$

$$\begin{bmatrix} \dot{x}_1 \\ \dot{x}_2 \end{bmatrix} = \begin{bmatrix} -1 & 1 \\ -1 & 0 \end{bmatrix}\begin{bmatrix} x_1 \\ x_2 \end{bmatrix} + \begin{bmatrix} 0 \\ 2 \end{bmatrix} u$$

51. By Mason gain formula

$$\frac{Y(s)}{U(s)} = \frac{P_k \Delta_k}{\Delta}$$
$$= \frac{P_1\Delta_1 + P_2\Delta_2}{\Delta}$$

Forward path gains

Here, $P_1 = 2\cdot\frac{1}{s}\cdot\frac{1}{s}\cdot 1(\cdot 0\cdot 5) = \frac{1}{s^2}$

$$P_2 = 2\cdot\frac{1}{s}\cdot 1\cdot 0\cdot 5 = \frac{1}{s}$$

Individual Loop gains,

$$L_1 = -\frac{1}{s}$$

$$L_2 = -\frac{1}{s^2}$$

$$\Delta = 1 - \left[-\frac{1}{s} - \frac{1}{s^2}\right]$$

$$= 1 + \frac{1}{s} + \frac{1}{s^2}$$

$$\Delta_1 = 1, \Delta_2 = 1$$

$$\therefore \quad \frac{Y(s)}{U(s)} = \frac{\frac{1}{s^2}\cdot 1 + \frac{1}{s}\cdot 1}{1 + \frac{1}{s} + \frac{1}{s^2}}$$

$$= \frac{s+1}{s^2 + s + 1}$$

52. 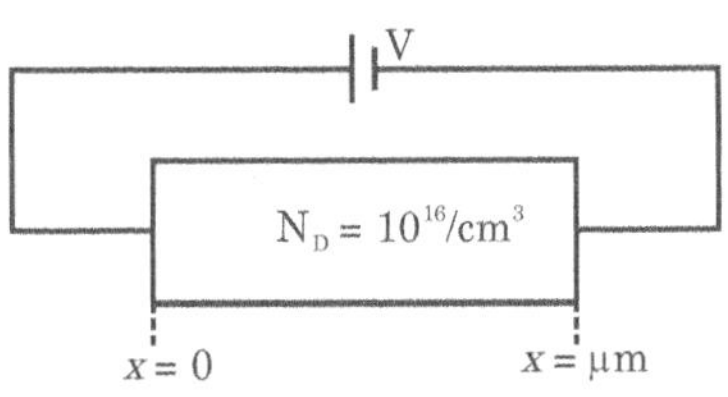

$$E = \frac{1}{1\mu} = 10K\,V/cm$$

It will remain same along the cross-section.

53. $J = N_D \times e \times \mu \times E$

Putting given data, we have

$$J = 10^{16} \times 1.6 \times 10^{-19} \times 1350 \times 10 \times 10^3$$

$$= 2.16 \times 10^4 \text{ A/cm}^2$$

54. Given : $\delta n(f) = \frac{N_0}{2}$

$$R(\tau) = \frac{N_0}{2} \times \delta(\tau)$$

$$R(0) = \frac{N_0}{2}$$

$$\text{Power} = 2 \times \frac{N_0}{2} \times 10^6 = 2 \times 10^{-14}$$

$$\text{Variance} = \frac{2}{\alpha^2}$$

which is equal to power since mean is zero.

$$\therefore \quad \alpha = 10^7$$

55. Probability of bit error $= \int_{\infty/2}^{\infty} P_{N_k}(n)$

$$= \int_{\infty/2}^{\infty} 0.5\alpha\, e^{-|\alpha||n|}$$

$$= 0.5e^{-10^{+7}\cdot 10^{-6}}$$

$$= 0.5e^{-10}$$

60. 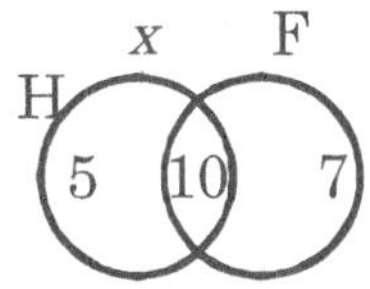

$\therefore \quad 5 + 10 + 7 + x = 25$

$\Rightarrow \quad x = 3$

62. By observation of summation, it can be seen that numbers are not in decimal. Let they have base n. Then converting all number in decimal

$$(137)_n = \left[1\times n^2 + 3n + 7n^0\right]_{10}$$

$$(276)_n = \left[2n^2 + 7n + 6n^0\right]_{10}$$

$$(435)_n = \left[4n^2 + 3n + 5n^0\right]_{10}$$

$$(731)_n = \left[7n^2 + 3n + 1n^0\right]_{10}$$

$$(672)_n = \left[6n^2 + 7n + 2n^0\right]_{10}$$

So $(137)_n + (276)_n = (435)_n$

$\Rightarrow n^2 + 3n + 7 + 2n^2 + 7n + 6n = 4n^2 + 3n + 5$

$\Rightarrow n^2 - 7n - 8 = 0$

$n = -1, 8$

Possible base $n = 8$ (+ ve)

So $(435)_n = (435)_8 = (4 \times 8^2 + 3 \times 8 + 5)_{10} = (285)_{10}$

$(731)_n = (731)_8 = (7 \times 8^2 + 3 \times 8 + 1)_{10} = (473)_{10}$

$(672)_n = (672)_8 = (6 \times 8^2 + 7 \times 8 + 2)_{10} = (442)_{10}$

So $(731)_8 + (672)_8 = (473)_{10} + (442)_{10} = (915)_{10}$

$(915)_{10} = (1623)_8$

63. 5 skilled workers build wall in 20 days

1 skilled worker build wall in 20 × 5 days

Hence in 1 day, part of work done by skilled work

$$= \frac{1}{100}$$

Similarly in 1 day part of work done by semi-skilled workers

$$= \frac{1}{25\times 8}$$

and in 1 day part of work done by un-skilled worker

$$= \frac{1}{30 \times 10}$$

So part of work done in 1 day by 2 skilled, 6 semi-skilled and 5 unskilled

$$= \frac{2}{100} + \frac{6}{200} + \frac{5}{300}$$

$$= \frac{1}{15}$$

So work done by given workers in days = 15

64. **Case 1** : First digit = 3 $\Rightarrow$ 3 ...

Rest 3 digit may be combination of

234 $\rightarrow$ 3!
223 $\rightarrow$ 3!/2!
224 $\rightarrow$ 3!/2!
332 $\rightarrow$ 3!/2!
334 $\rightarrow$ 3!/2!
442 $\rightarrow$ 3!/2!
443 $\rightarrow$ 3!/2!
444 $\rightarrow$ 1

So total combination = 25

Case 2 : First digit = 4 $\Rightarrow$ 4 ....

Rest 3 digit may be combination of

234 $\rightarrow$ 6
223 $\rightarrow$ 3
224 $\rightarrow$ 3
332 $\rightarrow$ 3
334 $\rightarrow$ 3
442 $\rightarrow$ 3
443 $\rightarrow$ 3
444 $\rightarrow$ 1
333 $\rightarrow$ 1

So total combination = 26

$\therefore$ Required answer = 25 + 26 = 51

65. (*i*) $H + G > I + S$

$\Rightarrow$ $I - H < G - S$ ...(1)

and (*ii*) $G - S = 1$ ...(2)

From equations (1) and (2), we get

$I - H < 1$

$\Rightarrow$ $H > I$

So order is SGHI

because G is not oldest and S is not youngest.

■■

SOLVED PAPER – 2009

ELECTRONICS AND COMMUNICATION ENGINEERING

Time : 3 Hours | Maximum Marks : 100

Q.1–20 carry one mark each.

1. The order of the differential equation

$$\frac{d^2y}{dt^2} + \left(\frac{dy}{dt}\right)^3 + y^4 = e^{-t} \text{ is}$$

(*a*) 1 (*b*) 2
(*c*) 3 (*d*) 4

2. The Fourier series of a real periodic function has only

P. cosine terms if it is even
Q. sine terms if it is even
R. cosine terms if it is odd
S. sine terms if it is odd

Which of the above statements are correct ?

(*a*) P and S (*b*) P and R
(*c*) Q and S (*d*) Q and R

3. A function is given by $f(t) = \sin^2 t + \cos 2t$. Which of the following is true ?

(*a*) f has frequency components at 0 and $\frac{1}{2\pi}$ Hz.

(*b*) f has frequency components at 0 and $\frac{1}{\pi}$ Hz.

(*c*) f has frequency components at $\frac{1}{2\pi}$ and $\frac{1}{\pi}$ Hz.

(*d*) f has frequency components at 0, $\frac{1}{2\pi}$ and $\frac{1}{\pi}$ Hz.

4. A fully charged mobile phone with a 12 V battery is good for a 10 minute talk-time. Assume that, during the talk-time, the battery delivers a constant current of 2 A and its voltage drops linearly from 12 V to 10 V as shown in the figure. How much energy does the battery deliver during this talk-time ?

(*a*) 220 J
(*b*) 12 kJ
(*c*) 13.2 kJ
(*d*) 14.4 kJ

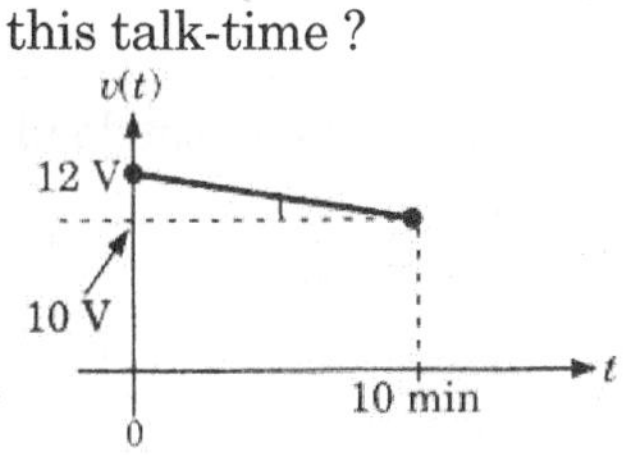

5. In an n-type silicon crystal at room temperature, which of the following can have a concentration of 4×10^{19} cm^{-3} ?

(*a*) Silicon atoms
(*b*) Holes
(*c*) Dopant atoms
(*d*) Valence electrons

6. The full forms of the abbreviations TTL and CMOS in reference to logic families are

(*a*) Triple Transistor Logic and Chip Metal Oxide Semiconductor
(*b*) Tristate Transistor Logic and Chip Metal Oxide Semiconductor
(*c*) Transistor Transistor Logic and Complementary Metal Oxide Semiconductor
(*d*) Tristate Transistor Logic and Complementary Metal Oxide silicon

7. The ROC of Z-transform of the discrete time sequence

$$x(n) = \left(\frac{1}{3}\right)^n u(n) - \left(\frac{1}{2}\right)^n u(-n-1) \text{ is}$$

(*a*) $\frac{1}{3} < |z| < \frac{1}{2}$ (*b*) $|z| > \frac{1}{2}$
(*c*) $|z| < \frac{1}{3}$ (*d*) $2 < |z| < 3$

8. The magnitude plot of a rational transfer fuction G(*s*) with real coefficients is shown below. Which of the following compensators has such a magnitude plot ?

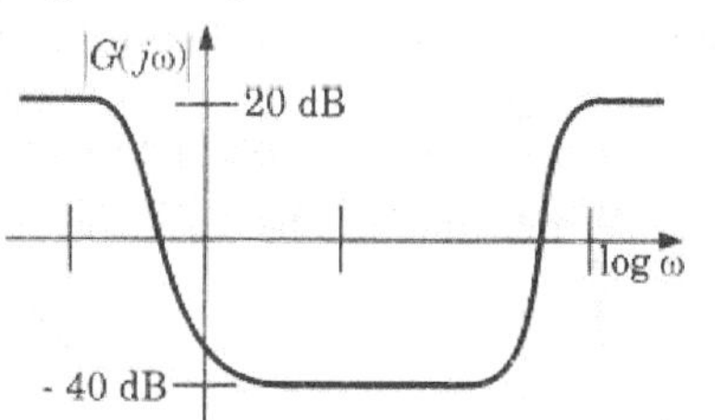

(*a*) Lead compensator
(*b*) Lag compensator
(*c*) PID compensator
(*d*) Lead-lag compensator

9. A white noise process $X(t)$ with two-sided power spectral density 1×10^{-10} W/Hz is input to a filter whose magnitude squared response is shown below.

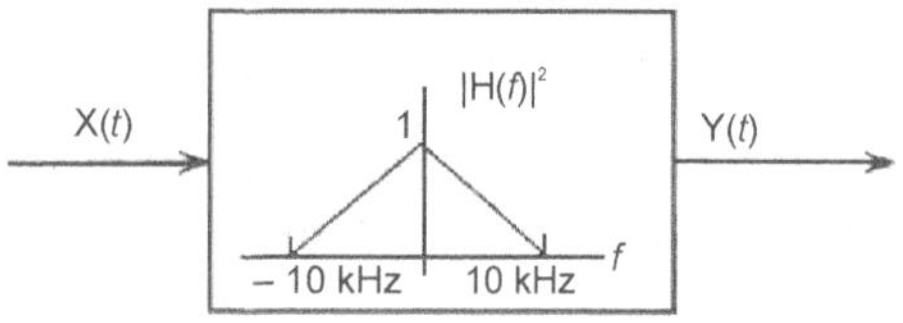

The power of the output process $Y(t)$ is given by

(a) 5×10^{-7} W (b) 1×10^{-6} W

(c) 2×10^{-6} W (d) 1×10^{-5} W

10. Which of the following statements is true regarding the fundamental mode of the metallic waveguides shown ?

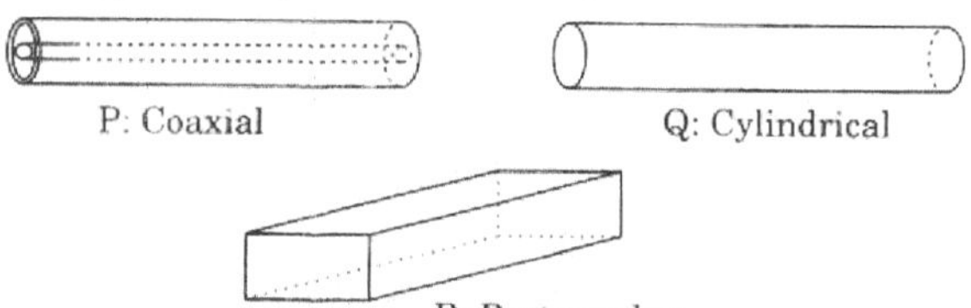

(a) Only P has no cut-off frequency

(b) Only Q has no cut-off frequency

(c) Only R has no cut-off frequency

(d) All three have cut-off frequencies

11. A fair coin is tossed 10 times. What is the probability that ONLY the first two tosses will yield heads ?

(a) $\left(\frac{1}{2}\right)^2$ (b) ${}^{10}C_2 \left(\frac{1}{2}\right)^3$

(c) $\left(\frac{1}{2}\right)^{10}$ (d) ${}^{10}C_2 \left(\frac{1}{2}\right)^{10}$

12. If the power spectral density of stationary random process is a sinc-squared function of frequency, the shape of its autocorrelation is

(a)
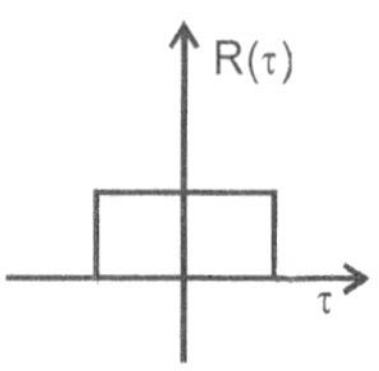

(b)
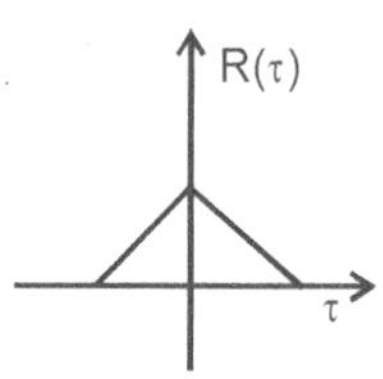

(c)
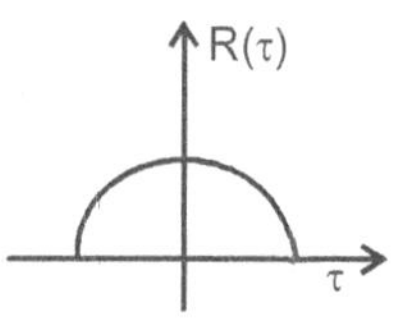

(d)
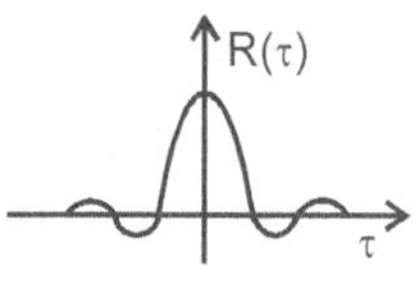

13. If $f(z) = C_0 + C_1 z^{-1}$, than $\oint_{\substack{unit \\ circle}} \frac{1+f(z)}{z} dz$ is given by

(a) $2\pi c_1$ (b) $2\pi(1+c_0)$

(c) $2\pi j c_1$ (d) $2\pi j(1+c_0)$

14. In the interconnection of ideal sources shown in the figure, it is known that the 60 V source is absorbing power.

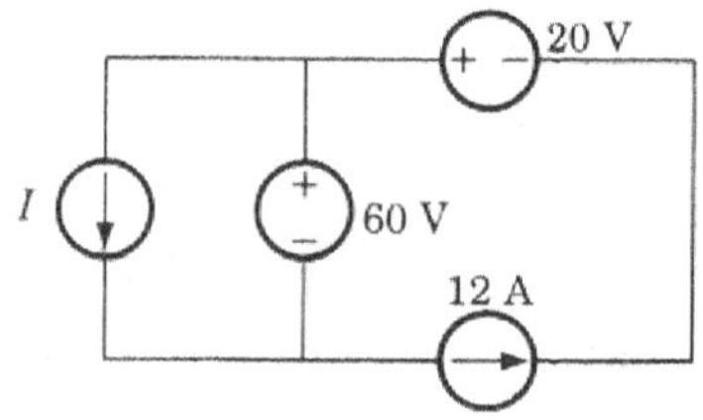

Which of the following can be the value of the current source I?

(a) 10 A (b) 13 A

(c) 15 A (d) 18 A

15. The ratio of the mobility to the diffusion coefficient in a semiconductor has the units

(a) V^{-1} (b) cm. V^{-1}

(c) V. cm^{-1} (d) V.s

16. In a microprocessor, the service routine for a certain interrupt starts from a fixed location of memory which cannot be externally set, but the interrupt can be delayed or rejected. Such an interrupt is

(a) non-maskable and non-vectored

(b) maskable and non-vectored

(c) no-maskable and vectored

(d) maskable and vectored

17. If the transfer function of the following network is

$$\frac{V_o(s)}{V_i(s)} = \frac{1}{2 + CRS}$$

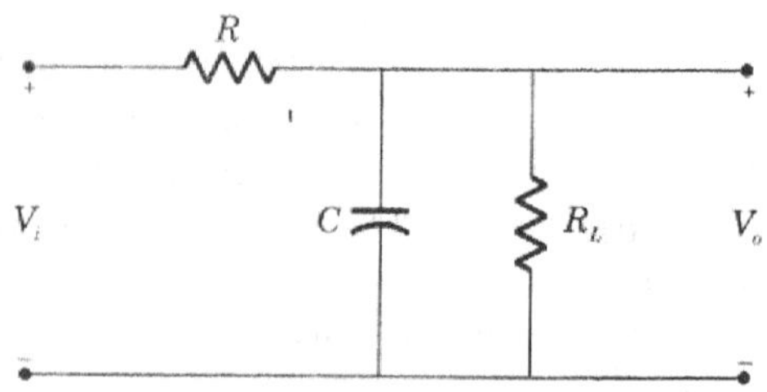

the value of the load resistance R_L is

(a) $\frac{R}{4}$ (b) $\frac{R}{2}$

(c) R (d) 2R

18. Consider the system $\frac{dx}{dt} = Ax + Bu$ with

$$A = \begin{bmatrix} 1 & 0 \\ 0 & 1 \end{bmatrix}, \text{ and } B = \begin{bmatrix} p \\ q \end{bmatrix}$$

where p and q are arbitrary real numbers.

Which of the following statements about the controllability of the system is true ?

(a) The systems is completely state controllable for any nonzero values of p and q

(b) Only $p = 0$ and $q = 0$ result in controllability

(c) The system is uncontrollable for all values of p and q

(d) We cannot conclude about controllability from the given data

19. For a message signal $m(t) = \cos(2\pi f_m t)$ and carrier of frequency f_o which of the following represents a single side-band (SSB) signal ?

(a) $\cos(2\pi f_m t)\cos(2\pi f_o t)$

(b) $\cos(2\pi f_c t)$

(c) $\cos[2\pi(f_m + f_c)t]$

(d) $[1+\cos(2\pi f_m)t]\cos(2\pi f_c t)$

20. Two infinitely long wires carrying current are as shown in the figure below. One wire is in the y-z plane and parallel to the y-axis. The other wire is in the x-y plane and parallel to the x-axis. Which components of the resulting magnetic field are non-zero at the origin ?

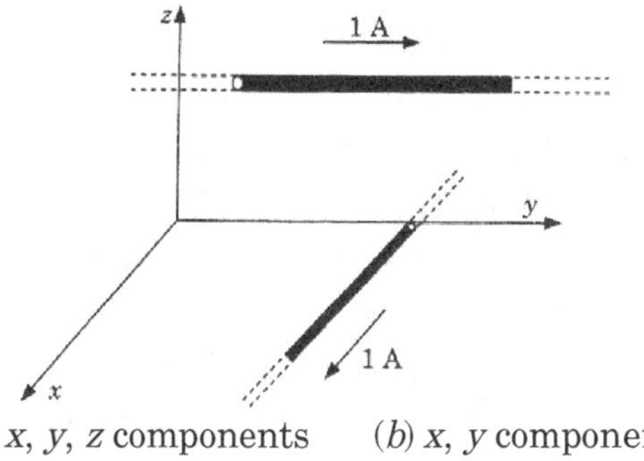

(a) x, y, z components (b) x, y components

(c) y, z components (d) x, z components

Q. 21 – 60 carry two marks each.

21. Consider two independent random variables X and Y with identical distributions. The variables X and Y take values 0, 1 and 2 with probabilities $\frac{1}{2}, \frac{1}{4}$ and $\frac{1}{4}$ respectively. What is theconditional probability

$P(X + Y = 2 \mid X - Y = 0)$?

(a) 0 (b) $\frac{1}{16}$

(c) $\frac{1}{6}$ (d) 1

22. The Taylor series expansion of $\frac{\sin x}{x-\pi}$ at $x = \pi$ is given by

(a) $1 + \frac{(x-\pi)^2}{3!} + \ldots$ (b) $-1 - \frac{(x-\pi)^2}{3!} + \ldots$

(c) $1 - \frac{(x-\pi)^2}{3!} + \ldots$ (d) $-1 + \frac{(x-\pi)^2}{3!} + \ldots$

23. If a vector field $\vec{V}$ is related to another vector field $\vec{A}$ through $\vec{V} = \nabla \times \vec{A}$, which of the following is true ?

Note: C and Sc refer to any closed contour and any surface whose boundary is C.

(a) $\oint_C \vec{V}.\vec{dl} = \iint_{S_c} \vec{A}.\vec{dS}$

(b) $\oint_C \vec{A}.\vec{dl} = \oint_S \nabla \times \vec{A}.\vec{ds}$

(c) $\oint_C \nabla \times \vec{V}\,\vec{dl} = \iint_{S_c} \nabla \times \vec{A}.\vec{dS}$

(d) $\oint_C \nabla \times \vec{A}.\vec{dl} = \iint_{S_c} \vec{V}.\vec{dS}$

24. Given that F(s) is the one-sided Laplace transform of $f(t)$, the Laplace transform of $\int_0^t f(\tau)d\tau$ is

(a) $sF(s) - f(0)$ (b) $\frac{1}{s}F(s)$

(c) $\int_0^s F(\tau)d\tau$ (d) $\frac{1}{s}[F(s) - f(0)]$

25. Match each differential equation in Group I to its family of solution curves from Group II.

Group I	Group II
A. $\frac{dy}{dx} = \frac{y}{x}$	1. Circles
B. $\frac{dy}{dx} = -\frac{y}{x}$	2. Straight lines
C. $\frac{dy}{dx} = \frac{x}{y}$	3. Hyperbolas
D. $\frac{dy}{dx} = -\frac{x}{y}$	

Codes :

	A	B	C	D
(a)	2	3	3	1
(b)	1	3	2	1
(c)	2	1	3	3
(d)	3	2	1	2

26. The eigen values of the following matrix are

$$\begin{bmatrix} -1 & 3 & 5 \\ -3 & -1 & 6 \\ 0 & 0 & 3 \end{bmatrix}$$

(a) 3, 3 + 5 j,– 6 – j (b) – 6 + 5j, 3 + j,3 –j

(c) 3 + j, 3 – j, 5 + j (d) 3, –1 + 3j, – 1– 3j

27. An AC source of RMS voltage 20 V with internal impedance Z_s = (1 + 2 j)Ω feeds a load of impedance $Z_L = (7+4j)\Omega$ in the figure below. The reactive power consumed by the load is

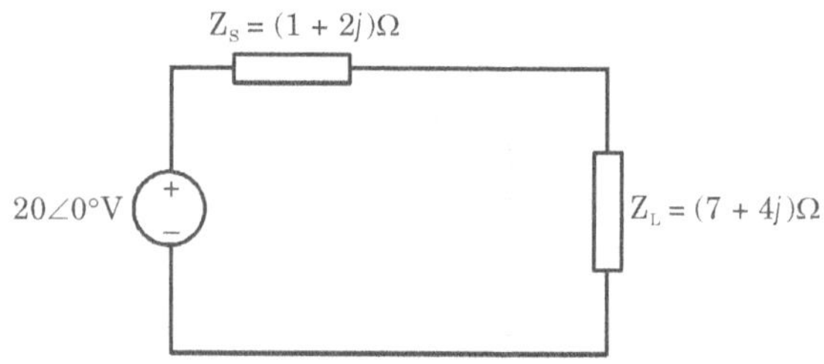

(a) 8 VAR (b) 16 VAR

(c) 28 VAR (d) 32 VAR

28. The switch in the circuit shown was on position *a* for a long time, and is moved to position *b* at time $t = 0$. The current $i(t)$ for $t > 0$ is given by

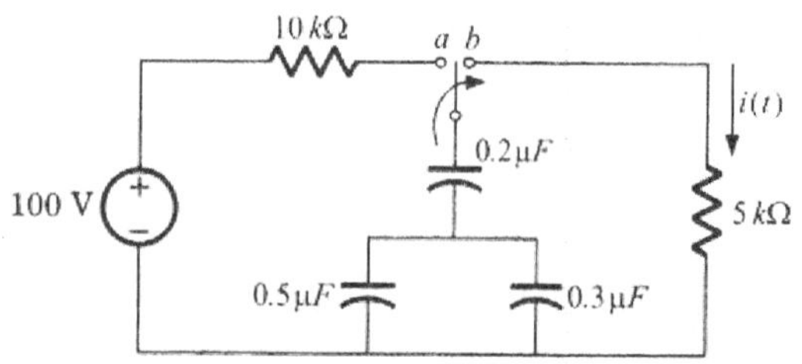

(a) $0.2e^{-125t}u(t)$ mA (b) $20e^{-1250t}u(t)$ mA

(c) $0.2\ e^{-1250t}\ u(t)$ mA (d) $20e^{-1000t}u(t)$ mA

29. In the circuit shown, what value of R_L maximizes the power delivered to R_L ?

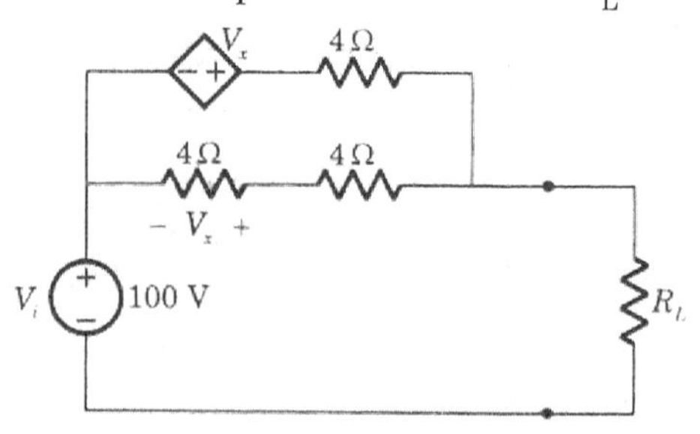

(a) 2.4 Ω (b) $\frac{8}{3}\Omega$

(c) 4 Ω (d) 6Ω

30. The time domain behaviour of a RL circuit is represented by

$$L\frac{di}{dt}+Ri=V_0(1+Be^{-\frac{Rt}{L}}\sin t)u(t).$$

For an initial current of $i(0)=\frac{V_0}{R}$, the steady state value of the current is given by

(a) $i(t)\to\frac{V_o}{R}$ (b) $i(t)\to\frac{2V_o}{R}$

(c) $i(t)\to\frac{V_o}{R}(1+B)$ (d) $i(t)\to\frac{2V_o}{R}(1+B)$

31. In the circuit below, the diode is ideal. The voltageV is given by

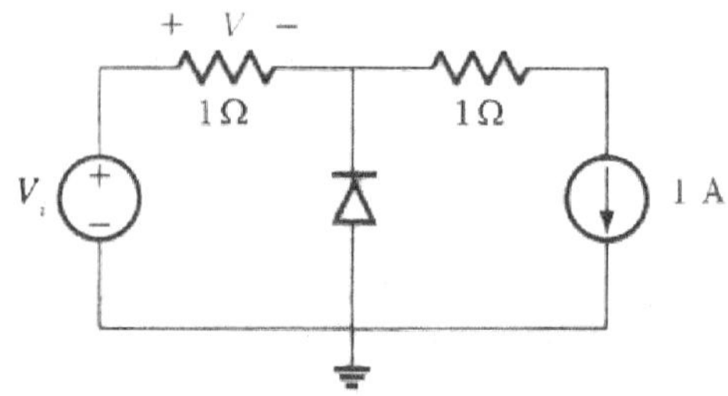

(a) min $(V_i,1)$ (b) max$(V_i,1)$

(c) min$(-V_i,1)$ (d) max$(-V_i,1)$

32. Consider the following two statements about the internal conditions in an n-channel MOSFET operating in the active region.

S1 : The inversion charge decreases from source to drain

S2 : The channel potential increases from source to drain

Which of the following is correct ?

(a) Only S2 is true

(b) Both S1 and S2 are false.

(c) Both S1 and S2 are true, but S2 is not a reason for S1

(d) Both S1 and S2 are true, and S2 is a reason for S1

33. In the following astable multivibrator circuit, which properties of $v_o(t)$ depend on R_2?

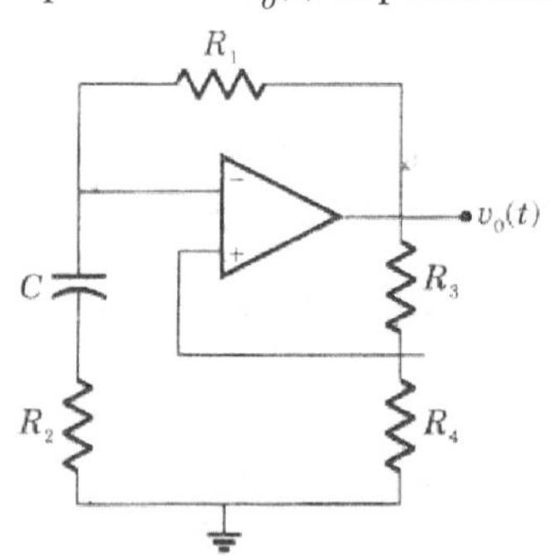

(a) Only the frequency

(b) Only the amplitude

(c) Both the amplitude and the frequency

(d) Neither the amplitude nor the frequency

34. In the circuit shown below, the op-amp is ideal, the transistor has V_{BE} = 0.6V and β =150. Decide whether the feedback in the circuit is positive or negative and determine the voltage V at the output of the op-amp.

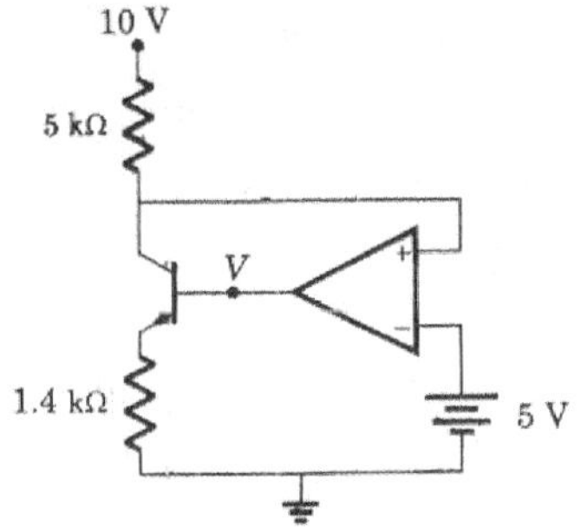

(*a*) Positive feedback, V = 10 V

(*b*) Positive feedback, V = 0 V

(*c*) Negative feedback, V = 5 V

(*d*) Negative feedback, V = 2 V

35. A small signal source

$$v_i(t) = \mathrm{A}\cos 20t + \mathrm{B}\sin 10^6 t$$

is applied to a transistor amplifier as shown below. The transistor has β = 150 and $h_{ie} = 3k\Omega$. Which expression best approximates $v_o(t)$?

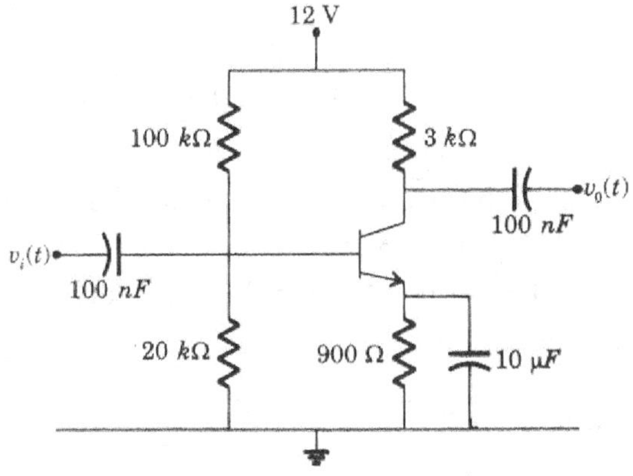

(*a*) $v_o(t) = 1500(\mathrm{A}\cos 20t + \mathrm{B}\sin 10^6 t)$

(*b*) $v_o(t) = -150(\mathrm{A}\cos 20t + \mathrm{B}\sin 10^6 t)$

(*c*) $v_o(t) = -1500\mathrm{B}\sin 10^6 t$

(*d*) $v_o(t) = -150\mathrm{B}\sin 10^6 t$

36. If X =1 in the logic equation

$$\left[\mathrm{X} + \mathrm{Z}\left\{\overline{\mathrm{Y}} + \left(\overline{\mathrm{Z}} + \mathrm{X}\overline{\mathrm{Y}}\right)\right\}\right]\left\{\overline{\mathrm{X}} + \overline{\mathrm{Z}}(\mathrm{X} + \mathrm{Y})\right\} = 1, \text{ than}$$

(*a*) Y = Z (*b*) $\mathrm{Y} = \overline{\mathrm{Z}}$

(*c*) Z =1 (*d*) Z = 0

37. What are the minimum number of 2-to-1 multiplexers required to generate a 2-input AND gate and a 2- input Ex-OR gate ?

(*a*) 1 and 2 (*b*) 1 and 3

(*c*) 1 and 1 (*d*) 2 and 2

38. Refer to the NAND and NOR latches shown in the figure. The inputs (P_1, P_2) for both the latches are first made (0, 1) and then, after a few seconds, made (1, 1). The corresponding stable outputs (Q_1, Q_2) are

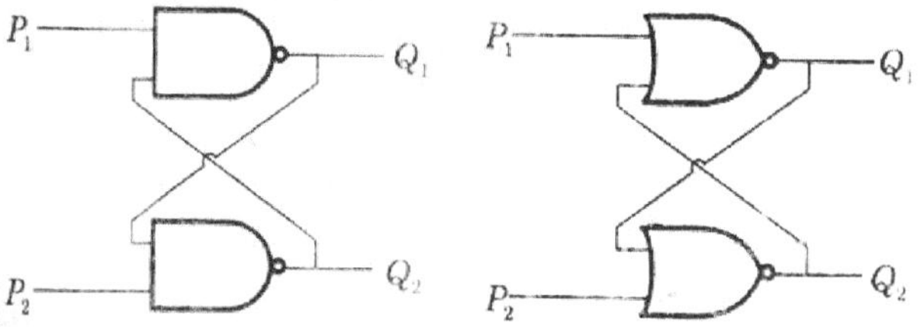

(*a*) NAND: first (0, 1) then(0,1) NOR: first (1,0) then (0,0)

(*b*) NAND: first (1,0) then (1,0) NOR: first (1,0) then (1,0)

(*c*) NAND: first (1,0) then (1,0) NOR: first (1,0) then (0,0)

(*d*) NAND: first ((1,0) then (1,1) NOR: first (0,1) then (0,)

39. What are the counting states (Q_1, Q_2) for the counter shown in the figure below ?

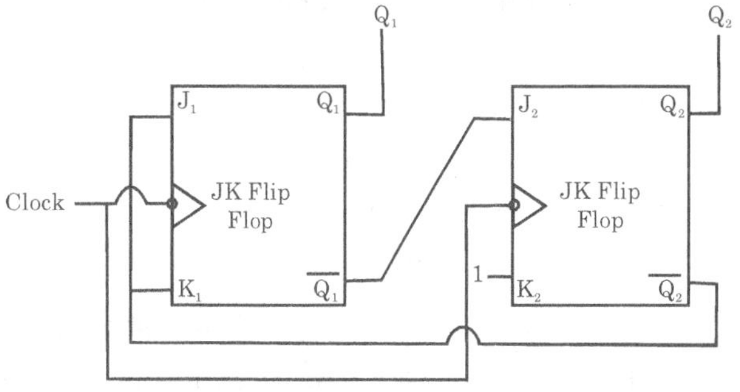

(*a*) 11, 10, 00, 11, 10,...

(*b*) 01, 10, 11, 00, 01,...

(*c*) 00, 11, 01, 10, 00, ...

(*d*) 01, 10, 00, 01, 10,...

40. A system with transfer function H(*z*) has impulse response *h*(×) defied as *h* (2) = 1, *h* (3) = 1 and *h*(*k*) = 0 otherwise. Consider the following statements.

S1 : H(*z*) is a low-pass filter.

S2 : H(*z*) is an FIR filter.

Which of the following is correct ?

(*a*) Only S2 is true

(*b*) Both S1 and S2 are false

(*c*) Both S1 and S2 are true, and S2 is a reason for S1

(*d*) Both S1 and S2 are true, but S2 is not a reason for S1

41. Consider a system whose input x and output y are related by the equation

$$y(t)\int_{-\infty}^{\infty} = x(t-\tau)h(2\tau)d\tau$$

where $h(t)$ is shown in the graph.

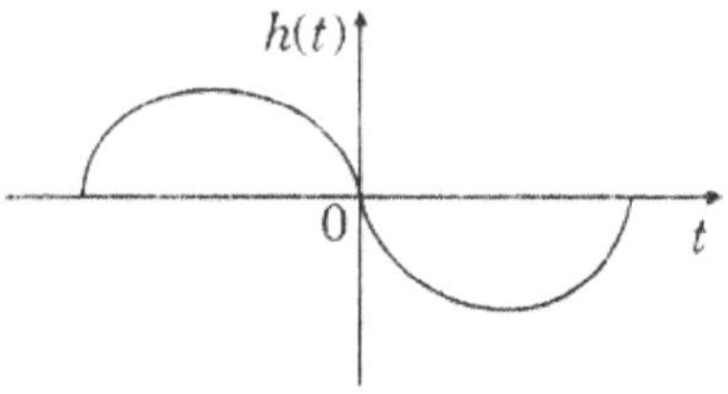

Which of the following four properties are possessed by the system ?

BIBO : Bounded input gives a bounded output.

Causal : The system is causal.

LP : The system is low pass.

LTI : The system is linear and time-invariant.

(*a*) Causal, LP (*b*) BIBO, LTI

(*c*) BIBO, Causal, LTI (*d*) LP, LTI

42. The 4-point Discrete Fourier Transform (DFT) of a discrete time sequence {1, 0, 2, 3} is

(*a*) [0, – 2 + 2*j*, 2, – 2 – 2*j*]

(*b*) [2, 2 + 2*j*, 6 – 2 – 2 *j*]

(*c*) [6, 1 – 3*j*, 2, 1 + 3*j*]

(*d*) [6, – 1 + 3*j*, 0, – 1 – 3*j*]

43. The feedback configuration and the pole-zero locations of $G(s) = \dfrac{s^2-2s+2}{s^2+2s+2}$ are shown below. The root locus for negative values of k, i.e for $-\infty < k < 0$, has breakway/break-in points and angle of departure at pole P (with respect to the positive real axis) equal to

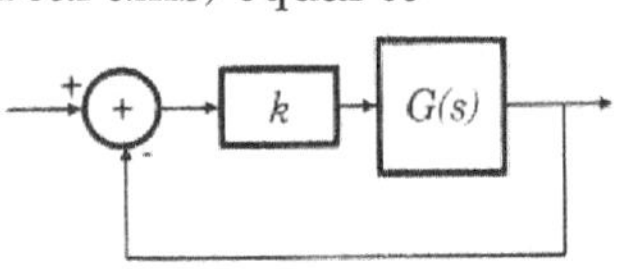

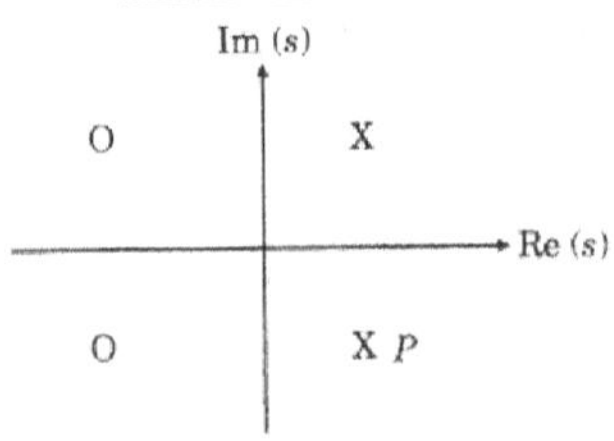

(*a*) $\mp\sqrt{2}$ and 0°

(*b*) $\pm\sqrt{2}$ and 45°

(*c*) $\pm\sqrt{3}$ and 0°

(*d*) $\pm\sqrt{3}$ and 45°

44. An LTI system having transfer function $\dfrac{s^2+1}{s^2+2s+1}$ and input $x(t) = \sin(t+1)$ is in steady state. The output is sampled at a rate ω_s rad/s to obtain the final output $\{y(k)\}$. Which of the following is true ?

(*a*) $y(x)$ is zero for all sampling frequencies ω_s

(*b*) $y(x)$ is nonzero for all sampling frequencies ω_s

(*c*) $y(x)$ is nonzero for $\omega_s > 2$, but zero for $\omega_s < 2$

(*d*) $y(x)$ is zero for $\omega_s > 2$, but nonzero for $\omega_s < 2$

45. The unit step response of an under-damped second order system has steady state value of –2. Which one of the following transfer functions has these properties ?

(*a*) $\dfrac{-2.24}{s^2+2.59s+1.12}$ (*b*) $\dfrac{-3.82}{s^2+1.91s+1.91}$

(*c*) $\dfrac{-2.24}{s^2-2.59s+1.12}$ (*d*) $\dfrac{-3.82}{s^2-1.91s+1.91}$

46. A discrete random variable X takes values from 1 to 5 with probabilities as shown in the table. A student calculates the mean of X as 3.5 and her teacher calculates the variance of X as 1.5. Which of the following statements is true ?

k	1	2	3	4	5
P(X) = k	0.1	0.2	0.4	0.2	0.1

(*a*) Both the student and the teacher are right

(*b*) Both the student and the teacher are wrong

(*c*) The student is wrong but the teacher is right

(*d*) The student is right but the teacher is wrong

47. A message signal given by

$$m(t) = \left(\frac{1}{2}\right)\cos\omega_1 t - \left(\frac{1}{2}\right)\sin\omega_2 t$$

is amplitude-modulated with a carrier of frequency ω_c to generate

$$s(t) = [1 + m(t)]\cos\omega_c t$$

What is the power efficiency achieved by this modulation scheme ?

(*a*) 8.33% (*b*) 11.11%

(*c*) 20% (*d*) 25%

48. A communication channel with AWGN operating at a signal to noise ratio SNR >> 1 and bandwidth B has capacity C_1. If the SNR is doubled keeping B constant, the resulting capacity C_2 is given by

(*a*) $C_2 \approx 2C_1$ (*b*) $C_2 \approx C_1 + B$

(*c*) $C_2 \approx C_1 + 2B$ (*d*) $C_2 \approx C_1 + 0.3B$

49. A magnetic field in air is measured to be

$$\vec{B} = B_0\left(\frac{x}{x^2+y^2}\hat{y} - \frac{y}{x^2+y^2}\hat{x}\right)$$

Which current distribution leads to this field ?

[**Hint :** The algebra is trivial in cylindrical coordinates.]

(*a*) $\vec{J} = -\frac{B_0\hat{z}}{\mu_0}\left(\frac{1}{x^2+y^2}\right), r \neq 0$

(*b*) $\vec{J} = -\frac{B_0\hat{z}}{\mu_0}\left(\frac{2}{x^2+y^2}\right), r \neq 0$

(*c*) $\vec{J} = 0, r \neq 0$

(*d*) $\vec{J} = \frac{B_0\hat{z}}{\mu_0}\left(\frac{2}{x^2+y^2}\right), r \neq 0$

50. A transmission line terminates in two branches, each of length λ/4 as shown. The branches are terminated by 50Ω loads. The lines are lossless and have the characteristic impedances shown. Determine the impedance Z_i as seen by the source.

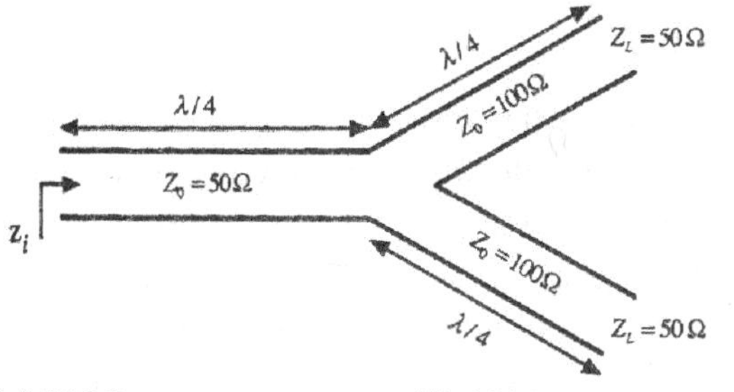

(*a*) 200Ω (*b*) 100Ω

(*c*) 50Ω (*d*) 25Ω

COMMON DATA QUESTIONS

Common Data for Q. 51 and 52 :

Consider a silicon p-n junction at room temperature having the following parameters:

Doping on the n-side = 1×10^{17} cm^{-3}

Deletion width on the-side = 0.1 μm

Depletion width on then p-side = 1.0 μm

Intrinsic carrier concentration = 1.4×10^{10} cm^{-3}

Thermal voltage = 26 mV

Permittivity of free space = 8.85×10^{-14} F. cm^{-1}

Dielectric constant of silicon = 12

51. The built-in potential of the junction

(*a*) is 0.70 V

(*b*) is 0.76 V

(*c*) is 0.82 V

(*d*) cannot be estimated from the data given

52. The peak electric field in the device is

(*a*) 0.15 MV. cm^{-1}, directed from p-region to n-region

(*b*) 0.15 MV. cm^{-1}, directed from n-region to p-region

(*c*) 1.80 MV. cm^{-1}, directed from p-region to n-region

(*d*) 1.80 MV. cm^{-1}, directed from n-region to p-region

Common Data for Q. 53 and 54:

The Nyquist plot of a stable transfer function G(*s*) is shown in the figure. We are interested in the stability of the closed loop system in the feedback configuration shown.

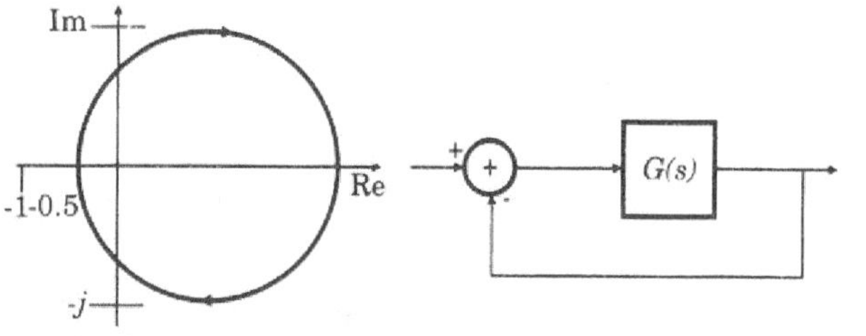

53. Which of the following statements is true ?

(*a*) G(s) is an all-pass filter

(*b*) G(s) has a zero in the right-half plane

(*c*) G(s) is the impedance of a passive network

(*d*) G(s) is marginally stable

54. The gain and margins of G(*s*) for closed loop stability are

(*a*) 6 dB and 180° (*b*) 3 dB and 180°

(*c*) 6 dB and 90° (*d*) 3 dB and 90°

Common Data for Q. 55 and 56:

The amplitude of a random signal is uniformly distributed between –5 V and 5 V.

55. If the signal to quantization noise ratio required in uniformly quantizing the signal is 43.5 dB, the step size of the quantization is approximately

(*a*) 0.0333 V (*b*) 0.05 V

(*c*) 0.0667 V (*d*) 0.10 V

56. If the positive values of the signal are uniformly quantized with a step size of 0.05 V, and the negative values are uniformly quantized with a step size of 0.1 V, the resulting signal to quantization noise ratio is approximately

(*a*) 46 dB (*b*) 43.8 dB

(*c*) 42 dB (*d*) 40 dB

LINKED ANSWER QUESTIONS

Linked Answer for Q. 57 and 58 :

Consider the CMOS circuit shown, where the gate voltage V_G of the n-MOSFET is increased from zero, while the gate voltage of the p-MOSFET is kept constant at 3 V. Assume that, for both transistors, the magnitude of the threshold voltage is 1 V and the product of the transconductance parameter and the $\left(\frac{W}{L}\right)$ ratio,

i.e. the quantity $\mu C_{ox}\left(\frac{W}{L}\right)$ is 1 mA.V^{-2}.

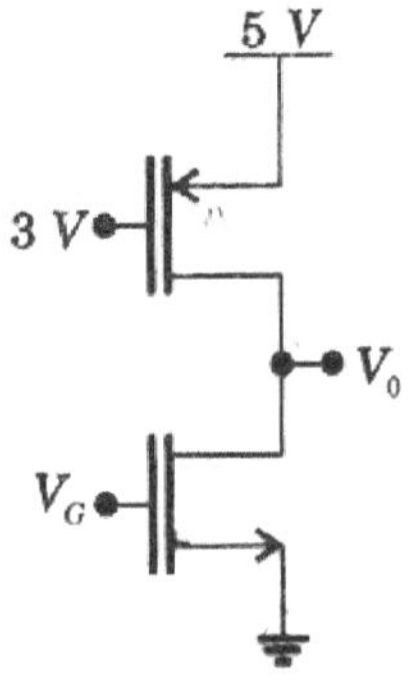

57. For small increase in V_G beyond 1 V, which of the following gives the correct description of the region of operation of each MOSFET ?
    (*a*) Both the MOSFETs are in saturation region
    (*b*) Both the MOSFETs are in triode region
    (*c*) n-MOSFET is in triode and p-MOSFET is in saturation region
    (*d*) n-MOSFET is in saturation and p-MOSFET is in triode region

58. Estimate the output voltage V_0 for V_G = 1.5 V. [Hint : Use the appropriate current-voltage equation for each MOSFET, based on the answer to Q. 57.]
    (*a*) $4-\frac{1}{\sqrt{2}}$ (*b*) $4+\frac{1}{\sqrt{2}}$ V
    (*c*) $-\frac{\sqrt{3}}{2}$ V (*d*) $4+\frac{\sqrt{3}}{2}$ V

Linked Answer for Q. 59 and 60 :

Two products are sold from a vending machine, which has two push buttons P_1 and P_2. When a button is pressed, the price of the corresponding product is displayed in a 7-segment display.

If no buttons are pressed, '0' is displayed, signifying 'Rs. 0'.

If only P_1 is pressed, '2' is displayed, signifying 'Rs. '2'.

If only P_2 is pressed, '5' is displayed, signifying 'Rs. 5'.

If both P_1 and P_2 are pressed, 'E' is displayed, signifying 'Error'.

The names of the segments in the 7-segment dis play, and the glow of the display for '0', '2' '5' and 'E', are shown below

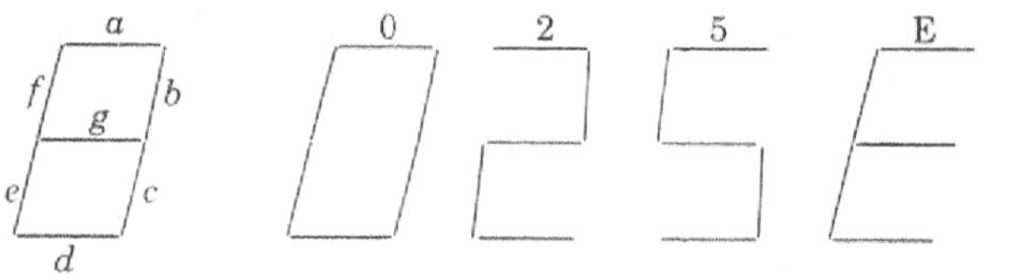

Consider

(*i*) push button pressed/not pressed is equivalent to logic 1/0 respectively.

(*ii*) a segment glowing/not glowing in the display is equivalent to logic 1/0 respectively.

59. If segments *a* to *g* are considered as functions of P_1 and P_2, then which of the following is correct ?
    (*a*) $g=\overline{P_1}+P_2,\ d=c+e$ (*b*) $g=P_1+P_2,\ d=c+e$
    (*c*) $g=\overline{P_1}+P_2,\ e=b+c$ (*d*) $g=P_1+P_2,\ e=b+c$

60. What are the minimum numbers of NOT gates and 2-input OR gates required to design the logic of the driver for this 7-segment display ?
    (*a*) 3 NOT and 4 OR (*b*) 2 NOT and 4 OR
    (*c*) 1 NOT and 3 OR (*d*) 2 NOT and 3 OR

ANSWERS

1. (*b*)	2. (*a*)	3. (*b*)	4. (*c*)	5. (*c*)	6. (*c*)	7. (*a*)	8. (*d*)	9. (*b*)	10. (*a*)
11. (*c*)	12. (*b*)	13. (*d*)	14. (*a*)	15. (*a*)	16. (*d*)	17. (*c*)	18. (*c*)	19. (*c*)	20. (*d*)
21. (*c*)	22. (*d*)	23. (*b*)	24. (*b*)	25. (*a*)	26. (*d*)	27. (*b*)	28. (*b*)	29. (*c*)	30. (*a*)
31. (*a*)	32. (*d*)	33. (*a*)	34. (*d*)	35. (*a*)	36. (*d*)	37. (*a*)	38. (*c*)	39. (*a*)	40. (*a*)
41. (*b*)	42. (*d*)	43. (*b*)	44. (*a*)	45. (*b*)	46. (*b*)	47. (*c*)	48. (*b*)	49. (*c*)	50. (*d*)
51. (*b*)	52. (*b*)	53. (*c*)	54. (*c*)	55. (*c*)	56. (*b*)	57. (*d*)	58. (*d*)	59. (*b*)	60. (*d*)

EXPLANATIONS

1. The order of a differential equation is the order of the highest deriavative involving in equation, so answer is 2.

2. Fourier series for periodic function

$$f(t) = a_0 + \sum_{n=1}^{\infty} (a_n \cos n\omega_0 t + b_n \sin n\omega_0 t)$$

Here, $a_0 = \frac{1}{T}\int_{-T/2}^{T/3} f(t)dt$

$b_0 = 0$

$a_n = \frac{2}{T}\int_{-T/2}^{T/2} f(t) \cos n\omega_0 t \, dt$

$b_n = \frac{2}{T}\int_{-T/2}^{T/2} f(t) \sin n\omega_0 t \, dt$

For even function, $f(t) = f(-t)$

so $b_n = 0$

Hence even function has only cosine terms.

For odd function , $f(t) = -f(-t)$

so $a_n = 0$

Hence odd function has only sine terms.

3. $f(t) = \frac{1}{2} - \frac{\cos 2t}{2} + \cos 2t$

$f(t) = \frac{1}{2} + \frac{\cos(2t)}{2}$

Thus $f(t)$ has frequency components

$f_1 = dc = 0$

$f_2 = \frac{2}{2\pi} = \frac{1}{\pi}$ Hz

4.

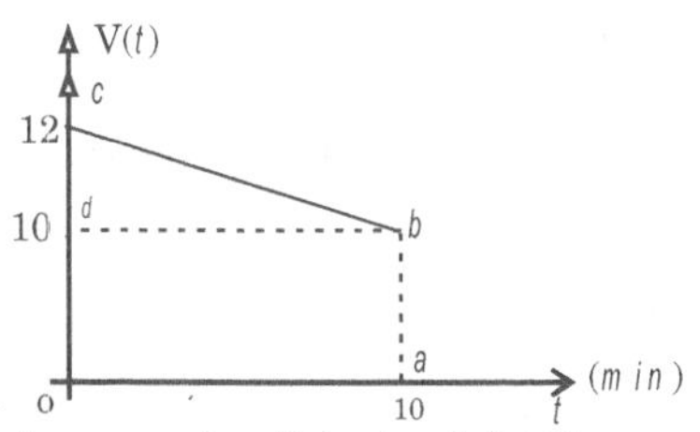

Area under the straight line

$$\Delta = \frac{1}{2}.(cd)\ (b\,d) + (d\,b).\ od$$

$$= \frac{1}{2}.\ (2.\ 10.\ 60)\ + 10.10.60)$$

Energy = Area × I

= 6600 × 2

= 13200 J

= 13.2 KJ

5. The concentration of Dopand atoms in n–type silicon is 4×10^{19} cm^{-3}.

7. $x(n) = \left(\frac{1}{3}\right)^n u(n) - \left(\frac{1}{2}\right)^n u(-n-1)$

Let, $x_1(n) = \left(\frac{1}{3}\right)^n u[n]$

$$x_1[Z] = \sum_{n=-\infty}^{\infty} \left(\frac{1}{3}\right)^n u(n)_z^{-n}$$

$$= -\sum_{n=0}^{\infty}\left(\frac{1}{3}z^{-1}\right)^n = \frac{1}{1-\frac{1}{3}z^{-1}}$$

$X_1[Z]$ will converge where $\left|\frac{1}{3}z^{-1}\right| < 1$ or, $|Z| > \frac{1}{3}$

and, $x_2[n] = -\left(\frac{1}{2}\right)^n u(-n-1)$

$$X_2[z] = -\sum_{n=-\infty}^{\infty}\left(\frac{1}{2}\right)^n u(-n-1)_z^{-n} - \sum_{n=-\infty}^{\infty}\left(\frac{1}{2}\right)^n z^{-n}$$

$$= -\sum_{n=1}^{\infty}\left(\frac{1}{2}z^{-1}\right)^{-n} = -\sum_{n=1}^{\infty}(2z)^n$$

$$= 1 - \sum_{n=0}^{\infty}(2z)^n$$

For convergence of $X_2(z)$ $|2z| < 1$ or $|z| < \frac{1}{2}$

Hence, ROC for $x(n)$ should be in the range of $\frac{1}{3} < |z| < \frac{1}{2}$

8. G (s) for lead compensator $= \frac{(s+z_1)(s+z_2)}{(s+p_1)(s+p_2)}$

and the Bode -plot for $G(j\omega)$ is shown below.

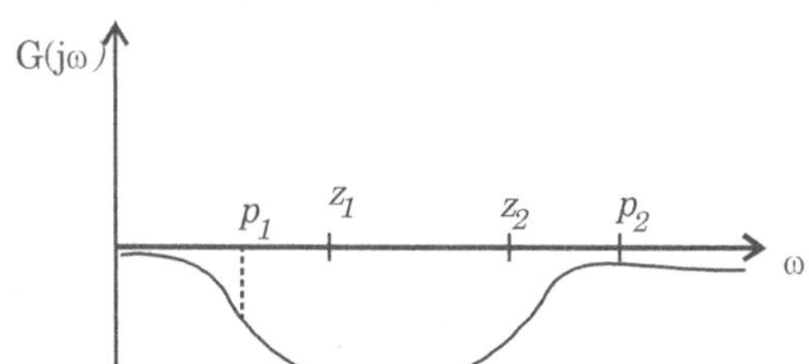

9. Power of the output proccss Y(t),

$Y(t) = |H(t)|^2 \cdot X(t)$

$X(t) = \frac{N_0}{2} = 1\times 10^{-10}$ W/HZ

$|H(t)|^2$ = area under curve $|u(t)|^2$

$= 2\left(\frac{1}{2}\times 10\times 10^3 \times 1\right) = 10^4$ Hz

then, $V(t) = 10^{-6}$ W

10. Rectangular wave guides for TE_{mn} modes have cut-off frequency,

$$f_c = \frac{1}{2\sqrt{\mu \in}}\sqrt{\frac{m^2}{a^2}+\frac{n^2}{b^2}}$$

here , a and b are dimension of waveguide.

Cut-off frequency for cylindrical waveguide

$$= \frac{K_{mn}}{2\pi\sqrt{\mu \in}} Hz$$

$$\therefore \quad K_{mn} = \frac{(K_a)_{mn}}{a}$$

where, a = radius

Coaxial waveguide has no cutt off frequency.

11. Probability of getting head in first toss = $\frac{1}{2}$

Probability of getting head in second toss = $\frac{1}{2}$

and in all other 8 tosses there should be tail alway.

So probability of getting head in first two tosses

$$= \left(\frac{1}{2}\right)^{10}$$

13.

$$f(z) = c_0 + c_1 z^{-1}$$

$$f(z) = \oint \frac{1 + c_0 + c_1 z^{-1}}{z}\, dz$$

$$= \oint\left(\frac{(1+c_0)z + c_1}{z^2}\right)dz$$

$$\therefore \quad f(z) = 2\pi j(\text{ residue of } f(z))$$

Residue at $z = 0$ (2 – order)

$$= \frac{1}{L_1}\frac{d}{dz}\left(z^2 - \frac{(1+c_0)z + c_1}{z^2}\right)$$

$$= \frac{1}{L_1}(1 + c_0).1$$

$$= (1 + c_0)$$

$$\therefore \quad f(z) = 2\pi j(1 + c_0)$$

14. If value of I is greater than 12 A, then both voltage sources are radiating power. There is no source for consuming because direction of current is same, so value of current I should be less than 12 ampere.

$$12 = I' + I$$

$$\Rightarrow \quad I = 12 - I'$$

$$I = 10 \text{ A}$$

$$I < 12 \text{ A}$$

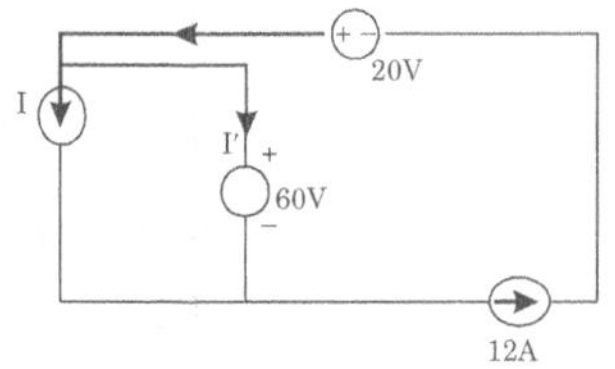

15. Unit for D : m^2/s

Unit for μ : m^2/V– s

$\therefore$ Ratio = V^{-1}

Alternately

$$J = e\,\mu nE$$

$$J = eD\,\frac{dn}{dx}$$

$$\frac{\mu}{D} = \frac{\frac{dn}{dx}}{nE}$$

$$= \frac{\text{number of } e^-}{cm} \times \frac{1}{\text{number of } e^-} \times \frac{\text{meter}}{\text{volt}}$$

$$= V^{-1}$$

16. If a interrupt has an address, then it is vectored interrupt, and if a interrupt can be rejected, then it is maskable interrupt.

17. Transfer function $\dfrac{\frac{R_L}{1+CsR_L}}{R + \frac{R_L}{1+CsR_L}} = \dfrac{V_o}{V_i}$

$$\Rightarrow \quad \frac{V_o}{V_i} = \frac{R_L}{R + CsR_L R + R_L}$$

$$= \frac{1}{\left(1 + \frac{R}{R_L}\right) + CsR}$$

According to question

$$1 + \frac{R}{R_L} = 2$$

$$\therefore \quad R = R_L$$

18. The system is uncontrollable for all values of p and q.

For controllable function, rank of matrix

$$[B \quad AB] = 2$$

$$\begin{bmatrix} p & p \\ q & q \end{bmatrix}$$

Rank of this matrix is not 2 but 1 for all values of p and q. So it is not controllable.

19. Modulated signal,

$$s(t) = m(t) \cdot \cos(2\pi f_c t)$$

$$= \cos(2\pi f_m t) \cdot \cos(2\pi f_c t)$$

$$= \frac{1}{2}[\cos 2\pi(f_c + f_m)\,t + \cos 2\pi(f_c - f_m)\,t]$$

First term represents upper side-band signal and second one represents lower side-band signal.

20. Magnetic field at origin due to wire in $y-z$ plane numbering parallel to the $y-$ axis

$$\overline{B}_1 = (k_1B_0)\cdot \overline{a}_x$$

and , the same due to wire in $x-y$ plane parallel to the $x-$ axis,

$$\overline{B}_2 = (k_2B_0)\,\overline{a}_z$$

Resultant field, $\overline{B} = \overline{B}_1 + \overline{B}_2$

$$= B_0\left[k_1\overline{a}_x + k_2\overline{a}_z\right]$$

Alternately

The direction of magnetic field according to Bio-Savart law is perpendicular to plane of $\vec{Idl}$ and $\vec{R}$. In first wire , $\vec{Idl}$ is y direction and $\vec{R}$ is in z direction In second wire, $\vec{Idl}$ is in x direction and $\vec{R}$ is in y direction.

21. Conditional probability

$$P\left(\frac{A}{B}\right) = \frac{P(A\cap B)}{P(B)}$$

$$P(x+y=2) = \frac{1}{2}\times\frac{1}{4}+\frac{1}{4}\times\frac{1}{4}+\frac{1}{4}\times\frac{1}{2} = \frac{5}{16}$$

Since condition : $x = 0,\quad y = 2$

$x = 1,\quad y = 1$

$x = 2,\quad y = 0$

$$P(x-y=0) = \frac{1}{2}\times\frac{1}{2}+\frac{1}{4}\times\frac{1}{4}+\frac{1}{4}\times\frac{1}{4} = \frac{6}{16}$$

Since condition: $x = 0,\quad y = 0$

$x = 1,\quad y = 1$

$x = 2,\quad y = 2$

$$P[(x+y=2)/(x-y=0)] = \frac{\frac{5}{16}-\frac{6}{16}}{\frac{6}{16}} = \frac{-1}{6}$$

But probability should not be negative , so answer is $\frac{1}{6}$

22. Since $\sin x = x - \frac{x^3}{3!} + \frac{x^5}{5!} + ----$

$$\therefore\quad \sin(x-\pi) = (x-\pi) - \frac{(x-\pi)^3}{3!} + \frac{(x-\pi)^5}{5!} + \ldots$$

or $$-\sin x = (x-\pi) - \frac{(x-\pi)^3}{3!} + \frac{(x-\pi)^5}{5!} + \ldots$$

or $$-\frac{\sin x}{(x-\pi)} = 1 - \frac{(x-\pi)^2}{3!} + \frac{(x-\pi)^4}{5!} + \ldots..$$

or $$\frac{\sin x}{(x-\pi)} = -1 + \frac{(x-\pi)^2}{3!} - \frac{(x-\pi)^4}{5!} + \ldots$$

23. It S be an oriented smooth surface that is bounded by a simple, closed, smooth boundary curve C with positive orientation.

Also set $\overline{F}$ be the vector field, then by stokes's theorem,

$$\oint_c \overline{F}\cdot d\overline{r} = \iint_S \text{curl}\,\overline{F}\cdot d\overline{s} \quad \ldots(A)$$

Here $\overline{F} = \overline{A}$ as were $\overline{F} = \overline{\nabla}\times\overline{A} = V$

then equation (A) given,

$$\oint_c \overline{A}\cdot dr = \iint_S (\overline{\nabla}\times\overline{A})\cdot d\overline{s}$$

24. $\int_0^t f(\tau)\,d\tau = \frac{1}{s}f(s)$ ....(Laplace formulae)

25. A – 2, B – 3, C – 3, D – 1

(*i*) $\frac{dy}{dx} = \frac{y}{x}$

$\therefore\ \log y = \log x + \log c$

or $y = cx$, which is straight line equation

(*ii*) $\frac{dy}{dx} = \frac{-y}{x}$

$\therefore\ \log y + \log x = \log c$

or $c = xy$, which is Hyperbola equation

(*iii*) $\frac{dy}{dx} = \frac{x}{y}$

$\therefore\ ydy = xdx$

or $y^2 = x^2 + c$, which is Hyperbola equaiton

(*iv*) $\frac{dy}{dx} = \frac{-x}{y}$

$\therefore\ y^2 + x^2 = c$, which is equation of a circle

26. Eigen values are given by

$$\begin{vmatrix} -1-\lambda & 3 & 5 \\ -3 & -1-\lambda & 6 \\ 0 & 0 & 3-\lambda \end{vmatrix} = 0$$

$$\Rightarrow\quad (3-\lambda)\left((-1-\lambda)^2+9\right) = 0$$

$$\Rightarrow\quad \lambda = 3,\ \lambda = -1+3j,\ -1-3j$$

27. $I = \frac{20}{z_{total}} = \frac{20}{\sqrt{8^2+6^2}} = 2$ Amp

$[\because Z_{total} = Z_s + Z_c = 8 + 6j]$

$\therefore$ Reactive power = Power on reactive element of load

$= 4\times 2^2 = 16$ VAR

28. When the switch is in the position (*a*), the circuit will be as shown,

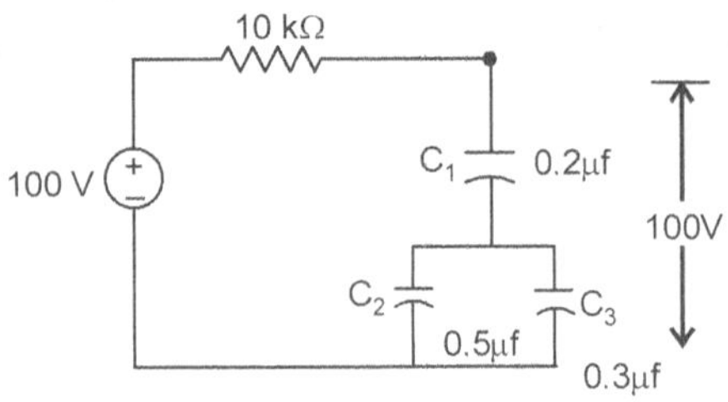

$C_1 = 0.2\quad F$

$C_2 = 0.5\quad F$

$C_3 = 0.3\quad F$

In steady state all the capacitors are open circuited,

⇒

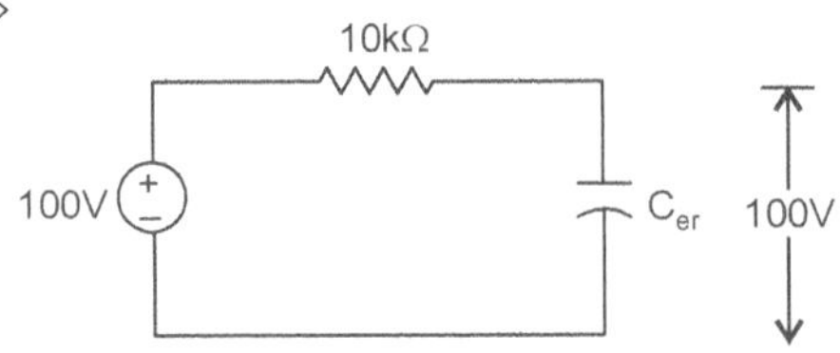

$$C_{eg} = \frac{C_1(C_2+C_3)}{C_1+C_2+C_3} = 0.16\quad F$$

When the switch is thrown to position (*b*)

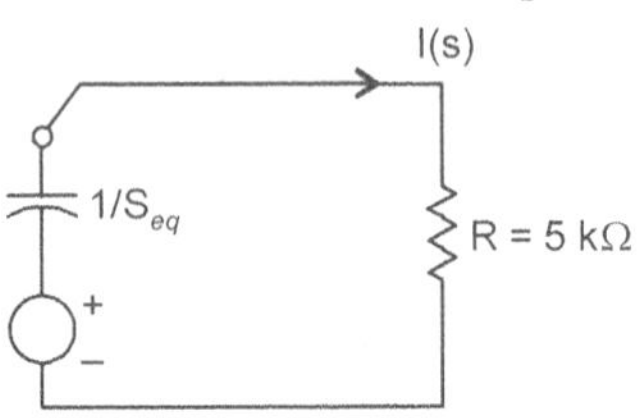

$$\frac{V(0+)}{S} = \frac{100}{S}$$

$$I(S) = \frac{V(0+)}{S}\cdot\frac{1}{R+\dfrac{1}{SC_{eq}}}$$

$$= \frac{V(0+)}{R}\cdot\frac{1}{S+\dfrac{1}{RC_{eq}}}$$

Inverse laplace given,

$$i(t) = V_F+(V_I-V_F)e^{-\frac{T}{R}} = \frac{V(0+)}{R}.e^{-\frac{t}{RC_{eq}}}.V(t)$$

$$= \frac{100}{5k}e^{-1250t}\cdot V(t) = 20e^{-1250t}V(t)\,mA$$

29.

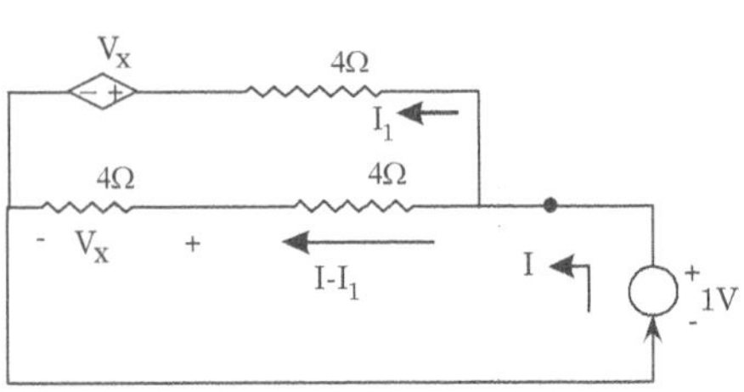

Using KVL in both loops, we have

$$1 = 8(I-I_1) = 2V_x \quad ...(i)$$

and $$V_x = 4(I-I_1) \quad ...(ii)$$

By equations (*i*) and (*ii*), we have

$$V_x = \frac{1}{2}\text{ volts}$$

and $$1 = 4I_1 + V_x$$

or $$I_1 = \frac{1}{8}$$

and $$1 = 8\left(I-\frac{1}{8}\right)$$

∴ $$I = \frac{1}{4}$$

Thus $$R_{TH} = \frac{V}{I} = 4\,\Omega$$

For max^m power to be delivered to R_L,

$$R_L = R_{TH} = 4\,\Omega$$

30. Taking Laplace transform, we have

$$sLI(s) - Li(0) + RI(s) = V_0\left(\frac{1}{s} + B\frac{1}{\left(s+\frac{R}{L}\right)^2 + L}\right)$$

For steady state current ,

$$\lim_{t\to\infty} i(t) = \lim_{s\to 0} sI(s)$$

∴ $$I(s) = \frac{V_0}{sL+R}\left(\frac{1}{s} + \frac{B}{\left(s+\frac{R}{L}\right)^2+1}\right) + \frac{LV_0}{R}\frac{S}{(R+LS)}$$

or $$\lim_{s\to 0} sI(s) = \frac{V_0}{R}$$

31. Voltage at 1 Ω is in negative direction of V_i and it is the maximum value of $(-V_i, 1)$ because diode is in reverse bias.

32.

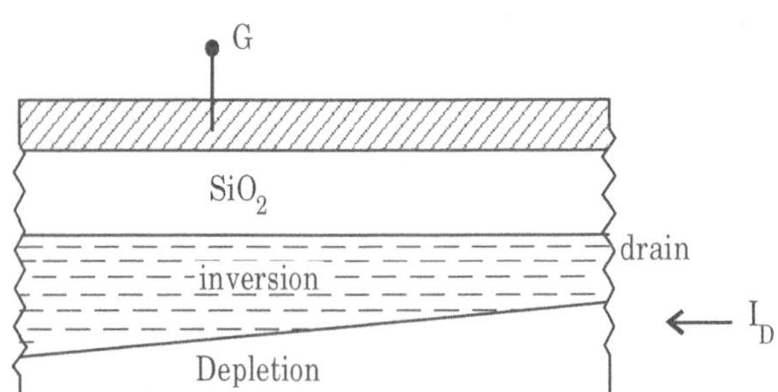

Both S_1 and S_2 are true, and S_2 is reason for S_1

33.

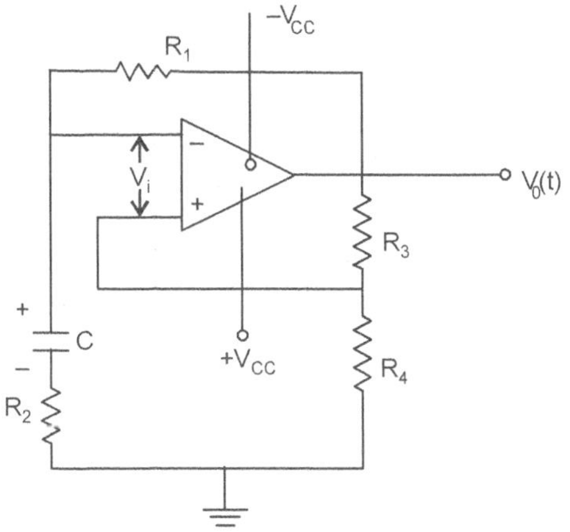

Feedback factor, $\beta = \dfrac{R_4}{R_3 + R_4}$

First assume output voltage at $+V_0 (\approx V_{CC})$

$$\text{then, } V_+ = V_- = \frac{V_{CC}R_4}{R_3 + R_4} = \beta V_{CC}$$

Now capacitor starts changing exponentially toward $+\beta V_{CC}$ thrugh $(R_2 + R_1)$ C

When capacitor voltage reaches $+\beta V_{CC}$, output voltage jumps from $+V_{CC}$ to $-V_{CC}$. Then cpacitor starts discharging towards $-\beta V_{CC}$

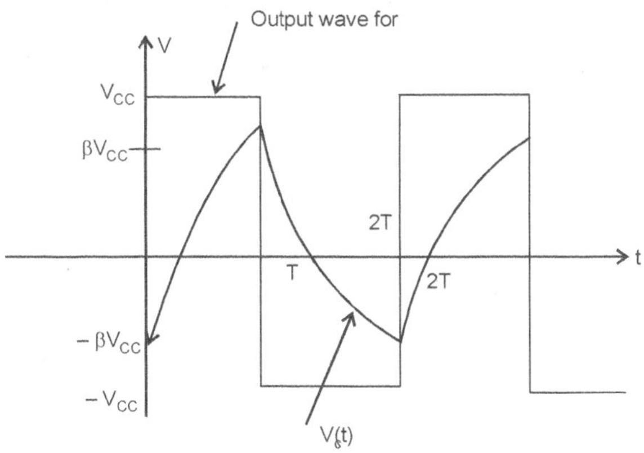

Now, $\quad V_C(t) = V_F + (V_1 - V_F)\, e^{-\frac{T}{RC}} \quad$ ...(A)

where $\quad R = R_2 + R_1$

V_1 = (tuntial voltage) = $-\beta V_{CC}$

$V_F = V_{CC}$

at $\quad t = T$ (half-period)

$V_C(t) = \beta V_{CC}$

then, Equation (A) given

$$\beta V_{CC} = V_{CC} + (-\beta V_{CC} - V_{CC})\, e^{-\frac{T}{RC}}$$

or, $\quad T = RC \ln\left[\dfrac{1+\beta}{1-\beta}\right]$

It can be seen that only time period frequency depends in R_2.

34. For ideal op-amp, voltage at non inverting point is equal to inverting point, so current from 5kΩ resistor is

$$\frac{10-5}{5k} = 1\ mA$$

Ideal op-amp has no current in input, so

$$I_c = 1\text{ mA} \approx I_E$$
$$V_o = 1.4 \times 1\text{ mA} + 0.6 = 2\text{ volts}$$

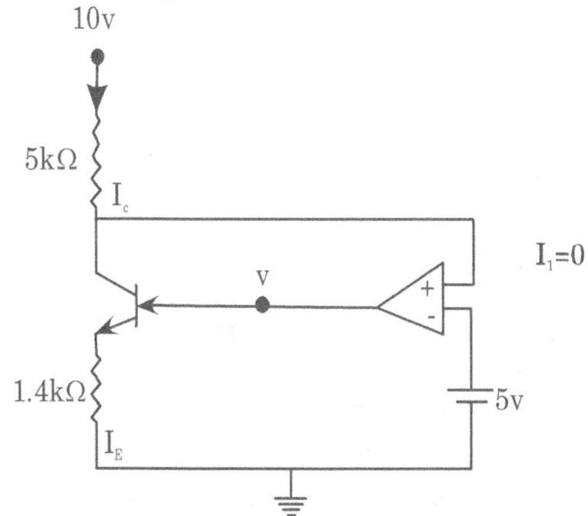

35. Drawing AC equivalent circuit, for AC sources, capacitors are short circuited

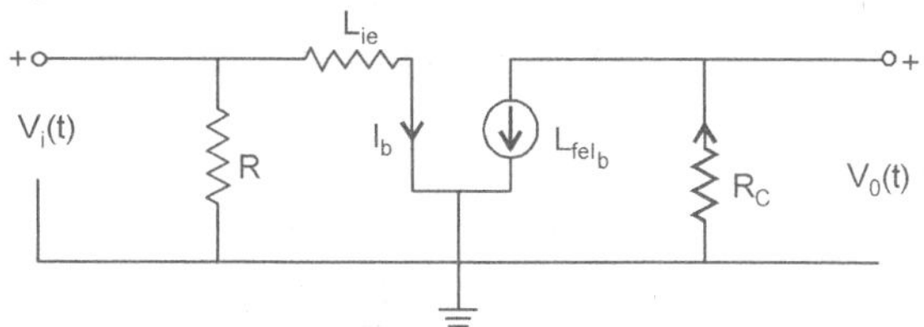

$$R = 100\ K\Omega \parallel 20\ K\Omega$$
$$= \frac{50}{3} K\Omega \text{ from the circuit}$$
$$V_i(t) = +\, h_{ie} I_b$$

and, $\quad V_0(t) = -\,h_{fe} I_b R_C$

$$= \frac{-h_{fe}R_C}{h_{ie}} v_i(t)$$
$$= \frac{-150(3K\Omega)}{(3K\Omega)} (A\cos 20t + B\sin 10^6 t)$$
$$= -150\,[A\cos 20t + B\sin 10^6 t]$$

36. $\left[X + Z\{\overline{Y} + (\overline{Z} + X\overline{Y})\}\right]\{\overline{X} + \overline{Z}(X+Y)\} = 1$

$\Rightarrow \quad \left[X\,\overline{Z}(X+Y) + \overline{X}Z\left[\overline{Y} + (\overline{Z} + X\overline{Y})\right]\right] = 1$

$\Rightarrow \quad \left[X\,\overline{Z} + XY + \overline{X}\,Z\,\overline{Y}\right] = 1$

For $\quad X = 1, \qquad \overline{Z} + Y = 1, \qquad Z = 0$

[since 1.1 = 1, and 0 + 1 = 1]

37. A 2 to 1 MUX has 1 selector line, so

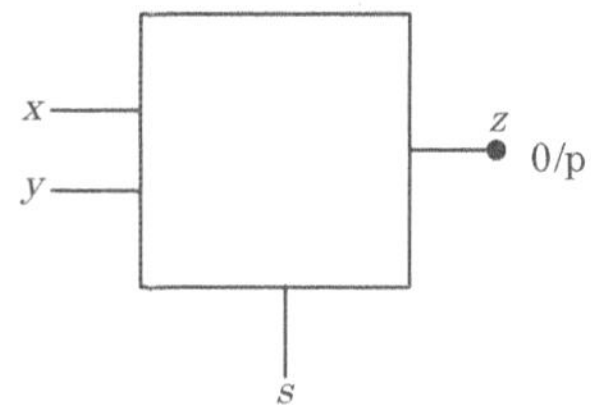

$$Z = X\bar{S} + YS$$

For generate and gate, we have

$$X = 0$$

$\therefore$ O/P = YS for that *a* AND gate output

For generate EX- OR gate, we have

$$Y = \bar{X}$$

and generate a NOT gate.

We again have a 2 –1 MUX having S = 1

So , total two (2 to 1) MUX required.

38. For NAND latches,

assuming $Q_1 = Q_2 = 0$ initially

$$P_1 = 0, \quad Q_1 = \overline{P_1 Q_2} = 1$$

$$P_2 = 1, \quad Q_2 = \overline{P_2 Q_1} = 0$$

Then after a few seconds

output , $Q_1 = \overline{P_1 . Q_2} = 1$

$$Q_2 = \overline{P_2 . Q_1} = 0$$

For NOR latches,

$$Q_1 = \overline{P_1 + Q_2}$$

and $Q_2 = \overline{P_2 + Q_1}$

$Q_1 = Q_2 = 0$ (Initially)

$P_1 = 0$, $Q_1 = 1$

$P_2 = 1$, $Q_2 = 0$

and , after a few seconds,

$P_1 = 1$, $Q_1 = 0$

$P_2 = 1$, $Q_2 = 0$

39. The counting sequence is shown here

	$(\bar{Q})$	$(\bar{Q}_2)$	$(\bar{Q}_1)$	$(\bar{Q}_1)$		
CLK	J_1	K_1	J_2	K_2	Q_1	Q_2
0					0	0
1	1	1	1	1	1	1
2	0	0	0	1	1	0
3	1	1	0	1	0	0
4	1	1	1	1	1	1
5	0	0	0	1	1	0

Truth-table

J	K	Q_{n+1}
0	0	Q_n
0	1	0
1	0	1
1	1	$\bar{Q}_{n+1}$

The output sequence is 11, 10, 00, 11, 10.......

40.

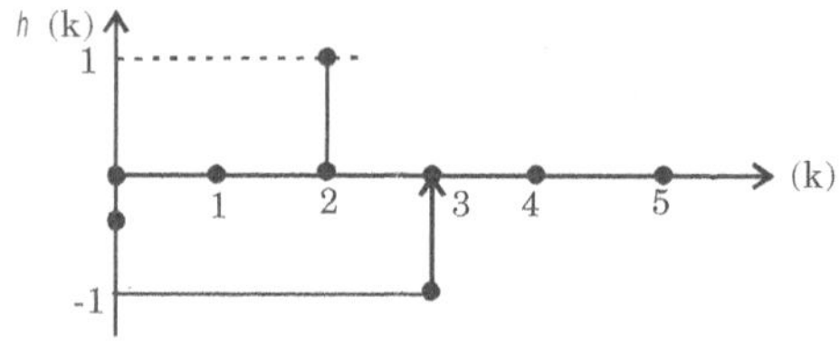

$\therefore$ $H(e^{j\omega}) = e^{-j2\omega} - e^{-j3\omega}$

So, it is FIR high pass filter.

41. System is LTI and BIBO system.

43. Characteristics equation is given by,

$$1 + G(s)\,H(s) = 0$$

$$\Rightarrow \quad 1 + \frac{K(s^2 - 2s + 2)}{s^2 + 2s + 2} = 0$$

$$K = -\frac{s^2 + 2s + 2}{s^2 - 2s + 2}$$

Breakway points are given by,

$$\frac{dx}{ds} = 0$$

and $\frac{dx}{ds} = \frac{(s^2 - 2s + 2)(2s + 2)(s^2 + 2s + 2)(2s - 2)}{(s^2 - 2s + 2)^2}$

then, $(s^2 - 2s + 2)(s + 1) = (s^2 + 2s + 2)(s - 1)$

or, $2s^2 - 4s^2 + 4 = 0$

or, $s = \pm\sqrt{2}$

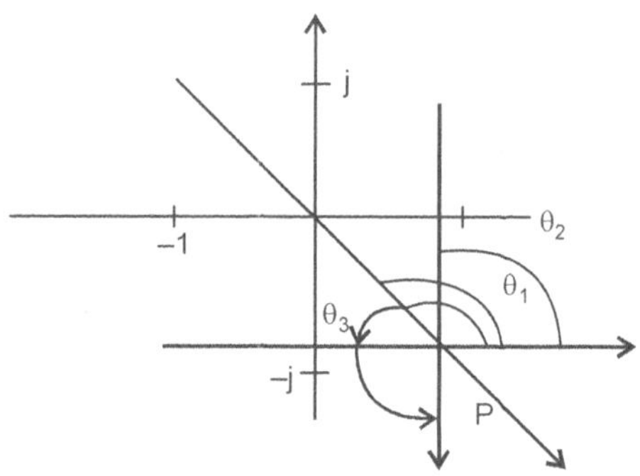

Angle of departure from Pole, Pj is, given by,

$$\phi_0 = \sum_{i=1}^{m} \angle(P_j - Z_i) - \sum_{j=1, i \neq j}^{n} (P_j - P_i)$$

m = number of zeros,

n = number of poles

$= 0° + 135° - 90° = 45°$

44. Input $x(t) = \sin(\omega t + 1)$

$$= \sin(t + 1)$$

$$\Rightarrow \quad \omega = 1$$

$$T(s) = \frac{s^2 + 1}{s^2 + 2s + 1}$$

$$T(j\omega) = \left.\frac{-\omega^2 + 1}{-\omega^2 + 1 + 2j\omega}\right|_{\omega=1} = 0$$

So $y(0)$ is zero for all sampling frequency ω_s.

45. For steady state value of –2,

$$\lim_{s\to 0}(\text{transfer function}) = -2$$

and for under damped, $0 < \delta < 1$

By checking option, we see that (*b*) option satisfies.

Alternately

$$T(s) = \frac{-3.82}{s^2+1.91s+1.91}$$

$$C(s) = T(s) - R(s)$$

$$= \frac{-3.82}{s(s^2+1.91s+1.91)}$$

Then, $C(\infty) = \underset{s\to 0}{lt}\ sC(s) = -2$

As for under damped system, $\delta < 1$

$\therefore \quad 1.91 = 2.\delta.\ \omega_n$

$$\Rightarrow \quad \delta = \frac{1.91}{2\sqrt{1.91}}$$

$$= \frac{\sqrt{1.91}}{2} = 0.69$$

46. Both the teacher and student are wrong

$$\overline{X} = (\text{Mean}) = \sum_{i=1}^{n} p_i k_i$$

$$= 0.1 + 0.4 + 1.2 + 0.8 + 0.5$$

$$= 3.0$$

$$\overline{X}^2 = \sum_{i=1}^{n} p_i k_i^{-2}$$

$$= 0.1\times 1 + 0.2\times 4 + 0.4\times 9 + 0.2\times 16 + 0.1\times 25$$

$$= 0.1 + 0.8 + 3.6 + 3.2 + 2.5$$

$$= 10.2$$

$$\sigma^2\ (\text{Variance}) = x^2 - x^{-2} = (-1.2)$$

47. Power efficiency, $\eta = \dfrac{P_{SB}}{P_{TOT}}$

where, P_{SB} = power in side bands

P_{TOT} = total power transmitted

$$s(t) = \left[1+\frac{1}{2}\cos\omega_1 t - \frac{1}{2}\sin\omega_2 t\right]\cos\omega_c t$$

$$= \cos\omega_c t + \frac{1}{4}\cos[(\omega_1+\omega_c)t] + \frac{1}{4}\cos[(\omega_1-\omega_c)t]$$

$$+\left\{\frac{1}{4}\sin[(\omega_2+\omega_c)t] + \frac{1}{4}\sin[(\omega_2+\omega_c)t]\right\}$$

Power in side bands, $P_{SB} = \dfrac{1}{2}\left[\dfrac{1}{16}+\dfrac{1}{16}+\dfrac{1}{16}+\dfrac{1}{16}\right]$

$$= \frac{1}{8}$$

$$P_{TOT} = \frac{1}{2} + \frac{1}{8}$$

$$= \frac{5}{8}$$

$$\therefore \quad \eta = \frac{1/8}{5/8}$$

$$= \frac{1}{5}$$

$$= 0.20$$

$$= 20\%$$

Alternately

$$m_1 = \frac{1}{2}$$

$$m_2 = \frac{1}{2}$$

$$m = \sqrt{m_1^2 + m_2^2}$$

$$= \frac{1}{\sqrt{2}}$$

Power efficiency $\eta = \dfrac{m^2}{2+m^2}$

$$= \frac{\frac{1}{2}}{2+\frac{1}{2}}$$

$$= \frac{1}{5} = 20\%$$

48. According to Shannon-Hartely law,

channel capacity, $C_1 = B\log_2\left(1+\dfrac{S}{N}\right) \approx B\log_2\left(\dfrac{S}{N}\right)$

$$C_2 = B\log_2\left(\frac{2S}{N}\right)$$

$$= B\log_2^2 + B\log\frac{S}{N}$$

$$C_2 = B + C_1$$

49. Conversion of cartesian co-ordinates into cylindrical co-ordinates is given as,

$$x = r\cos\phi,$$

$$\hat{a}_x = \cos\phi\hat{a}_i - \sin\phi\hat{a}_\phi$$

$$y = r\sin\phi,$$

$$\hat{a}_y = \sin\phi\hat{a}_n + \cos\phi\hat{a}_\phi$$

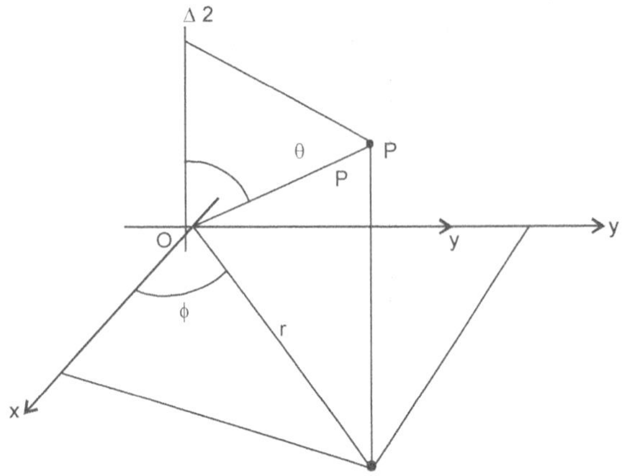

Given, $\bar{B} = B_0\left(\frac{x}{x^2+y^2}\hat{a}_y - \frac{y}{x^2+y^2}\hat{a}_x\right)$

Using the expressions given above, we get

$$\bar{B} = B_0\left(\frac{\cos\phi}{r}(\sin\phi\hat{a}_r + \cos\phi\hat{a}_\phi) - \frac{\sin\phi}{r}(\cos\phi\hat{a}_r - \sin\phi\hat{a}_\phi)\right)$$

$$\frac{B_0}{r}\hat{a}_\phi = \text{constant}$$

$$\bar{H} = \frac{B_0}{\mu_0 r}\hat{a}_\phi = \text{constant}$$

then, $\bar{J} = \nabla \times \bar{H}$

$= \text{zero}$

[curl of a constant quantity is zero]

and $r \# 0$

50. $\tan\beta l = \tan\left(\frac{2\pi}{\pi}\times\frac{\pi}{4}\right) = \infty$

Then, $Z_{in} = Z_0\left(\frac{Z_L + jZ_0\tan\beta l}{Z_0 + jZ_L\tan\beta l}\right)$

$$= Z_0\left(\frac{Z_L\cot\beta l + jZ_0}{Z_0\cot\beta l + jZ_L}\right)$$

$$Z_{in} = \frac{Z_0^2}{Z_L}$$

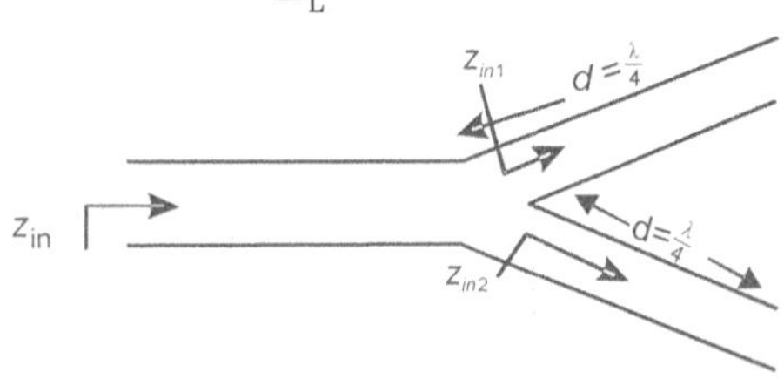

Impedance been by source,

$$Z_{in} = \frac{{Z_0}^2}{Z_L}$$

$$Z_i = \frac{{Z_0}^2}{Z_{L1}}$$

$$Z_{L1} = Z_{i1} \,||\, Z_{i2}$$

$$Z_{i1} = \frac{100^2}{50} = 200\Omega$$

$$Z_{i2} = 200\,\Omega$$

$$Z_{L1} = 100\Omega$$

$$Z_i = \frac{50^2}{100} = 25\,\Omega$$

51. $N_d = 1\times10^{17}\times10^6\, m^{-3}$

$X_n = 0.1\,\mu m$

$X_p = 1.0\,\mu\, m$

$N_i = 1.4\times10^{10}\times10^6\, m^{-3}$

$N_a = ?$

Now $X_pN_a = X_nN_d$

$\therefore$ $N_a = 10^{16}\times10^6\, m^{-3}$

Built in potential,

$$V_0 = \frac{kT}{q}\, ln\left(\frac{N_AN_D}{n_i^2}\right)$$

where, $\frac{kT}{q}$ = Thermal voltage = 26 mV

$$\therefore \quad V_0 = 26\times10^{-3}\, ln\frac{10^{16}\times10^{17}}{(1.4\times10^{10})^2}$$

$= 0.76$ V

Alternately

Built in potential

$$V_j = \frac{qN_DW^2}{2\varepsilon}$$

$$= \frac{1.6\times10^{-19}\times10^{17}\times10^{-6}}{2\times8.85\times10^{-14}}$$

$= 0.76$V

52. $e = q.N_D$ $(w = w_n + w_p)$

$$E = -\int\frac{e}{\varepsilon}dx$$

$$= \int_0^{1.1\mu m}\frac{-e}{\varepsilon}dx$$

$$= \frac{1.6\times10^{-19}}{12\times8.25\times10^{-14}}\times1\times10^{17}\times1.1\times10^{-6}\times10^2$$

$= 0.146$ (all dimension in cm)

$= 0.15$ MV. cm^{-1}

directed from n to p direction from positive charges of transistion region dipole toward the negative charges.

53. G(s) is the impedance of a passive network.

54.

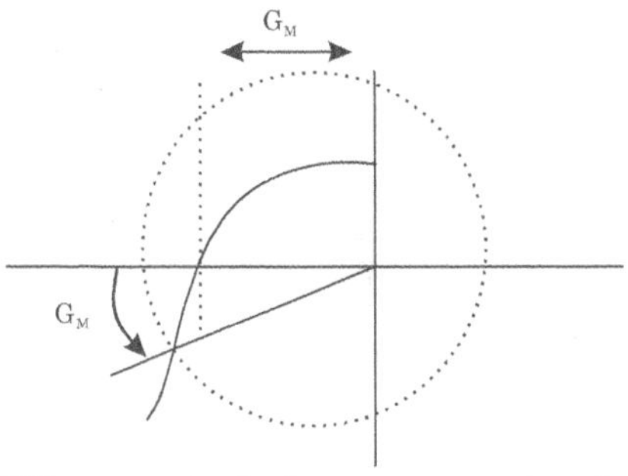

Where unit circle cross nyquist plot at that point the angle from negative real axis is phase margin = 90

Point at which Nyquist plot cut real axis in negative side is

$$\text{gain margin} = -20 \log 5$$
$$= 20 \log 2$$
$$= 20 \times 0\ 3010$$
$$= 6 \text{ dB}$$

55. $$\left(\frac{S}{N_q}\right)_{0dB} = 1\ 76 + 6\ 02u$$

$$\Rightarrow \quad 43\ 5 = 17\ 6 + 6\ 024$$
$$\Rightarrow \quad n \approx 7$$

$$\Delta\ (\text{Step size}) = \frac{2A}{2^n}$$
$$= \frac{10}{2^7} = 0\ 07 \text{ Volts}$$

Which is close to 0 0067 Volts

56. For positive value of Step Size, $\Delta = 0\ 05$ V

$$L = 2^n \geq \frac{5}{0\ 05} = 100$$

Then, $(S/N_q) = 1\ 76 + 6\ 02\ x$

or, $n = 7$

$$= 1\ 76 + 6\ 02 \times 7$$
$$= 43\ 76 \text{ dB}$$

For –ve value of step size,

$$\Delta = 0\ 1\text{V},$$
$$L = 2^n \geq \frac{5}{0\ 1} \geq 50$$
$$(S/N_q) = 1\ 76 + 6\ 02 \times 6$$
$$= 37\ 88 \text{ dB}$$
$$n = 6$$

Hence, the resulting (SNR) is about 43 8 dB for the positive signals

57-58.

Assuming both transitors in saturation, then,

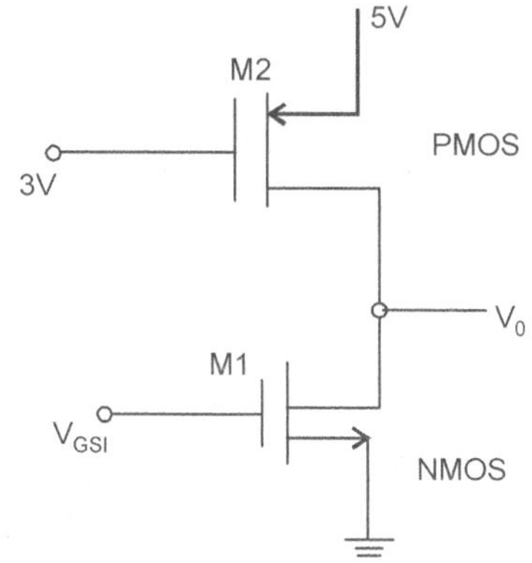

Drain current

$$I_{D_1} = \frac{\mu_n C_0 \times W}{2L}(V_{GS1} - V_{Tn})^2$$
$$= \frac{1}{2}(1\ 5 - 1)^2$$
$$= 0\ 125 \text{ mA} \quad [\text{Assuming } V_{US1} = 1\ 5\text{V}]$$

and $$I_{D_2} = \frac{\mu_p C_0 \times W}{2L}(V_{GS2} - V_{TP})^2$$
$$= \frac{1}{2}(3 - 5 + 1)^2$$
$$= 0\ 50 \text{ mA}$$

Since the gate currents being zero, the drain currents have to be equal to satisfy KCL Further saturation currents are not equal both the MOSFET's, both can't be in saturation simultaneously Thus NMOS which has saturation current of 0 125 mA, can never carry 0 50 mA Hence PMOS is not in saturation

Now, assume that NMOS is in saturation and PMOJ is in triode then,

$$I_{D_1} = 0\ 125\text{mA}$$
$$= I_{D_2} = \frac{\mu_p C_0 \times W}{L}\left(V_{GS_2} - V_{TP} - \frac{1}{2}V_{DS_2}\right)V_{DS_2}$$

$$\Rightarrow \quad 0\ 125 = 1\left(3 - 5 + 1 - \frac{V_D - 5}{2}\right)(V_D - 5)$$
$$\Rightarrow \quad 0\ 125 = \left(-1 - \frac{V_D - 5}{2}\right)(V_D - 5)$$
$$\Rightarrow \quad 0\ 25 = (-V_D + 3)(V_D - 5)$$
$$= \left(-V_D^2 + 8V_D - 15\right)$$
$$\Rightarrow \quad V_D^2 - 8V_D + 15 + \frac{1}{4} = 0$$
$$\Rightarrow \quad V_D^2 - 8V_D + \frac{61}{4} = 0 \qquad \text{(A)}$$

Solution of this quadratic equation (A) is given by

$$V_D = \frac{8 \pm \sqrt{64 - 4\frac{61}{4}}}{2}$$

$$= \left(4 \pm \frac{\sqrt{3}}{2}\right) \text{volts}$$

For $V_D = 4 - \frac{\sqrt{3}}{2}$

PMOS will be in saturation as

$$V_{DS} \leq V_{GC} - V_{TP} \qquad \begin{bmatrix} V_{DS} = V_D - 5 \\ V_{GS} = 3 - 5 = -2 \end{bmatrix}$$

Hence, $V_D = 4 + \frac{\sqrt{3}}{2}$ which will cause PMOS to put in triode region as

$$V_{DS} \geq V_{GC} - V_T$$

59-60

Truth-table for the given programme is shown below

P_1 P_2	a	b	c	d	e	f	g
0 0	1	1	1	1	1	1	0
0 1	1	0	1	1	0	1	1
1 0	1	1	0	1	1	0	1
1 1	1	0	0	1	1	1	1

K-maps for the truth-table are

P_1 \ P_2	0	1
0	1	1
1	1	1

$a = 1$

P_1 \ P_2	0	1
0	1	0
1	1	0

$a = \overline{P}_2$

P_1 \ P_2	0	1
0	1	1
1	0	0

$c = \overline{P}_1$

P_1 \ P_2	0	1
0	1	1
1	1	1

$d = 1$

P_1 \ P_2	0	1
0	1	0
1	1	1

$e = P_1 + \overline{P}_2$

P_1 \ P_2	0	1
0	1	1
1	0	1

$f = \overline{P}_1 + P_2$

P_1 \ P_2	0	1
0	0	1
1	1	1

$g = P_1 + P_2$

Also, $d = 1 = c + e$

Now, two NOT gate required for $b = \overline{P}_2$ and $c = \overline{P}_1$ and three OR gate required for $e = P_1 + \overline{P}_2$

$$t = \overline{P}_1 + \overline{P}_2 \text{ and } g = P_1 + P_2$$

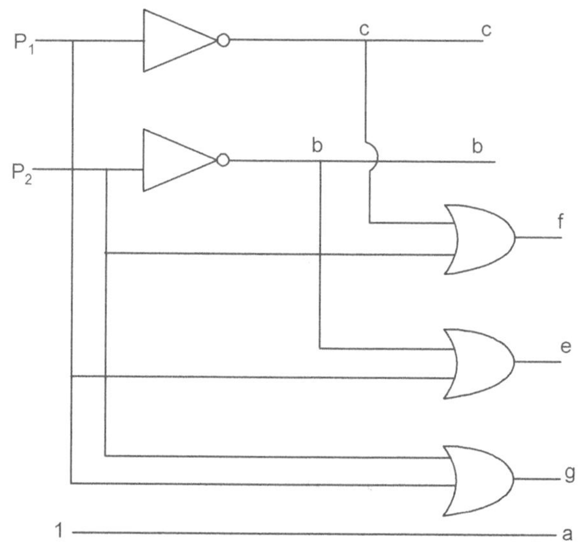

■■

SOLVED PAPER – 2008

ELECTRONICS AND COMMUNICATION ENGINEERING

Time : 3 Hours **Maximum Marks : 150**

Q.1 – Q.20 carry one mark each.

1. All the four entries of the 2 × 2 matrix $P = \begin{bmatrix} P_{11} & P_{12} \\ P_{21} & P_{22} \end{bmatrix}$ are nonzero, and one of its eigenvalues is zero. Which of the following statements is true?

   (*a*) $P_{11}P_{22} - P_{12}P_{21} = 1$

   (*b*) $P_{11}P_{22} - P_{12}P_{21} = -1$

   (*c*) $P_{11}P_{22} - P_{12}P_{21} = 0$

   (*d*) $P_{11}P_{22} + P_{12}P_{21} = 0$

2. The system of linear equations

$$4x + 2y = 7$$
$$2x + y = 6$$

   has

   (*a*) a unique solution

   (*b*) no solution

   (*c*) an infinite number of solutions

   (*d*) exactly two distinct solutions

3. The equation sin (z) = 10 has

   (*a*) no real or complex solution

   (*b*) exactly two distinct complex solutions

   (*c*) a unique solution

   (*d*) an infinite number of complex solutions

4. For real values of x, the minimum value of the function $f(x) = \exp(x) + \exp(-x)$ is

   (*a*) 2 (*b*) 1

   (*c*) 0.5 (*d*) 0

5. Which of the following functions would have only odd powers of x in its Taylor series expansion about the point x = 0?

   (*a*) $\sin(x^3)$ (*b*) $\sin(x^2)$

   (*c*) $\cos(x^3)$ (*d*) $\cos(x^2)$

6. Which of the following is a solution to the differential equation $\frac{dx(t)}{dt} + 3x(t) = 0$?

   (*a*) $x(t) = 3e^{-1}$

   (*b*) $x(t) = 2e^{-3t}$

   (*c*) $x(t) = -\frac{3}{2}t^2$

   (*d*) $x(t) = 3t^2$

7. In the following graph, the number of trees (P) and the number of cut-sets (Q) are

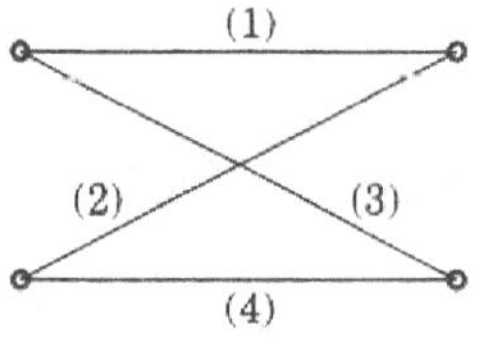

   (*a*) P = 2, Q = 2 (*b*) P = 2, Q = 6

   (*c*) P = 4, Q = 6 (*d*) P = 4, Q = 10

8. In the following ciruit, the switch S is closed at t = 0. The rate of change of current $\frac{di}{dt}(0^+)$ is given by

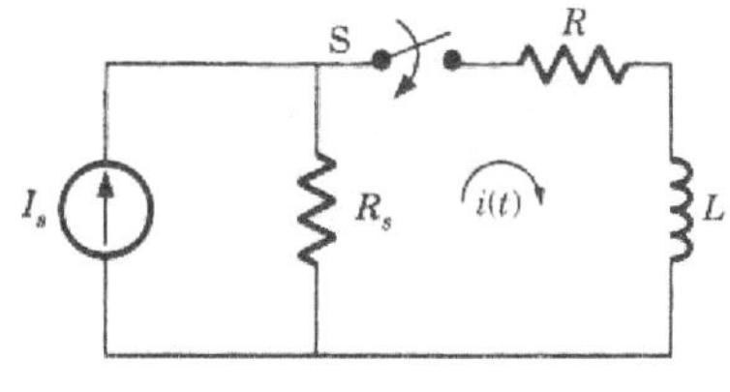

   (*a*) 0 (*b*) $\frac{R_s I_s}{L}$

   (*c*) $\frac{(R + R_s)I_s}{L}$ (*d*) ∞

9. The input and output of a continuous time system are respectively denoted by x(t) and y(t). Which of the following descriptions corresponds to a causal system?

   (*a*) $y(t) = x(t - 2) + x(t + 4)$

   (*b*) $y(t) = (t - 4)\,x(t + 1)$

   (*c*) $y(t) = (t + 4)\,x(t - 1)$

   (*d*) $y(t) = (t + 5)\,x(t + 5)$

10. The impulse response h(t) of a linear time-invariant continuous time system is described by $h(t) = \exp(\alpha t)u(t) + \exp(\beta t)u(-t)$, where u(t) denotes the unit step function, and α and β are real constants. This system is stable if

   (*a*) α is positive and β is positive

   (*b*) α is negative and β is negative

   (*c*) α is positive and β is negative

   (*d*) α is negative and β is positive

11. The pole-zero plot given below corresponds to a

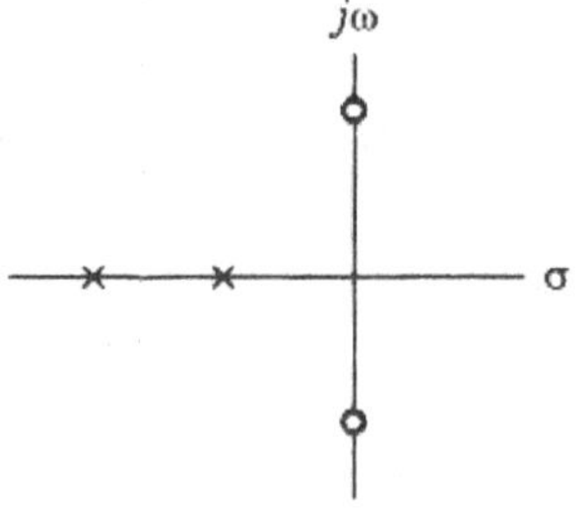

(*a*) Low pass filter (*b*) High pass filter
(*c*) Band pass filter (*d*) Notch filter

12. Step responses of a set of three second-order underdamped systems all have the same percentage overshoot. Which of the following diagrams represents the poles of the three systems?

(*a*)
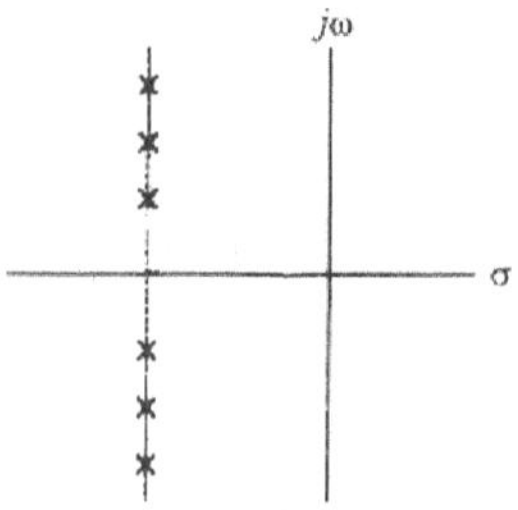

(*b*)
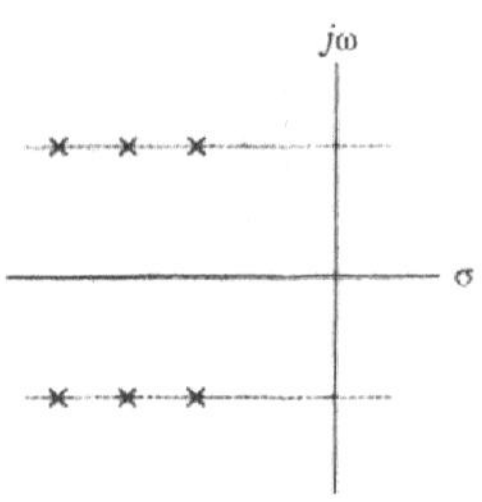

(*c*)
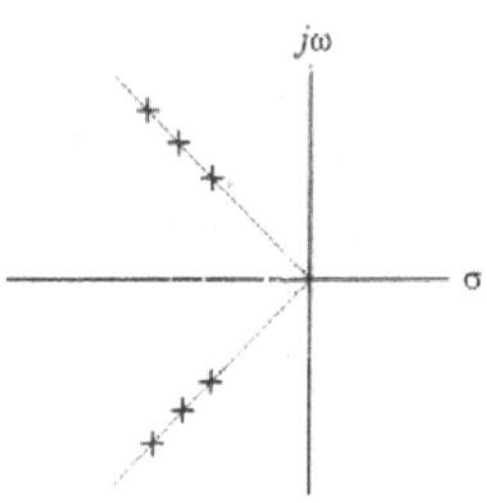

(*d*)
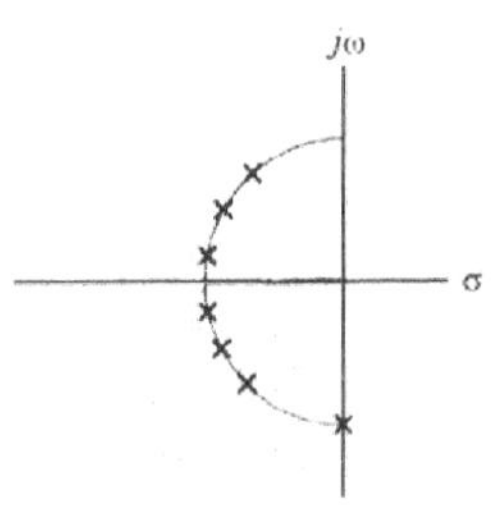

13. Which of the following is NOT associated with a p-n junction?
(*a*) Junction Capacitance
(*b*) Charge Storage Capacitance
(*c*) Depletion Capacitance
(*d*) Channel Length Modulation

14. Which of the following is true?
(*a*) A silicon wafer heavily doped with boron is a p^+ substrate
(*b*) A silicon wafer lightly doped with boron is a p^+ substrate
(*c*) A silicon wafer heavily doped with arsenic is a p^+ substrate
(*d*) A silicon wafer lightly doped with arsenic is a p^+substrate

15. For a Hertz dipole antenna, the half power beam width (HPBW) in the E-plane is
(*a*) 360° (*b*) 180°
(*c*) 90° (*d*) 45°

16. For static electric and magnetic fields in an inhomogeneous source-free medium, which of the following represents the correct form of two of Maxwell's equations?
(*a*) $\nabla.E = 0$ $\nabla \times B = 0$
(*b*) $\nabla.E = 0$ $\nabla.B = 0$
(*c*) $\nabla \times E = 0$ $\nabla \times B = 0$
(*d*) $\nabla \times E = 0$ $\nabla.B = 0$

17. In the following limiter circuit, an input voltage $V_i = 10 \sin 100\pi t$ is applied. Assume that the diode drop is 0.7 V when it is forward biased. The Zener breakdown voltage is 6.8V.

The maximum and minimum values of the output voltage respectively are

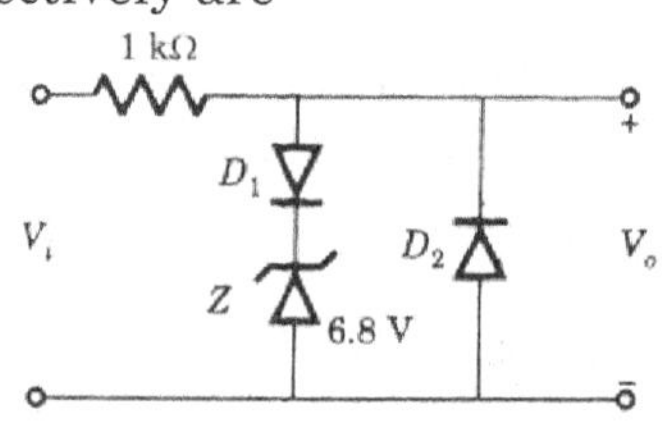

(*a*) 6.1 V, –0.7 V
(*b*) 0.7 V, –7.5 V
(*c*) 7.5 V, –0.7 V
(*d*) 7.5 V, –7.5 V

18. A silicon wafer has 100 nm of oxide on it and is inserted in a furnace at a temperature above 1000°C for further oxidation in dry oxygen. The oxidation rate
(*a*) is independent of current oxide thickness and temperature
(*b*) is independent of current oxide thickness but depends on temperature
(*c*) slows down as the oxide grows
(*d*) is zero as the existing oxide prevents further oxidation

19. The drain current of a MOSFET in saturation is given by $I_D = K(V_{GS} - V_T)^2$ where K is a constant. The magnitude of the transconductance g_m is

(*a*) $\dfrac{K(V_{GS} - V_T)^2}{V_{DS}}$ (*b*) $2K(V_{GS} - V_T)$

(*c*) $\dfrac{I_d}{V_{GS} - V_{DS}}$ (*d*) $\dfrac{K(V_{GS} - V_T)^2}{V_{GS}}$

20. Consider the amplitude modulated (AM) signal $A_c \cos\omega_c t + 2\cos(\omega_m t)\cos(\omega_c t)$. For demodulating the signal using envelope detector, the minimum value of A_c should be

(*a*) 2 (*b*) 1

(*c*) 0.5 (*d*) 0

Q.21 – 75 carry two marks each.

21. The Thevenin equivalent impedance Z_{th} between the nodes P and Q in the following circuit is

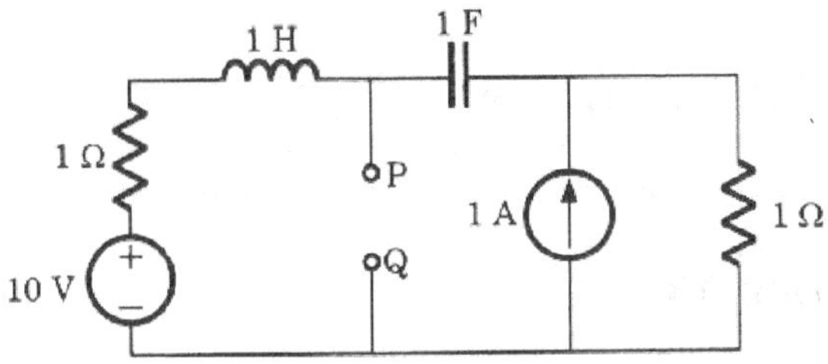

(*a*) 1

(*b*) $1 + s + \frac{1}{s}$

(*c*) $2 + s + \frac{1}{s}$

(*d*) $\frac{s^2 + s + 1}{s^2 + 2s + 1}$

22. The driving point impedance of the following network is given by $Z(s) = \frac{0.2s}{s^2 + 0.1s + 2}$. The component values are

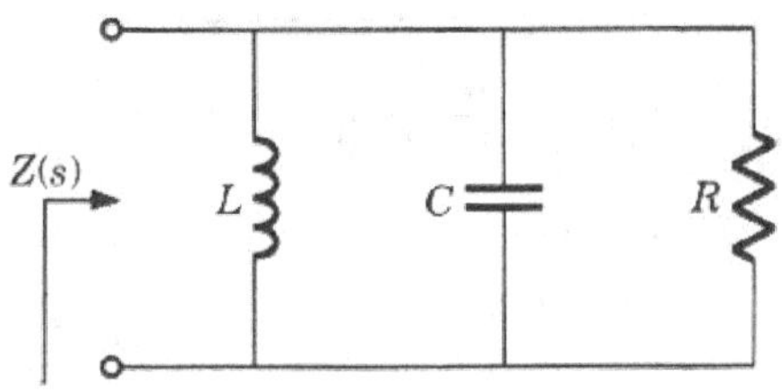

(*a*) L = 5H, R = 0.5Ω, C = 0.1F

(*b*) L = 0.1H, R = 0.5Ω, C = 5F

(*c*) L = 5H, R = 2Ω, C = 0.1F

(*d*) L = 0.1H, R = 2Ω, C = 5F

23. The circuit shown in the figure is used to charge the capacitor C alternately from two current sources as indicated. The switches S1 and S2 are mechanically coupled and connected as follows:

For $2nT \le t < (2n+1)T$, (n = 0, 1, 2...)

S1 to P1 and S2 to P2.

For $(2n+1)T \le t < (2n+2)T$, (n = 0, 1, 2...)

S1 to Q1 and S2 to Q2.

Assume that the capacitor has zero initial charge. Given that u(t) is a unit step function, the voltage $V_c(t)$ across the capacitor is given by

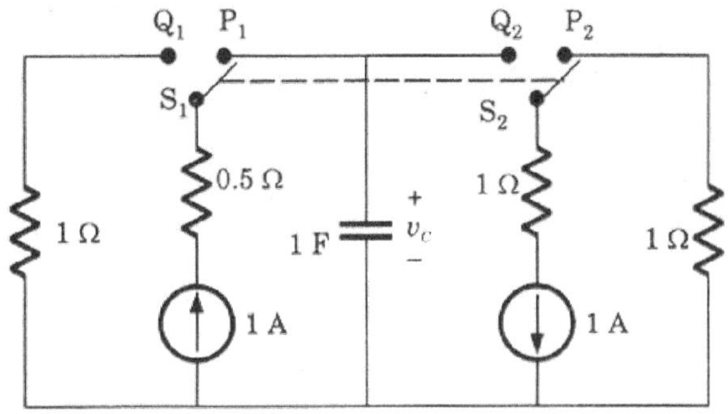

(*a*) $\sum_{n=0}^{\infty} (-1)^n \, t u(t - nT)$

(*b*) $u(t) + 2\sum_{n=1}^{\infty} (-1)^n \, u(t - nT)$

(*c*) $t u(t) + 2\sum_{n=1}^{\infty} (-1)^n \, (t - nT)\, u(t - nT)$

(*d*) $\sum_{n=0}^{\infty} \left[0.5 - e^{-(t-2nT)} + 0.5e^{-(t-2nT-T)}\right]$

24. The probability density function (PDF) of a random variable X is as shown below.

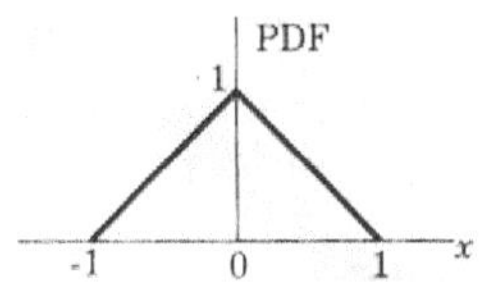

The corresponding cumulative distribution function (CDF) has the form

(*a*)

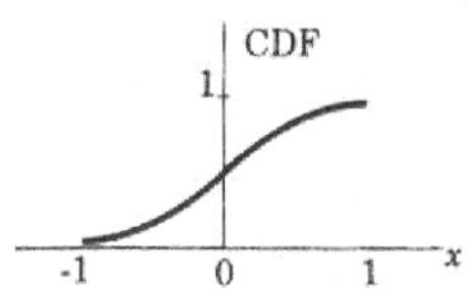

(*b*)

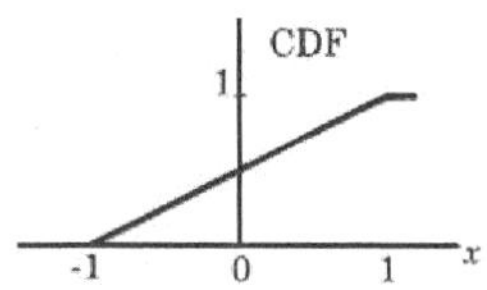

(*c*)

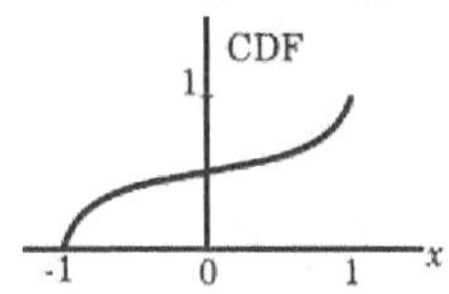

(*d*)

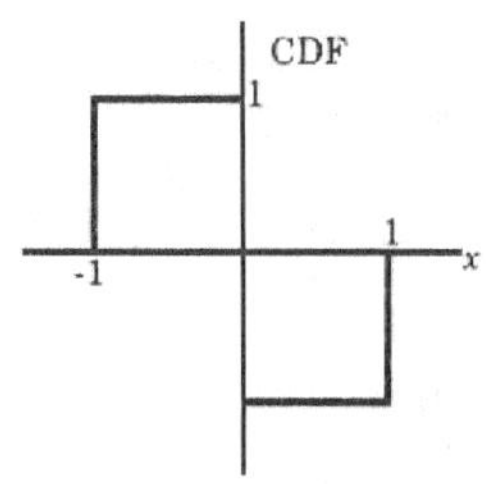

25. The recursion relation to solve $x = e^{-x}$ using Newton-Raphson method is

(*a*) $X_{n+1} = e^{-X_n}$

(*b*) $X_{n+1} = X_n - e^{-X_n}$

(*c*) $X_{n+1} = (1+X_n)\frac{e^{-X_n}}{1+e^{-X_n}}$

(*d*) $X_{n+1} = \frac{X_n^2 - e^{-X_n}(1+x_n) - 1}{x_n - e^{-X_n}}$

26. The residue of the function

$$f(z) = \frac{1}{(z+2)^2(z-2)^2} \text{ at } z = 2 \text{ is}$$

(*a*) $-\frac{1}{32}$ (*b*) $-\frac{1}{16}$

(*c*) $\frac{1}{16}$ (*d*) $\frac{1}{32}$

27. Consider the matrix $P = \begin{bmatrix} 0 & 1 \\ -2 & -3 \end{bmatrix}$. The value of e^p is

(*a*) $\begin{bmatrix} 2e^{-2} - 3e^{-1} & e^{-1} - e^{-2} \\ 2e^{-2} - 2e^{-1} & 5e^{-2} - e^{-1} \end{bmatrix}$

(*b*) $\begin{bmatrix} e^{-1} + e^{-2} & 2e^{-2} - e^{-1} \\ 2e^{-1} - 4e^{-2} & 3e^{-1} + 2e^{-2} \end{bmatrix}$

(*c*) $\begin{bmatrix} 5e^{-2} - e^{-1} & 3e^{-1} - e^{-2} \\ 2e^{-2} - 6e^{-1} & 4e^{-2} + e^{-1} \end{bmatrix}$

(*d*) $\begin{bmatrix} 2e^{-1} - e^{-2} & e^{-1} - e^{-2} \\ -2e^{-1} + 2e^{-2} & -e^{-1} + 2e^{-2} \end{bmatrix}$

28. In the Taylor series expansion of exp(x) + sin(x) about the point x=π, the coefficient of $(x-\pi)^2$ is

(*a*) exp(π) (*b*) 0.5 exp(π)

(*c*) exp(π)+1 (*d*) exp (π) –1

29. $P_x(x) = M \exp(-2|x|) - N \exp(-3|x|)$ is the probability density function for the real random variable X, over the entire x axis. M and N are both positive real numbers. The equation relating M and N is

(*a*) $M + \frac{2}{3}N = 1$ (*b*) $2M + \frac{1}{3}N = 1$

(*c*) M +N =1 (*d*) M +N =3

30. The value of the integral of the function $g(x, y) = 4x^3 + 10y^4$ along the straight line segment from the point (0, 0) to the point (1, 2) in the x-y plane is

(*a*) 33 (*b*) 35

(*c*) 40 (*d*) 56

31. A linear, time -invariant, causal continuous time system has a rational transfer function with simple poles at s =–2 and s =–4, and one simple zero at s =–1. A unit step u(t) is applied at the input of the system. At steady state, the output has constant value of 1. The impulse response of this system is

(*a*) [exp(–2t) + exp(–4t)]u (t)

(*b*) [–4 exp(–2t) –12 exp(–4t) – exp(–t)]u(t)

(*c*) [–4 exp(–2t) +12 exp(–4t)]u(t)

(*d*) [–0.5 exp(–2t) +1.5 exp(–4t)]u(t)

32. The signal x(t) is described by

$$x(t) = \begin{cases} 1 & \text{for} - \leq t \leq +1 \\ 0 & \text{otherwise} \end{cases}$$

Two of the angular frequencies at which its Fourier transform becomes zero are

(*a*) π, 2π (*b*) 0.5 π, 1.5π

(*c*) 0, π (*d*) 2π, 2.5π

33. A discrete time linear shift-invariant system has an impulse response h[n] with h[0] =1, h[1] =–1. h[2] =2, and zero otherwise. The system is given an input sequence x[n] with x[0] = x[2] = 1, and zero otherwise. The number of nonzero samples in the output sequence y[n], and the value of y[2] are respectively.

(*a*) 5,2 (*b*) 6,2

(*c*) 6,1 (*d*) 5,3

34. Consider points P and Q in the x-plane, with P=(1,0) and Q=(0,1). The line integral $2\int_P^Q (x\,dx + y\,dy)$ along the semicircle with the line segment PQ as its diameter

(*a*) is –1

(*b*) is 0

(*c*) is 1

(*d*) depends on the direction (clockwise or anti-clockwise) of the semicircle

35. Let x(t) be the input and y(t) be the output of a continuous time system. Match the system properties P1, P2 and P3 with system relations R1, R2, R3, R4

Properties	Relations
P1: Linear but NOT time-invariant	R1: $y(t) = t^2x(t)$
P2: Time-invariant but NOT linear	R2: $y(t) = t\|x(t)\|$
P3: Linear and time-invariant	R3: $y(t) = \|x(t)\|$
	R4: y(t) = x(t–5)

(*a*) (P1, R1), (P2, R3), (P3, R4)

(*b*) (P1, R2), (P2, R3), (P3, R4)

(*c*) (P1, R3), (P2, R1), (P3, R2)

(*d*) (P1, R1), (P2, R2), (P3, R3)

36. A memoryless source emits n symbols each with a probability p. The entropy of the source as a function of n

(*a*) increases as log n (*b*) decreases as log (1/n)

(*c*) increases as n (*d*) increases as n log n

37. $\{x(n)\}$ is a real-valued periodic sequence with a period N. x(n) and X(k) form N-point Discrete Fourier Transform (DFT) pairs. The DFT Y(k) of the sequence

$$y(n) = \frac{1}{N}\sum_{r=0}^{N-1} x(r)\, x(n+r) \text{ is}$$

(*a*) $|X(k)|^2$ (*b*) $\frac{1}{N}\sum_{r=0}^{N-1} X(r)\, X(k+r)$

(*c*) $\frac{1}{N}\sum_{r=0}^{N-1} X(r)\, X(k+r)$ (*d*) 0

38. Group I lists a set of four transfer functions. Group II gives a list of possible step responses y(t). Match the step responses with the corresponding transfer functions.

Group I

$$P = \frac{25}{s^2 + 25}$$

$$Q = \frac{36}{s^2 + 20s + 36}$$

$$R = \frac{36}{s^2 + 12s + 36}$$

$$S = \frac{49}{s^2 + 7s + 49}$$

Group II

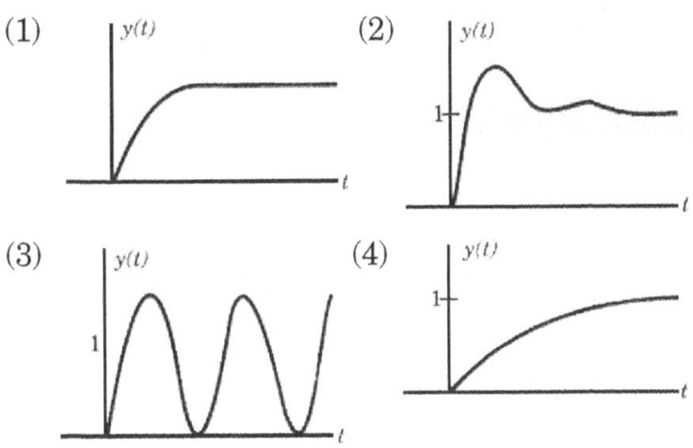

(*a*) P – 3, Q – 1, R – 4, S – 2

(*b*) P – 3, Q – 2, R – 4, S – 1

(*c*) P – 2, Q – 1, R – 4, S – 3

(*d*) P – 3, Q – 4, R – 1, S – 2

39. A certain system has transfer function

$$G(s) = \frac{s+8}{s^2 + \alpha s - 4}, \text{ where } \alpha \text{ is a parameter.}$$

Consider the standard negative unity feedback configuration as shown below.

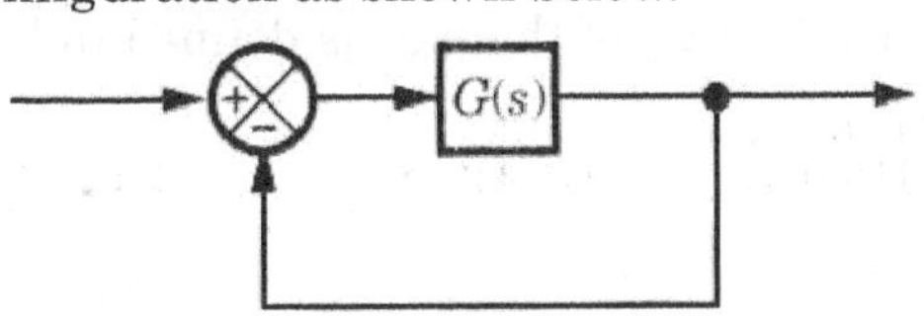

Which of the following statements is true?

(*a*) The closed loop systems is never stable for any value of α.

(*b*) For some positive values of α, the closed loop system is stable, but not for all positive values.

(*c*) For all positive values of α, the closed loop system is stable.

(*d*) The closed loop systems stable for all values of α, both positive and negative.

40. A signal flow graph of a system is given below.

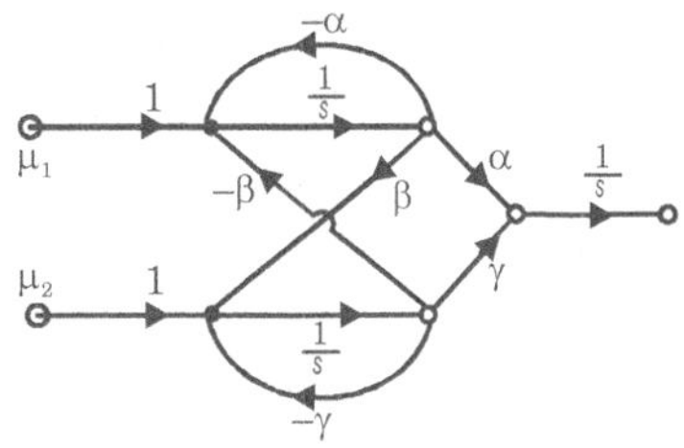

The set of equations that correspond to this signal flow graph is

(*a*) $\frac{d}{dt}\begin{pmatrix} x_1 \\ x_2 \\ x_3 \end{pmatrix} = \begin{bmatrix} \beta & -\gamma & 0 \\ \gamma & \alpha & 0 \\ -\alpha & -\beta & 0 \end{bmatrix}\begin{pmatrix} x_1 \\ x_2 \\ x_3 \end{pmatrix} + \begin{bmatrix} 0 & 0 \\ 0 & 1 \\ 1 & 0 \end{bmatrix}\begin{pmatrix} u_1 \\ u_2 \end{pmatrix}$

(*b*) $\frac{d}{dt}\begin{pmatrix} x_1 \\ x_2 \\ x_3 \end{pmatrix} = \begin{bmatrix} 0 & \alpha & \gamma \\ 0 & -\alpha & -\gamma \\ 0 & \beta & -\beta \end{bmatrix}\begin{pmatrix} x_1 \\ x_2 \\ x_3 \end{pmatrix} + \begin{bmatrix} 1 & 0 \\ 0 & 1 \\ 0 & 0 \end{bmatrix}\begin{pmatrix} u_1 \\ u_2 \end{pmatrix}$

(*c*) $\frac{d}{dt}\begin{pmatrix} x_1 \\ x_2 \\ x_3 \end{pmatrix} = \begin{bmatrix} -\alpha & \beta & 0 \\ -\beta & -\gamma & 0 \\ \alpha & \gamma & 0 \end{bmatrix}\begin{pmatrix} x_1 \\ x_2 \\ x_3 \end{pmatrix} + \begin{bmatrix} 1 & 0 \\ 0 & 1 \\ 0 & 0 \end{bmatrix}\begin{pmatrix} u_1 \\ u_2 \end{pmatrix}$

(*d*) $\frac{d}{dt}\begin{pmatrix} x_1 \\ x_2 \\ x_3 \end{pmatrix} = \begin{bmatrix} -\gamma & 0 & \beta \\ \gamma & 0 & \alpha \\ -\beta & 0 & -\alpha \end{bmatrix}\begin{pmatrix} x_1 \\ x_2 \\ x_3 \end{pmatrix} + \begin{bmatrix} 0 & 1 \\ 0 & 0 \\ 1 & 0 \end{bmatrix}\begin{pmatrix} u_1 \\ u_2 \end{pmatrix}$

41. The number of open right half plane poles of

$$G(s) = \frac{10}{s^5 + 2s^4 + 3s^3 + 6s^2 + 5s + 3} \text{ is}$$

(*a*) 0

(*b*) 1

(*c*) 2

(*d*) 3

42. The magnitude of frequency response of an underdamped second order system is 5 at 0 rad/sec and peaks to $\frac{10}{\sqrt{3}}$ at $5\sqrt{2}$ rad/sec. The transfer function of the system is

(*a*) $\frac{500}{s^2 + 10s + 100}$ (*b*) $\frac{375}{s^2 + 5s + 75}$

(*c*) $\frac{720}{s^2 + 12s + 144}$ (*d*) $\frac{1125}{s^2 + 25s + 225}$

43. Group I gives two possible choices for the impedance Z in the diagram. The circuit elements in Z satisfy the condition $R_2C_2 > R_1C_1$. The transfer function $\frac{V_o}{V_i}$ represents a kind of controller. Match the impedances in Group I with the types of controllers in Group II.

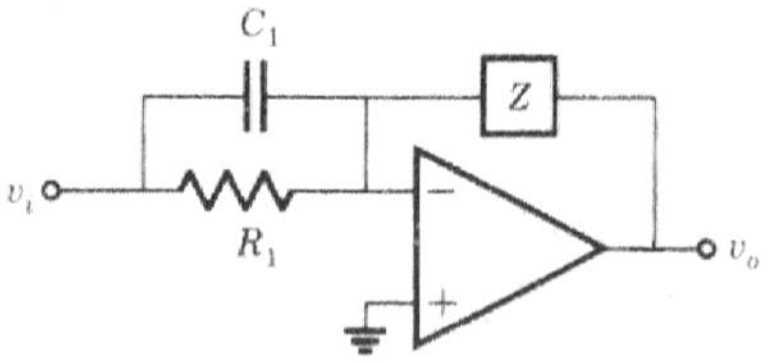

Group I	Group II
Q. 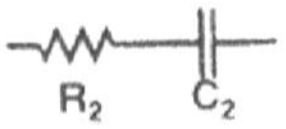	1. PID controller
R. 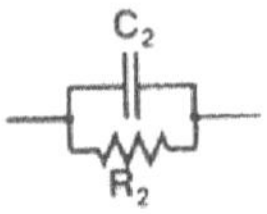	2. Lead compensator
	3. Lag compensator

(*a*) Q–1, R–2 (*b*) Q–1, R–3

(*c*) Q–2, R–3 (*d*) Q–3, R–2

44. For the circuit shown in the following figure, transistors M1 and M2 are identical NMOS transistors. Assume the M2 is in saturation and the output is unloaded.

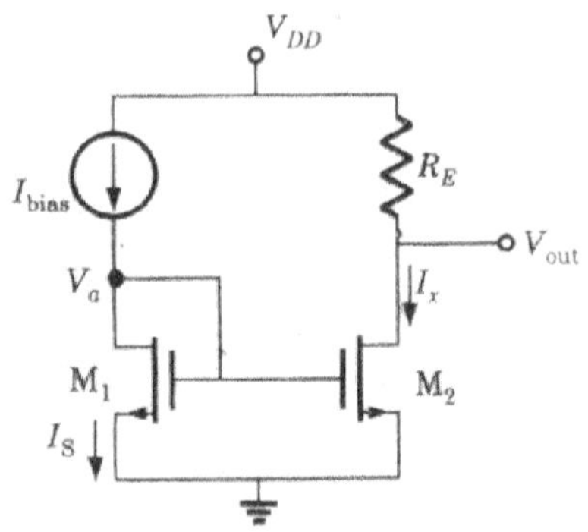

The current I_x is related to I_{bias} as

(*a*) $I_x = I_{bias} + I_s$

(*b*) $I_x = I_{bias}$

(*c*) $I_x = I_{bias} - I_s$

(*d*) $I_x = I_{bias} - \left(V_{DD} - \frac{V_{out}}{R_E}\right)$

45. The measured transconductance g_m of an NMOS transistor operating in the linear region is plotted against the gate voltage V_G at a constant drain voltage V_D. Which of the following figures represents the expected dependence of g_m on V_G?

(*a*)
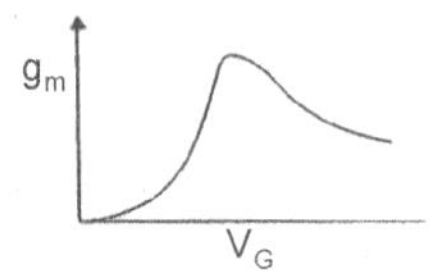

(*b*)
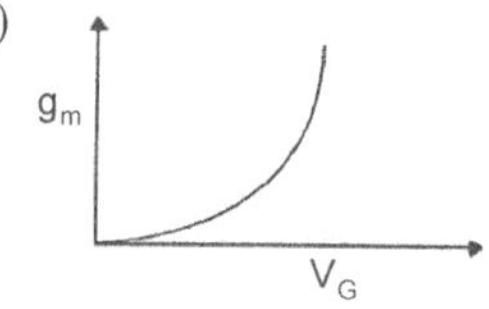

(*c*)
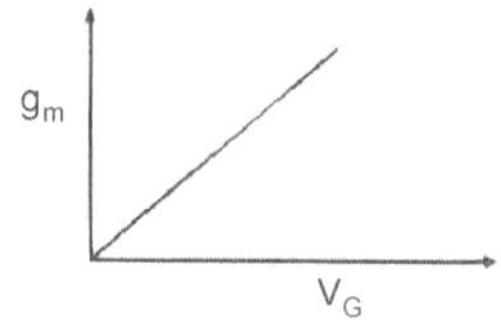

(*d*)
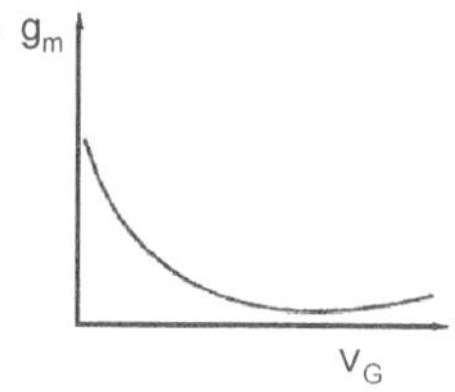

46. Consider the following circuit using an ideal OPAMP. The I–V characteristics of the diode is described by the relation $I = I_0\left(e^{\frac{V}{V_T}} - 1\right)$ where $V_T = 25$ mV, $I_0 = 1\mu A$ and V is the voltage across the diode (taken as positive for forward bias)

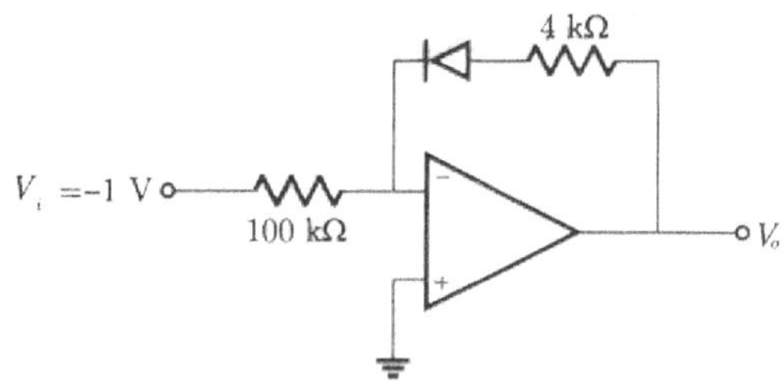

For an input voltage V_T = – I V, the output voltage V_0 is

(*a*) 0 V (*b*) 0.1 V

(*c*) 0.7 V (*d*) 1.1 V

47.

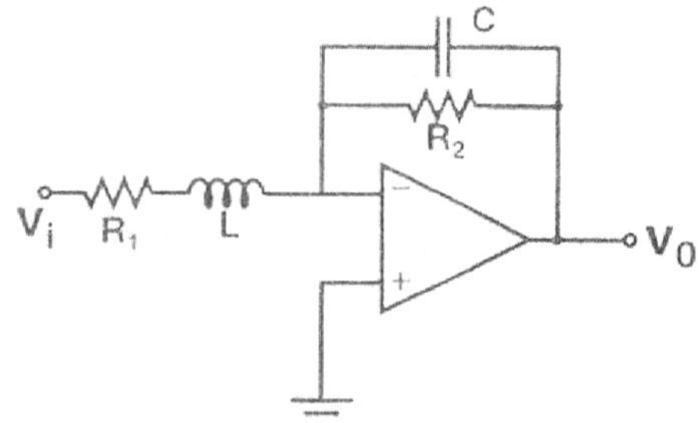

The OPAMP circuit shown above represents a

(*a*) high pass filter

(*b*) low pass filter

(*c*) band pass filter

(*d*) band reject filter

48. Two identical NMOS transistors M_1 and M_2 are connected as shown below . V_{bias} is chosen so that both transistors are in saturation. The equivalent g_m of the pair is defined to be $\frac{\partial I_{out}}{\partial V_i}$ at constant V_{out}

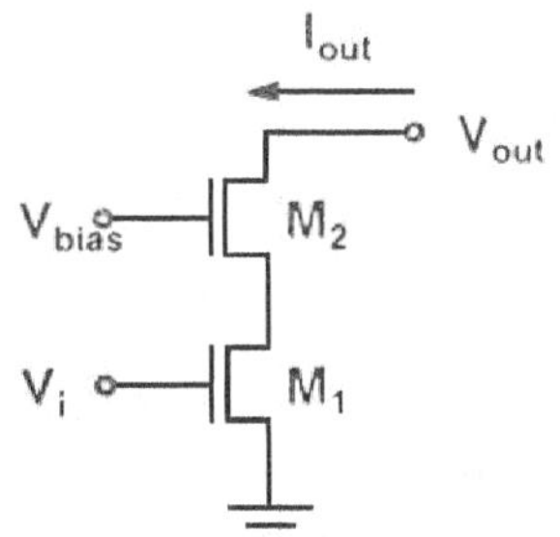

The equivalent g_m of the pair is

(*a*) the sum of individual g_m's of the transistors

(*b*) the product of individual g_m's of the transistors

(*c*) nearly equal to the g_m of M_1

(*d*) nearly equal to g_m/g_o of M_2

49. An 8085 executes the following instructions

2710 LXI H, 30A0H

2713 DAD H

2714 PCHL

All addresses and constants are in Hex. Let PC be the contents of the program counter and HL be the contents of the HL register pair just after executing PCHL.

Which of the following statements is correct ?

(*a*) PC = 2715H   HL = 30A0H

(*b*) PC = 30A0H   HL = 2715H

(*c*) PC = 6140H   HL = 6140H

(*d*) PC = 6140H   HL = 2715H

50. An astable multivibrator circuit using IC 555 timer is shown below. Assume that the circuit is oscillating steadily.

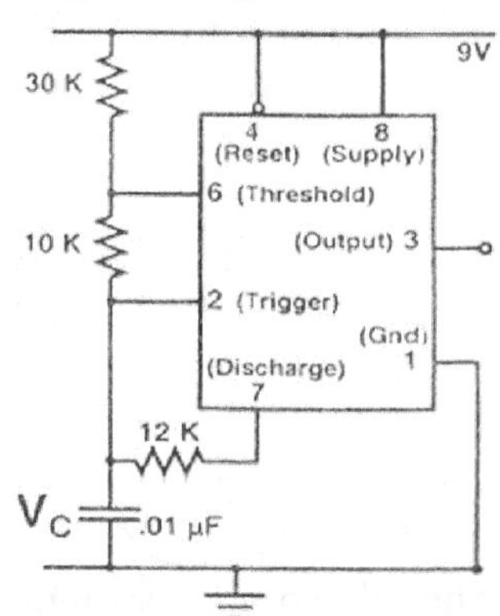

The voltage V_c across the capacitor varies between

(*a*) 3V to 5V

(*b*) 3V to 6V

(*c*) 3.6 V to 6V

(*d*) 3.6 V to 5V

51. Silicon is doped with boron to a concentration of 4×10^{17} atoms cm^3. Assume the intrinsic carrier concentration of silicon to be $1.5 \times 10^{10}/cm^3$ and the value of $\frac{kT}{q}$ to be 25 mV at 300 k.

Compared to undoped silicon, the Fermi level of doped silicon

(*a*) goes down by 0.13 eV

(*b*) goes up by 0.13 eV

(*c*) goes down by 0.427 eV

(*b*) goes up by 0.427cV

52. The cross section of a JFET is shown in the following figure. Let V_G be –2V and let V_p be the initial pinch-off voltage. If the width W is doubled (with other geometrical parameters and doping levels remaining the same), then the ratio between the mutual transconductances of the initial and the modified JFET is

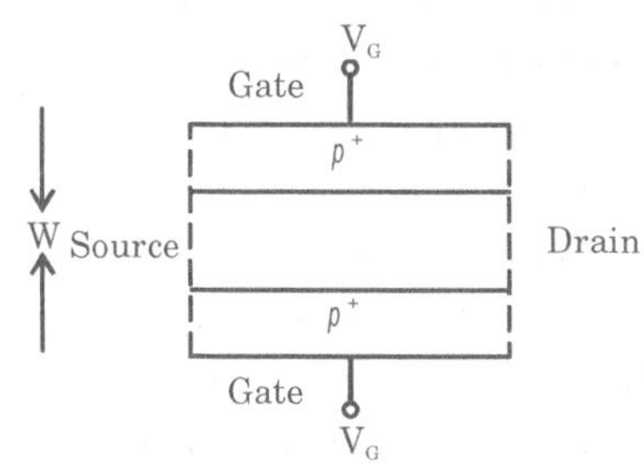

(*a*) 4

(*b*) $\left(\frac{1-\sqrt{2/V_p}}{1-\sqrt{1/(2V_p)}}\right) \times \frac{1}{2}$

(*c*) $\left(\frac{1-\sqrt{2/V_p}}{1-\sqrt{1/(2V_p)}}\right)$

(*d*) $\frac{1-(2/\sqrt{V_p})}{1-(1(2\sqrt{V_p}))}$

53. Consider the Schmidt trigger circuit shown below

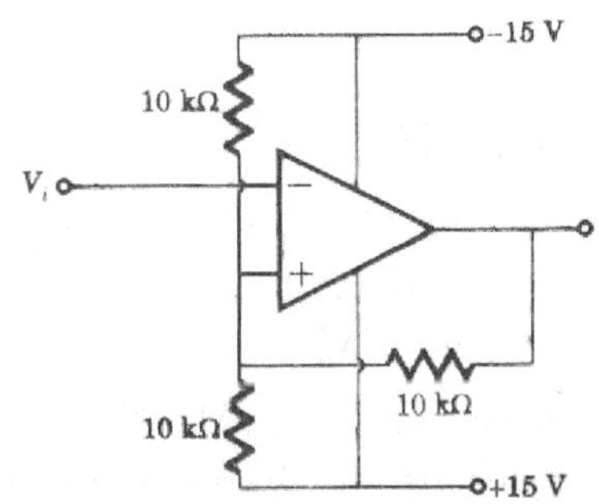

A triangular wave which goes from –12 V to 12 V is applied to the inverting input of the OPAMP. Assume that the output of the OPAMP swings from +15V to –15V. The voltage at the non-inverting input switches between

(*a*) –12V and +12 V

(*b*) –7.5V and –7.5V

(*c*) –5V and + 5V

(*d*) 0V and 5V

54. The logic function implemented by the following circuit at the terminal OUT is

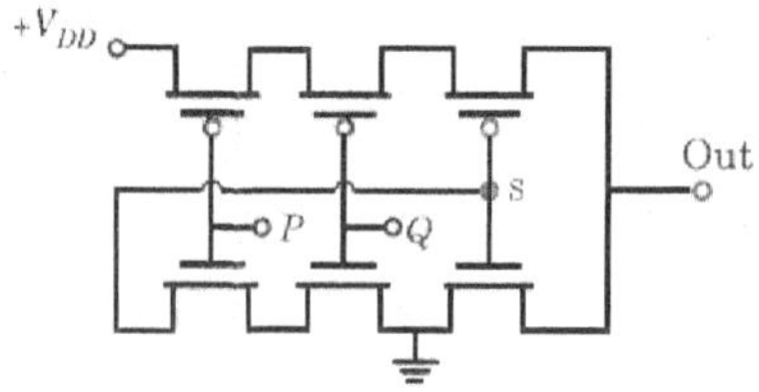

(a) P NOR Q (b) P NAND Q

(c) P OR Q (d) P AND Q

55. Consider the following assertions.

S1 : For Zener effect to occur, a very abrupt junction is required.

S2 : For quantum tunneling to occur, a very narrow energy barrier is required.

Which of the following is correct?

(a) Only S2 is true

(b) S1 and S2 are both true but S2 is not a reason for S1

(c) S1 and S2 are both true and S2 is a reason for S1

(d) Both S1 and S2 are false

56. The two numbers represented in signed 2's complement form are P =11101101 and Q = 11100110. If Q is subtracted from P, the value obtained in signed 2's complement form is

(a) 100000111 (b) 00000111

(c) 11111001 (d) 111111001

57. Which of the following Boolean Expression correctly represents the relation between P, Q, R and M_1?

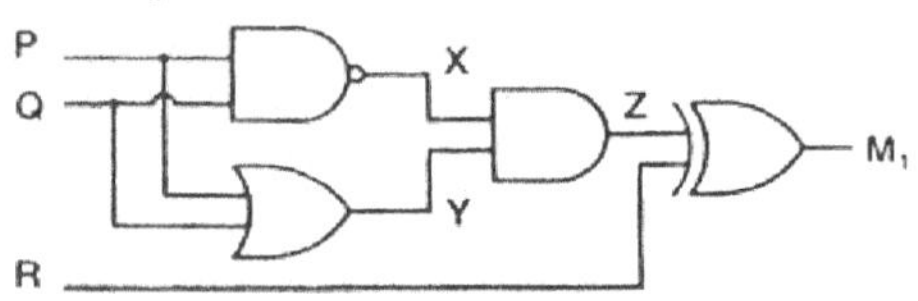

(a) M = (P OR Q) XOR R

(b) M_1 = (P AND Q) XOR R

(c) M = (P NOR Q) XOR R

(d) M_1 = (P XOR Q) XOR R

58. For the circuit shown in the following figure, I_0 –I_3 are inputs to the 4 : 1 multiplexer. R (MSB) and S are control bits.

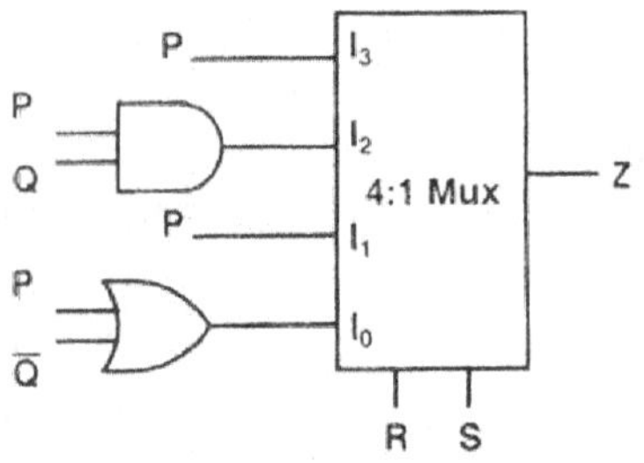

The output Z can be represented by

(a) $PQ + P\overline{Q}S + \overline{Q}\,\overline{R}\,\overline{S}$

(b) $P\overline{Q} + PQ\overline{R} + \overline{P}\,\overline{Q}\,\overline{S}$

(c) $P\overline{Q}\,\overline{R} + \overline{P}QR + PQRS + \overline{Q}\,\overline{R}\,\overline{S}$

(d) $PQ\overline{R} + PQR\overline{S} + P\overline{Q}\,\overline{R}S + \overline{Q}\,\overline{R}\,\overline{S}$

59. For each of the positive edge-triggered J-K flip flop used in the following figure, the propagation delay is ΔT .

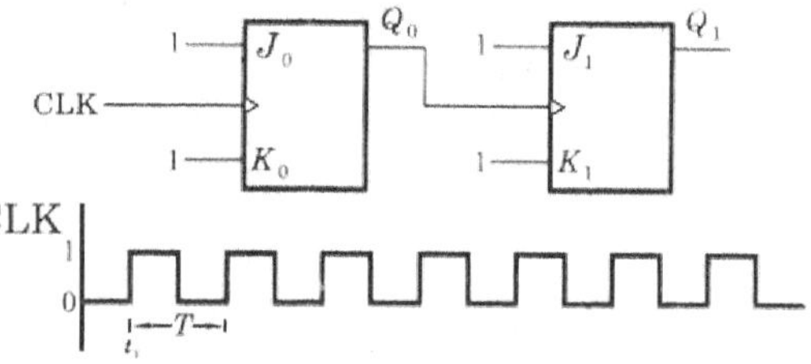

Which of the following waveforms correctly represents the output at Q_1?

(a)

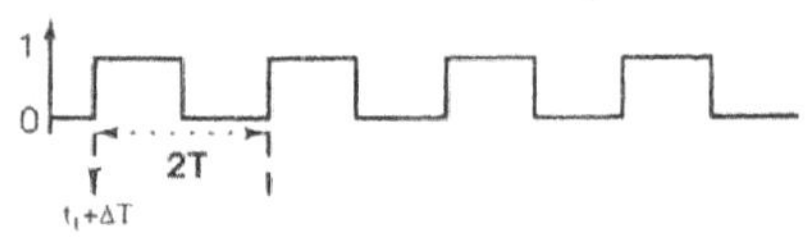

(b)

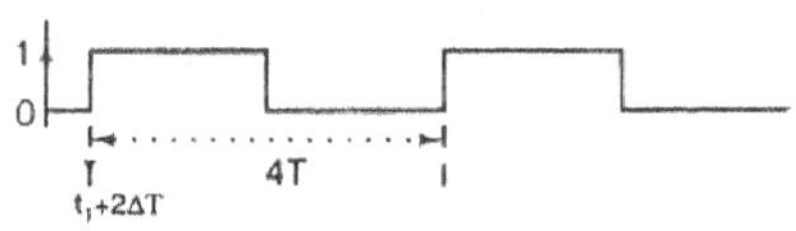

(c)

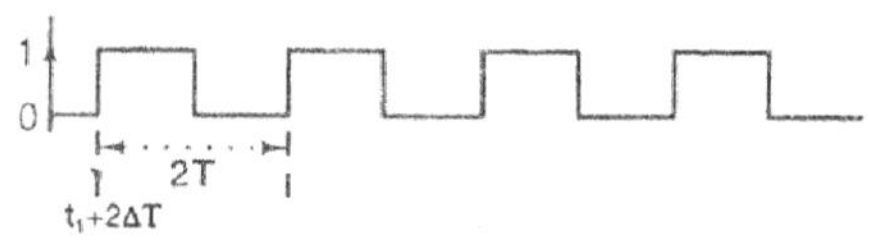

(d)

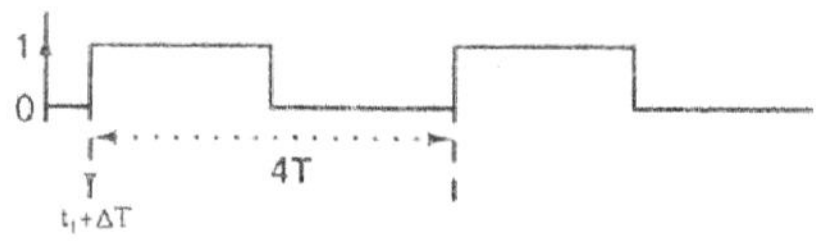

60. For the circuit shown in the figure, D has a transition from 0 to 1 after CLK changes from 1 to 0. Assume gate delays to be negligible .

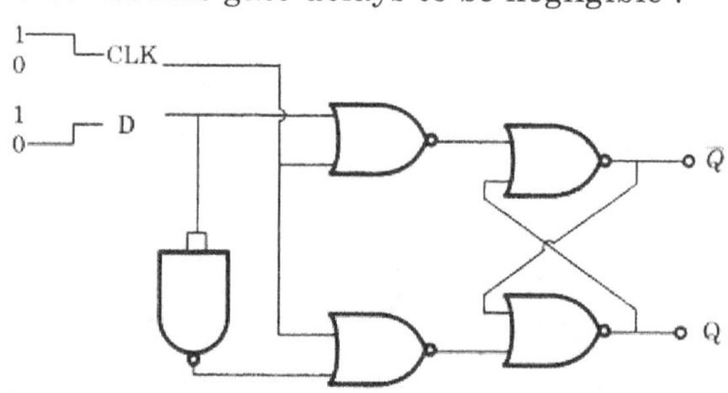

Which of the following statements is true?

(a) Q goes to 1 at the CLK transition and stays at 1.

(b) Q goes to 0 at the CLK transition and stays at 0.

(c) Q goes to 1 at the CLK transition and goes to 0 when D goes to 1.

(d) Q goes to 0 at the CLK transition and goes to 1 when D goes to 1.

61. A rectangular waveguide of internal dimensions (a = 4 cm and b = 3 cm) is to be operated in TE_{11} mode. The minimum operating frequency is

(*a*) 6.25 GHz (*b*) 6.0 GHz
(*c*) 5.0 GHz (*d*) 3.75 GHz

62. One end of a loss-less transmission line having the characteristic impedance of 75Ω and length of 1 cm is short-circuited. At 3 GHz, the input impedance at the other end of the transmission line is

(*a*) 0 (*b*) Resistive
(*c*) Capacitive (*d*) Inductive

63. A uniform plane wave in the free space is normally incident on an infinitely thick dielectric slab (dielectric constant ε = 9). The magnitude of the reflection coefficient is

(*a*) 0 (*b*) 0.3
(*c*) 0.5 (*d*) 0.8

64. In the design of a single mode step index optical fiber close to upper cut-off, the single-mode operation is NOT preserved if

(*a*) radius as well as operating wavelength are halved
(*b*) radius as well as operating wavelength are doubled
(*c*) radius is halved and operating wavelength is doubled
(*d*) radius is doubled and operating wavelength is halved

65. At 20 GHz, the gain of a parabolic dish antenna of 1 meter diameter and 70% effi-ciency is

(*a*) 15 dB (*b*) 25 dB
(*c*) 35 dB (*d*) 45 dB

66. Noise with double -sided power spectral density of K over all frequencies is passed through a RC low pass filter with 3dB cut-off frequency of f_c. The noise power at the filter output is

(*a*) K (*b*) K f_c
(*c*) K πf_c (*d*) ∞

67. Consider a Binary Symmetric Channel (BSC) with probability of error being p. To transmit a bit, say 1, we transmit a sequence of three 1s. The reveiver eill interpret the received sequence to represent1 if a least two bits are1. The probability that the transmitted bit will be received in error is

(*a*) $p^3 + 3p^2(1–p)$ (*b*) p^3
(*c*) $(1–p)^3$ (*d*) $p^3 + p^2(1–p)$

68. Four message band limited to W, W, 2W and 3W respectively are to be multiplexed using Time Division Multiplexing (TDM). The minimum band width required for transmission of this TDM signal is

(*a*) W (*b*) 3W
(*c*) 6W (*d*) 7W

69. Consider the frequency modulated signal $10 \cos[2\pi \times 10^5 t + 5\sin(2\pi \times 1500t) + 7.5\sin(2\pi \times 1000t)]$ with carrier frequency of 10^5 Hz. The modulation index is

(*a*) 12.5 (*b*) 10
(*c*) 7.5 (*d*) 5

70. The signal $\cos\omega_c t + 0.5 \cos\omega_m t \sin\omega_c t$ is

(*a*) FM only (*b*) AM only
(*c*) both AM and FM (*d*) neither AM nor FM

COMMON DATA QUESTIONS

Common Data for Questions 71, 72 and 73:

A speech signal, band limited to 4 kHz and peak voltage varying between +5V and –5V, is sampled at the Nyquist rate. Each sample is quantized and represented by 8 bits.

71. If the bits 0 and 1 are transmitted using bipolar pulses, the minimum band width required for distortion free transmission is

(*a*) 64 kHz (*b*) 32 kHz
(*c*) 8 kHz (*d*) 4 kHz

72. Assuming the signal to be uniformly distributed between its peak to peak value, the signal to noise ratio at the quantizer output is

(*a*) 16 dB (*b*) 32 dB
(*c*) 48 dB (*d*) 64 dB

73. The number of quantization levels required to reduce the quantization noise by a factor of 4 would be

(*a*) 1024 (*b*) 512
(*c*) 256 (*d*) 64

Common Data for Questions 74 and 75 :

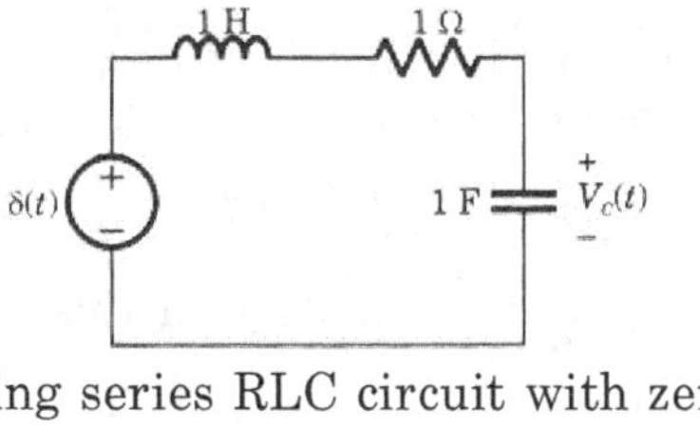

The following series RLC circuit with zero initial conditions is excited by a unit impulse functions δ(t)

74. For t >0, the output voltage $V_c(t)$ is

(a) $\frac{2}{\sqrt{3}}\left(e^{-\frac{1}{2}t} - e^{\frac{\sqrt{3}}{2}t}\right)$ (b) $\frac{2}{\sqrt{3}} te^{-\frac{1}{2}t}$

(c) $\frac{2}{\sqrt{3}} e^{-\frac{1}{2}t} \cos\left(\frac{\sqrt{3}}{2}t\right)$ (d) $\frac{2}{\sqrt{3}} e^{-\frac{1}{2}t} \sin\left(\frac{\sqrt{3}}{2}t\right)$

75. For t >0, the voltage across the resistor is

(a) $\frac{1}{\sqrt{3}}\left(e^{-\frac{\sqrt{3}}{2}} - e^{-\frac{1}{2}t}\right)$

(b) $e^{-\frac{1}{2}t}\left[\cos\left(\frac{\sqrt{3}t}{2}\right) - \frac{1}{\sqrt{3}}\sin\left(\frac{\sqrt{3}t}{2}\right)\right]$

(c) $\frac{2}{\sqrt{3}} e^{-\frac{1}{2}t} \sin\left(\frac{\sqrt{3}t}{2}\right)$

(d) $\frac{2}{\sqrt{3}} e^{-\frac{1}{2}t} \cos\left(\frac{\sqrt{3}t}{2}\right)$

LINKED ANSWER QUESTIONS

Q. 76 – 85 carry two marks each.

Statement for Linked Answer Questions 76 -77:

A two-port network shown below is excited by external dc scource. The voltages and the currents are measured with voltmeters V_1, V_2 and ammeters A_1, A_2(all assumed to be ideal), as indicated.

Under following switch conditions, the readings obtained are:

(i) S_1 – Open, S_2– Closed

$A_1 = 0$ A, $V_1 = 4.5$ V, $V_2 = 1.5$ V, $A_2 = 1$A

(ii) S_1– Closed, S_2 – Open

$A_1 = 4$ A, $V_1 = 6$ V, $V_2 = 6$V, $A_2 = 0$A

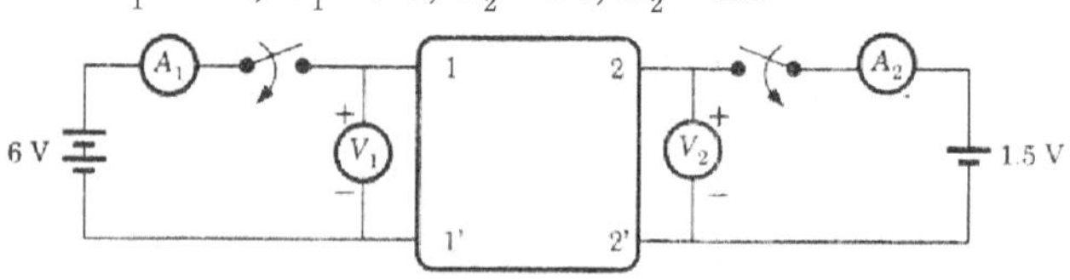

76. The z-parameter matrix for this network is

(a) $\begin{bmatrix}1.5 & 1.5\\4.5 & 1.5\end{bmatrix}$ (b) $\begin{bmatrix}1.5 & 4.5\\1.5 & 4.5\end{bmatrix}$

(c) $\begin{bmatrix}1.5 & 4.5\\1.5 & 1.5\end{bmatrix}$ (d) $\begin{bmatrix}4.5 & 1.5\\1.5 & 4.5\end{bmatrix}$

77. The h-parameter matrix for this network is

(a) $\begin{bmatrix}-3 & 3\\-1 & 0.67\end{bmatrix}$ (b) $\begin{bmatrix}-3 & -1\\3 & 0.67\end{bmatrix}$

(c) $\begin{bmatrix}3 & 3\\1 & 0.67\end{bmatrix}$ (d) $\begin{bmatrix}3 & 1\\-3 & -0.67\end{bmatrix}$

Statement for Linked Answer (Q. 78- 79):

In the following network, the switch is closed at t=0– and the sampling starts from t = 0. The sampling frequency is 10 Hz.

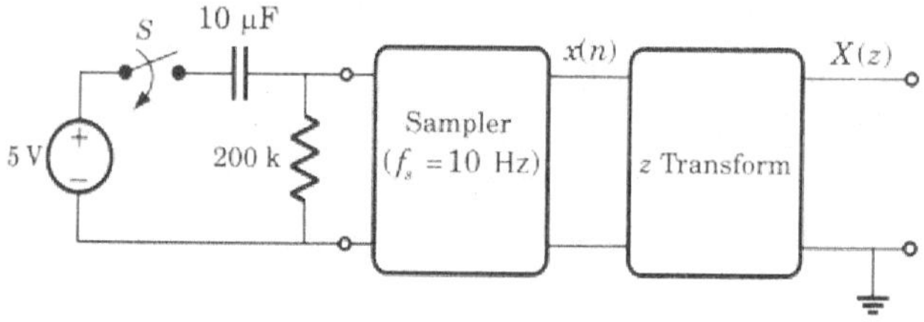

78. The samples x(n) n =(0,1, 2,.....) are given by

(a) $5(1- e^{-0.05n})$

(b) $5e^{-0.05n}$

(c) $5(1- e^{-5n})$

(d) $5e^{-5n}$

79. The expression and the region of convergence of the z-transform of the sampled signal are

(a) $\frac{5z}{z - e^{5}}$, $|z| < e^{-5}$

(b) $\frac{5z}{z - e^{-005}}$, $|z| < e^{-0.05}$

(c) $\frac{5z}{z - e^{-005}}$, $|z| > e^{-0.05}$

(d) $\frac{5z}{z - e^{-5}}$, $|z| > e^{-5}$

Statement for Linked Answer (Q. 80- 81):

In the following transistor circuit, $V_{BE} \approx 0.7$V, $r_e = 25$ m V/I_E, and β and all the capacitances are very large.

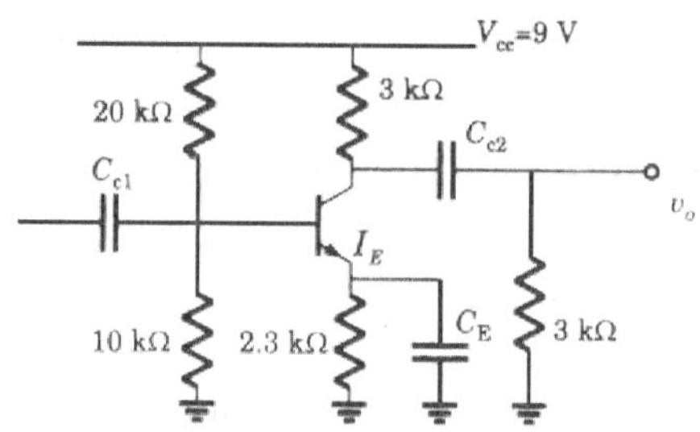

80. The value of DC current I_E is

(a) 1 mA

(b) 2 mA

(c) 5 mA

(d) 10 mA

81. The mid-band voltage gain of the amplifier is approximately

(a) –180

(b) –120

(c) –90

(d) –60

Statement for Linked Answer (Q 82-83):

In the following circuit, the comparator output is logic "1" if $V_1 > V_2$ and is logic "0" otherwise. The D/A conversion is done as per the relation $V_{DAC} = \sum_{r=0}^{3} 2^{n-1} b_n$ Volts, where b_3(MSB), b_2, b_1 and b_0 (LSB) are the counter outputs.

The counter starts from the clear state.

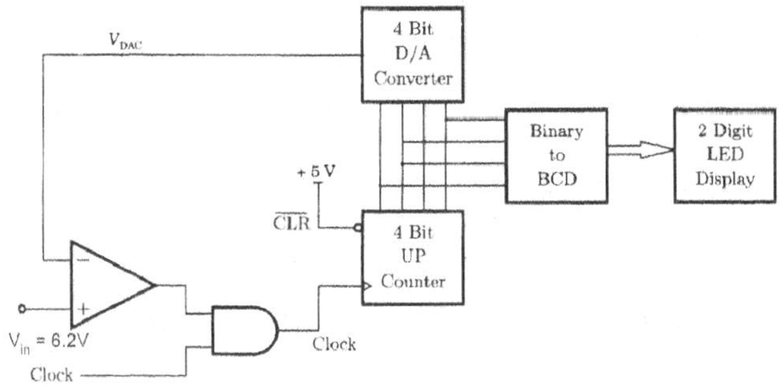

82. The stable reading of the LED display is
   (*a*) 06 (*b*) 07
   (*c*) 12 (*d*) 13

83. The magnitude of the error between V_{DAC} and V_{in} at steady state in volts is
   (*a*) 0.2 (*b*) 0.3
   (*c*) 0.5 (*d*) 1.0

Statement for Linked Answer(Q. 84– 85):

The impulse response h(t) of a linear time - invariant continuous time system is given by h(t) = exp(–2t) u(t), where u(t) denotes the unit step function.

84. The frequency response $H(\omega)$ of this system in terms of angular frequency ω, is given by $H(\omega)$
   (*a*) $\dfrac{1}{1+j2\omega}$
   (*b*) $\dfrac{\sin(\omega)}{\omega}$
   (*c*) $\dfrac{1}{2+j\omega}$
   (*d*) $\dfrac{j\omega}{2+j\omega}$

85. The output of this system, to the sinusoidal input x(t) = 2cos(2t) for all time t, is
   (*a*) 0
   (*b*) $2^{-0.25}\cos(2t - 0.125\pi)$
   (*c*) $2^{-0.5}\cos(2t - 0.125\pi)$
   (*d*) $2^{-0.5}\cos(2t - 0.25\pi)$

ANSWERS

1. (*c*)	2. (*b*)	3. (*a*)	4. (*a*)	5. (*a*)	6. (*b*)	7. (*c*)	8. (*b*)	9. (*c*)	10. (*d*)
11. (*d*)	12. (*c*)	13. (*d*)	14. (*a*)	15. (*c*)	16. (*b*)	17. (*c*)	18. (*c*)	19. (*b*)	20. (*a*)
21. (*a*)	22. (*d*)	23. (*c*)	24. (*a*)	25. (*c*)	26. (*a*)	27. (*d*)	28. (*b*)	29. (*a*)	30. (*a*)
31. (*c*)	32. (*a*)	33. (*d*)	34. (*b*)	35. (*a*)	36. (*a*)	37. (*a*)	38. (*d*)	39. (*c*)	40. (*d*)
41. (*c*)	42. (*a*)	43. (*b*)	44. (*b*)	45. (*c*)	46. (*b*)	47. (*b*)	48. (*c*)	49. (*c*)	50. (*b*)
51. (*c*)	52. (*c*)	53. (*c*)	54. (*d*)	55. (*b*)	56. (*b*)	57. (*d*)	58. (*a*)	59. (*b*)	60. (*c*)
61. (*a*)	62. (*d*)	63. (*c*)	64. (*d*)	65. (*d*)	66. (*c*)	67. (*a*)	68. (*d*)	69. (*d*)	70. (*a*)
71. (*b*)	72. (*c*)	73. (*b*)	74. (*d*)	75. (*b*)	76. (*c*)	77. (*a*)	78. (*b*)	79. (*c*)	80. (*a*)
81. (*d*)	82. (*d*)	83. (*b*)	84. (*c*)	85. (*d*)					

EXPLANATIONS

1. $P - \lambda I = 0$

For $\lambda = 0, \quad P = 0$

$$\Rightarrow \begin{bmatrix} P_{11} & P_{12} \\ P_{21} & P_{22} \end{bmatrix} = 0$$

$$\Rightarrow p_{11}\, p_{22} - p_{12}\, p_{21} = 0$$

2. Writing equations in matrix form,

$$\begin{bmatrix} 4 & 2 \\ 2 & 1 \end{bmatrix}\begin{bmatrix} x \\ y \end{bmatrix} = \begin{bmatrix} 7 \\ 6 \end{bmatrix}$$

Where $[A] = \begin{bmatrix} 4 & 2 \\ 2 & 1 \end{bmatrix}, B = \begin{bmatrix} 7 \\ 6 \end{bmatrix}$

The augmented matrix is [A | B] given as

$$\left[\begin{array}{cc|c} 4 & 2 & 7 \\ 2 & 1 & 6 \end{array}\right]$$

Using $R_2 - \dfrac{R_1}{2}$

We have $\left[\begin{array}{cc|c} 4 & 2 & 7 \\ 0 & 0 & 5/2 \end{array}\right]$

Now, rank of [A | B] = 2

and rank of [A] = 1

Since, Rank [A| B] ≠ Rank [A], no solution exists

3. Since $-1 \le \sin\theta \le 1$

therefore, it has no real or complex solution exists.

4. $$f(x) = e^x + e^{-x}$$

$$= e^x + \frac{1}{e^x} = \left(e^{\frac{x}{2}} - e^{\frac{1}{\frac{x}{2}}}\right)^2 + 2$$

The minimum value of $\left(e^{\frac{x}{2}} - e^{\frac{1}{\frac{x}{2}}}\right)$ is zero at $x = 0$ then min $|f(x)| = 2$

5. Taylor's series about x = 0 is

$$f(x) = f(0) + xf'(0) + \frac{x^2}{2!} f''(0) + \ldots\ldots$$

For $f(x) = \sin x^3$

$\Rightarrow t(0) = 0$

$f'(x) = 3x^2 \cos x^3$

$\Rightarrow f'(0) = 0$

$f''(x) = 6x \cos x^3 - (3x^2) \sin x^3$

$\Rightarrow f''(0) = 0$

$f'''(x) = 6 \cos x^3 - 36x^3 \sin x^3 - (3x^2)^2$

$\Rightarrow f'''(0) = 6$

Similarly for $f^{1x}(x)$ = constant we get the Taylor's series

$$f(x) = \frac{x^3}{3!}.6 + k.\frac{x^9}{9!} + \ldots\ldots$$

6. $\therefore \quad \dfrac{dx(t)}{dt} + 3x(t) = 0$

$$\Rightarrow \frac{dx(t)}{x(t)} = -3dt$$

$$\Rightarrow \ln[x(t)] = -3t + c$$

$$\Rightarrow x(t) = ke^{-3t}$$

7. Different trees are shown here, in figure (*a*)

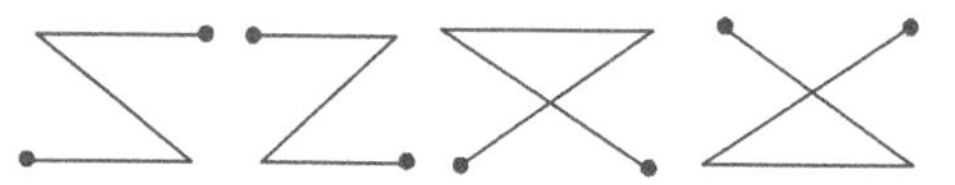

Fig. (*a*)

Different cut set are as shown in figure (*b*)

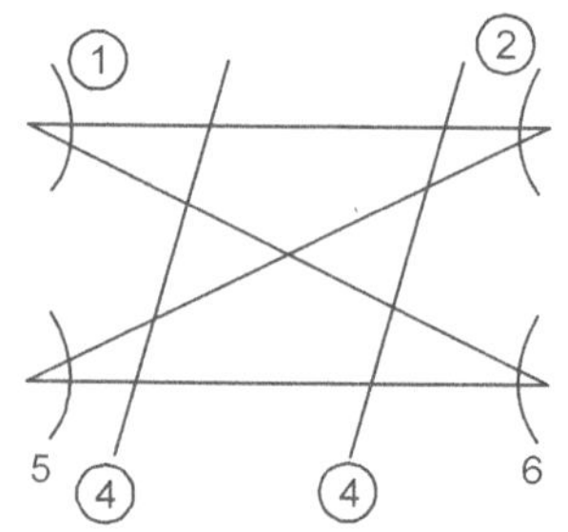

Fig. (*b*)

8. Initially $i(0^+) = 0$ due to inductor

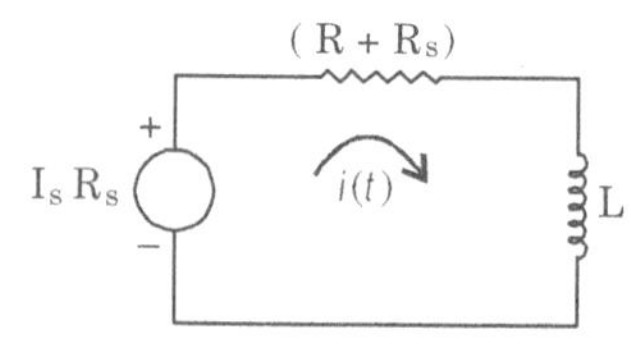

$$I_s R_s = (R + R_s)\, i(t) + \frac{L di(t)}{dt}$$

For $t = 0^+, \; I_s R_s = (R + R_s)\, i(0^+) + \dfrac{L di(0^+)}{dt}$

$i(0^+) = 0$

$$\Rightarrow \frac{di(0^+)}{dt} = \frac{I_s R_s}{L}$$

9. A system is said to be causal if output depends only on present and past states only.

10. $$u(t) = e^{\alpha t} u(t) + e^{\beta t} u(-t)$$

For the system to be stable,

$$\int_{-\infty}^{\infty} u(t) \cdot dt < \infty$$

or, $$\int_{-\infty}^{\infty} \left(e^{\alpha t} \cdot u(t) + e^{\beta t} u(-t)\right) \cdot dt < \infty$$

or, $$\int_{0}^{\infty} e^{\alpha t} \cdot dt + \int_{-\infty}^{0} e^{\beta t} \cdot dt < \infty \qquad \ldots(A)$$

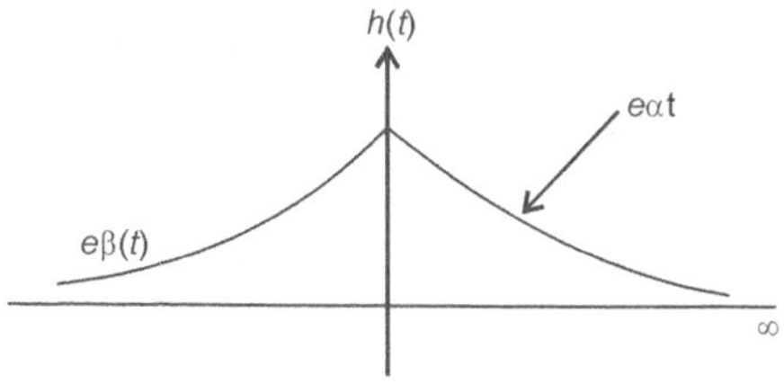

For, the conditon (A) to be satisfied,

$\alpha < 0$ and $\beta > 0$

11. Transfer functin, $T(S) = \dfrac{s^2 + as + b}{(s+p)(s+q)}$

At $S = 0, T(s) \neq 0$

and $S = \infty, T(s) \neq 0$

Frequency response of T(S) is shown here

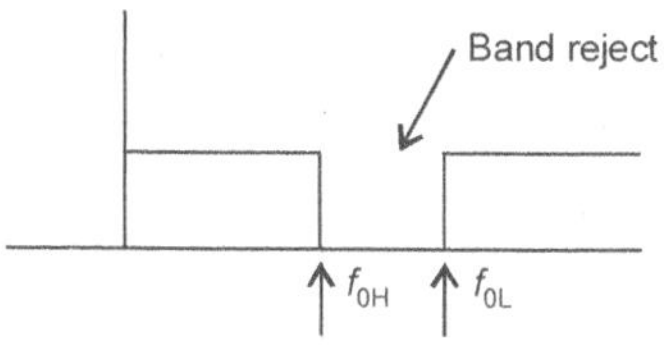

Hence, this is band reject a Notch filter.

12. Percentage overshoot is $M_i = \dfrac{-\pi r}{e\sqrt{1-r^2}}$

Where r is damping constant and is given by $r = \cos\theta$, θ is angle made w.r.t. real axis

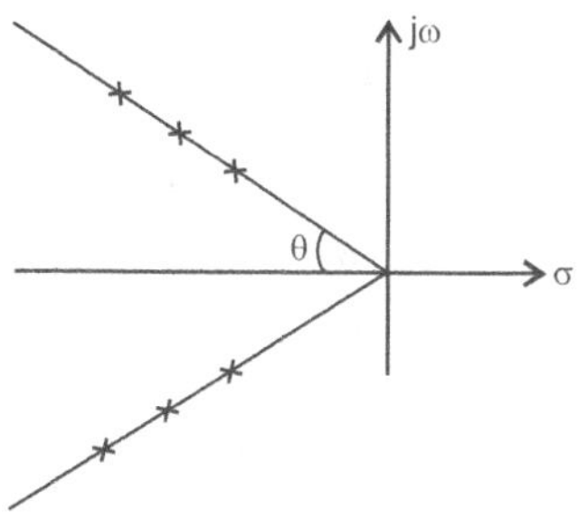

Since for all the poles in figure (*c*), cos θ is same, therefore *r* is same for all and percentage overshoot also.

13. Channel length modulation is related to FET.

14. Boron is an acceptor impurity high concentration of which makes p^+ substrate.

15.

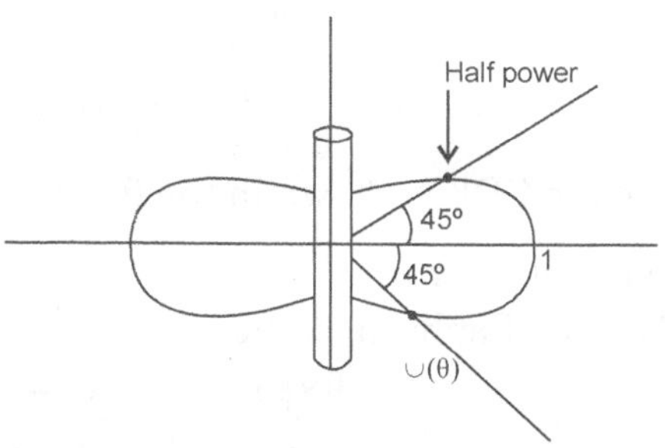

As shown above in the radiation intensity distribution for the Hertzian dipole the half-power points are at θ = 45° and θ = 135° and half-power beam width is therefore 90°.

16. Maxwell equations :

$$\nabla \times \vec{E} = -\vec{B}$$
$$\nabla \cdot \vec{E} = \rho/E$$
$$\nabla \cdot \vec{B} = 0$$

For static electric and magnetic fields

$$\nabla \times \vec{H} = \vec{J}$$
$$\nabla \times \vec{E} = 0$$
$$\nabla \cdot \vec{E} = \rho/E$$
$$\nabla \cdot \dot{B} = 0$$

For source-free medium,

$$\rho = 0$$

Then, $\nabla \cdot \overline{E} = 0$

17. In positive half cycle

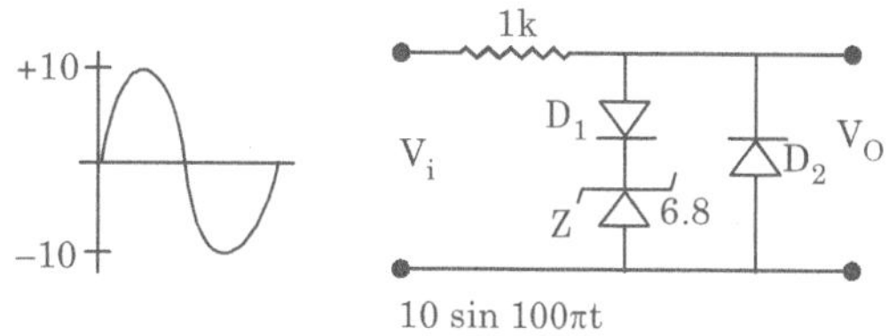

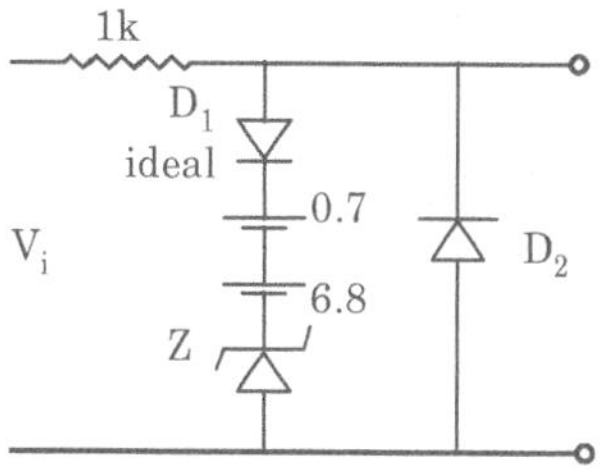

D_1 is forward bias

D_2 is reverse bias

$\therefore \quad (V_0)_{\text{max imum}} = 0.7 + 6.8$

$= 7.5\,V$

In negative half cycle

D_2 is Forward bias

D_1 is Reverse bias

$\therefore \; (V_0)\text{ min.} = V_{D2} = -0.7$ Volt

Alternately

During positive cycle, D_1 is ON and D_2 is OFF,

$$V_0 = V_Z + V_{D_1}$$
$$= 6.8 + 0.7$$
$$= 7.5V$$

During negative cycle,

D_1 is OFF and D_2 is ON

$$V_0 = -V_{D_2} = -0.7V$$

19. $$g_m = \frac{\partial I_D}{\partial I_{GS}}$$
$$= 2K(V_{GS} - V_T)$$

20.
$$x(t) = A_c \cos \omega_c t + 2 \cos (\omega_m t) \cos \omega_c t$$
$$= A_c \cos\omega_c t\left[1+\frac{2}{A_c}\cos\omega_m t\right]$$

Comparing with standard equation for AM wave,
$$x(t) = A_c \cos \omega_c t\,[1 + k_a m(t)]$$
(k_a is the amplitude sensitivity)

We get that to avoid over modulation, $|k_a m(t)|_{max} \le 1$
$$\Rightarrow \quad \left|\frac{2}{AC}\right| \le 1$$
or, minimum value of $A_C = 2$.

21. To find out Z_{th} make all the voltage sources short circuit and make all current sources open.

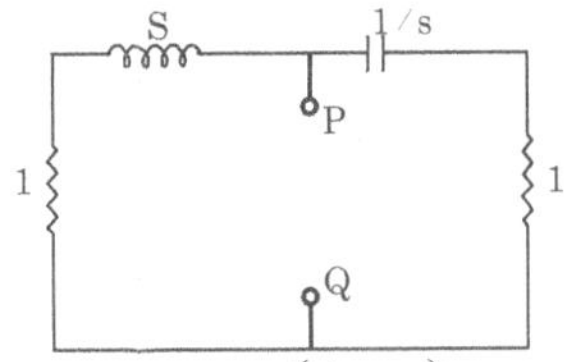

$$Z_{th} = (s + 1) \parallel \left(\frac{1}{s}+1\right) = 1$$
$$\text{or,} \quad Z_{TH} = \frac{(s^2+2s+1)}{s^2+2s+1} = 1$$

22. $Z(s) = R \parallel \frac{1}{sC} \parallel SL$

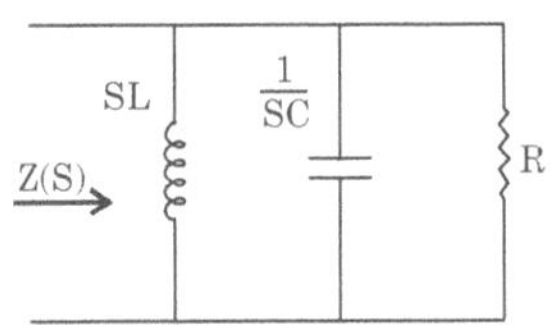

$$= \frac{\frac{S}{C}}{s^2 + \frac{s}{RC} + \frac{1}{LC}}$$

Comparing with $Z(s) = \dfrac{0.2\,s}{s^2 + 0.1\,s + 2}$
$$\frac{1}{C} = 0.2,$$
$$\Rightarrow \quad C = 5\text{ F}$$
$$\frac{1}{RC} = 0.1,$$
$$\Rightarrow \quad R = 2\ \Omega$$
$$\frac{1}{LC} = 2,$$
$$\Rightarrow \quad L = 0.1\text{ H}$$

23. For $n = 0$, $0 \le t \le T$, S_1P_1 and S_2P_2 are connected as shown
$$V_c(t) = \frac{1}{C}\int_0^t c\,dt = t, 0 \le t \le T$$

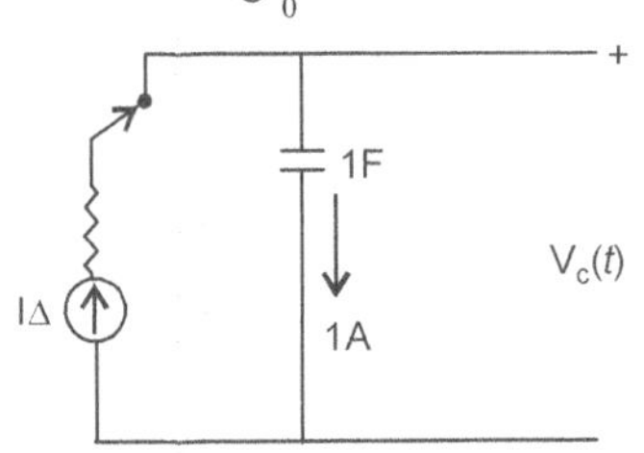

For $T \le t \le 2T$, $\quad S_1Q_1$ and S_2Q_2 are connected
$$V_c(t) - V_c(T) = \frac{1}{C}\int_T^t i \cdot dt = -\int_T^t dt$$
[$i = -1$A as flowing in opposite direction]

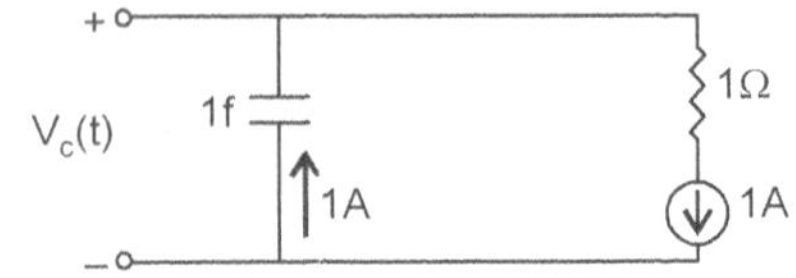

or, $\quad V_c(t) - T = -(t - T)$

or, $\quad V_c(t) = 2T - t$, for $T \le t \le 2T$.

Hence, $V_c(t)$ can be expressed as,
$$V_c(t) = t\,u(t) + 2[-(t - T)\,u(t - T)]$$
In generalised terms,
$$V_i(t) = t\,u(t) + 2\sum_{n=1}^{\infty}(-1)^n (t - nT)\,u(t - nT)$$

24. CDF : $F(x) = \int_{-\infty}^{x} (\text{PDF})\,dx$

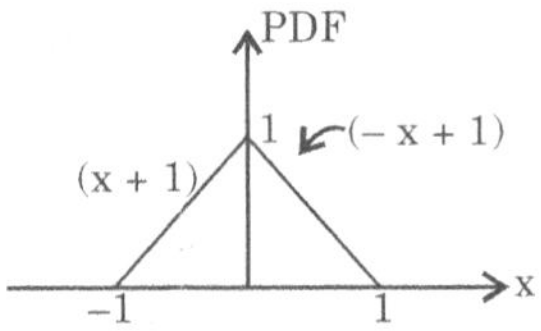

Fig. (a)

For $x < 0$, $F(x) = \int_{-1}^{x} (x+1)\,dx$
$$= \frac{x^2}{2} + x + \frac{1}{2} \text{ concave upwards}$$
$$\because \quad F(0) = \frac{1}{2}$$

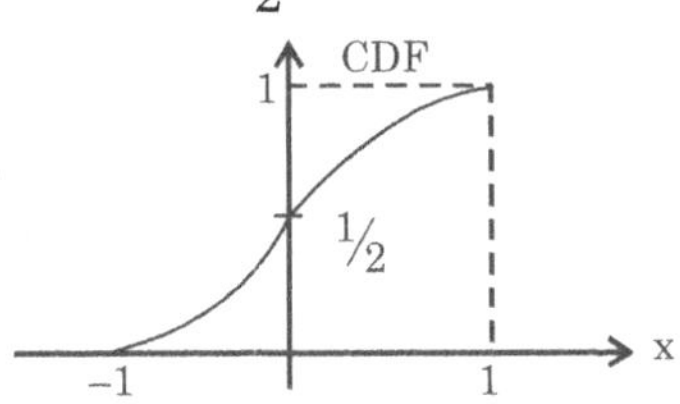

Fig. (b)

For $x > 0$, $F(x) = F(0) + \int_0^x (-x+1)\,dx$
$$= \frac{1}{2} + \left(\frac{-x^2}{2} + x\right) \text{ concave downwards}$$
Hence the CDF is shown in the figure (a).

25. Given: $\quad f(x) = x - e^{-x}$

By Newton Raphson method,
$$x_{n+1} = x_n - \frac{f(x_n)}{f'(x_n)} = x_n - \frac{x_n - e^{-x_n}}{1 + e^{-x_n}}$$
$$= (1 + x_n)\,\frac{e^{-x_n}}{1 + e^{-x_n}}$$

26. $f(z) = \dfrac{1}{(z+2)^2(z-2)^2}$

Residue at $z = a$ of order n is, defined as,

$$= \lim_{z\to 0} \frac{1}{|\underline{n-1}} \left[\frac{d^{n-1}}{dz^{n-1}} (z-a)^n f(z) \right]_{z=a}$$

$$= \text{Res}[f(z)]_{z=2} = \left[\frac{d}{dz} (z-2)^2 f(z) \right]_{z=2}$$

$$= \left[\frac{d}{dz} \frac{1}{(z+2)^2} \right]_{z=2} = \left. \frac{-2}{(z+2)^3} \right|_{z=2} = -\frac{1}{32}$$

27. $e^{P} = L^{1}\left[(sI - P)^{-1}\right]$

and $P = \begin{bmatrix} 0 & 1 \\ -2 & -3 \end{bmatrix}$

where $(sI - P)^{-1} = \begin{bmatrix} s & -1 \\ 2 & s+3 \end{bmatrix}^{-1}$

$$= \frac{1}{(s+1)(s+2)} \begin{bmatrix} s+3 & 1 \\ -2 & s \end{bmatrix}$$

$$= \begin{bmatrix} \dfrac{s+3}{(s+1)(s+2)} & \dfrac{1}{(s+1)(s+2)} \\ \dfrac{-2}{(s+1)(s+2)} & \dfrac{s}{(s+1)(s+2)} \end{bmatrix}$$

$$\therefore \quad e^{P} = L^{-1}\left\{ \begin{bmatrix} \dfrac{2}{s+1} - \dfrac{1}{s+2} & \dfrac{1}{s+1} - \dfrac{1}{s+2} \\ \dfrac{-2}{s+1} + \dfrac{2}{s+2} & \dfrac{2}{s+2} - \dfrac{1}{s+1} \end{bmatrix} \right\}$$

$$= \begin{bmatrix} 2e^{-1} - e^{-2} & e^{-1} - e^{-2} \\ -2e^{-1} + 2e^{-2} & 2e^{-2} - e^{-1} \end{bmatrix}$$

28. Let $f(x) = e^x + \sin x$

Taylor's series is

$$f(x) = f(a) + (x-a) f'(a) + \frac{(x-a)^2}{2!} f''(a)$$

where, $a = \pi$

$$\therefore \quad f(x) = f(\pi) + (x-\pi) f'(\pi) + \frac{(x-\pi)^2}{2!} f''(\pi)$$

Coefficient of $(x - \pi)^2$ is $\dfrac{f''(\pi)}{2}$

$$f''(\pi) = e^x - \sin x \Big|_{\text{at } x = \pi} = e^{\pi}$$

$\therefore$ Coefficient of $(x - \pi)^2 = 0.5 \exp(\pi)$

29. Given: $P_x(x) = Me^{-2|x|} + Ne^{-3|x|}$

For PDF< $P_x(x)$,

$$\int_{-\infty}^{\infty} P_x(x)\, dx = 1$$

$$\int_{-\infty}^{\infty} (Me^{-2|x|} + Ne^{-3|x|})\, dx$$

$$= 2\int_{0}^{\infty} (Me^{-2|x|} + Ne^{-3|x|})\, dx = 1$$

$$\Rightarrow \quad M + \frac{2}{3} N = 1$$

30. Given: $g(x, y) = 4x^3 + 10y^4$

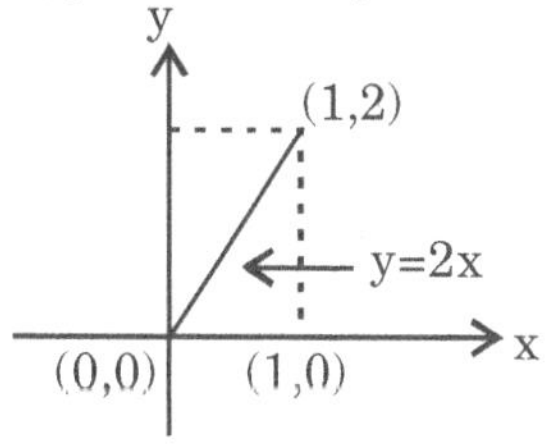

The straight line can be expressed as

$y = 2x$

Then $g(x, y) = 4x^3 + 10(2x)^4$

$$\therefore \quad I = \int_0^1 \left[4x^3 + 10(2x)^4\right] dx = \int_0^1 (4x^3 + 160x^4) dx$$

$$= \left. \frac{4x^4}{4} + \frac{160}{5} x^5 \right|_0^1 = 33.$$

31. $H(x) = \dfrac{k(s+1)}{(s+2)(s+4)}$

Output, $C(s) = H(s).R(s) = \dfrac{k(s+1)}{s(s+2)(s+4)}$

$\left[R(s) = \dfrac{1}{s}; \text{ unit step function} \right]$

At steady-state, final value theorem gives,

$$C(\infty) = \lim_{s\to 0} sc(s) = \lim_{s\to 0} \frac{k(s+1)}{(s+2)(s+4)} = \frac{k}{8}$$

given, $\dfrac{k}{8} = 1$

$\Rightarrow \quad k = 8$

Now, Impulse response,

$$h(t) = \mathcal{L}^{-1} H(s) = \lim_{s\to 0} \frac{8(s+1)}{(s+2)(s+4)}$$

$$= \frac{A}{s+2} + \frac{B}{s+4}$$

Where, $A = (s+2)H(s)\big|_{s=-2} = -4$

$B = (s+4)H(s)\big|_{s=-4} = 12$

then, $h(t) = \mathcal{L}^{-1}\left(\dfrac{-4}{s+2}\right) + \mathcal{L}^{-1}\left(\dfrac{12}{s+4}\right)$

$$= \left[-4e^{-2t} + 12e^{-4t}\right] u(t)$$

32. $$x(w) = \int_{-\infty}^{\infty} x(t)\cdot e^{-jwt}\cdot dt$$

$$= \int_{-1}^{1} e^{-jwt}\cdot dt = \frac{1}{jw}\left(e^{jw} - e^{-jw}\right)$$

$$x(w) = 0$$

$$\Rightarrow \quad e^{jw} - \frac{1}{e^{jw}} = 0$$

$$e^{2jw} - 1 = 0$$

$$\Rightarrow \quad e^{2jw} = 1$$

$$\Rightarrow \quad e^{jw} = \pm 1$$

For this equality to hold, $w = \pi, 2\pi$

33. Given $h(n) = [1, -1, 2]$

$x(n) = [1, 0, 1]$

$$y(n) = x(n) * h(n) = \sum_{k=-\infty}^{\infty} x(k)\; h(n-k)$$

$$y(2) = \sum_{k=-\infty}^{\infty} x(k)\; h(2-k)$$

$h(2-k) \Rightarrow$

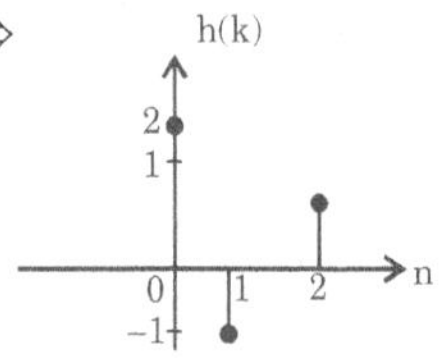

$x(k) \Rightarrow$

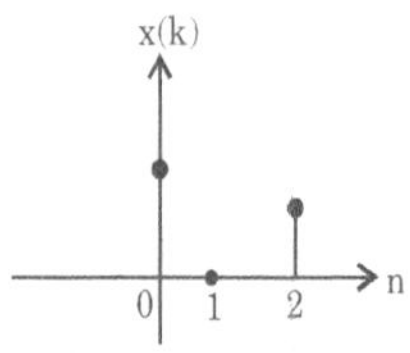

$$y(2) = 2 \times 1 + (-1) \times 0 + 1 \times 1 = 3$$

We find, $y(-1) = 0, y(0) = 1, y(1) = -1, y(2) = 3,$

$y(3) = -1, y(4) = 2, y(5) = 0$

Hence five non zero samples.

34.

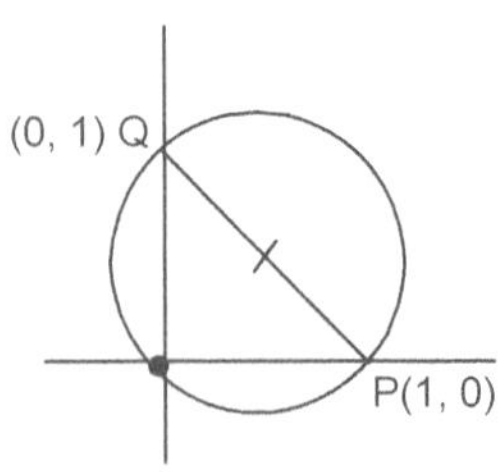

$$I = 2\int_P^Q xdx + ydy = 2\int_1^0 xdx + 2\int_0^1 ydy$$

$$= 2\left[\frac{x^2}{2}\Big|_1^0\right] + 2\left[\frac{y^2}{2}\Big|_0^1\right] = 0$$

35. R_1 : $y(t) = t^2x(t)$ Linear and time variant.

R_2 : $y(t) = t|x(t)|$ Non linear and time variant.

R_3 : $y(t) = |x(t)|$ Non linear and time invariant

R_4 : $y(t) = x(t-5)$ Linear and time invariant

36. For n symbols, $P = \frac{1}{n}$

$$\therefore \quad \text{Entropy} = \sum_{k=1}^{n} P_k \log_2 \frac{1}{P_k}$$

$$= \log_2\left(\frac{1}{P}\right) = \log_2(n)$$

37. Given: $$y(n) = \frac{1}{N}\sum_{r=0}^{N-1} x(r)\,x(n+r)$$

It is auto correlation

Hence $y(n) = r_{xx}(n)$

$$\therefore \quad r_{xx}(n) \xrightarrow[N]{DFT} |X(k)|^2$$

Alternately

$$X[n-r] = \sum_{n=0}^{N-1} x[n].W_N(N-r)n$$

$$\sum_{n=0}^{N-1} x(n).e^{-1}\left(\frac{2\pi}{N}\right)(N-r)n$$

Now, $e^{-1}\left(\frac{2\pi}{N}\right)(N-r) = e^{-12\pi n}e^{1}\left(\frac{2\pi}{N}\right)^{rn} = e^{1}\left(\frac{2\pi}{N}\right)^{rn}$

If x[n] is real, then x[n] = x[n]

and $$x[N-r] = \sum_{n=0}^{N-1} x[n].e^{j}\left(\frac{2\pi}{N}\right)^{rn}$$

$$= \left[\sum_{n=0}^{N-1} x[n].e^{-j}\left(\frac{2\pi}{N}\right)^{rn}\right] = X(k)$$

Hence, $Y[k] = [x(k)\{x^*[k]\} = |X(k)|^2$

38. $$P = \frac{25}{s^2 + 2s}, \zeta = 0,$$

$\omega_n = 0$ sinusoidal function option 3.

$$Q = \frac{36}{s^2 + 20s + 36}, \quad \omega_n = 6$$

$$\zeta = \frac{20}{2 \times 6} > 1$$

overdamped (hyperbolic function) option 4.

$$R = \frac{36}{s^2 + 12s + 36}, \quad \zeta = 1,$$

$\omega_n = 6$, Critically damped option 1.

$$S = \frac{49}{s^2 + 7s + 49}, \zeta = \frac{1}{2},$$

$\omega_n = 7$, underdamped option 2.

39. Given $G(s) = \dfrac{s+8}{s^2+\alpha s-4}$

Close loop transfer function $= \dfrac{G(s)}{1+G(s)}$

Characteristic equation is, $s^2 + \alpha s - 4 + s + 8 = 0$

$\Rightarrow \quad s^2 + (\alpha + 1)\, s + 4 = 0$

This will be stable if $(\alpha + 1) > 0$ and real.

40. We get

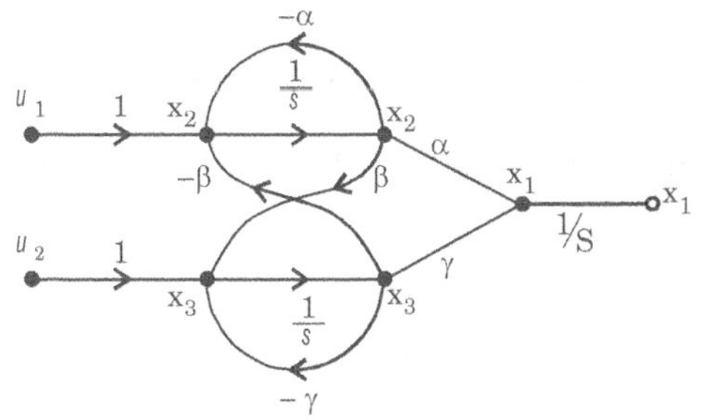

$$\dot{x}_1 = \alpha x_2 + \gamma x_3$$
$$\dot{x}_2 = \mu_1 - \alpha x_2 - \beta x_3$$
$$\dot{x}_3 = \mu_2 - \gamma x_3 + \beta x_2$$
$$\Rightarrow \begin{bmatrix}\dot{x}_1\\ \dot{x}_2\\ \dot{x}_3\end{bmatrix} = \begin{bmatrix}0 & \alpha & \gamma\\ 0 & -\alpha & -\beta\\ 0 & \beta & -\gamma\end{bmatrix}\begin{bmatrix}x_1\\ x_2\\ x_3\end{bmatrix} + \begin{bmatrix}0 & 0\\ 1 & 0\\ 0 & 1\end{bmatrix}\begin{pmatrix}u_1\\ u_2\end{pmatrix}$$

Depending upon position of x_1, x_2 and x_3 and u_1 and u_2, the rows and columns may interchange but the corresponding entries will remain the same, hence the solution is (d).

41. $1 + G(s) = 0$

or $s^5 + 2s^4 + 3s^3 + 6s^2 + 5s + 13 = 0$

Constructing Routh -array, we have

s^5	1	3	5
s^4	2	6	13
s^3	$0 \to \in$	$\dfrac{-3}{2}$	0
s^2	$\dfrac{6\in+3}{\in}$	13	0
s	$\dfrac{-3}{2}$	0	
s^0	13		

As $\in \to 0^+$, $\lim\limits_{\in\to 0^+}\left(\dfrac{6\in+3}{\in}\right) = +ve$

hence two sign changes, so two positive poles in RHS plane

43.

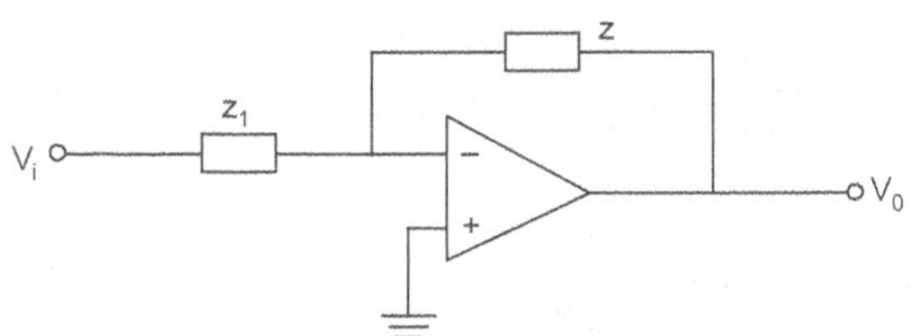

$$Z_1 = \frac{R_1}{I\dfrac{1}{C_1S}} = \frac{R_1}{R_1C_1S+1}$$

Using KCL at investing terminal,

$$\frac{V_i(S)}{Z_1} = \frac{0-V_0(S)}{Z}$$

$$\Rightarrow \quad \frac{V_0(S)}{V_i(S)} = \frac{-Z}{Z_1} = \frac{-Z(R_1C_1S+1)}{R_1}$$

Q : When $\quad Z = R_2 + \dfrac{1}{C_2S}$

$$\frac{V_0(S)}{V_i(S)} = -\left(\frac{R_1C_1S+1}{R_1}\right)\left(R_2+\frac{1}{C_2S}\right)$$
$$= -\left[R_2C_1\,S+\frac{R_2}{R_1}+\frac{C_1}{C_2}+\frac{1}{R_1C_2S}\right]$$
$$= -\left[\left(\frac{R_2}{R_1}+\frac{C_1}{C_2}\right)+R_2C_1S+\frac{1}{R_1C_2S}\right]$$

Thus represent PID controller

Q : When $\quad Z = \dfrac{R_2}{1\dfrac{1}{C_2S}} = \dfrac{R_2}{R_2C_2S+1}$

$$\frac{V_0(S)}{V_i(S)} = -\frac{R_2(R_1C_1S+1)}{R_1(R_2C_2S+1)} \quad \text{As } R_1C_1 < R_2C_2$$

$$\angle\frac{V_0(S)}{V_i(S)} = \tan^{-1} R_1C_1\omega - \tan^{-2}\omega R_2C_2 < 0$$

This represents lag compensator.

44. In satmation region, drain current for M1

$$I_{D1} = \frac{\mu_n C_{ox} W}{2L}(V_{GS_1} - V_T)^2 = I_{Bias}$$

and for M2,

$$I_{D2} = \frac{\mu_n C_{ox} W}{2L}(V_{GS_2} - V_T)^2$$

As $V_{GS1} = V_{GS2}$ for this circuit and the two transistors are identical,

$$I_{D1} = I_{D2} = I_{bias}$$

or, $\quad I_x = I_{bias}$

45. In linear region of NMOS transistor, diode current is given by,

$$I_D = \frac{\mu_n C_{ox} W}{L}\left(V_{GS} - V_T - \frac{1}{2}V_{DS}\right)V_{DS}$$

and $\quad V_{DS} \le V_{GS} - V_T$

Now, $\quad g_m = \left.\dfrac{\partial I_n}{\partial v_{gs}}\right|_{V_{DS=const.}}$

$$\frac{\partial I_D}{\partial v_{gs}} = \frac{\mu_n C_{ox} W}{L}\cdot V_{DS} = K \text{ [Constant].}$$

Hence, it can be infered that $\dfrac{\partial I_n}{\partial V_{gs}}$ follows a constant Linear variation.

46. $$I = \frac{1}{100k} = 10\ \mu A$$

$$I_0 = 1\ \mu A$$

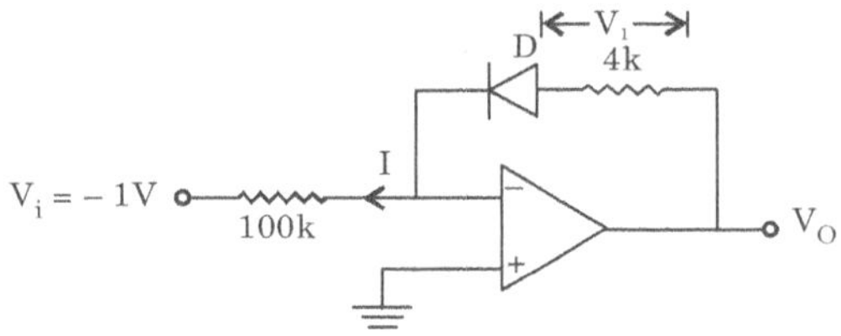

$\Rightarrow \quad 10\ \mu A = 1\ \mu A \left[e^{\frac{V_D}{25}} - 1\right]$

$\Rightarrow \quad V_D = 60$ mV

Voltage drop across 4K resistance

$$V_1 = 4K \times 10\ \mu A = 40\ mV$$

Total voltage output $= V_1 + V_D = 1$

$= (40 + 60)$ mV

$= 0.1$ volts

47. $$\frac{V_o}{V_i} = -\frac{R_F}{R_i}$$

$$R_F = R_2 \parallel \frac{1}{sC} = \frac{R_2}{R_2Cs + 1}$$

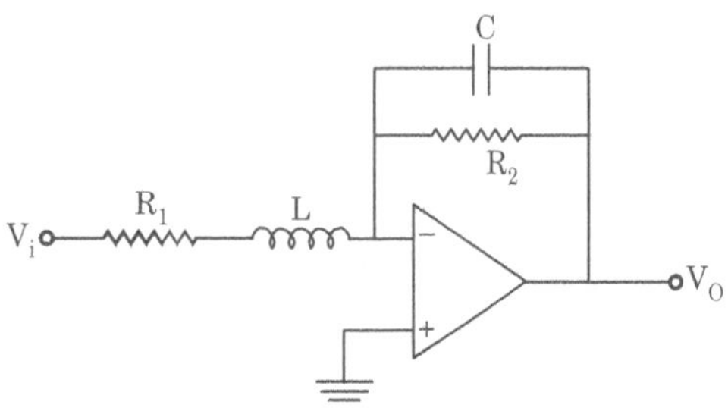

$$R_i = R_1 + sL$$

$$\frac{V_o}{V_i} = \frac{K}{(R_1 + sL)(R_2Cs + 1)}$$

$$= \frac{K}{\left(1 + j\frac{f}{f'_H}\right)\left(1 + j\left(\frac{f}{f''_H}\right)\right)}$$

As $\left.\frac{V_o}{V_i}\right|_{\omega \to 0} = 0$, it is a low pass filter.

48. $$\frac{1}{g_m} = \frac{1}{g_{m1}} + \frac{1}{g_{m2}}$$

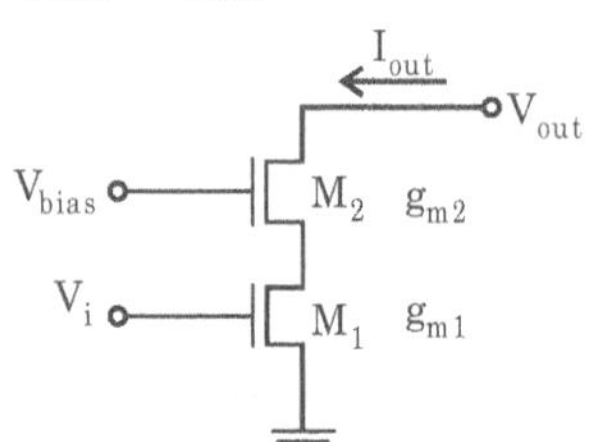

$\therefore$ M_2 is always saturated due to bias but g_{m1} changes according to V_i

$$g_m = \frac{g_{m1}\, g_{m2}}{g_{m1} + g_{m2}} = \frac{g_{m1}\, g_{m2}}{g_{m2}\left(1 + \frac{g_{m_1}}{g_{m_2}}\right)}$$

But $g_{m2} >> g_{m1}$

$\therefore \quad g_m = g_{m1}$

49. LXI H, 30AOH $\longrightarrow$ H 30 40 L

DAD H $\longrightarrow$ H 60 40 L

Adds the contents of H-L pair to itself

PCHL $\longrightarrow$ PC $\rightarrow$ 60 40 H

Loads the contents of H-L pair to PC.

50. Capacitor voltage and output waveform are shown below :

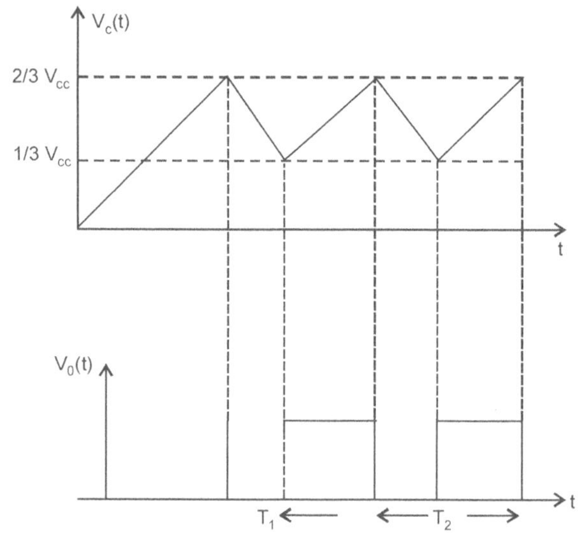

Hence $V_{cc} = 9$ volts

therefore, $V_c(t)$ varies between 3 volts to 6 volts.

51. $$N_A = 4 \times 10^{17}\ \text{atom/cm}^3$$

$$n_i = 1.5 \times 10^{10}/\text{cm}^3$$

For P-type semiconductor,

$$E_f - E_v = KT \ln \frac{N_A}{N_i}$$

$$= 0.025 \ln\left(\frac{4 \times 10^{17}}{1.5 \times 10^{10}}\right) ev$$

$$= 0.427\ ev$$

52. Transconductance,

$$g_{m1\,(Sat)} = G_o\left[1 - \left(\frac{-V_G}{V_P}\right)^{\frac{1}{2}}\right]$$

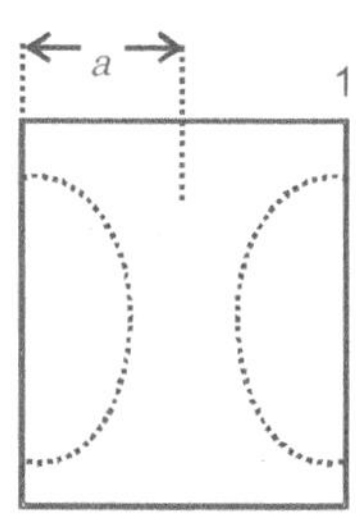

$$g_{m1\,(Sat)} = G_o\left[1 - \left(\frac{2}{V_P}\right)^{\frac{1}{2}}\right]$$

$$\because \quad V_P = \frac{qN_d\, W^2}{2\in}$$

By doubling W, we will increase V_P by four times.

$\Rightarrow \quad V_{P'} = 4V_P$

$$g_{m2} = G_o\left[1-\left(-\frac{V_G}{V_P'}\right)^{\frac{1}{2}}\right]$$

$$= G_o\left[1-\left(\frac{1}{2V_P}\right)^{\frac{1}{2}}\right]$$

$$\therefore \quad \frac{g_{m1}}{g_{m2}} = \left(\frac{1-\sqrt{\frac{2}{V_P}}}{1-\sqrt{\frac{1}{2V_P}}}\right)$$

Hint: $\dfrac{\text{Streetman}}{\text{P} - 294}$

$$G_o = \frac{2az}{PL} = \text{constant}$$

53. When output is + 15V, let voltage at non inverting terminal is V_0

$$\therefore \quad \left(\frac{V_0-15}{10}\right)2+\left(\frac{V_0+15}{10}\right)=0$$

$\Rightarrow \quad V_0 = +5$ volts

When output is – 15V

$$\frac{V_0-15}{10}+2\left(\frac{V_0+15}{10}\right)=0$$

$\Rightarrow \quad V_0 = -5$ volts

54. The output Out is 1 only when the mode (S) is 0, since then the UPPER MOSFET is shorted and connected to V_{dd}.

Node S is zero when P = Q =1 as it shorts then lower MOSFET's

Hence Output logic Y = P AND Q

55. S1 →True, in linearly graded function aralanche breakdown occurs.

S2 → For tunneling phenomenon to occur narrow energy barrier is required to charge tunnel from filled state in one side to empty state in other side.

Also, tunneling is independent on the doping profile of junction.

56.

P = 11101101

Q = 11100110

Subtract Q, hence find out Q's 2's complement = 00011010

P – Q = P + 2's complement of Q

$$\begin{array}{r} 0001011 \\ 1110010 \\ \hline 11111001 \end{array}$$

2's complement of P – Q = 00000111

57.

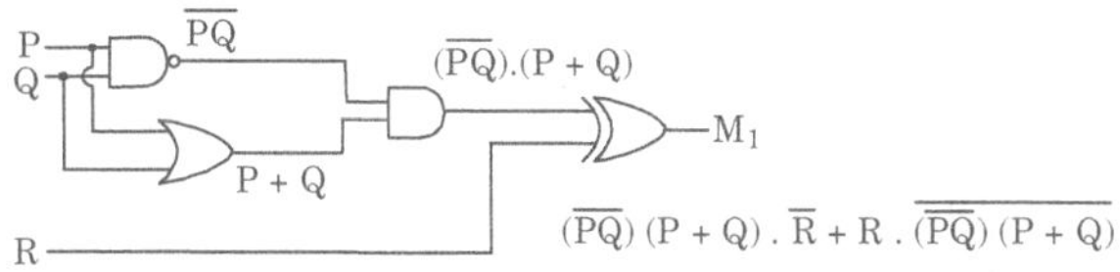

$$M_1 = (\overline{PQ})(P+Q)\overline{R}+R\left[\overline{\overline{PQ}+\overline{P+Q}}\right]$$

$$= (\overline{P}+\overline{Q})(P+Q)\overline{R}+R\left[PQ+\overline{P}.\overline{Q}\right]$$

$$= (P\overline{Q}+\overline{P}Q)\overline{R}+R(PQ+\overline{P}\overline{Q})$$

$$= (P\overline{Q}+\overline{P}Q)\overline{R}+R(\overline{P\overline{Q}+\overline{P}Q})$$

Output M_1 = (PX OR Q) XOR R

58. $Z = (P+\overline{Q})\ \overline{R}\,\overline{S} + P\overline{R}S + PQR\overline{S} + PRS$

PQ \ RS	$\overline{R}\overline{S}$	$\overline{R}S$	RS	$R\overline{S}$
$\overline{P}\overline{Q}$	1			
$\overline{P}Q$				
PQ	1	1	1	1
$P\overline{Q}$	1	1	1	

Hence, $Z = PQ + P\overline{Q}S + \overline{Q}\overline{R}\overline{S}$

59. At Q, frequency will be $\frac{1}{4}$th of clock frequency and a time delay of 2ΔT due to two flip flops.

<u>**Alternately**</u>

Assume initially $Q_0 = 0$

J	K	Q_{n+1}
1	1	$\overline{Q}_n$

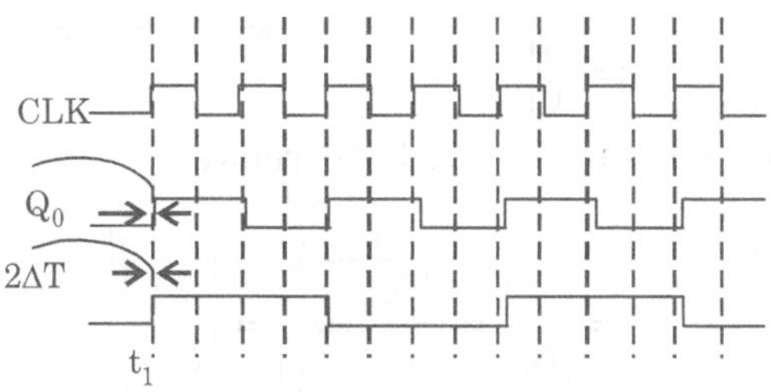

Q_0 acts as clock for second J.K flip-flop.

60. When C goes from $1 \rightarrow 0$

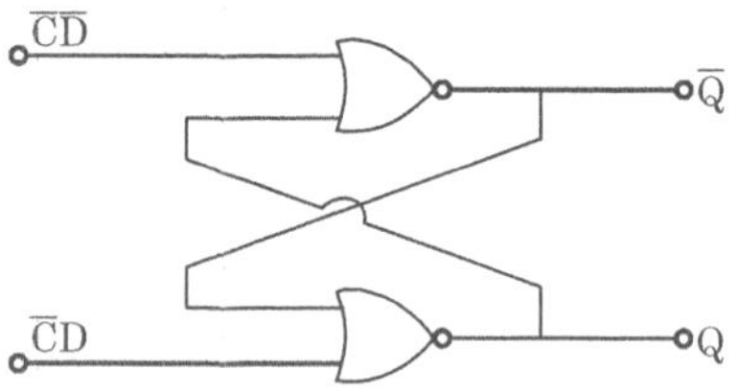

$\overline{C}$ becomes 1

Hence, output depends on D and $\overline{D}$

When D = 0, Q = 1

After sometime $D \rightarrow 1$

Then output Q = 0

61. For TE_{11} mode,

$$f_c = \frac{1}{2\sqrt{\mu \in}} \sqrt{\left(\frac{m}{a}\right)^2 + \left(\frac{n}{b}\right)^2};$$

where, a and b in cm.

and $$C = \frac{1}{\sqrt{\mu t}}$$

$$f_c = 1.5 \times 10^{10} \sqrt{\left(\frac{1}{4}\right)^2 + \left(\frac{1}{3}\right)^2}$$

$$= 6.25 \text{ GHz}$$

62. $$\beta = \frac{2\pi}{\lambda} = \frac{2\pi f}{c}$$

Path length, $$\beta l = \frac{2\pi f l}{c} = \frac{2\pi \times 3 \times 10^9 \times 0.01}{3 \times 10^8} = \frac{\pi}{5}$$

Input inpedance,

$$Z_{in} = Z_0 \left(\frac{Z_L + jZ_0 \tan \beta l}{Z_0 + jZ_L \tan \beta l} \right)$$

given, $Z_L = 0$ [short-circuited end]

$$Z_{in} = \frac{Z_0 jZ_0 \tan \beta l}{Z_0} = jZ_0 \tan \beta l$$

$$= j75 \tan \frac{\pi}{5}$$

$$= j54.5\Omega \text{[Inductive]}$$

63. $$\frac{E_{reflected}}{E_{incident}} = \frac{E_r}{E_i} = \frac{\eta_2 - \eta_1}{\eta_2 + \eta_1}$$

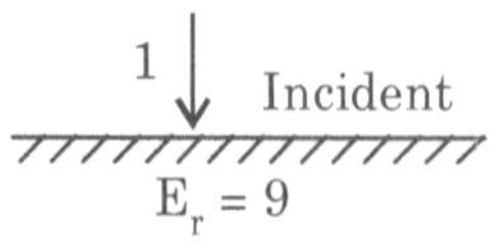

$$\eta = \sqrt{\frac{\mu}{E}}$$

But μ is same for both, therefore

$$K = \left|\frac{E_r}{E_i}\right| = \left|\frac{\sqrt{\frac{1}{E_r}} - 1}{\sqrt{\frac{1}{E_r}} + 1}\right| = \left|\frac{\frac{1}{3} - 1}{\frac{1}{3} + 1}\right| = \frac{2}{4} = 0.5$$

Alternately

Reflection coefficient,

$$T_L = \frac{\eta_2 - \eta_1}{\eta_2 + \eta_1}$$

$$T_L = \frac{\frac{\mu_2}{\epsilon_2} - \frac{\sqrt{\mu_1}}{\epsilon_1}}{\sqrt{\frac{\mu^2}{\epsilon_2}} + \sqrt{\frac{\mu_1}{\epsilon_1}}}$$

$$= \frac{\frac{1}{\sqrt{\epsilon_r}} - 1}{\frac{1}{\sqrt{\epsilon_r}} + 1}$$

$$= \frac{\frac{1}{3} - 1}{\frac{1}{3} + 1} = -0.5$$

$$|T_L| = 0.5$$

64. In optical fiber, the mode volume is given by,

$$V = \frac{nd}{\lambda} \sqrt{n_1^2 - n_2^2}$$

d = diameter of fiber, λ = wavelength

When V < 2.405, then fiber supports single mode for V > 2.405, then fiber supports multi mode

So single mode is NOT preserved if radius is doubled and wavelength is halved became is that case V > 2.405 and fibre will support multi mode.

65. Gain $\simeq 6 \left(\frac{D}{\lambda}\right)^2$

Here, D = 1 m.

$$\lambda = \frac{3 \times 10^8}{20 \times 10^9} = 1.5 \text{ cm.}$$

Effciency = 70%

$$\therefore \quad \text{Gain} = 0.7 \times 6 \left[\frac{100}{1.5}\right]^2$$

$$\therefore \quad \text{Gain (dB)} = 10 \log_{10} \left[4.2 \left(\frac{100}{1.5}\right)^2\right] \approx 45 \text{ dB}$$

66.

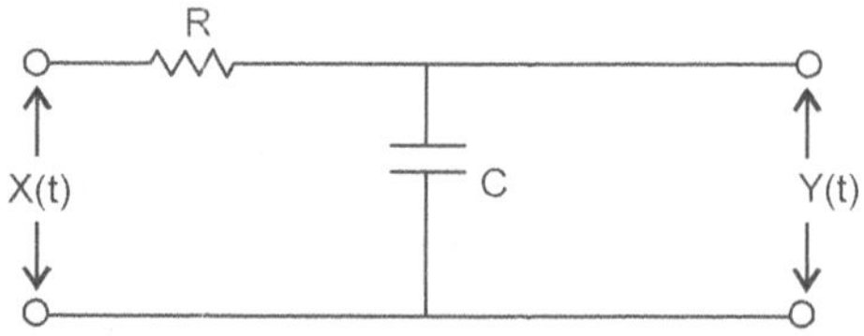

Frequency response of the RC filter is,

$$H(\omega) = \frac{1}{1 + j\omega RC}$$

$$\omega = 2\pi fc = \frac{1}{RC}$$

$S_{xx}(\omega) = k$ [While noise process]

$S_{yy}(\omega) = |H(\omega)|^2 \cdot S_{xx}(\omega)$

$$= \frac{1}{1+(\omega RC)^2} \cdot k$$

$$S_{yy}(\omega) = k\frac{\left(\frac{1}{RC}\right)^2}{\omega^2 + \left(\frac{1}{RC}\right)^2}$$

$$= \frac{k}{2(RC)} \cdot \frac{2/RC}{\omega^2 + \left(\frac{1}{RC}\right)^2}$$

Inverse fourier transform gives

$$R_{yy}(J) = \frac{k}{2RC} e^{-|J|/RC}$$

Noise power at the output,

$$E(Y^2(t)) = R_{YY}(0)$$

$$= \frac{k}{2RC} = \pi f_c k$$

67. p : transmitted bit will be received in error if at receiver, the bits are

000 or 001, 010, 100

$\downarrow$ $\rightarrow 3p^2(1-p)$

p^3

$p_e = p^3 + 3p^2(1-p)$

Alternately

The probability of ib being received in error is,

$$\binom{n}{i} p^i (1-p)^{n-i}$$

Hence the probability of error P_e is

$$p_e = \sum_{i=2}^{3} \binom{3}{i} p^i (1-p)^{3-i}$$

or $$p_e = \binom{3}{2} p^2 (1-p) + \binom{3}{3} p^3 (1-p)^0$$

or $$p_e = 3p^2(1-p) + p_3$$

68.

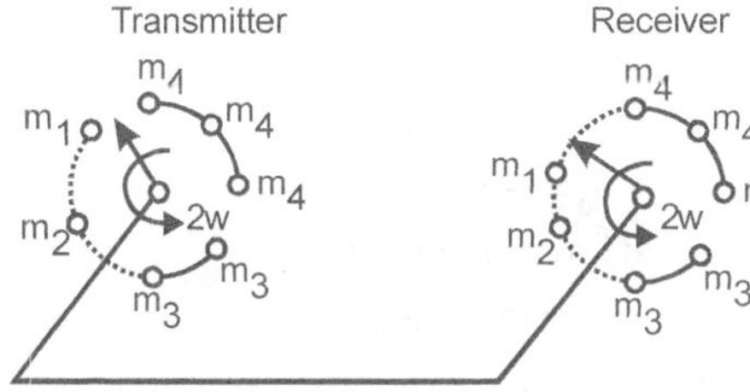

Bitrate

$R_b = (m_1 + m_2 + m_3 + m_3 + m_4 + m_4 + m_4) \times 2$

$= (\omega + \omega + \omega + \omega + \omega + \omega + \omega) \times 2$

$R_b = 7\omega \times 2$

$$\text{Bandwidth required} = \frac{R_b}{2} = \frac{14\omega}{2} = 7\omega$$

69. $s(t) = 10\cos[2\pi 10^5 t + 5\sin(2\pi \times 500t) + 7.5\sin(2\pi \times 1000t)]$

Let equation FM signal of type :

$$A\cos\left[\omega_c t + \frac{K_f V_m}{\omega_m} \sin\omega_{mt}\right]$$

then $\frac{K_f V_{m1}}{\omega_{m1}} = 5$; $\frac{K_f V_{m2}}{\omega_{m2}} = 7.5$

$\Rightarrow$ $K_f V_{m1} = 5 \times 2\pi \times 1500$

and $K_f V_{m2} = 7.5 \times 2\pi \times 1000$

we have $\Delta\omega$ (frequency deviation) same in both, so

$$\text{modulation index } m_f = \frac{\Delta\omega}{[\omega]_{max}}$$

$$= \frac{5 \times 2\pi \times 1500}{2\pi \times 1500} = 5$$

70. Ordinary AM is expressed as

$$x_{AM}(t) = m(t)\cos\omega_c t + A\cos\omega_c(t)$$
$$= [A + m(t)]\cos\omega_c t$$

Where as, angle-modulated signal is expressed

or $x_c(t) = A\cos[\omega_c(t) + \phi(t)]$

where $\phi(t) = k_p m(t)$ for PM

and $\frac{d\phi(t)}{at} = k_f m(t)$ for FM.

71. Nyquist rate = 8kHz

Number of bits transmitted per second = 8k × 8 = 64k bits/sec.

Minimum bandwidth required for distortion free transmission is one half of bit rate

$$= \frac{64}{2} = 32 \text{ kHz}$$

72. $$SNR = \frac{m^2(t)}{\frac{\Delta^2}{12}}$$

$$= \frac{3L^2}{(m_p)^2} \cdot m^2(t)$$

Let the modulating signal be sinusoidal.

$$\therefore \quad SNR = \frac{3L^2}{m_p^2} \times \frac{m_p^2}{2} = \frac{3L^2}{2} \cong 49.9 \text{ dB}$$

which is closest to 48 dB.

73. Noise $\propto \dfrac{1}{L^2}$

To reduce noise by facfor 4, L is two times, hence quantization levels = 2L = 512

Alternately

$$(SNR)_0 = \frac{3}{2}.2^{2n}$$

For $(SNR)_0$ to reduce by a factor 4, then

$$(SNR)_0' = \frac{(3NR)_0}{4} = \frac{3}{2}$$

or $\quad \dfrac{2^{2n}}{2^{2n'}} = \dfrac{1}{4}$

or $\quad n' = 9$

The number of quantization levels,

$$L = 2^n = 2^9 = 512$$

74.

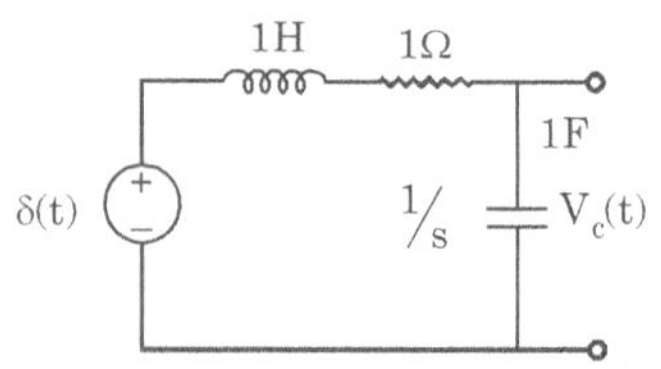

$$V_C(s) = \frac{1/s}{\left(\frac{1}{s}+s+1\right)}$$

$$= \frac{1}{\left(s^2+s+1\right)}$$

$$= \frac{2}{\sqrt{3}}\left[\frac{\frac{\sqrt{3}}{2}}{\left(s+\frac{1}{2}\right)^2+\frac{3}{4}}\right]$$

$$= \frac{2}{\sqrt{3}}\sin\left(\frac{\sqrt{3}}{2}t\right).e^{-\frac{t}{2}}$$

75. Let voltage across resistor be V_0

$$V_0(s) = \frac{1}{\left(\frac{1}{s}+s+1\right)} = \frac{s}{\left(s^2+s+1\right)}$$

$$= \frac{s}{\left(s+\frac{1}{2}\right)^2+\frac{3}{4}}$$

$$= \frac{\left(s+\frac{1}{2}\right)}{\left(s+\frac{1}{2}\right)^2+\frac{3}{4}} - \frac{\frac{1}{2}}{\left(s+\frac{1}{2}\right)^2+\frac{3}{4}}$$

$$= e^{-\frac{t}{2}}\cos\frac{\sqrt{3}}{2}t - \frac{1}{2}\times\frac{2}{\sqrt{3}}e^{-\frac{t}{2}}\sin\frac{\sqrt{3}}{2}t$$

$$= e^{-\frac{t}{2}}\left[\cos\frac{\sqrt{3}}{2}t - \frac{1}{\sqrt{3}}\sin\frac{\sqrt{3}}{2}t\right]$$

76. $\begin{bmatrix} V_1 \\ V_2 \end{bmatrix} = \begin{bmatrix} z_{11} & z_{12} \\ z_{21} & z_{22} \end{bmatrix}\begin{bmatrix} i_1 \\ i_2 \end{bmatrix}$

$$z_{11} = \left.\frac{V_1}{I_1}\right|_{I_2=0} = \frac{6}{4} = 1.5\ \Omega$$

$$z_{12} = \left.\frac{V_1}{I_2}\right|_{I_1=0} = \frac{4.5}{1} = 4.5\ \Omega$$

$$z_{21} = \left.\frac{V_2}{I_1}\right|_{I_2=0} = \frac{6}{4} = 1.5\ \Omega$$

$$z_{22} = \left.\frac{V_2}{I_2}\right|_{I_1=0} = \frac{1.5}{1} = 1.5\ \Omega$$

Thus, z-parameter will be expressed as

$$\begin{bmatrix} 1.5 & 4.5 \\ 1.5 & 1.5 \end{bmatrix}$$

77. $\begin{bmatrix} V_1 \\ I_2 \end{bmatrix} = \begin{bmatrix} h_{11} & h_{12} \\ h_{21} & h_{22} \end{bmatrix}\begin{bmatrix} I_1 \\ V_2 \end{bmatrix}$

$$h_{12} = \left.\frac{V_1}{V_2}\right|_{I_1=0}$$

$$= \frac{4.5}{1.5} = 3$$

$$h_{22} = \left.\frac{I_2}{V_2}\right|_{I_1=0}$$

$$= \frac{1}{1.5} = 0.67$$

From z matrix, we have

$$V_1 = z_{11}i_1 + z_{12}i_2 \quad ...(1)$$

$$V_2 = z_{21}i_1 + z_{22}i_2 \quad ...(2)$$

If $\quad V_2 = 0$

Then $\quad \dfrac{I_2}{I_1} = \dfrac{-z_{21}}{z_{22}} = \dfrac{-1.5}{1.5}$

$$= -1 = h_{21}$$

$\Rightarrow \quad I_2 = -I_1$

Put in equation (1), we get

$$V_1 = (z_{11} - z_{12})\,I_1$$

$$\left.\frac{V_1}{I_1}\right|_{V_2=0} = h_{11} = z_{11} - z_{12} = 1.5 - 4.5 = -3$$

Hence h-parameter will be $\begin{bmatrix} -3 & 3 \\ -1 & 0.67 \end{bmatrix}$

78. Input to the sampler

$$V_R(t) = S - V_c(t)$$

$$V_C(t) = 5\left(1 - e^{\frac{-t}{RC}}\right)$$

$$J = RC = 2$$

Then, $V_R(t) = 5e^{-\frac{t}{J}} = 5e^{-\frac{t}{2}}$

Sampling occurs at $T_s = \frac{1}{10}$ sec.

Then samples are, $V_R(n) = 5 \cdot e^{-\frac{n}{20}}$

$$= 5e^{-0.05n}$$

79. From 2-transform pair,

$$ka^n \longleftrightarrow \frac{k}{1 - az^{-1}}, |z| > a$$

then, for $x(n) = 5e^{-0.05n}$

$$X[Z] = \frac{5}{1 - e^{-0.05} \cdot z^{-1}}$$

$$= \frac{5z}{z - e^{-0.05}} \text{ for } |z| > e^{-0.05}$$

80. Drawing DC signal model, using Thevenin's theorem,

$$R_{TH} = 2011\ 10\ k = 6.67\ k$$

$$V_{TH} = \left(\frac{10}{10 + 20}\right) 9 = 3 \text{ volts}$$

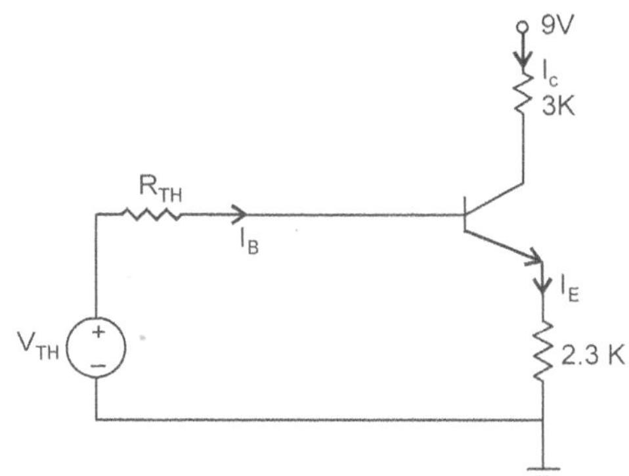

Using KVL in B-E loop.

$V_{TH} - R_{TH} \cdot I_B - V_{BE} - R_E I_E = 0$ ...(1)

and $I_E = I_B(\beta + 1)$

as, $\beta >> 1, I_E \approx \beta I_B = I_C$

then from equation (1),

$$I_E = \frac{V_{TH} - V_{BE}}{R_E} \qquad [I_B \approx 0]$$

$$= \frac{3 - 0.7}{2.3k} = 1mA$$

81. Using β-re AC model

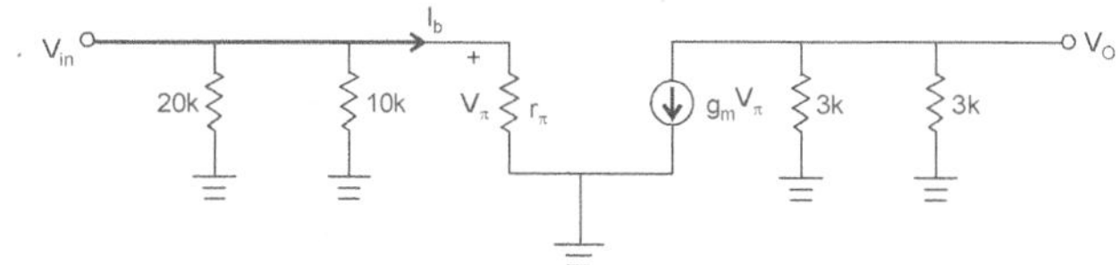

$$r_e = \frac{V_T}{I_E} = \frac{25mV}{1mA} = 25\Omega$$

$$g_m = \frac{I_C}{V_T} = \frac{1}{25} mA/V$$

$$V_O = -g_m V_\pi \times (1.5\ k) \qquad [\text{As } V_{in} = V_\pi]$$

$$= -\frac{1500}{25} V_{in} = -60 V_{in}$$

⇒ A_v(mid-band voltage gain)

$$= \frac{V_O}{V_{in}} = -60$$

82.
$$V_{DAC} = \sum_{n=0}^{3} 2^{n-1} b_n$$

$$= 2^{-1} b_0 + 2^0 b_1 + 2^1 b_2 + 2^2 b_3$$

Counter output will start from 0000 and will increase till $V_{in} > V_{DAC}$. When V_{DAC} = 6.5V, the comparator output will be zero and the counter will be stable at that reading and the number displayed at LED will be 13.

The counting sequence is shown below.

	b_3	b_2	b_1	b_0	V_{DAC}
0	0	0	0	0	0
1	0	0	0	1	0.5
2	0	0	1	0	1
3	0	0	1	1	1.5
4	0	1	0	0	2
5	0	1	0	1	2.5
6	0	1	1	0	3
7	0	1	1	1	3.5
8	1	0	0	0	4
9	1	0	0	1	4.5
10	1	0	1	0	5
11	1	0	1	1	5.5
12	1	1	0	0	6
13	1	1	0	1	6.5
14	1	1	1	0	7
15	1	1	1	1	7.5

83. Error voltage = $V_{DAC} - V_{in}$

$$= 6.5 - 6.2 = 0.3 \text{ volts}$$

84. $h(t) = e^{-2t}\,u(t)$

Fourier transform,

$$H(j\omega) = \int_{-\infty}^{\infty} h(t)e^{-j\omega t}\cdot dt$$

$$= \int_{0}^{\infty} e^{-2t}e^{-j\omega t}\cdot dt$$

$$= \int_{0}^{\infty} e^{-(j\omega+2)t}\cdot dt$$

$$= \frac{-1}{(2+j\omega)}\left(e^{-\infty} - e^{0}\right)$$

$$u(j\omega) = \frac{1}{2+j\omega}$$

85. $$H(s) = \frac{1}{(2+s)}$$

$$x(t) = 2\cos 2t$$

$$x(s) = \frac{2.s}{(s^2+4)}$$

$\therefore$ Output $= H(s)\,X(s)$

$$= \frac{1}{(2+s)} \times \frac{2s}{(s^2+4)}$$

$$= 2\left[\frac{s}{(s+2)(s^2+4)}\right]$$

$$= -\frac{1}{2(s+2)} + \frac{1}{2}\frac{s+2}{(s^2+4)}$$

↑

transient

$\therefore$ Output $= \frac{1}{2}\,[\cos 2t + \sin 2t]$

$$= \frac{1}{\sqrt{2}}\left[\cos\frac{\pi}{4}\cos 2t + \sin\frac{\pi}{4}\sin 2t\right]$$

$$= \frac{1}{\sqrt{2}}\cos\left[2t - \frac{\pi}{4}\right]$$

$$= \frac{1}{\sqrt{2}}\cos[2t - 0.25\pi]$$

SOLVED PAPER – 2007

ELECTRONICS AND COMMUNICATION ENGINEERING

Time : 3 Hours | Maximum Marks : 150

(Q. 1 – 20) carry one mark each

1. If E denotes expectation, the variance of a random variable X is given by
   (*a*) $E[X^2] - E^2[X]$ (*b*) $E[X^2] + E^2[X]$
   (*c*) $E[X^2]$ (*d*) $E^2[X]$

2. The following plot shows a function *y* which varies linearly with *x*. The value of the integral

$$I = \int_1^2 y\,dx \text{ is}$$

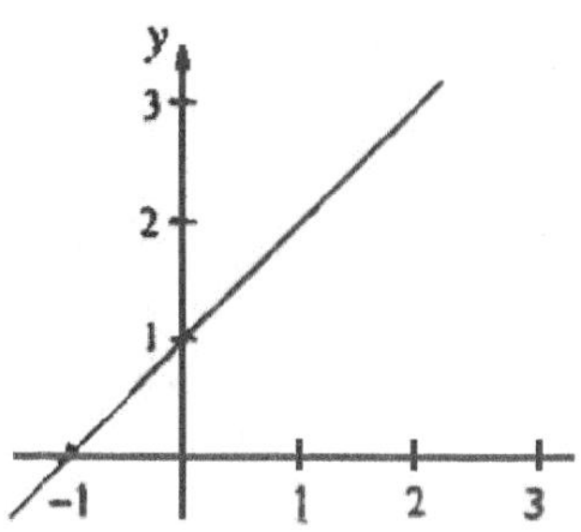

   (*a*) 1.0 (*b*) 2.5
   (*c*) 4.0 (*d*) 5.0

3. For $|x| << 1$, coth (*x*) can be approximated as
   (*a*) x (*b*) x^2
   (*c*) $\frac{1}{x}$ (*d*) $\frac{1}{x^2}$

4. $\lim_{\theta \to 0} \frac{\sin(\theta/2)}{\theta}$ is
   (*a*) 0.5 (*b*) 1
   (*c*) 2 (*d*) not defined

5. Which one of the following function is strictly bounded?
   (*a*) $\frac{1}{x^2}$ (*b*) e^x
   (*c*) x^2 (*d*) e^{-x^2}

6. For the function e^{-x}, the linear approximation around $x = 2$ is
   (*a*) $(3 - x)e^{-2}$
   (*b*) $1 - x$
   (*c*) $[3 + 2\sqrt{2} - (1 + \sqrt{2})x]e^{-2}$
   (*d*) e^{-2}

7. An independent voltage source in series with an impedance $Z_s = R_s + jX_s$ delivers a maximum average power to a load impedance Z_L when
   (*a*) $Z_L = R_S + jX_S$ (*b*) $Z_L = R_S$
   (*c*) $Z_L = jX_S$ (*d*) $Z_L = R_S - jX_S$

8. The RC circuit shown in the figure is

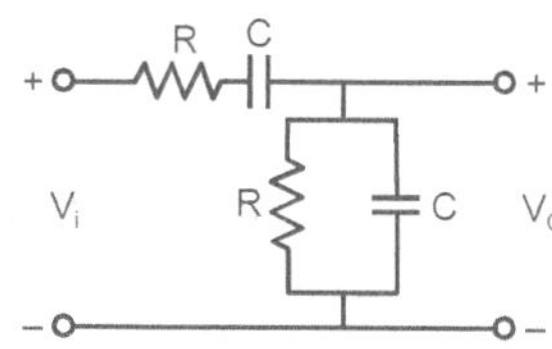

   (*a*) a low-pass filter (*b*) a high-pass filter
   (*c*) a band-pass filter (*d*) a band-reject filter

9. The electron and hole concentration in an intrinsic semiconductor are n_i per cm^3 at 300 K. Now, if acceptor impurities are introduced with a concentration of N_A per cm^3 (where $N_A >> n_i$), the electron concentration per cm^3 at 300 K will be
   (*a*) n_i (*b*) $n_i + N_A$
   (*c*) $N_A - n_i$ (*d*) $\frac{n_i^2}{N_A}$

10. In a $p^+ n$ junction diode under reverse bias, the magnitude of electric field is maximum at
   (*a*) the edge of the depletion region on the *p*-side
   (*b*) the edge of the depletion region on the *n*-side
   (*c*) the $p^+ n$ junction
   (*d*) the centre of the depletion region on the *n*-side

11. The correct full wave rectifier circuit is

(*a*)

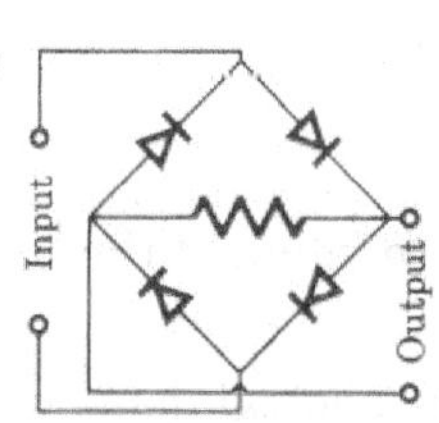

(*b*)

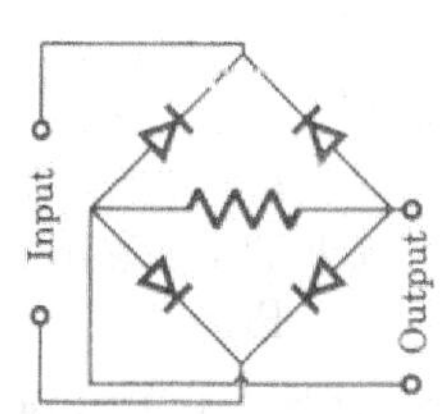

(*c*)

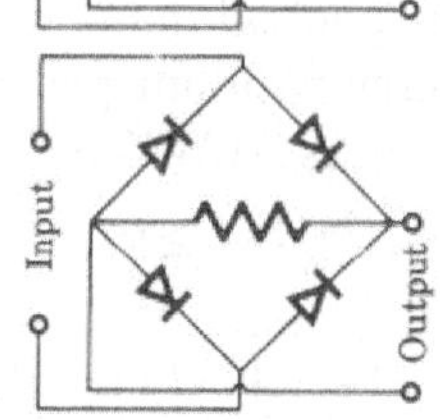

(*d*)

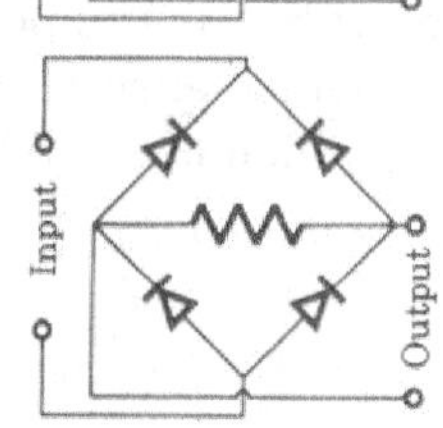

12. In a transconductance, amplifier, it is desirable to have
(a) a large input resistance and a large output resistance
(b) a large input resistance and a small output resistance
(c) a small input resistance and a large output resistance
(d) a small input resistance and a small output resistance

13. X = 01110 and Y = 11001 are two 5-bit binary numbers represented in two's complement format. The sum of X and Y represented in two's complement format using 6 bits is
(a) 100111 (b) 001000
(c) 000111 (d) 101001

14. The Boolean function Y = AB + CD is to be realized using only 2- input NAND gates.
The minimum number of gates required is
(a) 2 (b) 3
(c) 4 (d) 5

15. If the closed-loop transfer function of a control system is given as

$T(s) = \frac{s-5}{(s+2)(s+3)}$, then it is
(a) an unstable system
(b) an uncontrollable system
(c) a minimum phase system
(d) a non-minimum phase system

16. If the Laplace transform of a signal $y(t)$ is

$Y(s) = \frac{1}{s(s-1)}$, then its final value is
(a) –1 (b) 0
(c) 1 (d) Unbounded

17. If R(τ) is the autocorrelation function of a real, widesense stationary random process, then which of the following is NOT true?
(a) $R(\tau) = R(-\tau)$
(b) $|R(\tau)| \leq R(0)$
(c) $R(\tau) = -R(-\tau)$
(d) The mean square value of the process is R(0)

18. If $S(f)$ is the power spectral density of a real, wide-sense stationary random process, then which of the following is ALWAYS true?
(a) $S(0) \geq S(f)$ (b) $S(f) \geq 0$
(c) $S(-f) = -S(f)$ (d) $\int_{-\infty}^{\infty} S(f)df = 0$

19. A plane wave of wavelength λ is travelling in a direction making an angle 30° with positive x-axis and 90° with positive y-axis. The $\vec{E}$ field of the plane wave can be represented as (E_0 is a constant)
(a) $\vec{E} = \hat{y} E_0 e^{j\left(\omega t - \frac{\sqrt{3}\pi}{\lambda}x - \frac{\pi}{\lambda}z\right)}$
(b) $\vec{E} = \hat{y} E_0 e^{j\left(\omega t - \frac{\pi}{\lambda}x - \frac{\sqrt{3}\pi}{\lambda}z\right)}$
(c) $\vec{E} = \hat{y} E_0 e^{j\left(\omega t + \frac{\sqrt{3}\pi}{\lambda}x + \frac{\pi}{\lambda}z\right)}$
(d) $\vec{E} = \hat{y} E_0 e^{j\left(\omega t - \frac{\pi}{\lambda}x + \frac{\sqrt{3}\pi}{\lambda}z\right)}$

20. If C is a closed curve enclosing a surface S, then the magnetic field intensity $\vec{H}$, the current density $\vec{J}$ and the electric flux density $\vec{D}$ are related by
(a) $\iint_S \vec{H} \bullet d\vec{s} = \oint_c \left(\vec{J} + \frac{\partial \vec{D}}{\partial t}\right) \bullet \vec{dl}$
(b) $\int_C \vec{H} \bullet \vec{dl} = \oiint_S \left(\vec{J} + \frac{\partial \vec{D}}{\partial t}\right) \bullet d\vec{s}$
(c) $\oiint_S \vec{H} \bullet d\vec{s} = \int_C \left(\vec{J} + \frac{\partial \vec{D}}{\partial t}\right) \bullet d\vec{l}$
(d) $\oint_C \vec{H} \bullet d\vec{l} = \iint_S \left(\vec{J} + \frac{\partial \vec{D}}{\partial t}\right) \bullet d\vec{s}$

(Q. 21 – 75) carry two marks each.

21. It is given that $X_1, X_2, \text{-----} X_M$ are M non-zero, orthogonal vectors. The dimension of the vector space spanned by the 2M vector $X_1, X_2, \ldots X_M, -X_1, -X_2, \text{----} -X_M$ is
(a) 2M
(b) M + 1
(c) M
(d) dependent on the choice of $X_1, X_2, \text{---} X_M$

22. Consider the function $f(x) = x^2 - x - 2$. The maximum value of $f(x)$ in the closed interval [–4, 4] is
(a) 18 (b) 10
(c) –2.25 (d) indeterminate

23. An examination consists of two papers, Paper 1 and Paper 2. The probability of failing in Paper 1 is 0.3 and that in Paper 2 is 0.2. Given that a student has failed in Paper 2, the probability of failing in Paper 1 is 0.6. The probability of a student failing in both the papers is
(a) 0.5 (b) 0.18
(c) 0.12 (d) 0.06

24. The solution of the differential equation

$$k^2 \frac{d^2y}{dx^2} = y - y_2$$ under the boundary conditions

(*i*) $y = y_1$ at $x = 0$ and

(*ii*) $y = y_2$ at $x = \infty$,

where k, y_1 and y_2 are constants, is

(*a*) $y = (y_1 - y_2)\exp(-x/k^2) + y_2$

(*b*) $y = (y_2 - y_1)\exp(-x/k) + y_1$

(*c*) $y = (y_1 - y_2)\sin h(x/k) + y_1$

(*d*) $y = (y_1 - y_2)\exp(-x/k) + y_2$

25. The equation $x^3 - x^2 + 4x - 4 = 0$ is to be solved using the Newton-Raphson method. If $x = 2$ is taken as the initial approximation of the solution, then the next approximation using this method will be

(*a*) $\frac{2}{3}$ (*b*) $\frac{4}{3}$

(*c*) 1 (*d*) $\frac{3}{2}$

26. Three functions $f_1(t)$, $f_2(t)$ and $f_3(t)$, which are zero outside the interval [0, T] are shown in the figure. Which of the following statements is correct ?

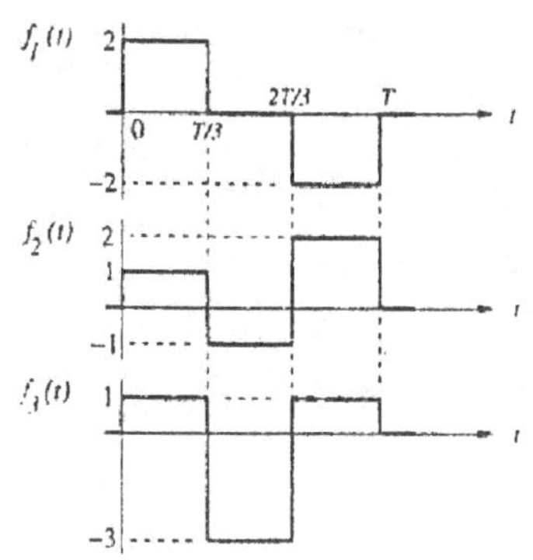

(*a*) $f_1(t)$ and $f_2(t)$ are orthogonal

(*b*) $f_1(t)$ and $f_3(t)$ are orthogonal

(*c*) $f_2(t)$ and $f_3(t)$ are orthogonal

(*d*) $f_1(t)$ and $f_2(t)$ are orthonormal

27. If the semi-circular contour D of radius 2 is as shown in the figure, then the value of the integral $\oint_D \frac{1}{(s^2 - 1)} ds$ is

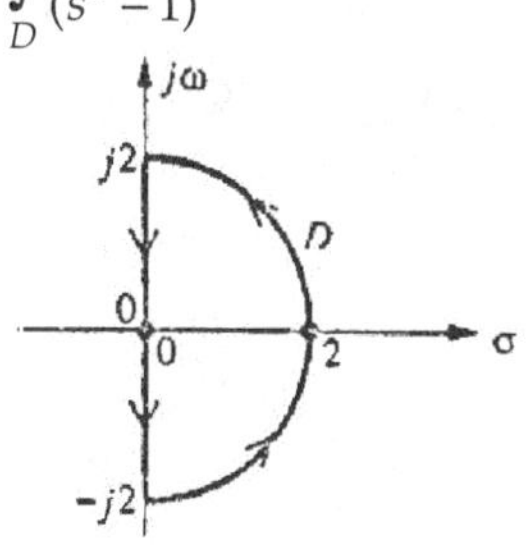

(*a*) $j\pi$ (*b*) $-j\pi$

(*c*) $-\pi$ (*d*) π

28. Two series resonant filters are as shown in the figure. Let the 3-dB bandwidth of Filter 1 be B_1 and that of Filter 2 be B_2. The value of $\frac{B_1}{B_2}$ is

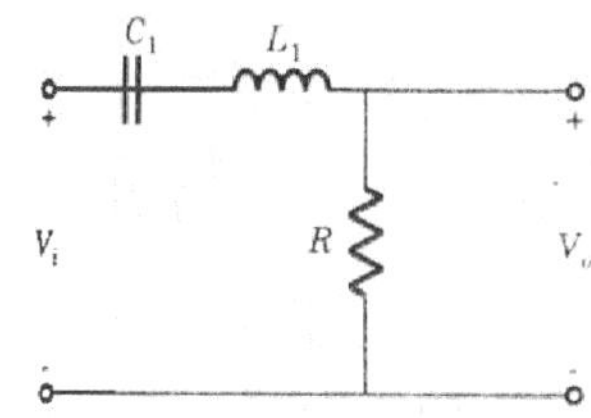

Filter 1

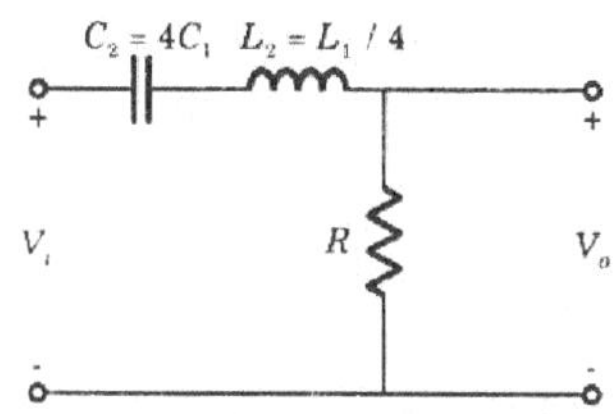

Filter 2

(*a*) 4

(*b*) 1

(*c*) $\frac{1}{2}$

(*d*) $\frac{1}{4}$

29. For the circuit shown in the figure, the Thevenin voltage and resistance looking into X-Y are

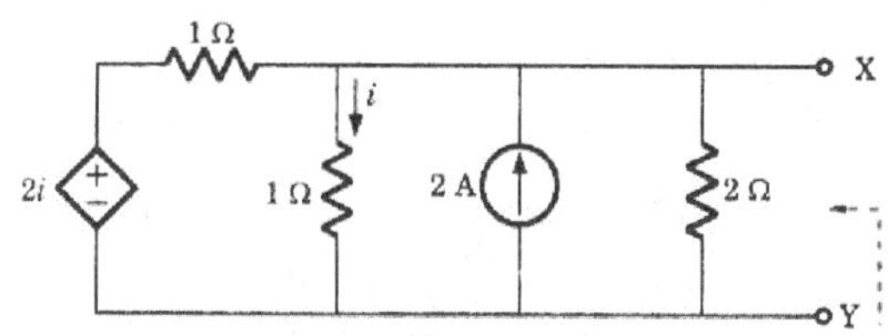

(*a*) 4/3 V, 2 Ω

(*b*) 4 V, 2/3 Ω

(*c*) 4/3 V, 2/3 Ω

(*d*) 4 V, 2 Ω

30. In the circuit shown, V_C is 0 volts at $t = 0$, sec. For $t > 0$, the capacitor current $i_c(t)$, where t is in seconds, is given by

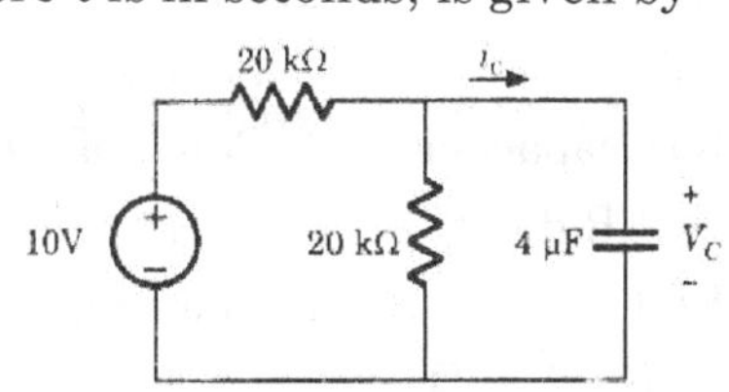

(*a*) 0.50 exp(–25*t*)mA

(*b*) 0.25 exp(–25*t*)mA

(*c*) 0.50 exp(–12.5*t*)mA

(*d*) 0.25 exp(–6.25*t*)mA

31. In the AC network shown in the figure, the phasor voltage V_{AB} (in Volts) is

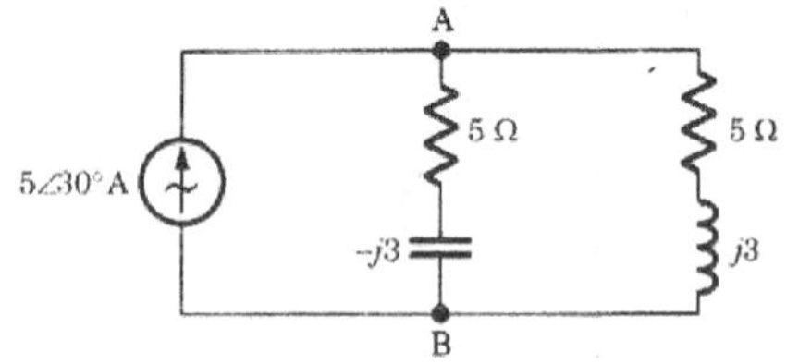

(a) 0 (b) 5 ∠30°

(c) 12.5 ∠30° (d) 17 ∠30°

32. A p^+ n junction has built-in potential of 0.8 V. The depletion layer width at a reverse bias of 1.2 V is 2 μm. For a reverse bias of 7.2 V, the depletion layer width will be

(a) 4 μm (b) 4.9 μm

(c) 8 μm (d) 12 μm

33. Group I lists four types of *p-n* junction diodes. Match each device in Group I with one of the options in Group II to indicate the bias condition of that device in its normal mode of operation.

Group I	Group II
A. Zener Diode	1. Forward bias
B. Solar cell	2. Reverse bias
C. LASER diode	
D. Avalanche Photodiode	

	A	B	C	D
(a)	1	2	1	2
(b)	2	1	1	2
(c)	2	2	1	1
(d)	2	1	2	2

34. The DC current gain (β) of a BJT is 50. Assuming that the emitter injection efficiency is 0.995, the base transport factor is

(a) 0.980 (b) 0.985

(c) 0.990 (d) 0.995

35. Group I lists four different semiconductor device. Match each device in Group I with its characteristic property in Group II.

Group I	Group II
A. BJT	1. Population inversion
B. MOS capacitor	2. Pinch-off voltage
C. LASER diode	3. Early effect
D. JFET	4. Flat-band voltage

	A	B	C	D
(a)	3	1	4	2
(b)	2	4	3	2
(c)	3	4	1	2
(d)	3	2	1	4

36. For the Op-Amp circuit shown in the figure, V_o is

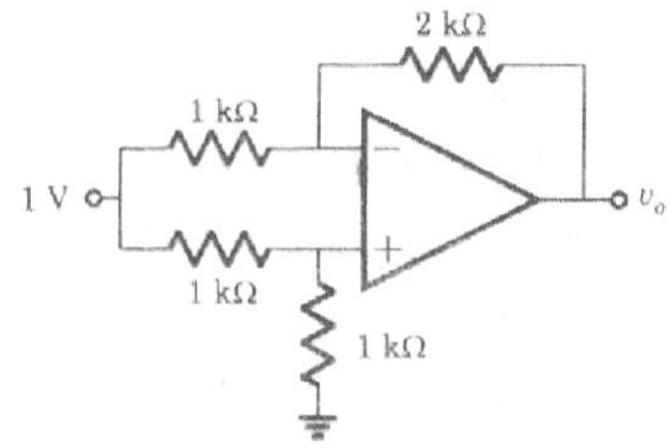

(a) –2 V (b) – 1 V

(c) – 0.5 V (d) 0.5 V

37. For the BJT circuit shown, assume that the β of the transistor is very large and V_{BE} = 0.7 V. The mode of operation of the BJT is

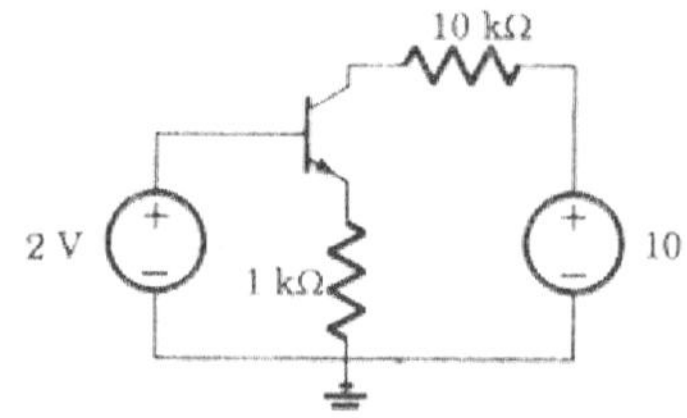

(a) cut-off (b) saturation

(c) normal active (d) reverse active

38. In the Op-Amp circuit shown, assume that the diode current follows the equation $I = I_s \exp(V/V_T)$. For V_i = 2 V, $V_0 = V_{01}$, and for V_i = 4 V, $V_0 = V_{02}$. The relationship between V_{01} and V_{02} is

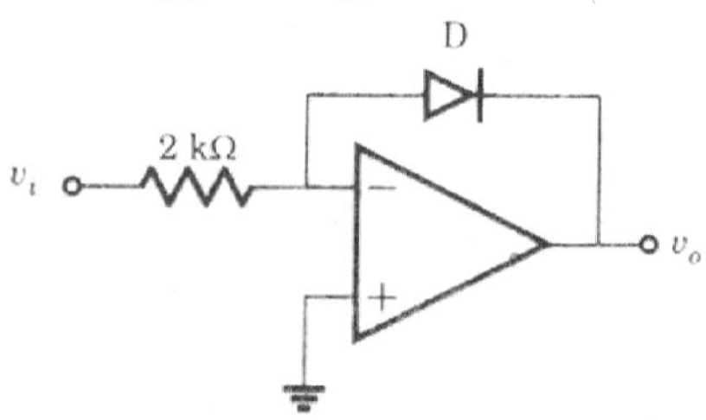

(a) $V_{02} = \sqrt{2}\, V_{01}$ (b) $V_{02} = e^2 V_{01}$

(c) $V_{02} = V_{01} \ln 2$ (d) $V_{01} - V_{02} = V_T \ln 2$

39. In the CMOS inverter circuit shown, if the transconductance parameters of the NMOS and PMOS transistor are $k_n = k_p = \mu_n C_{ox} \dfrac{W_n}{L_n} = \mu_p C_{ox} \dfrac{W_p}{L_p}$ = 40 μA/V² and their threshold voltages are $V_{THn} = |V_{THp}|$ = 1 V, the current I is

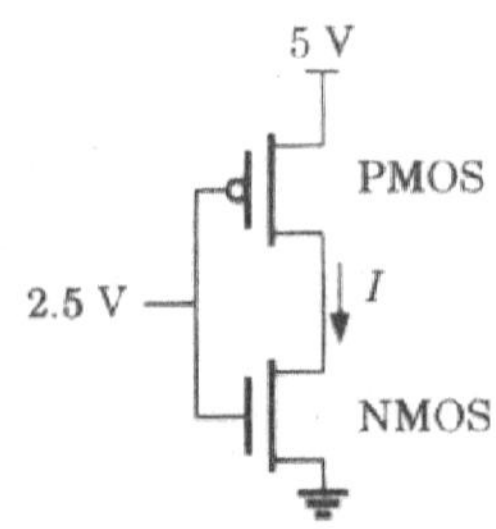

(a) 0 A

(b) 25 μA

(c) 45 μA

(d) 90 μA

40. For the Zener diode shown in the figure, the Zener voltage at knee is 7 V, the knee current is negligible and the Zener dynamic resistance is 10 Ω. If the input voltage (V_i) range is from 10 to 16 V, the output voltage (V_0) range from

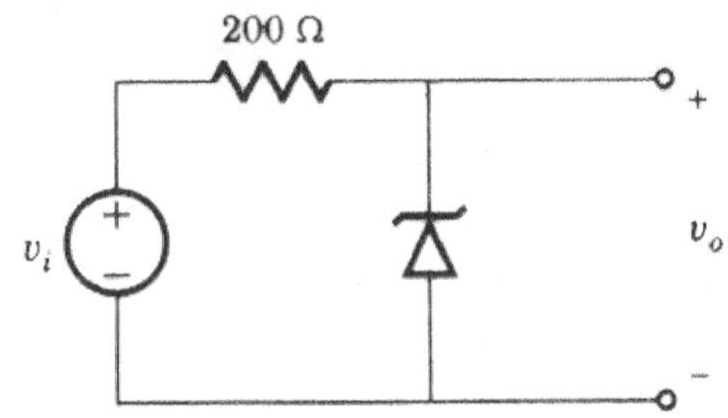

(*a*) 7.00 to 7.29 V

(*b*) 7.14 to 7.29 V

(*c*) 7.14 to 7.43 V

(*d*) 7.29 to 7.43 V

41. The Boolean expression $Y = \bar{A}\bar{B}\bar{C}D + \bar{A}BC\bar{D} + A\bar{B}\bar{C}D + AB\bar{C}\bar{D}$ can be minimized to

(*a*) $Y = \bar{A}\bar{B}\bar{C}D + \bar{A}B\bar{C} + A\bar{C}D$

(*b*) $Y = \bar{A}\bar{B}\bar{C}D + BC\bar{D} + A\bar{B}\bar{C}D$

(*c*) $Y = \bar{A}BC\bar{D} + \bar{B}\bar{C}D + A\bar{B}\bar{C}D$

(*d*) $Y = \bar{A}BC\bar{D} + \bar{B}\bar{C}D + AB\bar{C}\bar{D}$

42. Circuit diagram of a standard TTL NOT gate is shown in the figure. When V_i = 2.5 V, the modes of operation of the transistor will be

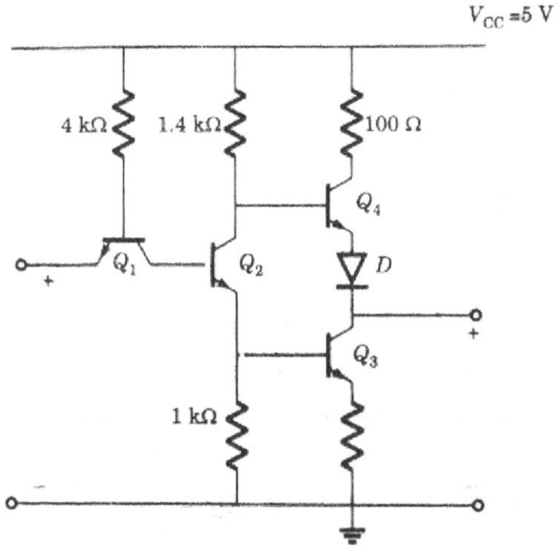

(*a*) Q_1 : reverse active; Q_2 : normal active; Q_3 : saturation; Q_4 : cut-off

(*b*) Q_1 : reverse active; Q_2 : saturation; Q_3 : saturation; Q_4 : cut-off

(*c*) Q_1 : normal active; Q_2 : cut-off; Q_3 : cut-off; Q_4 : saturation

(*d*) Q_1 : saturation; Q_2 : saturation; Q_3 : saturation; Q_4 : normal active

43. In the following circuit, X is given by

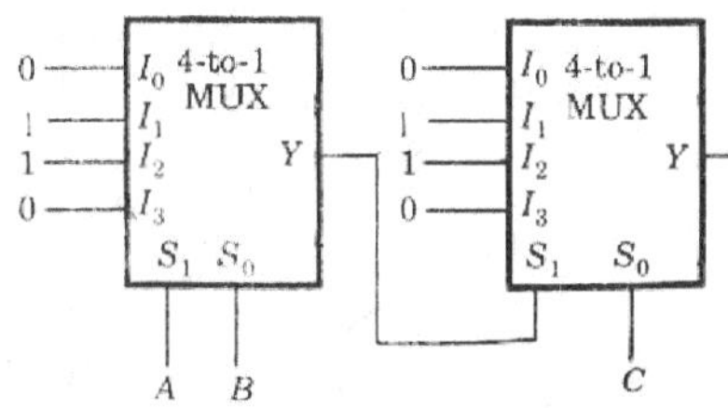

(*a*) $X = A\bar{B}\bar{C} + \bar{A}B\bar{C} + \bar{A}\bar{B}C + ABC$

(*b*) $X = \bar{A}BC + A\bar{B}C + AB\bar{C} + \bar{A}\bar{B}\bar{C}$

(*c*) X = AB + BC + AC

(*d*) $X = \bar{A}\bar{B} + \bar{B}\bar{C} + \bar{A}\bar{C}$

44. The following binary values were applied to the X and Y inputs of the NAND latch shown in the figure in the sequence indicated below :

X = 0, Y = 1; X = 0; Y = 0; X = 1, Y = 1

The corresponding stable P, O outputs will be

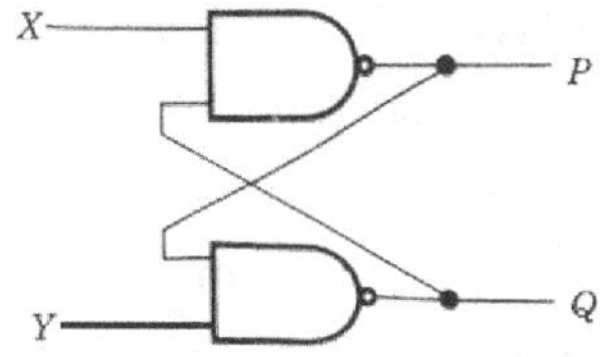

(*a*) P = 1, Q = 0; P = 1, Q = 0
P = 1, Q = 0 or P = 0, Q = 1

(*b*) P = 1, Q = 0; P = 0, Q = 1 or P = 0, Q = 1;
P = 0, Q = 1

(*c*) P = 1, Q = 0; P = 1, Q = 1;
P = 1, Q = 0 or P = 0, Q = 1

(*d*) P = 1, Q = 0; P = 1, Q = 1;
P = 1, Q = 1

45. For the circuit shown, the counter state (Q_1 Q_0) follows the sequence

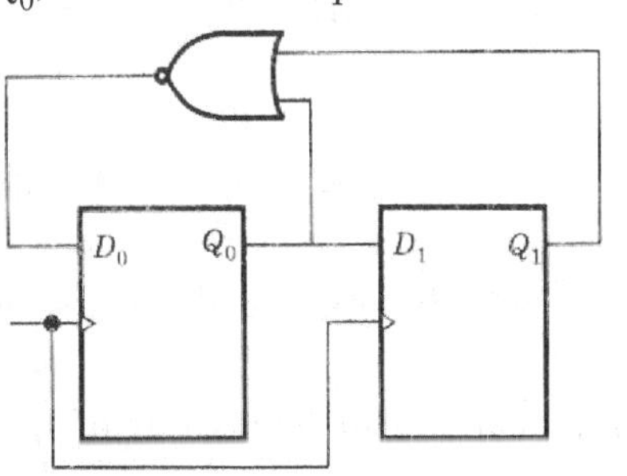

(*a*) 00, 01, 10, 11, 00...

(*b*) 00, 01, 10, 00, 01....

(*c*) 00, 01, 11, 00, 01...

(*d*) 00, 10, 11, 00, 10.....

46. An 8255 chip is interfaced to an 8085 microprocessor system as an I/O mapped I/O as shown in the figure. The address lines A_0 and A_1 of the 8085 are used by the 8255 chip to decode internally its three ports and the Control register.

The address lines A_3 to A_7 as well as the $IO/\overline{M}$ signal are used for address decoding. The range of addresses for which the 8255 chip would get selected is

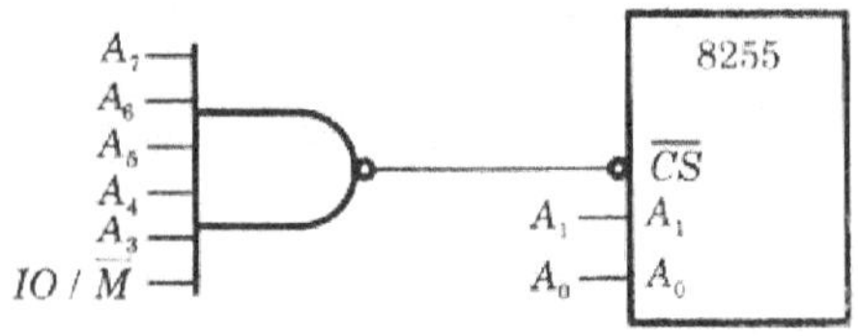

(*a*) F8H – FBH (*b*) F8H – FCH

(*c*) F8H – FFH (*d*) F0H – F7H

47. The 3–dB bandwidth of the low-pass signal $e^{-t}\,u(t)$, where $u(t)$ is the unit step function, is given by

(*a*) $\frac{1}{2\pi}$ Hz (*b*) $\frac{1}{2\pi}\sqrt{\sqrt{2}-1}$ Hz

(*c*) ∞ (*d*) 1 Hz

48. A Hilbert transformer is a

(*a*) non-linear system

(*b*) non-causal system

(*c*) time-varying system

(*d*) low-pass system

49. The frequency response of a linear, time-invariant system is given by

$$H(f) = \frac{5}{1 + j10\pi f}.$$

The step response of the system is

(*a*) $5\left(1 - e^{-5t}\right)u(t)$ (*b*) $5\left(1 - e^{-\frac{t}{5}}\right)u(t)$

(*c*) $\frac{1}{5}\left(1 - e^{-5t}\right)u(t)$ (*d*) $\frac{1}{5}\left(1 - e^{-\frac{t}{5}}\right)u(t)$

50. A 5-point sequence $x[n]$ is given as

$x[-3] = 1,\ x[-2] = 1,\ x[-1] = 0,\ x[0] = 5,\ x[1] = 1.$

Let $X(e^{j\omega})$ denote the discrete-time Fourier transform of $x[n]$. The value of $\int_{-\pi}^{\pi} X\left(e^{j\omega}\right) d\omega$ is

(*a*) 5 (*b*) 10π

(*c*) 16π (*d*) $5 + j10\pi$

51. The *z*-transform $X[z]$ of a sequence $x[n]$ is given by $X[z] = \frac{0.5}{1 - 2z^{-1}}$. It is given that the region of convergence of $X[z]$ includes the unit circle. The value of $x[0]$ is

(*a*) – 0.5 (*b*) 0

(*c*) 0.25 (*d*) 0.5

52. A control system with a PD controller is shown in the figure. If the velocity error constant $K_v = 1000$ and the damping ratio $\zeta = 0.5$, then the values of K_P and K_D are

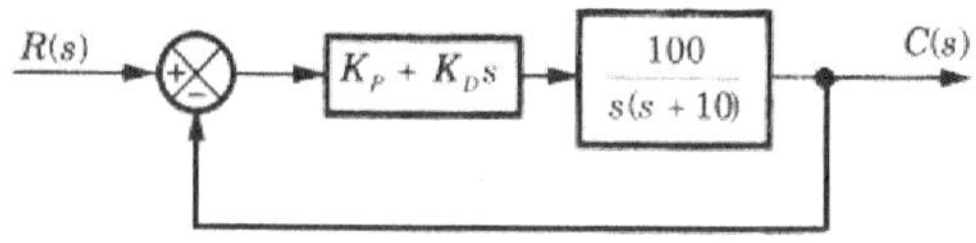

(*a*) $K_P = 100,\ K_D = 0.09$

(*b*) $K_P = 100,\ K_D = 0.9$

(*c*) $K_P = 10,\ K_D = 0.09$

(*d*) $K_P = 10,\ K_D = 0.9$

53. The transfer function of a plant is $T(s) = \frac{5}{(s+5)\left(s^2+s+1\right)}$. The second-order approximation of $T(s)$ using dominant pole concept is

(*a*) $\frac{1}{(s+5)(s+1)}$ (*b*) $\frac{5}{(s+5)(s+1)}$

(*c*) $\frac{5}{s^2+s+1}$ (*d*) $\frac{1}{s^2+s+1}$

54. The open-loop transfer function of a plant is given as $G(s) = \frac{1}{s^2-1}$. If the plant is operated in a unity feedback configuration, then the lead compensator that can stabilize this control system is

(*a*) $\frac{10(s-1)}{s+2}$

(*b*) $\frac{10(s+4)}{s+2}$

(*c*) $\frac{10(s+2)}{s+10}$

(*d*) $\frac{1(s+2)}{s+10}$

55. A unity feedback control system has an open-loop transfer function $G(s) = \frac{K}{s\left(s^2+7s+12\right)}$.

The gain K for which $s = -1 + j1$ will lie on the root locus of this system is

(*a*) 4 (*b*) 5.5

(*c*) 6.5 (*d*) 10

56. The asymptotic Bode plot of a transfer function is as shown in the figure. The transfer function $G(s)$ corresponding to this Bode plot is

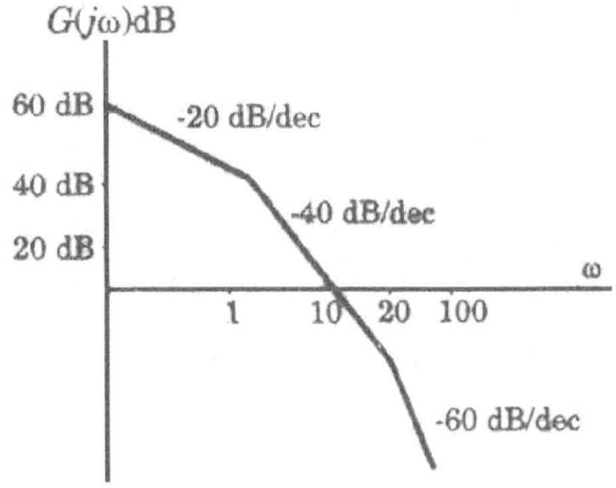

(a) $\frac{1}{(s+1)(s+20)}$ (b) $\frac{1}{s(s+1)(s+20)}$

(c) $\frac{100}{s(s+1)(s+20)}$ (d) $\frac{100}{s(s+1)(1+0.05s)}$

57. The state space representation of a separately excited DC servo motor dynamics is given as

$$\begin{bmatrix} \frac{d\omega}{dt} \\ \frac{di_a}{dt} \end{bmatrix} = \begin{bmatrix} -1 & 1 \\ -1 & -10 \end{bmatrix} \begin{bmatrix} \omega \\ i_a \end{bmatrix} + \begin{bmatrix} 0 \\ 10 \end{bmatrix} u$$

where ω is the speed of the motor, i_a is the armature current and u is the armature voltage. The transfer function $\frac{\omega(s)}{U(s)}$ of the motor is

(a) $\frac{10}{s^2+11s+11}$ (b) $\frac{1}{s^2+11s+11}$

(c) $\frac{10s+10}{s^2+11s+11}$ (d) $\frac{1}{s^2+s+1}$

58. In delta modulation, the slope overload distortion can be reduced by

(a) decreasing the step size

(b) decreasing the granular noise

(c) decreasing the sampling rate

(d) increasing the step size

59. The raised cosine pulse $p(t)$ is used for zero ISI in digital communications. The expression for $p(t)$ with unity roll-off factor is given by

$$p(t) = \frac{\sin 4\pi Wt}{4\pi Wt(1-16W^2t^2)}.$$

The value of $p(t)$ at $t = \frac{1}{4W}$ is

(a) – 0.5 (b) 0

(c) 0.5 (d) ∞

60. In the following scheme, if the spectrum M(f) of $m(t)$ is as shown, then the spectrum Y(f) of $y(t)$ will be

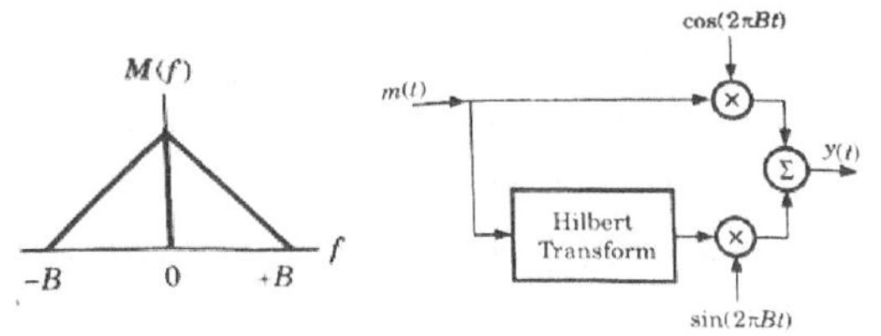

(a)

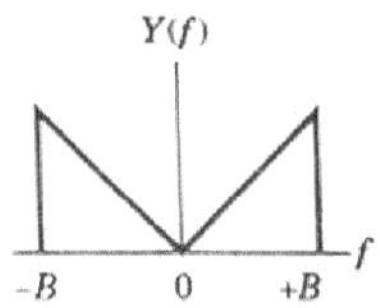

(b)

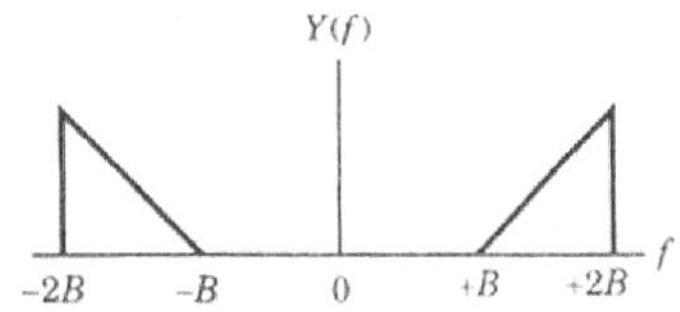

(c)

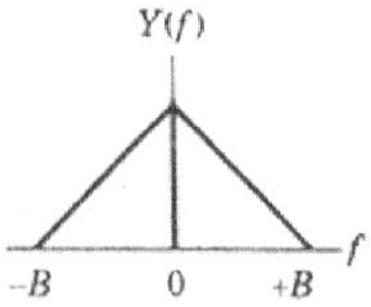

(d)

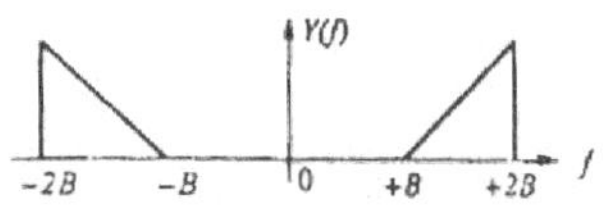

61. During transmission over a certain binary communication channel, bit errors occur independently with probability p. The probability of AT MOST one bit in error in a block of n bits is given by

(a) p^n

(b) $1 - p^n$

(c) $np(1-p)^{n-1} + (1-p)^n$

(d) $1 - (1-p)^n$

62. In a GSM system, 8 channels can co-exist in 200 kHz bandwidth using TDMA. A GSM based cellular operator is allocated 5 MHz bandwidth.

Assuming a frequency reuse factor of $\frac{1}{5}$, i.e. afive cell repeat pattern, the maximum number of simultaneous channels that can exist in one cell is

(a) 200 (b) 40

(c) 25 (d) 5

63. In a Direct Sequence CDMA system the chip rate is 1.2288×10^6 chips per second. If the processing gain is desired to be AT LEAST 100, the data rate

(a) must be less than or equal to 12.288×10^3 bits per sec (*b*) must be greater than 12.288×10^3 bits per sec

(*c*) must be exactly equal to 12.288×10^3 bits per sec

(*d*) can take any value less than 122.88×10^3 bits per sec

64. An air-filled rectangular waveguide has inner dimensions of 3 cm × 2 cm. The wave impedance of the TE_{20} mode of propagation in the waveguide at a frequency of 30 GHz is (free space impedance $\eta_0 = 377\ \Omega$)

(*a*) 308 Ω (*b*) 355 Ω

(*c*) 400 Ω (*d*) 461 Ω

65. The $\vec{H}$ field (in A/m) of a plane wave propagating in free space is given by

$$\vec{H} = \hat{x}\,\frac{5\sqrt{3}}{\eta_0}\cos(\omega t - \beta z) + \hat{y}\,\frac{5}{\eta_0}\sin\left(\omega t - \beta z + \frac{\pi}{2}\right).$$

The time average power flow density in watts is

(*a*) $\frac{\eta_0}{100}$ (*b*) $\frac{100}{\eta_0}$

(*c*) $50\,\eta_0^2$ (*d*) $\frac{50}{\eta_0}$

66. The $\vec{E}$ field in a rectangular waveguide of inner dimensions $a \times b$ is given by

$$\vec{E} = \frac{\omega\mu}{h^2}\left(\frac{\pi}{a}\right) H_0 \sin\left(\frac{2\pi x}{a}\right)\sin(\omega t - \beta z)\ \hat{y}$$

where H_0 is a constant, and a and b are the dimensions along the x-axis and the y-axis respectively. The mode of propagation in the waveguide is

(*a*) TE_{20} (*b*) TM_{11}

(*c*) TM_{20} (*d*) TE_{10}

67. A load of 50 Ω is connected in shunt in a 2-wire transmission line of Z_0 = 50 Ω as shown in the figure. The 2-port scattering parameter matrix (S – matrix) of the shunt element is

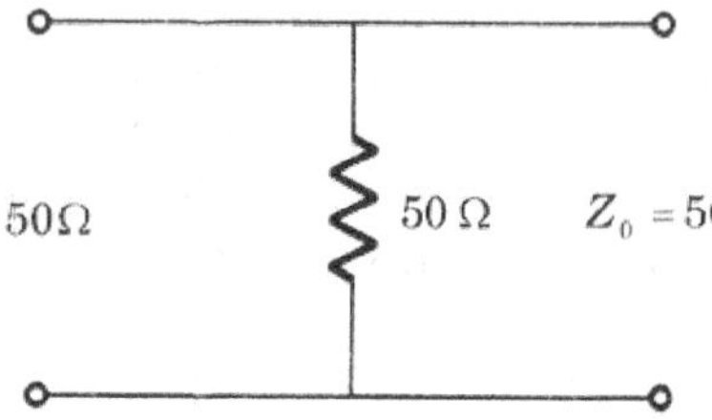

(*a*) $\begin{bmatrix} -\frac{1}{2} & \frac{1}{2} \\ \frac{1}{2} & -\frac{1}{2} \end{bmatrix}$ (*b*) $\begin{bmatrix} 0 & 1 \\ 1 & 0 \end{bmatrix}$

(*c*) $\begin{bmatrix} -\frac{1}{3} & \frac{2}{3} \\ \frac{2}{3} & -\frac{1}{3} \end{bmatrix}$ (*d*) $\begin{bmatrix} \frac{1}{4} & -\frac{3}{4} \\ -\frac{3}{4} & \frac{1}{4} \end{bmatrix}$

68. The parallel branches of a 2-wire transmission line are terminated in 100 Ω and 200 Ω resistors as shown in the figure. The characteristic impedance of the line is Z_0 = 50Ω and each section has a length of $\frac{\lambda_0}{2}$. The voltage reflection coefficient Γ at the input is

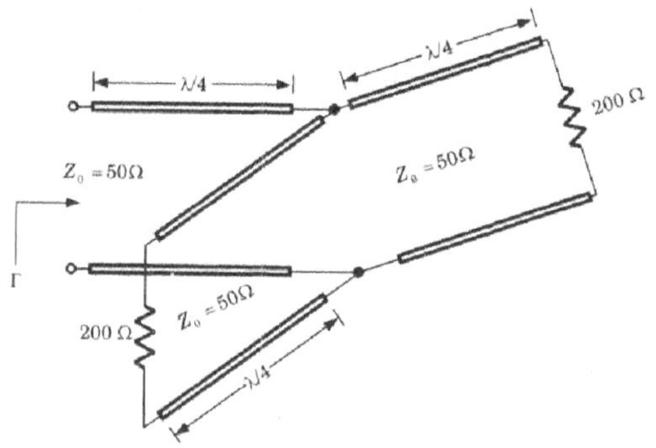

(*a*) $-j\frac{7}{5}$ (*b*) $\frac{-5}{7}$

(*c*) $j\frac{5}{7}$ (*d*) $\frac{5}{7}$

69. A $\frac{\lambda}{2}$ dipole is kept horizontally at a height of $\frac{\lambda_0}{2}$ above a perfectly conducting infinite ground plane. The radiation pattern in the plane of the dipole ($\vec{E}$ plane) looks approximately as

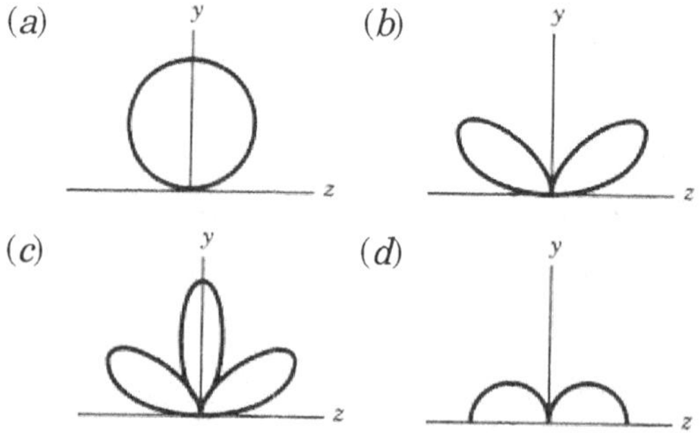

70. A right circularly polarized (RCP) plane wave is incident at an angle of 60° to the normal on an air-dielectric interface. If the reflected wave is linearly polarized, the relative dielectric constant ε_{r2} is

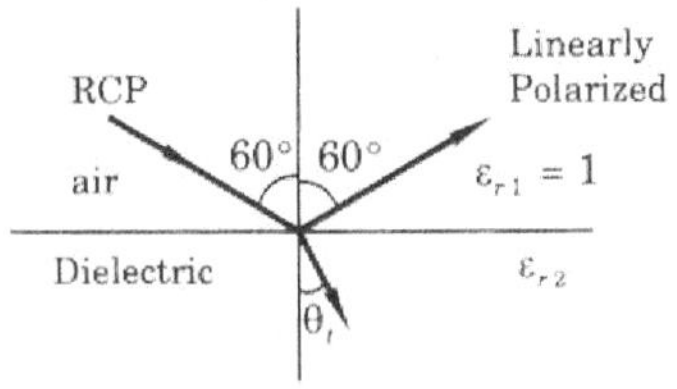

(a) $\sqrt{2}$ (b) $\sqrt{3}$

(c) 2 (d) 3

COMMON DATA QUESTIONS

Common Data for Questions 71, 72, 73 :

The figure shows the high-frequency capacitance voltage (C–V) characteristics of a Metal/SiO_2/ silicon (MOS) capacitor having an area of 1×10^{-4} cm^2. Assume that the permittivities ($\varepsilon_0\varepsilon_r$) of silicon and SiO_2 are 1×10^{-12} F/cm and 3.5×10^{-13} F/cm respectively.

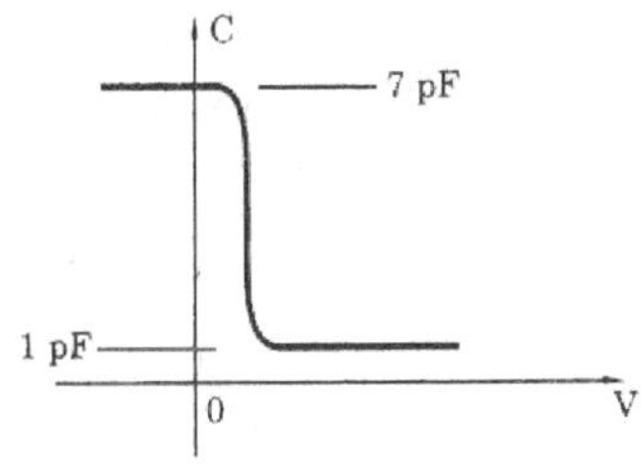

71. The gate oxide thickness in the MOS capacitor is

(a) 50 nm (b) 143 nm

(c) 350 nm (d) 1 μm

72. The maximum depletion layer width in silicon is

(a) 0.143 μm (b) 0.857 μm

(c) 1 μm (d) 1.143 μm

73. Consider the following statements about the C-V characteristics plot :

S1 : The MOS capacitor has an *n*-type substrate.

S2 : If positive charges are introduced in the oxide, the C-V plot will shift to the left.

Then which of the following is true ?

(a) Both S1 and S2 are true

(b) S1 is true and S2 is false

(c) S1 is false and S2 is true

(d) Both S1 and S2 are false

Common Data for Questions 74, 75 :

Two 4-ary signal constellations are shown. It is given that ϕ_1 and ϕ_2 constitute an orthonormal basis for the two constellations. Assume that the four symbols in both the constellations are equiprobable. Let $\frac{N_0}{2}$ denote the power spectral density of white Gaussian noise.

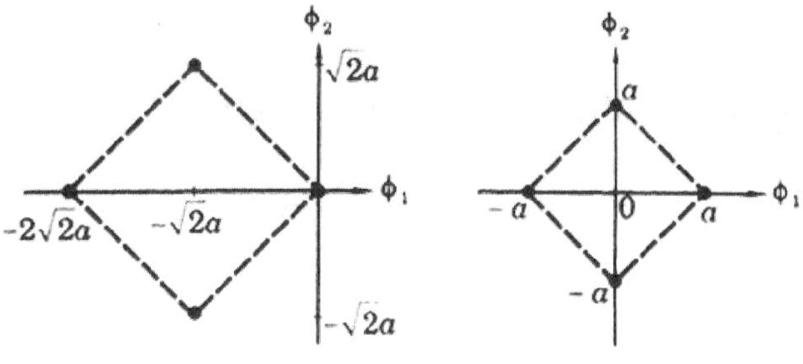

Constellation 1 Constellation 2

74. The ratio of the average energy of Constellation 1 to the average energy of Constellation 2 is

(a) $4a^2$ (b) 4

(c) 2 (d) 8

75. If these constellations are used for digital communications over an AWGN channel, then which of the following statements is true ?

(a) Probability of symbol error for Constellation 1 is lower

(b) Probability of symbol error for Constellation 1 is higher

(c) Probability of symbol error is equal for both the constellations

(d) The value of N_0 will determine which of the two constellations has a lower probability of symbol error

LINKED ANSWER QUESTIONS :

Q.76 – 85 carry two marks each

Statement for Linked Answer Questions 76 - 77

Consider the Op-Amp circuit shown in the figure.

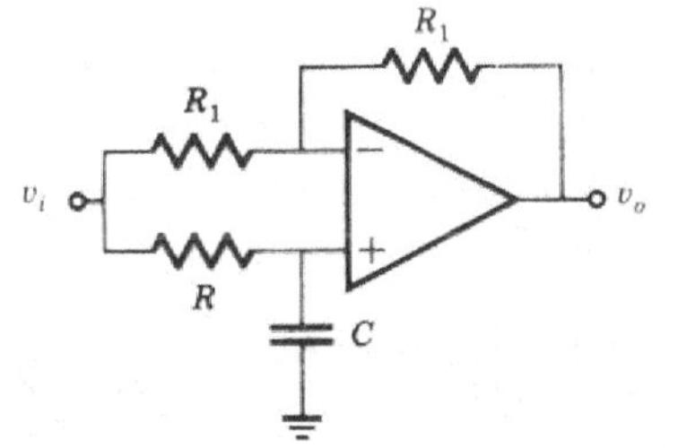

76. The transfer function $V_0(s) V_t(s)$ is

(a) $\frac{1 - sRC}{1 + sRC}$ (b) $\frac{1 + sRC}{1 - sRC}$

(c) $\frac{1}{1 - sRC}$ (d) $\frac{1}{1 + sRC}$

77. If $V_i = V_1 \sin(\omega t)$ and $V_0 = V_2 \sin(\omega t + \phi)$, then the minimum and maximum values of ϕ (in radians) are respectively

(a) $-\pi/2$ and $\pi/2$ (b) 0 and $\pi/2$

(c) $-\pi$ and 0 (d) $-\pi/2$ and 0

Statement for Linked Answer Questions 78 and 79

An 8085 assembly language program is given below

```
Line 1 : MVIA, B5H
     2 : MVI B, 0EH
     3 : XRI 69H
     4 : ADD B
     5 : ANI 9BH
     6 : CPI 9FH
     7 : STA 3010H
     8 : HLT
```

78. The contents of the accumulator just after execution of the ADD instruction in line 4 will be

(a) C3H (b) EAH

(c) DCH (d) 69H

79. After execution of line 7 of the program, the status of the CY and Z flags will be

(a) CY = 0, Z = 0 (b) CY = 0, Z = 1

(c) CY = 1, Z = 0 (d) CY = 1, Z = 1

Statement for Linked Answer Questions 80 and 81

Consider a linear system whose state space representation is $\dot{x}(t) = Ax(t)$. If the initial state vector of the system is $x(0) = \begin{bmatrix} 1 \\ -2 \end{bmatrix}$, then the system response is $x(t) = \begin{bmatrix} e^{-2t} \\ -2e^{-2t} \end{bmatrix}$

If the initial state vector of the system changes to $x(0) = \begin{bmatrix} 1 \\ -1 \end{bmatrix}$, then the system response becomes $x(t) = \begin{bmatrix} e^{-t} \\ -e^{-t} \end{bmatrix}$

80. The eigenvalue and eigenvector pairs (λ_i, v_i) for the system are

(a) $\left(-1, \begin{bmatrix} 1 \\ -1 \end{bmatrix}\right)$ and $\left(-2, \begin{bmatrix} 1 \\ -2 \end{bmatrix}\right)$

(b) $\left(-2, \begin{bmatrix} 1 \\ -1 \end{bmatrix}\right)$ and $\left(-1, \begin{bmatrix} 1 \\ -2 \end{bmatrix}\right)$

(c) $\left(-1, \begin{bmatrix} 1 \\ -1 \end{bmatrix}\right)$ and $\left(2, \begin{bmatrix} 1 \\ -2 \end{bmatrix}\right)$

(d) $\left(-2, \begin{bmatrix} 1 \\ -1 \end{bmatrix}\right)$ and $\left(1, \begin{bmatrix} 1 \\ -2 \end{bmatrix}\right)$

81. The system matrix A is

(a) $\begin{bmatrix} 0 & 1 \\ -1 & 1 \end{bmatrix}$ (b) $\begin{bmatrix} 1 & 1 \\ -1 & -2 \end{bmatrix}$

(c) $\begin{bmatrix} 2 & 1 \\ -1 & -1 \end{bmatrix}$ (d) $\begin{bmatrix} 0 & 1 \\ -2 & -3 \end{bmatrix}$

Statement for Linked Answer Questions 82 and 83

An input to a 6-level quantizer has the probability density function $f(x)$ as shown in the figure. Decision boundaries of the quantizer are chosen so as to maximize the entropy of the quantizer output. It is given that 3 consecutive decision boundaries are '–1', '0' and '1'.

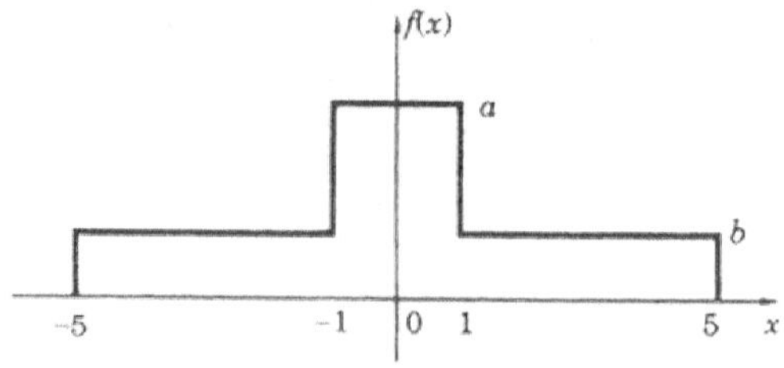

82. The values of a and b are

(a) $a = 1/6$ and $b = 1/12$

(b) $a = 1/5$ and $b = 3/40$

(c) $a = 1/4$ and $b = 1/16$

(d) $a = 1/3$ and $b = 1/24$

83. Assuming that the reconstruction levels of the quantizer are the mid-points of the decision boundaries, the ratio of signal power to quantization noise power is

(a) $\frac{152}{9}$ (b) $\frac{64}{3}$

(c) $\frac{76}{3}$ (d) 28

Statement for Linked Answer Questions 84 and 85

In the Digital-to-Analog converter circuit shown in the figure below, $V_R = 10$ V and R = 10 kΩ.

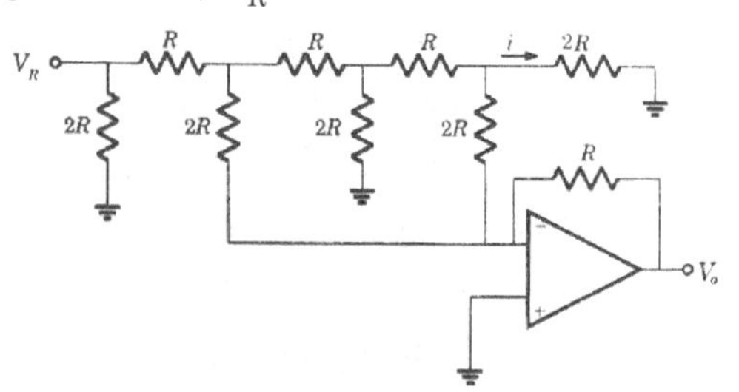

84. The current i is

(a) 31.25 μA (b) 62.5 μA

(c) 125 μA (d) 250 μA

85. The voltage V_o is

(a) – 0.781 V (b) – 1.562 V

(c) – 3.125 V (d) – 6.250 V

ANSWERS

1. (*a*) 2. (*b*) 3. (*c*) 4. (*a*) 5. (*d*) 6. (*a*) 7. (*d*) 8. (*c*) 9. (*d*) 10. (*c*)
11. (*c*) 12. (*a*) 13. (*c*) 14. (*b*) 15. (*d*) 16. (*d*) 17. (*c*) 18. (*b*) 19. (*a*) 20. (*d*)
21. (*c*) 22. (*a*) 23. (*c*) 24. (*d*) 25. (*b*) 26. (*b*) 27. (*a*) 28. (*d*) 29. (*d*) 30. (*a*)
31. (*d*) 32. (*a*) 33. (*b*) 34. (*b*) 35. (*c*) 36. (*c*) 37. (*b*) 38. (*d*) 39. (*d*) 40. (*c*)
41. (*d*) 42. (*b*) 43. (*a*) 44. (*c*) 45. (*b*) 46. (*c*) 47. (*a*) 48. (*a*) 49. (*b*) 50. (*b*)
51. (*b*) 52. (*b*) 53. (*d*) 54. (*a*) 55. (*d*) 56. (*d*) 57. (*a*) 58. (*d*) 59. (*c*) 60. (*a*)
61. (*c*) 62. (*b*) 63. (*a*) 64. (*c*) 65. (*d*) 66. (*a*) 67. (*b*) 68. (*d*) 69. (*c*) 70. (*d*)
71. (*a*) 72. (*b*) 73. (*c*) 74. (*c*) 75. (*a*) 76. (*a*) 77. (*c*) 78. (*b*) 79. (*c*) 80. (*a*)
81. (*d*) 82. (*a*) 83. (*d*) 84. (*b*) 85. (*c*)

EXPLANATIONS

1. $\text{var}[x] = \sigma_x^2 = E[(x-\mu_x)^2]$

where, $\mu_x = E[x]$

μ_x = expected or mean value of X

Defining $$E[X] = \int_{-\infty}^{\infty} x f_X(x)\,dx$$

$$= \int_{-\infty}^{\infty} x\left[\sum_i \rho(x_i)\delta(x-x_i)\right]dx$$

$$= \sum_i x_i \rho(x_i)$$

Variance σ_x^2 is a measure of the spread of the values of X from its mean μ_x.

Using relation, $E[X+Y] = E[X] + E[Y]$

and $E[CX] = C\,E[X]$

on $\text{var}[X] = \sigma_x^2 = E[(X-\mu_x)^2]$

$$\sigma_x^2 = E[X]^2 - \mu_x^2$$

$$= E[X^2] - E^2[X]$$

2. Two points on line are (–1, 0) and (0, 1)

Hence line equation is,

$$y = \left(\frac{y_2-y_1}{x_2-x_1}\right)x + c$$

$$y = x + c$$

$$y = x + 1 \qquad \text{....}(i)$$

$$I = \int_1^2 y\,dx = \int_1^2 (x+1)\,dx = \frac{5}{2} = 2.5$$

(since at $x = 1$, $y = 2$)

3. $$\coth(x) = \frac{\cos hx}{\sin hx} = \frac{e^x + e^{-x}}{e^x - e^{-x}}$$

$$e^x = 1 + \frac{x}{\lfloor 1} + \frac{x^2}{\lfloor 2} + \frac{x^3}{\lfloor 3} + \text{------}$$

$$e^{-x} = 1 - \frac{x}{\lfloor 1} + \frac{x^2}{\lfloor 2} - \frac{x^3}{\lfloor 3} + \text{------}$$

$$\therefore \quad \frac{e^x + e^{-x}}{e^x - e^{-x}} = \frac{1 + \frac{x^2}{\lfloor 2} + \frac{x^4}{\lfloor 4} + \text{----}}{\frac{x}{\lfloor 1} + \frac{x^3}{\lfloor 3} + \frac{x^5}{\lfloor 5} + \text{----}}$$

or $$\coth(x) = \frac{1}{x}$$

(Neglecting x^2 and higher order)

4. $$\lim_{\theta \to 0} \frac{\sin\theta/2}{\theta} = \lim_{0 \to 0} \frac{1}{2}\left[\frac{\sin\theta/2}{\theta/2}\right]$$

$$= \frac{1}{2}\left[\lim_{\theta\to 0}\frac{\sin\theta/2}{\theta/2}\right] = \frac{1}{2}$$

5. $y = f(x)$ $\quad x \leftarrow (0, \infty)$

For strictly bounded,

$$0 \le \left|\lim_{x\to 0} y\right| < \infty$$

or $$0 \le \left|\lim_{\theta\to\pm\infty} y\right| < \infty$$

So, $y = e^{-x^2}$ is strictly bounded

6. $f(x) = f(x_0) + \dfrac{(x-x_0)f'(x_0)}{\lfloor 1} + \dfrac{(x-x_0)^2 f''(x_0)}{\lfloor 2} + - - -$

$= e^{-2} + (x-2)(-e^{-2}) + \dfrac{(x-2)^2}{\lfloor 2}(+e^{-2}) - - - -$

$= e^{-2}\left(3 - x + \dfrac{(x-2)^2}{\lfloor 2} + - - - -\right)$

$= e^{-2}(3-x)$ (Neglecting higher power of x)

7.

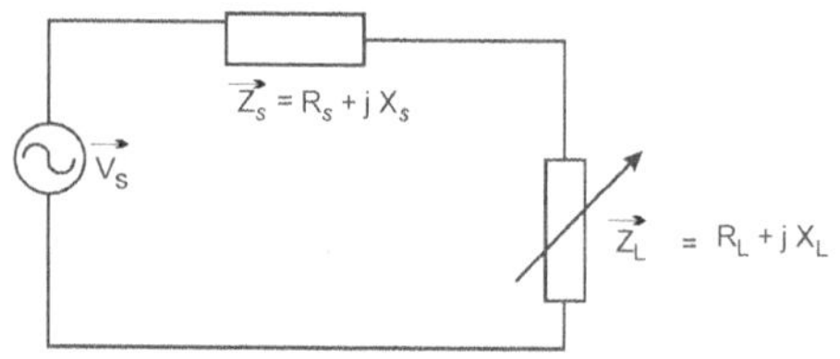

Current through the load,

$$\vec{I}_L = \frac{\vec{V}_S}{(R_S+R_L)+j(X_S+X_L)}$$

$$I_L = |\vec{I}_L| = \frac{V_S}{\sqrt{(R_S+R_L)^2+(X_S+X_L)^2}}$$

Average power to load,

$$P_L = I_L^2 R_L = \frac{V_S^2 R_L}{(R_S+R_L)^2+(X_S+X_L)^2}$$

$$\therefore \quad \frac{\partial P_L}{\partial X_L} = \frac{-2(X_S+X_C)V_S^2 R_L}{\left[(R_S+R_L)^2+(X_S+X_L)^2\right]^2} = 0$$

$\therefore \quad X_L = -X_S$ ..... (*i*)

Now putting $X_L = -X_S$ is P_L equation

$$P_L = \frac{V_S^2 R_L}{(R_S+R_L)^2}$$

But $\dfrac{\partial P_L}{\partial R_L} = 0$

$\therefore \quad R_L = R_S$ ..... (*ii*)

From equations (*i*) and (*ii*), $Z_L = R_S - jX_S$

8.

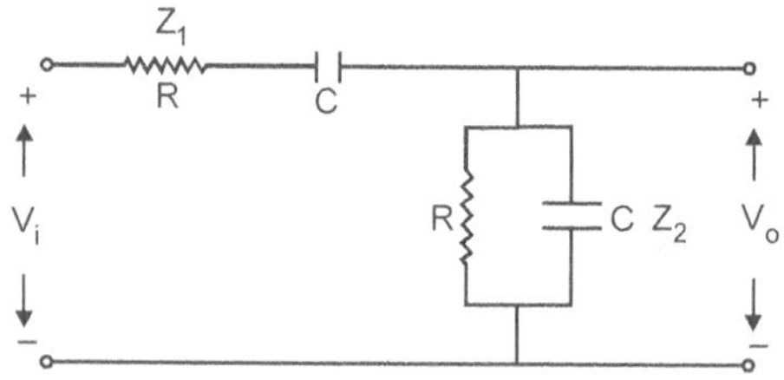

Laplace transform,

$$Z_1 = R + \frac{1}{sC} = \frac{1+sRC}{sC}$$

$$Z_2 = \frac{R \times \frac{1}{sC}}{\frac{1}{sC}+R} = \frac{R}{sCR+1}$$

$$V_0 = \frac{Z_2(s)}{Z_1(s)+Z_2(s)}$$

$$V_i = \frac{sRC \quad V_i}{s^2(RC)^2+3sRC+1}$$

$$G(s) = \frac{V_o}{V_i} = \frac{sRC}{s^2(RC)^2+3sRC+1}$$

$$G(j\omega) = \frac{j\omega RC}{-\omega^2(RC)^2+3j\omega RC+1} \quad s \to j\omega$$

$$= \frac{1}{3-j\left(\dfrac{1-\omega^2R^2C^2}{\omega RC}\right)}$$

For cut-off frequencies, $|G(j\omega)| = \dfrac{1}{\sqrt{2}}$

$$\therefore \quad 9+\left(\frac{1-\omega^2R^2C^2}{\omega RC}\right)^2 = 2$$

or $\quad (\omega RC)^4 + 5(\omega RC)^2 + 1 = 0$

It gives two cut-off frequencies ω_1 and ω_2

Band pass filter

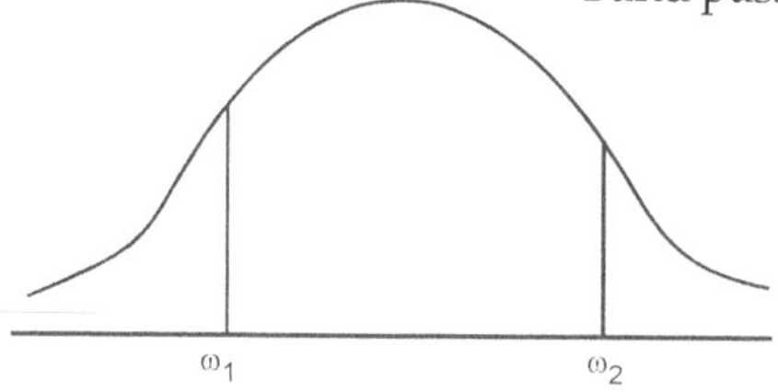

9. We know that at a particular temperature

$np = (n_i)^2$ (constant)

where, n = electron concentration

p = hole concentration

By doping acceptor impurities with a concentration of N_A per cm^2

$p \simeq N_A$

$\therefore \quad N_A \cdot n = n_i^2$

$$n = \frac{n_i^2}{N_A}$$

10. For PN junction

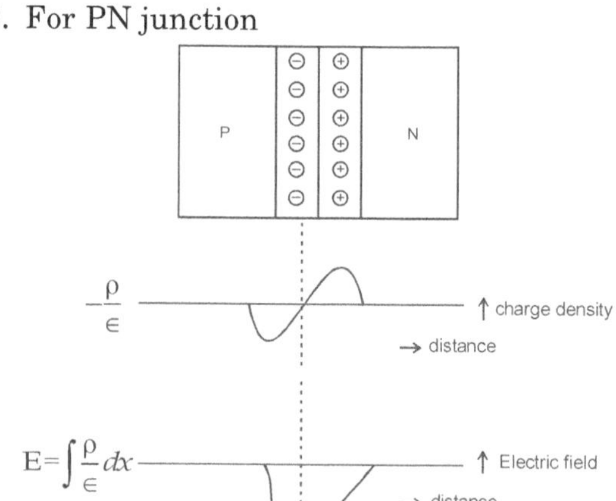

So maximum E field will be at the junction.

11. Full wave rectifier is

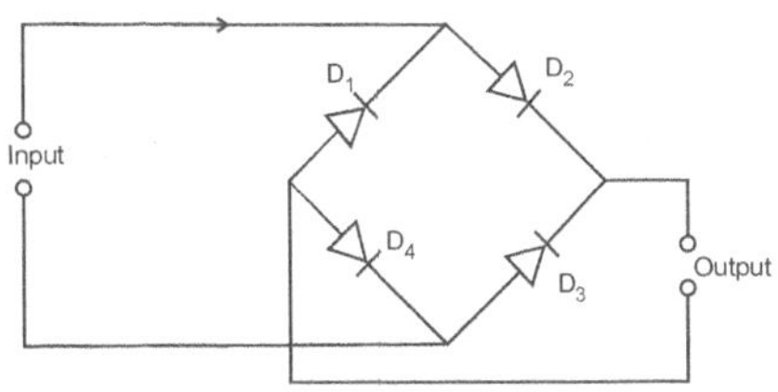

because

(1) during +ve cycle of input

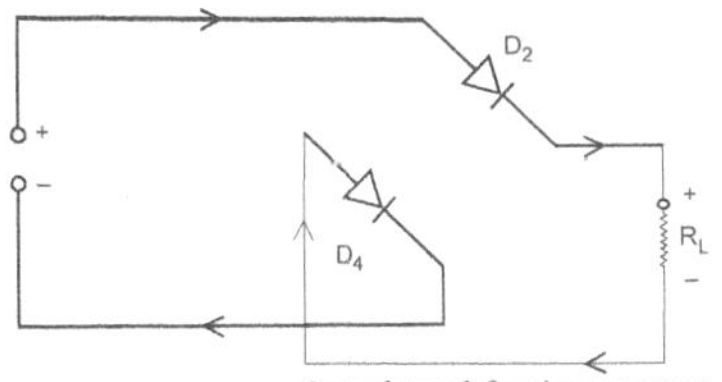

A complete path from input to output

(2) during – ve cycle of input

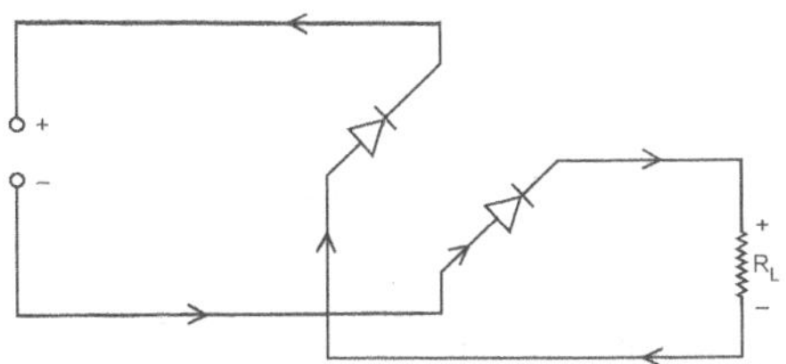

A complete path from input to output

In both cases current passes through R_L is in same direction (rectification)

12.

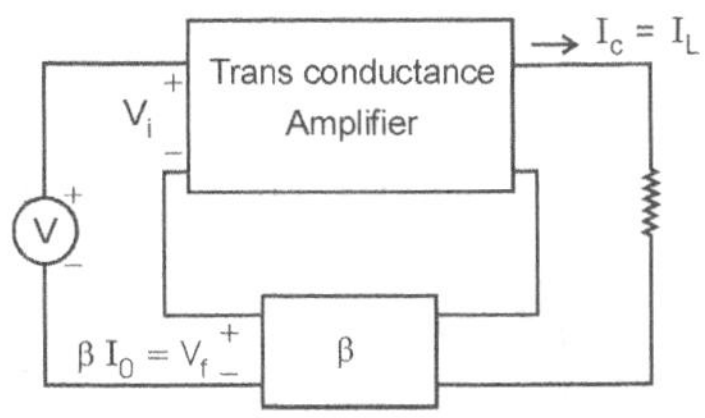

Input resistance, $R_{if} = R_i\,(1 + \beta\,Gm)$

So high input resistance

Output resistance, $R_{of} = Ro\,(1 + \beta\,Gm)$

So, Output resistance also will be high.

13. Given, X =01110 Y = 11001

In 2′s complement format MSB shows the sign of Number (0 for positive and 1 for negative) and negative number is represented in 2′s complement

X = 01110 Y = 11001

= + 14 00110

1

00111

Y = – 7

$\therefore$ X + Y = + 14 – 7 = + 7

S = + 7 in 6 bits representation

000111

It will remain same in 2′s complement because it is positive value

So answer is 000111

14. Y = AB + CD

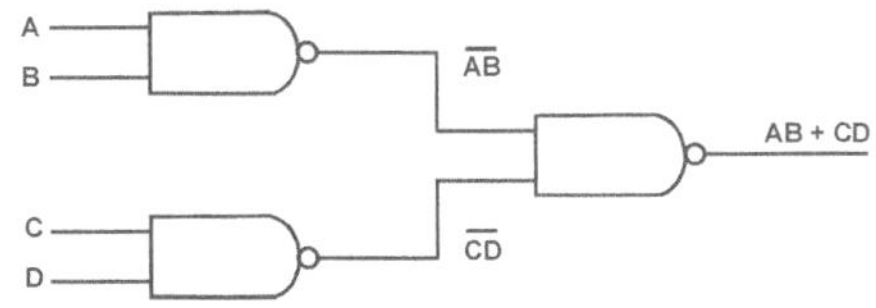

So, there is only 3 NAND gates will be required.

15. A minimum phase system is one which do not have any poles or zeros on right half s-plane where T(s) has on zero, s = 5 at right half s-place, so it is non-minimum phase system

16. Given, $Y(s) = \dfrac{1}{s(s-1)} = \dfrac{-1}{s} + \dfrac{1}{s-1}$

$Y(t) = -1 + e^t$

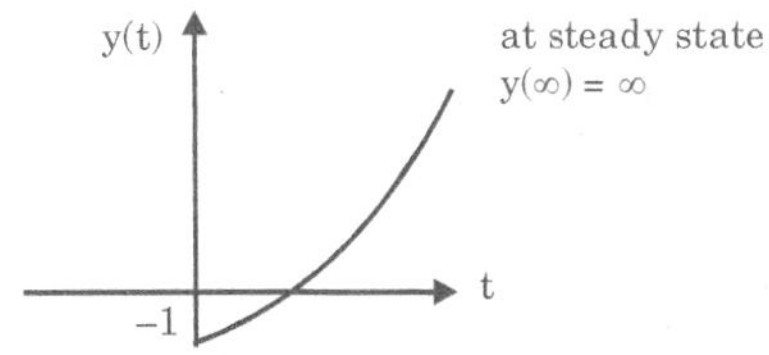

This graph show as $t \to \omega$, $y(t) \to \omega$

17. R(τ) is auto correlation function

$$R(\tau) = \frac{1}{T}\int_{-T/2}^{T/2} V(f)V(t+\tau)\,dt$$

and $$R(-\tau) = \frac{1}{T}\int_{-T/2}^{T/2} V(f)V(t-\tau)\,dt$$

Lett $-\tau = \sigma$

$$\therefore R(-\tau) = \lim_{T\to\infty}\frac{1}{T}\int_{-T/2}^{T/2} V(\tau+\sigma)V(\sigma)\,d\sigma$$

$R(-\tau) = R(\tau)$ even function

18. Power spectral density, $S(\omega) = \lim_{\tau\to\infty}\dfrac{|X_\tau(j\omega)|^2}{\tau}$

So for wide sense stationary random process, $S(f) \geq 0$

19.

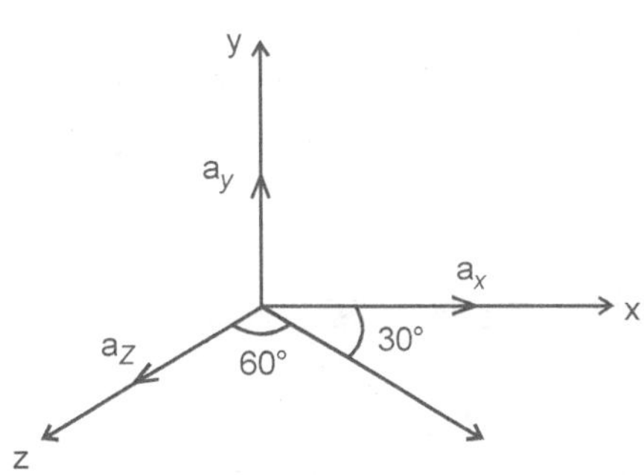

$a_x = a\cos 30 = \dfrac{\sqrt{3}}{2}a$

$a_y = a\cos 90 = 0$

$a_z = a\cos 60 = \dfrac{a}{2}$

$$\therefore \quad \vec{E} = \hat{y}\, E_0\, e^{j(\omega t - \frac{\sqrt{3}}{2}\beta x - \frac{\beta}{2} z)}$$

$$= \hat{y}\, E_0\, e^{j(\omega t - \frac{\sqrt{3}}{2}\left(\frac{2\pi}{\lambda}\right)x - \frac{1}{2}\frac{2\pi}{\lambda} z)}$$

$$= \hat{y}\, E_0\, e^{j(\omega t - \frac{\sqrt{3}\pi}{\lambda} x - \frac{\pi}{\lambda} z)}$$

20. From Ampere's law (or Maxwell's second equation)

$$\vec{\nabla} \quad \vec{H} = \vec{J} + \frac{\partial \vec{D}}{\partial t}$$

From Stoke's theorem, $\iint_S \overline{\nabla} \times \overline{H}.\overline{ds} = \oint_C \vec{H}.\vec{dl}$

$$\therefore \quad \oint_C \vec{H}.\vec{dl} = \iint_S \left(\vec{J} + \frac{\partial \vec{D}}{\partial t}\right).\overline{ds}$$

21. There are M non-zero, orthogonal vectors, so there is required M dimension to represent them.

22. Given, $f(x) = x^2 - x - 2$

$$\frac{df(x)}{dx} = 0$$

$$\Rightarrow \quad 2x - 1 = 0$$

$$\Rightarrow \quad x = \frac{1}{2}$$

$$\therefore \quad \frac{d^2 f(x)}{dx^2} = 2\ (+\text{ve}),$$

So it shows only minima for interval [–4, 4], it contains a maximum value that will be at

$x = -4$ or $x = 4$

$\therefore$ $f(-4) = 18$ and $f(+4) = 10$

So, maximum value is 18

23. Probability of failing in paper 1, P (A) = 0.3

Probability of failing in paper 2, P (B) = 0.2

Probability of failing in paper 1, when student has failed in paper 2, $P\left(\frac{A}{B}\right) = 0.6$

We know that, $P\left(\frac{A}{B}\right) = \frac{P(A \cap B)}{P(B)}$

$$\therefore \quad P(A \cap B) = P(B)\,.\,P\left(\frac{A}{B}\right)$$

$$= 0.6 \times 0.2 = 0.12$$

24.

$$k^2 D^2 y = y - y_2$$

$$\left(D^2 - \frac{1}{k^2}\right) y = \frac{-y_2}{k^2}$$

$$m_1 = \pm \frac{1}{k}$$

C.F. $\quad C_1 e^{x/k} + C_2 e^{-x/k}$

P. F. $\quad \frac{1}{D^2 - (1/k)^2} \left(\frac{-y^2}{k^2}\right)$

Let $\lambda \to \frac{1}{k} \quad \frac{1}{D^2 - \lambda^2}$

$$= \frac{1}{2\lambda}\left[\frac{-1}{\lambda}\left(1 - \frac{D}{\lambda}\right)^{-1} - \frac{1}{\lambda}\left(1 + \frac{D}{\lambda}\right)^{-2}\right]$$

$\therefore$ P.F. $\quad \frac{-1}{2\lambda^2} \times \frac{-y_2}{k^2}$

$$\left[1 - \frac{D}{\lambda} + \frac{D^2}{\lambda^2} - - - + 1 - \frac{D}{\lambda} - \frac{D^2}{\lambda^2} - -\right]$$

$$= \frac{y_2}{k^2 \lambda^2}\left(1 - \frac{D}{\lambda} + - -\right)$$

$$\text{P.F.} = y_2$$

$$y = C_1 e^{x/k} + C_2 e^{-x/k} + y_2$$

At $\quad y = y_1, \quad x = 0$

$\therefore \quad y_1 = C_1 + C_2 + y_2 \quad$ ....(*i*)

At $\quad y = y_2, \ x = \infty$ Hence C_1 must be zero

$\therefore \quad y_1 = C_2 + y_2$

$\Rightarrow \quad C_2 = y_1 - y_2$

$$\therefore \quad y = (y_1 - y_2)\exp\left(-\frac{x}{k}\right) + y_2$$

25.

$$y(t) = x^3 - x^2 + 4x - 4 = 0$$

$$x_0 = 2$$

Next approximation $x_1 = x_0 - \frac{f(x_o)}{f'(x_o)}$

$$x_1 = x_0 - \frac{x_o^3 - x_o^2 + 4x_o - 4}{3x_o^2 - 2x_o + 4}$$

$$x_1 = 2 - \frac{8}{12} = \frac{4}{3}$$

26. Two function will be orthogonal if

$$\int_0^T f_1(t) f_2(t)\, dt = 0$$

For $f_1(t)$ and $f_3(t)$ of given problem

$$\int_0^T f_1(t) f_3(t)\, dt = \int_0^{T/3} f_1(t) f_3(t)\, dt + \int_{T/3}^{2T/3} f_1(t) f_3(t)\, dt + \int_{2T/3}^{T} f_1(t) f_3(t)\, dt$$

$$= \int_0^{T/3} (2)(1)\, dt + \int_{T/3}^{2T/3} (0)(-3)\, dt + \int_{2T/3}^{T} (-2)(1)\, dt$$

$$= \frac{2T}{3} + 0 - \frac{2T}{3} = 0$$

27. $\oint_D \frac{1}{(s^2 - 1)} ds = 2\pi j\,(\text{sum of Residue})$

Singular points are $s = \pm 1$

Only $s = +1$ lies inside the given contour

Residue $s = +1 = \lim_{S \to 1} (s - 1) f(s)$

$$= \lim_{S \to 1} (s - 1) \frac{1}{(s^2 - 1)} = \frac{1}{2}$$

$$= 2\pi j \left(\frac{1}{2}\right) = \pi j$$

28. $Q = \frac{\omega_o}{B.\omega}$

$\therefore \quad B_1 = \frac{\omega_o}{Q_1}$

and $\quad B_2 = \frac{\omega_o}{Q_2}$

$Q_1 = \frac{\omega L_1}{R}$

and $\quad Q_2 = \frac{\omega L_2}{R} = \frac{\omega L_1}{4R} = \frac{Q_1}{4}$

$\therefore \quad B_1 = \frac{\omega_o}{Q_1}$

and $\quad B_2 = \frac{4\omega_o}{Q_1}$

$\therefore \quad \frac{B_1}{B_2} = \frac{1}{4}$

29.

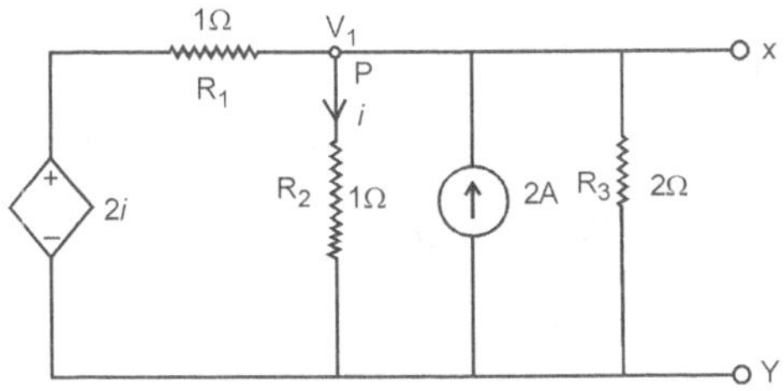

Apply Nodal equation at point P

$$\frac{V_1 - 2i}{R_1} + \frac{V_1 - 0}{R_2} + \frac{V_1 - 0}{R_3} = 2A$$

$$\Rightarrow \quad V_1 - 2i + V_1 + \frac{V_1}{2} = 2$$

But from diagram, $i = \frac{V_1}{R_2} = \frac{V_1}{1} = V_1$

$$\therefore \quad 2V_1 + \frac{V_1}{2} - 2V_1 = 2$$

or $\quad V_1 = 4$ volts

$\therefore \quad V_{TH} = 4$ volts

Determination of $R_{Th} = \frac{V_{Th}}{I_{Sc}}$

Where, I_{sc} = short circuit current throught X – Y

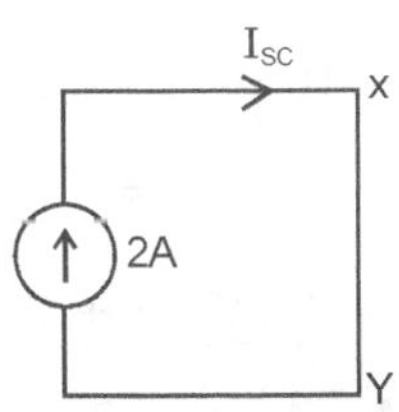

Short circuit at X and Y ; short the R_2 and R_3. Because of zero current at R_2, dependent source $2i$ also zero, and R_1 also short circuit.

$I_{sc} = 2A$

$\therefore \quad R_{Th} = \frac{4}{2} = 2\ \Omega$

30. Thevenin equivalent circuit across capacitor

$V_{th} = \frac{20}{40} \times 10 = 5$ volts

$R_{th} = 20\text{ k} \parallel 20\text{ k} = 10\text{ k}\Omega$

$5 = Ri + \frac{1}{C}\int i\,dt$

$\therefore \quad R\frac{di}{dt} + \frac{i}{C} = 0$

$\Rightarrow \quad \frac{di}{i} = -\frac{dt}{RC}$

$\int_o^t \frac{di}{i} = -\int_o^t \frac{dt}{RC}$

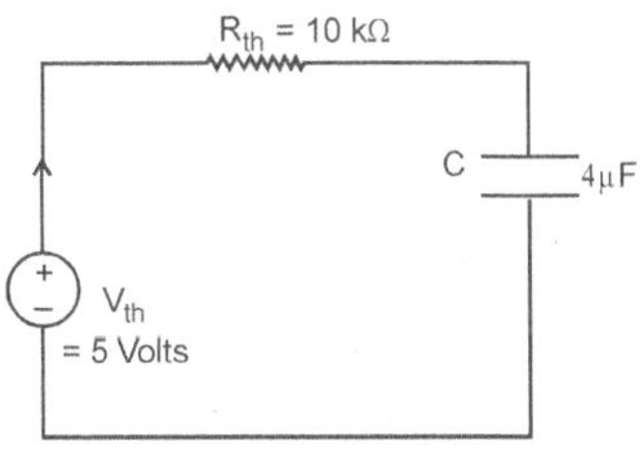

$\Rightarrow \quad \log(t) - \log i_o = -\frac{t}{RC}$

$\Rightarrow \quad i(t) = i_o\, e^{-t/RC}$

So, when $t = 0$ and $V_c = 0$

$\therefore \quad i_o = \frac{V_{th}}{R_{th}} = \frac{5}{10k} = 0.5$ mA

and $\frac{1}{RC} = \frac{1}{10k} \times \frac{1}{4\mu F} = \frac{1000}{10 \times 4} = 25$

So, current equation is,

$i(t) = 0.50\, e^{-25t}$ mA

31. $Z_{AB} = Z_1 \parallel Z_2 = (5 - 3j) \parallel (5 + 3j)$

$Z_{AB} = \frac{34}{10}$

$V_{AB} = I.\, Z_{AB}$

$= (5 \angle 30°)\left(\frac{34}{10} \angle 0°\right)$

$= 17 \angle 30°$ volts

32. Depletion capacitance,

$C_d \propto V^{-1/2}$

$\frac{\varepsilon A}{d} \propto V^{-1/2}$

$\therefore \quad d \propto V^{+1/2}$

$\therefore \quad \frac{d_2}{d_1} = \left(\frac{V_2}{V_1}\right)^{1/2}$

$\Rightarrow \quad d_2 = d_1\left(\frac{V_2}{V_1}\right)^{1/2}$

$= (2\mu m)\left(\frac{7.2 + 0.8}{1.2 + 0.8}\right)^{1/2} = 4\mu m$

33. A. Zener diode is used in reverse bias to give fixed zener voltage across it.

B. Solar cell operates in forward bias.

C. Laser diode operate in very high voltage forward bais to gives population insersion

D. Photo diode operate in reverse bias in avalanche region.

34. Given, $\beta = 50$

$\therefore \quad \alpha = \frac{\beta}{\beta+1} = \frac{50}{51}$

We know that, $\alpha = \beta^* \gamma$

where, β^* = base transport factor

γ = emitter injection efficiency (= 0.995)

$\therefore \quad \beta^* = \frac{\alpha}{\gamma} = \left(\frac{50}{51}\right)\frac{1}{0.995} = 0.9853 \approx 0.985$

35. BJT — Early effect

MOS capacitor — Flat-band voltage

LASER diode — Population inversion

JFET — Pinch-off voltage

36. In Op-Amp

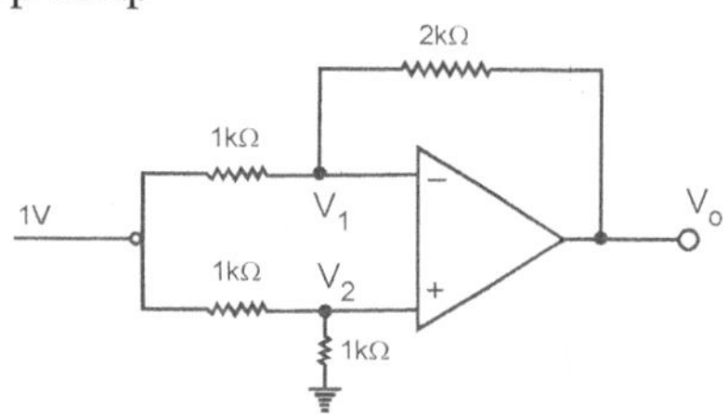

$V_1 \simeq V_2$

$V_2 = 1\left(\frac{1}{1+1}\right) = \frac{1}{2} = 0.5$ volts

Apply KCL at V_1,

$\frac{V_0 - V_1}{2k\Omega} = \frac{V_1 - 1}{1\,k\Omega}$

$\Rightarrow \quad 3V_1 - 2 = V_0$

$\Rightarrow \quad 3 \times 0.5 - 2 = V_0$

$\therefore \quad V_0 = -0.5$ volts

37. For saturation, $V_{BE} > 0.7$ volts

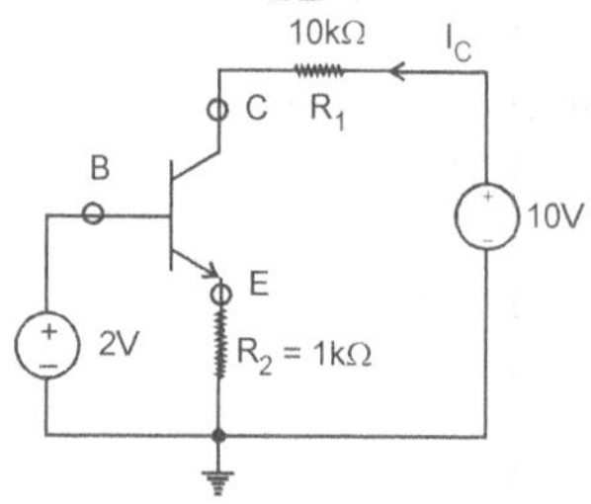

In saturation condition,

$I_c = \frac{V_{cc}}{R_1 + R_2} = \frac{10}{10+1} = \frac{10}{11}$

Apply KVL in base circuit

$V_{BB} = V_{BE} + I_c R_2$

$\Rightarrow \quad 2 = V_{BE} + \left(\frac{10}{11}\right)$

$\Rightarrow \quad V_{BE} = 1.09$

$\Rightarrow \quad V_{BE} > 0.7$

So, operate in saturation region.

38.

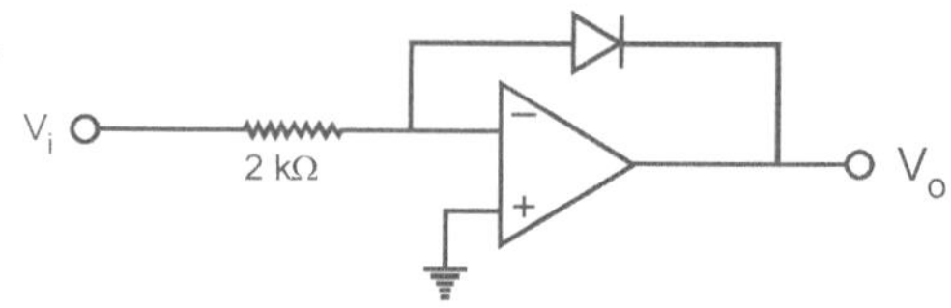

This is a circuit diagram of log amplifier

$V_0 = -V_T \ln V_i + C$

$V_{01} = -V_T \ln V_{i1} + C$

$\Rightarrow \quad V_{01} = -V_T \ln 2 + C$

and $\quad V_{02} = -V_T \ln V_{i2} + C$

$\Rightarrow \quad V_{02} = -V_T \ln 4 + C$

$\therefore \quad V_{01} - V_{02} = -V_T \ln 2 + V_T \ln 4$

$= V_T \ln 2$

39. Assuiming both PMOS and NMOS are in saturation

Then, $I_{D1} = \frac{\mu_{p_1} C_0 \times W}{2L}(V_{GS1} - V_{Tn})^2$ [For NMOS transistor]

$= \frac{40}{2}(2.5-1)^2 = 20 \times 1.5^2 = 45\,\mu A$

Now, For PMOS transistor

$I_{D2} = \frac{\mu_p C_0 \times W}{2L}(V_{GS2} - V_{TP})^2$

$= \frac{40}{2}(5 - 2.5 - 1)^2 = 45\mu A$

As $I_{D1} = I_{D2}$, both transistors are in saturation and

$I = I_{D1} = I_{D2} = 45 \quad A$

40.

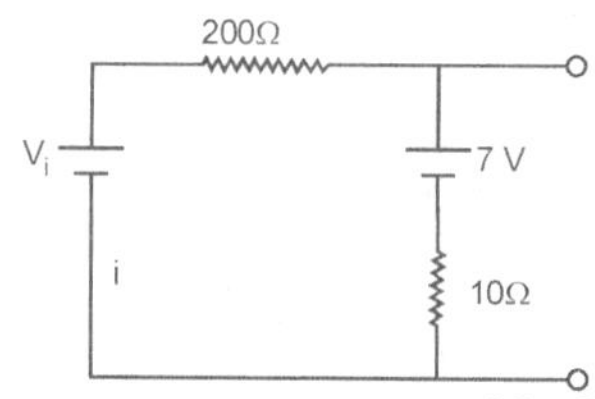

When $V_i = 10$ V, $i = \frac{10-7}{210} = \frac{3}{210}$ amp

$V_{01} = 7 + 10i = 7 + 10 \quad \frac{3}{210} = 7.14$ volts

When $V_i = 16$ V, $i = \frac{9}{210}$ amp

$V_{02} = 7 + 10\left(\frac{9}{210}\right) = 7.43$ volts

41. $Y = \overline{A}\,\overline{B}\,\overline{C}\,D + \overline{A}\,B\,C\,\overline{D} + A\overline{B}\,\overline{C}\,D + A\,B\,\overline{C}\,\overline{D}$

$= (\overline{A}\,\overline{B}\,\overline{C}\,D + A\overline{B}\,\overline{C}\,D) + \overline{A}\,B\,C\,\overline{D} + A\,B\,\overline{C}\,\overline{D}$

$= \overline{B}\,\overline{C}\,D\,(\overline{A} + A) + \overline{A}\,B\,C\,\overline{D} + A\,B\,\overline{C}\,\overline{D}$

$= \overline{B}\,\overline{C}\,D + \overline{A}\,B\,C\,\overline{D} + A\,B\,\overline{C}\,\overline{D}$

42.

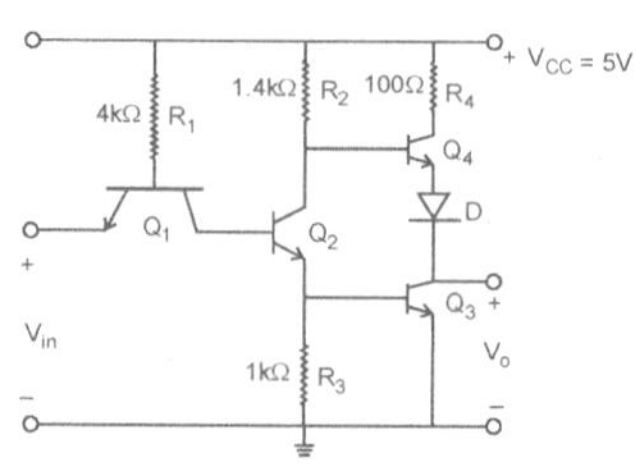

When V_{in} is at high voltage (2V – 5V), BE junction of Q_1 becomes reverse biased and current flows through R_1 and BC junction of Q_1 into the base of Q_2. So, Q_1 operate in reverse active.

Because of base current of Q_2 it drives into saturation because

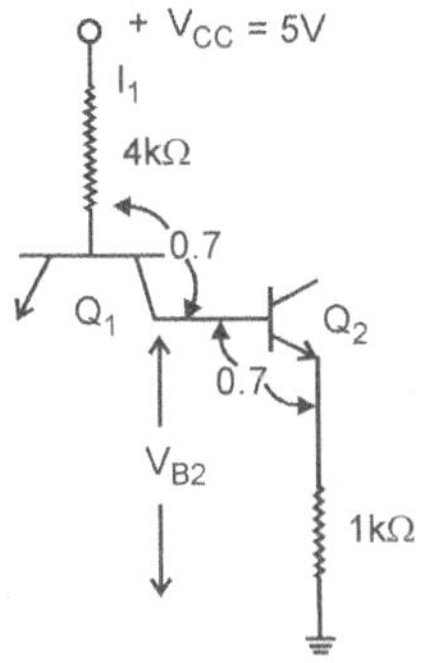

$$I_1 = \frac{5-0.7-0.7}{4+1} = 0.72 \text{ mA}$$

and $V_{B2} = 5 - 0.7 - I \,.\, 4k$

$= 5 - 0.7 - 0.72 \,.\, 4$

$= 1.42$ volts

$\therefore V_{B2} > 0.7$ volts so Q_2 operate in saturation region.

Because of saturation of Q_2, a voltage drop across R_3

$$I_2 = \frac{V_{CC}}{R_2+R_3} = \frac{5}{1.4+1} = \frac{5}{2.4} \text{ mA} = 2.03 \text{ mA}$$

$V_{B3} = (I_1 + I_3)R_3 = (0.72 + 2.03)\ 1k = 2.75$ volts

Since $V_{B3} > 0.7$ volts, so Q_3 also operates in saturation region.

Q_3 and Q_4 together form a totem pole Output, one transistor operate at a time, so Q_4 will be in cut off.

43.

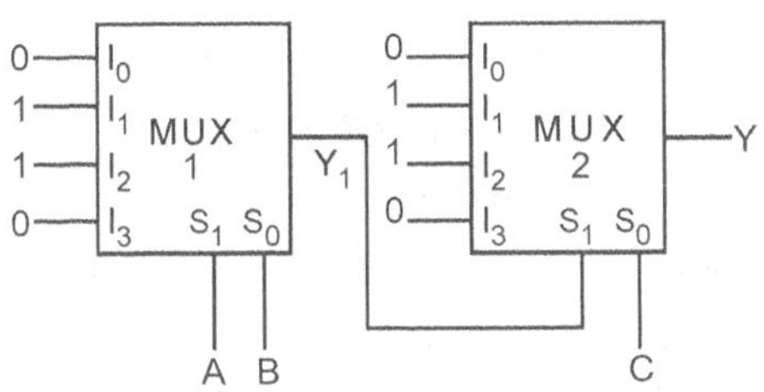

For MUX – 1

$A(S_1)$	$B(S_0)$	Y_1
0	0	0
0	1	1
1	0	1
1	1	0

$\therefore \quad Y_1 = \overline{A}B + A\overline{B}$

Similarly for MUX - 2

$Y_1(S_1)$	$C(S_0)$	Y
0	0	0
0	1	1
1	0	1
1	1	0

$Y = \overline{Y_1}C + Y_1\overline{C}$

$= \overline{\left(\overline{A}B + A\overline{B}\right)}\ C + \left(\overline{A}B + A\overline{B}\right) . \overline{C}$

$\therefore \quad Y = ABC + \overline{A}\,\overline{B}C + \overline{A}B\overline{C} + A\overline{B}\,\overline{C}$

44.

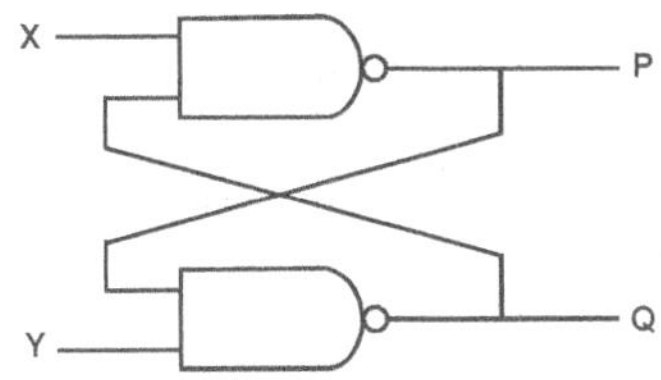

$P_{t+1} = \overline{X.Q_t} = \overline{X} + \overline{Q_t}$

$Q_{t+1} = \overline{Y.P_t} = \overline{Y} + \overline{P_t}$

X	Y	P_t	Q_t	P_{t+1}	Q_{t+1}
0	1	d	d	1	0
0	0	1	0	1	1
1	1	1	1	0	1
(Not stable)				1	0

Last condition is not stable, it continuously changes its state. Because changes of output again change its states.

45.

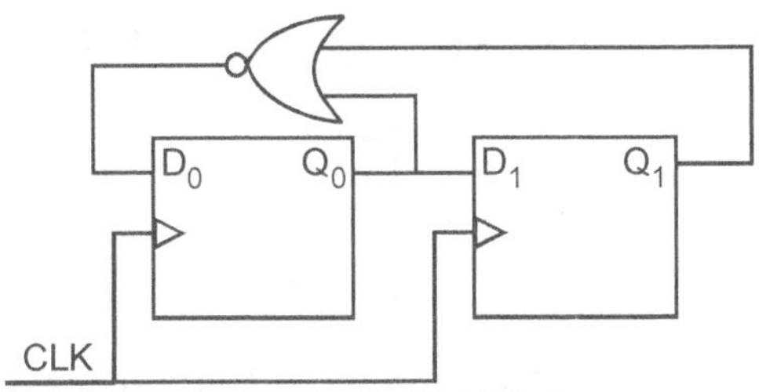

From diagram, $D_0 = \overline{Q_0 + Q_1}$

$= \overline{Q_0} \,.\, \overline{Q_1}$

and $D_1 = Q_0$

$Q_1(t)$	$Q_0(t)$	$Q_1(t+1) = D_1$ $= Q_0(t)$	$Q_0(t+1) = D_0$ $= \overline{Q_0 + Q_1}$
0	0	0	1
0	1	1	0
1	0	0	0
0	0	0	1
⋮	⋮	⋮	⋮

46. 8255 chip will select in I/O mapped when

$A_7\ A_6\ A_5\ A_4\ A_3$

1 1 1 1 1

there is A_2 in don't care condition

A_7	A_6	A_5	A_4	A_3	A_2	A_1	A_0	
1	1	1	1	1	0	0	0	= F8H
1	1	1	1	1	1	1	1	= FFH

So, range of chip selection is F8H – FFH

47. $f(t) = e^{-t}u(t)$

$$F(s) = L\,[f(t)] = L\,[e^{-t}\,u(t)] = \frac{1}{s+1}$$

In frequency domain, $s = j\omega$

$$F(j\omega) = \frac{1}{(1+j\omega)} = \frac{-1}{\sqrt{1+\omega^2}} \angle \tan^{-1}\omega$$

For 3 dB bandwidth,

$$\frac{1}{\sqrt{1+\omega^2}} = \frac{1}{\sqrt{2}}$$

$\therefore \quad \omega = +1$

$\omega = 2\pi f_c$

$\Rightarrow \quad f_c = \frac{1}{2\pi}$ Hz

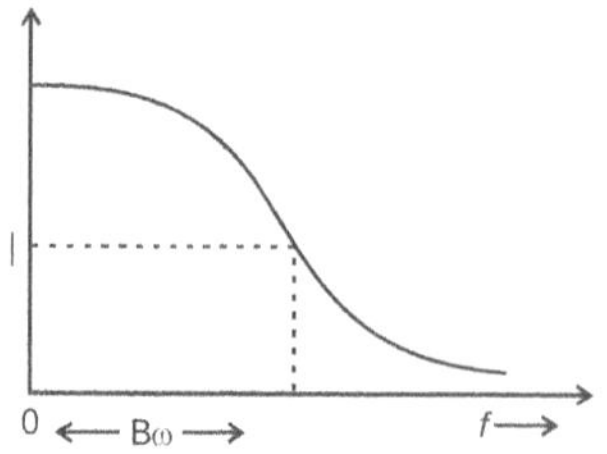

48. Hilbert transformer

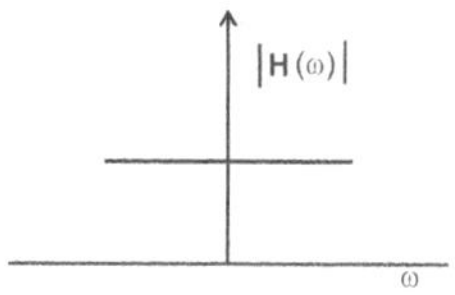

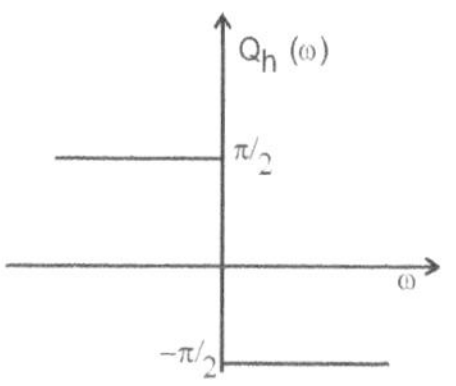

$$H(\omega) = -j\text{sgn}(\omega)$$

$$\begin{cases} -j = 1.e^{-\pi/2J} & \omega > 0 \\ +j = 1.e^{+\pi/2J} & \omega < 0 \end{cases}$$

For linear system, $Q_h(\omega) = -\omega t_d$

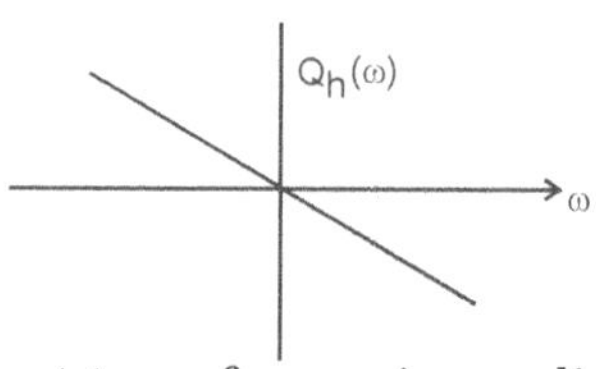

So, Hilbert transformer is non-linear and time invariant system.

49. $H(f) = \frac{5}{1+j\,10\pi f}$ $\Big|$ $\omega = 2\pi f$ and $s = j\omega$

$$\therefore \quad H(s) = \frac{5}{1+5s} = \frac{1}{s+0.2}$$

For step input,

$$y(s) = \frac{1}{s}.\ H(s) = \frac{1}{s(s+0.2)} = 5\left[\frac{1}{s} - \frac{1}{s+0.2}\right]$$

$$y(t) = \mathcal{L}^{-1}[y(s)] = 5\,[1 - e^{-0.2t}]$$

$$y(t) = 5[1 - e^{-t/5}]$$

50. For discrete time Fourier transform (DTFT)

$$x(n) = \sum_{k=n} \frac{1}{n} X(e^{(jk\omega_0)})\ e^{jk\omega_0 n}$$

$$\text{When } \lim N \to \infty, x[n] = \frac{1}{2\pi}\int_{-\pi}^{\pi} X(e^{j\omega})\ e^{j\omega n}\, d\omega$$

Putting $n = 0$,

$$x[0] = \frac{1}{2\pi}\int_{-\pi}^{\pi} X(e^{j\omega})\ e^{j\omega\, 0}\, d\omega$$

$$\therefore \int_{-\pi}^{\pi} X(e^{j\omega})\ d\omega = 2\pi\, x[0] = 2\pi \times 5 = 10\pi$$

51. For left- handed signal,

$$X[Z] = -\sum_{n=-\infty}^{\infty} a^n u(-n-1) z^{-n} = -\sum_{n=-\infty}^{-1} a^n z^{-n}$$

$$= -\sum_{n=1}^{\infty} (a^{-1}z)^n = 1 - \sum_{n=0}^{\infty} (a^{-1}z)^n$$

as $\quad \sum_{n=0}^{\infty} (a^{-1}z)^n = 1 + a^{-1}z + (a^{-1}z)^2 + \ldots$

$$= \frac{1}{1-a^{-1}z}$$

then, $X[Z] = 1 - \frac{1}{1-a^{-1}z} = \frac{1}{1-az^{-1}}$

and ROC is given by $|Z| < |a|$

Here $a = 2$,

then, ROC $|Z| < 2$ includies unit circle

Hence, $x[n] = -0.5[Z]^n u[-n-1]$

and $x[0] = 0$

52. $G = G_1G_2 = (K_P + K_D s)\left(\frac{100}{s(s+10)}\right)$

$$K_v = \lim_{s\to 0} s\,G(s) = \lim_{s\to 0} \frac{s[K_P + K_D s]100}{s(s+10)}$$

$\therefore \quad 1000 = 10\,K_P$

$\Rightarrow \quad K_P = 100$

$$T(s) = \frac{G(s)}{1+G(s)} = \frac{(K_P + K_D s)\left(\frac{100}{s(s+10)}\right)}{1 + (K_P + K_D s)\left(\frac{100}{s(s+10)}\right)}$$

$$= \frac{100(K_P + K_D s)}{s^2 + 10s + 100\,K_D s + 100 K_P}$$

$$= \frac{100(K_P + K_D s)}{s^2 + s\,(10 + 100 K_D) + 100\,K_P}$$

Compare characteristic equation with

$$s^2 + 2\xi\,\omega_n s + \omega_n^2$$

$$100\,K_D + 10 = 2\xi\omega_n \quad \text{and} \quad 100\,K_P = \omega_n^2$$

$\Rightarrow \quad \omega_n = \sqrt{100 \times 100} = 100$

$\Rightarrow \quad 100\,K_D + 10 = 2 \times 0.5 \times 100$

$\Rightarrow \quad K_D = 0.90$

53. Given, $\frac{C(s)}{R(s)} = T(s) = \frac{5}{(s+5)(s^2+s+1)}$,

$$r_{w_n} = 1, w_n = 1, P = 5$$

Unit step-reform to this transfer function is,

$$C_P(t) = 1 - k_1 e^{-5t} + k_2 e^{-t}\sin\left[\frac{\sqrt{3}}{2}t - \phi\right] \quad \ldots(A)$$

where $\phi = \tan^{-1}(-\sqrt{3}) + \tan^{-1}\left(\frac{\frac{\sqrt{3}}{2}}{4}\right)$

As here, $P >> rw_n$

The term k_1e^{5t} will loose its dominance more than the term given by k_2e^{-t}.....

i.e. $\underset{P\to 0}{lt}\ k_1 = 0$ and $\underset{P\to\infty}{lt}\ k_2 = \frac{2}{\sqrt{3}}$

Hence as the pole moves away from the origin, the reforme $C_p(t)$ is dominated by the complex poles at $-rw_n \pm jw_n\sqrt{1-r^2}$

Hence, the response of the system for $P >> rw_n$ is given by

$$T(s) = \frac{1}{s^2+s+1}$$

54. Given, $G(s) = \frac{1}{s^2-1} = \frac{1}{(s-1)(s+1)}$

It is an unstable system because $s = 1$ pole lies on right half s-plane.

To stabilise the unity gain must compensated by lead compensator that eliminate this pole.

From option $\frac{10(s-1)}{s+2}$

$$G_T(s) = \frac{1}{(s-1)(s+1)}\ \frac{10(s-1)}{s+2}$$

$$= \frac{10}{(s+1)(s+2)}$$ is a stable function.

55. Given, $G(s) = \frac{K}{s(s^2+7s+12)}$

For unity feedback, $\left|\frac{K}{s(s^2+7s+12)}\right| = 1$

$\Rightarrow\quad s = -1 + j1$

and $K = (-1+j)\,[(-1+j)^2 + 7(-1+j) + 12]$

$= (-1+j)\,[1 - 1 - 2j - 7 + 7j + 12]$

$= (-1+j)\,(5+5j)$

$= -5(1-j)\,(1+j)$

$= -5\,(1+1)$

$= -10$

$\therefore\quad K = |-10| = 10$

56. From given figure, corner frequencies at

$\omega_1 = 0,\quad \omega_2 = 1,\quad \omega_3 = 20$

$$\therefore\quad G(s) = \frac{k}{s(1+s\tau_1)(1+s\tau_2)}$$

$$\therefore\quad \tau_1 = \frac{1}{\omega_2} = 1,\qquad \tau_2 = \frac{1}{\omega_3} = 0.05$$

$$\therefore\quad G(s) = \frac{k}{s(s+1)(1+0.05\,s)}$$

From graph $|G\,z(j\omega)|\,dB = 60$ at $\omega = 0.1$

$$20\log_{10}|G(j\omega)| = 60$$

$$\Rightarrow\quad 20\log_{10}\left|\frac{k}{j\omega(1+j\omega)(1+0.005j\omega)}\right| = 60$$

$\Rightarrow\quad k = 100$

$$\therefore\quad G(s) = \frac{100}{s(s+1)(1+0.05\,s)}$$

57. From equation, $\frac{d\omega}{dt} = -\omega + i_a$ ....(1)

$\frac{di_a}{dt} = -\omega - 10\,i_a + 10u$ ....(2)

Taking laplace of (1)

$s\omega(s) = -\omega(s) + i_a(s)$

$\Rightarrow\quad (s+1)\,\omega(s) = i_a(s)$ ....(3)

Taking laplace of (2)

$s\,i_a(s) = -\omega(s) - 10\,i_a(s) + 10U(s)$

$\Rightarrow\quad (s+10)\,i_a(s) = -\omega(s) + 10U(s)$

Putting value of $i_a(s)$ from (3)

$(s+10)\,(s+1)\,\omega(s) = -\omega(s) + 10U(s)$

$$\frac{\omega(s)}{U(s)} = \frac{10}{s^2+11s+11}$$

58. For slope overload not to occur,

$$\frac{\Delta}{T_s} \geq \frac{d}{dt}m(t)$$

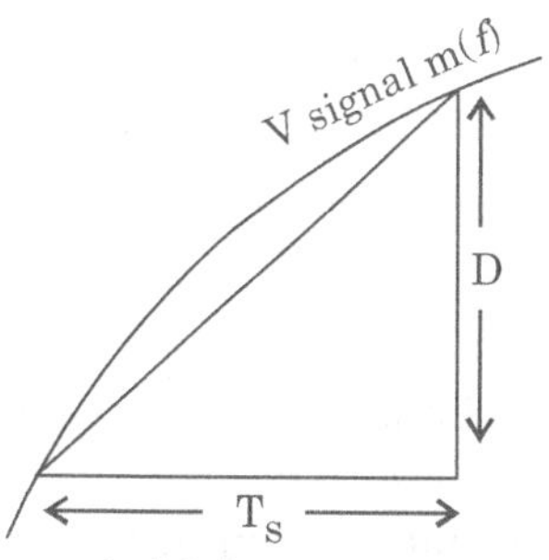

if $m(t)$ rate is high, Δ must increase

59. Given, $p(t) = \frac{\sin 4\pi Wt}{4\pi W\,t(1-16\,W^2t^2)}$

$$\lim_{t\to 1/4W} p(t) = \lim_{t\to 1/4W} \frac{\sin 4\pi Wt}{4\pi W\,t(1-16\,W^2t^2)}$$

$$= \lim_{t\to 1/4W} \frac{\frac{d}{dt}(\sin 4\pi Wt)}{\frac{d}{dt}[4\pi Wt\,(1-16W^2t^2)]}$$

$$= \lim_{t\to 1/4W} \frac{4\pi W\cos 4\pi Wt}{4\pi W\,(1-16\,W^2t^2) - 32W^2t(4\pi Wt)}$$

$$= \lim_{t\to 1/4W} \frac{4\pi W\cos 4\pi Wt}{4\pi W - 3\times 64\,W^3t^2}$$

$$= \frac{1}{2} = 0.5$$

60. $y(t) = m(t)\cos(2\pi Bt) + m_h(t)\sin(2\pi Bt)$

$m(t) = a\cos\omega_m t$

$m_h(t) = a\sin\omega_m t$

[Since $\frac{\pi}{2}$ shift by Hilbert transformer]

$y(t) = a\,[\cos \omega_m t \cos 2\pi Bt + \sin \omega_m t \sin 2\pi Bt]$
$= a \cos [(\omega_m - 2\pi Bt)]$

This is the equation of LSB.

$\omega_m = 2\pi B$

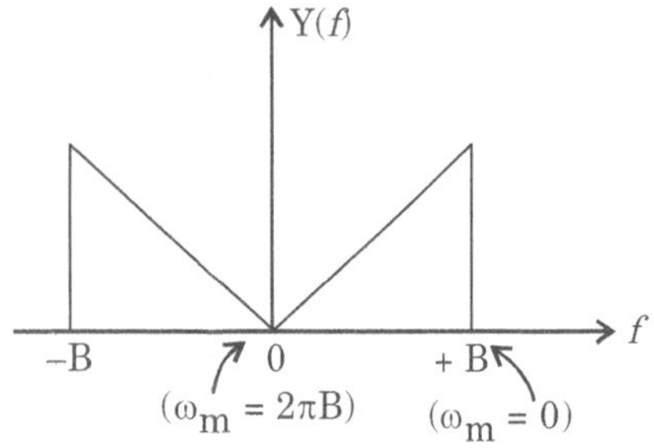

61. Atmost one-bit error

$= p(1 \text{ bit error}) + p(\text{no bit error})$
$= n_{C_1} \times p^1 \times (1-p)^{n-1} + (n_{C_0}\ p^o\ (1-p)^n)$
$= np(1-p)^{n-1} + (1-p)^n$

62. Total available band width = 5 MHz

$\Rightarrow$ Frequency reuse factor $\frac{1}{5}$, so five cell repeat pattern so available band width for each cell

$$(B.W)_{cell} = \frac{(BW)_{total}}{5} = \frac{5}{5}\text{MHz} = 1 \text{ MHz}$$

$\Rightarrow \quad (B.W)_{channel} = 200 \text{ kHz}$

Total number of channel in each cell,

$$(N)_{cell} = \frac{(BW)_{cell}}{(B.W)_{channel}} = \frac{1M}{200k} = 5$$

There is 8 channel coexist in same channel band width using TDMA

So, total number of simultaneous channel that conexist = 5 × 8 = 40

63. In Direct Sequence CDMA system,

$$\text{Process Gain} = G_P = \frac{f_{\text{chip rate}}}{f_{\text{data rate}}}$$

Given, $G_p \min = 100$

$G_p \geq 100$

$$\therefore \quad G_p = \frac{f_{\text{chip rate}}}{f_{\text{data rate}}} \geq 100$$

$$\Rightarrow \quad \frac{f_{\text{chip rate}}}{100} \geq f_{\text{data rate}}$$

$$\Rightarrow \quad \frac{1.2288 \times 10^6}{100} \geq f_{\text{data rate}}$$

So, $f_{\text{data rate}}$ must be less than $= 12.288 \times 10^3$ bit/sec

64. Given, $f_0 = 30$ GHz

$$\therefore \quad \lambda_0 = \frac{c}{f_o} = \frac{3 \times 10^8}{30 \times 10^9} = 1 \text{ cm}$$

$$\frac{1}{\lambda_c^2} = \left(\frac{m}{2a}\right)^2 + \left(\frac{n}{2b}\right)^2$$

For TE_{20} mode, $m = 2$ and $n = 0$

$\therefore \quad \lambda_c = a = 3$ cm

$$\eta = \frac{\eta_0}{\sqrt{1-\left(\frac{\lambda_0}{\lambda_c}\right)^2}} = \frac{377}{\sqrt{1-\left(\frac{1}{3}\right)^2}} = 400\ \Omega$$

65. Given, $\vec{H} = H_x \hat{x} + H_y \hat{y}$

So, plane wave propagating in free space will be more in z direction

We know that, pointing vector,

$$\vec{P} = \vec{E} \times \vec{H} = [\eta_0 H^2]\ \hat{k}$$

That shows the power flow in z-direction

Instantaneous power,

$$P = |\vec{P}| = \eta_0 H^2$$

Average power over an interval $(0, 2\pi)$ will be

$$P_{av} = \frac{1}{2\pi}\int_0^{2\pi} P d\theta = \frac{1}{T}\int_0^T P dt = \frac{1}{T}\int_0^T \eta H^2 dt$$

$$= \frac{\eta_0}{T}\int_0^T (H_x^2 + H_y^2)\, dt$$

$$= \frac{\eta_0}{T}\int_0^T \left[\frac{25 \times 3}{\eta_0^2}\cos^2(\omega t - \beta z) + \frac{25}{\eta_0^2}\sin^2(\omega t - \beta z + \frac{\pi}{2})\right] dt$$

$$= \frac{100}{T\eta_0}\int_0^T \cos^2(\omega t - \beta z)\ dt$$

$$= \frac{100}{T\eta_0}\left(\frac{T}{2}\right) = \left(\frac{50}{\eta_0}\right)$$

66. For TE_{mn} mode, magnetic field in z - direction is expened as,

$$\bar{H}_z \simeq H_{20} \cos\frac{m\pi x}{a} \cos\frac{n\pi y}{b} \cdot e^{-j\beta mnZ} \qquad ...(1)$$

then, $\bar{E}_y = J\frac{\omega\mu}{k^2}\frac{\partial \bar{H}2}{\partial x}$

$$= \frac{-j\left(\frac{m\pi\omega\mu}{axz}H_{20}\right)\sin\left(\frac{m\pi x}{a}\right)}{}$$

constant term $\quad \cos\left(\frac{n\pi y}{b}\right) \cdot e^{-j\beta_{mn}Z}$

then for TE_{20} mode $m = 2,\ n = 0$

We have

$$\bar{E}_y = -j\left(\frac{2\pi\omega\mu}{akz}H_0\right)\sin\left(\frac{2\pi x}{a}\right) \cdot e^{-j\beta_{mn}Z}$$

Comparing with given equation for $\bar{E}_y$, it can be inferred that the mode of propagation is TE_{20} mode.

67. S – matrix for 2 – port network

$$\begin{bmatrix} b_1 \\ b_2 \end{bmatrix} = \begin{bmatrix} S_{11} & S_{12} \\ S_{21} & S_{22} \end{bmatrix} \begin{bmatrix} a_1 \\ a_2 \end{bmatrix}$$

$$\bar{B} = \bar{S} \cdot \bar{A}$$

where, $\bar{S}$ = scattering matrix
$\bar{B}$ = scattered case matrix
$\bar{A}$ = incident case matrix

There is a shunt (R = 50Ω) is equal to the characteristic impedance ; so perfectly power condition occurs at both parts

$\therefore \quad S_{11} = S_{22} = 0$ (no reflection)

and $S_{12} = S_{21} = 1$ (complete power transfer)

$$\therefore \quad \bar{S} = \begin{bmatrix} 0 & 1 \\ 1 & 0 \end{bmatrix}$$

68. λ/4 section can be replaced by a

$$Z'_L = \frac{z_0^2}{z_L}$$

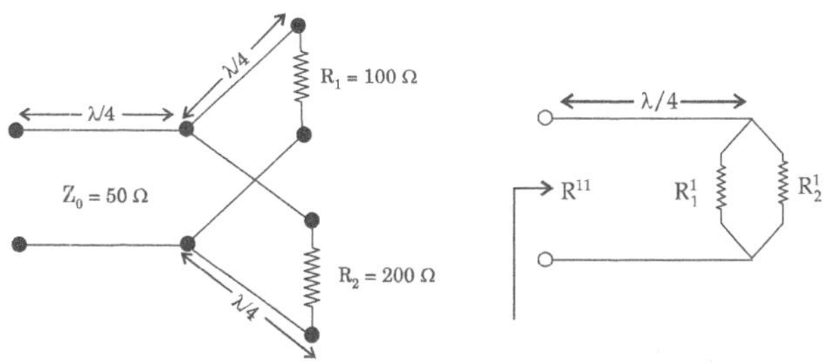

$$\therefore \quad R'_1 = \frac{Z_0^2}{R_1} = \frac{50 \times 50}{100} = 25\ \Omega$$

$$\therefore \quad R'_2 = \frac{50 \times 50}{200} = \frac{25}{2}\ \Omega$$

$$\therefore \quad R' = R'_1 \parallel R'_2 = 25 \parallel \frac{25}{2}$$

$$R' = \frac{25}{3}\ \Omega$$

$$\therefore \quad R'' = \frac{Z_0^2}{R_1} = \frac{50 \times 50 \times 3}{25} = 300\ \Omega$$

Voltage reflection coefficient,

$$\Gamma = \frac{Z_L - Z_0}{Z_L + Z_0} = \frac{300 - 50}{300 + 50} = \frac{5}{7}$$

69.

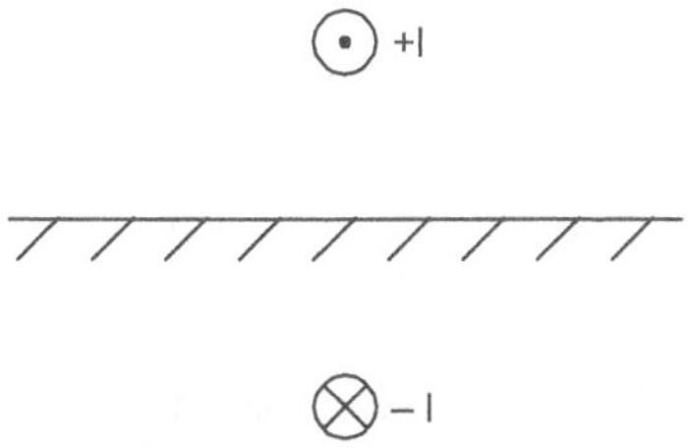

A λ/2 dipole kept λ/2 at a distance $\lambda_0/2$ above the grand equivalent to two λ/2 dipole at a distance of λ_0 between them with current flowing in them with a phase difference of π.

Now the array facts

$$A_f = \cos\left|\frac{\beta d \cos\phi + \alpha}{2}\right|$$

Here $\alpha = \pi$ and $d = \lambda_0$, $\beta = \frac{2\pi}{\lambda_0}$

$$A_f = \cos\left|\frac{2\pi\cos\phi + \pi}{2}\right| = \left|\sin(\pi\cos\phi)\right|$$

At $\phi = \frac{\pi}{2}$, array factor is zero and field strength is 0

At $\phi = 0$, field strength is also zero.

Hence, radiation field pattern will be as shown below

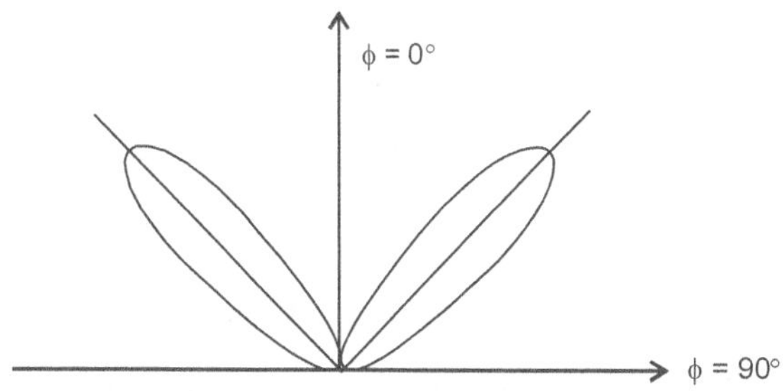

70. From Snell law, $n_1 \sin 60 = n_2 \sin\theta_t$

$n \propto \sqrt{\varepsilon}$; $\quad n_1 = 1$

$$\sqrt{\varepsilon_r}\ \sin\theta_t = \sin 60° \quad \text{....(1)}$$

Using boundary condition analysis

$$\frac{\tan 60°}{\tan\theta_t} = \varepsilon_r \quad \text{...(2)}$$

$$\sin\theta_t = \frac{\sqrt{3}}{2\sqrt{\varepsilon_r}}$$

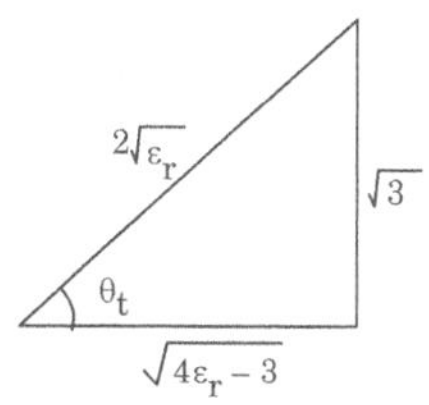

$$\frac{\sqrt{3} \times \sqrt{4\varepsilon_r - 3}}{\sqrt{3}} = \varepsilon_r$$

$\therefore \quad \varepsilon_r = 3$

71. **Case - 1** When no any gate voltage, there will not present any depletion layer, so all capacitance due to SiO_2.

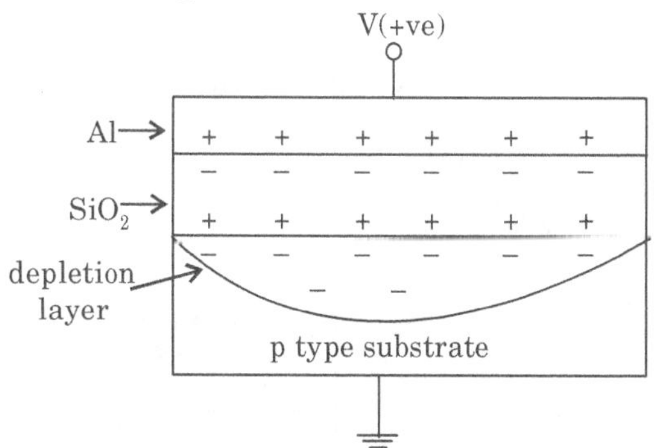

$$\frac{\varepsilon_1 A}{d_1} = 7\text{ pF} \quad (V = 0\text{ volt})$$

$$\therefore \quad \frac{3.5 \times 10^{-13}\text{ F/cm} \times 10^{-4}\text{ cm}^2}{d_1} = 7 \times 10^{-12}\text{ F}$$

$\Rightarrow \quad d_1 = 50$ nm

72. In maximum depletion layer width condition, there will be minimum capacitance (1pF) because both capacitance (SiO_2 and Si) comes in series

$$C_T = \frac{C_1C_2}{C_1+C_2} = 1 \text{ pF}$$

$$\therefore \quad \frac{7C_2}{7+C_2} = 1 \text{ pF}$$

$$\Rightarrow \quad C_2 = \frac{7}{6} \text{ pF}$$

$$\therefore \quad \frac{\varepsilon_2 A}{d_2} = C_2$$

$$\Rightarrow \quad d_2 = \frac{\varepsilon_2 A}{C_2}$$

$$\Rightarrow \quad d_2 = \left| \frac{10^{-12} \times 10^{-4}}{\frac{7}{6} \times 10^{-12}} \right|$$

$$= 0.857 \times 10^{-4} \text{ cm} = 0.857 \text{ μm}$$

73. • MOS capacitance has an *p*-type substrate
• If positive charges are introduced in the oxide, they increase the depletion layer and decrease the capacitance, so C–V curve will shift to left direction, because now small voltage is needed to decrease the capacitance.

74. $d_1^2 = \left[\left(\sqrt{2a}\right)^2 + \left(\sqrt{2a}\right)^2\right]$

$d_1 = \sqrt{4a^2}$ ...(*i*)

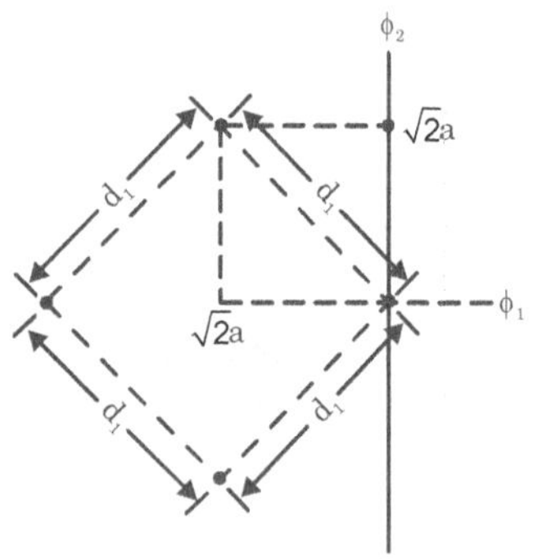

constellation - 1

$d_2 = \left[(a)^2 + (a)^2\right]^{\frac{1}{2}}$

$d_2 = \sqrt{2a^2}$ ...(*ii*)

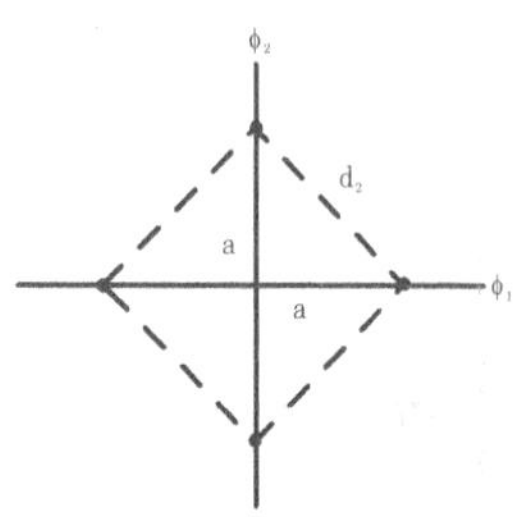

constellation - 2

we know that $E_b = \dfrac{d^2}{4}$

$$\text{Hence} \quad \frac{E_{b_1} = \frac{d_1^2}{4}}{E_{b_2} = \frac{d_2^2}{4}} = \frac{d_1^2}{d_2^2} = \frac{\left(\sqrt{4a^2}\right)^2}{\left(\sqrt{2a^2}\right)^2}$$

$$\frac{E_{b_1}}{E_{b_2}} = \frac{4a^2}{2a^2} = 2$$

75. $P(e) \propto \dfrac{1}{d}$

$d_1 > d_2$

Hence, $P_1(e) < P_2(e)$

where, $P_1(e)$ = probability of error in constellation-1

$P_2(e)$ = probability of error in constellation -2

76. $$V_+(s) = \frac{V_i(s) \times \frac{1}{sC}}{\frac{1}{sC} + R}$$

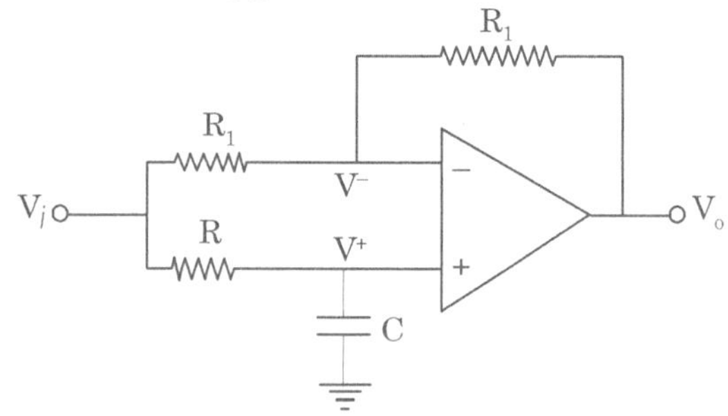

$$V_+(s) = \frac{V_i(s)}{1+sRC} \quad \text{......(1)}$$

For Op-Amp, $V^+ \approx V_-$

KCL at node V_-

$$\frac{V_0 - V_-}{R_1} = \frac{V_- - V_i}{R_1}$$

$$\therefore \quad 2V_- = V_o + V_i$$

$$\frac{2V_i(s)}{1+sRC} - V_i(s) = V_o(s) \quad \text{..... } (\because V_- \approx V_+)$$

$$\Rightarrow \quad \frac{2 - sRC - 1}{sRC + 1} = \frac{V_0(s)}{V_i(s)}$$

$$\Rightarrow \quad \frac{V_0(s)}{V_i(s)} = \frac{1 - sRC}{1 + sRC}$$

77. Given, $V_i = V_1 \sin \omega t$ and $V_o = V_2 \sin(\omega t + \phi)$

$$T(s) = \frac{V_o(s)}{V_i(s)} = \frac{1 - sRC}{1 + sRC}$$

$$\phi = \angle T(s) = -2 \tan^{-1}(sRC)$$

This ϕ phase is added by the system to the input signal (substitute $s = j\omega$)

When $\omega \to \infty$, $\phi = -2\left[\lim_{\omega \to \infty} \tan^{-1} \omega RC\right] = -\pi$

$$\therefore \quad \phi_{min} = -\pi$$

When $\omega \to 0, \phi = \lim_{\omega \to 0} -2\tan^{-1}(\omega RC) = -2 \times (0) = 0$

$\therefore \ \phi_{max} = 0$

Hence $\phi = -\pi$ to 0

78. After line (1) A contains B5 H

After line (2) B contains OEH

After line (3)

A is XOR with 69 H

A	10 11 01 01
69	01 10 10 01
DC	11 01 11 00

After XOR A contains DC

Now adding B to it

11 01 11 00

0000 11 10

11101010

So after line 4, A = EAH

79. Now line 5 ANI 9BH and immediate accumulator with 9BH

(A)	11 10 10 10
9BH	1001 1011
	10001010

Accumulator store 8A H in line 6 compare immediate with 9F H

Since 9FH is greater than 8AH so carry flag will be generated, while zero flage remain unaffected.

80. Given, $\dot{\vec{x}}(t) = A\,\vec{x}(t)$

Let $\vec{\phi}(t)$ is state transistion matrix

We know that,

$$\vec{x}(t) = \vec{\phi}(t)\ \vec{x}(o)$$

$$\therefore \begin{bmatrix} e^{-2t} \\ -2e^{-2t} \end{bmatrix} = \begin{bmatrix} \phi_{11} & \phi_{12} \\ \phi_{21} & \phi_{22} \end{bmatrix}\begin{bmatrix} 1 \\ -2 \end{bmatrix} \quad ......(1)$$

$$\text{and} \begin{bmatrix} e^{-t} \\ -e^{-t} \end{bmatrix} = \begin{bmatrix} \phi_{11} & \phi_{12} \\ \phi_{21} & \phi_{22} \end{bmatrix}\begin{bmatrix} 1 \\ -1 \end{bmatrix} \quad ......(2)$$

From equations (1) and (2),

$$\phi_{11} - 2\phi_{12} = e^{-2t} \text{ and } \phi_{11} - \phi_{12} = e^{-t}$$

$$\therefore \left.\begin{matrix} \phi_{12} = e^{-t} - e^{-2t} \\ \phi_{11} = 2e^{-t} - e^{-2t} \end{matrix}\right\} \quad ......(3)$$

Again from equations (1) and (2),

$$\phi_{21} - 2\phi_{22} = -2\,e^{-2t}$$

$$\phi_{21} - \phi_{22} = -\,e^{-t}$$

$$\therefore \left.\begin{matrix} \phi_{22} = 2e^{-2t} - e^{-t} \\ \phi_{21} = 2e^{-2t} - 2e^{-t} \end{matrix}\right\} \quad ......(4)$$

From equations (3) and (4),

$$\vec{\phi}(t) = \begin{bmatrix} 2e^{-t} & -e^{-2t} & e^{-t} & -e^{2t} \\ 2e^{-2t} & -2e^{-t} & 2e^{-2t} & -e^{-t} \end{bmatrix}$$

We know that $\vec{\phi}(t) = L^{-1}[(s\vec{I} - \vec{A})^{-1}]$

so $(s\vec{I} - \vec{A})^{-1} = L[\phi(t)]$

$$= \begin{bmatrix} \frac{2}{(s+1)} - \frac{1}{s+2} & \frac{1}{s+1} - \frac{1}{s+2} \\ \frac{2}{s+2} - \frac{2}{s+1} & \frac{2}{s+2} - \frac{1}{s+1} \end{bmatrix}$$

$$= \begin{bmatrix} \frac{s+3}{(s+1)(s+2)} & \frac{1}{(s+1)(s+2)} \\ \frac{-2}{(s+1)(s+2)} & \frac{s}{(s+1)(s+2)} \end{bmatrix} = \vec{M} \text{ (let)}$$

$$\therefore \ (s\vec{I} - \vec{A}) = [\vec{M}]^{-1} = \frac{1}{|\vec{M}|} \text{adj}[\vec{M}]$$

$$|\vec{M}| = \frac{1}{(s+1)(s+2)}$$

$$\text{adj } \overline{M} = \begin{bmatrix} \frac{s}{(s+1)(s+2)} & -\frac{1}{(s+1)(s+2)} \\ \frac{2}{(s+1)(s+2)} & \frac{s+3}{(s+1)(s+2)} \end{bmatrix}$$

$$[s\vec{I} - \vec{A}] = (s+1)(s+2)\begin{bmatrix} \frac{s}{(s+1)(s+2)} & -\frac{1}{(s+1)(s+2)} \\ \frac{2}{(s+1)(s+2)} & \frac{s+3}{(s+1)(s+2)} \end{bmatrix}$$

$$= \begin{bmatrix} s & -1 \\ 2 & s+3 \end{bmatrix}$$

Characteristic equation is

$$\Rightarrow \quad [s\vec{I} - \vec{A}] = (s+1)(s+2) = 0$$

$$s = 1, -2$$

So, eigen values are $-1, -2$

$$\lambda_1 = -1, \quad \lambda_2 = -2$$

For $\lambda_1 = -1$, $[\lambda_1 \vec{I} - \vec{A}][\vec{P_1}] = 0$

$$\begin{bmatrix} -1 & -1 \\ 2 & +2 \end{bmatrix}\begin{bmatrix} P_{11} \\ P_{12} \end{bmatrix} = 0$$

$$P_{11} = -P_{12} \qquad \therefore \left[-1, \begin{bmatrix} 1 \\ -1 \end{bmatrix}\right]$$

For $\lambda_2 = -2$,

$$\begin{bmatrix} -2 & -1 \\ 2 & +1 \end{bmatrix}\begin{bmatrix} P_{21} \\ P_{22} \end{bmatrix} = \begin{bmatrix} 0 \\ 0 \end{bmatrix}$$

$$P_{22} = -2P_{21} \qquad \therefore \left(-2, \begin{bmatrix} 1 \\ -2 \end{bmatrix}\right)$$

81. $s\vec{I} - \vec{A} = \begin{bmatrix} s & -1 \\ 2 & s+3 \end{bmatrix}$

$$\begin{bmatrix} s & 0 \\ 0 & s \end{bmatrix} - \begin{bmatrix} A_{11} & A_{12} \\ A_{21} & A_{22} \end{bmatrix} = \begin{bmatrix} s & -1 \\ 2 & s+3 \end{bmatrix}$$

$$\therefore \quad A = \begin{bmatrix} 0 & 1 \\ -2 & -3 \end{bmatrix}$$

82.

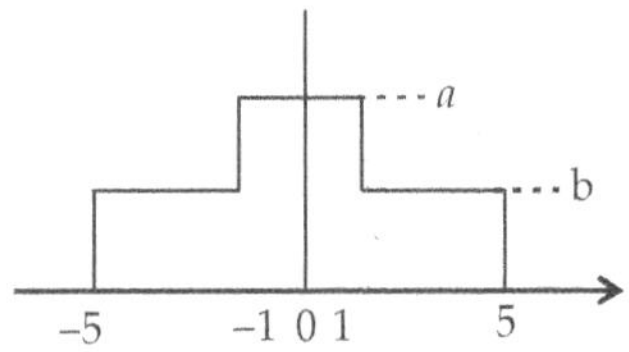

We know,

$$\int_{-\infty}^{\infty} f_X(x)\,dx = 1$$

$$2a + 8b = 1 \qquad ...(i)$$

$\because$ It is given that

$$P(-5 \le x \le -1) = P(-1 \le x \le 1) = P(1 \le x \le 5)$$

$$4b = 2a$$

$$a = 2b \qquad ...(ii)$$

Using equation (*i*) and (*ii*)

$$2 \times 2b + 8b = 1$$

$$4b + 8b = 1$$

$$12b = 1$$

$$b = \frac{1}{12}$$

$$a = \frac{2}{12} = \frac{1}{6}$$

83.

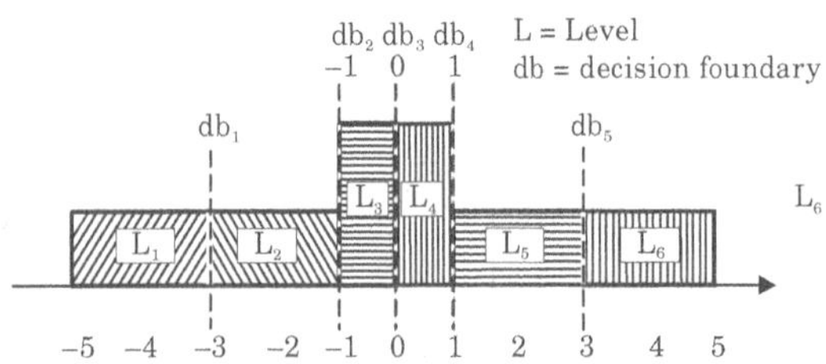

Signal power,

$$P_s = \frac{1}{12}\left[\frac{x^3}{3}\right]_{-5}^{-1} + \frac{1}{6}\left[\frac{x^3}{3}\right]_{-1}^{1} + \frac{1}{12}\left[\frac{x^3}{3}\right]_{1}^{5}$$

$$P_s = \frac{1}{12}\left[\frac{-1}{3} - \frac{-125}{3}\right] + \frac{1}{6}\left[\frac{1}{3} - \frac{-1}{3}\right] + \frac{1}{12}\left[\frac{125}{3} - \frac{1}{3}\right]$$

$$= \frac{2}{12}\left[\frac{125}{3} - \frac{1}{3}\right] + \frac{1}{6}\left[\frac{1}{3} + \frac{1}{3}\right]$$

$$= \frac{1}{6}\left[\frac{124}{3}\right] + \frac{1}{6}\left[\frac{2}{3}\right] = \frac{126}{18} = \frac{21}{3} = 7$$

$$P_s = 7$$

$$\text{Q.Noise} = \frac{2}{6} \times \left(\frac{1^2}{12}\right) + \frac{2}{6} \times \frac{2^2}{12} + \frac{2}{6} \times \frac{2^2}{12} = \frac{3}{12}$$

$$\text{Q.Noise} = \frac{3}{12} = \frac{1}{4},$$

$$\therefore \text{SNR} = \frac{P_s}{\text{Q.Noise}} = \frac{7}{\frac{1}{4}} = 28$$

84.

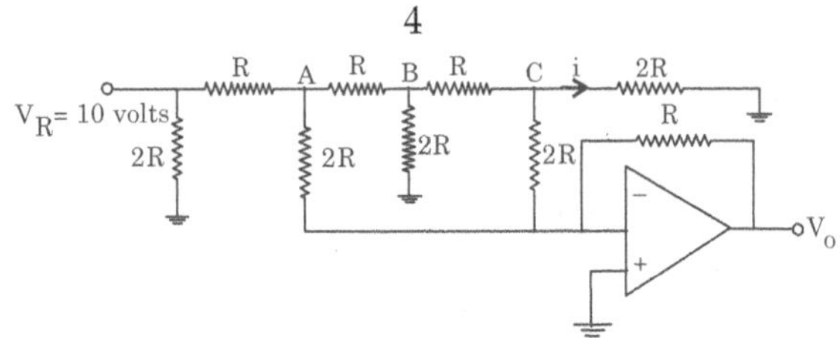

Given, R = 10 kΩ

Since, $V_+ = 0$, so in Op-Amp,

$$V_+ = V_-, \quad \text{so } V_- = 0$$

Using nodal analysis,

For node A,

$$\frac{V_A - V_R}{R} + \frac{V_A - V_B}{R} + \frac{V_A - V_{-1}}{2R} = 0$$

$$\Rightarrow \quad 5V_A - 2V_B - 2V_R = 0 \qquad ....(1)$$

For node B,

$$\frac{V_B - V_A}{R} + \frac{V_B - V_C}{R} + \frac{V_B}{2R} = 0$$

$$\Rightarrow \quad 5V_B - 2V_A - 2V_C = 0 \qquad ....(2)$$

For node C,

$$\frac{V_C - V_B}{R} + \frac{V_C - V_{-1}}{2R} + \frac{V_C}{2R} = 0$$

$$\Rightarrow \quad 4V_C - 2V_B = 0 \qquad ....(3)$$

Solving equations (1), (2), and (3), we get

$$V_A = 5 \text{ volts},$$

$$V_B = 2.5 \text{ volts}$$

$$V_C = 1.25 \text{ volts}.$$

$$\therefore \text{ Current, } i = \frac{V_C}{2R}$$

$$= \frac{1.25}{2 \times 10 \times 10^3} = 0.0625 \text{ mA}$$

$$= 62.5\ \mu A$$

85.

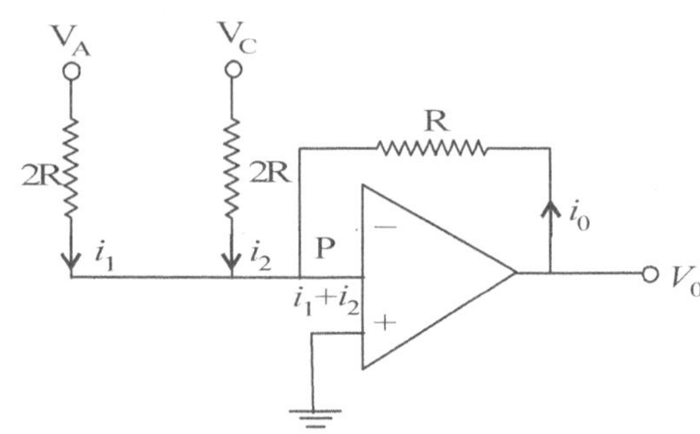

$$i_1 = \frac{V_A}{2R} = \frac{5}{20} \text{ mA};$$

$$i_2 = \frac{V_C}{2R} = \frac{1.25}{20} \text{ mA},$$

$$i_0 = \frac{V_0}{10} \text{ mA}$$

Apply KCL at node P

$$i_1 + i_2 + i_0 = 0$$

$$\Rightarrow \quad \frac{V_0}{10} = -[i_1 + i_2] = \frac{-5}{20} - \frac{2.5}{20}$$

$$\Rightarrow \quad V_0 = -3.125 \text{ volts}$$

■■

SOLVED PAPER – 2006

ELECTRONICS & COMMUNICATION ENGINEERING

Time Allowed : 3 Hours **Maximum Marks : 150**

Q. 1– 20 carry one mark each

1. The rank of the matrix $\begin{bmatrix} 1 & 1 & 1 \\ 1 & -1 & 0 \\ 1 & 1 & 1 \end{bmatrix}$ is

   (*a*) 0 (*b*) 1

   (*c*) 2 (*d*) 3

2. $\nabla \times \nabla \times P$, where P is a vector, is equal to

   (*a*) $P \times \nabla \times P - \nabla^2 P$ (*b*) $\nabla^2 P + \nabla (\nabla \bullet P)$

   (*c*) $\nabla^2 P + \nabla \times P$ (*d*) $\nabla (\nabla \bullet P) - \nabla^2 P$

3. $\iint (\nabla \times P) \bullet ds$, where P is a vector, is equal to

   (*a*) $\oint P \bullet dl$ (*b*) $\oint \nabla \times \nabla \times P \bullet dl$

   (*c*) $\oint \nabla \times P \bullet dl$ (*d*) $\iiint \nabla \bullet P\, dv$

4. A probability density function is of the form

   $p(x) = Ke^{-\alpha|x|}, x \in (-\infty, \infty)$

   The value of K is

   (*a*) 0.5 (*b*) 1

   (*c*) 0.5α (*d*) α

5. A solution for the differential equation

   $\dot{x}(t) + 2x(t) = \delta(t)$

   with initial condition $x(0-) = 0$ is

   (*a*) $e^{-2t} u(t)$ (*b*) $e^{2t} u(t)$

   (*c*) $e^{-t} u(t)$ (*d*) $e^{t} u(t)$

6. A low-pass filter having a frequency response $H(j\omega) = A(\omega)\, e^{j\phi(\omega)}$ does not produce any phase distortion, if

   (*a*) $A(\omega) = C\omega^2, \phi(\omega) = k\omega^3$

   (*b*) $A(\omega) = C\omega^2, \phi(\omega) = k\omega$

   (*c*) $A(\omega) = C\omega, \phi(\omega) = k\omega^2$

   (*d*) $A(\omega) = C, \phi(\omega) = k\omega^{-1}$

7. The values of voltage (V_D) across a tunnel-diode corresponding to peak and valley currents are V_P and V_V respectively. The range of tunnel-diode voltage V_D for which the slope of its $I–V_D$ characteristics is negative would be

   (*a*) $V_D < 0$ (*b*) $0 \le V_D < V_P$

   (*c*) $V_P \le V_D < V_V$ (*d*) $V_D \ge V_V$

8. The concentration of minority carriers in an extrinsic semiconductor under equilibrium is

   (*a*) directly proportional to the doping concentration

   (*b*) inversely proportional to the doping concentration

   (*c*) directly proportional to the intrinsic concentration

   (*d*) inversely proportional to the intrinsic concentration

9. Under low level injection assumption, the injected minority carrier current for an extrinsic semiconductor is essentially the

   (*a*) diffusion current

   (*b*) drift current

   (*c*) recombination current

   (*d*) induced current

10. The phenomenon known as "Early Effect" in a bipolar transistor refers to a reduction of the effective base-width caused by

    (*a*) electron-hole recombination at the base

    (*b*) the reverse biasing of the base-collector junction

    (*c*) the forward biasing of emitter-base junction

    (*d*) the early removal of stored base charge during saturation-to-cut-off switching

11. The input impedance (Z_i) and the output impedance (Z_0) of an ideal transconductance (voltage controlled current source) amplifier are

    (*a*) $Z_i = 0, Z_0 = 0$ (*b*) $Z_i = 0, Z_0 = \infty$

    (*c*) $Z_i = \infty, Z_0 = 0$ (*d*) $Z_i = \infty, Z_0 = \infty$

12. An *n*-channel depletion MOSFET has following two points on its I_D-V_{GS} curve

    (i) $V_{GS} = 0$ at $I_D = 12$ mA and

    (ii) $V_{GS} = -6$ Volts at $I_D = 0$

    Which of the following Q-points will give the highest trans-conductance gain for small signals?

    (*a*) $V_{GS} = -6$ Volts

    (*b*) $V_{GS} = -3$ Volts

    (*c*) $V_{GS} = 0$ Volts

    (*d*) $V_{GS} = 3$ Volts

13. The number of product terms in the minimized sum-of-product expression obtained through the following K-map is (where, "*d*" denotes don't care states)

1	0	0	1
0	*d*	0	0
0	0	*d*	1
1	0	0	1

(*a*) 2 (*b*) 3
(*c*) 4 (*d*) 5

14. Let $x(t) \longleftrightarrow X(j\omega)$ be Fourier Transform pair. The Fourier Transform of the signal $x(5t-3)$ in terms of $X(j\omega)$ is given as

(*a*) $\frac{1}{5} e^{-\frac{j3\omega}{5}} X\left(\frac{j\omega}{5}\right)$ (*b*) $\frac{1}{5} e^{\frac{j3\omega}{5}} X\left(\frac{j\omega}{5}\right)$

(*c*) $\frac{1}{5} e^{-j3\omega} X\left(\frac{j\omega}{5}\right)$ (*d*) $\frac{1}{5} e^{j3\omega} X\left(\frac{j\omega}{5}\right)$

15. The Dirac delta function $\delta(t)$ is defined as

(*a*) $\delta(t) = \begin{cases} 1, & t=0 \\ 0, & \text{otherwise} \end{cases}$

(*b*) $\delta(t) = \begin{cases} \infty, & t=0 \\ 0, & \text{otherwise} \end{cases}$

(*c*) $\delta(t) = \begin{cases} 1, & t=0 \\ 0, & \text{otherwise} \end{cases}$ and $\int_{-\infty}^{\infty} \delta(t)\, dt = 1$

(*d*) $\delta(t) = \begin{cases} \infty, & t=0 \\ 0, & \text{otherwise} \end{cases}$ and $\int_{-\infty}^{\infty} \delta(t)\, dt = 1$

16. If the region of convergence of $x_1[n] + x_2[n]$ is $\frac{1}{3} < |z| < \frac{2}{3}$, then the region of convergence of $x_1[n] - x_2[n]$ includes

(*a*) $\frac{1}{3} < |z| < 3$ (*b*) $\frac{2}{3} < |z| < 3$

(*c*) $\frac{2}{3} < |z| < 3$ (*d*) $\frac{1}{3} < |z| < \frac{2}{3}$

17. The open-loop transfer function of a unity-gain feedback control system is given by

$$G(s) = \frac{K}{(s+1)(s+2)}$$

The gain margin of the system in dB is given by

(*a*) 0
(*b*) 1
(*c*) 20
(*d*) ∞

18. In the system shown below, $x(t) = (\sin t)\, u(t)$. In steady-state, the response $y(t)$ will be

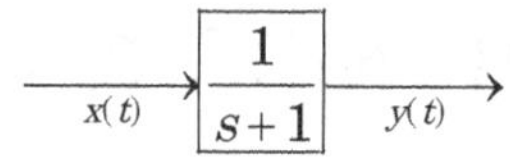

(*a*) $\frac{1}{\sqrt{2}} \sin\left(t - \frac{\pi}{4}\right)$ (*b*) $\frac{1}{\sqrt{2}} \sin\left(t + \frac{\pi}{4}\right)$

(*c*) $\frac{1}{\sqrt{2}} e^{-t} \sin t$ (*d*) $\sin t - \cos t$

19. The electric field of an electomagnetic wave propagating in the positive *z*-direction is given by

$$E = \hat{a}_x \sin(\omega t - \beta z) + \hat{a}_y \sin(\omega t - \beta z + \frac{\pi}{2})$$

The wave is

(*a*) linearly polarized in the z-direction
(*b*) elliptically polarized
(*c*) left-hand circularly polarized
(*d*) right-hand circularly polarized

20. A transmission line is feeding 1 Watt of power to a horn antenna having a gain of 10 *d*B. The antenna is matched to the transmission line. The total power radiated by the horn antenna into the free-space is

(*a*) 10 watts (*b*) 1 watt
(*c*) 0.1 watt (*d*) 0.01 watt

Q.21 –75 carry two marks each.

21. The eigenvalues and the corresponding eigen vectors of a 2 × 2 matrix are given by

Eigenvalue	Eigenvector
$\lambda_1 = 8$	$v_1 = \begin{bmatrix} 1 \\ 1 \end{bmatrix}$
$\lambda_2 = 4$	$v_2 = \begin{bmatrix} 1 \\ -1 \end{bmatrix}$

The matrix is

(*a*) $\begin{bmatrix} 6 & 2 \\ 2 & 6 \end{bmatrix}$ (*b*) $\begin{bmatrix} 4 & 6 \\ 6 & 4 \end{bmatrix}$

(*c*) $\begin{bmatrix} 2 & 4 \\ 4 & 2 \end{bmatrix}$ (*d*) $\begin{bmatrix} 4 & 8 \\ 8 & 4 \end{bmatrix}$

22. For the function of a complex variable W = ln Z (where, W = $u + jv$ and Z = $x + jy$), the u = constant lines get mapped in Z-plane as

(*a*) set of radial straight lines
(*b*) set of concentric circles
(*c*) set of confocal hyperbolas
(*d*) set of confocal ellipses

23. The value of the contour integral $\oint_{|z-j|=2} \frac{1}{z^2+4} dz$ in positive sense is

(a) $\frac{j\pi}{2}$ (b) $-\frac{\pi}{2}$

(c) $-\frac{j\pi}{2}$ (d) $\frac{\pi}{2}$

24. The integral $\int_0^{\pi} \sin^3\theta \, d\theta$ is given by

(a) $\frac{1}{2}$ (b) $\frac{2}{3}$

(c) $\frac{4}{3}$ (d) $\frac{8}{3}$

25. Three companies X, Y and Z supply computers to a university. The percentage of computers supplied by them and the probability of those being defective are tabulated below

Company	% of computers supplied	Probability of being defective
X	60%	0.01
Y	30%	0.02
Z	10%	0.03

Given that a computer is defective, the probability that it was supplied by Y is

(a) 0.1 (b) 0.2

(c) 0.3 (d) 0.4

26. For the matrix $\begin{bmatrix} 4 & 2 \\ 2 & 4 \end{bmatrix}$, the eigen value corresponding to the eigenvector $\begin{bmatrix} 101 \\ 101 \end{bmatrix}$ is

(a) 2 (b) 4

(c) 6 (d) 8

27. For the differential equation $\frac{d^2y}{dx^2} + k^2 y = 0$, the boundary conditions are

(i) $y = 0$ for $x = 0$, and

(ii) $y = 0$ for $x = a$

The form of non-zero solutions of y (where m varies over all integers) are

(a) $y = \sum_m A_m \sin \frac{m\pi x}{a}$ (b) $y = \sum_m A_m \cos \frac{m\pi x}{a}$

(c) $y = \sum_m A_m x^{\frac{m\pi}{a}}$ (d) $y = \sum_m A_m e^{-\frac{m\pi x}{a}}$

28. Consider the function $f(t)$ having Laplace transform

$$F(s) = \frac{\omega_0}{s^2 + \omega_0^2} \quad \text{Re}[s] > 0$$

The final value of $f(t)$ would be

(a) 0 (b) 1

(c) $-1 \le f(\infty) \le 1$ (d) ∞

29. As x is increased from $-\infty$ to ∞, the function

$$f(x) = \frac{e^x}{1+e^x}$$

(a) monotonically increases

(b) monotonically decreases

(c) increases to a maximum value and then decreases

(d) decreases to a minimum value and then increases

30. A two-port network is represented by ABCD parameters given by

$$\begin{bmatrix} V_1 \\ I_1 \end{bmatrix} = \begin{bmatrix} A & B \\ C & D \end{bmatrix} \begin{bmatrix} V_2 \\ -I_2 \end{bmatrix}$$

If port-2 is terminated by R_L, then input impedance seen at port-1 is given by

(a) $\frac{A + BR_L}{C + DR_L}$ (b) $\frac{AR_L + C}{BR_L + D}$

(c) $\frac{DR_L + A}{BR_L + C}$ (d) $\frac{B + AR_L}{D + CR_L}$

31. In the two port network shown in the figure below, z_{12} and z_{21} are, respectively

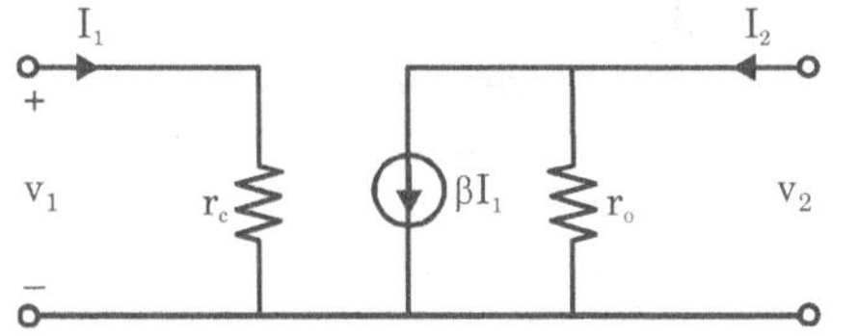

(a) r_e and βr_o

(b) 0 and $-\beta r_o$

(c) 0 and βr_o

(d) r_e and $-\beta r_o$

32. The first and the last critical frequencies (singularities) of a driving point impedance function of a passive network having two kinds of elements, are a pole and a zero respectively. The above property will be satisfied by

(a) RL network only

(b) RC network only

(c) LC network only

(d) RC as well as RL networks

33. A 2 mH inductor with some initial current can be represented as shown below, where s is the Laplace Transform variable. The value of initial current is

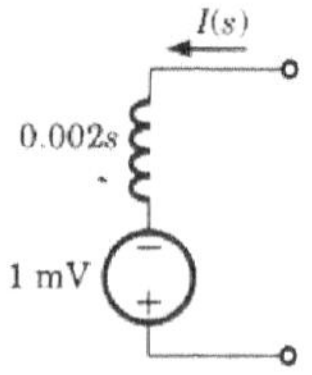

(a) 0.5 A (b) 2.0A

(c) 1.0 A (d) 4.0A

34. In the figure shown, assume that all the capacitors are initially uncharged.

If $V_i(t) = 10u(t)$ Volts, then $V_0(t)$ is given by

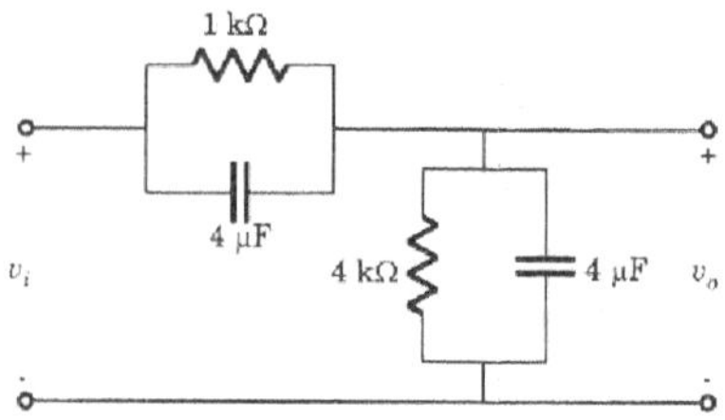

(a) $8e^{-0.004t}$ Volts

(b) $8(1 - e^{-0.004t})$ Volts

(c) $8u(t)$ Volts

(d) 8 Volts

35. Consider two transfer functions

$$G_1(s) = \frac{1}{s^2 + as + b}$$

and
$$G_2(s) = \frac{s}{s^2 + as + b}$$

The 3-dB bandwidths of their frequency responses are, respectively

(a) $\sqrt{a^2 - 4b}, \sqrt{a^2 + 4b}$

(b) $\sqrt{a^2 + 4b}, \sqrt{a^2 - 4b}$

(c) $\sqrt{a^2 - 4b}, \sqrt{a^2 - 4b}$

(d) $\sqrt{a^2 + 4b}, \sqrt{a^2 + 4b}$

36. A negative resistance R_{neg} is connected to a passive network N having driving point impedance $Z_1(s)$ as shown below. For $Z_2(s)$ to be positive real,

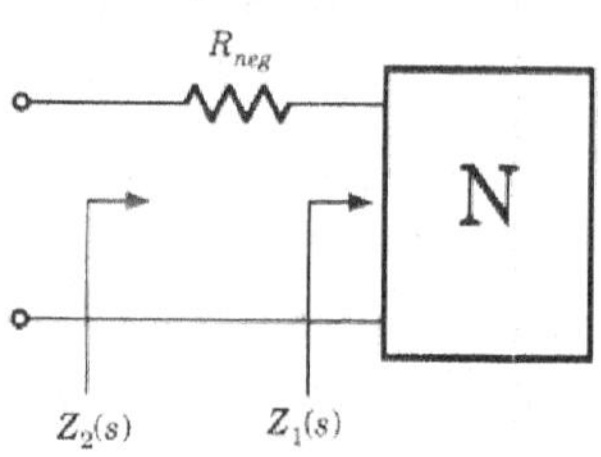

(a) $|R_{neg}| \le \text{Re}\, Z_1(j\omega), \ \forall\, \omega$

(b) $|R_{neg}| \le |Z_1(j\omega)|, \ \forall\, \omega$

(c) $|R_{neg}| \le \text{Im}\, Z_1(j\omega), \ \forall\, \omega$

(d) $|R_{neg}| \le \angle Z_1(j\omega), \ \forall\, \omega$

37. In the circuit shown below, the switch was connected to position 1 at $t < 0$ and at $t = 0$, it is changed to position 2. Assume that the diode has zero voltage drop and a storage time t_s. For $0 < t \le t_s$, V_R is given by (all in Volts)

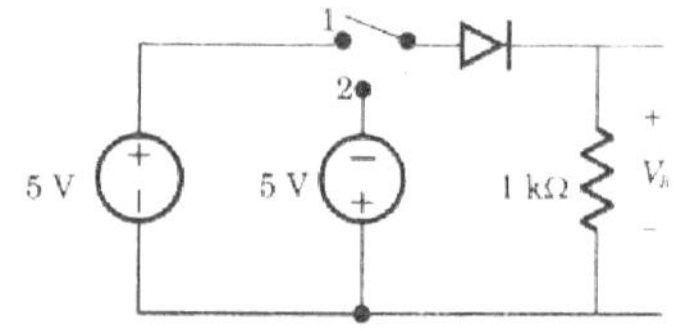

(a) $V_R = -5$

(b) $V_R = +5$

(c) $0 \le V_R < 5$

(d) $-5 < V_R < 0$

38. The majority carriers in an n-type semiconductor have an average drift velocity V in a direction perpendicular to a uniform magnetic field B. The electric field E induced due to Hall effect acts in the direction

(a) $\frac{V}{B}$ (b) $B \times V$

(c) along V (d) opposite to V

39. Find the correct match between Group-1 and Group-2

Group- 1	Group-2
A. Varactor diode	1. Voltage reference
B. PIN diode	2. High-frequency switch
C. Zener diode	3. Tuned circuits
D. Schottky diode	4. Current controlled attenuator

	A	B	C	D
(a)	4	2	1	3
(b)	2	4	1	3
(c)	3	4	1	2
(d)	1	3	2	4

40. A heavily doped n-type semiconductor has the following data

Hole-electron mobility ratio : 0.4

Doping concentration : 4.2×10^8 atoms/m^3

Intrinsic concentration : 1.5×10^4 atoms/m^3

The ratio of conductance of the n-type semiconductor to that of the intrinsic semiconductor of same material and at the same temperature is given by

(a) 0.00005 (b) 2,000

(c) 10,000 (d) 20,000

41. For the circuit shown in the following figure, the capacitor C is initially uncharged. At $t = 0$, the switch S is closed. The voltage V_C across the capacitor at $t = 1$ millisecond is

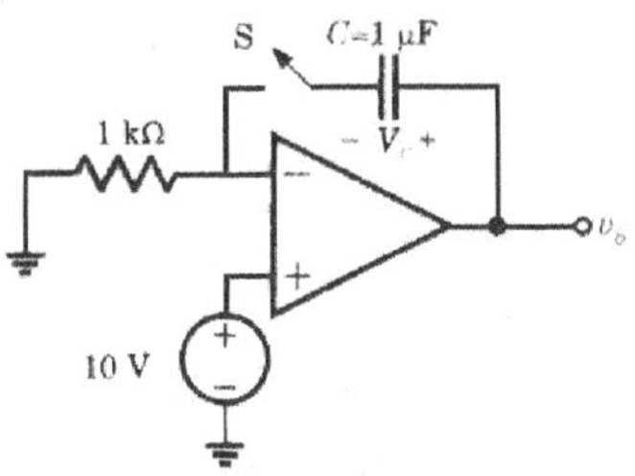

In the figure shown above, the OP AMP is supplied with ± 15V and the ground has been shown by the symbol ∇.

(*a*) 0 Volts (*b*) 6.3 Volts

(*c*) 9.45 Volts (*d*) 10 Volts

42. For the circuit shown below, assume that the zener diode is ideal with a breakdown voltage of 6 Volts. The waveform observed across R is

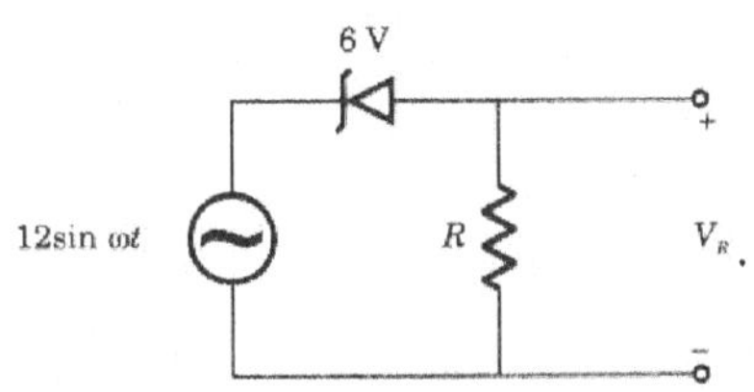

(*a*)

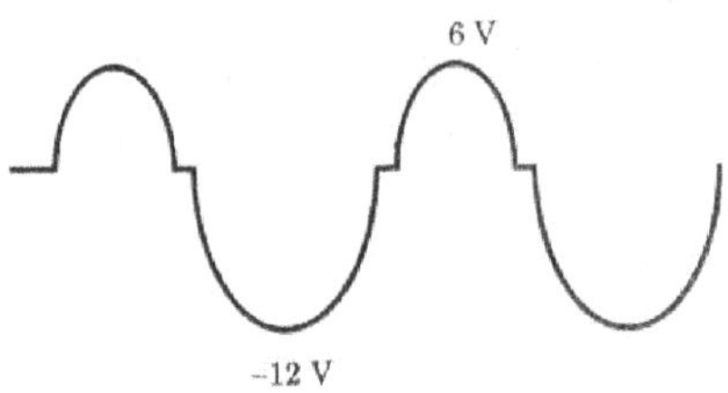

(*b*)

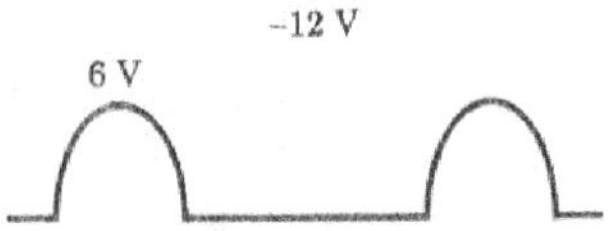

(*c*)

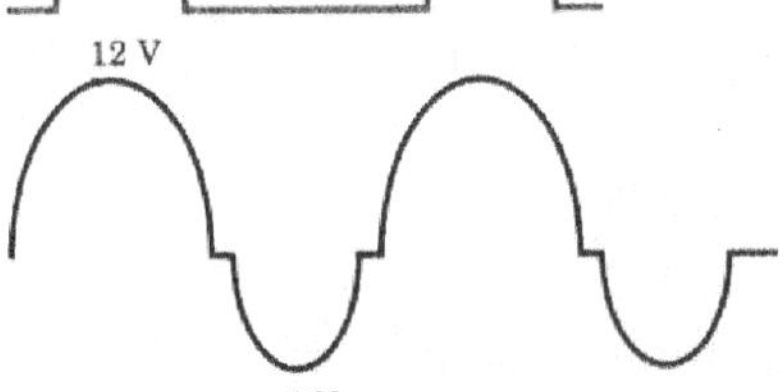

(*d*)

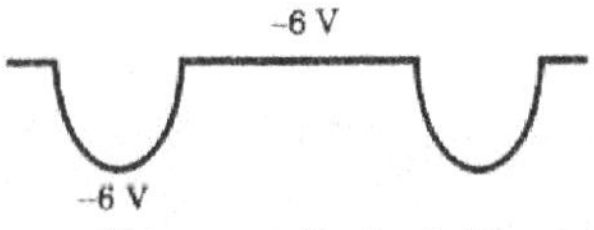

43. A new Binary Coded Pentary (BCP) number system is proposed in which every digit of a base-5 number is represented by tis corresponding 3-bit binary code. For example, the base-5 number 24 will be represented by its BCP code 010100. In this numbering system, the BCP code 100010011001 corresponds to the following number in base-5 system

(*a*) 423 (*b*) 1324

(*c*) 2201 (*d*) 4231

44. An $\frac{I}{O}$ peripheral device shown in Figure (*b*) below is to be interfaced to an 8085 microprocessor.

To select the $\frac{I}{O}$ device in the $\frac{I}{O}$ address range D4 H – D7 H, its chip-select $(\overline{CS})$ should be connected to the output of the decoder shown in Figure (*a*) below

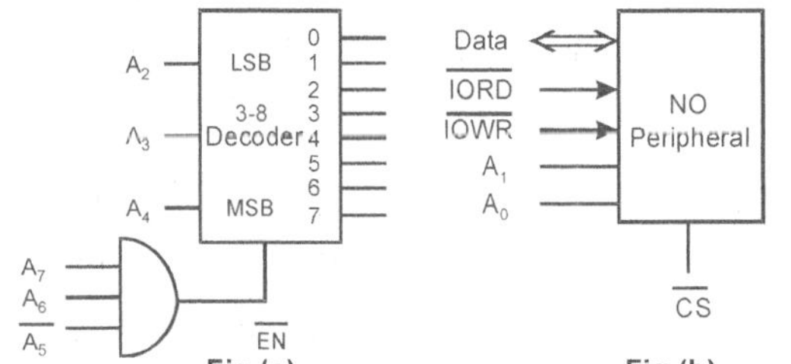

Fig.(a) Fig.(b)

(*a*) output 7 (*b*) output 5

(*c*) output 2 (*d*) output 0

45. For the circuit shown in figure below, two 4-bit parallel-in serial-out shift registers loaded with the data shown are used to feed the data to a full adder. Initially, all the flip-flops are in clear state. After applying two clock pulses, the outputs of the fulladder should be

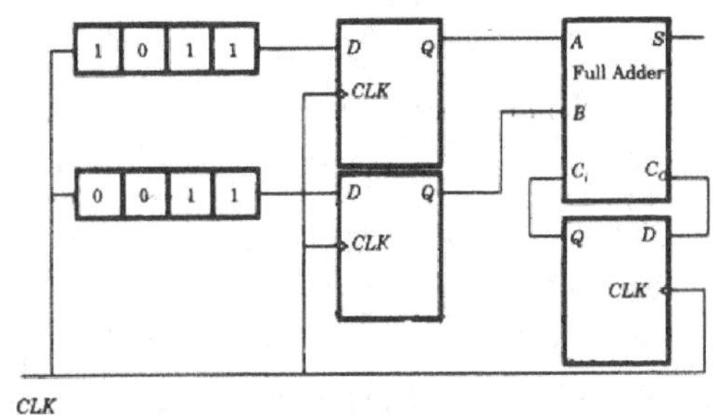

(*a*) $S = 0, C_0 = 0$ (*b*) $S = 0, C_0 = 1$

(*c*) $S = 1, C_0 = 0$ (*d*) $S = 1, C_0 = 1$

46. A 4-bit D/A converter is connected to a free-running 3-bit UP counter, as shown in the following figure. Which of the following waveforms will be observed at V_o?

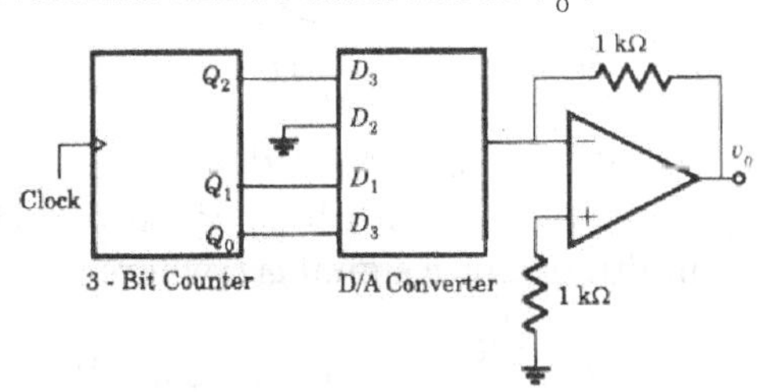

In the figure shown above, the ground has been shown by the symbol ∇

(*a*)

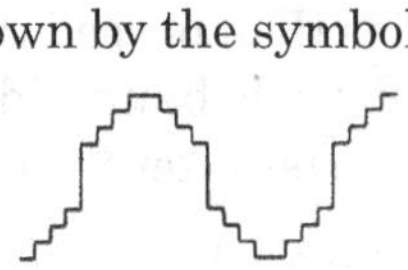

(*b*)

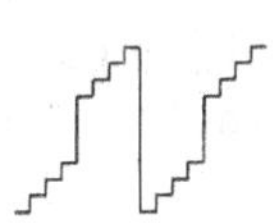

(*c*)

(*d*)

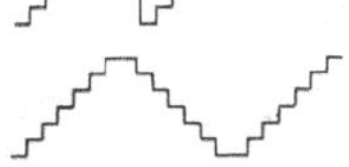

47. Two D-flip-flops, as shown below, are to be connected as a synchronous counter that goes through the following Q_1Q_0 sequence

$00 \longrightarrow 01 \longrightarrow 11 \longrightarrow 10 \longrightarrow 00 \longrightarrow \ldots\ldots$

The inputs D_0 and D_1 respectively should be connected as

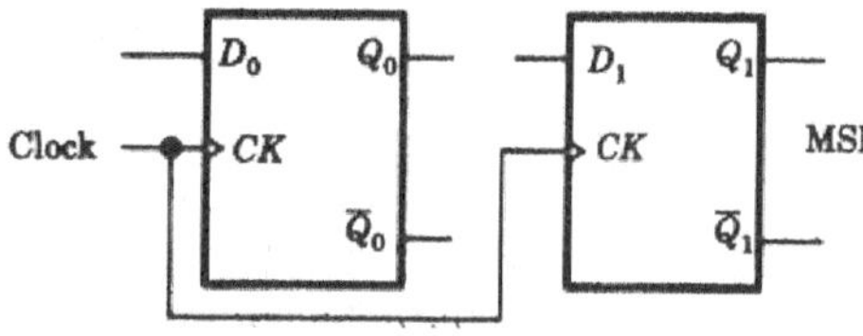

(*a*) $\overline{Q_1}$ and Q_0 (*b*) $\overline{Q_0}$ and Q_1

(*c*) $\overline{Q_1}\, Q_0$ and $\overline{Q_1}\, Q_0$ (*d*) $\overline{Q_1}\, \overline{Q_0}$ and $Q_1\, Q_0$

48. Following is the segment of a 8085 assembly language program

```
LXI SP, EFFFH
CALL 3000 H
:
:
3000 H : LXI H, 3CF4 H
PUSH PSW
SPHL
POP PSW
RET
```

On completion of RET execution, the contents of SP is

(*a*) 3CFO H (*b*) 3CF8 H

(*c*) EFFD H (*d*) EFFF H

49. The point P in the following figure is stuck-at-1. The output f will be

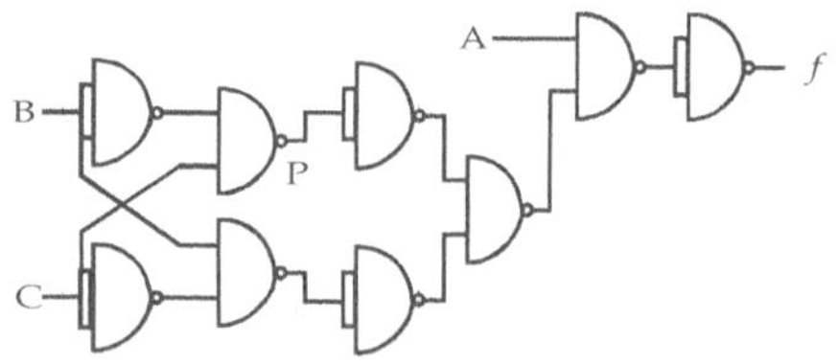

(*a*) $\overline{AB\overline{C}}$ (*b*) $\overline{A}$

(*c*) $AB\overline{C}$ (*d*) A

50. A signal $m(t)$ with bandwidth 500 Hz is first multiplied by a signal $g(t)$ where

$$g(t) = \sum_{K=-\infty}^{\infty} (-1)^k\, \delta(t - 0.5 \times 10^{-4} k)$$

The resulting signal is then passed through an ideal lowpass filter with bandwidth 1kHz. The output of the lowpass filter would be

(*a*) $\delta(t)$ (*b*) $m(t)$

(*c*) 0 (*d*) $m(t)\,\delta(t)$

51. The minimum sampling frequency (in samples/sec) required to reconstruct the following signal from its samples without distortion

$$x(t) = 5\left(\frac{\sin 2\pi\, 1000t}{\pi t}\right)^3 + 7\left(\frac{\sin 2\pi 1000t}{\pi t}\right)^2$$

would be

(*a*) 2×10^3 (*b*) 4×10^3

(*c*) 6×10^3 (*d*) 8×10^3

52. A uniformly distributed random variable X with probability density function

$$f_x(x) = \frac{1}{10}\,(u(x+5) - u(x-5))$$

where $u(.)$ is the unit step function is passed through a transformation given in the figure below. The probability density function of the transformed random variable Y would be

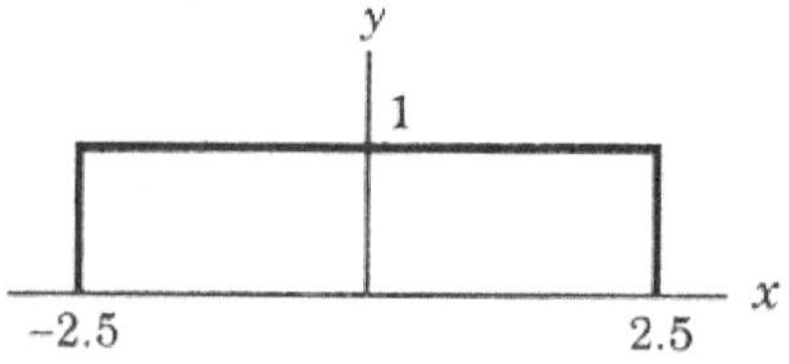

(*a*) $f_Y(y) = \frac{1}{5}\,(u(y+2.5) - u(y-2.5))$

(*b*) $f_Y(y) = 0.5\,\delta(y) + 0.5\,\delta(y-1)$

(*c*) $f_Y(y) = 0.25\,\delta(y+2.5) + 0.25\,\delta(y-2.5) + 0.5\,\delta(y)$

(*d*) $f_Y(y) = 0.25\,\delta(y+2.5) + 0.25\,\delta(y-2.5) + \frac{1}{10}\,(u(y+2.5) - u(y-2.5))$

53. A system with input $x[n]$ and output $y[n]$ is given as $y[n] = \left(\sin \frac{5}{6}\pi n\right) x(n)$.

The system is

(*a*) linear, stable and invertible

(*b*) non-linear, stable and non-invertible

(*c*) linear, stable and non-invertible

(*d*) linear, unstable and invertible

54. The unit-step response of a system starting from rest is given by

$$C(t) = 1 - e^{-2t} \text{ for } t \geq 0$$

The transfer function of the system is

(*a*) $\frac{1}{1+2s}$ (*b*) $\frac{2}{2+s}$

(*c*) $\frac{1}{2+s}$ (*d*) $\frac{2s}{1+2s}$

55. The Nyquist plot of G $(j\omega)$ H $(j\omega)$ for a closed loop control system, passes through $(-1, j0)$ point in the GH plane. The gain margin of the system in dB is equal to

(*a*) infinite (*b*) greater than zero

(*c*) less than zero (*d*) zero

56. The positive values of "K" and "*a*" so that the system shown in the figure below oscillates at a frequency of 2 rad/sec respectively are

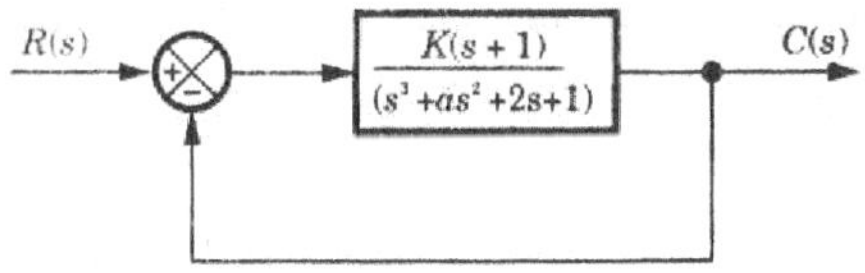

(*a*) 1, 0.75 (*b*) 2, 0.75

(*c*) 1, 1 (*d*) 2, 2

57. The unit impulse response of a system is

$$h(t) = e^{-t},\ t \geq 0$$

For this system, the steady-state value of the output for unit step input is equal to

(*a*) –1 (*b*) 0

(*c*) 1 (*d*) ∞

58. The transfer function of a phase-lead compensator is given by

$$G_c(s) = \frac{1 + 3Ts}{1 + Ts} \text{ where } T > 0$$

The maximum phase-shift provided by such a compensator is

(*a*) $\frac{\pi}{2}$ (*b*) $\frac{\pi}{3}$

(*c*) $\frac{\pi}{4}$ (*d*) $\frac{\pi}{6}$

59. A linear system is described by the following state equation

$$X(t) = AX(t) + BU(t), A = \begin{bmatrix} 0 & 1 \\ -1 & 0 \end{bmatrix}$$

The state-transition matrix of the system is

(*a*) $\begin{bmatrix} \cos t & \sin t \\ -\sin t & \cos t \end{bmatrix}$ (*b*) $\begin{bmatrix} -\cos t & \sin t \\ -\sin t & -\cos t \end{bmatrix}$

(*c*) $\begin{bmatrix} -\cos t & -\sin t \\ -\sin t & \cos t \end{bmatrix}$ (*d*) $\begin{bmatrix} \cos t & -\sin t \\ \cos t & \sin t \end{bmatrix}$

60. The minimum step-size required for a Delta-Modulator operating at 32 K samples/sec to track the signal (here $u(t)$ is the unit-step function)

$x(t) = 125\, t(u(t) - u(t-1)) + (250 - 125t)\,(u(t-1) - u(t-2))$

So that slope-overload is avoided, would be

(*a*) 2^{-10} (*b*) 2^{-8}

(*c*) 2^{-6} (*d*) 2^{-4}

61. A zero- mean white Gaussian noise is passed through an ideal lowpass filter of bandwidth 10 kHz. The output is the uniformly sampled with sampling period $t_s = 0.03$ m/sec.

The samples so obtained would be

(*a*) correlated

(*b*) statistically independent

(*c*) uncorrelated

(*d*) orthogonal

62. A source generates three symbols with probabilities 0.25, 0.25, 0.50 at a rate of 3000 symbols per second. Assuming independent generation of symbols, the most efficient source encoder would have average bit rate as

(*a*) 6000 bits/sec (*b*) 4500 bits/sec

(*c*) 3000 bits/sec (*d*) 1500 bits/sec

63. The diagonal clipping in Amplitude demodulation (using envelope detector) can be avoided if RC time-constant of the envelope detector satisfies the following condition, (here W is message bandwidth and ω_c is carrier frequency both in rad/sec)

(*a*) $RC < \frac{1}{W}$ (*b*) $RC > \frac{1}{W}$

(*c*) $RC < \frac{1}{\omega_c}$ (*d*) $RC > \frac{1}{\omega_c}$

64. In the following figure the minimum value of the constant 'C', which is to be added to $y_1(t)$ such that $y_1(t)$ and $y_2(t)$ are different, is

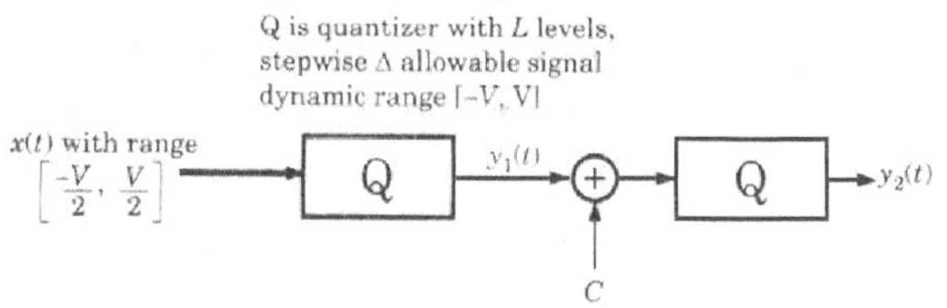

(*a*) Δ (*b*) $\frac{\Delta}{2}$

(*c*) $\frac{\Delta^2}{12}$ (*d*) $\frac{\Delta}{L}$

65. A message signal with bandwidth 10 kHz is Lower-Side Band (LSB) modulated with carrier frequency $f_{c1} = 10^6$ Hz. The resulting signal is then passed through a Narrow-Band Frequency Modulator with carrier frequency $f_{c2} = 10^9$ Hz.

The bandwidth of the output would be

(*a*) 4×10^4 Hz

(*b*) 2×10^6 Hz

(*c*) 2×10^9 Hz

(*d*) 2×10^{10} Hz

66. A medium of relative permittivity $\varepsilon_{r2} = 2$ forms an interface with free-space. A point source of electromagnetic energy is located in the medium at a depth of 1 meter from the interface. Due to the total internal reflection, the transmitted beam has a circular cross-section over the interface. The area of the beam cross-section at the interface is given by

(a) $2\pi\ m^2$ (b) $\pi^2\ m^2$

(c) $\frac{\pi}{2}\ m^2$ (d) $\pi\ m^2$

67. A medium is divided into regions I and II about $x = 0$ plane, as shown in the figure below. An electromagnetic wave with electric field $E_1 = 4\hat{a}_x + 3\hat{a}_y + 5\hat{a}_z$ is incident normally on the interface from region-I. The electric field E_2 in region-II at the interface is

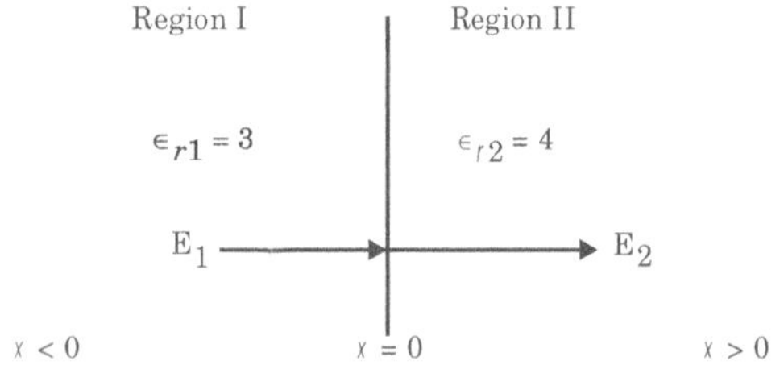

(a) $E_2 = E_1$

(b) $4\hat{a}_x + 0.75\hat{a}_y - 1.25\hat{a}_z$

(c) $3\hat{a}_x + 3\hat{a}_y + 5\hat{a}_z$

(d) $-3\hat{a}_x + 3\hat{a}_y + 5\hat{a}_z$

68. When a plane wave travelling in free-space is incident normally on a medium having $\varepsilon_r = 4.0$, then fraction of power transmitted into the medium is given by

(a) $\frac{8}{9}$ (b) $\frac{1}{2}$

(c) $\frac{1}{3}$ (d) $\frac{5}{6}$

69. A rectangular waveguide having TE_{10} mode as dominant mode is having a cut-off frequency of 18-GHz for the TE_{30} mode. The inner broad-wall dimension of the rectangular waveguide is

(a) $\frac{5}{3}$ cms (b) 5 cms

(c) $\frac{5}{2}$ cms (d) 10 cms

70. A mast antenna consisting of a 50 meter long vertical conductor operates over a perfectly conducting ground plane. It is base-fed at a frequency of 600kHz. The radiation resistance of the antenna in ohms is

(a) $8\pi^2$ (b) $\frac{\pi^2}{5}$

(c) $\frac{4\pi^2}{5}$ (d) $20\,\pi^2$

Common Data for Questions 71, 72, 73 :

In the transistor amplifier circuit shown in the figure below, the transistor has the following parameters :

$$\beta_{DC} = 60,\ V_{BE} = 0.7\ V,\ h_{ie} \longrightarrow \infty,\ h_{fe} \longrightarrow \infty$$

The capacitance C_c can be assumed to be infinite.

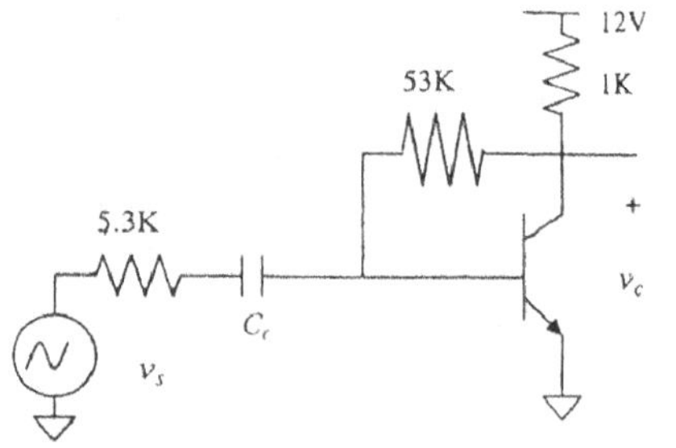

In the figure above, the ground has been shown by the symbol ∇

71. Under the DC conditions, the collector-to-emitter voltage drop is

(a) 4.8 Volts (b) 5.3 Volts

(c) 6.0 Volts (d) 6.6 Volts

72. If β_{DC} is increased by 10%, then collector-to-emitter voltage drop

(a) increases by less than or equal to 10%

(b) decreases by less than or equal to 10%

(c) increases by more than 10%

(d) decreases by more than 10%

73. The small-signal gain of the amplifier v_c/v_s is

(a) – 10 (b) – 5.3

(c) 5.3 (d) 10

Common Data for Questions 74, 75 :

Let $g(t) = p(t) * p(t)$, where * denotes convolution and $p(t) = u(t) - u(t-1)$ with $u(t)$ being the unit step function

74. The impulse response of filter matched to the signal $s(t) = g(t) - \delta(t-2) * g(t)$ is given as

(a) $s(1-t)$ (b) $-s(1-t)$

(c) $-s(t)$ (d) $s(t)$

75. An Amplitude Modulated signal is given as

$$x_{AM}(t) = 100\,(p(t) + 0.5\,g(t))\cos\omega_c t$$

in the interval $0 \le t \le 1$. One set of possible values of the modulating signal and modulation index would be

(a) t, 0.5 (b) t, 1.0

(c) t, 2.0 (d) t^2, 0.5

Linked Answer Questions : Q.76 – 85 carry two marks each.

Statement for Linked Answer Questions 76 and 77

A regulated power supply, shown in figure below, has an unregulated input (UR) of 15 Volts and generates a regulated output V_{out}. Use the component values shown in the figure.

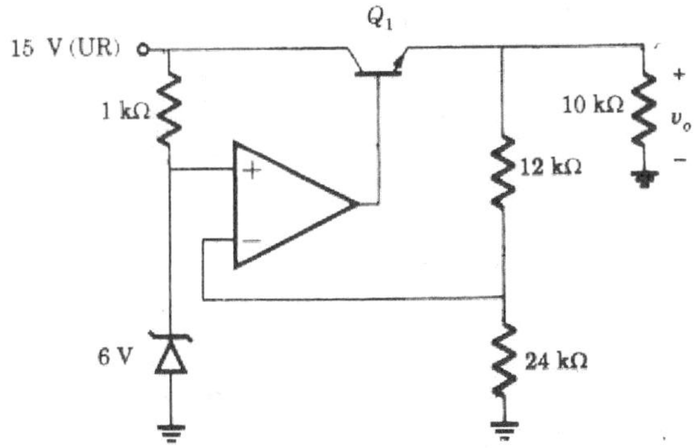

In the figure above, the ground has been shown by the symbol ∇

76. The power dissipation across the transistor Q_1 shown in the figure is

(a) 4.8 mW (b) 5.0 mW

(c) 5.4 mW (d) 6.0 mW

77. If the unregulated voltage increases by 20%, then power dissipation across the transistor Q_1

(a) increases by 20%

(b) increases by 50%

(c) remains unchanged

(d) decreases by 20%

Statement for Linked Answer Questions 78 and 79

The following two questions refer to wide sense stationary stochastic processes

78. It is desired to generate a stochastic process (as voltage process) with power spectral density

$$S(\omega) = \frac{16}{16 + \omega^2}$$

by driving a Linear-Time-Invariant system by zero mean white noise (as voltage process) with power spectral density being constant equal to 1. The system which can perform the desired task could be

(a) first order lowpass R-L filter

(b) first order highpass R-C filter

(c) tuned L-C filter

(d) series R-L filter

79. The parameters of the system obtained in Q. 78 would be

(a) first order R-L lowpass filter would have R = 4Ω, L = 4H

(b) first order R-C highpass filter would have R = 4Ω, C = 0.25 F

(c) tuned L-C filter would have L = 4H, C = 4F

(d) series R-L-C lowpass filter would have R = 1Ω, L = 4H, C = 4F

Statement for Linked Answer Questions 80 and 81

Consider the following Amplitude Modulated (AM) signal, where $f_m < B$

$$x_{AM}(t) = 10\,(1 + 0.5 \sin 2\pi f_m t) \cos 2\pi f_c t$$

80. The average side-band power for the AM signal given above is

(a) 25 (b) 12.5

(c) 6.25 (d) 3.125

81. The AM signal gets added to a noise with Power Spectral Density $S_n(f)$ given in the figure below. The ratio of average side-band power to mean noise power would be

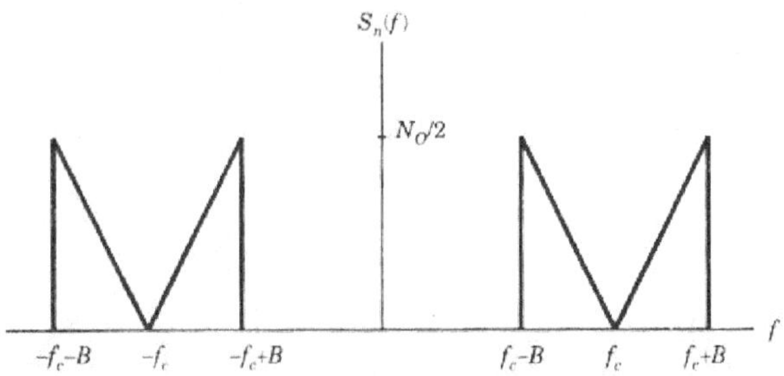

(a) $\frac{25}{8N_0B}$ (b) $\frac{25}{4N_0B}$

(c) $\frac{25}{2N_0B}$ (d) $\frac{25}{N_0B}$

Statement for Linked Answer Questions 82 and 83

Consider a unity-gain feedback control system whose open-loop transfer function is

$$G(s) = \frac{as + 1}{s^2}$$

82. The value of 'a' so that system has a phase-margin equal to $\frac{\pi}{4}$ is approximately equal to

(a) 2.40

(b) 1.40

(c) 0.84

(d) 0.74

83. With the value of '*a*' set for a phase-margin of $\frac{\pi}{4}$, the value of unit-impulse response of the open-loop system at $t = 1$ second is equal to

(*a*) 3.40 (*b*) 2.40
(*c*) 1.84 (*d*) 1.74

Statement for Linked Answer Questions 84 and 85

A 30-Volts battery with zero source resistance is connected to a coaxial line of characteristic impedance of 50 Ohms at $t = 0$ second and terminated in an unknown resistive load. The line length is such that it takes 400μs for an electromagnetic wave to travel from source end to load end and vice-versa. At $t = 400$ μs, the voltage at the load end is found to be 40 Volts.

84. The load resistance is

(*a*) 25 Ohms
(*b*) 50 Ohms
(*c*) 75 Ohms
(*d*) 100 Ohms

85. The steady-state current through the load resistance is

(*a*) 1.2 Amps
(*b*) 0.3 Amps
(*c*) 0.6 Amps
(*d*) 0.4 Amps

ANSWERS

1. (*c*)	2. (*d*)	3. (*a*)	4. (*c*)	5. (*a*)	6. (*b*)	7. (*c*)	8. (*b*)	9. (*a*)	10. (*b*)
11. (*d*)	12. (*d*)	13. (*a*)	14. (*a*)	15. (*d*)	16. (*d*)	17. (*d*)	18. (*a*)	19. (*c*)	20. (*a*)
21. (*a*)	22. (*b*)	23. (*d*)	24. (*c*)	25. (*d*)	26. (*c*)	27. (*a*)	28. (*c*)	29. (*a*)	30. (*d*)
31. (*b*)	32. (*b*)	33. (*a*)	34. (*c*)	35. (*c*)	36. (*a*)	37. (*a*)	38. (*b*)	39. (*c*)	40. (*d*)
41. (*c*)	42. (*a*)	43. (*d*)	44. (*b*)	45. (*d*)	46. (*b*)	47. (*a*)	48. (*b*)	49. (*d*)	50. (*b*)
51. (*c*)	52. (*b*)	53. (*c*)	54. (*b*)	55. (*d*)	56. (*b*)	57. (*c*)	58. (*d*)	59. (*a*)	60. (*b*)
61. (*b*)	62. (*b*)	63. (*a*)	64. (*b*)	65. (*b*)	66. (*d*)	67. (*c*)	68. (*a*)	69. (*c*)	70. (*c*)
71. (*c*)	72. (*b*)	73. (*a*)	74. (*c*)	75. (*a*)	76. (*c*)	77. (*b*)	78. (*a*)	79. (*a*)	80. (*c*)
81. (*b*)	82. (*c*)	83. (*c*)	84. (*d*)	85. (*b*)					

EXPLANATIONS

1. Given,
$$\Delta = \begin{bmatrix} 1 & 1 & 1 \\ 1 & -1 & 0 \\ 1 & 1 & 1 \end{bmatrix}$$
$$\det \Delta = \begin{vmatrix} 1 & 1 & 1 \\ 1 & -1 & 0 \\ 1 & 1 & 1 \end{vmatrix} = \begin{vmatrix} 0 & 0 & 0 \\ 1 & -1 & 0 \\ 1 & 1 & 1 \end{vmatrix}$$
$\therefore$ Rank = 2

4.
$$\int_{-\infty}^{\infty} P(x).dx = 1$$
$$\Rightarrow \int_{-\infty}^{\infty} Ke^{-\alpha|x|}.dx = 1$$
or
$$\int_{-\infty}^{0} Ke^{\alpha x}dx + \int_{0}^{\infty} Ke^{-\alpha x}.dx = 1$$
$$\left[\because |x| = x, \text{ for } x > 0 \atop \quad = -x \text{ for } x < 0\right]$$
$$\Rightarrow \frac{K}{\alpha} + \frac{K}{\alpha} = 1$$
$$\Rightarrow K = 0.5\,\alpha$$

5.
$$\dot{x}(t) + 2x(t) = \delta(t)$$
Taking Laplace transform of both sides, we have
$sX(s) - X(0) + 2X(s) = 1$
$$X(s) = \frac{1}{s+2}$$
From Inverse Laplace transform gives, we get
$$x(t) = L^{-1}[X(s)] = e^{-2t}.\,u(t)$$

6. For distortionless transmission, it requires
$$\frac{d.\phi(\omega)}{d\omega} = \text{constant}$$

7.

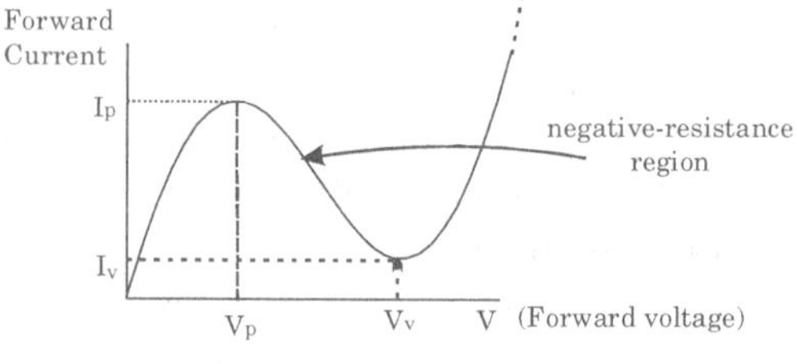

8.
$$np = n_i^2$$
where n_i = intrinsic concentration,
N_d = doping concentration for a *n*-type material

$$n \cong N_d$$

$$\Rightarrow \quad p = \frac{n_i^2}{N_d}$$

$$\Rightarrow \quad p \propto \frac{1}{N_d}$$

where p = Minority carrier concentration

9. $$J_n(x) = \underbrace{q\,u_n(x)\,E(x)}_{\text{Drift term}} + \underbrace{q\,D_n \frac{dn(x)}{dx}}_{\text{Diffusion term}}$$

For minority carrier current, low level injection, $n(x)$ is very less.

i.e. negligible but $\frac{d}{dx}$ of it is significant

Hence, the current is essentially due to diffusion term.

10. In a bipolar transistor, the increase in the depletion region of the reverse biased junction, tends to decrease the width of the base with increased reverse voltage. This in turn causes the current gain to increase with increased voltage resulting in a nearly linear increase in current even with the device in the saturation.

11. 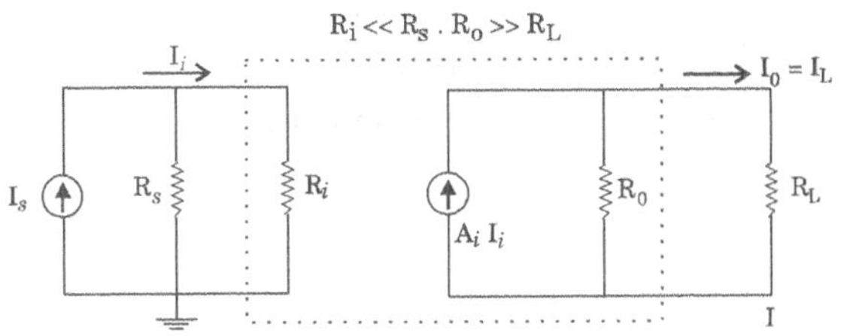

Noton's equivalent circuit of a current amplifier

For ideal transconductance amplifier, output current is proportional to the signal voltage independent of the magnitudes of R_S and R_L.

12. $$g_m = \left.\frac{\partial i_D}{\partial V_{GS}}\right|_{V_{DS}}$$

and, $$I_D = I_{DSS}\left(1 - \frac{V_{GS}}{V_P}\right)^2$$

$$g_m = \frac{\partial I_D}{\partial V_{GS}} = -\frac{2I_{DSS}}{V_p}\left(1 - \frac{V_{GS}}{V_P}\right)$$

g_m will be maximum when $V_{GS} = 0$ and given by,

$$g_{m0} = -\frac{2I_{DSS}}{V_p}$$

13.

AB \ CD	00	01	11	10
00	1	0	0	1
01	0	d	0	0
11	0	0	d	1
10	1	0	0	1

$$\therefore \quad Y = AC\overline{D} + \overline{B}\,\overline{D}$$

14. $$X(\omega) = \int_{-\infty}^{\infty} x(5t-3).e^{-j\omega t}.dt$$

Let, $5t - 3 = p$
or, $dp = 5dt$

$$\therefore \quad X(\omega) = \frac{1}{5}\int_{-\infty}^{\infty} x(p).e^{-\frac{j(p+3)\omega}{5}}.dp$$

$$= \frac{1}{5}\int_{-\infty}^{\infty} x(p).e^{\frac{-jp\omega}{5}}.e^{\frac{-j3\omega}{5}}.dp$$

$$= \frac{1}{5}.e^{-\frac{j3\omega}{5}}\int_{-\infty}^{\infty} x(p).e^{-\frac{jp\omega}{5}}.dp$$

$$= \frac{1}{5}.e^{-\frac{j3\omega}{5}}X\left(\frac{j\omega}{5}\right)$$

15. The heaviside function $H(x)$ is defined.

$$H(x) = \begin{cases} 0 & \text{for } x < 0 \\ 1 & \text{for } x > 0 \end{cases}$$

The derivative of the Heaviside function is zero for $x \neq 0$. At $x = 0$ the derivative is undefined. The derivative of the Heaviside function is the Dirac delta function, $\delta(x)$. The delta function is zero for $x \neq 0$ and infinite at the point $x = 0$. Since the derivative of $H(x)$ is undefined, $\delta(x)$ is not a function in the conventional sense of the word.

The Dirac delta function is defined by the properties.

$$\delta(x) = \begin{cases} 0 & \text{for } x \neq 0, \\ \infty & \text{for } x = 0, \end{cases} \quad \text{and} \int_{-\infty}^{\infty} \delta(x)\,dx = 1$$

The second property comes from the fact that $\delta(x)$ represents the derivative of $H(x)$. The Dirac delta function is conceptually pictured in figure A.

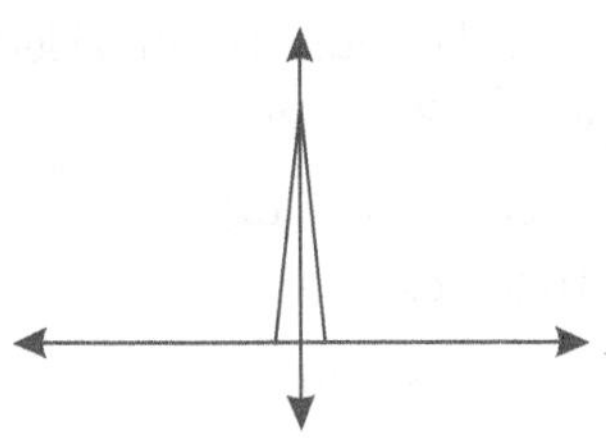

Figure A : The Dirac Delta Function.

16. For **right sided** exponential sequence,

$$x_1[n] = a^n u[n]$$

$$X_1[Z] = \sum_{n=-\infty}^{\infty} a^n u[n] Z^{-n} = \sum_{n=0}^{\infty} (az^{-1})^n$$

Convergence requires that $\sum_{n=-0}^{\infty} |az^{-1}|^n < \infty$

This is possible only or, if $|az^{-1}| < 1$

or, $|z| > |a|$

For **left-sided** exponential sequence

$$x_2[n] = b^{-n} u[-n-1]$$

$$X_2[Z] = \sum_{n=-\infty}^{\infty} -b^n u[-n-1] z^{-n} = 1 - \sum_{n=0}^{\infty} (b^{-1}z)^n$$

$X_2(n)$ converges only if $|b^{-1}z| < 1$ or, $|z| < |b|$

Hence, $x_1[n] + x_2[n]$ have ROC given by

$$\frac{1}{3} < |Z| < \frac{2}{3}, a = \frac{1}{3}, b = \frac{2}{3}$$

For, $x_1[n] - x_2[n]$;

$$X_3[z] = X_1[z] - X_2[z] = \sum_{n=0}^{\infty} (az^{-1})^n - 1 + \sum_{n=0}^{\infty} (b^{-1}z)^n$$

$X_2[z]$ will be converge for $|az^{-1}| < 1$ and $|b^{-1}z| < 1$

or, $|z| < |b|$ and $|z| > |a|$

ROC remains same $\frac{1}{3} < |z| < \frac{2}{3}$

18.

$$H(s) = \frac{1}{s+1}$$

$$H(j\omega) = \frac{1}{i\omega + 1}$$

$$\Rightarrow \quad H(j\omega)|_{\omega=1} = \frac{1}{1+i} = \frac{1}{\sqrt{2}}.\left(\angle -\frac{\pi}{4}\right)$$

For $x(t) = \sin t \,.\, U(t)$

$$y(t) = \frac{1}{\sqrt{2}} \sin\left(t - \frac{\pi}{4}\right)$$

19. Given, $E = \hat{a}_x \sin(\omega t - \beta z) + \hat{a}_y \sin(\omega t - \beta z + \frac{\pi}{2})$

If the phase difference between E_x and E_y components is $\frac{\pi}{2}$, then wave is circularly polarised and when $\hat{E}_y$ leads the wave is left hand circularly polarised and when $\hat{E}_x$ leads the wave is right hand circularly polarised.

20. Input power, $P_i = 1$ W

given, $|G|_{dB} = 10\ dB$

$\Rightarrow \quad 10 \log G = 10$

gain, $G = 10$

Output power, $P_0 = G.P_i = 10$ watts

21. For matrix, $A = \begin{bmatrix} 6 & 2 \\ 2 & 6 \end{bmatrix}$

We know $|\lambda I - A| = 0$

$$\Rightarrow \begin{vmatrix} \lambda - 6 & -2 \\ -2 & \lambda - 6 \end{vmatrix} = 0$$

$\Rightarrow \lambda^2 - 12\lambda + 32 = 0,$

$\Rightarrow \quad \lambda = 4, 8$ (Eigen values)

For $\lambda_1 = 4$, $(\lambda_1 I - A) = \begin{bmatrix} -2 & -2 \\ -2 & -2 \end{bmatrix}$, $v_1 = \begin{bmatrix} 1 \\ 1 \end{bmatrix}$

For $\lambda_2 = 8$, $(\lambda_2 I - A) = \begin{bmatrix} 2 & -2 \\ -2 & 2 \end{bmatrix}$, $v_2 = \begin{bmatrix} 1 \\ -1 \end{bmatrix}$

22. Given, $W = \log_e z$

$\Rightarrow \quad u + iv = \log_e(x + iy)$

$$= \frac{1}{2} \log (x^2 + y^2) + i \tan^{-1}\left(\frac{y}{x}\right)$$

Since, u is constant, therefore

$$\frac{1}{2} \log (x^2 + y^2) = c$$

$\Rightarrow \quad x^2 + y^2 = c'$

which is represented set of concentric circles.

23. Given, $\frac{1}{z^2 + 4} = \frac{1}{(z+2j)(z-2j)}$

Pole (0, 2) lies inside the circle $|z - j| = 2$

By Cauchy's Integral formula,

$$I = \frac{2\pi j}{(2j + 2j)} = \frac{\pi}{2}$$

24. $I = \int_0^{\pi} \sin^3 \theta . d\theta$

$$= \int_0^{\pi} (1 - \cos^2 \theta) \sin \theta . d\theta = -\int_1^{-1} (1 - t^2).dt$$

$$= \int_1^{-1} (t^2 - 1) dt = \left\{ \frac{t^3}{3} - t \Big|_1^{-1} \right\} = \frac{4}{3}$$

26. Given, Matrix $M = \begin{bmatrix} 4 & 2 \\ 2 & 4 \end{bmatrix}$

$$\therefore \quad M - \lambda I = \begin{bmatrix} 4-\lambda & 2 \\ 2 & 4-\lambda \end{bmatrix}$$

For the eigen vector $\begin{bmatrix} 101 \\ 101 \end{bmatrix}$,

$$(4 - \lambda)(101) + 2(101) = 0$$

$\Rightarrow \quad 4 - \lambda + 2 = 0$

$\Rightarrow \quad \lambda = 6$ (eigen value)

27. Given, Differential equation,

$$\frac{d^2y}{dx^2} + k^2 y = 0$$

Let $y = A \cos kx + jB \sin kx$

At $x = 0$, $y = 0$

$\Rightarrow \quad A = 0$

$y = jB \sin kx$

At $x = a, \quad y = 0$

$\Rightarrow \quad B \sin ka = 0$

or $B \neq 0$ otherwise $y = 0$ always

$\Rightarrow \quad \sin ka = 0$

$\Rightarrow \quad k = \frac{m\pi x}{a}$

$\therefore \quad y = \sum_m A_m \sin\left(\frac{m\pi x}{a}\right)$

28. $f(t) = L^{-1} f(x)$

$= \sin w_0 t$

As, $-1 \le \sin\theta \le 1$

Thus, $-1 \le f(\infty) \le 1$

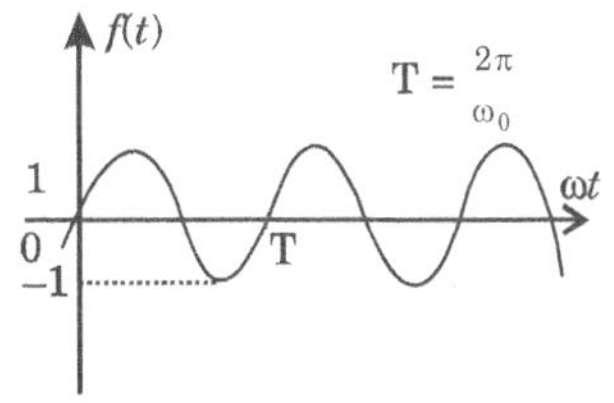

29. Given, $\quad f(x) = \frac{e^x}{1+e^x}$

$\therefore \quad f'(x) = \frac{(1+e^x).e^x - e^{2x}}{(1+e^x)^2} = \frac{e^x}{(1+e^x)^2}$

which is always +ve, so monotonically increases

30. $V_1 = AV_2 - BI_2$

$I_1 = CV_2 - DI_2$

From $\quad V_2 = -I_2R_1$

we get $\quad \frac{V_1}{I_1} = Z_{in} = \frac{AR_L + B}{D + CR_L}$

31. $V_1 = r_e I_1 \quad$ ...(*i*)

$I_2 = \beta I_1 + \frac{V_2}{r_0} \quad$ ...(*ii*)

Rearranging equation (*ii*), we get

$V_2 = -\beta r_0 I_1 + r_0 I_2 \quad$ ...(*iii*)

By equations (*i*) and (*iii*), we get

$Z_{12} = 0, Z_{21} = -\beta r_0$

32.

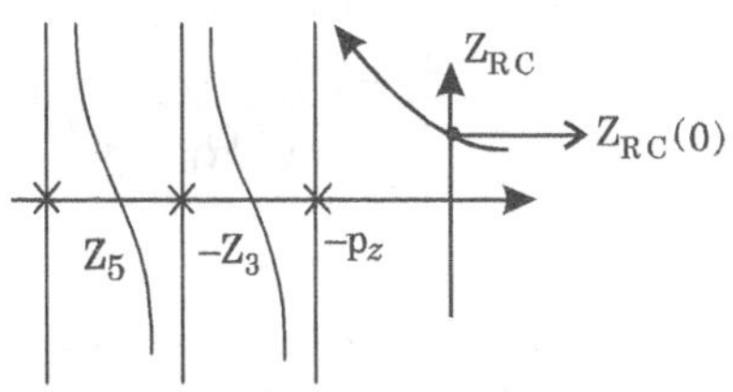

From the plot of Z_{RC} it is clear that first and last critical frequency are a pole and zero.

33. $I_L = \frac{1}{L}\int_0^t v_L \, dt$

$I_L(s) = \frac{v_L(s)}{Ls} + \frac{v_L(0+)}{L}$

Initial value of current, $I_L(0+) = \underset{s \to \infty}{I_L(s)}$

$\Rightarrow \quad I_L(0+) = \frac{v_L(0+)}{L} = \frac{1 \times 10^{-3}}{2 \times 10^{-3}} = 0.5\,A$

34. $\frac{V_0(s)}{V_i(s)} = \frac{\frac{R_0}{R_0C_0s+1}}{\frac{R_0}{R_0C_0s+1} + \frac{R_i}{R_iC_is+1}}$

$= \frac{R_0(R_iC_is+1)}{R_0R_i(C_i+C_0)s + R_0 + R_i}$

$= \frac{4K(4s+1)}{4K^2 5s + 5K} = \frac{4}{5}$

$V_0(s) = \frac{8}{s}$

$\Rightarrow \quad V_0(t) = 8u(t)$

35. Transfer functions $G_1(s) = \frac{1}{s^2 + as + b}$

and $\quad G_2(s) = \frac{s}{s^2 + as + b}$

3-*d*B bandwidths depends on the denominator and is specified by the discriminant

$D = \sqrt{a^2 - 4b}$

$\therefore \quad$ Bandwidth (BW) $= \sqrt{a^2 - 4b}$

36. For positive real function,

$R_e|Z_2(s)| \ge 0$

As, $\quad Z_2(1) = R_{neg} + R_e(Z_1(j\omega))] + jI_m[Z_1(j\omega)]$

$\therefore \quad R_{neg} + R_e[Z_1(j\omega)] \ge 0$

or $\quad R_e[Z_1(j\omega)] \ge -R_{neg}$

or $\quad R_e[Z_1(j\omega)] \ge |R_{neg}|$

37.

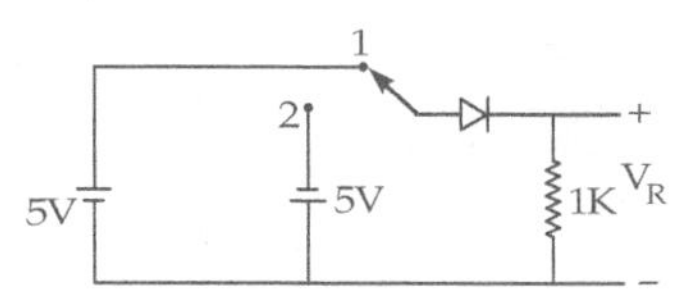

The injected charge carriers across the junction can not be removed instantaneously, but takes certain time (t_s). Thus for time t_s, diode will remain forward biased and will conduct.

Since, voltage drop across diode is zero, hence

$V_R = -5V$ for $0 \le t < t_s$.

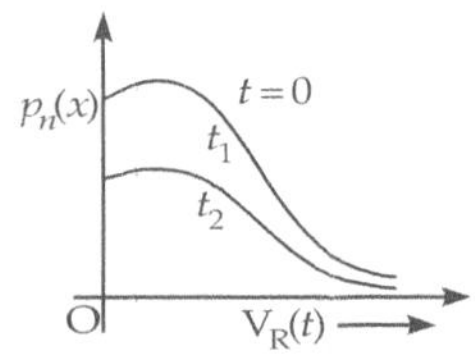

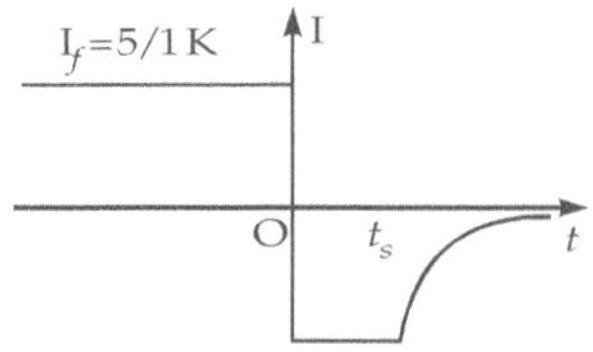

38. Lorentz force, $F = q(\overline{V} \times \overline{B})$

$= -e(VB)\overline{a}_x \times \overline{a}_z$

The *n*-type carriers will be carried towards back face. Thus clearly an electric field along negative *y*-axis, i.e. in the direction of $\overline{B} \times \overline{V}$.

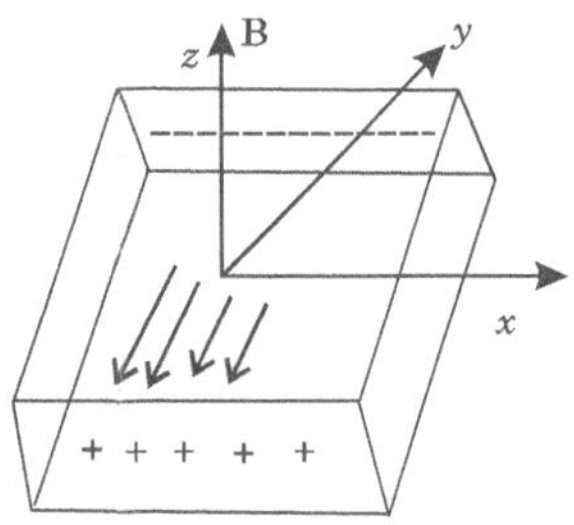

40. $$\text{Ratio} = \frac{qn\mu_u}{qn_i(\mu_u + \mu_p)} = \frac{n}{n_i\left(1 + \frac{\mu_p}{\mu_u}\right)}$$

$$= \frac{4.2 \times 10^8}{1.5 \times 10^4(1 + 0.4)} = 20{,}000$$

41. Virtual grand is Not applicable, became op-amp is in saturation

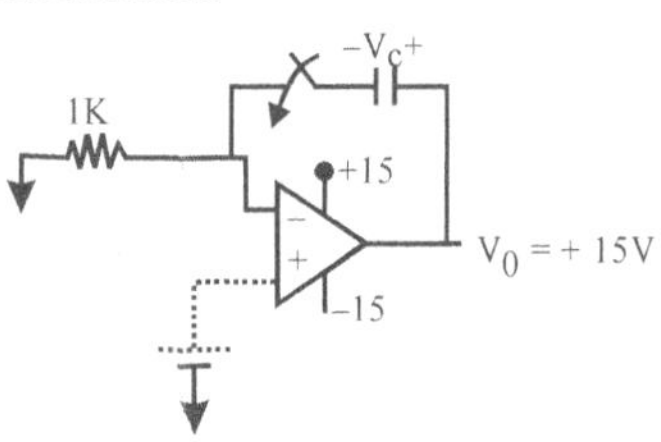

$V_0 = +15 = +V_{sat}$

$V_c = 15(1 - e^{-\frac{t}{R_C}})$

$V_c = 15(1 - e^{-1000t})$

at $t = 1$ ms

$V_c = 9.45$ Volts

42.

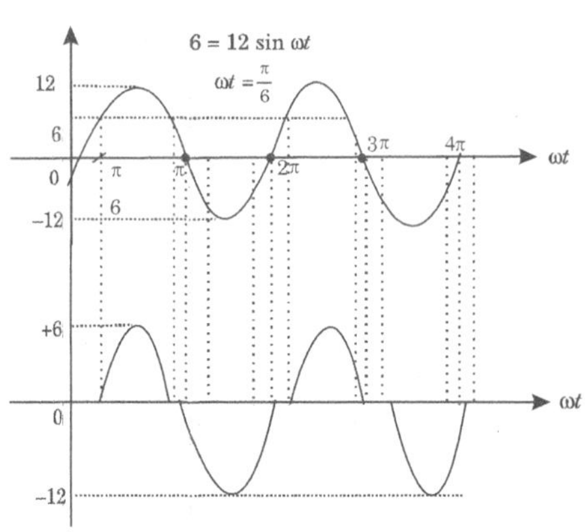

During $0 < \omega t < \frac{\pi}{6}$

Diode is OFF, No conduction

then, $V_R = 0$ volts

In the interval, $\frac{\pi}{6} < \omega t < \pi$

Zener diode voltage, $V_Z = 6$ volts

then, $V_0 = 12 \sin \omega t - V_Z$

$= 12 \sin \omega t - 6$

In the interval, $\pi < \omega t < 2\pi$, zener diode is in conducting state, $V_2 = 0$

then, $V_0 = 12 \sin \omega t$

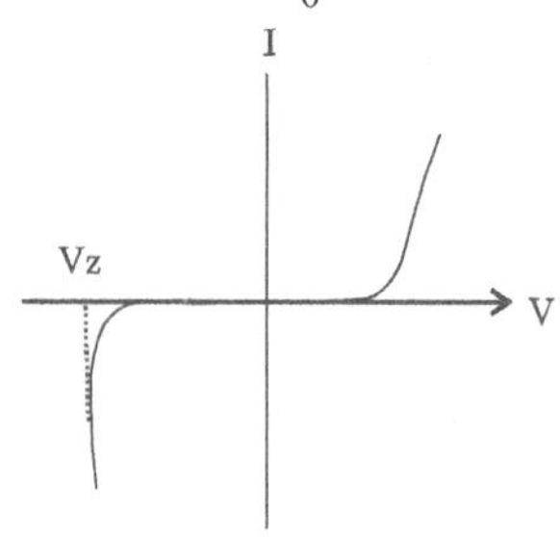

43. According to Binary Coded Pentary (BCP) number system,

24 is represented by

32 16 8   4 2 1
010   100
↓   ↓
2   4

Thus, 100 010 011 001 corresponds to 4231
(100 → 4, 010 → 2, 011 → 3, 001 → 1)

44.

A_7	A_6	A_5	A_4	A_3	A_2	A_1	A_0	
1	1	0	1	0	1	0	0	– D_4 H
1	1	0	1	1	1	1	1	– D_7 H

A_4	A_3	A_2	
1	0	1	= 5

Thus, to start to interface with address range $D_4H - D_7H$ to $\overline{CS}$ should connected with output 5.

45. Initially, $S_0 = 0$, $C_0 = 1$

After the first clock pulse,

$S_1 = Q_A * Q_B = 0$

$C_1 = Q_A \cdot Q_B = 1$

After the second clock pulse,

$S_2 = [Q_A * Q_B] + S_1 = 1$

$C_2 = Q_A \cdot Q_B = 1$

46.

C_K	D_3	D_2	D_1	D_0
1	0	0	0	0
2	0	0	0	1
3	0	0	1	0
4	0	0	1	1
5	0	0	0	0
6	0	0	0	1
7	0	0	1	0
8	0	0	1	1
9	1	0	0	0
10	1	0	0	1
11	1	0	1	0
12	1	0	1	1
13	1	0	0	0
14	1	0	0	1
15	1	0	1	1

From $C_K - 5$ to $C_K - 8$, count will be constant and output will rise.

48.(*b*) Location Insturction Operation

L X I SP, EFFFH SP → EFFFH

program transfers to memory location 3000H

CALL 3000H

3000H : L X IH, 3CFGH HL → 3CF4

PUSH PSW SP → EFFF → ///////

SP-1 → EFFE → X

SP-2 → EFFD → Y

X and Y are accumulators contents and flag register contents loaded into memory location EFFE and EFFD respectively

SPHL → SP → 3CF4 contents of H-L register loaded into SP.

POP PSW SP → 3CF4 Flag register → Z

SP → 3CF5 Accumulator → W

SP → 3CF6

Z and W are contents of memory locate 3CF4 and 3CF5 respectively

RET

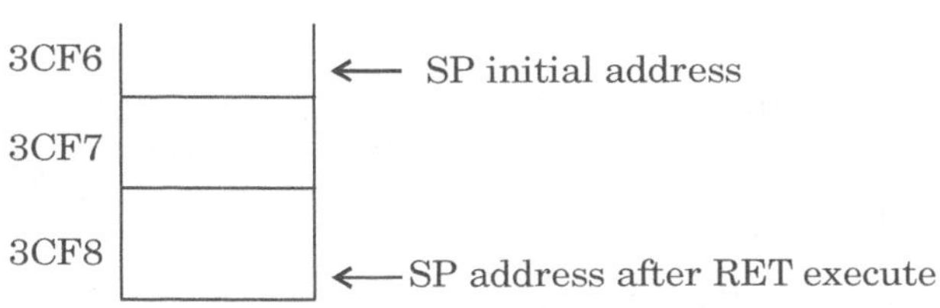

Hence SP → 3CF8H

49.

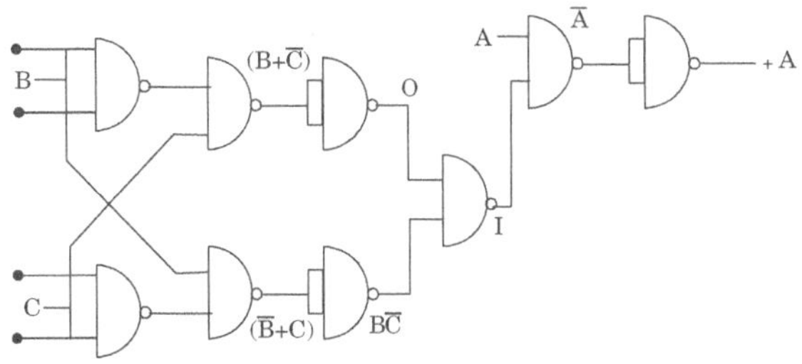

50. Given, $g(t) = \sum_{k=-\infty}^{\infty} (-1)^k \delta(t - 0.5\times10^{-4} k)$

Fourier series expansion of $g(t)$,

$$g(t) = \frac{a_0}{2} + \sum_{n=1}^{\infty} \left(a_n \cos 2\pi n f_0 t + b_n \sin 2\pi n f_0 t\right)$$

where $f_0 = 20\times10^3$Hz, $T_0 = 0.5\times10^{-4}$T

as $g(t)$ is even function of time, then $b_n = 0$

$$a_0 = \frac{2}{T}\int_{-T_0/2}^{T_0/2} g(t)\,dt - \frac{4}{T_s}\int_0^{T_0/2}\left[g(t) - g\left(t - \frac{T_0}{2}\right)\right]dt$$

$$= \frac{4}{T_s}\int_0^{T_0/2} g(t)\cdot dt - \frac{4}{T_s}\int_0^{T_0/2} g\left(t - \frac{T_0}{2}\right)dt = \frac{4}{T_0} - \frac{4}{T_0} = 0$$

$$\left[\int_{-\infty}^{\infty} g(t)\,dt = 1\right]$$

$$a_n = \frac{4}{T_0}\int_0^{T_0/2} g(t)\cos\frac{2\pi n}{T_0}\cdot t\,dt - \frac{4}{T_s}\int_0^{T_s/2} g\left(t - \frac{T_0}{2}\right)\cos\frac{2\pi n}{T_0}\cdot t\,dt$$

$$[g(t)\,x(t) = x(0)\,g(t)]$$

$$= \frac{4}{T_0}\int_0^{T_s/2} g(t)\,dt - \frac{4}{T_0}\cos n\pi \int_0^{T/2} g\left(t - \frac{T_0}{2}\right)dt = \frac{4}{T_0} - \frac{4}{T_0}\cos\pi n$$

$$[g(t - t_0)\,x(t) = x(t_0)\cdot g(t - t_0)]$$

then, $g(t) = \sum_{n=1}^{\infty} \frac{4}{T_0}\left[1 - (-1)^n\right]\cos(2\pi f_0 n\, t)$

$$g(t) = \frac{8}{T_0}\cos(40\pi\times10^3 t) + \frac{8}{T_0}\cos(60\pi\times10^3 t) + \ldots$$

Frequency spectrum of M(*f*) =

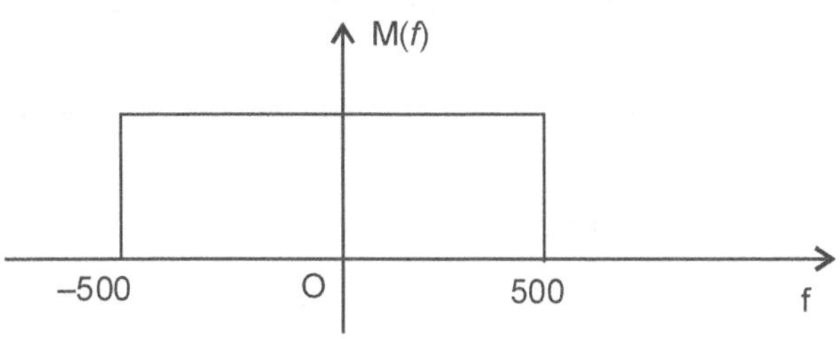

Frequency spectrum of G(*f*)

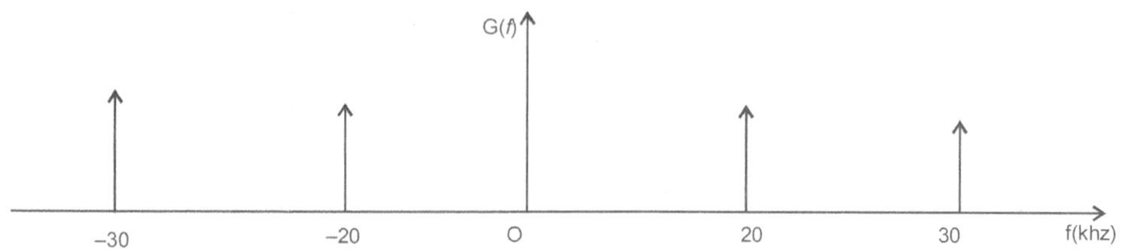

As $m(t)\,g(t) = M(f) \times G(f)$

Multiplication in time domain is equivalent to convolution in frequency domain

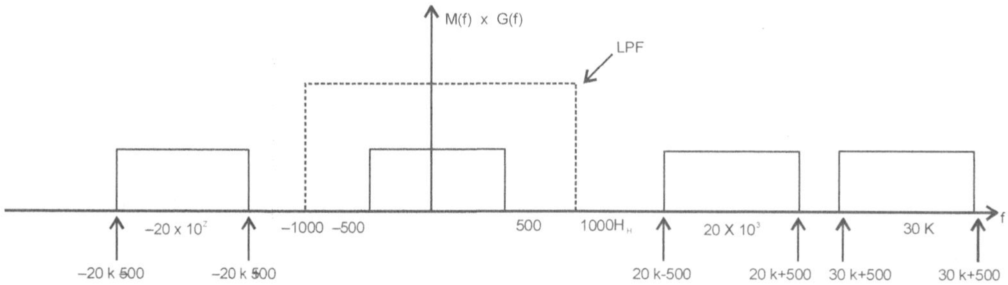

When the resulting signal passes through LPF having BW of 1 KHz, then only message signal $m(t)$ produces at the output.

51. The Bandwidth of $x(t)$ is determined by the highest frequency component of $[\sin c\,(2000\,t)]^3$ or $[\sin c\,(2000\,t)]^2$, whichever is the largest.

Since multiplication in the time domain compounds to convolution in the frequency domain, signal $[\sin c\,(2000t)]^3$ has a bandwidth equal to thrice to that of $[\sin c\,(2000\,t)]$. Further signal $[\sin c\,(2000\,t)]^2$ has a bandwidth equal to twice to that of $[\sin c\,(2000\,t)]$.

as $$\sin c(2wt) \rightleftharpoons \frac{1}{2w}\text{rect}\left(\frac{f}{2w}\right)$$

Signal $[\sin c\,(2000t)]$ has a bandwidth 1000 Hz and Nyquist rate = 2000 Hz

Hence, $x(t)$ has a sampling rate 3×2000 Hz $= 6 \times 10^3$

52.

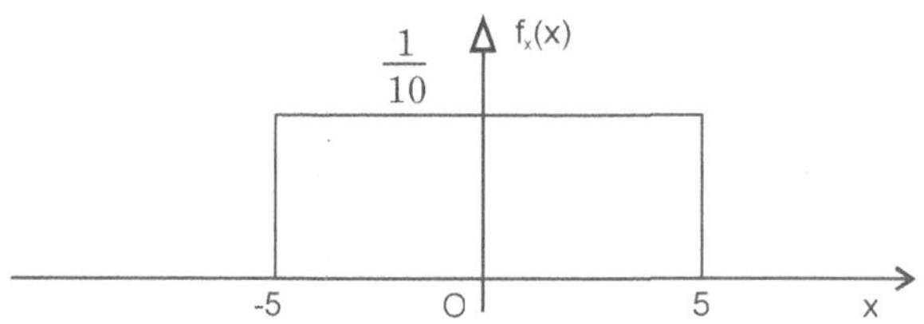

Now, $$f_Y(y) = \sum_{i=1}^{2} P(Y = y_i)\delta(y - y_i)$$

For $$i = 1,\ y_i = 0$$

$$P(Y = 0) = P(X < 2.5) + P(2.5 \le X < \infty)$$

$$= \int_{-\infty}^{-2.5} f_X(x)\cdot dx + \int_{2.5}^{-\infty} f_X(x)\cdot dx = \int_{-5}^{-2.5}\frac{1}{10}\,dx + \int_{2.5}^{5}\frac{1}{10}\,dx = 0.5$$

For $$i = 2,\ Y_2 = 1,\ P(Y = 1) = P(-2.5 \le X \le 2.5)$$

$$= \int_{-2.5}^{2.5} f_x(x)\cdot dx = \int_{-2.5}^{2.5}\frac{1}{10}\,dx = 0.5$$

Hence, $$f_Y(y) = \sum_{i=1}^{2} P(Y = y_i)\delta(y - y_i)$$

$$= P(Y = 0)\,\delta(y - 0) + P(Y = 1)\,\delta(Y - 1)$$

$$= 0.5\,\delta(Y) + 0.5\,\delta(Y - 1)$$

53. $y[n] = \left(\sin\frac{5}{6}\pi n\right) x(n)$

As $-1 \le \sin\frac{5}{6}\pi n \le 1$,

Then, for bounded $x[n]$, output $Y[n]$ is also bounded and the system is stable.

If any system to be invertible, input can be determined from the output of the system. For this to be true two different input signals should produce two different outputs. If some different input signals produce the same output signal then by processing output it can not be said which input produced the output.

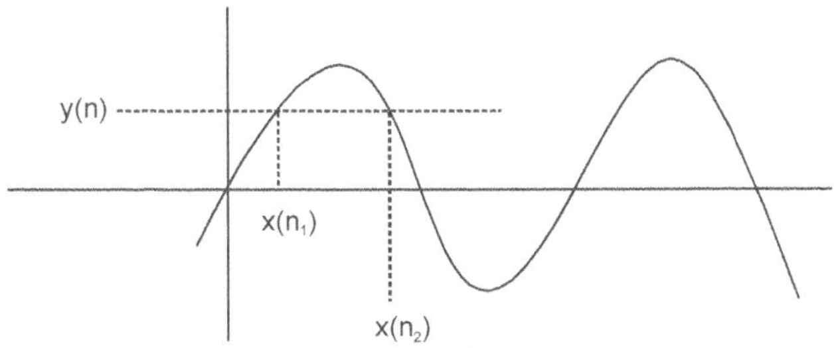

For different input $x(n_1)$ and $x(n_2)$, the output remains same. Therefore, it can not be ascentained which input produce the output and system is non-invertible.

54.
$$C(t) = 1 - e^{-2t}$$
$$\therefore \quad C(s) = \frac{1}{s} - \frac{1}{s+2} = \frac{2}{s(s+2)}$$
and
$$R(s) = L\{v(t)\} = \frac{1}{s}$$
$\therefore$ Transfer function, $H(s) = \dfrac{C(s)}{R(s)} = \dfrac{2}{s+2}$

55. GM $= 20\log\frac{1}{a} = 20 \log 1 = 0\ dB$

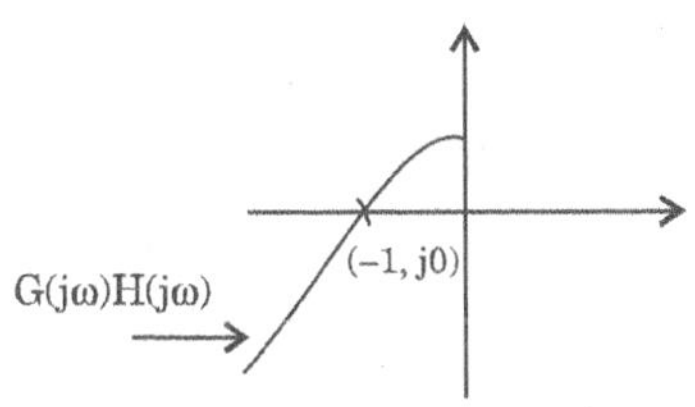

56. Characteristic equation is given by,
$$1 + G(s) = 0$$
$$\Rightarrow \quad s^3 + as^2 + s(K+2) + K + 1 = 0$$
Constructing Routh-Kuruoritz array,

$$\begin{array}{c|cc} s^3 & 1 & K+2 \\ s^2 & a & K+1 \\ s^1 & \dfrac{a(K+2)-K-1}{a} & 0 \\ s^0 & K+1 & \end{array}$$

For oscillations,
$$as^2 + K + 1 = 0$$
or
$$-\omega^2 a + K + 1 = 0$$
or
$$K + 1 = 4a \quad ...(i)$$
and
$$\frac{a(K+2)-K-1}{a} = 0$$
or
$$a(K+2) = K+1 \quad ...(ii)$$
Using equations (*i*) and (*ii*), we get
$$K = 2,\ a = 0.75$$

57.
$$H(s) = L\{h(t)\} = \frac{1}{s+1}$$
$$C(s) = H(s) \cdot R(s)$$
$$= \frac{1}{(s+1)} \cdot \frac{1}{s} = \frac{1}{s} - \frac{1}{s+1}$$
$$C(t) = 1 - e^{-t}$$

At steady state, $C(\infty) = 1$

58. $\phi = \angle G_c(s)\big|_{s=j\omega} = -\tan^{-1} T\omega + \tan^{-1} 3T\omega$

For maximum phase-shift;
$$\frac{d\phi}{d\omega} = 0 = \frac{-T}{1+(T\omega)^2} + \frac{3T}{1+(3T\omega)^2}$$
$$\Rightarrow \quad \frac{1}{1+(T\omega)^2} = \frac{3}{1+(3T\omega)^2}$$
$$\Rightarrow \quad \omega T = \frac{1}{\sqrt{3}}$$
$$\therefore \quad \phi_{max} = \tan^{-1}\sqrt{3} - \tan^{-1}\frac{1}{\sqrt{3}} = \frac{\pi}{6}$$

59. $\phi(t) = L^{-1}[SI - A]^{-1} = L^{-1}\begin{bmatrix} s & -1 \\ 1 & s \end{bmatrix}^{-1}$
$$= L^{-1}\begin{bmatrix} \dfrac{s}{s^2+1} & \dfrac{1}{s^2+1} \\ \dfrac{-1}{s^2+1} & \dfrac{s}{s^2+1} \end{bmatrix} = \begin{bmatrix} \cos t & \sin t \\ -\sin t & \cos t \end{bmatrix}$$

62. Entropy, $H(x) = -\sum_{i=1}^{m} P(x_i)\log_2 P(x_i)$

For three independent sources,
$$H(X) = -\sum_{i=1}^{3} P(x_i)\log_2 P(x_i)$$
$$= -\left[P(x_1)\log_2 P(x_1) + P(x_2)\log_2 P(x_2) + P(x_3)\cdot\log_2 P(x_3)\right]$$
$$= -\left[0.25\log_2 0.25 + 0.25\log_2 0.25 + 0.50\log_2 0.50\right]$$
$$= \frac{3}{2} \text{ bits/symbol}$$

Symbol rate, $r = 3000$ symbols/sec

Average bit rate $R = rH(X)$

$$= \frac{3}{2} \times 3000 = 4500 \text{ bits/sec}$$

64. The minimum value that can make $y_1(t)$ different from $y_2(t)$ is

$$\pm \frac{\text{LSB}}{2} = \frac{\Delta}{2}$$

65.

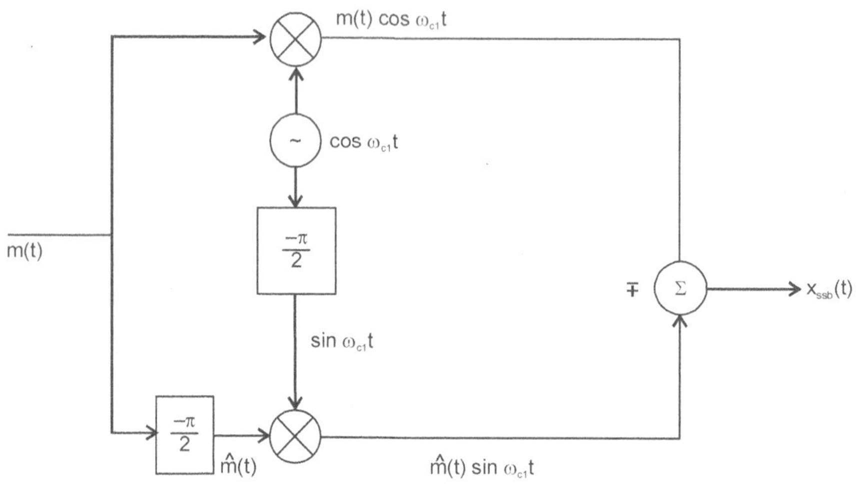

$\hat{m}(t)$ is the Hilbert transform of $m(t)$

$$\omega_{c1} = 2\pi f c_1$$
$$m(t) = A_m \sin(2\pi f_m t)$$

SSB-modulator

$$x_{\text{SSB}}(t) = m(t)\cos\omega_{c_1} t \mp \hat{m}(t)\sin\omega_{c_1} t$$

Lower-side-band SSB signal is represented as,

$$x_{\text{SSB}}(t) = m(t)\cos\omega_{c_1} t + \hat{m}(t)\sin\omega_{c_1} t \quad \text{...(A)}$$

$$= A_m \cos\omega_{c_1} t \sin(2\pi f_m t) - A_m \sin\omega_{c_1} t \cdot \cos(2\pi f_m t)$$

Now, Narrow band FM modulator is shown below to which SSB signal is being applied–

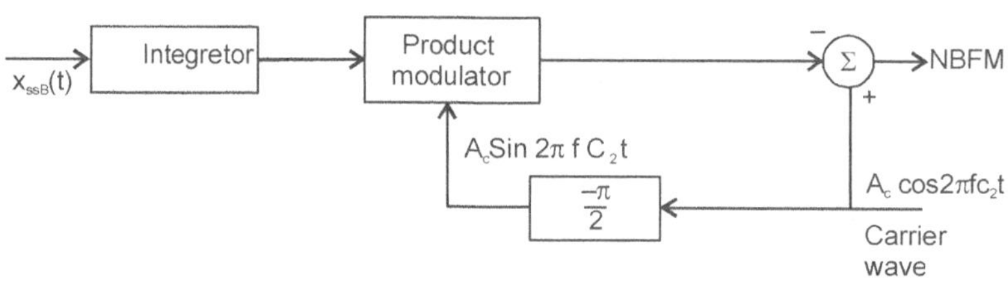

NBFM wave is represented as

$$x_{\text{NBFM}}(t) = A_c \cos(2\pi f_{c_2} t) - \beta A_c \sin(2\pi f_{c_2} t)\sin(\omega_{c_1} - \omega_m)t$$

$$= A_c \cos\omega_{c_2} t - \frac{\beta A_c}{2}\left[\cos(\omega_{c_2} - \omega_{c_1} + \omega_m)t - \cos(\omega_{c_2} + \omega_{c_1} - \omega_m)t\right]$$

BW of x_{NBFM} is

$$= 2(\omega_{c_1} - \omega_m)$$

$$\simeq 2\omega_{c_1} \approx 2f_{c_1} \approx 2 \times 10^6 \text{ Hz}$$

66. $\sin\theta = \frac{1}{\sqrt{\varepsilon_r}} = \frac{1}{\sqrt{2}}$

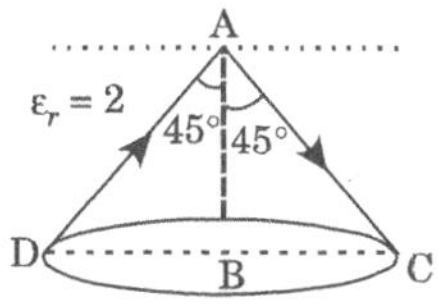

$\therefore$ θ = 45° BD = AB = 1 m

Area = π × DB² = π m²

67.
$$\bar{E}_{t_1} = \bar{E}_{t_2}$$
$$\bar{D}_{x_1} = \bar{D}_{x_2}$$

then, $\bar{E}_{y_1} = \bar{E}_{y_2},\ \bar{E}_{z_1} = \bar{E}_{z_2}$

and $\bar{D}_{x_1} = \bar{D}_{x_2}$

or $\varepsilon_{r_1}.\bar{E}_{x_1} = \varepsilon_{r_2}.\bar{E}_{x_2}$

or $\bar{E}_{x_2} = \frac{\varepsilon_{r_1}}{\varepsilon_{r_2}}.\bar{E}_{x_1} = \frac{3}{4}4\hat{a}_x = 3\hat{a}_x$

then $\bar{E}_2 = 3\hat{a}_x + 3\hat{a}_y + 5\hat{a}_z$

68. For free space, intrinsic impedance

$$\eta_1 = \sqrt{\frac{\mu_0}{\epsilon_0}} = 120\pi\Omega$$

and for medium having $\in_r = 4$, $_r = 1$

$$\eta_2 = \sqrt{\frac{\mu_0\mu_r}{\epsilon_0\epsilon_r}} = \sqrt{\frac{\mu_0}{\epsilon_0}}\cdot\frac{1}{\sqrt{\epsilon_r}} = 60\pi = \frac{\eta_1}{2}$$

Power transmitted,

$$P_t = P_i\cdot\frac{\eta_1}{\eta_2}\left(\frac{2\eta_2}{\eta_1+\eta_2}\right)^2$$

$$= P_i\cdot\frac{2\eta_2}{\eta_2}\left(\frac{2\eta_2}{2\eta_2+\eta_2}\right)^2 = P_i\cdot\frac{8}{9}$$

69. Cut-off frequency, f_c for TE_{mn} mode is given by,

$$f_c = \frac{C}{2}\sqrt{\left(\frac{m}{a}\right)^2 + \left(\frac{n}{b}\right)^2}$$

For, TE_{30} mode, $m = 3$, $n = 0$, $C = 3 \times 10^8$ m/s

$$f_c = 18\times10^9 = \frac{3\times10^8}{2}\sqrt{\left(\frac{3}{a}\right)^2}$$

$$\Rightarrow \frac{3}{a}\cdot\frac{3\times10^8}{2} = 18\times10^9$$

$$\Rightarrow a = \frac{5}{2}\text{ cm}$$

70.
$$R_{rad} = 80\pi^2\left(\frac{d}{\lambda}\right)^2$$

$$\therefore \quad \lambda = \frac{C}{V} = \frac{3\times10^8}{6\times10^5} = (0.5\times10^3\text{m})$$

$$\therefore \quad R_{rad} = 80\pi^2\left(\frac{50}{500}\right)^2 = 0.8\pi^2$$

71. Under DC conditions, capacitors get opened and the circuit becomes as,

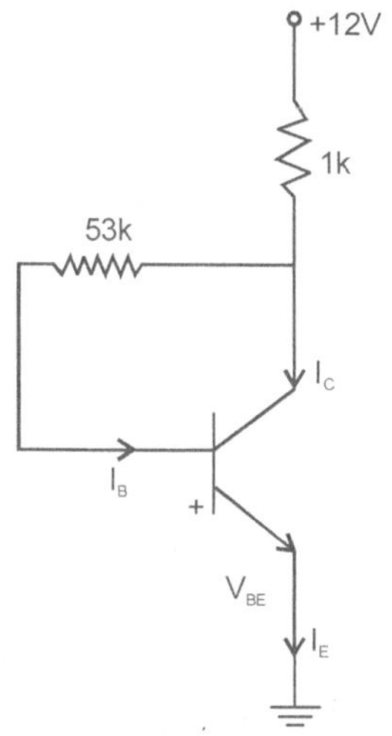

Using KVL in B-E loop of transistor

We get,

$12 - 10^3(I_B + I_C) - 53 \times 10^3 I_B - 0.7 = 0$ ...(*i*)

and $I_C = \beta_{DC} \times I_B$ ...(*ii*)

Using equation (*ii*) in eqn (*i*), we get,

$$11.3 - 61\times10^3 I_B - 53\times10^3 I_B = 0$$

$\Rightarrow I_B = 9.912\times10^{-5}\text{ A} \approx 0.1\text{ mA}$

and $V_{CE} = 12 - (I_C + I_B)\,10^3$

$= 12 - 61 \times 0.1 \approx 6$ volts

72. $\beta'_{dc} = 1.1 \times \beta_{dc} = 66$

then $I'_C = \beta'_{dc}\cdot I'_B$

from equation (1),

$$11.3 = 53\times10^3 I'_B + 10^3\times67\times I'_B$$

$$I'_B = \frac{11.3}{(53+67)\times10^3} = 9.41\times10^{-5}\text{A}$$

and,

$$V_{CE} = 12 - \beta'_{dc}I'_B = 12 - 67\times9.41\times10^{-5} = 5.69\text{ volt}$$

$$\%\text{ change in voltage} = \frac{5.69-5.95}{5.95}\times100$$

$= -4.369\%$

[decreases by less than 10%]

73. For $h_{ie} \to \infty$, $h_{fe} \to \infty$, drawing small-signal gain circuit,

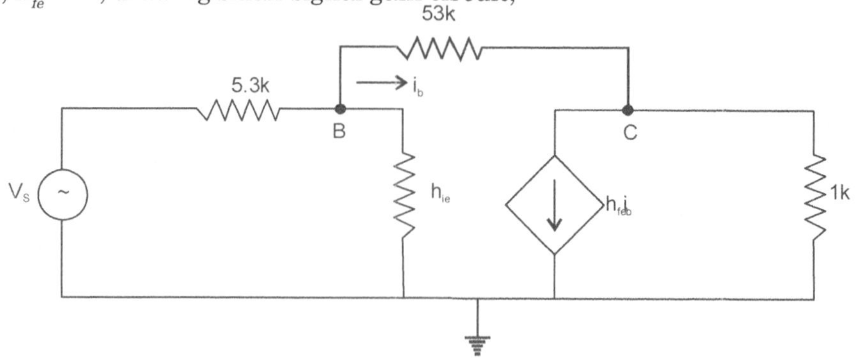

As h_{ie} and h_{fe} tends to ∞, the circuit will be same as the ideal OPAMP with $R_i = \infty$ and $A_V = \infty$.

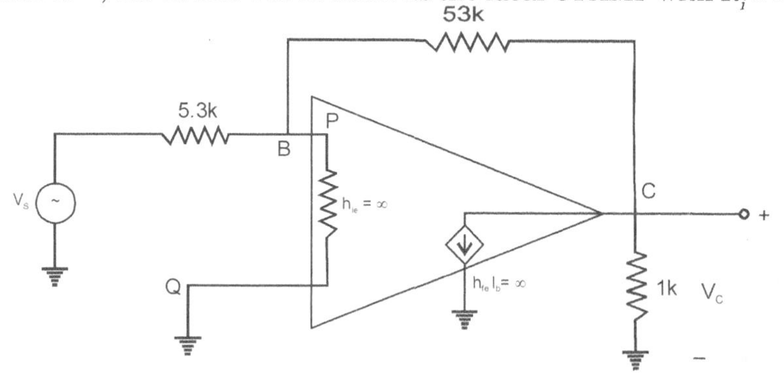

Applying virtual ground condition at terminal P and Q, and Using KCL at node P, we get

$$\frac{V_S}{5.3K} + \frac{V_c}{53K} = 0$$

$$\Rightarrow \qquad \frac{V_C}{V_S} = -10$$

74.(*c*) $g(t) = P(t) \cdot P(t)$

Taking laplace transform both sides.

$\Rightarrow \quad G(s) = P(s) \cdot P(s)$

$$P(s) = \mathcal{L}[U(t) - U(t-1)] = \frac{(1-e^{-s})}{s}$$

then,
$$g(t) = \mathcal{L}^{-1}\left(\frac{1-e^{-s}}{s}\right)^2 = \mathcal{L}^{-1}\left[\frac{1-2e^{-s}+e^{-2s}}{s^2}\right]$$
$$= r(t) - 2r(t-1) + r(t-2)$$

Now,
$$s(t) = g(t) - g(t-2) \cdot g(t) = g(t) - g(t-2) \qquad [\text{Using } x(t)\,\delta(t-t_0) = x(t-t_0)]$$
$$= [r(t) - 2r(t-1) + r(t-2) - r(t-2) + 2r(t-3) - r(t-4)]$$
$$s(t) = [r(t) - 2r(t-1) + 2r(t-3) - r(t-4)]$$

$s(t)$ is depicted in the figure below:

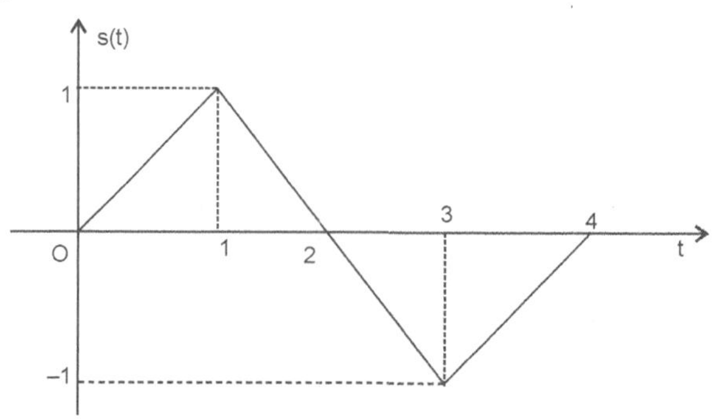

Matched filter response, $h(t) = s(T - t) = s(4 - t)$

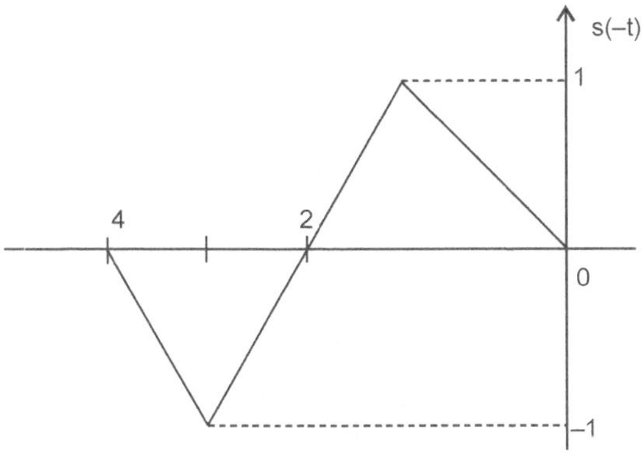

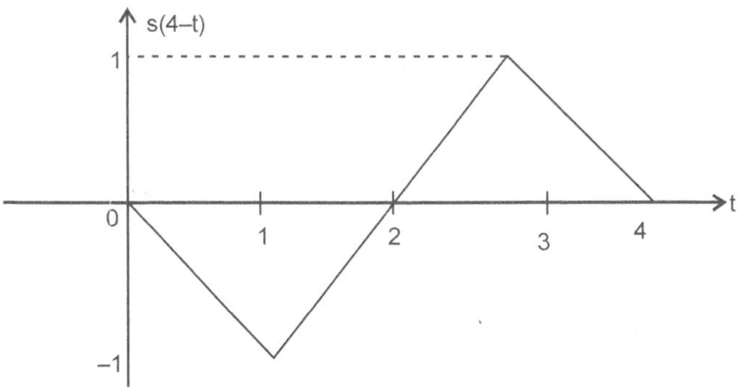

It can be seen that, $h(t) = -s(t)$

75.(*a*) $x(t)$ is depicted in the figure below:

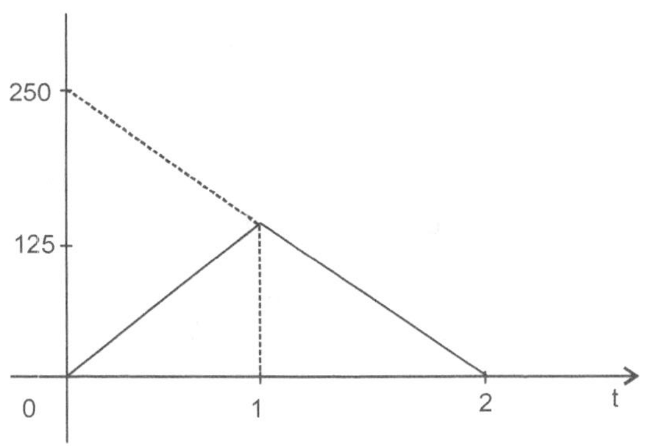

Minimum slope of $x(t)$, $\left|\frac{dx(t)}{dt}\right|_{max} = 125$

To avoid slope-overload direction,

$$\left|\frac{dx(t)}{dt}\right|_{max} \leq \frac{\Delta}{T_s}$$

$\Rightarrow \quad 125 \leq \Delta f_s$

$\Rightarrow \quad \Delta \geq \frac{125}{32 \times 10^3}$

$\Rightarrow \quad \Delta \geq 2^{-8}$

Minimum value of step, $\Delta_{min} - 2^{-8}$

75.(*b*)

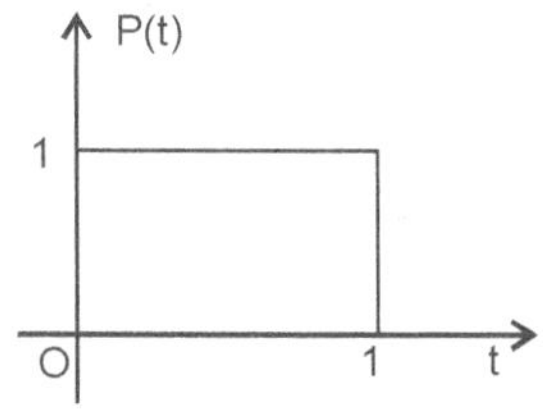

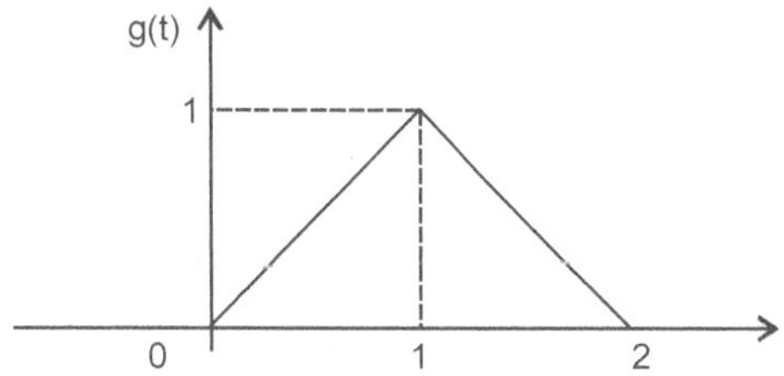

$g(t) = r(t) - 2r(t-1) + r(t-2)$

$x_{AM}(t) = 100[P(t) + 0.5\,g(t)] \cos \omega_c t$

In the interval $0 < t \leq 1$

$x_{AM}(t) = 100[1 + 0.5t] \cos \omega_c t$

Comparing with standard equation for the AM

$$x_{AM}(t) = A_C(1 + \quad \cos \omega_m t) \cos \omega_c t$$
$$= (A_C + m(t)) \cos \omega_c t$$

We have, $\quad = 0.5$ (modulation index)

and, $m(t) = t$ (modulating signal)

76.

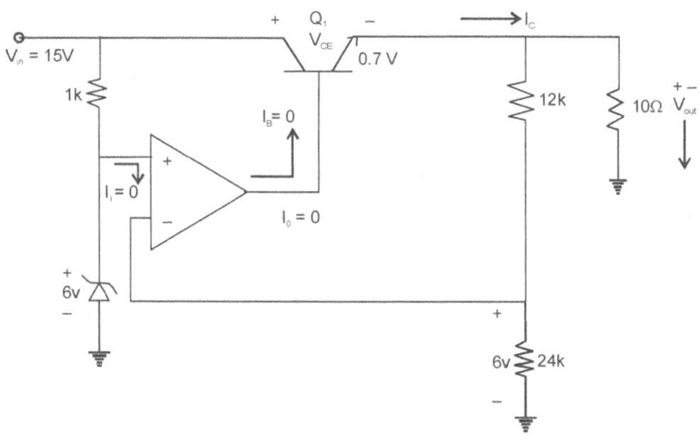

Voltage accross 24k resistor = $V_2 = 6V$

then, $\quad V_2 = 6 = \frac{V_0 \times 24}{12 + 24}$

$\Rightarrow \quad V_{out} = 9$ volts

and $\quad I_C = \frac{V_0}{(12k + 24k) \,||\, 10\Omega} = 900.25\ mA$

Using KVL in transister circuit, we have

$V_{in} - V_{CE} - V_0 = 0$

$\Rightarrow \quad V_{CE} I_c = V_{in} - V_0 = 15 - 9 = 6$ volts

Power discipation across transistor Q_1

$= V_{CE}I_C = 9 \times 900.25 \times 10^{-3}$

$= 5.4$ watts

77. If unregulated voltage V_{in} increased by 20%, then

$$V'_{in} = 15 \times 1.2 = 18 \text{ volts}$$

As output voltage is regulated, therefore it remains same.

Then, $V_{CE} = V'_{in} - V_0 = 18 - 9 = 9$ volts

Further collectors current also remains unchanged.

Then, power discipation accross Q_1 is

$= V_{CE} \times I_C = 9 \times 900.25$ mn $= 8.1$ wats

Hence, % change in power discipation

$$= \frac{8.1 - 5.4}{5.4} \times 100 = 50\%$$

78. $S_{YY}(\omega) = |H(\omega)|^2 . S_{XX}(\omega)$

Given, $S_{XX}(\omega) = 1$ [while noise]

$$S_{YY}(\omega) = \frac{16}{16+\omega^2} = |H(\omega)|^2$$

$$\Rightarrow \quad |H(\omega)| = \frac{4}{\sqrt{16+\omega^2}}$$

$\therefore$ $H(s) = \frac{4}{s+4}$ (First-order R–L filter)

79. $$H(s) = \frac{4}{s+4} = \frac{R}{R+Ls} = \frac{R/L}{s+R/L}$$

Comparing, we get R = 4 Ω, L = 1 H

80. Average side-band power, $P_{av} = \frac{\mu^2 A_c^2}{4}$

$$= \frac{(0.5)^2 \times 100}{4} = 6.25$$

81. Noise power (Area endored by the spectrum) $= N_0B$

Ratio of average side band honour to mean noise honer $= \frac{25}{4N_0B} = \frac{6.25}{N_0B}$

82. Phase Margin (PM) $= 180° + \angle G(i\omega)\big|_{\omega=\omega_c}$

$= 180° + \tan^{-1}\omega_c a - 2.90°$

$$\Rightarrow \quad \frac{\pi}{4} = \tan^{-1}\omega_c a$$

$$\Rightarrow \quad \omega_c a = 1$$

By definition, $|G(j\omega)|_{\omega=\omega_c} = 1$

$$\Rightarrow \quad \frac{\sqrt{(a\omega_c)^2+1}}{\omega_c^2} = 1$$

$$\Rightarrow \quad \omega_c = \frac{1}{2^{1/4}}$$

then, $a = 0.84$

83. $$G(s) = \frac{0.84s+1}{s^2}$$

Unit-Impulse response, we get

$$C(s) = G(s) . L . [\delta(t)]$$

or, $C(t) = L^{-1}\{C(s)\} = 0.84\,U(t) + t\ [\because L\{\delta(t)\} = 1]$

At $t = 1$, $C(t) = 1.84$

84.

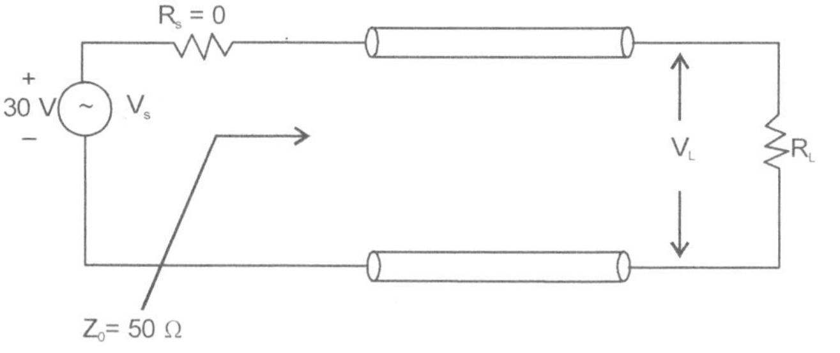

Voltage at receiver end,

$$V_L = V_S(1 + J_L)$$

$$\Rightarrow \quad 40 = 30(1 + J_L)$$

$$\Rightarrow \quad J_L = \frac{1}{3}$$

and Reflection co-efficient,

$$J_L = \frac{R_L - R_O}{R_L + R_O}$$

$$\Rightarrow \quad \frac{1}{3} = \frac{R_L - 50}{R_L + 50}$$

$$\Rightarrow \quad R_L = 100\ \Omega$$

85. $$I_{max} = \frac{V_{max}}{R_L} = \frac{40}{100} = 0.4\text{A}$$

$$I_{min} = \frac{V_{min}}{R_L} = \frac{20}{100} = 0.2\text{A}$$

Source current, $I_3 = \frac{I_{max} + I_{min}}{2} = 0.3\text{A}$

SOLVED PAPER – 2005

ELECTRONICS & COMMUNICATION ENGINEERING

Time Allowed : 3 Hours **Maximum Marks : 150**

Question 1 - 30 Carry One Mark Each

1. The following differential equation has

$$3\frac{d^2y}{dt^2} + 4\left(\frac{dy}{dt}\right)^3 + y^2 + 2 = x$$

(a) degree = 2, order = 1
(b) degree = 2, order = 2
(c) degree = 4, order = 3
(d) degree = 2, order = 3

2. Choose the function $f(t); -\infty < t < \infty$, for which a Fourier series cannot be defined.

(a) $3 \sin(25\, t)$
(b) $4 \cos(20\, t + 3) + 2 \sin(710\, t)$
(c) $\exp(-|t|) \sin(25\, t)$
(d) 1

3. A fair dice is rolled twice. The probability that an odd number will follow an even number is

(a) $\frac{1}{2}$ (b) $\frac{1}{6}$
(c) $\frac{1}{3}$ (d) $\frac{1}{4}$

4. A solution of the following differential equation is given by

$$\frac{d^2y}{dx^2} - 5\frac{dy}{dx} + 6y = 0$$

(a) $y = e^{2x} + e^{-3x}$ (b) $y = e^{2x} + e^{3x}$
(c) $y = e^{-2x} + e^{3x}$ (d) $y = e^{-2x} + e^{-3x}$

5. The function $x(t)$ is shown in the figure. Even and odd parts of a unit-step function $u(t)$ are respectively,

(a) $\frac{1}{2}, \frac{1}{2}x(t)$
(b) $-\frac{1}{2}, \frac{1}{2}x(t)$
(c) $\frac{1}{2}, -\frac{1}{2}x(t)$
(d) $-\frac{1}{2}, -\frac{1}{2}x(t)$

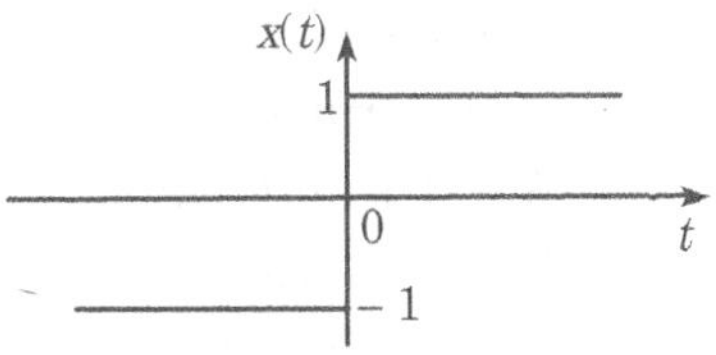

6. The region of convergence of Z-transform of the sequence

$$\left(\frac{5}{6}\right)^n u(n) - \left(\frac{6}{5}\right)^n u(-n-1) \text{ must be}$$

(a) $|z| < \frac{5}{6}$ (b) $|z| > \frac{6}{5}$
(c) $\frac{5}{6} < |z| < \frac{6}{5}$ (d) $\frac{6}{5} < |z| < \infty$

7. The condition on R, L and C such that the step response $y(t)$ in the figure has no oscillations, is

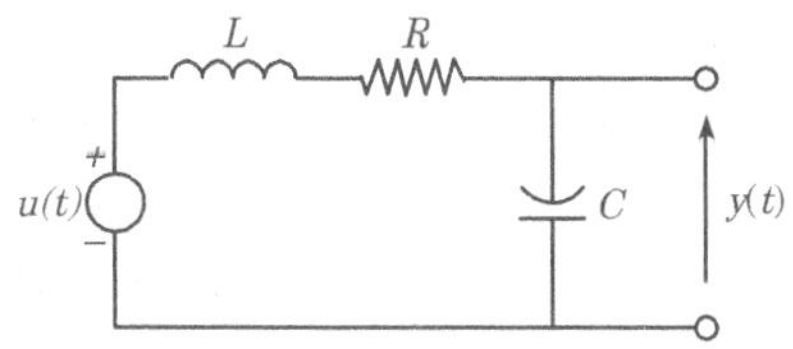

(a) $R \geq \frac{1}{2}\sqrt{\frac{L}{C}}$ (b) $R \geq \sqrt{\frac{L}{C}}$
(c) $R \geq 2\sqrt{\frac{L}{C}}$ (d) $R = \frac{1}{\sqrt{LC}}$

8. The ABCD parameters of an ideal n : 1 transformer shown in the figure are $\begin{bmatrix} n & 0 \\ 0 & X \end{bmatrix}$. The value of X will be

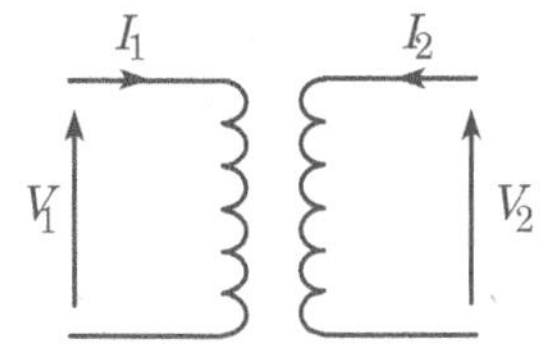

(a) n (b) $\frac{1}{n}$
(c) n^2 (d) $\frac{1}{n^2}$

9. In a series RLC circuit, R = 2 kΩ, L = 1 H, and $C \frac{1}{400} = \mu F$. The resonant frequency is

(a) 2×10^4 Hz (b) $\frac{1}{\pi} \times 10^4$ Hz
(c) 10^4 Hz (d) $2\pi \times 10^4$ Hz

10. The maximum power that can be transferred to the load resistor R_L from the voltage source in the figure is

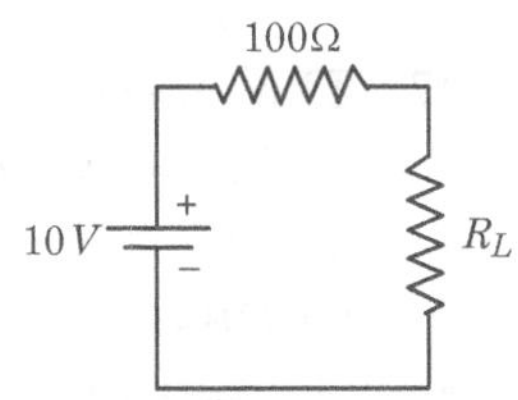

(a) 1 W
(b) 10 W
(c) 0.25 W
(d) 0.5 W

11. The bandgap of Silicon at room temperature is

(a) 1.3 eV (b) 0.7 eV
(c) 1.1 eV (d) 1.4 eV

12. A Silicon PN junction at a temperature of 20°C has a reverse saturation current of 10 pico-Amperes (pA). The reverse saturation current at 40° C for the same bias is approximately

(*a*) 30 pA. (*b*) 40 pA.
(*c*) 50 pA. (*d*) 60 pA.

13. The primary reason for the widespread use of Silicon in semiconductor device technology is

(*a*) aboundance of Silicon on the surface of the Earth.
(*b*) larger bandgap of Silicon in comparison to Germanium.
(*c*) favourable properties of Silicon-dioxide (SiO_2).
(*d*) lower melting point.

14. The effect of current shunt feedback in an amplifier is to

(*a*) increase the input resistance and decrease the output resistance.
(*b*) increase both input and output resistances.
(*c*) decrease both input and output resistances.
(*d*) decrease the input resistance and increase the output resistance.

15. The input resistance R_i of the amplifier shown in the figure is

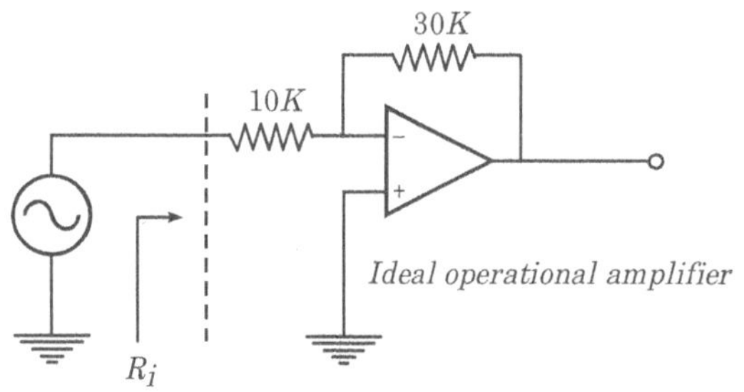

(*a*) $\frac{30}{4}$ kΩ (*b*) 10 kΩ
(*c*) 40 kΩ (*d*) infiinte

16. The first and the last critical frequency of an RC-driving point impedance function must respectively be

(*a*) a zero and a pole (*b*) a zero and a zero
(*c*) a pole and a pole (*d*) a pole and a zero

17. The cascode amplifier is a multistage configuration of

(*a*) CC - CB (*b*) CE - CB
(*c*) CB - CC (*d*) CE - CC

18. Decimal 43 in Hexadecimal and BCD number system is respectively

(*a*) B2, 0100 0011 (*b*) 2B, 0100 0011
(*c*) 2B, 0011 0100 (*d*) B2, 0100 0100

19. The Boolean function *f* implemented in the figure using two input multiplexers is

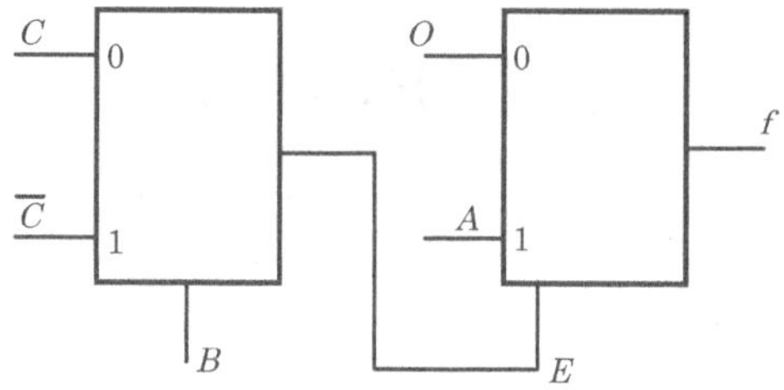

(*a*) $A\overline{B}C + AB\overline{C}$ (*b*) $ABC + A\overline{B}\,\overline{C}$
(*c*) $\overline{A}BC + \overline{A}\,\overline{B}\,\overline{C}$ (*d*) $\overline{A}\,\overline{B}C + \overline{A}B\overline{C}$

20. Which of the following can be impulse response of a causal system?

(*a*)
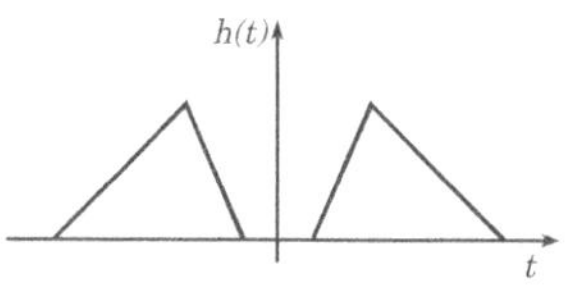

(*b*)
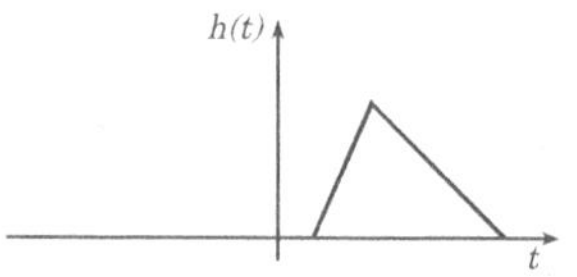

(*c*)
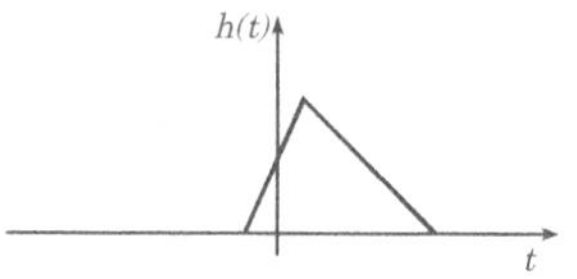

(*d*)
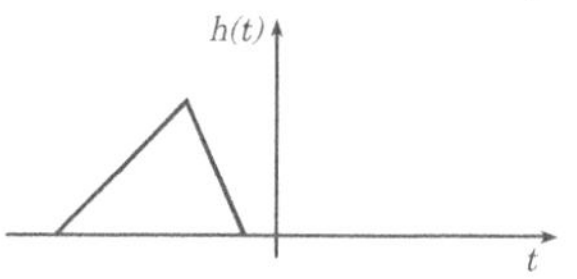

21. Let $x(n) = \left(\frac{1}{2}\right)^n u(n)$, $y(n) = x^2(n)$ and $Y\left(e^{j\omega}\right)$ be the Fourier transform of $y(n)$. Then $Y\left(e^{j0}\right)$ is

(*a*) $\frac{1}{4}$ (*b*) 2
(*c*) 4 (*d*) $\frac{4}{3}$

22. Find the correct match between group I and group II.

Group I

A – $\{1 + km(t)\}\, A \sin(\omega_c t)$
B – $km(t)\, A \sin(\omega_c t)$
C – $A \sin\{\omega_c t + km(t)\}$
D – $A \sin\left(\omega_c t + k\int_{-\infty}^{t} m(t)dt\right)$

Group II

1 – Phase modulation
2 – Frequency modulation
3 – Amplitude modulation
4 – DSB-SC modulation

	A	B	C	D
(*a*)	4	3	2	1
(*b*)	1	2	3	4
(*c*)	2	1	4	3
(*d*)	3	4	1	2

23. The power in the signal

$$s(t) = 8\cos\left(20\pi t - \frac{\pi}{2}\right) + 4\sin(15\pi t) \text{ is}$$

(*a*) 40 (*b*) 41

(*c*) 42 (*d*) 82

24. Which of the following analog modulation scheme requires the minimum transmitted power and minimum channel bandwidth?

(*a*) VSB (*b*) DSB-SC

(*c*) SSB (*d*) AM

25. A linear system is equivalently represented by two sets of state equations.

$$\dot{X} = AX + BU$$

and $$\dot{W} = CW + DU.$$

The eigenvalues of the representations are also computed as $\{\lambda\}$ and $\{\mu\}$. Which one of the following statements is true?

(*a*) $[\lambda] = [\mu]$ and $X = W$

(*b*) $[\lambda] = [\mu]$ and $X \neq W$

(*c*) $[\lambda] \neq [\mu]$ and $X = W$

(*d*) $[\lambda] \neq [\mu]$ and $X \neq W$

26. Which one of the following polar diagrams corresponds to a lag network?

(*a*)

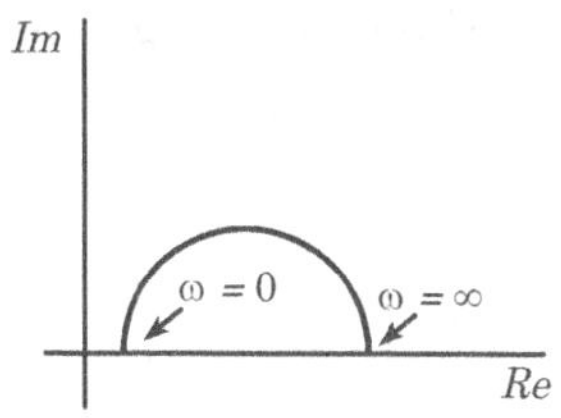

(*b*)

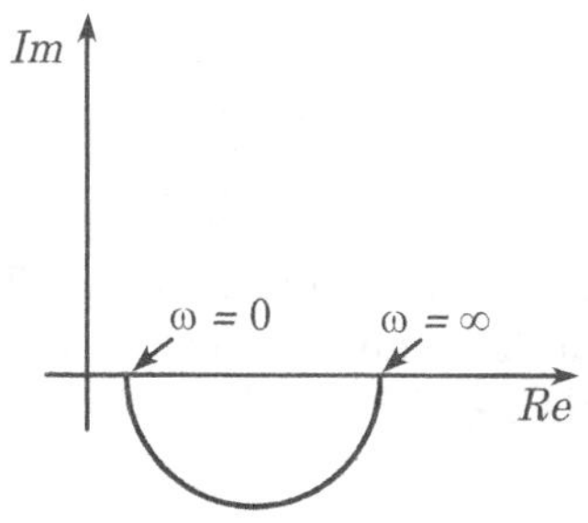

(*c*)

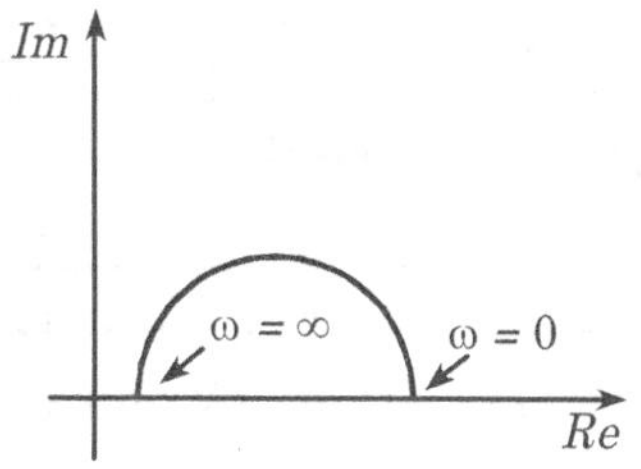

(*d*)

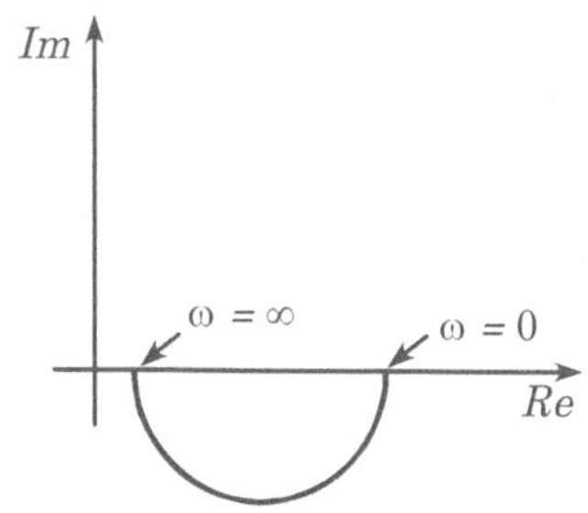

27. Despite the presence of negative feedback, control systems still have problems of instability because the

(*a*) components used have nonlinearities.

(*b*) dynamic equations of the subsystems are not known exactly.

(*c*) mathematical analysis involves approximations.

(*d*) system has large negative phase angle at high frequencies.

28. The magnetic field intensity vector of a plane wave is given by

$$\overline{H}(x, y, z, t) = 10\sin(50000t + 0.004x + 30\,\hat{a}_y),$$

where $\hat{a}_y$ denotes the unit vector in y direction. The wave is propagating with a phase velocity

(*a*) 5×10^4 m/s. (*b*) -3×10^8 m/s.

(*c*) -1.25×10^7 m/s. (*d*) 3×10^8 m/s.

29. Many circles are drawn in a Smith chart used for transmission line calculations. The circles shown in the figure represent

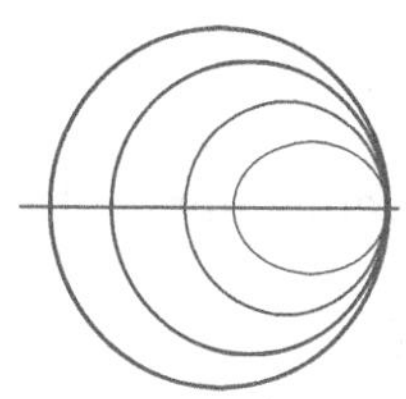

(*a*) unit circles.

(*b*) constant resistance circles.

(*c*) constant reactance circles.

(*d*) constant reflection coefficient circles.

30. Refractive index of glass is 1.5. Find the wavelength of a beam of light with a frequency of 10^{14} Hz in glass. Assume velocity of light is 3×10^8 m/s in vacuum.

(*a*) 3 μm (*b*) 3 mm

(*c*) 2 μm (*d*) 1 μm

Q. 31 - 80 Carry Two Marks Each

31. In what range should Re(s) remain so that the Laplace transform of the function $e^{(a+2)t+5}$ exits?

(*a*) Re (s) > a + 2 (*b*) Re (s) > a + 7

(*c*) Re (s) < 2 (*d*) Re (s) > a + 5

32. Given the matrix $\begin{bmatrix} -4 & 2 \\ 4 & 3 \end{bmatrix}$, the eigenvector is

(a) $\begin{bmatrix} 3 \\ 2 \end{bmatrix}$ (b) $\begin{bmatrix} 4 \\ 3 \end{bmatrix}$

(c) $\begin{bmatrix} 2 \\ -1 \end{bmatrix}$ (d) $\begin{bmatrix} -2 \\ 1 \end{bmatrix}$

33. Let, $A = \begin{bmatrix} 2 & -0.1 \\ 0 & 3 \end{bmatrix}$ and $A^{-1} = \begin{bmatrix} \frac{1}{2} & a \\ 0 & b \end{bmatrix}$.

Then $(a + b) =$

(a) $\frac{7}{20}$ (b) $\frac{3}{20}$

(c) $\frac{19}{60}$ (d) $\frac{11}{20}$

34. The value of the integral

$$I = \frac{1}{\sqrt{2\pi}} \int_0^{\infty} \exp\left(-\frac{x^2}{8}\right) dx \text{ is}$$

(a) 1 (b) π

(c) 2 (d) 2π

35. The derivative of the symmetric function drawn in given figure will look like

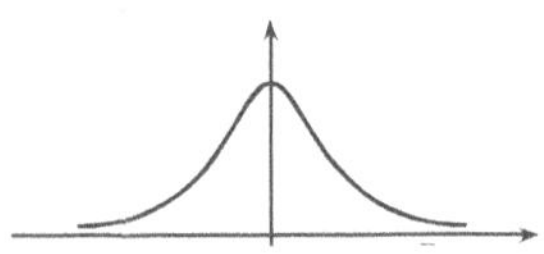

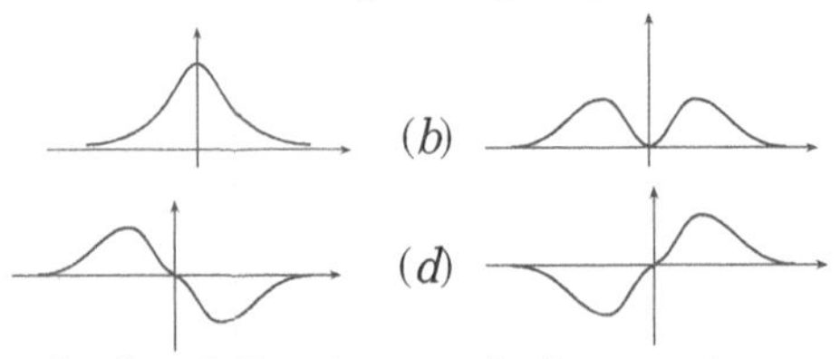

(a) (b) (c) (d)

36. Match the following and choose the correct combination:

Group I	Group II
A Newton-Raphson method	1. Solving nonlinear equations
B. Runge-kutta method	2. Solving linear simultaneous equations
C. Simpson's Rule	3. Solving ordinary differential equations
D. Gauss elimination	4. Numerical integration method
	5. Interpolation
	6. Calculation of Eigenvalues

	A	B	C	D
(a)	6	1	5	3
(b)	1	6	4	3
(c)	1	3	4	2
(d)	5	3	4	1

37. Given an orthogonal matrix

$$A = \begin{bmatrix} 1 & 1 & 1 & 1 \\ 1 & 1 & -1 & -1 \\ 1 & -1 & 0 & 0 \\ 0 & 0 & 1 & -1 \end{bmatrix} [AA^T]^{-1} \text{ is}$$

(a) $\begin{bmatrix} \frac{1}{4} & 0 & 0 & 0 \\ 0 & \frac{1}{4} & 0 & 0 \\ 0 & 0 & \frac{1}{2} & 0 \\ 0 & 0 & 0 & \frac{1}{2} \end{bmatrix}$ (b) $\begin{bmatrix} \frac{1}{2} & 0 & 0 & 0 \\ 0 & \frac{1}{2} & 0 & 0 \\ 0 & 0 & \frac{1}{2} & 0 \\ 0 & 0 & 0 & \frac{1}{2} \end{bmatrix}$

(c) $\begin{bmatrix} 1 & 0 & 0 & 0 \\ 0 & 1 & 0 & 0 \\ 0 & 0 & 1 & 0 \\ 0 & 0 & 0 & 1 \end{bmatrix}$ (d) $\begin{bmatrix} \frac{1}{4} & 0 & 0 & 0 \\ 0 & \frac{1}{4} & 0 & 0 \\ 0 & 0 & \frac{1}{4} & 0 \\ 0 & 0 & 0 & \frac{1}{4} \end{bmatrix}$

38. For the circuit shown in the figure, the instantaneous current $i_1(t)$ is

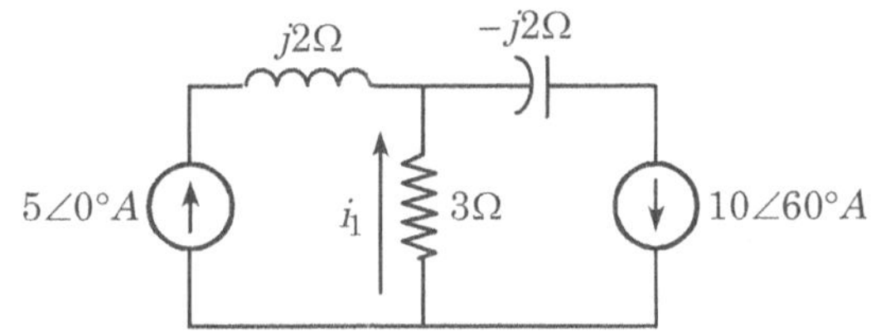

(a) $\frac{10\sqrt{3}}{2}\angle 90°$ (b) $\frac{10\sqrt{3}}{2}\angle -90°$

(c) $5\angle 60°$ (d) $5\angle -60°$

39. Impedance Z as shown in the given figure is

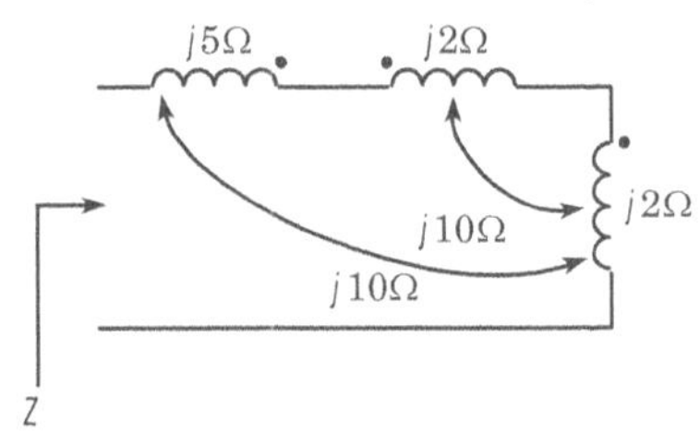

(a) $j29\Omega$ (b) $j9\Omega$

(c) $j19\Omega$ (d) $j39\Omega$

40. For the circuit shown in the figure, Thevenin's voltage and Thevenin's equivalent resistance atterminals a - b is

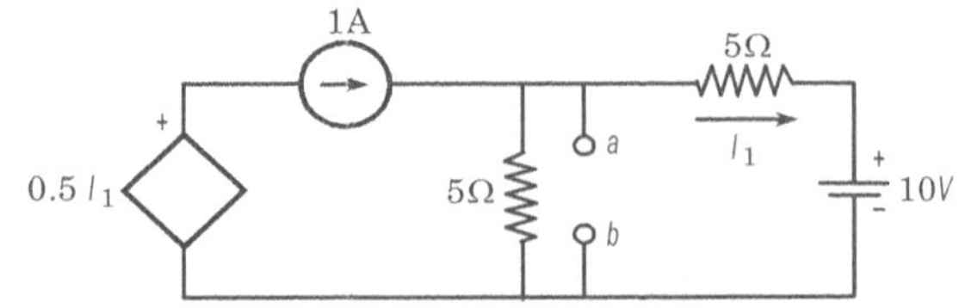

(a) 5 V and 2 Ω. (b) 7.5 V and 2.5 Ω.

(c) 4 V and 2 Ω. (d) 3 V and 2.5 Ω.

41. If $R_1 = R_2 = R_4 = R$ and $R_3 = 1.1R$ in the bridge circuit shown in the figure, then the reading in the ideal voltmeter connected between *a* and *b* is

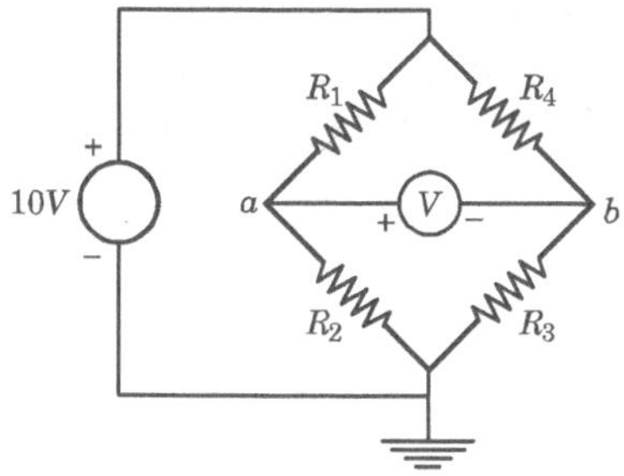

(*a*) 0.238 V (*b*) 0.138 V

(*c*) – 0.238 V (*d*) 1 V

42. The *h*-parameters of the circuit shown in the figure are

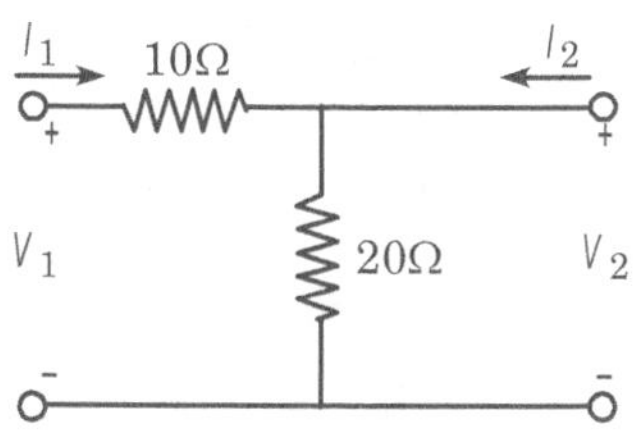

(*a*) $\begin{bmatrix} 0.1 & 0.1 \\ -0.1 & 0.3 \end{bmatrix}$ (*b*) $\begin{bmatrix} 10 & -1 \\ 1 & 0.05 \end{bmatrix}$

(*c*) $\begin{bmatrix} 30 & 20 \\ 20 & 20 \end{bmatrix}$ (*d*) $\begin{bmatrix} 10 & 1 \\ -1 & 0.05 \end{bmatrix}$

43. A square pulse of 3 volts amplitude is applied to C-R circuit shown in the figure. The capacitor is initially uncharged. The output voltage V_0 at time $t = 2$ sec is

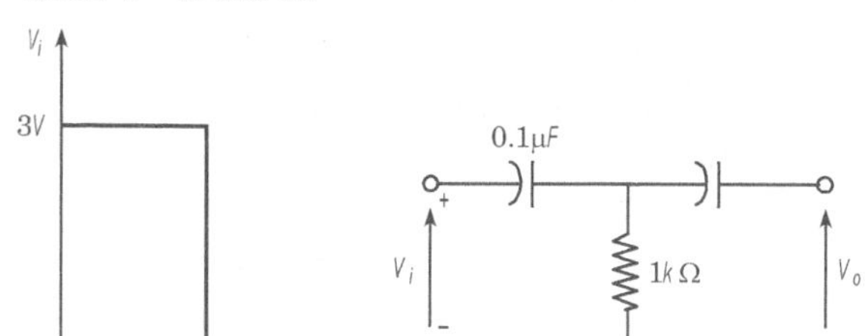

(*a*) 3V (*b*) – 3V

(*c*) 4V (*d*) – 4V

44. A Silicon sample A is doped with 10^{18} atoms/cm^3 of Boron. Another sample B of identical dimensions is doped with 10^{18} atoms/cm^3 of Phosphorus. The ratio of electron to hole mobility is 3. The ratio of conductivity of the sample A to B is

(*a*) 3 (*b*) $\frac{1}{3}$

(*c*) $\frac{2}{3}$ (*d*) $\frac{3}{2}$

45. A Silicon PN junction diode under reverse bias has depletion region of width 10μm. The relative permittivity of Silicon, $\varepsilon_r = 11.7$ and the permittivity of free space $\varepsilon_0 = 8.85 \times 10^{-12}$ F/m. The depletion capacitance of the diode per square meter is

(*a*) 100 μF. (*b*) 10 μF.

(*c*) 1 μF. (*d*) 20 μF.

46. For an npn transistor connected as shown in the figure, $V_{BE} = 0.7$ volts. Given that reverse saturation current of the junction at room temperature 300°K is 10^{-13} A, the emitter current is

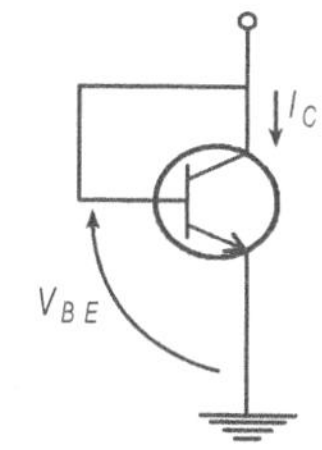

(*a*) 30 mA.

(*b*) 39 mA.

(*c*) 49 mA.

(*d*) 20 mA.

47. The voltage e_0 indicated in the figure has been measured by an ideal voltmeter. Which of the following can be calculated?

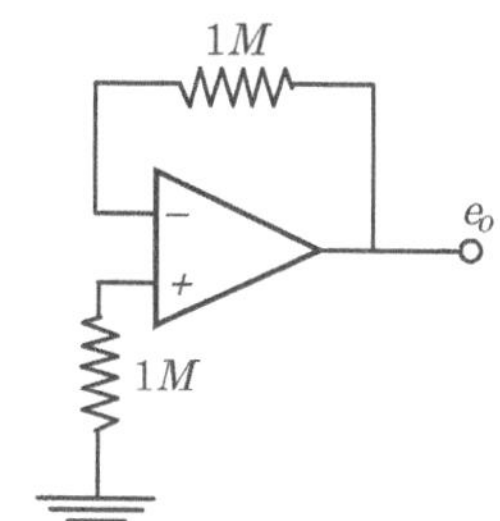

(*a*) Bias current of the inverting input only

(*b*) Bias current of the inverting and non-inverting inputs only

(*c*) Input offset current only

(*d*) Both the bias currents and the input offset current.

48. The OP-amp circuit shown in the figure is a filter. The type of filter and its cut-off frequency are respectively

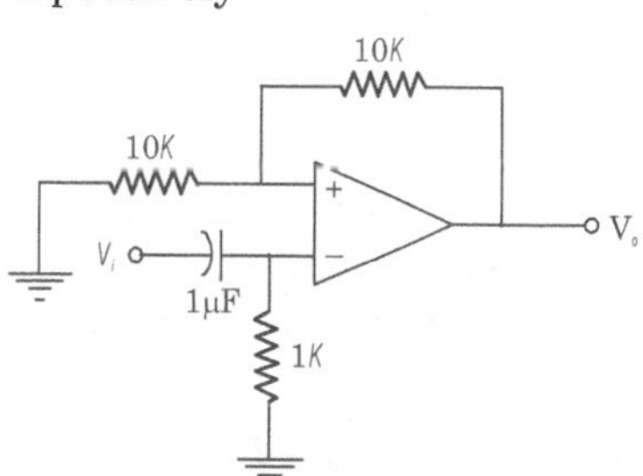

(*a*) high pass, 1000 rad/sec.

(*b*) low pass, 1000 rad/sec.

(*c*) high pass, 10000 rad/sec.

(*d*) low pass, 10000 rad/sec.

49. In an ideal differential amplifier shown in the figure, a large value of(R_E).

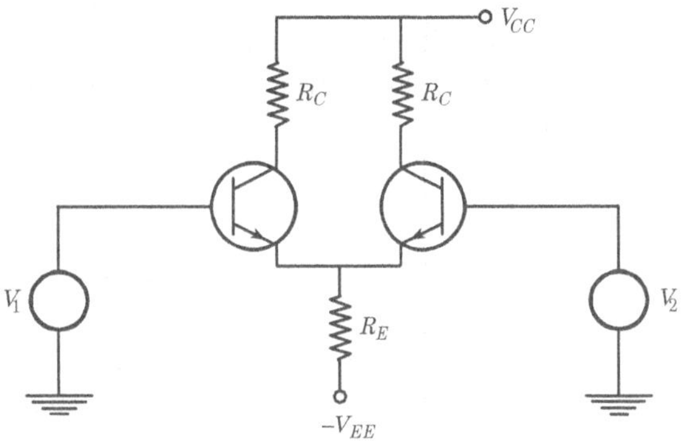

(*a*) increases both the differential and common-mode gains.

(*b*) increases the common-mode gain only.

(*c*) decreases the differential-mode gain only

(*d*) decreases the common-mode gain only.

50. For an n-channel MOSFET and its transfer curve shown in the figure, the threshold voltage is

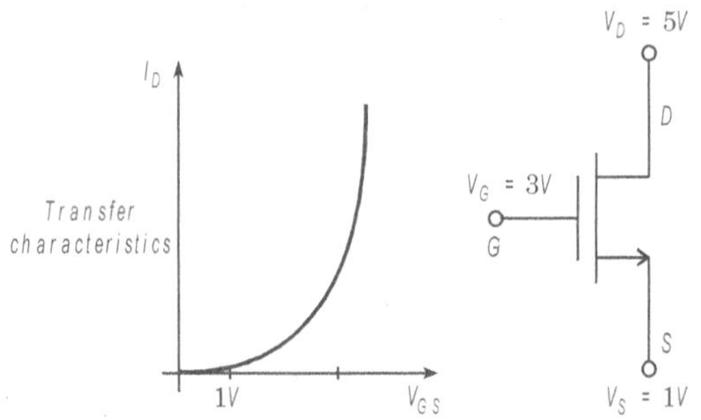

(*a*) 1 V and the device is in active region.

(*b*) – 1 V and the device is in saturation region.

(*c*) 1 V and the device is in saturation region.

(*d*) – 1 V and the device is in active region.

51. The circuit using a BJT with β = 50 and V_{BE} = 0.7 V is shown in the figure The base current I_B and collector voltage V_C are respectively

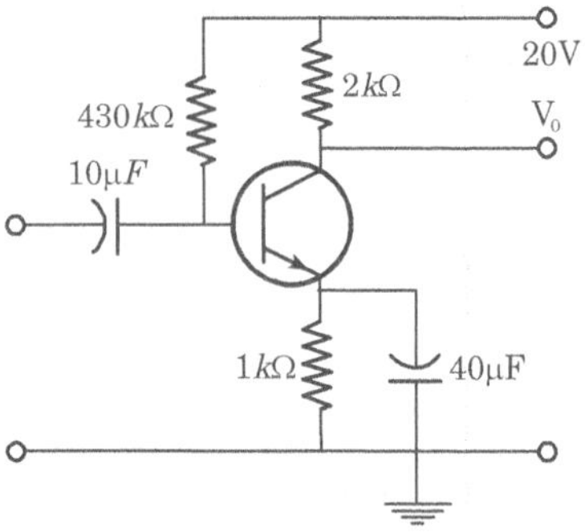

(*a*) 43 μA and 11.4 Volts.

(*b*) 40 μA and 16 Volts.

(*c*) 45 μA and 11 Volts.

(*d*) 50 μA and 10 Volts.

52. The zener diode in the regulator circuit shown in the figure has a Zener voltage of 5.8 volts and a Zener knee current of 0.5 mA. The maximum load current drawn from this circuit ensuring proper functioning over the input voltage range between 20 and 30 volts, is

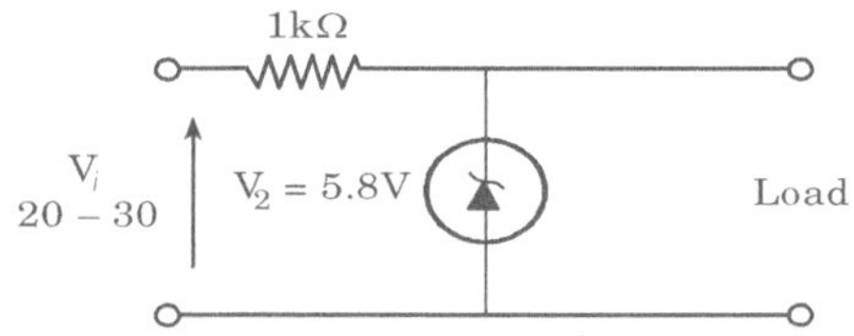

(*a*) 23.7 mA. (*b*) 14.2 mA.

(*c*) 13.7 mA. (*d*) 24.2 mA.

53. The transistors used in a portion of the TTL gate shown in the figure have a β = 100. The base-emitter voltage of is 0.7V for a transistor in active region and 0.75V for a transistor in saturation. If the sink current I = 1mA and the output is at logic 0, then the current I_R will be equal to

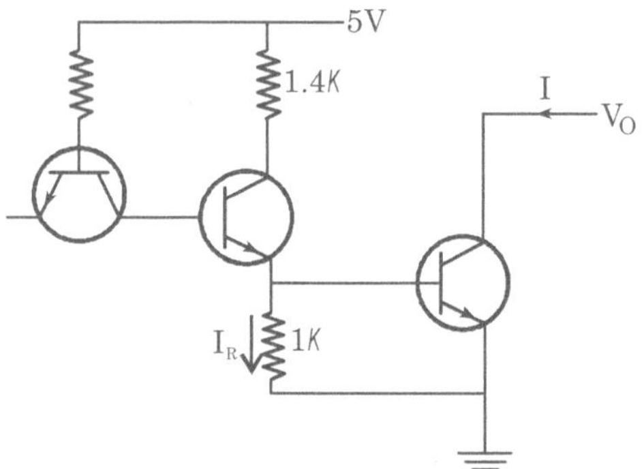

(*a*) 0.65 mA. (*b*) 0.70 mA

(*c*) 0.75 mA (*d*) 1.00 mA

54. The Boolean expression for the truth table shown is

A	B	C	f
0	0	0	0
0	0	1	0
0	1	0	0
0	1	1	1
1	0	0	0
1	0	1	0
1	1	0	1
1	1	1	0

(*a*) $B(A+C)(\overline{A}+\overline{C})$ (*b*) $B(A+\overline{C})(\overline{A}+C)$

(*c*) $\overline{B}(A+\overline{C})(\overline{A}+C)$ (*d*) $\overline{B}(A+C)(\overline{A}+\overline{C})$

55. Both transistors T1 and T2 shown in the figure, have a threshold voltage of 1 Volts. The device parameters K_1 and K_2 of T1 and T2 are, respectively, 36 μA/V^2 and 9μA/V^2. The output voltage V_0 is

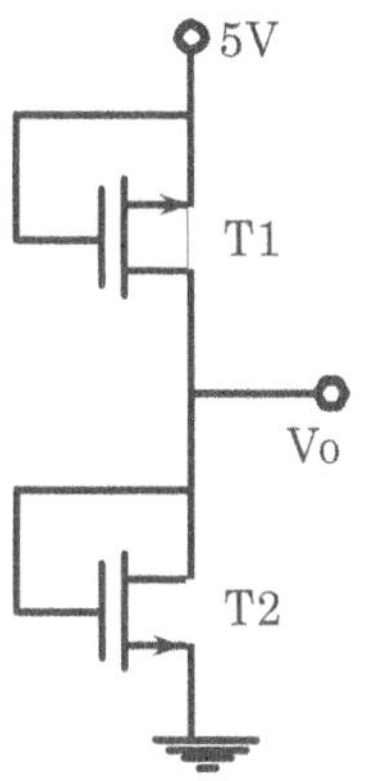

(*a*) 1 V (*b*) 2 V

(*c*) 3 V (*d*) 4 V

56. The present output Q_n of an edge triggered JK flip-flop is logic 0. If J = 1, then Q_{n+1}

(*a*) cannot be determined

(*b*) will be logic 0

(*c*) will be logic 1

(*d*) will race around

57. The given figure shows a ripple counter using positive edge triggered flip-flops. If the present state of the counter is $Q_2\,Q_1\,Q_0$ = 011, then its next state ($Q_2\,Q_1\,Q_0$) will be

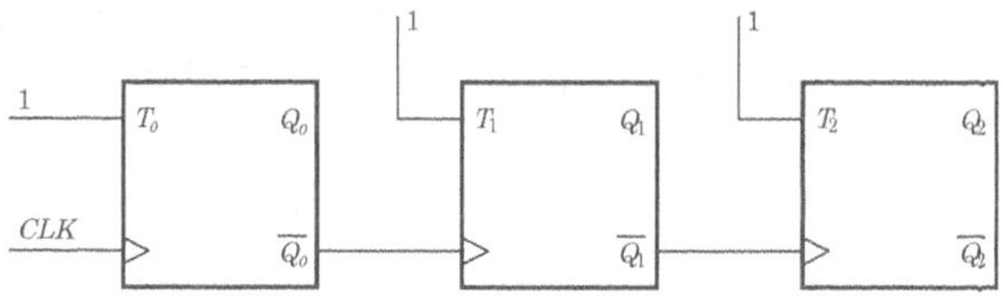

(*a*) 010 (*b*) 100

(*c*) 111 (*d*) 101

58. What memory address range is NOT represented by chip # 1 and chip # 2 in the figure A_0 to A_{15} in this figure are the address lines and CS means Chip Select.

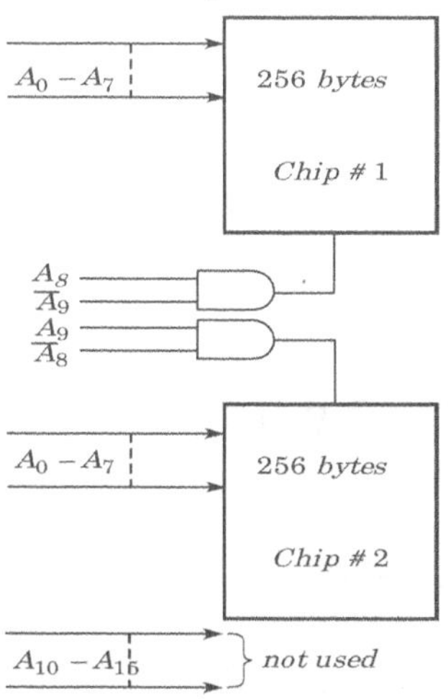

(*a*) 0100-02FF (*b*) 1500-16FF

(*c*) F900-FAFF (*d*) F800-F9FF

59. The output $y(t)$ of a linear time invariant system is related to its input $x(t)$ by the following equation:

$$y(t) = 0.5x(t - t_d + T) + x(t - t_d) + 0.5x(t - t_d - T).$$

The filter transfer function H(ω) of such a system is given by

(*a*) $(1 + \cos \omega T)e^{-j\omega t_d}$

(*b*) $(1 + 0.5 \cos \omega T)e^{-j\omega t_d}$

(*c*) $(1 - \cos \omega T)e^{-j\omega t_d}$

(*d*) $(1 - 0.5 \cos \omega T)e^{-j\omega t_d}$

60. Match the following and choose the correct combination.

Group I	Group II
A. continuous and aperiodic signal	1. Fourier representation is continuous and aperiodic
B. continuous and periodic signal	2. Fourier representation is discrete and aperiodic
C. Discrete and aperiodic signal	3. Fourier representation is continuous and periodic
D. Discrete and periodic signal	4. Fourier representation is discrete and periodic

	A	B	C	D
(*a*)	3	2	4	1
(*b*)	1	3	2	4
(*c*)	1	2	3	4
(*d*)	2	1	4	3

61. A signal $x(n) = \sin(\omega_0 n + \phi)$ is the input to a linear time-invariant system having a frequency response $H(e^{j\omega})$. If the output of the system $Ax(n - n_0)$, then the most general form of $\angle H(e^{j\omega})$ will be

(*a*) $-n_0\omega_0 + \beta$ for any arbitrary real β.

(*b*) $-n_0\omega_0 + 2\pi k$ for any arbitrary integer *k*.

(*c*) $n_0\omega_0 + 2\pi k$ for any arbitrary integer *k*.

(*d*) $-n_0\omega_0\ \phi$.

62. For a signal $x(t)$ the Fourier transform is $X(f)$. Then the inverse Fourier transform of $X(3f + 2)$ is given by

(*a*) $\frac{1}{2}x\left(\frac{t}{2}\right)e^{j3\pi t}$ (*b*) $\frac{1}{3}x\left(\frac{t}{3}\right)e^{-j4\pi t/3}$

(*c*) $3x(3t)e^{-j4\pi t}$ (*d*) $x(3t + 2)$

63. The polar diagram of a conditionally stable system for open loop gain K = 1 is shown in the figure. The open loop transfer function of the system is known to be stable . The closed loop system is stable for

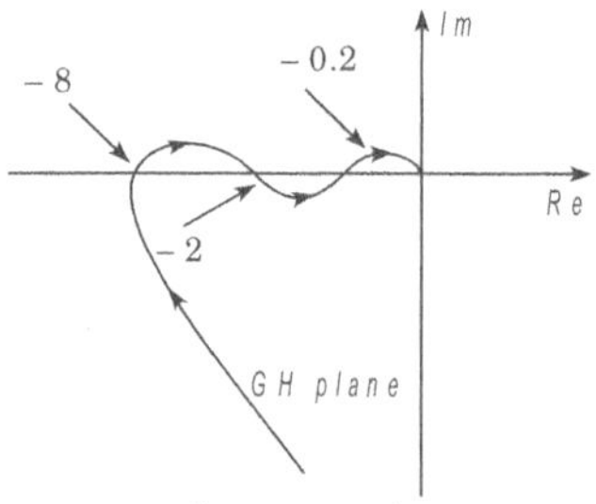

(*a*) $K < 5$ and $\frac{1}{2} < K < \frac{1}{8}$.

(*b*) $K < \frac{1}{8}$ and $\frac{1}{2} < K < 5$.

(*c*) $K < \frac{1}{8}$ and $5 < K$.

(*d*) $K > \frac{1}{8}$ and $K < 5$.

64. In the derivation of expression for peak percent overshoot,

$M_p = \exp\left(\frac{-\pi\zeta}{\sqrt{1-\zeta^2}}\right) \times 100\%$, which one of the following conditions is NOT required?

(*a*) System is linear and time invariant.

(*b*) The system transfer function has a pair of complex conjugate poles and no zeroes.

(*c*) There is no transportation delay in the system.

(*d*) The system has zero initial conditions.

65. Given the ideal operational amplifier circuit shown in the figure indicate the correct transfer characteristics assuming ideal diodes with zero cut-in voltage.

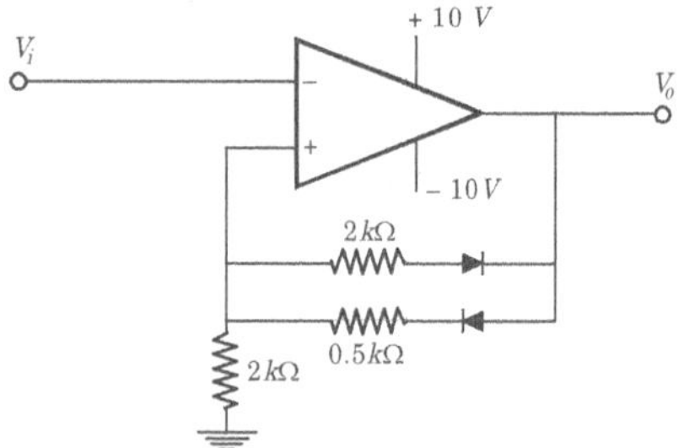

(*a*)

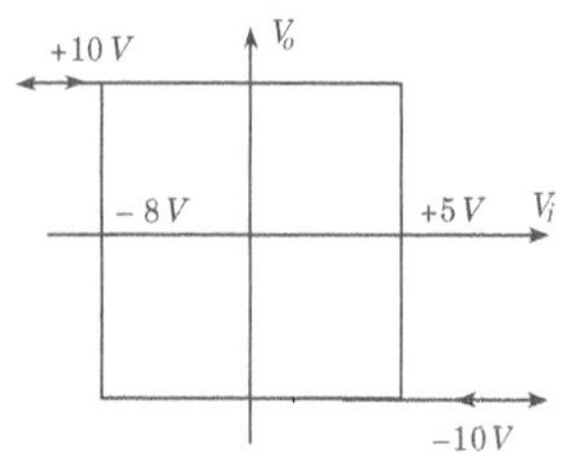

(*b*)

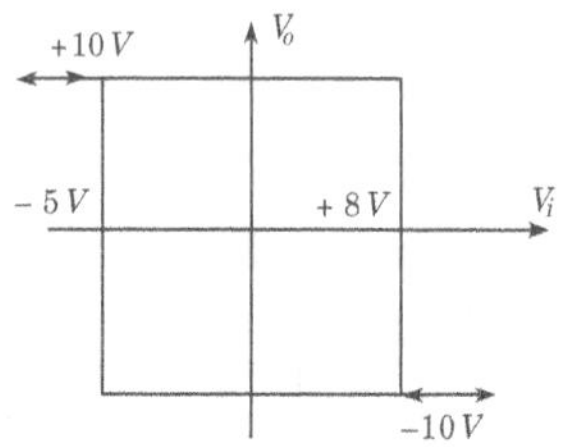

(*c*)

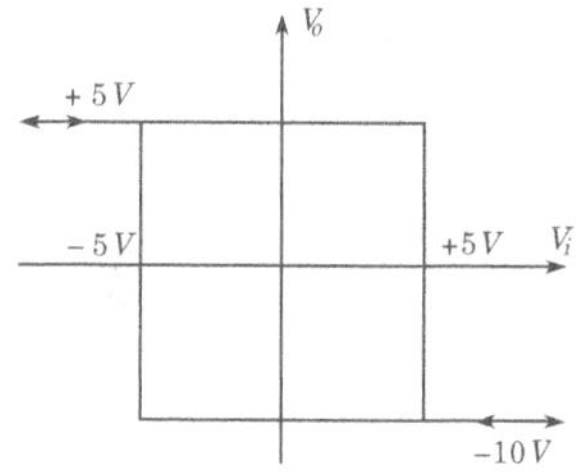

(*d*)

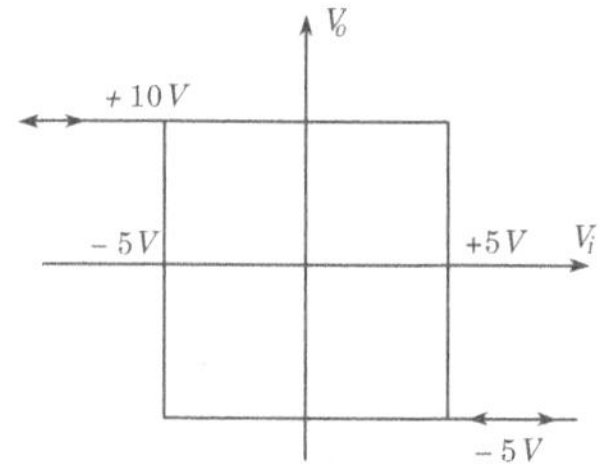

66. A ramp input applied to an unity feedback system results in 5% steady state error. The type number and zero frequency gain of the system are respectively

(*a*) 1 and 20 (*b*) 0 and 20

(*c*) 0 and $\frac{1}{20}$ (*d*) 1 and $\frac{1}{20}$

67. A double integrator plant, $G(s) = \frac{K}{s^2}$, $H(s) = 1$ is to be compensated to achieve the damping ratio $\zeta = 0.5$, and an undamped natural frequency, $\omega_n = 5$ rad/s. Which one of the following compensator $G_c(s)$ will be suitable?

(*a*) $\frac{s+3}{s+9.9}$ (*b*) $\frac{s+9.9}{s+3}$

(*c*) $\frac{s-6}{s+8.33}$ (*d*) $\frac{s-6}{s}$

68. An unity feedback system is given as,

$G(s) = \frac{K(1-s)}{s(s+3)}$

Indicate the correct root locus diagram.

(*a*)

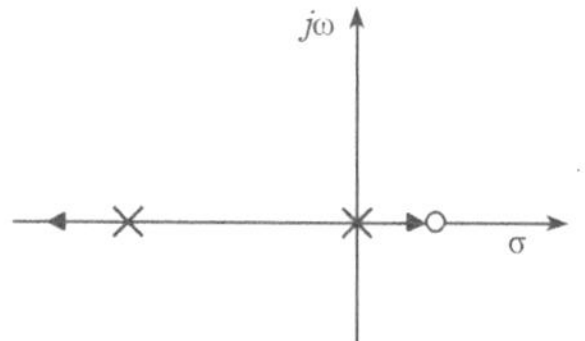

(*b*)

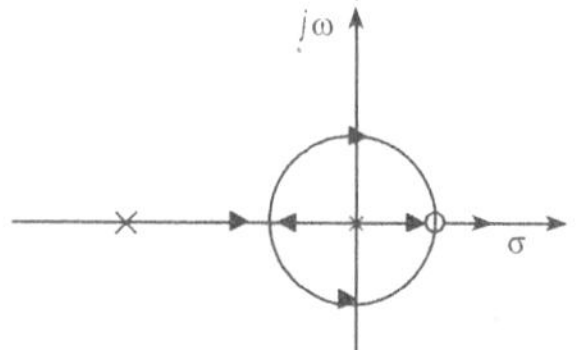

(*c*)

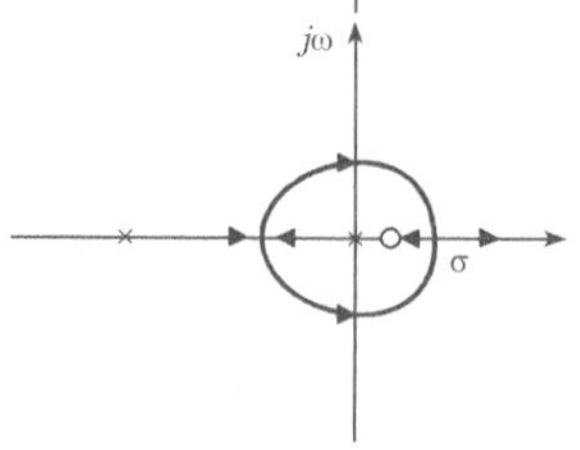

(*d*) 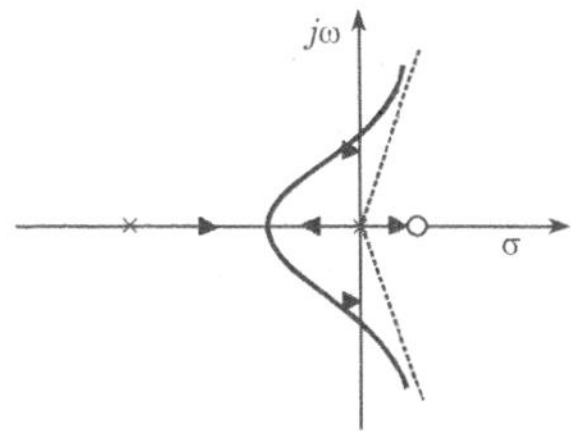

69. A MOS capacitor made using p type substrate is in the accumulation mode. The dominant charge in the channel is due to the presence of

(*a*) holes

(*b*) electrons

(*c*) positively charged ions

(*d*) negatively charged ions

70. A device with input $x(t)$ and output $y(t)$ is characterized by: $y(t) = x^2(t)$.

An FM signal with frequency deviation of 90 kHz and modulating signal bandwidth of 5 kHz is applied to this device. The bandwidth of the output signal is

(*a*) 370 kHz (*b*) 190 kHz

(*c*) 380 kHz (*d*) 95 kHz

71. A signal as shown in the figure is applied to a matched filter. Which of the following does represent the output of this matched filter?

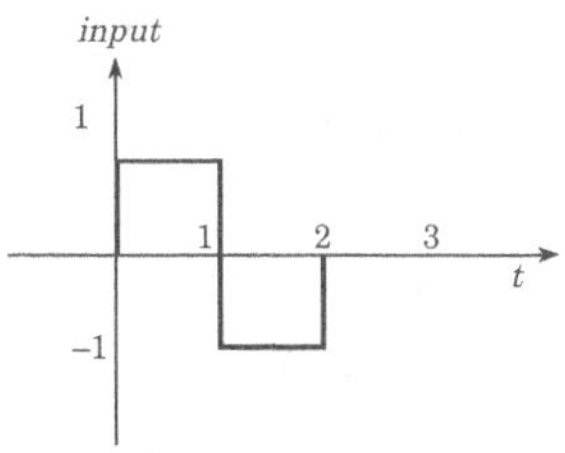

(*a*) 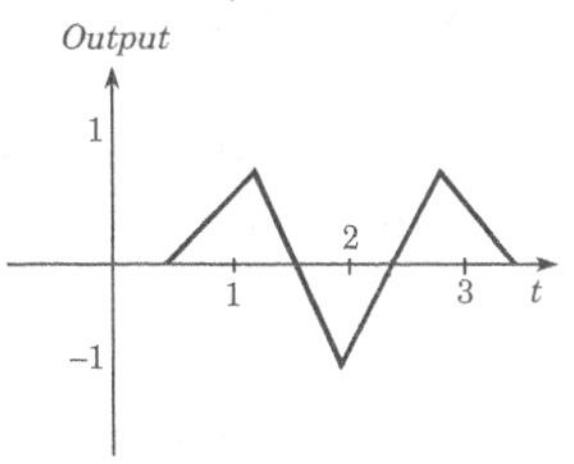

(*b*) 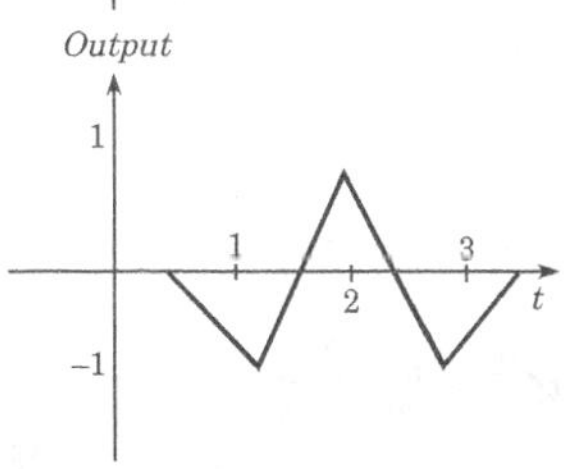

(*c*) 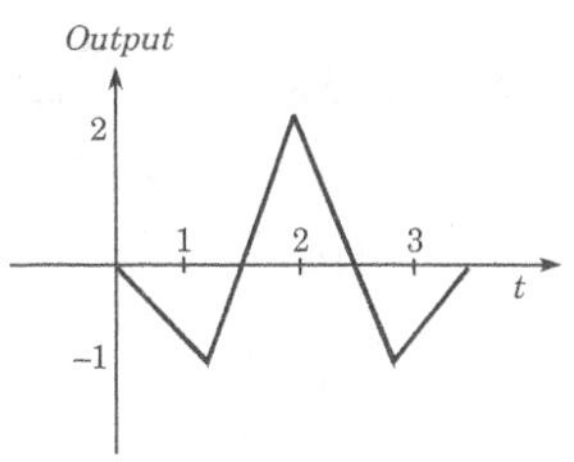

(*d*) 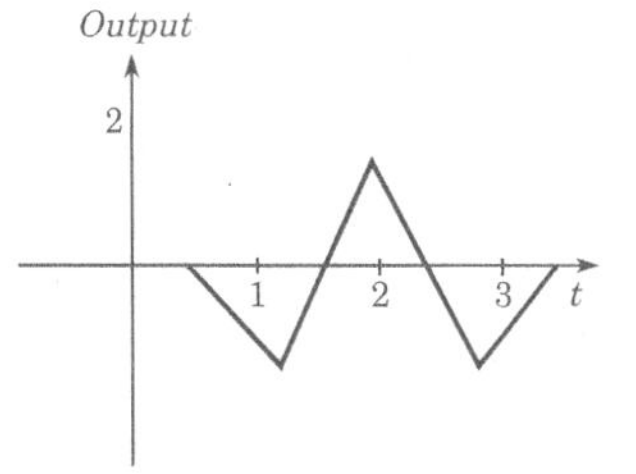

72. Noise with uniform power spectral density of N_0 W/Hz is passed through a filter $H(\omega) = 2\exp(-j\omega t_d)$ followed by an ideal low pass filter of bandwidth B Hz. The output noise power in Watts is

(*a*) $2N_0B$ (*b*) $4N_0B$

(*c*) $8N_0B$ (*d*) $16N_0B$

73. A carrier is phase modulated (PM) with frequency deviation of 10 kHz by a single tone frequency of 1 kHz. If the single tone frequency is increased to 2 kHz, assuming that phase deviation remains unchanged, the bandwidth of the PM signal is

(*a*) 21 kHz. (*b*) 22 kHz

(*c*) 42 kHz (*d*) 44 kHz

74. An output of a communication channel is a random variable ν with the probability density function as shown in the figure The mean square value of ν is

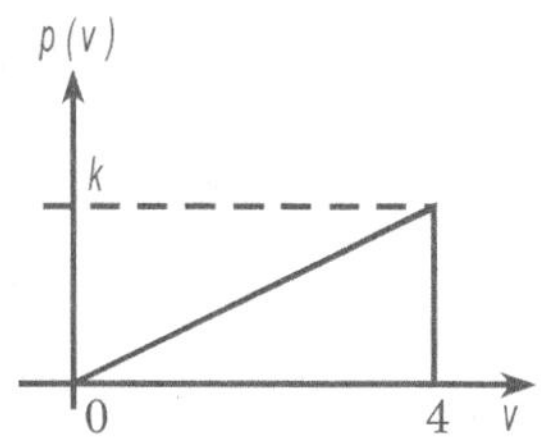

(*a*) 4

(*b*) 6

(*c*) 8

(*d*) 9

75. Which one of the following does represent the electric field lines for the TE_{02} mode in the cross-section of a hollow rectangular metallic waveguide?

(*a*) 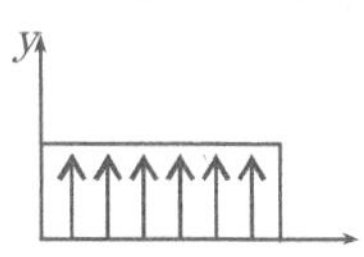

(*b*) 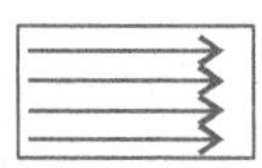

(*c*) 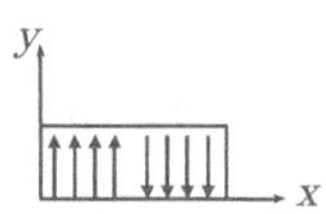

(*d*) 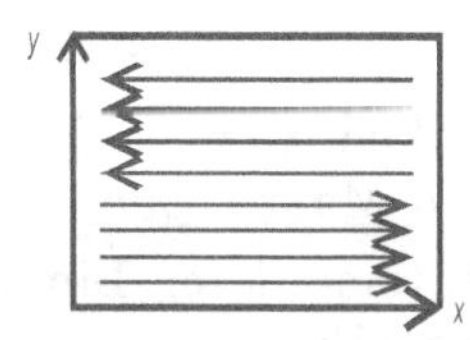

76. Characteristic impedance of a transmission line is 50Ω. Input impedance of the open-circuited line is $Z_{0c} = 100 + j150\ \Omega$. When the transmission line is short- circuited, then value of the input impedance will be

(*a*) 50Ω (*b*) $100 + j150\Omega$

(*c*) $7.69 + j11.54\Omega$ (*d*) $7.69 - j11.54\Omega$

77. Two identical and parallel dipole antennas are kept apart by a distance of $\frac{\lambda}{4}$ in the H-plane. They are fed with equal currents but the right most antenna has a phase shift of + 90°. The radiation pattern is given as

(a) 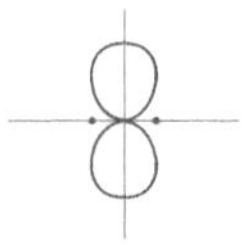
(b) 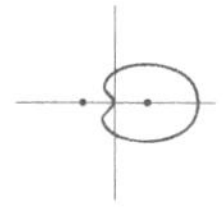

(c) 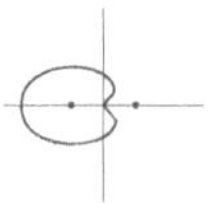
(d) 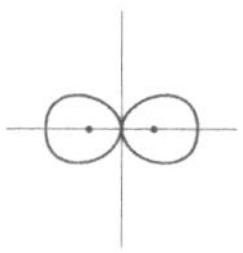

Common Data Questions 78, 79, 80

Given, $r_d = 20\ \text{k}\Omega$, $I_{DSS} = 10$ mA, $V_p = -8$V

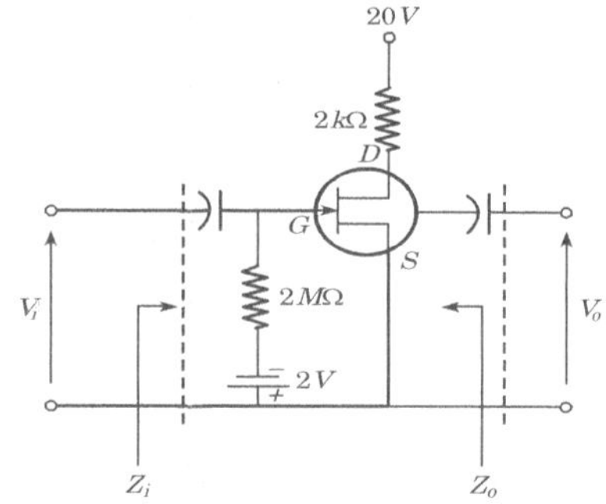

78. Z_i and Z_0 of the circuit are respectively

(a) 2 MΩ and 2 kΩ (b) 2 MΩ and $\frac{20}{11}$ kΩ

(c) infinity and 2 kΩ (d) infinity and $\frac{20}{11}$ kΩ

79. I_D and V_{DS} under DC conditions are respectively

(a) 5.625 mA and 8.75 V

(b) 7.500 mA and 5.00 V

(c) 4.500 mA and 11.00 V

(d) 6.250 mA and 7.50 V

80. Transconductance in milli-Siemens (mS) and voltage gain of the amplifier are respectively

(a) 1.875 mS and 3.41

(b) 1.875 mS and – 3.41

(c) 3.3 mS and – 6

(d) 3.3 mS and 6

Linked Answer Questions : Q. 81a to 85b Carry Two Marks Each

Statement for Linked Answer Questions 81a and 81b:

Consider an 8085 microprocessor system

81a. The following program starts at location 0100H.

LXI SP, 00FF

LXI H, 0107

MVI A, 20H

SUB M

The content of accumulator when the program counter reaches 0109H is

(a) 20H (b) 02H

(c) 00H (d) FFH

81b. If in addition following code exists from 0109H onwards,

ORI 40 H

ADD M

What will be the result in the accumulator after the last instruction is executed?

(a) 40H (b) 20H

(c) 60H (d) 42H

Statement for Linked Answer Questions 82a and 82b:

The open loop transfer function of a unity feedback system is given by

$$G(s) = \frac{3e^{-2s}}{s(s+2)}$$

82a. The gain and phase crossover frequencies in rad/sec are, respectively

(a) 0.632 and 1.26 (b) 0.632 and 0.485

(c) 0.485 and 0.632 (d) 1.26 and 0.632

82b. Based on the above results, the gain and phase margins of the system will be

(a) – 7.09 and 87.5°

(b) 7.09 and 87.5°

(c) 7.09dB and – 87.5°

(d) – 7.09dB and – 87.5°

Statement for Linked Answer Questions 83a and 83b

Asymmetric three-level midtread quantizer is to be designed assuming equiprobable occurence of all quantization levels.

83a. If the probability density function is divided into three regions as shown in the figure, the value of a in the figure is

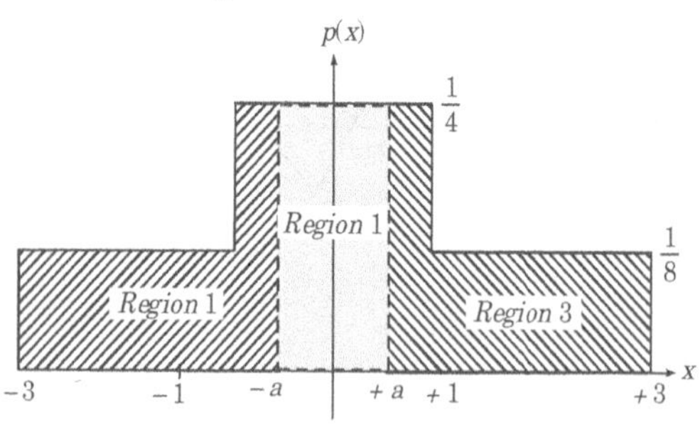

(a) $\frac{1}{3}$ (b) $\frac{2}{3}$

(c) $\frac{1}{2}$ (d) $\frac{1}{4}$

83b. The quantization noise power for the quantization region between - a and + a in the figure is

(*a*) $\frac{4}{81}$ (*b*) $\frac{1}{9}$

(*c*) $\frac{5}{81}$ (*d*) $\frac{2}{81}$

Statement of Linked Answer Questions 84a and 84b

Voltage standing wave pattern in a lossless transmission line with characteristic impedance 50Ω and a resistive load is shown in the figure.

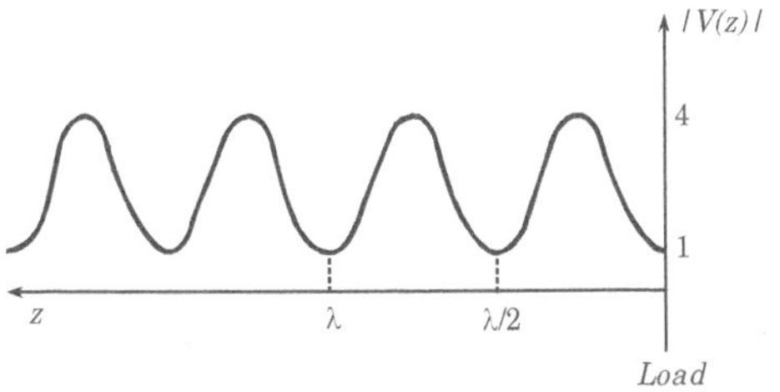

84a. The value of the load resistance is

(*a*) 50 Ω

(*b*) 200 Ω

(*c*) 12.5 Ω

(*d*) 0 Ω

84b. The reflection coefficient is given by

(*a*) – 0.6

(*b*) – 1

(*c*) 0.6

(*d*) 0

Statement of Linked Answer Questions 85a and 85b

A sequence $x(n)$ has non-zero values as shown in the figure

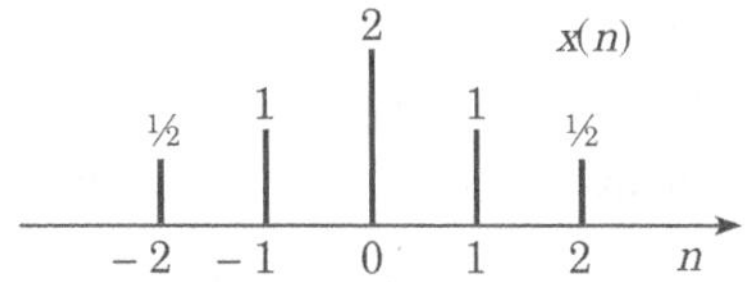

85a. The sequence $y(n) = \begin{cases} x\left(\frac{n}{2}-1\right) & \text{for n even} \\ & \text{for n odd} \end{cases}$ will be

(*a*)

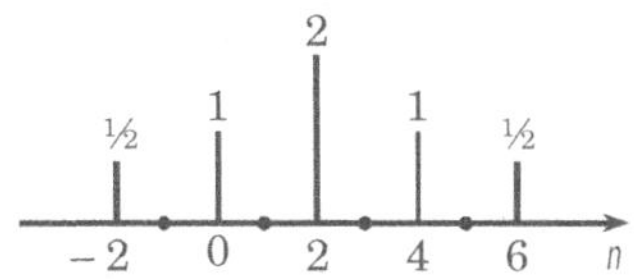

(*b*)

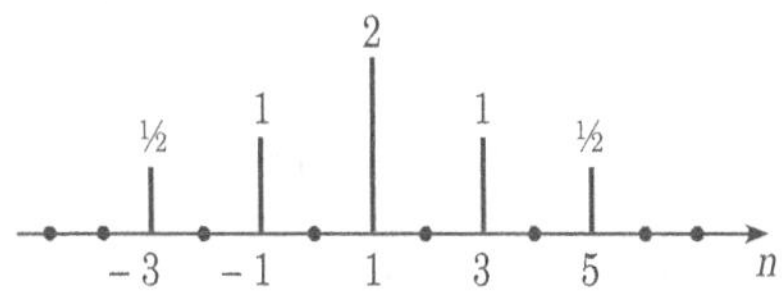

(*c*)

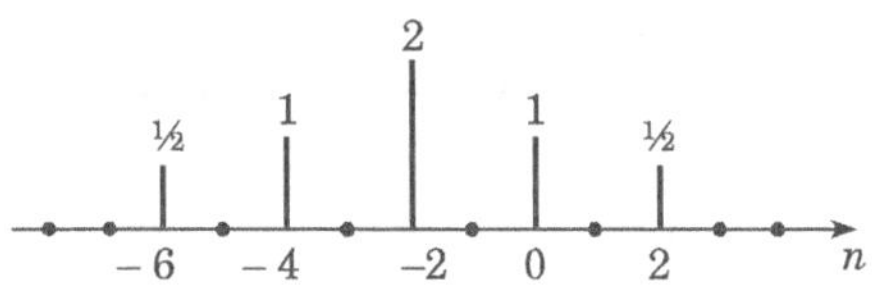

(*d*)

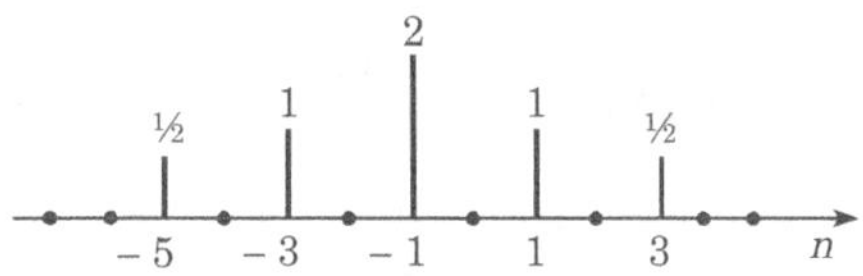

85b. The Fourier transform of $y(2n)$ will be

(*a*) $e^{-j2\omega}$ [cos 4ω + 2 cos 2ω + 2]

(*b*) [cos 2ω + 2 cos ω + 2]

(*c*) $e^{-j\omega}$ [cos 2ω + 2 cos ω + 2]

(*d*) $e^{-j2\omega}$ [cos 2ω + 2 cos ω + 2]

ANSWERS

1. (*b*)	2. (*c*)	3. (*d*)	4. (*b*)	5. (*a*)	6. (*c*)	7. (*c*)	8. (*b*)	9. (*b*)	10. (*c*)
11. (*c*)	12. (*b*)	13. (*c*)	14. (*d*)	15. (*b*)	16. (*d*)	17. (*b*)	18. (*b*)	19. (*a*)	20. (*b*)
21. (*d*)	22. (*d*)	23. (*a*)	24. (*c*)	25. (*c*)	26. (*d*)	27. (*a*)	28. (*c*)	29. (*b*)	30. (*c*)
31. (*a*)	32. (*c*)	33. (*a*)	34. (*a*)	35. (*c*)	36. (*c*)	37. (*c*)	38. (*a*)	39. (*b*)	40. (*b*)
41. (*c*)	42. (*d*)	43. (*b*)	44. (*b*)	45. (*b*)	46. (*c*)	47. (*c*)	48. (*a*)	49. (*d*)	50. (*c*)
51. (*b*)	52. (*a*)	53. (*c*)	54. (*a*)	55. (*c*)	56. (*c*)	57. (*b*)	58. (*d*)	59. (*a*)	60. (*c*)
61. (*b*)	62. (*b*)	63. (*b*)	64. (*c*)	65. (*b*)	66. (*a*)	67. (*a*)	68. (*c*)	69. (*b*)	70. (*a*)
71. (*c*)	72. (*c*)	73. (*d*)	74. (*c*)	75. (*d*)	76. (*d*)	77. (*a*)	78. (*b*)	79. (*a*)	80. (*b*)
81a. (*c*)	81b. (*c*)	82a. (*d*)	82b. (*d*)	83a. (*b*)	83b. (*a*)	84a. (*c*)	84b. (*a*)	85a. (*a*)	85b. (*c*)

EXPLANATIONS

1. $3\frac{d^2y}{dt^2}+4\left(\frac{dy}{dt}\right)^3+y^2+2=x$

Order of highest derivative = 2

Hence, most appropriate answer is (*b*)

2. All other f's are periodic oconstant.

3. Outcomes in which an odd number follows an even number,

(2, 1), (2, 3), (2, 5) (4, 1), (4, 3), (4, 5), (6, 1), (6, 3), (6, 5)

$$\text{Required probability} = \frac{\text{desired outcomes}}{\text{total possible outcomes}} = \frac{9}{36} = \frac{1}{4}$$

4. $\frac{d^2y}{dx^2}-5\frac{dy}{dx}+6y=0$

$\therefore$ A.E. is $D^2-5D+6=0$

$\Rightarrow$ $D = 2, 3$

Hence, solution is $y = e^{2x}+e^{3x}$

5. $\text{Even part} = \frac{u(t)+u(-t)}{2}$

Now $u(t) = 0;\ t<0$

$= 1,\ t \geq 0$

$\therefore$ $u(-t) = 0,\ -t<0$

$= 1,\ -t\geq 0$

i.e., $u(-t) = 1,\quad t \leq 0$

$= 0,\quad t>0$

$\therefore$ $\frac{u(t)+u(-t)}{2} = \frac{1}{2};\quad t\leq 0$

$= \frac{1}{2},\quad t>0$

$\therefore$ $\text{Even}[v(t)] = \frac{1}{2}$

Odd $(u(t)) = \frac{u(t)+u(-t)}{2}\begin{bmatrix}-\frac{1}{2}, & t\leq 0\\ \frac{1}{2}, & t>0\end{bmatrix}$

$= \frac{x(t)}{2}$ from given figure

6. $X(A) = \left(\frac{5}{6}\right)^n u(n)-\left(\frac{6}{5}\right)^n u(-n-1)$

$$X(z) = \sum_{n=-\infty}^{\infty}\left(\frac{5}{6}\right)^n u[n]\,z^{-n}-\sum_{n=-\infty}^{\infty}\left(\frac{6}{5}\right)^n u[-n-1]\cdot z^{-n}$$

$$= \sum_{n=0}^{\infty}\left(\frac{5}{6}\right)^n z^{-n}-\left(1+\sum_{n=-\infty}^{\infty}\left(\frac{6}{5}\right)^n u(-n)\,z^{-n}\right)$$

$$= \sum_{n=0}^{\infty}\left(\frac{5}{6}\right)^n z^{-n+1}+L-\sum_{n=0}^{\infty}\left(\frac{6}{5}\right)^{-n} z^{n}$$

The first term will converge when $\left|\frac{5}{6}Z^{-1}\right|<1$

or, $|Z|>\frac{5}{6}$

The second term will converge, when, $\left|\frac{6}{5}Z^{-1}\right|>1$

or, $|Z|<\frac{6}{5}$

Thus, the region of convergence is

$$\frac{5}{6}<|Z|<\frac{6}{5}$$

7. Transfer function is T(*s*)

Now $y(t) = \frac{\int i\,dt}{C}$

$\therefore$ $Y(s) = \frac{I}{sC}$

$u(t) = \left(\frac{1}{sC}+R+Ls\right)$

$\therefore$ $\frac{Y(s)}{u(s)} = \frac{1/sC}{1/sC+R+Ls}$

$= \frac{1}{LCs^{2+}+RCs+1}$

$T(s) = \frac{1/LC}{s^2+\frac{Rs}{L}+\frac{1}{LC}}$

$\therefore$ $\omega_n{}^2 = \frac{1}{LC}$

$\Rightarrow$ $\omega_n = \frac{1}{\sqrt{LC}}$

$\Rightarrow$ $2\xi\omega_n = \frac{R}{L}$

$\therefore$ $2\xi = \frac{R}{L}\sqrt{LC}$

$= R\sqrt{\frac{C}{L}}$

$\Rightarrow$ $\xi = \frac{R}{2}\sqrt{\frac{C}{L}}$

For no oscillations, $\xi \geq 1$

$\therefore$ $\frac{R}{2}\sqrt{\frac{C}{L}}\geq 1$

$\Rightarrow$ $R\geq 2\sqrt{\frac{L}{C}}$

8. For transform given

$$\frac{V_1}{V_2} = \frac{N_1}{N_2} = n$$

or, $V_1 = V_2 n$

and $$\frac{I_1}{I_2} = -\frac{N_2}{N_1} = \frac{-1}{n}$$

or, $$I_1 = -\frac{I_2}{n}$$

In terms of ABCD parameters,

$$\begin{bmatrix} V_1 \\ I_1 \end{bmatrix} = \begin{bmatrix} n & 0 \\ 0 & \frac{1}{n} \end{bmatrix} \begin{bmatrix} I_1 \\ -I_2 \end{bmatrix}$$

Comparing with the given matrix,

$$X = \frac{1}{n}$$

9. Resonant frequency $= \frac{1}{2\pi\sqrt{LC}}$

$$= \frac{1}{2\pi\sqrt{1 \times \frac{10^{-6}}{400}}}$$

$$= \frac{1}{\pi} \times 10^4 \text{ Hz}$$

10. From maximum power transfer theorem

$$P_{max} = \frac{E^2}{4R}$$

When $R = R_L$,

$$P_{max} = \frac{10^2}{4 \times 100} = 0.25 \text{ W}$$

11. Band gap of *Si* at any temperature T

$= 1.21 - 3.6 \times 10^{-4} T$

For $T = 300° K$

Bandgap $= 1.21 - 3.6 \times 10^{-4} \times 300 = 1.1 \; eV$

12. Reverse saturation current, at temperature, T_2,

$$I(T_2) = I(T_1) \times 2^{\frac{T_2 - T_1}{10}}$$

$$\text{or, } I(40° C) = I(20°C) \times 2^{20/10}$$

$= 10 \times 4 = 40$ pA

14.

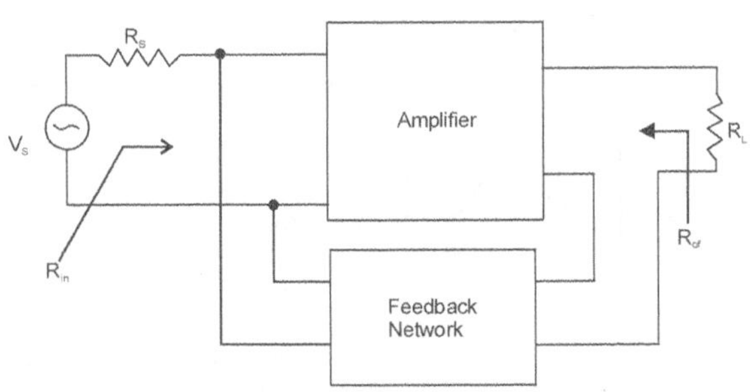

current shunt feedback network

As feedback network is connected in series, with output. Output resistance (R_{of}) increases, and the same is connected in shunt with the source resistance R_s, intput resistance decreases.

15. For an ideal OPAMP, $R_{in} = \infty$, $R_o = 0$, $A_V = \infty$

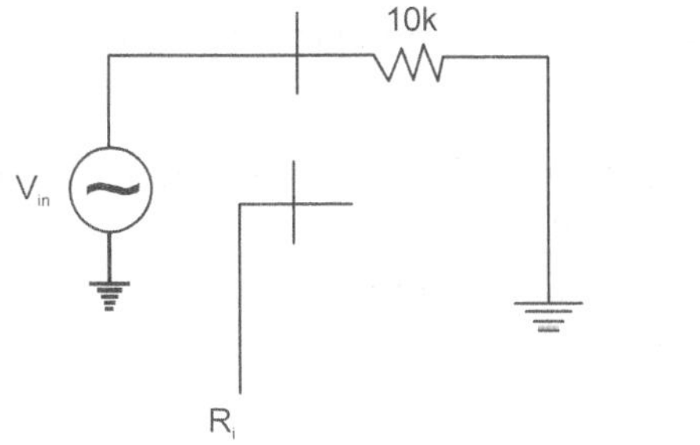

From the equivalent circuit

$R_i = 10K\Omega$

Equivalent circuit for an ideal OPAMP

16. For Series RC network

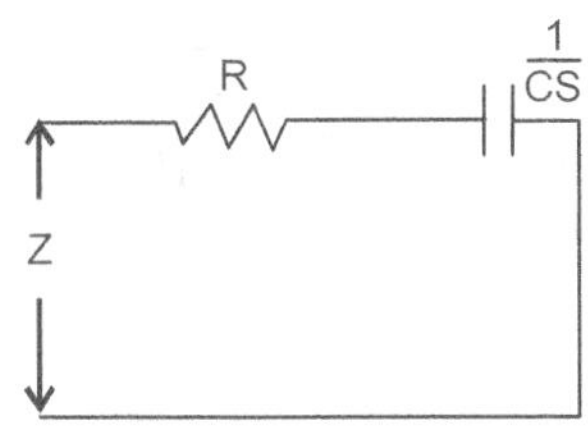

Fig - (a)

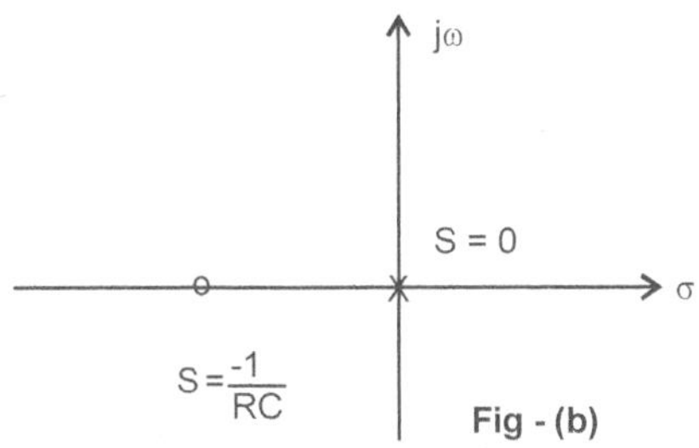

Fig - (b)

$$Z = R + \frac{1}{CS}$$

$$Z = \frac{RCS + 1}{CS}$$

Z has pole at origin and zero at $S = \frac{-1}{RC}$

From figure (*b*), first critical frequency is at S = 0

and last critical frequency is at $S = \frac{-1}{RC}$

For parallel RC network

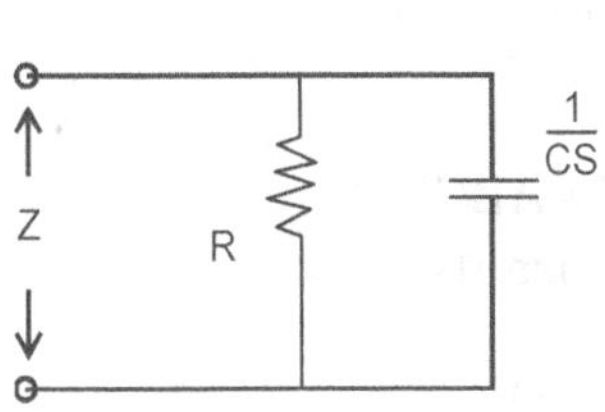

Fig - (c)

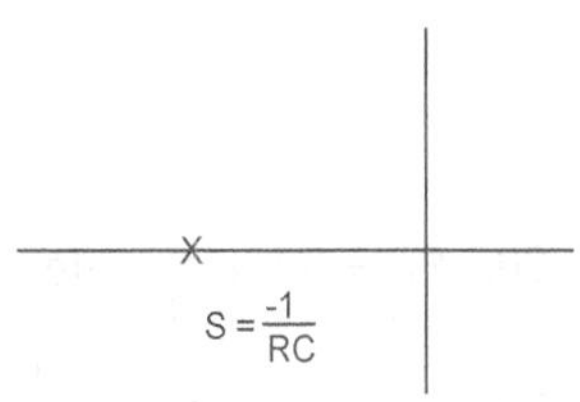

Fig - (d)

$$Z = \frac{R \cdot \frac{1}{CS}}{R + \frac{1}{CS}} = \frac{R}{RCS+1}$$

From figure (*d*), first critical frequency is at $S = \frac{-1}{RC}$ (pole) and last critical frequency is zero.

17. CE – CC,

Because of the impedance match of the configuration

$R_{(out)CE}$ = high,

and $R_{(in)CC}$ = high

18. 43 becomes 0100 0011 in BCD (4 = 0100, 3 = 0011)

When converted to binary,

43 ≡ 0010 1110 (0010 = 2, 1110 = B)

∴ Hexadecimal code is 2 B

19. First multiplexer

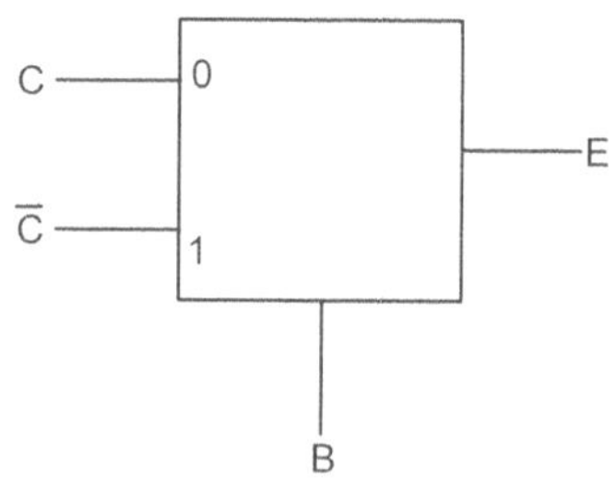

K-map for E

B \ C	0	1
0	0	1
1	1	0

$$E = \bar{B}C + \bar{C}B$$

For second multiplexer

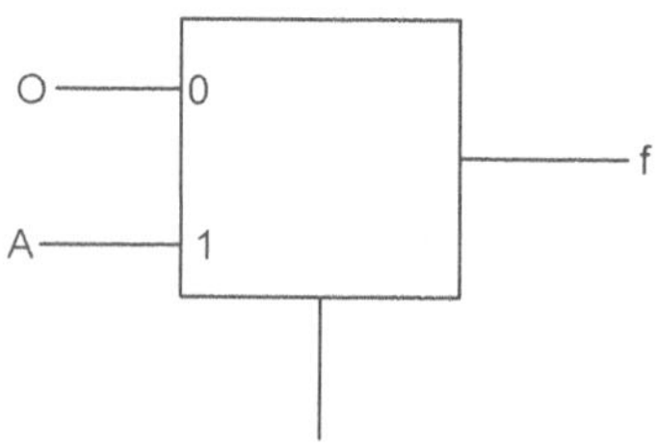

$$f = AE$$

$$= A\bar{B}C + AB\bar{C}$$

20. $f(t) = 0$ for $t < 0$ for causality

21. $$x(n) = \left(\frac{1}{2}\right)^n u(n)$$

$$\therefore \quad y(n) = \left(\frac{1}{2}\right)^{2n} u^2(n)$$

$$y(n) = \left(\frac{1}{2}\right)^{2n} u(n) = \left(\left(\frac{1}{2}\right)^2\right)^n u(n)$$

$$\therefore \quad y(n) = \left(\frac{1}{4}\right)^n u(n)$$

$$\therefore \quad y(z) = \frac{1}{1 - \frac{1}{4}z^{-1}}$$

$$y(e^{j\omega}) = \frac{1}{1 - \frac{1}{4}e^{-iw}}$$

$$y(e^0) = \frac{4}{3}$$

23. Power of $s(t) = 8\cos\left(20\pi t - \frac{\pi}{2}\right) + 4\sin(15\pi t)$

$$= \frac{8^2}{2} + \frac{16}{2}$$

$$= \frac{64+16}{2} = 40$$

26. Lag network is represented as,

$$H(s) = \frac{1+ST}{1+S\beta T}, \ \beta > 1 \ \&zT > 0$$

$$H(j\omega) = \frac{1+j\omega T}{1+j\omega\beta T}$$

$$\Rightarrow \quad |H(j\omega)| = \frac{\sqrt{1+(\omega T)^2}}{\sqrt{1+(\omega\beta T)^2}},$$

$$\angle H(j\omega) = \tan^{-1}\omega T - \tan^{-1}(\omega\beta T)$$

At $\omega = 0$, $|H(j\omega)| = 1$, $\angle H(j\omega)|_{\omega=0} = 0$

At $\omega = \frac{1}{T}, |H(j\omega)| = \frac{\sqrt{2}}{\sqrt{1+\beta^2}}$

$$\angle H(j\omega)|_{\omega=\frac{1}{T}} = \frac{\pi}{4} - \tan^{-1}(\beta)$$

At $\omega = \infty, |H(j\omega)|_{\omega=\infty}$

$$= \frac{1}{\beta}, \angle H(j\omega)|_{\omega=\infty} = 0$$

Hence, the polar diagram will be

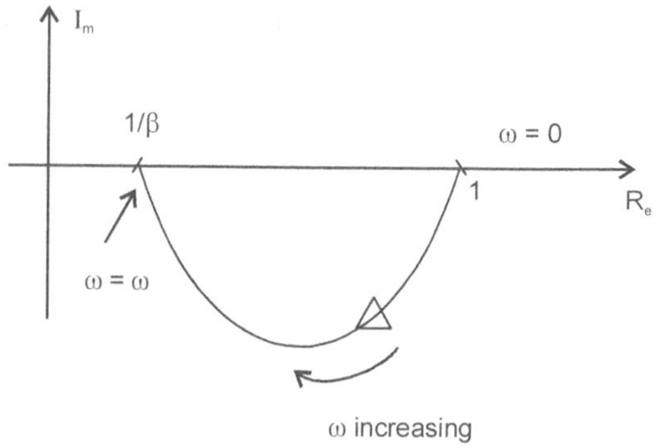

28. Magnetic field intensity vector is exprened as,

$$H(x, y, z, t) = H_0 \sin(\omega t + Kx + \phi)\, \bar{a}_y$$

Here $\omega = 5 \times 10^4$, K = 00.04

Since the wave is travelling in negative-x direction, phaswe velocity,

$$v_P = \frac{-\omega}{K} = -\frac{5000}{0.004}$$

$$= -1.25 \times 10^7 \text{ m/s}$$

30. Here, $\mu = 1.5$, $f = 10^{14}$ Hz

Now,

$$\therefore \quad f\lambda = \frac{C_0}{\mu}$$

$$\therefore \quad \lambda = \frac{C_0}{\mu f}$$

$$= \frac{3 \times 10^8}{1.5 \times 10^{14}} = 2 \times 10^{-6} = 2\mu\text{m}$$

31.

$$X(s) = \int_0^\infty x(t) \cdot e^{-st} \cdot dt$$

$$= \int_0^\infty e^5 e^{(a+2)t} \cdot e^{-st} \cdot dt$$

$$= e^5 \int_0^\infty e^{-(s-a-2)} \cdot dt$$

$$= \frac{e^5}{a+2-s}\left[e^{-t(s-a-2)}\right]_0^\infty$$

$$= \frac{e^5}{a+2-s}\left[e^{-\infty(s-a-2)} - 1\right]$$

For $e^{-\infty(s-a-2)}$ to be zero

Re $(s - a - 2) > 0$

or, $\text{Re}(s) > a + 2$

32.

$$A = \begin{bmatrix} -4 & 2 \\ 4 & 3 \end{bmatrix}$$

$$\therefore \quad (A - \lambda I) = 0$$

$$\Rightarrow \quad -(4-\lambda)(3-\lambda) - 2 \times 4 = 0$$

$$\Rightarrow \quad \lambda^2 + \lambda - 20 = 0$$

$$\lambda = -5, 4$$

Putting $\lambda = -5, \begin{bmatrix} 1 & 2 \\ 4 & 8 \end{bmatrix}\begin{bmatrix} x_1 \\ x_2 \end{bmatrix} = 0$

$$\Rightarrow \quad x_1 + 2x_2 = 0$$

$$x_1 = -2x_2$$

$$\Rightarrow \quad \frac{x_1}{2} = \frac{x_2}{-1}$$

Hence, $\begin{bmatrix} 2 \\ -1 \end{bmatrix}$ is eigen vector.

33. We know, $AA^{-1} = I \begin{bmatrix} 2 & -0.1 \\ 0 & 3 \end{bmatrix}\begin{bmatrix} \frac{1}{2} & a \\ 0 & b \end{bmatrix}$

$$= \begin{bmatrix} 1 & 2a - 0.1b \\ 0 & 3b \end{bmatrix}$$

$$\therefore \quad 3b = 1$$

$$\Rightarrow \quad b = \frac{1}{3}$$

or $2a - 0.1b = 0,$

$$\Rightarrow \quad 2a = \frac{b}{10},$$

$$\Rightarrow \quad a = \frac{1}{60}$$

$$\therefore \quad a + b = \frac{1}{3} + \frac{1}{60} = \frac{21}{60} = \mathbf{\frac{7}{20}}$$

34.

$$I = \frac{1}{\sqrt{2\pi}}\int_0^\infty \exp\left(\frac{-x^2}{8}\right)dx$$

Let $\frac{x}{2\sqrt{2}} = Z,$

then $\frac{dx}{2\sqrt{2}} = dZ$

$$\Rightarrow \quad dx = 2\sqrt{2}\,dZ$$

$$\therefore \quad I = \frac{2\sqrt{2}}{\sqrt{2\pi}}\int_0^\infty \exp(-Z^2)dZ$$

$$= \frac{2}{\sqrt{\pi}}\int_0^\infty \exp(-Z^2)dZ$$

$$= erf(\infty) = \mathbf{1}$$

35. $\frac{dy}{dx} > 0$ for $x < 0$

$\frac{dy}{dx} < 0$ for $x > 0$

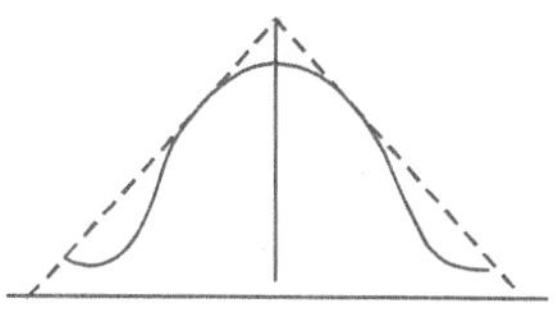

37. Since, $[AA^T]^{-1} = I$

38.

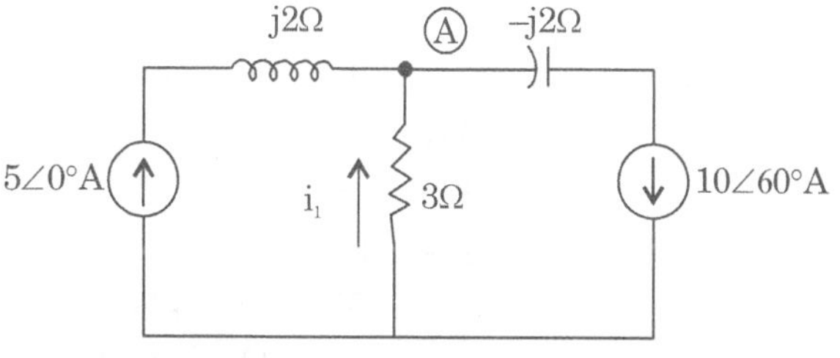

Using superposition theorem at node (A), we get

$$i_1 + 5\angle 0^\circ = 10\angle 60^\circ$$

$$\Rightarrow \quad i_1 = 10\angle 60^\circ - 5$$

$$= 10(\cos 60^\circ + j \sin 60^\circ) - 5$$

$$= 5 + 5\sqrt{3}j - 5$$

$$= 5\sqrt{3}j = \frac{10}{2}\sqrt{3}\angle 90^\circ \text{ Amp}$$

39. 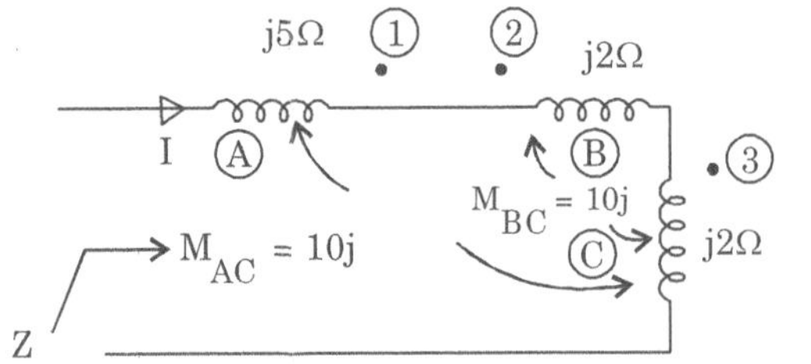

At node (1) current I is leaving and at node (2) and (3) current I is entering.

Then, for coil AC, $M_{AC} = -10j$

For coil BC, $M_{BC} = +10j$

Impedance $Z = j5 + j2 + j2 + jM_{BC} - jM_{AC}$

$= j9 + j10 - j10 = j9$

41. 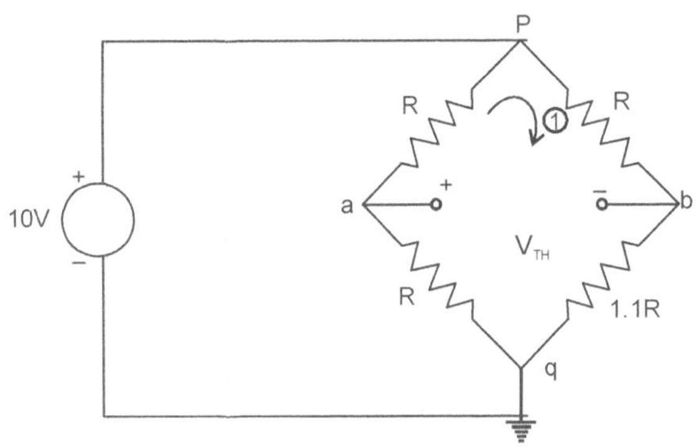

$$V_{ap} = \frac{10R}{R+R} = 5V$$

$$V_{pb} = \frac{10R}{1.1R+R} = \frac{10}{2.1} = 4.7519 \text{ V}$$

Using KVL in loop (1)

$V_{TH} + V_{ap} - V_{pb} = 0$

(Voltmotor reading) $V_{TH} = V_{pb} - V_{ap}$

$= 4.7519 - 5$

$= -0.248$ volts

42. $$\begin{bmatrix} V_1 \\ I_2 \end{bmatrix} = \begin{bmatrix} h_{11} & h_{12} \\ h_{21} & h_{22} \end{bmatrix} \begin{bmatrix} I_1 \\ V_2 \end{bmatrix}$$

For given circuit

$V_1 - 10I_1 - 20(I_1 + I_2) = 0;$

$V_2 - 20(I_1 + I_2) = 0$

$\Rightarrow$ $V_1 - 30I_1 + 20I_2 = 0;$

$V_2 - 20I_1 + 20I_2 = 0$

$V_1 = 30I_1 + 20I_2; V_2 = 20I_1 + 20I_2$

$\therefore$ $V_1 - V_2 = 10I_1$

$\Rightarrow$ $V_1 = V_2 + 10I_1$

Now, $V_1 = h_{11}I_1 + h_{12}V_2$

$\therefore$ $h_{11} = 10,$

and $h_{12} = 1$

$20I_2 = V_1 - 30I_1$

$= V_2 + 10I_1 - 30I_1$

$= V_2 - 20I_1$

$\Rightarrow$ $I_2 = \frac{1}{20}V_2 - I_1$

$\therefore$ $h_{21} = -1$ and $h_{22} = 0.05$

Hence, h-parameter are $\begin{bmatrix} 10 & 1 \\ -1 & 0.05 \end{bmatrix}$

43. Transfer function,

$$T(s) = \frac{R}{R + \frac{1}{Cs}}$$

or, $$\frac{V_O(s)}{V_i(s)} = \frac{R}{R + \frac{1}{Cs}} \quad ...(1)$$

$V_i(t) = 3[u(t) - u(t-2)]$

$V_i(s) = \frac{3}{s}(1 - e^{-2s})$

Their from equation. (1),

$$V_O(s) = \frac{3}{s}(1 - e^{-2s}) \cdot \frac{R}{R + \frac{1}{Cs}}$$

$$= \frac{3}{s}(1 - e^{-2s}) \frac{s}{s + 10^4}$$

$$V_O(s) = \frac{3}{s + 10^4} - \frac{3e^{-2s}}{s + 10^4}$$

Inverse Laplace Transform gives,

$V_O(t) = 3u(t)e^{-10^4 t} - 310^{-4(t-2)}u(t-2)$

At $t = 2$ sec

$V_O(2) = 3 \times 10^{-4} \times 2 - 3 \cong -3$ volts

44. Since, $\frac{\mu_n}{\mu_p} = 3$

and $N_A = N_D$

$\therefore$ $\frac{\sigma_A}{\sigma_B} = \frac{N_A \mu_P}{N_D \mu_n} = \frac{\mu_p}{\mu_e} = \frac{1}{3}$

45. $$C_T = \frac{\varepsilon_r \varepsilon_0 A}{d}$$

$\therefore$ $\frac{C_T}{A} = \frac{\varepsilon_r \varepsilon_0}{d} = \frac{11.7 \times 8.85 \times 10^{-12}}{10 \times 10^{-6}}$

$= \mathbf{10\mu F}$

46. Since base-collector terminals are short-circuited, therefore *n-p-u* transistor will function as P-N junction diode having forward voltage equal to base-emitter voltage (V_{BE}).

Then, emitter current

$$I_E = I_0\left(e^{\frac{V}{\eta V_T}} - 1\right) = I_0\left(e^{\frac{V_{BE}}{\eta V_T}} - 1\right)$$

I_0 reverse saturation current

$$V_{BE} = 0.7V,\ \eta = 1 \text{ for } S_i,$$

$$V_T = \frac{T}{11{,}600} = \frac{300}{11{,}600} = 0.026V$$

Then, $$I_E = 10^{-13}\left(e^{\frac{0.7}{0.026}} - 1\right) \cong 49 \times 10^{-3}\text{mA}$$
$$= 49 \text{ mA}$$

47. Only input offset current can be measured.

48. Using KCL at riverting terminal,

$$\frac{V_- - V_o}{10k} + \frac{V_- - 0}{10k} = 0$$

$$\Rightarrow \quad V_- = \frac{V_O}{2} \quad \text{...(A)}$$

At the non-inverting terminal,

$$\frac{V_+}{R} + \frac{V_+ - V_i}{\frac{1}{CS}} = 0$$

Since $$V_+ = V_-$$

then, $$V_+\left[\frac{1}{R} + CS\right] = CS\,V_i$$

$$\frac{V_O}{2}\left[\frac{1}{R} + CS\right] = CS\,V_i$$

$$\Rightarrow \quad \frac{V_O(s)}{V_i(s)} = \frac{2RCS}{RCS + 1}$$

$$\Rightarrow \quad \frac{V_O(j\omega)}{V_i(j\omega)} = T(j\omega) = \frac{2RCj\omega}{RCj\omega + 1}$$

When $\omega \to 0,\ |T(j\omega)| \to 0$

When $\omega \to \infty,\ |T(j\omega)| = 2$

Therefore this is high-pass filter. At cut-off frequency

$$|T(j\omega)|_{\omega_c = \omega_c} = \frac{1}{\sqrt{2}}|T(j\omega)|_{\omega = \infty}$$

$$\Rightarrow \quad \left|\frac{j2\omega_c RC}{jRC\omega_c + 1}\right| = \frac{1}{\sqrt{2}}$$

$$\Rightarrow \quad \sqrt{\frac{(2\omega_c RC)^2}{1 + (RC\omega_c)^2}} = \frac{1}{\sqrt{2}}$$

$$\Rightarrow \quad \frac{2\omega_c}{\omega_c^2 + 10^6} = \frac{1}{\sqrt{2}}$$

$$\Rightarrow \quad \omega_c = 1000 \text{ rad/sec}$$

49. Common mode gain is given by,

$$A_{cc} = \frac{-\beta R_C}{r_k + 2(\beta + 1)R_E} \approx \frac{R_C}{2R_E}$$

Differential-mode gain,

$$A_{dd} = -g_m R_C \quad [\text{Independent of } R_E]$$

Common mode gain will decrease while R_E increases and differential-mode gain stays constant.

50.
$$V_{GS} = V_G - V_S = 3 - 1 = 2 \text{ volts}$$
$$V_{DS} = V_D - V_S = 5 - 1 = 4 \text{ volts}$$

From the given figure,

$$V_T = 1V(\text{Threshold voltage})$$

Then, $$V_{GS} - V_T = 2 - 1 = 1V$$

As $$V_{DS} \geq V_{GS} - V_T,$$

n-channel will operate in saturation voltage.

51.
$$I_C \simeq \beta I_B \simeq I_E$$

Also, $$I_E = I_C + I_B$$
$$= (\beta + 1)\, I_B$$

In loop① $$20 = -0.7 + I_E \times 1\text{ k}\Omega$$

$$\Rightarrow \quad I_E = 20.7 \text{ mA}$$

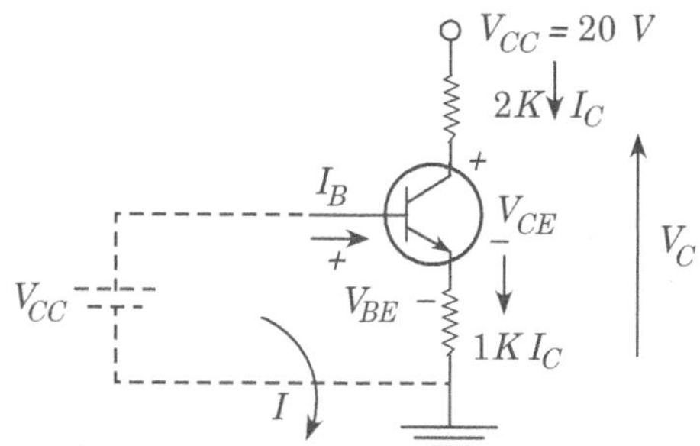

$$\therefore \quad I_B = \frac{I_E}{\beta + 1} = \frac{20.7}{51} = 40\mu A$$

$$V_C = V_{CC} - I_C \times 2K - I_E \times 1K - V_{CE}$$
$$= 20 - 3 \times 20\text{mA} \times 10^3 - 0.7 \simeq \mathbf{16\ V}$$

52.

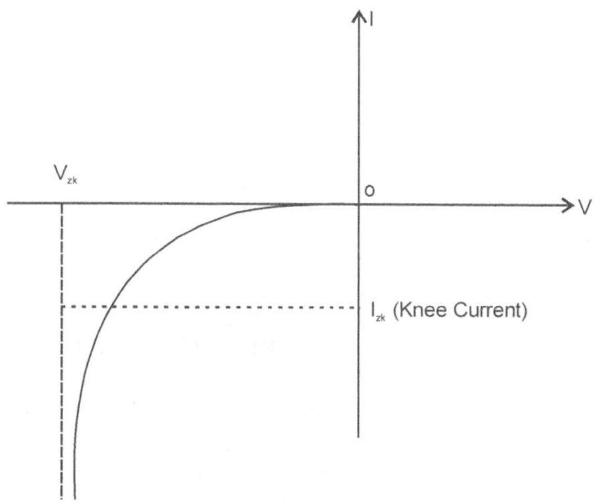

V–I characteristics of zener diode

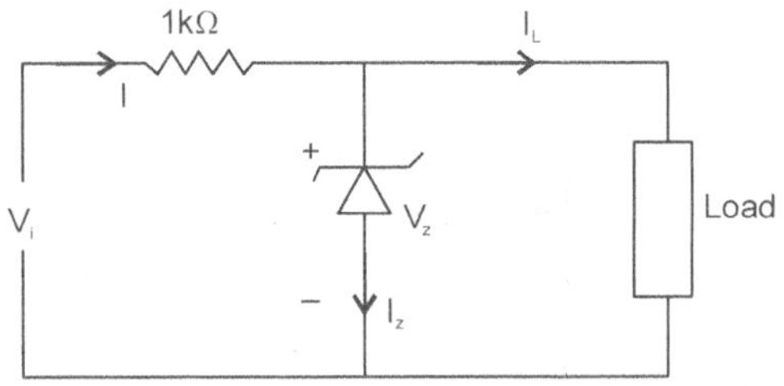

For $V_i = 20$ volts, minimum vlaue of current I

$$I_{min} = \frac{20-58}{1k} = 14.2 \text{ mA}$$

For $V_i = 30$ V, and maximum value of current I

$$I_{max} = \frac{30-5.8}{1k} \; 24.2 \text{ mA}$$

Maximum value of load current

$$I_{L,max} = I_{max} - I_{zk}$$

Where I_{zk} is the knee current

$$= 24.2 - 0.5 = 23.7 \text{ mA}$$

53.

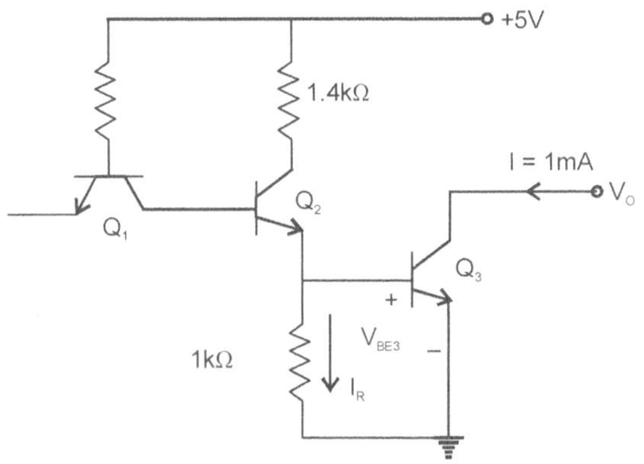

For output is at logic O, $V_O = 0$ V_OH

$$V_O = 0$$

only when Q_3(transistor) is in saturation.

then $\quad V_{BE3} = 0.75$ volts

Using KVL in B – E loop of transistor Q_3,

$$I_R \times 1 \times 10^3 - V_{BE3} = 0$$

$$\Rightarrow \quad I_R = \frac{V_{BE3}}{10^3}$$

$$\Rightarrow \quad I_R = 0.75 \times 10^{-3}$$

$$= 75\text{mA}$$

54. $$f = \overline{A}BC + AB\overline{C}$$

$$= B\left(\overline{A}C + A\overline{C}\right)$$

$$= B(A + C)\left(\overline{A} + \overline{C}\right)$$

$$\left\{\because \text{XOR} = \overline{A}C + C\overline{A} \equiv (A + C)\left(\overline{A} + \overline{C}\right)\right\}$$

55.

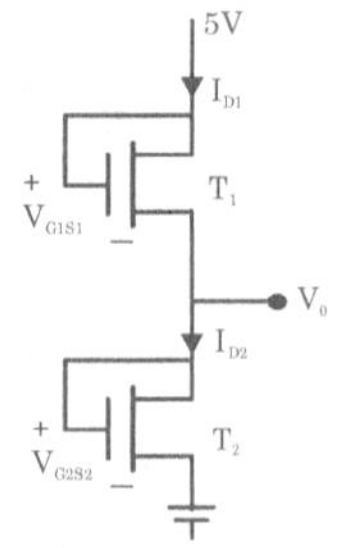

$$I_{01} = I_{02}$$

$$K_1 (V_{G1S1} - V_t)^2 = K_2 (V_{G2S2} - V_t)^2$$

$$\Rightarrow \quad 36\,[(5 - V_0) - 1]^2 = 9[(V_0 - 0) - 1]^2$$

$$V_0 = 3\text{V}$$

56. Truth table for JK FF

S.N.	J	K	Q_n	Q_{n+1}
1	0	0	0	1
2	0	0	1	1
3	0	1	0	0
4	0	1	1	0
5	1	0	0	1
6	1	0	1	1
7	1	1	0	1
8	1	1	1	0

It can be seen from truth-table

For $\quad J = 1,$

$K = X,$

$Q_{n+1} = 1 \qquad (\text{for } Q_n = 0)$

57. Truth table for T Flip-Flop

T	Q_n	Q_{n+1}
0	0	0
1	0	1
0	1	1
1	1	0

Here $\overline{Q}_0$ acts as clock for T1FF and $\overline{Q}_1$ acts as clock for T2FF

Now, the output waveforms for the given circuit is depicted below:

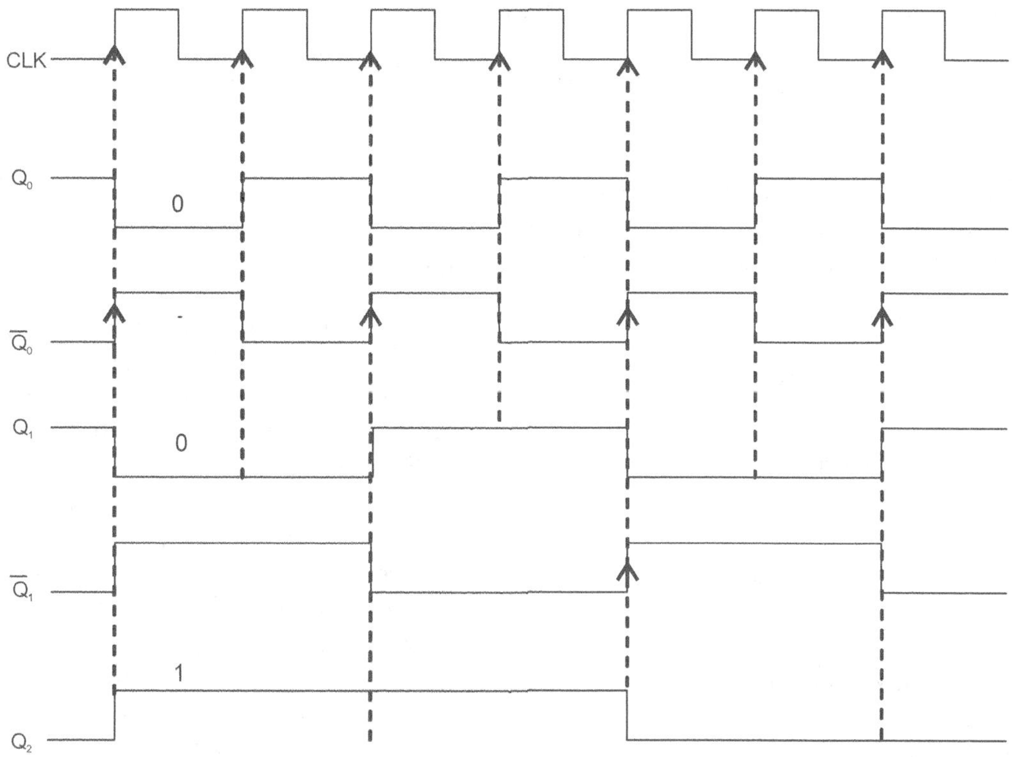

It can be seen from the nonform at the end of first state,

$$Q_2 = 1, Q_1 = 0, Q_0 = 0$$

58. Chip-1 selected for $A_8 = 1, A_9 = 0$

Chip-2 selected for $A_8 = 0, A_9 = 1$

A_{15}	A_{14}	A_{13}	A_{12}	A_{11}	A_{10}	A_9	A_8	A_7	A_6	A_5	A_4	A_3	A_2	A_1	A_0
X	X	X	X	X	X	0	1	X	X	X	X	X	X	X	X
X	X	X	X	X	X	1	0	X	X	X	X	X	X	X	X

Where X are the don't have states.

Now, for the following states of A_0 to A_{15}

	A_{15}	A_{14}	A_{13}	A_{12}	A_{11}	A_{10}	A_9	A_8	A_7	A_6	A_5	A_4	A_3	A_2	A_1	A_0
0100H =	0	0	0	0	0	0	0	1	0	0	0	0	0	0	0	0
02FFH =	0	0	0	0	0	0	1	0	1	1	1	1	1	1	1	1

The address range is 0100H – 02FFH and (0100 –02FF) comes within this address range

Again, for

	A_{15}	A_{14}	A_{13}	A_{12}	A_{11}	A_{10}	A_9	A_8	A_7	A_6	A_5	A_4	A_3	A_2	A_1	A_0
0100H =	0	0	0	0	0	0	0	1	0	0	0	0	0	0	0	0
FEFFH =	1	1	1	1	1	1	1	0	1	1	1	1	1	1	1	1

(01500 – 16FF) address range comes within this representative

Again, for the following states:

	A_{15}	A_{14}	A_{13}	A_{12}	A_{11}	A_{10}	A_9	A_8	A_7	A_6	A_5	A_4	A_3	A_2	A_1	A_0
F900H =	1	1	1	1	1	0	0	1	0	0	0	0	0	0	0	0
FAFFH =	1	1	1	1	1	0	1	0	1	1	1	1	1	1	1	1

Now for address range F800 – F9FFH, states are given by

	A_{15}	A_{14}	A_{13}	A_{12}	A_{11}	A_{10}	A_9	A_8	A_7	A_6	A_5	A_4	A_3	A_2	A_1	A_0
F800H	1	1	1	1	0	1	0	0	0	0	0	0	0	0	0	0
F9FFH	1	1	1	1	1	0	0	1	1	1	1	1	1	1	1	1

In the first case $A_9 = A_8 = 0$ [No Chip is selected]

for the second case, $A_9 = 0$ $A_8 = 1$ Chip-1 is selected

Hence, this address range is not addressed by the given figure.

59. Taking Fourier transform,

$$Y(\omega) = 0.5\, e^{j\omega(-t_d + T)} \times(\omega) + 0.0.5\, e^{-t_d j\omega} X(\omega) + 0.5\, X(\omega) e^{(-t_d - T) j\omega}$$

By time shifting property,

$$\frac{Y(\omega)}{X(\omega)} = H(\omega)$$

$$= (1 + \cos\omega t) e^{-j\omega t_d}$$

61. As, $x(n - n_0) \xleftrightarrow{FT} e^{-j\omega_0 n_0} \times (e^{j\omega})$

then, $Y(e^{j\omega}) = Ae^{-j\omega_0 n_0} \times (e^{j\omega})$

$$\Rightarrow \quad H(f\omega) = \frac{Y(e^{p\omega})}{X(e^{j\omega})}$$

$$= Ae^{-j\omega_0 n_0} \quad ...(A)$$

Given, frequency reforms

$$H(e^{j\omega}) = |H(e^{j\omega})| e^{j\angle H(e^{j\omega})} \quad ...(B)$$

Comparing equation. (A) with (B),

$$A = |H(e^{j\omega})|$$

and $\angle H(e^{j\omega}) = -\omega_0 n_0$

For LTI system, phase and frequency reform are periodic with 2π. The general form of $\angle H(e^{j\omega})$ is $-\omega_0 n_0 + 2\pi k$.

62. Time-scaling property

$$x(at) \xleftrightarrow{FT} \frac{1}{|a|} \times \left(\frac{f}{a}\right)$$

$$\text{or, } |a|\, x(at) \xleftrightarrow{FT} \times \left(\frac{f}{a}\right) \quad ...(A)$$

and frequency-shifting property

$$e^{-j2\pi f_0 t} x(t) \xleftrightarrow{FT} \times (f + f_0) \quad ...(B)$$

Using equation (B) in eqn (A), we get

$$|a|\, e^{-j2\pi f_0 at} x(at) \longleftrightarrow X\left\{\frac{1}{a}(f + af_0)\right\}$$

Putting $a = \dfrac{1}{3}$ and $f_0 = 2$, we get

$$\frac{1}{3} x\left(\frac{t}{3}\right) e^{-j4\pi t/3} \longleftrightarrow X(3f + 2)$$

63. As the open loop transfer function is stable, therefore number of poles in RHS of s-plane is zero. Now, according to Nyquist criteria, the closed-loop system is stable if number of encirclements of $(-1 + j0)$ point in anti-clockwise direction is equal to P-Z, where P and Z are the number of poles and zeros in the RHS of s-plane.

Here P = 0, therefore N = –Z

Hence, for the closed-loop system to be stable number of encirclements in anti-clockwise direction should be zero which will make number of zeros in RH-side of s-plane is zero.

Now, Drawing Nyquist plot.

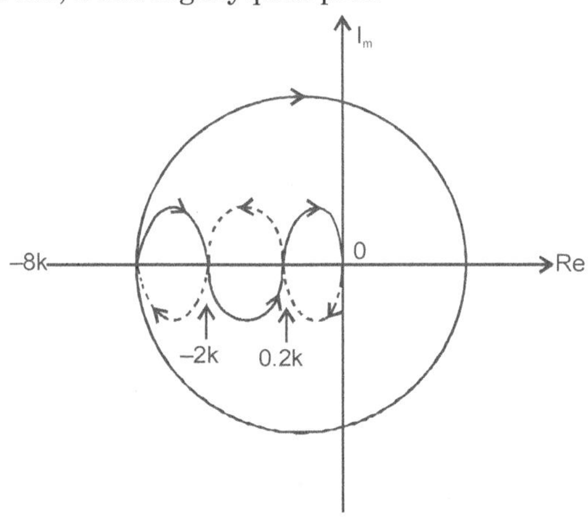

If $(-1 + j0)$ point lie between $-0.2k$ and $-2k$, then number of encirclements is zero. i.e. $2k > 1$ and $0.2k < 1$. Aslo, if $-8k > -1$, then there is no encirclements of $(-1 + j0)$.

Hence, the condition for stability is

$$k < \frac{1}{8} \text{ and } \frac{1}{2} < k < 5$$

64. Even if there is some transportation delay in the system, various time parameters, *e.g.*, t_r, t_s, t_d, t_p are affected but M_p remains same.

65.

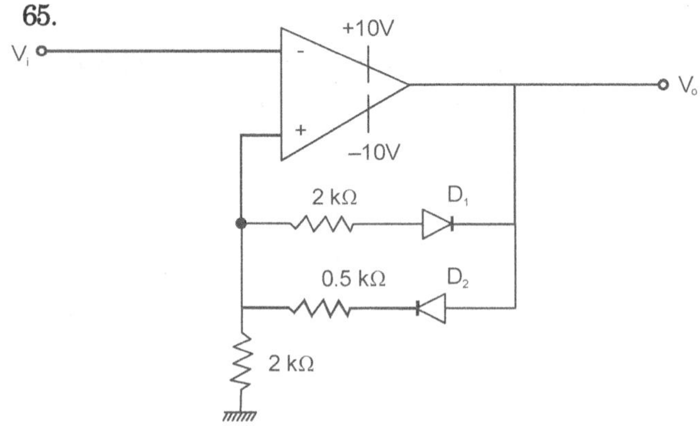

OPAMP output varies between $+V_{sat} = 10V$ to $-V_{sat} = -10V$

When $V_o > 0$, D_2 is forward-biased and D_1 is reverse-biased.

Then, V_+ = (Upper threshold voltage)

$$= +V_{sat}\frac{2}{2+0.5} = 10 \times \frac{2}{2.5} = 12 \text{ volts}$$

When $V_o < 0$, D_2 is reverse-biased and D_1 is forward-biased.

Then V_- = lower-threshold voltage

$$= -V_{sat} \cdot \frac{2}{2+2} = -5 \text{ volts}$$

Hence, the right option is (*b*)

66. $R(s) = \dfrac{1}{s^2}$

For type-*n* system,

open loop gain function is given as

$$G(s) = \frac{k(s+Z_1)(s+Z_2)(s+Z_3)\ldots(s+Z_n)}{s^n(s+P_1)(s+P_2)\ldots(s+P_n)}$$

$$\frac{C(s)}{R(s)} = \frac{G(s)}{1+G(s)}$$

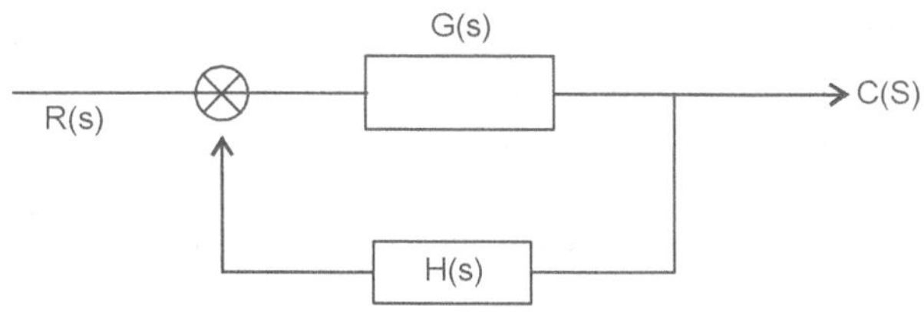

$H(s) = 1$ [Unity feedback system]

Steady-state error is given by

$$e_{ss} = \underset{s\to 0}{Lt}\, sE(s) = \underset{s\to 0}{Lt}\, s[R(s) - C(s)]$$

$$= \underset{s\to 0}{Lt}\frac{sR(s)}{1+G(s)}$$

For $R(s) = \dfrac{1}{s^2}$ (Ramp-input)

$$e_{ss} = \underset{s\to 0}{Lt}\frac{1}{s+sG(s)} = \underset{s\to 0}{Lt}\frac{1}{sG(s)} = \frac{1}{k_v}$$

For finite value of e_{ss}, type of the system should be 1, i.e. $n = 1$

and $\dfrac{1}{k_v} = e_{ss} = 0.05$

$\Rightarrow$ $k_v = 20$ (zero frequency gain)

68. Poles of the given system are at $s = 0, -3$ zero is at $s = 1$

Root locus will originate from open- loop poles at $s = 0, -3 (k = 0)$ and terminate at $s = 1$ $(k = \infty)$.

Characteristics equation is $1 + G(s) = 0$

or, $$1+\frac{k(1-s)}{s(s+3)} = 0$$

or, $$k = \frac{s(s+3)}{s-1}$$

For breakaway points

$$\frac{dk}{ds} = 0 = \frac{(s-1)(2s+3)-(s^2+3s)}{(s-1)^2}$$

$\Rightarrow$ $(s+1)(s-3) = 0$

$\Rightarrow$ $s = -1, 3$

Hence, the root-locus will be as shown in figure given below.

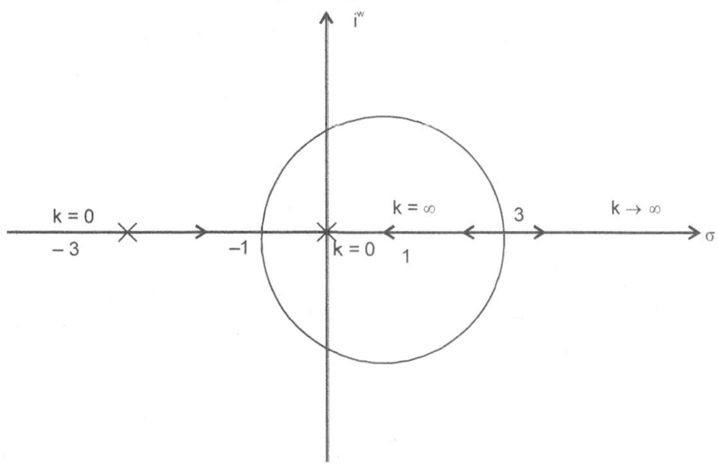

Thus figure corresponds to figure give in option (*c*)

69. Accumulation occurs when we set $V_{GB} < V_{FB}$, pulling the gate to a potential below the substrate. From the graph of potential accross the MOS structure (see figure 1), we know that $\frac{dV}{dx}$ is positive, meaning E(x), the electric field, is negative, or pointing in the negative x direction. This causes holes to accumulate at the top of the substrate and electrons to accumulate at the bottom of the n + gate.

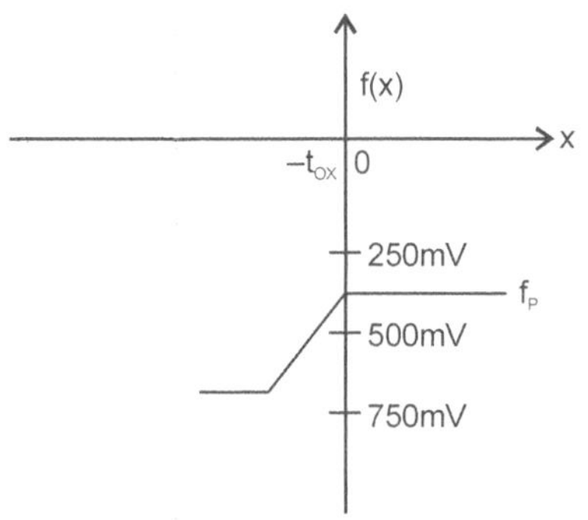

Fig 1 : Potential across as a MOS capacitor in accumulation

70. FM signal is expressed as

$$x(t) = A_C \cos(\omega_C t + \beta \cos \omega_m t)$$

Output of the given device,

$$Y = x^2(t)$$

$$= \frac{A_C^2}{2}\{1 + \cos 2(\omega_c t + \beta \cos \omega_m t)\}$$

Bandwidth, $BW = 2(\beta' + 1) f_m$

Here $\beta' = 2\beta = 2\frac{\Delta f}{f_m} = \frac{2 \times 90}{5} = 36$

Then $BW = 2(36 + 1) \times 5$ KHz

$= 370$ KHz

71. Transfer function of matched filter

$$h(t) = S(T - t)$$

Output of matched filter having input $s(t)$

$$y(t) = h(t) * s(t) = \int_{-\infty}^{\infty} S(J)\, h(t - J)\, dJ$$

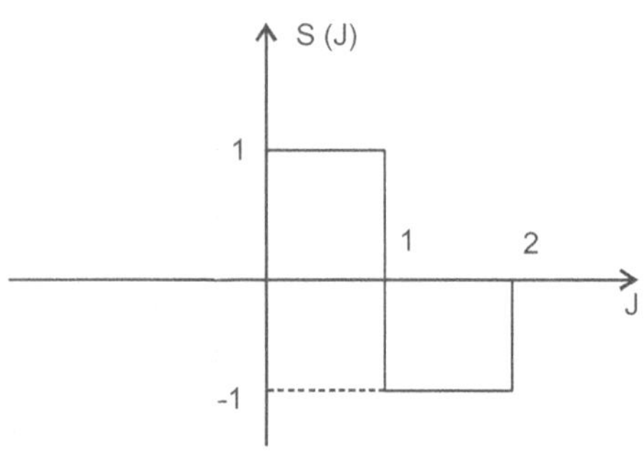

Fig (a)

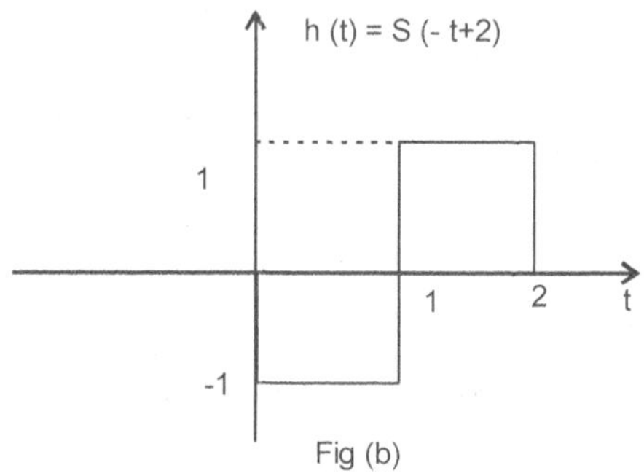

Fig (b)

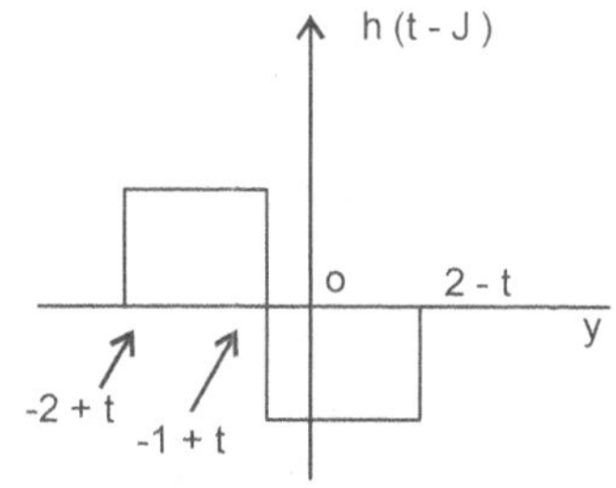

Fig (c)

For, $t = 1$,

$$P(t) = S(J)\, h(1 - J)$$

is shown in figure (d)

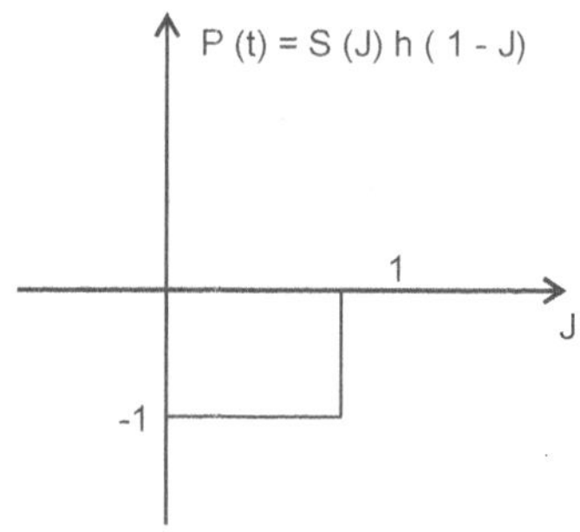

Fig (d)

$$Y(1) = \int_{-\infty}^{\infty} S(J)\, h(1 - J)\, dJ$$

$$= \int_0^1 -1 \cdot dJ = -1$$

For $t = 2$

S(J) [h(2 – J)] is shown in figure (e)

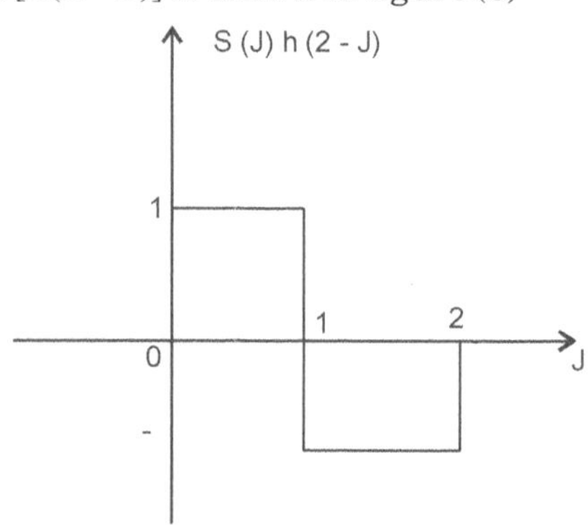

Fig (e)

$$Y(2) = \int_{-\infty}^{\infty} S(J)\,h(2-J)\cdot dJ$$

$$= \int_{-\infty}^{\infty} S(T)\,h(t-J)\cdot dJ$$

For $t = 0$

$S(J)h(-J) = 0$

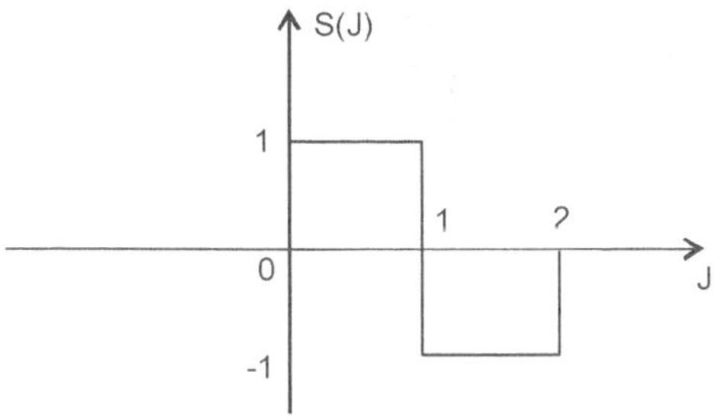

Then, $Y(0) = \int_{-\infty}^{\infty} S(J)\,h(-J)\cdot dJ = 0$

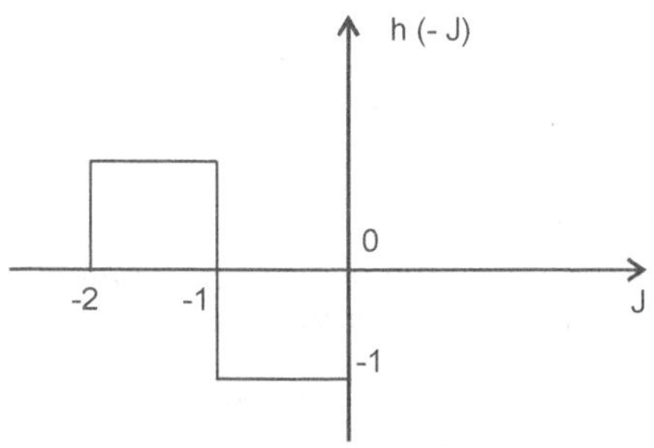

Hence, it can be inferred that right plot is in figure (*c*).

72. $H(\omega) = 2\exp(-j\omega t_d)$

or, $H(f) = 2\exp(-j2\pi f t_d)$

$$N_i(f) = N_o\,\frac{W}{Hz}$$

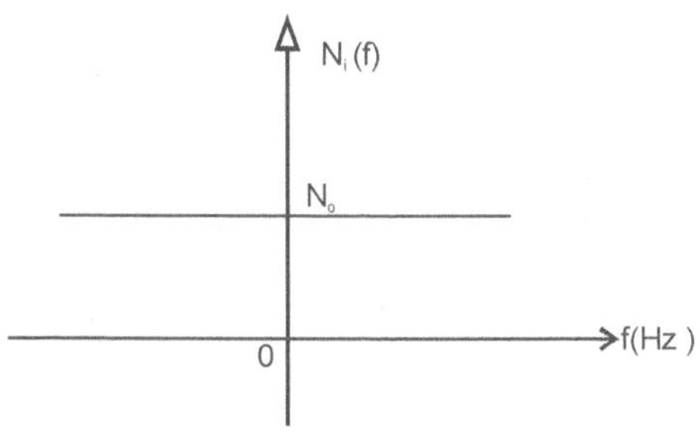

Ouput none power spectral density

$$N_O(f) = |H(f)|^2 Ni(f)$$

$$= 4\,N_O\,\frac{W}{Hz}$$

PSD of LPF is shown below

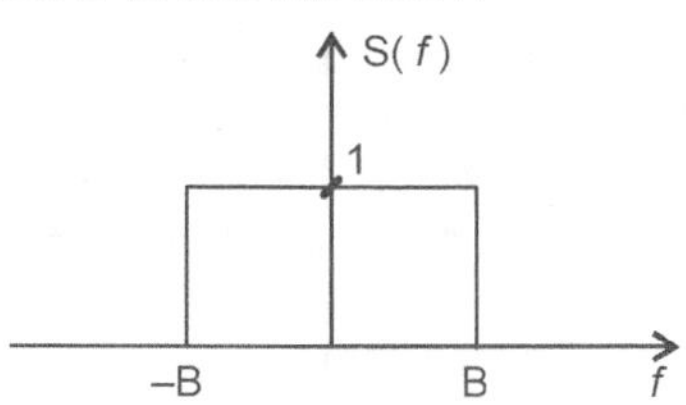

If output non passes through LPF, then PSD at the output of LPF will be as shown here

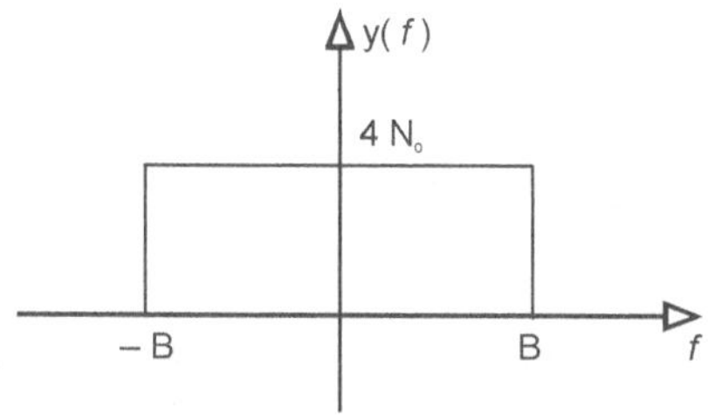

Output noise power is the area under the curved

$$= 4N_o \times 2B = 8N_oB$$

73. Modulation index, $\beta = \frac{\Delta f}{f_m} = \frac{10\text{ KHz}}{1\text{ KHz}} = 10$

When f_m increased to $f_m' = 2$ KHz

$$\text{Bandwidth} = 2(\beta+1)\,f_m'$$

Now, instantaneous angle is defined as,

$$Q_i(t) = 2\pi f_c t + K_p m(t)$$
$$= Q_c(t) + K_p m(t)$$

Phase deviation

$$Q(t) = |Q_i(t) - Q_c(t)|$$
$$= K_p m(t)$$

as phase deviation remains same

$$|K_p m(t)|_{max} = \text{constant}$$

$\Rightarrow$ $K_p a_m = \text{constant}\ [m(t) = a_m \cos\omega_m t]$

modulation index for PM is $\beta = K_p a_m$

Which is constant

then Bandwidth $= 2(\beta+1)\,f_m'$

$$= 2(10+1)2 = 44\text{ KHz}$$

74. $$\int_{-\infty}^{\infty} P(v)dv = 1$$

$\Rightarrow$ $\frac{1}{2}4 \times k = 1$

$\Rightarrow$ $k = \frac{1}{2}$

$\therefore$ $P(v) = \frac{v}{8}$

Now, mean square value

$$= \int_{-\infty}^{\infty} v^2 P(v)\cdot dv = \int_0^4 v^2 P(v)\cdot dv$$

$$= \int_0^4 \frac{v^2 v}{8}\cdot dv = \frac{1}{8}\left[\frac{v^4}{4}\right]_0^4 = 8$$

75. X-component of electric field for TE_{mn} nodes is given as

$$E_x = \frac{j\omega\mu}{h^2}\left(\frac{n\pi}{b}\right)H_0\cos\left(\frac{m\pi x}{a}\right)\sin\left(\frac{n\pi y}{b}\right)e^{-xz}$$

Where a, b are the dimensions of the rectangular waveguide.

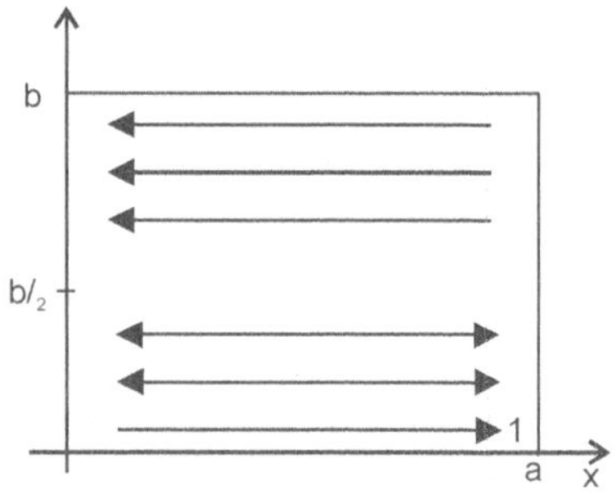

For TE_{02} mode, $m = 0$, $n = 2$

then, $E_x \propto \sin\left(\frac{2\pi y}{b}\right)$

At, $y = 0$, E_x is zero

At $y = \frac{b}{4}$,

E_x is maximum positive.

At $y = \frac{b}{2}$, E_x is zero again.

When $y > \frac{b}{2}$, then direction of E_x gets reversed and at $y = \frac{b}{4}$, E_x is negative maximum

At $y = b$, E_x is zero.

76. $Z_o = 50\Omega$

$Z_{oC} = 100 + j150\Omega$

then, Z_o(characteristics impedance) $= \sqrt{Z_{OC}Z_{SC}}$

$\Rightarrow \quad 50 = \sqrt{Z_{SC}(100 + j150)}$

$\Rightarrow \quad Z_{SC} = \frac{2500}{100 + j150} = 7.69 - j11.54\Omega$

77. The far-field approximation of the two-element array problem:

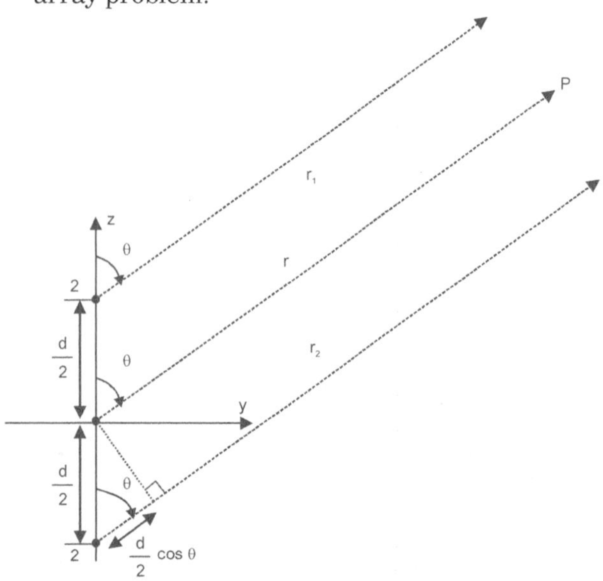

The assumptions as given are:

1. the array elements are identical, i.e.,
   $E_{n1}(\theta, \phi) = E_{n2}(\theta, \phi) = E_n(\theta, \phi)$,
2. they are oriented in the same way in space (they have identical polarizations), i.e.,
   $\hat{\rho}_1 = \hat{\rho}_2 = \hat{\rho}$,
3. their excitation is of the same amplitude, i.e.,
   $M_1 = M_2 = M$

Then, the total electric field,

$$E = E_1 + E_2,$$

$$E = \hat{\rho} M E_n(\theta, \phi)$$

$$x \frac{1}{r}\left[e^{-jk\left(r - \frac{d}{2}\cos\theta\right) + j\frac{\beta}{2}} + e^{-jk\left(r + \frac{d}{2}\cos\theta\right) - j\frac{\beta}{2}}\right],$$

$$E = \hat{\rho}\frac{M}{r} e^{-jkr} E_n(\theta, \phi)$$

$$x\left[e^{-j\left(\frac{kd}{2}\cos\theta + \frac{\beta}{2}\right)} + e^{-j\left(\frac{kd}{2}\cos\theta + \frac{\beta}{2}\right)}\right],$$

$$\boxed{E = \hat{\rho} M \frac{e^{-jkr}}{r} E_n(\theta, \phi) \times \underbrace{2\cos\left(\frac{kd\cos\theta + \beta}{2}\right)}_{AF}} \quad \text{...(A)}$$

The total field of the array is equal to the product of the field created by a single element located at the origin and the ***array factor***, AF:

$$AF = 2\cos\left(\frac{kd\cos\theta + \beta}{2}\right) \quad \text{...(B)}$$

Using the normalized fild pattern of a single element $E_n(\theta, \phi)$, and the normalized AF.

$$AF_n = \cos\left(\frac{kd\cos\theta + \beta}{2}\right) \quad \text{...(C)}$$

Given, $d = \frac{\lambda}{4}, \beta = \frac{\pi}{2}$ and $k = \frac{2\pi}{\lambda}$

then, $AF_n = \cos\left(\frac{\pi}{4}\cos\theta + \frac{\pi}{2}\right)$

[equation (C)]

Null-occurs for the resultant field

at $\frac{\pi}{4}\cos\theta + \frac{\pi}{2} = \frac{\pi}{2}$

$\Rightarrow \quad \cos\theta = 0$

or, $\theta = \frac{\pi}{2}$

Hence the resultant field pattern will be as shown below:

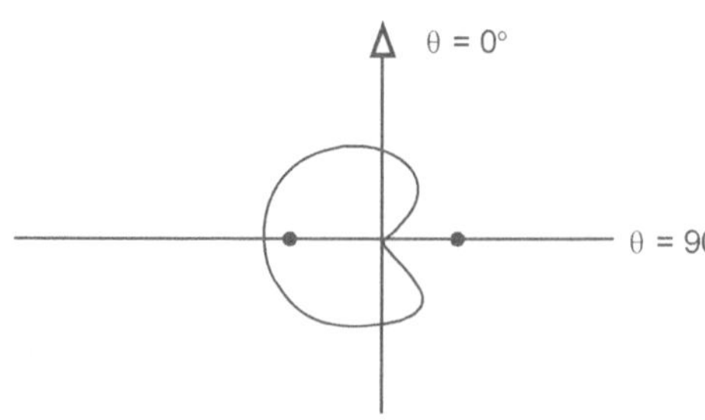

78. Drawing small signal model for the given circuit,

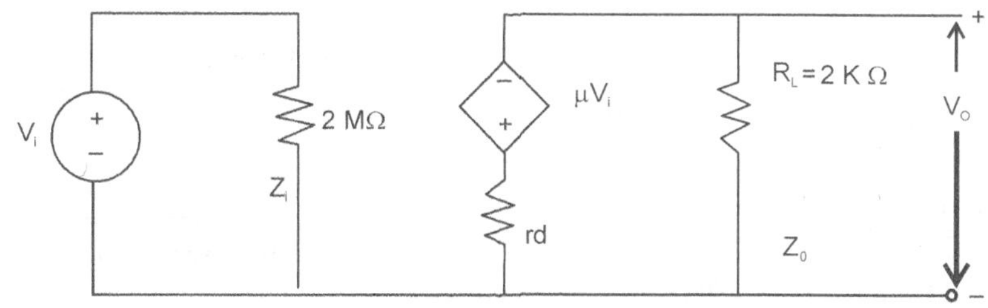

Fig (A)

Z_i = 2 mΩ

$$Z_o = R_L \| rd = 20\ K\Omega \| 2K\Omega = \frac{(20K\Omega)(2K\Omega)}{22K\Omega} = \frac{20}{11} K\Omega$$

79. Under DC conditions, capacitors gets open-circuited

$$V_{GS} = -2 \text{ volts}$$
$$V_P = -8 \text{ volts}$$

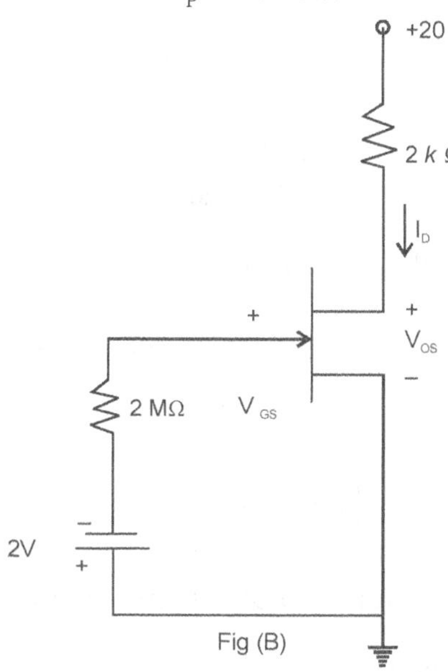

Fig (B)

N-channel MOSFET requires zero or negative gate bias and +ve drain voltage for to operate in active region.

Now, drain current

$$I_D = I_{DSS}\left(1 - \frac{V_{GS}}{V_P}\right)^2$$

$$\Rightarrow I_D = 10 \times 10^{-3}\left(1 - \frac{2}{8}\right)^2 = 5.625 \text{ mA}$$

80. Trans conductance

$$g_m = g_{m0}\left(1 - \frac{V_{GS}}{V_P}\right) \quad ...(A)$$

Drain to source current

$$I_{DS} = I_{DSS}\left(1 - \frac{V_{GS}}{V_P}\right)^2 \quad ...(B)$$

$$g_m = \frac{\partial I_{DS}}{\partial V_{GS}} = \frac{-2I_{DSS}}{V_P}\left(1 - \frac{V_{GS}}{V_P}\right) = \frac{-2I_{DSS}}{V_P}\sqrt{\frac{I_{DS}}{I_{DSS}}}$$

$$g_m = \frac{2\sqrt{I_{DSS} \cdot I_{DS}}}{V_P}$$

$$\Rightarrow g_m = \frac{2\sqrt{I_{DSS} \cdot I_{DS}}}{|V_P|} = \frac{2}{8}\sqrt{10 \times 10^{-3} \times 5.625 \times 10^{-3}}$$

$$= 1.875 \text{ mS}$$

From figure (A),

$$V_o = \frac{-\mu V_i R_L}{R_L + r_d} = \frac{-g_m r_d V_i R_L}{R_L + r_d}$$

$$\Rightarrow \frac{V_o}{V_i} = \frac{-g_m r_d R_L}{R_L + r_d} = -1.875 \times 10^{-3} \times \frac{20}{11} \times 10^3 = -3.41$$

81. (*a*)

Location	Muemonics	Operation
0100H	LXI SP, 00FF	SP → 00FF H
0103H	LXIH,0107	H-L → 0107 H
0106	MVI A, 20H	A → 20H
0108	SUB M	A → 00 H
0109		

Address of M specified by contents of H-L memory and H-L → 0107 H , and 0107H → 20H

When PC reaches 0109 H, contents of accumlater reduces to zero after having operation of instruction SUB M.

81. (*b*)

0109H	ORI 40H	A → $\boxed{40H}$
010BH	ADD M	A → $\boxed{40H + 20H}$ = $\boxed{60H}$

A → $\boxed{0000\ \ 0000}$H = 00H

M → $\boxed{0107H}$ 0100 0000 H = 40H

ORI → 0100 0000 H = 40H

$\boxed{0107H}$ → $\boxed{20H}$ → 0010 0000 H = 20H

ADDM → 0110 0000 H = 60H

82.(*a*) $G(j\omega) = \dfrac{3e^{-2j\omega}}{j\omega(1+j\omega)}$

If $|G(j\omega)|_{\omega=\omega_c=1}$,

then ω_c is gain-cross over frequency.

$$|G(j\omega)|_{\omega=\omega_c} = \frac{|3e^{-2j\omega_c}|}{|j\omega_c(2+j\omega_c)|} = \frac{3}{\omega_c\sqrt{\omega_c^2+4}} = 1$$

or, $\omega_c^4 + 4\omega_c^2 - 9 = 0$

$\omega_c^4 = 1.6$, $\omega_c = 1.26$ rad/sec

For phase-cross over frequency, $\omega = \omega_p$

$$\angle G(j\omega)|_{\omega=\omega_p} = -180^\circ$$

or, $(-2\omega_p) - \dfrac{\pi}{2} - \tan^{-1}\dfrac{\omega_p}{2} = -180^\circ = -\pi$

or, $+2\omega_p + \tan^{-1}\dfrac{\omega p}{2} = \dfrac{\pi}{2}$

or, $2\omega_p + \left(\dfrac{\omega_p}{2} - \dfrac{\left(\dfrac{\omega_p}{2}\right)^2}{3} + \cdots\right) = \dfrac{\pi}{2}$

Neglecting higher order terms, we get

$$\frac{5\omega_p}{2} - \frac{\omega_p^3}{24} = \frac{\pi}{2}$$

or, $\omega_p\left(5 - \dfrac{\omega_p^2}{12}\right) = \pi$

Trial and error procedures gives

$\omega_p \approx 0.632$ rad/sec

82. (*b*) At $\omega = \omega_p$

$$a = |G(j\omega)|_{\omega=\omega_p} = \frac{3}{\sqrt{\omega_p^2(\omega_p^2+4)}}$$

$$= \frac{3}{\sqrt{(0.632)^2 + (0.632)^2 + 4}} = 2.263$$

Gain margin $= 20\log_{10}\dfrac{1}{a}$

$$= 20\log_{10}\frac{1}{2.263} = -7.09\ dB$$

At $\omega = \omega_c$

$$\angle G(j\omega)|_{\omega=\omega_c} = -2\omega_c - \frac{\pi}{2} - \tan^{-1}\left(\frac{\omega_c}{2}\right)$$

$$= -2\times 1.26 - \frac{\pi}{2} - \tan^{-1}(0.63)$$

$$= -2.52 - 1.57 - 0.562$$

$$= -4.662 \text{ rad}$$

$\angle G(j\omega)|_{\omega=\omega_c} = -267.5^\circ$

then, PM $= \angle G(j\omega)|_{\omega=\omega_c} + 180^\circ$

$= -267.5^\circ + 180^\circ = -87.50^\circ$

83. (*a*) $\displaystyle\int_{-\infty}^{\infty} P(x)dx = 1$

Since, equiprobable, hence each has same probability $= \dfrac{1}{3}$

∴ $2a \times \dfrac{1}{4} = \dfrac{1}{3}$

⇒ $a = \dfrac{2}{3}$

83. (*b*) Quantization noise power $= \displaystyle\int_{-a}^{a} f(x)x^2 dx$

$$= \frac{1}{4}\frac{\left[x^3\right]_{-a}^{a}}{3}$$

$$= \frac{1}{12}\left[a^3 + a^3\right]$$

$$= \frac{a^3}{6} = \frac{\left(\frac{2}{3}\right)^3}{6} = \frac{4}{81}$$

84.(*a*) Standing-wave ratio,

$$\rho(\text{VSWR}) = \frac{V_{max}}{V_{min}} = 4$$

and
$$\rho = \frac{1+|\rho_e|}{1-|\rho_e|}$$

$$\Rightarrow \quad 4 = \frac{1+|\rho_e|}{1-|\rho_e|} \Rightarrow |\rho_e| = 0.6$$

Now reflection coefficeint, $\rho_e = |\rho_e| \cdot e^{jQ_e}$

Where Q_i is the phase-angle between incident and reflected wave and equal to π or 180°.

or, $\rho_e = 0.6\, e^{j\pi} = -0.6$

84. (*b*) As,
$$\rho_e = \frac{Z_L - Z_o}{Z_L + Z_o}$$

or,
$$-0.6 = \frac{Z_L - Z_o}{Z_L + Z_o}$$

or,
$$Z_L = \frac{Z_\omega}{4} = \frac{50\Omega}{4} = 12.5\Omega$$

85. (*a*)
$$y(x) = x\left(\frac{n}{2} - 1\right); \quad n \text{ even}$$
$$= 0 \quad ; \quad n \text{ odd}$$

$\therefore$ $y(x)$ is shifted 1 towards left and axis expanded by 2

85. (*b*) From the known $y(n)$,

$$y(2n) = x(n-1)$$

$\therefore$ Fourier transform

$$= \frac{1}{2}e^{j\omega} + 1 + 2e^{-j\omega} + e^{-2j\omega} + \frac{1}{2}e^{-3j\omega}$$

$$= e^{-j\omega}\left[\frac{1}{2}\left(e^{2j\omega} + e^{-2j\omega}\right) + 2 + \left(e^{j\omega} + e^{-j\omega}\right)\right]$$

$$= e^{-j\omega}\left[\cos 2\omega + 2 + 2\cos\omega\right]$$

SOLVED PAPER – 2004

ELECTRONICS & COMMUNICATION ENGINEERING

Time Allowed : 3 Hours | **Maximum Marks : 150**

Q. 1- 30 Carry One Mark Each

1. Consider the network graph shown in the figure. Which one of the following is NOT a 'tree' of this graph?

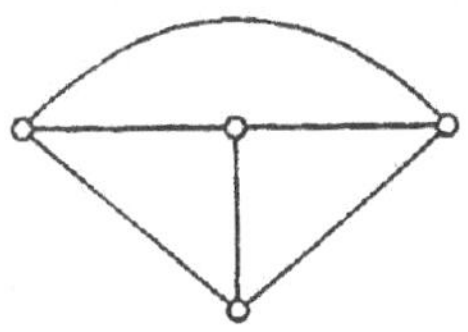

(*a*) 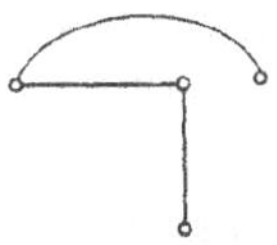
(*b*) 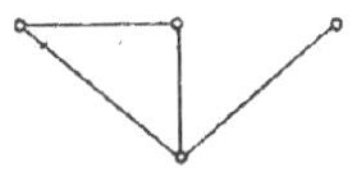
(*c*) 
(*d*) 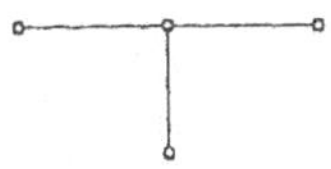

2. The equivalent inductance measured between the terminals 1 and 2 for the circuit shown in the figure is

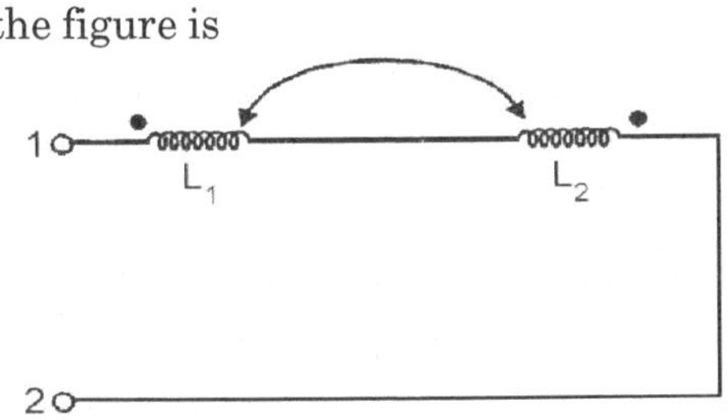

(*a*) $L_1 + L_2 + M$ (*b*) $L_1 + L_2 - M$
(*c*) $L_1 + L_2 + 2M$ (*d*) $L_1 + L_2 - 2M$

3. The circuit shown in in the figure, with $R = \frac{1}{3}$ Ω, $L = \frac{1}{4}$ H, C = 3 F has input voltage $v(t) = \sin 2t$. The resulting current $i(t)$ is

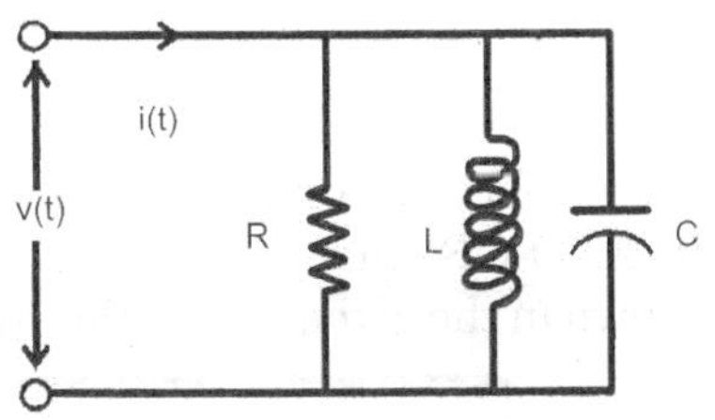

(*a*) $5 \sin (2t + 53.1°)$ (*b*) $5 \sin (2t - 53.1°)$
(*c*) $25 \sin (2t + 53.1°)$ (*d*) $25 \sin (2t - 53.1°)$

4. For the circuit shown in in the figure, the time constant RC= 1 ms. The input voltage is

$$v_i(t) = \sqrt{2} \sin 10^3 t.$$

The output voltage $v_0(t)$ is equal to

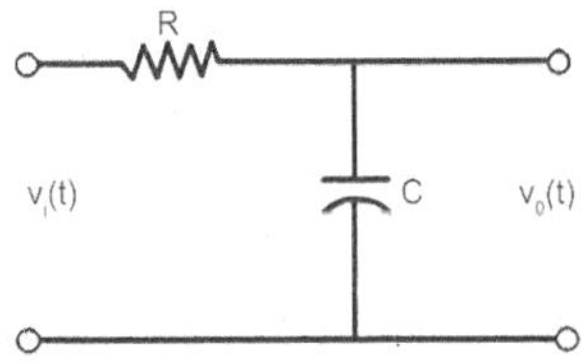

(*a*) $\sin (10^3t - 45°)$ (*b*) $\sin (10^3t + 45°)$
(*c*) $\sin (10^3t - 53°)$ (*d*) $\sin (10^3t + 53°)$

5. For the R-L circuit shown in the figure, the input voltage $v_i(t) = u(t)$. The current $i(t)$ is

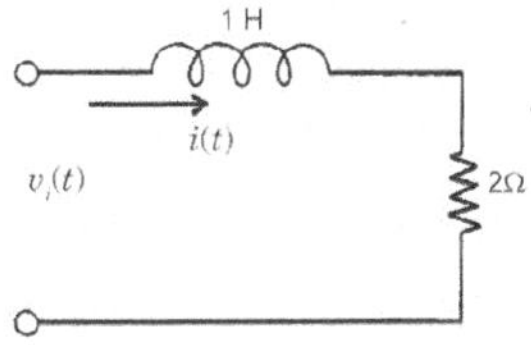

(*a*) 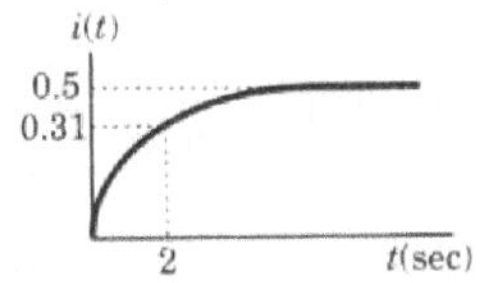

(*b*) 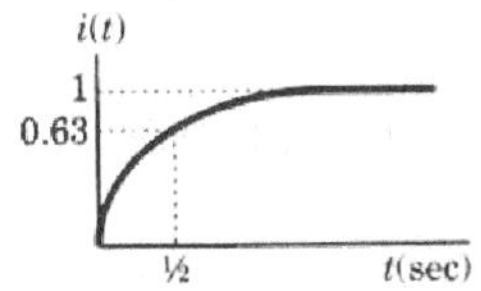

(*c*) 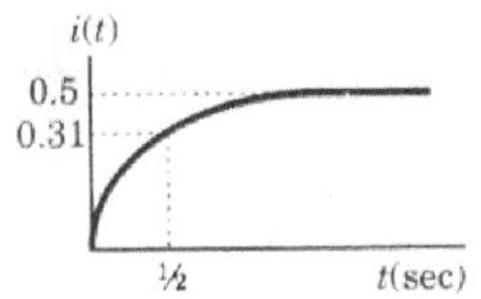

(*d*) 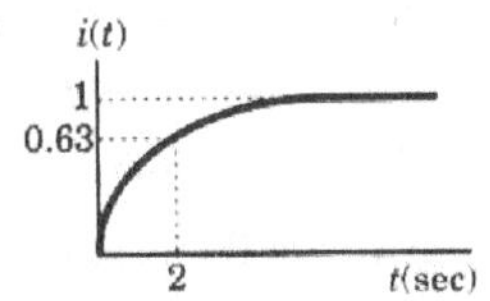

6. The impurity commonly used for realizing the base region of a silicon *n-p-n* transistor is
   (*a*) Gallium (*b*) Indium
   (*c*) Boron (*d*) Phosphorus

7. If for a silicon *n-p-n* transistor, the base-to-emitter voltage (V_{BE}) is 0.7 V and the collector-to-base voltage (V_{CB}) is 0.2 V, then the transistor is operating in the
   (*a*) normal active mode
   (*b*) saturation mode
   (*c*) inverse active mode
   (*d*) cut-off mode

8. Consider the following statements S1 and S2.

S1 : The β of a bipolar transistor reduces if the base width is increased.

S2 : The β of a bipolar transistor increases if the doping concentration in the base is increased.

Which one of the following is correct?

(*a*) S1 is FALSE and S2 is TRUE

(*b*) Both S1 and S2 are TRUE

(*c*) Both S1 and S2 are FALSE

(*d*) S1 is TRUE and S2 is FALSE

9. An ideal op-amp is an ideal

(*a*) voltage controlled current source

(*b*) voltage controlled voltage source

(*c*) current controlled current source

(*d*) current contolled voltage source

10. Voltage series feedback (also called series-shunt feedback) results in

(*a*) increase in both input and output impedances

(*b*) decrease in both input and output impedances

(*c*) increase in input impedance and decrease in output impedance

(*d*) decrease in input impedance and increase in output impedance

11. The circuit in the given figure is a

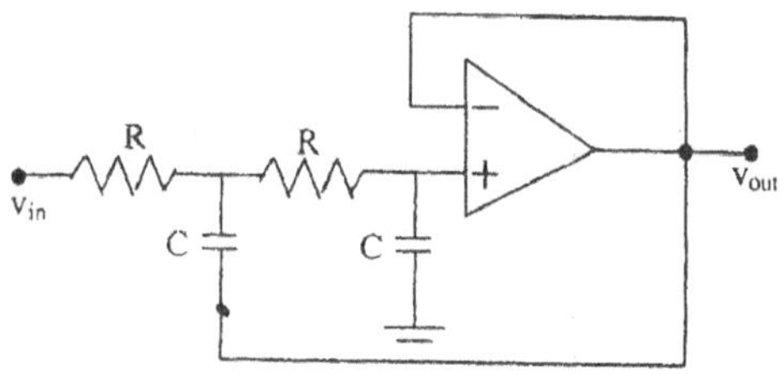

(*a*) low-pass filter (*b*) high-pass filter

(*c*) band-pass filter (*d*) band-reject filter

12. Assuming V_{CEsat} = 0.2 V and β = 50, the minimum base current (I_B) required to drive the transistor in the given figure to saturation is

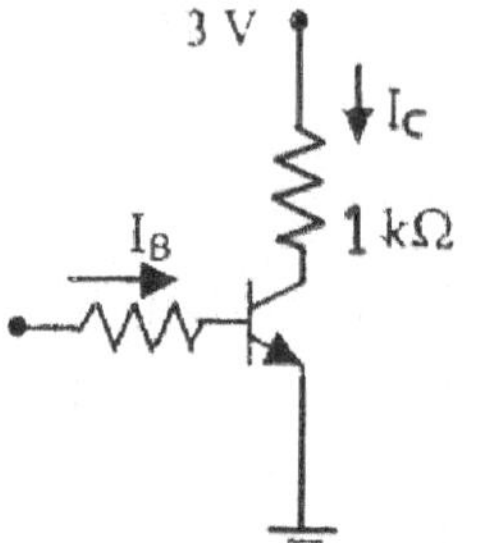

(*a*) 56 μA (*b*) 140 mA

(*c*) 60 μA (*d*) 3 mA

13. A master-slave flip-flop has the characteristic that

(*a*) change in the input immediately reflected in the output

(*b*) change in the output occurs when the state of the master is affected

(*c*) change in the output occurs when the state of the slave is affected

(*d*) both the master and the slave states are affected at the same time

14. The range of signed decimal numbers that can be represented by 6-bit 1's complement number is

(*a*) – 31 to + 31 (*b*) – 63 to + 64

(*c*) – 64 to + 63 (*d*) – 32 to + 31

15. A digital system is required to amplify a binary-encoded audio signal. The user should be able to control the gain of the amplifier from a minimum to a maximum in 100 increments. The minimum number of bits required to encode, in straight binary, is

(*a*) 8 (*b*) 6

(*c*) 5 (*d*) 7

16. Choose the correct one from among the alternatives A, B, C, D after matching an item from Group I with the most appropriate item in Group II.

Group I	Group II
A : Shift register	1 : Frequency division
B : Counter	2 : Addressing in memory chips
C : Decoder	3 : Serial to parallel data conversion

	A	B	C
(*a*)	3	2	1
(*b*)	3	1	2
(*c*)	2	1	3
(*d*)	1	2	2

17. Figure given below shows the internal schematic of a TTL AND-OR -Invert (AOI) gate. For the inputs shown in the given figure, the output Y is

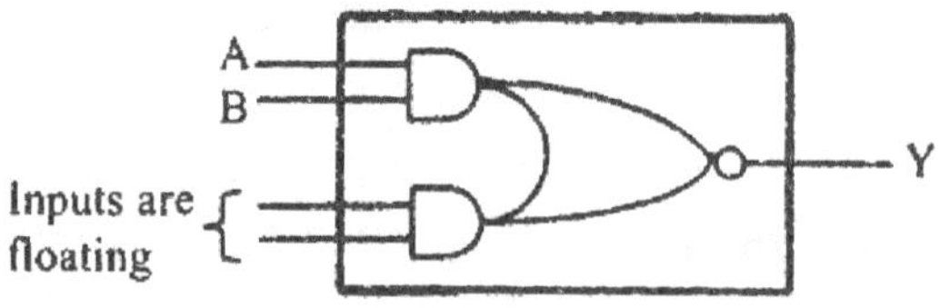

(*a*) 0 (*b*) 1

(*c*) AB (*d*) $\overline{AB}$

18. Given figure is the voltage transfer characteristic of

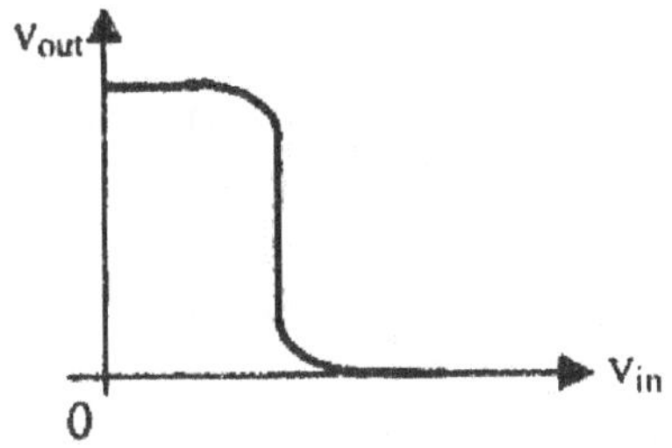

(*a*) an NMOS inverter with enhancement mode transistor as load
(*b*) an NMOS inverter with depletion mode transistor as load
(*c*) a CMOS inverter
(*d*) a BJT inverter

19. The impulse response h[n] of a linear time-invariant system is given by

$$h[n] = u[n+3] + u[n-2] - 2u[n-7]$$

where u[n] is the unit step sequence. The above system is
(*a*) stable but not causal
(*b*) stable and causal
(*c*) causal but unstable
(*d*) unstable and not causal

20. The distribution function $F_x(x)$ of a random variable X is shown in the figure. The probability that X = 1 is

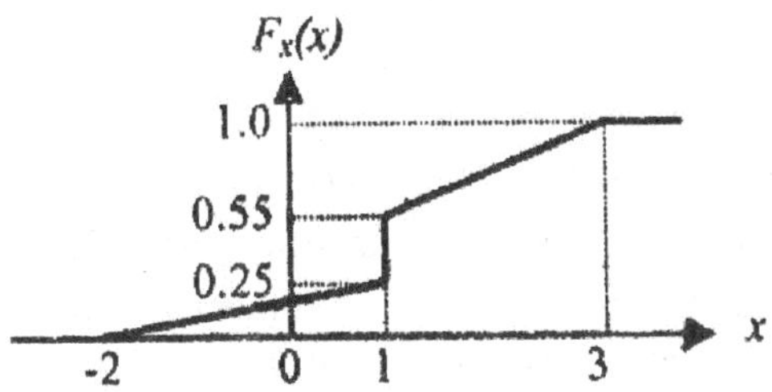

(*a*) zero (*b*) 0.25
(*c*) 0.55 (*d*) 0.30

21. The *z*-transform of a system is $H(z) = \frac{z}{z-0.2}$
If the ROC is $|z| < 0.2$, then the impulse response of the system is
(*a*) $(0.2)^n u[n]$ (*b*) $(0.2)^n u[-n-1]$
(*c*) $-(0.2)^n u[n]$ (*d*) $-(0.2)^n u[-n-1]$

22. The Fourier transform of a conjugate symmetric function is always
(*a*) imaginary
(*b*) conjugate anti-symmetric
(*c*) real
(*d*) conjugate symmetric

23. The gain margin for the system with open-loop transfer function $G(s)H(s) = \frac{2(1+s)}{s^2}$, is
(*a*) ∞ (*b*) 0
(*c*) 1 (*d*) – ∞

24. Given $G(s)H(z) = \frac{K}{s(s+1)(s+3)}$, the point of intersection of the asymptotes of the root loci with the real axis is
(*a*) – 4 (*b*) 1.33
(*c*) – 1.33 (*d*) 4

25. In a PCM system, if the code word length is increased from 6 to 8 bits, the signal to quantization noise ratio improves by the factor
(*a*) 8/6 (*b*) 12
(*c*) 16 (*d*) 8

26. An AM signal is detected using an envelope detector. The carrier frequency and modulating signal frequency are 1 MHz and 2 kHz respectively. An appropriate value for the time constant of the envelope detector is
(*a*) 500 μsec (*b*) 20 μsec
(*c*) 0.2 μsec (*d*) 1 μsec

27. An AM signal and a narrow-band FM signal with identical carriers, modulating signals and modulation indices of 0.1 are added together. The resultant signal can be closely approximated by
(*a*) broadband FM
(*b*) SSB with carrier
(*c*) DSB-SC
(*d*) SSB without carrier

28. In the output of a DM speech encoder, the consecutive pulses are of opposite polarity during time interval $t_1 < t < t_2$. This indicates that during this interval
(*a*) the input to the modulator is essentially constant
(*b*) the modulator is going through slope overload
(*c*) the accumulator is in saturation
(*d*) the speech signal is being sampled at the Nyquist rate

29. The phase velocity of an electromagnetic wave propagating in a hollow metallic rectangular waveguide in the TE_{10} mode is
(*a*) equal to its group velocity
(*b*) less than the velocity of light in free space
(*c*) equal to the velocity of light in free space
(*d*) greater than the velocity of light in free space

30. Consider a lossless antenna with a directive gain of + 6dB. If 1 mW of power is fed to it the total power radiated by the antenna will be

(*a*) 4 mW (*b*) 1 mW

(*c*) 7 mW (*d*) 1/4 mW

Q. 31- 90 Carry Two Marks Each

31. For the lattice circuit shown in the figure, $Z_a = j2\Omega$ and $Z_b = 2\Omega$. The values of the open circuit impedance parameters $Z = \begin{bmatrix} z_{11} & z_{12} \\ z_{21} & z_{22} \end{bmatrix}$ are

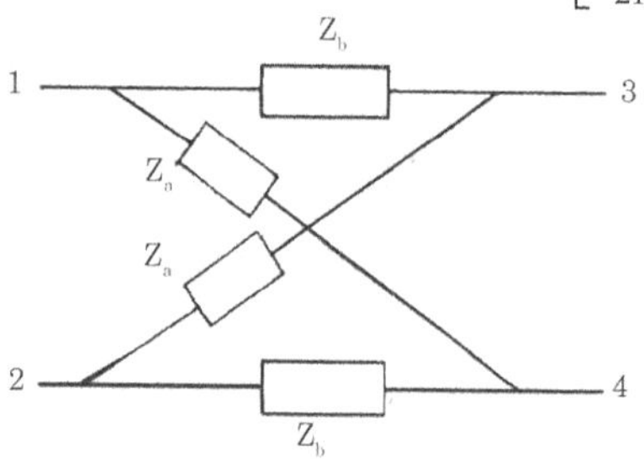

(*a*) $\begin{bmatrix} 1-j & 1+j \\ 1+j & 1+j \end{bmatrix}$ (*b*) $\begin{bmatrix} 1-j & 1+j \\ -1+j & 1-j \end{bmatrix}$

(*c*) $\begin{bmatrix} 1+j & 1+j \\ 1-j & 1-j \end{bmatrix}$ (*d*) $\begin{bmatrix} 1+j & -1+j \\ -1+j & 1+j \end{bmatrix}$

32. The circuit shown in the figure has initial current $i_L(0^-) = 1$ A through the inductor and an initial voltage $v_C(0^-) = -1$V across the capacitor. For input $v(t) = u(t)$, the Laplace transform of the current $i(t)$ for $t \geq 0$ is

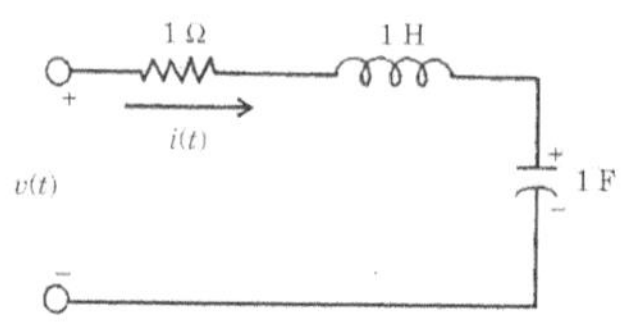

(*a*) $\dfrac{s}{s^2+s+1}$ (*b*) $\dfrac{s+2}{s^2+s+1}$

(*c*) $\dfrac{s-2}{s^2+s+1}$ (*d*) $\dfrac{s-2}{s^2+s+1}$

33. Consider the Bode magnitude plot shown in the given figure. The transfer function H(*s*) is

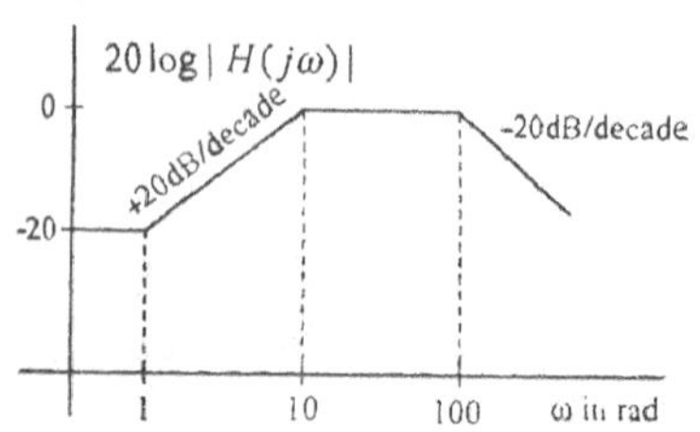

(*a*) $\dfrac{(s+10)}{(s+1)(s+100)}$ (*b*) $\dfrac{10(s+1)}{(s+1)(s+100)}$

(*c*) $\dfrac{10^2(s+1)}{(s+10)(s+100)}$ (*d*) $\dfrac{10^3(s+100)}{(s+1)(s+10)}$

34. The transfer function $H(s) = \dfrac{V_0(s)}{V_i(s)}$ of an R-L-C circuit is given by

$$H(s) = \frac{10^6}{s^2+20s+10^6}$$

The Quality factor (Q-factor) of this circuit is

(*a*) 25 (*b*) 50

(*c*) 100 (*d*) 5000

35. For the circuit shown in the figure , the initial conditions are zero. Its transfer function

$$H(s) = \frac{V_c(s)}{V_i(s)} \text{ is}$$

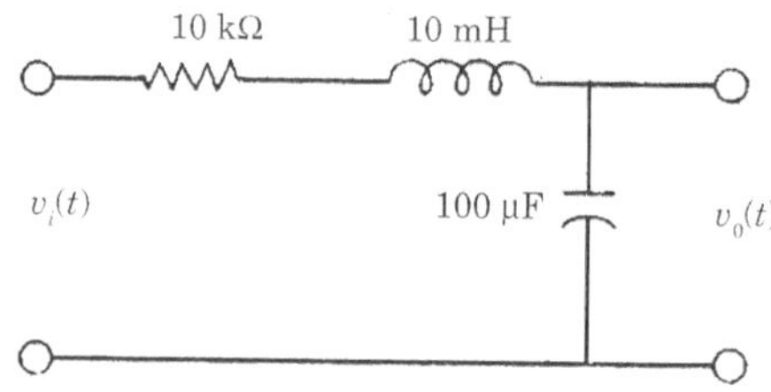

(*a*) $\dfrac{1}{s^2+10^6s+10^6}$ (*b*) $\dfrac{10^6}{s^2+10^3s+10^6}$

(*c*) $\dfrac{10^3}{s^2+10^3s+10^6}$ (*d*) $\dfrac{10^6}{s^2+10^6s+10^6}$

36. A system described by the following differential equation $\dfrac{d^2y}{dt^2}+3\dfrac{dy}{dt}+2y = x(t)$ is initially at rest.

For input $x(t) = 2u(t)$, the output $y(t)$ is

(*a*) $(1-2e^{-t}+e^{-2t})\,u(t)$

(*b*) $(1+2e^{-t}-2e^{-2t})\,u(t)$

(*c*) $(0.5+e^{-t}+1.5e^{-2t})\,u(t)$

(*d*) $(0.5+2e^{-t}+2e^{-2t})\,u(t)$

37. Consider the following statements S1 and S2

S1: At the resonant frequency the impedance of a series R-L-C circuit is zero.

S2: In a parallel G-L-C circuit, increasing the conductance G results in increase in its Q factor.

Which one of the following is correct?

(*a*) S1 is FALSE and S2 is TRUE

(*b*) Both S1 and S2 are TRUE

(*c*) S1 is TRUE and S2 is FALSE

(*d*) Both S1 and S2 are FALSE

38. In an abrupt p-n junction, doping concentrations on *p*-side and *n*-side are $N_A = 9 \times 10^{16}/cm^3$ and $N_D = 1 \times 10^{16}/cm^3$ respectively. The *p-n* junction is reverse biased and the total depletion width is 3 μm. The depletion width on the *p*-side is

(*a*) 2.7 μm (*b*) 0.3 μm

(*c*) 2.25 μm (*d*) 0.75 μm

39. The resistivity of a uniformloy doped *n*-type silicon sample is 0.5 Ω-cm. If the electron mobility (μ_n) is 1250 cm²/V-sec and the charge of an electron is 1.6×10^{-19} coulomb, the donor impurity concentration (N_D) in the sample is

(*a*) 2×10^{16}/cm³ (*b*) 1×10^{16}/cm³

(*c*) 2.5×10^{15}/cm³ (*d*) 2×10^{15}/cm³

40. Consider an abrupt *p-n*. Let V_{bi} be the built-in potential of this junction and V_R be the applied reverse bias. If the junction capacitance (C_j) is 1 pF for $V_{bi} + V_R = 1V$, then for $V_{bi} + V_R = 4V$, C_j will be

(*a*) 4 pF (*b*) 2 pF

(*c*) 0.25 pF (*d*) 0.5 pF

41. Consider the following statements S1 and S2.

S1 : The threshold voltage (V_T) of a MOS capacitor decreases with increase in gate oxide thickness

S2 : The threshold voltage (V_T) of a MOS capacitor decreases with increase in substrate doping concentration

Which one of the following is correct?

(*a*) S1 is FALSE and S2 is TRUE

(*b*) Both S1 and S2 are TRUE

(*c*) Both S1 and S2 are FALSE

(*d*) S1 is TRUE and S2 is FALSE

42. The drain of an n-channel MOSFET is shorted to the gate so that $V_{gs} = V_{DS}$. The threshold voltage (V_T) of MOSFET is 1 V. If the drain current (I_D) is 1 mA for V_{GS} = 2 V, then for V_{GS} = 3 V, I_D is

(*a*) 2 mA (*b*) 3 mA

(*c*) 9 mA (*d*) 4 mA

43. The longest wavelength that can be absorbed by silicon, which has the bandgap of 1.12 eV, is 1.1 μm. If the longest wavelength that can be absorbed by another material is 0.87 μm, then the bandgap of this material is

(*a*) 1.416 eV (*b*) 0.886 eV

(*c*) 0.854 eV (*d*) 0.706 eV

44. The neutral base width of a bipolar transistor , biased in the active region , is 0.5 μm. The maximum electron concentration and the diffusion constant in the base are 10^{14}/cm³ and D_n = 25 cm²/sec respectively. Assuming negligible recombination in the base, the collector current density is (the electron charge is 1.6×10^{-19} coulomb)

(*a*) 800 A/cm² (*b*) 8 A/cm²

(*c*) 200 A/cm² (*d*) 2 A/cm²

45. Assuming that the β of the transistor is extremely large and V_{BE} = 0.7 V, I_C and V_{CE} in the circuit shown in the figure are

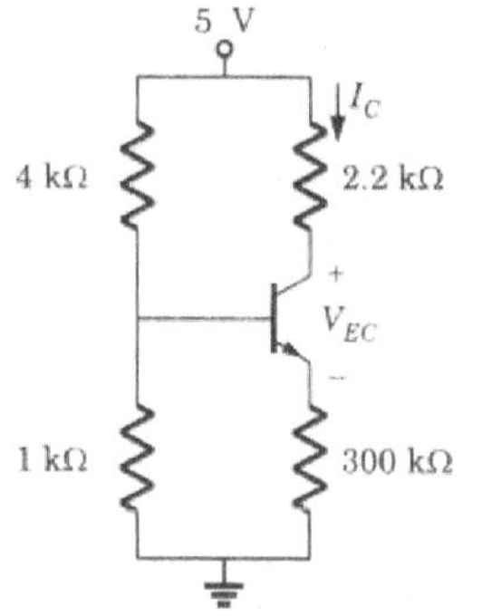

(*a*) I_C = 1 mA, V_{CE} = 4.7 V

(*b*) I_C = 0.5 mA, V_{CE} = 3.75 V

(*c*) I_C = 1 mA, V_{CE} = 2.5 V

(*d*) I_C = 0.5 mA, V_{CE} = 3.9 V

46. A bipolar transistor is operating in the active region with a collector current of 1 mA. Assuming that the β of the transistor is 100 and the thermal voltage (V_T) is 25 mV, the transconductance (g_m) and the input resistance (r_π) of the transistor in the common emitter configuration, are

(*a*) g_m = 25 mA/V and r_π = 15.625 kΩ

(*b*) g_m = 40 mA/V and r_π = 4.0 kΩ

(*c*) g_m = 25 mA/V and r_π = 2.5 kΩ

(*d*) g_m = 40 mA/V and r_π = 2.5 kΩ

47. The value of C required for sinusoidal oscillations of frequency 1 kHz in the circuit of given figure is

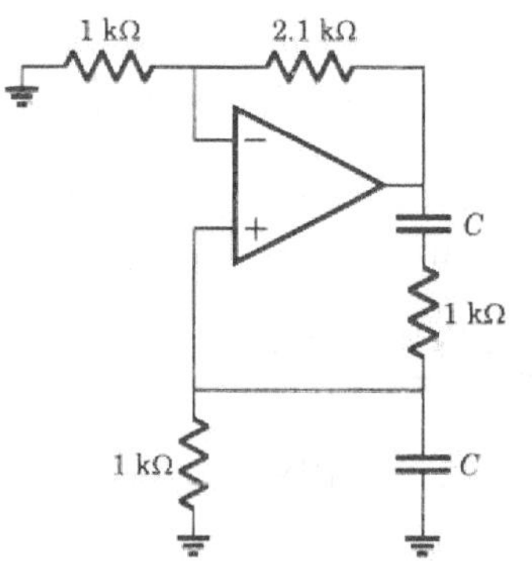

(*a*) $\frac{1}{2\pi}$ μF

(*b*) 2π μF

(*c*) $\frac{1}{2\pi\sqrt{6}}$ μF

(*d*) $2\pi\sqrt{6}$ μF

48. In the op-amp circuit given in the figure, the load current i_L is

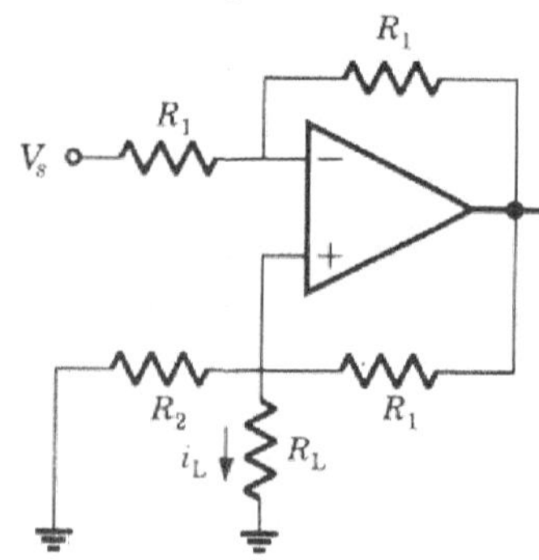

(a) $-\frac{V_s}{R_2}$ (b) $\frac{V_s}{R_2}$

(c) $-\frac{V_s}{R_L}$ (d) $\frac{V_s}{R_1}$

49. In the voltage regulator shown in the given figure, the load current can vary from 100 mA to 500 mA. Assuming that the Zener diode is ideal (i.e., the Zener knee current is negligibly small and Zener resistance is zero in the breakdown region), the value of R is

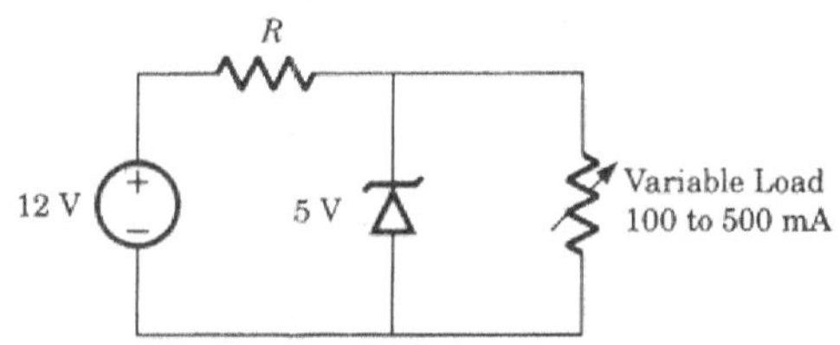

(a) 7 Ω (b) 70 Ω

(c) $\frac{70}{3}$ Ω (d) 14 Ω

50. In a full-wave rectifier using two ideal diodes, V_{dc} and V_m are the dc and peak values of the voltage respectively across a resistive load. If PIV is the peak inverse voltage of the diode, then the appropriate relationships for this rectifier are

(a) $V_{dc} = \frac{V_m}{\pi}$, PIV = $2V_m$

(b) $V_{dc} = 2\frac{V_m}{\pi}$, PIV = $2V_m$

(c) $V_{dc} = 2\frac{V_m}{\pi}$, PIV = V_m

(d) $V_{dc} = \frac{V_m}{\pi}$, PIV = V_m

51. The minimum number of 2-to -1 multiplexers required to realize a 4-to 1 mulitiplexer is

(a) 1 (b) 2

(c) 3 (d) 4

52. Boolean expression $AC + B\overline{C}$ is equivalent to

(a) $\overline{A}C + B\overline{C} + AC$

(b) $\overline{B}C + AC + B\overline{C} + \overline{A}C\overline{B}$

(c) $AC + B\overline{C} + \overline{B}C + ABC$

(d) $ABC + \overline{A}B\overline{C} + AB\overline{C} + \overline{A}BC$

53. 11001, 1001 and 111001 correspond to the 2's complement representation of which one of the following sets of number ?

(a) 25, 9 and 57 respectively

(b) –6, –6 and –6 respectively

(c) – 7, – 7 and – 7 respectively

(d) – 25, – 9 and –57 respectively

54. The 8255 Programmable Peripheral Interface is used as described below.

(I) An A/D converter is interfaced to a microprocessor through an 8255. The conversion is initiated by a signal from the 8255 on Port C. A signal on Port C causes data to be stored into Port A.

(II) Two computers exchange data using a pair of 8255s. Port A works as a bidirectional data port supported by appropriate handshaking signals.

The appropriate modes of operation of the 8255 for (I) and (II) would be

(a) Mode 0 for (I) and Mode 1 for (II)

(b) Mode 1 for (I) and Mode 0 for (II)

(c) Mode 2 for (I) and Mode 0 for (II)

(d) Mode 2 for (I) and Mode 1 for (II)

55. The number of memory cycles required to execute the following 8085 instructions

(I) LDA 3000 H

(II) LXI D, F0F 1 H

would be

(a) 2 for (I) and 2 for (II)

(b) 4 for (I) and 3 for (II)

(c) 3 for (I) and 3 for (II)

(d) 3 for (I) and 4 for (II)

56. In the modulo-6 ripple counter shown in the given figure, the output of the 2-input gate is used to clear the J-K flip-flops.

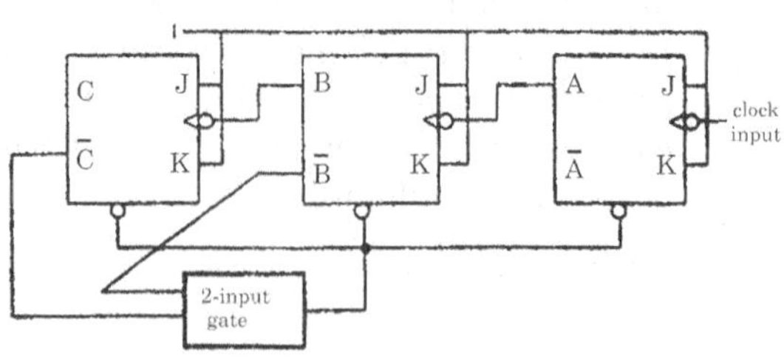

The 2-input gate is

(*a*) a NAND gate (*b*) a NOR gate

(*c*) an OR gate (*d*) an AND gate

57. Consider the sequence of 8085 instructions given below.

LXI H, 9258, MOV A, M, CMA, MOV M, A

Which one of the following is performed by this sequence ?

(*a*) Contents of location 9258 are moved to the accumulator

(*b*) Contents of location 9258 are compared with the contents of the accumulator

(*c*) Contents of location 9258 are complemented and stored in location 9258

(*d*) Contents of location 5892 are complemented and stored in location 5892

58. A Boolean function f of two variables x and y is defined as follows:

$$f(0,0) = f(0,1) = f(1, 1) = 1; \quad f(1, 0) = 0$$

Assuming complements of x and y are not available, a minimum cost solution for realizing f using only 2-input NOR gates and 2-input OR gates (each having unit cost) would have a total cost of

(*a*) 1 unit (*b*) 4 unit

(*c*) 3 unit (*d*) 2 unit

59. It is desired to multiply the numbers 0AH by 0BH and store the result in the accumulator. The numbers are available in registers B and C respectively. A Part of the 8085 program for this purpose is given below:

MVI A, 00H

Loop; - - - - - - - - - - - - -

HLT END

The sequence of instruction to the complete the program would be

(*a*) JNZ LOOP, ADD B, DCR C

(*b*) ADD B , JNZ LOOP, DCR C

(*c*) DCR C, JNZ LOOP, ADD B

(*d*) ADD B, DCR C, JNZ LOOP

60. A 1 kHz sinusoidal signal is ideally sampled at 1500 samples/sec and the sampled signal is passed through an ideal low-pass filter with cut-off frequencey 800 Hz. The output signal has the frequency

(*a*) zero Hz (*b*) 0.75 kHz

(*c*) 0.5 kHz (*d*) 0.25 kHz

61. A rectangular pulse train $s(t)$ as shown in the figure is convolved with the signal $\cos^2 (4\pi \times 10^3\, t)$. The convolved signal will be a

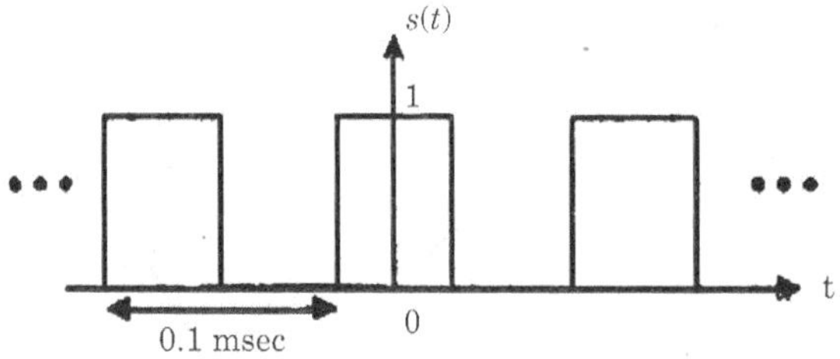

(*a*) DC (*b*) 12 kHz sinusoid

(*c*) 8 kHz sinusoid (*d*) 14 kHz sinusoid

62. Consider the sequence $x[n] = [-4 - j5 \quad 1 + j2 \quad 4]$ (arrow under $1 + j2$)

$\uparrow$

The conjugate anti-symmetric part of the sequence is

(*a*) $[-4 - j2.5 \quad j2 \quad 4 - j2.5]$

(*b*) $[-j2.5 \quad 1 \quad j2.5]$

(*c*) $[-j5 \quad j2 \quad 0]$

(*d*) $[-4 \quad 1 \quad 4]$

63. A causal LTI system is described by the difference equation

$$2y[n] = \alpha y[n-2] - 2x[n] + \beta x[n-1]$$

The system is stable only if

(*a*) $|\alpha| = 2, \quad |\beta| < 2$

(*b*) $|\alpha| > 2, \quad |\beta| > 2$

(*c*) $|\alpha| < 2$, any value of β

(*d*) $|\beta| < 2$, any value of α

64. A causal system having the transfer function $H(s) = \dfrac{1}{s+2}$ is excited with $10u(t)$. The time at which the output reaches 99% of its steady state value is

(*a*) 2.7 sec (*b*) 2.5 sec

(*c*) 2.3 sec (*d*) 2.1 sec

65. The impulse response $h[n]$ of a linear time invariant system is given as

$$h[n] = \begin{cases} -2\sqrt{2} & n = 1, -1 \\ 4\sqrt{2} & n = 2, \; 2 \\ 0, & \text{otherwise} \end{cases}$$

If the input to the above system is the sequence $e^{j\pi n/4}$, then the output is

(*a*) $4\sqrt{2}\, e^{j\pi n/4}$

(*b*) $4\sqrt{2}\, e^{-j\pi n/4}$

(*c*) $4e^{j\pi n/4}$

(*d*) $-4e^{j\pi n/4}$

66. Let $x(t)$ and $y(t)$ (with Fourier transforms $X(f)$ and $Y(f)$ respectively) be related as shown in the given figure.

Then $Y(f)$ is

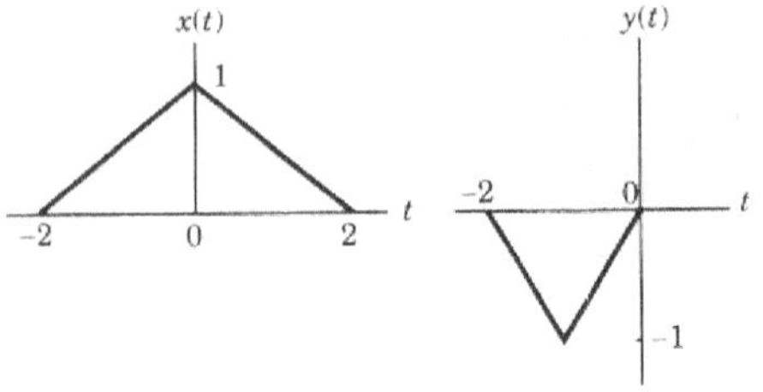

(a) $-\frac{1}{2}X(f/2)\,e^{-j2\pi f}$ (b) $-\frac{1}{2}X(f/2)e^{j2\pi f}$

(c) $-X(f/2)\,e^{j2\pi f}$ (d) $-X(f/2)\,e^{-j2\pi f}$

67. A system has poles at 0.01 Hz, 1 Hz and 80 Hz; zeros at 5 Hz, 100 Hz and 200 Hz. The approximate phase of the system response at 20 Hz is

(a) $-90°$ (b) $0°$

(c) $90°$ (d) $-180°$

68. Consider the signal flow graph shown in the figure below. The gain $\frac{x_5}{x_1}$ is

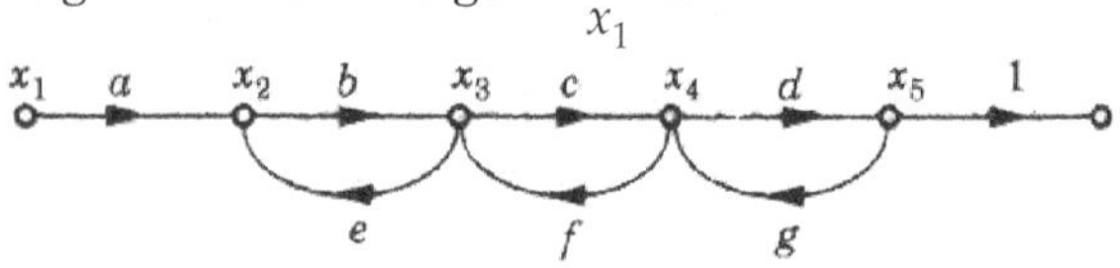

(a) $\frac{1-(be+cf+dg)}{abc}$

(b) $\frac{bedg}{1-(be+cf+dg)}$

(c) $\frac{abcd}{1-(be+cf+dg)+bedg}$

(d) $\frac{1-(be+cf+dg)+bedg}{abcd}$

69. If $A=\begin{bmatrix}-2 & 2\\ 1 & -3\end{bmatrix}$, then sin At is

(a) $\frac{1}{3}\begin{bmatrix}\sin(-4t)+2\sin(-t) & -2\sin(-4t)+2\sin(-t)\\ -\sin(-4t)+\sin(-t) & 2\sin(-4t)+\sin(-t)\end{bmatrix}$

(b) $\begin{bmatrix}\sin(-2t) & \sin(2t)\\ \sin(t) & \sin(-3t)\end{bmatrix}$

(c) $\frac{1}{3}\begin{bmatrix}\sin(4t)+2\sin(t) & 2\sin(-4t)-2\sin(-t)\\ -\sin(-4t)+\sin(t) & 2\sin(4t)+\sin(t)\end{bmatrix}$

(d) $\frac{1}{3}\begin{bmatrix}\cos(-t)+2\cos(t) & -2\cos(-4t)+2\sin(-t)\\ -\cos(-4t)+\sin(-t) & -2\cos(-4t)+\cos(-t)\end{bmatrix}$

70. The open-loop transfer function of a unity feedback system is $G(s)=\frac{K}{s(s^2+s+2)(s+3)}$

The range of K for which the system is stable is

(a) $\frac{21}{4}>K>0$ (b) $13>K>0$

(c) $\frac{21}{4}<K<\infty$ (d) $-6<K<\infty$

71. For the polynomial $P(s)=s^5+s^4+2s^3+2s^2+3s+15$, the number of roots which lie in the right half of the s-plane is

(a) 4 (b) 2

(c) 3 (d) 1

72. The state variable equations of a system are:

1. $x_1=-3x_1-x_2+u$
2. $x_2=2x_1$

$y=x_1+u$

The system is

(a) controllable but not observable

(b) observable but not controllable

(c) neither controllable nor observable

(d) controllable and observable

73. Given $A=\begin{bmatrix}1 & 0\\ 0 & 1\end{bmatrix}$, the state transition matrix e^{At} is given by

(a) $\begin{bmatrix}0 & e^{-t}\\ e^{-t} & 0\end{bmatrix}$ (b) $\begin{bmatrix}e^{t} & 0\\ 0 & e^{t}\end{bmatrix}$

(c) $\begin{bmatrix}e^{-t} & 0\\ 0 & e^{-t}\end{bmatrix}$ (d) $\begin{bmatrix}0 & e^{t}\\ e^{t} & 0\end{bmatrix}$

74. Consider the signal $x(t)$ shown in the figure. Let $h(t)$ denote the impulse response of the filter matched to $x(t)$, with $h(t)$ being non-zero only in the interval 0 to 4 sec. The slope of $h(t)$ in the interval $3<t<4$ sec is

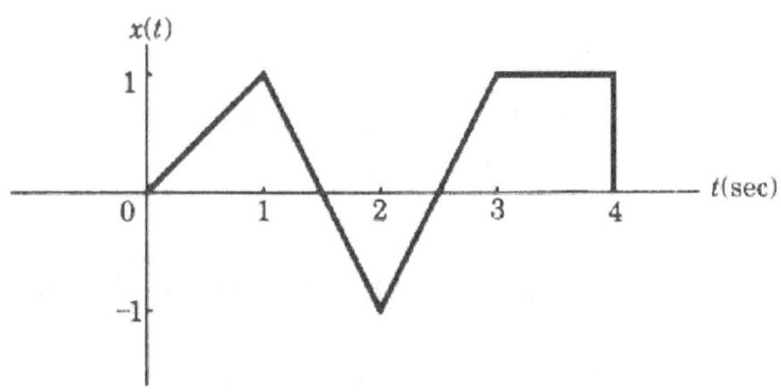

(a) $\frac{1}{2}$ sec^{-1} (b) -1 sec^{-1}

(c) $-\frac{1}{2}$ sec^{-1} (d) 1 sec^{-1}

75. A 1 mW video signal having a bandwidth of 100 MHz is transmitted to a receiver through a cable that has 40 dB loss. If the effective one-sided noise spectral density at the receiver is 10^{-20} watt/Hz, then the signal-to-noise ratio at the receiver is

(*a*) 50 dB (*b*) 30 dB
(*c*) 40 dB (*d*) 60 dB

76. A 100 MHz carrier of 1 V amplitude and a 1 MHz modulating signal of 1 V amplitude are fed to a balanced modulator. The output of the modulator is passed through an ideal high-pass filter with cut-off frequency of 100 MHz. The output of the filter is added with 100 MHz signal of 1V amplitude and 90° phase shift as shown in the given figure. The envelope of the resultant signal is

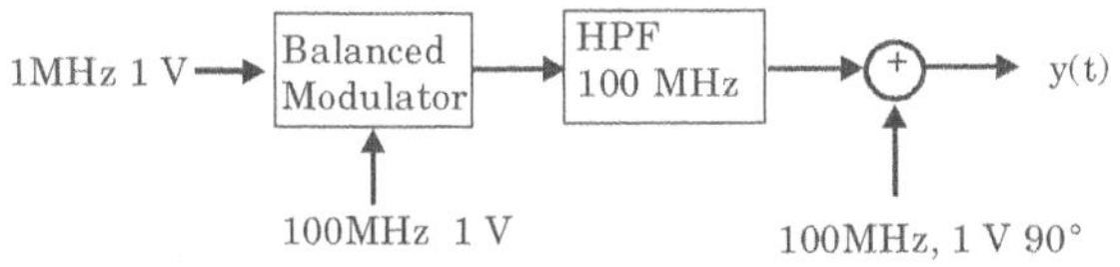

(*a*) constant

(*b*) $\sqrt{1+\sin\left(2\pi \times 10^{6} t\right)}$

(*c*) $\sqrt{5/4-\sin\left(2\pi \times 10^{6} t\right)}$

(*d*) $\sqrt{5/4+\cos\left(2\pi \times 10^{6} t\right)}$

77. Two sinusoidal signals of same amplitude and frequencies 10 kHz and 10.1 kHz are added together. The combined signal is given to an ideal frequency detector. The output of the detector is

(*a*) 0.1 kHz sinusoid
(*b*) 20.1 kHz sinusoid
(*c*) a linear function of time
(*d*) a constant

78. Consider a binary digital communication system with equally likely 0's and 1's. When binary 0 is transmitted the voltage at the detector input can lie between the levels – 0.25 V and + 0.25 V with equal probability: when binary 1 is transmitted, the voltage at the detector can have any value between 0 and 1 V with equal probabilty. If the detector has a threshold of 2.0 V (i.e., if the received signal is greater than 0.2 V, the bit is taken as 1), the average bit error probability is

(*a*) 0.15 (*b*) 0.2
(*c*) 0.05 (*d*) 0.5

79. A random variable X with uniform density in the interval 0 to 1 is quantized as follows:

If $0 < X < 0.3$, $x_q = 0$
If $0.3 < X < 1$, $x_q = 0.7$

where x_q is the quantized value of X.

The root-mean square value of the quantization noise is

(*a*) 0.573 (*b*) 0.198
(*c*) 2.205 (*d*) 0.266

80. Choose the correct one from among the alternatives A, B, C, D after matching an item from Group I with the most appropriate item in Group II.

Group I	Group II
A. FM	1. Slope overload
B. DM	2. μ-law
C. PSK	3. Envelope detector
D. PCM	4. Capture effect
	5. Hilbert transform
	6. Matched filter

	A	B	C	D
(*a*)	5	1	6	4
(*b*)	4	6	1	5
(*c*)	4	1	6	2
(*d*)	6	3	4	2

81. Three analog signals, having bandwidths 1200 Hz, 600 Hz and 600 Hz, are sampled at their respective Nyquist rates, encoded with 12 bit words, and time division multiplexed. The bit rate for the multiplexed signal is

(*a*) 115.2 kbps (*b*) 28.8 kbps
(*c*) 57.6 kbps (*d*) 38.4 kbps

82. Consider a system shown in the figure. Let X(*f*) and Y(*f*) denote the Fourier transforms of *x*(*t*) and y(*t*) respectively. The ideal HPF has the cut-off frequency 10 kHz.

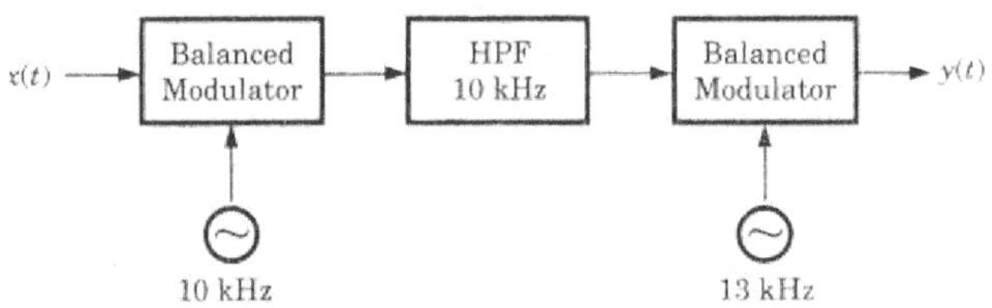

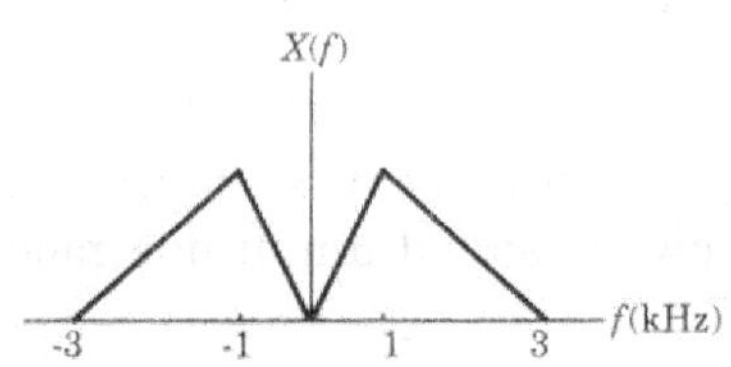

The positive frequencies where Y(*f*) has spectral peaks are

(*a*) 1 kHz and 24 kHz
(*b*) 2 kHz and 24 kHz
(*c*) 1 kHz and 14 kHz
(*d*) 2 kHz and 14 kHz

83. A parallel plate air-filled capacitor has plate area of 10^{-4} m^2 and plate separation of 10^{-3} m. It is connected to a 0.5 V, 3.6 GHz source. The magnitude of the displacement current is ($\varepsilon_0 = 1/36\pi \times 10^{-9}$ F/m)

(*a*) 10 mA (*b*) 100 mA
(*c*) 10 A (*d*) 1.59 mA

84. A source produces binary data at the rate of 10 kbps. The binary symbols are represented as shown in the figure given below.

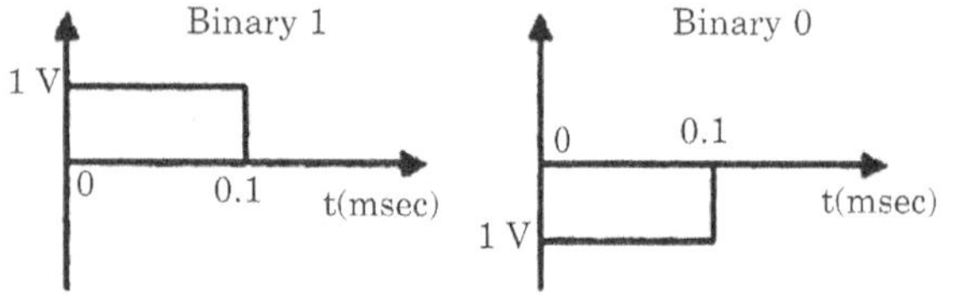

The source output is transmitted using two modulation schemes, namely Binary PSK (BPSK) and Quadrature PSK (QPSK). Let B_1 and B_2 be the bandwidth requirements of BPSK and QPSK respectively. Assuming that the bandwidth of the above rectangular pulses is 10 kHz, B_1 and B_2 are

(*a*) B_1 = 20 kHz, B_2 = 20 kHz
(*b*) B_1 = 10 kHz, B_2 = 20 kHz
(*c*) B_1 = 20 kHz, B_2 = 10 kHz
(*d*) B_1 = 20 kHz, B_2 = 10 kHz

85. Consider a 300 Ω, quarter-wave long (at 1 GHz) transmission line as shown in the figure. It is connected to a 10 V, 50 Ω source at one end and is left open circuited at the other end. The magnitude of the voltage at the open circuit end of the line is

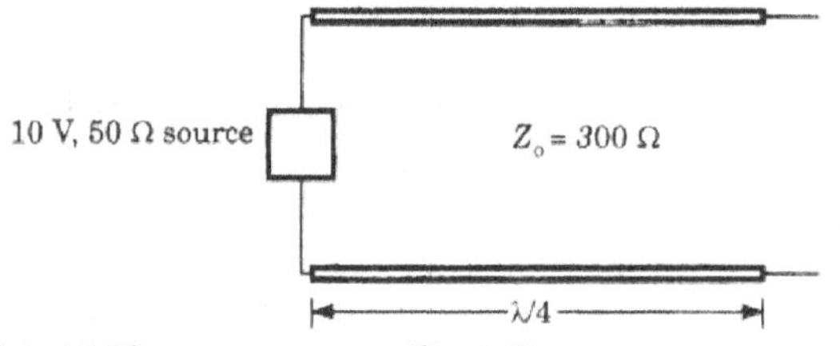

(*a*) 10 V (*b*) 5 V
(*c*) 60 V (*d*) 60/7 V

86. In a microwave test bench, why is the microwave signal amplitude modulated at 1 kHz?

(*a*) To increase the sensitivity of measurement
(*b*) To transmit the signal to a far-off place
(*c*) To study amplitude modulation
(*d*) Because crystal detector fails at microwave frequencies

87. If $\vec{E} = (\hat{a}_x + j\hat{a}_y)e^{jkz - j\omega t}$ and $\vec{H} = \left(\frac{k}{\omega\mu}\right)(\hat{a}_y + j\hat{a}_x)e^{jkz - j\omega t}$, the time averaged Poynting vector is

(*a*) null vector (*b*) $\left(\frac{k}{\omega\mu}\right)\hat{a}_z$

(*c*) $\left(\frac{2k}{\omega\mu}\right)\hat{a}_z$ (*d*) $\left(\frac{k}{2\omega\mu}\right)\hat{a}_z$

88. Consider an impedance Z = R + *j*X marked with point P in an impedance Smith chart as shown in the figure. The movement from point P along a constant resistance circle in the clockwise direction by an angle 45° is equivalent to

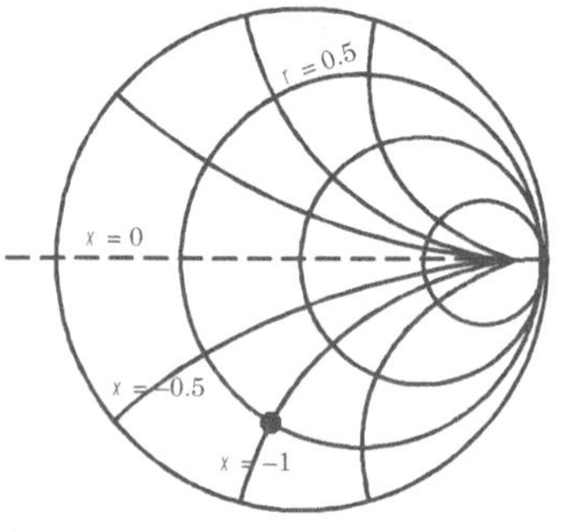

(*a*) adding an inductance in series with Z
(*b*) adding a capacitance in series with Z
(*c*) adding an inductance in shunt across Z
(*d*) adding a capacitance in shunt across Z

89. A plane electromagnetic wave propagating in free space in incident normally on a large slab of loss-less, non-magnetic, dielectric material with $\varepsilon > \varepsilon_0$. Maxima and minima are observed when the electric field is measured in front of the slab. The maximum electric field is found to be 5 times the minimum field. The intrinsic impedance of the medium should be

(*a*) 120 π Ω
(*b*) 60 π Ω
(*c*) 600 π Ω
(*d*) 24 π Ω

90. A lossless transmission line is terminated in a load which reflects a part of the incident power. The measured VSWR is 2. The percentage of the power that is reflected back is

(*a*) 57.73
(*b*) 33.33
(*c*) 0.11
(*d*) 11.11

ANSWERS

1. (*b*)	2. (*d*)	3. (*a*)	4. (*a*)	5. (*c*)	6. (*c*)	7. (*a*)	8. (*d*)	9. (*b*)	10. (*c*)
11. (*a*)	12. (*a*)	13. (*b*)	14. (*a*)	15. (*d*)	16. (*b*)	17. (*a*)	18. (*c*)	19. (*a*)	20. (*d*)
21. (*d*)	22. (*c*)	23. (*b*)	24. (*c*)	25. (*c*)	26. (*b*)	27. (*b*)	28. (*a*)	29. (*d*)	30. (*a*)
31. (*d*)	32. (*b*)	33. (*c*)	34. (*b*)	35. (*d*)	36. (*a*)	37. (*d*)	38. (*b*)	39. (*b*)	40. (*d*)
41. (*d*)	42. (*d*)	43. (*a*)	44. (*b*)	45. (*c*)	46. (*d*)	47. (*a*)	48. (*a*)	49. (*d*)	50. (*b*)
51. (*b*)	52. (*d*)	53. (*c*)	54. (*d*)	55. (*b*)	56. (*c*)	57. (*c*)	58. (*d*)	59. (*d*)	60. (*c*)
61. (*a*)	62. (*a*)	63. (*c*)	64. (*c*)	65. (*d*)	66. (*b*)	67. (*a*)	68. (*c*)	69. (*a*)	70. (*a*)
71. (*b*)	72. (*d*)	73. (*b*)	74. (*b*)	75. (*a*)	76. (*c*)	77. (*a*)	78. (*a*)	79. (*b*)	80. (*c*)
81. (*c*)	82. (*b*)	83. (*d*)	84. (*c*)	85. (*c*)	86. (*d*)	87. (*a*)	88. (*a*)	89. (*d*)	90. (*d*)

EXPLANATIONS

1. Options (*b*) contains a loop and for a tree there is no loop exist

2. $$V = L_1\frac{di}{dt} - M\frac{di}{dt} + (L_2 - M)\frac{di}{dt}$$
$$= (L_1 + L_2 - 2M)\frac{di}{dt} = L\frac{di}{dt}$$
$$L = L_1 + L_2 - 2M$$

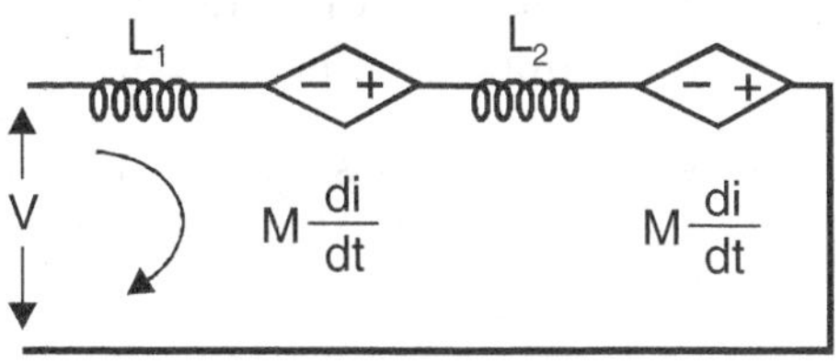

3. Here, $$i(t) = i_R(t) + i_L(t) + i_c(t)$$
$$i(t) = \frac{v(t)}{R} + \frac{v(t)}{(j\omega L)} + v(t) \times (j\omega C)$$
$$\Rightarrow \quad i(t) = \sin 2t\,[3 - 2j + 6j]$$
$$\Rightarrow \quad i(t) = 5\sin(2t + 53.1°)$$

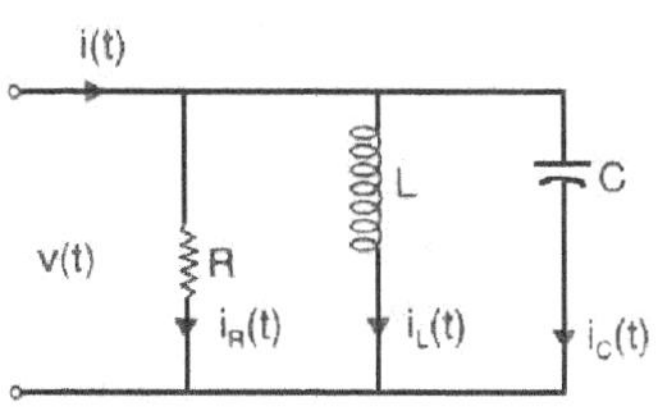

4. $$v_0(t) = \frac{v_i(t)}{R + \frac{1}{j\omega C}} \cdot \left(\frac{1}{j\omega C}\right)$$
$$= \frac{\sqrt{2}\sin 10^3 t}{j+1} = \sin(10^3 t - 45°)$$

5. $$v_i(t) = Ri(t) + L\frac{di(t)}{dt}$$

With Laplace transformation, we have
$$V_i(s) = [R + Ls]\,I(s)$$
or, $$I(s) = \frac{V_i(s)}{R + Ls} = \frac{1}{s(2+s)}$$
...[since, $V_i(s) = \frac{1}{s}$]

In time-domain, $i(t) = \frac{u(t)}{2}\,[1 - e^{-2t}]$

At $t = \frac{1}{2}$, $i(t) = 0.31$

and at $t = \infty$, $i(t) = 0.5$

6. Boron is trivalent atom. So it acts as an acceptor type impurity.

7. $V_{BE} = 0.7$ V [forward - biased]
and $V_{BC} = -V_{CB} = -0.2$V [Reverse - biased]
Hence the transistor operates in normal active mode.

8. Reducing base width causes decreasing recombination base current and therefore I_C increases as well as β, whereare, if doping concentration increases the result is vice-versa.

9. Ideal op-amp is an Ideal Voltage controlled voltage source.

10. Voltage series feedback results in increase in *i*/*p* impedance and decrease in *o*/*p* impedance.

11. The circuit is second - order low-pass filter.

12. The minimum value of base current to drive the transister into saturation is,
$$(I_B)_{min} = \frac{I_C}{h_{FE}}$$
From the figure given, $$I_C = \frac{3 - V_{CE,\,sat}}{1\,k\Omega}$$
$$= \frac{30 - 0.2}{1k} = 2.8 \text{ mA}$$
then, $$(I_B)_{min} = \frac{2.8\,mA}{50} = \mathbf{56\ \mu A.}$$

13. Change in *o*/*p* occurs when the state of the master is affected
∵ State of slave is depend on state of master.
Hence, the output is also depend on master.

14. The first complement range of numbers is $-2^{n-1}+1$ to $2^{n-1}-1$.
When $n = 6$, so -31 to $+31$

15. $$2^7 = 128$$
Hence for 100 increments 7 bits are required

16. Shift register → Serial to parallel data conversion
Counter → Frequency division
Decoder → Addressing in memory chips

17. In TTL AND-OR invertor gate propagation delay of transistor depend on RC when i/p is floating then $R \to 0$, hence o/p is zero.

18. Its CMOS invertor voltage transfer characteristic

19. For random variable X, the probability density function of the quantization error Q is given as,

$$f_X(x) = \begin{cases} 1, & 0 \le x \le 1 \\ 0, & \text{otherwise} \end{cases}$$

and $$\int_0^1 f_X(x) = 1$$

Now, Mean square value of qunatisation error is given by

$$\sigma_Q^2 = E[Q^2] = \int_0^1 q^2 f_x(x).dx$$

$$\sigma_Q^2 = \int_0^{0.3} x^2 f_x(x)\,dx + \int_{0.3}^1 (x - x_q)^2 \cdot f_x(x) \cdot dx$$

Where $q = x - x_q$

$$= \left.\frac{x^3}{3}\right|_0^{0.3} + \int_{0.3}^1 (x - 0.7)^2 \cdot dx$$

$$\sigma_Q^2 = 0.039$$

Root-mean square value $= \sqrt{\sigma_Q^2} = 0.198$

20. Probability distribution function of x is the function $F_X(x) = P\ (X \le x)$ for every x from $-\infty$ to ∞.

and $P\{x_1 < X \le x_2\} = F_X(x_2) - F_X(x_1)$

then, at $X = 1$,

$$P\{1^- < X \le 1^+\} = F_X(1^+) - F_X(1^-)$$
$$= 0.55 - 0.25 = 0.30$$

21. Using the following transform pair,

$$-A_k\,(d_k)^n\,u[-n-1] \overset{z}{\longleftrightarrow} \frac{A_k}{1 - d_k Z^{-1}}$$

with ROC $1\,|Z| < d_k$

We have $\frac{z}{z - 0.2} \overset{z}{\longleftrightarrow} -(0.2)^n\,u[-n-1]$.

22. If discrete sequence $x(n)$ is conjugate symmetry then, $x(-n) = x(n)$ and

$$X(w) = f(\underset{\infty}{X^0}\ (-n)) \qquad ...(A)$$

And, then fourier transform of

$$x(-n) = \sum_{n=-\infty}^{\infty} x^\phi(-n)\,e^{j\omega n}$$

put $-n = k$

then, $$f[X(-n)] = \sum_{k=-\infty}^{\infty} x^*(k)\,e^{j\omega k}$$

$$\Rightarrow \quad X(\omega) = \sum_{k=-\infty}^{\infty} x(k)\,e^{j\omega k}$$

$$\Rightarrow \quad X(\omega) = \sum_{k=-\infty}^{\infty} x(k)\,e^{-j\omega k}$$

and, $$X(\omega) = \sum_{k=-\infty}^{\infty} x(n)\,e^{-j\omega n} \quad \text{[equation. (A)]}$$

then $X(\omega) = X(\omega)$

Hence, It can be inferred than if complex conjugate of any function is the function itself, then function must be real.

23. Here, $\angle G(s)\,H(s)\big|_{s = j\omega} = -180° + \tan^{-1}\omega$
Now, $\angle G(s)\,H(s)\big|_{s = j\omega_c} = -180° + \tan^{-1}(\omega_c)$
$= -180°$

or, phase cross-over frequency, $\omega_c = 0$ rad/sec

Then, $GM = \frac{1}{a}$,

where $a = 1\ G(j\omega)\,H(j\omega)\big|_{\omega=\omega_c} = \infty = \frac{1}{\infty} = 0$

24. $$\text{Centroid}, (-\sigma_A) = \frac{\sum_{i=1}^{m} P_i - \sum_{j=1}^{n} Z_j}{m - n}$$

where m and n are the number of poles and zero respectively.

$$\therefore \quad \text{Centroid} = \frac{-4-0}{3} = -1.33$$

25. $$\rho = \frac{(SNR)_2}{(SNR)_1} = \frac{2^{2n_2}}{2^{2n_1}}$$

Here $n_1 = 6$, $n_2 = 8$, then

ratio, $\rho = 2^{2(n_2 - n_1)} = 16$.

26. $$\frac{1}{f_c} \le RC \le \frac{1}{\omega_m}$$

where, ω_m = message bandwidth
$= 2 \times 2 \times 10^3 = 4$ kHz
$f_c = 1 \times 10^6$ Hz

$$\therefore \quad 10^{-6} \le RC \le \frac{1}{4} \times 10^{-3}$$

i.e. $10^{-6} \le RC \le 25$ s.

27. Narrow-band FM signal is given as,

$x_{NBFM}(t) = A_C \cos(2\pi f_c t) - \beta A_C \sin(2\pi f_m t) \sin(2\pi f_c t)$ ...(A)

where $m(t) = A_m \cos(2\pi f m t)$

Carrier wave, $s(t) = A_C \cos(2\pi f_c t)$

AM signal is given as,

$S_{AM}(t) = A_C[1 + \ \cos(2\pi f_m t)] \cos(2\pi f_c t)$...(B)

Given $m(t)$ is same per both the signals and $\beta =$

Adding equation. (A) and equation (B), we have

$x_{NBFM}(t) + S_{AM}(t) =$

$$\underbrace{A_C \cos(2\pi f_c t)}_{\text{carrier wave}} + \underbrace{\frac{A_C\mu}{2}\cos 2\pi(f_m + f_c)t}_{\text{SSB wave}}$$

28. When signal is constant, then in Delta Modulation consecutive pulses are of opposite polarity.

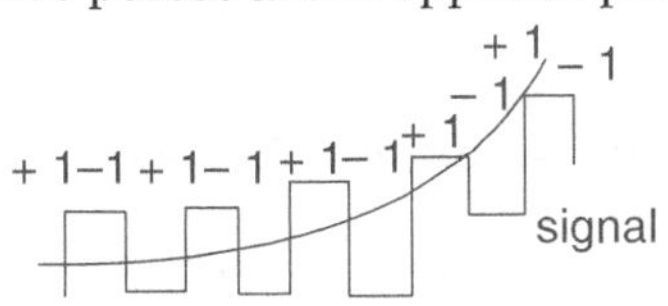

29. For TE_{10} mode, phase velocity,

$$v_g = \frac{v_p}{\left[1-(f_c/f)^2\right]^{1/2}}$$

and $$f_c = \frac{1}{2\sqrt{\mu\varepsilon}}\sqrt{\frac{m^2}{a^2}+\frac{n^2}{b^2}}$$

$$= \left(\frac{1}{2\sqrt{\mu\varepsilon}}\right)\cdot\frac{1}{a}$$

$m = 1$; $n = 0$ for TE_{10} mode]

$$f_c = \frac{c}{2a}$$

Now, $\frac{f_c}{f} < 1$, then $v_g > v_p$ (velocity of light).

30. Directive gain

$$G_d(\theta, \phi) = \frac{4\pi U(\theta,\phi)}{P_{rad} + P_L} = \frac{4\pi U(\theta,\phi)}{P_{rad}} \quad [P_L = 0]$$

For loss-less antenna,

(Input form) $P_{in} = P_{rad}$ (radiated power)

Directive gain, $10 \log_{10} G_d(\theta, \phi) = 6$

$\Rightarrow$ $G_d(\theta, \phi) = 3.981$

Then $P_{in} = P_{rad} = 1$ mW

then time-arranged radiated power/solid angle any u,

$$U(\theta, \phi) = \frac{P_{rad}\cdot G_d(\theta,\phi)}{4\pi}$$

Total power radiated $= P_{rad} \cdot G_d(\theta, \phi)$

$= 3.981 \times 1$ mW $= 4$ mW

31. $$V_1 = I_1 Z_{11} + I_2 Z_{12}$$
$$V_2 = I_1 Z_{21} + I_2 Z_{22}$$

$$Z_{11} = \left.\frac{V_1}{I_1}\right|_{I_2=0}$$

$$\frac{V_1}{(2+2j)} = i\,; \qquad I_1 - i = \frac{V_1}{2+2j}$$

$$I_1 = \frac{V_1}{2+2j} + \frac{V_1}{2+2j} = \frac{V_1}{1+j}$$

$$Z_{11} = 1 + j$$

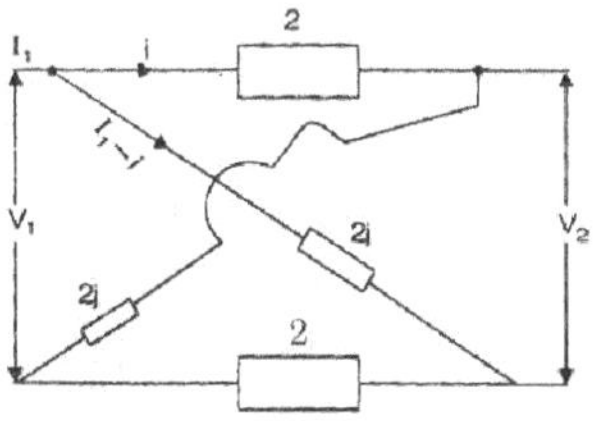

Likewise $$Z_{22} = \left.\frac{V_2}{I_2}\right|_{I_1=0} = 1 + j$$

$$\frac{V_2}{2j} - \frac{V_2}{2} = I_1$$

$$\frac{V_2}{I_1} = -1 + j = Z_{12}$$

Likwise $Z_{21} = -1 + j$

$$\begin{bmatrix} Z_{11} & Z_{12} \\ Z_{21} & Z_{22} \end{bmatrix} = \begin{bmatrix} 1+j & -1+j \\ -1+j & 1+j \end{bmatrix}$$

32. Using KVL around the loop, we have

$$v(t) = Ri(t) + \frac{L\,di(t)}{dt} + \frac{1}{C}\int_0^\infty i(t).dt$$

Taking Laplace transformation both of sides, we have

$$V(s) = RI(s) + LsI(s) - LI(0+) + \frac{I(s)}{sC} + \frac{v_c(o+)}{s}$$

$$\Rightarrow \quad \frac{1}{s} = I(s) + sI(s) - 1 + \frac{I(s)}{s} - \frac{1}{s}$$

$$\Rightarrow \quad \frac{2}{s} + 1 = \frac{I(s)}{s}\left[s^2 + s + 1\right]$$

$$\Rightarrow \quad I(s) = \frac{s+2}{s^2+s+1}.$$

33. At $\omega = 1$, slope changes from OdB/decade to 20 dB/decade, and, at $\omega = 10$, slope changes from 20dB/decade to 0dB/decade and at $\omega = 100$, slope changes from 0dB/decade to – 20 dB/decade.

Hence, there is a zero at $\omega = 1$ and poles at $\omega = 10, 100$.

Now, $$T(s) = \frac{k(s+1)}{(s+10)\,(s+100)}$$

$$T(j\omega) = \frac{k(j\omega+1)}{(j\omega+10)(j\omega+100)}$$

$$= \frac{k(j\omega+1)\cdot 10^{-3}}{\left(\frac{j\omega}{10}+1\right)\left(\frac{j\omega}{100}+1\right)}$$

Then, $20\log(10^{-3}k) = -20$

or, $k = 10^2$

Hence, $T(s) = \frac{10^2(s+1)}{(s+10)(s+100)}$.

34. $\left(R + sL + \frac{1}{Cs}\right) I(s) = V_i(s)$

$$\frac{I(s)}{V_i(s)} = \frac{s}{L\left(s^2 + \frac{Rs}{L} + \frac{1}{LC}\right)}$$

$$\frac{I(s)}{sC V_i(s)} = \frac{1/LC}{\left(s^2 + \frac{R}{L}s + \frac{1}{LC}\right)}$$

$$= \frac{\omega_n^2}{(s^2 + 2\omega_n s + \omega_n^2)}$$

$$\omega_n = \frac{1}{\sqrt{LC}}$$

$$\xi = \frac{R}{2L}\sqrt{LC} = \frac{R}{2}\sqrt{\frac{C}{L}}$$

Given transfer function $= \frac{10^6}{s^2 + 20s + 10^6}$

$$Q = \frac{\omega L}{R} = \frac{1}{R\omega C}$$

$$Q = \frac{1}{R}\sqrt{\frac{L}{C}}$$

$$= \frac{1}{2\xi} = \frac{1}{2\times 10^{-2}} = 50.$$

35. $$V_i(s) = RI(s) + Ls.I(s) + \frac{1}{Cs}I(s)$$

and $$V_c(s) = \frac{1}{Cs}I(s)$$

then $$\frac{V_c(s)}{V_i(s)} = \frac{1}{LCs^2 + RCs + 1}$$

or $$\frac{V_c(s)}{V_i(s)} = \frac{10^6}{s^2 + 10^6 s + 10^6}.$$

36. Given, $\frac{d^2y}{dt^2} + 3\frac{dy}{dt} + 2y = x(t)$

Taking Laplace transformation both sides, we have

$$[s^2 + 3s + 2]\, Y(s) = X(s) = \frac{2}{s}$$

or $$Y(s) = \frac{2}{s(s+1)(s+2)}$$

$$= \frac{1}{s} - \frac{2}{s+1} + \frac{1}{s+2}$$

Inverse Laplace transformation gives,

$$y(t) = (1 - 2e^{-t} + e^{-2t})\, u(t).$$

37. The impedance, $Z = R + j\left(\omega L - \frac{1}{\omega C}\right)$, at resonance,

$$\omega L - \frac{1}{\omega C} = 0$$

thus $Z = R$ [purely resistive]

In parallel G - R - C circuit, Q – factor varies inversely to G.

38. Since the net, charge must be zero, then

$$N_A W_p = N_D W_n \quad \text{... } (i)$$

and $$W_p + W_n = 3\ \mu m \quad \text{... } (ii)$$

Solving equations (*i*) and (*ii*) with given values, we have

$$W_p = 0.3\ \mu m$$

39. Conductivity, $\sigma = e(N_D \mu_n + N_A \mu_p)$

For, n – type silicon sample, $N_D >> N_A$

Then, $\sigma \approx eN_D\mu_n$

or, $$N_D = \frac{\sigma}{e\mu_n} = \frac{1}{\rho e\mu_n}$$

$$= \frac{1}{0.5\times 1.6\times 10^{-19}\times 1250} = 1\times 10^{16}/\text{cm}$$

40. Junction capacitance $C_j = \frac{tA}{W}$

and $$W = \sqrt{\frac{2t}{qN_D}V_j}$$

N_D = donor density

and V_j = function potential

where

Now, $C_j = V_j^{-1/2}$

or, $$\frac{C_{j1}}{C_{j2}} = \sqrt{\frac{V_{j2}}{V_{j1}}} \Rightarrow C_{j2} = C_{j1}\sqrt{\frac{V_{j1}}{V_{j2}}} = 0.5\text{PF}$$

41. S1: $C_{ox} = \frac{E_{ox}}{t_{ox}}$, t_{ox} = Gate oxide thickness

$$V_T = \phi_{GC} - 2\phi_F - \frac{Q_B}{C_{ox}} - \frac{Q_{ox}}{C_{ox}}$$

if $C_{ox}\uparrow$, then $V_T\uparrow$, $\frac{Q_B}{C_{ox}}\downarrow$, $\frac{Q_{ox}}{C_{ox}}\downarrow$

If $C_{ox} \downarrow$, then t_{ox} increases.

Hence, S1 is true.

S 2 : V_T of MOS increases by increase in substrate doping concentration.

Hence, S2 is false.

42. $$I_{DS} = I_{DSS}\left(1-\frac{V_{GS}}{V_T}\right)^2$$

Given, $V_{GS} = V_{DS}$

Then, $$I_{DS} = I_{DSS}\left(1-\frac{V_{GS}}{V_T}\right)^2$$

For $V_{GS} = 2V$

$$I_{DS} = 1 = I_{DSS}\left(1-\frac{2}{1}\right)^2 \quad \dots (i)$$

and for $V_{GS} = 3V$,

$$I_{DS} = I_{DSS}\left(1-\frac{3}{1}\right)^2 \quad \dots (ii)$$

From equations (*i*) and (*ii*), we have

I_{DS} For $V_{GS} = 3V$, $\mathbf{I_{DS} = 4mA}$.

43. We know $E = h\nu = \frac{hc}{\lambda}$

$$\therefore \quad \frac{E_2}{E_1} = \frac{\lambda_1}{\lambda_2}$$

$$\Rightarrow \quad E_2 = \frac{1.1}{0.87} \times 1.12\, eV = \mathbf{1.41609}$$

44. $$J_p = D_n q\frac{\partial N}{\partial x} = \frac{25\times1.6\times10^{-19}\times10^{14}}{0.5\times10^{-6}\times10^{2}} = 8\ \text{A/cm}^2$$

45. By voltage division, we have

$$\text{Here, } V_{BB} = 5\times\left(\frac{1}{4+1}\right) = 1 \text{ volt}$$

$$\text{and, } R_{BB} = \frac{4}{5}k\Omega$$

Now, using KVL in base-emitter loop, we have

$V_{BB} - R_B I_B - V_{BE} = R_E I_E \cong R_E I_C$ [since β is large]

or $0.3 \cong (0.3\text{ K})\, I_C$

or $I_C = 1\ mA$

and $V_{CE} = 5 - I_C R_C - R_E I_E$

$= 5 - 2.2 - 0.3 = \mathbf{2.5\ V}$.

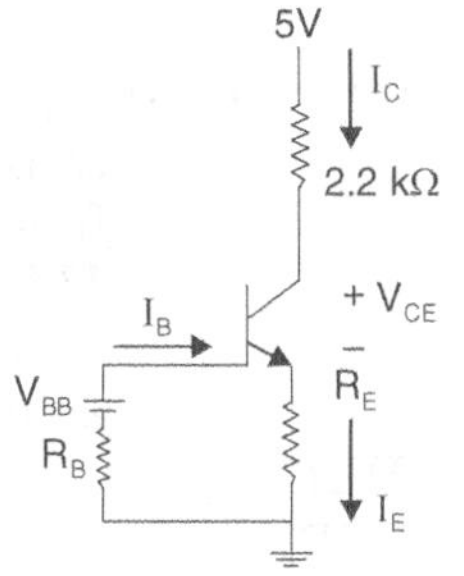

46. $$g_m = \frac{|I_C|}{V_T} = \frac{1\text{m A}}{25\text{mV}} = 0.04\ \text{A/V} = 40\ \text{mA/V}$$

$$r_\pi = \frac{\beta}{g_m} = \frac{100}{40\,mA/V} = 2.5\ k\Omega$$

47. $$\frac{V_o - V_1}{X_C + R} = \frac{V_1}{X_C} + \frac{V_1}{R}$$

$$V_0 = (X_C + R)\left[\frac{1}{X_C} + \frac{1}{R}\right]V_1 + V_1$$

$$\frac{V_0}{V_1} = \left(\frac{(X_C+R)}{X_C R}\right)^2 + 1 = \text{feedback gain}$$

$$X_C = -\frac{j}{\omega C}$$

and $R = 1\times10^3$

$$\frac{V_0}{V_1} = \frac{\frac{+1}{\omega^2C^2} + R^2 - \frac{2jR}{\omega C}}{\frac{+R}{j\omega C}} + 1$$

$$= +j\omega C\left(\frac{\frac{+1}{\omega^2C^2} + R^2 - \frac{2jR}{\omega C}}{R}\right) - 1$$

$$= \frac{+j\omega C\left(R^2 - \frac{1}{\omega^2C^2}\right) + 2R}{R}$$

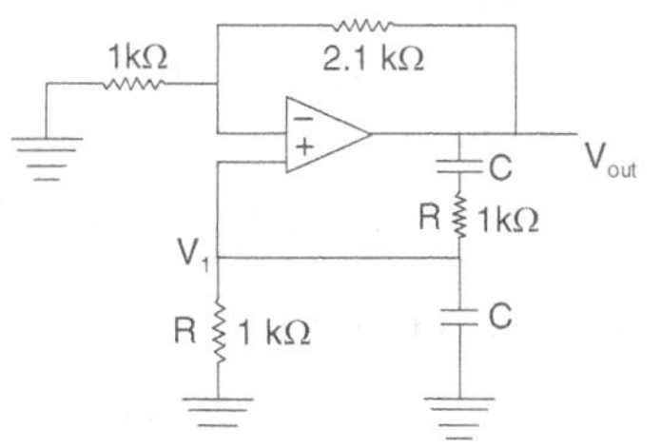

For oscillation imaginary part = 0

i.e. $R^2 - \frac{1}{\omega^2C^2} = 0$

$$\Rightarrow \quad C = \frac{1}{\omega R} = \frac{1}{2\pi\times1\times10^3\times1\times10^3} = \frac{1}{2\pi}\ \mu F$$

48. $$\frac{V_s - V_1}{R_1} = \frac{V_1 - V_o}{R_1}$$

$$\Rightarrow \quad \frac{V_s + V_o}{2} = V_1$$

$$I_L = \frac{V_1}{R_L}$$

$$\frac{V_o - V_1}{R_1} = \frac{V_1}{R_2} + \frac{V_1}{R_L}$$

$$V_o = \left(2 + \frac{R_2}{R_L}\right)V_1 = (2V_1 - V_s)$$

$$\frac{R_2 V_1}{R_L} = -V_s$$

$$\frac{V_1}{R_L} = \frac{-V_s}{R_2}.$$

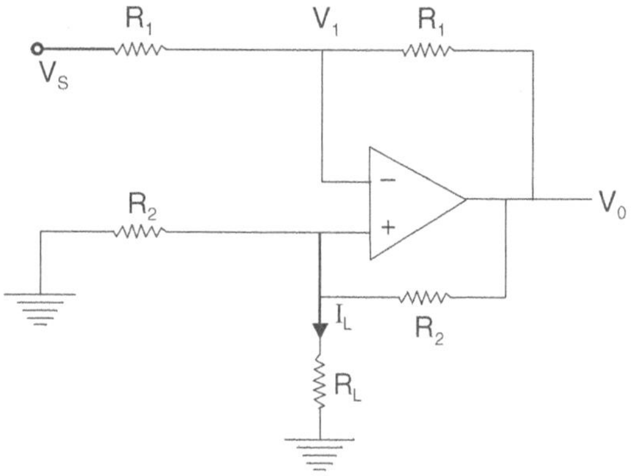

49. From the circuit given, we have

$$\frac{12 - V_z}{R} = I_L + I_z$$

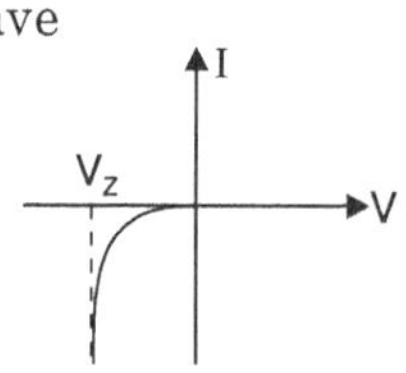

or, $$I_L = \left(\frac{12 - V_z}{R}\right) - I_z$$

Given : $100 \text{ mA} < I_L < 500\text{m A}$,

then, $$100 \times 10^{-3} < \left(\frac{7}{R} - I_z\right) < 500 \times 10^{-3}$$

or $$100 \times 10^{-3} < \frac{7}{R} < 500 \times 10^{-3}$$

{Zener knee current is negligibly small }

or $14\,\Omega < R < 70\,\Omega$

Hence required value of $R = 14\,\Omega$

[choosing minimum value]

50. In full wave rectifier, $V_{DC} = \dfrac{2V_m}{\pi}$

$\Rightarrow$ $PIV = 2V_m$

51.

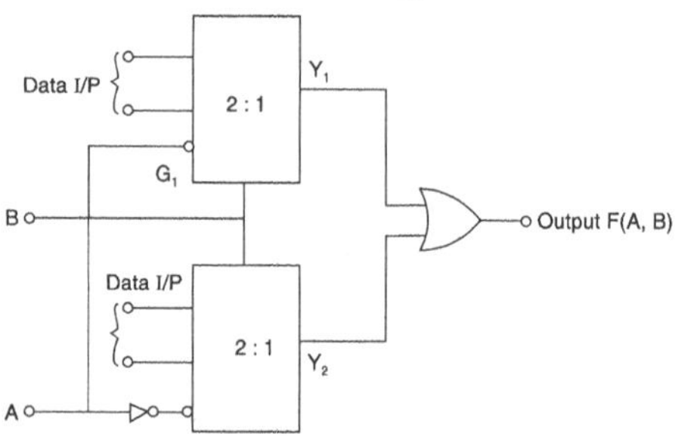

53. 11001 $\Rightarrow$ -00111 $\Rightarrow -7$
1001 $\Rightarrow$ -0111 $\Rightarrow -7$
111001 $\Rightarrow$ -000111 $\Rightarrow -7$

54. Mode 1 operation is for bidirectional data
Mode 2 operation causes data to be stored

55. For LDA, 4 memory cycles
For LXI, 3 memory cycles

56.

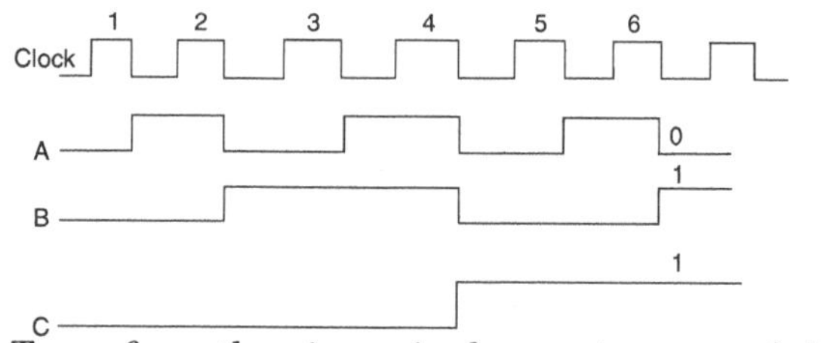

To perform the given ripple counter as module-6, the flip-flop should be cleared at the end of sixth pulse.

Since at 6th pulse,

$$B = C = 1,$$

then $$\overline{B} = \overline{C} = 0.$$

Hence, the 2 - input gate, should be OR gate so that CLR = 0 for all the three flip-flops.

57.

LXI H, 9258	⟶	memory location 9258
MOV A, M	⟶	data moved from memory to accumulated
CMA	⟶	data complement in accumulator
MOV M, A	⟶	complemented data moved to A

Best option is *c* as data is moved from memory location 9258 to the accumulator.

58.

x	*y*	F
0	0	1
0	1	1
1	0	0
1	1	1

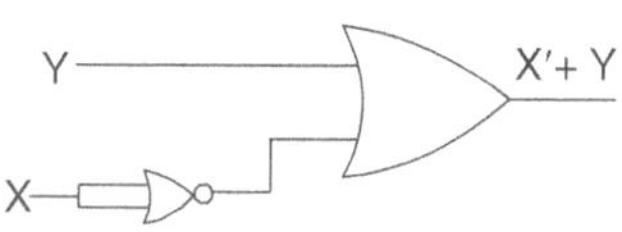

$$F = \overline{X} + Y$$

Hence 2 units are required.

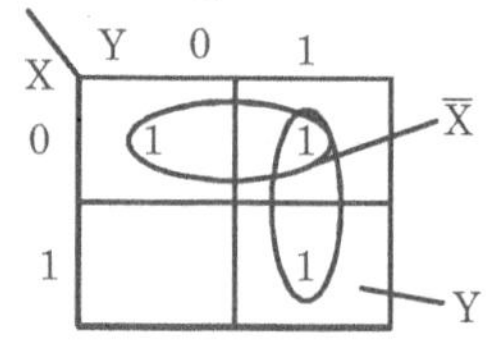

from k-map

$$f = \overline{X} + Y$$

59.

			Comments
	MVI A, OOH	⟶	$A \rightarrow 00$ H
Loop	ADD B	⟶	$A \rightarrow 0A$ H
	DCR C		Decreases the contents of register C OBH times, therefore adding OAH OBH = 11 times
	JNZ LOOP		Thus instruction causes the functioning of loop 11 times. Thus multiplication accomplishes.
	HLT		
	END		

60. Here, f_s = 1500 samples/sec, f_m = 1 kHz
The sampled frequency are 2.5 kHz, 0.5 kHz,
Since LPF has cut-off frequency 800 Hz, then only output signal of frequency 0.5 kHz would pass through it.

61.

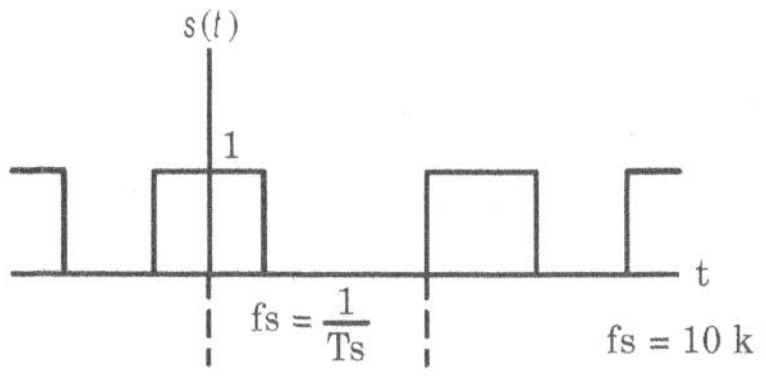

Using Fourier series expansion

$$s(t) = \frac{1}{2} + \frac{2}{\pi}\sum_{n=1}^{\infty}\frac{(-1)^{n-1}}{2n-1}\cos 2\pi\,(2n-1)\,f_s t$$

$$s(t) = \frac{1}{2} + \frac{2}{\pi}\cos 2\pi\, f_s t - \frac{2}{3\pi}\cos 2\pi \times 3\, f_s t + \ldots$$

and let $h(t) = \cos^2 (2\pi \times 2 \times 10^3)t$

$$= \frac{1}{2} + \frac{\cos(2\pi\times 4\times 10^3)t}{2}$$

Then, $s(t) \otimes h(t) \rightleftharpoons S(f)\,.\,H(f)$

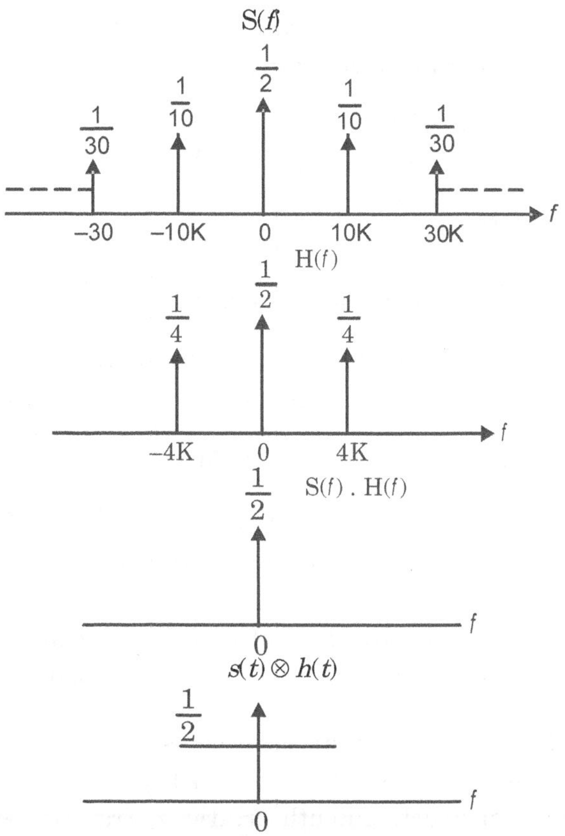

Hence, option (*a*) is correct.

62. $x[n] = [-4 - j5,\ 1 + 2j,\ 4]$
$\uparrow$ (at $-4-j5$)

$x[-n] = [4,\ 1 + 2j,\ -4 - j5\ ]$
$\uparrow$ (at $-4-j5$)

$x^*[-n] = [4,\ 1 - 2j,\ -4 + j5]$
$\uparrow$ (at $-4+j5$)

$$X_{CAS}[n] = \frac{x[n] - x^*[-n]}{2}$$

$$X_{CAS}[n] = [-\,4 - j2.5,\ j2,\ 4 - j2.5]$$

63.

$$2y(z) = \alpha z^{-2} y(z) + \beta z^{-1} x(z) - 2x(z)$$

$$(2 - \alpha z^{-2})\, y(z) = (\beta z^{-1} - 2)\, x(z)$$

$$\frac{y(z)}{x(z)} = \frac{(\beta z^{-1} - 2)}{(2 - \alpha z^{-2})}$$

For system to be stable, the ROC should include unit circle.

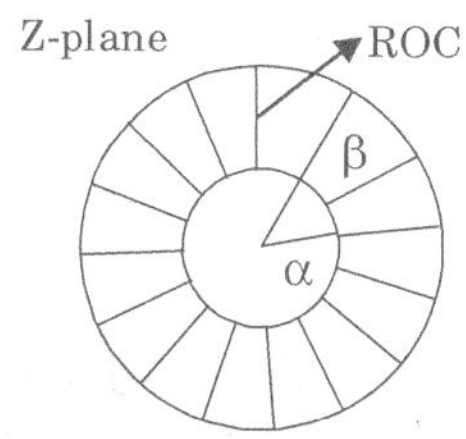

$\Rightarrow \quad 2 - \alpha z^{-2} > 0$

$\Rightarrow \quad 2 > \alpha z^{-2}$

$\Rightarrow \quad z > \sqrt{\frac{\alpha}{2}}$

$\beta z^{-1} > 2$

$$\frac{\beta}{2} > z > \sqrt{\frac{\alpha}{2}}$$

Hence $|\alpha| < 2$ and $|\beta|$ for any value.

64.

$$C(s) = H(s)\,.\,R(s)$$

$$\{R(t) = 10U(t)\}$$

$$C(s) = \frac{1}{s+2}\cdot\frac{10}{s}$$

$$= \frac{5}{s} - \frac{5}{s+2}$$

Then, $c(t) = 5[1 - e^{-2t}]$

Steady state value = 5

Hence $5[1 - e^{-2t}] = 0.99 \times 5$

or $\quad e^{-2t} = 0.01$

or $\quad t = 2.30$ sec.

65. $h[n] = [4\sqrt{2},\ -2\sqrt{2},\ 0,\ -2\sqrt{2},\ 4\sqrt{2}\ ]$
$\uparrow$ (at 0)

$x[n] = e^{\frac{j\pi n}{4}}$

$y[n] = x[n] \otimes h[n]$

Using property $x[n] \otimes \delta[n - n_0] = x[n - n_0]$

$y[n] = e^{\frac{j\pi n}{4}}\,[0 - 2\sqrt{2} \times 2\cos \pi/4]$

$y[n] = -\,4\,e^{\frac{j\pi n}{4}}$

66.

$$y(t) = x[\,2(t + 2)\,]$$

Now, Using $x(t - t_0) \longleftrightarrow X(f)\,.\, e^{-j\omega\pi f t_0}$

and $\quad x(at) \longleftrightarrow \frac{1}{|a|}X\left(\frac{f}{a}\right)$

We have $y(t) \longleftrightarrow -\frac{1}{2}X\left(\frac{f}{2}\right).e^{j2\pi f}$

67. Transfer function = $K\dfrac{\left(\frac{s}{5}+1\right)\left(\frac{s}{100}+1\right)\left(\frac{s}{200}+1\right)}{\left(\frac{s}{.01}+1\right)\left(\frac{s}{1}+1\right)\left(\frac{s}{80}+1\right)}$

$= \tan^{-1}\left(\frac{\omega}{5}\right)+\tan^{-1}\left(\frac{\omega}{100}\right)+\tan^{-1}\left(\frac{\omega}{200}\right)$

$-\tan^{-1}\left(\frac{\omega}{.01}\right)-\tan^{-1}\omega-\tan^{-1}\left(\frac{\omega}{80}\right)$

At $\omega = 20$, transfer function.

$= \tan^{-1}(4)+\tan^{-1}\left(\frac{1}{5}\right)+\tan^{-1}\left(\frac{1}{10}\right)-\tan^{-1}(2000)$

$-\tan^{-1}20-\tan^{-1}\left(\frac{1}{4}\right)=-90°.$

68. Forward path gain, $P_1 = abcd$

and, as it touches all the three loops,

$\Delta_1 = 1.$

Now, Three individual loops with loop gains.

$P_{11} = be,$
$P_{21} = cf,$
$P_{31} = dg$

and combination of two non-touching loops with loop gains products,

$P_{12} = bedg$

Then, $T = \dfrac{P_1\Delta_1}{1-P_{11}-P_{21}-P_{31}+P_{12}}$

$= \dfrac{abcd}{1-(be+cf+dg)+bedg}$

69. $A = \begin{bmatrix}-2 & 2\\ 1 & -3\end{bmatrix}$

Characteristics equation is given by $|\lambda I - A| = 0$

or, $\begin{bmatrix}\lambda+2 & -2\\ -1 & \lambda+3\end{bmatrix} = 0$

or, $\lambda^2 + 5\lambda + 4 = 0$

or, $(\lambda + 9)(\lambda + 1) = 0$

Hence, the eigen values are

$\lambda_1 = -4, \lambda_2 = -1$

Eigen vectors for

$\lambda_1 = -4,$

$(\lambda_1 I - A)\,M_1 = 0$

or, $\begin{bmatrix}\lambda_1+2 & -2\\ -1 & \lambda_1+3\end{bmatrix}\begin{bmatrix}m_{11}\\ m_{21}\end{bmatrix} = 0$

or, $\begin{bmatrix}-2 & -2\\ -1 & -1\end{bmatrix}\begin{bmatrix}m_{11}\\ m_{21}\end{bmatrix} = 0$

or, $m_{11} + m_{21} = 0$

For $\lambda_1 = -1,$

Eigen vector is $\begin{bmatrix}m_{11}\\ m_{21}\end{bmatrix} = \begin{bmatrix}-k\\ k\end{bmatrix} = k\begin{bmatrix}-1\\ 1\end{bmatrix}$

Now, eigen vector for $\lambda_2 = -1$, is given by

$(\lambda_2 I - A)M_2 = 0$

or, $\begin{bmatrix}1 & -2\\ -1 & 2\end{bmatrix}\begin{bmatrix}m_{12}\\ m_{22}\end{bmatrix} = 0$

or, $m_{12} - 2m_{22} = 0$

The eigen vector will than be given by

$\begin{bmatrix}m_{12}\\ m_{22}\end{bmatrix} = \begin{bmatrix}2k\\ k\end{bmatrix} = k\begin{bmatrix}2\\ 1\end{bmatrix}$

Diagonalization matrix,

$M = \begin{bmatrix}m_{11} & m_{12}\\ m_{21} & m_{22}\end{bmatrix} = \begin{bmatrix}-1 & 2\\ 1 & 1\end{bmatrix}$

$M^{-1} = \left(\frac{-1}{3}\right)\begin{bmatrix}1 & -2\\ -1 & -1\end{bmatrix}$

Digonal matrix of Sin At is

$D = \begin{bmatrix}\sin(\lambda_1 t) & 0\\ 0 & \sin(\lambda_2 t)\end{bmatrix} = \begin{bmatrix}\sin(-4t) & 0\\ 0 & \sin(-t)\end{bmatrix}$

matrix $B = \sin At = MDM^{-1}$

$= \left(\frac{-1}{3}\right)\begin{bmatrix}-1 & 2\\ 1 & 1\end{bmatrix}\begin{bmatrix}\sin(-4t) & 0\\ 0 & \sin(-t)\end{bmatrix}\begin{bmatrix}1 & -2\\ -1 & -1\end{bmatrix}$

$= \frac{1}{3}\begin{bmatrix}\sin(-4t)+2\sin(-t) & -2\sin(-4t)+2\sin(-t)\\ -\sin(-4t)+\sin(-t) & 2\sin(-4t)+\sin(-t)\end{bmatrix}$

70. The characteristic equation is, $1 + G(s) = 0$

or $s^4 + 4s^3 + 5s^2 + 6s + K = 0$

Constructing Routh – Hurwitz array, we have

s^4	1	5	K
s^3	4	6	0
s^2	$\frac{7}{2}$	K	
s^1	$(21-4K)\frac{2}{7}$		0
s^0	K		

For the system to be stable, we have,

$$21 - 4K > 0 \quad \text{and} \quad K > 0$$

Combining on we get, $\frac{21}{4} > K > 0.$

71. Constructing Routh array,

s^5	1	2	3
s^4	1	2	15
s^3	ε	-12	0
s^2	$\frac{2\varepsilon+12}{\varepsilon}$	15	0
s^1	$\frac{-12(2\varepsilon+12)}{\varepsilon} - 15\varepsilon$	0	0
s^0	15	$\frac{2\varepsilon+12}{\varepsilon}$	

Here ε is small positive quantity

Now, as number of sign changes are 2.

Therefore number of roots in RHS-plane is also 2.

72. Rewriting state equations in matrix form, we get

$$\begin{bmatrix}\dot{x}_1\\ \dot{x}_2\end{bmatrix} = \begin{bmatrix}-3 & -1\\ 2 & 0\end{bmatrix}\begin{bmatrix}x_1\\ x_2\end{bmatrix} + \begin{bmatrix}1\\ 0\end{bmatrix}u$$

$$y = [1 \;\; 0]\begin{bmatrix}x_1\\ x_2\end{bmatrix} + \begin{bmatrix}1\\ 0\end{bmatrix}u$$

Then, $A = \begin{bmatrix}-3 & -1\\ 2 & 0\end{bmatrix},$

$B = \begin{bmatrix}1\\ 0\end{bmatrix},$

$C = [1 \;\; 0]$

Now, $Q_c = [B : BA] = \begin{bmatrix}1 & -3\\ 0 & 2\end{bmatrix},$

$\det Q_c = \begin{vmatrix}1 & -3\\ 0 & 2\end{vmatrix} \neq 0$

Also, $Q_0 = [C^T \quad C^TA^T] = \begin{bmatrix}1 & -3\\ 0 & -1\end{bmatrix},$

$\det Q_c = \begin{bmatrix}1 & -3\\ 0 & -1\end{bmatrix} \neq 0$

Hence the system is controllable and observable.

73. Here $[sI - A]^{-1} = \begin{bmatrix}s-1 & 0\\ 0 & s-1\end{bmatrix}^{-1}$

Then, $\mathcal{L}^{-1}[sI - A]^{-1} = \mathcal{L}^{-1}\left\{\begin{bmatrix}\frac{1}{s-1} & 0\\ 0 & \frac{1}{s-1}\end{bmatrix}\right\}$

$$= \begin{bmatrix}e^t & 0\\ 0 & e^t\end{bmatrix}$$

74. $h(t) = X(T - t)$

So the response of $h(t)$ will be slope between $3 < t < 4$ is -1.

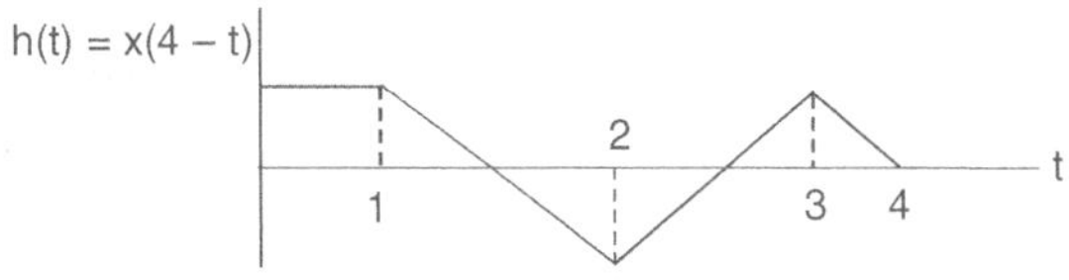

75. Total noise-power at the receiver is,

$$E[Y^2(t)] = \frac{1}{2\pi}\int_{-\infty}^{\infty} S_{YY}(\omega)\, d\omega$$

$$= N_0 W = 10^{-20} \times 100 \times 10^6$$

$$= 10^{-12} \text{ watt.}$$

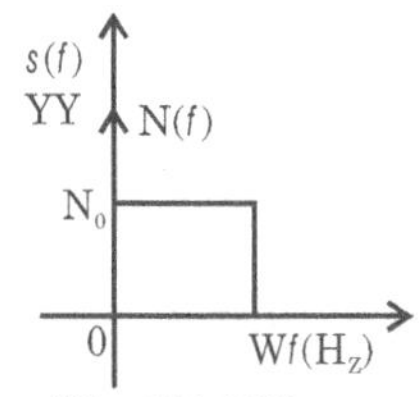

W = 100 MHz

Signal - to - noise ratio = $\frac{1 \times 10^{-3}}{10^{-12}} = 10^9$

In decibel, $10 \log_{10} 9 = 90$ dB

Considering loss of 40 dB, the SNR at the receiver is $(90 - 40)$ dB = 50 dB.

76. Output of the balanced modulator,

$$s(t) = \cos(2\pi\, 10^6 t) \cdot \cos(2\pi \cdot 10^8 t)$$

$$= \frac{1}{2}\left\{\cos\left[2\pi(10^8 + 10^6)t\right] + \cos\left[2\pi(10^8 - 10^6)t\right]\right\}$$

After high-pass filtering,

$$g(t) = \frac{1}{2}\cos\left[2\pi(10^8 + 10^6)t\right]$$

After adding another signal at the output, we have

$$y(t) = \frac{1}{2}\cos\left\{(10^8 + 10^6)2\pi t\right\} + \cos(10^8 \cdot 2\pi t)\frac{1}{2}$$

Its Hilbert transform,

$$\hat{y}(t) = \frac{1}{2}\sin\left\{(10^8 + 10^6).2\pi t\right\} - \cos(10^8 .2\pi t)$$

Then, pre-envelope,

$$y_+(t) = y(t) + j\,\hat{y}(t)$$

or, $$y_+(t) = \frac{1}{2}e^{j2\pi(10^8+10^6)t} - j.e^{j2\pi.10^8.t}$$

and, complex-envelope

$$\tilde{y}(t) = y_+(t).\exp(-j2\pi\,.\,10^8.\,t)$$

$$= \frac{1}{2}e^{j2\pi 10^6.t} - 1$$

$$\tilde{y}(t) = \frac{1}{2}[\cos(2\pi.10^6 t)] + j\left[\frac{1}{2}\sin 2\pi\,.\,10^6\,t - 1\right]$$

Hence envelope of $y(t)$ would be,

$$a(t) = \left[y_I^2(t) + y_Q^2(t)\right]^{1/2}$$

$$= \sqrt{\frac{5}{4} - \sin(2\pi \times 10^6 t)}$$

77. $y(t) = \sin(10^4 \times 2\pi t) + \sin(10.1 \times 10^3\, 2\pi t)$

After high-pass filtering, we have output, $s(t) = \sin(10.1 \times 10^3 \;\; 2\pi t)$

78. Signal is greater than 0.2 V

Hence probability of error

That 0 is transmitted and 1 is received

$= 0.25 - 0.2 = 0.05.$

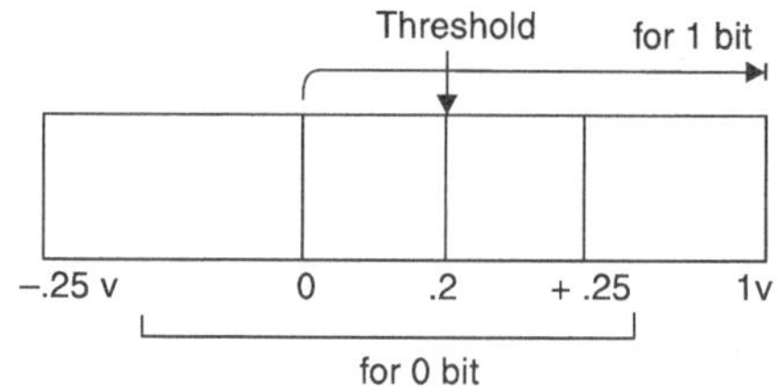

80. FM → capture effect

DM → slope overload

PSK → matched filter

PCM → μ– law

81. The three analog signals having band widths 1200 Hz, 600 Hz, 600 Hz have samples/sec of 2400, 1200 and 1200 respectively. Hence the total of 4800 samples /sec.

Then bit rate = 4800 sample/sec × 12

= 57.6 KbPs

82. For a message signal $m(t)$ and carrier wave $\cos \omega_c t$, frequency spectrum of an output signal is depicted below

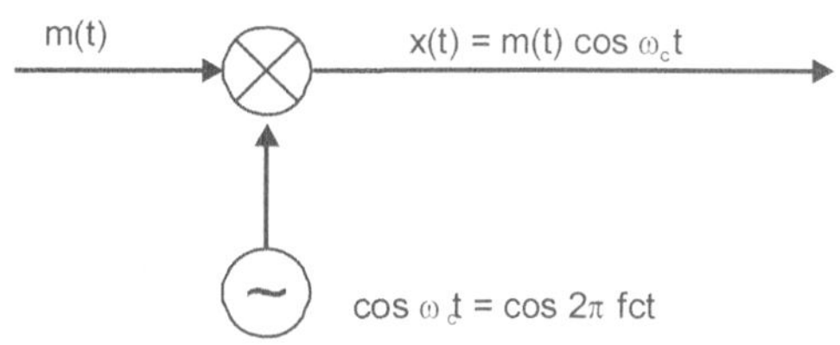

Balanced modulator

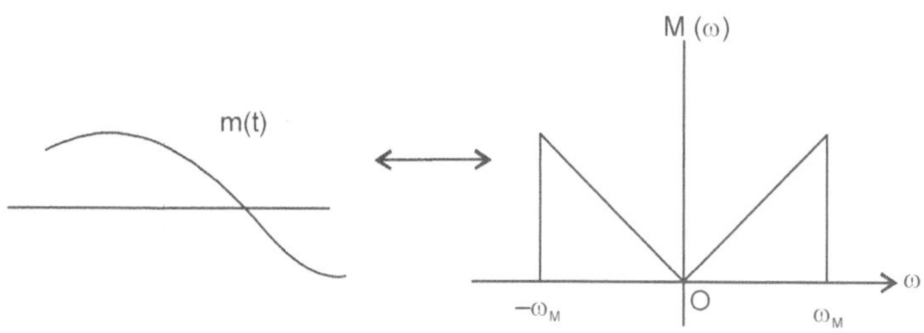

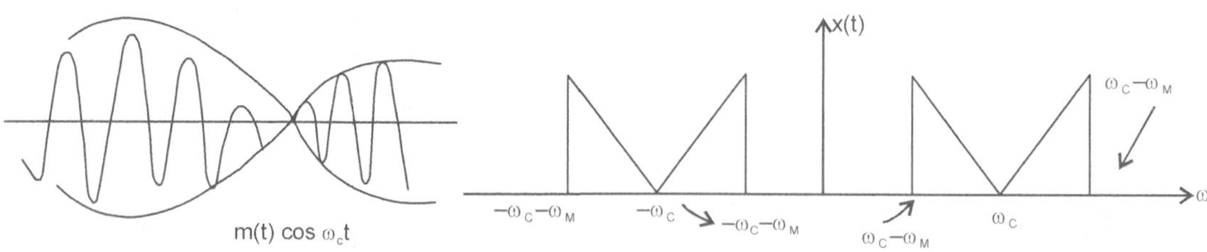

Frequency spectrum

Now, for f_c= 10KHz and $x(t)$

Frequency spectrum of the output at the first balanced modulator, is shown below

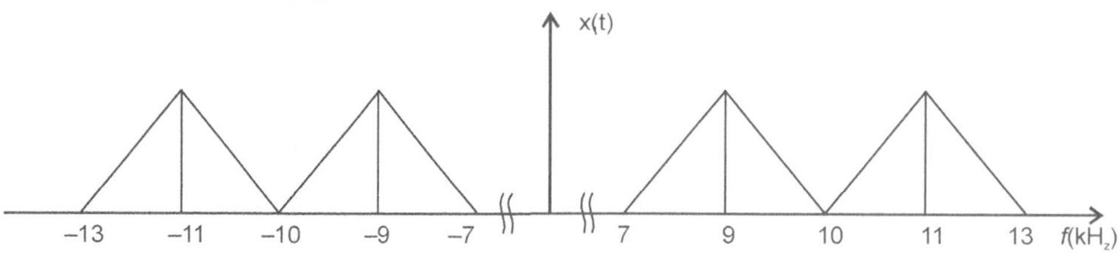

After high pass filtering, only signals having frequency greater than 10 KHz passes through

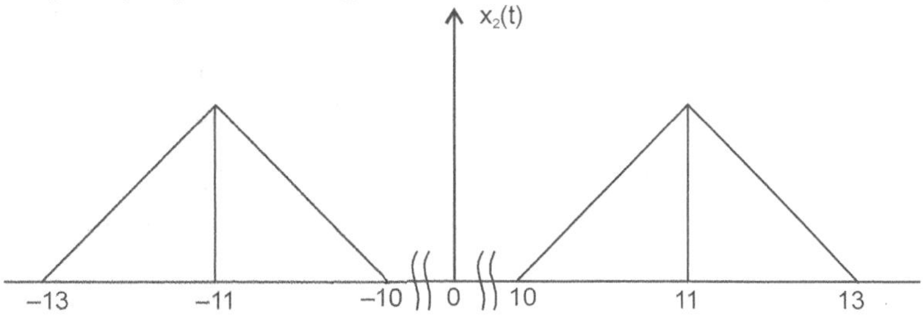

Now the output of the second balanced modulator is,

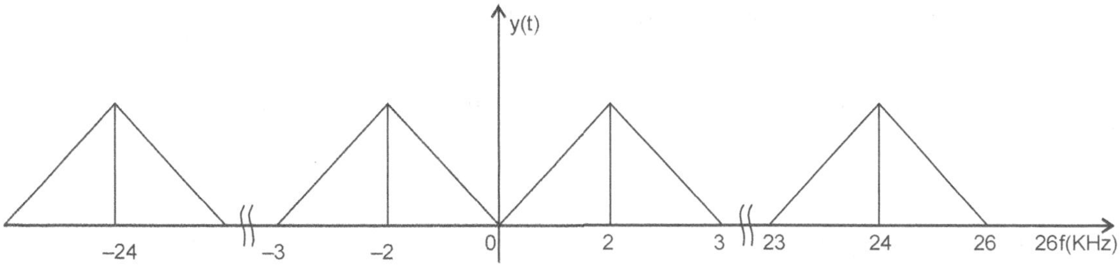

Hence positive peaks are at $f = 24$ KHz and $f = 2$KHz

83. Given : $A = 10^{-4}\ m^2$,

$d = 10^{-3} m$,

$v = 0.5$ V,

$f = 3.6 \times 10^9$ Hz

Displacement current density $= \frac{I_d}{A} = \varepsilon \cdot \frac{\partial E}{\partial t}$

$$E = E_o e^{j(\omega t + z\beta)}$$

$$E = \frac{-V}{d} e^{j(\omega t + z\beta)};$$

$$E_o = \frac{-V}{d}$$

$$\frac{\partial E}{\partial t} = +j\omega \frac{V}{d} e^{j(\omega t + z\beta)}$$

At $z = 0$, $\left|\frac{I_d}{A}\right| = \varepsilon_o \cdot \frac{V}{d}\omega$

$$I_d = \frac{\varepsilon_o V\omega}{d} A = \frac{1}{36\pi} \times 10^{-9} \frac{\times 0.5 \times 3.6 \times 10^9 \times 2\pi \times 10^{-4}}{10^{-3}} = \mathbf{1.59\ mA.}$$

84. For mary PSK bandwidth required is,

$BW = \frac{2f_b}{N}$ where $M = 2^N$ and f_b is the bit rate.

For BPSK, $M = 2 = 2^N$

$\Rightarrow$ $N = 1$

Then, BW (Bandwidth) $= 2f_b = 20$ KHz

$\Rightarrow$ $B_1 = 20$ KHz

For QPSK, $M = 4 = 2^N$

$\Rightarrow$ $N = 2$

then, Bandwidth, $B_2 = \frac{2f_b}{2} = 10$KHz

85. $\frac{V_L}{V_{in}} = \frac{Z_0}{Z_{in}}$

$$V_L = \frac{Z_0 . V_{in}}{Z_{in}}$$

$$V_L = \frac{10 \times 300}{50}$$

$$V_L = 60 \text{ V}$$

86. Microwave signals are amplitude modulate because crystal detector fails of microwave frequencies.

87. Time - averaged poynting vector,

$$P_{avg} = \frac{1}{2} Re[\bar{E} \times \bar{H}^*]$$

$$= \frac{1}{2} Re\left[\left(\bar{a}_x - j\bar{a}_y\right).e^{jkz-j\omega t} \times \left(\frac{k}{\omega\mu}\right)\left(\bar{a}_y - j\bar{a}_x\right)e^{-jkz+j\omega t}\right]$$

$$= \frac{k}{\omega\mu} Re\left[\left(\bar{a}_x + j\bar{a}_y\right) \times \left(\bar{a}_y - j\bar{a}_x\right)\right] = 0$$

=null vector.

88. Inductance is added in series when moving in clockwise direction.

89. $\frac{\eta_1}{\eta_2} = \frac{E_{min}}{E_{max}}$

$$\frac{\eta_2}{\eta_0} = \frac{E_0}{5E_0}$$

$$\eta_2 = \frac{\eta_0}{5} = 24\,\pi$$

90. VSWR, $\rho = \frac{1+\Gamma}{1-\Gamma}$

or, $2 = \frac{1+\Gamma}{1-\Gamma}$

or, $\Gamma = \frac{1}{3}$

Then, percentage of the power that is reflected back

$$\frac{P_{ref}}{P_{inc}} = \Gamma^2 = \frac{1}{9} \times 100$$

$$= \mathbf{11.11.}$$

SOLVED PAPER – 2003

ELECTRONICS & COMMUNICATION ENGINEERING

Time Allowed : 3 Hours **Maximum Marks : 150**

Q. 1–30 Carry One Mark Each

1. The minimum number of equations required to analyze the circuit shown in the figure is

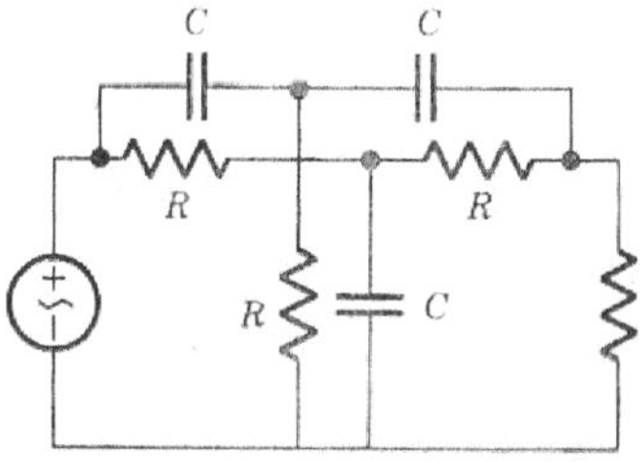

(a) 3 (b) 4

(c) 6 (d) 7

2. A source of angular frequency 1 rad/sec has a source impedance consisting of 1Ω resistance in series with 1 H inductance. The load that will obtain the maximum power transfer is

(a) 1 Ω resistance

(b) 1 Ω resistance in parallel with 1 H inductance

(c) 1 Ω resistance in series with 1 F capacitor

(d) 1 Ω resistance in parallel with 1 F capacitor

3. A series RLC circuit has a resonance frequency of 1 kHz and a quality factor Q = 100. If each of R, L and C is doubled from its origional value, the new Q of the circuit is

(a) 25 (b) 50

(c) 100 (d) 200

4. The Laplace transform of $i(t)$ is given by

$$I(s) = \frac{2}{s(1+s)}$$

As $t \to \infty$, the value of $i(t)$ tends to

(a) 0 (b) 1

(c) 2 (d) ∞

5. The differential equation for the current $i(t)$ in the circuit of the figure is

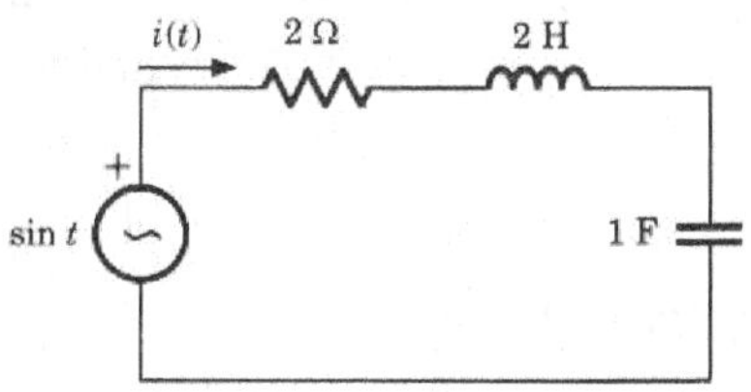

(a) $2\frac{d^2i}{dt^2} + 2\frac{di}{dt} + i(t) = \sin t$

(b) $\frac{d^2i}{dt^2} + 2\frac{di}{dt} + 2i(t) = \cos t$

(c) $2\frac{d^2i}{dt^2} + 2\frac{di}{dt} + i(t) = \cos t$

(d) $\frac{d^2i}{dt^2} + 2\frac{di}{dt} + 2i(t) = \sin t$

6. n-type silicon is obtained by doping silicon with

(a) Germanium (b) Aluminium

(c) Boron (d) Phosphorus

7. The bandgap of silicon at 300 K is

(a) 1.36 eV (b) 1.10 eV

(c) 0.80 eV (d) 0.67 eV

8. The intrinsic carrier concentration of silicon sample at 300 K is $1.5 \times 10^{16}/m^3$. If after doping, the number of majority carriers is $5 \times 10^{20}/m^3$, the minority carrier density is

(a) $4.50 \times 10^{11}/m^3$ (b) $3.33 \times 10^{4}/m^3$

(c) $5.00 \times 10^{20}/m^3$ (d) $3.00 \times 10^{-5}/m^3$

9. Choose proper substitutes for X and Y to make the following statement correct

Tunnel diode and Avalanche photodiode are operated in X bias and Y bias respectively.

(a) X : reverse, Y : reverse

(b) X : reverse, Y : forward

(c) X : forward, Y : reverse

(d) X : forward, Y : forward

10. For an n-channel enhancement type MOSFET, if the source is connected at a higher potential than that of the bulk (i.e. $V_{SB} > 0$), the threshold voltage V_T of the MOSFET will

(a) remain unchanged (b) decrease

(c) change polarity (d) increase

11. Choose the correct match for input resistance of various amplifier configurations shown below

Configuration	Input resistance
CB : Common Base	LO : Low
CC : Common Collector	MO : Moderate
CE : Common Emitter	HI : High

(a) CB-LO, CC-MO, CE-HI

(b) CB-LO, CC-HI, CE-MO

(c) CB-MO, CC-HI, CE-LO

(d) CB-HI, CC-LO, CE-MO

12. The circuit shown in the figure is best described as a

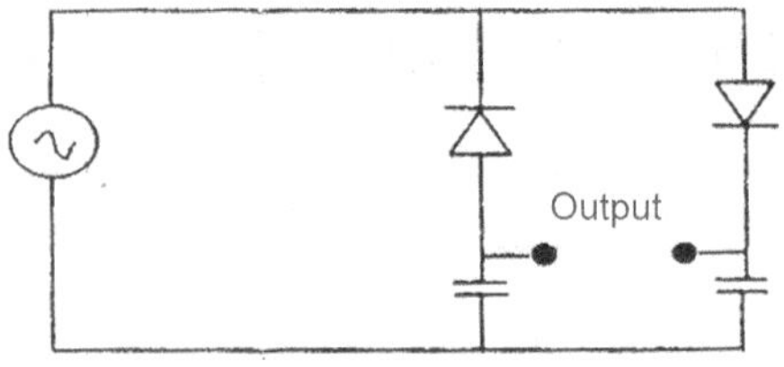

(a) bridge rectifier
(b) ring modulator
(c) frequency discriminatory
(d) voltage doubler

13. If the input to the ideal comparator shown in the figure is a sinusoidal signal of 8V (peak to peak) without any DC component, then the output of the comparator has a duty cycle of

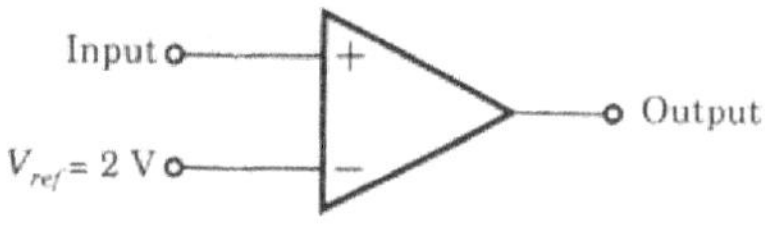

(a) 1/2 (b) 1/3
(c) 1/6 (d) 1/12

14. If the differential voltage gain and the common mode voltage gain of a differential amplifier are 48 dB and 2 dB respectively, then its common mode rejection ratio is

(a) 23 dB (b) 25 dB
(c) 46 dB (d) 50 dB

15. Generally, the gain of a transistor amplifier falls at high frequencies due to the

(a) internal capacitances of the device
(b) coupling capacitor at the input
(c) skin effect
(d) coupling capacitor at the output

16. The number of distinct Boolean expressions of 4 variables is

(a) 16 (b) 256
(c) 1024 (d) 65536

17. The minimum number of comparators required to build an 8 bit flash ADC is

(a) 8 (b) 63
(c) 255 (d) 256

18. The output of the 74 series of TTL gates is taken from a BJT in

(a) totem pole and common collector configuration
(b) either totem pole or open collector configuration
(c) common base configuration
(d) common collector configuration

19. Without any additional circuitry, an 8 : 1 MUX can be used to obtain

(a) some but not all Boolean functions of 3 variables
(b) all functions of 3 variables but none of 4 variables
(c) all functions of 3 variables and some but not all of 4 variables
(d) all functions of 4 variables

20. A 0 to 6 counter consists of 3 flip flops and a combination circuit of 2 input gate(s). The combination circuit consists of

(a) one AND gate
(b) one OR gate
(c) one AND gate and one OR gate
(d) two AND gates

21. The Fourier series expansion of a real periodic signal with fundamental frequency f_0 is given by

$$g_p(t) = \sum_{n=-\infty}^{\infty} c_n e^{j2\pi n f_0 t}$$

It is given that $c_3 = 3 + j5$. Then c_{-3} is

(a) $5 + j3$ (b) $-3 - j5$
(c) $-5 + j3$ (d) $3 - j5$

22. Let $x(t)$ be the input to a linear, time-invariant system. The required output is $4x(t-2)$. The transfer function of the system should be

(a) $4\,e^{j4\pi f}$ (b) $2\,e^{-j8\pi f}$
(c) $4\,e^{-j4\pi f}$ (d) $2\,e^{j8\pi f}$

23. A sequence $x(n)$ with the z-transform $X(z) = z^4 + z^2 - 2z + 2 - 3z^{-4}$ is applied as an input to a linear, time-invariant system with the impulse response $h(n) = 2\delta(n-3)$ where

$$\delta(n) = \begin{cases} 1, & n = 0 \\ 0, & \text{otherwise} \end{cases}$$

The output at $n = 4$ is

(a) -6 (b) zero
(c) 2 (d) -4

24. The figure shows the Nyquist plot of the open-loop transfer function $G(s)H(s)$ of a system. If $G(s)H(s)$ has one right-hand pole, the closed-loop system is

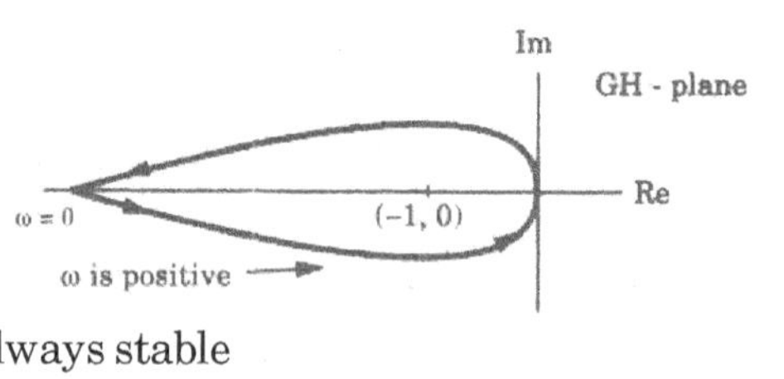

(a) always stable
(b) unstable with one closed-loop right hand pole
(c) unstable with two closed-loop right hand poles
(d) unstable with three closed-loop right hand poles

25. A PD controller is used to compensate a system. Compared to the uncompensated system, the compensated system has
(*a*) a higher type number
(*b*) reduced damping
(*c*) higher noise amplification
(*d*) larger transient overshoot

26. The input to a coherent detector is DSB-SC signal plus noise. The noise at the detector output is
(*a*) the in-phase component
(*b*) the quadrature-component
(*c*) zero
(*d*) the envelope

27. The noise at the input to an ideal frequency detector is white. The detector is operating above threshold. The power spectral density of the noise at the output is
(*a*) raised-cosine (*b*) flat
(*c*) parabolic (*d*) Gaussian

28. At a given probability of error, binary coherent FSK is inferior to binary coherent PSK by
(*a*) 6 dB (*b*) 3 dB
(*c*) 2 dB (*d*) 0 dB

29. The unit of $\nabla \times H$ is
(*a*) Ampere (*b*) Ampere/meter
(*c*) Ampere/meter^2 (*d*) Ampere-meter

30. The depth of penetration of electromagnetic wave in a medium having conductivity σ at a frequency of 1 MHz is 25 cm. The depth of penetration at a frequency of 4 MHz will be
(*a*) 6.25 cm (*b*) 12.50 cm
(*c*) 50.00 cm (*d*) 100.00 cm

Q. 31-90 Carry Two Marks Each

31. Twelve 1 Ω resistance are used as edges to form a cube. The resistance between two diagonally opposite corners of the cube is
(*a*) $\frac{5}{6}\,\Omega$ (*b*) $1\,\Omega$
(*c*) $\frac{6}{5}\,\Omega$ (*d*) $\frac{3}{2}\,\Omega$

32. The current flowing through the resistance R in the circuit in the figure has the form P cos 4*t*, where P is

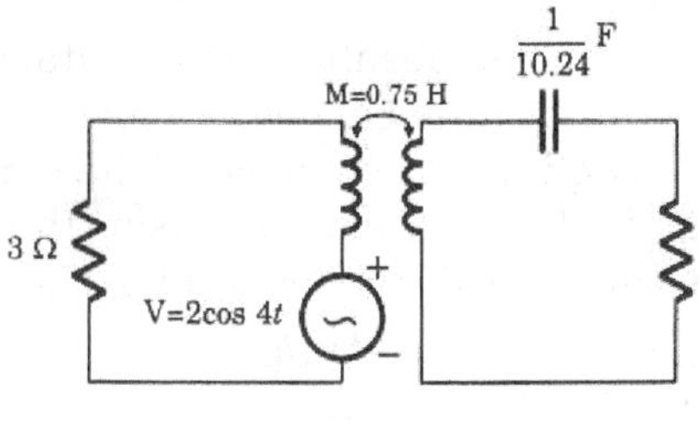

(*a*) $(0.18 + j0.72)$ (*b*) $(0.46 + j1.90)$
(*c*) $-(0.18 + j1.90)$ (*d*) $-(0.192 + j0.144)$

The circuit for Q. 33-34 are given in the figure. For both are the questions, assume that the switch S is in position 1 for a long time and thrown to position 2 at $t = 0$.

33. At $t = 0+$, the current i_1 is

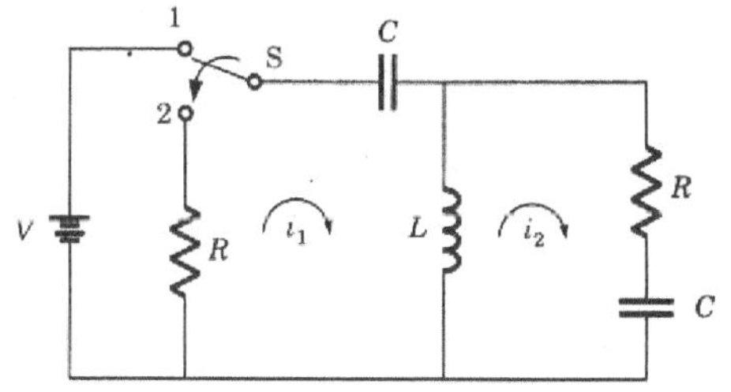

(*a*) $\frac{-V}{2R}$ (*b*) $\frac{-V}{R}$
(*c*) $\frac{-V}{4R}$ (*d*) zero

34. $I_1(s)$ and $I_2(s)$ are the Laplace transforms of $i_1(t)$ and $i_2(t)$ respectively. The equations for the loop currents $I_1(s)$ and $I_2(s)$ for the circuit shown in the figure after the switch is brought from position 1 to position 2 at $t = 0$, are

(*a*) $\begin{bmatrix} R + Ls + \frac{1}{Cs} & -Ls \\ -Ls & R + \frac{1}{Cs} \end{bmatrix} \begin{bmatrix} I_1(s) \\ I_2(s) \end{bmatrix} = \begin{bmatrix} \frac{V}{s} \\ 0 \end{bmatrix}$

(*b*) $\begin{bmatrix} R + Ls + \frac{1}{Cs} & -Ls \\ -Ls & R + \frac{1}{Cs} \end{bmatrix} \begin{bmatrix} I_1(s) \\ I_2(s) \end{bmatrix} = \begin{bmatrix} -\frac{V}{s} \\ 0 \end{bmatrix}$

(*c*) $\begin{bmatrix} R + Ls + \frac{1}{Cs} & -Ls \\ -Ls & R + Ls + \frac{1}{Cs} \end{bmatrix} \begin{bmatrix} I_1(s) \\ I_2(s) \end{bmatrix} = \begin{bmatrix} -\frac{V}{s} \\ 0 \end{bmatrix}$

(*d*) $\begin{bmatrix} R + Ls + \frac{1}{Cs} & -Ls \\ -Ls & R + Ls + \frac{1}{Cs} \end{bmatrix} \begin{bmatrix} I_1(s) \\ I_2(s) \end{bmatrix} = \begin{bmatrix} \frac{V}{s} \\ 0 \end{bmatrix}$

35. An input voltage $v(t) = 10\sqrt{2}\cos(t + 10°) + 10\sqrt{3}\cos(2t + 10°)$ V is applied to a series combination of resistance R = 1 Ω and an inductance L = 1H. The resulting steady-state current $i(t)$ in ampere is
(*a*) $10\cos(t + 55°) + 10\cos(2t + 10° + \tan^{-1} 2)$
(*b*) $10\cos(t + 55°) + 10\sqrt{\frac{3}{2}}\cos(2t + 55°)$
(*c*) $10\cos(t - 35°) + 10\cos(2t + 10° - \tan^{-1} 2)$
(*d*) $10\cos(t - 35) + 10\sqrt{\frac{3}{2}}\cos(2t - 35°)$

36. The driving-point impedance Z(s) of a network has the pole-zero locations as shown in the figure. If Z(0) = 3, then Z(s) is

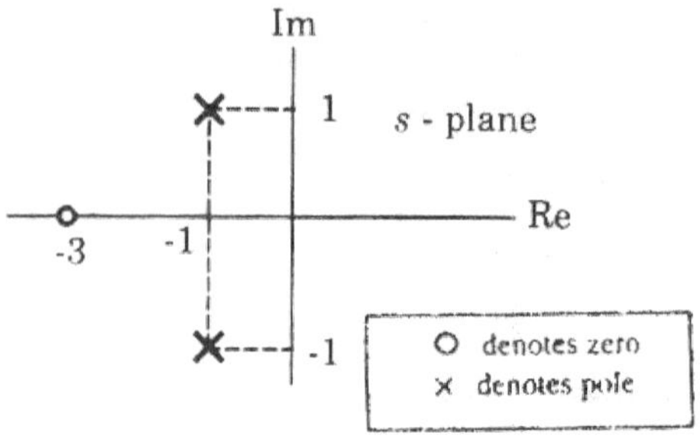

(a) $\frac{3(s+3)}{s^2+2s+3}$ (b) $\frac{2(s+3)}{s^2+2s+2}$

(c) $\frac{3(s-3)}{s^2-2s-2}$ (d) $\frac{2(s-3)}{s^2-2s-3}$

37. The impedance parameters Z_{11} and Z_{12} of the two-port network in the figure are

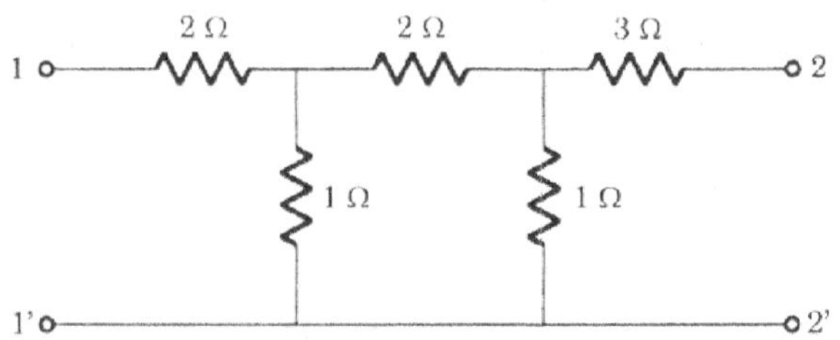

(a) Z_{11} = 2.75 Ω and Z_{12} = 0.25 Ω

(b) Z_{11} = 3 Ω and Z_{12} = 0.5 Ω

(c) Z_{11} = 3 Ω and Z_{12} = 0.25 Ω

(d) Z_{11} = 2.25 Ω and Z_{12} = 0.5 Ω

38. An n-type silicon bar 0.1 cm long and 100 μm² in cross-sectional area has a majority carrier concentration of $5 \times 10^{20}/\text{m}^3$ and the carrier mobility is 0.13 m²/V-s at 300 K. If the charge of an electron is 1.6×10^{-19} coulomb, then the resistance of the bar is

(a) 10^6 ohm (b) 10^4 ohm

(c) 10^{-1} ohm (d) 10^{-4} ohm

39. The electron concentration in a sample of uniformly doped n-type silicon at 300 K varies linearly from $10^{17}/\text{cm}^3$ at $x = 0$ to $6 \times 10^{16}/\text{cm}^3$ at $x = 2$ μm. Assume a situation that electrons are supplied to keep this concentration gradient constant with time. If electronic charge is 1.6×10^{-19} coulomb and the diffusion constant D_n= 35 cm²/s, the current density in the silicon, if no electric field is present, is

(a) zero

(b) – 506 A/cm²

(c) – 560 A/cm²

(d) – 1120 A/cm²

40. Match items in Group I with items in Group II, most suitably.

Group I	Group II
A. LED	1 Heavy doping
B. Avalanche photodiode	2 Coherent radiation
C. Tunnel diode	3 Spontaneous emission
D. LASER	4 Current gain

	A	B	C	D
(a)	1	2	4	3
(b)	2	3	1	4
(c)	3	4	1	2
(d)	2	1	4	3

41. At 300 K, for a diode current of 1 mA, a certain germanium diode requires a forward bias of 0.1435 V, whereas a certain silicon diode requires a forward bias of 0.178 V. Under the conditions stated above, the closest approximation of the ratio of reverse saturation current in germanium diode to that in silicon diode is

(a) 1 (b) 5

(c) 4×10^3 (d) 8×10^3

42. A particular green LED emits light of wavelength 5490 Å. The energy bandgap of the semiconductor material used there is (Planck's constant = 6.626 $\times 10^{-34}$ J-s)

(a) 2.26 eV (b) 1.98 eV

(c) 1.17 eV (d) 0.74 eV

43. When the gate-to-source voltage (V_{GS}) of a MOSFET with threshold voltage of 400 mV, working in saturation is 900 mV, the drain current is observed to be 1 mA. Neglecting the channel width modulation effect and assuming that the MOSFET is operating at saturation, the drain current for an applied V_{GS} of 1400 mV is

(a) 0.5 mA (b) 2.0 mA

(c) 3.5 mA (d) 4.0 mA

44. If P is Passivation, Q is n-well implant, R is metallization and S is source/drain diffusion, then the order in which they are carried out in a standard n-well CMOS fabrication process, is

(a) P-Q-R-S (b) Q-S-R-P

(c) R-P-S-Q (d) S-R-Q-P

45. An amplifier without feedback has a voltage gain of 50, input resistance of 1 kΩ and output resistance of 2.5 kΩ. The input resistance of the current-shunt negative feedback amplifier using the above amplifier with a feedback factor of 0.2 is

(a) 1/11 kΩ (b) 1/5 k Ω

(c) 5 kΩ (d) 11 kΩ

46. In the amplifier circuit shown in the figure is the values of R_1 and R_2 are such that the transistor is operating at $V_{CE} = 3$ V and $I_C = 1.5$ mA when its β is 150. For a transistor with β of 200, the operating point (V_{CE}, I_C) is

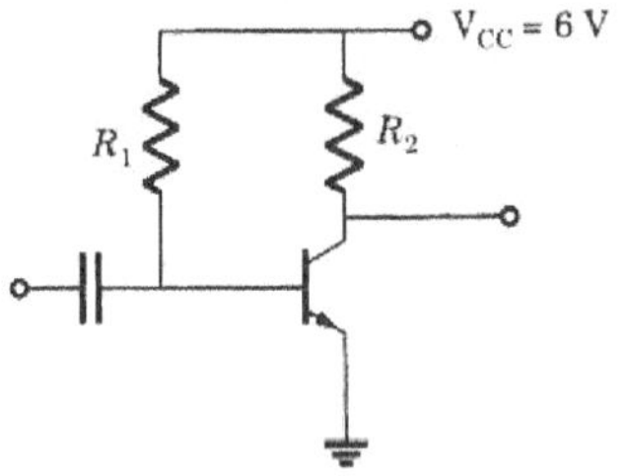

(*a*) (2 V, 2 mA) (*b*) (3 V, 2 mA)

(*c*) (4 V, 2 mA) (*d*) (4 V, 1 mA)

47. The oscillator circuit shown in the figure is has an ideal inverting amplifier. Its frequency of oscillation (in Hz) is

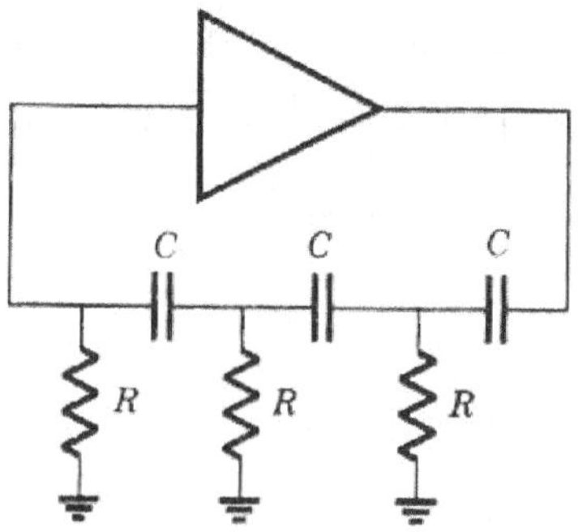

(*a*) $\dfrac{1}{(2\pi\sqrt{6}\,RC)}$ (*b*) $\dfrac{1}{(2\pi RC)}$

(*c*) $\dfrac{1}{(\sqrt{6}\,RC)}$ (*d*) $\dfrac{1}{\sqrt{6}\,(2\pi RC)}$

48. The output voltage of the regulated power supply shown in the figure is

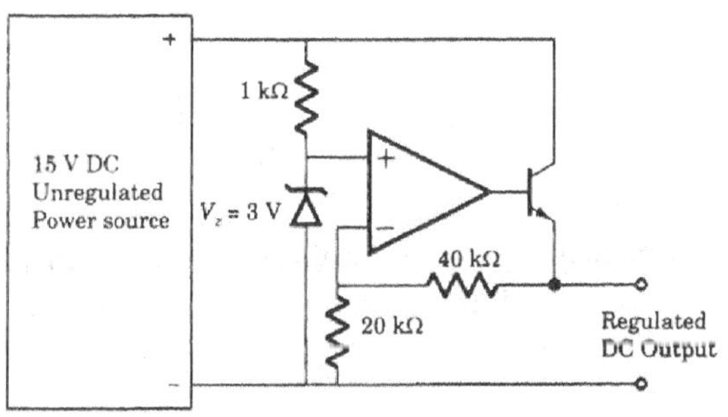

(*a*) 3 V (*b*) 6 V

(*c*) 9 V (*d*) 12 V

49. The action of a JFET in its equivalent circuit can best be represented as a

(*a*) Current Controlled Current Source

(*b*) Current Controlled Voltage Source

(*c*) Voltage Controlled Voltage Source

(*d*) Voltage Controlled Current Source

50. If the op-amp in the figure is idea, the output voltage V-out will be equal to

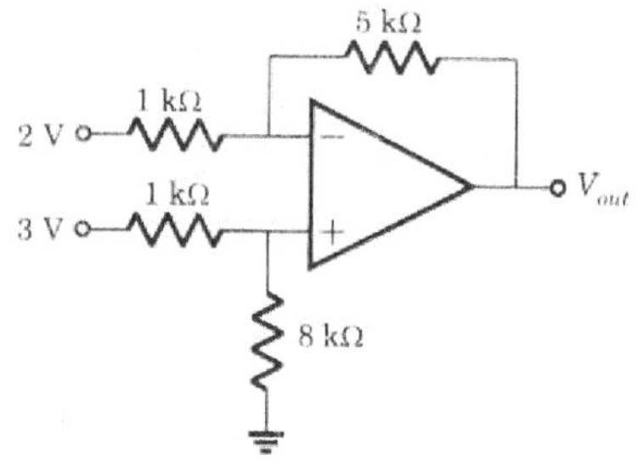

(*a*) 1 V (*b*) 6 V

(*c*) 14 V (*d*) 17 V

51. Three identical amplifiers with each one having a voltage gain of 50, input resistance of 1 kΩ and output resistance of 250 Ω, are cascaded. The open circuit voltage gain of the combined amplifier is

(*a*) 49 dB

(*b*) 51 dB

(*c*) 98 dB

(*d*) 102 dB

52. An ideal sawtooth voltage waveform of frequency 500 Hz and amplitude 3 V is generated by charging a capacitor of 2 μF in every cycle. The charging requires

(*a*) constant voltage source of 3 V for 1 ms

(*b*) constant voltage source of 3 V for 2 ms

(*c*) constant current source of 3 mA for 1 ms

(*d*) constant current source of 3 mA for 2 ms

53. The circuit shown in the figure has 4 boxes each described by inputs P, Q, R and outputs Y, Z with

$$Y = P \oplus Q \oplus R$$

$$Z = RQ + \overline{P}R + Q\overline{P}$$

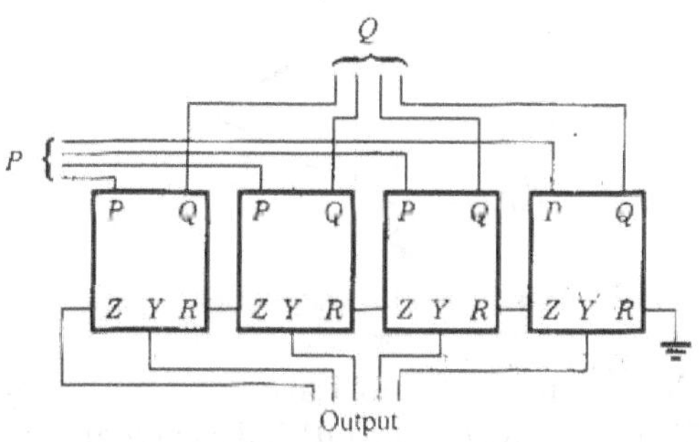

The circuit acts as a

(*a*) 4-bit adder giving P + Q

(*b*) 4-bit subtractor giving P – Q

(*c*) 4-bit subtractor giving Q – P

(*d*) 4-bit adder giving P + Q + R

54. If the functions W, X, Y and Z are as follows

$$W = R + \overline{P}Q + \overline{R}S$$

$$X = PQ\overline{R}\,\overline{S} + \overline{P}\,\overline{Q}\,\overline{R}\,\overline{S} + P\,\overline{Q}\,\overline{R}\,\overline{S}$$

$$Y = RS + \overline{PR + P\overline{Q} + \overline{P}\cdot\overline{Q}}$$

$$Z = R + S + \overline{PQ + \overline{P}\cdot\overline{Q}\cdot\overline{R} + P\overline{Q}\cdot\overline{S}}$$

Then

(*a*) $W = Z, X = \overline{Z}$ (*b*) $W = Z, X = Y$

(*c*) $W = Y$ (*d*) $W = Y = \overline{Z}$

55. A 4 bit ripple counter and a 4 bit synchronous counter are made using flip flops having a propagation delay of 10 ns each. If the worst case delay in the ripple counter and the synchronous counter be R and S respectively, then

(*a*) R = 10 ns, S = 40 ns

(*b*) R = 40 ns, S = 10 ns

(*c*) R = 10 ns, S = 30 ns

(*d*) R = 30 ns, S = 10 ns

56. The DTL, TTL, ECL and CMOS families of digital IC_s are compared in the following 4 columns

	(P)	(Q)	(R)	(S)
Fanout is minimum	DTL	DTL	TTL	CMOS
Power consumption is minimum	TTL	CMOS	ECL	DTL
Propagation delay is minimum	CMOS	ECL	TTL	TTL

The correct column is

(*a*) P (*b*) Q

(*c*) R (*d*) S

57. The circuit shown in the figure is a 4-bit DAC

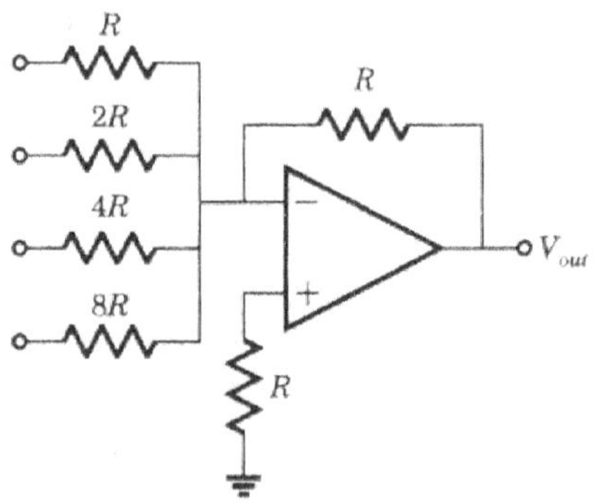

The input bits 0 and 1 are represented by 0 and 5V respectively. The OP AMP is ideal, but all the resistances and the 5 V inputs have a tolerance of ±10%. The specification (rounded to the nearest multiple of 5%) for the tolerance of the DAC is

(*a*) ± 35% (*b*) ± 20%

(*c*) ± 10% (*d*) ± 5%

58. The circuit shown in the figure converts

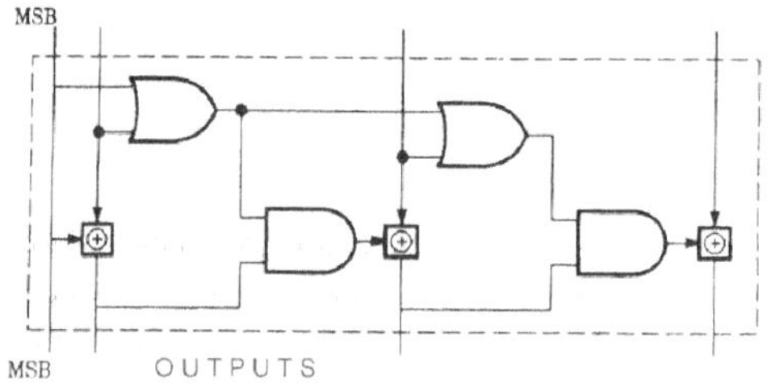

(*a*) BCD to binary code

(*b*) Binary to excess – 3 code

(*c*) Excess – 3 to Gray code

(*d*) Gray to Binary code

59. In the circuit shown in the figure A is a parallel-in, parallel-out 4 bit register, which loads at the rising edge of the clock C. The input lines are connected to a 4 bit bus, W. Its output acts as the input to a 16 × 4 ROM whose output is floating when the enable input E is 0. A partial table of the contents of the ROM is as follows

Address	0	2	4	6	8	10	11	14
Data	0011	1111	0100	1010	1011	1000	0010	1000

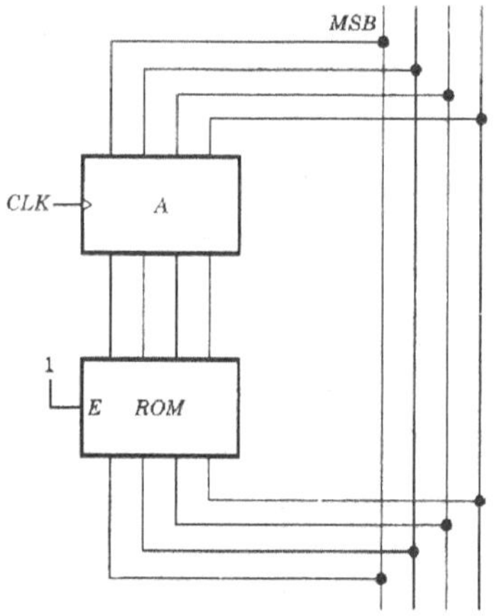

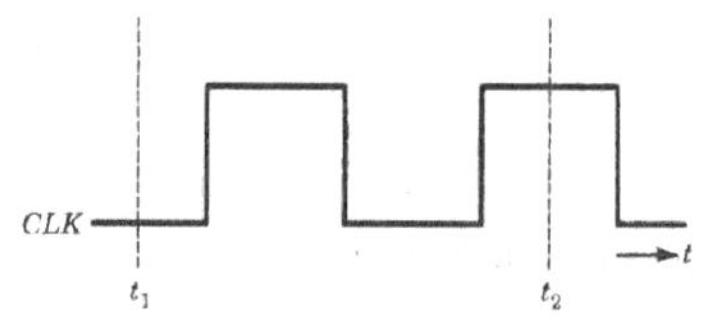

The clock to the register is shown, and the data on the W bus at time t_1 is 0110. The data on the bus at time t_2 is

(*a*) 1111 (*b*) 1011

(*c*) 1000 (*d*) 0010

60. In an 8085 microprocessor, the instruction CMP B has been executed while the content of the accumulator is less than that of register B. As a result

(*a*) Carry flag will be set but Zero flag will be reset

(*b*) Carry flag will be reset but Zero flag will be set

(*c*) Both Carry flag and Zero flag will be reset

(*d*) Both Carry flag and Zero flag will be set

61. Let X and Y be two statistically independent random variables uniformly distributed in the ranges (–1, 1) and (–2, 1) respectively. Let Z = X + Y. Then the probability that (Z ≤ – 2] is

(a) zero (b) $\frac{1}{6}$

(c) $\frac{1}{3}$ (d) $\frac{1}{12}$

62. Let P be linearity, Q be time-invariance, R be causality and S be stability. A discrete-time system has the input-output relationship,

$$y(n) = \begin{cases} x(n) & n \geq 1 \\ 0, & n = 0 \\ x(n+1), & n \leq -1 \end{cases}$$

where $x(n)$ is the input and $y(n)$ is the output. The above system has the properties

(a) P, S but not Q, R (b) P, Q, S but not R

(c) P, Q, R, S (d) Q, R, S but not P

Data for Q. 63-64 are given below. Solve the problems and choose the correct answers.

The system under consideration is an RC low-pass filter (RC-LPF) with R = 1.0 kΩ k and C=1.0μF

63. Let H(*f*) denote the frequency response of the RC-LPF. Let f_1 be the highest frequency such that $0 \leq |f| \leq f_1$, $\frac{|H(f_1)|}{H(0)} \geq 0.95$. Then f_1 (in Hz) is

(a) 327.8 (b) 163.9

(c) 52.2 (d) 104.4

64. Let $t_g(f)$ be the group delay function of the given RC-LPF and f_2 = 100 Hz. Then $t_g(f_2)$ in ms, is

(a) 0.717 (b) 7.17

(c) 71.7 (d) 4.505

For (Q. 65-66) : $x(t)$ is a random process with constant mean value of 2 and the auto corelation function $R_x(J) = 4[e^{-0.2[J]} + 1]$

65. Let X be the Gaussian random variable obtained by sampling the process at $t = t_i$ and let

$$Q(\alpha) = \int_a^{\infty} \frac{1}{\sqrt{2p}} e^{\frac{X^2}{2}} dy.$$

The probability that $[x \leq 1]$ is

(a) 1 – Q (0.5) (b) Q (0.5)

(c) $Q\left(\frac{1}{2\sqrt{2}}\right)$ (d) $1 - Q\left(\frac{1}{2\sqrt{2}}\right)$

66. Let Y and Z be the random variables obtained by sampling X(*t*) at *t* = 2 and *t* = 4 respectively. Let W = Y – Z. The variance of W is

(a) 13.36 (b) 9.36

(c) 2.64 (d) 8.00

67. Let $x(t) = 2\cos(800\pi t) + \cos(1400\pi t)$. $x(t)$ is sampled with the rectangular pulse train shown in the figure. The only spectral components (in kHz) present in the sampled signal in the frequency range 2.5 kHz to 3.5 kHz are

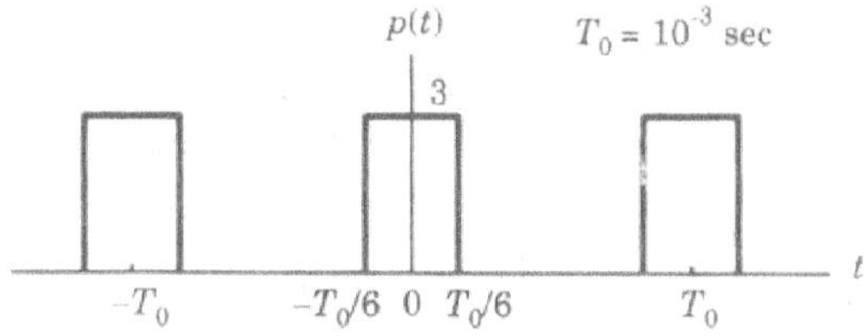

$T_0 = 10^{-3}$ sec

(a) 2.7, 3.4 (b) 3.3, 3.6

(c) 2.6, 2.7, 3.3, 3.4, 3.6 (d) 2.7, 3.3

68. The signal flow graph of a system is shown in the figure. The transfer function $\frac{C(s)}{R(s)}$ of the system is

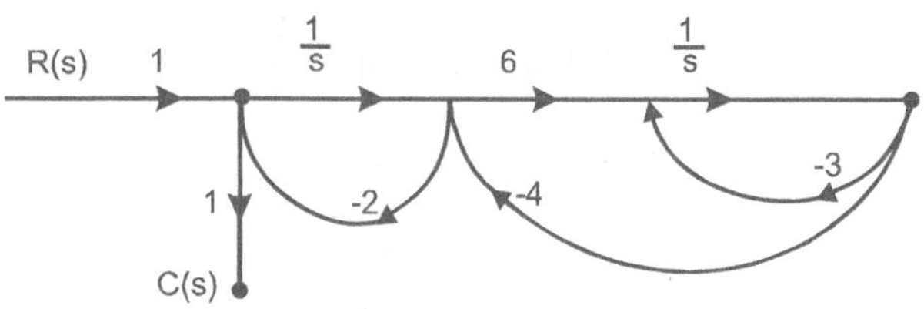

(a) $\frac{6}{s^2 + 29s + 6}$ (b) $\frac{6s}{s^2 + 29s + 6}$

(c) $\frac{s(s+2)}{s^2 + 29s + 6}$ (d) $\frac{s(s+27)}{s^2 + 29s + 6}$

69. The root locus of the system $G(s)H(s) = \frac{K}{s(s+2)(s+3)}$ has the break-away point located at

(a) (– 0.5, 0)

(b) (– 2.548, 0)

(c) (– 4, 0)

(d) (– 0.784, 0)

70. The approximate Bode magnitude plot of a minimum-phase system is shown in the figure. The transfer function of the system is

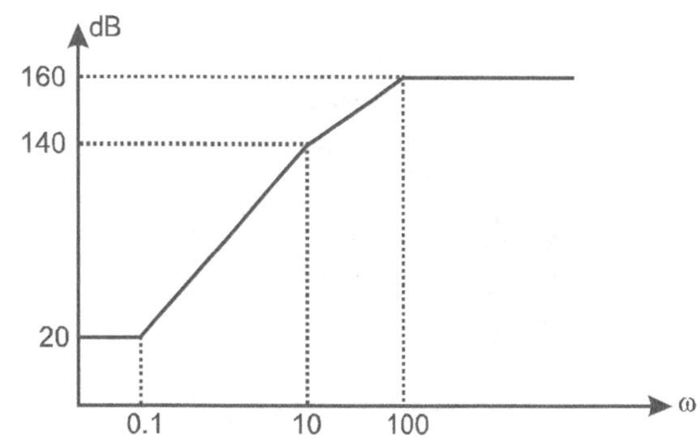

(*a*) $10^8 \dfrac{(s+0.1)^3}{(s+10)^2\,(s+100)}$

(*b*) $10^7 \dfrac{(s+0.1)^3}{(s+10)\,(s+100)}$

(*c*) $10^8 \dfrac{(s+0.1)^2}{(s+10)^2\,(s+100)}$

(*d*) $10^9 \dfrac{(s+0.1)^3}{(s+10)\,(s+100)^2}$ $v_i = A_c^i \cos(2\pi f_c^i t)$

71. A second-order system has the transfer function

$$\frac{C(s)}{R(s)} = \frac{4}{s^2 + 4s + 4}.$$

With $r(t)$ as the unit-step function, the response $c(t)$ of the system is represented by

(*a*)

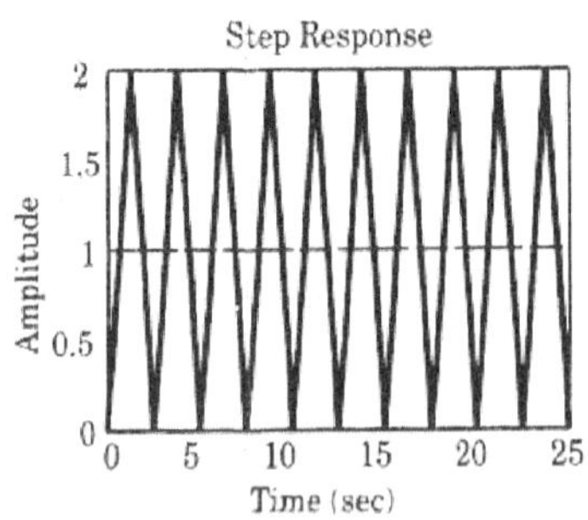

(*b*)

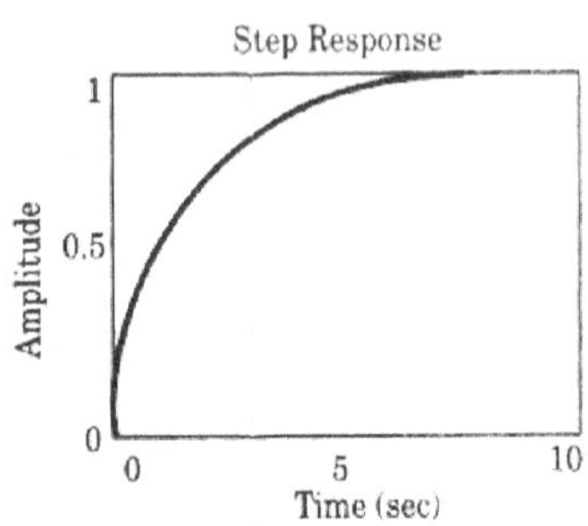

(*c*)

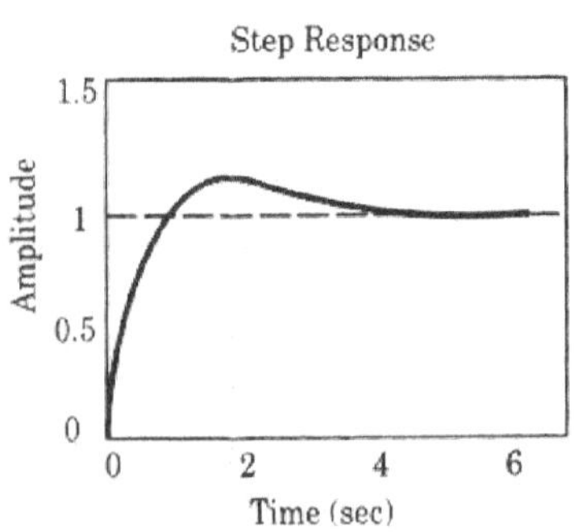

(*d*)

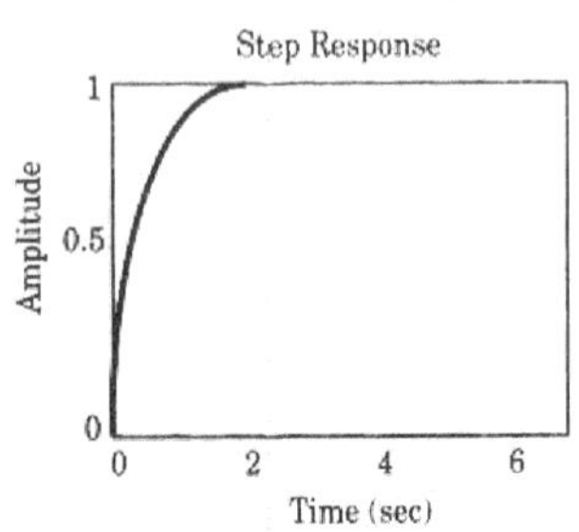

72. The gain margin and the phase margin of a feedback system with $G(s)H(s) = \dfrac{s}{(s+100)^3}$ are

(*a*) 0 dB, 0° (*b*) ∞, ∞

(*c*) ∞, 0° (*d*) 88.5 dB, ∞

73. The zero-input response of a system given by the state-space equation

$$\begin{bmatrix}\dot{x}_1\\ \dot{x}_2\end{bmatrix} = \begin{bmatrix}1 & 0\\ 1 & 1\end{bmatrix}\begin{bmatrix}x_1\\ x_2\end{bmatrix} \text{ and } \begin{bmatrix}x_1(0)\\ x_2(0)\end{bmatrix} = \begin{bmatrix}1\\ 0\end{bmatrix} \text{ is}$$

(*a*) $\begin{bmatrix}te^t\\ t\end{bmatrix}$ (*b*) $\begin{bmatrix}e^t\\ t\end{bmatrix}$

(*c*) $\begin{bmatrix}e^t\\ te^t\end{bmatrix}$ (*d*) $\begin{bmatrix}t\\ te^t\end{bmatrix}$

74. A DSB-SC signal is to be generated with a carrier frequency $f_c = 1$ MHz using a non-linear device with the input-output characteristic

$$v_0 = a_0 v_i + a_1 v_i^3$$

where a_0 and a_1 are constants. The output of the non-linear device can be filtered by an appropriate band-pass filter.

Let $v_i = A_c^i \cos(2\pi f_c^i t) + m(t)$ where $m(t)$ is the message signal. Then the value of f_c^i (in MHz) is

(*a*) 1.0 (*b*) 0.333

(*c*) 0.5 (*d*) 3.0

The data for Q. 75-76 are given below. Solve the problems and choose the correct answers.

Let $m(t) = \cos[(4\pi \times 10^3)t]$ be the message signal and $c(t) = 5\cos[2\pi \times 10^6)t]$ be the carrier

75. $c(t)$ and $m(t)$ are used to generate an AM signal. The modulation index of the generated AM signal is 0.5.

Then the quantity $\dfrac{\text{Total sideband power}}{\text{Carrier power}}$ is

(*a*) $\dfrac{1}{2}$ (*b*) $\dfrac{1}{4}$

(*c*) $\dfrac{1}{3}$ (*d*) $\dfrac{1}{8}$

76. $c(t)$ and $m(t)$ are used to generate an FM signal. If the peak frequency deviation of the generated FM signal is three times the transmission bandwidth of the AM signal, then the coefficient of the term $\cos[2\pi(1008 \times 10^3 t)]$ in the FM signal (in terms of the Bessel coefficients) is

(*a*) $5J_4(3)$ (*b*) $\dfrac{5}{2}J_8(3)$

(*c*) $\dfrac{5}{2}J_8(4)$ (*d*) $5J_4(6)$

77. Choose the correct one from among the alternative A, B, C, D after matching an item in Group I with the most appropriate item in Group II.

Group I	Group II
A Ring modulator	1 Clock recovery
B VCO	2 Demodulation of FM
C Foster-Seely discriminator	3 Frequency conversion
D Mixer	4 Summing the two inputs
	5 Generation of FM
	6 Generation of DSB-Sc

	A	B	C	D
(*a*)	1	3	2	4
(*b*)	6	5	2	3
(*c*)	6	1	3	2
(*d*)	5	6	1	3

78. A superheterodyne receiver is to operate in the frequency range 550 kHz-1650 kHz, with the intermediate frequency of 450 kHz. Let $R = \frac{C_{max}}{C_{min}}$ denote the required capacitance ratio of the local oscillator and I denote the image frequency (in kHz) of the incoming signal. If the receiver is tuned to 700 kHz, then

(*a*) R = 4.41, I = 1600 (*b*) R = 2.10, I = 1150

(*c*) R = 3.0, I = 1600 (*d*) R = 9.0, I = 1150

79. A sinusoidal signal with peak-to-peak amplitude of 1.536 V is quantized into 128 levels using a mid-rise uniform quantizer.

The quantization-noise power is

(*a*) 0.768 V (*b*) 48×10^{-6} V^2

(*c*) 12×10^{-6} V^2 (*d*) 3.072 V

80. If E_b, the energy per bit of a binary digital signal, is 10^{-6} watt-sec and the one-sided power spectral density of the white noise, $N_0 = 10^{-5}$ W/Hz, then the output SNR of the matched filter is

(*a*) 26 dB (*b*) 10 dB

(*c*) 20 dB (*d*) 13 dB

81. The input to a linear delta modulator having a step-size $\Delta = 0.628$ is a sine wave with frequency f_m and peak amplitude E_m. If the sampling frequency $f_s = 40$ kHz, the combination of the sine-wave frequency and the peak amplitude, where slope overload will take place is

	E_m	f_m
(*a*)	0.3 V	8 kHz
(*b*)	1.5 V	4 kHz
(*c*)	1.5 V	2 kHz
(*d*)	3.0 V	1 kHz

82. If S represents the carrier synchronization at the receiver and ρ represents the bandwidth efficiency, then the correct statement for the coherent binary PSK is

(*a*) ρ = 0.5, S is required

(*b*) ρ = 1.0, S is required

(*c*) ρ = 0.5, S is not required

(*d*) ρ = 1.0, S is not required

83. A signal is sampled at 8 kHz and is quantized using 8-bit uniform quantizer. Assuming SNR_q for a sinusoidal signal, the correct statement for PCM signal with a bit rate of R is

(*a*) R = 32 kbps, $SNR_q = 25.8$ dB

(*b*) R = 64 kbps, $SNR_q = 49.8$ dB

(*c*) R = 64 kbps, $SNR_q = 55.8$ dB

(*d*) R = 32 kbps, $SNR_q = 49.8$ dB

84. Medium 1 has the electrical permittivity $\varepsilon_1 = 1.5\,\varepsilon_0$ farad/m and occupies the region to the left of $x = 0$ plane. Medium 2 has the electrical permittivity $\varepsilon_2 = 2.5\,\varepsilon_0$ farad/m and occupies the region to the right of $x = 0$ plane. If E_1 in medium 1 is $E_1 = (2u_x - 3u_y + 1u_z)$ volt/m, then E_2 in medium 2 is

(*a*) $(2.0u_x - 7.5u_y + 2.5u_z)$ volt/m

(*b*) $(2.0u_x - 2.0u_y + 0.6u_z)$ volt/m

(*c*) $(1.2u_x - 3.0u_y + 1.0u_z)$ volt/m

(*d*) $(1.2u_x - 2.0u_y + 0.6u_z)$ volt/m

85. If the electric field intensity is given by

$$E = (xu_x + yu_y + zu_z) \text{ volt/m},$$

the potential difference between X (2, 0, 0) and Y (1, 2, 3) is

(*a*) +1 volt (*b*) –1 volt

(*c*) +5 volt (*d*) +6 volt

86. A uniform plane wave travelling in air is incident on the plane boundary between air and another dielectric medium with $\varepsilon_r = 4$. The reflection coefficient for the normal incidence, is

(*a*) zero (*b*) $0.5\angle 180°$

(*c*) $0.333\angle 0°$ (*d*) $0.333\angle 180°$

87. If the electric field intensity associated with a uniform plane electromagnetic wave travelling in a perfect dielectric medium is given by

$E(z, t) = 10\cos(2\pi \times 10^7\, t = 0.1\,\pi z)$ volt/m, then the velocity of the travelling wave is

(*a*) 3.00×10^8 m/sec

(*b*) 2.00×10^8 m/sec

(*c*) 6.28×10^7 m/sec

(*d*) 2.00×10^7 m/sec

88. A short-circuited stub is shunt connected to a transmission line as shown in the figure is. If Z_0 = 50 ohm, the admittance Y seen at the junction of the stub and the transmission line is

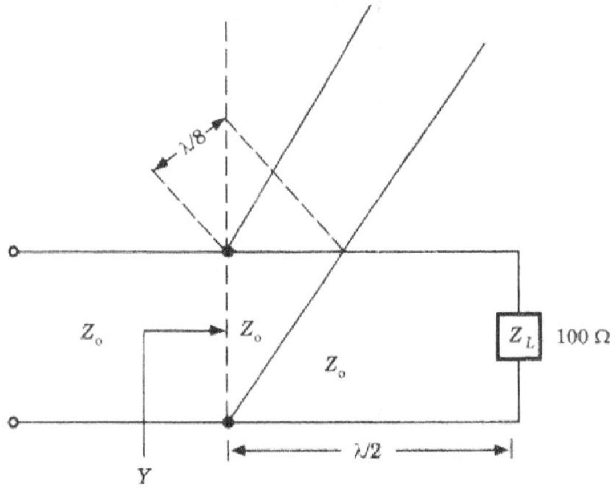

(*a*) (0.01 – *j*0.02) mho
(*b*) (0.02 – *j*0.01) mho
(*c*) (0.04 – *j*0.02) mho
(*d*) (0.02 + *j*0) mho

89. A rectangular metal wave guide filled with a dielectric material of relative permittivity $\varepsilon_r = 4$ has the inside dimensions 3.0 cm × 1.2 cm. The cut-off frequency for the dominant mode is

(*a*) 2.5 GHz (*b*) 5.0 GHz
(*c*) 10.0 GHz (*d*) 12.5 GHz

90. Two identical antennas are placed in the $\theta = \pi/2$ plane as shown in the figure. The elements have equal amplitude excitation with 180° polarity difference, operating at wavelength λ. The correct value of the magnitude of the far-zone resultant electric field strength normalized with that of a single element, both computed for $\phi = 0$, is

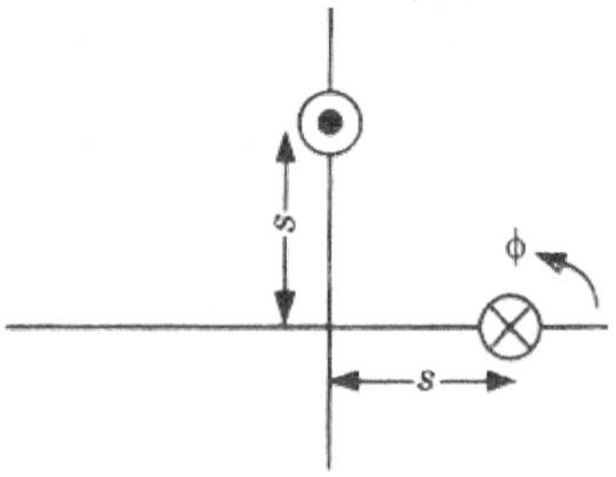

(*a*) $2\cos\left(\frac{2\pi s}{\lambda}\right)$

(*b*) $2\sin\left(\frac{2\pi s}{\lambda}\right)$

(*c*) $2\cos\left(\frac{\pi s}{\lambda}\right)$

(*d*) $2\sin\left(\frac{\pi s}{\lambda}\right)$

ANSWERS

1. (*b*)	2. (*c*)	3. (*b*)	4. (*c*)	5. (*c*)	6. (*d*)	7. (*b*)	8. (*a*)	9. (*c*)	10. (*d*)
11. (*b*)	12. (*d*)	13. (*b*)	14. (*c*)	15. (*a*)	16. (*d*)	17. (*c*)	18. (*b*)	19. (*d*)	20. (*d*)
21. (*d*)	22. (*c*)	23. (*b*)	24. (*a*)	25. (*c*)	26. (*a*)	27. (*c*)	28. (*b*)	29. (*c*)	30. (*b*)
31. (*a*)	32. (*)	33. (*a*)	34. (*c*)	35. (*c*)	36. (*b*)	37. (*a*)	38. (*a*)	39. (*d*)	40. (*c*)
41. (*c*)	42. (*a*)	43. (*d*)	44. (*b*)	45. (*a*)	46. (*a*)	47. (*a*)	48. (*c*)	49. (*d*)	50. (*b*)
51. (*c*)	52. (*d*)	53. (*b*)	54. (*a*)	55. (*b*)	56. (*b*)	57. (*a*)	58. (*a*)	59. (*c*)	60. (*a*)
61. (*d*)	62. (*a*)	63. (*c*)	64. (*a*)	65. (*b*)	66. (*c*)	67. (*d*)	68. (*d*)	69. (*d*)	70. (*a*)
71. (*b*)	72. (*b*)	73. (*c*)	74. (*c*)	75. (*d*)	76. (*d*)	77. (*b*)	78. (*a*)	79. (*c*)	80. (*b*)
81. (*b*)	82. (*a*)	83. (*b*)	84. (*c*)	85. (*c*)	86. (*d*)	87. (*b*)	88. (*a*)	89. (*a*)	90. (*d*)

EXPLANATIONS

1. Given, n = number of nodes = 4,
b = number of branches = 7
∴ Number of equations = $b - n + 1$
$= 7 - 4 + 1 = 4$

2. Source impedance, $R_s = R + j\omega L$
$= 1 + j$
For maximum power transfer, $R_L = R_S^* = 1 - j$ which indicates a 1 Ω resistance in series with 1F capacitor.

3. Resonant frequency, $\omega_0 = 2\pi f_0 = \frac{1}{\sqrt{LC}}$

and Quality factor, $Q = \frac{1}{\omega_0 RC}$

When, R, L, C all are doubled, then,

$$\omega'_0 = \frac{\omega_0}{2}$$

and, $Q' = \frac{1}{\omega'_0 R'C'} = \frac{2}{4(\omega_0 RC)}$

$= \frac{Q}{2} = 50$

4. From final value theorem,

$$\underset{t\to\infty}{\text{Lt}}\ i(t) = \underset{s'\to 0}{\text{Lt}}\ sI(s) = \underset{s\to 0}{\text{Lt}}\ \frac{2}{(1+s)} = 2$$

5. Using KVL around the loop, we have

$$\sin t = 2\,i(t) + 2\frac{di(t)}{dt} + \frac{1}{C}\int i(t)\,.dt$$

Differentiating $\cos t = \frac{2\,di(t)}{dt} + 2\frac{d^2 i(t)}{dt^2} + i(t)$

i.e. $2\frac{d^2 i}{dt^2} + 2\frac{di}{dt} + i(t) = \cos t$

6. Phosphorus is a pentavalent atom, therefore one free electron after making four bonds with Si atoms, thus making doping n–type

7. $E_g(t) = 1.27 - 3.60 \times 10^{-4}\,T$
For silicon, $T = 300$ K

8. Let majority carriers be electrons,
$n_i = 1.5 \times 10^{16}$, $n_e = 5 \times 10^{20}$
According to mass action law, $np = n_i^2$

or $p = \frac{n_i^2}{n} = \frac{(1.5\times10^{16})^2}{5\times10^{20}}$
$= 4.50 \times 10^{11}$ /m³

10. $V_T = V_{To} + \gamma\sqrt{|V_{SB} - 2\phi_F|} - \sqrt{|2\phi_F|}$
where, V_T = Threshold voltage
γ = substrate bias coefficient
V_{SB} = substrate bias voltage

11.

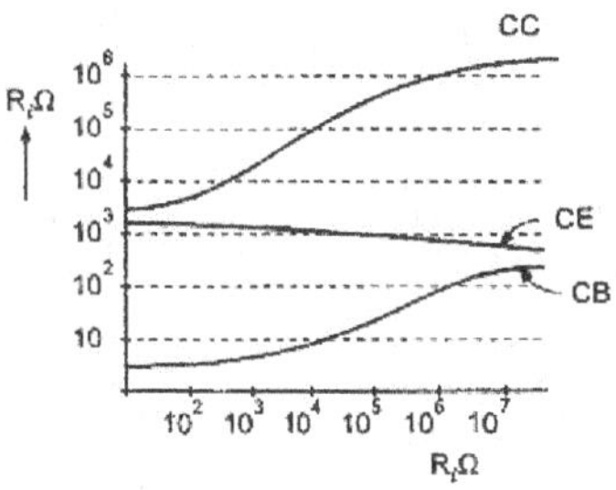

12.

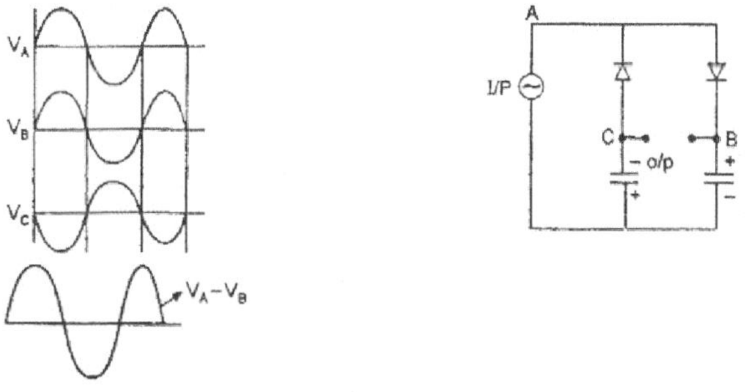

13. The output voltage waveform is shown in the figure,

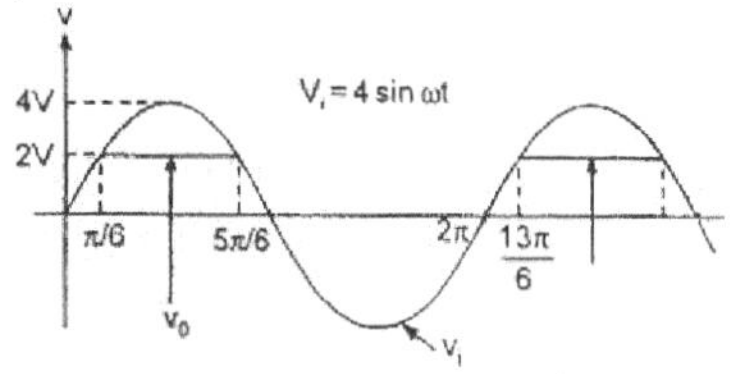

Here, at $V_i = 2$, $\sin\omega t = \frac{1}{2}$

or $\omega t = \frac{\pi}{6}$

Duty cycle $= \frac{T_{ON}}{T} = \frac{\pi - \frac{\pi}{6} - \frac{\pi}{6}}{2\pi} = \frac{1}{3}$

14. $\text{CMMR} = \frac{A_d}{A_c}$

Taking log on, $\log(\text{CMMR}) = \log\frac{A_d}{A_c}$

⇒ $20\log(\text{CMMR}) = 20\log A_d - 20\log A_c$
∴ $= 48 - 2 = 46$ dB

15. At high frequency strong internal capacitor come into existence which have considerable impedance.

16. Let ABCD be four variables.
⇒ Total number of variables
$= A, \overline{A}, B, \overline{B}\ C, \overline{C}, D, \overline{D}$ *i.e.* 8
Number of total elements in the 4 variable K Map
$= 2^4 = 16$
∴ Number of distinct Boolean expressions
$= 2^{16} = 65536$

17. Minimum number of comparators $= 2^N - 1$
$= 2^8 - 1 = 255$

19. Without any additional circuitry, i.e. without I/P only Selector expression will pass because of floating I/P will be assumed to be in high logic, i.e. 1

Ex : 1

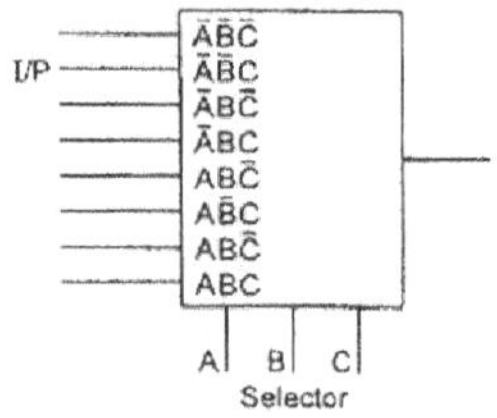

Ex : 2

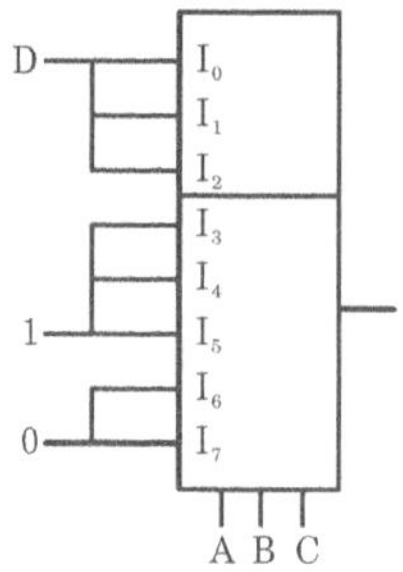

Ex : 3

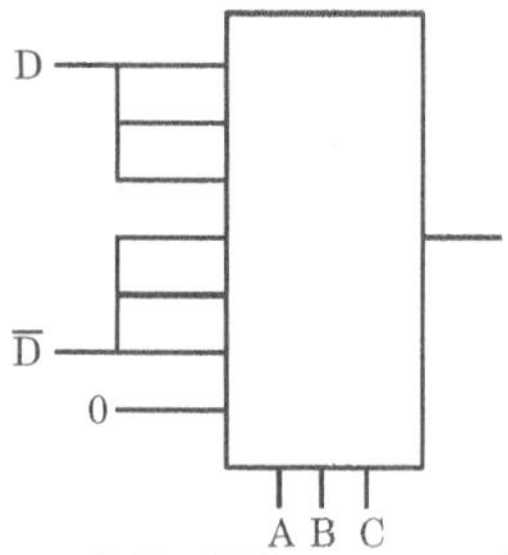

Convention : A B C D are available with their complements $\overline{A}$, $\overline{B}$, $\overline{C}$, $\overline{D}$.

20. Flip-Flop outputs

Counting sequence	Q_2	Q_1	Q_0
0	0	0	0
1	0	0	1
2	0	1	0
3	0	1	1
4	1	0	0
5	1	0	1
6	1	1	0
7	1	1	1

Counter should be reset at 7th state of counting sequence which can be done by following logic

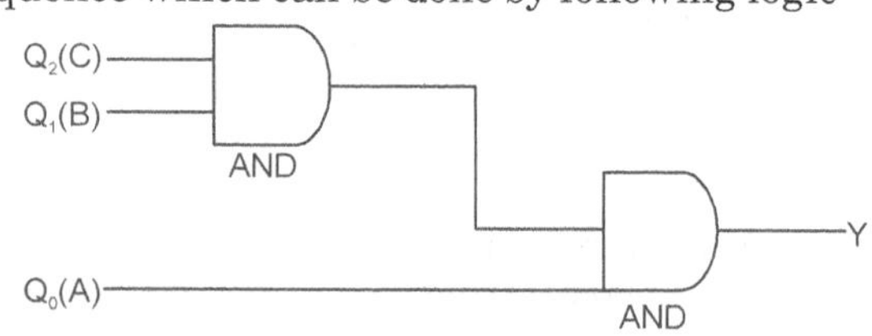

When A = B = C = 1, Y = 1 which can meet the flip-flops after being applied to reset function.

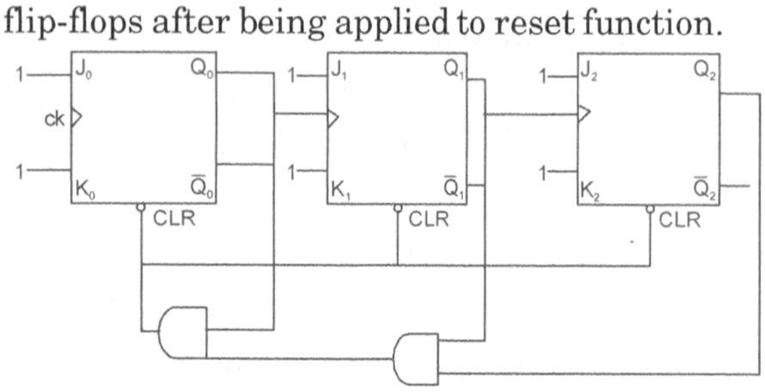

Circuit diagram for MOD-6 counter

21. Using relation between Trigonometric Fourier series coefficient and exponential Fourier series coefficient

$$c_n = a_n - jb_n \quad ..(i)$$

$$c_{-n} = a_n + jb_n \quad ...(A)$$

Given, $c_3 = 3 + j5$

$$c_3 = 3 - j(-5) \quad ...(ii)$$

Compare (*i*) with (*ii*)

$a_n = 3$, $b_n = -5$

Then using (A)

$$c_{-3} = 3 + j(-5) = 3 - j5$$

22. Output, $y(t) = 4x(t-2)$

Its, fourier transform, $Y(f) = 4x(f).\, e^{-j4\pi f}$

and the transfer function, $\dfrac{Y(f)}{X(f)} = 4e^{-j4\pi f}$

23. Output $Y(z) = H(z)X(z)$

$$= 2\,(z^4 + z^2 - 2z + 2 - 3z^{-4})z^{-3}$$

$$= 2\,(z + z^{-1} - 2z^{-2} + 2z^{-3} - 3z^{-7})$$

Taking inverse – z transform, we have

$$y(n) = 2\,[\delta(n+1) + \delta(n-1) - 2\,\delta(n-2) + 2\,\delta(n-3) - 3\delta(n-7)]$$

At $n = 4$, $y(4) = 0$

24. The encirclement of point (1, 0) is only once in anti-clockwise direction, then N = 1 and as P = 1 therefore N = P – Z = 0

and, the system is stable as there are no zeroes in right-hand s-plane.

26.

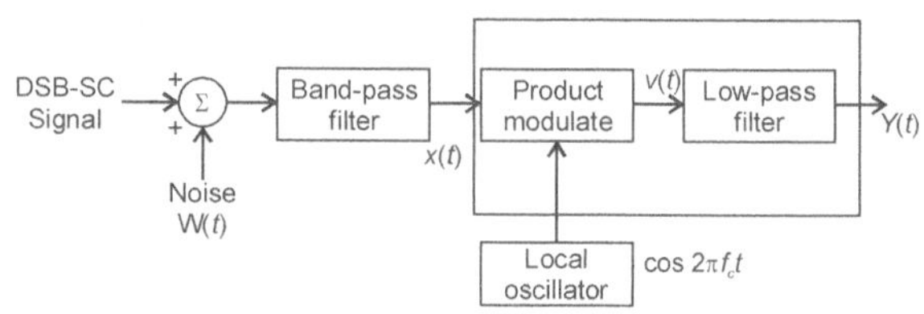

DSB – SC signal is expressed as $S(t) = CA_c \cos(2\pi f_c t).m(t)$ (here C is contant)

Average power of DSB-SC signal, $P_s = \dfrac{CA_c^2 P}{2}$

Hence P is the average power of message signal.

For noise spectral density of $N_0/2$, average noise power in the message bandwidth is WN_0.

then, $(SNR)_{C,\,DSB} = \dfrac{CA_C^2 P}{2WN_0}$

Using narrow band representation of the filtered noise $u(t)$, total signal at the coverent detector input,

$$x(t) = s(t) + n(t)$$
$$= CA_C \cos(2\pi f_c t)\, m(t) + n_I \cos(2\pi f_c t) - n_2(t) \sin(2\pi f_c t)$$

h_I = IN phase component of $n(t)$

u_Q = quadrative component of $u(t)$

Output of product modulator,

$$v(t) = x(t) \cos(2\pi f_c t)$$
$$= \frac{1}{2}CA_C m(t) + \frac{1}{2}n_I(t) + \frac{1}{2}[CA_C m(t) + n_I t] \cos(4\pi f_c t) - \frac{1}{2}n_Q(t)\sin(4\pi f_c t)$$

Low-pass filter given

$$y(t) = \frac{1}{2}CA_C m(t) + \frac{1}{2}n_I(t)$$

[Only in-phase component]

28. Probability of error, $P_e = \frac{1}{2}\, erfc\left(\sqrt{\frac{E_d}{2\eta}}\right)$

As P_e for FSK and PSK are same then E_d is also same for both

and then, $A_f^2 = 2A_P^2 \Rightarrow \frac{Af}{A_P} = \sqrt{2}$

Hence binary coherent FSK is interior by $20\log\sqrt{3} = 3dB$ to binary coherent PSK.

29. $$\nabla \times \overline{H} = \frac{\partial \overline{D}}{\partial t}$$

If $\overline{D}$ has unit of Coulomb/metre^2, then $\nabla \times \overline{H}$ have unit of Ampere/metre^2

30. Depth of penetration, $\delta = \frac{1}{\sqrt{\pi f \mu \sigma}}$

or $$\delta \propto \frac{1}{\sqrt{f}}$$

or $$\frac{\delta_1}{\delta_2} = \sqrt{\frac{f_2}{f_1}} = \sqrt{\frac{4}{1}}$$

or $$\delta_2 = \frac{\delta_1}{2} = 12.50 \text{ cm}$$

31. The current I will be distributed in the cube branches with the symmetricity.

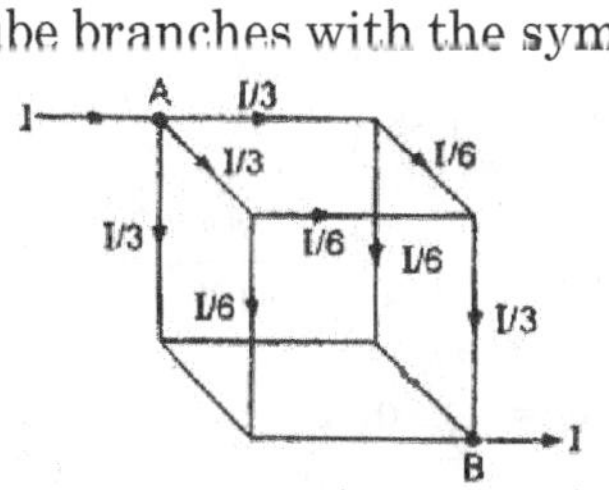

Then, $$V_{AB} = \left(\frac{I}{3} + \frac{I}{6} + \frac{I}{3}\right).1$$

or $$R_{AB} = \frac{V_{AB}}{I} = \frac{5}{6}\ \Omega$$

32. Here inductance has not been given, therefore ignoring inductance, we have, in the voltage source circuit,

$$I_1 = \frac{2\cos 4t}{3} = 0.67 \cos 4t$$

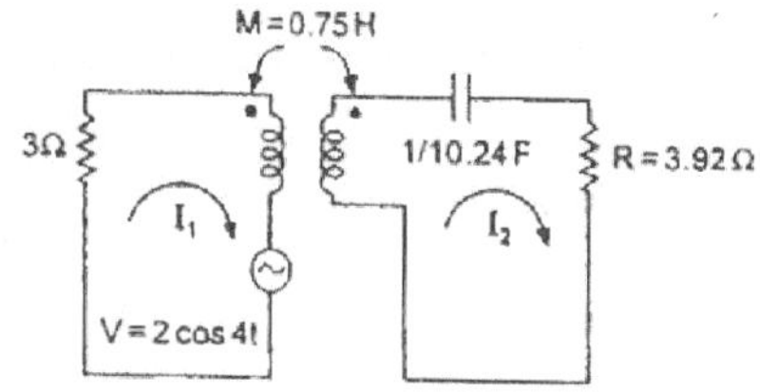

and, $$I_2 = \frac{-j\omega M I_1}{R - j/\omega C}$$
$$= \frac{-j4 \times 0.75 I_1}{3.92 - 2.56j}$$
$$= \frac{-j3 \times 0.67 \cos 4t}{3.92 - 2.56j}$$
$$= (0.23 - 0.35j)\cos 4t$$

Now, none of options seems to be true, it can be inferred from the question that something may be missing there.

33. At $t = 0^+$, the voltage across capacitors is $-V$

Capacitor in series with R is shorted, inductor open.

Then, resistance= R + R = 2R

and, $$i = \frac{-V}{2R}$$

34. Transformed circuit representation is as shown below

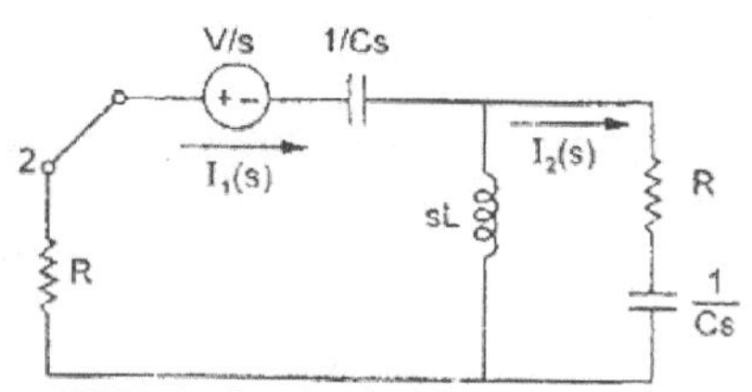

Using KVL in both the loops, we get

$$I_1(s).[1/Cs + sL + R] - I_2 . sL + \frac{V}{s} = 0$$

and $I_2(s)[R + 1/Cs + Ls] - I_1 Ls = 0$

Writing in matrix form,

$$\begin{bmatrix} R + sL + \frac{1}{Cs} & -Ls \\ -Ls & R + Ls + \frac{1}{Cs} \end{bmatrix} \begin{bmatrix} I_1(s) \\ I_2(s) \end{bmatrix} = \begin{bmatrix} -V/s \\ 0 \end{bmatrix}$$

35. The steady state-current,

$$i(t) = i_1(t) + i_2(t)$$

or $$i(t) = \frac{10\sqrt{2}\cos(t + 10°)}{R + j\omega_1 L} + \frac{10\sqrt{5}\cos(2t + 10°)}{R + j\omega_2 L}$$

here, $\omega_1 = 1$, $\omega_2 = 2$

$$= \frac{10\sqrt{2}\cos(t+10°)}{1+j} + \frac{10\sqrt{5}\cos(2t+10°)}{1+2j}$$

$$= \frac{10\sqrt{2}\cos(t+10°)}{\sqrt{2}.\angle\tan^{-1}1} + \frac{10\sqrt{5}\cos(2t+10°)}{\sqrt{5}.\angle\tan^{-1}2}$$

$$= 10\cos(t-35°) + 10\cos(2t+10° - \tan^{-1}2)$$

36. From the figure, $Z(s) = \frac{K(s+3)}{(s+1+i)(s+1-i)}$

$$= \frac{K(s+3)}{[(s+1)^2+1]}$$

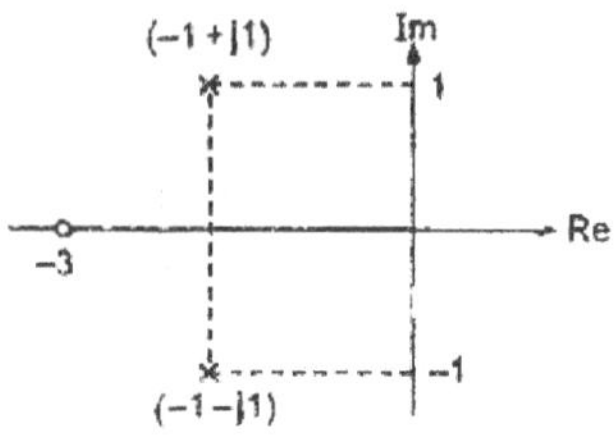

As, $Z(0) = 3$ then, $\frac{3K}{2} = 3$

or $K = 2$

then, $Z(s) = \frac{2(s+3)}{(s+1)^2+1}$

$$= \frac{2(s+3)}{s^2+2s+2}$$

37. Using $\nabla - Y$ conversion, we have

Here, $Z_{AP} = \frac{Z_{AB}Z_{AC}}{Z_{AB}+Z_{AC}+Z_{BC}} = \frac{2}{4} = 0.5$

$$Z_{PB} = \frac{Z_{AB}Z_{BC}}{Z_{AB}+Z_{BC}+Z_{AC}} = \frac{2}{4} = 0.5$$

$$Z_{PC} = \frac{Z_{AC}Z_{BC}}{Z_{AB}+Z_{BC}+Z_{AC}} = \frac{1}{4} = 0.25$$

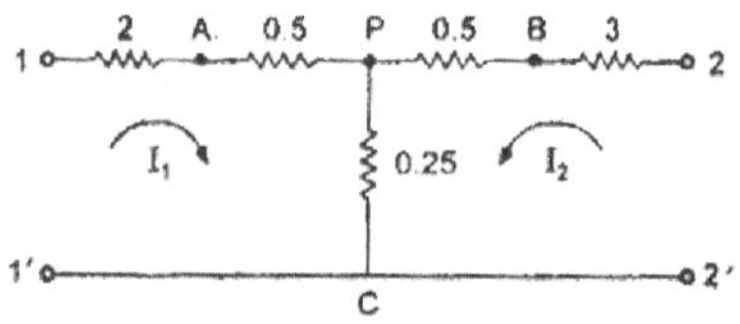

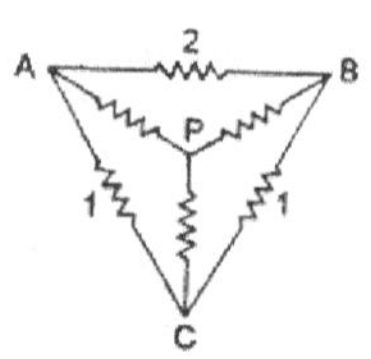

Now, from the resultant network,

$$V_1 = 2.75\ I_1 + 0.25\ I_2$$

$$= Z_{11}\ I_1 + Z_{12}\ I_2$$

Thus, $Z_{11} = 2.75\ \Omega$,

$Z_{12} = 0.25\ \Omega$

38. Conductivity, $\sigma = ne\ _n + pe\ _p$

As, silicon-bar is n–type, $\sigma \approx n\,e\ _n$

or Resistivity, $\rho = \frac{1}{ne\mu_n}$

and,resistance of the bar = $\frac{\rho l}{A} = \frac{l}{ne\mu_n A}$

or $R = \frac{10^{-3}}{5\times10^{20}\times1.6\times10^{-19}\times0.13\times100\times10^{-12}}$

$$= 0.96 \times 10^6\ \Omega$$

39. General formula is, $J_n = q\,u_n\,n\varepsilon + q.D_n\frac{dn}{dx}$

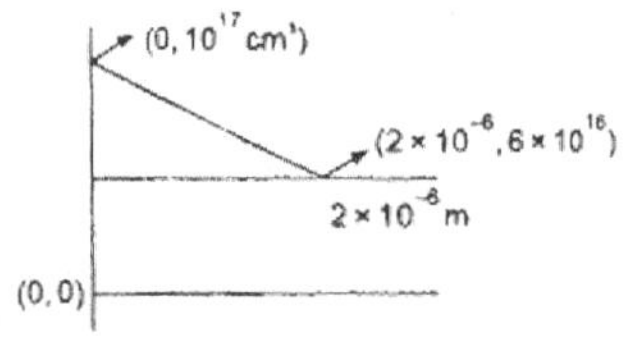

Here, $\varepsilon = 0$

$\Rightarrow$ $J_n = qD_n\frac{dn}{dx}$

where $\frac{dn}{dx} = \frac{-6\times10^{+16}}{2\times10^{-6}} = -3\times10^{22}$

$\Rightarrow$ $J_n = 1.6\times10^{-19}\times35\times10^{-4}\times(-3\times10^{22})$

$$= -1120\ \text{A/cm}^2$$

41. For Si, $I = I_{0,Si}\ (1-e^{0.718/\eta V_T})$

and for Ge, $I = I_{0,}\,ge(1-e^{0.1435/\eta V_T})$

Hence for equal currents, $\eta_{Si} = 2$; $\eta_{Ge} = 1$

$$\frac{I_{0,Si}}{I_{0,Ge}} = \frac{1-e^{-0.1435/V_T}}{1-e^{-0.359/V_T}} \approx e^{-0.2155/V_T}$$

or $\frac{I_{0,Ge}}{I_{0,Si}} \approx e^{(0.2155/26\times10^{-3})} = 4\times10^3$

42. $E = \frac{hc}{\lambda}$

$$= \frac{6.626\times10^{-34}\times3\times10^8}{5490\times10^{-10}} = 3.62\times10^{-19}\ \text{Joules}$$

In eV, the energy Band gap, $E_g = \frac{E}{e} = \frac{3.62\times10^{-19}}{1.6\times10^{-19}}$

$$= 2.26\ \text{eV}$$

43. $I_D = I_{DSS}\left(1-\frac{V_{GS}}{V_P}\right)^2$

or $1\text{mA} = I_{DSS}\left(1-\frac{900}{400}\right)^2 = \frac{25}{16}\ I_{DSS}$

or $I_{DSS} = \frac{16}{25}\ \text{mA}$

then for $V_{GS} = 1400$ mV $I_D = I_{DSS}\left(1-\frac{V_{GS}}{V_P}\right)^2$

$$= \frac{16}{25}\left(1-\frac{1400}{400}\right)^2 = 4\ \text{mA}$$

45. In current-shunt, negative feedback amplifier, the input resistance,

$$R'_i = \frac{R_i}{1+\beta A}$$

$$= \frac{1k\Omega}{1+50\times 0.2} = \left(\frac{1}{11}\right) k\Omega.$$

46.

$$V_{CC} - I_C R_2 = V_{CE}$$

or $$R_2 = \frac{3V}{1.5mA} = 2\ k\Omega.$$

$$I_B = \frac{I_c}{\beta} = 0.01\ mA$$

With change of β equal to 200,

$$I_C' = 0.01 \times 200 = 2\ mA$$

and $$V_{CE} = V_{CC} - I_C'\ R_2$$

$$= 6 - 2 \times 2$$

$$= 2\ \text{volts}.$$

Thus operating piont is (2V, 2mA)

47. The feedback network is shown below.

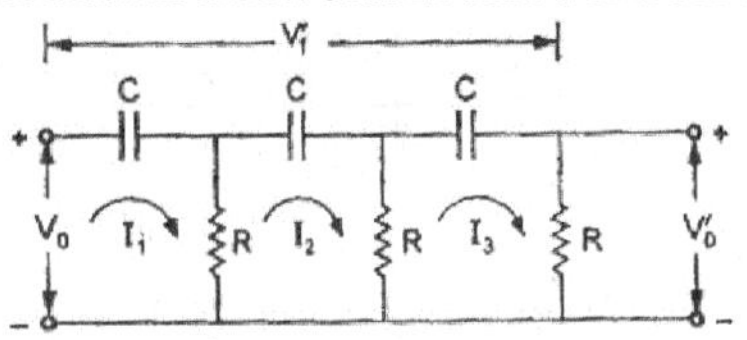

Let $X = \dfrac{1}{\omega C}$

then, using KVL, we have following set of linear equations

$$\begin{bmatrix} R - jX & -R & 0 \\ -R & 2R - jX & -R \\ 0 & -R & 2R - jX \end{bmatrix} \begin{bmatrix} I_1 \\ I_2 \\ I_3 \end{bmatrix} = \begin{bmatrix} V_0 \\ 0 \\ 0 \end{bmatrix}$$

Putting $\alpha = \dfrac{X}{R} = \dfrac{1}{\omega RC}$, we get

or $$R^3 \begin{bmatrix} 1 - i\alpha & -1 & 0 \\ -1 & 2 - i\alpha & -1 \\ 0 & -1 & 2 - i\alpha \end{bmatrix} \begin{bmatrix} I_1 \\ I_2 \\ I_3 \end{bmatrix} = \begin{bmatrix} V_0 \\ 0 \\ 0 \end{bmatrix}$$

Now, $$I_3 = \frac{\Delta_3}{\Delta}$$

where $$\Delta_3 = R^3 \begin{vmatrix} 1 - i\alpha & -1 & V_0 \\ -1 & 2 - i\alpha & 0 \\ 0 & -1 & 0 \end{vmatrix}$$

and, $$\Delta = R^3 \begin{vmatrix} 1 - i\alpha & -1 & 0 \\ -1 & 2 - i\alpha & -1 \\ 0 & -1 & 2 - i\alpha \end{vmatrix}$$

then, $$I_3 = \frac{\Delta_3}{\Delta} = \frac{V_0}{1 - 5\alpha^2 + j(\alpha^3 - 6\alpha)}$$

and, $$\frac{V_0'}{V_0} = \frac{-I_3 R}{V_0}$$

$$= \frac{-1}{(1 - 5\alpha^2) + j(\alpha^3 - 6\alpha)}$$

For 180° phase shift, $\alpha^3 - 6\alpha = 0$

or $$\alpha = \sqrt{6}$$

or $$\frac{1}{\omega RC} = \sqrt{6}$$

or $$f = \frac{1}{2\pi RC\sqrt{6}}$$

48. The voltage at OPAMP inputs, $V_+ = V_- = V_2 = 3$ volts

then, by voltage division rule, $\dfrac{V_0}{20+40} \times 20 = 3$

or $$V_0 = 9\ V$$

50. From the figure, by the voltage division

$$V_+ = V_- = \frac{3}{1+8} \times 8$$

$$= 8/3\ \text{volts}$$

and using KCL at inverting - end,

$$\frac{2 - V_-}{1} = \frac{V_- - V_0}{5}$$

or $$-10 + 6V_- = V_0$$

or $$V_0 = 6 \times 8/3 - 10 = 6\ \text{volt}$$

51. The voltage amplifier can be represented as

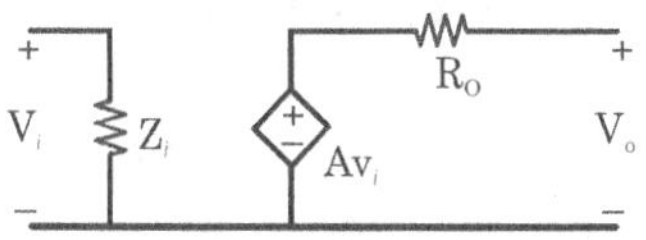

For three such cascaded amplifiers

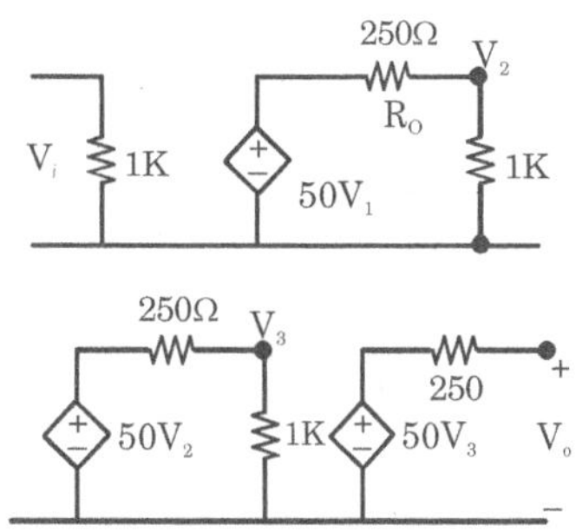

$$V_o = 50\ V_3$$

$$\downarrow$$

$$V_3 = \frac{1000}{1250} \times 50\ V_2 = 40\ V_2$$

$$\downarrow$$

$$V_2 = \frac{1000}{1250} \times 50\ V_1$$

$$= 40\ V_1$$

$$V_0 = 50 \times 40 \times 40\ V_1$$

$$A_v = \frac{V_o}{V_1} = 50 \times 40 \times 40$$

$$= 8 \times 10^4$$

$\therefore$ A_v in dB = 20 log (8×10^4) = 98 dB

52. Ideal sawtooth voltage waveform is shown below

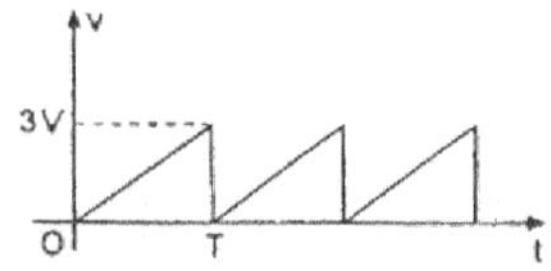

$$T = \frac{1}{f} = \frac{1}{500} = 2\text{ms}$$

and, $$I = C\frac{dV}{dt} = C\frac{3-0}{T} = \frac{3\times 2\times 10^{-6}}{2\times 10^{-3}} = 3\text{mA}$$

Thus charging requires a constant current source of 3 mA for 2ms.

53. Let $P = 1011$ and $Q = 1001$

Given, $Y_n = P_n \oplus Q_n \oplus R_n$

$Z = R_nQ_n + \overline{P}_nR_n + Q_n\overline{P}_n$

Now, constructing the truth - table

	P_n	Q_n	R_n	Z_n	Y_n
n = 1	1	1	0	0	0
n = 2	1	0	0	0	1
n = 3	0	0	0	0	0
n = 4	1	1	0	0	0
					0

From the circuit given, we have

$$R_{n+1} = Z_n, \quad 1 \le n \le 3$$

and $$Z_4 = Y_5\,(\text{MSB})$$

Hence, output is 00010 which show that it is a 4-bit subtractor giving $P - Q$.

54. $W = R + \overline{P}Q + \overline{R}S$

⇒

PQ \ RS	00	01	11	10
00		1	1	1
01	1	1	1	1
11		1	1	1
10		1	1	1

(i)

$X = PQ\overline{R}\overline{S} + \overline{P}\overline{Q}\overline{R}\overline{S} + P\overline{Q}\overline{R}\overline{S}$

⇒

PQ \ RS	00	01	11	10
00	1			
01				
11	1			
10	1			

(ii)

$$\begin{aligned} Y &= RS + \overline{PR + P\overline{Q} + \overline{P}\,\overline{Q}} \\ &= RS + \overline{PR}\,.\,\overline{P\overline{Q}}\,.\,\overline{\overline{P}\,\overline{Q}} \\ &= RS + (\overline{P}+\overline{R})(\overline{P}+Q)(P+Q) \\ &= RS + (\overline{P}+\overline{R}Q)(P+Q) \\ &= RS + \overline{P}Q + \overline{R}PQ + Q\overline{R} \end{aligned}$$

⇒

PQ \ RS	00	01	11	10
00			1	
01		1	1	1
11	1	1	1	
10			1	

(iii)

$$\begin{aligned} Z &= R+S+\overline{PQ+\overline{P}.\overline{Q}\,\overline{R}+P\overline{Q}\,\overline{S}} \\ &= R+S+\overline{PQ}\,.\,\overline{\overline{P}\,\overline{Q}\,\overline{R}}\,.\,\overline{P\,\overline{Q}\,\overline{S}} \\ &= R+S+(P+\overline{Q})(P+Q+R)(\overline{P}+Q+S) \\ &= R+S+(\overline{P}+\overline{Q}(Q+S))(P+Q+R) \\ &= R+S+(\overline{P}+\overline{Q}S)(P+Q+R) \\ &= R+S+\overline{P}Q+\overline{P}R+\overline{Q}PS+\overline{Q}RS \end{aligned}$$

⇒

PQ \ RS	00	01	11	10
00		1	1	1
01	1	1	1	1
11		1	1	1
10		1	1	1

(iv)

From Fig *(i)* and Fig *(iv)*, we see that

$$W = Z$$

From Fig *(ii)* and Fig *(iv)*, we see that

$$X = \overline{Z}$$

55. Propagation delay in 4-bit ripple counter $= 4\,T_d$ $= 40$ ns

and, since all the flip-flops in the synchronous counter are clocked simultaneously, then its worst delay will be equal to 10 ns.

57.

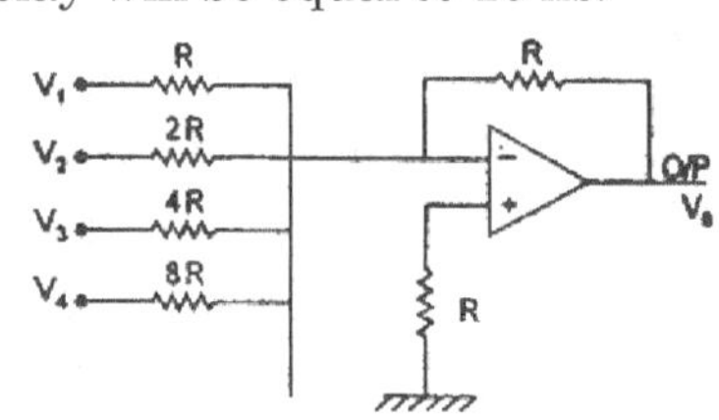

$$V_0 = -\left[\frac{R}{R}V_1 + \frac{R}{2R}V_2 + \frac{R}{4R}V_3 + \frac{R}{8R}V_4\right]$$

Exact value when $V_1 = V_2 = V_3 = V_4 = 5$ V

$$\Rightarrow \quad V_0 = -\left[5 + \frac{5}{2} + \frac{5}{4} + \frac{5}{8}\right]$$

$$= -5\,[1 + 0.5 + 0.25 + 0.125]$$

$$= -9.375$$

Maximum V_0 due to tolerance effect

$$= -\left[\frac{110\times 5.5}{90} + \frac{110\times 5.5}{2\times 90} + \frac{110\times 5.5}{4\times 90} + \frac{110\times 5.5}{8\times 90}\right]$$

$$= -\frac{110\times 5.5}{90}[1 + 0.5 + 0.25 + 0.125]$$

$$= 12.604$$

$$D = 34.44\% \approx 35\%$$

58. All BCD combination can be convert into binary code except 1000 & 1001.

$\therefore$ The best possible answer is (*a*).

59. At the first rising edge of clock after $t = t_2$, output of shift register is 0110 = 6 in decimal. At the address 6 data 1010 is stored which is applied to the input of shift register and at the next rising edge, the output of register is 1010 = 10. At the address 10 of ROM, it contains 1000.

60. CMPB subtracted the contents of register B from the accumulator A. Since accumulator has content less than that of register B, there CY = 1 as it shows negative result. Also, the result is being non-zero the zero flag will be reset.

61.

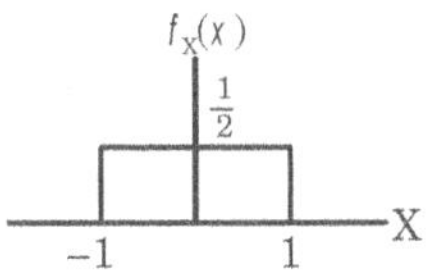

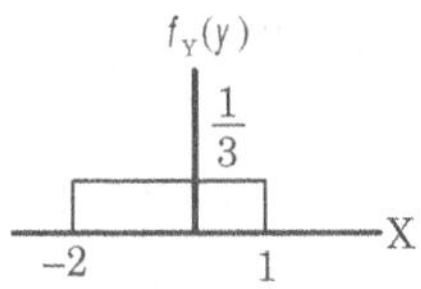

using central limit theorem

$$Z = X + Y$$

$$f_Z(z) = f_X(x) \otimes f_Y(y)$$

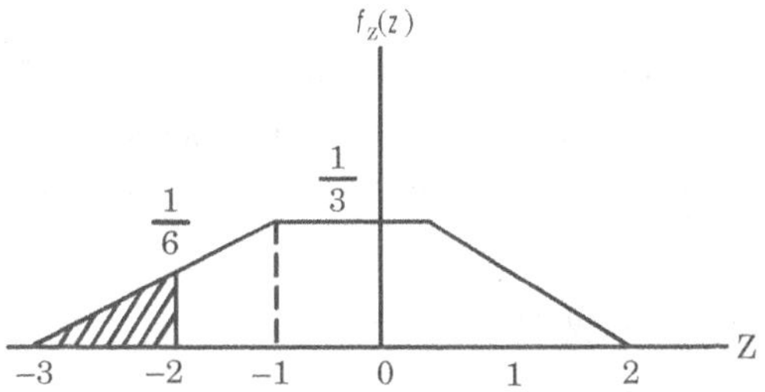

$$P(Z \le -2) = \frac{1}{6} \times 1 \times \frac{1}{2} = \frac{1}{12}$$

62. The equation is homogeneous

$\Rightarrow$ it is linear

$$y(n - n_0) \neq x(n - n_0)$$

$\Rightarrow$ not time invarient

$$y(n) = x(n + 1),\ n \le -1$$

not causal.

The system shows that not for bonded O/P but for bounded I/P

$\Rightarrow$ it is stable.

$\Rightarrow$ The system is linear and stable but not time invariant and causal.

63.

$$H(f) = \frac{1}{j2\pi R f C + 1}, \quad H(0) = 1$$

Now, as $\dfrac{|H(f_1)|}{H(0)} \ge 0.95$

then, $\dfrac{1}{\sqrt{4\pi^2 R^2 f_1^2 C^2 + 1}} \ge 0.95$

or $\quad 4\pi^2 R^2 f_1^2 C^2 \le 0.108$

or $\quad (f_1)_{max} = \dfrac{\sqrt{0.108}}{2\pi RC} = \dfrac{\sqrt{0.108}}{2 \times 3.14 \times 10^{-3}} = 52.2$ Hz.

64. Group Delay $t_g(f) = -\dfrac{1}{2\pi}\dfrac{d\phi(f)}{df}$

here, $\quad \phi(f) = -\tan^{-1}(2\pi RCf)$

then, $\quad t_g(f) = \dfrac{RC}{1 + 4\pi^2 R^2 C^2 f^2}$

$$tg(t)\big|_{f=f_2} = \frac{1 \times 10^3 \times 10^{-6}}{1 + 4\pi^2 (10^3)^2 \times (10^{-6})^2 \times (100)^2}$$

$$= 0.717 \text{ ms.}$$

65. Variance of $x(t)$, $\sigma x^2 = \bar{X}^2 - \mu^2$

μ is the mean value of $x(t)$ and $\mu = 2$

$$\overline{X^2} = R_x(0) = 8$$

then $\quad \sigma x^2 = 8 - 4 = 4$

Commulative distribution function,

Now, $\quad F(X) = 1 - F(-X)$ and $F(-X) = Q(X)$

Now, $P(x \le 1) = F_X(x)$

F(X) is expressed as, with mean μ and variance σx^2

$$F_X(x) = F\left(\frac{x - \mu}{\sigma_x}\right) = F\left(\frac{x - 2}{4}\right)$$

and $\quad P(X \le 1) = F_X(1) = F\left(\dfrac{1-2}{4}\right)$

$$= F\left(\frac{-1}{2}\right) = Q\left(\frac{1}{2}\right)$$

66.

$$Y = X[2],\ Z = X[4]$$

$$E[W] = E[Y - Z] = E[Y] - E[Z]$$

$$= E[X(2)] - E[X(4)] = 0$$

[As $x(t)$ is a random process having count at value of 2]

Now, $\quad E[W^2] = E[Y^2] + E[Z^2] - 2E[YZ] \quad \text{...(A)}$

$\quad E[X^2(t)] = R_X(0) = 4[e^{-0.2101} + 1] - 8$

And, $\quad E[Y^2] = E[X^2(2)] = 8$

$\quad E(Z^2) = E^2[X^2(4)] = 8$

$\quad E[XZ] = E[X(Z)\,X(4)] = R_X(4 - 2) = R_X(2)$

[As $E[X(t)\,X(t + J)] = R_X(J)$]

$\quad R_X(2) = [e^{-0.2|(4-2)|} + 1] = 6.68$

then from equation (A),

$$E[W^2] = 8 + 8 - (2 \times 6.68) = 2.64$$

67. Using Fourier series expansion,

$$C_n = \frac{1}{T_0}\int_{-T_0/6}^{T_0/6} A.e^{-Jnw_0 t}dt = \frac{A}{\pi n}\sin\left(\frac{n\pi}{3}\right)$$

all harmonics is present in C_n except integer multiple of 3.

Thus, frequencies of $p(t)$ for harmonics 1, 2, 4, 5, 7, 8, ... are 10^3, 2×10^3, 4×10^3, 5×10^3 .....

$\therefore$ $p(t) \times x(t)$ gives (1 ± 0.7)kHz, (2 ± 0.7)kHz, (4 ± 0.7)kHz ......

Thus, frequency present in range of 2.5kHz to 3.5kHz are, 2.7kHz, 3.3kHz.

68. Forword path gain, $P_1 = 1$

Since P_1 not touches two loops having gain of

$-\frac{3}{5}$ and $-\frac{24}{5}$

$$\Delta_1 = 1 - P_{11} - P_{12} = 1 + \frac{3}{5} + \frac{24}{5}$$

Indivdiual loop gain,

$$P_{11} = \frac{3}{5}, P_{12} = \frac{24}{5}, P_{13} = \frac{-2}{5}$$

Gain of pair of non-touching loops,

$$P_{21} = \frac{-2}{5} \times \frac{-3}{5} = \frac{6}{52}$$

then, transfer function,

$$T(s) = \frac{P_1\Delta_1}{1 - P_{11} - P_{12} - P_{13} + P_{21}}$$

or, $$T(s) = \frac{1\left(1 + \frac{3}{5} + \frac{24}{5}\right)}{1 + \frac{3}{5} + \frac{24}{5} + \frac{2}{5} + \frac{6}{5^2}}$$

$$= \frac{5(5+27)}{5^2 + 285 + 6}$$

69. The characteristic equations is $1 + G(s)H(s) = 0$

or $$1 + \frac{k}{s(s+2)(s+3)} = 0$$

or $$k = -(s^3 + 5s^2 + 6s)$$

For Break-away points,

$$\frac{dk}{ds} = 0 = -(3s^2 + 10s + 6)$$

or $$3s^2 + 10s + 6 = 0$$

or $$s = -0.784, -2.54$$

The root-locus is drawn below

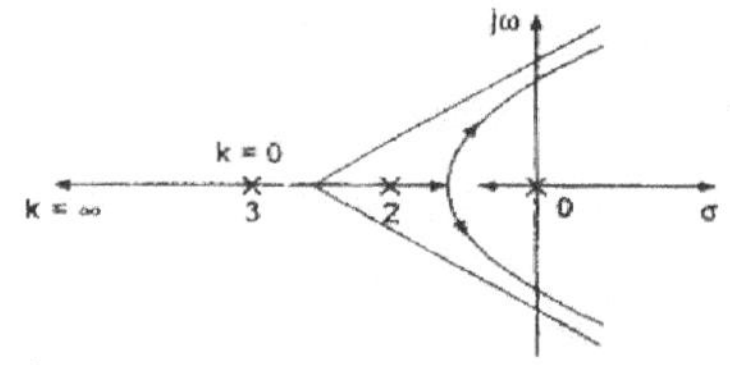

Since the point $s = -2.54$, doesn't lie on the root-locus, it can never be breakway point.

Hence $s = -0.784$.

70. From $\omega = 0.1$ to 10, i.e. in two decade dB change is + 120 dB which infers that there are three zeroes at $\omega = 0.1$.

Now from $\omega = 10$ to $\omega = 100$, i.e. in one decade dB change is – 20 dB, thus having a pole at $\omega = 10$.

Likewise at $\omega = 100$, slope changes to 0 dB/decade which infers a pole at $\omega = 100$.

Thus, $$T(s) = \frac{k(s+0.1)^3}{(s+10)^2(s+100)}$$

Given, $20\log[T(s)]|_{\omega=0.1} = 20 = 20\log(k \times 10^{-7})$

or $$k = 10^8$$

71.
$$2\xi\omega_n = 4$$
$$\xi\omega_n = 2$$
and $$\omega_n = 2$$
then, $$\xi = 1 \text{ (critically - damped system)}$$
and $$t_{ss}\text{ (Setting time)} = \frac{4}{\xi\omega_n} = 2 \text{ sec.}$$

Thus, only response is as shown in the figure. The figure is also critically-damped system, but its setting time is about 5 sec. Hence, the only choice is *(b)*.

72. $$G(j\omega)\ H(j\omega) = \frac{j\omega}{(100 + j\omega)^3}$$

$$= \frac{\left[3\times(100)^2\omega^2 - \omega^4\right] + j\left[(100)^3\omega - 3\times100\omega^3\right]}{\left[100^3 - 3\times(100)\omega^2\right]^2 + \left[3\times(100)^2\omega - \omega^3\right]^2}$$

imaginary term to become zero

$$\omega = \frac{100}{\sqrt{3}}$$

$$\text{Real axis cut} = a = +\frac{3}{80000}$$

$\because$ a is real axis cut at positive side

Hence, gain margin become ∞.

73. $$x(t) = e^{A(t-t_0)}X(t_0) + \int_{t_0}^{t} e^{A(t-s)}B\,U(\tau)ds$$

$$= e^{A(t-t_0)}X(t_0) + 0 \qquad (\because B = 0)$$

$$(sI - A) = \begin{bmatrix} s & 0 \\ 0 & s \end{bmatrix} - \begin{bmatrix} 1 & 0 \\ 1 & 1 \end{bmatrix}$$

where, $A = \begin{bmatrix} 1 & 0 \\ 1 & 1 \end{bmatrix}$ and $X(t_0) = \begin{bmatrix} 1 \\ 0 \end{bmatrix}$

$$\therefore (sI - A) = \begin{bmatrix} s-1 & 0 \\ -1 & s-1 \end{bmatrix}$$

$$(sI - A)^{-1} = \frac{\begin{bmatrix} (s-1) & 0 \\ 1 & (s-1) \end{bmatrix}}{(s-1)^2} = \begin{bmatrix} \frac{1}{s-1} & 0 \\ \frac{1}{(s-1)^2} & \frac{1}{(s-1)} \end{bmatrix}$$

$$L^{-1}[(sI - A)^{-1}] = e^{At} = \begin{bmatrix} e^t & 0 \\ te^t & e^t \end{bmatrix}$$

$$\Rightarrow \quad x(t) = e^{At} \times x[t_0] = \begin{bmatrix} e^t & 0 \\ te^t & e^t \end{bmatrix} \begin{bmatrix} 1 \\ 0 \end{bmatrix} = \begin{bmatrix} e^t \\ te^t \end{bmatrix}$$

74.

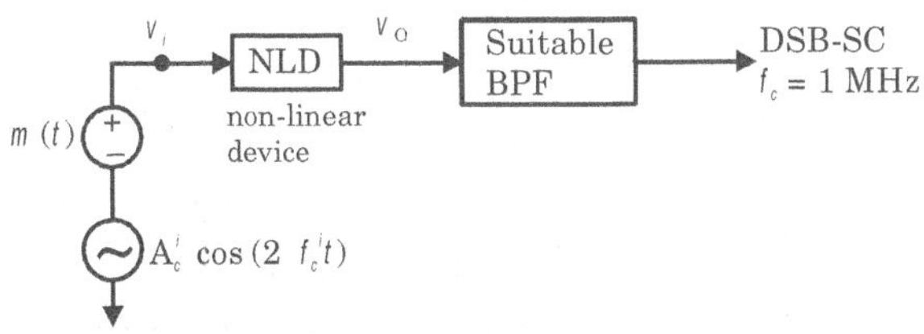

$$v_o = a_0 v_i + a_1 v_i^3$$

$$= a_0 [A_c^i \cos\omega_c^i t] + a_0\, m(t) + a_1 (A_c^i)^3 \cos^3 \omega_c^i t$$

$$+ a_1 m^3(t) + 3a_i\, m^2(t)\, A_c^i \cos\omega_c^i t$$

$$+ 3a_1 A_c^i \cos^2 \omega_c^i t \,.\, m(t)$$

$$v_o \rightarrow \boxed{BPF} \rightarrow \text{DSB-SC}$$

$$f_c = 1\text{ MHz}$$

The useful term for DSB - SC is

$3a_1\, A_c^i \cos^2 \omega_c^i t \,.\, m(t)$

$$\Rightarrow\ 3a_1\, A_c^i \left[\frac{1}{2} m(t) + \frac{\cos 2\omega_c^i t . m(t)}{2}\right]$$

For DSB - SC

$$\Rightarrow\ \frac{3a_1}{2} A_c^i \cos 2\omega_c^i t \,.\, m(t)$$

But f_c for DSB - SC is given which is 1 MHz.

Here compare $2 f_c^i$ with f_c = 1 MHz

$$f_c^i = 0.5\text{ MHz}$$

75.

$$P_t = P_c\left(1 + \frac{m^2}{2}\right)$$

$$\Rightarrow\ \frac{P_t}{P_c} = 1 + \frac{m^2}{2} = 1 + \frac{0.25}{2} = 1 + 0.125 = 1.125$$

$$\Rightarrow\ \frac{P_t - P_c}{P_c} = \frac{1.125 - 1}{1} = \frac{0.125}{1} = \frac{1}{8}$$

76. The FM signal is represented in terms of Bessel function as,

$$x_{FM}(t) = A \sum_{n=-\infty}^{\infty} J_n(\beta) . \cos(\omega_c \pm n\omega_m)t$$

where A is the amplitude of carrier signal.

then, $\Delta\omega = 6\,\omega_n$ and $\beta = \frac{\Delta\omega}{\omega_m} = 6$

Given $\omega_c + n\omega_m = (1008 \times 10^3).\ 2\pi$

or $2\pi \times 10^6 + n.4\pi \times 10^3 = (1008 \times 10^3).\ 2\pi$

or $n = 4$

thus, Bessel coefficient will be $5\, J_4(6)$.

78.

$$I = 700 + 450 \times 2$$

$$= 1600\text{ kHz}$$

$$f_{max} = 1650 + 450 = 2100\text{ kHz}$$

$$f_{min} = 550 + 450 = 1000\text{ kHz}$$

$$f = \frac{1}{2\pi}\ \frac{1}{\sqrt{LC}}$$

$$\Rightarrow\ \frac{2100}{1000} = \sqrt{\frac{LC_{max}}{LC_{min}}}$$

$$\Rightarrow\ \frac{LC_{max}}{LC_{min}} = (2.1)^2$$

$$\Rightarrow\ \frac{C_{max}}{C_{min}} = 4.41$$

79. Step-size, $\Delta = \frac{2m_p}{L} = 0.012$

then, quantization – noise power $\langle q_e^2 \rangle = \frac{\Delta^2}{12}$

$$= \frac{(0.012)^2}{12} = 12 \times 10^{-6}\,V^2$$

80. SNR for two sided PSD = $\frac{2E_b}{N_0}$

SNR for one sided PSD = $\frac{E_b}{N_0}$

$$\therefore\ 10 \log_{10} \frac{10^{-5}}{10^{-6}} = 10\text{ dB}$$

81. For slope overload to take place

$| \dot{x}(t) |_{max} > f_s \Delta ...(i)$

Now $x(t) = E_m \sin 2\pi f_m t$

$\Rightarrow | \dot{x}(t) |_{max} = E_m f_m \times 2\pi$ ...(ii)

$$= 6.28\, E_m f_m$$

$$f_s = 40\text{ kHz}$$

$$\Delta = 0.628$$

$$\Rightarrow\ \Delta f_s = 25.12\text{ kHz}$$

(a) → 6.28 × 0.3 × 8 = 15.07 < 25.12

(b) → 6.28 × 1.5 × 4 = 37.68 > 25.12

(c) → 6.28 × 1.5 × 2 = 18.44 < 25.12

(d) → 6.28 × 3 × 1 = 18.84 < 25.12

82. For coherent BPSK synchronization required for detection and efficiency is $\frac{1}{2}$.

83. Duration of each bit, $T_b = \frac{1}{f_s.8}$

then bit rate, $R = \frac{1}{T_b} = 8f_s = 64$ kbps

In dB, $(SNR)_q = 1.76 + 6.02\,n = 1.76 + (6.02) \times (8)$
$= 49.8$ dB

84.

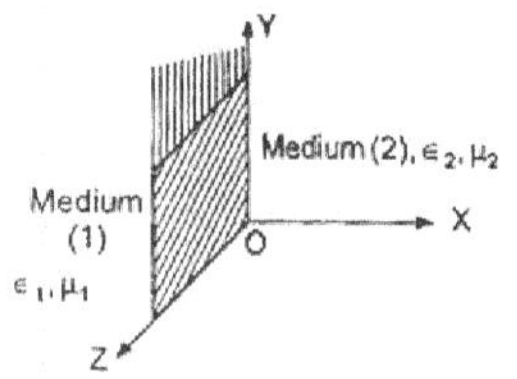

Since tangential components of electric field is same, therefore

$$\bar{E}_{x_1} = \bar{E}_{y_2} \quad \text{and} \quad \bar{E}_{z_1} = \bar{E}_{z_2}$$

and, as $\bar{D}_{x_1} = \bar{D}_{x_2}$

or $\bar{E}_{x_1} \in_1 = \bar{E}_{x_2} \in_2$

or $\bar{E}_{x_2} = \frac{(1.5)^2}{2.5}\bar{u}_x = 1.2\,\bar{u}_x$

thus $\bar{E}_2 = \bar{E}_{x_2} + \bar{E}_{y_2} + \bar{E}_{z_2} = \bar{E}_{x_2} + \bar{E}_{y_1} + \bar{E}_{z_1}$
$= (1.2\bar{u}_x - 3\bar{u}_y + 1\bar{u}_z)$ volt/m.

85.
$$V = -\frac{x^2}{2}\bar{u}_x - \frac{y^2}{2}\bar{u}_y - \frac{z^2}{2}\bar{u}_z$$

then, $V_x - V_y = -\frac{1}{2}\,[2^2 - 1^2 + 0^2 - 2^2 + 0^2 - 3^2]$
$= +5$ volts.

86. Reflection coefficient,

$$\Gamma = \frac{\eta_2 - \eta_1}{\eta_2 + \eta_1} = \frac{\sqrt{\frac{\mu_0}{\varepsilon_o t_r}} - \sqrt{\frac{\mu_0}{\varepsilon_o}}}{\sqrt{\frac{\mu_0}{\varepsilon_o t_r}} + \sqrt{\frac{\mu_0}{\varepsilon_o}}} = -\frac{1}{3} = 0.333\angle 180°$$

87. $E(z, t) = 10\cos(2\pi \times 10^7\,t - 0.1\,\pi z)$

Now $2\pi \times 10^7\,t - 0.1\,\pi\,z = \text{constant}$

Taking differentiation, we have

$$v = \frac{dz}{dt} = \frac{2\pi \times 10^7}{0.1\pi} = 2 \times 10^8 \text{ m/s}$$

88. $\beta = \frac{2\pi}{\lambda}$

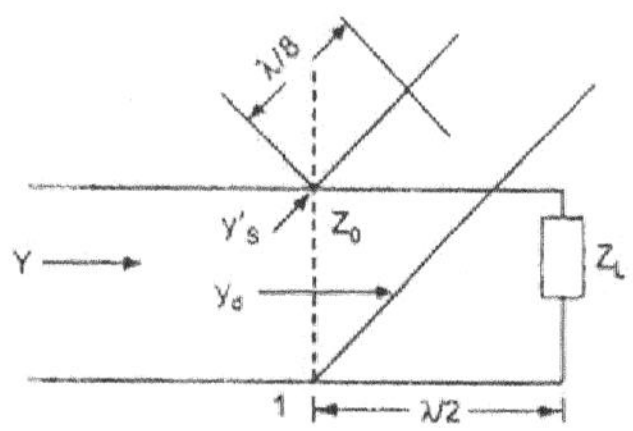

For y_d, $\beta d = \left(\frac{2\pi}{\lambda}\right)\frac{\lambda}{2} = \pi$

$$y_d = Z_0\left(\frac{Z_0 + jZ_L \tan\beta d}{Z_L + iZ_0 \tan\beta d}\right)$$

$$= \frac{50 + j100\tan\pi}{50(100 + j50\tan\pi)} = 0.01$$

For y_s, $\beta d = \frac{2\pi}{\lambda}\frac{\lambda}{8} = \frac{\pi}{4}$, $Z_L \to \infty$

then $y_s = \frac{Z_0 + jZ_L \tan\pi/4}{Z_0(Z_L + jZ_0\tan\pi/4)} = \frac{j}{Z_0} = 0.021$

Thus, $Y = y_d - y_s = (0.01 - 0.02\,j)$ mho

89. For TE_{10} mode

$\lambda_c = \sqrt{\varepsilon_r} \times 2 \times a = \sqrt{4} \times 2 \times 3$

$\lambda_c = 12$ cm.

$f_c = \frac{3 \times 10^{10}}{12} = 2.5$ GHz.

90. Normalized array factor $= 2\left|\cos\frac{\Psi}{2}\right|$

where $\Psi = \beta d\sin\theta\cos\phi + \delta$

Here $\theta = 90°$, $d = \sqrt{2}s$, $\phi = 45°$, $r = 180°$.

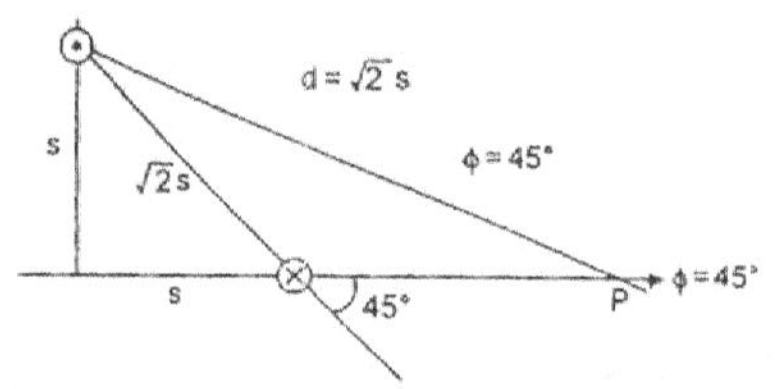

Then, $2\left|\cos\frac{\Psi}{2}\right| = 2\cos\left(\frac{\beta d\sin\theta\cos\phi + \delta}{2}\right)$

$$= 2\cos\left[\frac{2\pi}{\lambda.2}\sqrt{2s}\cos 45° + \left(\frac{180°}{2}\right)\right]$$

$$= 2\cos\left(\frac{\pi s}{\lambda} + 90°\right) = 2\sin\left(\frac{\pi s}{\lambda}\right)$$

SOLVED PAPER – 2002

ELECTRONICS & COMMUNICATION ENGINEERING

Time Allowed : 3 Hours | **Maximum Marks : 150**

SECTION - A (75 Marks)

1. This question consists of TWENTY FIVE sub-questions (1.1 - 1.25) of ONE mark each. For each of these sub-questions, four possible alternatives (*a, b, c* and *d*) are given, out of which only one is correct. Indicate the correct answer by darkening the appropriate bubble against the question number on the left hand side of the Objective Response Sheet (ORS). You may use the answer book provided for any rough work, if needed.

(1 × 25 = 25)

1.1 The dependent current source shown in given figure.

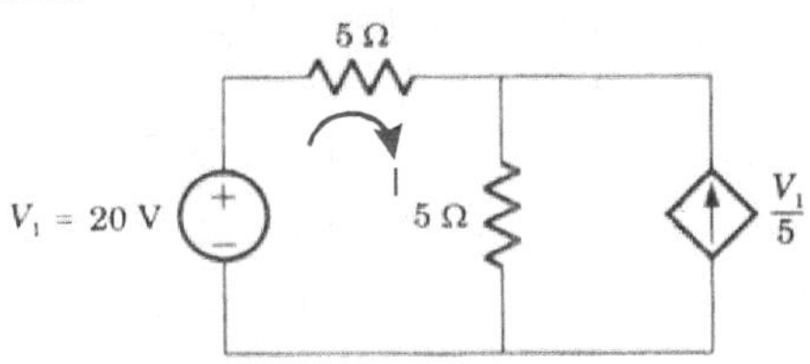

(*a*) delivers 80 W (*b*) absorbs 80 W
(*c*) delivers 40 W (*d*) absorbs 40 W

1.2 In given figure, the switch was closed for a long time before opening at $t = 0$. The voltage V_x at $t = 0^+$ is

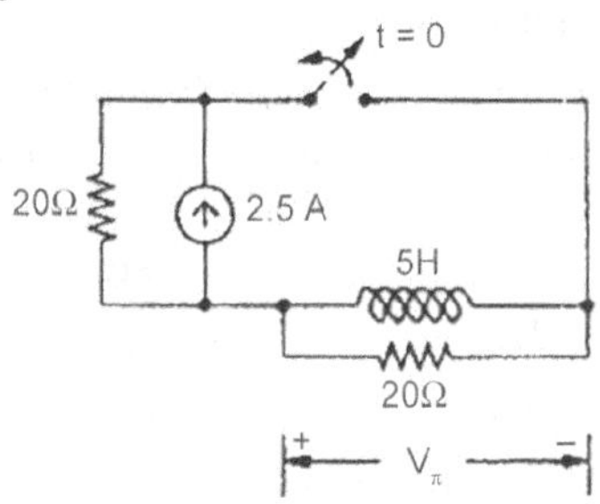

(*a*) 25 V (*b*) 50 V
(*c*) – 50 V (*d*) 0 V

1.3 Convolution of $x(t + 5)$ with impulse function $\delta(t - 7)$ is equal to

(*a*) $x(t-12)$ (*b*) $x(t+12)$
(*c*) $x(t-2)$ (*d*) $x(t+2)$

1.4 Which of the following cannot be the Fourier series expansion of a periodic signal?

(*a*) $x(t) = 2\cos t + 3\cos 3t$
(*b*) $x(t) = 2\cos \pi t + 7\cos t$
(*c*) $x(t) = \cos t + 0.5$
(*d*) $x(t) = 2\cos 1.5\pi t + \sin 3.5\pi t$

1.5 In given figure a silicon diode is carrying a constant current of 1 mA. When the temperature of the diode is 20° C, V_D is found to be 700 mV. If the temperature rises to 40° C, V_D becomes approximately equal to

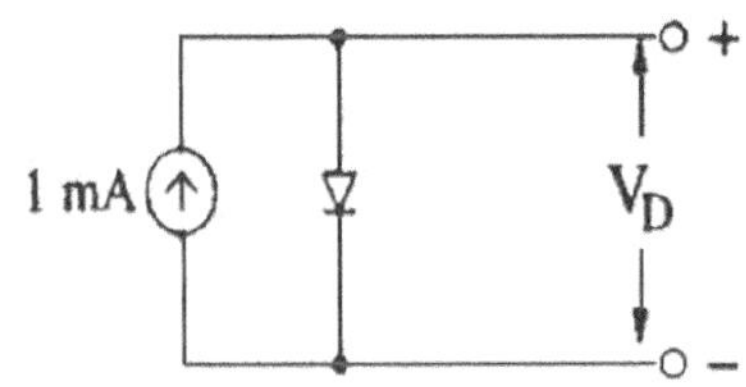

(*a*) 740 mV (*b*) 660 mV
(*c*) 680 mV (*d*) 700 mV

1.6 In a negative feedback amplifier using voltage-series (i.e. voltage-sampling, series mixing) feedback.

(*a*) R_i decreases and R_o decreases
(*b*) R_i decreases and R_o increases
(*c*) R_i increases and R_o decreases
(*d*) R_i increases and R_o increases

(R_i and R_o denote the input and output resistances respectively)

1.7 A 741-type op-amp has a gain-bandwidth product of 1 MHz. A non-inverting amplifier using this op-amp and having a voltage gain of 20 dB will exhibit a – 3-dB bandwidth of

(*a*) 50 kHz (*b*) 100 kHz
(*c*) $\frac{1000}{17}$ kHz (*d*) $\frac{1000}{7.07}$ kHz

1.8 Three identical RC-coupled transistor amplifiers are cascaded. If each of the amplifiers has a frequency response as shown in the figure is the overall frequency response is as given in

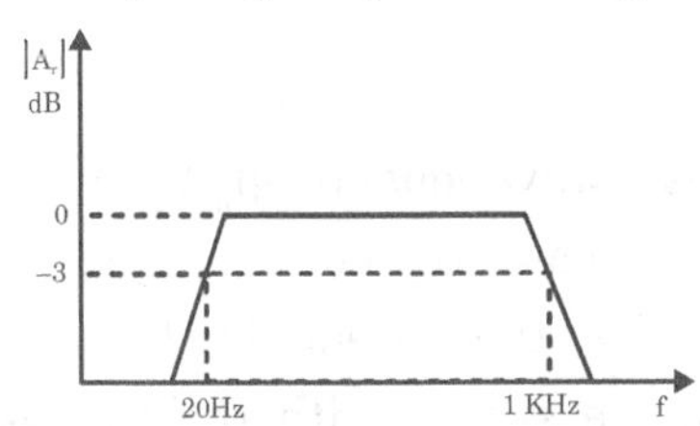

(*a*)

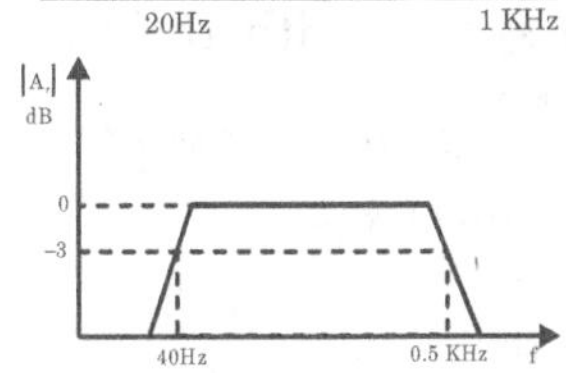

(*b*)

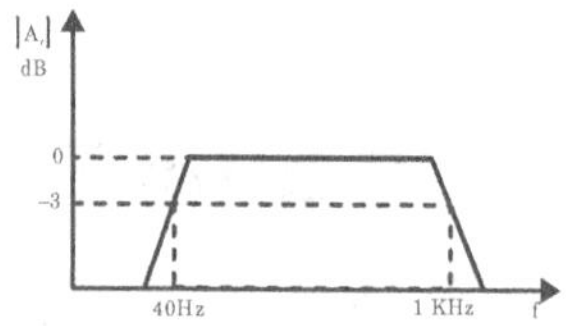

(*c*)

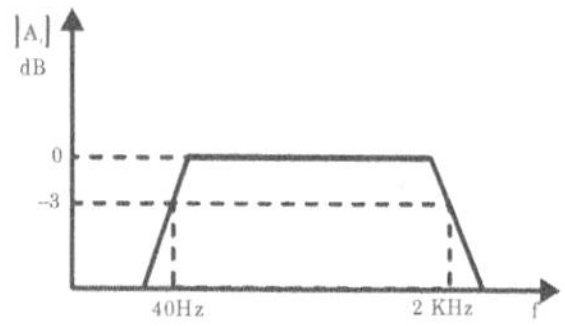

(*d*)

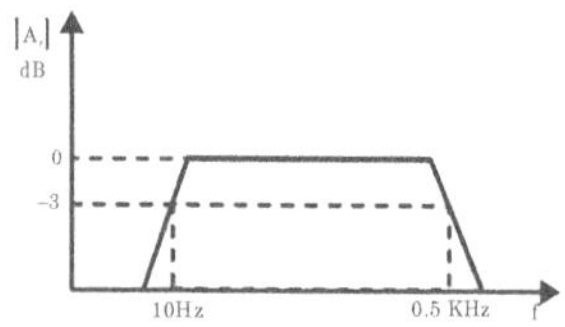

1.9 4-bit 2's complement representation of a decimal number is 1000. The number is

(*a*) + 8 (*b*) 0

(*c*) –7 (*d*) –8

1.10 If the input to the digital circuit the figure, consisting of a cascade of 20 XOR-gates is X, then the output Y is equal to

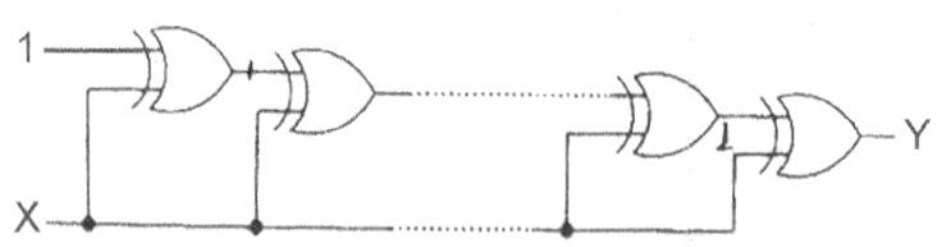

(*a*) 0 (*b*) 1

(*c*) $\overline{X}$ (*d*) X

1.11 The number of comparators required in a 3-bit comparator type ADC is

(*a*) 2 (*b*) 3

(*c*) 7 (*d*) 8

1.12 If the transistor in the figure is in saturation, then

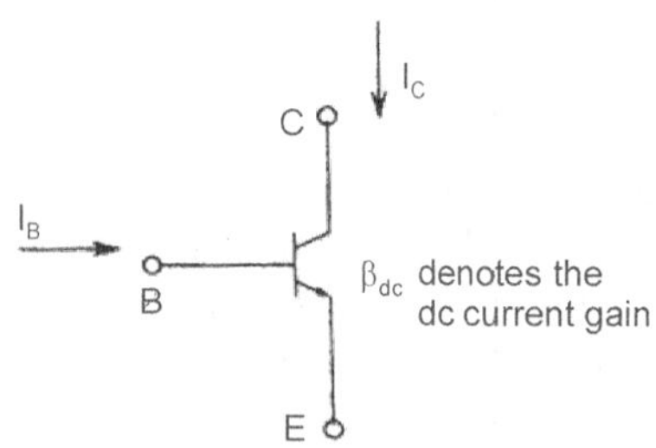

(*a*) I_C is always equal to $\beta_{dc}I_B$

(*b*) I_C is always equal to $-\beta_{dc}I_B$

(*c*) I_C is greater than or equal to $\beta_{dc}I_B$

(*d*) I_C is less than or equal to $\beta_{dc}I_B$

1.13 Consider a system with the transfer function $G(s) = \frac{s+6}{ks^2+s+6}$. Its damping ratio will be 0.5 when the value of k is

(*a*) 2/6 (*b*) 3

(*c*) 1/6 (*d*) 6

1.14 Which of the following points is NOT on the root locus of a system with the open-loop transfer function $G(s)H(s) = \frac{k}{s(s+1)(s+3)}$

(*a*) $s = -j\sqrt{3}$ (*b*) $s = -1.5$

(*c*) $s = -3$ (*d*) $s = -\infty$

1.15 The phase margin of a system with the open-loop transfer function $G(s)H(s) = \frac{(1-s)}{(1+s)(2+s)}$ is

(*a*) 0° (*b*) 63.4°

(*c*) 90° (*d*) ∞

1.16 The transfer function $Y(s)/U(s)$ of a system described by the state equations

$\dot{x}(t) = -2x(t) + 2u(t)$ and $y(t) = 0.5x(t)$ is

(*a*) $0.5/(s-2)$ (*b*) $1/(s-2)$

(*c*) $0.5/(s+2)$ (*d*) $1/(s+2)$

1.17 The Fourier transform $F\{e^{-t}u(t)\}$ is equal to $\frac{1}{1+j2\pi f}$. Therefore, $F\left\{\frac{1}{1+j2\pi t}\right\}$ is

(*a*) $e^{f}u(f)$ (*b*) $e^{-f}u(f)$

(*c*) $e^{f}u(-f)$ (*d*) $e^{-f}u(-f)$

1.18 A linear phase channel with phase delay T_p and group delay T_g must have

(*a*) $T_p = T_g$ = constant

(*b*) $T_p \propto f$ and $T_g \propto f$

(*c*) T_p = constant and $T_g \propto f$

(*d*) $T_p \propto f$ and t_g = constant

(f denotes frequency)

1.19 A 1 MHz sinusoidal carrier is amplitude modulated by a symmetrical square wave of period 100 μ sec. Which of the following frequencies will NOT be present in the modulated signal?

(*a*) 990 kHz (*b*) 1010 kHz

(*c*) 1020 kHz (*d*) 1030 kHz

1.20 Consider a sampled signal

$$y(t) = 5 \times 10^{-6}x(t)\sum_{n=-\infty}^{+\infty}\delta(t - nT_s)$$

where $x(t) = 10\cos(8\pi \times 10^3)t$ and $T_s = 100$ μsec. When $y(t)$ is passed through an ideal low pass filter with a cut-off frequency of 5 kHz, the output of the filter is

(*a*) $5 \times 10^{-6}\cos(8\pi \times 10^3)t$

(*b*) $5 \times 10^{-5}\cos(8\pi \times 10^3)t$

(*c*) $5 \times 10^{-1}\cos(8\pi \times 10^3)t$

(*d*) $10\cos(8\pi \times 10^3)t$

1.21 For a bit-rate of 8 kbps, the best possible values of the transmitted frequencies in a coherent binary FSK system are

(*a*) 16 kHz and 20 kHz (*b*) 20 kHz and 32 kHz

(*c*) 20 kHz and 40 kHz (*d*) 32 kHz and 40 kHz

1.22 The line-of-sight communication requires the transmit and receive antennas to face each other. If the transmit antenna is vertically polarized, for best reception the receive antenna should be

(*a*) horizontally polarized

(*b*) vertically polarized

(*c*) at 45° with respect to horizontal polarization

(*d*) at 45° with respect to vertical polarization

1.23 The VSWR can have any value between

(*a*) 0 and 1 (*b*) –1 and +1

(*c*) 0 and ∞ (*d*) 1 and ∞

1.24 In an impedance Smith chart, a clockwise movement along a constant resistance circle gives rise to

(*a*) a decrease in the value of reactance

(*b*) an increase in the value of reactance

(*c*) no change in the reactance value

(*d*) no change in the impedance value

1.25 The phase velocity for the TE_{10}-mode in an air-filled rectangular waveguide is

(*a*) less than *c* (*b*) equal to *c*

(*c*) greater than *c* (*d*) none of these

(*c* is the velocity of plane waves in free space)

2. This question consists of TWENTY FIVE sub-questions (2.1 – 2.25) of TWO marks each. For each of these sub-questions, four alternatives (*a, b, c* and *d*) are given, out of which only one is correct. Indicate the correct answer by darkening the appropriate bubble against the question number on the left hand side of the Objective Response Sheet (ORS). You may to use the answer book provided for any rough work, if needed. (2 × 25 = 50)

2.1 In the network of the figure is the maximum power is delivered to R_L if its value is

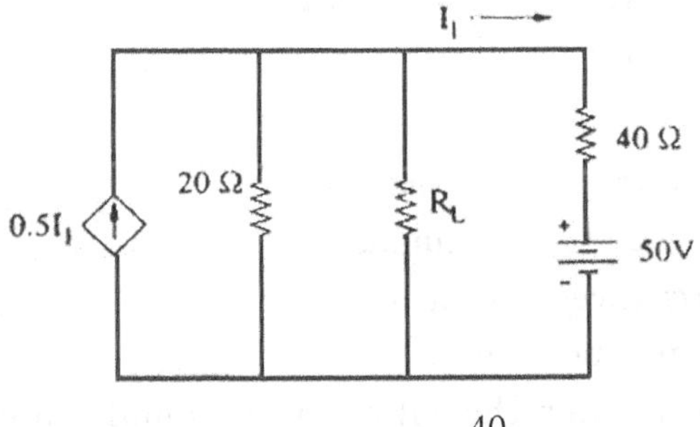

(*a*) 16 Ω (*b*) $\frac{40}{3}$ Ω

(*c*) 60 Ω (*d*) 20 Ω

2.2 If the 3-phase balanced source in the figure is delivers 1500 W at a leading power factor of 0.844, then the value of Z_L (in ohm) is approximately

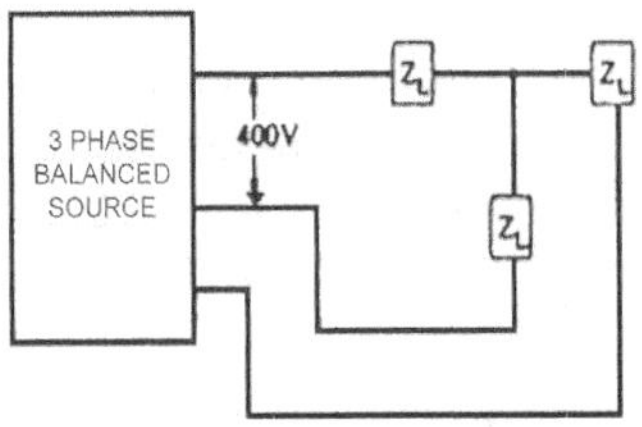

(*a*) 90 ∠32.44° (*b*) 80 ∠32.44°

(*c*) 80 ∠–32.44° (*d*) 90 ∠–32.44°

2.3 The Laplace transform of a continuous-time signal *x*(*t*) is $X(s) = \frac{5-s}{s^2-s-2}$. If the Fourier transform of this signal exists, then *x*(*t*) is

(*a*) $e^{2t}u(t) - 2e^{-t}u(t)$ (*b*) $-e^{2t}u(-t) + 2e^{-t}u(t)$

(*c*) $-e^{2t}u(-t) - 2e^{-t}u(t)$ (*d*) $e^{2t}u(-t) - 2e^{-t}u(t)$

2.4 If the impulse response of a discrete-time system is $h[n] = -5^n u[-n-1]$, then the system function $H(z)$ is equal to

(*a*) $\frac{-z}{z-5}$ and the system is stable

(*b*) $\frac{z}{z-5}$ and the system is stable

(*c*) $\frac{-z}{z-5}$ and the system is unstable

(*d*) $\frac{z}{z-5}$ and the system is unstable

2.5 An amplifier using an opamp with a slew-rate SR=1 V/μsec has a gain of 40 dB. If this amplifier has to faithfully amplify sinusoidal signals from dc to 20 kHz without introducing any slew-rate induced distortion, then the input signal level must not exceed.

(*a*) 795 mV (*b*) 395 mV

(*c*) 79.5 mV (*d*) 39.5 mV

2.6 The circuit in figure employs positive feedback and is intended to generate sinusoidal oscillation. If at a frequency f_0, $B(f) = \frac{V_f(f)}{V_0(f)} = \frac{1}{6}\angle 0°$, then to sustain oscillation at this frequency

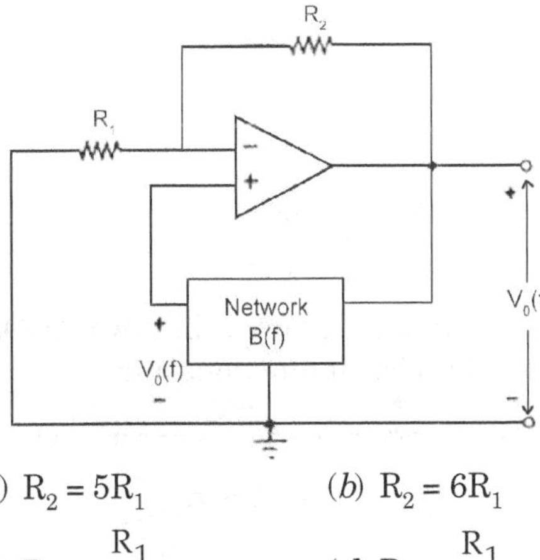

(*a*) $R_2 = 5R_1$ (*b*) $R_2 = 6R_1$

(*c*) $R_2 = \frac{R_1}{6}$ (*d*) $R_2 = \frac{R_1}{5}$

2.7 A zener diode regulator in the figure is to be designed to meet the specifications : $I_L = 10$ mA, $V_0 = 10$ V and V_{in} varies from 30 V to 50 V. The zener diode has $V_z = 10$V and I_{zk} (knee current) = 1 mA. For satisfactory operation.

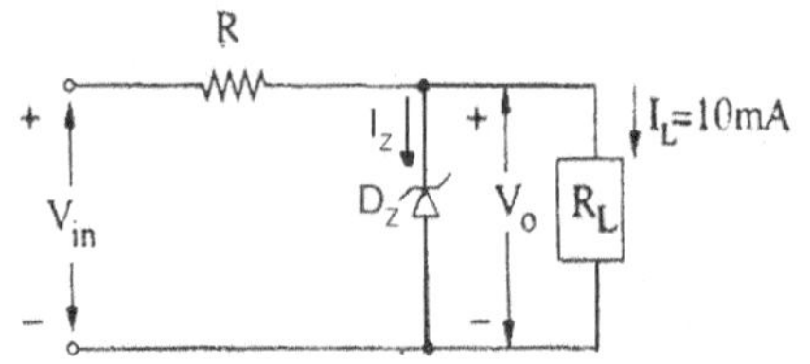

(*a*) $R \le 1800\,\Omega$ (*b*) $2000\,\Omega \le R \le 2200\,\Omega$

(*c*) $3700\,\Omega \le R \le 4000\,\Omega$ (*d*) $R > 4000\,\Omega$

2.8 The voltage gain $A_v = \frac{V_0}{V_i}$ of the JFET amplifier shown in the figure is

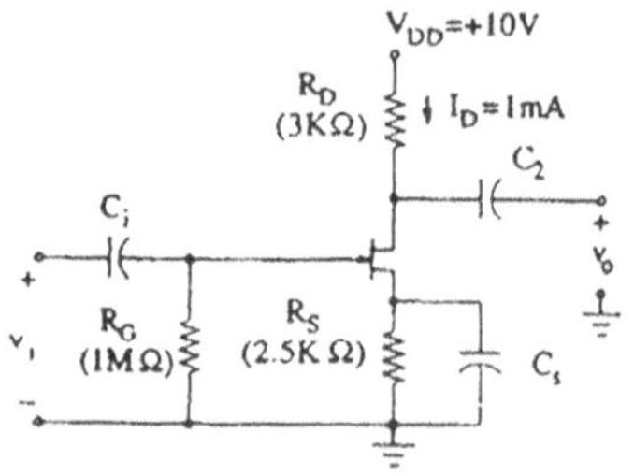

I_{DSS} = 10 mA; V_p = –5V (Assume C_1, C_2 and C_3 to be very large

(*a*) + 18 (*b*) – 18

(*c*) + 6 (*d*) – 6

2.9 The gates G_1 and G_2 in the figure is have propagation delays of 10 nsec and 20 nsec respectively. If the input V_i makes an abrupt change from logic 0 to 1 at time $t = t_0$, then the output wave form V_0 is.

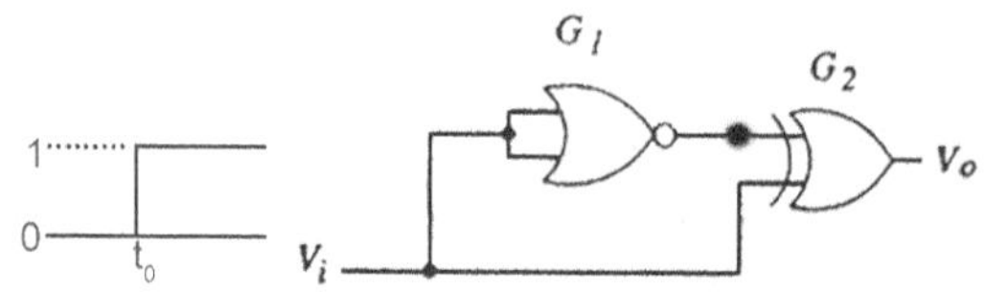

(*a*)

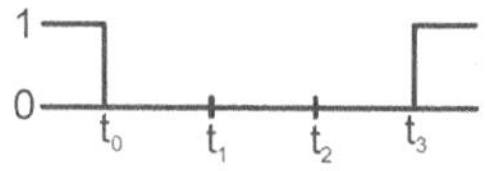

(*b*)

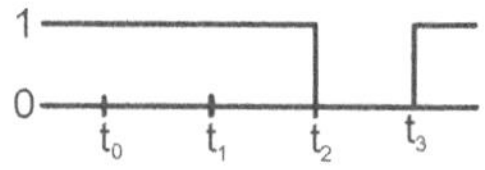

(*c*)

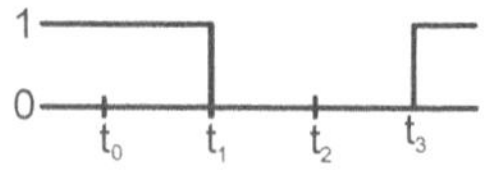

(*d*)

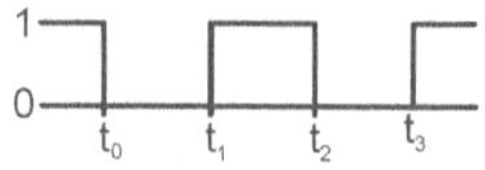

2.10 The circuit in the figure is has two CMOS NOR-gates. This circuit functions as a

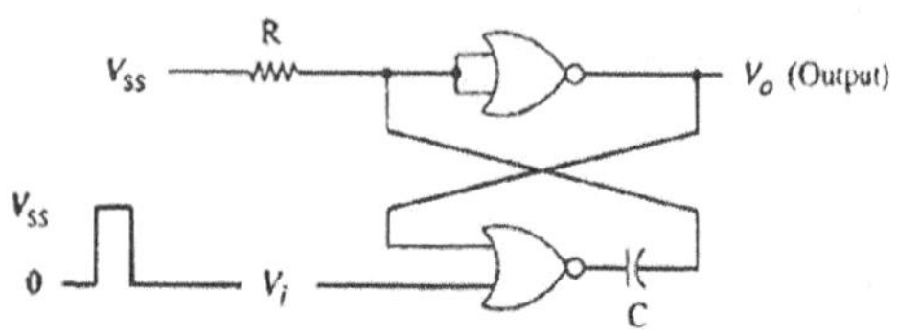

(*a*) flip-flop

(*b*) schmitt trigger

(*c*) monostable multivibrator

(*d*) astable multivibrator

2.11 Consider the following statements in connection with the CMOS inverter in the figure is where both the MOSFETs are of enhancement type and both have a threshold voltage of 2V.

Statement 1 : T_1 *conducts when* $V_i \ge 2$V.

Statement 2 : T_1 *is always in saturation when* $V_0 = 0$V.

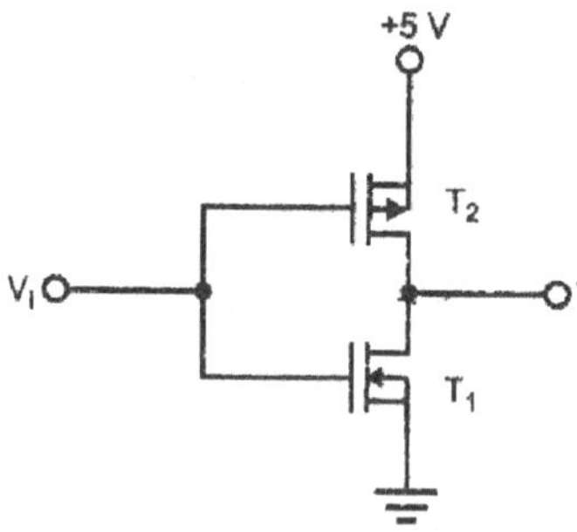

Which of the following is correct?

(*a*) Only statement 1 is TRUE

(*b*) Only statement 2 is TRUE

(*c*) Both the statements are TRUE

(*d*) Both the statements are FALSE

2.12 If the input X_3, X_2, X_1, X_0 to the ROM in the figure are 8–4–2-1 BCD numbers, then the outputs $Y_3Y_2Y_1Y_0$ are

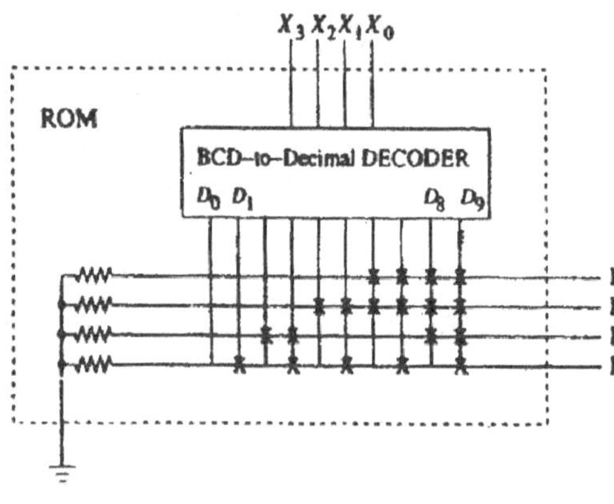

(*a*) gray code numbers

(*b*) 2421 BCD numbers

(*c*) excess-3 code numbers

(*d*) none of these

2.13 Consider the following assembly language program.

```
        MVI     B,87H
        MOV     A, B
START : JMP     NEXT
        MVI     B, 00H
```

```
        XRA     B
        OUT     PORT1
        HLT
NEXT :  XRA     B
        JP      START
        OUT     PORT 2
        HLT
```

The execution of the above program in an 8085 microprocessor will result in

(*a*) an output of 87H at PORT1

(*b*) an output of 87H at PORT2

(*c*) infinite looping of the program execution with accumulator data remaining at 00H.

(*d*) infinite looping of the program execution with accumulator data alternating between 00H and 87H.

2.14 The system shown in the figure is remains stable when

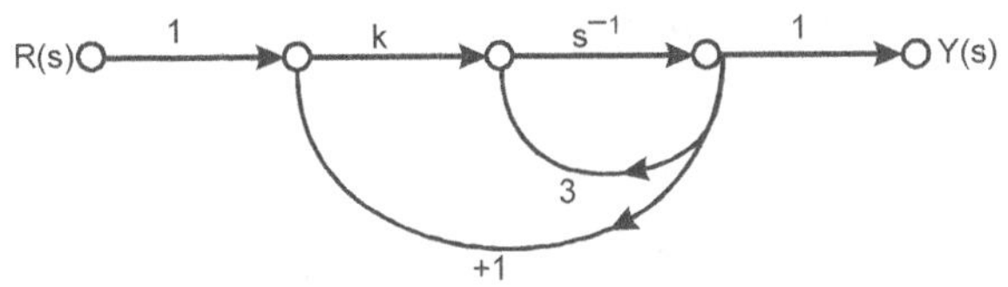

(*a*) $k < -1$ (*b*) $-1 < k < 1$

(*c*) $1 < k < 3$ (*d*) $k < -3$

2.15 The transfer function of a system is $G(s) = \frac{100}{(s+1)(s+100)}$.

For a unit-step input to the system the approximate settling time for 2% criterion is

(*a*) 100 sec (*b*) 4 sec

(*c*) 1 sec (*d*) 0.01 sec

2.16 The characteristic polynomial of a system is $q(s) = 2s^5 + s^4 + 4s^3 + 2s^2 + 2s + 1$.

The system is

(*a*) stable (*b*) marginally stable

(*c*) unstable (*d*) oscillatory

2.17 The system with the open loop transfer function

$G(s)H(s) = \frac{1}{s(s^2+s+1)}$ has a gain margin of

(*a*) –6 dB (*b*) 0 dB

(*c*) 3.5 dB (*d*) 6 dB

2.18 An angle-modulated signal is given by $s(t) = \cos 2\pi(2 \times 10^6 t + 30 \sin 150t + 40 \cos 150t)$.

The maximum frequency and phase deviations of $s(t)$ are

(*a*) 10.5 kHz, 140π rad (*b*) 6 kHz, 80π rad

(*c*) 10.5 kHz, 100π rad (*d*) 7.5 kHz, 100π rad

2.19 In the figure $m(t) = \frac{2\sin 2\pi t}{t}$, $s(t) = \cos 200\pi t$ and $n(t) = \frac{\sin 199\pi t}{t}$. The output $y(t)$ will be

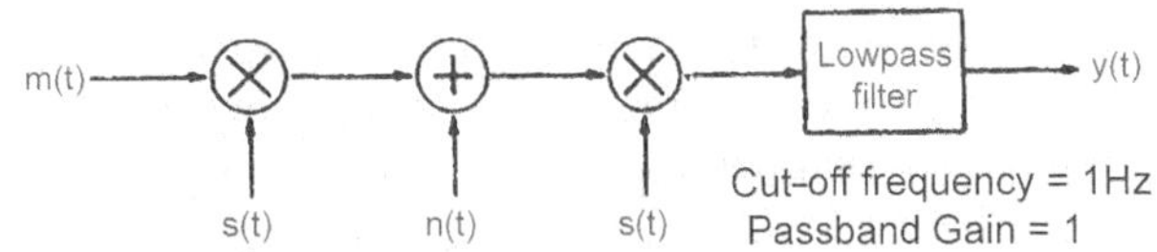

(*a*) $\frac{\sin 2\pi t}{t}$

(*b*) $\frac{\sin 2\pi t}{t} + \frac{\sin \pi t}{t}\cos 3\pi t$

(*c*) $\frac{\sin 2\pi t}{t} + \frac{\sin 0.5\pi t}{t}\cos 1.5\pi t$

(*d*) $\frac{\sin 2\pi t}{t} + \frac{\sin \pi t}{t}\cos 0.75\pi t$

2.20 A signal $x(t) = 100 \cos(24\pi \times 10^3)t$ is ideally sampled with a sampling period of 50μsec and then passed through an ideal lowpass filter with cut-off frequency of 15 kHz. Which of the following frequencies is/are present at the filter output?

(*a*) 12 kHz only (*b*) 8 kHz only

(*c*) 12 kHz and 9 kHz (*d*) 12 kHz and 8 kHz

2.21 If the variance σ_x^2 of $d(n) = x(n) - x(n-1)$ is one-tenth the variance σ_x^2 of a stationary zero-mean discrete-time signal $x(n)$, then the normalized autocorrelation function $R_{xx}(k)/\sigma_x^2$ at $k = 1$ is

(*a*) 0.95 (*b*) 0.90

(*c*) 0.10 (*d*) 0.05

2.22 A plane wave is characterized by $\vec{E} = (0.5\hat{x} + \hat{y}\, e^{j\pi/2})e^{jwt-jkz}$.

This wave is

(*a*) linearly polarized (*b*) circularly polarized

(*c*) elliptically polarized (*d*) unpolarized

2.23 Distilled water at 25°C is characterized by $\sigma = 1.7 \times 10^{-4}$ mho/m and $\epsilon = 78\epsilon_0$ at a frequency of 3 GHz. Its loss tangent tanδ is

(*a*) 1.3×10^{-5} (*b*) 1.3×10^{-3}

(*c*) $1.7 \times 10^{-4}/78$ (*d*) $1.7 \times 10^{-4}/(78\epsilon_0)$

($\epsilon = 10^{-9}/(36\pi)$ F/m)

2.24 The electric field on the surface of a perfect conductor is 2 V/m. The conductor is immersed in water with $\epsilon = 80\epsilon_0$. The surface charge density on the conductor is

(*a*) 0 C/m² (*b*) 2 C/m²

(*c*) 1.8×10^{-11} C/m² (*d*) 1.14×10^{-9} C/m²

($\epsilon_0 = 10^{-9}/(36\pi)$ F/m)

2.25 A person with a receiver is 5 km away from the transmitter. What is the distance that this person must move further to detect a 3-dB decrease in signal strength?

(*a*) 942 m (*b*) 2070 m

(*c*) 4978 m (*d*) 5320 m

SECTION - B (75 Marks)

This section consists of TWENTY questions (EC3 – EC22) of FIVE marks each. Attempt ANY FIFTEEN questions. Answers must be given in the answer book provided. Answer for each question must start on a fresh page and must appear at one place only (Answers to all parts of a question must appear together).
(5 × 15 = 75)

3. The switch in the figure is has been in position 1 for a long time and is, then moved to position 2 at $t = 0$.

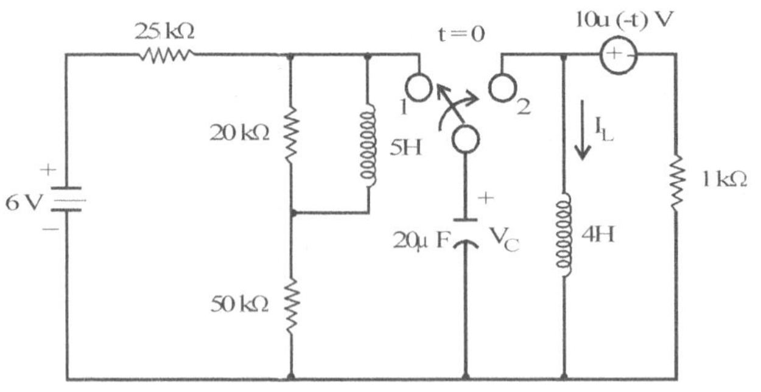

(*a*) Determine $V_C(0^+)$ and $I_L(0^+)$.

(*b*) Determine $\frac{dV_c(t)}{dt}$ at $t = 0^+$.

(*c*) Determine $V_c(t)$ for $t > 0$.

4. For the network shown in the figure is, $R = 1\ k\Omega$, $L_1 = 2$ H, $L_2 = 5$ H, $L_3 = 1$ H, $L_4 = 4$ H and $C = 0.2\ \mu F$. The mutual inductances are $M_{12} = 3$ H and $M_{34} = 2$ H.
Determine
(*a*) the equivalent inductance for the combination of L_3 and L_4.
(*b*) the equivalent inductance across the points A and B in the network.
(*c*) the resonant frequency of the network.

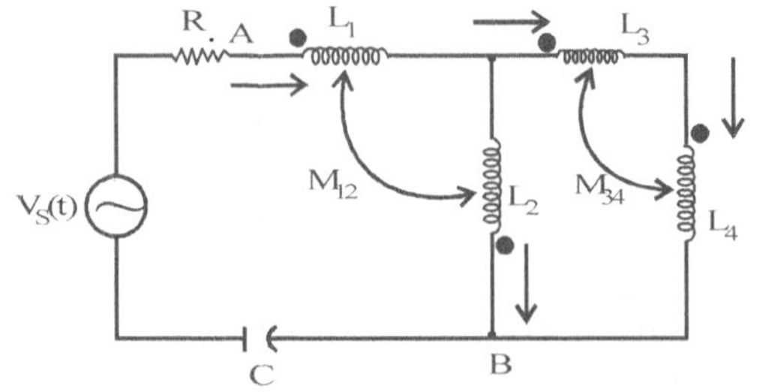

5. Consider the network in the figure is

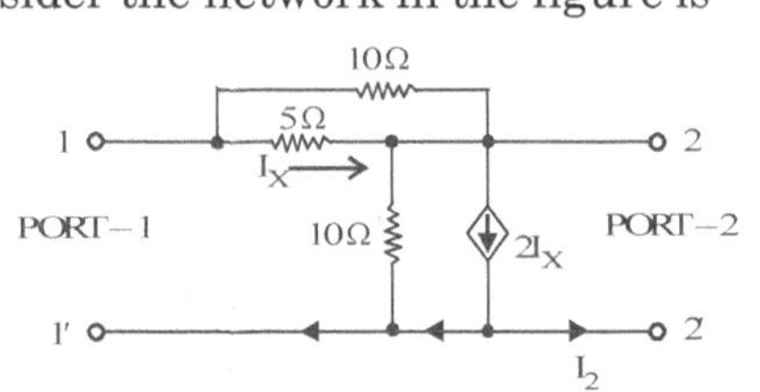

(*a*) Find its short-circuit admittance parameters.
(*b*) Find the open-circuit impedance Z_{22}.

6. A triangular voltage waveform $V_i(t)$ first figure applied at the input to the circuit of second figure. Assume the diodes to be ideal.

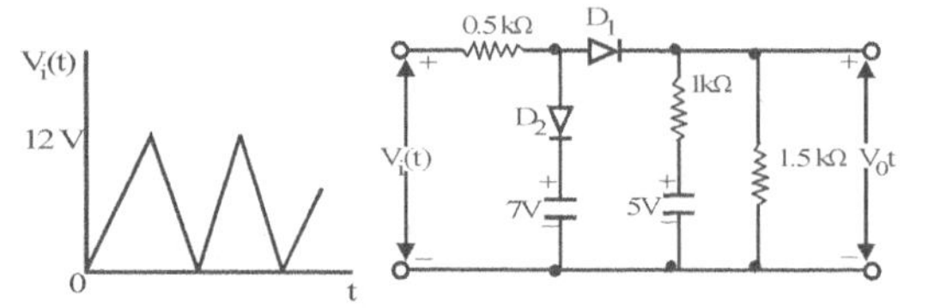

(*a*) Determine the output $V_0(t)$.
(*b*) Neatly sketch the output waveform superimposed on the input $V_i(t)$ and label the key points.

7. The given figure shows 2-stage amplifier. The transistors Q_1 and Q_2 are identical with current gain $\beta = 100$; further $\beta_{dc} = \beta_{ac} = \beta$. The zener diode D_z has a break-down voltage $V_z = 10.7$ volt. Assume that D_z is in break-down region and its dynamic resistance r_z is zero. The capacitors C_1 and C_2 are large and provide negligible impedance at signal frequencies.

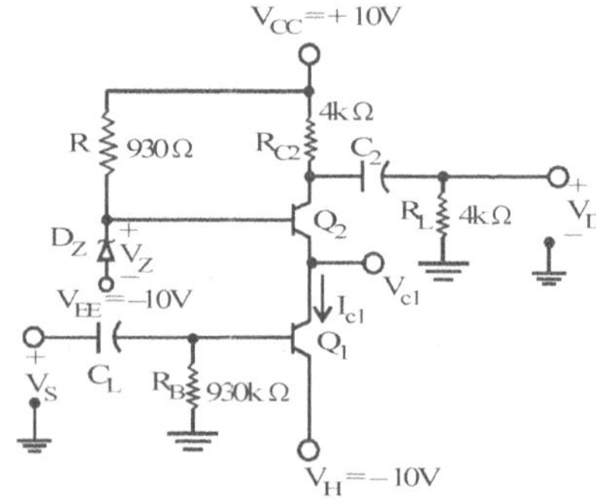

(*a*) Identify the configuration in each of the amplifier stages (*i.e.* whether CE, CC, CB etc.)
(*b*) Determine the quiescent quantities I_{C1} and V_{C1}.
(*c*) Derive an expression for the voltage gain

$A_V \Delta \frac{v_0}{v_s}$ and determine its value.

(Assume $V_{BE} = 0.7$ V, $r_0 = \infty$ and Thermal voltage $V_T = 25$ mV)

8. Consider the circuit of the figure. The opamp used is ideal.

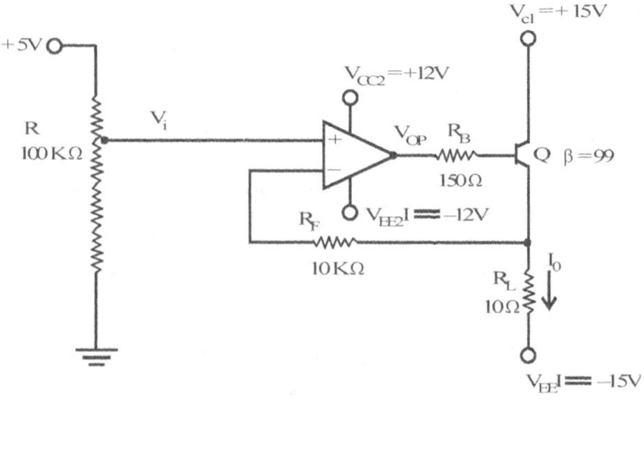

(*a*) In which mode is the BJT operating (*i.e.* active, saturation or cut-off)? Justify your answer.

(*b*) Obtain an expression relating the output current I_0 and the input voltage V_i.

(*c*) Determine I_0 and V_{OP} if $V_i = 2$ volt (V_{op} : output of op-amp)

(Assume $\beta = 99$ and $V_{BE} = 0.7$ V)

9. The inputs to a digital circuit shown in the figure is are the external signals A, B and C.

($\overline{A}$, $\overline{B}$ and $\overline{C}$ are not available). The +5 V power supply (logic 1) and the ground (logic 0) are also available. The output of the circuit is $X = \overline{A}B + \overline{A}\,\overline{B}\,\overline{C}$.

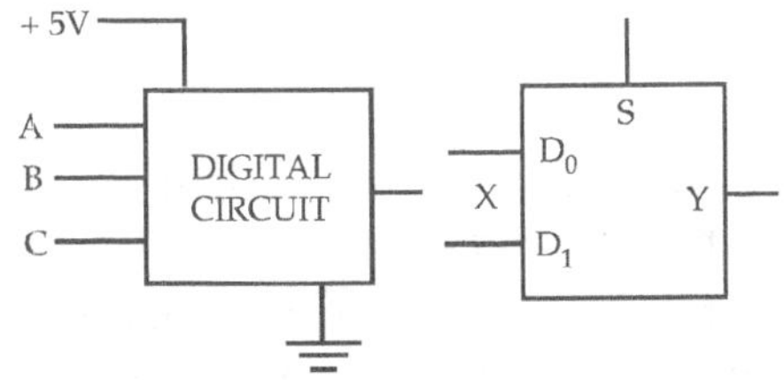

Table-9

S	Y
0	D_0
1	D_1

(*a*) Write down the output function in its canonical SOP and POS forms.

(*b*) Implement the circuit using only two 2:1 multiplexers shown in the figure where S is the data-select line, D_0 and D_1 are the input data lines and Y is the output line. The function table for the multiplexer is in given Table.

10. Each transistor in the figure is has *dc* current gain $\beta_{dc} = 50$, cut-in voltage $V\gamma = 0.65$ V and $V_{BE,\,sat} = 0.75$ V. The output voltage V_0 for T_2 in saturation can be as high as 0.2 V. Assume 0.7 V drop across a conducting *p-n* junction. Determine?

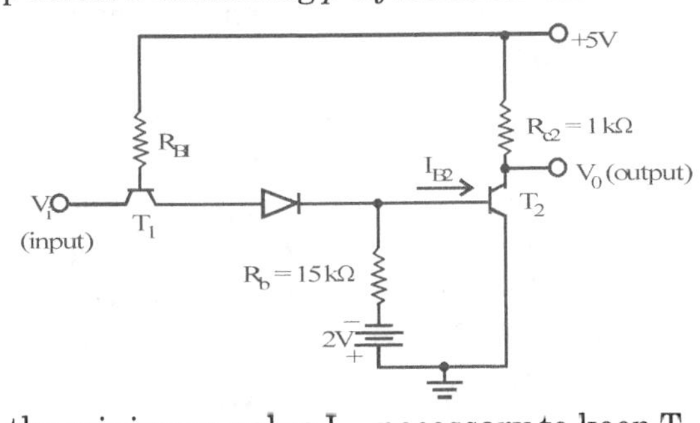

(*a*) the minimum value I_{B2} necessary to keep T_2 saturation.

(*b*) the maximum permissible value for the resistance R_{B1}.

(*c*) the worst-case high input (logic 1) and the worst-case low input (logic 0) for which T_2 will be either in saturation or in cut-off.

11. It is required to design a binary mod-5 synchronous counter using AB flip-flops such that the output $Q_2Q_1Q_0$ changes as $000 \rightarrow 001 \rightarrow 010$ .... and so on. The excitation table for the AB flip-flop is given in the table

A	B	Q_{n+1}
0	0	1
0	1	$\overline{Q}_n$
1	0	Q_n
1	1	0

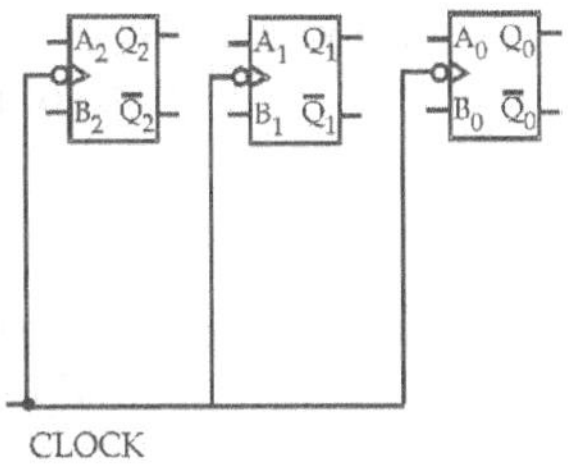

(*a*) Write down the state table for the mod-5 counter.

(*b*) Obtain simplified SOP expressions for the inputs A_2, B_2, A_1, B_1, A_0 and B_0 in terms of Q_2, Q_1, Q_0 and their complements.

(*c*) Hence, complete the circuit diagram for the mod-5 counter given in the figure using minimum number of 2-input NAND-gate only.

12. An 8085 microprocessor operating at 5 MHz clock frequency executes the following routine.

```
START : MOVE   A,B
        OUT    55H
        DCR    B
        STA    FFF8H
        JMP    START
```

(*a*) Determine the total number of machine cycles required to execute this routine till the JMP instruction is executed for the first time.

(*b*) Determine the time interval between two consecutive $\overline{MEMW}$ signals.

(*c*) If the external logic controls the READY line so that three WAIT states are introduced in the I/O WRITE machine cycle, determine the time interval between two consecutive $\overline{MEMW}$ signals.

13. A unity feedback system has the plant transfer function $G_p(s) = \dfrac{1}{(s+1)(2s+1)}$

(*a*) Determine the frequency at which the plant has a phase lag of 90°.

(*b*) An integral controller with transfer function $G_c(s) = \dfrac{k}{s}$ is placed in the feedforward path of the feedback system. Find the value of *k* such that the compensated system has an open-loop gain margin of 2.5.

(*c*) Determine the steady state errors of the compensated system to unit-step and unit-ramp inputs.

14. The Nyquist plot of an all-pole second order open-loop system is shown in the figure. Obtain the transfer function of the system.

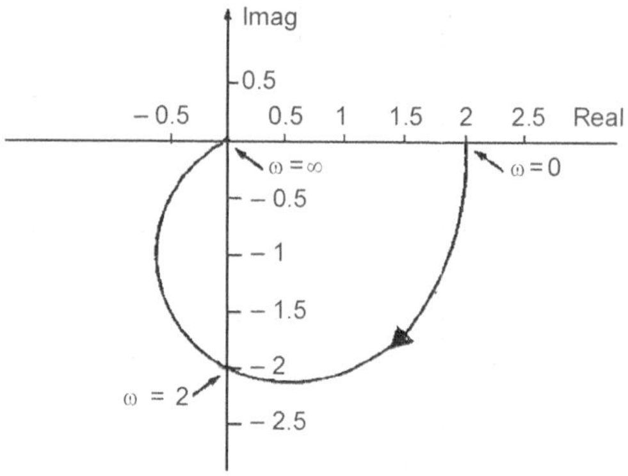

15. The block diagram of a linear time invariant system is given in the figure is

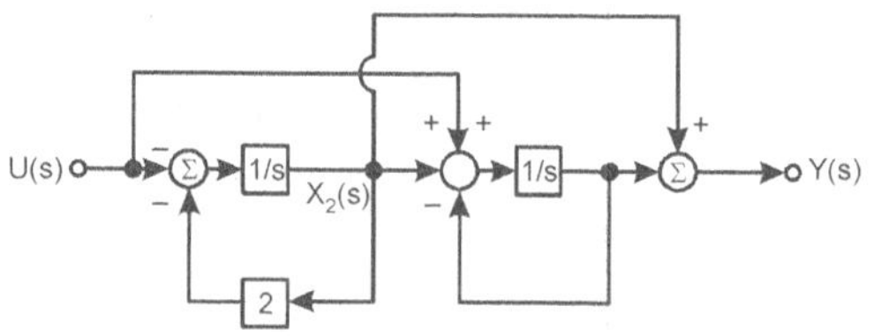

(*a*) Write down the state variable equations for the system in matrix form assuming the state vector to be $\left[x_1(t)\, x_2(t)\right]^T$

(*b*) Find out the state transition matrix.

(*c*) Determine $y(t)$, $t \geq 0$, When the initial values of the state at time $t = 0$ are $x_1(0) = 1$, and $x_2(0) = 1$.

16. A deterministic signal $x(t) = \cos 2\pi t$ is passed through a differentiator as shown in the figure is

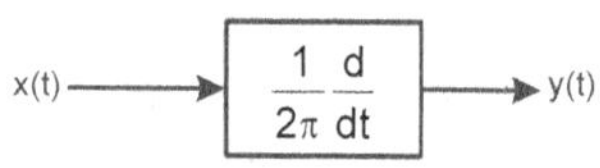

(*a*) Determine the autocorrelation $R_{xx}(T)$ and the power spectral density $S_{xx}(f)$.

(*b*) Find the output power spectral density $S_{yy}(f)$.

(*c*) Evaluate $R_{xy}(0)$ and $R_{xy}\left(\frac{1}{4}\right)$

17. A DSBSC modulated signal $s(t) = 10 \cos (2\pi \times 10^6 t + \phi) m(t)$ is corrupted by an additive white Gaussian noise of power spectral density 10^{-4} W/Hz. The message power spectral density $S_{mm}(f)$ is as shown in the figure and ϕ is uniformly distributed over the range 0 to 2π.

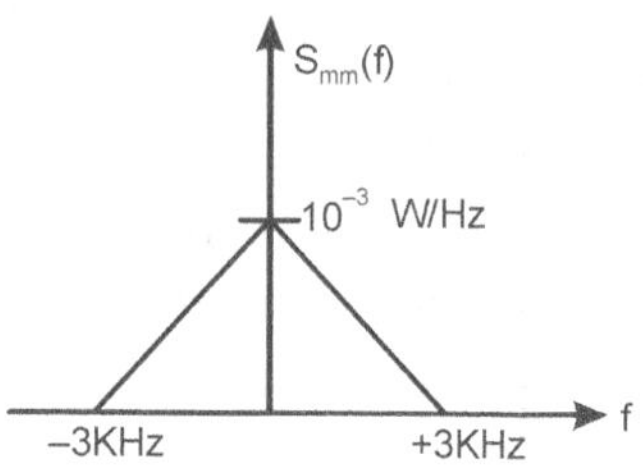

(*a*) Express the signal autocorrelation function $R_{ss}(T)$ in terms of the message autocorrelation function $R_{mm}(T)$. Clearly state the necessary assumptions.

(*b*) Determine the signal power spectral density $S_{ss}(f)$.

(*c*) Find the power of the modulated signal and the noise power in the transmission bandwidth.

18. A continuous-time signal with finite energy, band limited from 3 MHz to 5 MHz, is ideally sampled, encoded by a fixed length PCM coder and then transmitted over a digital channel of capacity 7 M*bps*. The probability density function (*pdf*) at the output of the sampler is uniform over the range –2V to + 2V.

(*a*) Determine the minimum sampling rate necessary for perfect reconstruction.

(*b*) Determine the maximum SNR (in dB) that may be achieved.

19. A discrete memoryless source generates either 0 or 1 at a rate of 160 K*bps*; 0 is generated three times more frequently than 1. A coherent binary PSK modulator is employed to transmit these bits over a noisy channel. The received bits are detected in a correlator fed with the basis function of unit energy (for this binary PSK scheme) as the reference signal. The receiver makes a decision in favour of 1 if the correlator output is positive, else decides in favour of 0. If 0 and 1 are represented as

$-(6\sqrt{2} \cos 640\pi \times 10^3 t)$V and

$+(6\sqrt{2} \cos 640\pi \times 10^3 t)$V respectively, then

(*a*) determine the transmitted signal energy per bit.

(*b*) determine the basis function of unit energy for this binary PSK scheme.

(*c*) determine the probability that the receiver makes a decision in favour of 1 when the channel noise is characterized as zero-mean AWGN with power spectral density $\frac{N_0}{2} = 3.125 \times 10^{-4}$ W/Hz

20. Transmission line transformation of a load Z_L to Z is given by

$$Z = Z_0 \frac{Z_L + jZ_0 \tan(\beta l)}{Z_0 + jZ_L \tan(\beta l)}$$

(*a*) Show that the above transformation implies that the impedance Z gets transformed to Z_L^* for real Z.

(*b*) What is the importance of the result derived in (*a*)?

21. Consider a parallel plate waveguide with plate separation d as shown in the figure electric and magnetic fields for the TEM-mode are given by

$E_x = E_0 e^{-jkz+j\omega t}$

$Hy = \frac{E_0}{n} e^{-jkz+j\omega t}$

where $k = \eta\omega\varepsilon$

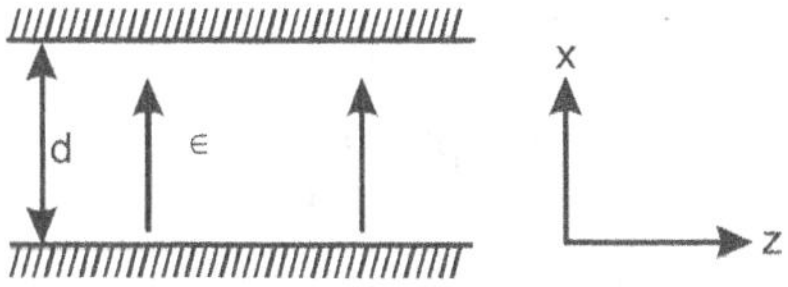

(a) Determine the surface charge densities p_s on the plates at $x = 0$ and $x = d$.

(b) Determine the surface current densities $\vec{J}_s$ on the same plates.

(c) Prove that p_s and $\vec{J}_s$ satisfy the current continuity condition.

22. Consider a linear array of two half-wave dipoles A and B as shown in the figure. The dipoles are $\frac{\lambda}{4}$ apart and are excited in such a way that the current on element B legs that on element A by 90° in phase.

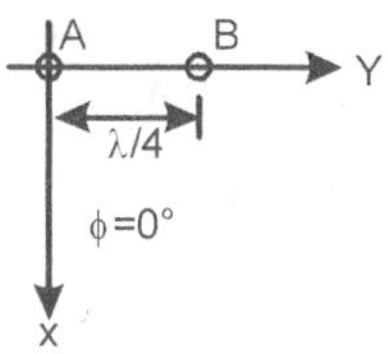

(a) Obtain the expression for the radiation pattern for E_θ in the XY plane (i.e.,θ =90°)

(b) Sketch the radiation pattern obtained in (a).

ANSWERS

1.1 (a)	1.2 (c)	1.3 (c)	1.4 (b)	1.5 (c)	1.6 (c)	1.7 (b)	1.8 (a)	1.9 (d)	1.10 (b)
1.11 (c)	1.12 (d)	1.13 (c)	1.14 (b)	1.15 (d)	1.16 (d)	1.17 (c)	1.18 (a)	1.19 (c)	1.20 (c)
1.21 (d)	1.22 (b)	1.23 (d)	1.24 (b)	1.25 (c)	2.1 (a)	2.2 (d)	2.3 (d)	2.4 (b)	2.5 (c)
2.6 (a)	2.7 (a)	2.8 (c)	2.9 (b)	2.10 (c)	2.11 (a)	2.12 (b)	2.13 (b)	2.14 (d)	2.15 (b)
2.16 (c)	2.17 (b)	2.18 (d)	2.19 (c)	2.20 (d)	2.21 (a)	2.22 (c)	2.23 (a)	2.24 (d)	2.25 (b)

EXPLANATIONS

SECTION - A

1.1 KVL gives, $20 = 5I + \left(\frac{V_1}{5} + I\right)5$

or $20 = 5I + (4 + I)\,5$

$I = 0$ amp.

$\therefore$ Power delivered $= 4^2 \times 5 =$ **80 W**

1.2 At $t = 0^+$, the circuit will be as shown,

$I_L(0^+) = 0,$

then $V_i = -50$ volts

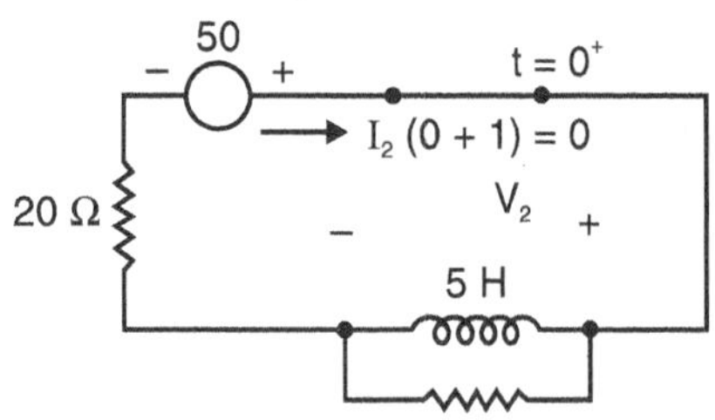

1.3 As, $g(t) * \delta(t - t_0) = g(t_0)$

then $x(t + 5) * \delta(t - 7) = x(t - 2)$

1.4 $x(t) = 2\cos \pi t + 7\cos t$, does not satisfy the **Dirichlet condition** $x(t)$ is not periodic Function, therefore, it can't be expanded as Fourier series.

1.5. Diode current, $I = I_0\left(e^{V_{D/n}V_{T-1}}\right)$

I_0 is the saturation current, For silicon, $\eta = 2$,

Now, At $T_1 = 273 + 20 = 293$ K, $V_{D1} = 0.70$ volts

$$I_1 = I_0\left(e^{\frac{V_D}{\eta V_{T_1}}} - 1\right) \qquad \text{...(A)}$$

$$V_{T1} = \frac{293.}{11600} = 0.025\text{V}$$

At $T_2 = 273 + 40 = 313$ K

and $I_{0_2} = I_{0_1} \cdot 2^{\Delta T/10} = I_{0_1} \cdot 2^2 = 4I_{0_1},\ V_{T_2} = \frac{373}{11600}$

$$I_2 = I_{02}\left(e^{V_{D2}/\eta V_{T2}} - 1\right) \qquad \text{...(B)}$$

As diode current remains unchanged,

$I_1 = I_2$

or, $I_{0_1}\left(e^{\frac{V_{D1}}{\eta V_{T1}}} - 1\right) = 4I_{0_1}\left(e^{\frac{V_{D_2}}{\eta V_{T2}}} - 1\right)$

or, $3 = 4e^{\frac{V_{D2}}{\eta V_{T2}}} - e^{\frac{V_{D1}}{\eta V_{T_2}}}$

$3 = 4e^{\frac{V_{D2}}{\eta V_{T2}}} - e^{\frac{0.70}{2 \times 0.025}}$

or, $V_{D_2} = \eta V_{T_2} \times 12.61$

$= 2 \times \frac{313}{11600} \times 12.61 = 680$ mV

1.6 $R_{if} = R_i(1 + \beta A)$

$$R_{of} = \frac{R_\delta}{1 + \beta A}$$

1.7 Gain-bandwidth product = BW × A

$\Rightarrow$ $10^6 = \text{BW} \times A$ ...(*i*)

Given, $20 \log A = 20$

$\Rightarrow$ $A = 10$

put A in equation (*i*)

then BW $= 10^5$ Hz = **100 KHz**

1.8 New lower 3-dB frequency,

$f_L' = f_L \cdot D$

and, new higher 3-dB frequency,

$$f_H' = \frac{f_H}{D}.$$

Where $D = \frac{1}{\sqrt{2^{1/n} - 1}}$

$n = 3$

1.9

1	0	0	0	| + 8
0	1	1	1	1's complement
+			1	
1	0	0	0	2's complement = –8

1.10.

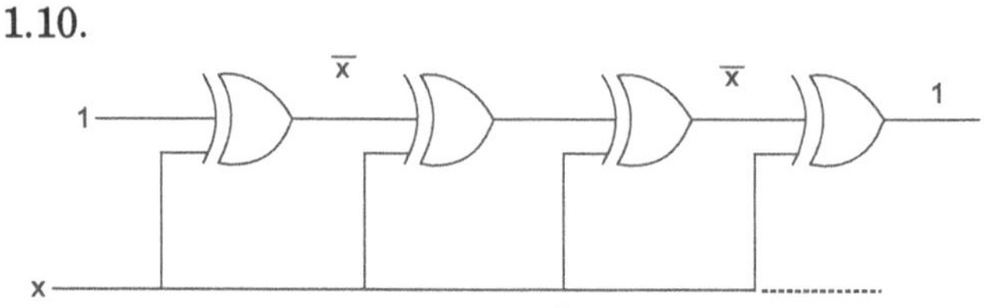

After 4-XOR gates, output is 1 and is repetitive therefor at the output of 20th X-OR gate is 1.

1.11 Number of comparators $= 2^N - 1 = 2^3 - 1 = 7$

1.12

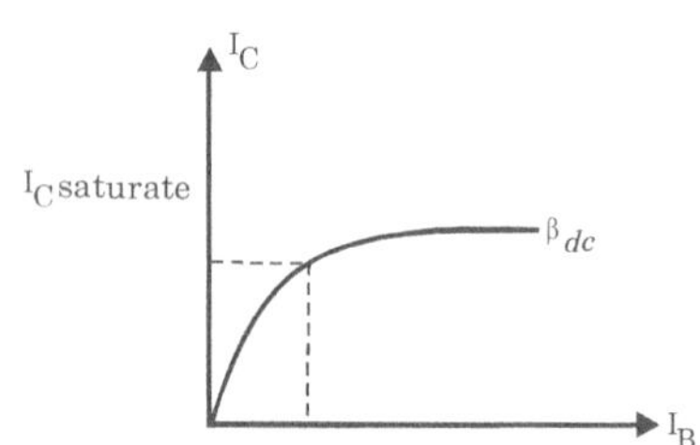

β_{dc} = constant

I_B = increases but I_C become saturates

1.13 Comparing G(*s*) with $G(s) = \frac{k}{s^2 + 2r\omega_n + \omega_n^2}$

or $2r\omega_n = \frac{1}{k}$ or $\omega_n = \frac{1}{k}$

and $\omega_n = \sqrt{\dfrac{6}{k}}$

then, $k = \dfrac{1}{6}$.

1.14

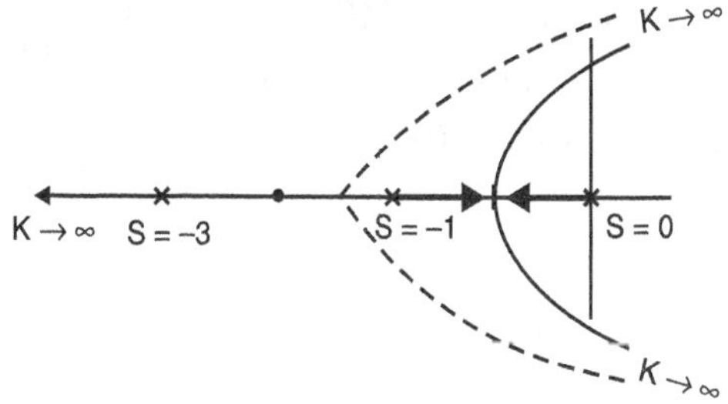

Root Locus lies where number. of poles and zeros to the right of the pole or zero is odd.

1.15 $|G(j\omega)\,H(j\omega)| \neq 1$, for any value of ω,

Therefore phase margin ∞.

or ω becomes imaginary

Thus,no gain crossover

$\therefore$ PM = ∞

1.16 $sX(s) = -2X(s) + 2U(s)$

$X(s)[s+2] = 2U(s)$

and $Y(s) = 0.5X(s)$

gives $\dfrac{Y(s)}{U(s)} = \dfrac{1}{s+2}$.

1.17. $f(t) = e^{-t}u(t)$

Fourier transform of $f(t) = \mathcal{L}[f(t)] = \dfrac{1}{1+j2\pi f}$

Duality property states that, if $8(t) \leftrightarrow G(f)$

then $G(t) \leftrightarrow g(-f)$

Here $F(f) = \dfrac{1}{1+j2\pi f}$

$F(t) = \dfrac{1}{1+j2\pi t}$ and $F(t) \leftrightarrow e^{f}u(-f)$

1.18. For linear phase channel,

$Q(\omega) = -\omega t_0$

$T_P(\text{phase delay}) = \dfrac{-\theta(\omega)}{\omega} = t_0$

and $T_g(\text{group delay}) = \dfrac{-d\theta(\omega)}{d\omega} = t_0$

Hence $T_P = T_g = t_0 = \text{constant}$

1.19

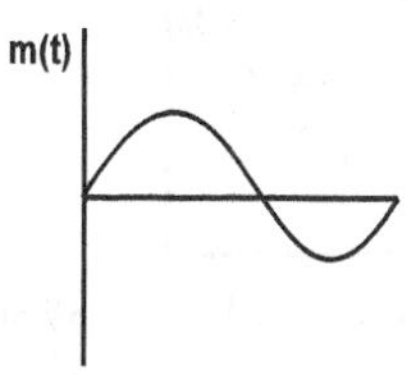

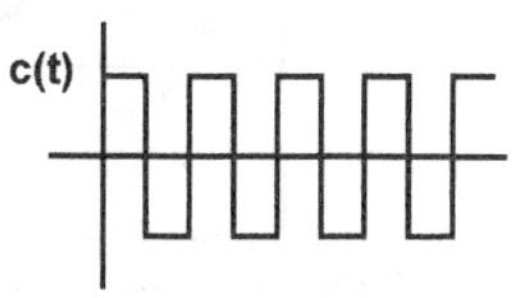

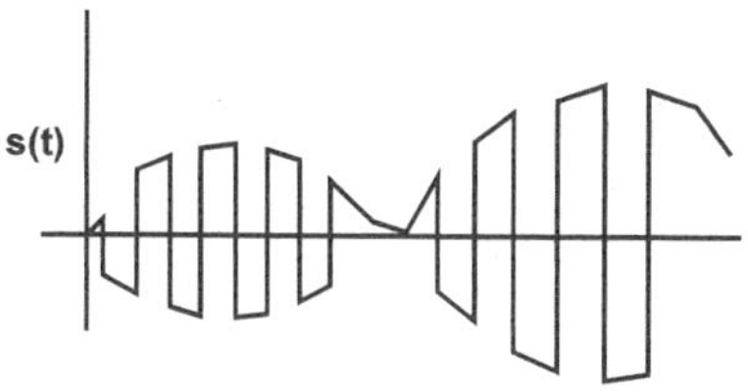

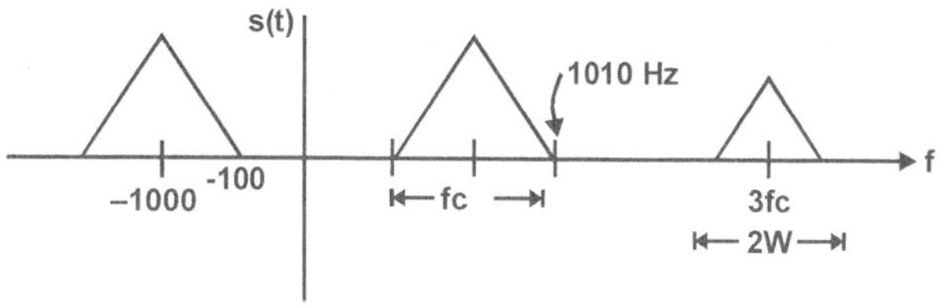

$s(t) = c(t)\,.\,m(t)$

Expressing square wave as,

$$c(t) = \frac{4}{\pi}\sum_{n=1}^{\infty}\frac{(-1)^{n-1}}{2n-1}\cos[2\pi f_c(2n-1)].$$

then, modulated output will be

$$s(t) = c(t)\,.\,m(t)$$

$$= \frac{4}{\pi}.\sum_{n=1}^{\infty}\frac{(-1)^{n-1}}{2n-1}\cos[2\pi f_c(2n-1)].m(t)$$

From, the spectrum it is clear than 1020 Hz signal will be absent.

1.20 $\Rightarrow \quad X(f) \otimes \dfrac{1}{T_s}\displaystyle\sum_{n=-\infty}^{\infty}\delta\left(f - \frac{n}{T_s}\right)$

$$\Rightarrow \quad \frac{1}{T_s} \times 5\times 10^{-6} \times 10 \sum_{n=-\infty}^{\infty}\left(\frac{\delta\left(f - f_m - \frac{n}{T_s}\right) + \delta\left(f + f_m - \frac{n}{T_s}\right)}{2}\right)$$

At $n = 0$

$$\Rightarrow \quad \frac{50\times 10^{-6}}{T_s}\;\frac{[\delta(f - f_m) + \delta(f + f_m)]}{2}$$

Taking inverse Fourier transform

$$= \frac{50\times 10^{-6}}{T_s}\cos\omega_m t$$

$$= \frac{50\times 10^{-6}}{100\times 10^{-6}}\cos\omega_m t = 5\times 10^{-1}\cos\omega_m t$$

1.21 Transmitted frequency,

$f_i = \dfrac{n_c + i}{T_b}$ for some fixed integer n_C and $i = 1, 2$

Given, $R_b = \dfrac{1}{T_b} = 8$ kbps,

then, $f_1 = (3+1)\,8 = 32$ kHz

and, $f_2 = (4+1)\,8 = 40$ kHz

1.23 $\text{VSWR} = \dfrac{1+|\tau_l|}{1-|\tau_l|}$

Since, $|\tau_l| \le 1$.

$\therefore \quad 0 < \text{VSWR} < \infty$

1.25 $V_p = \dfrac{V_c}{\sqrt{1+\left(\dfrac{f_c}{f}\right)^2}}$

For TE_{10} mode, $f_c < f$

$V_p > V_c = c$

2.1. Short-circuiting terminal across R_L.

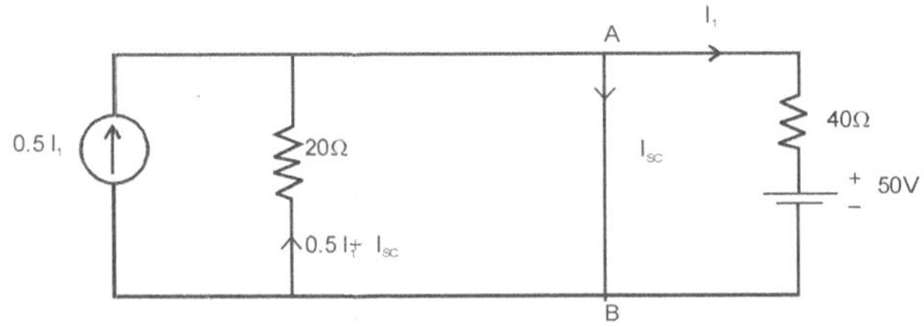

Using KVL we have

$\Rightarrow \quad 20(0.5\,I_1 + I_{SC}) - 40\,I_1 - 50 = 0$

$\Rightarrow \quad -3I_1 + 2I_{SC} = 5 \qquad \text{...(A)}$

and, $\quad -40I_1 - 50 = 0 \qquad \text{...(B)}$

$\Rightarrow \quad I_1 = \dfrac{-5}{4}$

then, from equation (A) and equation (B),

$I_{SC} = \dfrac{-5}{8}$ amp.

(Current flowing in the opposite direction to that of shown in figure)

Open circuiting terminal across R_L.

We get,

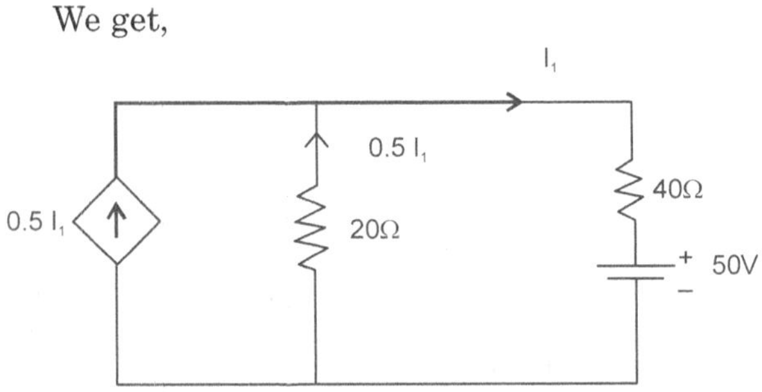

KVL gives, $-20 \times 0.5I_1 - 40\,I_1 - 50 = 0$

or, $\quad I_1 = -1$ amp.

then, $\quad V_{OC} = -20 \times 0.5I_1 = 10$ volts

Then, $\quad R_{TH} = \dfrac{V_{OC}}{I_{SC}} = \dfrac{10}{5/8} = 16\Omega$

For maximum transfer theorem,

Thevenin's resistance must be equal to R_L.

$R_L = R_{TH} = 16\Omega$.

2.2 $3\,V_p I_p \cos\theta = 1500$

or $\quad 3.\left(\dfrac{V_L}{\sqrt{3}}\right).\left(\dfrac{V_L}{\sqrt{3}\,Z_L}\right)\cos\theta = 1500$

or $\quad Z_L = \dfrac{V_L^2.\cos\theta}{1500} = \dfrac{(400)^2.(0.844)}{1500} = 90\,\Omega$

and $\quad \theta = \angle -\cos^{-1}(0.844)$

$= \angle -32.44^\circ$ [negative indicates lagging]

2.3 $X(s) = \dfrac{5-s}{(s+1)(s-2)} = \dfrac{A}{(s+1)} + \dfrac{B}{s-2}$

$A = \left.\dfrac{5-s}{s-2}\right|_{s=-1} = -2.$

and $\quad B = \left.\dfrac{5-s}{s+1}\right|_{s=2} = 1$

then, $\quad x(t) = L^{-1}X(s) = L^{-1}\left[\dfrac{-2}{s+1} + \dfrac{1}{s-2}\right]$

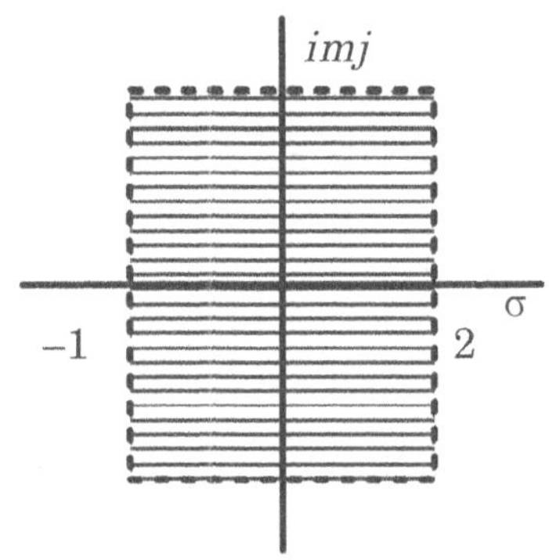

Given, ROC

$Re(s) < Re(a)$

$\dfrac{1}{s-a} \longleftrightarrow -e^{at}u(-t)$

$x(t)$ to be Fourier transformable

$x(t) = -2e^{-t}u(t) - e^{2t}u(-t)$

2.4 $\because \quad -a^n u[-n-1] \leftrightarrow \dfrac{z}{z-a} \qquad |z| < |a|$

$\therefore \quad -5^n u[-n-1] \leftrightarrow \dfrac{z}{z-5} \qquad |z| < |5|$

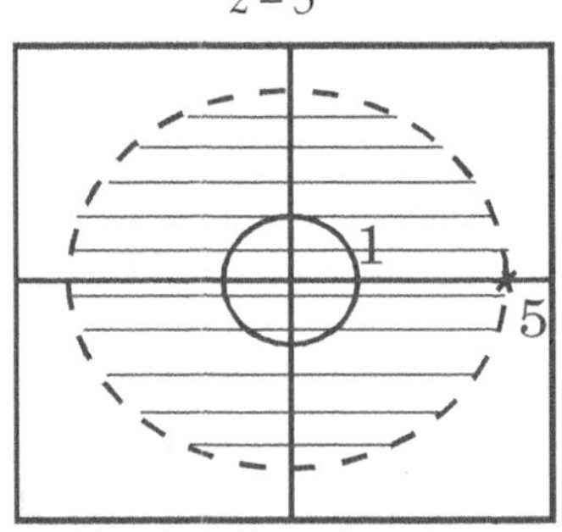

$\because$ ROC contains unit circle. Hence system is stable.

2.5 As, $20 \log A = 40$

$\therefore \quad A = 100$

Now, $v_0 = A\, v_m \sin \omega t$

or $$v_0 = \left| \frac{dv_o}{dt} \right|_{max} = AV_m\, 2\pi f$$

$$\therefore \quad v_m = \left| \frac{dv_o}{dt} \right|_{max} \cdot \frac{1}{2\pi A f}$$

$$= \frac{10^6}{2\pi \times 100 \times 20 \times 10^3} = \mathbf{79.5\ mV.}$$

2.6 Using KCL at inverting terminal, we get

$$\frac{\beta V_0 - 0}{R_1} + \frac{\beta V_0 - V_0}{R_2} = 0$$

or $$\beta\left(\frac{R_1 + R_2}{R_1 R_2}\right) = \frac{1}{R_2}$$

or $$\beta = \frac{R_1}{R_1 + R_2} = \frac{1}{6}$$

or $$R_2 = 5R_1$$

2.7 $$\frac{V_{in} - V_z}{R} \geq I_{2K} + I_L$$

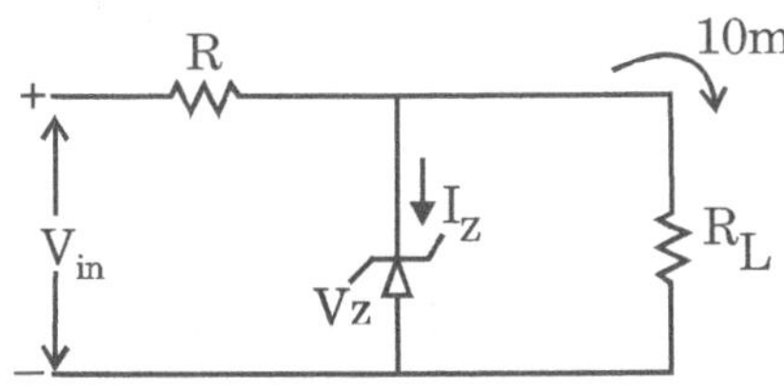

When $V_{in} = 20$ V, $\dfrac{20}{R} \geq 11 \times 10^{-3}$

or $R \leq 1818\ \Omega$ ...*(i)*

and when $V_i = 50$V

then $\dfrac{40}{R} \geq L \times 10^{-3}$

or $R \leq 3636$ ...*(ii)*

Combining equations *(i)* and *(ii)*, we get

$\mathbf{R \leq 1800\ \Omega}$

2.8 $$g_m = g_{m_0}\left(1 - \frac{V_{GS}}{V_p}\right) = \frac{2I_{DSS}}{[V_p]} \cdot \sqrt{\frac{I_D}{I_{DSS}}}$$

$V_G = 0,$

$V_s = I_D . R_s = 1\text{mA} \times 2.5\text{ k} = 2.5$ V

and $A_v \approx g_m R_s$

$V_{Gs} = V_G - V_s = -2.5$ V

$$\delta_m = \frac{2 \times 10\text{mA}}{5}\left(1 - \left(\frac{-2.5}{5}\right)\right) = 2\text{mS}$$

and $A_v = -gm\, R_D$

$= -2\text{mS} \times 3\text{k}\ \Omega = -6$

2.9

	t_0	t_1	t_2	t_3
Input to G_1	0	1	1	1
$\overline{G_1}$	1	1	0	0
Input to G_2	0	1	1	1
V_0	1	1	1	0

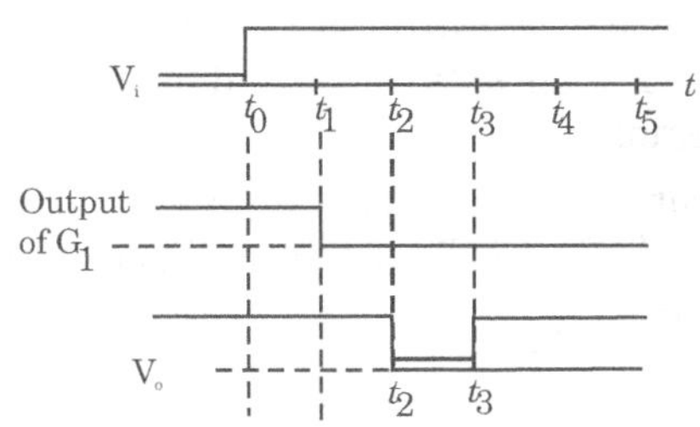

2.10. Working of two CMOS is depicted below:

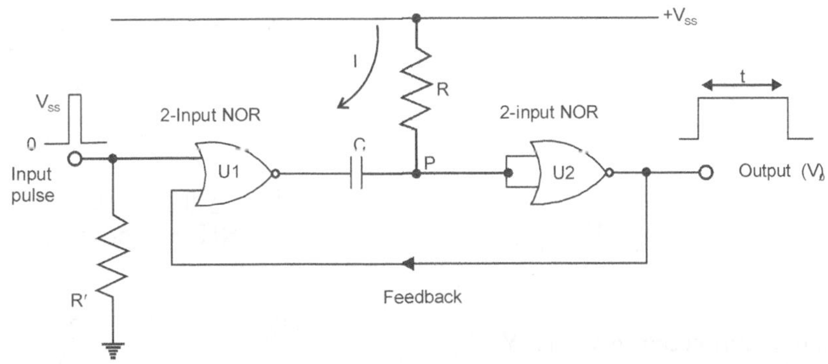

Suppose initially that the trigger input is LOW at a logic level "0" so that the output from the first NOR gate U1 is HIGH at logic level "1". The resistor, R is connected to the supply voltage so it is also equal to logic level "1", which means that the capacitor, C has the same charge on both of its plates. Junction P is therefore equal to this voltage so the output from the second NOR gate U2 will be LOW at logic level "0". This then represents the circuits "Stable State" with zero output.

When a positive trigger pulse is applied to the input at time t_0, the output of the first NOR gate U1 goes LOW taking with it the left hand plate of capacitor C1 there by discharging the capacitor. As both plates of the capacitor are now at logic level "0", so too is the input to the second NOR gate, U2 resulting in an output equal to logic level "1". This then represents the circuits second state, the "Unstable State" with an output voltage equal to $+V_{SS}$.

The second NOR gate, U2 will maintain this second unstable state until the timing capacitor now charging up through resistor, R reaches the minimum input Thershold voltage of U2 causing it to change state as a logic level "1" value has now appeared on its input. This causes the output to be rest to logic "0" which inturn is fed back (feed back loop) to one input of U2. This action automatically returns the monostable back to its original stable state and awaiting a second trigger pulse to restart the timing process once again.

2.11. For input voltage, $V_{IN} > V_T$, NMOS transistor (T_1) turns on, and current starts flowing through it. For V_{in} slightly greater than V_T, the PMOS (T_2)transistor stays in the triode region, and the NMOS transistor will be in saturation. For further increase of V_{IN}, drain current will increase and this will reduce voltage drop across T_2 thereby reducing V_0. Now both T_1 and T_2 enters into saturation and in that case further increase of V_{in} will cause larger drop in V_0 until low enough to pull out T_1 from saturation and to put in triode (linear region).

For saturation of NMOS (T_1), $V_{DS} \geq V_{GS} - V_T$ and for operation of NMOS (T_1) in triode (linear) region,

$$V_{DS} \leq V_{GS} - V_T$$

2.12. 8 – 4 2 – 1 BCD INPUT

X_3	X_2	X_1	X_0	Output
0	0	0	0	D_0
0	0	0	1	D_1
0	0	1	0	D_2
0	0	1	1	D_3
0	1	0	0	D_4
0	1	0	1	D_5
0	1	1	0	D_6
0	1	1	1	D_7
1	0	0	0	D_8
1	0	0	1	D_9

Now as D_0 is not connected to Y_3, Y_2, Y_1, Y_0

D_1 is connected to Y_0

D_2 is connected to Y_1

D_3 is connected to Y_1,Y_0

D_4 is connected to Y_2

D_5 is connected to Y_2,Y_0

D_6 is coneected to Y_3, Y_2

D_7 is coneected to Y_3, Y_2, Y_0

D_8 is coneected to Y_3, Y_2, Y_1

D_9 is coneected to Y_3, Y_2, Y_1, Y_0

Thus constructing truth-table for output Y,

2 – 4 – 2 –1

	Y_3	Y_2	Y_1	Y_0	Output
D_0	0	0	0	0	0
D_1	0	0	0	1	1
D_2	0	0	1	0	2
D_3	0	0	1	1	3
D_4	0	1	0	0	4
D_5	0	1	0	1	5
D_6	1	1	0	0	6
D_7	1	1	0	1	7
D_8	1	1	1	0	8
D_9	1	1	1	1	9

By the output shown

for $7 = 2Y_3 + 4Y_2 + Y_0$

$8 = 2Y_3 + 4Y_2 + 2Y_1$

$9 = 2Y_3 + 4Y_2 + 2Y_1 + 2Y_0$

Hence it can be inferred that out is 2 – 4 – 2 – 1 BCD code.

2.13 MVIB, 87 H ⟶ B ⟶ 87H

MOVA, B ⟶ A ⟶ 87 H

Start : ⟶ JMP NEXT

XRA B ⟶ A— OOH $s = 0$

JP START [will jump to start as $s = 0$]

JMP NEXT [Jump unconditionally to label Next]

XRA B ⟶ A ⟶ 87H

JP START [will not jump as $D_7 = 1$].

Out Port 2 ⟶ 87H

2.14 $$\frac{Y(s)}{R(s)} = \frac{K/s}{1-\left(\frac{3}{5}+K/s\right)} = \frac{K}{s-(K+3)}$$

The system will stable when, single pole lies in the RH-s plane, *i.e.*,

$$K + 3 > 0$$

or $$K < -3$$

2.15 Response, $$c(s) = \frac{100}{s(s+1)(s+100)}$$

[For $R(s) = \frac{1}{s}$ (unit-step input)].

then $C(t) = 1 - \frac{100}{90} e^{-t} + \frac{1}{99} e^{-t}$

[Using partial fraction]

the term $\frac{100}{99} e^{-t}$ will dominant on system response,

then $r\omega_n = 1$ and $t_s = \frac{4}{r\omega_n} = \frac{4}{1} = 4$ sec.

2.16 The characteristics equation is

$(2s + 1)(s^4 + 2s^2 + 1) = 0$

The roots are $s = -0.5$, $s = +1 \pm j$

Since, roots are present on the RHS and one root at negative real axis, therefore system will be unstable.

2.17 $\angle G(j\omega) H(j\omega) = -180°$

$$= -90° - \tan^{-1}\frac{\omega}{1-\omega^2}$$

$$\left[\text{Since } G(j\omega) = \frac{1}{j\omega\left(j\omega + 1 - \omega^2\right)}\right]$$

or $\tan^{-1}\frac{\omega}{1-\omega^2} = 90°$

which gives, $\omega = 1$ rad/sec.

then, [GM $= -\log |G(j\omega). H(j\omega)|_{\omega=1} = 0$ dB

2.18 Instantaneous frequency, ω_i, is given by,

$$\omega_i = \frac{d}{dt}[\theta(t)]$$

$$= \frac{d}{dt}\left[2\pi\left(2 \times 10^6 t + 30\sin 150t + 40\cos 150t\right)\right]$$

and $\Delta\omega = \omega_i - \omega_c$

$$= 3000\pi[3\cos 150t - 4\sin 150t]$$

$$= 15000\pi[\cos(150t + \alpha)],$$

where $\alpha = \tan^{-1}\frac{4}{3}$

Then maximum frequency deviation,

$$\Delta_f = \frac{|\Delta\omega|_{max}}{2\pi}$$

$$= \frac{15000\pi}{2\pi} = 7.5 \text{ kHz}$$

and $\phi(t) = 2\pi(30\sin 150t + 40\cos 150t)$

$$= 100\pi\left(\frac{3}{5}\sin 150t + \frac{4}{5}.\cos 150t\right)$$

$$= 100\pi . \sin(150t + \alpha')$$

where $\alpha' = \tan^{-1}\frac{4}{3}$.

Then, maximum phase deviation,

$$|\phi(t)|_{max} = 100\pi \text{ radian}$$

2.19

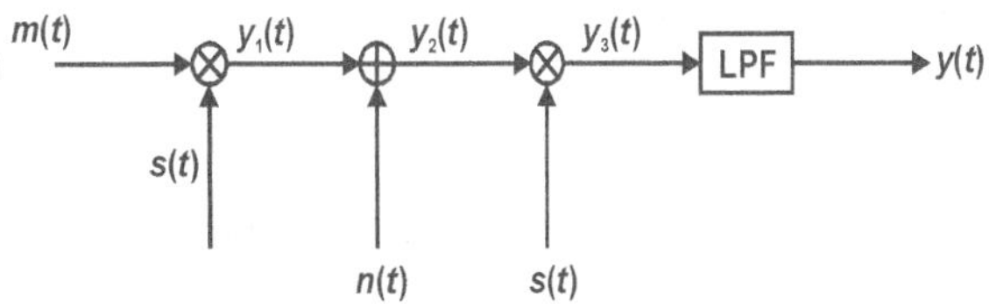

$$y_1(t) = m(t)\, s(t) = \frac{2\sin(2\pi t).\cos(200\pi t)}{t}$$

$$= \frac{\sin(202\pi t) - \sin(198\pi t)}{t} \quad y_2(t) = y_1(t) + n(t)$$

$$= \frac{\sin(202\pi t) - \sin(198\pi t) + \sin(199\pi t)}{t}$$

$$y_3(t) = y_2(t) . s(t)$$

$$= \frac{\sin(202\pi t).\cos(200\pi t) - \sin 198\pi t.\cos 200\pi t}{t} + \frac{\sin 199\pi t.\cos 200\pi t}{t}$$

$$= \frac{\sin 402\pi t + \sin 2\pi t - \sin 398\pi t + \sin 399\pi t + \sin(0.5\pi t).\cos(1.5\pi t)}{t}$$

After low-pass filterings, $y(t)$

$$= \frac{\sin 2\pi t + (\sin 0.5\pi t).(\cos 1.5\pi t)}{t}$$

2.20 $f_s = \frac{1}{T_s} = \frac{1}{50 \times 10^{-6}} = 20$ kHz

$$f_c = \frac{24\pi \times 10^3}{2\pi} = 12 \text{ kHz}$$

The frequencies passed through LPF are,

$f_c, f_s - f_c$ i.e., 12 kHz, 8 kHz

as LPF have cut-off frequencies is 15 kHz

2.21 $V_{ar}[d(n)] = V_{ar}[x(n) - x(n-1)]$

$\Rightarrow E[\{x(n) - x(n-1)\}^2]$

$\Rightarrow E[x^2(n)] + E[x^2(n-1)] - 2E[x(n).\, x(n-1)]$

$\Rightarrow V_{ar}[x(n)] + V_{ar}[x(n-1)] - 2R_{xx}(-1)$

[shifting does not effect variance]

$\sigma_d^2 = \sigma_x^2 + \sigma_x^2 - 2R_{xx}(1)$... even symmetric

$$\frac{\sigma_x^2}{10} = 2\sigma_x^2 - 2R_{xx}(1)$$

$$R_{xx}(1) = 0.95\,\sigma_x^2$$

$$\Rightarrow \frac{R_{xx}(1)}{\sigma_x^2} = 0.95$$

2.22

$$\overline{E}_X = (0.5).e^{j(wt-kz)}.\hat{x}$$

$$\overline{E}_Y = j.\, e^{j(\omega t - kz)}.\,\hat{y}$$

then, $$\frac{\overline{E}_x}{\overline{E}_y} = 0.5e^{-\frac{\pi}{2}} = \frac{1}{2}\angle -90^\circ$$

And, wave is elliptically polarized, as $\left|\frac{\overline{E}_x}{\overline{E}_y}\right| \neq 1.$

2.23 $$\tan\delta = \frac{\sigma}{\omega\,\varepsilon_0} = \frac{1.7\times10^{-4}}{78\,\varepsilon_0\times 2\pi\times f}$$

$$= \frac{1.7\times10^{-4}}{\frac{78\times10^{-9}}{36\pi}\times 2\pi\times3\times10^{9}} = 1.3\times10^{-5}$$

2.24 $$\overline{E} = \frac{\sigma}{t_0\varepsilon_0}$$

then, $$|\overline{\sigma}| = |\varepsilon_0\,\varepsilon\,\overline{E}|$$

$$= \frac{80\times2\times10^{-9}}{36\pi} = 1.41\times10^{-9}\ \text{C/m}^2$$

2.25. Signal strength $= \dfrac{P}{4\pi R^2}$

At $R = 5000$ m

$$P = P_1 = \frac{P}{4\pi(5000)^2}$$

Let at $R = x$ m,

$$(P_2)_{dB} = (P_1)_{dB} - 3$$

$$\Rightarrow \quad 10\log\frac{P_2}{P_1} = -3$$

$$\Rightarrow \quad 10\log\frac{P_1}{P_2} = 3$$

or, $$\frac{P_1}{P_2} = 2$$

then, $$\frac{P_1}{P_2} = 2 = \left(\frac{x}{5000}\right)^2$$

$$\Rightarrow \quad x = 5000\sqrt{2}$$

Required distance to be moved $= 5000(\sqrt{2}-1)$

$= 20 > 1$ m

SECTION - B

3. (*a*) At position 1, circuit will be in steady state as shown below :

$$U(-t) = 1, \quad t < 0$$
$$= 0, \quad t > 0$$

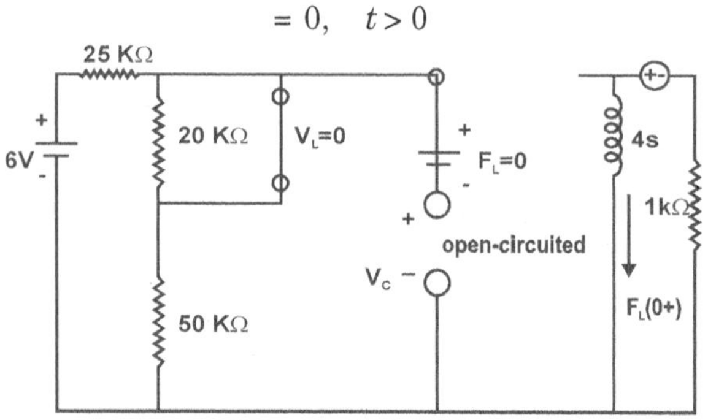

From the circuit,

$$V_c(0^+) = \frac{6}{75}.50 = 4\text{V}.,$$

$$I_L(0^+) = 10\text{ mA}$$

(*b*) When switch is thrown to position (2), the circuit will be as using KCL, we have

$$\frac{V_c}{R} + \frac{1}{L}\int_0^t V_c\,dt + C\frac{dV_c}{dt} = 0 \qquad (i)$$

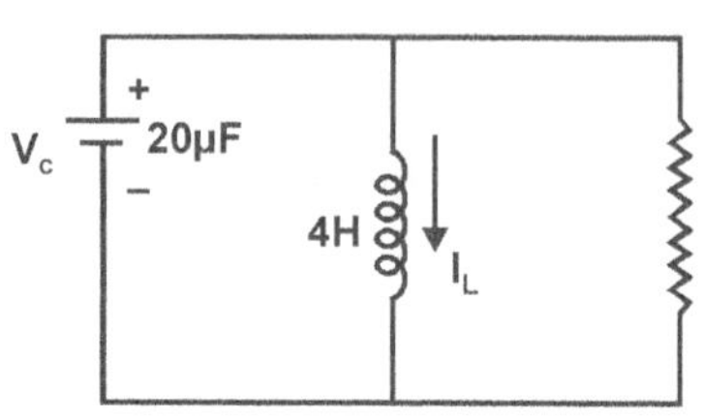

At $t = 0^+$,

$$\frac{dV_c}{dt} = \left[\frac{v_c(0+)}{R} + i_L(0^+)\right]\frac{1}{C}$$

$$= -[14].\frac{10^{-3}}{20\times10^{-6}}$$

$$= -\mathbf{700\ V/s}$$

(*c*) The solution of equation (*i*) is of the form,

$$V_c(t) = e^{-\alpha t}(A_1\cos\omega_d t + A_2\sin\omega_d t) \qquad ..(ii)$$

or $$\alpha = \frac{1}{2RC} = 25,$$

$$\omega_d = \sqrt{\omega_0^2 - \alpha^2} = 109$$

At $t = 0$, $v_C(0^+) = 4 = A_1$

Differentiating equation (*i*)

$$\frac{dV_c}{dt} = -\alpha\, e^{-\alpha t}(A_1\cos\ \omega_d t + A_2\sin\omega_d t)$$

$$4 + e^{-\alpha t}(-A_1\omega_d\cos\omega_d t + A_2\,\omega_d\,.\sin\omega_d t)$$

At $t = 0$

$$\left.\frac{dV_c}{dt}\right|_{t=0} = -A_1\alpha + \omega_d A_2 = -700$$

then, $A_2 = -5.50$

$$v_c(t) = e^{-25t}[4\cos(109\ t) - (5.5)\sin(109\ t)].$$

EC-4.

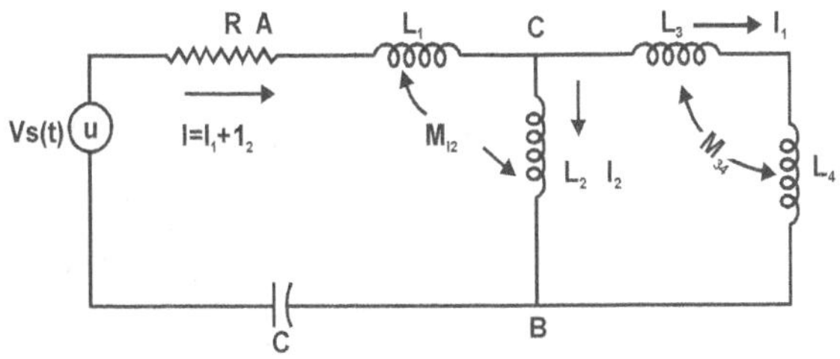

The equivalent circuit will be,

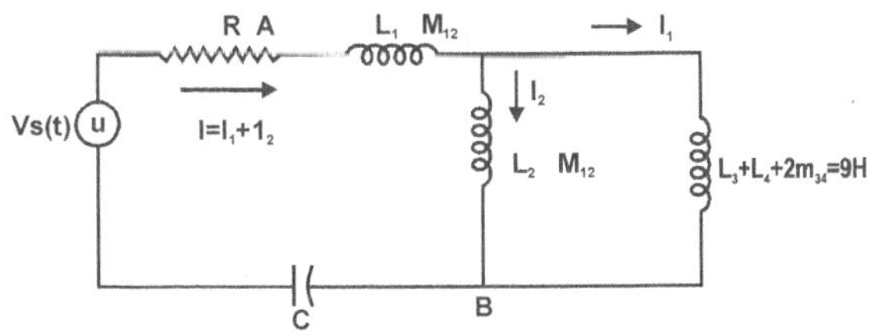

(*a*) Equivalent inductance for combination of L_3 and L_4

$$L = L_3 + L_4 + 2\,M_{34}$$
$$= 1 + 4 + 2.2 = 9H$$

(*b*) Equivalent inductance across the points A and B in the network

$$L_{eq} = \frac{9\,(L_2 - M_{12})}{9 + L_2 - M_{12}} + L_1 - M_{12}$$

$$= \frac{9 \times 2}{9 + 2} + 2 - 3 = \mathbf{0.634}$$

(*c*) $I = \dfrac{V_s\,(t)}{R + j\omega\,L_{eq} + \dfrac{1}{j\omega C}} = \dfrac{V_s\,(t)}{1 + j\left(\omega\,L_{eq} - \dfrac{1}{\omega C}\right)}$

Resonance occurs at frequency

$$\omega = \frac{1}{\sqrt{LC}} = \frac{1}{\sqrt{0.63 \times 0.2 \times 10^{-6}}}$$

$$= 2817.2 \text{ rad/sec}$$

$$f = \frac{\omega}{2\pi} = \frac{2817.2}{2\pi} = \mathbf{448.6\ Hz}$$

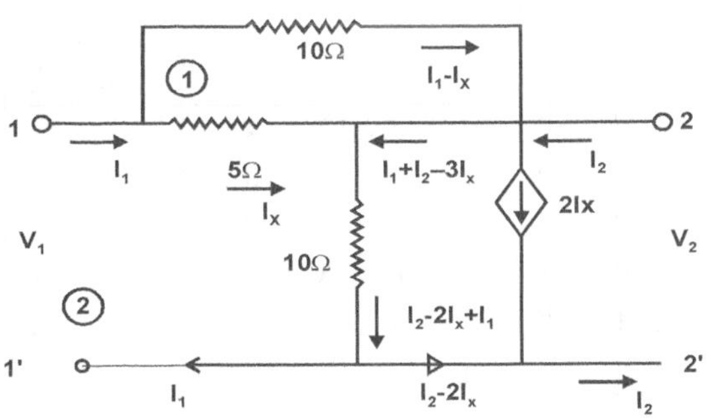

EC-5. (*a*) KVL in loop1 yields equation,

$$-10\,(I_1 - I_x) + 5\,I_x = 0$$

or $\quad 15\,I_x = 10\,I_1$

or $\quad 1.5\,I_x = I_1 \qquad \ldots(i)$

In loop (2),

$$V_1 = 5\,I_x + 10\,(I_1 + I_2 - 2I_x)$$
$$= -15\,I_x + 10\,I_1 + 10\,I_2 \qquad \ldots(ii)$$

Using equation *(i)*, we get $V_1 = 10\,I_2 \qquad \ldots(iii)$

Now, $\quad V_2 = 10\,(I_1 + I_2 - 2I_x) \qquad$ [from (*i*)]

$$= -3.33\,I_1 + 10\,I_2 \qquad \ldots(iv)$$

From equation (*iii*), $I_2 = \dfrac{V_1}{10} \qquad ..(v)$

and from equation (*iv*)

$$3.33\,I = -V_2 + V_1$$

or $\quad I_1 = +0.30\,V_1 - 0.30\,V_2 \qquad ..(vi)$

Hence, short-circuit admittance parameter,

$$Y_{SC} = \begin{bmatrix} 0.10 & 0 \\ 0.30 & -0.30 \end{bmatrix} \text{[Using equation } (v) \text{ and } (vi)\text{]}$$

(*b*) From equation (*iv*),

$$V_2 = -3.33\,I_1 + 10\,I_2$$

which gives, $Z_{22} = \left.\dfrac{V_2}{I_2}\right|_{I_1 = 0} = \mathbf{10}$

EC-6. (*a*)

Let, diode D_1 and D_2 are off and the circuit will be as shown

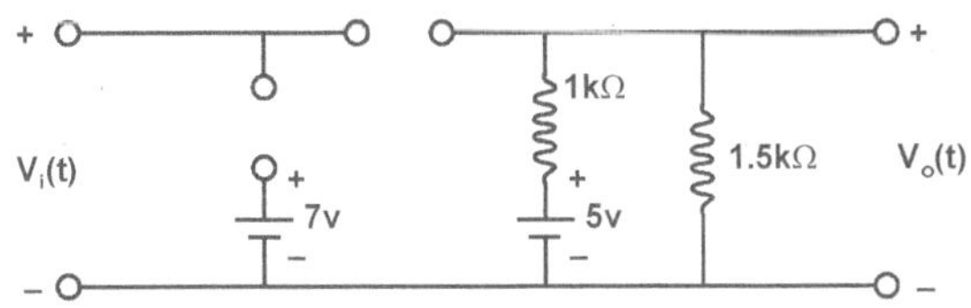

$$V_0 = \frac{5}{2.5}(1.5) = 3V$$

When $V_i > 5$ V, upper diode conducts and circuit will be as,

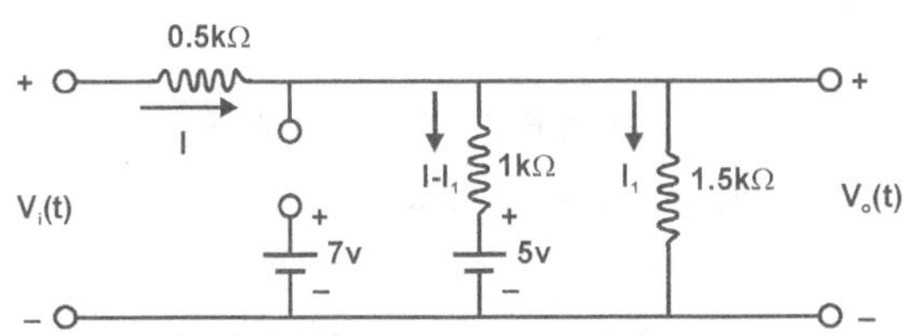

Using KVL, $V_i - 5 = 0.5\,I + (I - I_1)$

$$= 1.5\,I - I_1 \qquad \ldots(i)$$

and $\quad -1.5\,I_1 + 5 + (I - I_1) = 0$

or $\quad 5 + I = 2.5\,I_1 \qquad \ldots(ii)$

From (*i*) and (*ii*)

$$V_0 = 1.5\, I_1 = \frac{1.5\,(V_i + 2.5)}{2.75}$$

when, V_0 reaches 7 volt,

$V_i = 10.33$ volts

For $V_0 > 7$ volt, both diode conducts and

$V_0 = 7$ volts.

(*b*)

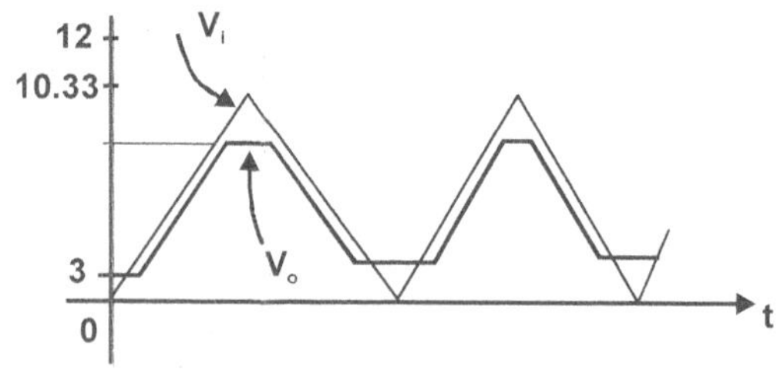

Output waveform superimposed on the input $V_i(t)$ shown below.

EC-8.

(*a*) Output of op-amp is '+'ve as, input to non-inverting terminal. Therefore B-E junction forward

biased and C-B junction reversed-biased, transistor operates in active region.

(*b*)
$$V_i = V_{EE1} + R_L I_0$$
$$= -15 + R_L I_0$$
$$= 10\, I_0 - 15$$

(*c*) If $V_i = 2V$

$I_0 = 1.7$ amp

Now, $I_0 = I_B(\beta + 1)$

$$\Rightarrow \quad I_B = \frac{I_0}{\beta + 1} = \frac{1.7}{100} = 0.017 \text{ amp.}$$

then, $V_{OP} = V_i + V_{BE} + R_B I_B$

$= 2 + 0.7 + (0.017) \times 150$

$=$ **5.25 volts.**

EC-9. Output, $X = \overline{A}B + \overline{A}\,\overline{B}\,\overline{C}$

$= \overline{A}B\,(C + \overline{C}) + \overline{A}\,\overline{B}\,\overline{C}$

$= \overline{A}BC + \overline{A}B\overline{C} + \overline{A}\,\overline{B}\,\overline{C}$ [in SOP form]

The K-map for X is,

A \ BC	00	01	11	10
0	1 (0)	0 (1)	1 (3)	1 (2)
1	0 (4)	0 (5)	0 (7)	0 (6)

From the K-map, output X is,

$X = \Pi M\,(1, 4, 5, 6, 7)$

$= (A + \overline{B} + C)(A + B + \overline{C})(A + \overline{B} + \overline{C})$
$(\overline{A} + \overline{B} + \overline{C})(\overline{A} + B + \overline{C})$

[in POS form]

(*b*) Preparing truth-table

Input			*Outout*
A	B	C	D
0	0	0	1
0	0	1	0
0	1	0	1
0	1	1	1
1	0	0	0
1	0	1	0
1	1	0	0
1	1	1	0

Now, comparing input C and output X, and preparing a truth-table, we have

A	*B*	*X*
0	0	$\overline{C}$
0	1	1
1	0	0
1	1	0

Now, complete circuit diagram will be as shown below :

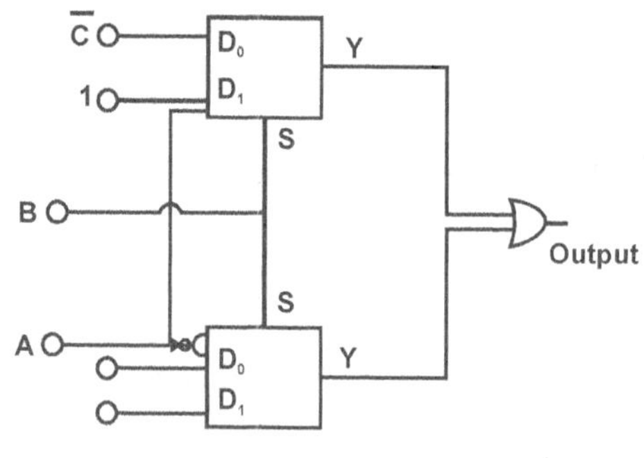

EC-11.

(*a*) Count-sequence

Q_2	Q_1	Q_0	A_2	B_2	A_1	B_1	A_0	B_0
0	0	0	1	X	1	X	0	X
0	0	1	1	X	0	X	X	1
0	1	0	1	X	X	0	0	X
0	1	1	0	X	X	1	X	1
1	0	0	X	1	1	X	1	X
0	0	0						

Excitation-table

Q_n	Q_{n+1}	A	B
0	0	1	X
0	1	0	X
1	0	X	1
1	1	X	0

(*b*) The K-maps are,

A_2

Q_2 \ Q_1Q_0	00	01	11	10
0	1	1	0	1
1	X	X	X	X

$A_2 = \overline{Q}_1 + \overline{Q}_0$

B_2

Q_2 \ Q_1Q_0	00	01	11	10
0	X	X	X	X
1	1	X	X	X

$B_2 = 1$

A_1

Q_2 \ Q_1Q_0	00	01	11	10
0	1	0	X	X
1	1	X	X	X

$A_1 = \overline{Q}_0$

B_1

Q_2 \ Q_1Q_0	00	01	11	10
0	X	X	1	0
1	X	X	X	X

$B_1 = Q_0$

A_0

Q_2 \ Q_1Q_0	00	01	11	10
0	0	X	X	0
1	1	X	X	X

$A_0 = Q_2$

B_0

Q_2 \ Q_1Q_0	00	01	11	10
0	X	1	1	X
1	X	X	X	X

$B_0 = 1$

The required circuit will be as shown below.

Mod-5 counter

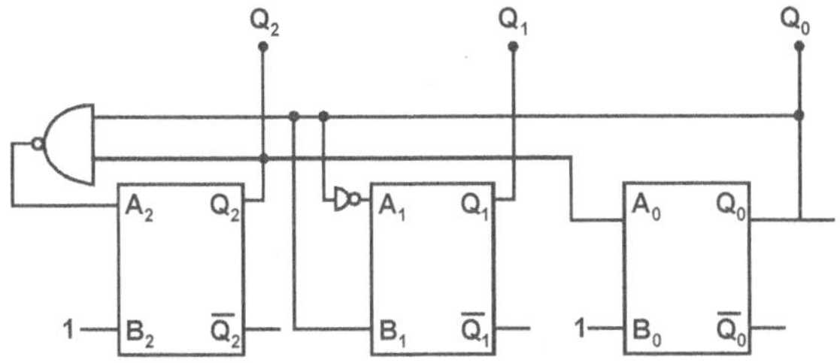

EC -12. Machine-cycles T-state

		Machine-cycles	T-state
MOV	A, B	2	7
OUT	55H	3	10
DCR	B	3	10
STA	FFF8H	4	13
JMP	START	3	10

(*a*) Number of machine cycles till the JMP executed $= 2 + 3 + 3 + 4 = \mathbf{12}$

(*b*) Timing of STA. FFF8H

M_1 (opcode fetch)				M_2 Memory read			M_3 Memory Read			M_4 Memory write		
T_1	T_2	T_3	T_4	T_1	T_2	T_3	T_1	T_2	T_3	T_1	T_2	T_3

$\overline{\text{MEMW}}$

From the diagram, time states required between two $\overline{\text{MEMW}}$ signals,

$$T = (7 + 10 + 10 + 13 + 10) \text{ state}$$
$$= 50 \text{ cycles}$$

and required time $= \dfrac{50}{5 \times 10^6} = \mathbf{10^{-5}}$ **sec**

(*c*) If three WAIT states are introduces, then 53 time-state required and then time interval

$$= \frac{53}{5 \times 10^6} = \mathbf{10.6 \times 10^{-6}\ sec.}$$

EC-13. Given, $G_p(j\omega) = \dfrac{1}{(j\omega+1)(2j\omega+1)}$

(*a*) $\angle G_p(j\omega) = -90° = -\tan^{-1}\omega - \tan^{-1}2\omega$

or $\tan^{-1}\omega + \tan^{-1}(2\omega) = 90°$

or $\tan^{-1}\dfrac{3\omega}{1-2\omega^2} = 90°$

which gives, $\omega = \dfrac{1}{\sqrt{2}}$ **rad/sec**

(*b*) New, open-loop transfer function

$$G(s) = G_p(s) \cdot G_c(s) = \frac{k}{s(s+1)(2s+1)}$$

$\angle G(j\omega) = -180°$

or $-90° - \tan^{-1}\omega - \tan^{-1}2\omega = -180°$

or $\tan^{-1}\dfrac{3\omega}{1-2\omega^2} = 90°$

or $\omega = \dfrac{1}{\sqrt{2}}.$

As given, $GM = 2.5 = \dfrac{1}{|G(j\omega)|_{\omega=\frac{1}{\sqrt{2}}}}$

or $\dfrac{1}{2.5} = \dfrac{k}{\left|\left(\frac{1}{\sqrt{2}}\right)\cdot\left(\frac{1}{\sqrt{2}}j+1\right)\left(\sqrt{2}j+1\right)\right|}$

or $\dfrac{1}{2.5} = \dfrac{2k}{3}$

or $k = \dfrac{3}{5}$

(*c*) Position error constant,

$k_p = \underset{s\to 0}{\text{Lt}}\, G(s) = \infty$

then, e_{ss} (due to step-input) $= \dfrac{1}{1+k_p} = 0$

Velocity error constant, $k_v = \underset{s\to 0}{\text{Lt}}\, S\,G(s)$

$$= \underset{s\to 0}{\text{Lt}} \frac{\left(\frac{3}{5}\right)}{(s+1)(2s+1)} = \frac{3}{5}$$

then, e_{ss} (due to ramp-input) $= \dfrac{1}{k_v} = \dfrac{5}{3}.$

EC - 14. The transfer function is of the form,

$$G(s) = \frac{k}{(S+P_1)(S+P_2)}$$

At $\omega = 0$,

$G(j\omega)\,|_{\omega=0} = \dfrac{k}{P_1P_2} = 2$ ...(*i*)

$$G(j\omega) = \frac{k}{(j\omega+P_1)(j\omega+P_2)} = \frac{k}{\left(-\omega^2 + P_1P_2 + j\omega(P_1+P_2)\right)}$$

For, Re $[G(j\omega)] = 0$, it requires, $P_1P_2 - \omega^2 = 0$

or $\omega = \sqrt{P_1P_2}$

Given, $\omega = 2,$

Then, $P_1P_2 = \omega^2 = 4$ ...(*ii*)

From equation (*i*),

$k = 2P_1P_2 = 8$

and $\dfrac{-jk}{\omega(P_1+P_2)} = -2j$

then, $P_1 - P_2 = \sqrt{(P_1+P_2)^2 - 4P_1P_2}$

$= \sqrt{4-16} = 2\sqrt{3}j$

thus, $P_1 = 1+\sqrt{3}j$

$P_2 = 1-\sqrt{3}j$

Hence, transfer function,

$$G(s) = \frac{8}{\left(s+1+\sqrt{3}j\right)\left(s+1-\sqrt{3}j\right)} = \frac{8}{\left[(s+1)^2+3\right]} = \frac{8}{(s^2+2s+4)}$$

EC-15. (*a*) From the given figure, we have

$$X_2(s) = \frac{1}{s}\left[-U(s) - 2X_2(s)\right]$$

or $sX_2(s) = [-2X_2(s) - U(s)]$ ...(*i*)

and $X_1(s) = [U(s) - X_1(s) + X_2(s)]\dfrac{1}{s}$

or $SX_1(s) = X_2(s) - X_1(s) + U(s)$ ..(*ii*)

and $X_1(s) + X_2(s) = Y(s)$ ...(*iii*)

Taking inverse laplace to all three equations (*i*), (*ii*) and (*iii*), we get

$\dot{x}_2(t) = [-2x_2(t) - U(t)]$ ...(*iv*)

$\dot{x}_1(t) = x_2(t) - x_1(t) + U(t)$ ..(*v*)

$Y(t) = x_1(t) + x_2(t)$ ..(*vi*)

From equations (*iv*), (*v*)

$$\begin{bmatrix}\dot{x}_1(t)\\ \dot{x}_2(t)\end{bmatrix} = \begin{bmatrix}-1 & 1\\ 0 & -2\end{bmatrix}\begin{bmatrix}x_1(t)\\ x_2(t)\end{bmatrix} + \begin{bmatrix}1\\ -1\end{bmatrix}[U(t)]$$

(*b*) $\phi(t) = L^{-1}[SI - A]^{-1}$

$$= L^{-1}\begin{bmatrix} s+1 & -1 \\ 0 & s+2 \end{bmatrix}^{-1}$$

$$= L^{-1}\begin{bmatrix} \frac{1}{s+1} & \frac{1}{(s+1)(s+2)} \\ 0 & \frac{1}{s+2} \end{bmatrix}$$

$$= \begin{bmatrix} e^{-t} & e^{-t} - e^{-2t} \\ 0 & e^{-2t} \end{bmatrix}$$

(*c*) From equation (*iv*)

$$y(t) = x_1(t) + x_2(t)$$

At $t = 0$, $x_2(0) = x_2(0) = 1$

then, $y(t) = 2$

EC-16. (*a*) Autocorrelation $R_{XX}(\tau) = E[X(t+\tau)X(t)]$

$$= \frac{1}{T}\int_{-T/2}^{T/2} \cos[2\pi(t+\tau)].\cos 2\pi t\, dt$$

$$= \frac{1}{2T}\int_{-T/2}^{T/2} [\cos 2\pi(2t+\tau) + \cos 2\pi\tau]\, dt$$

$$= \frac{1}{2T}\left[\frac{\sin 2\pi(2t+\tau)}{4\pi}\Bigg|_{-T/2}^{T/2} + (\cos 2\pi\tau).t\Bigg|_{-T/2}^{T/2}\right]$$

$$= \frac{\cos 2\pi\tau}{2}$$

and $S_{XX}(f) = \int_{-\infty}^{\infty} R_{XX}(\tau).e^{-j\omega\tau}.d\tau$

$$= \frac{1}{2}[\delta(f-1) + \delta(f+1)]$$

Using $\cos(2\pi f_c t) = \frac{1}{2}[\delta(f-f_c) + \delta(f+f_c)]$

(*b*) Now, $R_{YY}(\tau) = \frac{d}{dt}R_{XX}(\tau) = \pi\sin(2\pi\tau)$

and $S_{YY}(\phi) = H(\omega).S_{XX}(f)$

$$= j\omega.S_{XX}(f) = (j.2\pi f).\frac{1}{2}[\delta(f-1) + \delta(f+1)]$$

$$= (j\pi f).[\delta(f-1) + \delta(f+1)]$$

(*c*) $R_{xy}(\tau) = E[X(t)Y(t+\tau)]$

$$= \frac{2\pi}{T}\int_{-T/2}^{T/2} [\sin 2\pi(2t+\tau) + \sin(2\pi\tau)].d\tau$$

$$= \frac{\pi}{T}\left[\frac{\cos 2\pi(2t+\tau)}{4\pi}\Bigg|_{-T/2}^{T/2} + \sin 2\pi\tau.T\right]$$

$$= \pi\sin 2\pi\tau$$

then $R_{xy}(0) = 0$

$$R_{xy}\left(\frac{1}{4}\right) = \pi.\sin\frac{\pi}{2} = \pi$$

EC-17 (*a*) $s(t) = E[s(t)]$

$$= E[10\cos(2\pi\times10^6 t + \phi).m(t)]$$

$$= A\,E[m(t)].E[\cos(2\pi\times10^6 t + \phi)]$$

$$= 0$$

[$\because$ $m(t)$ and ϕ are independent and $E[m(t)] = 0$]

$R_{SS}(\tau) = R_{SS}(t, t+\tau) = E[s(t).s(t+\tau)]$

$$= E[100.m(t).m(t+\tau).\cos[\omega_c t + \phi].\cos[\omega_c(t+\tau) + \phi)]$$

$$= 50.E[m(t).m(t+\tau)]E[\cos\omega_c + \cos(2\omega_c t + \omega_c\tau + 2\phi)]$$

Here, $\omega_c = 2\pi\times10^6$ rad/sec

$$= 50.R_{mn}(\tau).\cos\omega_c\tau$$

$$= 50.R_{mn}(\tau).[\cos(2\pi\times10^6\tau)]$$

(*b*) $S_{ss}(f) = [R_{ss}(\tau)]$

$$= 50\,\tau[R_{mn}(\tau)\cos(2\pi\times10^6\tau)]$$

Given, $\tau[R_{mn}(\tau)] = S_{mn}(f)$.

$\tau[\cos(2\pi\times10^6\tau)]$

$$= \frac{1}{2}[\delta(f-f_c) + \delta(f+f_c)]$$

where [$f_c = 10^6$ Hz.]

Now, using convolutional theorem,

$$g_1(t) * g_2(t) = G_1(f).G_2(f)$$

and $g_1(t) * \delta(t-t_o) = g(t-t_0)$

We have, $S_{ss}(f) = 50[S_{mn}(f) * [\pi\delta[f-f_c] + \delta(f+f_c)]$

$$= 50[S_{mn}(f-f_c) + S_{mn}(f+f_c)]$$

(*c*) Power in the modulated signal $= (2f_c).50$

$$= 2\times10^6\times50 = 10^8 \text{ watt}$$

and, noise power $= (10^{-4}\text{ W/Hz}).2f_c$

$$= 10^{-4}\times2\times10^6$$

= 200 watt

EC-18.

(*a*) Minimum sampling rate

$$f_s \geq 2 \times f_m \;[f_m \text{ is higher-frequency}]$$
$$= 2.5 = \mathbf{10\ MHz}$$

(*b*) $(SNR)_{0,dB} = 6.02\,n + 1.76$ (n is the number of bits sample)

and, $nf_s = R_b$

$$n = \frac{R_b}{f_s} = \frac{7}{10}$$
$$= 0.7 \approx 1.$$

then, $(SNR)_{0,dB} = 6.02 \times 1 + 1.76 = \mathbf{7.78\ dB}$

EC-19.

(*a*) Two pair of signals as,

$$s_1(t) = \sqrt{\frac{2E_b}{T_b}}.\cos(2\pi f_c t) \quad \text{[represents 1]} \quad ....(i)$$

$$s_2(t) = \sqrt{\frac{2E_b}{T_b}}.\cos(2\pi f_c t) \quad \text{[represents 0]} \quad ....(ii)$$

Comparing with given,

$$\sqrt{\frac{2E_b}{T_b}} = 6\sqrt{2}$$

or Transmitted energy/bit, $E_b = 6T_b$

$$= 6 \cdot \frac{1}{R}$$

$$= 6.\frac{1}{160 \times 10^3} = \mathbf{0.375 \times 10^{-4}}$$

(*b*) From equation (*i*) and (*ii*), there is only one basis function, *i.e.*

$$\phi(t) = \sqrt{\frac{2}{T_b}}.\cos(2\pi f_c t)$$

where, $T_b = \frac{1}{160 \times 10^3}$ sec

and, $f_c = 640\,\pi \times 10^3$

$$\therefore \quad P_{eo} = \frac{1}{2} erf_c\left(\sqrt{\frac{E_b}{N_0}}\right)$$

$$= \frac{1}{2}\left[1 - erf\left(\sqrt{\frac{E_b}{N_0}}\right)\right]$$

$$= \frac{1}{2}[1 - erf\,(0.25)]$$

Since, $erf(0.25) = 0.276$

$$\therefore \quad P_{eo} = \frac{1}{2}(1 - 0.276) = \mathbf{0.362}$$

EC-22.

(a) $$|\overline{E}_\theta| = \frac{2E_m}{R_0}\left|\frac{\cos\left[\left(\frac{\pi}{2}\right)\cos\theta\right]}{\sin\theta}\right| \cdot \left|\cos\frac{\psi}{2}\right| \quad ....(i)$$

where $\psi = \beta d \sin\theta\cos\phi + r$

Given, $r = -90°$, $\theta = 90°$, which gives,

$$\Psi = \beta d - 90°$$
$$= \frac{\pi}{2} - \frac{\pi}{2} = 0°$$

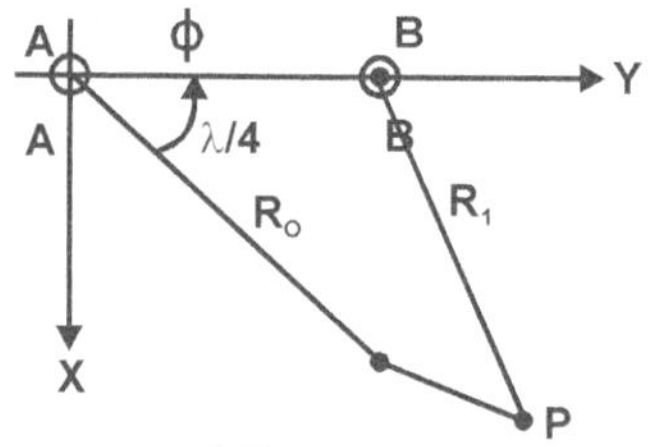

then, $|\overline{E}_\theta| = \frac{2E_m}{R_o}$ [Using equation (*i*)]

(*b*) Array factor,

$$|A(\phi)| = \left|\cos\frac{\Psi}{2}\right|$$

$$= \left|\cos\frac{\beta d\sin\theta\cos\phi + r}{2}\right|$$

$$= \left|\cos\frac{\pi}{4}(\cos\phi - 1)\right|$$

The radiation pattern is shown below,

H-plane radian pattern of two-element parallel dipole array

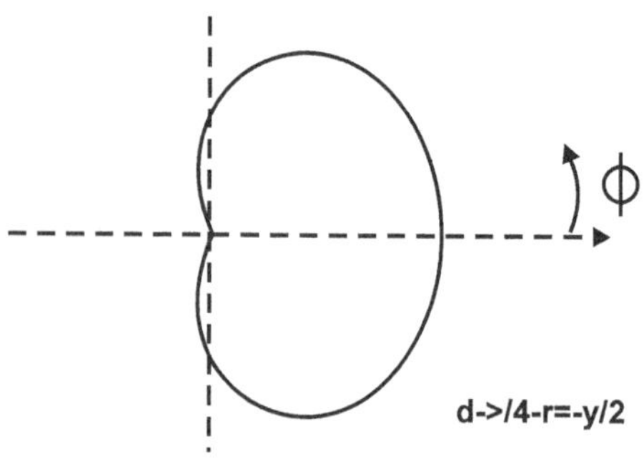

$$d = \frac{\lambda}{4}, \quad r = -\frac{\lambda}{2}$$

SOLVED PAPER – 2001

ELECTRONICS & COMMUNICATION ENGINEERING

Time Allowed : 3 Hours | **Maximum Marks : 150**

SECTION - A (75 Marks)

1. This question consists of TWENTY FIVE sub-questions (1.1 - 1.25) of ONE mark each. For each of these sub-questions, four possible answers (*a*, *b*, *c* and *d*) are given, out of which only one is correct. Answer each sub-question by darkening the appropriate bubble on the OBJECTIVE RESPONSE SHEET (ORS) using a soft HB pencil. Do not use the ORS for any rough work. You may like to use the Answer Book for any rough work, if needed. (1 × 25 = 25)

1.1. The Voltage e_0 in the figure,

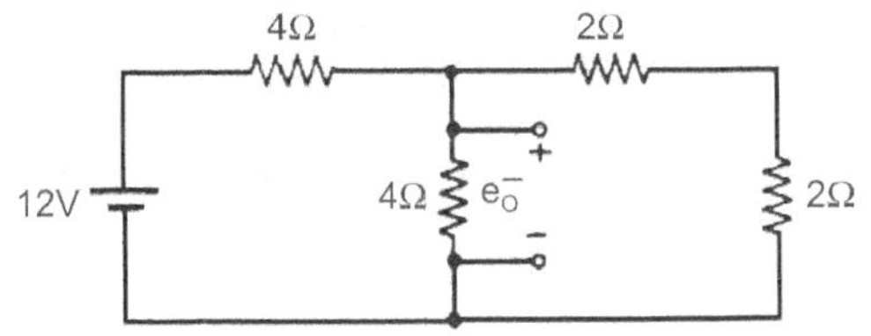

(*a*) 2 V (*b*) $\frac{4}{3}$ V

(*c*) 4 V (*d*) 8 V

1.2. If each branch of a Delta circuit has impedance $\sqrt{3}$ Z, then each branch of the equivalent Wye circuit has impedance.

(*a*) $\frac{Z}{\sqrt{3}}$ (*b*) 3 Z

(*c*) $3\sqrt{3}$ Z (*d*) $\frac{Z}{3}$

1.3. The transfer function of a system is given by H(*s*) $= \frac{1}{s^2(s-2)}$. The impulse response of the system is

(*a*) $(t^2 * e^{-2t})U(t)$ (* denotes convolution, and U(*t*) is unit step function)

(*b*) $(t * e^{2t})U(t)$

(*c*) $(te^{-2}t)U(t)$

(*d*) $(te^{-2t})U(t)$

1.4 The admittance parameter Y_{12} in the 2-port network in the figure,

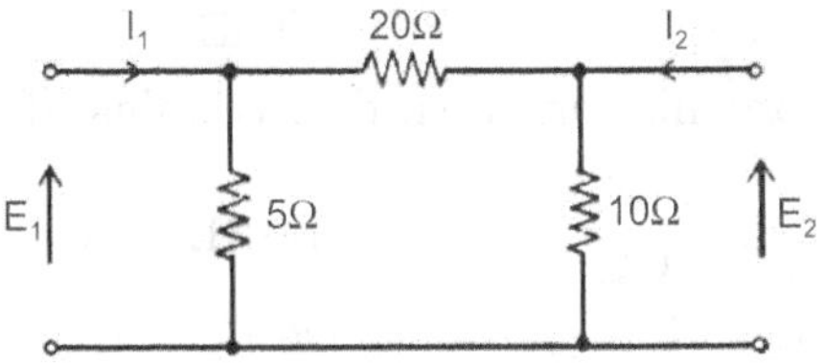

(*a*) – 0.2 mho (*b*) 0.1 mho

(*c*) – 0.05 mho (*d*) 0.05 mho

1.5 The region of convergence of the z-transform of a unit step function is

(*a*) $|z| > 1$ (*b*) $|z| < 1$

(*c*) (Real part of *z*) > 0 (*d*) (Real part of *z*) < 0

1.6 The current gain of a BJT is

(*a*) $g_m r_o$ (*b*) $\frac{g_m}{r_o}$

(*c*) $g_m r_\pi$ (*d*) $\frac{g_m}{r_\pi}$

1.7 MOSFET can be used as a

(*a*) current controlled capacitor

(*b*) voltage controlled capacitor

(*c*) current controlled inductor

(*d*) voltage controlled inductor

1.8 The effective channel length of a MOSFET in saturation decreases with increase in

(*a*) gate voltage (*b*) drain voltage

(*c*) source voltage (*d*) body voltage

1.9 The ideal OP-AMP has the following characteristics.

(*a*) $R_i = \infty$, $A = \infty$, $R_0 = 0$

(*b*) $R_i = 0$, $A = \infty$, $R_0 = 0$

(*c*) $R_i = \infty$, $A = \infty$, $R_0 = \infty$

(*d*) $R_i = 0$, $A = \infty$, $R_0 = \infty$

1.10 The 2's complement representation of –17 is

(*a*) 101110 (*b*) 101111

(*c*) 111110 (*d*) 110001

1.11 Consider the following two statements :

Statement 1 : A stable multivibrator can be used for generating square wave.

Statement 2 : B stable multivibrator can be used for storing binary information.

(*a*) Only statement 1 is correct

(*b*) Only statement 2 is correct

(*c*) Both the statements 1 and 2 are correct

(*d*) Both the statements 1 and 2 are incorrect

1.12 For the ring oscillator shown in the figure, the propagation delay of each inverter is 100 pico sec. What is the fundamental frequency of the oscillator output ?

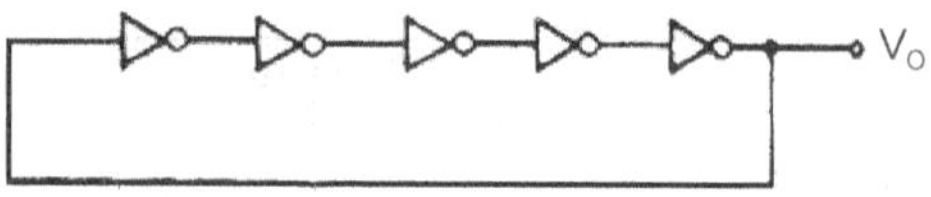

(*a*) 10 MHz (*b*) 100 MHz

(*c*) 1 GHz (*d*) 2 GHz

1.13 An 8085 microprocessor based system uses a 4K × 8 bit RAM whose starting address is AA00 H. The address of the last byte in this RAM is

(*a*) 0FFF H (*b*) 1000 H

(*c*) B9FF H (*d*) BA00 H

1.14 The equivalent of the block diagram in the figure, given in

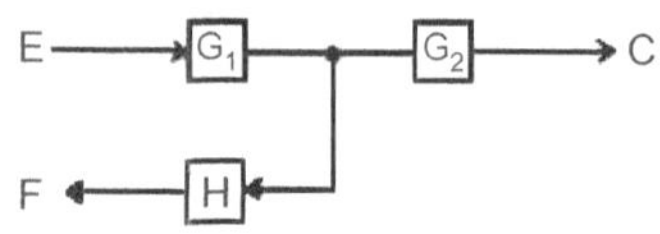

(*a*)

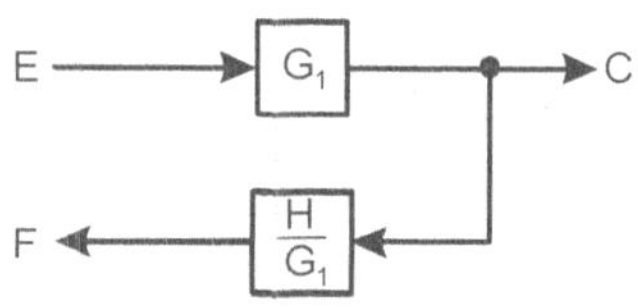

(*b*)

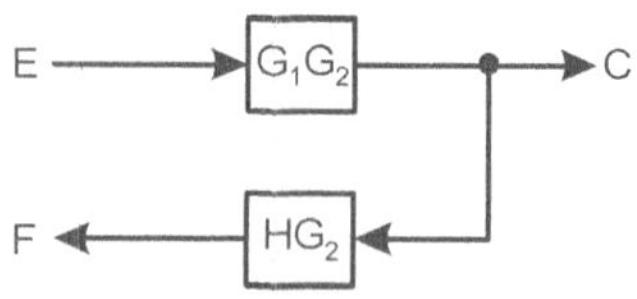

(*c*)

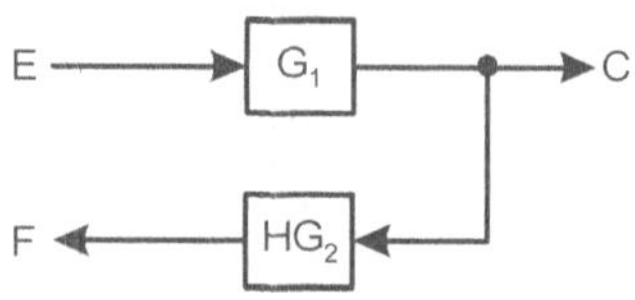

(*d*)

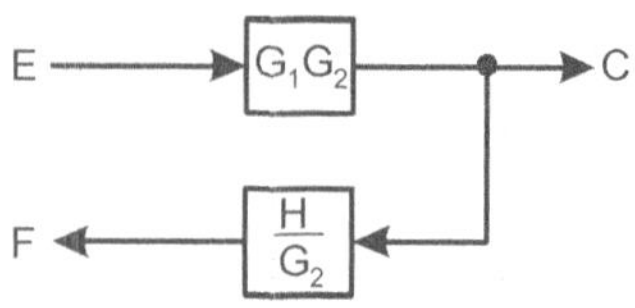

1.15 If the characteristic equation of a closed-loop system is $s^2 + 2s + 2 = 0$, then the system is

(*a*) overdamped (*b*) critically damped

(*c*) underdamped (*d*) undamped

1.16 The root-locus diagram for a closed-loop feedback system is shown in the figure is. The system is overdamped.

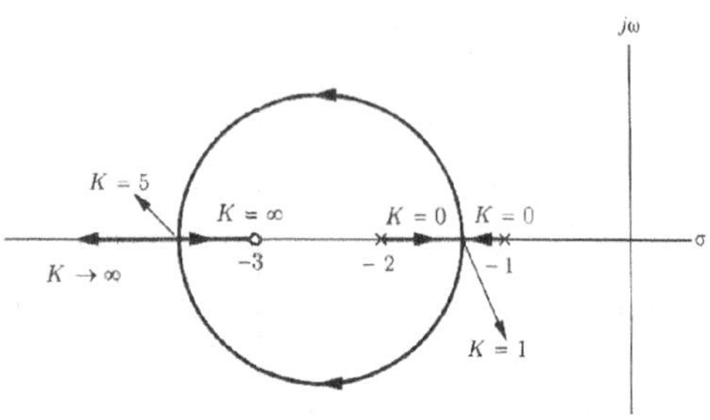

(*a*) only if $0 \le K \le 1$ (*b*) only if $1 < K < 5$

(*c*) only if $K > 5$ (*d*) if $0 \le K < 1$ or $K > 5$

1.17 The Nyquist plot for the open-loop transfer function G(*s*) of a unity negative feedback system is shown in the figure, if G(*s*) has no pole in the right-half of *s*-plane, the number of roots of the system characteristic equation in the right-half of *s*-plane is

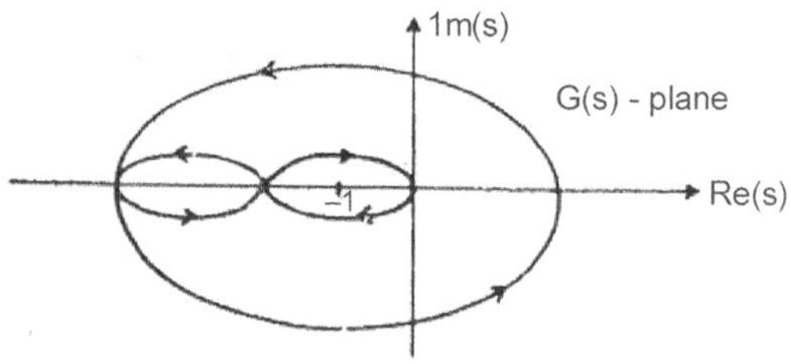

(*a*) 0 (*b*) 1

(*c*) 2 (*d*) 3

1.18 Let $\delta(t)$ denote the delta function. The value of the integral $\int_{-\infty}^{\infty} \delta(t)\cos\left(\frac{3t}{2}\right)dt$ is

(*a*) 1 (*b*) –1

(*c*) 0 (*d*) $\frac{\pi}{2}$

1.19 A bandlimited signal is sampled at the Nyquist rate. The signal can be recovered by passing the samples through

(*a*) an RC filter

(*b*) an envelope detector

(*c*) a PLL

(*d*) an ideal low-pass filter with the appropriate bandwidth.

1.20 The PDF of a Gaussian random variable X is given by $P(x) = \frac{1}{3\sqrt{2\pi}} e^{\frac{-(x-4)^2}{18}}$. The probability of the event {X = 4} is

(*a*) $\frac{1}{2}$ (*b*) $\frac{1}{3\sqrt{2\pi}}$

(*c*) 0 (*d*) $\frac{1}{4}$

1.21 If a signal *f*(*t*) has energy E, the energy of the signal *f*(2*t*) is equal to

(*a*) E (*b*) $\frac{E}{2}$

(*c*) 2E (*d*) 4E

1.22 A transmission line is distortionless if

(*a*) $RL = \frac{1}{GC}$ (*b*) RL = GC

(*c*) LG = RC (*d*) RG = LC

1.23 If a plane electromagnetic wave satisfies the equation $\frac{\partial^2 E_x}{\partial z^2} = c^2 \frac{\partial^2 E_x}{\partial t^2}$, the wave propagates in the

(*a*) *x*-direction

(*b*) *z*-direction

(*c*) *y*-direction

(*d*) *xy* plane at an angle of 45° between the *x* and *z*– directions

1.24 The phase velocity of waves propagating in a hollow metal waveguide is

(*a*) greater than the velocity of light in free space.

(*b*) less than the velocity of light in free space.

(*c*) equal to the velocity of light in free space.

(*d*) equal to the group velocity.

1.25 The dominant mode in a rectangular waveguide is TE10, because this mode has

(*a*) no attenuation

(*b*) no cut-off

(*c*) no magnetic field component

(*d*) the highest cut-off wavelength

2. This question consists of TWENTY FIVE sub-questions (2.1 – 2.25) of TWO marks each. For each of these sub-questions, four possible answers (*a*, *b*, *c* and *d*) are given, out of which only one is correct. Answer each sub-question by darkening the appropriate bubble on the OBJECTIVE RESPONSE SHEET (ORS) using a soft HB pencil. Do not use the ORS for any rough work. You may like to use the Answer Book for any rough work, if needed. (2 × 25 = 50)

2.1 The voltage e_0 in the figure,

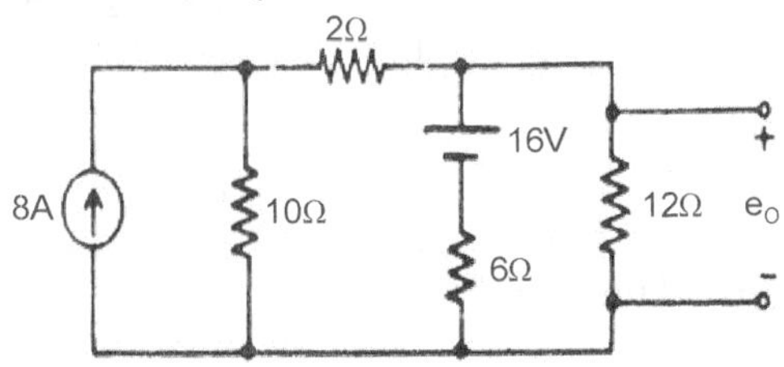

(*a*) 48 V (*b*) 24 V

(*c*) 36 V (*d*) 28 V

2.2 In the figure, the value of the load resistor R which maximizes the power delivered to it is

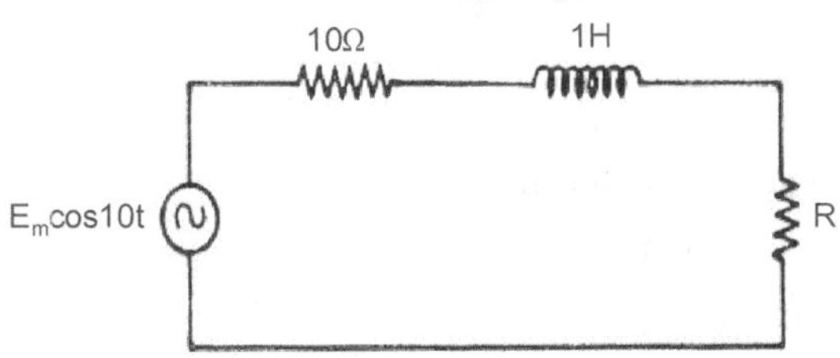

(*a*) 14.14 Ω (*b*) 10 Ω

(*c*) 200 Ω (*d*) 28.28 Ω

2.3 When the angular frequency ω in the figure, varied from 0 to ∞, the locus of the current phasor I_2 is given by

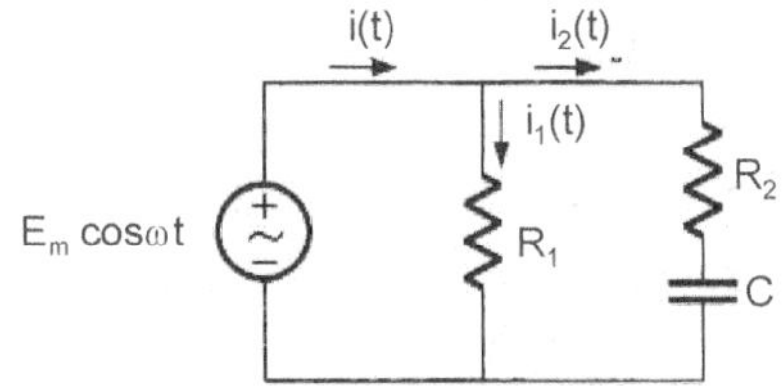

(*a*)

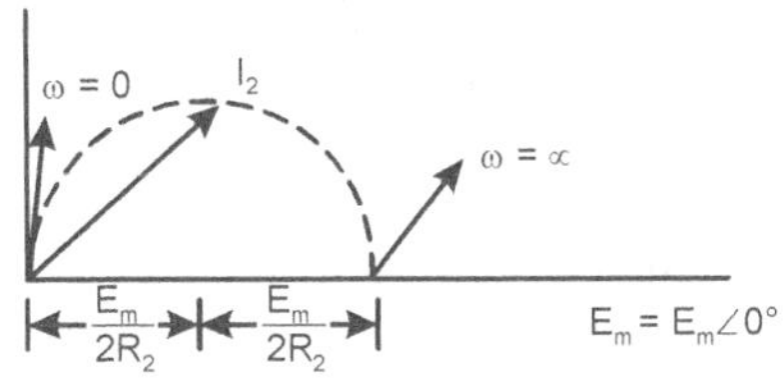

(*b*)

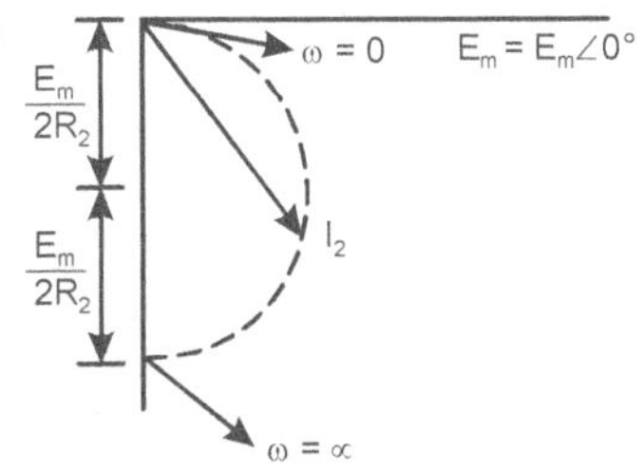

(*c*)

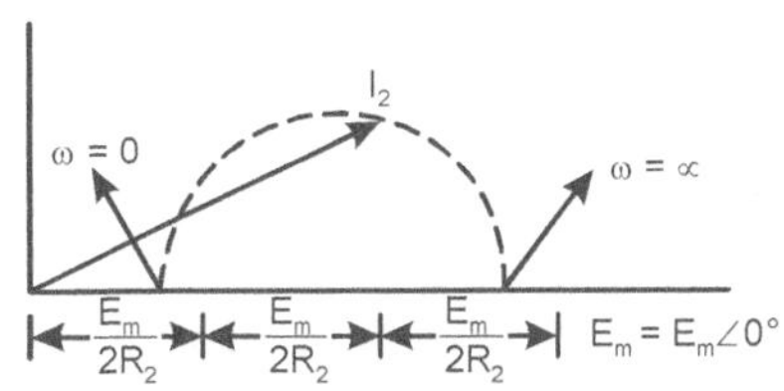

(*d*)

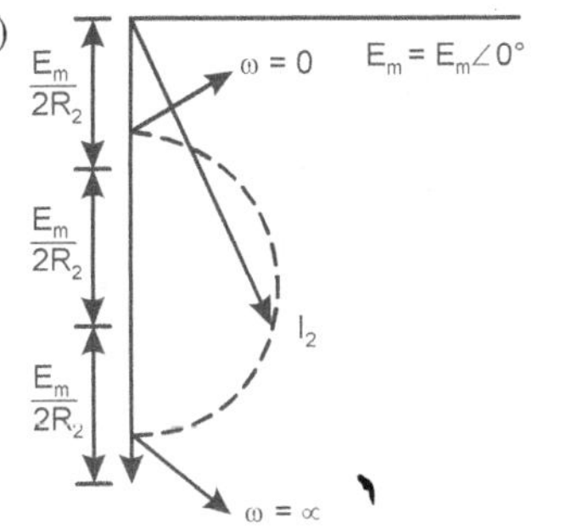

2.4 The Z-parameters Z_{11} and Z_{21} for the 2-port network in the figure,

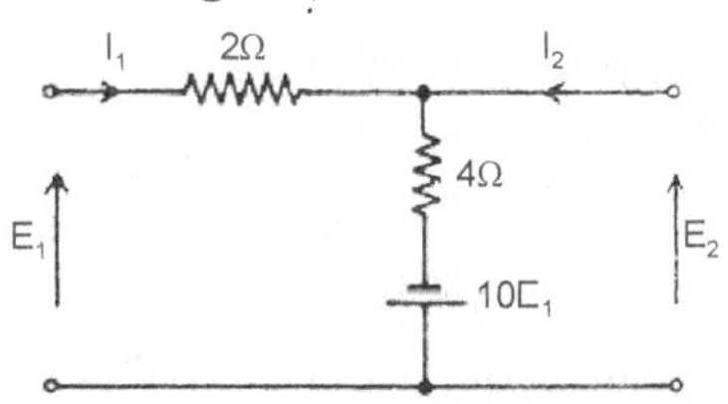

(*a*) $Z_{11} = -\frac{6}{11}\Omega; Z_{21} = \frac{16}{11}\Omega;$

(*b*) $Z_{11} = \frac{6}{11}\Omega; Z_{21} = \frac{4}{11}\Omega;$

(*c*) $Z_{11} = \frac{6}{11}\Omega; Z_{21} = -\frac{16}{11}\Omega;$

(*d*) $Z_{11} = \frac{4}{11}\Omega; Z_{21} = \frac{4}{11}\Omega;$

2.5 An npn BJT has gm = 38m A/V, $C_\mu = 10^{-14}$ F, $C_\pi = 4 \times 10^{-13}$ F, and DC current gain $\beta_0 = 90$. For this transistor f_T and f_β are

(*a*) $f_T = 1.64 \times 10^8$ Hz and $f_\beta = 1.47 \times 10^{10}$ Hz
(*b*) $f_T = 1.47 \times 10^{10}$ Hz and $f_\beta = 1.64 \times 10^8$ Hz
(*c*) $f_T = 1.33 \times 10^{12}$ Hz and $f_\beta = 1.47 \times 10^{10}$ Hz
(*d*) $f_T = 1.47 \times 10^{10}$ Hz and $f_\beta = 1.33 \times 10^{12}$ Hz

2.6 The transistor shunt regulator shown in the figure is has a regulated output voltage of 10 V, when the input varies from 20 V to 30 V. The relevant parameters for the zener diode and the transistor are : $V_Z = 9.5$, $V_{BE} = 0.3$ V, $\beta = 99$. Neglect the current through R_B. Then the maximum power dissipated in the zener diode (P_Z) and the transistor (P_T) are

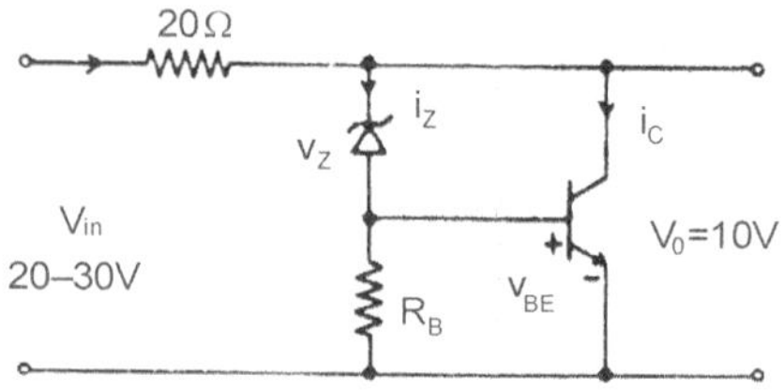

(*a*) $P_Z = 75$ mW, $P_T = 7.9$ W
(*b*) $P_Z = 85$ mW, $P_T = 8.9$ W
(*c*) $P_Z = 95$ mW, $P_T = 9.9$ W
(*d*) $P_Z = 115$ mW, $P_T = 11.9$ W

2.7 The oscillator circuit shown in the figure,

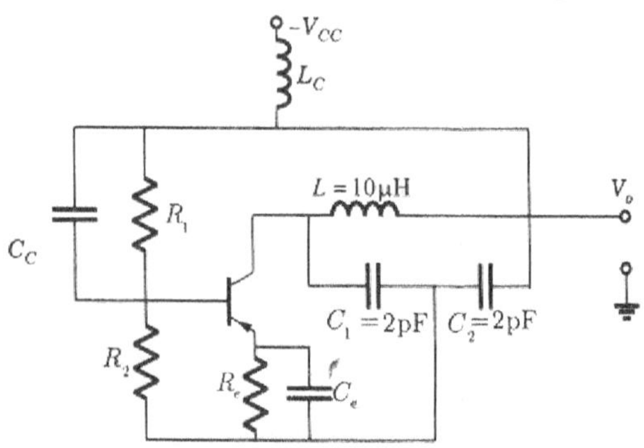

(*a*) Hartley oscillator with $f_{\text{oscillation}} = 79.6$ MHz
(*b*) Colpitts oscillator with $f_{\text{oscillation}} = 79.6$ MHz
(*c*) Hartley oscillator with $f_{\text{oscillation}} = 159.2$ MHz
(*d*) Colpitts oscillator with $f_{\text{oscillation}} = 159.2$ MHz

2.8 The inverting OP-AMP shown in the figure has an open-loop gain of 100. The closed-loop gain $\dfrac{v_0}{v_s}$ is

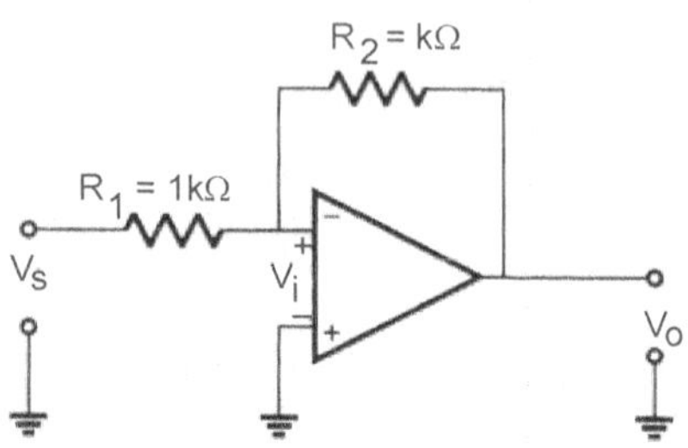

(*a*) – 8
(*b*) – 9
(*c*) – 10
(*d*) – 11

2.9 In the figure, assume the OP-AMPs to be ideal. The output v_0 of the circuit is

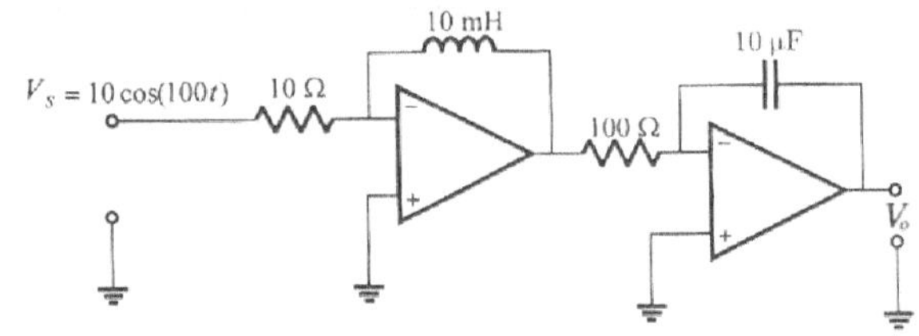

(*a*) $10\cos(100t)$
(*b*) $10\int_0^t \cos(100\tau)d\tau$
(*c*) $10^{-4}\int_0^t \cos(100\tau)d\tau$
(*d*) $10^{-4}\dfrac{d}{dt}\cos(100t)$

2.10 In the figure, the LED

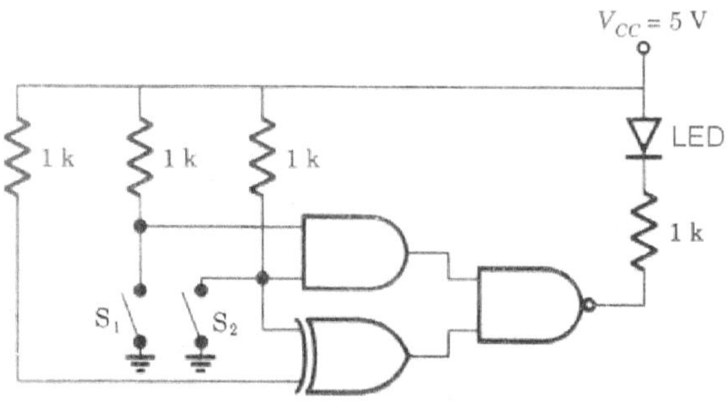

(*a*) emits light when both S_1 and S_2 are closed.
(*b*) emits light when both S_1 and S_2 are open.
(*c*) emits light when only of S_1 or S_2 is closed.
(*d*) does not emit light, irrespective of the switch positions.

2.11 In the TTL circuit in the figure, S_2 to S_0 are select lines and X_7 to X_0 are input lines. S_0 and X_0 are LSBs. The output Y is

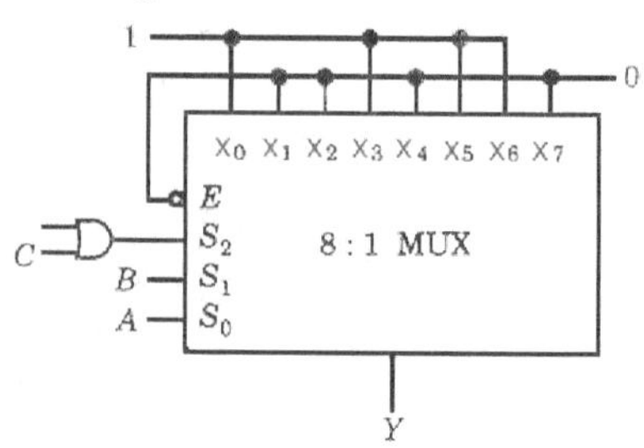

(*a*) indeterminate
(*b*) $A \oplus B$
(*c*) $\overline{A \oplus B}$
(*d*) $\overline{C}\cdot(\overline{A \oplus B}) + C\cdot(A \oplus B)$

2.12 The digital block in the figure, realized using two positive edge triggered D-flip-flops. Assume that for $t < t_0$, $Q_1 = Q_2 = 0$. The circuit in the digital block is given by

(*a*)

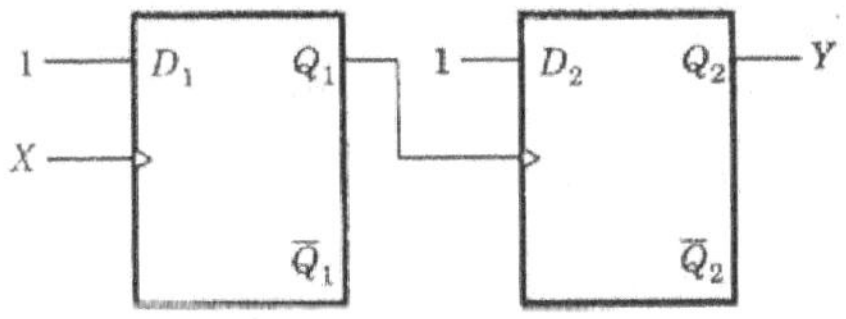

(*b*)

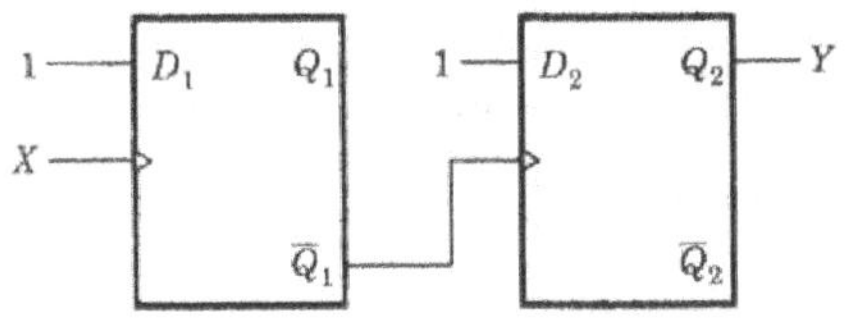

(*c*)

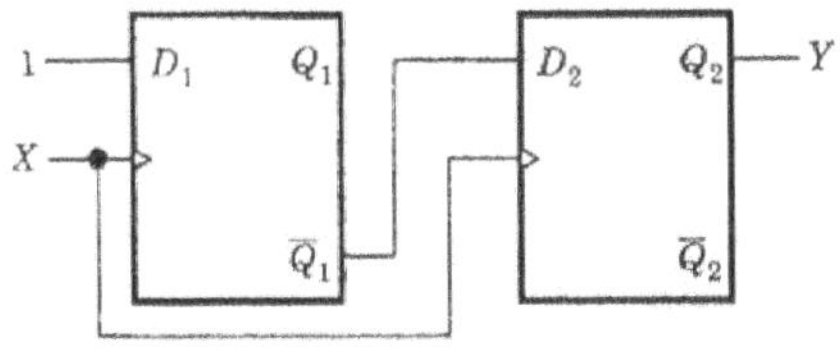

(*d*)

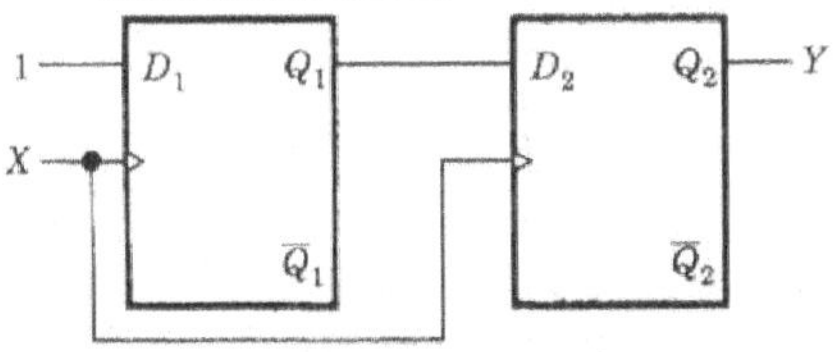

2.13 In the DRAM cell in the figure is the V_t of the NMOSFET is 1 V. For the following three combinations of WL and BL voltages.

WL = 5V, BL = 5V; WL = 5V, BL = 3V; WL = 5V, BL = 7V;

The corresponding voltages across the capacitor C are

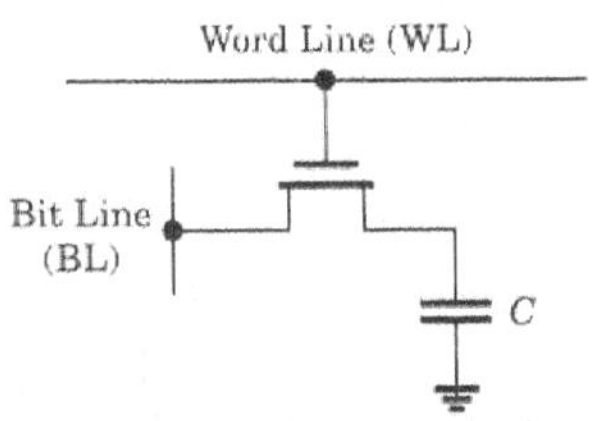

(*a*) 5 V; 3 V; 7 V (*b*) 4 V; 3 V; 4 V

(*c*) 5 V; 5 V; 5 V (*d*) 4 V; 4 V; 4 V

2.14 The impulse response functions of four linear systems S_1, S_2, S_3, S_4 are given respectively by

$h_1(t) = 1$ $h_2(t) = U(t)$

$h_3(t) = \dfrac{U(t)}{t+1}$ $h_4(t) = e^{-3t}U(t)$

Where U(*t*) is the unit step function. Which of these systems is time invariant, causal, and stable ?

(*a*) S_1 (*b*) S_2

(*c*) S_3 (*d*) S_4

2.15 An electrical system and its signal-flow graph representations are shown in the figure respectively. The values of G_2 and H, respectively, are

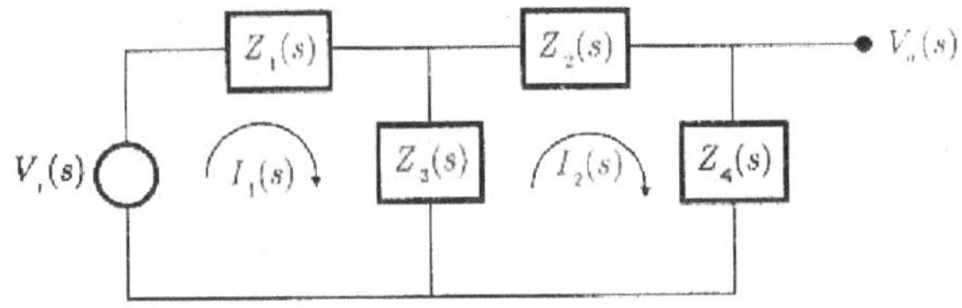

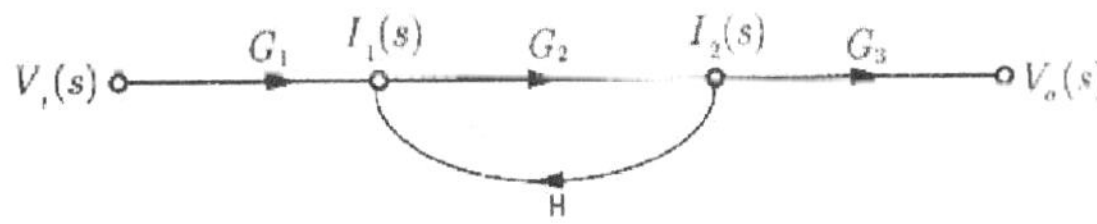

(*a*) $\dfrac{Z_3(s)}{Z_2(s)+Z_3(s)+Z_4(s)}, \dfrac{-Z_3(s)}{Z_1(s)+Z_3(s)}$

(*b*) $\dfrac{-Z_3(s)}{Z_2(s)-Z_3(s)+Z_4(s)}, \dfrac{-Z_3(s)}{Z_1(s)+Z_3(s)}$

(*c*) $\dfrac{Z_3(s)}{Z_2(s)+Z_3(s)+Z_4(s)}, \dfrac{Z_3(s)}{Z_1(s)+Z_3(s)}$

(*d*) $\dfrac{-Z_3(s)}{Z_2(s)-Z_3(s)+Z_4(s)}, \dfrac{Z_3(s)}{Z_1(s)+Z_3(s)}$

2.16 The open-loop DC gain of a unity negative feedback system with closed-loop transfer function $\dfrac{s+4}{s^2+7s+13}$ is

(*a*) $\dfrac{4}{13}$ (*b*) $\dfrac{4}{9}$

(*c*) 4 (*d*) 13

2.17 The feeback control system in the figure is stable.

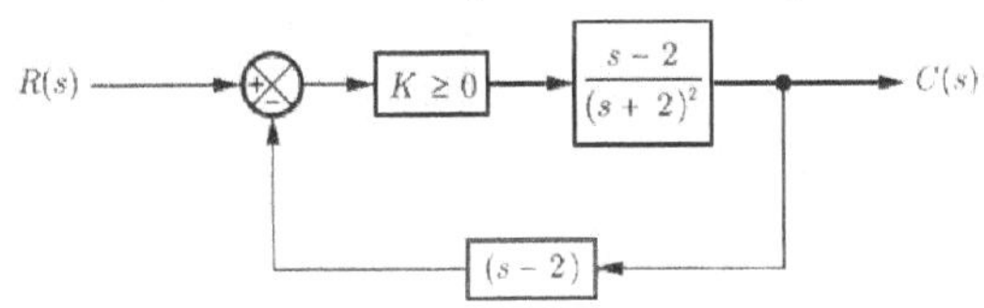

(*a*) for all $K \geq 0$ (*b*) only if $K \geq 1$

(*c*) only if $0 \leq K < 1$ (*d*) only if $0 \leq K \leq 1$

2.18 A video transmission system transmits 625 picture frames per second. Each frame consists of a 400 × 400 pixel grid with 64 intensity levels per pixel. The data rate of the system is

(*a*) 16 Mbps (*b*) 100 Mbps

(*c*) 600 Mbps (*d*) 6.4 Gbps

2.19 The Nyquist sampling interval, for the signal sin *c*(700*t*) + sin c (500*t*) is

(*a*) $\dfrac{1}{350}$ sec (*b*) $\dfrac{\pi}{350}$ sec

(*c*) $\dfrac{1}{700}$ sec (*d*) $\dfrac{\pi}{175}$ sec

2.20 During transmission over a communication channel, bit errors occur independently with probability p. If a block of n bits is transmitted, the probability of at most one bit error is equal to

(*a*) $1-(1-p)^n$ (*b*) $p+(n-1)(1-p)$

(*c*) $np(1-p)^{n-1}$ (*d*) $(1-p)^n + np(1-p)^{n-1}$

2.21 The PSD and the power of a signal $g(t)$ are, respectively, $S_g(\omega)$ and P_g. The PSD and the power of the signal $ag(t)$ are, respectively,

(*a*) $a^2S_g(\omega)$ and a^2P_g (*b*) $a^2S_g(\omega)$ and aP_g

(*c*) $aS_g(\omega)$ and a^2P_g (*d*) $aS_g(\omega)$ and aP_g

2.22 A material has conductivity of 10^{-2} mho/m and a relative permittivity of 4. The frequency at which the conduction current in the medium is equal to the displacement current is

(*a*) 45 MHz (*b*) 90 MHz

(*c*) 450 MHz (*d*) 900 MHz

2.23 A uniform plane electromagnetic wave incident normally on a plane surface of a dielectric material is reflected with a VSWR of 3. What is the percentage of incident power that is reflected ?

(*a*) 10 % (*b*) 25 %

(*c*) 50 % (*d*) 75 %

2.24 A medium wave radio transmitter operating at a wavelength of 492 m has a tower antenna of height 124 m. What is the radiation resistance of the antenna ?

(*a*) 25 Ω (*b*) 36.5 Ω

(*c*) 50 Ω (*d*) 73 Ω

2.25 In a uniform linear array, four isotropic radiating elements are spaced $\frac{\lambda}{4}$ apart. The progressive phase shift between the elements required for forming the main beam at 60° off the end-fire is

(*a*) $-\pi$ radians

(*b*) $-\frac{\pi}{2}$ radians

(*c*) $-\frac{\pi}{4}$ radians

(*d*) $-\frac{\pi}{8}$ radians

SECTION - B (75 Marks)

This section consists of TWENTY questions of FIVE marks each. ANY FIFTEEN out of these questions have to be answered in the Answer Book provided. (5 × 15 = 75)

3. For the circuit shown in the figure, determine the phasors E_2, E_0, I, and I_1.

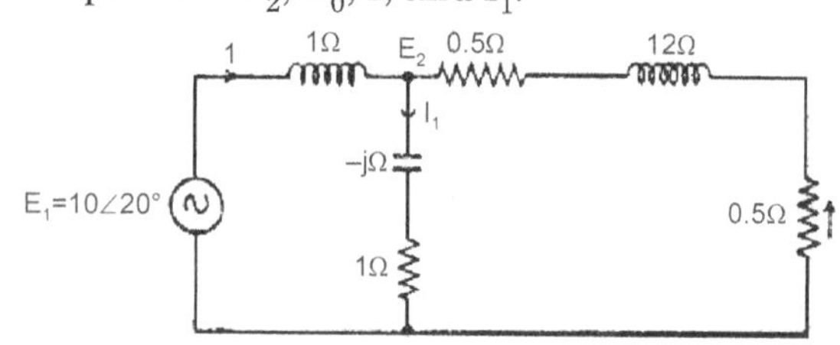

4. The circuit shown in the figure, operating in steady-state with switch S_1 colsed. The switch S_1 is opened at $t = 0$

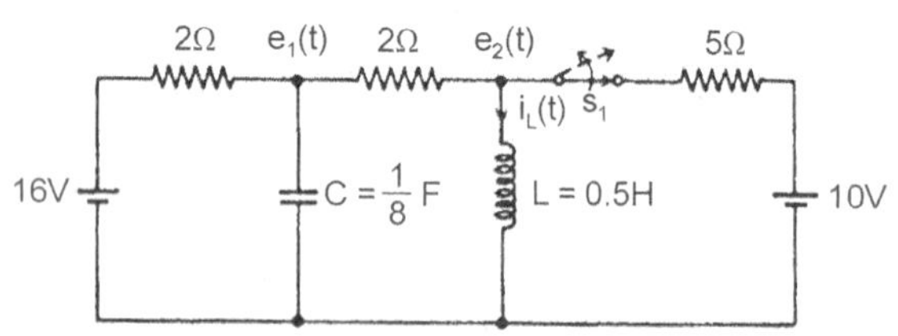

(*a*) Find $i_L(0^+)$.

(*b*) Find $e_1(0^+)$.

(*c*) Using nodal equations and Laplace transform approach, find an expression for the voltage across the capacitor for all $t > 0$.

5. The admittance parameters of a 2-port network shown in the figure, given by Y_{11} = 2 mho, Y_{12} = – 0.5 mho, Y_{21} = 4.8 mho, Y_{22} = 1 mho. The output port is terminated with a load admittance Y_L = 0.2 mho. Find E_2 for each of the following conditions

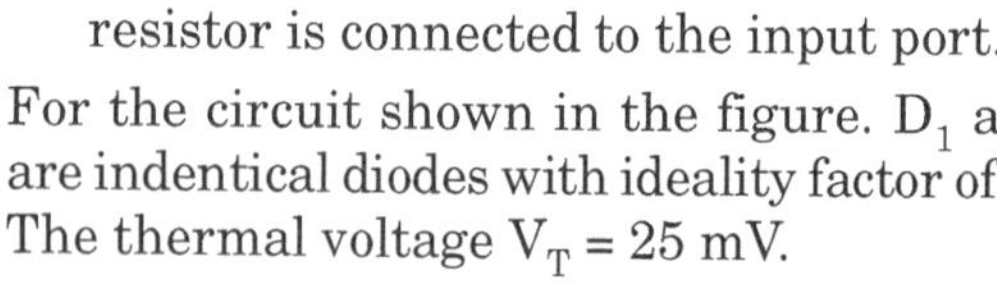

(*a*) $E_1 = 10\angle 0°$ V

(*b*) $I_1 = 10\angle 0°$ A

(*c*) A source $10\angle 0°$ V in series with a 0.25 Ω resistor is connected to the input port.

6. For the circuit shown in the figure. D_1 and D_2 are indentical diodes with ideality factor of unity. The thermal voltage V_T = 25 mV.

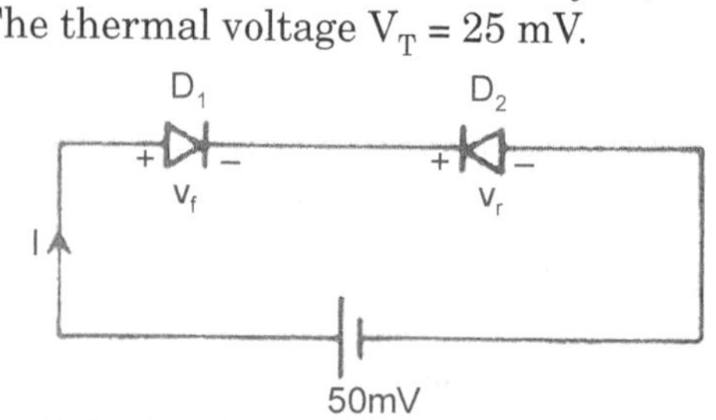

(*a*) Calculate V_f and V_r.

(*b*) If the reverse saturation current, I_s, for the diode is 1 *p*A, then compute the current I through the circuit.

7. An emitter-follower amplifier is shown in the figure, Z_i is the impedance looking into the base of the transistor and Z_0 is the impedance looking into the emitter of the transistor.

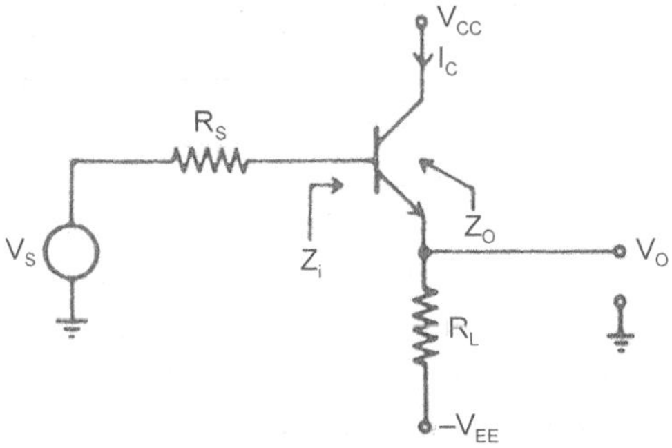

(*a*) Draw the small signal equivalent circuit of the amplifier.

(*b*) Obtain an expression for Z_i.

(*c*) Obtain an expression for Z_0.

(*d*) Determine Z_i and Z_0 if a capacitor C_L is connected across R_L.

8. Assume that the OP-AMP in the figure, ideal

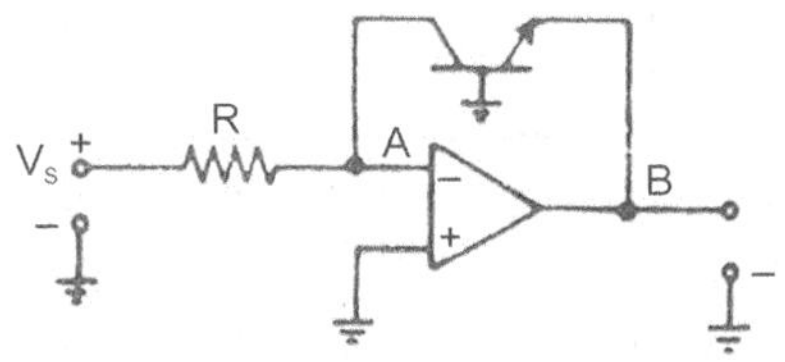

(*a*) Obtain an expression for v_o in terms of v_s, R, and the reverse saturation current I_s of the transistor.

(*b*) If R = 1Ω, I_s = 1*p*A and the thermal voltage V_T = 25 mV, then what is the value of the output voltage v_o for an input voltage v_s = 1V ?

(*c*) Suppose that the transistor in the feedback path is replaced by a *p-n* junction diode with a reverse saturation current of I_s. The *p*-side of the diode is connected to node A and the *n*-side to node B. Then what is the expression for v_o in terms of v_s, R and I_s?

9. A monochrome video signal that ranges from 0 to 8V, is digitized using an 8-bit ADC.

(*a*) Determine the resolution of the ADC in V/bit.

(*b*) Calculate the mean squared quantization error.

(*c*) Suppose the ADC is counter controlled. The counter is upcount and positive edge triggered with clock frequency 1 MHz. What is the time taken in seconds to get a digital equivalent of 1.59 V ?

10. In the figure, the output of the oscillator, V_1 has 10V peak amplitude with zero DC value. The transfer characteristic of the Schmitt inverter is also shown in the figure is. Assume that the JK flip-flop is reset at time $t = 0$.

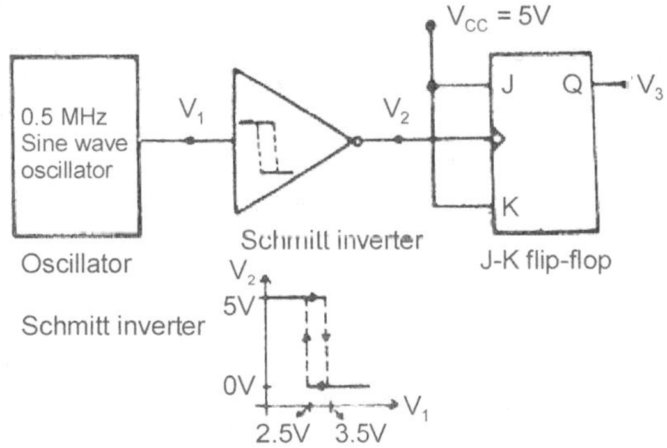

(*a*) What is the period and duty cycle of the waveform V_2?

(*b*) What is the period and duty cycle of the waveform V_3?

(*c*) Sketch V_1, V_2 and V_3 for the duration $0 \leq t \leq 6$ μs. Clearly indicate the exact timings when the waveforms V_2 and V_3 make high-to-low and low-to-high transitions.

11. For the digital block shown in the figure the output Y = *f*(S_3, S_2, S_1, S_0) where S_3 is MSB and S_0 is LSB. Y is given in terms of minterms as

$$Y = \sum m(1, 5, 6, 7, 11, 12, 13, 15) \text{ and its}$$

complement is $\overline{Y} = \sum m(0, 2, 3, 4, 8, 9, 10, 14)$.

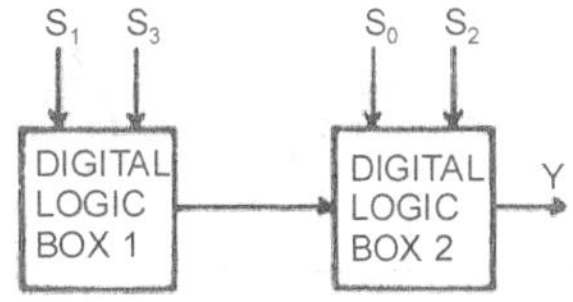

(*a*)

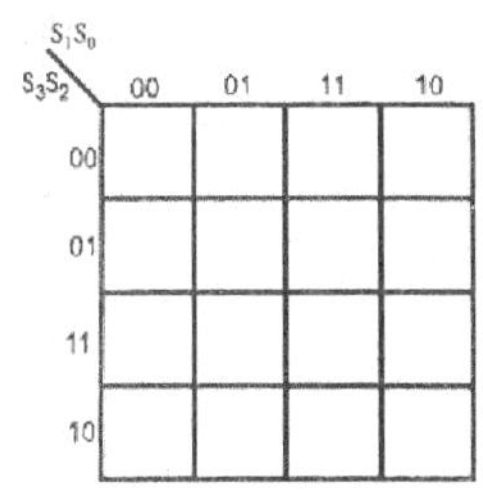

(*b*)

(*a*) Enter the logical values in the given Karnaugh map in the figure, for the output Y.

(*b*) Write down the expression for Y in sum-of-products from using minimum number of terms.

(*c*) Draw the circuit for the digital logic boxes using four 2-input NAND gates only for each of the boxes.

12. Consider the following sequence of instructions for an 8085 microprocessor based system.

Memory Address	Instructions	
FF00	MVI A,	FF H
FF02	INR A	
FF03	JC	FF0C H
FF06	ORI	A8H
FF08	JM	FF15 H
FF0B	XRA A	
FF0C	OUT	PORT 1
FF0E	HLT	
FF10	XRI	FF H
FF12	OUT	PORT2
FF14	HLT	
FF15	MVI A,	FF H
FF17	ADI	02 H
FF19	RAL	
FF1A	JZ	FF23 H
FF1D	JC	FF10 H
FF20	JNC	FF12 H
FF23	CMA	
FF24	OUT	PORT3
FF26	HLT	

(*a*) If the program execution begins at the location FF00 H, write down the sequence of instructions which are actually executed till a HLT instruction. (Assume all flags are initially RESET).

(*b*) Which of the three ports (PORT1, PORT2 and PORT3) will be loaded with data, and what is the bit pattern of the data ?

13. A feedback control system is shown in figure,

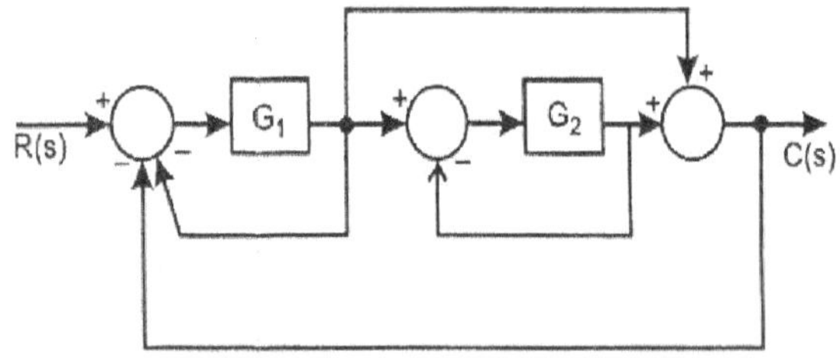

(*a*) Draw the signal-flow graph that represents the system.

(*b*) Find the total number of loops in the graph and determine the loop-gains of all the loops.

(*c*) Find the number of all possible combinations of non-touching loops taken two at a time.

(*d*) Determine the transfer function of the system using the signal-flow graph.

14. Consider the feedback control system shown in figure,

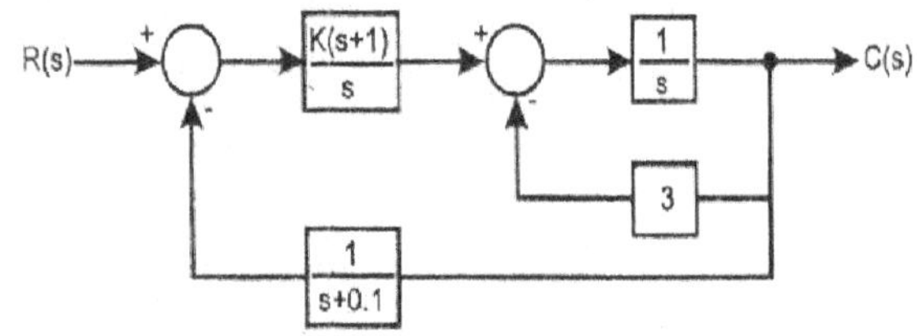

(*a*) Find the transfer function of the system and its characteristic equation.

(*b*) Use the Routh-Hurwitz criterion to determine the range of K for which the system is stable.

15. For the feedback control system shown in the figure, the process transfer function is $Gp(s) = \frac{1}{s(s+1)}$, and the complification factor of the power amplifier is $K \geq 0$. The design specifications required for the system, time constant is 1 sec and a damping ratio of 0.707.

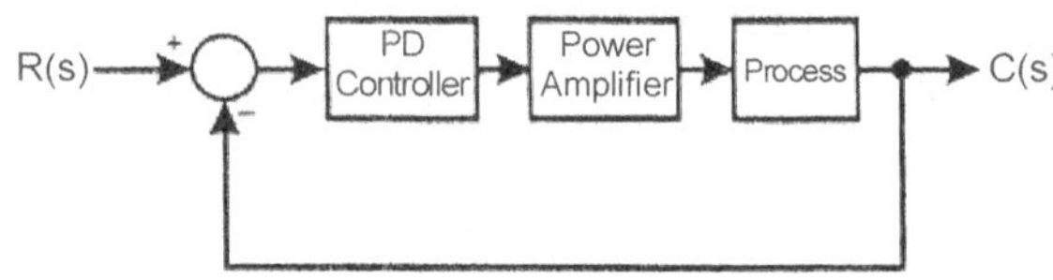

(*a*) Find the desired locations of the closed-loop poles.

(*b*) Write down the required characteristic equation for the system. Hence determine the PD controller transfer function $G_0(s)$ when K = 1.

(*c*) Sketch the root-locus for the system.

16. The Fourier transform G (ω) of the signal *g*(*t*) in the figure, given as $G(w) = \frac{1}{\omega^2}\left(e^{j\omega} - jwe^{j\omega} - 1\right)$. Using this information and the time-shifting and time-scaling properties, determine the Fourier transform of signals in the figure

(*a*)

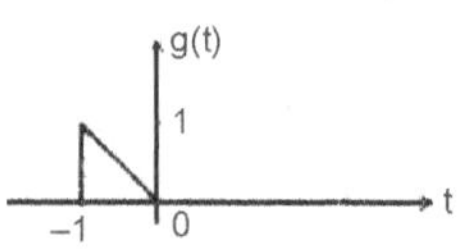

(*b*)

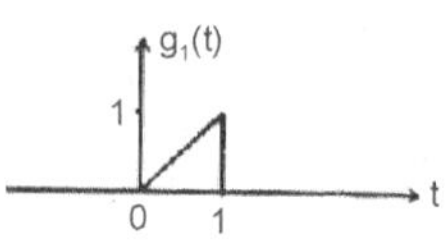

(*c*)

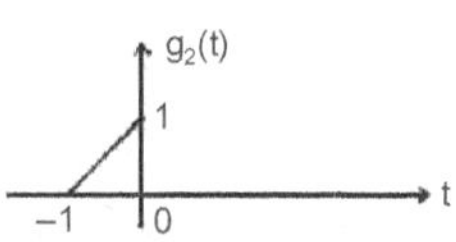

(*d*)

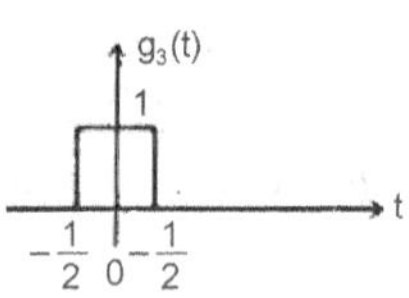

17. The periodic modulating signal $m(t)$ is shown in the figure. Using Carson's rule. estimate B_{FM} (bandwidth of the FM signal) and B_{PM} (bandwidth of the PM signal) for $kf = \pi \times 10^4$ and $kp = \frac{\pi}{4}$. Assume the essential bandwidth of $m(t)$ to consist only up to and including the third harmonic.

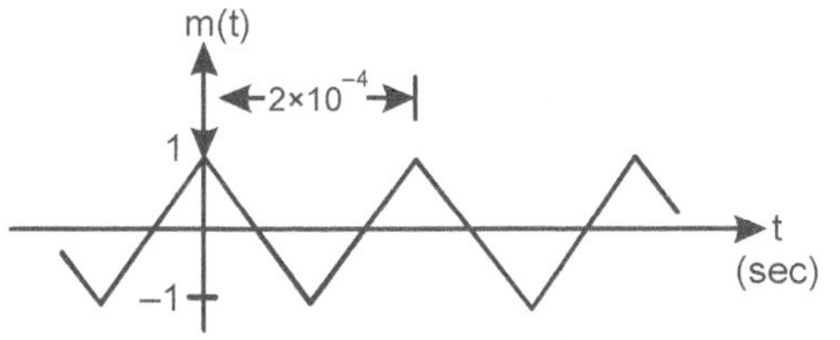

18. A baseband signal $g(t)$ bandlimited to 100 Hz modulates a carrier of frequency f_0 Hz. The modulated signal $g(t)\cos 2\pi ft$ is transmitted over a channel whose input x and output y are related by $y = 2x + x^2$. The spectrum of $g(t)$ is shown in the figure. Sketch the spectrum of the transmitted signal and the spectrum of the received signal.

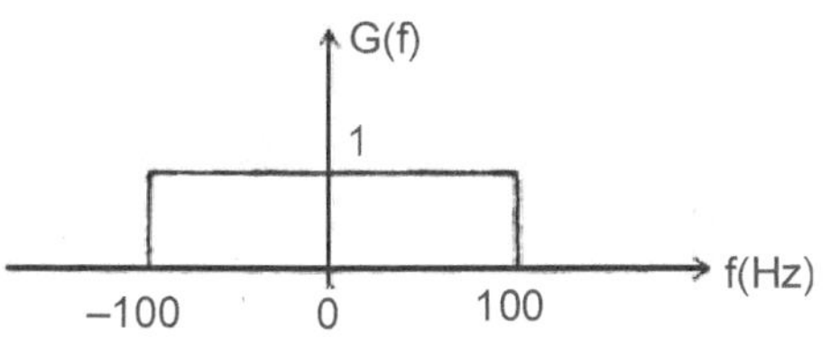

19. A periodic signal $g(t)$ is shown inthe figure. Determine the PSD of $g(t)$.

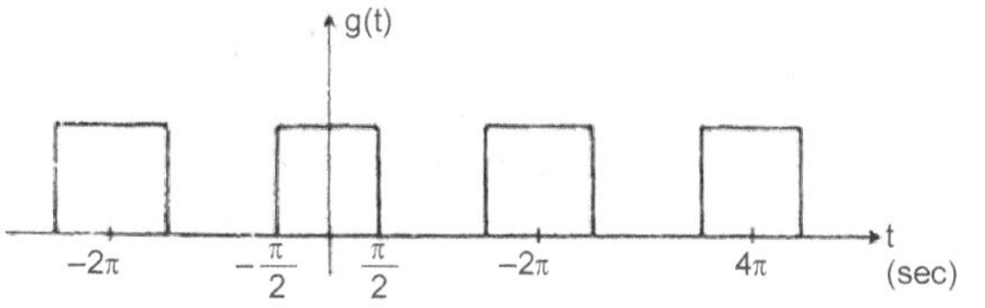

20. A system of three electric charges lying in a straight line is in equilibrium. Two of the charges are positive with magnitudes Q and 2Q, and are 50 cm apart. Determine the sign, magnitude and position of the third charge.

21. A medium has a breakdown strength of 16 KV/m r.m.s. its relative permeability is 1.0 and relative permittivity is 4.0 A plane electromagnetic wave is transmitted through the medium. Calculate the maximum possible power flow density and the associated magnetic field.

22. A rectangular hollow metal waveguide has dimensions a = 2.29 cm and b = 1.02 cm. Microwave power at 10 GHz is transmitted through the waveguide in the TE_{10} mode.

(*a*) Calculate the cut-off wavelength and the guide wavelength for this mode.

(*b*) What are the other (TE or TM) modes that can propagate through the waveguide ?

(*c*) If $a = b = 2.29$ cm, What are the modes which can propagate through the waveguidet ?

ANSWERS

1. 1 (*c*)	1. 2 (*a*)	1. 3 (*b*)	1. 4 (*c*)	1. 5 (*a*)	1. 6 (*c*)	1. 7 (*b*)	1. 8 (*b*)	1. 9 (*a*)	1. 10 (*b*)
1. 11 (*c*)	1. 12 (*c*)	1. 13 (*c*)	1. 14 (*d*)	1.15 (*c*)	1. 16 (*d*)	1. 17 (*a*)	1. 18 (*a*)	1. 19 (*d*)	1. 20 (*c*)
1. 21 (*b*)	1. 22 (*c*)	1. 23 (*b*)	1. 24 (*a*)	1. 25 (*d*)	2. 1 (*d*)	2. 2 (*a*)	2. 3 (*a*)	2.4 (*c*)	2. 5 (*b*)
2. 6 (*c*)	2. 7 (*b*)	2. 8 (*d*)	2. 9 (*d*)	2. 10 (*d*)	2. 11 (*d*)	2. 12 (*c*)	2. 13 (*)	2. 14 (*d*)	2. 15 (*c*)
2. 16 (*b*)	2. 17 (*c*)	2. 18 (*c*)	2. 19 (*c*)	2. 20 (*d*)	2. 21 (*a*)	2. 22 (*a*)	2. 23 (*b*)	2. 24 (*b*)	2. 25 (*c*)

EXPLANATIONS

1.1. Current through 4Ω resistor,

$$I = \frac{12}{4 + [4 \,||\, (2+2)]} = 2 \text{ Amp.}$$

Voltage drop, $e_0 = \left(\frac{I}{2}\right) \cdot 4 = 4 \text{ volts}$

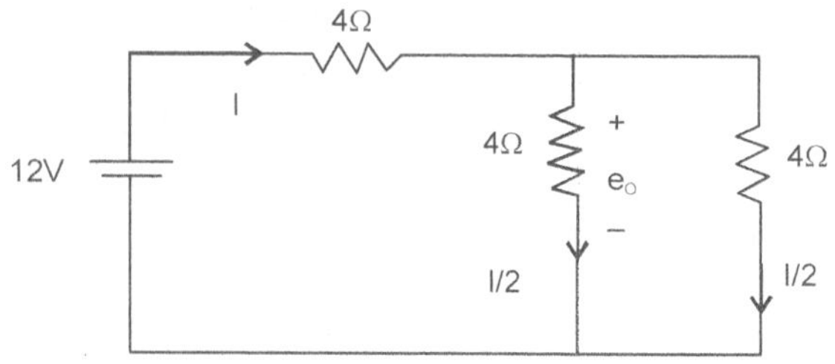

1.2 If the impedances are balanced then,

$$Z_\Delta = 3Z_Y$$

$$\Rightarrow \quad Z_Y = \frac{\sqrt{3}\,Z}{3} = \frac{Z}{\sqrt{3}}.$$

1.3 The impulse response,

$$C(s) = H(s)\,R(s)$$

$$\Rightarrow \quad C(s) = H(s)$$

For impulse signal

$$\mathcal{L}[\delta(t) = 1.$$

$$C(t) = \mathcal{L}^{-1}[H(s)] = \mathcal{L}^{-1}\left[\frac{1}{s^2} \cdot \frac{1}{(s-2)}\right]$$

$$f_1(t) = \mathcal{L}^{-1}[F_1(s)] = tU(t)$$

$$f_2(t) = \mathcal{L}^{-1}[F_2(s)] = e^{2t}U(t)$$

then $\quad e(t) = f_1(t) * f_2(t)$

$$= [e^{2t} * t]\,U(t)$$

1.4.

$$I_1 = E_1Y_1 + (E_1 - E_2)Y_2$$

$$Y_1 = \frac{1}{5}\mho, Y_2 = \frac{1}{20}\mho$$

$$I_2 = E_2Y_3 + (-E_1 + E_2)Y_2$$

$$Y_3 = \frac{1}{10}\mho$$

In matrix form,

$$\begin{bmatrix} I_1 \\ I_2 \end{bmatrix} = \begin{bmatrix} Y_1 + Y_2 & -Y_2 \\ -Y_2 & Y_2 + Y_3 \end{bmatrix} \begin{bmatrix} E_1 \\ E_2 \end{bmatrix}$$

$$Y_{12} = -Y_2 = -\frac{1}{20}\mho = -0.05 \text{ mho}$$

1.5.

$$H(z) = \sum_{n=0}^{\infty} u(n) \cdot z^{-n} = \sum_{n=0}^{\infty} 1 \cdot z^{-n}$$

For convergence, $\left|\sum_{n=0}^{\infty} z^{-n}\right| < \infty$

or, $\quad |z^{-1}| < 1$

or, $\quad |z| > 1$

1.12. Propagation delay of ring oscillator

$T_{pd} = 5 \times 100$ per second $= 500$ persecond

Frequency of oscillator,

$$f = \frac{1}{2T_{pd}}$$

$$= \frac{1}{2 \times 500 \times 10^{-12}} = 1\text{GHz}$$

1.13 4K × 8 –bit RAM

$4K = 2^{12}$ Address lines = 1000H

Starting Add.	AA00H	=	1010 1010 0000 0000
	+ OFFF	=	0000 1111 1111 1111
Last Add :		=	1011 1001 1111 1111
			B 9 F FH

1.14.

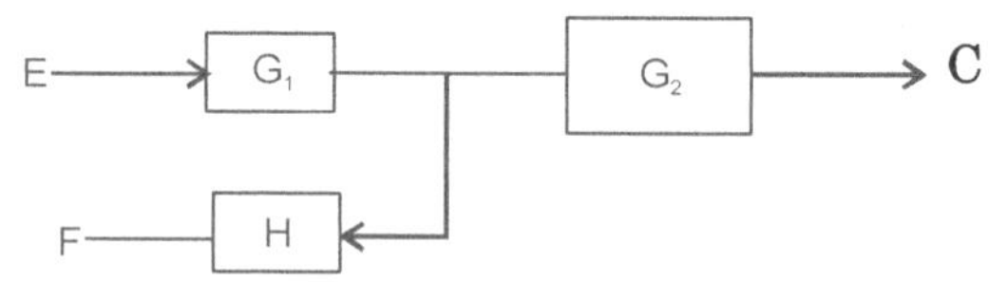

$F = EG_1H$

(*a*)

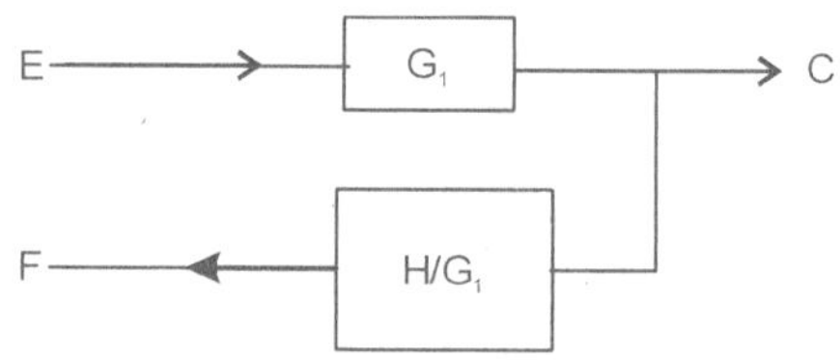

$F = EH$

(*b*)

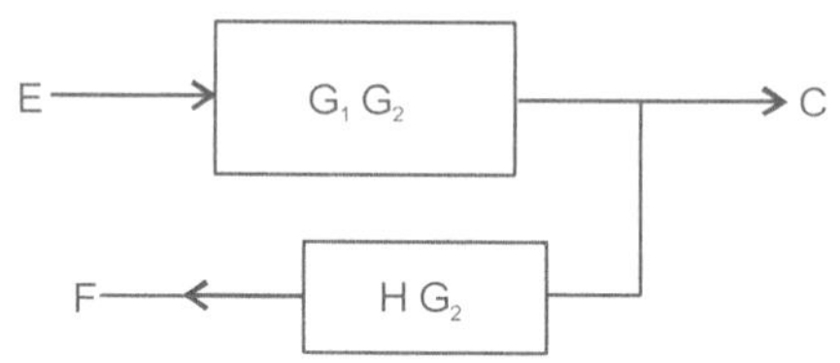

$F = G_1G_2^2H$

(*c*)

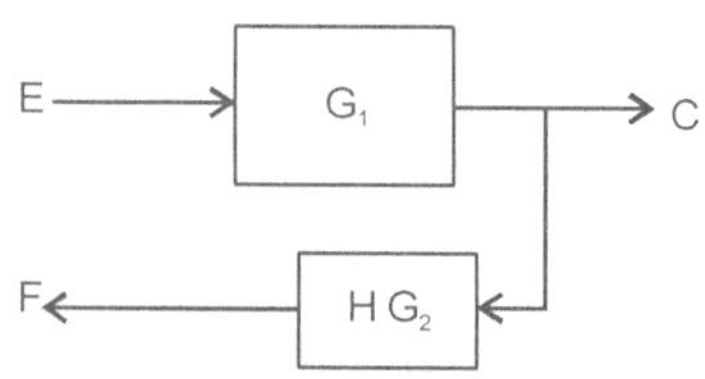

$F = EG_1G_2H$

(*d*)

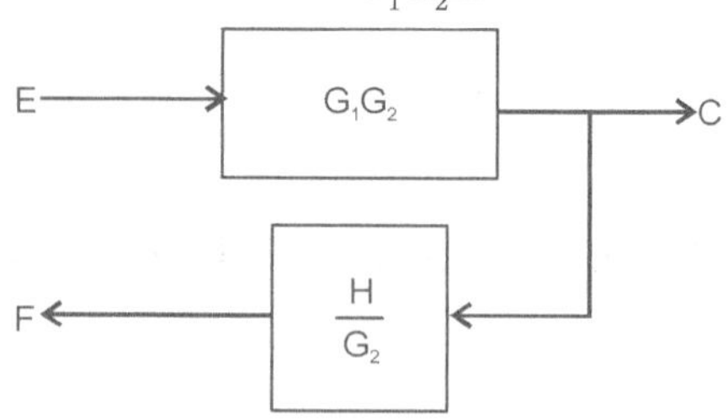

$F = EG_1H$

Hence the right option is (*d*).

1.15 Characteristic equation of a closed-loop system is

$$s^2 + 2s + 2 = 0$$

$$\omega_n = \sqrt{2} = 0$$

$$\delta = \frac{1}{\sqrt{2}} < 1, \quad \text{Hence underdamped system}$$

1.16. For 0 < K < 1, and K < 5, both the roots lie on the negative real axis which corresponds to overdamped system.

For, K = 1, system is critically damped.

For 1 < K < 5, the system is underdamped.

1.17 Pole in right half P = 0

Encirclement of $-1 + j0$

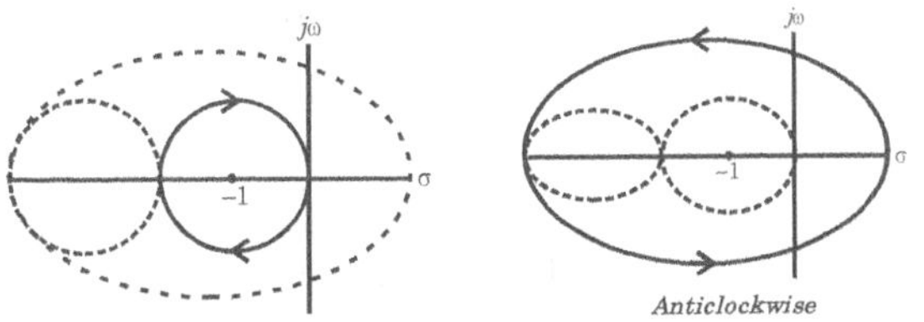

Clockwise Anticlockwise

N = (No. of anticlockwise encirclement – No. of clockwise encirclement)

N = (1 – 1) = 0

Thus, N = P – Z = 0

∴ Zero roots on RH of s-plane.

1.18 As $\int_{-\infty}^{\infty} \delta(at)\,\phi(t)\,dt = \frac{\phi(0)}{|a|}$ for any $a < 0$.

Thus $\int_{-\infty}^{\infty} \delta(t)\cos\left(\frac{3t}{2}\right)dt = \frac{\cos 0}{1} = 1.$

1.20 Since, $P_x(x)$ is continuous function

Hence, probability at point $[x = 4]$ tends to zero.

1.21 Energy content of a signal $f(t)$,

$$E = \int_{-\infty}^{\infty} |f(t)|^2\,dt \qquad ...(i)$$

Now, $E' = \int_{-\infty}^{\infty} |f(2t)|^2\,dt$ for signal $f(2t)$

Putting, $2t = z$, we get

$$E' = \frac{1}{2}\int_{-\infty}^{\infty} |f(z)|^2\,dz$$

$$= \frac{1}{2}E \text{ ......from equation } (i)$$

1.22 Characteristic equation of a transmission line

$$Z_0 = \sqrt{\frac{R + j\omega L}{G + j\omega C}}$$

$$\left[\text{At microwave frequencies, } R << \omega L, G << \omega C\right]$$

$$Z_0 \approx \sqrt{\frac{L}{C}}\left(1 + \frac{R}{2j\omega L}\right)\left(1 - \frac{G}{2j\omega C}\right)$$

$$\approx \sqrt{\frac{L}{C}}\left[1 + \frac{1}{2}\left(\frac{R}{j\omega L} - \frac{G}{j\omega C}\right)\right]$$

$$\approx \sqrt{\frac{L}{C}}\left[1 + \frac{1}{2j\omega LC}(RC - GL)\right]$$

Transmission line will be lossless when imaginary term is equal to zero.

i.e. RC = GL.

1.23 Equations for electric and magnetic field for a plane EM have propagating in the Z-direction

$$\frac{\partial^2 E_x}{\partial Z^2} = C^2\,\frac{\partial^2 E_x}{\partial t^2}$$

$$\frac{\partial^2 E_y}{\partial Z^2} = C^2 \cdot \frac{\partial^2 E_y}{\partial t^2}$$

1.25 The dominant mode in a particular waveguide is the mode having lowest cut-off frequency, below this cut-off frequency, EM will be attenuated to a negligible value. Lowest-cut-off frequency corresponds to highest cut-off wavelength.

2.1

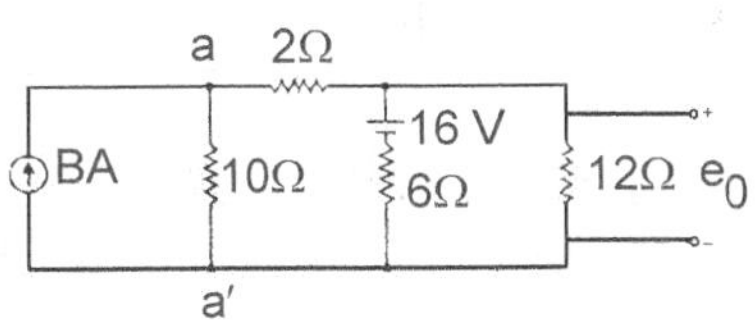

Applying Thevenin theorem at terminal a, a' we get circuit as follows

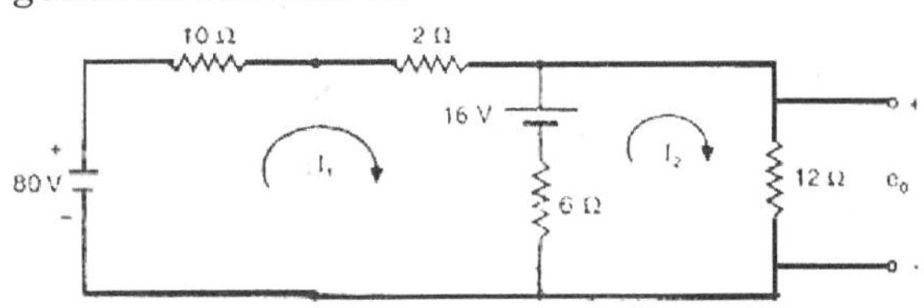

Applying Mesh rule to both the loops,

$$80 = 10I_1 + 2I_1 + 16 + 6(I_1 - I_2) \qquad ...(i)$$

$$-12I_2 - 6I_2 + 6I_1 + 16 = 0 \qquad ...(ii)$$

Solving we get, $I_1 = \frac{13}{3}$ amp, $I_2 = \frac{7}{3}$ amp.

$e_0 = 12 \times I_2 =$ **28 volts.**

2.2 Maximum power dissipation will be when

$$R = \sqrt{R_L^2 + (\omega L)^2}$$

As, ω = 10, $R_L = 10\,\Omega$, L = 1 H

$$\therefore \quad R = \sqrt{(10)^2 + (10)^2}$$

$$= \mathbf{14.14\,\Omega.}$$

2.3 $I_2(\omega) = E_m \cos \omega t \left[\frac{R_2}{R_2^2 + \left(\frac{1}{\omega C}\right)^2} - \frac{\frac{1}{j\omega e}}{R_2^2 + \left(\frac{1}{\omega C}\right)^2} \right]$

At $\omega = \infty$, $I_2(\omega) = \frac{E_m}{R_2}$.

Also, I_2 is leading with voltage phasor.

At $\omega = 0$, $I_2(0) = 0$.

Thus desired locus should be

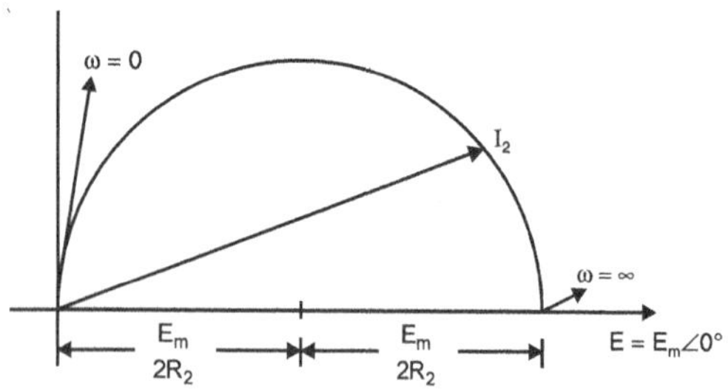

2.4

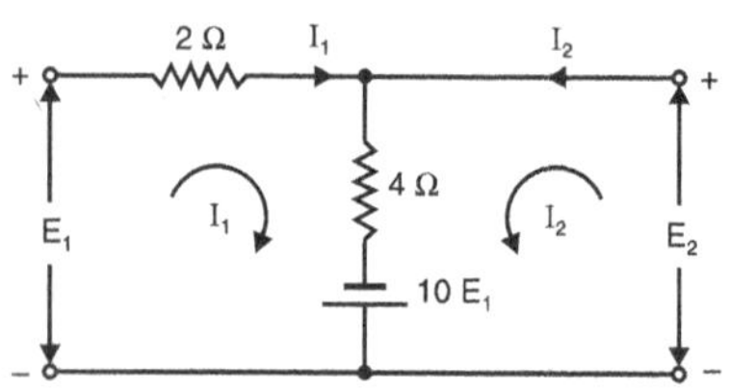

Loop law's to both the loops,

$$11E_1 = 6I_1 + 4I_2 \quad ...(i)$$

$$E_2 = (I_1 + I_2)\,4 - 10\,E_1 \quad ...(ii)$$

Putting the expression for E_1 from equation (*i*) in equation (*ii*), we get

$$E_2 = (I_1 + I_2)\,4 - \frac{10[2I_1 + 4(I_1 + I_2)]}{11}$$

$$= I_1\left[4 - \frac{20}{11} - \frac{40}{11}\right] + I_2\left[4 - \frac{40I_2}{11}\right]$$

Putting $I_2 = 0$ as, $Z_{21} = \left.\frac{E_2}{I_1}\right|_{I_2 = 0}$

$$E_2 = I_1\left[\frac{44 - 60}{11}\right] = -\frac{16}{11}I_1$$

From equation (*i*), we get

$$\therefore \quad Z_{11} = \left.\frac{E_1}{I_1}\right|_{I_2 = 0} = \frac{6}{11}\,\Omega.$$

2.5

$$f_T = \frac{g_m}{2\pi(C_\mu + C_\pi)}$$

$$f_\beta \,.\, h_{fe} = f_T$$

Putting the given values,

$$f_T = 1.47 \times 10^{10}\text{ Hz},$$

$$f_\beta = 1.64 \times 10^8\text{ Hz}.$$

2.6

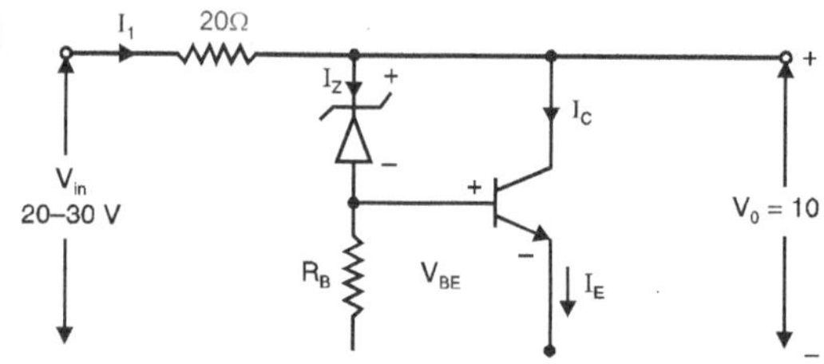

As, $I_1 \max = \frac{V_{in\,max} - V_0}{20}$

$$= \frac{30 - 10}{20} = IA$$

Also, $I_E = I_Z + I_C \quad ...(i)$

And $I_B = I_Z$

$$\frac{I_C}{I_B} = 99$$

From equation (*i*), $I_E = 100\,I_Z$

From the figure, $I_1 = I_Z + I_C = I_E = 100\,I_Z$

$$\Rightarrow \quad I_Z = \frac{1}{100} = 0.01\text{ A}$$

$$I_C = 99\,I_Z = 0.99\text{ A}$$

$$\therefore \quad P_Z = V_Z I_Z$$

$$= 9.5 \times 0.01\text{ W} = 95\text{ mW}$$

$$P_T = V_C\,I_C$$

$$= 10 \times 0.99 = \mathbf{9.9\ W}$$

2.7 This is Colpitts oscillator.

$$f_{oscillate} = \frac{1}{2\pi\sqrt{L\,.\,C_{eq}}}$$

where, $C_{eq} = \frac{C_1C_2}{C_1 + C_2} = \frac{2 \times 2}{4} = 1PF$

$$L = 10\,\mu F$$

$$f_{oscillate} = \frac{1}{2\pi\sqrt{10 \times 10^{-6} \times 10^{-12}}}$$

$$= 50.3\text{ MHz}$$

2.8

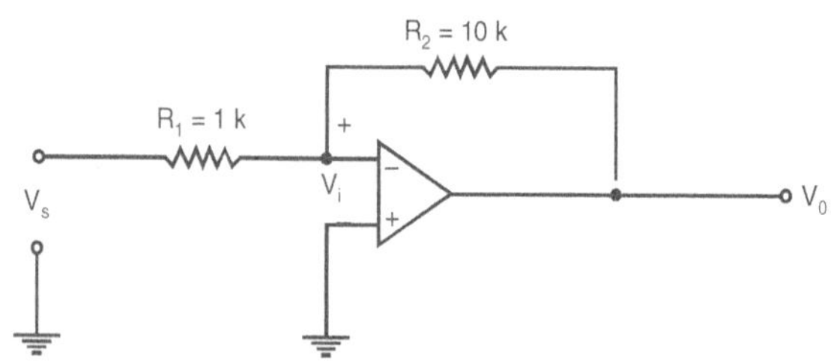

$$V_i A_v = V_0$$

KCL at inverting node,

$$\frac{V_s - V_i}{R_1} = \frac{V_i - V_0}{R_2}$$

$$\Rightarrow \quad \frac{V_s - \left(\frac{V_0}{A_v}\right)}{R_1} = \frac{\left(\frac{V_0}{A_v}\right) - V_0}{R_2}$$

$$\Rightarrow \quad \frac{V_0}{V_s} = \frac{A_v \cdot R_2}{(R_1 + R_2) - R_1 A_v}$$

$$= \frac{100.10}{11 - 100} \approx \mathbf{-11.}$$

2.9

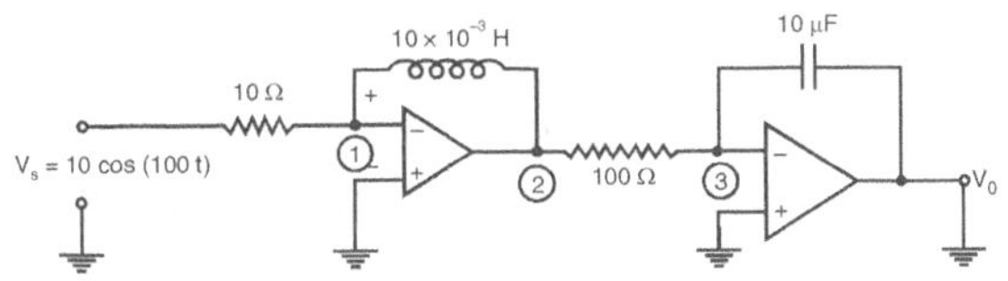

KCL at node (1),

$$\frac{V_s}{10} = -\frac{1}{L}\int V_2\,dt$$

$$\Rightarrow \quad V_s = \frac{-10}{L}\int V_2\,dt \qquad \ldots(i)$$

KCL at node (3),

$$\Rightarrow \quad \frac{V_3 - V_2}{100} = C\,\frac{dV_0}{dt}$$

$$\Rightarrow \quad -V_2 = 100\,C\,\frac{dV_0}{dt} \qquad [\because V_3 = 0]$$

$$\Rightarrow \quad V_0 = -\frac{1}{10^{-3}}\int V_2\,dt$$

From equation (*i*), we get

$$V_0 = 10 \cos 100\,t$$

2.10. Output at AND gate = S_1S_2

Output at XOR gate = $S_1\bar{S}_2 + \bar{S}_1S_2$

Output at NAND gate = $\overline{(S_1S_2)(S_1\bar{S}_2 + \bar{S}_1S_2)}$

$= \overline{S_1S_2\bar{S}_2 + S_1\bar{S}_1S_2} = 1$

As output is 1 irrespective of switch positions of S_1 and S_2, LED is reversed-biased and will not emit light.

2.11

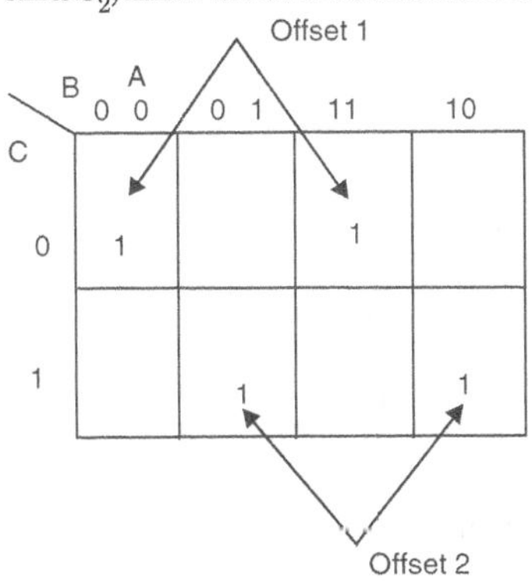

Offset 1. $F_1 = \bar{C}\,\overline{AB} + \bar{C}\,AB$

$= \bar{C} \cdot (AB + \bar{A}\,\bar{B})$

$= \bar{C} \cdot (\overline{A \oplus B})$

Offset 2. $F_2 = C\bar{B}A + CB\bar{A}$

$= C(A\bar{B} + \bar{A}B)$

$= C \cdot A \oplus B$

$\therefore \quad F = F_1 + F_2$

$= C \cdot (A \oplus B) + \bar{C} \cdot (\overline{A \oplus B})$

2.12. The wave four for the digital block given in figure (*c*) is depicted below

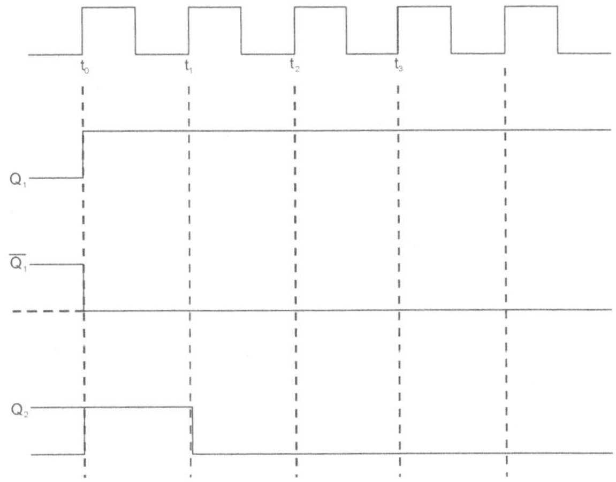

truth-table for D type FF

D	Q_n	Q_{n+1}
0	0	0
1	1	1
0	1	0
1	0	1

2.13.

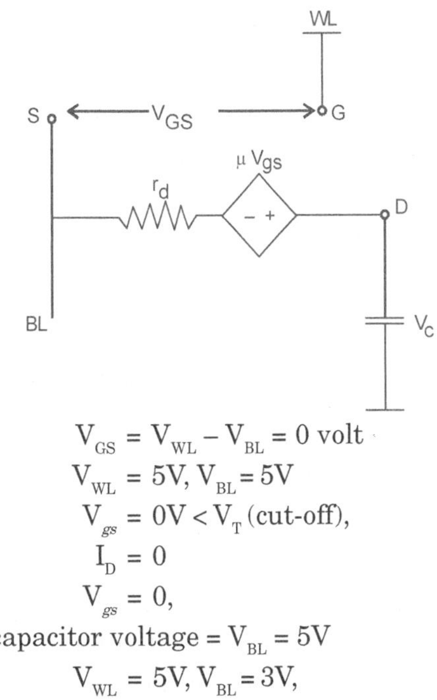

$V_{GS} = V_{WL} - V_{BL} = 0$ volt

(A) $V_{WL} = 5V, V_{BL} = 5V$

For, $V_{gs} = 0V < V_T$ (cut-off),

$I_D = 0$

and $V_{gs} = 0,$

the capacitor voltage = $V_{BL} = 5V$

(B) $V_{WL} = 5V, V_{BL} = 3V,$

$V_{GS} = V_{WL} - V_{BL} = 2$ volt

$V_{GS} > V_T$, therefore current I_D from through capacitor

and $V_C = -(V_{gs} + I_DR_D) + V_{BL}$

or $V_C < V_{BL}$

(C) $V_{WL} = 5V, V_{BL} = 7$ Volt

$V_{GS} = -2$ volt $< V_T$(cut-off)

Then, $I_D = 0$

and $V_{gs} = 0$

then capacitor voltage = $V_{BL} = 7$ volts

2.15 Applying KVL in Ist loop

$$V_i(s) = I_1(s)\,Z_1(s) + [I_1(s) - I_2(s)]\,Z_3(s)$$
$$V_i(s) = I_1(s)[Z_1(s) + Z_3(s)] - I_2(s)\,Z_3(s)$$
$$\frac{V_i(s)}{Z_1(s) + Z_3(s)} = I_1(s) - \frac{I_2(s)\,Z_3(s)}{Z_1(s) + Z_3(s)} \quad ...(i)$$

From IInd loop

$$[I_2(s) - I_1(s)]Z_3(s) + I_2(s)\,Z_2(s) + I_2(s)\,Z_4(s) = 0$$
$$I_2(s)[Z_2(s) + Z_3(s) + Z_4(s)] = I_1(s).\,Z_2(s)$$
$$G_2(s) = \frac{I_2(s)}{I_1(s)} = \frac{Z_3(s)}{Z_2(s) + Z_3(s) + Z_4(s)}$$

From SFG, $I_1(s) = V_1G_1(s) + I_2(s)\,H(s)$ ...(ii)

Compare (i) with (ii)

$$H = \frac{Z_3}{Z_1 + Z_3}$$

2.16

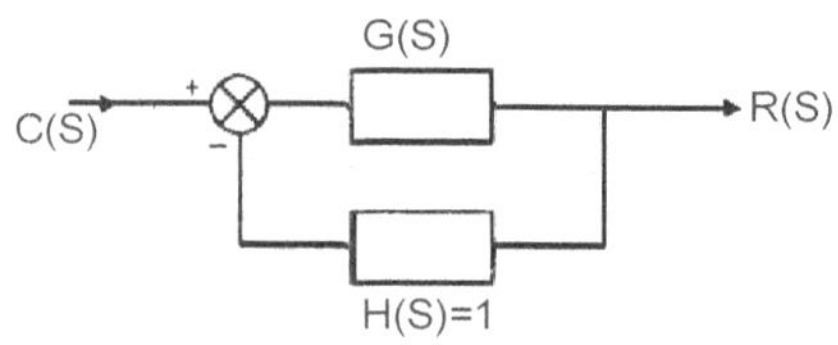

$$\frac{C(s)}{R(s)} = \frac{G(s)}{1 + G(s)\,H(s)}$$
$$\Rightarrow \quad \frac{G(s)}{1 + G(s)} = \frac{s+4}{s^2 + 7s + 13}$$
$$\Rightarrow \quad \frac{1 + G(s)}{G(s)} = \frac{s^2 + 7s + 13}{s+4}$$
$$\Rightarrow \quad \frac{1}{G(s)} = \frac{s^2 + 6s + 9}{s+4}$$
$$\Rightarrow \quad G(s) = \frac{s+4}{s^2 + 6s + 9}$$

DC gain will occur at $\omega = 0$

$$|G|_{DC} = \frac{4}{9}.$$

2.17 Characteristic equation is

$\Rightarrow \quad (k + 1)\,s^2 + s(4 - 4K) + 4 + 4K = 0$

Routh array drawn below,

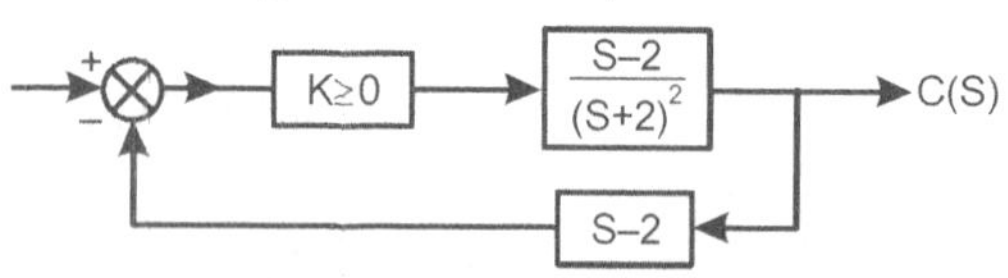

s^2	$K+1$	$4\,(1+K)$
s	$4\,(1-K)$	0

For stability, co-efficients in the first column is greater than zero, *i.e.* + ve

$$K + 1 > 0$$
$$1 > K \text{ and } K \geq 0.$$

or $$0 \leq K < 1$$

2.18. Given,

Pixel per frame = 400 × 400

Frames/second = 625

The number of bits required to produce 64 intensity lens is,

$$2^n = 64$$
$$\Rightarrow \quad n = 6$$

Hence, Data rate = 625 × 400 × 400 × 6

= 600 MbPS

2.19 $g(t) = \sin c\,(700\,t) + \sin c\,(500\,t)$

$$= \frac{1}{\pi t}[\sin(\pi\,700\,t) + \sin(\pi\,500\,t)]$$

$g(t)$ is band-limited with $f_M = 350$ Hz.

Hence the Nyquist rate will be 700 Hz

then Nyquist interval is $\frac{1}{700}$ sec.

2.20. Probability of error = P

Probability of no error = (1 – P)

For error in at most one bit, there should be either no error or error in only one bit.

Hence the required probability

$$= {}^nC_0\,(1-P)^n\,P^0 + {}^nC_1\,(1-P)^{n-1}\,P^1$$
$$= (1-P)^n + np(1-p)^{n-1}$$

2.21

$$(\text{PSD})\,S_g = \int_{-\infty}^{\infty} R_x(\tau)\,e^{-jW\tau}\,d\tau$$
$$R_x(\tau) = \underset{T\to\infty}{\text{Lt}}\,\frac{1}{T}\int_{-T/2}^{T/2} g(t)\,g(t+\tau)\,dt$$
$$P_g = \underset{T\to\infty}{\text{Lt}}\,\frac{1}{T}\int_{-T/2}^{T/2} |g(t)|^2\,dt$$

For $a\,g(t)$, $\quad S'_g = a^2\,S_g$

$P'_g = a^2\,P_g$

2.22

$$J = \sigma E = \omega D = \omega \epsilon E$$

or $$\frac{\sigma E}{D} = 2\pi f$$

or $$2\pi f = \frac{10^{-2}}{8.85 \times 10^{-12}}$$

or $$f = \mathbf{45\ MHz.}$$

2.23

$$\text{VSWR} = \frac{1 + |\tau|}{1 - |\tau|} = 3$$
$$|\tau| = 0.5$$

So reflected current $\quad I_{ref} = I_{inc} \times 0.5$

Power reflected $\quad \propto I_{ref}^2 \propto I_{ref}^2 \propto 0.25 I_{inc}^2 \propto 0.25 P_{inc}$

2.24 This is the monopole quarter-wave Antenna, and

for $\frac{L}{2} = \frac{\lambda}{4}$, $R_{rad} = 36.45\ \Omega$.

2.25 $$\Psi = \beta d(\cos\phi - 1)$$

$$= \frac{2\pi}{\lambda} \times \frac{\lambda}{4}(0.5 - 1) = -\frac{\pi}{4}.$$

4.

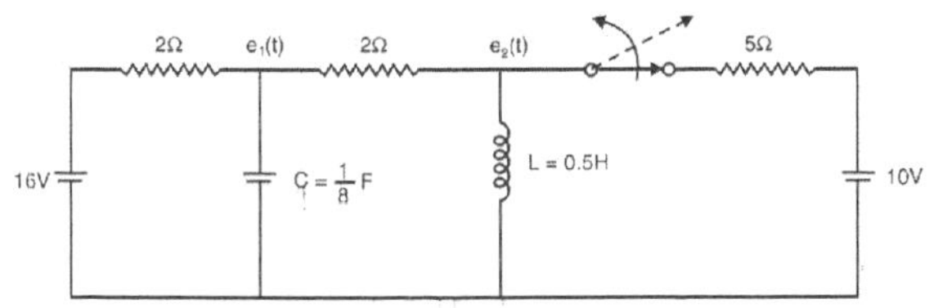

At steady state, $v_c(0^-) = \infty$, capacitor open-circuited for $t < 0$.

$v_2(0-) = 0$, inductor short-circuited for $t < 0$

Thus circuit at $t = 0$

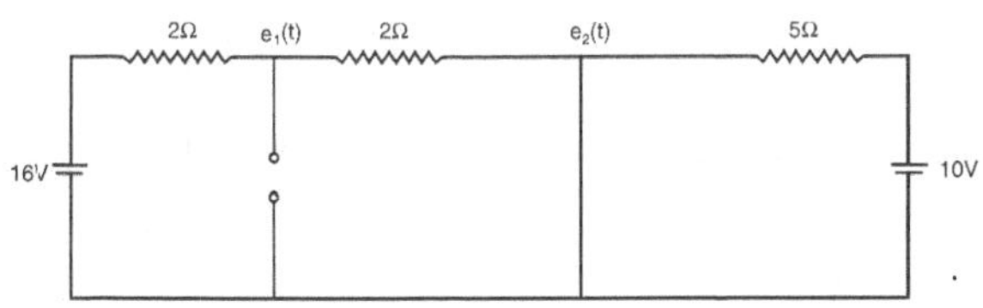

(*a*) $i_L(0^-) = i_L(0^+) = \frac{16}{4} = 4$ amp. [at $t = 0^+$, s_1 opened]

(*b*) $e_1(0^-) = e_1(0^+) = 16 - 4 \times 2 =$ **8 volt.**

(*c*) Applying KCL at nodes (1) and (2), we have

$$c\frac{de_1(t)}{dt} + \frac{e_1(t) - 16}{2} + \frac{e_1(t) - e_2(t)}{2} = 0 \quad ...(i)$$

$$\left[\frac{1}{L}\int_0^t e_2(t)dt + i_L(0^+)\right] + \frac{e_2(t) - e_1(t)}{2} = 0 \quad ...(ii)$$

Taking Laplace transform to both sides of the equation,

From equation *(i)*,

$$\frac{e_1(s)}{2} + \left[sCe_1(s) - CV_c(0^+)\right] + \frac{e_1(s) - e_2(s)}{2} = \frac{8}{s} \quad ...(iii)$$

$$\frac{e_2(s)}{Ls} + \frac{i_L(0+)}{s} + \frac{e_2(s) - e_1(s)}{2} = 0 \quad ...(iv)$$

From equation *(iii)*

$$\left[\frac{s}{8} + 1\right]e_1(s) - e_2(s) \cdot \frac{1}{2} = \frac{8}{s} + 1$$

$$-\frac{1}{2} \cdot e_1(s) + e_2(s)\left[\frac{2}{s} + \frac{1}{2}\right] = \frac{-4}{s}$$

Writing in matrix form,

$$\begin{bmatrix} e_1(s) \\ e_2(s) \end{bmatrix} \begin{bmatrix} \frac{s}{8} + 1 & -\frac{1}{2} \\ -\frac{1}{2} & \frac{2}{s} + \frac{1}{2} \end{bmatrix} = \begin{bmatrix} \frac{8}{s} + 1 \\ -\frac{4}{s} \end{bmatrix}$$

$$\Rightarrow \quad e_1(s) = \frac{\begin{bmatrix} \frac{8}{s} + 1 & -\frac{1}{2} \\ -\frac{4}{s} & \frac{4+s}{2s} \end{bmatrix}}{\frac{(s+8)(s+4)}{8 \times 2s} - \frac{1}{4}}$$

$$= \frac{\frac{(s+8)(s+4)}{2s^2} - \frac{2}{s}}{\frac{(s+8)(s+4) - 4s}{16s}}$$

$$= \frac{8}{s}\left[\frac{(s+8)(s+4) - 4s}{(s+8)(s+4) - 4s}\right] = \frac{8}{s}.$$

$e_1(t) = 8$ volt for all $t > 0$.

5.

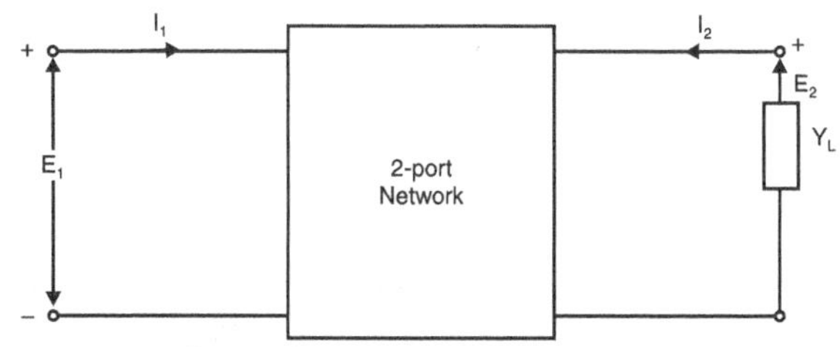

(*a*) From the definition of Y-parameter,

$$I_1 = E_1 Y_{11} + E_2 Y_{12} \quad ...(i)$$

$$I_2 = E_1 Y_{21} + E_2 Y_{22} \quad ...(ii)$$

As $I_2 = -E_2 Y_L$

$\therefore \quad -E_2 Y_L = E_1 \cdot Y_{21} + E_2 \cdot Y_{22}$

or $$E_2 = \frac{-E_1 Y_{21}}{Y_{22} + Y_L} \quad ...(iii)$$

$$= -\frac{10\angle 0° (4.8)}{(1.2)} = -40\angle 0$$

(*b*) Eliminating E_1 from equation *(ii)* with equation *(i)*, we have

$$I_2 = Y_{21} \cdot \left[\frac{I_1 - E_2 Y_{12}}{Y_{11}}\right] + E_2 Y_{22}$$

or $$-E_2 Y_L = \frac{Y_{21}}{Y_{11}}\left[I_1 - E_2 Y_{12}\right] + E_2 Y_{22}$$

or $$E_2\left[-Y_L + \frac{Y_{21} \cdot Y_{12}}{Y_{11}} - Y_{22}\right] = \frac{Y_{21}}{Y_{11}} \cdot I_1$$

or $$E_2 - \frac{Y_{21} I_1}{-Y_{11} \cdot Y_{22} + Y_{21} \cdot Y_{12} - Y_L \cdot Y_{11}}$$

$$= \frac{10\angle 0°}{2 - 2.4 - 0.4} = \mathbf{-12.5\angle 0}$$

(*c*)

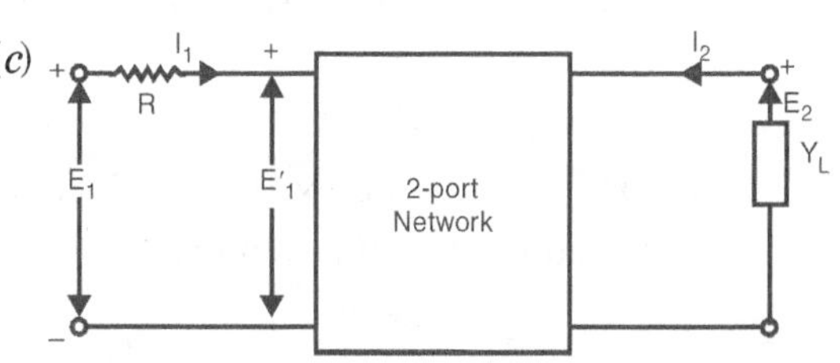

From the circuit, $E_1 - I_1 R = E_1'$

$$E_2 Y_L = -I_2$$

Now, $I_1 = E_1' Y_{11} + E_2 . Y_{12}$ ....(*i*)

$\Rightarrow$ $I_1 = (E_1 - I_1 R) Y_{11} + E_2 Y_{12}$

or $I_1 [1 + RY_{11}] = E_1 . Y_{11} + E_2 . Y_{12}$

$\Rightarrow$ $I_1 = \dfrac{E_1 Y_{11} + E_2 . Y_{12}}{1 + R Y_{11}}$ ....(*ii*)

$I_2 = E_1' Y_{21} + Y_{22} E_2$ ....(*iii*)

$$\Rightarrow I_2 = \left[E_1 - \frac{(E_1 Y_{11} + E_2 Y_{12})R}{1 + R Y_{11}}\right] Y_{21} + Y_{22} E_2$$

$$E_2 \left[\frac{Y_L}{\dfrac{Y_{12} . R . Y_{21}}{1 + R Y_{11}} + Y_{22}}\right]$$

$$= E_1 \left[- Y_{21} + \frac{Y_{11} . Y_{21} R}{1 + R Y_{11}}\right]$$

$$\Rightarrow E_2 = \frac{\dfrac{E_1 [-(1 + R Y_{11}) Y_{21} + Y_{21} Y_{21} R]}{(1 + R Y_{11}) Y_L}}{Y_{21} . Y_{12} . R + Y_{22} (1 + R Y_{11})}$$

$$\Rightarrow E_2 = \frac{\dfrac{- E_1 Y_{21}}{(1 + R Y_{11})(Y_L + Y_{22})}}{Y_{21} . Y_{12} . R}$$

Putting the value $E_2 = 20 \angle 0°$

6. 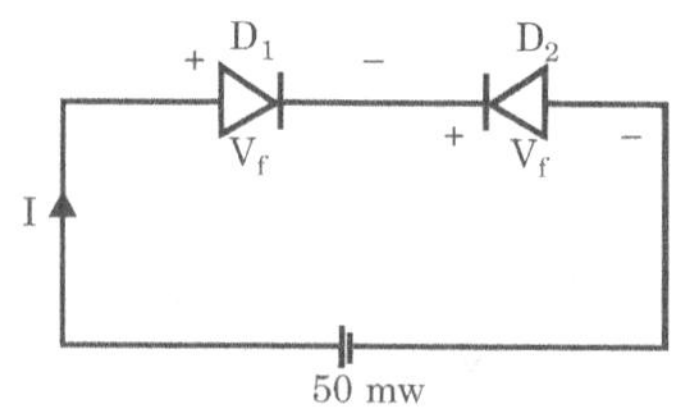

(*a*) The current I is given by

$$I = I_0 (e^{\frac{V}{\eta V_T}} - 1)$$

where V_T = volt - equivalent of temperature

Given, $\eta = 1$

Now, for diode D_1,

$$I = I_0 (e^{\frac{V_f}{V_T}} - 1) \quad ...(i)$$

For diode D_2, $I = - I_0 (e^{-\frac{V_r}{V_T}} - 1)$ ... (*ii*)

Equating equations (*i*) and (*ii*), we get

$$e^{\frac{V_f}{V_T}} + e^{\frac{V_r}{V_T}} = 2 \quad ...(iii)$$

Also $V_f + V_r = 50 \times 10^{-3}$ V = 50 mV ...(*iv*)

From equation (*iii*), we have

$$e^{\frac{V_f}{V_T}} + e^{\frac{-(50 - V_f)}{V_T}} = 2$$

$$\Rightarrow e^{\frac{V_f}{V_T}} [1 + e^{-2}] = 2$$

$$\Rightarrow e^{\frac{V_f}{V_T}} = \frac{2}{1 + e^{-2}}$$

$\Rightarrow$ $V_f = 14.15$ mv

And V_r = **35.85 mv**

(*b*) Current through the circuit,

$$I = I_s (e^{\frac{V_f}{V_T}} - 1)$$

$= 1 [e^{\frac{14.15}{25}} - 1]$ PA = **0.76 PA**

7. 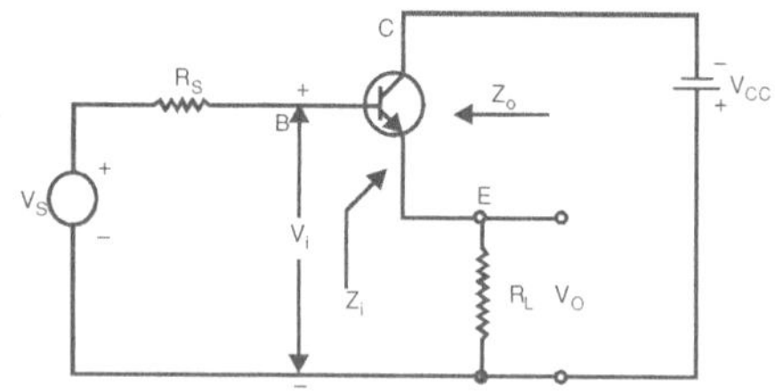

Its small-signal equivalent depicted in the following circuit :

(*a*) 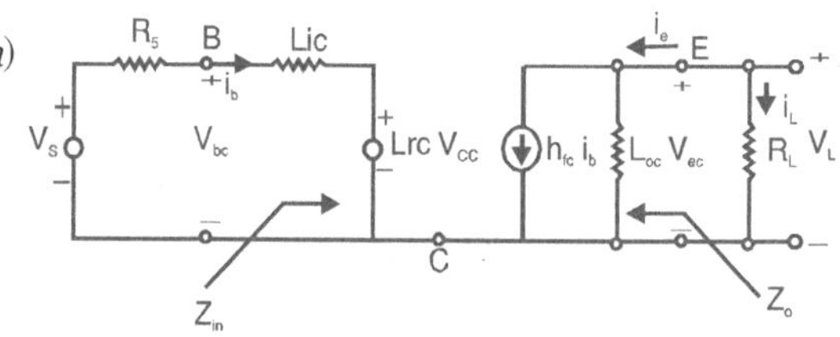

(*b*) From the above circuit,

$V_{bc} = i_b h_{ic} + h_{rc} V_{ec}$ ...(*i*)

$i_e = h_{oc} V_{ec} + h_{fc} i_b$ ...(*ii*)

$i_e = - h_{oc} R_L i_e + h_{fc} i_b$

$\Rightarrow$ $i_e (1 + h_{oc} R_L) = h_{fc} i_b$

$\Rightarrow$ $i_e = \dfrac{h_{fc} i_b}{1 + h_{oc} R_L}$ ...(*iii*)

From equation (*i*), $V_{bc} = i_b h_{ic} - h_{rc} R_L i_e$

$$= i_b h_{ic} - \frac{h_{rc} R_L i_b}{1 + h_{oc} R_L}.$$

$$Z_i = \frac{V_{bc}}{i_b} = h_{ic} - \frac{h_{rc} R_L h_{fc}}{1 + h_{oc} R_L}$$

(*c*) $Y_0 = \dfrac{I_e}{V_0}$ [Output admittance]

$= \dfrac{h_{fe} i_b + h_{oc} V_0}{V_0}$ [Setting the load R_L to ∞].

$= \dfrac{h_{fe} i_b}{V_{ec}} + h_{oc}$ ...(*iv*)

Setting the source voltage zero.

$R_s i_b + h_{rc} V_{ec} + h_{ic} i_b = 0$

$\Rightarrow$ $R_s i_b + h_{rc} V_{ec} + h_{ic} i_b = 0$

$$\Rightarrow \quad \frac{i_b}{v_{ec}} = \frac{-h_{rc}}{h_{ic}+R_s}$$

From equations *(iv)*, $Y_0 = \dfrac{-h_{fc}h_{rc}}{h_{ic}+R_s} + h_{oc}$

$$Z_0 = \frac{1}{Y_0} = \frac{1}{h_{oc} - \dfrac{h_{fc}\,h_{rc}}{h_{ic}+R_s}}$$

(*d*) $Z_L = \dfrac{R_L}{R_L\, j\omega c_L + 1}$

From equation *(ii)*, substituting Z_L for R_L, we get

$$Z_i = h_{ic} - \frac{h_{rc}\,R_L\,h_{fc}}{R_L\,j\omega C_L + 1 + h_{oc}R_L}$$

$$Z_0 = \frac{1}{h_{oc} - \dfrac{h_{fc}\,h_{rc}}{h_{ic}+R_s}} \text{ output impedance unaffected.}$$

8.

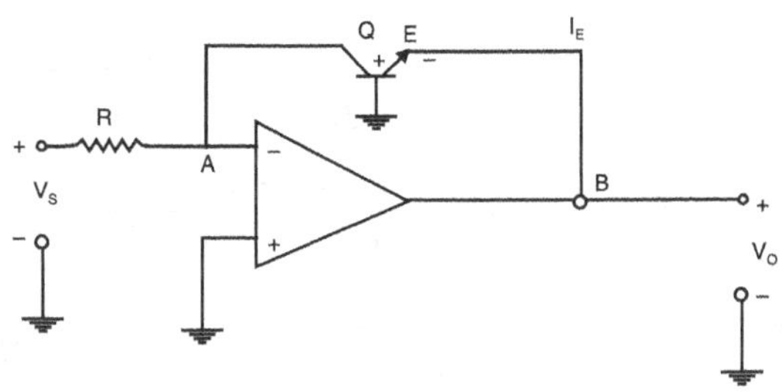

(*a*) For transistor Q, its emitter-base voltage and emitter current is related by,

$$I_E = I_s\, e^{\frac{v_{BE}}{V_T}} \quad \ldots.(i)$$

where I_s = reverse saturation current

Also, KCL at node A, $\dfrac{V_s - 0}{R} = I_E$

$$\Rightarrow \quad \frac{V_s}{R} = I_s\, e^{\frac{v_{BE}}{V_T}}$$

$$\Rightarrow \quad I_E = \frac{V_s}{R}$$

Putting this in equation *(i)*, we get

$$\Rightarrow \quad \frac{V_s}{R} = I_s\, e^{\frac{v_{BE}}{V_T}}$$

$$\Rightarrow \quad e^{\frac{v_{BE}}{V_T}} = \frac{V_s}{I_sR}$$

Taking logrithm both sides, we have

$$v_{BE} = V_T \ln\frac{V_s}{I_sR}$$

$$\Rightarrow \quad V_0 = -V_T \ln\frac{V_s}{I_sR} = V_T \ln\frac{I_sR}{V_s}$$

(*b*) From part *(a)*, we have

$$V_0 = V_T \ln\frac{I_sR}{V_s}$$

$$= 25\times10^{-3}\ln\frac{10^{-12}\times1}{1}$$

$$= -300\times10^{-3}\,\text{V}$$

$$= \mathbf{-0.3\ volt}$$

(*c*) As, $\quad I_D = I_s\,(e^{\frac{V_D}{\eta V_T}} - 1) \approx I_s e^{\frac{V_D}{\eta V_T}}$

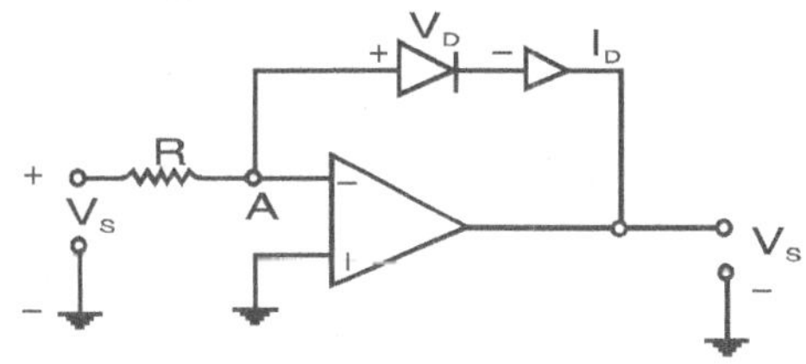

(*d*) $\quad V_0 = V_D = 0$

Also, $\quad i_D = \dfrac{V_s}{R}.$

Then from equation *(i)*, we have

$$\frac{V_s}{R} = I_s\, e^{\frac{V_D}{\eta V_T}}$$

$$\Rightarrow \quad e^{V_D/\eta V_T} = \frac{V_s}{I_sR}$$

Thus, $\quad V_0 = -\eta V_T \ln\dfrac{V_s}{I_sR}$

$$= \eta V_T \ln\frac{I_sR}{V_s}$$

10.

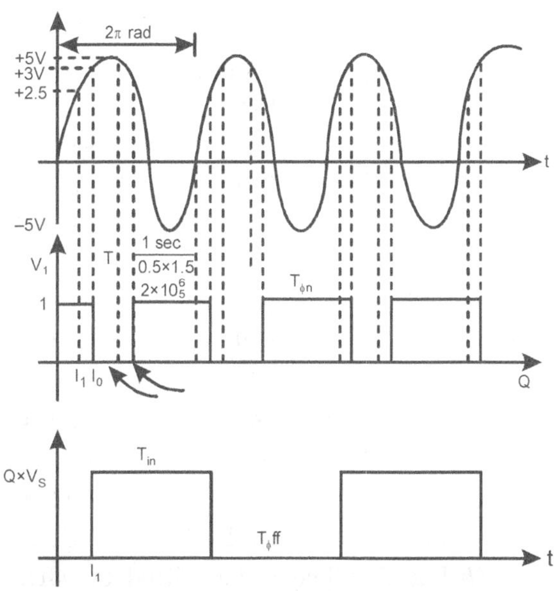

Given, J-K Flip-flop is -ve edge-triggering Flip-flop

$$5\sin\omega t_1 = 2.5$$

$$t_1 = \frac{1}{\omega}\sin^{-1}\left(\frac{2.5}{5}\right) = \frac{T}{12}.$$

And $\quad 5\sin\omega t_2 = 3.5, \quad t_2 = \dfrac{T}{8}.$

For V_2,

(*a*) Period $= \frac{1}{f} = T = 2\ \mu\text{sec}$

$$\text{Duty cycle} = 100.\frac{T_{ON}}{T} = 100.\frac{\frac{T}{2}+\frac{T}{8}+\frac{T}{12}}{T}$$

$$= \frac{17}{24} \times 100 = \mathbf{70.83}$$

(*b*) Duty cycle for waveform $V_3 = 100.\frac{T_{ON}}{T_{ON}+T_{off}}$

$$= 100.\frac{T}{2T} = \mathbf{50.}$$

Period = 2T = 4 sec.

11. (*a*)

$S_3S_2 \backslash S_1S_0$	00	01	11	10
00	0 (0)	1 (1)	0 (3)	0 (2)
01	0 (4)	1 (5)	1 (7)	1 (6)
11	1 (12)	1 (13)	1 (15)	0 (14)
10	0 (8)	0 (9)	1 (11)	0 (10)

(*b*) The minimized expression will be

$$Y = s_3s_2\bar{s}_1 + s_0s_1s_3 + \bar{s}_1s_0\bar{s}_3 + \bar{s}_3s_2s_1$$

$$= s_0[s_1s_3 + s_1\bar{s}_3] + s_2[s_1\bar{s}_3 + \bar{s}_1s_3].$$

12. (*a*)

```
          MVI A,   FFH
          INR A
          JC       FFOCH
          ORI      A8H
          JM       FF15H
address
FFISH     MVI A,   FFH
          ADI      02H
          RAL
          JZ       FF234
          JC       FF104
          JNC      FF124
FF124     OUT      PORT2
```

(*b*) Port 2 will be loaded with data 03H.

13. (*a*)

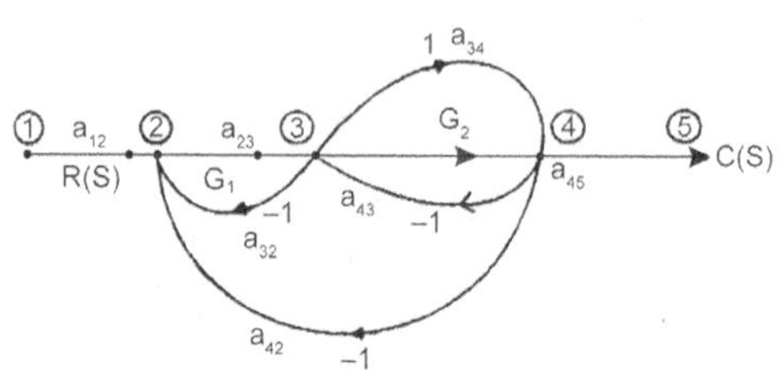

(*b*) There are five loops, the loop gains are

$$L_{11} = a_{23}a_{32} = -G_1$$
$$L_{21} = a_{34}a_{43} = -1$$
$$L_{31} = G_2 \,.\, a_{43} = -G_2$$
$$L_{41} = a_{23}.\ G_2 \,.\, a_{42} = -G_1G_2$$
$$L_{51} = a_{23}\ a_{34}\ a_{42} = -G_1.$$

(*c*) There is no combinations of non-touching loops.

(*d*) There are two forward paths

$$\mu_1 = a_{12}\ a_{23}\ a_{34} = G_1$$
$$\mu_2 = G_1\ G_2$$

Since both the forward paths touches all the loops, therefore

$$\Delta_1 = \Delta_2 = 1.$$

Hence, T(*s*) = (transfer function)

$$= \frac{\mu_1\Delta_1 + \mu_2\Delta_2}{\Delta}$$

$$\Delta = 1 - (L_{11} + L_{21} + L_{31} + L_{41} + L_{51})$$
$$= 2 + 2G_1 + G_2 + G_1G_2$$

Then, $T(s) = \frac{G_1 + G_1G_2}{2 + 2G_1 + G_2 + G_1G_2}$

14.

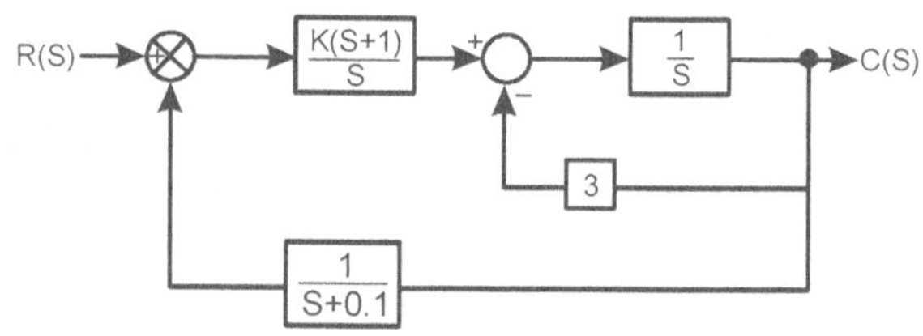

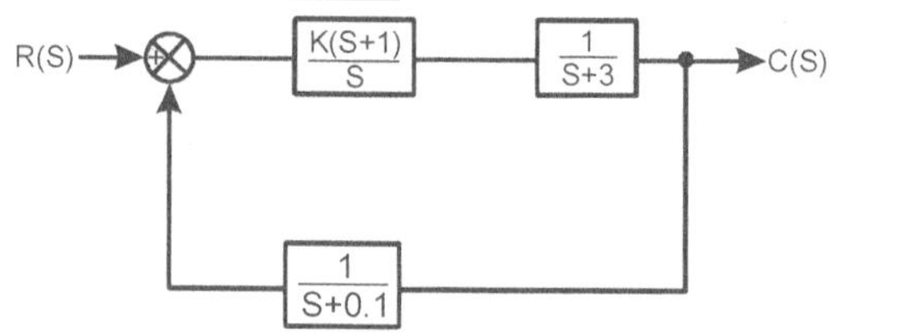

$$T(s) = \frac{C(s)}{R(s)}$$

$$= \frac{K(s+1)(s+0.1)}{s^3 + 3.1s^2 + s(K+0.3) + K}$$

Characteristic equation will be,

$$s^3 + 3.1\ s^2 + s(K + 0.3) + K = 0.$$

It south-away will be,

s^3	1	K+0.3	0
s^2	3.1	K	0
s^1	$\frac{2.1\,K+0.93}{3.1}$	0	0
s	K	0	0

For stability co-efficients of first column should be + ve,

$\therefore \quad K > 0$

And $\quad 2.1\,K + 0.93 > 0$

or $\quad K > -0.44$

Hence, K > 0 for stability. This will be the required range.

15. (*a*)
$$T(s) = \frac{K(K_D s + K_P)}{s^2 + s(KK_D + 1) + KK_P}$$

Given, $\quad \delta\omega_n = \frac{1}{1} = 1$

$$\Rightarrow \quad \omega_n = \frac{1}{0.707} = 1.41$$

The poles are,

$$\omega_1, \omega_2 = -\delta\omega_n \pm \omega_n\sqrt{1-\delta^2}$$
$$= \omega_n\left[-\delta \pm \sqrt{1-\delta^2}\right]$$
$$= -1 \pm 1j$$

(*b*)

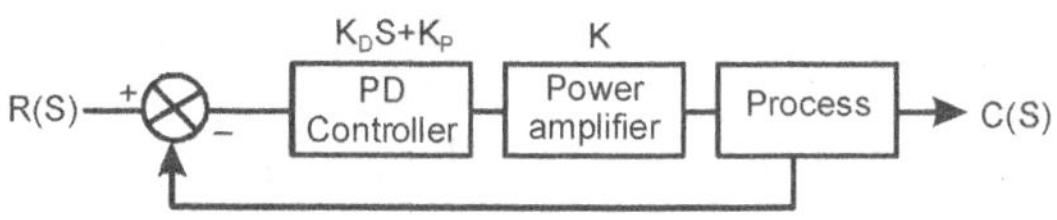

Characteristic equation, is

$$s^2 + s(KK_D + 1) + KK_P = 0.$$

when $\quad K = 1,$

$$K_P = \omega_n^2 = 2.$$

and, $\quad K_D + 1 = 2\delta\omega_n = 2$

$$K_D = 1$$

So PD controller equation will be,

$$K_DS + K_P = s+2$$

(*c*)

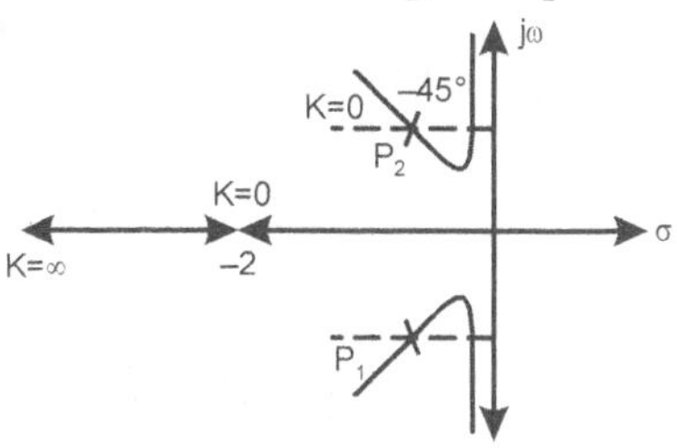

$$P_1 = -1 + j$$
$$P_2 = -1 - j$$

Angle of departure at pole P_1

$$= 180° - 135°$$
$$= -45°$$

Angle of departure, at pole P_2

$$= -180° + 135°$$
$$= 45°$$

The desired root locus is shown above.

16. The Fourier transform of any signal $x(t)$

$$X(\omega) = \int_{-\infty}^{\infty} x(t)\, e^{-j\omega t}\, dt$$

Then for signal $\quad x_1(t) = x(-t)$, it will be

$$X_1(\omega) = \int_{-\infty}^{\infty} x(-t)\, e^{-j\omega t}\, dt$$

Putting $\quad -t = P = -\int_{+\infty}^{-\infty} x(P)\, e^{+j\omega p}\, dp$

$$X_1(\omega) = \int_{-\infty}^{\infty} x(p)\, e^{j\omega p}\, dp$$

$$= X(-\omega) = X^*(\omega). \qquad ...(i)$$

For time-shifting signal $\quad x_2(t) = x_1(t+1)$

$$= \int_{-\infty}^{\infty} x_1(t+1)\, e^{-j\omega t}\, dt$$

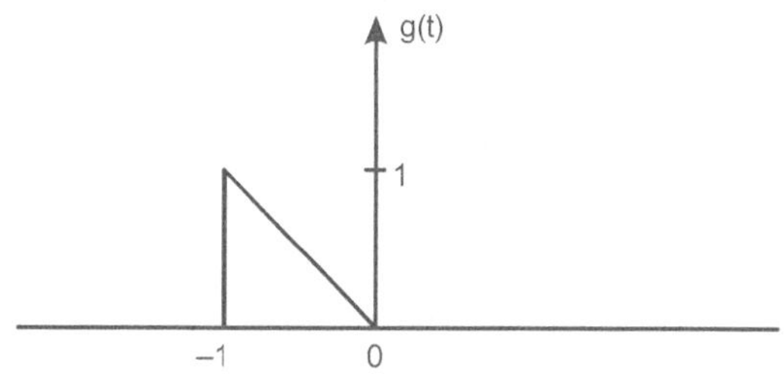

Put $\quad t + = p$

$$= +\int_{-\infty}^{\infty} x_1(P)\, e^{-j\omega p} . e^{j\omega}\, dp$$

$$= e^{j\omega} \int_{-\infty}^{\infty} x_1(p)\, e^{-j\omega p}\, d_p$$

$$= G_1(\omega) . e^{j\omega} \qquad ....(ii)$$

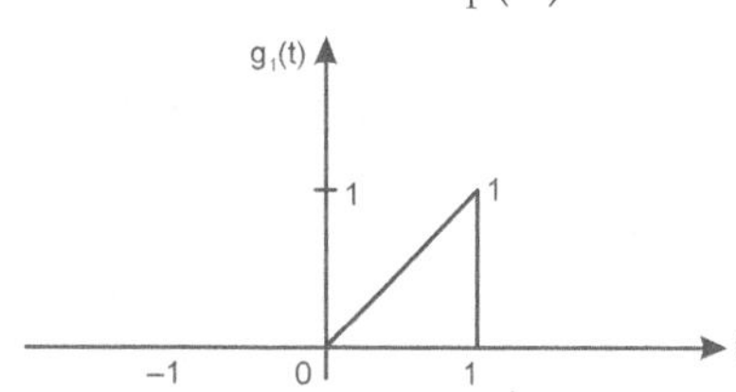

Given, for signal g (t), the Fourier transform

$$G(\omega) = \frac{1}{\omega^2}(e^{j\omega} - j\omega e^{j\omega} - 1).$$

$$g_1(t) = g(-t)$$

$\therefore \quad G_1(\omega) = G(-\omega)$ from equation (*i*)

$$= \frac{1}{w^2}(e^{-jw} + jwe^{-jw} - 1) \qquad ...(iii)$$

$$g_2(t) = g_1(t+1)$$

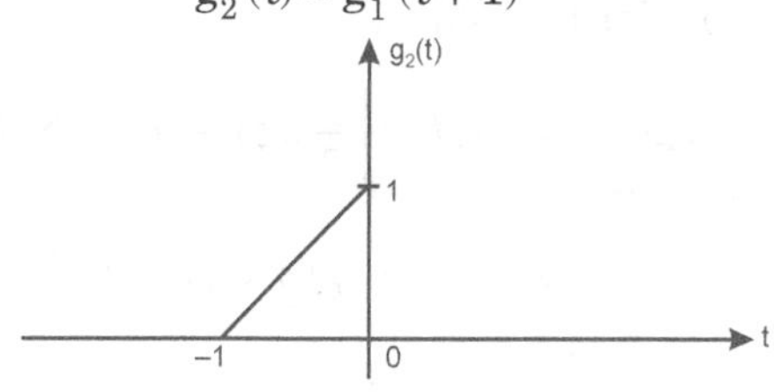

Then, $G_2(\omega) = G_1(j\omega) \cdot e^{j\omega}$

From equations *(ii)* and *(iii)*, we get

$$G_2(\omega) = \frac{1}{\omega^2}[1 + j\omega - e^{j\omega})$$

$$G_3(\omega) = \int_{-\infty}^{\infty} g_3(t)\, e^{-j\omega t}\, dt$$

$$= + \int_{-1/2}^{1/2} e^{-j\omega t} dt$$

$$= \sin C(f)$$

$$= \frac{\sin \pi f}{\pi f}$$

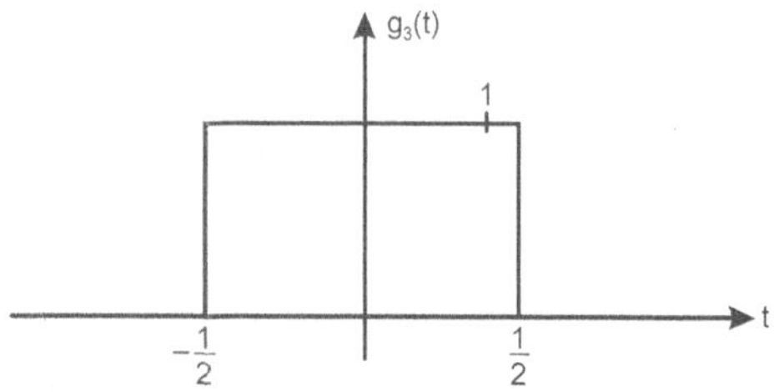

17.

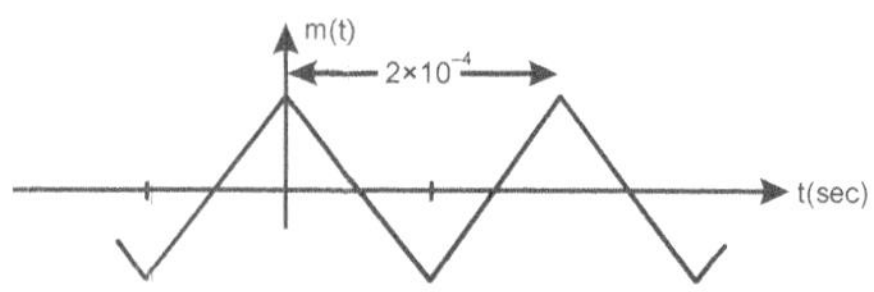

For FM signals the instantaneous frequency of the $f_i = f_c + k_f m(t)$

the frequency deviation,

$$\Delta_f = k_f A_m$$
$$= 2\, k_f$$

Then band width of the FM signal

$$= 2 \times 3 \times \Delta f$$
$$= 2 \times 3 \times 2\, k_f$$
$$= 12\,\pi \times 10^4 \text{ Hz.}$$

For PM, $\Delta f = K_p \left| \frac{dm(t)}{dt} \right|_{max}$

From figure $\frac{dm(t)}{dt} = 2 \times 10^4$

$$\Delta f = K_p\, 2 \times 10^4$$

$$\text{BW (for PM)} = 2 \times 3 \times K_p \times 2 \times 10^4$$
$$= 3\pi \times 10^4 \text{Hz.}$$

18. The modulating signal,

$$x = g(t) \cos 2\pi f_c t$$

Its Fourier transform

$$£[x(t)] = G(f) * \left[\frac{1}{2}\delta(f - f_c) + \delta(f + f_c)\right]$$

$$= [G(f - f_c) + G(f + f_c)] \cdot \frac{1}{2}$$

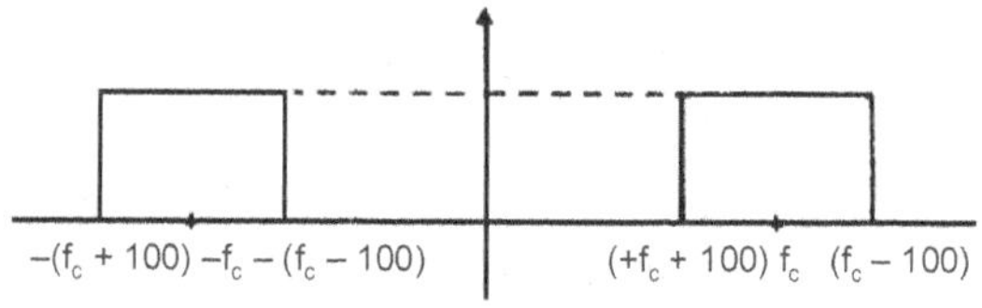

Spectrum for the modulating signal $x(t)$.

Now, $y = 2x + x^2$

$$= y_1(t) + y_2(t)$$

$$y_2(t) = x^2$$
$$= g^2(t) \cos^2 2\pi f_c t$$
$$= g^2(t) \times \frac{[1 + \cos 4\pi f_c t]}{2}$$

$$f[g^2(t)] = \int_{-\infty}^{\infty} V(f) V(f - \lambda)\, d\lambda$$

$$\Rightarrow \quad f[g^2(t)] = \int_{-100}^{100} V(f - \lambda)\, d\lambda$$

$$= \int_{f-100}^{f+100} V(\beta)\, d\beta,$$

where $V|\beta| = 1$

for, $|\beta| < 100$

i.e. $-100 < \beta < 100$

$V|\beta| = 0$, elsewhere

$$\Rightarrow \quad f[g^2(t)] = 200\left(1 - \frac{|f|}{200}\right) f$$

or $|f| < 200.$

then, $f[X^2] = \frac{1}{2}\left[f(g^2(t)\right]$

$$+ f\left[g^2(t) + \frac{1}{2}[\delta(f - 2f_c) + \delta(f + 2f_c)]\right]$$

$$= 100\left(1 - \frac{|f|}{200}\right) + 50\left(1 - \frac{|f - 2f_c|}{200}\right),$$

for $|f - 2f_c| \le 200$

$$= 100\left(1 - \frac{|f|}{200}\right) + 50\left(1 - \frac{|f + 2f_c|}{200}\right),$$

for $(f + 2f_c) \le 200$

Hence $Y = 2x + x^2$

$$= 2\, f[x] + f[x^2]$$

Its spectrum is shown below,

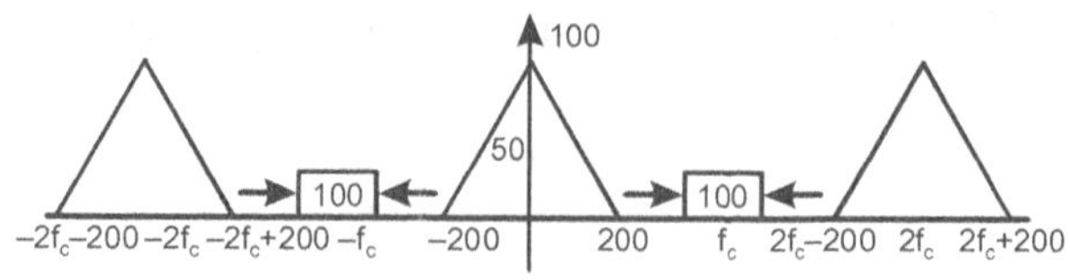

19.

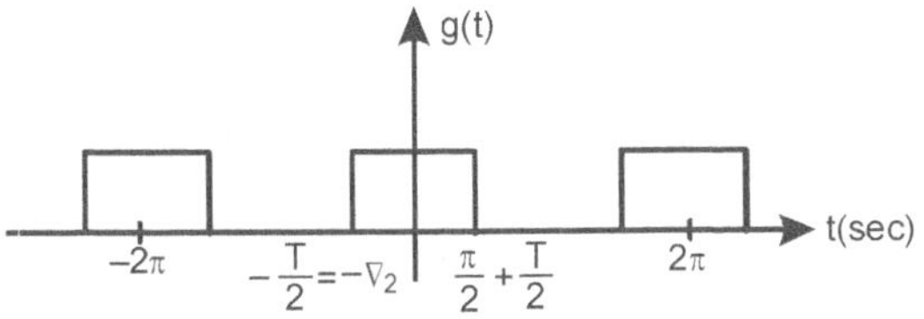

The Fourier series of the given signal is given by,

$$g(t) = \frac{1}{2} + \frac{1}{\pi}\sum_{n=-\infty}^{\infty} \frac{(-1)^{n-1}}{2n-1} \cdot e^{j\omega(2n-1)t}$$

It's correlation function R (τ) given by,

$$R(\tau) = \frac{1}{T}\int_{-T/2}^{T/2} g(t)\, g(t+\tau)\, dt$$

$$= \frac{1}{4} + \sum_{n=-\infty}^{\infty}\left[\frac{1}{\pi(2n-1)}\right]^2 e^{\frac{j\omega(2n-1)\tau}{T}}$$

$$= \frac{1}{4} + 2\sum_{n=1}^{\infty}\left[\frac{1}{\pi(2n-1)}\right]^2 \frac{\cos 2\pi(2n-1)\tau}{T}$$

The PSD of $g(t)$, $G(f) = \tau[R(\tau)]$

$$= \int_{-\infty}^{\infty}\left[\frac{1}{4} + \sum_{n=-\infty}^{\infty}\left[\left\{\frac{1}{\pi(2n-1)}\right\}^2 e^{\frac{j\omega(2n-1)\tau}{T}}\right] e^{-j2\pi f\tau}.d\tau\right]$$

$$= \frac{1}{4}\int_{-\infty}^{\infty} e^{-j2\pi f\tau}.\, d\tau + \sum_{n=-\infty}^{n}\left\{\frac{1}{\pi(2n-1)^2}\right\}^2 \int_{-\infty}^{\infty} e^{-2\pi\left(f - \frac{2n-1}{T}\right)^{\tau}} d\tau$$

$$= \frac{1}{4}.\delta(f) + \sum_{n=-\infty}^{\infty}\left\{\frac{1}{\pi(2n-1)^2}\right\}^2 \delta\left(f - \frac{2n-1}{T}\right).$$

20.

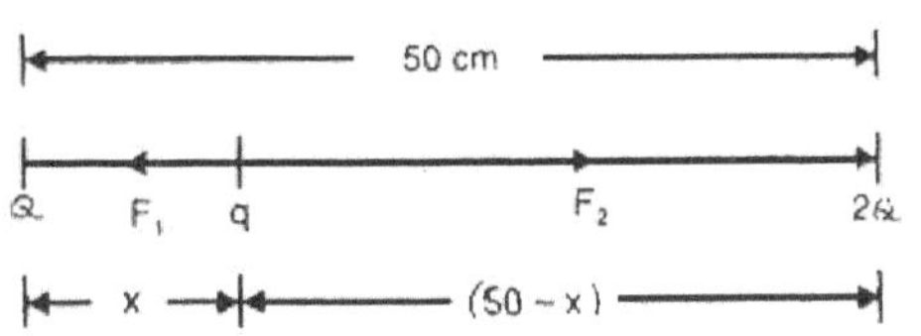

Let a charge q place between charges at distance x from charge Q,

For equilibrium $\bar{F}_1 + \bar{F}_2 = 0$

$$\Rightarrow \quad \bar{F}_1 = \frac{Qq}{4\pi\epsilon_0 x^2}\frac{\vec{x}}{|\vec{x}|}$$

$$\Rightarrow \quad \bar{F}_2 = \frac{2Qq}{4\pi\epsilon_0(50-x)^2}\cdot\frac{\overrightarrow{(50-x)}}{|\overrightarrow{50-x}|}$$

Now as, $\bar{F}_1 + \bar{F}_2 = 0$

$$\Rightarrow \quad \frac{Qq}{4\pi\epsilon_0 x^2}\frac{\vec{x}}{|\vec{x}|} + \frac{2Qq}{4\pi\epsilon_0(50-x)^2}\cdot\frac{\overrightarrow{(50-x)}}{|\overrightarrow{50-x}|} = 0$$

$$\Rightarrow \quad \frac{\vec{x}}{|\vec{x}|}\cdot\frac{1}{x^2} + \frac{\overrightarrow{(50-x)}}{|\overrightarrow{(50-x)}|}\cdot\frac{2}{(50-x)^2} = 0$$

If both the unit vectors are oppositely directed, then both the forces will cancel each other. It means that third charge will be '–'ve.

$$\Rightarrow \quad \frac{1}{x^2} = \frac{2}{(50-x)^2}$$

$$\Rightarrow \quad \frac{50-x}{x} = \sqrt{2}$$

$$\Rightarrow \quad x = \frac{50}{\sqrt{2}+1} = \mathbf{20.71\ cm.}$$

21. Given, $E = 16$ KV/m

$$E_{max} = 16\sqrt{2}\ \text{KV/m}$$

$$\eta = \sqrt{\frac{\mu_0\mu_r}{\epsilon_0\epsilon_r}}$$

$$= 120\pi \times \sqrt{\frac{1}{4}} = 60\pi\,\Omega$$

Maximum power flow density

$$= \frac{E^2}{\pi} = \frac{(16\sqrt{2})^2}{60\pi} = 2.71\ \text{MW.}$$

$$\text{Magnetic field} = \frac{E^2}{\eta} = \mathbf{0.12\ k\Omega m.}$$

22. (*a*) For TE_{mn} mode λ_c(cut-off wavelength)

$$= \frac{2}{\sqrt{\frac{m^2}{a^2} + \frac{n^2}{b^2}}}$$

For TE_{10} mode, $m = 1$, $n = 0$

Then, $\lambda_c = \dfrac{2}{\sqrt{\frac{1}{a^2} + 0}} = 2a$

λ_g(Guide wavelength for this mode,

$$\lambda_g = \frac{\lambda}{\sqrt{1 - \left(\frac{f_c}{f}\right)^2}}$$

As, $f_c = \dfrac{c}{2a} = \dfrac{3\times10^8}{2.29\times10^{-2}\times2} = 6.55$ GHz

Now, $\lambda_g = \dfrac{\frac{c}{f}}{\sqrt{1 - \left(\frac{f_c}{f}\right)^2}}$

$$= \frac{3\times10^8}{10\times10^9\times0.75} = \mathbf{0.04\ m}$$

(*b*) The other modes are TE_{11}, TE_{21}, TM_{21}, TM_{11}, TM_{21}, TE_{01}, TE_{22}, TE_{22}, TM_{01}, TM_{22}

This wavelength (cut-off) are lower than dominant wave cut-off wavelength.

■■

SOLVED PAPER – 2000

ELECTRONICS & COMMUNICATION ENGINEERING

Time Allowed : 3 Hours | **Maximum Marks : 150**

SECTION - A (75 Marks)

1. This question consists of 25 (Twenty Five) multiple choice questions, each carrying one mark. For each question (1.1 – 1.25), four alternatives (*a, b, c* and *d*) are given, out of which only one is correct. Write the correct answers in the boxes corresponding to the questions only on the FIRST sheet of the answer book. (1 × 25 = 25)

1.1. In the given circuit, the voltage $v(t)$ is

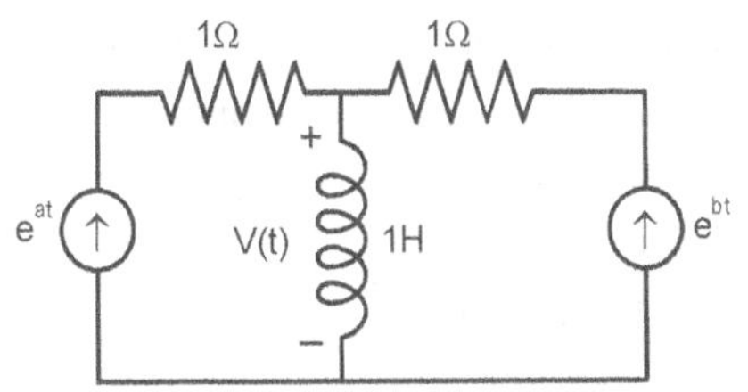

(*a*) $e^{at} - e^{bt}$ (*b*) $e^{at} + e^{bt}$

(*c*) $ae^{at} - be^{bt}$ (*d*) $ae^{at} + be^{bt}$

1.2. In the given circuit, the value of the voltage source E is

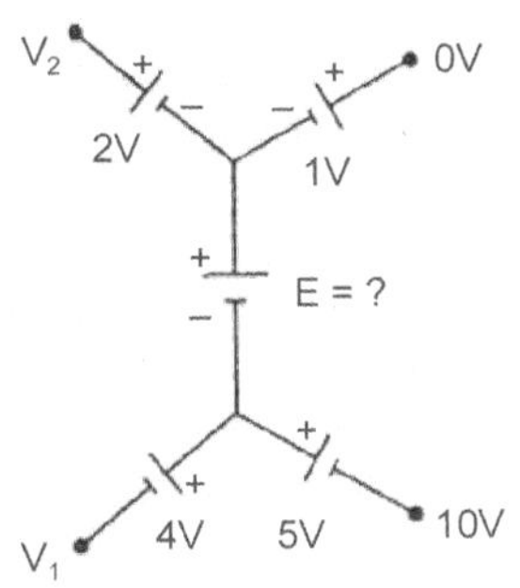

(*a*) – 16 V (*b*) 4 V

(*c*) –6 V (*d*) 16 V

1.3. Given that

$$\mathcal{L}[f(t)] = \frac{s+2}{s^2+1}, \mathcal{L}[g(t)] = \frac{s^2+1}{(s+3)(s+2)},$$

$$h(t) = \int_0^t f(\tau)g(t-\tau)d\tau. \ \mathcal{L}[h(t)] \text{ is}$$

(*a*) $\frac{s^2+1}{s+3}$ (*b*) $\frac{1}{s+3}$

(*c*) $\frac{s^2+1}{(s+3)(s+2)} + \frac{s+2}{s^2+1}$ (*d*) None of these

1.4. In the differential amplifer of the figure, if the source resistance of the current source I_{EE} is infinite, then the common-mode gain is

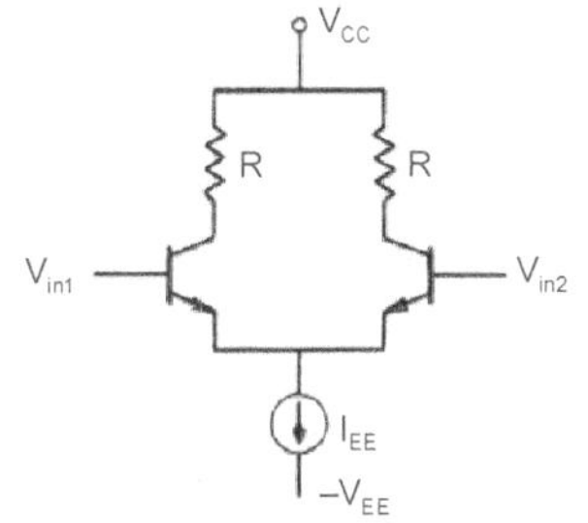

(*a*) zero

(*b*) infinite

(*c*) indeterminate

(*d*) $\frac{V_{in1} + V_{in2}}{2V_T}$

1.5. In the given circuit, V_o is

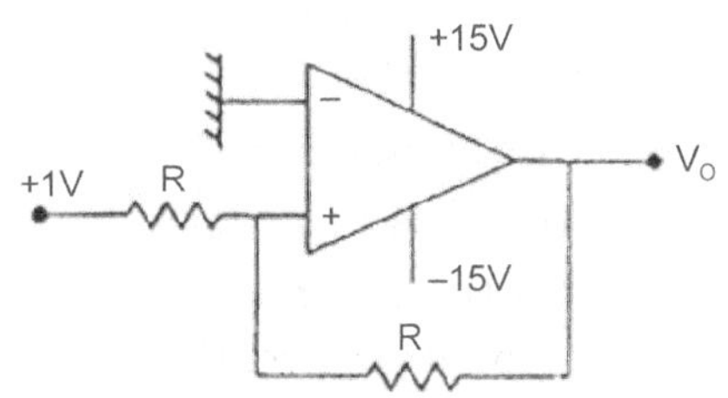

(*a*) –1 V (*b*) 2 V

(*c*) +1 V (*d*) +15 V

1.6. Introducing a resistor in the emitter of a common amplifier stabilizes the dc operating point against variations in

(*a*) only the temperature

(*b*) only the β of the transistor

(*c*) both temperature and β

(*d*) none of these

1.7. The current gain of a bipolar transistor drops at high frequencies because of

(*a*) transistor capacitances

(*b*) high current effects in the base

(*c*) parasitic inductive elements

(*d*) the Early effect

1.8. An amplifier with resistive negative feedback has two left half plane poles in its open-loop transfer function. The amplifier

(*a*) will always be unstable at high frequencies

(*b*) will be stable for all frequencies

(*c*) may be unstable, depending on the feedback factor

(*d*) will oscillate at low frequencies

1.9. If the op-amp in figure is ideal, then v_o is

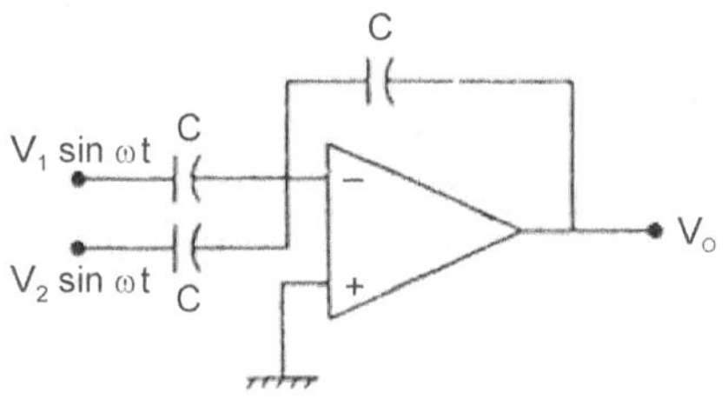

(*a*) zero (*b*) $(V_1 - V_2)\sin\omega t$

(*c*) $-(V_1 + V_2)\sin\omega t$ (*d*) $(V_1 + V_2)\sin\omega t$

1.10. The configuration of given figure is a

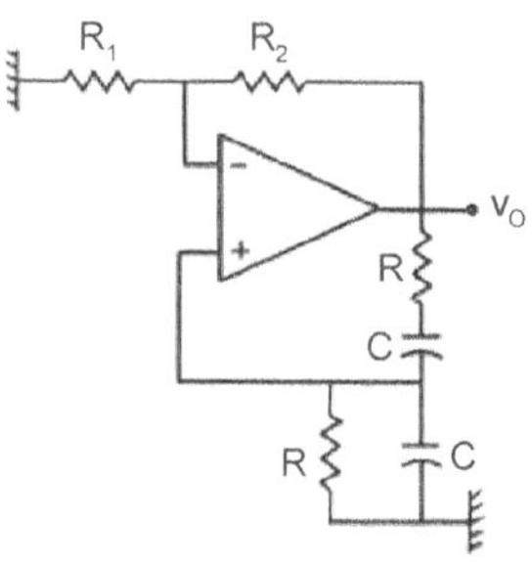

(*a*) precision integrator

(*b*) Hartley oscillator

(*c*) Butterworth high pass filter

(*d*) Wien-bridge oscillator

1.11. Assume that the op-amp of the figure is ideal. If v_i is a triangular wave, then vo will be

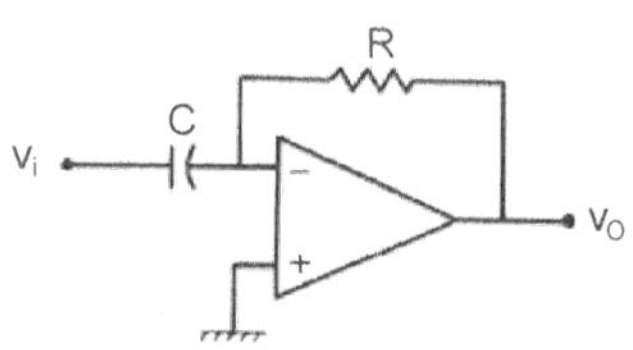

(*a*) square wave (*b*) triangular wave

(*c*) parabolic wave (*d*) sine wave

1.12. The Fourier Transform of the signal $x(t) = e^{-3t^2}$ is of the following form, where A and B are constants

(*a*) $A_e^{-B|f|}$

(*b*) A_e^{-Bf2}

(*c*) $A + B|f|^2$

(*d*) A_e^{-Bf}

1.13. A system with an input $x(t)$ and output $y(t)$ is described by the relation. $y(t) = tx(t)$. This system is

(*a*) linear and time-invariant

(*b*) linear and time varying

(*c*) non-linear and time-invariant

(*d*) non-linear and time-varying

1.14. The amplitude modulated wave form $s(t) = A_c[1 + K_a m(t)]\cos\omega_c t$ is fed to an ideal envelope detector. The maximum magnitude of $K_a m(t)$ is greater than 1. Which of the following could be the detector output?

(*a*) $A_c m(t)$ (*b*) $A_c^2[1 + K_a m(t)]^2$

(*c*) $|A_c[1 + K_a m(t)]|$ (*d*) $A_c[1 + K_a m(t)]^2$

1.15. An 8-bit successive approximation analog to digial converter has full scale reading of 2.55V and its conversion time for an analog input of 1V is 20 μs. The conversion time for a 2V input will be

(*a*) 10 μs (*b*) 20 μs

(*c*) 40 μs (*d*) 50 μs

1.16. The number of hardware interrupts (which require an external signal to interrupt) present in an 8085 microprocessor are

(*a*) 1 (*b*) 4

(*c*) 5 (*d*) 13

1.17. The most commonly used amplifier in sample and hold circuits is

(*a*) a unity gain inverting amplifier

(*b*) a unity gain non-inverting amplifier

(*c*) an inverting amplifier with a gain of 10

(*d*) an inverting amplifier with a gain of 100

1.18. The number of comparators in a 4-bit flash ADC is

(*a*) 4 (*b*) 5

(*c*) 15 (*d*) 16

1.19. For the logic circuit shown in the figure is the required input condition (A, B, C) to make the output (X) = 1 is

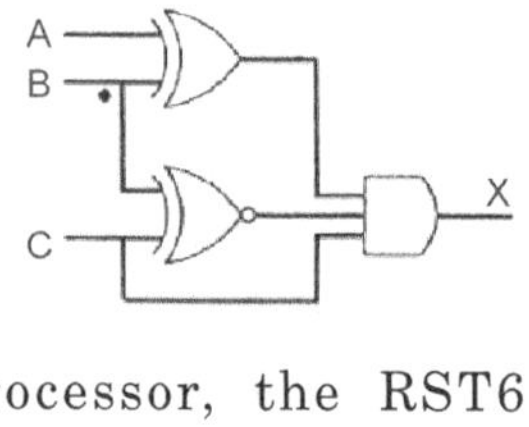

(*a*) 1, 0, 1

(*b*) 0,0,1

(*c*) 1,1,1

(*d*) 0,1,1

1.20. In the 8085 microprocessor, the RST6 instruction transfers the program execution to the following location:

(*a*) 30 H (*b*) 24 H

(*c*) 48 H (*d*) 60 H

1.21. The magnitudes of the open-circuit and short-circuit input impedances of a transmission line are 100Ω and 25Ω respectively. The characteristic impedance of the line is,

(*a*) 25 Ω (*b*) 50 Ω

(*c*) 75 Ω (*d*) 100 Ω

1.22. A TEM wave is incident normally upon a perfect conductor. The E and H fields at the boundary will be, respectively,

(*a*) minimum and minimum

(*b*) maximum and maximum

(*c*) minimum and maximum

(*d*) maximum and minimum

1.23. The frequency range for satellite communication is

(*a*) 1 kHz to 100 kHz (*b*) 100 kHz to 10 kHz

(*c*) 10 MHz to 30 MHz (*d*) 1 GHz to 30 GHz

1.24. If the diameter of a $\frac{\lambda}{2}$ dipole antenna is increased from $\frac{\lambda}{100}$ to $\frac{\lambda}{50}$, then its

(*a*) bandwidth increases

(*b*) bandwidth decreases

(*c*) gain increases

(*d*) gain decreases

1.25. The circuit of the figure represents a

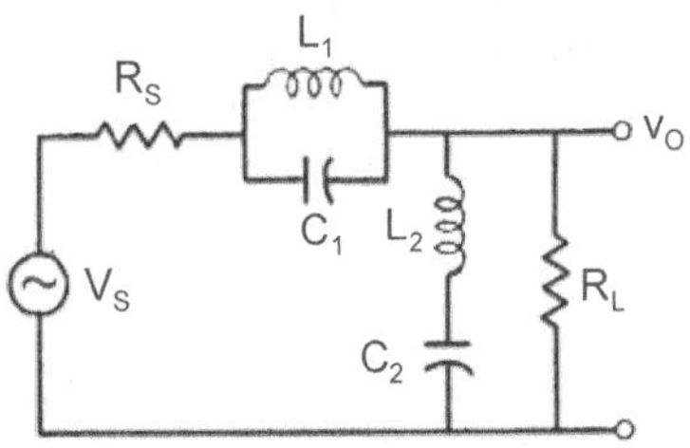

(*a*) low pass filter (*b*) high pass filter

(*c*) band pass filter (*d*) band reject filter

2. This question consists of 25 (Twenty Five) multiple choice questions, each carrying two marks. For each question (2.1 – 2.25), four alternatives (*a, b, c* and *d*) are given, out of which only one is correct. Write the correct answers in the boxes corresponding to the questions only on the SECOND sheet of the answer book. $(2 \times 25 = 50)$

2.1. The eigen values of matrix $\begin{bmatrix} 2 & -1 & 0 & 0 \\ 0 & 3 & 0 & 0 \\ 0 & 0 & -2 & 0 \\ 0 & 0 & -1 & 4 \end{bmatrix}$ are

(*a*) 2, –2, 1, –1

(*b*) 2, 3, – 2, 4

(*c*) 2, 3, 1, 4

(*d*) None of these

2.2. Use the data of the figure. The current i in the circuit of the figure is

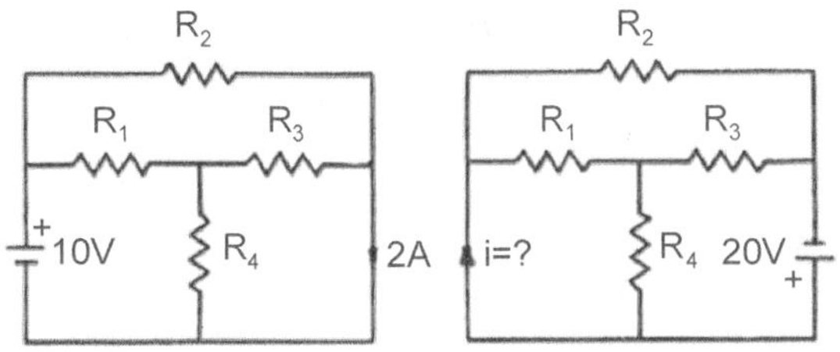

(*a*) – 2 A (*b*) 2 A

(*c*) – 4 A (*d*) + 4 A

2.3. For the circuit in the figure, the voltage v_o is

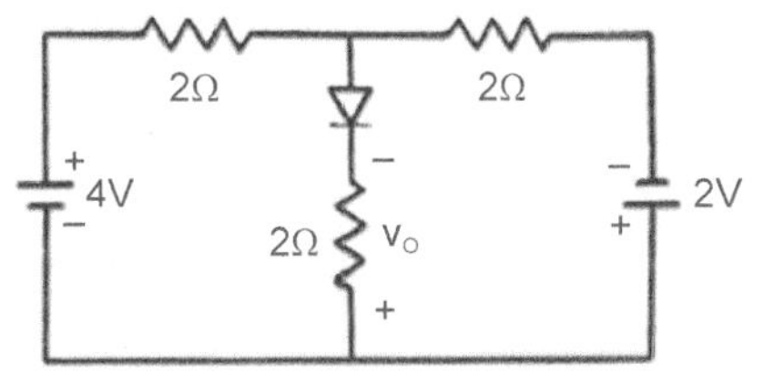

(*a*) 2 V (*b*) 1 V

(c) –1 V (*d*) None of these

2.4. A linear time invariant system has an impulse response e^{2t}, $t > 0$. If the initial conditions are zero and the input is e^{3t}, then output for $t > 0$ is

(*a*) $e^{3t} - e^{2t}$

(*b*) e^{5t}

(*c*) $e^{3t} + e^{2t}$

(*d*) None of these

2.5. In the circuit of the figure, assume that the transistor is in active region. It has a large β and its base-emitter voltage is 0.7 V. The value of I_c is

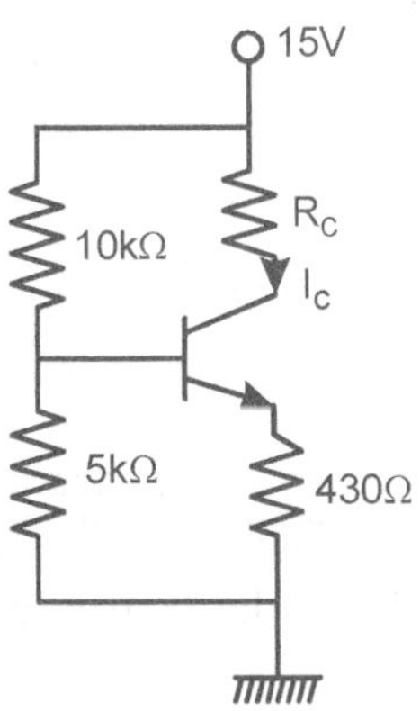

(*a*) Indeterminate since R_c is not given

(*b*) 1 mA

(*c*) 5 mA

(*d*) 10 mA

2.6. If the op-amp in given figure, has an input offset voltage of 5 mV and an open-loop voltage gain of 10, 000, then v_o will be

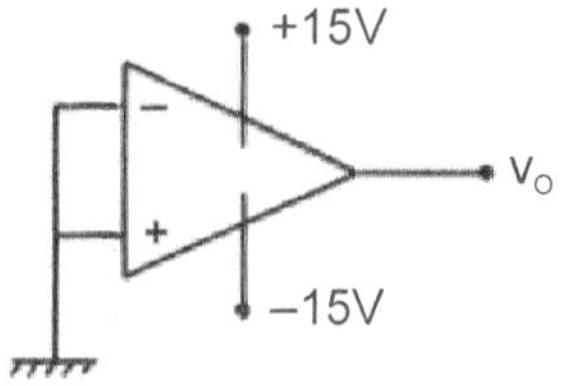

(*a*) 0 V (*b*) 5 mV

(*c*) +15 V or –15 V (*d*) +50 V or –50 V

2.7. For the logic circuit shown in, the simplified Boolean expression for the output Y is

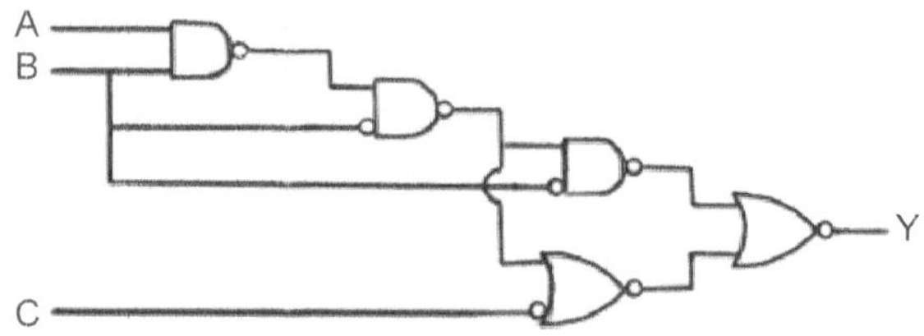

(*a*) A + B + C (*b*) A

(*c*) B (*d*) C

2.8. For the 4-bit DAC shown, the output voltage v_o is

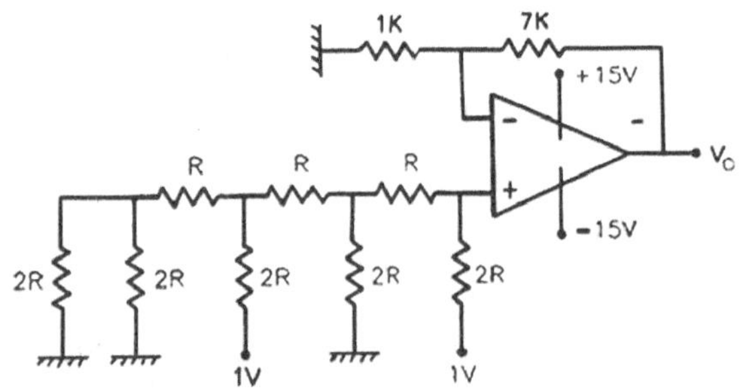

(*a*) 10 V (*b*) 5 V

(*c*) 4 V (*d*) 8 V

2.9. A sequential circuit using D Flip-Flop and logic gates is shown in figure, where X and Y are the inputs and Z is the output. The circuit is

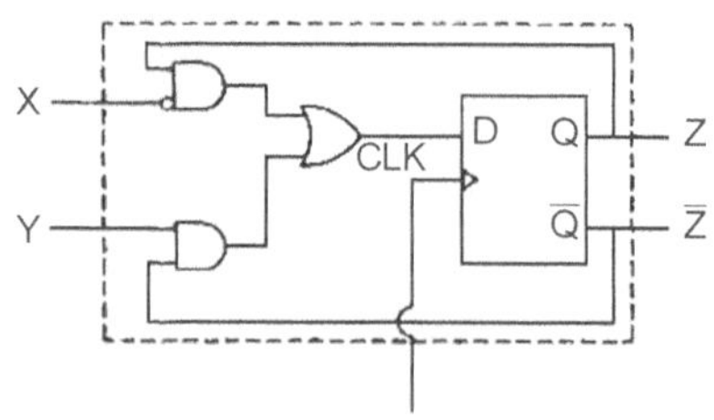

(*a*) S – R Flip-Flop with inputs X = R and Y = S

(*b*) S – R Flip-Flop with inputs X = S and Y = R

(*c*) J – K Flip-Flop with inputs X = J and Y = K

(*d*) J – K Flip-Flop with inputs X = K and Y = J

2.10. The contents of Register (B) and Accumulator (A) of 8085 microprocessor are 49H and 3AH respectively. The contents of A and the status of carry flag (CY) and sign flag (S) after executing SUB B instructions are

(*a*) A = F1, *CY* = 1, S = 1

(*b*) A = 0F, *CY* = 1, S = 1

(*c*) A = F0, CY = 0, S = 0

(*d*) A = 1F, CY = 1, S = 1

2.11. In given figure, the J and K inputs of all the four Flip-Flips are made high. The frequency of the signal at output Y is

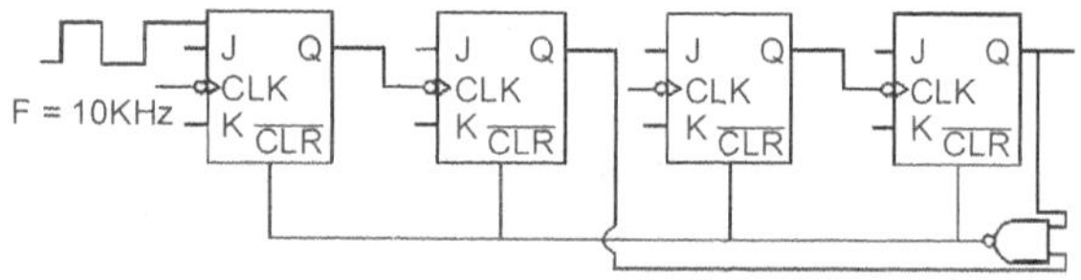

(*a*) 0.833 kHz (*b*) 1.0 kHz

(*c*) 0.91 kHz (*d*) 0.77 kHz

2.12. One period (0, T) each of two periodic waveforms W_1 and W_2 are shown in figure . The magnitudes of the *n*th Fourier series coefficients of W_1 and W_2. for $n \geq 1$, *n* odd, are respectvely proportional to

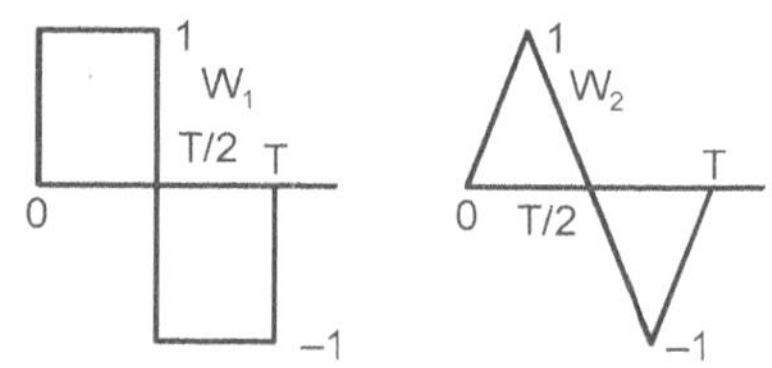

(*a*) $|n^{-3}|$ and $|n^{-2}|$

(*b*) $|n^{-2}|$ and $|n^{-3}|$

(*c*) $|n^{-1}|$ and $|n^{-2}|$

(*c*) $|n^{-4}|$ and $|n^{-2}|$

2.13. Let *u*(*t*) be the step function. Which of the waveforms in given figure, (*a*)-(*d*) corresponds to the convolution of $u(t) - u(t-1)$ with $u(t) - u(t-2)$?

(*a*)

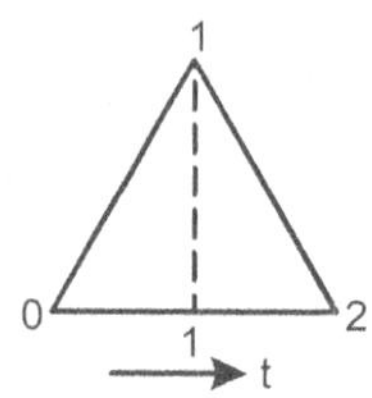

(*b*)

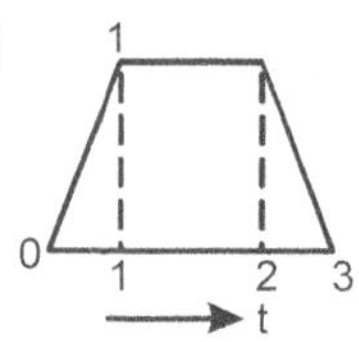

(*c*)

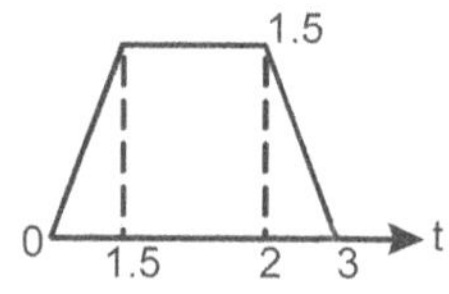

(*d*)

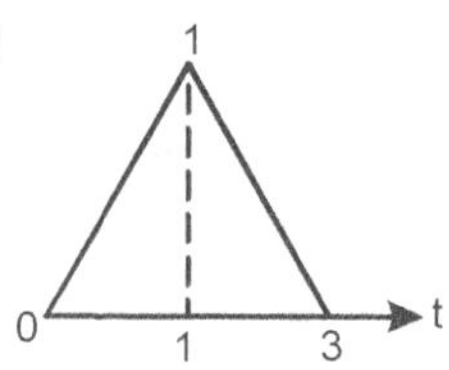

2.14. In given figure, the steady state output voltage corresponding to the input voltage $3 + 4 \sin 100t$ V is

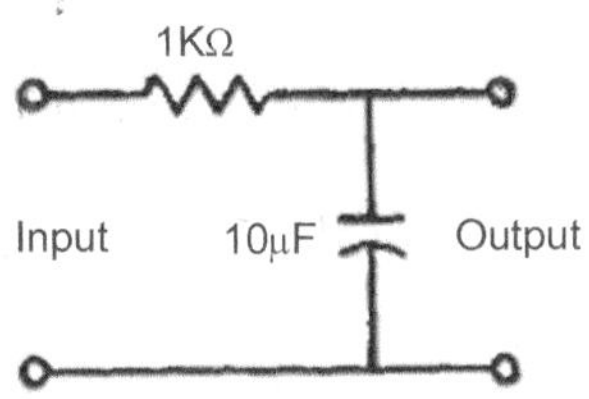

(a) $3 + \frac{4}{\sqrt{2}} \sin (100t - \frac{\pi}{4})$ V

(b) $3 + 4\sqrt{2} \sin (100t - \frac{\pi}{4})$ V

(c) $\frac{3}{2} + \frac{4}{\sqrt{2}} \sin (100t + \frac{\pi}{4})$ V

(d) $3 + 4 \sin (100t + \frac{\pi}{4})$ V

2.15. In a digital communication system employing Frequency Shift Keying (FSK), the 0 and 1 bit are represented by sine waves of 10 kHz and 25 kHz respectively. These waveforms will be orthogonal for a bit interval of

(a) 45 μsec (b) 200 μsec

(c) 50 μsec (d) 250 μsec

2.16. A message $m(t)$ bandlimited to the frequency f_m has a power of P_m. The power of the output signal in given figure is

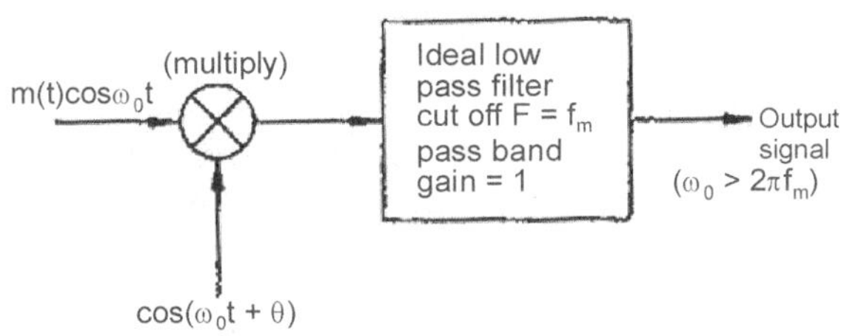

(a) $\frac{P_m \cos\theta}{2}$ (b) $\frac{P_m}{4}$

(c) $\frac{P_m \sin^2\theta}{4}$ (d) $\frac{P_m \cos^2\theta}{4}$

2.17. The Hilbert transform of $\cos \omega_1 t + \sin \omega_2 t$ is

(a) $\sin \omega_1 t - \cos \omega_2 t$ (b) $\sin \omega_1 t + \cos \omega_2 t$

(c) $\cos \omega_1 t - \sin \omega_2 t$ (d) $\sin \omega_1 t + \sin \omega_2 t$

2.18. A system has a phase response given by $\phi(\omega)$, where ω is the angular frequency. The phase delay and group delay at $\omega = \omega_0$ are respectively given by

(a) $-\frac{\phi(\omega_o)}{\omega_o}, \ -\left.\frac{d\phi(\omega)}{d\omega}\right|_{\omega=\omega_o}$

(b) $\phi(\omega_o), \ -\left.\frac{d^2\phi(\omega_o)}{d\omega^2}\right|_{\omega=\omega_o}$

(c) $\frac{\omega_o}{\phi(\omega_o)}, \ -\left.\frac{d\phi(\omega)}{d(\omega)}\right|_{\omega=\omega_o}$

(d) $\omega_o\phi(\omega_o), \int_{-\infty}^{\omega_o} \phi(\lambda)d\lambda$

2.19. A system described by the transfer function

$$H(s) = \frac{1}{s^3 + \alpha s^2 + ks + 3}$$ is stable.

The constraints on α and k are

(a) $\alpha > 0, \alpha k < 3$ (b) $\alpha > 0, \alpha k > 3$

(c) $\alpha < 0, \alpha k > 0$ (d) $\alpha > 0, \alpha k < 0$

2.20. In an FM system, a carrier of 100 MHz is modulated by a sinusoidal signal of 5 kHz. The bandwidth by Carson's approximation is 1 MHz. If $y(t)$ = (modulated waveform)^3, then by using Carson's approximation, the bandwidth of $y(t)$ around 300 MHz and the spacing of spectral components are, respectively.

(a) 3 MHz, 5kHz (b) 1 MHz, 15kHz

(c) 3 MHz, 15 kHz (d) 1 MHz, 5 kHz

2.21. A uniform plane wave in air impinges at 45° angle on a lossless dielectric material with dielectric constant ε_r. The transmitted wave propagates in a 30° direction with respect to the normal. The value of ε_r is

(a) 1.5 (b) $\sqrt{1.5}$

(c) 2 (d) $\sqrt{2}$

2.22. For an 8 feet (2.4 m) parabolic dish antenna operating at 4 GHz, the minimum distance required for far field measurement is closest to

(a) 7.5 cm (b) 15 cm

(c) 15 m (d) 150 m

2.23. A rectangular waveguide has dimensions 1cm × 0.5 cm. Its cut-off frequency is

(a) 5 GHz (b) 10 GHz

(c) 15 GHz (d) 20 GHz

2.24. Two coaxial cables 1 and 2 are filled with different dielectric constants ε_{r1} and ε_{r2} respectively. The ratio of the wavelengths in the two cables, (λ_1/λ_2) is

(a) $\sqrt{\varepsilon_{r1}/\varepsilon_{r2}}$ (b) $\sqrt{\varepsilon_{r2}/\varepsilon_{r1}}$

(c) $\varepsilon_{r1}/\varepsilon_{r2}$ (d) $\varepsilon_{r2}/\varepsilon_{r1}$

SECTION - B (75 Marks)

The section consists of TWENTYquestions of FIVE marks each. ANY FIFTEEN out of them have to be answerd. If more number of questions are attempted, score off the answers not to be evaluated, else only the first fifteen unscored answers will be considered.$(5 \times 15 = 75)$

3. For the circuit in given figure,

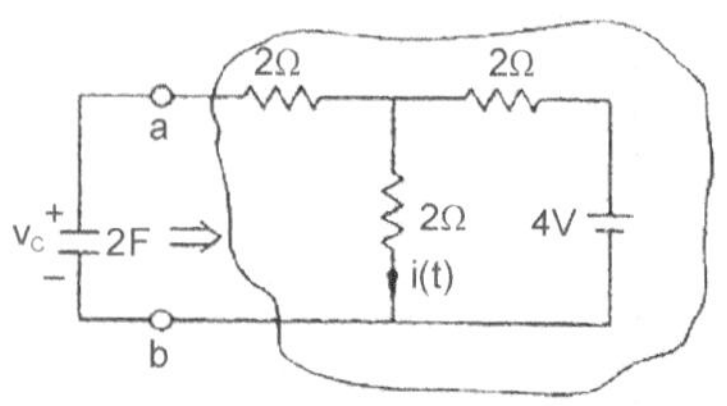

(*a*) Find the Thevenin equivalent of the sub circuit faced by the capacitor across the terminals *a*, *b*.

(*b*) Find $v_c(t)$, $t > 0$, given $v_c(0) = 0$

(*c*) Find $i(t)$, $t > 0$.

4. For the given circuit, which is in steady state

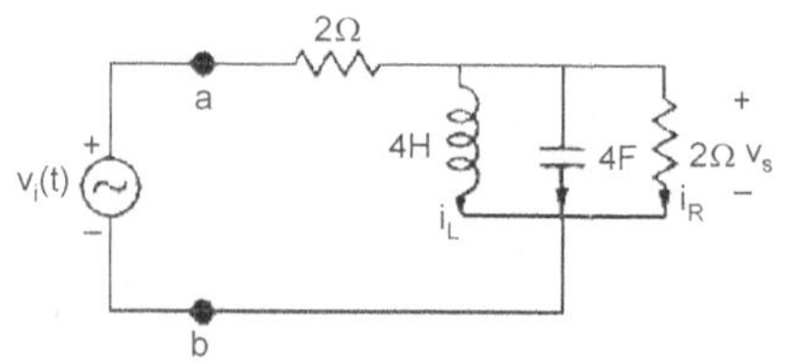

(*a*) Find the frequency ω_0 at which the magnitude of the impedance across terminals *a*, *b* reaches a maximum.

(*b*) Find the impedance across *a*, *b* at the frequency ω_0.

(*c*) If $v_s(t) = V \sin(\omega_0 t)$, find $i_L(t)$, $i_R(t)$.

5. For the given circuit, write the state equations using v_c and i_L as state variables.

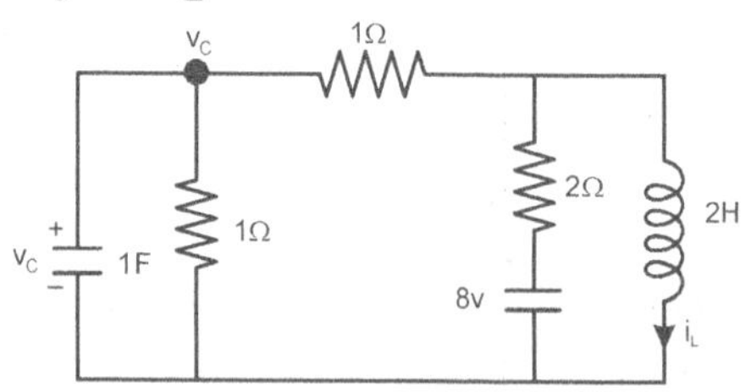

6. The network N in given figure consists only of two elements: a resistor of 1 Ω and an inductor of L henry. A 5 V source is connected at the input at $t = 0$ seconds. The inductor current is zero at $t = 0$. The output voltage is found to be $5e^{-3t}$ V, for $t > 0$.

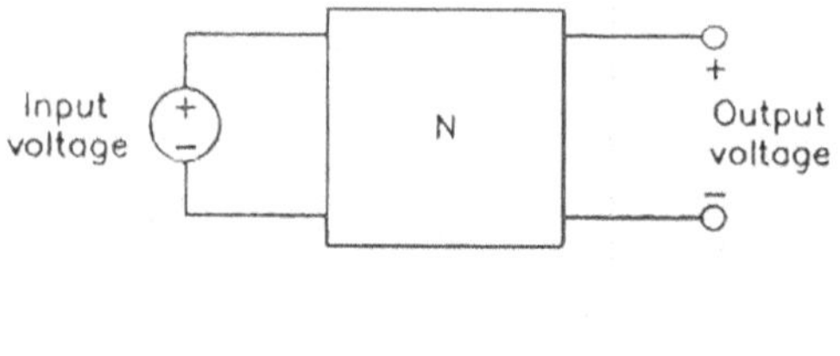

(*a*) Find the voltage transfer function of the network.

(*b*) Find L, and draw the configuration of the network.

(*c*) Find the impulse response of the network.

7. For the linear, time-invariant system whose block diagram is shown in figure with input $x(t)$ and output $y(t)$,

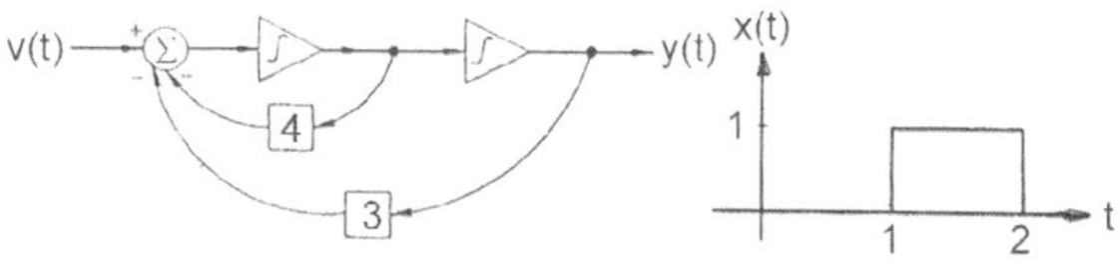

(*a*) Find the transfer function.

(*b*) For the step response of the system [*i.e.* find $y(t)$ when $x(t)$ is a unit step function and the initial conditions are zero.]

(*c*) Find $y(t)$, if $x(t)$ is as shown in figure and the initial conditions are zero.

8. A certain linear, time-invariant system has the state and output representation shown below:

$$\begin{pmatrix} \dot{x}_1 \\ \dot{x}_2 \end{pmatrix} = \begin{pmatrix} -2 & 1 \\ 0 & -3 \end{pmatrix} \begin{pmatrix} x_1 \\ x_2 \end{pmatrix} + \begin{pmatrix} 1 \\ 0 \end{pmatrix} u$$

$$y = (1\ 1) \begin{pmatrix} x_1 \\ x_2 \end{pmatrix}$$

(*a*) Find the eigenvalues (natural frequencies) of the system.

(*b*) If $u(t) = \delta(t)$ and $x_1(0_+) = x_2(0_+) = 0$, find $x_1(t)$, $x_2(t)$ and $y(t)$, for $t > 0$.

(*c*) When the input is zero, choose initial conditions $x_1(0_+)$ and $x_2(0_+)$ such that $y(t) = Ae^{-2t}$ for $t > 0$.

9. The block diagram of a feedback system is shown in the figure.

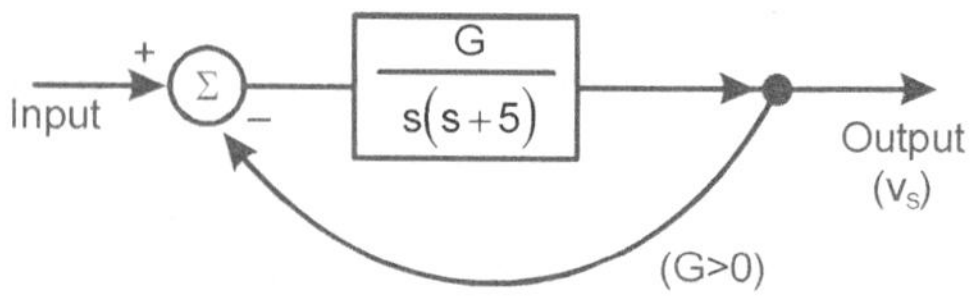

(*a*)

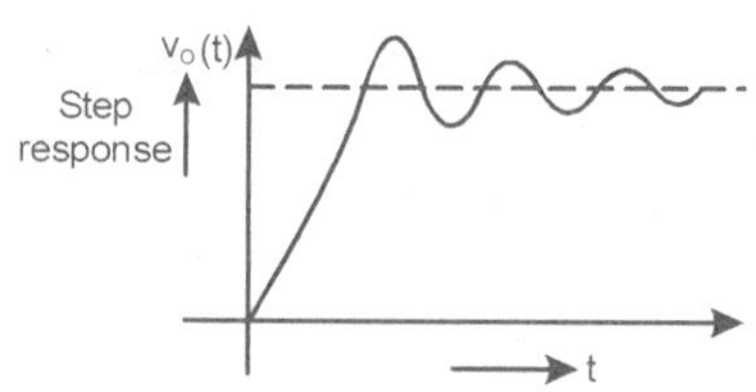

(*b*)

(*a*) Find the closed loop transfer function.

(*b*) Find the minimum value of G for which the step response of the system would exhibit an overshoot, as shown in figure.

(*c*) For G equal to twice this minimum value, find the time period T indicated in the figure.

10. (*a*) For given figure, Plot v_o under steady state conditions, with and without *C*. Assume that the diode is ideal.

(*b*) Design a circuit using two ideal diodes, one resistor and two voltage sources that would convert the input voltage of figure is to the output voltage of fourth figure. The resistor value need not be specified.

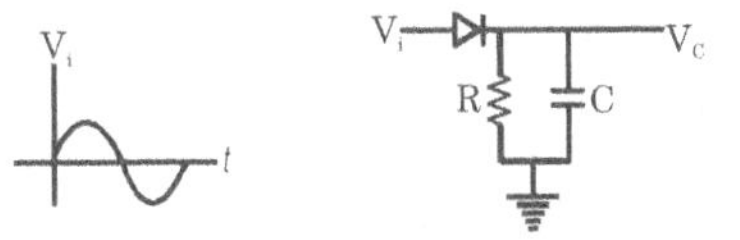

11. For the amplifier of given figure, $I_C = 1.3$ mA, $R_C = 2\ k\Omega$, $R_E = 500\ \Omega$, $V_T = \frac{KT}{q} = 26$ mV, $\beta = 100$, $V_{CC} = 15$ V $v_s = 0.01 \sin(\omega t)$ V and $C_b = C_e = 10\ \mu$F.

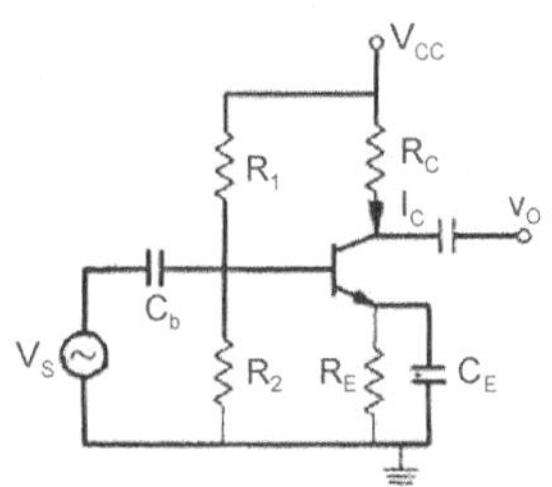

(*a*) What is the small-signal voltage gain, $A_v = v_o/v_s$?

(*b*) What is the approximate A_v, if C_e is removed ?

(*c*) What will v_o be if C_b is short circuited ?

12. For a feedback amplifier, the open loop transfer function has three poles at 100 *k* rad/s, 1 M rad/s and 10 M rad/s. The low frequency open loop gain is 1000 and the feedback factor (β) is 1. Use Bode plots to determine the phase margin of the amplifier. Is the amplifier stable ?

13. The figure shows a common base amplifier.

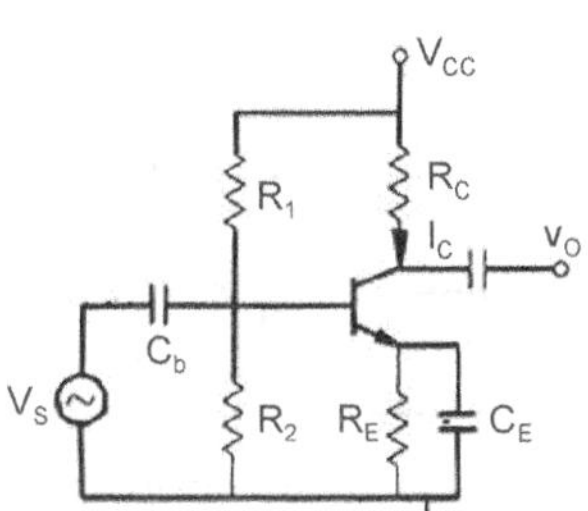

(*a*) Write expressions for the time-constants associated with the capacitors, C_B and C_S.

(*b*) What is the approximate lower cut-off frequency of the amplifier ?

14. For the CMOS monostable multivibrator of given figure, R = 50 *k*W, C = 0.01 μF, $V_{DD} = 5$V, and the CMOS NOR gates have a threshold voltage (V_T) of 1.5 V. v_{in} is a trigger pulse ($\tau_P << RC$) as shown in the figure.

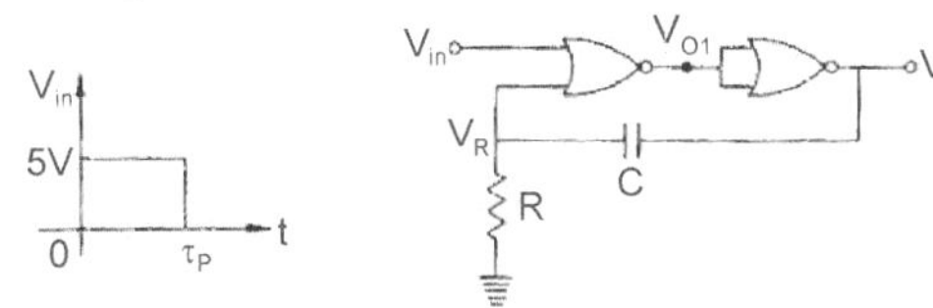

(*a*) Plot v_{01} and v_R as functions of time.

(*b*) Write the equation for $v_R(t)$, for $t > 0$.

(*c*) Find the time period of the output pulse.

15. The operating conditions (ON = 1, OFF = 0) of three pumps (*x*, *y*, *z*) are to be monitored. *x* = 1 implies that pump X is on. It is required that the indicator (LED) on the panel should glow when a majority of the pumps fail.

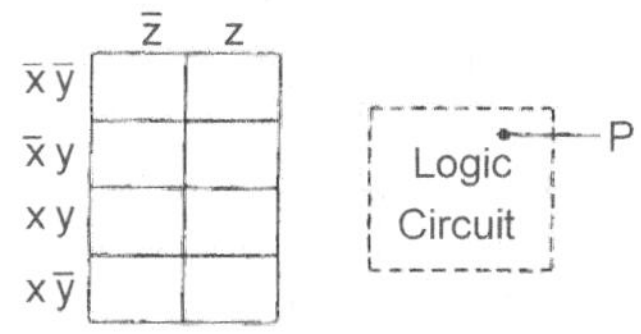

(*a*) Enter the logical values in the K-map in the format shown in figure. Derive the minimal Boolean sum-of- products expression whose output is zero when a majority of the pumps fail.

(*b*) The above expression is implemented using logic gates, and point P is the output of this circuit, as shown in given figure, P is at 0 V when a majority of the pumps fails and is at 5 V otherwise. Design a circuit to drive the LED using this output.

The current through the LED should be 10 mA and the voltage drop across it is 1V. Assume that P can source or sink 10 mA and a 5 V supply is available.

16. A one-bit full adder is to be implemented using 8-to-1 multiplexers (MUX).

(*a*) Write the truth table for sum (S) and carry to the next stage (C_N) in terms of the two bits (A, B) and carry from the previous stage (C_P), The truth table should be in the ascending order of (A, B, C_P), *i.e.* (000, 001, 010, ... etc.).

(*b*) Implement S and C_N using 8-to-1 multiplexers.

17. (*a*) The program and machine code for an 8085 microprocessor are given by

3E	MVI	A	C3
C3			
00	NOP		
80	ADD	B	
3D	DEC	A	
C2	JNZ	800A	
0A			
80			
C3	JMP	800C	
0C			
80			
D3	OUT	10	
10			
76	HLT		

The starting address of the above program is 7FFF H. What would happen if it is executed from 8000 H ?

(*b*) For the instructions given below, how many memory operations (read/write) are performed during the execution in an 8085 μP?

(*i*) Call 2000 H

(*ii*) LDA 2000 H

(*c*) Write an instruction which takes the minimum possible time to clear the accumulator of the 8085.

18. A bandlimited signal $x(t)$ with a' spectrum X (f) as shown in first figure is processed as shown in second figure is $p(t)$ is a periodic train of impulses as in third figure. The ideal bandpass filter has a passband from 26 kHz to 34 kHz.

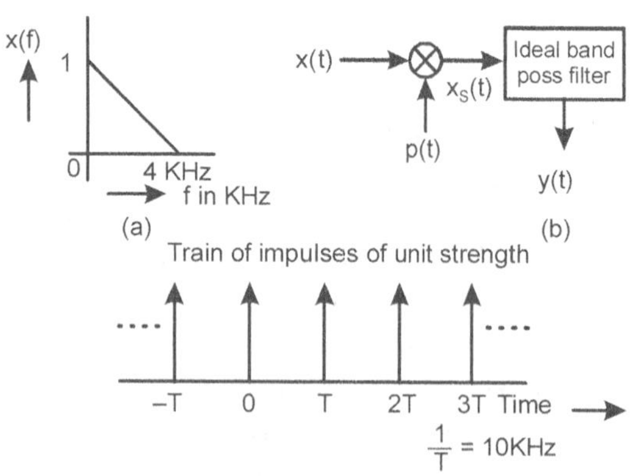

(*a*) Calculate the Fourier series coefficients C_n in the Fourier expansion of $p(t)$ in the form

$$p(t) = \sum_{n=-\infty}^{+\infty} C_n exp\,(jn2\pi t/T)$$

(*b*) Find the Fourier Transform of $p(t)$.

(*c*) Obtain and sketch the spectrum of $x_s(t)$.

(*d*) Obtain and sketch the spectrum of $y(t)$.

19. Zero mean white Gaussian noise with a two-sided power spectral density of 4 W/kHz is passed through an ideal lowpass filter with a cut-off frequency of 2 kHz and a passband gain of 1, to produce the noise output $n(t)$.

(*a*) Obtain the total power in $n(t)$.

(*b*) Find the autocorrelation function E $[n(t)\, n(t+\tau)]$ of the noise $n(t)$ as a function of τ.

(*c*) Two noise samples are taken at times t_1 and t_2. Find the spacing $|t_1 - t_2|$ so that the product $n(t_1)n(t_2)$ has the most negative expected value and obtain this most negative expected value.

20. Given $E = 10e^{-j(4x-kt)}\,\hat{y}$ V/m in free space.

(*a*) Write all the four Maxwell's equations in free space.

(*b*) Find $\nabla \times E$.

(*c*) Find H.

21. The three regions shown in figure are all lossless and non-magnetic. Find.

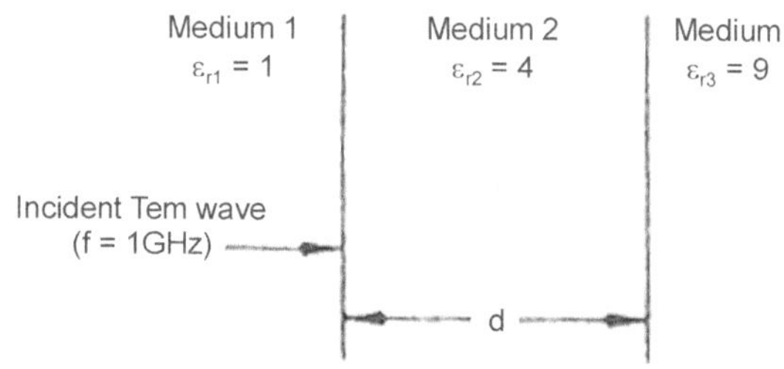

(*a*) Wave impedance in mediums 2 and 3.

(*b*) d such that medium 2 acts as a quarter wave ($\lambda/4$) transformer.

(*c*) Reflection coefficient (Γ) and voltage standing wave ratio (VSWR) at the interface of the mediums 1 and 2, when $d = \lambda/4$.

22. Design a lossless impedance matching network shown in figure to transform $Z_L = 10 + j10\ \Omega$ to $Z_{in} = 50\ \Omega$. Find the values of L, C and quality factor (Q) of the circuit at $f = 1$ GHz.

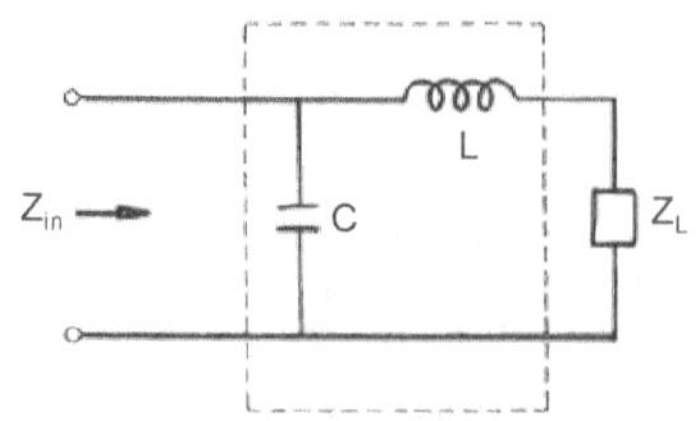

ANSWERS

1.1. (*d*)	1.2. (*a*)	1.3. (*b*)	1.4. (*a*)	1.5. (*d*)	1.6. (*c*)	1.7. (*a*)	1.8. (*b*)	1.9. (*c*)	1.10. (*d*)
1.11. (*a*)	1.12. (*b*)	1.13. (*b*)	1.14. (*a*)	1.15. (*b*)	1.16. (*c*)	1.17. (*b*)	1.18. (*c*)	1.19. (*d*)	1.20. (*a*)
1.21. (*b*)	1.22. (*a*)	1.23. (*d*)	1.24. (*b*)	1.25. (*d*)	2.1. (*b*)	2.2. (*c*)	2.3. (*d*)	2.4. (*a*)	2.5. (*d*)
2.6. (*c*)	2.7. (*c*)	2.8. (*b*)	2.9. (*d*)	2.10. (*a*)	2.11. (*b*)	2.12. (*c*)	2.13. (*b*)	2.14. (*a*)	2.15. (*b*)
2.16. (*d*)	2.17. (*a*)	2.18. (*a*)	2.19. (*b*)	2.20. (*a*)	2.21. (*c*)	2.22. (*d*)	2.23. (c)	2.24. (*b*)	

EXPLANATIONS

SECTION – A

1.1. $V(t) = \dfrac{Ldi}{dt} = \dfrac{d}{dt}(e^{at} + e^{bt}) = ae^{at} + be^{bt}$

1.2. $10 + 5 + E + 1 = 0$

$\Rightarrow \quad E = -16V$

1.3. $H(s) = F(s)\,G(s) = \dfrac{(s+2)}{(s^2+1)} \times \dfrac{(s^2+1)}{(s+3)(s+2)}$

$= \dfrac{1}{(s+3)}$

1.4. Both arms of differential amplifier are symmetrical. So terminal currents and collector voltages are equal. Characteristics of differential pair with common-mode input are similar to those of a C-E amplifier with large emitter resistor.

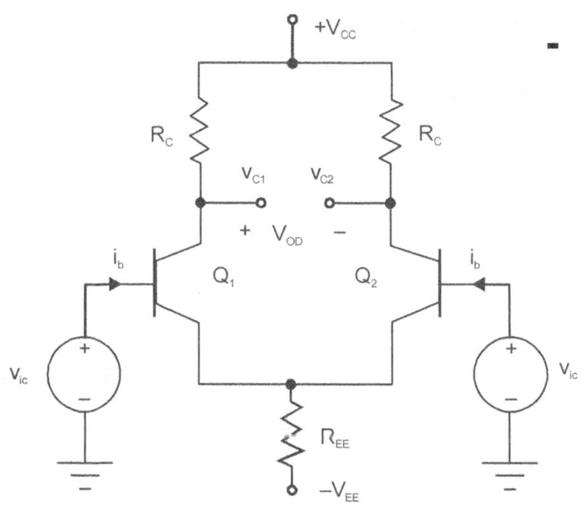

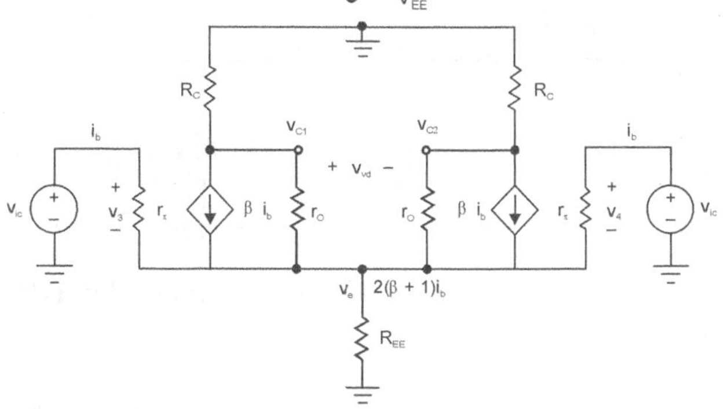

$$i_b = \frac{V_{ic}}{r_\pi + 2(\beta+1)R_{EE}}$$

Output voltages are:

$$V_{c1} = V_{c2} = -\beta i_b R_C = \frac{-\beta R_C}{r_\pi + 2(\beta+1)R_{EE}} V_{ic}$$

$$V_e = 2(\beta + 1)i_b R_{EE}$$

$$= \frac{2(\beta+1)R_{EE}}{r_\pi + 2(\beta+1)R_{EE}} V_{ic} \cong V_{ic}$$

Common-mode gain,

$$A_{CC} = \left.\frac{V_{0c}}{V_{ic}}\right|_{V_{i}=0} = \frac{-\beta R_C}{r_\pi + 2(\beta+1)R_{EE}}$$

For $R_{EE} \to \infty$, $A_{CC} \to 0$

1.5. Because of positive feedback

$$V_o = +V_{sat} = +15V$$

with positive feedback op-amp operates in saturation region $V_o = \pm V_{sat}$. Here +1 volt is applied at non inverting termal, hence

$$V_o = +V_{sat} = +15V$$

1.9.

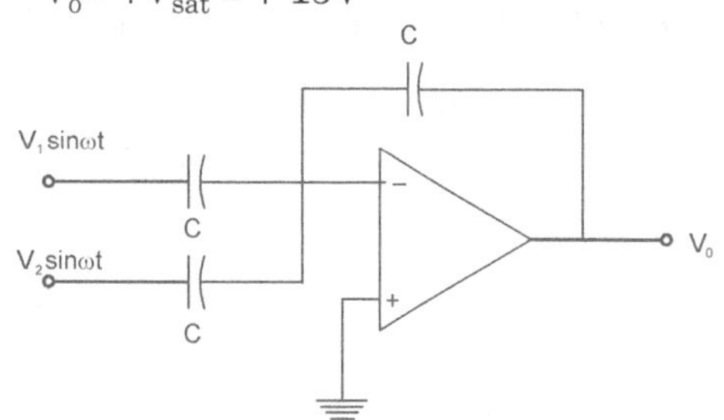

Using KCL at inverting terminal,

$$\frac{(V_1 \sin wt - 0)}{1/wc} + \frac{(V_2 \sin wt - 0)}{1/wc} = \frac{0 - V_0}{1/wc}$$

$\Rightarrow \quad V_0 = -(V_1 + V_2)\sin wt$

1.10.

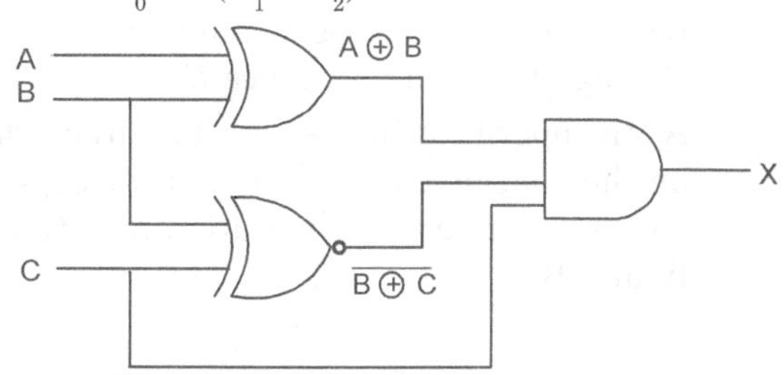

X =

$$(A \oplus B)(\overline{B \oplus C}) \cdot C \;(A\bar{B}+\bar{A}B)\left(\overline{\bar{B}C+B\bar{C}}\right).C$$

$$= (A\bar{B}+\bar{A}B)(B+\bar{C})\cdot(\bar{B}+C)C$$

$$X = (A\bar{B}+\bar{A}B)(\bar{B}\bar{C}+BC)C = \bar{A}BC$$

for $X = 1, A = 0, B = 1, C = 1$

1.11. $V_o = -R\omega CV_i = -RC\dfrac{dv_i}{dt}$

1.12. $e^{-\pi t^2} \rightleftharpoons e^{-\pi f^2}$

$$e^{-\pi\left(\sqrt{\frac{3}{\pi}}t\right)^2} \rightleftharpoons \sqrt{\frac{\pi}{3}}\, e^{-\pi\left(\frac{f}{\sqrt{3/\pi}}\right)^2}$$

$$e^{-3t^2} \rightleftharpoons Ae^{-Bf^2}$$

1.13. For the system H to be linear, the condition to be satisfied is

Here, $ay_1(t) = at\, x_1(t)$

$ay_2(t) = at\, x_2(t)H(\alpha f_1(x) + \beta f_2(x))$

$= \alpha H(f_1(x) + \beta H(f_2(x))$ .....(A)

$a[y_1(t) + y_2(t)) = at[x_1(t) + x_2(t)]$

Hence system is linear as condition (A) is satisfied.

If $g(x) = H[f(x)$ then for time-invariance system

$g(x + x_0) = H[f(x + x_0)$ .... (B)

Here $y(t - t_0) = (t - t_0)\, x(t - t_0)$

Therefore system is time-variant as condition (B) is not satisfied.

1.14. The AM signal, when not over modulated allows the recovery of $m(t)$ from its envelope. By using envelope detector whose operation is depicted below:

Consider the circuit show in figure (A)

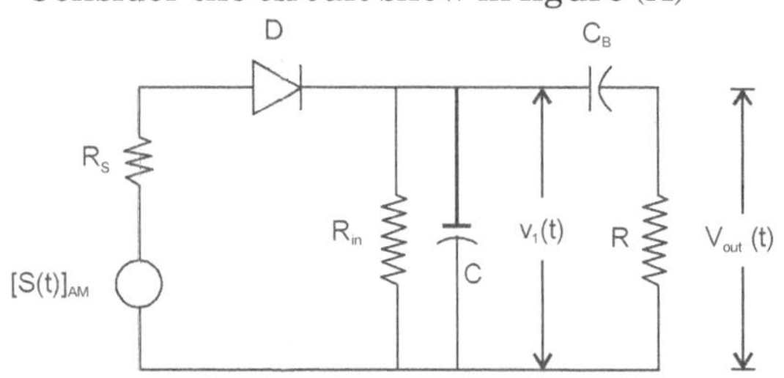

Fig. (A) The envelope detector circuit

We assume the diode D to be ideal. When it is forward biased, it acts as a short circuit and thereby, making the capacitor C charge through the source resistance R_s. When D is reverse biased, it acts as an open circuit and C discharges through the load resistance R_L.

As the operation of the detector circuit depends on the charge and discharge of the capacitor C, we shall explain this operation with the help of figure (B)

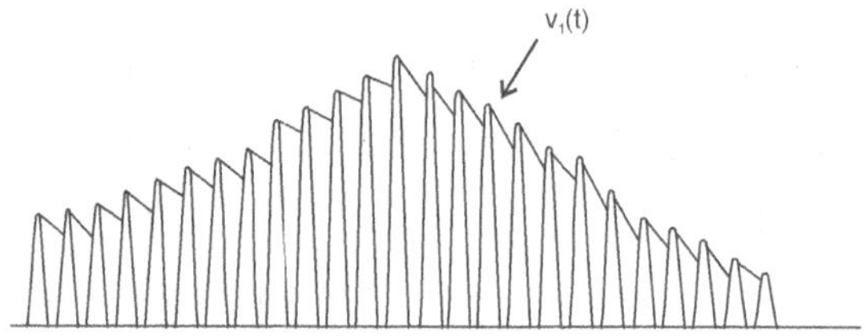

Fig.(B)

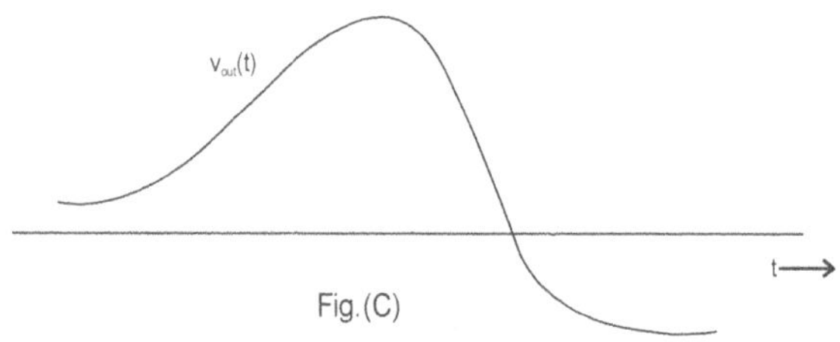

Fig.(C)

If the time constants R_sC and R_LC are properly chosen, $v_1(t)$ follows the envelope os $s(t)$ fairly closely. During the conduction cycle of D, C quickly charges to the peak value of the carrier at that time instant. It will discharge a little during the next off cycle of the diode. The time constants of the circuit will control the ripple about the actual envelope. C_B is a blocking capacitor and the final $v_{out}(t)$ will be proportional to $m(t)$, as shown in figure 2(b).

Hence, the output $V_{out}(t) = A_C m(t)$

1.18. Number of comparators $= 2^n - 1 = 2^4 - 1 = 15$

1.21. $Z = \sqrt{100 \times 25} = 50\Omega$

1.25. At $\omega \to 0$

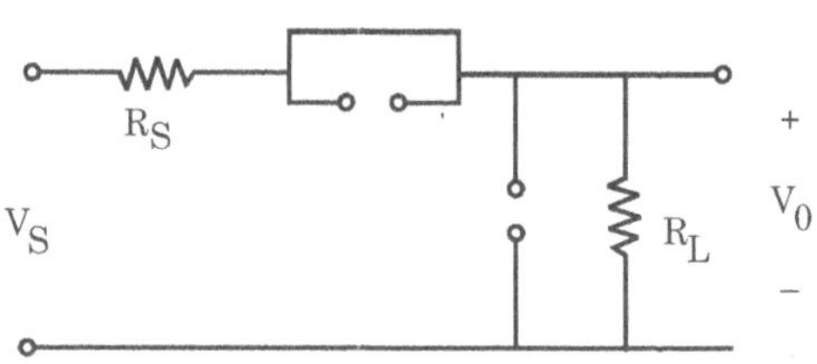

$V_o = AV_i$... Pass Band (P.B)

$A =$ constant

At $\omega \to \infty$

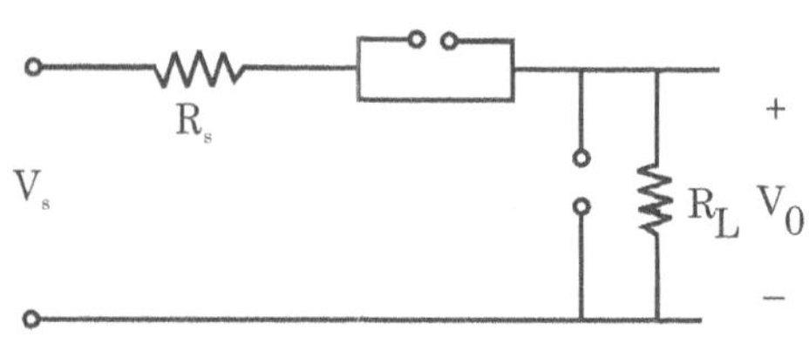

$V_o = AV_i$ ...Pass Band (P.B)

At $\omega \to \frac{1}{\sqrt{LC}}$

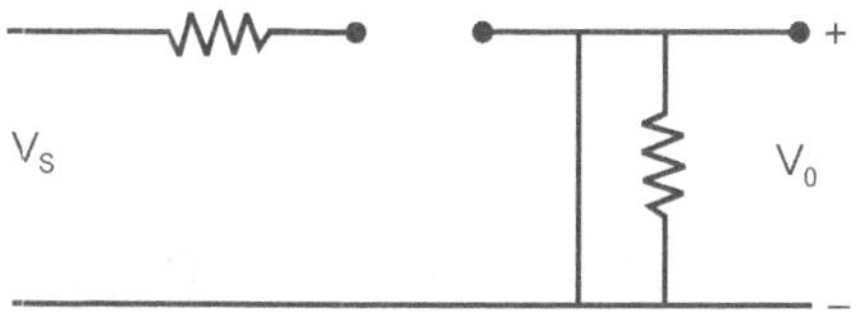

$V_o = 0$ ...Stop Band (SB)

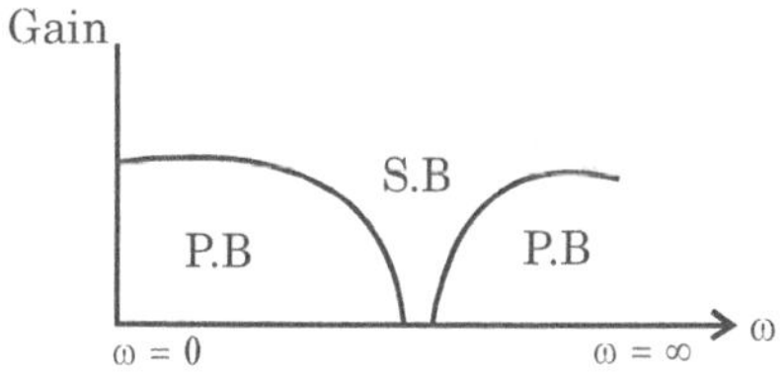

2.1. $(\lambda - 2)(\lambda + 2)(\lambda - 4)(\lambda - 3) = 0$

$\lambda = 2, 3, -2, 4$

2.2. According to reciprocitiy theorem,

The ratio of excitation to response remains invariant in a reciprical network with respect to an interchange between the points of application of excitation and measurement of response. This is illustrated here

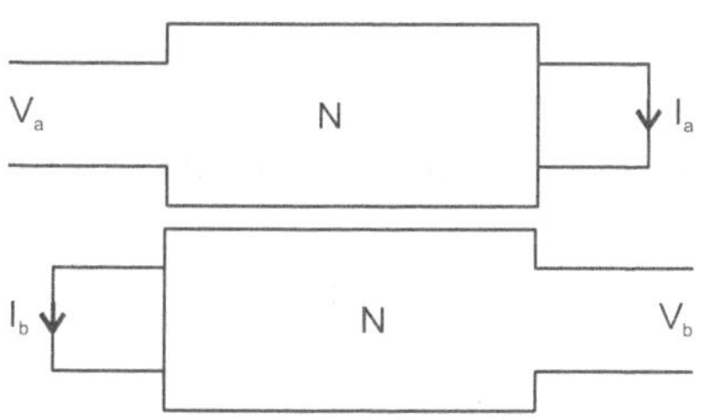

If the network is reciprocal,

$$\frac{I_a}{V_a} = \frac{I_b}{V_b}$$

From the given figure,

we have, $\frac{2A}{10V} = \frac{I}{20V}$

$\Rightarrow$ $I = 4A$

2.3. Diode is forward biased under given circuit condtion

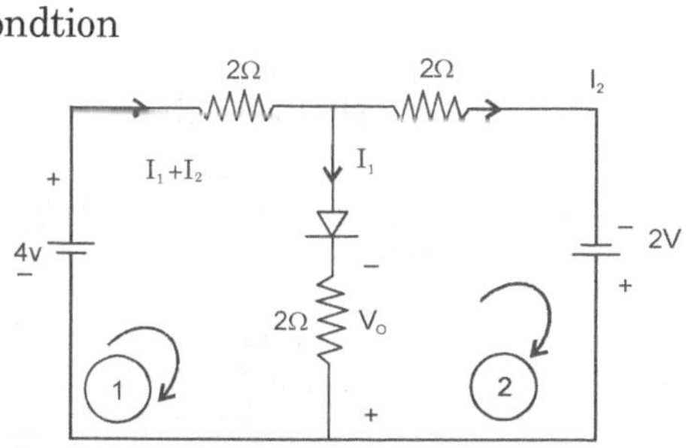

In loop(1), KVL gives,

$4 - 2(I_1 + I_2) - 2I_1 = 0$

or, $4 = 4I_1 + 2I_2$

or, $2 = 2I_1 + I_2$ ...(A)

and in loop (2), KVL gives,

$2 + 2I_1 = 2I_2 = 0$

or, $I_1 - I_2 = -1$ ...(B)

Equation (A) and equation (B) gives,

$$3I_1 = 1,\ I_1 = \frac{1}{3}$$

Then, $V_0 = -2I_1 = -\frac{2}{3}$ volts

2.4. $H(s) = \frac{1}{(s-2)}$, $R(s) = \frac{1}{(s-3)}$

$$\text{Output} = \frac{1}{(s-2)} \quad \frac{1}{(s-3)}$$

$$= \frac{1}{(s-3)} - \frac{1}{(s-2)}$$

$$C(t) = e^{3t} - e^{2t}$$

2.5. Using Theremin's theorem, we have,

$$R_{TH} = 10\ K\Omega \parallel 5\ K\Omega$$

$$= \frac{10}{3}\ K\Omega$$

$$V_{TH} = 15 \times \frac{5}{15} = 5 \text{ volts}$$

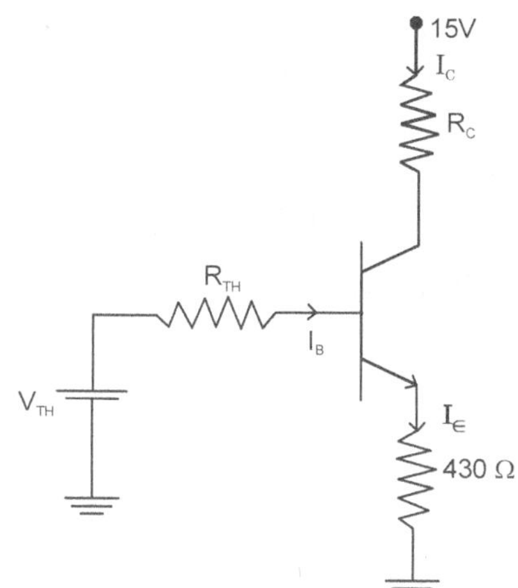

KVL in B – E loop gives,

$$V_{TH} - R_{TH} \cdot I_B - V_{BE} - (\beta + 1)I_D + 0.430 = 0$$

for $\beta >> 1$

We have, $I_C = \frac{V_{TH} - V_{BE}}{0.430} = \frac{5 - 0.7}{0.430} = 10$ mA

2.6. $V_o = V_i \,.\, A_v$

But V_o oscillates behei + V_{sat} to $-V_{sat}$

Therefore, $Vo = \pm 15$ Volts

$V_o = 5 \times 10^{-3} \times 10{,}000 = \pm 50V$

2.7.

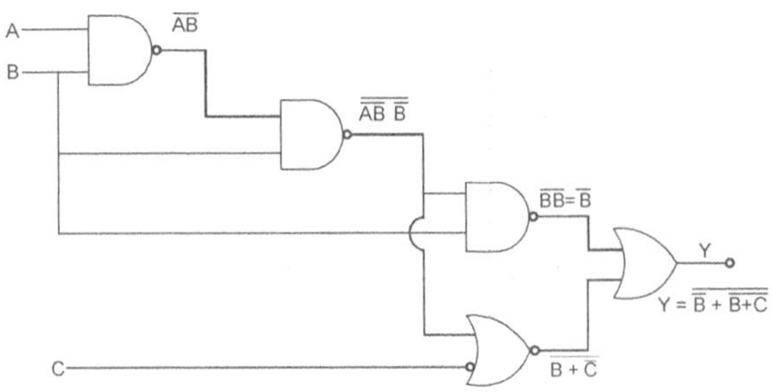

$$y = \overline{\overline{B} + \overline{B+C}}$$
$$= B(B + \bar{C})$$
$$= B + B\bar{C} = B$$

2.8

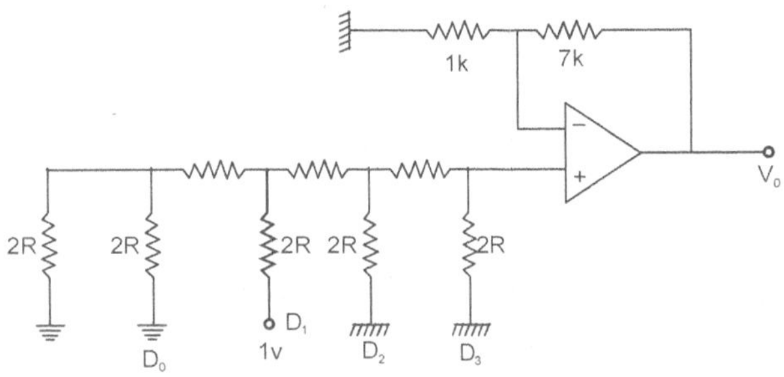

Voltage at non-inverting terminal,

$$V_T = \frac{V_R}{2^n}\left[D_0 2^0 + D_1 2^1 + D_2 2^2 + \cdots + D_{n-1} 2^{n-1}\right]$$
$$= \frac{V_R}{2^n}\left[0 + 2 \times 1 + 0 + 2^3 \times 1\right] = \frac{5}{8}\text{volts}$$

Using KCL at inverting terminal, we have

$$\frac{V_- - 0}{1K} + \frac{V_- - V_0}{7} = 0$$

$$\Rightarrow \qquad V_0 = 8V_- = 8V_+ = 5 \text{ volts}$$

2.9. From given circuit,

$$D = \bar{X}Z + Y\bar{Z}$$

X	Y	Z_n	D	Z_{n+1}
0	0	0	0	0
0	0	1	1	1
0	1	0	1	1
0	1	1	1	1
1	0	0	0	0
1	0	1	0	0
1	1	0	1	1
1	1	1	0	0

Truth table for X – Y

X	Y	Z_{n+1}
0	0	Z_n
0	1	1
1	0	0
1	1	$\bar{Z}_n$

Comparing with J-K truth-table,

We have X = K, Y = J

J	K	Q_{n+1}
0	0	Q_n
0	1	0
1	0	1
1	1	$\bar{Q}_n$

2.11. Output of NAND is zero when $Q_3Q_2Q_1Q_0$ have state $1010 = (10)_{Dec}$. Therefore, the given figure represents mod-10 counter. And, frequency of the signal will be

$$f = \frac{10\text{KHz}}{10} = 1\text{KHz}$$

2.12. Magnitude $(W_1) \propto \frac{1}{n} \times n^{-1}$

$$(W_2) \propto \frac{1}{n^2} \times n^{-2}$$

2.13. Given, Laplace transform of

$$f_1(t) = \mathcal{L}\, f_1(t) = \mathcal{L}\,[U(t) - U(t-1)]$$

$$F_1(S) = \frac{1}{S} - \frac{e^{-s}}{s}$$

and $\qquad \mathcal{L}\, f_2(t) = F_2(s) = \mathcal{L}\,[U(t) - U(t-2)]$

$$= \frac{1}{s}(1 - e^{-2s})$$

As convolution in time-domain is multiplication in s-domain *i.e.*

$$f_1(t) \otimes f_2(t) = F_1(1)F_2(1)$$

then, $\qquad F(S) = \frac{1}{s^2}(1 - e^{-3s} + e^{-s} + e^{-2s})$

Inverse laplace transform gives,

$$f(t) = \mathcal{L}^{-1}\, F(S)$$
$$= t - tu(t-2) - tu(t-1) + tu(t-3)$$
$$= t - t[u(t-1) - u(t-2) + tu(t-3)]$$

Which is dapicted in figure of option (*b*).

2.14. For DC input, capacitor gets open

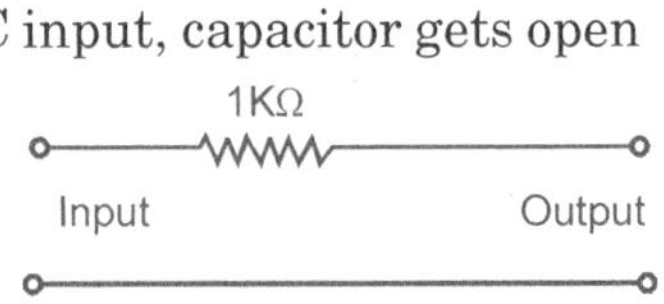

$$V_{in}(DC) = V_0(DC) = 3 \text{ volts}$$

For AC input

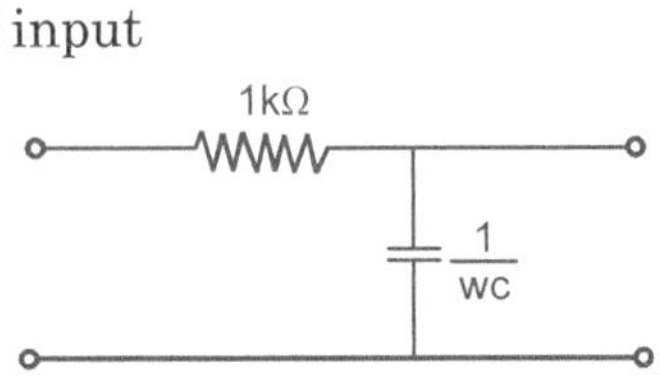

$$Z_C = \frac{1}{jw_c} = \frac{-j}{100 \times 10^{-5}}\,\Omega$$
$$Z_C = -110^{-3}\,\Omega$$

$$V_0 = V_i(AC)\frac{\frac{1}{j\omega c}}{R+\frac{1}{j\omega c}}$$

$$= V_i(AC)\cdot\frac{1}{j\omega RC+1}$$

$$= V_i(AC)\frac{1}{1+j100\times10^3\times10^{-5}}$$

$$= V_i(AC)\frac{1}{1+j}$$

$$V_0(AC) = 4\sin100t\cdot\frac{1}{\sqrt{2}}\angle-45°$$

$$= \frac{4}{\sqrt{2}}\sin\left(100t-\frac{\pi}{4}\right)$$

Hence, steady - state output,

$$V_0 = V_{i0}(DC)+V_0(AC)$$

$$= 3+\frac{4}{\sqrt{2}}\sin\left(100t-\frac{\pi}{4}\right) \text{ volts}$$

2.15. In a bit interval each wave should contain integral number of cycles for orthogonal to each other i.e.,

$$\Delta W = \frac{1}{2T}$$

2.16.

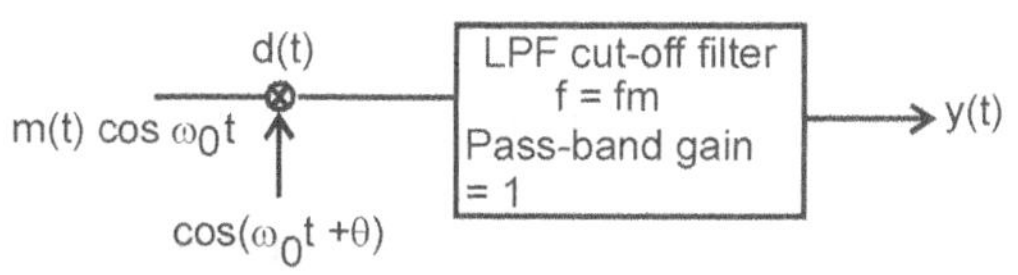

Output at product modulation

$$d(t) = m(t)\cos\omega_0 t * \cos(\omega_0 t+\theta)$$

$$= \frac{m(t)}{2}\left[\cos(2\omega_0 t+\theta)+\cos\theta\right]$$

since $\omega_0 > 2\pi f_m$

output at LPF, $y(t) = \frac{m(t)}{2}\cos\theta$ [higher frequency component rejected

[power in the output signal

$$P = E\left[y^2(0)\right]$$

$$= \frac{\cos^2\theta}{4}E[m^2(0)]$$

$$= \frac{\cos^2\theta}{4}\cdot P_m$$

2.17. Hilbert transform does change the phase by an angle $\frac{-\pi}{2}$ or 90°

If $g(t) = \cos\omega_1 t + \sin\omega_2 t$

then, Hilbert transform,

$$\hat{g}(t) = \cos\left(\omega_1 t-\frac{\pi}{2}\right)+\sin\left(\omega_2 t-\frac{\pi}{2}\right)$$

$$= \sin\omega_1 t - \cos\omega_2 t$$

2.18.

$$t_{group} = -\frac{1}{2\pi}\frac{d\phi(f)}{df}\bigg|_{f=fo}$$

$$t_{phase} = -\frac{1}{2\pi}\frac{\phi(f)}{(f)}\bigg|_{f=fo}$$

2.19. Characteristic equation

$$1 + n(s) = 0$$

$$\Rightarrow s^3 + \alpha s^2 + Ks + 3 = 0$$

Constructing routh array, we have

s^3	1	k
s^2	α	3
s^1	$\frac{\alpha k-3}{3}$	0
s^0	3	0

For system to be stable, terms in the first column of routh-array should be positive i.e., $\alpha > 0$, $\alpha k - s > 0$ $\alpha > 0$ and $\alpha k > 3$

2.20. Since $m(t)$ is sinusoidal

$$y_{FM}(t) = A_c \cos[\omega_c t + \beta\sin\omega_m t]$$

$$y(t) = [y_{FM}(t)]^3 = A_c^{\ 3}\cos^3[\omega_c t + \beta\sin\omega_m t]$$

Result. Modulation index β get multiplied by factor 3 and f_m remain constant.

$$f_m' = f_m' \ \beta' = 3\beta$$

Bandwidth, $B'_T = 2f_m(1+\beta') \approx 2f_m(3\beta)$ $\because$ $3\beta \gg 1$

hence Band width

$$B'_T = 3\ BT$$

$$= 3\times1\ M = 3\ MHz$$

$$f_{m'} = f_m = 5\ KHz$$

2.21.

$$\frac{\sin\theta_t}{\sin\theta_i} = \frac{\mu_1\varepsilon_1}{\mu_2\varepsilon_2}$$

$$\Rightarrow \quad \frac{\sin30°}{\sin45°} = \sqrt{\frac{1}{\varepsilon_r}}$$

$$\Rightarrow \quad \varepsilon_r = 2$$

2.22.

$$R \geq \frac{2d^2}{\lambda} \qquad d = 2.4\ m$$

$$\lambda = \frac{3\times10^8}{4\times10^9} = 7.5\ cm$$

$$R = \frac{2\times(2.4)^2}{7.5\times10^{-2}} = 153\ m$$

SECTION – B

3. (*a*)

$$V_{th} = \frac{4}{4} \times 2 = 2V$$

$$R_{th} = 2 + \frac{2\times 2}{4} = 3\Omega$$

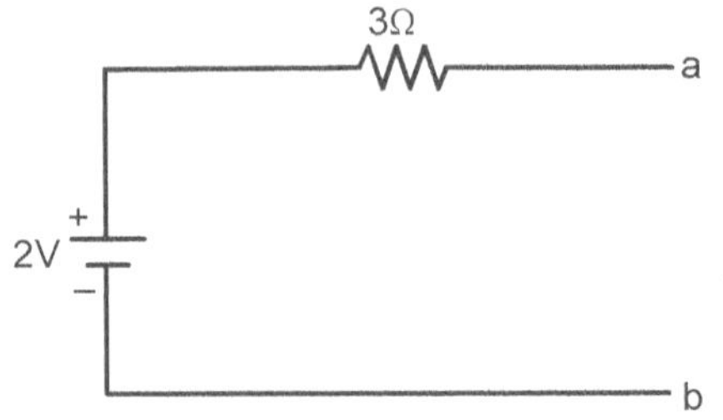

(*b*) $$\frac{2-V_c}{3} = 2\frac{dV_c}{dt}$$

Taking Laplace transform and solving

$$V_c(s) = \frac{1/3}{s(s+1/6)}$$

$$= \frac{2}{s} - \frac{2}{(s+1/6)}$$

or $$V_c = 2\,(1 - e^{-t/6})\,V$$

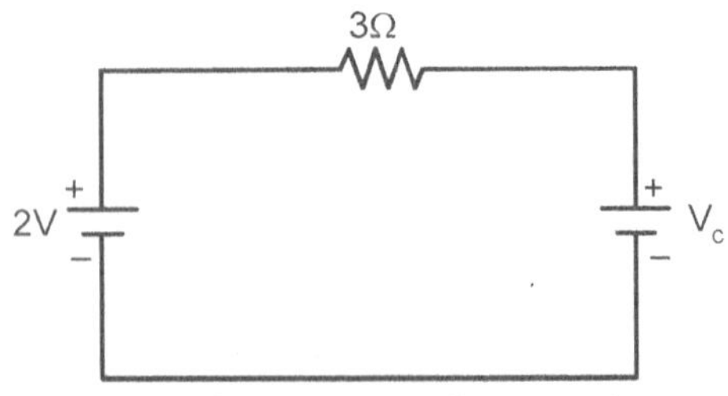

(*c*) Writing KCL in original network, assuming node voltage V

$$\frac{V-V_c}{2} + \frac{V}{2} + \frac{V-4}{2} = 0$$

or $$3V = V_c + 4$$

$$= 4 + 2\,(1 - e^{-t/6})$$

or $$V = 2 - \frac{2}{3}\,e^{-t/6}$$

But $$i(t) = \frac{V}{2}$$

$\therefore$ $$\boldsymbol{i(t) = \left(1 - \frac{1}{3}e^{-t/6}\right)A}$$

4. (*a*) Impedance across terminals *a, b* i.e.

$$Z = 2 + \frac{X_L X_C}{j(X_L - X_c)} \times \frac{2}{2+\dfrac{X_L X_C}{j(X_L - X_C)}}$$

where $X_L = \omega L\,\Omega$,

$$X_C = \frac{1}{\omega C}\Omega$$

or $$Z = 2 + \frac{2X_L X_C}{j2(X_L - X_C) + X_L X_C}$$

$$= 2 + \frac{2X_L X_C}{\sqrt{[2(X_L - X_C)]^2 + (X_L X_C)^2}}$$

From above expression, we can have for maximum Z

$$X_L - X_C = 0$$

or $$\omega = \omega_0 = \frac{1}{\sqrt{LC}}$$

$$= \frac{1}{\sqrt{4\times 4}} = 0.25 \text{ rad/sec.}$$

(*b*) $$Z_{max} = 2 + 2 = 4\Omega$$

(*c*) Total current through network

$$i(t) = \frac{V_i(t)}{4} = \frac{V}{4}\sin(\omega_o t)\,A$$

$$V_O = 2 \times \frac{V}{4}\sin(W_o t)$$

$$= \frac{V}{2}\sin(\omega_o t)\,V$$

$$i_L = \frac{V_o}{X_L}$$

$$= \frac{V}{2}\sin(\omega_o t - 90°)\,A$$

$$i_C = \frac{V_o}{X_c} = \frac{V}{2}\sin(\omega_o t + 90°)\,A$$

(since $X_L = X_C = 1\Omega$).

$$i_R = \frac{V_o}{R} = \frac{V}{4}\sin\omega_0 t$$

5.

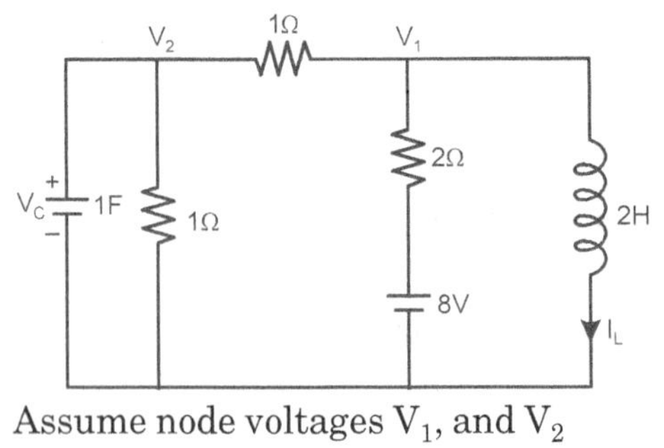

Assume node voltages V_1, and V_2

Writing node equations,

$$\frac{V_1 - 8}{2} + \frac{V_1 - V_C}{1} + i_L = 0$$

or $$\frac{3V_1}{2} = V_C - i_L + 4 \quad \ldots (i)$$

$$V_1 = L\,di_L = 2di_L \quad \ldots (ii)$$

$$V_2 = V_C \quad \ldots (iii)$$

$$\frac{V_2 - V_1}{1} = \frac{V_C}{1} + \dot{V}_C = 0$$

$$\dot{V}_C = -2V_C + V_1 \quad \ldots (iv)$$

From equations (*i*), (*ii*), (*iii*) and (*iv*), we get

$$\dot{V}_C = -\frac{4}{3}V_C - \frac{2}{3}i_L + \frac{8}{3} \quad \ldots (v)$$

$$\dot{i}_L = \frac{V_c}{3} - \frac{1}{3}i_L + \frac{4}{3} \quad \ldots (vi)$$

Equations (*v*) and (*vi*) can be written in matrix form,

$$\begin{bmatrix} \dot{V}_C \\ \dot{i}_L \end{bmatrix} = \begin{bmatrix} -\frac{4}{3} & -\frac{2}{3} \\ \frac{1}{3} & -\frac{1}{3} \end{bmatrix} \begin{bmatrix} V_C \\ i_L \end{bmatrix} + \begin{bmatrix} \frac{1}{3} \\ \frac{1}{6} \end{bmatrix} 8$$

6. (*a*)

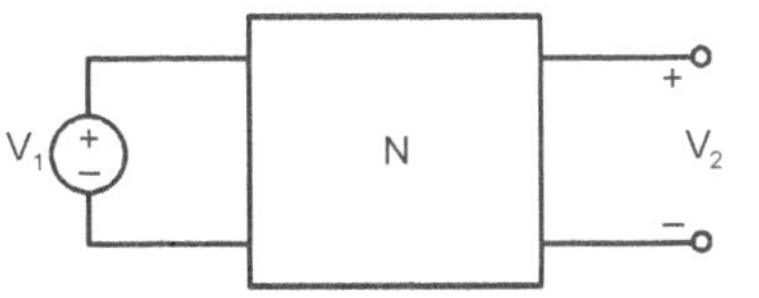

$$V_1 = 5V \quad \ldots (i)$$

or $$V_1(s) = \frac{5}{s}$$

$$V_2 = 5e^{-3t} \quad \ldots (ii)$$

or $$V_2(s) = \frac{5}{(s+3)}$$

$$\text{Voltage transfer function} = \frac{V_2(s)}{V_1(s)} = \frac{s}{(s+3)}$$

(*b*) Consider network N as shown below

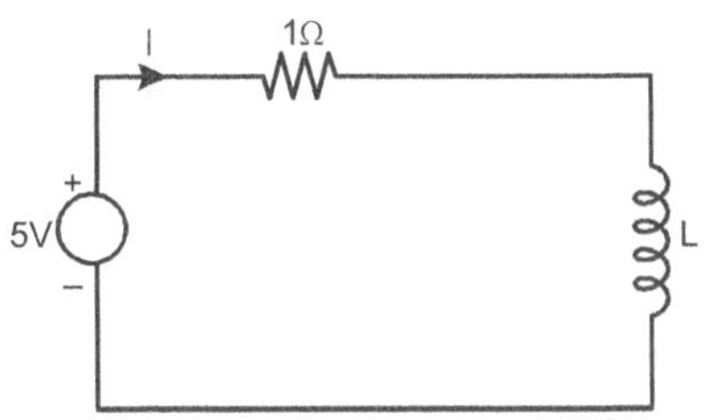

$$i = 5(1 - e^{-t/L}) \quad \ldots (iii)$$

It is obvious from above expression that output is across inductance

$$V_2 = L\frac{di}{dt} = (5e^{-t/L}) \quad \ldots (iv)$$

From equations (ii) and (iv), we get

$$\mathbf{L = \frac{1}{3} H}$$

(*c*) $V_1(s) = 1$ (since input voltage is an impulse)

$$V_2(s) = \frac{s}{(s+3)}$$

$$= 1 - \frac{3}{s+3}$$

$$V_2(t) = \delta(t) - 3e^{-3t} \text{ Volts}$$

At $t = 0^+$

$$V_2(t) = -3e^{-3t}V$$

7. (*a*) Transfer Function of inner loop,

$$G_1(s) = \frac{\frac{1}{s}}{1+\frac{4}{s}} = \frac{1}{(s+4)}$$

Transfer function of overall system,

$$G(s) = \frac{G_1(s)/s}{1+\frac{G_1(s)}{s}\times 3} = \frac{G_1(s)}{s+3G_1(s)}$$

$$= \frac{1/(s+4)}{s+3/(s+4)} = \frac{1}{s(s+4)+3}$$

$$= \frac{1}{s^2+4s+3}$$

(*b*) $$X(t) = 1$$

$\Rightarrow$ $$X(s) = \frac{1}{s}$$

$$\frac{Y(s)}{X(s)} = G(s)$$

$$Y(s) = \frac{1}{(s^2+4s+3)}\cdot\frac{1}{s}$$

$$= \frac{1}{s(s+1)(s+3)}$$

$$= \frac{1/3}{s} - \frac{1/2}{s+1} + \frac{1/6}{s+3}$$

$$Y(s) = \frac{1}{3} - \frac{1}{2}e^{-t} + \frac{1}{6}e^{-3t}$$

(*c*) $$X(t) = u(t-1) - u(t-2)$$

or $$X(s) = \frac{1}{s}(e^{-s} - e^{-2s})$$

$$Y(s) = \frac{1}{(s^2+4s+3)}\cdot\frac{1}{s}(e^{-s} - e^{-2s})$$

$$Y(t) = \left[\frac{1}{3}u(t-1)\right] - \frac{1}{2}e^{-(t-1)}u(t-1)$$

$$+\frac{1}{6}e^{-3(t-1)}u(t-1) - \frac{1}{3}u(t-2)$$

$$+\frac{1}{2}e^{-(t-2)}u(t-2) - \frac{1}{6}e^{-3(t-2)}u(t-2)$$

8. (*a*) $$[\lambda I - A] = 0$$

$$\begin{bmatrix} \lambda & 0 \\ 0 & \lambda \end{bmatrix} + \begin{bmatrix} 2 & -1 \\ 0 & 3 \end{bmatrix} = \begin{bmatrix} \lambda+2 & -1 \\ 0 & \lambda+3 \end{bmatrix}$$

or $$|\lambda I - A| = \begin{vmatrix} \lambda+2 & -1 \\ 0 & \lambda+3 \end{vmatrix}$$

$$(\lambda + 2)(\lambda + 3) = 0$$

$\Rightarrow$ $$\lambda = -2, -3$$

(*b*) $u(t) = \delta(t)$, initial conditions are zero (given)

$$\dot{x}_1 = -2x_1 + x_2 + \delta(t) \quad \ldots(i)$$

$$\dot{x}_2 = -3x_2 \quad \ldots(ii)$$

Taking Laplace transform of equations (*i*) and (*ii*) solving for $X_1(s)$ and $X_2(s)$.

$$X_1(s) = \frac{1}{(s+2)} \Rightarrow x_1(t) = e^{-2t}$$

$$X_2(s) = 0 \Rightarrow x_2(t) = 0$$

$\Rightarrow$ $$y(t) = x_1 + x_2 = e^{-2t}$$

(*c*) From equations (*i*) and (*ii*) we get

$$\dot{x}_1 = -2x_1 + x_2 \quad \ldots(iii)$$

$$\dot{x}_2 = -3x_2 \quad \ldots(iv)$$

Taking laplace transform of equations (*iii*) and (*iv*) considering initial conditions $x_1(0^+)$ and $x_2(0^+)$ respectively.

$$x_1(t) = x_1(0^+)e^{-2t} + (e^{-2t} - e^{-3t})\, x_2(0^+)$$

$$x_2(t) = x_2(0^+)\, e^{-3t}$$

$$y(t) = x_1(0^+)\, e^{-2t} + e^{-2t} x_2(0^+)$$

$$= [x_1(0^+) + x_2(0^+)]\, e^{-2t}$$

i.e. $x_1(0^+) + x_2(0^+) = A$

9. (*a*) Closed loop transfer function

$$= \frac{G/s(s+3)}{1 + \dfrac{G}{s(s+3)}} = \frac{G}{s(s+3)+G}$$

$$= \frac{G}{s^2 + 3s + G}$$

(*b*) Damping ratio (ξ) can be taken as 0.6 for the given response comparing above transfer function with second order standard one.

$$2\xi w_n = 3 \text{ and } \omega_n^2 = G$$

$$2\sqrt{G} = \frac{3}{0.6}$$

or $$G_m = 6.25$$

(*c*) $$G' = 2G_m$$

$$= 2 \times 6.25$$

$$= 12.5$$

$$\omega_n' = \sqrt{12.5}$$

$$= 3.53$$

$$\text{Damping ratio } (\xi) = \frac{3}{2 \times 3.53} = 0.424$$

$$\text{Damping frequency, } \omega_d = \omega_n' \sqrt{1-\xi^2}$$

$$\frac{2\pi}{T} = 3.53\sqrt{1-0.424^2}$$

$$= 3.197$$

$$T = 1.96 \text{ Sec.}$$

10. (*a*)

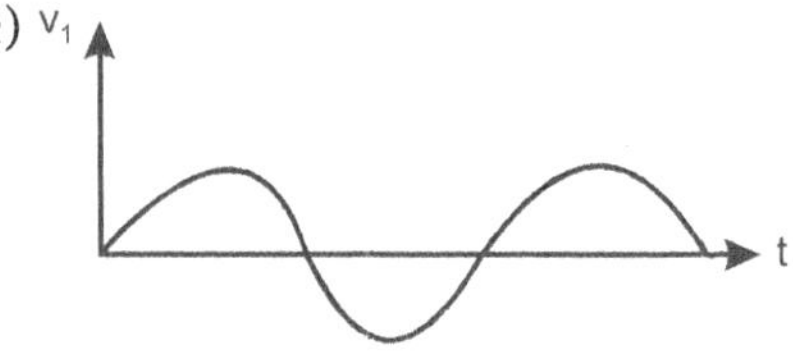

(a) Sinusoidal input

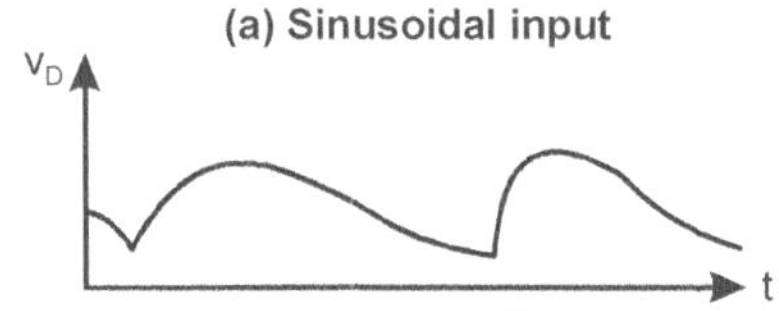

(b) Output with C

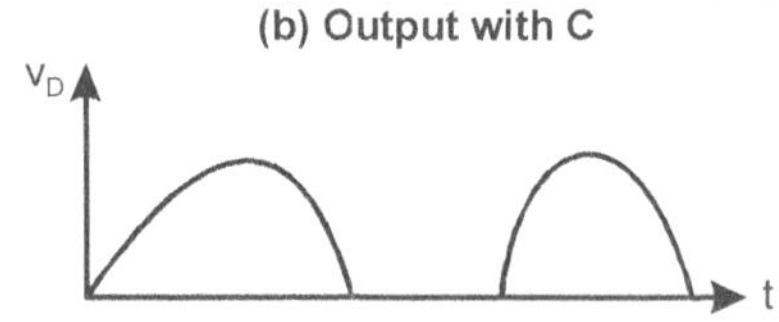

(c) Outputwithout C

(*b*)

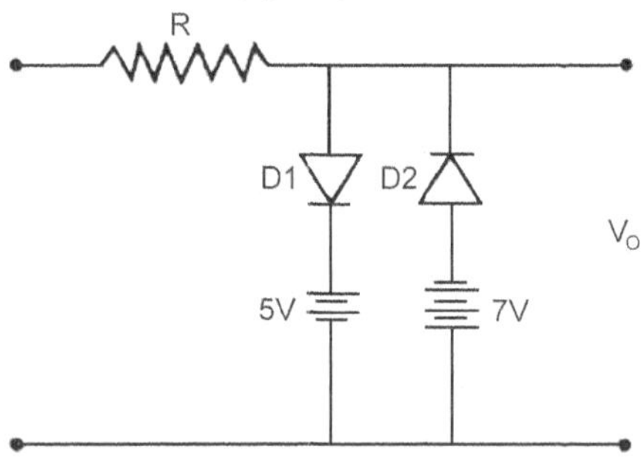

$V_1 = 5V$
$V_2 = 7V$ batteries

11. (*a*) $$A_v = \frac{V_o}{V_s} = -\frac{R_C}{r_e}$$

$$r_e = \frac{26mV}{I_E}$$

$$I_E = \frac{I_C}{\beta} + I_C = \frac{1.3}{100} + 1.3 = 1.313 \text{ mA}$$

$$r_e = \frac{26}{1.313} = 19.80$$

$$A_v = \frac{-2000}{19.80} = -101$$

(*b*) $$A_v \approx -\frac{R_C}{R_E} = -\frac{2000}{500}$$

$$= -4$$

(*c*) $$V_0 = A_v V_s$$

$$= -101 \times 0.01 \sin(\omega t)$$

$$= -1.01 \sin(\omega t) \text{ V}$$

12. (*a*) Bode plot is drawn on semilog paper.

Phase margin (*i.e.* phase of transfer function at which gain is zero dβ) –213.80° + 180° = –33.8°.

Since the phase marging of the system is negative, so amplifier is unstable.

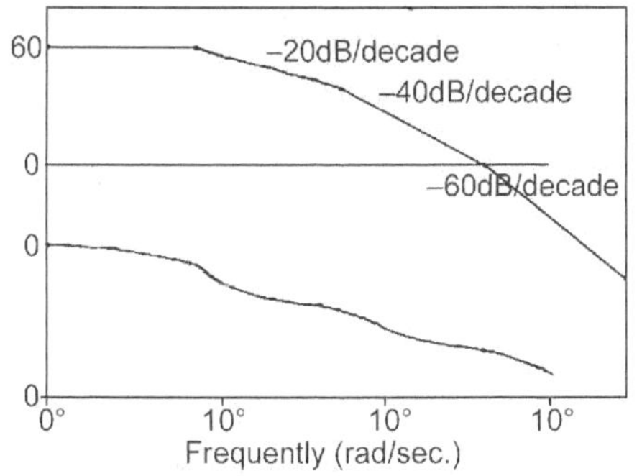

13. (*a*) Time constant associated with C_B, is $\tau_1 = RC_B$

where $R = h_{ib} \mid\mid R_2$

and h_{ib} = input impedance of circuit

Time constant associated with C_s, $\tau_2 = C_s (R_S + h_{ib})$

(*b*) Approximate lower cut-off frequency of the amplifier,

$$F_{LC} = \frac{1}{2\pi(R_s + h_{ib})C_S}$$

14. (*a*)

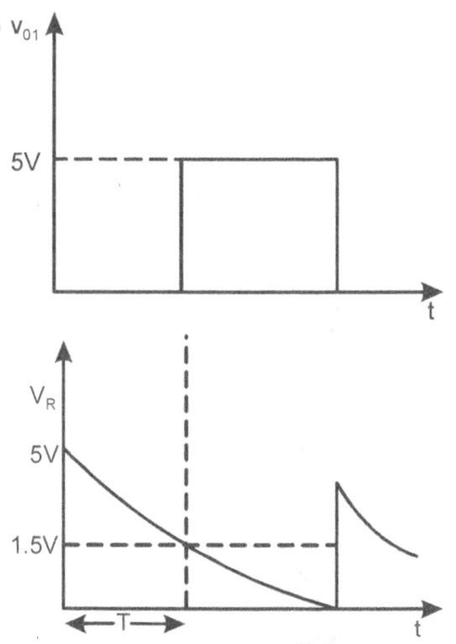

(*b*) $V_R(t) = 5e^{-t/RC} V,\ t > 0$

$RC = 50 \times 10^3 \times 0.01 \times 10^{-6}$

$= 0.5$ msec

$V_R(t) = 5e^{-2000t} V,\ t > 0$

(*c*) $5e^{-t/0.5 \times 10^{-3}} = 1.5$

$$-2000\, t' = \ln\left(\frac{1.5}{5}\right)$$

$\Rightarrow$ $t' = 0.6$ msec

T = t' = 0.6 msec

15. (*a*) Following are combination of pumps fail

x	*y*	*z*
0	0	1
0	1	0
1	0	0
1	1	0

Above entry can be made in given K-map

	$\overline{Z}$	Z
$\overline{X}\,\overline{Y}$	1	1
$\overline{X}Y$	1	0
XY	0	0
$X\overline{Y}$	1	0

P = XY + YZ + ZX (Output = 0)

(*b*) Above expression (p) can be implemented as

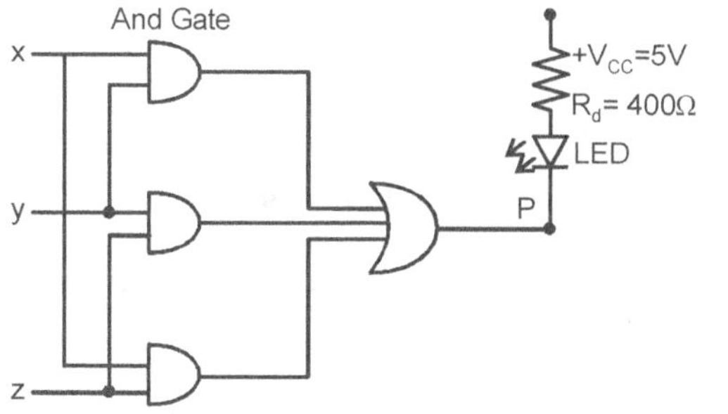

$$R_d = \frac{5-1}{10\text{mA}} = \mathbf{400\Omega}$$

16. (*a*) Truth table

	A	B	C_p	Sum(S)	Carry (C_N)
0	0	0	0	0	0
1	0	0	1	1	0
2	0	1	0	1	0
3	0	1	1	0	1
4	1	0	0	1	0
5	1	0	1	0	1
6	1	1	0	0	1
7	1	1	1	1	1

(*b*)

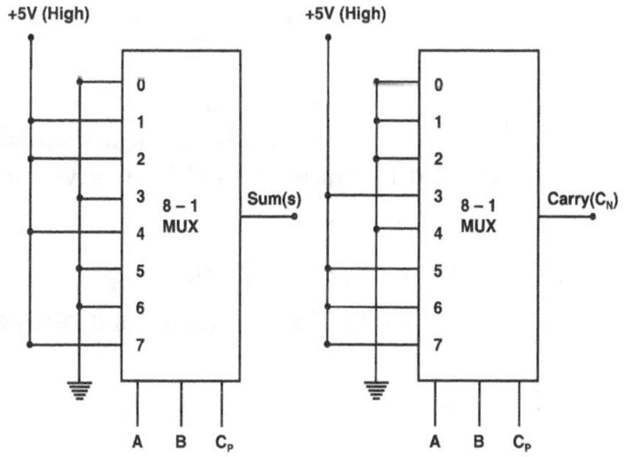

Fig. p16. (i) Implement for S (*ii*) Implement for C_N

17. (*a*) Assuming content of register B is 00H. After ADD B operation, content of Accumulator is 00H after DEC A operation, content of Accumulator is FFH. Hence content of output port 10 is FF H.

(*b*) (*i*) For call 2000 H memory operation (read/write) are 5.

(*ii*) For LDA 2000 H memory operations (read/write) are 4.

(*c*) XOR A

When this instruction is executed content of A becomes Zero (i.e. clear).

18. (*a*)
$$p(t) = \sum_{n=-\infty}^{\infty} C_n \exp\left(jn\frac{2\pi t}{T}\right)$$

or
$$C_n = \frac{1}{T}\int_{-T/2}^{T/2} p(t)\, e^{-j\frac{n2\pi t}{T}}\, dt$$

Function $p(t)$ in the interval $\left(-\frac{T}{2}, \frac{T}{2}\right)$ is simply $\delta(t)$, hence is 1.

so
$$C_n = \frac{1}{T}\int_{-T/2}^{T/2} e^{-j\frac{n2\pi t}{T}}\, dt$$

or
$$C_n = \frac{1}{T}$$

Hence
$$p(t) = \frac{1}{T}\sum_{n=-\infty}^{\infty} \exp\left(jn\frac{2\pi t}{T}\right).$$

(*b*) $f[p(t)]$ i.e., fourier transform of $p(t)$

$$= 2\pi \sum_{n=-\infty}^{\infty} \frac{1}{T}\delta(\omega - n\omega_0) = \frac{2\pi}{T}\sum_{n=-\infty}^{\infty}\delta(\omega - n\omega_0)$$

$$= \omega_0 \sum_{n=-\infty}^{\infty}\delta(\omega - n\omega_0) = \omega_0\delta\omega_0(\omega) = \frac{2\pi}{T}\delta\omega_0(\omega).$$

19. (*a*) Total power in $n(t) = \frac{1}{2\pi}\int_{-2\pi B}^{2\pi B} \eta\, d\omega$

where η is spectral power density of $n(t)$, here

$$\eta = 4\text{W/KHz},\ B = 2\text{ KHz}$$

$$P_n = \frac{1}{2\pi}\int_{-2\pi B}^{2\pi B} 4d\omega = 8 \times 2 = \mathbf{16W}$$

(*b*) Auto correlation function of $n(t)$ whose spectral power density η ω/Hz and low pass filter with a cut off frequency of B Hz is given as

$$\phi_{nn}(\tau) = \eta B\, S_a(2\pi B\tau)$$
$$\phi_{nn}(\tau) = 8S_a(4\pi \times 10^3 \tau)$$
$$= 8S_a(4\pi\tau), \text{ where } \tau \text{ is in milliseconds.}$$

(*c*) Time internal $= \frac{1}{2B}$

$$= \frac{1}{2\times 2\times 10^3} = \mathbf{0.25\ msec.}$$

20. (*a*) Maxwell's equations in free space

(*i*) $\nabla . E = 0$

(*ii*) $\nabla . H = 0$

(*iii*) $\nabla \times E = -\mu_0 \frac{\partial H}{\partial t}$

(*iv*) $\nabla \times H = \varepsilon_0 \frac{\partial E}{\partial t}$

(*b*)
$$\nabla \times E = \begin{vmatrix} \hat{x} & \hat{y} & \hat{z} \\ \frac{\partial}{\partial x} & \frac{\partial}{\partial y} & \frac{\partial}{\partial z} \\ 0 & 10e^{-j(4x-kt)} & 0 \end{vmatrix}$$
$$= z[10e^{-j(4x-kt)}.(j4)]$$
$$= -j40e^{-j(4x-kt)} z \text{ U/m}^2$$

(*c*) Since $\nabla \times E = -j\omega\mu_0 H$ (Here ω = K)

$$H = \frac{\nabla \times E}{-jk\mu_0} = \frac{-j40e^{-j(4x-kt)}}{-jk\mu_0}$$
$$= \frac{40}{k\mu_0} e^{-j(4x-kt)}$$

21. (*a*) Wave impedance in medium 2

$$Z_2 = \sqrt{\frac{\mu_0}{\varepsilon_{r2}\varepsilon_0}}$$
$$= \sqrt{\frac{4\pi\times 10^{-7}}{4\times 8.85\times 10^{-12}}} = 188.41\ \Omega$$

wave impedance in medium 3

$$Z_3 = \sqrt{\frac{\mu_0}{\varepsilon_{r3}\varepsilon_0}}$$
$$= \sqrt{\frac{4\pi\times 10^{-7}}{9\times 8.85\times 10^{-12}}}$$
$$= 125.61\Omega$$

(*b*)
$$d = \frac{\lambda_0}{2\sqrt{\epsilon\, r_z}}$$
$$\lambda_0 = \frac{c}{f}$$
$$d = \frac{c}{2f\sqrt{4}}$$
$$d = \frac{3\times 10^{10}}{1\times 10^9 \times 4} = \mathbf{7.5\ cm.}$$

(*c*)
$$Z_{in} = Z_2 \frac{Z_3 + jZ_2 \tan B_1, d}{Z_2 + jZ_3 \tan B_1, d}$$

where $B_1 = \frac{2\pi}{\lambda}$,

$$d = \frac{\lambda}{4}$$

$$Z_{in} = \frac{Z_2^2}{Z_3}$$

$$= \frac{188.41^2}{125.61} = 282.60\Omega$$

Reflection coefficient

$$\Gamma = \frac{282.60 - 377}{282.60 + 377}$$

$$= -\ 0.143$$

Voltage standing wave ratio

$$(VSWR) = \frac{1 + |\Gamma|}{1 - |\Gamma|} = \frac{1 + 0.143}{1 - 0.143}$$

$$= 1.33$$

22. 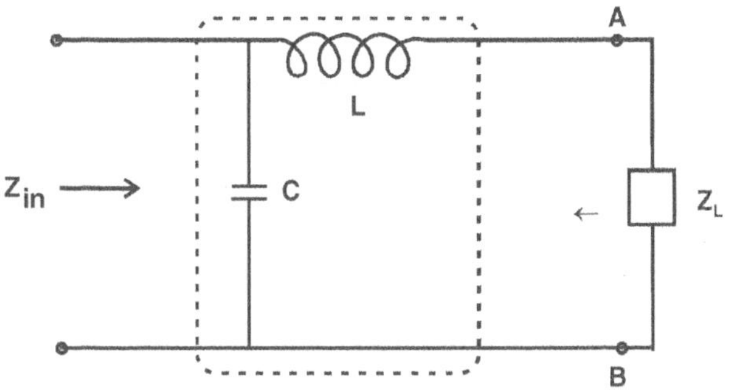

Since $Z_{in} = 50\Omega$

Input impedance from terminals A, B should be Z_L^*,

Hence $$j\omega L + \frac{50 \cdot \frac{1}{j\omega C}}{50 + \frac{1}{j\omega C}} = j\omega L + \frac{50}{1 + j50\omega C}$$

$$= 10 - j10$$

or $$\frac{j\omega_L\left[1+(50\omega C)^2\right]+50(1-j50\omega C)}{(1+50\omega C)^2} = 10 - j10$$

Comparing real and imaginary parts in above expression.

$$\frac{50}{1+(50\omega C)^2} = 10$$

$\Rightarrow$ $$\omega C = 0.04$$

or $$C = \frac{0.04}{2 \times 3.14 \times 1 \times 10^9}$$

$$= 6.37 \text{ pF}$$

and $$\frac{50\omega L - 2500 \times 0.4}{5} = -\ 10$$

or $$\omega L = 10$$

$\Rightarrow$ $$L = \frac{10}{2 \times 3.14 \times 1 \times 10^9}$$

$$= 1.59 \times 10^{-9} H$$

$\therefore$ $$\text{Q. Factor} = \frac{L}{CR_L^2}$$

$$= \frac{1.59 \times 10^{-9}}{6.37 \times 10^{-12} \times 100}$$

$$= 2.49$$